INSTRUMENT ENGINEERS' Handbook

PROCESS MEASUREMENT

INSTRUMENT ENGINEERS' Handbook

BÉLA G. LIPTÁK
EDITOR-IN-CHIEF
KRISZTA VENCZEL
ASSOCIATE EDITOR

WITHDRAWN
UTSA LIBRARIES

Revised Edition

CHILTON BOOK COMPANY RADNOR, PENNSYLVANIA

Copyright © 1969, 1982 by Béla G. Lipták
Revised Edition All Rights Reserved
Published in Radnor, Pennsylvania, 19089, by Chilton Book Company
and simultaneously in Canada by VNR Publishers,
1410 Birchmount Road, Scarborough, Ontario M1P 2E7

Library of Congress Cataloging in Publication Data
Main entry under title:

Instrument engineers' handbook.

Includes bibliographies and index.
1. Measuring instruments—Handbooks, manuals, etc. 2. Process control—Handbooks, manuals, etc. I. Lipták, Béla G. II. Venczel, Kriszta.
TS156.8.I56 1982 681'.2 81-70914
ISBN 0-8019-6971-9 AACR2

Designed by William E. Lickfield
Drawings by Adrian J. Ornik
Manufactured in the United States of America

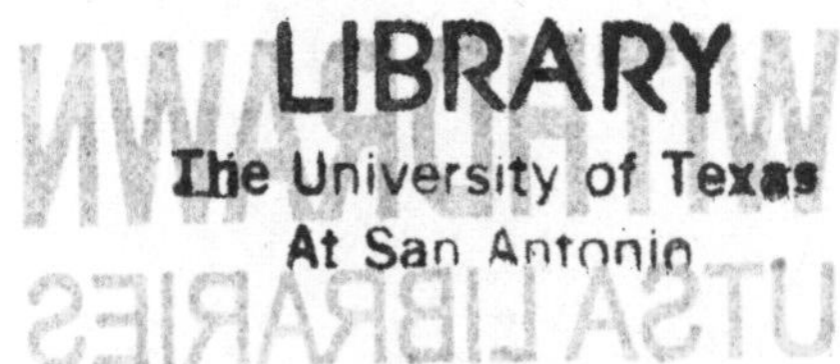

1 2 3 4 5 6 7 8 9 0 1 0 9 8 7 6 5 4 3 2

In honor of the 25th Anniversary
of the Hungarian Revolution
and
the Solidarity movement
in Poland

CONTENTS

CONTRIBUTORS

ROSS C. AHLSTROM, JR.
BS
Upper Level Research Specialist
The Dow Chemical Company
(Section 10.12)

JAMES B. ARANT
BSChE
Senior Consultant
E.I. du Pont de Nemours & Company
(Sections 2.8, 12, 23)

ARTHUR AUSTIN
BS, PE
Senior Research Engineer
Chevron Research Company
(Section 11.26)

CHRISTOPHER P. BLAKELEY*
BSChE
Marketing Manager, Water Management
Honeywell, Inc.
(Sections 10.19 and 11.20)

L. JOSEPH BOLLYKY*
PhD, PE
President
Bollyky Associates
(Sections 11.22 and 11.23)

AUGUST BRODGESELL*†
BSEE
President
CRB Systems, Inc.
(Sections 9.9, 11, 13; 10.6, 7, 17; 11.10, 14, 15, 16)

JAMES E. BROWN†
BSME, PE
Manager of Engineering
Union Carbide Corporation
(Sections 10.4, 5, 15, 16, 18)

THOMAS J. CLAGGETT†
BSEE
Application Specialist (Retired)
Honeywell, Inc.
(Sections 4.2 through 4.10 and 4.13, 14)

WILSON A. CLAYTON
BSChE, MSME
Chief Engineer, Thermal
Hy-Cal Engineering
(Section 4.11)

VINCENT B. CORTINA
BSChE, MS Indust. Mg't.
Business Element Manager
EG&E, Environmental Equipment Division
(Section 11.10)

GILES M. CRABTREE
BSEE, PE
Manager Customer Services
American Meter Division of Singer Company
(Section 2.14)

LOUIS D. DINAPOLI
BSEE, MSEE
Manager, Flowmeter Development
Leeds & Northrup Company
(Section 2.21)

ALBERT P. FOUNDOS
BSChE, MBA
President
Fluid Data, Inc.
(Sections 2.3; 10.2; 11.3)

WALTER F. GERDES*
BSEE, PE
Technical Specialist
The Dow Chemical Company
(Section 11.28)

*These authors' contributions have been updated from *Environmental Engineers' Handbook*, Béla G. Lipták, ed., Chilton Book Company, Radnor, Pa., 1974.

†These authors' contributions have been updated from the previous edition of *Instrument Engineers' Handbook*.

RICHARD GILBERT
PhD
Assistant Professor
University of South Florida
(Section 10.8)

ANTHONY C. GILBY
PhD
Research Coordinator
The Foxboro Company
(Section 10.9)

PAUL M. GLATTSTEIN†
BSEE
Senior Electrical Engineer
Crawford & Russell, Inc.
(Section 9.4)

JOHN D. GOODRICH, JR.†
BSME
Engineering Supervisor
Bechtel Corporation
(Section 5.6)

ROBERT J. GORDON*
PhD
Environmental Division Manager
Global Geochemistry Corporation
(Sections 11.5, 11, 18)

JOHN T. HALL
BS
Senior Technical Editor
Instruments & Control Systems
(Section 5.1)

CHARLES E. HAMILTON*
BSChE
Senior Environmental Specialist
The Dow Chemical Company
(Section 11.17)

ROBERT A. HERRICK*
BSChE, PE
Consulting Engineer
(Sections 11.4 and 11.29)

HEROLD I. HERTANU
MSEE, PE
Principal Engineer
Crawford & Russell, Inc.
(Section 8.1)

CONRAD H. HOEPPNER
BSEE, MSEE
Consultant
Simmonds Precision Products, Inc.
(Chapter VI)

MICHAEL F. HORDESKI
BSEE, MSEE, PE
Control Systems Engineering Consultant
Siltran Digital
(Section 9.2)

JOEL O. HOUGEN
PhDChE, PE
Consultant, Professor Emeritus
University of Texas
(Section 2.13)

WALTER D. HOULE
BSEE
Manager, Facilities Automation Engineering
IBM Corporation
(Section 9.3)

WILFRED H. HOWE†
BSEE, MBA, PE
Chief Engineer (Retired)
The Foxboro Company
(Sections 2.5, 18, 25)

DAVID L. HOYLE*
BSChE
System Design Engineer
The Foxboro Company
(Section 10.14)

JAY S. JACOBSON*
PhD
Plant Physiologist
Boyce Thomson Institute for Plant Research
(Section 11.9)

ROBERT F. JAKUBIK†
BSChE
Manager, Process Control Applications
Digital Applications, Inc.
(Section 12.4)

ROBERT H. JONES
BS
Principal Application Engineer
Beckman Instruments, Inc.
(Sections 10.13 and 10.14)

RICHARD K. KAMINSKI
BA
Senior Instrument Designer
Dravo Engineers and Constructors
(Sections 9.6; 11.8, 21)

DAVID S. KAYSER
BSEE
Senior Instrument Engineer
Texas City Refining, Inc.
(Sections 2.7, 16; Chapter III; 9.1)

THOMAS J. KEHOE
BSChE, PE
Manager, Technical Services
Beckman Instruments, Inc.
(Section 11.30)

CHANG H. KIM†
BSChE
Manager, Technical Services
ARCO Chemical Company
Division of Atlantic Richfield Co.
(Chapter VII)

JOHN G. KOCAK, JR.
BA
Consultant
(Section 10.9)

JOHN G. KOPP
BSME, PE
Senior Product Marketing Manager
Fischer & Porter Company
(Sections 2.9, 22)

BÉLA G. LIPTÁK
ME, MME, PE
President
Lipták Associates, P.C.
(Introduction, Sections 1.2; 2.2, 4, 6, 15, 17; 4.1, 12; 5.2 through 5.14; 8.1; 9.14; 10.1, 4)

DAVID H. F. LIU
PhD
Senior Planning Associate
Rhône-Poulenc, Inc.
(Section 10.3)

HARRY E. LOCKERY†
BSEE, MSEE, PE
President
Hottinger-Baldwin Measurements, Inc.
(Section 8.6)

DAVID J. LOMAS
Marketing Support Executive
Kent Process Control Limited
(Sections 2.1, 20, 24)

ORVAL P. LOVETT, JR.†
BSCE
Consulting Engineer, Instrumentation and Control Systems
Retired from E.I. du Pont de Nemours & Company
(Section 12.4)

DAVID C. MAIR†
BCE, PE
Manager, Sales Services
Wallace & Tiernan Division, Pennwalt Corp.
(Section 2.19)

THOMAS A. MAYER
BSE, MSE, PE
Senior Development/Research Engineer
PPG Industries
(Sections 9.7, 10; 12.3, 4; A.8)

GERALD F. MCGOWAN
BSEE, MSEE
Vice President of Engineering
Lear Siegler, Inc., Environmental Technology Division
(Section 11.27)

HUGH A. MILLS†
ME
President
Marcan Products
(Sections 8.2, 4, 5)

CHARLES F. MOORE†
BSChE, MSChE, PhDChE
Professor of Chemical Engineering
University of Tennessee
(Section 12.4)

THOMAS J. MYRON, JR.*
BSChE
Senior Systems Design Engineer
The Foxboro Company
(Section 10.14)

S. NISHI*
DSc
Research Scientist
National Chemical Laboratory for Industry, Ibaraki, Japan
(Section 11.13)

ROBERT NUSSBAUM†
BSEE
Senior Instrument Engineer
Crawford & Russell, Inc.
(Sections 9.5 and 11.7)

RICHARD T. OLIVER*
BSChE, MSChE, PhDChE
Senior Analog Systems Design Engineer
The Foxboro Company
(Section 10.10)

WILLIAM H. PARTH*
BS, MS
Senior Instrument Specialist
The Dow Chemical Company
(Section 11.25)

KURT O. PLACHE
BSChE, PE
Vice-President, Marketing
Micro-Motion, Inc.
(Section 2.10)

GEORGE PLATT
BSChE, PE
Staff Engineer
Bechtel Power Corp.
(Sections 12.1 and 12.2)

JAMES B. RISHEL
BSME
President
Corporate Equipment
(Section 9.12)

HOWARD C. ROBERTS
BAEE, PE
Consultant
(Sections 8.3 and 9.8)

LEWIS B. ROOF
BS, MS
Senior Measurement Engineer
Applied Automation, Inc.
(Section 10.11)

ERIC J. SCHAFFER
BSEE, MSEE
Project Engineer
MTS Systems Corporation
(Section 1.3)

DONALD J. SIBBETT*
PhD
Vice President
Geomet, Inc.
(Section 11.12)

ROBERT SIEV†
BSChE, MBA, CE
Engineering Specialist
Bechtel Corporation
(Sections 2.11, 19)

EDWARD TELLER
Professor-at-Large
University of California
(Preface)

AMOS TURK*
PhD
Professor, Dept. of Chemistry
The City College of the City University of New York
(Section 11.19)

WILLIAM H. WAGNER†
BSChE, PE
Staff Engineer
Union Carbide Corporation
(Section 10.1)

NORMAN S. WANER
BSME, MSME, ME, PE
Manager of Training and Development
Bechtel Corporation
(Section 11.26)

ALAN L. WERTHEIMER
PhD
Principal Scientist
Leeds & Northrup Company
(Section 11.24)

GEORGE P. WHITTLE*
BSChE, MSChE, PhDChE, PE
Associate Professor
University of Alabama
(Section 11.6)

THEODORE J. WILLIAMS
BSChE, MSChE, PhDChE, PE
Professor of Engineering, Director of Purdue Laboratory for Applied Industrial Control
Purdue University
(Section 1.3)

ROBERT W. WORRALL†
BA, PE
Principal Instrument Engineer
Catalytic, Inc.
(Sections 4.2 through 4.10 and 4.13, 14)

IRVING G. YOUNG*
BS, MS, PhD
Chemist, Advanced Technology Staff
Honeywell, Inc.
(Sections 11.1 and 11.2)

PREFACE

EDWARD TELLER

It is often said that the proliferation of knowledge and the accelerating development of technology have rendered our environment so complex that no one can understand even a small portion of it. The statement would be undoubtedly correct except for the circumstance that new knowledge can be better organized and that automation can relieve us of a considerable number of chores, as a consequence of which activity can be concentrated on subjects having technological meaning.

In this evolution, clear presentation of modern instruments is of the greatest importance. As machines are being increasingly applied, not only to replace human muscle, but also to develop a good substitute for the routine functions of human brains, it becomes more and more important to have an up-to-date summary of technological aids as they become available.

The difficulty is, of course, connected with the words "up-to-date." Techniques of measurement, processing of these measurements and regulating mechanisms which utilize the results are being improved from decade to decade and almost from year to year. This results in the immensely complicated structure known as automation, which has a deep influence on most phases of our industry.

One of the most rapidly developing aspects of this broad field is the employment of computers. Since they are designed to perform anything that the human brain can perform, as long as precise instructions are available, and since, indeed, industrial systems have to be planned in a precise manner, the increasing role of computers in industrial control is unavoidable. It is equally unavoidable that these computers be adjusted to the varying and complex needs of the systems which they serve.

It is understandable that the resulting, involved procedures should be viewed with some worry, which is accentuated whenever malfunction occurs on a big scale. A well-remembered event occurred on the 9th of November, 1965, when 30 million people in the northeastern part of the United States were left without electricity. What one should notice in connection with that event is that the cause for the breakdown remained obscure for a period of five days. While it is clear that the breakdown was connected with the complexity of electric interconnections, it should also be clear that the breakdown could be much more rapidly understood and that similar events can be more easily avoided by a more complete application of the principles of automation to information gathering, systems engineering and the connected aspects of safety devices.

Another widespread worry connected with the development of automation is the question whether the machines will create widespread unemployment among new strata of the working population. I do not believe that this will happen. Indeed, such a development would not be different from any broad phase of the industrial and scientific revolution. Human needs have a habit of growing faster than our ability to satisfy them. Human opportunities are growing as fast as our imagination and organizational abilities can create them.

The prospect is not that human intelligence will be replaced by automation and electronic brains. I would venture to predict that the future holds a fruitful symbiosis of man and machine in which increasingly creative contributions will have to be made by the human partner. Thus, I think that the next generation will have to be more ingenious, but will have to put up with much less boredom.

None of this, of course, can be executed successfully unless information on automation as a whole, and on its components, is readily available. One of the most destructive of contemporary ideas is that of the "black box." The required skill of the operator demands a thorough understanding of the apparatus. The same understanding is needed if malfunctions are to be avoided and, in case they do occur, corrected. It is for these reasons that a handbook for instrument engineers is of the greatest importance.

A work of this kind has the obvious property that it can be improved by better adaptation to the many facets of practical needs. This improvement will be speeded by constructive criticism from the users. It is hoped that this second edition will serve the broad purpose of technology even better.

INTRODUCTION

BÉLA G. LIPTÁK

The INSTRUMENT ENGINEERS' HANDBOOK was first published in 1969. Since that time, it has been added to the bookshelves of tens of thousands of instrument engineers. As the years have passed, the contents have become partially outdated.

This revised edition updates the information contained in the previous volumes and incorporates the many new developments that have occurred in the last decade. In this volume, we have concentrated our efforts on the subjects of measurement and analysis because these are the areas where most of the changes have occurred and where professional level technical literature is least available. Topics such as control valves and control systems are covered in the other volume of this HANDBOOK and in other works, such as *Instrumentation in the Processing Industries*.

It is hoped that this HANDBOOK will also contribute to the professional standing of the community of instrument engineers. We know that the social benefit of a competent and advanced instrument engineering profession can be substantial. We also know that our greatest national resource is our combined knowledge and professional dedication. The productivity of many industries could be doubled in some cases through the exploitation of state-of-the-art instrumentation and control. Similarly, the energy cost of many processes could be cut in half merely through optimized controls. We have the tools, the knowledge, and the need to reach these goals.

Chapter I

PERFORMANCE SPECIFICATIONS

INSTRUMENT SOCIETY OF AMERICA,
B. G. LIPTÁK,
E. J. SCHAFFER
and
T. J. WILLIAMS

CONTENTS OF CHAPTER I

1.1 INSTRUMENT TERMINOLOGY AND PERFORMANCE

The following section was reprinted with format change only from the work titled: *Standards and Practices for Instrumentation*, 6th ed., Instrument Society of America, 1980. ISA's permission to abstract from their standard S51.1 titled: "Process Instrumentation Terminology" is gratefully acknowledged.

The purpose of this Standard is to establish uniform terminology in the field of process instrumentation. The generalized test procedures described in the section titled "Test Procedures" are intended only to illustrate and clarify accuracy-related terms. It is not intended that they describe specific and detailed test procedures.

This Standard is intended to include all specialized terms used to describe the use and performance of the instrumentation and instrument systems used for measurement, control or both, in the process industries. Process industries include chemical, petroleum, power generation, air conditioning, metallurgical, food, textile, paper, and numerous other industries.

The terms of this Standard are suitable for use by people involved in all activities related to process instrumentation; including research, design, manufacture, sales, installation, test, use and maintenance.

The Standard consists of terms selected primarily from Scientific Apparatus Makers Association (SAMA) Standard PMC20.1 and American National Standards Institute (ANSI) Standard C85.1. Additional terms have been selected from other recognized standards. Selected terms and definitions have not been modified unless there was a sufficiently valid reason for doing so. New terms have been added and defined where necessary.

This Standard is primarily intended to cover the field of analog measurement and control concepts, and makes no effort to develop terminology in the field of digital measurement and control.

Introduction

Italicized Terms

Defined terms, where used as a part of other definitions, are set in italics to provide a ready cross reference.

Introductory Notes

In defining certain performance terms, the context in which they are used has been considered. It is fitting, therefore, that the philosophy of performance evaluation on which these terms are based be explained.

Ideally, instruments should be designed for realistic operating conditions, those they are likely to meet in service, and they should be evaluated under the same conditions. Unfortunately, it is not practical to evaluate performance under all possible combinations of operating conditions. A test procedure must be used which is practical under laboratory conditions and, at the same time, will make available with a reasonable amount of effort, sufficient data on which a judgment of field performance can be made.

The method of evaluation envisioned is that of checking significant performance characteristics such as accuracy rating, dead band, and hysteresis under a set of reference operating conditions, these having a narrow range of tolerances.

Reference performance is, therefore, to be evaluated and stated in terms of reference operating conditions.

Generally, reference performance under reference operating conditions represents the "best" performance that can be expected under ideal conditions.

The effect of change in an individual operating condition, such as ambient temperature, atmospheric pressure, relative humidity, line voltage, and frequency will be determined individually throughout a range defined as normal operating conditions. These can logically be expected to be encountered above and below the values of reference operating conditions during field operation.

While this approach does not duplicate all actual conditions, where many operating variables may vary simultaneously in random fashion, it does develop data from which performance may be inferred from any given set of operating conditions.

The effect of changes in an individual operating condition, all other operating conditions being held within the reference range, is herein called operating influence. There may be an operating influence corresponding to a change in each operating condition. In some cases the effect may be negligible, while in others it may have significant magnitude.

Tabulations of operating influences will usually denote the performance quality level of a given design. Comparisons of reference performance and operating influences for instruments of a given design, or for different designs, will show clearly their relative merits and probable performance under actual operating conditions.

Operating Conditions vs. Performance

Operating Conditions	*Performance*
Reference (narrow band)	Reference (Region within which accuracy statements apply unless indicated otherwise.)
Normal (wide band)	Conditional (Region within which the influence of environment on performance is stated.)
Operative Limits (extreme band)	Indefinite (Region within which influences are not stated and beyond which damage may occur.)

Sources and References

In the preparation of this Standard of Terminology, many standards and publications sponsored by technical organizations such as ASME, IEEE, and ISA were studied by the committee in addition to those listed as principal source documents. These are listed as References.

Existing terms and definitions have been used wherever considered suitable. In many cases terms have been extracted from source documents with verbatim definitions and in such cases permission to quote from the respective source document has been obtained from the organization concerned, as indicated below. Terms defined verbatim are followed by the reference number in parentheses. For example: (4) after a defined term indicates that this term is quoted verbatim from ANSI C85.1 "Terminology for Automatic Control."

In other cases definitions have been modified in varying degrees in conformity with current practice in process instrumentation. These have been noted in parentheses as "Ref." followed by the reference number. For example: (Ref. 8) indicates that this term is a modified definition of the referenced term in SAMA-PMC 20.1-1973 "Process Measurement and Control Terminology."

An omission or alteration of a note following a definition is not considered a modification of the definition and is not identified by the prefix, "Ref."

Principal source documents used from the many reviewed are as follows:

1. American National Standard C39.4-1966 "Specifications for Automatic Null-Balancing Electrical Measuring Instruments," published by American National Standards Institute, Inc.; Copyright 1966 by ANSI.
2. American National Standard C42.100-1972 "Dictionary of Electrical and Electronics Terms," published by the Institute of Electrical and Electronics Engineers, Inc.; Copyright 1972 by IEEE.
3. American National Standard C85.1-1963 "Terminology for Automatic Control," published by the American Society of Mechanical Engineers; Copyright 1963 by ASME.
4. SAMA Standard PMC20.1-1973 "Process Measurement and Control Terminology," published by Scientific Apparatus Makers Association, Process Measurement and Control Section, Inc.; Copyright 1973 by SAMA-PMC.

Copies of the American National Standards referred to above may be purchased from the American National Standards Institute, 1430 Broadway, New York, New York 10018. Copies of the SAMA Standard may be purchased from Process Measurement and Control Section, Inc., SAMA, 370 Lexington Avenue, New York, New York 10017.

Definition of Terms

*accuracy**—Degree of *conformity* of an indicated value to a recognized accepted standard value, or ideal value. (Ref. 4, Ref. 8)

See *accuracy rating*.

accuracy, measured—The maximum positive and negative *deviation* observed in testing a *device* under specified conditions and by a specified procedure.

Note 1: It is usually measured as an inaccuracy and expressed as accuracy.

Note 2: It is typically expressed in terms of the measured variable, percent of span, percent of upper range-value, percent of scale length or percent of actual output reading.

See *test procedure*.

accuracy rating—A number or quantity that defines a limit that *errors* will not exceed when a *device* is used under specified *operating conditions*. See Figure 1.1a

Note 1: When operating conditions are not specified *reference operating conditions* shall be assumed.

Note 2: As a performance specification, *accuracy* (or reference accuracy) shall be assumed to mean

*Throughout this handbook the term "inaccuracy" has been used instead of "accuracy." This appears to be the correct choice since this term relates to the error in a measurement. In the following paragraphs, we have allowed the term "accuracy" to be used, not because we agree with its use, but because this entire section is being quoted from an ISA Standard and, as such, it would not be appropriate for us to modify the text. (ed.)

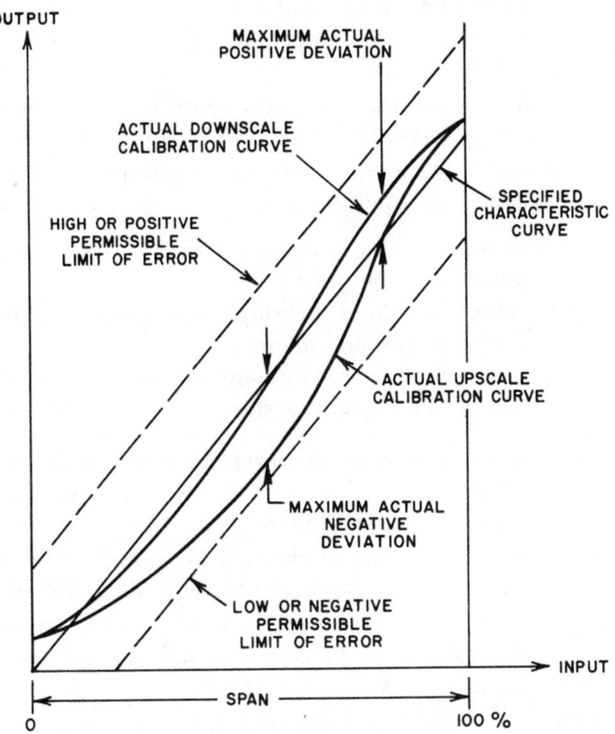

Fig. 1.1a Accuracy rating

accuracy rating of the *device*, when used at *reference operating conditions.*

Note 3: Accuracy rating includes the combined effects of *conformity, hysteresis, dead band* and *repeatability* errors. The units being used are to be stated explicitly. It is preferred that a ± sign precede the number or quantity. The absence of a sign indicates a + and a − sign.

Accuracy rating can be expressed in a number of forms. The following five examples are typical:

a. accuracy rating expressed in terms of the *measured variable.* Typical expression: The accuracy rating is ± 1°C, or ± 2°F.
b. accuracy rating expressed in percent of *span.* Typical expression: The accuracy rating is ±0.5% of span. (This percentage is calculated using scale units such as degrees F, psig, etc.)
c. accuracy rating expressed in percent of the upper range-value. Typical expression: The accuracy rating is ±0.5% of upper-range value. (This percentage is calculated using scale units such as kPa, degrees F, etc.)
d. accuracy rating expressed in percent of scale length. Typical expression: The accuracy rating is ±0.5% of scale length.
e. accuracy rating expressed in percent of actual output reading. Typical expression: The accuracy rating is ±1% of actual output reading.

accuracy, reference—see *accuracy, rating.*

actuating error signal—see *signal, actuating error.*

adaptive control—see *control, adaptive.*

adjustment, span—Means provided in an instrument to change the slope of the input-output curve. See *span shift.*

adjustment, zero—Means provided in an instrument to produce a parallel shift of the input-output curve. See *zero shift.*

air conditioned area—see *area, air conditioned.*

air consumption—The maximum rate at which air is consumed by a *device* within its operating *range* during *steady-state signal conditions.*

Note: It is usually expressed in cubic feet per minute (ft³/min) or cubic meters per hour (m³/h) at a standard (or normal) specified temperature and pressure. (8)

ambient pressure—see *pressure, ambient.*

ambient temperature—see *temperature, ambient.*

amplifier—A *device* that enables an *input signal* to control power from a source independent of the *signal* and thus be capable of delivering an output that bears some relationship to, and is generally greater than, the *input signal.* (3)

analog signal—see *signal, analog.*

area, air conditioned—A location with temperature at a nominal value maintained constant within narrow tolerance at some point in a specified band of typical comfortable room temperature. Humidity is maintained within a narrow specified band.

Note: Air conditioned areas are provided with clean air circulation and are typically used for instrumentation, such as computers or other equipment requiring a closely controlled environment. (Ref. 18)

area, control room—A location with heat and/or cooling facilities. Conditions are maintained within specified limits. Provisions for automatically maintaining constant temperature and humidity may or may not be provided.

Note: Control room areas are commonly provided for operation of those parts of a *control system* for which operator surveillance on a continuing basis is required. (18)

area, environmental—A basic qualified location in a plant with specified environmental conditions dependent on severity.

Note: Environmental areas include: *air conditioned areas; control room areas,* heated and/or cooled; *sheltered areas* (*process* facilities); *outdoor areas* (remote field sites). See *specific definitions.*

area, outdoor—A location in which equipment is exposed to outdoor ambient conditions; including temperature, humidity, direct sunshine, wind and precipitation. (Ref. 18)

area, sheltered—An industrial *process* location, area, storage or transportation facility, with protection against direct exposure to the elements, such as direct sunlight, rain or other precipitation or full wind pressure. Minimum and maximum temperatures and humidity may be the same as outdoors. Condensation can occur. Ventilation, if any, is by natural means.

Note: Typical area shelters for operating instruments, unheated warehouses for storage, and enclosed trucks for transportation. (18)

attenuation—1) A decrease in *signal* magnitude between two points, or between two frequencies. 2) The reciprocal of *gain*.

Note: It may be expressed as a dimensionless ratio, scalar ratio, or in decibels as 20 times the $\log_{10}$ of that ratio. (Ref. 4)

auctioneering device—see *signal selector*.

automatic control system—see *control system, automatic*.

automatic/manual station—A *device* which enables an operator to select an automatic *signal* or a manual *signal* as the input to a controlling element. The automatic *signal* is normally the output of a *controller* while the manual *signal* is the output of a manually operated *device*.

backlash—A relative movement between interacting mechanical parts, resulting from looseness when motion is reversed. (Ref. 4)

Bode diagram—A plot of log amplitude ratio and phase angle values on a log frequency base for a *transfer function*. See Figure 1.1b. (8, Ref. 4)

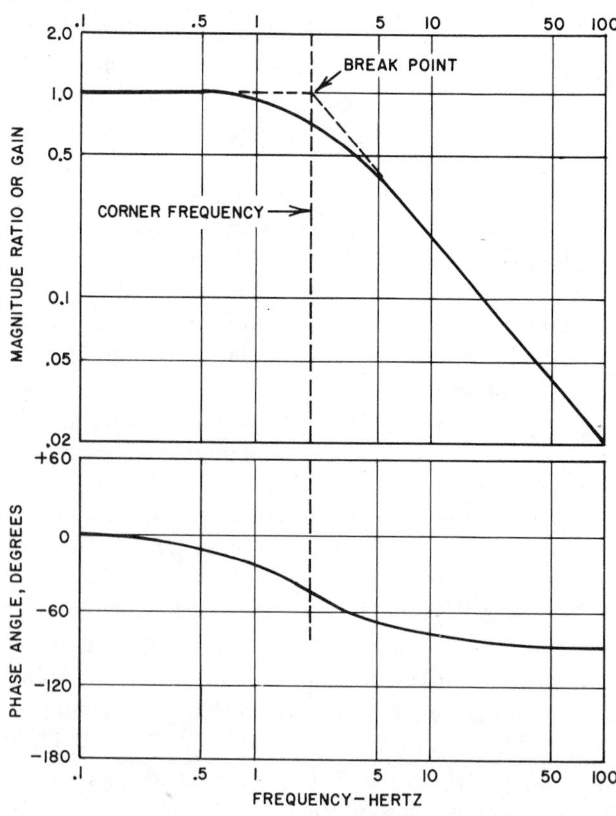

Fig. 1.1b Typical bode diagram

break point—The junction of the extension of two confluent straight-line segments of a plotted curve. See Figure 1.1b. (4, 8)

calibrate—To ascertain outputs of a *device* corresponding to a series of values of a quantity which the *device* is to measure, receive, or transmit. Data so obtained are used to:

1. determine the locations at which scale graduations are to be placed;
2. adjust the output, to bring it to the desired value, within a specified tolerance;
3. ascertain the *error* by comparing the *device* output reading against a standard. (Ref. 3)

calibration curve—A graphical representation of the *calibration report*. (Ref. 11)

For example, see Figure 1.1ee.

calibration cycle—The application of known values of the *measured variable* and the recording of corresponding values of *output* readings, over the *range* of the instrument, in ascending and descending directions. (Ref. 11)

calibration report—A table or graph of the measured relationship of an instrument as compared over its *range* against a standard. (Ref. 8)

For example, see Table 1.1ff.

calibration traceability—The relationship of the calibration of an instrument through a step-by-step process to an instrument or group of instruments calibrated and certified by a national standardizing laboratory. (Ref. 11)

cascade control—see *control, cascade*.

characteristic curve—A graph (curve) which shows the ideal values at *steady-state*, or an output variable of a system as a function of an input variable, the other input variables being maintained at specified constant values.

Note: When the other input variables are treated as *parameters*, a set of characteristic curves is obtained. (Ref. 17)

closed loop—see *loop, closed*.

closed loop gain—see *gain, closed loop*.

coefficient, temperature/pressure etc.—see *operating influence*.

cold junction—see *reference junction*.

common mode interference—see *interference, common mode*.

common mode rejection—The ability of a circuit to discriminate against a *common mode voltage*.

Note: It may be expressed as a dimensionless ratio, a scalar ratio, or in decibels as 20 times the $\log_{10}$ of that ratio.

common mode voltage—see *voltage, common mode*.

compensation—Provision of a special construction, a supplemental *device*, circuit, or special materials to counteract sources of *error* due to variations in specified *operating conditions*. (Ref. 11)

compensator—A *device* which converts a *signal* into some function which, either alone or in combination with other *signals*, directs the *final controlling element* to reduce *deviations* in the directly controlled variable.

See Figures 1.1j and 1.1k *for application of "set point compensator" and "load compensator."*

compliance—The reciprocal of *stiffness*.

computing instrument—see *instrument, computing*.

conformity—Of a curve, the closeness to which it approximates a specified curve (e.g., logarithmic, parabolic, cubic, etc.).

Note 1: It is usually measured in terms of nonconformity and expressed as conformity; e.g., the maximum deviation between an average curve and a specified curve. The average curve is determined after making two or more full *range* traverses in each direction. The value of conformity is referred to the output unless otherwise stated.

Note 2: As a performance specification, conformity should be expressed as *independent conformity, terminal-based conformity*, or *zero-based conformity*. When expressed simply as conformity, it is assumed to be *independent conformity*. (8, Ref. 4)

See *linearity*.

conformity, independent—The maximum *deviation of the calibration curve* (average of upscale and downscale readings) from a specified *characteristic* curve so positioned as to minimize the maximum *deviation*. See Figure 1.1c. (8)

conformity, terminal-based—The maximum *deviation* of the *calibration curve* (average of upscale and downscale readings) from a specified *characteristic curve* so positioned as to coincide with the actual *characteristic curve* at *upper* and *lower range-values*. See Figure 1.1d. (8)

conformity, zero-based—The maximum *deviation* of the *calibration curve* (average of upscale and downscale readings) from a specified *characteristic curve* so positioned as to coincide with the actual *characteristic curve* at the *lower range-value*. See Figure 1.1e. (Ref. 8)

control action—Of a *controller* or of a *controlling system*, the nature of the change of the output effected by the input.

Note: The output may be a *signal* or a value of a *manipulated variable*. The input may be the control loop *feedback signal* when the set point is constant, an *actuating error signal*, or the output of another *controller*. (Ref. 4, Ref. 8)

control action, derivative (rate) (D)—*Control action* in which the output is proportional to the rate of change of the input. (8, Ref. 4)

control action, floating—*Control action* in which the rate of change of the output variable is a predetermined function of the input variable.

Note: The rate of change may have one absolute value, several absolute values, or any value between two predetermined values.

(Ref. 17 "floating action")

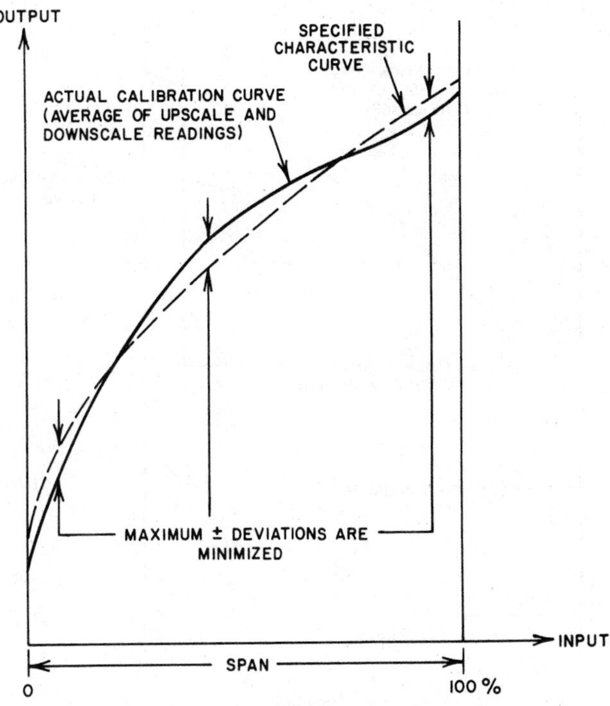

Fig. 1.1c Independent conformity

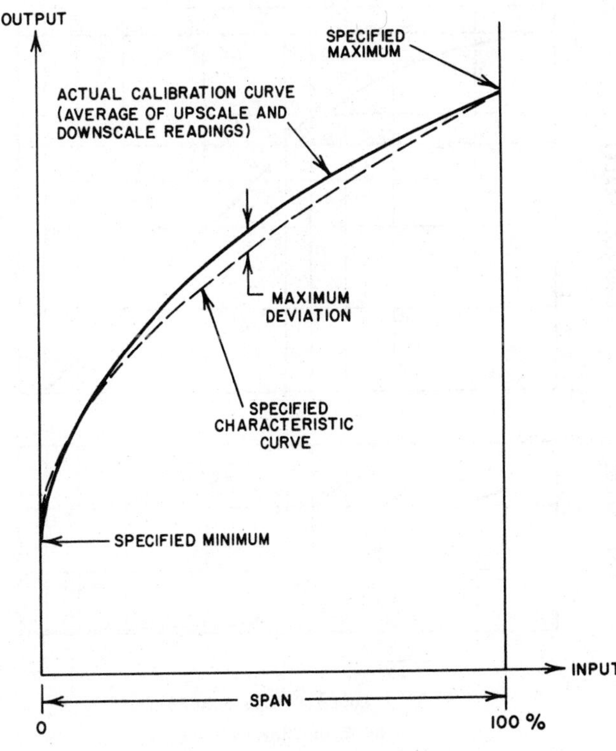

Fig. 1.1d Terminal-based conformity

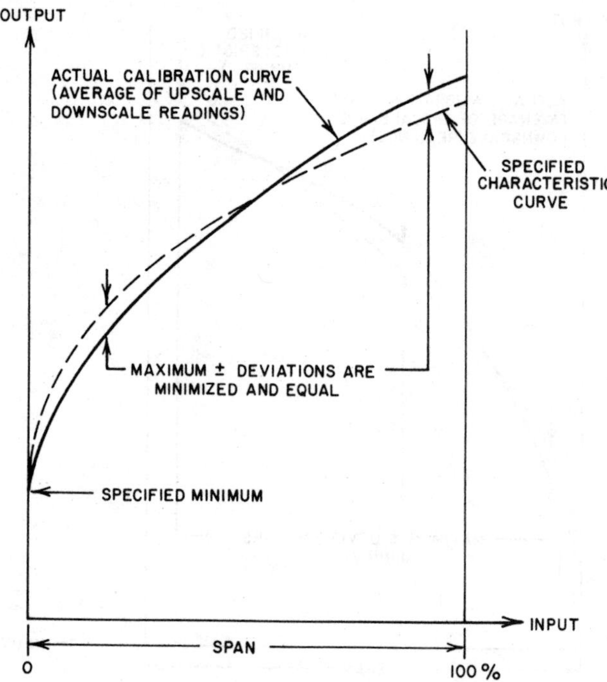

Fig. 1.1e Zero-based conformity

control action, integral (reset) (I)—*Control action* in which the output is proportional to the time integral of the input; i.e., the rate of change of output is proportional to the input. See Figure 1.1f.

Note: In the practical embodiment of integral control action, the relation between output and input, neglecting high frequency terms is given by

$$\frac{Y}{X} = \pm \frac{I/s}{bI/s + 1}, \text{ where } 0 \leqq b \ll 1$$

and

b = reciprocal of *static gain*
$^{I}/_{2\pi}$ = *gain crossover frequency* in hertz
s = complex variable
X = input transform
Y = output transform
I = *integral action rate* (4, 8)

See note under *control action.*

control action, proportional (P)—*Control action* in which there is a continuous linear relation between the output and the input.

Note: This condition applies when both the output and input are within their normal operating *ranges* and when operation is at a frequency below a limiting value. (4, 8)

See note under *control action.*

control action, proportional plus derivative (rate) (PD)—*Control action* in which the output is proportional to a linear combination of the input and the time rate-of-change of input. See Figure 1.1g.

Note: In the practical embodiment of proportional plus derivative control action, the relationship between output and input, neglecting high frequency terms is

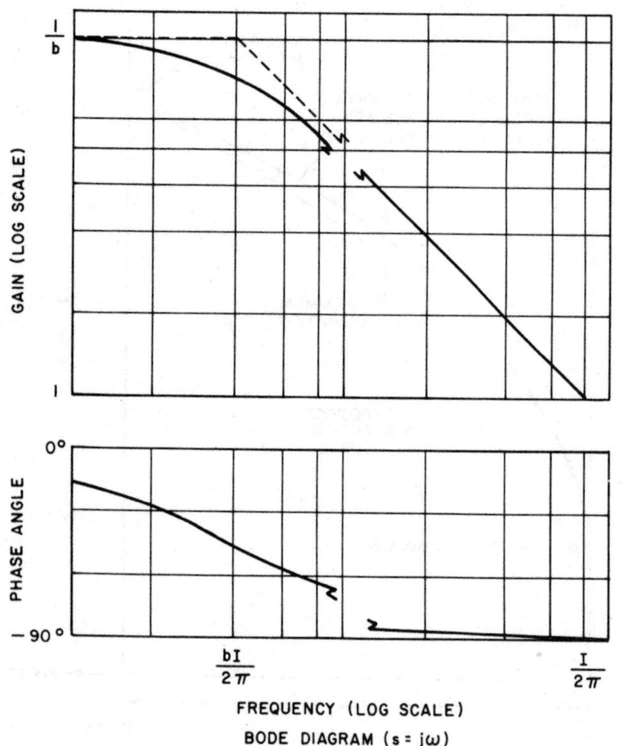

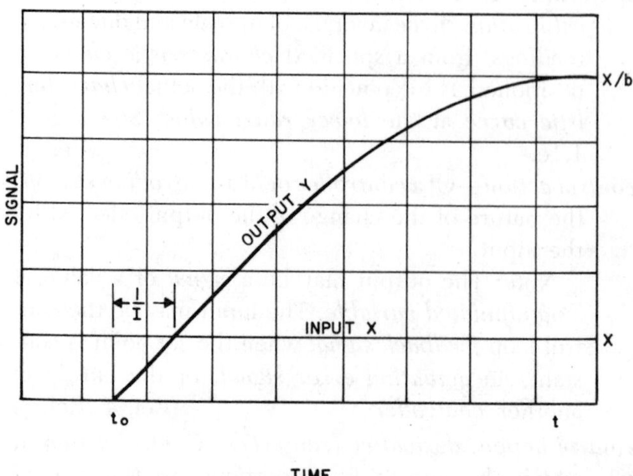

Fig. 1.1f Integral control action

$$\frac{Y}{X} = \pm P \frac{1 + sD}{1 + sD/a}, \text{ where } a > 1$$

and

a = *derivative action gain*
D = *derivative action time constant*
P = *proportional gain*
s = complex variable
X = input transform
Y = output transform (4, 8)

See note under *control action.*

control action, proportional plus integral (reset) (PI)—Control action in which the output is proportional to a linear combination of the input and the time integral of the input. See Figure 1.1h.

Note: In the practical embodiment of proportional plus integral control action, the relationship between output and input, neglecting high frequency terms, is

$$\frac{Y}{X} = \pm P \frac{I/s + 1}{bI/s + 1}, \text{ where } 0 \leqq b \ll 1$$

and

b = *proportional gain/static gain*
I = *integral action rate*
P = *proportional gain*
s = complex variable
X = input transorm
Y = output transform (4, 8)

See note under *control action.*

control action, proportional plus integral (reset) plus derivative (rate) (PID)—Control action in which the output is proportional to a linear combination of the input, the time integral of input and the time rate-of-change of input. See Figure 1.1i.

Note: In the practical embodiment of proportional plus integral plus derivative control action, the relationship of output to input, neglecting high frequency terms, is

$$\frac{Y}{X} = \pm P \frac{I/s + 1 + Ds}{bI/s + 1 + Ds/a},$$
$$\text{where } a > 1, 0 \leqq b \ll 1$$

and

a = *derivative action gain*
b = *proportional gain/static gain*
D = *derivative action time constant*
I = *integral action rate*
P = *proportional gain*
s = complex variable
X = input transform
Y = output transform (4, 8)

See note under *control action*.

control, adaptive—Control in which automatic means are used to change the type or influence (or both) of control *parameters* in such a way as to improve the performance of the *control system* (8, Ref. 4 "control system, adaptive")

control, cascade—Control in which the output of one *controller* is the *set point* for another *controller*. (Ref. 8 "control action, cascade")

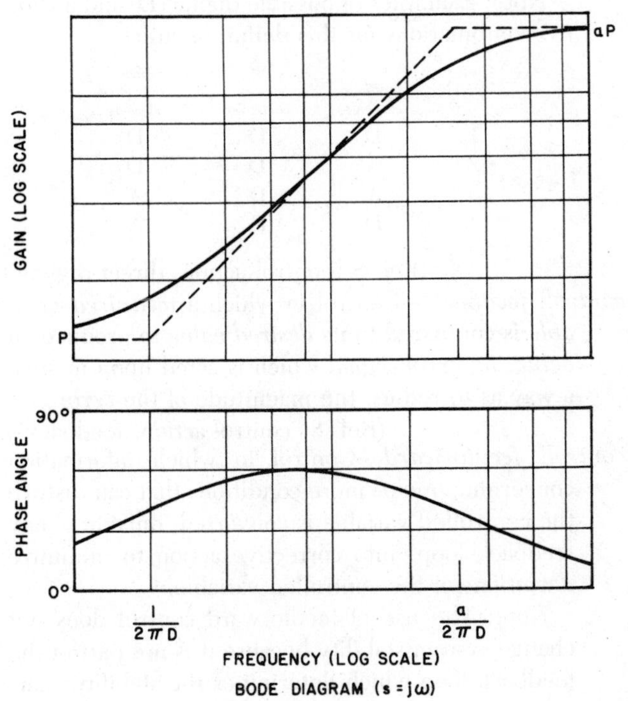

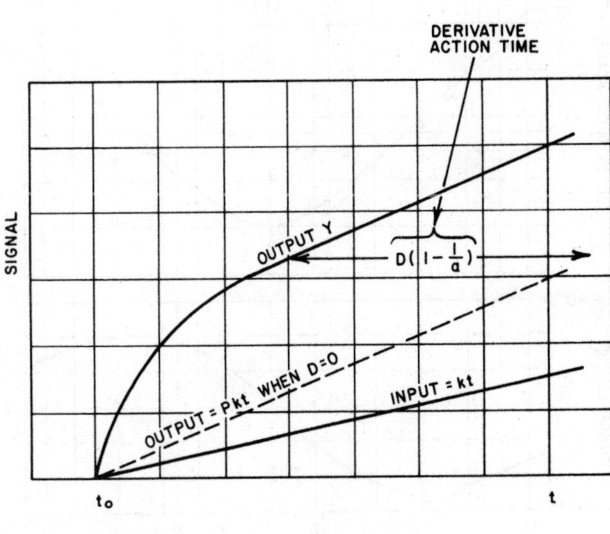

Fig. 1.1g Proportional plus derivative control action

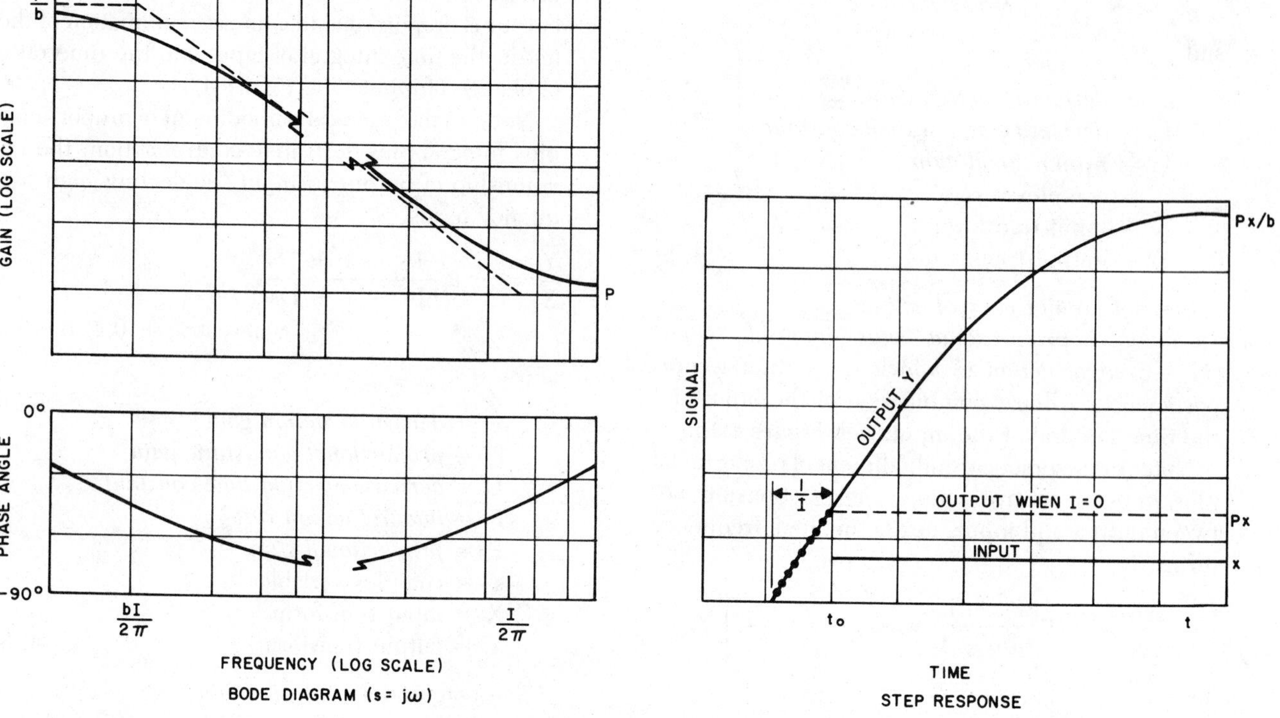

Fig. 1.1h Proportional plus integral control action

control center—An equipment structure, or group of structures, from which a *process* is measured, controlled and/or monitored. (Ref. 12)

control, differential gap—Control in which the output of a *controller* remains at a maximum or minimum value until the controlled variable crosses a band or gap, causing the output to reverse. The controlled variable must then cross the gap in the opposite direction before the output is restored to its original condition.

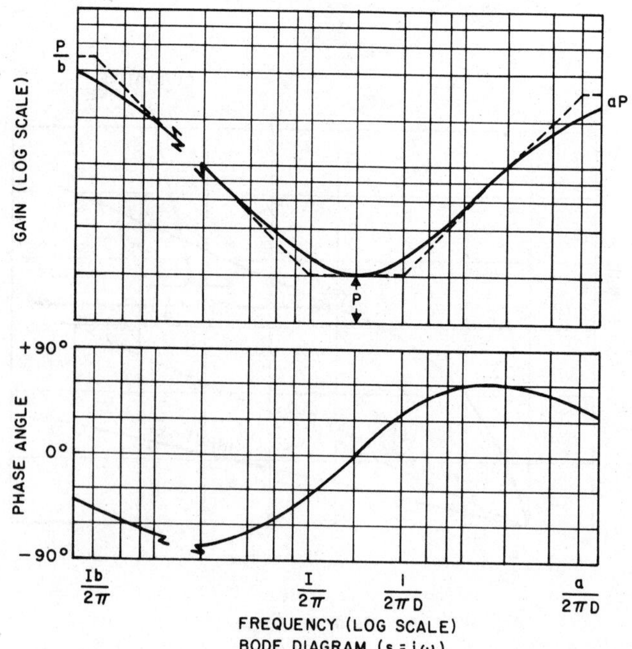

Fig. 1.1i Proportional plus integral plus derivative control action

control, direct digital—Control performed by a digital *device* which establishes the *signal* to the *final controlling element.*

Note: Examples of possible digital (D) and analog (A) combinations for this definition are:

	feedback elements	*controller*	*final controlling element*
1.	D	D	D
2.	A	D	D
3.	A	D	A
4.	D	D	A

(Ref. 8 "control action, direct digital")

control, feedback—Control in which a *measured variable* is compared to its *desired value* to produce an *actuating error signal* which is acted upon in such a way as to reduce the magnitude of the *error.*
(Ref. 8 "control action, feedback")

control, feedforward—Control in which information concerning one or more conditions that can disturb the controlled variable is converted, outside of any feedback loop, into corrective action to minimize *deviations* of the controlled variable.

Note: The use of feedforward control does not change system stability because it is not part of the feedback loop which determines the stability characteristics. See Figures 1.1j and 1.1k.
(Ref. 8 "control action, feedforward")

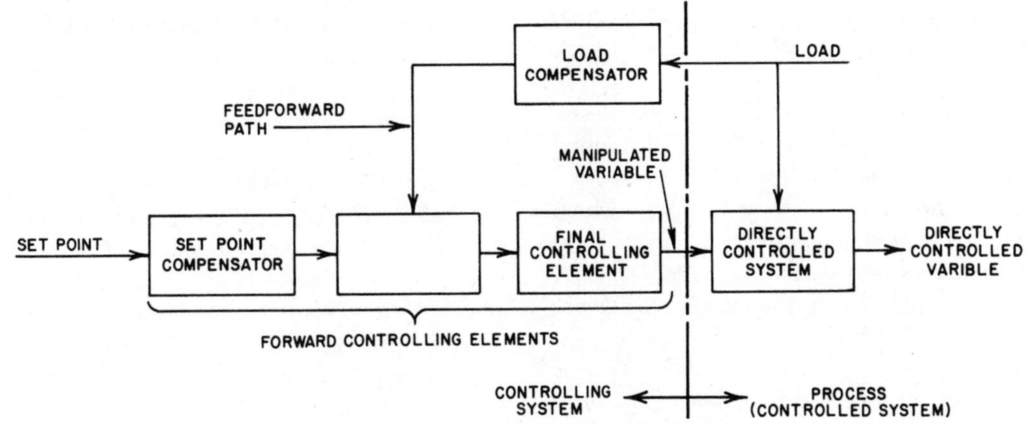

Fig. 1.1j Feedforward control without feedback

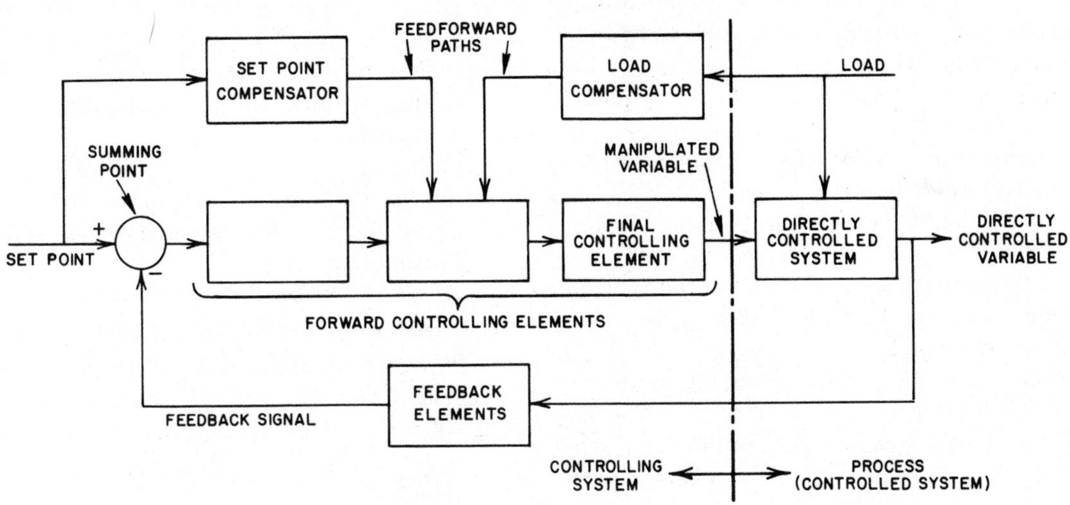

Fig. 1.1k Feedforward control with feedback

control, high limiting—Control in which the output *signal* is prevented from exceeding a predetermined high limiting value.

(Ref. 8 "control action, high limiting")

controlled system—see *system, controlled*.

controller—A *device* which operates automatically to regulate a controlled variable.

Note: This term is adequate for the *process* industries where the word "controller" always means "automatic controller." In some industries, "automatic" may not be implied and the term "automatic controller" is preferred.

(8, Ref. 4 "automatic controller")

controller, derivative (D)—A *controller* which produces *derivative control action* only.

controller, direct acting—A *controller* in which the value of the *output signal* increases as the value of the input (*measured variable*) increases. See *controller, reverse acting*. (Ref. 8)

controller, floating—A *controller* in which the rate of change of the output is a continuous (or at least a piecewise continuous) function of the *actuating error signal*.

Note: The output of the *controller* may remain at any value in its operating *range* when the *actuating error signal* is zero and constant. Hence the output is said to float. When the *controller* has *integral control action* only, the mode of control has been called "proportional speed floating." The use of the term *integral control action* is recommended as a replacement for "proportional speed floating control." (8)

controller, integral (reset) (I)—A controller which produces *integral control action* only.

Note: It may also be referred to as controller, proportional speed floating. (8)

controller, multiple-speed floating—A *floating controller* in which the output may change at two or more rates, each corresponding to a definite range of values of the *actuating error signal*.

(8, Ref. 4 "control system, multiple-speed floating")

controller, multi-position—A *controller* having two or

more discrete values of output. See Figure 1.1l.(8)

controller, on-off—A *two-position controller* of which one of the two discrete values is zero. See Figures 1.1n and 1.1o. (Ref. 8)

controller, program—A *controller* which automatically holds or changes *set point* to follow a prescribed program for a *process*.

controller, proportional (P)—A *controller* which produces *proportional control action* only. (8)

controller, proportional plus derivative (rate) (PD)—A *controller* which produces *proportional plus derivative (rate) control action*. (8)

controller, proportional plus integral (reset) (PI)—A *controller* which produces *proportional plus integral (reset) control action*. (8)

controller, proportional plus integral (reset) plus derivative (rate) (PID)—A *controller* which produces *proportional plus integral (reset) plus derivative (rate) control action*. (8)

controller, proportional speed floating—see *controller, integral (reset) (I)*. (8)

controller, ratio—A *controller* which maintains a predetermined ratio between two variables. (Ref. 4 "control system, ratio," Ref. 8)

controller, reverse acting—A *controller* in which the value of the *output signal* decreases as the value of the input *(measured variable)* increases. See *controller, direct acting*. (Ref. 8)

controller, sampling—A *controller* using intermittently observed values of a *signal* such as the *set point* signal, the *actuating error signal*, or the *signal* representing the controlled variable to effect *control action*. (8, Ref. 4 "control system, sampling")

controller, self-operated (regulator)—A *controller* in which all the energy to operate the *final controlling element* is derived from the *controlled system*. (Ref. 4, Ref. 8)

Fig. 1.1l Multi-position controller

controller, single-speed floating—A *floating controller* in which the output changes at a fixed rate increasing or decreasing depending on the sign of the *actuating error signal*. See *controller, floating*.

Note: A neutral zone of values of the *actuating error signal* in which no action occurs may be used. (Ref. 4 "control system, single speed floating," Ref. 8)

controller, three-position—A *multi-position controller* having three discrete values of output. See Figure 1.1m.

Note: This is commonly achieved by selectively energizing a multiplicity of circuits (outputs) to establish three discrete positions of the *final controlling element*. (Ref. 8)

controller, time schedule—A *controller* in which the *set point* or the *reference-input signal* automatically adheres to a predetermined time schedule. (8, Ref. 4)

controller, two-position—A *multi-position controller* having two discrete values of output. See Figures 1.1n and 1.1o. (8)

controlling system—see *system, controlling*.

control, low limiting—Control in which output *signal* is prevented from decreasing beyond a predetermined low limiting value. (Ref. 8 "control action, low limiting")

control mode—A specific type of *control action* such as *proportional*, *integral*, or *derivative*. (8)

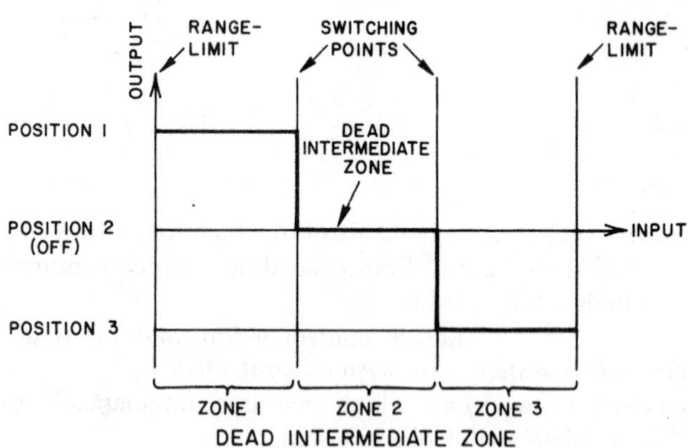

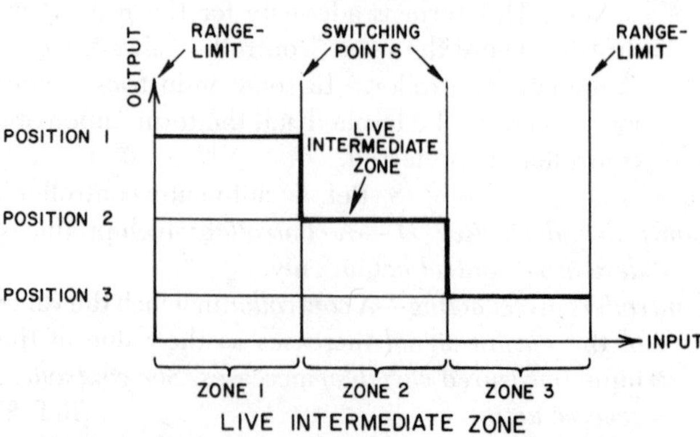

Fig. 1.1m Three-position controller

control, optimizing—Control that automatically seeks and maintains the most advantageous value of a specified variable, rather than maintaining it at one set value. (Ref. 4 "control action, optimizing")

control room area—see *area, control room.*

control, shared time—Control in which one *controller* divides its computation or control time among several control loops rather than by acting on all loops simultaneously. (Ref. 4 "control action, shared time")

control, supervisory—Control in which the control loops operate independently subject to intermittent corrective action; e.g., *set point* changes from an external source. (Ref. 4 "control action supervisory")

control system—a system in which deliberate guidance or manipulation is used to achieve a prescribed value of a variable. See Figure 1.1p.

Note: It is subdivided into a *controlling system* and a *controlled system.* (4, 8)

control system, automatic—A *control system* which operates without human intervention. (4) See also *control system.*

control system, multi-element (multi-variable).—A *control system* utilizing *input signals* derived from two or more *process* variables for the purpose of jointly affecting the action of the *control system.*

Note: Examples are *input signals* representing pressure and temperature, or speed and flow, etc. (Ref. 8)

control system, non-interacting—A *multi-element control system* designed to avoid disturbances to other controlled variables due to the *process* input adjustments which are made for the purpose of controlling a particular *process* variable. (8)

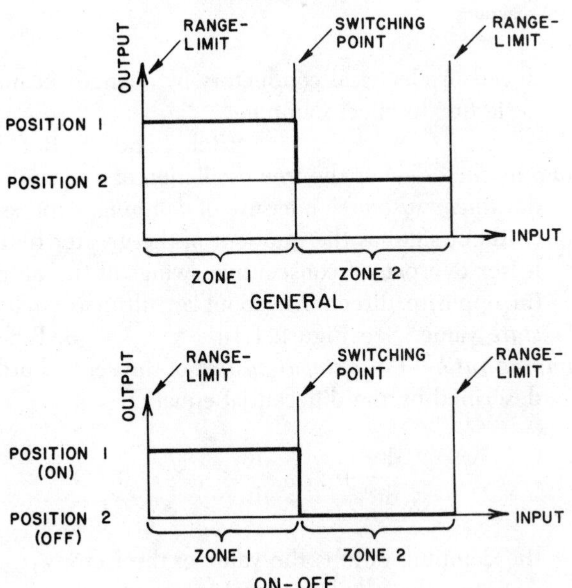

Fig. 1.1n Two-position controller

control, time proportioning—*Control* in which the output *signal* consists of periodic pulses whose duration is varied to relate, in some prescribed manner, the time average of the output to the *actuating error signal.* (Ref. 4 "controller, time proportioning")

control valve—A *final controlling element,* through which a fluid passes, which adjusts the size of flow passage as directed by a *signal* from a *controller* to modify the rate of flow of the fluid. (Ref. 17 "valve")

control, velocity limiting—Control in which the rate of change of a specified variable is prevented from exceeding a predetermined limit. (Ref. 8 "control action, velocity limiting")

corner frequency—In the asymptotic form of *Bode diagram,* that frequency indicated by a *break point,* i.e. the junction of two confluent straight lines asymptotic to the log *gain* curve. (4)

correction—The algebraic difference between the *ideal value* and the indication of the *measured signal.* It is the quantity which added algebraically to the indication gives the *ideal value.*

Note: A positive correction denotes that the indication of the instrument is less than the *ideal value.*

correction = *Ideal value* − indication.

See *error.* (Ref. 4, Ref. 8)

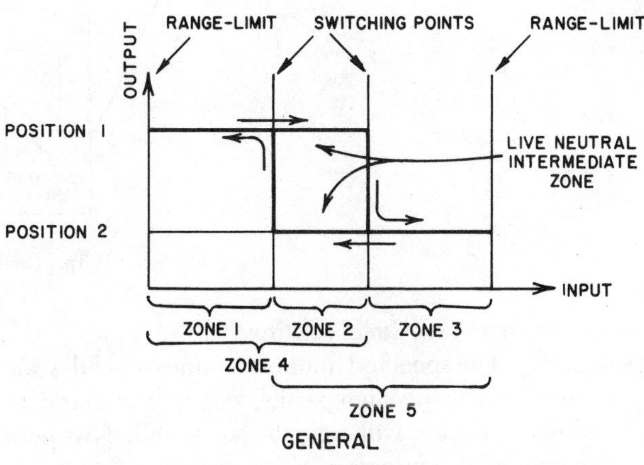

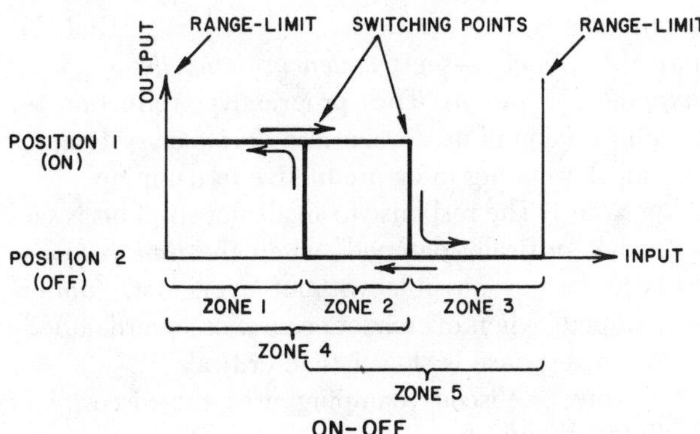

Fig. 1.1o Two-position controller with neutral intermediate zone

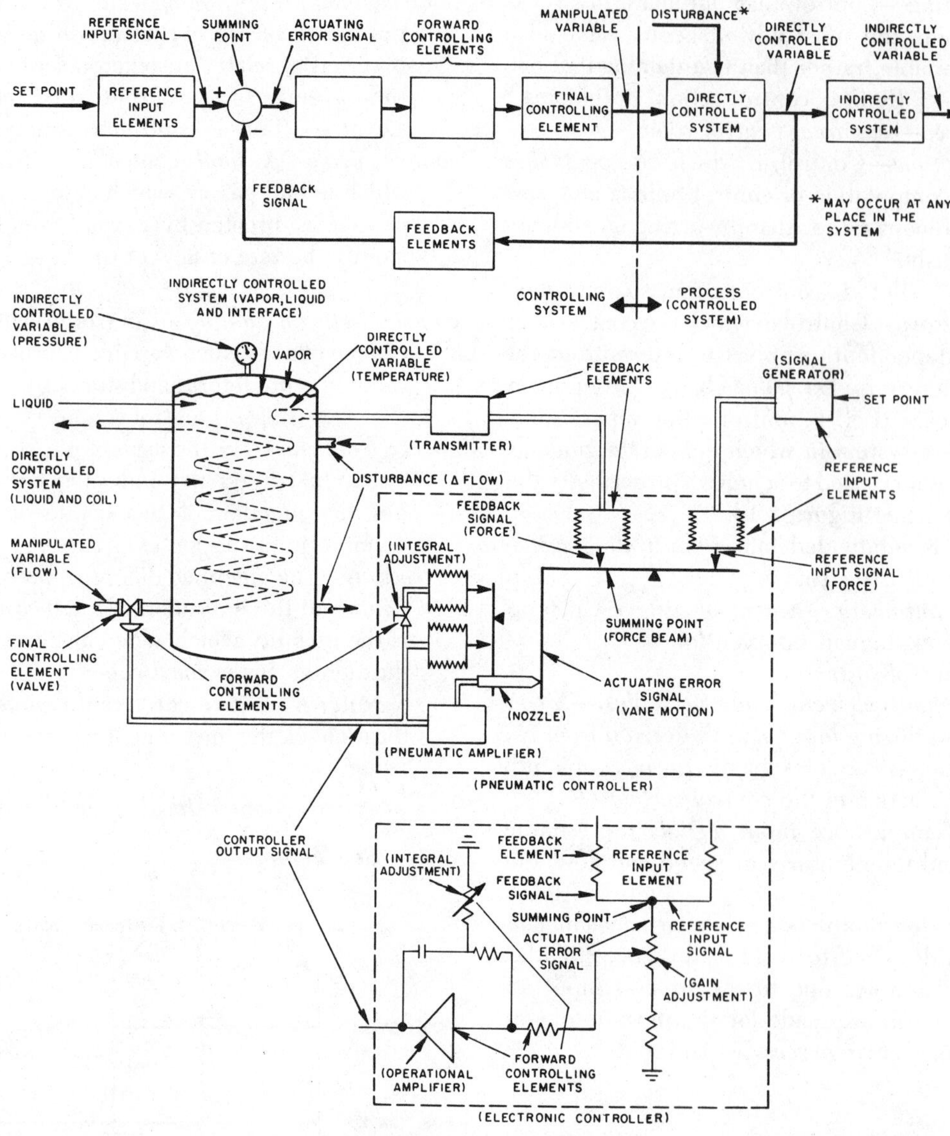

Fig. 1.1p Control system diagrams

correction time—see *time, settling.*

cycling life—The specified minimum number of full scale excursions or specified partial *range* excursions over which a *device* will operate as specified without changing its performance beyond specified tolerances. (Ref. 11)

damped frequency—see *frequency, damped.*

damping—(1) (noun) The progressive reduction or suppression of oscillation in a *device* or system. (2) (adj) Pertaining to or productive of damping.

Note 1: The response to an abrupt stimulus is said to be "critically damped" when the *time response* is as fast as possible without overshoot; "underdamped" when overshoot occurs, or "overdamped" when response is slower than critical.

Note 2: Viscous damping uses the viscosity of fluids (liquids or gases) to effect damping.

Note 3: Magnetic damping uses the current induced in electrical conductors by changes in magnetic flux to effect damping.

(Ref. 4, Ref. 8, Ref. 11)

damping factor—For the free oscillation of a second order *linear system*, a measure of *damping*, expressed (without sign) as the quotient of the greater by the lesser of a pair of consecutive swings of the output (in opposite directions) about an ultimate *steady-state* value. See Figure 1.1q. (8, Ref. 4)

damping ratio—For a *linear system* of the second order described by the differential equation

$$\frac{d^2x}{dt^2} + 2\zeta\omega_o \frac{dx}{dt} + \omega^2_o = 0$$

the damping ratio is the value of the factor ζ.

Note: ω_o is called the angular *resonance* frequency of the system. (4)

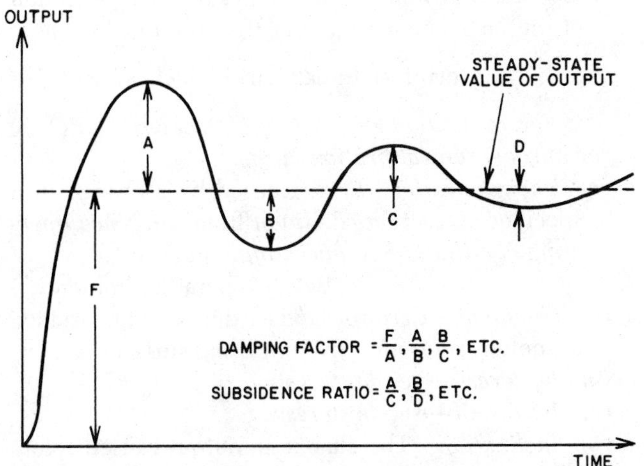

Fig. 1.1q Underdamped response of system with second-order lag

damping, relative—For an underdamped system, a number expressing the quotient of the actual *damping* of a second-order *linear system* or *element* by its critical *damping*.

Note: For any system whose transfer function includes a quadratic factor $s^2 + 2\zeta\omega_n s + \omega^2_n$, relative damping is the value of ζ, since $\zeta = 1$ for critical damping. Such a factor has a root $-\sigma + j\omega$ in the complex s-plane, from which $\zeta = \sigma/\omega_n = \sigma/(0^2 + \omega^2)$. (4)

d controller—see *controller, derivative (D)*.

dead band—The *range* through which an input can be varied without initiating observable response. See Figure 1.1r.

Note: Dead band is usually expressed in percent of *span*. (Ref. 4, Ref. 8)

See *zone, dead*.

See *test procedure*.

dead time—see *time, dead*.

dead zone—see *zone, dead*.

delay—The interval of time between a changing *signal* and its repetition for some specified duration at a downstream point of the *signal* path; the value L in the transform factor exp (− Ls).

See *time, dead*. (4)

derivative action gain—see *gain, derivative action (rate gain)*.

derivative action time—see *time, derivative action*.

derivative action time constant—see *time constant, derivative action*.

derivative control—see *control action, derivative (D)*.

derivative control action—see *control action, derivative*.

derivative controller—see *controller, derivative*.

design pressure—see *pressure, design*.

desired value—see *value, desired*.

detector—see *transducer*. (11)

deviation—Any departure from a *desired value* or expected value or pattern. (4, 8)

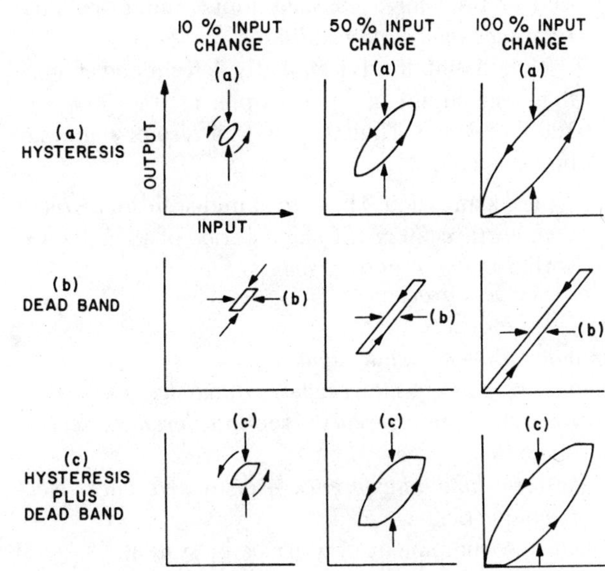

Fig. 1.1r Hysteresis and dead band

deviation, steady-state—The *system deviation* after *transients* have expired. (4, 8)

See also *offset*.

deviation, system—The instantaneous value of the *directly controlled variable* minus the *set point*. (8, Ref. 4)

See also *signal, actuating error*.

deviation, transient—The instantaneous value of the *directly controlled variable* minus its *steady-state* value. (Ref. 4)

device—An apparatus for performing a prescribed function. (8)

differential gap control—see *control, differential gap*.

differential mode interference—see *interference, normal mode*. (2, 8)

digital signal—see *signal, digital*.

direct acting controller—see *controller, direct acting*.

direct digital control—see *control, direct digital*.

directly controlled system—see *system, directly controlled*.

directly controlled variable—see *variable, directly controlled*.

distance/velocity lag—A delay attributable to the transport of material or to the finite rate of propagation of a *signal*. (Ref. 4, Ref. 17)

disturbance—An undesired change that takes place in a *process* which tends to affect adversely the value of a controlled variable. (8, Ref. 4)

dither—A useful oscillation of small magnitude, introduced to overcome the effect of friction, *hysteresis*, or recorder pen clogging. See Also *hunting*. (Ref. 4)

drift—An undesired change in the output-input relationship over a period of time. (Ref. 8)

drift, point—The change in output over a specified period of time for a constant input under specified *reference operating conditions.*

Note: Point drift is frequently determined at more than one input, as for example: at 0%, 50% and 100% of *range.* Thus, any *drift* of zero or *span* may be calculated.

typical expression: The *drift* at mid-scale for *ambient temperature* (70 ± 2°F) for a period of 48 hours was within 0.1% of output *span.* (8)

See *test procedure.*

droop—see *offset.*

dynamic gain—see *gain, dynamic.*

dynamic response—see *response, dynamic.*

electromagnetic interference—see *interference, electromagnetic.*

electrostatic field interference—see *interference, electromagnetic.*

element—A component of a *device* or system. (8)

element, final controlling—The *forward controlling element* which directly changes the value of the *manipulated variable.*
(8, Ref. 4 "controlling element, final")

element, primary—The system *element* that quantitatively converts the *measured variable* energy into a form suitable for measurement.

Note: For *transmitters* not used with external primary elements, the sensing portion is the primary element.
(Ref. 2 "detecting means," Ref. 8)

element, reference-input—The portion of the *controlling system* which changes the *reference-input signal* in response to the *set point.* See Figure 1.1p.
(8, Ref. 4)

element, sensing—The *element* directly responsive to the value of the *measured variable.*

Note: It may include the case protecting the sensitive portion. (Ref. 8)

elements, feedback—Those *elements* in the *controlling system* which act to change the *feedback signal* in response to the *directly controlled variable.* See Figure 1.1p. (Ref. 4, Ref. 8)

elements, forward controlling—Those *elements* in the *controlling system* which act to change a variable in response to the *actuating error signal.* See Figure 1.1p. (Ref. 4, Ref. 8)

elevated range—see *range, suppressed-zero.*
(4, Ref. 1 "Range, suppressed-zero")

elevated span—see *range, suppressed-zero.*

elevated-zero range—see *range, elevated-zero.*

elevation—see *range, suppressed-zero.*

environmental area—see *area, environmental.*

environmental influence—see *operating influence.*

error—The algebraic difference between the indication and the *ideal value* of the *measured signal.* It is the quantity which algebraically subtracted from the indication gives the *ideal value.*

Note: A positive error denotes that the indication of the instrument is greater than the *ideal value.*

error = Indication − *ideal value.*

See *correction.* (Ref. 2, Ref. 8)

error curve—see *calibration curve.*

error, environmental—*Error* caused by a change in a specified *operating condition* from *reference operating condition.* See *operating influence.*
(Ref. 8 "operating influence")

error, frictional—*Error* of a *device* due to the resistance to motion presented by contacting surfaces.

error, hysteresis—see *hysteresis.*

error, hysteretic—see *hysteresis.*

error, inclination—The change in output caused solely by an inclination of the *device* from its normal operating position. (Ref. 1 "influence, position")

error, mounting strain—*Error* resulting from mechanical deformation of an instrument caused by mounting the instrument and making all connections.

See also *error, inclination.*

error, position—The change in output resulting from mounting or setting an instrument in a position different from that at which it was *calibrated.*

See also *error, inclination.*

error signal—see *signal, error.*

error, span—The difference between the actual *span* and the ideal *span.*

Note: It is usually expressed as a percent of ideal *span.* (8)

error, systematic—An *error* which, in the course of a number of measurements made under the same conditions of the same value of a given quantity, either remains constant in absolute value and sign or varies according to a definite law when the conditions change.

error, zero—*Error* of a *device* operating under specified conditons of use, when the input is at the *lower range-value.* (Ref. 8)

Note: It is usually expressed as percent of ideal *span.*

excitation—The external supply applied to a *device* for its proper operation.

Note: It is usually expressed as a range of supply values.

See also *excitation, maximum.* (Ref. 11)

excitation, maximum—The maximum value of excitation *parameter* that can be applied to a *device* at rated *operating conditions* without causing damage or performance degradation beyond specified tolerances. (Ref. 11)

feedback control—see *control, feedback.*

feedback elements—see *elements, feedback.*

feedback loop—see *loop, closed (feedback loop)*

feedback signal—see *signal, feedback.*

feedforward control—see *control, feedforward.*

final controlling element—see *element, final controlling*.
floating control action—see *control action, floating*.
floating controller—see *controller, floating*.
flowmeter—A *device* that measures the rate of flow or quantity of a moving fluid in an open or closed conduit. It usually consists of both a primary and a secondary device.

Note: It is acceptable in practice to further identify the flowmeter by its applied theory; as differential pressure, velocity, area, force, etc. or by its applied technology as orifice, turbine, vortex, ultrasonic, etc. Examples include turbine flowmeter, magnetic flowmeter, fluidic pressure flowmeter, etc.

flowmeter primary device—The *device* mounted internally or externally to the fluid conduit which produces a signal with a defined relationship to the fluid flow in accordance with known physical laws relating the interaction of the fluid to the presence of the primary *device*.

Note: The primary device may consist of one or more *elements* necessary to produce the primary device *signal*.

flowmeter secondary device—The *device* that responds to the *signal* from the primary *device* and converts it to a display or to an output signal that can be translated relative to flow rate or quantity.

Note: The secondary *device* may consist of one or more *elements* as needed to translate the primary device *signal* into standardized or nonstandardized display or transmitted units.

forward controlling elements—see *elements, forward controlling*.
frequency, damped—The apparent frequency of a damped oscillatory *time response* of a system resulting from a non-oscillatory stimulus. (4)
frequency, gain crossover—(1) On a *Bode diagram* of the *transfer function* of an *element* or system, the frequency at which the *gain* becomes unity and its decibel value zero. (2) Of *integral control action*, the frequency at which the *gain* becomes unity.

See Figure 1.1f. (4)

frequency, phase crossover—Of a *loop transfer function*, the frequency at which the phase angle reaches ± 180°. (Ref. 4)
frequency response characteristic—The frequency-dependent relation, in both amplitude and phase, between *steady-state* sinusoidal inputs and the resulting fundamental sinusoidal outputs.

Note: Frequency response is commonly plotted on a *Bode diagram*. See Figure 1.1b.

(8, Ref. 4 "frequency-response characteristics")

frequency, undamped (frequency, natural)—1. Of a second-order *linear system* without *damping*, the frequency of free oscillation in radians or cycles per unit of time.

2. Of any system whose *transfer function* contains the quadratic factor $s^2 + 2z\omega_n s + \omega_n^2$ the value ω_n.

where:

s = complex variable
z = constant
ω_n = natural frequency in radians per second

3. Of a *closed-loop control system* or *controlled system*, a frequency at which continuous oscillation *(hunting)* can occur without periodic stimuli.

Note: In *linear systems*, the undamped frequency is the phase crossover frequency. With *proportional control action* only, the undamped frequency of a *linear system* may be obtained in most cases by raising the *proportional* gain until continuous oscillation occurs. (Ref. 4, Ref. 8)

frictional error—see *error, frictional*.
gain, closed loop—The *gain* of a *closed loop* system, expressed as the ratio of the output change to the input change at a specified frequency. (8, Ref. 4)
gain, crossover frequency—see *frequency, gain crossover*.
gain, derivative action (rate gain)—The ratio of maximum *gain* resulting from *proportional plus derivative control action* to the *gain* due to *proportional control action* alone. See Figures 1.1g and 1.1i. (4, 8)
gain, dynamic—the magnitude ratio of the *steady-state* amplitude of the *output signal* from an *element* or system to the amplitude of the *input signal* to that *element* or system, for a sinusoidal *signal*. (8)
gain, loop—The ratio of the change in the *return signal* to the change in its corresponding *error signal* at a specified frequency.

Note: The *gain* of the loop *elements* is frequently measured by opening the loop, with appropriate termination. The *gain* so measured is often called the open loop gain. (8, Ref. 5)

gain (magnitude ratio)—For a *linear system* or *element*, the ratio of the magnitude (amplitude) of a *steady-state* sinusoidal output relative to the causal input; the length of a phasor from the origin to a point of the transfer locus in a complex plane.

Note: the quantity may be separated into two factors: (1) a proportional amplification often denoted as K which is frequency-independent, and associated with a dimensioned scale factor relating to the units of input and output; (2) a dimensionless factor often denoted as G (jω) which is frequency-dependent. Frequency, conditions of operation, and conditions of measurement must be specified. A *loop gain characteristic* is a plot of log gain vs. log frequency. In nonlinear systems, gains are often amplitude-dependent. (4, 8)

gain, open loop—see *gain, loop.*

gain, proportional—The ratio of the change in output due to *proportional control action* to the change in input.

Illustration: Y = ± PX

where:

P = proportional gain
X = input transform
Y = output transform

See *proportional band.* (4, 8)

gain, static (zero-frequency gain)—Of *gain* of an *element*, or *loop gain* of a system, the value approached as a limit as frequency approaches zero.

Note: Its value is the ratio of change of *steady-state* output to a step change in input, provided the output does not saturate. (4, Ref. 8)

gain, zero frequency—see *gain, static (zero-frequency gain).*

hardware—Physical equipment directly involved in performing industrial *process* measuring and controlling functions.

hazardous (classified) location—see *location, hazardous (classified).*

high limiting control—see *control, high limiting.*

hunting—An undesirable oscillation of appreciable magnitude, prolonged after external stimuli disappear.

Note: In a *linear system*, hunting is evidence of operation at or near the stability limit; non-linearities may cause hunting of well-defined amplitude and frequency. See also *dither.* (4)

hysteresis—That property of an *element* evidenced by the dependence of the value of the output, for a given excursion of the input, upon the history of prior excursions and the direction of the current traverse.

Note 1: It is usually determined by subtracting the value of *dead band* from the maximum measured separation between upscale going and downscale going indications of the *measured variable* (during a full *range* traverse, unless otherwise specified) after *transients* have decayed. This measurement is sometimes called hysteresis error or hysteretic error. See Figure 1.1r.

Note 2: Some reversal of output may be expected for any small reversal of input; this distinguishes hysteresis from *dead band.*

See *test procedure.* (4)

I controller—see *controller, integral (reset) (I).*

idealized system—see *system, idealized.*

ideal value—see *value, ideal.*

impedance, input—Impedance presented by a *device* to the source. (3, 8)

impedance, load—Impedance presented to the output of a *device* by the load. (8, Ref. 3)

impedance, output—Impedance presented by a *device* to the load. (8, Ref. 3)

impedance, source—Impedance presented to the input of a *device* by the source. (8, Ref. 3)

inclination error—see *error, inclination.*

independent conformity—see *conformity, independent.*

independent linearity—see *linearity, independent.*

indicating instrument—see *instrument, indicating.*

indicator travel—The length of the path described by the indicating means or the tip of the pointer in moving from one end of the scale to the other.

Note 1: The path may be an arc or a straight line.

Note 2: In the case of knife-edge pointers and others extending beyond the scale division marks, the pointer shall be considered as ending at the outer end of the shortest scale division marks. (2, 8)

indirectly controlled system—see *system, indirectly controlled.*

indirectly controlled variable—see *variable, indirectly controlled.*

inherent regulation—see *self-regulation.*

input impedance—see *impedance, input.*

input signal—see *signal, input.*

instrumentation—A collection of instruments or their application for the purpose of observation, measurement or control.

instrument, computing—A *device* in which the output is related to the input or inputs by a mathematical function such as addition, averaging, division, integration, lead/lag, signal limiting, squaring, square root extraction, subtraction, etc.

instrument, indicating—A *measuring instrument* in which only the present value of the *measured variable* is visually indicated. (Ref. 8)

instrument, measuring—A *device* for ascertaining the magnitude of a quantity or condition presented to it. (Ref. 8)

instrument, recording—A *measuring instrument* in which the values of the *measured variable* are recorded.

Note: The record may be either analog or digital and may or may not be visually indicated. (Ref. 8)

insulation resistance—The resistance measured between specified insulated portions of a *device* when a specified direct currect voltage is applied, at *reference operating conditions* unless otherwise stated.

Note: The objective is to determine whether the leakage current would be excessive under *operating conditions.* (Ref. 11)

insulation voltage breakdown—The voltage at which a disruptive discharge takes place through or over the surface of the insulation. (3)

integral action limiter—A *device* which limits the value of the *output signal* due to *integral control action,* to a predetermined value. (8)

integral action rate (reset rate)—1. Of *proportional plus integral* or *proportional plus integral plus derivative control action devices;* for a step input, the ratio of the initial rate of change of output due to *integral control action* to the change in *steady-state* output due to *proportional control action.*

Note: Integral action rate is often expressed as the number of repeats per minute because it is equal to the number of times per minute that the proportional response to a step input is repeated by the initial integral response.

2. Of integral control action devices; for a step input, the ratio of the initial rate of change of output to the input change. (8, Ref. 4)

integral action time constant—see *time constant, integral action.*

integral control action—see *control action, integral.*

integral controller—see *controller, integral (reset).*

interference, common mode—A form of interference which appears between measuring circuit terminals and ground. (3, 8)

interference, differential mode—see *interference, normal mode.*

interference, electromagnetic—Any spurious effect produced in the circuits or *elements* of a *device* by external electromagnetic fields.

Note: A special case of interference from radio transmitters is known as "Radio Frequency Interference (RFI)."

interference, electrostatic field—see *interference, electromagnetic.*

interference, longitudinal—see *interference, common mode.*

interference, magnetic field—see *interference, electromagnetic.*

interference, normal mode—A form of interference which appears between measuring circuit terminals. (2, 8)

interference, transverse—see *interference, normal mode.*

intermediate zone—see *zone, intermediate.*

intrinsically safe equipment and wiring—Equipment and wiring which are incapable of releasing sufficient electrical or thermal energy under normal or abnormal conditions to cause ignition of a specific hazardous atmospheric mixture in its most easily ignited concentration. (14)

Laplace transform, unilateral—Of a function F(t), the quantity obtained by performing the operation

$$F(s) = \int_{\sigma}^{\infty} F(t)e^{-st}\,dt$$

where:

F(s) = function of s

s = complex variable, $\sigma + j\omega$

F(t) = function of t

t = time, seconds

σ = real part of the complex variable s

j = $\sqrt{-1}$

ω = angular velocity, radians per second. (8, Ref. 4)

leak pressure—see *pressure, leak.*

linear system—see *system, linear.*

linearity—The closeness to which a curve approximates a straight line.

Note 1: It is usually measured as a nonlinearity and expressed as linearity; e.g., a maximum *deviation* between an average curve and a straight line. The average curve is determined after making two or more full *range* traverses in each direction. The value of linearity is referred to the output unless otherwise stated.

Note 2: As a performance specification linearity should be expressed as *independent linearity, terminal-based linearity* or *zero-based linearity.* When expressed simply as linearity it is assumed to be *independent linearity.*

See *conformity.*

(8, Ref. 4 "Linearity of a signal.")

linearity, independent—The maximum *deviation* of the *calibration curve* (average of upscale and downscale readings) from a straight line so positioned as to minimize the maximum *deviation.* See Figure 1.1s.

See *test procedure.* (Ref. 8)

linearity, terminal-based—The maximum *deviation* of the *calibration curve* (average of upscale and downscale readings) from a straight line coinciding with the *calibration curve* at *upper* and *lower range-values.*

See Figure 1.1t.

See *test procedure.* (Ref. 8)

linearity, zero-based—The maximum *deviation* of the *calibration curve* (average of upscale and downscale readings) from a straight line so positioned as to coincide with the *calibration curve* at the *lower range-value* and to minimize the maximum *deviation.* See Figure 1.1u.

See *test procedure.* (Ref. 8)

linear system—see *system, linear.*

live zone—see *zone, live.*

load impedance—see *impedance, load.*

load regulation—The change in output (usually speed or voltage) from no-load to full-load (or other specified load limits). See *offset.*

Note: It may be expressed as the percentage ratio of the change from no-load to full load divided by the no-load value. (8)

location, hazardous (classified)—That portion of a plant where flammable or combustible liquids, vapors,

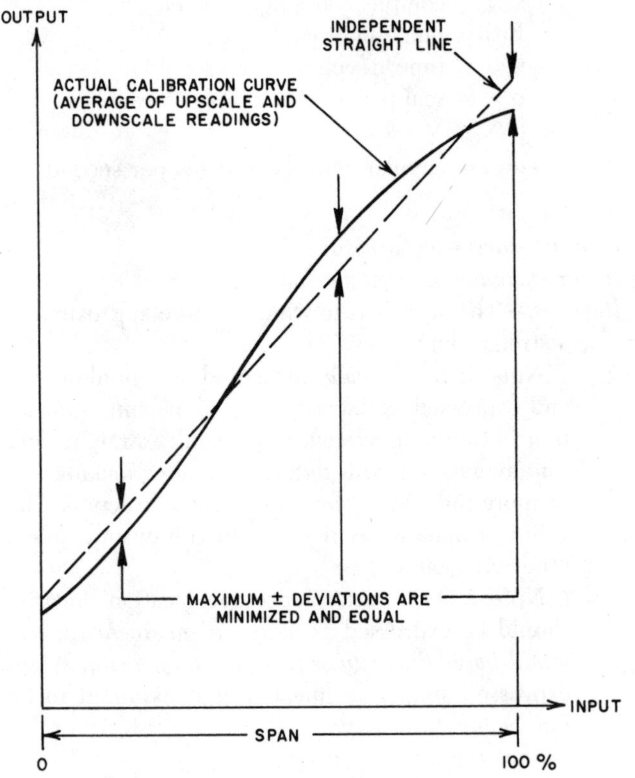

Fig. 1.1s Independent linearity

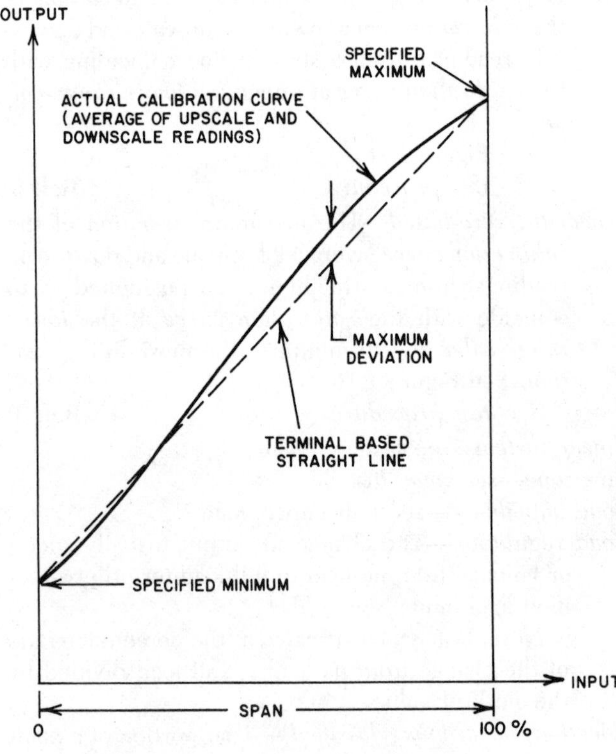

Fig. 1.1t Terminal-based linearity

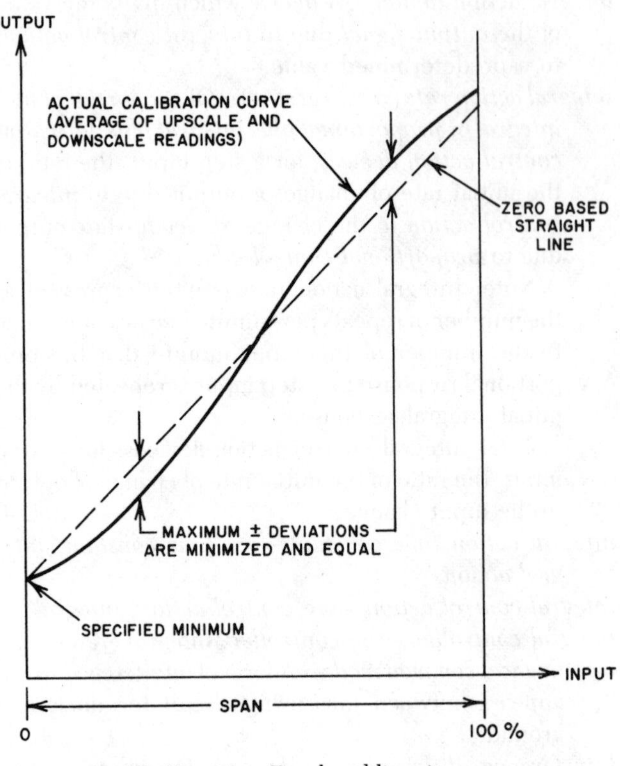

Fig. 1.1u Zero-based linearity

gases or dusts may be present in the air in quantities sufficient to produce explosive or ignitable mixtures. (Ref. 9)

longitudinal interference—see *interference, common mode.*

loop, closed (feedback loop)—A *signal* path which includes a forward path, a *feedback* path and a *summing point,* and forms a closed circuit. (4, 8)

loop, feedback—see *loop, closed (feedback loop).* (4, 8)

loop gain—see *gain, loop.*

loop gain characteristics—Of a *closed loop,* the *characteristic curve* of the ratio of the change in the *return signal* to the change in the corresponding *error signal* for all real frequencies. (8)

loop, open—A *signal* path without *feedback.* (4, 8)

loop transfer function—Of a *closed loop,* the *transfer function* obtained by taking the ratio of the *Laplace transform* of the *return signal* to the *Laplace transform* of its corresponding *error signal.* (4, 8)

lower range-limit—see *range-limit, lower.*

lower range-value—see *range-value, lower.*

low limiting control—see *control, low limiting.*

magnetic field interference—see *interference, electromagnetic.*

magnitude ratio—see *gain (magnitude ratio).*

manipulated variable—see *variable, manipulated.*

maximum excitation—see *excitation, maximum.*

maximum working pressure—see *pressure, maximum working (MWP).*

measurand—see *variable, measured.*
measured accuracy—see *accuracy, measured.*
measured signal—see *signal, measured.*
measured value—see *value, measured.*
measured variable—see *variable, measured.*
measuring instrument—see *instrument, measuring.*
mechanical shock—The momentary application of an acceleration force to a device.

Note: It is usually expressed in units of acceleration of gravity (g). (Ref. 8)

modulation—The process, or result of the process, whereby some characteristic of one wave is varied in accordance with some characteristic of another wave. (4, 8)

module—An assembly of interconnected components which constitutes an identifiable *device*, instrument, or piece of equipment. A module can be disconnected, removed as a unit, and replaced with a spare. It has definable performance characteristics which permit it to be tested as a unit.

Note: A module could be a card or other subassembly of a larger *device*, provided it meets the requirements of this definition.

mounting position—The position of a *device* relative to physical surroundings. (8)

mounting strain error—see *error, mounting strain.*
multi-element control system—see *control system, multi-element (multi-variable).*
multi-speed floating controller—see *controller, multiple-speed floating.*
multi-position controller—see *controller, multi-position.*
multi-variable control system—see *control system, multi-element (multi-variable).*
natural frequency—see *frequency, undamped.*
neutral zone—see *zone, neutral.*
noise—An unwanted component of a *signal* or variable.

Note: It may be expressed in units of the output or in percent of output *span*.

See *interference, electromagnetic.* (Ref. 4, Ref. 8)

non-incendive equipment—Equipment which in its *normal operating condition* would not ignite a specific hazardous atmosphere in its most easily ignited concentration.

Note: The electrical circuits may include sliding or make-and-break contacts releasing insufficient energy to cause ignition. Wiring which under normal conditions cannot release sufficient energy to ignite a specific hazardous atmospheric mixture by opening, shorting or grounding, shall be permitted using any of the methods suitable for wiring in ordinary locations.

non-interacting control system—see *control system, non-interacting.*
normal mode interference—see *interference, normal mode.*
normal mode rejection—The ability of a circuit to discriminate against a *normal mode voltage.*

Note: It may be expressed as a dimensionless ratio, a scalar ratio, or in decibels as 20 times the $\log_{10}$ of that ratio.

normal mode voltage—see *voltage, normal mode.*
normal operating conditions—see *operating conditions, normal.*
offset—The *steady-state deviation* when the *set point* is fixed. See also *deviation, steady-state.*

Note: The offset resulting from a no-load to a full-load change (or other specified limits) is often called "droop" or "*load regulation.*"

See *load regulation.* (8, Ref. 4)

on-off controller—see *controller, on-off.*
operating conditions—Conditions to which a *device* is subjected, not including the variable measured by the *device.*

Examples of *operating conditions* include: *ambient pressure, ambient temperature,* electromagnetic fields, gravitational force, inclination, power supply variation (voltage, frequency, harmonics), radiation, *shock* and *vibration.* Both static and dynamic variations in these conditions should be considered. (Ref. 2, Ref. 8)

operating conditions, normal—The range of *operating conditions* within which a *device* is designed to operate and for which *operating influences* are stated. See Figure 1.1v. (Ref. 8)

operating conditons, reference—The range of *operating conditions* of a *device* within which *operating influences* are negligible. See Figure 1.1v.

Note 1: The range is usually narrow.

Note 2: They are the conditions under which reference is stated and the base from which the values of *operating influences are* determined. (8)

operating influence—The change in a performance characteristic caused by a change in a specified *operating condition* from *reference operating condition,* all other conditions being held within the limits of *reference operating conditions.*

Note: The specified *operating conditions* are usually the limits of the *normal operating conditions.*

Operating influence may be stated in either of two ways:

1. As the total change in performance characteristics from *reference operating condition* to another specified *operating condition.*

Example: Voltage influence on *accuracy* may be expressed as:

> 2% of *span* based on a change in voltage from reference value of 120 volts to value of 130 volts.

2. As a coefficient expressing the change in a performance characteristic corresponding to unit change of the *operating condition,* from *ref-*

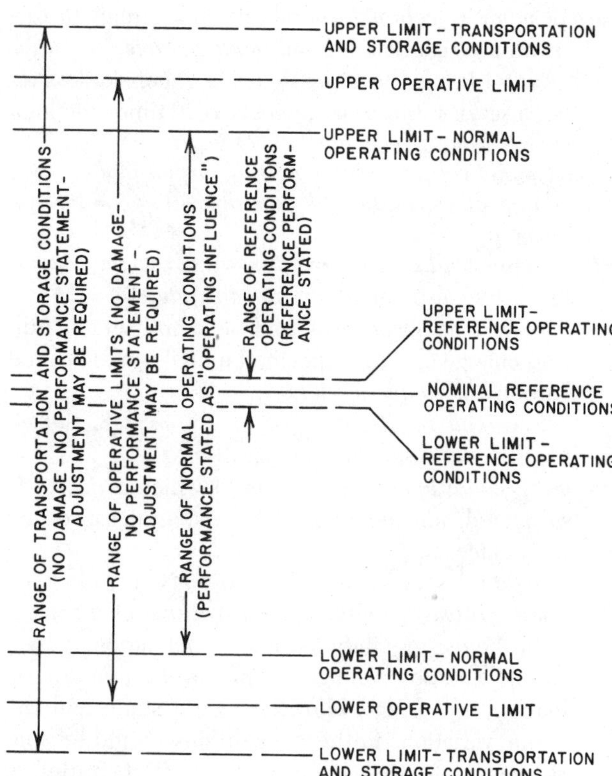

Fig. 1.1v Diagram of operating conditions

erence operating condition to another specified *operating condition*.

Example: Voltage influence on *accuracy* may be expressed as:

$$\frac{2\% \text{ of span}}{130\text{V-}120\text{V}} = 0.2\% \text{ of } \textit{span} \text{ per volt}$$

Note: If the relation between operating influence and change in *operating condition* is linear, one coefficient will suffice. If it is non-linear, it may be desirable to state more than one coefficient such as 0.05% per volt from 120 to 125V, and 0.15% from 125 to 130V. (8 Ref. 2)

operating pressure—see *pressure, operating*.

operative limits—The range of *operating conditions* to which a *device* may be subjected without permanent impairment of operating characteristics. See Figure 1.1v.

Note 1: In general, performance characteristics are not stated for the region between the limits of *normal operating conditions* and the operative limits.

Note 2: Upon returning within the limits of *normal operating conditions*, a *device* may require adjustments to restore normal performance.

(Ref. 2 "design limits", Ref. 8)

optimizing control—see *control, optimizing*.

outdoor area—see *area, outdoor*.

output impedance—see *impedance, output*.

output signal—see *signal, output*.

overdamped—see *damping*.

overrange—Of a system or *element*, any excess value of the *input signal* above its *upper range-value* or below its *lower range-value*. (8, Ref. 4)

overrange limit—The maximum input that can be applied to a *device* without causing damage or permenent change in performance.

overshoot—see *transient overshoot*.

parameter—A quantity or property treated as a constant but which may sometimes vary or be adjusted. (6)

P controller—see *controller, proportional*.

PD controller—see *controller, proportional plus derivative*.

pen travel—The length of the path described by the pen in moving from one end of the chart scale to the other. The path may be an arc or a straight line. (8)

phase crossover frequency—see *frequency, phase crossover*.

phase shift—(1) Of a *transfer function*, a change of phase angle with test frequency, as between points on a loop phase characteristic.

(2) Of a *signal* a change of phase angle with transmission. (4)

PI controller—see *controller, proportional plus integral*.

PID controller—see *controller, proportional plus integral plus derivative*.

pneumatic delivery capability—The rate at which a pneumatic *device* can deliver air (or gas) relative to a specified output pressure change.

Note: It is usually determined, at a specified level of input *signal*, by measuring the output flow rate for a specified change in output pressure. The results are expressed in cubic feet per minute (ft^3/min) or cubic meters per hour (m^3/h), corrected to standard (normal) conditions of pressure and temperature.

pneumatic exhaust capability—The rate at which a pneumatic *device* can exhaust air (or gas) relative to a specified output pressure change.

Note: It is usually determined, at a specified level of input *signal*, by measuring the output flow rate for a specified change in output pressure. The results are expressed in cubic feet per minute (ft^3/min) or cubic meters per hour (m^3/h), corrected to standard (normal) conditions of pressure and temperature.

point drift—see *drift, point*.

position—Of a *multi-position controller*, a discrete value of the *output signal*. See Figure 1.1l.

position error—see *error, position*.

power consumption, electrical—The maximum power used by a *device* within its operating *range* during *steady-state signal* condition.

Note 1: For a power factor other than unity, power

consumption shall be stated as maximum volt-amperes used under the above stated condition.

Note 2: For a *device* operating outside of its operating *range*, the maximum power might exceed that which is experienced within the operating *range*. (Ref. 8 "Power consumption")

power factor—The ratio of total watts to the total root-mean-square (rms) volt-amperes.

$$F_p = \frac{\Sigma \text{ watts per phase}}{\Sigma \text{ rms volt-amperes per phase}} = \frac{\text{active power}}{\text{apparent power}}$$

Note: If the voltages have the same waveform as the corresponding currents, power factor becomes the same as phasor power factor. If the voltages and currents are sinusoidal and for polyphase circuits, form symmetrical sets, $F_p = \cos(\alpha - \beta)$. (3)

pressure, ambient—The pressure of the medium surrounding a *device*. (8)

pressure, design—The pressure used in the design of a vessel or *device* for the purpose of determining the minimum permissible thickness or physical characteristics of the parts for a given *maximum working pressure* (MWP) at a given temperature.

pressure, leak—The pressure at which some discernible leakage first occurs in a *device*.

pressure, maximum working (MWP)—The maximum total pressure permissible in a *device* under any circumstances during operation, at a specified temperature. It is the highest pressure to which it will be subjected in the *process*. It is a designed safe limit for regular use.

Note: MWP can be arrived at by two methods: (1) Designed — by adequate design analysis, with a safety factor; (2) Tested — by rupture testing of typical samples.

See *pressure, design*.

pressure, operating—The actual pressure at which a *device* operates under normal conditions. This pressure may be positive or negative with respect to atmospheric pressure.

pressure, process—The pressure at a specified point in the *process* medium. (Ref. 8)

pressure, rupture—The pressure, determined by test, at which a *device* will burst.

Note: This is an alternate to the design procedure for establishing *maximum working pressure* (MWP). The rupture pressure test consists of causing the *device* to burst.

pressure, static—The *steady-state* pressure applied to a *device*; in the case of a differential pressure *device*, the *process pressure* applied equally to both connections.

pressure, supply—The pressure at the supply port of a *device*. (8)

pressure, surge—*Operating pressure* plus the increment above *operating pressure* to which a *device* may be subjected for a very short time during pump starts, valve closings, etc.

primary element—see *element, primary*.

process—Physical or chemical change of matter or conversion of energy; e.g., change in pressure, temperature, speed, electrical potential, etc.

process control—The regulation or manipulation of variables influencing the conduct of a *process* in such a way as to obtain a product of desired quality and quantity in an efficient manner.

process measurement—The acquisition of information that establishes the magnitude of *process* quantities.

process pressure—see *pressure, process*.

process temperature—see *temperature, process*.

program controller—see *controller, program*.

proportional band—The change in input required to produce a full *range* change in output due to *proportional control action*.

Note 1: It is reciprocally related to *proportional gain*.

Note 2: It may be stated in input units or as a percent of the input *span* (usually the indicated or recorded input *span*).

The preferred term is *proportional gain*.

See *gain, proportional*. (8)

proportional control action—see *control action, proportional*.

proportional controller—see *controller, proportional*.

proportional gain—see *gain, proportional*.

proportional plus derivative control action—see *control action, proportional plus derivative*.

proportional plus derivative controller—see *controller, proportional plus derivative*.

proportional plus integral control action—see *control action, proportional plus integral*.

proportional plus integral controller—see *controller, proportional plus integral*.

proportional plus integral plus derivative control action—see *control action, proportional plus integral plus derivative*.

proportional plus integral plus derivative controller—see *controller, proportional plus integral plus derivative*.

proportional plus rate control action—see *control action, proportional plus derivative*.

proportional plus rate controller—see *controller, proportional plus derivative*.

proportional plus reset control action—see *control action, proportional plus integral*.

proportional plus reset controller—see *controller, proportional plus integral*.

proportional plus reset plus rate control action—see *control action, proportional plus integral plus derivative*.

proportional plus reset plus rate controller—see *controller, proportional plus integral plus derivative.*

proportional speed floating controller—see *controller, integral.*

ramp response—see *response, ramp*.

ramp response time—see *time, ramp response.*

range—The region between the limits within which a quantity is measured, received, or transmitted, expressed by stating the *lower* and *upper range-values*.

Note 1: For example:

a. 0 to 150°F
b. −20 to + 200°F
c. 20 to 150°C

Note 2: Unless otherwise modified, input *range* is implied.

Note 3: The following compound terms are used with suitable modifications in the units: *measured variable range, measured signal range,* indicating scale range, chart scale range, etc. See Tables 1.1w and 1.1x.

Note 4: For multi-range *devices*, this definition applies to the particular range that the *device* is set to measure. (3, 8)

range, elevated-zero—A *range* in which the zero value of the *measured variable, measured signal,* etc., is greater than the *lower range-value*. See Table 1.1w.

Note 1: The zero may be between the *lower* and *upper range-values*, at the *upper range-value*, or above the *upper range-value*.

Note 2: Terms *suppression, suppressed range* or *suppressed span* are frequently used to express the condition in which the zero of the *measured variable* is greater than the *lower range-value*. The term "elevated-zero range" is preferred. (Ref. 8)

range-limit, lower—The lowest value of the *measured variable* that a *device* can be adjusted to measure.

Note: The following compound terms are used with suitable modifications to the units: *measured variable* lower range-limit, *measured signal* lower range limit, etc. See Tables 1.1w and 1.1x. (Ref. 8)

range-limit, upper—The highest value of the *measured variable* that a *device* can be adjusted to measure.

Note: The following compound terms are used with suitable modifications to the units: *measured variable* upper range-limit, *measured signal* upper range-limit, etc. See Tables 1.1w and 1.1x. (Ref. 8)

range, suppressed-zero—A *range* in which the zero value of the *measured variable* is less than the *lower range value*. (Zero does not appear on the scale.) See Table 1.1w.

Note 1: For example: 20 to 100

Note 2: Terms *elevation, elevated range* or *elevated span* are frequently used to express the condition in which the zero of the *measured variable* is less than the *lower-range-value*. The term "suppressed-zero range" is preferred. (Ref. 2, Ref. 8)

range-value, lower—The lowest value of the *measured variable* that a *device* is adjusted to measure.

Note: The following compound terms are used with suitable modifications to the units: *measured variable* lower range-value, *measured signal* lower range-value, etc. See Tables 1.1w and 1.1x. (Ref. 8)

Table 1.1w
ILLUSTRATIONS OF THE USE OF RANGE AND SPAN TERMINOLOGY

Typical Ranges	*Name*	*Range*	*Lower Range Value*	*Upper Range Value*	*Span*	*Supplementary Data*
0 +100	—	0 to 100	0	+100	100	—
20 +100	Suppressed Zero Range	20 to 100	20	+100	80	Suppression Ratio = 0.25
−25 0 +100	Elevated Zero Range	−25 to +100	−25	+100	125	—
−100 0	Elevated Zero Range	−100 to 0	−100	0	100	—
−100 −20	Elevated Zero Range	−100 to −20	−100	−20	80	—

Table 1.1x
U.S.A. DOMESTIC UNITS
ILLUSTRATIONS OF THE USE OF THE TERMS "MEASURED VARIABLE" AND "MEASURED SIGNAL"

Typical Ranges	*Type of Range*	*Range*	*Lower Range Value*	*Upper Range Value*	*Span*
(1) THERMOCOUPLE					
0 — 2000°F TYPE K T/C	Measured Variable	0 to 2000°F	0°F	2000°F	2000°F
−0.68 — +44.91 mV	Measured Signal	−0.68 to +44.91 mV	−0.68 mV	+44.91 mV	45.59 mV
0 — 20 ×100 = °F	Scale and/ or Chart	0 to 2000°F	0°F	2000°F	2000°F
(2) FLOWMETER					
0 — 10,000 lb/h	Measured Variable	0 to 10,000 lb/h	0 lb/h	10,000 lb/h	10,000 lb/h
0 — 100 in H_2O	Measured Signal	0 to 100 in H_2O	0 in H_2O	100 in H_2O	100 in H_2O
0 — 10 ×1000 = lb/h	Scale and/ or Chart	0 to 10,000 lb/h	0 lb/h	10,000 lb/h	10,000 lb/h
(3) TACHOMETER					
0 — 500 rpm	Measured Variable	0 to 500 rpm	0 rpm	500 rpm	500 rpm
0 — 5V	Measured Signal	0 to 5V	OV	5V	5V
0 — 80 ft/s	Scale and/ or Chart	0 to 80 ft/s	0 ft/s	80 ft/s	80 ft/s

range-value, upper—The highest value of the *measured variable* that a *device* is adjusted to measure.

Note: The following compound terms are used with suitable modifications to the units: *measured variable* upper range-value, *measured signal* upper range-value, etc. See Tables 1.1w and 1.1x. (Ref. 8)

rate—see *control action, derivative.*

rate control action—see *control action, derivative.*

rate gain—see *gain, derivative action.*

ratio controller—see *controller, ratio.*

recording instrument—see *instrument, recording.*

reference accuracy—see *accuracy rating.*

reference-input element—see *element, reference input.*

reference-input signal—see *signal, reference-input.*

reference junction—That thermocouple junction which is at a known or reference temperature.

Note: The reference junction is physically that point at which the thermocouple or thermocouple extension wires are connected to a *device* or where the thermocouple is connected to a pair of lead wires, usually copper. (ref. 7)

reference junction compensation—A means of counteracting the effect of temperature variations of the *reference junction*, when allowed to vary within specified limits.

reference operating conditions—see *operating conditions, reference.*

reference performance—Performance attained under *reference operating conditions.*

Note: Performance includes such things as *accuracy, dead band, hysteresis, linearity, repeatability,* etc. (8)

regulator—see *controller, self-operated (regulator).*

relative damping—see *damping, relative.*

reliability—The probability that a *device* will perform its objective adequately, for the period of time specified, under the *operating conditions* specified. (Ref. 4) *repeatability*—The closeness of agreement among a number of consecutive measurements of the output for the same value of the input under the same *operating conditions*, approaching from the same direction, for full *range* traverses. See Figure 1.1y.

Note: It is usually measured as a non-repeatability and expressed as *repeatability* in percent of *span*. It does not include *hysteresis*.

See *test procedure*. (8, Ref. 2)

reproducibility—The closeness of agreement among repeated measurements of the output for the same value of input made under the same *operating conditions* over a period of time, approaching from both directions.

Note 1: It is usually measured as a nonreproducibility and expressed as reproducibility in percent of *span* for a specified time period. Normally, this implies a long period of time, but under certain conditions the period may be a short time during which *drift* may not be included.

Note 2: Reproducibility includes *hysteresis, dead band, drift* and *repeatability*.

Note 3: Between repeated measurements the input may vary over the *range* and *operating conditions* may vary within *normal operating conditions*.

See *test procedure*. (8)

reset control action—see *control action, integral (reset)*.

reset rate—see *integral action rate*.

resolution—The least interval between two adjacent discrete details which can be distinguished one from the other. (4)

resonance—Of a system or element, a condition evidenced by large oscillatory amplitude, which results when a small amplitude of periodic input has a frequency approaching one of the natural frequencies of the driven system. (4, 8)

response, dynamic—The behavior of the output of a *device* as a function of the input, both with respect to time. (8, Ref. 2)

response, ramp—The total (transient plus *steady-state*) *time response* resulting from a sudden increase in the rate of change from zero to some finite value of the input stimulus.

(Ref. 4 "response, ramp-forced")

response, step—The total (*transient* plus *steady-state*) *time response* resulting from a sudden change from one constant level of input to another.

(4, "response, step-forced")

response, time—An output expressed as a function of time, resulting from the application of a specified input under specified *operating conditions*. See Figure 1.1z. (4, 8)

return signal—see *signal, return*.

reverse acting controller—see *controller, reverse acting*.

rise time—see *time, rise*.

rms value—see *value, rms*.

rupture pressure—see *pressure, rupture*.

sampling controller—see *controller, sampling*.

sampling period—The time interval between observations in a periodic sampling *control system*. (4, 8)

scale factor—The factor by which the number of scale divisions indicated or recorded by an instrument

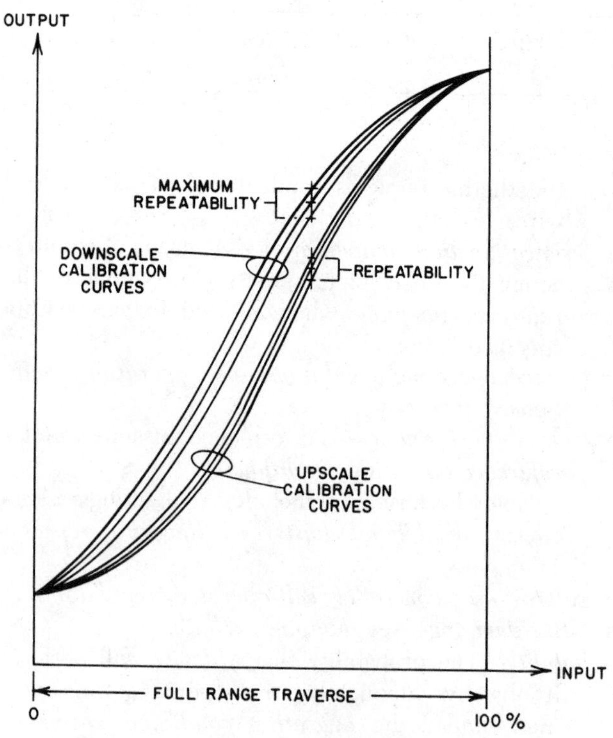

Fig. 1.1y Repeatability

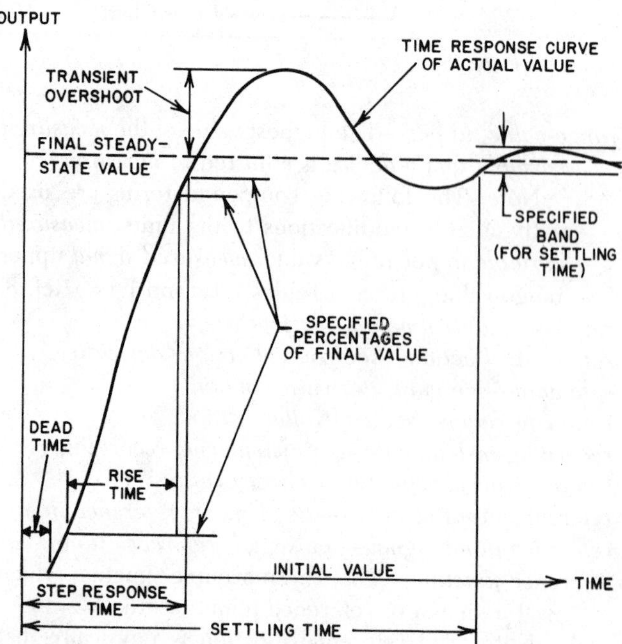

Fig. 1.1z Typical time response of a system to a step increase of input

should be multiplied to compute the value of the *measured variable.* (Ref. 8)

Note: Deflection factor is a more general term than scale factor in that the instrument response may be expressed alternatively in units other than scale divisions. (Ref. 3)

self-heating—Internal heating resulting from electric energy dissipated with a *device.* (Ref. 11)

self-operated controller—see *controller, self-operated (regulator).*

self-regulation (inherent regulation)—The property of a *process* or machine which permits attainment of equilibrium, after a *disturbance,* without the intervention of a *controller.* (4)

sensing element—see *element, sensing.*

sensing element elevation—The difference in elevation between the *sensing element* and the instrument.

Note: The elevation is considered positive when the *sensing element* is above the instrument. (8)

sensitivity—The ratio of the change in output magnitude to the change of the input which causes it after the *steady-state* has been reached.

Note 1: It is expressed as a ratio with the units of measurement of the two quantities stated. (The ratio is constant over the *range* of a linear *device.* For a nonlinear *device* the applicable input level must be stated.)

Note 2: Sensitivity has frequently been used to denote the *dead band.* However, its usage in this sense is deprecated since it is not in accord with accepted standard definitions of the term. (8, Ref. 4)

sensor—see *transducer.* (11)

servomechanism—An automatic *feedback control device* in which the controlled variable is mechanical position or any of its time derivatives. (Ref. 4)

set point—An input variable which sets the desired value of the controlled variable.

Note 1: The input variable may be manually set, automatically set, or programmed.

Note 2: It is expressed in the same units as the controlled variable. (8, Ref. 4 "command")

settling time—see *time, settling.*

shared time control—see *control, shared time.*

sheltered area—see *area, sheltered.*

signal—Physical variable, one or more *parameters* of which carry information about another variable (which the signal represents). (17)

signal, actuating error—The *reference-input signal* minus the *feedback signal.* See Figure 1.1p.

See also *deviation, system.* (8, Ref. 4 "signal, actuating")

signal amplitude sequencing (split ranging)—Action in which two or more *signals* are generated or two or more *final controlling elements* are actuated by an *input signal,* each one responding consecutively, with or without overlap, to the magnitude of that *input signal.* See Figure 1.1aa. (8)

signal, analog—A *signal* representing a variable which may be continuously observed and continuously represented.

signal converter—see *signal transducer.*

signal, digital—Representation of information by a set of discrete values in accordance with a prescribed law. These values are represented by numbers.

signal, error—In a *closed loop,* the *signal* resulting from subtracting a particular *return signal* from its corresponding *input signal.*

See also *signal, actuating error.* (4, 8)

signal, feedback—The *return signal* which results from a measurement of the *directly controlled variable.* See Figure 1.1p. (8, Ref. 4)

signal, feedforward—see *control, feedforward.*

signal, input—A *signal* applied to a *device, element* or system. (8, Ref. 4)

signal, measured—The electrical, mechanical, pneumatic or other variable applied to the input of a *device.* It is the analog of the *measured variable* produced by a *transducer* (when such is used.)

Example 1: In a thermocouple thermometer, the measured signal is an emf which is the electrical analog of the temperature applied to the thermocouple.

Example 2: In a *flowmeter,* the measured signal may be a differential pressure which is the analog of the rate of flow through the orifice.

Example 3: In an electric tachometer system, the measured signal may be a voltage which is the electrical analog of the speed of rotation of the part coupled to the tachometer generator.

See *variable, measured.* (8, Ref. 2)

signal, output—A *signal* delivered by a *device, element* or system. (8, Ref. 4)

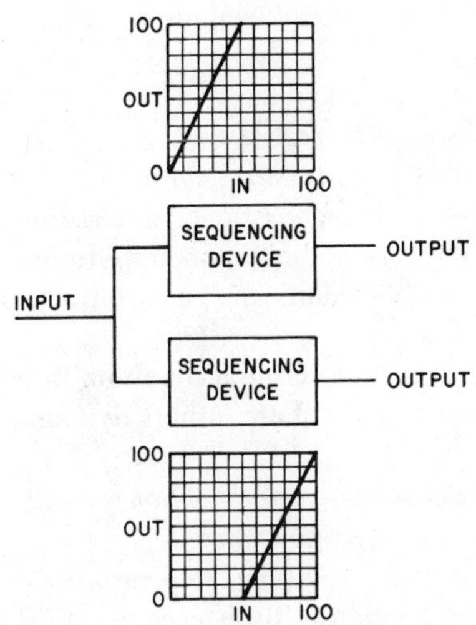

Fig. 1.1aa Signal amplitude sequencing

signal, reference-input—One external to a control loop, serving as the standard of comparison for the *directly controlled variable*. See Figure 1.1p. (4, 8)

signal, return—In a *closed loop*, the *signal* resulting from a particular *input signal*, and transmitted by the loop and to be subtracted from the *input signal*. See also *signal, feedback*. (4, 8)

signal selector—A *device* which automatically selects either the highest or lowest *input signal* from among two or more *input signals*.

Note: This device is sometimes referred to as a signal auctioneer. (Ref. 8 "auctioneering device")

signal to noise ratio—Ratio of *signal* amplitude to *noise* amplitude.

Note: For sinusoidal *signals*, amplitude may be peak or rms. For non-sinusoidal *signals*, peak values should be used. (8)

signal transducer (signal converter)—A *transducer* which converts one standardized transmission *signal* to another. (Ref. 8)

single speed floating controller—see *controller, single speed floating*.

source impedance—see *impedance, source*.

span—The algebraic difference between the *upper* and *lower range-values*.

Note 1: For example:

a. *Range* 0 to 150°F, *Span* 150°F

b. *Range* −20 to 200°F, *Span* 220°F

c. *Range* 20 to 150°C, *Span* 130°C

Note 2: The following compound terms are used with suitable modifications to the units: *measured variable span, measured signal span*, etc.

Note 3: For multi-range *devices*, this definition applies to the particular *range* that the *device* is set to measure. See Tables 1.1w and 1.1x. (8, Ref. 3)

span adjustment—see adjustment, span.

span error—see *error, span*.

span shift—Any change in slope of the input-output curve. See Figure 1.1bb.

split ranging—see *signal amplitude sequencing (split ranging)*.

static friction—see *stiction*.

static gain—see *gain, static*.

static pressure—see *pressure, static*.

steady-state—A characteristic of a condition, such a value, rate, periodicity, or amplitude, exhibiting only negligible change over an arbitrary long period of time.

Note: It may describe a condition in which some characteristics are static, others dynamic. (8, Ref. 4)

steady-state deviation—see *deviation, steady-state*.

step response—see *response, step*.

step response time—see *time, step response*.

stiction (static friction)—Resistance to the start of motion, usually measured as the difference between the driving values required to overcome static friction upscale and downscale. (4)

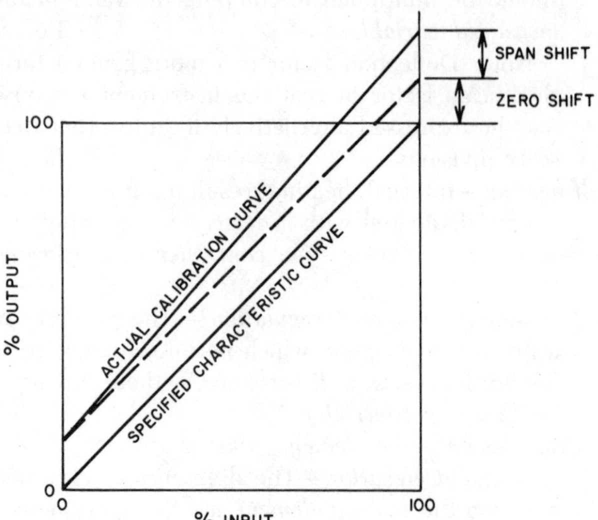

Fig. 1.1bb Shift, span and zero

stiffness—The ratio of change of force (or torque) to the resulting change in deflection of a spring-like *element*.

Note: Stiffness is the opposite of compliance.

subsidence—see *damping*, also *subsidence ratio*.

subsidence ratio—The ratio of the peak amplitudes of two successive oscillations of the same sign, the numerator representing the first oscillation in time. See Figure 1.1q. (16)

summing point—Any point at which *signals* are added algebraically. See Figure 1.1p. (4, 8)

supervisory control—see *control, supervisory*.

supply pressure—see *pressure, supply*.

suppressed range—see *range, elevated-zero*.

suppressed span—see *range, elevated-zero*.

supressed-zero range—see *range, suppressed-zero*.

suppression—see *range, elevated-zero*.

supression ratio—(of a *supressed-zero range*)—the ratio of the *lower range-value* to the *span*.

Note: For example:

Range 20 to 100,

$$\text{Suppression Ratio} = \frac{20}{80} = 0.25$$

See Table 1.1w. (8, Ref. 2)

surge pressure—see *pressure, surge*.

switching point—A point in the input *span* of a *multi-position controller* at which the *output signal* changes from one *position* to another. See Figure 1.1l. (8)

systematic error—see *error, systematic*.

system, control—see *control system*.

system, controlled—The collective functions performed in and by the equipment in which the variable(s) is (are) to be controlled.

Note: Equipment as embodied in this definition should be understood not to include any automatic control equipment. (Ref. 8 "process")

system, controlling—(1) Of a *feedback control system*, that portion which compares functions of a *directly controlled variable* and a *set point*, and adjusts a *manipulated variable* as a function of the difference. It includes the *reference-input elements; summing point; forward* and *final controlling elements*, and *feedback elements* (including *sensing element*).
(2) Of a *control system* without *feedback*, that portion which manipulates the *controlled system*. (Ref. 4, Ref. 8)

system deviation—see *deviation, system*.

system, directly controlled—The body, *process*, or machine directly guided or restrained by the *final controlling element* to achieve a prescribed value of the *directly controlled variable*. (4, 8)

system, idealized—An imaginary system whose ultimately controlled variable has a stipulated relationship to a specified *set point*.
Note: It is a basis for performance standards. (8, Ref. 4)

system, indirectly controlled—The portion of the *controlled system* in which the *indirectly controlled variable* is changed in response to changes in the *directly controlled variable*.
See Figure 1.1p. (Ref. 4, Ref. 8)

system, linear—One of which the *time response* to several simultaneous inputs in the sum of their independent *time responses*.
Note: It is represented by a linear differential equation, and has a *transfer function* which is constant for any value of input within a specified *range*. A system not meeting these condtions is described as "nonlinear". (Ref. 4)

tapping—see *dither*.

temperature, ambient—The temperature of the medium surrounding a *device*.
Note 1: For *devices* which do not generate heat this temperature is the same as the temperature of the medium at the point of *device* location when the *device* is not present.
Note 2: For *devices* which do generate heat this temperature is the temperature of the medium surrounding the *device* when it is present and dissipating heat.
Note 3: Allowable ambient temperature limits are based on the assumption that the *device* in question is not exposed to significant radiant energy sources. (8)

temperature, process—The temperature of the process medium at the sensing element. (8)

terminal-based conformity—see *conformity, terminal-based*.

terminal-based linearity—see *linearity, terminal-based*.

thermal shock—An abrupt temperature change applied to a *device*. (8)

three-position controller—see *controller, three-position*.

time constant—The value T is an exponential response term A exp $(-t/T)$ or in one of the transform factors $1 + sT$, $1 + j\omega T$, $1/(1 + sT)$, $1/(1 + j\omega T)$.

where:

s = complex variable
t = time, seconds
T = time constant
$j = \sqrt{-1}$
ω = angular velocity, radians per second.

Note: For the output of a first-order system forced by a step or an impulse, T is the time required to complete 63.2% of the total rise or decay; at any instant during the process, T is the quotient of the instantaneous rate of change divided into the change still to be completed. In higher order systems, there is a time constant for each of the first-order components of the process. In a *Bode diagram*, break points occur at $\omega = 1/T$. (Ref. 4, Ref. 8)

time constant, derivative action—Of *proportional plus derivative control action*, a *parameter* the value of which is equal to $1/2_\pi f_d$ where f_d is the frequency (in hertz) on a *Bode diagram* of the lowest frequency gain corner resulting from *derivative control action*. (4, 8)

time constant, integral action—(1) Of *proportional plus integral control action*, a *parameter* whose value is equal to $1/2_\pi f_i$ where f_i is the frequency (in hertz) on a *Bode diagram* of the highest frequency gain corner resulting from *integral control action*. (2) It is the reciprocal of *integral action rate*.
Note: The use of integral action rate is preferred. (8, Ref. 4)

time, correction—see *time, settling*.

time, dead—The interval of time between initiation of an input change or stimulus and the start of the resulting observable response. See Figure 1.1z. (Ref. 4, Ref. 8)

time, derivative action—In *proportional plus derivative control action*, for a unit ramp *signal* input, the advance in time of the *output signal* (after *transients* have subsided) caused by *derivative control action*, as compared to the *output signal* due to *proportional control action* only. (Ref. 5, Ref. 8)

time proportioning control—see *control, time proportioning*.

time, ramp response—The time interval by which an output lags an input, when both are varying at a constant rate.
(4, Ref. 3 "response time, ramp-forced")

time response—see *response, time.*

time, rise—The time required for the output of a system (other than first order) to change from a small specified percentage (often 5 or 10) of the *steady-state* increment to a large specified percentage (often 90 to 95), either before or in the absence of overshoot. See Figure 1.1z.

Note: If the term is unqualified, response to a unit step stimulus is understood; otherwise the pattern and magnitude of the stimulus should be specified. (Ref. 4, Ref. 8)

time schedule controller—see *controller, time schedule.*

time, setting—The time required, following the initiation of a specified stimulus to a system, for the output to enter and remain within a specified narrow band centered on its *steady-state* value. See Figure 1.1z.

Note: The stimulus may be a step impulse, ramp, parabola, or sinusoid. For a step or impulse, the band is often specified as ±2%. For nonlinear behavior both magnitude and pattern of the stimulus should be specified. (8, Ref. 4)

time, step response—Of a system or an *element*, the time required for an output to change from an initial value to a large specified percentage of the final *steady-state* value either before or in the absence of overshoot, as a result of a step change to the input. See Figure 1.1z.

Note: Usually stated for 90, 95 or 99 percent change.

See *"time constant" for use of 63.2% value.* (Ref. 4 "time response", Ref. 8)

transducer—An *element* or *device* which receives information in the form of one quantity and converts its to information in the form of the same or another quantity.

Note: This is a general term and definition and as used here applies to specific classes of *devices* such as *primary element, signal transducer,* and *transmitter.*

See *primary element, signal transducer, and transmitter.* (Ref. 4, Ref. 8)

transfer function—A mathematical, graphical, or tabular statement of the influence which a system or *element* has on a *signal* or action compared at input and at output terminals. (4, 8)

transient—The behavior of a variable during transition between two *steady-states.* (17)

transient deviation—see *deviation, transient.*

transient overshoot—The maximum excursion beyond the final *steady-state* value of output as the result of an input change. (Ref. 8)

transient overvoltage—A momentary excursion in voltage occurring in a *signal* or supply line of a *device* which exceeds the maximum rated conditions specified for that *device.*

transmitter—A *transducer* which responds to a *measured variable* by means of a *sensing element,* and converts it to a standardized transmission *signal* which is a function only of the *measured variable.* (Ref. 8)

transportation and storage conditions—The conditions to which a *device* may be subjected between the time of construction and the time of installation. Also included are the conditions that may exist during shutdown.

See Figure 1.1v.

Note: No permanent physical damage or impairment of operating characteristics shall take place under these conditions, but minor adjustments may be needed to restore performance to normal. (8)

transverse interference—see *interference, normal mode.*

two-position controller—see *controller, two-position.*

undamped frequency—see *frequency, undamped.*

underdamped—see *damping.*

upper range-limit—see range-limit, upper.

upper range-value—see range-value, upper.

value, desired—The value of the *controlled variable* wanted or chosen.

Note: The desired value equals the *ideal value* in an *idealized system.* (8)

value, ideal—The value of the indication, output or ultimately controlled variable of an idealized *device* or system.

Note: It is assumed that an ideal value can always be defined even though it may be impossible to achieve. (8)

value, measured—The numerical quantity resulting, at the instant under consideration, from the information obtained by a measuring *device.* (Ref. 16)

value, rms (root-mean-square value)—The square root of the average of the squares of the instantaneous values.

$$\text{Note: rms value} = \left[\frac{1}{T}\int_{t_o}^{t_o + T} x^2 dt\right]^{1/2}$$

where:

x is the instantaneous value,
t_o is any value of time,
T is the observation period (4)

variable, directly controlled—In a control loop, the variable the value of which is sensed to orginate a *feedback signal.* (4, 8)

variable, indirectly controlled—A variable which does not originate a *feedback signal,* but which is related to, and influenced by, the *directly controlled variable.* (4, 8)

variable, manipulated—A quantity or condition which is varied as a function of the *actuating error signal* so as to change the value of the *directly controlled variable*. (4, 8)

variable, measured—A quantity, property, or condition which is measured.

Note 1: It is sometimes referred to as the measurand.

Note 2: Common measured variables are temperature, pressure, rate of flow, thickness, speed etc. (Ref. 8)

velocity limit—A limit which the rate of change of a specified variable may not exceed. (Ref. 8)

velocity limiting control—see *control, velocity limiting*.

vibration—A periodic motion or oscillation of an *element, device*, or system.

Note 1: Vibration is caused by any excitation which displaces some or all of a particular mass from its position of equilibrium. The resulting vibration is the attempt of the forces, acting on and within the mass, to equalize.

Note 2: The amplitude and duration of any vibration is dependent on the period and amplitude of the excitation and is limited by the amount of *damping* present.

voltage, common mode—A voltage of the same polarity on both sides of a differential input relative to ground. See Figure 1.1cc.

voltage, normal mode—A voltage induced across the input terminals of a *device*. See Figure 1.1dd.

warm-up period—The time required after energizing a *device* before its rated performance characteristics apply. (8)

zero adjustment—see *adjustment, zero*.

zero-based conformity—see *conformity, zero-based*.

zero-based linearity—see *linearity, zero-based*.

zero elevation—For an *elevated-zero range*, the amount the *measured variable* zero is above the *lower range-value*. It may be expressed either in units of *measured variable* or in percent of *span*. (8)

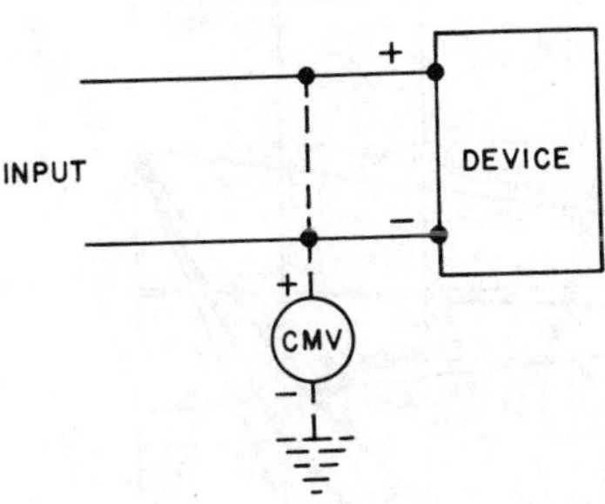

Fig. 1.1cc Common mode voltage

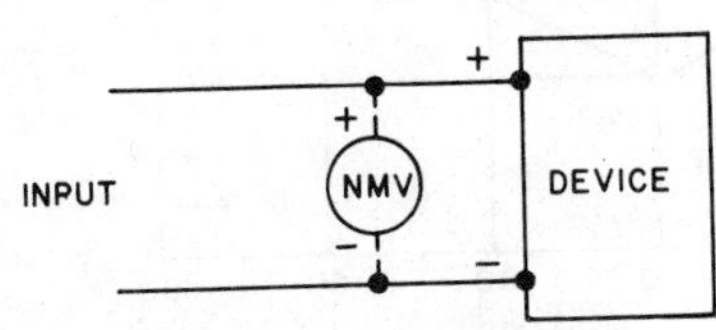

Fig. 1.1dd Normal mode voltage

zero error—see *error, zero*.

zero frequency gain—see *gain, zero frequency*.

zero shift—Any parallel shift of the input-output curve. See Figure 1.1bb.

zero suppression—For a *suppressed-zero range*, the amount the *measured variable* zero is below the *lower range-value*. It may be expressed either in units of the *measured variable* or in percent of *span*. (8)

zone—On a *multi-position controller*, the *range* of input values between selected switching points or any *switching point* and range-limit. See Figure 1.1l. (8)

zone, dead—(1) For a *multi-position controller*, a *zone* of input in which no value of the output exists. It is usually intentional and adjustable. (2) Dead zone is sometimes used to denote *dead band*. (Ref. 8)

zone, intermediate—Any *zone* not bounded by a range-limit (8)

zone, live—A *zone* in which a value of the output exists. (8)

zone, neutral—A predetermined *range* of input values in which the previously existing output value is not changed. See Figure 1.1o. (Ref. 4)

Test Procedures

Scope and Purpose

The purpose of the test procedures, as described herein, is to illustrate and clarify accuracy-related terms. It is intended only that the procedures indicate a generalized method of test.

The test procedures that follow are for the following terms:

accuracy, measured
dead band
drift, point
hysteresis
linearity, independent
linearity, terminal-based
linearity, zero-based
repeatability
reproducibility

Introduction

Tests described are for determination of static performance characteristics, not dynamic characteristics. When relating performance characteristics such as values of *accuracy* to values of other terms such as *linearity, hysteresis, dead band*, and *repeatability*, equivalent units must be used.

The *accuracy rating* of reference measuring means that relates to the characteristics being tested shall preferably be no greater than one tenth the tolerance allowed on the test device, but in any case not greater than one third the allowed tolerance.

> Example: *dead band*
> Test *device*, allowed *dead band* 0.2%
> Measuring *device*, preferred *dead band* 0.02%
> Measuring *device*, allowed *dead band* 0.06%

When the *accuracy rating* of the reference measuring means is one tenth or less than that of the *device* under test, the *accuracy rating* of the reference measuring means may be ignored. When the *accuracy rating* of the reference measuring means is one third or less but greater than one tenth that of the *device* under test, the *accuracy rating* of the reference measuring means shall be taken into account.

The *device* under test and the associated test equipment shall be allowed to stabilize under *steady-state operating conditions*. All testing shall be done under these conditions. Those *operating conditions* which would influence the test shall be observed and recorded. Where the performance characteristic being determined requires *reference operating conditions,* the conditions of test shall be maintained at *reference operating conditions*.

The number of test points to determine the desired performance characteristic of a *device* should be distributed over the *range*. They should include points at or near (within 10%) the *lower* and *upper range-values*. There should not be less than five points and preferably more. The number and location of these test points should be consistent with the degree of exactness desired and the characteristic being evaluated.

Prior to recording observations the *device* under test shall be exercised by a number of full range traverses in each direction.

At each point being observed the input shall be held steady until the *device* under test becomes stabilized at its apparent final value.

Tapping or vibrating the *device* under test is not allowed unless the performance characteristic under study requires such action.

Calibration Cycle

Maintain test conditions and precondition the test device as indicated in the introduction. Observe and record output values for each desired input value for one full range traverse in each direction starting near mid-range value. The final input must be approached from the same direction as the initial input. Apply the input in such a way as to not overshoot each input value.

Calibration Curve

For the purpose of the following test procedures, the *calibration curve* will be prepared as a "deviation plot." Determine the difference between each observed output value and its corresponding ideal output value. This difference is the *deviation* and may be expressed as a percent of ideal output *span*. The *deviation* is plotted versus input or ideal output. Figure 1.1ee illustrates percent

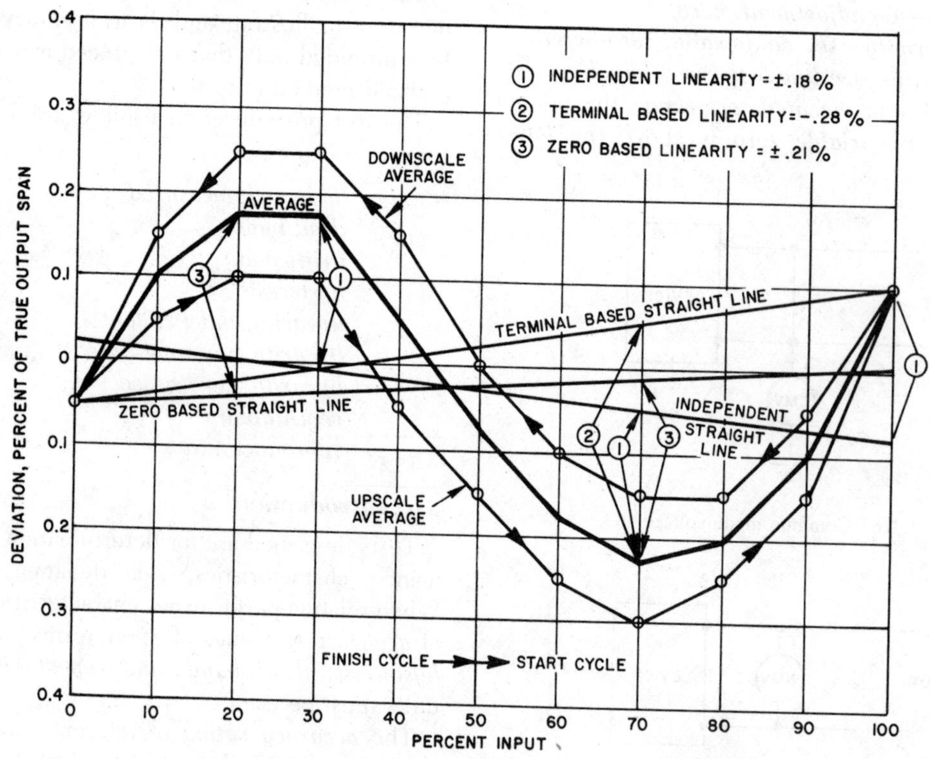

Fig. 1.1ee Calibration curve (of data from Table 1.1ff)

deviation plotted versus percent input. A positive *deviation* denotes that the observed output value is greater than the ideal output value.

Test Procedures

accuracy, measured—*Measured accuracy* may be determined from the *deviation* values (Table 1.1ff) of a number of calibration cycles. It is the greatest positive and negative *deviation* of the recorded values (from both an upscale and a downscale output traverse) from the reference or zero *deviation* line. *Measured accuracy* may be expressed as a plus and minus percent of ideal output *span*.

Example: The *measured accuracy* is +0.26%, to −0.32% of output *span*.

dead band—Maintain test conditions and precondition the test *device* as indicated in the introduction and proceed as follows:

1. Slowly vary (increase or decrease) the input to the *device* being tested until a detectable output change is observed.
2. Observe the input value.
3. Slowly vary the input in the opposite direction (decrease or increase) until a detectable output change is observed.
4. Observe the input value.

The increment through which the *input signal* is varied (difference between steps 2 and 4) is the *dead band.* It is determined from a number of cycles (steps 1 through 4). The maximum value is reported. The *dead band* should be determined at a number of points to make certain that the maximum *dead band* has been observed.

Dead Band may be expressed as a percent of input *span.*

Example: The *dead band* is 0.10% of input span.

drift, point—Maintain test conditions and precondition the test *device* as indicated in the introduction and proceed as follows:

1. Adjust the input to the desired value without overshoot and record the output value. Note: The test *device* should be permitted to warm up (if required) before recording the initial output value.
2. Maintain a fixed *input signal* and fixed *op-*

Table 1.1ff
CALIBRATION REPORT (See Fig. 1.1ee)

	Error										
Input	*Up Actual*	*Down Actual*	*Up Actual*	*Down Actual*	*Up Actual*	*Down Actual*	*Up Actual*	*Up Average*	*Down Average*	*Up Average*	*Average Error*
%	%	%	%	%	%	%	%	%	%	%	%
0		−0.04		−0.05		−0.06			−0.05		−0.05
10		+0.14	+0.04	+0.15	+0.05	+0.16	+0.06		+0.15	+0.05	+0.10
20		+0.23	+0.08	+0.26	+0.09	+0.26	+0.13		+0.25	+0.10	+0.175
30		+0.24	+0.09	+0.25	+0.10	+0.26	+0.11		+0.25	+0.10	+0.175
40		+0.13	−0.07	+0.15	−0.04	+0.17	−0.04		+0.15	−0.05	+0.05
50	−0.18	−0.02	−0.16	+0.01	−0.13	+0.01	−0.13	−0.15	0	−0.15	−0.075
60	−0.27	−0.12	−0.25	−0.10	−0.23	−0.08		−0.25	−0.10		−0.175
70	−0.32	−0.17	−0.30	−0.16	−0.28	−0.12		−0.30	−0.15		−0.225
80	−0.27	−0.17	−0.26	−0.15	−0.22	−0.13		−0.25	−0.15		−0.20
90	−0.16	−0.06	−0.15	−0.05	−0.14	−0.04		−0.15	−0.05		−0.10
100	+0.09		+0.11		+0.10			+0.10			+0.10

Measured Accuracy = +0.26% −0.32%

Hysteresis plus Dead Band = +0.22%

Repeatability = 0.05%

Note: Accuracy of reference measuring means was not considered in the determination of the average error.

erating conditions for the duration of the test.

3. At the end of the specified time interval observe and record the output value.

In evaluating the results of this test it is presumed that *dead band* is either negligible or of such a nature that it will not affect the value of *drift*.

Point drift is the maximum change in recorded output value observed during the test period. It is expressed in percent of ideal output *span* for a specified time period.

Example: The point drift is 0.1% of output span for a 24 hour test.

hysteresis—*Hysteresis* results from the inelastic quality of an *element* or *device*. Its effect is combined with the effect of *dead band*. The sum of the two effects may be determined directly from the *deviation* values (Table 1.1ff) of a number of test cycles, and is the maximum difference between corresponding upscale and downscale outputs for any single test cycle. *Hysteresis* then is determined by subtracting the value of *dead band* from the corresponding value of *hysteresis* plus *dead band* for a given input. The maximum difference is reported.

The difference may be expressed as a percent of ideal output *span*.

Example: The hysteresis is 0.12% of output span.

linearity, independent—*Independent linearity* may be determined directly from the *calibration* curve, (Figure 1.1ee) using the following procedure:

1. Plot a *deviation* curve which is the average of corresponding upscale and downscale output readings.
2. Draw a straight line through the average *deviation* curve in such a way as to minimize the maximum *deviation*. It is not necessary that the straight line be horizontal or pass through the end points of the average *deviation* curve.

Independent linearity is the maximum *deviation* between the average *deviation* curve and the straight line. It is determined from the *deviation* plots of a number of calibration cycles. It is measured in terms of independent nonlinearity as a plus or minus percent of ideal output *span*.

Example: The ideal *independent linearity* is ±0.18% of output *span*.

Note: The average *deviation* curve is based on the average of corresponding upscale and downscale readings. This permits observation of *independent linearity* independent of *dead band* or *hysteresis*. This concept assumes that if no *hysteresis* or *dead band* were present the *deviation* curve would be a single line midway between upscale and downscale curves.

linearity, terminal-based—*Terminal-based linearity* may be determined directly from the *calibration curve* (Figure 1.1ee) using the following procedure:

1. Plot a *deviation* curve which is the average of corresponding upscale and downscale output readings.
2. Draw a straight line such that it coincides with the average *deviation* curve at the *upper range-value* and the *lower range value*.

Terminal-based linearity is the maximum *deviation* between the average *deviation* curve and the straight line. It is determined from the *deviation* plots of a number of calibration cycles. It is measured in terms of terminal-based nonlinearity as a plus and minus percent of ideal output *span*.

Example: The *terminal-based linearity* is 0.28% of output *span*.

Note: The average *deviation* curve is based on the average of corresponding upscale and downscale readings. This permits observation of *terminal-based linearity* independent of *dead band* or *hysteresis*. This concept assumes that if no *hysteresis* or *dead band* were present, the *deviation* curve would be a single line midway between upscale and downscale readings.

linearity, zero-based—*zero-based linearity* may be determined directly from the calibration curve (Figure 1.1ee) using the following procedure:

1. Plot a *deviation* curve which is the average of corresponding upscale and downscale output readings.
2. Draw a straight line such that it coincides with the average *deviation* curve at the *lower range-value* (zero) and minimizes the maximum deviation.

Zero-based linearity is the maximum *deviation* between the average *deviation* curve and the straight line. It is determined from the *deviation* plots of a number of calibration cycles. It is measured in terms of zero-based linearity as a plus or minus percent of ideal output *span*.

Example: The *zero-based linearity* is ±0.21% of output *span*.

Note: The average *deviation* curve is based on the average of corresponding upscale and downscale readings. This permits observation of *zero-based linearity* independent of *dead band* or *hysteresis*. This concept assumes that if no *hysteresis* or *dead band* were present, the *deviation* curve would be a single line midway between upscale and downscale readings.

repeatability—Repeatability may be determined directly from the *deviation* values (Table 1.1ff) of a number of calibration cycles. It is the closeness of agreement among a number of consecutive measurements of the output for the same value of input approached from the same direction. Fixed *operating conditions* must be maintained.

Observe the maximum difference in percent *deviation* for all values of output considering upscale and downscale curves separately. The maximum value from either upscale or downscale curve is reported.

Repeatability is the maximum difference in percent *deviation* observed above and is expressed as a percent of output *span*.

Example: The *repeatability* is 0.05% of output *span*.

reproducibility—may be determined using the following procedure:

1. Perform a number of calibration cycles as described under "Calibration Cycle."
2. Prepare a calibration curve based on the maximum difference between all upscale and downscale readings for each input observed. The *deviation* values are determined from the number of calibration cycles performed for step 1 above. See section titled "Calibration Cycle."
3. Maintain the test device in its regular operating condition, energized and with an *input signal* applied.
4. At the end of the specified time repeat steps 1 and 2.

The test *operating conditions* may vary over the time interval between measurements providing they stay within the *normal operating conditions* of the test *device*. Tests under step 4 above must be performed under the same *operating conditions* that existed for the initial tests.

Reproducibility is the maximum difference between recorded output values (both upscale and downscale) for a given input value. Considering all input values observed, the maximum difference is reported. The difference is expressed as a percent of output *span* per specified time interval.

Example: The *reproducibility* is 0.2% of output *span* for a 30 day test.

REFERENCES

1. American National Standard C39.2-1964, "Direct-Acting Electrical Recording Instruments (Switchboard and Portable Types)."
2. American National Standard C39.4-1966, "Specifications for Automatic Null-Balancing Electrical Measuring Instruments."
3. American National Standard C42.100-1972, "Dictionary of Electrical and Electronics Terms."
4. American National Standard C85.1-1963, "Terminology for Automatic Control."
5. American National Standard C85.1a-1966, "Supplement to C85.1-1963 "Automatic Control Terminology."
6. American National Standard C85.1b-1966, "Supplement to C85.1-1963 "Automatic Control Terminology."
7. American National Standard MC96.1-1975, "Temperature Measurement Thermocouples."
8. Scientific Apparatus Makers Association Standard PMC20.1-1973, "Process Measurement and Control Terminology."
9. Instrument Society of America ISA-RP12.1-1960, "Electrical Instruments in Hazardous Atmospheres."
10. Instrument Society of America ISA-RP12.2-1965, "Intrinsically Safe and Non-Incendive Electrical Instruments."
11. Instrument Society of America ISA-S37.1-1969, "Electrical Transducer Nomenclature and Terminology."
12. Instrument Society of America ISA-S60.7-1975, (Draft Standard), "Control Center Construction."
13. Institute of Electrical and Electronics Engineers IEEE 279.
14. National Fire Protection Association NFPA 493.
15. National Fire Protection Association NFPA 501.
16. International Electrotechnical Commission, International Electrotechnical Vocabulary Publication 50(37)-1966, "Automatic Controlling and Regulating Systems."
17. International Electrotechnical Commission, International Electrotechnical Vocabulary Draft Chapter 351 (1037)-1972, "Automatic Control and Regulation-Servomechanisms."
18. International Electrotechnical Commission Technical Committee 65-Working Group 2 (IEC/TC65/WG2), "Service Conditions."

1.2 SYSTEM ACCURACY

Accuracy

There are a large number of technical subjects which would deserve separate discussion and elaboration as part of the introduction to this handbook. No topic, however, is more deserving than the subject of accuracy.* The reasons for this are multiple.

- Accurate measurement of process variables is an essential prerequisite to good control.
- The term itself is poorly defined and is widely misunderstood.
- The interrelationships between accuracy, rangeability, calibration and maintenance are not always recognized.
- There is a tendency on the part of some manufacturers to use misstatements of accuracy as a sales gimmick whereby their products appear in a more favorable light than those of more responsible suppliers.

Terminology

"Accuracy" is defined by Webster as "freedom from error or the absence of error." This, to start with, is contrary to the widespread use of this term. When an accuracy statement is given as "± 1 percent accuracy," in almost all cases the intended meaning is "± 1 percent inaccuracy." This mistake illustrates the carelessness (lack of accuracy) which prevails in dealing with this subject.

The purpose of all measurement is to obtain the true value of the quantity being measured, and error is thought of as the difference between the measured and the true quantity. Because it is impossible to measure a value without some uncertainty, it is equally impossible to know the exact error. All that can be stated in connection with the accuracy of a measurement, therefore, is the limits within which the true value will fall.

The accuracy-related terminology can be illustrated by an example of target shooting, as shown in Figure 1.2a. The spread in the nine shots fired in a tight pattern into the upper right hand corner of the target represents the *random error* of the shooter. His shooting is repeatable and precise, but precision alone does not represent accuracy; it is only the measure of random error.

The deviation between the mean impact and the bullseye represents the *systematic error.* This error (caused by the wind or faulty adjustments of the sights) is repeatable and can be eliminated, because it is not related to the shooter's inability to duplicate his shots. Systematic error is also referred to as *bias,* which is the displacement of the measured or observed value from the true one.

The shot in the lower left corner of the target represents the *illegitimate error* which is caused by blunders, and such should be totally eliminated.

Accuracy of measurement can thus be defined as the *sum of random and systematic error.* If the purpose of an installation is to maintain process conditions at previously experienced levels, without having an interest in the true values of these conditions, then it is desirable to reduce the random error, without paying much at-

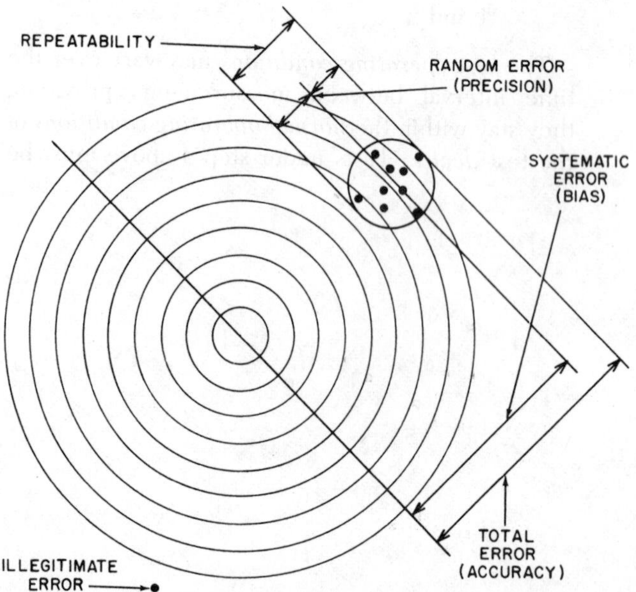

Fig. 1.2a Accuracy terminology

*Also see Section 1.1.

tention to the remaining bias. In such an installation, in other words, a precise, repeatable, but inaccurate measurement is sufficient.

Inversely, if the interest is in approaching the true value of the measurement because the installation serves such absolute purposes as accounting or quality control, then a repeatable detection is insufficient, and attention must be concentrated on accuracy, which will be achieved only through the reduction of both random *and systematic* errors.

If it is impossible to determine the systematic error, it can still be corrected by calibration against a fixed standard, such as standard thermal elements, analytical samples or weights.

It is often suggested[1] to elaborate on accuracy statements, beyond the point of stating some percentage values. A good example of this is the following quotation taken from a National Bureau of Standards Calibration Certificate for a turbine flowmeter:

> The results given are the arithmetic mean of ten separate observations, taken in groups of five successive runs on two different days. The reported values have an estimated *overall uncertainty* of ±0.13 percent, based on a standard error of ±0.01 percent and an allowance of ±0.1 percent for possible systematic error.

Flow Measurement Example

In order to bring this discussion into an area of specific relevance to the process industry, an example of flow measurement will be used. Figure 1.2b shows three flow detection installations all serving the purpose of totalizing a process stream. All flow sensors are for a full range of 100 GPM (380 l/m) instantaneous flow rate, and it will be our purpose to evaluate their accuracy at the flows of 20 GPM (76 l/m) and 80 GPM (304 l/m). (In all cases in this section, the GPM totals shown in the figures may be multiplied by 3.8 to obtain l/m.) The evaluation will consider two different basic assumptions—one, that component errors are *additive* and two, that system accuracy is likely to approach the accuracy of the least precise component in the system. Errors introduced by counter-totalizers will be neglected; devices having their accuracy based on the full scale (FS) range will be assumed to be inaccurate to ±0.5 percent FS; orifice plates will be treated as having a ±0.5 percent *of rate* inaccuracy; and a value of ±0.25 percent of rate will be used for turbine flowmeters. (Refer to the Orientation Table in Chapter II for details on flow sensor characteristics.)

Analog, Linear

The ideal behavior for a linear flow sensor, such as a magnetic flowmeter, is shown in Figure 1.2c. The line marked "actual" represents the relationship between the true flow and the output signal generated by a specific flow sensor. This deviation from the ideal is plotted in Figure 1.2d in terms of error, as a percentage of *full scale*, with the error limits being ±0.5 percent FS.

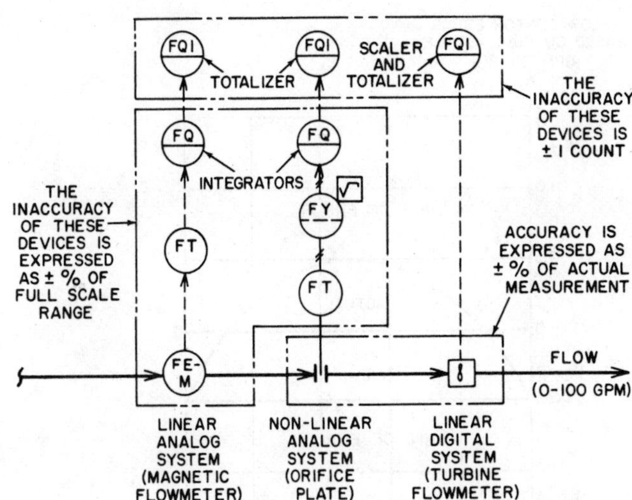

Fig. 1.2b Variations on possible flow totalization loops

The same performance, if plotted as a function of the actual flow rate reading (instead of full scale), results in the relationship shown in Figure 1.2e. The accuracy limits shown are conservative, in the sense that the specific detector performance is better, at most points of its range, than what the error limits would imply. It is also true that the meter inaccuracy increases with dropping flow rates if the inaccuracy is expressed as a percentage of actual reading, as shown in Figure 1.2f.

Analog, Non-Linear

In case of an orifice-type flow sensor, the actual measurement (pressure drop across the orifice plate) has a square relationship to the desired measurement (flow).

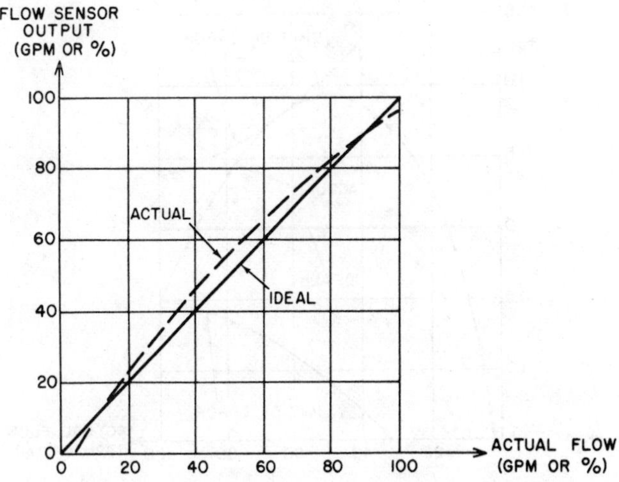

Fig. 1.2c Performance of a linear analog flow sensor

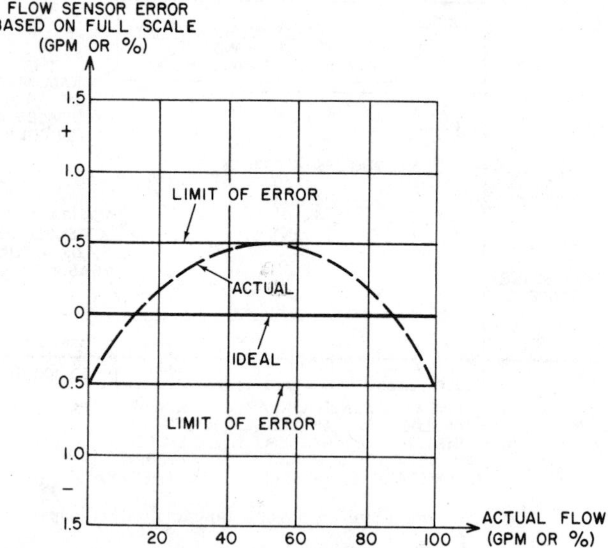

Fig. 1.2d Flow sensor error as a percentage of full-scale flow rate

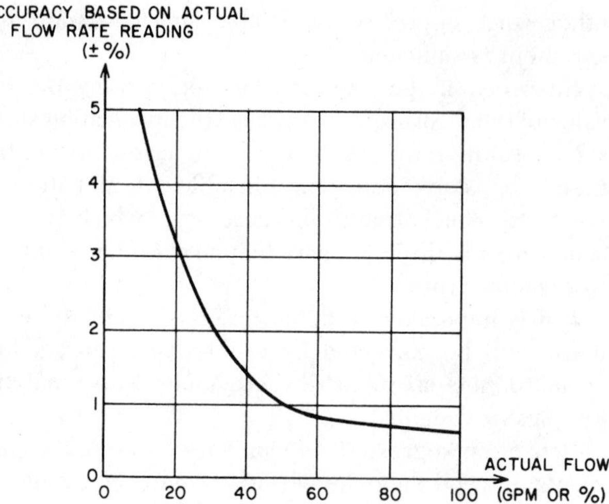

Fig. 1.2f Linear flowmeter inaccuracy as a percentage of actual flow reading

Figure 1.2g illustrates both this ideal non-linear relationship and the actual performance of a specific instrument. Because in most cases the square root must be extracted before the signal becomes useable (in our example of Figure 1.2b, a square-root extractor is installed before the integrator), the "gain effect" of this extraction must be recognized. As shown by Figure 1.2h, the extraction of the square root improves the accuracy at the higher flow rates, but degrades it as the flow rate is reduced.

Figure 1.2i depicts the relationship between the pipeline Reynolds number and the discharge coefficient for various head type flow elements. The Reynolds number is determined by the following equation:

$$Re = \frac{3.160 G_f Q_f}{D\mu} \quad 1.2(1)$$

where

G_f = process fluid specific gravity (at 60°F or 15.5°C),
Q_f = liquid flow in GPM (at 60°F or 15.5°C),
D = pipe inside diameter (in inches or mm), and
μ = process fluid viscosity (in centipoises).

As shown by Figure 1.2i the orifice plate discharge coefficient is constant within ±0.5 percent over a Reynolds number range of 2×10^4 to 10^6. This consistency of the discharge coefficient guarantees a corresponding limitation of inaccuracies based on *actual* flow, over a range that is wider than what the d/p flow transmitter can handle. This capability of the orifice plate element is illustrated in Figure 1.2j.

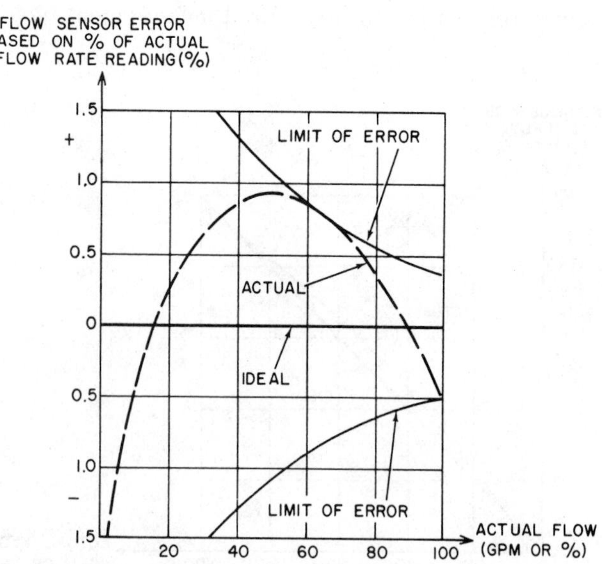

Fig. 1.2e Flow sensor error as a percentage of the actual flow rate reading

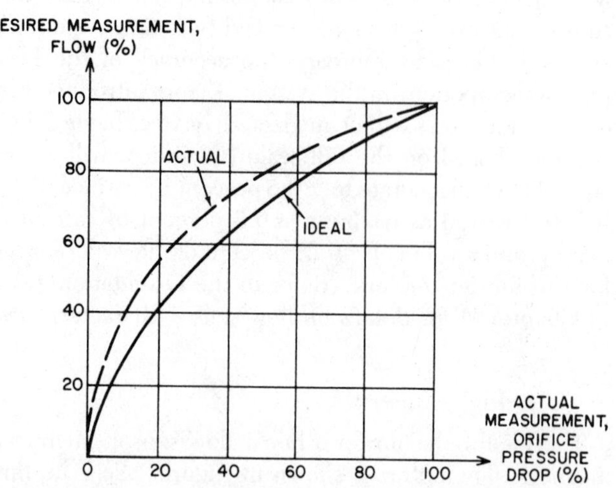

Fig. 1.2g Performance of a non-linear analog flow sensor

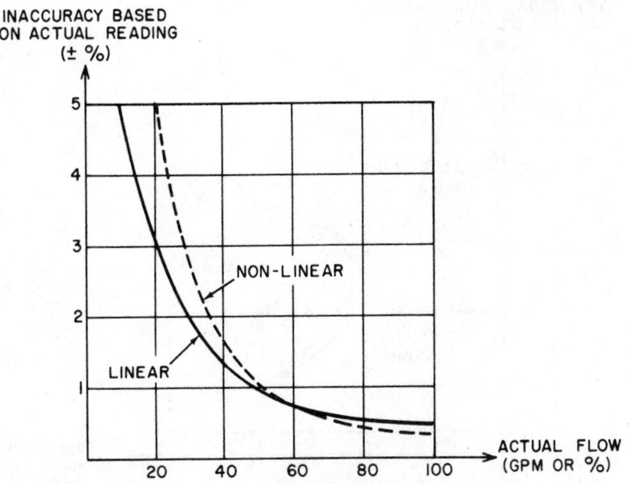

Fig. 1.2h Relative inaccuracies of linear and non-linear flowmeters

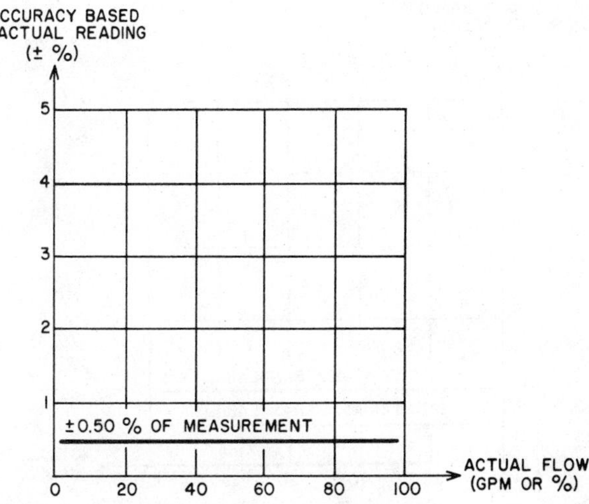

Fig. 1.2j Inaccuracy of orifice plate alone

Digital, Linear

Figure 1.2k illustrates the calibration of a turbine meter in terms of the K factor (pulses per gallon), which is rather similar to the calibration curve of an orifice plate (Figure 1.2i). The inaccuracy of a turbine meter is also stated as a percentage of the *actual* flow reading and not of full scale. As shown in Figure 1.2l, turbine meter accuracy can be improved by reducing the rangeability requirement of the unit.

System Accuracy

Having reviewed the inaccuracies of the various components in the loops shown in Figure 1.2b, the next step is to evaluate the resulting overall system accuracies. It is important to emphasize here that there is no proven basis for evaluating the accumulative effect of component inaccuracies, and only an actual system calibration can reliably establish the total inaccuracy. It should also be emphasized that probability favors the minimum number of components in a measurement loop to result in the best accuracy system. The only exception to this is the fact that in digital systems no additional error is introduced by the addition of functional modules.

Without actual system calibration the evaluation of loop accuracy must be based on assumptions. Table 1.2m summarizes system inaccuracies that can be expected under various conditions. The accumulated effect of component inaccuracies has been based on one of two assumptions:

Basis 1—Here it is assumed that the inaccuracy of each component is *additive*, and therefore system inaccuracy is the sum of component inaccuracies (very conservative basis).

Basis 2—Here the assumption is that all component inaccuracies can be neglected except that of the least accurate component, and therefore system inaccuracy is the same as the inaccuracy in this one component (very optimistic assumption).

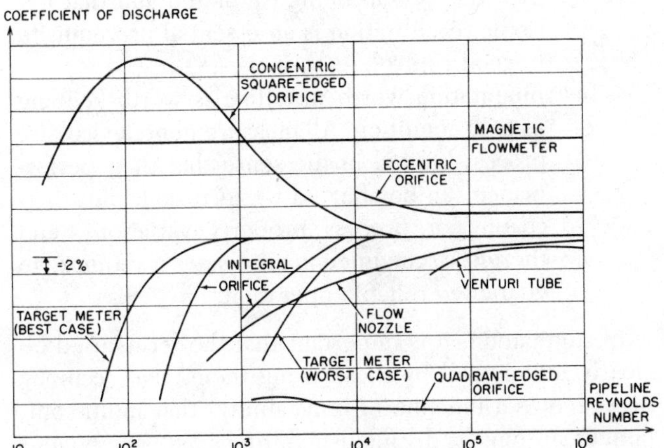

Fig. 1.2i Discharge coefficients as a function of sensor type and Reynolds number (Courtesy of The Foxboro Company)

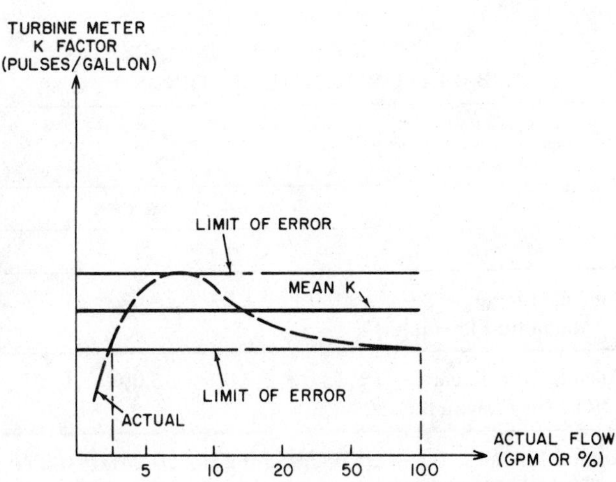

Fig. 1.2k Turbine flowmeter calibration curve

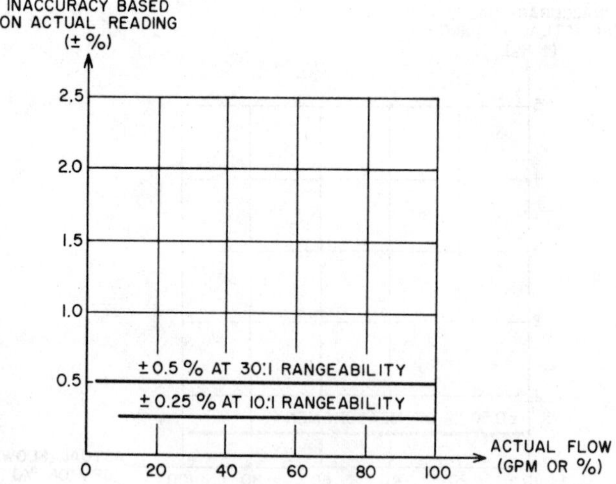

Fig. 1.2 l Turbine flowmeter inaccuracy as a function of rangeability

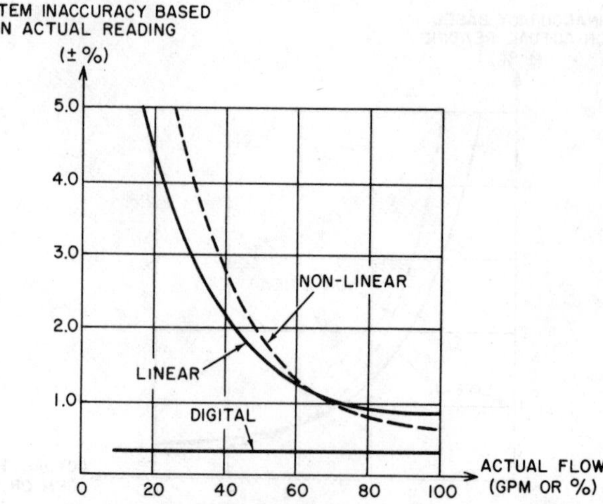

Fig. 1.2n System inaccuracies as a function of sensor type and flow rate (Accuracy in simple flow measurement, TI-1-30a, The Foxboro Company)

From Table 1.2m, it is seen that if Basis 1 is accepted for system accuracy evaluation, an orifice type installation operating at 20 percent of full scale will be inaccurate to ±12 percent of the reading, although no component inaccuracy in the loop exceeds ±0.5 percent FS.

Figure 1.2n illustrates the system inaccuracies if they were evaluated on a basis slightly more conservative than Basis 2 and much less conservative than Basis 1.

From the data in Table 1.2m and Figure 1.2n, it can be concluded that accuracy is by no means a clearly defined single number when one wishes to evaluate the performance of a multi-component system under varying load conditions.

From the foregoing discussion, therefore, only qualitative conclusions can be drawn. These include:

Accuracy is likely to be improved by reducing the number of components in a measurement loop.

Table 1.2m
SYSTEM INACCURACY SUMMARY
BASED ON ACTUAL READINGS

Assumption Used to Estimate Accumulated System Inaccuracy	*Basis 1*		*Basis 2*	
	Operating Flow Rate (GPM)			
Type of Flow Detection Loop	*20*	*80*	*20*	*80*
Analog, Linear (Magnetic Flowmeter)	±9.0%	±1.5%	±3.0%	±0.5%
Analog, Non-linear (Orifice Flowmeter)	±12.0%	±2.0%	±5.0%	±0.5%
Digital Linear (Turbine Flowmeter)	±0.25%	±0.25%	±0.25%	±0.25%

Accuracy is meaningful only in combination with rangeability. The wider the rangeability required (expected load variations), the more inaccurate the measurement is likely to be. Furthermore, the linear analog system accuracies are less affected by rangeability requirements than are non-linear analog systems, and the rangeability effect on digital systems is less than on any other type loop.

On non-accounting systems, the interest is focused on repeatability (random error) and not on accuracy. The repeatability of most measurement loops is several-fold better than their accuracy.

Instrumentation worth installing is worth calibrating. In this regard, several points should be made; namely, that the accuracy of a multi-component system is unknown unless calibrated as a system, that the calibration equipment must be *at least three times* more accurate than the system being calibrated and that periodic recalibration is an essential prerequisite to good control.

Instrumentation worth installing is worth keeping in good condition. All measurement devices are process-limited in the sense that their performance can be affected by corrosion, plugging, coating or process property variations, and therefore scheduled maintenance is required to guarantee reliable operation.

In summation, it is important that the terms used be clearly understood by those using them, that accuracy be stated as a function of rangeability, that multi-component systems be distinguished from single-component ones, and that the prerequisites of calibration and scheduled maintenance be emphasized.

REFERENCE

1. Kemp, R.E., "Accuracy for Engineers," *Instrumentation Technology*.

BIBLIOGRAPHY

Englund, D.R., "Loading Effects in Measurement Systems," *Instrument and Control Systems*, February, 1970, pp. 63–68.

Shinskey, F.G., "Estimating System Accuracy," Foxboro Publication #413-5.

1.3 REDUNDANT AND VOTING SYSTEMS

What is reliability in process control computing? It can be defined as the correct operation of a system up to a time t = T given that it was operating correctly at the starting time t = O.[1] Correct operation can have many meanings, though, depending upon the requirements previously established for the system. A common attitude today is that single or multiple failures can be accepted as long as the system does not go down or the desired operation is not interrupted or disturbed. Reliability is therefore a goal set by the users to be expected of a system.

In order to obtain a certain measure of reliability, the term fault-tolerant computing can be used. It may be defined as "the ability to execute specified algorithms correctly regardless of hardware errors and program errors."[2] Since different computers in different applications have widely different requirements for reliability, availability, recovery time, data protection and maintainability, an opportunity exists for the use of many different fault-tolerant techniques.[3]

The understanding of fault-tolerance can be helped first by understanding faults. Faults can be defined as "the deviation of one or more logic variables in the computer hardware from their design-specified values."[1] A logic value for a digital computer is either a zero or a one. A fault is the appearance of an incorrect value such as a logic gate "stuck on zero" or "stuck on one." The fault causes an "error" if it in turn produces an incorrect operation of correctly functioning logic elements. Therefore, the term "fault" is restricted only to the actual hardware that fails.

Faults can be classified in several ways. Their most important characteristic is a function of their duration. They can be either permanent (solid) or transient (intermittent). Permanent faults are caused by solid failures of components.[4] They are easier to diagnose, but usually require the use of more drastic correction techniques than do transient faults. Transient faults cause 80 to 90 percent of faults in most systems.[5] Transient faults or intermittents can be defined as random failures which prevent the proper operation of a unit for only a short period of time, not long enough to be tested and diagnosed before they disappear or become permanent. Permanent fault-tolerant techniques must then be used for system recovery.

The goal of systems reliability or of fault-tolerant computing is either to prevent or be able to recover from faults and continue correct system operation. This also includes immunity to software faults induced into the system. In order to achieve high reliability, it is essential that component reliability be as high as possible. "As the complexity of computer systems increases, almost any level of guaranteed reliability of individual elements becomes insufficient to provide a satisfactory probability of successful task completion."[6] Therefore, successful fault-tolerant computers must use a judicious selection of protective redundancy to help meet the reliability requirements. The three redundancy techniques are:

1. Hardware redundancy
2. Software redundancy
3. Time redundancy

These three techniques cover all methods of fault-tolerance. Hardware redundancy can be defined as any circuitry in the system not necessary for normal computer operation should no faults occur. Software redundancy, similarly, is additional program instructions present solely to handle faults. Any retrial of instructions is known as time redundancy.

Hardware Redundancy

Hardware redundancy can be classified as the set of all of those hardware components that need to be introduced into the system to provide fault-tolerance with respect to operational faults.[1] These components would be superfluous should no faults occur, and their removal would not diminish the computing power of the system in the absence of faults.

In achieving hardware fault-tolerance, it is clear that one should use the most reliable components available.[7] However, increasing component reliability has only a small impact on increasing system reliability. Therefore, it is "more important to be able to recover from failures than to prevent them."[8] Redundant techniques allow recovery, and thus are very important in achieving fault-tolerant systems. The techniques used in achieving hardware redundancy can be divided into two categories: static or masking redundancy and dynamic redundancy.

Static techniques are equally effective in handling both transient and permanent failures. Many different techniques of static redundancy can be applied. The simplest or lowest level of complexity is by a massive replication of the individual components of the system.[1] For example, four diodes connected as two parallel pairs which are themselves connected in series will not fail if any one diode fails "open" or "short." Logical gates in similar quadded arrangements [9,10] can also guard against single faults and even some multiple faults for largely replicated systems.

More sophisticated systems use replication at higher levels of complexity to mask failures. Masking is virtually instantaneous and automatic. It can be defined as any computer error correction method that is transparent to the user and often to the software. Redundant components serve to mask the effect of hardware failures of other components. Instead of using a mere massive replication of components configured in fault-tolerant arrangements, identical nonredundant computer sections or modules can be replicated and their outputs voted upon. Examples are triple modular redundancy (TMR) and more massive, N-modular redundancy (NMR) where N can stand for any odd number of modules.

In addition to component replication, coding can be used to mask faults as well as to detect them. With the use of some codes, data that has been garbled, i.e., bits changed due to hardware errors, can sometimes be recovered instantaneously with the use of redundant hardware. Dynamic recovery methods are, however, better able to handle many of these faults.

Higher levels of fault-tolerance can be achieved more easily through dynamic redundancy, implemented through the dual actions of fault detection and recovery. This often requires software help in conjunction with hardware redundancy. Many dynamic methods are extensions of static techniques.

More effective use of massive redundancy in components often results from dynamic control. Redundant modules or spares can have a better fault-tolerance when left unpowered until needed, because they will not degrade while awaiting use. This technique—stand-by redundancy [11]—can use dynamic voting techniques to achieve a high degree of fault-tolerance. The union of dynamic and stand-by redundancy is referred to as hybrid redundancy.[12] Hybrid redundancy requires additional hardware to detect and switch out faulty modules and switch in good spares.

Error detecting and error correcting codes [13] can be used dynamically to achieve fault-tolerance in a computing system. Coding refers to the addition and rearrangement of the bits in a binary word that contains information. The strategy of coding is to add a minimum number of check bits (the additional bits) to the message in such a way that a known degree of error detection or correction is achieved.[4] Error detection and correction is accomplished by comparing the new word, which should be unchanged after transmission, storage, or processing, with a set of allowable configurations of bits. Discrepancies thus discovered signal the existence of a fault which then can be corrected if enough of the original information remains intact—that is, the original binary word can be reconstructed provided a set number of bits in the coded word have not changed. Encoding and decoding words with the use of redundant hardware can be very effective in detecting errors. Often, through hardware or software algorithms, incorrect data also can be reconstructed. Otherwise, the detected errors can be handled by module replacement and software recovery actions. The actions taken are dependent upon the extent of the fault and of the recovery mechanisms available to the computing system.

Software Redundancy

Software redundancy refers to all additional software which would not be needed for a fault-free computer. It plays a major role in most fault-tolerant computers. Even computers that recover from failures mainly by means of hardware use software to control their recovery and decision making processes. The level of software used depends upon recovery system design, and recovery design is dependent upon the expected type of error or malfunction. Different schemes have been found to be appropriate for the handling of different errors. Some types of error correction are most efficiently accomplished with hardware. Others need only software, but most use a mixture of the two.

For a functional system, i.e., one without hardware design faults, errors can be classified into two varieties:

1. Software design errors
2. Hardware malfunctions

The first category can be corrected mainly by means of software. It is extremely difficult for hardware to be designed to correct for programmers' errors. Software methods, though, are often used to correct hardware faults, especially transient ones.

There are several software errors that computers may be designed to detect,[14,15] including the use of illegal instructions (i.e., instructions that do not exist), the use of privileged instructions without authorization, and address violations (i.e., reading or writing into locations that are beyond those of usable memory). These limits can often be set physically on the hardware. Computers capable of error detection cause interrupts, which route the program to specific locations in memory. The programmer, knowing these locations, can add his own code to branch to his specific subroutines which can handle each error specifically as he sees fit.

Chapter II

FLOW MEASUREMENTS

J.B. Arant, G.M. Crabtree,
L.D. DiNapoli, A.P. Foundos,
J.O. Hougen, W.H. Howe,
D.S. Kayser, J.G. Kopp,
B.G. Lipták, D.J. Lomas,
D.C. Mair, K.O. Plache, R. Siev

CONTENTS OF CHAPTER II

Table II ORIENTATION TABLE FOR FLOW SENSORS

SECTION	Type of Design	Applicable to Detect the Flow of: Clean Liquids	Viscous Liquids	Slurry	Gas	Solids	Direct Mass—Flow Sensor	Volumetric Flow Detector	Flow Rate Sensor	Inherent Totalizer	Direct Indicator	Transmitter Available	Linear Output	Rangeability	Pressure Loss Thru Sensor	① Approx. Straight Pipe-Run Requirement (Upstream Diam./Downstream Diam.)	Accuracy * ± % Full Scale ** ± % Rate *** ± % Registration	FLOW RANGE
2.5	Elbow Taps	√	L	L	√			√	√			√	SR	3:1②	N	⑦ 25/10	5-10*	gpm—m³/hr; SCFM—Sm³/hr
2.7	Jet Deflection				√			√	√			√	√	25:1	M	⑦ 20/5	2*	SCFM—Sm³/hr
2.8	Laminar Flowmeters	√	√		√			√	√			√	√	10:1	H	15/5	½-5*	gpm—m³/hr; SCFM—Sm³/hr
2.9	Magnetic Flowmeters	√⑤	√⑤	√⑤				√	√			√	√	10:1⑥	N	N	½**-2*	gpm—m³/hr
2.10	Mass Flowmeters	√	√	√	√	SD	√	√	√	SD	SD	√	√	100:1	A	N	½**	lbm/hr; SCFM—Sm³/hr
2.11	Metering Pumps	√	√	√				√		√		SD	√	20:1	–	N	⅒-1*	gpm—m³/hr
2.12	Orifice (Plate or Integral Cell)	√	L		√			√	√			√	SR	3:1②	H	⑦ 20/5	½-2*	gpm—m³/hr; SCFM—Sm³/hr
2.13	Pitot Tubes	√			√			√	√			√	SR	3:1②	L	⑦ 40/10	2-5*	gpm—m³/hr; SCFM—Sm³/hr
2.14	Positive Displacement Gas Meters				√			√		√	√	SD	√	200:1	M	N	½-1***	SCFM—Sm³/hr
2.15	Positive Displacement Liquid & Steam Meters	√	√					√		√	√	SD	√	10:1	H	N	¼-1**	gpm—m³/hr
2.17	Solids Flowmeters		SD	SD		√	SD	SD	√	√	SD	√	√	20:1	–	5/3	½**-4*	lbm/hr
2.18	Target Meters	√	√	L	√			√	√		SD	√	SR	4:1	H	15/5	1-5*	gpm—m³/hr; SCFM—Sm³/hr
2.19	Thermal Meters	√	L	L	√		√		√			√	L	20:1	A	5/3	1-2*	gpm—m³/hr; SCFM—Sm³/hr
2.20	Turbine Flowmeters	√	L		SD			√	√			√	√	10:1	H	10/5	¼**	gpm—m³/hr
2.21	Ultrasonic Flowmeters																	
	Transit	√	L	L				√	√			√	√	20:1	N	⑦ 15/5	1**	gpm—m³/hr③
	Doppler	√	L	√				√	√			√	√	20:1	N	⑦ 15/5	2-3*	gpm—m³/hr③
2.22	Variable Area Flowmeters	√	L	L	√			√	√		√	√	√	5:1	A	N	½*-10**	gpm—m³/hr; SCFM—Sm³/hr
2.23	Venturi Tubes and Flow Nozzles	√	L	L	√			√	√			√	SR	3:1 ②	M	⑦ 20/5	½-3*	gpm—m³/hr③; SCFM—Sm³/hr
2.24	Vortex Flowmeters																	
	Shedding	√		L	√			√	√			√	√	10:1 ⑧	H	15/5	½-1**	gpm—m³/hr
	Precession	L		L	√			√	√			√	√	10:1 ⑧	H	15/5	1**	ACFM—Am³/h
2.25	Weirs, Flumes	√	L	L				√	√			√	SD	100:1	M	See Text	2-5*	gpm—m³/hr③

FLOW RANGE scales:

Solids Flow Units: 0.1, 1.0, 10, 10², 10³, 10⁴ kgf/hr; 0.1, 1.0, 10, 10², 10³, 10⁴, 10⁵ lbm/hr

Gas Flow Units: 0.03, 0.3, 2.8, 28.3 cc/min; Sm³/hr or Am³/hr: 10⁻⁶, 10⁻⁵, 10⁻⁴, 10⁻³, 10⁻², 0.1, 1.0, 10, 10², 10³, 10⁴; SCFM or ACFM: 10⁻⁶, 10⁻⁵, 10⁻⁴, 10⁻³, 10⁻², 0.1, 1.0, 10, 10², 10³, 10⁴, 10⁵

Liquid Flow Units: .004, 0.04, 0.4, 3.8, 38, 379 cc/min; 10⁻⁶, 10⁻⁵, 10⁻⁴, 10⁻³, 10⁻², 0.1, 1.0, 10, 10², 10³, 10⁴ m³/hr; 10⁻⁶, 10⁻⁵, 10⁻⁴, 10⁻³, 10⁻², 0.1, 1.0, 10, 10², 10³, 10⁴ gpm 10⁵

– – – – – = Non Standard Range
L = Limited
SD = Some Designs
H = High
A = Average
M = Minimal
N = None
SR = Square Root

① = The data in this column is for general guidance only.
② = Inherent Rangeability of Primary Device is substantially greater than shown. Value used reflects limitation of differential pressure sensing device, when maximum readout accuracy is desired.
③ = Pipe size establishes the upper limit.
④ = Practically unlimited with the probe type design.
⑤ = Must be conductive.
⑥ = Can be re-ranged.
⑦ = Varies with upstream disturbance.
⑧ = Depends on application conditions.

2.1 APPLICATION AND SELECTION

Meter Selection

The variety of choices facing an engineer confronted with a flow measurement application is vast. The Orientation Table (See Table II) lists eighteen different categories of flowmeters. In nearly every case each category further subdivides into several distinctly different variants. For example, the positive displacement principle types include rotary piston, oval gear, sliding vane and reciprocating piston. If these sub-variants are included, the engineer faces a confusing list of well over 50 different meter types. Each type has advantages and limitations and no one type combines all the features and all the advantages.

A brief summary of the relevant advantages and limitation of each basic category is given below. Comparison tables on some of the key parameters, such as accuracy, flow range, and operating temperature, are also provided. It is hoped that this information, together with the basic selection philosophy outlined, will assist the engineer in finding his way through the maze of meter types and selecting suitable meters for his applications.

Having narrowed the list of choices, the engineer can then consult the other chapters for detailed information on the meters concerned.

Differential Pressure

This is undoubtedly the most widely used method of industrial flow measurement. A restriction is introduced into the pipe. The resulting pressure decrease is proportional to flow rate in accordance with the formula:

$$Q\text{ (flow)} = K\text{ (constant)}\sqrt{\frac{h\text{ (differential head)}}{d\text{ (fluid density)}}}$$

Differential pressure meters have the advantage that they offer the widest applicational coverage of any meter type. They are suitable for gas and liquid, viscous and corrosive fluids. There are no significant pipe size or flow rate limitations and there are no moving components. All differential pressure meters exhibit a square law relationship between head and flow rate which severely limits the usable flow range (typically 4:1 maximum). A separate transmitter is required; manifold and cocks are required (causing potential leakage points) and the fluid density must be known or measured. Types of differential pressure meters include:

Orifice Plates

Orifice plates are the simplest and cheapest form of differential pressure meter. The system cost is relatively independent of pipe diameter because the differential pressure transmitter remains at a fixed price regardless of main size. Orifice installations are consequently fairly expensive in small pipe sizes but very economical in pipe sizes above 6 in. (150 mm). Availability in a variety of materials and types (concentric, chord, eccentric) enables use on a wide range of applications. One advantage is that even when the orifice plate is badly worn or damaged, it will provide a reasonably repeatable output, albeit significantly inaccurate. The orifice plate can be isolated and the transmitter changed. Continuity of flow measurement to some degree is, therefore, virtually assured.

The limitations include high irrecoverable pressure loss (40 to 80 percent of generated head); deterioration in accuracy and long term repeatability with edge wear or deposition; maintenance required due to potential manifold leakage, tapping blockage, etc.; and limited flow range and accuracy. (See Figure 2.1a.)

Venturi and Proprietary Tubes

These tubes use specially designed and engineered shapes to give a low head loss. They are also more resistant to abrasion than an orifice plate and can be used on dirty fluids and slurries. They are, however, considerably larger, heavier and more expensive than the orifice plate. Installation is also more difficult. These tubes are largely used to enable lower capital expenditure on pumping equipment and save pumping energy costs.

Pitot Tubes

A pitot tube is a small tube inserted into the pipe. The differential pressure between the measuring point on the tube and the static pipeline pressure is measured. Pitot tubes offer a very low cost measuring system (for large diameter pipes) with negligible pressure loss. They are

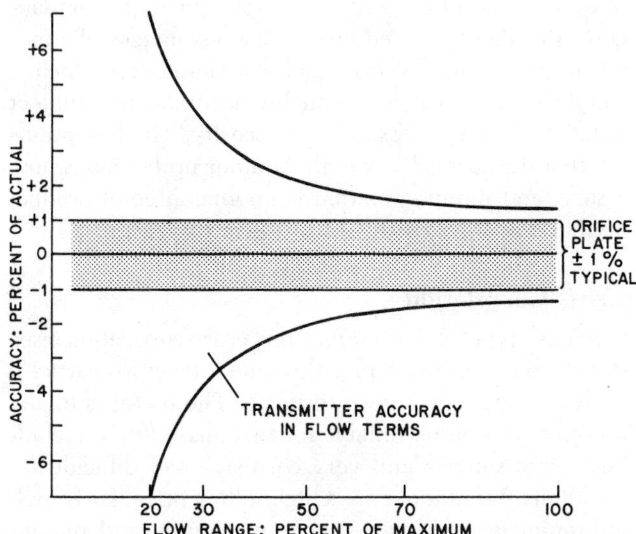

Fig. 2.1a Effect of ½ percent F.S.D. transmitter accuracy on overall differential pressure flow measurement system accuracy

also convenient for temporary measurements and pipe velocity traverses. Their principal limitation is that they measure a point velocity and, due to velocity profile changes, only provide very low accuracy volumetric readings. They are also subject to contamination and fouling.

Elbow Taps

Elbow taps provide flow rate by measuring the differential pressure across an elbow. This results in an economical installation using only one transmitter regardless of the pipe size. The system suffers from very poor accuracy; additionally, a relatively high flow velocity and long upstream straight pipe lengths are required.

Target (or Impact) Meters

In this system, deflection of a plate supported in the fluid flow is measured and an electronic or pneumatic analog output provided. The target meter is more expensive than a comparable orifice installation but since there are no tappings to block up, the target meter is more suited to fluids with suspended solids or "sticky" types of liquid. There are no moving parts. Operating flow range is limited (3:1) but it can be re-ranged.

Electromagnetic Flowmeters

By application of Faraday's Law, these meters measure the velocity of an electrically conductive liquid as it cuts the magnetic field produced across the metering tube. The principal advantages include no moving components; completely unobstructed bore and, therefore, no pressure loss and no components to become damaged; chemical compatibility with virtually all liquids; indifference to viscosity, pressure, temperature and density, linear analog output; suitability for bi-directional flow; extensive range of sizes; and ease and rapidity of reranging on site. The major limitation to the use of electromagnetic flowmeters is that the fluid must be electrically conductive. (This also excludes all gases.) The purchase price is high compared with alternative systems. A separate converter is required together with a power supply at the measurement point. Electromagnetic flowmeters are used for corrosive liquids, slurries and low pressure loss applications.

Turbine Meters

The speed of rotation of a bladed turbine meter, driven by the fluid, provides a digital output linear with flow rate. Turbine meters are available for liquids and gases, very low flow rates and as insertion designs. The liquid turbine meter is one of the most accurate meters available for low- to medium-viscosity products. Turbine meters offer good flow range capability (10:1) and are suitable for virtually unlimited pressures and extremes of high and low temperatures. They are easy to install and have a small size and weight relative to the pipeline diameter. The meter has a very fast response and can be made hygenic. The principal limitations are incompatibility with high viscosity liquids, possible damage due to overspeeding during a liquid/gas phase, and the necessity for secondary readout equipment. Filtration is also recommended. Turbine meters are widely used for high accuracy product sale, blending, test rig duty and general measurement.

Vortex Meters

The three types of vortex meters are vortex shedding, vortex precession and fluidic oscillation. All three types utilize fluid oscillation to provide meters with no moving components, suitable for gas or liquids. They generally have wide flow range capability, minimal maintenance, frequency or analog linear output, good accuracy, and long term repeatability. The liquid vortex meters are unsuitable for use on viscous or dirty fluids. The available size range is limited and Reynolds number must be greater than 10,000. Vortex shedding meters offer an attractive combination of technical features and low installed cost. Vortex precession meters are mainly used on gas and have a good performance capability. They are considerably more expensive than the others. Vortex shedding meters are gaining increasing acceptance and use as a general purpose, low cost alternative to the orifice plate and dp transmitter, and they are also used for specialized and demanding applications in the chemical industry.

Variable Area Meters

Variable area meters use a vertical tapered glass tube containing a float. The equilibrium position of the float is proportional to flow rate. The major features are very low cost, direct reading, very low flow rate capability, no power supply required, and suitability for both gases and liquids, including high viscosity products. Installation pipework is minimal, but the meter must be installed vertically. Limitations include limited size and pressure capability, high pressure loss, and limited accuracy. The transmission signal is not available as standard. Alternative designs feature metal tube gravity-loaded types with electrical sensors for higher flow rates, and pressure- and spring-loaded types. The latter meter offers good rangability and, potentially, a linear output. Variable area meters are widely used for small flow rate measurement where local indication is required, in test rigs and in general industry.

Positive Displacement Meters

Positive displacement meters trap a known fixed volume of fluid and transfer it from the inlet to outlet side of the meter. The number of "packages" of fluid is a measure of the flow. Designs include rotary piston, oval gear, sliding vane and reciprocating piston. Liquid PD meters offer good accuracy and flow range and are particularly suited to high viscosity products. The meters provide a local readout and do not require a power supply. They can be fitted with a transmitter to provide a linear transmission signal. The PD meter relies on close meshing surfaces and is, therefore, only suitable for clean fluids. It requires regular maintenance on non-lubricating liquids. Large sizes are bulky, heavy and more difficult to install. PD meters are used for viscous product sales and general purpose industrial metering.

Ultrasonic Meters

Ultrasonic meters are available in two forms: Doppler and transit time. With Doppler meters, an ultrasonic pulse is beamed into the pipe and reflected by inclusions, such as air or dirt. The Doppler meter is a "clamp on" device which can be fitted to existing pipelines. It is, effectively, a point velocity meter and as such its accuracy is not good. It is not suitable for clean fluids or gas. The meter cost is low and is independent of pipe size.

The single or multi-beam transit time flowmeters project an ultrasonic beam right across the pipe at an acute angle, first with the flow and then in opposition to the flow direction. The difference in transit time is proportional to flow rate. This type of ultrasonic meter is considerably more expensive but offers better accuracy. Unlike the Doppler meter, it requires a relatively clean fluid. There is no obstruction to flow and hence no pressure loss. It is ideally suited to very corrosive liquids. The approach pipework conditions are fairly critical. This type of meter is gaining limited acceptance in the water and chemical industries.

Metering Pumps

These units generate pressure and meter at the same time. The two basic types are the plunger pump for reasonable accuracy and the diaphragm type for dangerous or contaminated fluids. The advantages of a metering pump are that it is a self-contained unit which is simple to install; it is reasonably accurate, but subject to variation with pressure and viscosity. The limitations are that the cost is fairly high, strainer protection is necessary, and pump operation is no guarantee of product discharge.

Cross-Correlation

A new type of flowmeter, the cross-correlation uses statistical means to average the time it takes for particles in fluid to travel a known distance. The meter is totally non-invasive and is suitable for two phase flows, including heavy slurries and very corrosive and difficult applications. Limitations are a high selling price, fairly high minimum Reynolds number requirement, and reasonable to poor accuracy.

Sonic Venturi

This relatively new design is illustrated in Figure 2.1b. The meter body is that of a multiport digital control valve, where the area of each port is twice the size of a smaller one. The on/off ports are opened through binary manipulation and, therefore, the meter rangeability is only a function of the number of ports used. With 8 ports the rangeability is 255:1; with 10 it is 1023:1; with 12 it is 4095:1, and so on. The digital control valve is converted into a flowmeter through the insertion of a sonic velocity venturi into each of the ports. A sonic velocity venturi element passes a constant flow rate when the flow ve-

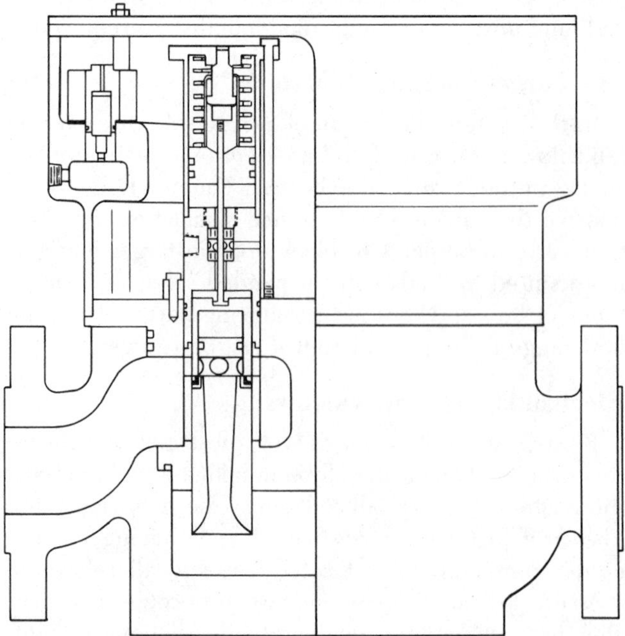

Fig. 2.1b Sonic venturi digital flowmeter featuring extremely wide rangeability

locity reaches the sonic velocity at its throat. Therefore, this flowmeter requires the continuous presence of sonic velocity of its throat, which in turn requires that the meter ΔP be greater than 15 percent of the upstream absolute pressure. Because of the inherent requirement for high pressure drops, this meter is ideal for applications where a combination device is required to both meter the flow and control the downstream pressure.

The meter accuracy of ½ to 1 percent of actual flow is the same at high or low flow rates throughout the meter range. With the addition of inlet gas pressure, temperature and/or density sensors, the meter can be converted for mass flow measurement. It can also meter liquids. This flowmeter is available in sizes from 1 to 8 in. (25 to 200 mm). Units have been built for up to 10,000 PSIG (69 MPa) pressure services and for temperatures from cryogenic to 1200°F (650°C).

Specialized Meters

The other meters listed in the orientation table serve more specialized functions and, as such, they do not tend to interact with the meters already mentioned. The individual sections contain details on these meters.

Flowmeter Selection Procedure

Meter selection falls into two main stages. First, identify the meters which are technically capable of performing the required measurement and are available in acceptable materials of construction; then, make the best choice from those available. A list should be made of the key parameters which the meter must be capable of accommodating. By comparing these requirements with the information given in Table II and Table 2.1c, a first pass elimination of technically unsuitable meters can be made. The list can then be further refined by a more detailed consideration of the applicational requirements against the "Features Summary" at the start of each appropriate meter section.

In order to cover special features such as reverse flow, pulsating flow, response time, and so on, it is necessary to study the individual meter specifications in detail and/or obtain the manufacturer's comments and advice.

Although the above steps will eliminate technically unsuitable meters, it does not necessarily follow that the available meters will be technically suitable for the application. The meter may possess the individual features required, but it may not be possible to find a combination of all of the desired features in one meter. Electromagnetic flowmeters, for example, are readily available for operating pressures of 1500 PSIG (10.3×10^6 N/m^2). They are also readily available for flow rates of 500,000 g.p.m. (31.5 m^3/sec.) but the availability of an electromagnetic flowmeter for use at 1500 PSIG *and* a flow rate of 500,000 g.p.m. is an entirely different proposition. Although such an applicational requirement may not exist, the example does illustrate the danger of assuming that all of a meter's features can automatically be incorporated in one individual meter.

The length of the list of technically suitable meters will depend on the complexity of the application. On an extreme application such as a highly corrosive, non-conductive liquid with large solid content, the list will probably consist of one meter at most (cross-correlation meter). On a straightforward clean water application, the list will consist of nearly all the flowmeters listed in the Orientation Table.

In order to make his subsequent selection, the engineer should concentrate on the reasons for measuring the flow. The key requirements should be identified and a weighting applied to them. For example, is high accuracy the most important requirement, or is long term repeatability, low installed cost, or easy maintenance?

It is essential that the requirements are objectively specified. Otherwise, a utopian but unavailable flowmeter will be required. Comments on various key topics follow.

Accuracy

The one parameter which is over-specified more than any other is accuracy. Statements such as "best possible," "better than one-quarter percent," and the like are frequently made. If taken at face value, they will severely limit the meter choice and often result in an unnecessarily high installation cost. A realistic accuracy level should, therefore, always be set. In many instances—for example, certain process control measurements, or repetitive batch dispensing—true accuracy is of no consequence at all, provided the meter gives a stable, repetitive, long term output. On such applications accuracy should be ignored and the emphasis placed on long term repeatability. Where true accuracy is required, it is necessary to be very specific about the meter specification and also to establish the likely on-site accuracy, which may well be different from the manufacturer's performance figures.

The difference between accuracies quoted as a percent of F.S.D. (full scale deflection) and those quoted as a percent of actual (or point or reading) is often overlooked. Figure 2.1d compares the two alternative specifications against a common scale. The rapid deterioration in true accuracy at low flow rates with the percent of F.S.D. specification is very apparent.

Likewise, the terms "linearity" and "accuracy" are often taken as one and the same thing. This is not the case. Accuracy is the closeness to the truth, whereas linearity is the deviation from a straight line which has a constant number of pulses per unit volume value. Consequently, a meter can be specified as being linear to ± 0.5 percent of point and the statement proved by an initial factory calibration, but unless the nominal calibration constant is known under the actual site operating

Table 2.1c
FLUID DUTY SELECTION TABLE

	Section	2.4	2.5	2.8	2.9	2.10	2.11	2.12	2.13	2.14	2.15	2.17	2.18	2.19	2.20	2.20	2.21	2.21	2.22	2.23	2.24	2.24	2.24
	Fluid Details / Meter Type	Correlation	Elbow Taps	Laminar	Electro-Magnetic	Angular Momentum	Metering Pumps	Orifice	Pitot	Gas Displacement	Liquid Displacement	Solids flowmeter	Target	Thermal	Liquid Turbine	Gas Turbine	Doppler U-Sonic	Transit U-Sonic	V.A.	Venturi	Vortex Shedding	Vortex Precision	Fluidic Oscillation
Liquid	Clean	X	√	√	*√	√	√	√	√	X	√	X	√	√	√	X	X	√	√	√	√	X	√
Liquid	Dirty	√	?	√	*√	√	√	?	?	X	X	?	√	√	?	X	√	?	√	√	?	X	?
Liquid	Slurries	√	X	?	*√	?	√	X	X	X	X	SD	?	?	X	X	?	X	X	?	X	X	X
Liquid	Low Viscosity	√	√	√	*√	√	√	√	√	X	?	X	√	√	√	X	√	√	√	√	√	X	√
Liquid	High Viscosity	√	?	?	*√	?	√	?	X	X	√	SD	?	?	X	X	?	?	?	?	X	X	X
Liquid	Corrosive	√	√	?	*√	√	?	√	√	X	?	X	?	?	?	X	√	√	√	?	?	X	?
Liquid	Very Corrosive	√	?	X	*√	X	X	?	?	X	X	X	X	?	X	X	√	√	√	X	X	X	X
Gas	Low Pressure	X	√	√	X	√	X	√	√	√	X	X	√	√	X	√	X	X	√	√	√	√	X
Gas	High Pressure	X	√	√	X	√	X	√	√	√	X	X	√	√	X	√	X	X	X	√	√	√	X
	Steam	X	X	?	X	X	X	√	X	X	X	X	√		X	SD	X	X	√	√	SD	X	X
	Reverse Flow	X	√	X	√	X	X	SD	X	X	X	X	X		SD	SD	√	√	X	X	X	X	X
	Pulsating Flow	?	X	√	√	X	X	?	X	X	X	X	X		X	X	√	√	?	?	X	X	X

* = Must Be Electrically Conductive
√ = Generally Suitable
? = Worth Consideration
X = Not Suitable
SD = Some Designs

conditions, the true accuracy of the meter is also unknown.

The published meter performance specification is generally based upon stipulated installation and operating conditions. It is an indication of the performance which the meter can achieve but it is no guarantee that the same performance will be achieved on site under actual operating conditions. Substandard approach pipework, resulting in a high degree of swirl in the liquid, will cause many meter types to experience a dramatic deterioration in the linearity tolerance and a shift in the nominal calibration constant. Consequently, the manufacturer's installation recommendations should be closely adhered to, or, if this is not possible, the resulting consequences on the meter's performance should be established.

Changes in the fluid characteristics can also alter the meter's performance. Figure 2.1e illustrates the effect of viscosity on two of the most accurate flowmeters, the

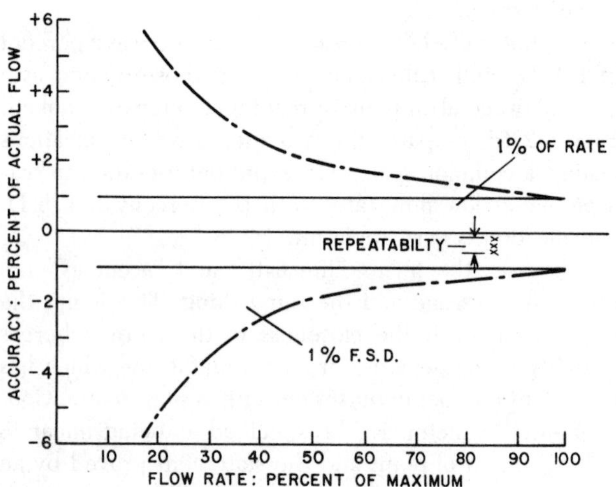

Fig. 2.1d Comparison of 1 percent F.S.D. accuracy with 1 percent of flow accuracy

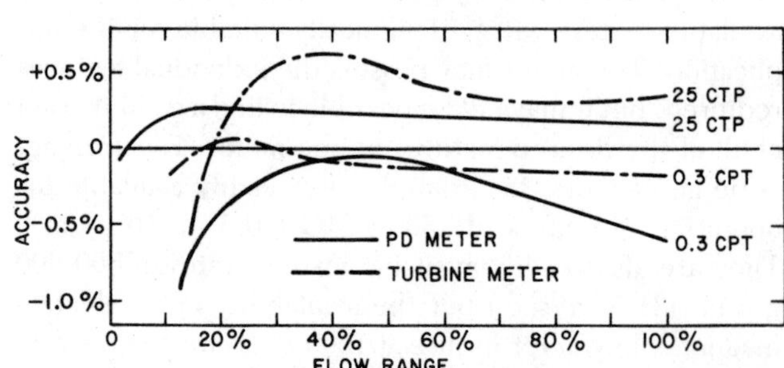

Fig. 2.1e Differing effects of viscosity variation on a turbine and a positive displacement meter

turbine meter and the positive displacement meter. With the turbine meter, an increase in viscosity causes a deterioration in the performance specification, whereas with the PD meter, a reduction in viscosity causes a deterioration in the performance specification. Acceptability of the proposed operating conditions should always be verified for any application.

Wear, analog drift, calibration shift, etc., should also be investigated for their differing effects on the alternative meter types being considered. Figure 2.1f illustrates an inline ballistic prover that can be used to recalibrate flowmeters, without requiring the interruption of the process flow.

Safety

Safety is arguably the most important consideration of all in the selection of any equipment. The meter and associated equipment must, obviously, be certified as being suitable for the electrical safety classification of the area in which it will be used. This may be achieved by use of purely mechanical or pneumatic equipment, or, more commonly, by use of certified intrinsically safe flameproof or explosion proof units.

Other safety aspects, which are more frequently overlooked, are the influence of the materials of construction on safety and the avoidance of leakage. Fluids such as oxygen or liquid chlorine can present a most dangerous explosive potential because of their chemical reaction with certain materials. If the heat of the reaction cannot be transferred away and, especially, if pressure is confined, a violent explosion can result. Various organic and inorganic substances, including ordinary lubricants such as oil, grease, and wax, can cause an explosion in the presence of oxygen or chlorine. It is, therefore, essential that any flowmeter for such duties does not have any of these elements present and that it has been thoroughly cleaned and degreased.

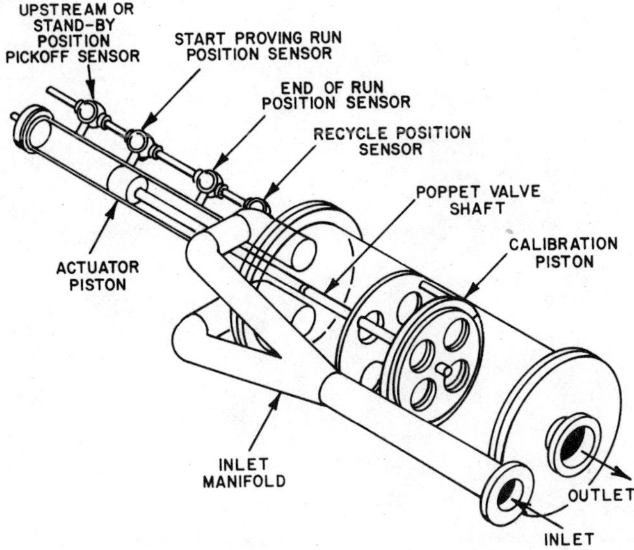

Fig. 2.1f Inline ballistic flow prover

The choice of the materials of construction is also critical for any oxygen or oxygen-enriched application. Steels, for example, present an increasing explosion hazard as the oxygen velocity and pressure increase. The cleanliness and surface finish affect the safe velocity and pressure limitations and, consequently, the point at which steels become unacceptable. This critical situation can be avoided by the use of alternative materials, such as phosphor bronze, gunmetal, brass, beryllium, copper, etc.

Where noxious or dangerous fluids are being metered, it is essential, from an operator safety standpoint, that any possibility of leakage is kept to an absolute minimum. The use of meter types which do not protrude into the pipe or feature solid metal housings without any manifolds, tappings, or fragile components help to ensure security and freedom from leakage.

Installation

Installation requirements vary dramatically between the various meter types (see Table II). Installation is often the deciding factor in meter selection. An extreme example is an operational pipeline which cannot be shut down or the measurement point bypassed. In this case, the user's choice is severely curtailed and is limited to clamp-on meters, such as Doppler and cross-correlation, or hot tap insertion meters.

When the above special conditions do not apply, installation requirements are still an important factor both with regard to cost and plant acceptability. In some locations, the requisite straight pipe lengths cannot be provided and it is then necessary to accept a de-rated performance or consider an alternative meter type, such as an electromagnetic meter which has minimal straight pipe requirements. Specific application requirements affect different meters in different ways. Unavailability of an electricity supply at the measurement point eliminates the electromagnetic flowmeter from consideration. If a vertical location cannot be found for a variable area meter, it cannot be used. A positive displacement meter requires a strainer, and so on. Even if the meter installation conditions can be met, their effect on the overall system cost should still be considered and quantified.

Cost

Cost is a critical factor in the selection of any equipment. To arrive at a reasoned decision on the subject, it is essential to consider cost in its widest possible context and not limit cost analysis to purchase price. Other factors, such as operating cost, maintenance, required spare parts inventory, the effect of down time—these and many others should all be considered if a "reasoned" decision is to be reached. On some applications, the most significant aspect related to cost is the effect which an improved measurement accuracy, system repeatability or operating flow range can have on plant efficiency and/or the quality of the end product.

The initial purchase cost comparison should be based upon system cost and not merely the flowmeter price. If, for example, the meter requires a separate converter or transmitter to produce the required pneumatic output, digital or analog transmission signal, linear output, etc., then these costs should be included. Likewise, the cost of ancillary items, such as recommended upstream and downstream pipe requirements, flow conditioning elements, filters, power supplies and so on, should all be considered. So too should the cost of the installation itself. Installation complexity varies significantly between meter types, and, with current high labor rates, installation cost can be a significant factor.

Operating cost is a complex and variable subject. The types of questions which should be asked about the various aspects include:

Routine Service and Maintenance—how much, if any, routine service is recommended? What level of personnel is involved? Are special flow simulator units available?

Versatility—Can the secondary units be used on other meters? Can the meter be easily re-ranged? Can the meter be used on other applications?

Spare Parts Inventory—What level and value of inventory is required? Are the spares interchangeable with other sizes and models?

Reliability—Are there moving components? What is the guarantee and estimated meter life? What are the likely failure modes?

Even the meter pressure loss can have cost ramifications. If the choice, for example, were between an orifice plate and a low loss differential pressure tube, the orifice plate would obviously be far cheaper but its irrecoverable head loss would be far greater. Pumping cost is a function of flow rate, electricity costs, running efficiency and pressure loss or head. Consequently, the higher pressure loss of the orifice plate would result in increased pumping costs throughout the life of the installation.

BIBLIOGRAPHY

Hayward, A.J., "Choose the Flowmeter Right for the Job," *Processing Journal*, 1980

Laskaris, E.K., "The Measurement of Flow," *Automation*, 1980

Lomas, D.J., "Selecting the Right Flowmeter," *Instrumentation Technology*, 1977.

Watson, G.A., "Flowmeter Types and Their Usage," *Chartered Mechanical Engineer Journal*, 1978.

Flow Sheet Symbol

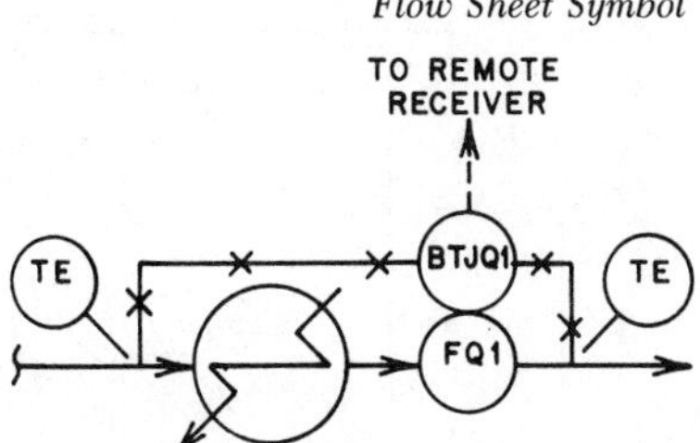

2.2 BTU FLOWMETERS FOR HEAT EXCHANGERS

BTU Flowmeter Types:	A—All mechanical design B—Electronic BTU computer
Approximate Cost:	A—About 50% more than the cost of the positive displacement flowmeter. If for example a 6 in. (150 mm) propeller meter is the flow sensor, the total cost is about \$3,000 B—\$2,100 in addition to the cost of the flow and temperature transmitters
Inaccuracy:	A— ± 2 to 5% of full scale B— ± 0.5% of full scale
Minimum ΔT:	A—5°F (2.8°C) B—1°F (0.56°C)
Partial List of Suppliers:	American Meter Co. (A); Hersey Products, Inc., Industrial Measurement Div. (A); ITT Barton (B). The analog or digital BTU computer can be made from the standard components of all major instrument suppliers

The first step toward energy conservation and toward the goal of energy efficient plant design is a reliable energy audit around the plant site. An overall heat balance around the plant can only be established if the individual loads are accurately measured. This is illustrated in Figure 2.2a.

The efficiency of the boiler is measured by the ratio of integrated energy flows at points 1 and 2. This is usually done by totalizing the fuel and steam flows over some period of time. The total fuel consumed over that period is multiplied by its heating value to obtain the total energy input into the boiler. Multiplying the totalized steam flow by the difference between the enthalpy of the steam and of the feedwater gives the total useful energy obtained from the boiler. The ratio of the two is the boiler's efficiency.

Similarly, the coefficient of performance of the chiller is measured by the ratio of energy flows at points 10 and 11. The efficiency of individual heat exchangers is also detected by measuring the energy flows on the utility and on the process sides, such as points 3 and 4 or points 12 and 13 in Figure 2.2a.

The efficiency of the utility distribution system is determined by comparing the sum of the individual loads with the total supply at the source. The difference between the two represents the losses due to insufficient thermal insulation of the pipe lines, bad steam traps and other causes.

When various optimization techniques are being considered, one of the first steps in the cost-benefit analysis is to empirically measure the energy consumption with or without optimization. The measured saving is then used on the benefit side to calculate the payback period for the installation.

BTU flowmeters are required in many of the energy flow sensor locations noted in Figure 2.2a. These BTU computing units are available either in mechanical or electronic designs.

Mechanical BTU Meters

In the mechanical BTU meters, flow is detected by positive displacement or propeller-type sensors and is mechanically transmitted through gear trains. The temperature difference is sensed by filled thermal bulbs, which are connected to bourdon springs as illustrated in Figure 2.2b. Dual cam rollers are used in the computing mechanism which produces the digital displays of both total BTUs and total flow.

The advantages of this design include its simplicity, low cost and the fact that it does not require any power supply. Transmitting attachments can also be provided where remote readouts are needed.

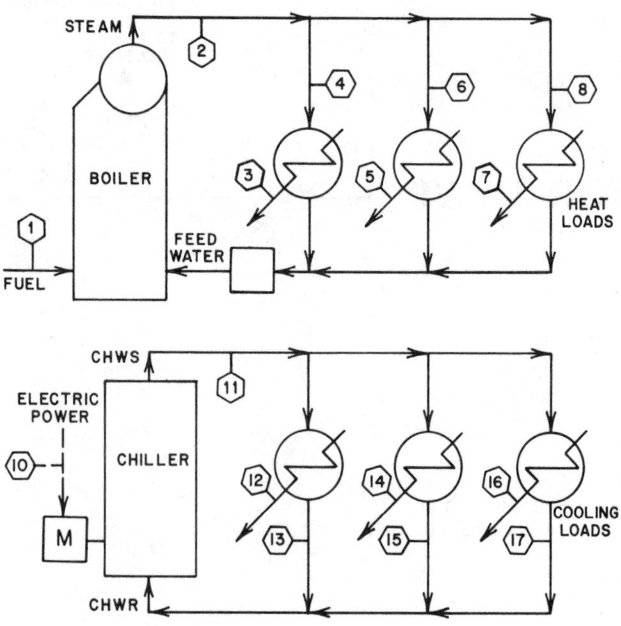

Fig. 2.2a Plant-wide energy audit

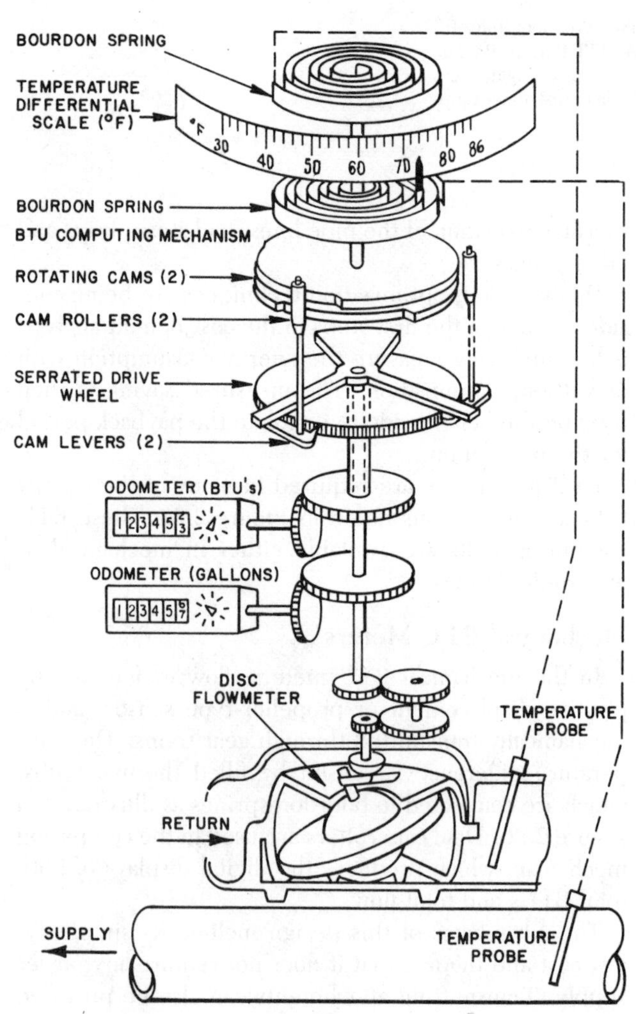

Fig. 2.2b Mechanical BTU Meter

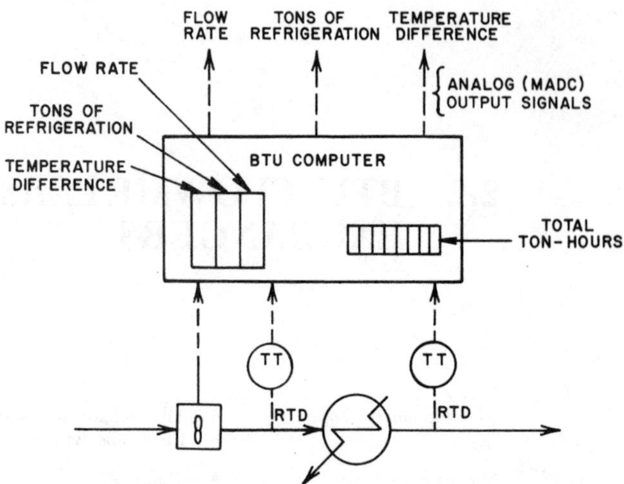

Fig. 2.2c Electronic BTU meter

The limitations of this design include its relatively low accuracy: about ± 2 percent of full scale error at temperature differences of 15°F (8.3°C) or higher. As the temperature difference decreases, the error tends to increase. These units are not recommended for temperature differences under 5°F (3°C).

The typical applications for the mechanical BTU meters are in the heating, ventilating, and air conditioning (HVAC) industry and in heat exchanger efficiency monitoring.

Electronic BTU Meters

In the electronic BTU computer packages, the flow sensor is usually a high accuracy turbine flowmeter. The two temperatures are usually detected by resistance bulb-type temperature transmitters. Both the flow and the temperature sensors are accurate devices with high repeatability and turndown.

Therefore, the main advantage of electronic BTU computers is their relatively high accuracy. The error is held under ± 0.5 percent of full scale. As illustrated in Figure 2.2c, the BTU computer digitally displays the accumulated total ton-hours and provides analog indication of flow rate, tonnage rate and temperature difference. In addition, analog electronic retransmission signals are provided to facilitate the remote display of flow rate, BTU rate and temperature difference.

While these units are more expensive than the mechanical BTU meters, the added cost can frequently be justified on larger or more critical installations, where accuracy is a prime concern.

BIBLIOGRAPHY

"BTU Computer," Bulletin BTU-1, published by Barton-ITT.

Reese, W.M. Jr., "Factor the Energy Costs of Flow Metering," *InTech*, July 1980.

Shinskey, F.G., *Energy Conservation Through Control*. New York: Academic Press, Inc., 1978.

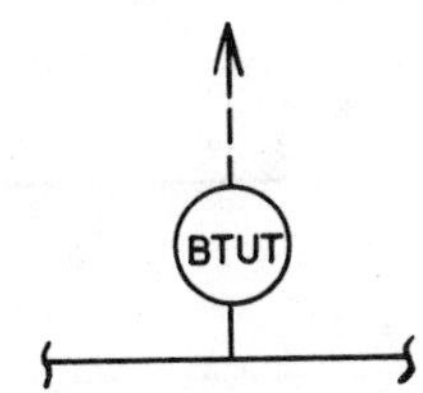

2.3 BTU FLOWMETERS FOR GASEOUS FUELS

Design Basis:	Conventional head-type flowmeter and heat release (Wobbe Index) measurements
Fluids:	Gaseous fuels
Applications:	Feedforward control loops for burning processes; BTU rate integration
BTU Flow Range:	From 100 BTU/min. to very large flow rates and as limited only by pipe sizes
Inaccuracy:	± 1.0% to ± 2.0% of full scale based on accuracy of head meter
Cost:	\$10,000 to \$12,000 — general purpose \$16,000 to \$18,000 — explosion-proof
Partial List of Suppliers:	Bailey Controls Company; Fischer & Porter Co.; Fluid Data, Inc.; Foxboro Co.; Honeywell Inc., Process Control Div.; Leeds and Northrup a unit of General Signal; Taylor Instrument Co., Div. of Sybron Corp.

Measurement of heat flow rate has become necessary as a result of the significant changes in sources of gaseous fuel (see Table 2.3a) and cost as dictated by supply and demand. Heretofore, standard procedure was simply to measure gas flow and use constants for the heat value and specific gravity of the gas, but this method is no longer satisfactory for efficient operation of burning processes.

Until recently, instruments to measure heat value of gaseous fuels have not been practical for continuous process conditions or for use in hazardous areas. These limitations have in the past required that specific gravity be used to determine inferentially the heat value of the fuel gas, in addition to the use of the specific gravity measurement as a means of compensation to arrive at a more efficient burning method. With heat value measuring instruments now available (see Section 11.3), it is possible to measure the potential heat of gaseous fuels and compensate for both variables — specific gravity and heat value — in combination with a gas flow measurement using an orifice meter, resulting in a BTU flow rate measurement.

Measuring Heat Flow

The heat flow rate of a gaseous fuel is defined as the volume of gas being delivered and the quality, or composition, of the gas which affects both its flow rate through an orifice as well as its heating value. The heat flow through an orifice can be expressed by the following simplified equation:

$$Q = V_0 \times CV = \frac{SCF}{HR.} \times \frac{BTU}{SCF} = \frac{BTU}{HR.} \qquad 2.3(1)$$

the flow through an orifice is expressed as:

$$V_0 = K\sqrt{\frac{\Delta P}{SG}} \qquad 2.3(2)$$

substituting equation 2.3(2) into 2.3(1):

$$Q = K\frac{\sqrt{\Delta P}}{SG} \times CV \qquad 2.3(3)$$

For the definition of Wobbe Index, the pressure is held constant, thus resulting in:

$$\text{Wobbe Index} = \frac{CV}{\sqrt{SG}} \qquad 2.3(4)$$

Rearranging equation 2.3(2):

$$\sqrt{\Delta P} = KV_0\sqrt{SG} \qquad 2.3(5)$$

Table 2.3a
COMBUSTION CONSTANTS AND COMPOSITION OF REPRESENTATIVE MANUFACTURED AND NATURAL GASES

	Blast Furnace Gas	*Coal Gas*	*Coke Oven Gas*	*Natural Gas Residual, Follansbee, W. Va.*	*Natural Gas Sandusky, Ohio*	*SNG Green Springs, Ohio*	*LNG Columbia Gulf Coast*	*NG Columbia Gulf Coast*	*Refinery Gas*	*Producer Gas*
% Methane, CH_4		34.0	28.5		83.5	98.914	85.136	97.528	27.0	2.6
% Ethane, C_2H_6				79.4	12.5	.01	10.199	1.238		
% Propane, C_3H_8				20.0			3.06	.241		
% Ethylene, C_2H_4		6.6	2.9				.0016		2.7	0.4
% Carbon monoxide CO	26.2	9.0	5.1			.025			10.6	22.0
% Carbon dioxide, CO_2	13.0	1.1	1.4		0.2	.439	.018	.487	2.8	5.7
% Hydrogen, H_2	3.2	47.0	57.4			.61			53.5	10.5
% Nitrogen, N_2	57.6	2.3	4.2	0.6	3.8	.002	.201	.224	3.4	58.8
% Oxygen, O_2			0.5				.007			
% Other*							1.37	.192		
BTU per cu. ft., high (gross) 60°F, 30 in Hg, satd. H_2O	93	634	536	1868	1047				516	136
BTU per cu. ft., low (net) 60°F, 30 in Hg, satd. H_2O	91.6	560	476	1711	946				461	128
Flame Temp. °F	2660	3910	3430	3830	3740				3970	3050

*Heavier Hydrocarbons and traces of compounds including sulfurs.

Multiplying equations 2.3(4) and 2.3(5):

$$\text{Wobbe Index} \times \sqrt{\Delta P} \qquad 2.3(6)$$

$$\frac{CV}{\sqrt{SG}} \times K\, V_0 \sqrt{SG} \qquad 2.3(7)$$

Eliminates the effect of specific gravity and provides an on-line measurement of heat flow as defined in equation 2.3(1)

where

V_0 = Volume at standard conditions
CV = Calorific Value
K = Constants
ΔP = Differential Pressure across orifice
SG = Specific Gravity

The BTU Flowmeter Loop

The BTU flowmeter for gaseous fuels consists of a heat value measuring instrument (see Section 11.3) and a typical flow measuring loop with an orifice element. When the signals are properly conditioned and multiplied as defined above in equation 2.3 (7), a typical BTU flowmeter loop providing a BTU flow rate signal results, as shown in Figure 2.3b.

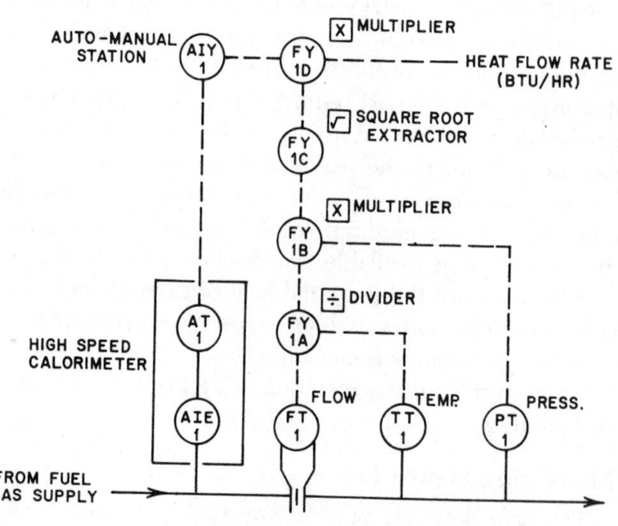

Fig. 2.3b BTU flowmeter loop

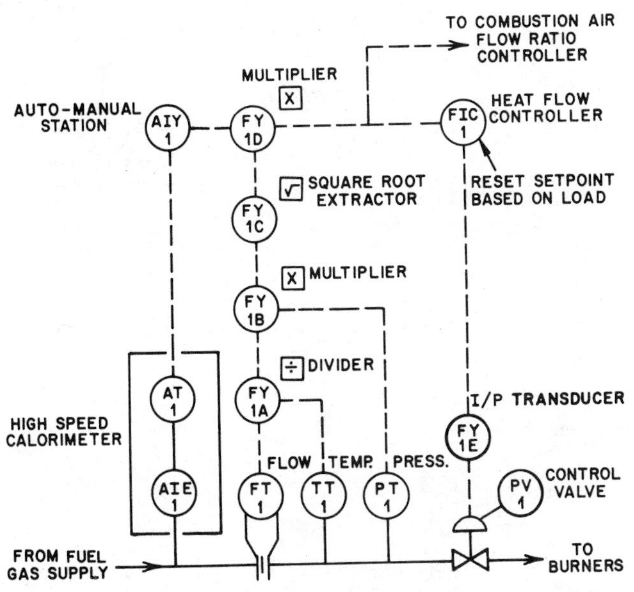

Fig. 2.3c Combustion control loop using BTU rate signal

Applications

As shown in Figure 2.3b, the BTU measuring loop can be used to generate a true indication of the BTU flow rate for use in a typical combustion control system. It will meter the air in a direct ratio to actual BTU flow rate in a feedforward mode as well as respond more directly to the feedback loop signal generated by the load requirements, as depicted in Figure 2.3c. This feedback signal modulates the fuel gas control valve to deliver the desired BTU flow rate rather than volumetric gas flow rate. This is one of several modes of application of the instrument, and perhaps the most important in view of its wide applicability in programs of improvement. It should be pointed out that in boilers or heaters using multi-fuel systems such as oil and gas, the measurement takes on even more significance since delivering the actual BTU flow rate from the gaseous fuel makes an excess air measurement more meaningful in controlling the other fuel parameters.

Another common application of BTU flowmetering is for blending an enriching or diluent gas into varying composition fuels to control the BTU to a constant level. The fuel gas thus produced can be distributed through a plant supply network to various boilers, heaters, and gas-fired processes. In this case, however, conventional flow control is directly related to the BTU flow rate. Figure 2.3d shows such a flow schematic of a typical blending control loop.

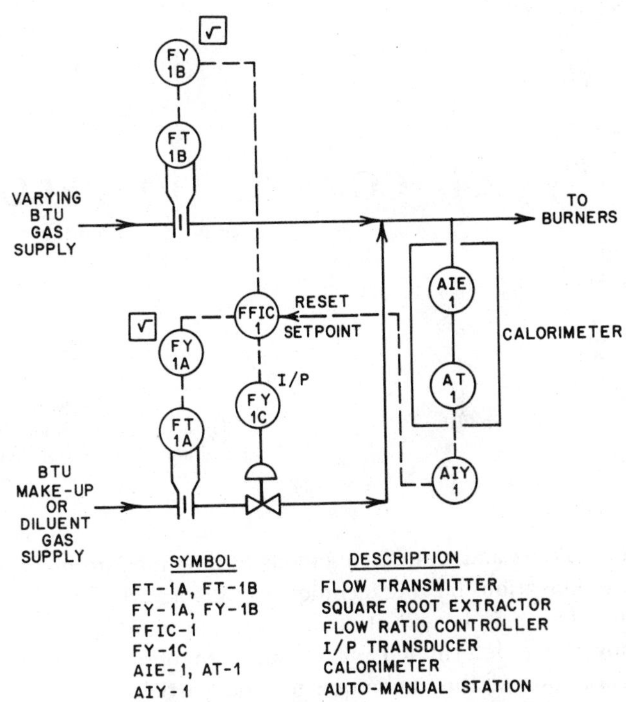

SYMBOL	DESCRIPTION
FT-1A, FT-1B	FLOW TRANSMITTER
FY-1A, FY-1B	SQUARE ROOT EXTRACTOR
FFIC-1	FLOW RATIO CONTROLLER
FY-1C	I/P TRANSDUCER
AIE-1, AT-1	CALORIMETER
AIY-1	AUTO-MANUAL STATION

Fig. 2.3d Mixing of two gas streams to supply controlled BTU gas

Conclusions

Using proven, available technology, a BTU flow rate loop can be designed for gaseous fuels to provide a direct signal for use in control applications. The basic elements of such a loop are an orifice flowmeter and a heat value measuring instrument with relatively good speed of response for typical burning processes. Measuring BTU flow rate provides a direct, accurate method of controlling combustion process variables to optimize controllability and, therefore, the rate of consumption of gaseous fuel.

BIBLIOGRAPHY

Foundos, A.P., "Measuring Heat Release Rate from Fuel Gases," *Instrumentation Technology*, Instrument Society of America, 1977.

Foundos, A.P., "On-Line Optimizing Fuel Gas Composition Variations," *Instrumentation Technology*, Instrument Society of America, 1980.

Perry, J., *Chemical Engineers Handbook*, 3rd Edition. New York: McGraw-Hill, 1950, p. 1577.

Reineke, H.F., Patent No. 1,055,259, Federal Republic of West Germany, 1957.

Flow Sheet Symbol

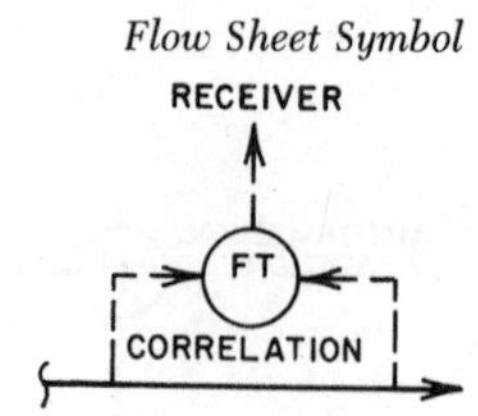

2.4 CORRELATION FLOWMETERING

Status: These devices are under development. No commercial units are yet available.

The oldest and simplest methods of flow measurement are the various tagging techniques. Here a portion of the flowstream is tagged at some upstream point and the flow rate is determined as a measurement of transit time. Variations of this technique include particle tracking, pulse tracking, dye or chemical tracing, including the radioactive types. The advantages of tagging techniques include the ability to measure the velocity of only one component in a multicomponent flowstream without requiring calibration or pipeline penetration. For example, electromagnetic tagging of gas-entrained particles allows for the determination of their speed through the detection of their time of passage between two points that are a fixed distance from each other.

Flowmetering based on correlation techniques[1,2] is similar in concept to the tagging or tracing techniques because it also detects transit time. As illustrated in Figure 2.4a, any measurable process variable which is noisy (displays localized variations in its value) can be used to build a correlation flowmeter. The only requirement is that the noise pattern must persist long enough to be seen by both detectors "A" and "B" as the flowing stream travels down the pipe. Flow velocity is obtained by dividing the distance (between the identical pair of detectors) by the transit time. In recent years, the required electronic computing hardware, with fast pattern recognition capability, has become available and consequently it is feasible to build on-line flowmeters using this technique.[3]

Fig. 2.4a Correlation flowmeter

The following process variables display persistent-enough noise patterns (or local fluctuations) so that correlation flowmeters can be built by using an identical pair of these sensors:

Density
Pressure
Temperature
Ultrasonics
Gamma Radiation
Capacitive Density
Conductivity

Several of the above process variables, such as temperature, [4,5] gamma radiation [6] and capacitive density [6] have been investigated as potential sensors for correlation flowmeters. At least one instrument has actually been developed, utilizing the principle of ultrasonic cross-correlation to measure heavy water flow.[3]

When fully developed, correlation flowmetering can extend the ability to measure flow not only into the most hostile process environments, but also in the areas of multiphase flow or when three-dimensional flow vector data is required.

REFERENCES

1. Porges, K.G., "On-line Correlation Flowmetering in Coal Utilization Plants," 1980 Symposium on Instrumentation and Control

of Fossil Energy Processes, June 9–11, 1980, Virginia Beach, Virginia.
2. Porges, K.G., "Correlation Flowmetering Review and Application," 1979 Symposium on Instrumentation and Control for Fossil Energy Processes, August 20-22, 1979, Denver, Colorado.
3. Flemans, R.S., "A New Non-Intrusive Flowmeter," Transactions, Flow Measurement Symposium, NBS, February 23–25, 1977.
4. Ashton, M.W., Bentley, P.G., "Design Study For On-Line Flow Measurement by Transit Time Analysis of Temperature Fluctuations," Conference on Industrial Measurement Techniques for On-Line Computers, June 11–13, 1968, London.
5. Boonstoppel, F., Veltman, B., Vergouwen, F., "The Measurement of Flow by Cross-Correlation Techniques," Conference on Industrial Measurement Techniques for On-Line Computers, June 11–13, 1968, London.
6. O'Fallon, N.M., "Review of the State-of-the-Art of Flow and Analysis Instruments," 1977 Symposium on Instrumentation and Control for Fossil Demonstration Plants, July 13–15, 1977, Chicago, Illinois.

BIBLIOGRAPHY

Beck, M.S., Calvert, G., Hobson, J.H., Lee, K.T., Mendies, P.J., "Flow Measurement in Industrial Slurries and Suspensions Using Correlation Techniques," Transactions of the Institute of Measurements and Control, Volume 4, Number 8. August, 1971, England.

Flow Sheet Symbol

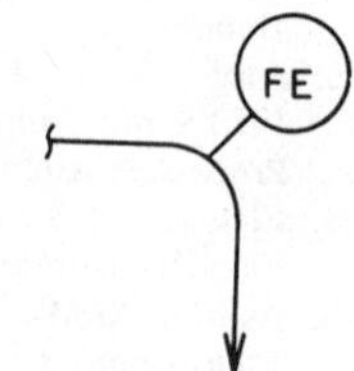

2.5 ELBOW TAPS

Design Pressure, Temperature:	Function of readout device and of the rating of the process pipe.
Fluids:	Liquids or gases
Flow Range:	Function of pipe size
Inaccuracy:	±5% to ±10% of full scale
Materials of Construction:	Same as process pipe
Cost:	Approximately that of a pipe elbow
Partial List of Suppliers:	Usually field fabricated by user

Flow measurement using elbow taps depends on a measurement of the differential pressure developed by centrifugal force as the direction of fluid flow is changed in a pipe elbow.* Taps are located at opposite ends of a diameter in the plane of the elbow; the diameter which passes through the taps is at either 45 degrees or 22½ degrees from the inlet face of the elbow (Figure 2.5a).

Elbow taps have an advantage in that most piping configurations already contain elbows in which taps can be located. This allows economical installation and results in no added pressure loss. The measurement introduces no obstructions in the line. With normal precaution against accumulation of extraneous material in the differential pressure connections, elbow taps may be used to measure flow of almost any fluid.

As with other head-type primary flow measurement devices, the differential pressure developed by a given flow is precisely repeatable. However, the coefficient of an elbow tap calculated from the physical dimensions is generally considered reliable to only ±5 to ±10 percent. This is quite satisfactory for many flow control applications, where repeatability is the primary consideration. If absolute accuracy is essential, a more precise type of meter is recommended, or an actual flow calibration of the system can be performed, preferably in place and using the working fluid. Data on elbow tap measurement is insufficient to establish precise correction factors for effects of upstream disturbances, viscosity, and roughness in pipe, elbow surfaces, etc.

Elbow taps develop a relatively low differential pressure. For this reason, their use is questionable for measurement of streams with low velocity. Typically, water flowing at an average velocity of 5 feet per second (roughly 200 gpm in a 4-inch pipe or 1.5 meters per sec-

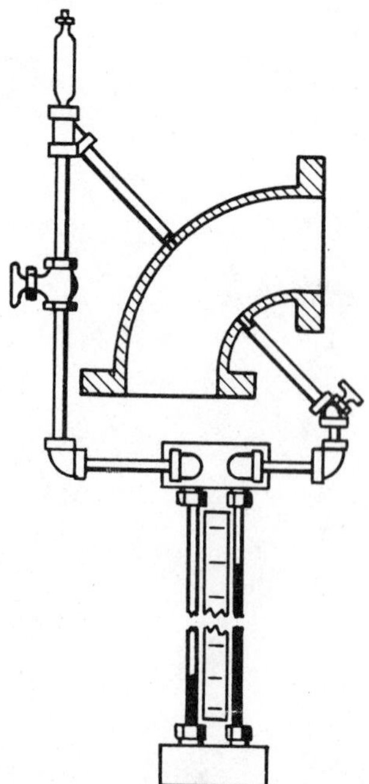

Fig. 2.5a Flow detection with elbow taps

*Bulletin #289—Wallace M. Lansford, Engineering Experiment Station, University of Illinois, Urbana, Illinois.

ond, roughly 45 cubic meters per hour in a 100 mm pipe) through a conventional elbow with a center line radius equal to the pipe diameter develops about 10 inches of water differential pressure. This approaches minimum full scale value recommended for reliable measurement. Taps in long radius pipe or tube bends do not develop sufficient differential pressure for good flow measurement at low flow velocities.

Care in installation of elbow taps is required, as it is with other head-type meters. Straight runs of pipe at least 25 pipe diameters upstream and 10 diameters downstream are recommended. The tap holes should be perpendicular to the surface of the elbow and slightly rounded at the pipe surface with no burrs or protrusions. Tap hole diameter should not exceed ⅛ of the pipe diameter. Elbows should be of the flange type with the elbow diameter equal to the pipe diameter. An elbow of smaller diameter than the pipe with a reducer between pipe and elbow has the advantage of higher differential for a given flow. Threaded elbows with the flow section larger than the pipe develop less differential pressure and introduce major uncertainty in calculated coefficient. Best results, particularly as to reliability of coefficient, are obtained with elbows with smooth inside surfaces. The elbow should be precisely aligned with the pipe with no projecting surfaces or gaskets protruding into the flowing stream either at the inlet or outlet of the elbow.

For elbow tap measurement, an elbow located between two horizontal pipes is preferable. This results in horizontal pressure taps. Second choice is an elbow between a vertical pipe with the flow upward and a horizontal pipe. Generally, except for an elbow between two horizontal pipes, one differential tap will slope upward and the other will slope downward. Especially with the low differential pressure developed in most elbow taps, any accumulation of liquid in supposedly gas or vapor-filled differential pressure connections, or accumulation of gas or vapor in supposedly liquid-filled connections can result in unsatisfactory performance.

Some tests indicate that the 22½-degree location for elbow taps is more stable, reliable, and less subject to approach conditions. However, these data are far from conclusive for the full range of operating conditions. For reversing flow measurement, the 45-degree location is advantageous.

Calculations of flow based on dimensions of the elbow are approximate with an uncertainty in absolute value of up to ±10 percent. A simple expression is

$$W = 244\sqrt{rhD^3\rho} \qquad 2.5(1)$$

where

W = mass flow in pounds per hour
r = elbow center line radius in inches
D = elbow (pipe) diameter in inches
h = differential pressure in inches of water*
ρ = operating density in pounds per cubic foot

It is apparent that the differential pressure which will be developed by a given flow in a known elbow can be estimated from a transposition of this equation.

Several special applications of centrifugal force for flow measurement are in use. Included are the Winter-Kennedy taps installed in the scroll case of hydraulic turbines. Another application is a full circle loop of pipe available as a proprietary product. Taps are located at the midpoint of the loop. High accuracy and minimum effects from upstream disturbances are claimed.

*Water density is assumed to be 62.32 pounds per cubic foot (one kilogram per liter), corresponding to 68°F (20°C).

BIBLIOGRAPHY

Hauptmann, E.G., "Take a Second Look at Elbow Meters for Flow Monitoring," *Instruments and Control Systems*, October, 1978, pp. 47–50.

Moore, D.C., "Easy Way to Measure Slurry Flowrates," *Chemical Engineering*, October 2, 1972, p. 96.

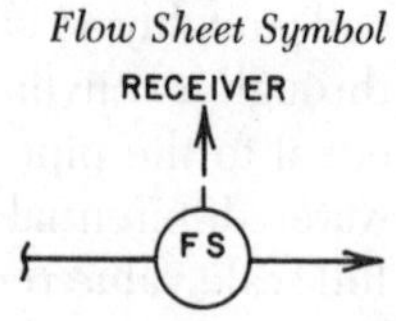

2.6 FLOW SWITCHES

Flow Switch Types:	A—Bypass B—Capacitance or Capacitance Noise C—Hot Wire Anemometer D—Paddle E—Thermal F—Ultrasonic G—Valve Body H—Variable Area
Applications:	0—Liquids (Flow–no flow) 1—Liquids (High or low flow) 2—Gases, Vapors (High or low flow) 3—Solids (High or low flow)
Approximate Cost:	$100 to $300. Some paddle designs are below this range and some ultrasonic and variable area units are above
Partial List of Suppliers:	W.E. Anderson, Inc. (D-1); Bestobell Meterflow Ltd. (F-1); Brooks Instrument Div. of Emerson Electric (F-1, H-1,2); DeLaval Turbine Co. (A-1); Fischer & Porter Co. (H-1,2); Flow Technology, Inc. (E-1,2); Fluid Components, Inc. (E-1,2); Hersey Products, Inc. (F-1); Kent Cambridge Instrument Co. (B-3); Kurz Instruments, Inc. (C-2); Magnetrol, Inc. (G-1); Moyno Pump Div. of Robbins & Myers (B-0); Penn Control, Div. of Johnson Controls (D-1,2); Power Engineering & Equipment Co. (D-1,2); TSI, Inc. (C-2); Universal Filters, Inc. (H-1,2)

Flow switches are used to determine if the flow rate is above or below a certain value. This value (the setpoint) can be fixed or adjustable. When the setpoint is reached, the response can be the actuation of an electric or pneumatic circuit. When the flow switch is actuated, it will stay in that condition until the flow rate moves back from the setpoint by some amount. This difference between the "setpoint" and the "reactivation point" is called the switch "differential." The differential can be fixed or adjustable. If the differential is small, the switch is likely to cycle its circuit as the flow fluctuates.

In certain applications a manual reset feature is desirable. This will guarantee that once the switch is actuated, it will not be allowed to return to its normal state until manually reset by the operator. This feature is designed to require the operator to review and eliminate the cause of the abnormal flow condition before resetting the switch.

All instruments that can measure flow can also be used as flow switches. On the other hand, if only a flow switch is required for a particular application, the installation of indicating or transmitting devices cannot be economically justified. Therefore, in this section only the direct flow switches will be discussed. Indirect devices, such as differential pressure switches piped around orifice plates, or receiving switches connected to the output signals of transmitters, are not covered.

Design Variations

The least expensive and therefore the most widely used are the various paddle type devices. At "no flow" the paddle hangs loosely, vertical to the pipe in which it is installed. As flow is initiated, the paddle begins to swing upward in the direction of the flow stream. This deflection of the paddle is translated into mechanical motion by a variety of techniques including a pivoting cam, a flexure tube or a bellows assembly. The mechanical motion causes the switch to open or close. If a mercury switch is used the mechanical motion drives a magnetic sleeve into the field of a permanent magnet which trips the switch. A hermetically sealed switch will be directly actuated by the permanent magnet as it moves up or down according to the paddle movement. If a micro-switch is used the translated motion will directly cause switch actuation.

The range and actuation point of paddle switches can be changed and adjusted by changing the length of the paddle. For any given pipe size, the flow rate at which switch actuation occurs decreases as the paddle length increases.

Paddle-type flow switches are sensitive to pipeline turbulence, pipeline vibration and installation configuration. For these reasons, it is advisable to provide them with the equivalent of a ten pipe diameter straight upstream run, to use dampers if pipe vibration or pulsing flow is expected and to readjust their settings if they are to be mounted in vertical upward flowlines. The conventional paddle-type designs are incapable of distinguishing low flow velocities from no-flow conditions. Therefore, if low flows are to be detected, the folding circular paddle should be used (Figure 2.6a) which permits the full diameter paddle to fold back upon itself, to minimize pressure drop.

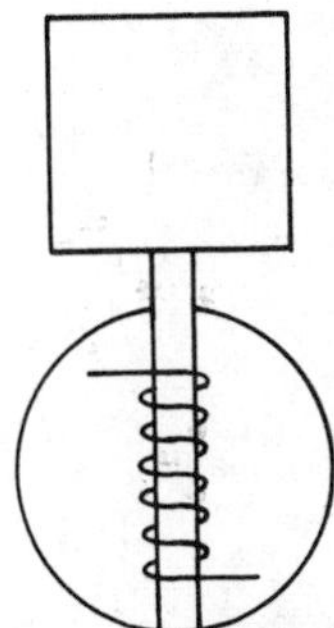

Fig. 2.6a Folding paddle switch

In smaller sized pipelines where it is desired to provide local flow indication, in addition to the flow switch action, the variable area-type flow switches can be considered. If the vertical upward flow configuration of the rotameter design is not convenient from a piping layout point of view, the circular swinging vane design, illustrated in Figure 2.6b, can be considered. In existing systems, the clamp-on-type ultrasonic liquid flow switch can be a convenient solution because it does not require pipe penetration. If the purpose of the flow switch is to protect pumps from running dry, the capacitance wafer-type flange insert unit is a good choice.

Flow switch reliability is increased by the elimination of moving parts, so that pipe vibrations or fluid flow pulses will not cause erroneous switch actuation. One of the most popular solid state designs is the thermal flow switch. All heat actuated flow switches sense the

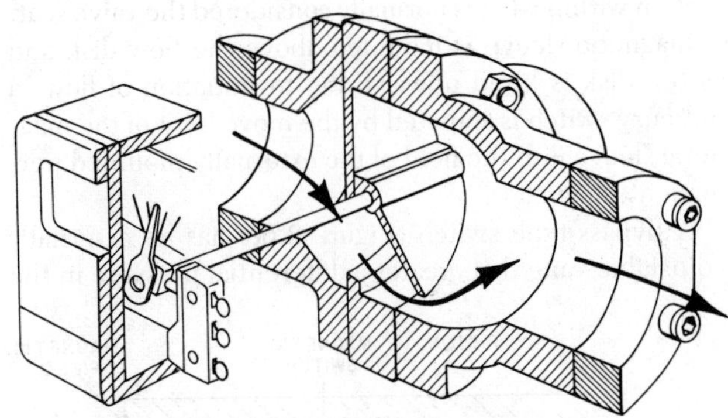

Fig. 2.6b Swinging vane flow switch

movement or stoppage of the process stream by detecting the cooling effect (temperature change) on one or more probes. One design consists of a heater probe and two sensor probes, connected in a Wheatstone bridge. When the flow stops, an inbalance in the bridge circuit occurs, as illustrated in Figure 2.6c. The main advantage of this

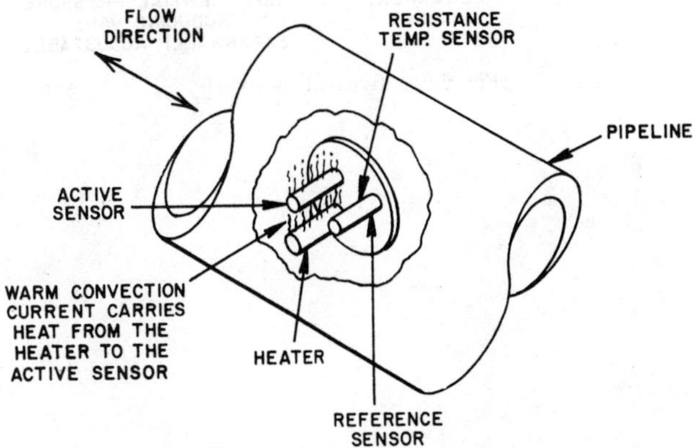

Fig. 2.6c Thermal flow switch

design is its ability to detect very low flow velocities (Table 2.6d). Its main limitation is that it cannot respond instantaneously to flow changes. Depending on switch adjustments and on type of process fluid, the speed of response will vary from two seconds to 2 minutes.

Table 2.6d
MINIMUM SETTINGS FOR FLOW SWITCHES

Flow Switch Type	*Minimum Velocity FPM (m/min)* Air	Water
Thermal	10 (3)	0.5 (0.15)
Variable Area	300 (91)	20 (6)
Ultrasonic	—	60 (18)
Paddle	300 (91)	60 (18)

The valve body-type flow switches are built into a pipe fitting that resembles the body of a single seated globe valve. A flow disk is allowed to move in a vertical direction within what is normally considered the valve seat. A magnetic sleeve is mounted above the flow disk and as the disk is lifted upward due to initiation of flow, a mercury switch is actuated by the movement of the magnetic sleeve into the field of the externally mounted permanent magnet.

A bypass-type switch (Figure 2.6e) has an externally adjustable vane that creates a differential pressure in the flow stream. This differential pressure forces a proportional flow through the tubing that bypasses the vane. A piston retained by a spring is in the bypass tubing and will move laterally as flow increases or decreases; the piston's movement actuates a switch. Bypass flow switches can be used for fairly low flow rates and their ability to be externally adjusted is a very desirable feature.

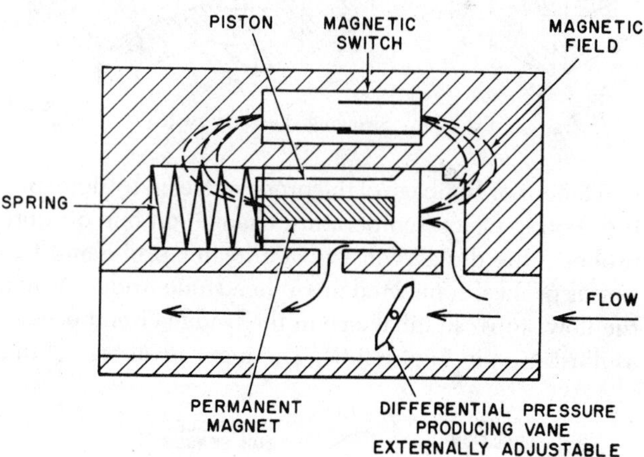

Fig. 2.6e Bypass flow switch

BIBLIOGRAPHY

Clark, W.J., *Flow Measurement.* Pergamon Press, 1967.

Spink, L.K., *Principles and Practices of Flow Meter Engineering.* Ninth Edition. The Foxboro Co., 1967.

Flow Sheet Symbol

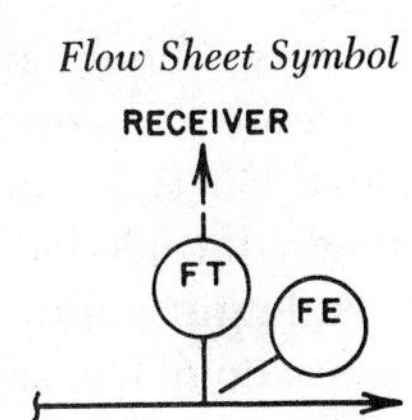

2.7 JET DEFLECTION FLOW DETECTORS

Design Pressure:	10 PSIG (69 kPa)
Design Temperature:	350°F (175°C) and higher up to 1200°F (650°C)
Materials:	Standard 316SS
Inaccuracy:	± 2% full scale
Range:	0–100 fps (0–30 mps)
Cost:	$3,300 to $7,500 depending on materials and accessories
Partial List of Suppliers:	Fluidynamic Devices, Ltd.; Lear Siegler, Inc.

Flow rate may be inferred from a measurement of gas velocity in a pipe or duct. The velocity measurement can be made with the jet deflection flow detector (Figure 2.7a). Air or other gas that is compatible with the process is expelled from the nozzle, forming a jet with a pressure profile as shown. Under conditions of no process flow, the profile is symmetrical in relation to the two receiver ports and the differential pressure across the ports is zero. As process flow increases, the jet is deflected downstream by an amount that is related to the velocity of the process stream. Deflection of the jet causes increased pressure at the downstream port and decreased pressure at the upstream port. The geometry of the ports is such that the change in differential pressure is linearly proportional to process stream velocity over the useful range of the element. Figure 2.7a also shows how the pressure profile of the jet shifts as the velocity of the process stream increases.

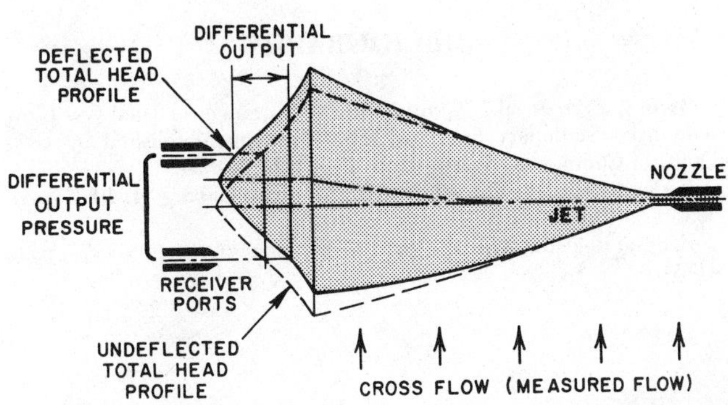

Fig. 2.7a Pressure profile of jet deflection flow detector

The relationship between stream velocity and flow rate is:

$$Q = 60VA$$

where:

Q = flow rate, actual cubic feet per minute
V = velocity, feet per second
A = pipe area, square feet

The velocity measurement of a gas in a rectangular or circular duct will not necessarily be related to the flow rate because the velocity profile is not uniform. For the case of laminar flow in a circular duct the profile is parabolic with the maximum velocity at the center and zero velocity at the walls. The maximum velocity is twice the average; therefore, one-half of the reading taken at the center will be an average for the laminar flow case. When there is turbulent flow in a circular duct the average velocity will be located at approximately 25 percent of a radius as measured from the duct wall. Determining the position of average velocity in rectangular ducts is not any easy matter and no rules of thumb can be given.

In many cases the point of average velocity will not be where it is expected because of disturbances introduced by the upstream piping configuration. The upstream piping should be straight for at least 20 pipe diameters to allow for the disturbances to smooth out. When this installation requirement cannot be met, an average velocity point can sometimes be established by using the flow element to make a traverse of the pipe, taking readings at a number of points in the traverse. A traverse should always be made on rectangular ducts if accurate measurement is required. Traversing rectangular ducts is complicated by the fact that a traverse might be needed on two or more planes, thereby necessitating multiple entries. Of course, an average velocity point can shift as flow rates vary in the rectangular or circular duct.

The jet deflection element, and other parts such as the pilot tube, can be installed so that they can be removed for inspection without shutting the process down. They can also be hot tapped into a line; that is, a new installation can be made without shutting the process down. Figure 2.7b shows the detail of an installation that enables removal of the probe when the pipe is under pressure. To remove the element, the gland nut is loosened just enough to allow the shaft of the element to be withdrawn until the element is in the chamber outside of the gate valve. The valve is then closed, the chamber is vented, and the gland nut is removed, allowing safe removal of the element. Normally the jet deflection element is installed in low pressure or vacuum piping and thus the risk of having the element blow out during removal is slight. Nonetheless, some users install stop rods or safety chains in order to eliminate the possibility of a blow out altogether.

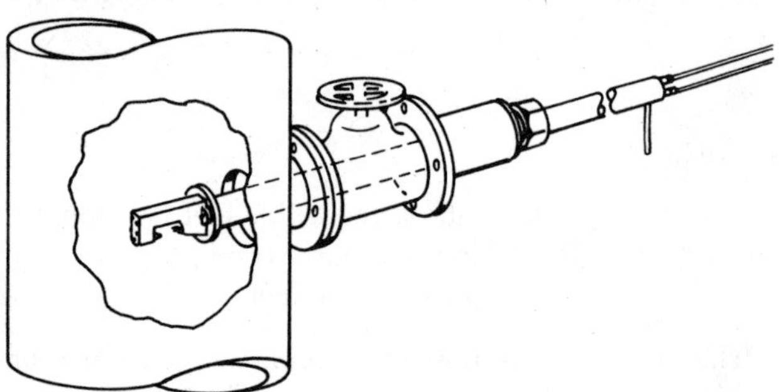

Fig. 2.7b Installation that allows removal of the element under pressure

The hot tapping procedure is shown step by step in Figure 2.7c. First a weld neck flange is welded to the pipe that is to be tapped. Next a flanged gate valve is made up to the flange. The third step is to bolt a hot tap machine to the downstream side of the gate valve and open the valve. The hot tap machine is used to drill the opening into the process pipe. The pressure seal on a hot tap machine is similar to that of a control valve packing box, allowing taps to be made in pipes that contain relatively high pressures. The machine shown in the figure is hand operated, but units are available with pneumatic or electric drives. A scale on the machine enables the operator to know how far he has drilled into the pipe. After the hole is made, the bit is retracted, the valve is closed, and the hot tap machine is removed. The tap is now ready for the installation of the flow element, gland nut, and chamber, as shown in Figure 2.7c. Rigid safety precautions must be used during hot tapping operations, particularly if the process is flammable or hazardous. Hot tapping generally is not done if the piping specification requires that welds be stress-relieved, nor is the weld made if there is no flow in the pipe.

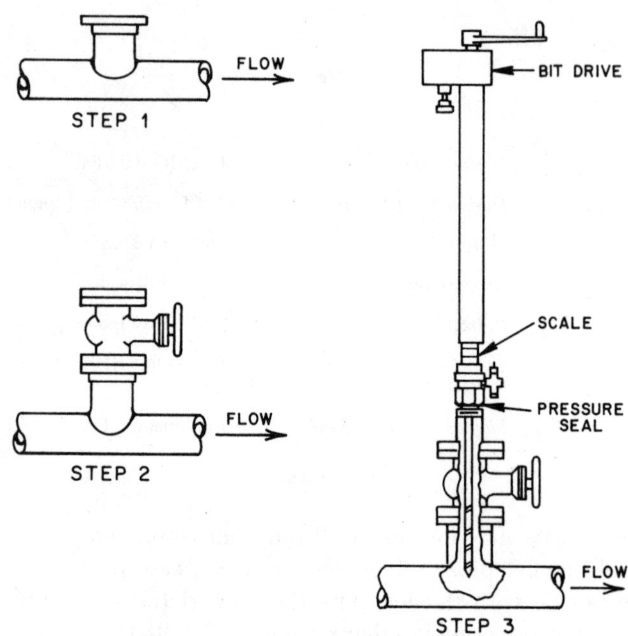

Fig. 2.7c Hot tap procedure

Conclusion

Jet deflection flow detectors are appropriate for flow measurement in low pressure circular and rectangular ducts. They can be purged or flushed and can be removed for inspection and cleaning; they are suitable for dirty, abrasive, corrosive, or plugging services. The accuracy of ± 2 percent of full scale is generally acceptable for measurements around flare headers, stacks, and air ducts although it should be remembered that overall installation accuracy is dependent on locating the element at a point of average velocity.

BIBLIOGRAPHY

Brooks, E.F. et. al., "Continuous Measurement of Total Gas Flow Rate from Stationary Sources," TRW Systems, Published by U.S. Dept. of Commerce, Washington, DC, P.B. 241894.

Federal Register, Vol. 42, No. 20, Monday, January 31, 1977, page 5936.

Federal Register, Vol. 42, No. 160, Thursday, August 18, 1977, page 41754.

Flow Sheet Symbol

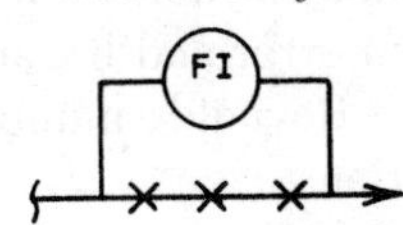

2.8 LAMINAR FLOWMETERS

Design Pressure:	Up to 5000 PSIG (34 MPa).
Design Temperature:	Up to 300°F (150°C) normally, but can be higher with special designs
Material of Construction:	Stainless steel, aluminum or any alloy available in small bore tubing
Fluids:	Liquids and gases
Flow Range:	0.0001 to 2000 scfm for gases (3 cm^3/min. to 57 m^3/ min.) 0.0003 to 10 gpm for liquids (1 cm^3/min. to 38 l/min.)
Flow Turndown:	10:1 minimum
Flow Characteristic:	Linear to approximately linear
Relative Cost:	Medium to high. A function of materials of construction, type of device, and whether user or vendor designed and built. Typical commercial units will range from \$400 to \$1800
Partial List of Suppliers:	Meriam Instrument Division of the Scott and Fetzer Co.; National Instrument Laboratories, Inc. Units are for gas flow only

Laminar flowmeters fill a special need in flow measurement where the requirements might include low to extremely low flow rates, linear calibration and low noise, the ability to measure high-viscosity liquids, or steady low flow repeatability and control accuracy. Laminar flowmeters are intended for very low flow rates where other types of meters are either marginal in performance or cannot be used at all. Laminar flowmeters can be constructed by various methods, but the most common method is with capillary tubes. Hence, the terms "laminar flowmeter" and "capillary flowmeter" are virtually synonomous. Proprietary commercial units use other matrix shapes and are intended for use with gases. Where gas is the metered flow, it is preferable to first investigate the commercial units before undertaking the design of a laminar flowmeter.

Similar to variable head meters (described in Section 2.12), the flowmeter consists of the laminar flow element and a differential pressure measuring instrument. While the flow is theoretically linear with pressure drop, in practice some nonlinearities are often encountered. In most cases, these are of little consequence.

The theory for laminar flowmeters is based upon the Hagen-Poiseuille Law for laminar flow and Reynolds number as a means of defining the type of flow. Both are required to investigate and design a laminar flow element. More detailed explanations and discussions of theory can be found in any standard textbook on fluid mechanics.

Theory

Reynolds Number

Fluid flow in pipes and tubes is characterized by a nondimensional number called the Reynolds number. Up to approximately Reynolds number 2,000, the flow is called laminar, viscous or streamline flow. Above 10,000 the flow is called fully developed turbulent. The region between 2,000 and 10,000, where the flow is shifting from laminar to turbulent, is not clearly defined but is called transitional. Generally, laminar flow elements are restricted to numbers under 2,000 and most commonly below 1,200. There are some methods that will enable a capillary element to be used satisfactorily up to a Reynolds number of 15,000 with a modest sacrifice in error and linearity.

Reynolds number is defined by the following equations:

For liquid flow

$$Re = \frac{50.7\,\rho Q}{D\mu} \text{ or } Re = \frac{6.32\,W}{D\mu} \qquad 2.8(1)$$

where

- Re = Reynolds number
- ρ = Density (lb./cu.ft.) at flowing temperature
- Q = Flow rate (gal./min.)
- D = Internal tube diameter (in.)
- μ = Viscosity of flowing temperature (centipoise)
- W = Flow rate (lb./hr.)

For gas flow

$$Re = \frac{6.32\,\rho Q}{D\mu} \text{ or } Re = \frac{6.32\,W}{D\mu} \qquad 2.8(2)$$

where

- ρ = Density at standard conditions (lb./cu.ft.)
- Q = Flow rate (scfh)

and other units are defined the same as for liquid

Hagen-Poiseuille Law

Once we have defined the tube inside diameter required to give laminar flow per the Reynolds number calculation, we need the length of the capillary to design our laminar flowmeter system. These equations are as follows:

For liquid flow

$$L = 1.5876 \times 10^{3}\,\frac{\Delta P D^{4}}{\mu Q} \qquad 2.8(3)$$

or

$$L = \frac{\Delta P D^{4}\,\rho}{7.86 \times 10^{-5}\,\mu W} \qquad 2.8(4)$$

where

- L = Length of tube (in.)
- ΔP = Differential pressure drop (in. water)
- D = Tube internal diameter (in.)
- μ = Viscosity at flowing temperature (centipoise)
- ρ = Density at flowing temperature (lbm./cu. ft.)
- Q = Flow rate (gal./min.)
- W = Flow rate (lbm./hr.)

Equation 2.8(4) can also be used for calculating a gas flow capillary element provided the value of ΔP is no greater than 10 percent of the inlet pressure. Otherwise, changes in gas density, specific volume and flow velocity cause too many calculation complications. While the calculation is in weight units, this can be easily converted to read in any desired scale units.

Design Parameters

There are a number of guidelines for successful design of a laminar flowmeter.

1. The differential pressure drop can range from 5 to 800 inches of water (1.24 kPa to 200 kPa).
2. (L/D)/Re should be a minimum of 0.3; for best linearity, a value of 0.6 or greater is preferable. Large L/D ratios and/or lower Reynolds numbers contribute to accuracy. For example, the entrance effect for laminar flow is negligible if (L/D)/Re $>$ 0.3 and Re $<$ 500.
3. The area of the flow conduit preceding the capillary should be a minimum of 20 times the capillary area.
4. The differential pressure instrument pressure connections should be located 100 to 200 capillary diameters from the capillary ends.
5. A filter capable of removing particles 0.1 in. (2.54 mm) or larger than the capillary internal diameter should be installed upstream of the system.
6. The metering system should be sloped up for liquids to permit gas venting and sloped down for gases to permit liquid draining.
7. Examination of the Hagen-Poiseuille equation shows that viscosity is a primary variable; changes in viscosity can result in large flow measurement errors. With a known fluid or composition, the only thing that affects viscosity is temperature. For this reason, the temperature must be known and held essentially constant. This can be done by immersing the metering system and measuring capillary in a constant temperature bath as shown in Figure 2.8a. If the flow is measured in weight units such as lbs/hr, then fluid density must be known. Fluid density also varies with temperature, but controlling the temperature to fix viscosity will also fix density. With some fluids, cooling may be required instead of heating, but the overall principle is the same.

Design Calculations

Based upon the flow rate calibration and the viscosity of the fluid, select a tube internal diameter that will result in a Reynolds number within the laminar range

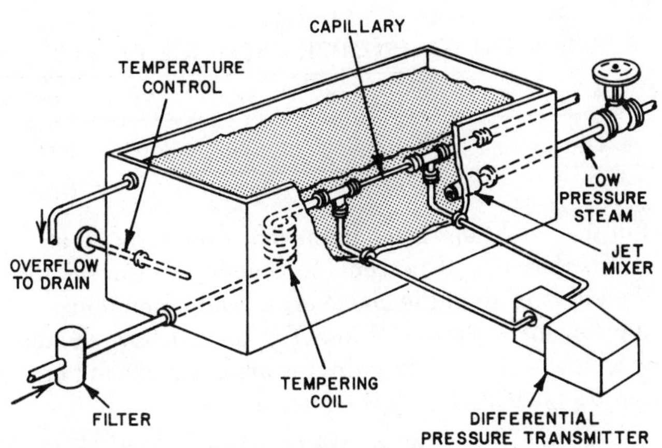

Fig. 2.8a Typical capillary with constant temperature bath

and preferably less than 1,200. Calculate the length of tubing required using the selected tube diameter to ensure that it is a reasonable length and that it meets the (L/D)/Re criteria. By working back and forth between the various equations, the system can be tailored to meet almost any design criteria. For example, let us assume that it is desired to design a capillary flowmeter to measure a small liquid catalyst stream, and the basic data for the catalyst flow is as follows:*

Maximum flow capacity:	50 lb. m/hr.
Viscosity:	20 cps at 100°F
Density:	53.8 lb. m/cu. ft.
Desired instrument ΔP:	100 in. of water

Small diameter standard stainless steel tubing that is readily available should be used. To design as linear and accurate a flowmeter as possible, a tube bore that provides a large (L/D)/Re is desirable. To minimize plugging problems and to enable the use of a filter that won't clog easily, start by looking at a 3⁄16 × 0.032 in. wall thickness tubing with a nominal internal diameter of 0.1235 in. From equation 2.8(1):

$$\mathrm{Re} = \frac{6.32\ \mathrm{W}}{\mathrm{D}\mu} = \frac{6.32 \times 50}{0.1235 \times 20} = 128$$

This is well into the laminar range, so the length of the flow element can be calculated to determine if it will make a reasonable design. From Equation 2.8(4):

$$\mathrm{L} = \frac{\Delta \mathrm{PD}^4\ \rho}{7.86 \times 10^{-5}\ \mu \mathrm{W}} = \frac{100 \times 0.1235^4 \times 53.8}{7.86 \times 10^{-5} \times 20 \times 50} = 15.7 \text{ in.}$$

$$(\mathrm{L/D})/\mathrm{Re} = (15.7/0.1235)/128 = 0.993$$

This is an easy length to work with in fabricating a meter element and a constant temperature bath and looks like a reasonable design based upon the criteria.

*See Appendix A.1 for SI units.

Errors

Viscosity and Density

As noted in the discussion of design parameters (and by examination of equation 2.8(4)), changes in viscosity and density can result in flow measurement errors. Viscosity changes in liquid due to temperature are often substantial, while density changes are more moderate. With gases, the reverse is usually true, with temperature having more influence on density and less on viscosity. The need for careful control of measurement temperature to minimize these effects on flow rate accuracy cannot be overemphasized.

Internal Diameter of Tubing

From equations 2.8(3) and 2.8(4), it can be seen that internal diameter of the tube is very important, because it is multiplied to the fourth power. While high-quality tubing will be very close to published specifications, manufacturing tolerances will result in variations from these dimensions both laterally and longitudinally. If the actual effective internal diameter of the capillary tube differs by 1 percent from the value used in the calculation for a given ΔP, an error of about 4 percent will result. For this and other reasons to be discussed, the laminar flowmeter should be calibrated on a known fluid before use, and appropriate design adjustments made as necessary.

Installation Design Losses

To measure the true capillary differential pressure drop per the Poiseuille equation, it would be necessary to put the pressure taps into the capillary at the calculated L dimension. This is impractical as a design because of the small size of tubing normally used. A pressure tap must be perfectly flush with the inside of the tube and must be clean with no burrs or other projections into the tube. Otherwise, considerable differential pressure measurement error will result. Using practical methods of constructing a capillary flowmeter, there are three additional sources of pressure drop in addition to the capillary loss. These are all additive and will give a greater indicated pressure drop than the capillary flow alone. These three sources of error are inlet loss, exit loss, and capillary entrance loss. These losses also contribute to nonlinearity.

Inlet Loss

There is very little loss from the entrance fitting into the capillary tube if laminar flow conditions exist. But if the piping cavity ahead of the capillary is extremely large relative to the capillary (approximating a reservoir) and the fluid velocity is thus extremely low (approaching zero), there can be an inlet effect and pressure loss.[1] This is due to the sudden contraction from the large reservoir to the small tube bore, forming a bell-mouth shape approach flow. This loss can be expressed by:

$$\Delta P = \frac{2.8 \times 10^{-7} W^2}{D^4 \rho} \qquad 2.8(5)$$

This equation is derived from Bernoulli's equation for flow out of a reservoir.

Exit Loss

When the fluid exits the capillary, the flow path enlarges. If the piping is similar to that described under inlet loss, the loss can be calculated by:

$$\Delta P = \frac{5.6 \times 10^{-7} W^2}{D^4 \rho} \qquad 2.8(6)$$

Entrance Loss

Entrance loss occurs in addition to the normal capillary pressure drop in the initial fluid path distance, or to state it in another way, for a short distance the pressure drop is higher than that predicted by the Poiseuille equation.[2,3] The additional loss is due to the work expended in the formation of the parabolic velocity distribution profile characteristic of laminar flow. It can be expressed in terms of an equivalent length of capillary, L_{eq}, added to that calculated by the Poiseuille equation. Refer to Figure 2.8b for determining the L_{eq}.

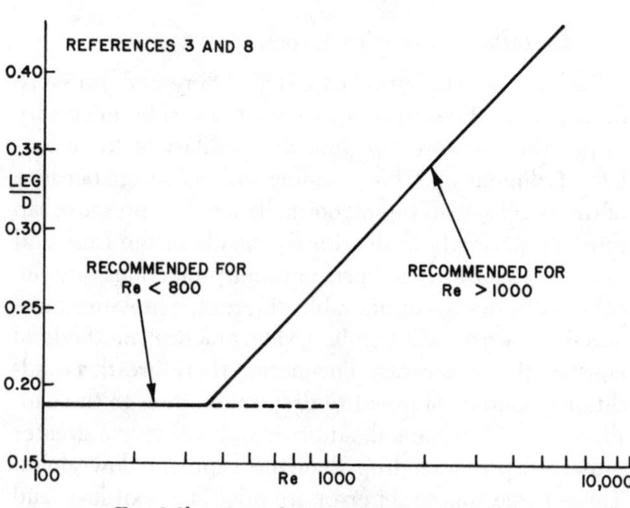

Fig. 2.8b Equivalent length of capillary (L_{eq})

The following equation can be used for the pressure drop:

$$\Delta P = \frac{1.96 \times 10^{-7} W^2}{D^4 \rho} \qquad 2.8(7)$$

Table 2.8c can be used as a quick guide for judging the design factors that will minimize overall entrance effects.

Table 2.8c
L/D RATIO TO MINIMIZE ENTRANCE EFFECT

Re	*10*	*50*	*100*	*500*	*1000*	*2000*
L/D>	15	75	150	750	1500	3000

For the conditions given in the table, the error involved will be less than 1 percent. In general, we can say that the effect of all of the above errors will be minimized if the Reynolds number is low, the laminar flow element is long and the pressure drop is high. The overall error can be calculated by this equation:

$$\text{Percent Error} = \frac{\Delta Pi + \Delta Pe + \Delta Pen \times 100}{\Delta P} = \frac{0.367\ W \times 100}{\mu L} = \qquad 2.8(8)$$

Range Extension Techniques

There are two techniques to expand the range capability of laminar flow elements. One is to use a number of capillary tubes in parallel. The other is to use a tight helical coil capillary. The choice of technique depends upon such factors as desired flow rate, nonlinearity requirements, Reynolds number, capillary length, and system space design limitations.

Parallel Capillaries

If the amount of flow desired is greater than can be conveniently handled by a single capillary, the flow can be split into as many smaller units as necessary.[4] Units with over 900 individual capillary tubes have been successfully built and used. The mechanical construction of multi-parallel capillaries can be a problem. Tube packing voids may not affect meter operation, but add considerable difficulty to calculating the meter range. Normally, it is best to eliminate the voids by filling the spaces with solder, braze material, or plastic resin; the filler material chosen will depend upon fluid compatability and operating conditions. Overall, it is a tricky mechanical design.

Coiled Capillary

Coiling a length of straight capillary results in a flow phenomenon called the Dean effect. When a fluid flows through a curved pipe or coil, a secondary circulation of fluid, known as a double eddy, takes place at right angles to the main direction of flow. This circulation accounts for the fact that the pressure drop in curved pipe is greater than in a corresponding length of straight pipe. The Dean effect stabilizes laminar flow and raises the Reynolds number at which turbulent flow starts. It has been established that this will allow properly designed coiled capillaries to be operated up to a Reynolds number of 15,000.[5] The Reynolds number at which laminar flow can be sustained for various coil curvature ratios is called the critical Reynolds number. It is a function of the internal diameter of the tube and the coil tightness or diameter. Table 2.8d gives the approximate critical Reynolds number at which laminar flow can be sustained for various coil curvature ratios.

Table 2.8d
CRITICAL REYNOLDS NUMBER VERSUS COIL CURVATURE RATIO

Coil Curvature Ratio (Dc/D)	*Critical Reynolds Number* $(Re)_c$
Straight Pipe	2,100
2,000	2,700
1,000	2,900
500	3,200
100	4,600
50	5,700
10	10,000
9	15,000

In this table, D is the tube inside diameter and D_c is the mean coil diameter, centerline to centerline. From a practical viewpoint, the ratio of $D/D_c = 1/9$ is equivalent to the maximum allowable critical Reynolds number of 15,000 and can be used as a safe design in most cases.

The pressure drop of laminar flow through coils can be expressed in terms of an equivalent length, L_e, of straight pipe of the same diameter and shape which will have the same friction loss as the curved pipe. The ratio of the equivalent to actual coil length, L_e/L, is a function

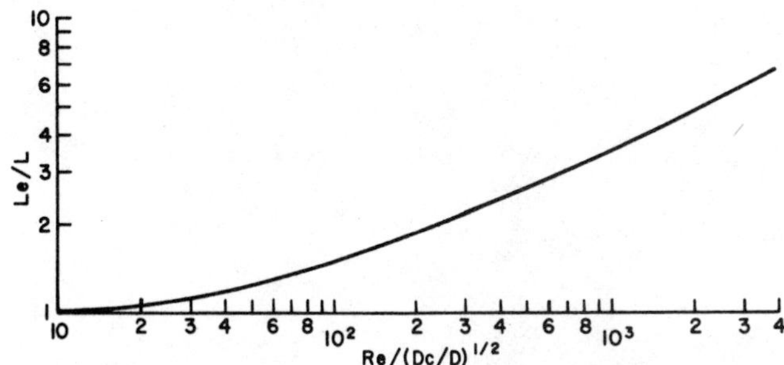

Fig. 2.8e Equivalent lengths for curved pipe

of the Dean number or $Re/(D_c/D)^{1/2}$ as shown in Figure 2.8e. This curve is accurate to about ± 5 percent.

The equation for calculating the length of a coiled capillary required to meet a specific metering design is expressed by:

$$L = \frac{\Delta P\, D^4\, \rho}{7.86 \times 10^{-5}\, \mu W C} \qquad 2.8(9)$$

where C = the coil factor correction

The coil factor correction is a function of the term $Re/(D/D_c)^{1/2}$. Refer to Figure 2.8f for C versus $Re/(D/D_c)^{1/2}$ or to Figure 2.8g for C versus Re for various D/D_c ratios. In very small capillaries, the coil diameter can be the

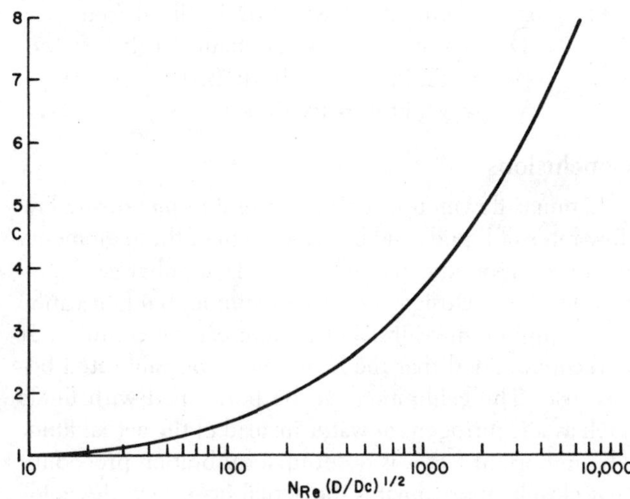

Fig. 2.8f Correction factors for coiled capillary flowmeter (data adapted from reference 5)

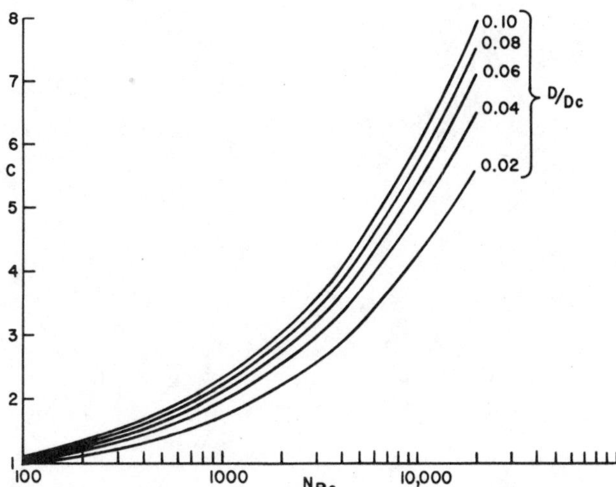

Fig. 2.8g Correction factors for coiled capillary flowmeters (data adapted from reference 5)

nominal value, since exact centerline measurement is insignificant.

Alternate Calculation Method

In laminar flow, the friction factor is a function of Reynolds number only and is independent of surface roughness. The friction factor can be expressed as:

$$f = 16/Re$$

Therefore, the Fanning Equation, 2.8(10), can be used as an alternate means of calculating the capillary element as shown by:

$$L = \frac{2\, \Delta P\, g_c\, D}{4 f\, \rho\, V^2} \qquad 2.8(10)$$

where

L = Capillary length (ft.)
ΔP = Pressure drop (lb. f/ft.^{2})

g_c = Gravity constant 32.17 (lb.-ft./sec.^{2})
D = Capillary internal diameter (ft.)
ρ = Fluid density (lb. m/ft.^{3})
V = Fluid velocity (ft./sec.)

Conclusions

Laminar flowmeters are very useful in measuring low flow rates of liquids and gases. Design of these elements is based upon the use of Reynolds number and Poiseuille's Law. Design is relatively simple, but fabrication of a complete measurement system can be complex. It is recommended that the final system be calibrated before use. The calibration can be performed with fluids such as air, nitrogen, or water instead of the actual fluid. The important factor is to set up a calibration procedure that closely approximates the actual in-service Reynolds number. Sources of error can be minimized by following design guidelines as closely as possible.

REFERENCES

1. Kreith, F. and Eisenstadt, R., Transactions of the ASME, July 1957, pages 1070–1078.
2. Willoughby, D.A. and Kittle, P.A., *Industrial and Engineering Fundamentals*, Vol. 6, No. 2, May 1967, pages 304–306.
3. Rivas, Jr., M.A. and Shapiro, A.H. Transactions of the ASME, April 1956, pages 489–497.
4. Greeff, C.E. and Hackman, J.R., ISA Journal, August 1965, pages 75–78.
5. Powell, H.N. and Browne, W.G., *The Review of Scientific Instruments*, Vol. 28, No. 2, February 1957, pages 138–141.

BIBLIOGRAPHY

Gann, R.G., *Journal of Chemical Education*, Vol. 51, No. 11, November 1974,pages 761–62.

Instruments and Control Systems, November, 1976, pages 75–76.

Mahood, R.F. and Littlefield, R., Private Communications, March 1952.

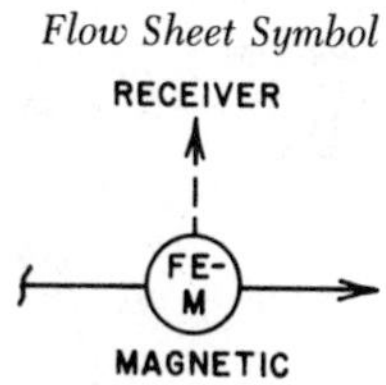

2.9 MAGNETIC FLOWMETERS

Standard Design Pressure:	275 PSIG (1.898 MPa) maximum
Standard Design Temperature:	Up to 360°F (180°C)
Materials:	Liners: fiberglass, neoprene, penton, polyurethane, rubber, Teflon, vitreous enamel, Kynar Electrodes: alloy 20, Hastelloy "C", platinum, stainless steel, tantalum, titanium, Monel, nickel
Flow Detection:	Volumetric flowrate
Fluid.	Liquid including slurries (Newtonian and non-Newtonian)
Flow Range:	0.01 through 100,000GPM (0.038 through 378,500 1/m)
Inaccuracy:	± 0.5% of rate to ± 2% of full scale; depends on type of meter (AC or pulsed DC excitation) and the calibration of the flowmeter and the type of receiver being used
Sizes:	0.1 through 96 in. (2.5 mm through 2.4 m)
Costs:	\$3,200 through \$110,000 as a function of size and construction. A 2 in. (50 mm) size, 304SS, Teflon-lined magnetic flowmetering tube only is about \$2,000. With amplifier-transmitter included it costs about \$4,500
Partial List of Suppliers:	Brooks Instrument Div., Emerson Electric Co.; Endress & Hauser Inc.; Fischer & Porter Co.; Flowmetering Instruments, Ltd.; Foxboro Co.; G. Kent, Ltd., Div. of Brown-Bover; Krohne; Schlumberger, Ltd.; Taylor Instrument Companies, Div. of Sybron

Magnetic-type flowmeters use Faraday's Law of electromagnetic induction for making a flow measurement. Faraday's Law states that when a conductor moves through a magnetic field of given field strength, a voltage level is produced in the conductor that is dependent on the relative velocity between the conductor and the field. This is the concept that is used in electric generators. Faraday foresaw the practical application of the principle to flow measurement, because many liquids are adequate electrical conductors. In fact, he attempted to measure the flow velocity of the Thames River using this principle. He failed because his instrumentation was not adequate, but 150 years later, we can successfully apply the principle in magnetic flowmeters.

Theory

Figure 2.9a shows how Faraday's Law is applied in the electromagnetic flowmeter. The liquid is the conductor which has a length equivalent to the inside diameter of the flowmeter, "D." The liquid conductor moves with an average velocity "V" through the magnetic field of strength "B." The induced voltage is "E." The mathematical relationship which applies is:

$$E = BDV/C$$

C is a constant to take care of the proper units.

When the pair of magnetic coils is energized, a magnetic field is generated which is in a plane that is mutuallly perpendicular to the axis of the liquid conductor

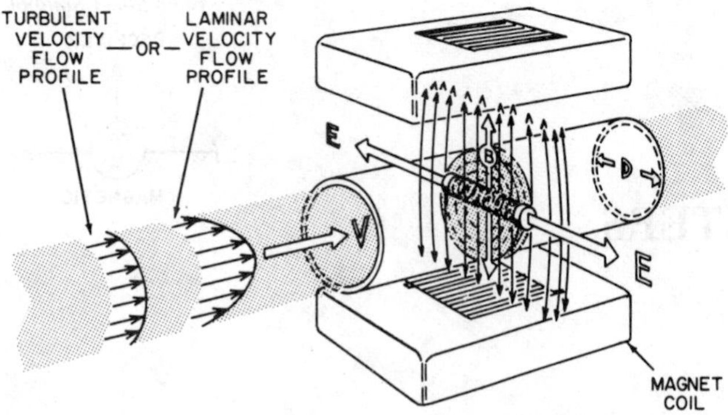

Fig. 2.9a Schematic representation of the magnetic flowmeter

and the plane of the electrodes. The velocity of the liquid is along the longitudinal axis of the flowmeter body; therefore, the voltage induced within the liquid is mutually perpendicular to both the velocity of the liquid and the magnetic field. The liquid should be considered as an infinite number of conductors moving through the magnetic field with each element contributing to the voltage that is generated. An increase in flow rate of the liquid conductors moving through the field will result in an increase in the instantaneous value of the voltage generated. Also, each of the individual "generators" is contributing to the instantaneous generated voltage. Whether the profile is essentially square (characteristic of a turbulent velocity profile), parabolic (characteristic of a laminar velocity profile), or distorted (characteristic of poor upstream piping), the magnetic flowmeter does an excellent job of averaging the voltage contribution across the metering cross section. The sum of the instantaneous voltages generated is therefore representative of the average liquid velocity because each increment of liquid velocity within the plane of the electrode develops a voltage proportional to its local velocity. The signal voltage generated is equal to the average velocity almost regardless of the flow profile. As with most flow measurement devices, we infer volumetric flow rate. The magnetic flowmeter is sensing the linear velocity of the liquid.

The Equation of Continuity ($Q = VA$) is the relationship which enables us to convert a velocity measurement to volumetric flow rate providing the area is constant. The area must be known and constant. Therefore, as with most other types of flowmeters, the pipe must be full for a fixed and known cross sectional area in order to obtain a correct measurement.

AC- and Pulsed DC-type Excitations

AC-type magnetic flowmeters apply line voltage to the magnet coils. The signal generated is a low level AC signal in the high microvolt to low millivolt range. A more recent development is the pulsed DC-type magnetic flowmeter. In this design, the magnet coils are periodically energized. There are many forms of excitation in use, but generally they can be categorized into two families: those which are on-off excitation and those which use plus-minus excitation. In either case, the principle is to take a measurement of the induced voltage when the coils are not energized and to take a second measurement when the coils are energized and the magnetic field has stablized. Figure 2.9b shows some of the types of excitation offered by various manufacturers.

In all of the pulsed DC approaches, the concept is to take a measurement when the coils are excited and store (hold) that information, then take a second measurement of the induced voltage when the coils are not excited. The voltage induced when the coils are energized is a combination of both noise and signal. The induced voltage when the coils are not energized is noise only. Subtracting the noise measurement from the signal plus noise yields signal only (see Figure 2.9c).

The pulsed DC-type systems establish zero during each on-off cycle. This occurs several times every second. Because zero is known, the end result is that pulsed DC systems are potential percent-of-rate systems. The AC-type systems must be periodically re-zeroed by stopping flow and maintaining a full pipe in order to zero out any voltage present at that time. The noise voltage can change with time, resulting in a potential offset; therefore, AC-type systems normally are percent-of-full scale systems.

Construction

Figure 2.9d is a cutaway view showing how the principle of electromagnetic induction is employed in a practical flowmeter. The basic elements of the flowmeter are a section of nonconducting pipe such as glass reinforced polyester or a nonmagnetic pipe section lined with an appropriate electrical conductor such as Teflon, Kynar, vitreous enamel, rubber, Neoprene, or polyurethane,

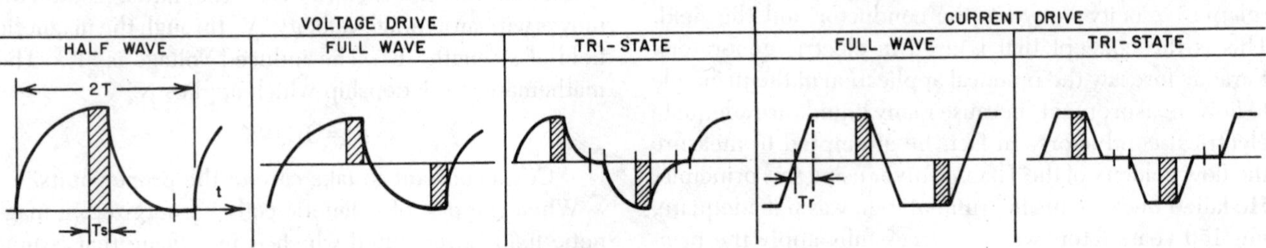

Fig. 2.9b Types of pulsed DC coil excitation

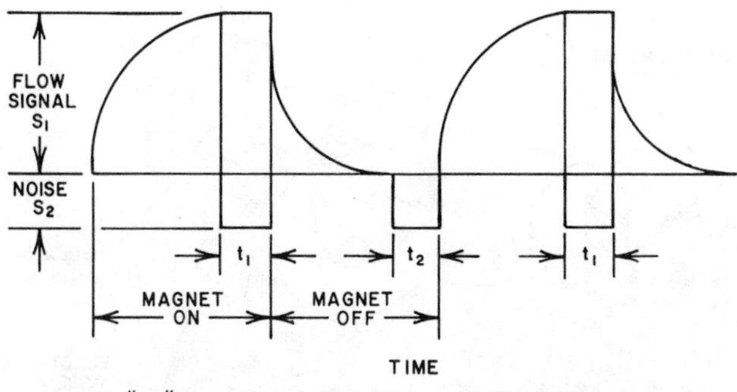

Fig. 2.9c Signal development of pulsed DC-type magnetic flowmeter with half wave excitation

As shown, the magnetic field is generated by a square wave which, in function, turns the magnet "on" and "off" in equal increments.

When "on," the associated signal converter measures and stores the signal which is a composite of flow plus a variable (non-flow related) residual voltage.

During the "off" period, the converter measures the variable (non-flow related) residual signal only. Since no field excitation is present, no flow signal will be generated.

The converter than subtracts the stored residual signal from the flow developed-plus residual signal, resulting in the display of a pure flow signal.

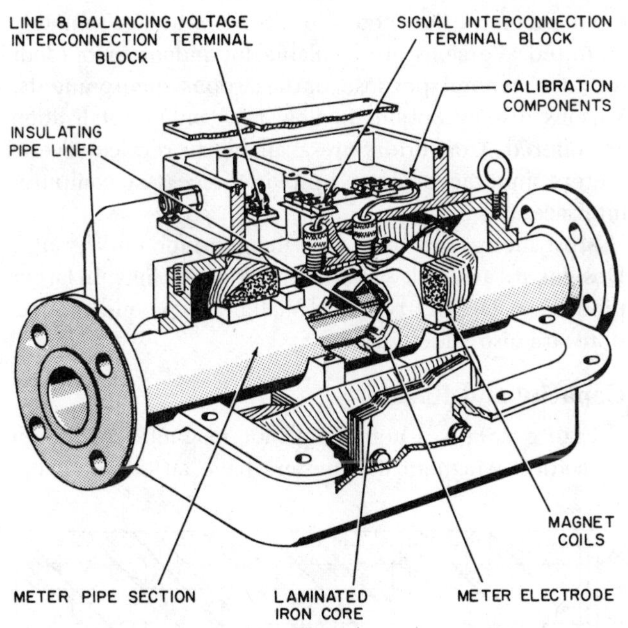

Fig. 2.9d Cutaway view of the magnetic flowmeter

among others. On alternate sides of the pipe section are magnet coils which produce the magnetic field perpendicular to the flow of liquid through the pipe. Mounted in the pipe, but insulated from it and in contact with the liquid, is a pair of electrodes which are located at right angles both to the magnetic field and the axis of the pipe. As the liquid passes through the pipe section, it also passes through the magnetic field set up by the magnet coils inducing a voltage in the liquid; the amplitude of the voltage is directly proportional to the liquid velocity. This voltage is conducted by the electrodes to a separate converter which in effect is a precision voltmeter (electrometer) capable of accurately measuring the voltage generated and converting that voltage to the desired control signals. These may be equivalent electronic analog signals, typically 4–20 mA DC, or a frequency or scaled pulse output.

Versions of magnetic flowmeters are available for periodic accidental submergence and for continuous submergence in water at depth of up to 30 feet (9 m). An outgrowth of the continuous submergence design is a sampling-type (pitot). The pitot-type magnetic flowmeter samples the flow velocity in large rectangular, circular, or irregularly shaped pipes or conduits. A typical design is shown in Figure 2.9e. A small size magnetic flowmeter is suspended in the flow stream. The magnet coils are completely encapsulated in the liner material, allowing submersion in the liquid to be measured. The short length of the meter body and the streamlined configuration are designed to minimize the difference of flow velocity through the meter and the velocity of the liquid passing around the meter. The velocity measurement of the liquid through the meter is representative of the pipe velocity. Repeatability of the system is typically 0.25 to 0.5 percent of full scale. As with any sampling-type flowmeter, the information from the flowmeter is representative of the flow through the flowmeter only. It is the user's responsibility to relate that "sampled" velocity to the average velocity that is proportional to volumetric flow rate. When applying any sampling-type flowmeter, including the pitot-type magnetic flowmeter, care must be taken in applications where there can be a profile change due to change in Reynolds number or due to piping configurations. The relationship between the measured velocity and the average velocity representative of whole body flow will change with profile change and would be reflected in a poor quality measurement.

Modern magnetic flowmeters are constructed so that the flux density of the magnetic field generated by the magnet coils is constant. The distance between the electrodes is also constant. Induced voltage is dependent only on liquid velocity and is not affected by temperature, viscosity, turbulence, or electrical conductivity, provided that electrical conductivity is above a minimum threshold.

Most manufacturers construct their flowmeters with coils external to the meter pipe section. Some designs place the coils within the flowmeter body, which is made from carbon steel to provide the return path for the magnetic field (see Figure 2.9f). In this design the meters can be shorter in length and have reduced weight and lower power consumption. Lowest power consumption occurs with the pulsed DC form of magnetic flowmeter

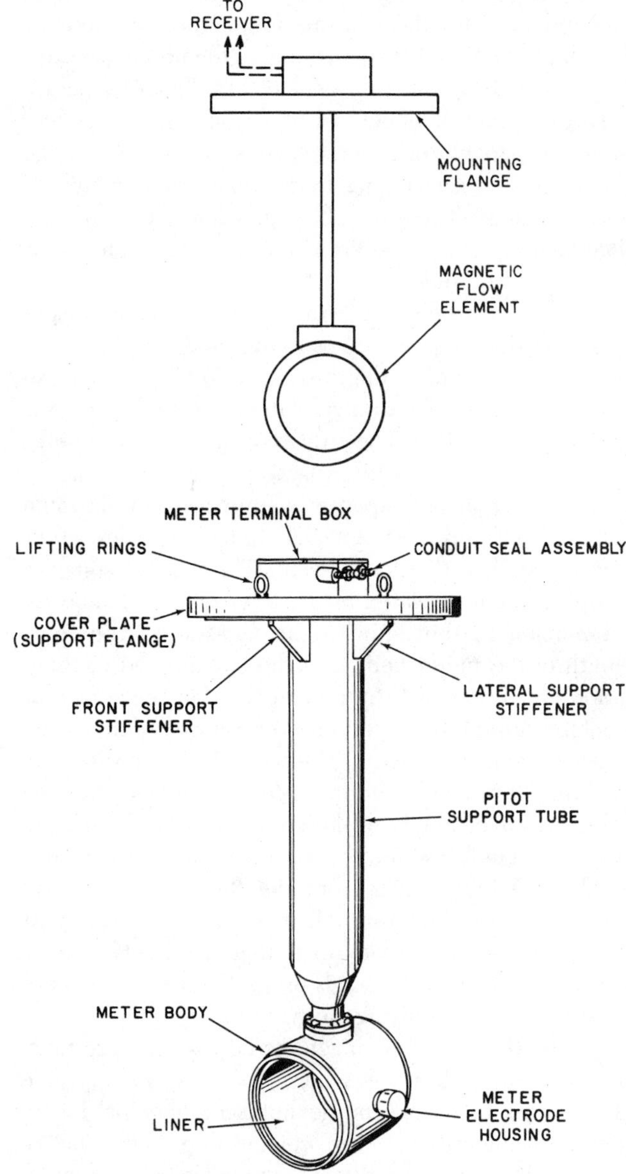

Fig. 2.9e Pitot-type magnetic flowmeter

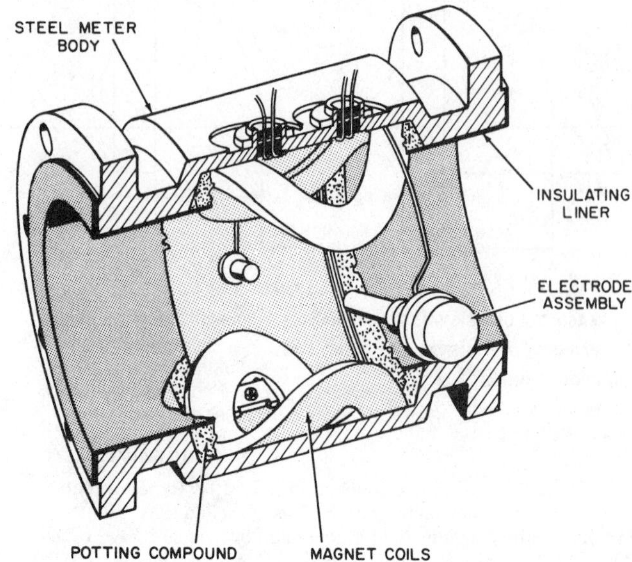

Fig. 2.9f The short-form magnetic flowmeter

because the coils are energized only part of the time. An additional saving with pulsed DC-types is that the power factor approaches one.

Each magnetic flowmeter requires an associated instrument to convert low level voltage signal into an analog or digital process signal. The converter can be mounted on the flowmeter, locally mounted, or remotely mounted. Versions are available for indoor or outdoor use, for general purpose or hazardous environments. Various display options of flow rate and/or totalization are offered. Converters are available to service several meters simultaneously and provide necessary computer interface.

Most meters are built with flanged-type end fittings. Designs are available with sanitary-type fittings. In larger pipe sizes, Dresser-type and Victaulic-type end connections are also widely used.

Capacity and Range

Figure 2.9g is a nomograph for magnetic flowmeter capacities. Magnetic flowmeters have an excellent op-

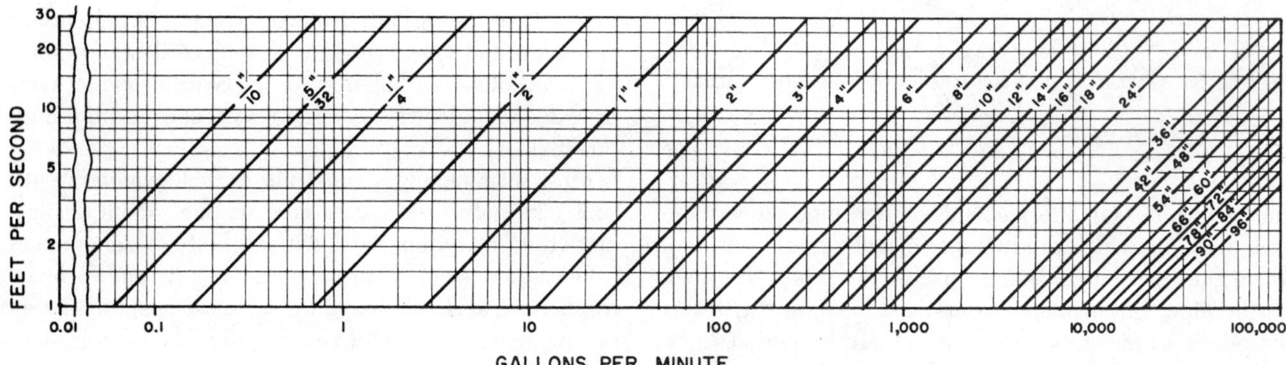

Fig. 2.9g Magnetic flowmeter capacity nomograph

erating range, at least 100:1. For AC types, typical inaccuracy is ±1 percent of full scale; to improve performance, range is usually divided into two portions and automatically switched between the two. Pulsed DC-types have typical inaccuracy of ±1 percent of rate applicable to a 10:1 range; operation of flow rates below 10 percent of maximum is on the order of ±0.1 percent of full scale. For magnetic flowmeters, regardless of type, the converter can be set for 20 mA output for any flow from 10 to 100 percent of meter capacity and still have at least a 10:1 operating range. This ability to field set or reset the meter for the actual operating conditions provides optimum performance.

Applications

Most liquids or slurries are adequate electrical conductors and can be measured by electromagnetic flowmeters. If the liquid conductivity is equal to 20 microSiemens per centimeter or greater, most of the conventional magnetic flowmeter systems can be used. Special systems are also available which will measure the flow of liquids with threshold conductivities as low as 0.1 microSiemens per centimeter. Some typical electrical conductivities are:

Liquid (at 25° C except where noted)	*Conductivity MicroSiemens/centimeter*
Acetic acid (up to 70% by weight)	250 or greater
Ammonium nitrate (up to 50% by weight)	360,000 or greater
Molasses (at 50°C)	5,000
Ethyl alcohol	0.0013
Formic acid (all concentrations)	280 or greater
Glycol	0.3
Hydrochloric acid (up to 40% by weight)	400,000 or greater
Kerosene	0.017
Magnesium sulphate (up to 25% by weight)	26,000 or greater
Corn syrup	16
Phenol	0.017
Phosphoric acid (up to 87% by weight)	50,000 or greater
Sodium hydroxide (up to 50% by weight)	40,000 or greater
Sulphuric acid (up to 99.4% by weight)	8,500 or greater
Vodka (100 proof)	4
Water (potable)	70

Above the threshold conductivity there is minimum effect for conductivity changes. The effect of liquid operating temperature upon the liquid conductivity should be considered. Most liquids have a positive temperature coefficient of conductivity; certain liquids which might be marginal at one temperature could become sufficiently nonconductive at a lower temperature so as to impair metering accuracy. At a higher temperature, the same liquid may be metered with good results. There are a few liquids that have a negative temperature-conductivity characteristic; these should be carefully investigated before applying magnetic flowmeters.

Magnetic flowmeters are not affected by viscosity or consistency (referring to Newtonian and non-Newtonian fluids respectively). The changes in flow profile due to Reynolds number change or due to piping do not affect magnetic flowmeter performance. The voltage generated is a summation of the incremental voltages across the entire area between the electrodes, resulting in a measure of the actual average fluid velocity. Nevertheless, conservative installation of the meter would be with 5 diameters of straight pipe before and 3 diameters of straight pipe following the meter.

Magnetic flowmeters are bi-directional. Most meters are used as uni-directional devices, but manufacturers offer specific converter options with an output signal for both direct and reverse flow conditions.

The magnetic flowmeter must always be used with a full pipe to assure accurate measurement because the flowmeter generates a voltage which is proportional to the average velocity of the fluid. To determine volumetric flow rate, we must have a known and constant cross sectional area and therefore the pipe must be full. Also, the meter will measure liquids with entrained gases as if the entire volume were filled with liquid; the measurement will be in error by the percent concentration of entrained gas.

The meter's electrodes must remain in electrical contact with the fluid being measured, and the electrodes should always be installed in the horizontal plane. In applications where a build-up or coating of the piping and the inside wall of the flowmeter occurs, periodic "flushing" or cleaning is recommended. Coatings can have conductivities that are the same, lower, or higher than the liquid. The effects are significantly different. Where the conductivity of the coating is essentially the same as that of the liquid, there is no effect on the accuracy of the measurement. This can be viewed as a specific profile condition and the meter will average the velocity to give the correct value for the particular flowrate. Fortunately, this is the most common coating problem. If the conductivity of the coating is significantly lower than the liquid being measured, the insulating coating can quickly disable the meter. If periodic cleaning is not possible, any one of the various types of electrode cleaning techniques offered by the manufacturers should be applied. Manufacturers also offer specifically shaped protruding electrodes to take advantage of the cleaning effect of adding turbulence to the local flow condition at the electrode. If the conductivity is higher, no corrective measure is needed.

Although magnetic flowmeters are not affected by temperature or pressure changes, there are specific limits of both temperature and pressure for particular designs. Temperature limitations for the particular lining material selected require special consideration.

Other Designs for Specific Applications

Sewage Sludge

Special systems for measuring sewage sludge are designed to prevent the build-up and carbonizing of sludge on the meter electrodes. Such systems use a self-heating principle to keep the metering body temperature at a level which prevents sludge and grease accumulation.

High Pressure

Special designs are available for high pressure services through 2500 PSIG (17MPa).

Permeable Slurries

Special systems are available for measuring slurries that contain permeable solids such as magnetic oxides and magnetic sulfides. Additional circuitry compensates for signal changes induced by the presence of permeable material that alters the slurry's magnetic field distribution.

DC-type

The magnetic flowmeters described thus far are of the alternating current-type. A DC magnetic flowmeter is available and is designed specifically for handling liquid metals such as liquid potassium and liquid sodium.

Bypass-types

The very low head loss of this family of flowmeters makes them suitable for use in measuring bypass flow around a mainline orifice, as shown in Figure 2.9h.

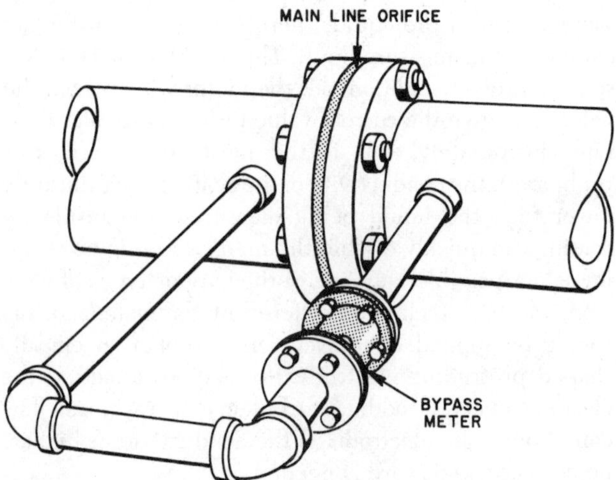

Fig. 2.9h Magnetic flowmeter used to measure bypass flow around an orifice plate to infer main line flow rate

Mass

Mass flowmetering systems are available using magnetic flowmeters in association with gamma radiation density gauges. The signal developed by the magnetic flowmeter relates to volumetric flow; the output signal from the gamma radiation density gauge is a signal proportional to density. The two signals are multiplied to obtain an output signal proportional to gravimetric flow rate for totalization, record or control. Some designs incorporate the flowmeter and the density gauge into one housing for installation ease and simplicity.

Installation

The signal generated by magnetic flowmeters is in the high microvolt to low millivolt range. Proper electrical installation and grounding is mandatory. Individual manufacturer recommendations for installation are the result of extensive experience and should be scrupulously followed (see Figure 2.9i).

It has been observed that AC-type magnetic flowmeters occasionally have shifted their no-flow indication after some operating time, requiring a zero reset. The shift occurs despite the proper application and correct installation of the meter. One of the most important installation considerations with electromagnetic flowmeters is a proper "bonding" of the flowmeter to the adjacent piping to minimize zero shifts. The intent of this bonding, or "jumpering," is to prevent stray current from passing through the flowmeter near the electrodes. Magnetic flowmeters are lined with an insulating material; generally, this lining covers the flange face of the meter, making the meter an electrical discontinuity in the system. The flange bolts should not be used for bonding since rust, corrosion, paint or other insulating materials will create an insulating barrier between the bolts and the flanges. Manufacturers supply and insist upon the installation of copper braid jumpers from the meter flange to the pipe flange at either end of the meter installation. Installing the flowmeter using the jumpers provides a continuous path to the stray current and a more stable zero can be expected. It is also essential to install a suitable ground strap to a grounded piece of structural steel, a grounding rod, or a cold water pipe.

The conservative installation of magnetic flowmeters recommends that five diameters of straight pipe, the same size as the flowmeter, be installed prior to the meter and two or three diameters of the same size pipe follow the meter. Meters can be installed in horizontal pipelines, vertical pipelines, or sloping lines. It is essential to keep the electrodes in the horizontal plane to assure uninterrupted contact with the liquid or slurry being metered. In gravity feed systems, the meter must be kept continually full; therefore, the meter should be installed in a "low point" in the system or, preferably, in a vertical upflow line.

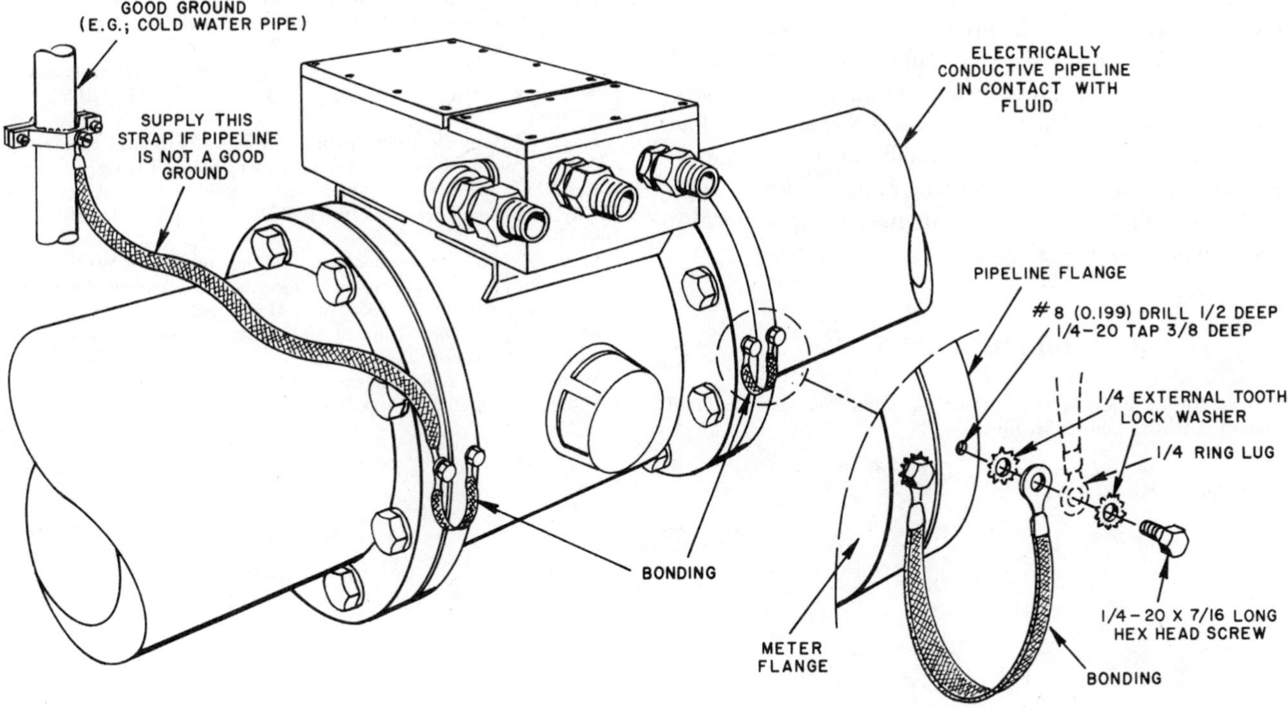

Fig. 2.9i Typical bonding and grounding procedure

Advantages

1. The basic advantage of the magnetic flowmeter is that it is totally obstructionless and, therefore, also has no moving parts. Pressure loss of the flowmeter is no greater than that of a piece of pipe of the same length. Pumping costs are minimized because there is no obstruction and because the fluid is not being manipulated in order to get a flow measurement.
2. Electric power requirements are extremely low, particularly with the pulsed DC-types. Electric power requirements as low as 15 or 20 watts are not uncommon.
3. The meters are suitable for most acids, bases, waters, and aqueous solutions because the lining materials selected are not only good electrical insulators but also are excellent materials for corrosion resistance. Only a small amount of electrode metal is required, and stainless steel, Alloy 20, the Hastelloys, nickel, Monel, titanium, tantalum, tungsten carbide, and even platinum are common.
4. The meters are widely used for slurry services not only because they are completely obstructionless but also because selected liners such as polyurethane, neoprene and rubber have superb abrasion or erosion resistance.
5. The meters are capable of handling extremely low flows with minimum sizes less than ⅛ in. (3.175 mm) inside diameter. The meters are also suitable for very high volume flowrates with sizes as large as 10 feet (3.04 m) offered.
6. The meters can be used as bi-directional meters.

Limitations

The meters do have some specific application limitations:

1. The meters work only with fluids which are adequate electrical conductors. Pure substances, including some hydrocarbons and all gases, cannot be measured. Most acids, bases, water, and aqueous solutions can be metered.
2. The meters are relatively heavy; especially in larger sizes.
3. Electrical installation care is essential.
4. The price of magnetic flowmeters ranges from moderate to expensive. They are generally selected where corrosion resistance, abrasion resistance, or high performance over wide turndown range is required. In such applications, the higher price can be justified.
5. The vast majority of designs presently available require AC power. DC powered versions are expected to gain acceptance in the future.
6. To periodically check the zero on AC-type magnetic flowmeters, block valves are required on either side to bring the flow to zero and keep the meter full.

Magnetic flowmeters are traditionally the first type of

flowmeters to be considered for very corrosive applications and for applications involving measurement of abrasive and/or erosive slurries. They are widely applied to pulp and paper stock measurement and other non-Newtonian fluids. They can be used for very low flow rates; pipe inside diameters as small as 0.1 in. (2.5 mm) are offered which can handle flow ranges as low as 0.01 to 0.1 gpm (0.038 to 0.38 l/m). Magnetic flowmeters also are available in pipe sizes up to 120 in. (3 m).

BIBLIOGRAPHY

Bean, H.S., "Fluid Meters," *ASME Research Committee on Fluid Meters*, Sixth Edition, 1971, pp. 125–128.

Eastman, P.C., Brodie, D.E. and Sawyer, D.J., "A Magnetic Flowmeter with Concentric Electrodes," *Instrumentation Technology*, June, 1970, pp. 52–55.

Elrod, H.J. and Fouse, R.R., "An Investigation of Electromagnetic Flowmeters," *Transactions ASME*, Volume 74, May 1952, pp. 589–594.

Head, V.P., "Electromagnetic Flowmeter Primary Elements," *AMSE Journal of Basic Engineering* Paper 58-A-126, 1958.

Kolin, A., "An Alternating Field Induction Flowmeter of High Sensitivity," *Review of Scientific Instruments*, Volume 16, May 1945, pp. 109–116.

Shercliff, J.A., "Experiments on the Dependence of Sensitivity of Velocity Profile on Electromagnetic Flowmeters," *Journal of Scientific Instruments*, Volume 32, 1955, pp. 441 and 442.

Shercliff, J.A., *The Theory of Electromagnetic Flow Measurement*, Cambridge University Press, 1962.

Webb, A.S., "Electromagnetic Flowmetering," *Instrumentation Technology*, March 1974, pp. 29–33.

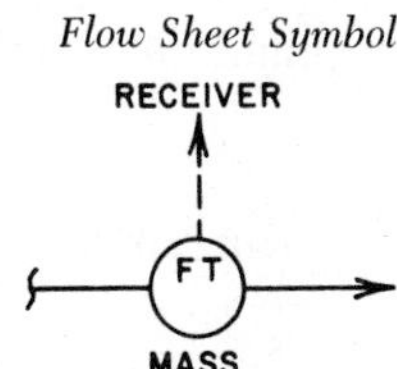

2.10 MASS FLOWMETERS

Design Pressure:	Up to 1500 PSIG (10 MPa) Special designs to 20,000 PSIG (140 MPa)
Design Temperature:	To 500°F (260°C)
Flow Rate Ranges:	Liquids: 1 lb/hr to 500,000 lbs/hr (0.13×10^{-3} kg/s to 65 kg/s) Gases: 1 lb/hr to 50,000 lbs/hr (0.13×10^{-3} kg/s to 6.5 kg/s)
Materials for Construction:	Stainless steel, Hastelloy, titanium
Inaccuracy:	½–1% actual flow rate
Cost:	\$2500 to \$6000
Partial List of Suppliers:	Agar Instrumentation, Inc.; Black, Sivalls & Bryson, Inc.; Flo-Tron, Inc.; Foxboro Co.; General Electric Co., Industrial Sales Div.; Micro Motion, Inc.

As an element of process control, the measurement of mass flow is not new. But until recently these measurements were not made directly; they were made by batch weighing or, in the case of continuous processes, by inference from simpler volumetric flow measurements. To extrapolate mass flow measurements from volumetric flow is a complex procedure, necessitating the application of various corrections based on parameters related to density such as pressure, temperature and specific gravity.

The most commonly cited examples of continuous processes that require mass flow measurements are those in which a hydrocarbon fuel combines with another compound to produce a reaction that results in the release of energy for heat or propulsion. In the case of rocket engines and combustion engines, the efficiency of the reaction—and thus of the engine itself—is related to the pounds, not gallons or cubic feet, of fuel consumed. Power plants rely on the combining of hydrocarbon fuels, either liquid or gas, with air, on a molecular weight basis, to provide heat for conversion into electrical energy.

Obviously, there are countless other processes that involve either the combination or separation of chemicals, in which mass flow measurements play a vital role.

Today, there is no longer any need to start with volumetric flow figures to eventually obtain the necessary mass flow measurements. Several techniques have been developed for measuring mass flow directly. These techniques offer greater accuracy and reliability; because they are direct and are independent of pressure, temperature and specific gravity, they eliminate many of the measurement and calculation steps required by the old, indirect method of obtaining mass flow measurements, thus vastly reducing the margin of potential error.

Theory

The principle of angular momentum can best be described by referring to Newton's second law of angular motion and the definition of angular momentum, using the following notation:

H = angular momentum (lbf-ft-sec)*
I = moment of inertia (lbf-ft²)
ω = angular velocity (rad/sec)
α = angular acceleration (rad/sec²)
Y = torque (ft-lbf)
r = radius of gyration (ft)
m = mass (slugs)
t = time (sec)

Newton's second law of angular motion states that

$$Y = I\alpha \qquad 2.10(1)$$

*For multipliers to convert to SI units, refer to Appendix A.1.

and defines that

$$H = I\omega \tag{2.10(2)}$$

But, since by definition

$$I = mr^2 \tag{2.10(3)}$$

equation 2.10(1) becomes

$$Y = mr^2\alpha \tag{2.10(4)}$$

and equation 2.10(2) becomes

$$H = mr^2\omega \tag{2.10(5)}$$

Since

$$\alpha = \frac{\omega}{t} \tag{2.10(6)}$$

equation 2.10(4) becomes

$$Y = \frac{m}{t}r^2\omega \tag{2.10(7)}$$

Solving for mass flow rate, $\frac{m}{t}$, (lbm/sec), we get

$$\frac{m}{t} = \frac{Y}{r^2\omega} \tag{2.10(8)}$$

Also, dividing both sides of equation 2.10(5) by t:

$$\frac{H}{t} = \frac{m}{t}r^2\omega \tag{2.10(9)}$$

Since torque is in terms of force, the right hand side of equation 2.10(8) must be multiplied by g (32.2 ft/sec^2 or 9.8 m/s^2) to obtain a dimensionally correct equation. Therefore, since r^2 is a constant for any given system, the mass flow of fluid can be determined if an angular momentum is introduced into the fluid stream and measurements are made of the torque produced by this angular momentum and of the fluid's angular velocity.

Impeller-Turbine

The impeller-turbine-type mass flowmeter uses two rotating elements in the fluid stream, an impeller and a turbine (see Figure 2.10a). Both elements contain channels through which the fluid flows. The impeller is driven at a constant speed by a synchronous motor through a magnet coupling and imparts an angular velocity to the fluid as it flows through the meter. The turbine located downstream of the impeller removes all angular momentum from the fluid and thus receives a torque proportional to the angular momentum. This turbine is restrained by a spring which deflects through an angle proportional to the torque exerted upon it by the fluid, thus giving a measure of mass flow.

Constant-Torque-Hysteresis Clutch

Another mass flowmeter version based upon the angular momentum principle eliminates the necessity of

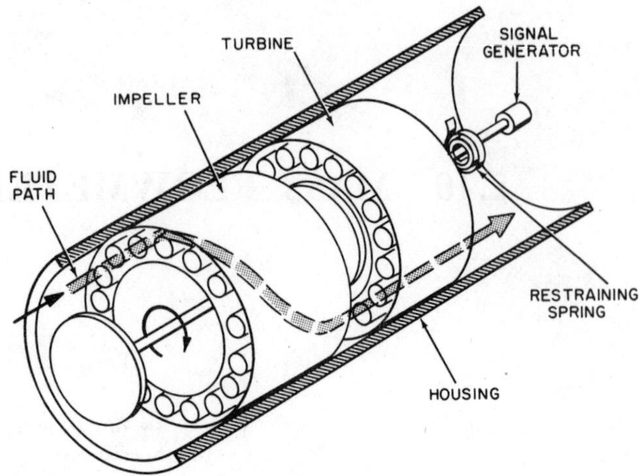

Fig. 2.10a Impeller-turbine mass flowmeter (illustration reproduced by permission of the General Electric Co.)

making a torque measurement by imparting a constant torque to the fluid stream. The relationship between mass flow and torque, as was derived previously, is

$$\frac{m}{t} = \frac{Y}{r^2\omega}$$

Therefore, if Y is held at a constant value, and since r^2 is a physical constant of any given system,

$$\frac{m}{t} = \frac{k}{\omega}$$

This relationship is translated into a mass flowmeter as follows. A synchronous motor is placed in the center of the flowmeter assembly. This motor is magnetically coupled to an impeller through which the fluid to be measured is allowed to flow. The magnetic coupling between the motor and the impeller is accomplished by means of a hysteresis clutch which transmits a constant torque from the motor to the impeller. Thus, a measurement of the rotational speed of the impeller is inversely proportional to the mass flow rate.

Twin-Turbine

Another approach to mass flowmetering, using the same basic principles of angular momentum and measurement by means of torque, has been used in the twin-turbine mass flowmeter. In this instrument, two turbines are mounted on a common shaft (see Figure 2.10b). They are connected with a calibrated torsion member. A reluctance-type pickup coil is mounted over each turbine, and the turbines are designed so that a very sharp pulse is generated in each of the coils by each of the corresponding turbines for each revolution of the aggregate twin-turbine assembly.

Each turbine is designed with a different blade angle, so that there is a tendency for the turbines to turn at different angular velocities. However, since the turbines,

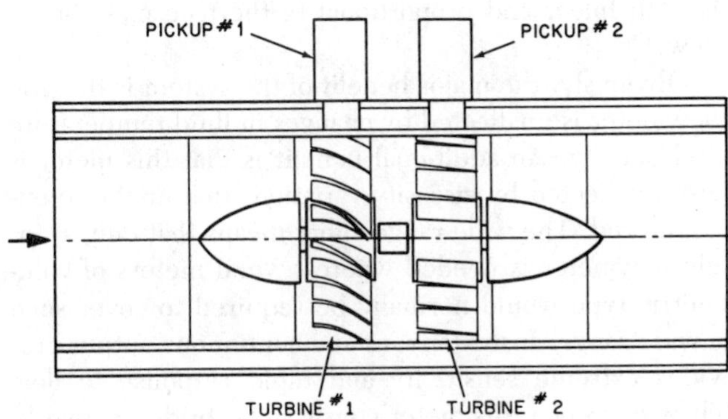

Fig. 2.10b Twin-turbine mass flowmeter

motion is restricted by the coupling torsion member, the entire assembly rotates in unison at some average velocity, and an angular phase shift is developed between the two turbines. This angle is a direct function of the angular momentum of the fluid. As was previously shown, angular momentum can be measured by torque, and angular momentum is a function of mass flow. In the twin-turbine assembly, the turbines are not restrained by a spring, but the torsion member which holds them together is twisted. This torsion member has a well-established torsion-spring rate (ft-lbf/rad). Therefore, the angle developed between the two turbines is a direct function of the twist or torque exerted by the system.

This angle is measured by a unique method. As each turbine magnet passes its pickup coil, the coil generates a pulse. The pulse from the upstream turbine is used to open a so-called electronic gate, while the pulse from the downstream turbine closes this electronic gate. An oscillator is placed in the electronic circuit, and the oscillations are counted while the gate is opened. The number of oscillations is thus a function of the angle between the two turbines. Knowledge of the angle gives the value of torque which, in turn, is proportional to mass flow rate.

Coriolis

The Coriolis-type of mass flowmeter (see Figure 2.10c) consists of a centrifugal-pump impeller wheel and a vaned sensing wheel which acts as a turbine wheel to extract the angular momentum imparted to the fluid by the impeller. The sensing (or turbine) wheel is contained in the same housing as the impeller and is attached to the latter by a strain gauge; the combination is driven at a known constant speed. The power applied to the impeller is merely that required to overcome the frictional drag of the system.

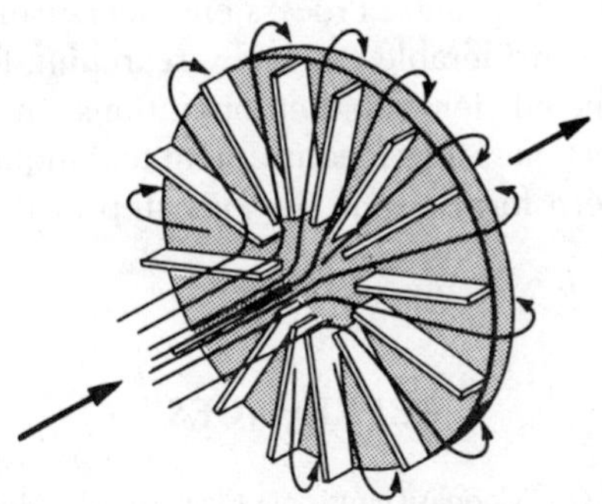

Fig. 2.10c Coriolis mass flowmeter

The torque measured is that required to impart to the fluid stream a Coriolis acceleration, and is given by the expression

$$Y = \omega\,(R_2^2 - R_1^2)\,\frac{m}{t}$$

where R_2, R_1 = outer and inner radii (ft)

Comprehensive analysis and laboratory testing of this type of flowmeter are reported in Reference 1.

Gyroscopic

Another angular momentum mass flowmeter (Figure 2.10d) operates on the principle of a gyroscope. It consists of a pipe shaped in the form of a circle or a square. A motor introduces an oscillating vibration at a constant angular velocity ω about the A axis. When the fluid passes through it, a precession-type moment is produced about the B axis and is measured by the deflection of a sensing element. This deflection can be shown to be directly proportional to mass flow.

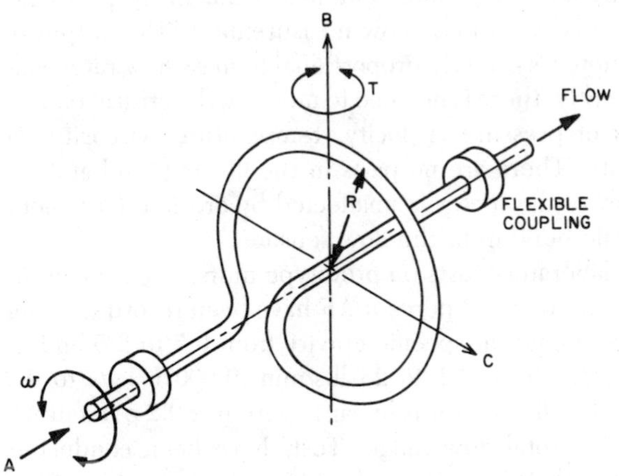

Fig. 2.10d Gyroscopic mass flowmeter

The gyroscopic mass flowmeter can handle slurries in the medium pressure and temperature ranges but its industrial use is very limited due to its high cost and inability to handle high flow rates.

Coriolis/Gyroscopic

The Coriolis/gyroscopic mass flowmeter (Figure 2.10e) employs a C-shaped pipe and a T-shaped leaf spring as opposite legs of a tuning fork. An electromagnetic forcer excites the tuning fork, thereby subjecting each moving particle within the pipe to a Coriolis-type acceleration.

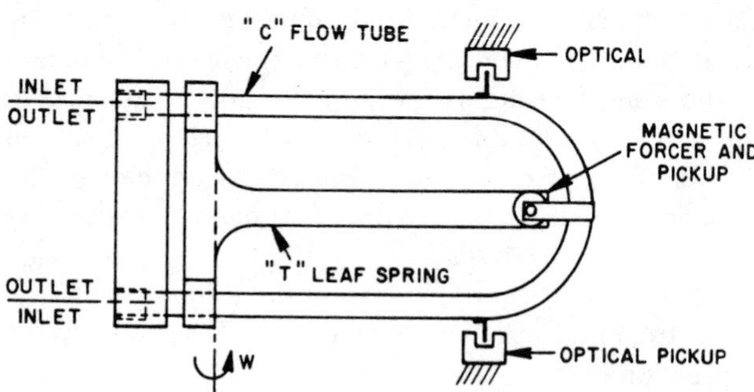

Fig. 2.10e Coriolis/gyroscopic flowmeter

The resulting forces angularly deflect the C-shaped pipe an amount that is inversely proportional to the stiffness of the pipe and proportional to the mass flow rate within the pipe.

The angular deflection of the pipe is optically measured twice during each cycle of the tuning-fork oscillation. The output of the optical detector is a pulse that is width-modulated proportional to the mass flow rate. An oscillator/counter digitizes the pulse width and displays a numerical indication of mass flow rate.

The total mass flow over a given time interval is obtained using a digital integrator to sum the pulses of the flow rate indicator. In this way totalized flow is updated with each oscillation of the tuning fork.

This mass flowmeter eliminates the many problems associated with mass flow measurement. The output of the meter is directly proportional to mass flow rate; consequently, there is no need to measure the critical parameters of pressure, velocity, temperature, viscosity, or density. There are no parts in the flowing fluid and accuracy of the meter is unaffected by erosion, corrosion, or scale buildup in the flow sensor.

In laboratory tests on prototype units, inaccuracies in the order of ± 0.2 percent AF have been recorded using fluids ranging in specific gravity from 0.5 to 2.5 and in flow rates from 0.1 to 25 lbs/min (0.00076 kg/s to 0.2 kg/s). The flowmeter is linear to within ± 0.2 percent AF over the total flow range. Tests have been conducted using both Newtonian and non-Newtonian fluids with a wide range of viscosities; the results indicate the meter is totally insensitive to viscosity. With the meter it is now possible to measure pulsating flow and two-phase flow.

Linear

The linear mass flowmeter is, in principle, a hydraulic equivalent of the electrical Wheatstone Bridge. Four matched orifices make up the bridge and an integral constant flow recirculating pump establishes the internal reference flow. Sensing a situation of unbalance generated by external flow through the meter, the hydraulic bridge produces an output of differential pressure which is both linear and proportional to the true mass liquid flow.

Obviously, the major benefit of the system is that the flowmeter is unaffected by changes in fluid temperature and density. An additional benefit is that this meter is also unaffected by viscosity variations that might be encountered. The wide rangeability means that only a single flowmeter is needed where several meters of volumetric type would normally be required to cover such a wide range. Instantaneous and continuous output provides extreme sensitivity and rapid response to flow changes. Since this meter consists of a bridge network, it also has the inherent capability to sense very low flow rates.

Conclusions

In addition to increased accuracy (plus or minus 0.25 percent of reading) and reliability, other significant advantages may be gained from direct mass flow measurement. Techniques are now available to accommodate all types of fluids, including gases, multi-phase materials, liquids of any viscosity, slurries, and so on. Some mass flowmetering systems offer corrosion resistance and intrinsic safety for use in combustible or explosive environments. Almost all provide a fast, dynamic response and are fully suitable for continuous industrial operation.

The advantages of direct mass flow measurement are self-apparent. Directly tied to those advantages are an array of tangible benefits which can be summed up in one word: efficiency. It has been pointed out that many processes requiring measurement involve fuels. In fact, a fuel is a form of energy which is represented by the BTU content of the fuel. In the case of almost all fuels, the measurement of the BTU content is far more constant on a per pound basis than on either a per cubic foot or per gallon basis. For example, comparison of a cubic foot of methane with a cubic foot of iso-butane reveals a 322 percent difference in BTU content. However, there is only a 12 percent change in the BTU/lb. of these gases.

Today's economy has sharply underlined the need for maximum precision and accuracy in the processing of fuels and other chemicals. Anything less is wasteful and, as raw materials costs continue to spiral, such waste finds its way to the bottom line in the form of increased expenses and low profits. Process engineers can offset this trend to a considerable extent by searching for ways to optimize the efficiency to their operations, and the adoption of direct mass flow measurement techniques, in their most modern forms, is an obvious step in the right direction.

REFERENCES

1. Plache, K.O., "Coriolis/Gyroscopic Flow Meter." Mechanical Engineering, March, 1979, p. 36.

BIBLIOGRAPHY

ASME Research Committee on Fluid Meters, "Fluid Meters, Their Theory and Application," fifth edition, 1959.

Close, D.L., "Cryogenic Mass Flowmeter," ISA Symposium in Pittsburgh, May 1971, Paper No. 1-6-34.

Hayward, A.J., "Choose the Flowmeter Right for the Job," Processing Journal, 1980.

Laskaris, E.K., "The Measurement of Flow," *Automation*, 1980.

Li, Y.T., and Lee, S.Y., "Fast-Response True-Mass-Rate Flow-Meter," Trans ASME, Vol. 75, July 1953, pp. 835–41.

Lomas, D.J., "Selecting the Right Flowmeter," *Instrumentation Technology*, 1977.

Snell, C.C., "Relative Volume Fraction Measurement of Gas/Liquid or Solid/Liquid Flow with a Rotating Field Conductance Gauge," 1979 ISA Symposium, ISBN 87664-432-9.

Spink, L.K., "Principles and Practice of Flow Meter Engineering," The Foxboro Co., 1958.

Watson, G.A., "Flowmeter Types and Their Usage," Chartered Mechanical Engineer Journal, 1978.

Flow Sheet Symbol

RECEIVER

ST

2.11 METERING PUMPS

Type of Hardware:	A. Peristaltic B. Piston C. Diaphragm
Flow Rate:	A. 0.005 cc/m to 250 GPH (18 l/m) B. 0.1 to 17,000 GPH (.38 l/hr to 1072 l/m) C. 50 to 2,500 GPH (3 to 158 l/m)
Inaccuracy:	A. 1/10% B. ½–1% C. ½–1%
Maximum Pressure:	A. 20 PSIG (138 kPa) B. 100,000 PSIG (690 MPa) C. 1,500 PSIG (10.3 MPa) with plastic diaphragm; 45,000 PSIG (283.5 MPa) with metal diaphragm
Maximum Temperature:	A. −70 to +600°F (−57 to +315°C) B. Approx. 1000°F (538°C) C. 400°F (204°C)
Materials of Construction:	A. Neoprene, Tygon, Viton, silicone B. Cast iron, steel, stainless steel, Hastelloy, plastics C. Polyethylene, Teflon, various metals
Cost:	\$300–1200 (A) \$500–3000 (B,C)
Partial List of Suppliers:	American Instrument Co., Travenol Laboratories Inc.(C); American Lewa Inc. (A, B, C); Amtec Products Co., W. S. Shamban Co., (A); Flo-Tron Inc. (B); Fluidyne Instrumentation (B); Jaeco Pump Co. (B); Milton Roy Co. (B); Neptune Chemical Pump Co. (B, C); Randolph Austin Co. (A, B); Viking Pump Div., Houdaille Industries, Inc. (B); Wallace & Tiernan Div., Pennwalt Corp. (B, C)

A metering pump is a positive displacement pump providing a predictable and accurate rate of process fluid flow. Normally the application-design, specification, and use of pumps are the concern of mechanical engineers and machinery designers. Metering pumps, however, are sometimes used to actually measure flow-rate, and in many cases they are the final control elements in an instrumentation loop. Therefore, the instrument engineer should be familiar with their operations and applications.

A great variety of controlled-volume pumps (as metering pumps are sometimes called) are commercially available. Many of these pumps are designed to meet the needs of just one particular application, such as adding sodium hypochlorite to a swimming pool or providing chemical reagents to a chromatograph. Thus, each industry appears to have its own particular types of metering pumps, and they could be classified by their application. A better way of classifying metering pumps, however, is to distinguish them by their basic mode of operation. Any positive displacement pump, due to is volumetric mode of fluid transfer, can be used as a metering pump. In practice, however, only those positive displacement pumps that have none or only very little internal and/or external leakage can provide the precision and accuracy that are normally required of a metering pump. These positive displacement pumps are the peristaltic, reciprocating piston, and diaphragm pumps. Their operation and application are described below.

Peristaltic Pumps

Peristaltic is defined in Webster's New Collegiate Dictionary as designating the peculiar wormlike wave motion of the intestines and other hollow muscular structures, produced by the successive contraction of the muscular fibers of their walls, forcing their content forward. In the peristaltic pump the fluid is moved forward by progressively squeezing a flexible container from the entrance to discharge. This container is usually a tube that can be made out of any material that possesses sufficient resiliency to allow it to recover its original shape immediately after compression. There are a variety of methods employed for squeezing the tube (or container) to produce flow-rate. Some of these are:

1. Rollers that are connected to a rotating body squeeze the tubing against a circular housing (see Figure 2.11a).

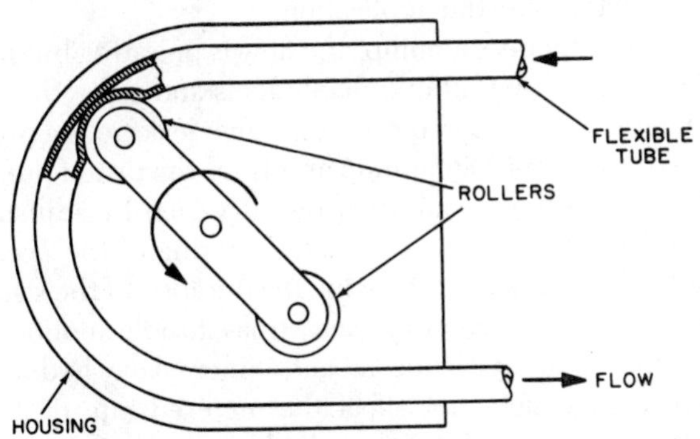

Fig. 2.11a Peristaltic pump utilizing rollers

2. Rollers that are driven by a chain drive squeeze tubes against a flat plate.
3. Cam operated fingers successively squeeze the tubing against a flat surface (see Figure 2.11b).

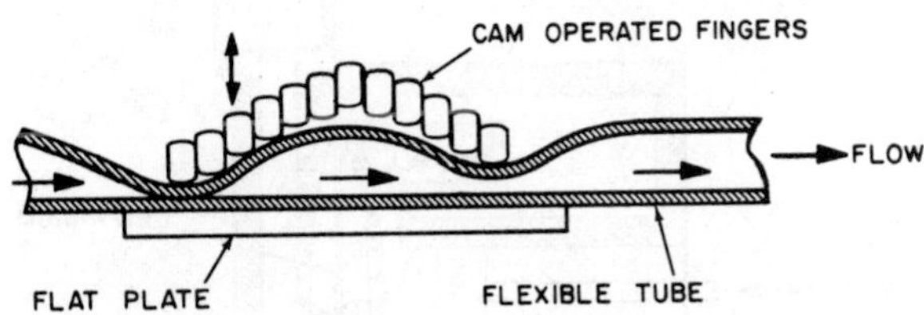

Fig. 2.11b Peristaltic pump with cam-operated fingers

4. A rotating wobbling cam squeezing a tube against a flat plate.

The plastic hose or tubing provides an external leak-tight, sanitary, and an easily cleanable and replaceable container. It must be remembered that the tube is the only component of the pump that comes into contact with the fluid. A plastic material can usually be found that will be suitable for even the most corrosive and abrasive application. However, the use of plastic tubing also places severe limits on the capability of the peristaltic pump. These pumps can furnish only low flow rates and low pressure heads.

The peristaltic pump has found particularly large acceptance in the medical and biochemical fields where high accuracy, low flow rates, inherent enclosure of the fluid, and sterilization are prime requirements. The flow-rate of the peristaltic pump can be adjusted by changing the speed of the squeezing mechanism. Power to these mechanisms is usually provided by 60-Hz, 110-volt electrical motors. Air operated and electrical explosion-proof motors are also available.

Another type of peristaltic pump employs a plastic liner separating a rotating cam from the pumped fluid. This allows the use of a cam positive displacement pump for metering service, but prevents any external leakage. Higher pressures and flow rates can be obtained from this type of pump, but the advantages of the flexible tubing are lost.

Piston Pumps

The piston pump employs a piston or plunger which moves with a reciprocating motion within a chamber. A fixed volume of liquid is delivered with each stroke. The flow rate is a function of piston diameter, chamber length, and piston speed. Check valves located at the pump inlet and outlet are required to prevent backflow. A schematic of a typical piston pump is shown in Figure 2.11c. The piston produces pressure in only one direction; therefore, the flow produced by plunger pumps (as piston pumps are sometimes called) is pulsating. If the pulsating flow characteristics are undesirable, a dampening reservoir (accumulator) should be installed

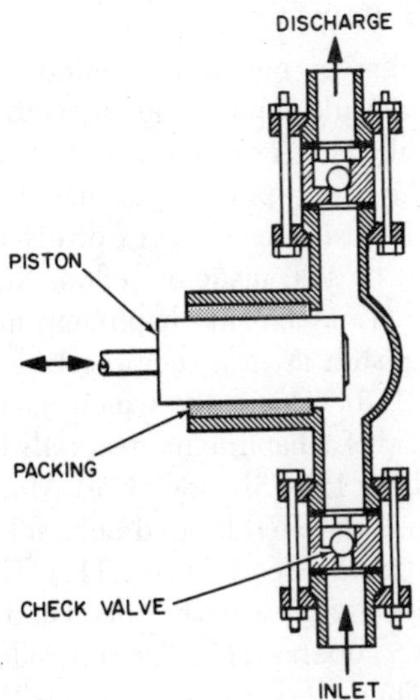

Fig. 2.11c Piston pump schematic

in the discharge line of the pump. Another method available to reduce pulsation is to use a pump that employs more than one chamber/piston combination in parallel. Pumps having as many as four chambers (cylinders) are commercially available. These multiple-piston pumps are called duplex pumps if they have two pistons, triplex, if they have three, etc.

The construction materials of the piston pump components must be selected with care since its housing, piston, piston packing, valve body and valve seat all come into intimate contact with the process fluid.

The main disadvantage of piston pumps with regard to metering service is that leakage occurs past the piston packing and the valve seats, which decreases accuracy and precision. However, in many instances the ability of the piston pump to deliver high head pressures or large flow rates outweighs the leakage problem. Leakage of the check valves is often minimized by using two valves in series (double valves) on both the pump suction and discharge ports.

Flow rate control in piston pumps is usually accomplished by varying the piston stroke length through alteration of the eccentricity of the piston crank. In many pumps this can be done without stopping the operation of the unit. Another method that is sometimes used to alter flow rate is to change the piston stroke duration by means of variable idle periods.

Piston pumps are usually driven by electric motors; air drives are also available.

An interesting derivative of the piston pump that can be used if extremely high pressures and low flow rates are required is the non-reciprocating "single-shot" pump. In this unit, the fluid is driven from the cylinder with a lead screw driven piston, usually employing a gear drive assembly.

The piston pump finds its main use in the heavy chemical and manufacturing industry.

Diaphragm Pumps

The diaphragm or membrane pump uses a flexible member to transmit a pulsating force to the pumped fluid without allowing external leakage such as that occurring past the piston pump's packing. Like the piston pump, inlet and exit check valves direct the flow. The diaphragm may be soft, made of Teflon, Neoprene, etc., or hard, made of metal. The diaphragm may be moved directly by a piston as in a reciprocating piston pump (see Figure 2.11d). This type, which may also employ a bellows instead of a diaphragm, generally has a pressure limitation of about 125 PSIG (862 kPa). More commonly, the diaphragm is pulsated by hydraulic oil from a reciprocating piston pump (see Figure 2.11e). The diaphragm is thus hydraulically pressure balanced and, with the aid of support plates, operates in a low controlled magnitude of deflection. Thus, the endurance limit of the diaphragm material is not exceeded, allowing long life. For better assurance of liquid sealing, double diaphragms may be provided. The materials exposed to the pumped fluid—diaphragm, housing and valves—must be carefully selected for the application.

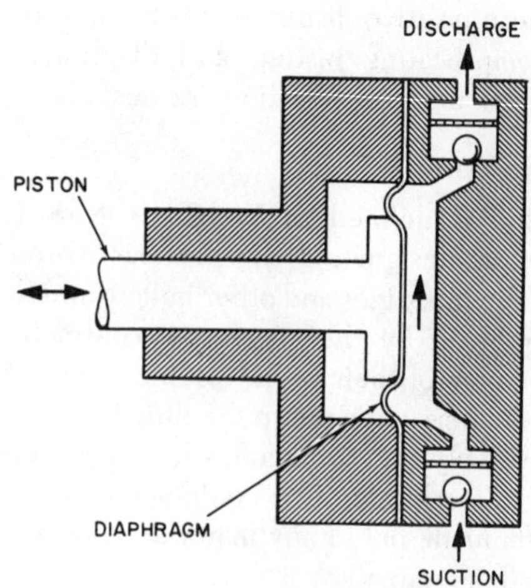

Fig. 2.11d Direct-driven diaphragm pump

As with the piston pump, the flow is pulsating but may be smoothed by multiple diaphragms and/or by the use of a dampening reservoir. Check valve leakage is a problem for the diaphragm pump as it is for the piston pump.

The slave fluid's pulsation rate is varied by adjusting the piston pump's stroke length, altering of the eccentricity of the crank, or changing the duration of the stroke by diverting a portion of each stroke to idle motion by mechanical or hydraulic means. Since slave hydraulic fluid leakage does not affect the metered rate of flow, many variations of the standard reciprocating piston pump are used.

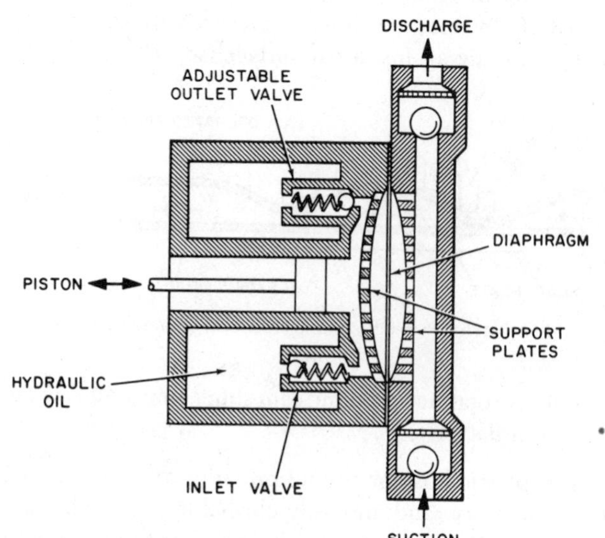

Fig. 2.11e Oil-driven diaphragm pump

Proportioning Pumps

At times multiple fluids need to be mixed at a set proportion regardless of the rate of flow, giving rise to proportioning pumps.

As a natural product of the finger- or chain-driven rollers of a flat plate peristaltic pump, multiple tubes of the same size or various sizes (up to 23) may be laid in parallel to proportion the fluid flow in each tube. This allows a set ratio of liquids to be pumped independent of the pumping rate as set by the motor speed. On piston and diaphragm pumps, multiple heads are set on the same drive with proportions set by their respective piston or diaphragm size. On some idle motion pumps, the stroke duration may be varied on individual pistons for proportion adjustments. Of course, in a more complex manner, several metering pumps set in parallel with the same speed control adjustment could accomplish the same function.

Conclusions

Each metering pump discussed has its particular application. Piston pumps are used to deliver high pressures. They require check valves and normally produce pulsating flows that can be dampened by various methods. Diaphragm pumps are utilized in the medium pressure range. The membrane serves as a moving partition between the mechanical or hydraulic drive and the process fluid. Rotary pumps furnish pulsation-free high flow rates, and are suitable for high-viscosity service. Their accuracy is a function of the clearances between the rubbing surfaces. This generally results in low precision and therefore the rotary pumps are not considered to be metering devices. Peristaltic pumps are very accurate, they can handle extremely small flows, are self-priming, and require no seals or check valves.

If the instrument engineer is responsible for the operation and maintenance of a metering pump, he must be aware that a pump differs in many aspects from other flowmeters. For example, the pump motor must be lubricated periodically. A control system must be used which makes it impossible for the pump to operate without liquid in it. The inlet piping must be designed to prevent cavitation. Running a pump dry or cavitating it will cause damage.

A metering pump should be calibrated, not only prior to first use, but also periodically during its operation. The calibration should duplicate fluid properties, suction and discharge pressures, and inlet and outlet piping configuration.

BIBLIOGRAPHY

"Metering Pump Survey," *Instruments and Control Systems*, April, 1971, pp. 103–109.

Payne, D.C., "Reliable Viscous Output with Positive-Displacement Pumps," *Chemical Engineering*, February 23, 1970.

2.12 ORIFICES

Flow Sheet Symbol

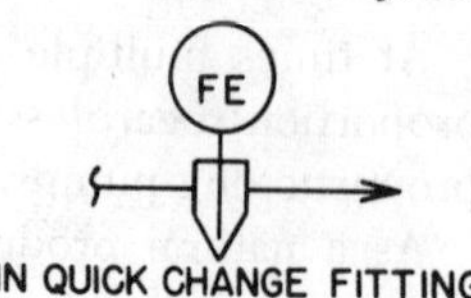

IN QUICK CHANGE FITTING

FE FLANGE TAPS

FO FIXED RESTRICTION

FE VENA CONTRACTA TAPS OR RADIUS TAPS

FT INTEGRAL ORIFICE TRANSMITTER

Design Pressure:	For plates it is limited by readout device only. Integral orifice transmitter to 1500 PSIG (10.3 MPa)
Design Temperature:	Function of associated readout system, only when the differential pressure unit must operate at the elevated temperature. For integral orifice transmitter the standard range is −20 to +250°F (−29 to +121°C)
Sizes:	Maximum size is function of related pipe size
Fluids:	Liquids and gases
Flow Range:	From a few cc/minute using integral orifice transmitters to any maximum flow limited only by pipe size
Inaccuracy:	± ¼ to ± 2% of full scale. This does not consider errors in bore diameter calculation or inaccuracies of the readout device used
Materials of Construction:	There is no limitation on plate materials. Integral orifice transmitter wetted parts can be obtained in steel, stainless steel, Monel, nickel, and Hastelloy
Cost:	\$45 to \$180 for plate only. For steel orifice flanges from 2 to 12′, the cost range is \$140 to \$820. For flanged meter runs in the same size range, the cost is \$300 to \$2400. The integral orifice transmitter costs vary from \$700 to \$800
Partial List of Suppliers:	Fischer & Porter Co.; Foxboro Co.; Honeywell Industrial Div.; Taylor Instrument Companies. Orifice plates can be obtained from specialized manufacturers and all major instrument manufacturers. Suppliers specializing on orifice plate accessories include Daniel Industries, Inc.; Fluidic Techniques, Inc.; Tech Tube Corporation

Head-Type Meters

Head-type flow meters comprise a class of devices for fluid flow measurement including orifice plates, venturi tubes, weirs, flumes, and many others. Their structure changes the velocity or direction of the flow, creating a measurable differential pressure or "pressure head" in the fluid.

Head metering is one of the most ancient of flow detection techniques. There is evidence that the Egyptians used weirs for measurement of irrigation water in the days of the Pharaohs, and that the Romans used orifices to meter water to households in Caesar's time. In the 18th century, Bernoulli established basic relationships between pressure head and velocity head, and Venturi published on the flow tube bearing his name. However, it was not until 1887 that Clemens Herschel developed the commercial venturi tube. Work on the conventional orifice plate for gas flow measurement was commenced by Weymouth in this country in 1903. Present activities include improved primary elements, refinement of data, more accurate and versatile test and calibrating equipment, better differential pressure sensors, and many others.

Head-type flow measurement is used extensively. It

is simple, reliable, relatively inexpensive, and capable of high, sustained accuracy. As a direct measurement of flow rate, it is particularly suited to automatic flow control. Total volume flow and total mass flow are readily obtained by integration of volume flow and mass flow rates. Head flow metering is unique in that the flow coefficient of the primary device is readily determined from simple measurements of dimensions; actual fluid flow calibration is not required, except where highest absolute accuracy is desired.

Six sections of this chapter (2.5, Elbow Taps; 2.8, Laminar Flowmeters; 2.12, Orifices; 2.13, Pitot Tubes; 2.18, Target Meters; 2.23, Venturi Tubes and Flow Nozzles; 2.25, Weirs and Flumes) deal with head-type measurement in various forms. Rotameters, discussed in Section 2.22, also depend upon the basic head-type principle.

Theory

Head-type flow measurement derives from Bernoulli's Theorem which states that in a flowing stream, the sum of the pressure head, the velocity head and the elevation head at one point is equal to their sum at another point removed in the direction of flow from the first point plus the loss due to friction between the two points. Velocity head is defined as the vertical distance through which a liquid would fall to attain a given velocity. Pressure head is the vertical distance which a column of the flowing liquid would rise in an open-end tube as a result of the static pressure.

This principle is applied to flow measurement by altering the velocity of the flow stream in a predetermined manner, usually by a change in the cross-sectional area of the stream. Typically, the velocity at the throat of a flow nozzle inserted in a pipe is increased relative to the velocity in the pipe. There is corresponding increase in velocity head. Neglecting friction and change of elevation head, there is an equal decrease of pressure head. This difference between the pressure in the pipe just upstream of the nozzle and the pressure at the throat of the flow nozzle is measured. Velocity is determined from the ratio of the cross-sectional areas of pipe and flow nozzle, and the difference of velocity heads given by differential pressure measurements. Flow rate derives from velocity and area. The basic equations are:

$$V = k\sqrt{\frac{h}{\rho}} \qquad 2.12(1)$$

$$Q = kA\sqrt{\frac{h}{\rho}} \qquad 2.12(2)$$

$$W = kA\sqrt{h\rho} \qquad 2.12(3)$$

where V is velocity, Q is volume flow rate, W is mass flow rate, A is cross-sectional area of the pipe, h is differential pressure between points of measurement, ρ is the density of the flowing fluid, and k is a constant which includes ratio of cross-sectional area of pipe to cross-sectional area of nozzle or other restriction, units of measurement, correction factors, etc., depending on the specific type of head meter. (For a more complete derivation of the basic flow equations, based on considerations of energy balance and hydrodynamic properties, consult references 1, 2, and 3.*)

Head Meter Characteristics

Two fundamental characteristics of head-type flow measurements are apparent from the basic equations. First, there is the square root relationship between flow rate and differential pressure. Second, the density of the flowing fluid must be taken into account both for volume and for mass flow measurements.

The square root relationship has two important consequences. Both are primarily concerned with readout. The primary sensor (orifice, venturi tube, etc.) develops a head or differential pressure. A simple linear readout of this differential pressure expands the high end of the scale and compresses the low end in terms of flow. Fifty percent of full flow rate produces 25 percent of full differential pressure. At this point, a flow change of 1 percent of full flow results in a differential pressure change of 1 percent of full differential. At 10 percent flow, differential pressure is only 1 percent; a change of 1 percent of full scale flow (10 percent relative change) results in only 0.2 percent full scale change in differential pressure. Both accuracy and readability suffer. Readability can be improved by a transducer which takes the square root of the differential pressure to give a signal linear with flow rate. However, errors in the more complex square root transducer tend to decrease overall accuracy.

For a large proportion of industrial processes which seldom operate below 30 percent capacity, a device with pointer or pen motion linear with differential pressure is generally adequate. Readout directly in flow can be provided by a square root scale. Where maximum accuracy is important, it is generally recommended that the maximum-to-minimum flow ratio shall not exceed 3 to 1 or at the most 3½ to 1 for any single head-type flow meter. The high repeatability of modern differential pressure transducers permits a considerably wider range for flow control, where constancy and repeatability of flow rate are the primary concern. However, where flow variations approach 10 to 1, use of two primary flow units of different capacities, or two differential pressure sensors with different ranges, or both, is generally recom-

*A large proportion of the material in Section 2.12 is updated and expanded material based upon the work of W.H. Howe in the original edition. A large part of the original material was abstracted from "Principles and Practice of Flow Meter Engineering" (Ref. 1). The "Shell Flow Meter Engineering Handbook" (Ref. 3, but now out of print) and the ASME report "Fluid Meters, Their Theory and Applications" (Ref. 2) are also highly recommended.

mended. It should be emphasized that the primary head meter devices produce a differential pressure which corresponds accurately to flow over a wide range. Difficulty arises in the accurate measurement of the corresponding extremely wide range of differential pressure. A 20 to 1 flow variation results in a 400 to 1 variation in differential pressure.

The second consequence of the square root relationship is also a matter of readout. Some computations require an input signal linear with flow. This is obvious when flow rates are integrated or when two or more flow rates are added or subtracted. In general, this is not necessarily true for multiplication and division; specifically, flow ratio measurement and control do not require an input signal linear with flow. A given flow ratio will develop a corresponding differential pressure ratio over the full range of the measured flows.

Density of the flowing fluid is involved in the determination of either mass flow rate or volume flow rate. In other words, head type meters *do not* read out directly in either mass or volume flow (weirs and flumes are an exception, as discussed in Section 2.25). The fact that density appears as a square root gives head type metering an actual advantage, particularly in applications where measurement of mass flow is required. Due to this square root relationship, any error which may exist in the value of the density used to compute mass flow is substantially reduced—a 1 percent error in the value of the fluid density results in a ½ percent error in calculated mass flow. This is particularly important in gas flow measurement, where the density may vary over a considerable range and where operating density is not easily determined with high accuracy.

β (Beta) Ratio. Most head meters depend on a restriction in the flow path to produce a change in velocity. For the usual circular pipe and circular restriction, the β ratio is the ratio between the diameter of the restriction and the inside diameter of the pipe. The ratio between the velocity in the pipe and the velocity at the restriction is equal to the ratio of areas or β^2. For noncircular configurations, β is defined as the square root of the ratio of area of the restriction to area of the pipe or conduit.

Reynolds Number

The basic equations of flow assume that the velocity of flow is uniform across a given cross section. In practice, flow velocity at any cross section approaches zero in the boundary layer adjacent to the pipe wall, and varies across the diameter. This flow velocity profile has a significant effect on the relationship between flow velocity and pressure difference developed in a head meter. In 1883, Sir Osborne Reynolds, an English scientist, presented a paper before the Royal Society, proposing a single, dimensionless ratio, now known as Reynolds number, as a criterion to describe this phenomenon. This number, Re, is expressed as

$$\mathrm{Re} = \frac{VD\rho}{\mu} \qquad 2.12(4)$$

where V is velocity, D is diameter, ρ is density, and μ is absolute viscosity. Reynolds number expresses the ratio of inertial forces to viscous forces. At a very low Reynolds number, viscous forces predominate, and inertial forces have little effect. Pressure difference approaches direct proportionality to average flow velocity and to viscosity. At high Reynolds numbers, inertial forces predominate and viscous drag effects become negligible.

At low Reynolds numbers, flow is laminar and may be regarded as a group of concentric shells; each shell reacts in a viscous shear manner on adjacent shells; the velocity profile across a diameter is substantially parabolic. At high Reynolds numbers, flow is turbulent, with eddies forming between the boundary layer and the body of the flowing fluid, and propagating through the stream pattern. A very complex, random pattern of velocities develops in all directions. This turbulent, mixing action tends to produce a uniform average axial velocity across the stream. Change from the laminar flow pattern to the turbulent flow pattern is gradual with no distinct transition point. For Reynolds numbers above 10,000, flow is definitely turbulent.

The value for k in the basic flow equations includes a Reynolds number factor. References 1 and 2 provide tables and graphs for Reynolds number factor. For head meters, this single factor is sufficient to establish compensation in coefficient for changes in ratio of inertial to frictional forces, and the corresponding changes in flow velocity profile; a gas flow with the same Reynolds number as a liquid flow has the same Reynolds number factor.

Compressible Fluid Flow

Density in the basic equations is assumed to be constant upstream and downstream from the primary device. For gas or vapor flow, the differential pressure developed results in a corresponding change in density between upstream and downstream pressure measurement points. For accurate calculations of gas flow, this is corrected by an "expansion factor" which has been empirically determined. Values are given in references 1 and 2. When practical, the full scale differential pressure should be less than 0.04 times normal minimum static pressure (differential pressure, stated in inches of water, should be less than static pressure stated in PSIA). Under these conditions, the expansion factor is quite small.

Choice of Differential Pressure Range

The most common differential pressure range for orifices, venturi tubes, and flow nozzles is 0–100 inches of water (0–25 kPa) for full scale flow. This range is high enough to minimize errors due to liquid density differences in the connecting lines to the differential pressure sensor or in seal chambers, condensing chambers, etc., caused by temperature differences. Most differential pressure responsive devices develop their maximum accuracy in or near this range, and the maximum pressure loss*—3.5 PSI (24 kPa)—is not serious in most applications. The 100-inch range permits a 2 to 1 flow rate change in either direction to accommodate changes in operating conditions; most differential pressure sensors can be modified to cover the range from 25 to 400 inches of water (6.2 to 99.4 kPa) or more, either by a simple adjustment, or by a relatively minor structural change. Applications, where the pressure loss up to 3.5 PSI is expensive or is not available, can be handled either by selection of a lower differential pressure range or by the use of a venturi tube or other primary element with high-pressure recovery. Some high velocity flows will develop more than 100 inches differential pressure with maximum acceptable ratio of primary element effective diameter to pipe diameter. For these applications, a higher differential pressure is indicated. Finally, for low static pressure (less than 100 PSIA) gas or vapor, a lower differential pressure is recommended to minimize the expansion factor. (See previous paragraph on compressible fluids.)

Pulsating Flow and Flow "Noise"

Short period (one second and less) variation in differential pressure developed from a head-type flow meter primary element arises from two distinct sources. First, reciprocating pumps, compressors, and the like may cause a periodic fluctuation in the rate of flow. Second, the random velocities inherent in turbulent flow cause variations in differential pressure even with a constant flow rate. Both have similar results and are often confused. However, their characteristics and the procedures to cope with them are distinct.

Pulsating Flow The so-called pulsating flow from reciprocating pumps, compressors, etc., may significantly affect the differential pressure developed by a head-type meter. For example, if the amplitude of instantaneous differential pressure fluctuation is 24 percent of the average differential pressure, a deviation from coefficient of as much as 1 percent may occur under normal operation conditions. For the pulsation amplitudes of 24, 48 and 96 percent values, the corresponding errors of ± 1, ± 4, ± 16 percent can be expected. In general terms the ratio between errors varies roughly as the square of the ratio between differential pressure fluctuations.†

For liquid flow, there is indication that the average of the square root of the instantaneous differential pressure (essentially average of instantaneous flow signal) results in a considerably lower deviation from coefficient than the measurement of average instantaneous differential pressure. However, for gas flow, extensive investigation has failed to develop any usable relationship between pulsation and deviation from coefficient beyond the estimate of maximum error.[4]

Operation at higher differential pressure is generally advantageous for pulsating flow. The only other valid approach to improve accuracy of pulsating gas flow measurement is the location of the meter at a point where pulsation is minimized.

Flow "Noise" Turbulent flow generates a complex pattern of random velocities. This results in a corresponding variation or "noise" in the differential pressure developed at the pressure connections to the primary element. The amplitude of the noise may be as much as 10 percent of the average differential pressure with a constant flow rate. This noise effect is a complex hydrodynamic phenomenon, and is not fully understood. It is augmented by flow disturbances from valves, fittings, etc., both upstream and downstream from the flow meter primary element and, apparently, by characteristics of the primary element itself.

Tests based on average flow rate accurately determined by static weight/time techniques, compared to accurate measurement of differential pressure including continuous, precise averaging of noise, indicate that the noise, when precisely averaged, introduces negligible (less than 0.1 percent) measurement error when average flow is substantially constant (change of average flow rate is not more than 1 percent per second).[5] It should be noted that average differential pressure, not average flow (average of the square root of differential pressure) is measured, since the noise is developed by the random and not the average flow.

Errors in determination of true differential pressure average will result in corresponding errors in flow measurement. For normal use, "damping" in one form or another in devices responsive to differential pressure is adequate. Where accuracy is a major concern, there must be no elements in the system that will develop a bias rather than a true average when subjected to the complex noise pattern of differential pressure.

Differential pressure noise can be reduced by use of two or more pressure sensing taps connected in parallel for both high and low differential pressure connections. Two taps provide the major noise reduction. Only minor improvement results from additional taps. Piezometer rings, a form of multiple connections, are frequently used with venturi tubes (see Section 2.23) but seldom with orifices or flow nozzles.

*Pressure loss may be as low as 50 percent of measured differential pressure for an orifice with $\beta = 0.75$.

†Results of work by the Joint ASME-AGA Committee on Pulsation Research.

The Orifice Meter

The orifice meter is the most common head-type flow measuring device. An orifice plate is inserted in the line, and the differential pressure across it is measured. This section is concerned with the primary device (the orifice plate, its mounting, and the differential pressure connections). Devices for measurement of the differential pressure are covered in Chapters III and V.

The Concentric, Thin Plate, Sharp-Edged Orifice Plate

The orifice in general, and the conventional thin, concentric, sharp-edged orifice plate in particular, has important advantages, including inexpensive manufacture to very close tolerances and ease of installation and replacement. Orifice measurement of liquids, gases and vapors under a wide range of conditions enjoys a high degree of confidence based on a great deal of accurate test work.

The standard orifice plate itself is a circular disc, usually stainless steel, from ⅛ to ½ inch thick (3.175 to 12.70 mm), depending on size and flow velocity, with a hole (orifice) in the middle, and a tab projecting out to one side which is used as a data plate (Figure 2.12a). The thickness requirement of the orifice plate is a function of line size, flowing temperature and differential pressure across the plate. Some helpful guidelines are:

By Size:
- 2–12 in (50–304 mm): ⅛ in. (3.175 mm) thick
- 14 in. (355 mm) and larger: ¼ in. (6.35 mm) thick

By Temperature ≥ 600°F (316°C)
- 2–8 in. (50–203 mm): ⅛ in (3.175 mm) thick
- 10 in. (254 mm) and larger: ¼ in (6.35 mm) thick

All plates thicker than ⅛ in. (3.175 mm) should be beveled 45° on the downstream side to a ⅛ in. edge thickness. In addition, any orifice hole 1 in. (25 mm) or less in size should be beveled 45° to a maximum ⅛ in. edge thickness (1/16 in. or 1.6 mm in a ⅛ in. thick plate).

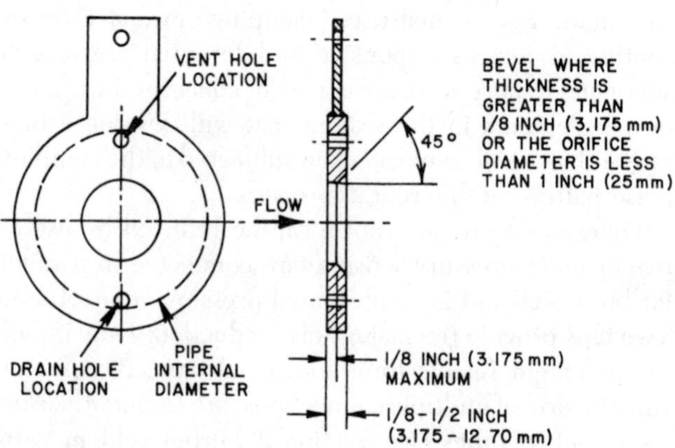

Fig. 2.12a Concentric orifice plate

Flow Through the Orifice Plate

The orifice plate inserted in the line causes an increase in flow velocity and a corresponding decrease in pressure. The flow pattern shows an effective decrease in cross section beyond the orifice plate, with a maximum velocity and minimum pressure at the vena contracta (Figure 2.12b). This location may be from .35 to .85 pipe diameters downstream from the orifice plate, depending on β ratio and on Reynolds number.

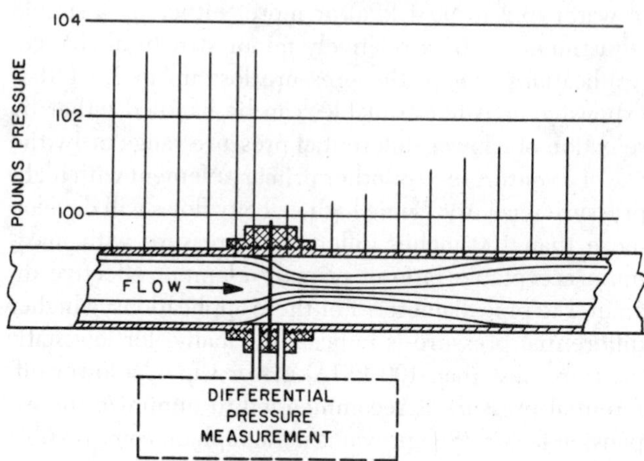

Fig. 2.12b Pressure profile with orifice plate

This flow pattern and the sharp leading edge of the orifice plate (Figure 2.12c) which produces it are of major importance. The sharp edge results in an almost pure line contact between the plate and the effective flow, with negligible fluid-to-metal friction drag at this boundary. Any nicks, burrs, or rounding of the sharp edge can result in surprisingly large errors in measurement.

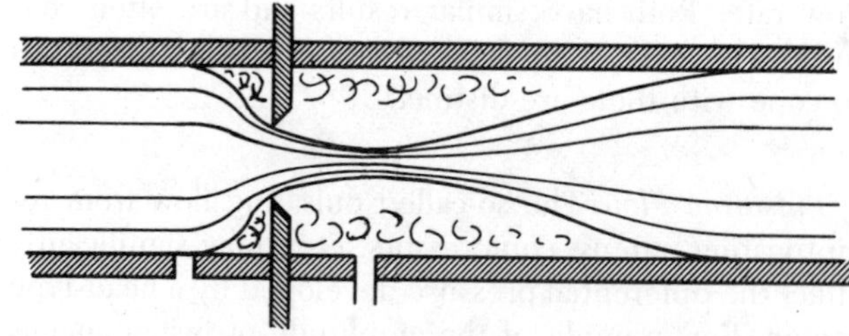

Fig. 2.12c Flow pattern with orifice plate

When the usual practice of measuring the differential pressure at a location close to the orifice plate is followed, friction effects between fluid and pipe wall upstream and downstream from the orifice are minimized so that pipe roughness has minimum effect. Fluid viscosity, as reflected in Reynolds number, does have a considerable influence, particularly at low Reynolds numbers. Since the formation of the vena contracta in an inertial effect, a decrease in the ratio of inertial to frictional forces (decrease in Reynolds number), and the corresponding change in the flow profile, results in less constriction of flow at the vena contracta and an increase of the coefficient. In general, the sharp edge orifice plate should not be used at pipe Reynolds numbers under 10,000. The minimum recommended Reynolds number will vary from 10,000 to 15,000 for 2 in (50 mm) through 4 in (102 mm) pipe sizes for β ratios up to 0.5 and 20,000 to 45,000 for higher β ratios. The Reynolds number requirement will increase with pipe size and β ratio and may range up to 200,000 for pipes 14 in. (355 mm) and larger. Maximum Reynolds numbers may be 10^6 through 4 in. (102 mm) pipe and 10^7 for larger sizes.

Location of Pressure Taps

For liquid flow measurement, gas or vapor accumulations in the connections between the pipe and the differential pressure measuring device must be prevented. Pressure taps are generally located in the horizontal plane of the center line of horizontal pipe runs. The differential pressure measuring device is either mounted close-coupled to the pressure taps, or connected through downward sloping connecting pipe of sufficient diameter to allow gas bubbles to flow up and back into the line. For gas, similar precautions to prevent accumulation of liquid are required. Taps may be installed in the top of the line, with upward sloping connections, or the differential pressure measuring device may be close-coupled to taps in the side of the line. For steam and similar vapors which are condensable at ambient temperatures, condensing chambers, or their equivalent, are generally used, usually with down-sloping connections from the side of the pipe to the measuring device. There are five common locations for the differential pressure taps: flange taps, vena contracta taps, radius taps, full flow or pipe taps, and corner taps.

Flange Taps

In the U.S.A., flange taps (Figure 2.12b) are predominantly used for pipe sizes 2 in. (50 mm) and larger. The manufacturer of the orifice flange set drills the taps into the flanges so that the center lines are 1 in (25 mm) from the orifice plate surface. This location also facilitates inspection and cleanup of burrs, weld metal, and so on, that may result from installation of a particular type of flange. Flange taps are not recommended below 2 in. (50 mm) pipe size and cannot be used below 1-½ in. (37.5 mm) pipe size, since the vena contracta may be closer than 1 in. (25 mm) from the orifice plate. Flow for a distance of several pipe diameters beyond the vena contracta tends to be unstable and is not suitable for differential pressure measurement.

Vena Contracta Taps

These taps use an upstream tap located one pipe diameter upstream of the orifice plate, and a downstream tap located at the point of minimum pressure. Theoretically, this is the optimum location. However, the location of the vena contracta varies with the orifice-to-pipe diameter ratio, and is thus subject to error if the orifice plate is changed. A tap location too far downstream in the unstable area may result in inconsistent measurement. For moderate and small size pipe, the location of the vena contracta is likely to lie at the edge of or under the flange. It is not considered good piping practice to use the hub of the flange to make a pressure tap. For this reason, vena contracta taps are normally limited to pipe sizes 6 in. (152 mm) or larger depending upon the flange rating and dimensions.

Radius Taps

Radius taps are similar to vena contracta taps, except that the downstream tap is located at one-half pipe diameter (one radius) from the orifice plate. This practically assures that the tap will not be in the unstable region regardless of orifice diameter. Radius taps today are generally considered superior to the vena contracta tap because they simplify the pressure tap location dimensions and do not vary with changes in orifice β ratio. The same pipe size limitations apply as to the vena contracta tap.

Full Flow or Pipe Taps

These taps are located 2-½ pipe diameters upstream and 8 diameters downstream from the orifice plate. Because of the distance from the orifice, exact location is not critical, but the effects of pipe roughness, dimensional inconsistencies, and so on, are more severe. Uncertainty of measurement is perhaps 50 percent greater with full flow taps than with taps close to the orifice plate. These taps are not normally used except where it is necessary to install an orifice meter in an existing pipe line and radius or vena contracta taps cannot be used.

Corner Taps

Corner taps (Figure 2.12d) are similar in many respects to flange taps, except that the pressure is measured at the "corner" between the orifice plate and the pipe wall.

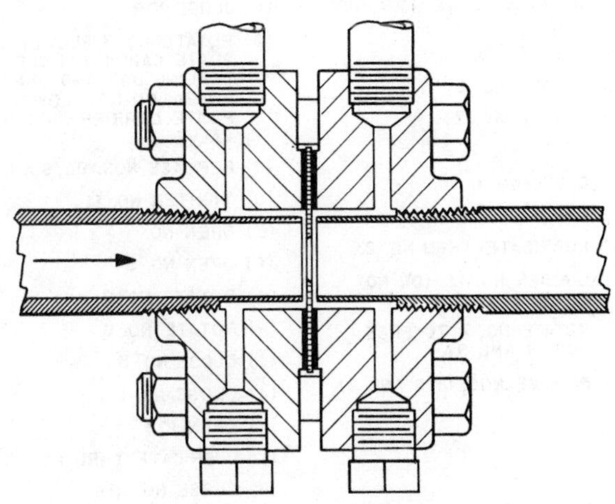

Fig. 2.12d Corner tap installation

Corner taps are very common for all pipe sizes in Europe. The relatively small clearances of the passages are a possible source of trouble. Also, some tests have indicated inconsistencies with high β ratio installations, which have been attributed to a region of flow instability at the upstream face of the orifice. For this situation, an upstream tap at one pipe diameter upstream of the orifice plate has been used. Corner taps are used in the U.S. primarily for pipe diameters of less than two inches (50 mm).

Installation

The orifice is usually mounted between a pair of flanges. Care should be exercised in installing the orifice plate to be sure that the gaskets are trimmed and installed so that they do not protrude across the face of the orifice plate beyond the inside pipe wall. A variety of special devices is commercially available for mounting orifice plates, including units by means of which orifice plates may be inserted and removed from flow lines under pressure without interrupting the flow (Figure 2.12e).

In order to avoid errors resulting from disturbance of the flow pattern due to valves, fittings, etc., a straight run of smooth pipe before and after the orifice is recommended. Required length depends on β ratio (ratio of diameter of orifice to inside diameter of pipe) and the severity of the flow disturbance.

For example, an upstream distance to the orifice plate of 45 pipe diameters with 0.75 β ratio is the minimum recommendation for a throttling valve. For a single elbow at the same β, the minimum distance would only be 17 pipe diameters. Figure 2.12f gives minimum values for a variety of upstream disturbances. Upstream lengths longer than the minimum are recommended. A downstream pipe run of 5 pipe diameters from the orifice plate is recommended in all cases, without interruption from thermowells or other devices inserted into the pipe.

Where it is not practical to install the orifice in a straight run of the desired length, the use of a straightening vane to eliminate swirls or vortices is recommended. Straightening vanes are manufactured in various configurations (Figure 2.12g) and are available from commercial meter tube fabricators. They should be installed so that there are at least 2 pipe diameters between

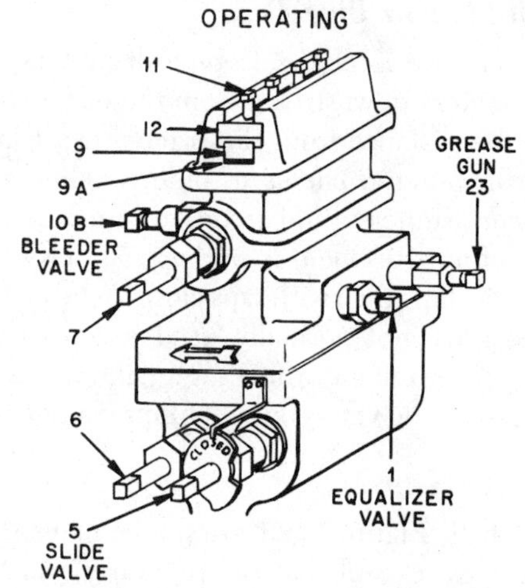

TO REMOVE ORIFICE PLATE

(A) OPEN NO. 1 (MAX. TWO TURNS ONLY)
(B) OPEN NO. 5
(C) ROTATE NO. 6
(D) ROTATE NO. 7
(E) CLOSE NO. 5
(F) CLOSE NO. 1
(G) OPEN NO. 10B
(H) LUBRICATE THRU NO. 23
(I) LOOSEN NO. 11 (DO NOT REMOVE NO. 12)
(J) ROTATE NO. 7 TO FREE NOS. 9 AND 9A
(K) REMOVE NOS. 12, 9 AND 9A

TO REPLACE ORIFICE PLATE

(A) CLOSE 10B
(B) ROTATE NO. 7 SLOWLY UNTIL PLATE CARRIER IS CLEAR OF SEALING BAR AND GASKET LEVEL. DO NOT LOWER PLATE CARRIER ONTO SLIDE VALVE.
(C) REPLACE NOS. 9A, 9 AND 12
(D) TIGHTEN NO. 11
(E) OPEN NO. 1
(F) OPEN NO. 5
(G) ROTATE NO. 7
(H) ROTATE NO. 6
(I) CLOSE NO. 5
(J) CLOSE NO. 1
(K) OPEN 10B
(L) LUBRICATE THRU NO. 23
(B) CLOSE NO. 10B

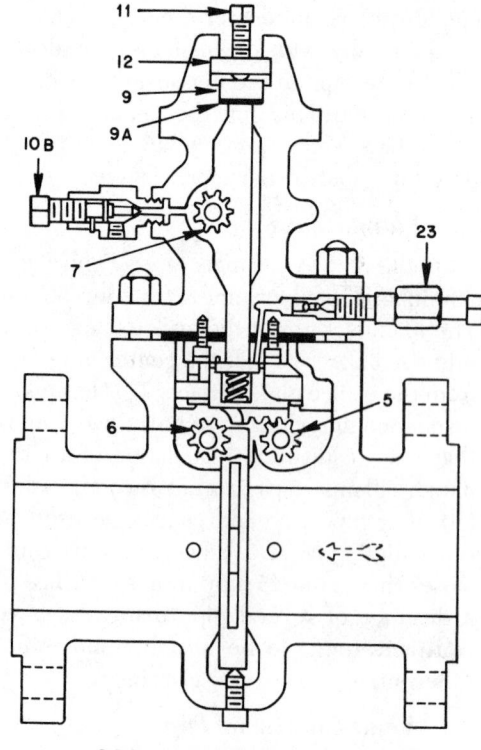

11
12
9
9A
10B
7
23
6
5
SIDE SECTIONAL ELEVATION

Fig. 2.12e Typical orifice fitting (courtesy Daniel Industries)

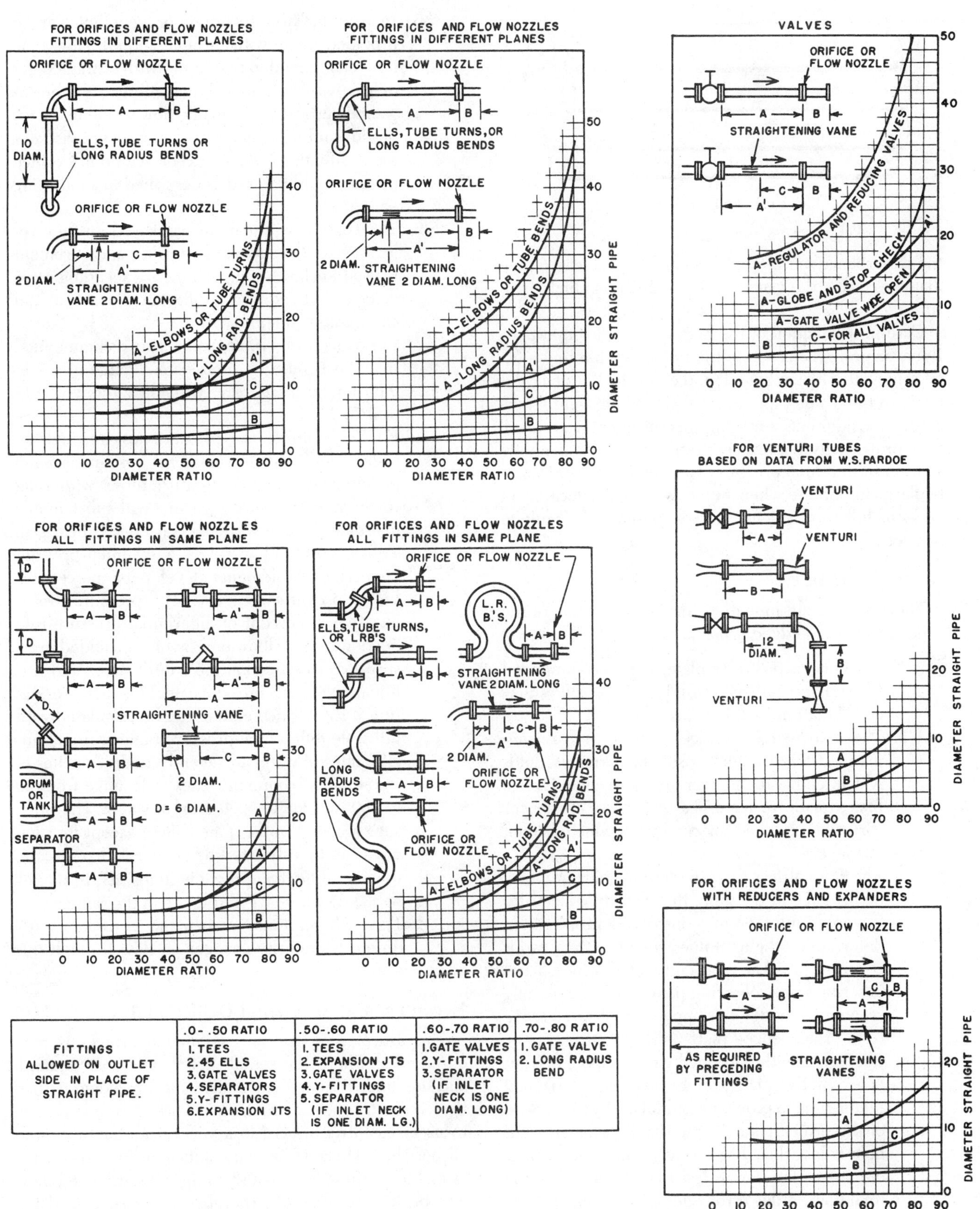

	.0-.50 RATIO	.50-.60 RATIO	.60-.70 RATIO	.70-.80 RATIO
FITTINGS ALLOWED ON OUTLET SIDE IN PLACE OF STRAIGHT PIPE.	1. TEES 2. 45 ELLS 3. GATE VALVES 4. SEPARATORS 5. Y-FITTINGS 6. EXPANSION JTS	1. TEES 2. EXPANSION JTS 3. GATE VALVES 4. Y-FITTINGS 5. SEPARATOR (IF INLET NECK IS ONE DIAM. LG.)	1. GATE VALVES 2. Y-FITTINGS 3. SEPARATOR (IF INLET NECK IS ONE DIAM. LONG)	1. GATE VALVE 2. LONG RADIUS BEND

Fig. 2.12f Orifice straight run requirements (Reprinted courtesy of The American Society of Mechanical Engineers)

Fig. 2.12g Straightening vane

the disturbance source and vane entry and at least 6 pipe diameters from the vane exit to the upstream high pressure tap.

Care in installation of the pressure taps is important. Burrs or protrusions where the tap enters the line must be removed. The tap hole should enter the line at a right angle to the inside pipe wall and should be slightly beveled. Considerable error in measurement can result from protrusions that react with stream flow and generate spurious differential pressure. Careful installation is particularly important when full flow taps are located in areas of full pipe velocity and in positions difficult to inspect.

Limitations

There are certain limitations in the application of the concentric, sharp-edged orifice.

1. The concentric orifice plate is not recommended for slurries and dirty fluids, where solids may accumulate near the orifice plate.
2. The sharp-edged orifice plate is not recommended for strongly erosive or corrosive fluids which tend to round over the sharp edge. Orifice plates made of materials that resist erosion or corrosion are used for conditions which are not too severe.
3. For flows at less than 10,000 Reynolds number, determined in the pipe, the correction factor for Reynolds number may introduce problems in determining the total flow when flow rate varies considerably. The quadrant-edged orifice plate, described in a following paragraph, is generally recommended for this application in preference to the sharp-edged plate.
4. For liquid with entrained gas or vapor, a "vent hole" in the plate can be used for horizontal meter runs to prevent accumulation of gas ahead of the orifice plate. If the diameter of the vent hole is less than 10 percent of the orifice diameter, then the flow is less than 1 percent of the total flow. If this error cannot be tolerated, then appropriate correction can be made to the orifice calculation. In general, vent or drain holes are considered of little value because they are subject to plugging, and are not recommended.
5. In a similar fashion, a drain or weep hole can be provided for gas with entrained liquid. However, it is recommended that liquid with entrained gas or gas with entrained liquid meter runs be installed vertically. Normally, flow would be up for liquids and down for gases. For severe entrainment situations, eccentric or segmental orifice plates should be used as described in a following paragraph.
6. The basic flow equations are based on flow velocities well below sonic. Orifice measurement is used for flows approaching or exceeding sonic velocity but requires a different theoretical and computational approach.
7. For concentric orifice plates, it is recommended that the β ratio be limited to a range of 0.2 to 0.65 for best accuracy. Outside limits are 0.15 to 0.75 β.
8. For large flows, the pressure loss through an orifice can result in significant cost in terms of power requirements. Venturi tubes with relatively large pressure recovery will substantially decrease the pressure loss. Lo-Loss Tubes, Dall Tubes, Foster Flow Tubes, and similar proprietary primary elements develop 95 percent or better pressure recovery. The pressure loss is less than 5 percent of differential pressure (see Section 2.23). Elbow taps involve no added pressure loss (see Section 2.5). Pitot tubes and Annubar* elements introduce neglible loss (see Section 2.13). Orifice meters can be calculated with full scale differential pressure ranging from 5 in. (127 mm) of water to several hundred inches of water. Most commonly they will range from 20 to 200 in. (508 to 5080 mm) of water. The pressure recovery ratio of an orifice (except for pipe taps) can be estimated by $(1 - \beta^2)$.
9. In general, for compressible fluids, $\Delta P/P_1$ should be ≤ 0.25 where ΔP and P_1 are in the same units. This will minimize errors and corrections required for density changes on flow through the orifice.

Eccentric and Segmental Orifice Plates

The use of eccentric and segmental orifices is recommended where horizontal meter runs are required and the fluids contain extraneous matter to a degree that the concentric orifice would not work. It is preferable to use concentric orifices in a vertical meter tube if at all possible. There is limited coefficient and other data available for these orifices and it must be recognized they will be less accurate. In the absence of specific data, concentric orifice data may be applied as long as it is recognized that a decrease in accuracy results.

*Proprietary device

The eccentric orifice plate, Figure 2.12h, is like the concentric plate except for the offset hole. The segmental orifice plate, Figure 2.12i, has a hole which is a segment of a circle. Both types of plates may have the hole bored tangent to the inside wall of the pipe or more commonly tangent to a concentric circle with a diameter no smaller than 98 percent of the pipe internal diameter. The segmental plate arc is parallel to the pipe wall. Care must be taken in the installation so that no portion of the flange or gasket interferes with the hole on either type plate. The equivalent β for a segmental orifice may be expressed as $\beta = \sqrt{a/A}$ where a is the area of the hole segment and A is the internal pipe area.

In general, the minimum line size for these plates is 4 in. (102 mm). However, the eccentric plate can be made in smaller sizes as long as the hole size does not require beveling. Maximum line sizes are unlimited and contingent only on calculation data availability. Beta ratio limits are limited to between 0.3 and 0.8. Lower Reynolds number limit is 2000D (D in inches), but not less than 10,000. For compressible fluids, $\Delta P/P_1 \leq 0.30$ where ΔP and P_1 are in the same units.

Flange taps are recommended for both types of orifices, but vena contracta taps can be used in larger pipe sizes. The taps for the eccentric orifice should be located in the quadrants directly opposite the hole. The taps for the segmental orifice should always be in line with the maximum dam height. The straight edge of the dam may be beveled if necessary using the same criteria as for a square edge orifice. In order to avoid confusion after installation, the tabs on these plates should be clearly stamped "eccentric" or "segmental."

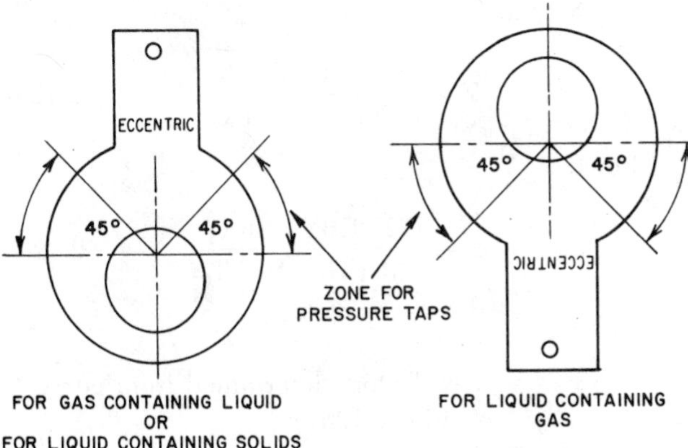

Fig. 2.12h Eccentric orifice plate

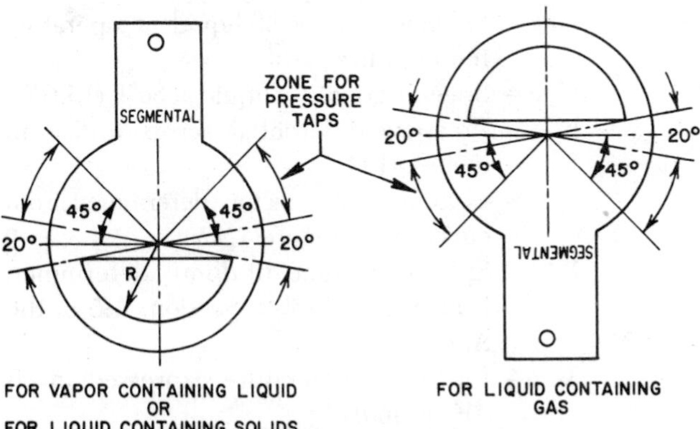

Fig. 2.12i Segmental orifice plate

Quadrant Edge and Conical Entrance Orifice Plate

The use of quadrant edge and conical entrance orifice plates is limited to lower pipe Reynolds numbers where flow coefficients for sharp-edged orifice plates are highly variable, with changes in the range of 50 to 10,000. With these special plates, the increased stability of the coefficients may be a factor of ten. The lower allowable Reynolds number is a function of β ratio and the allowable β ratio ranges are limited. Refer to Tables 2.12j and 2.12k for β ratio change and lower allowable Reynolds number. The higher allowable pipe Reynolds number ranges from $500{,}000 \times (\beta\text{-}0.1)$ for the quadrant edge to $200{,}000 \times (\beta)$ for the conical entrance plate. The conical entrance also has a minimum $D \geq 0.25$ in. (6.35 mm). For compressible fluids, $\Delta P/P_1 \leq 0.25$ where ΔP and P_1 are in the same units. Flange pressure taps are preferred for the quadrant edge but corner and radius taps can also be used with the same flow coefficients. For the conical entrance, reliable data is available for corner taps only. A typical quadrant edge plate is shown in Figure 2.12l and a typical conical entrance is Figure 2.12m. These plates are thicker and heavier than the normal sharp-edge-type. Because of the critical dimensions and shape, the quadrant edge is difficult to manufacture and it is

Table 2.12j
LOWER *Re* FOR QUADRANT EDGE ORIFICE PLATE

β	0.25	0.30	0.35	0.40	0.45	0.50	0.55	0.60
Re_D	250	300	400	500	700	1000	1700	3300

Table 2.12.k
LOWER *Re* FOR CONICAL ENTRANCE ORIFICE PLATE

β	0.10	0.11	0.12	0.13	0.14	0.15	0.16	0.17	0.18	0.19	0.20	0.21	0.22	0.23	0.24	0.25	0.26	0.27	0.28	0.29	0.30
Re_D	25	28	30	33	35	38	40	43	45	48	50	53	55	58	60	63	65	68	70	73	75

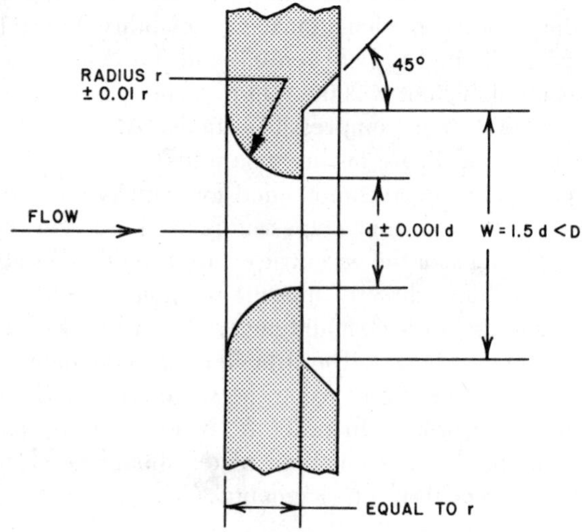

Fig. 2.12 l Quadrant edge orifice plate

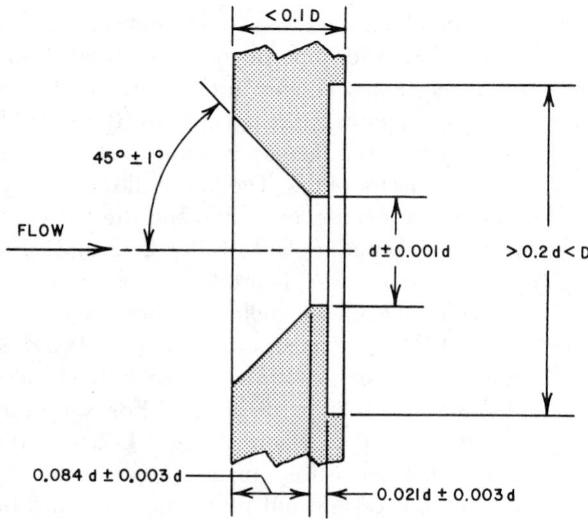

Fig. 2.12m Conical entrance orifice plate

recommended that it be purchased from skilled commercial fabricators. The conical entrance is much easier to make and could be made by any qualified machine shop. While these special orifice forms are very useful for lower Reynolds numbers, it is recommended that for a pipe Re > 100,000 the standard sharp-edge orifice be used. In order to avoid confusion after installation, the tabs on these plates should be clearly stamped "quadrant" or "conical."

Orifice Meter Calculations

Accurate flow calibration, traceable to recognized standards and using the working fluid under service conditions, is difficult and expensive. For large gas flows, it is nearly impossible. A major advantage of orifice metering lies in the facility with which flow can be accurately determined from a few simple, readily available measurements. In particular, for the concentric, sharp-edged orifice, measurement confidence is supported by a large body of experience and precise, painstaking tests.

Precise flow calculations are quite complex, although the calculation methods and equations have been well standardized. These calculation methods are thoroughly covered in the references 1 through 4 listed at the end of this section. Approximate calculations are practical where moderate accuracy is satisfactory or firming up the actual orifice design basis is being done. See Figure 2.12n for orifice bore determination and Table 2.12o for maximum flow capacities. The following equations are to be used with Figure 2.12n for the orifice bore determination.

For Liquid Flow:

$$Z = \frac{5.663\ ER\sqrt{h\ G_f}}{GPM\ G_t} \qquad 2.12(5)$$

*For Steam:

$$Z = \frac{358.9\ ERY}{lbm/hr.}\sqrt{\frac{h}{V}} \qquad 2.12(6)$$

*For Gas:

$$Z = \frac{7727\ ERY}{SCFH}\sqrt{\frac{h\ P_f}{G\ T_f}} \qquad 2.12(7)$$

where:

E = Area factor, determined from curve C on Fig. 2.12n
R = Pipe constant, determined from table on Fig. 2.12n
G = Specific gravity of gas (air = 1.0)
G_f = Specific gravity of liquid at operating temperature
G_t = Specific gravity of liquid at 60°F (15.6°C)
h = Pressure differential across orifice in inches H_2O
Y = Compressibility factor, determined from curve B on Figure 2.12n
V = Specific volume (ft³/lbm), determined from Steam Tables, Section, A.5 in the Appendix
T_f = Flowing temperature expressed in °R (°F + 460)
P_f = Flowing pressure in PSIA
X = Pressure loss ratio defined as $h/2P_f$.

A useful simplified form of the mass flow equation (equation 2.12(3)) is:

$$W = 359\ Cd^2\sqrt{\frac{h\rho}{1-\beta^4}} \qquad 2.12(8)$$

*For steam and gas, "h" expressed in inches H_2O should be equal to or less than "P_f" expressed in PSIA units.

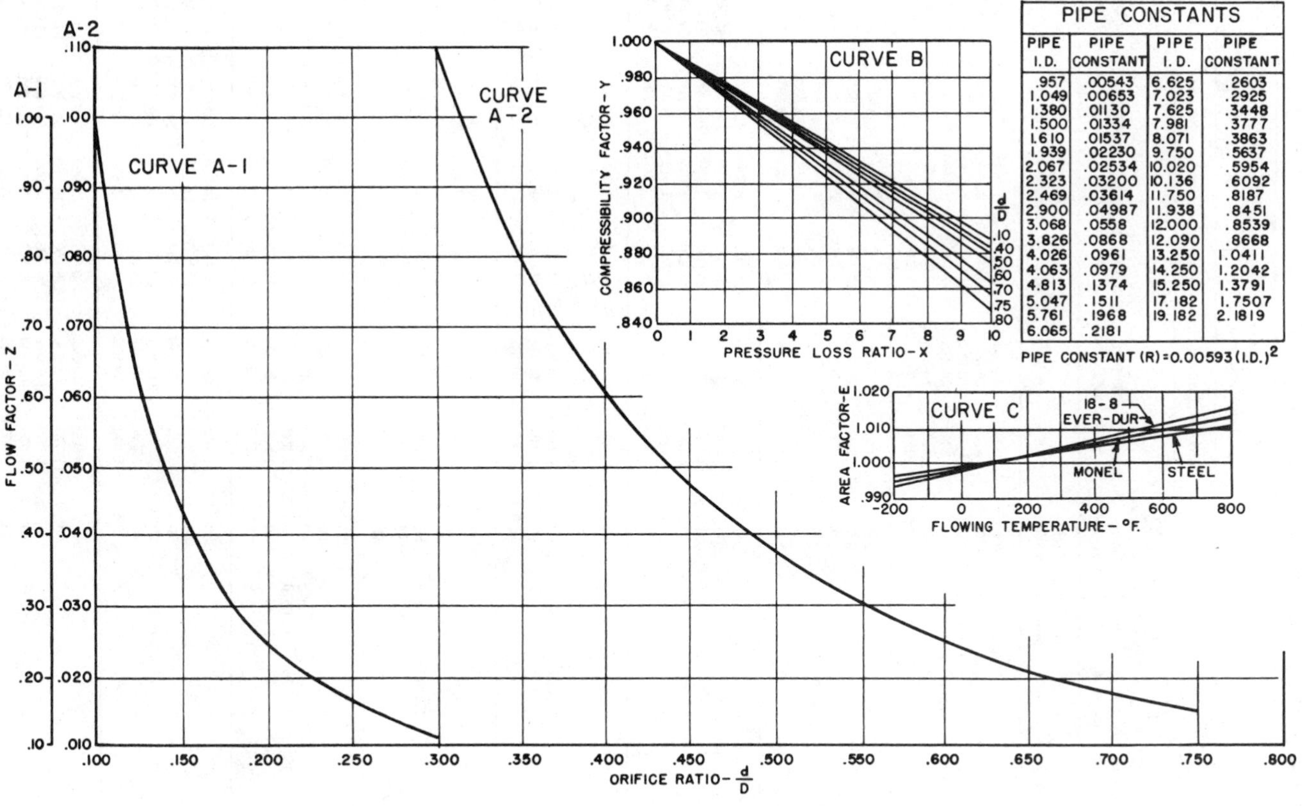

PIPE CONSTANTS

PIPE I.D.	PIPE CONSTANT	PIPE I.D.	PIPE CONSTANT
.957	.00543	6.625	.2603
1.049	.00653	7.023	.2925
1.380	.01130	7.625	.3448
1.500	.01334	7.981	.3777
1.610	.01537	8.071	.3863
1.939	.02230	9.750	.5637
2.067	.02534	10.020	.5954
2.323	.03200	10.136	.6092
2.469	.03614	11.750	.8187
2.900	.04987	11.938	.8451
3.068	.0558	12.000	.8539
3.826	.0868	12.090	.8668
4.026	.0961	13.250	1.0411
4.063	.0979	14.250	1.2042
4.813	.1374	15.250	1.3791
5.047	.1511	17.182	1.7507
5.761	.1968	19.182	2.1819
6.065	.2181		

PIPE CONSTANT (R) = 0.00593 (I.D.)^2

Fig. 2.12n Orifice bore determination chart (flange taps) © 1946 by Taylor Instrument Companies, Rochester, N.Y.

where

- W = mass flow in pounds per hour
- d = orifice diameter in inches
- h = differential pressure in inches of water*
- ρ = operating density in pounds per cubic foot
- β = ratio of orifice diameter to pipe diameter in pure number
- C = coefficient of discharge in pure number

This is a modification of the basic equation 2.12(3) for mass flow, substituting the $359\,Cd^2\sqrt{1-\beta^4}$ for kA. The figure 359 includes a factor for the chosen units of measurement. The coefficient of discharge is involved with the flow pattern established by the orifice, including the vena contracta and its relation to the differential pressure measurement taps. An average value of C = 0.607 can be used for flange and other close up taps, which gives a working equation:

$$W = 218d^2\sqrt{\frac{h\rho}{1-\beta^4}} \qquad 2.12(9)$$

*Water density is assumed to be 62.32 pounds per cubic foot, corresponding to 68°F (20°C).

For full flow taps, C = 0.715 and the equation becomes

$$W = 275d^2\sqrt{\frac{h\rho}{1-\beta^4}} \qquad 2.12(10)$$

These working equations can be used for approximate calculations of flow of liquids, vapors or gases through any type of sharp-edged orifice. When using orifices for measurement in weight units, errors in determination of ρ must be considered. (Refer to Chapter VI for density measurement and sensors.) Accurate determination of density under flowing conditions is difficult, particularly for gases and vapors. In some cases, even liquids are subject to density changes with both temperature *and* pressure (for example, pure water in high pressure boiler feedwater measurement).

For W, d, h and ρ given in dimensions other than those stated, simple conversion factors apply. Transfer of ρ in equations 2.12 (8), 2.12 (9), and 2.12 (10) from the numerator to denominator will give volume flow in actual cubic feet per hour at flowing conditions (see equations 2.12 (2) and 2.12 (3)).

Beta ratio, and hence orifice diameter, can be calculated from a transposed form of the mass flow equation 2.12 (8).

Table 2.12o
ORIFICE FLOWMETER CAPACITY TABLE*

Pipe Size	Actual Inside Diam. (I.D.) Sched. 40	Maximum Orifice Diam.	Meter Range	Flange and Vena Contracta Taps: Liquid, Water (SG = 1)	Flange and Vena Contracta Taps: Steam, 100 PSIG Saturated	Flange and Vena Contracta Taps: Gas, Air (SG = 1.0) @ 100 PSIG and 60°F	Pipe Taps: Liquid, Water (SG = 1)	Pipe Taps: Steam, 100 PSIG Saturated	Pipe Taps: Gas, Air (SG = 1.0) @ 100 PSIG and 60°F
Inches	Inches	Inches	Inches of Water	Gal./Min.	Lb./Hr.	Std. Cu. Ft/Min.	Gal./Min.	Lb./Hr.	Std. Cu. Ft./Min.
			200	10.6	338	119	15.7	506	178
			100	7.5	239	84	11.2	358	126
			50	5.3	170	59	7.9	253	89
½	0.622	0.435	20	3.3	107	37	5.0	160	57
			10	2.4	76	27	3.5	113	40
			2.5	1.17	38	13	1.7	56	20
			200	30	963	295	44.8	1 440	507
			100	21.2	682	239	31.7	1 017	358
			50	15.0	482	170	22.4	719	253
1	1.049	0.734	20	9.5	305	108	14.2	455	160
			10	6.7	216	76	10.1	323	113
			2.5	3.35	108	38	5.0	161	56
			200	70.7	2 270	796	105	3 380	1 190
			100	50.1	1 600	564	75	2 390	844
			50	35.1	1 135	399	52.7	1 690	596
1½	1.610	1.127	20	22.4	718	253	33.4	1 070	378
			10	15.8	683	178	23.6	758	267
			2.5	7.9	254	90	11.8	379	133
			200	116	3 740	1 313	174	5 580	1 966
			100	83	2 645	932	123	3 950	1 390
			50	58.5	1 870	658	87	2 790	983
2	2.067	1.448	20	37.0	1 183	417	55	1 768	623
			10	26.1	840	295	39	1 252	440
			2.5	13.1	420	148	19.4	625	220
			200	255	8 240	2 905	383	12 300	4 330
			100	181	5 830	2 080	271	8 700	3 070
			50	128	4 125	1 460	191	6 160	2 175
3	3.068	2.147	20	81.5	2 610	922	121	3 900	1 375
			10	57.5	1 843	653	86	2 760	975
			2.5	28.8	915	325	43	1 366	485
			200	512	16 400	5 780	764	24 500	8 630
			100	362	11 600	4 090	540	17 300	6 100
			50	255	8 170	2 890	382	12 200	4 310
4	4.026	3.02	20	162	5 180	1 830	242	7 730	2 730
			10	115	3 670	1 290	172	5 470	1 930
			2.5	57	1 820	647	85	2 710	965
			200	800	25 600	9 050	1 190	38 200	13 500
			100	557	18 200	6 410	845	27 100	9 560
			50	402	12 900	4 530	598	19 200	6 760
5	5.047	3.78	20	253	8 110	2 870	378	12 100	4 280
			10	180	5 750	2 020	268	8 580	3 020
			2.5	90	2 880	1 010	134	4 290	1 510
			200	1 158	37 100	13 100	1 730	55 300	19 500
			100	820	26 300	9 250	1 223	39 200	13 800
			50	580	18 600	6 540	866	27 700	9 760
6	6.065	4.55	20	367	11 700	4 140	547	17 500	6 180
			10	258	8 310	2 930	387	12 400	4 370
			2.5	129	4 150	1 460	193	6 200	2 180
			200	2 000	64 104	22 511	2 980	95 709	33 692
			100	1 413	45 320	15 952	2 110	67 682	23 853
			50	1 000	32 052	11 285	1 492	47 855	16 846
8	7.981	5.9858	20	634	20 275	7 156	943	30 263	10 674
			10	447	14 386	5 054	668	21 468	7 543
			2.5	223	7 186	2 534	333	10 719	3 772
			200	3 150	101 020	35 475	4 700	150 825	53 094
			100	2 230	71 481	25 138	3 325	106 658	37 589
			50	1 578	50 510	17 785	2 355	75 413	26 547
10	10.020	7.5150	20	998	31 950	11 277	1 487	47 691	16 821
			10	706	22 671	7 964	1 052	33 830	11 887
			2.5	352	11 324	3 994	525	16 891	5 944
			200	4 520	145 000	51 300	6 750	216 000	76 500
			100	3 200	103 000	36 200	4 775	153 000	45 100
			50	2 270	72 400	25 600	3 380	108 000	38 200
12	12.000	9.0000	20	1 430	46 000	16 200	2 135	68 600	24 200
			10	1 012	32 400	11 500	1 512	48 300	17 100
			2.5	507	16 200	5 740	757	24 200	8 560
			200	5 415	173 398	60 891	8 060	258 887	91 135
			100	3 830	122 588	43 148	5 720	183 076	64 520
			50	2 710	86 699	30 526	4 040	129 443	45 567
14	13.126	9.8445	20	1 715	54 842	19 356	2 555	81 860	28 873
			10	1 210	38 914	13 670	1 808	58 068	20 404
			2.5	603	19 437	6 855	900	28 994	10 202
			200	7 065	226 442	79 518	10 520	338 084	119 014
			100	5 000	160 089	56 347	7 460	239 081	84 258
			50	3 535	113 221	39 864	5 275	169 042	59 507
16	15.000	11.2500	20	2 240	71 619	25 277	3 335	106 902	37 705
			10	1 580	50 818	17 852	2 360	75 832	26 646
			2.5	788	25 383	8 952	1 175	37 865	13 323
			200	8 920	286 324	100 546	13 320	427 489	150 487
			100	6 330	202 424	71 248	9 270	302 305	106 539
			50	4 475	143 162	50 406	6 675	213 744	75 243
18	16.876	12.6570	20	2 830	90 558	31 962	4 220	135 172	47 676
			10	1 995	64 256	22 573	2 985	95 885	33 693
			2.5	995	32 095	11 320	1 485	47 876	16 847
			200	11 100	356 238	125 097	16 550	531 871	187 232
			100	7 870	251 352	88 645	11 720	376 121	132 554
			50	5 565	178 119	62 714	8 310	265 936	93 616
20	18.814	14.1105	20	3 520	112 671	39 766	5 250	168 177	59 318
			10	2 485	79 946	28 085	3 715	119 298	41 920
			2.5	1 240	39 932	14 084	1 850	59 566	20 960
			200	16 060	515 222	180 927	23 950	769 238	270 791
			100	11 375	364 250	128 206	16 960	543 978	191 710
			50	8 035	257 611	90 703	12 000	384 619	135 395
24	22.626	16.9695	20	5 090	162 954	57 513	7 585	243 233	85 790
			10	3 590	115 625	40 619	5 375	172 539	60 628
			2.5	1 795	57 753	20 369	2 675	86 150	30 314

*[illegible]

$$\beta^2 = \frac{W}{\sqrt{W^2 + 129{,}000\, h\rho\, C^2 D^4}} \qquad 2.12(11)$$

Where D is the internal diameter of the pipe in inches and all other units are the same as originally stated. For flange and other close up taps where C = 0.607, equation 2.12 (11) becomes:

$$\beta = \frac{1}{\sqrt[4]{1 + \frac{47{,}500\, h\rho\, D^4}{W^2}}} \qquad 2.12(12)$$

or

$$d = \frac{D}{\sqrt[4]{1 + \frac{47{,}500\, h\rho\, D^4}{W^2}}} \qquad 2.12(13)$$

For full flow or pipe taps, substitute 65,500 instead of 47,500.

Suppliers manufacture orifice plates to precise dimensions to develop a specified differential pressure under the stated flow conditions. The usual procedure is to make a plate to meet a set of conditions, and to supply with it a precise coefficient which can be used in the equations for exact flow calculations. By proper calibration of the differential pressure measuring instrument, flow can be read directly in engineering units for an assumed value of fluid density. Several suppliers of orifice flow meters have available special slide rules and other computing aids for orifice computations, useful for routine plant calculations (see Figure 2.12n and Table 2.12o).

The specific procedure that should be used depends upon the absolute accuracy of the flow measurement required. In many applications, constancy or repeatability of measurement is the major objective; approximate calculations are completely repeatable and quite adequate. At the other extreme, orifice measurement is almost universally used for measurement of the large quantities of natural gas transported in high-pressure gas transmission systems. This measurement can be the basis of purchase and sale. Large amounts of money are involved, so every precaution to assure ultimate accuracy is essential.

Where the ultimate in accuracy is required, actual flow calibration of the meter run (the orifice, assembled with the upstream and downstream pipe, including straightening vanes, if any) is recommended. Facilities are available for very accurate weighed water calibrations, in lines up to 24 in. (61 cm.) diameter and larger, and with a wide range of Reynolds number. For orifice meters, highly reliable data exists for accurate transfer of coefficient values for liquid, vapor, and gas measurement.

Miniature Flow Restrictors ("Integral Orifice" Cells)

Miniature flow restrictors provide a convenient primary element for measurement of small fluid flows. Units are available from several manufacturers that combine a plate with a small hole to restrict flow, its mounting and connections, and a differential pressure sensor, usually a pneumatic or electronic transmitter. Units of this type are often referred to as "integral orifice" flow meters. Each manufacturer has available interchangeable flow restrictors enabling his unit to cover a wide range of flows. A common minimum standard size is an 0.020 inch (0.5 mm) throat diameter which will measure water flow down to 0.0013 GPM (5 cm³/min.), air flow at atmospheric pressure down to 0.0048 SCFH (135 cm³/min.) and similar low values for other fluids.

Miniature flow restrictors are used in laboratory-scale processes, and pilot plants; to measure additives to major flow stream, and for other small flow measurements. Clean fluid is required, particularly for the smaller sizes, not only to avoid plugging of the small hole but also because of buildup of even a very thin layer on the surface of the element will introduce major error.

There is little published data on the performance of these small restrictors. These are proprietary products with performance data provided by the supplier. Where accuracy is important, direct flow calibration is recommended. Water flow calibration, using tap water, a stop watch, and a glass graduate or a pail and scale to measure

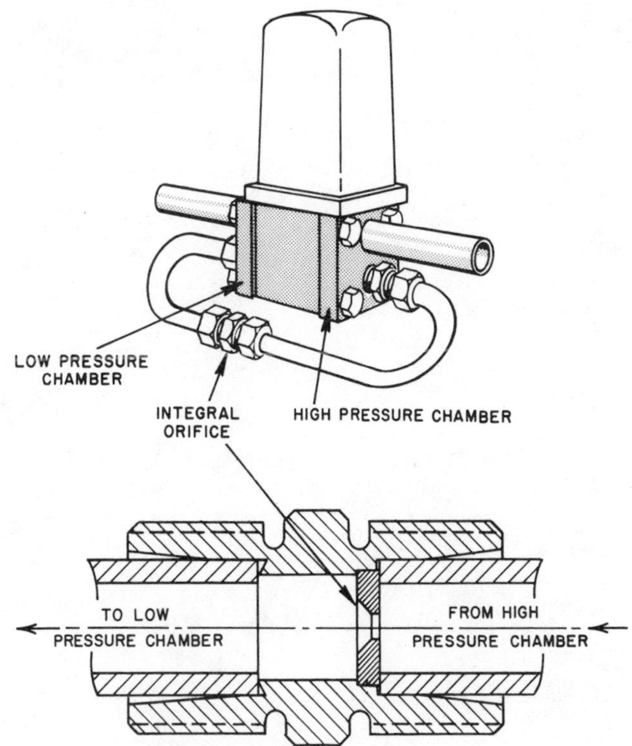

Fig. 2.12p Typical integral orifice meter

total flow, is readily carried out in the instrument shop or laboratory. For viscous liquids, calibration with the working fluid is preferable since viscosity has a substantial effect on most units. Calibration across the working range is recommended since precise conformity to the square law may not exist. Some suppliers are prepared to provide units with precision calibration for almost any specified range for an added fee. A typical unit is shown in Figure 2.12p.

REFERENCES

1. *Principles and Practice of Flow Meter Engineering.* L.K. Spink, The Foxboro Company, Foxboro, Massachusetts 02035—Ninth Edition, 1967. Chapter 5.
2. "Fluid Meters, Their Theory and Application." Report of ASME Research Committee on Fluid Meters, American Society of Mechanical Engineers, 345 East 47th Street, New York, New York, 10017.
3. *Shell Flow Meter Engineering Handbook.* Royal Dutch/Shell Group, Waltman Publishing Company, Delft, The Netherlands—1968.
4. *AGA Gas Measurement Manual.* American Gas Association, 605 Third Avenue, New York, New York 10016.
5. "Experimental Study of the Effects of Orifice Plate Eccentricity on Flow Coefficients." R.W. Miller and O. Kneisel, ASME Paper Number 68-WA/FM-1, Page 10, Conclusions 3, 4, 5. American Society of Mechanical Engineers, 345 East 47th Street, New York, New York 10017.

BIBLIOGRAPHY

Starrett, P.S., Nottage, H.B. and Halfpenny, P.F., "Survey of Information Concerning the Effects of Nonstandard Approach Conditions upon Orifice and Venturi Meters," presented at the annual meeting of the ASME, Chicago, November 7–11, 1965.

Flow Sheet Symbol

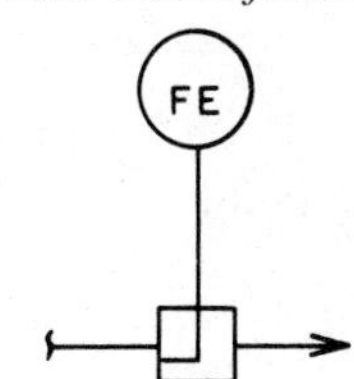

2.13 PITOT TUBES

Type of Design:	A—Standard B—Pitot Venturi* C—Multiple opening D—Area averaging
Design Pressure:	Usually up to 50 PSIG (345 kPa) with higher ratings on special or fixed position designs
Design Temperature:	Limited by sensing elements
Fluids:	Liquids and gases
Flow Range:	From 50 GPM or 100 SCFM (189 l/m or 170 m^3/hr) minimum flows to maximums, limited by pipe size only
Inaccuracy:	± ½% to 5% of full scale
Materials of Construction:	Standard materials are brass and stainless steel
Cost:	$450 to $675 for calibrated units
Partial List of Suppliers:	Air Monitor Corp. (D); Andersen Samplers Inc. (A); Brandt Industries Inc. (D); Cambridge Filter Corp. (D); Davis Instrument Mfg Co. Inc. (A); Dieterich Standard Corp. Subsidiary of Dover Corp. (C) Dwyer Instruments Inc. (A); Fischer & Porter Co. (A); Flow-Dyne Engineering Inc. (A); Foxboro Co. (A,B); GII Enterprises (A); Kent Process Control Inc. (A); Meriam Instrument Co. (A); Scott & Fetzer (C); Mid-West Instrument (C); No Loss Flow Tube (A); Permutit Co. Inc. Ranarax Instrument Dept. (A); Quik/ Industries Inc. (A); Sirco Products Ltd. (A); Sweet, J W Company (A); Taylor Instrument Co. Div. Sybron Corp. (B); Trimount Instrument Co. (A)

Since its conception by Henri de Pitot in 1732, the device which bears his name has, in various forms, been used as a means of measuring fluid velocity.

Theory

The stagnation or impact pressure on a body immersed in a moving fluid is the sum of the static pressure and the dynamic pressure. Thus,

$$Pt = P + Pv \qquad 2.13(1)$$

where:

Pt = total pressure which can be sensed by a fixed probe provided the fluid stagnates at the sensing point in an isentropic manner

P = the static pressure of the fluid whether in motion or at rest

Pv = the dynamic pressure equivalent to the kinetic energy of the fluid considered as a continuum

With respect to a static state, the energy relation at the isentropic stagnation point of an ideal probe is:

$$\int_{P}^{Pt} \frac{dp}{\rho} = \int_{o}^{Vp} \frac{VpdV}{gc} \qquad 2.13(2)$$

where:

Vp = the approach velocity at the probe location

*Registered Trademark

ρ = the fluid density
gc = constant

For a liquid of constant density, integration yields, at a point:

$$(Pt - P) = Pv = \frac{PVp^2}{2gc}$$

For a compressible perfect gas for which $\frac{P}{\rho\gamma}$ remains constant during an isentropic change, a similar relation emerges:

$$(Pt - P) = Pv = \rho\frac{(\gamma - 1)}{\gamma}\frac{Vp^2}{2gc}$$

where γ is the ratio of specific heats.

Assuming isentropic stagnation at the sensing point of the probe:

$$\int_{P}^{Pt}\frac{dp}{\rho} = \int_{o}^{Vp}\frac{VpdV}{gc} \qquad 2.13(3)$$

where, using English units*

Vp = velocity of approach, ft/sec.
P = pressure, Lbf/ft²
ρ = fluid density, lbm/ft³
gc = $32.2\frac{Lbm}{Lbf};\frac{ft}{sec^2}$

If density is constant, integration yields:

$$(Pt - P) = Pv = \frac{\rho(Vp^2)}{2\ gc}$$

For a compressible perfect gas, for which $\frac{P}{\rho\gamma}$ remains constant during an isentropic change, a similar relation is obtained:

$$(Pt - P) = Pv = \frac{(\gamma - 1)}{\gamma}\frac{p(Vp^2)}{2\ gc} \qquad 2.13(4)$$

where γ is the ratio of specific heats.

Specifications of Taps and Probe Design

In order to compute point velocities, it is necessary to measure both point values of static pressure (P) and total (Pt) pressures from whence:

$$Vp = C\frac{(Pt - P)}{\rho} \qquad 2.13(5)$$

where C is a dimensional constant.

For fluids flowing through conduits, the static pressure is commonly measured in one of three ways: through taps in the wall, from static probes inserted into the fluid stream, or from small apertures properly located on an aerodynamic body immersed in the flowing fluid.

The data of Shaw[1] as presented by Benedict[2] show that errors in the measurement of static pressure are minimal for velocities up to 200 ft/sec (60 m/s) if wall tap dimensions conform to those in Figure 2.13a for a tap diameter, d, of 0.0635 in. and if D ≅ 2d and 1.5 < l/d < 6.

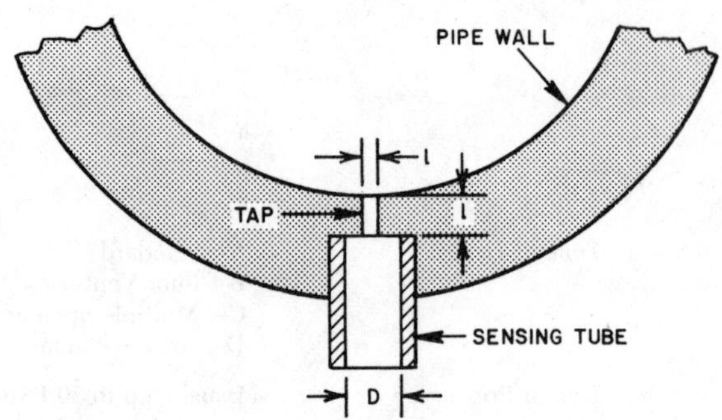

Fig. 2.13a Wall tap for static pressure measurement

Static pressure errors are also dependent upon fluid viscosity and velocity and whether the fluid is compressible or not. Shaw[3] states that for incompressible fluids flowing in a circular conduit with a pipe Reynolds number of 2×10^5, an error of about 1 percent of the mean dynamic pressure may occur using a wall tap having a diameter 1/10 that of the pipe. Rayle[4] mentions that a tap 0.03 in. (0.75 mm) in diameter with a conical countersink 0.015 in. (0.34 mm) deep will ensure nearly true static pressure sensing.

Static pressure may also be sensed through a tube inserted into the moving fluid. One configuration is shown in Figure 2.13b.

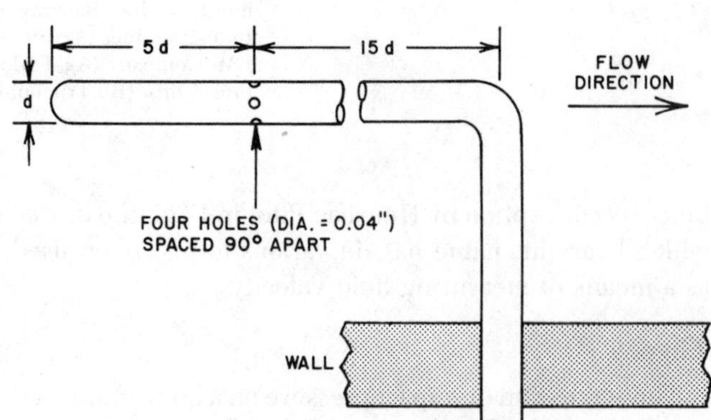

Fig. 2.13b Typical static pressure-sensing probe

Other static probe designs are described in the literature.[5] The aerodynamic probe is a bluff body inserted into the flowing fluid with appropriately located holes on its surface through which pressure signals are obtained. The probe is oriented so that the sensed pressure is a measure of the static pressure. Two configurations

*For SI units see Appendix A.1.

taken from Benedict[6] are shown in Figure 2.13c, the cylinder and the wedge. The probes are rotated until the pressure sensed from each hole is the same or, alternatively, the two taps may be manifolded to obtain an averaged pressure.

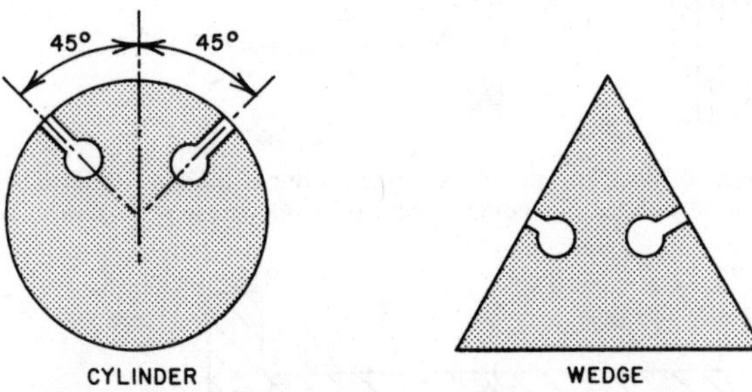

Fig. 2.13c Two shapes of aerodynamic probes used to sense static pressure

The total pressure develops at the point where the flow is stagnated isentropically which is assumed to occur at the tip of a Pitot tube or at a specific point on a bluff body immersed in the stream. Figure 2.13d is a typical Pitot tube which also shows the taps for sensing static pressure. Another variation is shown in Figure 2.13e.

Such probes must be carefully aligned with respect to the direction of fluid flow which is sometimes uncertain when turbulence exists. The Pitot tube is made less sensitive to flow direction if the impact aperture has an internal bevel of about 15 degrees extending about 1.5 diameters into the tube. Benedict[7] shows the behavior of various designs with orientation. Figure 2.13f shows

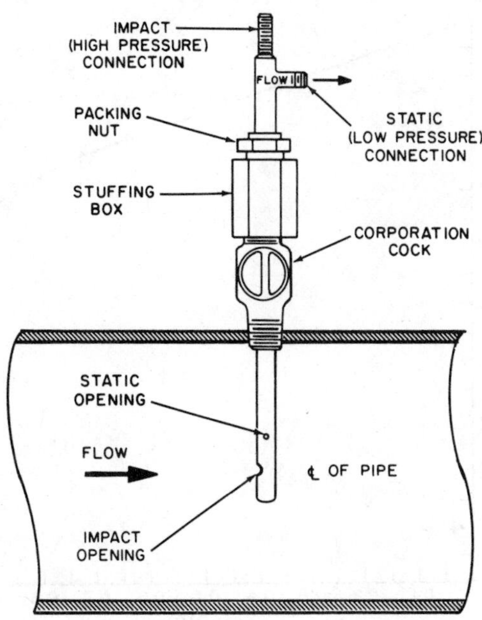

Fig. 2.13e Schematic of an industrial device for sensing static and dynamic pressures in a flowing fluid

typical performance data for a Pitot tube, although the fluid properties and flow conditions are not given.

To increase the magnitude of the pressure signals from in-stream velocity sensors, the Pitot Venturi and double Venturi have been developed as shown in Figures 2.13g and 2.13h. These elements are intended to remain in a fixed position in a duct and hence signals must be correlated with flow rate by calibration that accounts for fluid properties and flow conditions (e.g. Reynolds num-

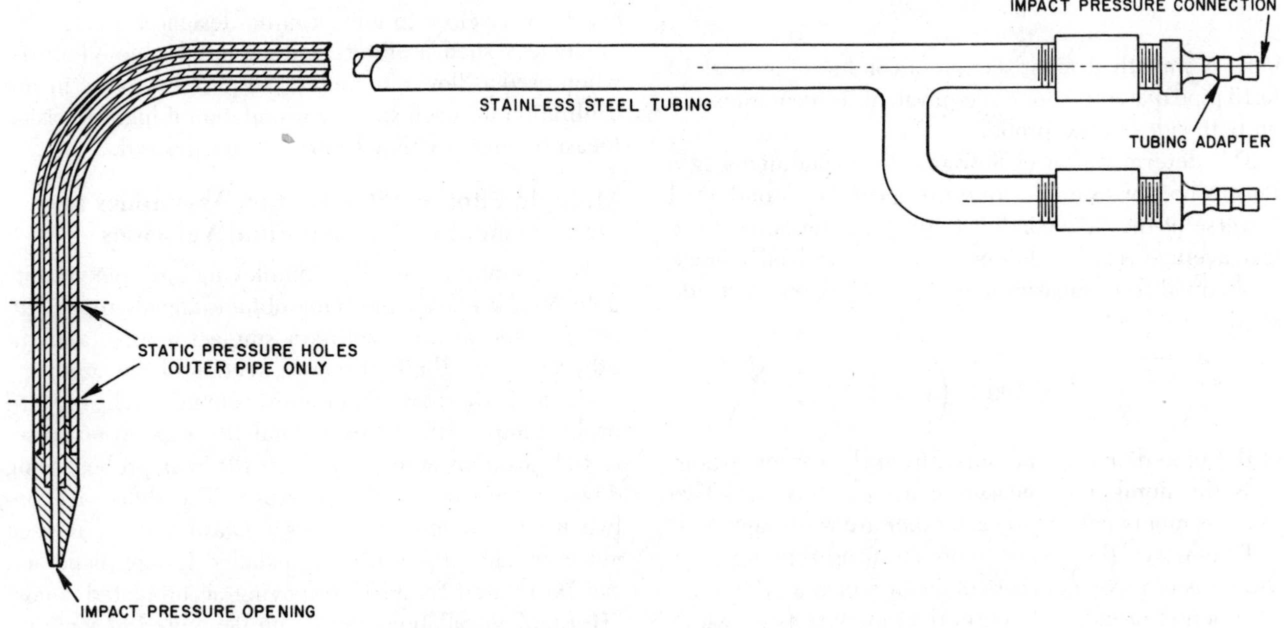

Fig. 2.13d Typical Pitot tube (Courtesy Andersen Samplers, Inc.)

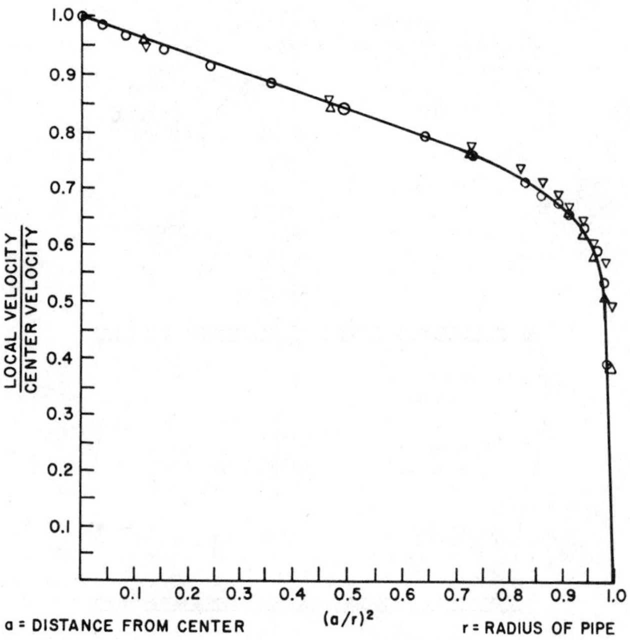

Fig. 2.13f A typical velocity distribution curve for a Pitot tube with ratio of local to center velocity plotted against the square of the ratio of the distance from center to the pipe radius

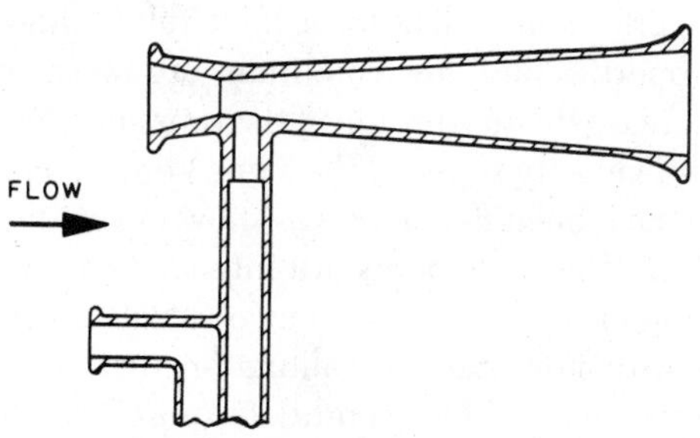

Fig. 2.13g A Pitot Venturi produces a higher differential pressure than the standard Pitot tube

ber). A smooth straight section of conduit of at least 10 to 15 pipe diameters, or the equivalent, is recommended on both sides of the probe.

The determination of flow rate in a conduit using a Pitot tube, for example, requires a carefully conducted traverse of the flow conduit to obtain a measure of the true average velocity. For circular pipes such an average is obtained from measurements of (Pt − P) on each side of the cross section at

$$\left(\sqrt{\frac{2n-1}{N}}\right) \times 100\%, \left(n = 1,2,3 \ldots \frac{N}{2}\right)$$

of the pipe diameter measured from the center, where N is the number of measurements per traverse. Two measurements normal to each other are recommended.

To enhance the ability to obtain measurements near the walls of pipes in excess of about 6 inches (150 mm), a reduction nozzle inserted in the pipe line as shown in Figure 2.13i is suggested.

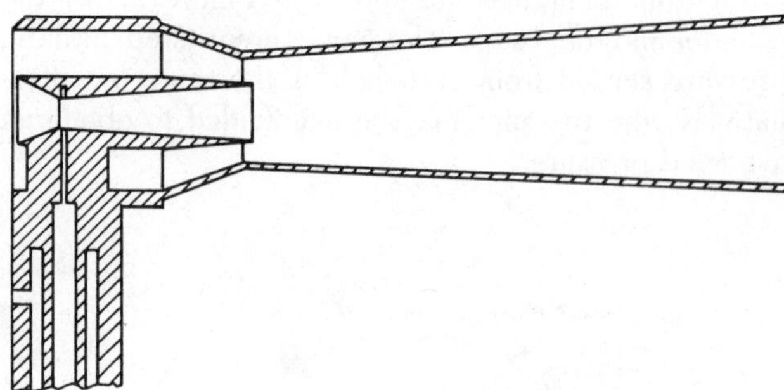

Fig. 2.13h The double Venturi produces a higher differential pressure than the standard Pitot tube (Courtesy of Foxboro Co.)

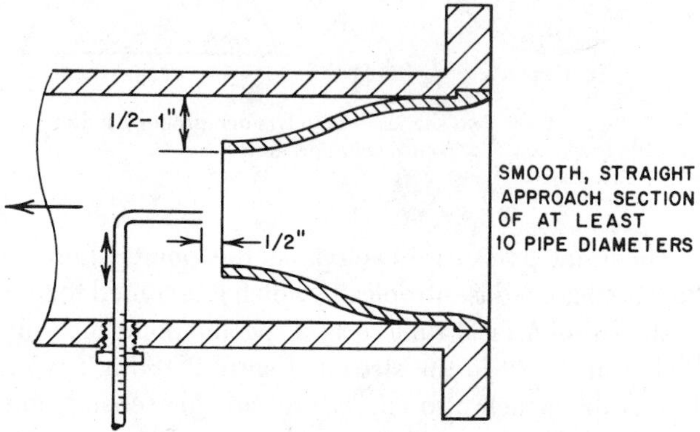

Fig. 2.13i Reduction nozzle used to expedite velocity traverses

Calibration of Pitot Tubes

The (absolute) method of calibrating a Pitot tube, as demonstrated by the National Bureau of Standards, is to mount the device on a carriage which can be drawn through stagnant air at a known velocity. Smoke is introduced into the room to verify the absence of turbulence. Such tests have shown that Pitot tubes with coefficients very close to unity can be designed.

Devices such as Pitot-Venturis or double Venturis, when used as flow rate meters, require extensive in situ calibration for each specific installation if high precision measurements (within 1 percent) are desired.

Multiple Pitot or Pitot-Venturi Assemblies for Measurement of Average Fluid Velocities

Two configurations, both employing multiple velocity detectors, are available. One obtains signals representative of the average across a single traverse, and the other averages the entire cross section of the conduit.

Figure 2.13j shows the Annubar element. The forward probe samples the average total pressure from preselected positions across the duct; the rear probe, facing downstream, senses the pressure. The difference between the two figures supplies a measure of the average velocity. The probe may be installed in any plane and can be rotated to aid in removing accumulated solids. "Hot tap" installations can be implemented in some instances.

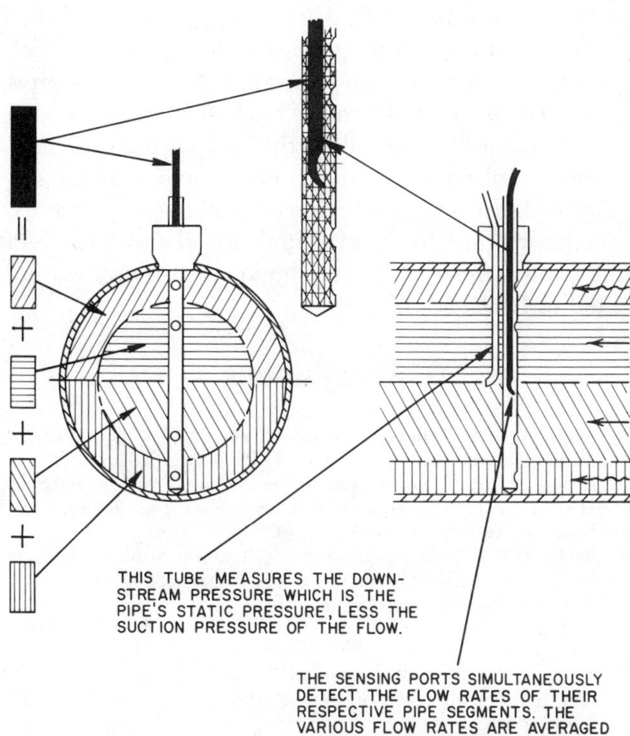

THIS TUBE MEASURES THE DOWN-STREAM PRESSURE WHICH IS THE PIPE'S STATIC PRESSURE, LESS THE SUCTION PRESSURE OF THE FLOW.
THE SENSING PORTS SIMULTANEOUSLY DETECT THE FLOW RATES OF THEIR RESPECTIVE PIPE SEGMENTS. THE VARIOUS FLOW RATES ARE AVERAGED BY THE PLENUM OF THE UPSTREAM ELEMENT. THE INTERNAL INTERPOLATING TUBE ENSURES THAT BOTH HALVES OF THE PIPE ARE EQUALLY REPRESENTED.

Fig. 2.13j A description of the function and operation of the Annubar element (Courtesy of Ellison Instrument Division of Dieterich Standard Corp.)

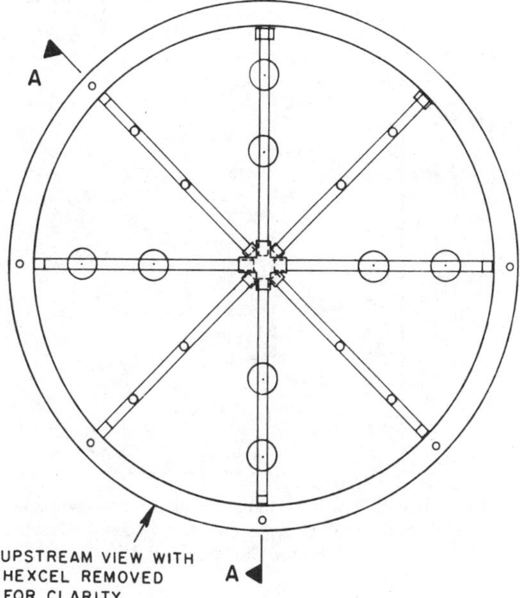

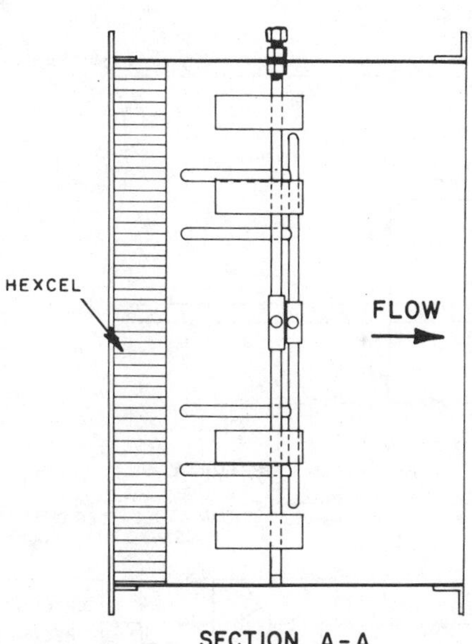

Fig. 2.13k Installation in circular duct of area-averaging Pitot tube ensembles for metering the flow rate of gases (Courtesy of Brandt Industries Inc.)

Area averaging ensembles are available from several sources. Configurations for circular and rectangular ducts are shown in Figures 2.13k and l. Straightening vanes, such as the "hexcel" (figure 2.13k) are recommended by some vendors. Note the shrouds surrounding the probes.

Measurement of Pressure Signals from Velocity Detectors

The devices described above characteristically develop output pressure signals of relatively small magnitude, commonly a maximum of a few inches of water. Thus precision of flow rate measurements may be poor unless appropriate signal conditioning and readout components are used.

With non-corrosive fluids, strain gauge pressure or differential pressure transducers can be used. Transducers may be obtained with a variety of operating ranges, typically ± 1 PSI (6.9 kPa) at a level of 5000 PSI (34.4 MPa). The outputs are about 5 mv/volt of excitation which can be AC or DC. When excited with 10 V DC and directed to a high-gain high-impedance amplifier with oscillograph readout, transducer output may be resolved within about 2 microvolts. A filter may be required to suppress high frequency fluctuations in the amplifier input caused by fluid turbulence at the probe. Such a measuring system is indicated in Figure 2.13 m. The use of such data systems for industrial process measurements is described in Hougen.[8] Manometric and dial-type indicators are typically offered by most manufacturers.

Isokinetic Sampling through Pitot Tube-type Probes

The procurement of representative samples of particulates suspended in gas streams demands that the velocity at the entrance to the sampling probe be precisely equal to the stream velocity at that point. This is accomplished by regulating the rate of sample withdrawal so that the static pressure within the probe is equal to the

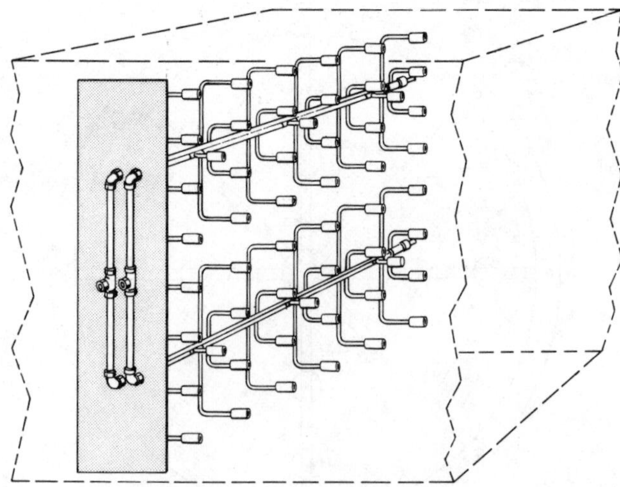

Fig. 2.13 l Installation in rectangular duct of area-averaging Pitot tube ensembles for metering the flow rate of gases (Courtesy Air Monitor Corporation)

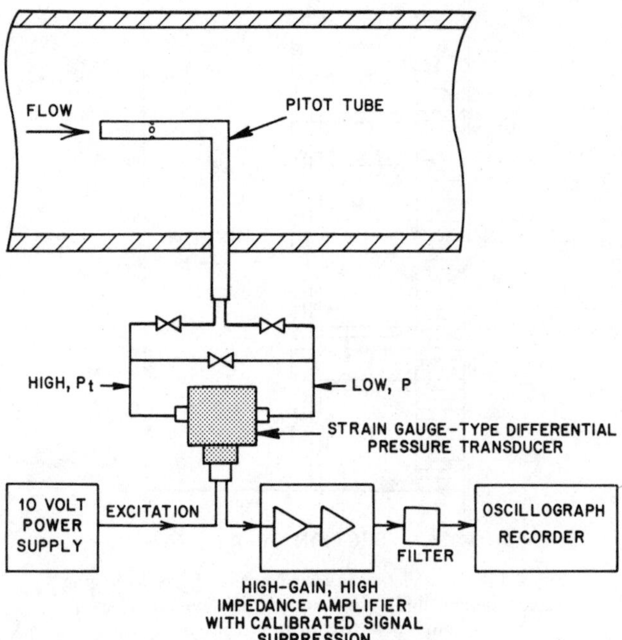

Fig. 2.13m Instrumentation for precise measurement of pressure and differential pressure

static pressure in the fluid stream at the point of sampling. A Pitot tube of special design is used for such purposes with means for measuring the pertinent pressures. The pressure difference can be maintained at zero by automatically controlling the sample draw-off rate. Fricke et. al.[9] describe such an isokinetic sampling system used successfully to retrieve samples in rapid succession from a duct transporting pulverized coal. Details of the probe construction may be found in the literature cited in Reference 9.

REFERENCES

1. Shaw, R., "The Influence of Orifice Geometry on Static Pressure Measurements," *Fluid Mechanics*, 7, Pt. 4, April 1960, p. 550.
2. Benedict, R.P., *Fundamentals of Temperature, Pressure, and Flow Measurements*, New York: John Wiley & Sons, Inc., 1969, p. 237.
3. Shaw, R., *Orifice Geometry*, p. 240.
4. Rayle, R.E., "Influence of Orifice Geometry on Static Pressure Measurements," ASME Paper 59-A-234, Dec. 1959.
5. Benedict, *Fundamentals*, p. 241.
6. Ibid, p. 245.
7. Ibid, p. 248.
8. Hougen, J.O., *Measurements and Control Applications*, 2nd edition, Instrument Society of America, Research Triangle Park, NC, 1979, p. 212.
9. Fricke, L.H., J.O. Hougen and O.R. Martin, "Isokinetic Sampling of Gas-Solid Mixtures by Analog Computer Control," Instrumentation in the Chemical and Petroleum Industries, Vol. 3, Plenum Press, 1967.

BIBLIOGRAPHY

"The Accuracy of the Pitot Tube Traverse Method of Measuring Pipe Flow at Various Distances up to 30 Diameters Downstream of a Smooth Right-Angled Bend," National Engineering Laboratory Flow Measurement Memo, No. 37, 1969.

Andrew, W.G. and Williams, H.B., *Applied Instrumentation in the Process Industries*, Vol. I, 2nd ed., Gulf Publishing Co., 1979.

Beitler, S.R., "Present Status of the Art of Flow Measurement in the Power Industry," ASME Paper No. 68-WA/PTC-7, Dec. 1968.

"Flow Meter Survey," *Instruments and Control Systems*, 42, No. 3 pp. 115–130, March 1969 and 42, No. 7, pp. 100–102, July 1970.

London, A.V., "Less Traditional Methods of Flow Measurements," *Process Engineering, Plant & Control*, Aug. 1968, pp. 47–50.

Malherbe G. and Silberberg, S., "Device for Measuring the Flow of Pulverized Control," Central Electricity Generating Board, Translation CE 4938 form *Automatisme* 13, No. 3, pp. 114–122, March 1968.

Ower, E. and Pankhurst, R.D., *The Measurement of Air Flow*, London: Pergamon Press, 4th ed., 1966.

Spencer, E.A., "Flow Measurements at the National Engineering Laboratories," *Process Engineering, Plant and Control*, pp. 53–57, August 1968.

Flow Sheet Symbol

FQI

2.14 POSITIVE DISPLACEMENT GAS METERS

Types of Designs:	A—Positive displacement B—Shunt flow
Design Pressure:	Low pressure, 5 to 100 PSIG (34 to 689 kPa); high pressure, to 1440 PSIG (10 MPa)
Design Temperature:	−30 to +140°F (−34 to +60°C)
Materials of Construction:	Aluminum, steel, plastics, synthetic elastomers
Inaccuracy:	± ½ to 1% registration
Cost:	\$100 to \$16,000 depending on size and materials
Partial List of Suppliers:	B.I.F., a division of General Signal (B); Dresser Measurement Div., Dresser Industries, Inc. (A); Rockwell International, Municipal and Utility Div. (A); Singer, American Meter Div. (A); Sprague Meter Div., Textron, Inc. (A)

Positive displacement gas meters measure by internally passing isolated volumes of gas that successively fill and empty compartments with a fixed quantity of gas. The filling and emptying process is controlled by suitable valving and is translated into rotary motion to operate a calibrated register or index that indicates the total volume of gas passed through the meter.

The liquid sealed drum meter is the oldest commercial positive displacement gas meter (see Fig. 2.14a). Developed in the early 1800's, it was used for many years during the gaslight era. This type of meter is still available today and remains one of the most accurate of the displacement-type meters. Applications of the liquid sealed drum meter today include laboratory work, appliance testing, pilot plant measurements, and as a calibration standard for other meter types.

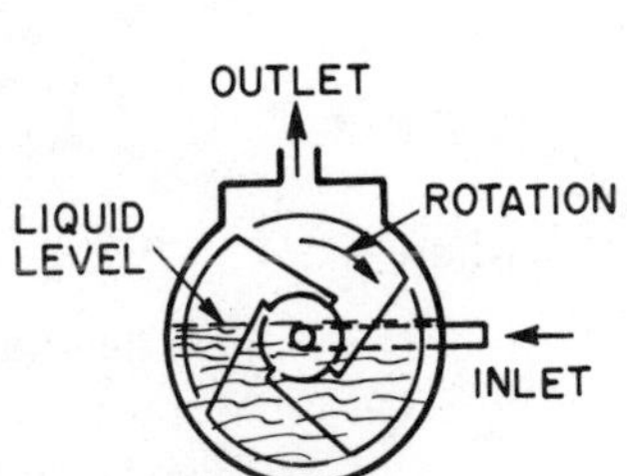

Fig. 2.14a The liquid sealed drum meter

Some of the inherent difficulties with the liquid sealed meter, such as changes in liquid level and freezing, were overcome in the 1840's with the development of the diaphragm-type positive displacement meter. The early meters were constructed with sheepskin diaphragms and sheet metal enclosures. Today, meters are made of cast aluminum with synthetic rubber-on-cloth diaphragms. The principle of operation, however, has remained the same for almost 150 years.

The operating principle of the four chamber diaphragm meter is illustrated in Figure 2.14b. The measurement section consists of four chambers formed by the volumes between the diaphragms and the center partition and between the diaphragms and the meter casing. Differential pressure across the diaphragms extends one diaphragm and contracts the other, alternately filling and emptying the four compartments. The control for the process is through the "D" slide valves which are synchronized with the diaphragm motion and timed to produce a smooth flow of gas by means of a crank mech-

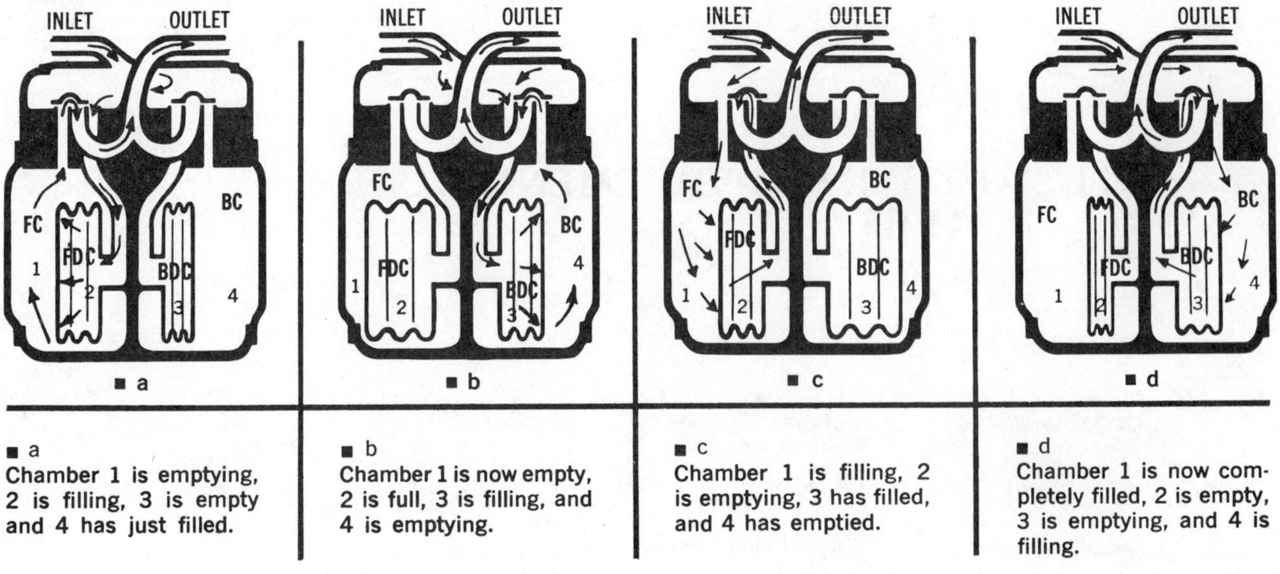

INLET
OUTLET
FC
BC
FDC
BDC
1
2
3
4
a
b
c
d
a
Chamber 1 is emptying, 2 is filling, 3 is empty and 4 has just filled.
b
Chamber 1 is now empty, 2 is full, 3 is filling, and 4 is emptying.
c
Chamber 1 is filling, 2 is emptying, 3 has filled, and 4 has emptied.
d
Chamber 1 is now completely filled, 2 is empty, 3 is emptying, and 4 is filling.

Fig. 2.14b The four chamber diaphragm meter. FC = front chamber; BC = back chamber; FDC = front diaphragm chamber; BDC = back diaphragm chamber.

anism. The crank and valve mechanism is designed and adjusted with no "top-dead-center" to prevent the meter from stalling. The rotating crank mechanism is connected through suitable gearing to the index which registers the total volume passed by the meter.

The rating of small diaphragm meters is usually specified in cubic feet per hour (0.03 m³/hr) of 0.6 specific gravity gas which result in a pressure drop of 0.5 inches water column (0.13 kPa). Larger meters are often rated for flow at 2 inches water column (0.5 kPa) differential.

Since most meters are sold to gas utility companies which serve natural gas with a specific gravity of approximately 0.6, it may be necessary to determine the flow rating of a diaphragm for other gases. This is accomplished by:

$$Qn = Qc \sqrt{\frac{(S\,G)c}{(S\,G)n}}$$

where

Qn = new flow rating, (ft³/hr)*
Qe = Meter rating, (ft³/hr)
(S G)c = Specific gravity for which meter is rated (usually 0.6)
(S G)n = Specific gravity of new gas

The inaccuracy of diaphragm positive displacement meters is typically ± 1 percent of registration over a range in excess of 200 to 1. This accuracy is maintained over many years of service. Deterioration of meter accuracy is rare unless unusual conditions of dirt, wear, or moisture in the gas are present.

*For SI units refer to Appendix A.1.

The lobed impeller meter (described in Section 2.15) is used for high volume measurement up to 100,000 ft³/hr (up to 3000 m³/hr). In this meter the close clearance of moving parts requires the use of upstream filters to prevent deterioration of accuracy performance. Typically the inaccuracy of lobed impeller meters is ± 1 percent over a 10 to 1 flow range at pressure drops of approximately 0.1 psi (0.7 kPa).

An improvement on the lobed impeller meter is the rotating vane meter illustrated in Figure 2.14c. Here four compartments formed by the vanes rotate in the same direction as a rotating gate. The fixed volumes of gas are swept through the meter by the vanes which are passed from inlet side to outlet side through the gate.

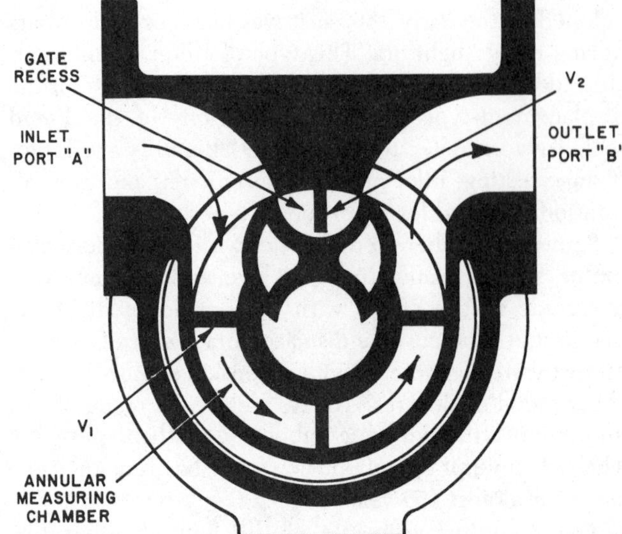

Fig. 2.14c The rotating vane meter

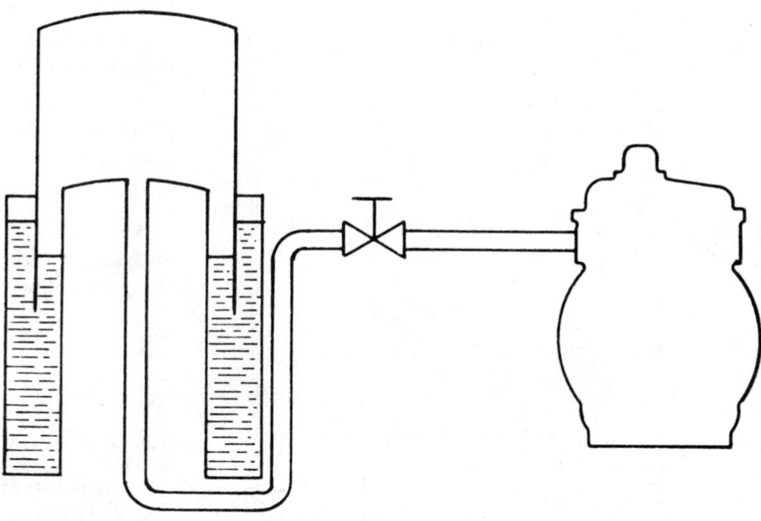

Fig. 2.14d The construction of a meter prover

The motion of the vanes and gate is synchronized by gears. Typical inaccuracy for the rotating vane meter is ± 1 percent over a 25 to 1 range at pressure drops of 0.05 inches of water column (0.013 kPa).

Application Notes

All displacement gas meters can be used to measure any clean dry gas that is compatible with the meters' construction materials and flow and pressure ratings. Dirt and moisture are the worst enemies of good meter performance; inlet filtering should be used when indicated. Since all gases change volume with pressure and temperature changes, these sources of possible error should be controlled or compensated. The national standard cubic foot of fuel gas is at 14.73 PSIA and 60° F and significant deviation from these values should be accounted for in measuring standard gas volumes. At elevated pressures and lower temperatures a deviation from the ideal gas laws occurs requiring the application of a compressibility factor to the measured volumes.

Testing and Calibration of Gas Meters

The testing, or proving as it is called in the gas utility industry, of gas meters is usually done using a special type of gasometer referred to as a prover. The construction of a meter prover is shown in Figure 2.14d. An accurately calibrated "bell" of cylindrical shape is sealed over a tank by a suitable liquid. The lowering of the bell discharges a known volume of air through the meter under test to compare the volumes indicated. Meter provers are typically supplied to discharge volumes of 2, 5 and 10 cubic feet (0.06, 0.15 and 0.3 m^3) and larger provers of several hundred cubic foot capacity are in use by meter manufacturers and gas utility companies. The volumetric inaccuracy of meter provers is on the order of ± 0.1 percent as determined by physical measurement and comparison with more accurate volumetric standards.

Other standards used to calibrate gas meters are calibrated orifices and critical flow nozzles. These devices compare rates of flow rather than fixed volumes and have inaccuracy ratings typically from ± 0.15 to ± 0.5 percent.

Gas Turbine Meters

While not a positive displacement type, gas turbine meters find application in fuel and other gas measurement applications because of their simplicity and wide rangeability. These are discussed in more detail in Section 2.20. Figure 2.14e shows the principle of the axial flow gas turbine meter. A flow diffuser increases the flowing gas velocity and directs it to a multi-bladed rotor mounted in precision bearings. The calibrated index is driven by the rotor through suitable gearing. Gas turbine meters are available in sizes from 2 to 12 inch pipe diameter (50 to 305 mm) and flow ratings up to 150,000 ft^3/hr (4500 m^3/hr). A desirable characteristic of gas turbine meters is their increase in rangeability at elevated

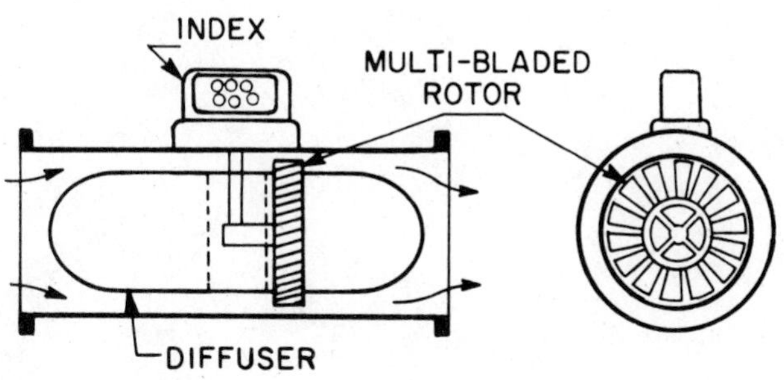

Fig. 2.14e The axial flow gas turbine meter

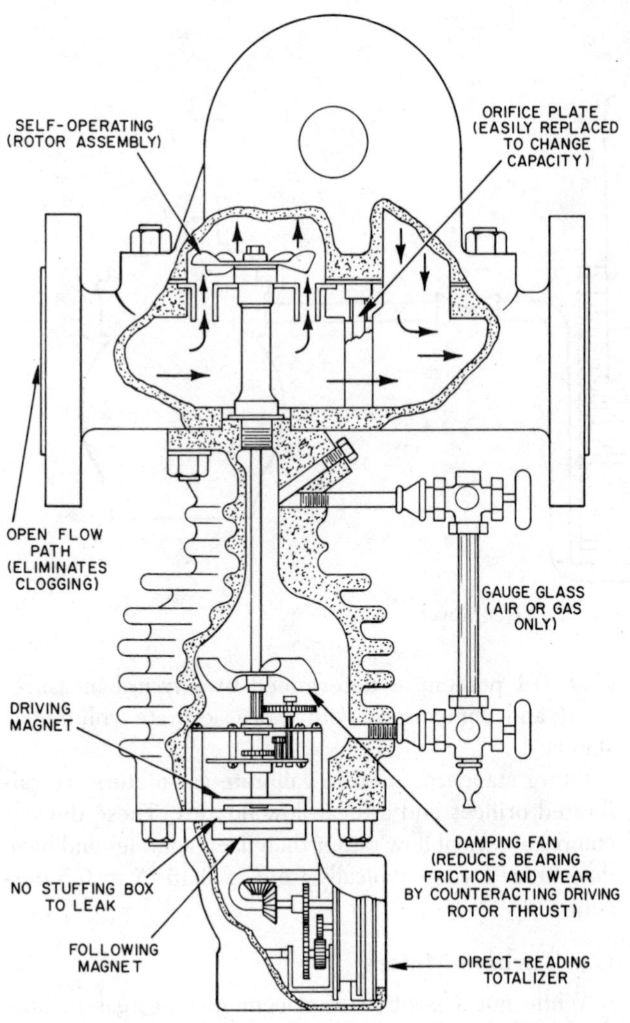

Fig. 2.14f Shunt flowmeter

operating gas pressures. Rangeabilities in excess of 100 to 1 are attainable in the large size meters operating at 1400 PSIG (9.7 MPa).

Shunt Flowmeters

Another flowmeter which is widely used in steam and gas flowmetering and totalizing applications is the shunt flowmeter illustrated in Figure 2.14f.

This is not a positive displacement design. It consists of an orifice plate in the main flow line and a self-operating rotor assembly in the bypass.

As gas flows through meter body, a portion of flow is diverted to drive the fan shaft assembly, rotating on a jewel bearing. A second set of blades on the fan shaft, rotating in damping fluid, acts as a damper or governor.

Rotational speed of the shaft is proportional to the rate of flow at all rates within the normal range of the meter.

These flowmeters are available in sizes of 2 in. (50 mm) and larger. Their inaccuracy is around ± 2 percent of the actual flow, and their rangeability is about 10:1

BIBLIOGRAPHY

Beck, H.V., *Displacement Gas Meters*, 1970, Singer American Meter Div., Philadelphia, Pa. 19116.

Consodine, D.M., *Encyclopedia of Instrumentation and Control.* New York: McGraw-Hill.

Evans, H.J., *Turbo-Meters—Theory and Application*, 1968, Rockwell International, Pittsburgh, Pa. 15208.

"Fluid Meters, Their Theory and Application," 6th ed., 1971, American Society of Mechanical Engineers, New York, NY 10017.

Hall, John, "Solving Tough Flow Monitoring Problems," *Instruments and Control Systems*, February, 1980.

Lief, A., *Metering for America.* New York: Appleton-Century-Crofts, 1961.

Lomas, D.J., "Selecting the Right Flowmeter," *Instrumentation Technology*, May, 1977.

Flow Sheet Symbol

FQI

2.15 POSITIVE DISPLACEMENT LIQUID METERS

Types of Designs:	A—Impeller, propeller, turbine B—Nutating disc C—Oval gear D—Piston E—Rotating vane F—Specialized, low-flow, etc. G—Viscous helix
Design Pressure:	To 300 PSIG (2.1 MPa) or greater
Design Temperature:	To 400°F (204°C)
Strainer Required:	Yes
Materials of Construction:	Bronze, cast iron, aluminum, steel, stainless steel, Monel, Hastelloy
Size Range:	1 to 16 in. (25 to 406 mm)
Flow Range:	0.01 GPH to 20,000 GPM. (0.04 l/hr to 4,542 l/hr)
Rangeability:	15:1
Inaccuracy:	± 1/10 to ± 2% of flow
Cost:	$1,100 for 2 in. (50 mm) aluminum meter with counter only. $3,800 for same unit in all-stainless construction
Partial List of Suppliers:	American Meter Div., Singer, (A,B); Astro Dynamics, Inc. (D); Badger Meter, Inc. (B,D); Brooks Instrument Div., Emerson Electric Co. (A,C,D); Conameter Corp. (F); Dresser Industries, Inc. (A); Fluidyne Instrumentation (C,D,G); Hays Republic Div., Milton Roy Co. (B); Hersey Products, Inc. (A,B,D); ITT Barton (A); Liquid Controls Corp. (E); Neptune Measurement Co. (B); Signet Scientific Co. (F); Tokheim Corp. (E)

Positive displacement meters split the flow of liquids into separate known volumes based on the physical dimensions of the meter, and count them or totalize them. They are mechanical meters in that one or more moving parts, located in the flow stream, physically separate the fluid into increments. Energy to drive these parts is extracted from the flow stream and shows up as pressure loss between the inlet and the outlet of the meter. The general accuracy of these meters is dependent upon minimizing clearances between the moving and stationary parts and maximizing the length of this leakage path. For this reason meter accuracy tends to increase as size increases.

Nutating Disk

This meter, also known as the disk meter, is used extensively for residential water service. The moving assembly, which separates the fluid into increments, consists of an assembly of a radially slotted disk with an integral ball bearing and an axial pin (see Figure 2.15a). This part fits into and divides the metering chamber into four volumes, two above the disk on the inlet side and two below the disk on the outlet side. As the liquid attempts to flow through the meter, the pressure drop from inlet to outlet causes the disk to wobble or nutate, and for each cycle to display a volume equal to the volume of the metering chamber minus the volume of the

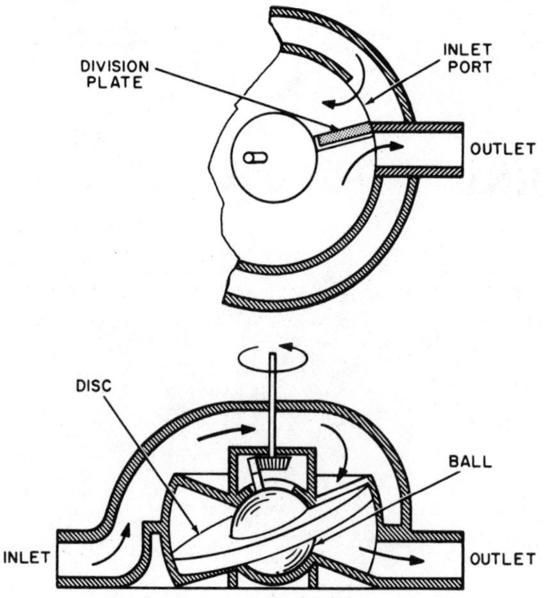

Fig. 2.15a Nutating disk meter

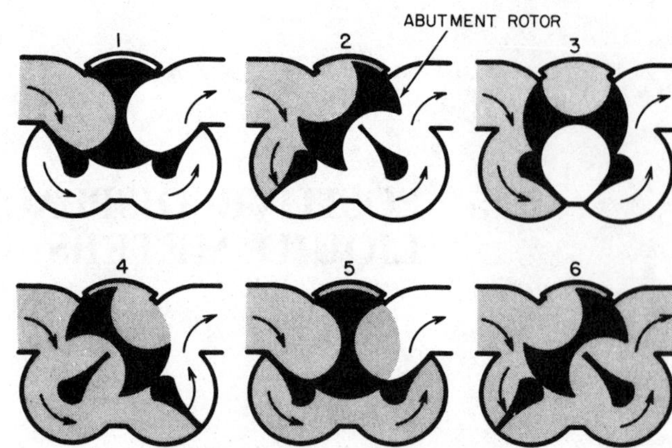

Fig. 2.15c Six phase metering cycle of rotary displacement-type flowmeter

disk assembly. The end of the axial pin, which moves in a circular motion, drives a cam that is connected to a gear train and the totalizing register. This flowmeter has an inaccuracy of about ±1–2 percent. It is built for small pipe sizes, and its maximum capacity is about 150 GPM (570 l/m).

Rotating Vane

This flowmeter has spring-loaded vanes that seal increments of liquid (Figure 2.15b) between the eccentrically-mounted rotor and the casing and transport it from the inlet to the outlet where it is discharged due to the decreasing volume. This type of meter is the most widely used in the petroleum industry and is used for such varied service as gasoline and crude oil metering, with ranges from a few GPM of low viscosity clean liquids to 17,500 GPM (66.5 l/m, or 25,000 bbl per hr) of viscous particle-ladened crudes. Accuracies of ±0.1 percent are normal, and ±0.05 percent has been achieved in the larger meters.

Fig. 2.15b Rotating vane meter

This instrument is built from a variety of materials of construction, and can be used for fairly high temperature and pressure service. Upper limits are approximately 350°F and 1000 PSIG (177°C and 6.9 MPa).

Another rotary design is illustrated in Figure 2.15c. Here an abutment rotor operates in timed relation with two displacement rotors and at half their speed.

Oscillating Piston

The moving portion of this meter consists of a slotted cylinder that oscillates about a dividing bridge that separates the inlet port from the outlet port. Spokes connect this cylinder to a pin located on the axis of the cylinder. As the cylinder oscillates about the bridge (Figure 2.15d) the pin makes one rotation per cycle. This rotation is transmitted to the gear train and registers either directly or magnetically through a diaphragm. This meter, in addition to being in common usage for the measurement of domestic water, has the capability of handling clean viscous and/or corrosive liquids. Inaccuracies are in the range of ±1 percent. This type of flowmeter is normally used in small pipe lines (2 in./50 mm or smaller) to measure low flowrates. Cost of this instrument depends upon size and materials of construction.

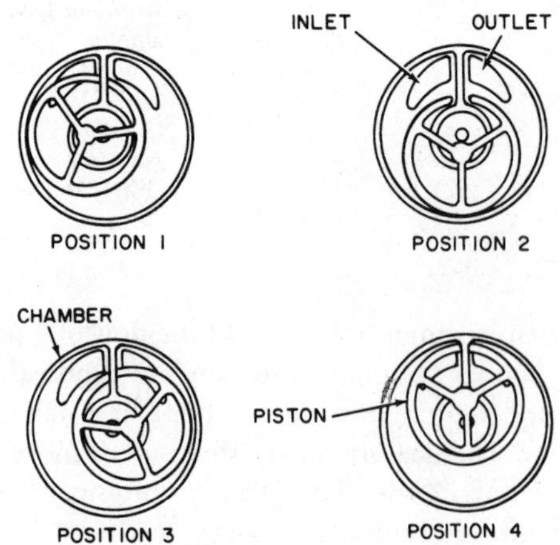

Fig. 2.15d Oscillating piston meter

Reciprocating Piston

The oldest of the positive displacement meters, this meter is available in many forms—multi-piston meters, double-acting piston meters, rotary valves, horizontal slide valves. Figure 2.15e shows a schematic of one va-

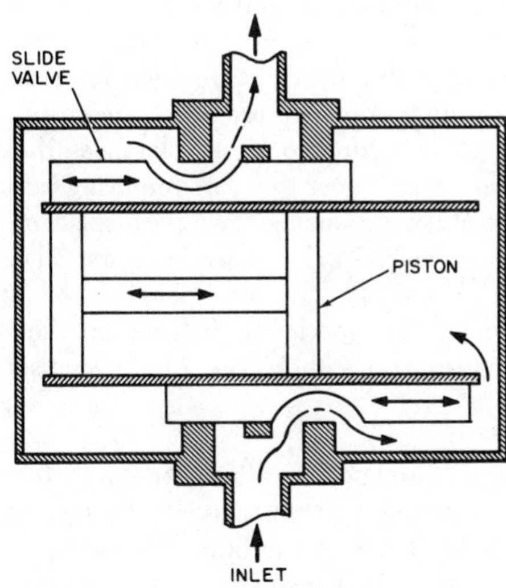

Fig. 2.15e Reciprocating piston meter

riety of this meter. A crank arm actuated by the reciprocating motion of the pistons drives the register. These meters are widely used in the petroleum industry and can achieve accuracies of ±0.2 percent.

Another version of this meter is shown in Figure 2.15f. The liquid enters the cylinder on the left forcing the piston down through lever action of the control plate. The piston on the right is forced up, discharging liquid through the port into the inner portion of the valve, down through the center of the meter, and out the meter discharge.

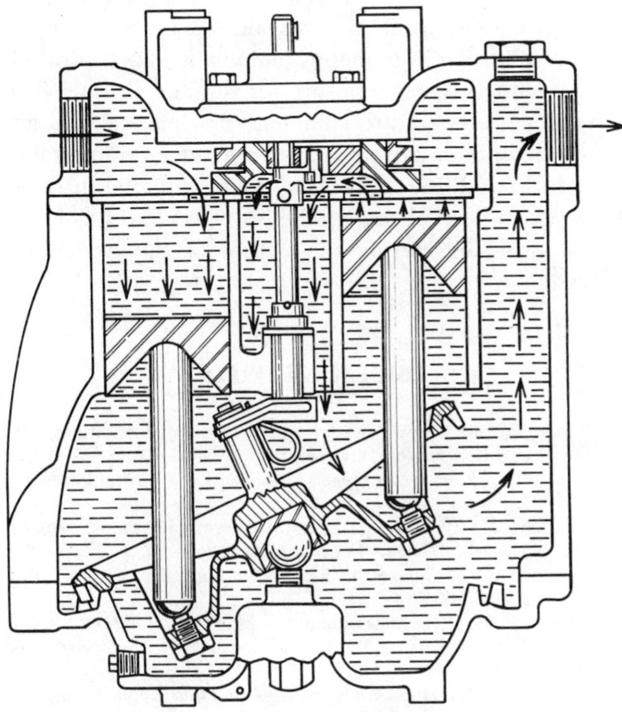

Fig. 2.15f Cutaway of reciprocating piston meter with two opposing pistons

Rotating Lobe

In this type of meter, two lobed impellers, which are geared together to maintain a fixed relative position, rotate in opposite directions within the housing (Figure 2.15g). A fixed volume of liquid is displaced for each revolution. A register is geared to one of the impellers. They are normally built for service in pipe sizes from 2 to 24 inches (50 to 610 mm) and their maximum capacities range from 8 to 17,500 GPM (30.4 to 66,500 l/m).

A recent addition to this general type of meter is one using oval-geared rotors in place of the lobed rotors.

At low flow rates (0.2 to 40 GPH or 0.8 to 152 l/hr) where inaccuracy due to clearance slippage can be substantial, the servo version of this meter can be considered. The concept behind this design is that if there is no pressure differential across the meter, then there will be no driving force to cause slippage. Elimination of this pressure differential is done by detecting up- and downstream pressures and by automatically adjusting a servo motor which varies the speed of the driven rotor so that up- and downstream pressures are the same.

These meters are increasingly accurate at higher flows where leakage or "slip" is decreased and though normally rated at ±0.5 percent they will operate at an accuracy of ±0.1 percent over a limited range. These flowmeters can be used for corrosive service. Their capacities range from 1 to 1600 GPM (3.8 to 6.080 l/m).

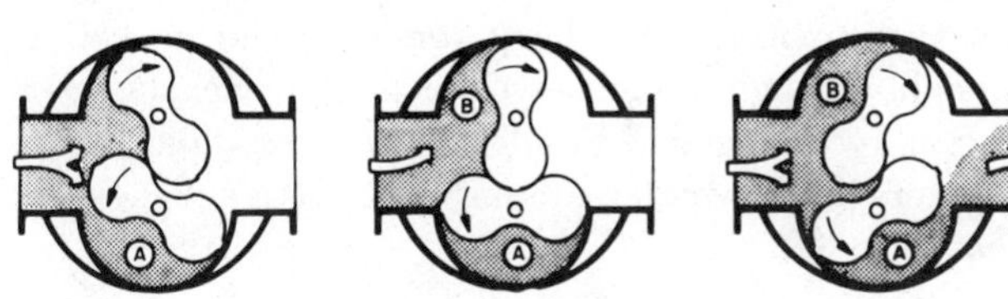

Fig. 2.15g Rotating lobe meter

Viscous Helix

The helix flow transducer (Figure 2.15h) is a positive displacement device utilizing two uniquely nested, radically-pitched helical rotors as the measuring elements. Close machining tolerances ensure minimum slippage and thus high accuracy. The design of the sealing surfaces provides the optimum ratio of longitudinal to lateral sealing for minimum pressure drop, especially with high-viscosity liquids.

This flow sensor is available in sizes up to 10 in. (250 mm) and can operate at temperatures up to 600°F (315°C) and at pressures 1000 PSIG (6.9 MPa) or higher. It is a high pressure drop device requiring at least 10 PSIG (69 kPa) for its operation at full flow. Its turndown can reach 100:1, while its metering inaccuracy is claimed to be under 0.5 percent of actual flow.

Available design variations include the heated versions to maintain line temperatures for meltable solids or polymer services. Also available are units with sanitary con-

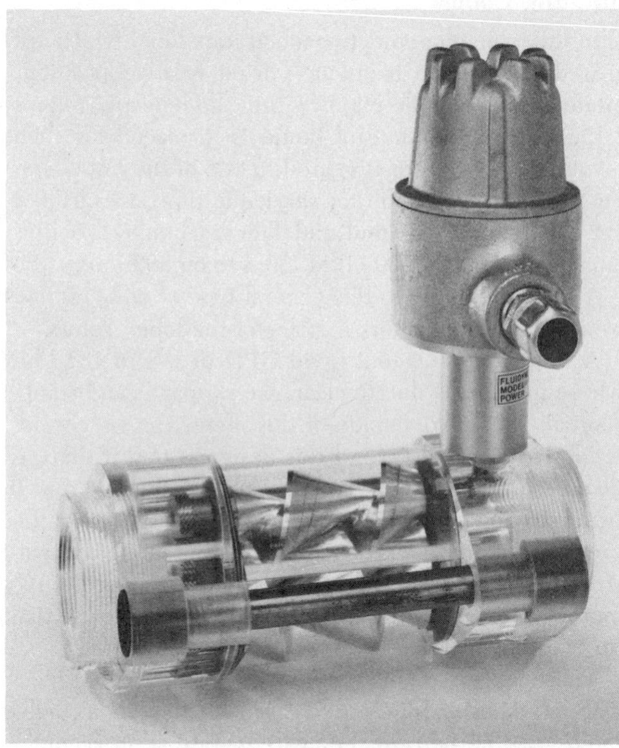

Fig. 2.15h Viscous helix flowmeter (Courtesy of Fluidyne Instrumentation)

struction. This meter is suited for high viscosity (over 1000 centipoises) and slurry services. The pocketless straight-through design serves to simplify cleanout.

It is recommended to filter the process fluid before it enters the flowmeter. US mesh size 30 is sufficient for that purpose.

Advantages

Positive displacement meters provide good accuracy (± 0.25 percent of flow) and high rangeability (15:1). They are repeatable to ± 0.05 percent of flow. Some designs are suited for high or variable viscosity services. They require no power supplies and are available with a wide variety of readout devices. Their performance is virtually unaffected by upstream piping configuration. Positive displacement meters are excellent for batch processing, mixing, or blending systems since actual quantities of liquid are measured.

These meters are simple and easy to maintain using regular maintenance personnel and standard tools. No specially trained crews or special calibration instruments are needed.

Disadvantages

Positive displacement meters require more expensive precision machined parts to achieve the small clearances upon which their accuracy depends. From this it follows that the liquids metered must be clean, for wear rapidly destroys accuracy. Contaminant particle size must be kept below 100 microns, and most of these meters are not adaptable to the metering of slurries. The moving parts require maintenance at frequent intervals; where corrosive liquids are metered this may result in high costs. In general, due to close tolerances the moving components are subject to wear, requiring periodic recalibration and maintenance. Positive displacement flowmeters are expensive in large sizes or in special materials. They can be damaged by overspeeding and require a high head loss for operation. In general, they are not suited for dirty, non-lubricating, or abrasive services.

Conclusions

As a class, liquid positive displacement meters are one of the most widely used instruments for measuring volumetric flow whenever the fluid is bought and sold on a contract basis. As a result, a large variety of meters, covering a broad spectrum of requirements, are available. Their good accuracy and ready availability warrant their primary consideration when selecting a volumetric meter.

These flowmeters are especially useful when the fluid to be measured is free of any entrained solids. A typical example is the measurement of water delivered to homes, factories, office buildings, etc.

Wear of parts introduces the major source of error over meter service life. Leakage error increases with lower viscosity fluids but remains relatively constant with time for its calibrated range. In the large meter range, temperature effects on fluid density and viscosity must be taken into consideration.

Standard available accessories include strainers, air release assemblies to remove all vapors prior to the stream entering the meter, automatic batch shut-off valves with two-stage closure for full and dribble flow operation, temperature compensators, manual and automatic printers, and pulse generators for remote totalization. In addition to the totalizer-type digital readout, flow rate indication can also be provided, but this is not economical considering the nature of the design involved.

BIBLIOGRAPHY

Blasso, Leonard, "Flow Measurement Under Any Conditions," *Instruments and Control Systems,* February, 1975.

Clark, W.J., *"Flow Measurement,"* Elmsford, New York: Pergamon Press, 1967.

Hall, John, "Solving Tough Flow Monitoring Problems," *Instruments and Control Systems,* February, 1980.

Hayward, A.J., "Choose the Flowmeter Right for the Job," Processing Journal, 1980.

Laskaris, E.K., "The Measurement of Flow," *Automation,* 1980.

Lomas, D.J., "Selecting the Right Flowmeter," *Instrumentation Technology,* 1977.

Spink, L.K., *Principles and Practices of Flow Engineering,* The Foxboro Company, 9th edition.

Watson, G.A., "Flowmeter Types and Their Usage," *Chartered Mechanical Engineer Journal,* 1978.

Flow Sheet Symbol

FG

2.16 SIGHT FLOW INDICATORS

Design Pressure:	To 600 PSIG (4 MPa) standard
Design Temperature:	To 450°F (230°C) standard
Materials of Construction:	Glass, plastics, carbon steel, stainless steel, and higher alloy metals
Sizes:	½–16 in. (12–406 mm)
Cost:	\$75 to \$150 per inch diameter for standard materials
Partial List of Suppliers:	Brooks Instrument Div., Emerson Electric Co.; Fischer & Porter Co.; Foxboro Co.; Jacoby-Tarbox Corp.; Schutte & Koerting Div., Ametek Inc.; Wallace & Tiernan Div., Pennwalt Corp.

Sight flow indicators are used when a visual inspection of the process is necessary. Figure 2.16a shows three designs. The flapper design is used in transparent or slightly opaque solutions and in gas services. Flow direction is vertically up or horizontally. Some indication of flow variations may be made by observing the position of the flapper. The drip tube design is used where the flow is vertically down and is particularly suited for intermittent flows. The paddle design is used in dark processes since the motion of the paddle can be easily detected. Flow through the paddle design can be vertical or horizontal.

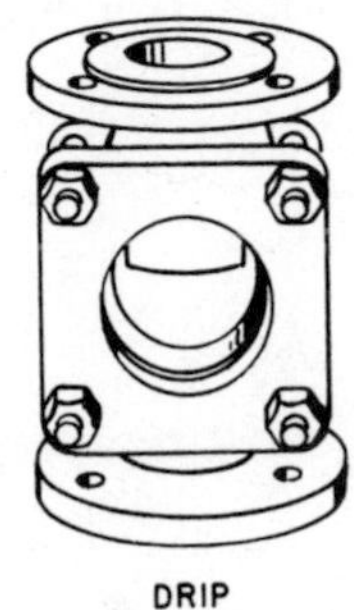

Fig. 2.16a Various designs of sight flow indicators

Figure 2.16b shows the cross section of a sight flow indicator that has flanged connections. The assembly consists of the chamber, glasses, gaskets, covers and bolts. It is similar in many respects to the transparent level gauge discussed in Section 3.11. The glass is normally a borosilicate which can be rated to 450°F (230°C) for sight glass applications, and has good resistance to mechanical and thermal shock. The glasses can also be made of silica glass or quartz thus enabling temperature ratings in excess of 1000°F (535°C). The chamber can be obtained in a wide variety of materials and can also be furnished with a number of different plastic liners. This enables a properly specified sight flow indicator to be used in almost all corrosive services. The bolts and covers are normally steel and iron respectively but these too may be obtained in different materials depending on atmospheric and temperature requirements.

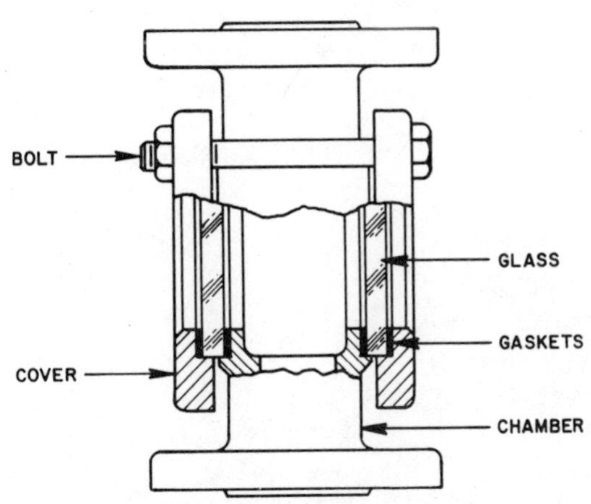

Fig. 2.16b Cross section of sight flow indicator

Figure 2.16c shows the cross section of a double window assembly. This assembly improves the safety of a sight flow indicator in several ways. In high temperature services the gradient across each glass is reduced. The outer glass protects the inner glass from thermal shock caused by splashes of cold water. If the outer or inner glass breaks there is a chance that the remaining glass will contain the process until the assembly can be repaired. Special designs are available that can be used in sanitary services such as food processing. Accessories such as jacketed indicators, illuminators, and spray rings for cleaning the glass in place are also available.

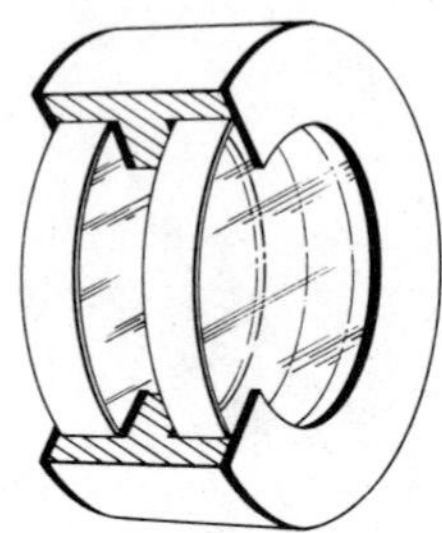

Fig. 2.16c Cross section of double window assembly

Sight flow indicators offer an inexpensive means of viewing the process to assure that the stream is flowing or to note process characteristics such as color, turbidity, or other properties that might indicate process deterioration or equipment malfunction. In fact, several types of analyzers require flow glasses so that the sensing element can see the process. They are sometimes used in secondary services, such as condensate pot installations as described in Section 3.7. Nonetheless, their use is limited in the industrial process area. It is difficult to estimate flow rate through a sight glass and a hazard is created if the glass breaks. They are used more commonly outside of the industrial processing area.

BIBLIOGRAPHY

Green, C.R., "Tank Sight Glasses," *Chemical Engineering*, September 25, 1978.

Flow Sheet Symbols

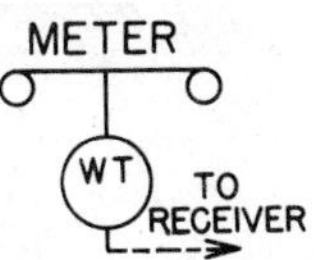

BELT TYPE

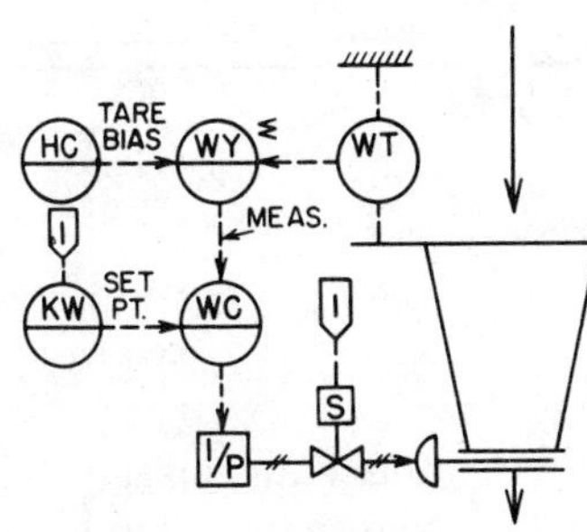

LOSS IN WEIGHT

2.17 SOLIDS FLOWMETERS AND FEEDERS

Types of Designs:	A—Accelerator B—Belt-type gravimetric feeder C—Volumetric D—Impulse sensor E—Loss-in-weight F—Vertical gravimetric feeder
Capacities:	A: 1,000–80,000 lbm/hr (450–36,000 Kgm/hr) B: Up to 180,000 lbm/hr (80,000 Kgm/hr) or up to 3,600 ft³/hr (100 m³/hr) C: Up to 3,600 ft³/hr (100 m³/hr) D: 3,000–3,000,000 lbm/hr (1,400–1,400,000 Kgm/hr) E: Determined by hopper or duct size F: Up to 40,000 lbm/hr (18,000 Kgm/hr)
Costs:	Under $3,000 (C) $2,000–$5,000 (A,D) $5,000–$10,000 (B,F) Over $10,000 (E)
Inaccuracy:	± 1% of rate over 10:1 range (B,E) ± ½% to ± 1% of full scale (A,D,F) ± 2% to ± 4% of full scale (C)
Partial List of Suppliers:	Acrison, Inc. (B); Bailey Controls Co. (C); B.I.F., a division of General Signal (B); Cutler-Hammer, Inc. (E); DeZurik, A division of General Signal (A); Howe Richardson Scale Co. (E); Leeds & Northrup, a division of General Signal (E); Merrick Scale Mfg. Co. (B); Milltronics, Inc. (D); Moore Products Co. (E); Ramsey Engineering Co. (B,F); Thayer Scale, Hyer Industries, Inc. (B); Vibra Screw, Inc. (B); Wallace & Tiernan Div., Pennwalt Corp. (B)

Belt Feeders

Belt feeders are compact factory-assembled devices utilizing belts to transport the material across a weight sensing mechanism. In the case of meters, an uncontrolled solids flow passes across a constant speed belt and the belt load signal is thus a function of gravimetric flowrate.

The feeder in its most basic form consists of a meter to which a controller and volumetric solids flow regulator is added. The flow regulator is normally a simple gate, but may be in the form of a rotary gate, screw, or other volumetric control device capable of being fitted with a suitable actuator. Other methods of control are based on varying belt speed or both belt speed and belt load.

Basic Belt Feeder Types

Figure 2.17a illustrates a simple feeder which is perhaps the forerunner of most modern belt feeders. It incorporates a constant speed belt coupled with a gate to modulate the solids flow rate such that belt load is balanced by an adjustable poise weight. The feeder, which is still used in some industrial applications today, is unique in its simplicity but includes a number of disadvantages relative to more modern designs as follows:

1. The entire feeder is weighed rather than only a portion of the belt; consequently, the low ratio of live load to tare coupled with mechanical fric-

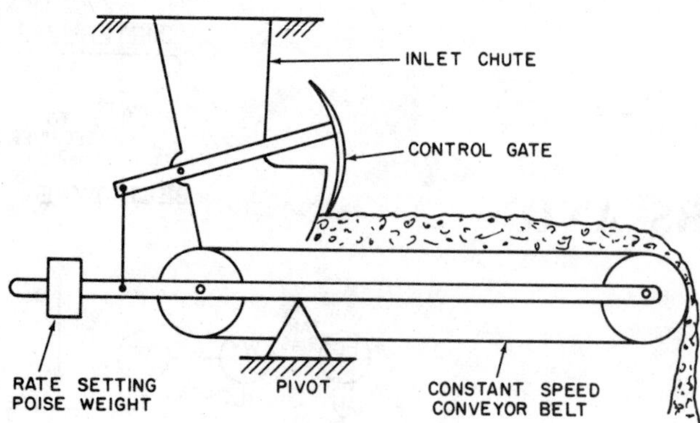

Fig. 2.17a Early belt-type mechanical gravimetric feeder

tion in the linkage pivots results in relatively low sensitivity in the belt load detection system.

2. The position of the gate control element is proportional to the belt load error. In the same manner that a float-operated level control valve cannot maintain level at set-point if valve supply pressure or tank drawoff vary, this feeder cannot maintain set gravimetric rate if the bulk density of the solids varies.

It should be noted that the basic principle involving the weighing of the entire feeder has been applied in modern designs. Successful operation of these versions has been achieved by adding belt load error detecting instrumentation and by actuating the control gate from an external power source. A controller with reset function eliminates the set-point error.

Figure 2.17b describes the basic construction of the electromechanical gravimetric feeder. Here the belt load is balanced by a mechanical beam and poise weight system which energizes one or the other of two clutches via a pair of mercury switches energized by a magnet attached to the beam. These clutches actuate and establish the direction of travel of the gate-positioning mechanism. The gate modulates as required to maintain the desired belt load as established by the position of the poise weight on the balance beam. It can be seen that this feeder will maintain belt load regardless of changes in material density, subject, of course, to the volumetric control limits of the gate. Belt load set-point is indicated by a mechanical counter geared to the beam poise weight drive. A second counter geared to the belt drive totalizes feet of belt. By varying drive gears, these counters can be provided to read direct in terms of pounds per foot of belt and belt travel in terms of total feet. Total weight fed can thus be calculated by multiplying the readings of the two counters. Remote belt load set-point and readout functions are available as well as a belt travel contact switch which may be used to operate a remote counter or to shut down the feeder via a predetermining counter after the desired total weight of material has been fed. Adjustable microswitches actuated by gate position may be utilized to activate alarms indicating either a stoppage of the material supply to the feeder or overtravel of the control gate resulting from abnormally low material density. These feeders are designed to meet Class II, Group G requirements for hazardous locations when equipped with similarly rated drive motor.

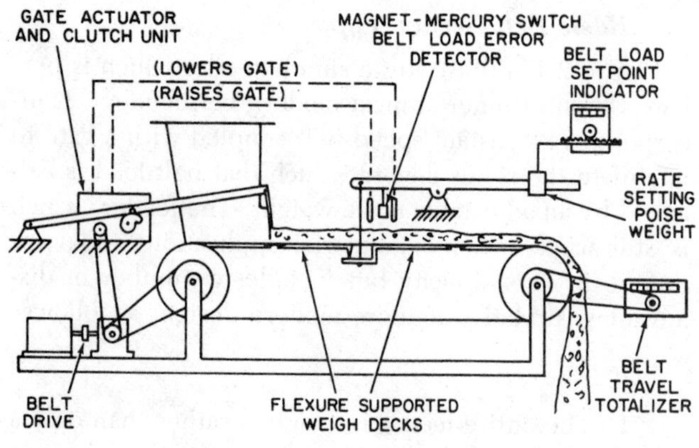

Fig. 2.17b Belt-type electromechanical gravimetric feeder

Figure 2.17c illustrates the gravimetric belt meter which is available with either pneumatic or electronic weight detection and transmission system. In the case of the pneumatic version shown, the preliminary calibration procedure involves adjustment of the tare weight with the beam in center position, and location of the nozzle relative to the flapper. This establishes a condition such that balance is achieved when balancing piston pressure is 3 PSIG (20.6 kPa). When material crosses the belt, beam movement throttles the nozzle. Nozzle backpressure is imposed on the pneumatic relay which in turn increases its output pressure until the balancing piston rebalances the beam. The balancing pressure is thus proportional to belt load and, since the belt speed is constant, balancing pressure is proportional to measured weight-rate. Also shown is an optional ball and disc integrator. The disc is driven by the front belt roll of the feeder and the ball is positioned by a pneumatic positioner. Unlike cam-operated sampling type integrators, the ball and disc type integrates continuously. It

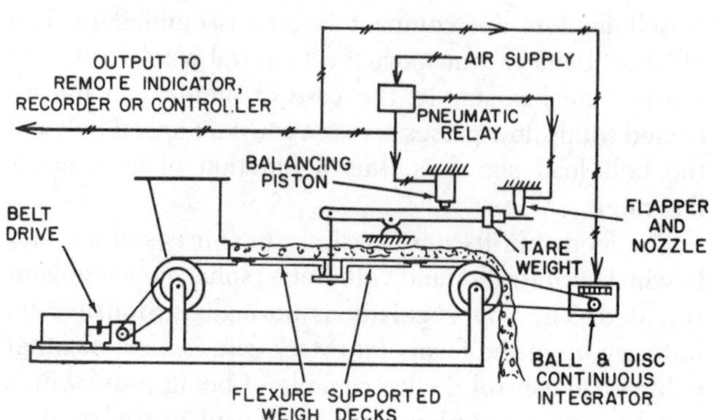

Fig. 2.17c Belt-type gravimetric meter

is especially recommended for use with gravimetric meters in applications involving the measurement of rapidly varying instantaneous flow rates. The integrator is supplied with a digital totalizer and can be furnished with a pulse transmitting switch to operate a remote counter.

Volumetric Belt Feed Sections

A gravimetric feeder consists of a weight-rate measuring mechanism coupled with a volumetric control device. The vertical gate volumetric regulator, which is perhaps the most popular, is not always suitable for controlling materials having large particle size, which are fibrous or irregularly shaped, or which tend to flow like a fluid due to fine particle size or because they become aerated. Unlike fluids that exhibit predictable flow behavior, solids flow characteristics are extremely difficult to evaluate on any basis other than an actual trial. For this reason, most manufacturers maintain a test and demonstration facility in which samples of a potential customer's material are fed by test feeders which may be equipped with various volumetric feed sections. Recognizing that a wealth of experience with commonly used materials can very often permit a feed section recommendation without the necessity of testing, it should be noted also that even a minor change in the properties of a material can drastically change its feeding characteristics. These changes might be in the form of a particle size or shape variation, entrainment of air during pneumatic conveying prior to entry to the feeder, or the addition of a relatively minor percentage of an additive to produce a preblend.

Vertical Gate

The vertical gate gravimetric feeder is available in a variety of sizes to produce typical material ribbon widths of 2 to 18 in. (50 to 457 mm) and to regulate up to 6 in. (152 mm) depth of material on the weigh belt. Gate actuators may be electromechanical or pneumatic, or in the form of electric servomotors or stepping motors. Manually adjustable gates are also available. The vertical gate has a typical depth control range of 10:1, and is generally suitable for materials which are not fluidized and which have a particle size not larger than about 1/8 in. (3.175 mm). Larger particles will not flow smoothly under the lip of the gate, thus resulting in an irregular belt load. This may necessitate the excessive dampening of the belt load transmitter output, which will have an undesirable effect on both control accuracy and sensitivity. In addition to producing undesirable control characteristics, rangeability will be decreased as particle size increases. As a rule of thumb, the minimum gate opening should be approximately three times the maximum particle size for materials which have irregularly shaped particles of random size. This factor may be reduced somewhat if the material is homogeneous and particles do not tend to interlock and tumble while in motion (typically, if particle shape approaches that of a sphere).

Rotary Vane

Figure 2.17d shows a rotary vane feeder with pneumatically-controlled variable speed drive and a belt-type gravimetric meter and controller. The rotary vane feeder is used as a volumetric feed section in instances where the material is aerated or has a low bulk density. Rotary feeders are generally not suitable for handling materials which have large particle sizes, and in some instances are undesirable if process requirements dictate that the material must not be abraded by the feeding device. Like the vertical gate, the rotary vane feeder is not suitable for handling fibrous or stringy materials, because sticky or hygroscopic materials tend to clog the pockets of the rotor. Pocket shape and depth selection are based on required volumetric flow rate and material characteristics. Care must be taken in determining a maximum practical rotor speed. Volumetric capacity of a given arrangement is regulated by rotor speed, but if the speed is too high, rotor pockets won't completely fill as they pass under the material inlet opening and volumetric output may decrease above a certain optimum rotor speed. Essentially, then, the rotary vane feeder is limited to use in connection with free-flowing powders or materials having small particle size but, unlike the vertical gate, it can control low density or aerated materials. The

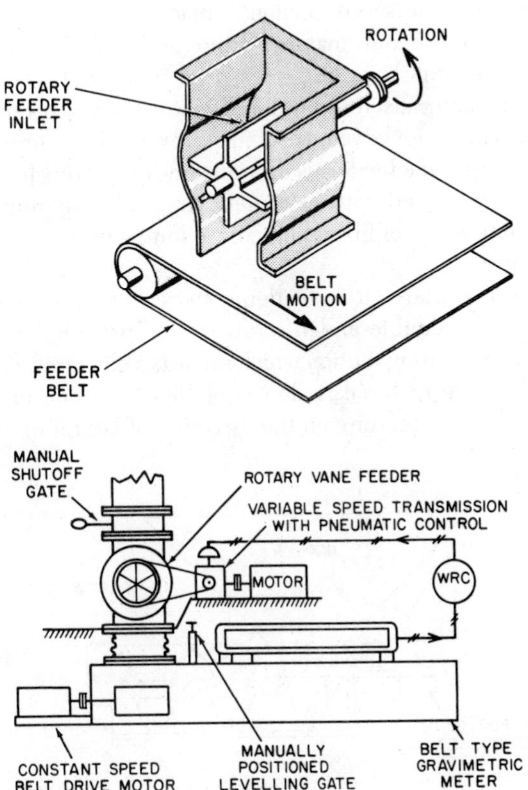

Fig. 2.17d Gravimetric feeding system utilizing a rotary valve volumetric feeder controlled by a belt-type gravimetric meter

gravimetric meter and rotary feeder should be separately mounted and interconnected by means of a flexible connection to prevent transmittal of vibration from the rotary feeder to the weight sensing mechanism. Figure 2.17d also shows a cutoff gate installed in the meter inlet ahead of the weighing section. This device drags across the irregular feed pattern created by a rotary feeder, levelling the ribbon of material and producing a more consistent feed to the weigh section and eventually to the process. The shutoff gate at the feeder inlet permits isolation of the feeder from the material supply when inspecting or servicing the feeder.

It is possible to operate several feeders in parallel or in cascade from the same set-point, and this mode of operation can be used effectively for blending solids.

Screw Feeder

The feeder element in this device is a screw whose rotary motion delivers a fixed volume of material per revolution (Figure 2.17e). The screw is located at the bottom of a hopper so that the feed element is always flooded with material. Rotation of the screw discharges material, at one or both ends of the screw, into the receiving vessel. Screws grooved in one direction discharge material at one end only. Screws grooved in opposite directions from the middle deliver material at both ends.

A gravimetric feeding system may also utilize a variable speed screw feeder to control low density or aerated materials. The screw section is made as long as is necessary to prevent material from flooding through it. Screw feeders have also been used successfully to feed fibrous materials and powdered materials which tend to cake. The major advantage of the screw feeder, compared to a rotary vane feeder, is that custom-built screw feeders can be provided with extremely large inlet openings to facilitate entry of fibers and coarse lumps to the conveying screw.

For materials with a tendency to cake or clog the feed screw, the double-ended screw can be provided with a lateral oscillating motion which imparts a cleaning action. In this case material is alternately fed from one end or the other, depending on the direction of lateral motion.

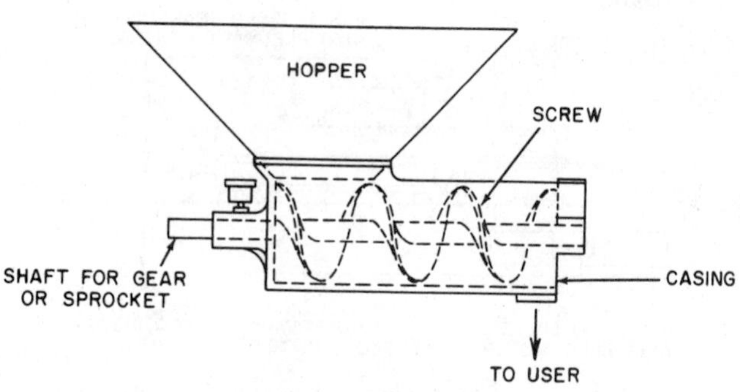

Fig. 2.17e Screw feeder

In order to assure an accurate feed, the hopper on the inlet side of the feeder must be designed to provide a uniform supply of material at the feed screw. Vibrators can be added to the hopper to keep the solids agitated and to prevent caking or bridging.

Feeder drives are usually electric motors. With a constant-speed drive, operating feed is adjustable over a 20 to 1 range by means of a mechanical clutch that varies the operating time per cycle. At 75 percent feed rate setting, the screw will be operating over 75 percent of a clutch revolution.

Additon of a variable-speed drive can extend the operating range to 200 to 1. The variable-speed drive can be electric or mechanical. The electric type will accept any standard milliampere signal; the mechanical type will operate on a 3–15 PSIG signal (20.6–103.4 kPa).

Vibratory Feeder

Vibratory feeders have been used in gravimetric feeding systems to handle materials consisting of particles which are too large to be handled by screw, rotary vane, or vertical gate feed sections, or in instances where the physical characteristics of the material particles may be adversely affected by passage through other types of volumetric feeding devices. The discharge flow pattern of a vibrating feeder is extremely smooth and thus is ideal for continuous weighing.

The vibratory feeder (Figure 2.17f) consists of a feed chute (which may be an open pan or closed tube) which is moved back and forth by the oscillating armature of an electromagnetic driver. Material transfer rate to the treatment process can be controlled by adjusting the current input to the electromagnetic driver, which controls the pull of the electromagnet and the length of the stroke. This feature also permits flow rate adjustment or control from a distant control panel.

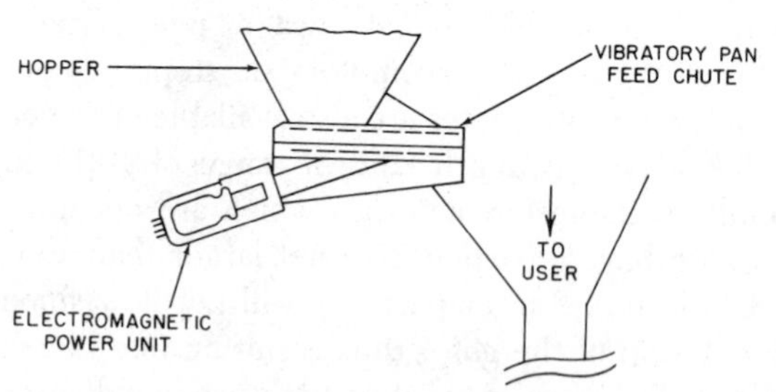

Fig. 2.17f Vibratory feeder

The feed chute can be jacketed for heating or cooling, and the tubular chutes can be made dust-tight by flexible connections at both ends. The vibratory feeders can resist flooding (liquid-like flow) and are available for a wide capacity range (from ounces to tons per hour).

Shaker Feeder

The shaker feeder design (Figure 2.17g) consists of a shaker pan mounted beneath a hopper. The back end of the shaker is mounted on hanger rods. The front end is carried on wheels, and can be moved by a crank. As the pan oscillates, the material is moved forward and dropped into the feed chute.

In most units the number of strokes is kept constant, while the length of the stroke is varied. The angle of inclination of the pan is chosen for the particular material to be fed and varies from about 8 degrees for freely flowing materials to about 20 degrees for sticky materials. If arching is expected in the hopper, special agitator plates are installed in the hopper to break up the arches. The shaker feeder is rugged and self-cleaning, and it can handle most types of material regardless of particle size or condition.

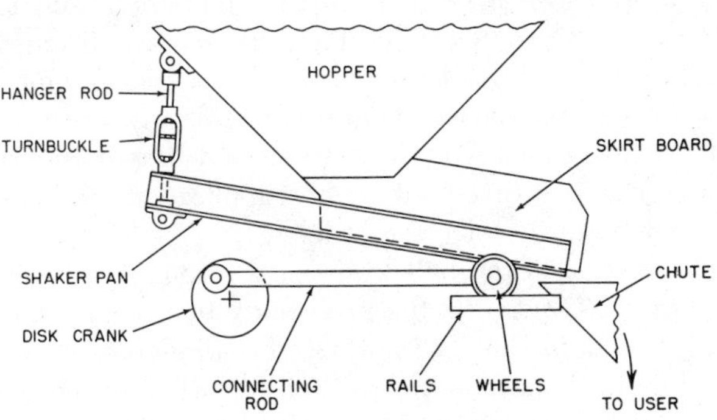

Fig. 2.17g Shaker feeder

Roll Feeder

Roll feeders are low-capacity devices for handling dry granules and powders (Figure 2.17h). The feeder consists of a feed hopper, two feed rolls and a drive unit. Guide vanes in the hopper distribute the material and provide agitation by oscillating. The feed rolls form the material into a uniform ribbon, and feed rate is controlled either by means of a slide that varies the width of the ribbon or by means of a variable-speed drive. Operating ranges are typically 6 to 1 with the feed slide and 10 to 1 with mechanical or electrical variable-speed drive. For materials that tend to cake or bridge in the hopper, agitators can be provided to maintain the material in a free-flowing state.

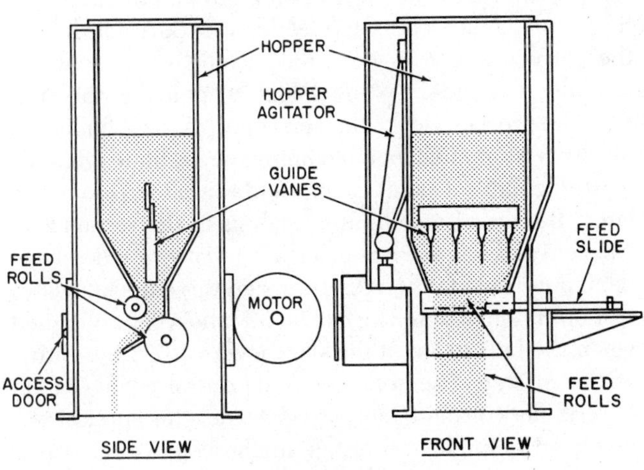

Fig. 2.17h Roll feeder

Revolving Plate Feeder

Revolving plate feeders (Figure 2.17i) consist of a rotating disk or table located beneath the hopper outlet. This (usually horizontal) table is driven by gears from above or below. As the table rotates, material is drawn from the hopper and is scraped off by the skirt board. Feed rate is controlled by adjusting the height of the adjustable gate or the position of the skirt board.

Revolving plate feeders handle both coarse and fine materials. Sticky materials are also handled satisfactorily because the skirt board is able to push them into the chute. This type of unit cannot handle materials that tend to flood. A variation of the revolving plate feeder utilizes rotating fingers to draw feed material from the bin. Revolving plate feeders can be equipped with arch-breaker agitators in the conical throat section of the hopper.

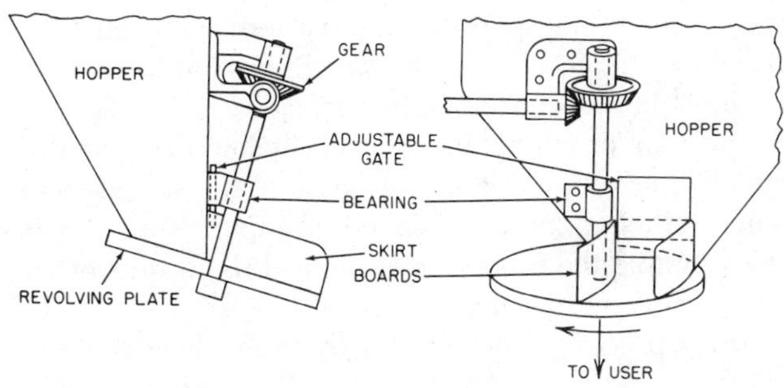

Fig. 2.17i Revolving plate feeder

Feed Rate Control

The feed rate of all belt-type gravimetric feeders is a function of the belt speed and the belt load. Belt speed is normally expressed in terms of feet per minute while belt load is defined as pounds per foot of belt.

$$\text{Rate} = \text{belt speed} \times \text{belt load} = \text{lbm/minute}$$

In the case of the constant speed belt feeders previously discussed, rate is directly proportional to belt load. Rate set-point is thus in terms of belt load, and the belt load signal generated by the device can be read out as rate.

Another method of rate adjustment utilizes belt speed variation with belt load a controlled constant. Still another involves variation of both belt speed and belt load wherein the rate signal is the multiplicant of the belt speed and belt load measurement signals generated by the feeder.

Figure 2.17j illustrates a standard constant speed belt feeder with pneumatic gate actuator. The length of the weight section and distance from end of weigh section to end of belt have been idealized to some extent but are approximately the same as those in an actual feeder. The response data illustrated in the curves have been simplified somewhat by disregarding the controller response. It is thus not precise since it assumes instantaneous gate response and does not include controller lags, but this is minor relative to the effect of belt transportation lag which is the major source of concern in the application of constant speed belt feeders.

The uppermost curve represents the belt load signal response of a feeder having a 12-feet-per-minute belt speed after a step change in belt loading. The curve immediately below it—as represented by a dashed line—represents the instantaneous feeder discharge rate leaving the end of the feeder belt; i.e., the feed rate as viewed by the downstream process. The effect of the change in belt loading is sensed almost immediately after the step change since the control gate is located at the upstream edge of the weigh section. At 12 fpm belt speed the weigh section has been completely covered by the material at the new rate in ⅛ minute after the step change. At this instant, however, the feeder is still discharging at the initial rate and an additional 1/24 minute is required to transport the material at the new rate from the downstream edge of the weight section to the end of the belt, a distance of 6 inches.

The pair of curves immediately below the pair discussed above describe the response of the same feeder when arranged for a belt speed of 2 fpm. Note that in this example the process does not feel the results of the step change until 1 minute after it is made.

Many processes, particularly those involving a single

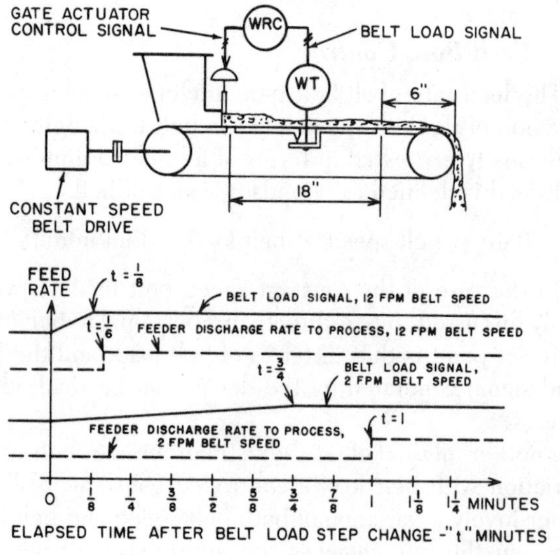

Fig. 2.17j Open loop response to belt load step change

feeder, can tolerate relatively long response times. In continuous blending operations, however, maintenance of instantaneous blend ratio is normally the major performance criterion, and it is in these situations that selection and application of constant speed feeders require careful study. If two feeders having belt speeds of 12 fpm and 2 fpm were controlled from a common belt load signal, the curves in Figure 2.17j indicate that the previously established ratio of materials entering the process would be disrupted for a period beginning 10 seconds after the belt load set-point change and continuing for a period of 50 seconds, at which time the original ratio would be restored.

The curves illustrate the basic fact that two or more constant speed gate feeders cannot maintain blend ratio after a common step change in belt load unless the belt speeds of all feeders are equal. Because of the normal feed rate variations encountered in blending applications, it is rarely possible to size a number of feeders for different rates so that they all have the same belt speed. It is true that if material flow characteristics permit, a narrow feed section can be utilized to decrease the width of the material ribbon on the belt and thus permit an increase in belt speed but, in most applications, this does not satisfactorily solve the problem.

Inspection of the belt load signal response curves in Figure 2.17j suggests the probability that maintenance of blend ratio can be improved by arranging the two feeders in a master-slave relationship wherein the belt load step change is applied to the gate actuator of the master feeder, and its belt load signal is applied to control the gate actuator of the slave feeder. The belt load signal and discharge rate response curves indicate that slaving the low speed feeder to the high speed feeder will only increase the duration of blend error. Because of transport time lag, the 2 fpm slave feeder will deliver material to the process at the initial feed rate for a period of one minute after the step change is applied to the 12 fpm feeder. Since the gate of the slave is controlled by the belt load signal of the master, the gate of the 2 fpm slave unit will not be at its new setting until ⅛ minute after the step change. The slave feeder will thus not deliver the corrected rate to the process until ⅛ + 1 or 1⅛ minutes after the set-point change is applied to the master. The process "sees" the correction of the 2 fpm slave feeder as a ramp change beginning one minute after the master feeder set-point change and completed ⅛ minute later. Results of a computer analysis of the response of two constant belt feeders in a master-slave situation have been published in the literature, and this study indicates that optimum maintenance of blend ratio will be obtained when the belt speed of the slave feeder is 1½ times that of the master. The selection and control configuration criteria for constant belt speed feeders in applications involving a single feeder, or in those utilizing two or more feeders for continuous blending, can be summarized as follows:

1. In single feeder applications, optimum response is obtained by utilizing the maximum possible belt speed commensurate with the characteristics of the material being fed and subject to the belt load limits established by the feeder manufacturer.
2. In applications involving two or more feeders for continuous blending when belt speeds of the feeders can be made equal, blend ratio will be maintained during a total throughput stepchange when the feeders are controlled in parallel from a common system rate setting signal. If the belt speeds are equal, the feeders should not be operated in a master-slave arrangement.
3. In blending applications utilizing constant speed belt feeders having different belt speeds, blend ratio cannot be exactly maintained during a total throughput change. Blend ratio error can be minimized, however, by arranging the individual feeders in a master-slave control relationship rather than in parallel and with the feeder having the lowest belt speed as the master. Minimum blend ratio disturbance can be obtained by proper belt speed selection such that the ratio of slave to master feeder belt speed is 1.5.

Belt Speed Control

The advantage of belt speed control over belt load control is that a feed rate change as viewed by the process is almost simultaneous with a change in belt speed set-point. The use of speed control in multi-feeder blending applications eliminates blend error due to differential transport lag which is typical in blending systems utilizing constant speed feeders. These systems are controlled via a common speed signal applied in parallel to all feeders and which increases or decreases the total throughput to the process. Ingredient ratio can be adjusted by varying either the belt loads of the individual feeders or by adjustment of individual speed signal ratio stations. The latter method is preferred in blending situations requiring ratio change while the system is operating; changing belt loading during operation will cause a blend error because of the transport lag between control gate and process. The use of a continuous integrator will provide an accurate totalization of the material fed regardless of changes in belt load or belt speed.

While the variable speed arrangement provides high rate response to set-point change and eliminates error in blending situations, it includes some undesirable features as far as the operator is concerned:

1. Unlike the constant speed feeder, the variable speed arrangement does not provide a readout of feed rate. Feed rate must be calculated by multiplying the speed set-point signal by the measured belt loading. In addition, the feed rate thus calculated is inferential since the device does not generate a signal proportional to actual belt speed.
2. If this configuration is used in a multifeeder blending system, a blend ratio change will require a change in belt loading or speed ratio set-point to one or more of the feeders. This will change the total throughput to the process unless a master speed adjustment is made to compensate.

What is needed, therefore, is an arrangement that can accept a single set-point proportional to feed rate, generate an output that maintains feed rate and at the same time controls the basis of speed change.

Figure 2.17k illustrates an arrangement that fulfills the above requirements. In the pneumatic version shown, the speed changing mechanism has a typical range of 10:1; electronic versions utilizing silicone controlled rectifier (SCR) drives provide a speed variation of at least 20:1. In the example shown, the feeder is equipped with a fixed gate for use in applications in which the material has a relatively constant density such that the speed actuator can accommodate variations in both density and gravimetric rate while operating within its working range. If the material density varies or if the feeder is to be used on a variety of materials having different bulk densities, such that the total range calculated by multiplying the required gravimetric range times density variations exceeds the range of the speed actuator, internal closed loop belt load control must be added.

As mentioned previously, weighing accuracy is optimized by utilizing the highest possible belt loading in which the ratio of live to dead load is maximum. The arrangement described here is usually sized by the manufacturer to handle the maximum gravimetric feed rate with the belt drive near maximum speed, the design belt loading at about 90 percent of maximum and based on

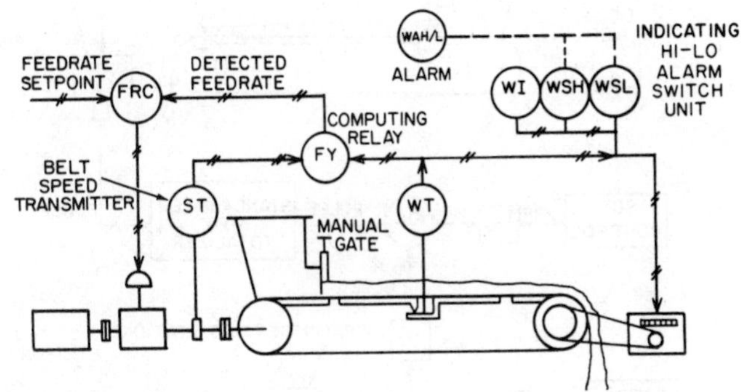

Fig. 2.17k Speed controlled belt feeder with feed rate setpoint and readout

the minimum expected material density. In order to assist the operator in setting the manual gate position, a belt load indicator is desirable. This is indicated in Figure 2.17k, combined with a pair of pressure switches in a commercially available unit. A belt load alarm is desirable to warn the operator in the event that unforeseen density variations dictate a readjustment of the feeder gate or that material flow to the belt has been interrupted.

Belt loading can also be measured by detecting the radiation absorption of a discrete length of material. A source of radiation is located under the belt and a pickup cell located above the material measures the radiant energy which has passed through the material. The difference between the energy transmitted by the source and that measured by the cell is a function of the mass of the material on the belt. The pickup cell converts the measured energy into an electric analog signal proportional to belt load. The balance of the feeder and control accessories is essentially the same as shown in Figure 2.17k.

Digital Control

A study of Figure 2.17k indicates that if the continuous integrator was provided with a feed rate transmitter, the belt speed transmitter and feed rate computer could be eliminated, and the feed rate signal from the integrator could be fed directly to the feed rate controller.

Figure 2.17l describes this arrangement which has been developed for use with commercial digital control systems. The control system is available in two basic arrangements; one for batching systems, the other for continuous systems.

In the batching version, the master oscillator in conjunction with a timer delivers a total number of pulses

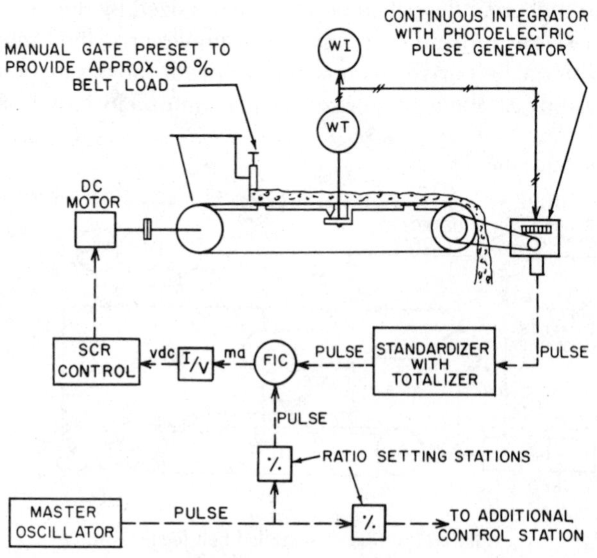

Fig. 2.17 l Belt-type gravimetric feeder with digital control

proportional to the total weight of material desired. The pulse frequency is adjusted to vary the duration of the batch preparation period. The pulses are applied as the set-point to the feed rate controllers via ratio setting stations for ingredient ratio. Pulses proportional to material feed rates are generated by photoelectric pulse generators driven by the feeder integrators and applied to the feed rate controllers via scaling and totalizing units. The controllers compare set and measured pulse frequencies and adjust feed rates as required by varying belt drive speed. The batch controller includes a memory feature by which the feeder is kept running until it has generated a total number of pulses equal to the total applied as a set-point to the controller via its ratio station. In a multifeeder batching system, this may result in feeders shutting down at different times but the batch blend ratio will be correct.

Continuous systems utilize another version of controller which includes a pacing feature. If one of the feeders experiences a decrease in feed rate which cannot be corrected by the controller, this controller gates the ouput of the master oscillator and thus paces down the feed rates of the other feeders to maintain blend ratio. If the faulty feeder corrects, the entire system is returned to control from the master oscillator. If the condition persists for a predetermined period, an alarm will be activated.

The digital control system is theoretically without error; the pulses generated by the master oscillator must be equalled by those generated by the integrator transmitter on the feeder. Laboratory evaluations confirmed by field tests have indicated that the feeder accuracy based on weighed samples versus total integrator pulses is better than 0.5 percent of feed rate over a 10 to 1 feed rate range.

These systems are utilized in situations involving a number of materials which must be blended in a wide variety of formulations which may be frequently changed. High accuracy, speed and ease of formula change, and centralized control characterize the digital control system. Although the cost of the feeder and its associated digital control is perhaps twice the cost of a feeder with conventional analog control, digital control systems are gaining acceptance in continuous blending systems, particularly in the food industry.

Feeder Accessories

Figure 2.17m illustrates a surge hopper which, when located between the storage hopper and feeder inlet, provides a means of dearating material and thus conditioning it so that it can be fed by a gate controlled belt feeder without flooding. The material supply to the surge hopper is controlled by bin switches to maintain a level zone in the hopper by on-off control of the hopper supply device. The supply control device may be a rotary vane

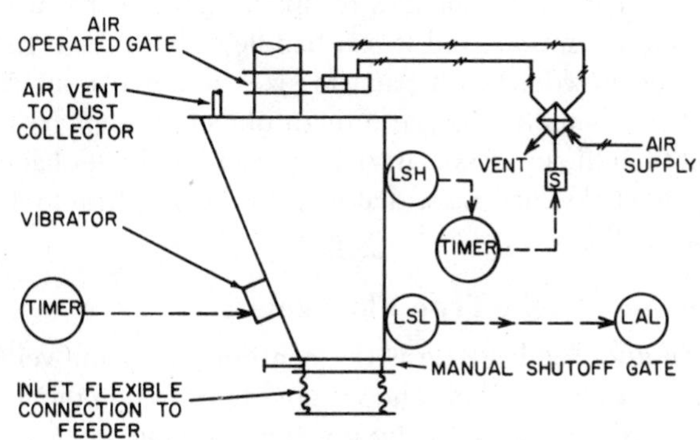

Fig. 2.17m Deaerating surge hopper

feeder, screw conveyor, or a knife gate with suitable actuator. If process requirements are such that the required feed rate is constant or nearly so, the bin switches are located to provide a hopper capacity equivalent to about 2 minutes retention time when operating at the design feed rate. If feed rates must be varied, it can be seen that the retention time will vary if the supply control device is directly actuated by the high and low bin switches. Excess retention time may be undesirable in some cases in which the material may compact in the hopper and interrupt the supply to the feeders. In situations where the feed rate will be varied, an adjustable timer is incorporated in the level control circuit to control the running time of the hopper supply device. This timer is controlled by the upper bin switch and shuts off the hopper supply when the material contacts the probe. The shutoff condition is maintained for a predetermined and adjustable period, after which the timer starts the hopper supply system which continues to feed material to the hopper until the upper bin switch level is reached. In this arrangement, the lower bin switch serves as a low level alarm actuator and is usually also utilized to shut down the feeder. Shutdown is usually desirable to prevent loss of the plug of material ahead of the belt feeder. With some materials, particularly those that aerate easily, loss of the plug of deaerated material can cause production to be delayed while steps are taken to re-establish the supply of deaerated material. Deaeration of some materials takes place in the surge hopper without the need of vibration. Other materials require some assistance through vibration, and hoppers can be furnished with suitable electric or pneumatic vibrators. The frequency and duration of vibration required will vary with the characteristics of the material and for this reason vibrators are supplied with accessory devices to permit adjustment of these variables.

A number of common materials, of which sulfur is an example, will compact unless kept in almost continuous motion, and in some instances will compact even while in motion under the pressure of a relatively low head of material. In these applications, small surge hoppers with bin switches are utilized to maintain a low head of material on the feeder belt. Retention time is in the order of seconds and external vibration is rarely if ever desirable.

The discharge flow pattern of a belt feeder varies with belt speed and material characteristics. A granular free-flowing material such as sugar will flow smoothly off the belt even at low belt speeds. Other materials having a high angle of repose coupled with a tendency to compact will drop off the end of the belt in lumps, especially at low belt speeds. The feed rate as viewed by the process is thus erratic and will result in short term blend error in a multifeeder system. The discharge flow pattern can be markedly improved by equipping the feeder with a material distributor. This device consists of a blade located across the full width of the belt at the discharge end of the feeder and vibrated by an electric or pneumatic vibrator. The blade is located so that it almost touches the belt and the material is directed across it. The vibration imparted to the material ribbon causes it to smoothly stream off the belt.

Virtually all manufacturers base their guarantee of performance on sample weight versus set-point in the case of feeders or sample weight versus totalizer reading in the case of the meters. Sample valves can be provided which facilitate performance checkout procedures. These are available either as a sample tray which can be inserted into the feeder discharge stream for a predetermined period and then weighed, or as a flap valve which diverts the discharge stream from the process duct into a sampling container. The flap-type is generally preferred since the tray-type is only suitable for low feed rates; sampling normally involves the taking of 10 consecutive 1 minute samples and comparing the average sample weight to the set-point. The flap-type is also faster acting than the tray-type and sample weights are thus more accurate.

Many installations involve feeding directly into processes which may be under low pressure or which may discharge corrosive vapors back through the feeder discharge ducting. If pressures are very low, the feeder can be purged with inert gas or a rotary valve can be installed in the ducting. The rotary valve body should be vented to remove process vapors from the valve pockets prior to their reaching the inlet or feeder discharge side of the valve. If the valve is not vented, blowback resulting from the release of pressure in the rotor pockets can cause discharge flow pattern disturbances and, in extreme cases, affect the feeder weigh section. The valve is vented into a dust or vapor collecting system via a vent port in the side of the valve rotor housing.

Each feeder or meter is usually supplied with a test weight or drag chain which may be used to check the calibration of the device without actually running material. The value of the weight is such that when applied to the weight sensing mechanism, a full scale output signal will be generated. The test weight is also useful in aligning the control set-points in multifeeder master-slave systems prior to running any material. With the test weight applied to the weighing section of the master feeder, the resultant output signal is applied to the ratio stations of the slave feeders. The ratio stations can thus be preset to provide the desired slave feeder set-points.

All manufacturers recommend that a feeder or meter be isolated from sources of vibration and some include shock mounts with each machine. Inlet and discharge flexible connections to isolate the equipment from vibration in the material inlet and outlet ducting are also available.

Vertical Gravimetric Feeder

The vertical gravimetric feeder is illustrated in Figure 2.17n. An agitator rotor within the supply bin guarantees a "live" bin bottom. The process material enters through a hole in the top cover of the pre-feeder and is swept through a 180 degree rotational travel by the rotor vanes, where it is dropped out through the discharge pipe. The solids are weighed together with the rotary weight feeder as it transports the solids to the outlet.

The advantages of this feeder include its convenient inlet-outlet configuration, its sealed, dust-tight design, and its self-contained nature wherein all associated control instruments are also furnished. After calibration, a ± 0.5 percent of full scale inaccuracy can be expected if a 5:1 rangeability is sufficient. At a 20:1 rangeability, the calibrated inaccuracy is ± 1 percent of full scale.

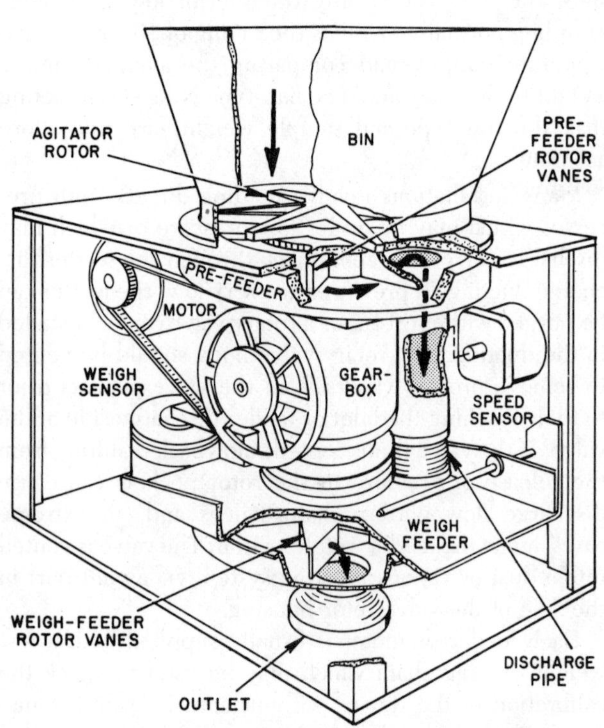

Fig. 2.17n Vertical gravimetric feeder

The main disadvantages of this design are that it has a limited capacity and it can handle only dry and free-flowing powders with particle sizes under 0.1 in. (2.5 mm) in diameter. Large foreign objects cannot be tolerated in the process material, nor can this feeder handle damp, sticky materials, that might cake or refuse to flow freely.

Loss-in-Weight-Type Flowmeters

Another feeder type is the continuous loss-in-weight feeder (Figure 2.17o). The weight of material in the hopper is counterbalanced by a retracting poise on a scale beam. The controller adjusts the speed of the rotary feeder to maintain a constant rate of poise retraction. This system can employ rotary screw or vibratory feeders as the modulated control elements. The poise is retracted by a lead screw along the scale beam at a preset rate,

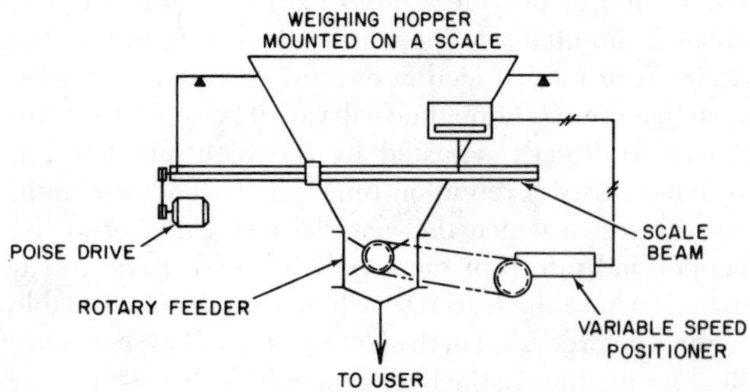

Fig. 2.17o Continuous loss-in-weight feeder

and any unbalance of the scale beam is counteracted by the controller by changing the rate of discharge.

In loss-in-weight systems, the weight sensing section of the system is a tank or silo rather than a horizontal belt surface open on all sides, and thus loss-in-weight feeding systems are suitable for handling liquids and slurries as well as solids. Manufacturers of such systems claim that their systems provide greater accuracy of delivery over short time increments than other continuous feeding systems because the weight sensing function is located ahead of the material discharge regulating device; that is, if a rate error exists, it is corrected before the material enters the process. These systems are available with either electronic or pneumatic weight sensing and control functions.

Operation

A material hopper or supply tank is supported by or suspended from one or more load cells. In order to eliminate side effect errors due to non-symmetrical hopper loading, tension-type cells are preferred. The weight signal generated by the load cell system is applied as a measured signal to a weight rate controller. The set-point of the controller is the output of a loss-in-weight signal generator or programmer. The difference between the weight of material in the hopper and the programmed weight is continually sensed and the flow rate of the material exiting from the hopper is regulated to counteract any deviations. In other words, the programmer generates a signal corresponding to a fixed reduction rate of the total weight in the hopper and this signal becomes the set-point for the controller.

The supply of material in the hopper must be replenished from time to time, and the filling cycle is initiated before the hopper is completely empty. A supply of material referred to as the "heel" always remains in the hopper and serves to minimize the effects of filling shock on the weight sensing cells and the material exit flow regulator. The filling operation is controlled by a separate subsystem which consists of a differential gap controller and a material supply valve, gate, or feeder. When the weight of material in the hopper reaches a predetermined low value equal to the desired weight of "heel," the differential gap controller starts the filling cycle and at the same time locks the weight controller output signal into the actuator of the material exit flow regulating device. When hopper weight reaches a filled condition, the differential gap controller stops the filling cycle and starts the feeding cycle by returning control of the material exit flow regulator to the loss-in-weight control system. It should be noted that the feeding system is operating on a volumetric rather than gravimetric basis during the filling cycle; hence filling is accomplished as rapidly as possible. It is desirable to design the system such that the filling cycle is a small portion of the total cycle time.

Equipment

The loss-in-weight programmer and weight controller are normally incorporated in a single package suitable for panel mounting. Accessories include loss-in-weight set-point indicator, set/measured deviation indicator, optional manual controls to regulate filling and feeding cycles, and alarm circuits to detect discrepancy between programmed and actual loss-in-weight, as well as malfunction of the automatic refilling cycle.

Load cells are sealed to withstand dust and corrosion, and include temperature compensation. Hermetically sealed cells also include compensation for barometric pressure changes. Because of the probability of shock loading, cells are normally designed to withstand overloads to 150 percent of rating or more. Since they are voltage sensitive, strain gauge load cell power supplies should be closely regulated and compensated for supply voltage variations. As mentioned previously, compression-type strain gauge load cells are generally sensitive to side load forces generated by non-symmetrical hopper loading or thermal expansion of the hopper and load transfer structure. For these reasons, tension-type cells are normally preferred.

Weigh hoppers can be built in a variety of materials of construction and are often supplied by the user rather than by the feeding system vendor. Maximum sensitivity of the weighing system will be obtained when the ratio of live load to tare is maximum. For this reason, the hopper or tank must be designed not only on the criteria of capacity and structural strength but also for minimum weight.

The material exit flow regulator may be a control valve if the material is a liquid or slurry. Solids may be controlled by a rotary vane feeder, belt feeder, vibrating feeder or a knife gate valve with positioner. The selection of the specific regulator to be utilized is based on required feed rate as well as on the physical characteristics of the material in question.

System Sizing Criteria

Design of these systems involves selection of the proper exit flow regulating device, but perhaps the most critical item is the hopper or tank. Theoretically, the hopper should be as large as possible commensurate with the space available to accommodate it. The larger the hopper, the longer the running cycle and the less frequent the filling cycle. It is obvious, however, that the required load sensitivity of the weighing mechanism must be increased as the hopper weight increases. For a given feed rate (loss-in-weight rate), the system accuracy will decrease as the weight of the material hopper and its contents increase. At least one manufacturer of loss-in-weight systems has established some basic design criteria. These state that the hopper should be sized to hold approximately 15 times the maximum pounds per minute flow rate. The differential gap controller is set to provide a feeding supply equivalent to ⅔ of the total supply thus leaving ⅓ of the total hopper capacity as the "heel." The refilling cycle is preferably accomplished in one minute or less.

Because of their total cost relative to cost of other continuous gravimetric feeding systems, loss-in-weight systems are generally not used to control the weight rate of easy-to-handle, free-flowing solids.

The loss-in-weight systems described in this section are not truly continuous weight rate control systems because the gravimetric rate control is interrupted during the refill cycle. Because of this, it also follows that accurate totalization is not possible, although optional counters are available to indicate the number of times that the hopper supply has been replenished.

Loss-in-weight systems are considered to be most useful for services involving hard-to-handle liquids and slurries. Flow rates of liquids which are highly viscous, nonconductive, corrosive, or contain abrasive solids cannot in many instances be measured or controlled by proportioning pumps or conventional flow control systems. In these instances, gravimetric control on the basis of loss-in-weight has been utilized and many highly satisfactory systems have been installed.

Accelerator-type Flowmeter

As illustrated in Figure 2.17p, the solids stream enters the "accelerator" section of the meter by gravity. The accelerator is driven by an electric motor at constant speed. As the flow stream is accelerated, it causes a torque on the motor. The change in torque is sensed by a torque transducer. The amplified pneumatic transmission signal is thus directly proportional to the mass flow rate of solids and can be used as the input to any pneumatic receiver instrument.

This flow sensor error is ±½ percent and can detect flow rate over a range of 25 to 1.

The unit is designed for use on a wide range of materials including powders, granules, pellets, and irregular solids as well as liquid slurries.

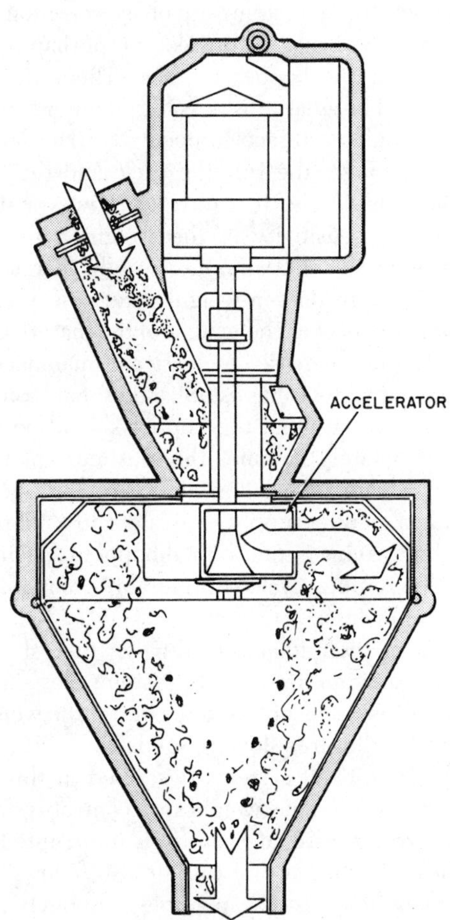

Fig. 2.17p Accelerator-type solids flowmeter

Volumetric Flowmeters

Positive displacement screw impellers provide volumetric flow measurement, but have to be used on uniform size solids, such as lead shot, if any reasonable accuracy is to be obtained. This type of instrument is similar to the turbine- or propeller-type flow meter used for volumetric liquid flow measurement. In the solids flow meter, a helical vane is used instead of the turbine. As the vane rotates, driven by the flow of granular material, a flexible cable transmits the rotation to a counting mechanism mounted outside of the pipe or duct (see Figure 2.17q). This counter can be a mechanical counter mounted directly against the piping, or it can be an electrical or pneumatic transmitter so that flow rate and/or total flow can be recorded at a remote location. In the transmitter, a rotary motion produced by a synchronous motor opposes the motion brought in by the flexible cable. This balancing of motions is utilized to position a slotted cam which determines the transmitter output signal.

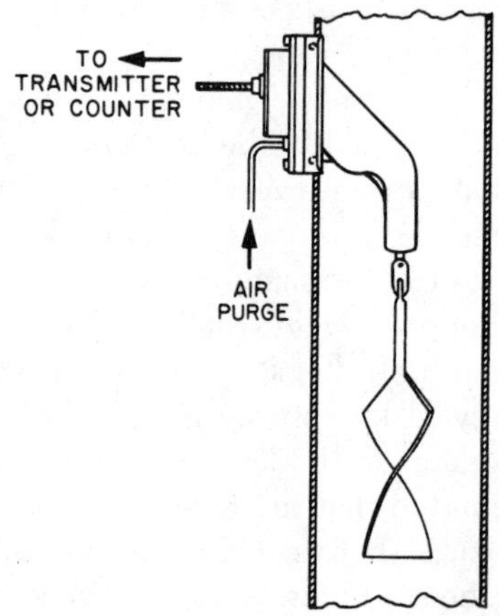

Fig. 2.17q Volumetric solids flow detector

The vane is normally installed in a vertical position and its bearing surfaces are protected with an air purge. In order to obtain accurate solids flow measurement, the instrument must be calibrated with the material for which its final use is intended.

This unit is capable of detecting volumetric solids flows with ±3 percent full scale accuracy if the flow rate is between 10 and 100 percent of design rate.

Impulse Flowmeter

The principles of impulse and momentum have long been used in liquid flowmeters such as the target or drag body flowmeters and the angular momentum flowmeter. They are based on Newton's second law of motion and the concept of conservation of momentum. These principles have now been successfully applied to solids flow measurement. The solid particles are allowed to fall by gravity on a calibrated spring-loaded resistance, the displacement of which, caused by the force of the falling particles, is a function of the mass flow rate of the solids. It is measured with a position transducer or transmitter. Figure 2.17r shows this type of instrument schematically.

These solids flow transmitters have a variety of appli-

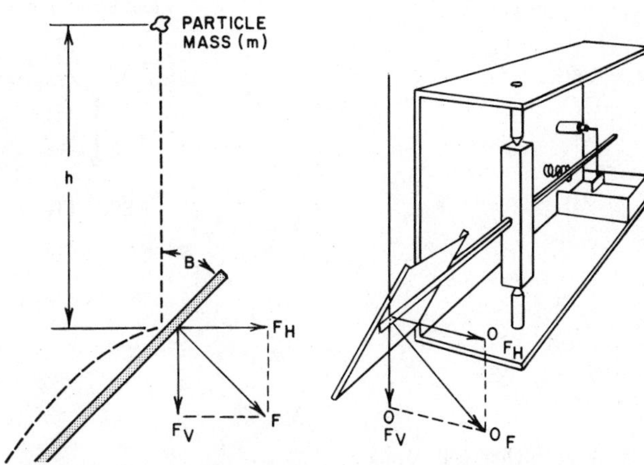
PARTICLE
MASS (m)
h
B
F_H
F_V
F
O
F_H
O
F_V
O F
THE HORIZONTAL COMPONENT OF THE IMPACT FORCE ON THE PLATE IS DIRECTLY PROPORTIONAL TO THE FLOW RATE OF MATERIAL OVER THE PLATE.

Fig. 2.17r Impulse flowmeter

cations such as continuous weighing, flow control or recording, or both, batching and alarming. Almost all conceivable types of solids can be measured, such as sugar, salts, cement, ores and the like.

BIBLIOGRAPHY

AWWA Standard for Quicklime and Hydrated Lime. New York: American Water Works Association, 1965.

Beck, M.S. and Plaskowski, A., "Measurement of the Mass Flow Rate of Powdered and Granular Materials in Pneumatic Conveyors using the Inherent Flow Noise," *Instrument Review*, Nov. 1967.

Colijn, H. and Chase, P.W., "How to Install Belt Scales to Minimize Weighing Errors," *Instrumentation Technology*, June 1967.

Grader, J.E., "Controlling the Flow Rate of Dry Solids," *Control Engineering*, March 1968.

Jenicke, A.W., "Storage and Flow of Solids," Bulletin 123. Utah Engineering Experiment Station. University of Utah, Salt Lake City, Utah, 1964.

Johanson, J.R. and Colijn, H., "New Design Criteria For Hoppers and Bins," *Iron and Steel Engineering* 41:85, Oct. 1964.

Nolte, C.B., "Solids Flow Meter," *Instruments and Control Systems*, May 1970.

"Solids Flowmeter Works Without Obstructing Flow," *Chemical Engineering*, September 18, 1972.

Stepanoff, A.J., *Gravity Flow of Bulk Solids and Transportation of Solids in Suspension.* New York: John Wiley & Sons, Inc., 1969.

Zanetti, R.R., "Continuous Proportioning for the Food Industry," *Instrumentation Technology*, March 1971.

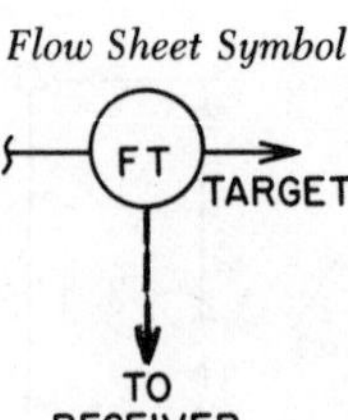

2.18 TARGET METERS

Design Pressure:	Up to 1500 PSIG (10.35 MPa), higher with strain gauge-type
Design Temperature:	Up to 750°F (400°C) with force balance and to 300°F (150°C) through 600°F (315°C) with strain gauge-type
Sizes:	½ to 8 in. (12.5 to 203 mm) with standard and up to 48 in. (1.2 m) pipes with the probe design
Fluids:	Liquids and gases
Flow Range:	From 1 GPM (3.785 l/m) or 1 SCFM (28 l/m) to practically any value using the probe design
Inaccuracy:	± ½ to ± 5 percent full scale
Materials of Construction:	Usually either carbon or stainless steel
Cost:	\$1,200 to \$3,700 as a function of size, design and materials of construction
List of Suppliers:	Delta Controls Corp., Automation Div.; Foxboro Co. (Target Meter, a proprietary flow meter); Ramapo Instrument Co., Inc. (Drag Body Flowmeter); West Coast Research Corp.

The target meter (Figure 2.18a) combines in a single unit an annular orifice and a force-balance transducer. Output is either an electric or pneumatic signal proportional to the square of the flow. Target meters are available in sizes from ½ to 8 in. (12.5 to 203 mm) pipe diameter. The annular orifice is formed by a circular disc supported in the center of a tubular section having the same diameter as schedule 80 pipe of the same nominal size. Flow through the open ring between disc and tube develops a force on the disc proportional to velocity head (the square of the flow). The disc is mounted on a rod passing out through a flexible seal. The force on the disc is measured from the rod outside the seal, using a standard force-balance transducer integrally mounted on the flow tube.

A similar operating principle applies to the drag body flowmeter (Figure 2.18b) which detects the impact forces by a strain gauge circuitry. This unit is also available in rectractable probe designs used in larger pipe sizes where it is desirable to withdraw the sensor periodically for cleaning, without opening the process line.

The target meter is applied in a number of fields for measurement of liquids, vapors and gases. It allows unimpeded flow of condensates and extraneous material along the bottom of a pipe and at the same time allows unimpeded flow of gas or vapor along the top of the pipe. It has given consistent dependable service on "difficult" measurements such as hot, tarry, sediment-bearing fuels to a pipe still where no other head-type meter has proved successful. There are no differential pressure connections to "freeze." This is useful in steam flow measurement in exposed locations, and for liquids that congeal at ambient temperature in pressure connections. Units are available for service up to 700°F (371°C).

Targets with diameters of 0.6, 0.7, and 0.8 times tube diameter are available. Combined with wide range force measurement transducers, a wide selection of full scale flow rates is provided.

In addition to providing an open flow both at the top and at the bottom of the pipe, the annular orifice has the advantage of being less sensitive to change in Reynolds number and to various upstream piping configurations than a concentric sharp-edged orifice. This is apparently because the main flow is directed outward toward the boundary layer. Greatest improvement occurs at 0.8 diameter ratio.

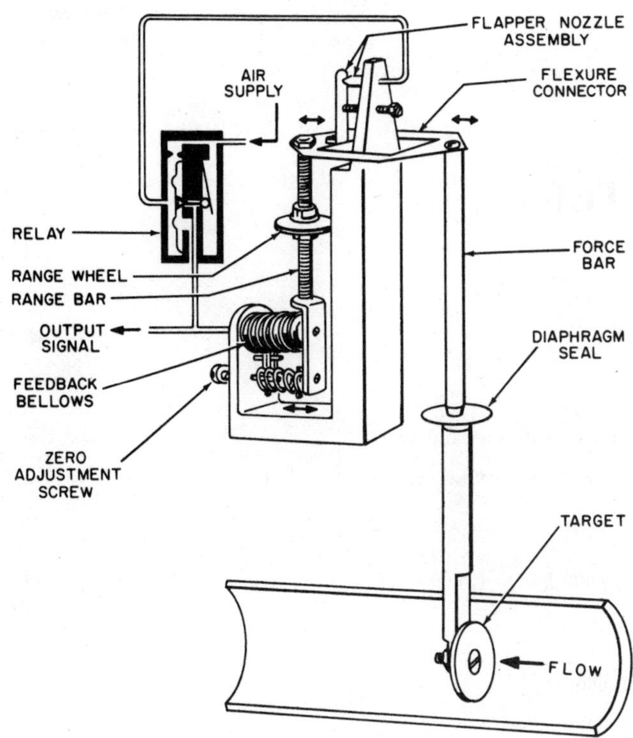

Fig. 2.18a Target meter

The important advantage of a single unit combining primary element and transducer also imposes a limitation. The transducer can be zeroed only under zero-flow condition. If zeroing under flowing conditions is desired, a bypass pipe and suitable valves are required. Modern force-balance transducers are sufficiently stable so that zero-check during flow is seldom demanded.

Calibration and Accuracy

Calibration data are provided by the manufacturers. A very wide choice of full scale flow values for each pipe size is available by selection of target size, transducer and by transducer adjustment.

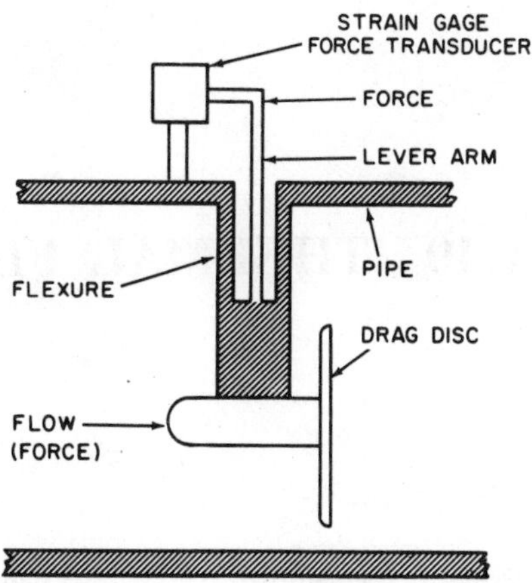

Fig. 2.18b The drag body flowmeter

Repeatability of output has proven excellent. Calibration accuracy includes not only the uncertainty of the primary element but also the characteristics of the transducer and the precision of the transducer adjustment. As with other proprietary devices, test data for determination of coefficient from physical dimensions over a wide range of fluids and operating conditions simply do not exist. Target meters with very accurate water flow calibration over almost any required range of Reynolds number can be obtained. Transfer characteristics to other fluids based on Reynolds number are highly reliable. Since calibration includes the characteristics of the transducer as well as the primary element, overall accuracy of flow calibrated target meters compares favorably with that of any other head metering system.

BIBLIOGRAPHY

Blasso, Leonard, "Flow Measurement Under Any Conditions," *Instruments and Control Systems*, February, 1975.

Hall, John, "Solving Tough Flow Monitoring Problems," *Instruments and Control Systems*, February, 1980.

Spink, L.K., *Principles and Practices of Flow Engineering*, The Foxboro Company, 9th edition.

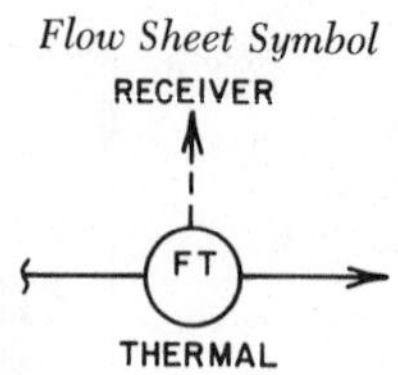

2.19 THERMAL FLOWMETERS

Design Temperature:	Up to 450° F (232° C), higher with special designs
Design Pressure:	Up to 1200 PSIG (8.3 MPa), higher with special designs
Fluids:	Gases or liquids
Flow Range:	From 0.5 cm^3/min to 40,000 lbm/hr (18,000 kg/hr) maximum
Inaccuracy:	±2% of full scale
Materials of Construction:	Stainless steel, glass, Teflon, etc.
Cost:	$3000 to $6000 per transmitting loop
Partial List of Suppliers:	Brooks Instrument Div., Emerson Electric Co.; Gould, Inc., Fluid Components Div.; Technology, Inc., Instruments and Controls Div.; Teledyne Hastings-Raydist; Thermal Instrument Co.; TSI, Inc.; Tylon Co. NOTE: For suppliers of thermal flow switches, see Section 2.6; for suppliers of thermal anemometers, see Section 9.1

Thermal flowmeters can be divided into two categories:

1. Flowmeters that measure the rise in temperature of the fluid after a known amount of heat has been added to it. They can be called heat transfer flowmeters.
2. Flowmeters that measure the effect of the flowing fluid on a hot body. These instruments are sometimes called hot wire probes or heated-thermopile flowmeters.

Both types of flowmeters can be used to measure flow rate in terms of mass, a very desirable feature especially on gas service.

Heat Transfer Flowmeters

The theory of the heat transfer flowmeter is based upon the specific heat equations.

$$Q = Wc_p (T_2 - T_1) \qquad 2.19(1)$$

where

Q = heat transferred (BTU/hour)
W = mass flow rate of fluid (lbm/hour)
c_p = specific heat of fluid (BTU/lbm°F)
T_1 = temperature of the fluid before heat is transferred to it (°F).
T_2 = temperature of the fluid after heat has been transferred to it (°F)

Solving for W we get

$$W = \frac{Q}{C_p(T_2 - T_1)} \qquad 2.19(2)$$

A simple flowmeter based upon this theory can be built in the plant and is shown schematically in Figure 2.19a. Heat is added to the fluid stream with an electric immersion heater. The power to the heater equals the heat transferred to the fluid (Q) and is measured by a wattmeter. T_1 and T_2 are thermocouples or resistance thermometers. Since we know the fluid, we also know the value of its specific heat. Thus, by measuring Q, T_1 and T_2 we can calculate flow rate (W). T_1 and T_2 do not have to be recorded separately; they can be connected together so that the temperature difference $(T_2 - T_1)$ is recorded directly.

A flowmeter of this type of construction has many problem areas. The temperature sensors and the heater

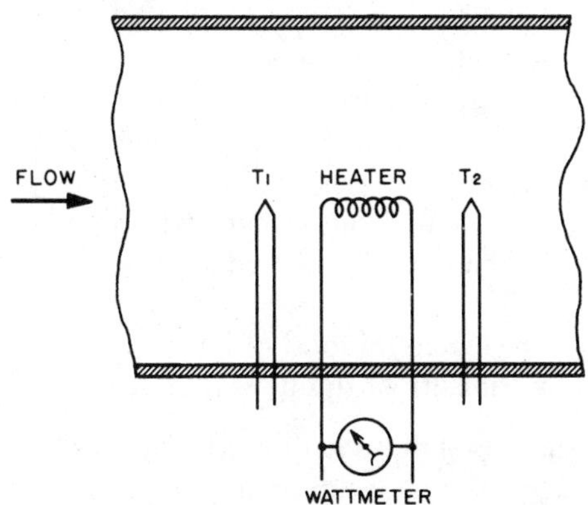

Fig. 2.19a Heat transfer flowmeter

must protrude into the fluid stream. Thus, these components (particularly the heater) are easily damaged by corrosion and erosion. Furthermore, the integrity of the piping is sacrificed by the protrusions into the fluid stream, increasing the danger of leakage.

To overcome these problems, the heater and the upstream and downstream temperature sensors can be mounted outside of the piping (see Figure 2.19b). In this type of construction the heat transfer mechanism from the heater to the fluid stream becomes more complicated and the relationship non-linear.

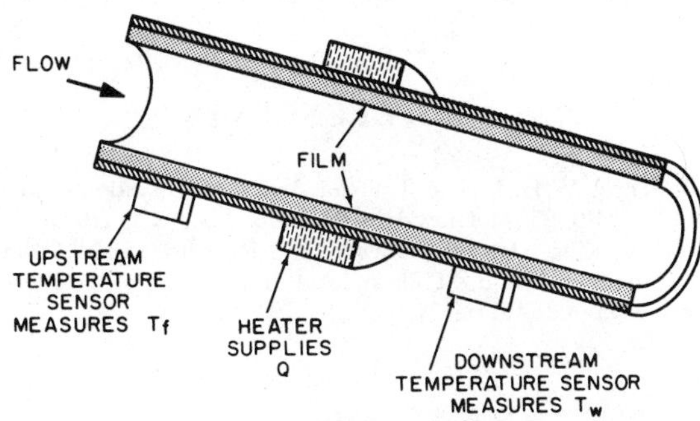

Fig. 2.19b Thermal flowmeter with external elements and heater

To understand the principle of operation of this flowmeter we must review two concepts of fluid mechanics and heat transfer. When a fluid flows in a pipe (turbulent or laminar) a thin layer (film) exists between the main body of the fluid and the pipe wall. When heat is passing through the pipe wall to the fluid, this layer offers a large resistance to the flow of heat and it must be considered in any heat transfer calculations. Now, if we insulate the heater sufficiently, and if the piping material is a good heat conductor, the heat transfer from the heater to the fluid can be expressed by the following equation:

$$Q = hA(T_{wall} = T_{fluid}) \qquad 2.19(3)$$

where

h = film heat transfer coefficient [BTU/(hr × ft² × °F)]

A = area of pipe through which heat is passing (ft²)

T_{wall} = temperature of wall (°F)

T_{fluid} = temperature of fluid (°F)

The film heat transfer coefficient h can be defined in terms of fluid properties and tube dimensions—one for laminar and another one for turbulent flow (reference 1).

$$h_{turbulent} = \frac{0.023K^{0.6}c^{0.4}W^{0.8}}{D^{1.8}\mu_1^{0.4}} \qquad 2.19(4)$$

$$h_{laminar} = \frac{1.75K^{0.67}c^{0.33}W^{0.33}}{DL^{0.33}} \qquad 2.19(5)$$

where

K = thermal conductivity of the fluid [BTU/(hr × ft × °F)]

c_p = specific heat of the fluid [BTU/(lbm × °F)]

D = pipe diameter (ft)

L = heated length (ft)

μ_1 = absolute viscosity of the fluid (lbf/hr-ft²)

W = flowrate (lbm/hr)

Using the turbulent flow condition as an example, let us solve equations 2.19(3) for "h" and 2.19(4) for "W"

$$h = \frac{Q}{A(T_{wall} - T_{fluid})} \qquad 2.19(6)$$

$$W^{0.8} = \frac{hD^{1.8}\mu_1^{0.4}}{0.023K^{0.6}c^{0.4}} \qquad 2.19(7)$$

Substituting equation 2.19(6) into 2.19(7) we get

$$W^{0.8} = \frac{QD^{1.8}\mu_1^{0.4}}{0.024K^{0.6}c^{0.4}A(T_{wall} - T_{fluid})} \qquad 2.19(8)$$

For any given fluid μ_1, K, and C will stay constant over a certain temperature range. D and A are constants for any given flowmeter and Q can he held constant. Letting all these constants = X, we get

$$W^{0.8} = \frac{X}{T_{wall} - T_{fluid}} \qquad 2.19(9)$$

The downstream temperature sensor is located near the heater so that it measures T_{wall}. The upstream temperature sensor is located where the wall and fluid temperatures are in equilibrium with each other. Thus, flow rate is obtained by measuring ΔT, knowing the geometry of the flowmeter, knowing thermal conductivity and thermal capacity and viscosity of the fluid, and keeping heater power constant. This type of flowmeter can also be operated by keeping the ΔT constant and measuring the power to the heater that is required.

When building and/or using a flowmeter based upon heat transfer principles, the instrument engineer must tread with caution. He must be sure that the heat is transferred and the fluid is flowing according to the mechanisms for which he is using various equations. He must ascertain that he is using the correct values for the various fluid properties. He must also understand that many relationships such as those given in equations 2.19(4) and 2.19(5) have very definite limitations such as range of Reynolds number, L/D ratio, etc. for which they are reliable. Finally, it is highly recommended that this type of instrument should be calibrated, either by the manufacturer or by the user, under conditions that duplicate as nearly as possible its final application.

Hot Wire Probes

The basic operating principle upon which flowmeters of this type are based is as follows. Two thermocouples are connected in series to form a thermopile. This thermopile is heated by passing an alternating current through it. A third thermocouple is placed in the direct current output circuit of the thermopile. Alternating current does not pass through this thermocouple and it is therefore not electrically heated. This assembly is now placed into the fluid (usually gas) stream. The gas will cool the heated thermopile by convection. Since the AC input power to the thermopile is held constant, the thermopile will attain an equilibrium temperature and produce an emf that will be a function of the temperature of the gas, velocity of the gas, and its density, specific heat, and thermal conductivity. The third thermocouple (not heated) will attain the ambient temperature of the gas, generating an emf that is proportional to the gas temperature and which cancels the effect of the ambient gas temperature on the output signal of the heated thermopile. A schematic of this type of flowmeter is shown in Figure 2.19c. A and B are the heated thermocouples, C is the unheated one.

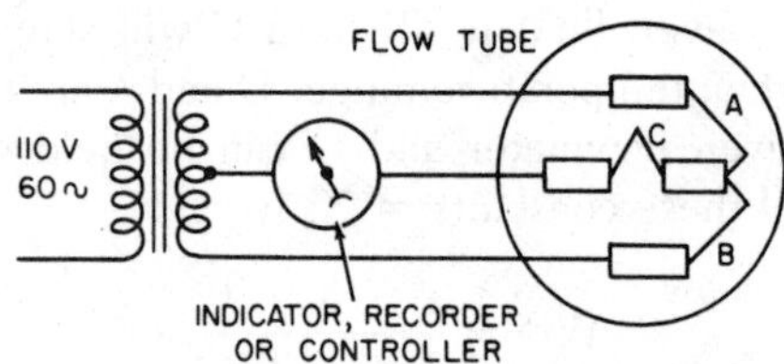

Fig. 2.19c Hot wire flow sensing probe

The output signal (voltage) of this instrument is given by the equation given below, as derived by King (reference 2).

$$e = \frac{C}{2(\pi K c_p \rho d v)^{1/2} + K} \qquad 2.19(10)$$

where

e = voltage generated

C = instrument constant

K = thermal conductivity of fluid (BTU/hr-ft-°F)

c_p = specific heat of fluid (BTU/lbm-°F)

d = diameter of heated thermocouple wire (ft)

v = velocity of fluid (ft/hr)

ρ = density of the fluid (lbm/ft³)

A closer analysis of this equation will show us that once the instrument has been calibrated for a certain gas, any change in gas temperature will have little effect on the gas properties, and thus on the output signal. For example, let us look at the properties of air, over a wide range of temperatures, as they apply to equation 2.19(10).

Temp (°F)	K (BTU/hr-ft-°F)	c_p (BTU/lbm°F)	ρ (lbm/ft³)	$(Kc_p\rho)^{1/2}$
70	0.0150	0.243	0.0753	0.0165
500	0.0246	0.245	0.0416	0.0159
1000	0.0359	0.263	0.0274	0.0161

Since K (in the denominator) is very small, and since the term $(Kc_p\rho)^{1/2}$ remains constant over a wide range of temperatures, this type of instrument can be used to measure the mass flow rate of gases.

REFERENCES

1. McAdams, W.H., "Heat Transmission" Third Edition, equations 9–10a and 9–23, McGraw-Hill Book Co. Inc., N.Y. (1954)
2. Louis V. King, "On the Convection of Heat from Small Cylinders in a Stream of Fluid," Philosophical Transactions of the Royal Society of London. Series A, Vol 214 (1914) pp 373–432

BIBLIOGRAPHY

Baker W.C., "Flow Without Fouling," *Measurements & Data*, Nov.–Dec., 1974.

Hayward, A.J., "Choose the Flowmeter Right for the Job," *Processing Journal*, 1980.

Laskaris, E.K., "The Measurement of Flow," *Automation*, 1980.

LeMay, D.B., "A Practical Guide to Gas Flow Control," *Instruments & Control Systems*, September, 1977.

Lomas, D.J., "Selecting the Right Flowmeter," *Instrumentation Technology*, 1977.

Watson, G.A., "Flowmeter Types and Their Usage," *Chartered Mechanical Engineer Journal*, 1978.

Flow Sheet Symbol

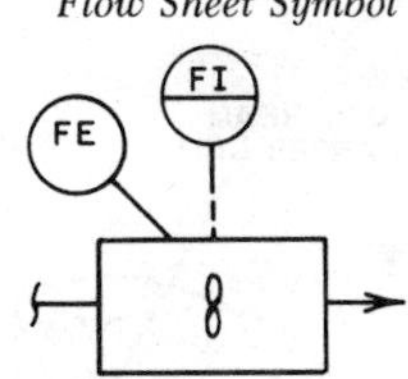

2.20 TURBINE FLOWMETERS

Sizes:	3/16 to 24 in. (4.7 to 610 mm)
Types:	Designs for liquid, gas, very low flow rates and insertion
Fluids:	Relatively clean liquids (special designs for gas)
Output Signal:	Linear frequency
Operating Pressure:	1500 PSIG (10.3 MPa) as standard; 5000 PSIG (34.5 MPa) as special
Operating Temperature:	−58 to +300°F (−50 to +150°C) as standard; −328 to +840°F (−200 to +450°C) as special
Materials of Construction:	Stainless steel with selected bearing materials; alternative materials available
Linearity:	± 0.25% of actual flow
Repeatability:	± 0.02% of point
Range:	10:1
Cost:	$2500 for 1 in. (25mm) diameter complete with flow rate indicator and totalizer, to $6000 for 8 in. (203mm) diameter complete with flow rate indicator and totalizer
Partial List of Suppliers:	Barton ITT Process Instruments & Control, Inc.; Bopp & Reuther GmbH; Brooks Instruments, Div. of Emerson Electric Co.; C-E Invalco, Div. of Combustion Engineering; Cox Instruments, Div. of Lynch Co.; Daniel Industries, Inc.; Faure-Herman; Flow Technology, Inc.; Fischer & Porter Co.; Foxboro Co.; Kent Process Control, Inc.; Simmonds Precision; Smith Meter Systems, Div. of Geosource, Inc.; Systecon, Div. of Corporate Equipment Co.

Introduction

Turbine meters are available for liquid, gas and very low flow rates in both full bore and insertion designs. The most widely used type is the full bore meter for liquid duty.

Operating Principles—Liquid Turbine Meters

A turbine meter consists of a multi-bladed rotor suspended in the fluid stream on a free running bearing (see Figure 2.20a). The axis of rotation of the rotor is perpendicular to the flow direction and the rotor blades sweep out virtually to the full bore of the meter. The fluid impinging on the rotor blades causes the rotor to revolve. Within the linear flow range of the meter, the angular speed of rotation is directly proportional to the volumetric flow rate. The speed of rotation is monitored by an electromagnetic pickup coil, which is fitted to the outside of the meter housing. Two types of pickup coil are primarily used: reluctance and inductance. Both operate on the principle of a magnetic field moving through a coil.

In the reluctance pickup coil system, the permanent magnet is the coil. The field produced is concentrated to a small point by the cone (see Figure 2.20b). The turbine rotor blades are made of a paramagnetic material, i.e., a material that is attracted by a magnet. As a blade

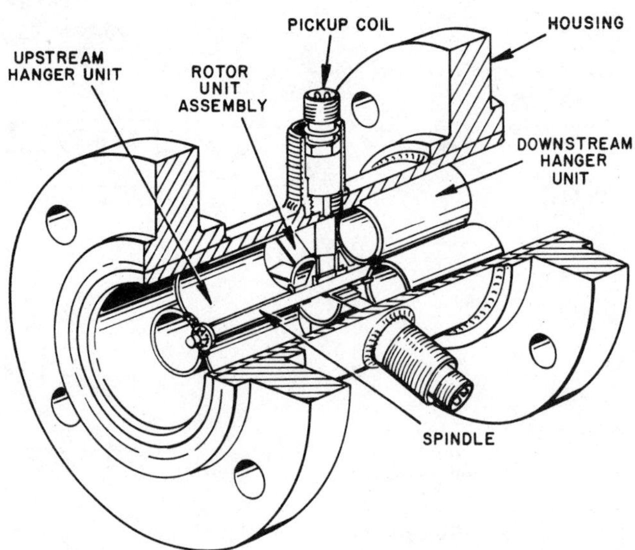

Fig. 2.20a Cutaway view of a typical turbine meter

approaches the cone point, its magnetic properties deflect the magnetic field. This deflection causes a voltage to be generated in the coil. As the blade passes under the cone point, the voltage decays, only to be built back up in the opposite polarity as the departing blade deflects the magnetic field in the opposite direction. Thus, each blade produces a separate and distinct voltage pulse as it passes the cone. Since each blade sweeps a discrete volume of fluid, each electrical impulse represents the same discrete volume of fluid.

With the inductance pickup coil system (see Figure 2.20b), the permanent magnet is embedded in the rotor. As the magnet rotates past the pickup coil position, it generates a voltage pulse for every complete revolution of the rotor.

The typical operating temperature range for standard pickup coils is −58 to +300°F (−50 to +150°C). Specially modified pickup coils are available, however, to cover operation at temperatures ranging from −328 to +840°F (−200 to +450°C). If the meter is located in a hazardous area, the pickup coil can be mounted in a flameproof or explosion-proof conduit box, or alternatively an intrinsically safe pickup coil can be used in conjunction with zener barrier to provide an inherently safe system.

Electronic Display Units

The output signal from the turbine meter is a continuous sine wave voltage pulse train with each pulse representing a small discrete volume of fluid. Associated electronic units then display total volumetric flow or flow rate and perform preset batching, control, automatic temperature correction and other functions.

Most turbine meter systems incorporate a totalizer unit with a factorizing and scaling function. The pulse output from the turbine meter is not in direct engineering units. For example, each pulse might represent 0.001231 gallons. The factorizer is set to this value and the incoming pulses are multiplied by 0.001231. The display presented is then in gallons.

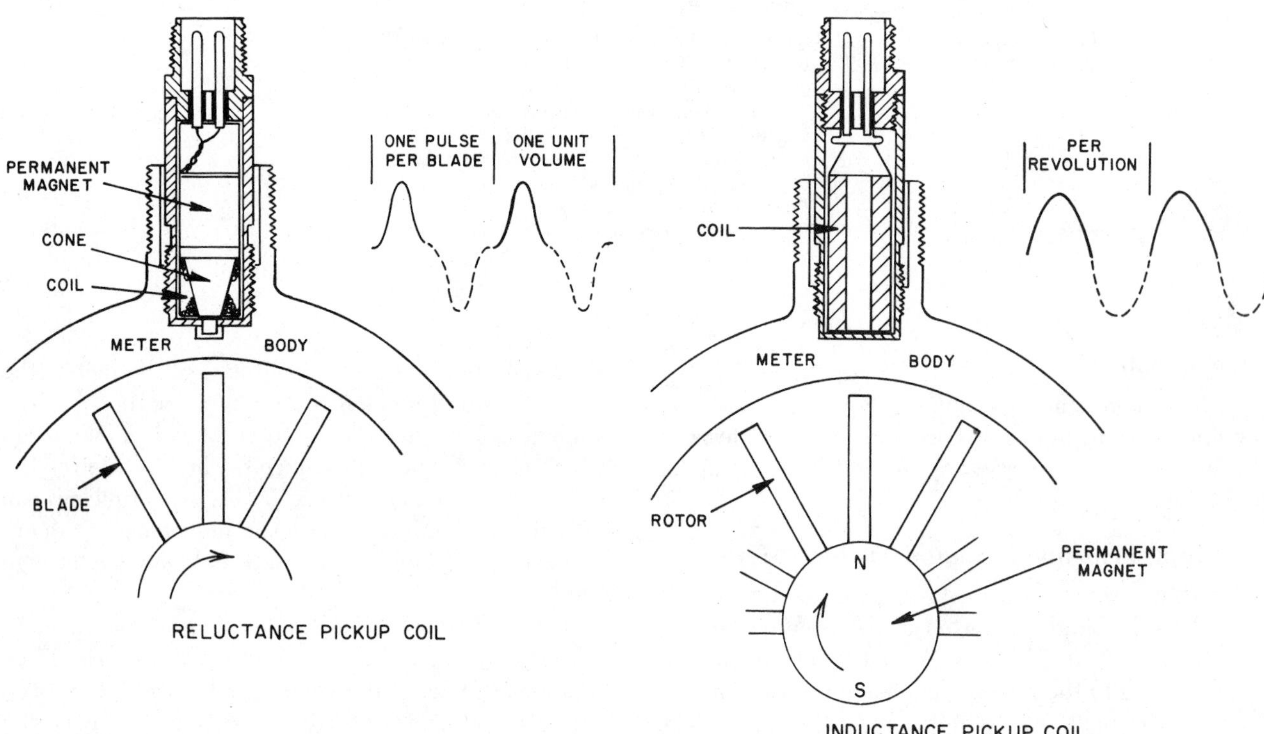

Fig. 2.20b Alternative signal generation systems

Alternatively, the totalizer can be a preset batch unit for automatically dispensing predetermined quantities of liquid. The required value is preset and the totalizer then counts down to zero and provides an output—that is, contact closure—to operate a valve and terminate the batch. In order to provide better system repeatability and avoid hydraulic shock, the preset batch unit can be fitted with an advance warning contact, or it can incorporate a ramp function. In the former case, an output is provided, typically 2 to 5 percent before batch completion. This output partially closes the valve and the batch is "topped off" at a low flow rate up to the final preset quantity. The latter system includes a ramp function in the preset batch unit providing an analog output signal at the start of the batch to open the valve at a predetermined rate. As the batch nears completion, the valve is progressively closed down to a low flow rate. The final valve closure signal is then given at the preset batch size.

Turbine meters measure volume flow at actual operating conditions. Consequently, if high accuracy is required and the fluid temperature is subject to variation, automatic temperature correction is necessary. This involves measuring the liquid temperature with a platinum resistance thermometer and providing an analog control signal proportional to temperature. The temperature/volume relationship for the metered liquid is built into the automatic temperature correction (A.T.C.) unit. Depending upon the measured temperature, the A.T.C. unit modifies the totalizer volume reading in accordance with the preset temperature coefficient of the liquid to give volume readout at the required reference temperature.

To safeguard against interference or lost pulses during signal transmission, a pulse comparitor is often used on high accuracy systems. This involves using two pickup coils (A and B) and taking two separate signal leads to the electronics. The pulse comparitor unit monitors the two signals for integrity. If any pulses are lost or picked up on either line, the correct pulse sequence (A, B, A, B, A, B, etc.) will be interrupted. Any such false pulses are logged and the associated totalizer reading corrected accordingly.

Most turbine meter systems require flow rate indication or an analog control signal. These options can generally be provided from the basic totalizer unit.

Linearity and Repeatability

The nominal "K" factor (the number of pulses per unit volume) is primarily determined by the size and type of turbine meter. In practice, the actual "K" factor varies slightly between apparently identical meters, due to manufacturing tolerances. Consequently, it is essential to calibrate each meter to establish its own specific "K" factor. A typical turbine meter calibration is shown in Figure 2.20c.

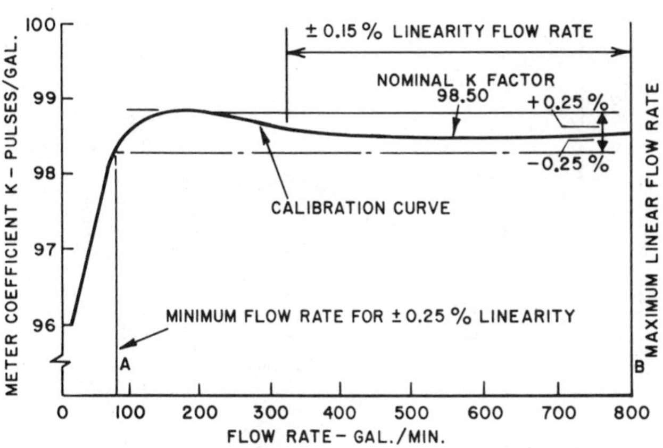

Fig. 2.20c Typical calibration curve for a turbine meter

The graph is a plot of "K" factor against flow rate. It will be noted that over the flow range "A" to "B" GPM, the "K" factor is a constant within the linearity tolerance band. The linearity tolerance band is typically ±0.25 percent of point over a 10:1 flow range for meters ¾ in. (20mm) and larger and ±0.5 percent of point over a 5 or 8:1 flow range on meters smaller than ¾ in. (20mm). It is important to note that the linearity is specified "of point" or "of actual reading" and is not "of full scale deflection."

The calibration in Figure 2.20c has a typical turbine meter hump in the low flow region (the lower 30 percent of the flow range). If this region is avoided, the turbine meter linearity can be improved to ±0.15 percent on the larger meters and ±0.25 percent on the smaller meters.

The repeatability of the turbine meter is typically ±0.02 percent of point at any flow rate within the linear range of the meter.

Viscosity and Density Effects

The principal fluid parameter which affects a turbine meter is viscosity. High viscosities change the nominal "K" factor and cause the calibration curve to fall away at a higher minimum flow rate (see Figure 2.20d). This causes a deterioration in the linearity tolerance over the full flow range or, alternatively, a shorter usable flow range at the standard linearity tolerance.

The effect of viscosity cannot be easily quantified because it is dependent upon the size and type of turbine meter. In general, larger meters are less affected by viscosity than smaller sizes. This does not imply that an oversize meter should be used on a viscous application. In fact, quite the reverse is true. On a high viscosity application it is advisable to size the meter so that its maximum permitted flow rate is as close as possible to the application flow rate. Thus, by tending to undersize the meter, the non-linear portion of the calibration is avoided and the best possible flow range is achieved.

The above comments about viscosity are applicable to

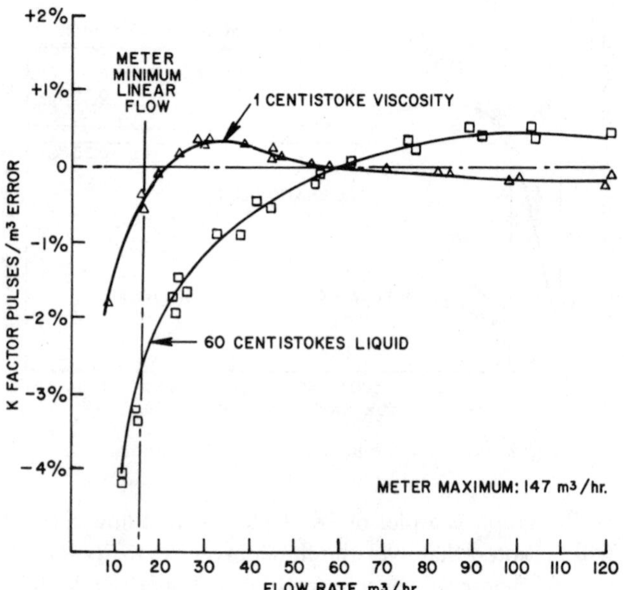

Fig. 2.20d Calibration curves illustrating the effect of high viscosity on meter performance

Table 2.20e
TYPICAL FLOW CAPACITY FOR A RANGE OF TURBINE METERS

Nominal Diameter		*Minimum Linear Flow*		*Maximum Linear Flow*	
Inches	*mm*	*G.P.M.*	*m³/hr.*	*G.P.M.*	*m³/hr.*
3/4	20	2.5	0.68	25	6.8
1	25	3.3	0.90	50	13.6
1½	40	7.2	1.96	108	29.5
2	50	19	5.17	187	51
3	75	54	14.7	540	147
4	100	104	28.4	1040	284
6	150	240	65.7	2410	657
8	200	415	113	4150	1130
10	250	715	195	6400	1750
12	300	1025	280	9160	2500
14	350	1210	330	10800	2950
16	400	1830	500	14650	4000
18	450	2310	630	18500	5050
20	500	2930	800	24000	6540

the linearity of the meter. Turbine meter repeatability will not be affected in this way and the standard repeatability tolerance will still be maintained at high viscosities. Consequently, a turbine meter can be used for such duties as on-off control on very viscous products. The control points can be determined impartially and the meter will then repeat these readings even though its calibration may be completely non-linear. To achieve reliable repeatability, the operating conditions must be constant.

Density has a small effect on the turbine meter's performance. On low density liquids, the meter's minimum flow rate is increased due to the lower driving torque, but the change in density has a minimal effect on the meter's calibration.

Meter Sizing

Turbine meters are sized by volumetric flow rate. Each meter size has a specified minimum and maximum linear flow figure and the meter should not normally be used outside these values. Typical flow capacities for a range of turbine meters from 3/4 in. (20mm) to 20 in. (508mm) are shown in Table 2.20e.

When sizing the meter, it is recommended that the application maximum flow rate is approximately 70 to 80 percent of the meter maximum flow rate. This results in a good flow range—7 or 8:1—and yet there is still approximately 25 percent spare capacity to allow for future expansion in production or increased metering requirements. Exceptions to this rule of thumb are applications which demand the best possible flow range capability and high viscosity applications.

In order to achieve optimium performance and flow range, most turbine meters are designed to a nominal maximum velocity of 30 ft./sec. (9.14 m/sec.). This velocity is higher than conventional pipework design velocities, which are typically 7 to 10 ft./sec. (2.13 to 3.05 m/sec.). Consequently, if the turbine meter is selected to be the same size as the pipeline, the meter flow range will be severely limited and will only be approximately 2 or 3:1. Hence, the importance of sizing on volumetric flow rate and not by pipe diameter cannot be overstressed. If the turbine meter is sized on volumetric flow rate, it will be smaller than the pipeline. This is a perfectly acceptable and normal practice, provided the meter is installed with the appropriate upstream and downstream straight pipe lengths and cone reducers (see Figure 2.20g).

Another aspect which must be considered when sizing the meter is available line pressure. Turbine meters have a typical pressure loss of 3 to 5 PSIG (20.7 to 34.5 kPa) at meter maximum flow rate. The pressure loss reduces on a square law with reducing flow rate. Consequently, if the meter is operating at 50 percent of maximum capacity, the pressure loss is 25 percent of that at maximum flow rate.

A typical pressure distribution through a turbine meter is shown in Figure 2.20f. As will be noted, the minimum pressure point occurs in the region of the rotor, with a substantial pressure recovery by the meter outlet occurring immediately thereafter. It is essential to provide sufficient line pressure to prevent liquid cavitation or gassing in the rotor region. To ensure that cavitation does not occur, the downstream line pressure must be at least 2 times the net meter pressure loss plus 1.25 times liquid vapor pressure. When the downstream line

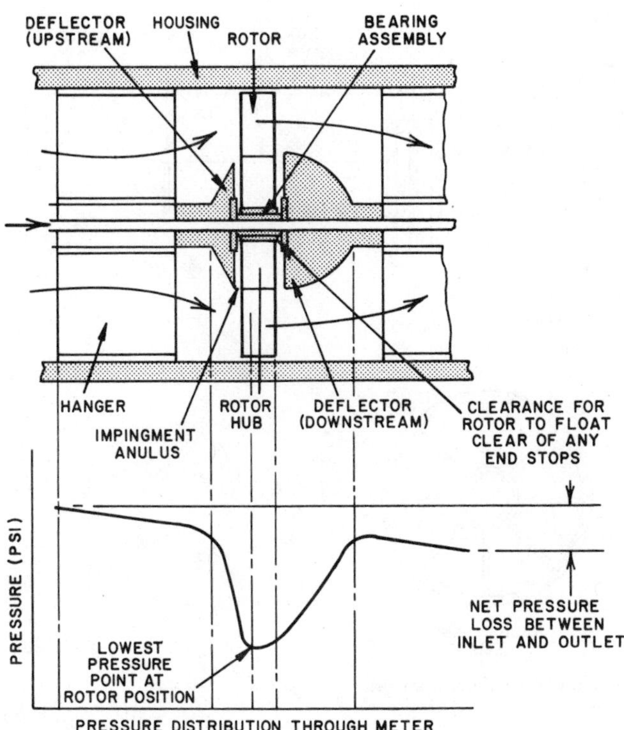

Fig. 2.20f Typical pressure distribution through a turbine meter

pressure is not sufficient to meet this requirement, a larger meter operating in a lower region of its flow range (with a resultant lower pressure loss) should be considered. The meter flow range will be reduced by this approach.

If cavitation occurs, it will cause an error in the meter output, and the meter will read high. If severe cavitation is present, it will cause serious over-speeding of the rotor, resulting in possible mechanical damage to the rotor and bearing.

Meter Characteristics and Features

The wetted materials of a turbine meter are generally stainless steel throughout except for the bearing. The most widely used bearing at present is a tungsten carbide sleeve bearing which offers exceptional reliability and immunity to wear. These materials provide good corrosion resistance capability on a wide range of process liquids. Where these materials are not suitable, other, more expensive possibilities, such as Hastelloy "C" with P.T.F.E. bearings, are feasible. On clean liquids, some meter designs use ball race bearings in order to achieve greater rangeability.

Turbine meters are suitable for extremes of temperature. When appropriate pickup coils and bearings are selected, turbine meters can operate at temperatures varying from −328°F (−200°C) to +840°F (+450°C). The turbine meter housing is a very good pressure vessel since there are no tappings or protrusions into the meter bore. Consequently, most small turbine meters are suitable for operating pressures up to 5000 PSIG (34.5 MPa), subject to the pressure limitation on the flanges or other end connections.

Another significant feature of the turbine meter is that it has a high throughput for a given size and is small in size and weight, relative to the pipeline. Consequently, turbine meters can handle large volume flow rates with a minimal requirement for space and no special mounting stands or pads. Other features of the turbine meter include fast response time, suitability for hygienic applications, linear digital output, ease of maintenance, and simple installation.

The principal limitations with a turbine meter are that it is not linear at high viscosities, it can be damaged by over-speeding or gassing, it requires secondary readout equipment, calibration is required (preferably at operating conditions), and it is relatively expensive, particularly in larger sizes.

Mechanical Installation

The turbine meter's high accuracy can easily be negated by a substandard installation. Upstream disturbances such as bends, valves, or filters may cause swirl and/or a non-uniform velocity profile which, in turn, affects both the linearity of the meter and the nominal "K" factor. The errors may be positive or negative, depending on the direction of the swirl. If there is sufficient straight pipe between the source of the disturbance and the meter, the fluid shear or internal friction between the liquid and the pipe wall will condition the flow to an acceptable degree. The length of straight pipe required depends upon the upstream disturbance and, in some instances, may have to be as long as 50 times the nominal meter diameter.

To avoid excessively long straight lengths of pipe, an internal flow-straightening element is generally used where good accuracy is required. The flow-straightening element may either be a bundle of thin-wall tubes or a series of radial vanes inserted longitudinally in the upstream section of the straight pipe. The location of the vane is important; the recommended position is shown in Figure 2.20g. When a flow-straightening element is used, the upstream straight pipe requirement is reduced to 10 times nominal meter diameter. The required downstream length is 5 times nominal meter diameter. It is, nevertheless, good practice to install the meter upstream of any severe source of disturbance, such as regulating control valves, whenever possible. If the meter is smaller in diameter than the process pipework, 15 degree included angle concentric cones should be fitted at either end of the metering pipework, as shown in Figure 2.20g. Care should be taken with the internal alignment of all flange joints in the metering section; no gaskets should protrude into the fluid path.

To avoid mechanical damage to the turbine meter and to ensure optimium life, a suitable mesh strainer should

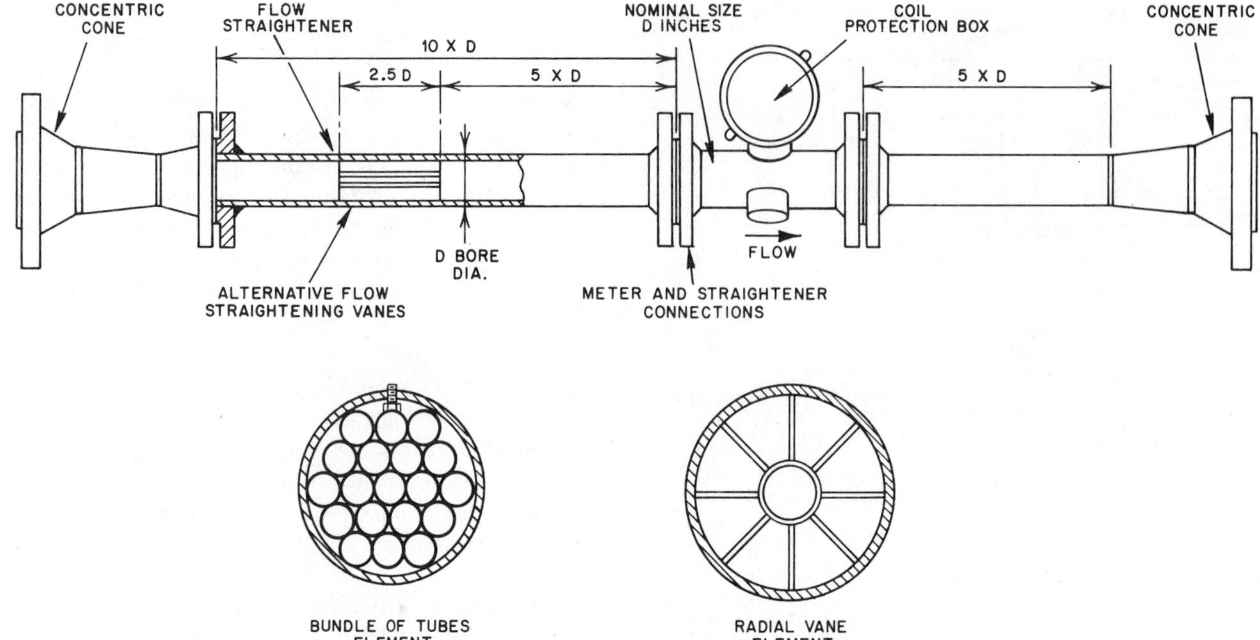

Fig. 2.20g Recommended turbine meter installation pipework

be fitted upstream of the meter. The recommended mesh size depends on the size and type of turbine meter, but typical guidelines are given in Table 2.20h. Close attention should be paid to any application where there are fibrous particles in the fluid. Contaminants of this type are frequently not removed by the strainer; the fibrous strands tend to wrap around the rotor and bearing causing the rotor to slow down and the calibration to change.

Table 2.20h
TYPICAL STRAINER RECOMMENDATIONS FOR TURBINE METER INSTALLATIONS

Turbine Meter Size Inches	*Recommended Strainer*			
	U.S. Sieve No.	*Wire Size (Inches)*	*Meshes/ Linear Inch*	*Opening (Inches)*
½ and smaller	120	0.0034	120.48	0.0049
¾ to 1½	45	0.0087	44.44	0.0138
2 and larger	18	0.0189	17.16	0.0394

Electrical Installation

The output frequency from a typical turbine meter pickup coil varies in frequency and amplitude with flow range. At low flows, the signal may be as small as 20 mV peak-to-peak. Consequently, if the turbine meter and electronic readout equipment are not from the same manufacturer, care must be taken to ensure that the two units are compatible with regard to pulse shape (sine wave or square wave), signal frequency and pulse amplitude and width.

Careful attention should also be given to the cable routing between the turbine meter and the electronics. Areas of electrical noise should be avoided, cable lengths should be kept as short as possible, impedance matching should be verified, and the appropriate shielded cable should be used. When long transmission distances are involved or the area is electrically noisy, a preamplifier should be fitted to the meter (see Figure 2.20i).

The preamplifier output signal amplitude is independent of flow rate and is typically a 12 volt square wave signal. This high level signal can be transmitted for great distances, typically 15,000 ft. (4,572 m) and is far more immune to electrical interference than an unamplified pickup signal. The limitations of a preamplifier include increased cost and the necessity for a DC power supply at the meter. In some designs, an additional cable is required (a three-wire system as opposed to a two-wire system) and the ambient temperature is typically limited to 212°F (100°C).

Applications

Due to its excellent performance characteristics, the turbine meter is widely used for high accuracy royalty and custody transfer of crude oil, refined hydrocarbons and other valuable liquids. Turbine meters are used throughout the petrochemical industry for many other applications, such as process control metering, blending, and pipeline leak detection. Turbine meters are also used in other industries for a broad range of applications, flow rates and duties. More specialized applications include measurement of cryogenic liquids (liquid oxygen and nitrogen), high pressure water injection to oil wells, aircraft fuel metering, test rig duty, and road tanker filling. Some of these applications require modified or special

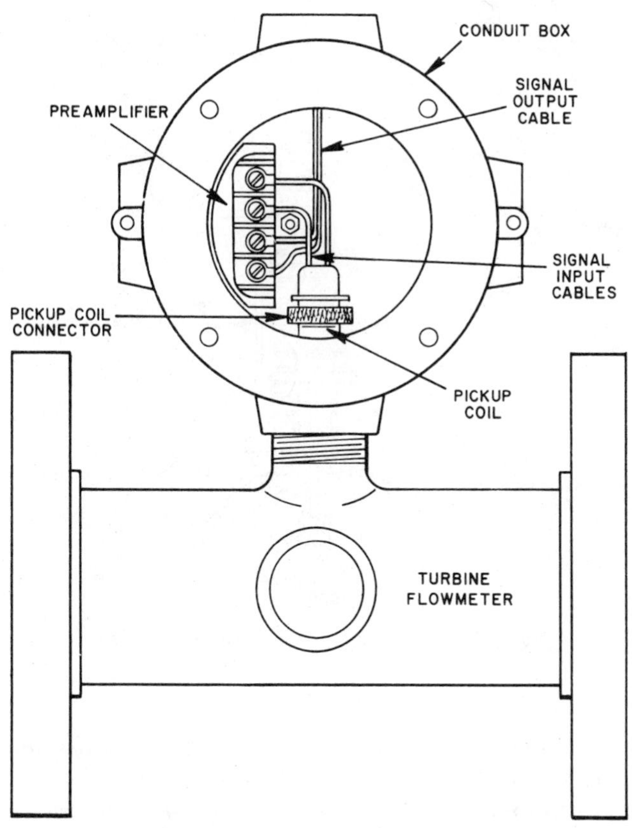

Fig. 2.20i Complete turbine flowmeter assembly showing pickup coil and preamplifier

meters—for example, aircraft meters are made from aluminum alloy to save weight—but fundamentally the same meter is used in all cases.

Gas Turbine Meters

The operating principle for the gas turbine meter is the same as that already described for the liquid turbine meter. The major difference is that due to the much lower density of the gas, the available fluid driving torque is greatly reduced. Consequently, gas turbine meters feature various design changes to enable the meter to operate at higher fluid velocities and to compensate for the lower driving torque. The principal changes are the use of larger hub diameters to give a smaller ratio of rotor annular area to pipe area (see Figure 2.20j), lightweight rotors, increased number of blades, modified blade angle, and alternative bearings. Some designs feature local mechanical volume flow indication, achieved by reduction gears in the rotor driving external gears via a magnetic coupling.

Due to the lower driving torque of the gas, it is essential to keep bearing frictional resistance to a minimum. The liquid turbine meter journal bearing is usually replaced by a ball race bearing. Any change in the bearing frictional resistance will result in a change in the meter calibration. Meters are frequently used in dust-

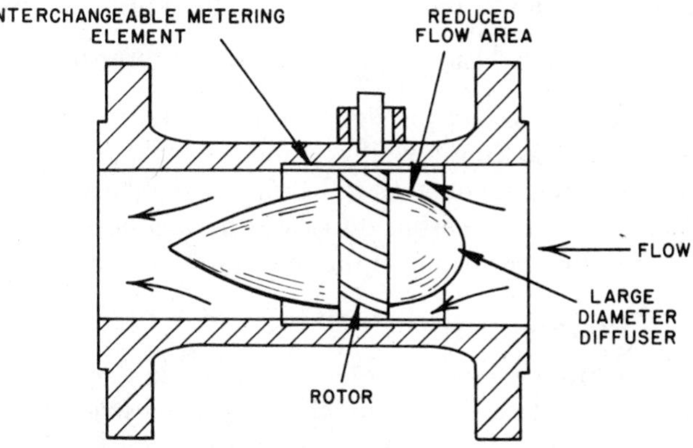

Fig. 2.20j Typical gas turbine meter showing low ratio rotor annular-to-pipe area

laden gases and the ball races are frequently of the sealed, self-lubricated type. Some designs, however, use gas bearings.

It is essential to calibrate the gas turbine meter initially, preferably under simulated operating conditions, to establish its own specific "K" factor. A typical calibration curve is shown in Figure 2.20k. Linearity is normally ±1 percent of point over a flow range of 20:1. Gas turbine meters have specific minimum and maximum volumetric flow rate values and it is essential to size the meter to these volumetric flow rates and not pipe size. It is also important to note that the meter must be sized on actual volume flow and not on reference or standard units.

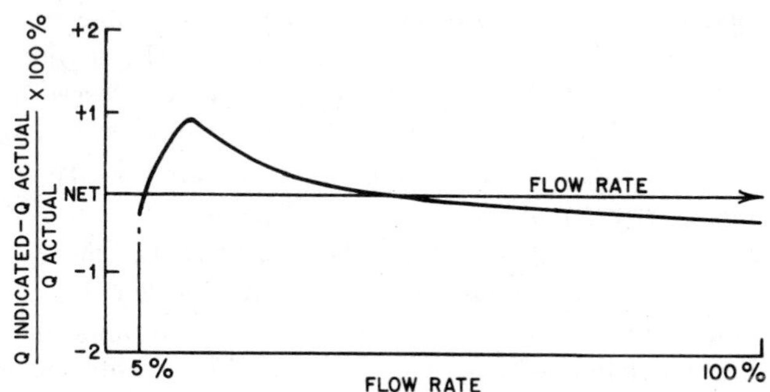

Fig. 2.20k Typical gas turbine flowmeter calibration

The turbine meter output frequency is proportional to the volumetric flow rate at the actual operating pressure and temperature. An appropriate pressure and temperature correction system is required to convert the meter output into a volume at reference conditions. If readout in mass units is required, either the above pressure and temperature correction system can be used—although it does not compensate for variations in the composition of the gas—or the meter reading can be combined with a density gauge reading to give true mass.

In any compensation system, the volume and pressure or density measurements should be related to the same point. The gas turbine meter has a typical pressure loss of one velocity head ($\frac{1}{2}\frac{\rho V^2}{g}$) and a similar pressure distribution to that of the liquid turbine meter shown in Figure 2.20f. Consequently, if the pressure or density measurement is not taken at the rotor, a slight correction factor may be necessary to relate the measured value back to that pertaining at the rotor position.

Features

Gas turbine meters are less sensitive to damage by grit and dust particles than other positive displacement meters. Gas turbine meters also can operate at higher pressures and have a high flow rate capacity for a given meter size. In addition, if the meter fails, the gas flow is not obstructed, ensuring continuity of flow.

Typical upstream pipe requirements are 20 times nominal meter diameter.

Due to possible variations in the meter bearing characteristics, calibration checks should be made at regular intervals if optimium performance is to be achieved.

Insertion Meters

Both of the turbine meters described above are full bore metering devices, and all flow passes through the meter. Their cost, however, increases proportionately with increasing pipe diameter. The insertion turbine meter is effectively a set of small turbine meter internals mounted on a probe in a larger diameter pipe (see Figure 2.20l). The meter operating principles are the same as described previously but the meter only measures the fluid velocity at a single point on the cross-sectional area of the pipe and does not "see" all the fluid. Total volumetric flow rate for the pipeline can then be inferred if certain assumptions are made about the velocity distribution across the pipe compared with the velocity at measurement point. The velocity distribution can either be established by "profiling" the line—that is, taking a series of measurements across the pipeline and establishing the fluid velocity profile—or by using impartial data which has been developed to establish the optimium compromise insertion depth for a range of pipe diameters.

The insertion meter cannot be as accurate as a full bore meter since it is only measuring velocity at one point on the cross-sectional area. It does, however, provide a very low cost metering system for large diameter gas or liquid pipelines where accuracy is not important.

Insertion meters can be hot-tapped into existing pipelines through a valving system without shutting down the pipeline. A flanged riser, complete with valve, is welded to the pipeline. A hot tap device is coupled to the valve, the valve opened, and the pipe penetrated. The hot tap unit is withdrawn and the valve closed. The insertion meter is then installed, the valve opened, and the meter screwed in to the appropriate depth.

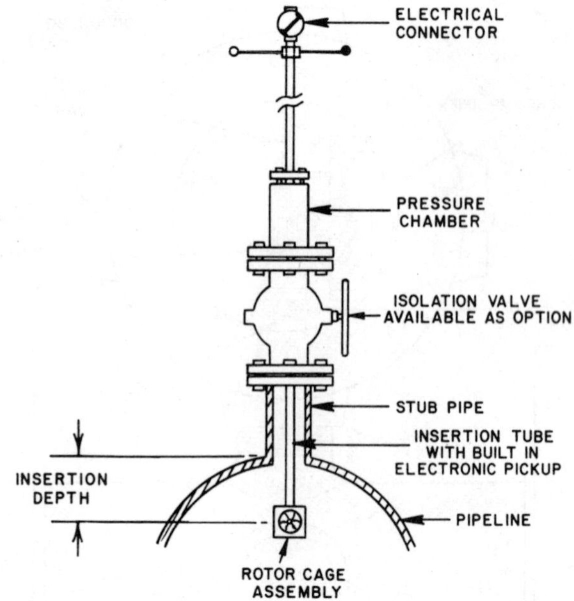

Fig. 2.20 l Insertion turbine flowmeter installed in large diameter pipe

Insertion meters can be used on pipelines above 4 in. (102 mm) and, due to the small cross-sectional area relative to the pipe area, their pressure loss is very low. Typical linearity and repeatability figures are ±1 percent and ±0.25 percent respectively. These are point velocity readings; in overall volumetric accuracy terms, the effects of changes in velocity profile must also be considered.

Optical Flow Sensor

A specialized version of the insertion-type turbine flowmeter is the optical photoflow sensor. The flow transducer consists of a probe supporting a low mass rotating element which interrupts a light ray traveling from a light source to a photo transistor. The result is a pulse train which is converted into a volumetric flow representation.

This flow transducer provides flow ranges as high as 100 to 1, bidirectional measurement without additional calibration, and extremely low pressure drop. The transmitter has only one moving part, the flow sensing element. The bearing for the element is not located directly in the flow stream, enabling the transducer to handle severe flow conditions such as heavy surging and pulsating flows.

The installation requirements include the need for 10 or more diameters of upstream straight run and the need to eliminate rotary valves (such as butterflies) at the ends of the measuring run.

Pelton Wheel Meters

It is not practical to make turbine meters for very low flow rates below 0.25 GPM (1.58×10^{-5} m³/sec.). Pelton wheel meters have been developed for these very low flow rates. The meter has a small orifice that projects the liquid onto a small Pelton wheel. The velocity of rotation is then measured electromagnetically and a frequency output signal produced. By varying the diameter of the orifice, a range of flow rates can be covered from 0.001 GPM through to 2 GPM (6.3×10^{-8} to 1.26×10^{-4} m^3/sec.). Flow range varies with meter type but is generally between 10 and 20:1. The meters offer good repeatability (±0.1 percent) but are generally non-linear and have a high pressure loss, typically 15 to 20 PSIG (103 to 138 kPa). Typical applications for this type of meter are metering internal combustion engine fuel flows in test rigs and additive dosing.

BIBLIOGRAPHY

American Petroleum Institute, "Measurement of Liquid Hydrocarbons by Turbine Meter Systems," A.P.I. Standard 2534.

Murphy, H.N., "Flow Measurement by Insertion Turbine Meters," Measurement Technology for the 80's, ISA Symposium, Delaware 1979.

Nichol, A.J., "An Investigation into the Factors Affecting the Performance of Turbine Meters," Conference on Fluid Flow Measurement in the Mid-1970's, East Kilbride.

Withers, V.R., Inkley, F.A., & Chesters, D.A., "Flow Characteristics of Turbine Flowmeters," Conference on Modern Developments in Flow Measurement, Harwell.

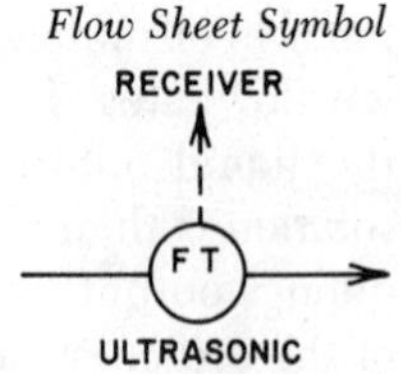

2.21 ULTRASONIC FLOWMETERS

Types:	A—Transit time B—Doppler
Process Temperature:	A: −76 to +392°F (−60 to +200°C) or higher B: −76 to +500°F (−60 to +260°C)
Design Pressure:	A: Up to 1000 PSI (6.9 MPa) or higher B: No limit; mounts outside the pipe
Materials:	A: Spool piece—cast iron, carbon steel, stainless steel B: Process pipe provided it conducts ultrasonic energy
Fluid:	A: Liquids—primarily clean but will tolerate a slight amount of aeration or solids B: Liquids—must be aerated or slurry with a concentration between 0.2% to 60% depending on particle size
Flow Range Velocity:	A: .1 ft/sec to 100 ft/sec (30 mm/sec to 30 m/sec) B: .2 ft/sec to 60 ft/sec (60 mm/sec to 18 m/sec)
Sizes:	A: .125 in. to 120 in. (3.1 mm to 3 m) B: .5 in. to 72 in. (12.5 mm to 1.8 m)
Inaccuracy:	A: 1% of rate to 2.5% of rate (depending upon velocity, pipe size and manufacturer) B: 2% to 5% of full scale (depending upon application, pipe size and manufacturer)
Cost:	A: 4 in. (100 mm)—$3,500, 10 in. (250 mm)—$5,000, 24 in. (500 mm)—$10,000 B: $2,000—independent of pipe size
Suppliers:	A: Badger Meter, Inc.; B.I.F. a division of General Signal; Controlotron Corp.; Manning Env. Corp.; Mapco, Inc.; ORE, Inc.; Sparling Div. Envirotech Corp.; Westinghouse Electric Corp. B: Andco Industries Inc.; Baird Controls, Inc.; Controlotron Corp.; Hersey Products, Inc.; Leeds & Northrup; Mapco, Inc.; Polysonics

Transit Time Flowmeters

Time Difference-type

As the name implies, these devices measure flow by measuring the time taken for ultrasonic energy to traverse a pipe section, both with and against the flow of the liquid within the pipe. Figure 2.21a is a diagram of a representative transit time flowmeter.

The time (t_{AB}) for energy to go from transducer A to transducer B is given by the expression:

$$t_{AB} = L/(C + V\cdot\cos\theta) \qquad 2.21(1)$$

The time (t_{BA}) to go from B to A is given by:

$$t_{BA} = L/(C - V\cdot\cos\theta) \qquad 2.21(2)$$

where C is the speed of sound in the fluid, L is the acoustic path length and θ is the angle of the path with respect to the pipe axis. By combining terms and simplifying it can be shown that:

$$\Delta t = t_{BA} - t_{AB} = 2\cdot L\cdot V\cdot\cos\theta/C \qquad 2.21(3)$$

Further simplification allows that:

$$V = L\cdot\Delta t/2\cdot\cos\theta\cdot t^2_{\bar{A}} = K\cdot\Delta t/t^2_{\bar{A}} \qquad 2.21(4)$$

where t_A is the average transit time between the transducers.

Since the cross-sectional area of the pipe section or "spool piece" is known, the product of area and velocity will yield volumetric flowrate.

Frequency Difference-type

In this type of meter, one oscillator is locked at a frequency, $f_{AB} = \frac{1}{t_{AB}}$ and a second oscillator is locked at a frequency $f_{BA} = \frac{1}{t_{BA}}$.

The difference in frequencies is related to the velocity as follows:

$$V = \Delta f \cdot L/2 \cdot \cos\theta \qquad 2.21(5)$$

Flowmeter Construction

The flowmeter usually consists of an electronics housing, transducers and a pipe section. Several options are open to the customer as to the construction of the transducers and pipe section. Some designs allow removal of the transducers without interrupting process flow. A spool piece with integral transducers is one of the most common types of construction and is shown in Figure 2.21a. The manufacturer mounts the transducers to a flanged pipe section (spool piece). Usually the unit is calibrated by the manufacturer to customer specifications. The spool piece thus becomes an integral part of the hydraulic system so it is not easily retrofitted into an existing system. Some manufacturers will supply a transducer assembly capable of being mounted outside of an existing pipe, as shown in Figure 2.21b. This type of system will be calibrated by the manufacturer only if detailed information such as pipe diameter, pipe wall thickness, process fluid, percent of solids concentration, process temperature, change of process temperature, and so on are defined by the customer. This type of flowmeter is easily retrofitted into an existing system, since no pipe section need be installed.

Some manufacturers provide transducers and mounting hardware that the user installs into an existing pipe. The user drills holes into the existing pipe and attaches

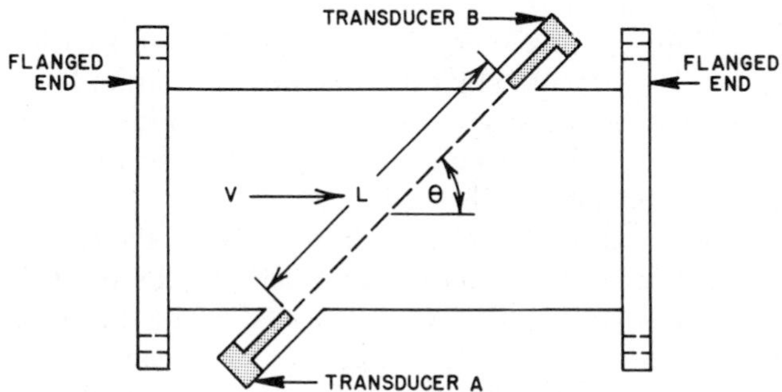

Fig. 2.21a Transit time flowmeter

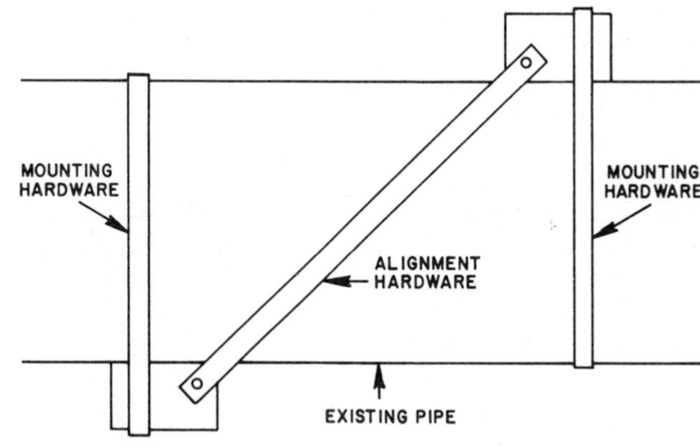

Fig. 2.21b External transducers

the transducer mounting hardware by welding or other suitable means. The transducers are then mounted and aligned. Usually this type of unit may be calibrated by the user only after measurements of transducer angle and spacing, and pipe diameter are made.

Application Considerations

As with most flowmeters, the spool piece or pipe section must always be full to assure proper operation and volumetric flow indication. Most manufacturers will specify the minimum distance from valves, tees, elbows, pumps, and so on that will ensure accurate flowmeter performance. Typically, 10–20 diameters upstream and 5 diameters downstream are required. The flowmeter relies upon an ultrasonic signal traversing across the pipe; therefore, the liquid must be relatively free of solids and air bubbles. Bubbles in the flowstream seem to cause more attenuation of the acoustic signals than solids do. The flowmeter can tolerate a few percent of solids but only a fraction of a percent of bubbles.

Depending on the process fluid, proper transducer materials and protection must be chosen to prevent transducer damage due to chemical action. Process temperature limitations must also be considered for proper flowmeter application.

Performance Specifications and Features

Accuracy is usually specified as a percent of rate. Typically it is 1–2.5 percent of rate, depending upon the manufacturer, velocity, pipe size and process. Some manufacturers calibrate each flowmeter at one or more points under actual flow conditions. **Repeatability** is usually specified as a percent of rate, typically better than 0.5 percent depending upon velocity range and manufacturer.

In an attempt to improve performance and accuracy for larger pipe sizes, some suppliers offer flowmeters with two, four, or more pairs of transducers arranged in multiple acoustic paths. Usually the cost of such units is higher than a single path flowmeter. Bi-directional flowmeters will measure flow in either direction. The display and output of such information depends upon the manufacturer. Totalizers are available in the form of an optional counter to indicate total flow through the flowmeter in some user-selected units. A current output (4 to 20mA) is usually standard. Voltage, pulse train or other digital outputs may be optionally available depending upon the manufacturer. Alarms for high or low flow are optionally available depending upon the manufacturer.

Doppler Flowmeters

Principle of Operation

As shown in Figure 2.21c, an ultrasonic wave is projected at an angle through the pipe wall into the liquid by a transmitting crystal in a transducer mounted outside the pipe. Part of the energy is reflected by bubbles or particles in the liquid and is returned through the pipe wall to a receiving crystal. Since the reflectors are traveling at the fluid velocity, the frequency of the reflected wave is shifted according to the Doppler principle.

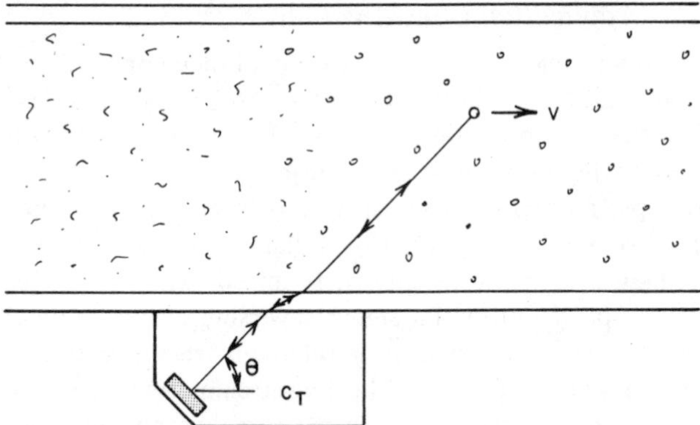

Fig. 2.21c Doppler flowmeter principle of operation

Combining Snell's Law and the classical Doppler equation, the velocity is as follows:

$$V = \Delta f \cdot Ct/(2 \cdot fo \cdot \cos\theta) = \Delta f \cdot K \qquad 2.21(6)$$

where Δf is the difference between transmitted and received frequency, fo is the frequency of transmission, θ is the angle of the transmitter and receiver crystal with respect to the pipe axis, and Ct is the velocity of sound in the transducer. As shown in equation 2.21(6), velocity is a linear function of Δf. Since the user can measure the inside diameter of the pipe, volumetric flowrate can be measured using equation 2.21(7) or a similar form:

$$GPM = 2.45 \cdot V \cdot (ID)^2 \qquad 2.21(7)$$

The single transducer is the most popular design. Both the transmitter and receiver crystal are contained in a single transducer assembly that mounts to the outside of the pipe. Alignment of the crystals is thus controlled by the manufacturer. This approach is shown in Figure 2.21c.

In the dual transducer design, the transmitter crystal and the receiver crystal are mounted separately on the outside of the pipe. Alignment is maintained by a mounting assembly between the transducers, as shown in Figure 2.21d.

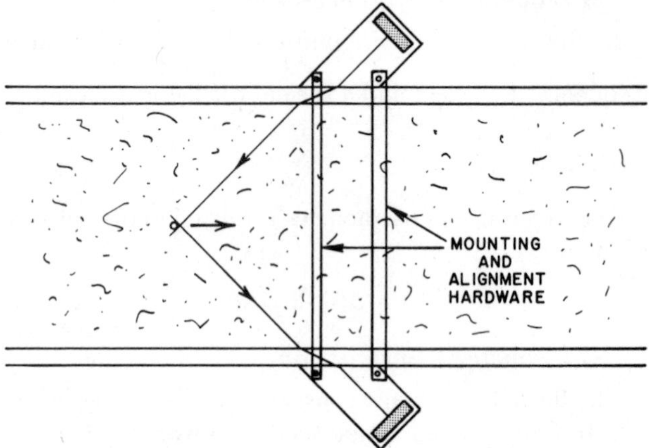

Fig. 2.21d Two transducer approach

Each manufacturer will have specific instructions on how to mount the transducer or transducers to the pipe. The acoustic coupling to the pipe and the alignment of the transducer to the pipe must be maintained in spite of pipe temperature changes and vibration. Therefore the manufacturer's instructions should be followed closely when mounting the transducer or transducers to assure a stable, reliable installation.

Application Considerations

As with transit time and other flowmeters, in order to properly indicate volumetric flow, the pipe must always be full. A Doppler will, however, indicate velocity in a partially full pipe as long as the transducer is mounted below the liquid in the pipe.

Most manufacturers will specify the minimum distance from valves, elbows, tees, pumps, and so on that will ensure accurate flowmeter performance. Typically 10–20 diameters upstream and 5 diameters downstream are required for relatively clean fluids, but this might change depending upon the process solids concentration or solids composition.

A Doppler flowmeter relies upon bubbles or particles in the flow stream to reflect the ultrasonic energy. Most manufacturers specify a lower limit of the concentration and size of solids or bubbles in the liquid for reliable, accurate operation. The flow must also be fast enough to keep the solids or bubbles in suspension, typically 6 ft/sec (1.8m/s) minimum for solids and 2.5 ft/sec. (0.75 m/s) for small bubbles. On horizontal pipes, the best place to locate the transducer around the circumference of the pipe is not always specified for all applications. The user should rely upon the manufacturer's empirical testing, application experience, and instructions for various applications.

Since energy need not go across the entire pipe, the single transducer Doppler can work with wide variations and high levels of solids concentration or aeration. In the Doppler with two transducers, ultrasonic energy must go across the pipe, so some effects on the flowmeter may occur due to wide variations and high levels of solids concentration or aeration.

The Doppler will operate independent of pipe material provided the pipe is sonically conductive. Such pipes as concrete, clay, and very porous cast iron absorb the ultrasonic energy and may not work with a Doppler. Depending upon the manufacturer, some Dopplers will work with lined pipes as long as the liner is well bonded to the inside wall of the pipe.

Transducer temperature limits must also be considered for proper flowmeter operation over the full process temperature range.

Performance Specifications and Features

Accuracy is usually specified as a percent of span. Typically, it is 2–3 percent depending upon manufacturer, velocity, pipe size, and process fluid. Some manufacturers calibrate the flowmeter at one or more points under actual flow conditions. As shown in equation 2.21 (7), to minimize the error in volumetric flow indication the inside diameter of the pipe must be measured very carefully because volumetric flowrate varies as the square of the diameter.

Repeatability is usually specified as a percent of span; typically it is better than .5 percent under simulated flow conditions. Under actual flow conditions it is typically 1 percent depending upon manufacturer, velocity, pipe size, and process conditions.

Bi-directional flowmeters will measure flow in either direction, but they only measure flow magnitude, not direction. Totalizers are available in the form of an optional counter to indicate total flow through the flowmeter in some user-selected units.

Pipe vibration at no-flow conditions can sometimes cause an upscale flow indication due to particle or bubble motion. Some manufacturers simply turn down the sensitivity of the detection circuitry, while others have proprietary circuitry to ensure a zero indication at no-flow conditions.

A current output (4 to 20mA) is usually standard. Voltage or pulse train outputs may be optionally available depending upon the manufacturer. Alarms for high or low flow are optionally available depending upon the manufacturer.

BIBLIOGRAPHY

Addie, G.R., Maffett, J.R., DiNapoli, L.D., Punis, G., "Doppler Flow Meter Tests at Georgia Iron Works," 1980 ISA Mining and Metallurgy Industries Division Symposium, Phoenix, AZ.

Faddick, R., Pouska, G., Connery, J., DiNapoli, L., Punis, G., "Ultrasonic Velocity Meter," Sixth International Conference on the Hydraulic Transport of Solids in Pipes, BRHA Fluid Engineering (Cranfield UK), Sept. 1979.

Lipták, B.G., "Ultrasonic Instruments," *Instrumentation Technology*, Sept. 1974.

Lynnworth, L.C., "Selected Alternatives to Conventional Ultrasonic Flowmeter," Ultrasonics International Conference of 1977 held in Guildford, England.

Lynnworth, L.C., "Clamp-on Ultrasonic Flowmeters," *Instrumentation Technology*, September 1975.

Raptis, A.C., "Accoustic Doppler Flowmeter," Fossil Energy I&C Briefs, March 1980.

Shane, J.L., "Ultrasonic Flowmeter Basics," *Instrumentation Technology*, July 1971.

Waller, J.M., "Guidelines for Applying Doppler Accoustic Flowmeters," *InTech*, October 1980.

Zachavias, E.M., "Sound Velocimeters Monitor Process Streams," *Chemical Engineering*, Jan. 22, 1973.

Flow Sheet Symbol

2.22 VARIABLE AREA FLOWMETERS

Standard Design Pressure:	350 PSIG (2.4 MPa) average maximum for glass metering tubes, dependent on size Up to 720 PSIG (5 MPa) for metal tubes and special designs to 6000 PSIG (41 MPa)
Standard Design Temperature:	Up to 400°F (204°C) for glass tubes and up to 1000°F (538°C) for some models of metal tube meters
End Connections:	Female pipe thread or flanged
Fluids:	Liquids, gases and vapors
Flow Range:	0.01 cc/minute to 4000 GPM (920 m^3/hr) of liquid 0.3 cc/minute to 1300 SCFM (2210 m^3/hr) of gas.
Inaccuracy:	±0.5% of rate to ±10% of full scale depending upon size, type, and calibration
Materials of Construction:	TUBE: Borosilicate glass, stainless steel, Hastelloy, Monel, alloy 20. FLOAT: *Conventional type*—brass, stainless steel, Hastelloy, Monel, alloy 20, nickel, titanium, or tantalum, and special plastic floats. *Ball Type*—glass, stainless steel, tungsten carbide, sapphire, or tantalum. END FITTINGS: Brass, stainless steel or alloys for corrosive fluids. PACKING: The generally available elastomers are used and "O" rings of commercially available materials, Teflon is also available
Cost:	$50 to $5,000 as a function of size, materials of construction, etc.
Suppliers:	Ametek, Inc., Schutte & Koerting Div.; Brooks Instrument Division of Emerson Electric Co.; Cox Instrument; Erdco Engineering Corp.; Fischer & Porter Co.; Flowmetrics, Inc.; Porter Instrument Co., Inc.; Process Systems, Inc. (digital valve binary weighted ports with sonic venturis); Universal Flow Monitors (spring and vane); Wallace & Tiernan Div., Pennwalt Corp.

The rotameter is a variable area-type flowmeter. It consists of a tapered metering tube and a float which is free to move up and down within the tube. The metering tube is mounted vertically with the small end at the bottom. The fluid to be measured enters at the bottom of the tube, passes upward around the float, and out at the top. Figure 2.22a is a representation of a rotameter.

When there is no flow through the rotameter, the float rests at the bottom of the metering tube where the maximum diameter of the float is approximately the same as the bore of the tube. When fluid enters the metering tube, the buoyant effect of the fluid lightens the float, but it has a greater density than the fluid and the buoyant effect is not sufficient to raise it. There is a small annular opening between the float and the tube. The pressure drop across the float increases and raises the float to increase the area between the float and tube until the upward hydraulic forces acting on it are balanced by its weight less the buoyant force. The metering float is "floating" in the fluid stream. The float moves up and down in the tube in proportion to the fluid flow rate and the annular area between the float and the tube. It reaches a stable position in the tube when the forces are in equilibrium. With upward movement of the float toward the larger end of the tapered tube, the annular opening between the tube and the float increases. As the area increases, the pressure differential across the float decreases. The float will assume a position, in dynamic equilibrium, when the pressure differential across the float plus the buoyancy effect balances the weight of the float. Any further increase in flow rate causes the float to rise higher in the tube; a decrease in flow causes the

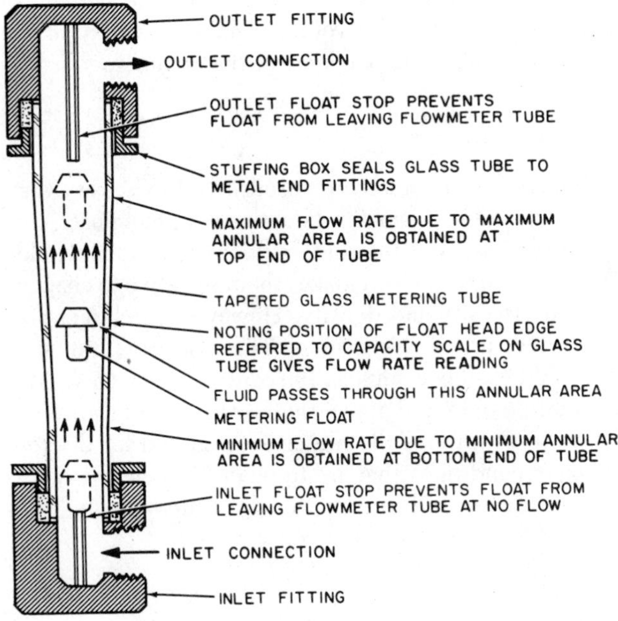

Fig. 2.22a The rotameter

float to drop to a lower position. Every float position corresponds to one particular flow rate and no other for a fluid of a given density and viscosity. It is merely necessary to provide a reading or calibration scale on the tube and flow rate can be determined by direct observation of the position of the float in the metering tube.

Metal metering tubes are used in applications where glass is not satisfactory. In this case, the float position must be indirectly determined by either magnetic or electrical techniques. The use of indirect float position sensors also provides functions other than direct visual indication. Rotameters are available which transmit pneumatic, electronic, or time pulse signals, or provide recording, totalizing, or control functions.

Sizing

To size a rotameter, it is customary to convert the actual flow to "standard flow." For liquid flows it is necessary to calculate the GPM (l/m or hr) water equivalent. For gases it is necessary to determine the SCFM (l/m or hr) air equivalent. Capacity tables are based on these "standard flows" of GPM or cc/min of water and SCFM or cc/min of air at standard conditions. The tables also are based on using stainless steel floats.

The equations necessary to calculate the water or air equivalent are:*

LIQUIDS

Volume Rate
GPM Water Equivalent

$$= \frac{(\text{GPM})(\rho)(2.65)}{\sqrt{(\rho_f - \rho)\rho}} \qquad 2.22(1)$$

Weight Rate
GPM Water Equivalent

$$= \frac{(\text{lbm/min})(0.318)}{\sqrt{(\rho_f - \rho)\rho}} \qquad 2.22(2)$$

Base or Contract Volume Rate
GPM Water Equivalent

$$= \frac{(\text{GPM}_b)(\rho_b)(2.65)}{\sqrt{(\rho_f - \rho)\rho}} \qquad 2.22(3)$$

GASES OR VAPORS

Standard Volume Rate
SCFM Air Equivalent

$$= \frac{(\text{SCFM})(\rho_g\ \text{std})(10.34)}{\sqrt{\rho_f(\rho_g\ \text{act})}} \qquad 2.22(4)$$

Weight Rate
SCFM Air Equivalent

$$= \frac{(\text{lbm/min})(10.34)}{\sqrt{\rho_f(\rho_g\ \text{act})}} \qquad 2.22(5)$$

Operating or Actual Volume Rate
SCFM Air Equivalent

$$= \frac{(\text{ACFM})(\rho_g\ \text{act})(10.34)}{\sqrt{\rho_f(\rho_g\ \text{act})}} \qquad 2.22(6)$$

GPM = maximum flow of liquid at metering condition in units of gallons per minute.
GPM_b = maximum flow of liquid at base or contract condition in units of gallons per minute.
lbm/min = maximum flow of fluid at metering condition in units of pounds per minute.
SCFM = maximum flow of gas referred to a base or standard condition in units of cubic feet per minute.
ACFM = maximum flow of gas at operating conditions in units of cubic feet per minute.

ρ = density of flowing liquid at metering conditions in units of grams per cubic centimeter.

*See Appendix A.1 for SI Units.

ρ_b = density of flowing liquid at base or contract conditions in units of grams per cubic centimeter.

ρ_f = density of float in units of grams per cubic centimeter.

ρ_g std = density of gas at 14.7 PSIA & 70°F or 14.4 PSIA & 60°F in units of pounds per cubic foot.

ρ_g act = density of gas at metering conditions in units of pounds per cubic foot.

To facilitate these computations, manufacturers offer slide rules or nomographs specifically designed for rotameter sizing.

Rotameter Characteristics

The rotameter has many characteristics which set it apart from other flowmeters.

Liquids

A very wide range of liquids can be handled by the rotameter. A wide choice of tube, float, end fitting, and packing or O-ring materials are available for the particular service being considered. Liquid metals, even very dense liquid metals like mercury and liquid lead, can be metered. Since these metals are more dense than the stainless steel float, they are metered by an inverted rotameter. In this case, the flow is from top to bottom. When the meter is full of the liquid metal but there is no flow, the stainless steel float is buoyed up by the heavier liquid and rests at the inlet which is at the top. When there is flow, the flow forces the float down against the net buoyant force and the float takes a position related to the flow rate.

Gases

The rotameter is an inexpensive flowmeter for gas flow measurement. The pressure drop across the meter is essentially constant over the full 10:1 operating range. Pressure drop is low, generally less than 1 PSI (6.89 kPa). Special designs are available for even lower pressure drop.

Range

The position of the float in the metering tube varies in a linear relationship with flow rate. This is true over ranges up to 10:1. Percent of maximum and direct reading scales are used. Rotameters can directly measure flows as high as 4000 GPM (920 l/hr). Higher flow rates can be economically handled using the bypass-type rotameter. The capacity of the rotameter can be changed by changing the float. Various float configurations are available for higher capacities and generally permit a 2:1 change in capacity. By using the same housing but changing both the metering tube and the float a gross change in capacity is possible. These changes can account for both a change in flow rate and a change in fluid density.

Slurry Service

The rotameter tends to be self-cleaning. The velocity of the flow past the float, and the freedom of the float to move vertically, enables the meter to clean itself of some buildup of foreign material. Liquids with fibrous materials are one of the exceptions and should not be metered with rotameters. Generally, the size of particle, type of particle whether fibrous or particulate, and the abrasiveness of the particle determine the suitability of the rotameter for a given service. Also, the percent of solids by weight or by volume and the density of the solids influence the selection of the rotameter for this service.

Viscosity Effects

Rotameters are relatively insensitive to viscosity variations. In the very small rotameters with ball floats, this is not the case and the meters do respond to Reynolds number changes, which makes them sensitive to changes in both viscosity and density. However, the larger size rotameters are less sensitive. The viscosity immunity threshold can be as high as 100 centipoises (1 Pa·s). Meters can be operated above the viscosity limit; however, for these conditions the meter is calibrated for discrete viscosity conditions that are to be encountered and correction curves are furnished to adjust the indicated flow to the actual flow for the given viscosity.

Mass Flow

The rotameter can also be used to measure mass flow rate since the float responds to changes in fluid density. For a fixed volumetric flow rate, the float position in the metering tube will change with changing fluid density. The effect of fluid density changes on float position is a function of the relative densities of the float and the fluid. The closer the float density approaches the fluid density, the greater the effect for a given fluid density change. It has been derived that if the float density is twice the fluid density then the compensation for fluid density change is exact and the rotameter is a mass flowmeter. However, fluid density normally varies and since the float density is not adjustable to follow the fluid density changes, a compromise is made. The mean fluid density is used to establish the float density. A 10 percent fluid density change from the reference causes only a 0.5 percent inaccuracy in mass flow measurement. The mass rotameter can only be used for low viscosity fluids such as raw sugar juice, gasoline, jet fuels, and other light hydrocarbons.

Accuracy

Although the vast majority of rotameters are 2 percent to 10 percent of full scale types, they are available with percent of rate performance. Logarithmic scale meters are designed to give the same percent of rate accuracy at all scale positions over the 10:1 range of the meter.

Accuracy statements of 0.5 percent of rate and 1 percent of rate are available. The high accuracy type rotameter finds greatest application in laboratory, testing, development, and production where best accuracy is mandatory.

Piping Effects

The meter is not affected by upstream piping effects. The meter can be installed with practically any configuration of piping prior to the meter entrance.

Accessories

The rotameter is a highly developed flowmeter. The meters are available with an extremely broad selection of alarms, indicators, transmitters, totalizers, controllers, and recorders. A choice of totalizers, controllers, recorders, indicators, and alarms are available locally at the flowmeter. Practically any combination of system requirements can be handled by the accessories and instruments associated with rotameters.

Types of Rotameters

Rotameters can be categorized into the following groups:

Purge

These are perhaps the most widely used flowmeters and certainly are the most widely used form of the rotameter. These meters take many forms, all of which are inexpensive, and are intended for low flow measurement. Most purge meters are selected to handle inert gases or liquids at low flow rates where these fluids are used as a purge; therefore, accuracy is not critical. Repeatability is normally the required performance characteristic. Purge meters are available with optional needle control valves. Figure 2.22b shows a typical purge-type rotameter with integral needle control valve.

General Purpose

This category is the widely used glass tube version of the rotameter. Figure 2.22c shows a cross section view of a representative general purpose rotameter. The meter is almost always used for flow indication only. A wide choice of materials is available for the float, packing, O-rings, and end fittings to handle the widest selection of fluids. The only fluids which cannot be handled are those that attack the glass metering tube. The meters also are limited to the pressure and temperature extremes of the glass metering tube, and by safety considerations.

Armored or Metal Tube Meters

These meters are used when the general purpose meters cannot be applied. They would be used for hot (above 100°F or 38°C) and strong alkalies (above 20 percent concentration) fluorine, hydrofluoric acid, hot water (above 200°F or 93°C), steam, slurries, or molten metals where glass cannot be used. This classification of meters is used where the operating temperature and pressure exceed the ratings of the glass tube or generally where transmission of electronic or pneumatic signals is needed. A typical metal tube meter is shown in Figure 2.22d.

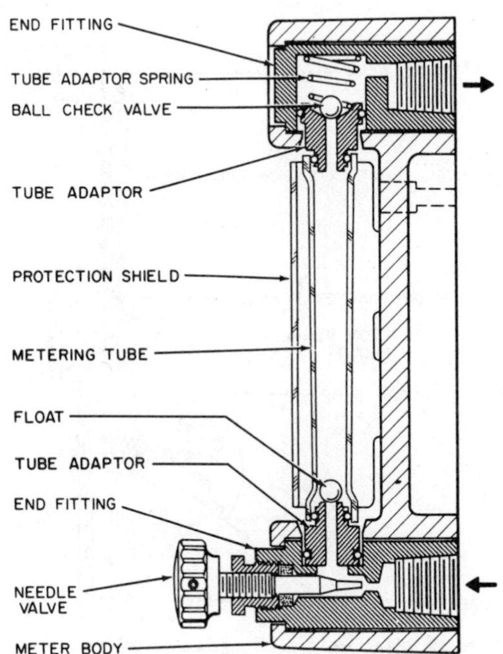

Fig. 2.22b Purge rotameter with integral needle valve

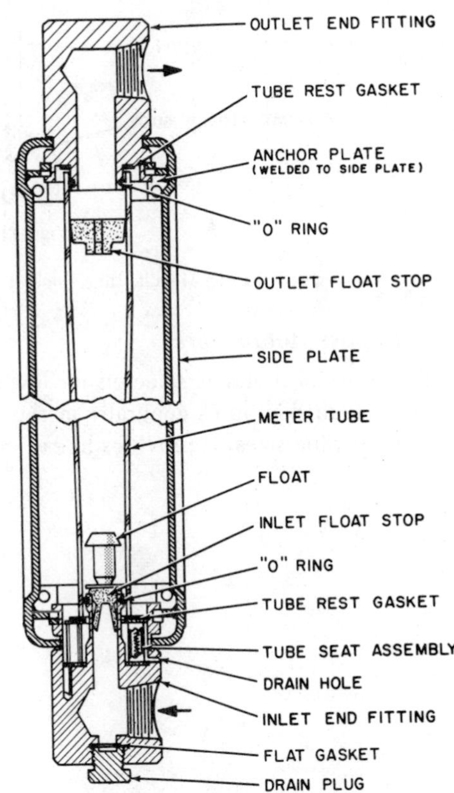

Fig. 2.22c Glass tube rotameter

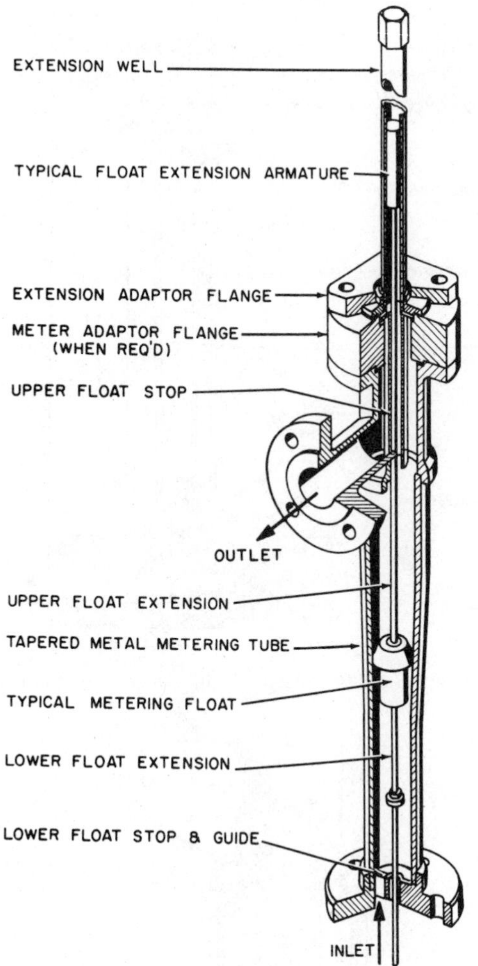

Fig. 2.22d Metallic tube rotameter

Bypass Rotameters

This class of meter is selected for low cost measurement of high flow rates generally in the 2 in. (50 mm) and larger pipe sizes. It provides linear measurement of fluid flow in conjunction with an orifice plate installed in the main line. The bypass rotameter measures the flow bypassing the main line orifice plate. The bypass rotameter is in parallel with the orifice plate in the main line. The rotameter is modified to include a range orifice which is sized so that the flow through the meter at maximum pressure drop across the orifice plate is equal to the flow rate necessary to lift the float to the maximum position. The flow through the range orifice is proportional to the instantaneous flow rate through the main line orifice; therefore; the bypass rotameter also measures the main line instantaneous flow rate. The indication is linear over the 10:1 range of the rotameter. The bypass meter is self-cleaning; therefore, there is generally no need for seal pots or water purges.

Accuracy

Rotameters can also be grouped based on accuracy of measurement.

a. 4 to 10 percent of full scale for all purge meters and the bypass rotameter.
b. 1 to 2 percent of full scale for the general purpose and metal tube meters in most industrial applications.
c. 0.5 to 1 percent of rate. The percent of rate performance is generally used for laboratory, development, and test type applications.

BIBLIOGRAPHY

Blasso, Leonard, "Flow Measurement Under Any Conditions," *Instruments and Control Systems*, February, 1975.

Cross, D.E., "Rotameter Calibration Nomograph for Gases," *Instrumentation Technology*, April, 1969, pp. 53–56.

Hall, John, "Solving Tough Flow Monitoring Problems," *Instruments and Control Systems*, February, 1980.

Lomas, D.J., "Selecting the Right Flowmeter," *Instrumentation Technology*, May, 1977.

Flow Sheet Symbol

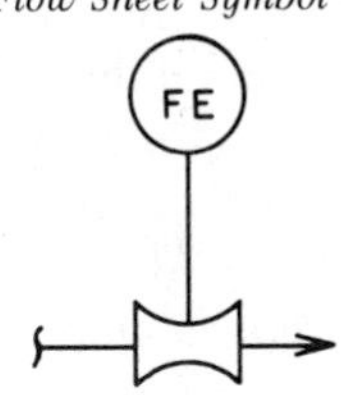

2.23 VENTURI TUBES, FLOW NOZZLES AND FLOW TUBES

Design Pressure:	Usually limited only by readout device or by pipe pressure ratings
Design Temperature:	Limited only by readout device, if operation is at very low or high temperature
Sizes:	Venturi tubes—1 in. (25 mm) up to 120 in. (3,000 mm) Flow nozzles—1 in. (25 mm) up to 60 in. (1,500 mm) Flow tubes—4 in. (100 mm) up to 48 in. (1,200 mm)
Fluids:	Liquids, gases and steam
Flow Range:	Limited only by minimum and maximum Beta (β) ratio and available pipe size range
Inaccuracy:	Venturi tubes—± 0.75% of rate uncalibrated, to ± 0.25% of rate calibrated in a flow laboratory Flow nozzles—± 1% uncalibrated to ± 0.25% calibrated Flow tubes—May range from ± 0.5% to ± 3% depending upon the particular design and variations in fluid operating conditions.
Materials of Construction:	Virtually unlimited. Cast venturi tubes are usually cast iron but fabricated venturi tubes can be made from carbon steel, stainless steel, most available alloys and fiberglass plastic composites. Flow nozzles are commonly made from alloy steel and stainless steel
Cost:	Dependent upon type of element, size and materials of construction. In general, flow nozzles are more economical than venturi tubes or flow tubes. Prices may range from low hundreds to several thousand dollars
Partial List of Suppliers:	Venturi tubes and flow nozzles—Daniel Industries; Flow-Dyne Engineering, Inc.; Fluidic Techniques, Inc.; Permutit Co., Inc.; Tech Tube Corp.; Vickery-Simms, Inc. Flow tubes—Badger Meter Co.; B.I.F., a division of General Signal; Fluidic Techniques, Inc.; ITT, Hammel-Dahl Div. (Gentile Patent Flow Tube); Vickery-Simms, Inc.

Theory

Venturi tubes, flow nozzles and flow tubes, like all differential pressure producers, are based upon Bernoulli's Theorem. General performance and calculations are similar to orifice plates. In these devices, however, there is continuous contact between the fluid flow and the surface of the primary device in contrast to the pure line contact between the orifice plate edge and main flow. Surface finish of the devices can have some effect on the meter coefficient, although the venturi tube has a relatively constant coefficient, seldom varying more than a fraction of 1 percent. Today's modern precision manufacturing techniques allow much greater accuracy of the coefficient for venturi tubes and flow nozzles computed from dimensions, and the coefficients are only moderately less reliable than those for orifice plates. The "C" or meter coefficient values for venturi tubes and flow nozzles have been well established with years of test data and are tabulated in reference sources such as the *ASME Fluid Meters Handbook*. In general, this is not true of the proprietary flow tubes and flow calibration is required to establish the actual meter coefficient. Meter coefficients for venturi tubes and flow nozzles are approximately 0.98 to 0.99 and for orifice plates average about 0.62. Therefore, almost 60 percent (98/62) more flow can be obtained through these elements for the same differential pressure.

The Venturi Tube

The venturi tube (Figure 2.23a) consists of (1) a straight inlet section of the same diameter as the pipe and in which the high pressure tap is located; (2) a converging conical inlet section in which the cross section of the stream decreases and the velocity increases with a consequent increase of velocity head and decrease of pressure head; (3) a cylindrical throat which provides for the low pressure tap location of this decreased pressure in an area where flow velocity is neither increasing nor decreasing; and (4) a diverging recovery cone in which velocity decreases and the decreased velocity head is recovered as pressure head. The pressure taps are located one-quarter to one-half pipe diameter upstream of the inlet cone and at the middle of the throat section. A piezometer ring is sometimes used for differential pressure measurement. This consists of several holes in the plane of the tap locations. Each set of holes is connected together in an annulus ring to give an average pressure.

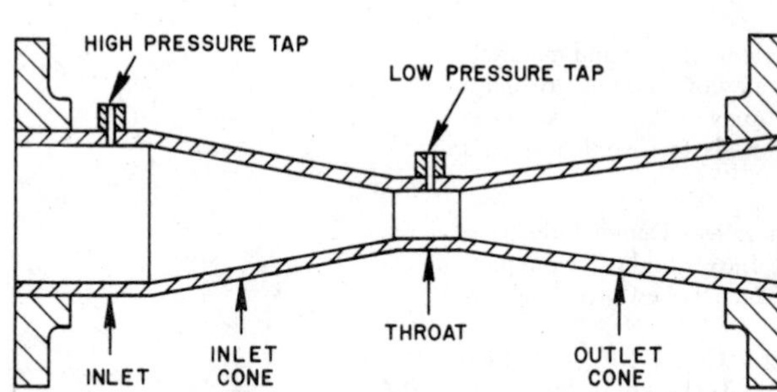

Fig. 2.23a Classic venturi tube

Venturis with piezometer connections are unsuitable for use with purge systems used for slurries and dirty fluids since the purging fluid tends to short circuit to the nearest tap holes. Piezometer connections are normally used only on very large tubes or where the most accurate average pressure is desired to compensate for variations in the hydraulic profile of the flowing fluid. Therefore, when it is necessary to meter dirty fluids and use piezometer taps, sealed sensors which mount flush with the pipe and throat inside wall should be used. These sensors function as independent measuring devices at each tap connection, yet function together to read differential pressure only, while automatically compensating for static pressure changes within the pipe. Single pressure tap venturis can be purged in the normal manner when used with dirty fluids. Because the venturi tube has no sudden changes in contour, sharp corners, projections or stagnant areas, it is often used to measure slurries and dirty fluids which tend to build up on or clog other primary devices.

Venturis are built in several forms. These include the standard long-form or classic venturi; a modified short form where the outlet cone is shortened; an eccentric form (Figure 2.23b) to handle mixed phases or to minimize buildup of heavy materials; and a rectangular form (Figure 2.23c) used in ductwork. If a rectangular venturi is substantially square, it is customary to converge-diverge all four sides with angles the same as for the circular form. Where duct width is different from height, the short sides are kept parallel with the long sides converging-diverging. A converging angle of 21 degrees and a diverging angle of 15 degrees gives satisfactory operation. Throat length should be equal to minimum throat height or width, whichever is smaller. Tap locations are the same as for the circular form.

The angle of convergence, which may range from 19 to 23 degrees, is the classical value established by Herschel in 1887. This angle is not particularly critical and

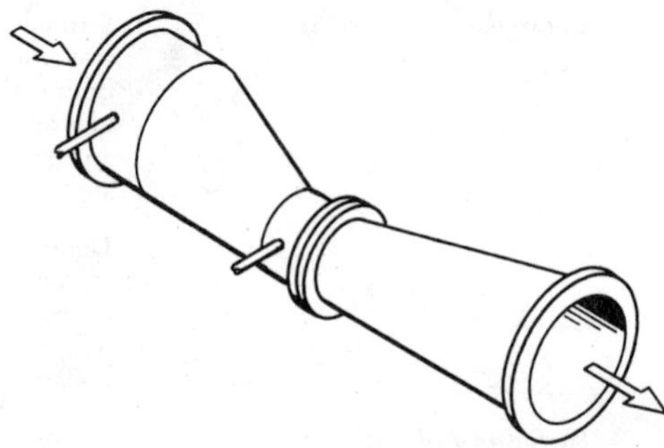

Fig. 2.23b Eccentric venturi tube

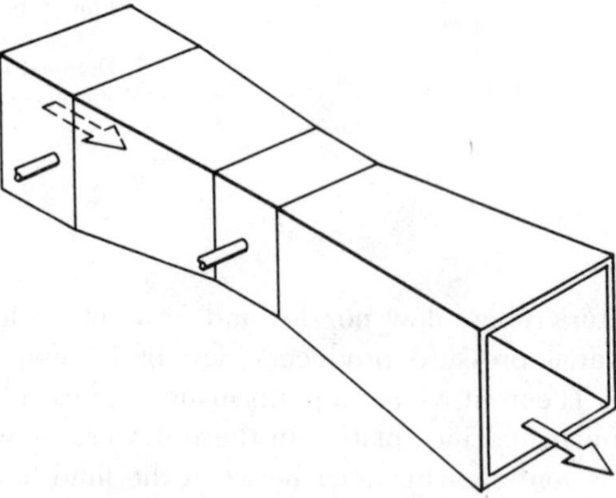

Fig. 2.23c Rectangular venturi tube

21 ± 1 degrees is commonly used. The recovery cone provides pressure recovery with its smooth flow transition. The classic long cone form is $7\frac{1}{2} \pm \frac{1}{2}$ degrees on the divergence, but up to 15 degrees is allowed, and the sharper angle allows the short-form version to be fabricated. The 15-degree outlet cone sacrifices a modest amount of pressure recovery (Figure 2.23d). The venturi pressure loss of 10 to 25 percent is the lowest of the standard primary head measurement elements. The long-cone form develops up to 89 percent pressure recovery at 0.75β ratio, decreasing to 86 percent at 0.25β ratio. The short-cone form develops up to 85 percent recovery at 0.75β, decreasing to 75 percent at 0.25β ratio. As an example of the power savings to be obtained in an energy-short era, an added pressure recovery of 50 inches of water (1270 mm of water) differential pressure can represent a 10-horsepower savings in a 24-inch (610 mm) waterline flowing at a velocity of 6 feet per second (1.829 meters per second). For a comparison of various head-meter elements from the pressure recovery point of view, see Figure 2.23e.

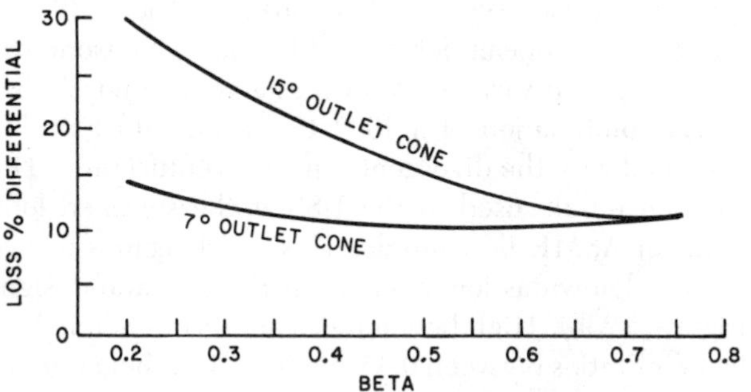

Fig. 2.23d Venturi pressure loss

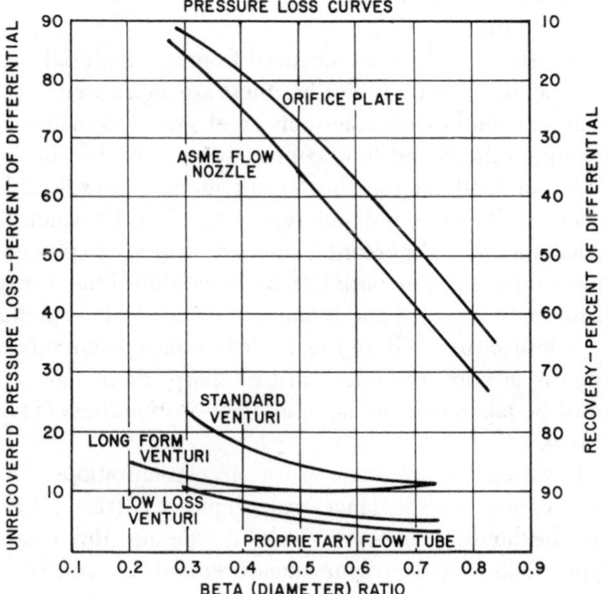

Fig. 2.23e Pressure loss curves

Limitations

The main limitation of venturi tubes is cost, both for the tube itself and often for the piping layout required for the length necessary in the larger sizes. However, the energy cost savings attributable to its higher pressure recovery and reduced pressure loss may easily justify a venturi tube in certain applications. Another restriction is the general limitation of a lower Reynolds number of 150,000, although some data is available down to a Re of 50,000 in some sizes. Also, due to its construction, a venturi is more difficult to inspect. This may not be a significant problem except in those applications subject to unusual erosion or corrosion; on the other hand, its hydraulic shape lends itself to greater dimensional reliability than devices such as orifice plates which depend upon a sharp edge for coefficient stability.

Accuracy

Operation and calibration of venturi tubes over a period of many years has resulted in extensive documentation. As a result, most manufacturers will guarantee a standard design inaccuracy of ± 0.75 percent of actual flow. This can be reduced to ± 0.25 percent by calibration at a recognized hydraulics laboratory. Modern manufacturing techniques have led to predictable discharge coefficients and a repeatability of ± 0.2 percent for venturi tubes of the same size and design.

For very small (under 4 inches or 100 mm) and very large (over 32 inches or 813 mm) venturi tubes and for very high (over 2,000,000) or very low (under 150,000) Reynolds numbers, flow calculations for venturi tubes have about a 50 percent greater uncertainty than a corresponding sharp-edged orifice plate. However, fluid flow calibration, particularly when made under conditions closely approximating service values, can provide a coefficient with practically the same accuracy as that of the calibration facilities.

Installation

A venturi tube may be installed in any position to suit the requirements of the application and piping. The only limitation is that with liquids the venturi is always full. In most cases, the valved pressure taps will follow the same installation guidelines as for orifice plates.

Upstream piping should be as long as needed to provide a proper velocity profile (Figure 2.23f). However, in most installations shorter upstream piping is required than for orifices, nozzles or pitot tubes because the venturi hydraulic shape itself provides some flow conditioning. Often, the combined length of a venturi and its upstream piping is less than the overall amount of piping required for an orifice or nozzle. Figure 2.23g shows typical upstream pipe diameters required for various elements at 0.7 Beta ratio and one elbow upstream. Straightening vanes can be used upstream to reduce the inlet pipe length. The vane installation should have a

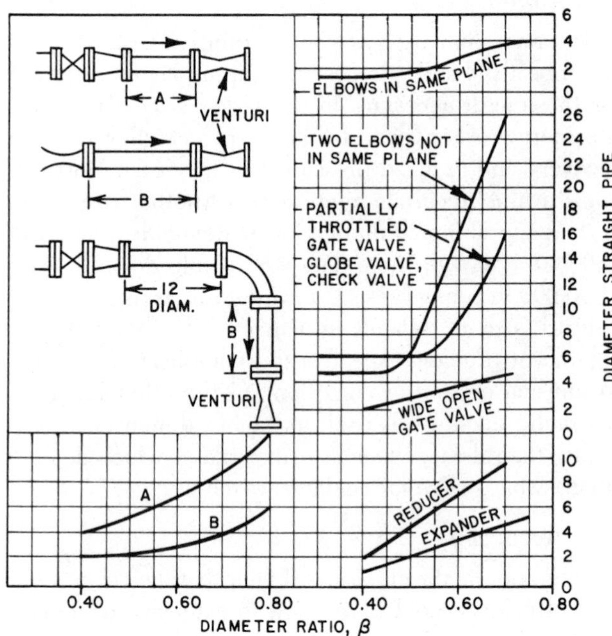

Fig. 2.23f Venturi piping requirement

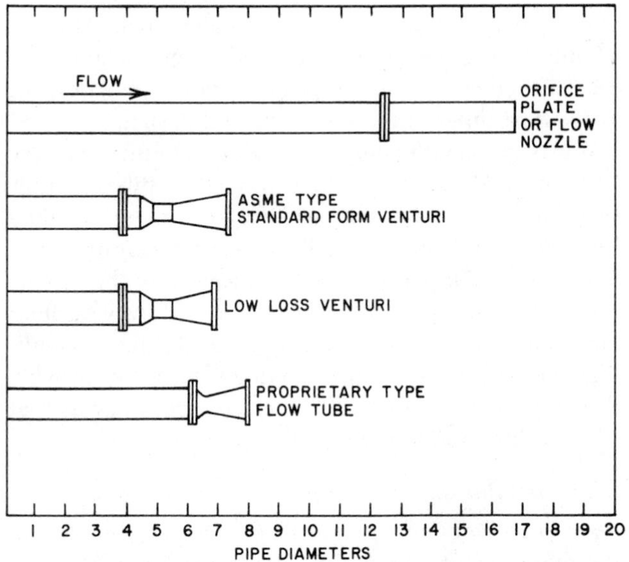

Fig. 2.23g Typical installation piping comparison

minimum of two pipe diameters upstream and the same distance downstream before entering the venturi. There is no limitation on piping configuration downstream of the venturi except a straight run of two pipe diameters before a valve.

Flow Calculations

The American Society of Mechanical Engineers Fluid Research Committee has adopted a general coefficient of discharge of 0.984 for the classic rough-cast entrance cone venturi tube from 4 inches (100 mm) through 32 inches (813 mm) and for beta ratios between 0.3 and 0.75. For tubes with machined entrance cones, the general coefficient is 0.995. Reynolds number must be 200,000 or greater. Approximate flow rates can be calculated from a working equation:

$$W = 353d^2 \sqrt{\frac{h\rho}{1 - \beta^4}} \qquad 2.23(1)$$

and, for approximate venturi tube design:

$$\beta = \frac{1}{\sqrt[4]{\frac{1 + 125{,}000\, h\rho D^4}{W^2}}} \qquad 2.23(2)$$

For Reynolds numbers between 50,000 and 200,000 substitute 344 instead of 353. Below 50,000, reliable data is not available. It should be noted that, in contrast to an orifice, a decrease in Reynolds number results in a decrease of flow corresponding to a given differential pressure. Other correction factors such as temperature coefficient of expansion, gas expansion factor, and so on are similar to those for orifices and flow nozzles.

Flow Nozzles

There are two types of flow nozzles. The 1932 ISA nozzle is a European design that has not seen use in the USA. A special variation known as a venturi nozzle is a hybrid combination of a 1932 ISA nozzle inlet profile combined with the divergent cone of a venturi tube. The common nozzle used in the USA is the so-called long radius or ASME flow nozzle. This nozzle comes in two versions known as low beta and high beta ratio designs (Figure 2.23h). High beta nozzles are recommended for diameter ratios between 0.45 and 0.80. Low beta nozzles are recommended for diameter ratios between 0.20 and 0.50. For beta values between 0.25 and 0.5, either design may be used. Both types of nozzles consist of a convergent inlet whose shape is a quarter ellipse and a cylindrical throat.

Nozzles may be manufactured from any material that can be machined; typically, they are fabricated from stainless steel or chrome-moly steel. Modern manufacturing methods and fluid contact surface finishes in the order of 6-10 microinches result in more predictable nozzle coefficients and high repeatable data. Flow nozzle inaccuracy of ± 1 percent is standard with ± .25 percent flow calibrated. The outlet or discharge side of the nozzle is normally beveled and is one of the more critical points of manufacture. Where the 10-degree back angle meets the throat bore, the edge must be sharp. Particular care must be taken to avoid taper and out-of-roundness of the throat.

Flow nozzles are made in various configurations. The most common is the flange-type (Figure 2.23i) but others are the holding ring-type, weld-in-type and throat-tap-type. Differential pressure measurement taps are commonly located one pipe diameter upstream and one-half pipe diameter downstream from the inlet face (USA prac-

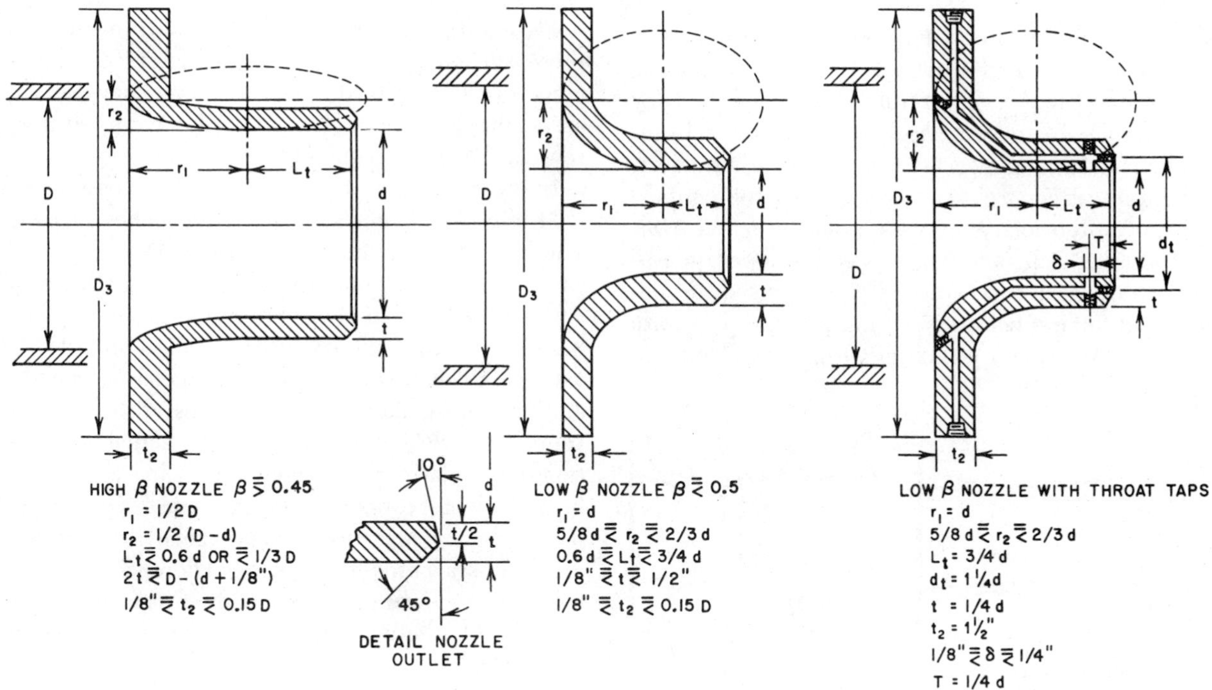

Fig. 2.23h ASME nozzle construction

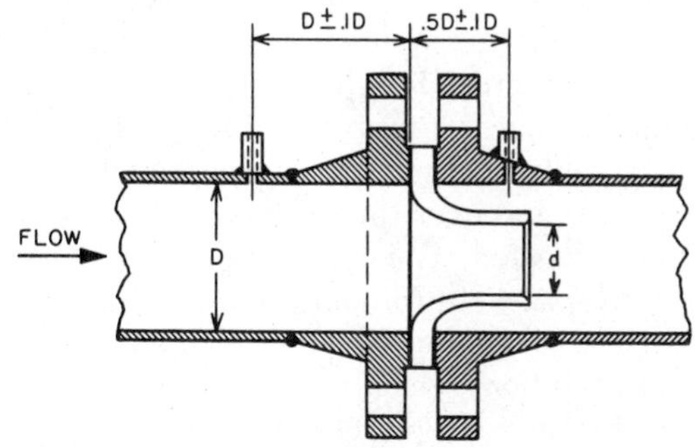

Fig. 2.23i Typical nozzle installation

tice), except for the throat-tap-type which has a special downstream construction. Tap installation precautions are the same as for orifice plates. The preferred installation position for flow nozzles is horizontal. However, a vertical downflow position is preferred for wet steam or gases and liquids with suspended solids. In general, upstream and downstream piping requirements are similar to those required for orifices. Because of width, nozzles installed between flanges are difficult to remove. Common practice is to provide a flange in the downstream piping to allow the nozzle to be removed as part of a spool section.

Flow nozzles are particularly suited for measurement of steam flow and other high velocity fluids, fluids with some solids, wet gases, and similar materials. Since the exact contour is not critical, the flow nozzle can be expected to retain good calibration for a long time under erosion or other hostile conditions. Because of its streamlined contour, it tends to sweep solids or moisture through the throat and is far superior to orifice plates in these services.

A flow nozzle will pass about 60 percent more flow than an orifice plate of the same diameter and differential pressure. It also has the advantage of operating acceptably over the wide beta ratio range of 0.2 to 0.8. For the same flow and differential pressure, the flow nozzle has a similar but slightly lower pressure loss than an orifice plate. This becomes apparent when it is recognized that the area of the throat and the velocity in the throat of a flow nozzle must be approximately the same as the area of flow and velocity at the vena-contracta following an orifice in order to develop the same differential pressure from the same flow. The slightly lower pressure loss of the flow nozzle is due to its streamlined entrance.

While nozzles should be used at Reynolds numbers of 50,000 or above, data is available for Re down to 6,000 so it is possible to use nozzles with more viscous fluids. Flow nozzles have very high coefficients of discharge, typically 0.99 or greater. Using a typical value of 0.993, approximate flow rates can be calculated from a working equation:

$$W = 358d^2\sqrt{\frac{h\rho}{1-\beta^4}} \qquad 2.23(3)$$

and, for approximate flow nozzle design:

$$\beta = \frac{1}{\sqrt[4]{1+\frac{128{,}000h\rho D^4}{W^2}}} \qquad 2.23(4)$$

Flow Tubes

There are several proprietary primary-head-type devices which have a higher ratio of pressure developed-to-pressure lost than a venturi tube (Figure 2.23j). They are all considerably more compact than the classical venturi tube with its long recovery cone, although the short-form venturi can come close to some types of these tubes. All of these proprietary units are available in the USA except for the Dall tube which was developed in England. All of these tubes vary in contour used, tap locations and differential pressure and pressure loss for a given flow. All have a laying length less than four diameters long.

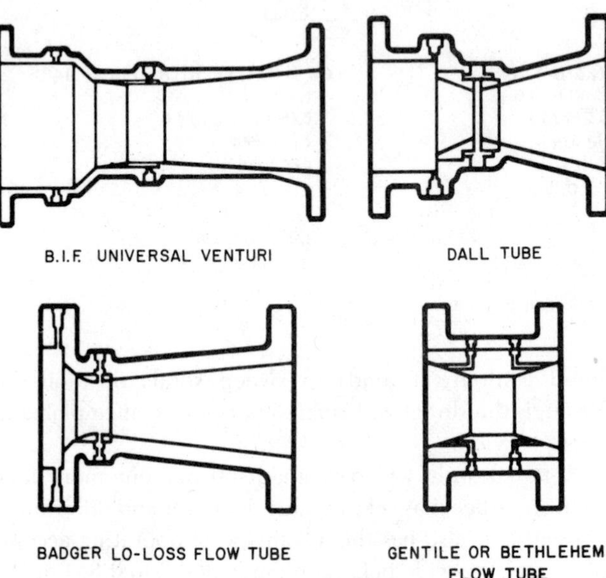

Fig. 2.23j Proprietary flow tubes

The B.I.F. Universal Venturi is the product that most closely approaches the Herschel design classic venturi. The inlet cone has two vena-contracta angles which condition the fluid as it enters the throat that is claimed to reduce the sensitivity to upstream piping configuration and give higher accuracy. Also claimed are a stable coefficient (0.9797) unaffected by internal surface roughness, lower Reynolds number application (90,000), low head loss (4–18 percent) and extensive documentation including expansion factors.

While these devices can be useful in larger sizes because of their shorter lay length, they may also require longer upstream pipe runs than the venturi for proper performance and thus lose any real advantage. They can be subject to coefficient change with viscosity and Reynolds number; the manufacturers can provide data on these effects. None has the smooth contour and resistance to clogging of the venturi meter; however, some are claimed to operate satisfactorily on wastewater and sewage flow measurement.

In general, these devices are available only in 4-inch (100 mm) and larger sizes up to 48 inches (1219 mm). There is little justification for their use in small flow, small pipe applications. In the larger sizes, the installed cost may be less than the venturi tube. Accuracy depends basically on the manufacturer's calibration data. Derivation of the flow coefficient by extrapolation from theory and tests on smaller sizes is much less direct than in the simple structure of the venturi tube; actual flow calibration, particularly in sizes above 24 inches (610 mm), can be difficult and expensive. While these devices generally have a better pressure recovery than the venturi (expressed as a percentage of the differential), most flow tubes have a lower coefficient of discharge (less efficient). As a result, there is often very little difference in the actual head loss.

In selecting a primary flow element, the possible advantages of slightly lower pressure loss and shorter laying length of the flow tubes should be carefully weighed against the metering accuracy and well established flow data available on the Herschel-form venturi. The ASME recommends that if a proprietary flow tube is used, it should be calibrated with the piping section in which it is to be used and over the full range of flows to which it will be subjected. The only possible exception to this is the Universal Venturi, and it must be carefully evaluated. The background of extensive tests under a wide range of conditions which supports orifice meters does not exist for these proprietary devices.

BIBLIOGRAPHY

Bean, H.S., ed., "Fluid Meters—Their Theory and Application," ASME Research Committee on Fluid Meters, 6th ed., 1971.

ISO/TC 30 231 E. Draft ISO Recommendation No. 1–157, November, 1966.

Lipták, B.G., ed., *Instrument Engineer' Handbook,* Volume I. Radnor, PA.: Chilton, 1969.

Shell Flow Meter Engineering Handbook, Royal Dutch/Shell Group. Delft, The Netherlands: Waltman Publishing Co., 1968 (now out of print).

Spink, L.K., *Principles and Practice of Flow Meter Engineering,* 9th ed., The Foxboro Co., 1967.

Flow Sheet Symbol

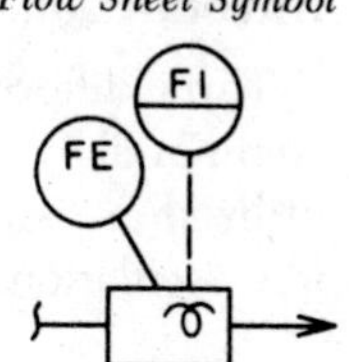

2.24 VORTEX FLOWMETERS

Type:	A—Vortex shedding B—Vortex precession C—Fluidic
Size:	A: 1 to 8 in. (25 to 203 mm) (larger as special design) B: 1 to 6 in. (25 to 152 mm) C: 1 to 4 in. (25 to 100 mm)
Flow Range:	A: Liquids: 8.8 to 2200 GPM (5.5×10^{-4} to 0.14 m^3/S) Gases: 2.9 to 2500 AGFM (1.4×10^{-3} to 1.11 m^3/S) B: 1.76 to 1500 ACFM (8.3×10^{-4} to 0.722 m^3/S) C: 1–1000 GPM (3×10^{-3} to 3 m^3/S)
Type of Fluid Metered:	A: Liquid, gas or steam B: Gas C: Liquid
Output Signal:	A: Linear digital (analog optional) B: Linear digital C: Linear digital
Type of Flow Measurement:	Volume flow rate
Design Pressure:	A: 1500 PSIG (10.3 MPa; higher as special design) B: 1500 PSIG (10.3 MPa) C: 600 PSIG (4.13 MPa)—1, 1½ in. (25, 38 mm) 150 PSIG (1.03 MPa)—2–4 in. (50–100 mm)
Design Temperature:	A: Typically −40 to +300°F (−40 to +150°C) B: 250°F (121°C) C: 0–250°F (−18 to +121°C)
Materials of Construction:	A: Carbon steel or stainless steel (other materials as special design) B and C: Stainless steel
Inaccuracy:	A: ± 0.75% of rate B: ± 1.25% of rate C: ± 1% of rate
Repeatability:	A: ± 0.15% B: ± 0.25% C: ± 0.2%
Cost:	A: $1300 for 2 in. (50 mm) diameter through $2000 for 6 in. (152 mm) diameter. Prices subject to variation depending on material and specification B: $2200 for 1 in. (25 mm) diameter through $4400 for 6 in. (152 mm) diameter C: $1600 for 1 in. (25 mm) diameter through $2000 for 4 in. (100 mm) diameter
Partial List of Suppliers:	Bopp & Reuther GmbH (A); Brooks Instrument Div. of Emerson Electric Co. (A); Fischer & Porter (A,B); Foxboro Co. (A); Kent Process Control, Inc. (A); Moore Products (C); Neptune/Eastech Inc. (A); Yokagawa Corp. of America (A)

Introduction

Three different types of vortex flowmeter are currently commercially available. The three operating principles utilized are vortex shedding, vortex precession and fluidic oscillation.

Vortex Shedding Flowmeters—Operating Principles

When a fluid flows past an obstacle, boundary layers of slow moving fluid are formed along the outer surfaces. If the obstacle is unstreamlined—that is, it is a bluff body—the flow cannot follow the contours of the obstacle on the downstream side and the separated layers become detached from the main stream of fluid and roll into eddies or vortices in the low pressure area behind the body (see Figure 2.24a). The vortices are shed from alternate sides of the body. The frequency at which the vortices are formed is directly proportional to the fluid velocity, thus providing the basis of a flowmeter. A perfect example of vortex shedding is a flag fluttering in the breeze.

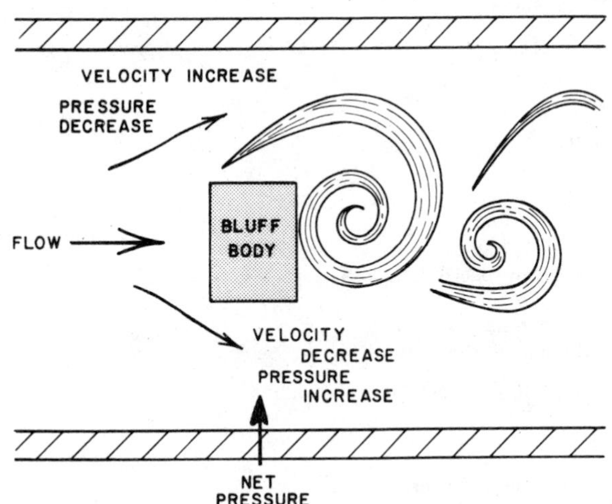

Fig. 2.24a' Vortex shedding from a bluff body causes velocity and pressure adjacent to the body to change

As a vortex is shed from one side of the bluff body, the fluid velocity on that side increases and the pressure decreases. On the opposite side the velocity decreases and the pressure increases, thus causing a net pressure change across the bluff body. The entire effect is then reversed as the next vortex is shed from the opposite side. Consequently, the velocity and pressure distribution adjacent to the bluff body change at the same frequency as the vortex shedding frequency.

Various detectors can be used to measure one of the following (see Figure 2.24b):

a. The oscillating flow across the face of the bluff body.
b. The oscillating pressure difference across the sides of the bluff body.
c. A flow through a passage drilled through the bluff body.
d. The oscillating flow or pressure at the rear of the bluff body.
e. The presence of free vortices in the downstream to the bluff body.

A flow-sensitive detector can be either a heated thermistor element or a spherical magnetic shuttle (with the movement of the shuttle measured inductively). Detectors that are sensitive to pressure use either metal diaphragms or vanes. Pressure exerted on diaphragms can be converted into a variable capacitance, or a variable strain on a piezo-resistive, piezoelectric or inductive sensor. Pressure exerted on vanes can similarly be converted into an electrical signal through any of the aforementioned sensors. Alternatively, the velocity components in the free vortices downstream of the bluff body can be used to modulate an ultrasonic beam diametrically traversing the meter housing. Depending on the characteristics of the sensing system, the flowmeter will be suitable for liquid or gas or both. The fundamental meter output is a frequency signal in all cases, which can be fed directly into digital electronic units for totalization and/or preset batching, computers or data loggers. The frequency signal also can be converted into a conventional 4–20mA DC analog signal for flow rate indication, recording or control purposes. Most meters are available in either a standard form or a design to satisfy Division 1 safety requirements.

Features

The vortex shedding meter provides a linear digital (or analog) output signal without the use of separate transmitters or converters, simplifying equipment installation. Meter accuracy is good over a potentially wide flow range although this range is dependent upon operating conditions. The shedding frequency is a function of the dimensions of the bluff body and is a natural phenomenon, ensuring good long term stability of calibration and repeatability of better than ± 0.15 percent of point. There is no drift because this is a frequency system.

The meter does not have any moving or wearing components, providing improved reliability and reduced maintenance. Maintenance is further reduced by the fact that there are no valves or manifolds to cause leakage problems. The absence of manifolds and valves results in a particularly safe installation, an important consideration when the process fluid is hazardous or toxic.

If the sensing system utilized is sufficiently sensitive, the same vortex shedding meter can be used on both gas and liquid. In addition, the calibration of the meter is virtually independent of the operating conditions (viscosity, density, pressure, temperature, and so on) whether the meter is being used on gas or liquid (see Figure 2.24c).

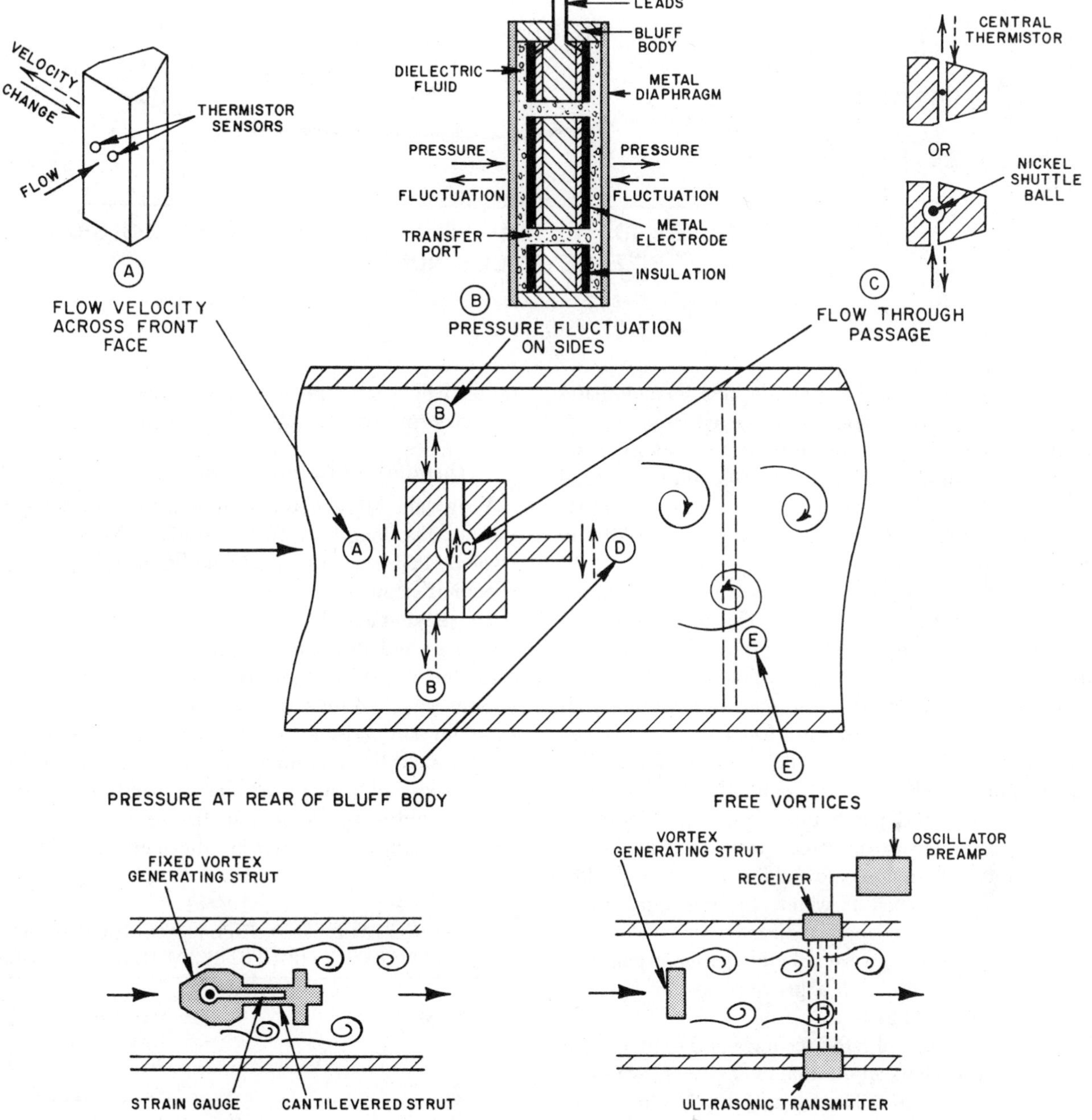

Fig. 2.24b A selection of different positions and methods of detecting vortices

The vortex shedding meter also offers a low installed cost, particularly in pipe sizes below 6 in. (152 mm) diameter, which compares competitively with the installed cost of an orifice plate and differential pressure transmitter.

The principal application limitations with the vortex shedding meter are covered in the "Sizing and Selection" section. Another limitation with the meter is available size range. Meters below 1 in. (25 mm) diameter are not practical and meters above 8 in. (203 mm) have limited application due to their high cost compared to an orifice system and their limited output pulse resolution. The number of pulses generated per unit volume decreases on a cube law with increasing pipe diameter. Consequently, a 24 in. (610 mm) diameter vortex shedding meter with a typical blockage ratio of 0.3 would only have a full scale frequency output of approximately 5 Hz at 10 ft./sec. (3 m/sec.) fluid velocity.

Selection and Sizing

Initially, the operating conditions (process fluid temperature, ambient temperature, line pressure, and so on) should be compared with the meter specification. The meter wetted materials, including bonding agents, and sensors should then be checked for compatability with the process fluid both with regard to chemical attack

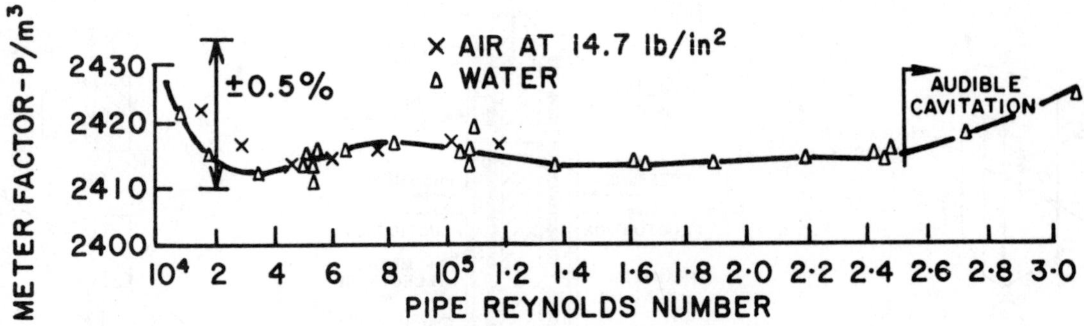

Fig. 2.24c Typical calibration curves for a 3 inch (76 mm) vortex meter showing the close correlation between water and atmospheric air calibrations

and safety. On oxygen, for example, non-ferrous materials should be used due to the reactive nature of oxygen. Applications where there are large concentrations of solids, two phase flow or pulsating flow should be avoided or approached with extreme caution. The meter minimum and maximum flow rates for the given application should then be established.

A typical performance curve for a vortex shedding flowmeter is shown in Figure 2.24c. The meter minimum flow rate is established by a Reynolds number of 10,000, fluid density and a minimum acceptable shedding frequency for the electronics. The maximum flow rate is governed by the meter pressure loss (typically two velocity heads), the onset of cavitation with liquids and compressibility with gases. Consequently, the flow range for any application depends totally upon the operating fluid viscosity, density and vapor pressure, and the application's maximum flow rate and line pressure. On low viscosity products such as water, gasoline, and liquid ammonia, and with an application maximum velocity of 15 ft./sec. (0.45 m/sec.), vortex shedding meters can have a rangeability of 40:1 with a pressure loss of approximately 4 PSIG (27.4 kPa).

The meter's good "of rate" accuracy and digital linear output signal make its application over wide flow ranges a practical proposition. The rangeability declines proportionally with increases in viscosity or reductions in the maximum flow velocity of the process. Vortex shedding meters are therefore unsuitable for high viscosity liquids.

On liquid applications, it is necessary to verify that sufficient line pressure exists to prevent cavitation in the vortex meter. The maximum pressure drop in a vortex shedding meter is in the region of the bluff body and there is a considerable pressure recovery by the meter outlet. Upstream line pressure requirements vary from one meter design to another, but a typical minimum acceptable upstream pressure requirement is given by the expression: Upstream pressure ≥ 1.3 (vapor pressure plus 2.5 times net pressure loss). Although cavitation conditions should be avoided, vortex shedding meters do offer the advantage that if the liquid "gasses," the meter will not be mechanically damaged (although the meter output will be seriously in error).

Installation Requirements

Vortex shedding meters require a fully developed flow profile. The length of upstream pipework necessary to ensure satisfactory approach conditions depends on the specific design of meter, the type of upstream disturbance present and the level of accuracy required. Typical upstream and downstream pipework requirements for a variety of disturbances are given in Table 2.24d. The lengths required for a typical orifice installation are also listed for comparison. Where there is a severe upstream disturbance, the resulting long straight lengths of pipe can be reduced by fitting a radial vane or bundle-of-tubes flow-straightening element in the upstream pipework. Wherever possible, however, the meter should be in-

Table 2.24d
TYPICAL INSTALLATION PIPEWORK REQUIREMENTS FOR A VARIETY OF UPSTREAM DISTURBANCES

	Upstream Straight Pipe Diameters		
		ISO R541 standard Orifice Dia. Ratio	
	Vortex Meters	*0.6*	*0.7*
Single bend (medium sweep 90°)	15	18	28
Elbow (90°)	15	*	*
Double bend (180°)	15	*	*
Two or more 90° bends in same plane	25	26	36
Two or more 90° bends in different planes	25	48	61
Reducing cone (15°)	15	9	14
Expansion cone (15°)	15	22	30
Butterfly valve fully open	5	*	*
Butterfly valve 20° open	25	*	*
Gate valve fully open	5	*	*
Gate valve 30% open	30	*	*
10° swirl	50	*	*
20° swirl	100	*	*
30° swirl	150	*	*

*Information not available in ISO R541

stalled upstream of any severe source of disturbance such as regulating control valves. The downstream straight pipe requirement is 5 times nominal meter diameter. The meter can be installed in any attitude (horizontal or vertical) but it is not suitable for reverse-flowmetering.

Vortex Precession (Swirl) Meters—Operating Principles

Construction of a typical vortex precession (swirl) meter and the operating principles are illustrated in Figure 2.24e. The fixed swirl-inducing helical vanes at the entrance to the meter introduce a spinning or swirling motion to the fluid. After the exit of the swirl vanes the bore of the meter contracts progressively causing the fluid to accelerate, but with the axis of rotation still on the center line of the meter. The swirling fluid then enters an enlarged section in the meter housing which causes the axis of fluid rotation to change from a straight to a helical path. The resulting spiraling vortex is known as vortex precession. The frequency of precession is proportional to velocity and, hence, volumetric flow rate above a given Reynolds number.

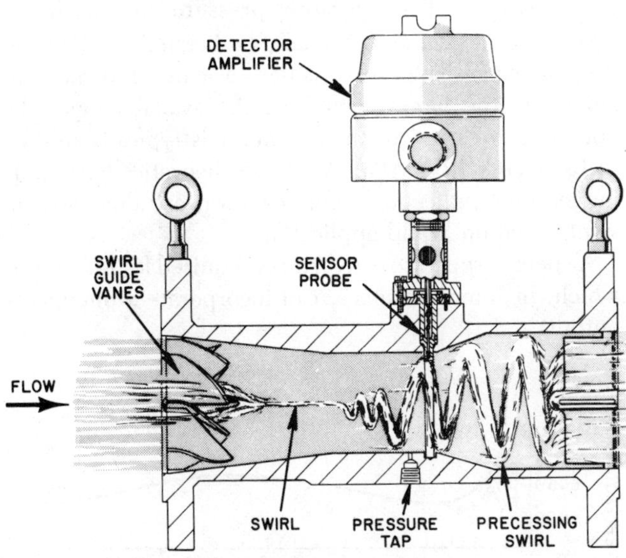

Fig. 2.24e Construction of a typical vortex precession (swirl) meter

The fluid velocity in the vortex is higher than that of the surrounding fluid. Consequently, as each vortex passes the sensor, there is a change in the local fluid velocity. The frequency at which the velocity changes occur is proportional to volumetric flow rate and can be detected by piezoelectric or thermistor sensors. The thermistor, which changes its resistance with temperature, is heated by a constant electrical current. The amount of heat extracted from the thermistor by the passing fluid is dependent upon the fluid velocity. Consequently, each "high velocity" vortex past the thermistor changes the resistance and, since a constant current is applied, the resistance change is detected as a voltage change. After amplification and filtering, a square wave voltage output signal is provided, linear with flow rate.

A flow-straightener is fitted at the meter outlet in order to isolate the meter from downstream piping effects which might otherwise impair the development of the precessing vortex.

General Characteristics

The general characteristics of the swirl meter are similar to those of the vortex shedding meter: no moving components, linear digital output signal, good flow range capability and long term stable calibration. The swirl meter does, however, require calibration and it also has a high pressure loss, namely ten velocity heads ($10\rho V^2$). The pressure loss is generally excessive for most liquid applications. On water, for example, at 15 ft./sec. (4.5 m/sec.) the pressure loss is approximately 20 PSIG (137.8 kPa). Swirl meters are primarily used in gaseous applications, where a very much lower density results in a significantly lower pressure loss.

The flow range which can be achieved with the meter depends upon the type of sensor and the gas density. Typical flow ranges and pressure losses for a range of different size swirl meters on air at atmospheric pressure conditions are given in Table 2.24f. It is important to note that sizing a vortex precession meter (or any other form of vortex meter) must be based upon actual operating conditions. The meter output is in actual volumetric units and if the pressure and/or temperature varies, a pressure and temperature compensation system is required to give readout in standard volumetric units or mass units.

Table 2.24f
TYPICAL SWIRL METER CAPACITIES

Size		*Overall Flow Range*		*Pressure Loss**	
Inches	*mm*	*ACFM*	*m³/hr*	*PSIG*	*N/m²*
1	25	1.76/17.6	3/30	0.36	2500
2	50	17.6/176	30/300	0.51	3500
3	80	4/410	70/700	0.58	4000
4	100	65/650	110/1100	0.72	5000
6	150	150/1500	260/2600	0.87	6000

*Pressure loss is for gas at 15°C and 14.7 PSIA pressure

A typical vortex precession meter calibration is shown in Figure 2.24g. The similarity of form with a vortex shedding meter will be noted.

The built-in swirl-inducer at the meter entry and flow-straightener at the meter outlet enable a vortex precession meter to be installed with a minimum of upstream and downstream straight pipe lengths. Normally, 5 times nominal meter diameter upstream and 3 times nominal meter diameter downstream are sufficient.

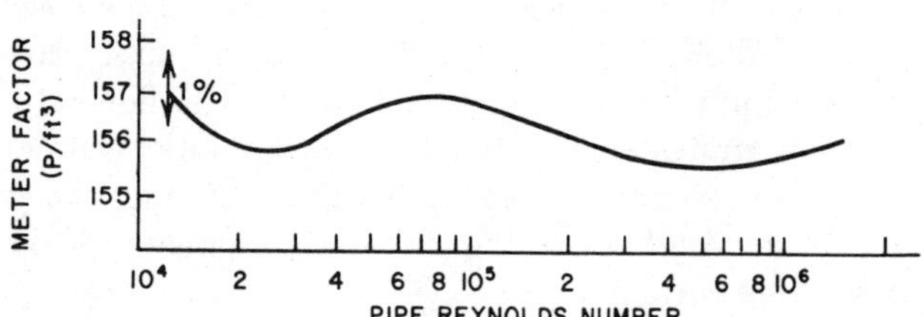

Fig. 2.24g Typical calibration of a swirl meter

The internal components of the swirl meter require a significant amount of complex machining; thus, it is more expensive than some other meter types.

Fluidic (Coanda Effect) Meter—Operating Principles

Fluid entering the meter is entrained into a turbulent jet from its surroundings, causing a reduction in pressure. The internal geometry of the meter body causes the jet to be deflected from its central position and initially attach itself to one of the side walls. The jet curvature is sustained by the pressure differential across the jet. If sufficient volume of fluid is then introduced into the control port on that side, it will cause the jet to switch to the opposite side wall. This is known as a Coanda Effect. The jet can be made to oscillate by one of two methods. The simplest method is a relaxation oscillator. In this system, the two ports are connected together. Fluid is sucked from the high pressure side to the low pressure side causing the jet to switch to the other wall.

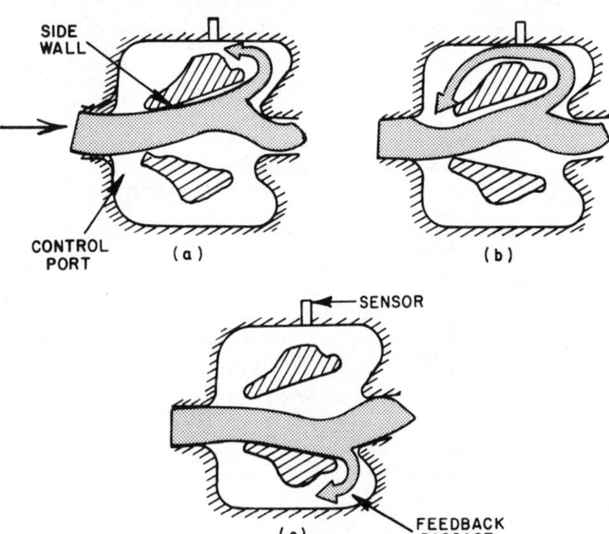

Fig. 2.24h Diagram of the mode of operation of a feedback oscillator

The jet thus continues to oscillate as the fluid is sucked alternately from one side to the other. The more commonly used system is the feedback oscillator (see Figure 2.24h). The deflected jet causes a low pressure area at the control port. At the upstream feedback passage the pressure is higher due to a combination of the jet expanding and the stagnation pressure. Thus, a small portion of the main stream of fluid is diverted through the feedback passage to the control port. The feedback flow intersects the main flow and diverts it to the opposite side wall. The whole feedback operation is then repeated, resulting in a continuous self-induced oscillation of the flow between the side walls of the meter body. The frequency of oscillation is linearly related to the volumetric flow rate above a minimum Reynolds number. As the main flow oscillates between the side walls, the flow in the feedback passages oscillates between zero and a maximum value. This frequency is detected by means of a thermistor sensor, providing a frequency output signal.

Characteristics

The principal features include no moving components, fixed calibration based on the geometry of the housing, linear digital or analog output and good rangeability. One advantage over other forms of vortex meter is that fluidic meters can operate down to a Reynolds number of 3,000. The maximum flow range (dependent on size and viscosity) is 30:1. The operating pressure and maximum practical pipe diameter are largely dictated by the complex housing shape. In practice, a 4 in. (100 mm) diameter is the largest commercially available, and the operating pressure in this diameter is typically limited to 150 PSIG (1.03 MPa). Although theoretically suitable for gaseous applications, fluidic meters are almost exclusively used on liquid applications.

A special, separate converter is required for the meter, which, in some instances, can incorporate a pneumatic output.

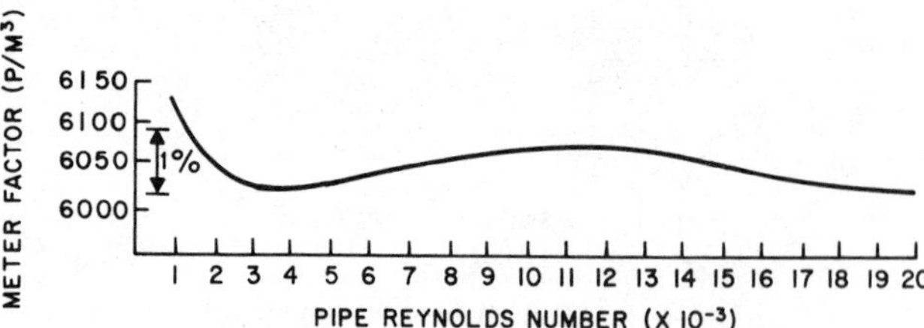

Fig. 2.24i Typical calibration for a 1 inch (25mm) fluidic meter

BIBLIOGRAPHY

Cousins, T., "The Performance and Design of Vortex Meters," Fluid Flow Conference, East Kilbride, 1975.

Herzl, J., "New Sensing Techniques and Modular Construction as applied to the Swirl Meter," ISA 28th Annual Conference, Pittsburgh.

Lomas, D.J., "Vortex Meters—A Practical Review," Measurement Technology for the 80's, ISA Symposium, Delaware 1979.

Medlock, R.S., "Vortex Shedding Meters," Liquified Gas Symposium, London 1978.

Within, W.G., "Theory, Design and Application of Vortex Shedding Flowmeters," Measurement Technology for the 80's, ISA Symposium, Delaware 1979.

Flow Sheet Symbol

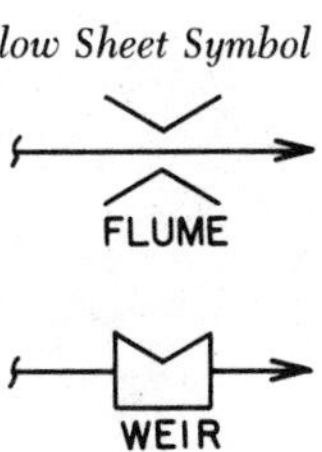

2.25 WEIRS AND FLUMES

Operating Pressure and Temperature:	Atmospheric
Fluids:	Liquids
Flow Range:	From 1 GPM (3.785 l/m) to any maximum flow
Inaccuracy:	± ½% to ± 5% of full scale
Cost:	Usually field fabricated
Partial List of Suppliers:	Atec, Inc.; B.I.F. Industries; Badger Meter Inc., Precision Product Div.; Fischer & Porter Co.; Industrial Technology, Inc.; Instrumentation Specialties Co., Environmental Div.; Inventron Industries, Inc.; F.B. Leopold Co., Inc.; Leupold & Stevens, Inc.; Manning Environmental Corp.; Marsh-McBirney, Inc.; Milltronics, Inc.; N.P. Industries, Inc.; Taylor Instrument Co., Div. of Sybron Corp.; Teledyne Gurley; Vickery-Simms, Inc., a Standco Co.

General

Weirs, flumes and similar devices develop a liquid head which is used to measure flow rate. Their application is primarily in water works, including irrigation, waste and sewage systems, and similar installations where flow is handled in open channels or in pipes and conduits that are generally not completely filled with liquid. While based on the same principles as other head-type flow meters, they differ in structure, characteristics and readout mechanisms.

Characteristics

Measurement characteristics of weirs and flumes can be established from physical dimensions without actual flow calibration. Theoretical derivations are supported by test data over a wide range of operating conditions. This is of particular importance in applications to very large flows—10,000 GPM (630 l/s) and up—where flow calibration would be difficult and expensive. Weir measurements in millions of gallons per day are not unusual.

For rectangular and trapezoidal weirs and Parshall flumes, flow is approximately proportional to the three-halves power of the measured head. For "V" notch weirs, flow is proportional to the five-halves power of head. This results in a wide range of flow measurement in a single device—75 to 1 for rectangular and trapezoidal weirs and Parshall flumes, and up to 500 to 1 for "V" notch weirs.

Head on weirs and flumes is customarily measured in terms of level or head of the flowing fluid. This results in a direct relation between head and actual volume flow with no correction for liquid density. Level measuring instruments are described in Chapter III.

Weirs

Weirs are apertures in the top of a dam across a channel through which flows the liquid to be measured (Figure 2.25a). The aperture may be rectangular (Figure 2.25b), trapezoidal (Figure 2.25c) or "V" notch (Figure 2.25d). The special case of a trapezoidal weir with side slopes of 1:4 (Figure 2.25c) is known as a Cippoletti weir; this form leads to a simplified flow calculation. "V"-notch weirs generally have a notch angle from 30 degrees to 90 degrees depending on required flow capacity.

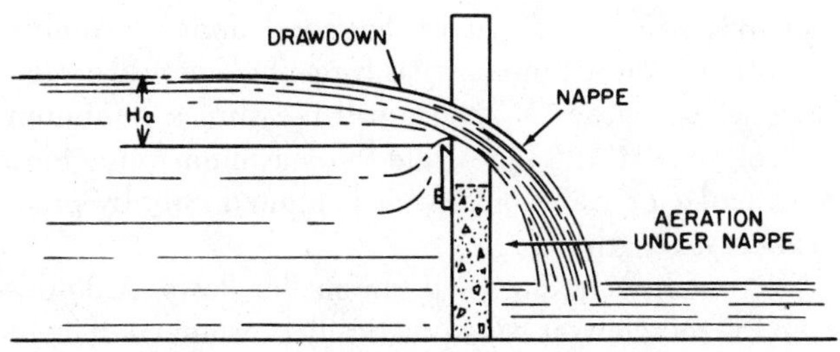

Fig. 2.25a Flow over a weir

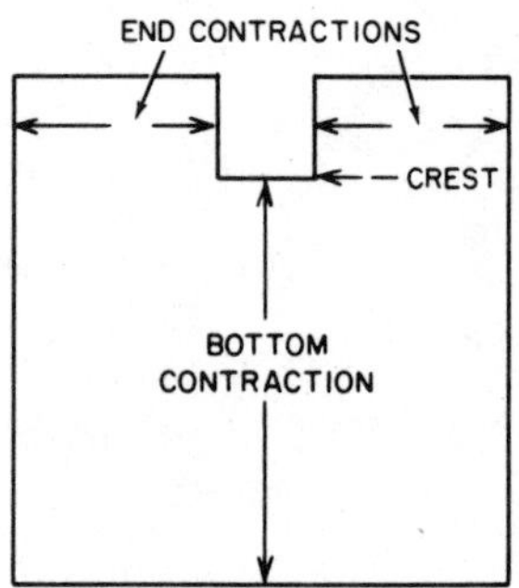

Fig. 2.25b Rectangular weir

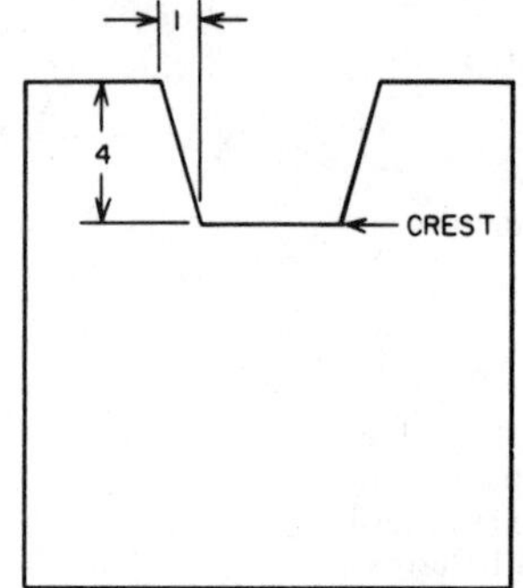

Fig. 2.25c Cippoletti (trapezoidal) weir

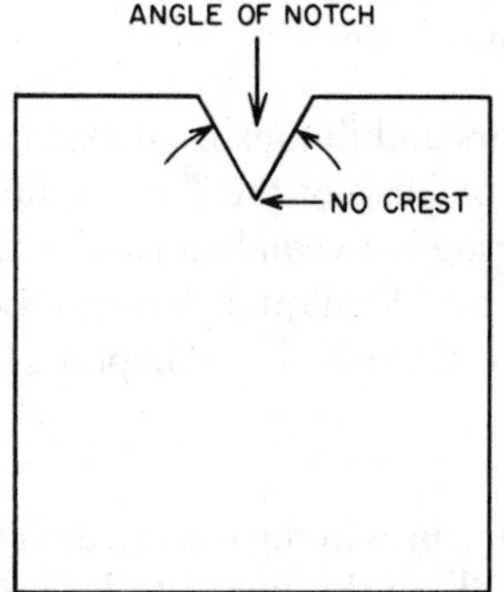

Fig. 2.25d Vee notch weir

The head is measured as the difference in level of the pool at an adequate distance upstream from the weir, compared to the horizontal crest of a rectangular or trapezoidal weir, or the bottom point of the "V" of a "V"-notch weir. Heads less than 0.1 foot (30 mm) for minimum measured flow or more than 1.0 foot (300 mm) for maximum flow are generally to be avoided although 1.25-foot (380 mm) head can be tolerated under favorable conditions. These limits are easily met by practical design since a 30-degree "V" notch will measure a minimum flow of 1 GPM (3.8 l/m) while the maximum value for a rectangular or trapezoidal weir is limited only by practical crest length.

"V"-notch weirs are used for smaller flows. A 30-degree "V"-notch weir has a practically constant coefficient from 3.0 to 300 GPM (11.4 to 1140 l/m) with flow proportional to the five-halves power of the head. Coefficient increases roughly 2 percent for flow down to 1 GPM (3.8 l/m) and changes relatively little for flow up to 500 gpm (1893 l/m). For notch angle up to 90 degrees, flow varies as the tangent of half the notch angle. Notch angle exceeding 90 degrees is not recommended.

Rectangular or Cippoletti weirs are used for larger flows. A rectangular weir with a crest two feet (0.6 m) long develops a head of about 0.2 foot (60 mm) for 250 GPM (946 l/m) and 1.0 foot (305 mm) for 2,700 GPM (10,221 l/m). For this weir, flow is directly proportional to crest length and to the three-halves power of the head.

The weir plate may be located in a dam in a natural channel or in a weir box (Figure 2.25e). The stilling basin ahead of the weir should be large enough so that the upstream velocity does not exceed ⅓ foot per second (.01 m/s). Width and depth immediately ahead of the weir should be sufficient so that the wall effect of the bottom and sides of the channel has negligible effect on the pattern of flow through the notch. It is important that the flow should break clear from the sharp edge of the notch with an air pocket maintained immediately beyond and below the weir plate. The channel downstream from the weir must be sufficiently wide and deep so that at maximum flow there is ample clearance between flow through the notch to downstream liquid level so that this air pocket is maintained (Figure 2.25a). The upstream edge of the weir should be sharp and straight. It is usual practice to bevel the downstream edge of the weir at 45 degrees to about a 1/32-inch (.8 mm) edge. For rectangular and Cippoletti weirs, the crest must be carefully leveled.

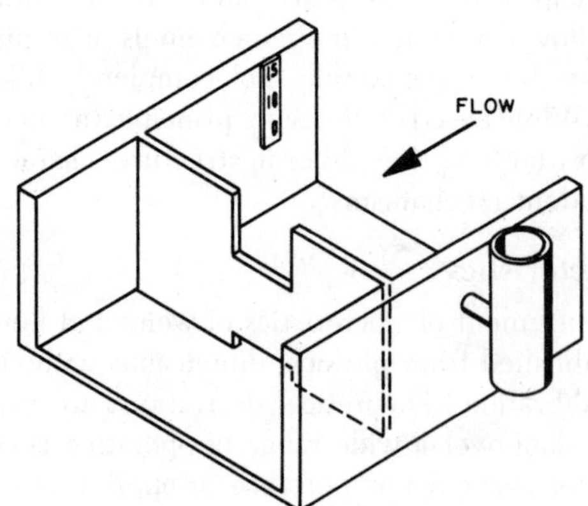

Fig. 2.25e Weir box

Accuracy

It is beyond the scope of this treatment to cover all the requirements for accurate weir measurement. With suitable precautions, accuracy of the relation between flow and head (level) to ± 2 percent is attainable, based on the dimensions of the primary device. Reference 1 gives full data on installation and operation of weirs.

Computations

The following equations establish the relationships between flow and measured head, provided that the installation and operation of the weir are as recommended in this section and also in the cited references.

For a "V"-notch weir:

$$Q = 2.48 \tan \tfrac{1}{2}\,\theta\, H^{2.5} \qquad 2.25(1)$$

For a rectangular weir:

$$Q = 3.33\,(L - 0.2H)\, H^{1.5} \qquad 2.25(2)$$

For a Cippoletti weir:

$$Q = 3.367\, LH^{1.5} \qquad 2.25(3)$$

Where
- Q = rate of flow in cubic feet per second
- θ = "V" notch angle in degrees
- H = head (note) in feet of flowing liquid
- L = crest length in feet

Note: Head is measured between level in stilling pond and crest of a rectangular or Cippoletti weir, or bottom of "V" of a "V"-notch weir.

For conditions other than exactly as recommended, see references for correction factors.

The Parshall Flume

Developed by R.L. Parshall at the Colorado Experiment Station of the Colorado Agricultural College in cooperation with the Division of Irrigation of the U.S. Department of Agriculture,[2] this device is a special type of Venturi flume (Figure 2.25f). The loss of head is about one-quarter of that for a weir of equal capacity. The effect of velocity of approach is practically eliminated so that a large upstream stilling basin is not required. The relatively high velocities in the system tend to flush away deposits of silt and other solids which might accumulate and alter measurement. There are no sharp edges, no pockets, few critical dimensions; the device can be locally fabricated from available materials. Calibration data based on physical dimensions are available from 3-inch (76 mm) throat width with minimum range of 0.03 second feet (13 GPM/49 l/m) up to 40-foot (12.2 m) throat width with maximum capacity of 2000 second feet (900,000 GPM/3,406,900 l/m). Flow is approximately proportional to the three-halves power of level with stated flow rangeability for a single unit of 35 to 1 or more, depending on size.

Extreme accuracy is not claimed for flow measurement using this device; however, measurement is very dependable with minimum maintenance and good repeatability. Accuracy is quite adequate for most applications to irrigation, waste and sewage flows.

Downstream level has no effect on the measurement so long as the level near the downstream end of the throat does not exceed 70 percent of the level measured near the upstream end of the converging section (Figure 2.25f). (Both levels are referred to the floor section of the flume.) For flumes less than 1 foot (305 mm) wide, the ratio of levels is 60 percent maximum. This is the preferred and more usual mode of operation. It provides best accuracy; only one measurement of level is required with flow computed directly from this upstream level measurement; direct, continuous readout of flow rate is readily provided.

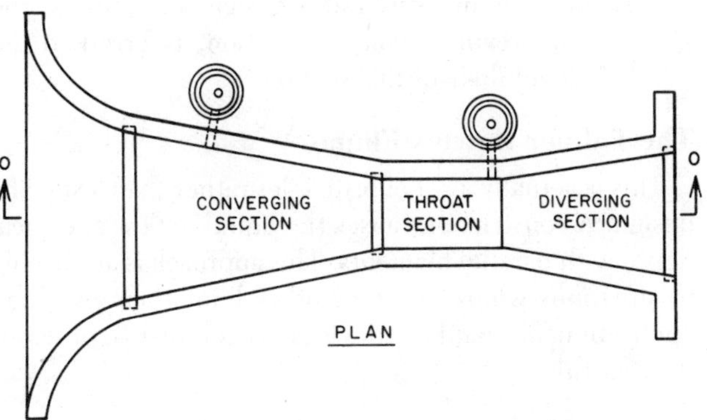

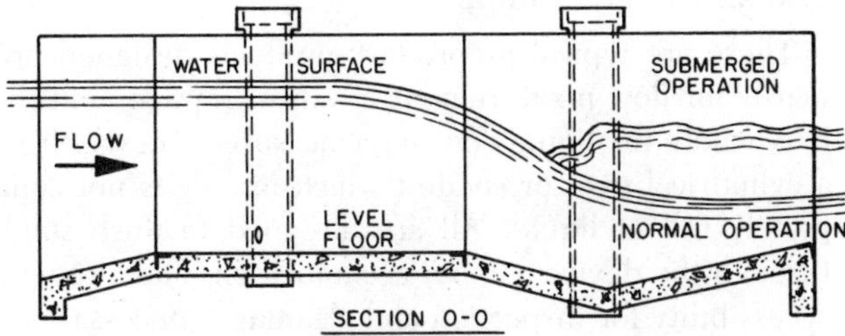

Fig. 2.25f Parshall flume

Where operating conditions (available head, maximum flow rate, weir size, etc.) result in a throat level greater than 70 percent of upstream level, so-called submersion results. Measurement can be obtained with a downstream level as great as 95 percent of upstream level. However, this requires a correction factor based on both upstream level and downstream level in the flow computation; accuracy suffers; standard equipment for direct readout in flow is not available.

The simplified equations based on a single measurement at the upstream location are as follows:

For L = ¼ foot

$$Q = 3.97\, L\, H^{1.547} \qquad 2.25(4)$$

½ foot

$$Q = 4.12\, L\, H^{1.58} \qquad 2.25(5)$$

¾ foot

$$Q = 4.10\, L\, H^{1.53} \qquad 2.25(6)$$

1 foot to 8 feet

$$Q = 4.0\ L\ H(1.522L)^{0.026} \qquad 2.25(7)$$

Over 8 feet

$$Q = (2.5 + 3.69\ L)H^{1.6} \qquad 2.25(8)$$

Where

L = width of throat section in feet
Q = volume flow rate in cubic feet per second
H = head in feet

H (head) is measured at a designated point in the upstream converging section, referred to the level floor of this section.

The Palmer Bowlus Flume

This is actually a set of principles rather than a specific design. A constriction causes the liquid to flow at critical depth with parallel filaments. This approach is applicable to situations where space or other limitations preclude application of standard devices. Reference 3 provides data on this.

The Kennison Nozzle, Parabolic Flume, and Leopold Lagco Flume

These are typical proprietary products designed primarily for flow measurement of waste, sewage, and the like, where the liquid flow to be measured emerges from a cylindrical pipe or conduit which usually is not completely full of liquid. All are designed to flush solids through the device without accumulations and also with accessibility for inspection and cleaning if necessary.

These devices develop heads which are a function of flow rate. In the Kennison Nozzle, head is almost linear with flow above 10 percent of maximum flow rate. Accuracy is stated as 2 percent in this range. For the parabolic flume and the Leopold Lagco Flume, flow varies approximately as the three-halves power of head.

These devices are available in medium to large sizes. Details as to structure, application and characteristics are available from the manufacturers.

REFERENCES

1. Streeter, V.L., "The Kinetic Energy and Momentum Corrections for Pipes and Open Channels of Great Width," *Civil Engineering*, April, 1942, Volume 12, Number 4, p. 212.
2. "Measuring Water in Irrigation Channels," *Farmers Bulletin 1682*, United States Dept. of Agriculture.
3. Paper Number 1948, *Proceedings of the Institute of Civil Engineers*, Volume 101 (1936), p. 1195.
4. Wells, E.A. and Gotaas, H.B., "Design of Venturi Tubes in Circular Conduits," *Proceedings of the American Society of Civil Engineers, Journal of the Sanitary Engineering Division*, Volume 82, April 1956. American Society of Civil Engineers, 345 E 47th St., New York 10017.

BIBLIOGRAPHY

Shinskey, G., "Characterizers for Flume and Weirs," *Instrument and Control Systems*, September, 1974, p. 111.

Thorsen, T. and Oen, R., "How to Measure Industrial Wastewater Flow," *Chemical Engineering*, February 17, 1975, pp. 97–100.

Chapter III

LEVEL MEASUREMENTS

D. S. KAYSER

CONTENTS OF CHAPTER III

Table III
ORIENTATION TABLE FOR LEVEL DETECTORS

Section	Instrument Type \ Service	LIQUIDS								SOLIDS	
		Switches		Local Indicators		Self-Contained Local Controller		Transmitters			
		Clean Fluid	*Foam*	*Accounting*	*Standard*	*Clean*	*Hard-to-Handle*	*Clean*	*Hard-to-Handle*	*Switches*	*Transmitters*
3.2	Antenna Level Sensors	G	—	—	—	—	—	P	P	G	P
3.3	Bubblers	P–F	—	P–F	P–F	P–F	P–F	P–F	P–F	—	—
3.4	Capacitance Probes	G	P–F	P–F	F	—	—	F	P–F	F	P–F
3.5	Conductivity Probes	F	P	—	—	—	—	—	—	P–F	—
3.6	Diaphragm Level Detectors	F	—	P	F	F	F	P–F	P	G	—
3.7	Differential Pressure Level Detectors	G	—	F	G	G	F	G	F	—	—
3.8	Displacer Level Detectors	E	—	F	G	E	P–F	E	P–F	—	—
3.9	Float Level Devices	G	—	F	F	—	—	P–F	P	—	—
3.10	Impedance Probes	G	P–F	P–F	F	—	—	G	G	F	F
3.11	Level Gauges	—	—	F	G	—	—	—	—	—	—
3.12	Optical Level Switches	G	—	—	—	—	—	—	—	—	—
3.13	Radiation Level Sensors	G	—	G	—	—	—	G	E	E	G
3.14	Resistance Tapes	—	—	P	F	—	—	P	P	—	F
3.15	Rotating Paddle Switches	—	—	—	—	—	—	—	—	G	—
3.16	Slip Tubes	—	—	P–F	P–F	—	—	—	—	—	—
3.17	Tape Level Devices	G	—	E	G	—	—	G	P–F	—	F–G
3.18	Thermal Level Sensors	G	—	—	—	—	—	—	—	—	—
3.19	Time Domain Reflectometry	—	—	—	—	—	—	—	—	—	—
3.20	Ultrasonic Level Detectors	G	P–F	F	G	—	—	F	G	G	F
3.21	Vibrating Reed or Tuning Fork Switches	G	F	—	—	—	—	—	—	G	—

E—Excellent G—Good F—Fair P—Poor

3.1 APPLICATION AND SELECTION

This chapter presents twenty-one basic principles used for level detection, such as buoyant force, hydraulic head, float position, radiation attenuation and ultrasonic wave reflection. Within each category a family of instruments has been designed to match process and output requirements insofar as possible. In order to help the reader correspond a design to a particular application, Orientation Table III has been prepared.

Selection

To use this table, the particular service is first defined. The service is divided into liquids and solids applications and is subdivided for switches, local indicators, local controllers, and transmitters. Further divisions are made to cover the nature of the material to be measured, such as foam, clean, or hard-to-handle. With the service defined the reader can scan down the column to find a letter indication (E—excellent, G—good, F—fair, P—poor) of the suitability of the various designs. The gradings are based on such factors as inaccuracy, reliability and ease of maintenance but do not take economics into account. Therefore, an instrument that is rated good or excellent for a particular service may not be an economical selection.

When the possible selections have been narrowed down to a few, the reader should refer to the corresponding sections of this chapter. In the front of each section there is a summary of basic features, such as inaccuracy, cost, range, material of construction, pressure and temperature ratings, and so on. By brief inspection, it can be determined if the instrument meets the general requirements of the application under consideration. If so, additional information may be obtained from the text in the section. If some of the features are unacceptable, the reader should examine the next instrument listed in the Orientation Table.

Application

Level measurement applications can be broadly grouped by service as follows: atmospheric vessels, pressurized vessels, and accounting grade. Accounting grade measurements are made in both atmospheric and pressurized vessels. The need for accuracy in accounting grade installations can be demonstrated as follows. A typical 750,000 barrel API storage tank has a diameter of 345 ft. (105m) and it takes some 8000 gallons (30 kl) to raise the level 1 in. (25 mm). A level measurement error of 1 in. (25 mm) would therefore indicate that 8000 gallons (30 kl) have been gained or lost. This is no small matter, particularly if the level measurement is used as a basis for custody transfer of the material. Consequently, substantial effort has been put into the development of storage tank gauging systems that have good reliability, high accuracy and high resolution. These efforts have been relatively successful, and the user can be confident of obtaining satisfactory results if adequate attention is given to installation details. Instruments for accounting grade measurements are covered in Section 3.17.

Atmospheric Vessels

Liquid level detection in atmospheric vessels rarely presents any serious problems. Moreover, several desirable features can be readily built into an atmospheric vessel level system. Instrumentation generally can be selected and installed so that it can be removed from the vessel for calibration or repair without draining the vessel. With few exceptions, an eye height indicator can be obtained, eliminating the necessity for the operator to climb on the vessel. A number of top mounted designs are available, eliminating the possibility of a spill if the instrument or nozzle corrodes or breaks off. And many vessels can be manually gauged. It is always comforting to know that such a simple procedure as manual gauging is available to verify an instrument reading.

Solids level gauging also is generally done in atmospheric tanks, but in this case the engineer has fewer available level detecting devices and less flexibility in installation methods. The devices that are suitable for point level detection of solids are noted in the table. Point detection units must be located at the actuation point and this can lead to accessibility problems. Except for the radiation type, it also means that a new connection must be cut into the vessel if the actuation point is changed. Solids that behave unpredictably can cause serious measurement problems. Therefore, if the solid is sticky or if it can bridge or rathole, the engineer should take particular care in the location and installation of the detector.

Continuous level measurement of solids can be made, with varying degrees of success, by using units graded in the "Solids, Transmitters" column of the table. The surface sensor design discussed in Section 3.17 is probably the most often used. All designs require top mounting, but since all can be equipped with ground or remote readouts, this is not a major limitation. As with the switches, performance of these instruments may not be good when the solids are not free flowing. If an inventory grade measurement is required, some users prefer to make a weight measurement by using load cells. Load cell installations are covered in Chapter VIII.

Pressurized Vessels

Point level detection of liquids in a pressurized vessel can be made using one of 14 or 15 principles of measurement, successfully solving most problem applications. For clean services in industrial processing plants, preference is given to the externally mounted displacer switch. This unit is rugged and reliable, it has above average resistance to vibration and, depending on the design, its actuation point can be changed easily. There are a number of cases where the capacitance switch or the float switch may be considered if they are arranged so that they may be removed for repair without depressurizing the vessel. And, as noted in Section 3.5, conductivity switches are used in water services to 700°F (370°C) and 3000 PSIG (21 MPa), although this does not imply that they should be automatically considered for hydrocarbon or chemical services. The balance of the switches, while less expensive, are generally used in noncritical secondary applications and atmospheric vessels.

Continuous liquid level detection in pressurized vessels is subdivided into clean and hard-to-handle processes. For clean services requiring local indication only, the clear choice is the armored level gauge. Even when a transmitted signal is required, many users specify that the transmitter be backed up with a level gauge so that the process can continue while the transmitter is valved out for service. The level gauge has an additional advantage in that it can be used to determine whether or not the transmitter is performing properly. Nevertheless, the need for a level gauge installation should be evaluated carefully since the level gauge can be the component most apt to fail in a pressure vessel system. Of course, level gauges will not work properly in hard-to-handle services.

Preferences for clean service transmitters vary from industry to industry. Petroleum refiners invariably think in terms of the externally mounted displacer transmitter, for several reasons. First, the displacer transmitter has a long record for reliability and accuracy. Second, since the installation can be made with 2 in. (50 mm) flanged connections, it is very rugged. This strength is important in the petroleum industry because a break at the instrument connection could cause a large spill of hydrocarbons that might be above their autoignition temperature. Third, the low side (vapor phase) connection does not require a seal, reducing maintenance requirements and eliminating possible inaccuracies. Finally, most refinery processes are compatible with carbon or low alloy steel, materials readily available in displacer design.

In other chemical processing industries, first consideration usually goes to the d/p cell when a transmitted level signal is required. This unit has a lower initial cost than the displacer transmitter, is reliable and accurate, and can be modified for some services that fall into the hard-to-handle category. The major problem with the d/p cell, when used for level measurement on pressurized vessels, is in handling the low side tap. If the low side leg can be arranged to run dry the problem is eliminated, but this is rarely the case. Normally the low leg must be filled with a seal oil or with the process material. If a seal oil is used, the oil must be compatible with, and heavier than, the process. If the leg is filled with the process material, the process fill must not boil away at high ambient temperatures. In either case, ambient temperature variations will change the density of the fill, which can cause inaccuracies in the level reading. The liquid seal also requires frequent inspection. Low side pressure repeaters and chemical seals are available that eliminate the seal problem but introduce inaccuracies of their own and increase the purchase cost. Despite this, d/p cells are successfully used in a wide range of applications and should be considered whenever the span to be measured is greater than 60 in. (1.5m).

Other devices such as capacitance probes and ultrasonic detectors are also used for pressurized vessel level measurement if the d/p cell or displacer is not satisfactory. Section 3.9 covers a number of float operated and magnetically coupled level detectors that are used in low volume storage tanks, underground tanks, transport tankers, and many other applications outside of the continuous processing area. Although these devices are normally thought of as local reading, almost all can be equipped with transmitters.

Reliable means are available for the point level detection of hard-to-handle liquids in pressurized vessels. The primary consideration is whether a penetrating or non-penetrating design will be required. Because of the process, it is probable that a penetrating design cannot be devised for removal without depressurizing and possibly draining the vessel. Devices are also available to continuously detect the level of hard-to-handle materials in pressurized vessels. Although the radiation detector might seem to be an obvious selection, high cost and licensing and regulatory requirements cause many engineers to consider other devices first, using the radiation detector only as a last resort. Many times the d/p cell can be equipped for a satisfactory installation in hard-to-handle service by specifying such options as flush or extended diaphragms, chemical seal assemblies and/or pressure repeaters. The capacitance probe or its cousin, the impedance probe, will also perform acceptably in some cases. Other devices, rated poor to fair in the Orientation Table, may be adequate for certain specific applications.

Point and continuous level detection of solids in pressure vessels are probably the most difficult of measurement problems. Within their pressure and temperature limitations, the devices used for solids detection in atmospheric vessels may also be used on pressure vessels. If these limitations are exceeded or if the solid is particularly abrasive or fouling, the radiation gauge becomes a possible choice.

BIBLIOGRAPHY

Akeley, L.T., "Eight Ways to Measure Liquid Level." Control Engineering, July 1967.

Andreiev, N., "Survey and Guide to Liquid and Solid Level Sensing." Control Engineering, May 1973.

API Guide for Inspection of Refinery Equipment, Chapter XV, Instruments and Control Equipment.

API Recommended Practice 550, Manual on Installation of Refinery Instruments and Control Systems. Part I—Process Instrumentation and Control, Section 2—Level.

Bailey, S. J., "Level Sensors '76, A Case of Contact or Non-Contact." Control Engineering, July 1976.

Belsterling, C.A., "A Look at Level Measurement Methods," Instruments and Control Systems, April 1981.

Buckley, P. S., "Liquid Level Measurement in Distillation Columns." ISA Trans 12 No 1:45–55, 1973.

Caldwell, A. B., "Process Control Series; Liquid and Solid Level Sensors." Engineering and Mining Journal, May 1967.

Considine, D. M., "Process Instrumentations; Liquid Level Measurement Systems; Their Evaluation and Selection." Chemical Engineering, February 12, 1968.

Cusick, C. F., "Liquid Level Measurement." Instrumentation 22 No. 1:22–7, 1969.

"Engineering Outline; Level Measurement." Engineering, October 6, 1967.

Hall, J., "Level Monitoring; Simple or Complex." Instruments and Control Systems, October 1979.

Hall, J., "Measuring Interface Levels," Instruments and Control Systems, October 1981.

ISA Directory of Instrumentation.

LaPadula, E. J., "Level Measuring Methods." I.S.A. Journal, February 1965.

Lawford, V. N., "How to Select Liquid-Level Instruments." Chemical Engineering, October 15, 1973.

Mariam, P. L., "Measuring Level in Hostile or Corrosive Environments." Instrumentation Technology, April 1979.

Morris, H. M., "Level Instrumentation from Soup to Nuts." Control Engineering, March 1978.

Morris, H. M., "Sensing Interface Level Poses Many Challenges." Control Engineering, August 1978.

Picker, S., "Measuring Levels of Freezing Liquids." Chemical Engineering, February 19, 1973.

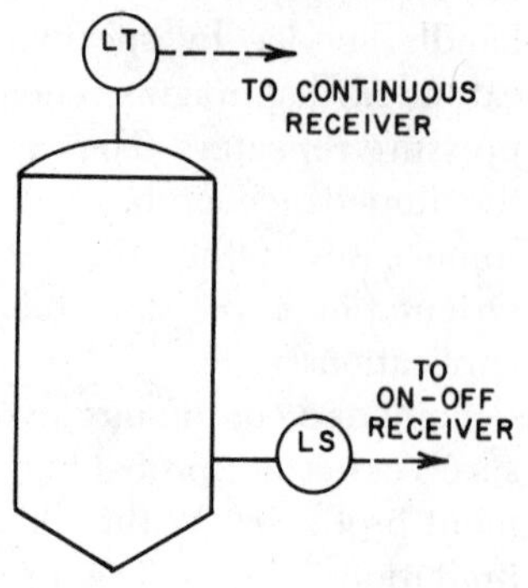

3.2 ANTENNA LEVEL SENSORS

Design Pressure	To 1500 PSIG (10 MPa) for switch, 100 PSIG (0.7 MPa) for continuous
Design Temperature:	−40 to +185°F (−40 to +85°C)
Materials of Construction:	Bare or plastic coated stainless steel and other alloys
Inaccuracy:	±¼ in. (±6 mm) for switch. ±15% for continuous
Range:	To 100 ft. (30.5 m)
Cost:	$375 for switch; $650 for probe and transmitter, readout not included
Supplier:	ASI Keystone

An oscillator circuit is one in which the output is fed back to the input in such a way that the loop gain is greater than one and the input is in phase with the output. The circuitry is self-limiting so that a gain of greater than one does not cause the circuit to run away. The frequency of oscillation of the circuit can be made to vary if passive elements in the circuit are varied. For example, a variable capacitor can be incorporated in the circuit in such a way as to make the oscillating frequency proportional to the varying capacitance. If, in turn, the capacitance is proportional to level in a tank, the oscillation frequency will be proportional to level. The antenna level sensor works on this principle.

Normally, an antenna is designed to be installed in the air and therefore the dielectric of the surrounding material is assumed to be 1.0. If the antenna is installed in a bin that has a rising and falling process material level, the value will be different because the material will have a different dielectric than that of air. A more detailed discussion of this phenomenon is presented in Section 3.4; it will suffice here to say that the antenna-to-ground (bin wall) capacitance will vary with changing level. By wiring the antenna into the oscillator circuit, the frequency of oscillation can be made to vary with level. The frequency of oscillation of the circuit containing the antenna is compared against the frequency of oscillation of a constant frequency oscillator, and the difference is used for point level detector switch operation or for continuous level measurement. Figure 3.2a shows a block diagram of how this is accomplished.

Figure 3.2b shows a typical installation of an antenna probe for point detection. The antenna is installed horizontally so that a large portion of the antenna is covered or uncovered at once, thereby causing a large change in capacitance with an attendant large change in the frequency of oscillation. As shown in Figure 3.2c, the continuous detector is suspended from the top of the bin and hangs down into it. Capacitance changes linearly with increasing level. The antenna can be made of a solid rod or stranded cable of various materials and can be bare or coated with Teflon or other plastic materials; plastic coatings are used when the process is electrically

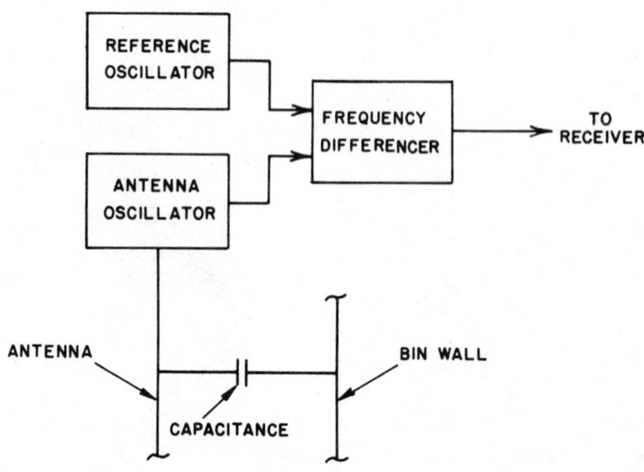

Fig. 3.2a Block diagram of antenna level detector

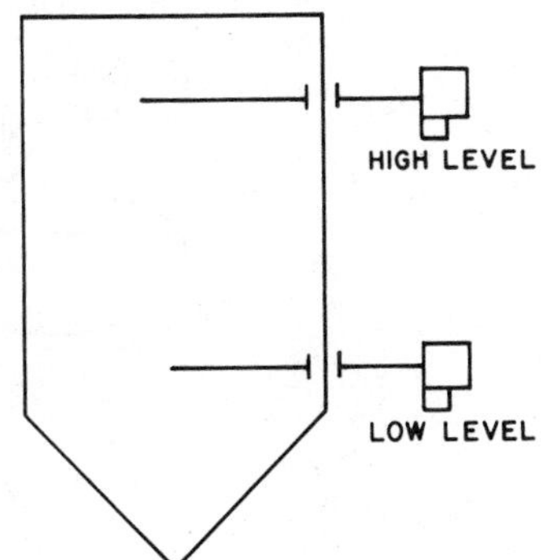

Fig. 3.2b Installation of antenna probes for point detection

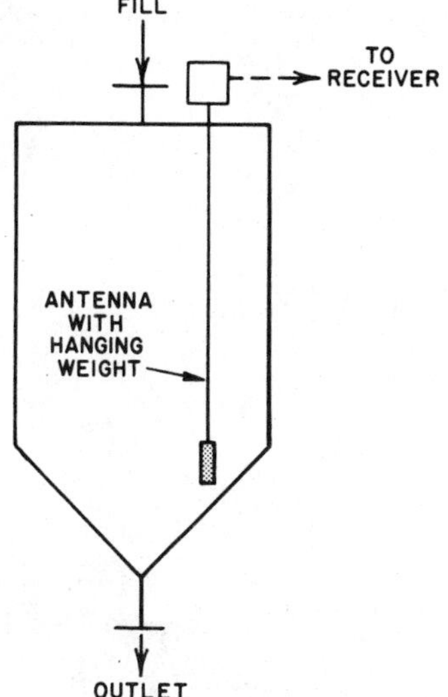

Fig. 3.2c Installation of antenna probe for continuous level detection

conductive. See Section 3.4 for a more detailed discussion of capacitance-type level detection in conductive processes.

Antenna probes are subject to certain problems and limitations. As with the capacitance probe, changes in the dielectric constant of the process material can affect accuracy. Radio frequency devices, such as walkie-talkies, also can interfere with proper operation. The bin wall must be grounded; thus, an antenna probe cannot be used in fiberglass bins. Conducting and non-conducting build-ups on the probe also may ruin the accuracy of the measurement. Most importantly, the system contains inherent non-linearities. While it is true that capacitance can be made to change linearly with rising or falling level, the fact is that the oscillation frequency will not change linearly with varying capacitance of one of the passive components of the oscillator circuit. This means, at least for the continuous level detector, that the antenna/oscillator circuit can only be calibrated for accurate performance at one point (one capacitive value) and that the performance of the system will become increasingly non-linear as the system operates farther away from that point.

Conclusion

The antenna level detector can be used selectively for point level detection. This application is subject to malfunction owing to process caking or build-up but does not suffer any problems with non-linearities. The continuous detector is not only subject to malfunction because of difficult processes but also can be non-linear as the level moves away from the calibration point.

BIBLIOGRAPHY

Andreiev, N., "Survey and Guide to Liquid and Solid Level Sensing," *Control Engineering*, May 1973.

Belsterling, C.A., "A Look at Level Measurement Methods," *Instruments and Control Systems*, April 1981.

Morris, H.M., "Level Instrumentation from Soup to Nuts," *Control Engineering*, March 1978.

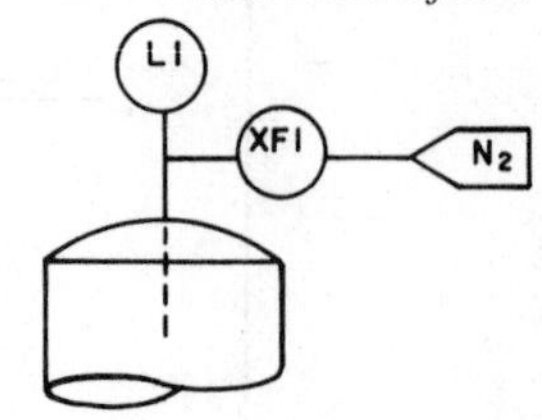

3.3 BUBBLERS

Pressure:	Low or moderate
Temperature:	Moderate
Materials:	Unlimited
Cost:	\$100–\$400
Inaccuracy:	±1–2%
Range:	Unlimited
Partial List of Suppliers:	Fisher Controls; Haywood and Haywood; King Engineering; Meriam Instrument Co.; Uehling Instrument Co.

The bubbler system has been used to measure liquid level ever since the advent of compressed air. A dip tube is installed in a tank with its open end about 3 in. (76 mm) from the bottom. A flow of air passes through the tube; when air bubbles escape from the open end, the air pressure in the tube equals the hydrostatic head of the liquid. As liquid head varies, the air pressure in the dip tube changes correspondingly. Pressure can be detected by gauges or manometers for continuous readout, or by pressure switches for on-off level control or alarm. Variations are available for pressurized or atmospheric tanks, for local or remote readouts and for situations where a compressed purge supply is not available.

Figure 3.3a illustrates an air bubbler installation for an open (atmospheric) tank with various purge controls. The enlargement shows how the tip of the dip tube may be formed to guarantee the continuous flow of small bubbles. The transmission line should be sloped toward the tank so that if the purge is lost and process vapors enter the transmission tube, the condensate will drain back into the vessel. If the readout device must be below tank level, a condensate trap can be installed as shown by the dotted line.

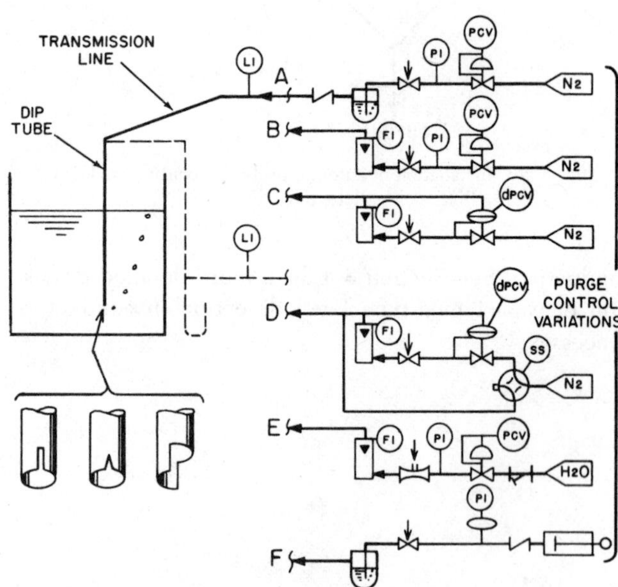

Fig. 3.3a Variations of air bubblers for atmospheric tanks

The dip tube is normally fabricated from ¼ or ½ in. (6 or 12mm) diameter tube of material compatible with the process. Plastic tubing, metal tubing runs longer than 6 ft. (1.8 m), and tubing in agitated tanks requires continuous support. If the tank is agitated, or if the installation of a top-entry dip tube would interfere with other equipment, the tube can be installed in the side of the vessel below the expected minimum level of the liquid.

The purge supply pressure should be at least 10 psi (69kPa) higher than the highest hydrostatic pressure to be gauged. The purge flow rate is kept small, about 1 SCFH (500 cc/min.), so that there is no significant pressure drop in the dip tube. Usually, the purge media is air or inert gas, although liquids can also be used. Several methods of gas purge controls are shown in the figure to illustrate some of the installation considerations. In system "A," nitrogen supply pressure is regulated to a value corresponding to a higher pressure than the hydrostatic head with the tank full. Purge flow is adjusted by a needle valve and is sent through a sight feed bubbler, which allows visual inspection of the actual flow. This system allows for detection of levels up to 10 ft. (3 m).

If higher levels are to be detected, system "B," which uses a rotameter instead of a sight feed bubbler, can be considered because a rotameter can withstand higher pressures. In systems "A" and "B," as the liquid level varies, the downstream pressure also will vary, thereby causing variations in the purge flow rate. Since the purge pressure at the readout device is the sum of hydrostatic head and dynamic dip tube pressure drop, variations in purge flow will cause non-linearities. To correct this condition, a differential pressure control valve can be installed across the fixed restriction of the needle valve as shown in system "C." This will cause the purge to flow at a uniform rate regardless of the liquid head.

If the process material can build up or plug the dip tube owing to loss of purge gas or because of the nature of the fluid, an aerator selector switch may be installed as shown on system "D" to allow for periodic blowing out of the transmission line. System "E" can be considered in remote locations where gas purge media is not available and a bubbler is desired instead of a liquid purge for level detection. Here water is jetted across a gap while air is aspirated into the stream and compressed. The air-water mixture enters the dip tube where the small amount of water runs down the inside of the bubbler tube while the pressure of the escaping air is detected as a measure of level. Such a setup would be in service only when the operator wanted to make a level reading, so that the water would not flow into the vessel continuously. System "F" shows a more common approach for bubbler installations in remote locations where a small hand pump is used to compress the purge air.

For tanks that operate under pressure or vacuum, the installation of a bubbler indicator becomes slightly more complex because the liquid level measurement is a function of the difference between the purge gas pressure and the vapor pressure above the liquid. Because of the differential measurement involved, the readout device normally is a manometer or other type of differential pressure detector. Figure 3.3b shows one of these installations. All of the previously discussed variations apply to pressure or vacuum installations. On pressurized tanks, a vapor pressure equalizing line is added. Purge control system "D" of Figure 3.3a is used to illustrate the bubbler installation for pressurized tanks in Figure 3.3b.

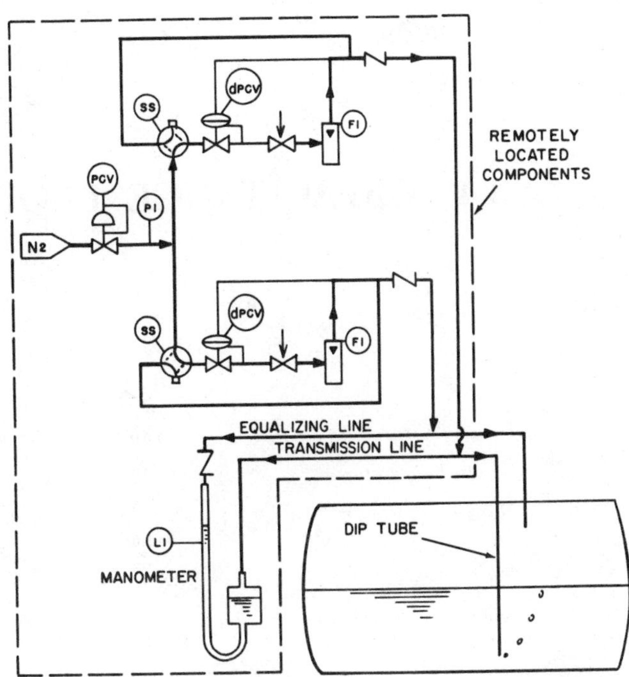

Fig. 3.3b Air bubbler installation for pressurized tanks

Conclusions

Bubbler systems are rarely used in industrial processing applications. In addition to limited accuracy, one major drawback is that bubblers must introduce foreign matter into the process. (Figure 3.6c illustrates a design which allows the use of a bubbler but does not introduce purge gases into the process.) Liquid purges can upset the material balance of the process and gas purges can overload the vent system on vacuum processes. If the purge media fails, not only is the readout lost, but the system is exposed to the process material, which can cause plugging, corrosion, freezing or safety hazards. At the same time, however, bubbler systems enable economical gauging in areas such as waste water handling, food processing and certain bulk storage applications. For remote tanks where compressed air is not available there is probably no simpler level indicator than a small hand pump and a gauge.

BIBLIOGRAPHY

"Air Pressure Tells Liquid Level," *Machine Design*, May 30, 1974.

Flemming, R.C., "Bubblers Determine Liquid Level when Density Varies," *Chemical Engineering*, February 24, 1969.

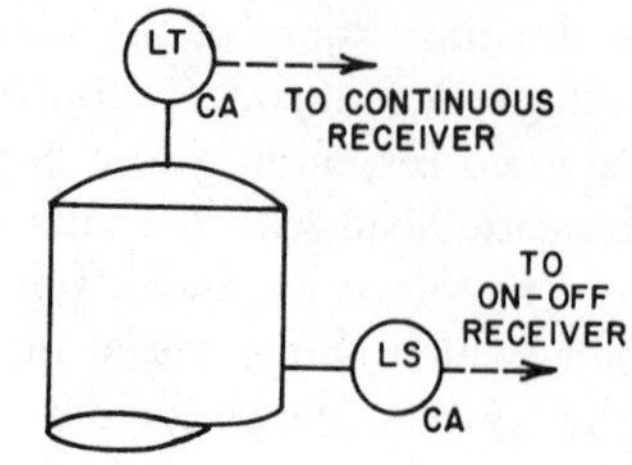

3.4 CAPACITANCE PROBES

Design Pressure:	To 1000 PSIG (6.9 MPa)
Design Temperature:	To 1500°F (815°C)
Materials of Construction:	Stainless steel, higher alloy metals, ceramic, Kel-F, Teflon
Inaccuracy:	Depends on process and probe sensitivity
Range:	To 20 ft. (6 m) for rod probe, to 150 ft. (46 m) for cable probe
Cost:	$500 for switch; $900 and up for continuous
Partial List of Suppliers:	Custom Controls Co.; Delavan Electronics, Inc.; Delta Controls, Drexelbrook Engineering Co.; Endress-Hauser, Inc.; Magnetrol International; Princo Instruments, Inc.; Robertshaw Controls

A capacitor consists of two conductors isolated by an insulator. The conductors are called plates and the insulator is referred to as the dielectric. The characteristic nature of a capacitor is its ability to accept and store an electric charge. When a capacitor is connected to a battery, as shown in Figure 3.4a, the negative electrons on plate #2 are attracted to the positive electrodes of the battery and the negative electrons on plate #1 are repelled by the negative electrode of the battery. The electrons flow until the battery potential is established across the capacitor. The larger the capacitor, the more current will flow to charge the unit. A perfect capacitor is one in which no current flows from one plate to the other. The magnitude of flow is directly proportional to the size of the capacitor.

Capacitor size is measured in farads. A capacitor has the capacitance of one farad if it stores a charge of one coulomb when connected to a one volt battery. Because this is a very large unit, one trillionth of it, noted as the picofarad (pf), is commonly used. The electric size of a capacitor is affected by its physical dimensions and by the material between the plates (dielectric). Figure 3.4a gives the capacitance in pf for cylindrical capacitors.

The term K in the equation represents the effect of the dielectric. When it is vacuum, K is unity. Replacing vacuum with Teflon doubles the capacitance. For pure substances, the dielectric constant is a fundamental property. The relationship betwen dielectric constant of a

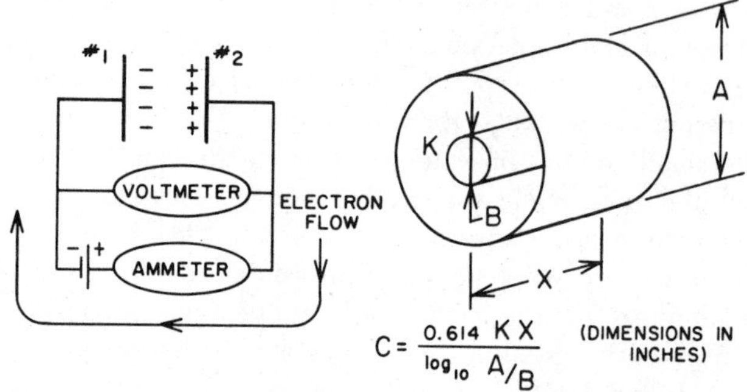

Fig. 3.4a The capacitance loop and the capacitance of cylindrical elements

binary mixture and the percentage of one of the ingredients can best be established experimentally. Table 3.4b lists dielectric constants for a selected group of substances. The capacitance of two or more capacitors connected in parallel is equal to the sum of the individual capacitances. When connected in series, the reciprocal of the total capacitance equals the sum of the reciprocals of the individual values. This relationship is the inverse of the behavior of electrical resistors. Another point worth noting is that the ideal capacitor definition is based on infinite parallel plates. Since there is no such thing in the physical world, the capacitance will be dependent on the geometry of the plates and will not perform linearly at the ends.

Table 3.4b
DIELECTRIC CONSTANTS FOR VARIOUS MATERIALS

Solids

	Dielectric Constant		*Dielectric Constant*
Acetic Acid	4.1	Phenol	4.3
Asbestos	4.8	Polyethylene	4.5
Asphalt	2.7	Polypropylene	1.5
Bakelite	5.0	Porcelain	5.7
Calcium Carbonate	9.1	Quartz	4.3
Cellulose	3.9	Rubber (Hard)	3.0
Ferrous Oxide	14.2	Sand	3.5
Glass	3.7	Sulphur	3.4
Lead Oxide	25.9	Sugar	3.0
Magnesium Oxide	9.7	Urea	3.5
Napthalene	2.5	Zinc Sulfide	8.2
Nylon	45.0	Teflon	2.0
Paper	2.0		

Liquids

	Temp. °F/°C	*Dielectric Constant*		*Temp. °F/°C*	*Dielectric Constant*
Acetone	71/22	21.4	Heptane	68/20	1.9
Ammonia	−27/−33	22.4	Hexane	68/20	1.9
Aniline	32/0	7.8	Hydrogen Chloride	82/28	4.6
Benzene	68/20	2.3	Iodine	224/107	118.0
Benzil	202/94	13.0	Kerosene	70/21	1.8
Bromine	68/20	3.1	Methanol	77/25	33.6
Butane	30/−1	1.4	Methyl Alcohol	68/20	33.1
Carbon Tetrachloride	68/20	2.2	Methyl Ether	78/26	5.0
Castor Oil	60/16	4.7	Mineral Oil	80/27	2.1
Chlorine	32/0	2.0	Napthalene	68/20	2.5
Chloroform	32/0	5.5	Octane	68/20	2.0
Cumene	68/20	2.4	Pentane	68/20	1.8
Cyclohexane	68/20	2.0	Phenol	118/47	9.9
Dimethylheptane	68/20	1.9	Phosgene	32/0	4.7
Dimethylpentane	68/20	1.9	Propane	32/0	1.6
Dowtherm	70/21	3.3	Pyridine	68/20	12.5
Ethanol	77/25	24.3	Styrene	77/25	2.4
Ethyl Acetate	68/20	6.4	Sulphur	752/400	3.4
Ethyl Benzene	68/20	2.5	Toluene	68/20	2.4
Ethylene Chloride	68/20	10.5	Urethane	74/23	3.2
Ethyl Ether	68/20	4.3	Vinyl Ether	68/20	3.9
Ethyl Benzene	76/24	3.0	Water	68/20	80.0
Formic Acid	60/16	58.5	Water	32/0	88.0
Freon 12	70/21	2.4	Water	212/100	48.0
Glycol	68/20	41.2	Xylene	68/20	2.4

Dielectric Measurements

A change in the characteristics of the material between the plates will cause a change in dielectric constant, which is often larger, more definite or more easily measured than changes in other properties. This makes the dielectric measurement suitable for detection not only of level, but also composition, moisture content or chemical structure of a substance. Because the dielectric constant of gases is nearly unity, gas composition cannot be measured by capacitance techniques. As can be assumed, changes in process material change the dielectric constant. While these changes are helpful in measuring various characteristics of the material, they also influence the accuracy of level measurement, and, therefore, must be evaluated carefully.

As material temperature increases, its dielectric constant tends to decrease. Temperature coefficients are in the order of 0.1 percent per degree Celius. Automatic temperature compensator units can be installed to cancel the effect of temperature variations. Chemical and physical composition and structure changes affect the dielectric constant. When the dielectric constant of solids is to be measured, it should be noted that variations in average particle size and changes in packing density will affect the dielectric constant detected. Current flow to ground through variable resistance tends to short out the capacitor. The short of the measured capacitance with a variable resistance can make the dielectric measurement very inaccurate if the resistance is low compared to the capacitive reactance. In this regard, there is a definite advantage to making the measurement at higher frequencies.

Level Measurement

Variations in process level cause changes in capacitance. Capacitance is measured by a bridge circuit excited by a high frequency oscillator (0.5–1.5 MHz). As shown in Figure 3.4c, the probe is insulated from the vessel and forms one plate of the capacitor; the vessel forms the other. (Process pressures and temperatures determine the type of seal used at the insulator, and corrosion conditions determine the type of probe material.) The material between the two plates is the dielectric. As level rises, vapors with low dielectric constant (1.0) are displaced by the higher dielectric process materials. Capacitance changes are detected with an instrument calibrated in units of level.

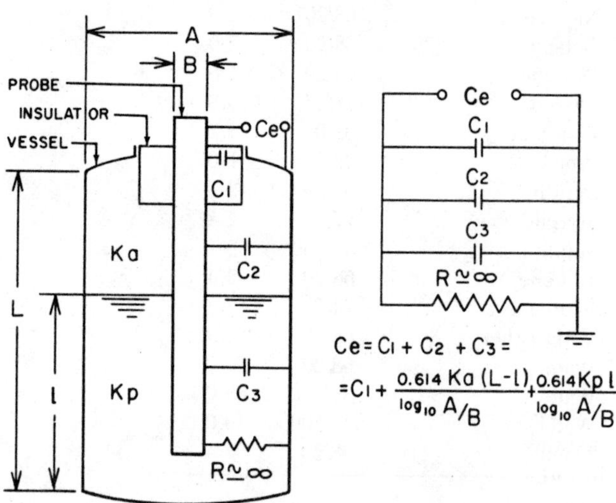

Fig. 3.4c Bare capacitance probe

For measurement of non-conductive materials, the bare metal probe shown in Figure 3.4c can be used. C_1 is the "dead capacitance" of the system which is unaffected by level changes. C_2 is the capacitance in the vapor phase and C_3 is for the process material. R is the effective resistance between the probe and vessel, which varies with the level in the vessel. If its value cannot be approximated as infinite ($C_e = C_1 + C_2 + C_3$), the measurement cannot be made with a bare probe. The capacitance of the system should be affected by changes in level only. For measuring the level of conductive materials, insulated (normally Teflon coated) probes are used. As shown in Figure 3.4d, this measurement is largely unaffected by the effective resistance and therefore this probe design is applicable to both conductive and nonconductive processes.

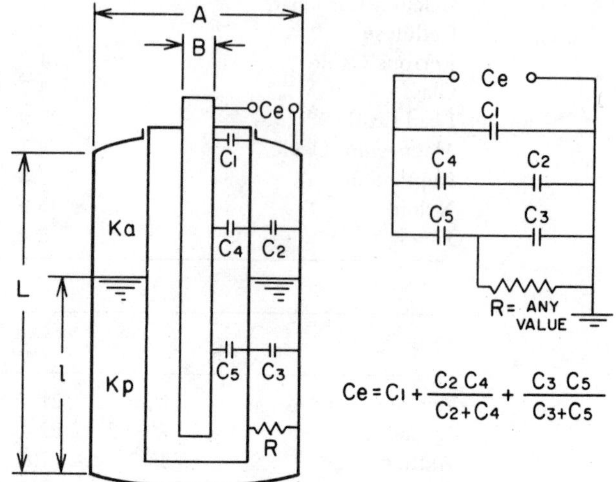

Fig. 3.4d Coated capacitance probe

If the process material adheres to the probe, a level reduction in the vessel will leave a layer of fluid on the probe. When this layer is conductive, the wet portion of the probe will be coupled to ground and the instrument will not read the new level but will register the level to which the probe is coated. Other than the changes in process material dielectric constant, this represents one of the most serious limitations of capacitance installations. It should be noted that if the probe coating is non-conductive, the interference with measurement accuracy is much less pronounced. One other problem frequently encountered with capacitance probe installations is grounding through the head assembly. Moisture can enter through capillary leakage or a cable entrance, causing a ground.

Sensitivity and drift in the measurement should be considered carefully. Some devices are sensitive to 0.5 pf and will drift 5 pf owing to a temperature change of 100°F (56°C). The best quality units available are sensitive to 0.1 pf and have a drift of 0.2 pf per 100°F (56°C).

Probe Sizing

For capacitance probe level measurement, differential capacitance has the same importance as differential pressure has to orifice flow detection. The differential capacitance over the level range determines the span of the instrument. It is desirable to keep this change in capacitance (ΔC) above 10 pf, although instruments with spans as narrow as 2 pf are available. A narrow span reduces measurement accuracy because of the limited sensitivity of the loop. Another important consideration is the ratio between differential and terminal capacitance (TC). The latter is the capacitance with the vessel empty. This ratio (TC/ΔC) should be kept between 4 and 0.25.

Sizing of the probe is illustrated by an example shown in Figure 3.4e, which indicates the dimensions of the vessel and the properties of the process fluid. Figure 3.4f relates the differential capacitance to the dielectric constant for several probe designs in a 5 ft. (1.5 m) tank. The dead, or gland, capacitance is estimated to be 12 pf. This plus capacitance in an empty tank gives the terminal capacitance. Initially, let us select a ½ in. (12.5 mm) diameter Teflon covered probe as described as Curve #1, Figure 3.4f. This, at a dielectric constant of unity, has a differential capacitance of 3 pf/1 ft. (305 mm). Therefore, the total terminal capacitance for the probe in Figure 3.4e is:

$$TC = 12 + (3 \times 7) = 33 \text{ pf.}$$

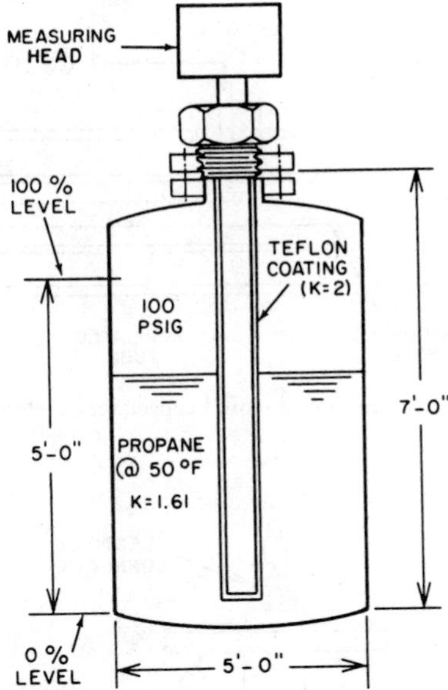

Fig. 3.4e Typical capacitance probe installation

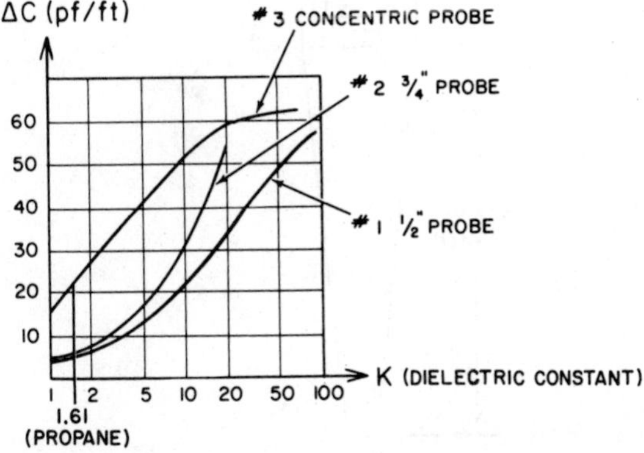

Fig. 3.4f Differential capacitance as a function of dielectric constant for several probes

If we enter the same graph for propane with a dielectric constant of 1.61, the differential capacitance for propane alone is 4.3 − 3 = 1.3 pf. The total change in capacitance over the level range is:

$$\Delta C = (4.3 - 3) \times 5 = 6.5 \text{ pf.}$$

The criteria established for a good installation was:

a.) $\Delta C > 10$ b.) $4 > (TC/\Delta C) > 0.25$

The selected probe fails on both counts and therefore is not satisfactory. From the equation in Figure 3.4c it can be seen that there are several alternatives to make this system work:

a.) The dielectric constant K cannot be changed, but it should be kept in mind that for K values greater than 2, the removal of the Teflon coating will increase differential capacitance.
b.) Increasing the sensitive probe length X would also increase the capacitance span, but this is limited by the process requirements.
c.) Increasing the probe diameter B will increase ΔC as shown on Figure 3.4f.
d.) Finally, reduction in the tank diameter A can also be considered. This does not imply that the vessel should be redesigned, but simply refers to installing a concentric shield around the probe, installing a grounded reference probe parallel to the capacitance probe, or locating the capacitance probe closer to the vessel wall.

Repeating the sizing procedure for the concentric probe design represented by Curve 3 on Figure 3.4f, we find the net *ΔC* to be 15 pf/ft.; therefore:

$$TC = 12 + (15 \times 7) = 117 \text{ pf and}$$

$$\Delta C = (22 - 15) \times 5 = 35 \text{ pf.}$$

This probe selection satisfies the installation since $\Delta C > 10$ and $TC/\Delta C < 4$.

Probe Configuration and Installation

Where the process tank cannot be opened in order to remove the probe for maintenance, the probe may be installed in an external chamber. For detection of liquid-liquid, liquid-vapor or solid-vapor interfaces in on-off service, the capacitance probe switches should be installed horizontally to provide a large change of wetted area due to a small change in level. Rugged probes, such as the one shown in Figure 3.4g are installed for detecting heavy solids, while knife edge designs are used for powders or fine solids. For continuous measurement, the probe is installed vertically and where the level span is large it is supported at both ends.

Capacitance measurements also can be made without contacting the process material. Installation of such a proximity probe is illustrated in Figure 3.4h. This design

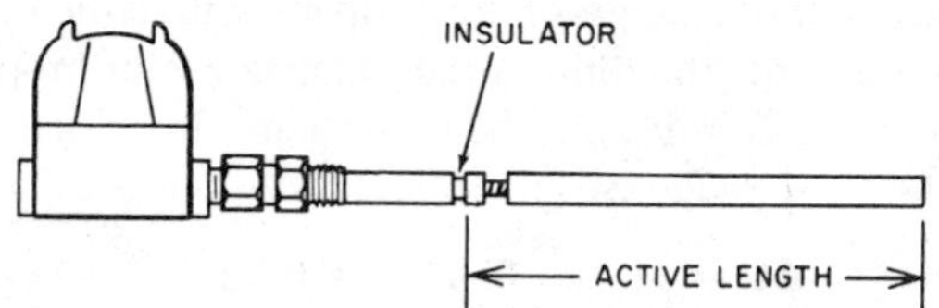

Fig. 3.4g Standard capacitance probe

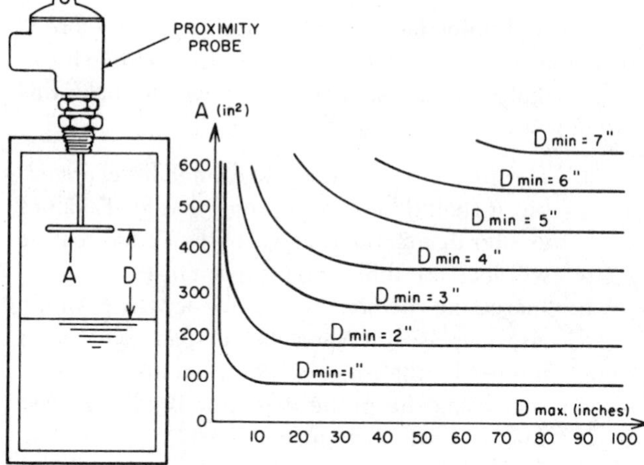

Fig. 3.4h Proximity capacitance probe

might be used where the process is hot—above 800°F (425°C)—or when it would otherwise coat the probe surface. Here, instead of the probe, a sensing plate is used that is spaced above the process material. Level variations change the geometry of the capacitor and thus its capacitance; this change can be detected and related to level change. The proximity probe is available as a high level switch or as a continuous level detector. It is not practical as a low level switch unless the distance between high and low settings is small.

The relationship between the plate to process distance and the plate area is very critical. Sufficient capacitance change must be developed by change in the process level. For continuous level measurement this should be at least 20 pf over the anticipated level range. The proximity plate area can be determined by using the following equation:

$$A = 4\frac{(\Delta pf)(D_{max})(D_{min})}{D_{max} - D_{min}} = 80\frac{(D_{max})(D_{min})}{D_{max} - D_{min}} \quad 3.4(1)$$

where

A = plate area, sq. in.,
Δpf = minimum required pf change,
D_{max} = maximum plate to process distance, in.,

and

D_{min} = minimum plate to process distance, in.

This equation is applicable for continuous proximity type capacitance level measurement if the process material is conductive. Any material with a moisture content of 15 percent or more may be considered conductive.

Capacitance measurements can also be made in pipelines. For example, petroleum products can be distinguished by their differing dielectric constants. By passing the pipeline product through an element similar to the one shown in Figure 3.4i, interface arrivals can be detected. Such in-line dielectric detectors can also be used to measure other process properties that vary with dielectric, such as moisture content. An interesting application of the capacitance probe is for flow measurement across weirs and flumes. As is discussed in Section 2.26, the flow across weirs and flumes is an exponential function of the upstream head. A special-purpose capacitance probe is available that is characterized such that the capacitance change versus head level is linearly proportional to flow rate. Figure 3.4j shows one method of doing this by characterizing the effective plate area.

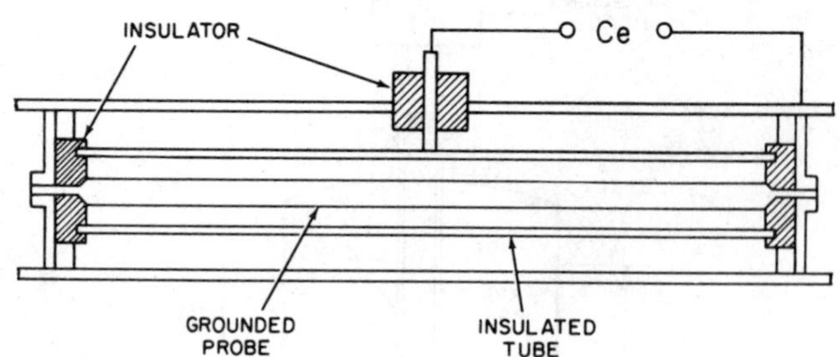

Fig. 3.4i In-line mounted capacitance element

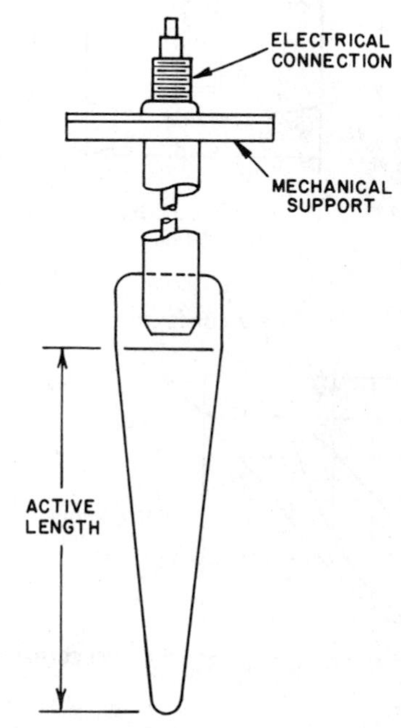

Fig. 3.4j Characterized probe for flow measurement over weir or flume

Conclusions

From the above discussion, the advantages and disadvantages of capacitance level measurement should be obvious. On the plus side, simplicity of design, absence of moving parts, and corrosion resistance can be noted, keeping in mind the proximity design which requires no contact with the process. The disadvantages are that accuracy is affected by changes in dielectric, and that process build-up on the probe can ruin the installation. Because of these, and recognizing that capacitance measuring installations are relatively expensive, there should be a compelling reason for selecting this method.

BIBLIOGRAPHY

"Capacitance Method for Liquid-Depth Measurement." *Electronics and Power*, December 1967.

Dinkel, J.A., "Universal Capacitance Probe Liquid Level Measuring System." *Review of Scientific Instruments*, November 1966.

Duncan, J. and Dutton, W., "Capacitance Probe Confirms Presence of Liquid NH_3 when Unloading." CIM Bulletin, January 1978.

Herbster, E.J. and Roth, J.H., "How to Gage by Capacitance." *I.S.A. Journal*, June 1965.

Mital, P.K., "Capacitor Sensor Monitors Stored Liquid Levels." *Electronics*, October 30, 1967.

Preshkov, V.P., "Capacitance Liquid Helium Level Indicator." *Cryogenics*, April 1969.

Schonfeld, S., "Capacitance Gaging Checks Spacecraft Fuel Level." *Hydraulics and Pneumatics*, April 1967.

Weiss, W.I., "Capacitance Level Control." *I.S.A. Journal*, November 1966.

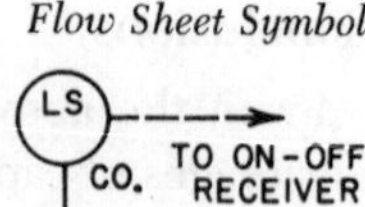

3.5 CONDUCTIVITY PROBES

Design Pressure:	To 3000 PSIG (21 MPa)
Design Temperature:	To 700°F (371°C)
Materials:	Wide selection of corrosion resistant materials
Cost:	\$40–\$400
Inaccuracy:	± ⅛ in. (± 3.175 mm)
Range:	On-off, actuation point determined by probe location
Partial List of Suppliers:	B/W Controls; Delavan Electronics, Inc.; Delta Controls; Electronics Corp. of America; Endress-Hauser, Inc.; National Controls Corp.; Zi-Tech Div., Aikenwood Corp

Figure 3.5a illustrates the principle of operation of the conductivity level probe. The electrode is shown above the liquid level on the left side of the sketch. The circuit, therefore, is open and no current is flowing through the relay coil to energize the relay and close the load contact. When the liquid level rises, as shown on the right side of the sketch, a conductive path between the electrode and the grounded tank is established, closing the circuit through the relay coil. Energization of the relay closes the load contact, operating pumps, solenoid valves or other processing equipment. In this system, the liquid in the tank acts as a switch in the relay circuit. The single probe installation described is normally used for high or low level detection. Although electro-mechanical relays are shown in the figure, solid state relays are commonly used. Also, if the tank is fabricated of fiberglass or other insulating material, the switching circuit can be made between the sensing probe and a reference probe.

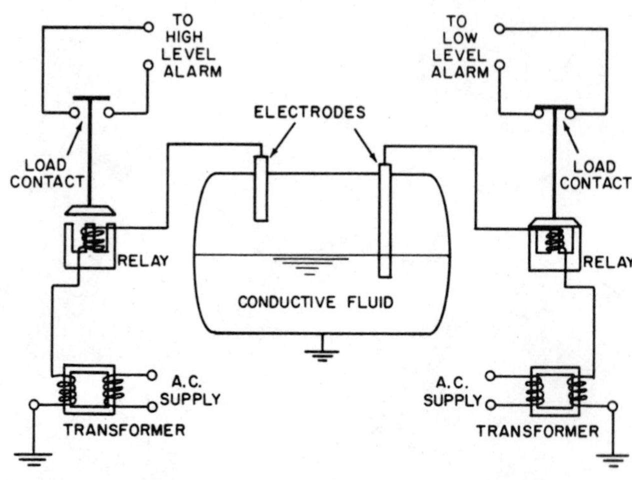

Fig. 3.5a Single point conductivity switch

Because these switches are available in a variety of configurations, they can be used for on-off control of one piece of equipment or for staged control of several pieces of equipment. When two pumps are installed in the same on-off service, it is sometimes necessary to automatically alternate the pumps so that wear is evenly distributed or so that pump motor hot starts are reduced. Figure 3.5b shows how one level switch with two conductivity probes can be used in conjunction with an electromechanical alternator to cycle the pumps. As level rises, LSL will close since this is the load contact operated by the lower probe. The control relay CR remains deenergized. When the level rises further, LSH closes, thereby energizing CR and closing contact CR-1. This is a hold in circuit identical to the one described in Section 3.9; CR will remain energized after LSH drops out. With CR energized, CR-3 closes and circuits are made through contact A of the position switch and contact 1 of the sequence switch. The first circuit starts the alternator switch motor which, in turn, moves the position switch to contact B. This deenergizes the switch motor, since CR-2 is held open when CR is energized. The second

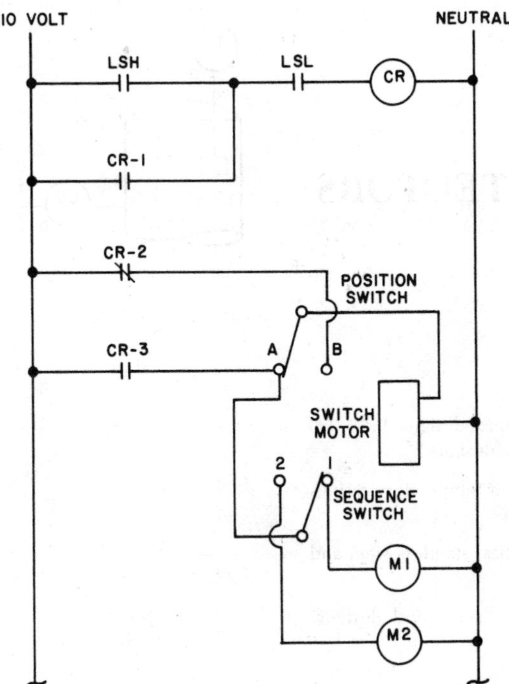

Fig. 3.5b Pump down alternator circuit

circuit through the sequence switch energizes M1, the starter coil for the motor associated with the first pump, and the first pump starts. When the level falls to LSL, relay CR is deenergized, dropping out CR-3 which stops the motor associated with the first pump. CR-2 closes and energizes the switch motor which steps to move the sequence switch to contact 2 and the position switch back to contact A. On a subsequent rise in level, M2 will be energized and the second pump will start. This is the most basic of the pump down alternator circuits. A similar circuit can be designed for pump up applications, and a wide variety of additional control requirements can be met by increasing the number of probes and relays.

The advantages of the conductivity switch include low cost, simple design and elimination of moving parts in contact with the process material. It can also be used to detect the level of moist bulk solids.

It has several disadvantages. In chemical processing equipment, the possibility of sparking when the liquid level is close to the probe can seldom be tolerated, although some of the solid state designs are rated for intrinsic safe operations. This switch is limited in application to conductive (below 10^8 ohm/cm resistivity) and non-coating processes. In most processes, electrolytic corrosion at the electrode can have harmful side effects. Electrolysis can be reduced, but not eliminated, by using AC currents. Finally, directly-connected level measuring devices are a poor choice for pressurized vessels.

Conclusion

Conductivity switches are rarely used in chemical processing services. However, they are used routinely in water level detection systems, including those on steam drums operated up to 3000 PSIG (21 MPa).

BIBLIOGRAPHY

Andreiev, N., "Survey and Guide to Liquid and Solid Level Sensing," *Control Engineering*, May 1973.

Belsterling, C.A., "A Look at Level Measurement Methods," *Instrument and Control Systems*, April 1981.

Hall, J., "Measuring Interface Levels," *Instruments and Control Systems*, October 1981.

Lawford, V.N., "How to Select Liquid-Level Instruments," *Chemical Engineering*, October 15, 1973.

Flow Sheet Symbol

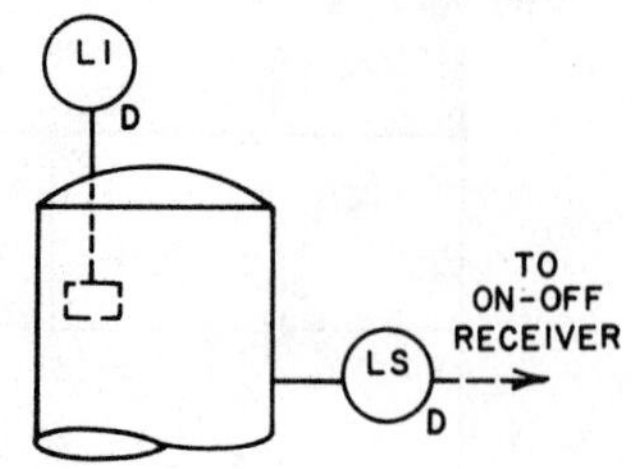

3.6 DIAPHRAGM LEVEL DETECTORS

Design Pressure:	Atmospheric or low for mechanical designs, to 15,000 PSIG (103 MPa) for electronic designs
Design Temperature:	0 to 130°F (−18 to 54°C), higher temperature ratings specially available
Materials:	Neoprene, Teflon, other plastics; stainless steel and high alloy metals
Inaccuracy:	±1–6 in. (±25–152 mm) for mechanical designs, ±0.3% for electronic designs
Range:	Practically unlimited
Cost:	\$75 to \$400 for mechanical, \$700 to \$1500 for electronic
Partial List of Suppliers:	Bindicator Co.; Delta Controls; ElectroSyn Corp.; Fisher Controls; King Engineering; Monitor Mfg.; Robertshaw; Viatran Corp

All diaphragm detectors operate on the simple principle of detecting the pressure exerted by the process material against the diaphragm. The designs discussed below include diaphragm switches for liquid and solid services and diaphragm devices for continuous liquid level detection.

Diaphragm Switches for Solids

For solid service, diaphragm switches can be selected from a number of design variations. Devices with mercury switches can be used with materials having a bulk density of more than 30 lb/ft^3 (0.5 g/m^3), while units with microswitches are used for lower density services. Some of the most sensitive diaphragm switches will actuate with as little as 6 oz. (170 g) of force on the diaphragm. The differential of a single diaphragm can be as high as 8 in. (203 mm), meaning that the switch will close its circuit when the solids rise to the top of the diaphragm and will open the circuit when they drop 8 inches. The lower the solid density, the larger the diaphragm area required for sensitive operation. Units are available with 4 to 10 in. (102 to 254 mm) diameter diaphragms. As illustrated in Figure 3.6a, there are three basic ways of installing these detectors. They can be suspended on a support pipe to provide for quick adjustment of the switch position, they can be mounted on the inside wall of thick-walled silos, or, as most commonly done, they can be externally mounted on thin-walled bins. The mounting location should always be selected to guarantee the free flow of solids to and from the diaphragm area.

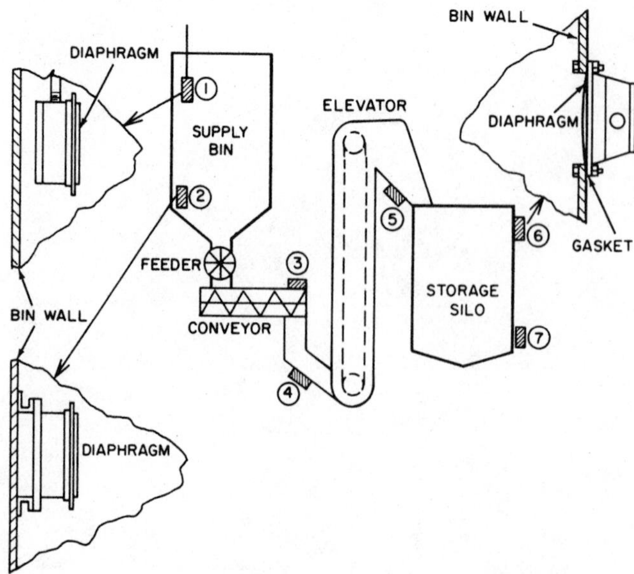

Fig. 3.6a Use of diaphragm switches in solids service

As shown in Figure 3.6a, diaphragm-type solid switches can serve several purposes. Switch #1 protects against overfilling, #2 signals low supply level, and #3 indicates choke-up in the screw conveyor. Diaphragm #4 detects overfeeding the elevator boot and #5 signals if the elevator discharge spout is choked -up. Diaphragm switches #6 and #7 in the storage silo will measure extreme level conditions and provide automatic shut off of the material flow when the desired level is reached.

Diaphragm Switches for Liquids

Figure 3.6b shows how diaphragm switches can be used to detect liquid level by sensing the pressure of a captive air column in a riser pipe beneath the diaphragm. An 8 in. (203 mm) head of liquid above the inlet of the riser pipe compresses the air sufficiently for switch actuation, and the unit can handle a maximum of 60 ft. (18 m) of liquid. The diaphragm is in contact with the captive air but not with the process. These units are limited in application to atmospheric tanks, and should be considered only for secondary applications where low cost is desired and accuracy is of no serious consideration.

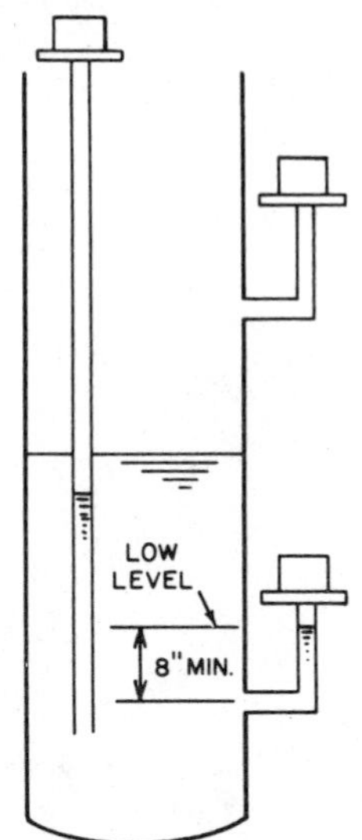

Fig. 3.6b Diaphragm switches in liquid service

Continuous Diaphragm Level Detectors for Liquid Service

Figure 3.6c illustrates two versions of the continuous detector, both limited to atmospheric tanks, and to applications where low cost is more important than quality or accuracy of measurement. The diaphragm box unit, shown on the left side of the sketch, is quite similar in operation to the previously discussed riser pipe diaphragm switches except that the diaphragm isolates the captive air from the process fluid. The unit consists of an air-filled diaphragm box connected to a pressure detector via capillary tubing. As the level rises above the diaphragm, the liquid head pressure compresses the captive air inside. The air pressure in the capillary tubing is sensed by a pressure element and interpreted as a level indication.

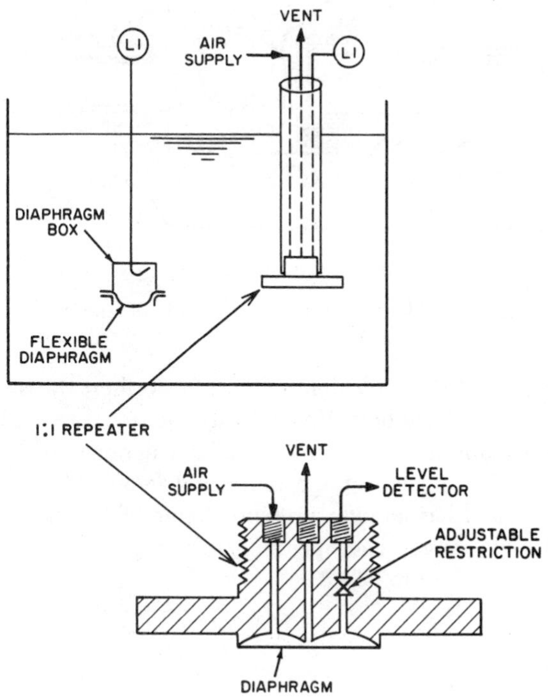

Fig. 3.6c Diaphragm devices for continuous detection of liquid levels

The one-to-one pressure repeater is illustrated on the right side of Figure 3.6c. This unit is submerged in the vessel; the static head of the liquid exerts an upward force on the diaphragm which increases as the level rises. The upward force is opposed by the air supply pressure on the other side of the diaphragm. The force due to rising level moves the diaphragm toward a bleed orifice, thus restricting its flow to atmosphere and causing the air pressure to build up until it equals the static head pressure. When the forces on the two sides of the diaphragm are equal, the unit is in equilibrium. The speed of response of the unit is changed by an adjustable restriction which, if opened, will increase sensitivity by allowing more air to flow onto the diaphragm. Air supply to the unit can be regulated at a pressure slightly in excess of the maximum hydraulic head to be repeated.

There are several electronic diaphragm pressure sensors available which may also be used to detect level. One is shown in Figure 3.6d. The unit consists of a diaphragm, a straight axis bourdon tube, a rotary differential transformer, and a housing. The bourdon tube is liquid-filled and is cantilevered from the process side. Pressure applied to the diaphragm is transferred to the liquid fill in the bourdon, causing the free end of the bourdon to rotate. The rotation of the bourdon is de-

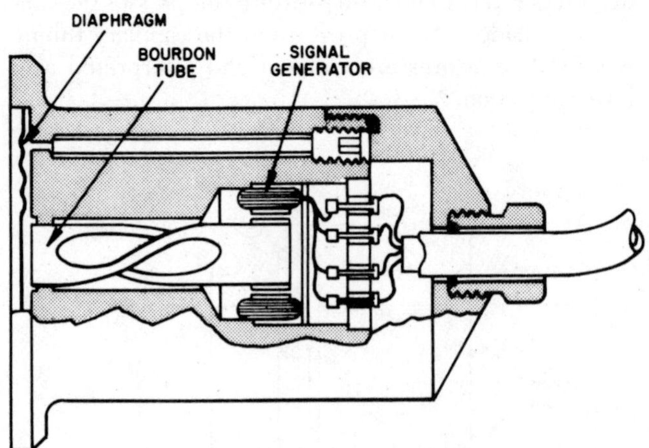

Fig. 3.6d Electronic diaphragm level sensor

tected by the rotary differential transformer; the signal so produced can be correlated with the pressure applied to the diaphragm. This unit is available in ranges of 0–100 in. (0–2.5 m) of water column and 0–300 PSIG (0–2 MPa), and has an over pressure rating of 1.5 times the full scale range. The operating temperature range is 0 to 130°F (−18 to 54°C).

Conclusions

Diaphragm level dectectors find a broad range of application as level switches in solid service and have proved to be reliable devices. The mechanical diaphragm devices for liquid service are discussed above more for the sake of completeness than to imply that they should be considered for critical applications where accuracy is of consequence. The electronic diaphragm devices for liquid service are precision instruments and can be considered for many moderate temperature applications.

BIBLIOGRAPHY

Belsterling, C.A., "A Look at Level Measurement Methods," *Instruments and Control Systems*, April 1981.

Hall, J., "Measuring Interface Levels," *Instruments and Control Systems*, October 1981.

Imsland, T., "Connecting d/p Elements for Level Sensing," *Instruments and Control Systems*, November 1975.

Lawford, V.N., "Differential Pressure Instruments: The Universal Measurement Tools," *Instrumentation Technology*, December 1974.

Slomiana, M., "Using Differential Pressure Sensors for Level, Density, Interface, and Viscosity Measurements," *Instrumentation Technology*, September 1979.

Flow Sheet Symbol

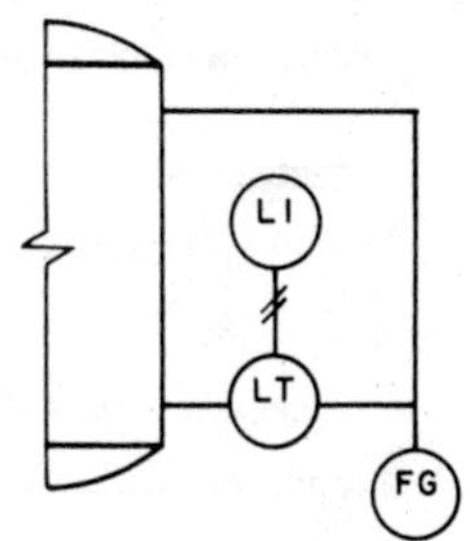

3.7 DIFFERENTIAL PRESSURE LEVEL DETECTORS

Design Pressure:	To 10,000 PSIG (69 MPa)
Design Temperature:	To 350°F (175°C) for d/p cell, to 1200°F (650°C) for filled systems, others to 200°F (93°C)
Materials of Construction:	Plastics, steel, stainless steel, higher alloys for d/p cells
Range:	d/p cell ranges available for all industrial applications
Cost:	\$600 to \$800 for transmitters in standard constructions, \$200 to \$500 for local indicators
Partial List of Suppliers:	Barton Instruments; Beckman Instruments; Foxboro Co.; Fischer and Porter Co.; Gould, Inc.; Hildebrandt Engineering Co.; Honeywell; Process Control Div, Rosemount, Inc.; Taylor Instruments

Liquid level can be measured by a differential pressure (d/p) instrument. For vessels operated at atmospheric pressure, the high side of the instrument is connected to the bottom of the vessel and the low side is vented to atmosphere. For pressurized vessels, the high side is connected to the vapor space in the vessel. This installation will give an accurate measurement of the liquid level provided that the density of the process does not change. Theoretically, differential pressure can be detected by sensing two pressures separately and taking the difference to obtain liquid level. In practice, it is desirable to use a single pressure difference sensor so that the static pressure levels are intrinsically balanced. The importance of this consideration can be visualized, for example, on a 0 to 100 in. (0 to 2.5 m) water column measurement where the expected accuracy will be $\pm\frac{1}{2}$ in. ($\pm$12.7 mm) water column. It would be impossible to approach this accuracy if the measurement were made at a static pressure level of 1000 PSIG (6.9 MPa) using two independent sensors.

Differential Pressure Devices

The categories of devices considered are: dry force balance designs, dry motion balance designs, and manometers. Generally, dry force balance designs are used when a pneumatic or an electronic transmission signal is required. The pneumatic transmitters are considered first; the options discussed are also available on electronic transmitters.

Figure 3.7a shows a schematic illustration of the pneumatic transmitter. A pair of diaphragms is welded to the opposite sides of the cell and the space between them is liquid-filled. This fill is used for damping out process noise, a common problem when the d/p cell is used on flow applications, but not normally a consideration for level installations. The differential pressure to be detected is applied to the two sides of this diaphragm capsule. The resulting force is then brought out of the d/p cell via the force bar. When a change of differential occurs, the force bar changes its position relative to the nozzle, causing a change in the pneumatic output signal. The change in air pressure is sensored by the feedback bellows and a new state of equilibrium (force balance) is established by the bar force being opposed by an equal force developed in the feedback bellows. As a result, the output signal is maintained proportional to differential pressure sensed by the cell.

Several important design features of the d/p cell merit attention. As shown in Figure 3.7b, the unit has zero, span, and elevation/depression adjustment. If the unit is piped so that the high and low sides can be equalized, the zero setting of the instrument can be checked by reading the output gauge. The span adjustment is large, generally in the range of 10 to 1. The elevation/depression adjustment can be used to "zero out" a constant head on either the low side or the high side. The pressure rating of the unit is determined by the rating of the enclosure and can be as high as 10,000 PSIG (69 MPa).

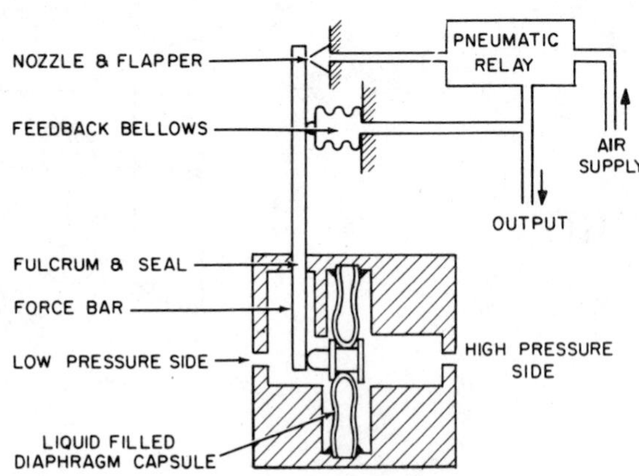

Fig. 3.7a Force balance d/p cell schematics

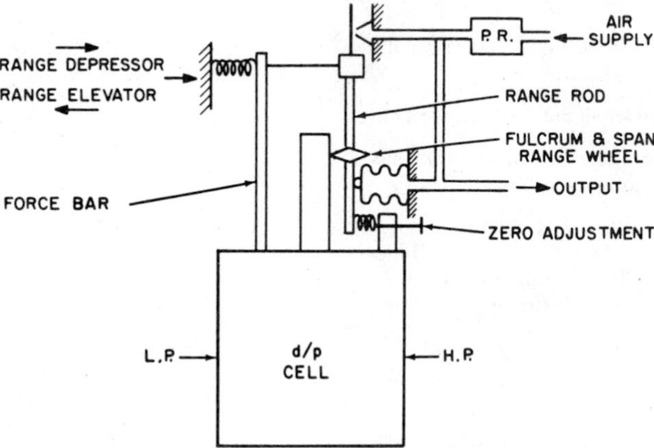

Fig. 3.7b Force balance d/p cell adjustments

Exposure of one side or the other to this pressure will not damage the instrument or necessitate recalibration because overpressure causes the diaphragms to bottom out on the capsule supports before the elastic limit of the diaphragms is reached. The displacement volume of the unit is small; thus, varying static pressures will not cause the level in the seal leg to drop. Also, because of limited motion, there is very little friction or wear in the unit. Finally, on some designs the low side and high side volumes are the same, so that liquid expansion or contraction due to temperature change has the same effect on both sides of the diaphragm. This is important if the unit is furnished with liquid-filled chemical seals. A wide range of materials are available for process-wetted parts. A common diaphragm material is Hastelloy. The body can be furnished in carbon steel, stainless steel, high alloys, and plastics such as Teflon, although plastic designs will have a considerably lower pressure rating. A common temperature rating for the metal body designs is 250°F (120°C).

Variations on Force Balance d/p Cells

The d/p cell can be modified for use on viscous, slurry, or other plugging applications. Figure 3.7c shows two such designs: the extended and the flat diaphragm d/p cell transmitters. Their principle of operation is the same as that of the conventional d/p cell illustrated in Figures 3.7a and b except that the high pressure side of the diaphragm capsule is in direct contact with the process, and the force detected by it is transmitted through a flexure to the lower end of the force bar.

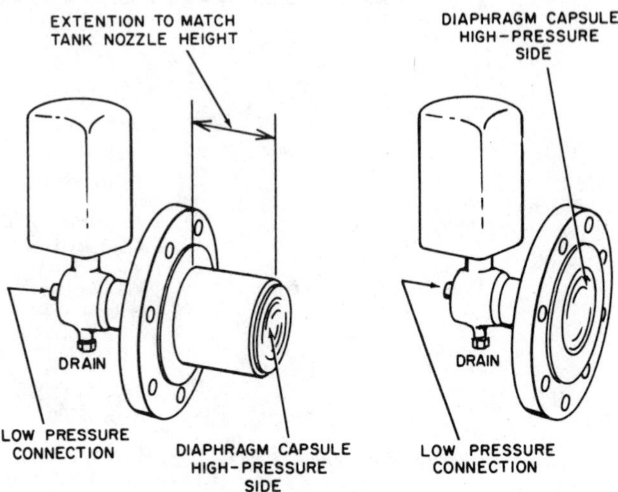

Fig. 3.7c Extended diaphragm d/p cell (left); flat diaphragm d/p cell

The extended diaphragm design is designed to bolt directly to the vessel nozzle; the protrusion can be sized to fill the space in the nozzle, placing the diaphragm flush with or slightly inside of the vessel wall. This design completely eliminates dead cavities and is used especially on materials that can freeze at high temperatures or that can deteriorate or discolor if pocketed. There is some resistance to this design because it cannot be serviced without depressurizing and draining the vessel.

The flat diaphragm design is normally installed by bolting it directly to the vessel block valve. Because the connection is large—3 in. (76 mm)—the process is less apt to bridge or plug the sensing connection. Flat diaphragm cells can be furnished with a solvent flush or steam-out connection.

Both the flat and the extended designs are available with ranges up to 850 in. (2160 mm) with an accuracy of $\pm\frac{1}{2}$ percent of span. The flat units can withstand 550 PSIG (3.8 MPa) operating pressures when the process is at 350°F (175°C). The maximum process temperature rating for the extended design is 750°F (400°C). Changes in process or ambient temperatures can cause zero shifts, as is the case with all similarly designed instruments. To minimize this effect, the d/p cell should be zeroed at the normal operating temperature, and the exposed body of the transmitter should be insulated.

Some suppliers offer the extended and flat diaphragm designs with a Teflon, Viton, or other plastic coating. This coating is intended as a slicking surface to minimize material build-up on the diaphragm. The plastic coatings should not be relied upon for corrosion protection unless the supplier states specifically that the coating is so designed. As a general rule, the engineer should not rely on coatings for corrosion protection of wetted parts of any process instruments. If the coating is nicked during installation, that assembly will be hazardous, particularly if exposed to pressure.

Another family of force balance d/p cell transmitters, also available with flat and extended diaphragms, has liquid-filled elements (chemical seals). The units shown in Figure 3.7c are used on atmospheric tanks or on pressurized tanks if the low side connection can be kept clean or sealed. The designs shown in Figure 3.7d and 3.7e are used on pressurized vessels where plugging or corrosive problems can occur on the high side and the low side. The chemical seal designs are available in a very broad range of materials including such metals as tantalum and zirconium. Vessel connection considerations are the same as already outlined, except they apply to both the high side and the low side. As can be seen from the figures, the process material contacts the diaphragm and the process side of the diaphragm flange. The instrument side of the diaphragm is filled with a liquid, and is connected by capillary to the high and low sides of the d/p cell. The differential pressure capabilities of these systems are dependent on the d/p cell selected. The accuracy of the system will always be worse than the accuracy of the d/p cell itself. The spring constant of the diaphragms at the chemical seals will cause some inaccuracy. This inaccuracy becomes more pronounced at small differential pressure ranges. A larger and less predictable error can result from the temperature-sensitive nature of the seal and capillary systems. Temperature differences between the low and high side systems will cause differing amounts of thermal expansion in the two systems; this will be interpreted by the d/p cell as a differential pressure (level) change. Because the unequal amounts of expansion can be caused by the process or by varying ambient conditions, it is not always possible to zero this error out.

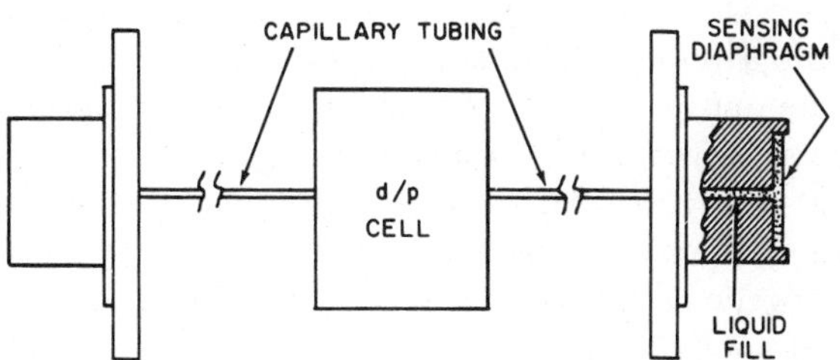

Fig. 3.7e d/p cell with extended chemical seal elements

The pressure and temperature ratings for these systems are dependent on seal design and filling liquid. Seals are readily available that are rated to 1500 PSIG (10 MPa). The fill material is normally a silicone oil which is good to 450°F (232°C). Other fill materials raise temperature ratings to 1200°F (650°C), allowing chemical seal designs to be considered for high temperature applications.

The Electronic d/p Cell

One common design of electronic d/p transmitter uses a force bar and fulcrum assembly similar to that shown in Figure 3.7a. Motion of the force bar is detected by use of a passive electrical element such as variable reluctance or bonded strain gauge. The output signal is fed back through a coil to rebalance the force bar and thus this instrument is classified as a force balance design. Figure 3.7f shows a section of a differential pressure sensing element that is used in a transmitter that does not have a force bar. Units of this kind detect position of a sensing diaphragm by measuring variable capacitance. For the unit shown, increasing pressure on the high side diaphragm causes filling oil to flow into the

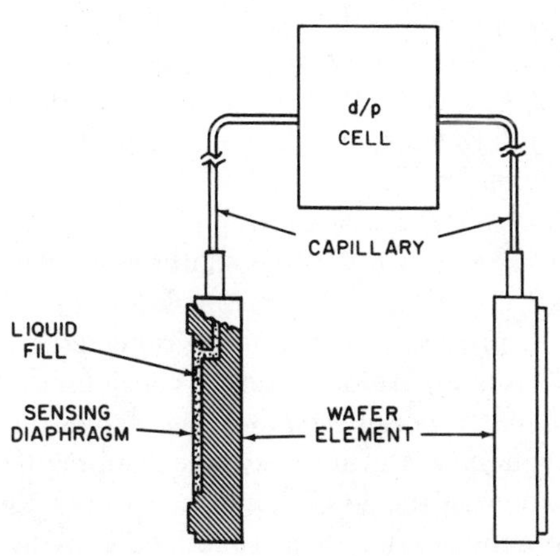

Fig. 3.7d d/p cell with wafer elements

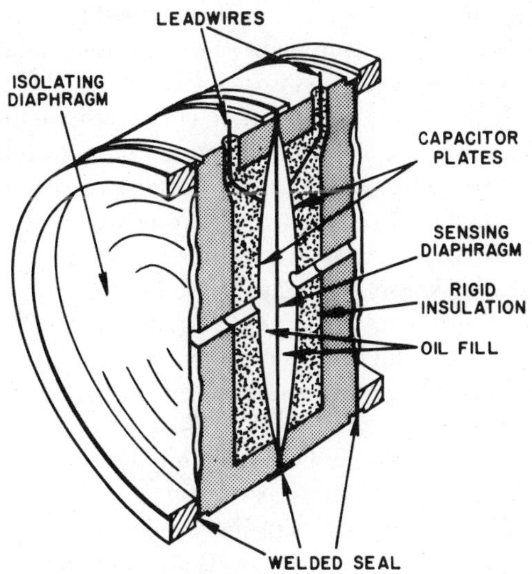

Fig. 3.7f Capsule for an electronic d/p cell

inner chamber, forcing the sensing diaphragm to the right. The capacitance between the high side capacitance plate and the sensing diaphragm increases and capacitance between the sensing diaphragm and the low side plate decreases. This change is detected by an electronic amplifier circuit which produces a milliampere current proportional to differential pressure. Because this unit does not have electro-mechanical feedback, it is classified as a motion balance design. However, the amount of motion is very small. As with the pneumatic d/p transmitter, the low side can be vented to atmosphere or it can be connected to the vapor space in the vessel in order to obtain static pressure compensation. Units are available that have ranges from 5 in. (127 mm) of water column to 1000 psi (6.9 MPa). Temperature ratings are in the range 200–250°F (93–121°C) depending on the unit selected.

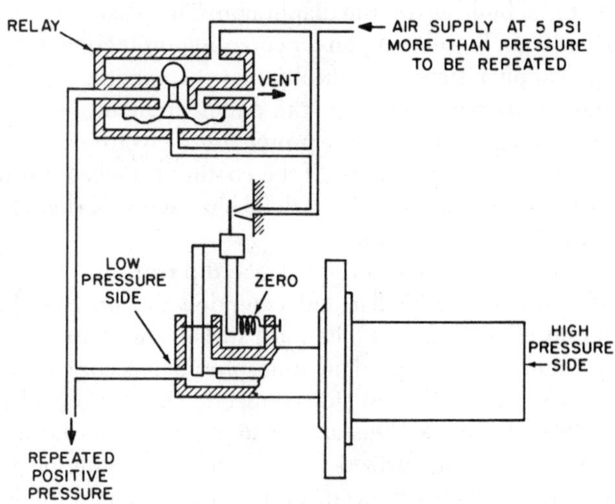

Fig. 3.7g Extended diaphragm positive pressure repeater

Force Balance Pressure Repeaters

When detecting the level in pressurized vessels, the low side pressure must be connected to the low side of the d/p cell to serve as a reference. On hard-to-handle materials, a one-to-one pressure repeater may be used to make this reference, and to simultaneously isolate the d/p cell from the process. These devices develop an air output pressure equal to the vapor pressure on the process side. Diaphragm level devices are discussed in Section 3.6 and illustrated in Figure 3.6c. These devices are inexpensive, but at the same time their accuracy is limited. The error in the repeated output pressure increases as the repeated pressure rises. At a pressure level of 40 PSIG (0.27 MPa) the error is 2 in. (51 mm) water column, while at a pressure of 400 PSIG (2.7 MPa), the error is 20 in. (508 mm) water column. Obviously, errors of this magnitude would not be acceptable for most process level measurements.

The repeater considered here is a modification of the d/p cell, is available with flat or extended diaphragm, and has an accuracy more in keeping with the requirement for level measurement. Designs are available for vacuum to positive pressures of 150 PSIG (1 MPa). Figure 3.7g illustrates the operation of the positive pressure repeater, which can reproduce pressures up to 100 PSIG (0.68 MPa) with high accuracy. A high pressure relay is required when the pressure to be repeated is above 100 PSIG. This device is similar to the d/p transmitter except that feedback from the relay is piped to the low pressure side of the d/p cell rather than to a bellows. The relay is modified for high pressure operation and the capsule is modified, but otherwise the repeater is identical to the transmitter.

Operation of the unit is as follows: As the vapor pressure in the vessel increases, it is detected by the high side of the diaphragm capsule, transmitting additional force to the force bar. This in turn causes the nozzle to be covered by the flapper, generating an increased back pressure to the relay. The relay responds by restricting the vent connection and opening the air supply to the relay outlet. This increased relay output pressure is then sent to the low pressure side of the d/p cell diaphragm capsule. When the pressures on the high and low sides are equal, the flapper is repositioned in relation to the nozzle in its throttling band and remains in that position until the process changes again. The air signal which is sent to the low pressure side of the capsule is equal to the vessel vapor pressure and is used to compensate the level transmitter in the liquid phase.

When it is desired to repeat vacuum instead of positive pressure, two techniques can be employed. One uses a suppression spring and the other uses an external vacuum source.

The range suppression (or depression) spring is illustrated in Figure 3.7b. If the suppression spring is set for 16.5 psi (113.6 kPa), the output signal of the repeater will be the sum of the pressure sensed by the high pressure side of the capsule plus the setting of the suppression spring. In other words, the output signal from the repeater will be 16.5 psi (113.6 kPa) higher than the pressure of the process vapors. If the process is at atmospheric pressure, then the output will be 16.5 PSIG (113.6 kPa); if it is at full vacuum, the output will be 1.8 PSIG (12.4 kPa). By the use of this bias, the need for a vacuum reference source is eliminated. The bias added in the repeater is eliminated at the d/p cell by proper setting of its suppression spring.

When a reference vacuum source is available that is at least 2 in. (50 mm) of water column lower than the minimum process pressure to be repeated, the design illustrated in Figure 3.7h can be considered. This unit incorporates a special vacuum relay. When the absolute pressure in the vessel increases, the diaphragm rod transmits motion to the force bar which draws the flapper away from the nozzle, causing a reduction in the pressure on the relay diaphragm. This moves the ball down, closing off the vacuum supply and opening the connection to the atmosphere. The net effect is an increase in the relay output pressure, which is piped to the low pressure side of the diaphragm capsule to balance the pressure on the process side. When the absolute pressure of the process decreases, the opposite sequence of events takes place. The force bar covers the nozzle, increasing the pressure on the relay diaphragm; the ball moves up, closing the connection to the atmosphere and opening it to the vacuum source. This causes a reduction in the relay output pressure that matches the lower process pressure on the process side of the capsule. Of course these changes are also transmitted to the low pressure side of the level transmitter.

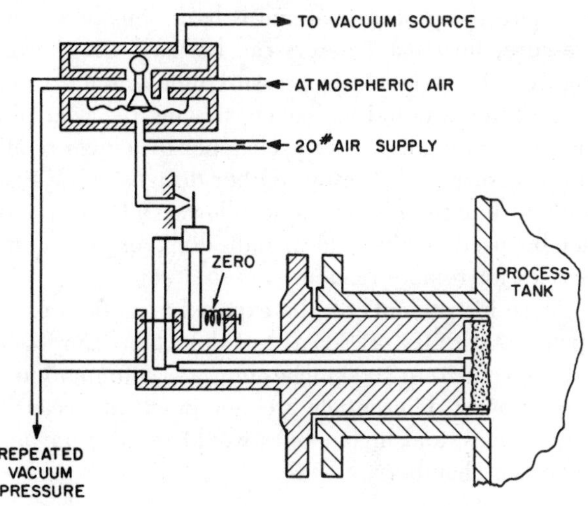

Fig. 3.7h Extended repeater for vacuum service

Dry, Motion Balance, Differential Pressure Devices

This family of differential pressure detectors is also referred to as bellows meters because it depends on liquid-filled, double opposed bellows. Bellows meters are most useful in areas where local indication or record is required and where compressed air or electric power is not available as an energy source. Their use as transmitters is limited because of the distinct advantages of the force balance units as discussed above.

Figure 3.7i illustrates the basic components of the unit, which are the high and low pressure chambers, the range spring, and the drive assembly to transfer bellows motion to the readout pointer. The bellows in both chambers and the passage between them are liquid-filled. When the unit is installed, the pressure in the high pressure chamber compresses the bellows, so that liquid flows from it into the low side bellows. When the low pressure (or range) bellows expands, it exerts a force against the range spring, which determines the span of the instrument. The linear motion of the "range" bellows moves the drive lever, mechanically transmitting a rotary

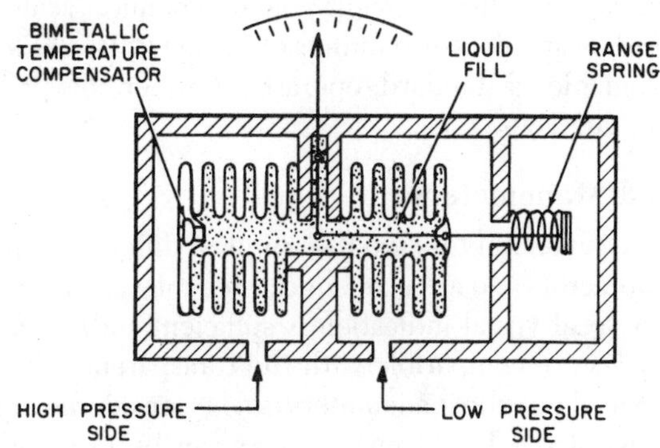

Fig. 3.7i Motion detector d/p indicator

motion through the sealed torque tube assembly to the indicator. The output motion from the torque tube assembly is limited to a few degrees of angular rotation. This is sufficient for most local indicators or recorders, but if the secondary device imposes a considerable load on the torque tube assembly, the accuracy and sensitivity of the unit are destroyed. For sustained accuracy, the bellows meter depends on the repeatability of its mechanical system, which has proven to be linear within ½ to 1 percent over its full range. A bimetallic temperature compensator inside the high pressure bellows automatically adjusts the capacity of the bellows assembly to the changing volume of the fill liquid due to ambient temperature variations. The device can be damaged by altering the restriction in the passage between the two bellows.

Bellows meters are provided with overrange protection. The operation of one of the protection mechanisms is as follows: The bellows move in proportion to the differential pressure applied across them. When they have moved over their calibrated travel, a valve mounted on the center stem seals against its seat, thereby trapping the fill liquid in the bellows. Because the liquid is noncompressible, the bellows are fully supported and cannot be ruptured regardless of the pressure applied. This overrange protection is furnished in both directions, protecting both bellows. Another design of overrange protectors involves the use of liquid-filled bellows with a number of diaphragm discs and spacer rings between them. As the bellows are subjected to overrange pressures, the diaphragms nest and the metallic spacer rings form a solid stop, thereby fully protecting the bellows from rupture.

Bellows meters can detect pressure differentials as low as 20 in. (508 mm) water column and as high as 400 PSIG (2.7 MPa). Measurement of low differentials is limited by the small forces available to actuate the motion detector mechanism. For very high differentials, the limitation is the mechanical strength of the bellows. Standard units are available with steel or stainless steel housings and stainless or beryllium copper bellows. For corrosive applications, other materials can be obtained or special high displacement volume chemical seals can be used. Static pressure ratings of 10,000 PSIG (69 MPa) are available as standard; operating temperature is limited to 200°F (93°C).

Liquid Manometers

Liquid manometers are the simplest differential pressure detectors and are economical, reliable and accurate. Where local visual indication is sufficient and the static pressures are compatible with the transparent tube design, the glass tube manometers can be considered. Float versions of the liquid manometers can be installed on high pressure services or where remote readout is desired.

Glass Tube Manometers

Figure 3.7j illustrates the most elementary U-tube manometer. The difference in level between the two columns of liquid is an indication of the pressure differential. There are no moving parts, no friction or inertia involved in the measurement, and therefore accuracy is limited only by scale visibility. As shown in Figure 3.7k, a more easily read scale can be attached to the well manometer. The liquid surface area on the high pressure side is 1000 or more times greater than the area on the low pressure side. Therefore, changes in differential pressure cause variations only in the low side column height and leave the high side level practically unaffected. Changes in well liquid level are compensated for by sliding the scale zero to match this level.

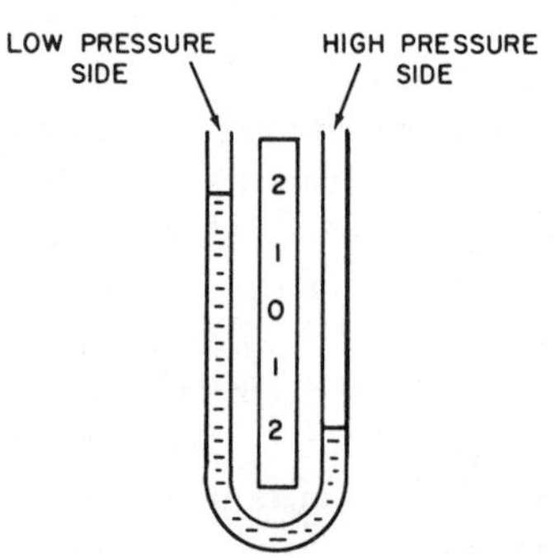

Fig. 3.7j U-tube manometer

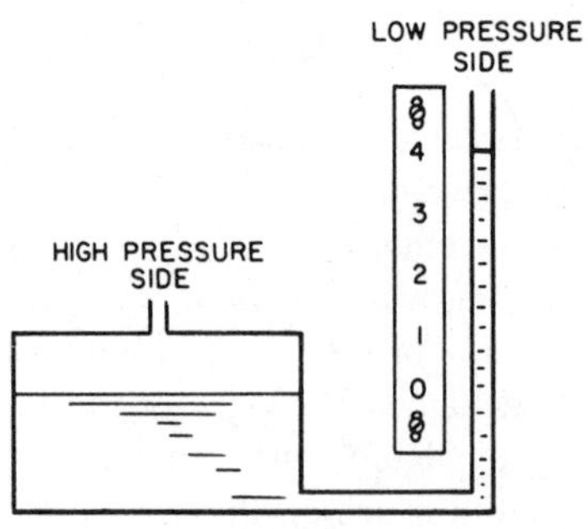

Fig. 3.7k Well manometer

The liquid fill for manometers must be inert and must be compatible with, immiscible with, and heavier than the process. Glass tube manometers are available with ranges up to 120 in. (305 cm) which is sufficient for most level applications if the filling fluid is mercury. Designs for pressures up to 1000 PSIG (6.9 MPa) can be obtained, but their use is discouraged because of the potential for danger if the tube is damaged. In industrial installations, the use of glass tube manometers is limited to locations where tube breakage will not endanger the operator either due to the exploding glass particles or because of the nature of the process material released.

Float Liquid Manometers

For installations where remote readouts are required, or where the process material is hazardous or at a high pressure, float manometers can be used. As shown in Figure 3.7l, variations in pressure differential cause the level of filling liquid to change, moving the float. Float motion is brought out of the chamber by a lever rotating in a pressure seal bearing. (Other designs are available using torque tube or magnetic followers.) Lever rotation can be used to drive local indicators, or transmitting devices for remote readout.

These instruments can be exposed to static pressures up to 6000 PSIG (41 MPa) and detect pressure differentials from 20 to 1000 inches of water column (5 to 250 kPa). Lower differentials are not practical because of inaccuracies, and higher ones would require unwieldly reference chambers.

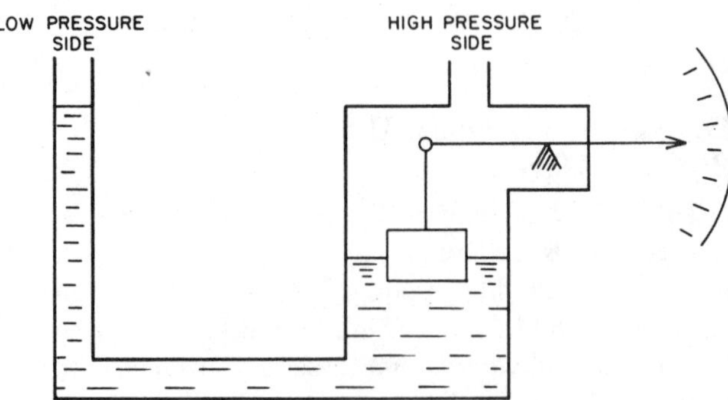

Fig. 3.7 l Float manometer

Stainless steel U-tube manometers having magnetic floats in both legs have been used for high precision at static pressures to 10,000 PSIG (69 MPa). The float positions in this design are detected by electric coils which provide a very precise measurement of differential pressure.

Conclusions

For industrial installations, where safety, initial cost, and low maintenance are important factors, the force balance type dry d/p cells warrant consideration. Where only local readouts are required and the cost of a transmitter cannot be justified, the dry motion detector bellows meters provide a good solution.

Variations of d/p-type Level Detection Loops

The application of pressure differential detectors as components in level measurement loops will be covered next. The requirements of atmospheric and pressurized tanks and the features of level loops on clean and hard-to-handle process fluids will be discussed separately. Figure 3.7m shows the symbols used for the various loop components.

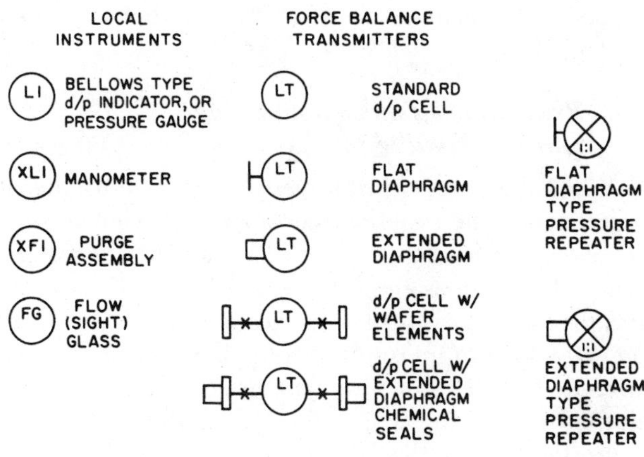

Fig. 3.7m Symbols for d/p level loops

Atmospheric Tanks Containing Clean Liquids

Unpressurized vessels containing clean liquids are the least demanding as far as level measurement is concerned, because the two most common sources of difficulties, vapor pressure compensation and plugging, are not present. Figure 3.7n shows five tanks equipped with five different types of level devices. The first two are for remote readout; the others are for local readout. On tank #1, a standard d/p transmitter with screwed connections is shown with its low pressure side open to the atmosphere. This installation can be made by using a pressure transmitter instead of a d/p transmitter. The pneumatic receiver gauge is normally calibrated for 0 to 100 percent level. The flat diaphragm-type d/p transmitter is shown on tank #2. Compared to the standard d/p cell, the flat diaphragm-type transmitter is simpler to install, and it is nozzle mounted, requiring no other means of support. It can also be less expensive, because only the diaphragm and the retaining ring are in contact with the process, so only these parts must be made of corrosion resistant materials; in the standard d/p cell, the entire body is exposed to the process fluids. A flat diaphragm-type pressure repeater can also be used in place of the d/p transmitter, in which case the receiver gauge will sense the actual hydraulic head instead of a 3 to 15 PSIG (21 to 103 kPa) transmitted signal. Tank #3 shows a motion

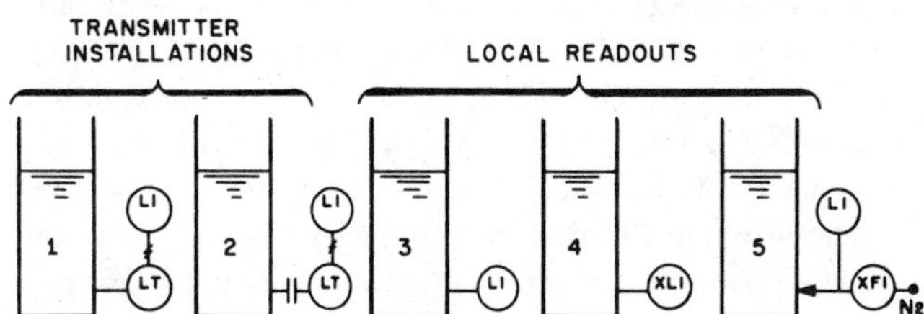

Fig. 3.7n Detection of clean liquid levels in atmospheric tanks by d/p instruments

balance local d/p indicator with the low pressure side vented to atmosphere. The same measurement can be made by using a standard pressure gauge. The level in tank #4 is detected by a manometer. Although this is one of the most accurate and economical devices to use for local readout, consequences of mechanical damage and proper selection of filling liquid must be considered. The installation on tank #5 is basically an air bubbler system which is covered in detail in Section 3.3.

Pressurized Tanks Containing Clean Liquids

When the level in a pressurized vessel is to be established by hydraulic head measurement, the instrument has to be compensated for the vapor pressures in the tank. This is done by exposing the low pressure side of the d/p cell to these vapor pressures. Compensation can be achieved by various means. Figure 3.7o shows seven variations of this installation. Tank #1 illustrates a wet leg application where the compensating leg is prefilled with a chemically inert liquid which will not freeze or vaporize under operating temperature conditions. The wet leg installations are used when the process vapors would otherwise condense into the compensating leg thereby exposing the low pressure side of the d/p cell to unpredictable hydraulic heads, or when the transmitter must be sealed from corrosive vapors. The prefilled wet leg creates a constant pressure on the low pressure side of the transmitter which is zeroed out by the range depression spring shown in Figure 3.7b. The leg is filled through a seal pot to provide excess capacity.

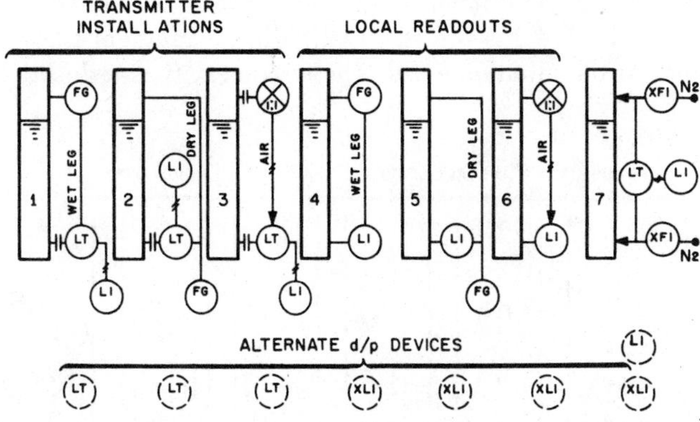

Fig. 3.7o Measurement of clean liquid levels in pressurized tanks by d/p instruments

It is desirable to make this seal pot out of a sight flow indicator so that the level of the filling liquid is visible to the operator. The d/p cell can be either the standard or the flat diaphragm design. On tank #2 the same d/p transmitter is installed in a dry leg system. This is acceptable when the process vapors are not corrosive and condensation at ambient temperatures is not expected. For such applications, a condensated pot is installed below the d/p cell, which should be made out of a sight flow indicator, so that the operator can visually determine if it is time to drain out the long term accumulation of condensate. Tank #3 illustrates the use of a flat diaphragm pressure repeater for vapor pressure compensation. It can repeat pressures between vacuum and 400 PSIG (2.7 MPa) with accuracies discussed above. The problems associated with range depressor adjustments, corrosion or condensate accumulation are eliminated by the use of repeaters. On tanks #4, 5, and 6, the same basic installations (wet leg, dry leg, and repeater) are shown in connection with local bellows indicators. Manometers can also be considered in place of the motion balance d/p indicators if mechanical damage, chemical inertness of the filling fluid, and its compatibility with the operating temperatures are previously established. Tank #7 illustrates a bubbler system with either transmitting or local indicating d/p devices. The limitations and drawbacks of such installations have been pointed out in Section 3.3.

Atmospheric Tanks Containing Hard-to-Handle Fluids

Level measurement is more difficult when the process fluid is highly viscous, is likely to freeze, contains solids that can settle out, or can gel or polymerize in dead-ended cavities. Figure 3.7p shows six installations that may be considered for these conditions. On tank #1, an extended diaphragm-type force balance transmitter is shown. The diaphragm motion is limited to a few thousandths of an inch and the nozzle cavity is completely eliminated by the diaphragm extension. The extended diaphragm transmitter is a good candidate for level measurement of hard-to-handle liquids, provided that the vessel can be drained when the transmitter needs service. Tank #2 is provided with a flat diaphragm transmitter mounted on a pad. The nozzle cavity is reduced,

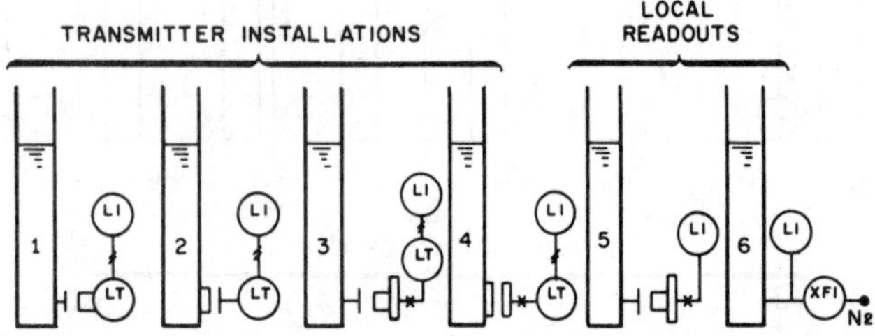

Fig. 3.7p Sensing of hard-to-handle liquid levels in atmospheric tanks by d/p devices

but not eliminated. Tank #3 is furnished with liquid-filled chemical seals such as the one shown in Figure 3.7e. This unit will perform as well as the extended diaphragm d/p cell, but owing to the liquid-filled capillary system it is subject to errors caused by temperature variations. A wafer-type liquid-filled element, such as the one illustrated in Figure 3.7d, is shown on tank #4. This sensing method combines the disadvantages of #2 and #3; the dead-ended cavity is not eliminated and it is subject to temperature errors. However, it can be used on extremely hot processes. On tank #5, the element is the same extended chemical seal as on #3, but the readout is a local pressure gauge. An air bubbler is shown on tank #6.

Pressurized Tanks Containing Hard-to-Handle Fluids

When the process in the vessel is hard-to-handle, it is frequently the case that the vapor space contains materials such as foam that can build up and plug the sensing line of the compensating leg. Figure 3.7q shows six methods for dealing with these applications. Tank #1 shows an extended d/p transmitter in the liquid region and an extended repeater in the vapor space. The use of these devices eliminates all possible plugging problems because the sensing diaphragms are flush with the inside of the vessel wall. This detecting system will function properly on all except the most difficult crystallizer applications, where the inside wall of the tank might be coated with a layer of solid crystals. The extended chemical seals shown on tank #2 will provide an installation similar to that on tank #1 and will perform as well as if ambient temperature variations do not cause inaccuracies. Tanks #3 and #4 are equipped with extended d/p transmitters, but a purge flow prevents the process vapors from entering the compensating legs. The purge medium can either be liquid or gas and, as shown, can be applied to both dry and wet leg installations. Such systems require additional maintenance and range depressor adjustments, and corrosion and condensate accumulation can cause calibration or reliability problems. Tank #5 shows a local indicator equipped with extended chemical seals. This device is subject to temperature effects and requires large displacement seals to match the displacement of the d/p indicator. For tanks #1 through #5, flat diaphragm elements can also be considered but it should be realized that they do not completely eliminate the dead-ended nozzle cavities in which material can accumulate. Such designs should only be considered where it is essential to have an isolating valve between the tank and the level device so that when maintenance is required, the vessel will not have to be drained prior to removal of the instrument. The bubbler system is illustrated on tank #6; either liquids or gases can be used as the purge media.

The extended diaphragm transmitter/repeater instal-

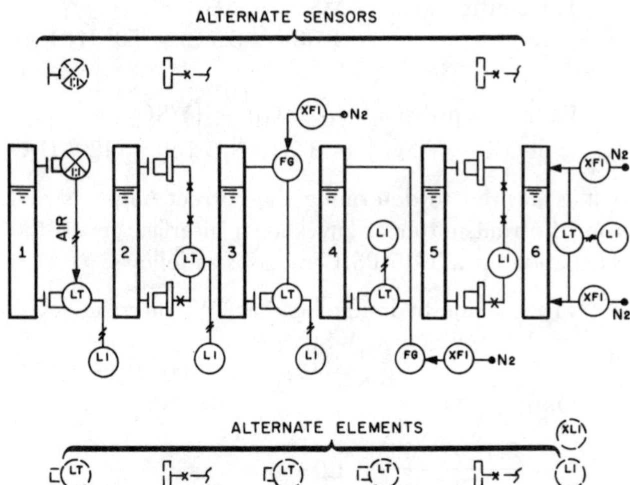

Fig. 3.7q Detection of hard-to-handle liquid levels in pressurized tanks by d/p instruments

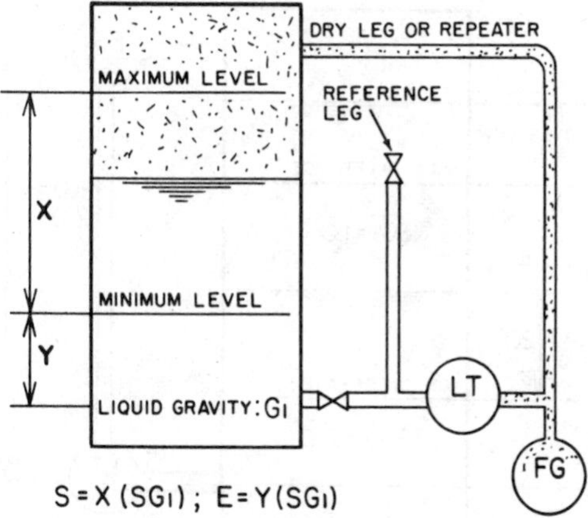

Fig. 3.7r Illustration for range elevation

lation is attractive for pressure vessels containing hard-to-handle materials if isolation valves are not required. Purged installations are also acceptable if the purge media is reliable and the process can tolerate the purge flow.

Determination of Span, Elevation and Depression Settings

As illustrated in Figure 3.7b, force balance d/p cells can be provided with zero, span, elevation, and depression adjustments. The tension in the elevation spring can be set to cancel out any initial pressure exerted on the high side of the diaphragm capsule. Similarly, the depression spring can be adjusted to compensate for initial forces on the low pressure side of the d/p cell. The amount of depression setting is limited to the full range of the capsule, while the sum of elevation setting and span cannot exceed the full range of the cell. These settings are normally adjusted in the factory if sufficient data are furnished to the manufacturer. If the setting is changed in the field, it will affect the span of the transmitter. A description is given below for calculating these settings.

Figure 3.7r shows a dry leg d/p cell installation with the desired minimum and maximum liquid levels noted. The output of the transmitter will be zero when the level is at minimum and 100 percent when it is at the predetermined maximum. The span (range) of the cell will be the product of liquid density and the distance between minimum and maximum levels desired. The elevation spring will be set for the product of density times distance between the minimum level desired and the cell datum. A reference leg is also shown on this sketch, which is convenient for establishing a check point for the transmitter. A reference figure is found by temporarily isolating the cell from the tank and filling the reference leg with a known gravity fluid. Once this figure is obtained, the repeatability of the unit can be checked periodically.

Figure 3.7s shows a wet leg installation. Span is determined the same way as before. Range depression is calculated as the difference between the hydraulic head in the wet leg and the range elevation desired. Difference between process and filling fluid densities must be considered so that the depression does not exceed the full range of the cell. For example, if the desired minimum level is at the cell datum line, the full range of the cell is 200 in. (5 m) of water column and the height of the wet leg is 100 in. (2.5 m), then the density of the filling liquid cannot be more than two. The actual span setting of the cell can be anywhere below the full range.

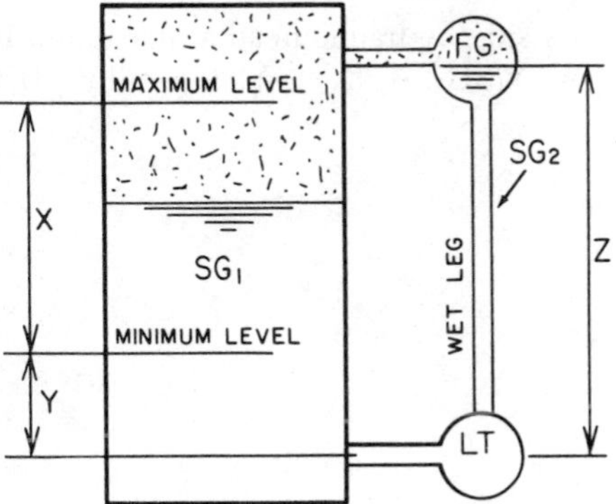

Fig. 3.7s Illustration for range depression

Figure 3.7t shows the settings for a liquid-liquid interface application. The span for this cell is the product of the density difference of the two liquids and the distance between the maximum and minimum interface levels. The range depression is the difference between the hydraulic head of the filling fluid and the sum of the range elevation plus the light liquid head over the range of minimum interface to overflow level. No depression is required if the minimum interface is at the cell datum, the height of the wet leg is the same as the maximum

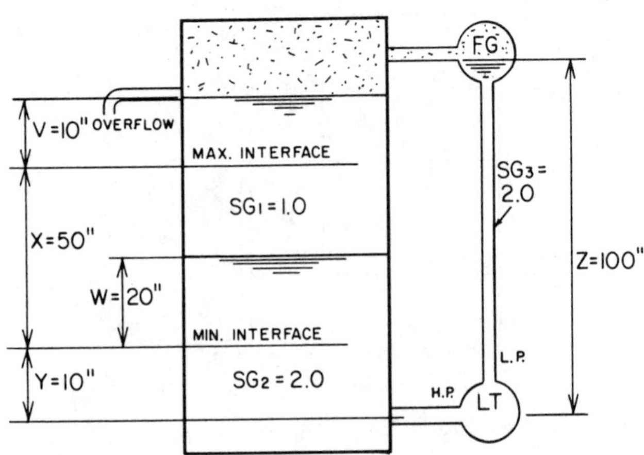

Fig. 3.7t Span and depression settings for interface detection

total level and the filling fluid density is the same as the light liquid. If the minimum interface is at the cell datum and a dry leg system is used, then instead of depression, the cell must be elevated by the hydraulic head of the light liquid over the range of cell datum to overflow level.

The following calculations, which are based on the data shown in Figure 3.7t, will serve as examples.

$$\text{Wet leg hydraulic head} = Z(SG_3) = 200\text{'' } H_2O \qquad 3.7(1)$$

Process side hydraulic head at minimum interface

$$= (V + X)SG_1 + Y(SG_2) = 80'' \ H_2O \qquad 3.7(2)$$

Process side hydraulic head at maximum interface

$$= V(SG_1) + (X + Y)SG_2 = 130'' \ H_2O \qquad 3.7(3)$$

$$\text{Transmitter span} = X(SG_1 - SG_2) = = 3.7(3) - 3.7(2) = 50'' \ H_2O \qquad 3.7(4)$$

$$\text{Range Depression} = Z(SG_3) - [Y(SG_2) + + (X + V)SG_1] = 3.7(1) - 3.7(2) = 120'' \ H_2O$$

If it is desired to determine the correct output signal from the transmitter at any known interface level, the calculation for a 3–15 PSIG range is as follows:

Process side hydraulic head at "W" interface level

$$= (Y + W)SG_2 + V(SG_1) + + (X - W)SG_1 = 100'' \ H_2O \qquad 3.7(5)$$

Output signal

$$= \frac{3.7(5) - 3.7(2)}{3.7(4)} 12 + 3 = = 12 \frac{100 - 80}{50} + 3 = 7.8 \text{ PSIG}$$

BIBLIOGRAPHY

Byrne, E.J., "Complete Guide to Liquid Handling; Measuring and Controlling." *Chemical Engineering*, April 14, 1969.

Duncan, W.J., "D/P Transmitters Handle Variety of Level Measurements." *Instruments and Control Systems*, August 1979.

Eman, J.F. and Gestrich, N., "Selecting Manometer-type Level Gauges." *Instruments and Control Systems*, July 1977.

Imsland, T., "Connecting d/p Elements for Level Sensing." *Instruments and Control Systems*, November 1975.

Lawford, V.N., "Differential Pressure Instruments; The Universal Measurement Tools." *Instrumentation Technology*, December 1974.

Slomiana, M., "Using Differential Pressure Sensors for Level, Density, Interface, and Viscosity Measurements." *Instrumentation Technology*, September 1979.

3.8 DISPLACER LEVEL DETECTORS

Flow Sheet Symbol

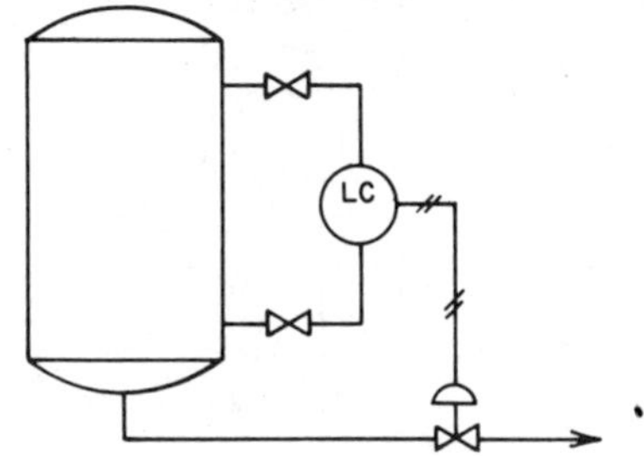

Design Pressure:	To 100 PSIG (0.7MPa) for flexible disc designs, to 600 PSIG (4MPa) or higher for other designs
Design Temperature:	To 400°F (200°C) standard, to 850°F (450°C) with finned extensions
Materials of Construction:	Carbon steel, stainless steel, Monel, Inconel, and other alloys
Inaccuracy:	± ½%
Range:	10 ft. (3m) for all designs except the flexible shaft which is narrow range
Cost:	\$400–\$600 for switch, \$1200 for 32 in. (812 mm) displacer with steel cage and pneumatic controller
Partial List of Suppliers:	Delta Controls; Fisher Controls; Foxboro Co.; Kieley-Muller; Magnetrol; Masoneilan; Moore Products; Robertshaw

Archimedes' Principle states that a body wholly or partially immersed in a fluid is buoyed up by a force equal to the weight of the fluid displaced. By detection of the apparent weight of an immersed displacer, a level instrument can be devised. If the cross sectional area of the displacer and the density of the liquid are constant, then a unit change in level will result in a reproducible unit change in displacer weight. The simplest level device of this type involves a displacer that is heavier than the process liquid and is suspended from a spring scale. When the liquid level is below the displacer, the scale shows the full weight of the displacer. As the level rises, the apparent weight of the displacer decreases, thereby yielding a linear and proportional relationship between spring tension and level. The spring scale can be calibrated 0 to 100 percent, or in other level units.

This simple device is limited to applications in open tanks. In actual industrial measurement, the basic problem is to seal the process from the spring scale or other force-detecting mechanism. This seal has to be frictionless and useful over a wide range of pressures, temperatures and corrosion conditions. The variations in the design of this seal distinguish the types of displacement detectors discussed below. They are the magnetically-coupled switch, the torque tube, the diaphragm and force bar, the spring balanced, the flexible disc, and the flexible shaft design. Each of these units operates on Archimedes' Principle, but are different as far as their seals are concerned. All of them can detect liquid-vapor interface, liquid-liquid interface and, if level is constant, they can be used to detect density changes. The flexible disc unit is available as a pneumatic transmitter, and the flexible shaft unit is available as a high gain pneumatic controller. The other designs are available with local pneumatic controllers and pneumatic or electronic transmitters.

Displacer Switch

The major difference between a float level switch and a displacer level switch is that a float stays on the surface whereas a displacer is partially or totally immersed. Switching arrangements and installation considerations are the same for the displacer and float switch, and since these are covered in Section 3.9, the following discussion will be limited to the design features of the displacer switch and its differences from the float switch. As shown in Figure 3.8a, the displacer is mounted on a flexible cable which is attached to a support spring. When the tank is empty, the spring is loaded with the full weight of the displacer. Changing buoyancy, resulting from immersion of the displacer, unloads the downward force on the support spring and provides the small stem move-

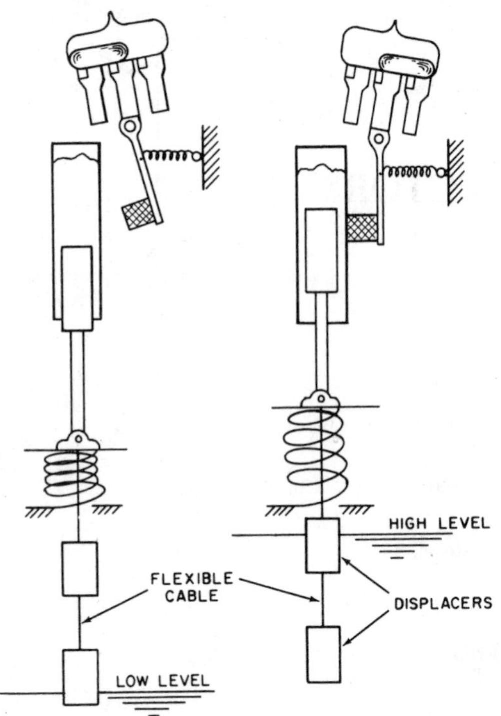

Fig. 3.8a Displacement level switches

ment required for switch action. The magnetically-coupled switching arrangement shown in Figures 3.9e through 3.9g and the installation sketches given in Figures 3.9h and 3.9i apply to the displacement design also, but the displacer switch has several advantages over the float switch:

a. The maximum differential between high and low settings can be as much as 50 ft. (15m).
b. Level settings or spans are easily adjusted by moving the displacer(s) to a new elevation on the cable.
c. Moderate surface turbulence is less apt to cause switch chatter because the cable is in tension.
d. Within broad ranges, fluid density has no effect on the displacer diameter, making units interchangeable between services of varying density simply by changing the support spring.
e. The displacer switch is less apt to cause spurious trips in vibrating services because the cable is always in tension. This is an important consideration for automatic shutdown systems such as may be used at compressor suction drums.

In addition to the above, the displacer switch is highly flexible. It can be provided with up to three displacers and so can be used for multiple control functions. For example, a single unit can be used to stage two pumps by actuating both pumps at high level, one at mid level and neither at low level. The one disadvantage of the displacer switch is that the support spring is exposed to the process. This limits the switch to applications that are clean, non-freezing and non-corrosive to the available spring materials.

Torque Tube Displacers

The torque tube shown in Figure 3.8b uses a torsion spring to support the displacer. By using a hollow torsion tube it is feasible to both support the displacer, which is always heavier than the process fluid, and to provide a frictionless pressure seal. This makes it possible to transfer the changes in the apparent weight of the displacer through the wall of the pressure vessel into a suitable measuring device. Figure 3.8b gives a schematic presentation of the displacer and torque tube. The displacers are cylindrical and can be furnished in a wide selection of plastic and alloy materials. Although any length displacer up to 10 ft. (3m) can be obtained, the most common lengths used are 14, 32, 48 and 60 in. (0.3, 0.8, 1.2 and 1.5m). The volume of the standard displacer is 100 cu. in. (1638cc) and consequently the diameter is reduced as the length increases.

The torque arm connects the displacer to the torque tube and absorbs lateral forces. Friction is minimized by use of a knife-edge bearing support. A limit stop prevents accidental over-stressing of the torque tube by limiting the downward motion of the torque arm. The angular displacement of the torque tube and torque arm are the same at the knife edge end of the tube. At the flange end, the tube does not rotate at all because it is solidly held, but the torque rod is free to rotate the same amount as it did at the knife edge. The angular displacement, which amounts to about 5 or 6 degrees, is linearly proportional to apparent weight of the displacer and thus level. With the difficult pressure sealing problem solved,

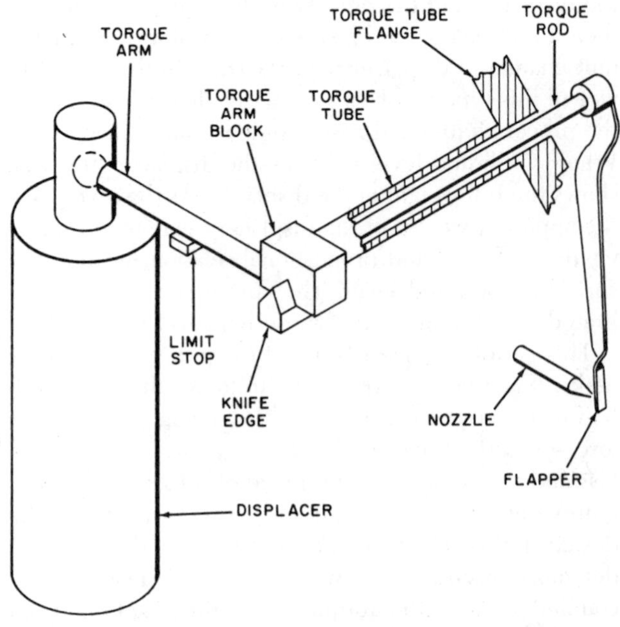

Fig. 3.8b Torque tube displacer level detector

it is a simple matter to convert the angular displacement to a usable electronic or pneumatic analog signal. The standard torque tube material is Inconel but it is also available in stainless steel, Hastelloy, Monel, nickel or Durimet.

Sizing of Displacers

The technique given below for determining the desirable displacer diameter is applicable for all types of buoyant force detectors, but is considered in conjunction with the torque tube design because that is the most common application. Determining the displacer diameter in effect sets the weight change of the displacer per level increment. The torque tube is designed to twist a fixed amount for each increment of buoyancy change. Therefore, in selecting the displacer diameter, the torque tube characteristics, the density of the process fluid and the level span must be considered. For purposes of this discussion, it will be assumed that the motion of the torque rod will be used to operate a proportional band controller. Proportional band (PB) refers to the response sensitivity of the controller, or, in other words, it determines how much corrective action will be generated by the controller in response to a change in the level. A 100 percent apparent proportional band setting means that the level in the tank has to cover the displacer completely to generate a full output signal, and that the level has to drop to the bottom of the displacer in order to generate the minimum output. It should be noted that a controller tuned this way can be used as a transmitter if the process fluid is water. At 50 percent apparent proportional band, the level has to vary over the middle half of the displacer length to produce minimum to maximum output, and at 25 percent setting, a level variation in the middle quarter of the displacer will generate the minimum to maximum output.

The term "apparent" proportional band has been used above. This was necessary to distinguish the "actual" band setting on the instrument based on water density from the resulting "apparent" band related to the density of the process fluid being measured. Figure 3.8c shows the relationship between the two terms as a function of process liquid density. It can be seen from Figure 3.8c that if the actual band setting on the instrument is 50 percent, the behavior of the controller will be such as to imply a 50 percent apparent setting with a liquid gravity of one, a 100 percent setting with a gravity of 0.5, and a 500 percent apparent band at 0.1. The process fluid density thus affects the apparent gain of the controller in an inverse linear fashion.

Because the weight change per unit level change generated by the displacer has to be balanced by the torsional spring, it is advisable to discuss the characteristics of the torque tube. The range of a standard torque tube will match the buoyant force generated by a 100 cu. in. (1638cc) displacer in water if the actual band setting is

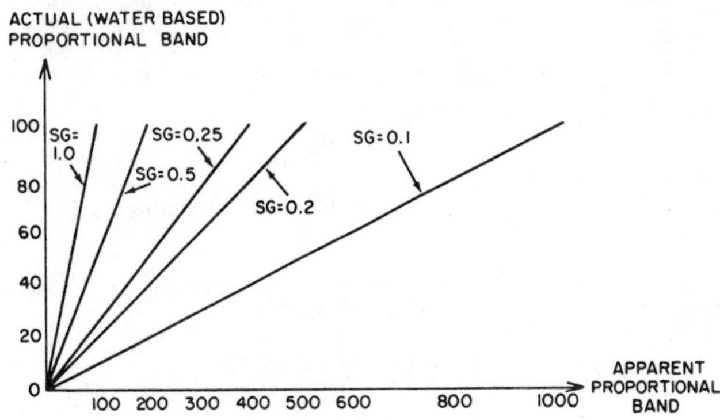

Fig. 3.8c Relationship between apparent and actual proportional band

100 percent. This is equivalent to a force range of 0–3.6 lbf. (0–16.02N). The table below lists the force ranges of standard and thin wall torque tubes at various actual proportional band settings. The thin wall tube requires one half the force for full range operation as the standard wall tube.

Actual PB Setting (percent)	*Force Range for Std. Wall lbf.(N)*	*Force Range for Thin Wall lbf.(N)*
100	0–3.60 (0–16.02)	0–1.80 (0–8.01)
50	0–1.80 (0–8.01)	0–0.72 (0–3.20)
20	0–0.72 (0–3.20)	0–0.36 (0–1.60)
10	0–0.36 (0–1.60)	0–0.18 (0–0.80)

In most liquid-vapor interface applications the density of the liquid is above 0.5, and this (or greater) density material will generate sufficient force on the torque tube using the standard volume displacer, unless a very narrow proportional band is required. The same cannot be said for liquid-liquid interface applications because the useful buoyant force is generated by the density difference between the two fluids (the displacer is completely submerged at all times). Therefore, the sizing of displacers for liquid-liquid interface will be discussed here in some detail.

On interface applications, it is advisable to select a displacer diameter to result in a 100 percent apparent band when the actual band setting is not less than 20 percent. This leaves additional adjustment capability to the operator, because if required he can reduce the actual band to 10 percent (the minimum recommended), achieving a reduction in apparent band to 50 percent. An example will illustrate this.

Assume an installation where the light fluid is 0.98 and the heavy fluid is 1.02 specific gravity, and the displacer required is 32 in. (812 mm) long. Further, it is desired to generate full controller output over a level interface change of 16 in. (406 mm). In order to select a displacer diameter suitable for these requirements, it is necessary to calculate the displacer volume that will generate the force range noted in the above table for a level variation of 16 in. (406 mm). A thin wall tube at an actual band of 20 percent requires a full range force of 0.36 lb. ft. (0.49 N·m). The calculation of displacer volume to generate this force is as follows:

$$\text{Volume for 16'' length} = \frac{\text{Torque Tube Force}}{(\Delta SG)\ (\text{Wt. of 1 cu. in. } H_2O)}$$

$$= \frac{0.36}{(0.04)\ (0.036)} = 250 \text{ cu. in.} \quad 3.8(1)$$

$$\text{Diameter} = \left(\frac{4V}{\pi L}\right)^{1/2} = \left[\frac{(4)\ (250)}{(\pi)\ (16)}\right]^{1/2} = 4.5 \text{ in.}$$

If the operator decides to broaden the actual proportional band to 40 percent, full controller output will result from a 32 in. (812 mm) change in the interface. If the band is narrowed to 10 percent, full controller output will correspond to a level change of 8 in. (203 mm). If a standard wall torque tube had been selected for the above example, the diameter of the tube would have worked out to about 6.5 in. The sizing method given can be used for all displacer applications including density detection.

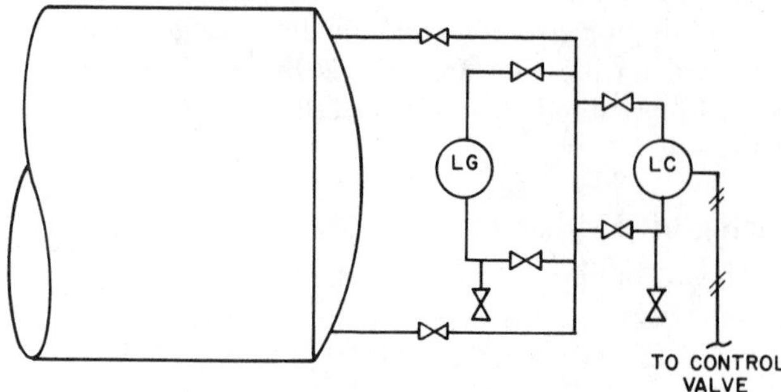

Fig. 3.8d Displacer controller and level gauge mounted on standpipe

Design Features and Accessories for Torque Tube Displacers

Torque tube level devices can be mounted internally or externally to the vessel. Internal displacers are used on applications where the tank can be drained when the level detector requires maintenance. If the displacer is to be internally mounted, it is good practice to install it inside of a stilling well which may be fabricated from a piece of pipe. The pipe should contain a number of vertical slots along its length and should have a stop bar welded across the bottom to prevent the displacer from sinking in the vessel in the event that it becomes disconnected from the torque tube.

For installations where the vessel cannot be depressurized and drained in order to perform maintenance on the displacer, the displacer should be installed in an external chamber mounted outside of the tank, isolated from the process by means of isolating valves. It is customary to cover the span of the displacer with a level gauge so that the operator can be assured of proper operation of the displacer unit. Figure 3.8d shows the installation of a displacer transmitter and level gauge mounted on a standpipe. Note that the two units have independent isolating and drain valves. Since the process is exposed to ambient temperatures when the external chamber is used, it is sometimes necessary to heat trace and insulate the installation for freeze protection. On liquid-liquid interface applications, the location of the chamber nozzles must be carefully considered to be sure that the interface in the vessel and chamber will be the same.

Special torque tube and displacer designs are available for operating pressures to 2500 PSIG (17 MPa). For high temperature installations, the torque tube material is the governing factor because the spring characteristics of the tube are affected. Low temperatures do not affect the spring characteristics of the tube. As far as the torque tube materials are concerned, Inconel is suitable for temperatures between −350° and 850°F (−212°C and 454°C). All other materials are limited to 500°F (260°C) except bronze which is rated at 300°F (150°C). Torque tube extensions are used to maintain the instrument case at or near ambient temperature. When the process is at a temperature above 500°F (260°C), finned extensions are recommended; when it is below 0°F (−18°C), plain extensions may be used. The extensions should not be insulated. Additional thermal insulation barriers can be used between the case and the torque tube flange. Jacketed displacer chambers are available for hard-to-handle services if field-applied tracing is judged to be inadequate. The process connections on external chamber displacers are normally 1½ or 2 in., so some consideration can be given to the use of these devices in mildly dirty services. They should not be used where there is a likelihood of plugging because of solids accumulation.

The torque tube displacer can be furnished with pneumatic or electronic transmitters, or with local pneumatic control. Controllers are available with gain, integral, and/or derivative control action. Although an output gauge can be installed on the transmitter signal to indicate level, it is always better to install an independent level gauge. The torque tube and displacer design has an excellent field record for accuracy and reliability. The torque tube seal is virtually trouble-free if properly specified to meet the process requirements. Because many processes are operated with infrequent changes in the level setpoints, these units are often used for local pneumatic level control regardless of how sophisticated the rest of the process controls may be. The unit may also be furnished with a pneumatic transmitter/controller combination, allowing the advantage of local control plus remote process indication.

Spring Balance Displacer

This instrument is similar to the torque tube unit with the exception of two basic features: one, the torsional spring of the torque tube is replaced by a conventional range spring; and two, the isolation of the process from the instrument is by means of a magnetic coupling. As illustrated in Figure 3.8e, the displacer is suspended in the liquid by means of an extension range spring. As the level in the vessel rises or falls, the effective weight of the displacer changes, causing the spring to extend to contract. A magnetic attracting ball attached to the displacer rod rises and falls in response to the displacer movement. The movement is about 1 in. (25 mm) full range. The ball is centered within the enclosing tube and therefore its movement is friction-free.

Other design features and accessories are similar to the ones discussed under the torque tube design. The spring balanced displacers are also available with corrosion-resistant wetted parts and are suitable for operating temperatures between −250°F (−157°C) and 600°F (316°C). Because interior to exterior motion is magnetically coupled, units with pressure ratings to 6000 PSIG (41 MPa) are available. The merits and disadvantages of the torque tube units discussed above apply to the spring balanced design as well. In addition, it should be noted that in this design the movement of the displacer is greater and that the range spring is exposed to the process. Greater movement almost always causes faster wear at pivot points. The range spring exposure creates difficulty on installations where vapor space condensation, polymerization, or crystallization is expected, because material build-up on the spring will cause the unit to function improperly. Inert gas purging of the spring chamber has been attempted to prevent the process vapors from entering the chamber, but the large flow rates required make this solution impractical.

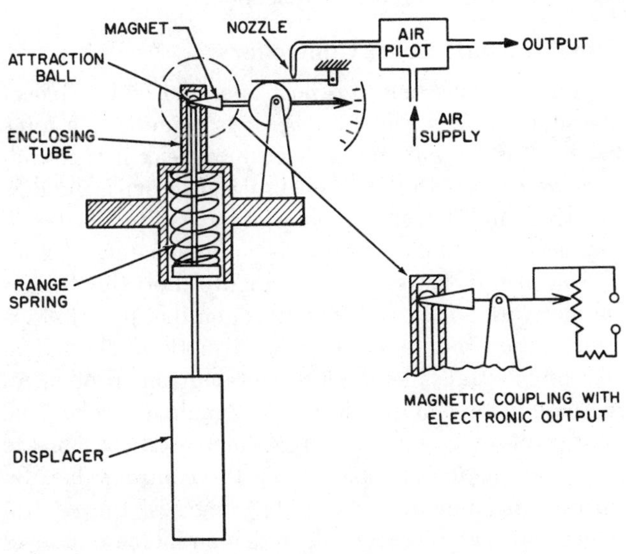

Fig. 3.8e Spring balance magnetically-coupled level displacer

Force Balance d/p Cell Adapted to Displacer Measurement

The basic mechanism of the differential pressure transmitter illustrated in Figures 3.7a and b can be altered to produce another family of displacer level devices. In terms of operating principle, these units are similar in design to the flexible disc design, discussed below, but are superior in adjustment flexibility and pressure rating. Figures 3.8f and 3.8g show the basic working and adjustment components of the top- and side-mounted designs. Level variations in the vessel cause a proportional change in the displacer buoyant force, reducing the apparent weight on the force bar as the level increases.

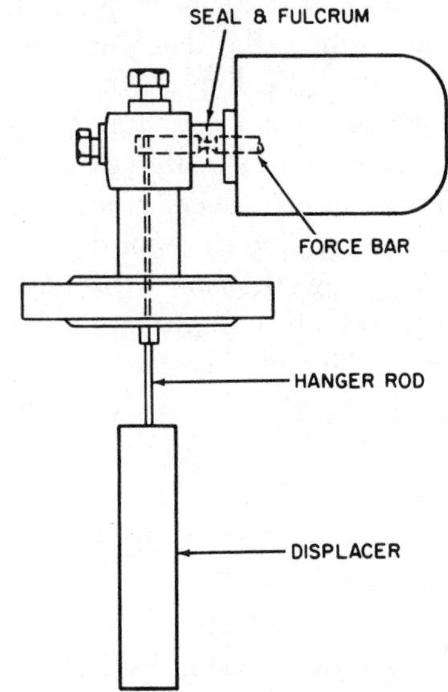

Fig. 3.8f Force balance, diaphragm sealed displacement level transmitter

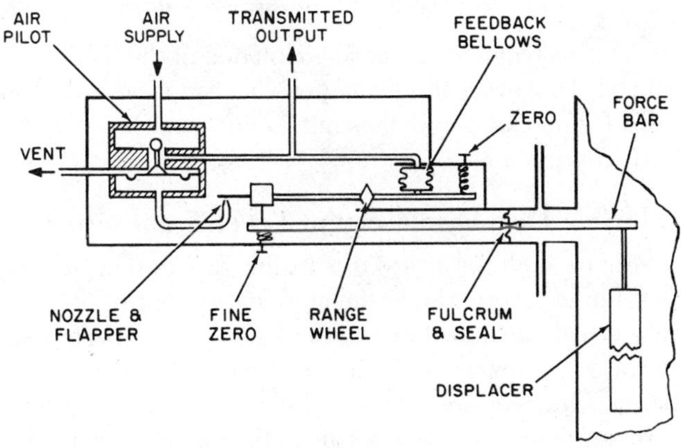

Fig. 3.8g Side-mounted, diaphragm sealed displacement transmitter

(See Section 2.18 for a similar modification of the d/p cell in which a measurement of force is used to infer flow rate.) The diaphragm serves as a fulcrum for the force bar and as a seal to contain the process material in the vessel. The buoyant force is transmitted to the change rod, which pivots on the range wheel. An increase in level causes a minute reduction of the clearance between the nozzle and flapper. This, in turn, increases the nozzle back-pressure, which closes the vent and opens the air supply of the air pilot, resulting in an increased transmitter output pressure. The rising output signal increases the force generated by the feedback bellows, which force-balances the system at a new equilibrium condition.

This instrument is available as either a pneumatic or as an electronic transmitter, and with the displacer mounted either inside the vessel or in an external chamber. The latter is used where maintenance must be performed while the tank is under pressure. The limitations of this device are similar to those of the torque tube design, but, in addition, the diaphragm seal is not as rugged as the torque tube, limiting operating pressures to 600 PSIG (4.1 MPa) and operating temperatures to 400°F (204°C). A variety of materials are available to meet requirements of corrosive services.

The displacer sizing procedure for density, interface, or level detection follows the same basic method outlined for the torque tube units, except that the buoyant force range involved is different. The buoyant force span for the standard unit is 2.90 lbf. (12.9N) or more; for the narrow design, it is 1.45 lbf. (6.45N) or more. The formula to calculate the buoyant force is:

$$F = (0.36)\,(SG)\,(V)\,(Lw)/L,$$

where

F = buoyant force in lbs.,
SG = specific gravity difference,
V = displacer volume in in.,
Lw = working length of displacer in in.,
L = total length of displacer in in.

This instrument is a good adaptation of the familiar d/p cell. However, the sealing mechanism is weaker than the torque tube and the unit is not available with an integral pneumatic controller.

Flexible Disc Displacement Level Transmitters

Figure 3.8h illustrates the flexible disc design. As rising liquid level reduces the apparent weight of the displacer, the float arm moves the bellows and nozzle seat toward the nozzle. This increases the nozzle back-pressure, which causes the air pilot to increase its output pressure until the feedback air to the bellows rebalances the system. Since the apparent weight of the displacer, the force of the zero adjustment spring, and the force generated by the feedback bellows must always be in equilibrium, the unit is classified as force balance. The output signal is directly related to the level in the vessel. Span adjustments are made by changing the total length of the float arm.

The float arm is supported by the flexible disc. Owing to the force balance principle of operation, all members maintain their predetermined position with essentially no movement, protecting the flexible disc from bending

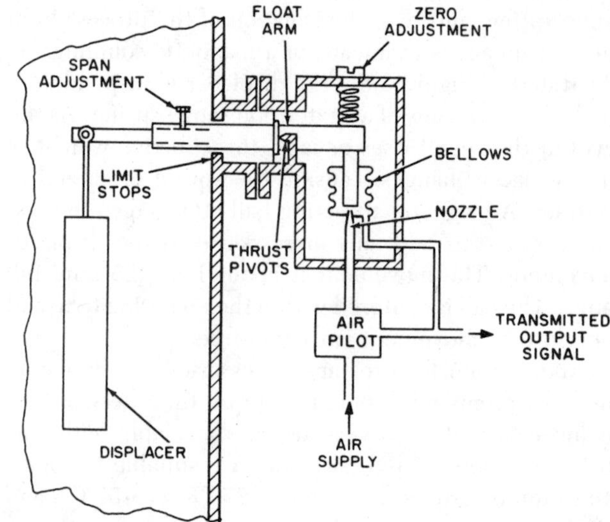

Fig. 3.8h Flexible disc sealed displacement transmitter

fatique. The weight of the displacer and float arm is carried by the flexible disc, and the static pressure in the vessel acts upon the thrust pivots. Limit stops keep the float arm motion within the elastic limits of the disc when the vessel is empty.

This design has the same limitations as the other displacer units but has a narrower range of application because the flexible disc seal limits use to 100 PSIG (0.7 MPa) operating pressure. At higher pressures, the unit first becomes inaccurate, and then unsafe. The flexible disc is normally made of stainless steel but it is also available in Monel or nickel. The displacer can be made of a wide variety of materials. From a performance and application standpoint this unit is inferior to some of the other designs covered in this section.

Flexible Shaft Level Controllers

This unit falls between float operated and displacer operated designs, but because it is essentially a force balance instrument with a mechanical seal, it is considered here. As with the other displacer designs, the flexible shaft unit detects the buoyant force of the float with practically no motion involved—approximately 1/32 in. (0.8 mm) travel. As shown in Figure 3.8i, the shaft is tubular with a flattened center section that permits vertical motion but resists horizontal motion. The extension tongue transmits the float motion to an air pilot. An increase in level moves the tongue, gradually closing the vent of the pilot and allowing an increased air signal to reach the pneumatic operator on the control valve. By turning the mounting flange 180 degrees, the control pilot action can be reversed, so that an increase in level will be accompanied by a decrease of control signal to

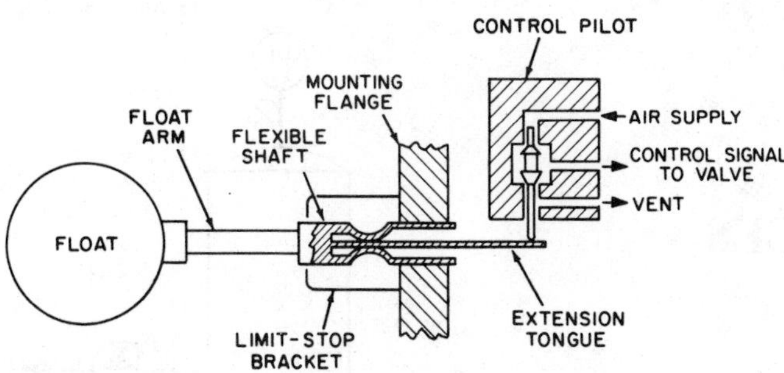

Fig. 3.8i Flexible shaft level controller

the valve. The unit can also be provided as an on-off switch with either pneumatic or electric output.

As with other displacer designs, this device can be used to control interface of fluid density, but it has some limitations in comparison with the torque tube design. As far as control quality is concerned, the flexible shaft unit is not available with adjustable proportional band or with integral action. It has a very narrow band action and virtually no adjustment for changes in density. The ball float has a throttling range of approximately ½ inch (12.7 mm), generating a full change in signal output from a very small level change. This high sensitivity is not desirable in most industrial level applications. The flexible shaft level controller can be furnished with pressure ratings to 1000 PSIG (6.9 MPa) and temperature ratings to 700°F (371°C). Construction materials include stainless steel, nickel, Monel, Hastelloy and others. The flexible shaft design is generally used in secondary services where on-off control action is acceptable.

Conclusion

Displacer instruments, particularly torque tube designs, should always be considered for clean industrial processing applications having span requirements up to 60 in. (1524 mm).

BIBLIOGRAPHY

Anderson, G., "Measuring Level with Displacers." *Instruments and Control Systems*, June 1979.

Flow Sheet Symbol

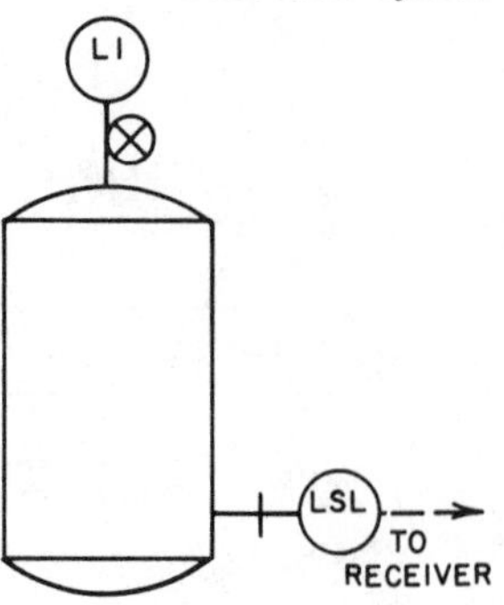

3.9 FLOAT LEVEL DEVICES

Design Pressure:	To 2000 PSIG (14 MPa)
Design Temperature:	To 500°F (260°C)
Float Materials:	Copper, stainless steel, Monel, Hastelloy, etc.
Cost:	\$100–\$400 for switches, \$200 and up for continuous indication
Inaccuracy:	$\pm\frac{1}{4}$ inch for switch and some continuous, $\pm 1\%$ for other continuous
Partial List of Suppliers:	Ball Mfg.; Bestobell-Mobrey, Chem-Tec Equipment Co.; Custom Controls Co., Inc.; Delta Controls; Flygt Corp.; Hersey Products, Inc.; Kessler-Ellis Products; Magnetrol, Inc.; Midland Mfg. Corp.; Reide Systems, Inc.; Robertshaw Controls Co.; Rochester Gauges, Inc.; Transamerica·Delaval, Inc.; Zi-Tech Division of Aikenwood Co.

Float level switches and indicators incorporate in their design a float which follows the liquid level or the interface level between liquids of differing specific gravities. Standard floats are normally spherical or cylindrical for top-mounted designs and spherical or oblong for side-mounted designs. Spherical floats are available from 3 to 7 in. (76 to 178 mm) in diameter. The small diameter floats are used in higher density materials whereas the larger floats are used for liquid-liquid interface detection, for lower density materials, or when the float must buoy up a long motion take-off assembly. Although there is a wide array of float operated switches and indicators, they all fall into one of three categories: 1) Direct connected for atmospheric tanks, 2) Sealed units for pressurized tanks, and 3) Tilt switches for liquids and solids. These three types will be discussed in the order given. No consideration will be given to designs that require stuffing boxes.

Atmospheric Tanks

Figure 3.9a shows one of the most direct and simple methods of float level measurement. The unit shown is basically a tape gauge, as discussed in detail in Section 3.17. A tape is connected to a float on one end and to a counterweight on the other to keep the tape under constant tension. The float motion results in the counterweight riding up and down a direct reading gauge

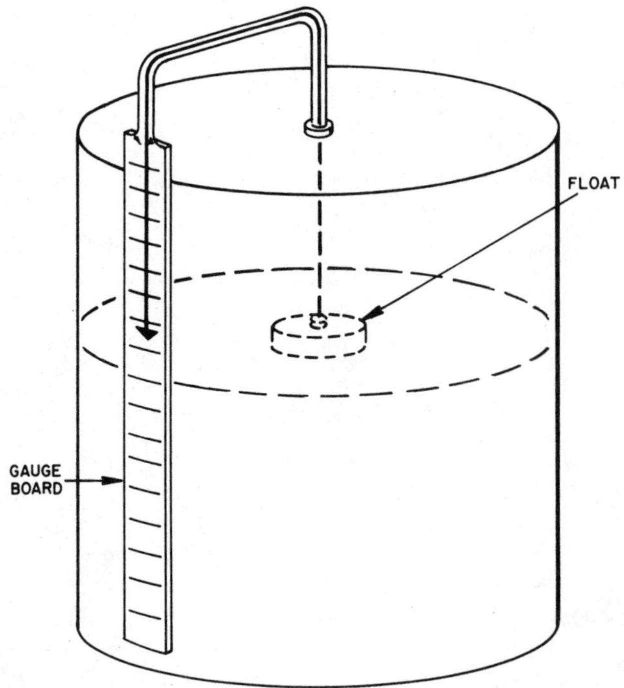

Fig. 3.9a Tape gauge using float and gauge board

board, thereby indicating level in the tank. The installation shown is normally used on water storage tanks, although it can be used in any process that can be left open to the atmosphere. Float and tape materials are selected to suit corrosion requirements. The instrument range is a function of tape length used, up to 100 ft. (30 m).

Figure 3.9b illustrates a second device for level indication or switching in atmospheric tanks or sumps. The ball float has a rod connected to it; rod motion may be used to indicate level on a gauge board or it may be used to trip high and low level switches for alarm or automatic lift pump start and stop. For lift pump applications the float and rod can be furnished and factory assembled by the pump manufacturer.

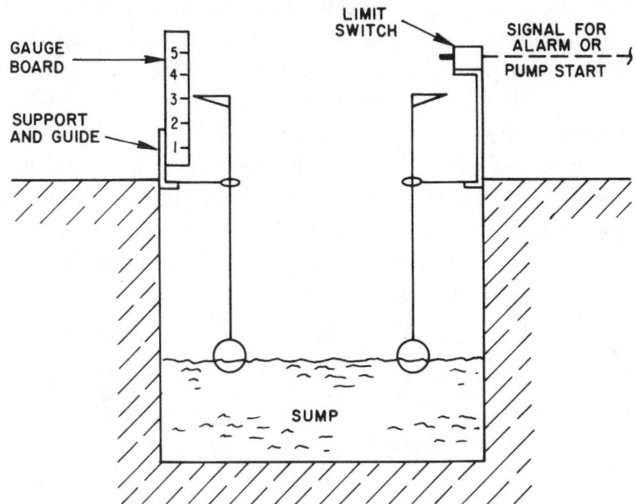

Fig. 3.9b Float operated point or switch trip for sump level

Pressurized Tanks

Float operated devices used in pressurized tanks require a seal between the process and the indicator or switch. In the majority of cases the float motion is transferred to the switch or indicator mechanism by magnetic coupling, but we will consider two other designs first.

In installations where no external power source is available, the float level device shown in Figure 3.9c can be used to obtain remote indication. When the level of the tank rises, the float moves up and the float arm turns on the float arm pivot, thus moving the push rod down, which operates the stroke lever. The stroke lever turns on its pivot which is sealed inside the bellows. This motion is carried by linkage to the tank side bellows, compressing bellows "A" and displacing some of its filling liquid into receiver bellows "C". At the same time, bellows "B" is expanded drawing some of the filling liquid from receiver bellows "D". The net effect is a level readout on the differential bellows remote indicator.

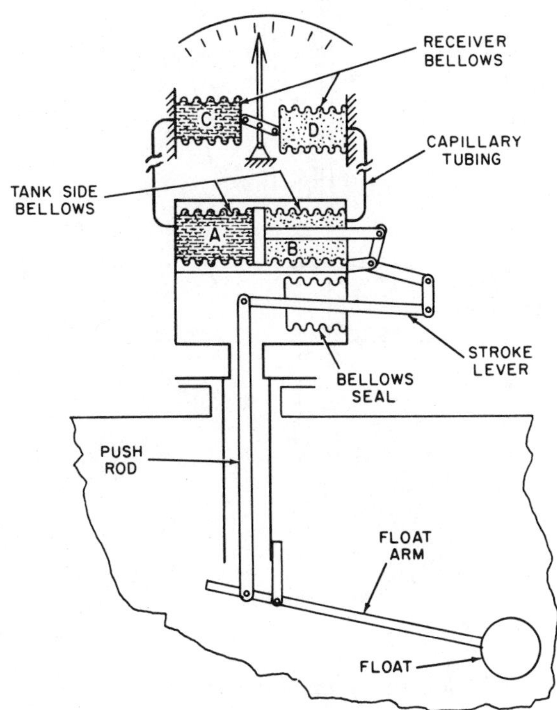

Fig. 3.9c Liquid-filled float gauge for remote readout

These liquid filled float gauges are available with temperature compensation to eliminate the effect of ambient temperature variations on the measurement. For installations where the process vapors are not corrosive to copper and the operating pressure is below 15 PSIG (0.1 MPa), the bellows seal shown in the sketch can be eliminated. The seal allows the unit to be exposed to 200 PSIG (1.4 MPa) operating pressures while isolating the process vapors from the majority of the working parts. The bellows seal and other process-wetted parts can be made out of stainless steel, Monel, aluminum, synthetic rubber or other materials. The capillary tubing between tank side and receiver bellows has been installed in lengths up to 250 ft. (76 m).

Rotameters for flow measurement are discussed in detail in Section 2.23. In these devices, the vertical position of the rotameter float is an indication of flow rate. Figure 3.9d shows how this instrument can be modified for level indication by installing a solid shaft on the float in the process fluid; the shaft moves the indicator in the rotameter housing with level change. The float can be placed directly into the tank or can be mounted in an external chamber furnished with isolating valves. The maximum range of this device is 15 in. (381 mm), which limits its use to narrow span applications. Materials of construction are either steel or stainless steel, with glass indicating tubes used on non-hazardous, low pressure services only. In other installations, metallic tubes are used with magnetic coupling to drive the scale indicator. This design can be used for operating pressures to 1200

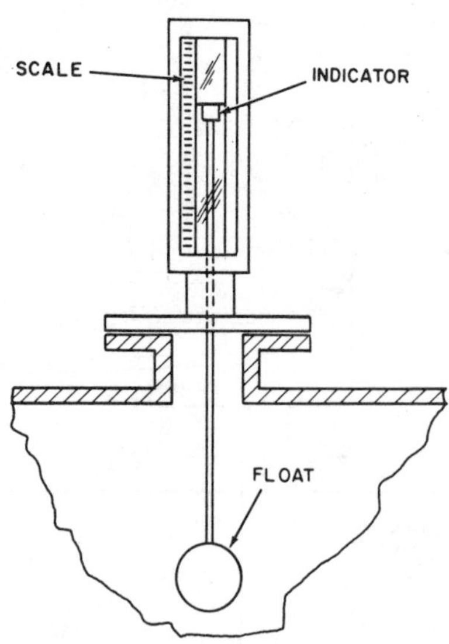

Fig. 3.9d Rotameter indicator adapted to level indication

PSIG (8 MPa) and temperatures to 800°F (427°C). The rotameter design can be used for local indication and can be equipped with high and low alarm switches and electronic or pneumatic transmitters. A similar magnetically-coupled level indicator is discussed in Section 3.11. The balance of the devices covered in this section use magnetic coupling as part of the process seal.

Figure 3.9e illustrates the operation of the spring loaded float switch. The magnet and switch are assembled on a swinging relay arm which operates on pivots. The figure shows the unit in a low level state. As liquid level rises, the float also rises, carrying the attracter up with it. When the attracter reaches its preset position, it pulls the magnet on the swing arm against the non-magnetic enclosing tube, thus tilting the mercury switch to the right. The spring is selected to assure snap action switching. When the switch tilts, the middle-to-left leg circuit is broken and the middle-to-right leg circuit is completed.

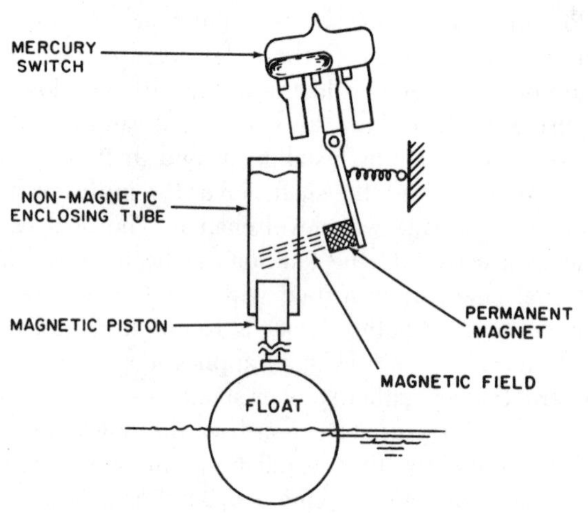

Fig. 3.9e Mercury float switch

The magnetic coupling across the enclosing tube eliminates contact between the process liquid and the switching elements of the instrument. The swing arm design can also be arranged to operate one or more dry contact switches. When two switches are used they can be spaced approximately 1 to 3 in. (25 to 76 mm) apart. It should be noted that time-consuming alterations are required to change the set point or span of this and other float switch designs.

Figure 3.9f shows how the float switch can be made to operate a two-position pneumatic switch. The position of the attracter determines the state of the magnetically actuated swinging carriage. The carriage operates the air valve through a lever and fulcrum assembly. In the figure, the attracter is in the field of the lower magnet and the pneumatic pilot air supply is open to flow to the detector or pneumatic valve operator. When the attracter rises, the upper magnet pulls in, rotating the carriage which, through a lever, closes the air supply port of the air relay and opens the exhaust port. Note in this illustration that the second magnet provides the switch snap action and therefore a loading spring is not required. Figure 3.9g illustrates how the same basic arrangement can be used to actuate a microswitch.

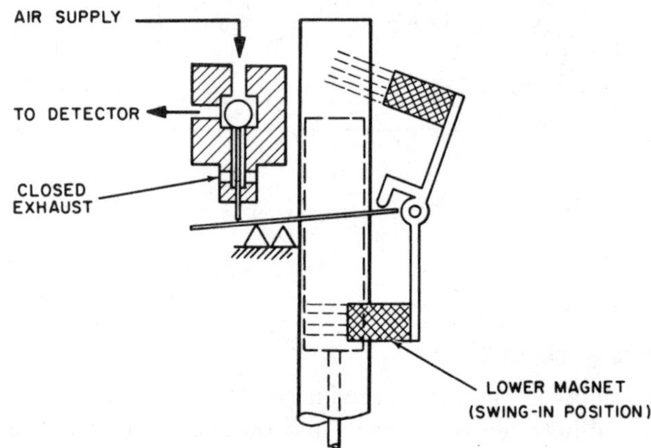

Fig. 3.9f Pneumatic float switch

Figure 3.9h shows top- and side-mounted float switch designs complete with the vessel flange connection and the switch housing. The vessel nozzle for the side-mounted switch must be at the same elevation at which the switch is to operate. These direct connected installations should be considered only if plant operations are such that the vessel can be depressurized and drained when the switch needs maintenance. If this would cause plant shutdown or interference with normal operations, then the external cage design illustrated in Figure 3.9i should be used. An external chamber can be isolated from the vessel by shutoff valves and can be maintained without draining the vessel. If ambient temperatures can

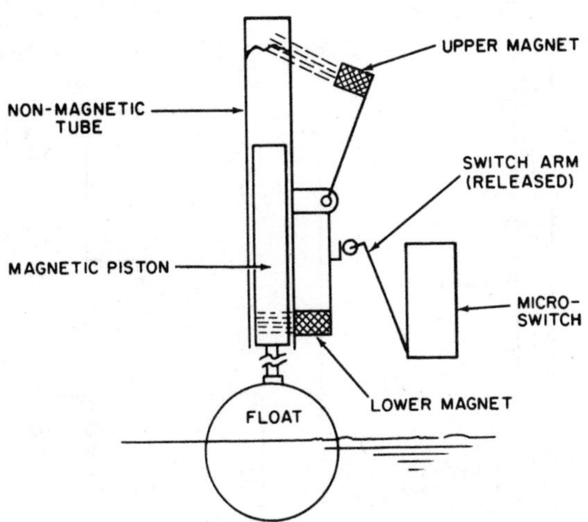

Fig. 3.9g Float switch with microswitch

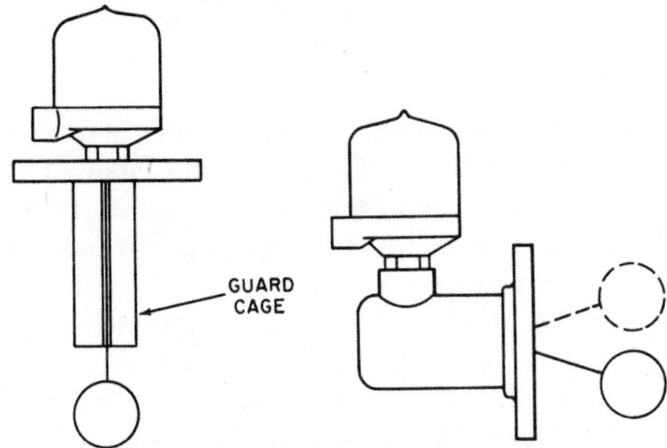

Fig. 3.9h Top- and side-mounted float switches

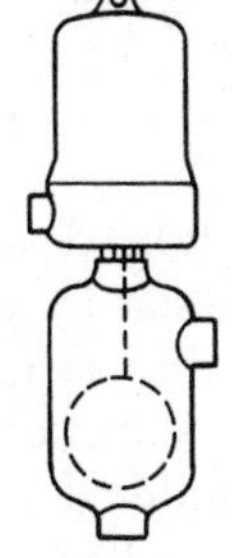

Fig. 3.9i External float chamber

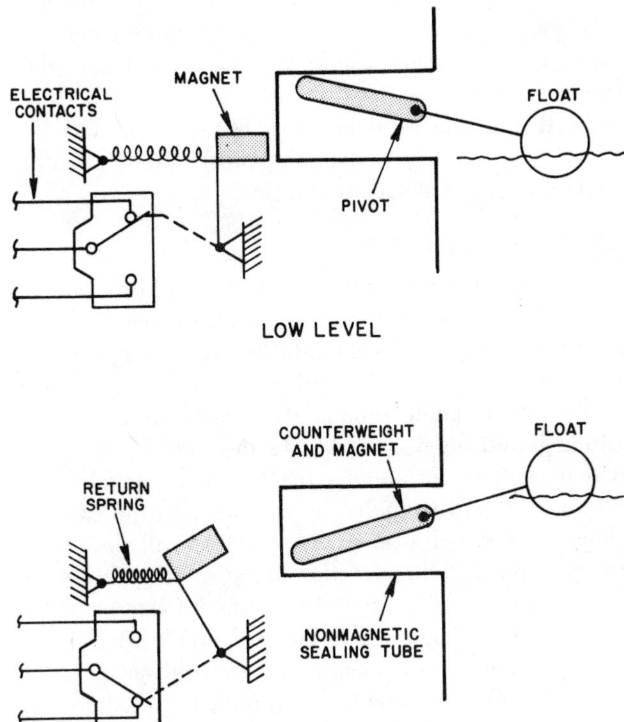

Fig. 3.9j Float and magnetically-coupled switch

cause the process fluid to freeze or gel, the external cage should be heat traced.

A second type of single point magnetically-coupled float switch is illustrated in Figure 3.9j. In this design, a float magnet and switch magnet are arranged to repel each other. As the process level rises, the float rises, the float magnet pivots down and the switch magnet is repelled up, thereby changing the state of the electric or pneumatic switch. This design is available with a plastic boot to cover the float magnet and float pivot point to prevent solids build-up or caking at the pivot. Use of the boot will require a reduction of the pressure and temperature rating of the switch. The previous discussion on direct versus external cage mounting also applies to this switch design.

A third type of magnetically-coupled switch is illustrated in Figure 3.9k. When the level rises, the float rises and rotates the cam counterclockwise. This moves the spring-opposed shuttle and magnet assembly into the switch housing and the magnet causes the switch to

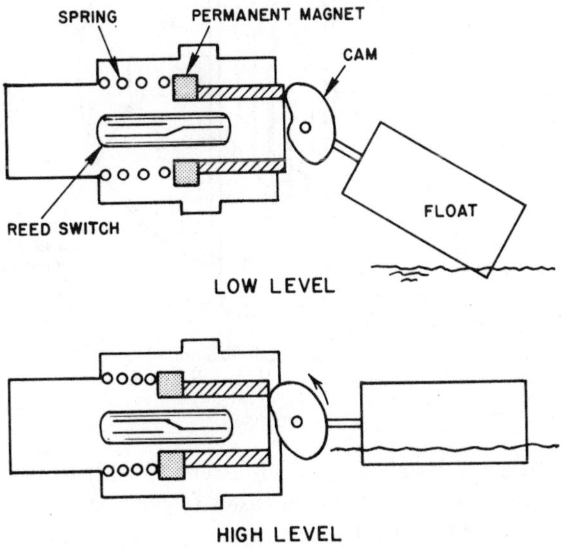

Fig. 3.9k Float and cam operated magnetically-coupled switch

change state. The mechanical coupling between the float and magnet may be somewhat better than other couplings in overcoming resistance to movement caused by the accumulation of foreign matter at the switch. However, this advantage is probably offset by the fact that more parts are exposed to the process than in other designs covered above.

Float and Guide Tube

The float and guide tube design shown in Figure 3.9l is a magnetically coupled float switch that can be modified for continuous level indication. In its simplest configuration, a reed switch is positioned inside a sealed and non-magnetic guide tube at the point where rising or falling liquid level is to actuate the switch. The float, which contains an annular magnet, rises or falls with liquid level and is guided by the tube. In the example shown, the switch is normally open and will close when the float reaches the elevation of the switch. The switch closure can be used to sound an alarm or to start a pump. The switch will reopen when the float falls, or if it continues to rise; the switch cannot know if it was closed by a falling float or a rising float. To remedy this shortcoming, a mechanical stop is placed on the guide to prevent the float from rising above the elevation of the switch. With the stop installed, the switch will stay closed whenever the level is at or above the switch elevation and will open only when the process level falls. Up to six mechanically-stopped floats can be placed along the tube to enable automatic multiple switching.

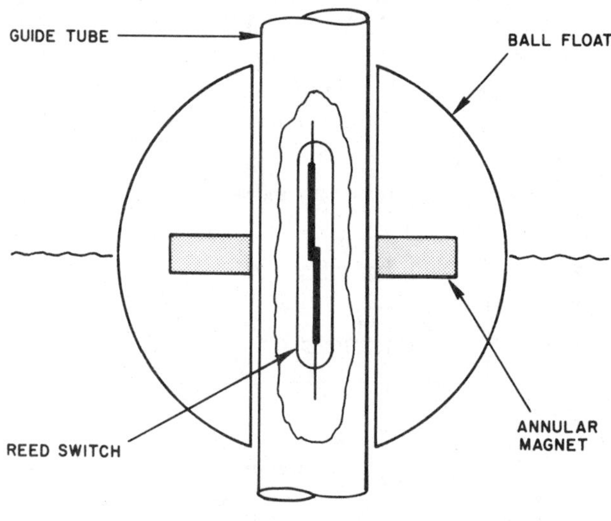

Fig. 3.9l Ball float and magnetically-coupled reed switch

Figure 3.9m shows how an assembly with two floats and two switches might be used to control level in a sump by automatically starting the sump pump on high level and stopping it on low level. When the level in the sump is below the elevation of LSL (level switch, low),

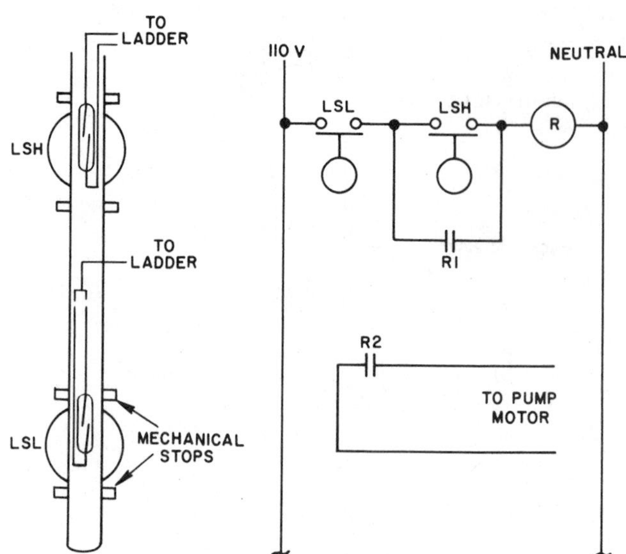

Fig. 3.9m Basic one pump pumpdown control scheme

LSL and LSH (level switch, high) are open and the relay coil R is de-energized. When level rises above LSL, it closes that switch, and continues to rise to LSH. When LSH closes, relay coil R is energized and relay contacts R_1 and R_2 are closed. Contact R_1 is a hold-in contact around LSH; R_2 closure automatically starts the pump motor. When the liquid level is pumped below LSH, it will open but the circuitry will remain energized because R_1 is still closed. When the level falls below LSL, that switch will open, thereby de-energizing the relay coil and allowing R_1 and R_2 to open. At this point, the pump is automatically shut down and the system is at the initially described state. By adding more floats and switches to the assembly, more complex control schemes can be devised.

The float and guide tube design can be modified for continuous level indication by placing a large number of closely spaced switches inside the guide tube and detecting which one is being held closed by the magnet in the ball float. A simplified scheme for doing this is shown in Figure 3.9n. A voltage divider network is made by connecting resistors R_1, R_2, . . . R_n in series across the power supply. Since the resistors have equal value, the voltage drop across each one will be equal. If the magnet in the ball float closes S_3, for example, the voltage between R_2 and R_3 will be impressed on R_L causing a very small current to flow through the ammeter. As the level falls, the magnet will drop S_3 and close S_4, impressing a somewhat smaller voltage on R_L. The ammeter can be scaled to read level in percent or in engineering units. Resistor R_L is made large so that current flow through the ammeter is small by comparison with that in the divider. The switches can be placed as close as $\frac{1}{4}$ in. (6 mm), the minimum resolution for this type of indicator. As can be expected, replacement of a faulty

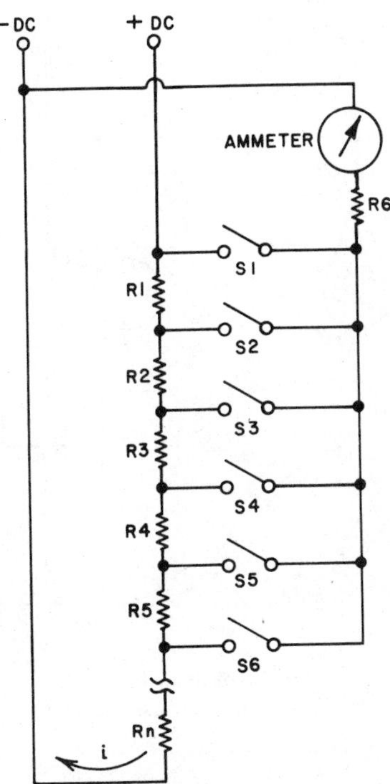

Fig. 3.9n Electrical schematic for float and guide tube indicator

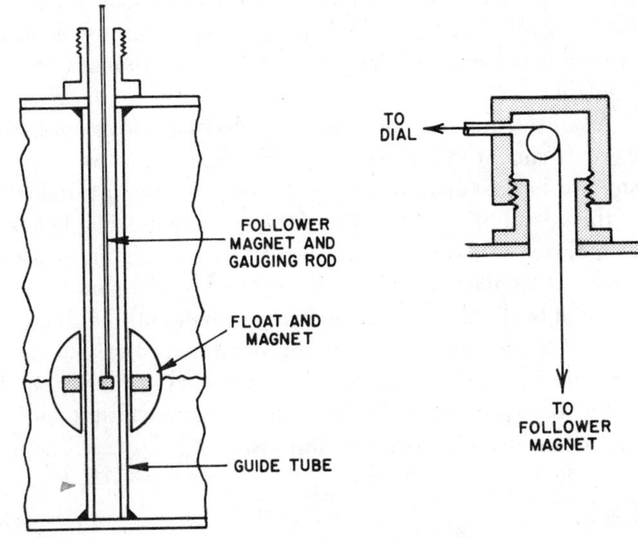

Fig. 3.9o Ground reading gauge using guide tube, float, and magnetic follower

switch would be an extremely time-consuming task and for this reason the indicator is normally equipped with a redundant switch at each point. Maximum length available for this design is 10 ft. (3 m) if the assembly is furnished with the metal guide tube. However, it is possible to obtain a flexible, plastic jacketed assembly that can be furnished in longer coils. These assemblies are field installed into a stainless steel or other non-magnetic guide tube. Another option available with this design is for solar powered circuitry, which reduces installation cost by eliminating the need for an electric power line to the indicator.

Figure 3.9o shows how the magnetically-coupled float and guide tube design may be used for manual gauging or direct dial reading. The sketch to the left shows the same magnet-carrying float and guide tube as described above. Instead of switch closures, the magnet is used to reposition a rod. As the float moves up with level, the rod projects farther out of the tank, thus indicating rising level. When the gauge is not in use, the rod can be pushed down to the bottom of the well and a cap put over the opening. This feature makes the gauge attractive for use on transportation tankers, and it is much safer than the slip tube design sometimes used for that purpose (see Section 3.16.). Given the right geometry, this system can be installed upside down, so that the reading can be made under the tank. As shown in the sketch to the right, this design is also available with a direct reading tape driven indicator. This unit is more expensive than the gauge rod, but its use eliminates the need for the operator to climb on top of the tank. Its major drawback is that the pipe that contains the tape is open to the process and can allow a substantial spill if ruptured.

Another magnetically-coupled float design is shown in Figure 3.9p. This gauge has been installed on a fair number of stationary and mobile liquified petroleum gas and anhydrous ammonia tanks. By means of gears and pivot, the vertical motion of the float is converted to rotary motion of the center shaft. The center shaft in turn positions a permanent magnet behind the dial pointer. There are no shaft connections or holes through the gauge head because the lines of force of the magnet on the center shaft pass through the non-magnetic gauge head and turn the pointer. This magnetic coupling guarantees leakproof operations and is applicable up to an operating pressure of 1000 PSIG (6.9 MPa). The readout dial can be calibrated for horizontal, vertical, or spherical tanks and, for good readout visibility, the dial diameter can be up to 8 in. (203 mm). The figure also illustrates some of the choices for locating the dial on the side or

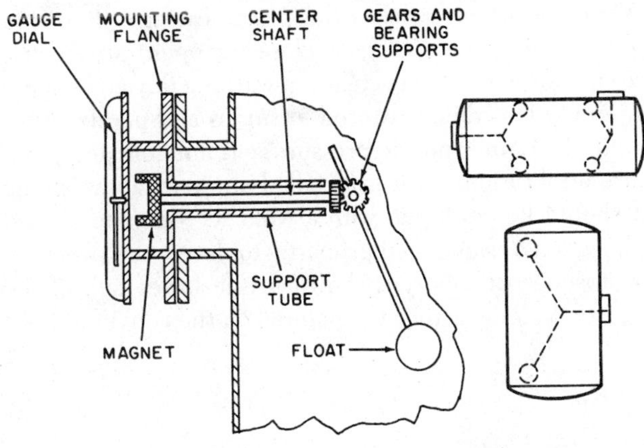

Fig. 3.9p Magnetically-coupled float level indicator

top of the tanks. The standard material of construction for the wetted parts is stainless steel. As with most float operated indicator designs, remote readout devices are available.

A slotted tube magnetically-coupled float level indicator is shown in Figure 3.9q. This design can be considered for installations where the shape or the internals of the vessel prevent the use of the float gauge. The float is enclosed in a tube which has a guide slot. As the float rises, its angular position is determined by the slot. The rotation of the float turns the magnet shaft and the magnet. The magnet repositions the pointer on the dial to indicate changing level. This device is used primarily on liquid transporters. It has a limited pressure rating and cannot be used in corrosive liquids.

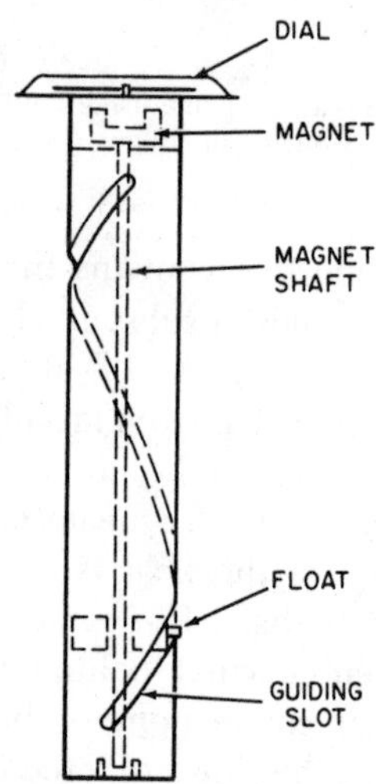

Fig. 3.9q Slotted tube and float gauge with magnetic coupling

Tilt Switch

Figure 3.9r illustrates two commercially available tilt switch designs. The one on the top is used for liquid services. A mercury switch enclosed in a plastic casing is freely suspended from a cable at the desired level. When the liquid reaches the plastic casing, the casing tilts, causing the switch to close (or open) an electric circuit, which in turn actuates a warning device or starts a pump. This device is used in sumps and ponds and is limited to atmospheric pressure and ambient temperature applications. If it is installed outdoors above grade it should be equipped with a wind screen. The second tilt switch design, used primarily to detect the presence or absence of solids, is shown in the lower half of the figure. As long as there is material on the conveyor belt,

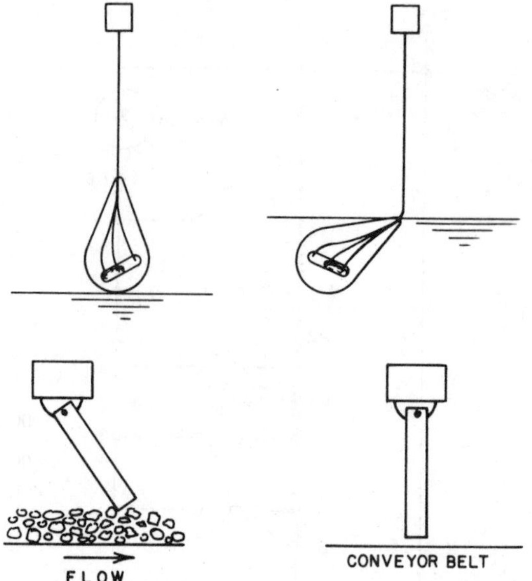

Fig. 3.9r Tilt switches for liquid and solids services

the switch is tilted up. If material feed to the belt is lost, the switch rotates to its vertical position and the switch contacts change state, thereby enabling alarm actuation, shutdown of the belt or other automated action.

Conclusions

Float operated devices are not normally considered for industrial processing applications. One drawback is that in pressure vessel services the wall of the float must be thick enough to prevent it from collapsing. In a number of cases this will result in an overly large float. Another limitation is that immersed magnets will attract pipe scale and other ferrous metal particles in the process that can interfere with proper switch operation. Many of the float operated designs have close tolerance moving parts exposed to the process and cannot be used in dirty or plugging services. Nonetheless, the ball float designs are relatively inexpensive and reliable, resulting in their selective use in industrial processes and routine use in a wide array of applications outside of the industrial area.

BIBLIOGRAPHY

"Flashlight, Sunlight Power Tank Level Indicator." *Design News*, July 9, 1979.

Stengel, R.F., "Solar Cells Power Tank-Gaging System." *Design News*, December 18, 1978.

"Tempkey Limited Introduces the Unique Kari Float Switch." *Ocean Engineering*, No. 6:617, 1979.

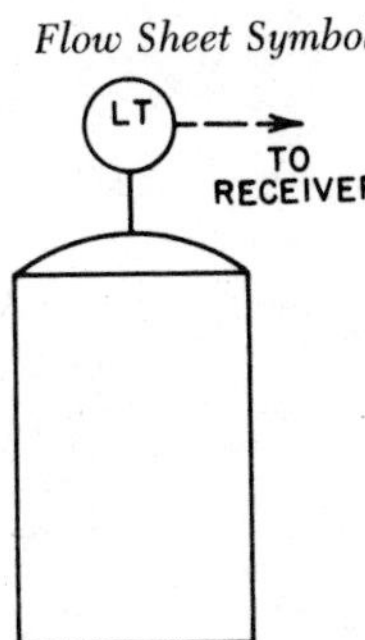

3.10 IMPEDANCE PROBES

Design Pressure:	Up to 1000 PSIG (6.9 MPa)
Design Temperature:	Up to 1500°F (815°C)
Materials of Construction:	Stainless steel, higher alloy metals, ceramic, kel-f, Teflon
Inaccuracy:	Depends on process and probe sensitivity
Range:	To 20 ft. (6 m) for rod probe, to 150 ft. (45 m) for cable probe
Cost:	\$700 for switch; \$1100 and up for continuous
Supplier:	Drexelbrook Engineering Co.

As discussed in Section 3.4, capacitance probes may not perform satisfactorily if the process builds up on the probe to any significant degree. The impedance probe, which is a modification of the capacitance probe, is designed to overcome this problem. Figure 3.10a shows a point detector probe that has been installed in a sticky conducting process. As material builds up on the probe, a low resistance path to ground (vessel wall) is formed. Since switch detector circuitry cannot discriminate between the current flow in the resistive path and the current flow in the capacitive path the switch will trip and stay tripped once the probe is coated. The method that is used to overcome this is to provide a bucking voltage in the resistive path to ground, thereby eliminating the resistive path current flow.

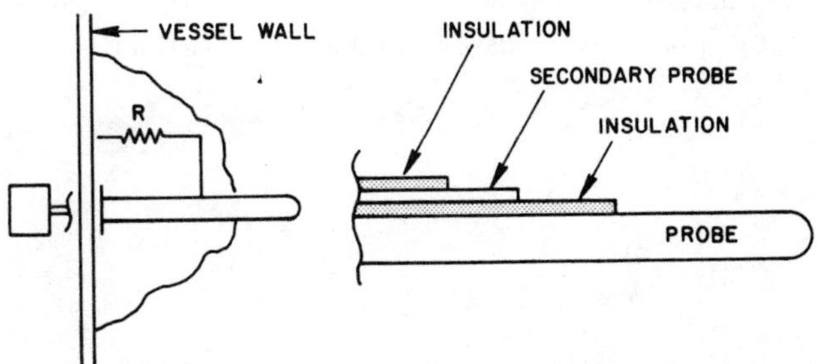

Fig. 3.10a Capacitance probe in sticky conducting material

A secondary probe, insulated both from ground and the primary probe, is driven in phase and at the same voltage as the primary probe. The secondary probe breaks the resistive path to ground when the probe is coated; the switch can be set to trip when liquid rises to cover the probe, thus completing the capacitive path.

For the case of non-conducting processes, a material build-up will not cause the formation of a resistive path to ground but will cause an unwanted capacitive path, thereby offsetting switch calibration and causing a high level trip in the absence of the process. By using the secondary probe assembly, the coupling effect of the build-up can many times be balanced out so that the effects of offset are minimized.

As shown in Figure 3.10b, continuous level measure-

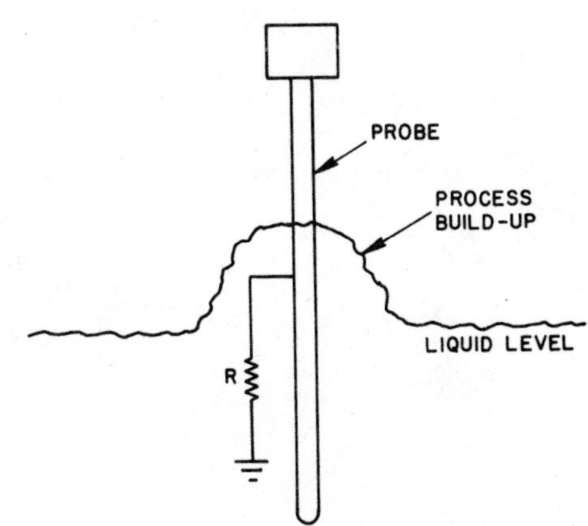

Fig. 3.10b Resistive path to ground caused by build-up of conducting process

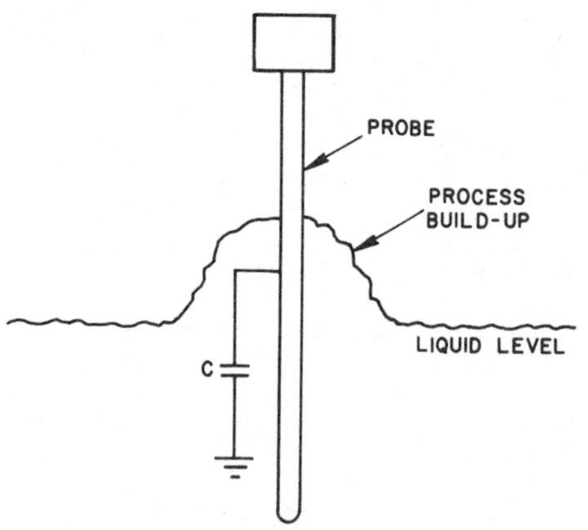

Fig. 3.10c Capacitive path to ground caused by build-up of non-conducting process

ment of sticky conducting processes introduces an additional complication. The resistive path to ground at the vessel connection can be broken in the same manner as was used for the switch. But since a portion of the probe is always covered by the process, a second resistive path to ground is formed through the liquid to the vessel wall. In order to make a measurement under these conditions, the electronic circuitry in the transmitter is designed so that it can discriminate between the resistive current flow and the capacitive current flow. Since these two are out of phase, it is possible to make this distinction in a number of cases. If continuous level measurement is to be made in a sticky non-conducting material, there may be unwanted capacitive coupling near the surface of the liquid as shown in Figure 3.10c. There is no way to correct for this, so some inaccuracies in the measurement must be expected.

Conclusion

The impedance probe, designed to overcome some of the weaknesses of the capacitance probe, is a candidate for a wider range of level detection applications. In addition, the strong features of the capacitance probe, such as no moving parts, have been preserved. On the other hand, the proper operation of the impedance probe and its accuracy can be affected by coating thickness and changes in process dielectric or conductivity. Impedance probe applications should be selected with care.

BIBLIOGRAPHY

Andreiev, N., "Survey and Guide to Liquid and Solid Level Sensing," *Control Engineering*, May 1973.

Belsterling, C.A., "A Look at Level Measurement Methods," *Instruments and Control Systems*, April 1981.

Hall, J., "Measuring Interface Levels," *Instruments and Control Systems*, October 1981.

Lawford, V.N., "How to Select Liquid-Level Instruments," *Chemical Engineering*, October 15, 1973.

Flow Sheet Symbol

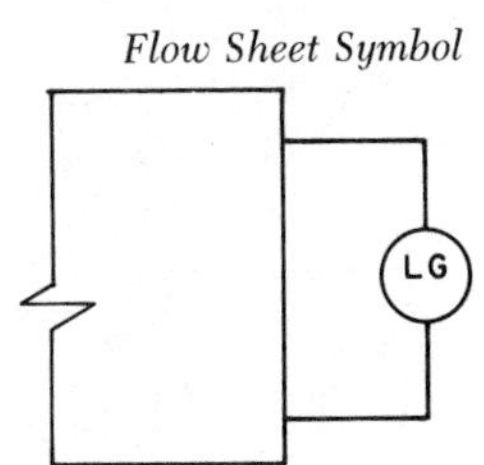

3.11 LEVEL GAUGES

Design Pressure:	To 10,000 PSIG (69 MPa)
Design Temperature:	To 700°F (371°C)
Material for Wetted Parts:	Steel, SS, most alloys all designs, plus glass for conventional designs
Cost:	Tubular: $10/ft. (300 mm) Reflex: $110/ft. (300 mm) Transparent: $175/ft. (300 mm) Magnetic: $360/ft. (300 mm) Remote: to $1500/ft. (300 mm) Valving, other accessories, and special materials extra
Inaccuracy:	See text
Range:	4 ft. (1.2 m) maximum recommended visible length
Partial List of Suppliers:	Clark Reliance; Daniel Industries; Diamond Power; Jerguson; Jogler; Kenco; Krohne; Penberthy; Strahman; Yarway

Level gauges have been used for many years to obtain a reliable and inexpensive indication of level in tanks and vessels. The original design was the tubular glass type. More recently flat glass, magnetically-coupled, remote reading, and bullseye designs have become available. Since these latter designs are safer, they are now used in the majority of applications. The level gauge can be equipped with a variety of accessories to meet special requirements.

Flat Glass Gauges

As illustrated in Figure 3.11a, the flat glass gauge is comprised of a metal chamber, gasket, glass, cushion, and cover; the assembly is held together by bolts. Glass sections are available in 4 to 20 in. (101 to 508 mm) visible lengths and the overall gauge assembly can have as many as five sections. There are two flat glass designs available: transparent and reflex. The transparent gauge has glass sections on opposite sides of the chamber, allowing the user to view the liquid level through the gauge. The reflex gauge has a single glass with prisms cut in the glass on the process side. Light striking the glass in the vapor phase is refracted back toward the viewer, making the vapor space appear silvery white. Light striking the glass covered with liquid is refracted into the liquid, making that portion appear black.

The reflex gauge is used on clean, clear, non-corrosive liquids. Transparent gauges are used on interface services and when the process is dirty or viscous. Transparent gauges are also used when internal mica or plastic shields are required to prevent chemical attack or discoloration of the glass. Such services include some caustics, hydrofluoric acid, and steam drums above 250 PSIG (1.7 MPa). Reflex gauge glass cannot be shielded because of the prisms that are cut in the glass. Transparent gauges should also be considered if the process is a solvent that can dissolve the internal coating of the chamber of the reflex gauge thereby reducing the effectiveness of the prisms.

Reference to Figure 3.11a will show that the chamber, glass, and gaskets are wetted by the process while the cushion, cover, and bolts are not. Material selection should be made with this in mind. Chambers, which are machined from one piece of barstock regardless of how many glass sections are used, are available in steel and a wide variety of alloys. For most applications, the glass used is a tempered borosilicate which has good resistance to most chemicals. It also has good resistance to thermal and mechanical shock. If the glass breaks, it will form an interlocking crystalline pattern which is supposed to prevent the formation of loose flying particles. Several gasket materials are available and they should be selected with regard to temperature and to the process media. Non-wetted metallic parts are generally forged carbon steel except for the cushion which is asbestos. However, for low temperature services or when atmospheric cor-

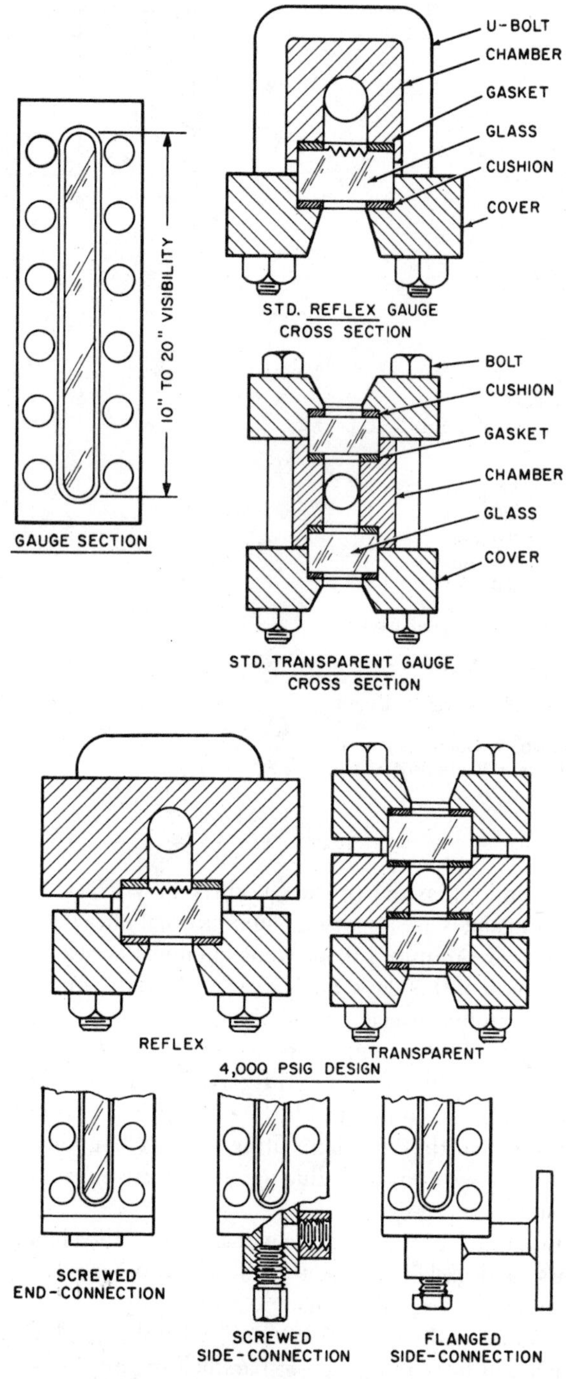

Fig. 3.11a Flat level gauge designs and cross sections

rosives are present, bolt, nut, and cover materials may be specified in stainless steel. Gaskets and cushions are installed in recesses that are machined into the chamber and cover. This insures proper assembly, stops gasket slippage, and prevents glass-to-metal contact. The bolts may be furnished with spring loaded washers to ensure gauge assembly tightness at high temperatures.

As shown in Figure 3.11b, commonly used reflex gauges can be rated to 4000 PSIG (27 MPa) and transparent gauges to 3000 PSIG (20 MPa), both at 100°F (38°C). These pressure ratings must be reduced as temperature is increased or if longer glass sections are used. There are some very heavily armored reflex and transparent designs available to 5000 PSIG (34 MPa) at 100°F (38°C), and there are transparent bullseye designs rated to 10,000 PSIG (69 MPa). Judgment is required for high temperature and pressure applications. For safety, one recognized industry standard recommends limiting gauge lengths to 5 ft. (1.5 m) between connections with a total of four sections for moderate process services, and places stricter limitations on gauges for higher temperature and pressure ratings. To cover larger visible lengths, two or more gauges may be overlapped as shown in Figure 3.11c.

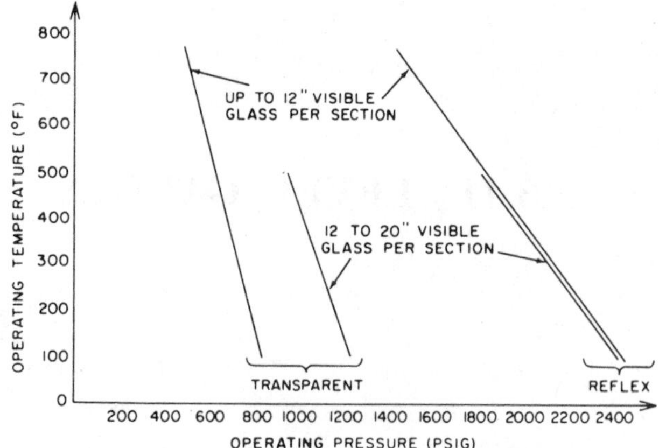

Fig. 3.11b Pressure-temperature ratings for standard level gauges

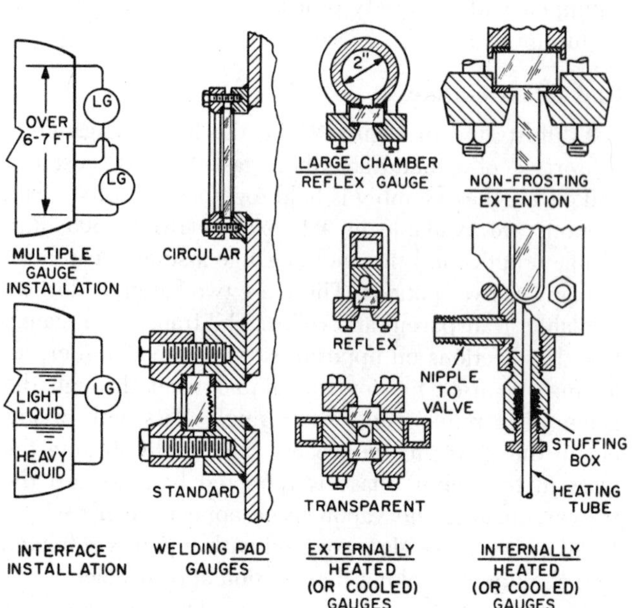

Fig. 3.11c Special level gauge designs and cross sections

There are several variations of the standard flat glass gauge design as shown in Figure 3.11c. One design is the large chamber reflex or transparent gauge. These gauges are used where a boiling or foaming liquid would ruin the reading in a smaller bore gauge. Since the chamber of this type of gauge is constructed of 2 in. (50 mm) pipe or tube, its pressure/temperature rating is limited to the low end of that shown in Figure 3.11b. Another variation is the welded pad design which is available in reflex only. This construction can be considered when vessel connections cannot be used. Such cases may involve liquids with suspended solids or where there are space problems around the vessel. It should be noted that this type of installation will weaken the vessel unless the vessel has additional reinforcing at the pad. Pad- and nozzle-mounted sight glasses are also available. Direct-mounted designs cannot be cleaned unless the vessel is depressurized and opened.

Gauging Inaccuracies

Flat glass gauges and the others described later can be inaccurate for several reasons. Reading inaccuracies may occur because of poor design, poor installation, lack of maintenance, or lack of proper lighting. High level indication caused by boiling or foaming liquids has already been mentioned in conjunction with use of the large chamber gauge. Heavy rates of condensation on the internal face of prismatic gauges can cause erroneous or blurred readings; this is most apt to occur when the gauge is operated at a temperature considerably below that at which the vessel operates. Gauge connection pluggage will cause inaccuracies; gauges should not be used in these services unless suitably flushed or heat traced. A serious but not so obvious cause of inaccuracy can arise in very high or very low temperature services. Considering high temperature first, since the liquid in the gauge is usually cooler than that in the vessel it will be more dense. This means that a shorter column of liquid in the gauge will balance a longer column of liquid in the vessel and the gauge indication will be lower than the liquid level in the vessel. Astounding inaccuracies of ½ inch per inch (12.5 mm per 25 mm) of visible length can be observed in services where the process is operated near its critical pressure, as in the case of high pressure steam drums. The inaccuracy can be even greater if the gauge piping is such that there is a cool vertical liquid leg below the bottom of the glass. In very cold services, the inaccuracy will be in the other direction: the gauge will indicate a higher level than that in the vessel. Manufacturers of very high and very low temperature equipment are aware of this problem and have developed solutions. One solution is to heat or cool the gauge. Others involve building offsetting errors into the gauge installation. Users should be cautious when modifying or retrofitting equipment in these services.

Level Gauge Accessories

The most common level gauge accessory is the gauge valve, several examples of which are illustrated in Figure 3.11d. Gauge valves are normally factory assembled to the gauge. They may be obtained with a number of different features depending on the service requirements. Standard valves for moderate services have ¾ inch (19.05 mm) NPT vessel connections and ½ inch (12.5 mm) NPT gauge and drain connections, although other sizes, both screwed and flanged, are available depending on application. Vessel side connections can be equipped with unions to facilitate gauge connection to the vessel. Similarly, gauge side unions are available to simplify gauge face orientation and to permit removal of the gauge with-

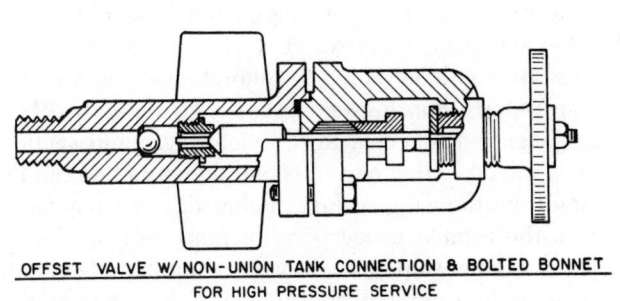

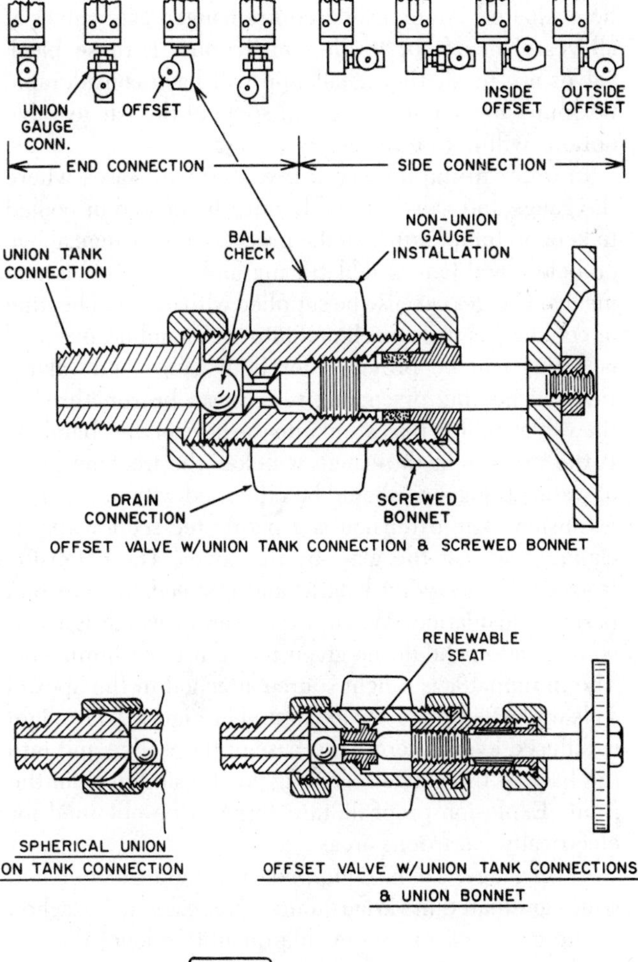

Fig. 3.11d Gauge valves

out depressurizing the vessel. Spherical unions are available for vessel side connections to correct for vertical or horizontal misalignment between the gauge connections and the vessel connections. Spherical unions should be avoided, if possible, because they can leak if they are disassembled and connected a second time. Gauge valves are routinely equipped with ball checks that minimize leakage to the atmosphere in the event a gauge section breaks. Valves can be furnished with back seating plugs so that the packing can be replaced while the valve is under pressure. Outside screw and yoke (O.S.&Y.) designs are available that are good for high temperature and corrosive services. Another option available is the offset design, allowing rod out and cleaning of the gauge without removal from the vessel. When considering gauge valve requirements, the entire installation should be evaluated. Many users require flanged gate valves at all vessel nozzle or bridle connections. If these block valves are used, the installation will have considerable flexibility for maintenance and some of the gauge valve options will not be necessary.

In order to improve accuracy, there are cases where the gauge and valve assembly must be heated or cooled to keep to temperature of the process in the gauge above or below ambient. Field tracing and insulation is one means. Gauges can also be supplied with external heating or cooling passages machined into the chamber, or metal jacketing can be provided for both gauge and valves. Internal heating or cooling tubing can be run through the valve stuffing boxes and into the bore of the chamber. If the process in the gauge is below the freezing point of water, the gauge should be equipped with a frost-free extension. The extension is a plastic tee section that is clamped against the glass by the cover. The extension is available in several lengths and is specified to extend past the insulation. When a transparent gauge is used, consideration should be given to the use of illuminator. The illuminator is a light source attached to the apex of a clear plastic wedge. As illustrated in Figure 3.11e, light is diffused evenly across the base of the wedge and into the back of the gauge, enabling good visibility from the front. Explosion-proof illuminators can be obtained for electrically hazardous areas.

Level gauges can be equipped with engraved or etched scales graduated in various units. The scales are attached to the gauge cover and enable quantitive level reading in the metering units selected. Tubular gauges with scales are commonly used to calibrate low volume metering pumps. For this service, as shown in Figure 3.11f, the gauge is connected to the metering pump suction and is valved so that once it is full, pump suction can be taken from the gauge column rather than the day tank. Since the tubular gauge bore tolerance can be closely held, a relationship between volume and liquid column travel can be accurately determined. If, for example, the scale is graduated in milliliters, an accurate calibration

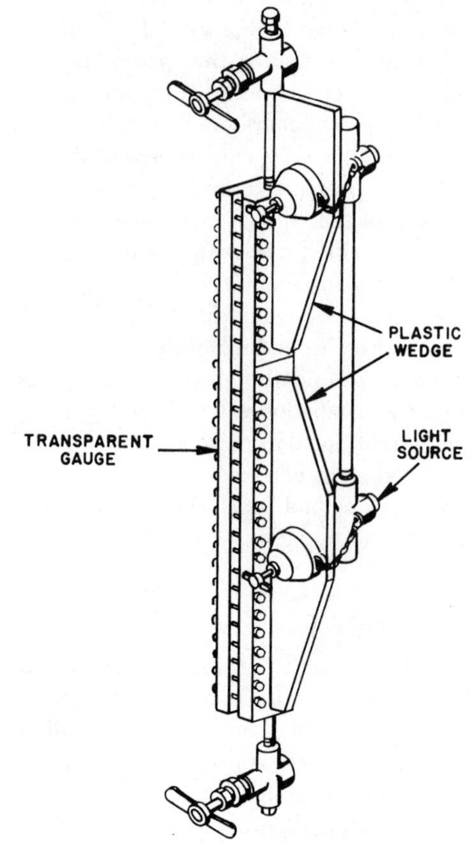

Fig. 3.11e Transparent gauge with illuminator

of the pump in ml./min. can be made by timing the calibration test and observing how many graduations the column has fallen during the test. This service is judged relatively safe for tubular gauges since the operation is attended, the process is at or near ambient temperature and atmospheric pressure, and the gauge can be drained and valved off or removed after the test. Except for this

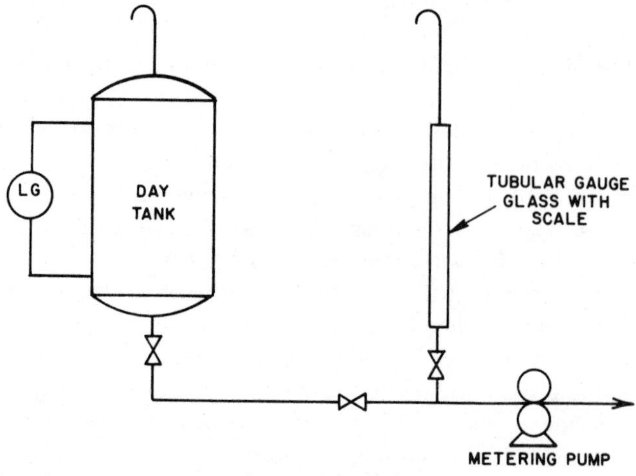

Fig. 3.11f Tubular gauge arranged for metering pump calibration

one special application, tubular gauges are rarely used for industrial services. One widely recognized industry recommended practice suggests that their use be limited to non-hazardous, non-toxic services below 15 PSIG (0.1 MPa) and 200°F (93°C).

Level gauging around steam boilers can have several unique requirements. One is for remote reading gauges, considered later in this section. Another is for special accessories, some of which are shown in Figure 3.11g. The major accessory is the water column, which is attached to the steam drum and, in turn, has level indicating and sensing devices attached to it. Because the steam drum is normally elevated, try cocks and gauges are available in quarter turn designs that may be chain operated. Also, to increase gauge visibility, tilted assemblies (that can be illuminated) are available. The column may be equipped with conductivity probes for actuation of alarm and interlock circuitry. Finally, the water column may be equipped with a float-actuated steam whistle to sound on high and/or low level. There are several designs of this whistle. In the design shown, a high and a low displacer are connected to a lever whose fulcrum is off center. If the high level displacer is immersed, the lever will rotate clockwise to uncap the steam whistle. Similarly, the lever will rotate clockwise if both displacers are uncovered. The reader should refer to applicable boiler codes for exact requirements for level measurement on steam drums. The ASME Boiler and Pressure Vessel Code, Section I, governs in many areas.

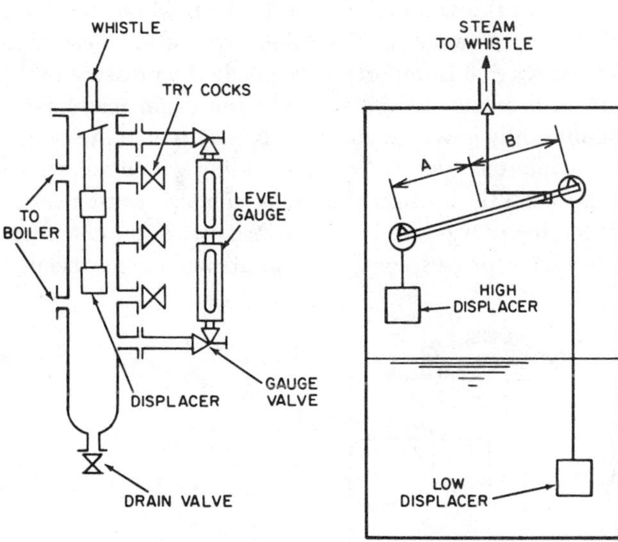

Fig. 3.11g The boiler water column

Installation

The level gauge installation should be designed so that its visible length covers the full operating range of interest, including any other level instruments on the vessel such as displacer transmitters and high and low level switches. In addition, consideration should be given to covering an extended range that may be of value during start-up and shutdown. When a transparent gauge is used for liquid/liquid interface indication, a middle connection should be made into the lighter liquid phase, in addition to the top and bottom connections. Where maximum visibility is important, side-side connected gauges can be used. When multiple gauges are used, the vessel or standpipe connections should be overlapped. A standpipe installation reduces the number of vessel nozzles required and allows greater flexibility in orienting and supporting the gauges. Gauges that are to be operated in high temperature service can be piped with an expansion loop. The expansion loop serves to minimize stress on the vessel nozzles caused by differential expansion of the vessel and gauge. Potential stress problems are particularly likely if the vessel and gauge chambers are made of different materials, if the gauge is to operate at a temperature considerably different from that of the vessel, or if the gauge can be valved out and allowed to cool to ambient temperature. Some of these installations are covered in Figures 3.11c and 3.11h.

Magnetically-Coupled Gauges

In certain toxic or corrosive services, gasketed glass designs may not be acceptable. For these applications, the magnetically-coupled level gauge may be considered.

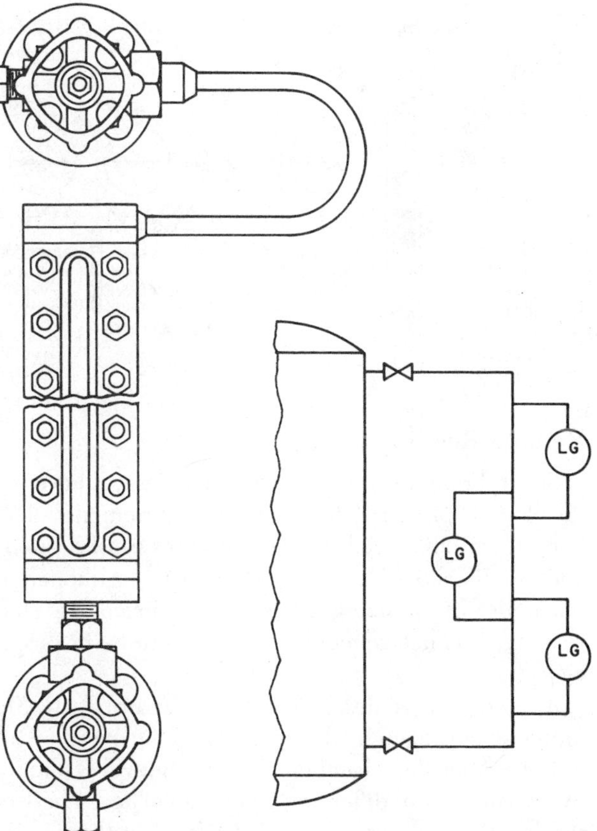

Fig. 3.11h Multiple gauges installed on standpipe and expansion loop for high temperature services

As shown in Figure 3.11i, this design consists of a stainless steel or other non-magnetic metal cage and an internal float that rides on the liquid level. A magnet in the float is coupled to an external indicator. There are two indicator designs. One is a magnetic follower that can be read against a scale. The other consists of a series of wafers that have one color on the front side and a contrasting color on the back. As the float and magnet pass by, the wafers are flipped over, thus indicating the level. Accuracy of the wafer design is limited to the width of the wafer, generally ¼ in. (6.3 mm); the follower type does not have this limitation.

Pressure and temperature ratings of these units vary depending on the manufacturer, but pressure ranges of full vacuum to 3500 PSIG (23 MPa) and temperature ranges of −320 to 750° (−196 to 400°C) are commonly available. The gauges can be used in processes having specific gravity of 0.45 or higher and may be used in lower gravity processes if other requirements for pressure, temperature, and corrosion resistance do not place undue restrictions on float design. Units are also available for indication of liquid/liquid interface provided that the difference in the specific gravities of the two is greater than 0.1. Accessories available are heating or cooling chambers, flanged connections, special materials of construction, alarm switches, and float extensions that enable the readout to be located above or below the vessel.

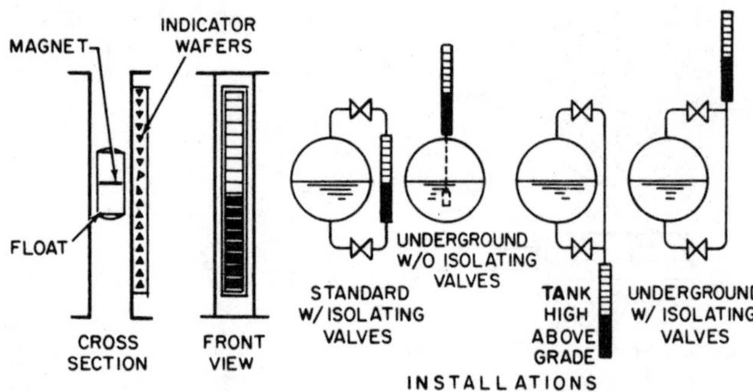

Fig. 3.11i Magnetic gauges

Remote Reading Gauges

Several remote reading level gauge designs have been developed, primarily to meet the requirements of various steam boiler codes. These designs can be rated to 3000 PSIG (20 MPa) at 700°F (371°C). Although the magnetically-coupled gauge can be arranged for remote reading, it is not generally considered to be in this category.

A remote gauge design is shown in Figure 3.11j. Maximum level is fixed by the elevation of the overflow tube in the condensing chamber. The actual level is obtained by measuring the difference in hydraulic heads between the level in the chamber and the level in the overflow tube, the latter being equal to the level in the steam drum. The legs are connected to a manometer assembly

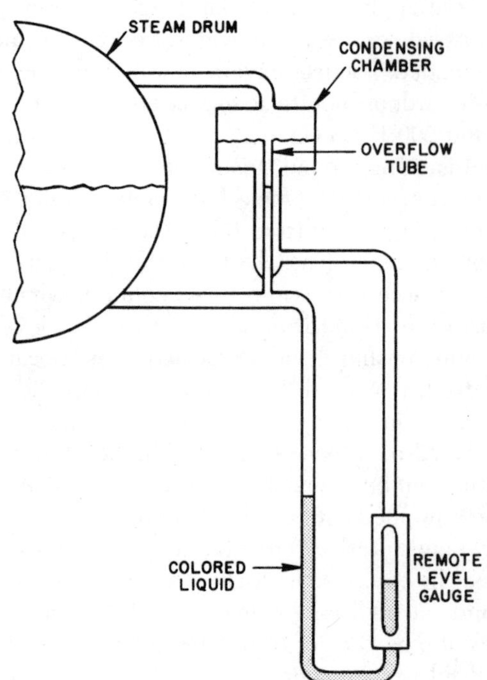

Fig. 3.11j Remote reading level gauge

filled with a colored liquid that is immiscible in water. A transparent gauge is used as a portion of the manometer and located so that the colored liquid in the gauge will rise as the level in the steam drum rises. For accurate level indication, it is important to keep both legs of the system at the same temperature. Also, when specifying this type of gauge, or any other type of level gauge in this service, it is important to specify the pressure of the steam drum, since the water in the drum may have a significantly lower specific gravity than that in the gauge.

A different type of remote reading gauge, shown in Figure 3.11k, consists of a metal chamber with a series of conductivity probes. As the water rises in the chamber, successive probes conduct, energizing their associated

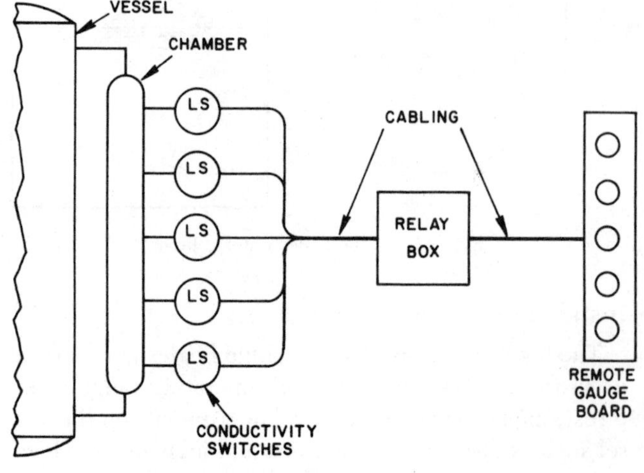

Fig. 3.11k Remote level gauge

switches and relays. The relays can be used to illuminate lights on a gauge board or to actuate a digital readout. This electric design is safer than the manometer design, particularly at higher pressures, because a glass gauge is not required.

A third type of gauge is used in steam boiler services, and although not specifically designed for the purpose, can be arranged for remote reading. As shown in Figure 3.11l, it consists of a series of backlighted and canted double bullseye assemblies mounted on a gauge chamber. By making use of the differing refractive indexes of steam and water, and by selecting the correct geometry, the bullseyes can be made to show one color if steam is in the section behind the bullseye and another color if water is present. By good fortune, if the section behind the bullseye is partially covered with water, the bullseye will divide its colors to show the exact water level. If the illuminators used are strong enough, the beams of the individual bullseyes can be transmitted with mirrors, thereby providing a remote reading.

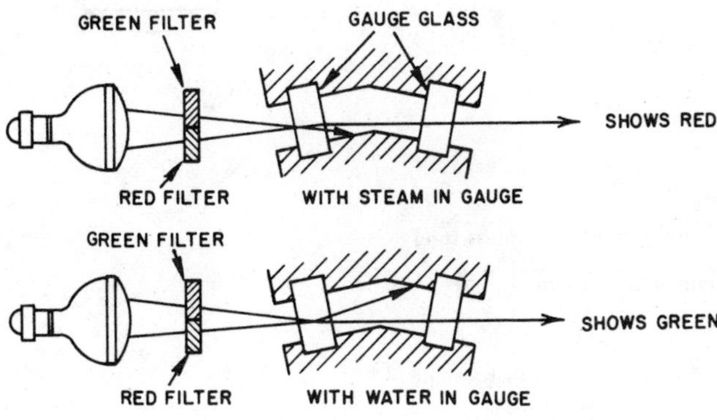

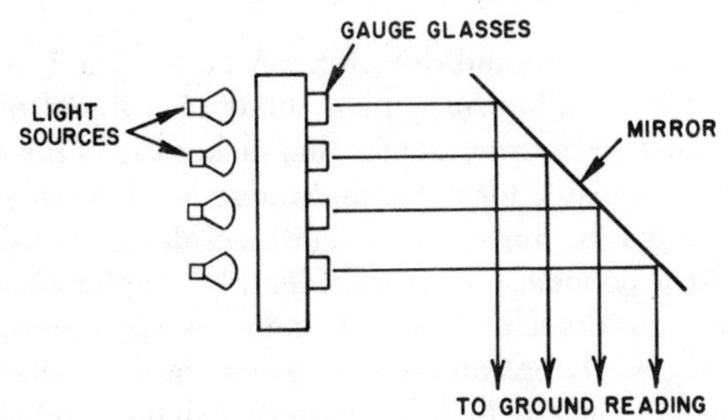

Fig. 3.11 l Bullseye level gauge arranged for remote reading

Closed circuit television systems enable remote monitoring of any important level indication.

Conclusions

Many factors are involved in obtaining safe and reliable level gauge installations, particularly where hazardous processes, high temperatures and/or high pressures are present. Tubular gauges are not suitable for continuous duty processes. The flat glass design is satisfactory in many cases, but magnetically-coupled designs and some of the remote reading designs are better for critical applications, since they reduce the risk of operator exposure and chemical spills.

BIBLIOGRAPHY

Cantieri, W.F., "Water Gauge Accuracy." Technical Publication by Diamond Power Specialty Corp.

Green, C.R., "Tank Sight Glasses." *Chemical Engineering*, September 25, 1978.

Flow Sheet Symbol

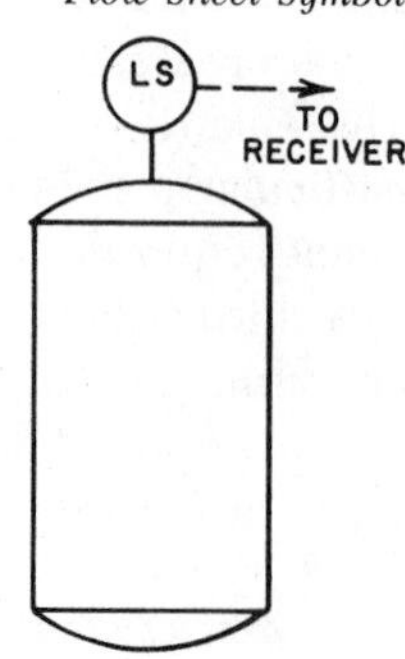

3.12 OPTICAL LEVEL SWITCHES

Design Pressure:	To 150 PSIG (1 MPa)
Design Temperature:	−60 to +160°F (−51 to +71°C)
Materials of Construction:	Contacting design: glass and plastic or stainless steel
Inaccuracy:	±1/16″ (±1.6mm) for contacting design
Cost:	\$150–\$300 for on-off
Partial List of Suppliers:	Courser, Inc.; Hi-G Electronics; Rexnord Instrument Products; Zi-Tech Div., Aikenwood Corp.

This section covers several devices that make use of a light beam to detect the presence of liquid. One design is illustrated in Figure 3.12a. A beam of light is aimed at the liquid and reflected back to a light sensitive transistor, located in the same holder as the light source. By adjusting the transistor sensitivity, the unit can be calibrated in the range of point level detection from ¼ in. (6.3mm) to 12 ft. (3.6m). The sensor can have several light sensitive detectors, permitting switching to occur at more than one point. The operating temperature range is −60° to 160°F (−51° to 71°C). Because of the noncontacting design of this switch, it is suitable for use on corrosive, sticking, or coating processes. The switch is adversely affected by changes in reflectivity of the process.

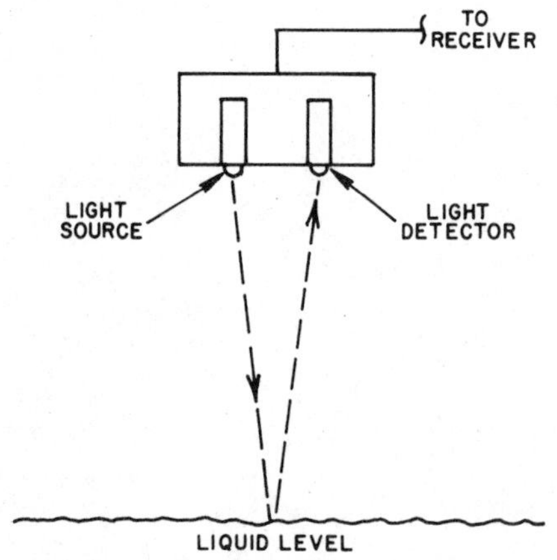

Fig. 3.12a Noncontacting level sensor

A light source and detector can be arranged to continuously monitor suspended solids in a liquid over a range of 0–4 percent. A unit that makes use of this principle is available for detecting sludge level, as shown in Figure 3.12b. Since the sludge level design is used to detect a boundary, it is considerably simpler than the continuous detector. The instrument uses an optical sensor to detect the interface between the low solids supernatant and the high solids settled sludge. In one design, the sensor and its electronics are portable, with the sensor furnished on a 30 ft. (9 m) cable. The operator lowers the sensor into the clarifier until the boundary is detected. Wetted-parts materials are available in nickel plated naval brass. The inaccuracy of the unit is 1 in. (25 mm) assuming that the cable is marked and read correctly.

Figure 3.12c illustrates a simple and inexpensive unit that also uses a principle of optics for level detection. A

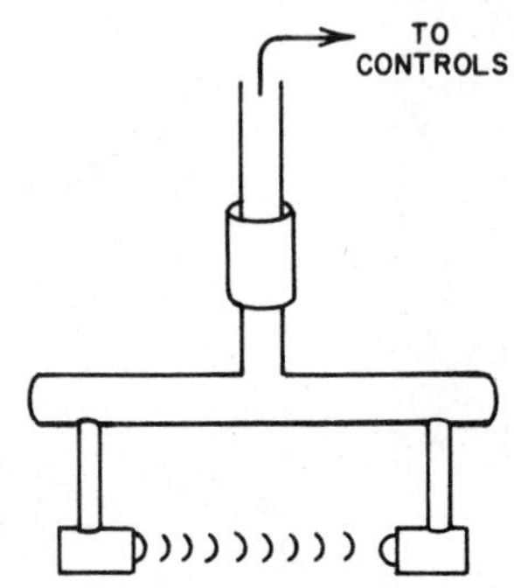

Fig. 3.12b Optical sludge level detector

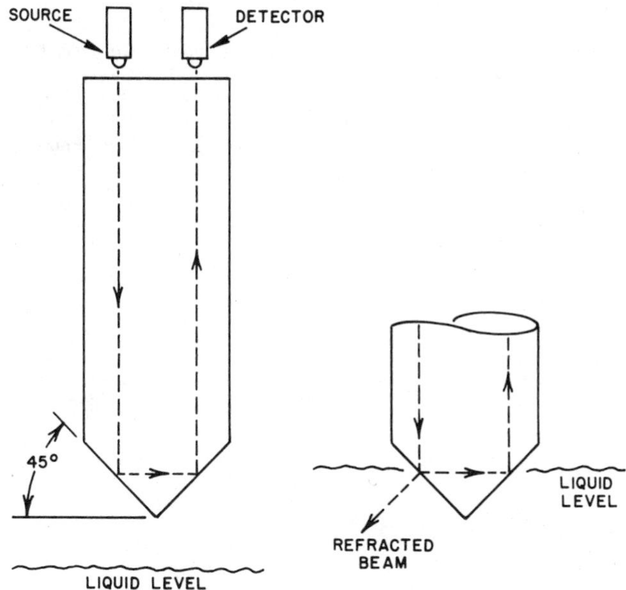

Fig. 3.12c Contacting level sensor

light beam is directed along a cylindrical translucent rod that has a 45 degree bevel at the base. When no liquid is present at the tip, the beam is reflected across the rod and up to a light sensitive transistor. As level rises to cover the tip of the probe, the index of refraction increases and light escapes to the liquid. This reduces the amount of light received by the transistor, causing it to switch. The unit is small, lightweight, and is available with a brass, aluminum, or stainless steel housing. It is sensitive to a change in level of 1/16 in. (1.6 mm), has a pressure rating of 100 PSIG (0.69MPa), and a temperature range of 15 to 250°F (−90 to 121°C). The switch cannot be used in caking or coating liquids. Also, if drops of liquid remain on the probe after the liquid level has dropped, the switch may continue to indicate a high level.

A laser beam and detector can be used for continuous level measurement under certain limited conditions. As shown in Figure 3.12d, a laser source is mounted on one side of the process at any angle between 15 and 60 degrees from horizontal. The detector is mounted on the other side of the process and the same angle as the source. As surface level changes, the beam is displaced as shown in the figure. The detector is arranged so that this displacement can be measured and the control unit is calibrated to convert the displacement to level. The controller produces an analog signal proportional to level. It can also be furnished with alarm or interlock contacts. In order for this unit to work properly, the surface of the process must be clean and reflective. The span is limited to approximately ½ in. (12.5 mm). This non-contacting instrument can be used to monitor the thickness of molten glass as it is formed into sheets. However, it is difficult to envision any application for the laser detector in the chemical processing industry.

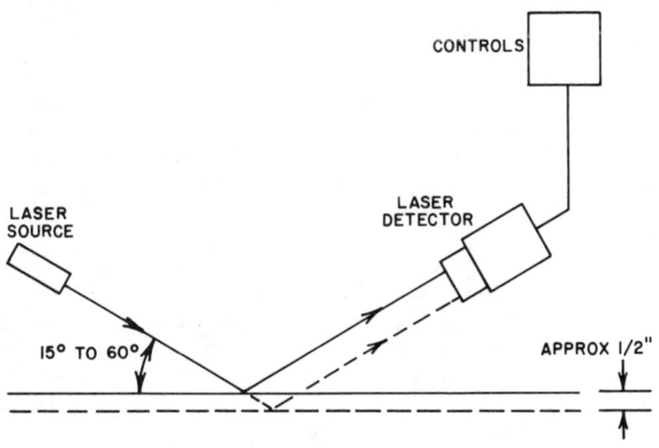

Fig. 3.12d Laser used to detect level

Figure 3.12e shows how a fiber optic system might be used for liquid level detection. A light beam travels through the fiber. When there is no liquid on the fiber, the return beam will have the same intensity as the source beam. As the liquid covers the fiber, the index of refraction increases, allowing light to escape into the liquid and reducing the strength of the return beam. This unit is not commercially available as of this writing.

Conclusion

Optical level devices are not generally used in continuous chemical processing applications because of their pressure and temperature limitation and because they will not operate properly in fouling and caking services. They are a good candidate for level detection in less severe environments.

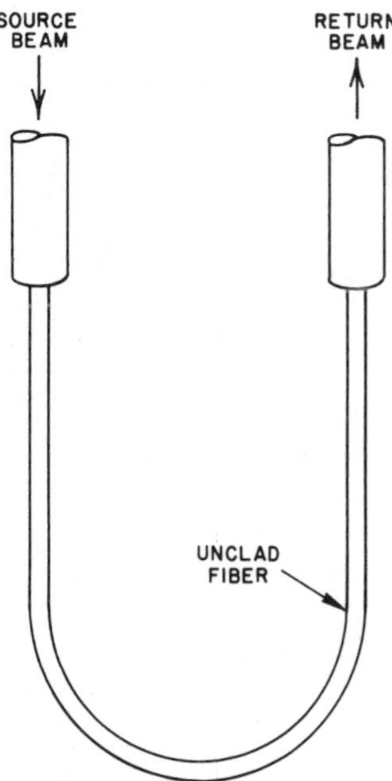

Fig. 3.12e Level detection using fiber optics

BIBLIOGRAPHY

"Fiber Optics Measure Liquid Level." *Machine Design*, June 28, 1973.

"Fiber Optics Monitor Oil-Tank Levels." *Machine Design*, October 21, 1976.

"Level Sensing with Fiber Optics." *Instruments and Control Systems*, January, 1977.

"Optic Liquid Level Sensor Guards Oil Tanks Against Overflow." *Control Engineering*, November 1976.

"Optical Switch Uses Reflection to Monitor the Level of Liquid." *Product Engineering*, September 23, 1968.

"Optical Unit Prevents Liquid Overfill." *Canadian Chemical Processing*, June 1976.

Passe, J., "Optical Probe for the Accurate Measurement of Liquid Levels." *Review of Scientific Instruments*, November, 1965.

"PFA's Optical Characteristics Put to Use in Liquid Level Sensor." *Modern Plastics*, May 1979.

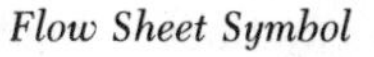

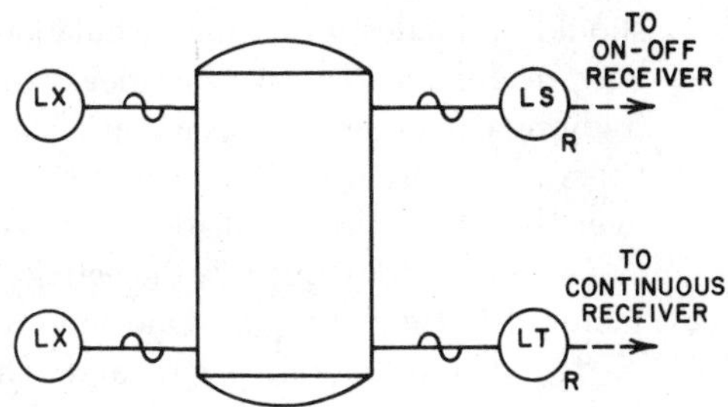

3.13 RADIATION LEVEL SENSORS

Design Pressure:	Not applicable
Design Temperature:	Detector and receiver −20 to +160°F (−30 to +70°C). Water cooled detectors available
Materials of Construction:	Not applicable
Cost:	\$1000 to \$1200 for on-off, \$2500 and up for continuous
Inaccuracy:	±¼ in. (±6.3 mm) for on-off, ±1 to ±2% for continuous
Range:	To 20 ft. (6.1 m) for stationary units. To 50 ft. (15.2 m) for motorized units
Partial List of Suppliers:	C-E Invalco; Endress-Hauser, Inc.; Kay-Ray; Ohmart Corp.; Texas Nuclear Div., Ramsey Engineering Co.

Radiation Phenomenon

Atoms with the same chemical behavior but with a different number of neutrons are called isotopes. Many elements have one or more naturally occuring stable isotopes. For example, the stable isotopes of oxygen are as follows:

Stable Isotopes	No. Protons	No. Neutrons	% Abundance
O 16	8	8	99.76
O 17	8	9	0.04
O 18	8	10	0.20

Most elements also have unstable (radioactive) isotopes. Oxygen's radioactive isotopes are O 15 and O 19 which have 7 and 11 neutrons respectively. The unstable isotopes disintegrate to form elements or stable isotopes. Most of the elements that are heavier than lead are also unstable and disintegrate to form lighter elements. Radioactive disintegration is accompanied by the emission of three different kinds of rays. Alpha (α) radiation consists of positively charged particles having two neutrons and two protons. Beta (β) radiation consists of electrons. Gamma (γ) radiation consists of electromagnetic waves which are comparable to X-rays. The relative penetrating powers of the three kinds of radiation are approximately in the range 1, 100, and 10,000 for the alpha, beta, and gamma rays respectively. The penetrating power of alpha rays is less than 8 in. (203 mm) of atmospheric pressure air. Alpha and beta rays carry an electrical charge and can be deflected by an electric or magnetic field. Because gamma rays have great penetrating power and cannot be deflected, gamma radiation sources are chosen for use in level detecting equipment.

The two most commonly used gamma sources are the radioactive isotopes Co 60 (cobalt) and Cs 137 (cesium). Co 60 is produced by bombarding the stable isotope Co 59 with neutrons. When Co 60 decays, it emits beta and gamma radiation to form the stable element Ni 60 (nickel). In similar fashion, when Cs 137 decays it emits beta and gamma radiation to form the stable element Ba 137 (barium). Cs 137 is one of a large number of fission products of uranium and is obtained when spent fuel rods from nuclear power plants are reprocessed. There are two important points to note about the decay phenomenon. One is that the decay produces electro magnetic energy which cannot induce other materials to become radioactive. This means that gamma sources can be used around such materials as food and food grade packaging materials. The second is that the source loses strength as it decays. The rate of decay is expressed as half-life, the period of time during which the source loses half of its strength. Co 60 has a half-life of 5.3 years; it will decay approximately 12.3 percent per year. The figures for Cs 137 are 30 years and 2.3 percent per year. For point or continuous level measurement, source decay does not affect accuracy, but the initial source size should be made so that the installation has a reasonably long useful life. In rare instances, the isotope Ra 226 (radium) can be used. This material has a half-life of 1602 years and therefore has no appreciable loss of strength over the life of the installation.

The unit used to quantify the activity of any radioactive material is the Curie (c). One gram of Ra 226 has 3.4 $\times 10^{10}$ disintegrations per second. This rate of activity is defined as one c whether it is produced by Ra or some other source. For most level detection applications source strengths of 100 millicuries (mc) or less are satisfactory. The unit of radiation is the Roentgen (r) which is defined as the quantity of radiation which will produce ionization equal to one electrostatic unit of charge in one cubic centimeter of dry air under standard conditions. A one Curie source will produce a dose of one Roentgen at a receiver placed one meter (3 ft.) away from the source for one hour. The dose rate unit is the Roentgen/hour (r/hr), a measure of the photons reaching the receiver at a defined distance. Radiation is attenuated when it penetrates liquids or solids, and the rate of attenuation is a function of the density of the material. This can be seen by reference to Figures 3.13a and 3.13b. The figures also illustrate what is sometimes called the half-value layer. For example, a 1 in. (25 mm) steel plate will reduce radiation from a Cs 137 source by half. An additional 1 in. plate will cause another 50 percent reduction so that the overall reduction caused by a 2 in. (50 mm) plate is $0.5 \times 0.5 = 0.25$.

As would be expected, the amount of radioactive material required to produce one c of activity depends on the material. One c is generated by one gram of Ra 226, by 0.88 milligram of Co 60 or by 0.115 milligram of Cs 137. The dose rate also will vary. Assuming a one mc source and a receiver 32 in. (812 mm) away, the dose rates will be 1.3 milliroentgens per hour (mr/hr) for Ra 226, 2.0 mr/hr for Co 60, and 0.60 mr/hr for Cs 137. Radiation field intensity in air can be calculated from the following equation:

$$D = 1000 \frac{K\,mc}{d^2} \qquad 3.13(1)$$

where

D = intensity, mr/hr,
mc = value of source in millicuries,
d = distance to source in inches, and
K = constant, 1.3 for Ra 226
0.6 for Cs 137
2.0 for Co 60.

Source Sizing

In actual installations, radiation must penetrate substances other than air, and it is of interest to determine the radiation field intensity after the gamma rays have

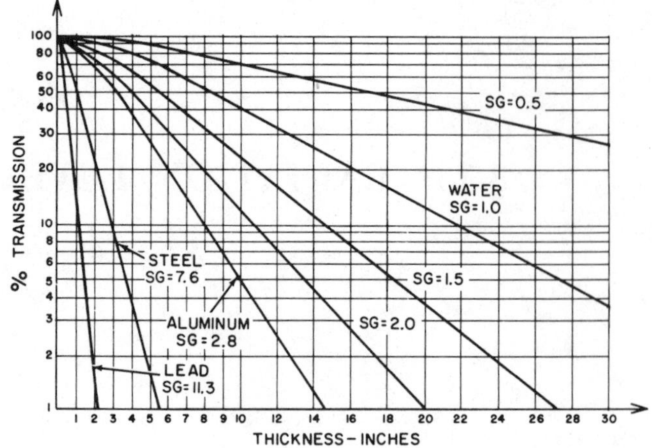

Fig. 3.13a Transmission of cesium-137 through various materials

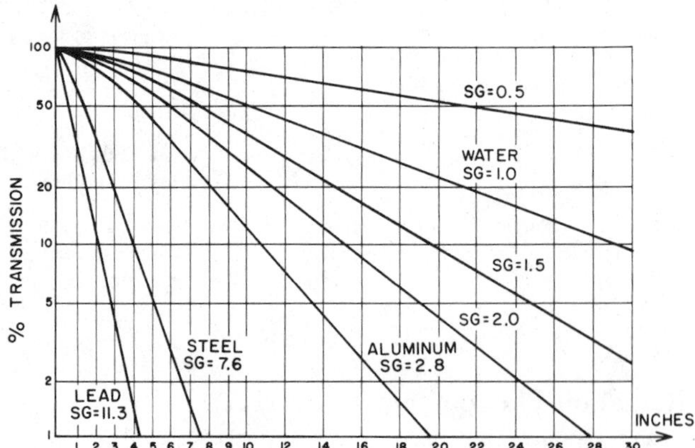

Fig. 3.13b Transmission of cobalt 60 (or radium) through various materials

passed through the vessel walls and process material. The previous equation and Figures 3.13a and 3.13b may be used for this purpose. Figure 3.13c shows how the radiation intensity at the receiver is determined and what levels of operator exposure to radiation may be expected. It will be assumed that the minimum radiation field intensity at the detector should be 2.0 mr/hr when the vessel is empty and that the field should be reduced at least 50 percent with the vessel full. The liquid in the vessel has a specific gravity of 1.0. The calculation is based on the use of a 50 mc source of Cs 137. With no vessel at all, field intensity at the detector would be:

$$D = 1000\frac{K\,mc}{d^2} = \frac{0.6 \times 50}{84^2} = 4.25 \text{ mr/hr}$$

With the empty vessel in place the attenuation through the two ½ in. steel walls will be $0.70 \times 0.70 = 0.49$ (see Figure 3.13a) and the resultant field intensity at the detector is $4.25 \times 0.49 = 2.1$ mr/hr. When the tank is full, the radiation will have to penetrate 72 in. (1.8 m) of material having a specific gravity of 1. The figure for Cs 137 does not cover thicknesses over 30 in. (0.7 m) but that does not matter since total attenuation will be the product of the attenuation in three 24 in. (0.6m) thicknesses. Each 24 in. (0.6 m) of material thickness has a transmission of 7.5 percent so the field intensity at the detector with the tank full is:

$$2.1 \times 0.075 \times 0.075 \times 0.075 = .0009 \text{ mr/hr}$$

This installation meets the given requirement for 2 mr/hr with the tank empty and at least a 50 percent reduction with the tank full. Cs 137 is the source that is normally selected for level measurements. Co 60 is considered only for applications where high penetration ability is required, as on vessels with thick walls. The main reason for this is that the Cs 137 has a much longer half-life. Even with this longer half-life, though, the 50 mc source selected for the example provides a field at the detector too close to the minimum 2 mr/hr required, and therefore would have to be made bigger to increase the useful life of the installation.

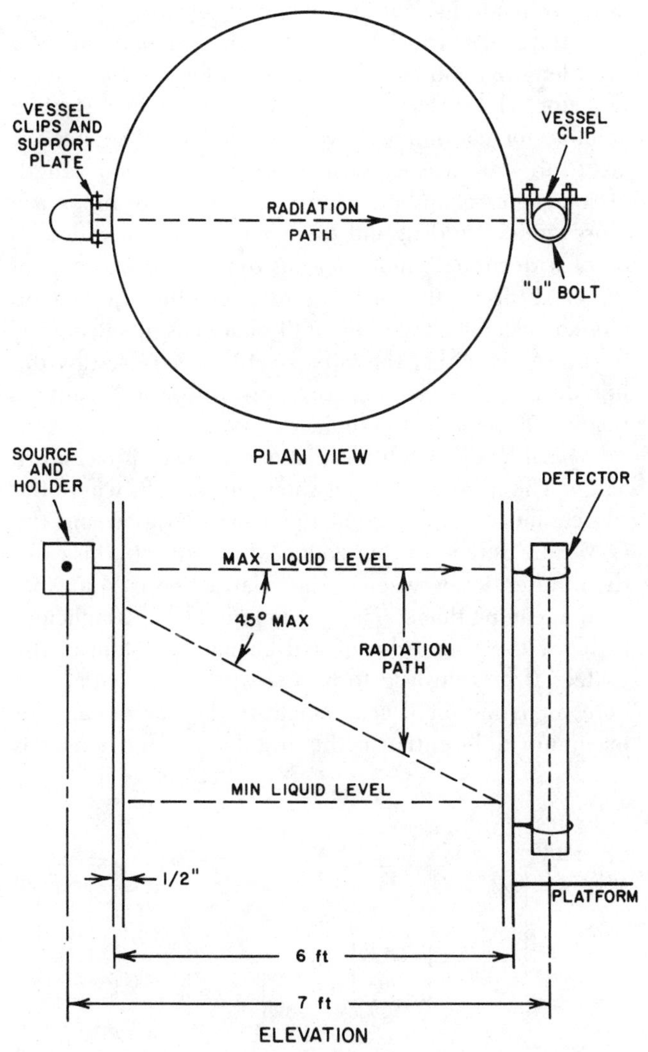

Fig. 3.13c Radiation instrument installation

Safety Considerations

A gamma source radiates electromagnetic energy in all directions just as a glowing ember radiates heat in all directions. Short term exposure to high intensity gamma radiation or long term accumulative exposure to lower intensity radiation is known to be hazardous. The degree of hazard, particularly to long term low intensity exposure, is a somewhat subjective determination and therefore if an error is made it should be made on the safe side.

Radiation sources are formed into ceramic pellets which are placed in a double-walled stainless steel capsule. The capsule is contained in a source holder that is constructed in such a way as to allow a radiation beam to escape through a very narrow window and to be blocked by lead shielding in all other directions. A shutter is provided for the window that can be closed when the source is being shipped or when it is out of service. Source shielding is thick enough to reduce the field intensity 1 ft (305 mm) from the source to 5 mr/hr or less. As was mentioned earlier, the disintegrations that produce gamma rays will also produce beta emissions. Because of their low penetrating power, beta rays cannot escape the stainless steel capsule that encloses the source.

Source holders are designed for a range of source sizes. For example, one holder may be used for sources in the range of 10 to 30 mc while the next heavier holder may be used for sources in the range of 31 to 90 mc. Obviously, if the source being used is at the low end of the range, the field intensity outside of the holder will be less, and will approach the maximum 5 mr/hr limit when the source being used is at the top of the range.

In the United States, rules governing safe exposure rates to radioactive materials have been established by the Nuclear Regulatory Commission (NRC). These rules are incorporated in the Occupational Health and Safety Act (OSHA). Exposure to external radiation is referred to in units of "rem" (Roentgen + equivalent + man). A person receives the dose of one rem when exposed to one roentgen of radiation in any time period. As illustrated in Figure 3.13d, a person should not receive more than 250 rems over his entire lifetime. The rate at which this exposure is accumulated is also important. It is desirable to keep the yearly dose below 5 rems, and it should definitely not exceed 12 rems per year or 3 rems per quarter. In most industrial chemical processing applications it is possible to keep operator exposure far below these levels.

For each industrial installation, it is essential to estimate the dosage received by personnel working in the vicinity of the source, using an assumed occupancy figure and a "worst case" figure for proximity to the source. Returning to the installation shown in Figure 3.13c, it is assumed that the occupancy is 25 hours per week and that the operator is within 12 in. (305 mm) of the tank

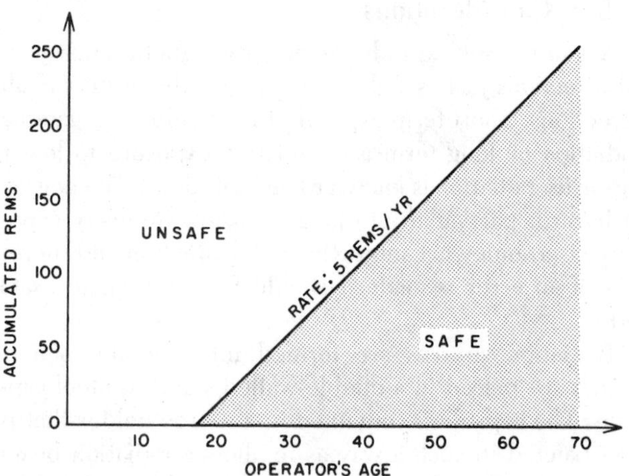

Fig. 3.13d Radiation exposure as a function of time and safety

during this period; the worst case would be if the operator were working next to the source. Assuming that the holder just meets the requirements for 5 mr/hr, operator exposure would be 5 × 25 = 125 millirem per week, or approximately 6.25 rem per 50-week year. This exposure would exceed set limits. The second worst case would be if the operator were working by the detector when the tank was empty. Here the field intensity would be approximately 2 mr/hr and operator's weekly and yearly exposure would be 50 millirem and 2.5 rem respectively. (The second worst case condition is cited to illustrate the point that the source shutter should always be closed when the tank is empty. The ultimate in bad practice is to allow a maintenance worker into the tank when the source shutter is open. Special interlock systems are available to prevent this from happening.)

After making the "worst case" calculations, the design engineer should determine what can be done to reduce operator exposure. In most cases, exposure can be minimized by positioning the detector and source so that the operator does not have to be in the area of high field intensity. If this is not possible, additional lead plate shielding can be permanently installed around the source holder and/or behind the detector. Changes in position cost nothing if done early in the project and additional shielding costs very little. In fact, for the majority of installations, the source and detector can be arranged or shielded so that the operator receives less than 10 percent of the safe levels, assuming a 40 hour per week occupancy. Referring to Figure 3.13c again, if a 1 in. (25 mm) thick lead shield were to be installed behind the source, operator exposure would be approximately 75 millirem per year (assuming a 40 hour per week exposure).

The Detector

There are a number of gamma radiation detectors available, but the two commonly used in conjunction with level detection are the Geiger-Mueller (G-M) tube and the gas ionization chamber. The G-M tube has a wire element anode in the center of a cylindrical cathode. The cathode tube is filled with inert gas and sealed. A voltage of 250 to 300 volts is impressed across the anode and the cathode. Incident gamma radiation ionizes the inert gas so that there is an electrical breakdown between the anode and cathode. The frequency of the breakdowns is related to the intensity of the gamma radiation and therefore field strength can be determined by counting the pulses so produced over a given time interval. The ionization chamber is likewise filled with inert gas and sealed, but rather than applying a breakdown voltage, a smaller voltage of 6 volts is applied across the chamber from end to end. When the chamber is exposed to gamma radiation, ionization occurs and a continuous current in the microampere range is caused to flow. This current is proportional to field intensity.

The G-M tube is always used for the level switch designs (point detection). The switch detector is arranged so that it sees the full field intensity (tank empty) or it sees little or no field (tank full). Both types of detectors are used for continuous level detection. Ionization chambers are made in continuous lengths up to 10 ft. (3 m). G-M tubes are furnished in 6 in. (152 mm) or 12 in. (305 mm) lengths, and can be stacked to form a continuous detector. There is some controversy as to which is more suitable for continuous level installations. The stacked G-M tubes are less expensive than an equivalent length of ionization chamber. However, the G-M tubes are more subject to drift and their performance can deteriorate with time. A more serious drawback to the use of the G-M tube is that an exposed (above the liquid level) element can fail altogether. If this happens, it will appear to the receiver that the failed section is covered by the liquid. In a five section assembly this would cause a reading 20 percent too high.

Typical level switch installations are shown in Figure 3.13e. The most common is shown at the left, where the source and detector are at the same elevation and the G-M tube detector is horizontally mounted. In this case, the differential between on-off relay action is ¼ in. (6.3 mm), meaning that a ¼ in. rise in liquid level is sufficient to block the source beam and change the state of the switch. If it is desired to have a wider differential, the detector is mounted at an angle to the horizontal. For maximum differential in this installation, the sensor is

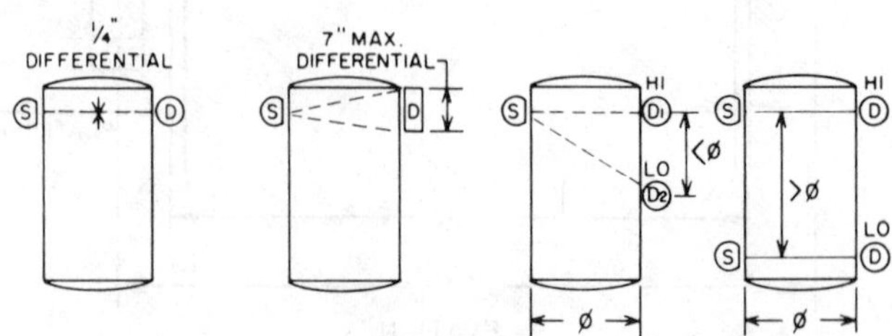

Fig. 3.13e On-off radiation switch installations

mounted vertically, producing a differential of 6 in. (152 mm). For even wider differentials, two detectors can be used with a single source. In this case, the maximum differential between high and low level settings can equal the tank diameter. Differentials greater than the tank diameter require two separate sets of souces and detectors.

In high level applications, the G-M tube and switch assembly is normally above the liquid level and therefore exposed to full field intensity. Pulses from the G-M are taken to a trigger circuit that, in turn, continuously resets a time-out relay much like the time-out relay used in a computer watch dog circuit. When rising level blocks the radiation beam, the relay is no longer reset and the switch changes state; for fail-safe operation, the switch would open. The switch circuitry is arranged so that the switch will open on failure of the G-M tube or failure of any of the switch components, and therefore the entire installation may be judged fail-safe. Low level switching applications are a different matter. In this case, the G-M tube or switch component failure would not be detected since exposure of the tube to the beam on falling level would not actuate the switch. Where fail-safe design for falling level applications is required, a test circuit can be installed in the G-M tube and switch assembly to test the switch when the level is high. There are several ways to do this, but in general, a small source is installed in the detector and used continuously or intermittently to test the integrity of the tube and switching circuitry.

Continuous Level Measurement

There are two methods for making continuous level measurements using fixed sources and detectors. One method, using a strip source and strip detector, is illustrated in Figure 3.13f. The second, shown in Figure 3.13c, uses a point source and a strip detector. The former, using the strip source, should be considered first because, although it is more expensive, it is more accurate and is suitable for a wider array of vessel geometries than the point source detector. As shown, the strip source radiates a long, narrow, uniform beam in the direction of the detector. As the level rises a small increment, a corresponding small increment of the detector is screened off. This incremental response is uniform and linear over the entire span, and therefore the signal produced is linear with level change over the entire span, except for small non-linear end effects near 0 and 100 percent of span. Moreover, this installation, unlike the point switch installation, is not sensitive to variations in the specific gravity of the material in the vessel. For example, if the specific gravity of the material varied from 0.40 to 0.70, the source would be sized for 0.40 specific gravity material; material of higher gravity would cause higher attenuation, and this would not affect the performance of the detector.

The point source and strip detector installation also works as a very small incremental on-off device, insofar as a small level rise blocks off the radiation beam of a corresponding increment of the detector. Unfortunately, as will be seen from an inspection of the high level-to-low level change in geometry, the incremental changes in level do not produce a uniform change in the coverage of the detector; consequently, this installation produces a non-linear signal with level change. Not only does the thickness of the material change as the level changes, but the geometry of the fixed portion of the system—the vessel walls, wall to detector distances, and the free space—also changes as the level changes. This system non-linearity can be rectified in most cases by correcting the receiver output electronically, if the ratio of level span to vessel diameter is no greater than 1 to 1. Note that this is the same limitation placed on the geometry that determines maximum spacing between two level switches and one source. As shown in Figure 3.13g, one way to reduce the non-linearity at the lower end is to use one strip detector and two or more point sources. This approach improves accuracy but increases the cost.

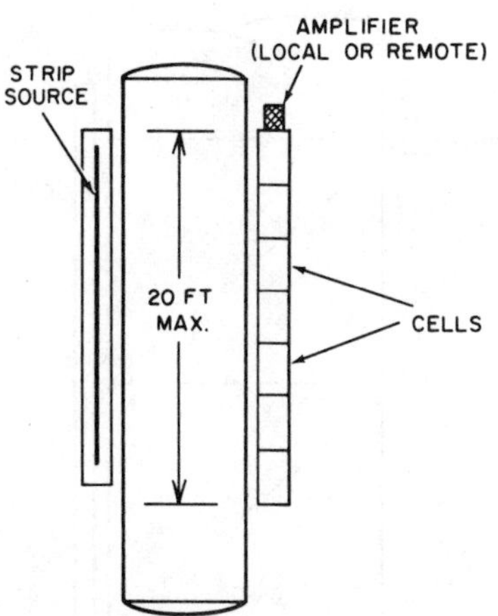

Fig. 3.13f Continuous level detection by use of strip source and electronic cell receivers

Radiation gauges can be arranged to detect, either at a point or continuously, solids levels and liquid/liquid interfaces. The accuracy of these installations depends on source size, detector sensitivity, material gravities and vessel geometry. Sometimes a system may be needed to continuously monitor a liquid/liquid interface, or a liquid or solid level over a long vertical straight side. In these cases, strip source or multiple source and strip detectors would be too expensive. Figure 3.13h shows how a point source and point detector may be motor-driven over a wide span to detect levels of the above type. The motor drive may be set up to continuously hunt the level—that is, undershoot and overshoot—or it may be set up to look for the level on operator demand, as might be required for an inventory. The exact location of the level is determined by the sprocket-driven position sensor.

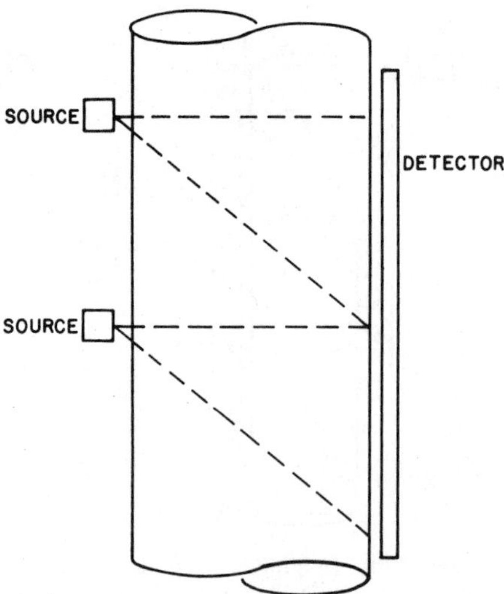

Fig. 3.13g Level detection using two sources and one detector

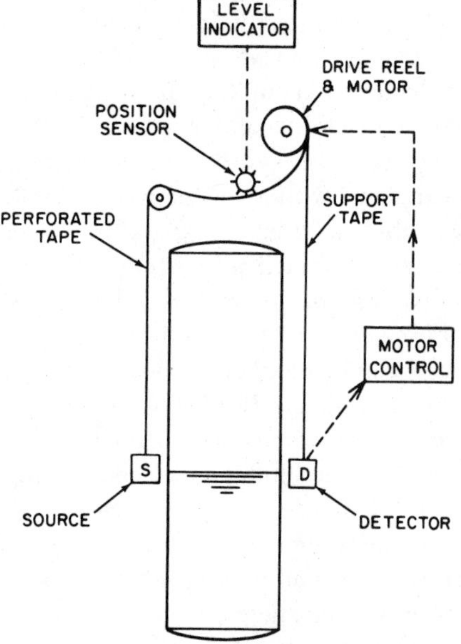

Fig. 3.13h Continuous high accuracy radiation detector system for accounting installations

Installation Notes

Source holders are furnished with a mounting flange on the window side of the holder. The detectors are supplied inside a piece of steel pipe that is capped on both ends. The preferred method of installing this equipment is to bolt it to clips that have been welded to the outside surface of the vessel. Figure 3.13c illustrates how this is done. The installation can be arranged so that the elevation of both the detector and the source can be changed easily. The system shown calls for the beam to pass through the center of the vessel. This is not necessary, and in some cases it would be undesirable. If the vessel has a center-mounted agitator, fill nozzle, or other internal obstruction, the source and detector should be located so that the beam radiates across a chord where there are no obstructions. A chord may also be selected on large diameter vessels in order to reduce source size. The cable run to the detector should be made in such a way that condensate from the conduit system will not flood the detector.

The technique for calibrating radiation level detectors is unique. A 100 percent full calibration reading may be made at any time by closing the source shutter, simulating a full vessel. The second calibration point, actual level, is found by running a portable G-M counter down the vessel wall between the vessel and the detector. The level is located when the counter reading drops off, because this is the point at which the liquid screens off the source.

Conclusions

Radiation level detection is very appealing for hard-to-handle, toxic, and corrosive processes because it does not require vessel wall penetrations. Safety and licensing requirements cannot be minimized, but are no serious impediment to a carefully designed system.

BIBLIOGRAPHY

McConnell, J.A. and Smuck, W.W., "Gamma Back Scatter Technique for Level and Density Detection." *Chemical Engineering Progress*, August 1967.

Rowe, S. and Cook, H.L., "Nuclear Gages for Density and Level Control." *Chemical Engineering*, January 27, 1969.

Thomason, E.M., "Design Procedure for Nuclear Level Gages." *Instrumentation Technology*, June 1968.

Flow Sheet Symbol

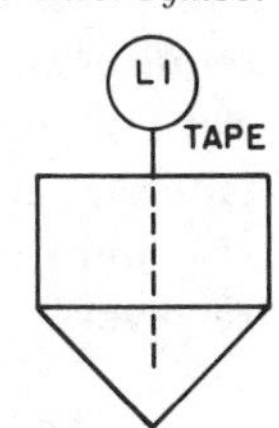

3.14 RESISTANCE TAPES

Design Pressure:	Atmospheric
Design Temperature:	5 to 160°F (−15 to +71°C)
Materials of Construction:	Outer tape sheath is corrosion-resistant plastic
Inaccuracy:	±1 to 6 in. (±25 to 152 mm)
Cost:	\$500–\$1000
Range:	To 100 ft. (30 m)
Supplier:	Metritape, Inc.

As illustrated in Figure 3.14a, the resistance tape detector responds to level changes in the tank with changing loop resistance. As the level of liquids or solids rises in the tank, the hydraulic pressure of the material compresses the jacket, causing progressive contact between the resistance element and the conducting base strip. The transverse sensitivity of the tape is 0.2 psi (1.4kPa), which is the pressure required to short out the tape section under the process material surface. Above the surface, the resistance element remains unshorted and it is this upper resistance leg that is measured to detect the material level. The resistance element winding provides two to four contact points per 1 in. (25 mm) of measuring tape. The jacket envelope serves to enclose and protect the electrical system, and to act as a pressure detecting diaphragm.

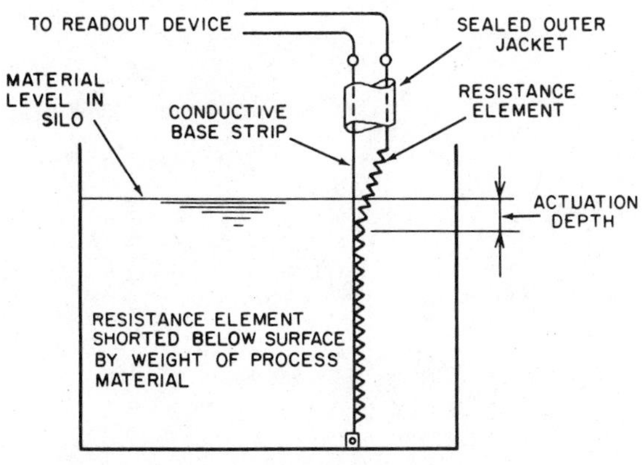

Fig. 3.14a Resistance level detector

Because 0.2 psi (1.4kPa) pressure is required to compress the jacket and thus short the resistance leg, the tape is never shorted out all the way up to the surface of the material. The uppermost electrical contact is made some distance below the surface, and this distance, "actuation depth," is determined by the density of the process material. For water, the actuation depth is approximately 6 in. (152 mm); for lighter materials, it is correspondingly greater. At the time of calibration, this offset is compensated by adjusting the zero setting of the instrument. The zero offset precludes any low level readings at or below the amount of the offset.

To maintain accuracy, the pressure inside the tape jacket must equal the pressure in the tank. For atmospheric tank applications, this is accomplished by venting the tape interior to the atmosphere through a small desiccant dryer. For tanks under pressure or vacuum, the tape internal and external pressures must be equalized. This can be done by installing a direct connected equalizing line, or by mounting a one-to-one pressure repeater on the tank and tubing the output of the repeater to the vent connection on the tape. (See Section 3.7 for a discussion of pressure repeaters.) However the equalization is made, care must be taken to prevent moisture or other contaminants from getting inside the tape jacket. As with many level detectors, changes in material specific gravity will affect accuracy of the resistance tape.

Resistance tape devices have been installed on both liquid and solid services. The tape may be gravity-suspended through an access hole in the tank roof with a weight on a lower end, or permanently attached to the bottom of the tank or silo. It may also be installed in a stilling well if the tank is to be agitated or if the tape requires mechanical protection for other reasons. A variety of readout devices is available for the resistance tape elements. They range from battery powered portable units for field use to single or multichannel remote receivers.

Conclusions

The unique construction of the resistance tape allows it to be used to measure the level of solids, slurries, and various liquids that are corrosive to more conventional level detection devices. However, limitations caused by "actuation depth," pressure equalization, and dryer maintenance suggest that applications for the resistance tape be made very selectively.

BIBLIOGRAPHY

Ehrenfried, A.D. "Resistive Metritape Level/Temp Gauge for Marine Closed-Tank Service," Marichem 79, Second International Conference on Marine Transportation, Handling and Storage of Bulk Chemicals, Monte Carlo, March 6–8, 1979.

Flow Sheet Symbol

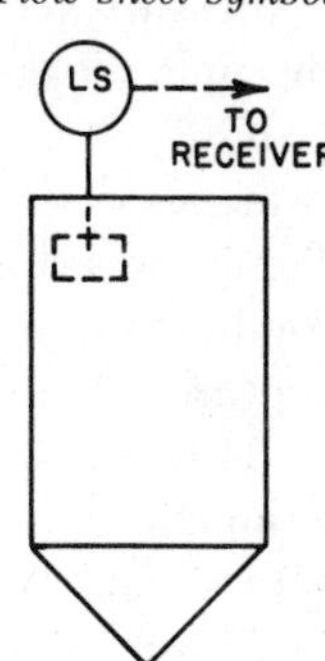

3.15 ROTATING PADDLE SWITCHES

Design Pressure:	To 30 PSIG (0.2 MPa)
Design Temperature:	To 250°F (121°C)
Materials of Construction:	Aluminum, steel, stainless steel
Inaccuracy:	±1 in. (±25 mm)
Cost:	$400
Partial List of Suppliers:	Bindicator; Monitor Manufacturing, Inc.; Semco, Inc.

The rotating paddle-type level switch is used to detect the presence or absence of solids in a silo. A small, geared, synchronous motor keeps the paddle in motion at very low speed. There is no torque on the paddle drive assembly when solids are absent. When level rises to the paddle, the paddle is stopped and torque is applied to the drive assembly; detection of the torque is used to actuate a switch which may, in turn, be used for alarm, or for control of silo filling or emptying equipment.

As shown in Figure 3.15a, one method for torque detection uses a modification of the displacer-type torque tube. (See Figure 3.8b for comparison.) When solids are not present (and thus there is no torque on the tube), the entire drive assembly rotates at the gear speed in a counterclockwise direction. With solids present the paddle stops, torque develops on the tube, and the torque rod is forced clockwise relative to the rest of the assembly. This clockwise motion rotates a beveled cam that rises to operate the switch. In some designs, the motor is rated to operate in a continuously stalled condition. In others, the motor is switched off and does not restart until the spring-opposed torque detector returns to the zero torque position.

A number of paddle designs are available; larger paddle areas are required to generate sufficient torque in lower bulk density materials. Conventional four blade paddles 6 in. (152 mm) wide by 2 in. (50 mm) high are used in materials with densities below 20 to 30 pounds per cubic foot (0.32 to 0.48 g/cm^3), while vane and wire designs are used in heavier materials. The latter types can be installed through a coupling on the silo wall. The larger paddle designs are furnished with a plate to cover the silo nozzle or cutout through which the paddles are inserted.

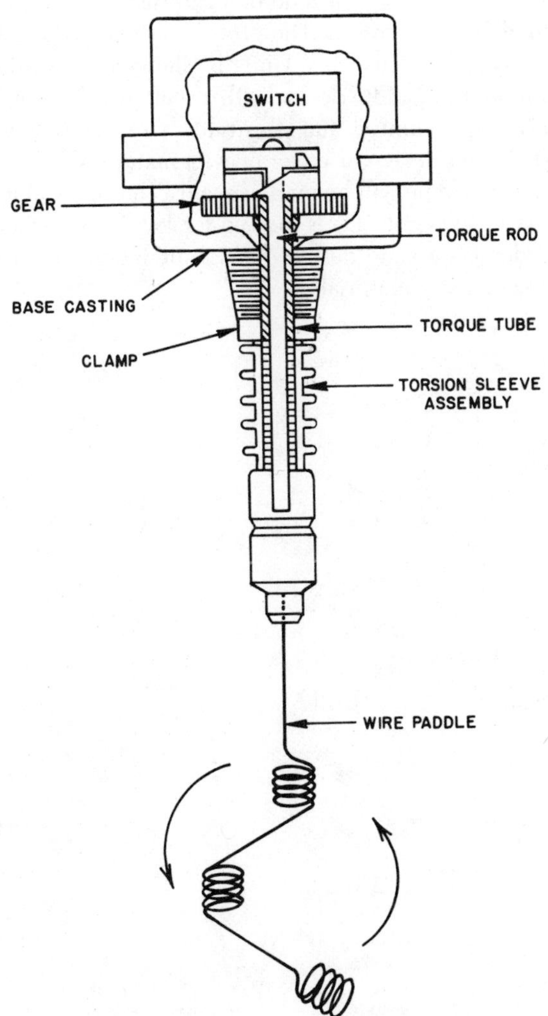

Fig. 3.15a Rotating paddle switch schematic

Pressure and temperature ratings vary, depending on the unit selected. Units have pressure ratings from 7.5 PSIG (52 kPa) to 30 PSIG (207 kPa). When the switch is installed on a pressurized silo, the electrical conduit should be sealed to prevent accumulation of dust in the conduit system in the event that the torque tube or seal ruptures. A common temperature rating is −30 to 200°F (−34 to 93°C) including the switch and motor housing, although higher ratings are available. Although the accuracy figure cited at the start of this section is ±1 in. (±25 mm), the user should be cautious because it only indicates repeatability in this application. More important for average level monitoring of a solid are such factors as location of the paddle versus angle of repose, repeatability of the angle of repose, and the possibility that the material will bridge around the switch and rathole elsewhere (or vice versa).

Figure 3.15b illustrates several paddle designs and some methods of installing these switches on silos. As noted, switch locations and immersion lengths are selected after determining the angle of repose of the material and how the solids fill and the empty nozzle orientation will affect average level. Unit "A" shows a wire design mounted in the side of a silo that may be in the path of falling material. The protective baffle is installed to prevent spurious trips. Unit "B" shows a conventional top-mounted paddle design with a shaft guard. Unit "C" is also top-mounted and illustrates a rectangular vane design. This design, used for heavier materials, is located so that the vane will be pinned against the silo wall upon rising level. Unit "D" is located for low level detection and depicts a vane design that is suitable for operation in low density materials.

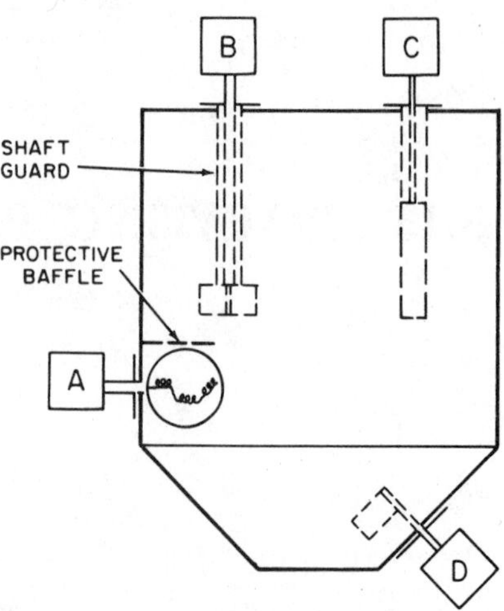

Fig. 3.15b Rotating paddle solids level switch

Conclusions

The principles underlying the operation of the paddle switch have always been good. The improvements and more conservative ratings available on the newer designs encourage one to believe that these switches can perform to expectations.

BIBLIOGRAPHY

Andreiev, N., "Survey and Guide to Liquid and Solid Level Sensing," *Control Engineering*, May 1973.

Belsterling, C.A., "A Look at Level Measurement Methods," *Instruments and Control Systems*, April 1981.

Hall, J., "Measuring Interface Levels," *Instruments and Control Systems*, October 1981.

Flow Sheet Symbol

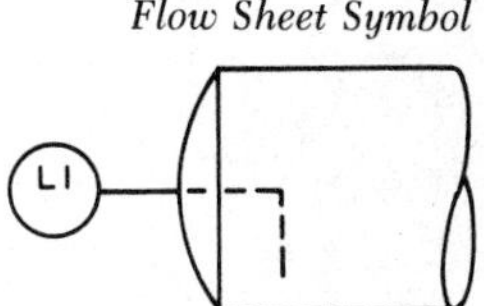

3.16 SLIP TUBES

Design Pressure:	To 300 PSIG (2 MPa)
Design Temperature:	Ambient
Materials of Construction:	Stainless steel
Inaccuracy:	±1–2 in. (±25–50 mm)
Cost:	\$50–\$200
Partial List of Suppliers:	Gas Equipment Co.; Metal Goods Manufacturing Co.

The two basic designs of slip tubes are the rotary and the vertical. The rotary slip tube is illustrated in Figure 3.16a. The unit is operated by opening the head assembly until the process vapor starts discharging through the bleed connection. The handle and slip tube assembly is then rotated to find the liquid/vapor interface. The liquid/vapor interface is established when liquid starts bleeding out. An approximate indication of tank level can be obtained by reading the position of the handle on the scale mounted behind it.

The vertical slip tube is operated similarly. It is installed on the top of a pressurized vessel. The tube is lowered into the tank until the tip makes contact with the liquid. At this point, liquid, instead of vapor, starts bleeding out of the cap assembly. For these units, the level scale is marked on the slip tube; the further it is pulled out after contacting the liquid/vapor interface, the higher the level in the tank.

Not only are these designs inaccurate, they are also dangerous. The rotary design will leak at the packing gland. The vertical design requires the operator to climb on top of the tank; once he is there, it can blow out in his face. These devices are mentioned only because they are still used on some chemical transport tankers.

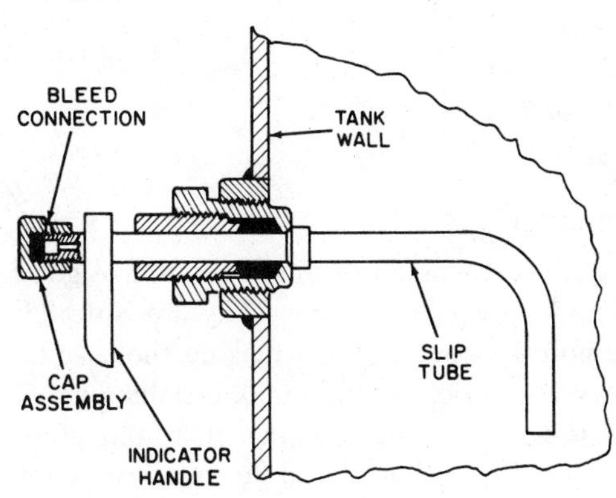

Fig. 3.16a Slip tube level indicator

BIBLIOGRAPHY

Andreiev, N., "Survey and Guide to Liquid and Solid Level Sensing," *Control Engineering*, May 1973.

Belsterling, C.A., "A Look at Level Measurement Methods," *Instruments and Control Systems*, April 1981.

Morris, H.M., "Level Instrumentation from Soup to Nuts," *Control Engineering*, March 1978.

Flow Sheet Symbol

3.17 TAPE LEVEL DEVICES

Design Pressure:	To 300 PSIG (2 MPa)
Design Temperature:	Cryogenic services to above 300°F (150°C)
Materials of Construction:	Plastics, steel, stainless steel, and high alloys
Cost:	\$1200 to \$3500 for sensor and transmitter
Inaccuracy:	To ±0.01 ft. (±2.5 mm)
Range:	To 100 ft. (30 m)
Partial List of Suppliers:	Bindicator; Delta Controls; Devtron Corp., Div. of Agar; Endress-Hauser; GPE Controls; Johnston and Jennings Co.; Kodata, Inc.; Monitor Manufacturing; Scientific Instruments, Inc.; Varec Div., Emerson Electric Co.

Introduction

The gauges discussed in this section have good absolute accuracy over long spans. They are routinely used with remote readout devices, making them suitable for inventory control of multiple tank installations. Others, such as the magnetically-coupled float and guide tube design, can also be considered for inventory control, as noted in the Orientation Table at the start of this chapter. There are several designs considered in this section. The first is an inductively-coupled float and fixed tape design. The second is the conventional wire-guided float detector. The third is a wire-guided capsule that is heavier than liquid and uses a thermal sensor principle for level detection. This group of detectors is used for liquid level measurements and can have a resolution approaching ±0.1 in. (±2.5mm). A fourth design is the surface sensor, which may be used in liquid or solid services. When used in liquids, the surface sensor can be as accurate as the first three types; for solids applications, its resolution approaches ±1 in. (±25 mm). For most solids applications, this resolution is adequate, since larger errors can be introduced in the correlation between level and volume by uneven solids level, bridging and ratholing.

These four devices are reviewed in the above order, followed by a brief description of a capacitance-type and a displacer-type level detector. The section concludes with a discussion of multiple tank wiring methods and a technique used to digitally encode an analog signal.

Inductively-coupled Float and Tape Detector

Figure 3.17a illustrates a fixed tape, float-actuated level measuring device. The tape is suspended from the roof of the tank and is anchored to the bottom. The tape is used to guide a float which contains an inductively-coupled transducer. The tape consists of a steel ribbon and a number of insulated conductors encapsulated in

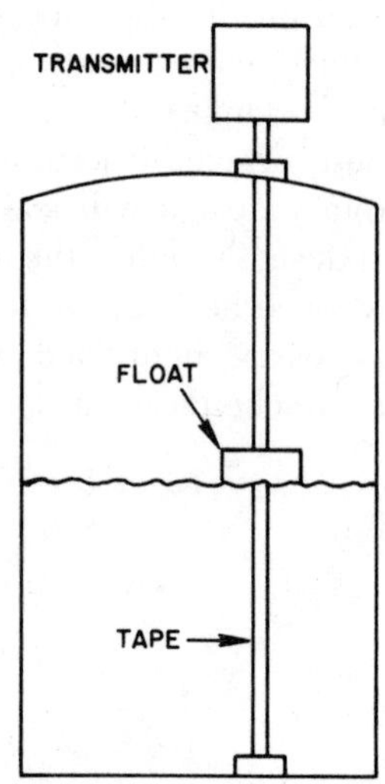

Fig. 3.17a Inductively-coupled float and tape

a Teflon jacket. In addition to providing mechanical strength, the steel tape is used to provide power to the transducer in the float through inductive coupling. At short intervals, this primary coupling is interrupted and a secondary inductive coupling from the transducer to the conductors on the tape is established. The conductors are arranged on the tape in a coded pattern so that each 0.1 in. (2.5 mm) increment has a unique code. The receiver mounted at the top of the tank reads which conductors have been inductively coupled, and from this information can determine where along the tape the float is located and thus the elevation of the liquid level. The float is Teflon coated and the tape-to-float clearance is approximately 1/4 in. (6.3 mm), to minimize float sticking and material build-up on the tape. The receiver can be furnished to transmit an analog signal proportional to level or it can transmit the digital signal that has already been produced by the tape and float assembly, which has a resolution of 0.1 in. (2.5 mm).

The conductors on the tape are arranged to produce a gray code digital word. In the gray code, only one digit in the word changes from one word to the next and, therefore, only one conductor must change its position from one 0.1 in. (2.5 mm) increment to the next. The number of conductors required increases with the span of the liquid level to be measured. The span covered is equal to 2^n where n is the number of conductors. Thus, if four conductors are used the span would be 16 increments, or 1.6 in. (40 mm). If 14 conductors are used the span would be 16,384 increments, or 135 ft. (41 m). In addition, each system requires a reference conductor and a return conductor. Figure 3.17b shows in schematic form how four conductors might be arranged on the steel tape to produce the gray code digital word for a 16 increment measuring system. If the conductor is on the right hand side of the tape, it is inductively-coupled to the transducer in the float; if on the left, it is not. The reference wire tells the receiver which side of the tape is the right hand side. The return conductor is common, completing the circuit for all conductors. As shown in the figure, if the float is at increment 7, conductors 1 and 2 will be inductively-coupled and conductors 3 and 4 will not. Thus, the gray code digital word produced is 0011, which is unique for the particular increment. As previously noted, each additional conductor doubles the preceding span; adding a fifth conductor to the arrangement shown in Figure 3.17b would enable measurement over 32 increments, a sixth conductor 64 increments, and so on.

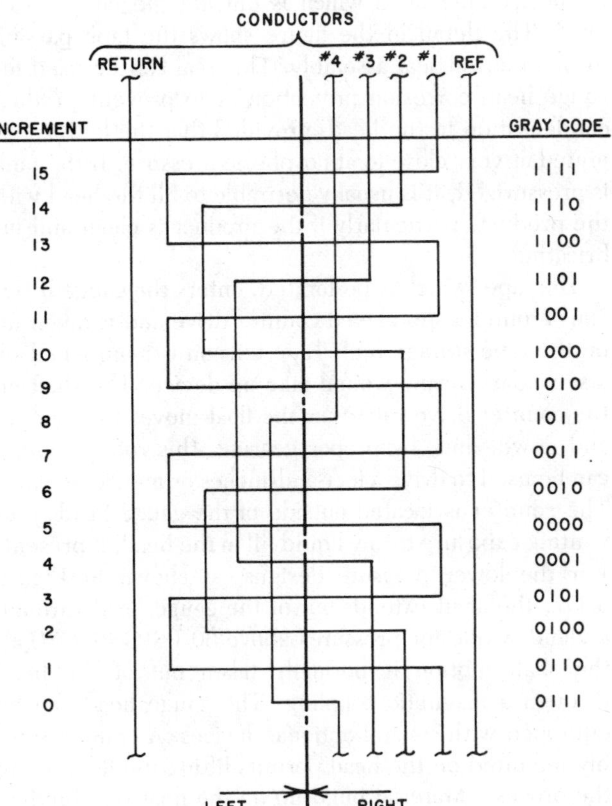

Fig. 3.17b Arrangement for 16 increment inductively-coupled system

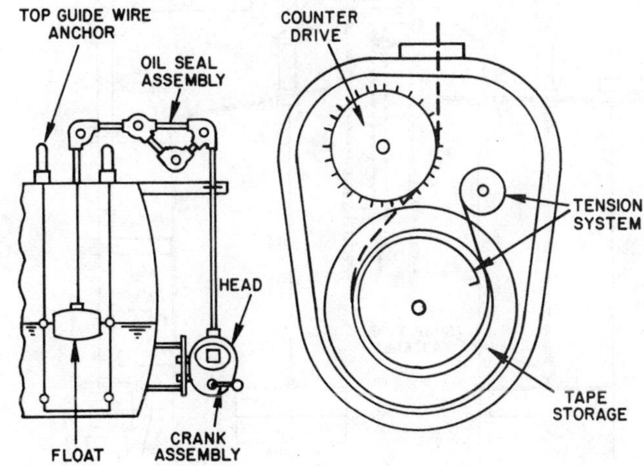

Fig. 3.17c Wire-guided float detectors and detail of head

Wire-guided Float Detectors

Figure 3.17c shows a wire-guided float detector that has a tape connection to a ground reading assembly. This installation can be compared with that shown in Figure 3.9a. The system shown here is suitable for tanks having an operating pressure to 30 PSIG (0.2 MPa) and a height of 60 ft. (18 m), although other designs are available that are rated to 300 PSIG (2 MPa). The float is about 15 in. (381 mm) in diameter and can be made of aluminum, stainless steel or other alloys; it can be hollow or filled with materials such as foam glass. In order to contain the float, guide wires are connected to top and bottom anchors. The top anchors are normally spring-loaded to maintain constant tension on the wires; the bottom anchors are tank clips. A tape runs from the connection on the top of the float, over sheave assemblies, and down to the gauging head which is outside the tank at eye level. The detail in the figure shows the tape passing through an oil seal assembly. This seal can be used for gauge head corrosion protection or to prevent product condensation in the head, provided that the tank is operated at very close to atmospheric pressure. If the tank is pressurized, it is usually desirable to fill the head with the product, particularly if the product is clean and lubricating.

The tape, which is perforated, enters the gauge head, runs around a sprocketed counter drive and is taken up on the tape storage reel. Tape tension is maintained by a secondary spring-wound take-up device. The shaft on the counter drive rotates as the float moves the tape up and down, and, by proper gearing, this rotary motion can be used to drive a feet-and-inches or metric readout. The counter is located outside of the gauge head, preventing exposure to any liquid fill in the head, if present. For the lower pressure designs, as shown in Figure 3.17c, the shaft extends out of the gauge head through a gland, while for pressures above 30 PSIG (0.2 MPa), the shaft motion is normally taken out of the head through a magnetic coupling. The gauge head can be equipped with several optional devices. A crank assembly mounted on the head permits lifting the float out of the process. Material build-up on the float that hinders smooth operation may make this necessary. Often the float can be freed by lifting and lowering it. The head also may be equipped with a variety of switch configurations for high and low level alarms or control circuitry actuation.

Figure 3.17d shows installation details for wire-guided floats in low pressure tanks. Figure 3.17e shows the installation detail for a high pressure installation rated to 300 PSIG (2 MPa). At the higher pressures, the flat circular float design is not suitable, and one or more spherical floats are used. Connections for a high pressure installation should be flanged, including those for the top anchor assemblies. Another feature to note in Figure 3.17e is the gate valve with rubber plug that is installed at the tank entry, allowing removal of the gauge head without tank depressurization.

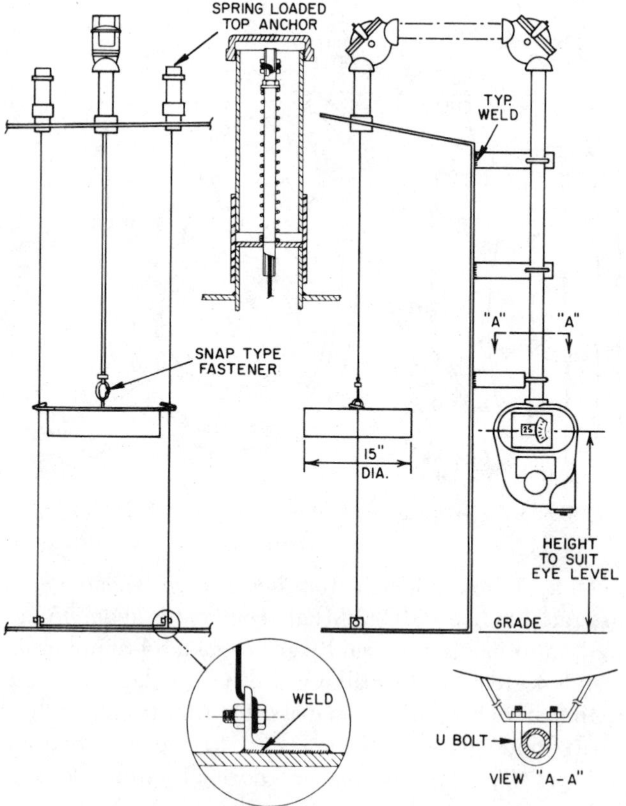

Fig. 3.17d Wire-guided float detector installation for low pressure tanks

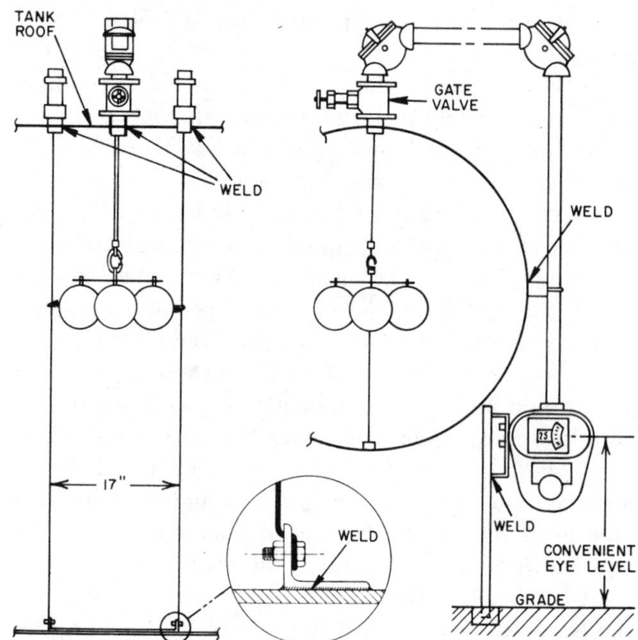

Fig. 3.17e Wire-guided float detector installation for high pressure tanks

Wire-guided Thermal Sensor

As discussed in Section 3.18, liquid level can be detected by using a pair of thermal sensors. Because liquid conducts heat better than vapor, the liquid level is bracketed by the two vertically displaced sensors when the lower one is cooler than the upper one. Figure 3.17f shows a schematic installation detail for a wire-guided thermal sensor. The sensor, which is heavier than the liquid being measured, is suspended from an armored control cable and guided by a wire attached to the top and bottom of the storage vessel. The control unit detects the position of the sensor relative to the liquid level, and issues step-up or step-down commands to the control cable take-up wheel until the lower sensor is in the liquid and the upper sensor is in the vapor. The system remains

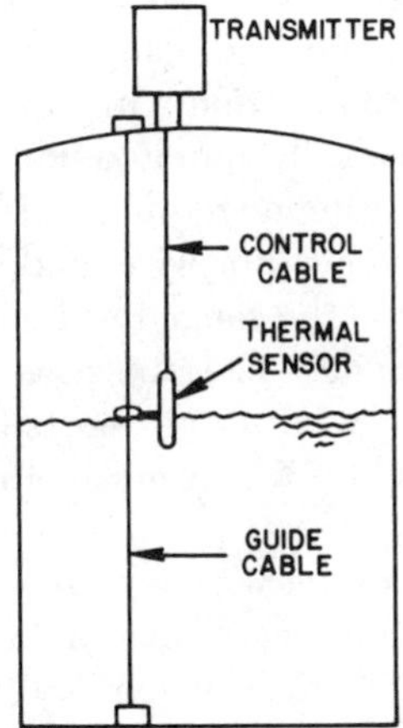

Fig. 3.17f Wire-guided thermal sensor

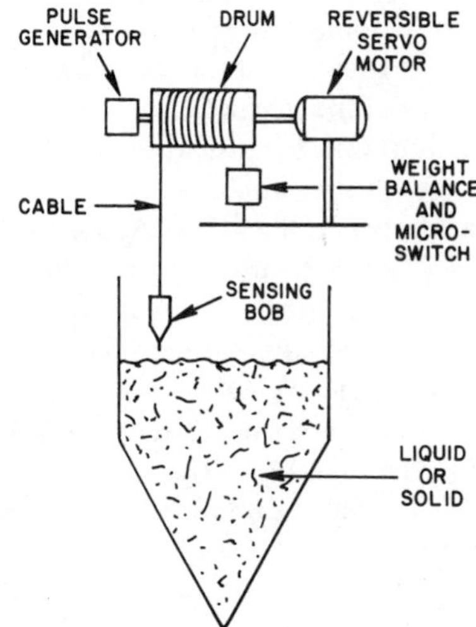

Fig. 3.17g Surface sensor

at rest as long as these conditions are met. When the sensor is moving, each stepping command adds or subtracts a length unit from the previous controller reading so that sensor position, and thus level, is accurately known.

The unique feature of this instrument is that the sensor is heavier than the liquid, allowing the unit to be lowered to the tank floor so that the control and counter circuitry can be zeroed. The controller contains a cable tension sensor to signal when the level sensor has hit bottom. A controller subroutine permits automatic zeroing. The sensor also can be equipped with a temperature detector to provide a thermal profile of the tank material. This is useful for accurate correction of level measurement, and can also be used to detect temperature inversions in cryogenic services.

Surface Sensor

Although this gauge was originally developed for solids level detection, it can be used for liquid level detection if equipped with a properly designed sounder. The gauging system is shown in Figure 3.17g; some sounder designs are shown in Figure 3.17h. As shown in Figure 3.17g, a sounder is suspended from the winding drum. Wire tension is continuously detected by the weight balance. A reversible servomotor rotates the drum on receipt of a starting signal, releasing wire until the sounder strikes the solid or liquid surface. When this occurs, the tension in the wire slackens, causing the weight balance to actuate a microswitch. After the momentary slackening of the cable, the microswitch reverses the motor, returning the sensing bob to its original reference position.

A pulse generator is coupled to the system to provide an input signal to the counter, which counts down from a preset maximum reference value in steps. When the sounder strikes the product surface, the solids level is displayed and, at the same time, the counter is automatically disconnected from the pulse generator. The reading stays on the counter until the next measurement. On receipt of a new start signal, the counter is reset to the maximum reference value and the measurement cycle is repeated. There are several ways in which the amount of cable paid out can be converted to pulses. In one design, the cable pays out over a measuring wheel with a six inch diameter. The measuring wheel drives a five-lobe cam that trips a stationary cam each time a lobe passes by. In this way, ten contact closures are produced for each foot of cable paid out. In a second high resolution design, the measuring wheel drives a disc that has fifty radial slots around its circumference. The slots are counted by a light beam and light sensitive transistor. One hundred pulses are generated for each foot of cable paid out. In either case, the level measurement reference is the top of the tank.

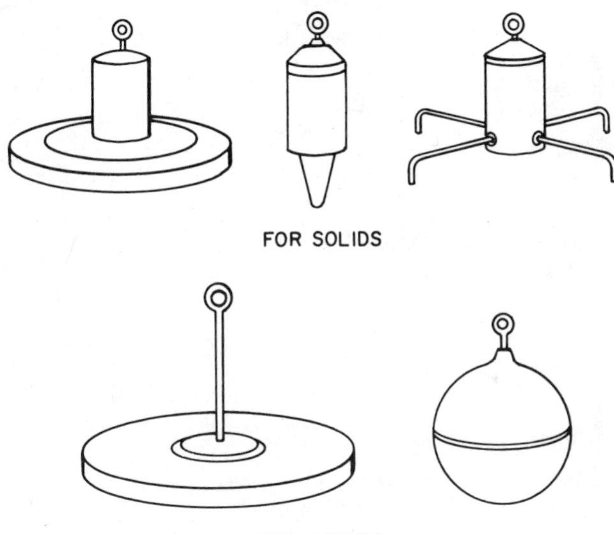

Fig. 3.17h Sounder designs

In dust-filled atmospheres, a solenoid operated pneumatic cleaning assembly is recommended to ensure reliable operation. As shown at the top of Figure 3.17i, the relative locations of the gauge nozzle and the solids inlet nozzle are important, because the sounder is used to take an average level reading. At no time should there be any contact between the filling stream and the sounder. If the inlet nozzle is in the center of the bin, the surface of granular products will tend to take the shape of a cone. The gauge, therefore, should be located to best obtain average level. As shown in the sketch, this requires the nozzle to be one sixth of the diameter from the bin wall for circular bins. The sketch on the bottom of Figure 3.17i shows the proper location for installations in rectangular bins.

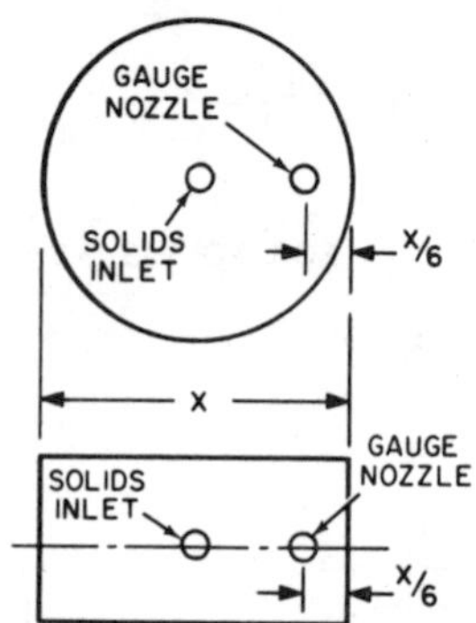

Fig. 3.17i Surface sounder installation for solids

There are at least two other tape level detector designs available. In one, the sensor is suspended on a cable and held a short distance above the liquid level. The distance is sensed by a proximity-type capacitance probe. (See Section 3.4 for a description of this mode of level measurement.) The control unit monitors the capacitance between the sensor and the liquid level, repositioning the sensor as the level changes. Sensor position, and thus level, is determined by measuring the amount of cable that has been paid out. In this respect it is the dame as the wire-guided thermal sensor already described. The second design uses a displacer mounted on the end of a cable. In this design, the displacer is continuously repositioned so that it is always immersed the same amount. Level is determined by the amount of cable paid out. The displacer design has cable weight compensation but is not compensated for changes in liquid density. Both the capacitance and the displacer designs are installed in stilling wells.

System Cabling

As previously mentioned, gauges covered in this section are used in conjunction with remote manually operated and automatically operated multiple tank gauging systems. Multiple gauging requires cables from each tank to the remote readout. For wire-guided float detectors, the shaft position on the gauge head must be transduced to an electrical signal. The objective in designing the cable system is to wire up all the tank gauges with as few wires as possible, which means that wires must be shared.

Figure 3.17j shows a wiring system used to obtain a level reading for any one of five tanks. Eight wires are used. By closing the tens switch #50 (at the top of the figure), one-half of the circuit to the relays at tanks 50 through 54 has been closed. Closure of any unit switch #00 through #04 will complete a circuit through the relay coil associated with the tank that is to be remotely metered. In the figure, switch #02 is closed; therefore relay R-52 is energized and relay contact R-52A is held closed. The remote readout can now obtain the level data that is available at the gauge head at tank 52. This technique allows a large number of tanks to be remotely monitored with relatively few wires. For example, a 100-tank installation can be made using 22 wires. The 22 wires would be comprised of 10 for the tens position, 10 for the units position, and two for the signal. A second group of 100 tanks can be picked up by adding only two more wires, one for the 100's series tanks and one for the 200's series tanks.

When the distance from the tanks to the remote readout is long, a satellite multiplexer may be considered.

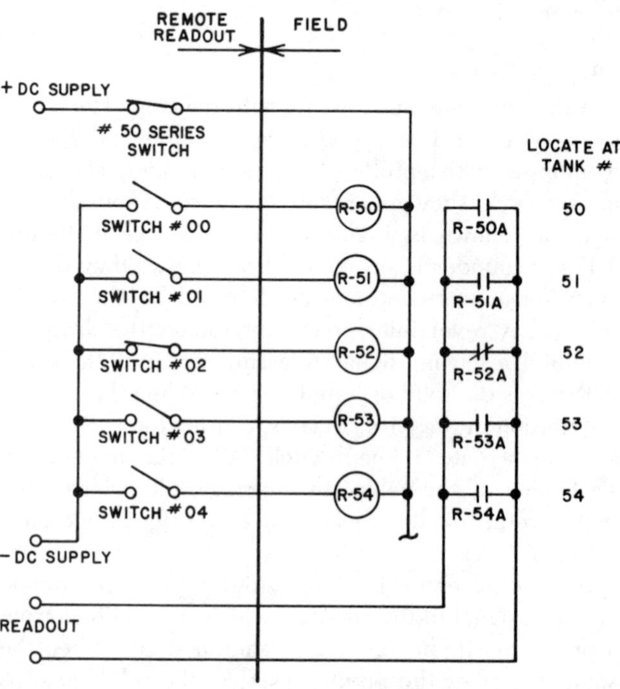

Fig. 3.17j Multiple tank system cabling

The satellite multiplexer collects level information for tanks in the immediate vicinity and transmits it, on demand, over two wires to the remote readout. The satellite multiplexing system might be used for a pipeline transmission installation where the various bulk storage facilities are hundreds of miles apart. The switches shown in Figure 3.17j constitute a manually operated multiplexer. These switches can be operated by a data logger or by a computer, making inventory monitoring completely automatic.

The reason for monitoring a tank farm with as few wires as possible is to reduce installation costs. It should be noted that in so doing the gauge head relays and much of the wiring are run in parallel. This means that a short to ground, an open wire or a malfunctioning gauge head can disrupt the entire system. Many tank farms are located in corrosive and/or humid environments. Therefore, particular care should be taken in the design and installation of the cable system, especially at the terminals in the gauge heads and junction boxes. Lightning strikes can be another source of trouble since most tank farm cabling systems are run overhead. Surge protection should be installed at each gauge head and at the remote readout. Associated loggers and computers should be electrically isolated. As an aid to troubleshooting a crippled system, isolating switches should be installed so that blocks of gauges can be separated from the system to enable a more rapid location of the fault.

Encoding

The wire-guided float detector level must be converted to an electrical signal by using the shaft rotation of the gauge head to drive encoding discs. Figure 3.17k shows one conversion method. The input shaft drives the "inches" wheel and the gear assembly at the left of the sketch drives the "foot" wheel. For purposes of this sketch, the level tape sheave and shaft are set up to rotate 180° for each one foot of level change. As the inch wheel completes one-half of a revolution, it steps the foot wheel up or down, corresponding to rising or falling level. Stepping of the foot wheel occurs when the notches on the inch wheel pass the gear. The wheels are coded such that a rotation corresponding to a 0.01 ft. level change presents a new and unique digital code to the code take-off assembly. Codes are available for foot, inch and fraction, or meter and millimeter readouts. Since the principle of operation is the same for all, only the foot, tenths, and hundredths will be covered.

The wheel has a number of concentric tracks on it, each track representing one digit of the digital word. The tracks are designed to produce the zero or one information needed for the digital word. This can be done in several ways. One way is to plate portions of the track with a conductor and allow a conducting brush to ride on the track. If the brush is on a conducting portion of the track, a current path will be formed and a one will

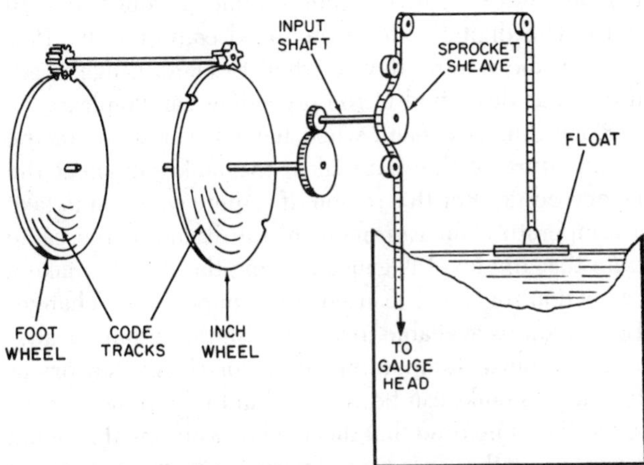

Fig. 3.17k Schematic of foot and inch wheel drives

be produced. If the brush is on a non-conducting part of the track, the current path will not be formed and a zero will be produced. Another encoding method is to use optical coupling. Portions of the tracks are plated with a reflecting material and a light is beamed on the tracks. If the beam hits a reflecting portion of the track, a light-sensitive transistor conducts, thereby producing a one. If the beam hits a nonreflecting portion, the transistor does not conduct and a zero is produced.

Figure 3.17l shows how a modified gray code could be used to produce a digital word unique for a given wheel position and thus a given level. (Only eight tracks are shown, instead of a 16-track arrangement that would be required for a ± 0.01 ft. resolution over an 80 ft. (24.3 m) span. Also, the tracks are shown as being linear, whereas in the actual configuration they would be on a closed circular track.) The shaded areas represent conducting portions of the tracks and the light portions are non-conducting. Thus, if the float were at a level cor-

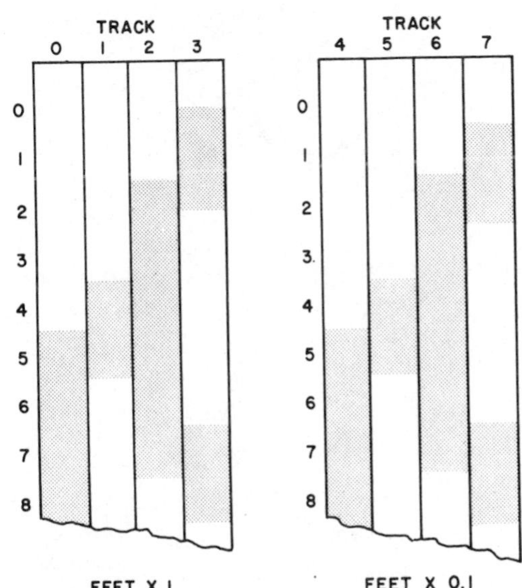

Fig. 3.17 l Encoding with a modified gray code

responding to 6.4 ft. the code produced would be 1010 0110. The digital code is produced continuously. It is read by the remote device when the tank gauge is addressed as described in the preceding paragraphs.

Since liquids expand when heated, variations in the temperature of the material in the tank will affect the level reading. For this reason, it is quite common to take a temperature measurement of the liquid at the same time that the level is gauged, using the data to make a correction of level caused by temperature change. Equipment is available to accomplish this automatically. The resistance temperature detector (RTD) sensors at the various tanks can be switched in to the remote readout at the same time that the level measurement is being made, using the same type of wiring arrangement shown in Figure 3.17j. There is a wide array of remote temperature and level readout equipment commercially available. The simplest is the manually operated unit with pushbutton random access to all tanks. These units generally display the number of the tank called, its level and temperature. The more complex systems are micro- or mini-computer based and can have automatic logging of temperature compensated level, and other features such as high and high high level alarm. Most systems can be readily interfaced with larger computers.

Conclusions

Automatic tank gauging systems are found in almost all tank farms of any size. They enable accurate inventory monitoring at a given time each day. The wire-guided float systems are most commonly used for liquid services. The design of the remote metering portion of these systems has been improved markedly over the past years, and these systems can be expected to perform satisfactorily if they are properly installed and maintained. The tape and float systems and the wire-guided capsule systems are more expensive and do not have as great a degree of field exposure as the wire-guided float types. However, the tape and float system has fewer moving parts. The wire-guided capsule can be zeroed, and can be used for temperature profiling. Some users regard the in-tank electrical circuitry required by these latter designs as a drawback. The surface sensor design has been used for some time for solids level measurement. There have been reports that the sensor can become detached from the take-up cable, contaminating product or ruining downstream equipment. A more rugged cable and a sound design should overcome this problem. The surface sensor would not appear to be the best selection for use on liquids stored in external floating roof tanks.

BIBLIOGRAPHY

"Floats on Pulleys Keep Track of Tank Levels." *Machine Design*, February 12, 1976.

Flow Sheet Symbol

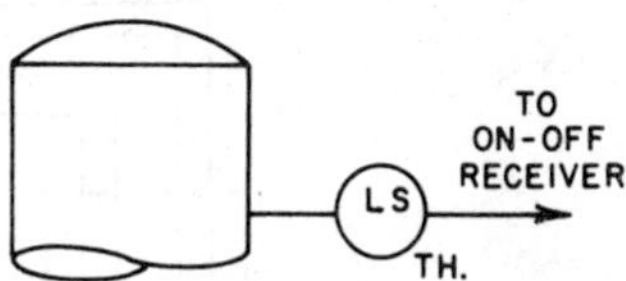

3.18 THERMAL LEVEL SENSORS

Design Pressure:	To 3000 PSIG (21 MPa)
Design Temperature:	Cryogenic to 300°F (149°C)
Materials of Construction:	Stainless steel standard, others available
Inaccuracy:	±¼ in. (±6.3 mm)
Cost:	\$150–\$600
Partial List of Suppliers:	Chromalox; Fluid Components, Inc.; Scientific Instruments

Thermal level sensors, used in on/off services, sense the difference in the thermal conductivity of the process materials. The simplest of these designs is shown in Figure 3.18a wherein the sensor is being used to detect low water level. The probe contains a resistive heater element that has a current flowing through it. A temperature switch is used to monitor the temperature of the probe. If the probe is submerged, heat generated by the resistive element will be carried into the water and the element will remain close to the water temperature. Once the level falls below the probe, the probe temperature will begin to rise because the water vapor or air above the water has a lower thermal conductivity. The temperature switch detects this rise and its associated switch interrupts power to the probe. This relay contact opening can be used for alarm or shutdown.

Figure 3.18b shows a second thermal level switch design. The measuring portion of the switch contains two temperature sensitive probes and a heater element. The probes are resistance elements that are located in a horizontal plane such that rising liquid level will contact them at the same time. The heater element is arranged so that the active probe is heated. When the probes are in the vapor phase, the active probe will be warmer than the reference probe and its resistance will be higher, causing the bridge circuit to unbalance. When the liquid covers both probes, their temperatures will be approximately equal to that of the liquid and their resistances will be essentially equal, balancing the bridge. The change in the balance of the bridge circuit can be detected and used for alarm or interlock actuation. The sensitivity of this design is quite good, so it may be used for liquid/liquid interface detection. This design may also be used as a flow switch, as described in Chapter 2.

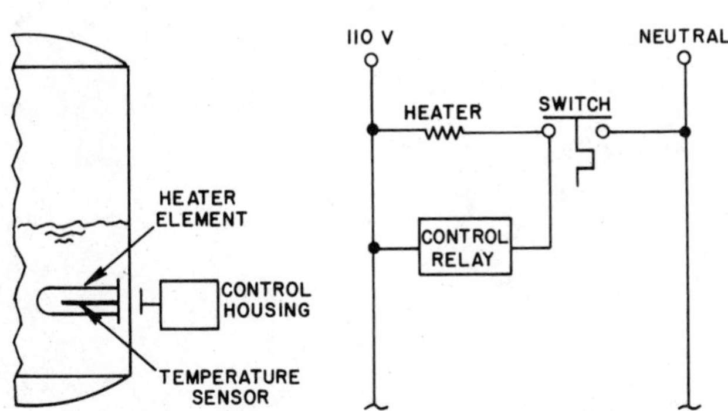

Fig. 3.18a Low level thermal conductivity switch

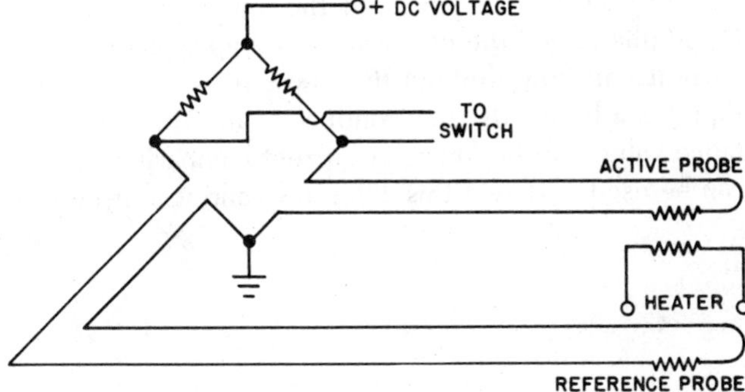

Fig. 3.18b Circuit schematic for a thermal conductivity level switch

Figure 3.18c illustrates a third design. The probe, which can be a stainless steel capsule, contains two resistive elements that are vertically displaced from each

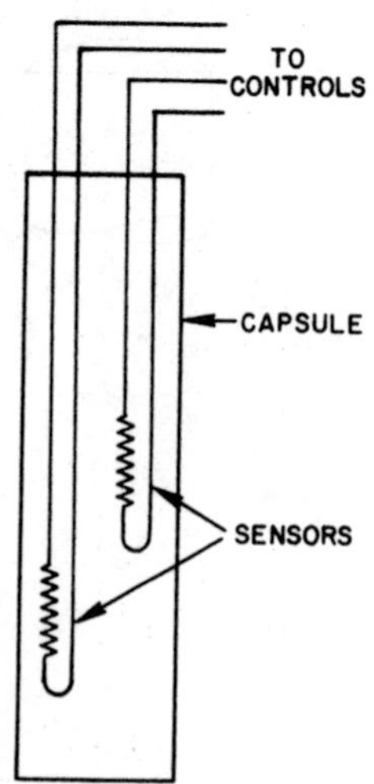

Fig. 3.18c Vertically displaced thermal sensors for liquid/vapor or liquid/liquid interface

other. A voltage is applied to both elements. If they are both in the vapor phase or both in the liquid phase, heat transfer from the resistors to the process will be the same, their resistances will be the same, and the current flow to each will be equal. If the lower one is in the liquid phase, and the upper one is in vapor phase, more current will flow through the lower resistor, since the liquid is a better thermal conductor, and thus its resistance value will be lower. A current comparitor circuit can be used to detect this difference, and its output can be used for alarm or control circuitry actuation. The unique feature of this design is that the capsule can be cable-suspended in a tank, and the sensor output can be used to drive a cable take-up motor. In this fashion the assembly may be used as a continuous level detector for liquid/liquid or liquid/vapor interface in a tank. This system can be used for high accuracy inventory monitoring in storage tanks; additional features are covered in Section 3.17, Tape Level Gauging.

Conclusions

Thermal level detectors cannot be used on caking or plugging materials, nor can they be used when additional heating will cause product degradation. Outside of these limitations, thermal detectors are appealing because they contain no moving parts, they are based on sound, easily understood principles, and they can be very sensitive. The unit shown in Figure 3.18a should only be used in secondary nonhazardous services, since it uses 120 volt power and depends on proper operation of a temperature switch in order to actuate the associated controls.

BIBLIOGRAPHY

Cantor, H.P. and Williams, O.M., "Hot Wire Sensor for Liquid Level Detection." *Scientific Instruments*, December 1976.

"Monitor Temperature and Measure Liquid Level Using a Thermistor." *Control Engineering*, July 1977.

"Thermistor Probes Measure Tank Level." *Control Engineering*, March 1975.

3.19 TIME DOMAIN REFLECTOMETRY

The principle behind Time Domain Reflectometry (TDR) is that a portion of an electrical signal will be reflected back toward its source by a discontinuity in the cable that is carrying the signal. By measuring the time that it takes for the signal to reach the discontinuity plus the time for the reflected return, the discontinuity can be located very accurately. Systems are available that can pinpoint a fault in several miles of cable to within a few feet. By observing the characteristics of the reflected signal it is also possible to determine the nature of the fault, such as open circuit, short to ground, moisture in the insulation, and so on. These units are used routinely by power and communications companies to locate problems in their transmission systems.

Some work has been done in an attempt to adapt TDR to level detection. One of the applications that has been studied is the detection of the interface between a vapor and a very hot bed of fluidized solids by looking for a discontinuity at the interface in a cable suspended in the vessel. These efforts apparently have not been successful. It is probable that the accuracy of the system would be adversely affected by temperature change and vessel geometry changes.

BIBLIOGRAPHY

Fuchs, Walter and Yavorsky, Paul M., "New Method for Measuring Bed Levels." *Chemical Engineering*, June 24, 1974.

Flow Sheet Symbol

TO CONTINUOUS RECEIVER

LT

S

TO ON-OFF RECEIVER

LS

S

3.20 ULTRASONIC LEVEL DETECTORS

Design Pressure:	To 3000 PSIG (21 MPa) for on-off, to 150 PSIG (1 MPa) for continuous
Design Temperature:	−100 to 300°F (−73 to 149°C) for on-off and −30 to 150°F (−34 to 66°C) for continuous
Materials:	Wide selection of corrosion resistant materials
Inaccuracy:	±⅛ in. (±3 mm) for on-off, ±2–3% for continuous
Range:	To 100 ft. (30 m) for in-air-type, to 2000 ft. (610 m) for in-liquid-type
Cost:	\$300–\$400 for on-off; \$1800 for continuous
Partial List of Suppliers:	Controlotron Corp.; Delavan Electronics, Inc.; Endress-Hauser, Inc.; Interface Detector Co.; National Sonics; Princo Instruments, Inc.; Pro-Tech Environmental Instruments; Rexnord Instrument Products; Sirco Control; Wesmar

Introduction

Ultrasonic level devices can be used for both continous and point measurement. The point detectors—for measurement of gas/liquid, liquid/liquid, liquid/foam or solid/gas interfaces—can be grouped by design into damped sensor and on-off transmitter categories; and by method of packaging as single element and two element units. The continuous level detector designs can be categorized as under-liquid sensors and above-liquid sensors. Most designs use a 20 kHz or higher oscillator circuit as the ultrasonic signal generator. Some designs incorporate filters or discriminatory circuitry in electronics to prevent false readings that might be caused by random noise. Each of these basic design variations will be discussed separately.

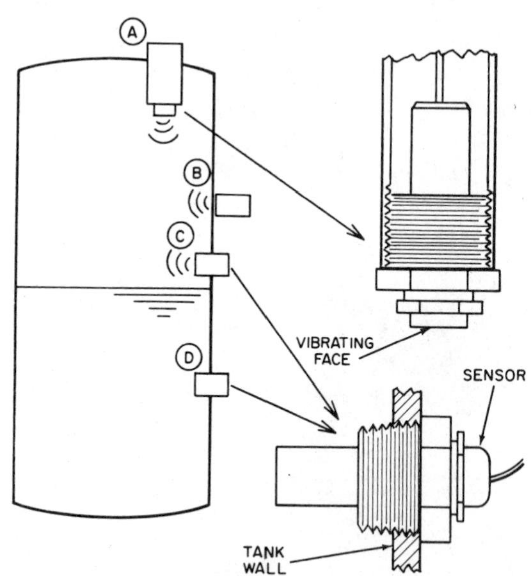

Fig. 3.20a Dampened ultrasonic sensors

Dampened Sensor-type Level Switches

The devices in this category operate on similar principle to that of the vibrating reed switch. As long as the sensor face is in the vapor space of the tank, it vibrates at its resonant frequency but is dampened out when the process material contacts it. Some designs incorporate a piezoelectric crystal in the vibrating tip. Figure 3.20a shows some of the elements and their installation. Four units are shown in the figure. "A" notes the top entry installation where the vibrating face is in the vapor space (and therefore undampened). This design can be repositioned manually or automatically for flexible adjustment of control point. Installation "B" is a unique design in that it does not penetrate the tank wall and thus is not in contact with the process fluid. When the liquid rises to the opposite side of the wall, the transducer is dampened and a switch action is indicated. Designs "C" and "D" show the side-mounted switch elements, one dampened, the other in undampened condition. These units are normally limited to liquid service because the dampening effect of solids is not sufficient in most cases. Design "B" can be used on any liquid, while the others are limited to clean, non-coating fluids.

On–Off Transmitter-type Level Switches

These ultrasonic switches contain transmitter and receiver elements. The transmitter generates pulses in the ultrasonic range and the receiver detects these pulses if they are transmitted through the media in which the probe is located. The transmitter and receiver can be mounted on the same probe or they can be located on the opposite sides of the tank. Figure 3.20b illustrates some of the variation of this design. Installation "A" shows a design where the transmitter and receiver are packaged separately. This design transmits in air; the switch will actuate when the ultrasonic sound beam is interrupted by the rising process material. Reflectors are installed to narrow the sonic beam angle when the distance between source and receiver is more than ten feet (3 m). Installation "B" is a single probe design in which the pulses generated by the transmitter will be sensed by the detector only if they are submerged in a noncompressible liquid. The pulses are not transmitted in the vapor space. Design "C," which is similar to "B," transmits only in liquid. When fluid is present in the gap of the single probe, the ultrasonic sound beam is received by the detector in the tip and signals the presence of liquids. Probe "D" is a multipoint variation of "C" allowing the measurement of both high and low levels by the same probe. Design "E" is mounted at about 10 degrees from the horizontal, and it can be used for detection of liquid/liquid interfaces. The ultrasonic sound beam generated by the transmitter crystal will be detected by the receiver crystal if the probe is in only one liquid. If an interface is present in the probe cavity, the interface will reflect the signal, preventing it from reaching the receiver.

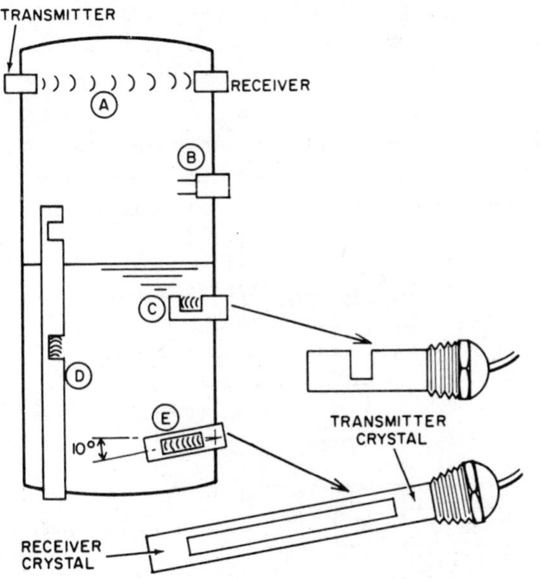

Fig. 3.20b Transmitting ultrasonic point sensors

All of the above discussed designs are applicable to clean liquid service, and none are particularly suitable for slurry or coating services. Only design "A" will detect the level of solids. In connection with solids level detection, it should be noted that these devices, being point sensors, will not take into consideration the angle of repose during filling or emptying, nor will they detect ratholes, arches or bridging.

Figure 3.20c illustrates a special application of ultrasonic level detection for filling nozzles. The nozzle itself contains a transmitting and receiving element to detect when liquid reaches the nozzle, at which point flow is shut off.

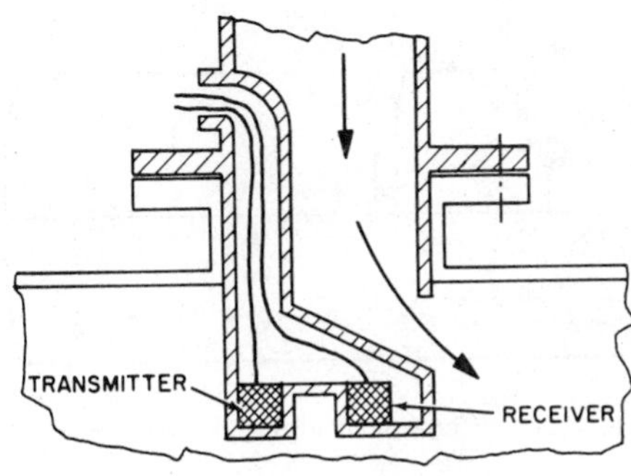

Fig. 3.20c Ultrasonic filler nozzle

Continuous Level Detectors

The principle of operation of these units is very similar to that of the well-known echometers used to measure the depth of wells. In that design, a blank shell is fired; the time needed for the echo to return is converted to an indication of the depth. The continuous ultrasonic level detector (SONAR) measures the time required for an ultrasonic pulse to travel to the process surface and back. The source is an oscillator-type ultrasonic speaker, and the receiver, in most designs, is a metal disc which is both electrically and mechanically resonant. The transducer can be mounted either under or above the liquid level.

Figure 3.20d illustrates some of the design features and possible installations. Installation "A" shows a two-element continuous detector, where the transmitting and receiving transducers are packaged separately. This device transmits in air. The time required to receive the ultrasonic reflection from the surface is the measure of the vapor depth of the space, which is an indirect indication of level. Another version of an "in-air," continuous level system is illustrated in "B." The transducer and receiver are packaged as a single unit. The transducer generates short bursts of ultrasonic energy, and while the acoustic energy is being produced, the receiver is blanked off. When the ultrasonic waves are on their way, the receiver gate is opened to detect the echo. Mounting the transducers in the vapor space has the advantage that the instrument does not contact the process materials, but it has the disadvantage that some energy is lost in traveling through the vapor space.

On liquid level applications, the aiming angle must be within ±2 degrees from the vertical. When measuring level of solids, the angle of repose should be tested. Unit "B" in Figure 3.20d is shown equipped with a calibration bar. In this design, the electronics package has a level detector and a calibration detector. Since the length of the bar is known, the unit is being calibrated continuously. Units so equipped can be very accurate, provided that the density of the material in the vapor space is uniform.

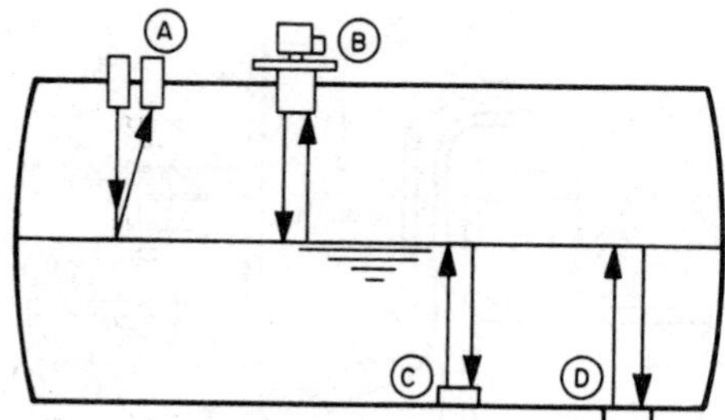

Fig. 3.20d Continuous ultrasonic level detectors

In installation "C," the time for the ultrasonic echo is a true indication of level. The transducer can also be mounted on the outside of the tank ("D"), with the added advantage that the sensing element does not penetrate the tank. Designs "C" and "D" are applicable to continuous detection of clean liquid levels, while designs "A" and "B" also can be used to measure the level of solids. By using several sensors in the same bin, a visual profile can be obtained, showing the angle of repose and indicating if the bin is being filled or discharged.

Figure 3.20e shows an interesting application of the continuous ultrasonic level detector. When hydrocarbons are stored in salt dome wells, the hydrocarbon rests on a brine layer. When additional hydrocarbons are pumped in, the brine is displaced; the hydrocarbons are recovered by displacing them with brine. For reasons of safety, inventory monitoring and cavity use, it is important to know where the hydrocarbon/brine interface is located. This device is suitable for finding the interface.

Ultrasonic level detectors have a good application history and have been used succesfully on many troublesome solids. The cost of a single, continuous ultrasonic level detector is relatively high, but when it is used to scan the level in many storage silos, the unit cost becomes more reasonable. Figure 3.20f shows a twenty-four bin system. In this installation, each of the transducers is connected to an automatic scanning console.

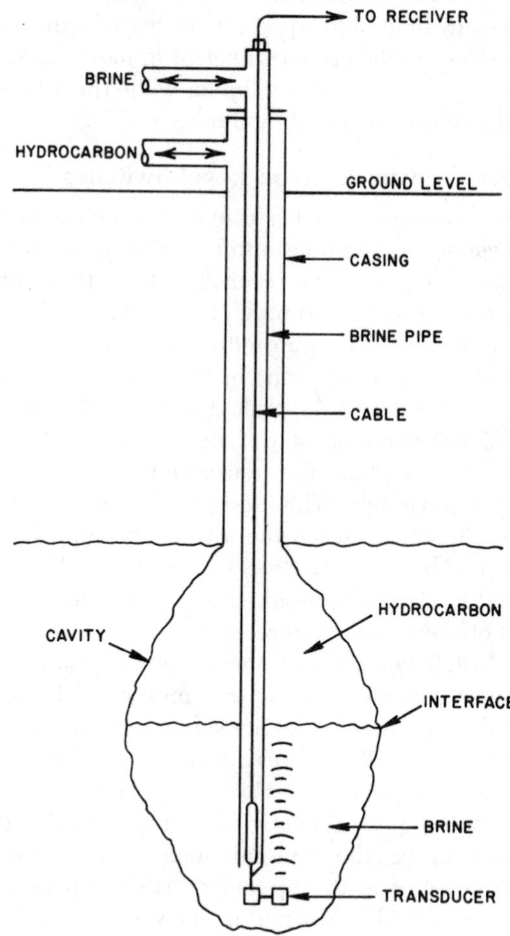

Fig. 3.20e Interface detector for hydrocarbon storage cavity

The output from the console is sent to individual readout devices, which can be indicators or recorders, with or without alarm functions. The scanning frequency can be adjusted to suit the requirements of the process conditions.

Conclusions

The ultrasonic technique for level detection is a proven, reliable method. Its advantages include the absence of moving parts and the capability for continuous

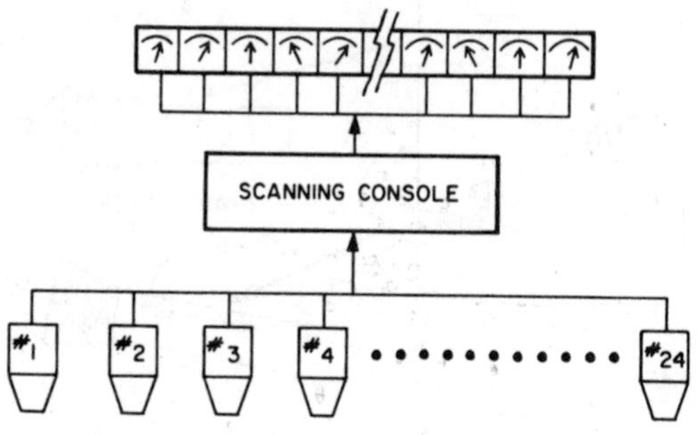

Fig. 3.20f Ultrasonic silo scanning system

measurement without contacting the process material. In some designs, penetration of the tank can also be avoided. The measurement does not depend on detecting various process properties and thus the reliability of the instrument is unaffected by variations in composition, density, thermal or electrical conductivity, capacitance or other variables. The limited accuracy of the continuous unit can be improved to ± 0.1 percent by the installation of a temperature compensator and subsequent calibration. The ultrasonic level detectors are second only to the radiation-type designs in flexibility and ease of installation, and they can approach the accuracy of some accounting-type level instruments. This does not imply that ultrasonic units should be considered where conventional devices have a good history of reliable operation, but it does suggest consideration where difficult applications are involved.

BIBLIOGRAPHY

Andsager, R.L. and Knapp, R.M., "Acoustic Determination of Liquid Levels in Gas Wells." *Petroleum Technology,* May 1967.

Kalmus, H.P., "New Ultrasonic Liquid Level Gauge." *Review of Scientific Instruments,* October 1965.

Kaminski, R.K., "Sonics Measure Level." *Instrumentation Technology,* December 1967.

"Liquid Levels are Measured by Ultrasonic Signal." *Product Engineering,* October 1975.

Reason, J., "Ultrasonics: Practical Plant Engineering Tool for Level Control." *Plant Engineering,* December 28, 1972.

"Ultrasonic Monitor/Sensor Controls Sludge Flow." *Water and Sewage Works,* June 1978.

"Ultrasonics Pinpoint Liquid Levels." *Automotive Engineering,* September 1975.

Weiss, W.I., "Ultrasonic Level Control." *I.S.A. Journal,* December 1966.

Wolff, J., "Ultrasonics in the Sewage Industry." *Water and Sewage Works,* June 1973.

3.21 VIBRATING REED SWITCHES

Flow Sheet Symbol

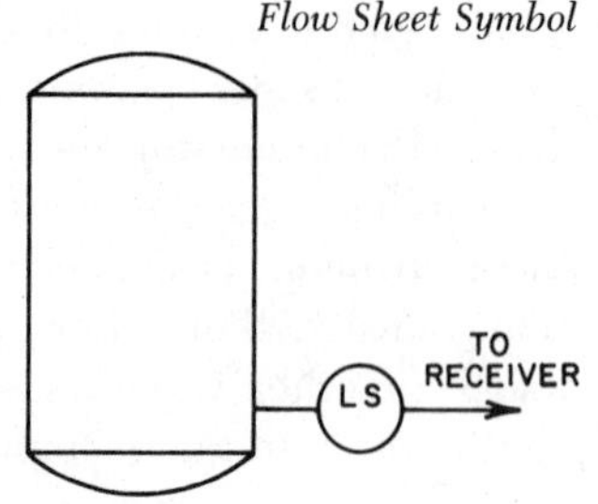

Design Pressure:	To 3000 PSIG (21 MPa) at 100°F (40°C)
Design Temperature:	− 150 to 300°F (− 100 to 149°C)
Materials of Construction:	Wide selection of corrosion resistant materials
Sensitivity:	Materials with bulk density 1.0 lbm/ft³ (0.016 g/cc) and up
Inaccuracy:	± ⅛ in. (±3 mm)
Cost:	\$300 to \$500
Partial List of Suppliers:	Automation Products, Inc.; Endress-Hauser, Inc.

One design of the vibrating reed level switch is shown in Figure 3.21a. The unit consists of a driver, paddle, and pickup. The driver coil induces a 120 cycle per second vibration in the paddle, that is dampened out when the paddle is covered with process material. The pickup end contains a permanent magnet and a coil that generates a millivolt output signal when the paddle is vibrating. When the paddle is covered, the signal decreases and a control relay is de-energized. A welded pressure seal is made at the node points. The device will detect liquid/liquid, liquid/vapor, or solid/vapor interfaces because the switch is sensitive enough to detect relatively small changes in the density of the surrounding material.

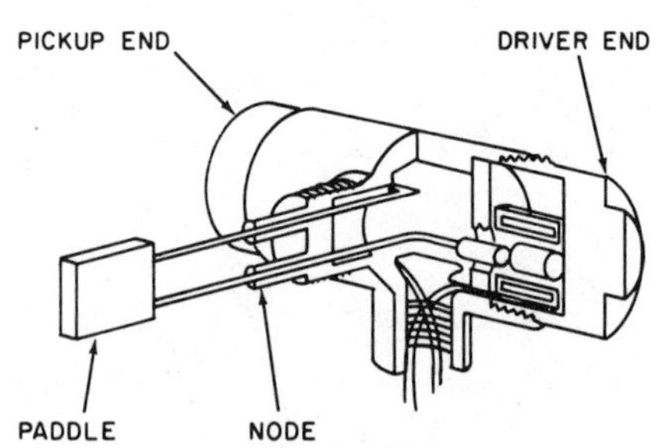

Fig. 3.21a Vibrating reed switch

The characteristics of this switch are given in Figures 3.21b and 3.21c. Figure 3.21b is a plot of millivolt signal strength versus paddle coverage. Curve #1 describes the switch behavior in flour, #2 in water, #3 in polyethylene pellets, and #4 in granular sugar. From the figure, it can be noted that in all four cases the switch

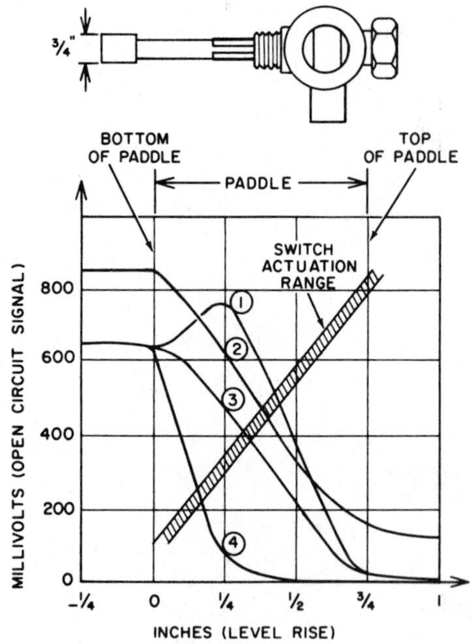

Fig. 3.21b Reed switch actuation ranges for liquid service

will actuate before the paddle is fully covered by the process material. For example, in the case of water (#2), switch actuation occurs when the water level is about 0.4 in. (10 mm) above the bottom of the paddle.

The curves in Figure 3.21c refer to granular powders at various densities. Curve #1 is for 60 lbm/ft³ (0.96g/cc), #2 for 50 lbm/ft³ (0.8 g/cc) and #3 for 40 lbm/ft³ (0.64 g/cc). As would be expected, the lighter the powder, the more has to build up before switch actuation

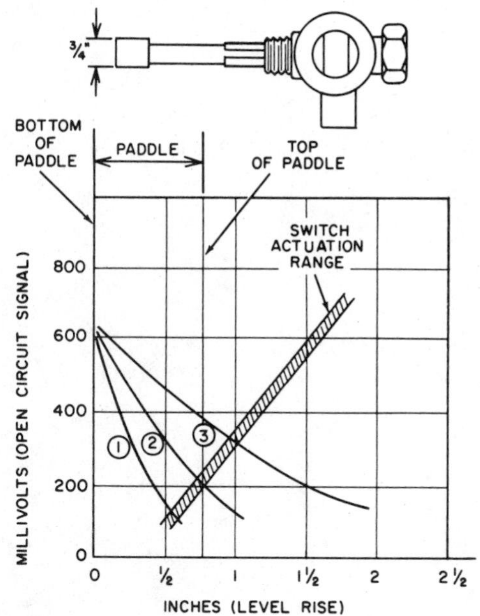

Fig. 3.21c Reed switch actuating ranges for heavy solids

takes place. In case of the 40 lbm/ft^3 powder (#3), the switch will actuate only when the level builds up to ¼ in. (6 mm) above the top of the paddle.

The vibrating reed level switch can be used to detect both rising and falling level, and it can be installed in tanks or pipelines. Its use for measuring density and viscosity are discussed in Sections 6.13 and 7.13 respectively. If the process material has a tendency to adhere to the paddle, the build-up can be removed by periodically purging the in-line assembly through a purge well, but normally the unit described is not recommended for applications where build-up is probable. The pair of sensing wires between probe and receiver should be shielded and grounded at both ends. Supply voltage variations between 105 and 125 volts will not have harmful effects.

When used on wet powders, the vibrating paddle has a tendency to create a cavity in the granular solids. If this occurs, false level detection will result, because the vibration amplitude will soon be the same as if the paddle were in the vapor space. Where the solids bins are purposely vibrated and the vibration frequency is close to 120 cycles per second, the reliability of level measurement can be destroyed.

An alternate design is the self-sustaining resonant frequency probe. When the probe is in the vapor space, an electro-mechanical oscillator is formed, causing the probe to vibrate at its natural mechanical resonant frequency of approximately 100 cycles per second. One coil drives the probe and a second coil monitors the vibration and generates a corresponding AC voltage signal. This feedback signal is amplified and reapplied to the drive coil, sustaining the mechanical vibration. When the process fluid rises to cover the probe, it dampens the vibration, causing the feedback signal and drive voltages to collapse, and oscillation ceases. A relay located in the control unit detects the state of oscillation and provides corresponding contact closures. The advantage of the self-sustaining resonant frequency probe design is that material build-up on the probe has no effect, because the coating layer changes the natural frequency of the probe but does not collapse the oscillation. Therefore, the sensor is suitable for slurry or similar services. This design is also applicable to low density solids level detection.

Conclusions

Vibrating reed switches are reliable devices that operate on a readily understood principle. Since they can detect the presence of materials with a bulk density as low as 1.0 lb./ft^3 (0.016 g/cc) they have a wide range of applications, particularly in solid level detection. Their major limitation is that the switch setting cannot be easily changed.

BIBLIOGRAPHY

"Tuning Fork Notes Level of Salt in CEGB Brine Tanks." *Process Engineering*, January 1975.

Chapter IV

TEMPERATURE MEASUREMENTS

T. J. Claggett
W. A. Clayton
B. G. Lipták
R. W. Worrall

CONTENTS OF CHAPTER IV

Table IV
ORIENTATION TABLE FOR TEMPERATURE SENSOR

Section	Type	Temperature Range (°C: −268, −184, −73, −18, 38, 93, 260, 538, 1094, 2760, 5538; °F: −450, −300, −100, 0, 100, 200, 500, 1,000, 2,000, 5,000, 10,000)	Available Span			Accuracy		Cost ($)			Sensor Size			Available With			Stability	Repeatability	Response Time	Sensitivity	① Interchangeable	Linear	② Complete System	③ Maximum Distance to Readout in Feet (0.305m)
			Under 100°F (38°C)	Between 100–1000°F	Above 1000°F (538°C)	% Full Scale	Best Attainable in °F (°C = 5°F/g)	Under $100	Between $100-1000	Above $1000	Small	Medium	Large	Recorder	Indicator	Controller								
4.2	Bi-Metallic Elements		√	√		1–2	0.2	√			√	√		√	√	√	E	F	G	G		N	√	N
4.3	Color Indicators		√				8	√							√					F			√	N
4.4	Filled Elements																							
	Liquid		√	√		0.5–2	0.1	√	√		√	√		√	√	√	E	F	F	G		√	√	40
	Vapor		√	√		0.5–2	0.6	√	√			√		√	√	√	E	F	G	F		N	√	200
	Gas			√	√	0.5–2	1.2	√	√				√	√	√	√	E	F	F	F		√	√	150
	Mercury		√	√	√	0.5–2	0.25	√	√		√	√		√	√	√	E	F	F	G		√	√	50
4.5	Glass Stem Therm.		√	√		0.1–2	0.01	√				√			√		E	E	G	G	√	√	√	N
4.6	Misc.—																							
	Spectroscopy				√	1				√			√		√		G	G,E	G			N	√	
	Paramagnetic Salts		√			1	0.005			√			√		√		G	G,E	E	F,G		N	√	
	Diodes		√			2			√		√	√			√		G	G	G	G		√	N	
	Transistors		√			2	1		√		√	√			√		G	G	G	G		√	N	
	Liquid Crystals		√				1	√					√		√		E	G,E	G	E,G			√	
	Carbon Resistors		√				0.5		√		√	√			√		F,G	G	G	G		N	N	
	Pneumatic Probe			√		2			√				√		√		F	F	F	F		N	N	
	Fluidic Sensor			√		2			√				√		√		F	F	F	F		N	N	
	Ultrasonic		√	√	√	5				√			√		√		G	F,G	E	E,G		N	N	
	Thermography			√			1	√	√						√					F			√	

Table IV *Continued*

ORIENTATION TABLE FOR TEMPERATURE SENSOR

4.7	Pyrometric Indicators		√				8	√			√	√			√					F			√	N
4.8	Pyrometers Manual Optical				√	1–2	10		√	√		√	√		√				F,G			N	√	N
	Infrared			√	√	1–2	5			√		√	√	√	√	√			E			N	√	1000
4.9	Pyrometers Radiation				√	0.5–1	5			√		√	√	√	√	√	G	G,E	G	F		N	N	2000
4.10	Quartz Crystals		√	√		0.1	0.2			√		√	√	√	√		E	G	G	E	√	√	N	1000
4.11	Resistance Bulbs Nickel		√	√		0.1	0.2	√	√	√	√	√		√	√	√	G,E	E	G,E	G,E	√	N	N	1000
	Platinum		√	√	√	0.1	0.05		√	√	√	√		√	√	√	G,E	E	G,E	G,E	√	√	N	3000
4.13	Thermistors		√	√		0.1	0.2		√		√			√	√	√	G,E	G	G,E	E	√	N	N	3000
4.14	Thermocouples Type T			√		0.3–1	1	√	√		√	√		√	√	√	G	G,E	E	G,E	√	N	N	3000
	Type J			√	√	0.3–1	2	√	√		√	√		√	√	√	G	G,E	E	G,E	√	N	N	3000
	Type K			√	√	0.3–1	2		√		√	√		√	√	√	G	G,E	E	G,E	√	√	N	3000
	Type R & S			√	√	0.3–1	3		√		√	√		√	√	√	G	G,E	E	G,E	√	N	N	3000

Terminology

N—No or None
E—Excellent
G—Good
F—Fair

(1) Interchangeable sensor, without recalibration of entire system.
(2) System is complete when sensor and readout is sold as single unit. When several readouts can be used with the same sensor, system is not considered to be complete.
(3) Without special compensation.

━━━━ Recommended

▬ ▬ ▬ ▬ Avaliable but not recommended

4.1 APPLICATION AND SELECTION

Temperature is an expression denoting a physical condition of matter, just as are mass, dimensions and time. Yet, the idea of temperature is a relative one, arrived at by a number of conflicting theories. The classical theory depicts heat as a form of energy associated with the activity of the molecules of a substance. These minute particles of all matter are assumed to be in continuous motion which is sensed as heat. Temperature is a measure of this heat.

To standardize on the temperature of objects under varying conditions, several scales have been devised. The Fahrenheit scale arbitrarily assigns the number 32 to the freezing point of water and the number 212 to the boiling point of water and divides the interval into 180 equal parts. The Centigrade or Celsius scale calls the freezing point of water 0 and its boiling point 100.

In line with the classic theory, some relation to the point where molecular motion is at a minimum had to be established, and the Kelvin scale in terms of absolute zero, related to thermodynamics and using Centrigrade divisions, was drawn. Zero Kelvin was determined to be −273.19°C. The Rankine scale places its zero at −459.61°F using Fahrenheit divisions in the same arbitrary way in which Lord Kelvin used the Celsius scale.

The International Practical Temperature Scale

The International Practical Temperature Scale is the basis of most present-day temperature measurements. The scale was established by an international commission in 1948 with a text revision in 1960.[1] A revision of the scale was formally adopted in 1968 and is reproduced in Table 4.1a. The scale is defined by reproducible temperature points established by physical constants of readily available materials. Interpolation between these fixed

Table 4.1a
PRIMARY TEMPERATURE POINTS AND VALUES OF IPTS OF 1968

Temperature °C	*Defining Fixed Point*	*Interpolating Instrument*	*Accuracy of Realizing (IPTS) °C*
−183.09	Oxygen, liquid-vapor equilibrium	Platinum resistance thermometer	0.02
0.00	Water, solid-liquid equilibrium	Platinum resistance thermometer	0.0002
0.01	Water, triple point	Platinum resistance thermometer	0.0002
100.00	Water, liquid-vapor equilibrium	Platinum resistance thermometer	0.0005
419.58	Zinc, solid-liquid equilibrium	Platinum resistance thermometer	0.002
444.67	*Sulfur, liquid-vapor equilibrium	Platinum resistance thermometer	0.002
961.92	Silver, solid-liquid equilibrium	Platinum-platinum 10% rhodium thermocouple	0.2
1,064.43	Gold, solid-liquid equilibrium	Platinum-platinum 10% rhodium thermocouple	0.2

*Zinc point preferred

points is made by several standard measuring instruments summarized in Table 4.1a.

For a complete graphical presentation of the conversion factors to bring temperatures based on the IPTS of 1948 into agreement with the IPTS of 1968, the reader is referred to NBS Special Publication 300 (Volume 11).[2]

Orientation Table

The range in temperature within the universe varies from the near zero of black space to the billions of degrees in the nuclear fusion process deep within stars. But the practical range on earth can be considered as extending from 1°R upward about five decades to around 20,000°R. This is still a tremendous range and no single sensor could possibly cover it. Therefore, one of the restrictions on the temperature sensor concerns the temperature range over which it can stay reasonably accurate. Table IV (at the front of this chapter) has been prepared to show the approximate temperature ranges of each of the sensor types. The many types of sensors are listed along the left hand vertical column while some of their characteristics are shown horizontally across the top. If it is not known what general type of sensor will do a specific job then the table will help to point the way to selection.

Once the class of sensor has been found, a number or descriptive indication in the table will give a rough idea of the applicability of that design. Letters refer to a relative consideration of the sensor for the category listed. In order to choose a sensor for a job all columns should be considered.

When the possible choices of selection have been narrowed down to a few instrument types, the reader should turn to the corresponding sections of this chapter. In the front of each section there is a summary of basic features, such as accuracy, cost, range, etc. Inspecting these briefly, he can determine if the instrument generally meets his requirements or not. If it does, he should read the section, which describes the design and its available variations in detail. If some of the features are unacceptable, he should proceed to the next choice noted in the orientation table. Additional hints on selecting a temperature sensor may be found in the conclusions.

Selection of Temperature Sensors

Temperature sensors should be selected to meet the requirements of specific applications. Table 4.1b should serve to assist the reader in this task.[3] If the application engineer determines the required temperature level, the nature of the information required (point or average temperature) and the nature of the process environment, this table can be used to determine the suitability of various sensors to that application. The most difficult temperature measurement applications are those where high temperatures are to be detected within a hostile environment, such as that which exists within a fluid-bed coal gasifier.

High Temperature Service

At high temperatures in the 2000–3000°F (1100–1650°C) range, the problem of thermowell drooping becomes a serious consideration. If direct temperature measurement is to be used (where the temperature sensor is physically inserted into the process), metallic thermowells are likely to droop, resulting in problems of replacement and maintenance. If ceramic thermowells are used instead of metallic thermowells, erosion becomes a problem in fluid bed-type environments due to the relatively low surface strength of the ceramic materials.

If the thermowell is flush with the inside diameter of the reactor, its life is likely to be increased, but only at the expense of having a less representative temperature measurement because of the greater influence of wall temperatures. Another way of increasing the reliability of the overall installation is to provide automatic, scheduled, preventive maintenance replacement of the thermowells. This can be done on a redundant basis, so that at any one time two or even three temperature sensors are in operation. A voting system can be arranged in critical locations so that the reading of the one sensor which disagrees with the other two is disregarded. While it is desirable to replace the thermowells without shutting the process down, and while it is desirable to replace them before they fail, all this effort is not inexpensive, and over a period of years of operation it can easily approach the initial cost of indirect sensors, such as infrared pyrometers.

The long range trend in high temperature measurement seems to favor the indirect, noncontacting-type measurements. The unit cost of such installations can be lowered by making multiple readings, using fiber optics to connect many sensing points to the same infrared source and the same receiving electronics package. This seems to be the most promising solution to the problems of high temperature measurements in hostile environments.

Factors Affecting Temperature Measurements

All temperature measurements fall into two basic categories including measurements of fluid temperature (liquid, gas, vapors and slurries) and measurement of temperature on or within solid bodies. The discussion immediately following has been limited to sensing devices employed by attaching to or immersing in the substance of interest.

Temperature Range

Specification of temperature range can identify the type of sensing device which must be used. This is apparent from Table 4.1c. For example, if the temperature range for a measurement is 1,000 to 3,000°F (538 to 1650°C), the choice of temperature sensors is narrowed

TABLE 4.1b
TEMPERATURE SENSOR SELECTION TABLE

Measured Temperature	Under 500°C								Above 500°C							
Reading[1]	Point				Average				Point				Average			
Hostile Environment[2]	No		Yes		No		Yes		No		Yes		No		Yes	
Interference[3]	No	Yes	No	Yes	No	Yes	No	Yes	No	Yes	No	Yes	No	Yes	No	Yes
Sensors[4] Color Indicators	G(L)	G(L)	G(L)	G(L)					G(L)	G(L)	G(L)	G(L)				
Bimetallic Units	G	F	G(P)	F(P)												
Filled Elements	G	F	G(P)	F(P)	G(L)	F	F(P)	F(P)								
Resistance Bulbs	E	F	E(P)	F(P)	E	F	E(P)	F(P)	E	F	G(P)	F(P)	G(L)	F	G(P,L)	F(P,L)
Thermistors	E(L)	F	G(P)	F(P)	E(L)	F	G(P)	F(P)								
Thermocouples	G	F	G(P)	F(P)	F(L)	F(L)	F(L)	F(L)	E	F	E(P)	F(P)	F(L)	F(L)	F(L,P)	F(L,P)
Quartz Crystals	E(L)	F														
Radiation Pyrometers									E(L)	G(L)	E(L)	G(L)	E(A,L)	G(A,L)	E(A,L)	G(A,L)
Infrared Pyrometers	E(L)	E(L)	E(L)	G(L)					E(L)	E(L)	E(L)	E(L)				
Spectroscopic (Fraunhofer) Sensors																
Thermopile					G(A)	F(A)							G(A)	F(A)		
Acoustic Time Domain Reflectometry (TDR)					D	D	D	D					D	D	D	D

CODE LETTERS: D–in development
L–limited
F–fair
G–good
E–excellent
P–protective well reduces speed of response
A–detects the average temperature of an area

EXAMPLE: G(D)–this combination of code letters refers to a device which is a good selection for the particular service, but is not yet commercially available.

NOTES:
1 This device either detects a point or the average temperature of some section of the process or of the refractory.
2 The term "hostile environment" here is used to mean processes such as fluid beds, where the sensor is likely to experience the mechanical impact of high velocity solid particles
3 "Interference" refers to need to overcome temperature interferences due to hot refractories or to temperature differences between the carrier gas and the solid particles in it.
4 For considerations of measurement error, span, cost stability, response time, linearity, materials of construction, etc., refer to the text.

to several types of commercially available thermocouples. If the temperature range is from 0 to 200°F (−18 to 93°C), the choice includes several types of thermocouples, liquid filled bulbs such as mercury thermometers, resistance thermometers, thermistors and bimetallic thermometers.

Accuracy

Accuracy requirements impose an additional restriction on the choice of sensing device. In considering accuracy two factors must be reviewed. First is the inherent accuracy of the measuring device, which is summarized in Table 4.1c. This generally represents the highest accuracy that can be achieved with the device under ideal conditions. The second consideration, which in general is a greater limitation than the inherent accuracy of the device, is the inaccuracy of actual measurement due to environmental effects, method of installation and use.

Environmental Influences

In considering the validity of a measurement, it should be remembered that temperature sensing devices respond only to the temperature which they experience. It may be considerably different from the temperature one is attempting to measure if the sensor is of improper size or configuration, or is installed in a manner such that it interferes with the temperature that would exist if the sensor was not present, or if the sensor is not adequately coupled thermally to the media whose temperature is being measured.

Measuring the Temperature of Solids

The next consideration is the allowable size and configuration of the sensor so as to assure a useful measurement. This requires some knowledge of the heating or cooling conditions together with an estimate of the magnitude of the temperature gradients that are likely to exist in the region in which the measurement is to be made. A simple, rule-of-thumb indicator to determine if significant gradients are likely to be present is the magnitude of the Biot modulus (hL/K), where h is the surface heat transfer coefficient, L is the smallest dimension of the solid and K is the thermal conductivity of the solid. If this modulus is hL/K>0.2, significant temperature gradients are likely to exist in the solid, and care should be exercised in choosing the size, location and orientation of the sensor within the solid. If the Biot modulus is less than 0.2, no significant gradient is ex-

Table 4.1c
TEMPERATURE SENSOR ACCURACY AND RANGE

		Inaccuracy (Error)	
Description of Sensors	*Temperature Range**	*ISA Standard Limits*	*ISA Special Limits*
I Thermocouples			
a. Chromel/Alumel	−300°F to +2500°F	±4°F, 0°F to +530°F	—
		±¾%, +530°F to +2500°F	
b. Iron/Constantan	−310°F to +1400°F	±4°F, 0°F to 530°F	±2°F, 0°F to +530°F
		±¾%, +530°F to +1400°F	±⅜%, +530°F to +1400°F
c. Copper/Constantan	−310°F to +750°F	±2%, −150°F to −75°F	±1%, −300°F to −75°F
		±1½°F, −75°F to +200°F	±¾°F, −75°F to +200°F
		±¾%, +200°F to +700°F	±⅜%, +200°F to +700°F
d. Platinum 10% Rhodium/Platinum	+32°F to +3200°F	±5°F, +32°F to +1000°F	—
		±½%, +1000°F to +2700°F	
e. Tungsten/Tungsten 26% Rhenium	+32°F to +5200°F	±8°F, +32°F to +800°F	—
f. Tungsten 5% Rhenium/Tungsten 26% Rhenium	+32°F to +5200°F	±1%, +800°F to +4200°F	—
II Resistance Thermometers			
a. Platinum (high purity strain free element)	−435°F to +900°F	±0.01°F limited range, ±0.05% full range	
b. Platinum (commercial purity element)	−320°F to +2000°F	±0.2% full range	
c. Nickel	−320°F to +800°F	±0.4% full range	
d. Balco	−320°F to +500°F	±0.4% full range	
e. Semiconductor (germanium)	−458°F to −400°F	±0.01°F	
f. Semiconductor (thermistors)	−40°F to +300°F	±0.05°F limited range, ±1°F full range	
III Bimetallic Thermometers	0°F to +1000°F	±2°F limited range, ±10° full range	
IV Liquid Filled Thermometer (glass stem)	−328°F to +800°F	±0.05°F limited range, ±1° full range	
V Liquid Filled Thermometer (metal bulb)	−150°F to +700°F	±2°F limited range, ±5°F full range	

*To convert from °F to °C use the following equations:

$$+°C = \frac{5(°F-32)}{9} \quad \text{and} \quad -°C = \frac{5(°F+32)}{9}$$

pected and a measurement anywhere on or within the solid should give identical results regardless of size or configuration of the sensor. If significant gradients are likely to exist, the maximum rate of heat transfer to the surface of the solid must be known or estimated, and the maximum gradient at the point of measurement must be determined. The following relationship allows the maximum gradient at the surface of a solid to be calculated:

$$\frac{\Delta T}{\Delta X} = \frac{q}{K} \qquad 4.1(1)$$

where

$\frac{\Delta T}{\Delta X}$ = temperature gradient at the surface
q = heat transfer rate per unit area at the surface
K = thermal conductivity of solid

Under certain conditions of heating or cooling, if measurements at points other than the surface are important, it may be necessary to evaluate anticipated heat transfer conditions and resulting temperature gradients.[4,5]

On the basis of this gradient, it is possible to establish limits on the size of the sensing device. For example, the length of any one of the three dimensions of the sensor (lead wires excluded) should not be greater than the distance between two points of the process that are different in temperature by more than the acceptable measurement error. It is assumed that the sensors are in satisfactory thermal coupling with the process material, which is not always the case. If the thermal coupling is poor, the sensor will not reflect the true temperature history experienced by the solid, a condition which can produce dynamic errors.[6] The best thermal coupling is achieved by direct bonding of the sensor, such as welding a thermocouple to the solid surface or into a cavity within the solid. The bond line between the sensor and the solid should be kept as thin as possible and should not fracture or fail during thermal cycling. Various epoxy and ceramic cements, with fillers to improve their conductivity, have been successfully used for such bonding. For example, a flat resistance thermometer bonded to a surface with an epoxy bond line 0.005 inch (0.1 mm) thick will produce a lag time of about 1 second, which will produce a dynamic error equal to the rate of temperature rise of the surface, i.e., a dynamic error of 25°F (14°C) for a rate of surface temperature change of 25°F per second.[6]

Methods of installing temperature sensors on or within solid bodies are shown in Figure 4.1d.

Measuring the Temperature of Fluids

The fundamental problem of measuring the temperature of a fluid is one of assuring strong thermal coupling. For a fluid temperature measurement to have meaning, the sensor must come to equilibrium with the temperature of the fluid. The difference between the equilibrium temperature of the sensor and the fluid temperature is a direct error. Consequently, with rapidly changing temperatures the rate of heat transfer between the sensor and process fluid must be sufficient to overcome the thermal capacity of the sensor in order that it can follow the fluctuations in fluid temperature. Under such conditions one can write the following equation which expresses the temperature, T, experienced by a sensor which was initially at temperature T_1:[4]

$$\frac{T_1 - T}{T_1 - T_0} = e^{-(hA/WC_p)\theta} \qquad 4.1(2)$$

where

T = temperature
e = base of natural logarithm
h = heat transfer coefficient at surface of sensor
A = surface area of sensor
W = weight of sensor
C_p = specific heat of sensor
θ = time after immersing sensor in fluid

A sensor, to respond rapidly to changes in fluid temperature, should have a large surface area-to-mass ratio. Furthermore, the heat transfer coefficient, which is a direct function of the fluid mass flow rate over the sensor, should also be as large as possible. Forced convection or rapid flow of the fluid over the sensor is therefore desirable.

A practical measure of sensor sensitivity is its time constant. Since the sensor approaches change in fluid temperature asymptotically, it is difficult to determine when it has reached that temperature. The time constant of a sensor is the time required to accomplish 63.2 percent of the step change. This point is reached when the exponent of equation 4.1 (2) is 1. Thus, the time constant of a fluid stream temperature sensor (τ) is defined as:

$$\tau = \frac{1}{hA/WC_p} \qquad 4.1(3)$$

When measuring the temperature of a gas stream in a heated duct or furnace where temperature differences between the sensor and its surroundings exceed 1,000°F (538°C), significant errors can occur due to radiation exchange between the sensor and its surroundings. Under such conditions the sensor must be shielded against thermal radiation exchange. This can cause disturbances in the flow of the fluid around the sensor and hence affect the directional response characteristics.

Another thermal effect in measuring the temperature of gas streams at high velocities is the recovery factor, which results from an increase in the temperature of the gas at the sensor due to compression heating as the gas is brought to stagnation against the sensor.[7,8] Fluid

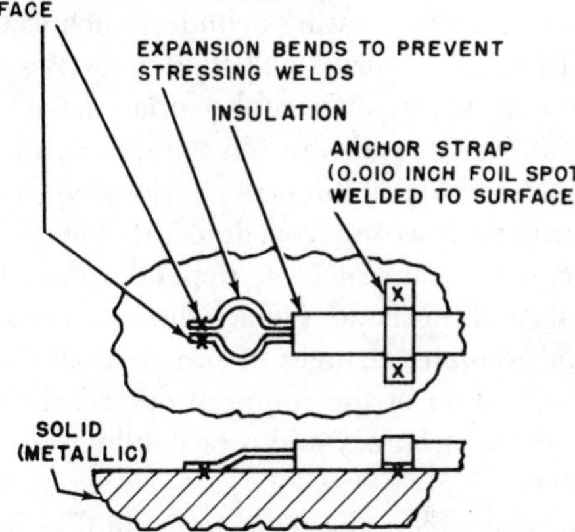

A-TYPICAL THERMOCOUPLE INSTALLATION FOR JUNCTION FORMED THROUGH SOLID BY DIRECT SPOT WELDING TO SURFACE

CHARACTERISTIC:
1. FAST RESPONSE
2. SIMPLE TO INSTALL
3. MAY CAUSE DISTURBANCE TO FLOW PATTERN UNDER HIGH SPEED FORCED CONVECTIVE, COMPRESSIBLE FLOW

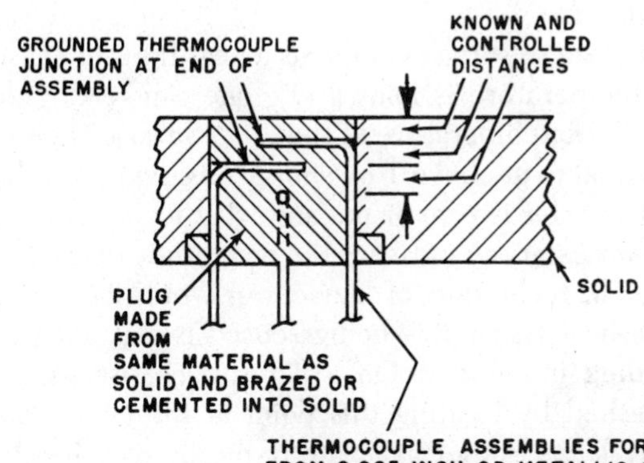

C-TYPICAL EMBEDDED THERMOCOUPLE INSTALLATION IN SOLID

CHARACTERISTICS:
1. ALLOWS MEASURING GRADIENTS WITHIN SOLID
2. NO DISTURBANCE TO FLUID FLOW PATTERN AT HEATED SURFACE

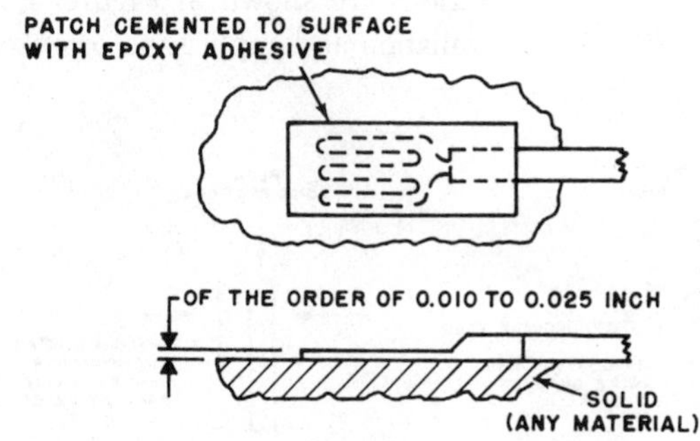

B-TYPICAL RESISTANCE THERMOMETER PATCH INSTALLATION ON SURFACE OF SOLID

CHARACTERISTICS:
1. SAME AS "A"

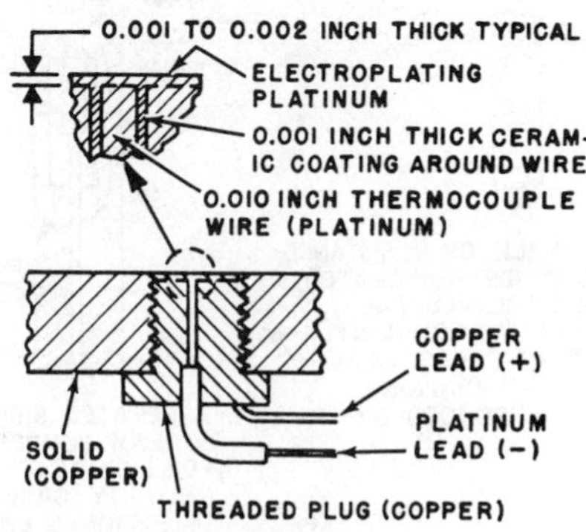

D-SURFACE THERMCOUPLE FOR HIGH SURFACE HEATING RATES

CHARACTERISTICS:
1. RESPONDS AT SAME RATE AS SOLID AT SAME DEPTH
2. NO DISTURBANCE TO FLUID FLOW PATTERN AT HEATED SURFACE
3. MINIMUM DISTURBANCE TO NATURAL HEAT FLOW IN SOLID
4. APPLICABLE UNDER EXTREMELY HIGH PRESSURE
5. USEFUL UP TO MELTING POINT OF SOLID

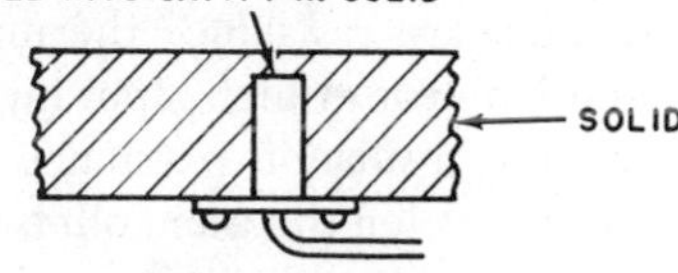

E-TYPICAL RESISTANCES THERMOMETER EMBEDDED IN SOLID

CHARACTERISTICS:
1. SLOW RESPONSE
2. INACCURATE KNOWLEDGE OF LOCATION OF TEMPERATURE MEASURED

Fig. 4.1d Some selected methods of installing temperature sensors on or within solids

stream sensor configurations are shown in Figure 4.1e together with some installation and application oriented guidance.

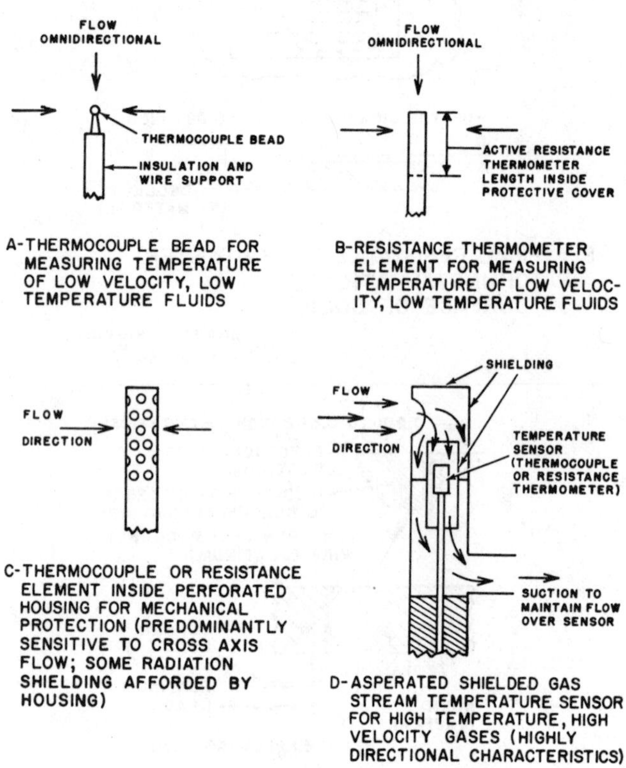

Fig. 4.1e Fluid stream sensor configurations

Atmospheric Effects

Adverse atmospheric conditions can cause problems with temperature sensors. For example, in a highly humid or very moist environment it is essential that the element of a resistance thermometer or the bead of a thermistor be well insulated electrically. If moisture contacts the resistance element or thermistor bead, it may cause a short. Thermocouples in general are less sensitive to moisture than are resistance thermometers. If null balance potentiometric or high input impedance readout devices are used, insulation resistance between legs of the couple as low as ten thousand ohms can be tolerated without serious error in the indicated temperature. If low input impedance current measuring readout devices are used, a high insulation resistance between legs of the couple becomes as important as with resistance thermometers and should be in the megohm range.

Corrosive, reducing or oxidizing environments also create problems. If a sensing device is exposed to such environments, it must be protected by some form of envelope or coating. It should be remembered that protection applied to a sensor will increase its mass and thus, in general, adversely affect its response characteristics.

Non-Contact Temperature Measurement

Non-contact temperature measurement of a moving surface is needed when measuring webs of paper, plastic, textile and metal or rotating cylinders such as calender rolls or drier cans. Very small thermocouples or resistance elements placed close to the surface have had some success but, in general, are too vulnerable to physical damage and ambient influences. Various types of radiation devices such as infrared detectors have also been employed with some success, depending on the environmental conditions under which the measurements are made. For example, changes in emissivity of the surface and contamination of the required optical elements directly affect the accuracy and repeatability of such measurements.

A new principle, the convective-null-heat-balance concept,[9,10] involves the null-balance of convective heat transfer between a sensing head of known temperature and a surface of unknown temperature, in which temperatures are compared rather than directly measured. If, for example, two surfaces are placed in close proximity to one another, their temperatures can be compared by measuring the rate and direction of heat transfer between them. The direction of heat transfer is a definite measure of which surface is at the higher temperature; if the heat transfer between them is zero, their temperatures must be equal.

If one of these surfaces contains a heat transfer sensor and its temperature is known (T), one can readily determine if the temperature of the other surface is either below, equal to or above T by simply observing whether the heat transfer is zero or not and what its direction is. It is not necessary to measure the magnitude of the heat transfer. The technique, of course, works only if the point at which heat transfer is being sensed is isolated from surrounding influences. On a flat surface, isolation is accomplished by locating this point at the center of a relatively large isothermal plate (typically 6 inches/150 mm in diameter) spaced approximately 1/16 to 1/8 (1.5 to 3 mm) inch from the surface being measured.

The basic concept of null-heat-balance can be used either to monitor or to control surface temperature. In the monitoring mode, the temperature of a sensing head is automatically varied to keep it at null-heat-balance (at the same temperature) as the moving surface. The head temperature, which is equal to the surface temperature, can then be measured.

In the control mode, the head is held at the desired set point temperature at which the surface is to be controlled and the signal from the heat flow sensor is employed to manipulate the surface temperature.

In another arrangement, the signal from the heat flow detector is calibrated for temperature difference between the sensor, which is at a constant known temperature, and the surface being measured.

REFERENCES

1. Stimson, H.F., "International Practical Temperature Scale of 1948, Text Revision of 1960." NBS Monograph 37, 1961.
2. NBS Special Publication 300, Volume II, January, 1969.
3. Lipták, B.G., "Overview of Coal Conversion Process Instrumentation," Argonne National Laboratory publication: ANL-FE-49628-TM01, May, 1980.
4. Giedt, W.H., "Principles Of Engineering Heat Transfer." New York: Van Nostrand, 1957.
5. Carslow, H.S. and Jaeger, J.C. "Conduction Of Heat In Solids." London: Oxford, 1947.
6. Rall, D.L. and Hornbaker, D.R., "A Rational Approach To The Definition Of A Meaningful Response Time For Surface Temperature Transducers." New York: 21st Annual ISA Conference, October 22–27, 1966.
7. Glawe, G.E., Simmons, F.S. and Stickney, T.M. "Radiation and recovery corrections and time constants of several chromel-alumel thermocouple probes in high-temperature, high-velocity gas streams." NACA TN-3766, October, 1956.
8. Scadron, M.D. and Warshawsky, I., "Experimental determinations of time constants and Nusselt numbers for bare-wire thermocouples in high-velocity air streams and analytic approximations of conduction and radiation errors." NACA TN-2599k, January, 1952.
9. U.S. and foreign patents have been issued and others are pending, Trans-Met Engineering, Inc., La Habra, California.
10. Hornbaker, D.R. and Rall, D., "The Convective Null-Heat-Balance Concept For Non-Contact Temperature Measurements Of Sheets, Rolls, Fibers And Wire." Fifth Symposium On Temperature—It's Measurement And Control In Science And Industry, June 21–29, 1971, Washington, D.C.

BIBLIOGRAPHY

Hall, J., "Applying Temperature Sensors," *Instruments and Control Systems*, June, 1980.

Hormuth, G.A., "Ways to Measure Temperature," *Control Engineering*, Reprint No. 948, 1971.

Jutila, J.M., "Temperature Instrumentation," *Instrumentation Technology*, February, 1980.

Lipták, B.G., "Overview of Coal Conversion Process Instrumentation," Argonne National Laboratories Publication: ANL-FE-49628-TM01.

Plumb, H.H., "Temperature, Its Measure and Control in Science and Industry," 5th Symposium on Temperature held in 1971 by NBS, API and ISA.

Schooley, J.F., "State of the Art of Instrumentation for High Temperature Thermometry," Argonne National Laboratories Publication Number: ANL-78-7, 1977 Symposium.

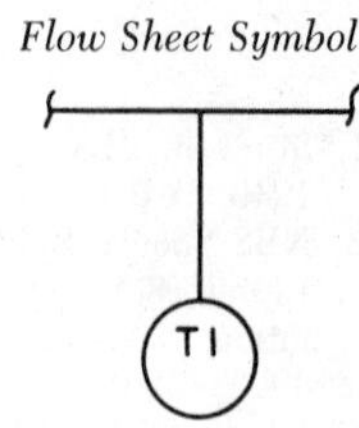

4.2 BIMETALLIC THERMOMETERS

Temperature Range:	Extreme, −100 to 1000°F (−73 to 538°C); Practical, −80 to 800°F (−62 to 427°C). Minimum span, 30°F (−1°C) for 90 angular degrees
Linearity:	Linear over most of range.
Cost:	\$25–\$110
Inaccuracy:	1–20°F (1–7°C)
Partial List of Suppliers:	Ametek, Inc., US Gauge Div.; Dresser Industries; Marsh Instrument Co.; Marshalltown Instruments; Moeller Instrument Co.; Rochester Gauges, Inc.; Weiss, Albert A. & Sons, Inc.; Weston Instrument Div. of Sangamo Weston, Inc.

Bimetallic thermometers make use of two fundamental principles: (1) that metals change volume with temperature and (2) this coefficient of change is not the same for all metals. If two different straight metal strips are bonded together and heated, the resultant strip will tend to bend toward the side of the metal with the lower expansion rate. Deflection is proportional to the square of the length and the temperature change, and inversely proportional to the thickness.

The motion is small; to amplify it in a reasonably sized space the bimetal strip may be wound in the form of a spiral or a helix. A classical example of an ambient air temperature thermometer is shown in Figure 4.2a. The outside edge of the spiral is pinned to the frame and a pointer is connected to the center. As the temperature increases the spiral winds up deflecting the pointer clockwise.

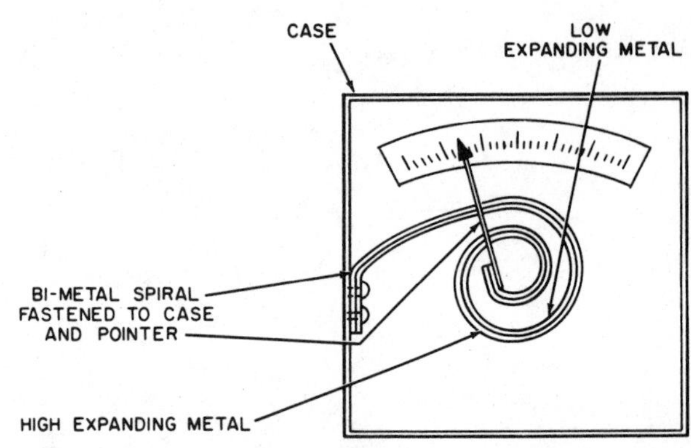

Fig. 4.2a Bimetallic ambient air thermometer

Knowing the coefficients of expansion of the two metals, their thickness, and the desired scale length and range, the total length of the spiral is easily computed. A favorite combination of metals is low expanding invar (64 percent Fe, 36 percent Ni) against high expanding nickel-iron alloy with chromium or manganese added.

Most industrial or residential bimetal thermometers use a helical coil which can be designed to fit into a stem more easily than the spiral. The element is surrounded by a protecting tube or well. The device can be mounted to measure the temperature of the gas or liquid inside a duct. The design is frequently used on domestic furnaces and can be competitive with glass stem thermometers in many other applications.

A single helix moves axially as it winds or unwinds with heat and cold. This requires clearance for a vertical movement of the pointer. The difficulty can be overcome, if desired, by using a multiple element, wound coaxially so as to form coils within coils. This construction is more costly but has an advantage in requiring less immersion depth (Figure 4.2b).

Bimetal thermometers are also made in types that have the dial face at any set angle with respect to the axis of the stem. This can even be a swivel type as shown on Figure 4.2c. Any of these constructions requires a bend in the motion transmission from coil to pointer. This is

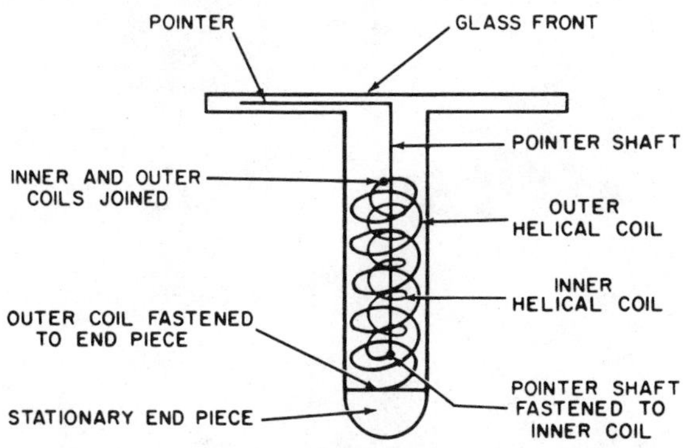

Fig. 4.2b Bimetal thermometer with dual helical coil

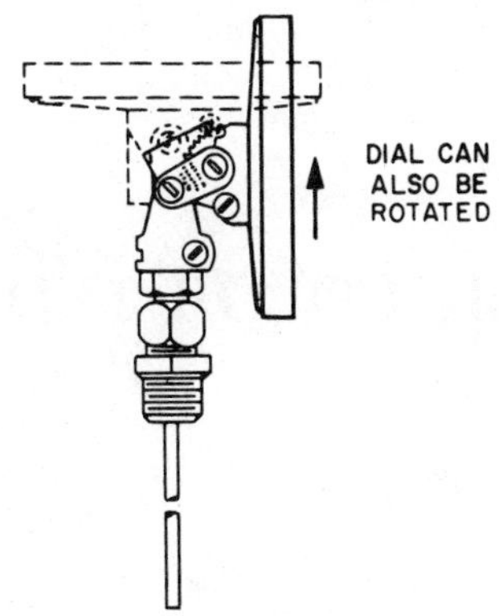

Fig. 4.2c Adjustable dial bimetallic thermometer

done with an edgewound helical spring which eliminates backlash and requires little torque to operate.

Another feature usually included is complete sealing. A dry gas is in the dial face portion of the assembly while silicone fluid fills the stem and surrounds the coil to dampen vibration and accelerate heat transfer.

Readout dials are available varying from 2 to 5 in. (50 to 125 mm) in diameter and with stem lengths up to 24 in. (600 mm). Wells, made of carbon steel, stainless steel or other materials are available to protect against corrosive environments.

Bimetallic elements can be made sufficiently sturdy to actuate a recording pen. A chart, driven by mechanical clockwork behind the pen, forms a complete measuring and recording system which is independent of outside electric power and very reasonably priced with about 2–5 percent accuracy.

Advantages over glass stems are that the bimetallic design is less subject to breakage and easier to read, plus it can be equipped with the recording feature. They are low in cost compared to thermal systems and electrical sensors but more expensive than the glass stem type.

A disadvantage is that rough handling changes their calibration. The overall accuracy is not as good as that of the glass stem design for the same reason. The bimetallic thermometers are confined to local measurement.

BIBLIOGRAPHY

Considine, D.M., *Encyclopedia of Instrumentation and Control,* McGraw-Hill, New York, 1971.

Hormuth, G.A., "Ways to Measure Temperature," *Control Engineering,* Reprint No. 948, 1971.

Plumb, H.H., "Temperature, Its Measure and Control in Science and Industry," 5th Symposium on Temperature, held in 1971 by NBS, AIP, and ISA.

4.3 COLOR INDICATORS

Temperature Range:	105–2500°F (41–1371°C)
Cost:	\$4–\$5 per crayon. \$2–\$4 per oz. of paint
Inaccuracy:	± 10°F (± 6°C)
Partial List of Suppliers:	Markal Co.; Tempil Div., Big Three Industries Inc.

Color indicators are a class of sensors which have the property of changing their original color when a certain temperature is reached. The change is distinct, not just an alteration in shade. For instance, it may change from yellow to grey or from light blue to light brown. Some can go through several color changes at different temperatures.

Paints and crayons are familiar forms of these indicators, which are applied directly to a solid object either when it is cold and about to be heated or when it is already hot. Some indicators can determine the temperature of solid objects immersed in oil. They are not recommended for use in hot gases.

Temperature is indicated by a chemical reaction, where a molecule of a gas, such as ammonia, CO_2, or water vapor, is driven off the basic stock (colorful salts of metals like nickel, cobalt or chromium), thus changing its color. The change is usually permanent after the object cools down. An exception occurs when the gas is water vapor; the indicator may slowly reabsorb this gas from the air and revert to the original color.

Change in color of these types of indicators is not only a function of temperature but also of time. For this reason, the immediate past temperature history of the indicator will influence the exact point at which it will change color. The indicators are usually rated for a specific temperature over a certain time period, for instance 140°F (60°C) in 30 minutes. This means that if held at a constant 140°F the color change will occur in 30 minutes. If the color change occurs in less than 30 minutes, the average temperature is higher than 140°F and vice versa. On such an indicator, if the temperature does not exceed 130°F (54°C) the change will never occur. It is stable below this temperature.

Figure 4.3a shows typical time-temperature relationships for two different crayons. In the examples shown, the temperature at which color change occurs is quite critical when exposure time is short. For lower soakout temperatures, the changeover will occur in a longer time.

Flatter curves than those shown are possible where changeover will occur within a few seconds after "operating" temperature is reached or it won't occur at all.

Many different temperature ratings are available. They can be obtained in a series for every few degrees to the maximum offered (about 2500°F or 1371°C). This class of indicators is quite inexpensive and is used in industry where only an end point is needed and someone can be present to watch for or interpret the results. The material adheres tightly to the object on which it is placed and presents a minor problem if it must be removed later.

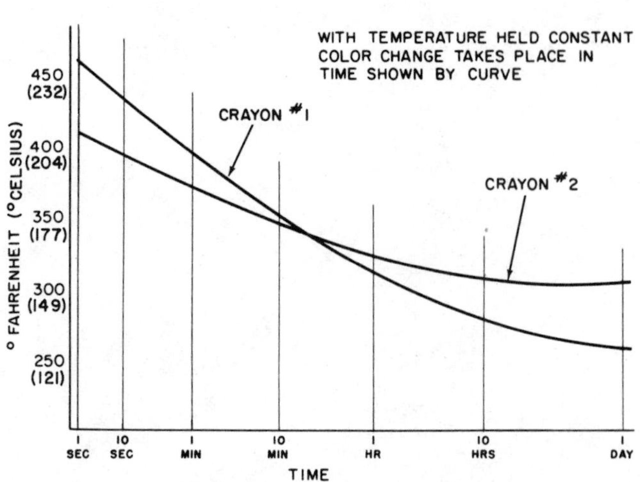

Fig. 4.3a Time-temperature relationship for color indicators

BIBLIOGRAPHY

Considine, D.M., *Encyclopedia of Instrumentation and Control*, McGraw-Hill, New York, 1971.

Hormuth, G.A., "Ways to Measure Temperature," *Control Engineering*, Reprint No. 948, 1971.

Plumb, H.H., "Temperature, Its Measure and Control in Science and Industry," 5th Symposium on Temperature, held in 1971 by NBS, AIP, and ISA.

Flow Sheet Symbol

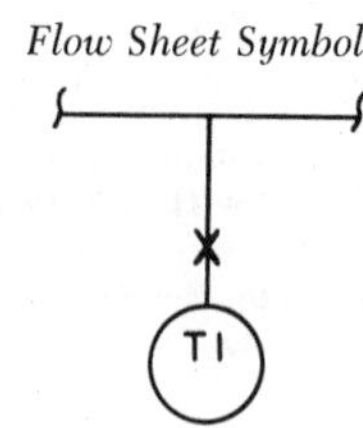

4.4 FILLED THERMAL ELEMENTS

Temperature Range:	Extreme, −450 to 1400°F (−268 to 760°C); practical, −300 to 1000°F (−184 to 538°C)
Linearity:	Linear, except vapor systems
Cost:	\$100–\$250 for local indicators; \$700–\$1100 for standard transmitters
Inaccuracy:	$\pm\frac{1}{2}$ to $\pm 2\%$ of full scale
Partial List of Suppliers:	Ametek Inc., US Gauge Div.; Bristol Co.; Duro Instrument Corp.; Fischer & Porter Co.; Foxboro Co.; Honeywell, Industrial Div.; G. Kent Ltd.; Moore Products Co.; Palmer Instruments; Taylor Instrument Companies; Thermometer Corp. of America; Weksler Instruments Corp.

A filled thermal system is basically a pressure gauge modulated in a readout instrument and connected by small bore tubing to a bulb acting as the temperature sensor. The whole system is gas-tight, and filled with an appropriate confined gas or liquid under pressure.

There are many different types of filled systems, each having certain peculiarities which give it advantages over others. These are some of the factors which should be considered in selecting for a specific application:

1. Ambient temperature compensation
2. Range limitations
3. Even vs. uneven scale divisions
4. Bulb designs available
5. Bulb size and tubing length
6. Bulb material
7. Bulb elevation
8. Overrange capacity
9. Cost factors
10. Torque requirements
11. Objection to presence of mercury

The Scientific Apparatus Makers Association (SAMA) has classified filled system thermometers into four major classes according to filling material. Table 4.4a has been prepared to compare some of the properties of the various thermal systems.

Ambient Temperature Compensation

Thermal expansion and increased pressure due only to an increase in ambient temperature would affect the reading slightly if the system did not have compensation for these effects. The spiral alone may be compensated (case compensation) or, if the range is narrow and/or the bulb is small and capillary long, then full compensation of both spiral and capillary is used.

Compensation for temperature changes along the capillary in case of a vapor pressure (Class II) system is unnecessary because the interface between the liquid and gas, which establishes the pressure, must always be in the bulb.

An uncompensated thermal system is shown in Figure 4.4b. It may be used in Class I systems only if range is wide and length of capillary short. Class III and V systems are never supplied without compensation.

Case compensation, shown in Figure 4.4c, is adequate when case and capillary are at the same temperature (near ambient) and the length of capillary not too long. A bimetallic compensator acts in a nearly equal and opposite direction on the pointer to that caused by the ambient temperature change.

When a thermal system must be fully compensated the simplest way is to supply a duplicate spiral and capillary, the latter being closed off just before the active bulb. The two capillaries are parallel to each other and the two spirals are mechanically connected so that am-

Table 4.4a
THERMAL SYSTEMS COMPARED

SAMA Classification	Type Compensation	Fill	Temperature °F (°C) Max.	Temperature °F (°C) Min.	Scale Linearity	Speed of Response (No Well)	Head Effect	Overrange Capabilities	Superior Features	Less Desirable Features
I	None	Liquid	700 (371)	−125 (−87)	Linear Except at Low Temps.	6–7 Sec.	Significant; Must Be Compensated	150%	Smaller Bulb Sizes; Narrower Spans; Lower Cost	Shortest Capillary; Compensation More Difficult
IA	Full									
IB	Case									
IIA	Not Required	Vapor	Only Above Ambient to 650 (343)		Scale Divisions Increase with Temp. Increase	4–5 Sec. Except When Passing Through Ambient	Appreciable Compensation	None	Long Capillary Lengths Available; Ambient Temperature Comp. Not Req.; Fast Speed of Response	Non-Linear Scale; Difficult to Provide Through Ambient Temps. No Overrange Capacity
IIB			Only Below Ambient to −300 (−184)				None			
IIC			Above and Below Ambient Not Through				Not Usable; Can't Compensate			
IID			Above, Through and Below Ambient				Same as IIA			
IIIA	Full	Gas	1400 (760)	−450 (−268)	Linear Except Very Low Temps.	4–7 Sec.	None	150–300%	No Head Effect; Greatest Overrange; Greatest Range of Temperature	Large Bulb Size; Widest Spans Req.; Least Power for Control
IIIB	Case									
VA	Full	Mercury	1200 (649)	−40 (−40)	Linear	4–5 Sec.	Not Significant Up to 30 Ft. (9.1m)	150%	Most Linear Scale; Least Difficult to Compensate; Most Power for Control Elements	Objection to Mercury on Accidental Breakage
VB	Case									

There is no SAMA CLASSIFICATION IV

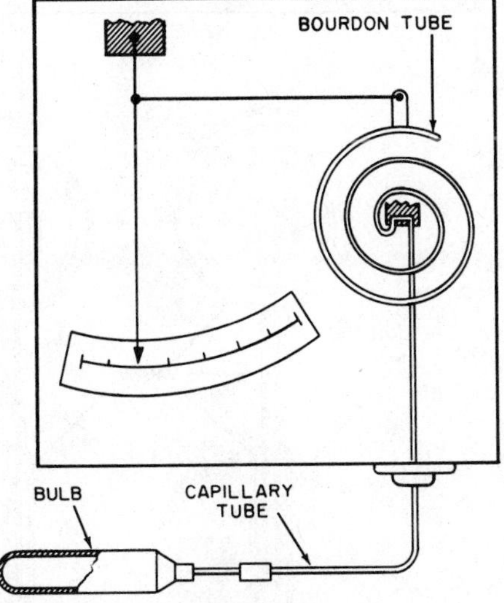

Fig. 4.4b Uncompensated filled thermometer

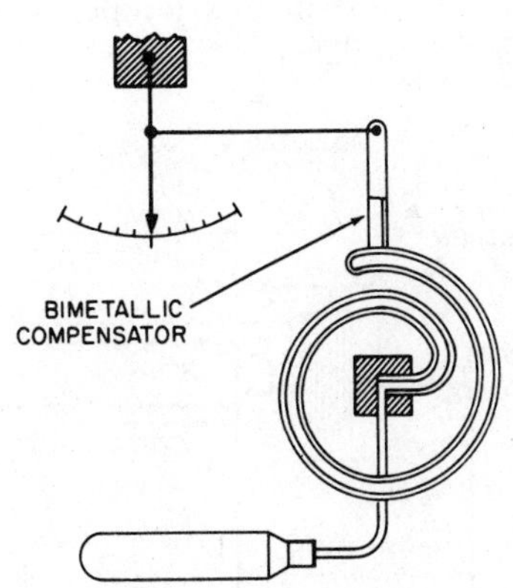

Fig. 4.4c Case compensation

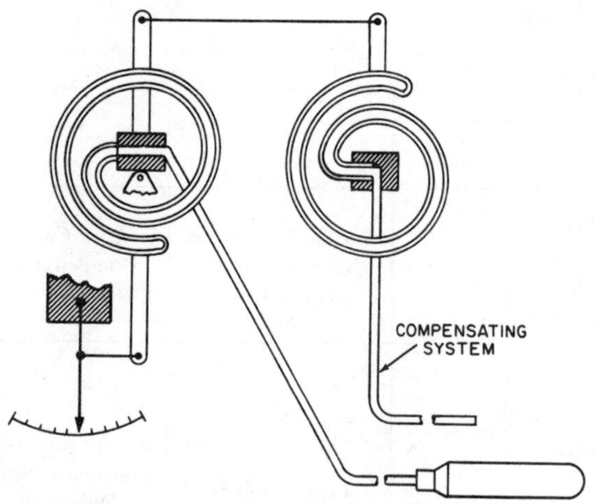

Fig. 4.4d Full compensation

bient temperature changes are cancelled out (see Figure 4.4d). Another compensating method is described in Figure 4.4h.

Bulbs, Wells and Capillaries

The temperature sensitive element, the bulb, comes in many sizes and shapes to handle the many different applications. It is good practice to use the largest bulb which will do the job. This will cut down on ambient temperature errors, permit smaller spans and longer capillaries.

Plain bulbs are used where the measured medium is not under pressure and will not harm the bulb material. If this is not the case then a separate well to protect the bulb from the measured medium is needed. This, of course, will slow down the response time slightly (see Figure 4.4e).

High speed of response can be obtained with a long, thin, bendable bulb used to sense the average temperature in large areas. Another way of presenting a long bulb is by coiling it. For use in low velocity gas flow temperature measurement the coil is set at the factory and cannot be uncoiled.

Most bulbs are made of stainless steel, which is relatively inert and will withstand high temperatures. Other materials are readily available.

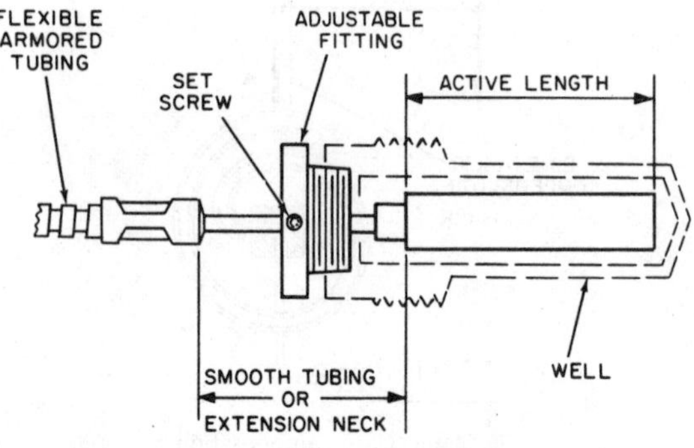

Fig. 4.4e Plain bulb with adjustable fitting and well

The relatively fragile thin-wall capillary should be protected by a flexible armored stainless steel or PVC-covered bronze tubing. An extension neck to the bulb prevents the tubing from being immersed directly in the measured medium. Bendable smooth steel tubing might also be used for capillary protection.

Class I, Liquid-Filled Systems

These systems are completely filled with a liquid (other than mercury) and operate on the principle of liquid expansion with increase of temperature. The filling fluid is usually an inert hydrocarbon, such as xylene (C_8H_{10}) which has a coefficient of expansion six times that of mercury and makes smaller bulbs possible. Other liquids (even water) are sometimes used. The criterion is that the pressure inside the system must be greater than the vapor pressure of the liquid to prevent bubbles of vapor from forming in the spiral. Also, the liquid should not be allowed to solidify even in storage, otherwise the calibration may be affected.

Class II, Vapor Systems

The pressure element, capillary and bulb of a Class II system have the filling medium in both the liquid and gaseous form. The interface between the two must occur in the bulb and this will move slightly with temperature, affecting the pressure. On Class IIA systems, the bulb would be mostly filled with gas while the capillary and spiral would contain liquid. On Class IIB systems the reverse would be true. Such systems are not suitable when the ambient temperature is the same or close to the measured temperature. Under such conditions the interface will not be defined.

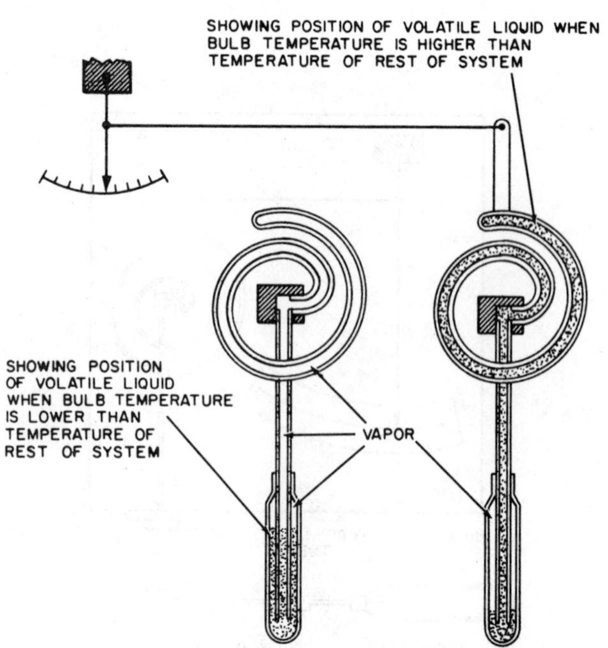

Fig. 4.4f Class II C system

A Class IIC system (see Figure 4.4f) permits measurement on both sides of ambient but msut use a larger bulb. The action across ambient is delayed and the system requires time to settle down as the interface between liquid and gas reverses.

The Class IID system (see Figure 4.4g) overcomes this limitation by using a second non-volatile fluid to act as a hydraulic transmitter of pressure to the measuring spiral.

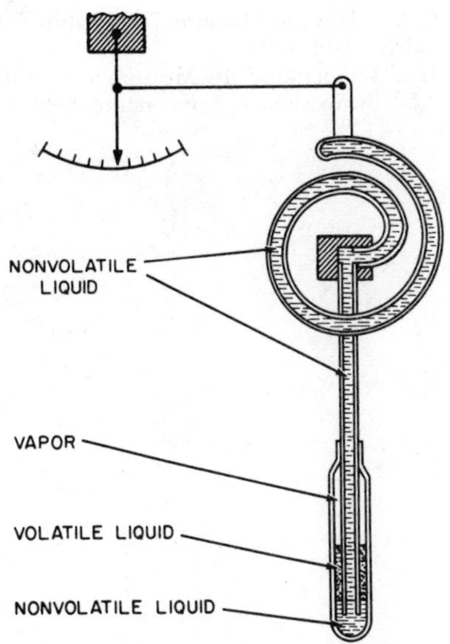

Fig. 4.4g Class II D system

Class III, Gas-Filled Systems

The operating principle for gas-filled systems is that in a perfect gas confined to a constant volume the pressure is proportional to the absolute temperature. The gas is not perfect and not all at the same temperature nor is the volume constant, but variances are small enough so that a measurement of pressure can be used to indicate temperature.

Nitrogen is the favorite fill for a Class III system because it is inert and inexpensive. It does react somewhat with the steel bulb material at temperatures exceeding 800°F (427°C), and it does act less like a perfect gas at extremely low temperatures. Under these conditions helium should be used. Different ranges are obtained by selecting the correct filling pressure.

In general, bulbs should be as large as practical to lessen the influence of temperature variations along the capillary. One way of avoiding long capillaries is to terminate a short capillary at a small diaphragm chamber. The force due to gas pressure on the diaphragm causes it to compress the spring. This motion is amplified and used to regulate another pressure which is transmitted to the spiral. This arrangement, though more expensive, permits much smaller bulbs than could otherwise be used.

Class V, Mercury-Filled Systems

Mercury is a liquid and in this respect Class V systems are similar to Class I. The two are separated because of the unique characteristics of mercury and its importance as a temperature measuring medium. Mercury provides rapid response, accuracy, and plenty of power for operating control elements. Pressures within the working system are relatively high—as much as 1200 PSIG (8.3 MPa) for the higher temperatures, dropping to 400 PSIG (2.8 MPa) at the low temperature end of the range. This high pressure cuts down on any head effect error (difference in elevation between bulb and measuring instrument).

The incompressible nature of mercury makes ambient temperature compensation less of a problem. Besides the standard method of full compensation (Figure 4.4d), another method is to place in the single capillary an "Invar" wire which has negligible coefficient of expansion. The diameters of the wire and the inside diameter of the capillary are selected so that an increase in ambient temperature will increase the size of the annular space between capillary and wire by just the amount of expansion of the mercury (see Figure 4.4h).

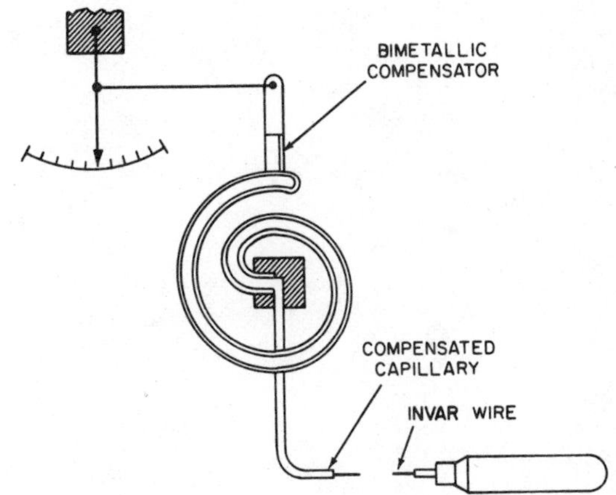

Fig. 4.4h Class V full compensation with internal wire

Conclusions

Filled thermal systems are, in general, intermediate in cost and performance between the simplest measuring devices like glass stem and bimetallic thermometers and the more complex electrical measuring element.

ADVANTAGES

1. Simplicity allows inexpensive design and rugged construction.
2. Response time, sensitivity and accuracy, though not as good as electrical instruments, are sufficient to meet most industrial requirements.
3. The system is self-contained. No auxiliary power is necessary. With a hand-wound clock for driv-

ing a chart, the system is naturally explosion-proof.

4. There is sufficient power to operate both electric and pneumatic controls.
5. Three (or more) separate systems can be put in a single instrument case.

DISADVANTAGES

1. Bulb size still tends to be larger than a small thermocouple or resistance thermometer bulb.
2. Minimum spans are not as narrow as the bimetallic thermometer or electric systems.
3. The thermal system, being under pressure, cannot be broken without destroying the calibration. Repair must usually be done at the manufacturer's factory.
4. Distance from measuring bulb to readout instrument is a maximum of 50 to 200 ft (15 to 60 m), far less than in electrical systems.

BIBLIOGRAPHY

Considine, D.M., *Encyclopedia of Instrumentation and Control*, McGraw-Hill, New York, 1971.

Hormuth, G.A., "Ways to Measure Temperature," *Control Engineering*, Reprint No. 948, 1971.

Plumb, H.H., "Temperature, Its Measure and Control in Science and Industry," 5th Symposium on Temperature, held in 1971 by NBS, AIP, and ISA.

Flow Sheet Symbol

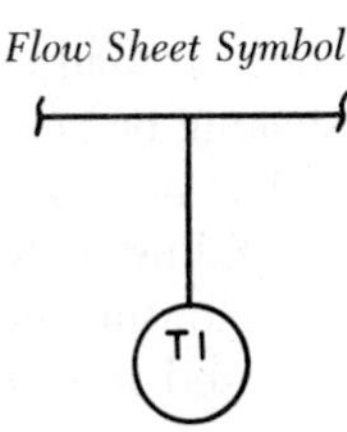

4.5 GLASS STEM THERMOMETERS

Temperature Range:	Extreme, −321 to 1100°F (−196 to 593°C) practical, −200 to 600°F (−129 to 316°C).
Linearity:	Linear
Cost:	\$11 to \$45 nominal
Inaccuracy:	0.1 to 2°F (0.1 to 1.1°C)
Partial List of Suppliers:	Ametek, Inc., US Gauge Div.; Duro Instrument Corp.; Economy Gauge and Instrument Supply; Moeller Instrument Co., Inc.; Palmer Instruments; Princo Instruments, Inc.; Taylor Instrument, Consumer/Industrial Products; Thermal Corp.; Thermometer Corp. of America; Weksler Instruments Corp.

The glass stem thermometer was the first closed thermal expansion system and has been known since Gabriel Daniel Fahrenheit investigated the expansion of mercury in the eighteenth century.

Mercury or some other liquid (alcohol, pentane) fills the glass bulb (see Figure 4.5a) and extends into the capillary bore of the stem. Generally, the space above the mercury column to the sealed top is evacuated, but occasionally it may be filled with an inert dry gas, such as nitrogen, to increase the temperature range.

The expansion of pure mercury is .01%/°F (0.005%/°C) and very linear; therefore, the volume of the bulb must be about 10,000 times the volume of the capillary between two marks 1°F (0.56°C) apart.

Although bulb and capillary could be made from the same type of glass, it is more convenient to make the bulb from glass with a good stability factor and the capillary from glass easier to work. For accurate measurements, the capillary must be properly annealed after it is drawn to the correct bore. Uniformity of bore is desirable but not absolutely necessary if the thermometer is calibrated at a sufficient number of points.

The glass stem thermometer can be made for a very narrow range in temperature. For instance, consider a clinical thermometer whose full range may be 96–102°F (35.6–38.9°C) and with a stem 4 in. (101.6 mm) long, active length. If the bulb volume is 0.03 cubic inch (.5 cc) the diameter of the bore is about .001 inch (.025 mm).

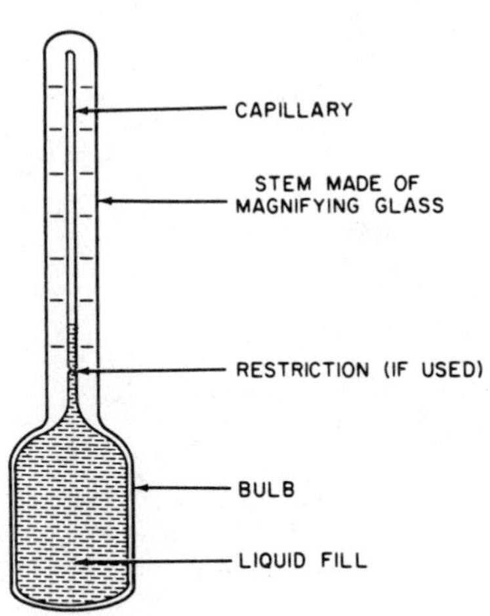

Fig. 4.5a Liquid-in-glass thermometer

The stem is frequently designed to magnify such a small column for easier readability.

Another design feature of some glass stem thermometers, notably the clinical type, is a restriction purposely placed in the capillary which prevents the liquid from returning toward the bulb when the thermometer is removed from the warmer object. This creates a separation of the column in the stem. In this case, it is a desirable feature because it permits a "highest point" reading or peak picker. If the separation occurs inadvertently in any glass stem thermometer, however, the result is an erroneous reading. The column can usually be rejoined by shaking or tapping.

The design of a glass stem thermometer requires that the filling material be a liquid over the entire range of temperature desired. Mercury is most suitable and can be used from its freezing point (−38°F or −39°C) up to nearly its boiling point (over 1000°F or 538°C). At this upper limit the space above the mercury column must be under great pressure with an inert gas to prevent evaporation of mercury from the top of the column. Alcohol and a few other hydrocarbons may be used for low temperatures. Colorfast dyes are usually added to these liquids to increase visibility.

To minimize accidental breakage a metallic thermowell is sometimes used to protect the bulb. This has no effect on the accuracy but may reduce the speed of response. Advantages of liquid-in-glass thermometers are low cost, simplicity and long life if treated properly. Disadvantages are difficult reading, confinement to local measurement and non-adaptibility to recording or automatic control. They also break very easily.

BIBLIOGRAPHY

Considine, D.M., *Encyclopedia of Instrumentation and Control,* McGraw-Hill, New York, 1971.

Hormuth, G.A., "Ways to Measure Temperature," *Control Engineering,* Reprint No. 948, 1971.

Plumb, H.H., "Temperature, Its Measure and Control in Science and Industry," 5th Symposium on Temperature, held in 1971 by NBS, AIP, and ISA.

4.6 MISCELLANEOUS SENSORS

In this section several unrelated methods of temperature measurement will be briefly discussed. Some of these are still in the experimental stage while others have not been exploited commercially, because they are not yet competitive with the more common sensors. This fast changing field is so broad that some types of temperature sensors will not have been mentioned at all and for this the authors make due apology.

Self Measuring Devices

Sometimes temperatures can be measured by allowing a material to serve as its own thermometer. This is done when determining the pressure of a confined gas to indicate its mean temperature, or the resistance of heater coils for the same purpose.

Actually, any property having a consistent rate of temperature variation will serve to indicate temperature. The frequency of the chirp of crickets on a summer night is an indication of their temperature environment. The rate at which a viscous substance such as oil drips through a small hole in the bottom of its container is just as much an indication of its temperature (provided the time-temperature relationship is known) as would be obtained by inserting a glass stem thermometer.

Spectroscopic Temperature Measurement

Spectroscopic methods are often used to measure the temperature of hot gases. They are, in fact, the only possible way to measure the surface temperature of stars.

The spectroscope in its simplest form is the familiar triangular glass prism which breaks up light from a hot object into its constituent colors (its spectrum). The chemical composition of glowing gas is determined from the pattern of dark (Fraunhofer) lines which appear across the spectrum.

Many procedures for temperature determination from the spectrum have been developed such as measurement of brightness and actual color, reversal temperatures, population temperature estimates, measurements made of spectral line shifts in ionized gases and many others. These are all laboratory techniques seldom employed industrially because of their complexity and relatively high costs. For further information on this subject the reader is referred to Chapter 56, "Spectroscopic Methods of Temperature Measurement," of the book, *Temperature—Its Measurement and Control in Science and Industry*, Volume III, Part I.

Paramagnetic Salts

Magnetic thermometry has been developed chiefly to measure temperatures near absolute zero (below −458°F or −272°C). The temperatures themselves are obtained by adiabatic demagnetization of a paramagnetic salt. An isothermal magnetization at the lowest attainable liquid Helium temperature (about −458°F or −272°C) followed by an adiabatic demagnetization is used. The entropy is decreased, with a simultaneous heat flow from the sample, when the magnetic ions are oriented parallel to the field. During subsequent adiabatic demagnetizations the entropy of the salt remains constant, if demagnetization is reversible, and temperature decreases.

To obtain the temperature, some temperature dependent quality of the salt under investigation is used such as the magnetic susceptibility.

If a sphere or rotational ellipsoid of an isotropic paramagnetic salt is located in the homogeneous part of the magnetic field of a coil of a mutual inductance or a self inductance, the inductance of the coil is a function of the temperature. Inductance can be measured with an ac bridge as shown in Figure 4.6a whose balance is inde-

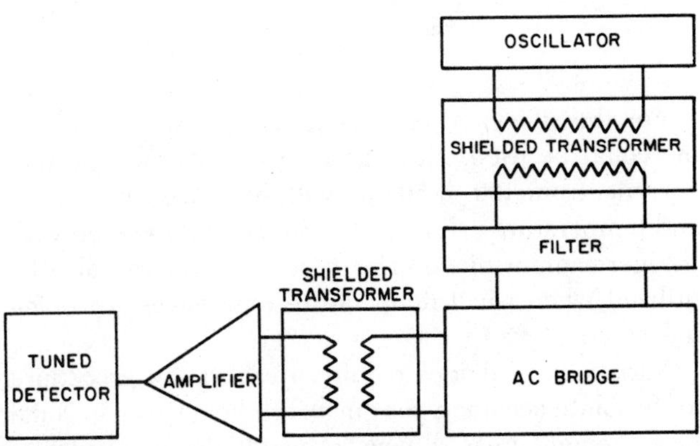

Fig. 4.6a Circuitry for temperature measurement by paramagnetic salts

pendent of frequency. A galvanometer can be used for detection. Effective shielding is a requirement.

In a paramagnetic salt with a coil surrounding it, self inductance is related to temperature. An Anderson ac bridge has been used to measure magnetic temperature in such a situation. The relationship between self inductance and susceptibility of a salt has been found to be linear when the ellipsoidal or spherical shaped salt piece is placed in the homogeneous part of the measuring field.

Accuracy of the magnetic method has been estimated on the order of .001°F (.00056°C). The method is the best available for measurements near absolute zero.

Diodes

Both silicon and germanium diodes have been used as temperature sensors. The silicon diode performs fairly well between the temperatures of −58 to 300°F (−50 to 149°C). Germanium types cover a range from −420 to 110°F (−251 to 43°C) with good linearity. Simple circuits for each type are illustrated in Figures 4.6b and c.

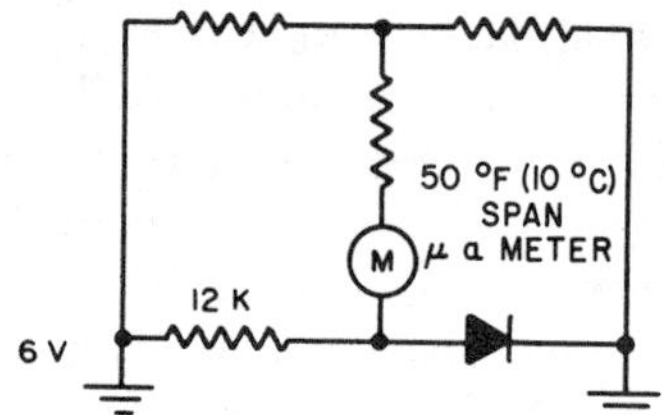

Fig. 4.6b Germanium diode thermometer circuit

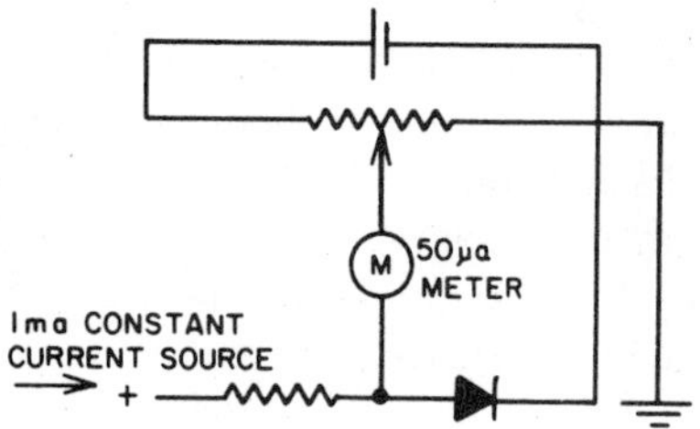

Fig. 4.6c Simple silicon diode thermometer circuit

For the silicon type, currents between 10μa and 10 ma could be used, although experiments indicate that currents from 100 to 500 μa will minimize self heating and temperature errors. A 0 to 50 μa meter is used with the germanium diode. This type of meter can also be utilized when small temperature differences are being measured.

Variations in diodes require a calibration procedure for absolute accuracy. Size may also be a factor in some measurement areas, although microdiodes are available. Low source impedances enable simple microampere indicating meters to be used. Diodes are inexpensive and their response to temperature is linear over a fair range.

Acoustic Time Domain Reflectometry

This thermometer operates on the principle that ultrasound pulses travel in solids at speeds which are a function of the temperature of the solids. The measurement is made by detecting the time needed for the acoustic pulses to travel from the transducer to the impedance demarcation point (which may be the junction between the wire and the wall of the tank) and back to the transducer. This device is in the development stage and shows good potential, although some drift in the measurement has been reported. The rhenium sensor used in the system will probably require protection with some sheathing material in the hostile environment of coal gasifiers.

Thermal Noise Thermometer

Another thermometer under development capitalizes on the phenomenon that electrical resistors provide a voltage related to thermal noise. These devices also use rhenium sensors. They are referred to as the Johnson noise detection-type thermometers.

Transistors

Temperature sensitive transistors have been used as sensors in an electronic thermometer circuit. As the temperature decreases, the base bias must be increased to maintain constant collector current. The base bias versus temperature is linear over a range of from near absolute zero to +130°F (54°C). With a constant collector current, the base bias can be calibrated as a temperature scale in degrees. Sensitivities vary, but with a limited span, differentials on the order of 0.1°F (.056°C) have been measured.

Advantages and disadvantages are similar to those of diodes. Accuracy may vary between sensors and circuitry must be home-made. Sensitive small size transistors can be easily obtained.

Liquid Crystals

Used in non-destructive testing for surface temperature measurements, liquid crystals undergo a series of color changes as temperature varies. They are an organic compound which is physically liquid, but which exhibits optical properties similar to those of a crystalline solid.

A number of solutions are available from the minimum temperature of about 68°F (20°C) to a maximum of approximately 340°F (170°C). The solutions are packaged in kit form for various ranges within these limits. Mixtures are made covering spans as narrow as 4°F (2.2°C) within the selected range.

Temperature is read by comparing the color exhibited when a thin coating is subjected to the conditions under question to a standard reference color. Response speed is less than one second. The indication is continuous and reversible. Cost is low. Disadvantages are manual preparation and limited range in addition to lack of automatic readout.

Carbon Resistors

Commercially available carbon resistors have been used as temperature sensors in the cryogenic temperature area near absolute zero (from about −424°F or −253°C downward to below −458°F or −272°C). The problems below −458°F (−272°C) have been noted in connection with paramagnetic salt measurements.

Resistor sizes of 0.1 to 1 watts and ambient resistance values up to 150 ohms exhibit a large increase in resistance below −424°F (−253°C). Reproducibilities on the order of 0.2 percent are obtainable.

Small size, low cost and general availability make their use attractive in cryogenic work.

The influence of stray radio interference and a loss of sensitivity are drawbacks. Variations between resistors make calibration difficult.

Pneumatic Probe Thermometer

Originally invented in 1893 for use in blast furnaces, the pneumatic probe has most recently been used to determine very high gas temperatures, such as in turbojet engines or rockets. It depends upon the principle that the mass flow rate of a gas through a known area restriction is a function of gas total temperature provided pressure and pressure drop across the restriction are known.

It can be used for local temperature measurement, but each case must be designed for the application. It also requires appreciable amounts of gas to be continuously withdrawn for the measurement (see Figure 4.6d).

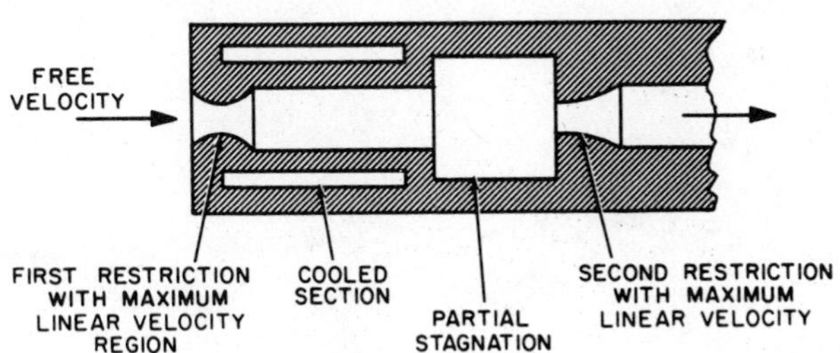

Fig. 4.6d Pneumatic thermometer

Factors affecting the reading include restriction area, friction, cooling in tube, downstream pumping pressure, approach velocity, downstream temperature, pressure loss, compressibility, and relaxation phenomena. Many of these are treated as random errors. Dynamic response is affected by a combination of pneumatic and thermometric lags.

Fluidic Sensors

The fluidic sensor is a device for converting gas temperatures into gas pressure. Actually, it is a beat-frequency detector system which contains no moving parts other than the gas.

One type of design is shown in Figure 4.6e. The oscillator is a two-chamber resonator in which the entering gases are split by a knife edge. The gases are reflected from one chamber into the other, setting up oscillations whose frequency is proportional to the square root of the absolute temperature.

A reference signal input from a temperature sensitive resonant oscillator is compared with the unknown in a beat-frequency detector made of beam-deflection fluid amplifiers. The frequency of its output is the beat frequency of the combined reference and oscillator signals. The components in the frequency converter create a steady pressure proportional to the beat frequency. Accuracies of about 2 percent and temperatures up to 2000°F (1093°C) are claimed for the system.

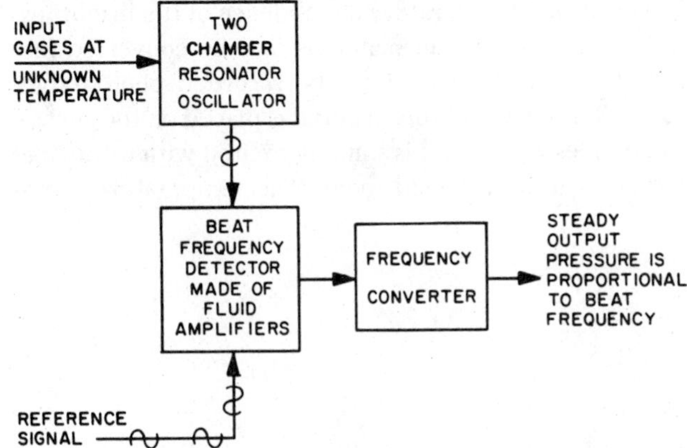

Fig. 4.6e Fluidic temperature-to-pressure converter

Ultrasonic Temperature Measurements

The relationship between sound velocity and temperature has been known for nearly one hundred years, but until recently it has been little used. Temperature dependence of velocity in an ideal gas is expressed as

$$v^2 = \frac{\gamma RT}{M_w} \qquad 4.6(1)$$

where v is sound velocity, γ is the ratio of specific heats, R is the gas constant per mole, M_w is the molecular weight and T is the absolute temperature.

A method of temperature measurement in a plasma jet involves the use of two quartz probes set a fixed distance apart. The sound velocity is determined by circuitry for the continuous measurement of the ultrasonic wave transit time (see Figure 4.6f).

Making the measurement one of time is an advantage resulting in high accuracy. Because the gas is the thermometer element, errors such as leakage are absent and fast changes can be followed. Disadvantages include cost, non-ideal gas behavior, pressure correction, accurate

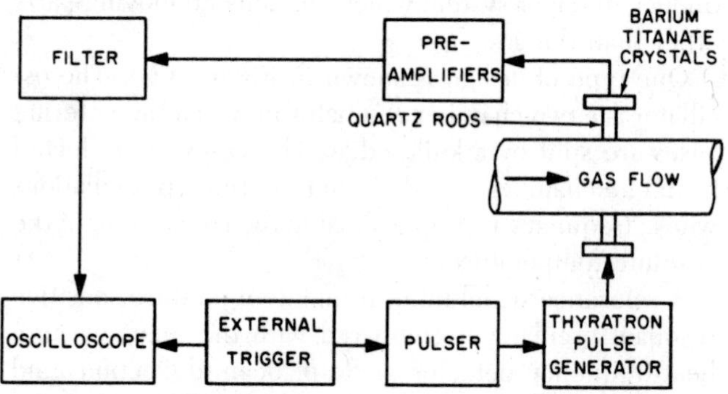

Fig. 4.6f Simplified ultrasonic temperature measuring apparatus

determination of γ and the inability to make point measurements.

Thermography

The strong temperature dependence of the brightness of certain luminescent materials may be converted into a pattern of color which can be recorded photographically. A thin layer of this material is placed on the surface to be investigated and is suitably excited with ultraviolet radiation in a darkened room. The temperature is implied by the brightness of the coating in comparison to the brightness of the same coat at a known temperature.

The sensitivity of the phosphors used gives a 10 percent brightness change per °F and this can be picked up with a relatively crude system of photometry. Temperature range for this type of measurement is from 32 to about 750°F (0 to about 400°C).

REFERENCE

1. Plumb, H.H., "Temperature, Its Measure and Control in Science and Industry," 5th Symposium on Temperature, held in 1971 by NBS, AIP, and ISA.

BIBLIOGRAPHY

Bliss, P. and Morgan, R.K., "Errors in Temperature Measurement," *Instrumentation Technology*, March, 1971.

Hormuth, G.A., "Ways to Measure Temperature," *Control Engineering*, Reprint No. 948, 1971.

Lynnworth, L.C., "Sound Ways to Measure Temperature," *Instrumentation Technology*, April 1969.

Schooley, James F., "State of the Art of Instrumentation for High Temperature Thermometry," Argonne National Laboratories' 1977 Symposium, Publication No. ANL-73-7.

4.7 PYROMETRIC INDICATORS

Temperature Range:	1085–3659°F (585–2015°C)
Cost:	\$2.50–\$4.50 per box of 50 cones; \$0.25–\$1.20 per plug
Inaccuracy:	Can be ±2°F usually ±5°F (±1.1°C usually ±2.8°C)
Partial List of Suppliers:	Bell Clay Co.; Orton Ceramic Foundation; Shell Oil Co.

This category includes those substances which imply temperature through exposure. They are small expendable plugs, chips or geometrically shaped solids whose purpose is to accompany work through a heating cycle. An indication of the temperature is obtained by noting the physical or metallurgical change which they incur during or after the heat program. Obviously, they are designed for one-time use.

Ceramic Industry

The indicator material is generally quite similar to the substance of the work under test. Pyrometric cones, used mainly in the ceramic industry, are actually composed of ceramic materials very carefully blended to soften at a certain temperature. The slender cone is slightly tilted from the vertical; when its softening point is reached the tip bends over and may actually touch the base. This action can be watched through the window of the firing furnace or its condition can be studied after cooling. Observation of a fired cone will show the experienced operator if the furnace atmosphere was oxidizing, reducing or carburizing. If the latter has taken place the cone will have formed a shell less dense than the interior. Presumably the work will have taken on the same characteristic.

Cones may be supplied singly or arranged in groups of four mounted on a plaque (see Figure 4.7a). In the latter case each cone in the series has a slightly different softening point and very fine temperature gradients can be obtained, sometimes within ±2°F (±1.1°C).

As an alternate to cones, the indicator may take the shape of a long cylindrical bar. The bars are supported at their ends with axes horizontal. On temperature rise they soften and sag at the middle under gravity. The deformation serves as a measure of temperature.

Another group of indicators operates by shrinkage rather than deformation. After removal from the furnace the diameter of a hole in the indicator, or perhaps the indicator's length, is measured and compared with the original dimension.

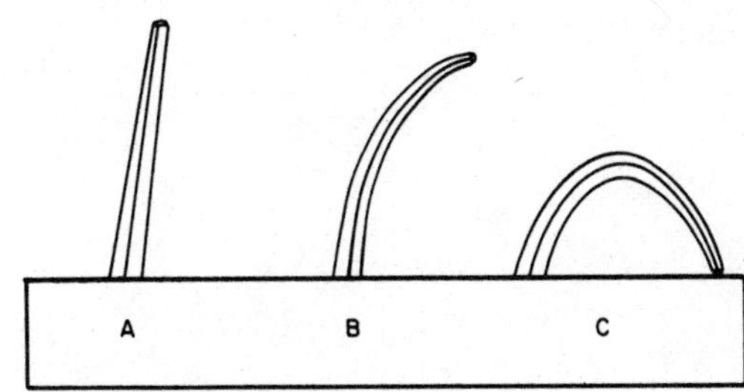

Fig. 4.7a Pyrometric cones

Like color indicators, pyrometric ceramics should not be considered exact temperature measuring devices. The fusion, bending and/or shrinking that they undergo is a *time-temperature* relationship and, as such, it is only useful to determine the end point of the specific job. This property is frequently more important than an exact measurement of the instantaneous temperature. The use of this type of indicator may almost be considered an art.

Engine Test Research

An entirely different material, used in a similar manner, is the metal test plug. This small device can tell temperature by a change in hardness that results from the heat treatment it has received. One use is to have it located carefully in an operating engine, in an otherwise inaccessible spot, where it will respond to the temperatures that occur during operation. When the test is over the plug is removed and carefully analyzed to determine the change in hardness along the horizontal axis.

Time is again a factor, but metal responds much faster than ceramic material. Exposures of less than one second duration can be detected.

Advantages of this class of temperature sensors are their relative economy and design for a very specific job. Their shortcomings are obvious.

BIBLIOGRAPHY

Considine, D.M., *Encyclopedia of Instrumentation and Control,* McGraw-Hill, New York, 1971.

Hormuth, G.A., "Ways to Measure Temperature," *Control Engineering,* Reprint No. 948, 1971.

Plumb, H.H., "Temperature, Its Measure and Control in Science and Industry," 5th Symposium on Temperature, held in 1971 by NBS, AIP, and ISA.

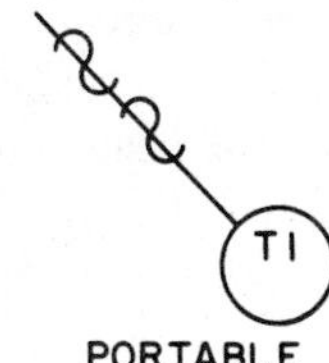

4.8 PYROMETERS—OPTICAL AND INFRARED TYPES

Types:	A—Manual optical pyrometers B—Automatic optical and infrared pyrometers
Temperature Range:	A—1400–6300°F (760–3500°C) B—0–6000°F (0–3300°C)
Temperature Span:	A—Not available B—As narrow as 100°F (56°C) or less
Linearity:	A and B. Non-linear, varies as 4th power
Cost:	A—$600–$3300 for complete system B—$1300–$11,000 for complete system, depending on required accessories
Inaccuracy:	A and B—±1–2% of full scale B—Can be ±0.1°F (0.05°C) on narrow spans
Partial List of Suppliers:	Barber-Colman Co. (B); Barnes Engineering Co. (B); Instrument Div., Honeywell, Industrial Div. (B); Irçon Inc. (B); Infrared Industries, Inc. (B); Leeds & Northrup (A and B); Pyrometer Instrument Co., Inc. (A); Raytek, Inc. (B); Vanzetti Infrared & Computer Systems (B)

Manual Optical Pyrometers

Optical pyrometers are narrow-band or two-color radiation pyrometers (see Section 4.9 for theory) which operate in the visible spectrum around the 0.65 micron point. The human eye, acting as the detector in the manually-balanced type, compares a source of known radiant energy generated within the instrument to the incoming unknown source. A filter interposed between the eye and both sources of energy cuts out the shorter wavelengths. This serves a dual purpose: (a) it minimizes the difference between eyes, permitting an easier color match, and (b) it permits an extension of the temperature range beyond the point where the eye could no longer tolerate the amount of energy if viewed directly.

The instrument is shaped to be held in the hand and up to the eye so that it may be sighted on the target. An adjustable focus permits the operator to focus an image of the source whose temperature is to be determined. The filament of the standard source is placed on the same plane as this image so that the two appear superimposed on one another when viewed through the eyepiece.

A null type of balance is usually used where a rheostat, moving against a calibrated dial, is manually rotated to vary the current through the standard source until it just disappears into the field of the unknown. A slight modification of this principle maintains the standard source constant and varies the amount of interposing absorbing material in the optical path.

The range of the manual optical pyrometer is limited on the low end to a minimum of 1400°F (760°C) since there is insufficient emission of visible light for an accurate comparison below this figure. At 2400°F (1316°C), the image would become too bright to look at directly but filters are usually interposed to permit readings as high as 6300°F (3500°C).

The use of the human eye as the detector restricts accuracy somewhat because the eye responds to both color and brightness rather than directly to energy, and no two eyes are exactly the same. However, it is possible to detect both a color and a brightness match by adjusting to the "minimum difference" between known and unknown.

The manual optical pyrometer is a self-contained unit with its own power supply for operating the current for the known radiant energy source. It can be mounted in place or hand-held by an operator as he takes a sighting.

Its advantages include that it is a light, portable self-contained unit of reasonable accuracy if sighted into a near black body furnace.

Its disadvantages are:

1. It requires the operator to adjust the temperature dial manually. It is not suitable for alarm or control functions.
2. It can only be used at relatively high temperatures where plenty of visible energy is given off.
3. It is subject to emissivity errors inherent to a narrow-band radiation pyrometer.

Automatic Optical and Infrared Pyrometers

The automatic optical pyrometer uses an electrical radiation detector rather than the human eye, and consequently it is not limited to the visible wave lengths of the spectrum. It can reach far into the infrared or the near ultraviolet using either a narrow band, a two color or a wide band selection in accordance with the optical system and detector (refer to Section 4.9 for theory).

Although there are many adaptations, the automatic optical and/or infrared pyrometer operates essentially by comparing the amount of radiation emitted by the target with that emitted by an internally-controlled reference source. The output is proportional to the difference in radiation between the variable source and the fixed reference. The system usually consists of two components, the optical head and the electronic amplifier.

In some models (Figure 4.8a), the optical head contains a temperature-controlled black body source, the required filters, a detector, a preamplifier and an optical chopper. The chopper, driven by a synchronous motor, alternately exposes the detector to incoming and cavity radiations at a frequency which might be 60–120 cycles per second range.

Another model (Figure 4.8b) uses the human eye to adjust the focus. Radiant energy passes through the front lens onto a dichroic mirror which allows visible light to pass through, while infrared radiation is reflected onto the detector. The operator adjusts the eyepiece for the correct focus by just filling the circle reticle plate with the visible image.

The calibrate flag is solenoid-operated from the amplifier and when actuated cuts off the radiation coming through the lens and focuses the calibrate lamp onto the

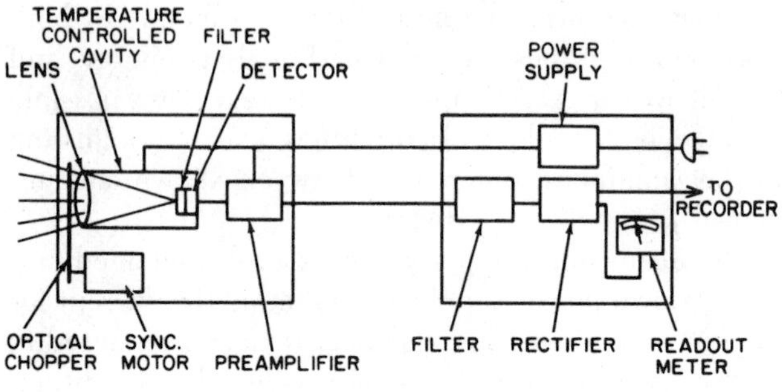

Fig. 4.8a Infrared system

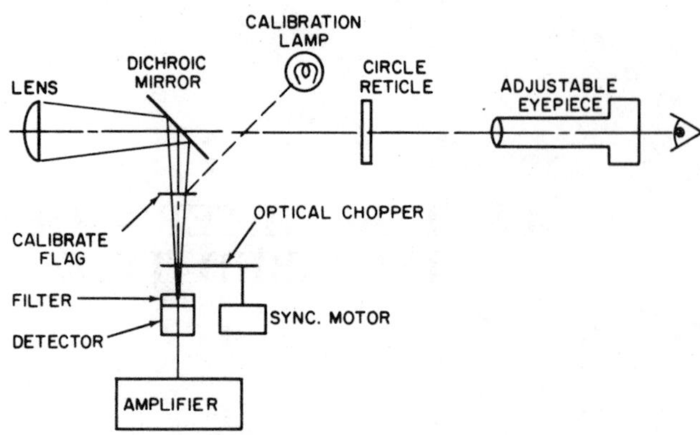

Fig. 4.8b System with dichroic mirror

detector. The known radiation from the lamp allows the entire system to be checked.

Automatic optical or infrared pyrometers are closely tied in with radiometers and other optical devices. Telescopic lenses are sometimes employed for their special effects. The instrument may have a wide-angle field of view for large area scanning or it may have a very narrow angle where the target is only .05 in. (1.27 mm) diameter at 8 in. (203 mm) distance.

The entire pyrometer, optics plus amplifier, has been packaged into a single, hand-held, battery-operated device for open field work. It is shaped somewhat like a pistol held in the hand and aimed at the target. Pulling the trigger energizes the standard reference source and the readout indicator.

A very practical application of the infrared pyrometer in the 8–14 micron range is as a hot spot detector. The operator can move quickly from one object to another looking for undesirable internal sources of heat generation.

It is possible to attach a camera to the unit and take a temperature profile as a picture for immediate or future analysis.

Table 4.8c shows some typical uses at various spectral responses for this versatile temperature measuring device.

Fiber Optic Thermometers

Fiber optic measurement systems consist of three elements: the fiber optic assembly, an infrared detector, and an electric console. A single fiber or bundle of several fibers gather infrared radiation from the target, transmit to the detector and convert the radiation to a voltage suitable to the required function.

Fibers are sensitive only to the infrared portion of the spectrum and filter out other radiation; thus they are not activated by flames and fumes. Even though infrared radiation becomes detectable at about 140°F (60°C), it is best to use 212°F (100°C) as the starting point in any monitoring application.

Fibers, whether single or in bundles, are always en-

Table 4.8c
USES OF AUTOMATIC OPTICAL AND INFRARED PYROMETERS

Spectral Response (Microns)	*Main Areas of Application*
0.2–0.3	Jet Engine Rotor Blades, Vacuum Deposition
0.653	Arc Furnaces, Hot Metal Filaments, Refractory Metals
0.6–1.0	Metal Processing Industry
2.0–2.6	Textiles, Plastics, Ceramics, Food Products
3.43 ± 0.14	Thin Film Polymers, Vinyl, Nylon
4.42	Carbon Dioxide Gas
4.8–5.6	Glass, Dry Chemicals, Oxidized Metals
8–14	Background Land, Sea, Clouds, Profiles for Hot Spot Detection

closed in metal or ceramic sheaths for protection. A unique characteristic of the special glass fibers is the ability to bend light and to transmit it for distances up to 30 ft. (9.1 m) without distortion or loss of definition.

Perhaps the one outstanding feature of fiber optic systems is the ability of the fibers to withstand and function in hostile environments, including intense heat. Fibers maintain resolution exceeding one degree change at temperatures over 2000°F (1100°C). At extreme high temperatures and under other severe conditions, the fibers are protected with air or inert gas purging or by water cooling.

The advantages of using fiber optics in non-contact temperature measurement applications include their inertness, relatively rugged design, small size and ability to "look around" opaque objects. With the addition of a telescopic lens system, the same fiber optic assembly can be made to monitor different target areas at various focal distances.

BIBLIOGRAPHY

Bulkley, D., "Temperature Measurement by Infrared Radiation," Instruments and Control Systems, September, 1970.

Greelish, W.T., "Non-Contact Temperature Measurement Systems," Paper Number 76–807 at the 1976 ISA Conference in Houston, Texas.

Intrieri, A.J., "Optical Fibers Look Around Obstacles to Measure Temperature," Control Engineering, December, 1977.

Schooley, James F., "State of the Art of Instrumentation for High Temperature Thermometry," Argonne National Laboratories, 1977 Symposium, Publication Number: ANL-78-7.

Flow Sheet Symbol

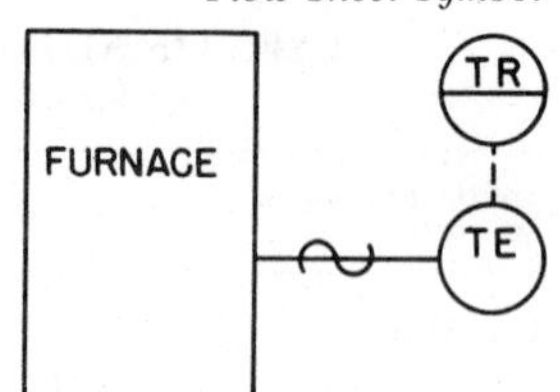

4.9 PYROMETERS—RADIATION

Temperature Range:	0–7000°F (0–3900°C)
Linearity:	Non-linear, approximate 4th power
Cost:	Sensor alone, \$600 to \$2500; complete system, \$1,000 to \$6,000
Inaccuracy:	± ½–1% of full scale
Partial List of Suppliers:	Bailey Controls Co.; Barber Colman Co., Industrial Instruments Div.; Barnes Engineering Co.; Bristol-Babcock Div., ACCO Industries, Inc.; Honeywell Process Control Div.; Hy-Cal Engineering; Ircon, Inc.; Leeds & Northrup Co.; Mikron Instrument Co.; Pyrometer Instrument Co.; Raytek, Inc.

Thermal radiation is a universal property of matter which is absent only when the material is an inert gas like helium, or is at absolute zero temperature. From this property of matter, the technique of radiation pyrometry has been developed so that it is possible to infer the temperature of an object without contacting it. This is done with an optical system which collects the visible and infrared energy, sometimes through a filter, and focuses it on a detector (see Figure 4.9a). The detector converts the concentrated energy to an electrical signal, a complex function of the absolute temperature.

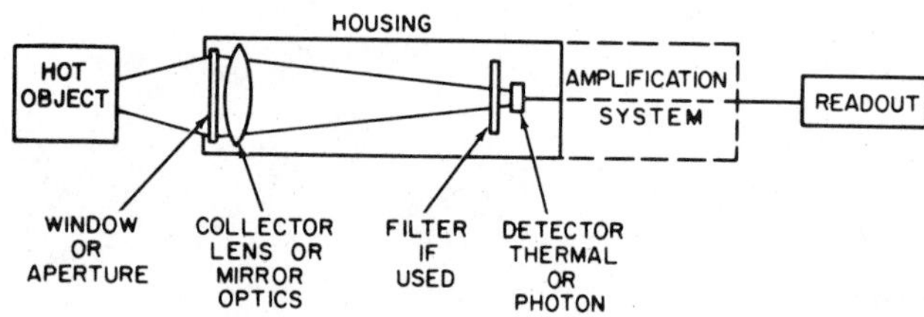

Fig. 4.9a General radiation pyrometer system

The development of radiation pyrometry stems from Planck's quantum theory, about 1900, and from Stefan-Boltzmann's law for total radiated energy, $W = KT^4$, where K is the Boltzmann constant and T is the absolute temperature.

Only the energy given off between about 0.3 and 20 microns is of sufficient magnitude to be of any consequence. This encompasses the visible spectrum (0.35–0.75 microns*) and the near infrared. The intensity and distribution of this energy from a substance may be compared with that of a black body* which radiates its energy in a theoretically predictable spectral distribution and intensity, such as that shown for several temperatures in Figure 4.9b. The area under the curve represents the total amount of energy radiated at all wavelengths.

Real targets, however, always deviate from an idealized black body to some degree. The ratio of energy radiated by a real body to that of a black body under similar conditions is termed the emittance* (E). Two other optical ratio characteristics of targets are reflectance* (R) and transmittance* (T) and for a body whose temperature is constant, $E + R + T = 1$ for any wavelength. The radiation pyrometer would gather in all three of these energy sources over the wavelength band to which it was sensitive. In Figure 4.9c, if the purpose were to measure the temperature of Object B, and if A were the same temperature as B, B would absorb, emit, reflect and transmit radiant energy and would appear to be a black body.

But frequently A is not uniform in temperature nor does it completely surround B. Furthermore, B might be cooler than A or have a high reflectance which causes it to reflect extraneous sources of radiant energy. If any of these conditions prevail, the measurement of the total

*Terms marked with an asterisk are defined at the end of this Section.

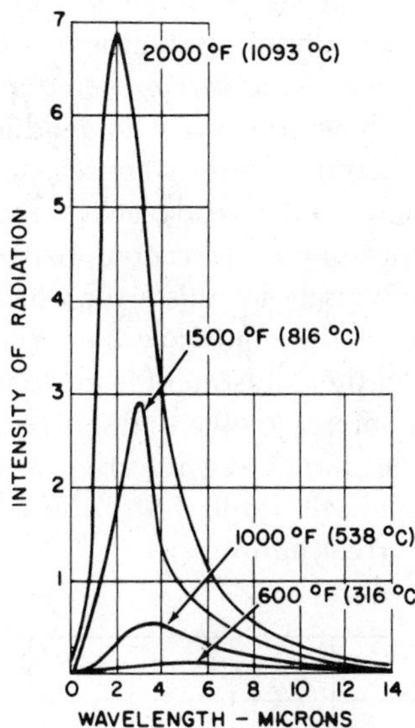

Fig. 4.9b Radiation vs. wavelength for various temperatures of black body

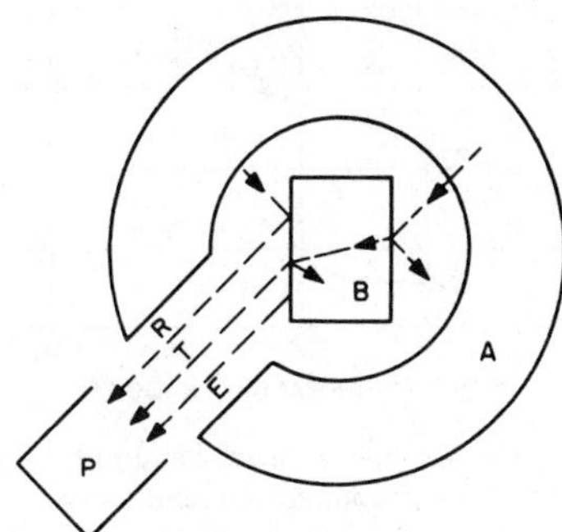

Fig. 4.9c Emission, reflection and transmission

energy radiated by B cannot be converted exactly into temperature with the Stefan-Boltzmann law.

For best results, emittance should be high and reflectance low. Transmittance of most solid objects will be near zero. If the process material is not solid, the radiant energy detector will actually "see" beneath the surface, or if the object is thin, right through it.

Emittance, reflectance and transmittance are not easy factors to derive and they vary considerably with wavelength. Materials like ferrous metals with a matte surface have a high emittance at the short end of the spectrum but become lesser emitters for the longer waves. On the other hand, glass acts in nearly the opposite way, being practically transparent to visible energy and almost opaque to wavelengths in the 5–7 micron region.

Pyrometer Design and Geometry

In designing or choosing the radiation pyrometer for a particular application, consideration must be given to the following characteristics:

1. The field of view, or the size-distance relationship.
2. The transmission qualities of the collector system and any windows or filters in the optical path.
3. The band pass* and sensitivity of the detector.

Figure 4.9d shows a typical wide-angle field of view. Note how the target size requirement necks down to a minimum at the focal length of a lens in such a system. The narrow-angle field of view, shown on Figure 4.9e, flares out more slowly. Cross-sectional area in either case can vary from circular to rectangular and even slot-shaped depending on apertures in the housing design. On some designs telescopic eye pieces can magnify the radiant energy so that much smaller targets at greater distances can be viewed. Targets 1/16 in. (1.6 mm) in diameter are feasible with the proper pyrometer design.

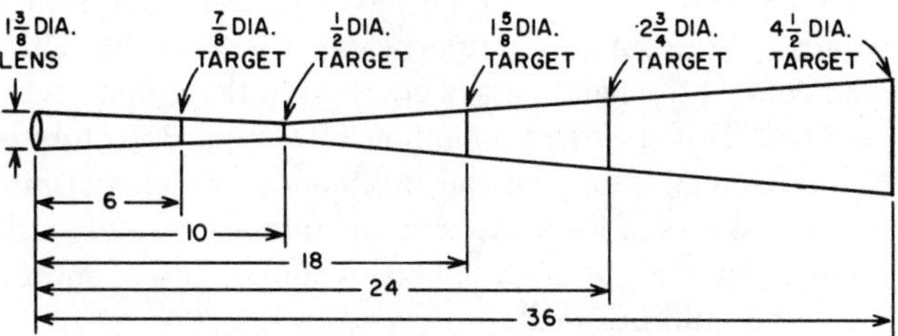

Fig. 4.9d Sighting path of the wide-angle total radiation pyrometer

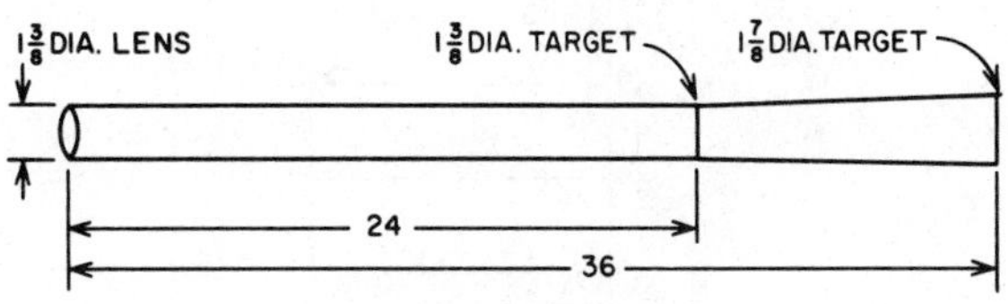

Fig. 4.9e Sighting path of the narrow-angle total radiation pyrometer

The physical shape of the optical system (lens or curved mirrors) and its mounting within the pyrometer housing control the sighting path while the material from which it is made determines the optical properties. Glass does not transmit well beyond 2.5 microns and is suitable only for the higher temperatures where plenty of output is available. Other popular optical materials are quartz (fused silica) to 4 microns and crystalline calcium fluoride out to about 10 microns. Lesser used (and more expensive) materials will increase the transmission even more.

Windows and filters in front of or behind the optical system can alter the transmission properties greatly. A plate glass window in front of a calcium fluoride lens, for instance, will very effectively stop the longer wavelengths which would have passed through the lens.

A band pass filter might be purposely placed in front of the detector to cut off unwanted wavelengths.

Detectors

The detector receives the radiant energy focused on it by the optical system and converts it into an electrical signal. There are two classes:

1. Thermal detectors produce an output because they absorb energy and become heated. A thermopile* is most often used in regular wide band pyrometers although bolometers* have also been used to a limited extent.
2. Photomultiplier tubes and photon detectors* produce an output because the radiant energy reaching them releases electrical charges in the body of the detector. Photon detectors are a semi-crystalline material such as silicon, lead sulphide or others.

While thermopiles and bolometers will respond to energy throughout the entire spectrum, photon detectors are very sensitive to the wavelength and are sometimes chosen for this property (see Figure 4.9f). They can control the band pass in addition to that imposed by a filter. The speed of operation of photon detectors is many times faster than the thermopile, which permits them to be used for measuring the temperature of small objects moving at high speed, where relative energy output would be small.

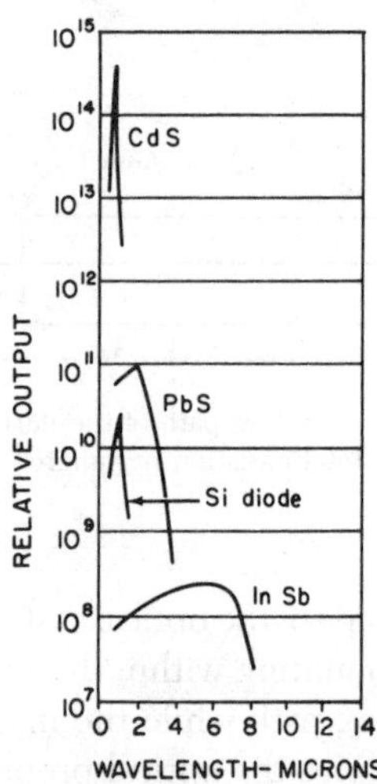

Fig. 4.9f Response to energy of some photon detectors

Radiation Pyrometer Types

Emittance is a very uncertain characteristic of any material and methods are continually sought to overcome this variable effect. Three groups of radiation pyrometers are recognized each having a slightly different attack on the emittance problem.

Broad band radiation pyrometers attempt to measure as much of the radiant energy coming from a hot body as possible. The simplest of the three, it has substantially no selectivity for specific wavelengths other than a cut-off inherent in the optical system and is dependent on the total emittance of the surface being measured.

Figure 4.9g shows the error in reading for various emissivities and temperatures when a typical instrument of this type is calibrated to read black body temperatures. A calibration control on the readout instrument permits the user to compensate for this error which corrects the reading as long as the emittance does not change. The proper setting of the adjustment is made with the aid of a second thermometer known to be correct; or, knowing the nature of the target and referring to a table giving total emissivities (see Table 4.9h), the adjustment can be set to the correct figure.

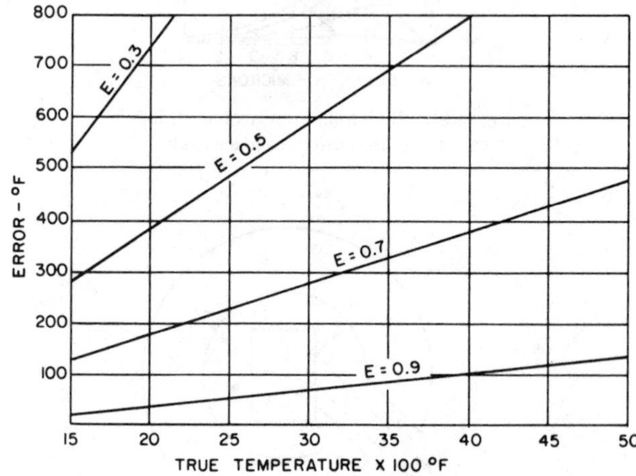

Fig. 4.9g Error in reading of broad band radiation pyrometer as function of temperature and emissivity

Table 4.9h
TOTAL EMITTANCES OF VARIOUS SUBSTANCES

Material	*Emittance*	*Material*	*Emittance*
Aluminum		Iron	
Unoxidized	0.06	Oxidized	0.89
Oxidized	0.19	Rusted	0.65
Brass (oxidized)	0.60	Lead (oxidized)	0.63
Calorized Copper	0.26	Monel (oxidized)	0.43
Calorized Copper (oxidized)	0.19	Nickel	
		Bright	0.12
Calorized Steel (oxidized)	0.57	Oxidized	0.85
		Silica Brick	0.85
Carbon	0.79	Silver (unoxidized)	0.03
Cast Iron		Steel (oxidized)	0.79
Oxidized	0.78	Steel Plate (rough)	0.97
Strongly oxidized	0.95	Tungsten (unoxidized)	0.07
Copper (oxidized)	0.60	Wrought Iron (dull oxidized)	0.94
Fire Brick	0.75		
Gold	0.03		

Broad band (sometimes called total) radiation pyrometers must have an unobstructed sighting path to the target. The presence of smoke or carbon dioxide will absorb some of the radiant energy and cause a low reading. The optical system must be kept clean, with a sighting window protected from any corrosive gases which otherwise would come in contact with it. In furnaces, a closed-end sighting tube is frequently used to protect the optics and to provide a clear sighting path. The back end of the tube assumes the same temperature as the furnace. Water or air cooling is sometimes used to keep the housing temperature cool on these installations.

Ambient temperature compensation is required for those total radiation pyrometers using a thermopile detector. Nickel wire, having a temperature coefficient opposite to that of the thermopile wires, is most often utilized. For low temperature work, a thermostatically-controlled housing is often employed to eliminate any ambient temperature fluctuations.

Single band pass pyrometers operate over a selected, usually narrow, band of the energy spectrum centered at a desired point. For high temperature measurement of metals, for instance, the band might be very narrow at the 0.65 micron point, the red end of the visible spectrum where metal emissivity is highest. At this visible point the instrument might be referred to as a brightness pyrometer.*

To measure gas temperatures, a band around 4.3 microns to pick up carbon dioxide might be chosen, while glass surface temperatures, previously mentioned, would be measured in the 5–7 micron wavelength.

Emissivity over a narrow wavelength band will not vary as much as it would over the total spectrum, but the limited band pyrometer will suffer somewhat from a lack of sensitivity because of the reduced energy available. The use of more sensitive photon detectors helps make up this deficiency as well as providing a desired band pass.

*The ratio** or *two-color pyrometer** measures the energy it receives from two rather narrow bands and divides one by the other. When the two bands are chosen so that there is very little change in emissivity from one to the other, such as would be the case if the bands were close together, the emissivity factor nearly cancels out. Thus, low emissivity bodies which create the most error for the broad and single band pyrometers do not have such a large effect on the ratio-type instrument.

The selection of the two wavelengths is not arbitrary but, like the single band pyrometers, is made for the particular application.

Summary

The broad band pyrometer is used generally in industry for readout and automatic control. It can cover all temperature ranges and is the least expensive of the three types. Narrow band and two-color pyrometers are used, where necessary, to minimize the emissivity effects and for special applications where it is desirable to select the particular band pass.

The prospective user of a radiation pyrometer should consider the following points.

1. Target temperature, low, normal and high limits.
2. Minimum target size and distance factors.
3. Target material and emittance.
4. Angle of observation.
5. Is target stationary or moving? If moving, will the speed of response of the pyrometer be fast enough?
6. Atmospheric conditions between target and detector.
7. Ambient temperature.
8. Can pyrometer sight directly on target or must it sight through a sealed auxiliary window such as required for vacuum or pressure?
9. Is scale to be read directly in temperature units or will an arbitrary reading be satisfactory?

ADVANTAGES

1. Does not require physical contact with material whose temperature is being measured.
2. Fast speed of response—can be used on moving targets.
3. Can look at small targets (1/16 in. or 1.6 mm in diameter) or measure the average temperature over a wide area.
4. Measures much higher temperatures than thermocouples.

DISADVANTAGES

1. More fragile and costlier than thermocouples, RTDs or thermistors.
2. Non-linear scale shape, approximating the 4th power of the temperature.
3. Emissivity of target may cause a low temperature reading if not corrected.
4. Relatively wide temperature span required.

Definition of Terms Used in Radiation Pyrometry

Absorbance—Ratio of radiant energy absorbed by a body to the corresponding absorption of a black body at the same temperature. Absorbance equals emittance on bodies whose temperature is not changing.

Band pass filter—An optical or detector filter which permits the passage of a narrow band of the total spectrum. It excludes or is opaque to all other wavelengths.

Black body—The perfect absorber of all radiant energy striking it. It radiates energy in a theoretically predictable spectral distribution and intensity (Figure 4.9b). Its emittance is unity.

Bolometer—Thermal detector which changes its electrical resistance as a function of the radiant energy striking it.

Brightness pyrometer—Uses the radiant energy on each side of a fixed wavelength of the spectrum. This band is quite narrow and usually centered at 0.65 microns in the orange-red area of the visible spectrum.

Emittance—Ratio of radiant energy emitted from a body to the corresponding radiation emission of a black body at the same temperature.

Gray body—Has an emittance less than unity but this emittance does not change with wavelength over that part of the spectrum in which interest is centered.

Infrared—That portion of the spectrum whose wavelength is longer than that of red light. Only the portion between 0.7 and 20 microns gives useable energy for radiation detectors.

Micron—.001 millimeters. 10,000 Angstrom units. A unit used to measure wavelengths of radiant energy.

Photon detector—Measures thermal radiation by producing an output through release of electrical changes within its body. They are small flakes of crystalline materials such as CdS or InSb which respond to different portions of the spectrum, consequently showing great selectivity in the wavelengths at which they operate.

Ratio pyrometer—See *two-color pyrometer*.

Reflectance—The percentage of the total radiation falling on a body which is directly reflected without entry. Reflectance is zero for a black body, and nearly 100 percent for a highly polished surface.

Spectral emissivity—The ratio of emittance at a specific wavelength or very narrow band to that of a black body at same temperature.

Thermopile—Measures thermal radiation by absorption to become hotter than its surroundings. It is a number of small thermocouples arranged like the spokes of a wheel with the hot junction at the hub. The thermocouples are connected in series and the output is based on the difference between the hot and cold junctions.

Total emissivity—The ratio of the integrated value of all spectral emittances to that of a black body.

Transmittance—The percentage of the total radiant energy falling on a body which passes directly through it without being absorbed. Transmittance is zero for a black body and nearly 100 percent for a material like glass in the visible spectrum region.

Two-color pyrometer—Measures temperature as a function of the radiation ratio emitted around two narrow wavelength bands. Also called ratio pyrometer.

BIBLIOGRAPHY

Hormuth, G.A., "Ways to Measure Temperature," *Control Engineering*, Reprint No. 948, 1971.

Instrumentation and Control Systems Engineering Handbook, TAB Books, 1978.

Leftwich, R.F., "Infrared Radiometry Applied to Critical Temperature," Argonne National Laboratories' 1977 Symposium, Publication No. ANL-78-7.

Schooley, James F., "State of the Art of Instrumentation for High Temperature Thermometry," Argonne National Laboratories' 1977 Symposium, Publication No. ANL-73-7.

Flow Sheet Symbol

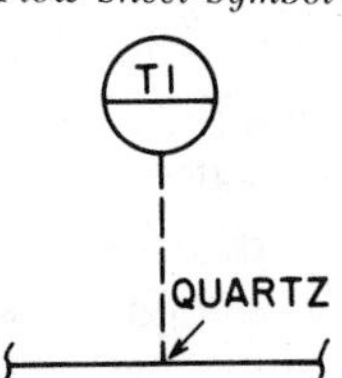

4.10 QUARTZ CRYSTAL THERMOMETRY

Temperature Range:	−40 to 450°F (−40 to 232°C)
Linearity:	±0.3°F (±0.2°C) over full range; ±0.04°F (±0.02°C) over 0 to 200°F (93°C) portion
Cost:	$5000 single probe; $7300 differential type
Inaccuracy:	±0.2°F (±0.1°C)
Partial List of Suppliers:	Bailey Controls Co.; Hewlett-Packard Co.

Temperature measurement using quartz thermometers is based on the change in resonant frequency in response to a temperature change. This temperature sensitivity is on the order of .0005°F (.0003°C) under ideal conditions. Temperature deviations on the order of 10^{-6}°F have been measured in the laboratory.

The effect of temperature on the frequency of quartz crystal controlled oscillators has been known for some time, work on this subject having been published in 1946. Work by various individuals through 1962 resulted in development of the sensitive laboratory device previously mentioned.

A commercial application of the quartz crystal thermometer includes a unique angle of cut, which exhibits a very linear relationship between resonant frequency and temperature. In this case a reference oscillator and a sensor oscillator are employed. The reference oscillator frequency is selected to provide zero beat with the sensor oscillator when the probe is at 32°F (0°C). Sensor oscillators operate to obtain a frequency sensitivity of 500 cycles per second per degree F. Dividers are arranged to obtain a sample period of .01 seconds. A resolution of 0.2°F (0.1°C) is obtained with a digital readout (Figure 4.10a). A digital-analog converter changes the output for use in a strip-chart recorder. A differential measurement can also be made using similar circuitry, with the principal difference being that gating circuits are added to enable the oscillator circuits to be heterodyned against each other.

Probes are provided with the quartz crystal hermetically sealed in a stainless steel cylinder, similar to a well for a thermocouple (Figure 4.10c) or a resistance thermometer capsule (Figure 4.10b). This makes the sensor unit larger than either the two sensors mentioned or the thermistor; a ⅜ in. (9.5 mm) outside diameter is typical. However, the probes can be used to pressures of 3000 PSIG (21 MPa) and can stand shocks of 10,000g without changing calibration.

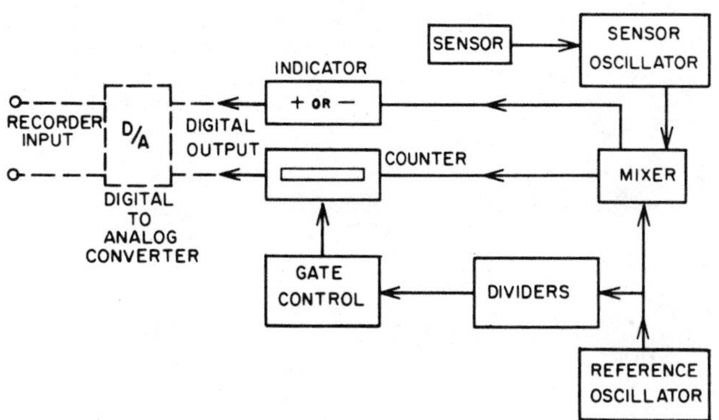

Fig. 4.10a Quartz thermometer system block diagram

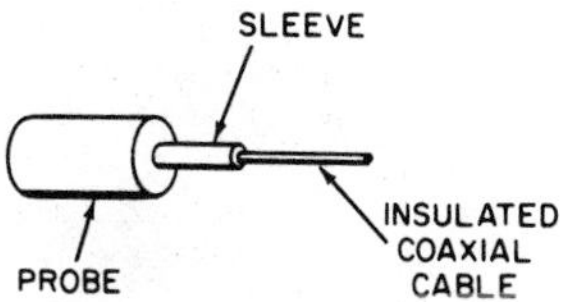

Fig. 4.10b Quartz crystal sealed in capsule

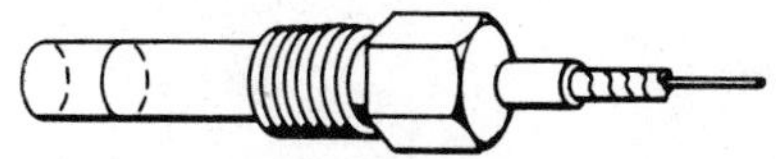

Fig. 4.10c Quartz crystal in stainless steel well

Response time is stated as being 1 second for a step change in a well-stirred water bath. A strong feature is stability, which for short term is in the order of .0001°F (.000056°C). Long term stability, for periods of a month or over, is in the order of .02°F (.01°C). These can be degraded by oscillator drift, but, in an environmentally controlled area, this is not much of a problem.

Advantages and disadvantages are summed up as follows:

ADVANTAGES

1. No lead resistance or noise problems because temperature is converted to frequency.
2. Excellent short term stability.
3. Good accuracy.
4. One second response time.
5. Accurate differential measurements possible.
6. Rugged—can withstand shocks without changing calibration.

DISADVANTAGES

1. Expensive.
2. Accuracy somewhat lower than resistance thermometer or thermistor.
3. Probe size larger than thermocouple of RTD.
4. Best used in laboratory environment.

BIBLIOGRAPHY

Hormuth, G.A., "Ways to Measure Temperature," *Control Engineering*, Reprint No. 948, 1971.

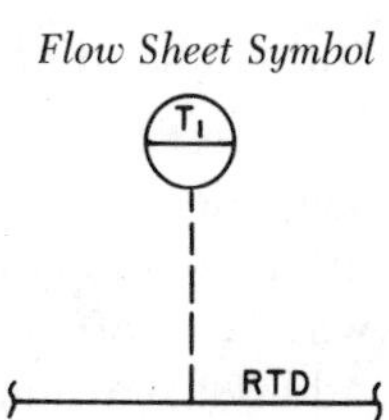

4.11 RESISTANCE THERMOMETERS

Application:	Industrial and lab temperature measurement
Temperature Range:	−430 to 1200°F (−257 to 649°C)
Inaccuracy:	.1°F (.05°C), platinum only
Stability:	Better than .4°F (.2°C) per year, platinum only
Interchangeability:	.5°F (.25°C) or .5% of span
Linearity:	Platinum and copper linear, nickel and balco nonlinear
Configurations available:	All, probe and surface types
Cost:	\$25 to \$80 industrial, to \$2000 labs standards
Partial List of Suppliers:	Action Instruments Co.; Burns Engineering Inc.; Foxboro Co.; Honeywell, Process Control Div.; Hy-Cal Engineering; Leeds & Northrup Co.; Minco Products, Inc.; RdF Corporation; Rosemount Inc.; Yellow Springs Instrument Co.

Resistance thermometry is based upon the increasing electrical resistance of conductors with increasing temperature. The application of this property using platinum was first described by Sir William Siemens at the Bakerian Lecture of 1871 before the Royal Society in Great Britain. The necessary temperature limitations and methods of construction were established by Callendar, Griffiths, Holborn and Wein between 1885 and 1900. Platinum resistance thermometers are now the international standard for temperature measurements between the triple point of hydrogen at 24.86R (13.81K) and the freezing point of antimony at 1167.35°F (630.75°C). The laboratory application of platinum resistance thermometers recognizes the unsurpassed stability and repeatability of this noble metal sensor. Platinum resistance thermometers of more rugged industrial construction retain this advantage over other conductors used in industrial resistance thermometers.

The conductors used for resistance thermometry include platinum, nickel of various purities, 70 percent nickel/30 percent iron (Balco[tm]) and copper, listed in order of decreasing temperature range capability. No other candidates are generally available in commercial thermometers. These conductors are all available as fine wire for sensor winding. Platinum is also available as a deposited film sensor, and nickel and Balco are available in foil-type sensors. All share in varying degrees the characteristics of repeatability, high temperature coefficient, long term stability, and linearity over a useful temperature range. While no sensor material surpasses platinum in overall performance, each has at least one characteristic that may encourage its selection.

The application of platinum as a temperature standard (SPRT), including the careful application of manual bridge resistance measurements, differs considerably from industrial practice.[1] The temperature standard is 25.5 ohms at the ice point to stay within the range of practical Mueller bridges while providing a nominal .1 ohm/°C sensitivity. Standard platinum thermometers are constructed in a manner to be almost totally strain-free, using very lightly supported wire of larger size (and cost) than typical in an industrial thermometer (RTD). Temperature coefficients near the theoretical maximum for pure platinum and maximum thermal stability are obtained at the expense of fragility and large size. The industrial platinum RTD offers ruggedness at a negligible loss in temperature coefficient compared to the standard SPRT, while retaining the best stability and repeatability available in industrial thermometers over a wide temperature range.

Industrial RTD Construction Requirements

All resistance thermometers require the following considerations in their manufacture. Wire-wound sensors must be supported on mandrels closely matching the wire in thermal expansion to minimize strain effects. Additional assembly materials, such as cements, should not introduce additional strain in the operating temperature range. The final assembly must be in a stable, annealed condition, trimmed to the required resistance tolerance. Only high purity materials and clean assembly methods should be used to avoid sources of contamination that might degrade the sensor. All internal connections should be welded, and connecting leads should be chosen for the required temperature capability and avoidance of thermoelectric junctions. To realize the ruggedness of fully-supported elements in the total sensor assembly, all internal connections should be anchored and isolated from effects of thermal and mechanical strains, including shock and vibration. The same requirements apply when deposited film or foil-type resistance elements are used.

For equivalent performance in their respective temperature ranges, base metal RTD's cost the same as 100 ohm platinum RTD's. Construction requirements and materials cost are similar. Base metal RTD's have a materials cost advantage at higher resistance values compared to wire-wound platinum sensing elements. Thin film platinum elements erase this advantage.

Industrial Platinum RTD's

In the case of platinum RTD's, the fully-supported rugged construction using "reference grade" wire results in a temperature coefficient (alpha) over the interval 32 to 212°F (0 to 100°C) between .00387 and .003915 ohms/ohm-°C, depending on the manufacturer. Compared to .003927 ohms/ohm-°C on a SPRT, the reduction in sensitivity is insignificant. For best accuracy, the user should be aware of or specify the actual temperature coefficient. One common value available from most manufacturers is .003902 ohms/ohm-°C. This is the result obtained for windings on pure alumina mandrels.

Standard platinum industrial RTD curves based on slightly doped platinum wire have been adopted by several countries. These curves are all substantially identical with an alpha of .00385 ohms/ohm-°C. This result is reproducible by manufacturers everywhere and the so-called "international grade" platinum curve is the most widely used curve, even in the U.S. An exception is temperatures below −320°F (−196°C) (LN_2) where only reference grade wire RTD's are well characterized.

Platinum RTD's using thick or thin films are also available to the same curve as international grade wire-wound sensors. Performance is often equivalent to wire-wound sensors except maximum temperature may be reduced. Wire-wound platinum RTD's are most common at 100 ohms ice point resistance, with 200 and 500 ohms available at additional cost. Using thin films, ice point resistances of 100 and 1000 ohms are available at the same cost with slightly lower alpha specified at 1000 ohms.

Base Metal RTD's

Second in usage to platinum is high purity nickel which offers the highest temperature coefficient, second highest temperature range and lower assembled cost than wire-wound platinum at high resistance values. 120 and 500 ohm resistances are the most common, with 1000 ohms available. Nickel is nonlinear, with an increasing temperature coefficient at increasing temperature. Nickel is also highly strain sensitive and requires great care by the manufacturer to obtain good interchangeability. The temperature coefficient of nickel is highly influenced by both purity and state of anneal. In addition, lower purity nickel such as 99 percent nickel (no longer available) or ballast nickel has been used and provides a somewhat lower temperature coefficient. The maximum temperature of nickel sensors is limited to near 500°F (260°C). There is no internationally recognized standard curve for nickel sensors, although there are national standards, and several manufacturers in the United States can provide sensors to a common curve characterized by an alpha temperature coefficient between 32 and 212°F (0 and 100°C) of .00672 ohms/ohm-°C.

Third in usage is the 70 percent nickel/30 percent iron alloy tradenamed Balco. The sole basis for its use is a very high specific resistance which makes possible very high resistance windings without increasing size. Ice point resistances are commonly 2000 or 10,000 ohms. It has the second highest temperature coefficient, third highest temperature capability and, like pure nickel, is nonlinear with an upward bending R vs. T curve. There is no recognized standard curve for Balco sensors.

Last in usage is pure copper, which is generally available only at 10 or 100 ohm ice point resistance values due to low specific resistance of the winding wire. Copper's temperature coefficient is almost the same as platinum and it is very linear above the ice point. Copper in bifilar windings is used in electrical machinery due to very low inductive or capacitive reactance, but platinum can also be used. Some traditional applications have also exploited the linearity of copper sensors in narrow range temperature difference measurements where two sensors are connected on opposed arms of a bridge. There is no internationally recognized standard curve for copper, although some national standards exist.

RTD Selection

Industrial RTD's are supplied in all of the same configurations as thermocouples. Figure 4.11a illustrates the most common process temperature assembly. Figure 4.11b shows two types of surface temperature sensors. Two-, three- and four-wire lead configurations are used

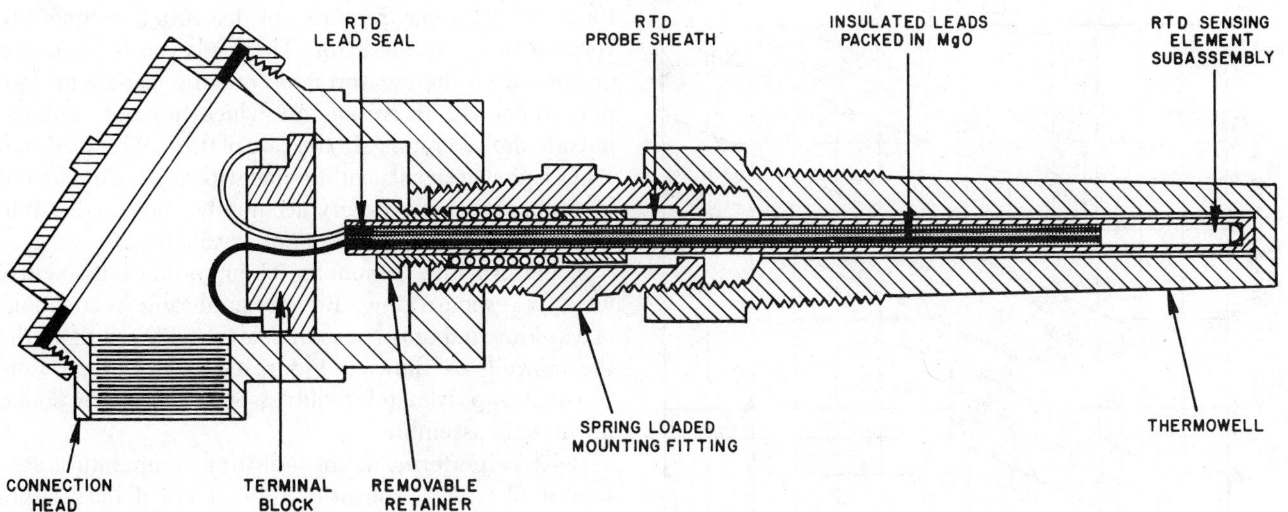

Fig. 4.11a Industrial RTD/thermowell assembly

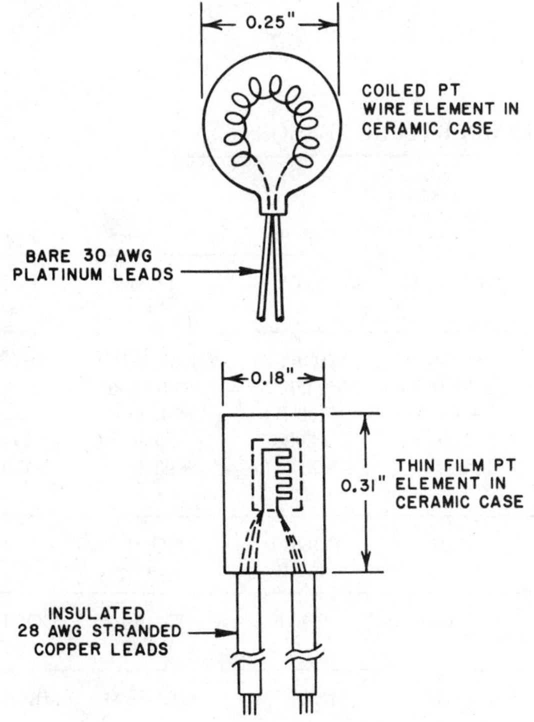

Fig. 4.11b RTD surface temperature sensors

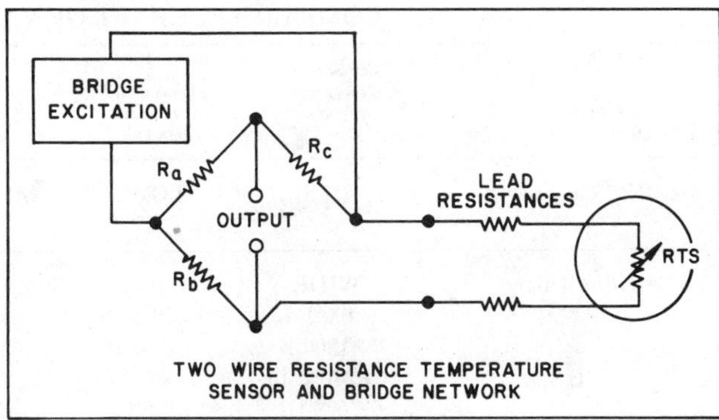

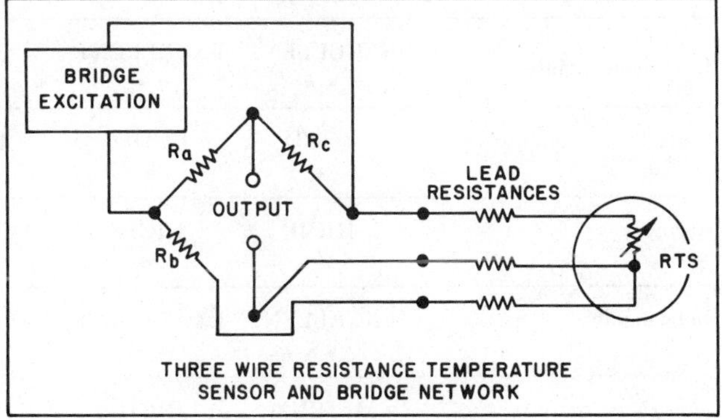

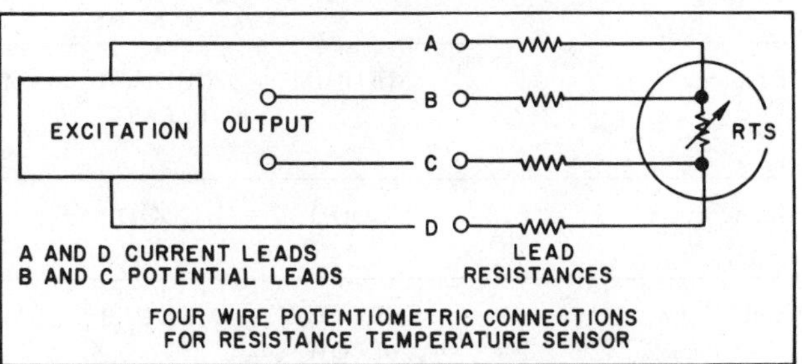

Fig. 4.11c RTD connections

in order of increasing possible precision in the temperature measurement. Two-wire sensors are common in applications where leads are short, such as machine installations. The resistance of the two leads is included with sensing element resistance. Three-wire sensors are the most widely used in process applications. Even relatively long leads are self compensating where one lead is included in each arm of the traditional bridge interface as shown in Figure 4.11c. The practical limitation in the compensation with three wire leads is that stranded lead wire is available to no better than ±10 percent resistance

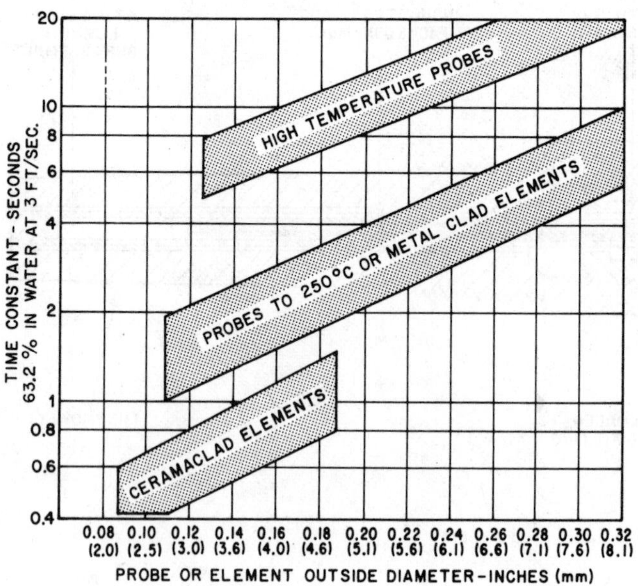

Fig. 4.11d Response time of typical resistance temperature sensor

tolerance. This may require a bridge zero adjustment to remove the resulting error. The leads can be removed entirely from the measurement only by using four-wire potentiometric measurements which are not common outside the laboratory. In the case of three-wire platinum RTD's, analog signal conditioning with voltage or current loop outputs that are very accurately linearized within .05 percent of span is commonly available.

A performance parameter of importance in control loops is response time. Broadly applicable correlations of experimental time constant data for RTD's without the thermowells are shown in Figure 4.11d. In general, time constant approximately doubles with each layer added in the final assembly.

Relative performance of industrial temperature sensors of all types is shown in Table 4.11e. RTD's, especially platinum RTD's, are clearly superior for temperature measurement requirements below 1000°F (540°C).

Table 4.11e
COMPARISON CHART OF VARIOUS TEMPERATURE SENSORS

Evaluation Criteria	*Platinum Rtd 100Ω wire wound and thin film*	*Platinum Rtd 1000Ω thin film*	*Nickel Rtd 1000Ω wire wound*	*Balco Rtd 2000Ω wire wound*	*Thermistor*	*Thermocouple*	*Semi-Conductor Devices*
Cost-OEM Quantity	HIGH	LOW ★	MEDIUM	MEDIUM	LOW ★	LOW ★	LOW ★
Temperature Range	WIDE −400°F to +1200°F (−240°C to +649°C) ★	WIDE −320°F to +1000°F (−196°C to +538°C) ★	MEDIUM −350°F to +600°F (−212°C to +316°C)	SHORT −100°F to +400°F (−73°C to +204°C)	SHORT to MEDIUM −100°F to +500°F (−73°C to +260°C)	VERY WIDE −450°F to +4200°F (−268°C to +2316°C) ★★	SHORT −57°F to +257°F (−49°C to +125°C)
Interchangeability	EXCELLENT ★★	EXCELLENT ★★	FAIR	FAIR	POOR to FAIR	GOOD ★	FAIR
Long Term Stability	GOOD ★	GOOD ★	FAIR	FAIR	POOR	POOR to FAIR	GOOD to FAIR
Accuracy	HIGH ★	HIGH ★	MEDIUM	LOW	MEDIUM	MEDIUM	MEDIUM
Repeatability	EXCELLENT ★★	EXCELLENT ★★	GOOD ★	FAIR	FAIR to GOOD	POOR to FAIR	GOOD ★
Sensitivity (output)	MEDIUM	HIGH ★	HIGH ★	VERY HIGH ★★	VERY HIGH ★★	LOW	HIGH ★
Response	MEDIUM	MEDIUM to FAST ★	MEDIUM	MEDIUM	MEDIUM to FAST ★	MEDIUM to FAST ★	MEDIUM to FAST ★
Linearity	GOOD ★	GOOD ★	FAIR	FAIR	POOR	FAIR	GOOD ★
Self Heating	VERY LOW to LOW ★	MEDIUM	MEDIUM	MEDIUM	HIGH	N/A	VERY LOW to LOW ★

Table 4.11e *Continued*
COMPARISON CHART OF VARIOUS TEMPERATURE SENSORS

Evaluation Criteria	*Platinum Rtd 100Ω wire wound and thin film*	*Platinum Rtd 1000Ω thin film*	*Nickel Rtd 1000Ω wire wound*	*Balco Rtd 2000Ω wire wound*	*Thermistor*	*Thermocouple*	*Semi-Conductor Devices*
Point (end) Sensitive	FAIR	GOOD ★	POOR	POOR	GOOD ★	EXCELLENT ★★	GOOD ★
Lead Effect	MEDIUM	LOW ★	LOW ★	LOW ★	VERY LOW ★★	HIGH	LOW ★
Physical Size/Packaging	MEDIUM to SMALL	SMALL to LARGE ★	LARGE	LARGE	SMALL to MEDIUM ★	SMALL to LARGE ★	SMALL to MEDIUM ★

★★ Best Rating ★ Good Rating

REFERENCE

1. Riddle, J.L., Furukawa, G.T., and Plumb, H.H., "Platinum Resistance Thermometry," NBS Monograph 126, 1973.

BIBLIOGRAPHY

American Petroleum Institute Manual Number API RP 550.

Baker, H.D., Ryder E.A., and Baker, M.A., *Temperature Measurement in Engineering*, Volume II, John Wiley and Sons, Inc., New York, 1961.

Corruccini, Robert J., *Interpolation of Platinum Resistance Thermometers, 10° to 273.15°K*, Temperature, ITS Measurement and Control in Science and Industry, Vol, 3, Part I, Van Nostrand Reinhold, New York, 1962.

Hormuth, G.A., "Ways to Measure Temperature," *Control Engineering*, Reprint No. 948, 1971.

Instrumentation and Control Systems Engineering Handbook, TAB Books, 1978.

The International Practical Temperature Scale of 1968, Amended Edition of 1975, Metrologia, 12, July 17, 1976.

Schooley, James F., "State of the Art of Instrumentation for High Temperature Thermometry," Argonne National Laboratories' 1977 Symposium, Publication No. ANL-73-7.

Standard DIN 43760, Fundamental values of measuring resistors for resistance thermometers, German Standards Committee (DNA).

Flow Sheet Symbol

ALARM — TS; DAMPER OR VALVE — TC

4.12 TEMPERATURE SWITCHES AND THERMOSTATS

Types of Designs:	A—Temperature Switches B—Thermostats
Approximate Costs:	From $75 for an HVAC quality pneumatic room thermostat to $300 for an industrial quality calibrated two-stage temperature switch
Range:	A—From 0 to 75°F (−18 to 24°C) to 50 to 650°F (10 to 343°C) B—Room thermostat setpoints are usually adjustable from 45 to 85°F (7 to 29°C) for heating and from 55 to 105°F (13 to 40°C) for cooling service
Inaccuracy:	A—Usually ±1% of span for industrial quality units B—Thermostats are usually uncalibrated and therefore their manufacturers usually do not guarantee their performance to any quantitatively defined value of maximum error
Partial List of Suppliers:	Automatic Switch Co. (A); Barber-Colman Co. (B); Dresser Industries, Inc. (A); Fenwal Inc., Div. Walter Kidde & Co. (A,B); Honeywell, Inc. (A,B); ITT General Controls (A,B); Johnson Controls, Inc. (A,B); MCC Powers, a Unit of Mark Controls Corp. (B); Mercoid Corp. (A,B); Robertshaw Controls Co. (A,B); Square D Co. (A,B); Thermo Electric Co., Inc. (A); United Electric Controls Co. (A)

Temperature switches are on-off devices while thermostats are narrow (frequently fixed) proportional band controllers. Their common characteristic is that measurement, setpoint and control functions are all combined into a single instrument.

Temperature Switches

Temperature switches are used to energize and de-energize electrical circuits as a function of the relationship between the process, temperature and a predetermined setpoint. They are available with accuracies up to ± ½ percent of span. The sensing elements are mostly the elastic types including filled and bimetallic elements. The electric switching assemblies are either snap-acting mechanical micro-switches or mercury switches. The latter contains no mechanical moving parts; it must be mounted on a vibration-free level surface.

Figure 4.12a illustrates some of the terminology used in connection with temperature switches. The temperature range within which the actuation point can be set is referred to as *adjustable range*. The switch may actuate at its setpoint on rising (*high*) or falling (*low*) temperature. The setpoint is the temperature that actuates the switch to open or close an electric circuit. The setpoint accuracy defines the band within which repetitive actuations will occur. Differential or dead band is the difference between setpoint and reactuation point. For example, if a switch is set to actuate (close) at 100°F (38°C) on rising temperature, it will close at that point; when the temperature drops, it may not open again until the temperature has fallen to 95°F (35°C). In this case, the differential is 5°F (2.8°C). Tolerance is the repeatable accuracy of the reactuation point.

Temperature switch elements should be selected with service life and maximum operating temperature in mind. Most elastic elements will have a service life of close to a million cycles if the cycle time is not less than five seconds. Selection of the adjustable range for a specific installation should consider both the setpoint actuation accuracy and the life factor. For greatest accuracy, the setpoint should fall in the upper half of the range, but for longest service life it should be in the lower half.

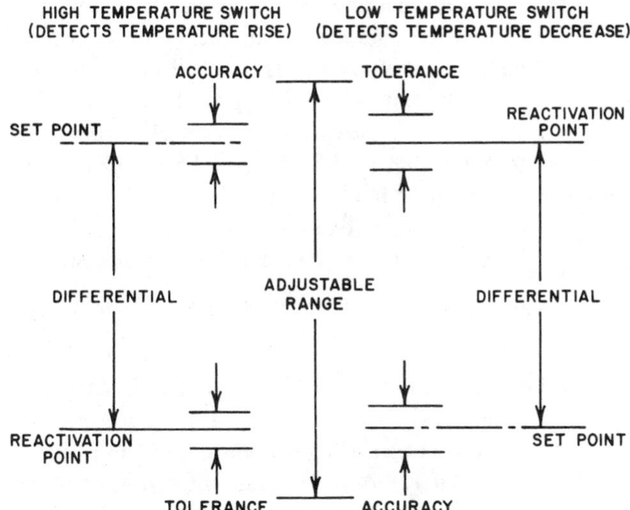

Fig. 4.12a Temperature switch terminology

The usually acceptable compromise is to locate the setpoint in the middle third of the range.

It is desirable to have an external calibrated knob provided on the temperature switch for setpoint adjustment. Uncalibrated or internal setpoint adjustments are generally undesirable on industrial installations. The "fixed differential" temperature switches are furnished with a single adjustment for setpoint. These units are factory set with close differentials amounting to ½ to 1 percent of span. On double adjustment-type designs, both set and reactuation points can be independently adjusted. The maximum differential in such designs is the range of the switch, while the minimum varies between 2 and 8 percent of span. Temperature switches with dual control are also available. Here, two independent switches mounted in the same housing are responding to the same process temperature in opening or closing two independent circuits.

The electrical rating of temperature switches at a 115 volt operating level varies from 0.3 to 10 amperes on ac or dc circuits. Generally, the dual control and the fixed differential switches have lower ratings, and the double adjustment type units higher ratings. The available circuit arrangements are very flexible. Some of the standard arrangements include single-pole-single-throw(SPST), single-pole-double-throw(SPDT), and double-pole-double-throw(DPDT); designs, but units are available with up to four poles.

There are basically three standard case designs: general purpose (NEMA 1), weather resistant (NEMA 2 and 3), and explosion proof (NEMA 7) cases.

In some designs, the control mechanism is an integral part of the explosion proof case, while in others it can be removed in the field for maintenance. As noted earlier, the setpoint should be externally adjustable so that the explosion proof case does not need to be opened in order to effect a setpoint change. For added convenience of the operator, some switches also provide continuous process temperature indication. Another optional feature is pushbutton reset, which must be manually operated before the circuit will be restored to its original state after an automatic operation. For example, the operation of a low temperature switch with "manual reset" would be such that if it is to start a heating device when the low temperature setting is reached, it will not stop that device when normal temperature is restored, but will keep it in operation until reset manually.

Conventional Room Thermostats

Room thermostats are simple, inexpensive, self-contained temperature controllers. Thermostats are distinguished from other controllers in that they provide proportional action only and that their proportional band (gain) is *narrow and fixed*. A conventional room thermostat might have a setpoint range of 55 to 85°F (13 to 29°C) and a *fixed sensitivity* (gain) of 2.5 PSI/°F (17 kPa/°C). Assuming that this thermostat operates a control valve having a 9 to 13 PSIG (62 to 90 kPa) spring range, we can convert the fixed sensitivity of 2.5 PSI/°F into a percent proportional band value that is better understood by instrument engineers.

Sensitivity (or gain) is stated as the ratio between a change in measurement and a change in the corresponding controller output. A 1°F (0.56°C) change in measurement is 3.3 percent of the measurement range of 30°F (55–85°F or 13–29°C), while the 2.5 PSI change in controller output is 62 percent of the controller output range of 4 PSI (9–13 PSIG) [or 28 kPa from a range of 62–90 kPa]. If a 3.3 percent change in input results in a 62 percent change in output, that is a gain of 62/3.3 = 19. Converting this typical thermostat gain to proportional band, we get:

$$P = \frac{100}{G} = 5.3 \text{ percent}$$

This controller, having a fixed proportional band of 5 percent, is very sensitive, one with "offset" but without tuning requirement. Let us examine the consequences of all three of these characteristics separately. A sensitive controller is suited for the control of very slow, large capacity processors. Temperature control in the HVAC industry usually fits that description and therefore thermostats with fixed, high gains can give acceptable results, if space temperature can change only very slowly. On the other hand, such thermostats cannot control spaces with fast dynamics (short time constants) and will cycle or lose control when applied to such service.

The "offset" inherent in all proportional controllers is not a serious drawback because the resulting error is small, due to the narrowness of the proportional band. Using our earlier example and assuming that the thermostat output is set at 11 PSIG (76 kPa) when the error is zero (thermostat on setpoint), we can calculate the maximum error due to offset. When there is no error, the control valve is 50 percent open (output is 11 PSIG or 76 kPa). As the load changes, an error must be allowed to develop in order for the valve opening to change. Assuming that the valve has been correctly sized and is large enough to handle all expected loads, the maximum error due to offset will be that deviation from setpoint which is required to move the valve from half to full opening. This 2 PSIG (14 kPa) change in thermostat output will occur when the deviation from setpoint is 0.8°F (0.4°C).

Therefore, this is the size of our permanent offset error for a gain of 19:1. There is an inverse linear relationship between gain and offset error, such that if the gain is cut in half, the offset error will be doubled.

A relative advantage of thermostats in comparison to PID controllers is that they have no dynamics and therefore they need not be tuned. They simply respond to the size of the deviation from setpoint (error); thus, the same thermostat can control both large and small rooms that have different dead times and time constants. The same cannot be said for PID controllers. If one were to attempt to use PID controllers as room thermostats, the controllers would have to be individually tuned to match the dynamics of each room, or some rooms would go off controls while the control valves of other rooms would be cycling.

From the above discussion it might be concluded that the conventional room thermostat is a good selection for the HVAC type applications and that more expensive instruments, such as PID controllers, would not necessarily improve the overall performance. While that is true, it should also be noted that conventional thermostats are usually uncalibrated devices and their manufacturers usually do not guarantee their accuracy. This is a very serious limitation, because it is possible to have some thermostats with 5–10°F (3–6°C) error in their measurement. The HVAC industry will benefit when a new generation of thermostats is introduced, with guaranteed measurement error limits of 1°F (0.6°C) or less.

Standard and Optional Design Features

Application engineers can choose from a fairly large variety of design features when specifying thermostats:

- Thermostats can be electric or pneumatic
- They can be indicating or blind
- They can have direct or reverse action
- Some thermostats will change from direct- to reverse-acting in response to a change in air supply pressure
- The measuring element can be bimetallic, filled or electronic
- The thermostat setpoint can be local or remote
- Local setpoints can be external, concealed or accessible only through the use of a key
- If the setpoint of a pneumatic thermostat is adjusted remotely by an air signal, each 1 PSIG (7 kPa) change in setpoint pressure will move the setpoint by an adjustable preset amount. The range of this adjustment is usually from 0.15 to 1.4°F (0.1 to 0.8°C) per 1 PSI (7 kPa)
- If the setpoint is to change as a function of the time of day, a timer can automatically operate a solenoid and thereby switch the setpoint signal

Some of the more recently developed and more advanced thermostat features include:

- Adjustable gains, or proportional bands. Another term used to describe the sensitivity of thermostats is *throttling range*. As shown in Figure 4.12b, it refers to the amount of temperature change that is required to change the thermostat output from 3 to 13 PSIG (21 to 90 kPa). The throttling range is usually adjustable from 2 to 10°F (1 to 5°C)

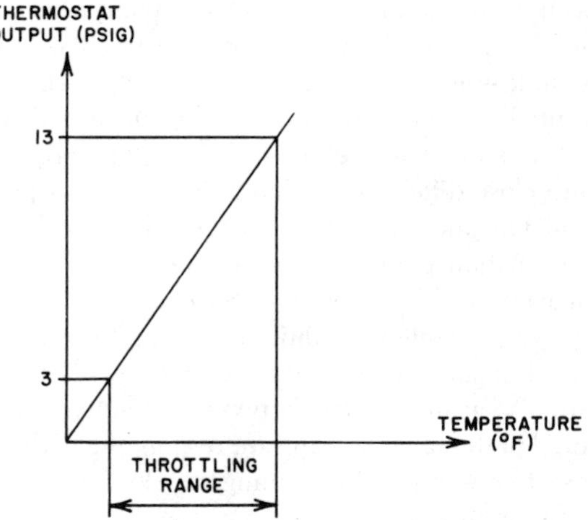

Fig. 4.12b Definitions of throttling range

- Dual setpoints. These thermostats will switch their settings in response to a change in air supply pressure. Both setpoints can be manually adjustable, with the day setting made by external thumbwheel and the night setting concealed internally
- "Limited control range." These thermostats allow the occupant of an office to move the setpoint to any value he wishes, but will disregard any setting that exceeds the limit value. For example, in heating applications, the limit could be 74°F (23°C). In this case the space temperature will be limited to a maximum of 74°F, regardless of the setting by the occupant. Similar limit values can be set for cooling

Zero Energy Band Control

A recent addition to the available thermostat choices is the Zero Energy Band (ZEB) design. The idea behind ZEB control is to conserve energy by not using any when the room is comfortable. As illustrated by Figure 4.12c, the conventional thermostat wastes energy by continuing to use it when the area's temperature is already comfortable. The "comfort gap" or "ZEB" is adjustable and can be varied to match the nature of the particular space involved.

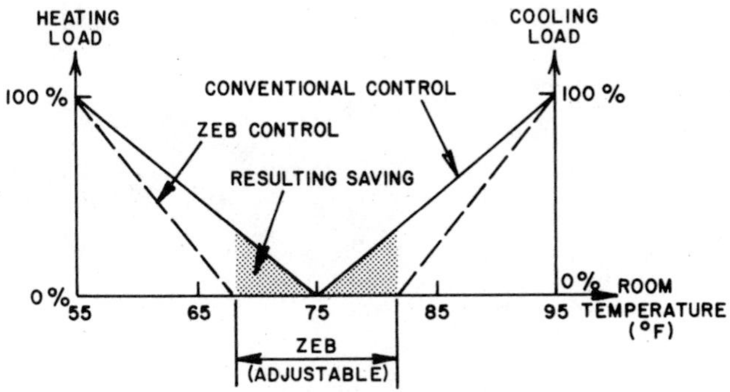

Fig. 4.12c Zero Energy Band (ZEB) control

ZEB control can be accomplished in one of two ways. The single setpoint and single output approach is illustrated on the left side of Figure 4.12d. Here the cooling valve fails closed and is shown to have an 8–11 PSIG (55–76 kPa) spring range, while the heating valve is selected to fail open and has a 2–5 PSIG (14–34 kPa) range. Therefore, between 5 and 8 PSIG (34 and 55 kPa) both valves are closed, and no pay energy is expended while the thermostat output is within this range. The throttling range is usually adjustable from 5 to 25°F (3 to 13°C). Thus, if the ZEB is 30 percent of the throttling range, it can be varied from a gap size of 1.5°F (.85°C) to 7.5°F (4.2°C) by changing the throttling range (or gain).

While the split-range approach is a little less expensive than the dual setpoint scheme (shown on the right of Figure 4.12d), it is also less flexible and more restrictive. The two basic limitations of the split range approach are:

1. The gap width can only be adjusted by also changing the thermostat gain; maximum gap width is limited by the minimum gain setting of the unit.
2. In this design, the heating valve must fail open which is undesirable from an energy conservation point of view.

These limitations are removed when a dual setpoint, dual output thermostat is used. Here both valves can fail closed and the band width is independently adjustable from the thermostat gain. The gains of the heating and cooling thermostats are also independently adjustable. In Figure 4.12d, the heating thermostat is reverse-acting and the cooling thermostat is direct-acting.

The most recent advances in thermostat technology are the microprocessor-based units. These are programmable devices with memory capability. They can be monitored and reset by central computers using pairs of telephone wires as the communication link. Microprocessor-based units can be provided with continuously recharged back-up batteries and with accurate room temperature sensors. They can also operate without a host

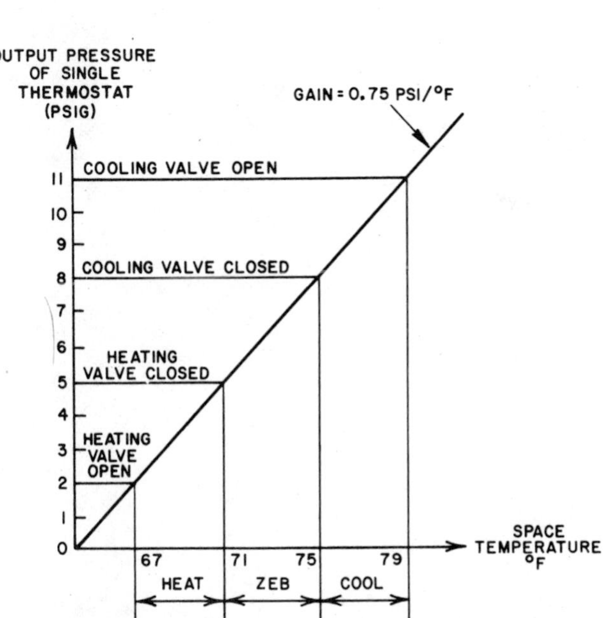

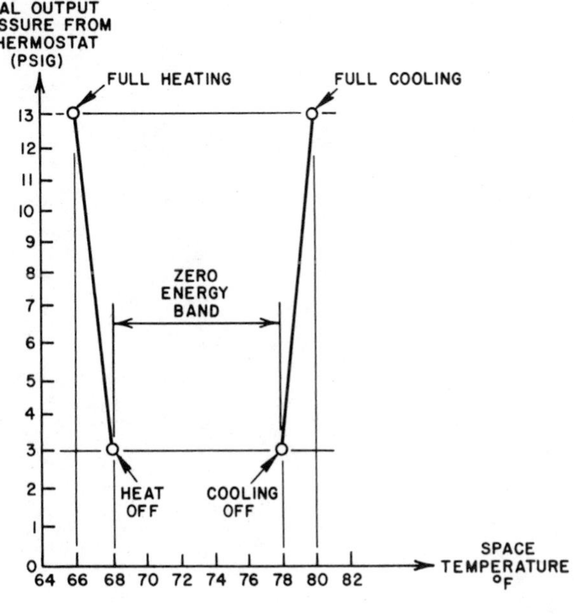

Fig. 4.12d Implementation of ZEB control schemes

computer (in the "stand alone" mode). In this case the user manually programs the thermostat to maintain various room temperatures as a function of the time of day and other considerations.

BIBLIOGRAPHY

ASHRAE Handbook and Product Directory, 1980 Systems, Chapter 34 on "Automatic Control." An ASHARE Publication.

Jutila, J.M., "Temperature Instrumentation," *Instrumentation Technology*, February, 1980.

Lipták, B.G., "Reducing the Operating Costs of Buildings by the Use of Computers," ASHRAE Transactions, 1977, Volume 83, Part 1.

Plumb, H.H., "Temperature, Its Measure and Control in Science and Industry," 5th Symposium on Temperature, Held in 1971 by NBS, API and ISA.

Roots, W.K., *Fundamentals of Temperature Control*, Academic Press, New York, 1969.

Spethmann, D.H., "Importance of Control in Energy Conservation," ASHRAE Journal, February, 1975.

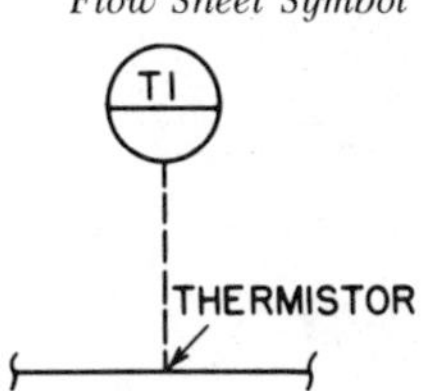

4.13 THERMISTORS

Temperature Range:	−150 to 600°F (−101 to 316°C)
Linearity:	Non-Linear
Cost:	$23 to $170
Inaccuracy:	0.1°F (0.05°C)
Partial List of Suppliers:	Atkins Technical, Inc.; Barber-Colman Co., Industrial Instruments Div.; Conax Corp.; Fenwall Electronics; Honeywell, Process Control Div.; Horiba Instruments; Hy-Cal Engineering; Omega Engineering, Inc.; Thermometrics, Inc.; Yellow Springs Instrument Co.

Thermistors are semiconductors made from specific mixtures of pure oxides of nickel, manganese, copper, cobalt, iron, magnesium, titanium and other metals sintered at temperatures above 1800°F (982°C). Their distinguishing characteristics are a high temperature coefficient and the fact that their resistance is a function of absolute temperature. Bell Laboratories manufactured them some fifty years ago, naming them from the term "*therm*ally sensitive res*istors*". While their use as temperature sensors was recognized over 100 years prior to the Bell investigation, the sensors were unstable and not reproducible. Therefore, they were seldom used outside the laboratory.

The Bell project resulted in the development of thermistors stable and reproducible enough to make their large scale use worthwhile in telephone work about 1940. However, industry in general did not accept these sensors until the 1950s. Bad experiences with commercially available thermistors hampered their acceptance. Variances in resistance at a given temperature and in rate of change of temperature made individual calibration a requirement. Overcoming this difficulty, one manufacturer patented a process for interchangeable thermistors, resulting in production of probes interchangeable to 0.05°F (0.03°C). The interchangeability feature enables them to be used in Wheatstone bridge circuits (Figure 4.13a) recorders and controllers to take advantage of their small size, great sensitivity and swift response.

Their temperature coefficient is usually negative, although it can be positive as well. In some cases, with the negative coefficient type, thermistor resistance decreases at the rate of over 3 percent for each degree F temperature rise. This makes them very good for narrow span measurement, but rather hard to handle for wide span applications.

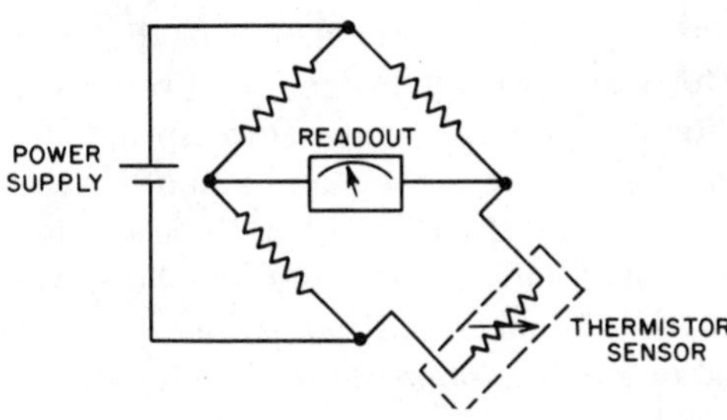

Fig. 4.13a Wheatstone bridge

A number of configurations are possible. Most familiar is the bead type (Figure 4.13b), usually glass coated. However, thermistors can be made into washers (Figure 4.13c), discs (Figure 4.13d) or rods (Figure 4.13e). They can also be encapsulated in plastic, cemented, soldered in bolts, encased in glass tubes, needles or a variety of other forms.

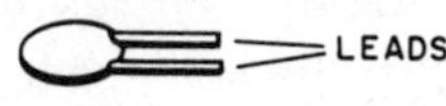

Fig. 4.13b Bead type

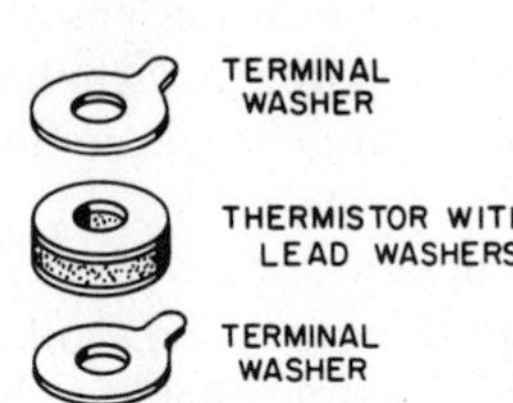

Fig. 4.13c Washer type

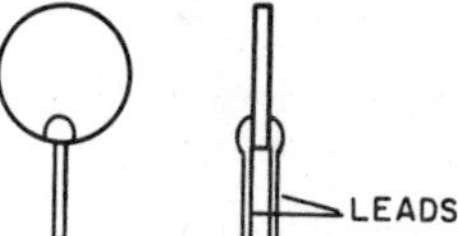

Fig. 4.13d Disc type

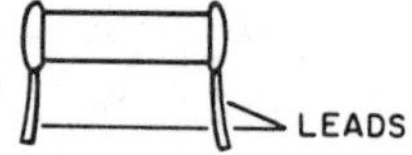

Fig. 4.13e Rod type

These assemblies serve to support the sensor, protect against damage to the wires, direct flow across the unit uniformly, permit sealing of conduits or flow lines, and provide for easier handling.

Thermistors can be used in a number of applications other than temperature measurement. Some of these are voltage regulation, flow meters, power level controls, gas analyzers and vacuum detectors (see Section 5.14).

The I^2R temperature increase is a direct function of the dissipation constant in its mounting environment, and, unlike resistance thermometers or thermocouples, thermistor resistance values are varied by varying their composition to suit the temperature span, range, and sensitivity desired for a given application. If a small current flows through the thermistor, there is a negligible increase in its temperature due to this flow. The resistance can thus be measured and, since resistance is proportional to absolute temperature, the temperature can be inferred.

However, if the current flow is slowly increased, the heat generated within the sensor will gradually begin to raise its temperature above that of its environment. This in turn lowers its resistance and more current will flow. Eventually, the current input will reach a level where it is balanced by the heat output of the thermistor. At this point, sensor temperature would be 300 or 500 degrees F (149 or 260 degrees C) above ambient and sensor resistance would have dropped to a value approximately 10^{-3} times its value at the original low current.

For most temperature measuring applications, self-heating is not a problem, since thermistor resistances are usually high and currents used are relatively low. If a current of higher magnitude is used, under fixed conditions, an offset allowance may be made for the self-heating effect. However, anything that affects the dissipation constant will change the offset. This could be the result of a number of things, including flow changes in the measured medium, changes in fluid composition and the like.

To measure temperature with a thermistor, a Wheatstone bridge circuit (Figure 4.13a) can be used as mentioned earlier. But, a simple circuit such as that shown in Figure 4.13f consisting of a battery, a sensor and a microammeter can also be used. In such a circuit, the sensor will have a very high resistance. As long as the voltage is constant, the current flow will be determined only by changes in resistance of the thermistor. The sensor can be mounted fairly far from the meter and ordinary copper wire can be used for transmission. With a sensor resistance on the order of 10^5 ohms, meter temperature changes and leadwire resistances can be neglected.

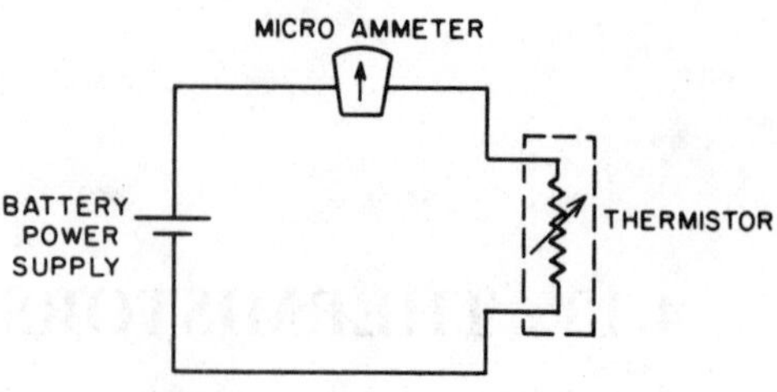

Fig. 4.13f Microammeter readout

Locating the thermistor in one leg of a bridge circuit (Figure 4.13g) with a center zero galvanometer enables very narrow temperature spans to be displayed relatively inexpensively. Range depends upon galvanometer sensitivity and can be as low as 2°F (1°C).

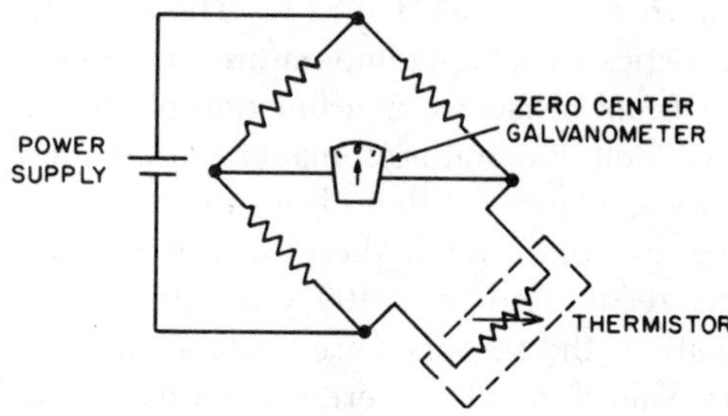

Fig. 4.13g Galvanometer readout

Very accurate temperature measurements can be made with a differential circuit (Figure 4.13h). With two thermistors in different locations in a bridge, the unbalance will be determined by resistance difference caused by temperature difference of the sensors. With a high gain amplifier, differentials of 10^{-3}F (6×10^{-5} C) can be measured.

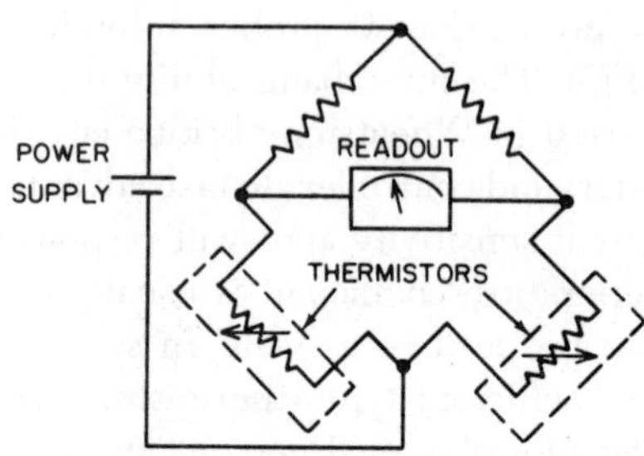

Fig. 4.13h Differential measurement

Combinations of thermistors and wire-wound resistors can be connected in networks (Figures 4.13i and j) to produce either a varying voltage or resistance which is linear with temperature. The basic equation for a divider network of R_1 and R_2 in series is $E_{out} = E_{in} R_1/(R_1 + R_2)$, where E_{out} is the voltage drop across R_1. When R_1 is a thermistor and E_{out} is plotted against temperature, the total curve is non-linear and "S" shaped. If R_1 is modified by the addition of another thermistor and resistor of proper values, linearity of the center section of the S curve can be extended to cover a relatively wide temperature range. This section then is considered to follow the general equation of a straight line. For the resistance mode, this would be $R_t = MT + b$, where M is slope in ohms per degree. T is temperature in degrees F* and b is value of R_t when T = 0°F. The advantages of such a system are obvious.

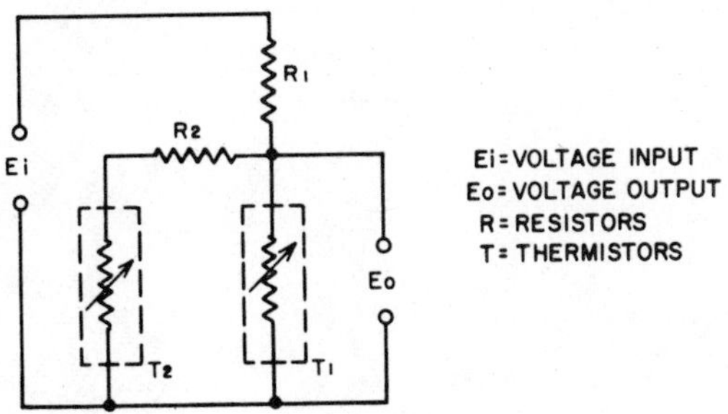

Fig. 4.13i Two-thermistor network

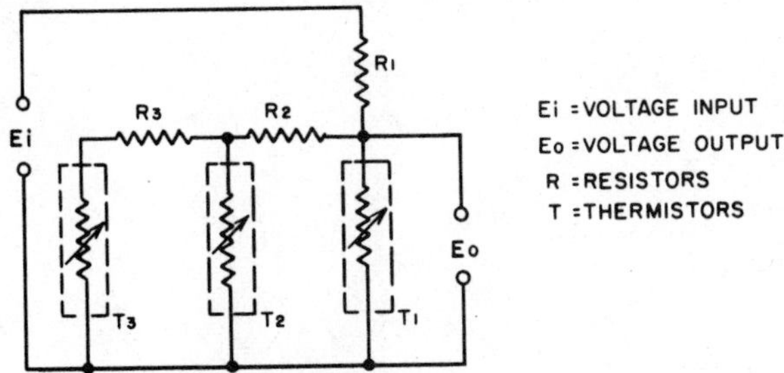

Fig. 4.13j Three-thermistor network

The National Bureau of Standards has offered a limited calibration service for thermistors over a temperature range of approximately −150 to 200°F (−101 to 93°C). Calibrations on the order of ±.02°F (±.01°C) are the rule. Users have claimed better than ±.002°F (±.001°C) stability over a two-year period.

The following should be checked to test the operating parameters of thermistors:

1. Zero power resistance. This test is done under conditions that produce negligible heating of the sensor by test current. Most common sources of error are measurement of ambient temperature, self-heating error, thermocouple effects at junctions of dissimilar metals and accuracy of test equipment. An ordinary ohmmeter is too crude. A five dial Wheatstone bridge and a good mirror galvanometer will do the job.
2. Temperature coefficient of resistance. The rate of change of thermistor resistance vs. temperature at the desired temperature.
3. Voltage developed across the thermistor under conditions of thermal equilibrium with a constant current. A current-limiting resistor should be used. Maximum current should not be exceeded, even for short periods. The sensor should not be moved to a medium of lower thermal conductivity during a test. The problem is similar to that of measuring voltage across ordinary resistors; the difference is that stabilization times are longer and small currents are used.
4. Time required for the thermistor to pass a certain current after the voltage is applied.
5. Dissipation constant measurements, which have been defined as the ratio at a certain temperature of a change in power dissipation in a thermistor to the resultant body temperature change.
6. Thermal time constant, which is the time for a 63.2 percent change from initial to final temperatures when subjected to a step change.

We have spoken mainly of negative coefficient thermistors, since they are most widely used. However, positive temperature coefficient thermistors are also used to protect motors from overheating, for very accurate measurements at high temperatures, for non-expendable current limiters and for flowmeters. In these applications, the positive increase in resistance with temperature and rapid change in sensitivity at switching temperatures are useful.

Advantages and disadvantages of thermistors can be listed as follows:

ADVANTAGES

1. Small size makes fast response and variety of element configurations possible
2. Temperature coefficient of resistance is several times that of meters
3. When temperature coefficient is negative, the sensor exhibits a greater sensitivity as the temperature drops
4. Good for narrow spans
5. Resistance is a function of absolute temperature, so cold junction compensation is unnecessary
6. Polarity means nothing
7. Large sensor resistance made contact or lead-wire resistance negligible

$$^*C = \frac{F - 32}{1.8}$$

8. 90 percent of aging occurs in first week
9. Stability increases with age
10. Low cost

DISADVANTAGES

1. Very non-linear temperature vs resistance curve
2. Interchangeability can be a problem
3. Stability at higher temperatures (over 600°F or 316°C) is somewhat of a problem
4. Not suitable for wide spans
5. Too much resistance requires shielded power lines, filters, or dc voltage

BIBLIOGRAPHY

Handbook of Thermistor Applications, Victory Engineering Corp., Springfield, N.J., 1968.

Hormuth, G.A., "Ways to Measure Temperature," *Control Engineering*, Reprint No. 948, 1971.

Instrumentation and Control Systems Engineering Handbook, TAB Books, 1978.

Flow Sheet Symbol

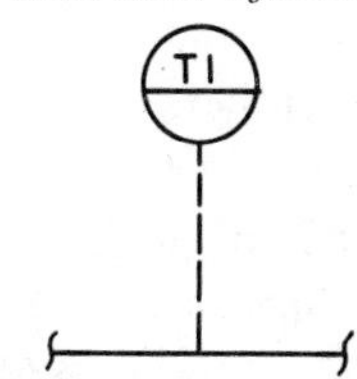

4.14 THERMOCOUPLES

Temperature Range:	−440 to +5,000°F (−262 to +2760°C)
Linearity:	Chromel-alumel (ISA Type K) most linear. Iron-constantan (ISA Type J) nearly linear
Cost:	\$56 to \$170
Inaccuracy:	0.2°F (0.1°C)
Partial List of Suppliers:	Action Instruments; BLH Electronics; Barber-Colman Co., Industrial Instruments Div.; Beckman Instruments, Inc.; Bristol-Babcock Div., ACCO Ind. Inc.; Burns Engineering, Inc.; Cal Temp Instrument, Inc.; Conax Corp.; Emerson Electric Co.; Fisher Controls Co.; Foxboro Co.; Honeywell, Process Control Div.; Hy-Cal Engineering; Kent Process Control, Inc.; Leeds & Northrup Co.; Omega Engineering, Inc.; Pyrometer Instrument Co., Inc.; Rochester Instrument Systems; Rosemount, Inc.; Taylor Instrument Co.; Tem Tex Co.; ThermoElectric Co., Inc.; Thermo-Couple Prod. Co.; United Electric Controls Co.; Weed Instrument Co., Inc.

Imparting heat to the junction of two dissimilar metals causes a small continuous emf to be generated. One of the simplest of all temperature sensors, the thermocouple depends upon the principle known as the Seebeck Effect. T. J. Seebeck discovered this phenomenon in 1821 and in the ensuing years the thermocouple has become the most widely used electrical temperature sensor. The word is a combination of *thermo* for the heat requirement and *couple* denoting two junctions.

An ordinary thermocouple consists of two different kinds of wires, each of which must be made of a homogeneous metal or alloy. The wires are fastened together at one end to form a *measuring junction*, normally referred to as the *hot* junction, since a majority of the measurements are made above ambient temperatures. The free ends of the two wires are connected to the measuring instrument to form a closed path in which current can flow. The point where the thermocouple wires connect to the measuring instruments is designated as *reference junction*, or the *cold* junction (see Figure 4.14a).

The emf developed at wire junctions is a manifestation of the *Peltier Effect* and occurs at every junction of dissimilar metals within the measuring system. This effect involves the liberation or absorption of heat at the junction when a current flows across it. The resultant heating or cooling depends upon the direction of current flow. Applications of this principle are becoming increasingly useful in electric heating and refrigeration.

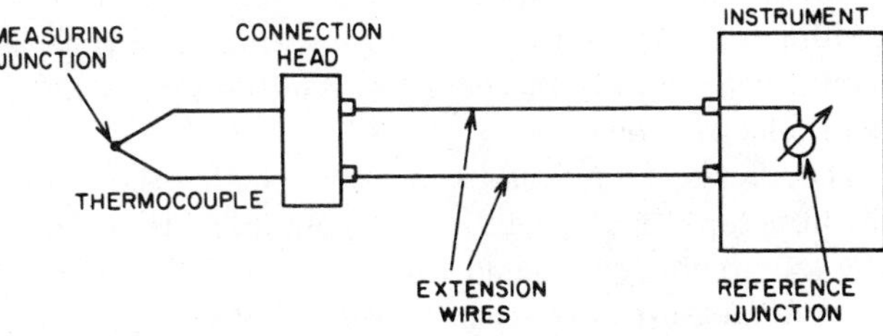

Fig. 4.14a Thermocouple terminology

A second emf develops along the temperature gradient of a single homogeneous wire. This is the *Thomson Effect*. It is most important that each section of wire in a given circuit be homogeneous, since, with no change in the composition or physical properties along its length, the circuit emf depends only upon the metals employed and the temperature of their junction; circuit emfs are independent of both length and diameter of wires. Another reason for requiring homogeneous wire is that thermal emfs within a single strand passing from a warmer to a cooler area, or vice versa, will cancel one another.

Further, if both junctions of a homogeneous metal are held at the same temperature, the metal does not contribute to the net emf of a circuit. Since some thermocouples are made of extremely expensive metals, this fact can be used to cut costs by supplying copper extension wire for long runs.

It follows, then, that by holding temperatures constant at all junctions except one within a given circuit, we can measure temperature as a function of the hot junction temperature with respect to the cold junction temperature.

When a readout device is employed, it converts the emf produced by the temperature difference between the hot and cold junctions to record or otherwise display the temperature of the hot junction. To prevent errors due to secondary emfs produced by variations of temperature at the cold junction and within the readout device, these emfs must be compensated for. One method is to hold the cold junction at a constant temperature (as can be done in laboratories with an ice bath). An oven can also be used, although keeping an oven temperature constant presents another set of problems. In practice, this compensation is usually accomplished by using resistors whose combined temperature resistance coefficient curves match those of the voltage-temperature curves produced by the secondary junctions, cancelling any variations in the cold junction temperature.

Since thermocouple emfs are of low level type, precautions must also be taken against stray currents resulting from proximity to electrical wiring. Best practice is never to run thermocouple wire in the same conduit with electric power wires.

The Seebeck emf can be measured with either a millivoltmeter (Figure 4.14b) or a potentiometer (Figure 4.14c) circuit. We should remember that the thermocouple measures only the difference between its reference and hot junctions. How closely it matches the accepted emf curve has a bearing on accuracy. Emf tables are usually based upon 32°F (0°C) reference temperatures for convenience. When the reference temperature is an ice bath and the hot junction is exposed to an unknown temperature, the emf developed will vary directly with changes in the unknown temperature.

To relieve the control engineer of the problem of compensating for temperature instability at the reference

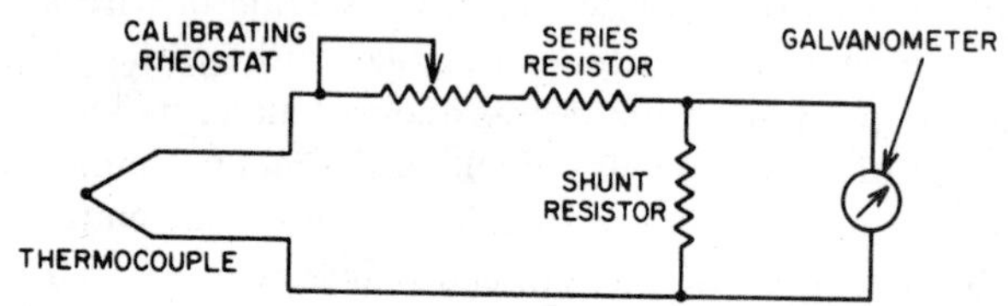

Fig. 4.14b Millivoltmeter circuit

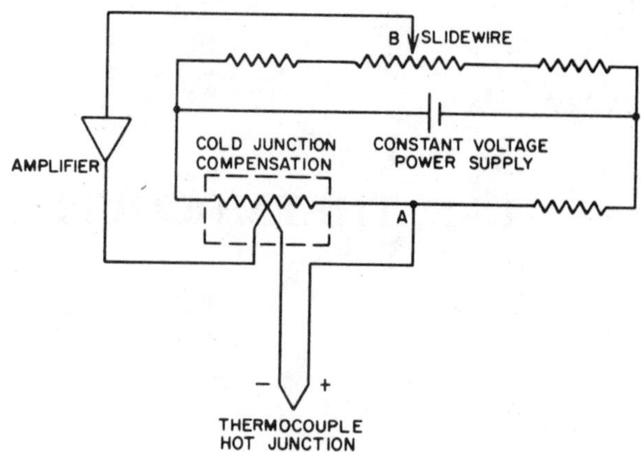

Fig. 4.14c Potentiometer circuit

junction, a copper or nickel resistor is placed in a bridge so that the thermocouple emf is opposed by an emf corresponding to the required ambient temperature correction. Operating on the null balance principle, the resulting potentiometer (Figure 4.15c) tends to reduce any voltage difference between points A and B to zero.

The Law of Intermediate Temperatures and Metals

The law of intermediate temperatures states that the sum of the emf's generated by two thermocouples, one with its junctions at 32°F (0°C) and some reference temperature, the other with its junctions at the same reference temperature and the measured temperature, is equivalent to that produced by a single thermocouple with its junctions at 32°F (0°C) and the measured temperature.

This is represented in Figure 4.14d where the measured temperature is 700°F (371°C). By adding an emf equal to that produced by thermocouple A in Figure 4.14d (with its junctions at 32°F (0°C) and the reference temperature) to that of thermocouple B, a total emf equivalent to that generated by the hypothetical thermocouple C results. In most pyrometers, this is done by a temperature-sensitive resistor, which measures the variations in reference junction temperature caused by ambient conditions, and automatically provides the necessary emf by means of a voltage drop produced across it. Thus, the instrument calibration becomes independent of reference temperature variations.

The law of intermediate metals states that the introduction of a third metal into the circuit will have no effect upon the emf generated so long as the junctions of the third metal with the other two are at the same temperature. Any number of different metals can be introduced, providing all the junctions are at the same temperature. Thus, in Figure 4.14e the circuits shown all generate the same electromotive force, even though the second and third circuit diagrams show materials C, D, E, and F inserted between A and B.

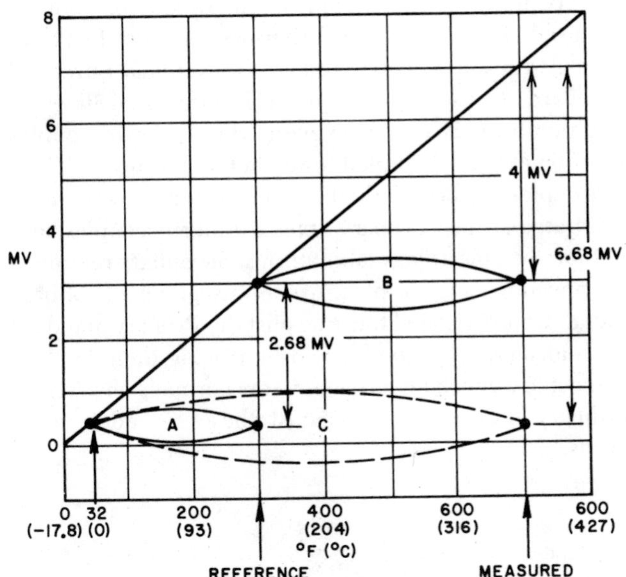

Fig. 4.14d According to the law of intermediate temperatures, the emf of thermocouple A plus the emf of thermocouple B is equal to the emf of thermocouple C

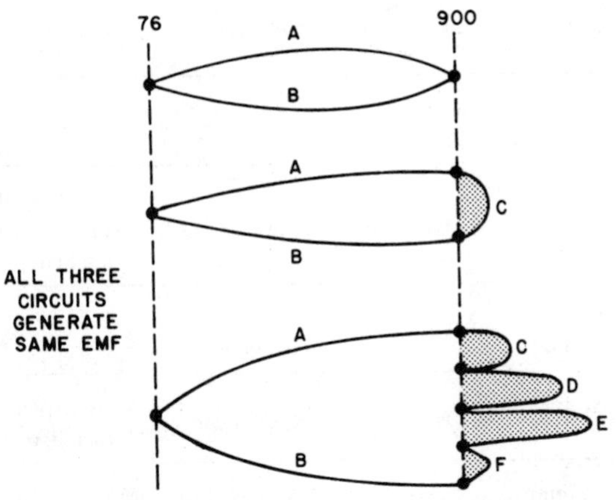

Fig. 4.14e No harmful effect is caused by introducing any number of metals at a thermocouple junction if all connections are at the same temperature

Thermocouple Types

Based on possible combinations of metals there could be countless numbers of thermocouples; but, as a matter of fact there are relatively few. Things that determine a metal's usefulness in thermocouple wire include:

1. Melting points
2. Reaction to various atmospheres
3. Thermoelectric output in combination
4. Electrical conductance
5. Stability
6. Repeatability
7. Cost
8. Ease of handling and fabrication

The platinum-platinum 90 percent/rhodium 10 percent (ISA Type S) thermocouple is most important, for it is used to define the International Temperature Scale between 1166.9°F (630.5°C), the point at which antimony freezes, and 1945.4°F (1063°C), the gold point. This thermocouple is not limited to the above range. It can be used from about 32°F to almost 2900°F (0°C to 1600°C) with excellent results. Industrial thermocouples of this material will match the standard calibration curve to better than ±.25 percent.

Several other thermocouples are commonly used (see Table 4.14f for a comparison). Platinum-platinum 13 percent rhodium (ISA Type R) and platinum 30 percent rhodium-platinum 6 percent rhodium (ISA Type B) are recommended for use in oxidizing atmospheres. They are relatively easily contaminated in other atmospheres.

Copper-constantan (ISA Type T) can be used in either oxidizing or reducing atmospheres. Thermocouples of this type exhibit a high resistance to corrosion from moisture, provide a relatively linear emf output and are good from the medium to the very low temperature range.

Iron-constantan (ISA Type J) can be used in reducing atmospheres. These thermocouples provide a very nearly linear emf output. They are the least expensive commercially available type.

Chromel-alumel (ISA Type K) can be used in oxidizing atmospheres. They are the most linear thermocouple in general use.

Chromel-constantan (ISA Type E) thermocouples provide the highest emf per degree of temperature change. However, they also tend to drift more than the others. They can be used in oxidizing atmospheres.

Tungsten-tungsten 26 percent rhenium thermocouples can be used to measure the highest temperatures. They cannot be used in oxidizing atmospheres, are brittle and hard to handle. They are used in a vacuum or in a clean inert gas.

To protect thermocouples from harmful atmospheres, corrosive fluids, mechanical damage, to support the thermocouple, or to permit entry into a pressurized system, protecting tubes or wells are supplied (Figure 4.14g). These tend to reduce the speed of response of the thermocouple, so small-mass, thin-wall, or needle-type installations are supplied where feasible (Figure 4.14h). Disposable-tip thermocouples are supplied in furnace applications (Figure 4.14i). They can also be peened or welded into a tube or tank well. Their low cost makes it feasible to place them in concrete beams while curing, or to be used in other single-time operations.

Thermocouples are not limited to single point measurements. They can be connected in parallel to provide the average temperature in a system (Figure 4.14j). They can also be used to measure the difference between two temperatures (Figure 4.14k). A single thermocouple can be utilized, with proper precautions, by two separate measuring instruments (Figure 4.14l). Generally speaking, a potentiometer draws no current from the thermocouple circuit at balance. With burnout, a small current circulates through the thermocouple. The control point setting also has a bearing. A thermocouple is now available with a built in current transmitter.

Since all thermocouples are subject to drift calibration, checks are done regularly in laboratories and industrial plants. A thermocouple is only as accurate as the wire from which it is made. Therefore, it is common practice for best accuracy to make all thermocouples from the same coil of wire. This assumes uniformity of the wire. Most manufacturers offer either standard or special calibrations which imply more care in selection of wire, handling and manufacturing. But the careful selection of materials, proper construction, installation and handling alone will not maintain highest accuracy; an adequate checking program is a must.

Depending upon the application, various procedures are used. Primary standard thermocouples of Pt-Pt, 10 percent Rh can be calibrated by the National Bureau of Standards to fixed points on the International Practical Temperature Scale. However, these thermocouples must be carefully handled to retain their accuracy. Thermocouples can be supplied by most major manufacturers calibrated against Primary Standard thermocouples kept in their own metrology laboratories. Secondary reference thermocouples for in-plant use are usually made of base metal. Comparison of these against the Primary Standard thermocouple is accomplished by placing them in close contact in a checking furnace. Users normally check their ordinary thermocouples against these secondary standards.

Table 4.14f
THERMOCOUPLE COMPARISON TABLE

ISA Type Designation	*Positive Wire* (*Numbers = Percentages*)	*Negative Wire* (*Numbers = Percentages*)	*Millivolts per °F*	*Recommended Temp °F** *Min.*	*Max.*	*Scale Linearity*	*Atmosphere Environment Recommended*	*Favorable Points*	*Less Favorable Points*
B	Pt70-RH30	Pt94-Rh6	.0003-.006	32	3380	Same as for type R couple	Inert or slow oxidizing	—	—
E	Chromel	Constantan	.015-.042	−320	1830	Good	Oxidizing	Highest emf/°F	Larger drift than other base metal couples
J	Iron	Constantan	.014-.035	−320	1400	Good; nearly linear from 300–800	Reducing	Most economical	—
K	Chromel	Alumel	.009-.024	−310	2500	Good; most linear of all T/C	Oxidizing	Most linear	More expensive than T or J
R	Pt87-Rh13	Platinum	.003-.008	0	3100	Good at high temps. Poor below 1000°F	Oxidizing	Small size, fast response	More expensive than type K
S	Pt90-Rh10	Platinum	.003-.007	0	3200	Same as R	Oxidizing	Same as R	More expensive than type K
T	Copper	Constantan	.008-.035	−310	750	Good but crowded at low end	Oxidizing or reducing	Good resis. to corrosion from moisture	Limited temp.
Y	Iron	Constantan	.022-.033	−200	1800	About same as type J	Reducing	—	Not industrial standard
—	Tungsten	W74-Re26	.001-.012	0	4200	Same as R	Inert or vacuum	High temp.	Brittle, hard to handle, expensive
—	W94-Re6	W74-Re26	.001-.010	0	4200	Same as R	Inert or vacuum	Same as above	Slightly less brittle than above
—	Copper	Gold-Cobalt	.0005-.025	−450	0	Reasonable above 60°K	—	Good output at very low temp.	Expensive lab. type T/C
—	Ir40-Rh60	Iridium	.001.-.004	0	3800	Same as R	Inert	—	Brittle, expensive

$*°C = \frac{°F - 32}{1.8}$

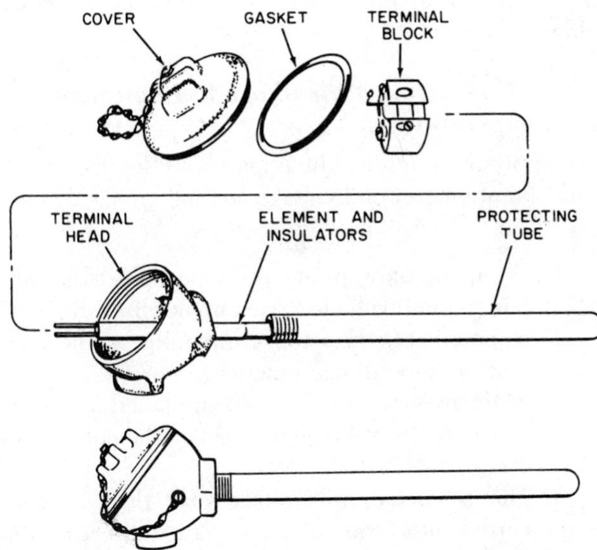

Fig. 4.14g Exploded view of T/C assembly and protecting tube (top); complete assembly with protecting tube (bottom)

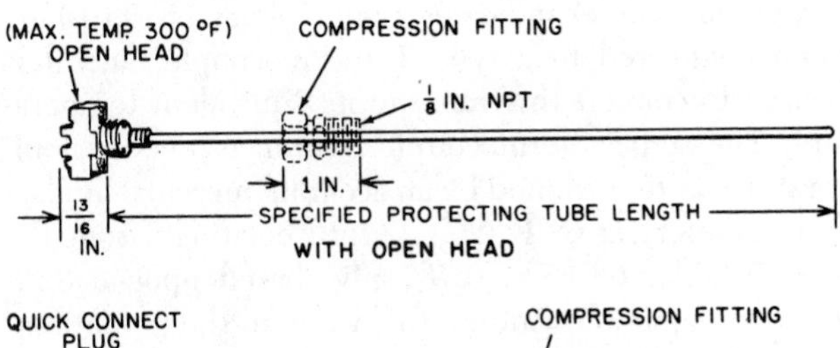

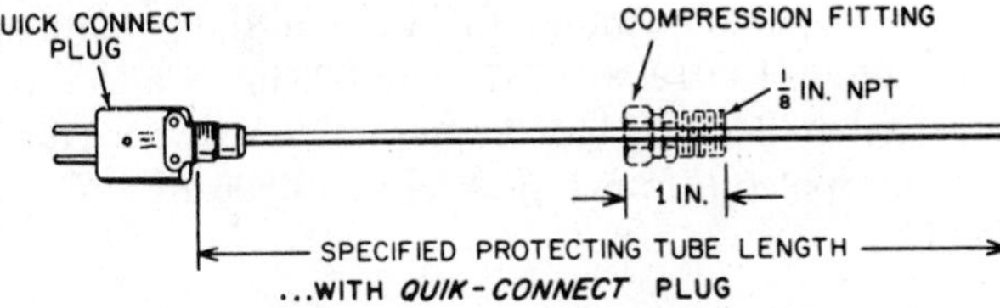

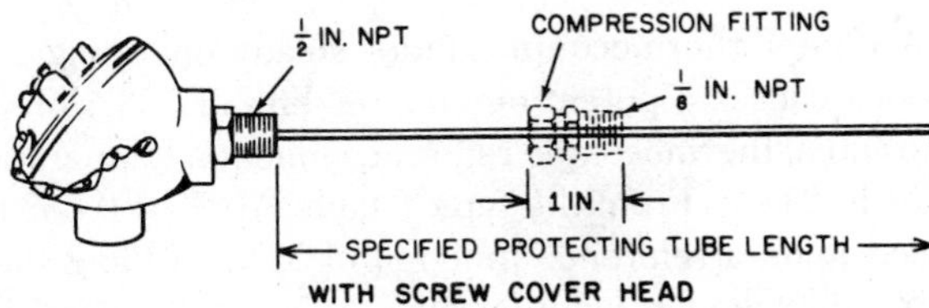

Fig. 4.14h High speed small O.D. T/C assemblies with stainless steel protecting tube

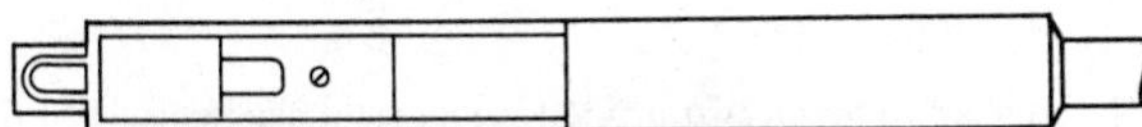

Fig. 4.14i Molten steel expendable thermocouple

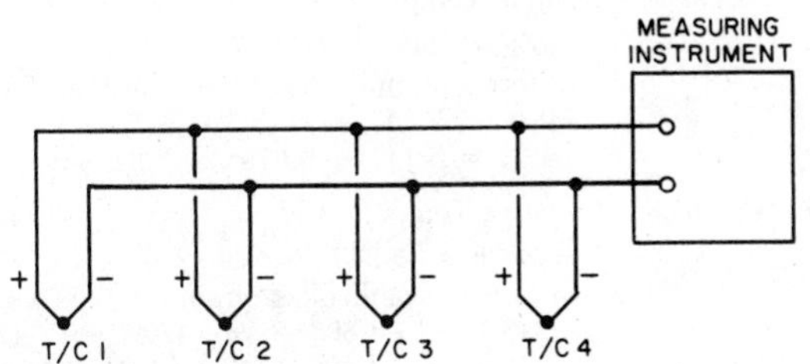

Fig. 4.14j Average temperature measurement

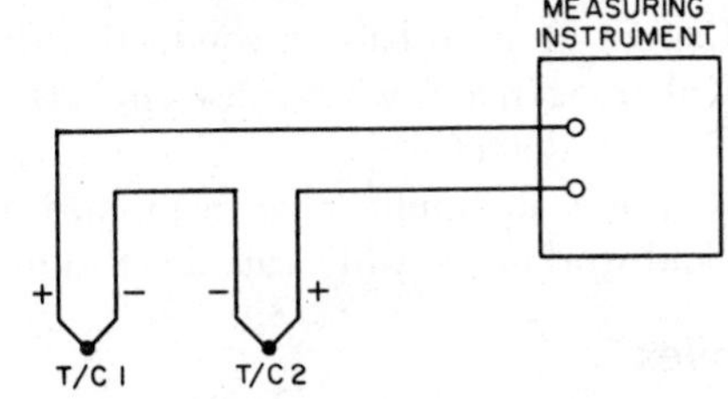

Fig. 4.14k Temperature difference

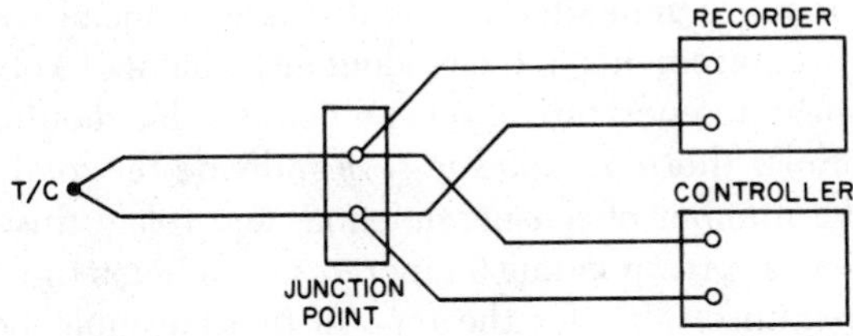

Fig. 4.14 l Parallel operation from common TC

Leadwire has been mentioned previously. With connections correctly made, copper extension wire can be used over long distances. However, it is recommended that iron-constantan and copper-constantan always be used with leadwire of the same material. To guard against mistakes in connection, industry practice is to color-code the wires, with the *negative* lead always *red*. Smaller gauge wire provides faster response; but, heavier gauge wires last longer and resist contamination or deterioration at high temperatures.

The advantages and disadvantages of thermocouples can be summed up as follows:

ADVANTAGES

1. Small size
2. Convenient for mounting—can be welded into a wall
3. Low cost—expendable
4. Rugged—can take abuse
5. Wide range—from near absolute zero to over 5,000°F (2,760°C)
6. Fairly accurate—calibration easily checked
7. Signal can be used by recording instrument
8. Long transmission distances are feasible
9. No danger of contaminating the process by filling fluid
10. Speedy response
11. Circuit emf's independent of length and diameter of wires (as long as homogeneous)
12. Reproducibility good for long term

DISADVANTAGES

1. Stray pickup a factor
2. Calibration must match temperature—emf relationship it is assumed to follow
3. Must avoid temperature gradients
4. Not as simple as direct reading thermometers

5. 70°F (39°C) nominal minimum span
6. Cannot be used bare in conducting fluid
7. Unhomogeneous wires cause emf's that are difficult to detect
8. Requires an amplifier for many measurements
9. Cold working of wires can affect calibration

Thermopiles

Thermopiles are thermocouples connected in series with electrically-insulated junctions (Figure 4.14m). Thermopiles generate large emf's, thus reducing sensitivity requirements in the readout instrument. To obtain the mean temperature at several points being monitored by similar thermocouples in series, divide the total emf by the number of sensing junctions and relate this emf value to a corresponding temperature reading in the emf-temperature table for the type of thermocouple being used.

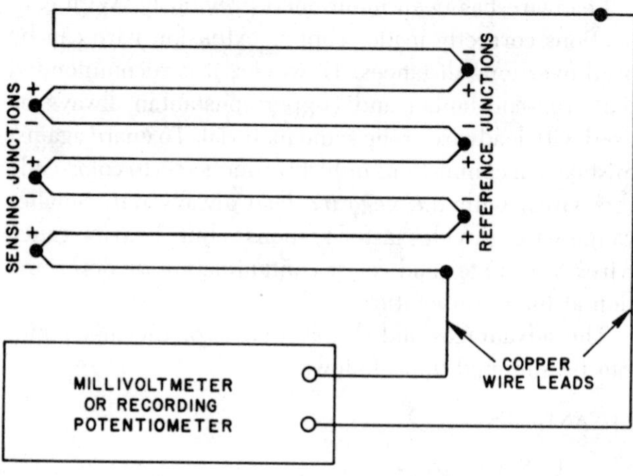

Fig. 4.14m Thermopile design

The principal objections to the use of thermopiles are the need for electrical isolation of individual thermocouples and the error that might go unnoticed when the output of one of the thermocouples is reduced by a short circuit. One satisfactory application for thermopiles is to use them as temperature differential detectors.

Thermocouple Tables

All thermocouple tables in this handbook are based upon a reference junction temperature of 32°F (0°C); therefore, direct conversion from the tables can be made only when an ice bath is used at the reference junction.

If it is not possible to maintain the reference junction temperature at 32°F (0°C), a correction factor must be applied to the millivolt values shown in the thermocouple tables. Note that the millivoltage produced by a given thermocouple is decreased when the temperature difference between the measuring junction and the reference junction is decreased. Correct for reference junction temperatures other than 32°F (0°C) as described below.

A. Converting Millivoltage to Equivalent Temperature

To apply the reference junction correction factor to a given potentiometer millivoltage reading, proceed as follows:

1. From the appropriate thermocouple table, obtain the millivoltage (based upon a 32°F R/J) corresponding to the actual temperature of the thermocouple reference junction.
2. ADD algebraically the value obtained in step 1, above, to the millivoltage read on the potentiometer.
3. The corrected millivoltage may then be converted into terms of temperature directly from the same table.

EXAMPLE 1

A potentiometer indicates a millivoltage of 13.019 mv when connected to a type T thermocouple, and it is desired to convert this value to its equivalent temperature. The actual thermocouple reference junction temperature, as determined by an accurate mercury-in-glass thermometer, is 68°F (20°C). Interpolating* from the type T table, 68°F = 0.787 mv, based upon a 32°F reference junction. Adding this value to the potentiometer reading, 13.019 + 0.787 = 13.806 mv, which is the corrected millivoltage based upon a 32°F reference junction. Interpolating from type T table, 13.806 mv = 539°F (282°C).

EXAMPLE 2

A type T thermocouple under steady operating conditions causes a potentiometer reading of −3.357 mv. The actual thermocouple reference junction temperature is 70°F (21°C). From the type T table, 70°F = 0.832 mv based upon a reference junction of 32°F. Adding these two millivoltages algebraically, −3.357 + 0.832 = −2.525 mv. Interpolating, −2.525 mv = −98°F (−72°C).

*To interpolate between two printed values, add algebraically to the smaller value a proportionate part of the difference between the two printed values, thus:

EXAMPLE 1 (Positive Temperature)

$248°F = 245° + 3/5\,(250° - 245°)$.
In terms of millivoltage from the type T table,
$248°F = 5.147 + 3/5\,(5.280 - 5.147)$
$= 5.147 + 0.0798 = 5.227$ mv

EXAMPLE 2 (Negative Temperature)

$-248°F = -245° + 3/5\,(-250° - (-245°))$
In terms of millivoltage from the type T table,
$-248°F = -4.688 + 3/5(-4.747 - (-4.688))$
$= -4.688 + 3/5(-4.747 + 4.688)$
$= -4.688 - 0.0354 = -4.723$ mv

B. Converting Temperature to Equivalent Millivoltage

To determine the proper millivolt input required to check the calibration of an instrument, proceed as follows:

1. From the appropriate table, obtain the millivoltage based upon a 32°F reference junction corresponding to the actual temperature at the input terminals of the instrument to be checked.
2. From the same table, obtain the millivoltage based upon a 32°F reference junction for the temperature to be checked.
3. Subtract algebraically the value obtained in step 1, above, from the value obtained in step 2.

EXAMPLE 1

It is desired to check the calibration of an instrument at 300°F (149°C). The instrument has a scale graduated in degrees Fahrenheit for a type T thermocouple. The actual temperature at the input terminals of the instrument to be checked, as determined by an accurate mercury-in-glass thermometer, is 70°F (21°C). From the type T table, 70°F = 0.832 mv and 300°F = 6.647 mv based upon a reference junction temperature of 32°F. Subtracting, the corrected millivoltage input required on the basis of a 70°F reference junction is 6.647 − 0.832 = 5.815 mv.

EXAMPLE 2

It is desired to determine the correct millivoltage input required to check the calibration of an instrument at −200°F (−129°C). The instrument scale is graduated in degrees Fahrenheit for a type T thermocouple. The actual temperature at the input terminals of the instrument is 68°F (20°C). From the type T table, 68°F = 0.787 and −200°F = −4.111 mv based upon a 32°F reference junction. Subtracting algebraically, the corrected millivolt input on the basis of a 68°F reference junction is −4.111 − 0.787 = −4.898 mv.

BIBLIOGRAPHY

Bartosiak, G., "Guide to Thermocouples," *Instruments & Control Systems*, November, 1978, pp. 53–55.

Coffee, M.B., "Common-Mode Rejection Techniques for Low-Level Data Acquisition," *Instrumentation Technology*, July, 1977, pp. 45–49.

Klipec, B., "How to Avoid Noise Pickup on Wire and Cable," *Instruments & Control Systems*, December 1977, pp. 27–30.

Park, R.M., "Applying the Systems Concept to Thermocouples," *Instrumentation Technology*, August, 1973, pp. 25–31.

Pederson, R., "Choosing Thermocouples or Resistance Thermometers," *Instruments & Control Systems*, June, 1975, p. 39.

Table 4.14n
TYPE J—IRON-CONSTANTAN THERMOCOUPLE

Degrees Fahrenheit vs. Millivolts. Temperatures are based on the International Temperature Scale of 1948. emf is expressed in absolute millivolts.

Reference Junction 32°F (0°C).

Note: Instruments calibrated to this curve have scales identified as "Type J" Thermocouple.

°F*	−300	−200	−100	−0	+0	100	200	300	400	500	600	700	800	900	1000	1100	1200	1300	1400	1500
										Millivolts										
0	−7.52	−5.76	−3.49	−0.89	−0.89	1.94	4.91	7.94	11.03	14.12	17.18	20.26	23.32	26.40	29.52	32.72	36.01	39.43	42.96	46.53
5	−7.59	−5.86	−3.61	−1.02	−0.75	2.09	5.06	8.10	11.18	14.27	17.34	20.41	23.47	26.55	29.68	32.89	36.18	39.61	43.14	46.71
10	−7.66	−5.96	−3.73	−1.16	−0.61	2.23	5.21	8.25	11.34	14.42	17.49	20.56	23.63	26.70	29.84	33.05	36.35	39.78	43.32	46.89
15	−7.73	−6.06	−3.85	−1.29	−0.48	2.38	5.36	8.40	11.49	14.58	17.64	20.72	23.78	26.86	30.00	33.21	36.52	39.96	43.50	47.07
20	−7.79	−6.16	−3.97	−1.43	−0.34	2.52	5.51	8.56	11.65	14.73	17.80	20.87	23.93	27.02	30.16	33.37	36.69	40.13	43.68	47.24
25		−6.25	−4.09	−1.56	−0.20	2.67	5.66	8.71	11.80	14.88	17.95	21.02	24.09	27.17	30.32	33.54	36.86	40.31	43.85	47.42
30		−6.35	−4.21	−1.70	−0.06	2.82	5.81	8.87	11.96	15.04	18.11	21.18	24.24	27.33	30.48	33.70	37.02	40.48	44.03	47.60
35		−6.44	−4.33	−1.83	+0.08	2.97	5.96	9.02	12.11	15.19	18.26	21.33	24.39	27.48	30.64	33.86	37.20	40.66	44.21	47.78
40		−6.53	−4.44	−1.96	+0.22	3.11	6.11	9.17	12.26	15.34	18.41	21.48	24.55	27.64	30.80	34.03	37.36	40.83	44.39	47.95
45		−6.62	−4.56	−2.09	+0.36	3.26	6.27	9.33	12.42	15.50	18.57	21.64	24.70	27.80	30.96	34.19	37.54	41.01	44.57	48.13
50		−6.71	−4.68	−2.22	0.50	3.41	6.42	9.48	12.57	15.65	18.72	21.79	24.85	27.95	31.12	34.36	37.71	41.19	44.75	48.31
55		−6.80	−4.79	−2.35	0.65	3.56	6.57	9.64	12.73	15.80	18.87	21.94	25.01	28.11	31.28	34.52	37.88	41.36	44.93	48.48
60		−6.89	−4.90	−2.48	0.79	3.71	6.72	9.79	12.88	15.96	19.03	22.10	25.16	28.26	31.44	34.68	38.05	41.54	45.10	48.66
65		−6.97	−5.01	−2.61	0.93	3.86	6.87	9.95	13.04	16.11	19.18	22.25	25.32	28.42	31.60	34.85	38.22	41.72	45.28	48.83
70		−7.06	−5.12	−2.74	1.07	4.01	7.03	10.10	13.19	16.26	19.34	22.40	25.47	28.58	31.76	35.01	38.39	41.90	45.46	49.01
75		−7.14	−5.23	−2.86	1.22	4.16	7.18	10.25	13.34	16.42	19.49	22.55	25.62	28.74	31.92	35.18	38.57	42.07	45.64	49.18
80		−7.22	−5.34	−2.99	1.36	4.31	7.33	10.41	13.50	16.57	19.64	22.71	25.78	28.89	32.08	35.35	38.74	42.25	45.82	49.36
85		−7.30	−5.44	−3.12	1.51	4.46	7.48	10.56	13.65	16.72	19.80	22.86	25.93	29.05	32.24	35.51	38.91	42.43	46.00	49.53
90		−7.38	−5.55	−3.24	1.65	4.61	7.64	10.72	13.81	16.88	19.95	23.01	26.09	29.21	32.40	35.68	39.08	42.61	46.18	49.70
95		−7.45	−5.65	−3.36	1.80	4.76	7.79	10.87	13.96	17.03	20.10	23.17	26.24	29.37	32.56	35.84	39.26	42.78	46.35	49.88
100		−7.52	−5.76	−3.49	1.94	4.91	7.94	11.03	14.12	17.18	20.26	23.32	26.40	29.52	32.72	36.01	39.43	42.96	46.53	50.05

$$*{}^\circ C = \frac{{}^\circ F - 32}{1.8}$$

Table 4.14o

TYPE K—CHROMEL-ALUMEL THERMOCOUPLE

Degrees Fahrenheit vs. Millivolts. Temperatures are based on the International Temperature Scale of 1948. emf is expressed in absolute millivolts

Reference Junction 32°F (0°C).

°F*	0	100	200	300	400	500	600	700	800	900	1000	1100	1200	1300	1400	1500	1600	1700	1800	1900	2000	2100	2200	2300
												Millivolts												
0	−0.68	1.52	3.82	6.09	8.31	10.57	12.86	15.18	17.53	19.89	22.26	24.63	26.98	29.32	31.65	33.93	36.19	38.43	40.62	42.78	44.91	47.00	49.05	51.05
5	−0.58	1.63	3.94	6.20	8.42	10.68	12.97	15.30	17.64	20.01	22.37	24.74	27.10	29.44	31.76	34.05	36.31	38.54	40.73	42.89	45.01	47.10	49.15	51.15
10	−0.47	1.74	4.05	6.31	8.54	10.79	13.09	15.41	17.76	20.13	22.49	24.86	27.22	29.56	31.88	34.16	36.42	38.65	40.84	42.99	45.12	47.21	49.25	51.25
15	−0.37	1.86	4.17	6.42	8.65	10.91	13.20	15.53	17.88	20.24	22.61	24.98	27.34	29.67	31.99	34.28	36.53	38.76	40.95	43.10	45.22	47.31	49.35	51.35
20	−0.26	1.97	4.28	6.53	8.76	11.02	13.32	15.65	18.00	20.36	22.73	25.10	27.45	29.79	32.11	34.39	36.64	38.87	41.05	43.21	45.33	47.41	49.45	51.45
25	−0.15	2.09	4.40	6.65	8.87	11.13	13.44	15.76	18.11	20.48	22.85	25.22	27.57	29.91	32.22	34.50	36.76	38.98	41.16	43.31	45.43	47.52	49.55	51.54
30	−0.04	2.20	4.51	6.76	8.98	11.25	13.55	15.88	18.23	20.60	22.97	25.34	27.69	30.02	32.34	34.62	36.87	39.09	41.27	43.42	45.54	47.62	49.65	51.64
35	+0.07	2.32	4.63	6.87	9.09	11.36	13.67	16.00	18.35	20.72	23.08	25.46	27.80	30.14	32.45	34.73	36.98	39.20	41.38	43.53	45.64	47.72	49.76	51.74
40	+0.18	2.43	4.74	6.98	9.21	11.48	13.78	16.12	18.47	20.84	23.20	25.57	27.92	30.25	32.57	34.84	37.09	39.31	41.49	43.63	45.75	47.82	49.86	51.84
45	+0.29	2.55	4.86	7.09	9.32	11.59	13.90	16.23	18.58	20.95	23.32	25.69	28.04	30.37	32.68	34.96	37.20	39.42	41.60	43.74	45.85	47.93	49.96	51.94
50	0.40	2.66	4.97	7.20	9.43	11.71	14.02	16.35	18.70	21.07	23.44	25.81	28.15	30.49	32.80	35.07	37.31	39.53	41.70	43.85	45.96	48.03	50.06	52.03
55	0.51	2.78	5.08	7.31	9.54	11.82	14.13	16.47	18.82	21.19	23.56	25.93	28.27	30.60	32.91	35.18	37.43	39.64	41.81	43.95	46.06	48.13	50.16	52.13
60	0.62	2.89	5.20	7.42	9.66	11.94	14.25	16.59	18.94	21.31	23.68	26.05	28.39	30.72	33.02	35.29	37.54	39.75	41.92	44.06	46.17	48.23	50.26	52.23
65	0.73	3.01	5.31	7.53	9.77	12.05	14.36	16.70	19.06	21.43	23.80	26.16	28.50	30.83	33.14	35.41	37.65	39.86	42.03	44.17	46.27	48.34	50.36	52.33
70	0.84	3.12	5.42	7.64	9.88	12.17	14.48	16.82	19.18	21.54	23.91	26.28	28.62	30.95	33.25	35.52	37.76	39.96	42.14	44.27	46.38	48.44	50.46	52.42
75	0.95	3.24	5.53	7.75	10.00	12.28	14.60	16.94	19.29	21.66	24.03	26.40	28.74	31.07	33.37	35.63	37.87	40.07	42.24	44.38	46.48	48.54	50.56	52.52
80	1.06	3.36	5.65	7.87	10.11	12.40	14.71	17.06	19.41	21.78	24.15	26.52	28.86	31.18	33.48	35.75	37.98	40.18	42.35	44.49	46.58	48.64	50.65	52.62
85	1.18	3.47	5.76	7.98	10.22	12.51	14.83	17.17	19.53	21.90	24.27	26.63	28.97	31.30	33.59	35.86	38.09	40.29	42.46	44.59	46.69	48.74	50.75	52.72
90	1.29	3.59	5.87	8.09	10.34	12.63	14.95	17.29	19.65	22.02	24.39	26.75	29.09	31.42	33.71	35.97	38.20	40.40	42.57	44.70	46.79	48.85	50.85	52.81
95	1.40	3.70	5.98	8.20	10.45	12.74	15.06	17.41	19.77	22.14	24.51	26.87	29.21	31.53	33.82	36.08	38.32	40.51	42.67	44.80	46.90	48.95	50.95	52.91
100	1.52	3.82	6.09	8.31	10.57	12.86	15.18	17.53	19.89	22.26	24.63	26.98	29.32	31.65	33.93	36.19	38.43	40.62	42.78	44.91	47.00	49.05	51.05	53.01

*$°C = \frac{°F - 32}{1.8}$

Table 4.14p
TYPE R—PLATINUM vs. PLATINUM PLUS 13% RHODIUM THERMOCOUPLE

Degrees Fahrenheit vs. Millivolts. Temperatures are based on the International Temperature Scale of 1948. emf is expressed in absolute millivolts.

Reference Junction 32°F (0°C).

°F*	0	100	200	300	400	500	600	700	800	900	1000	1100	1200	1300	1400	1500
								Millivolts								
0	−0.089	0.220	0.596	1.030	1.504	2.012	2.547	3.103	3.677	4.264	4.868	5.488	6.125	6.773	7.436	8.116
5	−0.076	0.237	0.616	1.052	1.529	2.038	2.575	3.132	3.706	4.294	4.899	5.519	6.156	6.805	7.470	8.150
10	−0.062	0.255	0.637	1.075	1.553	2.065	2.602	3.160	3.735	4.324	4.930	5.551	6.188	6.838	7.503	8.184
15	−0.049	0.272	0.657	1.098	1.578	2.091	2.630	3.188	3.764	4.354	4.960	5.582	6.220	6.871	7.537	8.218
20	−0.035	0.291	0.678	1.121	1.603	2.117	2.657	3.217	3.794	4.384	4.991	5.614	6.252	6.904	7.571	8.253
25	−0.021	0.308	0.700	1.144	1.628	2.144	2.685	3.245	3.823	4.413	5.022	5.645	6.285	6.937	7.605	8.287
30	−0.006	0.327	0.721	1.167	1.653	2.170	2.712	3.273	3.852	4.443	5.053	5.677	6.317	6.970	7.639	8.322
35	+0.009	0.345	0.742	1.191	1.678	2.197	2.740	3.302	3.882	4.473	5.084	5.709	6.349	7.003	7.672	8.356
40	+0.024	0.363	0.763	1.214	1.703	2.223	2.768	3.330	3.911	4.503	5.115	5.741	6.381	7.037	7.706	8.391
45	+0.039	0.381	0.785	1.238	1.729	2.250	2.796	3.359	3.941	4.533	5.146	5.773	6.414	7.069	7.740	8.426
50	0.055	0.400	0.807	1.261	1.754	2.277	2.823	3.387	3.970	4.563	5.176	5.805	6.446	7.103	7.774	8.460
55	0.071	0.419	0.828	1.285	1.779	2.303	2.851	3.416	3.999	4.593	5.208	5.837	6.479	7.136	7.808	8.495
60	0.086	0.438	0.850	1.309	1.805	2.330	2.879	3.445	4.029	4.624	5.238	5.869	6.511	7.169	7.842	8.530
65	0.103	0.457	0.872	1.333	1.831	2.357	2.907	3.473	4.058	4.654	5.270	5.901	6.544	7.202	7.877	8.565
70	0.119	0.476	0.894	1.357	1.856	2.384	2.935	3.502	4.087	4.685	5.301	5.933	6.577	7.235	7.911	8.599
75	0.135	0.496	0.917	1.381	1.882	2.412	2.963	3.531	4.116	4.715	5.332	5.964	6.609	7.269	7.945	8.634
80	0.152	0.516	0.939	1.406	1.908	2.438	2.991	3.560	4.146	4.746	5.363	5.996	6.642	7.302	7.979	8.669
85	0.169	0.536	0.962	1.430	1.934	2.466	3.019	3.589	4.175	4.776	5.394	6.028	6.674	7.336	8.013	8.704
90	0.186	0.556	0.984	1.455	1.960	2.493	3.047	3.618	4.205	4.807	5.426	6.060	6.707	7.369	8.047	8.739
95	0.203	0.576	1.007	1.480	1.986	2.520	3.075	3.647	4.235	4.837	5.457	6.092	6.740	7.403	8.081	8.774
100	0.220	0.596	1.030	1.504	2.012	2.547	3.103	3.677	4.264	4.868	5.488	6.125	6.773	7.436	8.116	8.809

*$°C = \frac{°F - 32}{1.8}$

Table 4.14p *Continued*
TYPE R–*Continued* PLATINUM vs.
PLATINUM PLUS 13% RHODIUM THERMOCOUPLE

°F*	1600	1700	1800	1900	2000	2100	2200	2300	2400	2500	2600	2700	2800	2900	3000
							Millivolts								
0	8.809	9.516	10.237	10.973	11.726	12.488	13.255	14.027	14.798	15.568	16.340	17.110	17.875	18.636	19.394
5	8.844	9.552	10.274	11.011	11.765	12.526	13.293	14.065	14.837	15.607	16.378	17.148	17.913	18.674	19.432
10	8.879	9.587	10.310	11.048	11.802	12.564	13.332	14.104	14.875	15.645	16.417	17.186	17.951	18.712	19.470
15	8.914	9.623	10.347	11.085	11.840	12.602	13.371	14.142	14.914	15.684	16.455	17.225	17.989	18.750	19.508
20	8.949	9.659	10.383	11.122	11.878	12.641	13.409	14.181	14.952	15.722	16.494	17.263	18.027	18.788	19.545
25	8.984	9.694	10.420	11.160	11.916	12.679	13.448	14.219	14.991	15.761	16.532	17.301	18.065	18.826	19.583
30	9.019	9.730	10.456	11.197	11.954	12.718	13.486	14.258	15.029	15.800	16.571	17.340	18.103	18.864	19.621
35	9.054	9.766	10.493	11.235	11.992	12.756	13.525	14.296	15.068	15.838	16.610	17.378	18.141	18.902	19.659
40	9.090	9.802	10.529	11.273	12.029	12.795	13.564	14.335	15.107	15.877	16.648	17.416	18.179	18.940	19.697
45	9.125	9.838	10.566	11.310	12.068	12.833	13.602	14.374	15.145	15.915	16.687	17.455	18.218	18.978	19.735
50	9.161	9.874	10.603	11.348	12.105	12.871	13.641	14.412	15.184	15.954	16.725	17.493	18.255	19.016	19.773
55	9.196	9.910	10.639	11.385	12.144	12.909	13.679	14.451	15.222	15.992	16.764	17.532	18.294	19.054	19.811
60	9.232	9.946	10.676	11.424	12.182	12.948	13.718	14.490	15.261	16.031	16.802	17.569	18.332	19.092	19.848
65	9.267	9.982	10.712	11.461	12.220	12.986	13.756	14.528	15.299	16.070	16.842	17.608	18.370	19.129	19.886
70	9.303	10.019	10.749	11.499	12.258	13.025	13.795	14.567	15.338	16.108	16.880	17.646	18.408	19.168	19.924
75	9.338	10.056	10.786	11.537	12.296	13.063	13.833	14.606	15.377	16.147	16.918	17.685	18.446	19.205	19.962
80	9.374	10.092	10.823	11.575	12.335	13.102	13.872	14.644	15.415	16.185	16.957	17.723	18.484	19.243	19.999
85	9.409	10.129	10.861	11.613	12.373	13.140	13.911	14.683	15.454	16.224	16.995	17.761	18.522	19.281	20.037
90	9.445	10.164	10.898	11.651	12.411	13.178	13.949	14.721	15.492	16.263	17.033	17.799	18.560	19.318	20.075
95	9.481	10.201	10.936	11.689	12.450	13.216	13.988	14.760	15.531	16.301	17.072	17.837	18.598	19.356	20.112
100	9.516	10.237	10.973	11.726	12.488	13.255	14.027	14.798	15.568	16.340	17.110	17.875	18.636	19.394	20.150

$*°C = \frac{°F - 32}{1.8}$

Table 4.14q
TYPE S—PLATINUM vs.
PLATINUM PLUS 10% RHODIUM THERMOCOUPLE

Degrees Fahrenheit vs. Millivolts. Temperatures are based on the International Temperature Scale of 1948. emf is expressed in absolute millivolts. Reference Junction 32°F (0°C).

°F*	0	100	200	300	400	500	600	700	800	900	1000	1100	1200	1300	1400	1500
								Millivolts								
0	−0.092	0.221	0.595	1.017	1.474	1.956	2.458	2.977	3.506	4.046	4.596	5.156	5.726	6.307	6.897	7.498
5	−0.078	0.238	0.615	1.039	1.498	1.981	2.484	3.003	3.533	4.073	4.623	5.184	5.755	6.336	6.927	7.529
10	−0.064	0.256	0.635	1.061	1.521	2.005	2.510	3.029	3.560	4.100	4.651	5.212	5.784	6.635	6.957	7.559
15	−0.050	0.274	0.655	1.083	1.545	2.030	2.535	3.056	3.587	4.128	4.679	5.241	5.813	6.394	6.987	7.589
20	−0.035	0.291	0.676	1.106	1.569	2.055	2.561	3.082	3.614	4.155	4.707	5.269	5.842	6.424	7.017	7.620
25	−0.021	0.309	0.696	1.128	1.593	2.080	2.587	3.108	3.640	4.182	4.735	5.298	5.871	6.453	7.046	7.650
30	−0.006	0.327	0.717	1.151	1.616	2.105	2.613	3.135	3.667	4.210	4.763	5.326	5.899	6.483	7.076	7.681
35	+0.009	0.346	0.738	1.173	1.640	2.130	2.638	3.161	3.694	4.237	4.790	5.354	5.928	6.512	7.106	7.711
40	+0.024	0.364	0.758	1.196	1.664	2.155	2.664	3.188	3.721	4.264	4.818	5.383	5.957	6.542	7.136	7.742
45	+0.040	0.383	0.779	1.219	1.688	2.180	2.690	3.214	3.748	4.292	4.846	5.411	5.986	6.571	7.166	7.772
50	0.056	0.401	0.800	1.242	1.712	2.205	2.716	3.240	3.775	4.319	4.874	5.440	6.015	6.601	7.196	7.803
55	0.071	0.420	0.822	1.264	1.736	2.230	2.742	3.267	3.802	4.347	4.902	5.469	6.044	6.630	7.226	7.834
60	0.087	0.439	0.843	1.287	1.761	2.255	2.768	3.293	3.829	4.374	4.930	5.497	6.073	6.660	7.257	7.864
65	0.104	0.458	0.864	1.311	1.785	2.281	2.794	3.320	3.856	4.402	4.959	5.526	6.102	6.689	7.287	7.895
70	0.120	0.477	0.886	1.334	1.809	2.306	2.820	3.347	3.883	4.430	4.987	5.555	6.131	6.719	7.317	7.925
75	0.136	0.496	0.907	1.357	1.833	2.331	2.846	3.373	3.910	4.457	5.015	5.583	6.161	6.749	7.347	7.956
80	0.153	0.516	0.929	1.380	1.858	2.357	2.872	3.400	3.937	4.485	5.043	5.612	6.190	6.778	7.377	7.987
85	0.170	0.535	0.951	1.404	1.882	2.382	2.898	3.426	3.964	4.512	5.071	5.640	6.219	6.808	7.407	8.018
90	0.187	0.555	0.973	1.427	1.907	2.407	2.924	3.453	3.991	4.540	5.099	5.669	6.248	6.838	7.438	8.048
95	0.204	0.575	0.994	1.450	1.931	2.433	2.951	3.480	4.019	4.568	5.128	5.698	6.277	6.867	7.468	8.079
100	0.221	0.595	1.017	1.474	1.956	2.458	2.977	3.506	4.046	4.596	5.156	5.726	6.307	6.897	7.498	8.110

$*{}^{\circ}C = \frac{{}^{\circ}F - 32}{1.8}$

Table 4.14q *Continued*

TYPE S—*Continued* PLATINUM vs. PLATINUM PLUS 10% RHODIUM THERMOCOUPLE

°F*	1600	1700	1800	1900	2000	2100	2200	2300	2400	2500	2600	2700	2800	2900	3000	3100	3200
								Millivolts									
0	8.110	8.732	9.365	10.009	10.662	11.323	11.989	12.657	13.325	13.991	14.656	15.319	15.979	16.637	17.292	17.943	18.590
5	8.141	8.764	9.397	10.041	10.695	11.356	12.022	12.690	13.358	14.024	14.689	15.352	16.012	16.670	17.324	17.975	18.622
10	8.172	8.795	9.429	10.074	10.728	11.389	12.055	12.724	13.391	14.058	14.722	15.385	16.045	16.702	17.357	18.008	18.655
15	8.203	8.827	9.461	10.106	10.761	11.423	12.089	12.757	13.425	14.091	14.755	15.418	16.078	16.735	17.389	18.040	18.687
20	8.234	8.858	9.493	10.139	10.794	11.456	12.122	12.790	13.458	14.124	14.789	15.451	16.111	16.768	17.422	18.073	
25	8.265	8.890	9.525	10.171	10.827	11.489	12.155	12.824	13.491	14.157	14.822	15.484	16.144	16.801	17.455	18.105	
30	8.296	8.921	9.557	10.204	10.860	11.522	12.189	12.857	13.525	14.191	14.855	15.517	16.177	16.834	17.487	18.137	
35	8.327	8.953	9.589	10.237	10.893	11.556	12.222	12.891	13.558	14.224	14.888	15.550	16.210	16.866	17.520	18.170	
40	8.358	8.984	9.621	10.269	10.926	11.589	12.256	12.924	13.591	14.257	14.921	15.583	16.243	16.899	17.552	18.202	
45	8.389	9.016	9.654	10.302	10.959	11.622	12.289	12.957	13.625	14.290	14.954	15.616	16.275	16.932	17.585	18.235	
50	8.420	9.048	9.686	10.334	10.992	11.655	12.322	12.991	13.658	14.324	14.988	15.649	16.308	16.965	17.618	18.267	
55	8.451	9.079	9.718	10.367	11.025	11.689	12.356	13.024	13.691	14.357	15.021	15.682	16.341	16.997	17.650	18.299	
60	8.482	9.111	9.750	10.400	11.058	11.722	12.389	13.058	13.725	14.390	15.054	15.715	16.374	17.030	17.683	18.332	
65	8.513	9.143	9.782	10.433	11.091	11.755	12.423	13.091	13.758	14.423	15.087	15.748	16.407	17.063	17.715	18.364	
70	8.545	9.174	9.815	10.465	11.124	11.789	12.456	13.124	13.791	14.457	15.120	15.781	16.440	17.095	17.748	18.396	
75	8.576	9.206	9.847	10.498	11.157	11.822	12.490	13.158	13.825	14.490	15.153	15.814	16.473	17.128	17.780	18.429	
80	8.607	9.238	9.879	10.531	11.190	11.855	12.523	13.191	13.858	14.523	15.186	15.847	16.506	17.161	17.813	18.461	
85	8.638	9.270	9.912	10.564	11.224	11.888	12.556	13.224	13.891	14.556	15.219	15.880	16.538	17.194	17.845	18.493	
90	8.670	9.302	9.944	10.597	11.257	11.922	12.590	13.258	13.924	14.589	15.253	15.913	16.571	17.226	17.878	18.526	
95	8.701	9.333	9.976	10.629	11.290	11.955	12.623	13.291	13.958	14.623	15.286	15.946	16.604	17.259	17.910	18.558	
100	8.732	9.365	10.009	10.662	11.323	11.989	12.657	13.325	13.991	14.656	15.319	15.979	16.637	17.292	17.943	18.590	

*°C = $\frac{°F - 32}{1.8}$

Table 4.14r

TYPE T—COPPER-CONSTANTAN THERMOCOUPLE

Degrees Fahrenheit vs. Millivolts. Temperatures are based on the International Temperature Scale of 1948. emf is expressed in absolute millivolts.

Reference Junction 32°F (0°C).

°F*	−300	−200	−100	−0	+0	100	200	300	400	500	600	700
						Millivolts						
0	−5.284	−4.111	−2.559	−0.670	−0.670	1.517	3.967	6.647	9.525	12.575	15.773	19.100
5	−5.332	−4.179	−2.645	−0.771	−0.567	1.633	4.096	6.786	9.674	12.732	15.937	19.269
10	−5.379	−4.246	−2.730	−0.872	−0.463	1.751	4.225	6.926	9.823	12.888	16.101	19.439
15		−4.312	−2.814	−0.973	−0.359	1.869	4.355	7.066	9.973	13.046	16.264	19.608
20		−4.377	−2.897	−1.072	−0.254	1.987	4.486	7.208	10.123	13.203	16.429	19.779
25		−4.441	−2.980	−1.171	−0.149	2.107	4.617	7.349	10.273	13.362	16.593	19.949
30		−4.504	−3.062	−1.270	−0.042	2.226	4.749	7.491	10.423	13.520	16.758	20.120
35		−4.566	−3.143	−1.367	+0.064	2.346	4.880	7.633	10.574	13.678	16.924	20.291
40		−4.627	−3.223	−1.463	+0.171	2.467	5.014	7.776	10.726	13.838	17.089	20.463
45		−4.688	−3.301	−1.559	+0.280	2.589	5.147	7.920	10.878	13.997	17.255	20.634
50		−4.747	−3.380	−1.654	0.389	2.711	5.280	8.064	11.030	14.157	17.421	20.805
55		−4.805	−3.457	−1.748	0.499	2.835	5.415	8.207	11.183	14.317	17.588	
60		−4.863	−3.533	−1.842	0.609	2.958	5.550	8.352	11.336	14.477	17.754	
65		−4.919	−3.609	−1.934	0.720	3.082	5.685	8.497	11.490	14.637	17.921	
70		−4.974	−3.684	−2.026	0.832	3.207	5.821	8.642	11.643	14.799	18.089	
75		−5.029	−3.757	−2.117	0.944	3.332	5.957	8.788	11.797	14.961	18.257	
80		−5.081	−3.829	−2.207	1.057	3.458	6.094	8.935	11.953	15.122	18.425	
85		−5.134	−3.901	−2.296	1.171	3.584	6.232	9.082	12.108	15.284	18.593	
90		−5.185	−3.972	−2.385	1.286	3.712	6.370	9.229	12.263	15.447	18.761	
95		−5.235	−4.042	−2.472	1.401	3.839	6.508	9.376	12.418	15.610	18.930	
100		−5.284	−4.111	−2.559	1.517	3.967	6.647	9.525	12.575	15.773	19.100	

*$°C = \frac{°F - 32}{1.8}$

Table 4.14s

CHROMEL-CONSTANTAN THERMOCOUPLE

°F*	−300	−200	−100	−0	+0	100	200	300	400	500	600	700
						Millivolts						
0	−8.30	−6.40	−3.94	−1.02	−1.02	2.27	5.87	9.71	13.75	17.95	22.25	26.65
10	−8.45	−6.62	−4.21	−1.33	−0.71	2.62	6.25	10.11	14.17	18.38	22.69	27.09
20	−8.60	−6.83	−4.47	−1.64	−0.39	2.97	6.62	10.51	14.59	18.81	23.13	27.53
30		−7.04	−4.73	−1.94	−0.07	3.32	7.00	10.91	15.00	19.23	23.57	27.97
40		−7.24	−4.98	−2.24	0.26	3.68	7.38	11.31	15.42	19.66	24.00	28.42
50		−7.44	−5.23	−2.54	0.59	4.04	7.76	11.71	15.84	20.09	24.44	28.86
60		−7.62	−5.48	−2.83	0.92	4.40	8.15	12.11	16.26	20.52	24.88	29.31
70		−7.80	−5.72	−3.11	1.26	4.77	8.54	12.52	16.68	20.95	25.32	29.75
80		−7.97	−5.95	−3.39	1.59	5.13	8.93	12.93	17.10	21.39	25.76	30.19
90		−8.14	−6.18	−3.67	1.93	5.50	9.32	13.34	17.52	21.82	26.20	30.64
100		−8.30	−6.40	−3.94	2.27	5.87	9.71	13.75	17.95	22.25	26.65	31.09

°F*	800	900	1000	1100	1200	1300	1400	1500	1600	1700	1800
						Millivolts					
0	31.09	35.57	40.06	44.56	49.04	53.50	57.92	62.30	66.63	70.90	75.12
10	31.54	36.02	40.51	45.01	49.49	53.94	58.36	62.74	67.05	71.32	75.53
20	31.98	36.47	40.96	45.46	49.93	54.38	58.80	63.17	67.48	71.75	75.95
30	32.43	36.92	41.41	45.91	50.37	54.83	59.24	63.60	67.91	72.17	76.37
40	32.87	37.37	41.86	46.36	50.82	55.27	59.68	64.04	68.34	72.60	
50	33.32	37.82	42.31	46.81	51.27	55.71	60.11	64.47	68.76	73.02	
60	33.77	38.26	42.76	47.26	51.72	56.15	60.55	64.90	69.19	73.44	
70	34.22	38.71	43.21	47.71	52.16	56.59	60.99	65.34	69.62	73.86	
80	34.67	39.16	43.66	48.15	52.61	57.03	61.43	65.77	70.05	74.28	
90	35.12	39.61	44.11	48.60	53.05	57.48	61.86	66.20	70.47	74.70	
100	35.57	40.06	44.56	49.04	53.50	57.92	62.30	66.63	70.90	75.12	

$*°C = \frac{°F - 32}{1.8}$

Chapter V

PRESSURE MEASUREMENT

J. D. GOODRICH JR.
J. T. HALL
B. G. LIPTÁK

CONTENTS OF CHAPTER V

Table V

ORIENTATION TABLE FOR PRESSURE DETECTORS

Applicable Pressure Ranges scales: *mm Hg absolute* (1mm Hg = 133 Pa): 10^{-14} 10^{-10} 10^{-6} 10^{-3} 10^{-1} 1 50 200 400 600 4 7 11 10^2 10^3 10^4 10^5 10^6; *"H_2O* (1"H_2O = 250 Pa): −300 −200 −100 −10 −5 −1 ±0.1 +1 +5 +10 +100 +200 +300; *PSIG* (1PSIG = 6.9 kPa)

Section		Type of Design	Inline Device	Laboratory or Pilot Plant Device	Local Readout (Gauge)	Remote Readout Trans.	Applicable Pressure Ranges (reference)
5.3	Bellows	Abs. Press. Motion Balance	√	√	√	√	Ⓥ
		Abs. Press. Force Balance	√	√		√	Ⓥ
		Atm. Press. Ref. Motion Bal.	√	√	√	√	Ⓐ
		Atm. Press. Ref. Force Bal.	√	√		√	Ⓐ
		Aneroid Manostats	√	√			Ⓥ
5.4	Bourdon	C.-Bourdon	√	√	√	√	Ⓐ
		Spiral Bourdon	√	√	√	√	Ⓐ
		Helical Bourdon	√	√	√	√	Ⓐ
		Quartz Helix		√	√	√	Ⓥ Ⓐ
5.5	Diaphragm	Abs. Press. Motion Balance	√	√	√	√	Ⓥ
		Abs. Press. Force Balance	√	√		√	Ⓥ
		Atm. Press. Ref. Motion Bal.	√	√	√	√	Ⓐ
		Atm. Press. Ref. Force Bal.	√	√		√	Ⓐ
5.7	Electronic	Strain Gauge	√	√	√	√	Ⓥ Ⓐ
		Electronic Transmitters	√	√		√	Ⓥ Ⓐ
		Capacitive Sensors	√	√		√	Ⓥ Ⓐ
5.8	High-Press. Sensors	Dead Weight Piston Gauge		√	√		Ⓐ
		Bulk Modulus Cell	√	√		√	Ⓐ
		Manganin Cell		√		√	Ⓐ
5.9	Manometers	Inverted Bell	√	√	√	√	Ⓥ Ⓐ
		Ring Balance	√	√	√	√	Ⓐ
		Float Manometer	√	√	√	√	Ⓐ
		Barometers	√	√	√		Ⓥ
		Visual Manometers	√	√	√		Ⓐ
		Micromanometers		√	√		Ⓥ Ⓐ
		Cartesian Divers	√	√			●
5.12	Pressure Repeaters	D/P Cell	√	√		√	●
		Std. Diaphragm	√	√		√	●
		Button Diaphragm	√	√		√	●
5.14 High-Vacuum	Ionization	Hot Cathode		√	√	√	●
		Cold Cathode		√	√	√	●
	Thermal	Thermocouple	√	√	√	√	●
		Thermopile	√	√	√	√	●
		Resistance Wire	√	√	√	√	●
	Mechanical	McLeod	√	√	√		●
		Molecular Momentum	√	√	√		●
		Capacitance	√	√	√	√	Ⓥ

Ⓥ Indicates that the device uses full-vacuum reference in its operation.
Ⓐ Indicates that the instrument uses atmospheric pressure reference.
● Indicates that the operating principle used does not involve the use of reference pressures.

5.1 APPLICATION AND SELECTION

Introduction

Pressure detection devices can be classified on the basis of the pressure levels they can measure, by the design principle involved in their operation, or by their application. In this chapter, the various categories are not separated in any strict manner. In general, low-pressure sensors are discussed in the earlier sections, and high-pressure detectors in the later parts of the chapter. Industrial instruments are discussed in reasonable detail, with emphasis on the most commonly used devices, and laboratory instruments are covered in less detail. Information on accessory equipment, such as repeaters, switches and vacuum sensors, will be found near the end of the chapter.

Each section starts with a brief summary of basic features applicable to the group of instruments discussed in that section. This information allows the reader to choose the material that is most suitable for his application.

Application and Selection

This chapter covers a wide range of pressure sensors. The elements described can measure pressures from ultrahigh vacuums, such as 10^{-13} mmHg (133×10^{-13} Pa), to ultrahigh gauge pressures approaching 400,000 PSIG (2,800 MPa). The principles of operation of these local indicating and transmitting devices are varied, and the instruments may be used in both the laboratory and industry.

The range of costs and inaccuracies of the instruments is equally broad. A simple, 1½-in. diameter, ±5 percent inaccurate gauge might cost only $2.00, while a 0.01 percent inaccurate fused quartz helix sensor with digital readout could cost $3,000. With so many types from which to choose, it might seem that making the proper selection for a particular installation would be difficult and time consuming. Actually, this is not the case. The multitude of devices covered is all-inclusive, and for a typical industrial installation the selection is fairly simple.

The reader should find Table V of value in narrowing his choices. The table indicates the overall pressure range that a particular design can detect. It also notes whether the unit is available for industrial in-line installation or for laboratory use only. In addition, it shows whether the instrument can furnish remote readout or is limited to direct local indication.

When local pressure indicators are required, most installations between 0–10 in. H_2O (2.6 kPa) and 0–100,000 PSIG (690 MPa) can be handled by conventional pressure sensors (Sections 5.3, 5.4, and 5.5). If small, near-atmospheric pressures are to be detected, the bellows and diaphragm sensors (Sections 5.3 and 5.5) or manometers (Section 5.9) can be considered. For local vacuum measurement down to 1 mmHg (0.13 kPa), both the diaphragm absolute sensors and the vacuum manometers (Sections 5.5 and 5.9) will give satisfactory performance.

Where remote readouts are required, the force or motion balance pneumatic or electronic transmitters with elastic elements (Sections 5.3, 5.4, and 5.5) will handle most applications. These include vacuums down to 1 mmHg (0.13 kPa) absolute and gauge pressures up to 100,000 PSIG (690 MPa). When small, near-atmospheric pressures, or pressures up to 200,000 PSIG (1,400 MPa) are to be transmitted, almost any of the electronic sensors described in Section 5.7 will be satisfactory. Above 200,000 PSIG (1,400 MPa), the high-pressure sensors going up to 400,000 (2,800 MPa) are used (Section 5.8).

Multiple pressure sensors, including scanners and multiplexers, are covered in Section 5.10. Pressure gauges for local readout with a range of 10 in. H_2O (2.6 kPa) up to 100,000 PSIG (690 MPa) are described in Section 5.11.

Various types of pressure repeaters ranging from full vacuum to 10,000 PSIG (69 MPa) are described in Section 5.12. Details on pressure switches for use up to 20,000 PSIG (138 MPa) will be found in Section 5.13. Virtually every type of vacuum sensor, ranging from 10^{-11} mmHg (1.3×10^{-9} Pa) to 760 mmHg (0.1 MPa), is discussed in Section 5.14.

As in level detection, a frequent problem in pressure sensing is maintenance. For this reason, careful consideration should be given to vibration, freezing, corrosion, temperature and hard-to-handle fluids. In this connection, the various protection devices discussed in Section

5.2 should be just as carefully evaluated as the sensing element to be used.

BIBLIOGRAPHY

Comber, J. and Hockman, P., "Pressure Monitoring: What's Happening?" *Instruments and Control Systems*, April, 1980.

Elliott, T.C., "Temperature, Pressure, Level, Flow—Key Measurements in Power and Process," *Power*, September, 1975.

Hall, J., "Monitoring Pressure with Newer Technologies," *Instruments and Control Systems*, April, 1979.

Herceg, E.E., "Handbook of Measurement and Control," Schaevitz Engineering (Pennsauken NJ), 1972.

Lawford, V.N., "Differential Pressure Instruments: The Universal Measurement Tools," *Instrumentation Technology*, December, 1974.

Moore, R.L., "Basic Instrumentation Lecture Notes and Study Guide—Measurement Fundamentals," Instrument Society of America, 1976.

Slomiana, Maria, "Selecting Differential Pressure Instruments," *Instrumentation Technology*, August, 1979.

Slomiana, Maria, "Using Differential Pressure Sensors," *Instrumentation Technology*, September, 1979.

Soisson, H.E., "Instrumentation in Industry," John Wiley and Sons (New York), 1975.

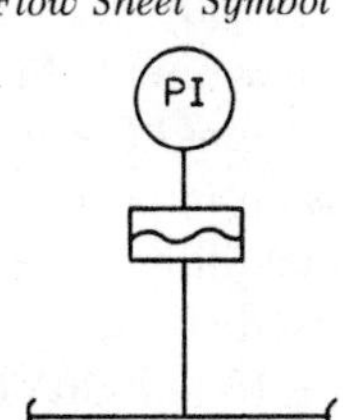

5.2 ACCESSORIES

Design Pressure:	Up to 2,500 PSIG (17 MPa) with standard units, specials to 10,000 PSIG (69 MPa) and higher
Design Temperature:	Up to 600°F (316°C) with standard units, specials up to 1,200°F (650°C)
Materials of Construction:	Wetted parts available in steel, stainless steel, bronze, aluminum, carpenter 20, Hastelloy, Inconel, Monel nickel, tantalum, titanium, Kel-F or Teflon lining; Saran, PVC, rubber or glass are also available for the bottom housing and noble (noncorrosive) metals as diaphragm materials
Cost:	$50 minimum; cost increase as a function of size and materials of construction
Partial List of Suppliers:	Ametek Inc., US Gauge Div.; Belfab, Bailey Controls Co.; Bellofram Corp.; Bristol–Babcock Div.; ACCO Industries, Inc.; Helicoid Gauge Div., ACCO; Marsh Instrument Co., A Unit of General Signal; Parker–Hannifin Corp.; Ronningen–Petter Div. of Dover Corp.

Pulsation Dampeners

On steam service, it is desirable to prevent live steam from entering the sensing element to protect against overtemperature damage. This is accomplished by installing a coil pipe syphon between the gauge and the process connection.

Where sudden shock pressures or rapidly fluctuating pressures are expected, snubbers or pulsation dampers should be installed between the gauge and the process. Some of the design variations for snubbers are illustrated in Figure 5.2a. One design consists of a fitting with a corrosion resistant porous metal filter disc. Such a device delays the equilibrium reading on the indicator by about 10 seconds. Another snubber design depends for its damping action on a small piston in the inlet fitting, which rises and falls with pressure impulses, and thereby absorbs shock and surge. Still another snubber design uses the adjustable restriction created by a microvalve in the inlet fitting to damp pulsations.

Where there is a possibility that the process fluid might freeze inside the pressure element, a small *electric heater* can be installed inside the case. The heater will provide for a temperature rise of about 80°F (27°C) and can be ordered with gauges having dial diameters 4½ in. (112 mm) or larger.

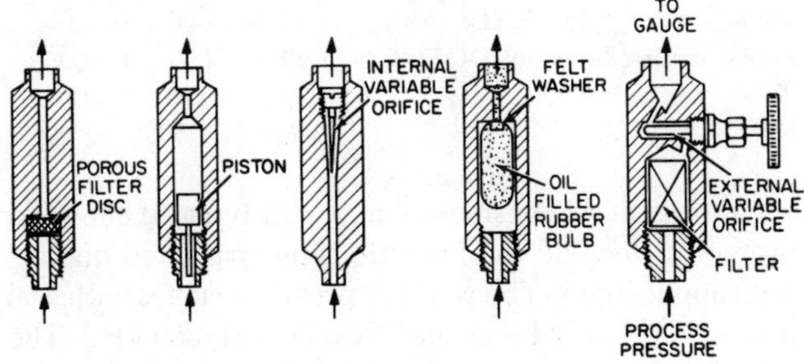

Fig. 5.2a Variations on snubber designs

Chemical Seals

Chemical seals or diaphragm protectors can be provided with most pressure sensors. These components (a.) keep the process fluid in the elastic pressure element from freezing, gelling or settling out due to temperature changes; (b.) prevent poisonous, noxious or corrosive process materials from entering the pressure sensor, which is of standard construction materials; and (c.) prevent slurries or viscous polymers from entering and plugging the detector element.

In the following discussion, several chemical seal designs will be described. All are noncompressible, liquid-filled elements with means of transmitting the process pressure through their filling media into the protected instrument. Some of the limitations are common to all designs. Because of the liquid filling, thermal expansion is an important consideration. Below 0°F (−18°C), special filling fluids, such as ethyl alcohol and kerosene, are used. Between the process temperatures of 0 and 400°F (−18 and 205°C), conventional filling fluids and Teflon or other plastic gaskets can be considered. Up to 600°F (316°C), asbestos gaskets give satisfactory performance. At temperatures above 600°F, special volumetric elements with NaK filling can be utilized. Even with the proper selection of filling fluids, the temperature effects will not be completely eliminated. Therefore, it is desirable to calibrate critical units at operating temperature to eliminate errors in the approximate area of 3 PSI per 100°F (21 kPa per 38°C). The table below lists the properties of some common filling fluids. These should not decompose at the highest process temperature and should have low enough viscosities to operate the process elements efficiently at the lowest process temperatures.

Filling Fluid	*Temperature Range (°F)**	*Expansion Coeff. (SG/°F)*
Toluene	−40 to +200	0.00063
Dow Corning Silicone (DC-200)	−30 to +300	0.00075
Kerosene	−30 to +350	0.00051
Hooker Chemical Fluorlube (FS-5)	−20 to +300	0.00049
Mercury	−30 to +700	
Dibutyl Phthalate	+20 to +300	0.00080
22% Na–78% K	+20 to +1,400	
70% Glycerin, 30% Water	+30 to +300	0.00051
Instrument Oil	+35 to +300	0.00035
Light Turbine Oil	+40 to +300	0.00048
Dow Corning Silicone (DC-550)	+40 to +500	0.00042
Dow Corning Silicone (DC-703)	+40 to +600	0.00055
96.5% Glycerin	+70 to +450	0.00039
56% Na–44% K	+70 to +1,500	

Other general considerations for all types of chemical protectors include the limitations on spans and on vacuum applications. The working pressure ranges with seal protection should be greater than 50 PSI (350 kPa). The use of diaphragm seals on vacuum service is questionable. Small amounts of entrapped air and the vapor pressure of the filling liquid limit such services because the noncompressible filling fluid will not accurately transmit tension, and the air or vapor pressure will interfere with the measurement. The higher the operating temperature, the more severe this limitation becomes, and it is suggested that devices other than liquid-filled elements on vacuum service be used.

The measuring instrument can be direct-connected to the seal or remote-connected with capillary tubing. In general, it is desirable to limit the tubing length to 25 ft. (7.5 m) and to maintain the ambient temperature around the capillary and the instrument case within 50 and 110°F (10 and 43°C).

$$*°C = \frac{°F - 32}{1.8}$$

Standard Seals

Figure 5.2b illustrates the three main components of a standard seal: the top and bottom housings and the oil-filled, nested diaphragm capsule. The upper housing is in contact with the filling fluid only; therefore, if the atmosphere is not corrosive, standard construction materials are acceptable. A filling screw permits the instrument to be assembled to the seal and be filled under vacuum as a unit.

The diaphragm capsule shown is one of several designs. Some seal manufacturers use a single diaphragm element welded to the upper housing. The illustrated capsule consists of a thin lower diaphragm bonded to a sturdy, corrugated backup plate, and the whole assembly is screwed into the top housing. Leakage is the most frequent source of diaphragm failure, and when it occurs, the diaphragm is pressed against the corrugated backup plate to prevent diaphragm distortion and to seal in the process pressure. The rupture pressure rating of conventional seals is 2,500 PSIG (17 MPa). The diaphragm capsule is in contact with the process fluid and is available in corrosion resistant metallic construction or with Kel-F, Teflon or a similar lining. The top section and the diaphragm capsule can be removed with the associated instrument without disconnecting the bottom housing. This enables the operator to clean the assembly without refilling or recalibrating the unit.

The bottom housing is in contact with the process media and is available in corrosion resistant metal or

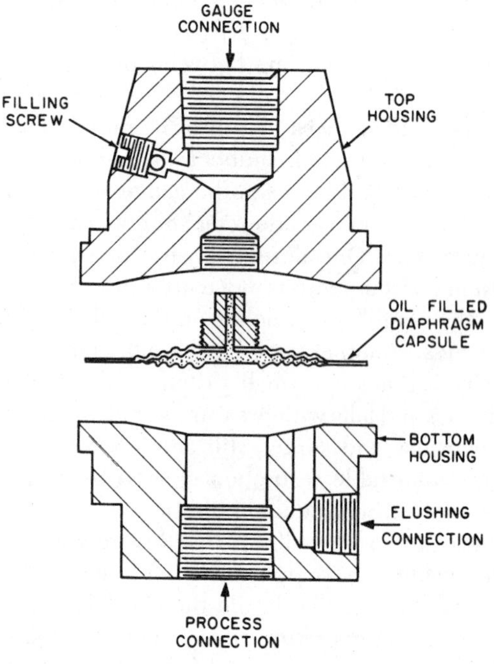

Fig. 5.2b Basic components of conventional chemical seals

plastic. A flushing connection can be provided on the lower housing, leading to the pressure chamber below the diaphragm. This allows for continuous or intermittent purge of cleaning fluid to be applied to remove material buildup that may have accumulated below the diaphragm or in the connecting piping.

Figure 5.2c shows standard off-line seals with screwed or flanged process connections. These units are utilized where the seal is to provide corrosion or freezing protection. Due to the dead-ended cavity between the process line and the seal, they are not suitable as protection against plugging.

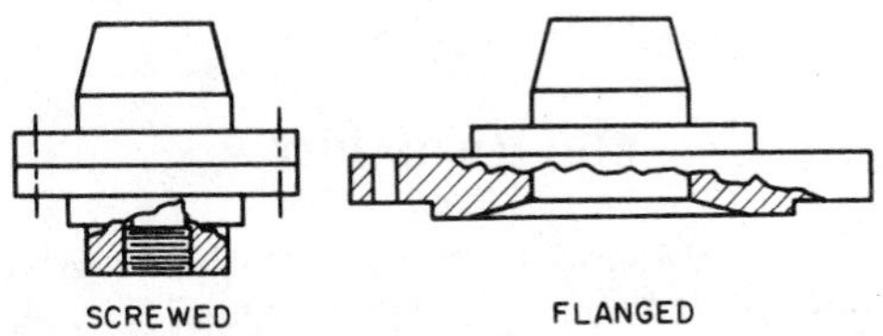

Fig. 5.2c Off-line chemical seals

Figure 5.2d illustrates some of the in-line or flow-through designs in which the dead-ended cavities have been minimized. These units can be considered for both plugging or corrosive services, but their removal for maintenance necessitates the draining of the process pipe.

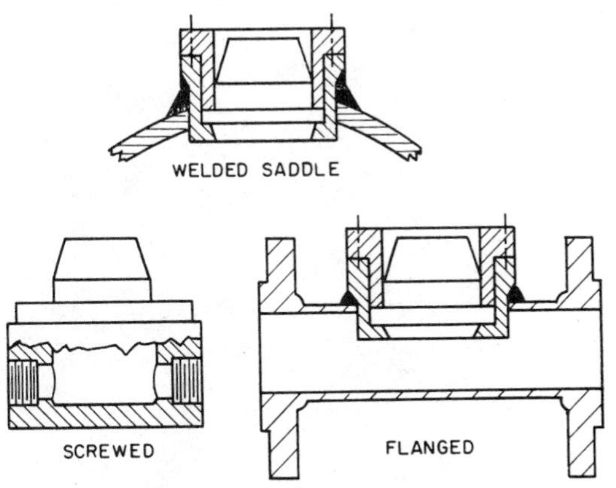

Fig. 5.2d In-line chemical seals

Because displacement of these standard seals is small, they are compatible with only small displacement-sensing elements, such as Bourdon tubes, force balance diaphragms, spiral and helical elements or small-diameter bellows. For instruments with bellows diameters larger than ¾ in. (19 mm), the capacity of the standard seal is not sufficient to match the displacement of the pressure element, and, therefore, large capacity seals are needed. Such design with a rolling diaphragm is illustrated in Figure 5.2e.

When material buildup is a very serious consideration, even the in-line seals shown in Figure 5.2d will not give satisfactory performance, and it is necessary to eliminate all cavities completely. This is achieved with the full-stream seal shown in Figure 5.2f. The three main components of this seal are the flexible cylinder, the cast iron housing and the end flanges. The space between the flexible sleeve and the housing is filled with the sensing fluid. The sleeve is available in neoprene or Teflon and the end flanges can be of stainless steel or other metals compatible with the process. This design eliminates all cavities and is applicable to low- and medium-pressure applications up to 200 PSIG (1.4 MPa).

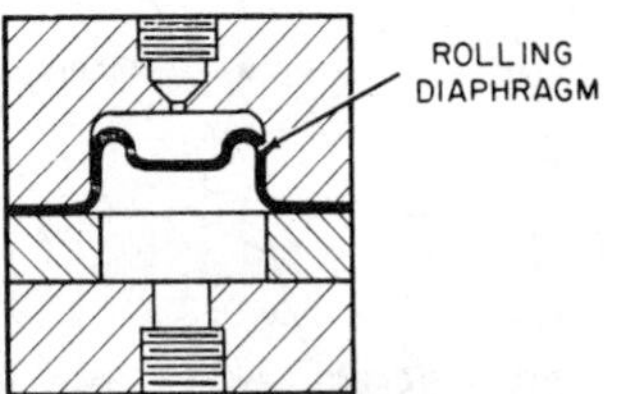

Fig. 5.2e Large displacement seal

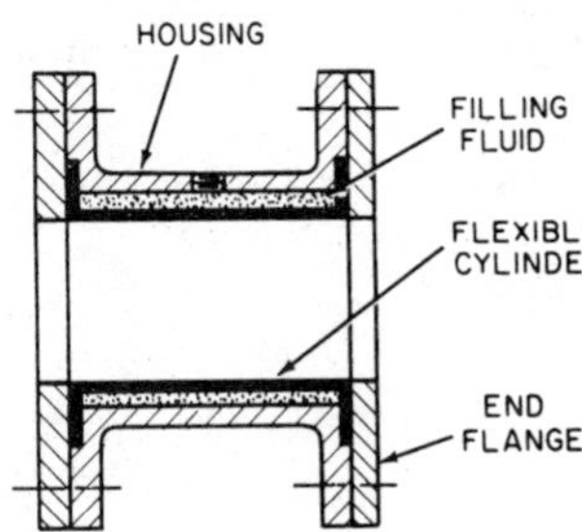

Fig. 5.2f Full-stream spool seal

Volumetric Seal Elements

Figure 5.2g shows a number of volumetric seal designs, all of them consisting of a flexible member, a housing and filling liquid to transmit the process pressure to the pressure instrument through capillary tubing. Each of these devices serves to minimize or eliminate the cavities in which process fluid could build up. In the diaphragm units, there is a definite relationship between the process pressure to be detected and the required diaphragm area. Extended or wafer elements with greater than 2 in.-diameter (50 mm) diaphragm surfaces can handle pressure spans from 0–50 PSIG to 0–1000 PSIG (0–3 MPa to 0–6.9 MPa). The small button diaphragm designs will work with a minimum span of a few hundred pounds per square inch and can handle spans as high as 10,000 PSIG (69 MPa) or greater. The bellows design is more sensitive than the same diameter button diaphragm and can handle spans from 75 to 1,000 PSI (0.5 to 6.9 MPa). This element, having flexibility in both directions, is more applicable to compound pressure applications than the others. The tube seal element is applicable to spans between 1,000 and 5,000 PSI (6.9 and 35 MPa).

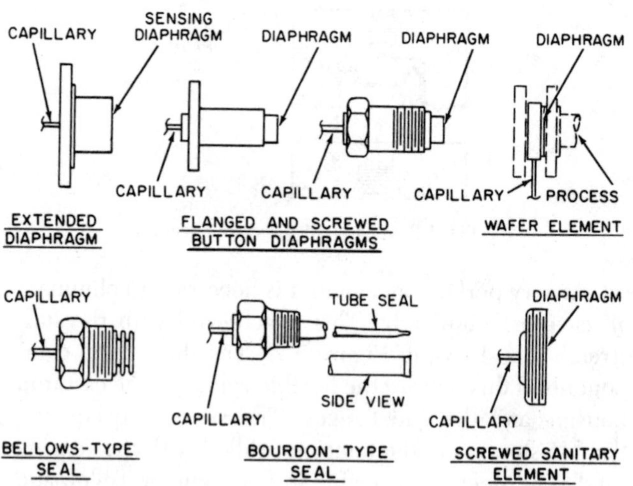

Fig. 5.2g Volumetric seal elements

The general remarks at the beginning of this section are also applicable to volumetric elements. The construction materials for these units are more limited than those for standard seals. Stainless steel is the standard material for wetted parts, but more corrosion resistant materials can also be obtained on special order. Liquid-filled, extended diaphragm seals can be obtained with extension lengths and diameters as needed to bring the diaphragm surface flush with the inside of the pipe or other equipment.

On applications in which low-pressure span, vacuum service or process temperature considerations would make the liquid-filled seals undesirable, but in which dead-ended cavities cannot be tolerated, the extended diaphragm d/p cells illustrated in Figure 5.6c or the extended diaphragm pressure repeaters shown in figure 5.12a can be considered.

BIBLIOGRAPHY

Comber, T. and Hackman, P., "Pressure Monitoring: What's Happening?" *Instruments and Control Systems*, April, 1980.

Krell, George H., "Sensing Pressure through Diaphragms," *Chemical Engineering*, July 1, 1968.

Reed, C.J., "How to Refill Chemical Seals," *Instrumentation Technology*, February, 1971.

Flow Sheet Symbol

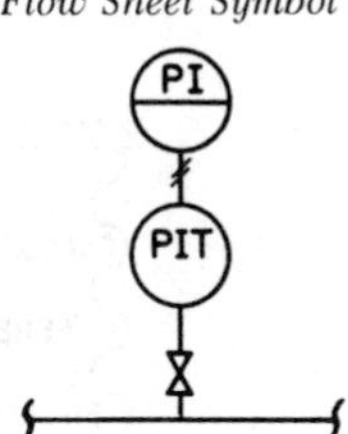

5.3 BELLOWS-TYPE PRESSURE SENSORS

Design Pressure:	Up to 20 PSIG (138 kPa) with evacuated motion balance bellows; up to 375 PSIG (2.6 MPa) with evacuated force balance bellows; up to 100 PSIG (690 kPa) with atmospheric reference motion balance bellows; up to 6,000 PSIG (41 MPa) with dual bellows element; up to 4,000 PSIG (27 MPa) with atmospheric reference force balance bellows
Design Temperature:	Up to 200°F (93°C) with brass bellows
Materials of Construction:	Brass, stainless steel, beryllium copper, Monel
Inaccuracy:	±1% of span; better with force balance designs
Range:	Absolute pressure ranges from 0–100 mmHg to 0–35 PSIA (0–13 kPa to 0–240 kPa); gauge pressure ranges from 0–5 in. H_2O to 0–2,000 PSIG (0–1.3 kPa to 0–14 MPa)
Cost:	$600 for direct recorder, $900 for transmitter; both in standard materials
Partial List of Suppliers:	Fischer & Porter Co.; Foxboro Co.; Honeywell, Industrial Div.; ITT Barton; Kent Process Control, Inc.; Taylor Instrument Co., Div. of Sybron Corp.; Wallace & Tiernan Div., Pennwalt Corp.

Bellows-type pressure elements have undergone the same improvements over the last few decades as the diaphragm capsules discussed in Section 5.5. Precision bellows are now available with minimum error due to drift, friction and elastic hysteresis. The most frequently used bellows materials are brass and stainless steel. Sensitivity of the bellows increases as a function of size. In general, it can be said that bellows elements will deliver higher forces and can detect slightly higher pressures than diaphragm capsules. The disadvantages of bellows elements include being subject to work hardening and sensitive to ambient temperature variations. In most cases, the elastic action of the bellows alone is not sufficient for accurate measurement, and a spring needs to be added for precise characterization.

Four basic designs of bellows pressure detectors will be discussed. They are classified by the reference used as absolute, or gauge, and differential pressure detectors, and can also be grouped as force or motion balance types, depending on the instrument design. Direct local indication is available only with motion balance units, but force balance sensors are more accurate. In addition, bellows manostats (self-contained vacuum controllers) will be described.

Motion Balance Absolute Pressure Sensors

When absolute pressure is to be sensed with bellows elements, it normally involves two bellows, one for measuring and the other for compensating. The compensating element is fully evacuated and sealed, while the sensing element is connected to the process being measured. Figure 5.3a illustrates the beam balance version of this design, and Figure 5.3b shows the opposed bellows design. In either arrangement, an increase in process pressure causes the measuring bellows to extend, which results in an increase of readout through the motion balance mechanism. If the process pressure is constant but the barometric pressure changes, force will be exerted equally on the outside of both bellows causing no change in the readout. The evacuated bellows are capable of compensating for barometric pressure variations as high as 100 mmHg (200 kPa).

Dual bellows detectors are available with spans from 0–100 mmHg to 0–30 PSIA (0–13 kPa to 0–200 kPa).

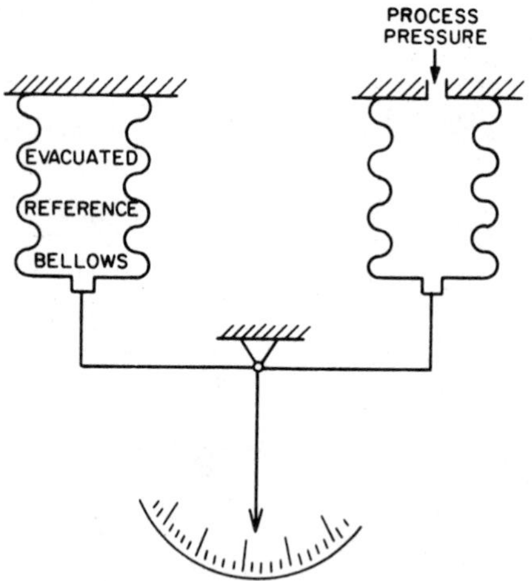

Fig. 5.3a Beam balance bellows sensor

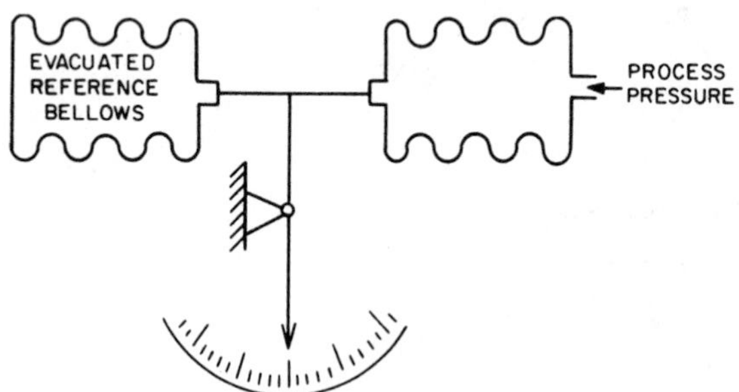

Fig. 5.3b Opposed bellows detector

The elements can be exposed to overpressures up to 35 PSIA (240 kPa). These motion balance absolute pressure sensors are inaccurate to ±1 percent of full-scale, and, therefore, can give reasonably reliable measurement of absolute pressures down to approximately 5 mmHg (0.7 kPa).

Force Balance Absolute Pressure Sensors

The force balance detector shown in Figure 5.3c is available as a pneumatic or electronic transmitter, but not as a direct indicator because there is no motion to drive a pointer. For local indication, an output gauge can be installed.

Because the force balance transmitter mechanism is illustrated in Figure 5.6b, it has not been repeated in Figure 5.3c. The pressure being sensed is applied to the inside of the bellows within the capsule. The space on the outside of the bellows is evacuated, thus providing a zero absolute pressure reference. The process pressure exerts a force on the capsule that is applied to the lower end of the force bar. Due to the force balance nature of the unit, the force bar is constantly balanced; therefore, the sensing bellows do not move as long as the pressure detected is within the range of the instrument. If the range of the capsule is exceeded, the bellows extend to the right where they are supported by the capsule backup plate.

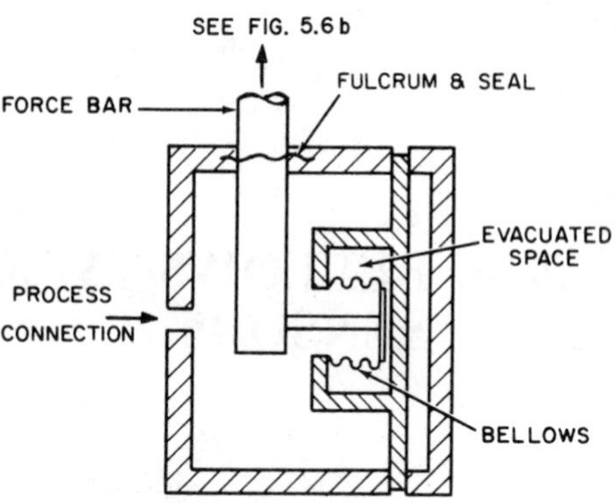

Fig. 5.3c Force balance absolute pressure sensor

The standard material for construction of this instrument is stainless steel. The inaccuracy is between ¼ and ½ percent of span. The minimum span for this unit is 25 PSIA (170 kPa) and the maximum is 250 PSIA (1,700 kPa). These spans are adjustable within the 0–350 PSIA (0–2.4 MPa) range limits of the capsule. The unit will withstand a maximum pressure overrange of 375 PSIG (2.6 MPa). The ambient temperature limits on this device are from −40 to +250°F (−40 to +120°C).

Motion Balance Pressure Sensors with Atmospheric Reference

When bellows are used as the pressure-sensing element it is desirable to add a spring for ranging and accurate characterization. Without the calibration, spring temperature effects and work hardening of the bellows would contribute to loss of accuracy. The bellows are either brass or stainless steel, and the calibration spring is Ni-Span C, which insures uniform motion throughout the range with practically no change due to ambient temperature variations.

Figures 5.3d and 5.3e illustrate designs using single-bellows elements. In both designs, the inside of the bellows is open to atmosphere, which represents the pres-

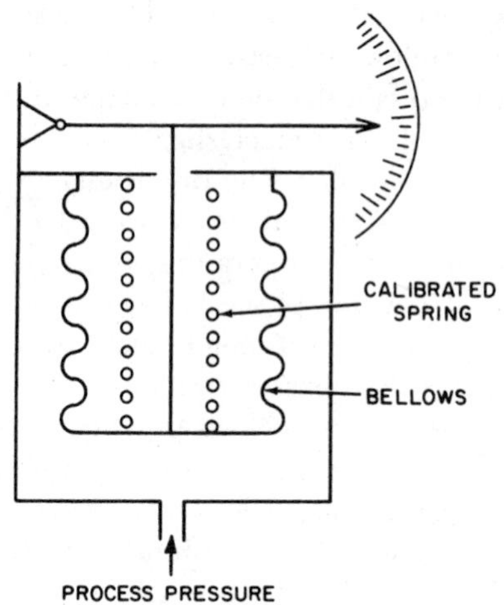

Fig. 5.3d Bellows sensor with calibrated spring

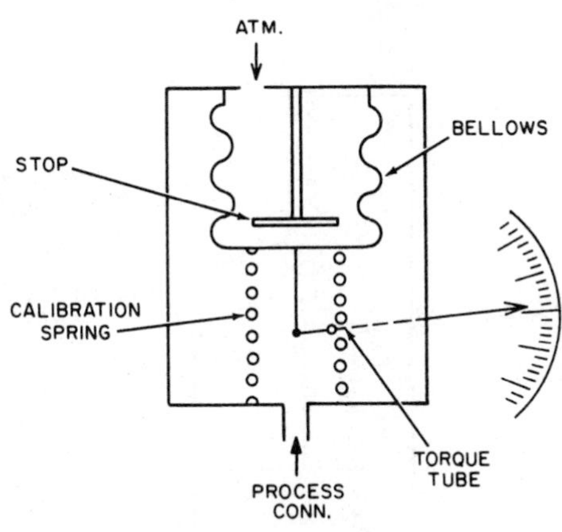

Fig. 5.3e Bellows detector with calibration spring

sure reference, and the outside is exposed to the process pressure. The differences in the two designs involve only the location of the calibration springs and the method applied in transmitting bellows motion to the readout pointer. The spring-loaded metal bellows are compressed by the process pressure forcing the lower end of the bellows upward against the opposing force of the spring. This vertical movement is transmitted through suitable linkage or a torque tube assembly to the pointer. These units can also act as differential pressure detectors if the inside of the bellows is connected to a process pressure instead of being left open to the atmosphere.

The spans of the single-bellows element vary with the diameter and construction materials of the bellows. Listed below are span ranges for some standard bellows elements. By selecting the proper spring, the span can be on the positive, or vacuum, side of the atmospheric reference pressure or can cover a compound range.*

Type of Bellows	*Minimum Span*	*Maximum Span*
Large diameter brass	0–5 in. H_2O	0–90 in H_2O
Small diameter brass	0–90 in. H_2O	0–450 in. H_2O
Small diameter stainless steel	0–15 PSIG	0–40 PSIG

The inaccuracy of these units is between ½ and 2 percent of span, depending on design, and the overpressure limitation varies from 25 to 100 PSIG (170 to 690 kPa).

Dual bellows elements are also available as differential pressure sensors. These elements will be discussed in detail in Section 5.6 (see Figure 5.6f). If either the low- or the high-pressure side is left open to atmosphere, the unit will detect gauge or vacuum pressures. The various features and operating principles of this instrument will not be repeated here, but a brief summary is given below of the various bellows available and of their span limitations.*

Type of Bellows Stainless Steel	*Minimum Span*	*Maximum Span*
3¾ in. diameter	0–20 in. H_2O	0–400 in. H_2O
2⅛ in. diameter	0–15 PSI	0–50 PSI
1⅝ in. diameter	0–30 in. H_2O	0–35 PSI
¾ in. diameter	0–35 PSI	0–400 PSI

These instruments can be designed for overpressures up to 6,000 PSIG (41 MPa), can operate between −60 and +200°F (−50 and +93°C) temperature and are inaccurate to ½ to 1 percent of span. The torque tube shaft in this unit can drive local indicators, recorders, controllers, switches or transmitters.

Force Balance Pressure Sensors with Atmospheric Reference

Figures 5.3f and 5.3g illustrate two of the bellows element gauge pressure sensors. When relatively low pressures need to be measured, the process pressure is introduced into the inside of the bellows, while for higher pressures the outside surface of the bellows is exposed to the process pressure. The force of the process pressure is applied to the lower end of the force bar. Due to the force balance nature of the unit, the force bar is constantly balanced; therefore, the sensing bellows do not move as long as the pressure detected is within the range of the instrument. These units are available as transmitters only, because they do not have enough motion

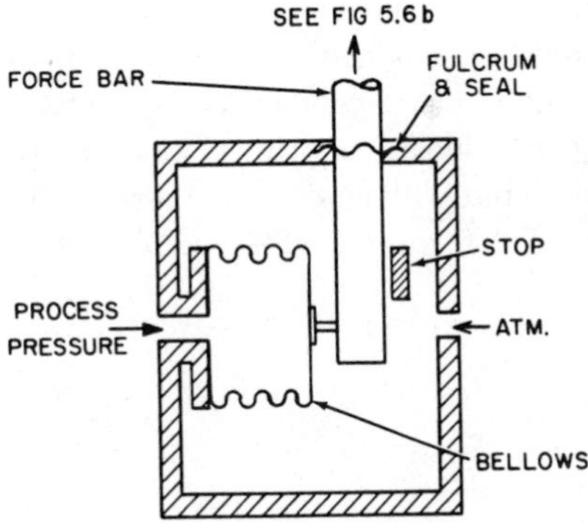

Fig. 5.3f Force balance low-gauge pressure sensor

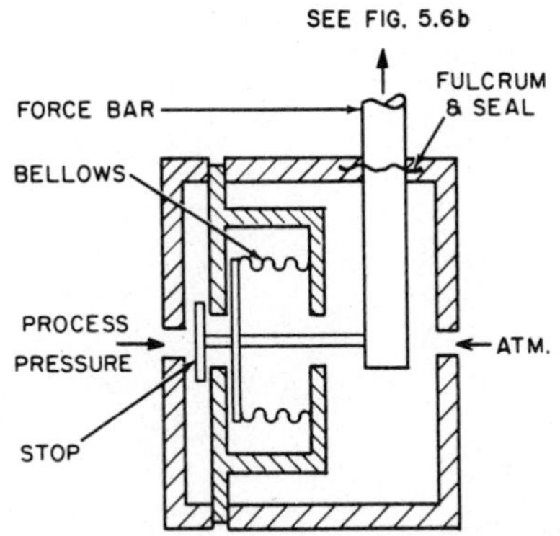

Fig. 5.3g Force balance gauge pressure detector

*See Appendix A.1 for SI Units.

to drive a local pointer (see Figure 5.6b). The bellows are available in brass, stainless steel or Monel, and inaccuracies are between ± ¼ and ½ percent of span. The tabulation below lists the spans and overpressures for some of the standard bellows elements.*

Type of Bellows	*Minimum Span*	*Maximum Span*	*Overpressure Limit*
Low-Pressure Bellows	0–10 PSI	0–60 PSI	100 PSIG
½ in. Diameter	0–25 PSI	0–250 PSI	500 PSIG
⅜ in. Diameter	0–50 PSI	0–500 PSI	1,000 PSIG
¼ in. Diameter	0–100 PSI	0–1,000 PSI	2,000 PSIG
11/64 in. Diameter	0–200 PSI	0–2,000 PSI	4,000 PSIG

Aneroid Manostats (Vacuum Controllers)

The aneroid manostat illustrated in Figure 5.3h is a self-contained bellows-type vacuum controller. The bellows is fully evacuated to provide a zero absolute pressure reference unaffected by barometric changes. The spring is temperature compensated, and its tension is adjustable over the entire range of the contorller, which is 1 to 60 in. Hg (3.4 to 200 kPa) absolute pressure. The sensitivity of setting is about ½ mmHg (0.07 kPa) and the inaccuracy of control is ±2 percent of set pressure.

When below-atmospheric pressures are to be maintained, connections are made both to the vacuum source (usually a vacuum pump) and to the controlled system. The spring tension is set for the desired set pressure, and expansion or contraction of the bellows moves the valve port to control the air flow through the manostat. The manostat is primarily designed for dead-ended service, but will handle small flows. For example, at a setting of 300 mmHg (2,000 kPa) without any restrictions in the vacuum source line, it will pass 6 l. of air each minute.

*See Appendix A.1 for SI Units.

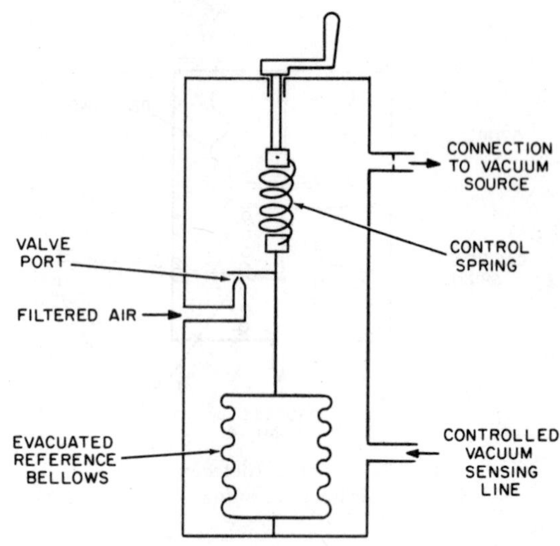

Fig. 5.3h Aneroid manostat

When above-atmospheric pressures are to be controlled, the filtered air supply is not atmospheric any longer, but connected to a pressurized supply, and the vacuum source connection is left open to vent the unit.

BIBLIOGRAPHY

Brombacher, W.G., "40 Years of Precise Pressure Measurement," *Instruments and Control Systems*, September, 1967.

Comber, J. and Hockman, P., "Pressure Monitoring: What's Happening?" *Instruments and Control Systems*, April, 1980.

Elliott, T.C., "Temperature, Pressure, Level, Flow—Key Measurements in Power and Process," *Power*, September, 1975.

Hall, J., "Monitoring Pressure with Newer Technologies," *Instruments and Control Systems*, April, 1979.

"Pressure Systems," *Chemical Engineering*, January 29, 1968.

Flow Sheet Symbol

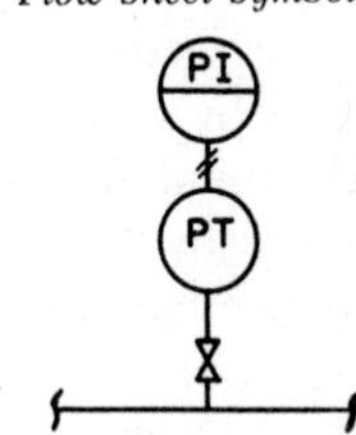

5.4 BOURDON-TYPE PRESSURE SENSORS

Design Pressure:	Up to 100,000 PSIG (690 MPa)
Design Temperature:	Up to 600°F (316°C) with stainless steel Bourdon tube; lower with others; ±1 percent zero shift per 50°F (28°C) ambient variation
Materials of Construction:	Phosphor bronze, alloy steel, 316 & 403 stainless steel, beryllium copper, Monel, Ni-Span C
Inaccuracy:	¼ to 5 percent of span
Range:	Minimum span 15 PSI (100 kPa); maximum span 100,000 PSI (690 MPa)
Cost:	$50 for local indicator; $700 for transmitter in conventional materials
Partial List of Suppliers:	Ametek Controls Div.; Automation Services; BIF; Bailey Controls Co.; Babcock & Wilcox, a McDermott Co.; Bristol-Babcock Div.; ACCO Industries, Inc.; C-E Invalco; Fischer & Porter Co.; Foxboro Co.; Hays-Republic Div., Milton Roy Co.; ITT Barton; Kent Process Control, Inc.; Mensor Corp.; Moore Products Co.; Taylor Instrument Co., Div. of Sybron Corp.

Bourdon tube pressure sensors are the elastic pressure elements most often used for industrial purposes. They have been in use for over a century. Their accuracy was substantially improved around 1930 when the friction effects, drift and elastic hysteresis were reduced, and ±0.1 percent inaccuracies became possible. These elements are not ideally suited for low-pressure, vacuum or compound measurements because the spring gradient in the Bourdon tube is too low for precision measurements at spans of 30 PSI (200 kPa) or below. Frequently used Bourdon tube materials include bronze, Monel, alloy and stainless steel. In this section some of the basic types of elastic Bourdon elements will be discussed, including the C–Bourdon tube, the spiral and helical elements and the fused quartz helix. The features and accessories of instruments using these elements will be covered in later sections (pressure gauges in Section 5.11 and switches in 5.13). (Chemical protectors were discussed in Section 5.2)

C-Bourdon Pressure Sensors

Figure 5.4a illustrates a C–Bourdon tube used in a direct indicating gauge, which usually has an arc of 250°. The process pressure is connected to the fixed socket end of the tube, while the tip end is sealed. Because of the difference between inside and outside radii, the Bourdon tube presents different areas to pressure, which causes the tube to tend to straighten when pressure is applied. The resulting tip motion is nonlinear because less motion results from each increment of additional

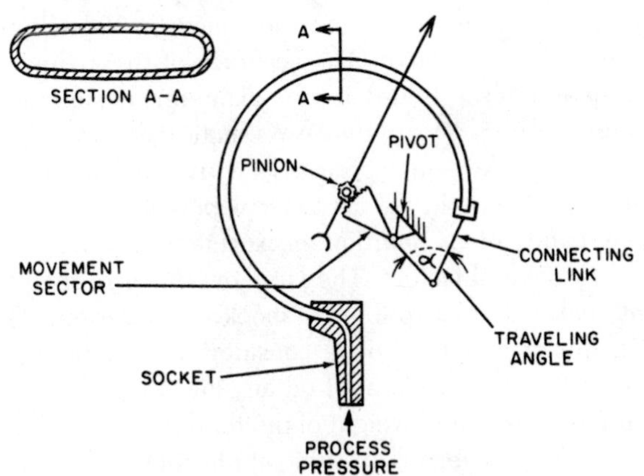

Fig. 5.4a C–Bourdon pressure element

pressure. This nonlinear motion has to be converted to linear rotational pointer response. This is done mechanically by means of a geared sector and pinion movement. The tip motion is transferred to the tail of the movement sector by the connector link. The angle between the connecting link and the sector tail is called the *traveling angle*. This angle changes with tip movement in a nonlinear fashion, compensating for the nonlinearity of the tip movement itself. It is designed to minimize backlash and provide smooth roll-on and roll-off characteristics in the geared sector and pinion movement by using fine pitch gears or by eliminating the gears altogether and using a cam sector that positions on a roller surface. This design eliminates the gears that eventually wear out and reduces the play that occurs when teeth are worn. On gear and pinion designs, the operation has been improved by the use of nylon and Teflon materials.

The following table lists Bourdon tube materials, noting some of the important characteristics by the letters P (poor), F (fair), G (good) and also indicating the maximum pressure range that the Bourdon tube can detect.

Tube Material	*Corrosion*	*Spring Rate*	*Temp. Coefficient*	*Hysteresis*	*Maximum Pressure*
Phosphorus Bronze	P	F	P	F	800 PSIG
Beryllium Copper	P	G	P	G	5,000 PSIG
316 Stainless Steel	G	P	P	P	10,000 PSIG
403 Stainless Steel	G	P	P	P	20,000 PSIG
Ni-Span C	G	G	G	G	12,000 PSIG
K-Monel	G	P	P	P	20,000 PSIG

This table shows that Ni-Span C seamless tubing, drawn to the profile in Figure 5.4a, is the best choice for Bourdon tube applications in all cases except those at very high pressure or those that are extremely corrosive. Direct indicators or motion balance transmitters using C–Bourdon elements are available with spans from 0–15 PSI to 0–20,000 PSI (0–100 kPa to 0–140 MPa) and can be used for positive, negative or compound pressure ranges, but the indication on the vacuum side will not be accurate or sensitive. The accuracy of these devices is a function of the Bourdon tube diameter, design quality, and calibration procedures. As such, it can vary from ±0.1 to ±5 percent inaccuracy, with the majority of these units falling in the area of ±1 percent.

The quality of C–Bourdon measurement can be influenced by several factors. The large overhang in the element makes it susceptible to shock or vibration. The filling media should also be considered. If a Bourdon element has been calibrated on air, and in the process it is filled with liquid, weight of the fluid in the overhang will introduce an error. In some applications, the tubes will need to be flushed, for which purpose tube tip bleed valves are available. On liquid service, air is likely to be trapped in the tube end, which acts as a cushion and results in sluggish performance. Bourdon designs are available with an internal capillary tube that allows entrapped air to escape from the tip.

Figure 5.4b illustrates the design that uses a C–Bourdon element in a force balance transmitter. (See Figure 5.6b for a detailed view of a force balance transmitter mechanism.) The pressure applied to the tube tends to straighten it. The flexure transmits the resulting force to the lower end of the force bar. Due to the force balance nature of the unit, the force bar is constantly balanced; therefore, the sensing Bourdon does not move as long as the pressure sensed is within the range of the instrument. If the range is exceeded, the lower end of the force bar moves to the left where it is supported by a limit stop. The Bourdon tube is made of Ni-Span C, and all other wetted components are stainless steel. Available spans vary from 1,000 to 12,000 PSI (6.9 to 83 MPa) with 150 percent overrange protection. The inaccuracy of this transmitter is ±0.5 percent in the lower and ±1 percent in the higher ranges. Maximum ambient temperature is limited to 250°F (120°C).

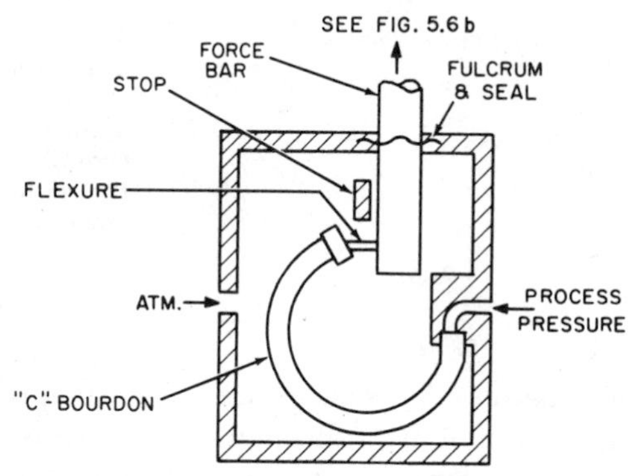

Fig. 5.4b Force balance C–Bourdon pressure sensor

Spiral Bourdon Pressure Sensors

The free end motion of the C–Bourdon tube is insufficient to operate some of the motion balance devices, such as the transmitters. The spiral element shown in Figure 5.4c is essentially a series of C–Bourdon tubes

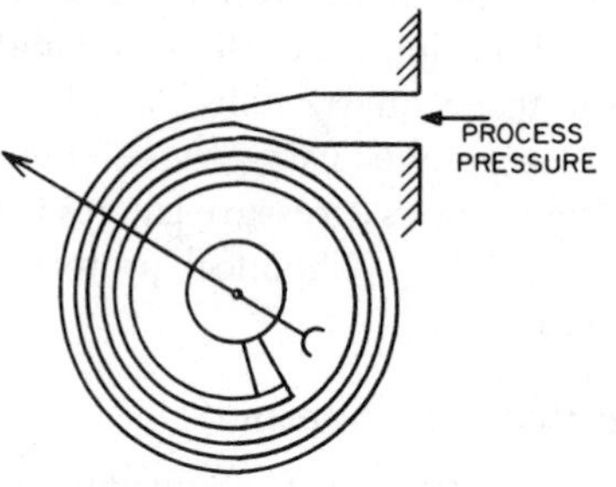

Fig. 5.4c Spiral Bourdon element

joined end to end. When pressure is applied, this flat spiral tends to uncoil and produces a greater movement of the free end requiring no mechanical amplification. This increases the sensitivity and accuracy of the instrument because no lost motion or friction is introduced through the links and levers.

Standard spiral materials include bronze, steel, stainless steel, beryllium copper, Monel and Ni-Span C. Spans as low as 10 PSI (69 kPa) are available for positive, negative or compound ranges. The various manufacturers use different break points between the spiral and the helical elements. Some start using helical elements at above 200 PSI (1.4 MPa) spans, others stay with the spiral element up to 4,000 PSI (27 MPa) and higher.

Unflattened, heavy-wall, spiral tube springs to detect pressures up to 100,000 PSIG (690 MPa) are also available in direct indicators.

Figure 5.4d is a simplified illustration of a motion balance transmitter operated by a spiral element. This sketch would also apply if the process pressure was detected by a C–Bourdon or a helical element. In this unit, the air supply passes through a restriction before being applied to the top of the relay diaphragm and the nozzle. An increase in process pressure tends to straighten out the spiral, which causes the flapper to move closer to the nozzle. This increases the nozzle back pressure sensed by the relay diaphragm, which will move down, opening up the air supply to the output. The increased output pressure is felt by the feedback bellows and restores the flapper to its throttling position. For each value of process pressure there is a corresponding definite flapper position and output pressure. Hence, the name *motion balance*. The inaccuracy of these motion balance pressure transmitters is generally placed at ±0.5 percent and, as shown in Figure 5.4d, they will read the process pressure even if the air supply is lost.

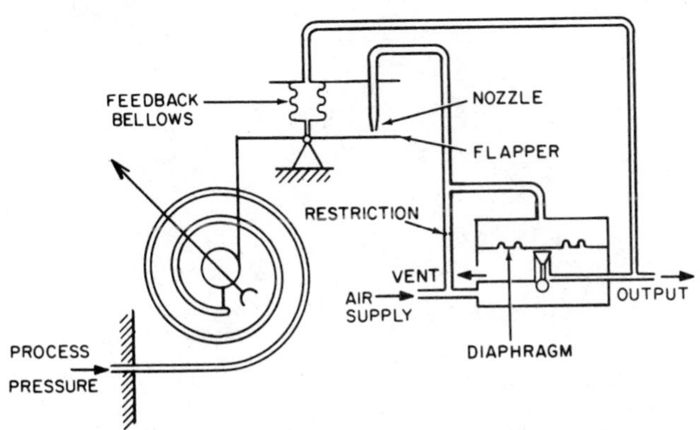

Fig. 5.4d Motion balance pressure transmitter with spiral element

Helical Bourdon Pressure Sensors

Figure 5.4e shows the construction of a helical Bourdon element. This sensor produces an even greater motion of the free end than the spiral element, eliminating

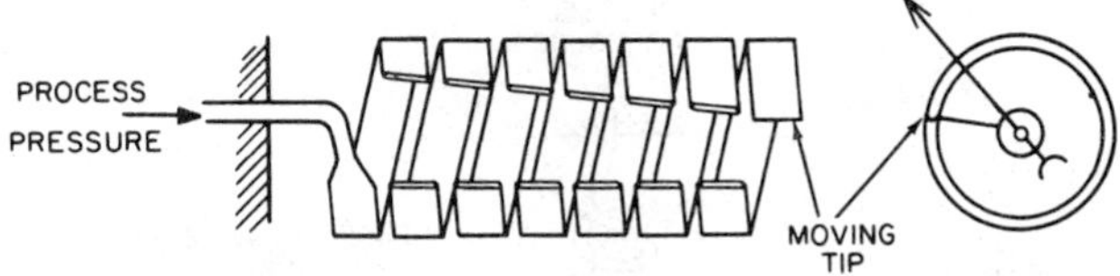

Fig. 5.4e Helical Bourdon pressure sensor

the need for mechanical amplification. Other advantages of this design include the high overrange protection available; for example, a 0–1,000 PSIG (0–6.9 MPa) element may be safely exposed to 10,000 PSIG (69 MPa) pressure, and it is suitable for pressure measurement on continuously fluctuating services. The range of the helical coil is affected by the diameter, wall thickness, number of coils used and the construction materials. High-pressure elements might have as many as twenty coils while low-span sensors can have only two or three coils. Available materials include bronze, beryllium copper, Ni-Span C and stainless steel. The various suppliers have different practices concerning the pressure levels at which they change from spiral to helical elements. For this reason, helical elements are available with spans from 0–30 PSI, to 0–80,000 PSIG (0–200 kPa up to 0–550 MPa), depending on supplier. For pressures exceeding 5,000 PSIG (35 MPa) heavy duty stainless steel is the standard element material.

Helical coils can also be used as the element in differential pressure sensors if one of the pressures is acting on the outside surface and the other on the inside of the coil.

Motion balance instruments with helical elements include direct indicators, recorders, controllers, switches and transmitters with either pneumatic or electronic outputs. Their inaccuracy is normally in the area of ±1 percent of span.

Fused Quartz Helix Pressure Sensors

As shown in Figure 5.4f, the process pressure causes the free end of the helix to rotate together with the mirror attached to it. The quartz helix is mounted above a large precision gear that can rotate concentrically around the pressure sensor. A lamp and a pair of photo cells are attached to the gear. The light from the lamp is reflected from the mirror and distributed equally on the two photo cells. When the process pressure changes, the free end of the helix rotates the mirror, unbalancing the photo cells. This signal causes the balancing motor to turn the gear until the light falling on the photo cells is equally distributed once more. Consequently, the gear follows the movement of the free end of the helix without mechanical contact. The rotation of the gear is registered by a digital counter that is calibrated in pressure units.

This instrument detect vacuum, absolute, gauge and differential pressures between 1 micron (10^{-3} mmHg)

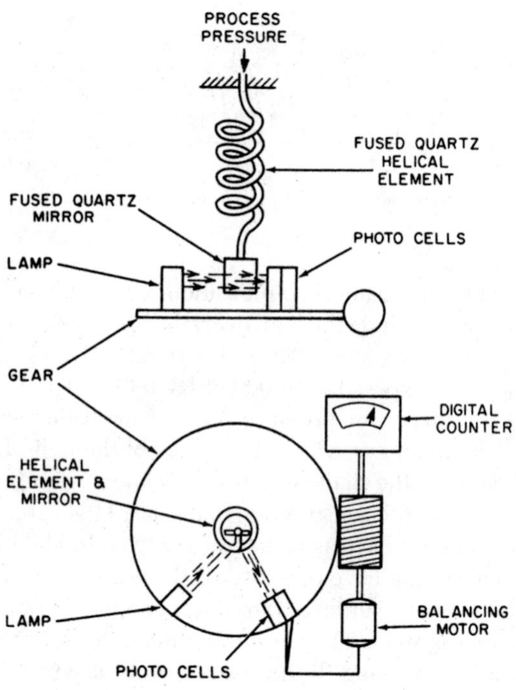

Fig. 5.4f Fused quartz helical pressure sensor

and 500 PSIG (3.5 MPa) in several ranges with a minimum span of 5 PSI (35 kPa). The device is repeatable to 2 parts in 100,000 or inaccurate to ±0.01 percent.

Its resolution and sensitivity are high while its hysteresis is negligible. It is presently used in laboratories, but the principle should make it suitable for industrial installations in which precision pressure detection is desired. The cost is high, approximately $4,500 for a loop that includes the digital counter readout. The unit is sensitive to, and thus should be protected from, vibration. One of its drawbacks is the slow response speed of 2 min. for full-scale travel.

BIBLIOGRAPHY

Brombacher, W.G., "40 Years of Precise Pressure Measurement," *Instruments and Control Systems,* September, 1967.

Buffenmyer, W.L., "Selecting Bourdon–Tube Gauges," *Instruments and Control Systems,* February, 1961.

Elliott, T.C., "Temperature, Pressure, Level, Flow—Key Measurements in Power and Process," *Power,* September, 1975.

Hall, J., "Monitoring Pressure with Newer Technologies," *Instruments and Control Systems,* April, 1979.

"Pressure Systems," *Chemical Engineering,* January 29, 1968.

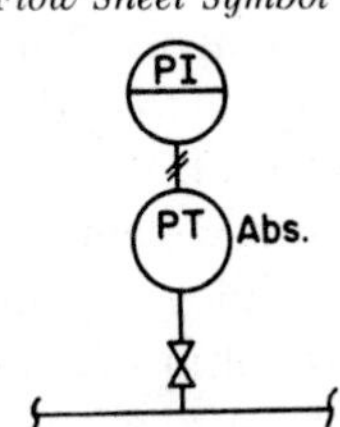

5.5 DIAPHRAGM-TYPE PRESSURE SENSORS

Design Pressure:	Up to atmospheric with evacuated motion balance capsule; up to 50 PSIG (344 kPa) or more with evacuated force balance capsule; up to 200 PSIG (1.4 MPa) with atmospheric reference motion balance capsule; up to 1,500 PSIG (10 MPa) with atmospheric reference force balance capsule
Design Temperature:	Phosphor bronze (−50 to +250°F or −46 to 120°C), Ni-Span C (−50 to +300°F or −46 to 149°C), 316 stainless steel (−400 to 600°F or −240 to 316°C), Inconel (−300 to 1,000°F or −184 to 538°C). The other components besides the diaphragm element can limit the operating temperature to 250°F (120°C) or less
Materials of Construction:	Buna-N, nylon, Inconel, Ni-Span C, phosphor bronze, 316 stainless steel, beryllium-copper, Monel, tantalum, Hastelloy, nickel, duranickel, Teflon, Kel-F
Inaccuracy:	±1% of span; some force balance designs with ±½%
Range:	Absolute pressure ranges from 0-5 mmHg to 0-50 PSIA (0–0.7 kPa to 0–350 kPa); gauge pressure ranges from 0-0.5 in. H_2O to 0-200 PSIG (0–0.12 kPa to 0–1.4 mPa)
Cost:	$600 for direct recorder; $900 for pressure transmitter, both in standard materials of construction
Partial List of Suppliers:	Ametek Controls Div.; Automation Services; BIF; Belfab, Bailey Controls Co.; Babcock & Wilcox, a McDermott Co.; Brandt Industries Inc.; Bristol-Babcock Div.; ACCO Industries, Inc.; Fischer & Porter Co.; Foxboro Co.; Hays-Republic Div., Milton Roy Co.; Honeywell Process Control Div.; ITT Barton; Kent Process Control, Inc.; Moore Products Co.; Taylor Instrument Co., Div. of Sybron Corp.

Pressure sensors that depend on the deflection of a diaphragm have been in use for over a century. In the last few decades, the elastic hysteresis, friction and drift effects have been reduced to approximately ±0.1 percent of span in the high-quality designs. New materials have been introduced with improved elastic qualities, such as beryllium copper, and with very low temperature coefficients of elasticity, such as Ni-Span C. Inconel and stainless steel have been used when extreme operating temperatures or the corrosive nature of the process demanded them. Even quartz diaphragms were introduced when minimum hysteresis and drift were desired.

The diaphragm is a flexible disc, either flat or with concentric corrugations, which is made of sheet metal with precise dimensions. The pressure deflection characteristics of both flat and corrugated diaphragms have been well investigated. Some instruments use the diaphragm as the pressure sensor, others use it as a basic component for a capsular element. Figure 5.5a shows a single corrugated diaphragm and also some capsule designs. The capsules consist of two diaphragms welded together at their peripheries. The sketch illustrates the two basic types of capsules: the convex and the nested. Evacuated capsules are used for absolute pressure detection and single diaphragms for highly sensitive measurements. The sensitivity of a capsule increases in proportion to its diameter, which in the conventional designs varies from 1 to 6 inches (25.4 to 152.4 mm). Multiple capsule elements can be built from either convex or nested capsules as shown on Figure 5.5b. These elements are useful in increasing the output motion resulting from a pressure change.

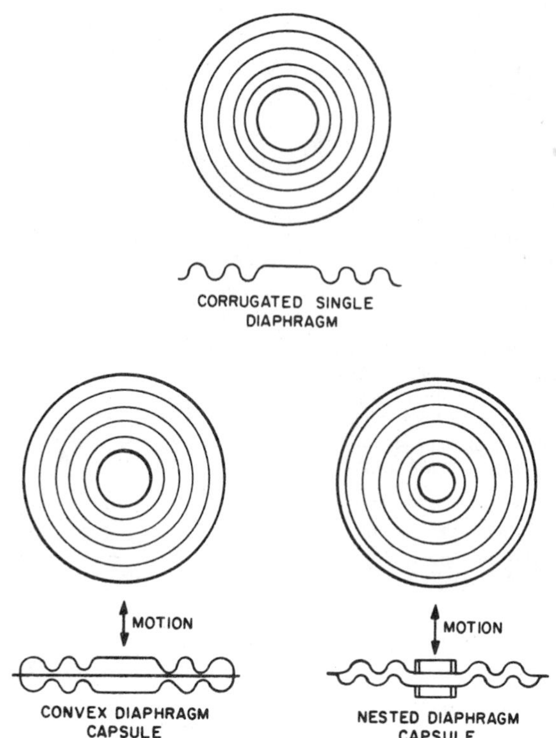

Fig. 5.5a Standard diaphragm elements

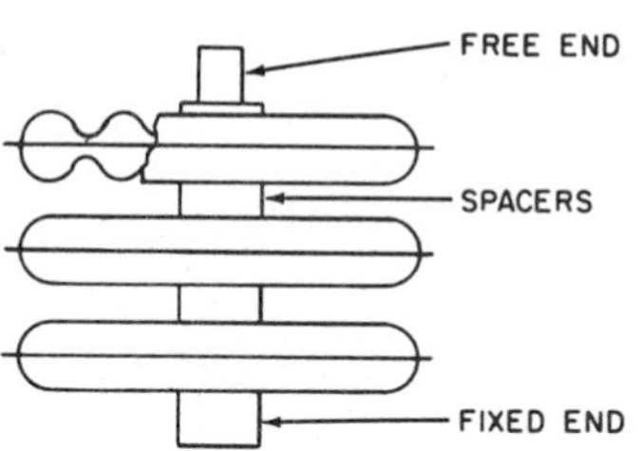

Fig. 5.5b Multiple capsule element

In this section, four basic designs of diaphragm pressure sensors will be discussed. They can be distinguished by the pressure reference used (full-vacuum or atmospheric) and by whether the design is force or motion balance. The output of sensors that work with a full-vacuum reference is related to absolute pressure, while the atmospheric reference units generate gauge or differential-pressure-related outputs. Motion balance units are capable of driving local, direct reading indicators, but are subject to hysteresis, friction and dead-band errors. Force balance designs are transmitting devices with high accuracy, but without local indication capability. In addition, a brief description of pressure repeaters will be given.

Motion Balance Absolute Pressure Sensors

These sensors are illustrated in Figure 5.5c. The capsular element is fully evacuated and changes its length as a function of the process pressure in its housing. Because its length depends on the difference between internal and external pressures and because the internals have been fully evacuated, the element length is a measure of the absolute pressure acting on the outside of the capsule. The capsular element is sealed inside the pressure-tight housing. The meter shaft transmits the capsule motion to the readout device or controller through a bellows seal. This seal protects against air leakage under vacuum and is fairly frictionless.

The capsule and the bellows seal are normally available in bronze, stainless steel, nickel and silver or gold plate. Because the evacuated diaphragm element would collapse as the pressure on its outside increases, the capsules have been designed to "bottom" on each other when they are exposed to pressures above the range of the instrument. For this reason, the unit can be exposed to pressures up to atmospheric without damage to the capsule element.

The ranges available with this instrument can be as low as 0 to 5 mmHg (0 to 0.7 kPa) or as wide as 0 to 760 mmHg (0 to 100 kPa). Inaccuracy can be better than ±1 percent full-scale for the wide range elements, and for the narrow range capsules, a fraction of a millimeter of mercury.

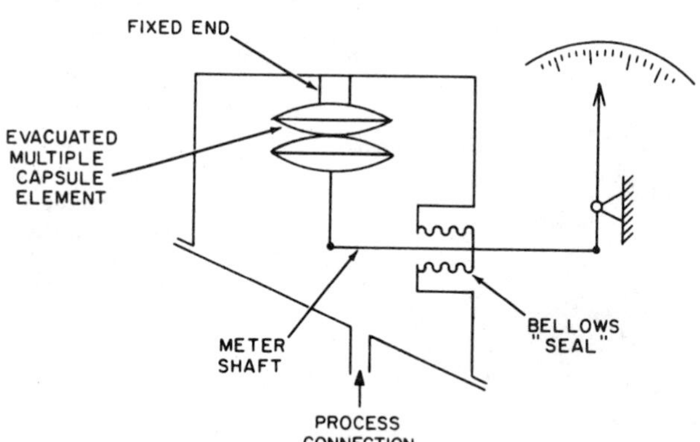

Fig. 5.5c Motion balance absolute pressure sensor

Force Balance Absolute Pressure Sensors

The force balance detector illustrated on Figure 5.5d is available only as a pneumatic or electronic transmitter, not as a direct indicator because there is no motion available to drive a pointer. If local indication is desired, an output gauge can be installed on the transmitter to provide it.

For a detailed sketch of the force balance transmitter mechanism see Figure 5.6b. The main sensing element

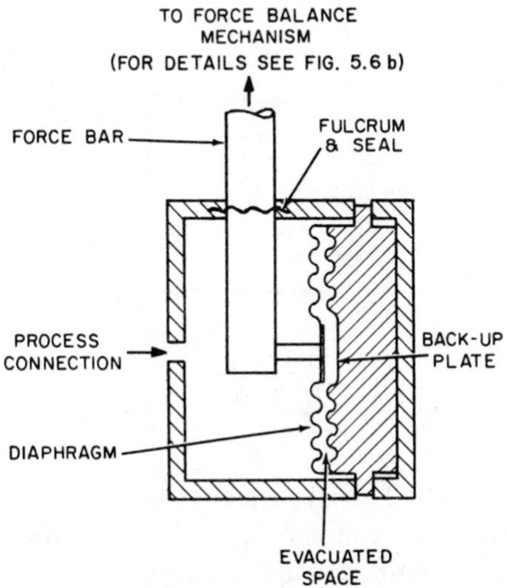

Fig. 5.5d Force balance absolute pressure detector

in this unit is the capsule. The pressure being detected is applied to the left side of the diaphragm in the capsule, while the space on the other side of the diaphragm is fully evacuated, thus providing a zero absolute pressure reference. The force felt by the force bar is related to the difference between full vacuum on the right side of the diaphragm and the process pressure on the other side or to the absolute pressure of the system measured. Due to the force balance nature of the unit, the force bar is constantly balanced; therefore, the sensing diaphragm does not move as long as the pressure detected is within the range of the instrument. If the range of the capsule is exceeded, the diaphragm moves to the right where it is supported by the backup plate with matching convolutions. For this reason, high overrange protection can be obtained with these transmitters.

The standard material of construction for these units is stainless steel. The following table lists some of the important features of this transmitter family.* The units are distinguished from one another by the diameter of the capsule. It should be noted that between the minimum and maximum ranges they are adjustable in the field.

Capsule Diameter:	5 in.	3½ in.	2 in.
Minimum Range:	0–10 mmHg	0–40 mmHg	0–375 mmHg
Maximum Range:	0–40 mmHg	0–400 mmHg	0–1,520 mmHg
Overpressure Protection up to:	50 PSIG	100 PSIG	150 PSIG
Inaccuracy (% Span):	1%	½–1%	½–1%
Operating Temperature:	−40 to 250°F	−40 to 250°F	−40 to 250°F

*See Appendix A-1 for SI Units.

Special transmitters are available in all-Monel construction or with Hastelloy body and tantalum capsule.

Motion Balance Pressure Sensors with Atmospheric Reference

In this pressure detector, shown in Figure 5.5e, the process pressure inside the capsule is balanced against the spring action of the element. The outside surface of the diaphragm assembly is exposed to atmospheric pressure, which is then the pressure reference for this instrument. This unit is available either as a direct readout recorder or as a transmitter for remote readout. Due to the straight-line expansion of the capsule, the force-to-friction ratio is reasonably high. The disphragm area is selected to produce a substantially greater force than that required to operate the movement.

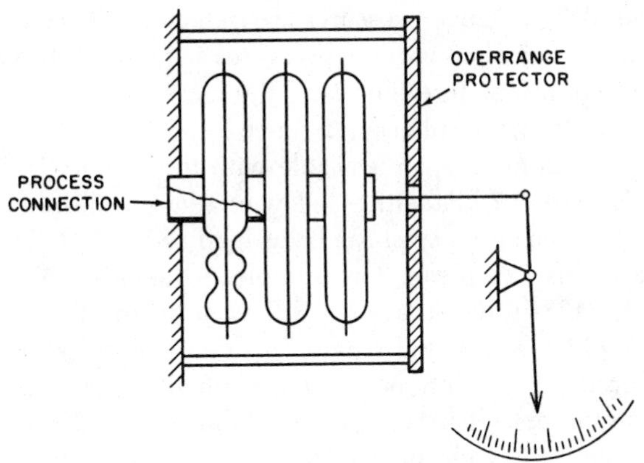

Fig. 5.5e Motion balance pressure sensor with atmospheric reference

The diaphragm capsules can be either convex or nested (see Figure 5.5a). The nested capsule can be exposed to higher overpressures than the convex ones, but all units are designed to withstand at least 100 percent overpressure over their full range. Figure 5.5f shows the nested capsule made of a convex and a concave diaphragm with the process pressure acting on the outside of the capsule rather than on the inside as illustrated in Figure 5.5c. This is the reason why high overpressures can be tolerated with the nested capsule. The range of this element is a function of the materials of construction, capsule diameter and of the particular design. Some of the frequently used diaphragm materials include: Cu-Ni-Mn (60 percent copper, 20 percent nickel, 20 percent manganese), phosphor bronze, Ni-Span C (constant modulus nickel alloy, 42 percent nickel, 2.4 percent titanium, 5.4 percent chrome, 50 percent iron), 316 stainless steel and Inconel. The available capsule diameters vary from 2 to 5 in. (50.8 to 127.0 mm), with the larger diameters being sensitive enough to detect the lower pressures. These elements can detect both gauge and vacuum pressures. It is important to distinguish between absolute and vacuum pressures. To make an absolute

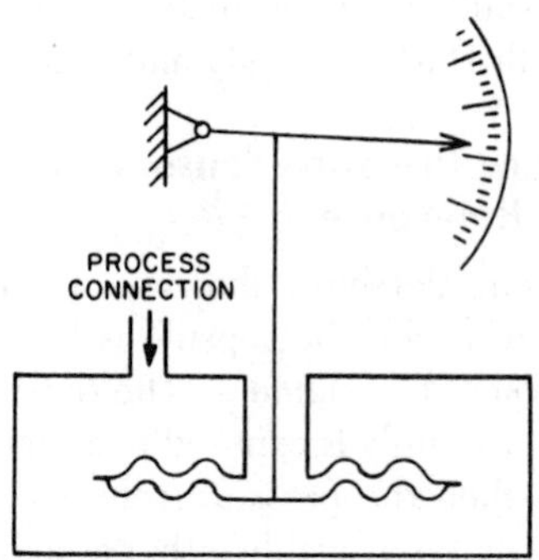

Fig. 5.5f Nested diaphragm sensor

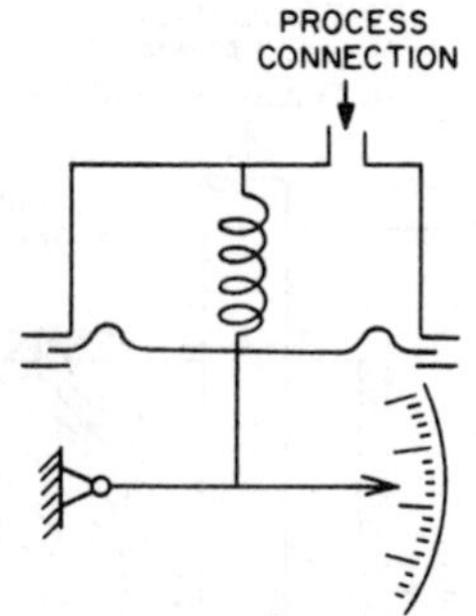

Fig. 5.5g Slack-diaphragm detector

pressure measurement with an elastic element, it is essential to have a full-vacuum reference point. The instruments discussed here can only evaluate the pressure differential between atmospheric and operating pressures. When gauge pressures are detected, the capsule expands and when vacuum pressures are sensed the element contracts. In either case, the measurement is relative to the atmospheric reference.

The minimum span available with these units is 0–3 in. H_2O (0–0.7 kPa) either below or above atmospheric, and the maximum span can be as high as 180 PSI (1,200 kPa). This can mean, for example, a range of 20–200 PSIG (138–1,400 kPa) or a full vacuum to 165 PSIG (1,100 kPa) compound range. This range of pressures is covered by a variety of capsular elements, each having a minimum and maximum span. For the reader's orientation, a few of the standard capsule spans are noted below. The zero adjustment on the instrument allows the range to be shifted to above or below atmospheric, and the span adjustment on the unit allows for narrowing or widening the range within the minimum and maximum span limits noted.*

Capsule Material	*Minimum Span*	*Maximum Span*
2 in. Diameter Cu-Ni-Mn	20 in. H_2O	5 PSI
3 in. Diameter Cu-Ni-Mn	8 in. H_2O	40 in. H_2O
Ni-Span C—large diameter	3 in. H_2O	5 PSI
Ni-Span C—small diameter	5 PSI	15 PSI
Ni-Span C—nested capsule	12 PSI	180 PSI

The inaccuracy of these units is normally ±1 percent of span or better.

Another design of motion balance pressure sensors with atmospheric pressure reference is illustrated in Figure 5.5g. This unit is also called the *slack-diaphragm* element. The process pressure detected by this unit is balanced against either the spring action of the diaphragm itself or against a calibrated spring as shown in Figure 5.5g. This device is the most sensitive pressure detector in the family of elastic element sensors, and as such, it is capable of measuring near-atmospheric draft pressures. If the housing encloses both sides of the diaphragm, the unit can act as a differential pressure detector. The diaphragms are available in metal, such as stainless steel, or can be made of elastomers, such as Buna-N or nylon. The slack-diaphragm elements can actuate local, direct readout devices or transmitters for remote readout. Although the pressures measured are small, the force-to-friction ratio is still satisfactory due to the large diaphragm areas in these designs. The available spans range from 0–0.5 in. H_2O to 0–120 in. H_2O (0–0.12 kPa to 0–30 kPa). These spans can be either on the positive or vacuum pressure side or can cover compound ranges. Inaccuracy is in the range of ±1 to ±2 percent of span, and the slack-diaphragm elements can withstand overpressures between 5 and 25 PSIG (35 and 173 kPa), depending on the particular design.

*See Appendix A-1 for SI Units.

Force Balance Pressure Sensors with Atmospheric Reference

The force balance differential pressure detectors that will be discussed in Section 5.6 can also be used as electronic or pneumatic pressure detectors by leaving the low-pressure side of the d/p cell open to atmosphere. In such installation, the output signal of the transmitter will be related to the process gauge pressure sensed on the high-pressure side of the d/p diaphragm capsule.

Figures 5.6a, b and c illustrate the design and operating components of these instruments. The miminum span available with these units is 0–2 in. H_2O (0–0.5 kPa) and the maximum span is 0–30 PSIG (0–200 kPa). This range is covered by three capsules for which the minimum-maximum spans and overpressure limits are noted in the following table.*

Capsule	*Minimum Span*	*Maximum Span*	*Overpressure Limit*
Low Range	0–2 in. H_2O	0–20 in. H_2O	500 PSIG
Medium Range	0–20 in. H_2O	0–200 in. H_2O	1,500 PSIG
High Range	0–200 in. H_2O	0–30 PSIG	1,500 PSIG

*See Appendix A-1 for SI Units.

These capsules can be built into the conventional d/p cell body shown in Figure 5.6a or into the flat and extended diaphragm arrangements illustrated in Figure 5.6c. In the flat diaphragm design, the low-range capsule requires a 6-in. (150 mm) flange, while the medium- and high-range capsules can be accommodated by a 3-in. (75 mm) flange. The extended diaphragm version is available with medium- or high-range capsules only and requires a 4-in. (100 mm) mounting flange.

These transmitters have inaccuracies between ±½ and ±1 percent of span and can withstand operating temperatures of 250°F (120°C) in the standard and 350°F (177°C) in the flanged construction. The maximum static pressure that they can be exposed to varies from 225 PSIG (1.6 MPa) for the 6-in. diameter flanged unit to 6,000 PSIG (41 MPa) for the high-pressure version of the standard unit shown in Figure 5.6a.

The parts exposed to the process fluid can be made of a large variety of materials, including carbon and stainless steel, Monel, duranickel, Hastelloy, tantalum and nickel, or provided with coatings such as Kel-F or Teflon.

Pressure Repeaters

Pressure repeaters are force balance devices capable of generating an air output signal having the same pressure as the process to which the repeater is attached. These units are useful for isolating the process media from the measuring element to prevent corrosion, hazards, the possibility of plugging in the sensing line and so on. Because these units will be discussed in Section 5.12, only a summary of their features is given here.

The flanged repeaters shown in Figures 5.12a, b and c are available in 2, 3 and 4-in. sizes (50, 75 and 100mm), with either flat or extended diaphragm construction. The extended design is used when the main consideration for installing a repeater is protection against plugging. The available materials for wetted parts include stainless steel, Hastelloy, nickel, Teflon, Kel-F, PVC, penton and others.

Available spans vary from 0–1 PSIG to 0–300 PSIG (0–6.9 kPa to 0–2.1 MPa) and the repeater inaccuracy is ±¼ percent or better. Repeater inaccuracy at 20 PSIG (138 kPa) is ±1 in. H_2O (±0.25 kPa) and at 200 PSIG (1,400 kPa) it is about ±10 in. H_2O (±2.5 kPa).

When positive pressures are to be repeated, the reference is the atmosphere and the air (or other gas) supply pressure has to be 125 percent of the highest pressure to be repeated. On vacuum service, the repeater inaccuracy of ±2 mmHg (±0.3 kPa) normally limits use to absolute pressures above 40 mmHg (5.2 kPa). The repeaters can work either on a vacuum reference (see Figure 5.12b), in which case the reference vacuum always has to be lower than the lowest absolute pressure to be repeated, or on atmospheric pressure reference with a suppression spring. If the spring is set at 16.5 PSI (114 kPa), the repeater output will be zero-shifted by this amount, representing the sum of the process pressure and the spring setting. When the process is at atmospheric pressure the output is 16.5 PSIG (114 kPa), and when the process is at full vacuum, the output is 1.5 PSIG (10 kPa).

The operating temperature and static pressure limitations of these units are similar to those of the force balance d/p cells.

BIBLIOGRAPHY

Brombacher, W.G., "40 Years of Precise Pressure Measurement," *Instruments and Control Systems*, September, 1967.

Comber, J. and Hockman, P. "Pressure Monitoring; What's Happening?" *Instruments and Control Systems*, April, 1980.

Elliott, T.C., "Temperature, Pressure, Level, Flow—Key Measurements in Power and Process," *Power*, September, 1975.

Hall, J., "Monitoring Pressure with Newer Technologies," *Instruments and Control Systems*, April, 1979.

"Pressure Systems," *Chemical Engineering*, January 29, 1968.

Flow Sheet Symbol

RECEIVER

PDT

5.6 DIFFERENTIAL PRESSURE INSTRUMENTS

Design Pressure:	Up to 5,000 PSIG, some as high as 10,000 PSIG (up to 35 MPa, some as high as 69 MPa)
Design Temperature:	Up to 350°F (177°C) for dry, force balance; up to 1,200°F (650°C) on liquid-filled force balance; others up to 200°F (93°C)
Materials of Construction:	Stainless steel, Kel-F, Teflon, Hastelloy, Monel, tantalum, titanium, duranickel for force balance types
Inaccuracy:	± ½ to 1% full range
Cost:	$600 to $1,200 for transmitters in standard materials of construction; $100 to $500 for local indicators in standard materials; $150 for switches only
Partial List of Suppliers:	Ametek Controls Div.; Automation Services BIF; Belfab, Bailey Controls Co.; Babcock & Wilcox, a McDermott Co.; Brandt Industries Inc.; Bristol-Babcock Div.; ACCO Industries, Inc.; Fairchild Industrial Products; Fischer & Porter Co.; Foxboro Co.; Hays-Republic Div., Milton Roy Co.; Honeywell, Process Control Div.; ITT Barton; Kent Process Control, Inc.; King Engineering Corporation; Moore Products Co.; Taylor Instrument Co., Div. of Sybron Corp.

The devices developed for the measurement and indication or transmission of pressure differentials can be divided into *dry* and *wet*. Dry units can be further subdivided into *force balance* and *motion detection* types. Similarly, wet units have two major categories: *liquid seal* (inverted bell) and *liquid manometer*. Liquid manometers are divided into the *visual* and the *float* designs. These designs will be discussed in the same order as listed, starting with the dry, force balance units.

Dry, Force Balance, Pressure Differential Detectors

In this section, the basic operating principles of these units will be discussed, together with a brief description of some design variations, such as flat and extended diaphragm versions, chemically sealed and wafer units and pressure repeaters. The latter are not d/p devices by design, but are used as elements in such applications and are obtained by a slight redesign of the force balance d/p instruments.

Force balance level detector d/p cells are widely used in applications where transmitters are required for remote readout. Figure 5.6a is a schematic of this unit. A pair of diaphragms is welded to opposite sides of the d/p capsule and the space between them is filled with liquid. The differential pressure to be detected is applied to the two sides of this diaphragm capsule. The resulting force is then brought out of the d/p cell via the force bar, which at its fulcrum point is sealed by a metallic diaphragm. When a change occurs in the steady-state conditions, the force bar changes its position relative to the nozzle, causing a change in the pneumatic output signal. The new air pressure is sensed by the feedback bellows, and a new state of equilibrium (force balance) is established by the bar force being opposed by an equal force developed in the feedback bellows. As a result, the output signal is maintained in proportion to the differential pressure sensed by the cell.

When electronic d/p transmitters are used, the bar force is balanced against the magnetic force developed in a coil. Here, the coil replaces the feedback bellows, and the current flow replaces the output air signal. As the flowing current is changed, the force balance is reestablished by matching each bar force with an equal

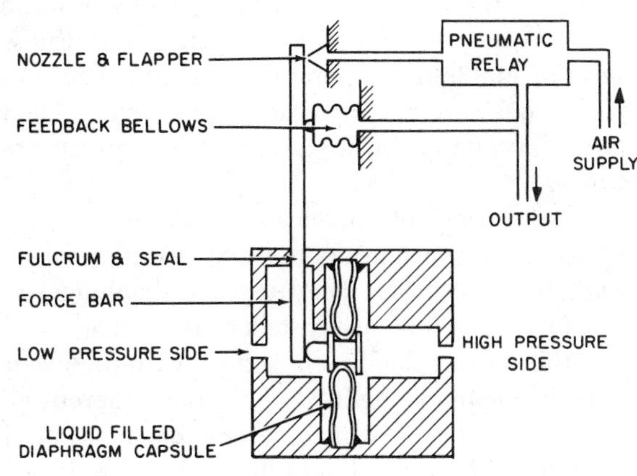

Fig. 5.6a Force balance d/p cell

magnetic force; thus, the current output signal is proportional to the pressure differential in the cell.

Most advantages of the force balance design are self-evident; for example, wear and friction are negligible in this unit when the motion is limited to a few thousands of an inch. The displacement volume of the capsule diaphragm is very small, and if exposed to higher differentials than designed for, the diaphragm is simply pressed against and rests on the capsule body without damage to the unit.

The adjustment capabilities of this instrument are noted in Figure 5.6b. By moving the span range wheel, the range of most of these units can be adjusted over a ten-to-one range. For example, a d/p cell having the full water column range of 200 in. (50 kPa) can be set for its full range to operate anywhere between 20 and 200 in. H_2O (5 and 50 kPa). Cells with full-scale ranges between 1 and 1,000 in. H_2O (0.25 and 250 kPa) or more are available and can withstand static pressures up to 5,000 or 6,000 PSI (35 or 41 MPa). As shown in Figure 5.6b, zero adjustments or range elevation and range depression adjustments are also provided.

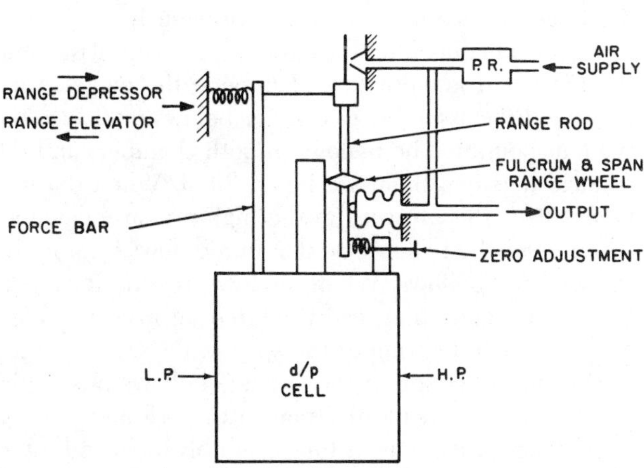

Fig. 5.6b Force balance d/p cell with adjustments shown

Design Variations on Force Balance d/p Cells

Force balance d/p cell designs were developed for use on viscous, corrosive or slurry applications. One of the major considerations was the elimination of dead-ended cavities, in which solids could accumulate and eventually plug the instrument.

In Figure 5.6c, two of these designs are shown: the extended and the flat diaphragm d/p cell transmitters. Their principle of operation is the same as for the conventional d/p cells illustrated in Figures 5.6a and b, except that the high-pressure side of the diaphragm capsule is in direct contact with the process, and the force detected by it is transmitted through a flexure to the lower end of the force bar.

When the major need is corrosion protection, the flat diaphragm design can be used with corrosion resistant lining (Kel-F, Teflon, etc.) or with the high-pressure side being made of corrosion resistant materials, such as Hastelloy or titanium. In such installations, an isolating valve can be installed between the d/p cell and the tank nozzle, which allows for maintenance on the instrument without draining the tank.

This unit is also widely used on installations in which corrosion is not a particular problem, but simplicity of installation is desired. The flat diaphragm transmitter, being mounted directly on the tank nozzle, provides a self-supporting installation without need for additional piping or support materials that tend to increase the installation cost of the conventional d/p cells.

When the main consideration is protection against plugging, the extended diaphragm version is considered. These units are also available in corrosion resistant materials, but their main design feature is elimination of the dead-ended cavity in which process material can collect and gel, polymerize or solidify. The unit is installed in the tank nozzle so that the diaphragm is flush with the inside of the tank or extends slightly into the tank.

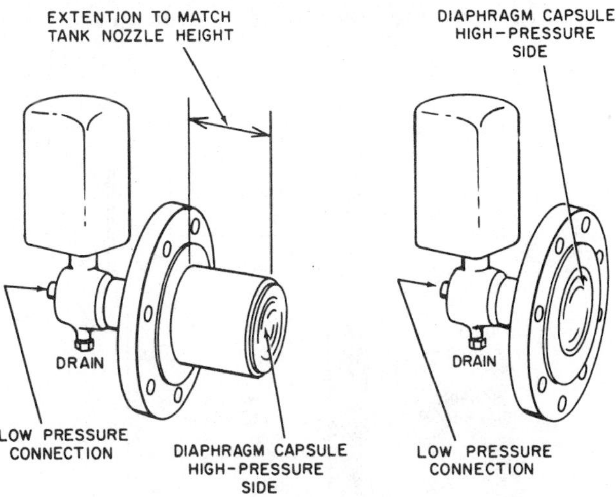

Fig. 5.6c Extended diaphragm d/p cell (left); flat diaphragm d/p cell (right)

Both the flat and the extended diaphragm designs are available with ranges up to 850 in. H_2O (213 kPa) with their inaccuracy being ± ½ percent up to a range of 450 in. H_2O (113 kPa) and less on higher spans. The units can withstand 550 PSIG (3.8 MPa) operating pressures when the process is at 350°F (177°C) temperature. Changes in process or ambient temperatures can cause zero shifts, as is the case with all similarly designed instruments. To minimize this effect, the d/p cell should be zeroed at the normal operating temperature, and the exposed body of the transmitter should be insulated. In more critical installations, the low-pressure side of the transmitter can be filled with a clean heat-transfer fluid that, in most cases, will eliminate the temperature gradient across the device. (Maximum process temperature for the extended design is 750°F or 400°C.)

Another family of force balance d/p cell transmitters has liquid-filled elements. Two basic versions of these designs will be discussed here: the wafer and the chemical seal element.

Wafer elements are illustrated in Figure 5.6d. These elements mount on pad connections, thereby reducing, but not eliminating, the dead-ended cavity on the process side. These units are available with ranges up to 800 in. H_2O (200 kPa) and with inaccuracies of ±1 percent. The filling liquid in the elements has a low thermal expansion coefficient, but temperature effects still cannot be disregarded. Actually, the pressure differential detected by the transmitter is the combined result of the hydraulic head in the tank and the differential thermal expansion in the two elements. For example, a 50°F (28°C) process temperature difference at the two elements will cause a zero shift of 4 in. H_2O (1 kPa). Even if the two wafers are at the same temperature, zero and span shifts will occur, as either the process or the ambient temperature changes. Wafer elements can tolerate up to 1,500 PSIG (10 MPa) in static pressure and can be exposed to 1,200°F (650°C) process temperatures. When used on vacuum service, the transmitter body should be mounted below the lower wafer element. In short, this design does not fully eliminate the possibility of plugging, and due to its liquid-filled design, it will drift with process or ambient temperature changes. The unit is normally considered for installations in which economical means of level detection are desired at high-process temperatures.

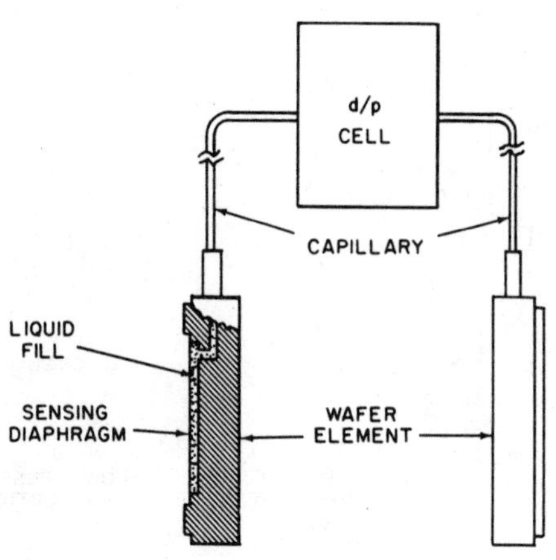

Fig. 5.6d d/p cell with wafer elements

A broad variety of chemical seal protectors can be attached to d/p transmitters to serve as either corrosion or plugging protectors. The extended-diaphragm seals shown in Figure 5.6e are installed in the tank nozzles so that the sensing diaphragm is flush with the inside of the vessel or extends slightly into the tank, thereby eliminating any cavities in which material could accumulate and solidify. These are liquid-filled elements in which the pressure sensed by the diaphragm is transmitted through the filling fluid via the capillary tubing into the d/p cell, in much the same way as with the wafer elements. Their features and limitations are also similar to those of the wafer design.

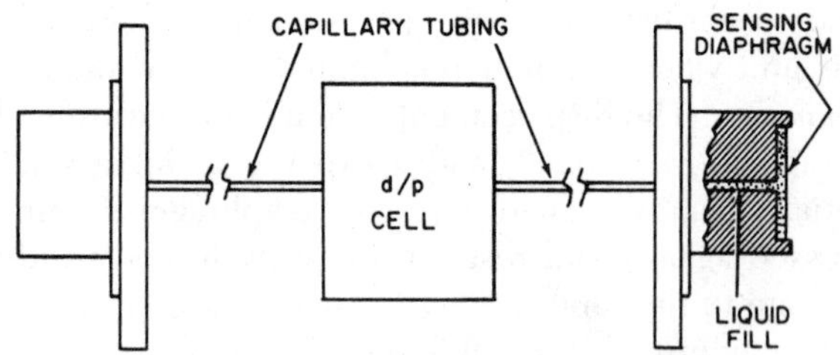

Fig. 5.6e d/p cell with extended chemical seal elements

Dry, Motion Balance, Differential Pressure Detectors

This family of differential pressure detectors is also called bellows meters because for their operation they depend on liquid-filled, double-opposed bellows. They are most useful when local indication or record is required and where compressed air or electric power is not available as the energy source. Their use as transmitters is limited because of the distinct advantages of the force balance units discussed previously.

Figure 5.6f illustrates the basic components of the unit: the high- and low-pressure chambers, the range spring and the drive assembly to transfer belows motion to the readout pointer. The bellows in both chambers and the passage between them are liquid-filled. When the unit is installed, the pressure in the high-pressure chamber compresses the bellows, so that liquid flows from it into the low-side bellows. When the low-pressure (or range) bellows expand, they exert a force against the range spring, which determines the span of the instrument.

The linear motion of the "range" bellows moves the drive lever, mechanically transmitting a rotary motion through the sealed torque tube assembly to the indicator. The output motion from the torque tube assembly is

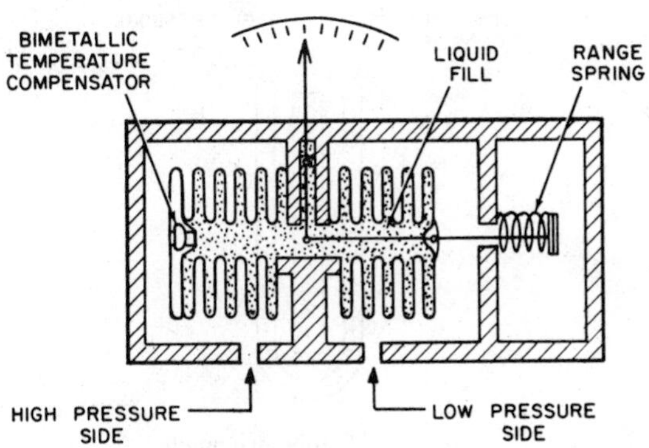

Fig. 5.6f Motion detector d/p indicator

limited to a few degrees of angular rotation. This is sufficient for most local indicators or recorders, but when the secondary responsive device imposes a considerable load on the torque tube assembly, the accuracy and sensitivity of the unit can be destroyed. For sustained accuracy, the bellows meter depends on the repeatability of its mechanical system, which has proven to be linear within ½ to 1 percent over its full range.

A bimetallic temperature compensator inside the high-pressure bellows automatically adjusts the capacity of the bellows assembly to the changing volume of the fill liquid due to ambient temperature variations. The device can be damaged by adjusting the restriction in the passage between the two bellows.

Each of the bellows meters is provided with overrange protectors. The operation of one of the protection mechanism is as follows: The bellows move in proportion to the differential pressure applied across them. When they have moved over their calibrated travel, a valve mounted on the center stem seals against its seat, thereby trapping the fill liquid in the bellows. Because the liquid is noncompressible, the bellows are fully supported and cannot be ruptured, regardless of the overpressure applied. This overrange protection is furnished in both directions, protecting both bellows.

Another design of overrange protectors involves the use of liquid-filled bellows with a number of diaphragm discs and spacer rings between them. As the bellows are subjected to overrange pressures, the diaphragms "nest" and the metallic spacer rings form a solid stop, thereby fully protecting the bellows from rupture.

Bellows meters can detect pressure differentials as low as 20 in. H_2O (5 kPa) and as high as 400 PSIG (2.8 MPa). Measurement of low differentials is limited by the small forces available to actuate the motion detector mechanism. For very high differentials, the limitation is in the mechanical strength of the bellows. Standard units are available with steel or stainless steel housings and stainless or beryllium copper bellows. On corrosive applications, other materials can be obtained, or special high-displacement volume chemical seals can be used. Static pressure ratings of 10,000 PSIG (69 MPa) are available as standard, and the operating temperature is limited to 200°F (93°C) maximum.

Liquid-Sealed Manometers

The desirable features of liquid-sealed d/p detectors are their self-powered nature and their strong outputs. Their disadvantages include the many moving parts, some of which are exposed to the process, their relatively high first and installation costs and their various limitations caused by process materials with which they are incompatible for reasons of corrosion, temperature, chemical reaction or plugging. In short, dry d/p detectors (force or motion balance types) can handle almost all installations with less investment and maintenance than the wet devices.

Liquid-sealed differential pressure detectors are seldom used as hydraulic head sensors on level measurement applications, but because the coverage of d/p devices would be incomplete without them, a brief description of their design is given below.

Inverted Bell d/p Detectors

These instruments are used chiefly to measure low-pressure differentials. The two well-known members of this family are the cylindrical and the Ledoux bell designs. Figure 5.6g illustrates the operation of the cylindrical bell unit. Here, the force developed by the differential pressure acting on the bell is opposed by a spring. The resulting bell motion is a measure of the pressure differential. Because a large bell acts as a force amplifier, even a small differential can generate a sizeable force to drive the readout mechanism. Depending on design and the type of seal fluid used, these units can respond to full-range differentials from 1 to 15 in. H_2O (from 0.25 to 3.8 kPa) and can withstand substantial static pressures.

The Ledoux inverted bell is different from the cylindrical design in two important aspects. Instead of a spring, the pressure differential on the bell is balanced against the weight of the heavy bell as it rises and falls in the seal fluid, which is normally mercury. Also, the

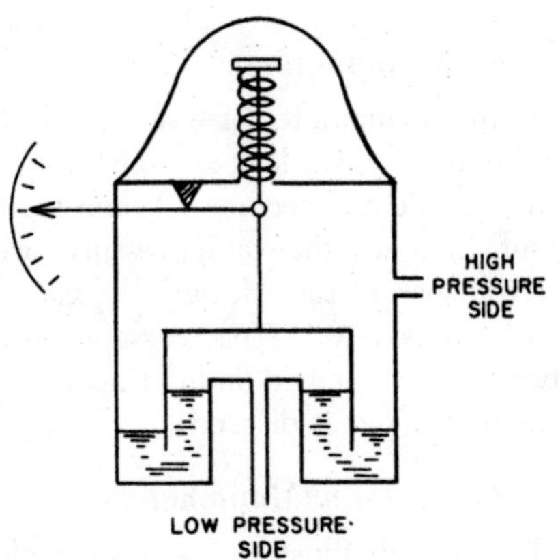

Fig. 5.6g Inverted bell d/p detector

bell is shaped so that its motion corresponds to the square root of the pressure differential acting upon it. This feature makes it ideally suitable to d/p flow measurement, in which an inherently linear output signal is desired. The unit can detect differentials up to 200 in. H_2O (50 kPa) and can be installed to operate at up to 6,000 PSIG (41 MPa) static pressure.

Ring Balance Meters

These meters (see Figure 5.6h) are widely used in Europe but have found limited acceptance in the United States. Their application and operating principle are the same as those of the inverted bell meters. A tubing ring formed in the vertical plane is pivoted at its central axis, and the pressure differential tends to rotate this ring. The opposing torque is developed by a weight attached to the ring. Depending on the design and the filling fluid used, differential spans of 5 to 500 in. H_2O (1.25 to 125 kPa) can be covered by this unit. The output signal can be made proportional to the square root of the differential pressure sensed, in which case it can act as a linear flow meter. Static pressures up to 6,000 PSIG (41 MPa) can be tolerated, but at high pressures the flexible connections at the sensing leads become a source of inaccuracy and need constant maintenance.

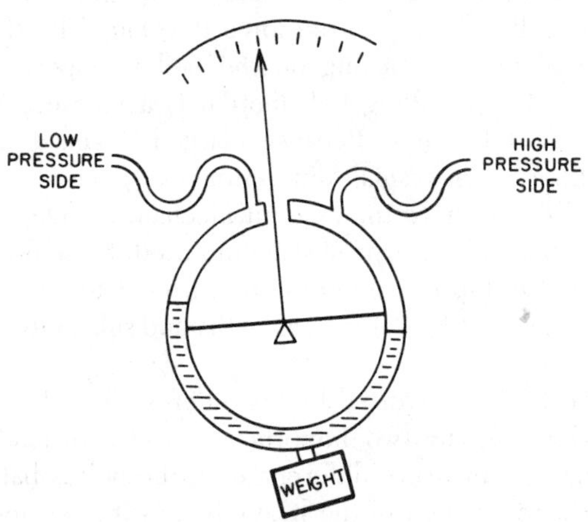

Fig. 5.6h Ring balance d/p sensor

Liquid Manometers

Liquid manometers are the simplest differential pressure detectors, while at the same time they are economical, reliable and accurate. When local visual indication is sufficient and the static pressures are compatible with the transparent tube design, the glass tube manometers can be considered. Float versions of the liquid manometers can be installed on high-pressure services or when remote readout is desired.

Glass Tube Manometers

Figure 5.6i illustrates the most elementary type of manometer, the U-tube. The difference between the two columns of liquid is an indication of the pressure differential. There are no moving parts and no friction or inertia involved in the measurement; therefore, its accuracy is limited only by the scale visibility.

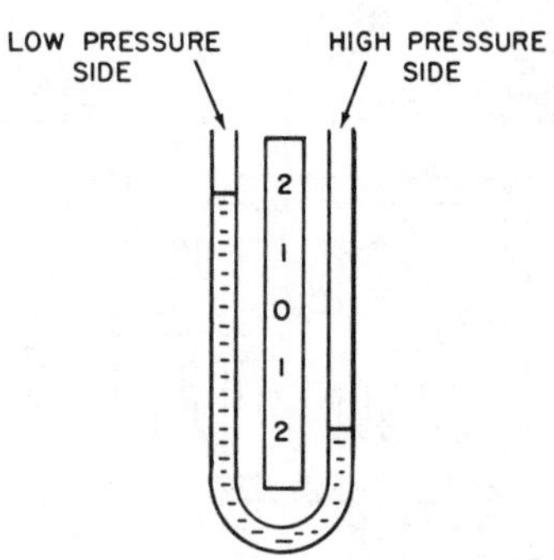

Fig. 5.6i U-tube manometer

An easier-to-read scale can be attached to the well manometer, when the liquid surface area on the high-pressure side is 1,000 or more times greater than the area on the low-pressure side. Therefore, changes in differential pressure cause variations only in the column height on the low side, as shown in Figure 5.6j, but leave the high side level practically unaffected. Changes in well liquid level are compensated for by sliding the scale zero to match this level.

The filling fluid is selected on the basis of its chemical inertness to the process materials it comes in contact with and its compatibility with the process temperatures expected so that it will not freeze or vaporize under operating conditions.

Glass tube manometers are available with up to 120-in. ranges (30 kPa), which is sufficient for most level applications, if the filling fluid is mercury. Design pressures up to 1,000 PSIG (6.9 MPa) can be obtained, but their use is discouraged because of mechanical hazards. For this reason, the use of glass tube manometers is limited to locations where tube breakage will not create hazardous conditions for the operator, as a result of either exploding glass particles or the nature of the process material released.

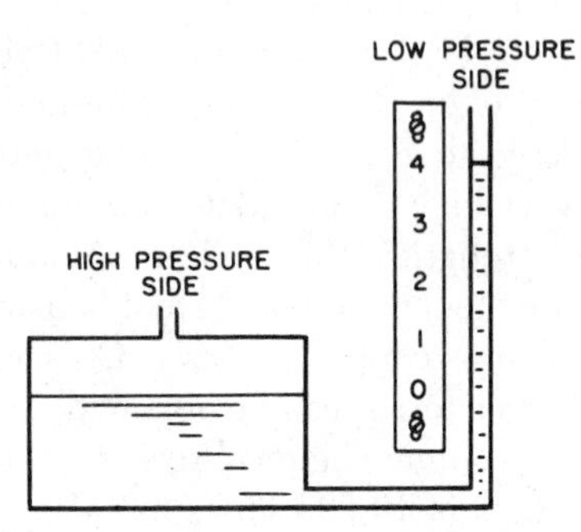

Fig. 5.6j Well manometer

Float Manometers

When remote readouts are required or when the process material is hazardous or at high pressures, float manometers are used. In these units (see Figure 5.6k), the variations in pressure differential cause the level of filling liquid to change, and the float moves with the changing level. The float motion is brought out of the chamber by a lever rotating in a pressure seal bearing.

Other designs are available using torque tube or magnetic followers. The lever rotation can be used to drive local indicators or transmitting devices for remote readout.

These instruments can be exposed to static pressures up to 6,000 PSIG (41 MPa) and can detect pressure differentials from 20 to 1,000 (in. H_2O (5 to 250 kPa). Lower differentials are not practical from an accuracy point of view and higher ones would require too-long reference chambers.

For high precision at static pressures up to 10,000 PSIG (69 MPa) stainless steel U-tube manometers have been used with magnetic floats in both legs. The float positions in this design are detected by electric coils to provide a very precise measurement of differential pressure.

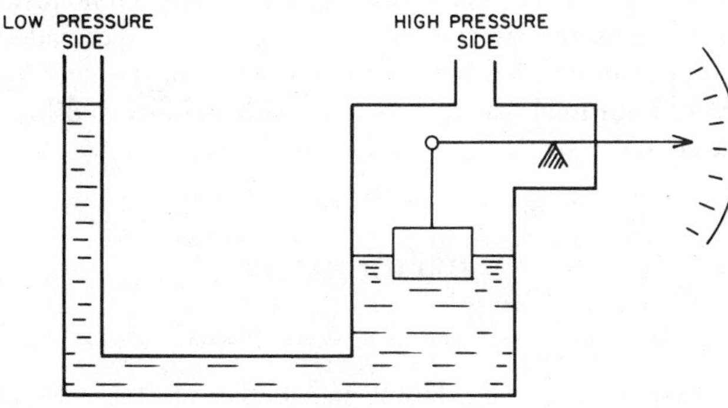

Fig. 5.6k Float manometer

Torque Tube Sensors

These instruments incorporate small torque tubes that function like the torque tubes in displacement level instruments. The tube is subjected to torsion that is proportional to the force (or movement) developed by the sensing element. For pressure instruments, a Bourdon or bellows element produces a linear movement proportional to pressure. For differential pressure, a diaphragm element produces a linear movement similar to that illustrated in Figure 5.6l. This movement is applied to the torque tube lever arm, which causes an angular displacement of the output shaft. This shaft rotation becomes the input of a motion transmitter that provides a milliampere or pneumatic output signal.

These are motion transmitting devices and are not supplied as direct-indicating types. All elements produce the same angular movement; consequently, all transmitters of a given type (i.e., pneumatic or electronic) are identical. Zero and span adjustments are made in the transmitter. Electronic transmitters use two wires that conduct both the signal and transmitter power. These transmitters meet Class I, Group D, Division I explosion-proof and intrinsic safety requirements.

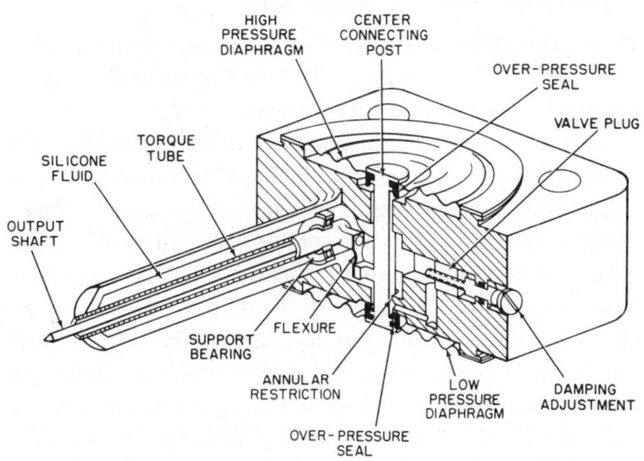

Fig. 5.6 l Differential pressure element and torque tube

Piston Sensors

Piston sensors (Figure 5.6m) are currently available as low-cost, moderately accurate indicators, switches or indicating switches. A common application is the monitoring of differential pressure across filters.

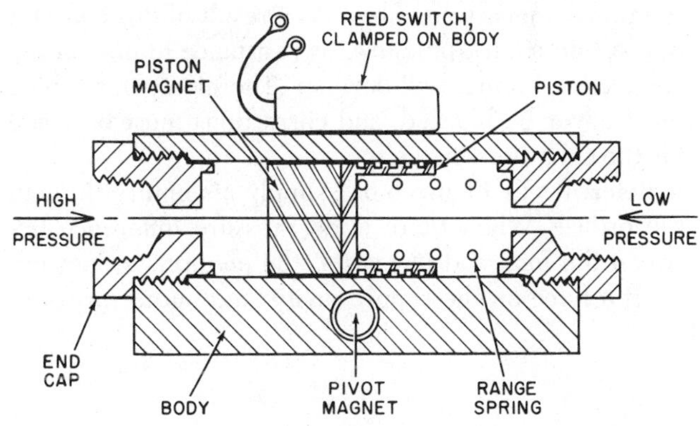

Fig. 5.6m Piston sensor

The high pressure is applied to the top of a piston and the low pressure to the underside of the piston against which a spring presses. A given differential pressure produces a corresponding spring movement. A magnet is attached to the piston which, through a nonmagnetic body, positions a pivot magnet, thereby driving the pointer or tripping as many as four switches, or both. Since piston movement is not transmitted mechanically, there are no associated moving seals. The only seals required are the piston seals and the static seals for the end caps.

Low-Differential Transmitters

Transmitters with differential pressure spans of 0.01 in. H_2O (2.5 kPa), 0.05 in. H_2O (12.5 kPa), and so on are available, which utilize the *pressure balance* principle. Pressure balance measurement is made possible by the unique characteristics of a device that senses and amplifies pressure by means of a membrane rather than a diaphragm. Since the membrane does not move or displace during measurement, the valve has no moving or mechanical parts to wear or contribute to hysteresis. Thus, the device will accurately sense very low pressures at its input ports, while being insensitive to shock, vibration and mechanical problems normally associated with high-mass pressure-sensing devices. It consumes less than ⅕ the amount of instrument air of a traditional d/p cell.

Because this transmitter does not have bellows or diaphragms that must fill with air to force balance, it responds much faster to an input change than any other device measuring pressure changes at very low levels. This means that in an application measuring turbulent air or gas flow, the transmitter will exhibit a higher output than manometers or similar devices because it will respond to true average flow, which has short-term peaks not "seen" by a manometer or mechanical device. This transmitter is not "wetted" by the process, but rather purges the high and low ports continuously with a low flow of supply air. This prevents the entrance of water, dirt or process gases into the sensitive measurement cavities of the transmitter, much as a purgemeter is used with a conventional d/p cell to purge contaminants from the process measuring lines. As a result of this constant purge, some *back pressure,* or resistance to flow in the connecting tubing, will develop. Therefore, this tubing must never be blocked, and corrections must be made for the tube pressure drop.

Referring to Figure 5.6n, supply air enters through the orifice. When there is no pressure difference between the high and low ports, the membrane does not exert a force on the supply cavity, and output pressure does not appear at the output port. If the output is dead-ended, all the supply flow is relieved through the ports.

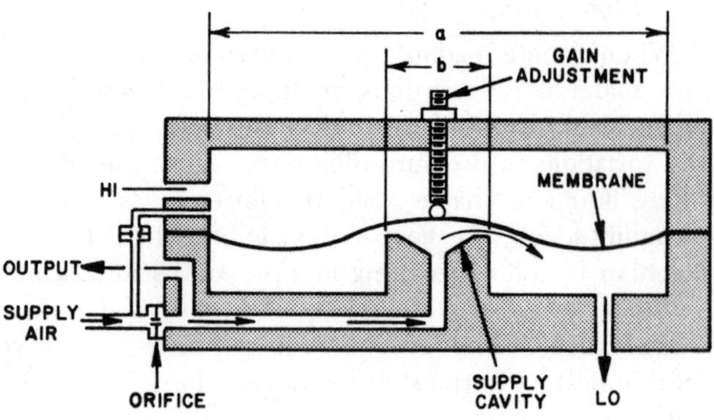

Fig. 5.6n Membrane low d/p transmitter

When a difference exists between the high and low ports, the membrane exerts a force on the supply cavity, which also appears at the output port. This pressure is a function of the gain, which is a function of dimensions *a* and *b*. When the pressure in the supply cavity satisfies the gain equation, supply air flows out of the supply cavity into the low port. At this point, the valve is in equilibrium. If the applied pressures decrease, the reverse occurs. The valve has very stable gain and no measurable hysteresis; therefore, the valves can be cascaded (output of one fed into the input of the next valve) for very high total gains, consistent with stability and high accuracy.

BIBLIOGRAPHY

"Capacitance Change Senses Differential Pressure," *Control Engineering*, June, 1970.

Lawford, V.N., "Differential Pressure Instruments: The Universal Measurement Tools," *Instrumentation Technology*, December, 1974.

Slomiana, Maria, "Selecting Differential Pressure Instruments," *Instrumentation Technology*, August, 1979.

Slomiana, Maria, "Using Differential Pressure Sensors," *Instrumentation Technology*, September, 1979.

Flow Sheet Symbol

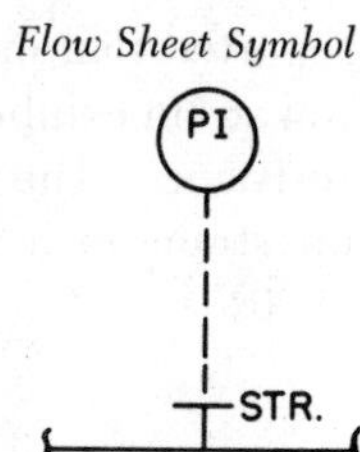

5.7 ELECTRONIC PRESSURE SENSORS

Design Pressure:	Up to 200,000 PSIG (1,400 MPa)
Design Temperature:	Generally up to 250°F (120°C); special designs to 600°F (316°C)
Materials of Construction:	Stainless steel wet parts for strain gauge and capacitance transducers; large variety of corrosion resistant materials for electronic transmitters
Inaccuracy:	Between ±0.1 and ±1% of span
Range:	From 3 in. H_2O to 200,000 PSIG (0.08 kPa to 1,400 MPa)
Cost:	$700 to $1,300 for strain gauge element; $700 to $900 for electronic transmitters; $1,000 to $2,000 for capacitance pressure sensors
Partial List of Suppliers:	Ametek Controls Div.; BIF; Bailey Controls Co.; Babcock & Wilcox, a McDermott Co.; Beckman Instruments Inc., CA; Bell & Howell Co., CEC Division; Bristol-Babcock Div.; ACCO Industries, Inc.; Dresser Industries; Dresser Ind. Valve & Instr. Oper.; Dynisco; Fairchild Industrial Products; Fischer & Porter Co.; Fisher Controls Co.; Foxboro Co.; Foxboro ICT, Inc.; Hays-Republic Div., Milton Roy Co.; Healy-Ruff Company; Honeywell, Process Control Div.; ITT Barton; Kent Process Control, Inc.; Kulite Semiconductor Prod. Inc.; Leeds & Northrup Co.; Leslie Co.; Moore Products Co.; Robertshaw Controls Co., Industrial Instrumentation Div.; Rochester Instrument Systems; Rosemount Inc.; Taylor Instrument Co., Div. of Sybron Corp.; Teledyne Taber; Telmar, Inc.; Three D Instruments; Validyne Engineering Corp.; Yellow Springs Instrument Co.

In this section various electronic pressure detectors and transmitters will be discussed. With few exceptions, these devices incorporate one of the pressure-sensing elements discussed in Section 5.6 and produce an electronic output related to the force or motion generated by the process pressure.

Electronic sensors are more accurate and have better response speeds than the previously discussed units, but at the same time they tend to be more expensive. The various electrical transducers developed during the last three decades include the bonded and unbonded resistance wire strain gauge, the differential transformer, the variable reluctance, the piezo-electric and the electrical capacitance elements.

Strain Gauge Pressure Transducers

The word strain means changes in dimensions of solid bodies due to the actions of forces exerted upon them. The earliest devices used to detect strain were the extensometers. These instruments simply determine the change in length of bodies, and the change in length divided by the original length is then a measure of the average strain. The true strain gauge does not measure extension, but changes its physical characteristics so that its electrical resistance changes as a function of the strain to which it is subjected.

The operating principle of strain gauges is more than 100 years old, and was discovered when Lord Kelvin reported that metallic conductors subjected to mechanical strain exhibited a corresponding change in electrical resistance. This principle was first put to practical use on strain measurement of concrete structures in the 1930s.

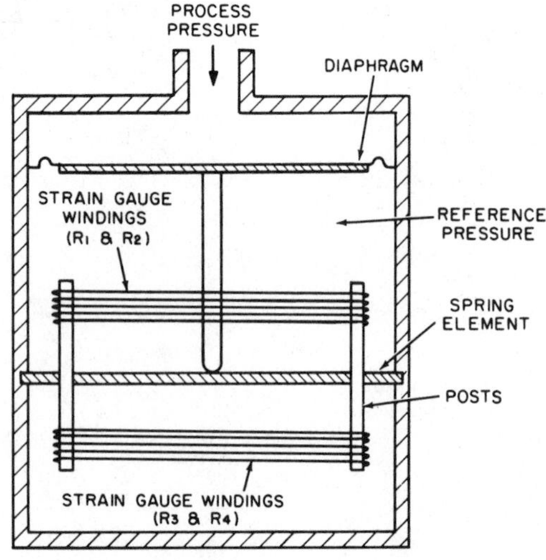

Fig. 5.7a Pressure transducer with unbonded strain gauge wires

These units are called unbonded wire gauges (see Figure 5.7a) because the wire elements are mounted on a mechanical frame whose parts can move in relation to each other, causing a change in wire tension as load changes. Therefore, the change in electrical resistance is a measure of strain.

The bonded strain gauge represented a major advance in strain gauge technology. This design eliminates the mechanical frame by attaching the sensing wire directly to the strained surface. The bond must be an insulator that forces the conductor wire to follow the strains on the surface without excessive stresses developing in the bond itself. This can be done when the conductor cross-sectional area is small in relation to the surface area per unit length. Strain gauge wires of less than 0.001-in. (0.025 mm) diameter have a surface area that is several thousand times more than the cross-sectional area. Therefore, the bond between the strained surface and the wires mounted on paper or plastic carriers is sufficiently strong. Foil gauges have also been used, in which the foil thickness can be as low as 0.0001 in. (0.025 mm).

Recently, another breakthrough occurred in strain gauge technology. This involved the use of semiconductors, such as silicon and germanium, as the gauge elements. The most attractive characteristic of semiconductors is their sensitivity, which is close to 100 times greater than that of metallic wires. When metallic wires are used, they are likely to be chrome-nickel alloys, although pure metals, such as platinum, are also used as temperature compensators. The desirability of a certain alloy used as a wire element is partially a function of its strain sensitivity. Strain sensitivity (S) is defined as the ratio between unit extension ($\Delta L/L$) and corresponding change in specific resistivity ($\Delta r/r$).

$$S = \frac{\Delta r/r}{\Delta L/L} \qquad 5.7(1)$$

where L is the initial length of the wire and r is the specific resistivity in that unstrained condition. If the strain occurs within a certain portion of the elastic range, the strain sensitivity is constant, meaning that by detecting resistance, we are sensing a value that is linearly related to strain. In actual installations, several factors, such as geometry of grid, wire size and material, direction of strain, type of bond, and so on, affect the relationship between axial strain and change in resistance. This is expressed by the constant called gauge factor:

$$\text{Axial Strain} = \frac{\Delta R/R}{\text{Gauge Factor}} \qquad 5.7(2)$$

When strain-gauge-sensing elements are used, process temperature variations must be compensated for because of the following reasons. Temperature changes will cause both the base material to which the element is bonded and the element itself to expand or contract. In addition, the coefficient of resistivity of the element will vary with temperature. There are several methods of temperature compensation.

a. In near-ambient temperature ranges (−100 to +150°F or −73°C to +66°C), thermal errors can be made negligible without compensation.
b. The element can be directly calibrated for the anticipated temperature range, and then by detecting process temperature, the reading can be corrected.
c. The element can be selected so that its thermal properties will match those of the mounting surface material.
d. Dual element gauges are also employed when, one having positive and the other negative response to temperature changes, they cancel out each other's effects within a set temperature range.
e. The most frequently employed method is the use of dummy elements. The dummy gauge is mounted on the same surface as the active element, and is exposed to the same temperature, but is not subject to the forces applied. If such a dummy is connected in a Wheatstone bridge arm adjacent to the active element, it will automatically compensate for temperature effects.

The resistance change of strain gauges being small, precise instrumentation is required to detect it with good accuracy. The Wheatstone bridge is one of the common configurations used for strain gauge measurement. Usually each arm of the bridge contains a strain sensitive element (see Figure 5.7b). Some of the elements can be active or dummies. The bridge will be balanced when $R_1R_3 = R_2R_4$. After an initial condition of balance, the change in output voltage is:

$$\Delta E = \frac{V}{4Ro}(\Delta R_1 + \Delta R_3 - \Delta R_2 - \Delta R_4), \quad 5.7(3)$$

where Ro is the initial, equal resistance of each element. From this equation it can be seen that if two elements form adjacent arms of the bridge (R_1 and R_2, or R_3 and R_4), the temperature effects will be minimized because their influence on the output is subtractive. If the active gauges are on opposite arms of the bridge, their effect is additive, and dummy elements are needed to achieve compensation. The Wheatstone circuit can detect both static and dynamic strains and is well suited for temperature compensation.

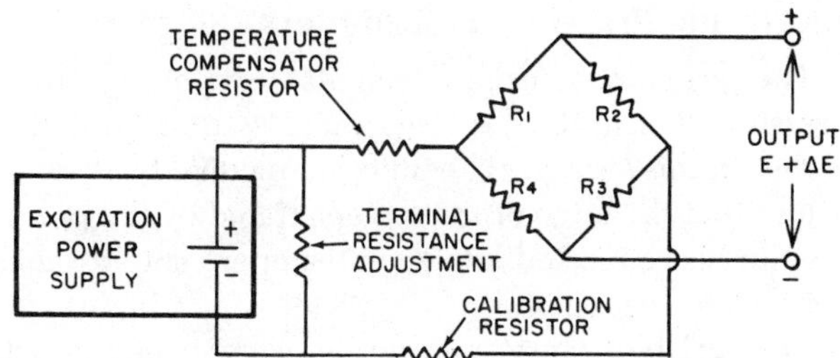

Fig. 5.7b Wheatstone circuit for strain gauges

The ballast or potentiometric circuit is arrived at by making R_2 = Infinity and R_3 = 0 in Equation 5.7(3). This circuit is simpler than the Wheatstone bridge and has the added advantage of possibility for common ground for the measuring instrument, amplifier and measuring circuitry. The drawbacks include the difficulty of temperature compensation, and the fact that the circuit is suited to dynamic, not static, strain sensing.

The strain gauge transducers normally have a bridge resistance that varies from 100 to 500 ohms. They can be excited by either ac or dc voltage from a power supply providing an output voltage in the range of 8 to 40 volts. The output generated by the bridge can be anywhere from 1 to 4 millivolts per volt excitation. Calibrated inaccuracy is 0.25 percent or better.

The output millivolt signal from the circuit can be converted to dc milliamperes or sensed directly by analog or digital readout devices, including typewriters or punched tape. Where large numbers of measurements are involved, a switching box with up to 50 channels is provided, which allows a single digital indicator to read any of the measurements. When fluctuating pressures are to be detected, the frequency of vibration has to be taken into account. For frequencies up to 50 cycles per second, conventional recorders are acceptable, up to 2,000 cps, galvanometer systems recording on light sensitive paper are preferred and at still higher frequencies the cathode ray oscilloscope is used.

In addition to pressure, strain gauges can be used to measure torque, weight, horsepower, velocity and acceleration. Here some of the more common strain gauge pressure transducers will be briefly described. The unbonded design is illustrated in Figure 5.7a. The pressure to be detected causes displacement of the diaphragm. The force applied to the diaphragm is transmitted by the force rod to the spring element. Motion of the spring center causes movement of the posts upon which the 0.0003-in. diameter (0.0075mm) resistive wires are mounted. The strain is increased in two of the windings and is decreased in the other two. These windings in the form of a Wheatstone bridge provide a millivolt output that is a linear function of process pressure.

One of the bonded designs is shown in Figure 5.7c. Here the process pressure is applied to a flat diaphragm. The strains resulting from the diaphragm deflection are sensed by four strain elements that are bonded directly to the underside of the diaphragm. The changes in resistance of these elements are measured as an indication of process pressure.

The working element of the strain gauge transducer shown in Figure 5.7d is a tube closed on one end, with the other end open to the process pressure. Four strain gauges are bonded to the outside of this tube. Two of

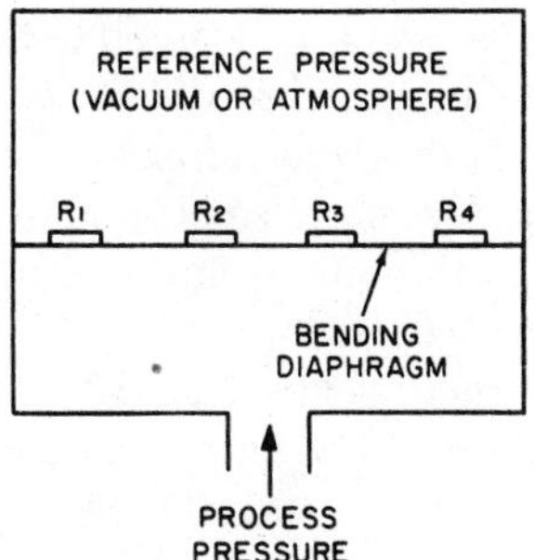

Fig. 5.7c Strain gauge transducer with diaphragm element

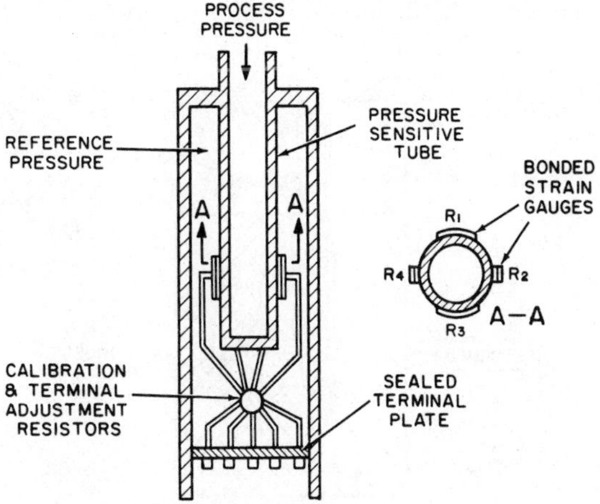

Fig. 5.7d Strain gauge transducer with elements bonded to tube surface

the elements are strained under pressure and two are not because they are mounted longitudinally and circumferentially. When the tube is pressurized, its minute expansion changes the resistance of the gauges, which are connected to a Wheatstone bridge (see Figure 5.7b). Calibration and terminal adjustment resistors are provided outside the tube (see Figures 5.7b and d).

In Figure 5.7e, two bellows are the working elements. When used to detect differential pressures, both bellows are connected to the process. Otherwise only one is connected, and the other provides atmospheric or full-vacuum reference for the unit. Here, the strain detector elements are bonded to the bending beam as shown.

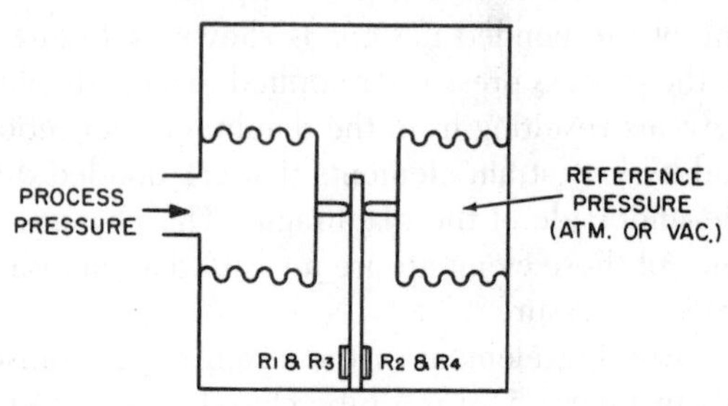

Fig. 5.7e Bonded strain gauge transducer with bellows elements

Still another design is illustrated in Figure 5.7f where the strain gauge sensors are adapted to the force balance d/p cell. Here, the pressure difference across the diaphragm transmits a corresponding force to the lower end of the force bar. This bar is connected to the strain tube, and the strain in the tube is in direct proportion to the pressure differential.

In summary, these devices can detect absolute, gauge and differential pressures with spans from 30 in. H_2O to 200,000 PSIG (7.5 kPa to 1,400 MPa). Each of these designs covers a segment of this overall range. Their inaccuracy falls between 0.2 and 0.5 percent of span, with thermal shift between 0 and 150°F (−18 and +66°C) amounting to less than 0.01 percent. Compensated units are readily available for temperatures between −60 and +250°F (−50 and +120°C). Special designs can handle process temperatures up to 600°F (316°C).

Most transducers are available with steel or stainless steel wetted parts and in very small sizes. They are stable devices with one of the highest speeds of response of any sensor, and generate strong electric output signals in the range of 15 to 100 ac or dc millivolts. A large variety of readout devices is available, some of which can be located up to 1,000 ft. (300 m) from the transducer. Most designs contain no moving parts, are suitable for high overloads and have good shock and vibration characteristics. All strain gauge installations necessitate regulated power supplies for the excitation voltage.

Their cost is relatively high, thus they should be limited to installations in which small size, fast speed of response under dynamic loads or high-static pressures are necessary.

Electronic Pressure Transmitters

The pressure-sensing elements used in electronic pressure transmitters are the same as those discussed earlier in this section. The units are modified only to the extent that the force or motion generated by the process pressure is converted into dc milliampere output signals for remote readout.

The variable transformer design consists of an elastic pressure element, such as a metallic bellows, with the core of the variable transformer connected to it. Process pressure variations result in core movement that generates a corresponding electrical output signal.

The differential transformer design is illustrated in Figure 5.7g in connection with a force balance pressure transmitter. The process pressure applied to the elastic element of the unit (not shown) exerts a proportional force on the lower end of the force bar, which pivots on the flexure seal. This force through the levers (shown) is transmitted to an electronic force balance system consisting of three main components: the detector, the feedback motor and the oscillator-amplifier. The detector primary is excited by the oscillator. A change in process

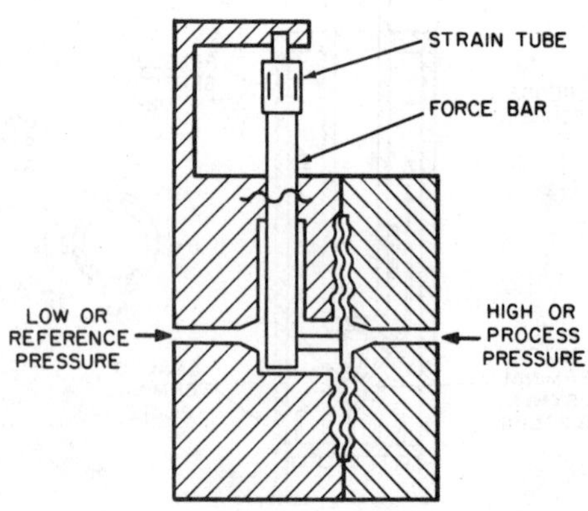

Fig. 5.7f Force balance d/p cell with strain gauge elements

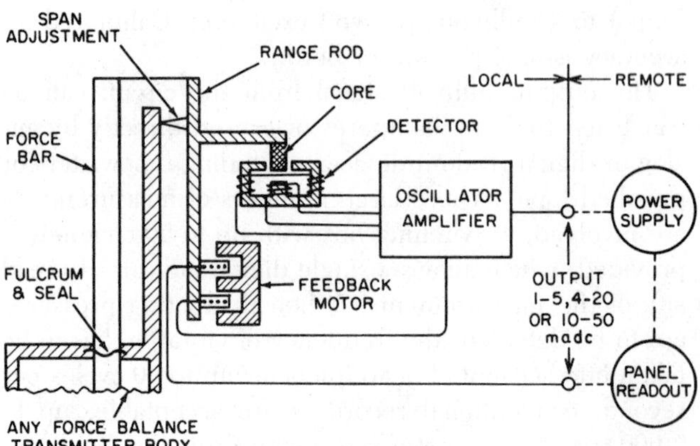

Fig. 5.7g Differential transformer force balance electronic pressure transmitter

pressure results in a slight movement of the laminated core, strengthening the inductive coupling, which increases the secondary voltage to the dc amplifier. The output from the amplifier is fed to the feedback motor in series with the remote readout. As the current in the feedback motor coil increases, an increased force is developed, which repositions the laminated core in the detector. The feedback coil applies an equal and opposite force to that produced by the change in process pressure, and, therefore, maintains the system in continuous force balance.

Figure 5.7h shows a similar electronic pressure transmitter design. Here the elastic sensing element deflects in proportion to the process pressure, thereby exerting a force on the balance beam through the input spring. The resulting motion changes the air gap in the detector assembly. The detector consists of two pieces of ferrite, one mounted on the force beam and the other on the chassis. As the air gap is changed, there is a resulting change in the inductance of the oscillator circuit. When this occurs, the oscillator, acting as a variable resistor, changes the output current correspondingly. The output current is fed through the magnet coil in the feedback motor, producing an equal and opposite force on the beam to balance the force produced by the change in process pressure.

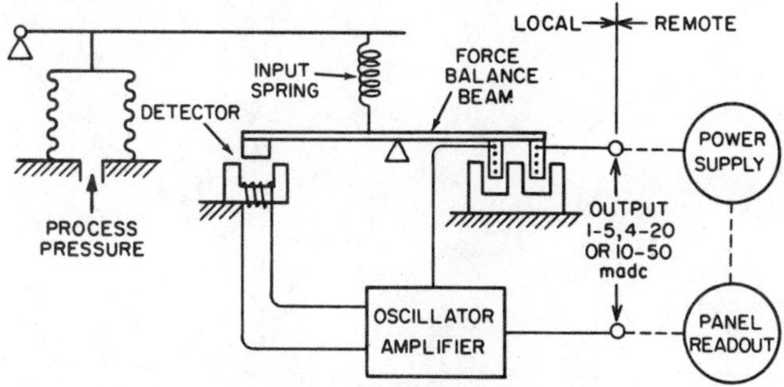

Fig. 5.7h Inductance electronic force balance pressure transmitter

The variable reluctance electronic transmitter is shown in Figure 5.7i. Electrical reluctance is the equivalent of electrical resistance in a magnetic circuit. A change in process pressure exerts a force on the metal diaphragm that moves the armature between two ferrite core coils. The air gap change causes a predetermined change in the inductance ratio of the two coils. The resultant change in core reluctance is detected by the amplifier bridge. The bridge, in conjunction with the amplifier rebalancing circuits, changes the transmitter output current to flow in direct proportion to the process pressure. The function of the amplifier rebalancing circuit is to change the capacitive reactance of the varactors (voltage sensitive capacitors) in the bridge a like amount. Therefore, the feedback adjusts the varactors so that the capacitance ratio of the two varactors equals the inductance ratio of the two inductors.

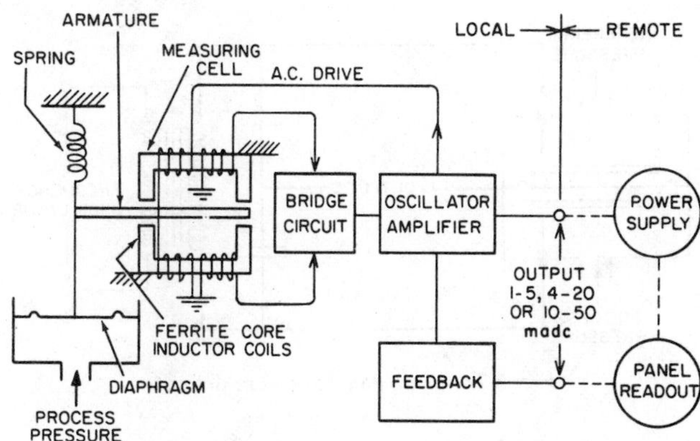

Fig. 5.7i Variable reluctance electronic force balance pressure transmitter

A simple schematic for the resistance pressure detector is given in Figure 5.7j. A change in process pressure sensed by the elastic element moves the connecting rod positioning a noble metal contact wiper over the precision potentiometer, thereby converting process pressure into electrical resistance in a manner somewhat similar to that of the strain gauge elements. Spans vary from 1 to 5,000 PSI (1 to 35 MPa) and accuracy is between 1 and 2 percent. Readout devices are similar to those used for strain gauges.

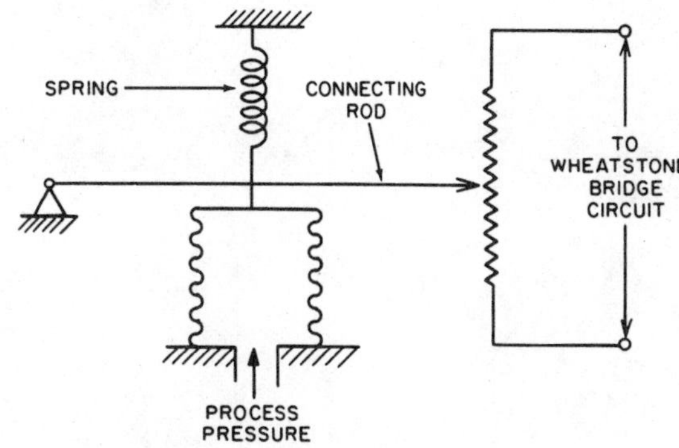

Fig. 5.7j Resistance pressure sensor

Capacitance Pressure Detectors

The basic operating principle involved in all capacitive pressure sensors is the measurement of change in capacitance resulting from the movement of an elastic element. The elastic element in most designs is a Ni-Span C or stainless steel diaphragm exposed to the process pressure on one side and to the reference pressure on the other. Depending on the reference pressure used, the unit can detect absolute, gauge or differential pressures.

The unit shown in Figure 5.7k incorporates two capacitor plates, while other designs have only one such plate. A high-voltage, high-frequency oscillator is used

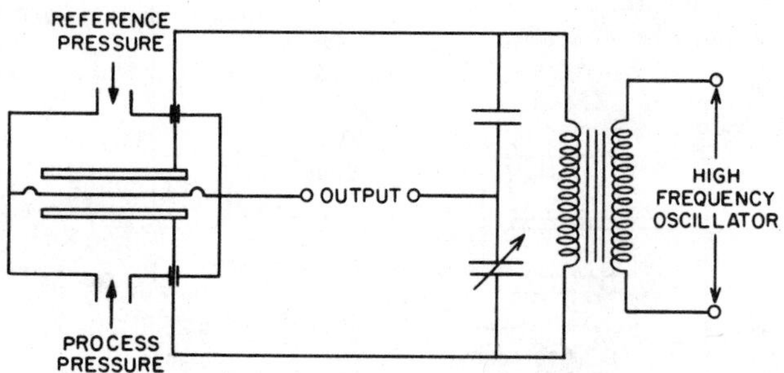

Fig. 5.7k Capacitance pressure detector

to energize the sensing element. Changes in process pressure deflect the diaphragm, and the resultant change in capacitance is detected by a bridge circuit. The two-plate design can be operated in balanced or unbalanced modes. If the circuit is operated in the balanced mode, the output voltage is fed to a null detector, and the capacitor arms are varied to maintain the bridge at null. In this mode, the null setting itself is a measure of process pressure. If the circuit operates in the unbalanced mode, the ratio between output voltage and excitation voltage is the indication of process pressure.

In the single capacitor design, the plate is positioned on one side of the sensing diaphragm. The capacitance of the element, being a function of diaphragm deflection, is a measure of process pressure. The element's capacitance is converted and amplified into a dc milliampere current signal. If desired, direct dc voltage output can also be used when the readout device preference is similar to a digital voltmeter.

Capacitive pressure sensors are inaccurate to ±0.1 to ±0.2 percent of span and with the proper selection of diaphragms can handle pressure ranges from 3 in. H_2O up to 5,000 PSIG (0.08 kPa to 35 MPa). Both their temperature sensitivity and hysteresis are low, while their speed of response is high. Their output is linear and can be temperature compensated for optimum results. (See Section 5.14 for the vacuum detector version of this unit.)

Capicitance sensors are relatively expensive and are sensitive to dielectric constant variations in the process fluid. For this reason, their use is limited to dielectric fluid and to dry, noncondensible gas services.

BIBLIOGRAPHY

Bradley, C.D., "Semiconductor Strain Gauges," *ISA Journal*, January, 1966.

"Capacitance Change Senses Differential Pressure,"*Control Engineering*, June, 1970.

Comber, J. and Hockman, P., "Pressure Monitoring: What's Happening?" *Instruments and Control Systems*, April, 1980.

Hall, J., "Monitoring Pressure with Newer Technologies," *Instruments and Control Systems*, April, 1980.

Harvey, G.F., "ISA Transducer Compendium," Instrument Society of America, 1969.

Flow Sheet Symbol

5.8 HIGH-PRESSURE SENSORS

Design Pressure:	Up to 400,000 PSIG (up to 2,800 MPa)
Materials of Construction:	Steel and stainless steel
Inaccuracy:	From better than one part in 10,000 to ±2% depending on design
Range:	From 0–25 PSIG to 0–425,000 PSIG (0–173 kPa to 0–2,900 MPa); special low-pressure units are also available
Cost:	$900 and higher
Partial List of Suppliers:	Bourns Instruments, Inc; Crosby Valve & Gauge Co.; Dresser Industries; Dynisco; Foxboro ICT, Inc.; Harwood Engineering Co., Inc.; Rosemount Engineering Co.; Ruska Instrument Corp.; Strainsert Co.; Tavis Corp.; Validyne Engineering Corp.

If we define high-pressure detectors as sensors that can measure pressures in excess of 20,000 PSIG (140 MPa), we will find that we have already discussed some of the high pressure detectors. The various Bourdon designs (Section 5.4) are capable of detecting pressures at this level, and the strain gauge elements (Section 5.7) can measure pressures up to 200,000 PSIG (1,400 MPa). This section will cover high-pressure sensors that were not discussed previously, such as dead weight piston gauges, bulk modulus cells, manganin cells and others. No industrial processes, with the exception of diamond synthesis, are known to operate above 200,000 PSIG. This is the main reason for the limited interest in pressure sensors in that range.

Dead Weight Piston Gauges

As illustrated in Figure 5.8a, these are piston gauges in which the test pressure is balanced against a known weight that is applied to a known piston area. The test pressure is applied by the secondary piston. The principal purpose of these free-piston gauges is as a primary standard to calibrate other pressure sensors. The National Bureau of Standards has been using these devices for the past 50 years.

Piston gauges, or dead weight testers, are normally provided with a number of interchangeable piston assemblies and NBS certified weights. They can be used to calibrate at pressure levels as low as 5 PSIG (35 kPa) or as high as 100,000 PSIG (690 MPa). The range has been extended to even greater pressures, but research on piston and cylinder materials and their treatment to withstand loads has uncovered a serious limitation. To illustrate the problems at higher pressures, let us consider some parameters of a standard piston gauge for the detection of 100,000 PSIG pressure. If the dead weight is to be kept under 1,000 lbs. (450 Kg), the piston area cannot exceed 0.01 in.^{2} (6.3 mm^2). This means that an approximately 0.1-in. diameter (2.5 mm) piston must support 1,000 lb. of weight and also be able to be rotated.

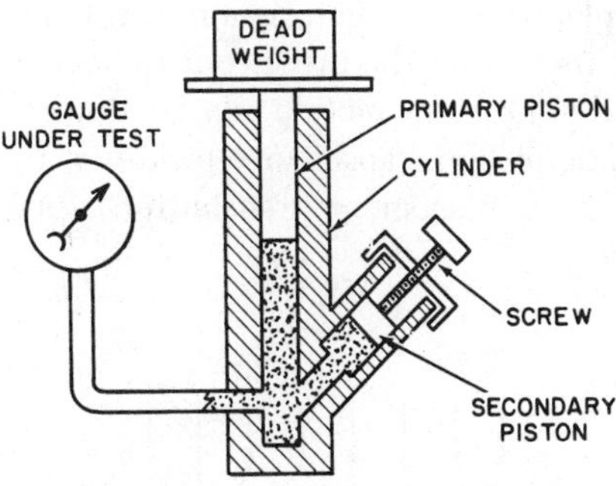

Fig. 5.8a Dead weight piston tester

The accuracy of dead weight piston testers has been improved over the years. For higher pressure services, the controlled piston-cylinder clearance design is a major improvement. This clearance is controlled by pressurizing the outside surface of the cylinder. Thus, the piston-cylinder clearance is kept constant, resulting in a slow rate of fall for the piston unaffected by pressure level. The laboratory piston gauges are standardized by NBS, calibrating the associated weights and measuring the piston diameter. NBS has found these dead weight testers to be inaccurate to 1.5 parts in 10,000 of the measured pressure at values greater than 40,000 PSIG (280 MPa) and to 5 parts in 100,000 at lower pressures. The inaccuracy of industrial dead weight testers is better than ±0.1 percent of span.

The free-piston gauge is limited to its principal purpose, a primary standard for calibrating other pressure sensors, because it is slow in response and is not practical for direct industrial installation.

Presently the utility of the high-accuracy piston gauges is being extended to the lower pressure ranges by the newly developed tilting-type, air-lubricated designs. With such design, pressures (and pressure differentials) in the mmHg range have been detected to one part in 100,000 full-scale inaccuracy.

Bulk Modulus Cells

These cells, shown in Figure 5.8b, are comprised of a hollow cylindrical steel probe closed at the inner end, and a stem that projects beyond the outer end of the probe. When subjected to process pressures, the active part of the probe contracts isotropically, causing its tip to be displaced to the right. As a result, the stem moves outward, increasing the distance it projects beyond the outer end. The stem motion can be detected by electromagnetic pickup, capacitance pickup or by the use of mechanical displacement transmitters (pneumatic or electronic).

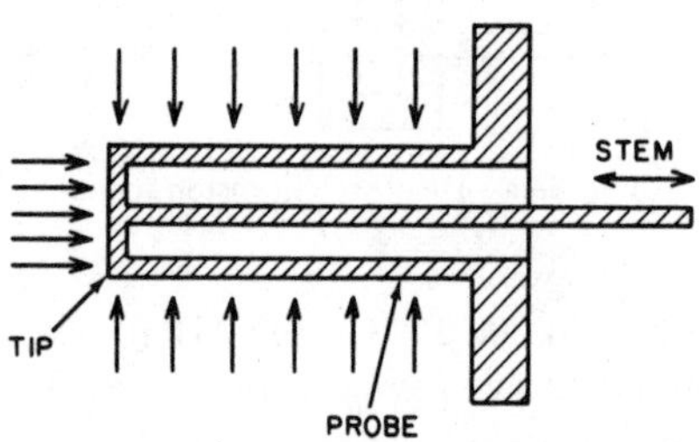

Fig. 5.8b Bulk modulus cell

The unit is available with ranges of 0–50,000 PSIG to 0–200,000 PSIG (0–350 MPa to 0–1,400 MPa), and its inaccuracy is ±1–2 percent of full-scale. Its advantages, when compared with other high-pressure sensors, include its relatively fast response, the fact that it is characteristically remote-reading and that the design is absolutely safe because the probe is not subject to fatigue. The hysteresis and temperature sensitivity of the bulk modulus cell are similar to that of other elastic element pressure sensors.

Manganin Cells

When a small coil of Manganin wire is subjected to high-process pressures, the coil resistance changes linearly with pressure. The pressure-resistance relationship for Manganin is substantial, positive, linear, and, therefore, can be detected by a bridge. Manganin is relatively insensitive to temperature variations.

These cells can be obtained with ranges from 0–50,000 PSIG to 0–425,000 PSIG, and their inaccuracy is between ±1/10 and ±1/2 percent of full-scale.

The main disadvantage of this cell is its delicate nature. Both the gauge coils and the coil protection bellows can be easily damaged by rapid pressure changes, liquid viscosity or other causes.

The pressure-resistance relationship of other materials, such as platinum, gold-chromium or lead, has the same desirable features as Manganin, and they have been used as elements in pressure-resistance cells.

Other Techniques of High-pressure Measurement

One other method for high-pressure sensing is to determine the pressure at which change-of-state occurs in various materials and then to apply that as a standard. Some of the change-of-state points have already been determined. For example, it has been established that the melting point of mercury at 0°C is 109,765 ± 30 PSIG (757 ± 0.2 MPa). Similarly, the first polymorphic transition point of bismuth has been found to be between 365,000 and 370,000 PSIG (2,519 MPa 2,553 MPa).

Dynamic Pressure Sensors

The interest in dynamic pressure measurement to detect blast pressures, rapid chemical reactions, combustion pressures of rocket propellants and so on has increased in recent years. Several electric transducers have been developed for use with elastic elements. Because these devices were covered in Section 5.7, only a brief listing will be given here.

Electrical transducers for dynamic pressure detection include the piezoelectric transducers, the bonded and unbonded strain gauge elements, variable reluctance, differential transformer, electrical capacitance and others.

Strain gauges bonded to diaphragm or bellows elements have given good performance in measuring blast pressures. In connection with underwater explosions and noises, piezoelectric crystals have been successfully used. These units are directionally sensitive to force, necessitating a seal interposed between the element and the process and converting pressure to force for optimum response.

BIBLIOGRAPHY

Budenberg, G.F., "Dead Weight Pressure Measurement," *Instruments and Control Systems*, February, 1971.

Comber, J. and Hockman, P., "Pressure Monitoring: What's Happening?" *Instruments and Control Systems*, April, 1980.

Hall, J., "Monitoring Pressure with Newer Technologies," *Instruments and Control Systems*, April, 1979.

Harvey, G.F., "ISA Transducer Compendium," Instrument Society of America, 1969.

Kaminski, R.K., "Measuring High Pressures Above 20,000 PSIG," *Instrumentation Technology*, August, 1968.

Flow Sheet Symbol

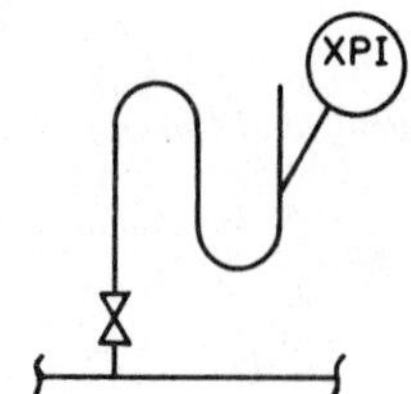

5.9 MANOMETERS

Design Pressure:	Up to 6,000 PSIG (up to 41 MPa)
Design Temperature:	Function of seal fluid; usually ambient
Materials of Construction:	Pyrex, brass, steel, aluminum, stainless steel
Inaccuracy:	±1% of span for most; ±0.1% for precision visual types
Range:	Minimum span is 0.15 in. H_2O (38 Pa); maximum span 60 PSIG (410 kPa)
Cost:	$60 for glass tube indicator and $900 for transmitter, both in standard materials of construction
Partial List of Suppliers:	Dwyer Instruments, Inc.; Fischer & Porter Co.; Manostat Corp.; Meriam Instrument Div. of Scott & Fetzer Co.; Petrometer Corp.; Princo Instruments, Inc.; Trimount Instrument Co.; Uehling Instrument Co.; Wallace & Tiernan Div., Pennwalt Corp.

This section describes various manometers, and also a self-contained vacuum controller called the cartesian diver. The pressure detector designs discussed in the previous sections are "dry" units, while the sensors in this section are "wet" detectors. There are two major groups of wet pressure instruments: the liquid-sealed and the liquid manometer. The liquid-sealed category includes various bell designs and the ring balance units. The liquid manometer group is subdivided into visual and float types. There are a variety of visual manometers, including barometers; well, U-tube and inclined tube manometers; and micromanometers.

Liquid-Sealed Designs

Because these instruments have been discussed in Section 5.6 from a level measurement point-of-view, here we shall concentrate on features that have not yet been covered in detail.

Inverted Bell Pressure Sensors

In all inverted bell devices, the bell dips into a sealing liquid and generates a vertical motion as a function of the pressure differential acting on the inner and outer surfaces of the bell. The bell motion might be balanced against a calibration spring or by the weight of the bell. Depending on the pressure on the reference side of the bell, this device can detect absolute, positive, negative or differential pressures.

Variations include single- and double-bell units, mercury- and oil-sealed types and designs in which the vertical bell motion is linearly or square-root-related (Ledoux) to the pressure differential sensed.

Figure 5.6g shows the operation of the single cylindrical bell design. If the underside of the bell is evacuated by a vacuum pump, with this full vacuum reference, the unit will detect absolute pressures. If the reference side is left open to the atmosphere, the instrument will measure pressures above or below atmospheric pressure. By changing the calibrated spring, which normally is a temperature-compensated helix, the range of the unit can also be changed.

Figure 5.9a illustrates a double-bell unit, one of the most sensitive pressure sensors when measurement of near-atmospheric pressures is required. It can also be used as a differential pressure detector with the two process pressures connected to the underside of the two bells. As a result of a change in differential pressure, one of the bells moves up and the other down in direct proportion to the differential pressure detected. In this design, there is no calibration spring because the weight of the bells balances the pressure differential.

When the bell is shaped so that its vertical motion

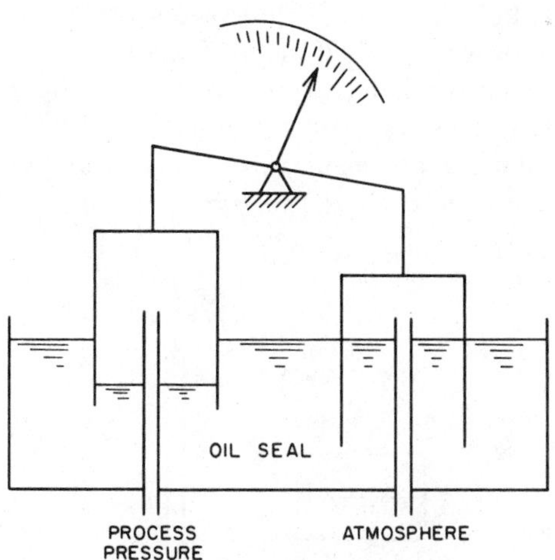

Fig. 5.9a Double inverted bell manometer

corresponds to the square root of the pressure differential acting on it, the resulting signal is linear with volumetric flow rate through orifice elements.

All inverted bell units have the common desirable feature that the large bell area generates ample power to operate the mechanical motion detector levers even at very low process pressures. These units are made of materials that are normally suitable for noncorrosive services only, and the process media has to be compatible with the seal fluid used. Their inaccuracy is ±1 percent of span (Ledoux is ±2 percent), and they are bulky units sensitive to ambient temperature variations.

The maximum differential pressure these units can withstand is determined by the depth of the sealing fluid. If that limitation is exceeded, the seal fluid will be blown out. The same limitation exists for the measurement of static pressures with vacuum or atmospheric references. When the unit is used for differential pressure sensing, the static pressure limitation is determined by the design of the housing. In the Ledoux bell design, housings can withstand up to 6,000 PSIG (41 MPa) service pressures. The spans available with the Ledoux bells are higher (up to 200 in. H_2O or 50 kPa) than the others. The minimum span available with the cylindrical bells is 0.15 in H_2O (0.038 kPa), and the maximum is about 15 in. H_2O (3.8 kPa). The actual range can be high-vacuum with vacuum-pump reference or near-atmospheric with open-vent reference. Compound ranges are also available.

Ring Balance Manometers

The minimum span for these units is 0.2 in. H_2O (0.05 kPa) and the maximum span is 500 in. H_2O (125 kPa). (See Figure 5.6h.) Depending on the reference pressure used, it can detect absolute or near-atmospheric pressures. Although the housings can withstand up to 6,000 PSIG (41 MPa), the pressure differential between reference and process pressure should be below the limit, which would cause the sealing fluid to be blown out.

Float Manometers

Figure 5.6k shows the features and the operation of these instruments. When used as a pressure sensor, the process pressure is connected to one side of the manometer and a reference pressure to the other. If the reference pressure is atmospheric, the float movement is related to the pressure differential between the process pressure and atmosphere. Spans are available from 20 to 1,000 in. H_2O (from 5 to 250 kPa). The range can be set for positive, negative or compound pressure detection. Housings are available for up to 6,000 PSIG (41 MPa). Measurement inaccuracy is ±1 percent of span. If the float housing is cylindrical, the float motion is linearly related to the pressure differential sensed. By forming the housing tube to variable parabolic cross-sections or other shapes, any desired nonlinear relationship between pressure differential and float motion can be achieved. When precision measurements are required in the area of ±0.05 percent of reading, the servomanometer illustrated in Figure 5.9b can be considered. The tube of this instrument can be either glass or stainless steel when mechanical strength is desired. The principle of operation is as follows. Process pressure variations result in float movement that repositions the magnetic armature. A differential transformer is continuously positioned by a servomotor, to be electrically centered about the armature. The shaft position in the servotransmission is used as a measure of the liquid column height in the manometer tube.

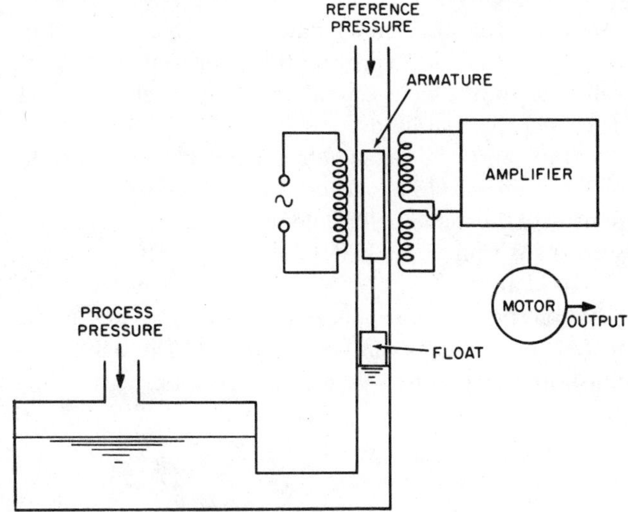

Fig. 5.9b Metallic tube servomanometer

Visual Manometers

Here the various visual manometers are covered, noting considerations and design features not already discussed (see Figures 5.6i and j).

Liquid Barometers

The liquid barometer is a fundamental instrument for detecting atmospheric pressure that is used for precise measurements and for calibration of other sensors. As shown in Figure 5.9c, a glass tube that is open on one end is filled with mercury and then inverted into an open mercury bath. The height of the resulting column in the tube is a measure of the barometric pressure in millimeters of mercury. Accuracy is affected only by the visibility of reading and by capillary effects in the tube. The barometer will act as an absolute pressure sensor if its well is connected to the process.

U-Tube, Well and Inclined Liquid Manometers

Figures 5.6i and j illustrate the U-tube and well designs. In addition the following features should be noted.

The standard construction materials for manometers include brass, steel, aluminum and stainless steel for the metallic wetted parts and Pyrex glass for the tube. The glass tube with $\frac{1}{4}$ in. (6.3 mm) inside diameter can be exposed up to 400 PSIG (2.8 MPa) pressure, while the extra heavy-wall tube with $\frac{1}{8}$ in. (3.1 mm) inside diameter is designed to withstand 2,000 PSIG (14 MPa). These pressure ratings must be qualified in two respects. First, they do not take mechanical damage into consideration. Therefore, if the process material is toxic, hazardous or should not be allowed to escape for other reasons, glass tube manometers should not be used, even at low operating pressures. Second, the static pressure limitations noted apply only when the manometer detects pressure differentials. When its purpose is straight pressure measurement, the manometer fluid will be blown out if its range is exceeded. To prevent this from occurring, check valves or return wells can be provided which will block or collect the seal fluid.

U-tube and well manometers are available in lengths of 6 to 140 in. (150 mm to 35 m). Depending on the density of the manometer fluid used, they can read pressure spans from 5 in. H_2O to about 60 PSIG (1.25 kPa to 410 kPa).

When below-atmospheric pressures are to be detected, two designs can be considered. The barometer illustrated in Figure 5.9c will give readings in absolute pressure units based on its full-vacuum reference. If the manometer well is raised the manometer will sense negative pressures relative to atmosphere (see Figure 5.9d). In some designs, the well elevation is adjustable, allowing for shifting the manometer range from vacuum to compound or to positive pressure detection.

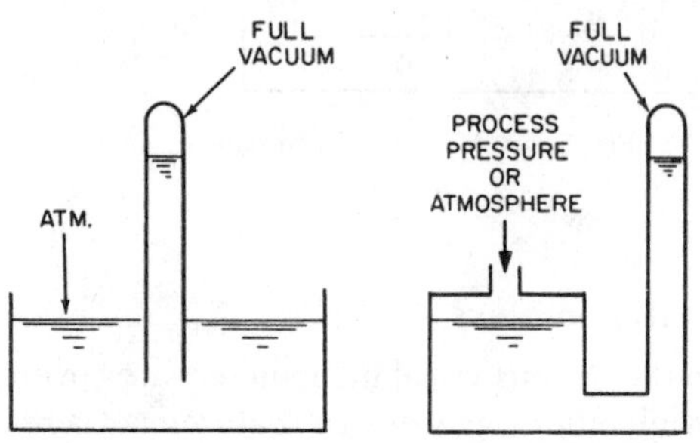

Fig. 5.9c Liquid column manometer

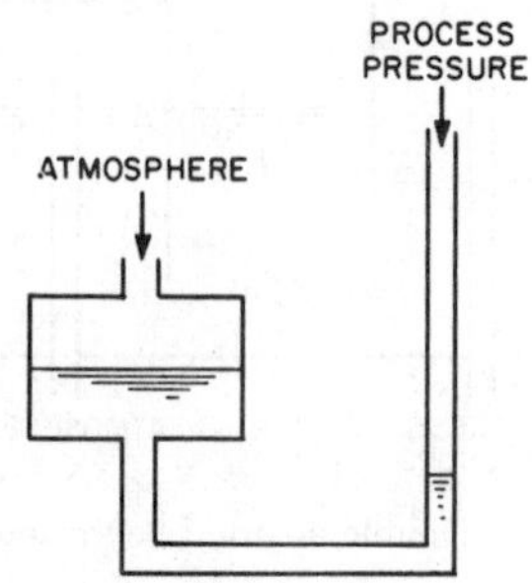

Fig. 5.9d Raised well manometer for positive or negative pressure detection

Figure 5.9e illustrates a multitube manometer with atmospheric reference. Up to 50 tubes can be packaged in a common bank; thus the same number of process pressures can be detected from the same reference well. If desired, some or all of the tubes can be provided with separate wells for flexibility.

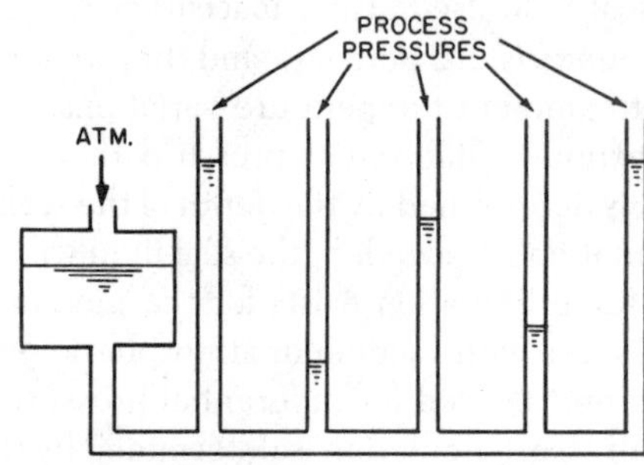

Fig. 5.9e Multitube manometer

The accuracy of measurement is largely a function of the operator's capability to read precisely the height of the liquid column. For this reason, designs have been developed which tend to amplify the reading. One of these units is shown in Figure 5.9f. This design is called the *double inverted-well manometer*. If the left leg is filled with a manometer fluid of 1.0 density and the right leg with a fluid of 0.9 density, the movement of interface between the two will be 10 times that of a U-tube manometer filled with water.

If the vertical tube of a well manometer is laid down in an inclined, almost horizontal position so that a slight change in process pressure will cause a large movement of liquid column, the reading visibility will be improved. With this design, illustrated in Figure 5.9g, full spans of 0.5 to 4 in. H_2O (0.13 to 1 kPa) are available.

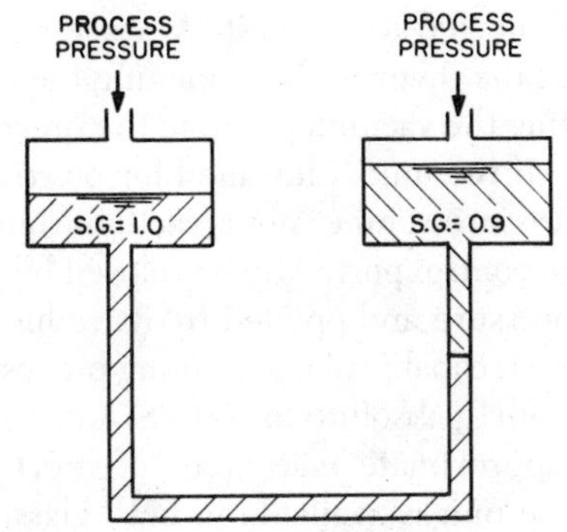

Fig. 5.9f Double liquid balance manometer

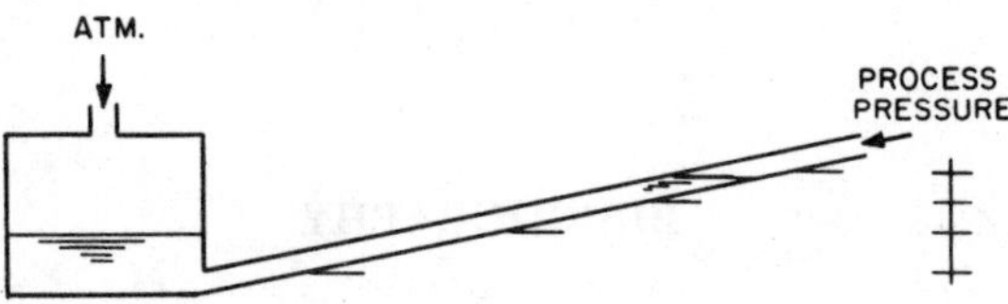

Fig. 5.9g Inclined manometer

The performance of any manometer installation is largely a function of the indicating fluid selected. It has to satisfy several criteria, which in many cases are hard to meet. The filling fluid has to be chemically inert and compatible with the process media and produce a clear, visible interface. The fluid should not coat the glass tube and should not be corrosive to standard materials, such as copper-bearing alloys, aluminum or steel. Its surface tension should be low to minimize capillary effects, and it should be stable so that no flashing occurs under high-operating temperature or vacuum conditions. Similarly, the filling fluid should not freeze due to low-ambient temperatures. In general, manometer fluids are available for an operating temperature range of −70 to +150°F (−57 to +66°C). The most important characteristic of a good manometer fluid is its ability to maintain density unaffected by temperature variations, so that the height of its column is a reproducible measure of process pressure. Standard filling fluids are available in a variation of colors covering a specific gravity range of 0.8 to 13.6. None of them satisfies all the requirements listed above; therefore, they have to be selected separately for each installation.

Manometers are normally mounted close to the point of measurement to reduce lag time. They are available for bench, wall or panel mounting, although in modern plants, remote panel-mounted manometers have been replaced by transmitter-receivers. This is desirable not only from space and time-lag point of view, but also for safety reasons in order to prevent process materials from entering the general purpose central panel. When the pressure to be measured is unsteady, pulsation dampeners are installed to steady the indicating column for better readability.

Visual manometers are normally used only as direct indicators, but they can also be provided with switches to detect the level of indicating fluid in the tubes. For this purpose, either photo cells or electric conductivity switches are installed that actuate alarms, solenoids or any other electrical devices.

Micromanometers

The inclined and the double liquid balance manometers are capable of amplifying the movement of the manometer fluid column. The micromanometer illustrated in Figure 5.9h achieves the same thing, but its amplification capability is much greater.

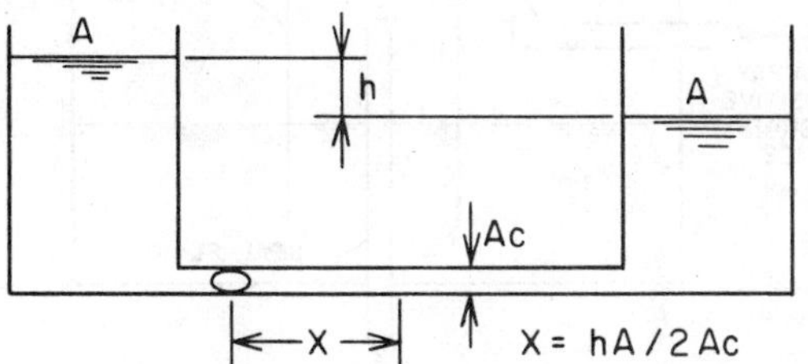

Fig. 5.9h Micromanometer

The unit consists of two large-diameter tubes connected by a small-bore capillary. The capillary either contains an air bubble to indicate filling fluid movement, or the two large-diameter tubes are filled with two non-mixing liquids of different colors, in which case fluid movement is indicated by interface travel. When both wells are exposed to the same pressure, the air bubble comes to rest in the center. Then if one of the wells is connected to the process, the difference between that and the atmospheric reference causes fluid movement from one well to the other, developing a head differential. If we call the well cross-sectional area A and the capillary bore area Ac, the resulting movement of the bubble in the capillary can be expressed as:

$$X = hA/2Ac \qquad 5.9(1)$$

Consequently, the head motion has been changed into air bubble motion and has been amplified by the factor A/Ac. This ratio is limited only by the selection of tubes and can be as high as 100:1 or 1000:1. If the operator is capable of determining the bubble position to an accuracy of 1 mm, the actual pressure can be read to an accuracy of $\pm 10^{-2}$ or $\pm 10^{-3}$ mmH_2O. Depending on the reference pressure, this device can then be used for the sensing of near-atmospheric, absolute or differential pressures. Its performance as a high-vacuum sensor is a function of the quality of the vacuum reference available.

The speed of response of this instrument is relatively slow because the smaller the bore of the capillary, the longer it takes for the bubble to assume its final position.

At the same time, small capillaries are desired to prevent fluid leakage between the bubble and the capillary.

Cartesian Diver Regulators

Cartesian divers are self-contained pressure, or vacuum, regulators operating on principles somewhat similar to those of the aneroid manostats discussed in Section 5.3. As shown in Figure 5.9i, the set pressure for this

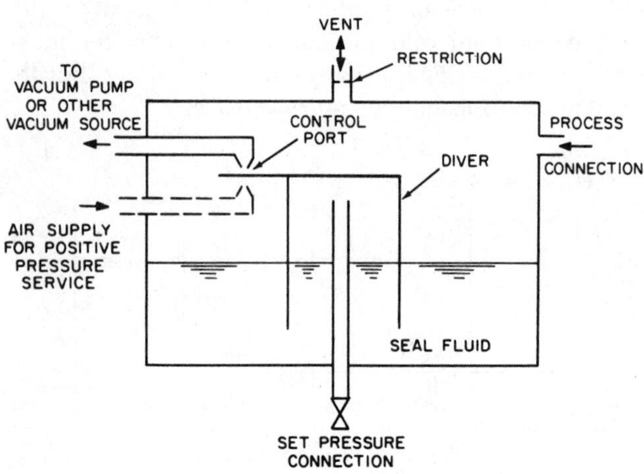

Fig. 5.9i Cartesian diver

controller is sealed in under the diver. The process pressure acts on the outside of the diver causing it to sink or rise as pressure varies. If the unit is to control a vacuum process, a vacuum pump (or other vacuum source) is connected to the unit. A process pressure increase causes the diver to sink, opening the control port and connecting the vacuum pump to the process to lower its pressure. If the unit is installed for positive pressure control, a pressure source, not a vacuum source, is connected to the control port, which is closed by an increase in process pressure and opened by its reduction.

This device is capable of maintaining process pressures between 1 mmHg absolute to 100 PSIA (0.13 kPa to 690 kPa) to an approximate inaccuracy of ± 0.1 percent of set-point. The unit is available in both glass and metal, requires no external power source, and is simple to operate or to change its set pressure. Because of its limited flow capacity, it can control small volume systems only.

BIBLIOGRAPHY

Burka, E.S., "Micromanometers," *Instruments and Control Systems*, September, 1964.

Comber, J. and Hockman, P., "Pressure Monitoring: What's Happening?" *Instruments and Control Systems*, April, 1980.

Hall, J., "Monitoring Pressure with Newer Technologies" *Instruments and Control Systems*, April, 1979.

Herceg, E.E., "Handbook of Measurement and Control," Schaevitz Engineering (Pennsauken NJ), 1972.

Meriam, J.B., "Manometers," *Instruments and Control Systems*, February, 1962.

Utterbach, N.G., "Reliable Submicron Pressure Readings with Capacitance Manometer," *Review of Scientific Instruments*, July, 1966.

Flow Sheet Symbol

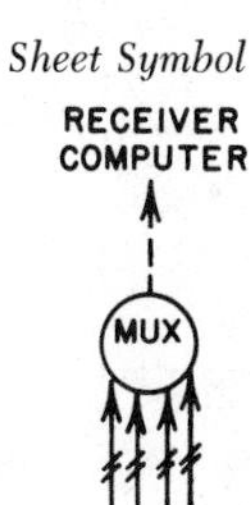

5.10 MULTIPLE PRESSURE SCANNERS

Design Variations: A—Rotary pressure scanners; B—high-speed pressure scanners; C—rotary air signal distributor; D—dedicated pressure multiplexers; E—air signal distributor manifold

Inaccuracy: (A) 0.1 to 0.2% of full-scale; (B) 0.25 to 0.5% of full-scale; (C) 0.25 to 0.5% of full-scale; (D) 0.15 to 0.3% of full-scale

Speed: (A) 0.1 to 0.2 sec. per point or 11 sec. to transduce all 64; (B) Scan 64 pressures in one second; (C) 3 sec. per point

Cost: (A) $4,000 or about $62 per point; (B) $7,500 or about $120 per point; (C) $3,000 or about $125 per point; (D) $2,400 or about $150 per point; (E) $1,500 or about $250 per point

Partial List of Suppliers: Computer Controls Corp. (E); Foxboro/ADEC Inc. (D); Scanivalve Corp. (A,B,C).

To computerize a pneumatically instrumented plant, a large number of analog pneumatic signals must be interfaced with this digital electronic device. Figure 5.10a illustrates two alternate methods of interfacing. In alternate A, pneumatic multiplexing is used, while in alternate B, the multiplexer is on the electronic side. Consequently, in alternate B a converter needs to be assigned to each pneumatic signal, while in alternate A the converter is shared between many pneumatic signals. Therefore, the same goal can be accomplished with a substantially lower investment, while providing better measure of inaacuracy (0.1 percent of full-scale), because a high-quality transducer can be justified on a time-shared basis.

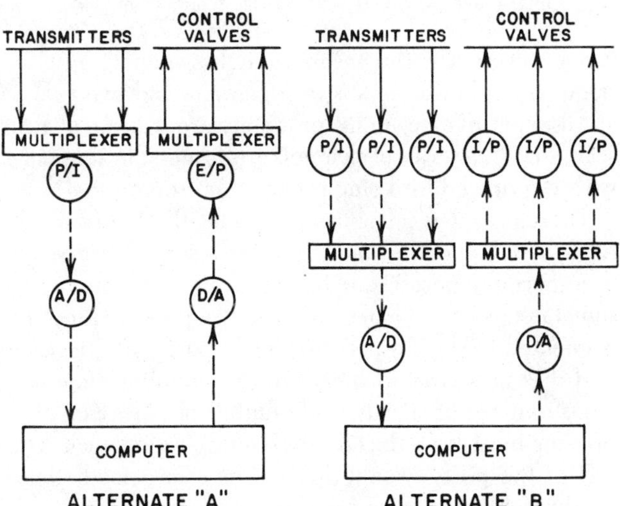

Fig. 5.10a Alternate methods of computer interfacing for pneumatic systems

Rotary Pressure Scanners

Figure 5.10b illustrates a rotary pneumatic selector with a 0.016 cm^3 switching volume. This is the volume of the rotating U-tube that connects the individual signals to the common transducer in the middle. This pneumatic multiplexer can scan 64 input signals, and if accurate zero or span reference signals are connected to some of the ports, the computer can also automatically recalibrate the system. Because of the small volume of the rotating U-tube, the response time is less than 50 ms, which theoretically allows a scanning speed of up to 20 samples per second.

The mechanical-drive mechanism is rated at six steps per sec. for continuous duty, in order to limit mainte-

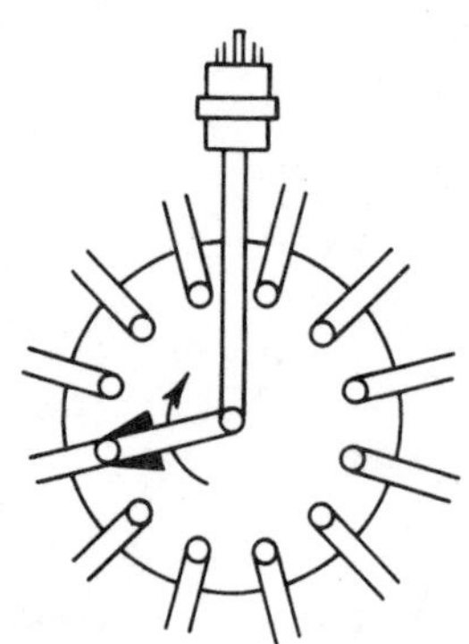

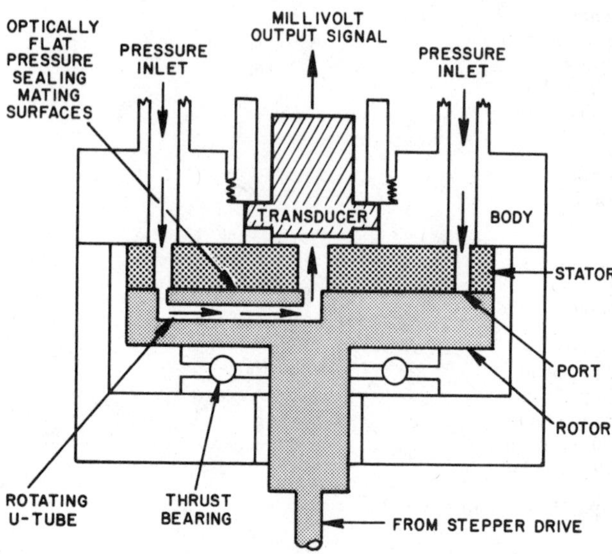

Fig. 5.10b Scanner module showing "traveling volume" between the various signal ports and the transducer membrane

nance. Presently the mean time between failure for pneumatic multiplexers is between one and two years.

All repetitive scanning programs are a form of ring scan; that is, a series of channels are scanned in sequence with the procedure being repeated on some time base.

Therefore, it is advantageous to let the hardware do the scheduling of the scan as the valve steps successively from port 0 to port 63 and back to 0. For example, if one signal needs to be "looked at" very frequently, it can be connected to several ports of the 64-port multiplexer.

If it is necessary to speed up the sampling time of a large number of signals, the number of ports per multiplexer must be reduced. Two multiplexers, each with half of the ports, would double the scan speed, three would treble it, and so on.

Other features of the rotary pressure multiplexers include a nine-track optical disc as a channel encoder. This provides a binary-coded computer address number for each of the pressure input ports and for the "Home" position. Another feature of these scanners is their capability to scan temperatures and pressures at the same address. This requires the addition of a thermocouple scanner module.

High-speed Pressure Scanners

In these scanners all pressure inputs are simultaneously compared to a reference pressure, which changes linearly with time as a linear ramp function. Each input signal will equal the moving reference pressure at different times, and these times will correspond to the signal pressures involved.

Therefore, this system does not use the conventional pressure transducer followed by an analog to digital converter. Instead, it uses 64 individual pressure switches and a precision single-slope pressure ramp that has a linear relationship between pressure and time. A binary counter is synchronized to this pressure ramp, providing a digital count proportional to pressure. The multiple pressure inputs are individually compared to this pressure ramp and the digitizing is accomplished as a function of the time at which each comparator trips its enable line. This enable line transfers the binary count into the appropriate memory cell. At the completion of each pressure ramp a 64-word memory will contain digital data representing the 64 pressures that are piped into the front panel of the system. The above ramp and update function is completed in less than 800 milliseconds. All or part of the stored data can be accessed, using the appropriate command(s). A two-page memory is used so that one page can be updated while the other is being asynchronously accessed by the host system.

Dedicated Pressure Multiplexers

In locations such as blow-down wind tunnels where even higher scan rates are required, or when the mean time between failures of mechanical multiplexers is considered to be excessive, as in direct control, dedicated pressure sensors can be used.

To add a high-quality transducer to each pneumatic signal tends to be too expensive. Therefore, if a microprocessor is available and 0.15 to 0.3 percent full-scale inaccuracy can be tolerated, mass-produced pressure sensors that have been partially temperature compensated are available in specialized lower cost packages.

One such assembly is available containing 16, 24 or 32 field replaceable solid-state pressure sensors, mounted on a printed circuit board. The basic element of the pressure sensor is an integrated circuit diaphragm that incorporates a piezoresistive sensor that changes resistance in proportion to the input pressure. The diaphragm is fabricated from a single silicon crystal on which the strain-sensitive elements arranged in the form of a four-active-arm Wheatstone bridge have been diffused.

The complete miniature module contains the pressure sensors, low-level multiplexer, instrumentation amplifier and a valving circuit that allows each sensor to be pneumatically switched from its input pressure to a common calibration manifold as often as the user desires. Channel selection is controlled by a four-bit binary address input to the assembly.

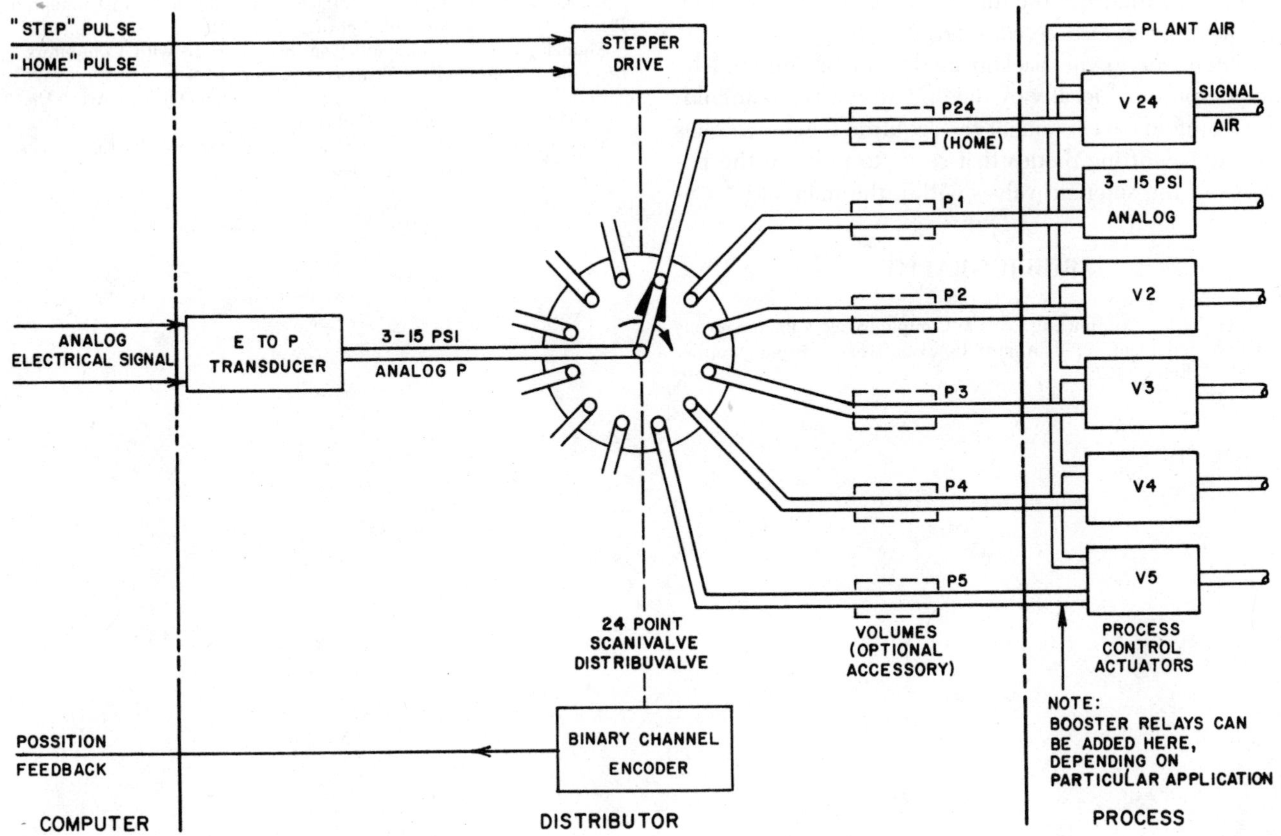

Fig. 5.10c Rotary signal distributor

Rotary Air Signal Distributors

As illustrated in Figure 5.10c, some 24 electronic analog signals can be converted to 3–15 PSIG (0.2–1.0 atmosphere) signals by time-sharing one transducer. Because the rotary distributor is remotely controlled by the computer, it can be installed in the field, near the actuators served. As a stepping motor sequentially connects the 24 outputs, an optical channel encoder provides the required channel position feedback for the computer. As shown in Figure 5.10c, it is desirable to equalize the volumes of the 24 output tubes and if some of the lines are long or if there is substantial leakage at the actuators, the addition of booster relays can also be justified, because the distributor will spend at least 3 sec. at each port. This means that with 24 outputs, the time between updates is one or two minutes.

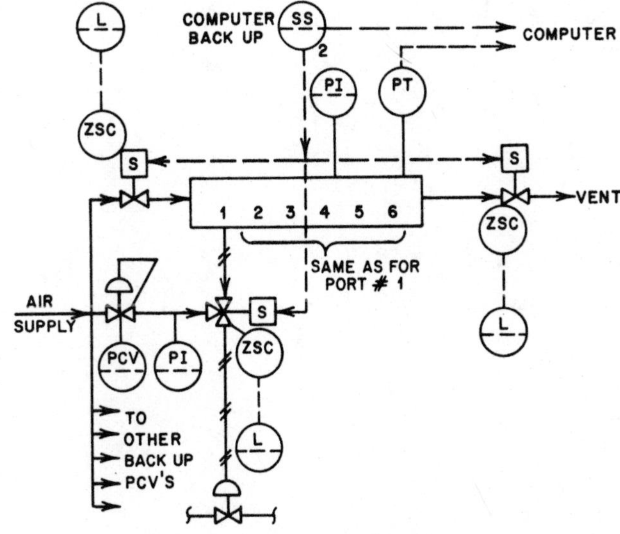

Fig. 5.10d Air signal distributing manifold

Air Signal Distributor Manifolds

If only a few pneumatic output signals are required or if a manual backup mode of operation is desired (when the computer is down), the unit illustrated in Figure 5.10d can be considered.

This multiplexer is capable of sequentially updating six (6) pneumatic output signals. Each line is sequentially pressurized or bled as required, while the solid-state transducer (PT) provides the needed feedback signal.

Similarly to Figure 5.10c, tube volume equalization and/or boosting can also be considered.

Switchover to the backup mode can be initiated by the computer or locally by the operator at the manifold. The manifold cover contains the required toggle switches and light-emitting diodes that serve to indicate the positions of the solenoid valves within the manifold.

BIBLIOGRAPHY

Baker, W.L., "Minimizing Costs in Computerizing a Pneumatically Instrumented Plant," 27th Annual ISA Conference, Paper Number 72–527, October, 1972.

Cooke, E.F., "Computer Interfaces in Pneumatically Instrumented Plants," *Oil & Gas Journal*, February 23, 1970.

Mamzic, C.L., "Pneumatic Controls Interface with Computers," *Instruments and Control Systems*, October, 1975.

Pemberton, J.C., "Pressure Scanning—State Of The Art," *Measurement and Data*, January/February, 1975.

Weyermuller, G.H., "Pneumatic Control Systems Play Key Process Role," *Chemical Processing*, October, 1973.

Flow Sheet Symbol

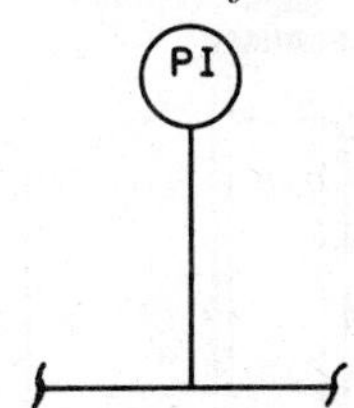

5.11 PRESSURE GAUGES

Design Pressure:	Up to 100,000 PSIG (690 MPa)
Design Temperature:	Sensing element normally kept at ambient by use of syphons or chemical seals
Materials of Construction:	See Sections 5.3, 5.4 and 5.5
Inaccuracy:	Between ±0.1 and ±5% of span
Range:	From 10 in. H_2O to 100,000 PSIG (from 2.5 kPa to 690 MPa) and 0.1 to 760 mmHg absolute (0.013 to 100 kPa)
Cost:	$10 to hundreds of dollars, depending on dial size and accuracy
Partial List of Suppliers:	Ametek Inc., U.S. Gauge Div., Ametek Controls Div.; Bailey Controls Co.; Babcock & Wilcox, a McDermott Co.; Barber-Colman Co.; Industrial Instruments Div.; Bristol-Babcock Div.; ACCO Industries, Inc.; Crosby Valve & Gage Co.; Dresser Industries; Dresser Ind. Valve & Instr. Oper.; Dwyer Instruments, Inc.; Dynamic Sciences; Dynisco; Ernst Gage Co.; Ernst, Eugene Products Co.; Fischer & Porter Co.; Foxboro Co.; Helicoid Gauge Div. ACCO; Hildebrandt Eng. Co.; Honeywell, Process Control Div.; Kent Process Control, Inc.; Marsh Instrument Co.; Marshalltown Instruments; Robertshaw Controls Co.; Ruska Instrument Corp.; Span Instruments Inc.; Taylor Instrument Co., Div. of Sybron Corp.; Wallace & Tiernan Div. of Pennwalt Corp.; Weiss, Alberta A & Sons Inc.; Weksler Instruments Corp.; Weston Instruments

This section will concentrate on the various features of pressure gauges. Because the bellows Bourdon, and diaphragm elements utilized in pressure gauges were discussed in sections 5.3, 5.4 and 5.5, the elastic elements and the gauge movements will not be included here. Accuracy has also been covered, so it is only noted here that large diameter precision indicators can be obtained with inaccuracies up to ±0.1 percent of span. Because readability can be a limiting factor on inaccuracy, only large-diameter dials are used for precision measurement. Dials are available with diameters ranging from 1 to 24 in. (25 to 600 mm).

The main parts of a pressure gauge are the retaining ring, glass lens, lens gasket, pointer, dial, movement, elastic element and the case.

Cases

Gauge cases are available in drawn steel, cast iron, aluminum, brass, polypropylene, or phenolic resin materials. Drawn steel represents the most economical selection, cast iron is generally used for ruggedness, brass is selected for decorative value and for antimagnetic applications, aluminum is utilized on limited-weight installations and phenolic resin or polypropylene case materials are used in corrosive atmospheres. Other materials, such as stainless steel, are available for special applications.

The standard mounting styles include back flange, front flange and turret types (see Figure 5.11a). The selection is based on the mounting requirements, such as surface, local or flush-mounted gauges.

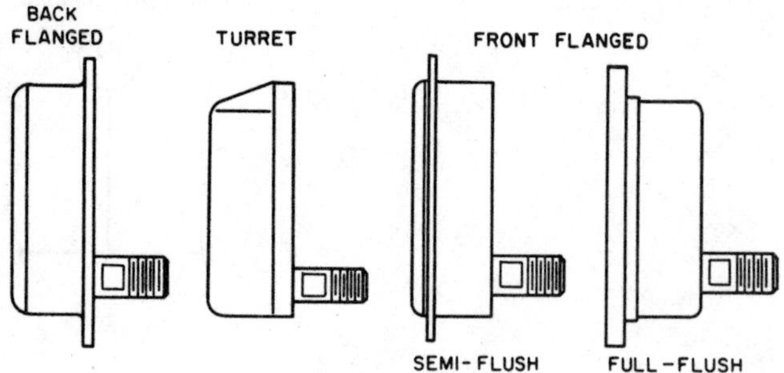

Fig. 5.11a Pressure gauge case design variations

The gasketed lens is held against the case front by rings. The ring can be attached to the case in several ways, including snap, friction, slip, threaded and hinged designs. The appearance of the gauge can also be affected by ring design variations, including peaked, flat-flared and beaded gauges.

The lens crystals are available in a variety of materials, including shatterproof, flat and beveled glass or cellulose acetate, acrylic and nonelectrostatic plastics.

The elastic pressure-sensing element in the gauge is exposed to the process pressure, and there is a possibility of rupturing this element due to overpressure. Cases are available with various degrees of safety. Almost all industrial gauges are provided at least with a blow-out disc that allows the internal pressure to escape. For severe or dangerous services, the solid-front design is recommended. Here, an integrally cast solid wall separates the dial and lens from the sensing elements, and a safety blow-out disc covers the entire back of the case (see Figure 5.11b). The disc is gasketed and spring mounted so that it releases internal pressures as soon as they reach about 1 PSIG (6.9 kPa). As an additional convenience, the full-area disc provides quick access to the element and gauge movement for inspection, recalibration or repair without need to remove the dial.

The process connections of most gauges are either ¼ or ½ in. (6.3 or 12.5 mm) in size and are located on

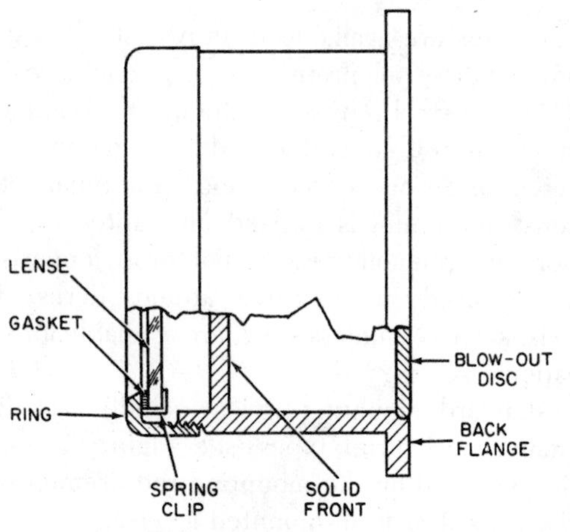

Fig. 5.11b Safety gauge case

either the back or the bottom of the case. Special connection sizes and locations are also available.

Dials and Pointers

Gauge dials are made of steel, brass, aluminum, solid or laminated plastic with black graduations on white background or in other color combinations. Numerals are placed at the major graduations. The number of minor graduations usually conforms to gauge inaccuracy. For example, a 1,000 PSIG (6.9 MPa) gauge with ±1 percent inaccuracy will have minor graduations at 20 PSI (138 kPa) intervals. On precision gauges, the major graduations are placed on the inner circle and the subdivisions on the outer circle for easier reading.

To eliminate parallax error, one of the following designs can be considered: raised dial, use of split pointer or use of mirror ring.

For improved readability, gauges are available with 360 or 720° dials. For example, on an 8½ in. (212.5 mm) dial with two revolutions corresponding to full-scale reading, the actual scale length is 45 in. (1,125 mm). Another common method of improving readability is the use of suppressed scales, which instead of being zero-based, are selected to cover ±50 percent of the average operating pressure.

The pointer adjustment methods are a function of gauge quality. On economy designs, the pointer is fixed, so that the hub has to be lifted from the shaft for adjustment and then repositioned. The standard industrial gauges are furnished with adjustable pointers with friction fit between pointer and shaft. Adjustment is made by turning the hub while the shaft is firmly held. Precision or test gauges are furnished with micrometer pointer adjustment.

When it is desired to have a permanent indication of the highest pressure reached, a *maximum* pointer can be added to the gauge. This pointer is useful, for example, for indicating the maximum pressure reached before a burst disc or a safety valve relieved the system. The maximum pointer is manually reset by the operator. On some installations it is desirable to provide a set pointer, which is permanently positioned for the operator's information at the normal operating pressure desired.

Special Features

When it is desired to read the pressure from a considerable distance, the gauge dial can be illuminated. These dials are normally jet black with white numerals and graduations. The translucent glass dial is illuminated by a number of electric lamps arranged in the case to eliminate all shadows. The lamps can provide standard white or ultraviolet (black) light, and the pointers and graduations can be phosphorescent.

On high-precision gauges, it is necessary to compensate for ambient temperature variations. Temperature changes affect the modulus of elasticity of the sensing element and also cause thermal growth of the linkage members. Compensation is achieved by inserting a bimetallic bar into the linkage between the elastic element and the gauge movement.

When it is desired to indicate two related process pressures on the same dial, duplex gauges can be used. These consist of two independent sensing element, movement, and pointer assemblies. When the differential between two pressures is of interest, a single pointer is available. Differential gauges are available with predetermined or interchangeable high- and low-pressure connections to read differentials in both directions. When severe vibrations exist, the gearless gauges can be considered (see Figure 5.11c).

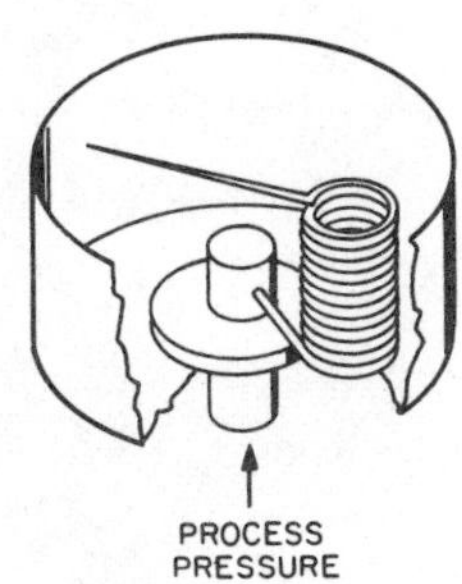

Fig. 5.11c Gearless gauge for heavy vibration service

Diaphragm Vacuum Gauges

The diaphragm sensors discussed in Section 5.5 are also used as elements in direct-reading vacuum gauges. The diaphragm gauge indicates the pressure difference between the process vacuum and the fully evacuated reference chamber. A thin, copper beryllium diaphragm isolates the reference chamber from the process, and the diaphragm movement is a measure of absolute process pressure. Mechanical linkage is used to trasmit diaphragm motion to scale readout. The measurement is independent of atmospheric pressure variations. The scale is nonlinear and is expanded for precise reading below 50 mmHg. (6.7 kPa) absolute. Above 50 mmHg, the scale permits rough readings only. The full-scale range is 0.1 to 760 mmHg. (0.013 to 100 kPa) on a scale length of approximately 8 in. (200 mm). The unit can tolerate overpressures up to 7 PSIG (48 kPa).

When the diaphragm vacuum gauge is considered for corrosive service, a purge flange accessory is recommended for installation between the gauge and the process. A small, controlled quantity of atmospheric air (or other protective gas) is fed through the purge flange into the process. The purge flow is adjusted to displace all corrosive gases and to prevent their diffusion into the gauge. Care should be taken in selecting the proper dimensions, so that the pressure drop through the purge flange will not affect the reading accuracy.

BIBLIOGRAPHY

Buftenmyer, W.L., "Selecting Bourdon-Tube Gauges," *Instruments and Control Systems*, February, 1961.

Comber, J. and Hockman, P., "Pressure Monitoring: What's Happening?" *Instruments and Control Systems*, April, 1980.

Hall, J., "Monitoring Pressure with Newer Technologies," *Instruments and Control Systems*, April, 1979.

Moore, R.L., "Basic Instrumentation Lecture Notes and Study Guide—Measurement Fundamentals," Instrument Society of America, 1976.

"Precision Pressure Gauges," *Instruments and Control Systems*, June, 1961.

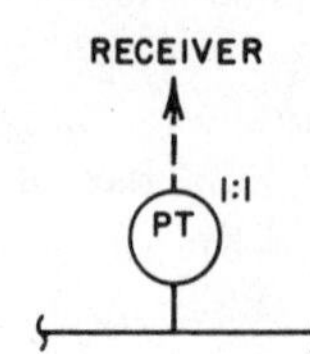

5.12 PRESSURE REPEATERS

Design Variations:	(A)—Force balance pressure repeaters; (B)—Conventional repeaters; (C)—Button diaphragm repeaters
Range:	(A, B) Full-vacuum to 100 PSIG (690 kPa); (C) 1,500 to 10,000 PSIG (10–70 MPa)
Inaccuracy:	(A) ±1 to 2 in. H_2O (0.25–0.5 kPa); (B) ±1 to 5 in. H_2O (0.25–1.25 kPa); (C) ± 1 percent of full-scale
Limitations:	(A, B) Limited by vacuum or air supply; (B) subject to air supply overpressure damage; (C) suited for high pressures only
Costs:	(B) Under \$500; (A, C) \$500 to \$1,500, depending on materials of construction
Partial List of Suppliers:	Fisher Controls (B); Foxboro Co. (A, B,); Moore Products Co. (A, B); Rosemount, Inc. (C).

When detecting the level in pressurized tanks, the low-pressure side of the level transmitter is exposed to the vapor pressure above the process liquid. On hard-to-handle materials, one of the methods used is installation of pressure repeaters. These devices develop an air output pressure equal to the vapor pressure on the process side.

Force Balance Pressure Repeaters

Although pressure repeaters are not d/p transmitters, they are used as one of the important elements in differential pressure level detection loops. The error in the repeated output pressure increases as the pressure to be repeated rises. At a pressure level of 40 PSIG (276 kPa), the error is 1 to 2 in., (25 to 50 mm) while at a pressure of 400 PSIG (2.7 MPa), the error is ±10 to 20 in. H_2O (±2.5 to 5 kPa).

Both the flat and the extended diaphragm differential pressure transmitters are available as repeaters, and they can be used to repeat both positive and vacuum pressures. The basic design features for these units are the same as those of their transmitter versions.

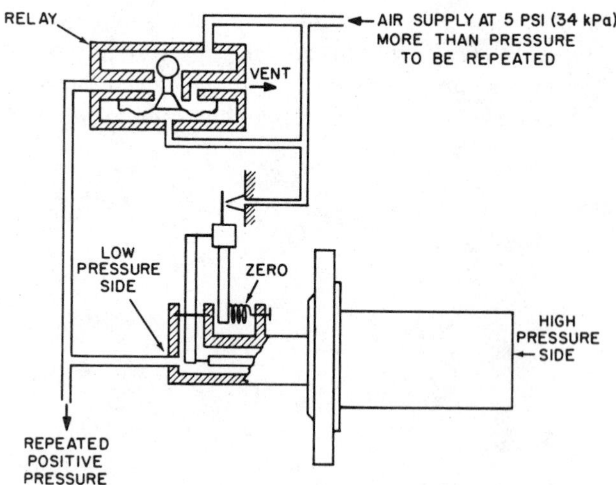

Fig. 5.12a Extended diaphragm positive pressure repeater

Figure 5.12a illustrates the operation of the positive pressure repeater, which can reproduce process pressures up to 100 PSIG (690 kPa) with high accuracy. To repeat higher pressures, a high-pressure relay, which can be obtained on special order, is required. This device is very similar to the transmitting unit, except that the feedback from the relay is piped to the low-pressure side of the d/p cell instead of to the feedback bellows, which have been eliminated. The relay itself is modified for the higher operating pressure and the capsule filling is different, but otherwise the repeater is identical to the transmitter.

The operation of the unit is as follows: When the vapor pressure in the tank increases, it is detected by the high side of the diaphragm capsule, transmitting additional force to the force bar. This in turn causes the nozzle to be covered by the flapper, generating an increased back pressure to the relay. The relay responds to this by restricting the vent connection and opening the air supply to the relay outlet. This increased relay output pressure is then sent to the low-pressure side of the d/p cell diaphragm capsule. When the pressures on the high and low side are equal, the flapper is repositioned in relation to the nozzle in its throttling band and remains in that position until the process pressure changes again. The air signal that is sent to the low-pressure side of the capsule is at the same time the air pressure equivalent of the tank vapor pressure that is used to compensate the level transmitter in the liquid phase.

When it is desired to repeat vacuum instead of positive pressures, two techniques can be employed. One involves the use of a suppression spring, the other an external vacuum source.

The range depression (or suppression) spring is illustrated in Figure 5.12b. If the suppression spring is set for 16.5 PSI (114 kPa), the output signal from the repeater will be the sum of the pressure sensed by the high-pressure side of the capsule and the setting of the suppression spring. In other words, the output signal from the repeater will be 16.5 PSI higher than the pressure of the process vapors. If the process is at atmospheric pressure, the output is 16.5 PSIG, and if it is full vacuum, the output is 1.5 PSIG (10 kPa). By the use of this bias, the need for a vacuum reference source is removed. This bias is *removed* in the d/p level transmitter by setting the same amount on *its* suppression spring.

When a reference vacuum source is available which

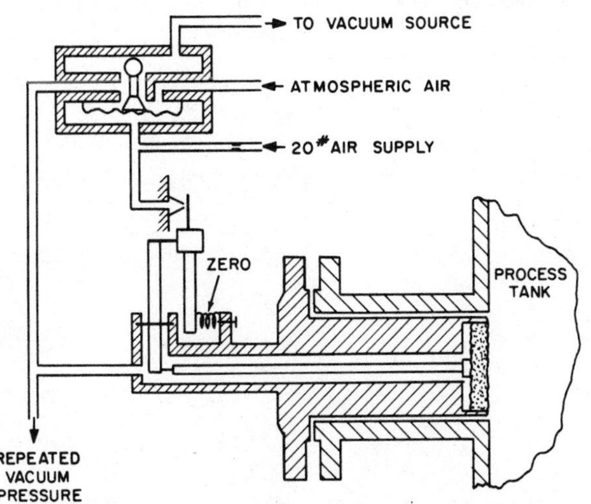

Fig. 5.12b Extended repeater for vacuum service

is at least 2 in. of water column lower than the minimum absolute pressure to be repeated, the design illustrated in Figure 5.12b can be considered. This unit incorporates a special vacuum relay. When the absolute pressure in the tank increases, a force is exerted on the force bar, which in turn tends to draw the flapper away from the nozzle, causing a reduction in the pressure on the relay diaphragm. This moves the ball down, closing off the vacuum supply and opening the connection to the atmosphere. The net effect is an increase in the relay output pressure, which is piped to the low-pressure side of the diaphragm capsule to balance the equal pressure on the process side. When the absolute pressure of the process decreases, the opposite sequence of events takes place; namely, the force bar tends to cover the nozzle, which increases the pressure on the relay diaphragm, pushing the ball up, thereby closing the connection to the atmosphere and opening it to the vacuum source. The net effect is a lowering of the relay output pressure which then matches the lower process pressure on the other side of the capsule.

As noted, pressure repeaters are available for most vapor pressure compensation applications. Their accuracy is sufficient for the average level detection loop between full vacuum and 100 to 150 PSIG (690 to 1,000 kPa) positive pressures. Therefore, in this range these units merit consideration when the process vapors are hard to handle. At higher pressures, the lower accuracy of the repeater has to be balanced against the advantages of having clean air on the low-pressure side of the level transmitter.

Conventional Repeaters

Figure 5.12c illustrates a simpler repeater design. This unit is submerged in the vessel so that the static head of the liquid exerts an upward force on the diaphragm, which increases as the level rises. This force is opposed by the supply air pressure on the other side of the diaphragm. The force caused by the rising level moves the diaphragm toward a bleed orifice, restricting the flow to atmosphere and causing the air pressure to build up until it equals the static head pressure. When the forces on the two sides of the diaphragm equal each other, the unit is in equilibrium. The speed of response of the unit is changed by an adjustable restriction, which if opened, will increase sensitivity by allowing more air to flow onto the diaphragm. The air supply to the unit can be regulated at a pressure slightly in excess of the maximum hydraulic head to be repeated.

Button Diaphragm Repeaters

These instruments have been developed for the plastic extrusion and synthetic fiber industries, in which high pressures and temperatures are encountered together with a need to eliminate dead-ended cavities to prevent plugging.

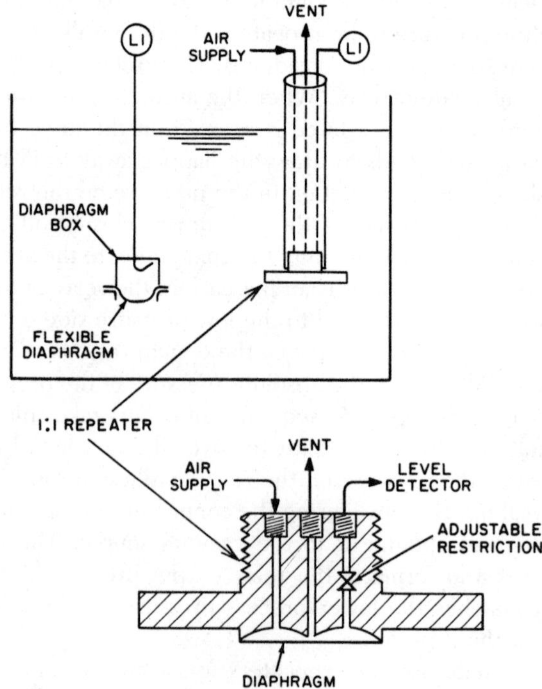

Fig. 5.12c Diaphragm devices for continuous detection of liquid levels

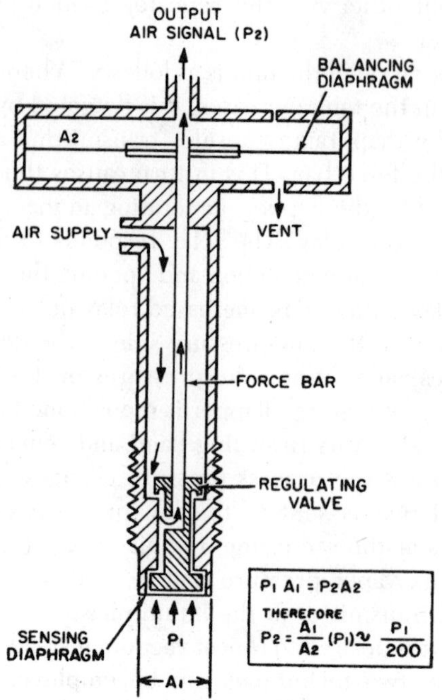

Fig. 5.12d Button diaphragm repeater

Figure 5.12d illustrates a button diaphragm that operates on a direct force balance principle. Basically, this is a pressure repeater that generates an air signal that is about 1/200th of the process pressure. The main components of the unit are the sensing and balancing diaphragms connected by the force bar and the regulating valve. The process pressure acting on the sensing diaphragm is in force balance with the air output acting on the balancing diaphragm. The air output is proportional to the process pressure by the inverse ratio of the two diaphragm areas. When the process pressure increases, the measuring diaphragm raises the force bar, thereby opening the regulating valve to the air supply. This causes a corresponding increase in the air output signal, which rebalances the system.

The unit is available with ranges from 0–1,500 PSIG to 0–10,000 PSIG (0–10 MPa to (0–69 MPa) and can be used on process temperatures up to 800°F (430°C). The temperature effect is relatively small, ±0.5 percent of span per 100°F (56°C) change, and inaccuracy is ±½–1 percent of full-scale. The wetted parts are made of 316 stainless steel, and the button diaphragm requires a process connection size of only ½ in. (12.5 mm).

BIBLIOGRAPHY

Elliott, T.C., "Temperature Pressure, Level, Flow—Key Measurements in Power and Process," *Power*, September, 1975.

Herceg, E.E., "Handbook of Measurement and Control," Schaevitz Engineering (Pennsauken NJ), 1972.

Moore, R.L., "Basic Instrumentation Lecture Notes and Study Guide—Measurement Fundamentals," Instrument Society of America, 1976.

Soisson, H.E., *"Instrumentation in Industry,"* John Wiley and Sons (New York), 1975.

Flow Sheet Symbol

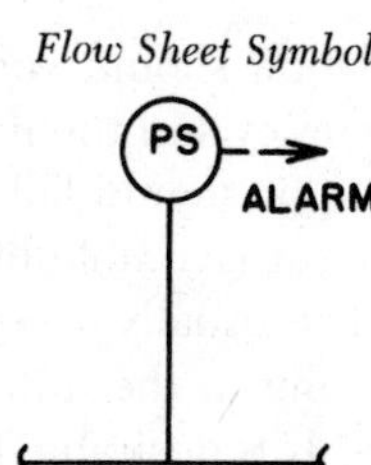

5.13 PRESSURE SWITCHES

Design Pressure:	Up to 20,000 PSIG (138 MPa)
Design Temperature:	Sensing element usually kept at ambient by use of syphons or chemical seals (−65 to 165°F) (−54 to +74°C)
Materials of Construction:	See Sections 5.3, 5.4 and 5.5
Inaccuracy:	Between ±½ and ±3 percent of span
Range:	0.1 in. H_2O to 12,000 PSIG (25 Pa to 83 MPa)
Cost:	$35 to $450, depending on quality and features
Partial List of Suppliers:	W. E. Anderson, Inc.; Automatic Switch Co.; Barksdale Valves; Custom Component Switches, Inc.; DeLaval Turbine, Inc.; Barksdale Control Div.; Dwyer Instruments, Inc.; Fairchild Industrial Products; Instruments, Inc.; International Controls Corp.; Mercoid Corp.; United Electric Controls Co.

Pressure switches are used to energize and de-energize electrical circuits as a function of the relationship between the process pressure and a predetermined set-point. They are available to detect absolute, compound, gauge and differential pressures, with inaccuracies up to ±½ percent of span. The sensing elements are mostly the elastic types discussed in Sections 5.3, 5.4 and 5.5. In addition, some designs depend on sealed piston or Belleville disc springs for their operation. The electric switching assemblies are either snap-acting, mechanical microswitches or mercury switches. The latter contain no mechanical moving parts and have to be mounted level and free of vibrations.

Figure 5.13a illustrates some of the terminology used in connection with pressure switches. The pressure range within which the actuation point can be set is called adjustable range. The switch may actuate at its set-point on rising or falling pressure as distinguished in the sketch. The set-point is the pressure that actuates the switch to open or close an electric circuit. The set-point accuracy defines the band within which repetitive actuations will occur. Differential or dead-band is the difference between set-point and reactuation point. For example, if a switch is set to actuate (close) at 100 PSIG (690 kPa) on rising pressure, it will close at that point, and when the pressure drops, it may not open again until the pressure has fallen to 95 PSIG (656 kPa). Here, the differential is 5 PSI (35 kPa). Tolerance is the repeatable accuracy of the reactuation point.

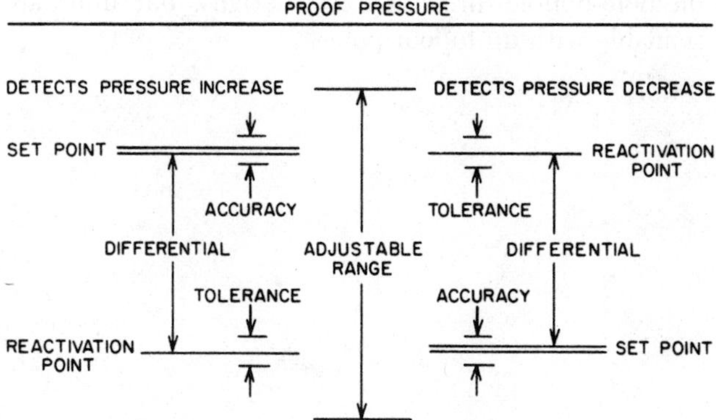

Fig. 5.13a Pressure switch terminology

Pressure switch elements should be selected on the basis of service life and proof pressures. Most elastic elements will have a service life of close to a million cycles, if the cycle time is not less than 5 seconds. If longer service life or higher cycle frequencies are required, metal fatigue tends to limit the usefulness of elastic elements and special designs, such as sealed piston elements, should be considered.

Selection of the adjustable range for a specific installation should consider both the set-point actuation accuracy and the life factor. For greatest accuracy, the set-point should fall in the upper half of the range, but for most favorable life factor, it should be in the lower half. The usually acceptable compromise is to locate the set-point in the middle third of the range.

It is desirable to have an external calibrated knob provided on the pressure switch for set-point adjustment. Uncalibrated or internal set-point adjustments are generally undesirable on industrial installations. The "fixed differential" pressure switches are furnished with a single adjustment for set-point. These units are factory set with close differentials amounting to ½ to 1 percent of span. On double-adjustment designs, both set- and reactuation points can be independently adjusted. The maximum differential in such designs is the range of the switch, while the minimum varies between 2 and 8 percent of span. Pressure switches with dual control are also available. Here, two independent switches mounted in the same housing are responding to the same process pressure in opening or closing two independent circuits.

The electrical rating of pressure switches at 115-volt operating level varies from 0.3 to 10 amp. on ac or dc circuits. Generally, the dual control and the fixed differential switches have the lower, and the double-adjustment units the higher ratings. The available circuit arrangements are very flexible. Some of the standard arrangements include the single-pole-single-throw (SPST), single-pole-double-throw (SPDT), and the double-pole-double-throw (DPDT) designs, but units are available with up to four poles.

The case variations for pressure switches are similar to those for pressure gauges, but here the electrical area classification also needs to be considered. There are three standard case designs: the general purpose (NEMA 1), the weather resistant (NEMA 2 and 3), and the explosion proof (NEMA 7) cases.

In some designs, the control mechanism is an integral part of the explosion-proof case, while in others, it can be removed in the field for maintenance. As noted earlier, the set-point should be externally adjustable so that the explosion-proof case does not need to be opened in order to effect a set-point change. For added convenience of the operator, some pressure switches also provide continuous process pressure indication.

BIBLIOGRAPHY

Elliott, T.C., "Temperature, Pressure, Level, Flow—Key Measurements in Power and Process," *Power,* September, 1975.

Godfrey, P.E., "Choosing the Right Pressure Switch," *Instruments and Control Systems,* April, 1979.

Harvey, G.F., "ISA Transducer Compendium," Instrument Society of America, 1969.

Herceg, E.E., "Handbook of Measurement and Control," Schaevitz Engineering (Pennsauken, NJ), 1972.

Moore, R.L., "Basic Instrumentation Lecture Notes and Study Guide—Measurement Fundamentals," Instrument Society of America, 1976.

Rhodes, T.J., *"Industrial Fundamentals for Measurement and Control,"* McGraw-Hill Book Co. (New York), 1972.

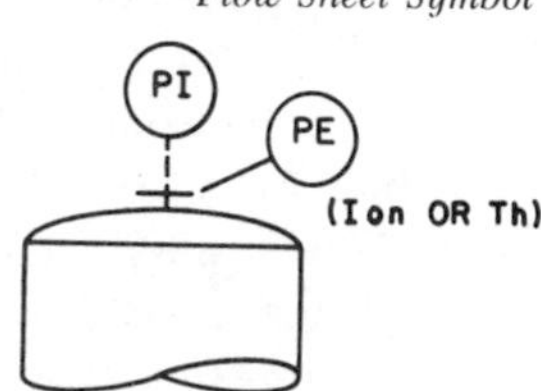

5.14 VACUUM SENSORS

Design Variations:	A—Ionization A1—Hot cathode A2—Cold cathode B—Thermal B1—Thermocouple B2—Thermopile B3—Resistance wire C—Mechanical C1—McLeod C2—Molecular momentum C3—Capacitance
Range:	(A1) 10^{-3} to 10^{-11} mmHg (0.13 to 1.3×10^{-9} Pa) with 6 ranges; (A2) 10^{-1} to 10^{-7} mmHg (13 to 1.3×10^{-5} Pa) with 3 ranges; (B1) 5 to 10^{-3} mmHg (650 to 0.13 Pa); (B2, C2) 20 to 10^{-3} mmHg (2,600 to 0.13 Pa); (C1) 10 to 10^{-6} mmHg (1,300 to 1.3×10^{-4} Pa)
Inaccuracy:	(A1) ±10 percent of reading at 10^{-5} mmHg (1.3×10^{-3} Pa); (A2) ±20 percent of reading at 10^{-5} mmHg (1.3×10^{-3} Pa); (B) ±2 percent to ±10 percent; (C) ±20 percent in the micron range; (C1) ±10 percent; (C2) ±5 to ±25 percent (higher error percent at higher vacuums)
Cost:	(A2, B1, B2, C1, C2) \$300 to \$600; (A1) \$1,000
Partial List of Suppliers:	Cooke Vacuum Products, Inc. (A,B,C); Edwards High Vacuum, Inc. (A,B,C); Electron Technology, Inc. (A); Frederiks Co. (A,B); General Electric Co., Industrial Sales Div. (A,B,C); Manostat Corp. (C); MKS Instruments, Inc. (C); Teledyne, Hastings-Raydist, Inc. (B); Vacuum Instrument Corp. (A); Varian Associates (A,C); Veeco Instruments, Inc. (A,B)

Ionization Vacuum Sensors

The ionization vacuum detectors have been available since 1916 and their detectable pressure level has been gradually lowered from 10^{-7} to 10^{-13} mmHg (10^{-5} to 10^{-11} Pa) since then. Some of the major improvements included the Bayard-Alpert modification, which reduced the effect of extraneous X-ray current produced in the tube. The various designs of ionization gauges can be grouped by the method used in generating the electrons, the number of which is related to the absolute pressure detected. These are the hot cathode and the cold cathode.

Hot Cathode Ionization Gauges

All ionization vacuum gauges detect the electric current that is generated from ionization of the gas whose pressure is being measured. They are distinguished by the method applied in producing the ions. To convert a gas molecule into a positive ion, an electron must be removed from it. This is done by supplying the molecule with energy equal to its ionization potential. The approximate energy level involved is 5 to 30 electron volts. If this energy is supplied at a constant rate, the ions will be produced at a similarly constant rate. The ion current produced is proportional to the pressure of the gas.

In the hot filament vacuum gauge illustrated in Figure 5.14a, the ionization energy is supplied by electron bombardment. The electrons are derived from thermionic emission from the hot filament. The electrons acquire their kinetic energy in being attracted to and passing through the grid. When they collide with the gas molecule from the vacuum system, positive ions are produced. These ions are then attracted to the negatively charged collector plate to form an ion current. At constant accelerating voltage, the number of ions formed is proportional to the gas pressure if it is below 10^{-3} mmHg (0.13 Pa). At higher pressures, the relationship between plate current and pressure is not linear because the mean free path becomes so short that an ionized molecule may pick up a free electron to become a molecule once more.

The accuracy of these devices is lower than that of the average instrument because the physical amounts being measured are very small. At an absolute pressure of 10^{-5} mmHg (1.3×10^{-3} Pa) inaccuracy would be about ± 10 percent. The vacuum range detectable by the hot cathode gauge is 10^{-3} to 10^{-11} mmHg (1.3×10^{-1} to 1.3×10^{-9} Pa). The minimum span of the readout device is one decade, and it can be furnished with five or six ranges. Range switching is by manual selector switch on the front of the instrument. The sensitivity of the hot cathode gauge is 100 microamperes per micron (10^{-3} mmHg) pressure. The readout device can be combined with thermocouple readout to extend its coverage to vacuums in the range of 1 to 10^{-3} mmHg (133 to 0.13 Pa). A protective relay circuit is also furnished to turn off the hot filament at pressures higher than 10^{-3} mmHg to protect against burn-out. Besides multirange indicators, the readout device can also record or actuate electrical control and alarm circuits. When several ionization sensors are involved, a switching console can be provided to monitor the detectors and thereby reduce system cost.

The hot filament vacuum detector provides the widest pressure reading range with fast response and fair accuracy. Its application is limited to gases that will not decompose on the hot filament. The ionization tube, by its nature, is subject to mechanical damage, and, therefore, is used mostly in laboratories and in pilot plants.

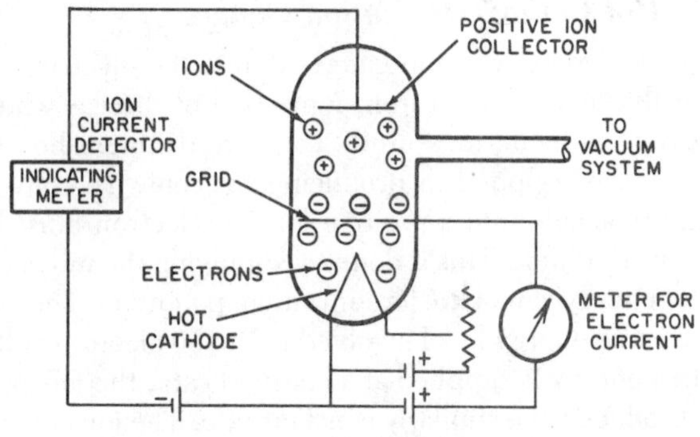

Fig. 5.14a Schematics for hot cathode ionization vacuum gauge

The filament current is controlled in such a way that a constant flow of electrons is emitted from it. In some designs, calibration is maintained by controlling the grid charge so that if emissivity of the filament is decreasing, the grid receives a correspondingly greater charge to maintain the entire circuit in equilibrium.

Cold Cathode Ionization Gauges

The cold cathode vacuum gauges are also called Philips detectors after their first manufacturer. The operating principle had been first outlined by F. M. Penning in 1937. The basic difference between cold and hot cathode gauges is in the method by which ions are produced. In the hot filament unit, the electrons are derived from thermionic emission. In the cold cathode design, the electrons are withdrawn from the cathode surface by a high-potential field. Because the rate of electron emission is lower in the cold cathode units, the collision frequency between gas molecules and electrons would also be lower if the electrons traveled in a straight path. To increase the path length of the electrons, a magnetic field is created around the tube to deflect the electrons. Thereby, the electrons emanating from the cathode are caused to spiral as they move across a magnetic field to the anode. This spiraling action greatly increases the path of travel; thus, the chance of collision with gas molecules is increased. The overall result is greater sensitivity of the cold cathode gauge than that of the hot cathode.

The inaccuracy of the cold cathode unit is about ± 20 percent at an absolute pressure of 10^{-5} mmHg (1.3×10^{-3} Pa). The detectable vacuum range is from 10^{-2} to 10^{-7} mmHg (1.3 to 1.3×10^{-5} Pa) and the minimum span is one decade. Readout devices are available with one, two or three ranges, which are selected by the manual switch at the front of the instrument. The sensitivity is 5 milliamperes per micron pressure. The available features and variations or combinations of the available readout devices are the same as those for hot cathode sensors.

The advantages of this design in comparison with the hot cathode tube include lower cost and more rugged basic design, while the drawbacks are lower accuracy, lower detectable range and nonlinear output at pressures below 10^{-4} mmHg (1.3×10^{-2} Pa). Both designs are sensitive to gauge tube contamination.

In recent years a combination ionization gauge has been designed. This gauge is basically a cold cathode unit, but is furnished with a hot cathode serving to "trigger" the discharge from the cold cathode. This design extends the detectable pressure range to 10 decades or from 10^{-4} to 10^{-14} mmHg (1.3×10^{-2} to 1.3×10^{-12}) absolute.

Thermal Vacuum Detectors

The thermal conductivity of a gas is a function of the gas pressure. Therefore, if a heated element is placed in the gas with a constant power input, the resulting surface temperature will be a function of the heat conductance of the gas, which is related to its pressure. Thermal vacuum gauges consist of three basic elements: a heater, a temperature sensor and a compensator for process temperature variations. There are two basic designs, depending on the type of temperature sensor used: the thermocouple and the resistance wire vacuum detectors. In general, the thermal vacuum detectors are limited in being able to measure down to 10^{-3} mmHg (0.13 Pa) absolute pressures. This coverage can be extended with special designs involving special amplifiers and liquid nitrogen cooling around the gauge tube to reduce radiation losses.

Thermocouple Vacuum Gauges

As shown in Figure 5.14b, the single thermocouple detector consists of a wire heated by the passage of constant ac or dc current. A thermocouple is welded to the center of this heated filament, thereby providing means to measure the temperature of the filament directly. The top of the tube is left open to be connected into the vacuum system being measured.

In operation, the constant current passing through the heater wire is in the order of 20 to 200 milliamperes, and the thermocouple sensor develops a full-scale output in the order of 20 millivolts dc. For any constant value of current through the filament, the heater wire temperature increases as the pressure in the tube is reduced. The temperature detected by the thermocouple depends on the thermal conductivity of the gas surrounding the junction. For the same gas, thermal conductivity is a measure of pressure. A sensitive millivolt meter or a potentiometer provides an indication of thermocouple output. The most common range is 1 to 10^{-3} mmHg (133 to 0.13 Pa) on a logarithmic output.

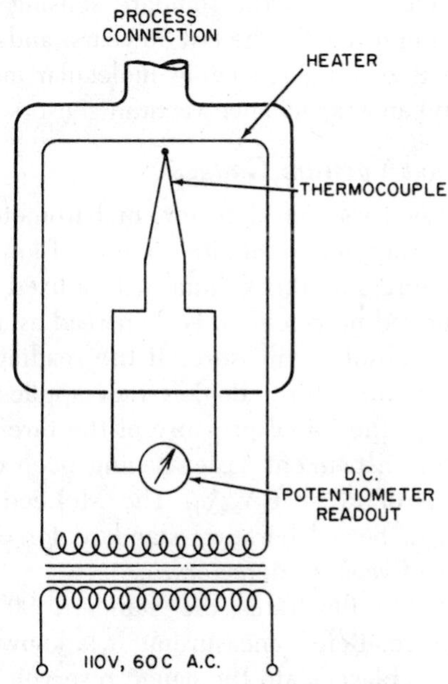

Fig. 5.14b Single thermocouple vacuum gauge

The filament temperature is kept below 400°F (205°C) to reduce the possibility of the sample gases decomposing or forming a deposit on the gauge tube elements. Compensation is provided for process temperature variations, which otherwise would introduce an error by affecting the filament temperature. This is achieved by the use of a reference gauge that has been fully evacuated and sealed. The same current and voltage are applied to both the measuring and the reference gauges. The temperature of the two heater wires is then compared, and the difference is used as a measure of process vacuum. The reference tube compensates for ambient temperature changes because the two cells are at the same temperature.

The available readout devices are as varied as are the receiving instruments to measure thermocouple output signals. They include plain indication to recording, on-off or throttling controllers, analog or digital readouts, and others. At higher ranges, such as 5 to 10^{-3} mmHg (665 to 0.13 Pa), dual scales are usually used, while for the standard range a single scale is sufficient. The reading inaccuracy at midscale is ±2 percent while ±10 percent covers the full range. On multistation installations, a single readout device can serve several thermocouple gauges through a manual selector switch. Because all filaments are on at the same time, the readings are instantaneous.

The detection tubes can be of one-piece metal construction rugged enough to withstand 150 PSIG (1,035 kPa) pressure and to be handled with a wrench.

Each thermocouple vacuum detector must be calibrated for the specific gas involved, so that the thermal conductivity of the particular process material is taken into consideration.

It is important that the filament current and voltage be carefully maintained for stable calibration. Filaments should not become dull or tarnished due to contamination because this would cause radiation losses affecting calibration. One way to overcome this problem is to precoat the filaments so that further contamination would have no effect. Unfortunately, in thermocouple gauges precoating tends to impair sensitivity. This is the main reason why the thermopile designs discussed below have been developed: several precoated thermocouples connected in series provide the extra potential output, and, therefore, sensitivity that has been lost as a result of precoating.

Thermopile Vacuum Gauges

To increase sensitivity, thermopiles (several thermocouples) can be used to detect heater temperature. This design is shown in Figure 5.14c. In this design, the thermocouples (A & B) are heated by low-voltage alternating current; therefore, the heater and the temperature-sensing functions are combined in the same noble (noncorrosive) metal thermopile. A change in process pressure results in a change of thermopile temperature

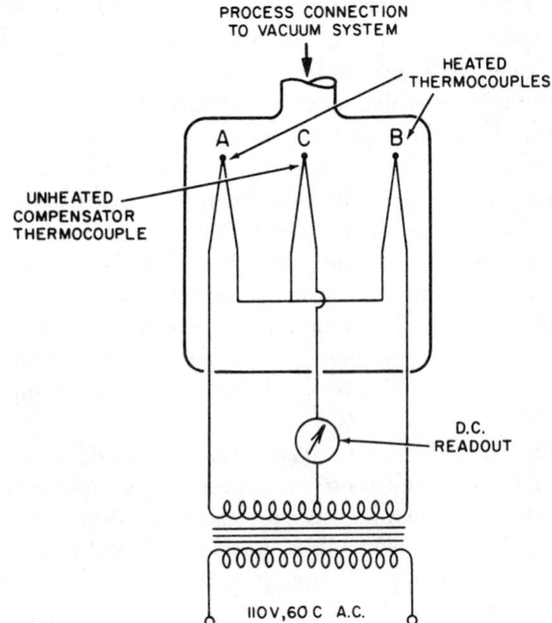

Fig. 5.14c Multiple thermocouple (thermopile) vacuum gauge

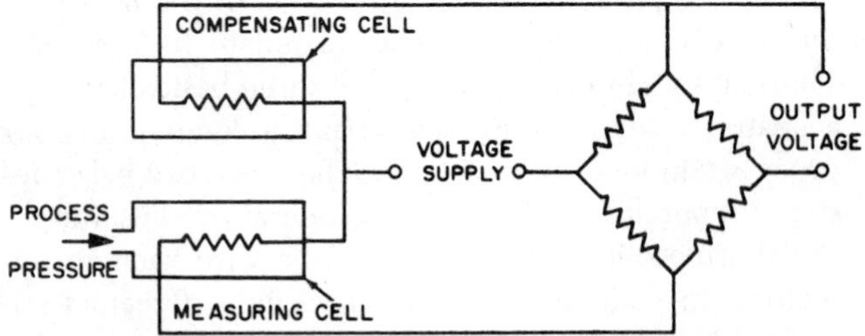

Fig. 5.14d Resistance wire (Pirani) vacuum detector

causing a new dc output from the thermocouples. A third unheated thermocouple (C) is included in the circuit to compensate for operating process temperature variations. This couple is the same size as the heated ones, but it is connected in opposite polarity. A change in process temperature develops voltages in all the thermocouples, but the transient effects are equal and opposite in the heated and unheated elements. Therefore, compensation is achieved.

The features of and accessories for the single thermocouple design also apply here. In addition, the following should be noted.

Because of the noble metals used in the thermopile, oxidation of the couples does not occur. The operating temperature of the heated thermopile is lower than that of the hot filament used in the single couple design. Therefore, the probability of sample gas decomposition or deposit formation is remote. Gauge tubes are available to withstand several thousand PSIG overpressures. The most accurate readings are obtained in the range of 10^{-1} to 10^{-3} mmHg (13 to 0.13 Pa), but extended coarse detection is feasible over a broader range up to 100 mmHg (13 kPa) and down to 10^{-4} mmHg (1.3×10^{-2} Pa) absolute pressures.

Resistance Wire Vacuum Gauges

The resistance wire, or Pirani, vacuum detector shown in Figure 5.14d operates on the principle that heat loss from a resistance wire filament carrying a constant electric current is related to the gas pressure surrounding the filament if the process temperature is constant. The Pirani gauge uses the same heating principle as the thermocouple detectors, but the temperature is sensed by the change in the resistance of the wire. As in the thermopile detectors, the heating and sensing elements are combined. Voltage is applied to the resistance wire, which is part of a Wheatstone bridge, causing self-heating of the resistance element.

A second resistance wire, which is enclosed in a reference vacuum, is used to compensate for process temperature variations. The readout device detects the amount of current or voltage that is necessary to return the Wheatstone bridge to balance after a change in the vacuum being measured.

The standard pressure-sensing range for this unit is between 10^{-3} and 1 mmHg (0.13 to 133 Pa) absolute. The inaccuracy of the gauge is about ± 2 percent at the calibration pressure and ± 10 percent over the operating range. Maximum output signal is about 0.1 milliamperes at full-scale reading.

Other features, accessories and readout devices are the same as those for thermocouple vacuum detectors.

Mechanical Vacuum Gauges

This section will cover high-vacuum detectors that operate on principles other than ionization or thermal conductivity. These include the well-known McLeod gauge, which has been the standard sensing device in the vacuum industry for the last 50 years, and some less frequently used designs, such as molecular momentum (or viscosity) and capacitance vacuum sensors.

McLeod Vacuum Gauges

The McLeod vacuum detector, or barometer gauge, operates by trapping a definite volume of low-pressure gas and compressing this volume with a fixed pressure. The pressurized new volume is then read as a measure of the initial absolute pressure. If the reading is taken at constant temperature, Boyle's Law applies: $V_1P_1 = V_2P_2$. Here P_1 the initial pressure of the rarefied gas is given on the instrument scale, having been calibrated on the basis of $P_1 = P_2V_2/V_1$. The McLeod gauge is considered to be a laboratory standard for calibrating other types of vacuum detectors.

Figure 5.14e illustrates the standard 90° rotation McLeod gauge. Before measurement, a known volume of mercury is placed into the gauge reservoir as shown

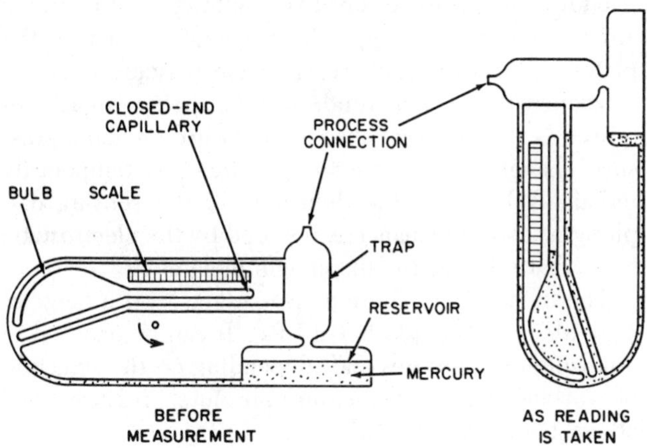

Fig. 5.14e 90° rotation McLeod vacuum gauge

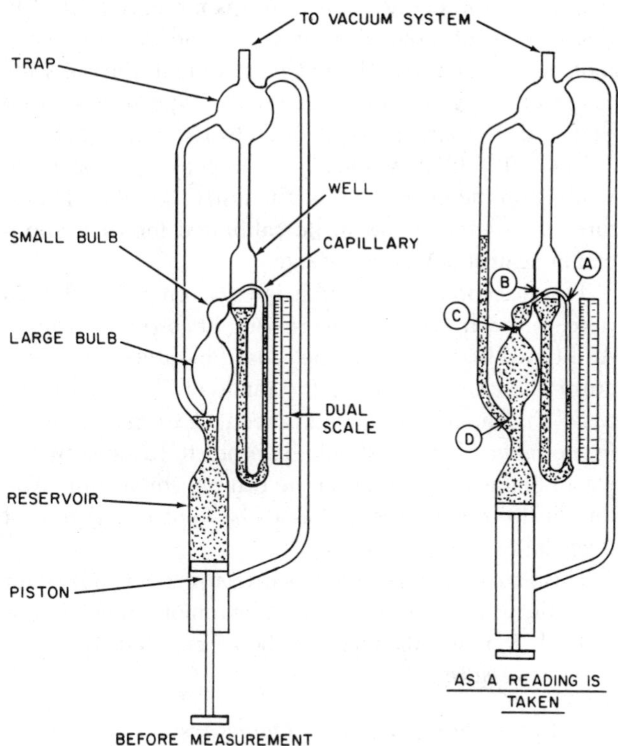

Fig. 5.14f Piston McLeod vacuum gauge

at the left side of the sketch. Then the gauge is connected to the vacuum system to be detected and is rotated 90°, thereby trapping a fixed volume of rarefied gas (the volume of the bulb and capillary), which is compressed by the column of mercury shown at the right side of the sketch. The level to which the mercury rises in the capillary is read on the scale that has been calibrated per Boyle's Law to read in units of initial pressure. This device has no moving parts so friction, inertia or hysteresis does not affect the measurement, therefore making it as basic a measuring instrument as a manometer. Assuming constant temperature and proper quantity of filling mercury, the only limitation of the unit is due to capillary effects. Below 10^{-4} mmHg, accuracy is affected because it is hard to obtain a flat mercury surface free from capillary effects even at the highest practical compression ratio.

Figure 5.14f shows an improved version of the McLeod gauge. In this design, the unit is stationary, and a piston is used to trap in the rarefied gas. One of the advantages of this unit is that it does not use a dead-ended capillary, and, therefore, the problems associated with keeping the capillary clean are eliminated. At the left side of the sketch, the unit is shown just prior to the taking of a measurement. As the piston with micrometer adjustment is moved up, the rarefied gas is trapped when the mercury reaches poing D; thus, the initial volume (V_1) is the volume between points A and D. The filling of the gauge with mercury is done through the process connection. It runs down into the well, up the capillary and syphons over through the bulbs into the reservoir. When the syphon breaks, the mercury level is at point A. When the instrument is connected to the vacuum system, the mercury level in the reservoir is below point D to allow trapped gases to be liberated. As the piston is moved up, the mercury fills thelarge bulb up to point C. Here a reading can be taken on the dual scale if the pressure to be detected is in the mmHg range. If the vacuum is higher, the piston is moved further up, increasing the compression ratio until the mercury reaches point B above the small bulb. In this case, the reading is taken on the micron side of the dual scale.

McLeod gauges can cover the vacuum range between 1 and 10^{-6} mmHg (133 and 1.3×10^{-4} Pa). At pressures below 10^{-4} mmHg (1.3×10^{-2} Pa), the reading accuracy is limited by capillary effects. This device is a laboratory instrument that measures on a "sampling" rather than on a continuous basis, but it is also available for industrial installations where a rugged case is required. No remote readout devices are available. Condensation as a result of compression of the gases can contribute to inaccuracy; therefore, it is desirable to condense and trap the condensible vapors before they are admitted into the gauge.

Molecular Momentum Vacuum Gauges

In these gauges there are two basic working parts: a rotating and a restrained cylinder. The gas molecules coming in contact with the rotating cylinder (at a constant speed of 3,600 rpm) are set in motion in the direction of rotation. These molecules, having obtained their energy from contact with the moving cylinder, then strike the restrained cylinder, transferring energy to it. This force moves the restrained cylinder a distance proportional to the energy transferred, which in turn is a function of the number of gas molecules in that space. The number of molecules is related to the absolute pressure of the gas. The pointer attached to the restrained cylinder indicates the gas pressure on the scale.

Because the energy transferred is not only related to the number of molecules (pressure) and the velocity of molecules, but also to themolecular weight, the full-scale range of the gauge depends on the type of gas to be detected. For air, the range is 20 to 10^{-3} mmHg (2.7 to 1.3×10^{-4} kPa), while for hydrogen, the maximum reading on the instrument is 280 mmHg (37 kPa). Therefore, this detector has to be calibrated for each application against a McLeod gauge.

The inaccuracy of the unit is between ± 5 and ± 25 percent, with accuracy decreasing at lower pressures. Additional inaccuracy can be caused by process temperature variations which, in the range of 50 to 100°F (28 to 56°C), can amount to ± 2 percent. External vibration in the range of 50 cps should be protected against by the use of bellows couplings on the process connection. The sample from the process has to be kept clean, free of dust, oil or other particles.

This gauge gives continuous direct readout, but is not available as a signal transmitter for remote indication or control. It is not damaged by being exposed to atmospheric pressure.

Capacitance Vacuum Detectors

These devices, also called micromanometers, are differential pressure detectors using a flexible diaphragm in an electric capacitance circuit. One plate of the capacitor is the pressure-sensitive diaphragm in the bridge circuit. A dc balancing voltage connected across the plates of the capacitor exerts an electrostatic force for rebalancing. The magnitude of rebalancing voltage required is a measure of absolute pressure. When a pressure change occurs, the bridge circuit is temporarily unbalanced due to the deflection of the sensing diaphragm, and the balance is restored by the electrostatic force that opposes the diaphragm movement.

The device gives continuous readings over a range of 1 to 10^{-3} mmHg (133 to 0.13 Pa). It can be exposed to atmospheric pressures and, depending on the selection of working parts, can measure absolute pressures well above 1 mmHg (133 Pa).

BIBLIOGRAPHY

Brombacher, W.G., "40 Years of Precise Pressure Measurement," *Instruments and Control Systems*, September, 1967.

Comber, J. and Hockman, P., "Pressure Monitoring: What's Happening?" *Instruments and Control Systems*, April, 1980.

Elliott, T.C., "Temperature, Pressure, Level, Flow—Key Measurements in Power and Process," *Power*, September, 1975.

Hall, John, "Measuring Negative Pressures," *Instruments and Control Systems*, April, 1980.

Harvey, G.F., "ISA Transducer Compendium," Instrument Society of America, 1969.

Chapter VI

DENSITY MEASUREMENT

C. H. Hoeppner
B. G. Lipták

CONTENTS OF CHAPTER VI

TABLE VI
ORIENTATION TABLE FOR DENSITY SENSORS

Section Number	*LIQUID* *Density Sensor Design*	*Applicable to* *Clean Process Streams*	*Applicable to* *Slurry Service*	*Applicable to* *Viscous or Polymer Streams*	*Minimum Span Based on Water SG = 1.0*	*Inaccuracy (% span)*	*Design Pressure and Temperature Limitations* (PSIG/°F)*	*Temperature Compensation Available*	*Direct Local Indicator*	*Transmitter*
6.2	Angular Position Type	√			0.1	0.5	1000/500	N.S.		√
6.3	Ball Type	√			Digital	0.1	5000/300		√	√
6.4	Capacitance Type	√	√	√	0.1	2	1000/200	√		√
6.5	Displacement Type									
	Buoyant Force Displacer	√			0.005	1	600/400	N.S.		√
	Chain Balance Float	√			0.01	1	125/400	√	√	√
	Electromagnetic Suspension	√			0.01	1	200/350	√		√
6.6	Fluid Dynamic Type	√			0.1	2	100/200		√	
6.7	Hydrometers	√			0.1	1	100/200	√	√	√
6.8	Hydrostatic Head Type	√	√	√	0.05	0.5-1	5000/350	N.S.	√	√
6.9	Oscillating Fork Deflection	√			0.05	2	1000/500	√		√
6.10	Radiation Type	√	√	√	0.05	1	Unlimited	√		√
6.11	Sound Velocity Type	√		√	0.2	5	50/200	√		√
6.12	Torsional Vibration	√			0.05	1	500/200			√
6.13	Vibrating Plate Type	√			0.1	1	500/200			√
6.14	Vibrating Spool Type	√			0.1	1	200/200			√
6.15	Vibrating Tube Type	√			0.05	1–3	1000/300	√		√
6.16	Weight of Fixed Volume Type	√			0.05	1	2400/500	√	√	√

Section Number	*GAS* *Density Sensor Design*	*Minimum Span Based on Air SG = 1.0 or in lbm/ft³ Units**	*Inaccuracy (% span)*	*Design Pressure and Temperature Limitations* (PSIG/°F)*	*Manually Operated Indicator*	*Continuous Indicator*	*Transmitter*
6.17	Sensors Operating at Actual Flowing Conditions						
	Centrifugal Type	1.0 lbm/ft³	0.1-0.5	2000/300		√	√
	Displacement Type	1.0 lbm/ft³	0.25	1500/200	√	√	√
6.18	Specific Gravity Detectors Operating at Near-Ambient Conditions			Near Ambient			
	Electromagnetic Suspension Type	0.01 SG	1				√
	Gas Column Balance Type	0.25 SG	0.002 SG			√	√
	Manual Displacement Type Elements	0.02 SG	1			√	
	Thermal Type	0.5 SG	0.01 SG				√
	Viscous Drag Type	0.1 SG	1-2			√	√

N.S.: Non-standard
*For SI units, see Appendix A·1

6.1 APPLICATION AND SELECTION

The density of a substance is one of its important characteristics, used to provide information concerning composition, concentration, mass flow and, in fuels, caloric content. Determination of specific gravity relative to air yields information on stream composition; the measurement is performed under near-ambient conditions. Direct density measurement at operating conditions eliminates the need for separate pressure, temperature, supercompressibility or humidity measurements when the ultimate purpose is to determine mass flow. It should be noted that all "Mass Flowmeters" (discussed in Chapter II) can be used as densitometers, if the volumetric flow rate through them is kept constant.

The reader, by referring to Orientation Table VI, will be able to determine service applicability such as clean, slurry, or polymer services, compensation capability for temperature variations, span and accuracy limitations, pressure and temperature limitations, and applicability for manual or continuous, local or remote measurements.

Instruments for the measurement of the density of solids are covered under "Solids Flowmeters." Solids are generally measured for density by weighing them suspended in air and then weighing them suspended in a liquid of known density. Density is given by:

$$D = \frac{W_a D_L - W_L D_a}{W_a - W_L}$$

Where:

D_a = Density of air
D_L = Density of the liquid
W_a = Weight in air
W_L = Weight in liquid

Alternately, a bin is filled with the solid material to a known volume and the whole mass is weighed.

Liquid Density Detectors

Liquids are measured in terms of their density and often expressed in terms of specific gravity, which is the ratio of the density of the measured substance to the density of water. The compressibility of most liquids is slight and the effects of pressure upon density measurement generally may be disregarded.

In this chapter, a number of liquid density gauges are described. When a selection is to be made for a particular installation, several considerations should influence the decision.

If the purpose of measurement is to obtain a transmitted signal, the most economical selections are the hydraulic head and the displacement-type sensors. If only local indication is required, then the above noted units and the hydrometers represent the least expensive choice.

The inaccuracy of these detectors is expressed as percent full span and, therefore, the narrower the range of the sensor the higher the measurement precision will be. The accuracies of practically all devices listed are acceptable for industrial applications, but where operating temperatures vary, it is necessary to compensate for these changes. All density detectors discussed can be furnished with some means of temperature compensation, but this feature is standard only for some designs and requires additional components in others as noted in Table VI.

Most density gauges are limited in their use to clean, non-viscous fluids. The selection for these applications can be based on economics and accuracy. If the process fluid is viscous or of the slurry type, then only the radiation, the hydrostatic head and the U- or straight-tube sensors can be considered. Of these three types, the U- or straight-tube gauges are seriously limited in their pressure and temperature ratings and will handle only moderately viscous or slurry-type streams. With the hydrostatic head-type measurement, it is difficult to detect narrow spans because the corresponding height of the required standpipe becomes excessive. In addition, on heavy slurry or viscous services, the operating pressure is limited by the rating of the pressure repeater or by the pressure at which purge media are available. This is the case because the low pressure side of the differential pressure detector is likely to be plugged if piped into the process without protection.

It should be fairly clear from the above discussion that a combination of slurry or high viscosity process with high operating pressures necessitates the use of radiation-type sensors.

Density can also be detected indirectly through the measurement of some other process property. Measurement of boiling point elevation is one of the common methods of indirect density gauging. Here resistance elements compare the temperature of the boiling process sample with that of boiling water at the same pressure. The differential temperature scale for a particular solution can be calibrated in terms of density. The above method is also used for end-point determination in evaporators.

Gas Density Detectors

The effect of pressure on gas density cannot be disregarded. Because gases are compressible materials, the same detector will seldom be able to detect their specific gravity (or molecular weight) and their density at operating conditions. For this reason, two families of gas sensors have been developed:

a. Stream composition determination, and
b. Mass flow rate calculation from orifice or volumetric flow data.

The first requires comparison of sample gas density to air at ambient conditions while the second necessitates the measurement of process gas density at operating conditions. Sensors have been developed for both of these measurements, and the two families of detectors will be discussed in separate sections.

When the purpose of measurement is stream composition determination, the sample gas gravity is detected at ambient conditions, disregarding the actual operating pressures and temperatures. The scales of these gauges are calibrated either in specific gravity units based on air, or in molecular weight units. The two scales are interchangeable, because:

$$SG = \frac{M_{wg}}{M_{wa}} \qquad 6.1(1)$$

This can be proven by considering that:

$$SG = \frac{\rho_g}{\rho_a} = \frac{P_g Z_a R_a T_a}{P_a Z_g R_g T_g} \qquad 6.1(2)$$

But the pressures and temperatures are the same:

$$SG = \frac{Z_a R_a}{Z_g R_g} \qquad 6.1(3)$$

and,

$$\frac{R_a}{R_g} = \frac{M_{wg}}{M_{wa}} \qquad 6.1(4)$$

therefore,

$$SG = \frac{Z_a\, M_{wg}}{Z_g\, M_{wa}} \qquad 6.1(5)$$

For gases (substantially superheater vapors), the ideal gas equations are quite accurate at ambient conditions, and, therefore, the compressibility factors can be considered as unity.

$$SG = \frac{M_{wg}}{M_{wa}} \qquad 6.1(6)$$

Where:

SG = Specific gravity based on air = 1.0
ρ_g, ρ_a = Gas and air densities respectively (lbm/ft.³)
P_g, P_a = Gas and air pressures respectively (lbf/in.² abs.)
Z_g, Z_a = Gas and air compressibility factor
R_g, R_a = Gas and air engineering gas constant (ft-lbf/lbm °R)
T_g, T_a = Gas and air temperature (°R)
M_{wg}, M_{wa} = Gas and air molecular weight

When the purpose of measurement is mass flow rate determination, the gas density is detected under actual operating conditions. The relationship between gas properties and density is as follows:

$$\rho_g = \frac{M_{wg}\, P_g}{10.73\, T_g\, Z_g} \qquad 6.1(7)$$

Volumetric flow rate at standard conditions is established by dividing the mass flow rate by the specific gravity. If the pressure drop across an orifice plate and the density at operating conditions are measured, then the mass and volumetric flow rates can both be determined using the following equations:

$$W = C_1 \sqrt{\Delta P\, \rho_g} \quad \text{and} \quad Q = \frac{C_2 \sqrt{\Delta P\, \rho_g}}{SG} \qquad 6.1(8) \text{ and } 6.1(9)$$

Where:

W = Mass flow rate (lbm/hr)
ΔP = Differential pressure (lbf/in.²)
C_1, C_2 = Constants containing appropriate orifice and other factors
ρ_g = Gas density (lbm/ft³)
Q = Volumetric flow rate (ft³/hr)

Density sensors which are capable of measuring the gas at operating conditions can also be used to determine the compressibility factor of the vapor. Supercompressibility is a measure of the *deviation* between ideal and actual gas behavior and is defined as the *ratio* between the actual specific weight and the theoretical based on the perfect gas law. If the gas pressure, temperature, specific gravity and flowing density are measured, then the supercompressibility factor is calculated as follows:

$$Z = \frac{P \times SG}{\rho . T . R_{air}} \qquad 6.1(10)$$

Where:

Z = Supercompressibility factor (dimensionless)
P = Operating pressure (lbf/in.^{2})
SG = Specific gravity (dimensionless)
ρ = Flowing density (lbm/ft^3)
T = Absolute temperature (°R)
R_{air} = Gas constant for air 53.3 (ft-lbf/lbm °R)

When selecting gas and vapor density gauges, the application engineer should first determine the purpose of the installation. If it is for composition or concentration gauging, then the information desired is specific gravity or molecular weight, which can be obtained by measuring the sample under near-ambient conditions. If the purpose of the measurement is direct density detection under operating conditions for mass flow rate determination, then the instruments described in Section 6.17 should be evaluated. The important difference between these two groups is that in case of mass flow rate determination there is no need for separate pressure, temperature, supercompressibility or humidity measurements, while when gas composition is the purpose of the installation, these variables must be taken into consideration and their effects compensated for.

In the category of gas-specific gravity sensors (composition detectors), economics would favor the displacement-type, manually operated indicators. If continuous indication or remote signal transmission is required, the gas column balance and the viscous drag-type instruments will be the most economic selections. When the overriding consideration is high sensitivity and accuracy, the electromagnetic suspension-type sensors will satisfy that requirement. If corrosive gas samples are involved and the sensing elements should not come in contact with the process stream, the thermal gauges can be considered. All of these units operate at near-ambient conditions and, therefore, in case of non-ambient samples it is necessary to regulate their pressures and temperatures and to establish the moisture content of both the sample and the reference gas, which is usually air.

If the purpose of measurement is direct density detection under operating conditions, the most economical choice is the manually operated displacement indicator. Continuous indication or remote readout can be provided by both the displacement and the centrifugal-type designs. Both of these units are accurate and capable of withstanding high operating pressures and temperatures. The most important consideration is to provide the gas sample under operating conditions without loss of pressure or temperature in the sampling system. This is achieved by close-coupling the instruments to the process and by insulating the sample lines. Both designs include some restrictions and moving parts which should be protected from deposits by filtering the sample.

Units and Definitions

Density is defined as the quantity of matter per unit volume. The most common unit is grams per cubic centimeter but the units of pounds per cubic foot are also used.

Specific gravity is defined as the ratio between the density of a process material to that of water or air at specified conditions. Being a ratio, this property has no units associated with it.

Both of these values characterize the same physical property of the process media and they have meaning only if defined at stated temperature levels. In case of specific gravity, the temperatures might be different for the process and the reference fluid, which is acceptable if clearly stated. For example, a specific gravity table might list a process fluid as having $0.87^{80/40}$ specific gravity, which means that this liquid at 80°F (27°C) will have a density of 0.87 times that of water at 40°F (4.4°C).

In cases of gases, the specific gravity is normally based on standard conditions meaning that both the process vapors and the reference air density are measured at 60°F (16°C) and atmospheric pressures. For perfect gases, the ratio of molecular weights yields the same data.

Some of the commonly used specific gravity units are defined below for the reader's assistance in interpreting the various standard scales.

Scale in *Baumé* degrees for light liquids including ammonia:

$$\frac{140}{SG\ @60°F} - 130 = °Be \qquad 6.1(11)$$

Scale in *Baumé* degrees for heavy liquids including acids:

$$145 - \frac{145}{SG\ @\ 60°F} = °Be \qquad 6.1(12)$$

Scale in *A.P.I.* hydrometer degrees for petroleum products:

$$\frac{141.5}{SG\ @\ 60°F} - 131.5 = °API \qquad 6.1(13)$$

Scale in *Sikes, Richter or Tralles* degrees reads volume percent of ethyl alcohol in water.

Scale in *Quevenne* degrees is used in the milk industry. Each degree is 1/1000th of a SG unit above 1.000. (40°Q means SG = 1.040) or (SG − 1.000) × 1000 = °Q.

Table 6.1a
DENSITY CONVERSION TABLE

SG	Pounds per Gallon	°Be	°API	% Proof	Vol. % of Alcohol	°Be	°Bk	Degrees Brix or Balling	°T_w
0.6087	5.066	100	100.96						
0.6364	5.296	90	90.86						
0.6667	5.549	80	80.75						
0.7000	5.827	70	70.64						
0.7368	6.134	60	60.54						
0.7778	6.475	50	50.43					LIQUIDS HEAVIER THAN WATER	
0.7955	6.623	46	46.39	199.36	99.68				
0.8235	6.857	40	40.32	186.00	93.00				
0.8750	7.286	30	30.21	151.72	75.86				
0.9333	7.772	20	20.11	101.00	50.50				
1.0000	8.328	10	10.00	0.00	0.00				
1.000	8.328					0	0	0	0
1.007	8.385					1	7	1.75	1.4
1.036	8.625					5	36	9.00	7.2
1.074	8.945					10	74	18.00	14.8
1.115	9.289					15	115	27.00	23.0
1.160	9.660					20		36.20	32.0
1.208	10.063					25		45.40	41.6
1.261	10.501					30		55.00	52.2
1.318	10.978		LIQUIDS LIGHTER THAN WATER			35		64.66	63.6
1.381	11.501					40		74.72	76.2
1.450	12.076					45		85.00	90.0
1.526	12.711					50		95.84	105.2
1.543	12.849					51		98.14	108.6
1.611	13.417					55			122.2
1.706	14.207					60			141.2
1.813	15.095					65			162.6
1.933	16.101					70			186.5
2.000	16.656					72.5			200.0

Values from Bureau of Standards Circular C-410

Scale in *Twaddell* degrees is applied to fluids heavier than water. Each degree is 5/1000th of a SG unit above 1.000. (10°Tw means SG = 1.050) or (SG − 1.000) × 200 = °Tw.

Scale in *Brix* degrees is common of the sugar industry representing percent sugar by weight in 60°F water solution. (°Br)

Scale in *Balling* degrees is typical of the brewing industry and indicates weight percent of dissolved solids in 60°F water. (°Ba)

Scale in *Barkometer* degrees is used in the tanning industry. Each degree is 1/1000th of a SG unit above or below 1.000. (+10°Bk means SG = 1.010) or (SG − 1.000) × 1000 = °Bk.

Proof means twice the volumetric percentage of ethyl alcohol in water.

Table 6.1a gives density conversion for fluids that are heavier and also for those that are lighter than water.

BIBLIOGRAPHY

Note: This bibliography applies to all sections of Chapter VI.

Agar, J., "Vibrating Spool Densitometer." *Instruments and Control Systems*, January, 1970.

Cameron, D., "An Instrument for Measurement of Liquid Density." *Industrial Electronics*, March, 1967.

Cook, H.L., "Slurry Measurement and Control." *ISA Journal*, June, 1964.

Frost & Sullivan, Inc., "On Stream Process Analyzer Market." *Report No. 669*, August, 1979.

Greene, G.Jr., "Measure and Sell Gas by the Pound." *Pipeline Industry*, January, 1967.

Hough-Grassby, A.W., "The Evaluation of Process Analyzers." *Advances in Instrumentation*, Volume 34, ISA, 1979.

Jutila, J.M., "On-Stream Analyzers." *InTech*, October, 1980.

Lipták, B.G., "Instruments to Measure Slurries." *Chemical Engineering*, February, 1967.

November, M.H., "Electronic Density Measuring System." ISA 1973 ASI 73207 (19–24).

November, M.H., "Measuring Fluid Density and Specific Weight." *Instrumentation Insight*, Volume 2, Number 3, August, 1975.

Plache, K.D., "Coriolis/Gyroscopic Flow Meter." ASME Publication 77-WA-M4.

Puzniak, T.J., "Analyzers—The Key to Advanced Control." Proceedings of the 1980 ISA Symposium on Analysis.

Smith, B.W., "Radioisotope Gauging in the Mining Industry." *The Canadian Mining and Metallurgical Bulletin*, January, 1964.

Torrance, J.W., "LPG Mass Flow Using a Densitometer of Resonating Twin Tube Design." 1976 ISA Conference, Paper No. 76-852.

Yokogawa, T., "Vibration Type Liquid Density Measuring System." A publication of Yokogawa Electric Works, Tokyo, Japan.

Zacharias, E.M., "The Sonic Interface Detector." *Oil & Gas Journal*, July 8, 1970.

Flow Sheet Symbol

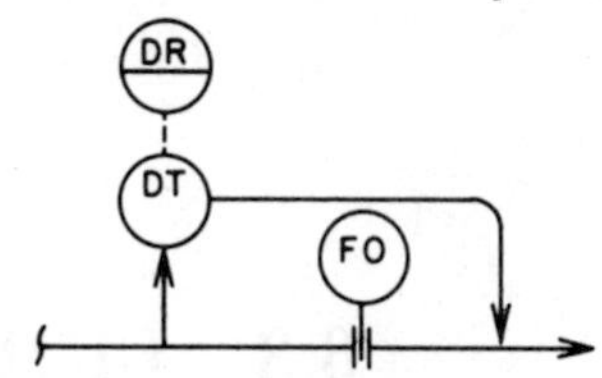

6.2 LIQUID DENSITY— ANGULAR POSITION TYPE

Design Pressure:	Up to 1000 PSIG (6.9 MPa)
Design Temperature:	Up to 500°F (260°C)
Materials of Construction:	Aluminum or stainless steel
Inaccuracy:	0.0005 SG
Minimum Span:	0.1 SG
Maximum Span:	1.0 SG
Approximate Cost:	$2000
Supplier:	Potter Aeronautical Corp.

As illustrated in Figure 6.2a, the process fluid sample flows continuously through the detector at a rate under 30 GPH (114 l/h). The gauge chamber contains three displacer floats, each of different density and volume. The solid displacers are spaced 90–100 degrees apart and assembled to a common shaft. Each fluid density sample positions the shaft and displacers at a precise angular position. The displacer moments are a function of float position and buoyant force. By having the three displacer moments in balance, the assembly is in equilibrium at all times. The angular position of the assembly is transmitted to the electrical components through magnetic coupling. The output signal to remote readout devices is available in either analog or digital form.

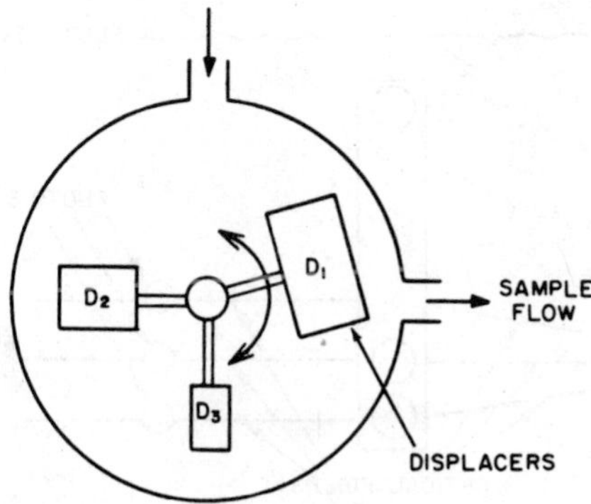

Fig. 6.2a Angular position density sensor

Because the displacers are made of solid materials, process pressure variations will have no effect on their volume. Similarly, the influence of process temperature variations is minor because the floats are made of materials with low coefficients of expansion. A change of 200°F (111°C) will result in a measurement shift of only 0.0001 specific gravity.

Detection accuracy is ± 0.0005 to 0.001 SG and it is unaffected by the span used. Due to the magnetic coupling utilized in transmitting the angular position of the shaft, the design pressure and temperature rating of the unit are high.

It is desirable to maintain the sample flow rate constant and to remove from it any particles in excess of 20 micron size. The unit is limited to clean, non-viscous services, although viscosities up to 500 centipoises (0.5 Pa·s) can be tolerated. Temperature compensation is not provided as standard but can be obtained through the installation of additional sensing and computing elements.

Flow Sheet Symbol

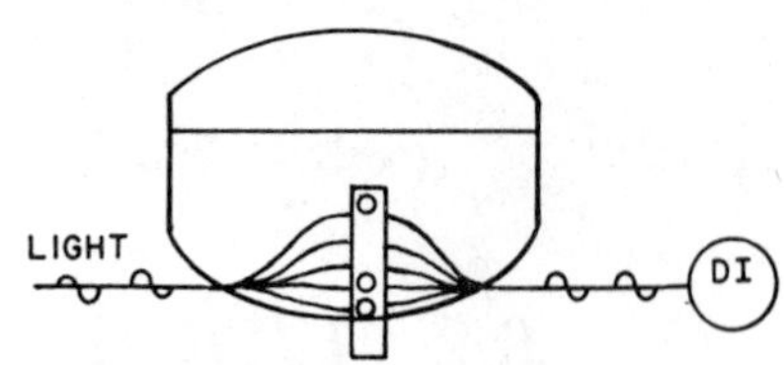

6.3 LIQUID DENSITY—BALL TYPE

Design Pressure:	600 PSIG (4 MPa)
Design Temperature:	−65 to 160°F (−54 to 71°C)
Materials of Construction:	Aluminum, silica glass balls and optical fibers
Inaccuracy:	To 0.01%
Range:	Available 0.7 to 0.9 specific gravity; can be supplied in other ranges
Approximate Cost:	0.01%—$2000 0.1%—$500
Limitations:	Limited to transparent or translucent fluids
Supplier:	Simmonds Precision Products, Inc.

The ball-type density meter consists of a number of ¼ in. (6.3 mm) diameter hollow opaque silica glass balls, typically 10 to 50 balls. Silica glass is chosen because of its low thermal expansion of 0.5 parts in 10^6 per degree F. The balls are free to move in tubes that are immersed in the fluid to be measured. Each ball has a different density; for example, in the range of 0.7 to 0.9 specific gravity, twenty balls are graded as 0.70, 0.71, 0.72, . . . to 0.89 specific gravity.

Fiberglass probes conduct light through one side of the tubes in which the balls are contained. Probes through the other side of the tubes pick up the light if the ball is not intervening. The last ball to float determines the density of the fluid. The tubes are usually located near the bottom of the tank and are only twice as long as the ball train which they enclose. Alternatively, however, the tubes may extend to the full height of the tank to detect density variations and density stratifications. In this case, fluid height is indicated by the ball of least density. The spherical shape of the floats reduces friction to the tube walls, permitting the density meter to be mounted as much as 85 degrees to the vertical.

Photo diodes are used to produce electrical signals from the light signals to enter a data bus or provide an analog readout.

A valuable feature of this density meter is its full intrinsic safety. It may be used in explosive mixtures without danger of causing an explosion. It is also not affected by electric or magnetic fields.

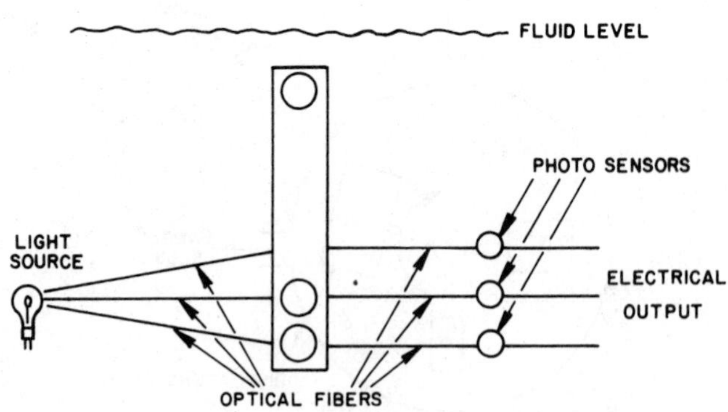

Fig. 6.3a Multiple free ball density meter

Flow Sheet Symbol

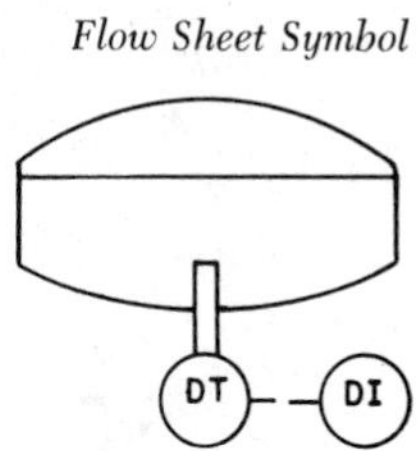

6.4 LIQUID DENSITY—CAPACITANCE TYPE

Design Pressure:	500 PSIG (3.45 MPa)
Design Temperature:	−65 to 160°F (−54 to 71°C)
Materials of Construction:	Fiberglass and aluminum
Inaccuracy:	1% of span
Range:	0.7 to 0.9 specific gravity (for fuel applications)
Approximate Cost:	$1200
Limitations:	Limited to electrically non-conducting fluids
Partial List of Suppliers:	Gull Engineering Inc.; Ragen Precision Industries; Simmonds Precision

The capacitance-type density meter consists of two concentric cylinders into which the fuel to be measured is introduced. A bridge circuit measures the capacitance between the cylinders. The capacitance is proportional to dielectric constant which is, in turn, proportional to density.

This method can also be applied to measure the total fuel in a tank in pounds. To do so, the probe, consisting of two concentric cylinders, extends from top to the bottom of the tank and fills to the fuel level in the tank. The product of density, inferred from dielectric constant, and volume is measured and displayed as number of pounds in the tank.

Two features of this system are of interest. First, for irregularly shaped tanks, the electrodes may be contoured to match the shape of the tank, thereby providing a linear output. Second, many probes may be connected in parallel to sum the mass in several tanks or to compensate for tilting of the tank, as is usual in aircraft applications.

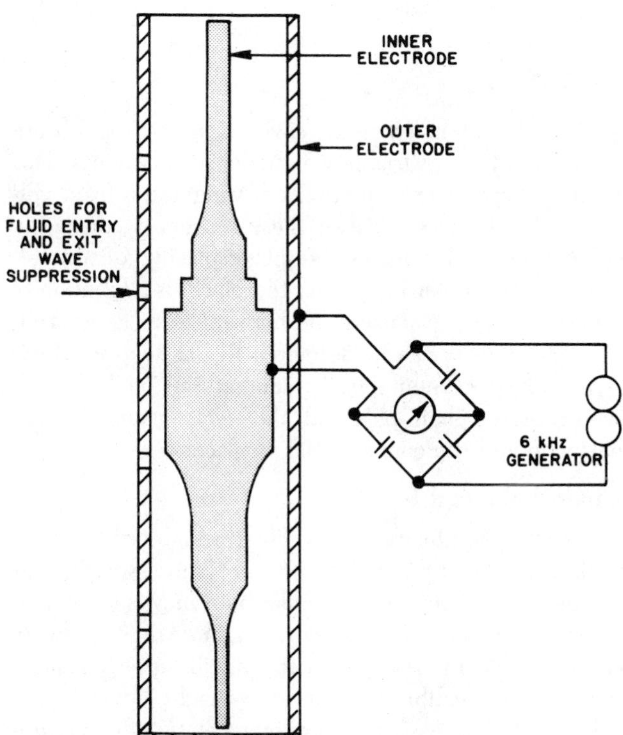

Fig. 6.4a Capacitance probe contoured to compensate for tank irregularities

Flow Sheet Symbol

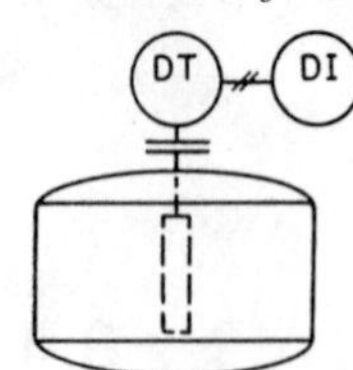

6.5 LIQUID DENSITY—DISPLACEMENT TYPE

Design Pressure:	Glass to 125 PSIG (861 kPa), others to 600 PSIG (4 MPa) or more
Design Temperature:	Up to 400°F (204°C), higher with special designs
Materials of Construction:	Glass, brass, stainless steel, Monel, Hastelloy, Teflon-lined, Durimet, plastics, epoxy
Inaccuracy:	½ to 3% of span
Range:	Minimum span is 0.005, maximum is 1.0 or more specific gravity units
Approximate Cost:	$450 for local indicator, $1400 or more for pneumatic transmitter. Electronic transmitter $2700 and higher
Partial List of Suppliers:	Black, Sivalls & Bryson, Inc.; Fischer & Porter Co.; Foxboro Co.; Kieley & Mueller, Inc.; Masoneilen Div. of Worthington Corp.; Precision Thermometer & Instrument Co.; Sangamo Weston, Inc.

The displacement elements covered in this section are constant-volume, variable-weight density elements. This represents one of the methods of weighing a fixed volume. The displacer float in these sensors is constantly submerged in the process liquid, displacing a fixed volume of fluid. Float buoyancy, therefore, is a function of liquid density; by sensing the buoyant force generated, the fluid density can be detected. In this section, three types of displacement density elements will be discussed: the calibrated chain-balanced, the electromagnetic force balanced and the conventional displacers.

Chain-balanced Float

Figure 6.5a illustrates this flow-through-type density indicator. The submerged float and chain assembly displaces a fixed fluid volume. Float buoyancy is a function of liquid density and therefore an increase in density causes the float to rise. As it rises, it will support a larger portion of the calibrated chain, the weight of which cancels out the increase in buoyancy so that a new equilibrium condition is achieved. The new float position is an indication of fluid density.

The wetted parts of this indicator are available in a large variety of corrosion-resistant materials including Pyrex and clear plastics for the chamber. Specific gravity spans from 0.01 to 0.5 can be obtained within the limits of 0.5 and 3.5 SG. The inaccuracy is about ±1 percent of full span and the detector can be exposed to pressures up to 125 PSIG (861 kPa) and temperatures to 400°F (204°C). The sample flow rate through the instrument should be kept between 5 and 30 GPH (19 and 114 l/h) with process streams having less than 50 centipoise (0.05 Pa·s) viscosities. Higher viscosities or flow rates will introduce velocity and friction effects which will reduce sensing accuracy.

The chain-balanced density sensor is also available as a transmitter (see Figure 6.5b). The density-related vertical position of the float is detected by the inductance pickup arrangement shown. This consists of a magnetic

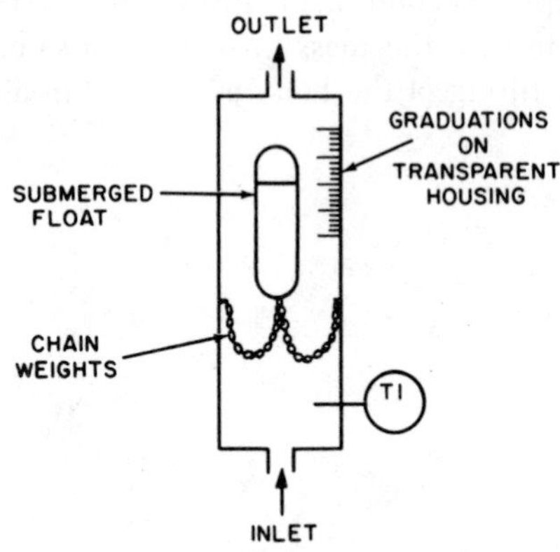

Fig. 6.5a Chain-loaded density indicator

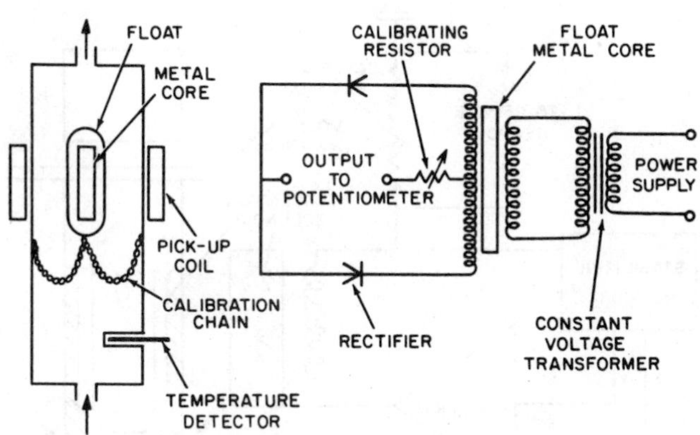

Fig. 6.5b Chain-balanced density transmitter

core inside the float and a three-winding differential transformer outside the chamber. Float movement causes an inductance change in the transformer with a resultant change in the differential voltage output. This output is rectified and detected by a potentiometer as a measure of density. When the process temperature varies, it is necessary to include automatic compensation, which corrects the apparent fluid density to a predetermined base temperature. A resistance-type temperature sensor is furnished for this purpose, and the readout instrument is provided with adjustment controls to permit selection of compensation values.

General features of the chain-balanced density transmitter are similar to those of the indicator, except that the chamber can be furnished in corrosion-resistant metallic or plastic-lined materials, and consequently pressure and temperature limits are increased to 500 PSIG (3.4 MPa) and 500°F (260°C) respectively. The specific gravity spans for the transmitters are the same as for the indicators, but accuracy is somewhat less. They can be calibrated to ±1 percent at the desired control point, but at extreme ends of the span only ±3 percent accuracy can be expected. The speed of response is a function of sample flow rate, and averages out to be around 30 seconds.

Conventional Displacer-type Density Sensors

A detailed description of the operation and features of displacement-type level devices has been given in Section 3.8. For that reason, the discussion here will be limited to those features which are particular to the application of these instruments for density detection. For information concerning construction materials, pressure rating, and so on, the reader is referred to Section 3.8.

The basic difference between a level and a density displacer is that the displacer float, in the case of level sensing, is only partially covered with the process fluid, while for density detection it is always totally immersed. Therefore, changes in buoyant force are only a function of density variations.

Figure 3.8b shows the basic components of a torque tube-type displacer instrument, and Figure 6.5c illustrates some of the modifications that are desirable when this device is used as a density sensor. The sample fluid enters around the center section of the cage, through a piezometer ring which eliminates the velocity effects of the flowing liquid. If the sample fluid velocity is below two feet (61 cm) per minute, the piezometer ring is not essential. The sample fluid leaves the cage through the top and bottom connections. It is recommended that these flows be maintained at equal rates by the use of purgemeters. The above system applies when the density measurement is made in a sample bypass of the process piping. If the density is to be detected in tanks or vessels, the flange mounted displacer illustrated in Figure 3.8f should be considered.

Sizing the displacer float for density applications involves a similar procedure to that discussed in Section 3.8 in connection with interface detection. At an actual proportional band setting of 20 percent the standard torque tube requires 0–0.72 lbf (0–3.20 N) buoyant force change for full range operation. The thin-wall tube requires half as much, or 0–0.36 lbf (0–1.60 N). Therefore, that volume of the displacer float has to be selected which will generate this buoyant force when the fluid density changes from the minimum limit of its range to the maximum. In equation form:

$$\text{Displacer volume (in.}^3) = \frac{\text{Torque Tube Force}}{(\text{SG Differential})(\text{Weight of 1 in.}^3 \text{ of } H_2O)}$$

If our density detector is to have a range of 0.990 to 1.000 and we are contemplating use of a thin-wall torsion tube, then the volume required is:

$$V = \frac{0.36}{(1.000\text{–}0.990)(0.36)} = 1{,}000 \text{ in.}^3$$

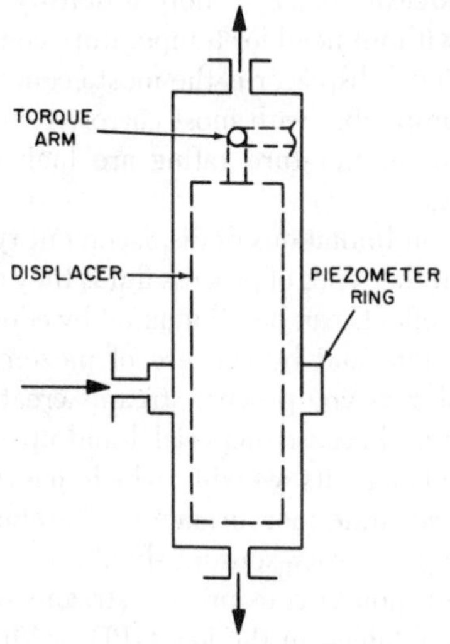

Fig. 6.5c Displacer density sensor

This would approximately correspond to a 36 in. long, 6 in. diameter (914 mm long, 152 mm diameter) displacer. The table lists some of the standard displacers with their corresponding volumes and the density spans that will generate the required buoyant force change for standard and thin-wall torque tubes at 20 percent actual proportional band settings.

Displacer Float			*Minimum Specific Gravity Span*	
Diameter (in./mm)	*Length (in./mm)*	*Volume (in.³/cm³)*	*Standard Tube*	*Thin-Wall Tube*
3/76	14/356	99/1622	0.202	0.101
3/76	32/813	226/3703	0.088	0.044
3/76	48/1219	340/5572	0.059	0.030
3/76	60/1524	425/6964	0.047	0.024
4/102	14/356	176/2884	0.114	0.057
4/102	32/813	402/6588	0.050	0.025
4/102	48/1219	602/9865	0.033	0.017
4/102	60/1524	753/12339	0.027	0.014
6/152	14/356	396/6489	0.051	0.026
6/152	32/813	905/14830	0.021	0.011
6/152	48/1219	1360/22286	0.015	0.008
6/152	60/1524	1700/27858	0.012	0.006

As can be seen from the table, there is no real minimum limit on the span of these density detectors because span is only a function of increasing displacer volume.

As a transmitter, the displacer type density sensor is available with both pneumatic and electronic output signals for remote readout. Temperature compensation is not available as a standard, but can be built into the system by use of computing relays.

The various displacement-type density sensors covered in this section are reliable and accurate instruments. When only local indication is desired, the chain-balance unit is the logical choice. Where a density transmitter is required without need for temperature compensation, the conventional displacer is the most economical selection. It is compatible with most corrosive fluids and its pressure and temperature rating are high enough for most applications.

The common limitations of displacement-type density sensors are in the types of process fluids they can handle. The velocity effects can be eliminated by controlling the sample flow rate and by the use of piezometer rings. Viscous or slurry-type process streams create more serious problems because material build-up on the displacer will change its weight, which has to be kept constant for accurate measurement. Therefore, the displacement-type density sensors should only be considered for clean, non-viscous process streams with sample flow rates maintained in the low GPH or l/h range.

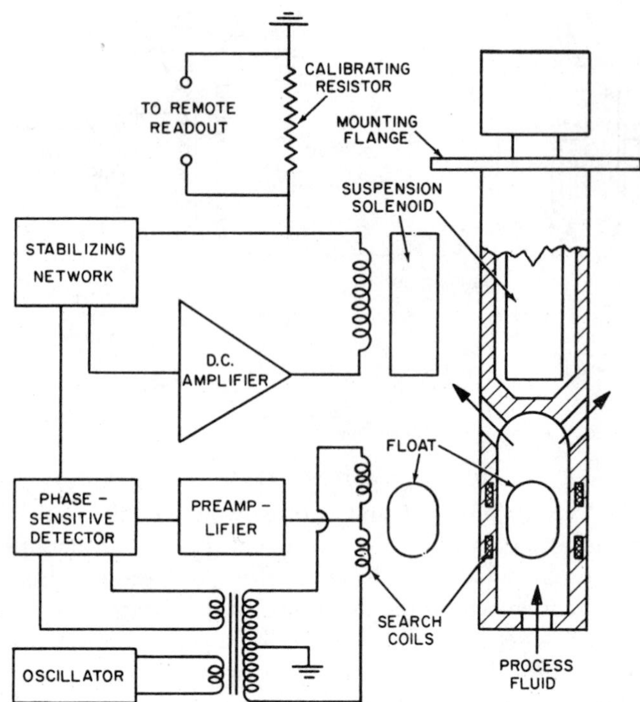

Fig. 6.5d Magnetic suspension density sensor

Electromagnetic Suspension

The instrument illustrated in Figure 6.5d includes a float, suspended electromagnetically, totally immersed in the process fluid. The position of the gold-plated ferrous alloy float is detected by a pair of search coils fed by a high frequency supply. The float selected should be slightly more dense than the maximum density of the fluid to be measured so that it always tends to sink. However, it is prevented from doing so by the electromagnetic force from the solenoid situated directly above the float. The search coils, controlled by the electronic circuits in the head of the instrument, allow just enough current to flow in the solenoid to maintain the float centrally between the two search coils.

If the density of the fluid decreases, the float will start to sink. The change in position is detected by the search coils and results in an unbalance signal to the amplifier and demodulator, causing an increase in the current to the solenoid. This increases the force of magnetic attraction on the float and restores it to its original centered position. Thus, a closed-loop servo system within the instrument accurately positions the float by variation of the solenoid current. This current is measured, using a stable calibrating resistor, as a measure of fluid density. Where operating temperatures are expected to vary, the density transmitter is furnished with a temperature compensation unit using platinum resistance wire elements. The adjustments on this unit allow for setting the temperature reference anywhere between 32 and 212°F (0 and 100°C). The temperature coefficient for the particular fluid is also adjustable between zero and 0.0009 SG/°F (°C).

The plummet floats inside the lower portion of the epoxy probe housing, touching nothing but the process liquid and, therefore, is insensitive to the effects of liquid surface tension. This density sensor can be mounted on tank or pipeline nozzles, directly immersed in the process stream, or provided with a sample chamber for mounting in a sample by-pass line. In either case, it is to be mounted within one degree of the vertical, and it is also important to avoid magnetic materials being placed closer than a foot (305 mm) from the float. The wetted parts are epoxy, stainless steel and gold-plated for corrosion resistance. Sample flow rates exceeding a few GPH or viscosities above a few centipoises can result in vertical forces on the float, causing error in the measurement. Deposits on the float will result in measurement error. For this reason the detector is not recommended for slurry service.

The available full spans for this instrument range from 0.01 to 0.4 specific gravity units within the limits of 0.4 and 2.0 SG. Measurement accuracies of ½ to 1 percent of full span can be obtained. One reason for this high accuracy is that the instrument detects only the difference between float and fluid densities. Consequently, the best results are obtained when the liquid density to be sensed is close to the density of the float. Pressure and temperature limitations are 200 PSIG (138 kPa) and 350°F (177°C) respectively.

Flow Sheet Symbol

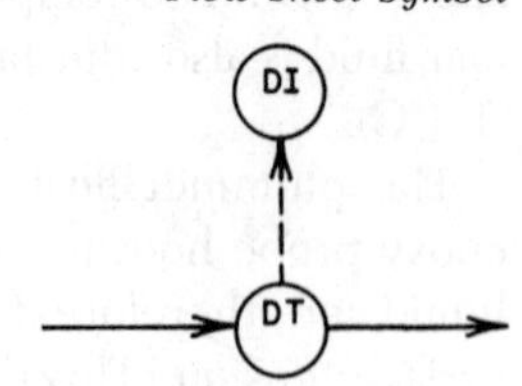

6.6 LIQUID DENSITY—FLUID DYNAMIC TYPE

Design Pressure:	50 PSIG (345 kPa)
Design Temperature:	1200°F (650°C)
Materials of Construction:	316 stainless
Inaccuracy:	2% of span
Range:	Molecular weight of 2 to 100+ at standard conditions of temperature & pressure
Approximate Cost:	\$4000 to \$8000
Partial List of Suppliers:	FluiDynamic Devices Ltd.

The fluid dynamic density meter (Fig. 6.6a) is used to measure the densities of gases and liquids. It is composed of two chambers (A and B) each having a supply nozzle (C and C[1]) and an opposing receiver port (D and D[1]). One chamber (A) is used as a reference chamber having only a small outlet port; it is filled with a suitable supply fluid (E) such that the dynamic pressure (P_r) of this jet on the receiver port (D) serves as a reference value. Directly adjacent to this reference chamber is a similar measuring chamber (B) which has large inlet and outlet ports through which the measured fluid is pumped by the action of the same supply fluid (E). The product whose density is to be measured, being entrained by this supply jet, affects the dynamic pressure (P_m) of this jet on the receiver port inversely to the density. A comparison of the pressure differential produced between the reference chamber and the measurement pressure is a measure of the density of the unknown product.

The fluid dynamic density meter can be used for measurement of density of process streams and effluents, both liquids and gases. It has no moving parts, a very high sensitivity and a high rate of response, but it is not particularly suitable for non-Newtonian fluids. It can be made of a wide variety of materials and can be mounted directly in a body or stream of fluid to give precise point measurements. Thus, petroleum and petrochemical refineries, natural gas processing plants, chemical process plants and the pulp and paper industry are typical applications. This densitometer is also useful for the manufacture of ethylene.

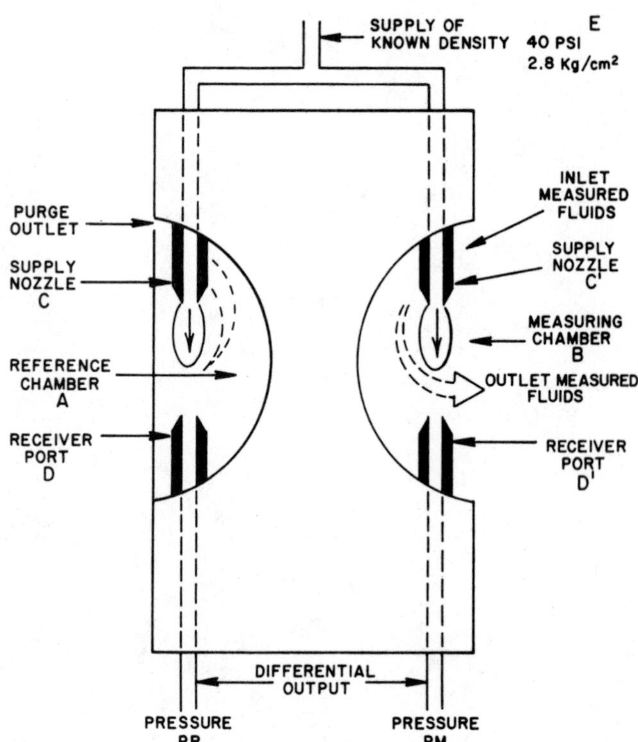

Fig. 6.6a Fluid dynamic density meter

Flow Sheet Symbol

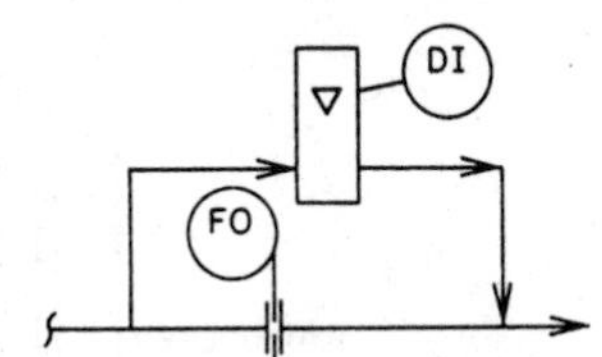

6.7 LIQUID DENSITY—HYDROMETERS

Design Pressure:	Glass design limited to 100 PSIG (690 kPa)
Design Temperature:	Up to 200°F (93°C)
Materials of Construction:	Glass, brass, stainless steel, Monel, Hastelloy, Teflon-lined, Durimet, nickel
Inaccuracy:	1% of span
Range:	Minimum span is 0.10 (maximum 0.50) specific gravity unit, within the values of 0.6 and 2.1
Approximate Cost:	\$450 for local indicator, \$1400 for transmitter
Partial List of Suppliers:	Black, Sivalls & Bryson, Inc.; Fischer & Porter Co.; Precision Thermometer & Instrument Co.

The hydrometer element consists of a weighted float with a small diameter indicator stem attachment at the top as shown on Figure 6.7a. The stem is graduated in any of the previously discussed density units. According to Archimedes' principle, when a body is immersed in a fluid it loses weight equal to the liquid weight displaced. The hydrometer element is a constant weight body which, if immersed in fluids with differing densities, will displace differing volumes of fluid. Therefore, the degree of stem scale submersion is an indication of fluid density. Readings are made at the point where the stem emerges from the liquid; the accuracy of measurement is a function of surface tension, turbulence and sample contamination, all of which affect reading accuracy.

One of the simplest in-line density indicators is illustrated in Figure 6.7b. It consists of a transparent glass tee with a hydrometer element. The process fluid sample enters from the bottom and overflows to maintain constant level in the tee. The sample flow rate is maintained at less than 1 GPM (0.004 m^3/min) rate to minimize velocity and turbulence effects. Where the process temperature is allowed to vary, a thermometer is added to the assembly for temperature correction capability.

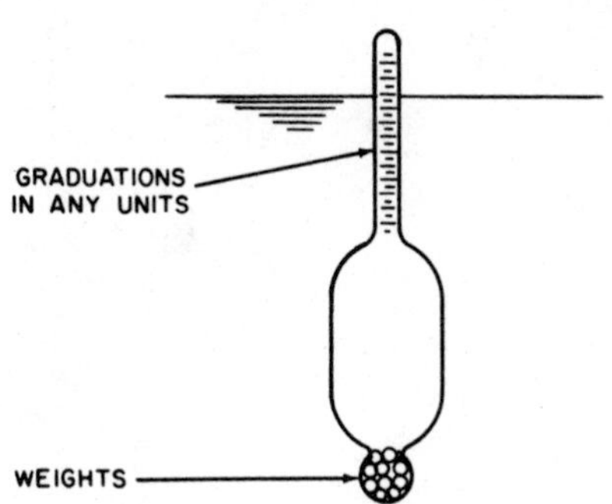

Fig. 6.7a Hand hydrometer

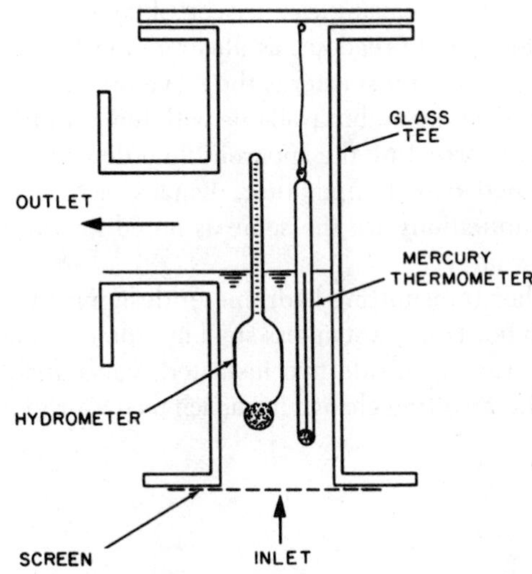

Fig. 6.7b In-line hydrometer indicator

The hydrometer element can also be mounted inside a rotameter housing. In such designs, overflow pipes are used to maintain constant level of the sample fluid inside the glass tube as shown in Figure 6.7c. Needle valves for sample rate control at about 15 GPH (0.057 m^3/min) and integral thermometers are standard accessories.

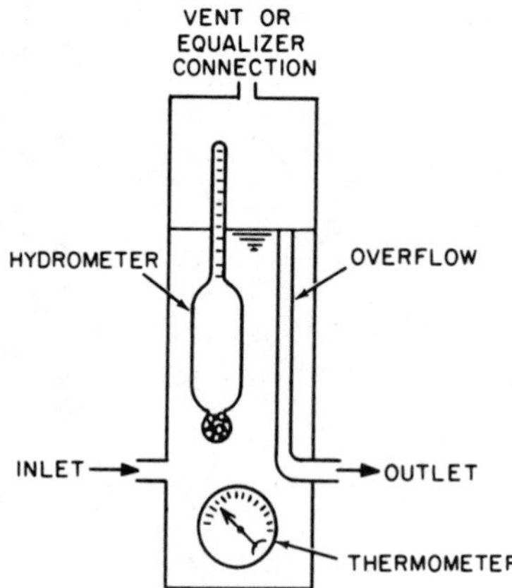

Fig. 6.7c In-line hydrometer in rotameter housing

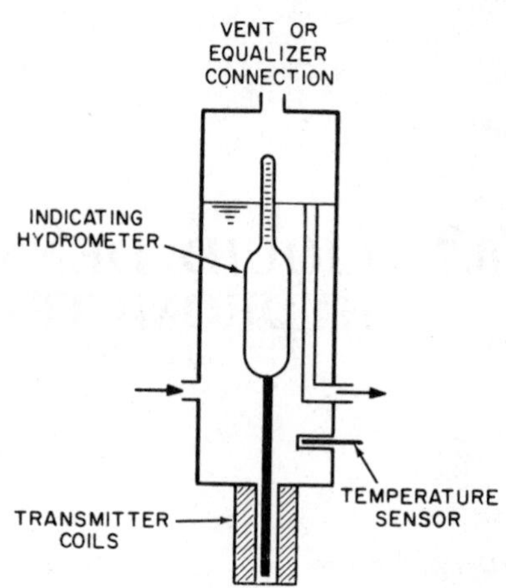

Fig. 6.7d Hydrometer transmitter in rotameter housing

Wetted parts are available in the same range of materials as conventional rotameters. These units can withstand up to 200°F (93°C) and 100 PSIG (690 kPa). Specific gravity spans of 0.1 to 0.5 can be obtained within the limits of 0.6 and 2.1 SG. Reading inaccuracy is 1 percent of span or the smallest division on the scale (0.001 to 0.005 SG).

The variable immersion hydrometer element mounted in a rotameter housing can also be obtained as a transmitter for remote readout, as illustrated in Figure 6.7d. This electronic transmitter is the servo-operated impedance bridge type and is available with temperature compensation correcting the apparent fluid density to a predetermined base temperature. Ranges, accuracies and design limitations are the same as noted for the in-line indicator above.

Another transmitting hydrometer design is the capacitance type. Here, a stainless steel hydrometer positions a dielectric cup inside two insulated, concentric cylinders. The resulting change in capacitance is proportional to density. Automatic temperature compensation is provided with the unit so that the transmitted output signal is referred to 60°F (16°C). The transmission can either be analog or digital and the specific gravity detected has to be between 0.7 and 0.9.

The hydrometers discussed above are basic to density measurement. They are accurate, frictionless, direct-indicating without need for mechanical linkages or external energy sources, and are compatible with most corrosive fluids. Their limitations are in the process fluids they can handle and in the pressures and temperatures they can be exposed to. Because the float position should not be affected by anything except the fluid density, the velocity, friction, turbulence, and viscosity effects must be minimized. Also, because the basis for their operation is the constant weight float, material build-up on the float cannot be tolerated. For these reasons, the above-discussed hydrometers should only be considered for clean, non-viscous process fluids with the sample flow rate controlled at the low GPH level.

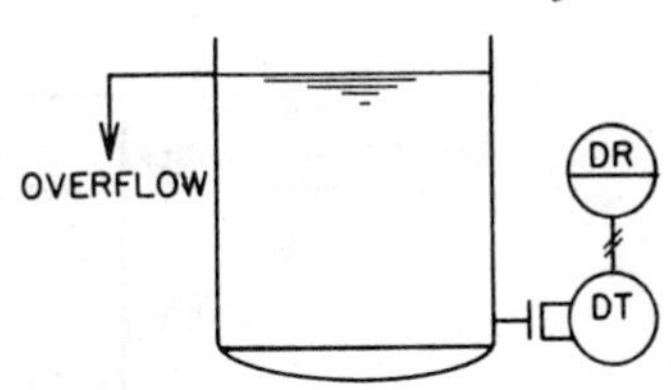

6.8 LIQUID DENSITY—HYDROSTATIC HEAD TYPE

Design Pressure:	Up to 5000 PSIG (34.5 MPa)
Design Temperature:	Up to 350°F (177°C), higher with filled volumetric elements
Materials of Construction:	See Sections 3.3 and 3.7
Inaccuracy:	±½% to ±1% of span
Range:	From 0.05 to any greater span
Cost:	\$100–\$450 for local indicators \$600–\$1700 for remote readout
Partial List of Suppliers:	Bailey Controls Co.; Fischer & Porter Co.; Fisher Controls Co.; Foxboro Co.; Hagan Controls Div., Westinghouse Electric Corp.; Honeywell, Industrial Div.; ITT Barton; G. Kent Ltd.; Meriam Instrument Co.; Petrometer Corp.; Taylor Instrument Companies; Uehling Instrument Co.

The hydrostatic head of a liquid column with fixed height will vary as a function of the process fluid density. The hydrostatic head approach to density measurement is identical to liquid interface level detection. The two distinguishing features between these measurements and conventional level detection are:

a. Both hydrostatic head sensing points can be located below the liquid surface.
b. The range of the detector instrument is suppressed.

The differential pressure approach to density detection requires the same hardware as is needed for level measurement. This instrumentation has been discussed in detail in Section 3.7 and, therefore, will not be repeated here. The reader should apply the same precautions in selecting his d/p detector for density as recommended for level measurement, keeping in mind the nature of the process fluid and that it is essential to protect the hydrostatic head sensing leads from corrosion, plugging, contamination, freezing or temperature effects.

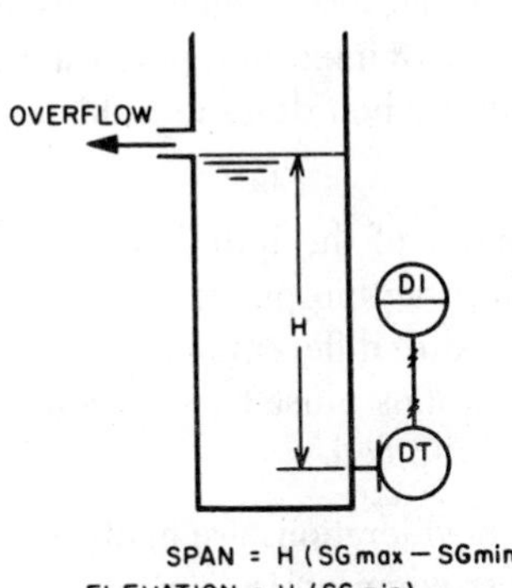

Fig. 6.8a Density detection in fixed level atmospheric tank

Determination of Span, Elevation or Depression Settings

Figures 6.8a and b illustrate some of the basic variations of hydraulic head density sensor installations. The letter H in these sketches designates the constant liquid column height being measured. If it is desired to measure the specific gravity from SG_{min} to SG_{max} the span (or range) of the d/p detector can be defined as H multiplied by (SG_{max} − SG_{min}). It should be noted that the span is always H (SG_{max} − SG_{min}) regardless of the type of installation.

When the specific gravity of the process fluid is SG_{min}, the d/p cell output should be zero. Therefore, it is necessary to "elevate" or "depress" the instrument "zero," allowing its full range to be devoted to the differential caused by density changes. In case of open tanks with fixed levels (Figure 6.8a) or any other type of installation

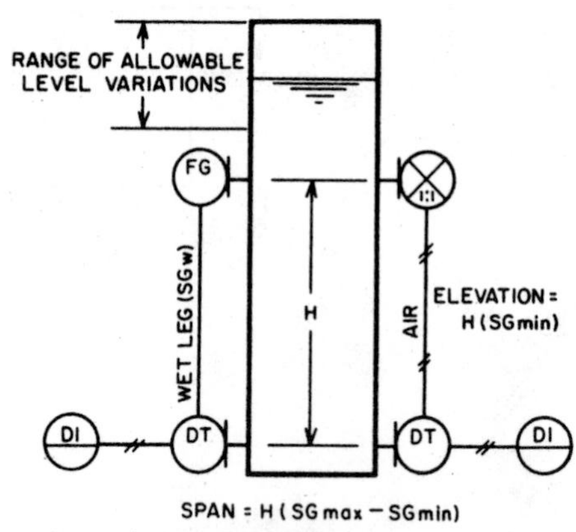

Fig. 6.8b Density detection in variable level open or pressurized tanks

with an air pressure repeater at the upper sensing point (right side of Figure 6.8b) the d/p cell zero is elevated by $H(SG_{min})$.

When the installation is of the "wet leg" type (left side of Figure 6.8b), the required shift in the instrument "zero" is a function of the filling fluid density. If the filling fluid is of the same density as the minimum on the instrument range (SG_{min}), then the d/p cell "zero" requires no adjustment. If it is greater, the "zero" must be depressed; if lower, elevation is required.

Besides the d/p hardware selection discussed in Section 3.7, there are two important practical considerations to be kept in mind when designing this type of density sensor loop.

a. The height of the liquid column detected has to be kept within practical limits.
b. The pressure differential generated due to density variations must be sufficient to match the selected d/p cell.

The above two considerations are in obvious conflict and selecting the best compromise can be illustrated by an example.

Let us assume that the process fluid involved requires an extended-diaphragm-type d/p cell (Figure 3.7c), which has a minimum differential range of 20 in. H_2O (5 kPa) and a maximum depression of 250 in. H_2O (63 kPa). We are to measure the specific gravity between 0.65 and 0.70 SG and the selected filling fluid for this wet leg installation has a gravity of 1.20. Calculate the minimum height of the fluid column and the zero depression required.

$$H = \frac{\text{Span}}{(SG_{max} - SG_{min})} = \frac{20}{0.05} = 400'' = 33'\ 4''$$

$$\text{Depression} = H(SG_w - SG_{min}) = \\ = 400\ (1.20 - 0.65) = 220''\ H_2O$$

The zero depression calculated in this example is compatible with available hardware. It is also clear that to detect narrow density spans with conventional d/p cells, very tall standpipes are required. If low range d/p detectors could be used for the particular installation, the column height could be reduced correspondingly, but when the process fluid is not compatible with low range d/p cell designs, the only alternate choice is to consider some other method of density detection. For the above noted reasons, it is not practical to use the hydrostatic head technique for specific gravity spans under 0.05.

The table below lists some of the relevant features of standard d/p cells.

d/p Cell Type	*Min. Range (in. H_2O)*	*Max. Range (in. H_2O)*	*Max. Elevation (in. H_2O)*	*Max. Depression (in. H_2O)*
Low	0–5	0–25	50–(Range)	50
Medium	0–20	0–250	250–(Range)	250
High	0–200	0–850	850–(Range)	850

Note: 1 in. H_2O = 0.25 kPa

Other Considerations

Temperature compensation is not a standard feature of hydrostatic head density loops. It can be provided by the addition of temperature sensors and computing relays, but is required only as a function of measurement accuracy desired. As shown in Figure 6.8c, the temperature effect of near-ambient temperature variations is relatively small for most liquids.

The choice of d/p detector for density measurement should be based on the same considerations discussed in connection with level sensors (see Section 3.7). The drawbacks and limitations of some of these d/p detectors are to be carefully considered. Purge systems, for example, are recommended only for open tank applications and even there it should be realized that the speed of response will be slow due to the volume of the purge system, and that in order to obtain reasonable accuracy, frequent maintenance will be required. Pressure repeaters are limited in their performance on vacuum service and require an independent vacuum source or biasing

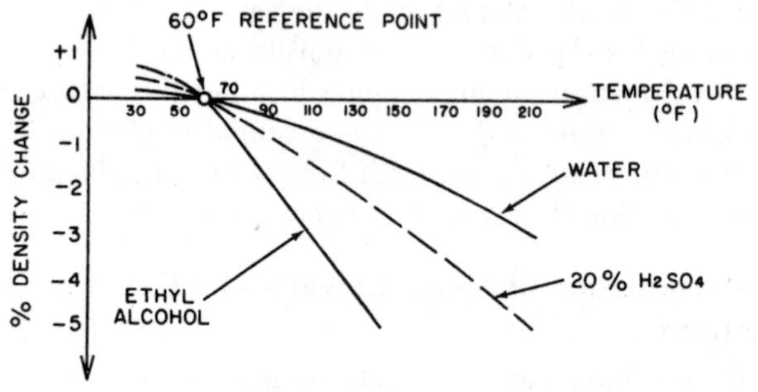

Fig. 6.8c Density change as a function of near-ambient temperature variations

the repeater signal. This places additional restrictions on allowable density spans and tank depths.

Conclusions

In hydrostatic head-type density loops, the force balance d/p cells are the most reliable devices to consider. The nature of the process fluid determines the particular design—such as extended-diaphragm-type—to be selected for the specific application.

The limitations of the d/p type density detectors include the maximum practical height of the liquid column and the resulting minimum span.

Flow Sheet Symbol

DT DR

6.9 LIQUID DENSITY— OSCILLATING CORIOLIS

Design Pressure:	0–200 PSIG (0–1.4 MPa)
Design Temperature:	−100 to 500°F (−73 to 260°C)
Materials of Construction:	Stainless steel
Inaccuracy:	0.2%
Range:	0.5 to 2.5 SG, 0.1 to 25 lbs. per minute
Approximate Cost:	$2500
Partial List of Suppliers:	Micro Motion, Inc.

The oscillating coriolis density detector consists of a C-shaped pipe and a T-shaped leaf spring as opposite legs of a tuning fork. An electromagnetic forcer excites the tuning fork, thereby subjecting each moving particle within the pipe to a Coriolis-type acceleration. The resulting forces angularly deflect the C-shaped pipe an amount that is inversely proportional to the stiffness of the pipe and proportional to the mass flow rate within the pipe. When volumetric flowrate is constant, deflection is proportional to density. The angular deflection of the pipe is optically measured twice during each cycle of the tuning fork oscillation. The output is proportionally modulated to the mass flow rate. An oscillator/counter digitally encodes the pulse width and displays a numerical indication of mass flow rate.

The total mass flow over a given time interval is obtained using a digital integrator to sum the pulses of the flow rate indicator. In this way, totalized flow is updated for each oscillation of the tuning fork.

This mass flowmeter eliminates the many problems associated with mass flow measurement. The output of the meter is directly proportional to mass flow rate. Consequently, there is no need to measure the critical parameters of pressure, velocity, temperature, viscosity or density. There are no parts in contact with the flowing fluid, and accuracy of the meter is unaffected by erosion, corrosion or scale build-up in the flow sensor.

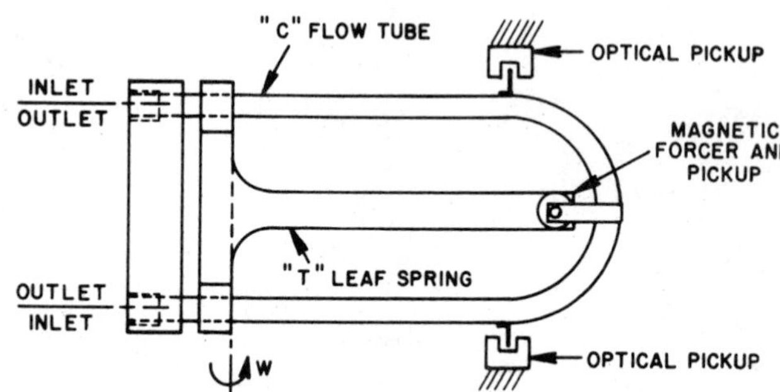

Fig. 6.9a Coriolis densitometer

Flow Sheet Symbol

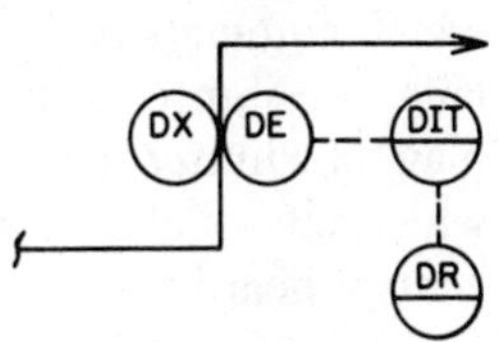

6.10 LIQUID DENSITY— RADIATION TYPE

Process Pressure:	Unlimited
Process Temperature:	Unlimited
Materials of Construction:	External device with no parts in contact with process
Inaccuracy:	±1% of span
Range:	Minimum span is 0.05, maximum is 1.0 or more specific gravity units
Approximate Cost:	\$7000–\$11,000 per complete recording loop, depending on type of amplifier, source size, area classification and accessories
Partial List of Suppliers:	Kay-Ray, Inc.; Ohmart Corp.; Texas Nuclear Div., Ramsey Engineering Co.

The features and limitations of radiation-type density sensors are very similar to those of radiation-type level detectors. For this reason, the reader is referred to Section 3.13, where some of the basic considerations for radiation instrumentation have been discussed. That material will not be repeated here, but will be assumed to have been read. The following is particular to density measurement only, although many of the points would also apply to liquid/liquid interface level measurement.

The basic components of the density gauge comprise a radioactive source beaming through a pipe and a detector system to measure the amount of transmitted radiation.

When gamma rays pass through a process fluid, they are absorbed in proportion to the material density (see Figure 6.10a). An increase in process density results in a reduced output current because a denser process fluid absorbs more of the gamma rays.

When the density inside large pipes or containers is measured, the radiation source and detector would be mounted as illustrated on the left side of Figure 6.10a. In smaller diameter pipe (under 6 in.) the radiation path is not adequate to provide high accuracy and sensitivity. Therefore, the installation shown on the right side of the figure would be used, which lengthens the radiation path and consequently increases accuracy.

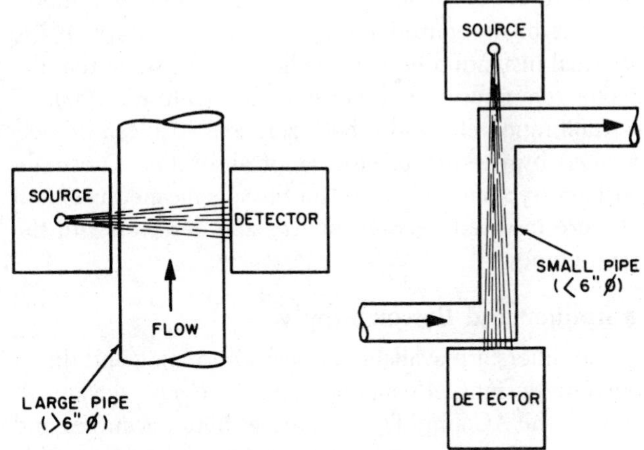

Fig. 6.10a Radiation density sensors

Radiation Source

For the majority of density applications, cesium 137 is used as the radioisotope. Source sizes normally vary between 200 and 2000 millicuries as a function of pipe diameter and specific gravity span. Gauge sensitivity is increased by the use of collimated (narrow) beam geometry which restricts radiation in all directions except for a direct path to the detector. This minimizes scatter and permits the use of larger sources with increased measurement sensitivity. Most designs are such that the radiation intensity at one foot from the gauge surface in any direction will not exceed five (5) milliroentgens per hour. This is a safe value for any process area where the operator's occupancy is twenty hours or less per week. The source holder is provided with a shutter mechanism to close the radiation beam port during installation.

Available shutter features include the fail-safe design which automatically closes the shutter whenever power fails, the shutter switch which permits remote light display of shutter position, and remote shutter control which allows closing the radiation beam from a central control board.

Radiation Detector

There are three basic types of gamma ray detectors: the Geiger tube, the ionization chamber, and the scintillation detector. The Geiger tube is a low accuracy device which measures radiation through the ionization of a halogen gas at about 500 volts DC potential. Ionization cells operate by the ionization of pressurized gas between two dissimilar metals under incident radiation, generating a low current signal (10^{-10} amps.). The scintillation detector senses the light photons resulting from gamma rays incident on certain crystal materials. This is the most sensitive but least stable detector. It is particularly affected by variations in environmental conditions such as temperature. For density gauging, the ionization chambers are used almost without exception. They require stable amplification, but when this is provided they supply simple, accurate and reliable measurements.

It is desirable to heat and thermostatically-control the detector chamber to eliminate temperature variation and moisture condensation problems. The process temperature is of no consequence, but it is necessary to use thermal insulation between cell and process, so that detector temperature will not rise above 140°F (60°C).

Calibration checks for both zero and span can be performed by the use of equivalent absorbers. These absorbers are accurately made for the specific measurement and are inserted between source and detector with the pipe empty.

Amplifier and Power Supply

Amplifiers are available in both alternating and direct current designs. Economics favor the DC design although the AC amplifier guarantees better accuracy and less drift. The DC amplifier needs weekly or bi-monthly standardization, while the solid-state, vibrating-capacitor type AC unit requires standardization much less frequently.

The minimum full scale span is about 0.05 specific gravity units with a corresponding accuracy of 0.0005 or better. When measuring small spans, the zero drift due to source decay becomes an important consideration. The source decay compensator unit is a must for such installations. For wider ranges, it is essential only if the source is cobalt 60. The source decay effect with cesium 137 is only 3 percent per year.

Process materials with high temperature expansion coefficients are provided with temperature compensation if variations in process temperature are expected.

The output signal from the detector to the amplifier is in micromicro amperes and therefore the signal should be protected by the use of coaxial shielded cable. The maximum distance between detector and amplifier is about a 1000 feet (300 m), but should be made shorter whenever possible to minimize noise pickup and time lag problems.

Beta Radiation Densitometers

Beta-ray absorption has been successfully used in cryogenic density applications. The source in this unit is strontium-90 and the receiver is a silicon surface-barrier detector.

The system counts the number of beta particles that strike the detector with an energy greater than the discrimination level.

The source and receiver are packaged in a small, light, corrosion-resistant probe. It can be used for level and other measurements besides density sensing.

Conclusions

Economics represents the only real limitation of the radiation-type density gauge. It performs reliably and with precision on all hard-to-handle process streams. The discussion in Section 3.13 concerning sizing, safety and other aspects of radiation instrumentation also applies to density gauges.

Flow Sheet Symbol

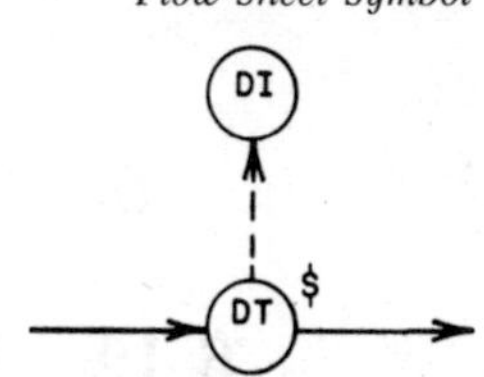

6.11 LIQUID DENSITY—SOUND VELOCITY TYPE

Design Pressure:	10,000 PSIG (69 MPa)
Design Temperature:	−20 to 390°F (−29 to 199°C)
Materials of Construction:	Stainless steel, epoxy, Hastelloy, titanium
Inaccuracy:	0.1% of span
Range:	0.5 to 1.5 SG
Approximate Cost:	\$2500 to \$5000
Limitations:	Relatively constant bulk modulus required
Partial List of Suppliers:	Agar Div. of Redland Automation Inc.

The sound velocity density detector consists of a device for measuring the speed of sound in the liquid while compensating for the effects of temperature and pressure. The speed of sound is given by $C = \sqrt{E/\rho}$, where E is the bulk modulus and ρ is the mass density.

The sensor is calibrated by immersing the sensor in distilled water whose temperature is closely controlled at intervals between 32 and 167°F (0 and 75°C), and by measuring cycle repetition frequencies to an accuracy of 0.001 percent.

Sound velocity meters have several limitations. The liquid must be clear; emulsions, dispersions and slurries scatter the sound waves and cannot be monitored. Similarly, undissolved gases and bubbles (even those as small as several thousandths of an inch in diameter) cause errors and occasionally render the instrument inoperable. Since the velocity of sound is a function of bulk modulus and density, variations in bulk modulus will cause significant measurement errors.

The instruments are generally used in pipelines to detect the interface between two different hydrocarbon products.

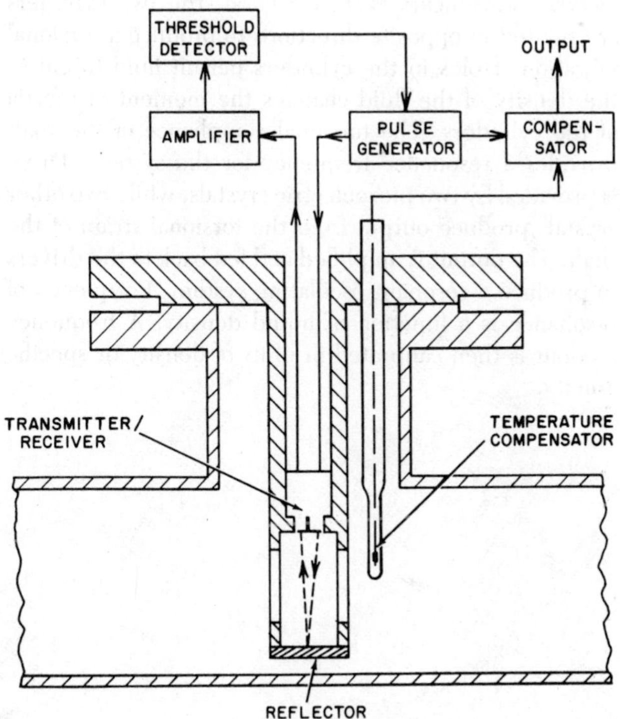

Fig. 6.11a Sound velocity meter

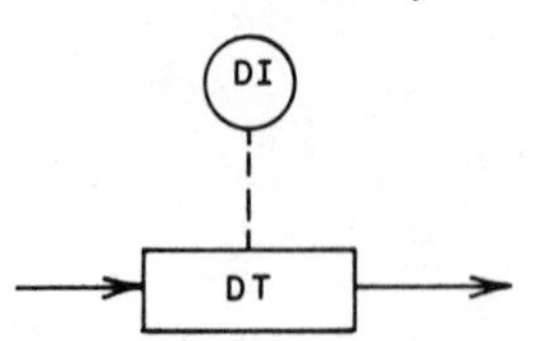

6.12 LIQUID DENSITY—TORSIONAL VIBRATION

Design Pressure:	500 PSIG (3.4 MPa)
Design Temperature:	−65 to 200°F (−54 to 93°C)
Materials of Construction:	Invar, Ni-Span C, ceramic
Inaccuracy:	0.1%
Range:	25% (typically 0.75 to 1)
Partial List of Suppliers:	General Electric Co.

The torsional vibration density detector consists of two colinear cylinders mounted on a single shaft with piezoelectric transducers. When excited, the two cylinders are rotated in opposite directions to produce a torsional vibration. Holes in the cylinders permit fluid to enter; the density of the fluid changes the moment of inertia of the cylinders. The torsional compliance of the shaft provides a resonance frequency for the system. Drive is provided by two piezoelectric crystals, while two other crystals produce output from the torsional strain of the shaft. The output is amplified and fed back to the drivers to produce a resonant oscillating system. Frequency of resonance is a function of liquid density. A frequency readout is then calibrated in units of density or specific gravity.

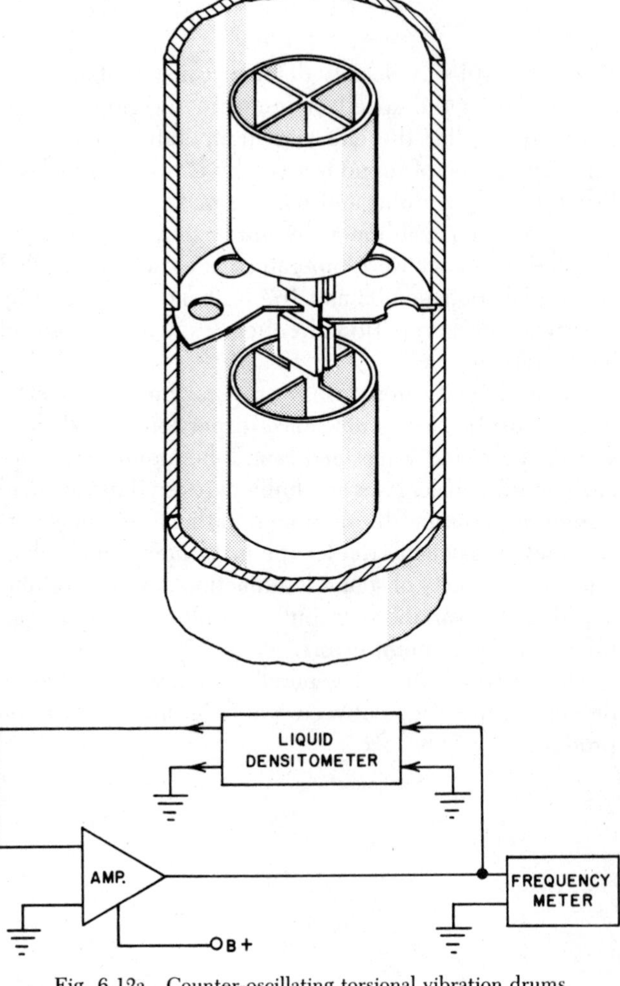

Fig. 6.12a Counter-oscillating torsional vibration drums

Flow Sheet Symbol

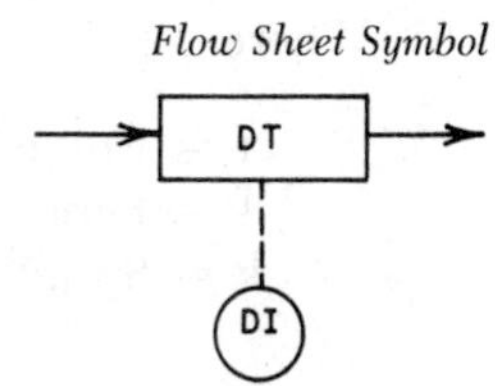

6.13 LIQUID DENSITY— VIBRATING PLATE TYPE

Design Pressure:	500 PSIG (3.4 MPa)
Design Temperature:	−425 to 250°F (−254 to 121°C)
Materials of Construction:	Stainless steel, Ni-Span C, Alnico
Inaccuracy:	±¼% of reading
Range:	0.5 to 1.5 SG
Approximate Cost:	$2500
Partial List of Suppliers:	ITT Barton

The vibrating plate-type detector consists of a flexible rectangular plate fixed at both ends within a tube through which the liquid is flowing. A distributed mass of fluid of defined volume interacts with the plate, causing it to vibrate with a simple harmonic motion. The system is permitted to oscillate at its resonant frequency. A simple measurement of "t" (period of oscillation) is made and related to the density of fluid substance in which it is immersed. The deflection is a half sine wave with nodal

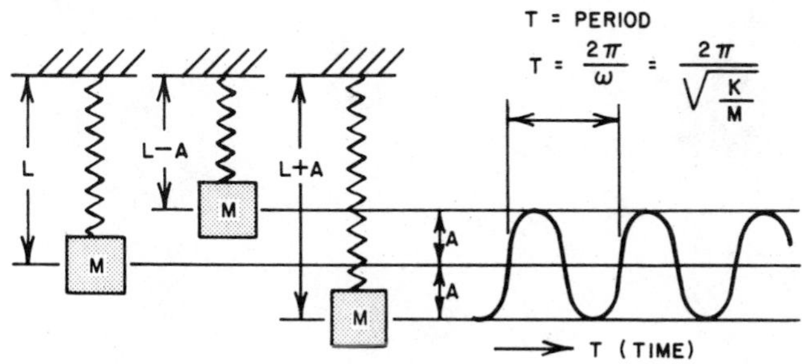

Fig. 6.13a Mechanical vibration

points at the point of fixation to its supporting structure. The motion described is simple and harmonic, and the laws governing such motion define the changes in velocity and acceleration of any point on the elastic member, and hence the imparting of an acceleration to the fluid particles distributed relative to the plate.

To illustrate elements of the principle by a simple example, a mass whose volume is readily established is suspended from a spring having a spring factor k (lbs/inch deflection) as shown in Figure 6.13b.

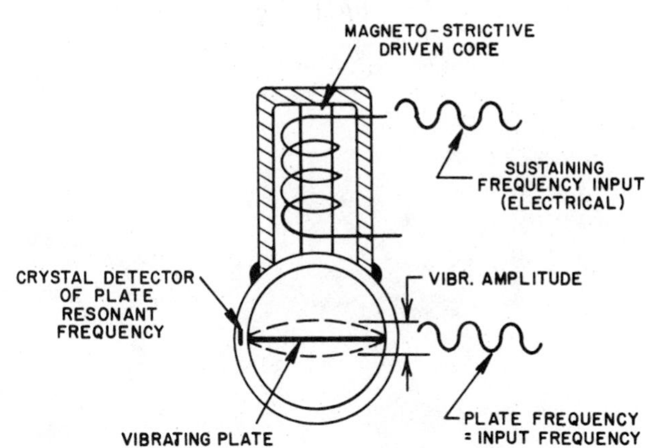

Fig. 6.13b Vibrating plate density meter

Assume, at first, that no motion exists and spring length "L" is established in which the tensile reaction of the (weightless) spring equals the suspended mass "M." An external force is now applied reducing L to L−a. The supporting force is removed and the mass is permitted to fall under the action of gravity to a position (L + a). In a frictionless undampened system, the mass "M" will oscillate with a deflection of ± "a" distance with simple harmonic motion and a period:

$$+ = 2\pi\left(\frac{k}{M}\right)^{-1/2} \qquad 6.13(1)$$

or:

$$M = \frac{k}{(2\ \pi)^2} \times t^2 \qquad 6.13(2)$$

Where

f = Frequency of oscillation
t = Period of oscillation
k = Spring constant

Equation 6.13(1) is derived from solution of classical case equation:

$$M \frac{d^2 x}{dt^2} + kx = O \text{ (Reference Figure 6.13b)}$$

Equation 6.13(2) suggests that a determination of mass "M" can be made by a measure of frequency or period of the oscillation.

For the case of a thin plate vibrating in a fluid, it can be shown that the fundamental plate frequency is defined by the equation:

$$f = \frac{f_0}{(1 + \beta)^{1/2}} \qquad 6.13(3)$$

$$\beta = \left[\frac{\rho}{\rho_p}\right] \times \left[\frac{a}{h}\right] \qquad 6.13(4)$$

Where

$\left[\frac{\rho}{\rho_p}\right]$ is the ratio of fluid density to vibrating plate density, and

$\left[\frac{a}{h}\right]$ is the plate length to thickness ratio.

The solution of equations 6.13(3) and (4) for fluid density in terms of plate resonant frequency, yields:

$$\rho = \frac{A^1}{f^2} + B^1 \qquad 6.13(5)$$

$$\rho = A^1 t^2 + B^1 \qquad 6.13(6)$$

Where

$A^1 = f_0\, 2 \left[\frac{h}{a}\right] \rho_p$

$B^1 = \rho_p \left[\frac{h}{a}\right]$

f_0 = Plate resonant frequency when $\rho = 0$ (perfect vacuum for gas measurement)

Equation 6.13(6) implies that the measured density is directly related to the "square" of the period "t" of vibration. Note the similarity to equation 6.13(1).

Flow Sheet Symbol

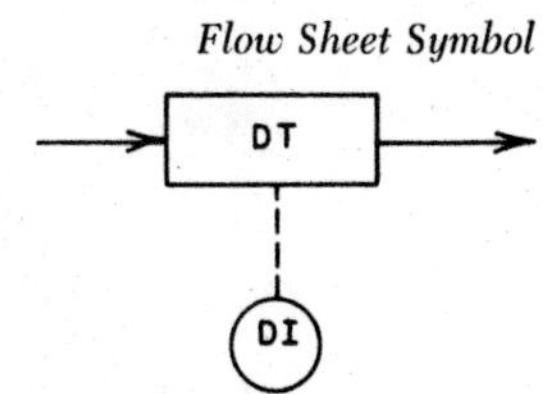

6.14 LIQUID DENSITY—VIBRATING SPOOL TYPE

Design Pressure:	10,000 PSIG (69 MPa)
Design Temperature:	−65 to 300°F (−54 to 149°C)
Materials of Construction:	Stainless steel, Ni-Span C, epoxy, copper, Alnico
Inaccuracy:	0.1 SG units
Range:	0.5 to 1.0 SG
Approximate Cost:	$4000
Partial List of Suppliers:	Agar Div. of Redland Automation, Inc.; Solartron Ltd.

If a cylindrical spool is immersed in a fluid and if circumferential oscillation normal to the spool is induced and sustained, the spool will vibrate at a frequency that is a function of its stiffness and the oscillating mass. Since the fluid surrounding the spool is caused to oscillate, the mass of the entire system in vibration consists of the mass of the spool plus that of the fluid. The system can be treated as a lumped parameter damped harmonic oscillator over a small range of frequencies. If a loop closed around the oscillator exhibits a 90 degree phase shift, vibration is sustained at the natural frequency irrespective of fluid viscosity. Thus, for a spool of a fixed stiffness and mass, variations in oscillating frequencies are due solely to variations in fluid densities.

In the instrument shown in Figure 6.14a, oscillations are induced and sustained by a feedback amplifier. A predetermined number of oscillations is counted and the elapsed time is measured by a high-frequency "clock." The signal is then developed from these data. Developing the signal in this manner provides a faster response than does measurement of frequency.

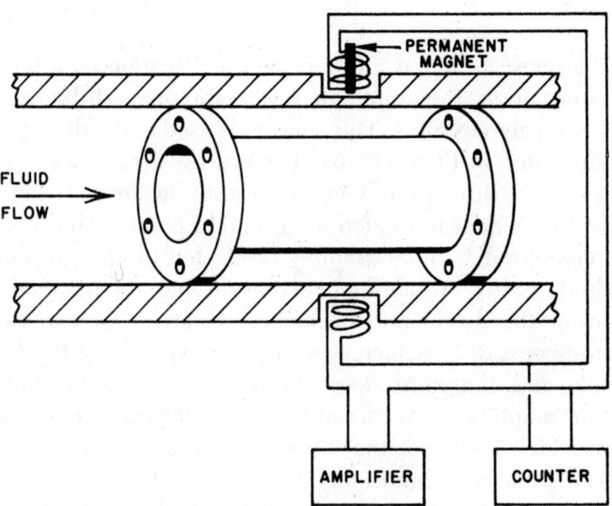

Fig. 6.14a Vibrating spool density sensor

Variations in spool stiffness and natural oscillating frequencies are minimized by several means. Temperature effects on spool dimensions and elastic modulus are kept as small as possible by alloys like Ni-Span C. Fluid pressure is exerted on both surfaces of the spool, so that pressure variations have no influence on oscillation. Material is stressed considerably below fatigue levels because oscillation amplitude is very slight.

Flow Sheet Symbol

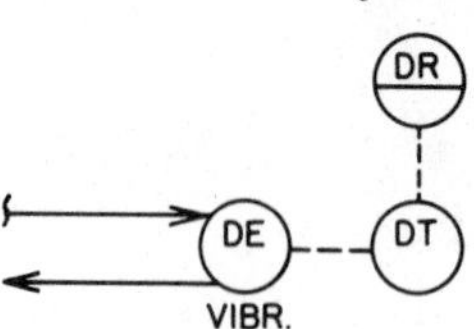

6.15 LIQUID DENSITY— VIBRATING TUBE TYPE

Design Pressure:	Up to 1000 PSIG (6.9 MPa)
Design Temperature:	Up to 300°F (149°C)
Materials of Construction:	Stainless steel
Inaccuracy:	± 1–3% of span
Minimum Span:	0.05 SG
Maximum Span:	0.5 SG
Approximate Cost:	$4500
Suppliers:	Automation Products, Inc.; Bell & Howell Ltd., CEC Div.

If a body is excited into mechanical vibration by a pulsating drive, the amplitude of its vibration will be proportional to its mass. This concept is used in the detector illustrated in Figure 6.15a. The process fluid flows continuously through a ½ in. (12.5 mm) diameter U-tube section which is welded at the node points. The total mass of the U-tube assembly is affected by the process fluid density. A pulsating current through the drive coil brings the U-tube into mechanical vibration. An increase in process density increases the effective mass of the U-tube and, therefore, decreases the corresponding vibration amplitude. An armature and coil arrangement is provided to detect the vibration at the "pickup" end. The armature vibrates together with the U-tube and induces an AC voltage proportional to fluid density in the pickup coil. This AC voltage is then converted into DC millivolts, which is more compatible with remote recorders or controllers.

In installations where the process temperature is expected to vary, an automatic temperature compensating circuit can be added. This consists of a resistance-type temperature element in the process stream and solid state circuitry in the converter to perform the required temperature correction.

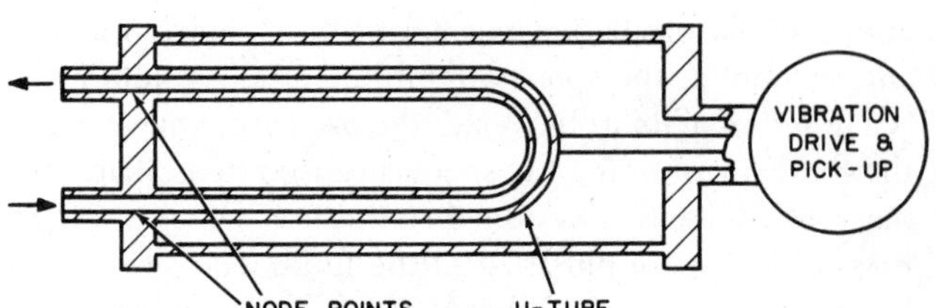

Fig. 6.15a Vibrating U-tube density detector

The main advantages of these sensors are that the design pressures and temperatures are not limited by flexible connectors and that ambient temperature, process pressure, sample flow rate or viscosity variations have practically no effect on the measurement.

If the process stream contains entrained gases, then it is important to maintain the process pressure and flow, because low velocities can cause separation and trapping of the gas, and pressure variations will affect the volume of the gas bubbles, influencing the density readout.

The main limitation of this unit is that it can handle only clean fluids with low or moderate viscosities. High viscosity streams or heavy slurries are likely to plug the small diameter U-tube.

Flow Sheet Symbol

6.16 LIQUID DENSITY—WEIGHT OF FIXED VOLUME TYPE

Design Pressure:	2400 PSIG max. (16.6 MPa) with solid connections. 200 PSIG max. (1.37 MPa) with flexible metallic connectors. 40 PSIG (276 kPa) with rubber connectors
Design Temperature:	500° F (260°C) max. with solid connections, 180 to 300° F (82 to 149°C) max. with various flexible hoses
Materials of Construction:	For wetted parts: Bronze, stainless steel, Carpenter 20, Pyrex, nickel, Monel, Karbate, PVC or ebonite lining For flexible hoses: Stainless steel, Monel, Neoprene, Hypalon, Teflon, Viton-A, butyl, silicone rubber
Inaccuracy:	±1% of span or 0.0003, whichever is greater
Minimum Span:	0.01 to 0.50 SG depending on design
Maximum Span:	0.4 to 3.5 SG depending on design
Approximate Cost:	\$2700 to \$9000 as a function of design, construction materials, features and accessories
Partial List of Suppliers:	Arcco Instrument Co.; Bell Corp.; Halliburton Co.; Honeywell, Industrial Div.

One of the methods of weighing a fixed volume has already been discussed in Section 6.5 where the displacement-type density sensors were covered. The other basic category of constant volume variable-weight density elements is discussed here. All of these units contain a constant volume flow-through chamber, which is continuously weighed. The design variations within this group involve the shape of the weighing chamber which can be a straight piece of pipe, a U-tube section or a small tank. Furthermore, the connection between the weighing chamber and the process can be solid or flexible, and the instrument can be a multi-purpose device or be limited to function only as a controller. Details of these design variations will be covered in the following paragraphs.

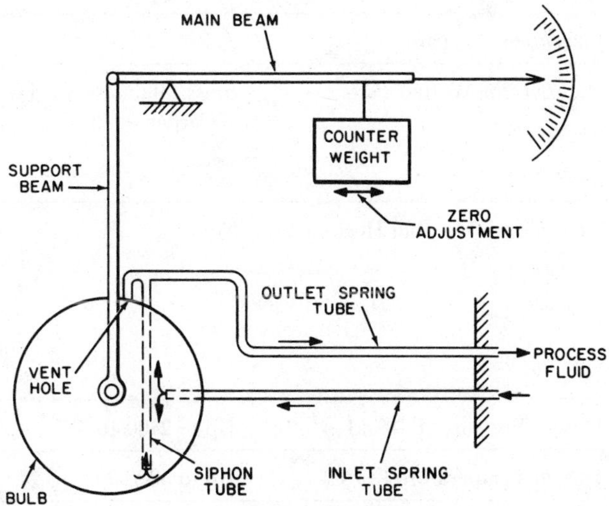

Fig. 6.16a Weighed-bulb density sensor

Weighed-bulb Density Sensor

The closed liquid system illustrated in Figure 6.16a consists of a bulb and sample tubing. These tubes also serve as spring elements of the weighing mechanism. The bulb is suspended from one end of the main beam, which is balanced by the counterweight on the other end. The counterweight is used to zero the instrument by balancing the dead weight of both the bulb and of the minimum density fluid within the bulb. (For a range of 0.65 to 0.75 SG, 0.65 represents the minimum density noted above.) As liquid density increases, the bulb will descend. Its descent is resisted by the spring action of the sample tubes. This bulb motion is directly proportional to the density variations in the process fluid and is transmitted to the main beam so that it, too, moves in proportion to the changes in specific gravity of the flowing sample. This movement can actuate an indicating or recorder pen. The unit is also available as a motion balance transmitter or controller of either the pneumatic or the electronic type. When the readout device is to be located several miles from the detector, telemeter transmission systems are considered, utilizing any two-wire circuit or its equivalent, such as telephone or radio channels.

This unit is provided with a thermostatic coil which detects the outlet process temperature and automatically corrects the reading for changes in specific gravity due to temperature variations. The compensator can correct for process temperatures between −100 to 500°F (−73 to 260°C).

Table 6.16b compares some features of the density detectors discussed in this section. From this table, it can be seen that the main advantage of the weighed-bulb design is that it contains no flexible hoses and, therefore, is applicable to high process pressures and temperatures.

The unit is designed for low sample flow rates with velocities in the laminar range. The advantage of such design is that the forces created by fluid turbulence are eliminated, while the disadvantage is that the small tubes and other openings are subject to plugging. For this reason, the sensor should be considered for only clean and nonviscous services.

As shown in Figure 6.16a, the use of the vent hole and the siphon tube guarantees that no gases or liquids are trapped in the bulb.

U-tube Density Gauge

The design involves a U-loop pipe section pivoted on flexures about the horizontal axis (see Figure 6.16c). Process fluid flows through this loop and its weight is transferred to a weight beam. A counterweight balances the pipe section and its minimum density contents on the weight beam. An increase in process density produces a proportional additional force on the weight beam, which is sensed by the nozzle-flapper section of a force balance transmitter. Increased density increases the back-pressure on the nozzle by moving the flapper closer to it and results in an increased output signal from the pneumatic amplifier relay. This signal, representing

Table 6.16b
COMPARISON BETWEEN DENSITY SENSORS WHICH DETECT THE WEIGHT OF FIXED VOLUMES

	Type of Design			
Features	*Weighed Bulb*	*U-Tube*	*Straight-Tube*	*Direct Controller*
Minimum Span (SG)	0.05	0.02–0.05	0.5	—
Maximum Span (SG)	0.5	3.5	1.0	—
Inaccuracy (% span)	±1%	±1–2% or 0.0003	±1%	±1% of set point
Materials for Wetted Parts	Brass, Stainless Steel, Carpenter 20	Stainless Steel, Monel, Nickel, Karbate, Glass or lined with PVC or Ebonite	Stainless Steel	Stainless Steel and Pyrex
Materials for Flexible Hoses	None	Neoprene, Hypalon, Viton-A, Butyl, Teflon or Silicone rubber, also Stainless Steel or Monel	Neoprene, Butyl, Viton-A	Neoprene, Butyl
Design Pressure (PSIG/kPa)	Up to 2400/16,547	40 to 200/276 to 1,379	50/345	10/69
Design Temperature (°F/°C)	Up to 500/260	180 to 300/82 to 149	180/82	180/82
Weigh Chamber Volume (Gallon/l)	0.265/1	0.2 to 1.0/1 to 3.8	0.5 to 1.0/1.9 to 3.8	5.0/18.9
Sample Flow Rate (GPM/lpm)	0.265/1	1 to 70/3.8 to 265	125/473	5 to 15/18.9 to 56.7
Sample Connection Size (in./mm)	¼/6.35	⅞ to 2/22.2 to 50.8	2/50.8	1½/38
Temperature Compensation	Automatic to 500°F/260°C	Automatic but limited	Non-Standard	No

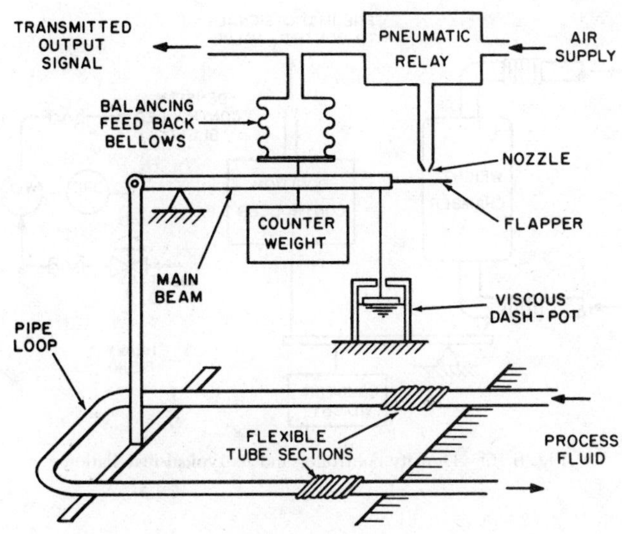

Fig. 6.16c U-tube density transmitter

the transmitter output, is also sent to the feedback bellows which exerts a counter force to rebalance the main beam. Because of the force balance nature of the design, the total motion of the flapper is a few thousanths of an inch. The U-tube sensors are also available as electronic transmitters.

Effects of process pressure or flow rate variations are minimized by locating the pivots on the vertical center line of the U-tube. This stays the loop against horizontal forces. The viscous dashpot is used to eliminate vibration effects or other sudden changes in vertical force components. This design has been successfully used on slurries with the flow velocity high enough to prevent settling. This normally requires velocities in the range of 5 to 8 feet per second (1.5 to 2.4 m/s).

The process fluid can also contain gas bubbles. If the process density without gas entrainment is desired, then a trap or separator is installed upstream to the density gauge.

If the specific gravity of the total stream is to be measured, then the process pressure must be controlled so that the gas bubbles maintain their volume, and the fluid velocity must be high enough to prevent separation or trapping of the gas within the U-tube.

As shown in Table 6.16b the design pressure and temperature limitations of these sensors are a function of the flexible connectors used, which also affect the accuracy of the measurement by their spring characteristics and hysteresis. For example, accuracy when using neoprene liners on the braided connectors is twice as good (±1 percent) as that when using Viton liners (±2 percent).

Small variations in process temperature (±20°F or ±11°C) can be compensated for automatically if the gauge has been factory calibrated with the particular process fluid and at the applicable process temperature. If large temperature variations are expected, it is necessary to install a temperature transmitter in the process stream and to add its signal to the output of the density transmitter. The maximum temperature compensation span is a function of the density span used. Determination of the temperature transmitter range is described in the example below.

Assume a density span for the particular installation of 0.08 SG, or a range of 0.60 to 0.68 SG. The coefficient of thermal expansion for the particular process fluid is 0.0004 SG/°F. Because the effect of temperature change on density should not exceed the density span of the detector, the temperature span is determined as density span divided by expansion coefficient, which in this case is 0.08/0.0004 = 200°F (93°C). Assuming that the density measurement is to be based on 60°F (16°C) the actual range of the transmitter can be 60–260°F (33–144°C) if only above ambient temperatures are expected. See Figure 6.16d which illustrates this compensation, assuming that the thermal expansion coefficient is linear for the above noted temperature range.

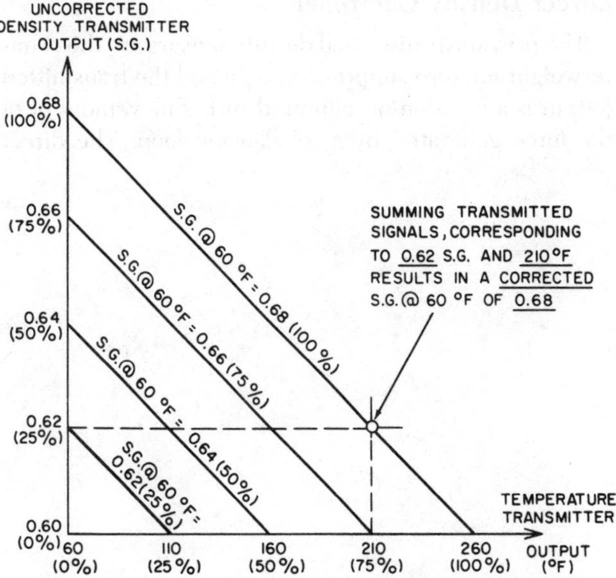

Fig. 6.16d Temperature compensation of density sensors by summing transmitter outputs

Straight-tube Density Transmitter

On applications where the process fluid is a heavy slurry or one that contains large solid particles which would not flow through a U-tube loop, the design illustrated in Figure 6.16e can be considered. The principle of operation is the same as for the U-tube design, but the performance is of lower quality. Its main advantage is that any process fluid that will flow through a 2 in. (50 mm) pipe will pass through this detector. The drawbacks as noted in Table 6.16b include the fact that narrow spans are not available and, therefore, the ± 1 percent inaccuracy based on span represents a substantially

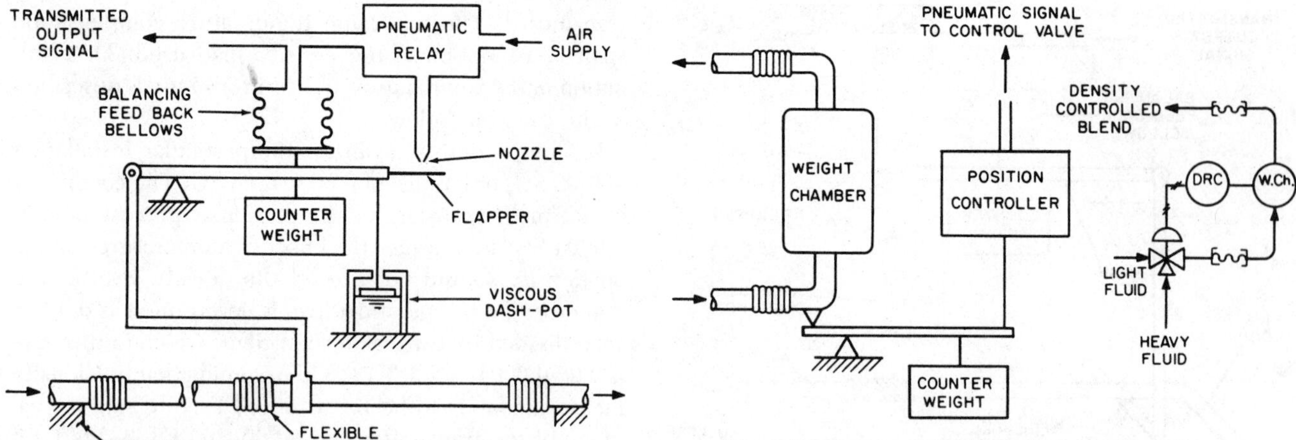

Fig. 6.16e Straight tube density transmitter

Fig. 6.16f Density controller and its typical installation

greater error than in the U-tube design. Temperature compensation can be achieved by the use of a temperature transmitter and summing relay.

Direct Density Controller

The previously discussed density sensors use the counterweight for zero suppression only and the transmitted output is a function of calibrated spring movement or of the force generated by a rebalancing loop. The direct density controller shown in Figure 6.16f is different. Here the counterweight on the beam scale is set to balance the sample chamber and the liquid it contains at the desired density. Therefore, the counterweight represents the set point of this controller and any deviation in density will deflect the beam, which will initiate a counteraction by the controller. In other words, the controller detects the position of the beam and counteracts any movement away from the horizontal.

Flow Sheet Symbol

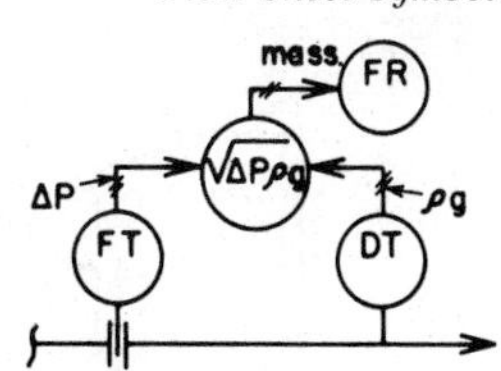

6.17 GAS DENSITY SENSORS OPERATING AT ACTUAL FLOWING CONDITIONS

Design Pressure:	Up to 2000 PSIG (13.8 MPa)
Design Temperature:	Up to 150°F (66°C); up to 300°F (149°C) with special designs
Materials of Construction:	Cast iron, steel, aluminum, stainless steel
Inaccuracy:	±0.1–0.5% full scale
Minimum Span:	1.0 lbm/ft³ (16 kg/m³)
Maximum Span:	25.0 lbm/ft³ (400 kg/m³)
Approximate Cost:	$1100–$2700 as a function of features required
Partial List of Suppliers:	Foxboro Co.; Rotron Inc.; UGC Industries, Inc.

In Section 6.18, gas density sensors are discussed that can determine the stream composition by detecting the molecular weight or specific gravity of the sample.

In this section, gauges are covered that detect the process gas density at operating conditions and generate a measurement signal which can be used for mass flow rate calculation from orifice or volumetric flow data.

There are only two basic designs of these detectors, the displacement and the centrifugal types. These instruments are described below. The liquid densitometers described in Sections 6.6, 6.11, 6.13 and 6.14 can also be adapted to gas density detection.

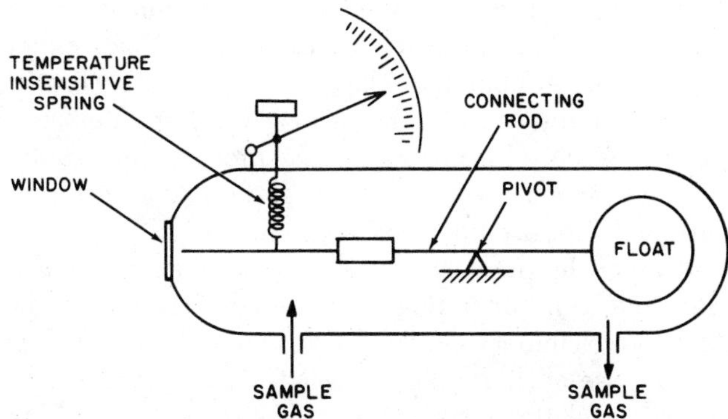

Fig. 6.17a Manual gas density sensor

Displacement Type Gas Density Sensors

This instrument measures the gas density at actual flowing conditions. The effects of process pressure, temperature, supercompressibility or specific gravity changes are automatically detected without need of measuring these properties individually. Archimedes' Principle states that the buoyant force on a float is a function only of the fluid density surrounding the float.

The manually operated version of this gauge is illustrated in Figure 6.17a. It consists of a buoyant float attached to one end of a pivoted beam and a temperature insensitive spring attached to the other end. The spring tension in this case is adjusted manually using a micrometer until the beam is brought into a null position. The readout linkage is attached to the spring and provides linear density data.

Besides the manual method described above, the spring tension can be adjusted by pneumatic relays or electric motors. Figure 6.17b shows a pneumatic-type. While the connecting rod is in equilibrium, both nozzles #1 and #2 are open to their corresponding chambers of the dual-acting cylinder. In this condition, the process pressure is exposed on both sides of the piston, and, therefore, no motion takes place. When the process density changes, a proportional change occurs in the buoyant force acting on the float which causes the connecting rod to rotate around the pivot. As a result, one of the nozzles is covered up and on the corresponding side of the dual-acting cylinder the pressure drops due to the continuous venting. This results in a pressure differential across the piston, causing it to move and thereby adjusting the tension of the spring until the connecting rod is balanced again. When it is rebalanced, the nozzle reopens and the

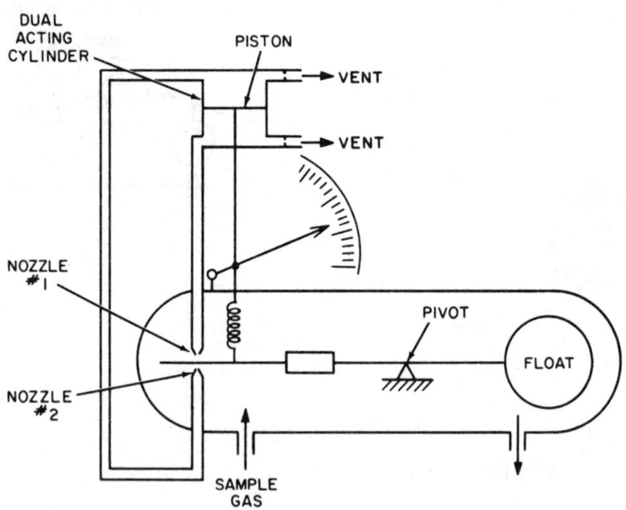

Fig. 6.17b Continuous gas density sensor with pneumatic pilot

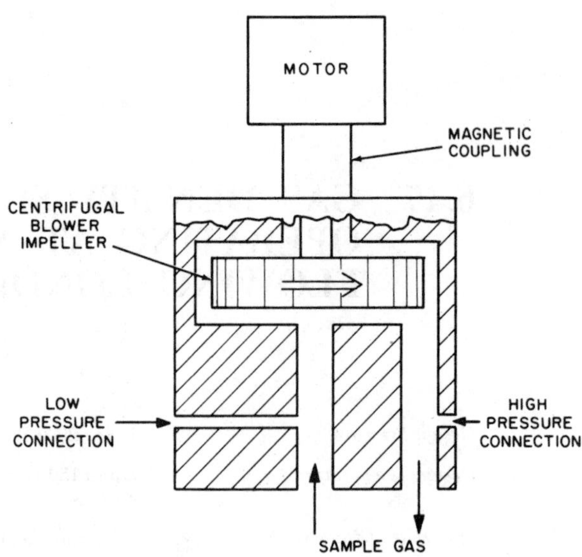

Fig. 6.17c Density detector with centrifugal element

pressures on both sides of the piston are equalized. The equilibrium piston position is proportional to the process density which is indicated, recorded or transmitted as the linear density readout. In cases where the process gas pressure is not sufficient (below 5 PSIG or 34 kPa) or the gas is dirty and might plug the pneumatic relay system, instead of the above described method, either instrument air-operated relays or electric motors can be used to adjust spring tension to balance the detector. The spring itself is not affected by process temperature changes and a great variety of springs is available to match any desired density range.

One of the prerequisites of accurate measurement is that the same operating conditions be present in the detector chamber as in the process line. This is achieved by the use of short (close-coupled) sample lines and by utilizing relatively high sample flow rates (10 SCFH or 280 l/hr).

Spans from 1.0 lbm/ft^3 to 25.0 lbm/ft^3 (16 to 400 kg/m^3) can be obtained for measurements at 0.25 percent of full scale accuracy.

Centrifugal Gas Density Sensors

The centrifugal detector also measures the gas density at flowing pressure and temperature conditions. As shown in Figure 6.17c, a small centrifugal blower operating at constant high speed (3000 to 13,000 RPM) extracts a small sample gas from a tank or pipe line. The impeller is driven by magnetic coupling to allow removal of the driver and to prevent gas leakage. Gas enters the impeller at the center and is thrown outward by centrifugal force. This action creates a pressure differential across the impeller which is directly proportional to the gas density. The differential pressure can be indicated locally or used as the input signal into a d/p transmitter for remote readout.

Some designs can be close-coupled around an orifice plate in the pipeline; others can be installed directly in tanks or pipe.

When the gas sample is dirty, deposits may accumulate in the unit, affecting the sample flow rate. This will not influence the measurement accuracy until the material build-up becomes substantial.

Errors can be introduced by temperature differences between the process and the density chamber. This temperature difference may be caused by the difference between ambient and process temperatures and also by the motor and friction heat developed in the sensor. The error can be as high as 0.1 or 0.2 percent per °ΔF. In most installations, close-coupling and insulating the detector are sufficient. In case of critical measurements,

Density Range		*Output Differential Range*		*Gas Line Pressure Corresponding to Maximum Density Shown (PSIG/kPa)*	
(lbm/ft³)	*(kg/m³)*	*("H₂O)*	*(kPa)*	*Natural Gas*	*Ethylene*
0.25–1.25	4.0–20.0	7–35	1.7–8.7	400/2758	—
0.5–2.5	8.0–40.0	9–44	2.2–10.9	800/5516	—
1.0–5.0	16.0–80.0	11–57	2.7–14.2	1500/10,343	—
2.0–10.0	32.0–160.2	14–68	3.5–16.9	—	600/4137
3.0–15.0	48.0–240.3	16–78	4.0–19.4	—	900/6206
4.0–20.0	64.0–320.4	18–90	4.5–22.4	—	1200/8274
5.0–25.0	80.0–400.5	18–90	4.5–22.4	—	1500/10,343

the process and chamber temperatures are detected and automatic correction is applied for the difference.

Available density spans vary from 1.0 to 20 lbm/ft^3 (16 to 300 kg/m^3) with full scale inaccuracies at ±½ percent or less. Repeatability can be as high as ±0.05 percent.

Wetted components are available in cast iron, aluminum, steel or stainless steel, and the detectors can be exposed to pressures up to 2000 PSIG (14 MPa) and temperatures up to 300°F (149°C).

The maximum pressure differentials developed by the blower are a function of the impeller size selected, speed of rotation and density span. These values vary between 30 and 100 in. H_2O (7.5 to 25 kPa). The relationship between density and pressure differential developed is linear within the range of 5:1. The preceding table lists some typical ranges and some related data.

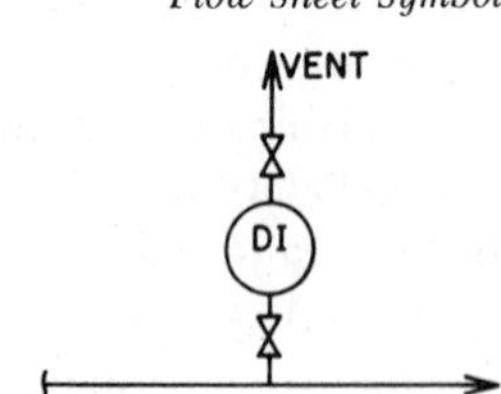
Flow Sheet Symbol

6.18 GAS SPECIFIC GRAVITY DETECTORS OPERATING AT NEAR-AMBIENT CONDITIONS

Design Pressure:	Normally atmospheric
Design Temperature:	Normally ambient
Materials of Construction:	Brass, stainless steel, Pyrex, Monel and/or nickel
Inaccuracy:	±1–2% of span up to ±0.0002 SG
Minimum Span:	0.01 SG (based on air = 1.0)
Maximum Span:	2.5 SG
Approximate Cost:	$1100 to $4500 depending on degree of automation and features
Partial List of Suppliers:	Arcco Instrument Co.; Beckman Scientific Instruments, Inc.; Fischer & Porter Co.; Gow-Mac Instrument Co.; Permutit Co. Inc., Ranarax Instrument Dept.

The purpose of the sensors discussed in this section is stream composition determination in most cases. The devices described are capable of indicating the relative density of gas sample whether this is an inherent quality of the gas or is due to its pressure and temperature.

$$\text{Specific Gravity} = \text{Gas density}\frac{(14.7\ \text{PSIG})\ (\text{Oper. Temp.})}{(\text{Oper. Press.})\ (520°\text{R})}$$

$$\text{or } \frac{(100\ \text{Pa})\ (\text{Oper. Temp.})}{(\text{Oper. Press.})\ (272°\text{K})}$$

Gas specific gravity or molecular weight is of interest in these measurements; consequently, they are usually performed with the sample gas at ambient conditions.

The liquid densitometers described in Sections 6.6, 6.11, 6.13 and 6.14 can also be adapted to gas density detection.

Gas Column Balance Sensors

As their name implies, these instruments operate on the principle of measuring the difference in weight between a column of gas and a column of air having equal heights and pressures.

As shown in Figure 6.18a, the gas sample flows continuously to the interior of the oil-sealed working bell on the right and then leaves the bell through a column of set height into the atmosphere. At the same time, dry air enters the interior of the reference bell and exits to the atmosphere through a column of equal height. The weight of gas and air in the columns exerts an upward force on the bells and the difference in weight between the two columns is the force which causes the beam movement. Beam movement is resisted by the weight of a pendulum, which allows the pen mechanism to move in direct ratio to the specific gravity of the gas flowing through the unit.

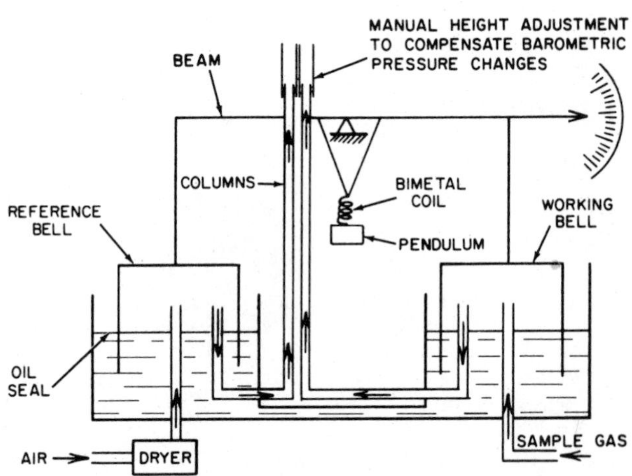

Fig. 6.18a Gas column balance density gauge

The sample flow rate is 2 SCFH (0.9 l/m) or less at a near atmospheric pressure limited by the oil seal in the unit. This corresponds to an approximate measurement time lag of three minutes. Available spans vary between 0.25 and 1.0 SG and ranges can be selected between 0 and 2.5 SG, based on air. Detection accuracy is ±0.002 SG.

The wetted parts can be made of brass or stainless steel and a large number of accessories and features are available, including direct recorders, controllers or transmitters of both electronic and pneumatic types.

Whenever the basis of operation for a density sensor is to compare the gas sample to air in arriving at a specific gravity reading, it is important to compensate for the following:

a. Ambient temperature variations
b. Barometric pressure variations
c. Moisture content of air

The sensor illustrated on Figure 6.18a is provided with an automatic temperature compensator. This is a bimetallic coil so calibrated that it shifts the pendulum weight center of gravity in direct ratio to the temperature effect on the density. Therefore, regardless of the ambient temperature variations, the specific gravity readout is always based on 60°F (16°C).

The forces acting on the working bell are very small. When the sample gravity is 0.5, the total resulting pressure head is 0.036 in. H_2O (9 Pa) and the instrument sensitivity of 0.002 SG corresponds to only 0.00014 in. H_2O (0.04 Pa). With such sensitivity, even barometric pressure variations are sufficient to introduce an error. As an example, a change of 0.5 in. Hg (0.065 kPa) in the barometric pressure can result in an error of 1 percent. Compensation can either be by manual adjustment of column height or by use of an aneroid barometer which automatically corrects for barometric pressure changes.

When air contains moisture, its density at the same barometric pressure is less than what it would be under dry conditions. Changes in humidity have a very definite effect on measurement precision. Therefore, a silica gel drier is installed to keep the reference air purge dry.

From the above discussion it is clear that this detector can automatically correct its readings for humidity, temperature and barometric pressure variations.

Manual Displacement-type Elements

The buoyant force exerted upon a body suspended in a gas is proportional to the density of that gas. If the gas is at ambient conditions, the buoyant force is an indication of the molecular weight or specific gravity of the sample.

One of these gauges is illustrated in the upper part of Figure 6.18b. It consists of a balance beam with a glass cylinder on one end and a counterweight on the other. The manual use of this device first requires filling the chamber with dry air until the beam just balances and noting the corresponding air pressure P_a on the manometer. After this, the chamber is evacuated and filled with

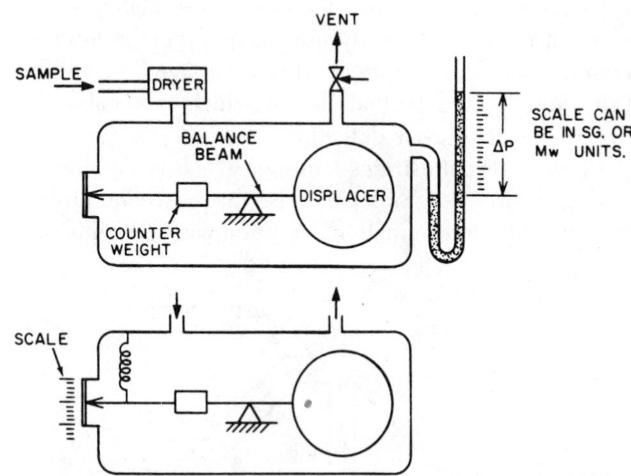

Fig. 6.18b Displacement-type manual density gauges

sample gas until the beam is once more balanced. The corresponding absolute gas pressure P_g is noted and the specific gravity of the sample gas is determined as the ratio of the two pressures:

$$SP_g = \frac{P_g(PSIA)}{P_a(PSIA)} \qquad 6.18(1)$$

It is not necessary to balance the unit in air each time the gauge is used; once it has been balanced, the manometer scale can be calibrated in either molecular weight or specific gravity units. The sample molecular weight is determined by:

$$M_{wg} = M_{wa}\frac{P_g(PSIA)}{P_a(PSIA)} \qquad 6.18(2)$$

At the lower part of Figure 6.18b, a slightly different version of this gauge is shown. Here the arc through which the beam swings is the measure of sample density as indicated on the calibrated scale. If the bulb and beam are balanced in air and the gas sample introduced is of the same pressure, temperature and humidity as the balancing air, then the gauge can read out either in specific gravity or in molecular weight units.

The sensors shown in Figure 6.18b operate on the buoyancy gas balance principle, where the buoyant force of air and gas is compared under identical conditions. The accuracy of measurement is largely a function of eliminating those factors (pressure, temperature, humidity) which can influence density. If the air and gas are compared under identical conditions then the accuracy can be as great as ±0.0002 SG. These sensors are mainly used for high precision laboratory measurements. Their main drawback is that they cannot be adapted for continuous measurement.

Thermal Density Gauges

Thermal conductivity-type sensors are discussed under Section 10.16 and also in Section 10.4 in connection with the chromatographs. Because the thermal-type elements are most frequently used as analyzers or as components in analyzer systems, their coverage here will be brief and limited to that design which is suitable for direct specific gravity detection.

Figure 6.18c illustrates a design which is suitable for both gas chromatography and also for on-stream molecular weight determination. A pneumatic Wheatstone bridge with two hot wire detectors is mounted in the vertical plane. The reference gas tends to cool the thermistors and as long as the flow is balanced the two detectors are equally cooled. When a sample gas is introduced, it will upset the system balance if its molecular weight is different from that of the reference gas. Zero balance is checked by introducing the reference gas as sample. If the sample gas is lighter than the reference gas, the sample molecules will tend to rise, partially obstructing the reference gas flow at detector #1 and, therefore, causing a temperature rise at that point. Simultaneously, a corresponding increase of reference gas flow will be experienced at detector #2, causing a drop in temperature at that point. The temperature difference between the two thermistors is a measure of the sample gas molecular weight.

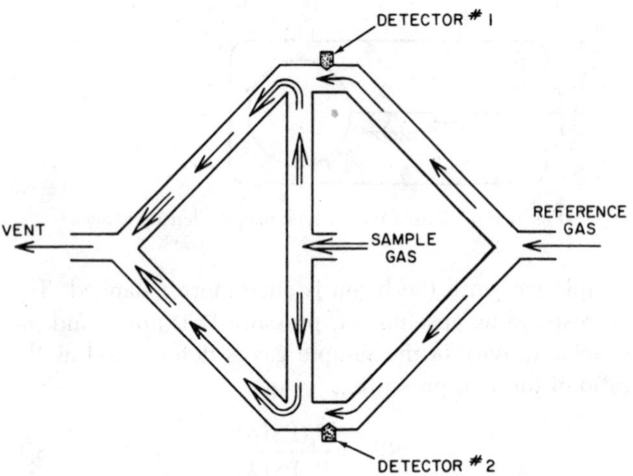

Fig. 6.18c Thermal gas density sensor

The reference gas selected should have sufficient difference in density from the sample for accurate measurement, and preferably have high molecular weight, high heat capacity and low viscosity. Nitrogen, argon or carbon dioxide tend to satisfy these requirements. The sample flow is normally maintained at about 10 cc/minute, while the reference flow is set at about ten times that to avoid back-diffusion.

The main advantage of this design is that the sample gas never comes in contact with the detectors, and, therefore, they cannot become coated, coked or contaminated by the sample. Another desirable feature is that the cell contains no moving parts.

Due to the tubular design, brass, stainless steel, Monel or nickel construction materials can be provided without substantial difference in cost.

Variations in ambient temperature can introduce errors and, therefore, the cell temperature is normally controlled within ±1°F (±0.6°C).

Viscous Drag-type Density Sensors

Another method of detecting the density of a gas relative to air (specific gravity or molecular weight) is illustrated in Figure 6.18d.

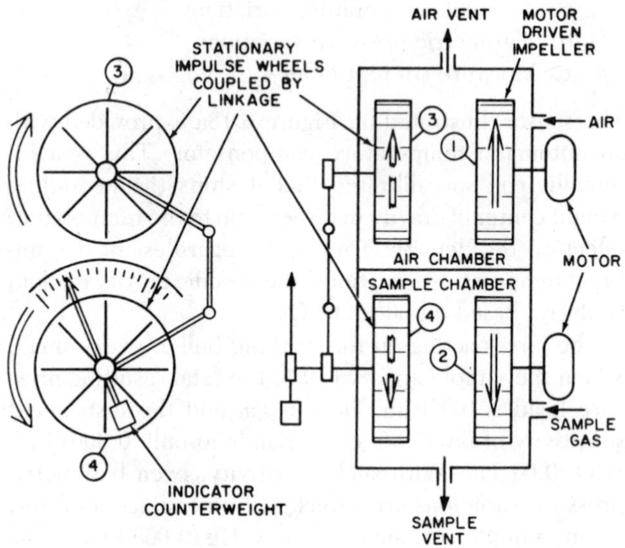

Fig. 6.18d Viscous drag density sensor

The unit consists of a sample chamber and an air chamber with a motor driven impeller and an impulse wheel in each. Impeller #1 draws in air and spins it against the vanes of impulse wheel #3, creating a clockwise torque proportional to the air density. Impeller #2 draws in the gas sample and sets it in a counterclockwise rotation. The whirling gas strikes the impulse wheel #4 and imparts a counterclockwise torque to its shaft proportional to the gas density. The difference between the opposing torques is a measure of the sample specific gravity. It is transmitted through a lever and linkage arrangement to the pointer which moves in front of the scale, which is calibrated in specific gravity, molecular weight or percent concentration units.

As in the case of the previously discussed detectors, here, too, the sample temperature, pressure, humidity and solid particle content will affect the precision of measurement.

The sample gas temperature should be the same as that of ambient air. This is normally guaranteed by the large surface area of the sample tubing that brings the gas temperature to ambience without need for auxiliary devices.

Because both the gas and air are discharged to atmosphere, the barometric pressure variations are automatically compensated for.

If the sample gas is under pressure, it has to be reduced to less than an inch of water before entering the instrument, and if it is under vacuum, the exhaust will have to be piped into the same vacuum system.

The relative humidity of air and gas should be the same. This can be achieved either by saturating both streams with a humidifier or by drying both streams through a reference dryer.

Sample filters are normally used to remove suspended particles or corrosive constituents because the wetted parts are not highly corrosion-resistant. Brass, aluminum and stainless steel are in contact with the sample.

The sample flow rate of about 20 SCFH (9 l/m) guarantees a response time of less than 20 seconds.

Spans from 0.1 to 2.0 SG are available within the limits of 0.1 and 3.0 SG. The measurement inaccuracy varies between 1 and 2 percent of full scale and the unit is normally calibrated on the actual sample gas.

This instrument can be obtained as an indicator, recorder, switch or transmitter. Both pneumatic and electronic versions are available.

Electromagnetic Suspension-type Sensors

The main working component of this detector is a small dumbbell supported on a quartz fibre. One of the spheres on the dumbbell is sensitive to buoyancy effects (changes in gas density), the other is not. Insensitivity to buoyancy is achieved by puncturing that sphere (see Figure 6.18e).

A mirror fixed to the dumbbell axis reflects a light beam to a dividing mirror which splits the beam equally between two photocells. A change in gas density tends to rotate the dumbbell causing the light to be unevenly divided between the photocells. Therefore, the signals produced by the photocells will differ; when this difference is amplified it can be applied to restabilize the dumbbell. This is achieved by applying a new electrical potential to the electrodes which generate the electrostatic field around one of the spheres. The sphere is made

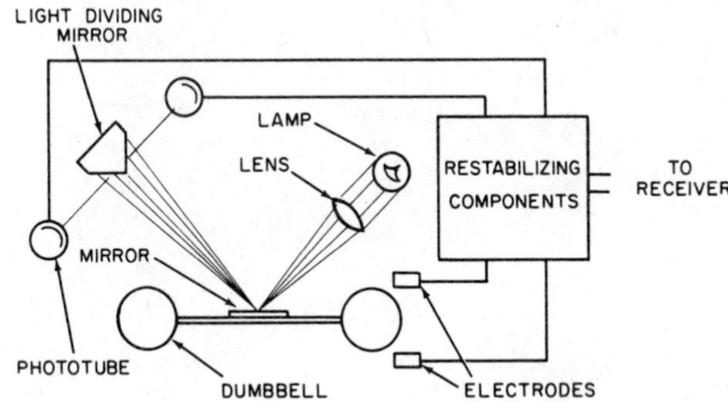

Fig. 6.18e Electromagnetic suspension-type sensor

conductive with rhodium coating. Measuring the electrical potential required to stabilize the dumbbell gives a linear indication of the torque created by the differential buoyancies, which in turn is an indication of sample density. If the sample is at ambient conditions, the instrument scale can be calibrated in specific gravity or molecular weight units.

In contrast with the manual displacement units, these detectors are adaptable for continuous measurements and do generate output signals for remote potentiometric readout. They are available with spans of 0.01 SG to 2.0 SG based on air with ±1 percent full scale accuracy. Single range units have a total span of 1.0 SG and multi-range designs are available with a 5:1 range ratio. The measurements are performed at near atmospheric pressures and ambient temperatures, utilizing sample flow rates in the 50 to 500 cubic centimeter per minute range. Depending on this flow rate, the 95 percent response time of the detector varies from a few seconds to one minute. Due to the small openings and the optical parts involved, the sample gas must be clean upon entering the sensing cell. This detector is compensated for barometric changes and is manually calibrated on air or other reference gases.

This device is selected for those installations where high precision overrides the cost considerations.

Chapter VII

VISCOSITY MEASUREMENT

C. H. Kim

CONTENTS OF CHAPTER VII

Table VII
ORIENTATION TABLE FOR VISCOMETERS

	Section	Type of Design	Features	Provides Continuous Signal	In-Line Device	Laboratory Device	Local Readout	Remote Readout Trans.	Temp. Compensation	Fluids: Gas	Fluids: Newtonian	Fluids: Non-Newtonian	Maximum Design Pressure, PSIG **	Maximum Design Temperature, °F**	Inaccuracy, % (1) Based on Full Scale (2) Based on Measurement	Minimum Sample Size or Flow Rate**	APPLICABLE VISCOSITY RANGES* Centipoises** (10^{-2} 10^{-1} 1 10^1 10^2 10^3 10^4 10^5 10^6 10^7 10^8 10^9 10^{10})
Laboratory	7.2	Bubble Time	Manual			√	√				√		ATM.	77	2.0(2)	13 CC	
	7.3	Capillary Tube	Manual Timing			√	√				√		ATM.	300	0.35(2)	20 CC	
			Auto. Timing		√	√	√	√			√		ATM.	300	0.01(2)	20 CC	
	7.4	Capillary Extrusion	Influx			√	√				√	√	100	300	2.0(2)	0.7 CC	
			Efflux			√	√	√			√	√	5,000	640	2.0(2)	30 CC	
	7.5	Efflux Cup	Saybolt			√	√				√		ATM.	250	0.1(2)	60 CC	
			Ford Cup			√	√				√		ATM.	80	2.0(2)	150 CC	
			Zahn Cup			√	√				√		ATM.	80	2.0(2)	44 CC	
			Auto Timing		√	√	√	√			√	√	ATM.	80	5.0(2)	—	
	7.6	Falling Ball	Manual			√	√			√	√		ATM.	300	0.1(2)	30 CC	
			Automatic		√	√	√				√	√	15,000	350	0.1(2)	70 CC	
	7.7	Rotational	Coaxial-Cylinder			√	√	√			√	√	ATM.	2,500	1.0(2)	1 CC	
			Cone & Plate			√	√	√			√	√	ATM.	750	0.5(2)	0.1 CC	
Industrial	7.8	Continuous Capillary	Differential Pressure	√	√	√	√	√			√		115	240	1.0(1)	1 GPH	
			Back Pressure	√	√	√	√	√			√		65	210	0.5(1)	1 GPH	
	7.9	Falling Piston			√	√	√	√	√		√	√	500	650	1.0(1)	—	
	7.10	Float	Single Float	√	√	√	√	√			√	√	650	450	4.0(2)	0.75 GPM	
			Two-Float	√	√	√	√				√		300	450	2.0(2)	0.25 GPM	
			Concentric	√	√	√	√	√	√		√	√	650	450	2.0(2)	2 GPM	
	2.11	Plastometer	Cone & Plate			√	√	√				√	100	400	0.5(1)	25 CC	Arbitrary Units Are Used / Mooney / 0–200 Points
			Kneader	√	√	√	√	√				√	ATM.	570	1.0(1)	80 CC	% Scale / 0–1,000 Division
			Capillary	√	√	√	√	√				√	5,000	570	2.0(1)	0.6 #/HR	MI & CIL / 0–200 & 0–100
	7.12	Rotational	Rotating Cone	√	√	√	√	√	√		√	√	100	650	1.0(1)	—	
			Agitation Power	√	√		√	√			√	√	125	200	~5.0(1)	15 GPM	
	7.13	Vibrational	Reed	√	√	√	√	√	√		√	√	3,000	300	1.0(1)	—	

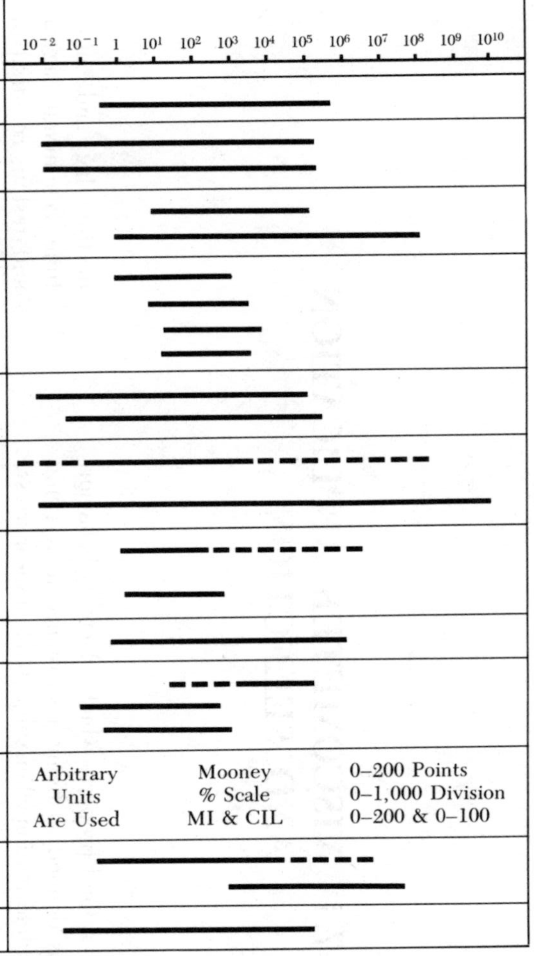

* ▬▬▬ Normal Range ▬ ▬ ▬ With Special Modifications.
**For SI units, see Appendix A.1

7.1 VISCOMETER APPLICATION AND SELECTION

Viscosity is a fluid property which affects its behavior. If a fluid is defined as being a substance undergoing continuous deformation when subjected to a shear stress, then the consistency can be termed the resistance offered by the fluid to such deformation. If static pressure and temperature are fixed, the consistency is constant for gases and Newtonian liquids, and is called absolute viscosity. The consistency of non-Newtonian fluids varies even though the static pressure and temperature are fixed, as a function of the applied shear stress. In some cases, the consistency may vary with duration of the applied shear stress. The consistency of non-Newtonian fluids is frequently expressed in terms of apparent viscosity. Thus the viscometer is an instrument which measures consistency of gases, and Newtonian and non-Newtonian fluids.

In modern science and in processing plants, viscosity measurements are used in determining (1) flowability of fluids, (2) concentration, size and shape of solids in a slurry, (3) molecular weight and its distribution of high molecular weight substances, and (4) color (of inks). Many designs of viscometers are available to characterize these properties in a laboratory or in a producing plant environment, or for both.

Viscometer Selection

In selecting a viscometer to do a specific task, the following should be known:

1. Is this instrument for laboratory use or for continuous measurement in the plant for control?
2. What type of materials will come in contact with the instrument?
   a. Highly volatile?—needs closed system?
   b. Newtonian or non-Newtonian fluids, or both?
   c. Rheological characteristics of material—plastic, thixotropic, dilatant, etc.
   d. Corrosiveness of the fluids.
   e. Does the fluid contain solids and what are the special characteristics of this slurry or emulsion?—settling-out, adhesive, fibrous, etc.
   f. Temperature and pressure of the fluids.
   g. Do the sample composition (due to reaction or time lag) and consistency change with time? Is minimum lag time between sampling and testing required or is on-stream type needed?
   h. Is viscosity vs. temperature relationship known?
3. Testing environment—needs explosion proofing?
4. Viscosity ranges to be measured.
5. Accuracy and sensitivity required (for continuous process viscometer, repeatability requirement).
6. Special features needed:
   a. Remote indication or recording.
   b. Automatic operation.
   c. Automatic closed-loop control.
   d. Temperature compensating system.
7. Viscometer response time requirement.
8. Flow Conditions—laminar or turbulent flow?

An orientation table (Table VII) has been prepared to highlight viscometer features. When, based on a prepared application requirement check list, several possible choices of selections exist, it is suggested to look for further details in the appropriate section for final selection.

This chapter has been so organized that the first seven sections cover the laboratory type designs and the last seven sections deal with industrial in-line detectors. The discussions on the theoretical aspects of viscosity are given in the text with the corresponding hardware description.

Terminology and Units

Here only a brief summary is given of terminology and equations:

Stress Force/Area—F/A

Velocity Gradient (Shear) Rate of change of liquid velocity across the stream—V/L for linear velocity profile, dV/dL for non-linear velocity profile. Units are $V/L = ft/sec/ft = sec^{-1}$.

Absolute (dynamic) Viscosity (μ) Constant of proportionality between applied stress and resulting shear velocity (Newton's Hypothesis).

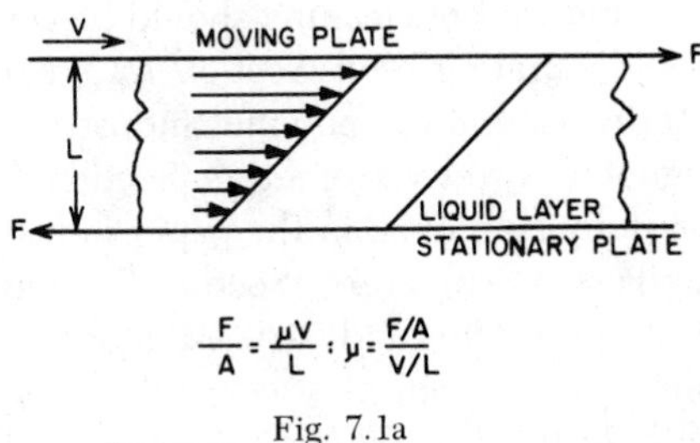

$$\frac{F}{A} = \frac{\mu V}{L};\ \mu = \frac{F/A}{V/L}$$

Fig. 7.1a

Poise (μ) Unit of dynamic or absolute viscosity—dyne-sec/cm^2.

Kinematic Viscosity (ν) Dynamic viscosity/density = μ/ρ.

Stoke Unit of kinematic viscosity (ν).—cm^2/sec.

Hagen-Poiseuille Law (flow through a capillary).

$$Q = \frac{\pi R^4}{8\mu L}(P_1 - P_2) \qquad 7.1(1)$$

Saybolt Viscometer (Universal, Furol) Measures time for given volume of fluid to flow through standard orifice; units are seconds.

Fluidity Reciprocal of absolute viscosity; unit in the cgs system is the rhe, which equals 1/poise.

Specific Viscosity Ratio of absolute viscosity of a fluid to that of a standard fluid, usually water, both at same temperature.

Relative Viscosity Ratio of absolute viscosity of a fluid at any temperature to that of water at 20°C (68°F). Since water at this temperature has a μ of 1.002 cp, the relative viscosity of a fluid equals approximately its absolute viscosity in cp. Since density of water is 1, kinematic viscosity of water equals 1.002 ctks at 20°C.

Apparent Viscosity Viscosity of a non-Newtonian fluid under given conditions. Same as consistency.

Consistency Resistance of a substance to deformation. It is the same as viscosity for a Newtonian fluid, and the same as apparent viscosity for a non-Newtonian fluid.

Saybolt Universal Seconds (SUS) Time units referring to the Saybolt viscometer.

Saybolt Furol Seconds (SFS) Time units referring to the Saybolt viscometer with a Furol capillary, which is larger than a universal capillary.

Shear Viscometer Viscometer that measures viscosity of a non-Newtonian fluid at several different shear rates. Viscosity is extrapolated to zero shear rate by connecting the measured points and extending curve to zero shear rate.

For viscosity conversion data, refer to the information below.

To convert centipoises into other units, multiply the centipoises by the following constants:

0.001 for Pascal seconds
0.01 for viscosity in poises
0.000672 for viscosity in (lb mass)/(sec)(ft) units
2.42 for viscosity in (lb mass)/(hr)(ft) units
3.60 for viscosity in (kg mass)/(hr)(m) units
5.60×10^{-5} for viscosity in (lb mass)/(sec)(in.) units
1.45×10^{-7} for viscosity in (lb force)(sec)/(in.^{2}) units

The value of the kinematic viscosity (in cm^2/sec units) can be obtained approximately from the indications in seconds t of various viscometers by the following equations:

Saybolt Universal,	
when $32 < t < 100$,	$\nu = 0.00226t - 1.95/t$
when $t > 100$,	$\nu = 0.00220t - 1.35/t$
Saybolt Furol,	
when $25 < t < 40$,	$\nu = 0.0224t - 1.84/t$
when $t > 40$,	$\nu = 0.0216t - 0.60/t$
Redwood No. 1 (English),	
when $34 < t < 100$,	$\nu = 0.00260t - 1.79/t$
when $t > 100$,	$\nu = 0.00247t - 0.50/t$
Redwood Admiralty (English),	$\nu = 0.027t - 20/t$
Engler (German),	$\nu = 0.00147t - 3.74/t$

Viscometer Application

A viscosity measurement can be of value for one of the following two reasons:

1. Viscosity is a direct measurement of fluid characteristics and behavior when in motion. It is very difficult to size a pump, pipe line, orifice meter, or agitator without knowing the viscosity of the process fluid. In any operation where liquids are applied (sprayed, coated, or dipping process), viscosity of the fluid determines its effectiveness and the quality of the finished product. In short, viscosity is one of the most important process properties.
2. Viscosity detection can be a very sensitive indirect measurement of other properties. Molecular weight and its distribution in polymers, lubrication oils, and others, as well as insoluble solids concentration, specific gravity, color, and solids size, shape and their distribution in a slurry or in an emulsion can all be reflected in viscosity variations.

The reader is encouraged to look for other applications of viscosity measurement, because it is one of the simplest, most accurate and reliable of measurements.

In reviewing the techniques available for measuring fluid viscosity, the viscometers can be regrouped based on their intended use: (1) finished product specification, (2) routine laboratory testing, (3) scientific research study, and (4) in-line process control.

Finished Product Specification

Here, the type of viscometer to be used has been already specified by industry-approved standard testing methods. One caution is that the testing procedures should be carefully followed and test results correctly reported.

Routine Laboratory Testing

Simple to operate, easy to clean and direct reading viscometers should be considered for this purpose. The coaxial-cylinder type viscometer (Section 7.7) is well suited for this purpose since it is inexpensive and meets most of the above requirements. The efflux-cup viscometer is recommended (Section 7.5) for field laboratory testing work. If sample size is a problem (less than 1 cc), then modified coaxial-cylinder or cone-and-plate rotating viscometers (Section 7.7) should be considered.

Scientific Research Study

For this purpose, accuracy and versatility should be considered first. Cone-and-plate rotational viscometers (Section 7.7) are the most versatile units but are also the most expensive. If extreme accuracy is desired, consider the automatic capillary-tube viscometer (Section 7.3). If recording of results is important for a permanent record, the previously noted viscometers have this feature. For measurement of viscosity of gases, the falling-ball viscometer (Section 7.6) is the only one that can be used.

In-Line Process Control

In selecting an in-process viscometer, cost, repeatability, sensitivity, construction materials, reliability, response time, and ease of cleaning should be considered. Ultrasonic and vibrating-reed (Section 7.13) viscometers are successfully used in the polymer industry. Rotating-cone and agitator power viscometers (Section 7.12) have been successfully employed in the paper industry. Continuous capillary viscometers (Section 7.8) are widely applied in the petroleum industry. Manufacture of synthetic rubbers and certain plastics would be almost impossible without the plastometers (Section 7.11).

It has been found that today's continuous viscometers are reliable. In modern production, viscosity measurement need not be an expensive and time-consuming operation. There are continuous in-line viscometers available today to satisfy most processes, including molten steel viscosity and there are automatic control systems in which viscosity is the primary variable. Viscometer signals can be readily accepted by a process computer to calculate other fluid properties from the viscometer data and to perform closed-loop control.

BIBLIOGRAPHY

Hallikainen, K. E., "Viscometry," *Instruments and Control Systems*, November, 1972.

Nelson, Robert C., "Automatic Force Viscometers," *Instruments and Control Systems*, April, 1963.

Nelson, R. C., "Manual Force Viscometers," *Instruments and Control Systems*, May, 1963.

"Viscometer Survey," *Instruments and Control Systems*, January, 1971.

Flow Sheet Symbol

7.2 LABORATORY VISCOMETERS—BUBBLE TIME

Operating Temperature:	77°F (25°C)
Operating Pressure:	Atmospheric or slightly higher
Materials of Construction:	Glass
Cost:	\$10–\$170
Inaccuracy:	±2 % to ±10% of reading
Range:	0.5 centistokes to 125,000 centistokes ($5 \times 10^{-7}m^2/s$ to $0.13m^2/s$)
Partial List of Suppliers:	L. C. Eitzen Co.; Gardner Laboratory, Inc.; Koehler Instrument Co., Inc.; Precision Scientific, Subs. of GCA Corp.

The time for a given volume of liquid to gravity flow through a capillary, or other restrictive orifice, is a measure of its kinetic viscosity. In a bubble time viscometer, a liquid streams downward in the annular zone between the glass wall of a sealed tube and the perimeter of a rising air bubble.

The bubble time viscometer is used to determine kinematic viscosity of transparent liquids, either in bubble seconds or approximate stokes, by timing or by comparison.

The comparison method determines the viscosity of a fluid by comparing the speed of bubble-rise in a tube, having the same dimensions as the standard tube, which is filled with fluids of known viscosities. A set of 36 standard tubes is available covering a viscosity range of 22 centistokes to 100,000 centistokes ($22 \times 10^{-6}m^2/s$ to $0.1m^2/s$) in uniform, logarithmically even increments of about 26 percent. The viscosity of the fluid under test is assumed to be equal with the viscosity of that fluid in the standard tube which had the nearest bubble speed. Another type, rather than requiring sets of standard tubes, consists of only one standard tube, and determines the viscosity of the fluid by reading a precalibrated scale in Saybolt seconds. The distance which the bubble travels in a fixed time period is related by the scale to viscosity.

The direct timing of the bubble speed method is based on the observation that the nominal viscosity (in stokes) of the liquid in the bubble tube (10.65 ± 0.025mm inside diameter) is numerically equal to the time (in seconds) required for the bubble to travel a distance of 73 millimeters.

Because of the influence of specific gravity, surface tension and thixotropy, the accuracies are low and depend upon the scale range being used.

For best results, particularly when the timing method is used to obtain kinematic viscosities, precautions must be taken to assure true verticality of the glass tube with a standard bore and to provide good temperature control. Care in preparation of the sample tube should be exercised to ensure a bubble of suitable and uniform size. A tube tipped by only one radius off the vertical axis will give an error of approximately 10 percent in the time of bubble travel. A temperature variation of only 1°F (0.6°C) will cause a 5 percent variation in the timed bubble travel. For low viscosity liquids with bubble speed below 5 seconds, it is advised to make comparisons against predetermined standard values.

One advantage of this type of viscometer is that it requires no calibration or recalibration. The bubble time viscometer is well suited for viscous solutions where evaporation losses must be avoided. It is not suitable for liquids containing crystal, fiber, or gel particles. Because of its simplicity and low cost, it is suited for routine industrial use by operators without any special training.

BIBLIOGRAPHY

Nelson, R. C., "Manual Force Viscometers," *Instruments and Control Systems*, May, 1963.

"Viscometer Survey," *Instruments and Control Systems*, January, 1971.

Flow Sheet Symbol

VI

7.3 LABORATORY VISCOMETERS—CAPILLARY

Operating Temperature:	100 to +300°F (−73 to +150°C)
Operating Pressure:	Atmospheric
Materials of Construction:	Glass
Cost:	\$25–\$250 for capillary tubes; \$340–\$2500 for thermostatic bath; \$6000 for auto-viscometer
Inaccuracy:	±0.35% of reading for manual clocking; ±0.01% of reading for auto-viscometer
Range:	0.2–120,000 centipoises (0.0002–120 Pa·s)
Partial List of Suppliers:	Cannon Instrument Co.; Chatas Glass Co.; Fish-Schurman Corp.; Gardner Laboratory, Inc.; Hewlett-Packard Co.; Koehler Instrument Co., Inc.; Precision Scientific Co., Subs. of GCA Corp.

The capillary-tube viscometer consists of a fluid reservoir to hold a specified volume of sample liquid and of a capillary tube. The hydro-static head of the fluid causes the liquid to flow through the capillary as illustrated in Figure 7.3a. A clock for measuring the efflux time of the fixed liquid volume and a thermostatic device complete the apparatus.

The capillary-tube viscometer gives kinematic viscosity in stokes from measurements of the pressure gradient and of the volumetric flow-rate in a cylindrical tube of precisely known dimensions as stated by the Hagen-Poiseuille Law for the flow of fluids through a capillary.

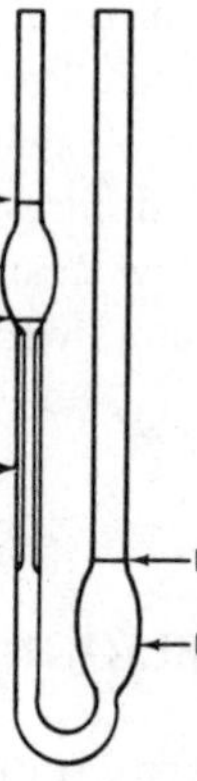

UPPER ETCHED INDEX LINE
EFFLUX VESSEL
LOWER ETCHED INDEX LINE
CAPILLARY TUBE
FILLING ETCHED LINE
RECEIVING VESSEL

Fig. 7.3a Ostwald viscometer

$$\nu = \frac{\mu}{\rho} = \frac{\pi g h R^4 t}{8VL} \qquad 7.3(1)$$

where

ν = kinematic viscosity, Stokes (cm^2/sec)
μ = absolute viscosity, Poises (dyne-sec/cm^2)
ρ = density of liquid, g/cm^3
g = acceleration due to gravity, cm/sec^2
h = vertical distance between ends of capillary, cm
R = radius of capillary, cm
L = length of capillary, cm
V = volume of liquid flowing, cm^3, in time t, sec.

The simplest type as illustrated in Figure 7.3a is the Ostwald viscometer. Various modifications of the classical Ostwald device are available to suit various application needs. The method of measuring kinematic viscosity is detailed in ASTM D-445 bulletin. In practice, a sample liquid of fixed volume is charged to the lower receiving vessel and the viscometer is placed in a thermostatic bath. After time is allowed for the sample liquid to reach thermal equilibrium (about 5 minutes), the sample is drawn up into the efflux vessel by suction until the level is above the upper etched index line. The fluid is then permitted to flow down through the capillary by releasing the suction. When the fluid surface passes the upper etched index line, a stopwatch is started. The watch is stopped when the surface passes the lower etched index line of the efflux vessel.

From this efflux time (t), the kinematic viscosity of the fluid is calculated by multiplying it by the viscometer calibration constant.

Because of the small driving force caused by the hydrostatic head of the fluid and because of the change in hydrostatic head with time, the capillary-tube viscometer is usually restricted to low-viscosity Newtonian fluids. Even so, the device is a simple and convenient instrument for measuring kinematic viscosity accurately in the range of 0.2 to 120,000 centistokes (0.0002 to 120 Pa·s). If the procedure of ASTM test is carefully followed, a repeatability of ±0.10% and a precision of about ±0.35% of reading may be achieved. To achieve this type of accuracy and repeatability, (1) the constant-temperature bath should be maintained with a uniformity of ±0.02°F (0.01°C), (2) efflux times of 100 to 700 seconds are desired, and (3) precise calibration should be performed at that temperature at which the device will operate. Besides measuring kinematic viscosity, capillary-tube viscometers are used for intrinsic viscosity determinations, molecular weight measurements by relating it to intrinsic viscosity, and study of molecular shapes of natural and synthetic polymers.

Since the unit is normally operated under vented (atmospheric) conditions and because the time lag from sample taking to measurement is large, its use on highly evaporative or hydroscopically deteriorative samples should be avoided. This type of viscometer is recommended for use under static and stable conditions. Also, the sample liquid should be filtered before use to prevent clogging of the capillary tube by solids.

The capillary-tube viscometer, with controlled temperature, is capable of very accurate measurements, and is relatively inexpensive, easy to operate, and needs little or no maintenance aside from cleaning.

Calibration

The first step in making kinematic or absolute viscosity measurements is to calibrate the capillary-tube viscometer. This is done by measuring the efflux time for calibrating liquids whose viscosity and density are known. Water or a standard ASTM calibrating liquid (see Tables 7.3b, c and d) may be used. The calibration constant is determined graphically by the following equation:

$$\frac{\mu}{\rho t} = A - \frac{m\beta}{t^2} \qquad 7.3(2)$$

where

μ = absolute viscosity, poises
ρ = density of liquid, g/cm²
t = efflux time, seconds
A = viscometer constant
m = kinetic energy coefficient
β = kinetic energy constant

By plotting $\mu/\rho t$ vs. $1/t^2$ for several known viscosities, and connecting the points by a straight line, the value of $m\beta$ is obtained from the negative slope of the line; the value of A is obtained from the intercept of the line with the ordinate. (See Figure 7.3e).

Table 7.3b
APPROXIMATE VISCOSITY OF ASTM VISCOSITY STANDARDS IN CENTISTOKES

ASTM Viscosity Standard	*Approximate Kinematic Viscosity in Centistokes,* at*							
	*−65°F***	*−40°F*	*68°F*	*77°F*	*86°F*	*100°F*	*122°F*	*210°F*
S-3	300	80	4.6	4.0	—	3.0	—	1.2
S-6	—	—	10	9.0	—	6.0	—	1.8
S-20	—	—	44	35	—	20	—	3.9
S-60	—	—	160	120	—	60	—	7.7
S-200	—	—	700	480	—	200	—	16
S-600	—	—	2,500	1,600	—	600	280	32
S-2,000	—	—	9,000	5,700	—	2,000	—	76
S-8,000	—	—	38,000	22,000	—	8,000	—	—
S-30,000	—	—	—	—	50,000	27,000	11,000	—

*1 centistoke = 10^{-6} m²/s

**$°C = \frac{°F - 32}{1.8}$

Table 7.3c
APPROXIMATE VISCOSITY OF ASTM VISCOSITY STANDARDS IN CENTIPOISES

ASTM Viscosity Standard	Approximate Viscosity in Centipoise,* at					
	65°F**	77°F	86°F	100°F	122°F	210°F
S-3	3.8	3.3	—	2.5	—	0.9
S-6	8.6	7.7	—	5.1	—	1.5
S-20	38	30	—	17	—	3.2
S-60	140	100	—	51	—	6.3
S-200	620	430	—	180	—	14
S-600	2,200	1,400	—	530	250	32
S-2,000	7,900	5,000	—	1,700	—	63
S-8,000	34,000	19,000	—	7,000	—	—
S-30,000	—	—	46,000	24,000	9,500	—

*1 Centipoise = 10^{-3} Pa·s

**$°C = \frac{°F - 32}{1.8}$

With viscometer factors A and mβ known, efflux time measurements can be made for unknown liquids to determine their kinematic viscosity.

$$\nu = At - \frac{m\beta}{t} \qquad 7.3(3)$$

When the efflux time is very large, the kinetic energy correction (mβ/t) in the viscosity equation will be small and can often be neglected.

With the kinematic viscosity known, the absolute viscosity is calculated by multiplying the kinematic viscosity by the fluid density. Measurement of liquid density and efflux time is to be made at the same temperature for accurate values of absolute and kinematic viscosity.

Automatic

Automatic capillary-tube viscometers are designed to operate using conventional glassware and to give precise kinematic viscosity measurements automatically. The system consists of a basic control console to provide programming of influx and efflux operations, measurement of the efflux time, automatic data display, print-out and precise temperature control units.

Precise results of ±0.01% inaccuracy and ±0.007% repeatability may be obtained because the temperature is maintained within ±0.01°F (0.006°C), automatic influxing (same level every time) eliminates drainage error, and the electronic timing unit is controlled to 0.01 second resolution.

Once the viscometers have been loaded with samples, the control circuitry starts the pump and valves and thus influxes the sample into the measuring bulb. Here a specified time (about 3 minutes) is allowed to reach bath temperature. At the end of the equilibrium period, the timing circuit signals the liquid to descend. When fluid passes the upper photocell detector, the electronic timing counter starts and continues to run until the fluid passes the lower photocell detector.

The timing counter is driven by a thermostatic crystal-controlled oscillator and displays centistoke values directly on a digital read-out device.

The photocells actuated by changes in light refraction rather than changes in intensity of transmitted light are preferred to avoid effects of sample color variation.

Table 7.3d
APPROXIMATE VISCOSITY OF ASTM VISCOSITY STANDARDS IN SAYBOLT UNIVERSAL SECONDS

ASTM Viscosity Standard	Approximate Viscosity in SUS, at	
	100°F*	210°F
S-3	36	—
S-6	46	—
S-20	100	—
S-60	290	—
S-200	930	—
S-600	—	150

*$°C = \frac{°F - 32}{1.8}$

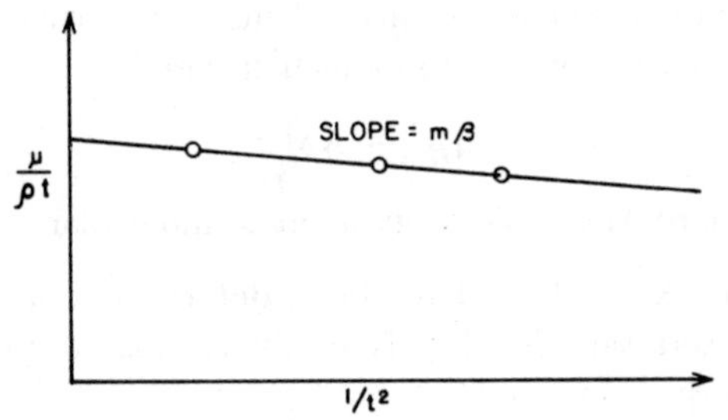

Fig. 7.3e Calibration of capillary tube viscometers

The unit requires an external cooling medium, a nitrogen gas supply, vacuum (20″ Hg [68 kPa] or better) and 115/230 volts ±10%, 50 or 60 Hz, of electricity. It can contain several capillary tube assemblies to permit continuous operation, by having one viscometer in the measurement cycle, while others can be in cleaning, drying, loading and prewarm cycle stages.

Because of the detector response time of 10 microseconds, minimum meniscus speeds of one inch per minute are required. The constant temperature bath is controlled between 40°F to 275°F (4.4°C to 135°C).

Use of photocell detectors and automatic cleaning methods suggest that highly viscous liquids and staining liquids can not be tested if they tend to cling to or stain the capillary tubes.

Because of accuracies obtainable with this type of instrument, it is well suited for research work, even though it is more expensive than other types.

The automatic capillary-tube viscometer can be converted to process viscometry by providing an automatic sampling system with adjustable influx pressure regulation. This will give intermittent viscosity measurements, but it is not suitable for fluids in processes undergoing rapid changes.

Intrinsic Viscosity (LVN) and Molecular Weight

Intrinsic viscosity (also called limiting viscosity number or LVN) is useful in determining the molecular weight and the general shape of polymer molecules in solution.

The intrinsic viscosity of dilute polymer solutions can be obtained from efflux time measurements. The polymer solution should be dilute enough to give Newtonian characteristics.

First, the value for viscosity ratio (μ_r = relative viscosity) is calculated:

$$\mu_r = \frac{t}{t_0} \qquad 7.3(4)$$

where

t = efflux time of the solution
t_0 = efflux time of the solvent

The specific viscosity (μ_{sp}) is derived:

$$\mu_{sp} = \mu_r - 1 \qquad 7.3(5)$$

Then intrinsic viscosity $[\mu_I]$ is calculated from either one of the following formulas:

$$[\mu_I] = \left(\frac{\mu_{sp}}{C}\right)_{C \to 0} \qquad 7.3(6)$$

$$[\mu_I] = \left(\frac{\log_e \mu_r}{C}\right)_{C \to 0} \qquad 7.3(7)$$

where C is concentration.

Either or both of the bracketed quantities are plotted against concentration and are extrapolated to infinite dilution to obtain the intrinsic viscosity as illustrated in Figure 7.3f.

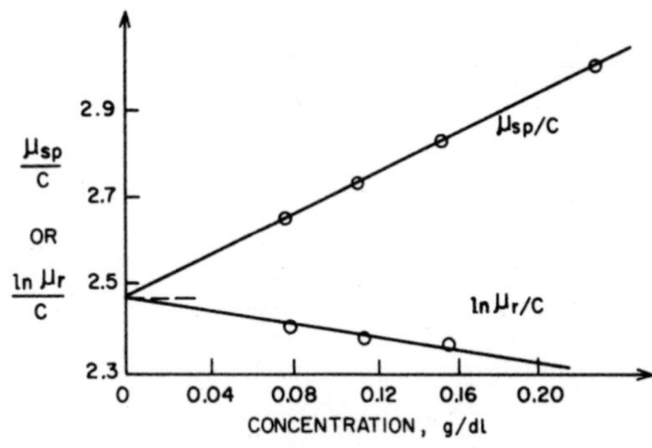

Fig. 7.3f Intrinsic viscosity determination from specific or relative viscosity

The Mark-Houwink equation expresses the relationship between intrinsic viscosity and molecular weight:

$$[\mu_I] = KM_w^a \qquad 7.3(8)$$

or, in logarithmic form:

$$\log[\mu_I] = \log K + a \log M_w \qquad 7.3(9)$$

where

M_w = molecular weight
K and a = constants for a given polymer-solvent system

Efflux time measurements are made of well-fractionated or mono-disperse systems and $[\mu_I]$ are calculated as noted above. M_w is measured osmotically or by light-scattering, and the values of $[\mu_I]$ are plotted against M_w on log-log paper as shown in Figure 7.3g.

From this plot, the value of "a" is calculated as the slope between two convenient points. The value for K is determined from the slope "a" and from a single point on the line.

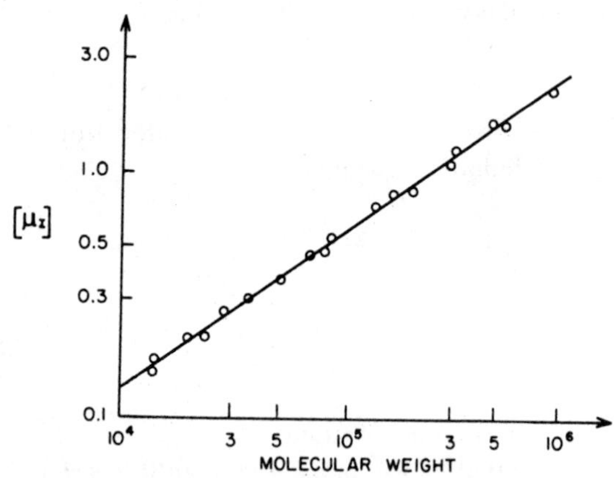

Fig. 7.3g Molecular weight determination from intrinsic velocity

For polydisperse systems, intrinsic viscosity takes the following relationship to molecular weight:

$$[\mu] = K\bar{M}_w{}^a \qquad 7.3(10)$$

where $\bar{M}_w$ = viscosity average molecular weight

Once "K" and "a" have been determined on a monodispersed sample, $\bar{M}_w$ is easily calculated from the above equation.

BIBLIOGRAPHY

Nelson, R.C., "Manual Force Viscometers," *Instruments and Control Systems*, May, 1963.

"Viscometer Survey," *Instruments and Control Systems*, January, 1971.

Flow Sheet Symbol

7.4 LABORATORY VISCOMETERS—CAPILLARY EXTRUSION

Operating Temperature:	50 to 640°F (10 to 338°C)
Operating Pressure:	0 to 5,000 PSIG (0 to 35 MPa)
Material of Construction:	Hardened stainless steel
Cost:	\$675–\$13,000 as a function of the degree of automation
Inaccuracy:	±2.0% of reading
Range:	Up to 10^8 centipoises (10^5 Pa·s)
Partial List of Suppliers:	Burrell Corp.; Instron Corporation; Seiscor Div., Seismograph Service Corp,; Techne (Princeton) Ltd.

As discussed in section 7.3, the capillary-tube viscometers are limited to low viscosity Newtonian fluids because of small driving force and inability to vary shear stress and shear rates. Non-Newtonian fluid flow properties are much dependent on shear rate.

Several manufacturers have designed capillary-extrusion viscometers to study flow behavior of non-Newtonian high viscosity fluids as illustrated in Figure 7.4a.

One type utilizes compressed nitrogen gas from a commercial cylinder to drive the sample through a capillary. The sample to be tested is charged to a thermostatically-controlled steel reservoir and a free floating follower plug rides on top of the sample to distribute the pressure evenly over the surface. A series of extrusions may be made through an orifice at varying pressures. The rate of flow in cubic centimeters per second is calculated from the weight extruded and the elapsed time of each extrusion. The driving force is measured by a pressure gauge. The viscosity in poises, the rate of shear and the stress are all calculated from the observed data using the formulas containing the viscometer parameters provided by the manufacturer.

Another type utilizes an automated tensile tester to extrude materials through a capillary with constant plunger speed. A sample is placed in a thermostated extrusion chamber and a piston is fastened to the moving crosshead to force the material through a selected capillary at a controlled fixed rate. The force required to drive the piston is detected by a compression load cell on the crosshead. The load cell signal is recorded on a strip chart recorder to detect the steady-state condition. The rate of flow can be determined by the piston speed or by measurement of extruded material weight and the elapsed time of the test. The piston speed can easily be changed and a series of tests at different shear rates can

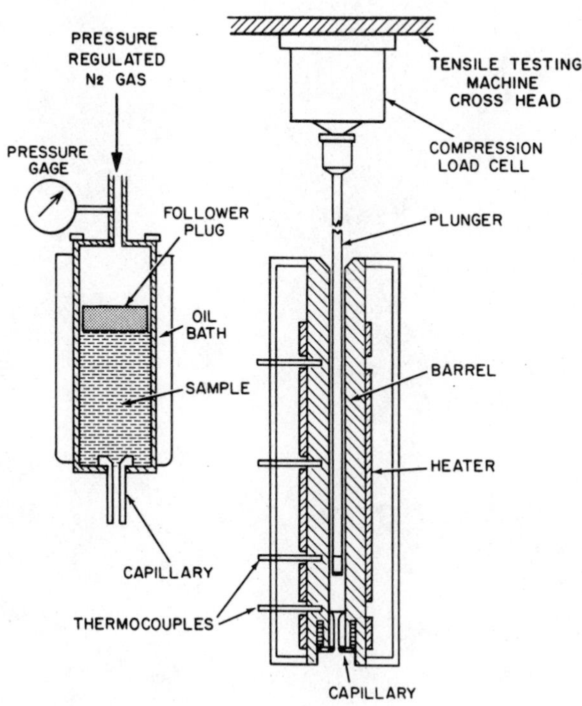

Fig. 7.4a Schematic diagram of capillary-extrusion viscometer

be made by loading one sample. The recorded load history makes it possible to study non-Newtonian materials having properties of dilatancy, thixotropy, etc. The flow properties—viscosity, shear stress and shear rate—are calculated from the measured data and from the viscometer geometry constants.

Still another type measures influx time of material through a capillary. The sample liquid is forced through an open ended capillary by a constant air pressure which is developed by a vibrating vertical piston. The influx time taken for the sample liquid to travel a marked distance along the capillary tube is measured as viscosity in poises.

For these viscometers, a variety of capillary lengths and diameters are available. The capillary-extrusion viscometer is useful for characterization of non-Newtonian fluids with viscosities up to 10^8 centipoise (10^5 Pa·s) over a shear rate range of 1 to 10^4 sec^{-1} with inaccuracies within $\pm 2\%$ of reading. This device is valuable for the study of materials that are processed through injection molding operations, since the tests are carried out under closely simulated processing conditions.

When using the capillary-extrusion viscometer to study non-Newtonian fluids, its sources of errors should be recognized. The following factors may have a significant effect on the results obtained with the capillary-extrusion viscometer:

1. Non-uniformity of sample shear rate.
2. Entrance effects—energy losses at the capillary entrance and discharge points.
3. Compressibility of fluids.
4. Pressure loss produced by the flow in the sample chamber.
5. Temperature gradient created by shear induced heat.

These factors are contrary to the assumptions made in defining Newtonian fluid behavior in a capillary tube. (See Equation 7.3(1).) For scientific research work, great care should be taken to correct for these errors. The unit has excellent reproducibility, and is well suited for routine industrial and scientific work.

BIBLIOGRAPHY

Nelson, R.C., "Manual Force Viscometers," *Instruments and Control Systems*, May, 1963.

"Viscometer Survey," *Instruments and Control Systems*, January, 1971.

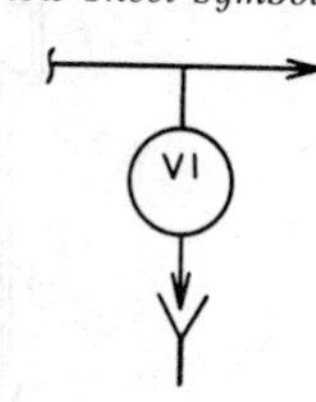

7.5 LABORATORY VISCOMETERS—EFFLUX CUP

Operating Temperature:	60 to 250°F (16 to 120°C)
Operating Pressure:	Atmospheric
Materials of Construction:	Aluminum, brass, stainless steel, paper and others
Cost:	\$35–\$225 for cups; \$350–\$1300 for thermostatic bath; \$675 for autoviscometer
Inaccuracy:	±0.1%–±5% of reading
Range:	1 to 1,200 centipoises (10^{-3} to 1.2 Pa·s)
Partial List of Suppliers:	L. C. Eitzen Co.; Gardner Laboratory, Inc.; General Electric (GE) Co., Industrial Sales Div.; Greiner Scientific Co.; Koehler Instrument Co., Inc.; Precision Scientific Co., Subs. of GCA Corp.; Shell Oil Co.; P. E. Tobias Associates

Efflux cup viscometers are most commonly used for field work to measure the viscosity of oils, syrups, varnish, lacquer, paints and bitumen emulsions, although they have some inherent inaccuracies. The testing procedure is quite similar to the capillary-tube viscometers where efflux time of a specified volume of fluid is measured through fixed orifice at the bottom of a cup to represent the viscosity of the fluid. Since the viscosities of Newtonian liquids are independent of dimensions of viscometers used, it is possible to convert efflux times to kinematic viscosities by conversion charts or by formulas suggested by the equipment manufacturers.

To obtain high accuracy, the liquid holding vessel and orifice are temperature controlled by immersing them in a thermostatically-controlled bath.

There are 40 or more variations of these viscometers available today. Some of the commonly used types are described below.

Saybolt

This viscometer is the standard instrument for testing petroleum products. The method is described in ASTM D88-56 standard method of testing for Saybolt viscosity. The thermostatic baths have a temperature control uniformity of better than ±1°F (±0.6°C) over a range of 70°F to 230°F (21°C to 110°C). The viscosity determinations should be conducted in a room free from drafts and rapid changes in temperature to attain the highest degree of accuracy. Accuracies within ±0.1% of reading are possible when the standard testing procedures are followed. Also, each orifice and cup assembly should be calibrated with ASTM viscosity standards (see Table 7.3d) to obtain the correction factor as described in Section 7.3 of this chapter.

There are three types of orifices available—universal, furol, and asphalt. The furol and asphalt orifice respectively have an efflux time of approximately one-tenth and one-hundredth that of the universal orifice. The cup-orifice combination should be selected to provide an efflux time within the range of 20 to 100 seconds.

Of these three types, the Universal orifice (Saybolt universal viscometer) is most commonly used and its efflux time is designated as Saybolt Universal Seconds (SUS). The Saybolt universal viscometer, as shown in Figure 7.5a, measures the time required for 60 cc of sample fluid to flow out through an orifice having dimensions of 0.176 cm in diameter and 1.225 cm in length.

Saybolt Universal Seconds (t) can be converted to kinematic viscosity (ν) by the following equations:

when $t < 100$ sec,

$$\nu = 0.226\,t - 195/t \text{ centistokes*} \qquad 7.5(1)$$

*1 centistoke = 10^{-6} m²/s

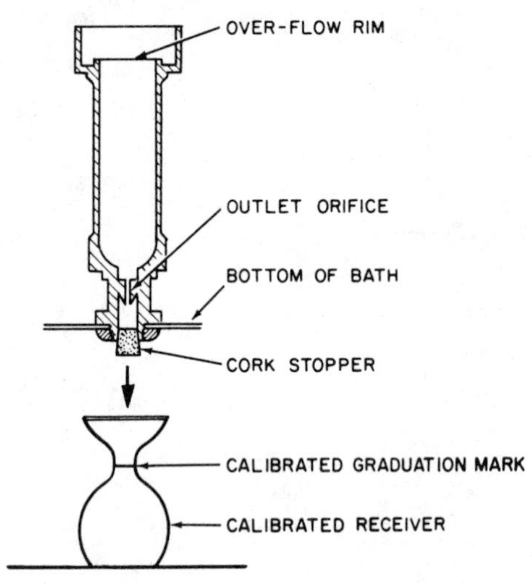

Fig. 7.5a Saybolt viscometer

when $t > 100$ sec,

$$\nu = 0.220\,t - 135/t \text{ centistokes} \qquad 7.5(2)$$

Ford Cups

Ford cup viscometers are used for the determination of low viscosity liquids that do not deviate much from the ideal (Newtonian) liquid behavior. ASTM D-1200-58 standard method defines the procedure for determining viscosity of paints, varnishes, lacquers and related liquid materials with the Ford cup viscometers. Ford cups have a conical bottom fitted with a standardized orifice. Results are expressed as efflux time, in seconds, for the liquid to flow through the orifice until the first break in the stream. Efflux time may be converted to the corresponding kinematic viscosity by the following equations:

Table 7.5b
CONVERSION OF FORD CUP SECONDS

Cup Number	*Orifice Diameter*	*Conversion Equation*
2	0.0995 inch*	$\nu = 0.013t\text{-}10/t$
3	0.134 "	$\nu = 0.023t\text{-}8/t$
4	0.162 "	$\nu = 0.037t\text{-}4/t$

*1 inch = 25 mm

where

ν = kinematic viscosity, stokes**
t = efflux time, seconds

The particular cup-and-orifice combination should be selected to provide an efflux time within the range of 20–100 seconds at a controlled temperature of 77°F (25°C), and temperature drift during the test should not exceed 0.5°F (0.28°C) as determined by a thermometer in the efflux stream. Wide ranges of materials of construction are available and, usually, the body is aluminum and orifices are brass. Cups and orifices are interchangeable, but the whole assembly must be calibrated each time the orifice is changed. It is always good practice to calibrate the Ford cups on a regular basis to correct for the errors caused by orifice enlargement due to cleaning or material deposits in the orifice.

**1 stoke = 10^{-4} m²/s

Ford viscosity cups No. 2, No. 3, and No. 4 are normally used for viscosity ranges of 10 to 1000 centistokes (10^{-5} to 10^{-3} m²/s) with inaccuracies of $\pm 2.0\%$ of reading.

Zahn Cups

The Zahn cup viscometer is widely used by paint manufacturers to standardize their products during manufacturing stages, because of its convenience and rapid use. It is also utilized for measuring the viscosity of many coatings such as varnishes and lacquers.

It is essentially a bullet-shaped container having a definite volume (44 cc), with an orifice in the bottom. To insure a level position, it is suspended from a ring and ball.

It provides readings in units called Zahn-seconds, which represents the efflux time of this definite volume of liquid to flow through the orifice. For best results, liquids should flow through the calibrated orifice in the bottom of the cup in approximately 20 to 40 seconds. Repeatability of $\pm 2\%$ is possible.

Five different orifice sizes are available.

Table 7.5c
CONVERSION OF ZAHN SECONDS

Cup Size	*Range in Zahn Seconds*	*Range in Relative Centipoises**
1	40–85	20–85
2	20–70	30–170
3	25–60	170–550
4	20–65	200–900
5	15–60	250–1200

*1 centipoise = 10^{-3} Pa·s

There is no general formula to convert Zahn-seconds to other viscosity units. As a result, Zahn cup viscometers are used for regulating the thickness of coatings and as a guide for coating quality determination.

Automatic

This is essentially an automated Zahn cup viscometer to be used on-line where filling, efflux timing and solvent washing operations are controlled by a cycle-time programmer (see Figure 7.5d).

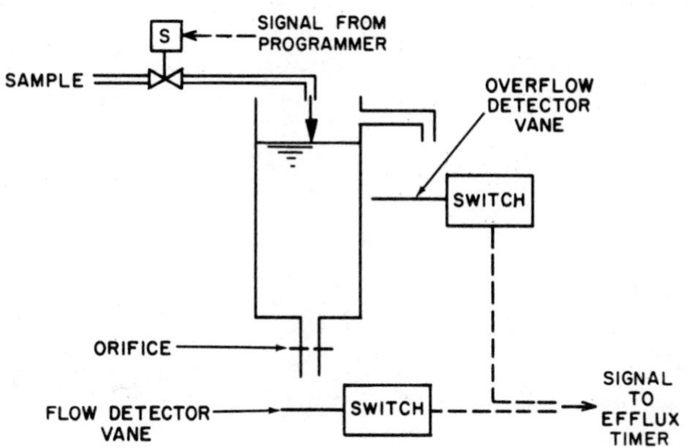

Fig. 7.5d Automatic efflux-cup viscometer

The programmer allows the liquid to enter the cup until the overflow detector actuates. Drainage of fluid from the bottom of the cup and the overflow are sensed by deflection of torsion wire-mounted vanes which operate sensitive magnetic switches to detect the efflux time.

It is a low cost on-line viscometer and is recommended where accuracy is not critical and intermittent measurements with substantial time lags between detections can be tolerated.

BIBLIOGRAPHY

Nelson, R.C., "Manual Force Viscometers," *Instruments and Control Systems,* May, 1963.

"Viscometer Survey," *Instruments and Control Systems,* January, 1971.

Flow Sheet Symbol

7.6 LABORATORY VISCOMETERS—FALLING BALL

Type of Hardware:	A. Manual Falling Ball B. Automatic Falling Ball
Operating Temperature:	A. −30 to +300°F (−34 to +150°C) B. Up to 350°F (up to 177°C)
Operating Pressure:	A. Atmospheric or slightly higher B. Up to 15,000 PSIG (104 MPa)
Materials of Construction:	Glass, corrosion-resistant steel alloys
Cost:	A. $800 B. $8000
Inaccuracy:	A&B. ± 0.1 to ± 1.0% of reading
Range:	A. 0.01 to 160,000 centipoises (10^{-5} to 160 Pa·s) B. 0.1 to 200,000 centipoises (10^{-4} to 200 Pa·s)
Partial List of Suppliers:	Coleman Tulsa Co. (B); Fish-Schurman Corp. (A); Gardner Laboratory, Inc. (A): Roger Gilmont Industries (A); Ruska Instrument Corp. (B)

Manual Falling Ball Viscometers

This viscometer operates on the falling ball principle, published by Stokes in 1851, based on his investigations on spheres falling through liquids.

Stokes' Law

$$\nu = \frac{2r^2(\rho_S - \rho_L)g}{9\mu} \qquad 7.6(1)$$

where

ν = terminal velocity of fall, cm/sec.
r = radius of sphere, cm
ρ_S = density of sphere, g/cm^3
ρ_L = density of liquid, g/cm^3
g = gravity, cm/sec^2.
μ = coefficient of viscosity, poises*

*1 poise = 0.1 Pa·s

Stokes' Law relates the viscosity of gases and liquids to the steady-state velocity of a known weight sphere falling through a fluid of known density at very low Reynolds numbers.

The falling ball viscometer consists essentially of a precision-bore glass tube and calibrated glass or special steel balls of optical precision, as shown in Figure 7.6a. The tube has a length of about 200 mm, a bore of about 16 mm, and is mounted in a glass water jacket held at a specified angle to the vertical.

A special capillary plug permits positive seal and prevents introduction of air into the column during the tube inversion process. This permits as many readings to be taken as are desired with a single sample, either as checks or to find the variation of viscosity with temperature. The closed system allows operation without effects of skin formation, surface tension, evaporation, or other losses of material.

The measurement is made by timing the fall of the ball through the accurately calibrated distance between two marks on the glass tube. The absolute viscosity of the fluid can be found by multiplying that time interval with a factor involving the temperature, the specific gravity of the fluid, and a constant characteristic of the ball used, in accordance with the following formula:

$$\mu = t \times (\rho_{SB} - \rho_{SF}) \times B \qquad 7.6(2)$$

where

μ = absolute viscosity, centipoises
t = time interval of the falling ball, seconds

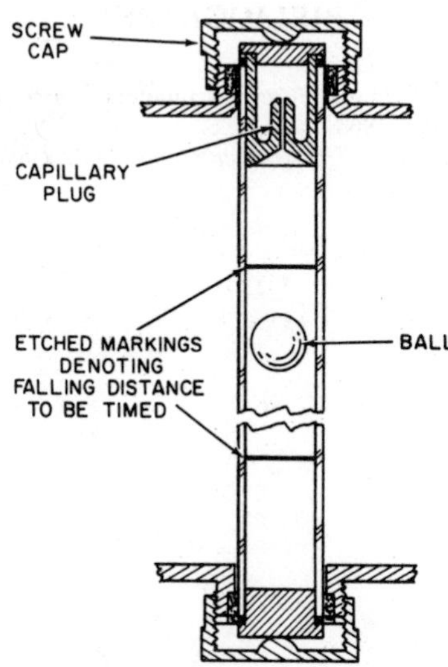

Fig. 7.6a Falling ball viscometer

ρ_{SB} = specific gravity of the ball
ρ_{SF} = specific gravity of the fluid at the measuring temperature
B = ball constant

The use of several calibrated balls of different sizes in the same glass tube covers visocity ranges of 0.01 to 1,000,000 centipoises (10^{-5} to 1,000 Pa·s)—a range which includes gases (hydrogen, air, carbon dioxide, etc.) and viscous liquids which will hardly pour through a 16 mm opening. If the temperature of the liquids in the glass jacket is maintained precisely at 68°F (20°C), the usual range of viscosities (10 to 600 centipoises [0.01 to 0.6 Pa·s]) can be determined with an inaccuracy of ±0.1% when the time interval is measured by a stopwatch of 0.02 seconds inaccuracy. Outside this viscosity range ±0.5% inaccuracy is expected.

To obtain this degree of accuracy, measurement of liquid density and correction factor on the ball density become very critical. For the metal ball, the density of the liquid under examination should be known within 0.02 SG, and for the glass ball, the density should be within 0.005 SG. The variation of the density of the balls with temperature is usually too small to have a measurable effect on the ball constants. However, for extremely accurate measurements, a correction factor should be used as recommended by the manufacturer.

The viscosity of gases at 68°F (20°C) can be calculated from the following relationship:

$$\mu = \frac{\text{period of fall in gas}}{\text{period of fall in air}} \times 0.01836 \qquad 7.6(3)$$

Absolute cleanliness of the tube and ball is essential to obtain any degree of accuracy when the instrument is used for gases since the clearance between the ball and the tube is very minute.

It is always good practice to calibrate the instrument with great care and as often as possible. The ball constant can be determined by rearranging equation 7.6(2):

$$B = \frac{\mu}{(\rho_{SB} - \rho_{SF})t} \qquad 7.6(4)$$

Water at 68°F or ASTM calibrating liquids may be used. The instrument should be carefully cleaned, and the temperature should be held constant within ±0.02°F (±0.01°C).

For laboratory instruments such as this, care should be taken to minimize subjective errors. A stopwatch of insufficient accuracy is often the cause of error. Lack of attention to temperature is another. Viscosities at room temperature may vary from 2.5% to 16% for each degree change in temperature, depending on the type of liquid to be tested. Maintenance of temperature within hundredths of degree at low temperatures, and within tenths above 100°F (38°C), is absolutely essential for accurate results.

Other possible sources of error are vibration during time of fall, inaccurate leveling of instrument, foreign matter or gas bubbles in the liquid, inaccuracy of thermometer, or internal heating of the liquid by absorption of infra-red radiation from direct sunlight or other sources.

Simplicity of operation, speed and virtual elimination of subjective variables render the instrument well suitable for routine industrial use as well as for precise scientific measurements of gases and Newtonian liquids, including dark or opaque colored fluids as long as they are free from crystal or gel particles.

Automatic Falling Ball Viscometer

These viscometers are provided with automatic controls and timers to measure viscosities of 0.1 centipoises (10^{-4} Pa·s) or higher, with a repeatability of 0.1%. Approximately 70 cc of sample is required to charge the viscometer, which can operate at pressures as high as 15,000 PSIG (104 MPa) and temperatures as high as 350°F (177°C).

The viscometer consists of two units: the viscosity test assembly (equipped with a precision bore tube, heating and temperature control components and the ball position detector), and the auxiliary control unit. The auxiliary control unit contains the temperature controller, electrical timer with reset and the electrical circuits. The viscosity test chamber, mounted on a trunnion, can be rotated to various angular positions. The electrical timer is activated by the ball release signal and when the ball closes the contact at the lower end of the tube, the timer stops. The timer is accurate to 0.001 minute. The fall time of the ball is proportional to the sample viscosity, and the readings are interpreted by calibration curves.

The automatic falling ball viscometer is simple and

safe to operate and can measure fluid viscosity under simulated process conditions. Low inaccuracy ($\pm 0.1\%$) is obtained by judicious selection of the proper calibrated ball size and by adjusting the angular position of the precision bore tube.

BIBLIOGRAPHY

Nelson, R.C., "Manual Force Viscometers," *Instruments and Control Systems*, May, 1963.

"Viscometer Survey," *Instruments and Control Systems*, January, 1971.

Flow Sheet Symbol

VI

7.7 LABORATORY VISCOMETERS—ROTATIONAL

Operating Temperature:	−10 to 2500°F (−23 to 1370°C)
Operating Pressure:	Atmospheric
Materials of Construction:	Stainless steel, nickel plated brass, plastic, Al-Mg light alloy, ceramic, platinum
Cost:	\$1,000–\$5,000 for the coaxial type and \$30,000–\$50,000 for the cone and plate type
Inaccuracy:	±0.5 to ±1.0% of measurement
Range:	10^{-4} to 10^{8} poise (10^{-5} to 10^{7} Pa·s)
Partial List of Suppliers:	Beckman Instruments, Inc.; C. W. Brabender Instruments, Inc.; Brookfield Engineering Laboratories, Inc.; Ferranti Electric, Inc.; Gardner Laboratory, Inc.; PolyScience Corp., International Div.; Sweets Martin, Co. Inc., Instr. Div.

The rotational viscometer is probably the most widely used rheometer. The operating principle of the rotational type viscometer is based on the fact that the torque necessary to overcome the viscous resistance to the induced movement (torque) by rotation of a spindle is directly proportional to the viscosity of the fluid. The entire sample is subjected to a uniform or nearly uniform rate of shear, and direct determination of viscosity by measurement of the corresponding shear stress is, therefore, possible.

The rotational viscometer is the most versatile laboratory rheometer, and is useful for measurement of fluid viscosities ranging from 10^{-4} to 10^{8} poise (10^{-5} to 10^{7} Pa·s), at a full range of shear rates from 10^{-4} sec^{-1} to 10^{4} sec^{-1} with varying spindle size and speed of rotation.

Ability to perform continuous measurement under varying conditions on the same sample or at a given set of boundary conditions for an extended time period are chief advantages of the rotational viscometer. They are particularly valuable for investigation of non-Newtonian fluids.

The rotational viscometers are classified in two major groups—coaxial-cylinder types and cone-and-plate types—because of their basic differences in spindle design. The cone-and-plate type viscometer is more suitable for study of non-Newtonian fluids than the coaxial-cylinder type viscometer, and is more versatile.

Coaxial-Cylinder

There are many types of coaxial-cylinder viscometers which are available commercially and are similar to the design shown in Figure 7.7a. All have three features in common: a mechanical means of driving a spindle at a constant speed, a torque measuring device, and a means of correlating shear rate (spindle diameter and speed of rotation).

One type measures the force required to rotate a spindle or bob in the sample at a specified speed by use of weights as the driving force. They are suspended at the free end of a cable, the other end of which is wound around a pulley geared to the shaft to which the spindle is attached. A standard one-pint friction-top can is used as a container. The weights are a measure of torque and the shear rate is calculated from the rotation speed. This unit is inexpensive and is used for measurement of viscosities ranging from 0.1 to 5,000 poise (0.01 to 500 Pa·s).

Another type rotates a flat disc-shaped stirrer at a constant speed and a sample container (half-pint size can) is firmly placed on a turntable which is equipped with torque measuring capability. A spring attached to the turntable spindle is used to counteract the torque transmitted through the sample. The angular deflection of the turntable is presented as the viscosity of the sample fluid. This type is used for low viscosity measurements—from 0.1 to 15 poise (0.01 to 1.5 Pa·s).

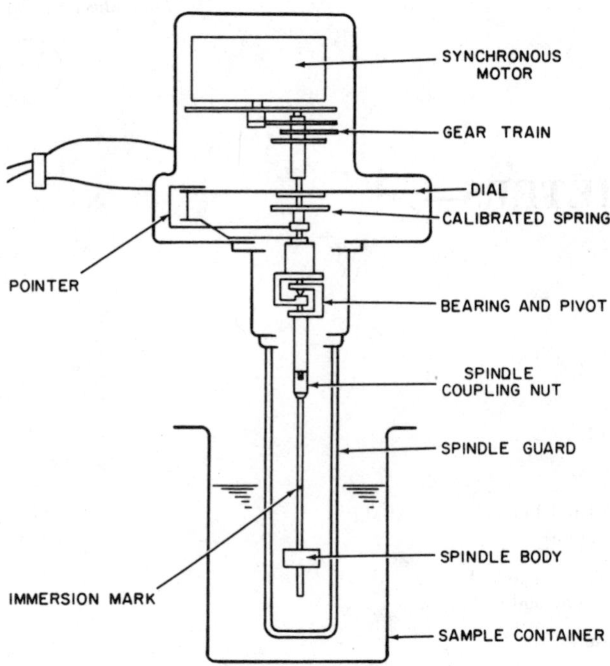

Fig. 7.7a Rotational viscometer

There is another type where the sample container is driven at a constant speed and measures the torque required to restrain the freely suspended spindle by spring-controlled angular deflection of the supporting torque wire. In some cases, the spring movement is converted to a current signal by use of poles of a magnet, and the current required to restrain the spring is detected as viscosity. This instrument is good for low to medium viscosity fluids at relatively low shear rates.

Still another type drives the spindle at a desired speed through a calibrated spring and the viscosity of the fluid is measured by the degree to which the spring is wound. Constant speed is maintained by a synchronous motor and speeds of rotation are changed by a gear train (up to 8 speeds). The spring torque can be converted to pneumatic or electrical signals by a flapper nozzle arrangement or by capacitance of a condenser plate assembly for remote transmission of measured data.

Perhaps one of the most versatile coaxial-cylinder viscometers available commercially is the one whose rotor speed is changed by line frequency manipulation. With this model, the torque on the rotor is measured by noting the deflection of the torsion spring, mounted between the rotor and the drive transmission, by means of potentiometer mounted on the spring.

For the latter two viscometers discussed above, shear diagrams can be plotted automatically on an X-Y recorder of shear stress vs. shear rate or shear stress vs. time of shear because of their ability to change speed of rotation continuously and their ability to transmit the measured torque data. Any X-Y recorder can be utilized providing enough sensitivity is available. A minimum frequency response of 1 second is recommended. This is particularly useful for study of non-Newtonian fluid properties such as: apparent viscosity relationship, dilatancy, thixotropy, rheopexy, yield value, plasticity.

For some types, the electrical components are housed in an explosion-proof housing and the spindle extension can be adapted for hazardous areas or for remote testing in a case of very high temperature environment. A wide variety of cups, bobs (spindles), and special attachments is available to extend the working ranges. These units will permit measurement of fluids having viscosities ranging from 5×10^{-3} to 16×10^{7} poise (5×10^{-4} to 16×10^{6} Pa·s) over a six-decade range of shear rates that are from 10^{-2} to 10^{4} $\sec^{-1}$.

These coaxial-cylinder viscometers are reasonably inexpensive and well suited for routine industrial and research work.

In the design of the coaxial-cylinder viscometer, many assumptions were made to give the appearance of Newtonian fluid behavior; i.e., steady laminar flow, isothermal flow, no slippage at the wall, constant temperature, viscosity unaffected by shear rate, etc. Thus the coaxial-cylinder viscometer provides a rapid means of obtaining reproducible absolute viscosity readings of Newtonian fluids by subjecting the sample to a shear stress.

Uncertainty does arise when non-Newtonian fluids are dealt with. The fundamental problem here is that the very property that we are interested in measuring is affected by the viscometer's shear. Because of this, it is important to provide well defined shearing conditions within the viscometer. A useful interpretation of the flow data is only possible when the limits of variations are known, even if the shear rate is not uniform.

The major sources of errors with the coaxial-cylinder viscometer when used on non-Newtonian fluids are as follows:

1. The shear stress, shear rate and therefore the viscosity varies across the gap between the sample cup and the bob used to measure the viscosity.
2. The end effect, which is the contribution to the torque that arises in the fluid between the end of the bob and the bottom of the cup.
3. The stress-induced heat generation within the fluid at high shear rate.

For these reasons, many different geometrical configurations of bob and sample holding devices were developed (see Figure 7.7b). The end effect is minimized by specifying the immersion depth, by use of bottomless cups, or by trapping air beneath the bob. When using the coaxial-cylinder viscometers, the following factors may have a significant effect on the results, and care should be taken to specify these conditions.

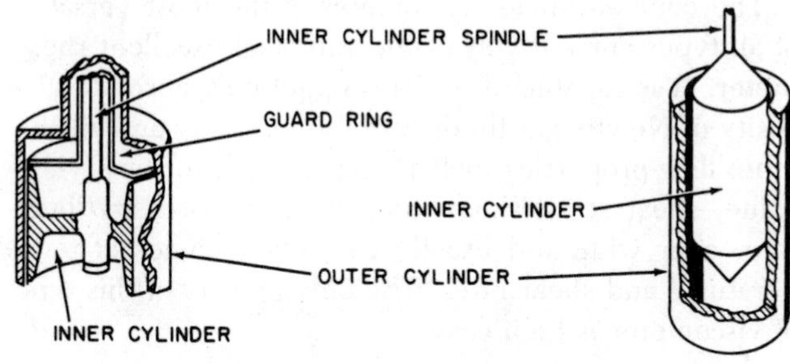

Fig. 7.7b Coaxial cylinder viscometers

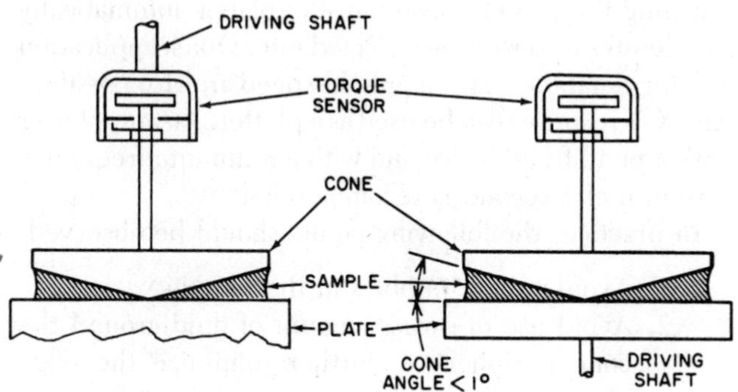

Fig. 7.7c Cone-and-plate viscometers

A. Conditions to be specified

1. Speed of rotation
2. Spindle size and shape
3. Temperature of liquid
4. Size and shape of sample container
5. Depth of liquid from bottom of the container
6. Spindle immersion depth
7. Elapsed time of rotation before reading is taken

B. Undesirable factors

1. Dirt on torque measuring device
2. Dirty, corroded, pitted or deformed spindle
3. Off-center placement of spindle in container
4. Wrong frequency of electric power
5. Lack of regular recalibration
6. Turbulent flow (low viscosity fluid with high speed)
7. Large air bubbles in the sample

In conclusion, undertainty arising from viscometric data obtained from non-Newtonian fluids is minimized by observing the two principal points: obtain a viscometer which provides the most uniform shear rate throughout the measured sample, and use a consistent experimental procedure.

Cone-and-Plate

The cone-and-plate type viscometer has been designed to eliminate some of the drawbacks of the coaxial-cylinder type viscometer and to provide rapid means of obtaining reproducible flow characteristics of non-Newtonian fluids.

The geometrical configuration of a cone and a plate provides uniform shear rate and stress throughout the fluid sample due to linear increase in both the sample thickness and tangential velocity as a function of distance from the center at a given angular velocity. The influence of shear-induced heat within the fluid at high shear rates is substantially reduced because the sample is a very thin layer.

The cone-and-plate viscometer essentially consists of a flat plate and a cone with a very small angle (less than 1°). As shown in Figure 7.7c, the apex of the cone almost touches the flat plate surface and the fluid sample fills the narrow gap between the cone and the plate. Capillary action keeps the materials in place during operation.

A continuously variable speed motor drives the rotating platen; the speed is precisely maintained by an electronically-controlled velocity servo-mechanism.

Some of these cone-and-plate types rotate the flat plate while the conical disc is held by a torsion measuring head. The measurement of torsion bar movement is made by a displacement transducer of a differential transformer type.

Another type rotates the conical disc through a dynamometer while the bottom flat plate is stationary. The viscous traction on the cone is measured on an electromechanical torque dynamometer (combination of torque spring and potentiometer).

For both types, temperature of the sample is measured by thermocouples embedded in the plate, and precise temperature control is maintained. The gap between cone and plate is kept constant by a servo system. The relative position of platen to cone is sensed by a ceramic proximity gauge or by a thermal detection unit.

By choice of platen diameter and cone angle, and by using the drive gearbox ratios, viscosities from 10^{-4} to 10^{8} poise (10^{-5} to 10^{7} Pa·s) can be measured within a full range of shear rates from 10^{-4} to 10^{4} sec^{-1}.

To determine the dynamic properties of visco-elastic materials, one manufacturer has incorporated a sinusoidal motion over a wide frequency range into his cone-and-plate viscometer. This oscillatory method of testing allows the moduli of elasticity and viscosity to be derived from the ratio of applied strain and resultant stress, and the phase difference between the two to be in an almost undisturbed state. This is important for materials with unstable structure or yield value or those that can be damaged by continuous rotation.

The chief advantage of the commercially available cone-and-plate viscometer is its ability to record rheograms automatically for non-Newtonian fluids which exhibit shear dependent and time dependent behavior over a wide range of shear stress and rates. The control unit gives uniform acceleration of the platen up to a selected maximum speed at a preselected acceleration rate. On reaching the preset maximum, the platen automatically decelerates to zero at the selected rate. Quick application and termination of strain or hold-speed are also possible. Any X-Y recorder can be used as a plotter. An autoplotter with a pen lifting device and with a minimum frequency response of 1 second is recommended.

In practice, the following points should be observed:

1. Avoid any air bubbles in the sample.
2. Avoid use of excess amount of fluid around the cone periphery to further minimize the edge effect.
3. Avoid testing high-viscosity materials such as polymers which tend to ball-up and leave a gap between the cone and plate unless processing temperature is high enough to melt the materials and assure proper contact.

The cone-and-plate viscometer is the most versatile of all types currently available and is an excellent rheometer. It is capable of measuring not only absolute viscosity of Newtonian fluids but also elasticity and all the other flow properties such as dilatancy, thixotropy, yield value, shear stress/shear rate, apparent viscosity/shear rate, over wide and exactly controlled ranges of temperatures and shear rates. The only penalty of this type of viscometer is high cost.

BIBLIOGRAPHY

Nelson, R.C., "Manual Force Viscometers," *Instruments and Control Systems*, May, 1963.

"Viscometer Survey," *Instruments and Control Systems*, January, 1971.

Flow Sheet Symbol

7.8 INDUSTRIAL VISCOMETERS—CAPILLARY

Design Pressure:	Up to 670 PSIG (60 to 125 normal) (up to 5 MPa, 0.4 to 0.9 MPa normal)
Design Temperature:	Up to 900°F (200°F normal) (up to 480°C, 93°C normal)
Materials:	Hardened stainless steel or other corrosion-resistant metals
Cost:	\$3,400 to \$15,750
Inaccuracy:	±0.5% to ±2.0% of full scale
Range:	0 to 2,500 centipoises (0 to 2.5 Pa·s); up to 1.5×10^6 centipoises (up to 1.5×10^3 Pa·s) with the high pressure design
Partial List of Suppliers:	GCA, Precision Scientific Group

Capillary viscometers utilize the flow of the process liquid through a capillary to measure viscosity. The technique of determining the viscosity of Newtonian fluids by measuring the pressure drop across a capillary tube during isothermal, laminar flow is well known. The Poiseuille law states that the pressure drop in a Newtonian liquid passing through a capillary tube is directly proportional to its viscosity if the flow rate is maintained constant.

$$\mu = K\frac{d^4P}{VL} \qquad 7.8(1)$$

where

μ = absolute viscosity, centipoise
K = a constant
d = inside diameter of a capillary tube, inches
P = pressure drop across the capillary tube, PSI
V = flow rate, GPH
L = length of the capillary tube, inches

Newton's hypothesis assumes that the viscosity of Newtonian fluids is independent of the rate of shear or shearing force if temperature and pressure are fixed. The Poiseuille law and Newton's hypothesis indicate that if the flow rate is held constant through a fixed capillary tube, the absolute viscosity is a linear function of the pressure drop.

The continuous capillary viscometers are primarily designed to measure the viscosity of Newtonian liquids. Because pressure transducers are used to transmit the measured viscosity, it is readily adaptable to the automatic control of processes. Temperature compensation is not practical with this type of viscometer. Rather, a thermostatic device is used to control the temperature of the sample at a reference temperature before metering through the capillary tube. This type of viscometer has been found successful at viscosities up to 15,000 poises (1,500 Pa·s) and at temperatures up to 900°F (480°C). The span of the instrument is determined by the bore and length of the capillary, and a large variety of viscosity ranges are possible. The use of large-diameter bores and long capillary tubes is recommended for minimum end effects (see Section 7.4).

Although this type of continuous viscometer is simple enough to be even field fabricated rather than purchased, one chief disadvantage is that the capillary tube must be kept absolutely clean for accurate and reliable measurements. Undetected fouling may occur since the capillary tube diameter is very small, in the range from 0.05 to 0.2 inches (1.25 to 5 mm). In addition, the low sample flow rates may limit the continuous capillary viscometer to by-pass installations in order to automatically control the processes with minimum time lag. Other sources of errors are: (1) fluctuating (consistently high or low) flow rate through the capillary tube, (2) dirty or plugged capillary, (3) leakage in the viscometer, (4) zero and span of transducer incorrectly set or drifting, (5) fluctuating fluid pressure, (6) insufficient sample supply pressure, and (7) fluid temperature fluctuations due to thermostat malfunctions or to temperature variation in sample supply temperature.

The continuous capillary viscometer can be calibrated by using the following equations:

$$d = \sqrt[4]{\frac{6\mu VL}{\Delta P \times 10^9}} \qquad 7.8(2)$$

$$\mu = \frac{\Delta P d^4 \times 10^9}{6VL} \qquad 7.8(3)$$

Equation 7.9(2) is used to calculate the diameter of the capillary tube with a known viscosity fluid. Equation 7.8(3) is used to construct the calibration curve of viscosity vs. pressure drop with several known viscosity fluids. It should produce a straight line through a zero viscosity-zero pressure drop point.

The continuous capillary viscometers are successfully used in oil refineries to control various products such as fuel oils, hydraulic oils, lubricating oils, fuels, and various grades of asphalts. It is also used in fuel oil viscosity control to optimize atomization in the power industry.

There are basically two different types of viscometers that utilize the continuous capillary principle. One type measures pressure drop across the capillary tube and the other measures upstream pressure as the sample flows through the capillary tube.

Differential Pressure

Figure 7.8a is a schematic flow diagram of a differential pressure type viscometer which requires a bypass installation. In operation, the externally mounted strainer provides the reasonably clean fluid sample to the instrument. A constant sample flow rate (at about 1 GPH or 3.8 l/h) is maintained with a precision metering pump driven by a synchronous motor. Two heat exchangers,

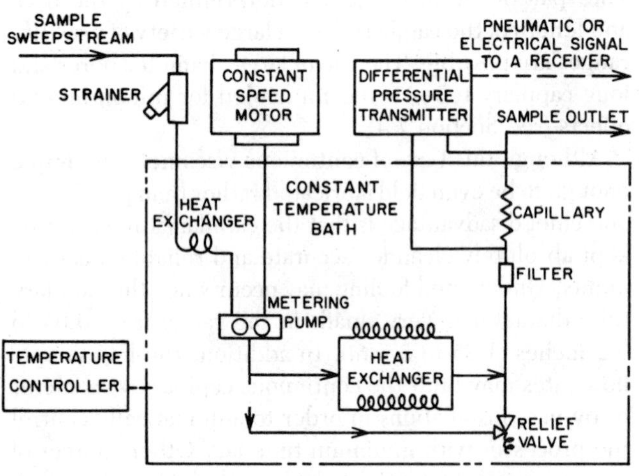

Fig. 7.8a Schematic flow diagram of a differential pressure continuous capillary viscometer

before and after the metering pump, are used to keep the sample fluid in equilibrium with the thermostatic bath. A relief valve protects the flow system against excessive pressures that may occur due to blockage of the capillary.

To measure the liquid viscosity, the pressure drop across the capillary is measured with a differential pressure transducer connected to the inlet and outlet sides of the capillary. The transducer outlet signal may be pneumatic or electric and is used for indicating, recording or controlling the process.

This type of viscometer has been found successful at viscosities up to 2,500 centipoises (2.5 Pa·s) and at temperatures up to 240°F (116°C). High pressure capillaries and high pressure metering pumps are used to measure viscosities up to 15,000 poises (1,500 Pa·s) and line pressures up to 670 PSIG at 900°F (4.6 MPa at 480°C). The range and span of the transmitter determine the range of a single capillary tube. Overall inaccuracy is about ±1% of full scale with repeatability of 1% of full scale. The average response time is about 2 minutes (0.6 min. time constant), which depends on the length of the sample loop.

To improve the response time of the instrument, to less than one second, the capillary tube may be housed in the main process stream and the liquid flow-rate may also be increased (about 4 GPH or 15.2 l/h), as shown in Figure 7.8b. A constant flow rate through the capillary

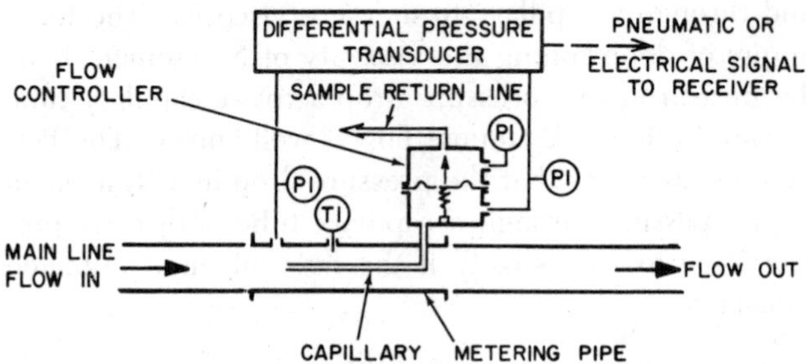

Fig. 7.8b Schematic diagram for an in-line, differential pressure continuous capillary viscometer

tube is maintained by a pressure-regulated, diaphragm-type flow controller. Inaccuracy is reduced to about ±2.0% of full scale at a viscosity range of 5 to 30 centistokes (5×10^{-6} to 30×10^{-6} Pa·s) in this fast-response-time design. Since viscosity measurements are made at the main stream temperature, this instrument is generally used to maintain constant fluid viscosity by adjusting temperature, in such applications as fuel oil viscosity control to maintain optimum atomization patterns for industrial furnaces, heating plants and steam power stations as well as for marine boilers.

Back Pressure

The operation of this instrument is quite similar to the differential pressure type except that it measures only

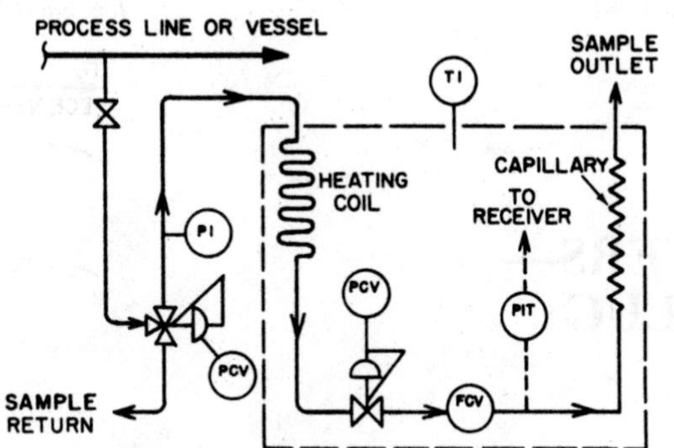

Fig. 7.8c Schematic system diagram for a back-pressure continuous capillary viscometer

upstream pressure to a capillary tube which discharges to atmosphere or returns the sample to a pressure-regulated process line. As shown in Figure 7.8c, the sample fluid is continuously fed to the instrument from the process line or from a vessel. The sample temperature is maintained by flowing through a heat exchanger immersed in a constant-temperature bath. Sample then passes through a pressure regulator to a constant-rate metering device. Under conditions of constant flow rate and temperature, the sample pressure at the entrance to the measuring capillary tube is proportional to the viscosity of the liquid. The inlet pressure is sensed by a strain gauge. The strain gauge signal is converted to convenient viscosity units to indicate, record, or control the process viscosity. Since it measures only the inlet pressure to the capillary tube, it is extremely important to maintain the outlet side at constant pressure by discharging to atmosphere or to a pressure-regulated vessel or pipe line.

The back-pressure type viscometer operates within the viscosity range of 5 to 500 centistokes (5×10^{-6} to 5×10^{-4} Pa·s) at temperatures up to 210°F (99°C) with overall inaccuracy and repeatability of ±1% of full scale. The response time is rather slow—3 to 6 minutes.

BIBLIOGRAPHY

Nelson, Robert C., "Automatic Force Viscometers," *Instruments and Control Systems*, April, 1963.

"Viscometer Survey," *Instruments and Control Systems*, January, 1971.

Flow Sheet Symbol

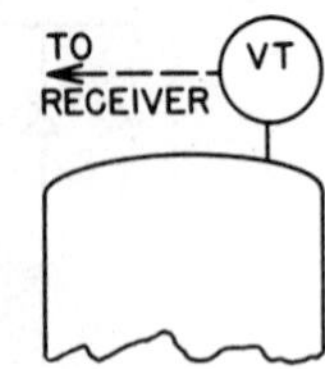

7.9 INDUSTRIAL VISCOMETERS—FALLING PISTON OR SLUG

Type of Hardware:	A. Piston B. Slug
Design Pressure:	A. Up to 500 PSIG (3.5 MPa) B. Up to 200 PSIG (1.4 MPa)
Design Temperature:	A. Up to 650°F (340°C) B. Up to 300°F (150°C)
Materials of Construction:	Stainless steel, and other non-corrosive and hardened alloys
Cost:	A. \$2,500 to \$5,600 B. \$6,750 to \$11,000
Inaccuracy:	±1% full scale (both types)
Range:	A. 0.1 to 10^6 centipoise (10^{-4} to 10^3 Pa·s) B. 10 to 10^6 centipoise (0.01 to 10^3 Pa·s)
Partial List of Suppliers:	Gam Rad, Inc. (B); Norcross Corp. (A)

Falling Piston Viscometer

The working principle of the falling piston viscometer is quite similar to the falling ball viscometer operation discussed in section 7.6. Its excellent reproducibility makes it possible to utilize it for the measurement of both Newtonian and non-Newtonian fluid viscosities as an in-process instrument.

The measuring element of the instrument consists of a piston in a measuring tube as shown in Figure 7.9a. The measuring element can be installed in a tank, open or closed, or it may be a in a liquid-filled line as long as the measuring tube is completely immersed in the fluid. During the filling phase, the piston at the bottom of the tube is automatically raised by an air lifting mechanism or by a motor-cam mechanism. As the piston is raised, a sample of the liquid to be measured is drawn in through openings in the sides of the tube, and fills the measuring tube as the piston is withdrawn.

For some in-line units, the piston is raised by the process fluid. The measuring tube is filled through an opening in the bottom of the tube. The filling and measuring cycle is controlled by the periodic opening and closing of a valve. The pressure drop across the measuring tube is about 5 PSI (35 kPa). During the measuring phase, the piston assembly is allowed to fall by gravity, expelling the sample out of the tube through the same route that it entered. The time of fall is a measure of viscosity using the clearance between the piston and the inside wall of the measuring tube as the measuring orifice. The timed interval is tben displayed on an indicator or recorded. For process control or hi-low alarm purposes, the intermittent time signal is converted to a continuous pneumatic or electronic signal by a simple clutch mechanism or by a sliding-wire mechanism which is linked to the recorder pen. To be able to measure high viscosity materials in a reasonable cycle time and to accommodate any position of mounting, a two-way air cylinder is used to lift and force down the piston.

The total viscosity range is 0.1 to 10^6 centipoise (10^{-4} to 10^3 Pa·s). Each piston has a range of 100 to 1. The full scale inaccuracy is about 1%; the reproducibility is about 0.25%; and sensitivity of 0.1% full scale has been noted. Because the frequency of measuring cycles is reasonably frequent—up to 2 cycles per minute, once every 2 minutes is standard—it is applicable to either batch or continuous installations with or without automatic process control capabilities. Pressure and temperature ratings on the standard in-line model are from full vacuum to 300 PSIG (2 MPa) and temperatures up to 650°F (340°C). A special unit can be constructed to operate under pressures up to 500 PSIG (3.5 MPa).

This instrument has been applied in paper sizing,

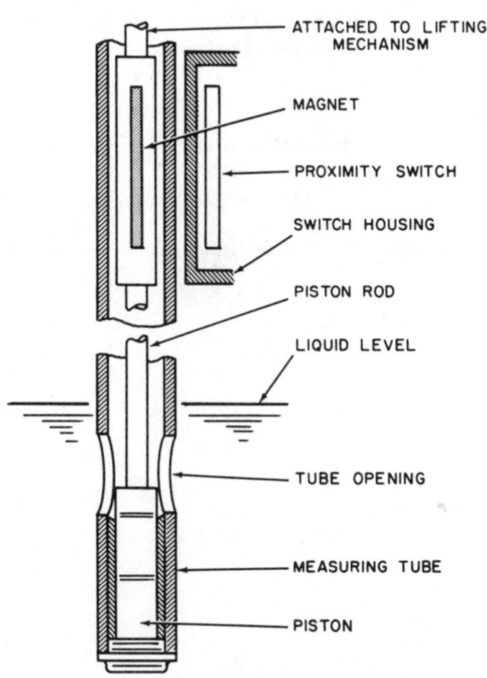

Fig. 7.9a Diagram of falling piston viscometer

printing, coating, polymerization of resins, starch conversion, textile sizing, and blending processes.

Like all other viscometers, operating temperature and pressure should be specified and strictly maintained to utilize fully the capabilities of the instrument. Depending on the fluid characteristics, the error caused by small variations in the process fluid temperature can be quite substantial. Use of a controlled sampling system is recommended for this service. A temperature compensating system should be used whenever possible. Viscosity vs. temperature relationship should be known to provide correct compensation. A pneumatic compensating system is generally effective over a range of ±25°F (±14°C) and over a viscosity range of 3 to 1. It requires a custom made cam to correct for any non-linearity.

Since it is a batch-type measuring instrument, use of this detector should be avoided where fast (less than one minute) response time is required. Where a sampling loop is used, the sample pumping rate should be sufficient to minimize the sample lag time.

To obtain reasonable reproducibility, the following points should be observed:

1. Avoid any vapor entrainment in the sample liquid due to agitation or boiling.
2. Avoid turbulent environment.
3. Calibrate the instrument regularly to take care of the measurement drift caused by gradual material build-up or wear of the piston and tube.
4. The measuring unit should be cleaned at regular intervals—duration is dependent upon the material being handled.
5. Size of any undissolved solids in the liquid to be tested should be small enough not to interfere with the measurement. An in-line filter should be used to remove any large foreign materials.
6. Sensitivity requirements should be known for blending process applications.
7. Avoid use of this instrument for liquids having poor flow characteristics.
8. Erratic readings may result if it is installed in severe vibration environment.

Falling Slug Viscometer

This instrument automatically measures the time required for a cylindrical slug of a specific density to fall a given distance in a vertical tube filled with the process liquid at a known, constant temperature.

The sample pump introduces the fresh sample and purges the system of the previous sample (Figure 7.9b).

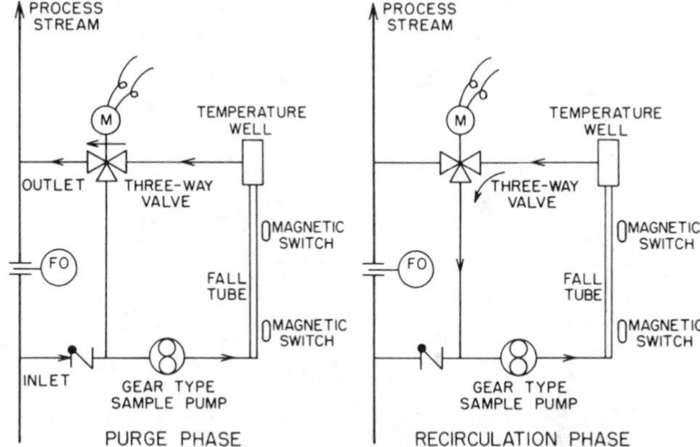

Fig. 7.9b Measuring cycle of falling-slug viscometer

Two separate thermostats in the temperature well control the purge and recirculation cycle by activating a three-way valve. The flow velocity raises the slug to the top of the fall tube, and when the sample temperature is reached, the pump and the sample stop, thereby permitting the slug to fall. As it does, it actuates a magnetic switch attached to the side of the fall tube, which in turn starts the recorder motor. The slug then actuates a second magnetic switch located at an adjustable distance (1 to 20 inches [25 to 500 mm]) below the first. This switch stops the recorder motor. The resultant time measurement is directly proportional to the viscosity of the sample. Actuation of the lower switch also initiates the system to the purging phase.

The total viscosity range is 10 to 1,000,000 centipoise (0.01 to 10^3 Pa·s). Specific ranges for a given process are field-selectable by adjusting the distance between the two magnetic switches on the fall tube. Also, recorder pen drive gears provide full scale indication from 10 to 250 seconds in 5 different steps. The full scale inaccuracy is about ±1%, and the reproducibility is about ±1% full scale, depending on the accuracies of the thermostat and the recorder. The viscometer is designed to operate at temperatures up to 300°F (150°C) and pressures up to 200 PSIG (1.4 MPa), and it is applicable to those continuous installations for which the three-minute maximum cycle frequency is sufficient. Depending on the fluid characteristics, the error from small variations in the process fluid temperature can be substantial. The falling slug viscometer is recommended only for clean streams that are not shear sensitive.

BIBLIOGRAPHY

Nelson, Robert C., "Automatic Force Viscometers," *Instruments and Control Systems*, April, 1963.

"Viscometer Survey," *Instruments and Control Systems*, January, 1971.

Flow Sheet Symbol

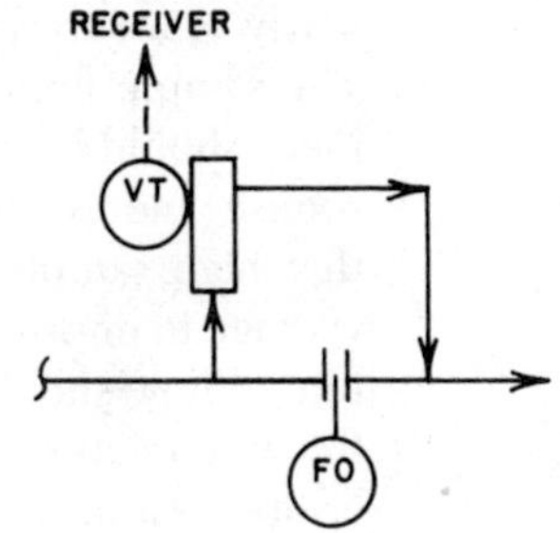

7.10 INDUSTRIAL VISCOMETERS—FLOAT

Type of Hardware:	Float
Design Pressure:	300 PSIG (2 MPa)
Design Temperature:	450°F (232°C)
Materials of Construction:	Body and fittings—wide selection; metering tube—stainless steel or glass; floats—wide selection
Cost:	Single and two-float designs: \$675 to \$1,500; concentric type \$2,250 to \$7,800
Inaccuracy:	±2% of reading below 35 centipoises (0.035 Pa·s). ± 4% of reading above
Range:	0.5–10,000 centipoises (5×10^{-4} to 10 Pa·s) total range, with a maximum span of 10:1 and a minimum span of 3:1
Partial List of Suppliers:	Fischer & Porter Co.

Float Viscometers

These viscometers are used industrially and in the pilot plant to measure the viscosity of process fluids and to continuously indicate, record, or control the process. This type of viscometer is used in a closed flow system. The operating principle is similar to that of the variable area flow meter (rotameter), where the viscous drag force on a float is proportional to the orifice opening required (between float and tapered tube) to move the fluid through that orifice at a constant flow rate.

In a rotameter-type flow meter the forces acting on the float are affected by the flow rate, float and liquid specific gravity, and viscosity of the fluid being metered. For flow meter applications, the floats are designed so that the viscous drag area is relatively small and the float is viscosity insensitive. Thus the float is sensitive only to flow rate and density changes of the fluid. In the viscometer, the flow rate through the variable area meter is held constant. Therefore, the position of the float is a measure of only fluid viscosity and density or of kinematic viscosity. To increase its sensitivity, the viscometer float is designed with a relatively large viscous drag area. For float-type viscometers, accurate viscosity measurement demands a carefully controlled flow rate. Based on their flow controller requirements, there are three different types: single float, two float and concentric.

For these viscometers, the following points should be carefully observed:

1. The float viscometer must be installed vertically, with the outlet at the top.
2. The instrument should be installed in a location that is reasonably vibration-free.
3. Process fluid temperature should be carefully controlled, or temperature compensation should be used. For non-compensating units, viscosity vs. temperature curves should be available for use by operating personnel.
4. For best results, the flow of the fluid should be smooth. Use of pulsating metering pumps should be avoided. The pressure of the fluid should be kept constant.
5. The process fluid should be free of foreign material to prevent plugging of the small orifice (0.1 inch [2.5 mm] diameter minimum) inside of the instrument. In-line filters are recommended to remove these foreign materials.
6. Install bypass lines to permit the flushing of pipelines and to facilitate instrument servicing.
7. Make certain that operating conditions are not exceeded.
8. It is suggested that recalibration with a known viscosity fluid should be made on a regular ba-

sis. The tube and float should be cleaned regularly to assure reliable service.

9. The sample fluid flow rate through the instrument should be sufficient to give maximum response and sensitivity. It has been observed that high sample rates often constitute an advantage in improving speed of response, but the rate of flow should not create turbulent flow in the viscometer, because that would result in erratic readings.
10. The sample liquid should be void of any air or vapor entrainment.
11. Use insulation or steam tracing when fluid temperature deviates considerably from room temperature.

The float viscometer has been successfully used to control viscosity of fuel oils for maximum combustion efficiency in marine and stationary boilers. Other areas of application have been in cement slurry, starch, glue and petroleum (motor oils) products.

Single Float

The single float viscometer is a direct reading viscosity instrument for continuous measurement. As illustrated in Figure 7.10a, a positive displacement pump (other flow control devices can also be used) provides the constant sample flow rate through the instrument. The recommended flow rate is between 0.75 and 2.0 GPM (2.9 and 7.6 l/m). Temperature compensation units are not available for this type of viscometer. If a metering pump is used to provide a constant flow-rate, the temperature rise through the pump should be known for proper correction by the use of a viscosity vs. temperature curve. Transmission of the float position is possible for recording or controlling purposes by the use of an armature attached to the float extension rod with a magnetic sensing device around its outer periphery.

The glass tube viscometer is rated at 450°F (232°C) temperature and 90 PSIG (621 kPa) pressure. For steel tube viscometers, the pressure rating depends on temperature limitations; 650 PSIG at 450°F (4.5 MPa at 232°C) is available.

The single float viscometer can be used with non-Newtonian fluids of viscosities less than 400 centipoises (0.4 Pa·s) and can handle Newtonian fluids up to 10,000 centipoises (10 Pa·s) with a maximum span of 6:1 and a minimum span of 3:1. Inaccuracy of the instruments is ±4% of indication, and reproducibility is ±1% of indication.

Two Float

This is a low cost viscometer designed to provide intermittent viscosity measurement in the laboratory, on the test bench, or in industry. It is only for local indication and viscosity signal transmission or automatic control is not possible. It incorporates two floats. The upper float is sensitive to fluid flow rate and the lower to viscosity. In operation, the fluid flow rate is manually adjusted to a predetermined value as indicated by the position of the upper float. By maintaining a constant reference flow rate as indicated by the flow rate float, the position of the other float indicates the viscosity of the fluid on a direct reading scale.

The recommended piping for a two float viscometer is shown in Figure 7.10b. A throttling valve or an orifice plate may be used in the main line between the instrument inlet and outlet connections to produce the pressure differential required for sufficient flow through the instrument. The required flow rate is 0.25 to 2.5 GPM (0.95 to 9.5 l/m) depending on size. The throttling valve or orifice plate is not required if the outlet line is discharged to a pressure lower than the main header or to a vented tank. A needle valve should be used on the instrument inlet line to provide sensitive and accurate flow rate adjustment. The unit is rated for pressures up to 300 PSIG at 450°F (2 MPa at 232°C) depending on size.

The two float viscometer is suitable for Newtonian

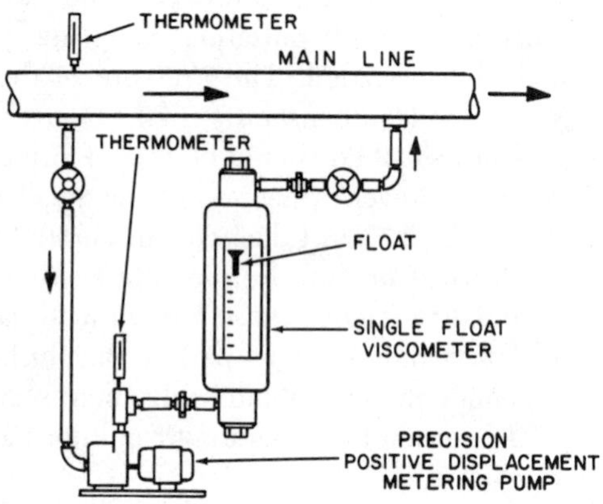

Fig. 7.10a Typical piping arrangement for single-float viscometer

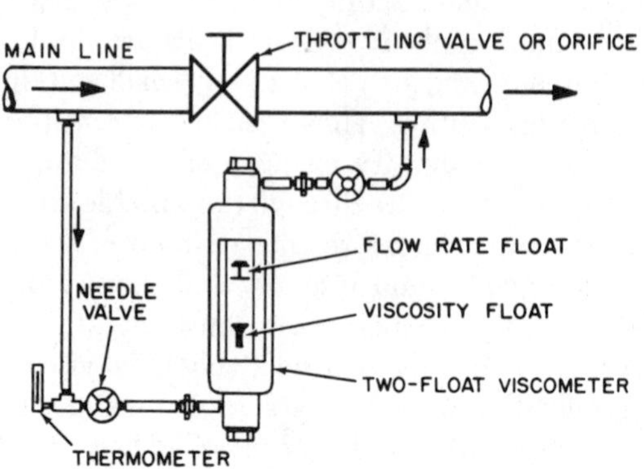

Fig. 7.10b Typical bypass piping arrangement for two-float viscometer

fluids with viscosities from 0.3 to 250 centipoises (10^{-4} to 0.25 Pa·s) and is available with a span range of 10:1. Inaccuracy of the instrument is within ±4% of indication for viscosities higher than 35 centipoises (0.035 Pa·s) and ±2% for lower viscosities. Reproducibility is ±1% of indication.

Concentric

As shown in Figure 7.10c, the concentric viscometer consists of a differential pressure regulator that maintains a constant pressure drop across the meter, and a variable area flowmeter with a viscosity-sensitive float. As the fluid enters the instrument, it splits into two streams. The portion of the fluid that flows upward around the differential pressure float is used to control the pressure drop across the meter. The upper end of the differential pressure float acts as a control valve, and when flow rate changes, it throttles the flow of fluid to maintain a constant pressure drop. The pressure drop across this portion of the meter is determined by the weight of the float. The portion of the fluid that flows downward enters the viscometer tube by way of an orifice and then passes the viscosity sensitive float. Constant flow rate through the viscometer tube is maintained since the fluid flows through an orifice at a fixed pressure drop. Thus, measurement of viscosity is possible under constant flow rate.

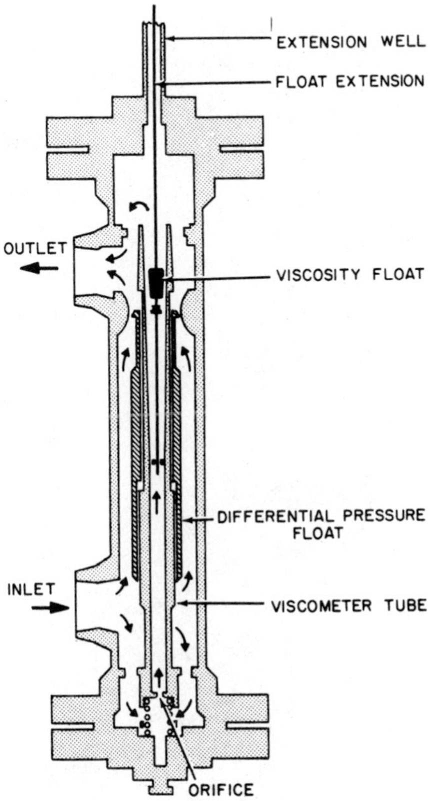

Fig. 7.10c Cross-sectional view of concentric viscometer

An extension attached to the viscosity float transmits its movement through a magnetic coupling to a receiver for display, recording or automatic process control, as desired.

The unit may be placed in either the main flow line or in a sample line depending on the process flow rates. Figure 7.10d shows the flow insensitive range of the instrument.

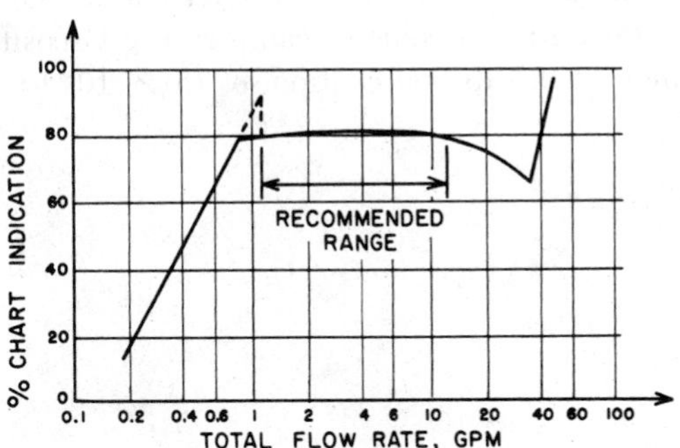

Fig. 7.10d Operating characteristics of concentric viscometer

The flow-rate through the instrument should not exceed 30 GMP (114 l/m). The recommended piping for bypass installation is shown in Figure 7.10e. The restriction orifice should be sized for approximately 4 PSI (28 kPa) drop. If a valve is used in place of an orifice, the correct valve position can be determined by slowly opening the valve until the viscometer reading levels off. Further opening will cause the reading to fall off and then rise sharply as illustrated in Figure 7.10d due to turbulent condition. The valve openings which produce steady viscosity readings are in the operating range. This procedure should be carried out with maximum main

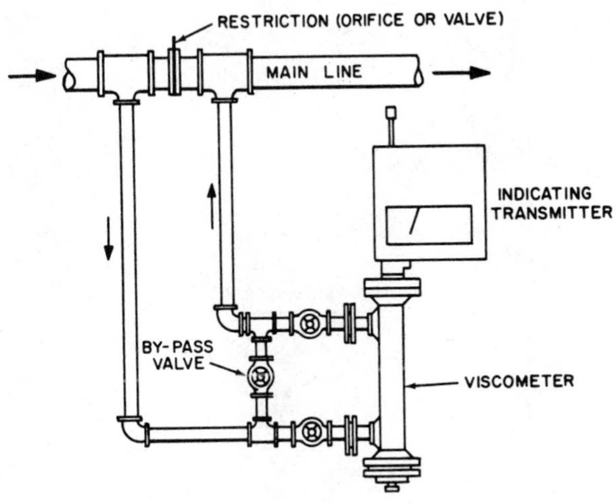

Fig. 7.10e Bypass installation of concentric viscometer

line flow. It is desirable to lock the valve in place to prevent unauthorized adjustments. If the main line flow-rate is lower than 2 GPM (7.6 l/m), use a booster pump to provide adequate flow-rate through the meter.

A pneumatic temperature compensating unit is available for this unit with a total cost of about $7,200. Its logarithmic compensation allows the unit to operate 50°F (28°C) higher than the reference temperature. The unit is rated at 650 PSIG at 450°F (4.5 MPa at 232°C).

The concentric viscometer can measure viscosities in the range of 0.5 to 550 centipoises (5×10^{-4} to 0.55 Pa·s) and is available with a span range of 10:1, with an inaccuracy of ±2% of indication for viscosities lower than 35 centipoises (0.035 Pa·s), and ±4% for higher viscosities.

BIBLIOGRAPHY

Nelson, Robert C., "Automatic Force Viscometers," *Instruments and Control Systems*, April, 1963.

"Viscometer Survey," *Instruments and Control Systems*, January, 1971.

Flow Sheet Symbol

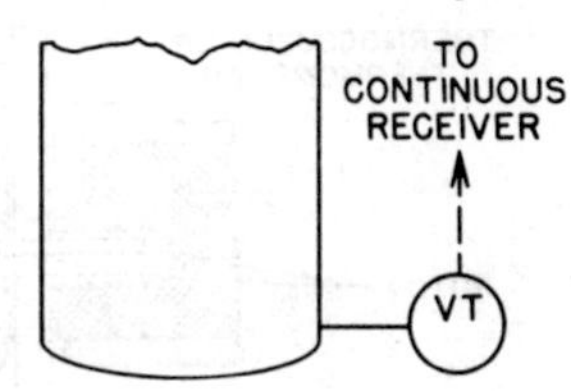

7.11 INDUSTRIAL VISCOMETERS— PLASTOMETERS

Design Pressure:	Up to 5,000 PSIG (35 MPa)
Design Temperature:	Up to 570°F. (299°C)
Materials:	Hardened stainless steel or other metals
Cost:	\$4,500 to \$45,000 as a function of degree of automation
Repeatability:	±1% to 2% of full scale
Range:	0–200 Mooney points, 0–200 MI, 0–100 CIL or 0–1000 division units
Partial List of Suppliers:	C. W. Brabender Instruments, Inc.; 6 am Rad, Inc.; Seiscor Div., Seismograph Service Corp.

Plastometers are utilized to study melt flow behavior (plastic behavior) of plastic materials or the molecular weight distribution of polymers.

A number of materials do not begin to flow unless the stress has exceeded their yield value. Such materials behave as elastic bodies with shears below the yield value, but will yield increasingly with time under greater stresses. This phenomenon is called plastic behavior. Bingham assumed that for some materials the rate of flow was zero for stresses below a critical value, and was linear with stresses above this value. Such a body is called a "Bingham solid" or ideal plastic. Extremely few solids show this behavior. In general, the yield value is not sharp and shows nonlinear behavior with a delayed and imperfect recovery. The plastic behavior differs for different types of plastic solids and at different temperatures.

For this reason, the plastometer gives a relative indication of polymer behavior, thus characterizing a plastic material. Since the shear stress/shear rate relation is based on the molecular behavior of materials, the plasticity can be related to the molecular weight of a material or to its distribution. The plasticity of plastic materials is expressed in arbitrary units such as ASTM melt index (MI), or Canadian Industries Limited (CIL) flow index, or Mooney points, or percent of full scale.

In general, the plastometer consists of a constant temperature sample body heating chamber, a mechanism to apply high shear force, and a device to measure torque or flow-rate of material in detecting the plasticity of the sample.

It can be used in a laboratory to study polymeric behavior with time under high shear stress or may be used on-line to control the process manually or automatically. This type of rheometer is an invaluable tool in manufacturing plastics and synthetic rubbers.

Cone-and-Plate Type

This plastometer incorporates the features of the Mooney plastometer and is designed to meet the requirements of ASTM standard test method D-1646. Basically, the working principle is the same as the one discussed in connection with cone-and-plate viscometers (see Section 7.7). It is designed to eliminate polymer "ball-up" and slippage tendencies by confining the sample in a disc-shaped cavity, as shown in Figure 7.11a. The cone-and-plate plastometer is successfully applied to conduct tests for the evaluation of crude rubber, rubber compounds or reclaims, control of mill breakdown of polymer molecules, determination of time to scorch, calculation of optimum cure time and the evaluation of the processing characteristics of plastics.

In operation, the sample is placed in the cylindrical test chamber and is allowed to reach the predetermined test temperature (up to 400°F [204°C]) by the use of integral heaters. Machined serrations on all of the die and rotor surfaces prevent slippage. After a warm-up period, the rotor is driven at a constant speed (normally

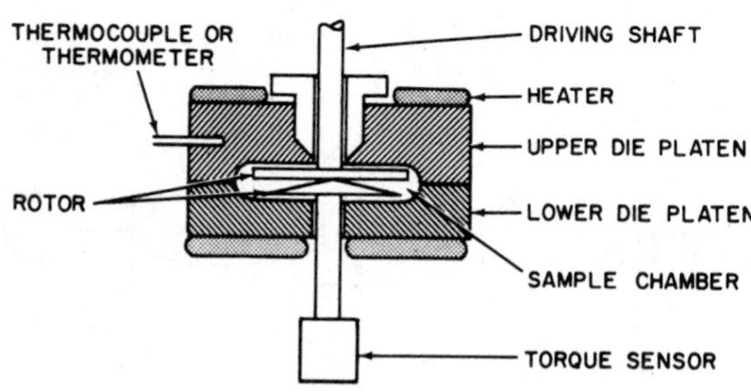

Fig. 7.11a Cone-and-plate plastometer

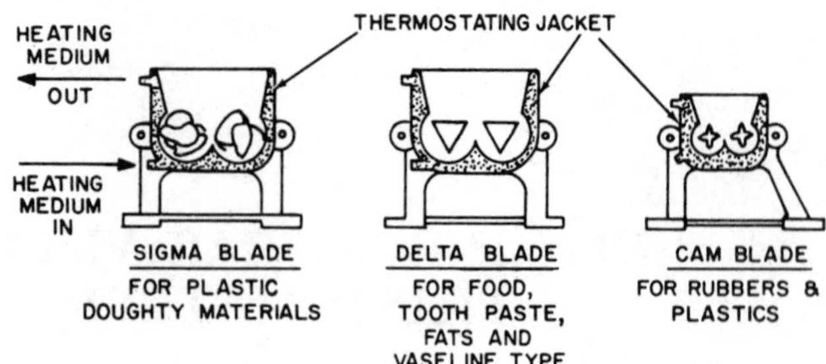

Fig. 7.11b Measuring heads of kneader plastometer

at 2 RPM) or at various speeds from 0.05 to 20 RPM with a continuously variable speed drive to test for relative molecular weight or to study plastic behavior. The shearing action which takes place between the rotor and the die cavity is measured by the deflection of a calibrated U-spring attached to the torque sensing rotor. The deflection of the U-spring is read from a dial indicator and is directly proportional to the shearing torque of the specimen being tested. An electronic strain gauge may be attached to the U-spring to transmit the signal for continuous recording.

The plastometer operates within a range of 0–200 (Mooney) units with repeatability of ±0.5 units. Dr. Mooney gives the following equation for the determination of average viscosity ($\bar{\mu}$) of solids.

$$\bar{\mu} = \frac{G\ 188.44g\theta}{2\pi\omega R^3(R/2a + h/b)} \qquad 7.11(1)$$

where

- G = gauge reading
- g = acceleration due to gravity
- θ = pitch radius of worm gear
- ω = angular velocity of rotor in radians/sec.
- R = radius of rotor, inches
- a = vertical clearance between rotor and stator above or below the rotor, inches
- h = thickness of rotor, inches
- b = effective radial clearance between rotor and stator

Kneader Type

This type of plastometer is equally suitable for both laboratory work and on-line indication or control of highly viscous flowing materials. The chief advantage of this instrument is that it measures plastic behavior or melt viscosity of plastic materials under very similar conditions to those prevailing in processing equipment. This instrument is widely used for plastics and rubber, food, pigment, cement, paint, cosmetics, and for coating products. There are many different shapes of measuring heads available (see Figure 7.11b) to accommodate wide ranges of viscosities.

In operation, the sample is kept at a constant temperature (up to 570°F [299°C]) by the jacketed heater. The kneader is driven by a dynamometer (available with variable speed as well as fixed speed drives). Resistance encountered by the mixing blades in the material under test is transmitted to the dynamometer housing which tends to rotate in a direction opposite to that of the driving shaft. The torque is transmitted on a direct reading balance system through levers and is recorded. The pen movement can be transmitted to a controller for automatic control of processes involving medium viscosity, free flowing fluids.

The reading and recording are on a 1000-unit division scale which is arbitrary for indicating shearing torque of the specimen being tested. Repeatability is ±1% of full scale. By the application of different weights on the scale system, the measuring range can be varied.

Capillary Type

It is designed for use in polymer manufacturing plants, and is based on the capillary tube viscometer principle. The instrument is calibrated to record either ASTM melt index (flow rate of polyethylene through an open ended capillary at 190°C [374°F] and 43.2 PSIG [298 kPa] pressure) or CIL flow index (190°C and 1500 PSIG [10 MPa] for polypropylene) or it can record both by alternating from one to the other (see Figure 7.11c). These two measurements together can be interpreted as a molecular weight parameter and as a molecular weight distribution parameter for a particular polymer. The capillary type plastometer is not as versatile as those previously covered.

The advantage of this instrument is that it can be used as an automatic on-line process control device for both viscous fluids and plastic solids. Especially for the solution polymerization processes, the polymer solution can be directly analyzed in the plastometer from the processing reactors through an auto sampling device which flashes the solvent and the unreacted monomers, and melts the polymer before feeding it to the plastometer. For plastic solids, the die unit (capillary) can be mounted to the pelletizing extruder of the process stream for measurement. The capillary type plastometer is ideally suited for study of plastic behavior of those materials that are processed through injection molding or finishing operations.

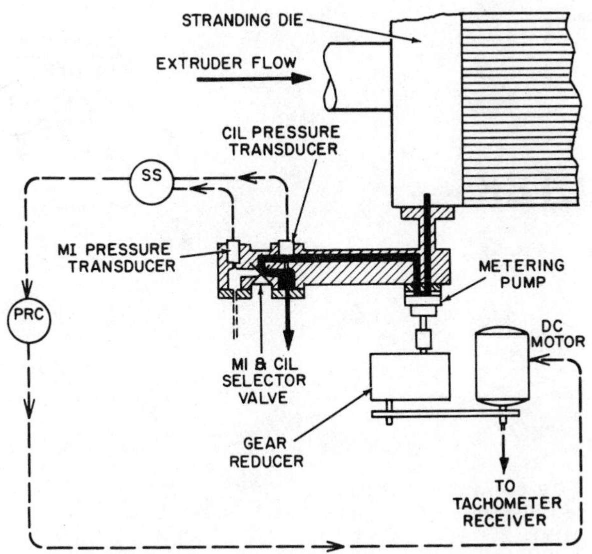

Fig. 7.11c Pressure-controlled capillary plastometer

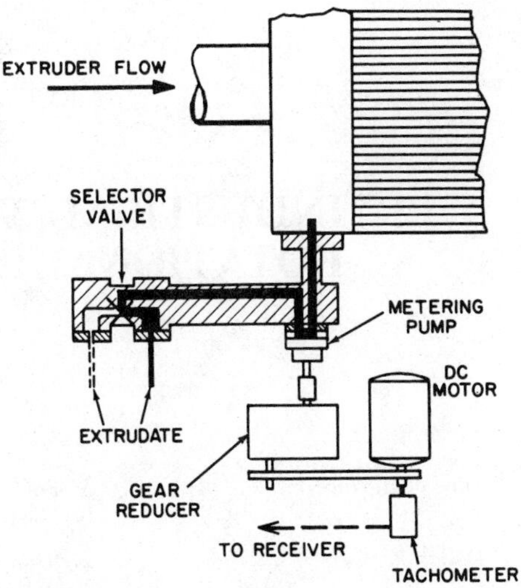

Fig. 7.11d Flow-sensing capillary plastometer

In feeding the capillary type plastometer, the polymer melt is first conditioned to be at some specific temperature and pressure. It is then extruded through a capillary of suitable dimensions. The rate of flow of the polymer through the capillary is measured (the output of DC generator tachometer on the metering pump) and recorded in units appropriate for the test (see Figure 7.11d).

The operating range of this plastometer is 0–200 MI and 0–100 CIL units with repeatability within ±2% of full scale. The unit is designed to operate at up to 570°F (299°C) and up to 5,000 PSIG (35 MPa) in pressure. Erratic results may arise from the same causes that have been pointed out in Section 7.8.

BIBLIOGRAPHY

McKennell, R., "The Influence of Viscometer Design on Non-Newtonian Measurements," *Analytical Chemistry*, Volume 32, Page 1458, October, 1960.

Nelson, Robert C., "Automatic Force Viscometers," *Instruments and Control Systems*, April, 1963.

Nelson, R.C., "Manual Force Viscometers," *Instruments and Control Systems*, May, 1963.

"Viscometer Survey," *Instruments and Control Systems*, January, 1971.

Flow Sheet Symbol

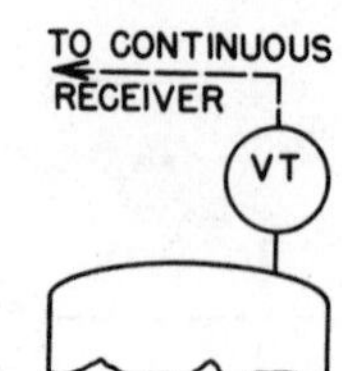

7.12 INDUSTRIAL VISCOMETERS—ROTATIONAL

Type of Hardware:	A. Rotating Cone B. Magnetically Coupled
Design Pressure:	A. 8 mmHg to 100 PSIG (1.0 to 690 kPa); up to 2000 PSIG (14 MPa) with special designs B. Up to 1,600 PSIG (11 MPa)
Design Temperature:	A. Up to 160°F (71°C) for head; up to 600°F (316°C) for mounting flange; up to 2,500°F (1370°C) for spindle B. Up to 570°F (299°C)
Materials of Construction:	Wide selection of materials for high pressure and corrosive applications
Cost:	A. \$2,250 to \$5,600 B. \$3,500 to \$8,500
Inaccuracy:	A. ±1% of full scale between 0–100 and 0–50,000 centipoises (0–0.1 and 0–50 Pa·s) B. ±0.25%
Range:	A. 0–10 centipoises (0–0.01Pa·s) minimum to 0–50,000 (0–50 Pa·s) centipoises standard and up into the 10^6 cps (10^3 Pa·s) range with special designs B. 1 to 4,000,000 centipoises (0.01 to 4,000 Pa·s)
Partial List of Suppliers:	Bristol Co. (A); Brookfield Engineering Labs., Inc. (A); Contraves Viscometer Div. (B); Eur-Control, U.S.A., Inc.; Foxboro Co. (A); Honeywell Industrial Div. (A); Olkon Corp. (A)

Rotating Cone

Rotating cone viscometers are designed to operate under industrial environment on a continuous basis. Viscosity is measured by sensing the torque required to rotate a spindle continuously in a liquid. In a process viscometer application, the sample is continuously replaced and is subjected to a constant shear rate, thus measurement of non-Newtonian apparent viscosity is possible, as well as absolute viscosity measurement of Newtonian fluids.

In operation, a synchronous induction motor (for safety) drives a cage coupled through a calibrated spring to a spindle arm which supports the spindle or cylinder in the fluid being measured. During measurement, the spring tends to wind up until its force equals the viscous drag on the spindle. At this point, the cage and spindle both rotate at the same speed but with a definite angular relationship to each other proportional to the torque on the spring. Two methods are used to convert the angular relationship into a viscosity reading. As shown in Figure 7.12a, one side of a variable capacitor is attached to the cage and the other to the spindle arm. A capacitance is thus made proportional to the angular relationship between cage and spindle.

The other method of signaling is based on a potentiometer mounted in the cage with its free member connected to the spindle arm. A resistance is made directly proportional to the angular relationship between cage and spindle. In either case, the measured signal is transmitted to a suitable receiver as an electrical impulse, and is converted into viscosity units. The variable capacitor type is preferred for most applications in the low viscosity ranges.

The capacitance type can operate within any range from 0–10 centipoises (0–0.01 Pa·s) to 0–50,000 centi-

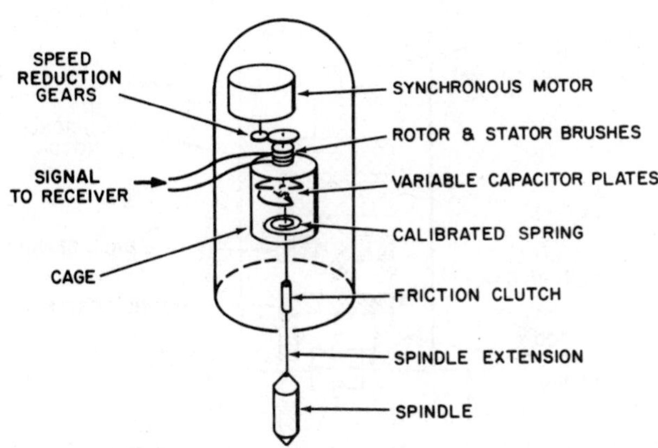

Fig. 7.12a Schematic diagram of the operation of an industrial rotational viscometer

poises (0–50 Pa·s) maximum, with an inaccuracy of ±1% full scale between 0–100 and 0–50,000 centipoises (0–0.1 and 0–50 Pa·s) and with a repeatability of ±0.3% of full scale for all ranges. Response time is less than 30 seconds for full scale change. The viscosity measurement range of any unit can be changed over a 30:1 ratio by changing spindle size or shape. The standard spindle speed is 50 RPM. Depending on installation, the maximum range can be increased to 0–100,000 centipoises (0–100 Pa·s).

The measuring head may be installed on a tank, open vessel, or in a sampling line (see Figure 7.12b). For in-line installation, the process liquid flow rate should be constant, non-pulsating and laminar (less than 3 GPM [11.4 l/m] for standard 4 in. [101.6 mm] pipe). Swirling motion in the measurement chamber may be reduced by installation of a spider type of deflector in the inlet port. The head is designed to operate at temperatures up to 160°F (71°C) and at pressure ranges of 8 mm Hg absolute to 100 PSIG (1 to 690 kPa). The head should be purged with dry air or other inert gases at a rate of 0.5 cubic feet per minute (2.4×10^{-4} m^3/s) or greater, and a positive pressure differential of 3 PSI (21 kPa) between the viscometer head and vessel should be maintained to prevent liquid or vapor penetration to the head.

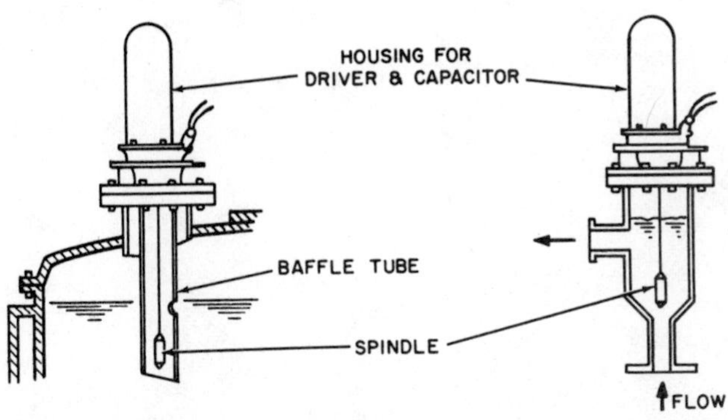

Fig. 7.12b Installation details

The temperature of the mounting flange on which the viscometer is placed should not exceed 600°F (316°C). Above 220°F (104°C) it is necessary to use a cooling pad placed between the flange and the viscometer. The process liquid temperature can be up to 2,500°F (1370°C).

The rotating cone viscometer is not affected by normal industrial vibration. Over-range protection is provided by a friction clutch on the spindle extension. It is simple to clean the spindle. The fluid should be maintained at a constant temperature or be compensated for temperature variations. Logarithmic temperature compensating units are available and are matched to the viscosity-temperature relationship shown by the particular process fluid. Logarithmic compensation to within 1% error over a span of 10°F (6°C) can be provided. For an agitated tank or reactor installation, always use a baffle tube to obtain laminar flow condition, but if very violent agitation or bubbling conditions exist, this viscometer should not be used. For both types of installation, the spindle should always be immersed and liquid level should be kept constant to obtain reproducible data.

The frequency of power supply should be checked. Wide deviations from the specified frequency will cause spindle speed changes, and can introduce errors in the viscosity measurement of non-Newtonian fluids since their apparent viscosity is related to shear rate.

The rotating cone viscometer is an electromechanical device. A monthly preventive maintenance and cleaning schedule is suggested. If the unit is operated continuously, the speed reduction gears will need replacement on a yearly basis.

Because of its simple design, ease of cleaning and non-clogging features, it has been successfully used for solid and liquid blending processes. The rotating cone viscometers are quite versatile in their application over wide ranges of viscosity and are equally applicable to Newtonian and non-Newtonian fluids with or without automatic process control features.

Rotating cone viscometers have been applied chiefly to measure non-Newtonian fluids, such as paints, printing ink, starch, size solution, and varnishes.

Agitator Power

Operation of this instrument is the same as that of the rotating cone viscometer except that the torque exerted on the process agitator blade is measured by a transmitting watt meter (thermal converter). It measures the power consumed in driving an agitator in a mixing tank. Since most of the industrial agitator motors are oversized, the torque response is poor at low viscosities, as shown in Figure 7.12c. The size of the motor or impeller size should be selected to operate in a region where the viscosity vs. torque relationship is linear. Keep in mind that the motor, reducer, bearing assembly, and pressure seal are all part of the viscometer and any change in characteristics of these components would affect the

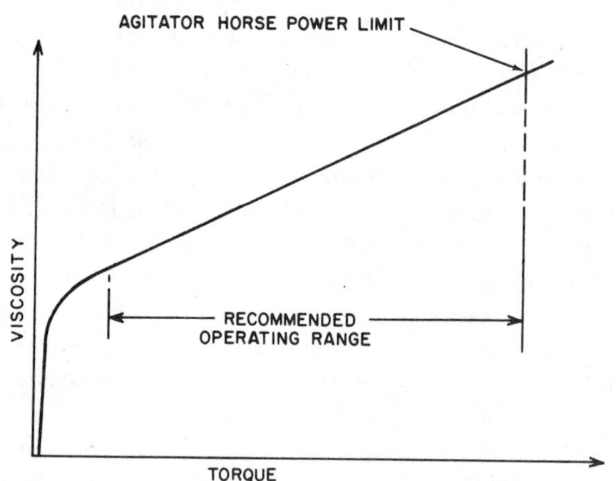

Fig. 7.12c Relationship between viscosity and torque in agitator power viscometer

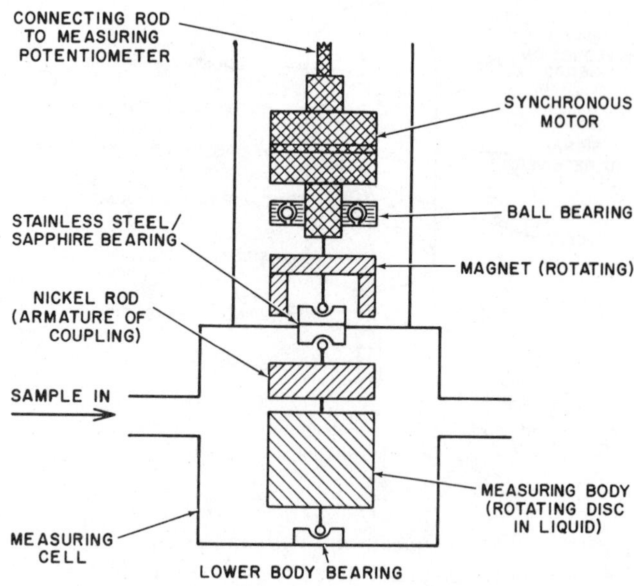

Fig. 7.12d Magnetically-coupled rotational industrial viscometer

power consumption of the motor and, in turn, the viscosity reading. This instrument is very simple and easy to install, but has a low ratio between changes in agitator power consumption and changes in viscosity of the fluid. Many different designs of impellers are available to improve the sensitivity and rangeability.

This type of instrument is widely used in the paper industry to control and measure the consistency of paper pulp slurries. Since it is a self cleaning and agitating design, it is ideal for materials that have a tendency to cling to slow moving parts or to settle out from suspensions.

This type of instrument should not be applied to fluids which are thixotropic or rheopectic (viscosity changes with duration of agitation or shear). However, it can be used in in-line applications for all non-Newtonian fluids since the process fluid is continuously replaced by a fresh sample at a reasonably constant flow rate.

The accuracy and sensitivity of the instrument are poor but its repeatability is at about ±1% of full scale.

Magnetically-Coupled Rotational Viscometer

A recent development in the design of the rotating cone viscometer is the introduction of a magnetic coupling between the electronic detector at atmospheric pressure and the rotating sensor, which is exposed to the process pressure (Figure 7.12d). With this separation between atmospheric and pressurized areas, no purging is required to keep the liquid away from the measuring instrument, and operating conditions at the levels of 2,850 PSIG (19.7 MPa) per 20°C or 1,620 PSIG (11.2 MPa) per 300°C can be tolerated. The magnetically-coupled viscometer should not be used to measure fluids containing fiber, ferrite or abrasive materials because of interference with the operation of the magnetic coupling and of the stainless steel-sapphire bearings.

BIBLIOGRAPHY

Minard, R.A., "Continuous Control of Viscosity," *Instruments and Automation*, Vol. 31, No. 7, July, 1958.

Nelson, Robert C., "Automatic Force Viscometers," *Instruments and Control Systems*, April, 1963.

"Viscometer Survey," *Instruments and Control Systems*, January, 1971.

Flow Sheet Symbol

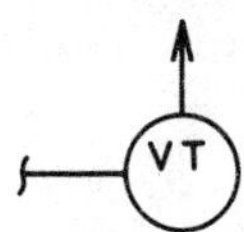

7.13 INDUSTRIAL VISCOMETERS—VIBRATIONAL

Type of Hardware:	A. Vibrating Reed B. Torsional vibration
Design Pressure:	A. Up to 3000 PSIG at 100°F (21 MPa at 38°C) B. Up to 15,000 PSIG (104 MPa)
Design Temperature:	A. From −150 to 300°F (−101 to 149°C) B. From −100 to 300°F (−73 to 149°C)
Materials of Construction:	Wide selection of corrosion-resistant materials and coatings
Cost:	A & B. $3,000
Inaccuracy:	A & B. ±1% of full scale
Range:	A. 0.1 to 10^5 centipoises (10^{-4} to 10^2 Pa·s) B. 0.1 to 10^6 centipoises (10^{-4} to 10^3 Pa·s)
Partial List of Suppliers:	Automation Products, Inc. (A); Nametre Co. (B)

Vibrating Reed

The vibrating reed viscometer is designed for continuous measurement of viscosity in process streams. It can be installed directly in process vessels or pipelines. As illustrated in Figure 7.13a, it consists of a frequency generator, vibrating spring rod, probe, and a pick-up unit to complete a measurement loop through the process material. The drive coil is excited at a frequency of 120 cycles per second (cps) from a 60 cps input frequency. This produces a pulsating magnetic field which causes the drive armature to vibrate at the same frequency.

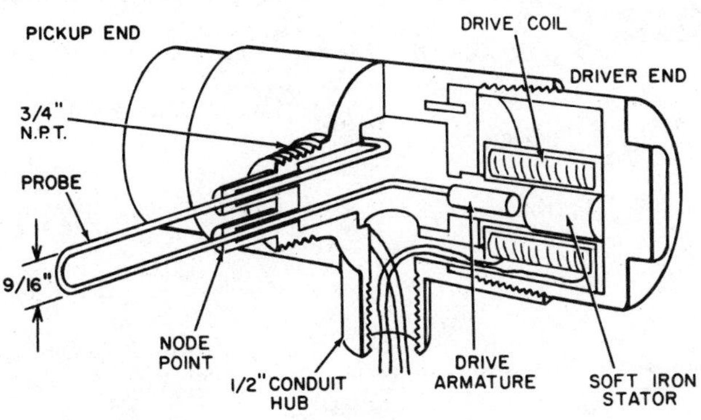

Fig. 7.13a Vibrating reed viscometer

Mechanical vibration of the drive armature is transmitted to the probe through an all-welded pressure sealed node, where amplitude of vibration is zero. The fundamental principle of operation is that the amplitude of probe vibration depends upon the viscosity of the process media. The resistance to the shearing action caused by the probe vibration increases with increase in the process media viscosity. The amplitude of probe vibration is transferred through a second welded node point along the upper spring rod to the pickup end of the detector. The pickup end is similar to that of the driver end, except that a permanent magnet is used to induce a 120 cps AC voltage in the pickup coil from the vibration of the pickup armature. The magnitude of the voltage generated in the pickup coil is a measure of the process viscosity since it is proportional to the pickup armature vibration. The 120 cps output signal from the detector is converted into a 0–10 mvdc signal to indicate, record or control viscosity of Newtonian or non-Newtonian fluids. Inaccuracy of the instrument is about ±1%, and reproducibility is little better. There are 5 different detector units to cover viscosity ranges of 10^{-1} to 10^5 centipoise (10^{-4} to 10^2 Pa·s) at relatively low shear rate. Typical response curves of different detector units are shown in Figure 7.13b.

Since viscosity of fluids is dependent on temperature and pressure, it is a must to maintain the process pres-

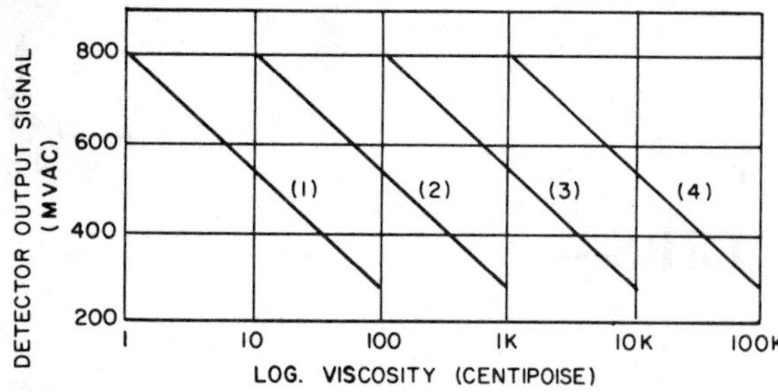

Fig. 7.13b Typical response curves

sure and temperature constant. If maintaining the temperature is difficult or the fluid temperature fluctuates, it is recommended that temperature compensation be used as shown in Figure 7.13c. Before ordering a viscometer for in-process use, the fluid should be characterized against temperature to incorporate a suitable compensating circuitry in the converter.

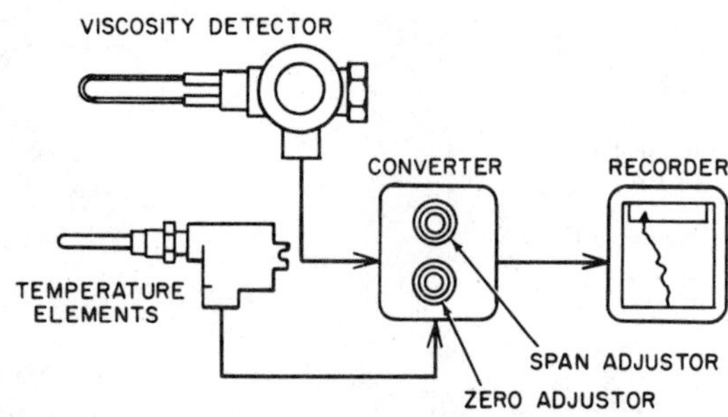

Fig. 7.13c Schematic diagram of recording system

For reliable measurements, care should be exercised to maintain *laminar* flow conditions, since turbulence of flow would give erratic readings. Installation of the viscometer directly in an agitated vessel is not recommended for continuous recording or control purposes. If it must done, the viscosity reading should be taken after turning off the agitator.

Complete immersion of the probe in the fluid is essential. For pipeline installation, mounting of the unit in vertical upward flow, as shown in Figure 7.13d, is recommended to assure complete immersion of the probe at all times. It is a very sound idea to install the viscometer in a sample loop where temperature, pressure and flow rate are controlled.

Amplitude of vibration will change with condition of the probe. It is good practice to calibrate the instrument on a regular basis when it is used for critical operations. Build-up on the probe can occur if the process material has a tendency to adhere or contains long fibrous materials. Slight or loose build-up can be removed by periodical purging of the in-line assembly through a purge well. If the build-up is severe, use of this type of viscometer should be avoided. Since the detector unit transmits a low voltage signal, shielded wires are recommended, and should be grounded at both ends to minimize noise pickup. Supply voltage variations between 105 and 125 volts will have no harmful effects. *Frequency of supply power should be checked* since the probe vibration frequency is controlled by the line frequency, and measurement of viscosity is dependent on the amplitude of this vibration. If the vessel or pipeline vibrates and this vibration frequency is close to 120 cps, it can interfere with the reliability of viscosity measurement by this instrument.

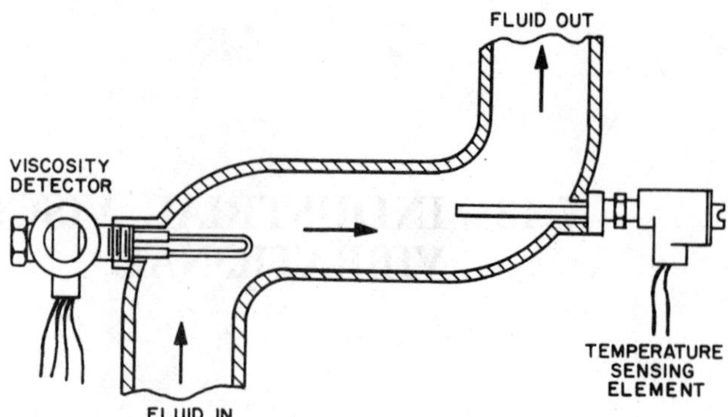

Fig. 7.13d Schematic diagram of recommended installation practice for in-line vibrating reed viscometer

Torsional Vibration

Torsional vibration viscometers are damped proportionally to the fluid friction and measure the viscosity of the fluid by the time of decay of the vibrations.

The amplitude of torsional vibration of a spherical, stainless steel tip immersed in a liquid is maintained constant, and the power required to do it is measured. The readings are converted to a viscosity-density product by chart. The chart reading is then divided by liquid density to obtain actual viscosity.

The spherical tip (Figure 7.13e) is firmly connected

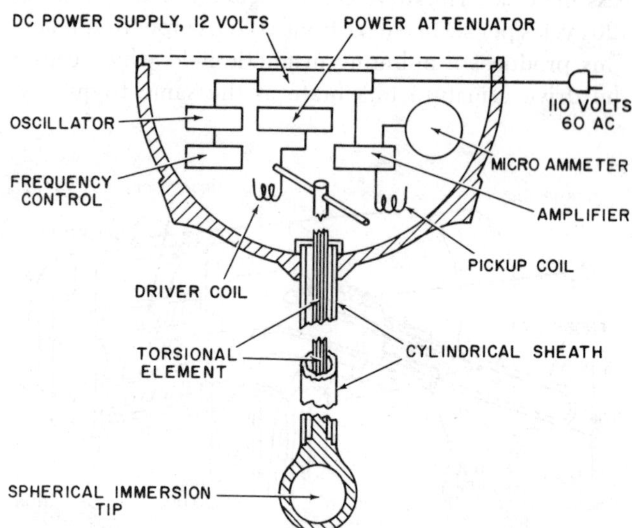

Fig. 7.13e Working components of torsional vibration viscometer

to the instrument housing by a cylindrical sheath. The center rod, surrounded by this sheath for the greater part of its length, is maintained in torsional vibration by an energized coil on a small magnet at one end of the cross-arm. The rod and sheath, joined at their lower ends by the spherical tip, are in torsional vibration about their concentric axis. Rod vibrations are controlled by the elasticity of the sheath. The power required to overcome fluid friction and to maintain a given amplitude of vibration is a measure of liquid viscosity. Amplitude is measured by a stationary coil at the other end of the cross-arm. The subsequent electrical signal is amplified and displaced on a microammeter. For each application there is a resonant frequency at which maximum vibrational amplitude is obtained for a given power input. Measurements are made at this resonant frequency, and readings to 1 part in 10,000 are possible.

One tip is used throughout the entire range of 0.1 to 1,000,000 centipoises (10^{-4} to 10^3 Pa·s). The viscometer is calibrated with standard liquids of known viscosity and density and is easy to clean and to operate. The unit is suitable for both laboratory and on-line industrial use.

BIBLIOGRAPHY

Nelson, Robert C., "Automatic Force Viscometers," *Instruments and Control Systems*, April, 1963.

"Viscometer Survey," *Instruments and Control Systems*, January, 1971.

Chapter VIII

WEIGHT MEASUREMENT

H. I. Hertanu
B. G. Lipták
H. E. Lockery
H. A. Mills
H. C. Roberts

CONTENTS OF CHAPTER VIII

Table VIII
ORIENTATION TABLE FOR WEIGHING SYSTEMS

	Features													
Type of Scale	*Range*	*Installed Cost*	*Accuracy*	*Sensitivity*	*Temperature Sensitivity*	*Ease of Installation*	*Amount of Site Preparation Required*	*Amount of Maintenance Required*	*Applicable to Outdoor Installation*	*Applicable to Dynamic Loads*	*Remote Readout is Conveniently Available*	*Inherently Explosion Proof*	*Applicable to Corrosive Atmospheres*	*Primary Measurement Signal Generated*
Mechanical Lever Scales	L—M	M–H	E	E	L	F	H	H	Y	F	N	Y	N	Displacement
Hydraulic Load Cells	M–H	L	G	G	M	E	L	M	Y	E	Y	Y	Y	Hydraulic Pressure
Pneumatic Load Cells	L	L–M	G	G	L	E	L	L	N	F	Y	Y	Y	Pneumatic Pressure
Strain Gauge Load Cells	Un-limited	M–H	E	E	H	E	M	L	Y	E	Y	N	Y	Electronic

L—Low
M—Medium
H—High

E—Excellent
G—Good
F—Fair

Y—Yes
N—No

8.1 APPLICATION AND SELECTION

Introduction

The greatest number of weighing installations involve the batch charging of various chemicals in accordance with a recipe. Because the use of weight scales necessitates the purchase of weight tanks and pumps, the cost of such installations can only be fairly evaluated by considering the price of all related equipment and building space. If a comparison is made on this basis with flow metering hardware that requires no additional equipment or support design, the relative merits of weighing versus metering techniques will be placed in a more realistic perspective. (See Chapter II for volumetric and mass flowmeters.)

In general, the weighing method is favored in all cases where the batch media is hard to handle, such as in solids, slurries, and very high viscosity or very hot fluids. Elsewhere, the choice might be made on the basis of installed cost, accuracy and other relevant considerations.

The orientation table (Table VIII) gives a brief summary of important weighing element features. The weight sensor should be selected considering some of these features and also keeping in mind the nature of the plant involved. For example, if the instrumentation concept is one of centralized electronic controls with computer supervision, then the logical selection could be the strain gauge element. In general, it is a good rule to consider that measurement technique which requires the least number of signal conversions, if detection precision is of importance.

General Considerations in Weighing

The old axiom, "A Pound's a Pound, the World Around," while hardly true today, nevertheless gives a fair indication of the faith that mankind has in the process of weighing.

What Is "Weighing"?

In the sense we are presently concerned with, to "weigh" an object is to measure the force of gravity on the mass of the object at the particular location in which it is situated. Thus, if we establish a standard of mass and agree on a particular location, we may compare objects with our standard and establish a relative relationship between them and arbitrarily call this ratio "weight." It is true, then, that in order to have an agreeable system of weighing for commercial or process control purposes, for all places and locations throughout the world, it is necessary to agree on a standard of mass and on a standard of acceleration due to the force of gravity. This has been done through convention of the various countries (Metric Treaty of 1875). Our system of measurement is based upon a cylinder of platinum-iridium alloy kept at the International Bureau of Weights and Measure at Sevres, France. Although this prototype is actually a standard kilogram, through the use of a factor (1 kgm = 2.2046 lbm), we may derive a corresponding standard mass equal to one pound.

Force and Weight

One frequently encounters the terms "force" and "weight." What is their fundamental relationship? To understand this, one must consider Newton's fundamental law of universal gravitation, which points out that all material objects attract all other material objects. Thus, everything on or near the earth's surface is attracted toward the earth's center. By agreement, this attractive force of gravity has been termed the object's "weight."

However, if we recognize that the gravitational attraction varies from place to place over the earth's surface (as much as 0.55%), we can readily see that in order for force and weight to be equal, we must assume standard conditions for gravity. In the United States, this has been done and is established as 980.0 cm/sec^2 or 32.15 ft/sec^2. All calibration standards used in this country and all standard expressions of weight are based on practices and standards established by the Department of Commerce, United States Bureau of Standards. Thus, one may use the terms "force" and "weight" interchangeably, provided one accepts the concept that the standard value for "gravity" is assumed in this discussion.

Historical Considerations

Although the discovery of the principle of the lever as a means of weighing is attributed to Archimedes

(287–212 B.C.), it is a fact that the common steelyard was in use throughout the Orient and India many centuries before Archimedes. While Archimedes may have first adroitly defined the mathematical principles involved, simple First Class lever balance systems were frequently referred to in the earliest records of civilization. Mankind thus acknowledged the principle of exchange based on weight. Other early standards and equipment may have been crude in comparison with those of the present.

It is interesting to note that nearly sixteen centuries went by without important addition to the fundamental teaching of Archimedes. Not until the time of Leonardo da Vinci (1452–1519) were the various classes of levers investigated. Da Vinci sketched out and described many styles and designs of lever weighing systems. The basic principle of the lever platform scale, the multiple lever (articulated) scale, and the first direct indicating scale are all attributed to the genius of da Vinci. Although the principle of the simple spring balance was later enumerated by Hooke in 1676, da Vinci had illustrated and described such a system of weighing nearly two centuries earlier.

From the 17th century to the present, while vast improvements have been made in methods of fabrication and available material, no fundamental changes in principle of the lever scales or spring balance have occurred. Even today, the majority of scales and weighing systems installed are based on the early principles set forth by Leonardo da Vinci.

Not until about 100 years ago, with the advent of the hydraulic load cell, and more recently the discovery and development of the electrical strain gauge, have there been many significant changes in weighing principles and machinery.

Application of Weighing Principles

As compared to volumetric measurement, weighing has several outstanding advantages. Perhaps most fundamental is the freedom from temperature effects on volume. In volumetric measurement, one must always take into consideration the temperature of the material being handled, and apply correction factors to obtain results based on mass. Weight is unaffected by temperature, and if the material can be physically handled, it can be weighed. Consider, for example, the weighing of molten metal where temperatures of 2400–2600°F (1300–1400°C) are commonplace. The difficulty of accurate volumetric measurement is easily visualized. Yet highly accurate (0.1%) weighing is possible with modern strain gauge weighing equipment.

In many manufacturing operations and in process control systems, the gravimetric (weight) control may provide the most direct and most accurate measurement. Consider, for example, the problem of maintaining constant proportions between solvent and solute where the solvent may enter the process at widely varying temperatures. If volumetric measurement of the liquid solvent is employed, constant temperature correction must be applied, whereas if both ingredients are controlled by weight, then direct control is maintained and more accurate results may be expected.

In the handling of liquids and solids, the weighing techniques provide commerce and industry with an accurate and direct method of control, relatively independent of process media characteristics, physical location, time, atmospheric conditions and temperature. Units of weight are easily convertible to various monetary values, thus facilitating commerce and exchange.

In the handling of such material as chlorine, liquified petroleum gases and other like materials, it should be noted that the weighing process provides a safe and convenient method of control without container penetration, external sight-glasses, etc. In the case of dangerous or highly corrosive materials, this is a very important advantage. The same may also hold true for cryogenic materials.

Methods of Weighing

Mechanical Lever Scales

By far the largest number of weighing installations are based on mechanical lever systems. These are available for such applications as motor truck scales, railroad track scales, hopper scales, tank scales, platform scales, crane scales, dynamometer scales and many other specialized types and applications.

They utilize the principles of the lever and involve countless combinations of load support arrangements. Although portable models are available in the lower capacities, most installations are permanent in nature and usually require some preparation of the installation area.

Mechanical lever scales have reached a high degree of perfection through many years of development by hundreds of individuals and manufacturers. One outstanding advantage is their basic simplicity, which allows them to be used even in those parts of the world where sophisticated maintenance service is not available.

Because of their accuracy and reliability, the performance of mechanical lever scales has been the basis for all laws and standards established in connection with weights and measures for the regulation of commerce on a local national and international level.

Spring Balance Scales

Perhaps the simplest of all methods of weighing, the spring balance scale consists basically of a calibrated spring and some means of reading deflection or elongation of the spring under load. This rudimentary principle can be very accurate when properly applied in correct design, with presently available high-grade constant temperature modulus alloys. Although spring balance scales are dependent on relatively large deflection of the weighed load, and are best suited for relatively light loads, they represent a most economical approach to accurate weighing.

Load Cell Weighing—Hydraulic, Pneumatic and Electrical

Although available in one form or another for over fifty years, the use of load cells as weighing devices has only recently become of widespread economic and engineering importance. Starting in the industrial expansion period following World War II, a number of important inventions and developments improved the level of performance of load cells of all types. Today, although trailing far behind mechanical lever scales in total number of installations, load cells have proven their accuracy and reliability for many types of installations that heretofore had proven most difficult. Load cells and load cell weighing systems have several outstanding advantages: they are relatively easy to install; they have fast response and short settling time; they are not subject to wear; they permit remote location of the readout module; and they yield an analog weight signal that is either directly compatible with, or easily converted to, standard control instrumentation.

Factors Affecting Weighing System Performance

Some of the common factors affecting weighing system performance are:

1. Temperature
2. Vibration
3. Ambient conditions
4. Maintenance

Temperature

In general, most scales or weighing systems are designed to operate under normal atmospheric temperature conditions. Operating personnel are usually present, at least part time. We shall assume average operating limits of +20°F to +120°F (−6.7 to +48.9°C). Operations outside these limits are commonplace, as, for example, wintertime operation of platform scales with the scale indicator head located in a heated control room. This example also infers operation at different temperature levels on a single piece of equipment.

All modern weighing systems are compensated for operation over varying temperatures. Performance and accuracy guarantees are usually given for operation at 70°F ±50°F (21°C ±10°C). For operation outside these limits, temperature correction curves are available from most manufacturers.

Mechanical Lever Scales. A mechanical lever scale system is largely self-compensating when subject to uniform gradual temperature changes. Inasmuch as we are here concerned with a lever ratio relationship, it can be readily seen that although the overall length of a lever may change due to expansion, the relative position of the fulcrum point with respect to the reaction points does not change and the lever ratio remains constant.

It is true, however, that under conditions of non-uniform temperature on a lever system, some error in reading or balance may be introduced.

It should be recognized that errors on direct reading scales due to temperature shift may be divided into two classifications: (1) zero shift, (2) span shift.

In the case of zero shift, the effect is translatory, and will represent a constant error over the weight indicator range. In the case of span shift, the error is rotational in nature, and one will find variable errors over the instrument range. Zero error or unbalance may be checked from time to time by direct observation. All standard systems are equipped with "zeroing" adjustment which can be manipulated as required.

In the case of span errors resulting from temperature change, a much more serious problem is faced. It is substantially impossible for an operator to compensate for such errors unless temperature correction factors or temperature correction curves are available. The use of such factors or charts involves careful determination of temperature. Where possible, it is most desirable that the manufacturer provide the highest degree of self-compensation within the instruments themselves, thus avoiding errors or inconvenience in use.

Load Cell Weighing Systems. Load cell weighing systems differ from mechanical lever systems in principle of operation and design. It is not unexpected, then, that the effect of temperature on load cell weighing systems is quite different from that on mechanical lever systems. The elimination of levers and mechanical linkages precludes errors (or compensations) arising from these members. The individual load cells must, therefore, be designed for minimum effect of temperature and contain compensating devices or networks that will assure performance within specification. Although some weighing systems are offered with a centralized temperature compensating device, this presumes uniform temperature at each individual load cell and at the compensator. This condition may not prevail at all times.

Electrical Load Cells. All types of load cells discussed contain some form of load supporting member and an associated electrical component designed to detect precisely deflection of the support member under load. Deflection may be in the order of 0.001" to 0.010" (0.025 to 0.25 mm) or more. The electrical signal may take the form of changing current, voltage or resistance. Load cells involving one or more bonded or unbonded strain gauges also contain appropriate compensation networks that correct not only for changing electrical characteristics due to temperature variation, but also in most instances for modulus changes in the support member and for other temperature related errors.

Hydraulic Load Cells. Hydraulic load cells are subject to zero and span errors when subjected to varying temperature. Temperature compensation must be provided to prevent errors in output signals.

Pneumatic Load Cells. Pneumatic load cells considered here are of the force-balance type. They are relatively free of temperature related errors, provided proper care has been given to the choice of diaphragm material. It is of fundamental importance that the operating medium, whether compressed air or other gas, shall be sufficiently dry to prevent condensation of water vapor and freezing at temperatures below 32°F (0°C).

Vibration

Vibration and dynamic disturbances may seriously affect weighing system performance. Accuracy, stability of reading and frequency of maintenance are factors that may be affected. Vibrational disturbances may arise from two sources. These can be either external to the weighing system, such as building vibration, or caused internally, such as through impact loading.

In the case of external sources, selection in the choice of scale location may be possible. Locations having maximum foundation stiffness and stability, away from known sources of vibration such as compressors, punch presses, etc., are most desirable. Where foundation stability is known to be poor, one should consider methods of improving stability, as by erecting additional bracing columns, gussets or concrete work. Consideration can also be given to the use of suitable vibration absorption or insulation material, such as "Fabreeka".

Vibrational disturbances internal to the weighing system, such as those caused by loading methods, or agitation of tank content etc., may also cause serious problems. Design consideration should be given to the loading method on all weighing systems, to eliminate or minimize impact or shock loading. Baffle plates are frequently helpful in reducing surging and gyration of fluid contents in weighing vessels.

Most weighing systems are provided with adjustable vibration dampers. These, of course, should be maintained in good working order, and adjusted for optimum scale performance under the operating conditions.

Ambient Conditions

Weighing system installations frequently have to be made in areas where the equipment may be continuously or intermittently exposed to high or low temperatures, wet or humid atmosphere, corrosive or abrasive contaminants, etc. In such areas, steps should be taken to afford as much protection as possible to the weighing system. Foundation pits should always be provided with adequate draingage that will allow quick run-off and thorough drainage of the pit. Attention should also be directed to the possibility of accidental back-flooding through the drain in the event of flooding in other areas.

Standard methods of protection such as corrosive-resistant paint, electroplating, etc., may be employed. It is often practical to specify that materials and alloys unaffected by the anticipated corrosive materials be used in the manufacture of the equipment.

When conditions are particularly severe, it may be desirable to provide a small atmosphere-controlled area to house the readout instruments for the weighing system. Prefabricated structures are available for this purpose.

Maintenance

Proper and timely maintenance is most important in the operation of any weighing system. Reference should be made to manufacturer's suggested practices, and adjustments in this program should be made to accommodate special conditions.

The weighing process is frequently the basis of control for manufacturing or chemical processes. It is, therefore, very important that calibration checks be made from time to time to assure accurate performance.

Where the weighing system is used for commercial purposes, i.e., buying or selling of goods or merchandise, it is the legal responsibility of the seller to maintain the accuracy and performance of the weighing equipment in conformance with local laws on weights and measures.

Load Cell Selection

Concepts and selection procedures for weigh systems based on load cells are focused on accuracy and repeatability of measurements of relatively large loads within a variety of constraints imposed by shape and vessel sizes, structural and mechanical arrangements, materials to be weighed, and environmental conditions. This section presents established criteria for load cell selection, design concepts used in load cell installation, analysis of the impact of piping arrangements, mechanical equipment, and structural deformation on total weigh system performance.

To obtain the best performance in designing a load cell based weighing system, consideration must be given both to the selection of the load cells and to the choice of load cell installation assemblies. A correctly designed weighing system must have the following characteristics:

1. Low deflection (typically 0.005 to 0.008 inches [0.125 to 0.2 mm]) permitting process piping to be attached to the weighed vessel.
2. Excellent repeatability: ±0.02% of full scale, or better within a prescribed temperature range.
3. Accuracy: Ranges from 1% to better than 0.05% of full scale.

The following factors shall be considered when selecting load cells:

Mode of loading—tension or compression. This choice

is determined by the type and capacity of the vessel and by structural and mechanical design criteria. It is proven by experience that tension support for very large tanks or hoppers is more difficult to design and is more costly than a compression support. In addition, tension load cells with female threads on both ends and the need for eyes and rods require greater vertical clearances. A single load cell (see Figure 8.1e) supporting a small tank in tension is stable and less expensive than a multiple compression support. A weight of maximum 20,000 lbs. (9,000 Kg) load usually limits the tension mounting applicability.

Ambient temperature. The compression load cell assembly will require low friction expansion assemblies in order to accommodate differential thermal expansion or contraction between vessel and supporting structure. The tension load cell assembly does not require additional accessories for expansion compensation. The adjustment of flexure rods which are part of the standard tension mounting assembly will compensate for differential thermal expansion between vessel and structure.

Lateral restraints. Lateral restraints on vessel movements are always required when a compression assembly is chosen. In a tension assembly, lateral restraints are not required for vented vessels which store, for instance, dry, non-hazardous materials, since a hanging mass is inherently stable.

Structure vibrations. Tension assemblies are more sensitive to vibration because of reduced structural stiffness and damping capability caused by tension linkages. The compression assemblies sensitivity to vibration are a function of the stiffness of the structure and vessel supports.

Number of load cells. The number of load cells required is determined by the plane view geometry of the supported structure—hopper, tank or silo. A vertical circular tank might require three cells (see Figure 8.1b) while the same tank in a horizontal positon might require four cells (see Figure 8.1c). Considerations must also be given to the strength and rigidity of the weighbridge structure. Horizontal dimensions in excess of 25 feet (7.5 m) may increase the number of load cells needed. Related to the number of load cells required are two additional factors: Capacity and degree of precision required. When a load is supported from more than one point, pivots or flexures can replace some of the cells, depending on the symmetry of the load, at the expense of overall accuracy expected. Vertical vessels with supports above the center of gravity may use three cells instead of one cell and two pivots. Three cells give more accuracy but cost more. Instead of four cells, a horizontal vessel can use a flexure and two cells or a flexure and one cell. The use of a flexure reduces the staying requirements but with some loss of accuracy.

Load cell capacity. The minimum load cell capacity can be calculated with the following formula:

$$C = 1.25\,K\,\frac{W_T + W_N}{N} \qquad 8.1(1)$$

where

C = minimum load cell capacity
1.25 = allowance factor for low tare estimates and unequal load distribution on the load cells as installed
W_T = tare weight of the empty vessel
W_N = net weight of projected vessel content (live load)
N = number of load cells
K = dynamic factor ($K = 1.25$ for certain dynamic loads, otherwise $K = 1$)

When calculating the tare weight of the empty vessel, one must include any additional equipment attached to the vessel such as agitators, valves, and filters which contribute to the weight that the empty vessel (including its accessories) will exert on the load cells.

Examples of anticipated dynamic loads are vessels with crane buckets and vessels with horizontal agitators. Assuming for these cases $K = 1.25$, one can provide an extra capacity for sizing, resulting in a higher capacity load cell selection. A higher capacity load cell will perform better under repeated impact loads or high cycle fatigue.

Type of load cells. A selection must be made between hydraulic and electric load cells. As a rule of thumb, hydraulic load cells are selected for large and very large vessels whenever required accuracy is within 0.25% to 1%. Hydraulic load cells' ruggedness is good and their cost is reasonable. On outdoor installations where temperature changes are drastic, special precautions are necessary. Maintenance is usually low. Electric load cells are more expensive, most accurate and most trouble free. Inaccuracy is as low as 0.1% and can be even lower depending on mounting, staying and piping.

Classes of load cells. Manufacturers offer the following classes of load cells:

general purpose
precision
high temperature environment
corrosive environment
rugged design

General purpose load cells may be used in any service (tension or compression) whenever weigh system accuracy required is not better than 1%. Precision load cells are specified in systems where accuracy is expected to be 0.1% or better. Specially designed load cells using adequate materials of construction are utilized in a high temperature environment (maximum 450°F [232°C]). Precision and high temperature load cells have temperature compensation accessories which make the operation unaffected by temperature variations within the compensation range (+15 to +115°F [−10 to +46°C] for general purpose and precision and +15 to +425°F [−10 to +218°C] for high temperature design). Load cells can be protected with a special coating in order to prevent deterioration due to the presence of corrosive chemicals. Rugged load cells are offered whenever mechanical shocks may affect their performance.

Load Cell Installation Assemblies

Load cells measure all vertical forces acting upon the vessel. Forces other than the vessel and contents must be kept small, elastic and repeatable so that their effect can be removed by field calibration.

Load Cell Arrangement

The general rules for load cell and vessel arrangements are:

1. The vessel structure in the area of the load cell mounting must be rigid.
2. The supporting structure or foundation depending upon the loading mode (tension or compression) must be rigid. If more than one vessel is to be supported on the same structure, the structure must be designed with sufficient rigidity to prevent interaction errors caused by large deflections.
3. On multiple load cell arrangements, the load cells must be positioned and should be installed so that after the vessel is fully loaded each cell will carry not more than 120% of rated capacity.
4. Optimal vessel stability requires flexibility in the vertical plane and rigidity in the horizontal plane. Figures 8.1a through 8.1e show typical load cell arrangements which include some design recommendations. However, each weight application must be individually analyzed and an optimal arrangement should be chosen by assessing stability vs. cost for an arrangement which will produce the desired weigh system performances.

Vessel Stabilizing Devices

Stabilizing devices are mechanical elements designed to secure a weigh vessel to the structure allowing free movement in a vertical plane and restraining lateral movements thereby maintaining an initial alignment throughout service life. The following disturbance factors may destabilize a vessel mounted on load cells:

fluid sloshing
violent chemical reactions
thrust and impact forces due to mass flow entering or leaving the vessel
vibration of live bottoms
agitators
thermal expansion of attached piping
structural support vibration and/or deflection
wind
seismic events

Two major considerations added to the above mentioned factors make stabilizing devices a requirement for most weigh applications using load cells:

Safety. Attached piping rupture or vessel rupture due to motion produced by one of the destabilizing factors may create a hazardous situation if dangerous materials are weighed.

Weigh system accuracy. Vessel vibration, translation or oscillation can apply side loads on load cells causing readout errors.

Types of Stabilizing Devices

Two types of stabilizing devices have emerged as design solutions for vessel lateral movement restraints:

1. Stay rods.
2. Safety check rods.

Stay rods are mechanical elements installed securely between a gusset on the vessel support bracket and a rigid floor bracket. Vessel translation and rotation is thus restricted, while radial thermal expansion is relatively free. Installation and sizing of stay rods in a perfect horizontal plane must insure a linear response to vertical movement. Figures 8.1a and 8.1g show typical installation of stay rods on vertical and horizontal vessels.

Safety check rods are back-up elements whose function is to prevent gross vessel tipping or wobbling (see Figure

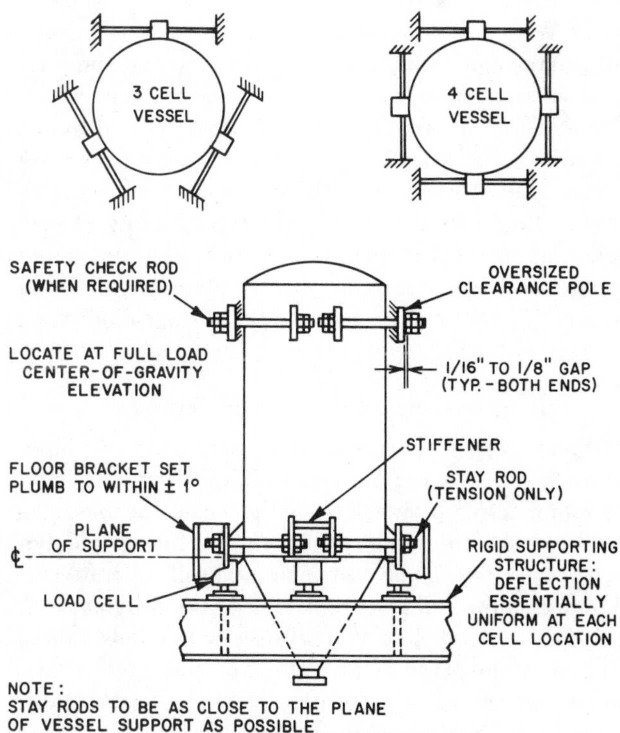

Fig. 8.1a Typical staying arrangement

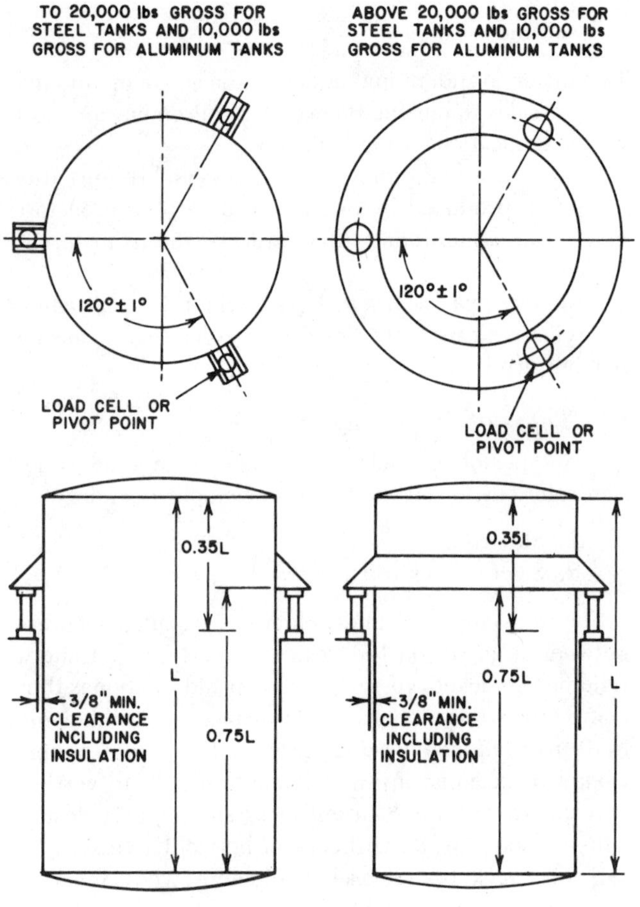

TO 20,000 lbs GROSS FOR STEEL TANKS AND 10,000 lbs GROSS FOR ALUMINUM TANKS
ABOVE 20,000 lbs GROSS FOR STEEL TANKS AND 10,000 lbs GROSS FOR ALUMINUM TANKS
120° ± 1°
120° ± 1°
LOAD CELL OR PIVOT POINT
LOAD CELL OR PIVOT POINT
0.35L
0.35L
L
0.75L
0.75L
L
3/8" MIN. CLEARANCE INCLUDING INSULATION
3/8" MIN. CLEARANCE INCLUDING INSULATION

Fig. 8.1b Load cell locations for vertical vessels

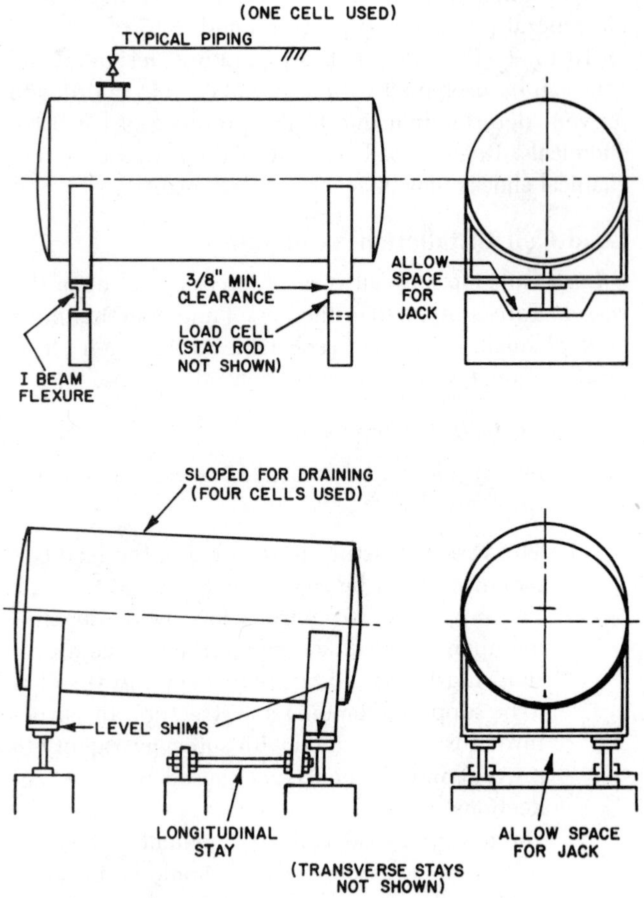

(ONE CELL USED)
TYPICAL PIPING
3/8" MIN. CLEARANCE
LOAD CELL (STAY ROD NOT SHOWN)
I BEAM FLEXURE
ALLOW SPACE FOR JACK
SLOPED FOR DRAINING (FOUR CELLS USED)
LEVEL SHIMS
LONGITUDINAL STAY
ALLOW SPACE FOR JACK
(TRANSVERSE STAYS NOT SHOWN)

Fig. 8.1c Load cell and stay locations for horizontal tanks

8.1a). Stay and check rod sizing and arrangements vary from application to application. The design criteria define the total acceptable vertical force generated by the deflection of all stabilizing devices expressed in percent of total load. A limit is established for the total rod elongation, thus minimizing significant sideloading of load cells and attached piping. The mechanical arrangement of stay and check rods to obtain optimal equilibrium depends on the center of gravity of the fully loaded vessel in relation to the staying plane.

Piping Associated with Weigh Vessels

Weigh vessels are generally connected to pipes through which materials are introduced and extracted. Any pipe directly attached to the vessel becomes an active part of the weigh system; any relative motion between the vessel and a particular pipe will generate vertical and horizontal reaction forces on the weigh vessel. The total vertical force V generated by the deflection of all piping connected to a weigh vessel (see Figure 8.1a) should not exceed a percentage of maximum live load, proportional to the required system accuracy A:

$$V \leq (30\,A)L \qquad 8.1(2)$$

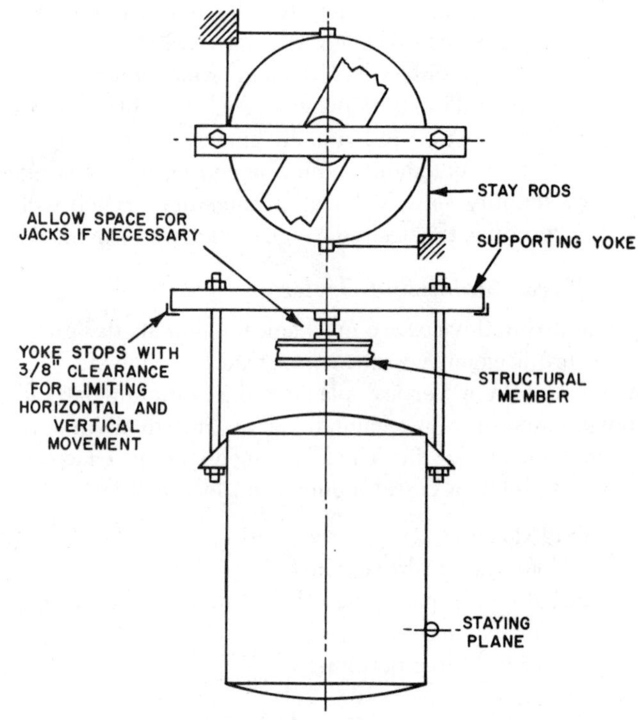

Fig. 8.1d Suspended tank installation detail, load cell in compression

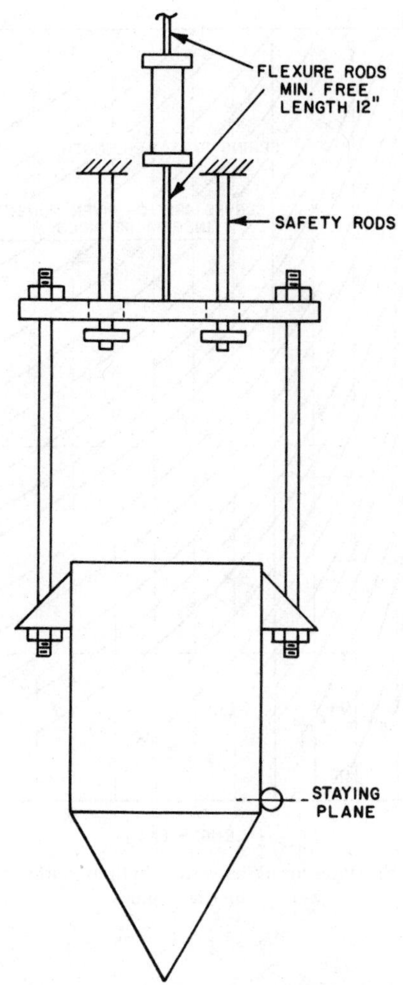

Fig. 8.1e Suspended tank installation detail, load cell in tension (electric cell only)

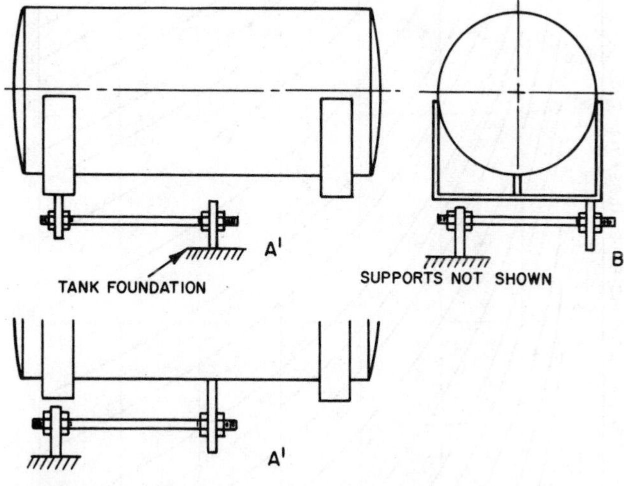

Fig. 8.1f Staying method for horizontal tanks

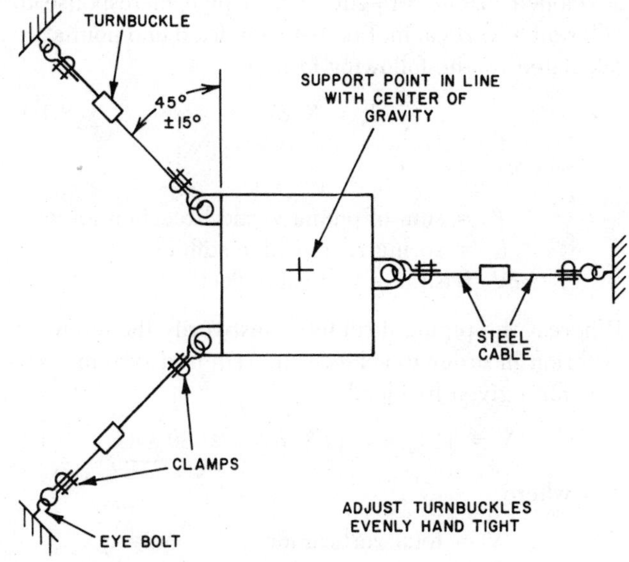

Fig. 8.1g Staying method for suspended hoppers

where

V = Total vertical force
A = System accuracy, percent
L = Maximum live load

There are two types of vertical forces generated due to piping:

1. Differential motion between the pipe connecting point on the vessel and the first pipe support.
2. Thermal expansion of the vertical run between the points mentioned above.

In order to minimize the vertical forces associated with piping, it is recommended to mount all first pipe supports to the vessel support structure using only horizontal runs between the vessel and the first piping supports. The closest pipe support to the weigh vessel is the first support. The total vertical piping force represents the sum of individual vertical piping reactions, generated as described above. The total vertical deflection, which must be evaluated for each pipe run, is expressed in the following formula:

$$\delta V = \delta s + \delta t_v + \delta t_p + \delta a \qquad 8.1(3)$$

where

δV = vertical deflection
δ_s = vessel support deflection
δt_v = vessel thermal expansion to point of pipe attachment
δt_p = pipe thermal expansion in vertical runs
δa = deflection of first pipe support or anchorage

Deflection tending to increase the vessel weight is given a positive sign; deflection tending to decrease the vessel weight is given a negative sign. The vertical reactions developed in one straight span of pipe in response to differential vertical motion between fixed end points are calculated by the following formula:

$$P_i = K_i \, \delta V_i \qquad 8.1(4)$$

where

P_i = sum of piping vertical reaction forces
K_i = spring rate or pipe stiffness
δV_i = sum of vertical deflections

Whereas the piping flexibility must verify the following criterion in order to achieve the required system accuracy for a given live load:

$$V = |\Sigma P_i| = |\Sigma K_i \, \delta V_i| \leq (30\ A)L \qquad 8.1(5)$$

where

V = total vertical force
P_i = vertical reaction force exerted by span pipe i
K_i = spring rate of span pipe i
δV_i = vertical deflection of span pipe i
A = system accuracy, percent
L = maximum live load
i = 1,n where n is the number of pipes connected to the weigh vessel. The span of pipe is assumed to be the portion of piping between the vessel connection and the first support.

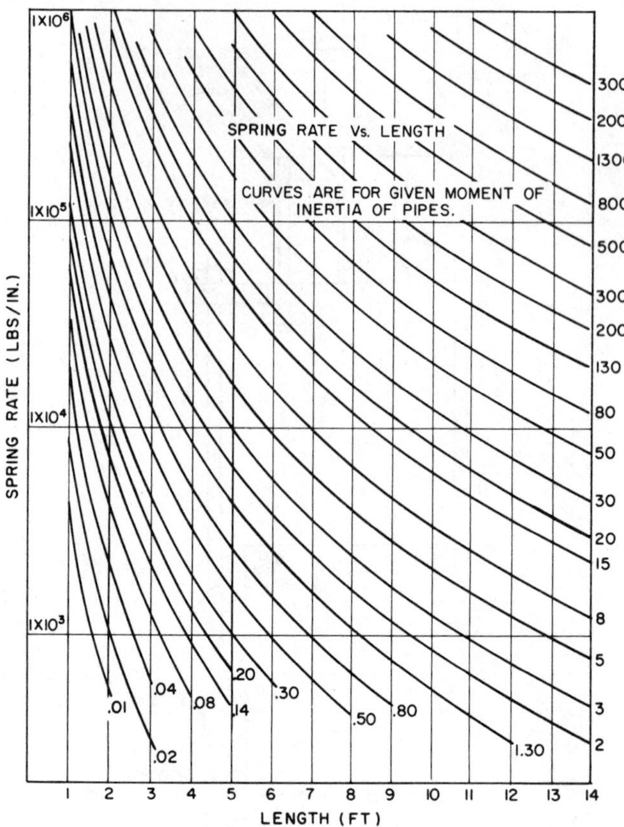

Fig. 8.1h Pipe spring rates vs. length for various moments of inertia of pipes

The spring rate K of any pipe is derived from the bending formula, taking into account the moment of inertia for the pipe section under consideration. Figure 8.1h shows pipe spring rates versus length for various moments of inertia of pipes. Figure 8.1i shows spring rate versus length for schedule 40 steel pipe. If the resultant vertical force exceeds the live load criterion for a required accuracy, design solutions are required to increase the piping flexibility. There are three practical ways available:

1. Reducing the pipe spring rate by choosing a different pipe schedule.
2. Introducing flexible devices whenever possible. Limitations are usually related to design pressure, temperature and material of construction.
3. Increasing the length of horizontal runs between the vessel connection and the first support. Table 8.1j presents calculated force reactions for several pipe runs and flexible devices.

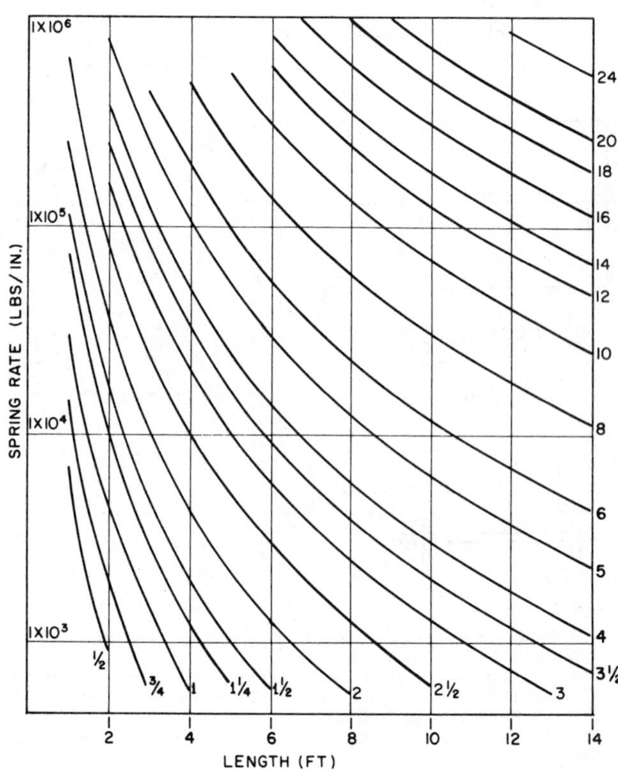

Fig. 8.1i Spring rates for schedule 40 steel pipe

Vessel and Structure Expansion

Temperature variations can cause the vessel or the supporting structure to contract or expand. Under these circumstances, load cells are subjected to horizontal loads resulting in weighing errors. Whenever the mode of loading is compression, one can minimize the expan-

Table 8.1j
VERTICAL REACTION GENERATED BY A 0.25 INCH (6.35 mm) END DEFLECTION (COMPARISON)

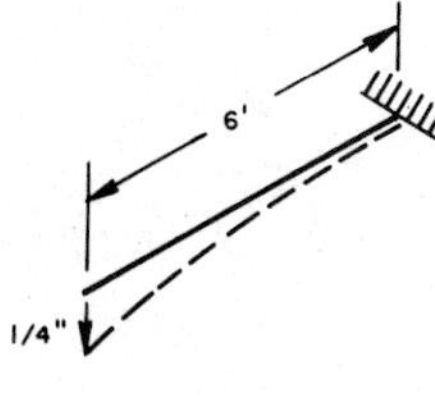

Nominal Pipe Size (mm/in.)	Steel Pipe Sch 40 (kg/lb)	Steel Pipe Sch 10S (kg/lb)
76.20 / 3.00	325 / 715	204 / 450
304.80 / 12.00	34,050 / 75,000	14,755 / 32,500

Nominal Pipe Size (mm/in.)	Steel Pipe Sch 40 (kg/lb)	Steel Pipe Sch 10S (kg/lb)
76.20 / 3.00	66 / 145	41 / 90
304.80 / 12.00	6810 / 15,000	2950 / 6500

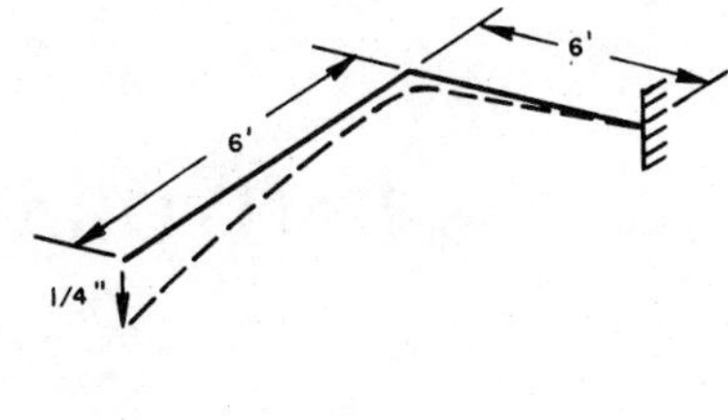

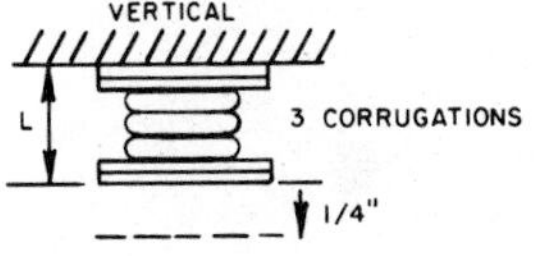

EXPANSION JOINT—VERTICAL

Minimum L (mm/in.)	Steel (kg/lb)	Minimum L (mm/in.)	Teflon (kg/lb)
238 / 9.375	123 / 270	92.10 / 3.625	27.20 / 60
356 / 14.00	141 / 310	200 / 7.875	43.10 / 95

EXPANSION JOINT—HORIZONTAL

Minimum L (mm/in.)	Steel (kg/lb)	Minimum L (mm/in.)	Teflon (kg/lb)
238 / 9.375	329 / 725	92.10 / 3.625	47.70 / 105
356 / 14.00	1105 / 2435	200 / 7.875	114 / 250

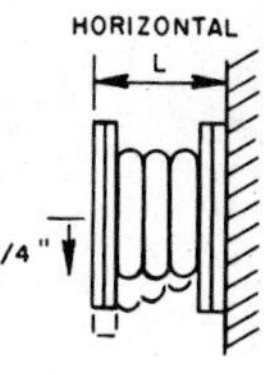

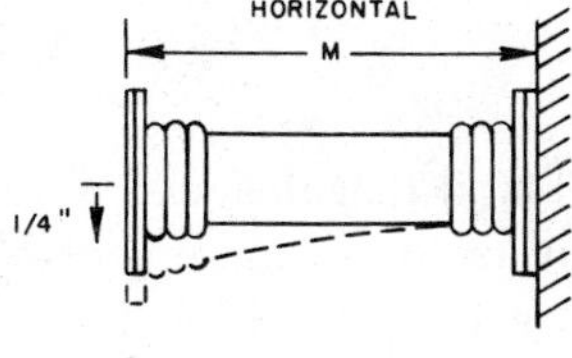

UNIVERSAL JOINT

Minimum M (mm/in.)	Steel (kg/lb)	Teflon (kg/lb)
689 / 27.13	2.72 / 6	1.82 / 4
883 / 34.75	20.43 / 45	8.17 / 18

sion/contraction effect by adapting the following solutions:

I-Beam flexure. I-Beam flexures are short lengths of standard I-Beams used to provide flexible support for weigh vessels. I-Beam flexures bend through very small angles about their web and allow slight motion perpendicular to the web. The flexures mounted as supports can accomodate lateral movement up to 0.010″ (0.25 mm). I-Beam flexures are utilized in weigh systems where load cells sense only a portion of the tank weight (see Figure 8.1c). This arrangement is commonly used for weighing liquids where an accuracy of 0.5% or less is acceptable.

Expansion assemblies. These assemblies are, in principle, sliding bearing units which have a low coefficient of friction and can move laterally within ±3/8″ (9.5 mm). Load cells used outdoors in areas subjected to large temperature variation should be provided with expansion assemblies. In cases where the mode of loading is tension, flexure rods are used (see Figure 8.1e). Flexure rods link the load cell with the structure in a tension weighing arrangement. The flexure rod has tensile strength of approximately 90,000 psi (621 MPa) and can accommodate deflection of ±3/32″ (2.3 mm).

BIBLIOGRAPHY

Considine, Douglas M., *Industrial Weighing*, Reinhold Publishing Co., New York, 1948.

"Electronic Weigh Systems Handbook," No. 002, BLH Electronics, Nov. 1979.

Elengo, J.J.,Jr., "Selecting and Applying Load Cells," *Instruments and Control Systems*, Sept., 1980.

Jensen, M.W. and Smith, R.W., *The Examination of Weighing Equipment*, NBS Handbook No. 94, 1965, U.S. Dept. of Commerce, Washington, DC.

Kauffman, J.H., "Tank Weighing by Load Cells," *Instruments and Control Systems*, February, 1966.

Lockery, H.E., "New Developments in Electronic Weighing Instrumentation," ISA Conference in Philadelphia, October, 1970.

"Review of Electronic Load Cells," *Instruments and Control Systems*, December, 1970.

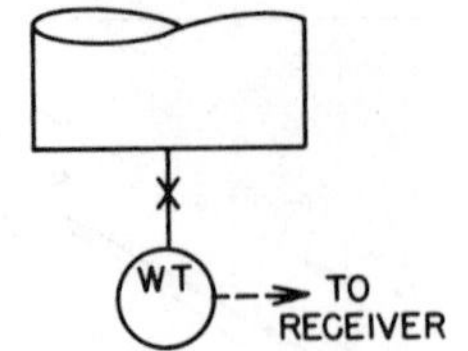

8.2 HYDRAULIC LOAD CELLS

Design Range:	100–1,000,000 lbm (45–45,000 kg)
Design Temperature:	0–125°F (−18–52°C)
Materials of Construction:	Aluminum, cast iron, steel and stainless steel; some designs contain elastic parts, rubber, neoprene, etc.
Inaccuracy:	Between 0.1% and 1% of range
Cost:	$1,100 and up depending on function and complexity. $1,100 covers single cell and pressure gauge readout only
Partial List of Suppliers:	A. H. Emery Co.; W. C. Dillon & Co.; ENERPAC Div., Applied Power, Inc.; Martin-Decker Corp.

All hydraulic load cells function on the principle of a force counterbalance. Weight imposed upon the load cell causes a change in internal fluid pressure. A wide variety of pressure detecting devices is employed to translate pressure into analog signal proportional to weight. The most popular readout device is a precision bourdon tube.

For practical application, hydraulic load cells must function without leakage, they must be relatively free of internal friction and, most desirably, they must be linear and precise in operation.

One approach to the above criteria has been through the use of a close fitting piston and cylinder, using an ordinary O-ring as a means of preventing leakage past the piston. While such devices will function rather satisfactorily under some conditions, one must guard against frictional losses due to the rubbing of the O-ring.

The Rolling Diaphragm Design

With the introduction of the so-called rolling diaphragm, a new and very successful design of hydraulic load cell appeared. This is illustrated in Figure 8.2a. Here we note that the hydraulic fluid is confined within the diaphragm chamber by means of the clamped seal between the cylinder wall and base plate. The piston or load bearing member is guided within the cylinder by two sets of ball guide rings. Thus, the piston is substantially limited to one degree of freedom; i.e., parallel to its major axis. The effective acting area of the piston equals the area of a circle whose diameter is the mean diameter of the diaphragm convolution. It has been found that this area is very constant over a wide range of piston displacements. Thus, the requirement of linearity and precise performance is satisfied. One limiting factor on this design is the ability of the elastomeric diaphragm to withstand pressure. Materials available limit the maximum internal pressure to 800 to 1000 PSIG (5.5 to 6.9 MPa). This limitation can be offset by increasing the size (area) of the load cell, within some practical limits. Here again, diaphragm molding techniques tend to limit size.

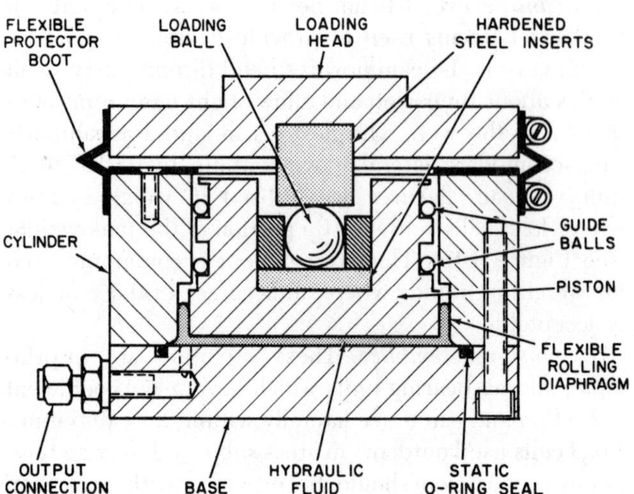

Fig. 8.2a Cross section of the rolling diaphragm hydraulic load cell

Performance of the rolling diaphragm type of hydraulic load cell is acceptable for most process weighing applications. One may expect measurement inaccuracies of ±0.25% FS or better on properly installed systems.

An outstanding feature of the rolling diaphragm type hydraulic load cell is its relative insensitivity to the amount of hydraulic filling. Thus, in making connections to gauges or other read-out equipment, high-pressure hoses rather than rigid tubing may be permitted, where desirable.

It is also easy to visualize that because of the relative insensitivity to filling, changes in hydraulic fluid volume due to ambient temperature variations have little effect on load cell performance. The system tends to be quite stable under varying temperature conditions provided, of course, that other factors, such as diaphragm stiffness variations, do not affect its performance.

All Metal Design

A more complex design, eliminating the flexible (elastomeric) diaphragm, is shown in Figure 8.2b. This design is characterized by all metal construction, with a very limited quantity of hydraulic fluid. One outstanding feature of this design is its ability to accept extremely heavy unit loads. Successful hydraulic load cells with individual capacities of 10,000,000 lbm (4,500,000 kg) have been constructed.

By eliminating all bearings, pivots and knife edges, hydraulic load cells offer high sustained accuracy. Displacement under capacity load, although dependent on connected auxiliary instrumentation, can usually be limited to 0.005 to 0.010″ (0.125 to 0.25 mm). Thus the natural frequency of hydraulic load cells is high, and on dynamic load applications, resonance is rarely, if ever, encountered. The viscous damping characteristics of the hydraulic medium tend to yield stable weight signals even under dynamic disturbances.

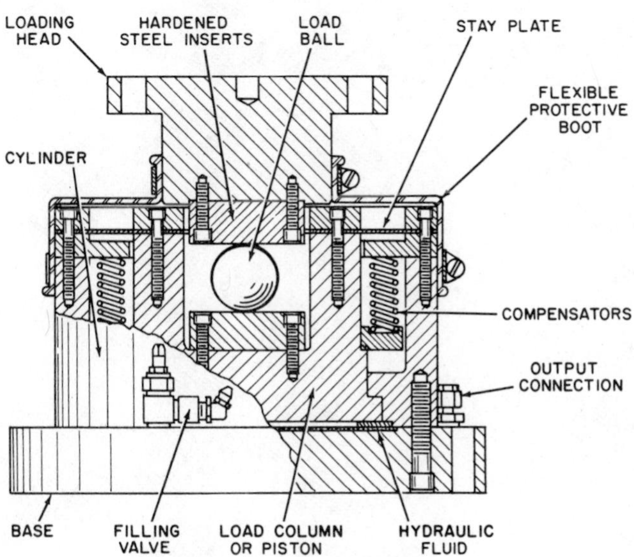

Fig. 8.2b Cross section of the all-metal hydraulic load cell

Hydraulic load cells are self-contained and require no outside power for operation. They are inherently explosion-proof and are available for both tension and compression force measurement. The hydraulic load cells illustrated in Figures 8.2a and b are applicable to tank, bin and hopper weighing. In both types, the supported load is borne by a top member, or head plate, which in turn rests upon a ball or rolling member. The rolling member is supported by the load sensitive piston or column of the load cell. Any tendency of the load head to move in a horizontal plane, as under the influence of an expanding or contracting vessel, is accommodated by a corresponding rolling action of the ball. The load cell itself is protected from heavy side forces that would tend to interfere with its free vertical displacement under varying load conditions.

Hydraulic Totalizers

In using hydraulic load cells for process weighing applications, a special problem arises when the vessel is supported on more than one load cell. In order to obtain the total weight of the supported body, the output of the support points must be added. If the load cells are simply interconnected, and an average pressure is obtained, the danger of "grounding" of one point may occur, especially under conditions of non-uniform support loading.

This problem is solved through the use of a hydraulic totalizer, as shown in Figure 8.2c. Here the output of each load cell is conducted to individual modules, which are, in effect, small pistons and cylinders. The output forces of the piston/cylinder combinations are collected on an output module, usually of larger acting area than the input modules. Provided this can be accomplished without serious internal losses, one pressure signal proportional to the several inputs may be developed. Units totalizing two, three and four inputs have been constructed with totalizing inaccuracy of ±0.1% FS. However, due to temperature sensitivity and other non-linear

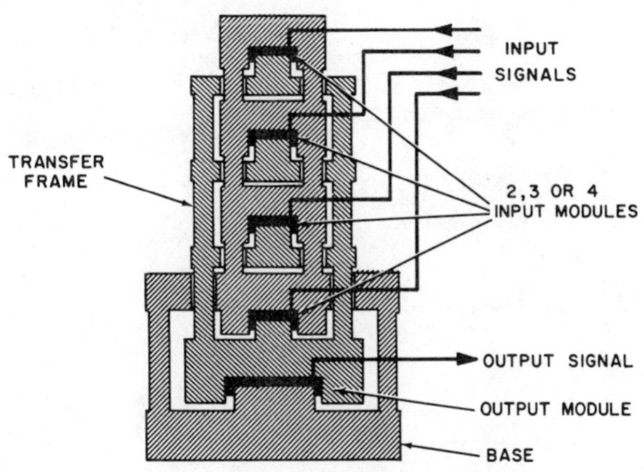

Fig. 8.2c The hydraulic totalizer

effects, hydraulic totalizing inaccuracy in the order of ±0.25% to ±0.50 FS is more commonly encountered.

Electronic Totalizers

Hydraulic load cells used in multiple may also be totalized by transducing the hydraulic pressure output into proportional DC voltage or current. Commercial transducers of high quality are available for this purpose. This method has the added capability of very long transmission without loss of accuracy.

Other Features

Hydraulic load cells are particularly well adapted for high impact loading applications and will withstand high overloads (300–400% in some instances), without loss of accuracy or zero shift.

Well designed hydraulic load cells do include some means of temperature compensation for both span and zero effect. Nevertheless, most manufacturers specify standard operating limits of 0 to 120°F (−18 to 49°C) as a basis for the performance guarantees. Operation outside these normal limits will necessitate the reference to temperature correction charts and graphs available from all suppliers.

Hydraulic load cells have found other applications in the force measurement and weighing field. The high natural frequency, low deflection and fast response rate made this device well adapted to web tension control, dynamometer torque measurement, jet engine and rocket thrust measurement and other similar highly dynamic installations.

BIBLIOGRAPHY

Fraade, D.J., "Load Cells for Batch Weighing," *Instrumentation Technology*, December 1977.

Kauffman, J.H., "Tank Weighing by Load Cells," *Instruments and Control Systems*, February 1966.

Flow Sheet Symbol

8.3 MECHANICAL LEVER SCALES

Range:	0–0.5 lbm to 0–500,000 lbm (0.2 kg to 225,000 kg)
Temperature Range:	−10 to +135°F (−23 to +57°C)
Materials of Construction:	Cast iron, aluminum, steel, stainless steel
Inaccuracy:	0.05%–0.1% of scale span
Cost:	$1,000 and up; but usually part of larger installation
Partial List of Suppliers:	Cardinal Co.; Cutler-Hammer Inc.; Detecto Scales, Inc.; Fairbanks Div. of Colt Industries; Hardy Co.; Howe-Richardson Scale Co.; Merrick Scale Mfg. Co.; Thayer Scale Div. Hyer Industries, Inc.; Toledo Scale Corp.

Principle of Operation

All mechanical lever scales employ lever systems which balance the weight of the unknown (gravity pull) against a known (calibrated) lever and mass: it is in fact a balancing of one moment against another. It is customary to adjust the lever system so that the pull from the unknown will fall within a convenient range—usually 25 to 50 pounds (11 to 23 kg). The unknown mass will include the mass of the bin, hopper, or platform holding the material to be weighed; this will be balanced out by a tare device. In Figure 8.3a, the hopper supports, hopper, gathering levers, hanger rods, and the pull rod leading to the counterbalancing means, or balance device, are shown for a typical industrial scale. In the same figure two widely-used balance devices are shown.

Balance Devices

The most widely used balance devices are the beam and the pendulum; these are sketched in principle in Figure 8.3a.

The beam is an array of (usually) three smooth bars, each marked off linearly, and each carrying a poise. One beam is for balancing out tare weight; it may not be calibrated. Another is for balancing out hundreds (or perhaps thousands) of pounds; it carries a rather heavy poise. The third is calibrated to balance out tens and units, and its poise is one-tenth the weight of the "hundreds" poise. The total moment exerted by the beam is the sum of the three. Balance is indicated by the position of the free end of the beam, usually guided within a trig loop.

The pendulum (often a double pendulum for greater stability and accuracy) employs a heavy mass swinging around a horizontal pivot; the moment is proportional to the displacement of the mass in a horizontal direction. This displacement is indicated by a pointer on a circular scale; linear calibration is made possible by a contoured cam and drawband. Tare corrections are made with a tare beam.

Types of Scales, Ranges

A few of the most widely used types of industrial scales, and typical ranges for them, are listed here:

Type	*Typical Capacity lbm/kg.*
Even-arm scale	5/2.26
Bench dial scale	200/90
Platform scale	1,500/680
Floor scale	6,000/2,720
Overhead scale	12,000/5,443
Suspended hopper scale	25,000/11,345
Truck scale	100,000/45,400
Railroad track scale	400,000/181,800

Applications

Mechanical lever scales are used in virtually every phase of industry, in development and in sales. The greatest scope for their application is probably in the weighing of stationary objects or quantities of material.

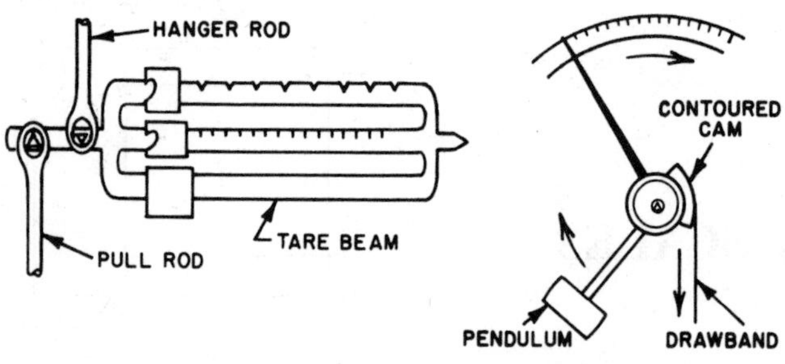

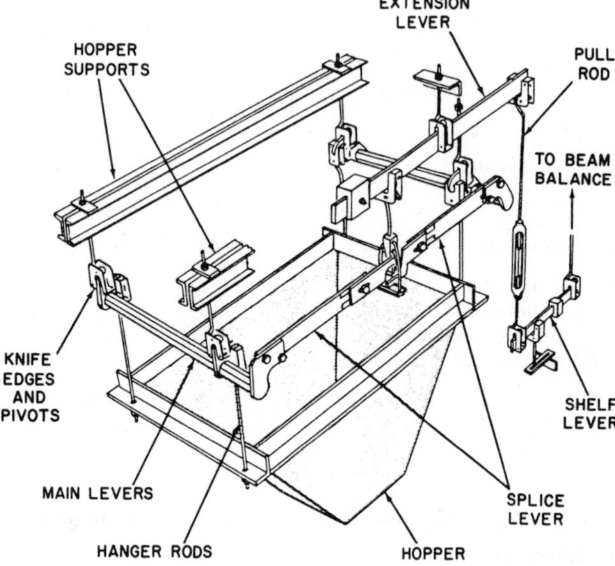

Fig. 8.3a Typical counterbalancing devices, and typical mechanical lever scale installation

Such scales may have almost any capacity, and can accommodate loads of almost any material. Overhead track scales, motor truck scales, and railroad track scales are all forms of mechanical lever scales.

A great many moving-body scales are also used. Vehicles in motion (trucks, railroad cars, etc.) can be weighed on mechanical lever scales if their velocity is low—usually less than 5 miles per hour (8 km/h). Such scales usually employ pendulum counterbalances, indicating on a dial for easy reading.

Granular materials are conveniently weighed on conveyor-belt scales. A short section of conveyor belt is built into the scale mechanism, with tare adjustment to balance the scale with the belt empty; the balancing device then indicates the weight of the material on the belt.

An extension of this is the integrating weighing device. The belt is driven at a known speed; the total amount of material delivered is readily computed from the duration of the operation and the average weight of material on the belt. Accuracy is improved if the amount of material on the belt is kept uniform and constant.

Supplying granular material at a known rate can be conveniently done by a conveyor-belt scale. The rate of feed to the belt is controlled by the balance device; the load on the belt is thus kept constant. The rate of feed of the material then may be controlled by adjusting the speed of the belt.

Many industrial processes require the weighing of batches of material individually; others require the weighing of a series of materials for later mixing or other treatment. Batches of constant size are readily weighed into a hopper, using beam scales, with the position of the beam indicating when the feed should be stopped. A series of quantities of materials can be weighed into the same hopper; this is most readily done with a dial-type scale (although a series of balance beams, dropped into position in succession, can also be used). The dial pointer positions for each added material are made to actuate gates to stop the flow of each material when the required weight is reached. Pointer positions are detected by various means: photoelectric pick-ups, reed switches, etc. Similar devices are used to sense balance-beam position.

Advantages and Limitations

Mechanical lever scales are notable for long-time accuracy, with proper maintenance; they are also quite resistant to most environmental conditions. They are available in an extremely wide range of capacities and forms; they are available in sizes ranging from quite small to a railroad-track scale more than 100 feet (30 m) long. They can readily be made a working part of other industrial devices; in fact most industrial scales are so used.

Their principal limitation is in speed of response. The mass and inertia of the lever system does not permit weighing speeds as great as strain-gauge load cells, which can be used to weigh vehicles moving at high speeds. The normal output of a mechanical lever scale is a visual signal, not readily coupled into other systems, but this may be handled quite easily.

Output Signals

While many scales provide only a visual indication of balance, many electrical output devices are available. The simplest of these are "cut-off" devices, which indicate only when a desired weight (or each one of a series of weights) has been reached. There are also transducers which are attached to dial-scale pointers; these provide a continuous electrical output which can be fed into computer-type controls to perform quite sophisticated functions.

BIBLIOGRAPHY

Colijn, Hendrik (ed.), *Weighing and Batching of Bulk Solids*, Trans Tech Publications, Bay Village, Ohio, 1975.

Considine, Douglas M., *Industrial Weighing*, Reinhold Publishing Co., New York, 1948.

Jensen, M.W. and Smith, R.W., *The Examination of Weighing Equipment*, NBS Handbook No. 94, 1965, U.S. Dept. of Commerce, Washington D.C.

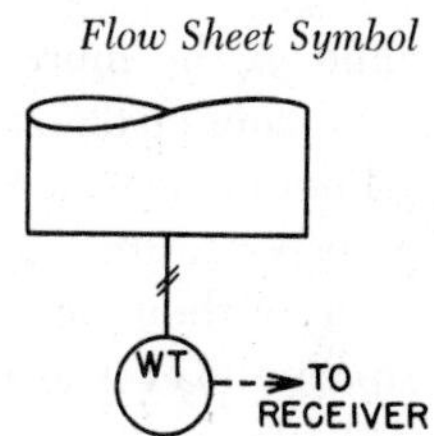

8.4 PNEUMATIC LOAD CELLS

Design Range:	0–10,000 lbm (0–4,500 kg)
Design Temperature:	0 to 125°F (−18 to 52°C)
Materials of Construction:	Aluminum, cast iron, steel, stainless steel. All contain flexible elastomeric diaphragms
Inaccuracy:	0.1% FS to 1% FS
Cost:	\$2,250 and up depending on function and complexity
Partial List of Suppliers:	Kane Air Scale Co.; Link Engineering Co.; Moore Products Co.

Pneumatic load cells are successfully applied in process weighing. Units presently available are all force-balance in principle and function with high accuracy. Most pneumatic weighing systems are offered with tare balancing chambers which enhance their overall performance.

The pneumatic output signal from the load cell may be read locally or transmitted by metal or plastic tubing to a remote point. Readout of weight is usually by precision bourdon tube gauge, although the pneumatic pressure signal may be transduced into electronic or digital form.

Pneumatic weighing systems have several outstanding advantages. They are inherently explosion-proof, are quite insensitive to temperature variation and in the event of rupture or leakage, they contain no contaminating fluid medium, such as hydraulic fluid. This feature is of particular interest to the food and drug industry.

For successful operation, pneumatic load cells and associated weighing equipment must have a carefully regulated source of clean, dry air. Although systems have been operated for short periods on inert gases such as dry nitrogen, such operation would, obviously, be too expensive and impractical for process application. Therefore, when installing a pneumatic weighing system, in addition to the system components themselves, attention must be directed to the energy source for the system—the air supply. An average requirement would be 10 SCFM (0.0047 m^3/s) of dry air (−40°F [−40°C] Dew Point) per load cell.

Figure 8.4a illustrates a pneumatic load cell cross section.

The natural frequency of pneumatic load cells is quite low. Under certain conditions of dynamic loading, resonance may occur. This problem has been largely over-

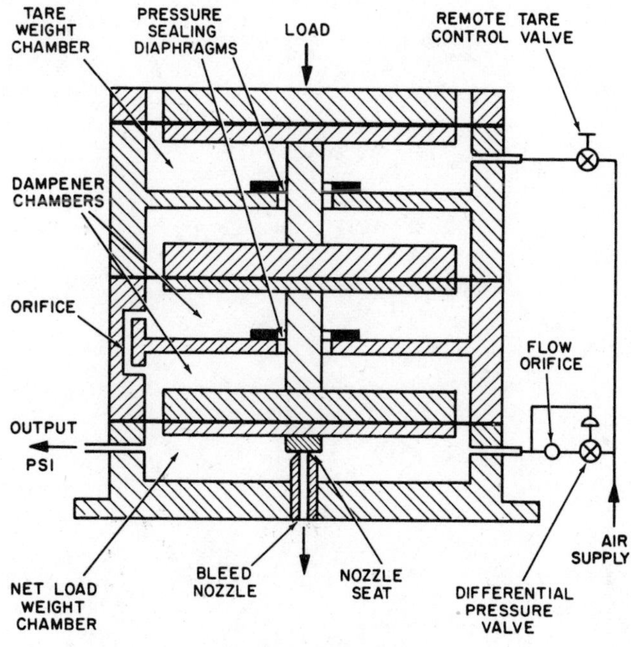

Fig. 8.4a Pneumatic load cell cross section

come in the more successful designs by incorporating stabilizing or dampener chambers (see Figure 8.4a).

Pneumatic weighing systems have relatively slow rates of response to incremental load changes; however, because of their force-balance principle, deflection under capacity load is also low—0.003 to 0.005 in. (0.075 mm to 0.125 mm).

BIBLIOGRAPHY

Beckett, R.H., "Control Considerations for Applying Pneumatic Load Cells to Gravimetric Feeders," ISA Conference in Philadelphia, October 1970.

Flow Sheet Symbol

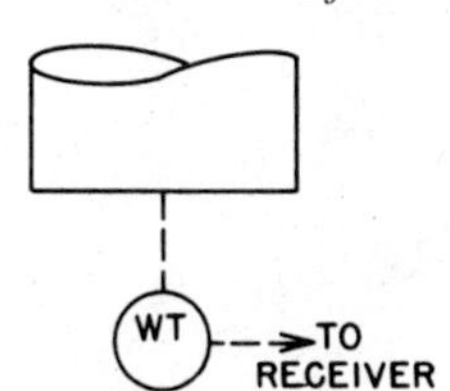

8.5 STRAIN GAUGE LOAD CELLS

Design Range:	0–1 lbm to 0–12,000,000 lbm (0–0.45 kg to 0–5,400,000 kg)
Design Temperature:	+15 to +115°F (−9 to +46°C)
Materials of Construction:	Steel, stainless steel
Inaccuracy:	±0.1% FS to ±1.0% FS
Cost:	$2,250 to $11,250 per remote readout installation
Partial List of Suppliers:	BLH Electronics; Fairbanks Weighing Div., Colt Industries; Hottinger Baldwin Measurements, Inc.; Howe Richardson Scale Co.; Revere Corp. of America; Streeter Amet Div., Mangood Corp.; Toledo Scale Div., Reliance Electric Co.

This section will consider the bonded strain gauge load cells.

One of the first uses of the bonded resistance wire strain gauge following its recognition in the early 1940s was in the development of an accurate and reliable load cell or force transducer. In fact, the strain gauge and its application, in general, have been one of the most intensely researched fields in recent technological history. As a result of this work, there is a wide variety of accurate, stable and reliable strain gauge load cells available for nearly all applications.

Strain gauge load cells are commonly used for weighing tanks, bins, hoppers and ladles. They are installed underneath platforms, rails, conveyors and monorails. Their applications include thrust measurement on rocket and jet engine test stands, launching pads and also wind tunnels and other branches of aeronautical research.

If a wire is bonded to a supporting member in such a manner that its cross section varies as the supporting member is strained, it is possible to establish a relationship between the electrical resistance of the wire and the force causing deformation of the supporting member. Strain gauge load cells are designed to permit controlled elastic deformation of the supporting member. The supporting member usually consists of one or more steel columns mounted upon a rigid base and hermetically sealed within a suitable chamber (see Figure 8.5a). Here we see a column loaded in the Z-axis. Bonded to the four sides of the column are grids of fine wire, a, b, c and d. As load increases, gauges "a" and "c" tend to decrease in length, and their resistance decreases. Gauges "b" and "d," mounted perpendicular to Z are placed in tension by the column tending to decrease in cross section (Poisson effect), and their resistance will increase. The four gauges are connected into a Wheatstone bridge circuit as shown in Figure 8.5b. By having gauges "b" and "d" strain opposite to "a" and "c," the bridge unbalance due to load variations is amplified and output voltage is greater than if "b" and "d" were strained in the same manner as "a" and "c."

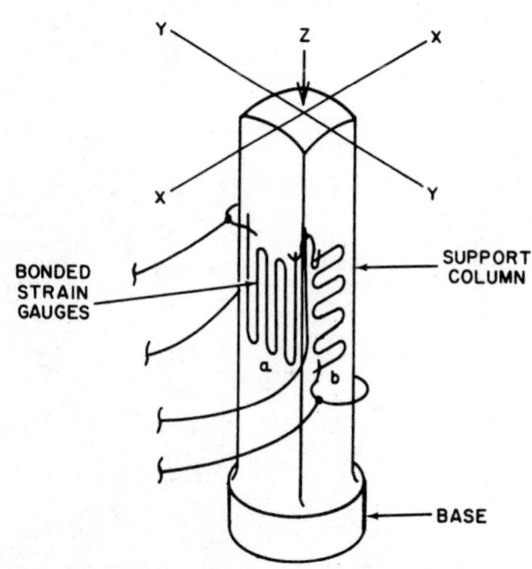

Fig. 8.5a Strain gauge element

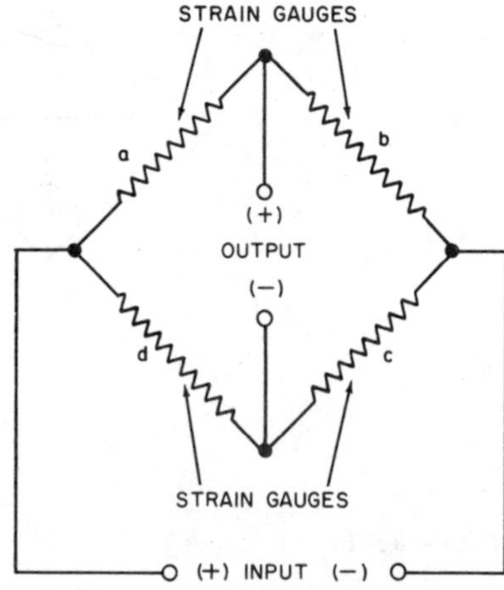

Fig. 8.5b Wheatstone bridge for strain gauge load cells

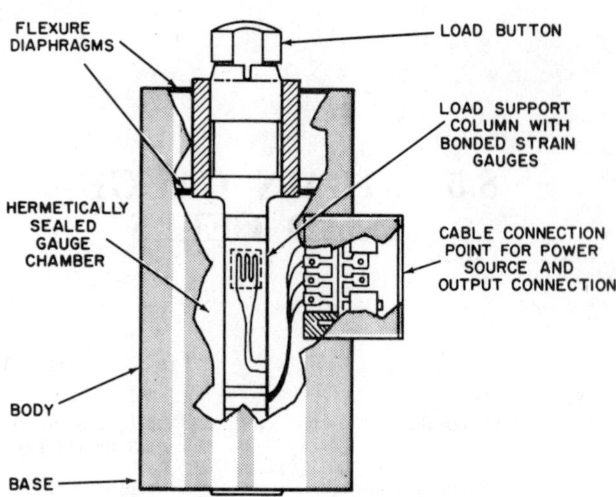

Fig. 8.5d Assembled strain gauge load cell

All strain gauge load cells are compensated for the effects of temperature on zero shift and span. This is accomplished by making the strain wires out of temperature-insensitive alloys and introducing suitable compensating resistors into the bridge network. A typical strain gauge circuit is shown diagrammatically in Figure 8.5c, and an assembled unit in Figure 8.5d.

In the application of strain gauge load cells, one may encounter loading conditions or forces which impose angular loads upon the cell. Such forces exert a bending moment upon the support column which the strain gauge can not discriminate from purely normal loads. Figure 8.5d shows the support column stabilized by two flexure diaphragms. The support column is thus limited to one degree of freedom, parallel to the Z-axis (Figure 8.5a). Other designs, involving more than one support column, can also successfully reduce this source of error.

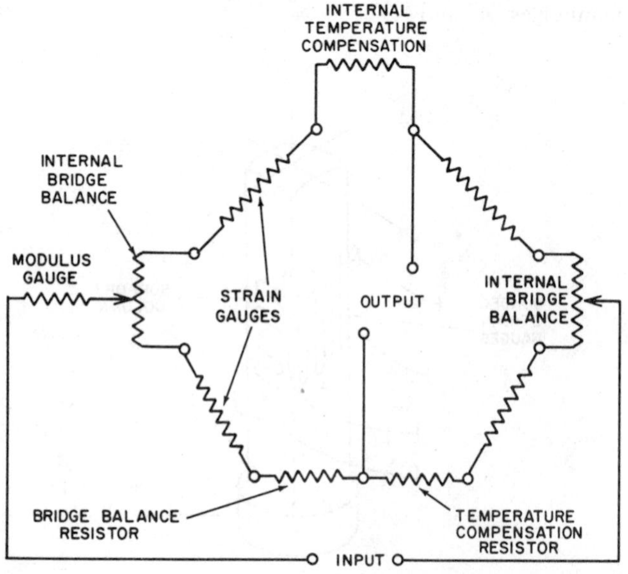

Fig. 8.5c Strain gauge circuit schematic

Design Variations

Strain gauge load cells are available for compression load measurement, tension load measurement and as combination, or universal, load cells that may be used in either direction.

Load cells with two sets of strain gauge elements ("double bridge") are also available. Thus two output signals can be generated in one load cell and may be used to supply two different readout systems. An example of such an installation would be when a plant is provided with both conventional and computer controls. Here it is possible to AC excite one set of elements in the load cell to generate the output for the conventional readouts and DC excite the other set to provide a second, computer compatible signal. Such systems are relatively expensive, however, and not often used on industrial applications.

The output signal of a strain gauge is relatively small and is related to the excitation. A common value is 2–3 millivolts per volt of excitation. The excitation voltage (AC or DC) is usually in the range of 5 to 20 volts, with values in the order of 12 volts recommended for average installations.

Successful use of strain gauge load cells in a weighing application requires that attention be paid to the design of the excitation power supply, output amplifier, summation network and other back-up equipment, such as filters. Most suppliers have developed compatible auxiliary instrumentation for use with their load cells, such as recorders, printers and data processors, to mention a few. The strain gauge load cell has been successfully applied to such weighing processes as automatic batching, automatic inventory recording and control, constant rate feeders, continuous (belt) weighers and many other devices involving the gravimetric principle.

Critical Evaluation of Strain Gauge Load Cells

Strain gauge load cells find extensive application in the chemical industry. As a result of increasing experience, and a widening background of empirical knowledge, it is practical to consider the use of strain gauge load cells on installations requiring inaccuracies of ±0.1% to ±0.25% of full scale.

Temperature compensation systems for span and zero shift are now an intrinsic part of all high quality strain gauge load cells. Nevertheless, for operation outside normal temperature limits, generally 15 to 115°F (−9 to +46°C), one should anticipate the use of correction factors or provide means of controlling temperatures around the load cells by auxiliary means. Strain gauge load cells should be protected from angular or non-axial loads. Any force other than normal or axial will tend to cause bending of the support column or columns. Inasmuch as a strain gauge can not discriminate between bending and axial loads, errors in output can result. Where strain gauge load cells are installed under tanks, bins or hoppers which are subject to excessive bending expansion or contraction, special mounting equipment is available to help isolate the load cell from undesirable external side forces. In extreme cases, specially designed mounting pedestals may be required.

Strain gauge load cells are designed for operation within specific capacity ratings. Overloads greatly in excess of rating may result in loss of accuracy or failure. In general, the load cells should not be subjected to more than 125% of their rated capacity. This includes impact or shock loading, as well as static loading.

BIBLIOGRAPHY

"Direct Force Measurements," *Control Engineering*, December 1969.

Electronic Weigh System Handbook, BLH Electronics, First Edition, December, 1979.

Fraade, D.J., "Load Cells For Batch Weighing," *Instrumentation Technology*, December 1977.

Lockery, H.E., "New Developments in Electronic Weighing Instrumentation," ISA Conference in Philadelphia, October 1970.

"Review of Electronic Load Cells," *Instruments and Control Systems*, December 1970.

Trevathan, V.L., "Weighing Silos with Strain Gauges," *Instrumentation Technology*, December 1970.

8.6 WEIGHING COMPONENT ADVANCES

New Sensing Techniques

For a more complete understanding and appreciation of the advantages and disadvantages of new weighing transducers or devices, it is advisable to be familiar with the basic sensing principles on which they are based.

Semiconductor Strain Gauge

The piezoresistive characteristics of germanium and silicon semiconductor materials were discovered by scientists at Bell Laboratories in the mid-1950's. It was discovered that the terminal resistance of these devices is highly sensitive to applied stress or strain. In fact, their gauge factors (unit change in resistance divided by unit strain) are more than fifty times that of their metallic wire or foil strain gauge counterparts.

While possessing very high strain sensitivity relative to that of metallic strain gauges, they also exhibit substantial non-linearity, and temperature effects on strain sensitivity and terminal resistances are also relatively high. The latter characteristics have limited their application. Nevertheless, semi-conductor strain gauges are used in force measuring devices in which high output signal level and low system cost are the primary objectives.

Nuclear Radiation Sensor

This form of weight sensing is generally applied to in-motion weighing of bulk materials. It utilizes a radioactive source of gamma rays which are imposed on or directed through a certain section of the moving material. The material absorbs some of the gamma rays and allows others to pass through. The amount of radiation transmitted through the bulk material depends on the amount of material on the conveyor. A radiation sensor converts the transmitted radiation to an electronic signal which bears a known relationship to the amount of material on the weighing section of the conveyor.

The nuclear radiation form of weight sensing is applicable when the weight sensor should not contact the material or the conveying devices. Certain shortcomings of conventional belt scales can be avoided with this technique.

Inductive Sensing Technique

Inductive weight sensors use the change in inductance of a solenoid coil with changing position of an iron core. Two forms of the inductive sensing principle are illustrated in Figure 8.6a. In configuration #1, motion of the iron core to the right increases the inductance of coil B and decreases the inductance of coil A. Arranging the two coils in a Wheatstone bridge with resistors completing the bridge network provides a means for developing a voltage signal proportional to the core position.

Configuration #2 utilizes three solenoid coils. Coils C and D are wound in opposite directions and surround an iron core, whereas coil E is placed between the two coils and is excited by an external AC voltage source. When the iron core is centrally located, voltages induced into the secondary coils (C and D) are equal and opposite, and no voltage appears across the output terminals (F and G). If the iron core is moved to the right, the voltage coupled into coil D is greater than that coupled into coil C and hence a voltage is developed at the output terminals. If the core were moved in the opposite direction by the same amount, a similar voltage of opposite phase would be developed. Other embodiments of inductive sensors are in current use. Those discussed here are for illustrative purposes only.

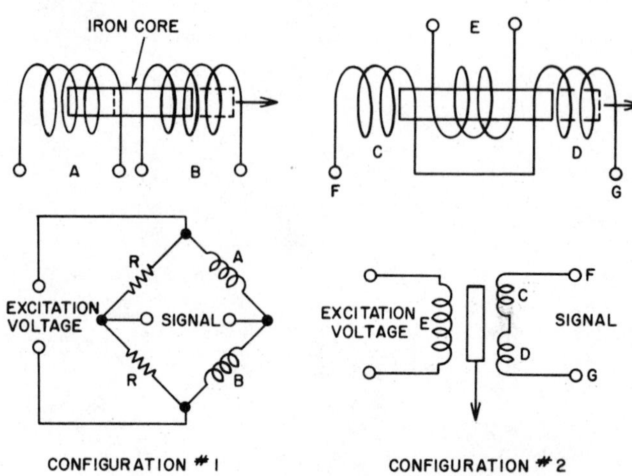

Fig. 8.6a Inductive sensing techniques

Inductive sensors furnish relatively high output signal levels and efficient null stability. Since their inertial masses are greater than strain gauge sensors, they are more subject to vibration.

Variable Reluctance Sensing Technique

Similar to the inductive sensing principle except that the inductance of one or more coils is usually changed by altering the reluctance of a very small air gap, this technique is illustrated in Figure 8.6b. Solenoid coils A and B are mounted on a structure of ferromagnetic material, and a U-shaped armature completes the magnetic circuit through air gaps 1, 2 and 3. Motion of the coil assembly to the right decreases air gap (2) while air gap (1) is increased. Air gap (3) remains constant during the translation of the coil assembly. As a result of horizontal translation, the inductance of coil B increases while that of coil A decreases. Incorporating the two coils in a Wheatstone bridge similar to that utilized in the inductive sensing principle permits development of a voltage proportional to the translation of the coil assembly.

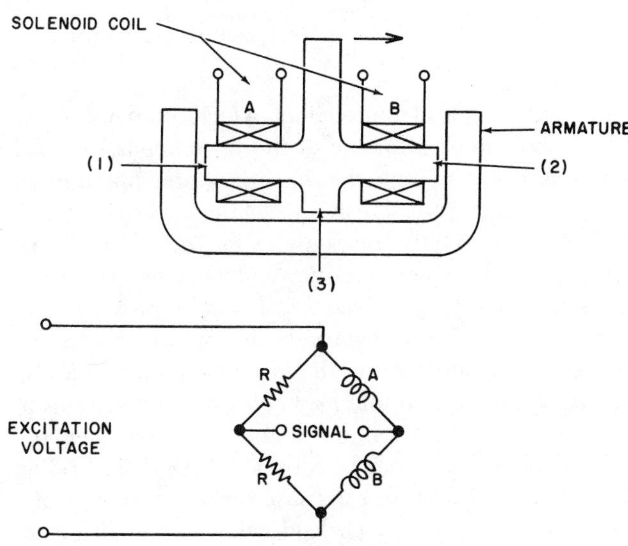

Fig. 8.6b Variable reluctance sensing technique

The reluctive sensing principle also offers a relatively high output voltage and efficient null stability with the higher vibration sensitivity due to the relatively high inertial masses of the mechanical structure.

Magnetostrictive Sensing Technique

Based on the Villari effect, this technique utilizes the change in permeability of ferromagnetic materials with applied stress. A stack of laminations forms a load-bearing column (Figure 8.6c), and primary and secondary transformer windings are wound on the column through holes oriented as shown. Coil A is excited with an AC voltage and coil B provides the signal voltage. In the unstressed condition, the permeability of the material is uniform

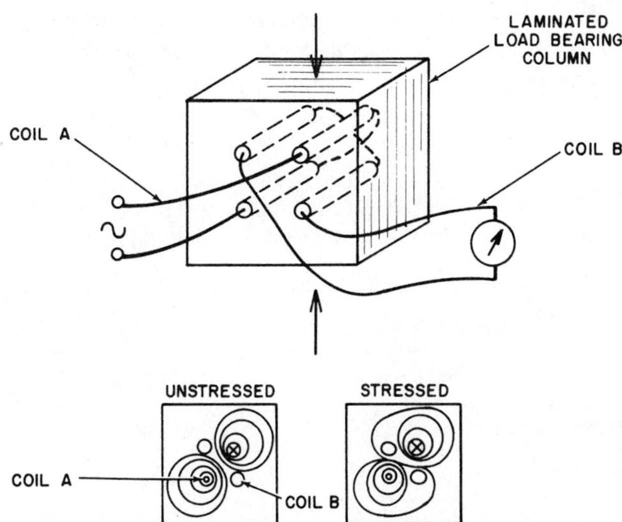

Fig. 8.6c Magnetostrictive sensing principle

throughout the structure and since the coils are oriented at 90° with respect to each other, little or no coupling exists between coil A and coil B. Hence, no output signal is developed. When the column is loaded, the induced stresses cause the permeability of the column to be nonuniform, resulting in corresponding distortions in the flux pattern within the magnetic material. Magnetic coupling now exists between the two coils and a voltage is induced in the signal coil, providing an output signal proportional to the applied load.

The magnetostrictive principle produces relatively high output signal levels and offers extreme ruggedness in load cells incorporating this sensing principle.

New Developments in Weight Transducers

Although many of the electromechanical force sensing transducers in current weighing systems have been available for some time, improvements have been made over the past several years facilitating their application to new and improved weighing systems. Some of these transducers as well as certain force transducers which were not treated in the previous sections will be discussed here.

Semiconductor Strain Gauge Load Cells

Metallic wire and foil strain gauge units have long been used in weighing systems, furnishing accuracies of better than 0.1% of reading. Within the past five years semiconductor strain gauge load cells have been used in those systems requiring low cost and moderate accuracies.

Semiconductor strain gauges in load cell configurations provide units with rated output capabilities of 1.0 volt at 15 volts bridge excitation. As a result of the high signal level, semiconductor units are used in simple weighing systems with simple regulated power supplies and direct meter readouts. Sometimes an amplifier is interposed between the transducer and the meter display.

Typical performance characteristics of semiconductor load cells are listed in Table 8.6d together with those of their metallic strain gauge counterparts.

Table 8.6d
PERFORMANCE CHARACTERISTICS OF SEMICONDUCTOR AND STRAIN GAUGE TYPE LOAD CELLS

Performance Characteristic	*Semiconductor Load Cell*	*Metallic Strain Gauge Load Cell*
Output (at 15 v)	1.0 v	30 mv
Terminal Linearity	0.25%	0.05%
Hysteresis	0.02%	0.02%
Temperature Effect on Zero Balance	±0.25%/100°F (38°C)	±0.15%/100°F
Temperature Effect on Output	±0.5%/100°F	±0.08%/100°F
Selling Price Index	1.3	1.0

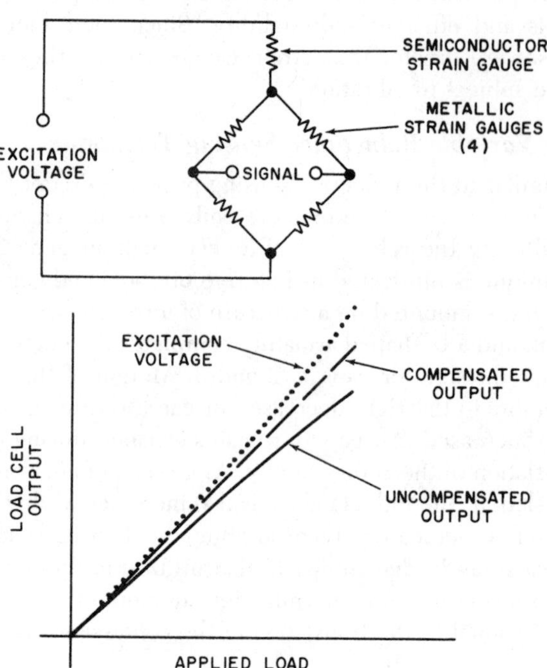

Fig. 8.6e Linearization of column load cells

The moderately high cost of semiconductor load cells and the dramatic cost reductions in linear integrated circuitry have limited the use of the semiconductor load cell in low cost weighing systems. In other words, the cost of linear amplification required to raise metallic strain gauge load cell signals to the levels offered by their semiconductor counterparts is now less than the additional cost for semiconductor load cells.

Linearization of Column Type Load Cells

Column type strain gauge load cells in capacities above 10,000 lbs. (4,500 kg) heretofore have suffered from a characteristic non-linearity of about 0.15% of rated capacity. The inherent non-linearity of these devices results from electrical bridge non-linearities caused by the fact that all strain gauges are not subjected to equal strain. Additional non-linearity also results from the column area change with increasing load. The characteristic column type load cell non-linearity is parabolic and lends itself to almost perfect compensation by utilizing a semiconductor strain gauge compensating element. Figure 8.6e shows a semiconductor strain gauge incorporated in series with the excitation terminals of the load cell bridge circuitry. From the curve of output voltage versus applied load, an uncompensated column type load cell exhibits a drooping concave downward characteristic when loaded in compression. The linearizing strain gauge senses column strain induced by the applied compressive load and, due to its piezoresistive characteristics, its terminal resistance decreases with increasing load. The decreasing resistance with load causes the net excitation voltage applied to the bridge circuitry to increase with increasing load (dotted line), which compensates for the drooping characteristic of the uncompensated load cell and results in improved linearity (interrupted line). Adjusting the terminal resistance of the linearizing strain gauge almost exactly compensates for the inherent parabolic drooping characteristic, and terminal linearities of better than 0.02% of rated load can be provided.

Linearities of this magnitude not only eliminate external linearization within the instrumentation, but also reduce errors in multiple load cell weighing systems in which unequal distribution of total load between the individual load cells may be substantial. Unequal loading on non-linear load cells can cause serious system errors, even in systems in which load cell non-linearity compensation is included in the display instrumentation.

Beam Type Strain Gauge Transducers

A new generation of force measuring transducers that will find wide application in the weighing field consists of a slotted bending beam construction (Figure 8.6f). Strain gauges at the locations shown are arranged in a Wheatstone bridge configuration so that the output of the sensing element is independent of the position of the applied load. Electrical compensation of the bridge circuitry can reduce the load position sensitivity to virtually zero. As will be observed, this is a very important feature.

Protection from environmental effects is by a simple bellows arrangement (Figure 8.6f), thereby making unnecessary the conventional diaphragm and cylindrical casing. Strain gauge location and beam design provide

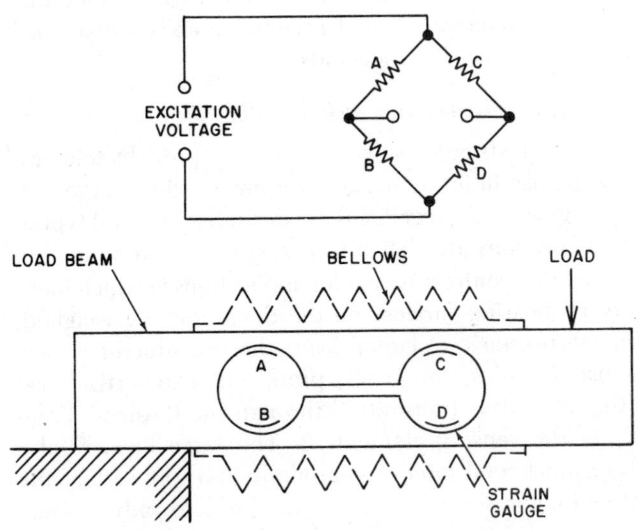

Fig. 8.6f Beam-type strain gauge transducer

inherent adverse loading sensitivity. This simplified load sensing configuration provides inherent linearity as well as very low creep and very high repeatability. Typical performance features are summarized in Table 8.6g.

Table 8.6g
PERFORMANCE OF BEAM TYPE STRAIN GAUGE TRANSDUCER

Rated Capacity Range	10 lbs to 5,000 lbs. (4.5 kg to 2270 kg)
Output	2 mv/v
Terminal Linearity	±0.03%
Hysteresis	0.015%
Creep (30 min)	0.02%
Temperature Effect on Zero	0.15%/100°F (38°C)
Temperature Effect on Output	0.08%/100°F (38°C)

Monorail Weighing Transducer

Another beam type weighing transducer is being used in monorail conveyor systems such as those in meat processing plants. Conventional monorail weighing systems support the "live" rail by load cell and flexure assemblies. The new monorail transducer replaces the live rail and is self-supporting.

A typical unit is shown in Figure 8.6h. Strain gauges at the designated location sense bending strains as the load traverses the transducer. The gauges interconnected in a bridge arrangement supply a constant output independent of load position on the transducer. The sloping arrangement on the upper edge of the transducer decouples the moving carrier from the "pusher" mechanism during the sensing period. That is, on the downward slope the carrier rolls free, eliminating contact with the pusher mechanism.

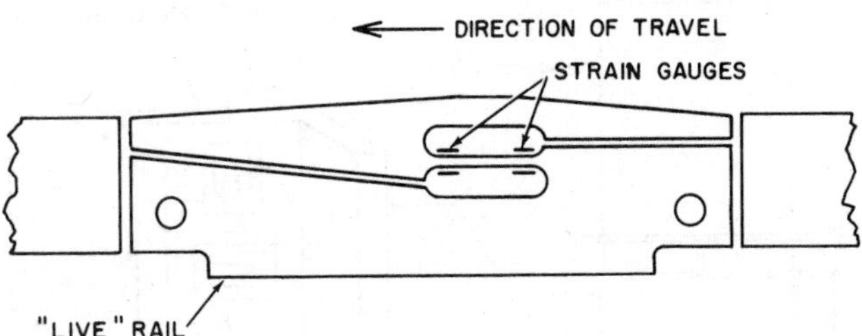

Fig. 8.6h Monorail weighing transducer

This simplified transducer not only provides greater measuring accuracy but also eliminates a second load cell and associated flexure and mounting arrangements, thereby reducing overall cost.

High Temperature Load Cells

As load cell weighing was applied to the metal processing industry, the need for devices to withstand high environmental temperatures became pressing. In recent years organic and inorganic bonded strain gauge backing and installation materials have become available and can withstand higher temperatures than conventional units. Bonded strain gauges with organic backings are now available for continuous operation at temperatures as high as 500°F (260°C).

On special applications high temperature strain sensing wire alloys have been installed with inorganic bonding materials such as ceramic cements and flame spray techniques, wherein molten aluminum oxide is sprayed on the sensing element and on the strain sensing grid to hold the latter firmly in place. These installations allow short-term operation at temperatures of 1,000°F (538°C) but with some degradation in performance.

Direct Weighing of Tank Legs

Frequently, structures do not lend themselves to support by load cells of any type. They are already fabricated and erected, and their support by load cells would require extensive field modification.

One solution to the weighing of these structures is installing strain gauges directly on the supporting legs. The legs become the transducer sensing element to which the strain gauges are applied in full bridge configurations.

In a typical installation (Figure 8.6i), a pair of gauges is applied longitudinally, sensing the compressive stresses in the tank legs. Another pair is applied in the transverse direction, sensing the tensile strains due to the Poisson effect. The four gauges are connected in a Wheatstone bridge arrangement and leads are brought out from each leg to a summing box and from there to the readout instrumentation. The installation is thoroughly protected with waterproofing materials.

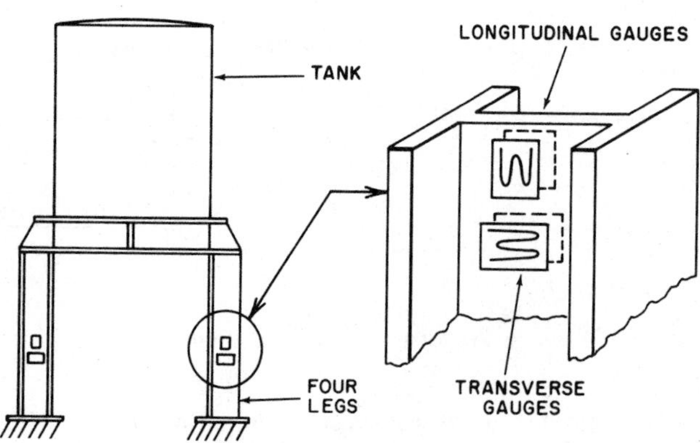

Fig. 8.6i Direct gauging of tank supports

Usually, the strains established in the supporting legs are very low and it is difficult to achieve perfect waterproofing permanently. As a result, the accuracy of such a weighing system tends to be relatively poor—3 to 5 % of rated capacity.

Inductive and Reluctance Load cells

Inductive and reluctance load cells incorporate the two basic sensing principles in the same way, i.e., the motion of a ferromagnetic core (inductive) or a coil assembly (reluctance) is converted to a voltage signal directly proportional to the displacement.

Various force sensing elements convert the applied force (weight) to a displacement to which the sensing element is coupled (Figure 8.6j).

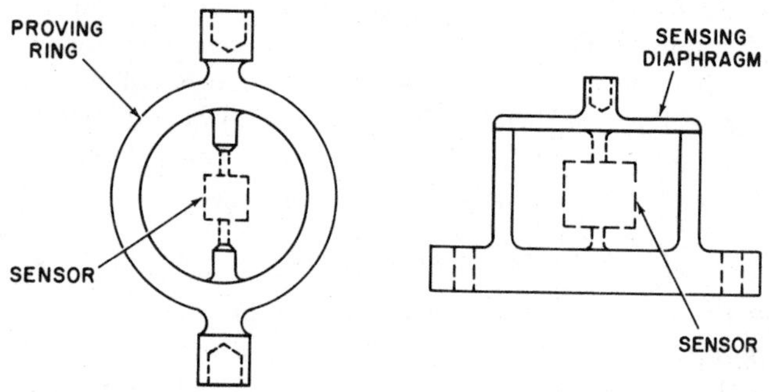

Fig. 8.6j Inductive and reluctive load cells

Table 8.6k
PERFORMANCE CHARACTERISTICS FOR INDUCTIVE OR RELUCTANCE LOAD CELLS

Capacity Range	0.01 to 100,000 lbs. (0.0045 to 45,000 kg)
Rated Output Range	5 to 200 mv/v
Linearity Range	0.1 to 0.5%
Repeatability	0.05%
Temperature Effect on Zero	1%/100°F (38°C)
Temperature Effect on Output	1%/100°F (38°C)

These transducers furnish relatively high output signal levels and moderate to high accuracies, and cover a broad range of measuring capacities.

Magnetostrictive Load Cells

Magnetostrictive sensing load cells (pressductors) are finding use in industrial applications in which large output signals and ruggedness are desirable. Several typical configurations are shown in Figures 8.6l and 8.6m.

The first configuration is for applications in which there are no bearing surfaces on the devices being weighed; in the presence of lateral loads the pressductor is very sensitive unless adequately protected. The vertical load (Figure 8.6l) is transmitted through the flexures, 1 and 2, to the sensing element, 3. The same flexures also transmit lateral forces to "ground" in a way so that the pressductor sensing unit is subjected to only a small portion of the adverse lateral loads.

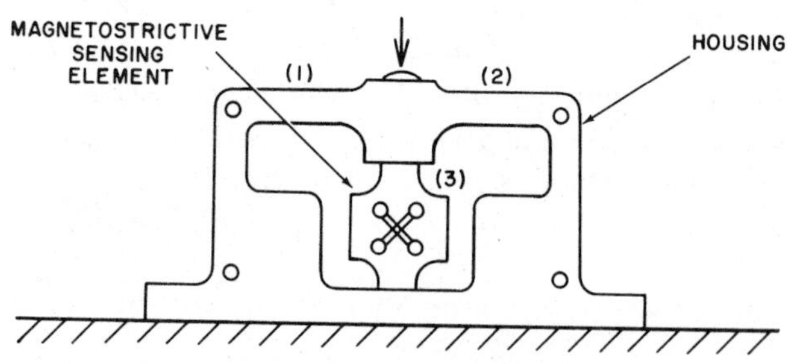

Fig. 8.6 l Magnetostrictive load cell

The second embodiment (Figure 8.6m), designed for weighing during coiling operations, uses a similar construction with an additional overhanging member, 4, which supports the coiler shaft, and continuous weighing during coiling operations is provided. All units are adequately protected with watertight covers to accommodate applications in industrial environments.

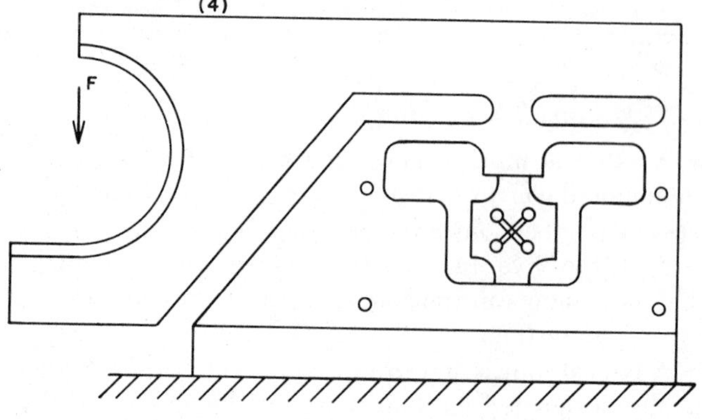

Fig. 8.6m Magnetostrictive load cell

New pressductor designs provide weighing inaccuracies of 0.1% of rated capacity. Output signal levels range from 1 to 20 VDC, with source impedances ranging from 0.5 to 25 ohms. Overload ratings as high as fifteen times the rated load are supplied. Although usable for weighing, the pressductor has greater applicability in the steel industry for the measurement of roll-forces in rolling mills and strip-tension in strip mills.

Load Cell Weighing Accessories

Several load cell weighing accessories required for satisfactory application of load cells to weighing installations were discussed the previous sections. Others will now be discussed.

A New Load Cell Adaptor

As already noted, many load cell weighing installations involve large differential expansions which can impose severe horizontal forces on the installed load cells. Also, in vehicle scales large horizontal forces can be applied owing to deceleration and acceleration forces associated with bringing the vehicle on and off the scale. Development of a new load cell adaptor virtually eliminates such forces.

Primarily a mechanical arrangement, the active weighing platform is suspended from the top of the load cell by three suspension links (Figure 8.6n), and an upper plate and adaptor ring contact the load cell at the desired loading point. The upper plate carries the three links by link pins projecting radially from the upper plate. Hanging on the opposite end of the links is the lower plate which includes three additional link pins for engaging the lower end of the links.

The lower plate is connected to the active weighing platform, thereby transmitting the weight through the links and upper plate to the top of the load cell. The load cell is supported by a base plate which rests on the foundation or ground structure. The base plate also serves to absorb heavy side loads when the horizontal deflection of the weigh-bridge exceeds the clearances provided between the base plate and the cutout portion of the lower plate. The height of the adaptor assembly can be adjusted by a center screw, enabling the equal distribution of total load among the several load cells in a given installation.

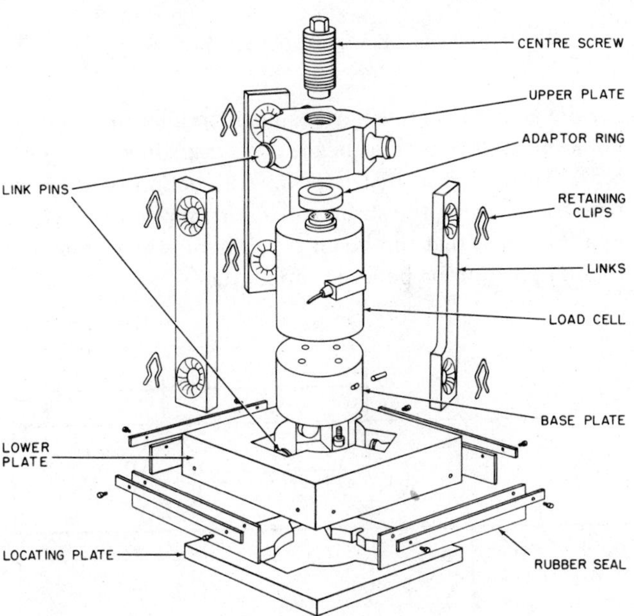

Fig. 8.6n A new load cell adaptor

The structure provides a highly flexible load cell adaptor assembly which transmits virtually no side loads to the load cell caused by differential expansion of the weighing structure relative to the ground structure. The side loads that are transmitted to the load cell are from weigh-bridge deflections, imposing angular loads on the load cell. These are minimized by appropriate structural design of the weigh-bridge.

Rocker Assembly

Another load cell adaptor commonly used in weighing systems is the rocker assembly (Figure 8.6o). An adaptor

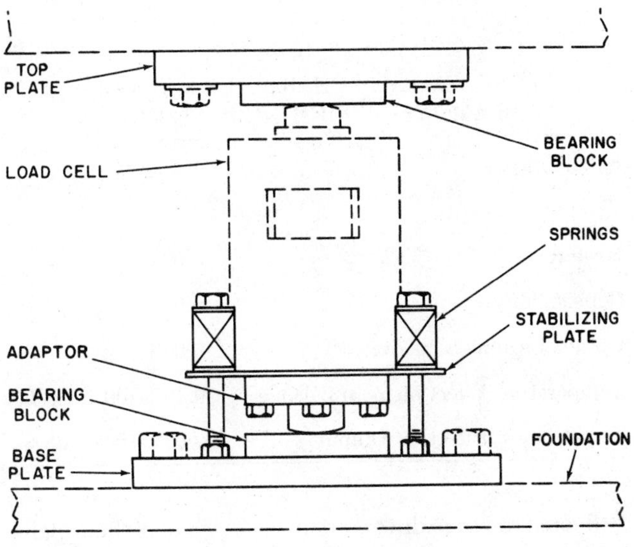

Fig. 8.6o Rocker assembly

is added to the bottom of the load cell which in effect provides a convex loading surface on the bottom as well as on the top of the unit. The load cell and adaptor are located in place by a stabilizer plate. Load is introduced to the load cell through the upper bearing block and transmitted through the load cell and the lower bearing block to the mounting plate. The stabilizer plate allows partial rotation of the load cell while at the same time restricting excessive lateral motion.

Differential expansion between the structure being weighed and the foundation causes slight rotation of the load cell, greatly reducing the magnitude of the horizontal forces which would have been present in the absence of the rocker assembly. The load cell is thus protected from the adverse effects of large lateral forces caused by differential expansion in multiple cell weighing systems.

Various rocker assemblies for load cell capacities range from 20,000 to more than 300,000 lbs. (from 9,000 to more than 135,000 Kg).

Integrated Weighing Devices

For several years a number of integrated weighing devices have become available combining force transducing elements and weigh-bridges or platforms in a single self-contained weighing structure.

Beam Type Weighing Platform

These transducers are now the basic sensing elements in medium capacity, low height, semiportable platform scales. Available in capacities from 500 to 14,000 lbs. (225 to 6,300 Kg), one can be located almost anywhere within the process area without the need for pits and attendant preparation costs. Typical units, depending on capacity, are from 2 to 35 ft² (0.19 to 3.3 m²) in platform area and 1.5 to 6 inches (37.5 to 152 mm) in overall height (Table 8.6q).

Table 8.6q
PERFORMANCE CHARACTERISTICS OF BEAM TYPE WEIGHING PLATFORM

Rated Output	1.5 mv/v
Terminal Linearity	0.05%
Hysteresis	0.03%
Repeatability	0.02%
Corner Loading Sensitivity	0.03%
Temperature Effect on Zero Balance	0.15%/100°F (38°C)
Temperature Effect on Output	0.08%/100°F (38°C)

Beam type transducers are in each corner of the weighing platform (Figure 8.6p). A steel cable or alternative flexible arrangements suspend the active weighing platform from the end of the transducer. The flexible connection allows the transducer to operate in its normal deflection mode, thereby capitalizing on its inherently high accuracy. The flexible connections also render the entire structure much less susceptible to uneven foundation conditions.

In addition to low height, high accuracy and relative insensitivity to foundation unevenness, the units are semiportable.

Smaller, lower capacity beam type platforms available in the range of 10 to 250 lbs. (4.5 to 114 Kg) use two beam type sensing elements (Figure 8.6r). Features and characteristics are similar to those of platforms already described. One significant difference, however, relates to the use of only two beam type sensing elements. In this configuration the reduction of corner loading sensitivity or load placement errors must be approached differently. It is no longer a matter of simply dividing the load among four linear transducers but rather involves a reduction of the effect of changing load position on two transducers.

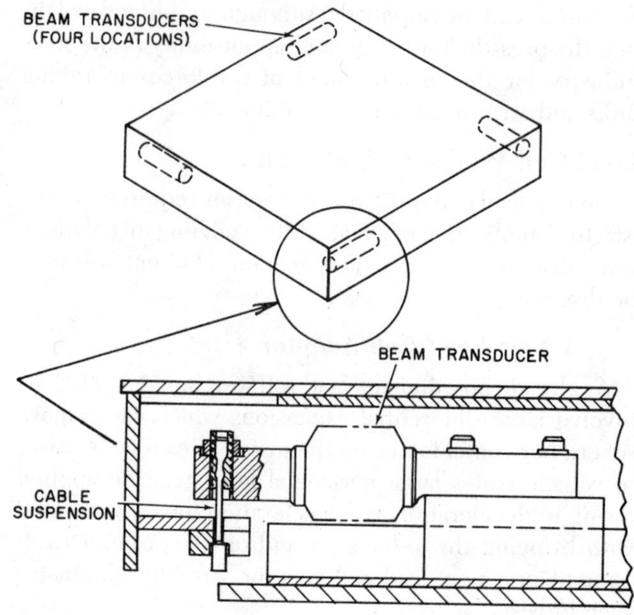

Fig. 8.6p Beam-type weighing platform

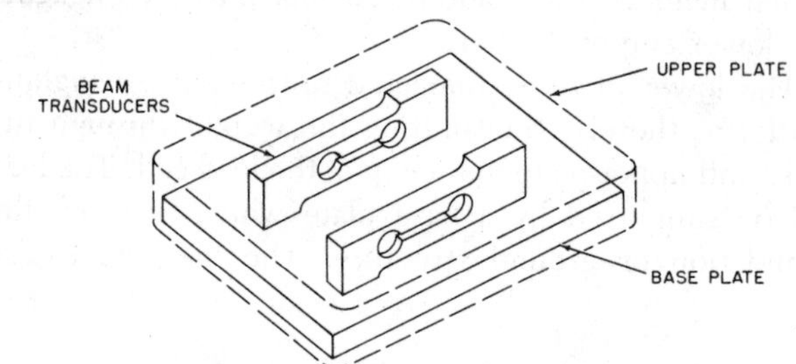

Fig. 8.6r Low capacity beam platform

A changing load position results in a different bending moment applied to the beam type transducers (Figure 8.6s). However, as discussed in connection with beam type transducers, the strain gauges are applied and wired in a way such that the beam type element is insensitive to load position or bending moment.

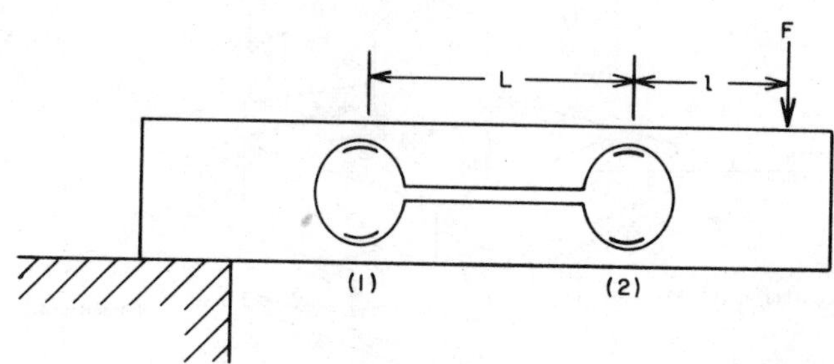

Fig. 8.6s Load position desensitization

This concept can be explained as follows. Strain gauges at location 1 sense the bending moment there (M_1). Gauges at position 2 sense the bending moment at that position (M_2). The strain gauge bridge is arranged so that the output is proportional to the difference $M_1 - M_2$, and this may be expressed by the relationship

$$M_1 - M_2 = F(L + 1) - F(1) \quad 8.6(1)$$

$$M_1 - M_2 = FL \quad 8.6(2)$$

Hence, the output of the beam type transducer is proportional to the applied force and the distance between the strain gauge locations (L). It is insensitive to the placement of the load (1). This relationship is theoretically correct, assuming perfect strain gauge location and zero machining tolerances. Since this relationship seldom exists, electrical "tuning" methods tailor the sensitivities at locations 1 and 2 to give virtually perfect moment rejection.

The lower capacity platform configurations are available in capacities of 10 to 250 lbs. (4.5 to 114 Kg) with roughly 15 by 18 in. (381 by 457 mm) weighing area and height of about 3.5 in. (88 mm).

Treadle Scales

Heretofore, scales for weighing vehicle axles on highways at toll booth locations utilized individual load cells, weigh-bridges and mechanical stabilization hardware, all of which required expensive pits and additional installation. An integrated treadle scale now eliminates some of these requirements and provides a self-contained unit which can be lowered into place in a shallow pit.

Conceptually, the treadle scale (Figure 8.6t) consists of two strain gauge sensing elements located at opposite ends of the bridging plate. The sensing elements measure shear forces by virtue of the sensing element configuration and gauge location. Strain gauges at each of the reduced sensing sections sense the principal tensile and compressive strains due to the applied shear forces.

Additional strain gauges on the leading and trailing ends of the sensing elements are connected in a bridge configuration which provides direction sensing information for the associated instrumentation. Invalid vehicle entries are thereby detected.

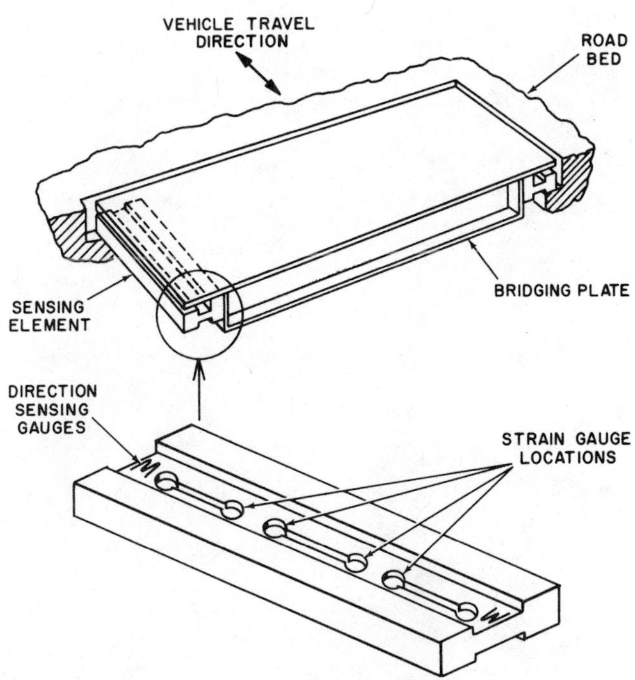

Fig. 8.6t Treadle scale

Table 8.6u
TREADLE SCALE PERFORMANCE CHARACTERISTICS

Rated Capacity	18,000 lbs. (8,100 kg)
Rated Output (10 volts excitation)	0.25 v
Terminal Linearity	0.75%
Hysteresis	0.25%
Temperature Effect on Zero Balance	1%/100°F (38°C)
Temperature Effect on Output	0.25%/100°F (38°C)
Size	15 in. × 60 in. (381 mm × 1,524 mm)
Weight	200 lbs. (90 kg)

Two units are generally installed end on end across a lane, spanning a distance of 10 ft. (3 m). Total capacity for the pair is 36,000 lbs. (16,300 Kg). Currently, the units are used to sense overweight vehicles for subsequent weighing by law enforcement scales. Another use is axle weight measurement for toll assessment by weight classification.

Portable Platform Scales

Another portable platform scale (Figure 8.6v) is composed of a honeycomb weigh-bridge and four load cells, one at each corner. Adjustable leveling pads level the platform. Each pad includes a swivel joint for accommodating sloping surfaces. An entire structure weighs less than 140 lbs. (63 Kg).

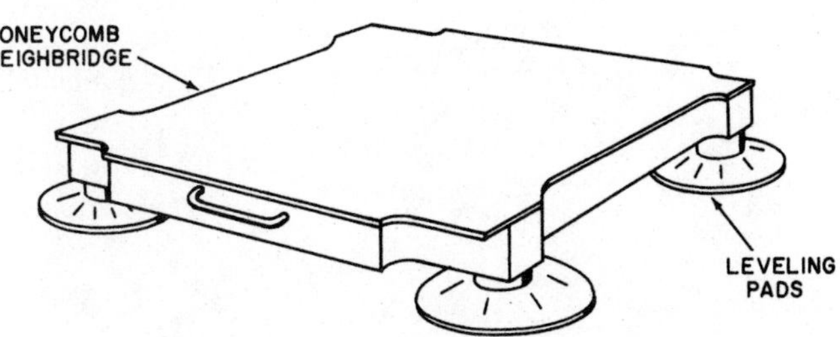

Fig. 8.6v Portable platform scale

Table 8.6w
PERFORMANCE CHARACTERISTICS OF PORTABLE PLATFORM SCALES

Rated Capacity	50,000 lbs. (22,500 kg)
Rated Output	1.5 mv/v
Terminal Linearity	0.05%
Hysteresis	0.05%
Temperature Effect on Zero	0.1%/100°F (38°C)
Temperature Effect on Output	0.15%/100°F (38°C)

The units are available with ramps and spacers for use with various vehicle wheel configurations. They are being used in portable aircraft and vehicle weighing.

BIBLIOGRAPHY

Fraade, D.J., "Load Cells for Batch Weighing," *Instrumentation Technology*, December 1977.

Lockery, H.E., "New Developments in Electronic Weighing Instrumentation," ISA Conference in Philadelphia, October 1970.

Chapter IX

MISCELLANEOUS SENSORS

A. Brodgesell
P. M. Glattstein
M. F. Hordeski
W. D. Houle
R. K. Kaminski
D. S. Kayser
B. G. Lipták
T. A. Mayer
R. Nussbaum
J. B. Rishel
H. C. Roberts

CONTENTS OF CHAPTER IX

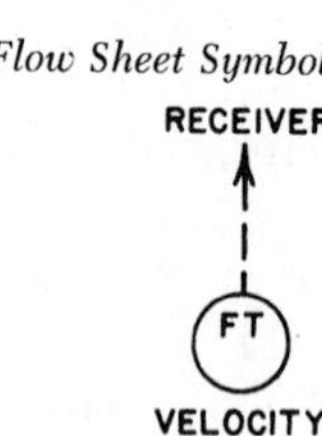

9.1 ANEMOMETERS

Types:	A.—Hot wire B.—Mechanical C.—Laser
Design:	All ambient conditions; some resistant to salt-air corrosion
Range:	To 200 mi/hr (320 km/hr)
Sensitivity:	0.5 mi/hr (0.8 km/hr)
Inaccuracy:	±1 percent
Cost:	To \$300 for local readout, to \$2000 for remote readout
Partial List of Suppliers:	Alnor Instrument Co. (A,B); Airflow Development Ltd. (A,B); Bendix Corp., Process Instruments Div.; Davis Instrument Mfg. Corp. Inc. (A,B); Dwyer Instruments, Inc. (B); J-Tec Associates, Inc. (C); Kurz Instruments, Inc. (A); TSI Inc. (A,C); Teledyne Hastings-Raydist (A); Weathermeasure Corp. (A,B)

An anemometer is an instrument that detects and indicates air velocity or wind speed. Several of the elements discussed in Chapter II can be arranged for flow measurement in ducts, flues, and free air; thus they also meet the definition of an anemometer. (See in particular Sections 2.7 and 2.12.) This section is limited to a description of the mechanical anemometers; the vane, cup, and impeller anemometer designs; and the hot wire anemometer.

Figure 9.1a illustrates the vane design. The vanes rotate in response to air flow with the angular velocity of the vanes being proportional to the wind speed. When a portable unit is required or when the local readout is satisfactory, vane motion is passed to the indicator through a gear and spring assembly. If the reading is to be remote, a magnetically- or capacitively-coupled pickup can be used to generate a transmission signal.

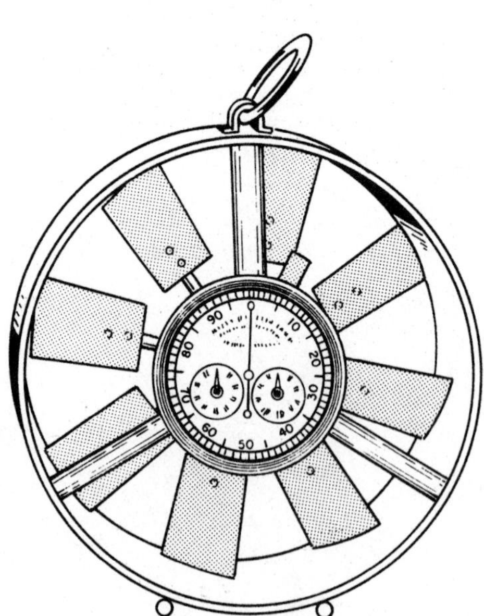

Fig. 9.1a Vane anemometer with local readout

Figure 9.1b shows a three-cup anemometer. In one design the shaft drives a direct current (dc) tachometer generator with an output voltage that is proportional to the wind speed. This signal may be used to drive a remote mounted indicator or recorder. The impeller design shown in Figure 9.1c, also has a shaft-driven tachometer. Since the tail on the impeller design always keeps the impellers pointed into the wind, this instrument can be used to detect both wind speed and wind direction.

The speed of response for anemometers is given in feet of wind and is known as the distance constant. One

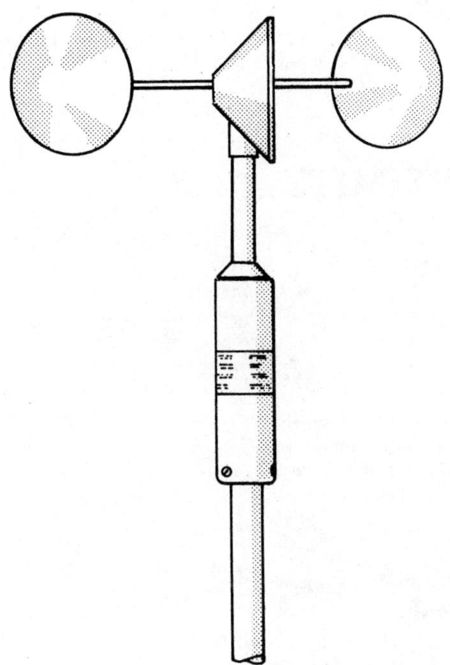

Fig. 9.1b Three-cup anemometer

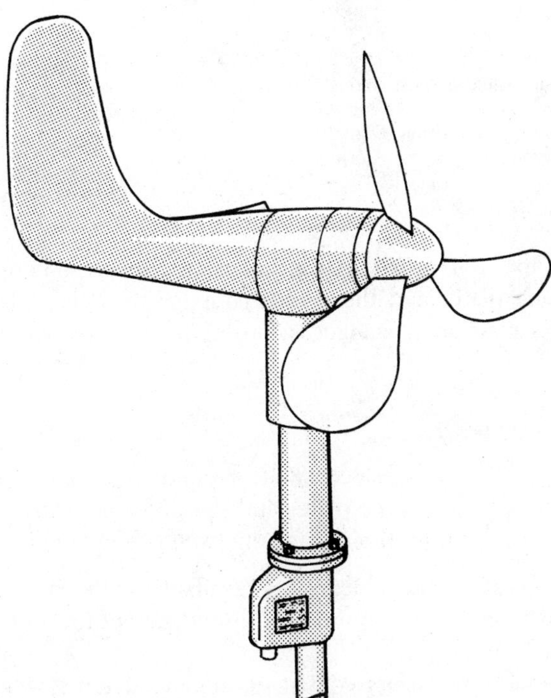

Fig. 9.1c Impeller anemometer

commercially available unit has a distance constant of 6 ft (1.8 m) which means that the instrument will recover 63 percent of a step change after 6 ft of wind has passed the sensor.

The hot wire anemometer (Figure 9.1d) operates as a heated thermopile which is cooled at a rate that is proportional to the air (or gas) velocity at the probe tip. It is available in ranges of 100–2000, 50–1000, and 20–500 feet per minute (0.5–10.0, 0.25–5.0, and 0.1–2.5 meters per second, respectively).

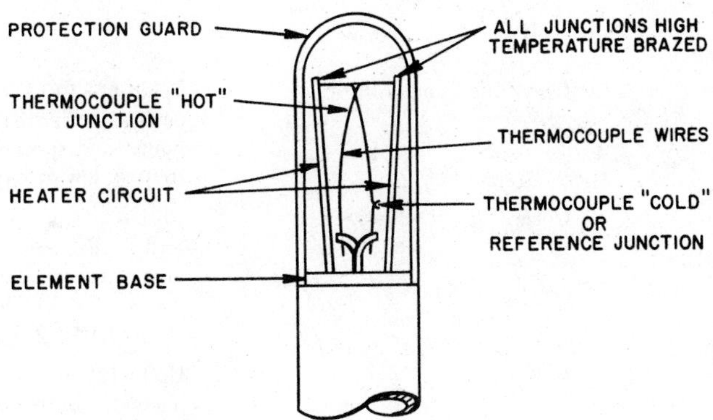

Fig. 9.1d Hot wire anemometer

When selecting a location for an anemometer, it should be remembered that the structure on which it is mounted will disturb the air flow. A rectangular building will disturb air flow up to about two times its height above grade, six times its size to the leeward, and two times to the windward.

It is always a good idea to have wind direction indicators in industrial processing units so that the operators will know in which direction to travel in the event of a spill. Knowledge of wind speed is of considerably less importance under these circumstances.

BIBLIOGRAPHY

Laskaris, E.K., "The Measurement of Flow," *Automation*, 1980.

LeMay, D.B. "A Practical Guide to Gas Flow Control," *Instruments of Control Systems*, September 1977.

Lomas, D.J., "Selecting the Right Flowmeter," *Instrumentation Technology*, 1977.

MacCready, P.B. and Jex, H.P., "Response Characteristics and Meteorology Utilization of Propeller and Vane Wind Sensors," *Journal of Applied Meteorology*, 1964.

Flow Sheet Symbol

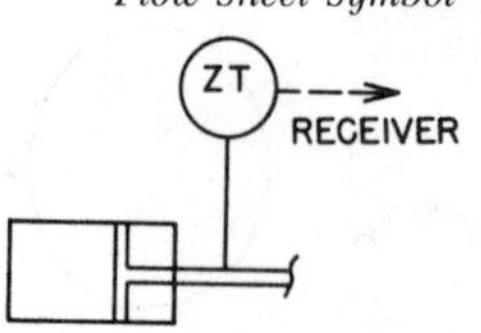

9.2 DISPLACEMENT AND PROXIMITY SENSORS

Operating Temperature:	Typical −50 to +150°F (−46 to +66°C); high-accuracy devices may require a restricted temperature range, and specialized sensors are available for extremely low or high temperature applications
Cost:	Cost is a function of sensor accuracy and range as well as the sensor type. A variable resistor displacement sensor or single-coil inductive proximity sensor may cost $20 or less whereas a high-accuracy shaft encoder may cost $2000
Range:	Up to 120 in. (3 m) for linear displacement sensors; photoelectric proximity sensors can be used in applications up to 100 ft (30 m) depending on the target size
Inaccuracy:	0.1 percent linear, 5 sec angular
List of Suppliers:	Columbia Research Labs, Inc.; Electro-Sensors, Inc.; Kaman Sciences Corp.; Moxon, Inc.; Pickering & Co., Inc.; Robinson-Halpern Co.; Siltran Digital; Skan-A-Matic Corp.; Teledyne Gurley; Zygo Corp.

Sensors for the measurement of position, displacement, and proximity, may use resistive, capacitive, inductive, or optical methods. Displacement sensors measure the linear or angular position and are connected mechanically between the point or object being sensed and a reference or fixed point or object. Proximity sensors also measure linear or angular motion but without any mechanical linkage.

The output from a displacement or proximity sensor may be an analog or digital function of the absolute distance being sensed or it may be a function of distance from a given starting point.

Analog Output

Variable resistors can be used as voltage or current dividers to provide displacement information. The wiper or movable arm of the resistor slides over the resistance element, which can be wire or conductive film. Sensors are available for linear and angular measurement including fractional and multiturn operation.

Capacitive sensors are generally used for linear rather than angular proximity measurements. Either the dielectric or one of the capacitor plates is movable for displacement measurement.

Capacitive proximity sensors use the measured object as one plate, and the sensor contains the other plate. The capacitance changes according to the equation

$$C = k/d \qquad 9.2(1)$$

where

k = is a constant, depending on the area of the plates and the dielectric constant
d = the distance between the plates

Capacitive transducers are available with packaged signal-conversion circuitry for direct current (dc) output operation.

Inductive sensors consist of single-coil units, which use a change in the self-inductance of the coil, and multiple-coil units, which rely on the change in magnetic coupling or reluctance between coils. Single-coil displacement sensors use a movable core to change the self-inductance whereas single-coil proximity sensors use the magnetic properties of the object itself to modify the self-inductance. The change in inductance is usually sensed with a bridge circuit or oscillator.

Multiple-coil inductive sensors consist of the differential transformer and its variations. The linear variable differential transformer (LVDT) uses three windings and a movable core to sense linear displacement.

A typical LVDT configuration is shown in Figure 9.2a. The transformer's secondary windings are wound to produce opposing voltages and connected in series. With the core in the neutral or zero position, voltages induced in the secondary windings are equal and opposite and the net output is a minimum. Displacement of the core increases the magnetic coupling between the primary coil and one of the secondary coils and decreases the coupling between the primary coil and the other secondary coil. The net voltage increases as the core is moved away from the center position, and the phase angle increases or decreases as a function of the direction in which the core is moved.

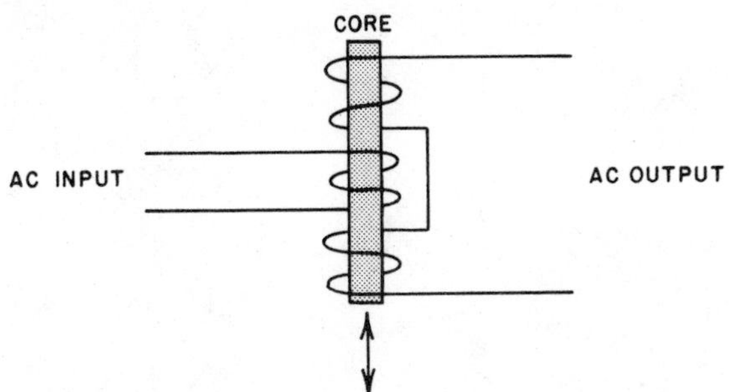

Fig. 9.2a An LVDT configuration

A demodulator circuit can be used to produce a dc output from this winding configuration. Differential transformers are also available for angular measurement in which the core rotates about a fixed axis.

Variations in winding configurations are used in synchros, resolvers, and microsyns. Inductance bridge sensors utilize two coils with a moving core to change the inductance of the coils that form one half of an alternating current (ac) bridge. These sensors are available in linear and angular configurations.

Digital Output

Digital outputs or pulses are produced in displacement and proximity sensors using changes in electrical conduction, induction, or photoelectric conduction.

Conducting encoders use brushes or wipers to detect the position of a coded disk or plate. If a single track is employed, a number of pulses are produced as the disk or plate is moved. Direction sensing is performed by adding another track which is offset to produce sequence logic. Electronic counting circuitry is used to count the number of pulses and perform the conversion to angular or linear measurement. Multiple-track encoders provide a digital or binary coded output, which is a function of the absolute angular or linear position.

Magnetic proximity sensors can be employed with gears of ferromagnetic material to produce pulses from a change in linear or angular position. Direction sensing can be obtained by shaping the gear teeth in an asymmetrical pattern in order to modify the output waveform.

Photoelectric encoders use a light source and the detector with disks or plates of transparent and opaque windows. Operation is similar to conducting encoders except that switching is accomplished by breaking the path of the light beam between the source and detector. Multiple-tracked encoders may use arrays of sources, such as light-emitting diodes, and detectors as shown in Figure 9.2b.

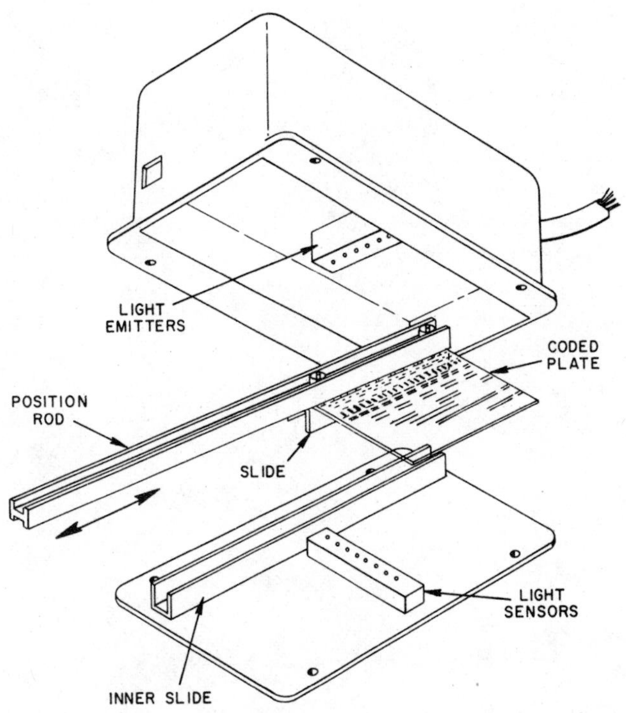

Fig. 9.2b Digital displacement sensor

The laser interferometer uses a laser beam which is directed at a reflector on the measured object. Changes in the linear displacement of the object produce interference fringes which are counted by electronic circuitry.

Application

Displacement and proximity sensor selection begins with the technical and economic requirements. In dc systems potentiometric transducers are simple to apply and can be used with output levels to 50 V or higher, and for displacements to 24 in. or from 5° or 3600°. Reluctive transducers with dc-to-dc conversion circuitry offer displacements between 0.01 and 120 in. (0.25 mm and 3 m) and between 0.05° and 90°. Capacitive and inductive proximity sensors as well as photoelectric sensors can be used to detect displacement changes as small as 1 μin. In ac systems multiple-coil inductive sensors are used more than all others. Maximum accuracy can be obtained with incremental and absolute digital displacement sensors along with photoelectric units such as interferometers. Shaft-angle encoders can provide a system accuracy of less than 5 seconds of arc.

BIBLIOGRAPHY

Ellis, J.F. "Using Eddy Currents for Proximity Measurements," *Instruments and Control Systems,* April 1973.

Hordeski, M.F., "Adapting Electric Actuators to Digital Control," *Instrumentation Technology,* March 1977.

Hordeski, M.F., "Digital Position Encoders for Linear Applications," *Measurements and Control,* July–August 1977.

Hordeski, M.F., "Digital Sensors Simplify Digital Measurement Systems," *Measurements and Data,* May–June 1976.

9.3 ENERGY MANAGEMENT DEVICES (PEAKLOAD SHEDDING)

Types:	A—Electromechanical B—Electronic C—Mini-computer based
Financial Savings Potential:	Up to approximately 25% of utility cost
Energy Savings Potential:	Up to approximately 25% of energy consumption
Controllable Loads:	8 and more
Effectiveness:	Widely variable
Cost:	\$500 to \$25,000
Partial List of Suppliers:	AMF Paragon Inc. (B); Barber-Coleman, Inc. (B,C); CSL Industries (B); Econ Systems (A,B); Honeywell Corp. (B,C); IBM Corp. (C); Johnson Controls Corp. (B,C); Lumenite Electronic Co. (A); MCC-Powers Inc. (B,C); Pacific Technology Corp. (B); Simplex Time Recorder Corp. (A); Square D Co. (A,B); Watt Master U.S.A. (B)

Introduction

Energy management devices are employed to regulate the "on" and "off" times of selected loads, such as fans, heaters, and motors, within a building in order to reduce electrical demand (kilowatts) and to regulate energy consumption (kilowatt hours). The energy management devices that are used for this purpose can be either electromechanical, electronic, or mini-computer based. All but the simplest continuously monitor energy consumption by means of an attachment to the electrical utility meter. The operation of one or more of the loads is interrupted for brief periods when it appears that the demand will reach some preselected critical value. More complex energy management systems can base control actions on complex algorithms using building-operating parameters, such as temperatures, air flow, or occupancy. The results obtained are almost entirely dependent on the imagination and skill of the user. Savings in electrical energy use and cost can be from zero to 50 percent or more.

Table 9.3a
FUNCTIONS OF THE VARIOUS TYPES OF SYSTEMS

	Mechanical	*Electronic*	*Mini-Computer*
Maximum number of loads	1–6	16–32	64–256
Peak demand limiting	No	Yes	Yes
Time-of-day control	Yes	Yes	Yes
Cycling control	Yes	Yes	Yes
User adjustable	Yes	Yes	Yes
User programmable	No	No	Yes
Remote adjustments	No	Some	Yes
Optimization calculations	No	No	Yes
Printed reports	No	Some	Yes

Electrical Demand

Electrical demand is defined as the average load connected by a user to an electrical generating system as measured over a short, fixed period of time, usually 15 to 30 minutes. The electrical demand is measured in kilowatts and recorded by the generating company meter for each measurement period during the billing month. The highest recorded electrical demand during the month is used to determine the cost of each kilowatt hour (kWh) of energy consumed. Electrical demand should not be confused with instantaneous load, which is also measured in kilowatts (kW) (Figure 9.3b).

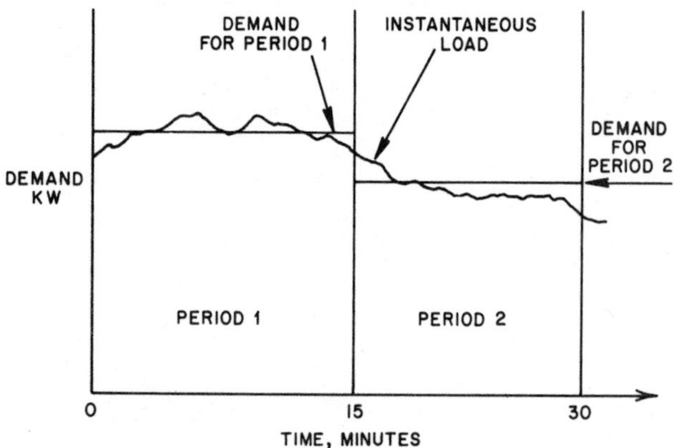

Fig. 9.3b Relationship between electrical demand and instantaneous load

Demand Load Shedding

Energy management devices, particularly load shedders, reduce the demand (i.e., average load) in critical demand periods by interrupting electrical service to selected motors, heaters, and so on for short periods. Since the load which has been turned off would normally have been operating continuously, the effect is to reduce the average load or demand for that period (Figure 9.3c).

$$\frac{\text{Energy Consumed (kWh)}}{\text{Demand Period (hr)}} = \text{Demand (kW)}$$

The instantaneous load when the load is operating, of course, remains the same. If the period involved was going to have the highest monthly demand, considerable money could be saved in rate reductions. In all periods other than the highest demand period, some energy is still saved. Originally, before the advent of high energy cost, the load shedder was used primarily to avoid demand cost penalties. Now it is used to limit energy consumption, by cycling loads on and off for brief periods, as well as to reduce demand. With the availability of inexpensive electronics, more exotic methods can be used to limit energy use, such as computer optimization of start times, set points, and other operating parameters based on weather, temperatures, or occupancy.

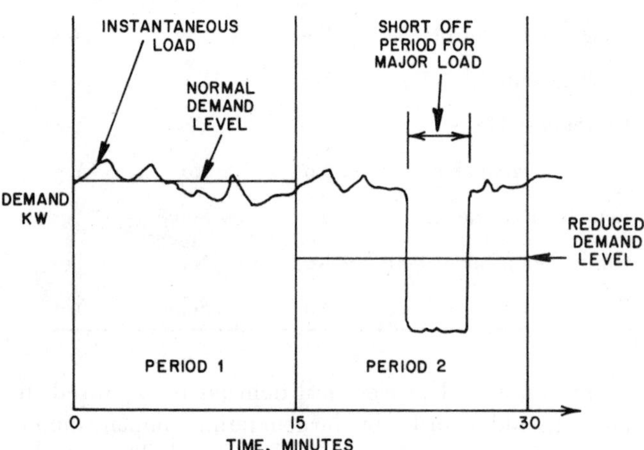

Fig. 9.3c Effect on demand of cycling a major load off for a brief period

Meter Connections

Energy management devices base all or most of their control actions on the demand level in each demand period. To accomplish this, they receive two signals from the conventional electrical utility meters. These signals are usually in the form of two momentary contact closures of approximately 500 ms duration which indicate the beginning of each demand period or the consumption of a unit of electrical energy (kWh). Meters can be modified by the utility company for a fee ranging from \$500 to over \$1000. In those cases in which the electrical utility cannot or will not supply the timing contact closure, external clock mechanisms can be substituted. The more sophisticated energy management systems are capable of generating their own timing signals with good results (Figure 9.3d). The energy (kWh) consumption pulse value can be converted to a unit of demand (kW) by dividing it by the demand period length in hours.

$$\frac{\text{kWh per Pulse}}{\text{Demand Period in Hours}} = \text{kW (Demand)}$$

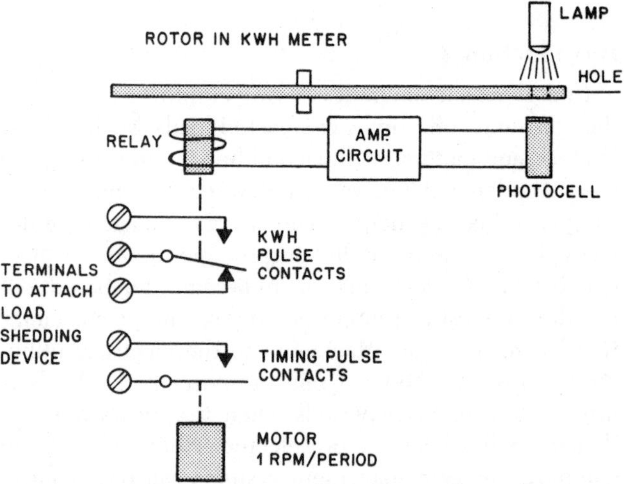

Fig. 9.3d Electrical connections for meter attachment. The timing pulse contact closes momentarily at the end of each demand period. The kWh pulse transfers once for every rotation of the meter rotor.

Electromechanical Devices

The least expensive and simplest power control device is the time clock. By means of a small electrical motor and an arrangement of cams and switches, several loads can be operated in a cyclical mode or simply turned on or off at preset times. Although their most popular application has been to operate advertising signs and display lighting, they have been adapted to control electric energy use and to limit demand. The time clock is set so that a load is turned off and on for brief intervals during each demand period. This has the effect of limiting energy consumption and reducing demand. Control is limited to those devices which can be cycled continuously.

Electronic Demand Limiters

A wide variety of electronic demand limiting devices are available. Electronic circuitry is used to monitor and measure the actual demand and provide control actions to limit the operation of the attached devices when the measured demand reaches a preset value. In operation these devices receive two signals from the utility meter, the kilowatt hour (kWh) or demand pulse, which indicates that a unit of electrical energy has been consumed, and the timing pulse, which indicates the end of one demand pulse and the start of the next one. Each demand period is controlled independently of all others. The timing pulse is used to reset the circuitry. The kWh or demand pulse is used to increment an electronic counter; each pulse from the meter is given a calibration kW weight corresponding to the meter being used. As these counts accumulate within a demand period, they are compared with a target value which corresponds to the maximum desired demand. When the value accumulated in the counter exceeds the target value, one or more of the attached loads will be turned off. If, when the next demand pulse is received, or after a 1-min delay, the demand still exceeds the demand target, a second load is turned off. This procedure is repeated until all available loads are turned off or until the demand no longer exceeds the target. Once turned off, the loads stay off until the end of the demand period. The timing or reset pulse from the meter restores the counter to zero and turns *all* the loads back on again. Some load shedders are somewhat more sophisticated in that the demand target is not fixed but increases at a steady rate. Energy usage is accumulated by a counter and compared with the target. This comparison is done at regular intervals throughout the demand period. When the difference between the accumulated usage count and the incrementing target reaches a set value, a load is turned off. The load will stay off until the difference decreases to a second set value or until the demand period ends. In some devices, the turn off–turn on set points can be adjusted independently for individual loads. In others, the maximum turn-off time is limited by a preset duration of time.

Mini-Computer Based Systems

The employment of computers and computer technology to the regulation of electrical energy opens up a wide range of control options. The computer's high speed memory and decision-making circuitry are combined with a customized program to maximize demand and energy savings with minimum impact on operating conditions. In operation, the computer receives inputs from the electric meter and from multiple sensors monitoring critical parameters within a building or facility. Loads can be cycled based on maximum demand target; time of day and day of week; rate of demand increase; process temperatures, pressures, rates, and schedules; inside and outside temperatures and humidity; wind direction and velocity; and combinations of all of these factors.

The maximum control durations are variable and can be changed automatically according to these same parameters. A second demand target is sometimes used so that, if the primary demand target is exceeded, extra control actions can be applied. The demand target can also be made to float upward and downward according to current demand rates, time of day, and all other parameters. Different targets and control strategies can be applied to different sections of buildings and plants. Printed reports showing actual consumption by demand period and by day, week, or month are also available. These systems are also being applied to manage the energy consumed by central air conditioning systems. Intake and exhaust dampers are controlled on the basis of air temperatures, so that the mix of air requiring the least energy is obtained at all times. The start-up and shut-down of air conditioning, heating, and lighting systems can be regulated according to inside and outside temperatures as well as occupancy. The intention is always to produce the conditions which consume the least energy.

BIBLIOGRAPHY

Energy Users News, Vol. 5, No. 33, August 18, 1980.

Hunt, V.D., *Energy Dictionary*, Van Nostrand Reinhold, New York, 1979.

Jennings, B.H., *Environmental Engineering*, International Textbook Company, Scranton, Pa., 1970.

Roose, R.W. (ed.), *Handbook of Energy Conservation in Mechanical Systems in Buildings*, Van Nostrand Reinhold, New York, 1978.

Stobaugh, R. and Yergin, D. (ed.), *Energy Future*, Random House, New York, 1979.

9.4 ELECTRICAL QUANTITIES

Types of Meters and Accessories:	A—Permanent magnet, moving coil, meter B—Rectifier-type meter C—Moving iron vane meter D—Electrodynamic meter E—Current transformer F—Shunt G—Potential transformer H—Resistor *Note:* In the summary below, the letters A to H refer to the listed equipment. A table with additional information will be found at the end of this section.
Meter Selection for Various Measurement Requirements:	ac current (C,B), high-range ac current (C,B with E) dc current (A), high-range dc current (A with F) ac voltage (C,B), high-range ac voltage (C,B with G) dc voltage (A), high-range dc voltage (A with H) 1 or 3 phase ac power (D with E,G) dc power (D), high-range dc power (D with H)
Inaccuracy:	±0.25 percent of rating (F) ±0.3 to ±0.6 percent of secondary rating (H) ±0.6 to ±1.2 percent of secondary rating (E) ±1 to ±2 percent of full scale (A,C,D) ±3 percent of full scale (B)
Approximate Cost:	\$30 (E,F), \$50 (A,C), \$70 to \$90 (B,H), \$160 (G), \$450 to \$700 (D)
Partial List of Suppliers:	Bi-Tronics Inc. (A,B,C,D,F), Compagnie Générale de Metrologie (A,B,C), Dowa Trading Co., Ltd. (A,B,C,D,F), Esterline-Angus Div., Esterline Corp. (All), Ferranti Ltd. (All), General Electric Co. (All), Hickok Electrical Instruments Co. (A,B,C,D,F), Simpson Electric Co. (All), Singer Co., American Meter Div. (All); Triplett Corp. (A,B,C,D,F), Westinghouse Electric Corp., Relay and Instrument Div. (All), Weston Instruments Division, Weston Instruments Inc. (All)

Basic Meter Movements

Four types of meter movements that commonly measure basic electric quantities include (1) permanent magnet moving coil; (2) rectifier; (3) moving iron vane; and (4) electrodynamic.

Although details of construction vary widely among the different meter movement designs, essentially the same principle underlies them all, i.e., rotation of the pointer is caused by a current-induced magnetic field.

Permanent Magnet Moving Coil Movements

More commonly known as the D'Arsonval type, these movements consist of a pointer attached to a coil of fine wire suspended between the poles of a permanent magnet (Figure 9.4a). Current through the coil creates a magnetic field which reacts with the field of the permanent magnet, causing a deflection of the coil proportional to the amount of current. The coil shaft is usually mounted on jeweled bearings, and hairsprings provide

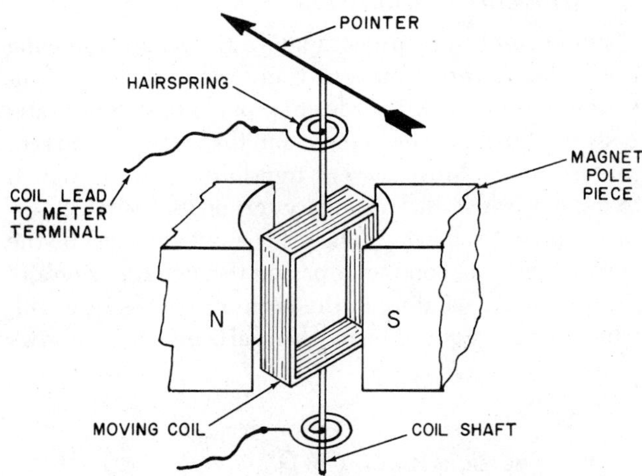

Fig. 9.4a Internal construction of a permanent magnet moving coil meter

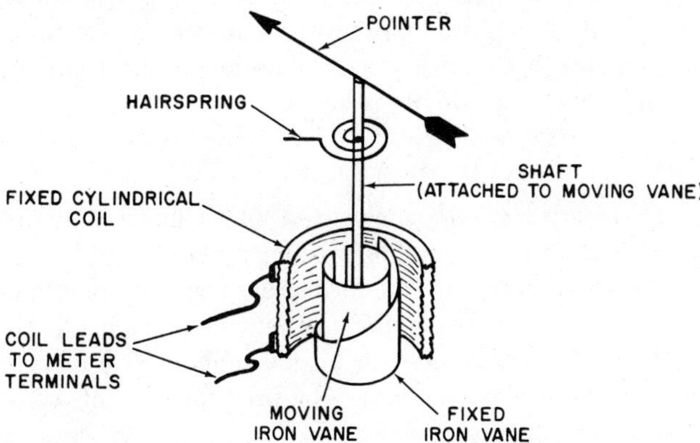

Fig. 9.4c Internal construction of a moving iron vane-type meter movement

the force necessary to restore the pointer to zero. An alternative method is to support the coil by flat metal bands attached to a supporting framework. The bands carry current to the coil and furnish the necessary restoring force. Permanent magnet, moving coil meters are accurate and sensitive, consume small amounts of power, possess linear display scales, and are widely used to measure dc current and voltage.

Rectifier-type Movements

These movements consist of a permanent magnet, moving coil-type sensor combined with silicon rectifiers arranged in a full wave bridge for use on ac circuits (Figure 9.4b). Rectifier current ratings limit these movements to relatively low ranges of ac current and voltage measurements. A wide band of frequencies can be accommodated because accuracy is not affected by frequency. Scales are essentially linear with some crowding at the lower end because of changes in rectifier values at very low currents. Meters must be calibrated to read effective or root mean square values since the movement responds to average ac values. Sensitivity is very high; power consumption is very low.

Moving Iron Vane Movements

These movements consist of two cylindrical soft iron vanes mounted within a fixed current-carrying coil (Figure 9.4c). One vane is held immobile and the other is free to rotate, carrying with it the pointer shaft. Current in the coil induces both vanes to become magnetized, and repulsion between the similarly magnetized vanes produces a proportional rotation. A hairspring provides restoring force. Only the fixed coil carries current, and therefore the movement may be constructed so as to withstand high current flows. Moving iron vanes may be used for dc current and voltage measurement, but small dc errors due to residual magnetism cause them to be more widely utilized for ac current and voltage detection even though sensitivity is low and power consumption is moderately high. These movements indicate effective or root mean square ac values and are subject to minor frequency errors only. Scales are nonlinear and somewhat crowded in the lower third, since pointer deflection is approximately proportional to the square of the coil current.

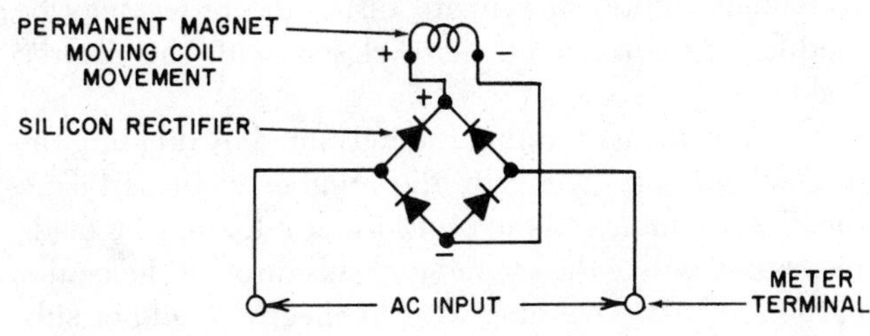

Fig. 9.4b Rectifier movement with full wave bridge circuit

Electrodynamic Movements

These movements are similar to the permanent magnet, moving coil-type elements except that the magnet is replaced by two fixed coils which produce the magnetic field when energized (Figure 9.4d). These movements can measure ac or dc current, voltage, and power. Cost and performance compared with other movements re-

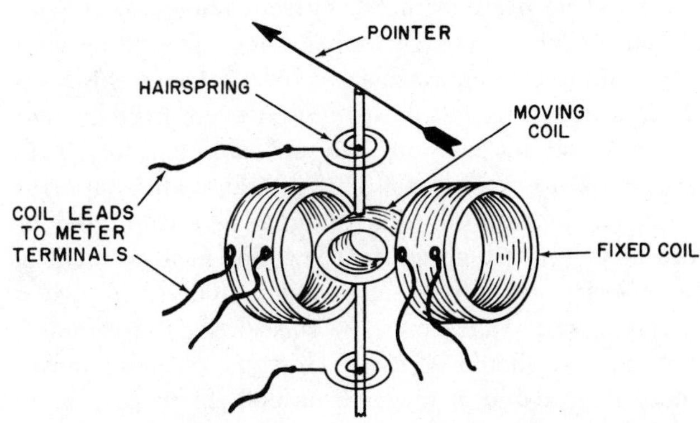

Fig. 9.4d Internal construction of an eletrodynamic meter movement

strict the use of this design to ac and dc power measurement. When used for power measurement, its scale is linear (calibrated in average values for ac). Its accuracy is high but its sensitivity is low.

Additional Meter Movements

These are less commonly used meter movements or those that are restricted to specific applications.[1–5]

Induction-type movements depend for their operation on the reaction between the magnetic field of a fixed coil and the current induced in a solid moving disk or cylinder. These movements may be used for current, voltage, or power measurement.

Electrothermic movements usually consist of one or more dissimilar metal thermocouple junctions in contact with a heater. Current through the heater causes the thermocouple to produce a dc voltage which may be measured with a permanent magnet moving coil meter. These instruments are suitable for measurements of current, voltage, and power.

Electrostatic movements utilize the forces of attraction or repulsion between electrically charged parts to deflect a pointer and measure voltage only.

Electronic meters with vacuum tube or transistor circuitry utilize a wide variety of circuit arrangements to permit measurement of current, voltage, and power.

Transducers permit the use of the permanent magnet moving coil meter for a broad range of ac and dc, voltage, and power measurements. Transducer designs vary widely; however, they generally are one of the following types: balanced bridge circuits, analog converters or multipliers, hall effect transducers, saturable reactors, or resistance divider networks.

Current Measurement

AC Current

Alternating current indicators are almost exclusively of the moving iron vane type owing to their wide range and low sensitivity to frequency variations (less than the electrodynamic type) and waveform errors.

This meter movement in ranges from 1 to 50 A full scale can be used for higher current ranges by the addition of external current transformers. Operating voltage is limited to approximately 750 V. Above this level a current transformer must be used even if the current is within the meter rating. Meters calibrated for 60 Hz are usable from 25 to 400 Hz, with an additional error of only ½ percent full scale, and may be recalibrated for use at levels as high as 1000 Hz. For measurement of small currents at frequencies above 400 Hz or, when variable frequencies must be measured, the rectifier-type meters should be used. This type of meter movement is available in ranges from 500 μA to 20 μA full scale and may be used from 20 to 10,000 Hz without loss of accuracy on sine wave circuits.

Current Transformers

For current measurement above the range of moving iron vane meters a current transformer should be inserted in the circuit in order to provide the proper ratio between the meter movement and the measured current (Figure 9.4e). Most current transformers are designed to deliver 5 A at the secondary terminals when full primary current is flowing. The primary winding can be the current carrying conductor passed through the center of the secondary winding in the form of a single-turn coil, which design requires no mechanical connection or break in insulation.

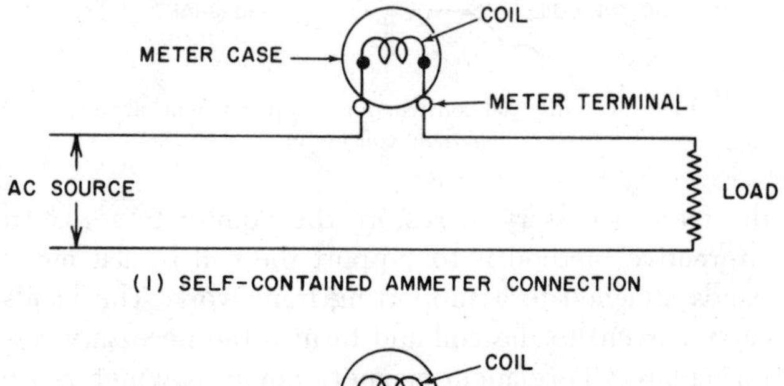

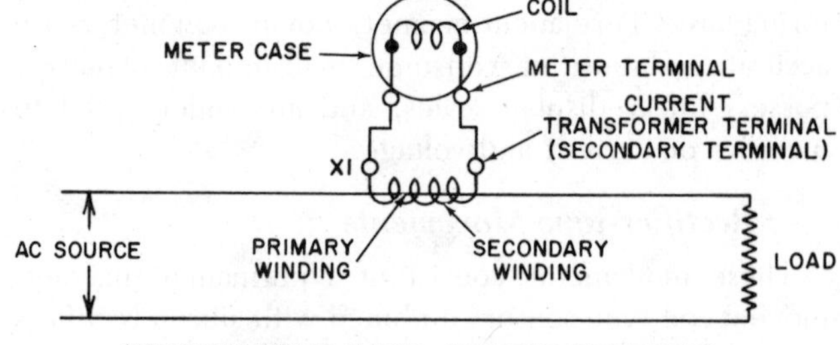

Fig. 9.4e Alternating current ammeter connections

Selection of the current transformer ratio should be based on the closest standard primary rating over the maximum current to be measured. Because there are large intervals between standard primary ratings for current transformers, it is often desirable to modify the transformer ratio by passing the primary conductor through the center of the secondary winding two or more times. A typical current transformer with a rating of 100 to 5 A has a normal ratio of 1 primary turn to 20 secondary turns or 1 to 20. Passing the primary conductor through the center of the secondary winding twice (Figure 9.4f) will increase the primary turns to two and the ratio will become 1 to 10, thus giving a new rating of 50 to 5 A. By taking additional primary turns, the rating may be modified to correspond more closely with the desired scale.

The new primary rating is determined by dividing the original primary rating by the number of turns taken. Small adjustments to the transformer ratio may be made by passing one of the secondary leads through the center of the secondary winding so as to effectively add or subtract secondary turns (Figure 9.4g). A current trans-

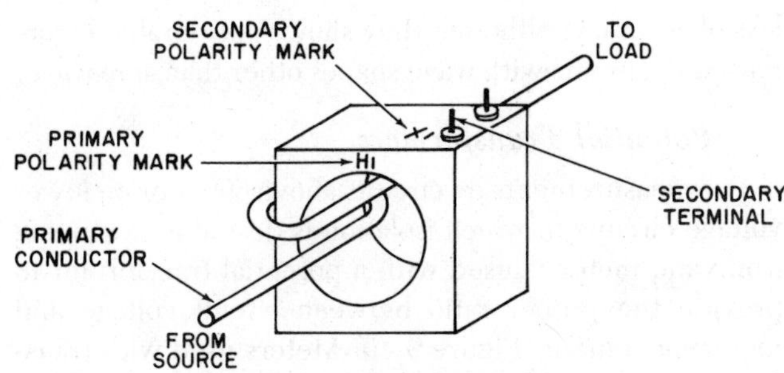

Fig. 9.4f Current transformer with two primary turns added

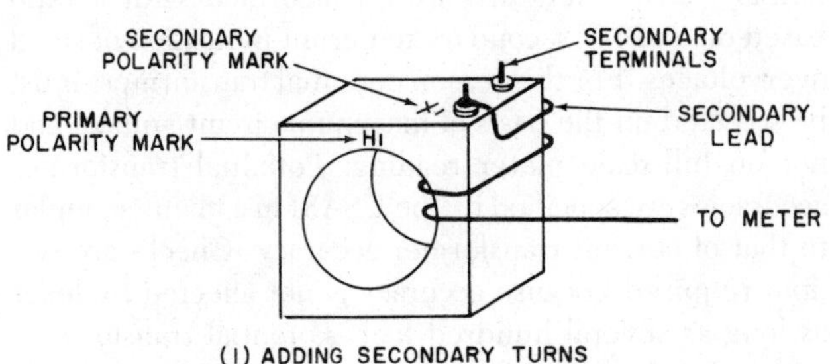

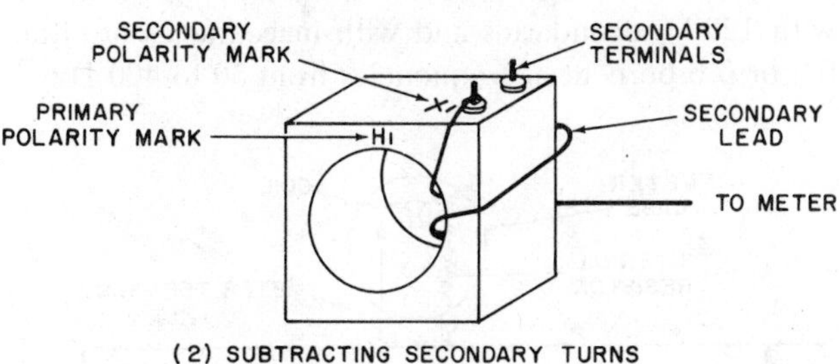

Fig. 9.4g Modifying the number of secondary turns on a current transformer

former with a 100 to 5 A rating has 20 secondary turns. Therefore, the addition of one secondary turn changes the turns ratio to 1 to 21, for a new rating of 105 to 5 A. By adding or subtracting the required number of secondary turns, the rating may be adjusted to almost any value. The new primary rating is determined by multiplying the total number of secondary turns by the secondary rating, which is 5 A in all cases. When both primary and secondary turns are modified, the new primary rating may be calculated by equations 9.4(1) and 9.4(2).

$$\text{Original secondary turns} = \frac{\text{Original primary turns}}{\text{Secondary rating}} \quad 9.4(1)$$

$$\text{Adjusted primary rating} = \frac{(\text{Total secondary turns})(\text{Secondary rating})}{\text{Primary turns}} \quad 9.4(2)$$

Primary and secondary polarity markings are provided on current transformers with the primary identified as H1 and the secondary terminal as X1. To add secondary turns, one should connect a lead to terminal X1 and pass it through the center of the secondary winding from the side opposite the H1 polarity mark, as indicated in Figure 9.4g. To subtract secondary turns, connect a lead to terminal X1 and pass it through the secondary winding from the same side as the H1 polarity mark. Current transformer accuracies are expressed as the maximum error for a particular load class as specified by the U.S.A. Standards Institute (USASI). Although load accuracy data are available for the meter manufacturers, it is seldom necessary to check the meters because if they are listed as transformer rated they are also designed so that current transformer accuracy will be maintained when connected with #14 AWG leads at lengths up to 150 ft (45 m) and #12 AWG leads at lengths up to 250 ft (75 m).

Current transformers are generally insulated for use on systems up to 600 V, but 600 V transformers may be used at higher voltages if the primary conductor is fully insulated. Current transformers designed for ammeters are available with 0.6 and 1.2 percent inaccuracy and in the following standard primary ratings: 100, 150, 200, 250, 300, 400, 500, 600, and 800 amperes.

DC Current

Direct current indicators are usually of the permanent magnet moving coil type to take advantage of the low power consumption and inherent linear scale factor while avoiding the residual magnetic errors of the iron vane types and the slower response of electrodynamic movements. Current carrying capacity of the internal connections to the coil limit the self-contained meters to a range of 20 μA to 1 A full scale. So-called self-contained meters are available for up to 50 A; however, they are actually low-range movements with an internal shunt. Above this range external shunts are employed. Meters are suitable for use on up to 600 V circuits. Above this value other types of meter movements or transducer-rated meters are generally used.

Ammeter Shunts

Direct current ammeter ranges can be greatly extended by connecting a shunt of the proper resistance in parallel with the meter movement so that a specific proportion of the current passes through the meter movement, the remainder being carried by the shunt (Figure 9.4h). Meters designed for shunts are rated either at 50 mV or at 100 mV, depending on the voltage drop across the meter terminals at full-scale deflection. When a shunt is used it must have a corresponding millivolt rating. The actual resistance value of the proper shunt does not have to be known since shunts are rated on the basis of full-scale meter current and are available from 5 to 20,000 A, calibrated to ±0.25 percent inaccuracy. Shunt-rated ammeters are calibrated for a particular shunt lead resistance, factors which vary from one manufacturer to another, and the correct calibrated shunt leads should be used for maximum accuracy or the meter recalibrated for use with different leads. Noncal-

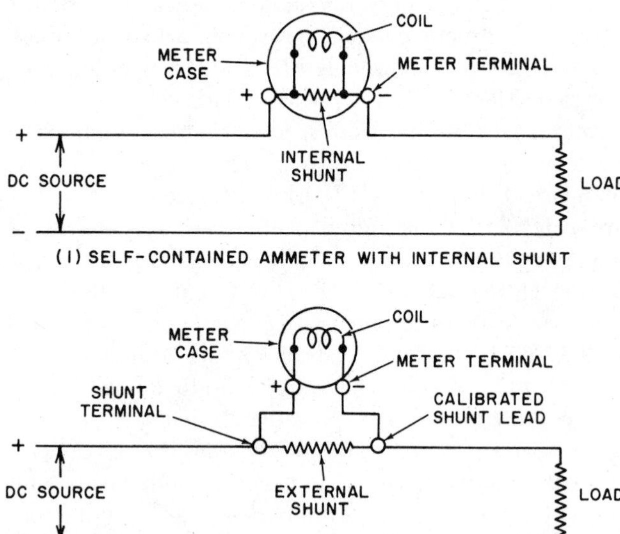

Fig. 9.4h Direct current ammeters with internal and external shunts

ibrated shunt leads of #14 AWG wire have an additional error of ½ percent for 50 mV meters at distances up to 25 ft (7.5 m) or for 100 mV meters at distances up to 40 ft (12 m).

Voltage Measurement

AC Voltage

Indicators for ac voltage measurement may be of the moving iron vane type or the rectifier type, since each movement has certain advantages for a particular voltage range. Moving iron vane voltmeters have the advantage of high accuracy and wide range, although their sensitivity is poor. Voltmeter sensitivity is determined by the coil current required for full-scale detection and is expressed in ohms per volt since the coil current is inversely proportional to the total resistance. A series resistor is generally installed within the meter case to provide the total resistance required by the movement so that full voltage equals full-scale meter current. Self-contained moving iron vane voltmeters are available in ranges from 3 to 600 V full scale with corresponding sensitivities ranging from 4 to 250 Ω/V. Meters are usually calibrated for 60 Hz; however, they may be used in circuits from 25 to 1000 Hz when calibrated for a specific frequency.

Rectifier-type meters have an advantage in that their high-resistance coils permit greatly increased sensitivity although they are not as accurate as the moving iron vane type, owing to varying rectifier losses. They are available in ranges from 3 to 800 V full scale with standard sensitivity of 1000 Ω/V at all ranges. Calibrated for 60 Hz, these meters may be used from 20 to 10,000 Hz without loss of accuracy, although they show considerable errors if used in circuits with wave shapes other than sine wave.

Potential Transformers

For measurements on circuits above 600 V or on lower voltage circuits in which isolation is desirable, a moving iron vane meter is used with a potential transformer to provide the proper ratio between circuit voltage and meter movement (Figure 9.4i). Meters used with transformers are usually provided with 150 V movements having a sensitivity of 100 Ω/V. The corresponding potential transformers are always furnished with a ratio based on a 120-V secondary to permit indication of small overvoltages. For this reason potential transformers must be selected on the basis of maximum circuit voltage and not on full-scale meter reading. Potential transformer accuracies are specified by the USASI in a manner similar to that of current transformer accuracy. Checks are seldom required because accuracy is not affected by leads as long as several hundred feet. Potential transformers with primary ratings from 120 to 14,400 V are available with 120-V secondaries and with inaccuracies of either 0.3 or 0.6 percent at frequencies from 50 to 400 Hz.

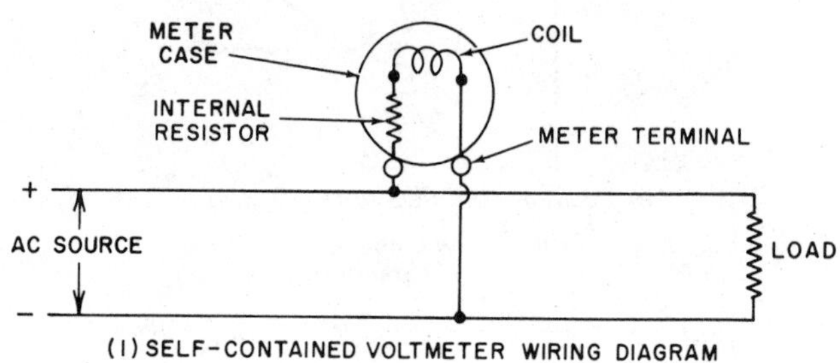

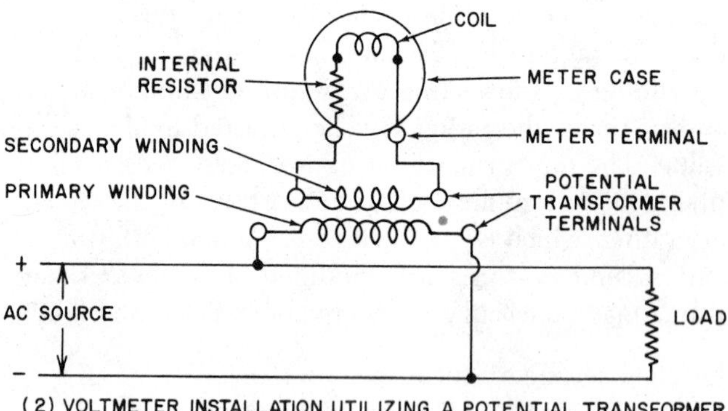

Fig. 9.4i Alternating current voltmeter installations

DC Voltage

Almost all meters for dc voltage measurement are the permanent magnet moving coil type because of its high sensitivity, linear scale, and wide range. Self-contained dc voltmeters are available in ranges from 1 to 600 V full scale with a standard sensitivity of 1000 Ω/V in all ranges. Above this range self-contained meters are not available because the internal resistors would become very high and resistor losses would result in excessive heating. External resistors can be used on circuits above 600 V to extend the measurement range of the permanent magnet moving coil meter. Circuit isolation is not possible in this arrangement, and the meter will be damaged if the external resistor is accidentally shorted.

Movements with sensitivities up to 20,000 Ω/V are available, although not in all ranges.

Voltmeter Resistors

Inserting an external resistor of the proper value in series with a permanent magnet moving coil voltmeter movement (Figure 9.4j) permits measurement of higher voltages or increased sensitivity at low ranges. Meters with external resistors generally have 1-mA movements and 125-Ω internal resistance. Standard resistors designed to afford a sensitivity of 1000 Ω/V may be selected on the basis of full-scale meter reading without the necessity of calculating resistance. Standard resistors are available in voltage ratings from 250 to 30,000 V, and resistors for nonstandard voltages may be calculated using equations 9.4(3) and 9.4(4).

$$\text{Total resistance} = \frac{\text{Full-scale meter reading}}{\text{Full-scale meter current}} \qquad 9.4(3)$$

$$\text{External resistor value} = \\ = \text{Total resistance} - \text{meter resistance} \qquad 9.4(4)$$

Movements requiring less current for full-scale deflection permit greater sensitivity and once the total resistance has been determined for a particular voltage range the new sensitivity may be calculated from equation 9.4(5).

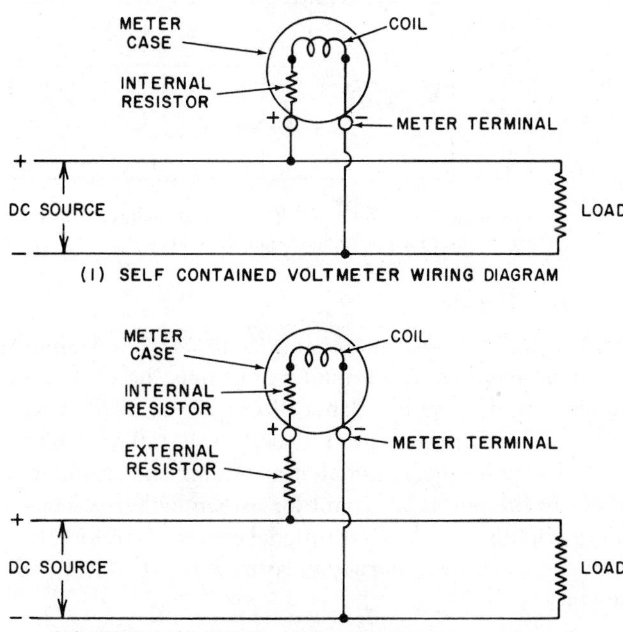

Fig. 9.4j Direct current voltmeter installations

$$\text{Sensitivity} = \frac{\text{Total resistance}}{\text{Full-scale meter reading}} \qquad 9.4(5)$$

Accuracy is not noticeably affected by lead length because the value of the external resistor is usually very high and calibrated inaccuracies of ±0.25 percent are generally available.

Power Measurement

AC Power

Indicators for ac power measurement are usually constructed with electrodynamic movements because the separate fixed and moving coils permit two different types of input signals for the same movement. The fixed coils are usually connected to measure currents, and the moving ones are connected to monitor voltages (Figure 9.4k). Connection of the coils in this way causes a deflection of the moving coil proportional to the instantaneous product of the circuit current and voltage. Inertia prevents the moving coil from responding quickly to current and voltage variations in ac circuits above 25 Hz, and the pointer will indicate the average ac power regardless of wave shape. The way wattmeters are wired introduces a certain amount of error which can be minimized by altering the wire connections as a function of the current and voltage ranges being measured.

In the wattmeter shown in Figure 9.4k the current in the current-carrying coils is the true load current, but the voltage across the potential coil is higher than the load voltage by an amount equal to the voltage drop across the current coils. This arrangement will result in a positive error in power indication by an amount equal to the power consumed by the current coils. This error will be lowest when the wattmeter is used on high-voltage circuits with low-current loads. If the external wiring is changed, as shown in Figure 9.4l, the voltage across the potential coil will be the true load voltage, but the current drawn by the potential coil will also pass through the current coil and the wattmeter reading will therefore be high by an amount equal to the power consumed in

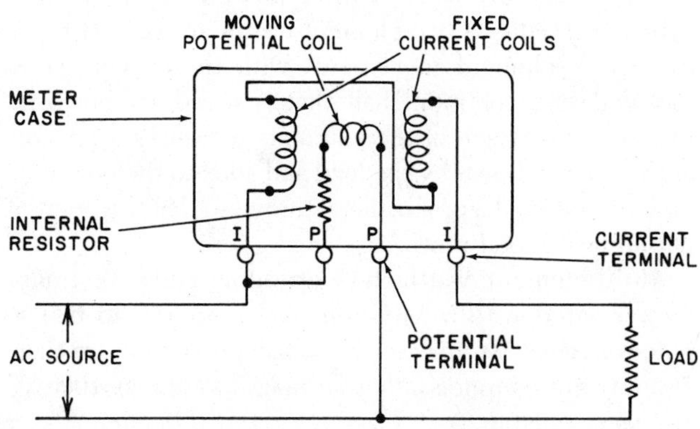

Fig. 9.4k Single-element wattmeter, wired for accurate detection of low-current and high-voltage loads

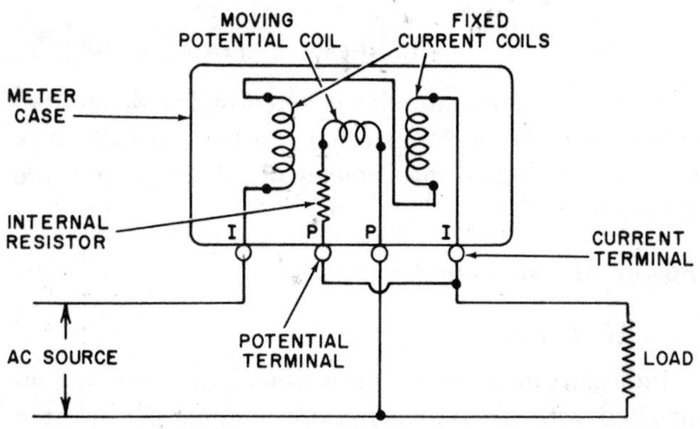

Fig. 9.4 l Single-element wattmeter, wired for accurate detection of high-current and low-voltage loads

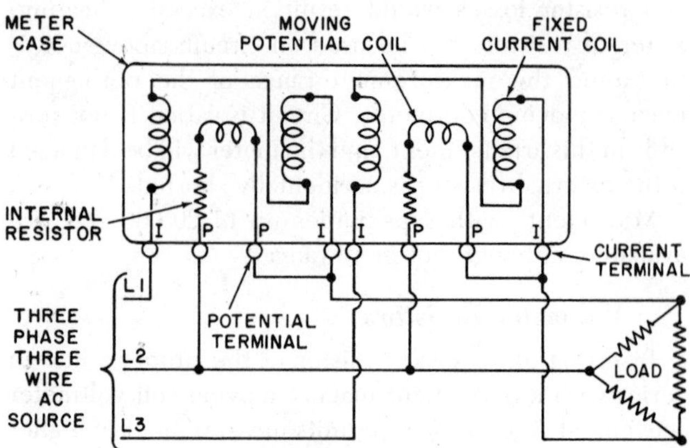

Fig. 9.4m Two-element wattmeter, wired for three-phase, three-wire loads

the potential coil. The error will be lowest when the meter is used in low-voltage circuits with high-current loads.

Meter scales are linear, accuracy is high, but low sensitivity is the result of the high power consumption. Calibration is usually for 60 Hz and recalibration is necessary for use at any frequency up to 400 Hz. Wattmeters are rated for maximum current and voltage in addition to maximum wattage because circuits with large phase angle differences between current and voltage (circuits with low power factor) can overload the current or voltage coils without causing large wattage readings. Power measurement in single-phase, two-wire circuits requires a single element wattmeter consisting of two fixed current coils and one moving potential coil.

The fixed current coils are wound with heavy wire as required by the meter current rating, whereas the moving coil is usually wound with fine wire and is provided with a series resistor to insure a high resistance–low inductance circuit for potential measurement. Self-contained meters have ranges from 125 to 1000 W full scale with current ratings from 1.25 to 10 A at 120 V. Transformer-rated meters are also available, with ratings of 5 A at 120 V and scale ranges from 1000 W to above 100 mw, depending on current transformer and potential transformer ratios.

Power measurement in three-phase, three-wire circuits necessitates a two-element wattmeter consisting of two single-element movements with the moving coils attached to a common shaft that is wired as shown in Figure 9.4m. Each of the elements measures a portion of the power drawn by the load and adds to the common moving coil shaft a proportional torque so that the pointer will indicate total power.

Multi-element wattmeters are almost always transformer rated with 5-A current coils and 120 to 600 V potential coils. Meter ranges from 5000 W to more than 100 mw are common although ranges as low as 1000 W are also available.

Power measurements can be made in three-phase, four-wire circuits by a three-element wattmeter. Such a design is seldom utilized because the two-element wattmeter movement can be modified to permit measurement on three-phase, four-wire systems by reconnecting one of the fixed coils for each element (Figure 9.4n). Meters of this type are known as 2½-element wattmeters and will correctly indicate power for three-phase, four-wire loads as long as the lines to neutral voltages are balanced for all three phases. Meter ranges from 10,000 W to more than 100 mw are available in this design, with ratings of 5 A at 120 or 240 V.

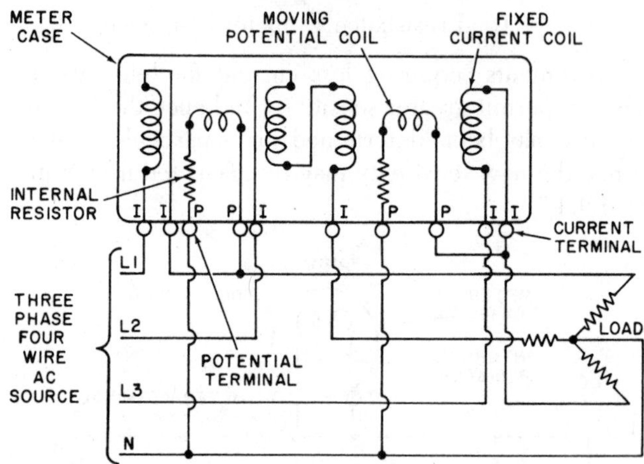

Fig. 9.4n Two-and-one-half-element wattmeter, wired for three-phase, four-wire loads

DC Power

Direct current wattmeters utilize the same movement as single-element ac wattmeters but must be calibrated for use on dc circuits. Ranges from 100 to 2000 W are available with ratings from 1 to 20 A at 120 V. Higher ranges may be accommodated by adding an external resistor to the potential circuit for use on higher voltages, although this introduces additional errors. Therefore, the higher ranges are generally measured by a dc watt transducer.

Meter Scales

In general, the best choice of meter-scale range is one in which the maximum anticipated value of current, voltage, or power falls to 80 percent of the full-scale reading. Scale selection on this basis provides reasonable utilization of the meter scale, furnishing good visibility with the capability to indicate moderate overloads.

When many meters are grouped in a small area, it may be desirable to have all normal readings at the midscale position in order to permit easier identification of an abnormal condition. The midscale pointer position on meters with nonlinear scales indicates a value of approximately 65 percent of full scale, whereas on linear displays it corresponds to a value of 50 percent of full scale.

Table 9.4o
ORIENTATION TABLE FOR AMMETERS, VOLTMETERS, AND WATTMETERS

Meters	Features: Type of Meter Movement	Accessories Required	Full-Scale Meter Range (A-Amperes V-Volts W-Watts)	Permissible Overload in Multiples of Full Scale and for Noted Time Duration	Recommended Applications
Ammeters ac	Rectifier	None	0.5–20 × 10^{-3}A	*For meters:* 1.2×: 8 hr 100×: 1 sec	Low-range, high-frequency
	Moving iron vane	None	1–50 A		General use up to 750 volts
	Moving iron vane	Transformer	10–8000 A	*For transformer:* 50×: 2 sec	High-range, over 750 volts, long meter leads
dc	Permanent magnet moving coil	None	0.02 × 10^{-3} – 50 A	1.2×: 8 hr	General use
	Permanent magnet moving coil	Shunt	20–20,000 A	100×: 1 sec	High-range
Voltmeters ac	Rectifier	None	3–800 V	*For meters:* 1.2×: continuous 100×: 1 sec	Low-range, high-frequency
	Moving iron vane	None	3–600 V		General use
	Moving iron vane	Transformer	150–18,000 V	*For transformer:* 1.1×: continuous 1.25×: 1 min	High-range, circuit isolation
dc	Permanent magnet moving coil	None	1–600 V	1.2×: continuous 100×: 1 sec	General use
	Permanent magnet moving coil	Resistor	250–30,000 V		High-range, high-sensitivity
Wattmeters Single-phase ac	1-element electrodynamic	None	125–1000 W	*For current:* 1.5×: continuous 10×: 1 min	Low power, single-phase 2-wire circuits
	1-element electrodynamic	Transformer	1000 – 100 × 10^6 W		General use, single-phase circuits
Three-phase ac	2-element electrodynamic	Transformer	1000 – 100 × 10^6 W	*For voltage:* 1.2×: continuous 10×: 1 min	General use, three-phase, three-wire
	2½-element electrodynamic	Transformer	1 × 10^4 – 1 × 10^8 W		General use, three-phase, four-wire
dc	1-element electrodynamic	None	100–2000 W		General use, low-power
	Permanent magnet moving coil	Transducer	400 – 100 × 10^6W	Varies with design	High-power

REFERENCES

1. Drysdale, C.V. and Jolley, A.C., "Electrical Measuring Instruments," 2 vols., Ernest Bern, London, 1924.
2. Edgcumbre, K. and Ockenden, F.E., "Industrial Electrical Measuring Instruments," Sir Isaac Pitman and Sons, London, 1933.
3. Golding, E.W., "Electrical Measurements and Measuring Instruments," Sir Isaac Pitman and Sons, London, 1948.
4. Harris, F.K., "Electrical Measurements," John Wiley and Sons, New York, 1952.
5. Laws, F.A., "Electrical Measurements," McGraw-Hill Book Co., New York, 1938.

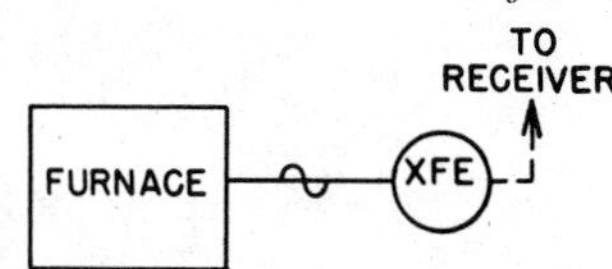

9.5 FLAME SENSORS

Design Temperature:	Up to 2400°F (1300°C) for flame rods Up to 150°F (66°C) for all other sensing devices
Cost:	\$55–\$400
Partial List of Suppliers:	Bailey Controls Co.; Electronic Development Labs, Inc.; Electronics Corp. of America, Photoswitch Div.; Honeywell, Process Control Div.; Ircon Inc.; Pyronics Inc.

In this section a brief description of the more widely used means of flame detection will be presented. After some basic theoretical discussions, the various devices will be described. The feature tabulation will enable the reader to select the most desirable sensor for the application at hand. Functions of flame sensors as part of burner control systems will be discussed later.

Among the many characteristics of flame, the following have been successfully used to detect its presence:

1. Heat generated
2. Ability to conduct electricity (ionization)
3. Radiation at various wavelengths, such as
   a. Visible
   b. Infrared
   c. Ultraviolet

Methods of detecting these characteristics are described below.

Heat Sensors

The earliest flame sensors utilized the most obvious characteristics of the flame: namely, the heat generated. These devices were thermocouples, bimetallic elements, etc. For small installations and domestic burners these devices were satisfactory and are still used. Their relatively slow response (2–3 min) renders them unsuitable and indeed dangerous for larger installations. Let us take a large reforming furnace as an example. If it requires 1000 scfm (0.47 m^3/sec) of fuel gas, and flame failure is detected only after 2–3 min, the 2000–3000 ft^3 (56–84 m^3) of unburned fuel gas admitted to the furnace will create an explosion hazard. A detector with a response of 4–6 sec or less will not permit sufficient amounts of gas to enter the furnace to cause an explosion. Another obvious disadvantage of these sensors is that, since they only sense heat, they will be unable to distinguish between heat generated by flame or that radiated by the hot refractory.

Electric Conduction-type Detectors

The first major breakthrough toward fast (but yet unreliable) detection of flame was the discovery that flame is capable of conducting electricity. This is true because the flame, being a chemical reaction between a fuel and oxygen, liberates a large number of electrons. Because of ionization, the flame can conduct both direct and alternating currents (dc and ac), which are utilized to establish an electrical circuit. A rod immersed in the flame (flame rod) acts as one electrode, the burner as the other. This proved to be fast but unreliable in that a high-resistance electrical short caused by, say, faulty insulation could "simulate" the presence of flame.

Rectification

Any electrical device which offers a low resistance to the current in one direction but a high resistance to the current in the opposite direction is called a rectifier. An ideal rectifier is one with zero resistance in one direction and infinite resistance in the opposite direction.

When ac is passed through a rectifier, the current thus obtained will be rectified current, consisting of only that portion of the input current to which the rectifier presented low resistance.

By making the area of one of the electrodes (the burner in this case) much larger than the other, conduction will essentially take place in one direction only (see Figure 9.5a).

The explanation for the above phenomenon is that

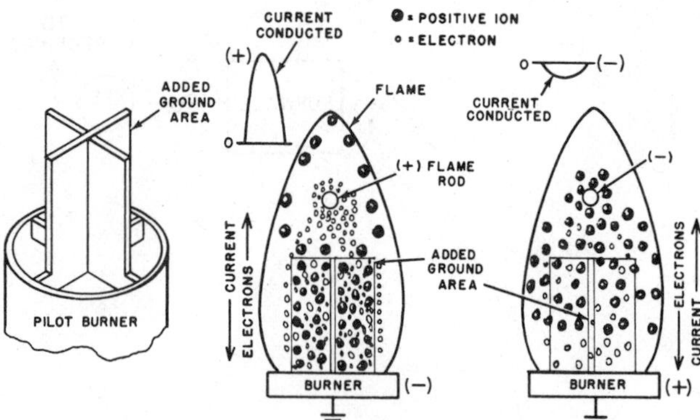

Fig. 9.5a Rectification by flame

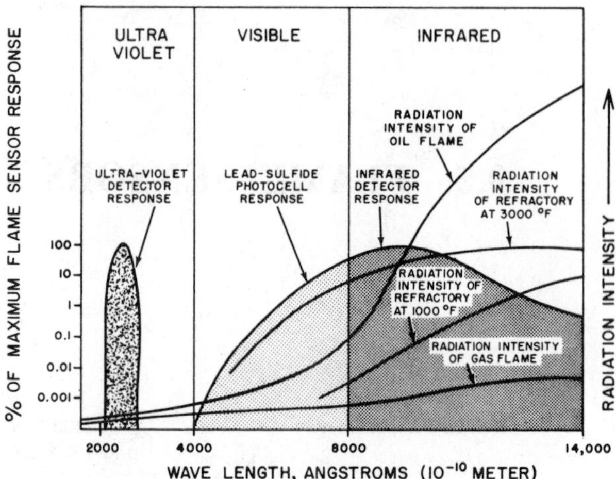

Fig. 9.5b Applicable ranges of selected flame sensors, and the range of hot refractory effect

when ionization takes place, electrons are liberated from the gas molecules and are free to move about, thus constituting electric current. In addition to the freed electrons, the negative electrode contains many surplus electrons acquired through the negative side of the external circuit; that is what makes the electrode negative. These surplus electrons repel each other and, given enough positive ions to attract them, will leave the negative electrode. The number of electrons leaving the negative electrode and entering the positive electrode determines the rate of current flow. It is apparent that the current flow depends on the number of positive ions that get near enough to the negative electrode. If the area of one electrode is made several times larger than the other, and that electrode is negative, it will accommodate a larger number of positive ions. This, in turn, will increase the flow of electrons to the other (positive) electrode.

If alternating current is applied, the current through the flame will be rectified. Indeed, using this arrangement and thereby obtaining a *half-wave rectified* current, the lack of safety due to simulated flame was eliminated, as the circuitry associated responded only to a half-wave rectified signal. Unrectified alternating current or direct current input to the detecting circuit will result in a safe shutdown. Because flame rods are in direct contact with the flame, they have to be cleaned or replaced often. Above approximately 2000°F (1100°C), very few metals can be used and even these become brittle. Fuels with high sulphur content burn with flames having low resistance, which results in low flame rod output and leads to nuisance shutdowns.

To avoid direct contact with the flame and the maintenance problems arising from this, another obvious characteristic of the flame, its ability to radiate energy, was used.

Radiation Types

Radiation emitted by the flame covers the energy spectrum for wavelengths corresponding to ultraviolet, visible, and infrared range (see Figure 9.5b).

Visible Radiation

Visible radiation occupies about 8 percent of the total band of wavelengths radiated by the flame. To detect this portion of the radiation, a rectifying phototube is used. It consists of a light-sensitive coated cathode of large surface, and an anode encapsulated in a vacuum (see Figure 9.5c).

The number of electrons emitted by the cathode is approximately proportional to the light intensity. Conduction will take place only when the anode is positive.

If ac is applied, the phototube will act as a half-wave rectifier. The associated electronic circuit will respond only to such a rectified signal and high-resistance shorts will not simulate flame. At high temperatures the refractory (noncombustible insulator used to line interiors of furnaces) emits visible radiation, which the detector might not be able to distinguish from visible radiation emitted by flames; therefore, caution must be exercised in installing the detector in such a position that it will not be able to sight hot refractory. Use of rectifying photocells is limited to oil-fired burners because visible radiation emitted by gas flame is insufficient to be detected by this type of flame sensors.

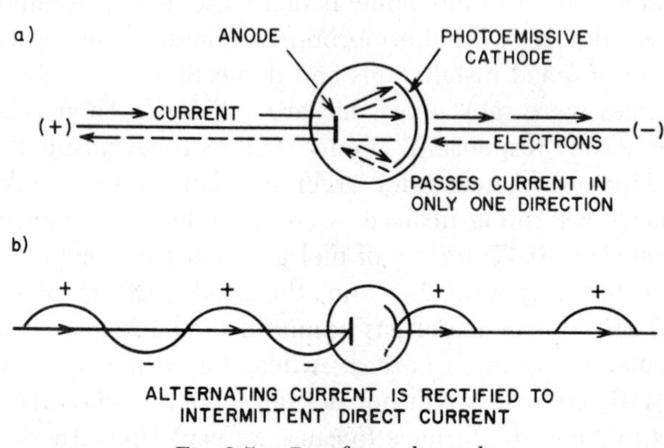

Fig. 9.5c Rectifying phototube

Cadmium-Sulfide Photocell

The single element of this type of photocell is coated with cadmium sulfide which is sensitive to radiation in the visible spectrum only. In the absence of light, cadmium sulfide offers high resistance to an electrical circuit but, when exposed to light, conducts freely and almost instantaneously. The electrical resistance of cadmium sulfide decreases directly with increasing intensity of light. Like lead sulfide, it acts as a variable resistor, and conducts current equally well in either direction.

This photocell is used in series with the coil of a low voltage relay in an ac circuit. When the cell sees sufficient light to pass a given current, the relay "pulls in." The sensitive region of the cadmium-sulfide cell is such that it will not respond to gas flames, and can, therefore, be used with oil flames only. The cell will not be affected by hot refractory or "shimmering" effects.

Infrared Radiation

Infrared radiation covers the largest portion, approximately 90 percent, of the band of wavelengths emitted by the flame, and its intensity is by far the strongest of all radiation emitted. The infrared radiation emitted is a more reliable means of detection than the visible radiation. It is emitted by gas as well as oil flames, and the intensity never drops to zero.

Lead-Sulfide Photocell

This is sensitive to infrared radiation and is the most widely used device for such detection. Similarly to the principle of the cadmium-sulfide photocell, the principle of the lead-sulfide photocell is to change its resistance inversely to the infrared radiation it is subjected to. It is a variable resistor and conducts electricity equally well in both directions without rectification. The current flow is a measure of flame strength.

The signal sensing circuit responds only to so-called flame frequencies. When looking at a flame, it seems to burn steadily. But if the eyes were sensitive enough to very rapid changes, it could be seen that the flame burns brightly at one instant and less brightly the next. The flame really flickers or pulsates, but much too rapidly for this to be detected by the human eye. The frequency of the flicker is very irregular but may be detected by an electronic circuit designed to accept only a proper band of frequencies. Such a tuned circuit is, therefore, selective and will not be affected by infrared radiation emitted by hot refractory, nor can a high-resistance short simulate flame.

Experiments, however, show that a "shimmering" effect caused by movement of hot gases between the refractory and photocell can "fool" the circuit and flame simulation can occur. This so-called shimmering can be demonstrated by viewing objects through the heated air over a candle.

Ultraviolet Radiation

Seemingly, the least significant portion of radiation is that of ultraviolet. It represents only about 1 percent of the total radiation and covers 10 percent of the emitted band of wavelengths. Ultraviolet (UV) radiation is emitted by gas as well as oil flame. The device popularly known as a UV detector is a gas-filled tube with voltage applied between its two electrodes (anode and cathode).

The tube is conductive only if both of the following conditions exist: voltage is applied at its terminals, and it is subjected to ultraviolet radiation.

It is readily seen that a high-resistance short cannot "simulate" flame, nor can it be affected by the hot refractory because no ultraviolet radiation will be emitted by refractories below approximately 2500°F (1370°C). Compared to the detectors so far mentioned, the UV cell is foolproof.

In addition to all the attributes, some UV detectors have an additional built-in safety feature. This consists of a shutter arrangement that blocks the "view" of the cell for a fraction of a second about 20 times each second, interrupting the circuit which can reestablish current flow again only if ultraviolet radiation still exists. This arrangement makes the detector responsive to flame failure within a fraction of a second, because it has to "convince" itself 20 times a second that flame exists.

Installation

Proper installation of flame sensors is essential to safe operation. Installations of these sensors fall into three categories: (a) pilot and main flames supervised simultaneously; (b) only pilot flame supervised; and (c) only main flame supervised.

a. Checking the pilot and main flames simultaneously is the most desirable method of flame supervision. When this arrangement is used, the pilot is capable of safely igniting the main flame (see Figure 9.5d). The flame rod in this case is located in the path of both the main flame and the pilot flame.

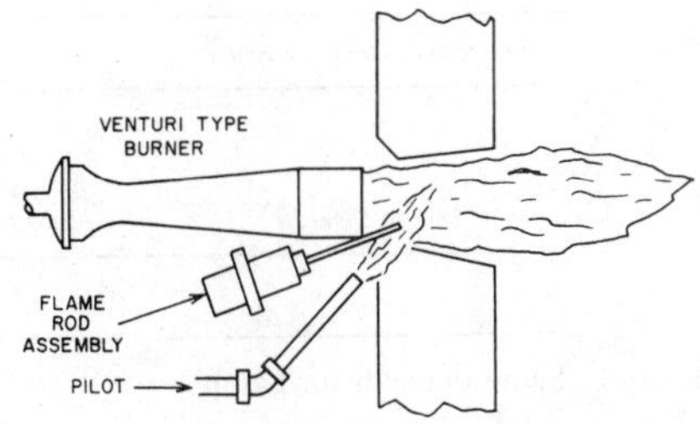

Fig. 9.5d Pilot and main flame monitored by flame rod

b. On some installations it may be impossible to prove the pilot and main flames simultaneously because variations of the main flame envelope at different firing rates. This condition prevails, for example, in mechanical fan-type burners. The recommended practice of installation in this case is to supervise the pilot flame (see Figure 9.5e).

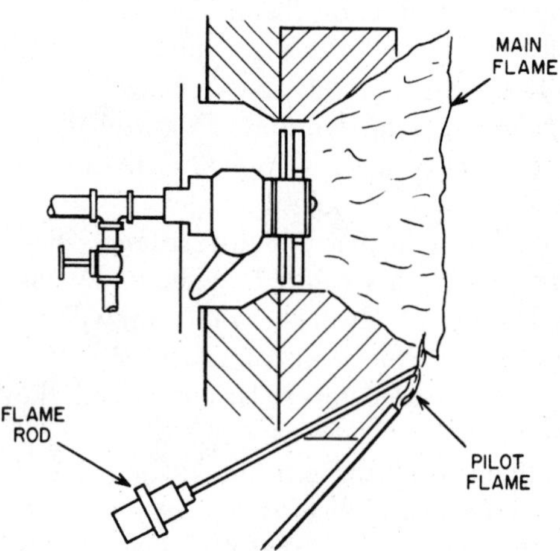

Fig. 9.5e Mechanical fan-type burner. Pilot flame only is proved

c. In applications where the pilot flame is not lit continuously but only on light-off, the main flame is supervised. It is essential that the sensor be installed in such a way that it will be able to "feel" or "see" the flame during all variations in firing rates and draft adjustments.

The above considerations are applicable for flame rods as well as for the different viewing devices.

Additional care must be exercised in the installation of rectifying phototubes and lead-sulfide photocells to insure that incandescence of hot refractory or shimmering of hot air will not simulate flame. This is accomplished by limiting the view of the devices through a restriction orifice.

Portholes for viewing devices must be cleaned regularly to insure an unobstructed view and thus prevent nuisance shutdowns.

Conclusions

The order of presentation for the different types of sensors was intended to show how the limitations of each design have been eliminated.

Starting with the heat sensors, it was seen that their response time rendered them unsuitable for industrial applications. The flame rods provided for fast response; however, their direct contact with the flame and the consequences, such as brittleness or oil and carbon deposits on the rod, limited their usefulness. Photocells utilizing the visible range of radiation were limited by their inability to reject radiation from the refractory, and their useful range covered only oil flames.

The infrared detector eliminated most of the above shortcomings; however, shimmering of hot air could still "fool" the detecting circuit. It is safe to say that UV detectors can be used reliably in any application. Their relatively high cost, however, demands that the situation be carefully investigated for possible application of other sensors.

In an oil-fired furnace, the cadmium-sulfide cell can be safely used if it is installed so that it will not "see" the hot refractory. For combination oil and gas furnaces, the lead-sulfide cell will work satisfactorily if it is installed with its view pointed toward the flame through a restricting orifice so that it will not "see" shimmering hot air.

All devices except the flame rods share the common advantage of not being in contact with the fuel and the flame. The temperature limitations imposed on them are easily satisfied because the temperature outside of the furnace is unlikely to reach 150°F (66°C). The purpose of these installations is safety and, therefore, cost considerations can be only secondary.

For a summary of flame sensor features, see Table 9.5f.

Table 9.5f
COMPARISON OF FLAME SAFEGUARDS

Principle of Flame Detection	*Rectification*		*Infrared*	*Visible Light*	*Ultraviolet*
Type of Detector	*Rectifying Flame Rod*	*Visible Light Rectifying Phototube*	*Lead-Sulfide Photocell*	*Cadmium-Sulfide Photocell*	*Ultra-Violet Detector Tube*
		Advantages			
Same detector for gas or oil flame			√		√
Can pinpoint flame in three dimensions	√				
Viewing angle can be orificed to pinpoint flame in two dimensions		√	√	√	√

Table 9.5f *continued*
COMPARISON OF FLAME SAFEGUARDS

Principle of Flame Detection	*Rectification*		*Infrared*	*Visible Light*	*Ultraviolet*
Type of Detector	*Rectifying Flame Rod*	*Visible Light Rectifying Phototube*	*Lead-Sulfide Photocell*	*Cadmium-Sulfide Photocell*	*Ultra-Violet Detector Tube*
		Advantages			
Not affected by hot refractory	√			√	√
Checks own components prior to each start	√	√	√	√	√
Can use ordinary thermoplastic covered wire for general applications, no shielding needed	√	√		√	√
No installation problem because of size			√	√	
		Disadvantages			
Difficult to sight at best ignition point			√		
Exposure to hot refractory may reduce sensitivity to flame flicker and require orificing			√		
Flame rod subject to rapid deterioration and warpage under high temperatures	√				
Not sensitive to extremely hot premixed gas flame			√	√	
Temperature limit too low for some applications	√	√	√	√	
Shimmering of hot gases in front of hot refractory may simulate flame			√		
Hot refractory background may cause flame simulation		√			
Electric ignition spark may simulate flame					√

BIBLIOGRAPHY

Factory Mutual Staff, *Handbook of Industrial Loss Prevention*, McGraw-Hill Book Co., New York, 1967.

Shinskey, F.G., *Energy Conservation Through Control*, Academic Press, New York, 1978.

9.6 LEAK DETECTORS

Methods of Detection:
A—Pressurization
B—Bubble emission or other simple methods
C—Analyzers
D—Acoustic emission
E—Other approaches
F—Thermography

Inaccuracy:
Very difficult to generalize; many methods require good techniques to obtain high level of accuracy

Cost:
Methods A and B can be less than $100
Methods C, D and E can be $500 and higher
Design F begins at about $4000

Partial List of Suppliers:
Acoustic Emission Leak Locators Corp. (D); AGA Corp. (F); American Gas & Chemical Co., Ltd. (B); Baghouse Accessories Co. (B); Barnes Engineering Co. (F); CRC Bethany International, Inc. (E); Dukane Corp. (D); Expander Seal Tools, Inc. (A); Formulabs, Inc. (B); Gator Hawk External Testors, Inc. (A); General Electric (C); Guy Speaker Co., Inc. (B); Heath Consultants, Inc. (D); Hughes Aircraft Co., Industrial Products Div. (F); Matheson (B); Metronics Associates (C); Mine Safety Appliances (C); MDA Scientific, Inc. (C); National Draeger, Inc. (C); Safety Pipe Line Stopper Co. (A); Scott Engineering (B); Techsonics (D); Telatemp Corp. (B); United Survey, Inc. (A); Uscon Corp. (A); Volumetrics (A)

Although special techniques and equipment permit quantitative determinations, this survey stresses those techniques that provide a qualitative answer to a problem. Chemical materials and fuels that are not retained represent both a financial loss and a resource loss. Energy goes into the shipment, as well as the processing of chemicals, so fluid leaks waste energy. Safety aspects of leak detection are extremely important because of the many materials that are toxic or hazardous because of their flammable or explosive characteristics.

Pressurization

Pressurization by pneumatic or hydraulic means is probably the most widely used method in industry. The fall of a pointer on a pressure gauge can indicate a leak.

Various standards and codes explain specific procedures for piping, pneumatic instrumentation systems, and vessels. The recommended practice of the National Fire Protection Association (NFPA) goes into considerable detail on how to handle underground leaks. American Petroleum Institute (API) Standard 527 describes a method for checking the seats of safety relief valves.

Test equipment is available for pressure testing, including special fittings for plugging off lines and testing joints.

Bubble Emission and Other Simple Methods

Commercial formulations that bubble or foam at the point of a leak improve the value of this technique, and safe materials are available for different chemical applications. Once a leak has been found by pressurization, the bubble emission technique can help to isolate it.

Aerosols, paints, and papers that change color because of chemical action are used in much the same way as the materials that bubble and foam. They must be applied to the exact point which is to be tested.

One vendor claims to have a proven way of detecting leaking steam traps. A tape is fastened to the outlet piping. If a spot on the tape remains silver, the trap is working. If live steam passes through the trap continuously, the spot turns black because of overheating.

Dyes can label sewage steams to check flows and seepage areas. One company has biodegradable, fluorescent dyes that are still visible after dilution to 1 ppm in water.

Phosphorescent powders can be useful for finding defective bags and other leaks in dust collectors. The powder is put into the dirty side of the collector, and it works its way through the system in a few minutes. Then the inside of the collector is checked with an ultraviolet light to find leakage points, which are easy to see because of the glowing powder.

White smoke can be used in sewers, ducts, and other places that cannot be pressurized. Candles, bombs, and generators with blowers produce various quantities of smoke.

Analyzers

If a liquid is toxic or flammable, a wide variety of analyzers can be considered for leak detection. These instruments can be infrared, colorimetric, or based on one of a dozen other principles. Chemical or physical properties can be measured. These instruments range from handheld to continuous, permanently installed monitors.

Portable detector tube units provide measurements of low concentrations of gases and vapors. Most of the tubes have direct reading scales. The length of the discoloration in the tube provides the concentration reading. Some tubes contain chemicals that change color; therefore, these require a color comparison for the reading.

Manufacturers have tubes for a hundred or more specific analyses. Tubes have been certified in some cases by the National Institute for Occupational Safety and Health (NIOSH).

For carbon monoxide, hydrogen sulfide, and a few other toxic materials, small badges and stickers (dosimeters) that change color as a function of total exposure are available.

Personal alarm units can provide considerable protection against toxic materials. These devices record or continuously integrate toxic levels to provide a time-weighted average (TWA) exposure number. An audible alarm occurs if a preset threshold limit value (TLV) is exceeded at any instant.

The most common type of combustible gas detector utilizes the catalytic combustion principle. A platinum wire filament (in a Wheatstone bridge circuit) is usually mounted in a diffusion-sensing head. Sulfur compounds and halides can prevent proper operation of the sensor. For flammable gases and vapors, the technique warrants serious consideration for intermittent and continuous monitoring.

Thermal Conductivity Leak Detectors

Thermal conductivity leak detectors have cells that contain coils in bridge circuits. Heat dissipation increases with the concentration of the gas or gases in the sample, and the cooling effect changes coil resistance. The instrument will detect many different gases and has a wide dynamic range.

A halogen-bearing sample produces ionization and current flow in a special cell. Cell life is shortened by high halogen concentrations. Although this instrument type is very valuable for checking refrigeration systems and for certain kinds of production testing, it is of less value in many industrial applications. The cell runs very hot, so the available detectors cannot be used in flammable atmospheres.

For leak detection, the mass spectrometer is usually made sensitive to helium. The helium acts as a tracer gas, inside or outside of the item that is being tested. These expensive instruments would be difficult to use in a chemical plant but have considerable value for production (assembly line) leak testing.

Acoustic Emission

Fluids escaping from openings generate sonic and ultrasonic waves. The acoustic emission technique has some advantages. Leaks can be detected at a distance. One vendor's explanation is that a 10-PSIG (68.9 kPa) gas leak might be detectable at 50 ft (15 m), but would be detected at only 10 ft (3 m) if the pressure was 1 PSIG (6.89 kPa). The method can be considered for the location of leakage from buried pipes and tanks. Although the technique has limitations for low-pressure gas problems, it can detect the "gurgling" of leaks from sewers and other low-pressure liquid lines.

Other Instrumented Approaches

A wide variety of other techniques have been developed:

1. Conductivity—Checking the integrity of the walls in glass-lined vessels and monitoring heat-exchanger tubes for leakage. (A nonconductive liquid is checked for contamination by a conductive fluid.)
2. Optical density—Detection of broken bags or other leaks in dust collectors.
3. Float switch—Seal leak detector.
4. Flow switches
5. Flow meters—Mass flow in and out of a system.
6. Moisture—RF power loss method to spot leaking or broken packages.
7. Special cable—Cable degrades and initiates an alarm when contacted by petroleum derivatives.
8. Computer—Flow balance algorithm.

Thermography

This technique involves the detection of temperature differences by scanning in the infrared region of the spectrum. The result is a thermal picture that is usually two colors, but sophisticated systems can convert the measurements into graded colors. If a hot fluid is escaping, it can be detected at a distance, above or below ground, or under insulation. Leaking valves and steam traps are easy to find.

Thermography has already seen many other applications that only involve thermal energy. It has great value for the inspection of insulation and refractory materials. As a maintenance tool, it can be used to check heater tubes and electrical equipment.

BIBLIOGRAPHY

Hammock, A.A., "Operate Large Plants Safely," *Hydrocarbon Processing,* April 1979, pp. 263–270.

Kaplan, H. and Leftwich, R.F., "A Guide to Infrared Temperature Measurements," *Instruments and Control Systems,* January 1978, pp. 33–35.

National Fire Protection Association, "Underground Leakage of Flammable and Combustible Fluids," NFPA-329, 1977.

Nielsen, C. and Powers, J., "Varied Uses Found in Chemical Plant for Infrared Thermography," *Chemical Processing,* December 1977, pp. 116–117.

Prellwitz, S.B., "Thermal Imaging Techniques Applied to Solving Steel Plant Problems," *Iron and Steel Engineer,* March 1976, pp. 59–63.

Schaeffer, J., "Use Flammable Vapor Sensors?" *Hydrocarbon Processing,* January 1980, pp. 211–220.

9.7 METAL DETECTORS

Operating Temperature:	−40 to 120°F (−40 to 49°C)
Operating Pressure:	Atmospheric
Materials of Construction:	Metallic support structure, aluminum and/or plastic housing, for search-head and electronics circuitry
Cost:	$30 to $400
Inaccuracy:	Variable; very strongly influenced by surrounding medium and skill of operator
Range:	1 to 7 in. (2.54 to 17.78 cm) for 0.3-oz (9.3 g) of nonferrous metal objects, and shorter distances and smaller weights for ferrous objects. The distance from the search coil to the object is in direct proportion to the size of the object.
Partial List of Suppliers:	Applied Electronics Inc.; Edmund Scientific Co.; Gardiner Electronics; Goring Kerr Inc.; Lafayette Radio Electronics Corp.; Radio Shack, a Tandy Corp.; White Electronics

Metal detection in a process stream is an unusual instrumentation concern. The more likely solutions to detecting aggregates in a process stream, depending on the size of the particulates, would be mechanical filtering, mass flow measurement, or densitometry. For the latter two, state-of-the-art electronics and/or microprocessors could facilitate rapid and accurate detection of small changes in the density (mass) of the process stream.

A quality metal detector would be used to locate underground piping or cabling needed for and those which may interfere with the construction of new plants or expansion of existing facilities (e.g., gas and oil pipelines or electric utility and telephone cables). The highest sensitivity detectors, also the more expensive ones described here, would be capable of indicating a pipeline 10 to 20 ft (3 to 6 m) or more below ground level.

Additional use for the metal detector is found in the industry by geologists seeking metal and mineral deposits, in the consumer market by hobbyists "beach combing" for coins, jewelry, and so on, and in science by archeologists.

Metal detectors operate with magnetic fields, where the magnetic circuit of the search coil is influenced by the differential conductance of the medium. The greater the difference between the conductance of the medium (sand, water, air, etc.) and that of the minerals or metallic objects, the better the sensitivity of the detector; that is, smaller objects are detectable from a greater distance.

Two types of detectors are used: the transmit–receive (T/R) and the beat frequency oscillator (BFO). Both utilize the magnetic field of a search coil and low or very low frequencies (VLF) of 1 kHz or lower.

The transmitter and receiver coils are located in the search head for T/R types. The transmitter coil—in reality an antenna—sets up a magnetic field which is altered by metallic objects near the head, and the perturbation detected by the receiving coil is processed for audio and analog output.

For the BFO type, the search coil in the head is driven at the frequency of operation and an internal reference frequency signal is maintained. The operating frequency changes if a metallic object is near enough to the coil and the change (compared with the reference frequency) is converted to the target identification signal.

All models are battery operated. Cost variations between models are due to the power output available from the driving amplifier, the size of the battery pack, water and/or saltwater proofing, oscillator stability, and miscellaneous features.

With considerable practice, differentiation is possible

between metal objects of various sizes as well as materials of different compositions.

The response of all models is deemed too slow to detect metal fragments in a flowing process stream, where other means, such as density measurements, may be faster and more accurate, but not as inexpensive.

BIBLIOGRAPHY

Consumer Research, Vol. 62, No. 4, April 1979, p. 17.

Edmund Scientific Catalog, 1979.

Mueller, K. von, *Treasure Hunter's Manual*, Dover, New York, 1978.

White's Electronics Corporation Catalog, 1980.

White's Manual Coinmaster, 5000/D, G.E.B. Instruction.

Table 9.7a
ORIENTATION TABLE FOR METAL DETECTORS

Operating Mode	*Detection Range**		
	Medium	*Object*	*Distance, in. (cm)***
Transmitter–receiver	air	dime	3–6 (7.5–15)
	water	dime	4–7 (10–17.5)
	sand	dime	6–7 (15–17.5)
Transmit–receive, very low frequency	air	dime	6–8 (15–20)
	water	dime	6–8 (15–20)
	sand	dime	6–7 (15–17.5)
Beat frequency oscillator	air	dime	6–7 (15–17.5)
	water	dime	6–7 (15–17.5)
	sand	dime	6–7 (15–17.5)

*Range or distance to target depends on operator skill.
**Instrument sensitivity, i.e., the maximum distance at which a specific target can be detected, is determined by the size of the object, the purity of the metal, the medium, and the output power of the unit.

9.8 NOISE SENSORS

Types:	A—Microphones, for sound measurement in air B—Hydrophones, for sound measurement in water
Ranges:	Usually from about 20 to 20,000 Hz; for special purposes, 0.1 to 40,000 Hz. Usually from 30 to 140 dB
Cost:	Wide range: general purpose, \$150 up; instrument grade, \$300 and up
Partial List of Suppliers:	A—Acoustic Instruments International Inc.; Altec Sound Products Division; Bolt, Beranek, and Newman; Brüel and Kjaer; Columbia Research Laboratories; Electro Voice, Inc. ENDEVCO, Dynamic Instrument Div.; Prod Inc.; B—Electrodynamics Div., Bendix Corp.; Massa Corp.

Nature of the Measurement

Acoustic noise (sound) is defined as an oscillation in an elastic medium, occurring within the frequency range to which the human ear is sensitive: that is, from approximately 20 to 20,000 cycles per second (Hz). The elastic medium is usually air, and occasionally a liquid such as water. Sound waves occur as pressure waves or as particle-velocity or particle-displacement waves; any of these three waveforms can be measured. In water, because of its limited compressibility, only pressure waves are usually measured.

The pressure variations to be measured are small. At the threshold of audibility for the normal human ear, the root mean square value of pressure is about 20 μN/m^2; at the threshold of feeling for the normal ear, the pressure will be perhaps 200 N/m^2—10 million times as great, yet still only one one-thousandth of "atmospheric pressure." Consequently, the effect of steady pressure must be eliminated; the sensing device does not respond to it.

Because of the extremely wide range of sound pressures which may be measured (far more than the 10 million-to-one ratio just mentioned), a scale of logarithmic units is used to describe sound pressure levels; that is, the measured pressure is compared with a standard pressure (that at the threshold of hearing, 20 μN/m^2) using the following formula:

$$SPL = 20 \log_{10} (P_m/P_{ref}) \text{ dB re } 20\ \mu\text{N/m}^2 \qquad 9.8(1)$$

The symbol *SPL* indicates *sound pressure level*, P_m is measured pressure, and P_{ref} is the reference pressure. The *decibel* (dB) is the logarithmic unit which describes sound levels.

Transducer Principles

Conversion of pressure variations into an electrical output may be accomplished by several means; the most popular of these are inductive, dynamic, electrostatic, piezoelectric (or electrostrictive), and resistance variation.

The inductive principle uses the variation of reluctance in a magnetic circuit; an iron reed (or a suspended iron armature) is made to vibrate by the sound waves and thus to change the air gap in a magnetic circuit. A coil of wire is placed around an element of that magnetic circuit, and an electromotive force appears at the terminals of the coil, its waveform represents the waveform of the sound. Inductive or "magnetic" microphones are used most for low-quality but high-level applications; they are not usually very precise devices.

The moving-conductor principle—a conductor moving (and cutting lines of force) in a magnetic field—is used effectively in microphones; one form is the ribbon microphone. Here a very thin, lightweight metal ribbon is suspended in a magnetic field (see A in Figure 9.8a). Sound waves striking the ribbon cause it to move, thus generating an electromotive force which is fed to an amplifier input. The ribbon is extremely thin and light; it effectively follows the motion of the air particles. Its velocity is thus approximately equal to the velocity of the air particle motion, since the entire area of the ribbon is exposed to the sound waves and moves in unison with them.

Dynamic microphones also employ the moving-conductor principle: a lightweight diaphragm carries a coil of wire which moves within a magnetic field, as shown at B in Figure 9.8a. Sound waves strike the diaphragm; their pressure causes the coil to move and an electromotive force is produced at the coil terminals and fed through a suitable transformer to an amplifier. These microphones are pressure actuated, since the diaphragm (and the coil) moves in response to sound-wave pressure.

Capacitor (condenser) microphones depend upon a fundamental principle in electricity: when some amount of electrical charge is held within a capacitor, the electrical potential between the condenser plates will change if the capacitance is changed. Thus, in a capacitor consisting of two plates, one of which is movable—as at C in Figure 9.8a—movement of one plate will change the capacitance, and the potential between the plates will change in response. A change in air pressure, caused by sound waves, actuates the capacitor microphone. The initial electrical potential may be supplied from a separate source through a high resistance; or the electrical charge may be provided by making either the backplate or the diaphragm of a material which has been given a permanent electrical charge (an electret).

The carrier-type capacitor microphone is structurally like that illustrated in C of Figure 9.8a, but its electrical circuitry is different. No polarizing potential is used; instead, the capacitance of the microphone is made a part of an electrical oscillating circuit, so that a change in capacitance causes a change in frequency—the circuitry converts this change to the kind of output desired. This permits operation down to steady pressures (zero frequency), which is not possible with the usual circuitry.

The piezoelectric microphone depends on a specific behavior of some crystalline materials—that a deformation of the crystal will cause electrical potentials to appear on the surfaces of the crystal; the magnitude of the potential is in proportion to the force that is causing the deformation. The assembly shown in diagram D of Figure 9.8a indicates a typical construction; a diaphragm receives the sound waves and applies their force to the crystal element, causing it to bend. The crystal element is usually an assembly of thin slices called a bimorph; this is an efficient and convenient form for the application.

In years past, most piezoelectric microphones used crystal elements made of Rochelle salt or of ammonium dihydrogen phosphate. Recently, various ceramic products have been developed with suitable characteristics for this use; they may be called *electrostrictive* rather than piezoelectric, but they serve the same function.

The most widely used microphone is the resistive device called the carbon telephone transmitter. Its characteristics make it unsuitable for most quantitative uses, but it is excellent for voice communication.

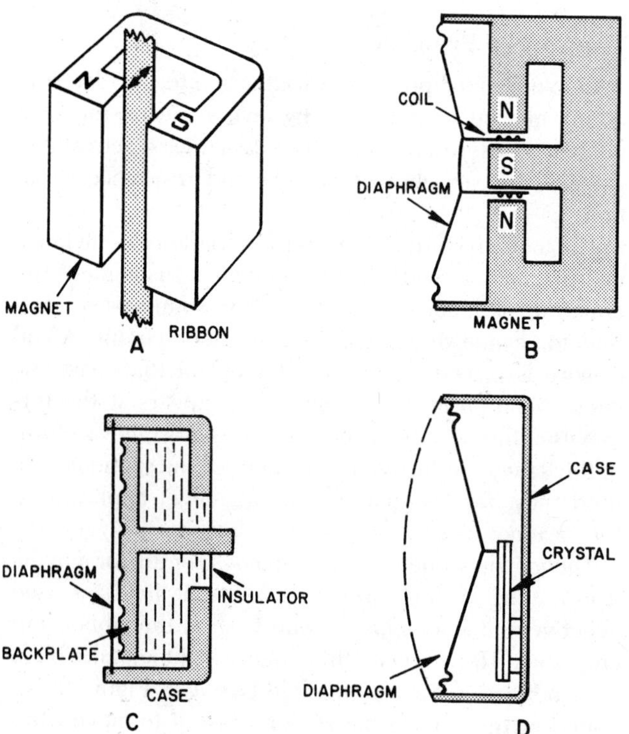

Fig. 9.8a Construction of microphones

Microphone Types

A useful way of classifying microphones is according to type of response: whether the instrument is actuated by velocity, displacement, or pressure of the sound waves. The distinction is not always entirely clear-cut.

The ribbon microphone (in its usual form) responds to velocity; the moving ribbon is very light and flexible and easily moved. Because of this, it needs no added mechanical damping and its motion is very nearly identical with the air particle motion, its electrical output is also proportional to velocity. But a ribbon microphone can be made into a pressure microphone. The sound waves may be made to pass through a pipe before they reach the ribbon, and this pipe may be so damped that it acts as an acoustical resistance, controlling microphone characteristics. This permits some desirable control over frequency sensitivity. The usual form of ribbon microphones does not offer this control; its damping is a result of its construction, and it is quite insensitive to even large changes in steady pressure.

Microphones that employ elastic diaphragms are nearly all pressure types, since the velocity of the coil motion (in the dynamic microphone), the diaphragm displacement (for the condenser microphone), and the crystal deformation (in the piezoelectric) are all in proportion to the applied force—the instantaneous sound pressure. For this to be true, the microphone structure must be properly designed; the elastic deformations must be linear, and there must be suitable damping. Damping is usually provided by introducing air leakages (through small tubes which add the proper acoustical resistance) from the interior of the microphone to the open air. Damping in the condenser microphone is difficult because the diaphragm is very close to the backplate; grooves in the backplate solve that problem.

Pressure-gradient microphones are sensitive to the difference in pressure between two points separated by a finite distance; thus, they provide a form of velocity response. They are usually special-purpose devices; they can be given some very specific directional characteristics.

When extremely high directivity is required, other special forms are available: the most popular of these are the parabolic reflector microphone and the line-type microphone. The line microphone uses an array of pickup tubes of varying length; the sound that reaches the microphone through these tubes adds vectorially, making high directivity possible, though uniformity of frequency response suffers. In Figure 9.8b the array is sketched and the general type of frequency response is shown. Microphones of these two types are often called wave microphones.

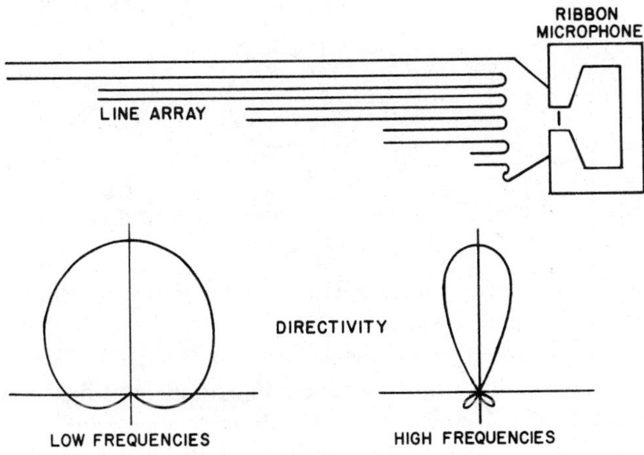

Fig. 9.8b A line microphone

Some special-purpose sound-measuring devices have been developed, which are sensitive to the energy of sound but not to delineating waveform: the hot-wire microphone, the Lindval glow microphone, and the Rayleigh disk are examples. They do not seem to be applicable, in general, to industrial use. The Rayleigh disk, however, is still a practical device for calibration.

Ultrasonic microphones form a special class; they respond only to frequencies above the hearing range, that is, about 25,000 to 45,000 Hz. They are widely used in remote-control devices, gas-leak detectors, guidance devices (as for the blind), and the like. They are usually piezoelectric units; the crystal may be made resonant at some frequency.

Microphone Characteristics

In selecting a microphone, the user is seldom particularly interested in the physical principles involved in its construction; he is interested in its operating characteristics. Table 9.8c serves as a general guide to these characteristics.

The sensitivity of microphones can be given in volts per unit of pressure; a typical condenser microphone might produce about 10 mV for a sound-pressure input of 1 N/m^2. Trade practice often describes microphone sensitivities in terms of decibels referred to 1 $V/dyne/cm^2$. (The example just given would be −40 dB.) One dyne per square centimeter is about 74 dB on the usual scale of sound-pressure levels. Amplification is always needed; amplifier characteristics must be suited to the microphone characteristics.

The direction from which sound waves reach a microphone usually has some effect on the output produced. A "pressure" microphone should measure pressure at a point; the smaller the microphone size the nearer the approach to this condition. At high frequencies, an error may appear; if, for example, the diameter of the microphone diaphragm is equal to one wavelength of the sound, a series of waves striking the diaphragm at nearly grazing incidence may cause almost no net effect—a far different result from that produced by the same series of waves striking normal to the surface (see Figure 9.8d). Typically, sound in a room is directed diffusely; that is, it comes from all directions. If high accuracy is required at frequencies above about 10,000 Hz, this must be considered.

In general, microphones are classified as omnidirectional, unidirectional, or bidirectional; these classifications are only approximately accurate, but the terms are useful guides. An omnidirectional microphone (see A in Figure 9.8e) displays nearly uniform sensitivity to sound coming from all directions; one would probably use such a microphone for measuring sound levels in a room. Bidirectional response (see B in Figure 9.8e) is of great value in such instances as broadcasting, and live stage performances. Bidirectional response is most often provided by a ribbon microphone with both sides of the ribbon exposed to the sound. If only one side is exposed, the response pattern becomes cardioidal, as shown in C of Figure 9.8e.

The possible frequency range of a microphone depends on both its operating principle and its construction. Nearly all modern microphones are quite flat over the usual range of the human ear (see Table 9.8c). Dynamic microphones and diaphragm-type crystal microphones depend on good design and construction, including proper damping, to insure uniform response with

Table 9.8c
A USER'S CLASSIFICATION OF MICROPHONES

Type of Microphone	*Frequency Response Range, Hz*	*Output, dB**	*Typical Characteristics*	*Typical Use*
Carbon telephone transmitter	300–4,000 (useful range)	−25 to −45	Inexpensive, with high output in the speech frequency range	Telephone
Capacitor microphone	12–15,000	−48	Extremely stable, wide, and flat frequency response	Measurement of sound level
Carrier-type capacitor microphone	0.1–20,000	depends on auxiliary unit	Widest possible frequency response; uses auxiliary electronics, with any of several microphone units	Measurement of sound in extreme conditions
Crystal microphone	30–12,000	−65	Good frequency response, usually semicardioid pattern. Often temperature sensitive	Public address, recording
Cardioid microphone	20–12,000	−80	Good frequency response, cardioid or "unidirectional" pattern	Public address, recording, etc.
Ribbon microphone	20–15,000	−85	Good frequency response; can be used either as pressure or velocity type	Live performance, recording
Wave microphone	80–8,000	−80	Can be highly directional, because of construction	Broadcasting, special uses

*Microphone sensitivity in terms of 0 dB = 1 V/dyne/cm².

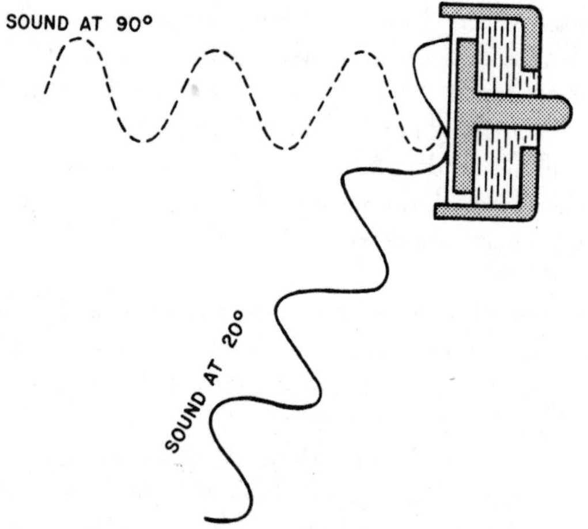

Fig. 9.8d Normal incidence versus grazing incidence for sound waves

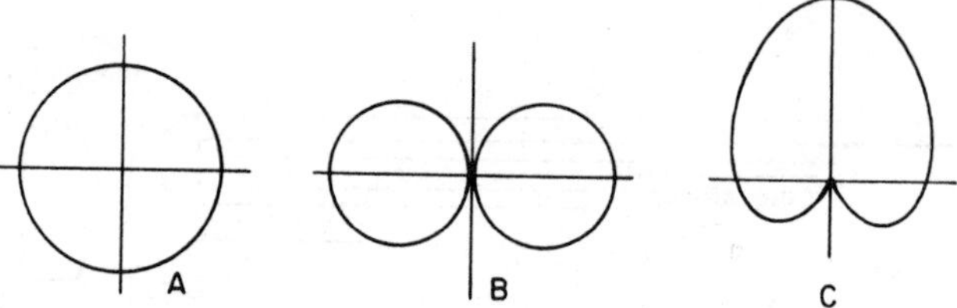

Fig. 9.8e Typical directivity patterns for microphones

no noticeable resonance peaks. Condenser microphones have a much smaller moving mass, as do ribbon microphones, and thus fewer mechanical problems; their frequency limitations are more likely to appear in the electronics associated with them. Manufacturers' data usually describe these factors adequately.

The amplitude range which a particular microphone can accept depends on its construction; pressures that are too high can cause mechanical damage. Most microphones can tolerate the maximum acceptable to the human ear—perhaps a 140-dB sound-pressure level—but special forms are available which can be used successfully at much higher levels.

For use under water, special forms are available. These are microphone units installed in strong housings arranged in such a way that the vibratory components of pressure may enter, but, because of the depth of immersion, the steady component does not. These are usually called hydrophones; they require special cables and often special electronics.

With modern microphones, the stability of microphones—the constancy of the rates of output to input at all frequencies and under all conditions—is usually more dependent on environmental considerations than on elapsed time.

Environmental Considerations

Microphones may be exposed to extremes of temperature and humidity, mechanical shock, magnetic fields, and other factors. Any of these can affect sensitivity and frequency response, either on a temporary or a permanent basis.

Protection from mechanical damage is obviously desirable; the rather delicate mechanical suspensions can be damaged, crystals cracked, etc. Dust can clog the equalizing openings between the interior of the microphone and the outside, thus impairing the frequency response. Reasonable care should be taken to protect this moderately delicate equipment from such things.

Dynamic microphones contain permanent magnets; exposure to strong magnetic fields could weaken these and change the sensitivity of the microphone. If the microphone is used in magnetic fields, there could be noise pickup by either the microphone or its cable. The same is true for ribbon microphones; they may be even more vulnerable to pickup of alternating current. Alternating current fields should be avoided in the use of any such equipment.

Crystal microphones are not usually greatly affected by temperature within reasonable ranges, though this was not true of earlier models. They can, however, pick up noise from electric fields.

Condenser microphones are usually extremely stable for most conditions, but moisture can seriously impair their response. They contain high-quality insulating materials, but they are not entirely unaffected by humidity. Their electronic auxiliaries also may be vulnerable to environmental effects.

It is always possible that microphone characteristics may change. However, modern microphones are extremely good. The manufacturer's information is usually a reliable guide.

One environmental effect—wind—is extremely troublesome. Wind produces microphone noise; this can be reduced by windscreens installed on the microphone.

Calibration at appropriate intervals will detect, and perhaps correct, most changes that occur over a period of time.

Calibration

Microphones may be supplied with nominal sensitivity values, but microphones intended for measurement are usually supplied with detailed calibration data. The electronic equipment used with the microphone becomes a part of the calibrated system.

Precision calibration of microphones may be done by several procedures; this is beyond the scope of this discussion. The user is interested in day-to-day checks to establish that no perceptible change has occurred. Microphones may be given precision calibration at the necessary intervals by the manufacturer.

Portable calibration devices are available from most suppliers of microphones. They are usually small sound sources which can be used on individual microphones; it is important that the correct procedure be used. Here, again, the recommendation of the manufacturer should be followed.

BIBLIOGRAPHY

Beranek, L.L., *Acoustics*, McGraw-Hill Book Co., New York, 1954.

Gaylord, M.L., *Electroacoustics: Microphones, Earphones and Loudspeakers*, Elsevier, New York, 1971.

Olson, H.F., *Acoustical Engineering*, Van Nostrand Co., Princeton, N.J., 1957.

Tremaine, Howard M., *Audio Cyclopedia*, Howard W. Sams and Co., Indianapolis, 1973.

Flow Sheet Symbol

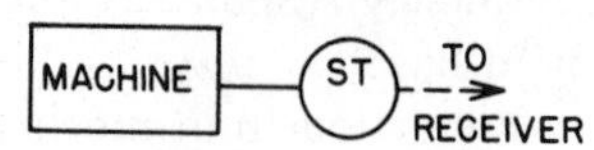

9.9 SPEED SENSORS

Operating Temperature:	To 200°F (93°C)
Cost:	\$500–\$800. Lower prices for switches
Range:	Virtually unlimited
Inaccuracy:	1–3%
Partial List of Suppliers:	Airpax Electronics, Inc.; Allen-Bradley Co.; J.G. Biddle Co.; Electric Tachometer Corp.; Delavan Electronics Inc.; Foxboro/Jordan Inc.; General Electric (GE) Co., Industrial Sales Div.; Jones Instrument Corp.; C.J. Maux Co.; Meriam Instrument Div. of Scott & Fetzer Co.; Metron Instruments, Inc.; Power Instruments Inc.; Weston Instruments, Div. of Sangamo Weston Inc.

Speed sensing devices are available in a multitude of designs using electromagnetic induction, voltage generation, light sensing, and magnetic force measurement to achieve their objective. Since devices employing the same principle vary in design from manufacturer to manufacturer, only the operating principles are discussed below.

Tachometer Generators

The ac tachometer generator is an electromechanical device very similar to a two-phase induction motor. It consists of a primary winding, a secondary winding placed at 90° (mechanically) to the primary, and a rotor. This arrangement is shown schematically in Figure 9.9a. With the rotor stationary and the primary winding excited by a sinusoidal voltage, the induced voltage in the secondary is almost zero due to the relative position of the two windings. However, as the rotor begins to turn, a sinusoidal voltage is induced in the secondary winding whose magnitude is in proportion to the rotor speed. Since the output signal is a voltage, the readout instrument used with this tachometer should have a high input resistance to give near-zero current flow in the secondary winding. Any current flow in the output winding will cause a voltage drop, which will be subtracted from the true measured voltage, and thus the speed measurement will be in error.

In the dc tachometer generator the magnetic flux is provided by a permanent magnet, and the output winding is located on the rotor. At zero speed there is no relative motion between magnetic field and winding, and the output voltage is zero. As rotor speed increases, so does the relative speed between magnetic field and winding and hence the output voltage. The voltage induced in the winding is sinusoidal; a commutator and brushes on the rotor convert the ac voltage in the winding to a dc output in the same manner as in the dc generator. The dc tachometer is shown schematically in Figure 9.9b.

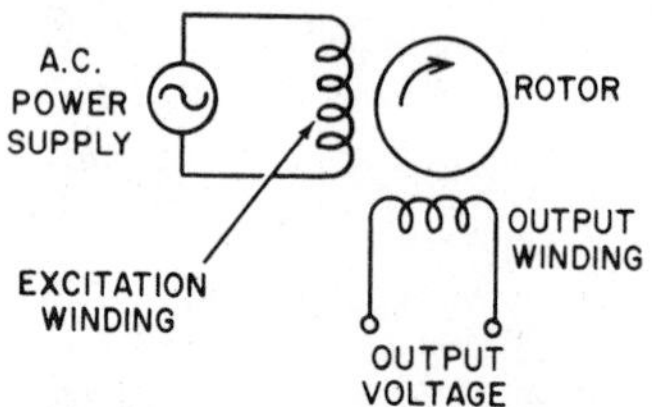

Fig. 9.9a Alternating current tachometer schematic

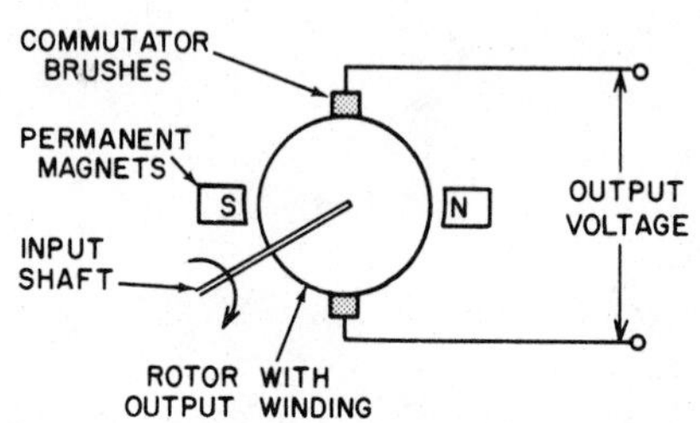

Fig. 9.9b Direct current tachometer schematic

Induction-type Sensors

Several configurations are possible with induction-type sensors. Basically, the unit consists of a coil located in a magnetic field. This magnetic field is made to vary as a function of speed. One design, shown schematically in Figure 9.9c, uses a coil wound on a permanent magnet and located in the proximity of a rotating toothed wheel of magnetic material. As the teeth of the wheel pass the magnet, they disturb the magnetic field, which in turn induces voltage pulses in the coil. The pulse frequency is related to rotational speed as expressed by the relationship rpm = $60f/N$, where rpm is rotational speed in revolutions per minute, f is the frequency in Hertz (cycles per second), and N is the number of teeth on the wheel. Figure 9.9d shows a design in which the magnet is mounted in the rotating member, thus inducing voltage pulses in the stationary coil.

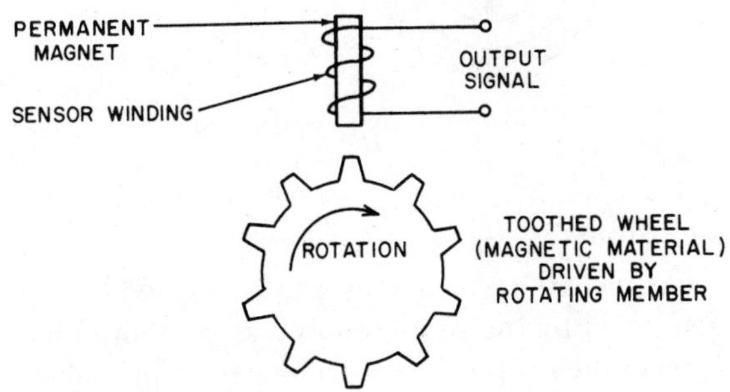

Fig. 9.9c Schematic of an induction-type speed sensor

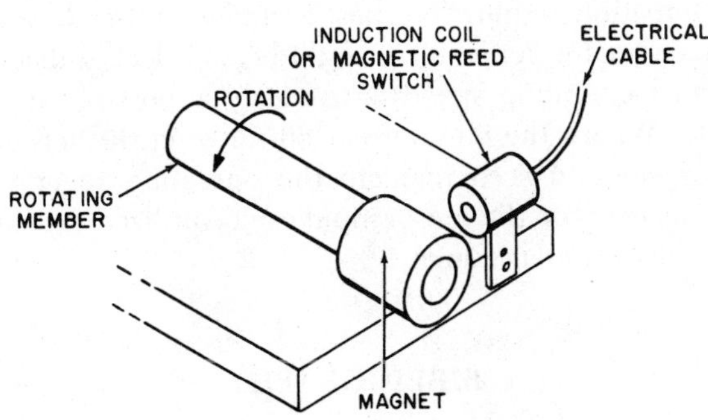

Fig. 9.9d Mounting arrangement for magnetic or inductive velocity sensor

Magnetic-type Sensors

Magnetic-type sensors (Figure 9.9e) utilize magnets to accomplish speed detection, as implied by the name. The sensor consists of a multipole magnet mounted on the rotating member and a stationary magnetic reed switch mounted in the proximity of the magnet. Passage of the magnet poles causes the contacts of the reed switch to open and close, thus making and breaking an electrical circuit. The output is thus a pulse train whose frequency is in proportion to the rotation speed.

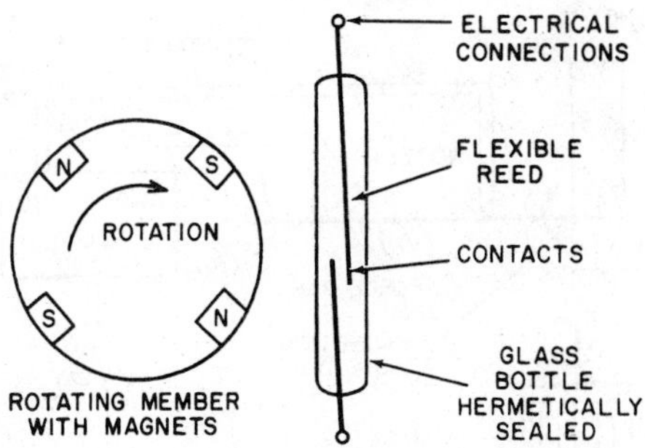

Fig. 9.9e Schematic of magnetic speed sensor

Photoelectric-type Sensors

In this design, a light beam is modulated by a rotating member, and the modulations produce current pulses in the output of a photocell. In the design shown in Figure 9.9f, dark and light areas on the rotating disk alternately absorb and reflect the light of the source beam. The variations in reflected light are sensed by a photocell which produces current pulses whose frequency is in proportion to the speed of rotation. Alternately, the light source and photocell can be located on opposite sides of a rotating, perforated disk.

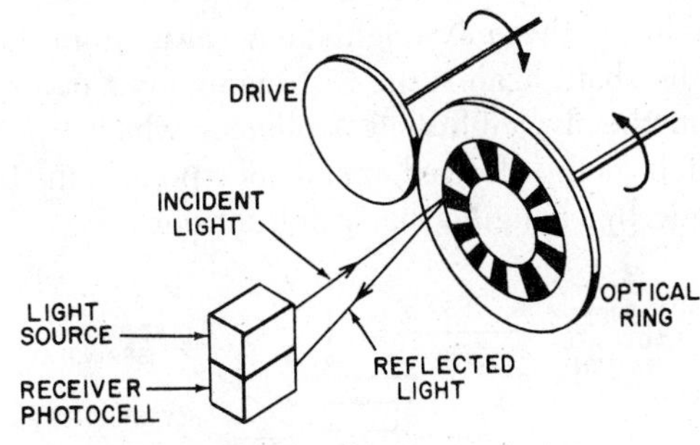

Fig. 9.9f Photoelectric speed sensor

Pneumatic Speed Transmitter

An eight-pole permanent magnet is mounted on the input shaft as shown in Figure 9.9g. Flexure strips hold the disk between the base plate and the magnet poles. As the input shaft rotates, magnetic lines of force cutting through the disk tend to rotate it. The force bar is subjected to two forces: the force resulting from the rotation of the magnet and the opposing force from the feedback unit. Imbalance of these forces results in motion of the force bar so that the air pressure at the nozzle changes. The pneumatic relay amplifies the nozzle pressure and produces a restoring force through the feedback unit until torque due to magnet rotation and restoring force of the feedback unit are balanced. Since the torque due to rotation of the magnet is in proportion to the speed of the magnet, feedback pressure is in proportion to speed. However, the feedback pressure is also the transmitter output signal, and hence the output pressure is in proportion to speed.

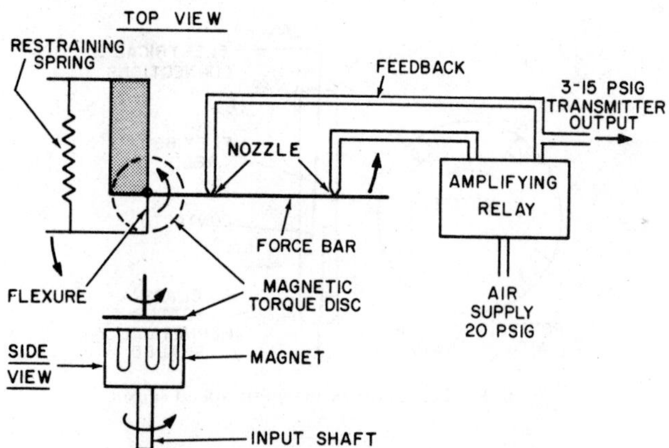

Fig. 9.9g Pneumatic speed transmitter

Speed Switches

Sensors with a continuous output can be provided with relays actuated by the measuring circuitry to provide a contact closure signal for an alarm or shutdown function. Where only a contact closure is required—to signal overspeed, low speed, or zero speed—devices with on-off output action may be sufficient.

One type of switch shown in Figure 9.9h consists of flyweights mounted on a rotating shaft coupled to the machine to be monitored. Under the influence of centrifugal force, these flyweights move away from the center of the shaft, against the restraining force of a spring. Toes on the flyweights lift a plunger which tips a microswitch at the desired speed. Set point adjustments are made by changing the spring tension.

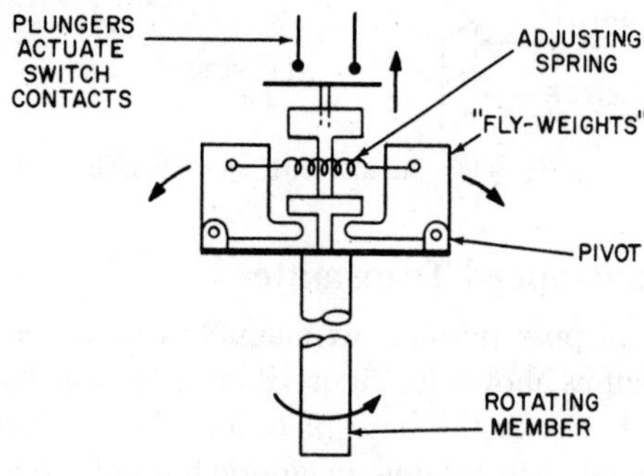

Fig. 9.9h Speed switch

However, the switch is nonindicating, and set point changes must be made on a bench at known speed or the speed of the machine must be measured independently. These switches can be provided with actuation on increasing or decreasing speed. A manual reset feature is available, which keeps the switch in the actuated position, regardless of machine speed, until it is manually reset. A switch to monitor zero speed is shown in Figure 9.9i. Relative motion between the coil and magnet produces a torque which holds the switch contacts in the operating position. As speed decreases, the electromagnetic torque decreases until the spring force is sufficient to overcome the torque and the switch contacts transfer. This switch will actuate only on decreasing speed.

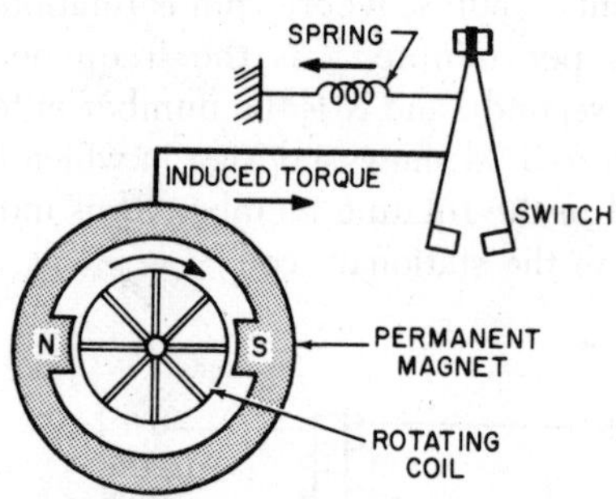

Fig. 9.9i Zero speed switch

Conclusions

Selection of the type of sensor to be used will generally be governed by the purpose of measurement. Tachometer generators or pneumatic tachometers for analog control have an advantage over other types since the output is already an analog signal. For digital control and readout, magnetic, inductive, and photoelectric sensors offer a compatible output. For best accuracy, full-scale speed of tachometer generators should be so selected that the normal operating speed is as close as possible to full scale. Where the inherent speed range of the sensor is insufficient to accommodate the operating speed, reducing gears on the sensor input can be utilized to extend the measurement range.

BIBLIOGRAPHY

"Accelerometer Survey," *Instruments and Control Systems*, November 1968.

Kaufman, A.B., "Velocity and Acceleration Transducers," *Instruments and Control Systems*, April 1971.

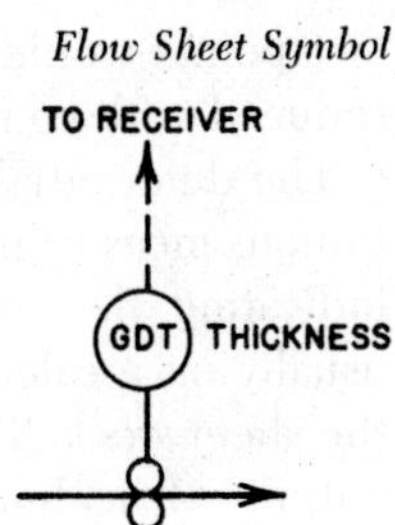

9.10 THICKNESS MEASUREMENTS

Applications: Metal foils, plastic films, web

Operating Temperature: 32 to 100°F (0 to 37°C)

Operating Pressure: Atmospheric

Cost: \$35 to \$45,000

Inaccuracy: ±10 μin. (±25 × 10^{-5} mm) for radiation-type gauges to ±0.001 in. (±0.025 mm) for micrometer calipers

Partial List of Suppliers: Bendix Corp., Automation and Measurement Div.; Electron Machine Corp.; Federal Product Corp., an Esterline Co.; Hitec Corp.; LFE Corp.; Lion Precision Corp.; Mechanical Technology Inc.; Nortec Corp.; Ohmart Corp.; L.S. Starrett Co.; Twin City Testing Corp.

The thickness of web, film, foil, and sheet materials is monitored and controlled during manufacturing by the use of thickness measuring devices. The devices may be those which compare the thickness to a reference length or those which indicate the thickness based on some physical relationship. (Compare specific versus nonspecific measurements.) The former devices are generally considered contacting types whereas the latter are typically noncontacting types. The final reading is an averaged indication due to either mechanical or electronic damping of spurious signals. The cost of a specific instrument depends on its sensitivity, which determines the complexity of the device. Table 9.10a reviews the various available instruments.

Contacting-type Gauges

The familiar micrometer or vernier caliper gauges are only suitable for noncontinuous sampling or batch quality control and calibration measurements, because the con-

Table 9.10a
THICKNESS MEASUREMENTS

Instrument	*Type**	*Sensitivity***	*Application*****
Micrometer caliper	C	0.0001 in	Sampling, quality control, calibration, foils, web, sheet, film materials (N)
Differential roller gauge	C	20 μ in.	Low-speed continuous foils, web, sheet, film materials, calibration, sampling (Y)
Sonic gauge	C	0.01 in.	Rigid, relatively thick sheets, accessible from one side only (N)
Capacitance gauge	NC	0.001 in.	Insulating sheets, films (Y)
Radiation gauge	NC	50 mg/cm²***	Metal foils and plastic films (Y)

*C = contacting; NC = noncontacting.
**1 in. = 25.4 mm; 1 μ in. = 2.54 × 10^{-5} mm.
***Actual sensitivity depends on the specified value for a given instrument divided by its density (g/cm³).
****Whether a device is suitable for continuous process instrumentation is indicated by N (no) or Y (yes).

tact points would wear out and the accuracy would be reduced with continuous use.

The differential dial gauge adapts the calipers to continuous measurement by using rolling contact points and indicating the difference between a reference wheel usually on a calender roll and the measuring wheel on the sheet stock. The thickness signal is derived from the output of a linear variable differential transformer (LVDT) (Figure 9.10b). The difference between the secondary voltages caused by displacement of the movable iron core (armature) is linearly proportional to the displacement. With the proper power source and input mechanism dimensional gauging from 0.001 in. to several inches (0.0254 mm to several centimeters) is possible. The accuracy is independent of the finish of the calender roll. A single roller dial gauge contacting the stock would rely on the accuracy of the backing roll for overall accuracy.

Sonic gauges are used to measure the thickness of materials accessible from one side only (i.e., the wall of a pressure vessel or a storage tank). The wall is made to resonate by varying the applied ultrasonic frequency. The resonant frequency is then used to calculate the thickness from the relation

$$t = 0.5\ v/f \qquad 9.10(1)$$

where

t = thickness of the material being gauged (ft) or (m)
v = velocity of sound in material (ft/sec) or (m/sec)
f = frequency of resonance (Hz)

The device is generally not suitable for dynamic reading of moving material because accurate dimensioning requires intimate contact between the generator crystal and the surface of the material.

Under static conditions thickness from 0.01 to >10 in. (0.254 to >254 mm) may be measured with very great accuracy.

Noncontacting Gauges

Certain properties of materials may be used to indirectly indicate the thickness of films or sheets of those materials.

For insulating films, the capacitance gauge may be used (see Figure 9.10c). Because the capacitance varies directly in relation to the thickness of the dielectric material between the capacitor plates, it is possible to measure thickness of insulating films if the dielectric constants are a great deal different from that of air. The capacitor and the material to be measured are part of an

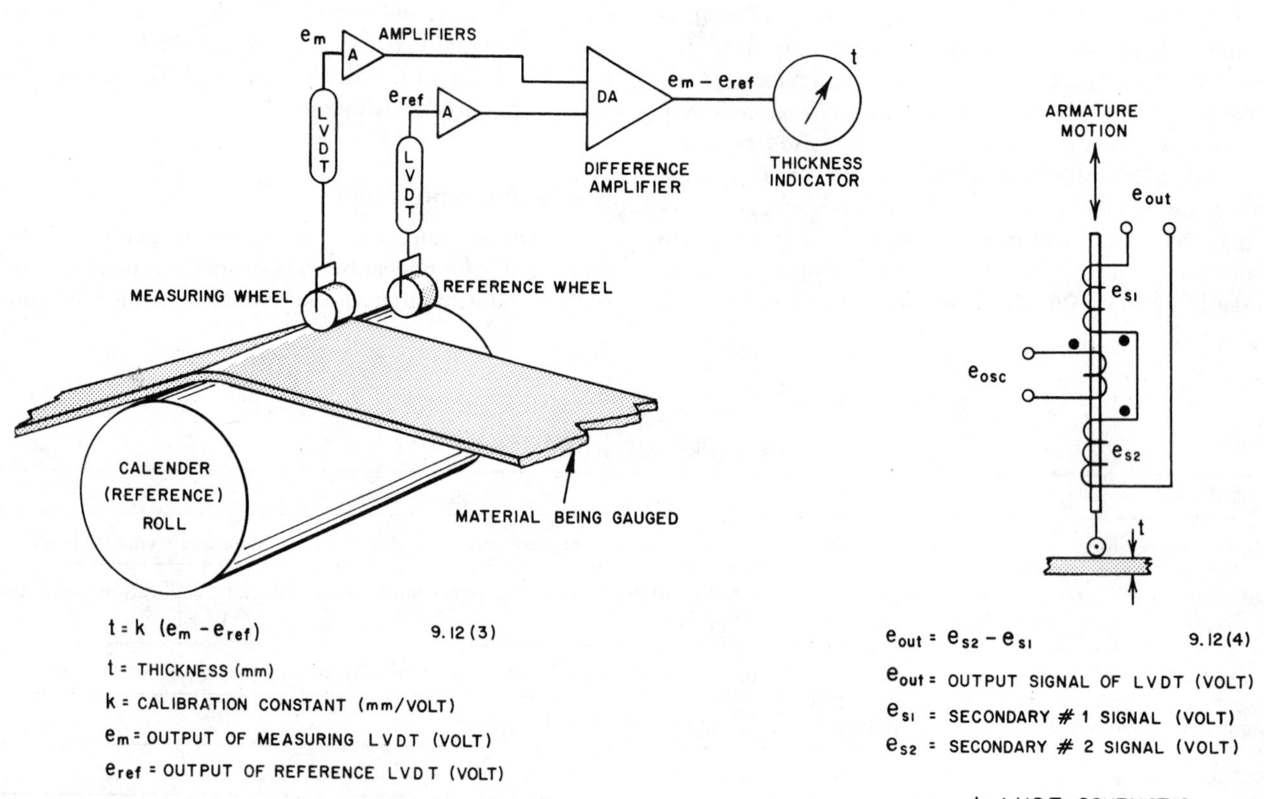

Fig. 9.10b Differential roller gauge with linear variable differential transformers (LVDT)

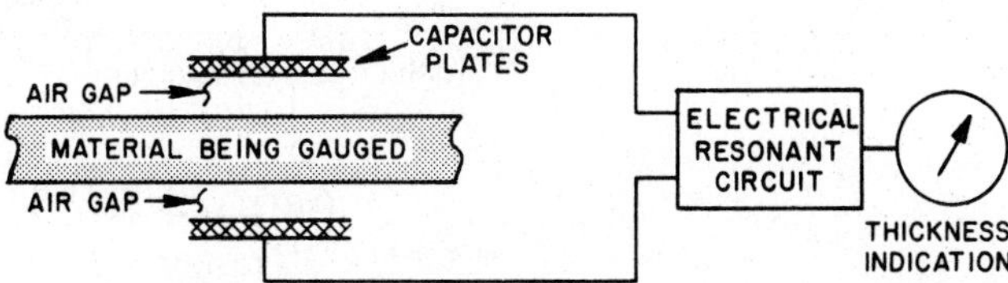

Fig. 9.10c Exaggerated sketch of the capacitance gauge

electrical resonant circuit whose output is calibrated to indicate thickness. Moisture content of the material and/or air gap between the plates and the material are the primary sources of error for this instrument. (The principle is applied to moisture measurement.) It also is not usable for conductive films.

The attenuation of radiation from x-rays or radioactive decay by matter is utilized in the radiation absorption gauge to measure the thickness of the material. The equation is

$$\Delta I = I_o [1 - \exp(-\mu t)] \qquad 9.10(2)$$

using averaged ionization current for signal, where

ΔI = change in ionization current when absorber is inserted
I_o = ionization current without absorber
μ = absorption coefficient ($cm^2/\mu g$)
t = thickness ($\mu g/cm^2$)

The display is calibrated to indicate thickness.

Radiation gauges are subject to errors from the statistical nature of radioactive decay and from the dependence of the absorption coefficient on the composition of the material being measured (see Figure 9.10d).

For all thickness gauges, but particularly for the nonspecific, noncontacting types, frequent calibration

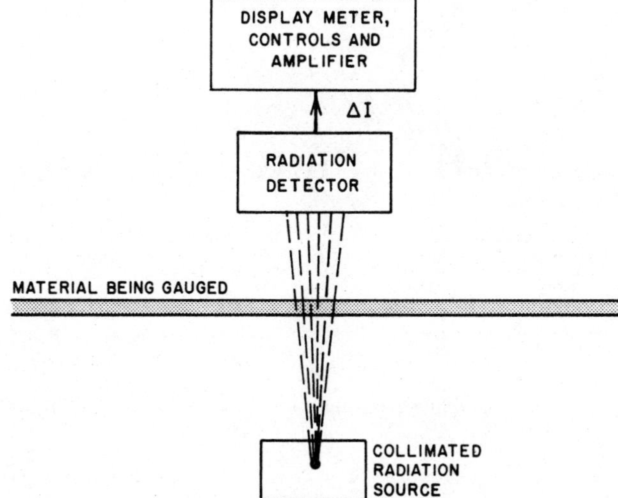

Fig. 9.10d Radiation absorption gauge

checks with samples of the material to be gauged are recommended. Specific or contacting-type gauges are usually calibrated against precision standards.

BIBLIOGRAPHY

Accoustics Handbook, Application Note 100, Hewlett-Packard Company, 1968.

Baumeister, T., ed., *Mark's Standard Handbook for Mechanical Engineers*, 8th ed., McGraw-Hill Book Co., New York, 1978.

Considine, D. M., ed., *Handbook of Applied Instrumentation*, McGraw-Hill Book Co., New York, 1964.

Fink, D. G., ed., *Electronics Engineers' Handbook*, 1st ed., McGraw-Hill Book Co., New York, 1975.

Chirons, N. P., ed., *Machine Devices and Instrumentation*, McGraw-Hill Book Co., New York, 1966.

Flow Sheet Symbol

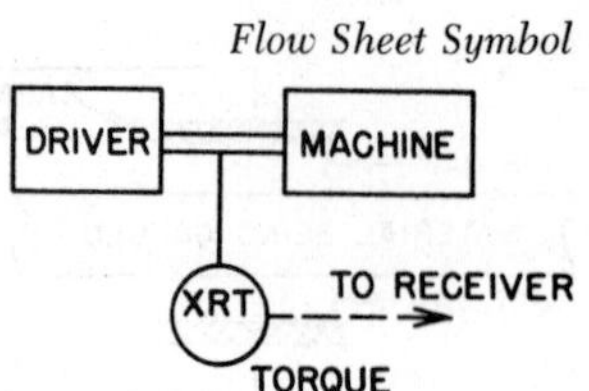

9.11 TORQUE TRANSDUCERS

Operating Temperature:	−60 to 300°F (−50 to 150°C)
Cost:	\$1000–\$6000; higher prices reflect extremely low or high ranges
Inaccuracy:	1–5 percent
Partial List of Suppliers:	Advanced Mechanical Tech. Inc.; W.C. Dillon & Co., Inc.; S. Himmelstein and Co.; Indicon Co., Inc.; Lebow Associates, Inc.; Tensitron, Inc.; Vibrac Corp.

Torque is defined as that force which tends to produce rotation. Specifically, it is the moment due to a tangential force, $T = LW$, where T is torque, L the length of the force arm, and W is the force.

All torque sensors consist of two basic components, a structure designed to deform microscopically under applied torque, and a device to convert the deformation to a measurement signal. Basically, there are two types of torque sensors available for torque measurement on process machines. These are the in-line rotating torque sensors and the in-line stationary sensors. Reaction torque sensors, those that measure the torque on a machine housing rather than on a shaft, are generally not suitable for process measurement because connecting power and pipe lines affect the measured torque.

Rotating Sensors

The in-line rotating sensor consists of a metal shaft with bonded strain gauges electrically connected in the form of a Wheatstone bridge. Figure 9.11a illustrates the stresses acting on a rotating shaft subject to torsion.

In one direction, at a 45° angle to the axis, pure tensile stress exists, whereas 45° in the other direction pure compressive stress is extant. The rotor shaft is elastic and will deflect minutely under the imposed stresses. The strain elements are located on the shaft to sense compressive and tensile deformation due to torsion. The Wheatstone bridge output is in proportion to torsion and hence to torque. Bridge power and output voltage are connected to the sensor through slip rings and brushes. This type of pickoff is limited to rotational speeds in the order of 100 ft/sec (30 m/sec) at the brush surfaces. For very high speeds, the rotating sensor can be provided with mist-lubricated bearings and air-cooled brush surfaces.

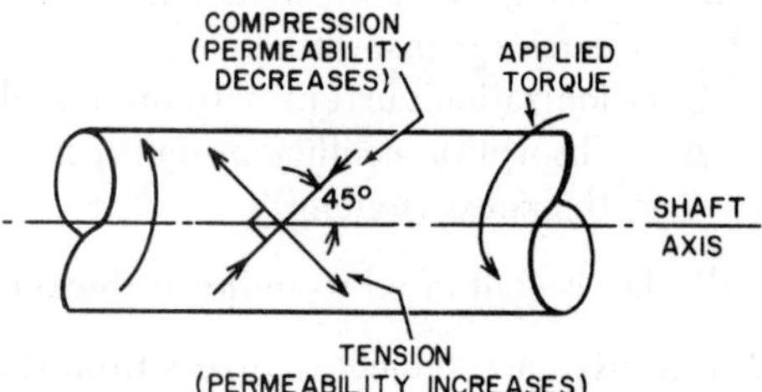

Fig. 9.11a Tension-compression stresses on surface of a circular shaft

The schematic drawing of a sensor with noncontacting power supply and signal pickoff is shown in Figure 9.11b. The bridge power and output signals are transmitted between the rotating and the stationary members through transformers. The bridge power is a constant amplitude, high-frequency sine wave, and the output is a sine wave of the power frequency but whose amplitude is a function of torque. Consequently, the power supply

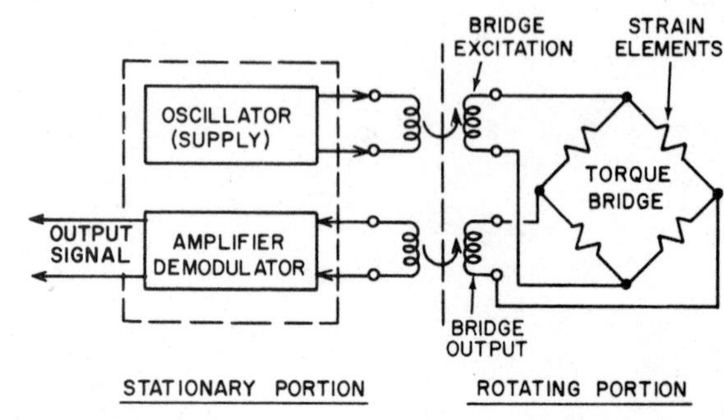

Fig. 9.11b In-line rotating torque sensor

electronics must include an oscillator to generate the carrier frequency. The output electronics includes a demodulator to produce a dc signal in proportion to the peak of the output sine wave. This sensor is superior to the slip ring design since contact resistance, contact friction, and heating are absent and cannot affect the measurement.

Stationary Sensors

Several designs of stationary sensors are possible. One type utilizes the Wiedemann Magnetostriction Effect to achieve a torque measurement. This effect causes changes in the permeability of the materials subjected to tensile or compressive stresses. Permeability is a magnetic property related to the ability of the material to concentrate magnetic flux. Permeability increases under tensile and decreases under compressive stress. The sensor consists of two primary and two secondary windings mounted near the rotating shaft as shown in Figure 9.11c. The secondary coils, S_1 and S_2 are coupled by magnetic lines of flux through the shaft to the primaries. With no loading on the shaft, the permeability of the shaft is uniform, and equal but opposite voltages are induced in S_1 and S_2. With the shaft under torsion, permeability and the number of magnetic lines of flux increase in the direction of tension and decrease in the direction of compression. The voltages induced in S_1 and S_2 do not cancel and their algebraic sum is in proportion to the torque. Magnetically, the windings are arranged to produce a circuit analagous to the Wheatstone bridge. Figure 9.11d shows the Wheatstone bridge analogy with the battery being equivalent to the primary windings, P_1 and P_2, in Figure 9.11c. Reluctances R_{P1}, R_{P2}, R_{S1}, and R_{S2} are analogous to electrical resistance and represent the reluctance of the corresponding air gap between winding and shaft in Figure 9.11c. Reluctance of the metal shaft is represented by R_A, R_B, R_C, and R_D. With no loading on the shaft, the bridge is in balance and the output voltage is zero.

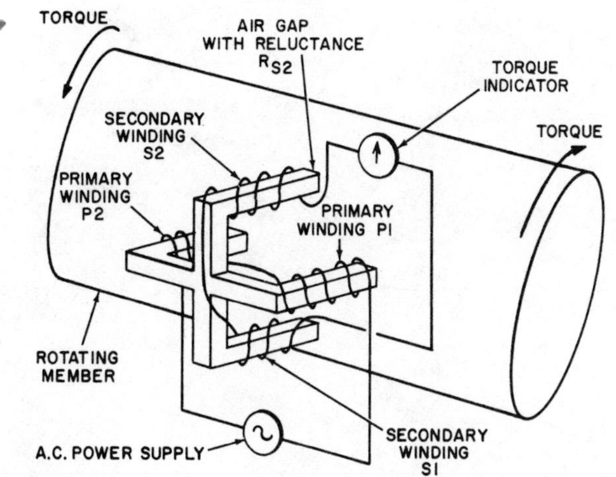

Fig. 9.11c Noncontacting torque sensor

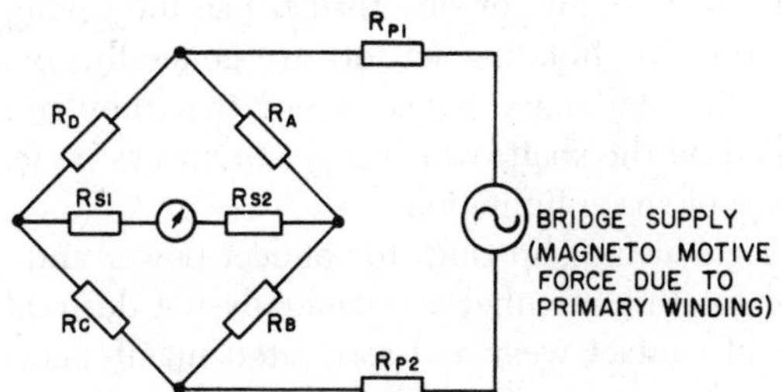

RPI, RP2 = PRIMARY WINDING AIR-GAP RELUCTANCE
RSI, RS2 = SECONDARY WINDING AIR-GAP RELUCTANCE
RA, RB, RC, RD = RELUCTANCE OF METAL BETWEEN ADJACENT AIR-GAPS

Fig. 9.11d Wheatstone bridge analogy of magnetic circuit for sensor in Fig. 9.11c

A variation of the basic design consists of many primary and secondary windings arranged on a ring fitted around the torsion shaft. This feature eliminates errors in output voltage due to residual magnetism which will induce an additional voltage in the output windings of the abovementioned unit. By locating many pole pieces around the shaft circumference, the stresses are integrated with respect to time so that the effect of instantaneous errors is minimized.

The angular displacement between two sections of a rotating shaft can also be related to torque. Two identical toothed wheels are fixed on the shaft a certain distance apart, as shown in Figure 9.11e. Two proximity sensors, one at each wheel, produce alternating output voltages whose phase difference is in proportion to torque. Alternatively, a light source and photocell can be mounted on opposite sides of the wheels and the amount of light falling on the photocell is related to torque.

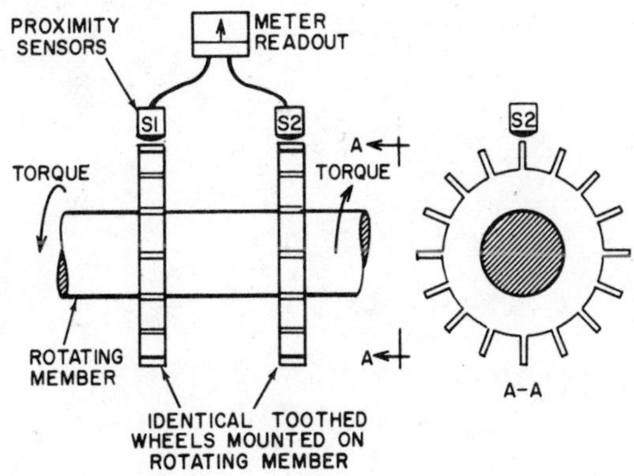

Fig. 9.11e Proximity sensors used for torque measurement

Conclusions

Of the designs available, stationary sensors measuring variations of permeability lend themselves very well to existing installations where modifications to accommodate a sensor must be minimized. The ring version of this design is preferred since it is less sensitive to speed variations, bending, or side thrust than the single point measurement. Rotating sensors are generally comparable to the stationary types except that they must be installed on the shaft, which may not always be feasible because of space limitations.

Designs using slip rings to connect power and signal to the rotating member are generally not desirable because of contact wear and associated maintenance and electrical noise introduced into the output. Slip ring designs should be avoided, especially in dirty environments, where dirt buildup can cause erratic output or complete loss of signal.

All designs discussed here are temperature sensitive and compensation is required.

BIBLIOGRAPHY

Foskett, R., "Torque Measuring Transducers," *Instruments and Control Systems,* November 1968.

"Measurement of Shaft Horsepower," *Instruments and Control Systems,* September 1962.

"Torque Measurement Without Touching," *Control Engineering,* May 1968.

"Torque Measuring Equipment," *Instruments and Control Systems,* February 1964.

Flow Sheet Symbol

9.12 VARIABLE SPEED DRIVES

Types of Drives:	A—Electrical drives: Variable frequency Wound rotor regenerative Direct current Variable voltage Two-speed motors B—Electromechanical drives: V-belt drives Hydraulic couplings Hydroviscous couplings Eddy current couplings
Partial List of Suppliers:	American Standard; Avtek Systems; Eaton Corp.; General Electric Co.; Ideal Electronics Corp.; Louis Allis Div., Litton Industrial Products, Inc.; Peerless Pump; Ramsey Controls, Inc.; Reliance Electric Co.; Robicon Corp.; Systecon Div., Corporate Equipment Co.; U.S. Motors Corp.

The centrifugal pump is used for many purposes and accounts for an appreciable amount of energy consumption, which results in an increased interest in variable speed drives as the cost of energy escalates. This chapter describes the current technology of variable speed drives for centrifugal pumps. The general equations and discussion will be limited to water at temperatures where the specific gravity is 1.0. Much of these data can be used for other liquids by applying specific gravity corrections to the equations. Equations will be developed for the calculation of energy requirements for the various drives, including development of wire-to-water efficiencies.

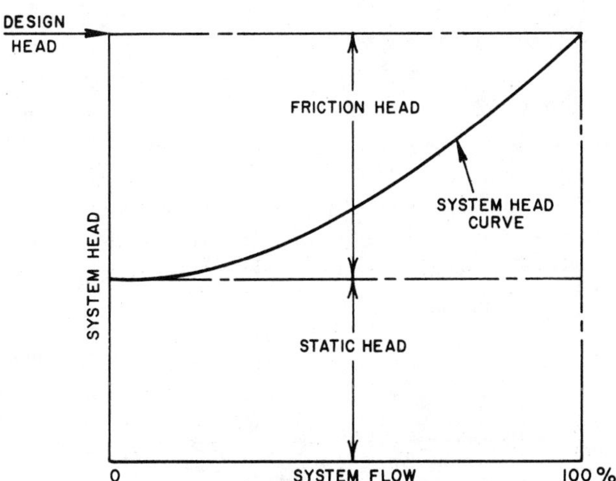

Fig. 9.12a Typical system, head curve

Water System Analysis

An analysis should be made of hydraulic systems to demonstrate the need for variable speed pumps before the drives themselves are evaluated. Any hydraulic system has what is defined as a system head curve. This curve demonstrates the need for pump head from zero to maximum flow. Figure 9.12a describes the general form of this curve; it indicates that the curve consists of static head plus friction head. The static head consists of static rise (elevation change) or terminal pressure such as boiler pressure. The friction head varies with flow, increasing exponentially as the flow increases. The exponent is usually between 1.85 and 2.0. The system head curve is all-important as it will demonstrate whether or not variable speed is economical.

A hydraulic system with a great amount of system friction has a system head curve such as the one shown in Figure 9.12b. The shaded area between the pump head capacity curve and the system head curve demonstrates the energy savings possible with variable speed drives. Contrarily, a hydraulic system with little system friction offers minimum energy savings possibilities for

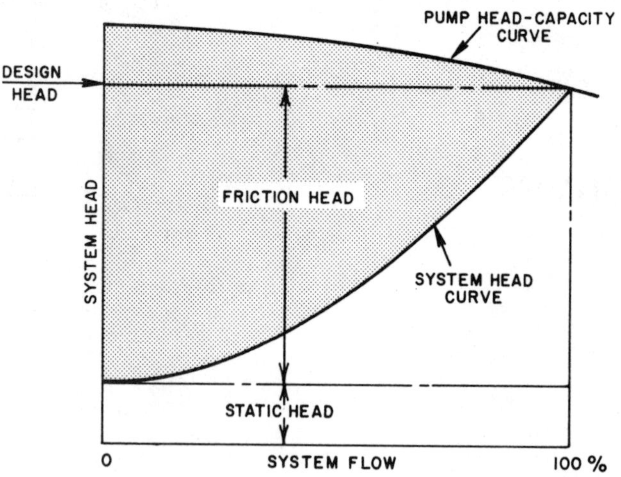

Fig. 9.12b Hydraulic system with high friction head

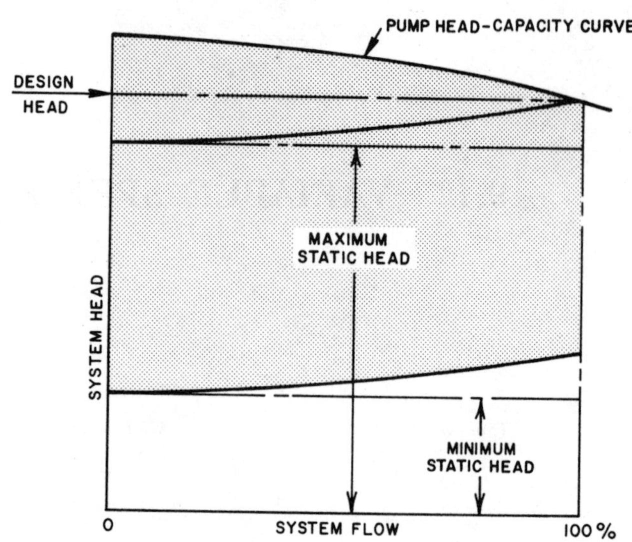

Fig. 9.12d Hydraulic system with variable static head

variable speed, as is shown in Figure 9.12c. Some hydraulic systems have a variable static head, which is caused by a change in supply pressure to the system. This is shown in Figure 9.12d. Again, the shaded area describes the energy savings possibilities with variable speed. In actual systems, a simplified system head curve may not exist; often what exists is a system head band, as shown in Figure 9.12e. This is because of load variations that change the system friction at the same flow in the system. Hydraulic systems can be open (noncirculating) or they can be loop-type (circulating). An open, noncirculating system is shown in Figure 9.12f and a loop-type, circulating system is shown in Figure 9.12g. Water supply and distribution systems in cities and buildings are typical open systems whereas hot-and-chilled-water heating and cooling systems are typical loop systems.

Hydraulic systems must also be evaluated as to whether flow is restricted or unrestricted. Restricted-flow systems are those that include valves to regulate the flow through the system. Plumbing systems or hot-and-chilled-water systems are restricted-flow systems since hand or automatic valves control the flow. Unrestricted-flow systems include sewage and storm water lift stations as well as municipal water flow into elevated storage tanks.

All of the above characteristics must be evaluated when variable speed is considered for centrifugal pumps. Generally, hydraulic systems that prove to be suitable for variable speed are those with (1) high system friction, (2) variable static head, (3) a loop-type configuration, and (4) restricted water flow. Typical applications are (1) public buildings with high system friction or variable static head, (2) municipal water booster stations that deliver water to suburban areas without storage tanks, and (3) circulating water systems such as hot, chilled, or condenser water as found in heating and cooling systems used for human comfort or industrial processes. Systems that normally do not need variable speed are systems

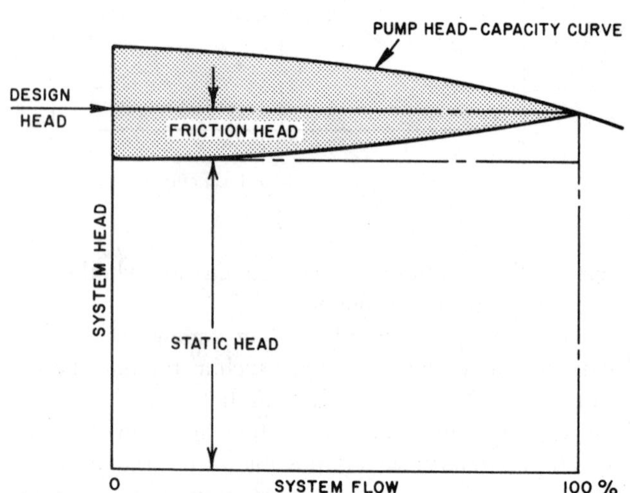

Fig. 9.12c Hydraulic system with low friction head

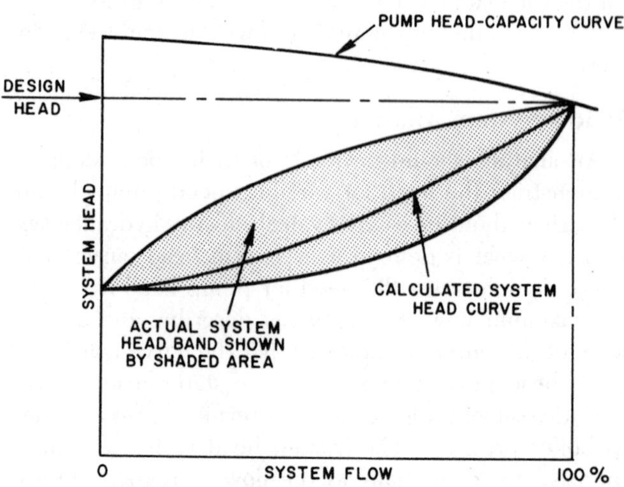

Fig. 9.12e System head band

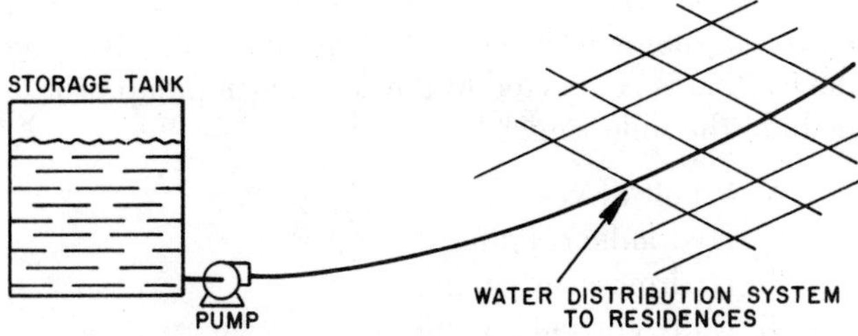

Fig. 9.12f Noncirculating or open water system

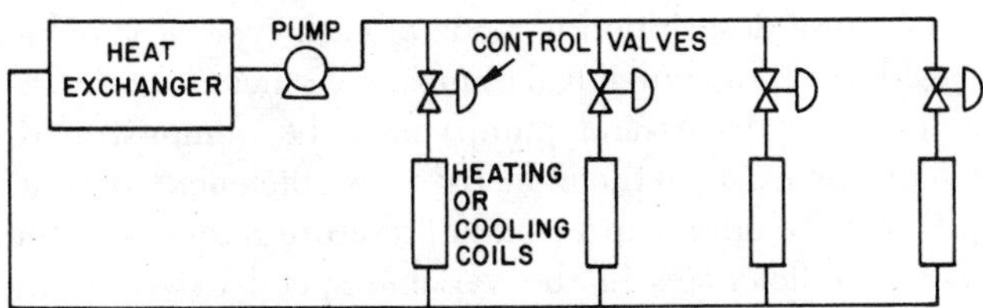

Fig. 9.12g Circulating or loop-type water system

with (1) low system friction head and constant static head or (2) unrestricted flow. These are generalizations and must be supplanted with actual system analysis for each specific system before variable speed is accepted or rejected.

Centrifugal Pump Performance

Along with hydraulic system characteristics, a brief analysis must be made of centrifugal pump performance before variable speed drives can be discussed in detail. Most centrifugal pumps have a head capacity curve at a constant speed, as shown in Figure 9.12h; the head rises as flow is reduced. As indicated in this curve, the efficiency drops rapidly as the flow is decreased. When the pump speed is also reduced, a series of curves is generated, with the peak efficiency following a parabolic curve toward zero. This is important as it indicates that pump efficiency is higher with variable speed pumps. These facts are shown in Figure 9.12i.

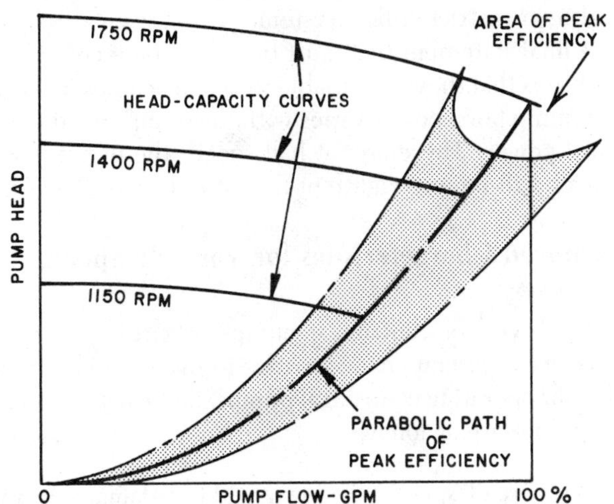

Fig. 9.12i Variable speed pump characteristics

The combination of the pump-capacity curve and the system head curve, as shown in Figures 9.12b, c, d, and e, enables the designer to evaluate system head requirements and pump performance. Figure 9.12j describes further considerations that must be made. This figure describes "steep" and "flat" pump head capacity curves as well as points of operation of pumps. Steep pump head capacity curves should be avoided wherever possible because of the overpressuring they cause. As to points of operation of pumps, on systems with unrestricted flow, the pump, if it is variable speed, will always

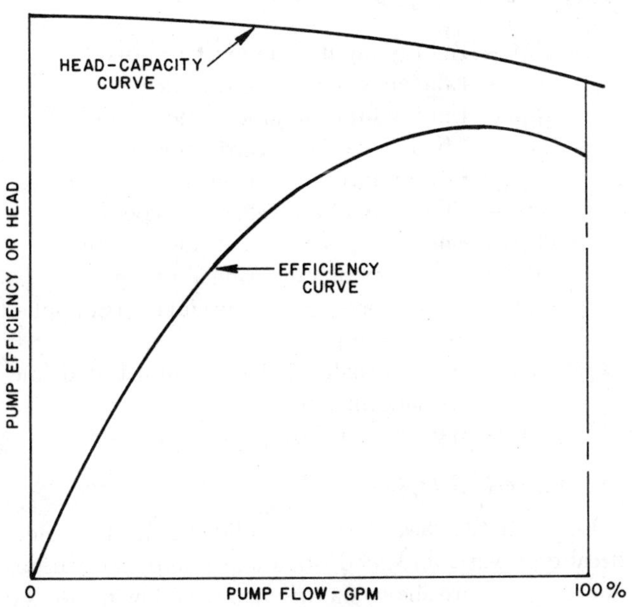

Fig. 9.12h Typical centrifugal pump characteristics at constant speed

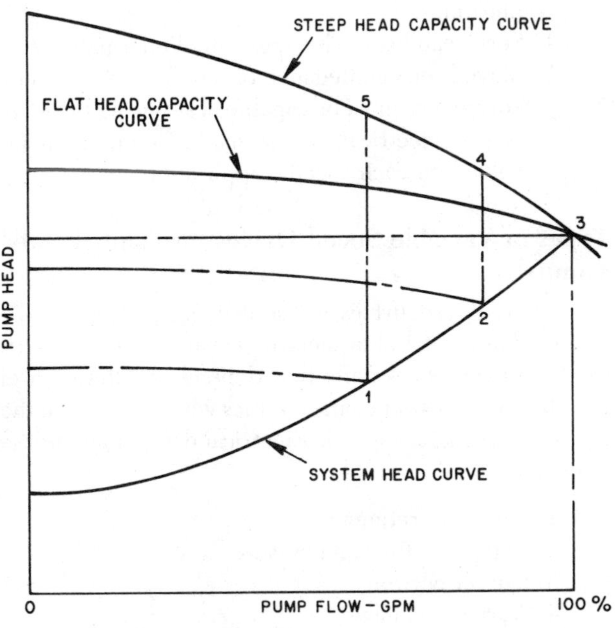

Fig. 9.12j Pump head capacity curves and points of operation

operate on the system head curve as shown at Points 1, 2, and 3 of Figure 9.12j. Contrarily, a constant speed pump on a restricted flow system will operate at Points 3, 4, and 5. A variable speed pump, with a properly located pressure transducer, will operate at Points 1, 2, and 3 on restricted flow systems.

A final statement that must be made about centrifugal pumps is that they are variable torque machines in which the pump horsepower varies with the pump speed to the third power. Therefore, at half speed, the pump horsepower will be one-eighth of that at full speed.

Required Characteristics for Variable Speed Drives

All drives for centrifugal pumps, regardless of type, must have certain characteristics to make them acceptable for operating centrifugal pumps. Some of these characteristics are as follows:

1. Broad speed turndown without damage to the motor or variable speed drive. The operating speed range for most variable speed pumps is from 50 to 100 percent of the synchronous speed. However, through misadjustment or momentary system dynamics, the controls may cause the pump to run at zero speed. The drive must be able to function with zero pump speed without damage or actuation of circuit protection devices such as thermal overloads.
2. Control repeatability, so that water system controls can achieve desirable pump speeds at all loads.
3. Reliability. The variable speed drive and motor must be designed to fit field operating conditions which vary appreciably in centrifugal pump applications.
4. Serviceability. The personnel normally employed for maintenance of motors and pumping equipment must be capable of servicing the electronic, electrical, or mechanical equipment involved on each specific application.

Types of Variable Speed Drives for Centrifugal Pumps

Variable speed drives for centrifugal pumps can be classified as electrical or electromechanical. The electrical drives consist of units that depend on alteration of the electrical current characteristics which will cause the motor to change its speed. Electrical drives include the following types:

1. Variable frequency
2. Wound rotor regenerative
3. Direct current
4. Variable voltage
5. Two-speed motors

Electromechanical drives include those that utilize an electrical motor to drive a mechanical, speed-changing device that is connected to the centrifugal pump. These include the following:

1. V-belt drives
2. Hydraulic couplings
3. Hydroviscous couplings
4. Eddy current couplings

Efficiency of Variable Speed Drives and Pumps

Before a description is made of each type of variable speed drive for centrifugal pumps, the efficiency for variable speed drives and pumps must be demonstrated. Confusion exists in the industry since efficiencies of components are often expressed in literature rather than the overall efficiencies of the variable speed drive or the efficiency of the combination of the drive and pump.

The following three efficiencies should be evaluated for any application of various speed pumps:

1. Wire-to-shaft efficiency of the variable speed drive and the motor. Obviously, this is the ratio of the useful energy applied to the pump shaft divided by the energy applied to the drive-motor combination.
2. Pump efficiency.
3. Wire-to-water efficiency of the drive, motor, and pump. This is the ratio of the useful energy applied to the water divided by the energy applied to the pump–motor–drive combination.

Following is a list of terms that will be used to describe the efficiencies of various pump–motor–drive combinations. All of the energy terms are in horsepower to simplify this discussion; they can easily be converted into watts.

Chp = Energy input to an electrical drive
Ce = Efficiency of an electrical drive
Ihp = Energy input to an electric motor
Me = Efficiency of an electric motor
Dhp = Energy input to a mechanical-type drive
De = Efficiency of a mechanical-type drive
Php = Energy input to a centrifugal pump
Pe = Efficiency of a centrifugal pump
Whp = Water horsepower or useful energy applied to the water
Ws = Wire-to-shaft efficiency (often called line-to-shaft efficiency)
Ww = Wire-to-water efficiency

Electrical Drives

Figure 9.12k describes the configuration of an electrical-type variable speed drive and pump; the various energy terms are shown on this figure. Following are the equations for this combination. The efficiencies are entered into these equations as fractions.

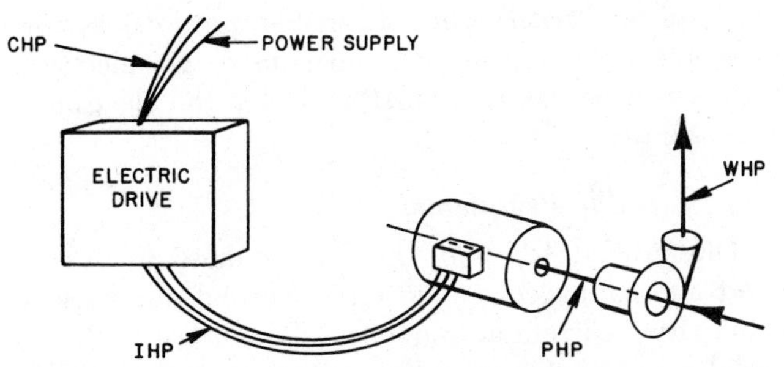

Fig. 9.12k Electrical variable speed drive arrangement

$$Whp = \text{Water horsepower} = \qquad 9.12(1)$$
$$= \frac{\text{GPM} \times \text{pump TDH (in ft of water)}}{3960}$$

$$Php = \text{Pump brake horsepower} = \qquad 9.12(2)$$
$$= \frac{Whp}{Pe}$$

$$Ihp = \text{Motor input horsepower} = \qquad 9.12(3)$$
$$= \frac{Php}{Me}$$

$$Chp = \text{Energy input to electrical drive} = \qquad 9.12(4)$$
$$= \frac{Ihp}{Ce}$$

$$Ws = \text{Wire-to-shaft efficiency} = \qquad 9.12(5)$$
$$= \frac{Php \times 100\%}{Chp} = Ce \times Me \times 100\%$$

$$WW = \text{Wire-to-water efficiency} = \qquad 9.12(6)$$
$$= \frac{Whp}{Chp} = Ce \times Me \times Pe \times 100\%$$

Electromechanical Drives

Figure 9.12l describes the configuration of the electromechanical drive and pump along with the various energy terms that are applicable to this combination. The following equations are similar but not necessarily equal to those for electrical drives. Equations 9.12(1) and 9.12(2), for water horsepower and pump brake horsepower, are the same.

$$Dhp = \text{Mechanical drive input horsepower} = \qquad 9.12(7)$$
$$= \frac{Php}{De} \text{(same as the electric motor shaft horsepower)}$$

$$Ihp = \text{Motor input horsepower} = \qquad 9.12(8)$$
$$= \frac{Dhp}{Me}$$

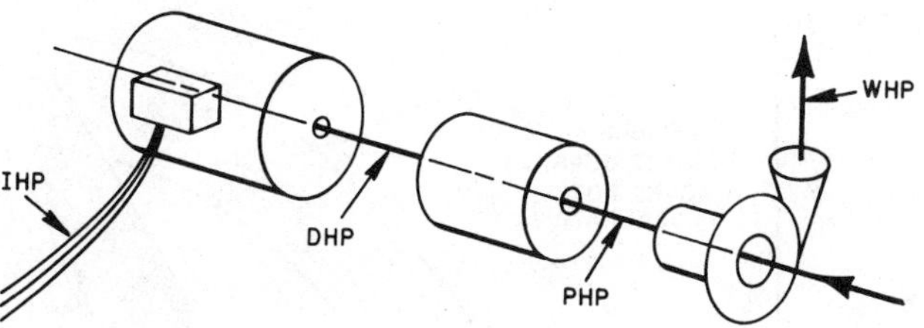

Fig. 9.12 l Electromechanical variable speed drive arrangement

$$Ws = \text{Wire-to-shaft efficiency} = \qquad 9.12(9)$$
$$= Me \times De \times 100\%$$
$$Ww = \text{Wire-to-water efficiency} = \qquad 9.12(10)$$
$$= Me \times De \times Pe \times 100\%$$

Evaluation of Variable Speed Drive Efficiencies

The above equations demonstrate that the wire-to-shaft efficiency for any drive–motor combination is dependent upon the efficiency of the electric motor and the drive, whether the drive is electrical or mechanical. As shown in equations 9.12(5) and 9.12(9) the efficiency of both the motor and the drive must be determined to achieve the true wire-to-shaft efficiency. These component efficiencies should be determined throughout the anticipated operating speed range, not just at full-speed condition. The advent of the high-efficiency electric motor has created the opportunity of achieving higher wire-to-shaft efficiencies for variable speed drives, but it is imperative that equivalent motors be used when comparing the efficiency of one type of drive with that of a different type. If high-efficiency motors are used, their efficiencies should have been secured from tests conducted in accordance with Nema Standard MG1-12.53a which is based on Institute of Electrical and Electronics Engineers (IEEE) Standard 112, Method B.

Figure 9.12m provides a general comparison of the efficiencies of various types of variable speed drives for centrifugal pumps. The curves shown in this figure *should not* be used for energy calculations for a specific application. Rather, efficiencies for drives and motors under consideration for that application should be certified by the manufacturers of that equipment.

As shown in Figure 9.12m, most electric drives are more efficient than electromechanical drives at reduced speeds. The exception is the variable voltage electric-type drive which is generally less efficient than mechanical drives. The mechanical drives are usually less efficient than the electric drives since most of them utilize slip between the input and output shafts of the mechanical drive. This results in a slip loss which is directly in proportion to the amount of slip. The equation for slip loss is

$$\text{Slip horsepower loss} = \frac{\text{Slip rpm} \times Php}{\text{Output rpm}} \qquad 9.12(11)$$

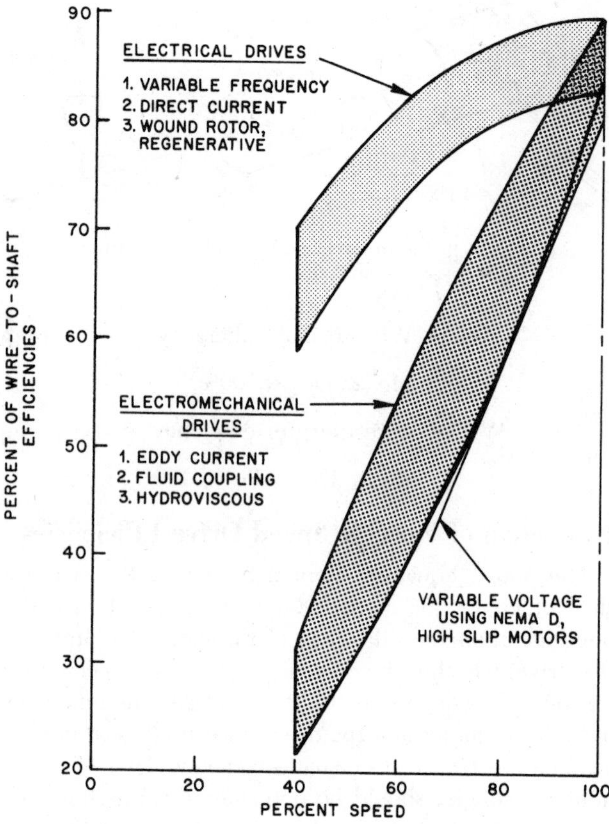

Fig. 9.12m Comparative wire-to-shaft efficiencies of variable speed drives for centrifugal pumps

The wire-to-shaft efficiency of the electromechanical drive is

$$Ws = \frac{\text{Php}}{\text{Ihp}} =$$

$$= \frac{\text{Php} \times \text{Me} \times 100\%}{\text{Php} \times \dfrac{\text{Slip rpm}}{\text{Output rpm}} + \text{Php} + \text{Circulation losses}} \qquad 9.12(12)$$

$$= \frac{\text{Php} \times \text{Me} \times 100\%}{\dfrac{\text{Input rpm}}{\text{Output rpm}} \times \text{Php} + \text{Circulation losses}} \qquad 9.12(13)$$

Circulation losses are those losses incurred by the constant rotation of the input shaft, along with energy for specific uses such as the loss imparted by oil pumps to circulate oil in fluid couplings and hydroviscous drives. Circulation losses must always be stated by mechanical drive manufacturers.

Evaluation of Variable Speed Drives

All of the foregoing discussion of hydraulic systems and centrifugal pumps, along with the definition of applicable equations, has been developed to aid the engineer in selecting the optimum drive for each application. To complete this work, an evaluation should be made of the various types of variable speed drives now available to the centrifugal pump industry. The electrical drives will be described first, and then the electromechanical drives.

Variable Frequency

Often called *Adjustable Frequency,* these drives convert alternating current to direct current and back to alternating current at frequencies from 0 to 120 Hz. Of solid state construction, these drives provide a highly reliable means of varying pump speed. They have wire-to-shaft efficiencies as high as 89 percent at full speed and 70 to 75 percent efficiencies at 40 percent speed when driving the new, high-efficiency motors. The motors are standard induction types, found in stock in most major cities. The primary objections to these drives are their high first cost and complex designs. The development of power transistors and other electronic advances is reducing the cost of these drives, and the ongoing training of field personnel in electrical and electronic service is making variable frequency more acceptable in fields that have, traditionally, used mechanical, variable speed drives. Variable frequency drives are available up to 2000 horsepower (1491 kW) with broad speed turndown ranges.

Wound Rotor Regenerative

The wound rotor motor can be adapted readily to variable speed; motor speed is changed by varying the resistance in the rotor. The wound rotor motor has been used for many years to vary the shaft speeds of cranes, machine tools, and similar heavy equipment. The energy loss due to heating of secondary resistance in these earlier applications has been recovered through the use of rectifiers and inverters and has resulted in a highly efficient drive for a centrifugal pump. The cost of the wound rotor motor and drive relegate this drive to larger pumps in the range of 500 to 10,000 horsepower (373 to 7455 kW).

Direct Current

Direct current motors are becoming more popular for pump drives because of the development of silicon controlled rectifiers (SCRs). The SCRs and direct current motor have resulted in a variable speed drive of modest first cost and high efficiency. The significant objection to direct current motors has been maintenance of commutators and brushes. Direct current motors have been used on elevators in public buildings for many years. Like centrifugal pumps, elevator loads are of the variable torque type. Brush life varies from 6 weeks to 18 months on elevator applications. At present, no extensive experience is available on brush and commutator wear and maintenance on centrifugal pump applications in the 1 to 100 horsepower (0.7457 to 74.57 kW) range. If trained maintenance personnel are available at the point of application, the direct current drive can provide a highly efficient, variable speed drive for centrifugal pumps. Cost of brushes and brush maintenance must be included in any economic evaluation. At the present, direct current motors are not as readily available as ac induction motors.

Variable Voltage

One of the least desirable drives for centrifugal pumps is the variable voltage drive utilizing Nema D, high-slip, induction motors. This is because of their poor efficiency, limited speed turndown, and need for special motors. This drive is not used extensively on centrifugal pumps.

Two-Speed Motor

The two-speed motor can be utilized on simple pumping systems in which accurate control of pump pressure is unnecessary. Typical applications of this drive are storm and sewage lift stations that have appreciable pipe friction, as shown in Figure 9.12b. They should not be used in systems with high static head and low friction, as shown in Figure 9.12c. This demonstrates that it is necessary to carefully calculate the system head curve before the two-speed motor is used. Where it is acceptable, an efficient, low-cost variable speed system can be achieved.

Eddy Current Couplings

The eddy current coupling has been the most acceptable drive for centrifugal pumps, particularly for large vertical turbine pump applications. It is rugged in design and available in both horizontal and vertical configurations. The eddy current coupling uses a standard, constant-speed motor which drives the eddy current coupling. The coupling consists of a drum connected to the electric motor and an electromagnetic pole-type rotor assembly which is connected to the centrifugal pump, inside of and free from the drum. The speed control system regulates the amount of flux that exists between the drum and rotor assemblies. The amount of slip or speed difference between motor and pump increases and decreases with the flux density in the coupling. Being a slip-type, variable speed device, the eddy current coupling does not have as high an efficiency curve as the variable frequency, direct current, and wound rotor regenerative, electrical types of variable speed drives.

Fluid Couplings

Fluid couplings consist of two coupling halves, one driven by a standard electrical motor and the other connected to the centrifugal pump. Oil is circulated in the coupling to regulate pump speed. A splitter or diverter assembly applies the oil to the coupling or bypasses it around the coupling. The pump shaft speed increases as the amount of oil supplied to the coupling is increased, thus producing a simple control system for varying pump speed. The fluid coupling has been one of the most popular variable speed devices for centrifugal pumps because of its relatively low first cost, high reliability, and ease of maintenance. It is a slip-type device, like the eddy current coupling, and has a lower efficiency curve than the variable frequency, direct current, and wound rotor regenerative types of variable speed drives.

Hydroviscous Drives

Hydroviscous drives are similar to fluid couplings in configuration and operation. Instead of an oil-filled coupling, one or more disk assemblies are used, with driving and driven members which are pressed together by oil pressure. Increasing the oil pressure increases the pump speed; likewise, reducing the oil pressure reduces the pump speed. The hydroviscous drive is, therefore, a slip-type device, like the eddy current coupling, and has a lower efficiency curve than most electrical-type drives. This drive often requires a more complex control system than other drives because of variations in oil pressure caused by oil viscosity or temperature.

Summary

The above brief descriptions provide general information on the various types of variable speed drives for centrifugal pumps. Actual selection of a drive for a specific application will require an extensive evaluation of (1) wire-to-shaft efficiency, (2) first cost, (3) reliability, (4) serviceability, (5) maintenance costs, (6) need for special motor, (7) speed range, and (8) control repeatability.

No variable speed drive should be selected for a centrifugal pump without careful evaluation of the water system itself to determine the feasibility of variable speed. This, along with consideration for the eight factors listed above, should result in the selection of the optimum drive for each centrifugal pump application.

BIBLIOGRAPHY

Bower, J.R. "The Economics of Variable-Speed Pumping with Speed-changing Devices," *Pump World*.

Gottliebson, M., "Explore the Use of Variable Speed Water Booster Pumps," *Water and Wastes Engineering*, May 1978.

Janki, C., "What's New in Motors and Motor Controls?" *Instruments and Control Systems*, November 1979.

Liu, T., "Controlling Pipeline Pumps for Energy Efficiency," *In Tech*, June 1979.

Merritt, R., "Energy Saving Devices for AC Motors," *Instruments and Control Systems*, March 1980.

Merritt, R., "What's Happening with Pumps," *Instruments and Control Systems*, September 1980.

Rishel, J.B., "The Case for Variable-Speed Pumping Systems," *Plant Engineering*, November 1974.

Rishel, J.B., "Matching Pumps to System Requirements," *Plant Engineering*, October 1975.

Rishel, J.B., "Water System Head Analysis," *Plant Engineering*, October 1977.

Schroeder, E.G., "Choose Variable Speed Devices for Pump and Fan Efficiency," *In Tech*, September 1980.

Flow Sheet Symbol

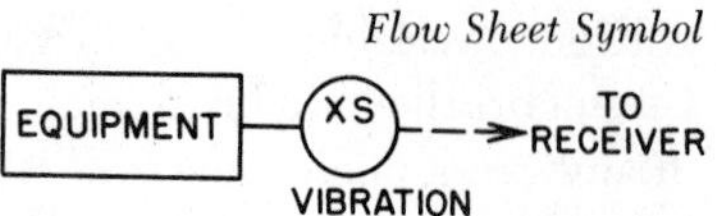

9.13 VIBRATION MONITORS

Design Temperature:	To 500°F (260°C)
Element Materials:	Alnico magnets, copper coils, piezoelectric crystals, semiconductors
Cost:	$200
Range:	3–50,000 Hz
List of Suppliers:	ADE Corp.; B & K Instruments; BBN Instruments Corp.; Bell & Howell Co., CEC Div.; Disa Electronics; DYMAC, a Division of Spectral Dynamics Corp.; Enderco, Dynamic Instrument Div.; Kaman Sciences Corp.; Ling Electronics Inc.; Robertshaw Controls Co.; Sprengnether Instruments Inc.; Unholtz-Dickie Corp.; Vitec, Inc.

Vibration can be defined in terms of displacement, velocity, or acceleration. Traditional mechanical tripout devices such as the spring-loaded magnetic holding types are all acceleration sensors. These are actuated by changes in the vibration force at the measurement. However, at low frequencies these devices are severely limited because acceleration decreases with frequency, and large displacement is required to produce sufficient acceleration to trip the device. In fact, the displacement becomes large enough to destroy the machine before tripout occurs. For example, at a frequency of 5 Hz, the displacement must be in excess of 0.75 in. (18.75 mm) to produce an acceleration of Ig (Ig is the acceleration due to gravity at the surface of the earth).

Magnetic electric sensors, using velocity pickup to sense displacement electronically, have been developed for low-speed machinery. Piezoelectric accelerometers are a relatively new development which can be used on very high speed machines to monitor vibration frequencies.

Mechanical-Magnetic Switches

The detector mechanism of this switch consists of an armature suspended on a flexure pivot and restrained from moving by a permanent magnet as shown in Figure 9.13a. A compression spring provides an adjustable force opposing the magnet. The armature acts as a lever where the adjusting spring force is balanced by the magnetic force. The armature is constrained to move in only one plane by the flexure pivot. When the entire assembly is subjected to vibration perpendicular to the base, the product of acceleration and armature mass produces an inertial force trying to pull the armature away from the magnet. Since the spring force aids the inertial force, the spring loading is selected to trip the mechanism at the desired acceleration. When the peak acceleration exceeds the preset level, the armature leaves the stop pin and moves to the latch magnet, which holds the armature in the excessive vibration position. Motion of the armature actuates a snap acting switch that opens or closes an electrical circuit. Once the switch has been tripped by excessive vibration, the armature must be returned to the stop pin. Resetting can be accomplished remotely by energizing the reset coil momentarily or by

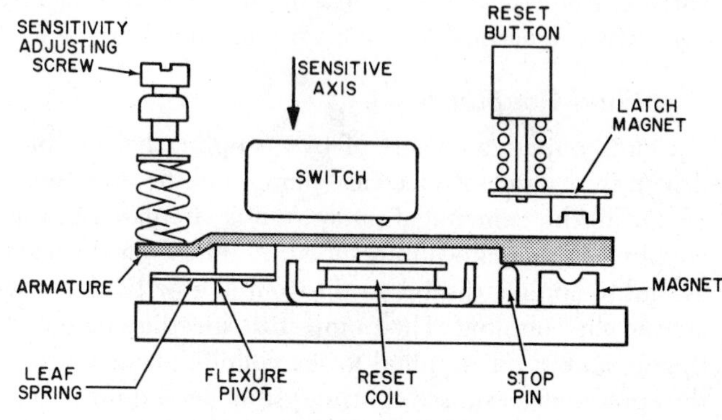

Fig. 9.13a Mechanical-magnetic switch

depressing the reset button mounted on the switch. The reset coil requires direct current for operation. If dc power is not available, a rectifier can be supplied with the switch. The rectifier output is sufficient to power several reset coils.

Another version of this switch uses a ball in a cup to provide acceleration sensing. Under excessive vibration, the ball leaves the cup and trips an electrical circuit. This switch is provided only with a local reset which returns the ball into the cup. The internals of this switch are shown in Figure 9.13b.

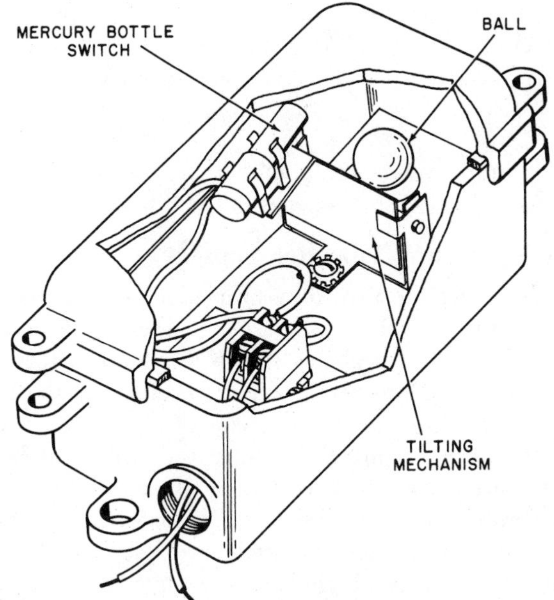

Fig. 9.13b Ball-in-a-cup switch

Piezoelectric Accelerometers

These devices are made of various types of crystals sandwiched between metal contacts and mounted on the equipment to be monitored. A separation of charge is produced on the opposite faces of the crystal when it is subjected to acceleration forces. The magnitude of the voltage produced is in proportion to mechanical deformation and, hence, acceleration. Generally, these units are provided as two components: the sensor (see Figure 9.13c) which mounts on the equipment and includes a solid-state amplifier and alarm switch, and a control unit that contains the readout meter, power supply, alarm reset, and facilities for vibration recording or waveform analysis by oscilloscope.

Piezoelectric crystals are affected in their output by temperature variations; however, quartz and some of the newer piezoelectric ceramics are superior to such ceramics as barium titanate or lead titanate-zirconate in their reduced sensitivity to temperature. In noisy electrical environments, electrical shielding of the crystal is recommended to reduce noise pickup.

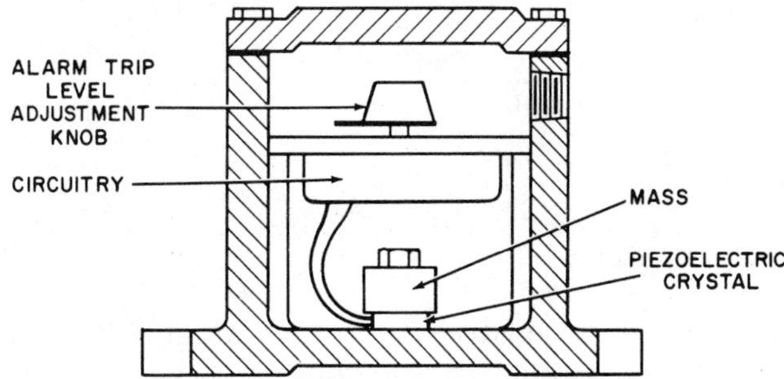

Fig. 9.13c Piezoelectric sensor

Strain Gauge Accelerometer

Strain gauge accelerometers measure a change in electrical resistance which is in proportion to the force due to acceleration. Although numerous configurations of this type of instrument are available, only two will be discussed here. The semiconductor strain gauge type consists of semiconductors bonded to a mass whose deformation under acceleration forces is reflected as a change of resistance. The resistance measurement is made by means of a Wheatstone bridge or half Wheatstone bridge with the elements so arranged on the mass that half of them sense tension and the other half compression. A bending type of accelerometer with circuit diagram is shown in Figure 9.13d.

At full-scale deflection of the beam, R_{G1} and R_{G2} are a maximum and minimum, respectively, since they are subject to equal but opposite deformations. For an initial resistance of $R_{G1} = R_{G2} = 1000\ \Omega$ and

$$\frac{\Delta R_G}{R_G} = 0.1$$

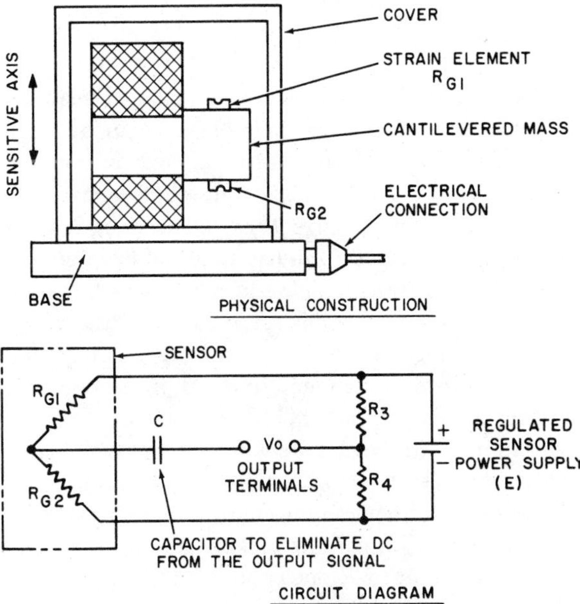

Fig. 9.13d Bending-type vibration sensor (acceleration sensitive)

the full-scale output is given by the relationship

$$V_0 = E\left[\frac{R_{G1} + \Delta R_G}{(R_{G1} + \Delta R_G) + (R_{G2} - \Delta R_G)} - \frac{R_4}{R_3 + R_4}\right]$$

$$V_0 = 10\left[\frac{1100}{2000} - \frac{1000}{2000}\right]$$

$$V_0 = 500 \text{ mV}$$

Of course, the readout meter must have a high resistance so that for practical purposes the current flow through the meter is zero. In the above example, the accelerometer contained only two active elements. Sensitivity can be improved by the addition of two more elements to complete the Wheatstone bridge.

The unbonded wire-type accelerometer is shown in Figure 9.13e. Mass m is attached to the base by cantilever springs, and the resistance elements are connected in the form of the Wheatstone bridge. Usually the entire assembly is contained in an oil-filled case for viscous damping.

Strain gauges are temperature sensitive and should be used with care where wide temperature fluctuations are anticipated.

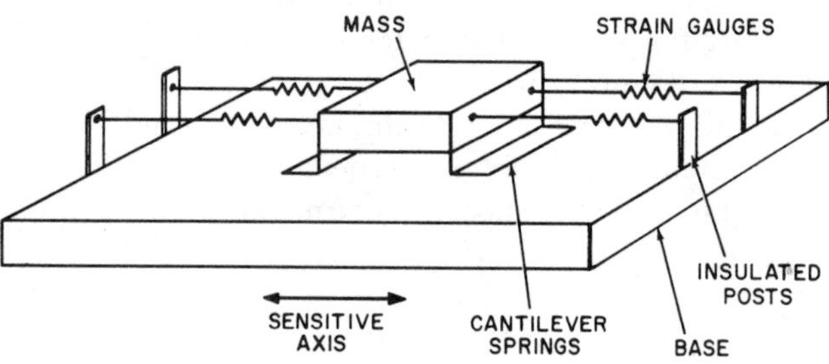

Fig. 9.13e Strain gauge vibration sensor (acceleration sensitive)

Velocity Sensors

Velocity sensors consist of an electrical coil suspended by a spring or springs in the field of a permanent magnet (see Figure 9.13f). The entire assembly may be contained in silicone oil or, for low-mass coils, the reactive forces caused by the induced currents can be used for damping. The spring coil mass system has a natural frequency below which it moves with the enclosure. Above the natural frequency, the coil remains stationary, and the relative motion between coil and magnet induces a voltage in the coil. This voltage has an amplitude in proportion to the vibration velocity and a frequency equal to the frequency of vibration. With velocity pickup, the output can be in the form of mils (thousandths of an inch) displacement by integration of velocity. A control unit mounted separately is required for these pickups and it contains the readout, amplifier, and power supply. Electronic filtering can be provided where it is desired to look at only specific frequency bands. These devices are fairly rug-

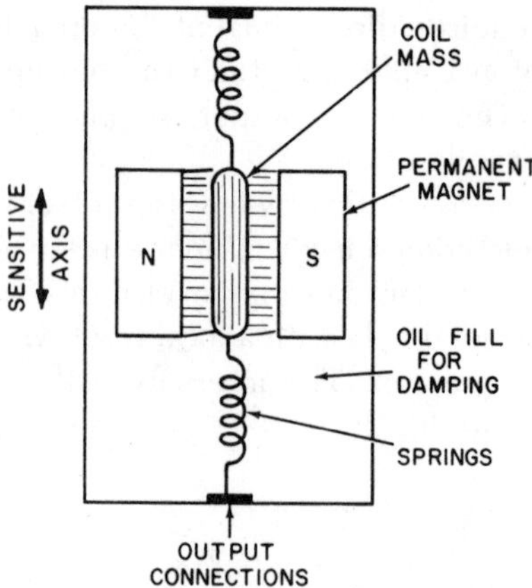

Fig. 9.13f Schematic representation of a velocity sensor

ged, but the oil-damped types are temperature sensitive because of viscosity changes of the filling oil.

Noncontacting Vibration Sensors

To detect the proximity of conducting materials, an eddy current probe can be used. The schematic arrangement of such a probe is shown in Figure 9.13g. Two identical coils are wound on the probe, and these, together with the resistances, complete a bridge circuit. With no conducting surface near the probe, the bridge is in balance. When a conducting object is brought near the probe, the bridge becomes unbalanced and the output signal is in proportion to the proximity of the object. The excitation is a high-frequency signal which induces eddy currents in the test object. These currents produce losses in the bridge circuit in such a way that bridge imbalance is related to the proximity of the object. The output signal amplitude is related to the vibration or displacement amplitude, and the frequency of amplitude variations is the frequency of vibration. The bridge is powered by a high-frequency (100 kHz to 2 MHz) oscillator which supplies the rapidly changing magnetic field required for sensor operation.

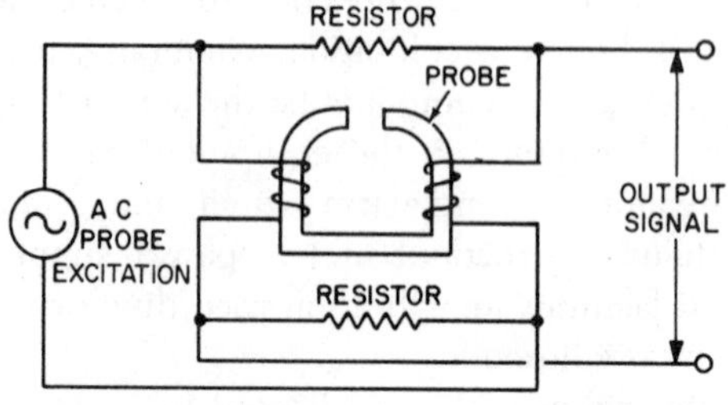

Fig. 9.13g Eddy current probe schematic

Conclusions

Mechanical vibration sensors are limited to detection of lower frequencies of vibrations. Output signals for waveform analysis are not available with these units.

They are used as trip devices to shut down equipment or alarm on excessive vibration. If vibration sensing is for the purpose of detecting impending malfunction, devices with continuous outputs are required. Acceleration sensors have one inherent advantage in that they filter out low-frequency background vibrations. Velocity sensors, however, offer a direct measurement of the energy dissipated in vibration, and thus indicate a quantity directly related to efficiency and the destructive forces. Where vibrations are poorly transmitted to the equipment housings, such as on turbine shafts, noncontacting sensors can be utilized to monitor the critical components of the machine.

BIBLIOGRAPHY

Kaufman, A.B., "Monitor Acceleration, Velocity or Displacement," *Instruments and Control Systems*, October 1975, pp. 37–40.

Kaufman, A.B., "Vibration—A Warning Signal That Shouldn't Be Ignored," *Instruments and Control Systems*, July 1976, pp. 23–25.

Verhoff, W.H., "Measuring Torsional Vibration," *Instrumentation Technology*, November 1977, pp. 61–66.

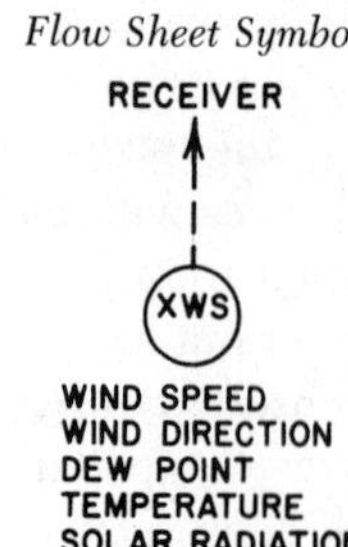

9.14 WEATHER STATIONS

Variable Ranges:	Barometric Pressure: 28–32 in. Hg (95–108 kPa) Dew Point: −40 to 120°F (−40 to 50°C) Precipitation: Each tip of bucket is 0.01 in. (0.25 mm) Relative Humidity: 0–100% RH Solar Radiation: 75 mV/cal/(cm²/min) Temperature: −30 to 120°F (−34 to 50°C) Wind Direction: 0–360° Wind Speed: 0.5–125 mph (0.2–56 m/sec)
Inaccuracies:	Barometric Pressure: ±0.3% to ±0.5% full scale Dew Point: ±0.9°F (±0.5°C) Precipitation: ±1% Relative Humidity: ±2% full scale Solar Radiation: ±1 to 2% Temperature: ±0.5°F (±0.2°C) Wind Direction: ±2.5° Wind Speed: ±1% or ±0.2 mph (± 0.1 m/sec)
Cost of Station:	$5000 to $15,000 depending on number and quality of sensors and accessories
Partial List of Suppliers:	Belfort Instrument Co.; Climatronics Corp.; G.M. Mfg. & Instrument Corp.; Kahl Scientific Instrument Corp.; Maximum, Inc.; Meteorology Research, Inc.; Science Associates, Inc.; Texas Electronics, Inc.; R. M. Young Co.

A typical weather station includes a supporting frame with a number of sensors mounted on it, transmitters, cables, and remote readout devices. All of this equipment can be purchased as a package or separately. In larger installations, the weather information is usually sent to the central computer, and therefore no separate readout devices are required for the weather station.

Many of the sensors that are used in weather stations have already been discussed in other parts of this handbook. Therefore, only those detectors that are not covered elsewhere will be discussed here in detail.

Solar Radiation Sensors

The intensity of solar radiation can be measured by (1) absorbing the incoming radiation and determining the rate of heat absorption and (2) by making use of photovoltaic or photoresistive transducers. Among the first accurate measurements of solar radiation were those made with miniature waterflow calorimeters in which the flow rate was kept constant and the temperature rise due to absorption of solar radiation was determined by high-accuracy mercury-in-glass thermometers and later by thermopiles.

Thermal Radiometers

The term *pyranometer* is applied to a solar radiometer which measures the shortwave radiation coming directly from the sun and diffusely from all parts of the sky. *Pyrheliometer* is the word used to identify an instrument which measures only the direct radiation from the solar disk itself.

Pouillet in France (1830) and Ericsson in the United States (1870) devised pyrheliometers which measured the rate of temperature rise of blackened metal masses located at the bottom of tubes that excluded all but the direct rays of the sun. C. G. Abbott of the Smithsonian Institution perfected this technique with his silver-disk pyrheliometers, which are still in use as secondary standards. He had earlier built a waterflow calorimeter in which a precisely controlled flow of distilled water carried away the heat absorbed by a blackened cone located at the bottom of a collimating tube. A platinum-resistance thermometer was used to measure the temperature rise and then an exactly equal rise was created by passing the water through an electrical heating coil. Precise measurement of the heating current and the coil resistance enabled the user of the instrument to make first-order determinations of the rate at which the incoming solar energy was being absorbed. This instrument was long used as the primary standard for measuring solar radiation.

The primary standard radiometer used today was originated in 1899 by K. J. Angström of Sweden. His instrument uses two blackened strips of Manganin, each of which can be heated either electrically or by the rays of the sun. The measurement is made by exposing one strip to the sun, measuring its temperature precisely, and then measuring the current required to heat the adjacent strip to exactly the same temperature. By reversing the operation of the strips several times, an average value of the radiation intensity within 1 percent of the absolute value can be obtained. The results may be expressed in langleys (1 ly = 1.0 g-cal/cm^2/min = 221.2 Btu/hr/ft^2) or in electrical units (1 ly = 69.7 mW/cm^2).

The solar radiometer which is in widest use today employs a multijunction thermopile to detect the difference in temperature between whitened and blackened segments of divided or concentric circles (Fig. 9.14a), which are generally protected by hemispherical glass domes. These instruments characteristically produce from 2.0 to 10.0 mV/ly of incoming radiation, and their output can be readily measured with indicating or recording millivoltmeters. Mechanical or electrical integrators can be used to total daily or monthly insolation. Since the response of the thermopile instruments is independent of wavelengths, they can be used with filters to determine radiation intensity in selected portions of the solar or longwave spectrums.

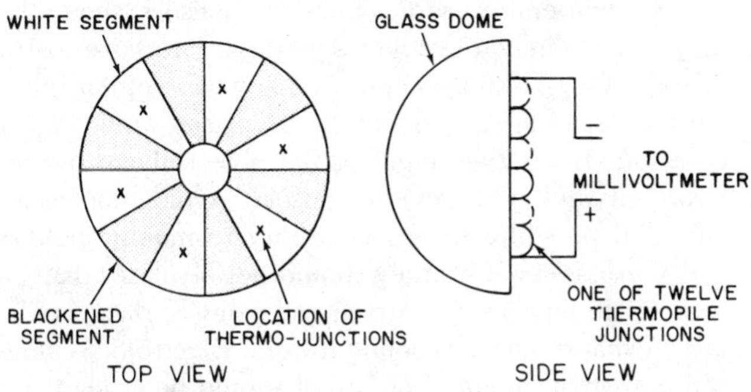

Fig. 9.14a Thermopile-type pyranometer

Photovoltaic Sensors

The newest form of solar radiometer uses silicon photovoltaic cells which produce a short-circuit current that is directly in proportion to the intensity of the solar radiation falling on them. Widely used in the space program to provide power for space probes and satellites, silicon cells are rugged, sensitive, and capable of producing currents large enough to operate self-powered integrators.

Wind Direction and Speed Transmitters

The wind direction and wind speed transmitters can be either separate units or combined into a single assembly, as shown in Figure 9.14b.

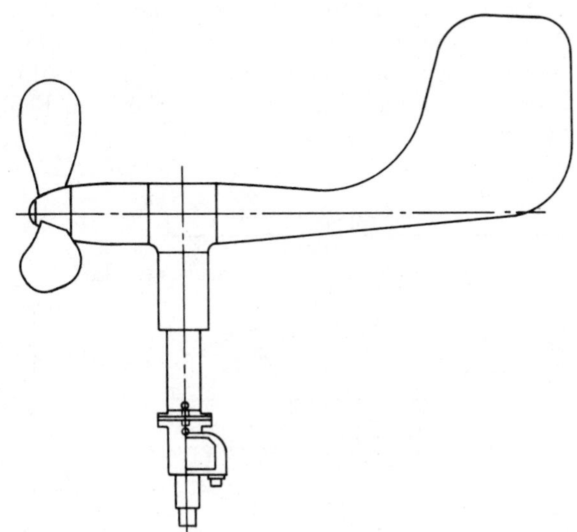

Fig. 9.14b Wind direction and speed transmitter

The wind direction is usually detected by a potentiometer, which is mechanically coupled to the wind vane shaft. The resistance at the potentiometer wiper contact is directly in proportion to the angular position of the vane.

Wind speed is sensed by 3- or 6-bladed impellers or by anemometers of 3-cup or 6-cup configurations. The anemometer shaft can be connected to a photochopper or to a dc or ac generator. If it is connected to the photochopper, a 60-slot disk is rotated to interrupt a light beam with a photosensitive transistor on the opposite side of the disk. This design is more maintenance-free than the dc-type tachometer generators, which require brush contacts for their operation. The contact between rotating and stationary components can also be eliminated through the use of ac generators.

Transmitters are usually selected to be sensitive to light winds and yet be strong enough to withstand winds of hurricane force.

Rain Gauge Transmitters

A collector and a tipping bucket mechanism is provided in this instrument such that each 0.01 in. (0.25 mm) of rain causes the alternate fill and tip of the mechanism. A magnet is attached to the bucket, and causes a momentary closure in a magnetic switch with each tip of the bucket. The resulting electrical impulse can be sent to a counter, an event recorder, to telemetering or data acquisition equipment.

Other Sensors

Barometric pressure is detected by bellows with sealed-in vacuum reference. These differential pressure sensing bellows are usually coupled to the core of a linear variable differential transformer (LVDT), which generates the required electronic output signal.

Ambient temperature can be detected by thermocouples, resistance bulbs, or thermistors. They are to be shielded from direct sun exposure as shown in Figure 9.14c.

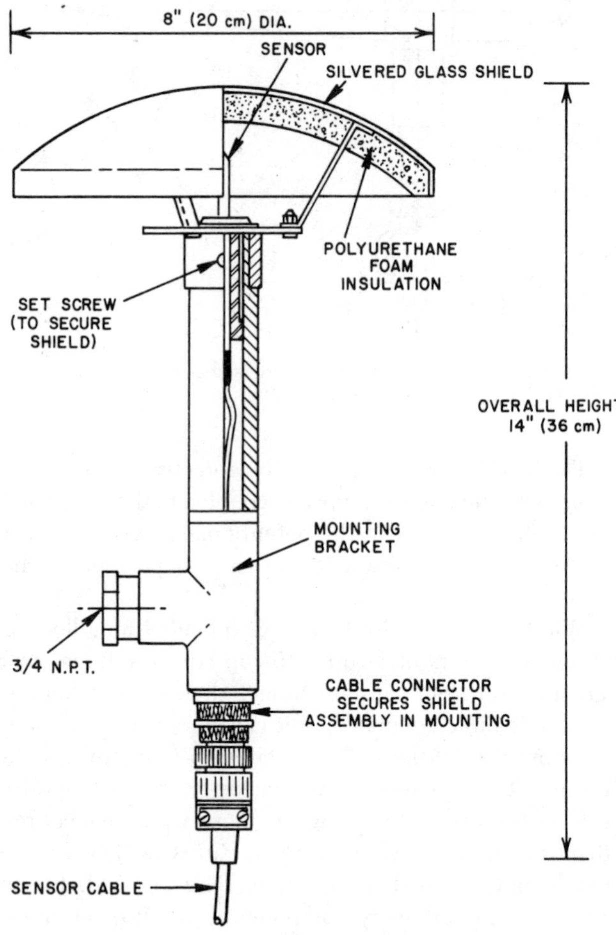

Fig. 9.14c Temperature radiation shield with natural ventilation

Dew point is usually detected by lithium chloride sensors or, where high precision is required, by cooled mirrors. For relative humidity measurements the elongation and contraction of hygroscopic organic or inorganic elements is usually utilized.

Readouts

The sensors can be 1000 ft (300 m) or more from the readout instruments. Some weather station packages are provided with multiplexers, so that the outputs of six or more sensors can all be read from the same wires. The connection between sensors and readout can be through hard wiring or through radio telemetering. The outputs of six or more sensors can all be recorded on the same chart.

Sensor Location

The detection of wind speed should ideally be made on an open terrain. If this is not possible, one can minimize the effect of single-story buildings by (1) locating the sensor a distance upwind of a building equal to the building height; (2) locating the sensor at least one building-height above the roof; or (3) locating the sensor a distance downwind of the building equal to five to ten building-heights. If taller buildings are involved, it is best to erect a small tower to clear any obstruction. If this is not practical, an alternative is to place the instrument on a corner of the building which is upwind the majority of the time or the corner which is exposed to the greatest frequency of wind. This will generally be the west corner of the building.

Before making a permanent installation, it would be wise to visually monitor a small flag on the end of a pole placed in various locations on the roof to help in determining the location which is most representative.

Horizontal-mount booms which extend from existing towers should be fabricated so that sensors will extend a distance of 5 to 10 ft (1.5 to 3 m) from the tower assembly (depending on tower thickness). Wind direction sensors are oriented at the time of installation in reference to either true north or magnetic north.

Instrument Shelters

Whenever possible, instrument shelters, as well as remote temperature and/or humidity sensors, should be installed at a height of 4 ft (1.2 m) (or greater) over earth or sod, at least 100 ft (30 m) from any concrete or other hard-surfaced area, and not closer to any other object than four times the height of the object above the instrument shelter or remote sensors. Avoid roof installations if possible. If it is necessary to mount shelters and sensors on a roof, they should not be closer than 30 ft (9 m) to any large, vertical reflecting surface (walls, etc.), exhaust fans, or cooling towers. Electronic remote sensors when mounted on a roof should be at least 9 ft (2.7 m) (or greater) above the roof surface. To minimize radiation effects from the roof, remote sensors can also be mounted on a horizontal boom so that they will extend from the side of a building roof or existing tower. Horizontal booms should extend approximately 5 to 10 ft (1.5 to 3 m) from the side of the building roof or tower assembly.

Precipitation Gauges

Rain gauges should be installed on a level plot of ground, at a distance from any object of at least two and preferably four times the height of the object above the top of the gauge. All types of gauges must be exposed with the rim of the receiver in a horizontal plane and at a level well above the average level of snow surfaces. Roof-mounting of rain gauges should be avoided when possible. Air currents at heights other than ground level have been observed to cause an apparent decrease in rainfall catch commensurate with the increase in mounting height above the ground level.

Objects which individually or in small groups constitute a "windbreak" reduce prevailing wind speed in the vicinity of the gauge. The consequence of this reduction of wind speed will be to reduce possible eddy currents and turbulence around the gauge. The presence of such objects is usually beneficial in providing a more accurate rainfall catch. Ideally, the "windbreak" objects (fences, bushes, etc.) should be generally uniform in height and in their distance from the gauge. Their height above the gauge should not exceed about twice their distance from the gauge.

Remote Barometric Pressure Sensors

Select a site where the instrument will not be subject to rapid fluctuations of temperature or to jarring and continuous vibration. Avoid exposing the instrument to direct sunlight or radiant heaters and to direct drafts such as from open windows and doors.

BIBLIOGRAPHY

MacDonald, T.H., "Some Characteristics of the Eppley Pyrheliometer," *Monthly Weather Review*, Vol. 79, No. 8, August 1951, pp. 153–159.

U.S. Department of Commerce, *National Weather Service Bulletin*, LS5927, January 1963.

Yellott, J.I., "Solar Radiation Sensors," in *Environmental Engineers Handbook*, B. G. Lipták (ed.), Vol. 2, 1974, p. 442.

Yellott, J.I., "Solar Radiation Measurement, Low Temperature Applications of Solar Energy," American Society of Heating, Refrigerating, and Air Conditioning Engineers. New York, 1967, pp. 19–25.

Yellott, J.I. and A.L. Pittinger, "Development of an Indicating and Integrating Solar Radiometer," ASME Paper 67-WA/Sol-3, Winter Annual American Society of Mechanical Engineers Meeting, 1967.

Chapter X

TYPES OF ANALYZERS

R. C. AHLSTROM, JR. • C. P. BLAKELEY
A. BRODGESELL • J. E. BROWN
A. P. FOUNDOS • R. GILBERT
A. C. GILBY • D. L. HOYLE
R. H. JONES • J. G. KOCAK, JR.
B. G. LIPTÁK • D. H. LIU
T. J. MYRON • R. T. OLIVER
L. P. ROOT • W. H. WAGNER

CONTENTS OF CHAPTER X

Table X

Grouping	Sensor Type \ Manufacturers of Online Analyzers	ACSI	American Instrument Co.	Anacon, Inc.	Applied Automation, Inc.	Bacharach Instruments Co.	Bailey Controls Co.	Barton ITT Process Instruments and Controls	Beckman Instruments, Inc.	Bell & Howell Co., CEC Div.	Bendix Corp.	Bristol Div. ACCO	Brookfield Engineering Labs, Inc.	Corning Glass Works	Dasibi Environmental Corp.	Delta Scientific, National Sonics Div. Enviroteck Corp.	DeZurik, a unit of General Signal	DuPont Co. Analytical Instrument Div.	EG&G, Inc. Environmetal Equipment Div.	Electron-Machine Corp.
Grouped by the methods of analysis applied	*Electromagnetic Radiant Energy*																			
	Ultrasonic																			
	Microwave																			
	Infrared			✓					✓		✓							✓		
	Visible																			
	Colorimeter		✓	✓					✓							✓		✓		
	Flame photometer															✓		✓		
	Refractive index			✓																✓
	Turbidity	✓	✓				✓		✓							✓		✓		
	Ultraviolet		✓	✓		✓	✓		✓									✓		
	Radiation (gamma, neutron)																			
	Electrochemical																			
	Electrolytic cells									✓										
	Galvanic cells			✓			✓									✓				
	pH								✓					✓		✓				
	Ion selective electrodes								✓					✓		✓				
	ORP (redox)															✓				
	Polarographic cells																			
	Conductivity								✓							✓				
	Inferential Property																			
	Conductivity-resistivity								✓							✓				
	Consistency												✓				✓			
	Density, thickness						✓				✓									
	Dielectic constant																			
	Flash, melt, pour point																			
	Flow index																			
	Molecular weight								✓		✓									
	Plastometry																			
	Thermal conductivity		✓				✓		✓		✓									
	Vapor pressure																			
	Viscosity										✓		✓							
	Combination, Miscellaneous																			
	Chromatography																			
	Adsorption				✓				✓	✓	✓									
	Partition				✓				✓	✓	✓									
	Gel-permeation																			
	Total carbon, hydrocarbon															✓				
	Ionization (mass spectrometry)																			
	Thermal & flame								✓	✓	✓									
	Electrostatic									✓										✓
	Radiation										✓									
	Paramagnetic										✓									
	Catalytic combustion					✓														
	Piezoelectric																	✓		
	Wet chemistry (autotitrators)															✓				
Grouped by component for which analysis is being made	Chlorine															✓				
	Combustibles					✓	✓				✓									
	Dissolved oxygen															✓				
	Flame						✓													
	Humidity		✓									✓							✓	
	H_2S							✓			✓									
	Moisture in gas or liquid samples			✓					✓	✓	✓	✓						✓		
	Moisture in solid samples			✓																
	Oxygen			✓		✓	✓		✓		✓					✓				
	Ozone														✓	✓				
	Sampling systems					✓			✓		✓									
	Smoke and dust density						✓											✓		
	SO_2 and SO_3							✓						✓						
	Water quality monitoring packages															✓				

Table X *continued*

Grouped	Sensor Type / Manufacturers of Online Analyzers	Electronic Associates, Inc.	Environmental Systems Corp.	Farrand Optical Co., Inc.	Fischer & Porter Co.	Foxboro Co.	Gam Rad, Inc.	GCA/Precision Scientific Group	General Electric	Hach Chemical Co.	Hays-Republican Div., Milton Roy Co.	Honeywell Inc.	Houston Atlas, Inc.	Industrial Nucleonics Corp.	Infrared Industries, Inc.	Jacoby-Tarbox Corp.	Kay Ray, Inc.	Kollmorgen Corp.	Leeds & Northrup Co., a unit of General Signal	Lockwood & McLorie, Inc.
Grouped by the methods of analysis applied	*Electromagnetic Radiant Energy*																			
	Ultrasonic																			
	Microwave																			
	Infrared								✓						✓				✓	
	Visible																			
	Colorimeter			✓	✓			✓		✓			✓					✓		
	Flame photometer												✓							
	Refractive index																			✓
	Turbidity				✓	✓	✓	✓		✓		✓				✓				
	Ultraviolet			✓																
	Radiation (gamma, neutron)								✓					✓			✓			
	Electrochemical																			
	Electrolytic cells																			✓
	Galvanic cells										✓									✓
	pH					✓						✓							✓	
	Ion selective electrodes					✓					✓	✓							✓	
	ORP (redox)					✓						✓							✓	
	Polarographic cells				✓							✓								
	Conductivity					✓						✓							✓	
	Inferential Property																			
	Conductivity-resistivity					✓						✓							✓	
	Consistency				✓	✓													✓	
	Density, thickness					✓								✓			✓			
	Dielectic constant							✓												
	Flash, melt, pour point							✓												
	Flow index																			
	Molecular weight																			
	Plastometry																			
	Thermal conductivity										✓									
	Vapor pressure					✓		✓												
	Viscosity				✓	✓														
	Combination, Miscellaneous																			
	Chromatography																			
	Adsorption																			
	Partition																			
	Gel-permeation																			
	Total carbon, hydrocarbon																			
	Ionization (mass spectrometry)																			
	Thermal & flame																			
	Electrostatic	✓																		
	Radiation																			
	Paramagnetic										✓									
	Catalytic combustion										✓									
	Piezoelectric																			
	Wet chemistry (autotitrators)									✓										
Grouped by component for which analysis is being made	Chlorine				✓															
	Combustibles										✓				✓					
	Dissolved oxygen										✓	✓								
	Flame																			
	Humidity																			
	H_2S												✓							
	Moisture in gas or liquid samples							✓											✓	
	Moisture in solid samples																✓			
	Oxygen										✓								✓	✓
	Ozone				✓															
	Sampling systems							✓			✓									✓
	Smoke and dust density		✓													✓				
	SO_2 and SO_3		✓																✓	
	Water quality monitoring packages		✓									✓								

Table X *continued*

Grouping	Sensor Type \ Manufacturers of Online Analyzers	*Mine Safety Appliances Co.*	*Moisture Systems Corp.*	*Monsanto Co.*	*Norcross Corp.*	*NUS Corp.*	*Ohmart Corp.*	*Panametrics, Inc.*	*Permutit Co., Inc.*	*Princo Instruments, Inc.*	*Process Analyzers, Inc.*	*Research Appliance Co.*	*Seiscor Div. Seismograph Service Corp.*	*Taylor Instrument Process Control Div.*	*Technicon*	*Teledyne Analytical Instruments*	*Texas Nuclear Div., Ramsey Engineering Co.*	*UCC Industries, Inc.*	*Union Carbide Corp.*	*Waters Associates, Inc.*	*Westinghouse Electric Corp.*
Grouped by the methods of analysis applied	*Electromagnetic Radiant Energy*																				
	Ultrasonic					✓															
	Microwave		✓																		
	Infrared	✓												✓		✓					
	Visible																				
	Colorimeter															✓					
	Flame photometer																				
	Refractive index																				
	Turbidity															✓			✓		
	Ultraviolet															✓					
	Radiation (gamma, neutron)						✓	✓				✓		✓			✓				
	Electrochemical																				
	Electrolytic cells										✓								✓		✓
	Galvanic cells										✓					✓					
	pH																		✓		
	Ion selective electrodes																				
	ORP (redox)																		✓		
	Polarographic cells																				
	Conductivity																		✓		
	Inferential Property																				
	Conductivity-resistivity																		✓		
	Consistency				✓																
	Density, thickness					✓	✓		✓	✓	✓			✓			✓	✓			
	Dielectic constant							✓													
	Flash, melt, pour point																				
	Flow index										✓										
	Molecular weight			✓									✓								
	Plastometry			✓									✓								
	Thermal conductivity	✓									✓					✓					
	Vapor pressure																				
	Viscosity			✓	✓								✓								
	Combination, Miscellaneous																				
	Chromatography																				
	Adsorption	✓									✓										
	Partition	✓									✓										
	Gel-permeation																			✓	
	Total carbon, hydrocarbon										✓								✓		
	Ionization (mass spectrometry)																				
	Thermal & flame	✓														✓					
	Electrostatic	✓																			
	Radiation	✓															✓				
	Paramagnetic	✓																			
	Catalytic combustion	✓																			
	Piezoelectric																				
	Wet chemistry (autotitrators)														✓				✓		
Grouped by component for which analysis is being made	Chlorine																				
	Combustibles	✓																			
	Dissolved oxygen																		✓		
	Flame																				
	Humidity							✓													
	H_2S										✓										
	Moisture in gas or liquid samples	✓						✓			✓		✓			✓					
	Moisture in solid samples	✓	✓		✓		✓							✓							
	Oxygen										✓					✓					✓
	Ozone																				
	Sampling systems	✓											✓								
	Smoke and dust density											✓							✓		
	SO_2 and SO_3	✓									✓				✓						
	Water quality monitoring packages														✓				✓		

10.1 ANALYZER APPLICATION AND SELECTION

In 1980, the total domestic purchase of instruments was about 20 billion dollars. Of this, analytical instrumentation amounted to over 2 billion dollars. This is a drastic increase, if one considers that ten years ago the total purchases of all instruments in the United States amounted to only 5 billion dollars.

This growth rate will change the nature of the analyzer industry as it presently exists because large production runs can now replace the custom-built units, and funds will become available for large-scale research and development activities. The demand for more and better analyzers is reinforced by the growing concern about environmental pollution and the associated need to enforce safety standards by detecting the concentrations of various pollutants.

Another reason why more and better analyzers are needed is our concern for improved productivity and the increasingly competitive and sophisticated nature of the petrochemical industry. A plant may operate at a profit or at a loss, depending on its ability to increase the efficiency (by only a few percentage points) of converting feeds into products. This can be done only if the operators know what is flowing in the pipelines and are aware of the rates of flow. This represents a higher level of sophistication compared to that of laboratory analysis of grab samples, which was previously considered to be sufficient.

Another factor in more and better analyzers is the new generation of control systems that replaced the regulatory and feedback concepts with adaptive, feedforward and optimizing modes of control, targeting energy conservation as one of its major goals.

That there is a need for more and better analyzers does not mean that their application is, or ever will be, a simple, routine task. An analyzer system will fulfill its expectations only if careful planning and evaluation precede its purchase, and if the users realize that if an expensive analyzer is worth purchasing, it is also worth calibrating and maintaining properly after installation. Another important factor is the operator's acceptance, which depends largely on training and familiarization. The recent introduction of self-diagnosing and self-calibrating analyzers is a major contribution to improved operators' acceptance. What follows is a discussion of selection and application in addition to a brief state-of-the-art review.

Analyzer Selection

The reasons for purchasing process analyzers include cost and safety, as well as improvement of product quality or quantity, reduction of by-products, decrease in analysis time, tightening of specifications and monitoring of contaminants, toxicants or pollutants. By continuing to ask how to accomplish these objectives, one arrives at the specific measurement requirement—to measure material A in the presence of material B, or to measure property X of material C.

Additional requirements and characteristics of the measurement should also be defined, such as the frequency of analysis, sample availability and the like. Distinguishing between the actual needs and the desirable characteristics can save both time and money.

A form is helpful in gathering data on needs and desires (Table 10.1a). In selecting an analysis system, one must ask whether the need for the measurement justifies the cost of determining what type of analysis system is required. If the answer is yes, the study progresses until several types of analyzers are selected, at which time the question of cost is again raised, and some analyzers may be eliminated from further consideration. After the complete system has been defined, the estimated costs can be compared with the expected return and a decision reached.

Some of the more important factors that must be weighed by the instrument engineer are noted in Table 10.1b, in which both the desirable and undesirable features are explained. Unfortunately, no single analyzer combines all the desirable features of providing on-stream, specific, continuous, unattended high-sensitivity readings without drift, noise or need of a sampling system. Therefore, the selection is always a compromise, which is likely to give satisfactory results only if it is preceded by careful evaluation.

If the problem has been defined as one requiring the measurement of one material in the presence of another, it is necessary first to look for a unique property of the

Table 10.1a

ANALYZER SPECIFICATION FORM

Project: .... Date: ....
Specification No.: .... Code: ....
Information Compiled by: ....

A. GENERAL INFORMATION:
   1. Plant: .... 2. Unit: ....
   3. Process: ....

B. CONDITIONS AT PLANNED ANALYZER LOCATION:
   1. Ambient Temperature Range: .... to .... Normal .... °C ☐ °F ☐
   2. Protected from Weather: Yes ☐ No ☐
   3. Unusual Ambient Conditions ....
      (Corrosive or explosive atmosphere, excessive moisture, dust, etc.)
   4. Power Available:................ Volts................ Hertz
      (a) Voltage Variation: .... to .... Volts
      (b) Frequency Variation: .... to .... Hertz
      (c) Grounding Facilities Available: Yes ☐ No ☐
   5. Lighting Level: *Good* *Average* *Poor*
      (a) Front of Instrument ....
      (b) Back of Instrument ....
      (c) Direct Sunlight Will Strike Instrument: Yes ☐ No ☐
   6. Steam Lines Near Location: ....
   7. Instrument Air Available:
      (a) Pressure Range: from .... to .... Normal .... PSIA
      (b) Temperature Range: from ...... to ...... Normal ...... °C ☐ °F ☐
      (c) Contaminants: ....
      (d) Size of Header: .... Volume: .... ft.^{3}/min.
   (Use Separate Page for Each Stream to be Analyzed)

C. SAMPLING INFORMATION
   1. Form of Sample: Gas ☐ Liquid ☐ Other ....
   2. Temperature Range: from ...... to ...... Normal ...... °C ☐ °F ☐
   3. Pressure Range: from .... to .... Normal .... PSIA
   4. Dew Point: .... °C At .... PSIG
   5. Quantity Available: .... Per Hour
   6. Low-pressure Return Line Available:
      Yes ☐ No ☐ Back-pressure .... PSIG
   7. Specific Gravity ....
   8. Contaminants in Sample:
      Oil ☐ Wax ☐ Solids ☐ Particle Size ....
      (Identity and concentration to be included in list of components below)
   9. Corrosive Nature: Acid ☐ Basic ☐ Other ....
   10. Other Data (Viscosity, Unusual Surges, etc.): ....
   11. Materials of Construction That May Be Used in Contact with Sample: ....
   12. Distance from Sample Tap to Analyzer Location: .... ft.
   13. Size of Tap at Process Line if Any: ....
   14. Concentration Ranges of All Components (even if only traces) in Stream: (Specify unit of measurement: % by volume, % by weight or ppm for each component)

Components to be Analyzed					*Other Components*				
	Max	*Min*	*Normal*	*Unit*	Water (liquid) (vapor)	*Max*	*Min*	*Normal*	*Unit*

   15. The Above Stream Composition Information Is Considered Proprietary:
       Yes ☐ No ☐
   16. Method of Lab Analysis Used to Measure Sample: ....
   17. Desired Response Time of Analyzer: .... Minutes .... Seconds
   18. Accuracy Required: % of full-scale reading

D. INSTALLATION REQUIREMENTS:
   1. Type of Installation: Permanent ☐ Temporary ☐ Portable ☐
   2. Type of Mounting: None ☐ Rack ☐ Panel ☐
      Other ....
   3. Electrical Code: Class .... Group .... Division ....
   4. Recorder or Indicator Required: ....
      (a) To Be Supplied with Analyzer: Yes ☐ No ☐
   5. Location of Recorder: At Analyzer ☐ Distance from Analyzer ... ft.
   6. Recorder Mounting: None ☐ Rack ☐ Panel ☐
   7. Accessories:
      (a) Alarms: High ☐ Low ☐ (b) Controls: ....
      (c) Others: ....
   8. Date Required: ....
   9. Sketch of System Indicating Sample Points and Distances to Analyzer:

Table 10.1b
FACTORS AFFECTING ON-STREAM ANALYZER PERFORMANCE

	Code	Desirable	Code	Undesirable
Selectivity	2,3,4,5,8,11, 12,14,16,23, 31,32,33,34, 35,36,37,38, 40,41	Responds only to specific component of interest; is unaffected by or can be compensated for pressure, temperature, other factors.	6,7,9,13,15,17, 20,21,22,24, 25,26,27,28, 29,39	Responds to a process property that is affected by both the process conditions and the presence of components other than the one of interest. Applications limited to compensated binary systems.
Operation	All except 23,31,32,33,34	Continuous, unattended	23,31,32,33,34	Semicontinuous or intermittent
Quality	2,3,6,8,11,12, 16,31,32,33, 35,36,37,38, 39,40,41	Good sensitivity, low drift, high signal-to-noise ratio.	1,4,5,7,9,13, 14,15,17,20, 21,22,23,24, 25,26,27,28, 29,34	Low sensitivity, substantial noise or drift.
Sample phase	3,4,8,21,22,25, 29	Handles both liquid and gas samples.	All except 3,4,8,21,22,25, 29,41	Limited to either liquid or gas samples.
	2,3,9,17,20,21, 22	Can analyze solids.	All except 2,3,9,17,20,21, 22,41	Not applicable to solid samples
Sample handling	2,9,17,21,22	No sampling system required. Sensor is not in contact with process stream.	4,5,8,11,23,24, 25,26,27,28, 31,32,33,34, 35,36,37,38, 39,40,41	Sample must be withdrawn from process and taken to analyzer. Results: transportation lag, possible need for altering.
	3,6,7	Sensor requires a window but is not in contact with process.		
	1,12,13,14,15, 16,17,22,29	Sensor has retractable probe or can be cleaned ultrasonically by air or by liquid jets.		This could involve flashing, condensing, filtering, diluting, drying, or other manipulation that could affect representativeness of sample.

*Explanation of Sensor Code Numbers**

Type: Electromagnetic radiant energy
- 1 Ultrasonic
- 2 Microwave
- 3 Infrared
- Visible
  - 4 Colorimeter
  - 5 Flame photometer
  - 6 Refractive index
  - 7 Turbidity
- 8 Ultraviolet
- 9 Radiation (gamma, neutron)

Type: Electrochemical
- 11 Electrolytic cells
- 12 Galvanic cells
- 13 pH
- 14 Ion selective electrodes
- 15 ORP (redox)
- 16 Polarographic cells
- 17 Conductivity

Type: Inferential property
- 17 Conductivity-resistivity
- 20 Consistency
- 21 Density
- 22 Dielectric constant
- 23 Flash, melt, pour point†
- 24 Flow index
- 25 Molecular weight
- 26 Plastometry
- 27 Thermal conductivity
- 28 Vapor pressure
- 29 Viscosity

Type: Combination, miscellaneous
- Chromatography
  - 31 Adsorption
  - 32 Partition
  - 33 Gel-permeation
- 34 Total carbon analyzer
- Ionization (mass spectrometry)
  - 35 Thermal & flame
  - 36 Electrostatic
  - 37 Radiation
- 38 Paramagnetic
- 39 Catalytic combustion
- 40 Piezoelectric
- 41 Wet chemistry

*Note: Codes 10, 18, 19, and 30 unassigned
†Physical properties sensors

material to be measured. Usually, a first step in determining the suitable methods of analysis is to investigate laboratory methods to determine the desired property. The ASTM has established certain test methods for the determination of various properties and materials, as have several other organizations. Suppliers of the material to be analyzed are possible sources of information for methods of analysis as are suppliers of on-line analyzers. Properties[1] that can be utilized for process composition analysis are shown in Table 10.1b.

Once prospective methods of analysis have been selected, the search for appropriate hardware begins. Manufacturers' names and addresses can be obtained from this handbook or from *Buyers' Guides* and the *Thomas Register*. Ordinarily, details of the desired measurement are given to the potential analyzer suppliers and estimates of performance are obtained. The analysis system can often be purchased on a guaranteed performance basis. For proprietary installations the user himself must determine how the general analyzer specifications apply to his needs, and since the user is buying hardware only, the supplier will guarantee only the quality of materials and workmanship.

Specificity (Selectivity)

Specificity is the characteristic of responding only to the property or component of interest. The specificity of the analyzer will not exceed that of the analysis. Specificity is often a function of the range to be measured, the sample background and the process conditions (solid, liquid or gas; pressure and temperature).

Inaccuracy and Repeatability

Frequently, absolute inaccuracy cannot be established owing to the lack of a suitable calibration standard. For this reason, other terms, such as repeatability, take on added significance. Often, the terms do not have industry-wide significance, so the definition should be agreed upon with each supplier. (For a detailed discussion of such terms as accuracy, see Section 1.1.)

For the purpose of this discussion, repeatability is defined as the ability of an analyzer to produce the same output each time the sample contains the same quantity of component or property being measured. The terms stability, reliability and reproducibility are sometimes used synonymously with repeatability. Thus a repeatable analysis system, properly calibrated, also becomes precise. Lack of repeatability may be caused by the analyzer, the sample system or the effects of temperature, voltage, composition, pressure and flow rate. Repeatability and inaccuracy are normally expressed either as a percentage of the full measurement range or as a percentage of the actual reading. If the analyzer error can be given as percent of reading, measurement reliability will not deteriorate toward the lower end of its range.

Calibration

The ability to calibrate an analyzer properly usually depends on the availability of a reliable reference sample, or on the ability to perform reliable laboratory analysis on the actual sample that is entering the analyzer. This ability to calibrate has been the subject of many articles.[2]

Interferences with the analysis are physical or chemical effects that cause a deviation in the analyzer output. If the effect of the interfering substance remains constant, a compensation factor can usually be applied in the calibration procedure. During checkout of the installed system, one should verify the effect of interference by calibrating the analyzer for each suspected substance.

When the interferences cannot be predetermined, it may be practical to purchase the instrument subject to a plant test or to send known samples to prospective suppliers for evaluation. In any case, known and suspected interferences and their concentrations should be made known to prospective suppliers when requesting quotations.

Whether to believe the analyzer or the calibration standard can become a "chicken or the egg" proposition. Both the method of calibration and the accuracy of the calibration procedure should be established before one purchases a process analyzer.

Analysis Frequency

One should first determine whether continuous analysis, automatic-repetitive batch sampling or an occasional "spot check" is required. The information from the analyzer and the rate of the dynamic changes in the sample are the main factors of consideration.

The need for process control suggests either a continuous analyzer or an analyzer with a continuous output signal because a continuous analyzer provides a better chance for the system to reach and maintain equilibrium with the sample. Also, the mechanical design of a continuous analyzer ordinarily is less complicated than that of a discontinuous system. However, the discontinuous analyzer may be more attractive if automatic zero checks are frequently needed, if reagents are blended with the sample or if the sample is corrosive. Some analyzers, for example, chromatographs, are inherently discontinuous.

The rate at which the measured variable changes in the process is an important factor in determining the frequency of discontinuous analyses. If the sample has to be withdrawn and transported to the analyzer, the time lag factor must also be considered. If several different samples are to share the analyzer, additional time allowances are required.

Radiant Energy Sensors

This family of sensors operates on either absorption or reflection principles. If radiation at different wavelengths (Figure 10.1c) is passed through a process ma-

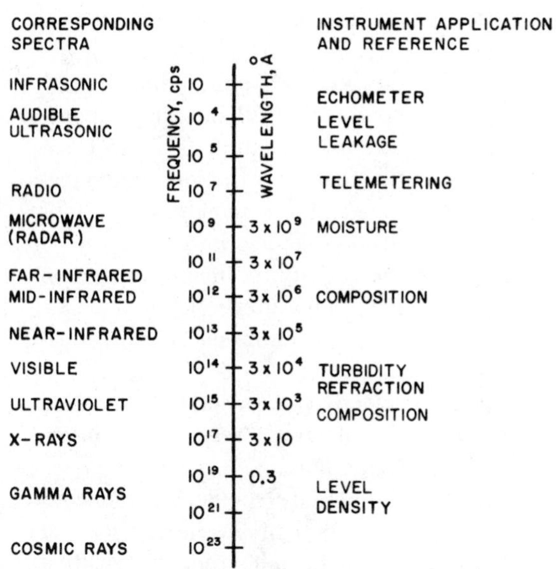

Fig. 10.1c Radiation spectrum pertaining to analytical instrumentation

terial, the amount of absorption is an indicator of sample identity or composition. The output of these analyzers is frequently nonlinear because Beer's law of radiation absorption is logarithmic:

$$A = abc = \ln\frac{I_0}{I} \qquad 10.1(1)$$

where

A = absorbance
a = molar absorption
b = sample path length
c = absorber (sample) concentration
I_0, I = radiation intensity entering and leaving sample

The microwave analyzer operates in both the absorption and the reflection modes. One of its applications is in the measurement of moisture in solids without requiring physical contact with the sample.

Infrared (IR) was one of the first analyzers to be moved from the laboratory to the pipeline and is available for composition detection of gas, liquid or solid streams. In the absorption mode of operation on liquid samples, the path length has to be very short. The noncontacting back-scatter designs detect the moisture in solids. The more recent probe version of the IR analyzer eliminates the need for a sampling system.

The refractive index (RI) is a unique property of a chemical compound and as such can be used for composition determination in binary systems. Snell's law of refraction expresses the relationship between RI, the angle of incidence (α) and the angle of refraction (β) as light passes through the interface between two materials:

$$RI = \sin\alpha/\sin\beta \qquad 10.1(2)$$

Differential refractometers utilize this relationship by keeping α constant; therefore, a measurement of β expresses RI. At a critical value of α the light is totally reflected; the measurement of this angle can also be related to RI as follows:

$$\alpha \text{ critical} = \text{arc sin}\left(\frac{\text{variable sample RI}}{\text{fixed prism RI}}\right)$$

Although less sensitive, this latter technique (critical angle refractometry) requires no sampling system and, therefore, is preferred for on-stream applications. (For a listing of refractive indexes, see Table 10.1d.)

As with refractometers, turbidity sensors also work with visible light, either in the absorption mode or in the reflectance (nephelometry), light-scattering mode. The first detects the sum of all light-absorbing effects, including color variations, whereas the second measures only the concentration of suspended solids.

Ultraviolet (UV) analyzers are less specific but more sensitive than IR and can handle only clean, single-phase samples. Recent studies indicate that the UV bands of some chemicals shift with changes in pH, a phenomenon that could also be used for composition analysis. (See Table 10.1e for a list of UV-absorbing compounds.)

The absorption or back-scatter of neutron or gamma radiation can also be correlated to composition in binary systems. Neutrons have been used to measure the moisture content of solids in processes in which hydrogen is present only in the free water and is not bonded to the other molecules. Gamma rays can penetrate metallic walls, thus giving composition data for binary systems without contacting the process stream. Scheduled re-

Table 10.1d
REFRACTIVE INDEX TABLE
All Data Based on 68°F (20°C)

Acetic acid	1.3718	Formic acid	1.3714
Acetone	1.3588	Glycerol	1.4729
Acrylic acid	1.4224	Glycol	1.4318
Amyl acetate	1.4012	Heptane	1.3876
Benzene	1.5011	Hexane	1.3749
Butyl acetate	1.3951	Hexanol	1.4135
Butyl alcohol	1.3993	Hydrazine	1.470
Butylene	1.3962	Hydrogen chloride	1.256
Carbon disulfide	1.6295	Lead tetraethyl	1.5198
Carbon tetrachloride	1.4631	Menthol	1.458
Chlorobenzene	1.5248	Methyl alcohol	1.3288
Chloroform	1.4464	Methyl-ethyl ketone	1.3807
Cycloheptane	1.4440	Nitric acid	1.397
Cyclohexane	1.4262	Nonane	1.4055
Cyclohexanone	1.4503	Octane	1.3975
Cyclopentane	1.4065	Pentane	1.3575
Decane	1.41203	Perchloroethylene	1.5053
Di-ethyl benzene	1.4955	Phenol	1.5425
Di-ethyl ether	1.3497	Pronanol(iso)	1.3776
Di-methyl benzene	1.4972	Propanol(n)	1.3851
Ethyl acetate	1.3722	Styrene	1.5434
Ethyl alcohol	1.3624	Toluene	1.4969
Ethylbenzene	1.4952	Water	1.3330

Table 10.1e
PARTIAL LIST OF UV-ABSORBING COMPOUNDS

Acetic acid	Formic acid
Acetone	Hydrogen peroxide
Ammonia	Hydrogen sulfide
Benzene	Iodine
Bromine	Isoprene
Butadiene (1,3)	Mercury
Carbon disulfide	Naphthalene
Carbon tetrachloride	Nitric acid
Chlorine	Ozone
Chlorobenzene	Perchloroethane
Crotonaldehyde	Phenol
Cumene	Phosgene
Cyclohexane	Styrene
Cycloheanane	Sulfur
Cyclohexanol	Sulfur dioxide
1,3, Cyclopentadiene	Sulfuric acid
Ethylbenzene	Toluene
Fluorine	Trichlorobenzene
Formaldehyde	Xylene (orthi, meta, and para)

calibration using calibrated absorber plates is necessary to compensate for source decay and drift.[3]

Electrochemical Sensors

In a recent development, the hydrochromatograph, the chromatographic column separates the water from the sample, and an electrolytic cell generates the electrolysis current, which is proportional to the quantity of water. This technique makes it possible to detect the moisture content of unsaturated hydrocarbons that otherwise could polymerize and plug the electrolytic cell.

In galvanic and polarographic cells in probe packaging, the electrolyte gel is separated from the process stream only by a membrane, a feature that eliminates the need for a sample system and performs well if the membrane is kept clean. The electrochemical reaction in the electrolyte is either spontaneous or is caused by the polarizing voltage, and the resulting current flow is an indication of composition.

The oxidation-reduction potential (ORP) sensors are also available in probe designs to detect the ratio of reducing agent to oxidizing agent—an important parameter in effluent treatment controls.

A family of analytical sensors that detects the electrical potential generated in response to the presence of dissolved ionized solids in a solution includes pH, conductivity and ion selective probes. Before discussing each of these devices, we will review the principle of their operation, which is based on the Nernst equation:

$$E = E_0 + \frac{F}{n} \log (\gamma c + s_1\gamma_1 c_1 + s_2\gamma_2 c_2 \ldots), \qquad 10.1(3)$$

where

E = potential difference between sensing and reference electrodes

E_0 = base potential, which is a function of the construction and characteristics of the electrodes

F = the Nernst factor, which is approximately 60 mv. if temperature is constant at 77°F (25°C)

n = the charge on the ion being measured (1 for monovalent, 2 for divalent and so forth)

γ = the activity coefficient of the ion being measured

c = the concentration being measured

s_1, s_2 = selectivity constants reflecting the electrode response to interfering ions

$\gamma_1, \gamma_2, c_1, c_2$ = the activity coefficient and concentration of interfering ions to which the electrode is sensitive with the selectivity of s_1 and s_2

Conductivity sensors measure a solution's ability to conduct electricity, which is a function of all dissolved ionized solids in the solution. These detectors are packaged either as probes (with isolating valves for removal, without opening up the process) or in the flow-through, inductive, designs.

Ion selective electrodes have received much attention recently, although the principles, which are based on the Nernst law, were known and applied (pH) for a long time. If total ionized solids (conductivity) are constant, a correlation can be drawn between the activity of a specific ion and its concentration in the process stream. The ideal reference electrode produces a constant potential that is independent of the composition of the solution. A "perfect" measuring electrode gives a 60-mv. change in potential for each tenfold change in the activity of a monovalent ion. It is important to emphasize that it is the *activity* of free ions that the electrodes respond to and *not* concentration. It is equally important to understand that according to the Nernst equation, concentration (c) can be determined by the measurement of activity (activity = γc) *only if* the other variables of the equation (F, s_1, s_2, γ_1, γ_2, c_1, c_2 and so forth) are *constant*. To achieve this involves scrupulous design; occasionally it also requires sample preparation.

Unless the interferences to which ion selective measurements are subject are recognized and eliminated in the potential installations, misapplications are likely. The available ion selective electrodes are listed in Table 10.1f, where they are also grouped by the type of membrane utilized. Coating or material buildup on the membranes calls for the same degree of maintenance as required for pH electrodes.

The pH sensors are ion selective electrodes, sensitive as they are to the activity of free hydrogen ions in the process stream, and, as such, reflect the acidity or alkalinity of the sample. Recently pH hardware has been improved in various ways. Local, integral preamplifiers have become available, reducing drift, instability and transmission distance limitations. New reference electrodes with "nonflowing junctions" have reduced maintenance and eliminated pressurization. Other improvements include the combination of the measuring and reference electrodes into a single probe and the development of various electrode-cleaning devices, such as air and water jets and mechanical and ultrasonic cleaners. Despite all this activity, pH remains a difficult measurement for which installation of standby spare sensors and scheduled periodic maintenance are likely to be necessary.

Table 10.1f
ION SELECTIVE ELECTRODES—
TYPES AND APPLICATIONS

Ion	*Type of Electrode*	*Lower Detectable Limit (ppm)*	*Principal Interferences*
Bromide	Solid-state	0.4	CN^-, I^-, S^-
Cadmium	Solid-state	0.01	Ag^+, Hg^{++}, Cu^{++}, Fe^{++}, Pb^{++}
Calcium	Liquid-ion exchange	0.4	Zn^{++}, Fe^{++}, Pb^{++}, Cu^{++}, Ni^{++}, Sr^{++}, Mg^{++}, Ba^{++}
Chloride	Liquid-ion exchange	0.4	ClO_4^-, I^-, OH^-, NO_3^-, Br^-, OAc^-, HCO_3^-, F^-, SO_4^-
Chloride	Solid-state	1.8	Br^-, CN^-, SCN^-, I^-, NH_3, S^-
Copper (II)	Solid-state	0.006	Ag^+, Hg^{++}, Fe^{+++}
Cyanide	Solid-state	0.3	S^-, I^-
Fluoride	Solid-state	0.02	OH^-
Fluoroborate (boron)	Liquid-ion exchange	0.11	I^-, HCO_3^-, NO_3^-, F^-, Br^-, OAc^-, OH^-, Cl^-, SO_4^-
Iodide	Solid-state	0.007	S^-, CN^-, $S_2O_3^-$
Lead	Solid-state	0.02	Ag^+, Hg^{++}, Cu^{++}, Cd^{++}, Fe^{++}
Nitrate	Liquid-ion exchange	0.6	ClO_4^-, I^-, ClO_3^-, S^-, Br^-, NO_2^-, CN^-, HCO_3
Perchlorate	Liquid-ion exchange	1.0	OH^-, I^-, NO_3^-, MnO_4^-, IO_4^-, $Cr_2O_7^-$
Potassium	Liquid-ion exchange	2.0	H^+, NH_4^+, Ag^+, Na^+, Li^+, Cs^+
Redox	Solid-state	Varies	All redox systems
Silver	Solid-state	0.01	Hg^{++}
Sodium	Solid-state	0.02	Ag^+, H^+, Li^+, K^+
Sulfide	Solid-state	0.003	
Thiocyanate	Solid-state	0.6	I^-, $S_2O_3^-$, Br^-, Cl^-, NH_3
Water hardness	Liquid-ion exchange	0.001	Zn^{++}, Fe^{++}, Cu^{++}, Ni^{++}, Ba^{++}, Sr^{++}

Inferential Property Sensors

Concerning property sensors and combination analyzers, the 1960s were a time of consolidation and improvement without major breakthroughs in the state of the art. Some of the more recent improvements and trends are noted below.

The detection of the dielectric constant has been used to measure the composition, or moisture content, of samples. Now instruments can measure the dielectric loss[4] of moving materials with all components external to the piping and with increased sensitivity. For example, the dielectric loss tangent of moist materials is frequently two orders of magnitude greater than the loss tangent of dry materials.

In plastometry, flow index or intrinsic viscosity measurement (the determination of molecular weight distribution and plastic behavior), the main development is that some of the Mooney, Kneader and capillary extrusion plastometers are now available for continuous, unattended on-stream service in polymer plants.

As with the plastometers, gel permeation chromatographs can also be equipped with automated sampling systems for on-stream measurements. The main drawback is the lengthy analysis; the advantage is that these chromatographs furnish data on molecular weight distribution in addition to molecular weight averages.

The piezoelectric effect is responsible for the phenomenon that an electric charge appears across crystals when they are exposed to a deforming force or that the frequency of crystal oscillation is affected by material deposits on the surface of the crystals. The first phenomenon has been used to detect pressure, acceleration, temperature, force and thickness or to generate ultrasonic waves; the second phenomenon has recently been applied to measure the moisture content of gases. The technique is moderately accurate because each microgram of moisture deposited in the hygroscopic coating of the crystal results in a frequency change of 2000 Hz.

In connection with pollution controls, one of the parameters of interest is the chemical oxygen demand (COD) of the effluent. The automatic total carbon analyzer gives close correlation to COD and is, therefore, an important analytical tool. The organic and inorganic carbon-bearing compounds in the sample are reacted catalytically, or otherwise, to form carbon dioxide and water. Carbon dioxide is then detected by an IR analyzer, and the entire cycle is completed in a few minutes.

The most powerful on-stream analyzer is the chromatograph, and its widespread use is reflected in the fact that 30 percent of all analyzer outlay is for this type of instrument. The chromatograph operates in two distinct steps. First, it separates the component(s) of interest from the rest of the sample; second, through elution by a carrier gas, it permits the use of binary sensors. These sensors are mainly of the thermal conductivity type; despite their higher cost, flame ionization detectors are the second most frequently used devices, owing to their superior sensitivity. In moisture measurement, for example, the phosphorus pentoxide cell (hydrochromatograph)[5] has been used successfully. Other chromatograph-detector combinations utilize the specific wavelength absorption characteristics of IR, visible and UV analyzers, the piezoelectric effect or, when both quantitative and qualitative analyses are desired, the ionization sensors, such as the mass spectrometer.

The orifice chromatographic detector is a recent development. This device operates at controlled constant flow rates to generate a pressure differential that responds to changes in the molecular weight of the eluting vapors. The simplicity and low cost of its design allows it to be dedicated to the measurement of a single component in a single process stream. This dedicated operation tends to increase the reliability of the whole installation. Another advantage is safety, due to the fact that the detector has no electrical components.

Improvements in chromatographic technology include parallel columns and programmed multiple temperature zones that contribute to the reduction of analysis time; signal-storing peak-pickers that close the control loop with an otherwise discontinuous analysis signal; solid-state circuitry that improves reliability; and new sampling valves. One improvement is in liquid sampling, in which peak-tailing and baseline separation have been minimized by preventing the unvaporized, or adsorbed, portion of the sample from entering the chromatographic column.

The cost of the chromatograph itself can be less than half, sometimes only 30 percent, of the total installation cost, and if the expense of maintenance is also included, the cost is reduced accordingly. This consideration has contributed to the emergence of two schools of thought, as have the reports indicating that for every successful chromatograph installation in the chemical industry, one has failed and been abandoned.

Adherents of the first school favor simple inexpensive chromatographs, monitoring one component in a single sample to increase reliability. Proponents maintain that an overburdened chromatograph is more a liability than an asset and that even when it is operating, the volume of information generated is more likely to swamp than to assist the operator. Therefore, their target is to achieve a degree of standardization and simplicity similar to that found in flow or temperature detectors. It is reported[6] that these simple chromatographs can accommodate the majority of present applications. The experience of about one hundred installations, including closed loop control systems, indicates that these standardized interchangeable units are easy to operate and maintain without the need for specialists.

Proponents of the second point of view feel that the answer lies in the opposite direction. They do not suggest reducing the number of samples or components analyzed but propose attacking complex problems with the tools of advanced technology. This means the use of dedicated minicomputers, each handling six or more chromatographs, at a total computer hardware-software package cost of $100,000 or more. For the most critical analyses, a standby manual mode of operation is frequently available to prevent plant shutdown during the expected annual downtime of five days. The computer can accurately integrate the chromatographic peaks, log or rearrange the measurement data, calculate control models or control functions, program all sequential or logic steps in the analysis cycle, optimize and, thereby, reduce cycle time and detect malfunctions. More important, it can anticipate maintenance needs before they occur and thus predict failures, such as sample valve leakage, and increase reliability by preventive maintenance.

Sampling Systems

If a sample has to be brought to the analyzer, a sample transportation delay and a potential for interference with the integrity of the sample will be introduced. The transportation lag can seriously deteriorate the closed loop control stability of the system.

Even more serious is the potential for interference with the integrity of the sample, due to the effects of filtration, condensation, leakage, evaporation and so on. As a result, the information on the composition of the process fluid is not only delayed, but also degraded.

Consequently, the best solution is to eliminate all sampling systems and place the analyzer directly into the process. Such "in-pipe" designs are becoming more and more available. Particularly well suited for this design are the various radiant energy analyzers and the probe sensors.

Sample systems are rarely duplicated, and thus each system must be "debugged" as a new entity. In deciding the design of a sampling system, the following questions should be raised:

1. Will the sample be adversely affected by sample transport and conditioning?
2. From what stage in the process will the sample be taken?
3. How can the sample be transported to the analyzer?
4. Is the sample solid, liquid, gas or a mixture?
5. In what phase must the sample be for analysis?
6. Must the sample be altered (filtered)?
7. Is sufficient sample available?
8. What will be the time lag introduced by transporting the sample?
9. Where is the excess sample returned?
10. Will the analyzer be shared by one or more samples?

These considerations are applicable both to the continuously flowing sample systems and to the less frequently used grab samples. Grab sampling is limited by variations in cleanliness of the sample containers, changes in the method of collecting the sample, delays in transporting the sample and deviations in withdrawing the sample from the sample container.

The piped-in sample system usually includes the hardware for calibrating the analysis system. It also provides a tap for obtaining samples for laboratory testing. Sample conditioning systems are usually costly to maintain.

As shown in Table 10.1a, some of the most powerful analyzers also require sampling systems. The installation price for the sampling system frequently exceeds the cost of the analyzer, but its importance overrides this economic consideration because a second-class analyzer can still furnish useable data if it operates with an efficient sample system, whereas a poor sample invalidates the entire measurement.

Therefore, the most important criterion requires keeping the samples representative, both in time (short sample lines guarantee minimum transportation delays) and in composition. Whenever possible, the sample should not be tampered with because the steps of sample preparation (drying, vaporizing, condensing, filtering and diluting) always degrade the representativeness of the sample. If there is no sampling system, the integrity of the sample is automatically guaranteed, and preference should be given to sensors that are external to the process pipe or penetrate it with a retractable, cleanable probe. Probe sensors, either solid or membrane, require periodic cleaning, which can be done manually, by withdrawing the probe through an isolating valve so that the process is not opened when the electrode is cleaned, or automatically. Automatic probe-cleaning devices may be pressurized liquid or gas jets, or thermal, mechanical or ultrasonic cleaning and scraping instruments (see Table 10.1g).

When a sample system is unavoidable, it should be made as simple as possible, with the minimum number

Table 10.1g
SELECTION OF AUTOMATIC PROBE CLEANERS

		Applicable Choice of Probe Cleaner						
		Mechanical		*Chemical*				
		Brush	*Rotary Scraper*	*Acid*	*Base*	*Emulsifier*	*Hydro-dynamic (self-cleaning)*	*Acoustical or Ultrasonic*
Service								
Oils, fats			√			√		√
Resins (wood, pulp)					√			√
Emulsions of Latex			√					
Fibers (paper, textile)							√	
Solid suspensions							√	
Crystalline precipitations (carbonates)		√	√	√				
Amorphous precipitations (hydroxides)		√	√	√				√
Material of construction		Stainless steel (brush pH 7–14)	Stainless Steel	PVC	PVC	PVC	Stainless steel	Polypropylene, stainless steel
Temperature	°F	40–140	40–140	40–140	40–140	40–140	40–250	40–195
	°C	4–60	4–60	4–60	4–60	4–60	4–120	4–90

of components. Even a well-designed multistream sampling system will most likely become expensive to maintain, owing to the number of components.

If the process is not pressurized, solid Teflon aspirators activated by water or air jets can withdraw the samples. They are easier to maintain than sample pumps. Similarly, the ball valves with block-bleed-block arrangements perform better than solenoids in sampling systems.

A frequent problem of sampling systems is plugging. There are two ways to eliminate this problem. The older, more traditional approach is filtering. Unfortunately, as the filters remove materials that might otherwise plug the system, they also remove process constituents and make the sample less representative.

The newer approach is to eliminate the potential for plugging by reducing the size of solid particles (homogenization) while maintaining the integrity of the sample. Thus, when pulverizers are used to replace filters, the analyzer samples become more representative.

If the material to be removed is dust, the self-cleaning bypass filter (Figure 10.1h) with automatic blowback constitutes a potential solution. In some instances liquid scrubbing or electrostatic precipitation should be considered, as well as cyclone separators. In the latter device (Figure 10.1i), the process stream enters tangentially to provide a swirling action, and the cleaned sample is taken near the center. Transportation lag can be kept to less than 1 minute, and the unit is applicable to both gas and liquid samples. This type of centrifuge can also separate sample streams by gravity into their aqueous and organic constituents.

Another good filter design is the rotary disc filter (Figure 10.1j). Here the filtered liquid enters through the small pores in the self-cleaning disc surfaces. The

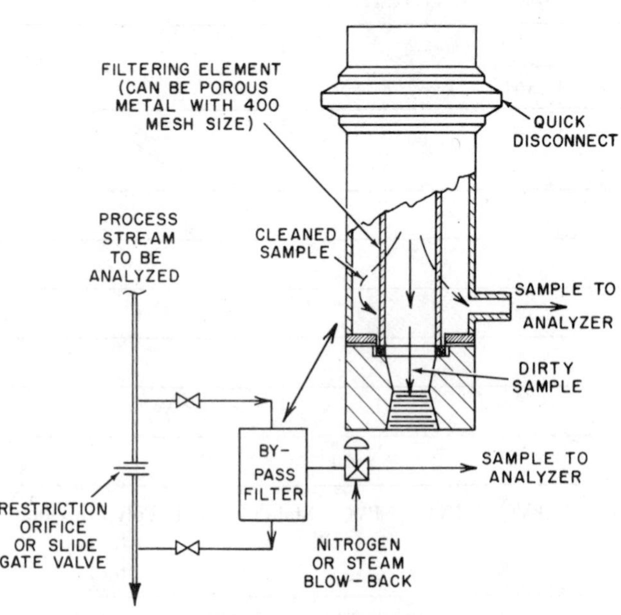

Fig. 10.1h Self-cleaning bypass filter and its installation

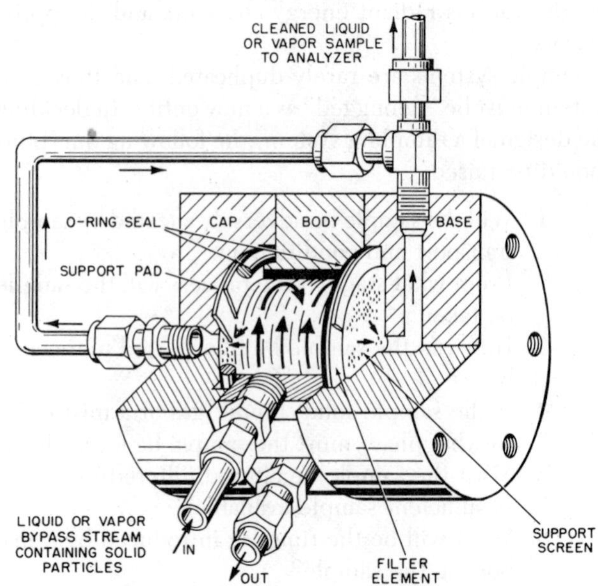

Fig. 10.1i Bypass filter with its cleaning action amplified by the swirling of the tangentially entering sample

sample liquid is drawn by the sample pump through the hollow shaft and is transported to the analzyer.

For the removal of small amounts of polymer dust in vapor samples there are melt filters with removable, heated metallic surfaces which melt and collect the polymer dust from the sample.

When the impurities in the process gas stream are both solids and liquids, such as in particulate matter and mist—carry-over problems in chlorine plants—the fiber mist eliminator[7] (Figure 10.1k) should be considered. The liquid particles form a film on the fiber surface, and the drag of the gas moves this film and the dissolved solids radically, while gravity causes them to move downward, resulting in self-cleaning action.

For dissolved solids or polymer-forming compounds in a process stream, which would leave a residue and eventually plug the liquid sample valve if not removed, the logical solution is to force the residue formation to

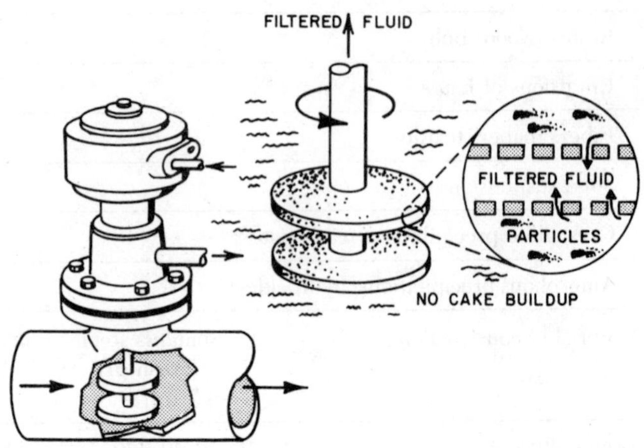

Fig. 10.1j Rotary disc filter

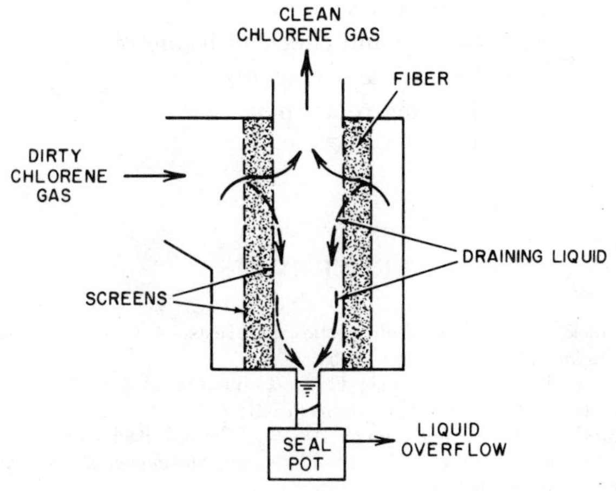

Fig. 10.1k Fiber mist eliminator

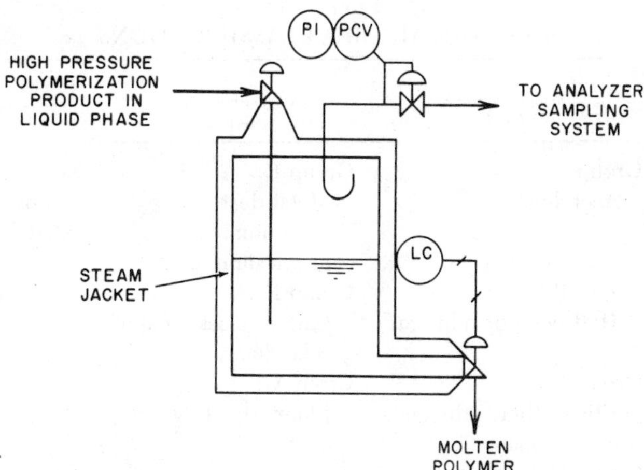

Fig. 10.1m Flash chamber makes the analysis of unreacted monomers possible

take place in a controlled area, such as the fiberglass filter in the spray-stripping chamber shown in Figure 10.1l. If polymers represent a substantial portion of the process stream, the need for filter replacements becomes excessive and, therefore, impractical. A better technique is to vaporize the unreacted monomers through pressure reduction while keeping the polymers in a molten state through heating. This technique (Figure 10.1m) not only discharges polymers continuously, but also provides a useable vapor sample.

When it is necessary to clean the windows on the various photometers operating on gas samples, a warm air purge can be used, keeping the window compartment isolated from the sample.

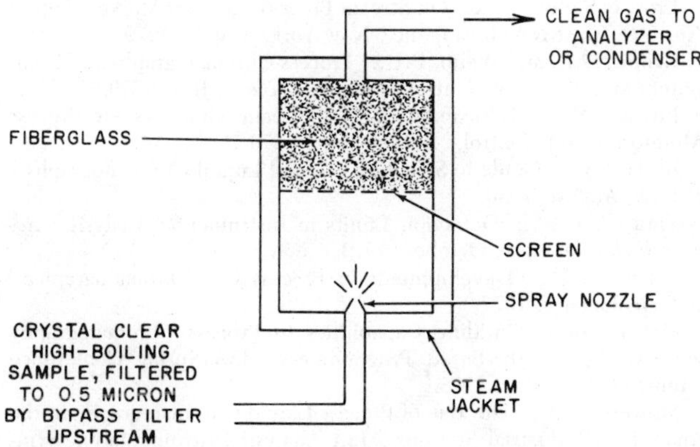

Fig. 10.1 1 When dissolved inorganic solids or polymer-forming compounds are present, stripping the liquid sample may be the answer

Analyzer Location

Location for the analyzer must be selected after considering ambient requirements; availability of utilities, space and readout location; safety, access for maintenance and analyzer response time.

An inferior quality utility can cause inaccuracies in the output or degradation of the analysis system. Problems are frequently caused by varying voltage, changes in electrical frequency, transients in the voltage, changing pressure in vents and drains and oil or moisture, or both, in plant or instrument air. The temperature and humidity of the atmosphere surrounding the analyzer may also contribute to inefficient performance.

Analyzers may be purchased in one housing or in several modules. A modular approach can often reduce space requirements, provide remote readout, improve safety, provide easy maintenance access and reduce sample transport time.

The major safety hazards are fire and explosion (see Table 10.1n).[8] The inability of laboratory analyzers to meet electrical requirements limits their use in on-line processing. Often, however, purging an analyzer that is not explosion-proof allows its use in a hazardous area of the plant. Samples or stored reagents containing toxic, flammable or noxious substances are also safety hazards.

Maintenance

Analyzer hardware is likely to receive better care if it is accessible and housed in pleasant surroundings, such as air-conditioned buildings. Spare parts and special testing components should be ordered at the same time as the system to avoid future delays; the availability and most effective use of qualified maintenance personnel are also factors of major importance.

An analyzer already in the plant is helpful in determining maintenance requirements. Major subassemblies not stocked as spare parts can also be interchanged, and help to obviate waiting for parts for the duplicated system.

Cost

The cost elements of an analyzer installation include the following:

Engineering study
Analyzer

Table 10.1n
ELECTRICAL AREA CLASSIFICATIONS

Flammable Gas Class I		*Combustible Dust Class II*		*Ignitable Fibers Class III*	
Group A: Acetylene		Group E: Metal dust, aluminum, magnesium, etc.		No groups listed	
Group B: Hydrogen or mfd. gas		Group F: Carbon black, coal or coke dust.			
Group C: Ethyl-ether, ethylene, cyclopropane		Group G: Flour, starch, grain dust			
Group D: Gasoline, hexane, naphtha, benzene, propane, butane, alcohol, acetone, benzol, lacquer solvent, natural gas					
Div. 1	Div. 2	Div. 1	Div. 2	Div. 1	Div. 2

CLASS I FLAMMABLE GASES

Division 1 Locations	*Division 2 Locations*
Locations in which: (1) Hazardous concentrations exist continuously, intermittently or periodically under normal operating conditions. (2) Gases exist frequently because of repair, maintenance, operations or leakage. (3) Breakdown or faulty operation might release gases that might also cause simultaneous failure of electrical equipment.	Locations in which: (1) Gases are handled, processed, or used, but will normally be confined to closed containers or systems from which they can escape only in accidental rupture or abnormal operation of equipment. (2) Gases are normally prevented by positive ventilation but might become hazardous through failure of ventilating equipment. (3) Adjacent to Class I, Division 1 locations in which gases might occasionally be communicated.

Recorder or other display hardware
Sample system
Startup and checkout in the laboratory
Calibration standard
Installation costs
Spare parts
Startup and checkout in the plant
Utilities and reagents
Training repairmen
Maintenance

REFERENCES

1. Mock, J.A., "Physical Properties and Tests—A to Z," *Materials Engineering*, 67:82, June, 1968.
2. Gray, T.A. and Kuczynsk, E.R., "Calibration of SO_2 Monitoring Instruments," *ISA Transactions*, 7:327, 1968.
3. McConnell, J.A. and Smuck, W.W., "Gamma Backscatter Technique for Level and Density Detection," *Chemical Engineering Progress*, August, 1967.
4. Wood, H.H., "An Instrument for On-Line Moisture Measurement Utilizing the Principle of Dielectric Loss," ISA 16th Analysis Symposium, May, 1970.
5. Penther, C.J. and Notter, L.J., "Hydro-Chromatography," *Analytical Chemistry*, February, 1964.
6. Topham, W.H., "A New Approach to Chromatograph Systems for Process Control," ISA 25th Conference, October 1970.
7. Nichols, J.H. and Brink, J.A., Jr., "Use of Fiber Mist Eliminators in Chlorine Plants," *Electrochemical Technology*, July–August, 1964.
8. Kuller, B.E., "Interpretation and Application of Article 500 of the National Electrical Code," presented at the Symposium on Limiting Electrical Losses, 63rd National Meeting of the American Institute of Chemical Engineers.

BIBLIOGRAPHY

Demming, P.L., "Process Chromatographs with Integral Microprocessors Upgrade Composition-Based Control," *InTech*, July, 1980.

Frost & Sullivan, Inc., On Stream Process Analyzer Market, Report No. 669; Frost & Sullivan, Inc., New York, August, 1979.

Griffin, D.E. and Webb, P.U., "Process Chromatographs and Computers in Optimizing Control Systems," *InTech*, July, 1979.

Jutila, J.M., "Multicomponent On-Stream Analyzers for Process Monitoring and Control," *InTech*, July, 1979.

Jutila, J.M., "Guide to Selecting Gas and Liquid Chromatographs," *InTech*, August, 1980.

Karasek, F.W., "Detection Limits in Instrumental Analysis," *Research/Development*, October, 1979.

Martin, F.D., "Developments in Process Gas Chromatography," *InTech*, January, 1977.

McCoy, R.D., "Adding Capabilities to Process Chromatography with Microprocessor-Based Programmers," ISA Spring Conference (Houston, TX), May, 1978.

Mowery, R.A., "The Use of Process Liquid Chromatography in the Analysis of Industrial Streams," ISA Analysis Instrumentation Symposium (Charleston, WV), May, 1977.

Oliver, B.M., "The Role of Micro-Electronics in Instrumentation and Control," *Scientific American*, September, 1977.

Villalobos, R; Porter, R.; LeBlanc, R. and Hearn, R. "New Concepts in Microprocessor-Based Process Chromatographs," 33rd Annual Texas A&M Symposium, January, 1978.

Wise, S.A. and May, W.E., "Unusual Experimental Detectors for LC," *Research/Development*, October, 1977.

Worthy, W., "Analytical Chemistry," *Chemical and Engineering News*, July 31, 1978.

Flow Sheet Symbol

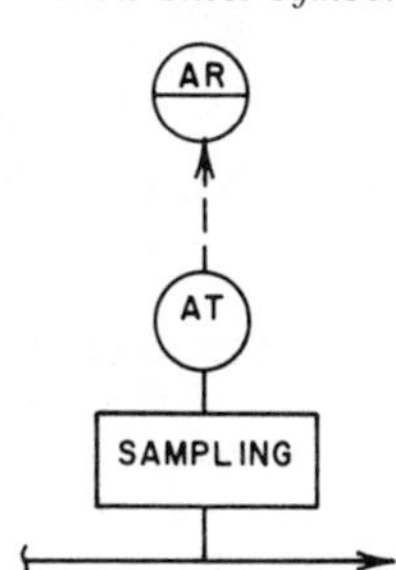

10.2 ANALYZER SAMPLING SYSTEMS—PROCESS

Cost:	Single-stream gas sample: \$1,500 to \$2,500 Single-stream liquid sample: \$2,000 to \$4,500 Multistream: 60 percent of single-stream cost multiplied by number of streams Specialty components: \$500 to \$10,000
Partial List of Suppliers:	Applied Automation, Inc.; Beckman Instruments, Inc.; Bendix Corp., Automation & Measurement Div.; Comsip Customline Corp.; Foxboro Co.; Leeds & Northrup, a unit of General Signal; Siemens Corp.; Tex-A-Mation; United Controls Div., Xertex Corp.

Processes are controlled by the application of automatic controls to produce higher quality, more uniform products, increase throughput and protect the safety of personnel and property. An analyzer system enhances process control by providing specific measurement of physical or compositional data of a process or ambient conditions. With the advent of computer technology, the need for such analysis has increased substantially to provide on-line data that the computers can use to optimize process control.

The sampling system is an integral and key part of an analyzer system and should be designed to obtain a representative sample, transport the sample to the analyzer system, condition the sample, accomplish sample stream switching if necessary, provide facilities for return and/ or disposal of the sample, and provide calibration facilities, preventive maintenance features and alarm functions for on-line reliability and operator alerts.

Sample Data Requirements

A comprehensive listing of the characteristics of each sample stream, including any abnormal conditions, should be prepared before an analyzer sample system is designed. Data are usually available from process conditions and laboratory analysis in existing plants and from design data for new plants. Table 10.2a is a form to be used in summarizing the characteristics of a sample.

Sample Take-off Point

Sample conditioning begins with the location of a suitable sample take-off point. To obtain a representative sample, the take-off is usually located at the side of a process line, especially in the case of liquid samples where there is the possibility of vapor on the top of a horizontal line and dirt or solids on the bottom of the line. For sampling vapors, the connection may be located in the side or top of the process line, but in both cases with due consideration to accessibility for maintenance.

Ideally, the sample at the appropriately selected take-off point will require little or no conditioning; however, it is good practice to install a sampling probe (Figure 10.2b) for most applications as a precautionary measure to prevent particulates from entering the sample transport system. Sampling processes that are still reacting chemically or pyrolysis gases may require reaction quenching, or fractionation, at the sample take-off. This is done by cooling or back-flushing with an inert gas or liquid to keep the sample take-off clean and reliably active (Figure 10.2c), while drawing off a reproducible sample for analysis.

With the advent of in situ analyzer detectors, the sample take-off becomes, in fact, the point of analysis, and for proper location of in situ analyzers the above considerations must be carefully evaluated.

Sample Transport

A representative sample extracted from a process line must be continuously conditioned while in transport to avoid compromising the sample integrity. Thus, provisions must be made to heat or cool the line as necessary for the specific condition. The sample is normally transported in one of three ways (Figure 10.2d).

Table 10.2a
SAMPLE ANALYZER DATA FORM

Stream Name or Identification ______________________________

STREAM COMPOSITION DATA

COMPONENT	CONCENTRATION IN MOL.%, WT.%, PPM	RANGE OF COMPONENT TO BE MEASURED

OPERATING PROCESS DATA

Temperature ______________ Pressure ______________

Phase: Liquid ______________ Vapor ______________

Corrosive Components/Solids ______________

Stability (Polymerize, decomposes, etc.) ______________

Sample Bubble Point ______________ Dew Point ______________

SAMPLING CONDITIONS

Maximum Distance—Tap to Analyzer ______________ Analyzer to Return ______________

Speed Loop Required: Yes ________ No ________

Sample Return Pressure Point ______________

Sample Probe Requirements: Connection Size ______________ Orientation ______________

Materials of Construction: SS ________ Teflon ________ Viton ________ Glass ________ Other ________

Electrical Area Classification ______________

Power Supply ______________

Output Signal ______________

Utilities Available: Steam ______________ Air ______________ Cooling Water ______________

(NOTE: *Specify Conditions and Units Where Applicable)*

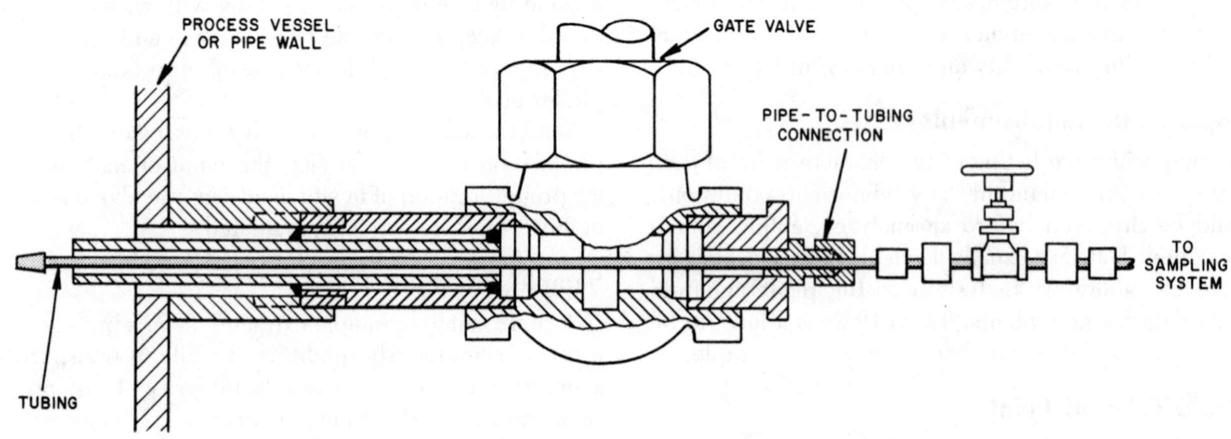

Fig. 10.2b Sample probe assembly with process shut-off valve

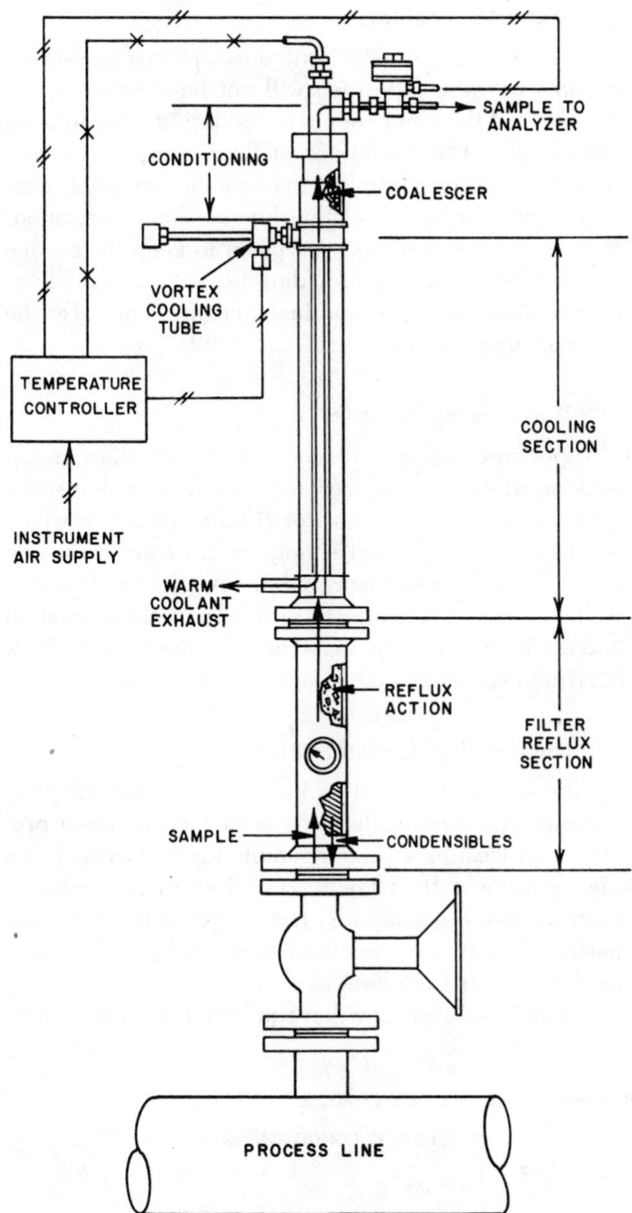

Fig. 10.2c Pyrolysis gas sample fractionation and conditioning unit

Single-line Transport

This is the most direct approach and is used when the sample line volume is small in relation to the analyzer sample consumption so that the transport time lag is reasonably short. It is usually used when the analyzer is field mounted close to the sample point and sample exhaust facilities are available.

Bypass Stream Transport

This is a commonly used method for maintaining a high-sample transport velocity that provides minimum transport lag. This method is used when samples are vaporized at the tap and no facilities exist for returning

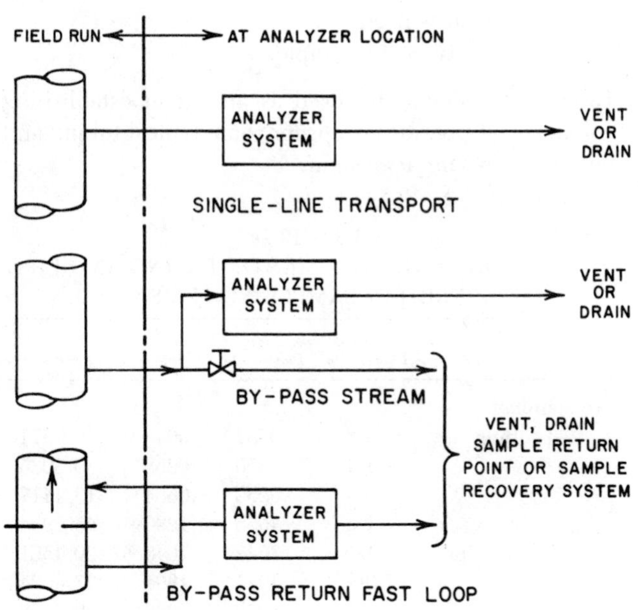

Fig. 10.2d Sample transport methods

the vapor to the process, or in similar situations. Consideration must be given to the fact that the sample bypass must be piped to a drain or vent. This procedure may have a negative impact on the environment, despite the fact that the cost of the sample being disposed of may economically justify a sample recovery system.

Bypass Return Fast Loop

This is the most commonly used transport loop for analyzers mounted away from the sample take-off point. By circulating a continuous high velocity across a device, to create a differential pressure, and drawing off a bypass stream to the analyzer, a fast loop is obtained, which is adjustable with no waste in the product. In most cases, such devices as pumps, control valves and process equipment exist in the process for this purpose. Otherwise it may be necessary to install an orifice in the process line or a circulating sample pump. With a circulation sample pump, care must be taken to prevent cavitation by locating the pump close to the sample take-off.

Calculating Sample Transport Lag

After selecting the appropriate sample transport method for each analyzer system, a calculation of the sample time lag should be made, using conventional flow equations based on the following:

1. Available differential pressure.
2. Total length of the fast loop from the sample take-off point to the analyzer location and back to the sample return point. Any restrictions on this loop should be excluded from available differential pressure.

3. Line sizes used.
4. Viscosity of the sample.

Table 10.2e should be used as an aid in establishing volumes and pressure drops for some typical tubing and pipe used in sample system.

Table 10.2e
DIMENSIONS AND VOLUMES OF TUBING AND PIPE USED IN SAMPLE SYSTEMS

	Nominal Diameter In.	*Inner Diameter In.*	*Internal Area In.²*	*Volume per Ft. cc*
316 stainless steel tubing	1/8	.0787	.0048	.9571
	1/4	.1850	.0268	5.3035
	3/8	.0253	.0684	13.4417
	1/2	.4055	.1290	25.2984
Schedule 40 pipe	1/4	.3642	.1040	20.4521
	3/8	.4921	.1891	37.4904
	1/2	.6220	.3038	59.7408
	3/4	.8268	.5363	106.6800

Sample Conditioning

The extracted sample begins conditioning at the take-off point, continues through the transport and finishes conditioning at the analyzer location prior to entering the analyzer. All samples require some form of conditioning to make them suitable for the analyzer and to assure reliable on-stream operation. The conditioning is done at the appropriate location in the sample system loop in order to maintain the integrity of the sample (Figures 10.2f,g).

Sample Washing

This washing is usually limited to dirty, particle-laden streams whose composition will not be affected by the solubility of the components in the liquid used to wash the sample. The conditions of flow, temperature and pressure must be controlled to maintain a relatively constant predetermined relationship of the composition. When washing, care must be taken to keep the sample in the vapor phase by providing heated transport lines or by making provisions for final moisture removal as the analyzer may require.

Vaporizing Samples

Vaporizing is frequently necessary for equilibrium liquids or when the analyzer requires it. This is usually done by a vaporizer regulator (Figure 10.2h) in which the sample is vaporized simply by pressure reduction across a capillary, or more often, by using a heater as well. Care must be exercised to avoid partial vaporization and/or fractionation by selecting a suitably heated vaporizing regulator to accommodate the sample.

Removal of Entrainments

This is a routine matter in liquid and gas sampling systems which normally starts with the minimum provisions of locating a proper sample tap and providing a sample probe at the take-off point. Filtration is normally used for gases and liquids. It removes both liquid and particulate entrainments from gases and primarily particulate matter from liquids.

For gases requiring further conditioning of heavy load-

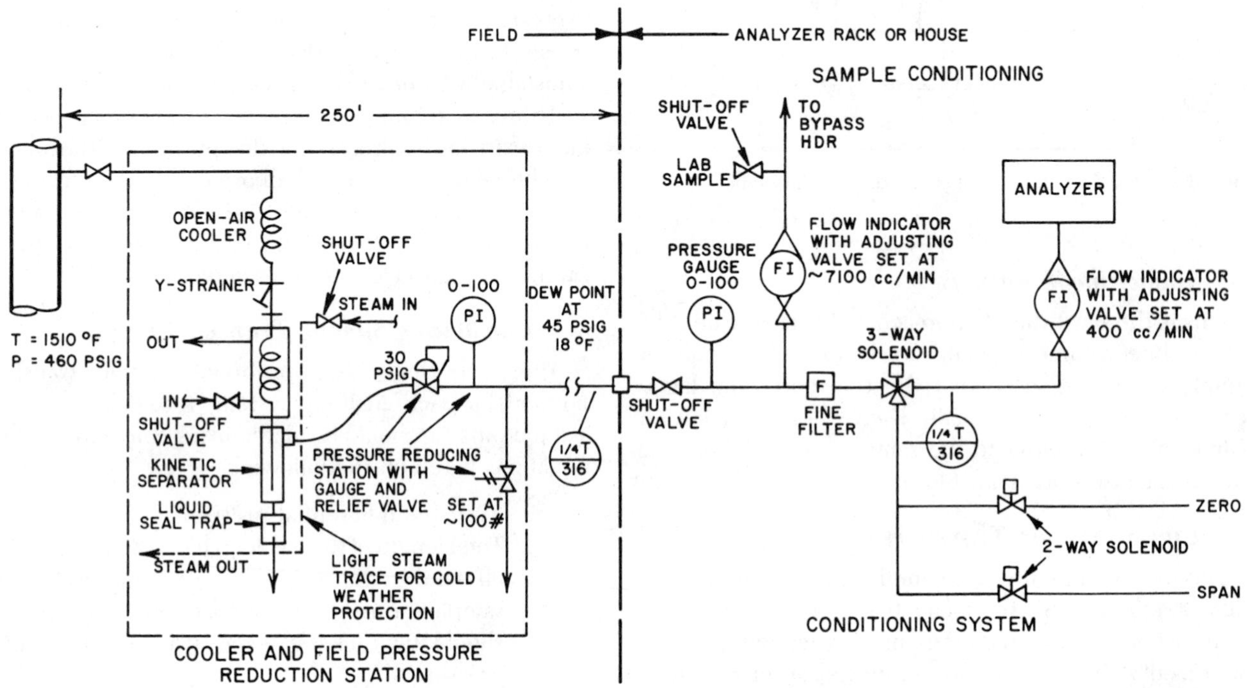

Fig. 10.2f Typical sample conditioning system with remote preconditioning unit

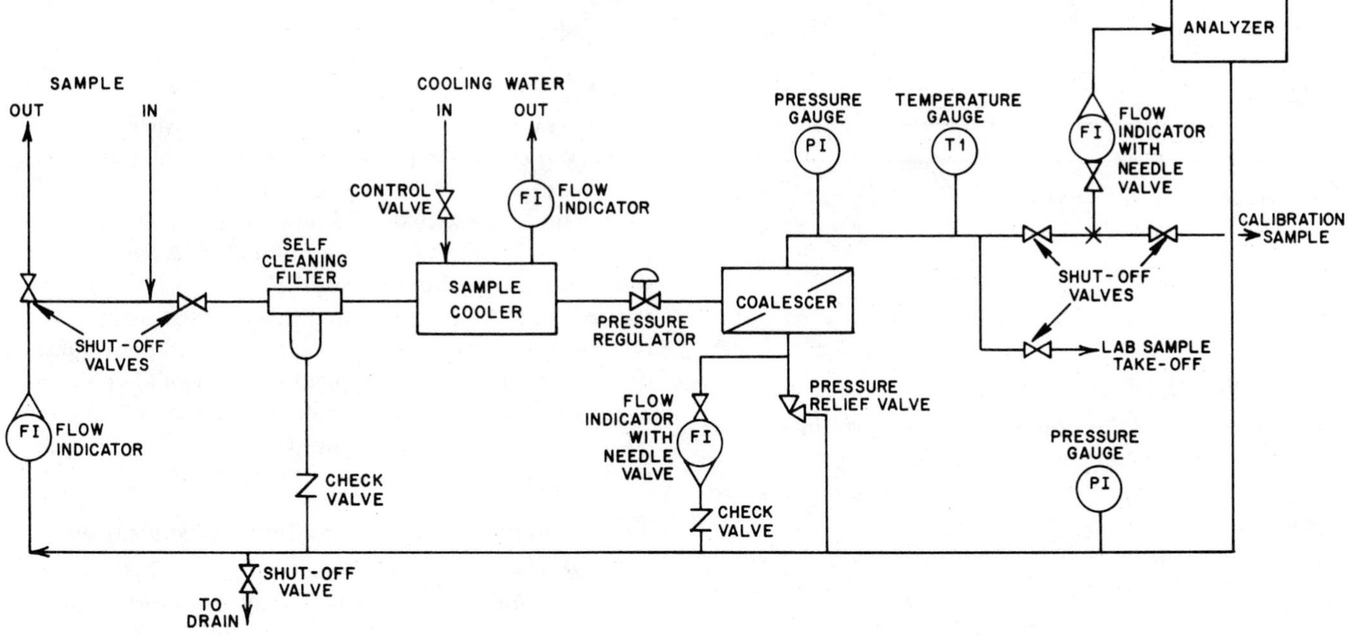

Fig. 10.2g Typical liquid product sample system for refinery applications

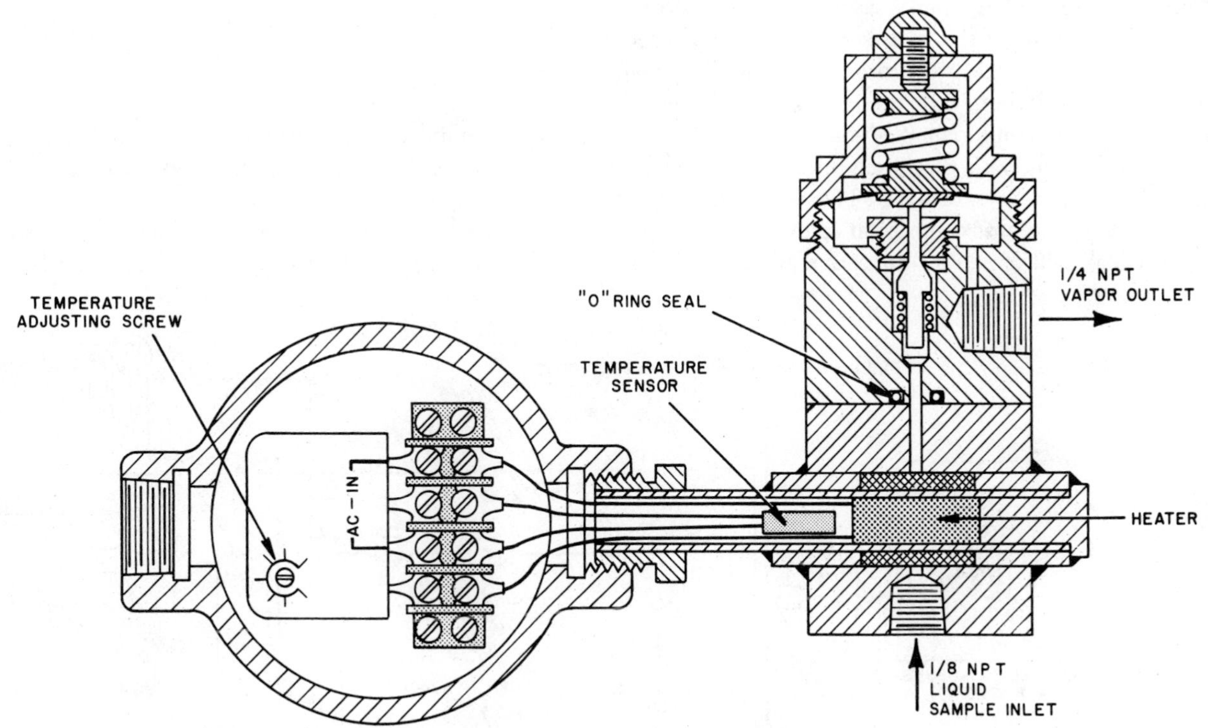

Fig. 10.2h Vaporizing regulator assembly (electrically heated)

ing, cyclone filters can be used if the sample has adequate velocity. For liquids, coalescers are frequently used to remove both undesirable gases and liquids by gravity. The removal of free water from a hydrocarbon stream is usually accomplished by passing the liquid through a hydrophilic element causing the water droplets to accumulate on the element. The hydrocarbon stream is then passed through a hydrophobic element that rejects the water, removing it from the bottom with a hydrocarbon bypass stream. Another method of removing moisture from a stream uses a selective permeation device with a drying medium that creates differential pressure to drive the water through the permeable material, thus removing it from the flowing stream (Figure 10.2i).

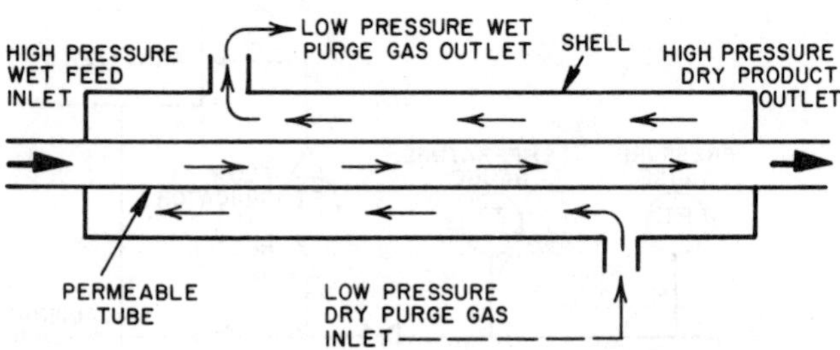

Fig. 10.2i Permeable tube sample dryer

Other Physical Properties

Adsorption of the sample components of interest on the walls or surfaces with which the sample comes in contact will affect the analysis, especially for measurement in the parts per million (ppm) range. Therefore, proper selection of materials and conditioning is essential for establishing an equilibrium that will make analysis reliable. Specifically, water vapor samples reach equilibrium more rapidly in stainless steel lines than in copper or plastic tubing. Diffusion is another consideration. Sample system design should assure that gases do not permeate the walls of the sample system. This is especially important in high-pressure systems and ppm analysis. Another problem with diffusion is leakage which can change the sample composition because gas molecules will flow in both directions of the leak and can significantly affect ppm analysis. This is best illustrated by the fact that in an oxygen ppm analyzer system, the slightest leak will create a full-scale reading on the instrument even though the leak is from a high-pressure sample to atmospheric pressure.

Chemical Treatment and Drying

Some sample streams may contain corrosive gases or water vapor that influence the accuracy of the measurement or potentially damage the analyzer. Removal of such undesirable matter or components can be accomplished by passing the sample through a packed bed of solid chemicals and/or desiccant. Further, a liquid treating agent may be applied with provisions that gas streams be broken up into small bubbles to assure proper contact between the liquid and gas phases. Care should be taken in such systems to avoid alteration of the sample and to condition the sample to a desired and reproducible form for analysis.

Trace Analysis Sampling System Considerations

Trace analysis sampling systems necessitate more stringent requirements than normal analyzers sample systems because of contamination, adsorption and desorption (Figure 10.2j). Care should be taken in the selection of construction materials and proper application of design criteria to avoid alteration of the sample. The following is a list of recommended practices for such systems.

1. Stainless steel seamless tubing is the preferred material because it provides inertness, smooth surfaces and low porosity.
2. All components should be thoroughly cleaned of oil, gas or other contaminating materials.

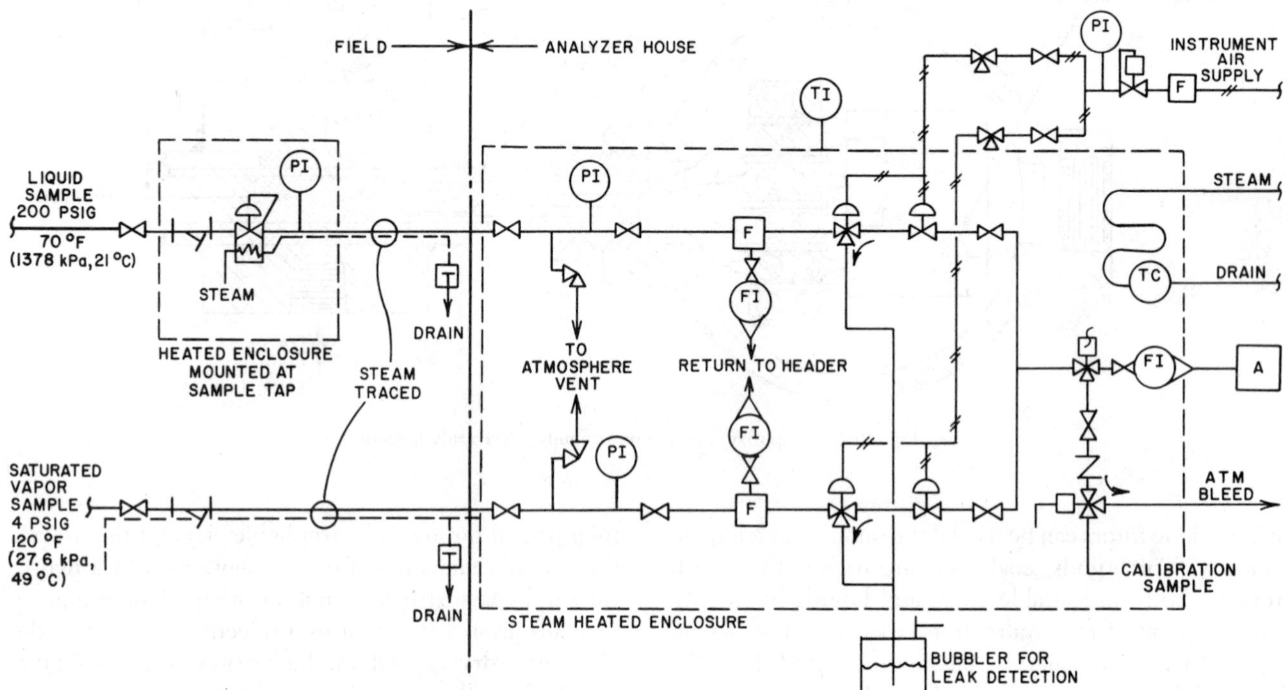

Fig. 10.2j Two-stream sampling system for trace analysis with double-block-double-bleed

3. Dense fluorocarbons, or other soft inert materials, may be used as diaphragms as required.
4. Tubing sizes are critical, especially where low flows are used because they limit the amount of increase in the adsorption/desorption phenomenon. A rule of thumb is to use the smallest possible tubing to achieve maximum flow to accommodate the sample loop design.
5. Packless shutoff valves with a diaphragm or bellows seals should be used; however, due to their high cost, serious consideration must be given to this area, and conventional valves are useable in most applications.
6. Filtration of the sample in ppm analysis can create significant problems unless the filter is totally inert. Therefore, stainless steel filters using a high flow rate, or dense fluorocarbon inert materials are recommended for such applications.
7. Conditioning of the lines and system to accommodate the ambient temperature requirements for preventing condensation of components of interest in the sample must be considered; if necessary, heat tracing of the lines must be furnished.
8. When it is necessary to provide an aid for transporting the sample from its sample point to the sample system through the analyzer, as is frequently done in ambient monitoring systems, a sample pump, ejector or aspirator is necessary. In such cases, the pump must be a diaphragm. If practical, an ejector to pressurize the sample or an aspirator to aspirate it through the measuring device can be used. Both are more desirable than a sample pump.

Multistream Switching

Multistream switching is usually used when it is practical to analyze several streams using one analyzer. However, because this system is more complex than single-stream sample conditioning, the following considerations should be reviewed to determine if multistreaming is feasible:

1. The contamination problem that can be encountered from stream to stream.
2. The importance of each analysis and frequency of analysis.
3. The loss of information of more than one analysis in the case of analyzer failure.
4. The cost of an additional analyzer vs the cost of multistreaming.
5. Maintenance requirements.

After reviewing the above, one can decide if multistreaming is feasible and whether it should be manual or automatic. Whether manual or automatic, multistreaming requires good quality valves for stream switching.

Typical systems are shown in Figure 10.2k, and one common and important requirement in all such systems is that a continuous bypass be provided for each sampling point to avoid dead-ended sample lines. The sample system should be laid out in such a manner that contamination between streams is avoided. This is best accomplished by arranging the solenoid valve in a double-block-double-bleed arrangement, which is rather expensive. More often a three-way solenoid valve is used for each stream with the venting port always at low pressure to create a relief in case of a leak. To prevent contamination, dead volume of the sample system should be

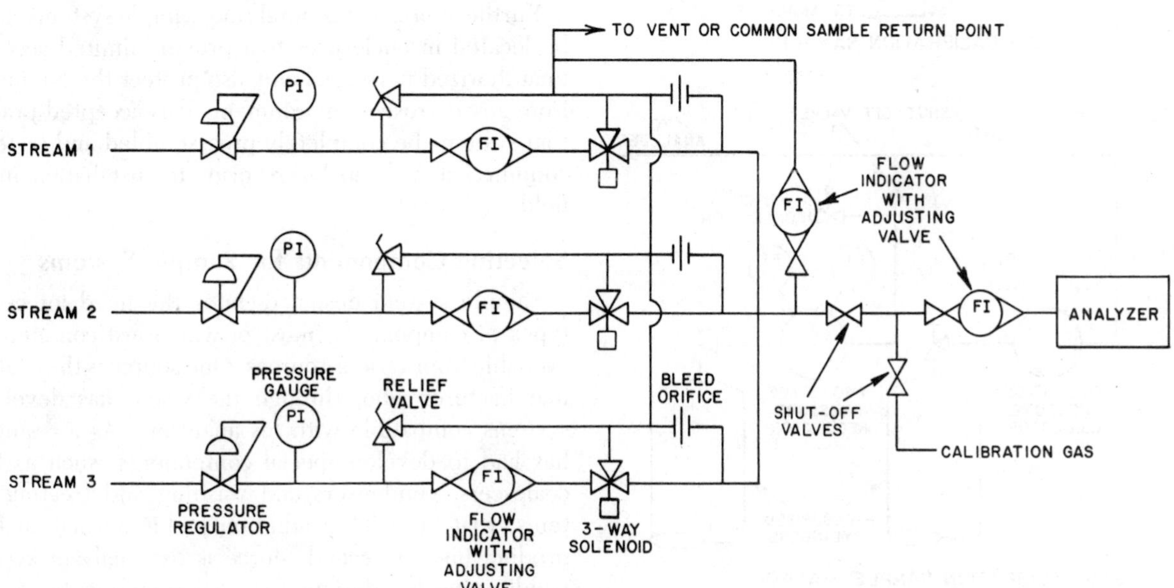

Fig. 10.2k Typical multistream automatic sampling system

considered, as well as equalization of the pressures upstream of the three-way valves. The problem is more severe in ppm sampling systems because of the adsorption/desorption effects, and careful consideration should be given to the design criteria described above.

Test and Calibration Provisions

The reliability of each analyzer is measured by its ability to check the analyzer calibration as recommended by the manufacturer. Therefore, each system must include provisions for testing and calibration. As is shown in Figure 10.2l, a means of isolating the inlet is needed in order to allow a calibration sample to be introduced manually or automatically.

Also necessary is a suitable calibration manifold with gas or liquid to furnish a reliable calibration sample to the analyzer. Storage of the test sample may be a consideration, especially for unstable liquids or gases with low dew points. Treatment of the containers is very important for trace analysis samples. In all cases, the calibration provisions must be incorporated into the systems and a sample provided that is compatible with the desired stream composition and suitable for analysis by the analyzer used for the system. It is desirable, but not essential, that the calibration sample be introduced automatically from a remote location so that the instrument can be periodically checked; however, in most systems the introduction of the sample is done by manual switching at the analyzer.

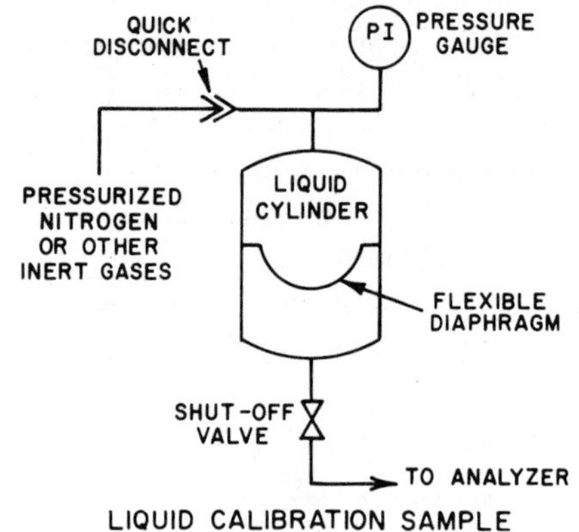

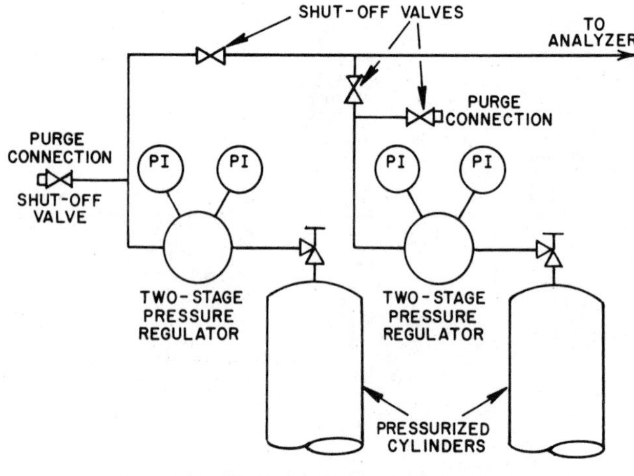

Fig. 10.2 1 Calibration provisions for liquid and gas samples

Sample Disposal

Sample disposal is a critical area, both from the economic point of view that would preserve any quantity of sample not used, and from the environmental side that would prevent the emission of most hydrocarbons into the air.

When there is an economic justification for saving the sample, for example, liquids in boiling point and viscometer analyzers, a sample collection and return system must be furnished to collect the sample at atmospheric pressure and pump it back at high pressure into the process. For gases with no sample return point, the sample can be pressurized back into the process, or as is most frequently done, the sample can be vented into the flare system. However, except in rare cases, venting is done directly into the atmosphere. When this is not possible, extreme caution should be taken to control the back-pressure when venting the sample into a flare or other collection system with varying pressures.

Ambient Considerations

Sample systems condition a sample suitable for introduction into an analyzer while maintaining the integrity of the sample. As has been emphasized throughout this section, once a sample is conditioned, it must be preserved in the conditioned state; therefore, provisions for heating, or in some cases cooling, the sample lines and system must be furnished for the integrity of the sample. Thus, the entire system must be protected from varying ambients that will condense or flash the sample.

Furthermore, as a general rule, sample systems should be located in enclosures that provide limited access to unauthorized personnel and also protect the equipment from any corrosive environment. It is accepted practice that systems be completely preassembled and tested in conjunction with analyzers prior to installation in the field.

Selecting Components for Sample Systems

Sample system design dictates the need for certain types of components, most of which are commercially available from various sources. One source is the analyzer manufacturer who, through the years, has developed systems compatible with his analyzers. As a result, he has had to develop special components, such as filter coalescers, condensers and washing and treating systems, that he sells in addition to his normal analyzer product line. A second source is the analyzer systems vendor who has designed special components, such as kinetic separators, filter probes and the like, for use with analyzers for applications in rather hard service. A third source is the specialty vendor, who has developed unique sampling components, such as pyrolysis gas sample conditioners, permeation devices for water removal systems, high-efficiency, self-cleaning filters and so on, that are a key to proper sample system design. Therefore, it is always desirable to seek out all the specialty items rather than trying to design new components. Most of the components have taken years of field testing to develop and properly modify for successful application. When a large number of analyzers is used, the components selected must be of the same type and manufacture for interchangeability and stocking of spare parts. Documenting a sample system with complete flow schematics, part identification and manufacture of various components is an essential part of being able to properly start a system and maintain it successfully over a long period.

In summary, sample systems require careful analysis of the physical and chemical conditions of the stream, as well as serious consideration of the ambient and transport conditions to ensure integrity of the sample. Therefore, care should be taken in evaluating the above considerations with respect to a given stream and in applying the correct analyzer to provide the desired measurement. Sample system design is based on experience, and whenever possible, previous experience should be given prime consideration in the selection and application of components. A successful sample system normally results in a successful analyzer system. Therefore, no effort must be spared in proper sample system design for effective implementation of analyzer systems.

BIBLIOGRAPHY

American Petroleum Institute, "Manual on Installation of Refinery Instruments and Control Systems: Part II—Process Stream Analyzers," May, 1976.

Clevett, Kenneth J., *Handbook of Process Stream Analyzers*. Chichester, Ellis Horwood, Ltd., 1973.

Considine, Douglas M., ed., *Process Instruments and Controls Handbook*. New York, McGraw-Hill Book Company, 1974.

Erk, G.F., "Engineering Analyzer Systems," *InTech*, August, 1979.

Houser, E.A., *Principles of Sample Handling and Sampling Systems Design for Process Analysis*. Pittsburgh, PA, Instrument Society of America, 1977.

Utterback, V.C., "Online Process Analyzers," *Chemical Engineering*, June 21, 1976.

Williams, A.O. and Green, P.W., "Applying Analyzers in the Process Industries," *Hydrocarbon Processing*, June, 1980.

10.3 ANALYZER SAMPLING SYSTEMS—STACKS

Type of Sample:	Gas containing particulates
Standard Design Pressure:	Generally atmospheric or near-atmospheric
Standard Design Temperature:	−25 to 1,500°F (−32 to 815°C)
Sampling Velocity:	400 to 10,000 ft. (120 to 3,000 m) per min.
Materials of Construction:	316 or 304 stainless steel for pitot tubes; 304 or 316 stainless, quartz or incoloy for sample probes
Cost:	$6,000 to $12,000 for a complete EPA particulate-sampling system (Reference Method 5)
Partial List of Suppliers:	Andersen Samplers, Incorporated; Bendix Corp., Environmental & Process Instrument Div.; Joy Manufacturing Company, Western Precipitation Div.; Research Appliance Co.; Scientific Glass & Instruments, Inc.

A complete EPA particulate-sampling system (Reference Method 5)[1,2,3,4] is comprised of four major subsystems:

1. A pitot tube probe or pitobe assembly for temperature and velocity measurements and for sampling.
2. A two-module sampling unit that consists of a separate heated compartment with provision for a filter assembly, and a separate ice-bath compartment for the impinger train and bubblers.
3. An operating/control unit with a vacuum pump and a standard dry gas meter.
4. An integrated, modular umbilical cord that connects the sample unit and pitobe to the control unit.

Figure 10.3a is a schematic of an EPA particulate-sampling train (Method 5). As shown in the figure, the system can be readily adapted for sampling sulfur dioxide (SO_2), sulfur trioxide (SO_3) and sulfuric acid (H_2SO_4) mist (Method 8).[1,2]

This section will give a detailed description of each of the four subsystems.

Pitot Tube Assembly

Description

Figure 10.3b shows the pitot tube manometer assembly for measuring stack gas velocity. The Type S (Stauscheibe or reverse) pitot tube consists of two opposing openings, one made to face upstream and the other downstream during the measurement. The difference between the impact pressure (measured against the gas flow) and the static pressure gives the velocity head.

Figure 10.3c illustrates the construction of the Type S pitot tube. The external tubing diameter is normally between 3⁄16 and 3⁄8 in. (4.8 and 9.5 mm). As can be seen, there is an equal distance from the base of each leg of the tube to its respective face-opening planes. This distance, (P_A and P_B), is between 1.05 and 1.50 times the external tube diameter. The face openings of the pitot tube should be aligned as shown.

Figure 10.3d shows the pitot tube in combination with the sampling probe. The relative placement of these components eliminates the major aerodynamic interference effects. The probe nozzle is of the bottom-hook or elbow design. It is made of seamless 316 stainless steel or glass with a sharp, tapered leading edge. The angle of taper should be less than 30°, and the taper should be on the outside to preserve a constant internal diameter. For probe lining either borosilicate or quartz glass probe liners are used for stack temperatures up to approximately 900°F (482°C); quartz liners are used for temperatures between 900 and 1,650°F (482 and 899°C). Although borosilicate or quartz glass probe linings are generally recommended, 316 stainless steel, Incoloy or other corrosion resistant metal may also be used.

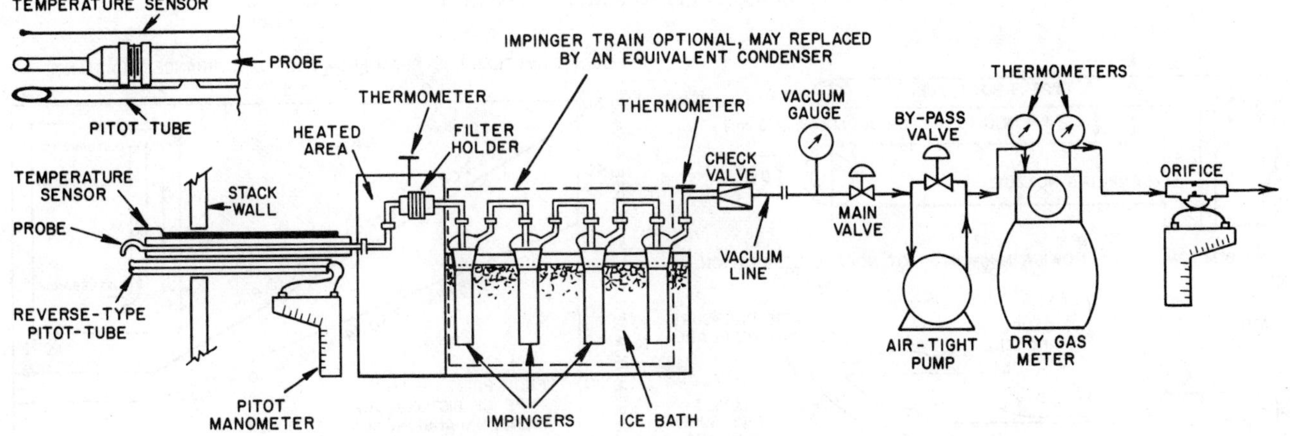

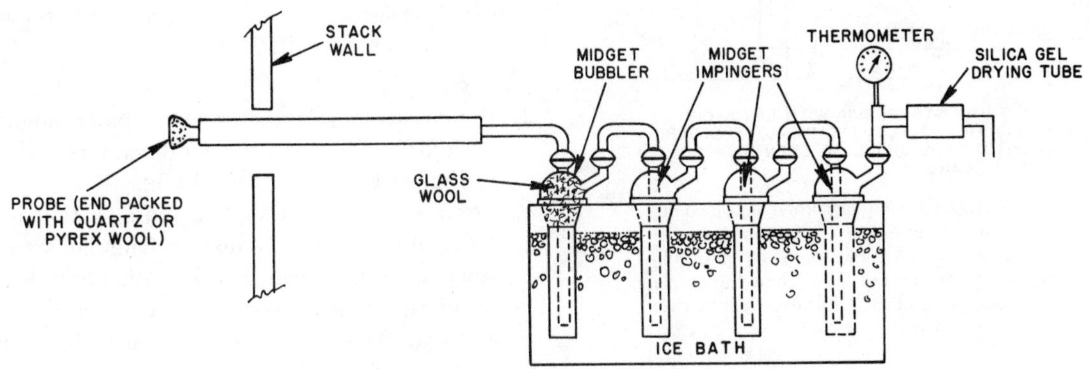

Fig. 10.3a (Top) EPA particulate-sampling train (Method 5) (Bottom) Sampling case for SO_2, SO_3 and H_2SO_4 mist (Method 8)

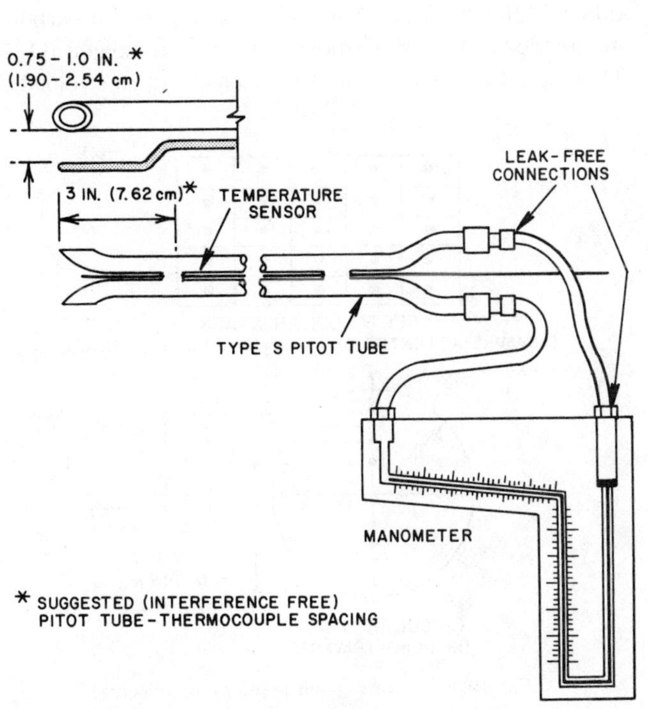

Fig. 10.3b Type S pitot tube manometer assembly

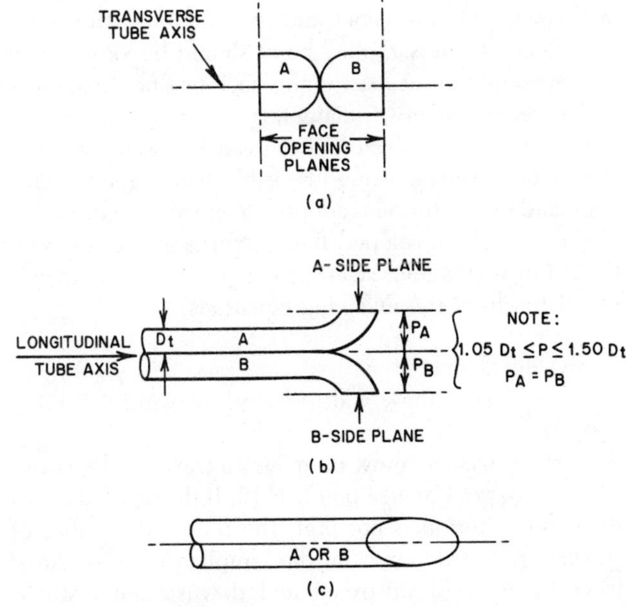

Fig. 10.3c Properly constructed Type S pitot tube: (A) End view; face-opening planes perpendicular to transverse axis; (B) top view; face-opening planes parallel to longitudinal axis; (C) side view; both legs of equal length and center lines coincident, when viewed from both sides. Baseline coefficient values of 0.84 may be assigned to pitot tubes constructed this way.

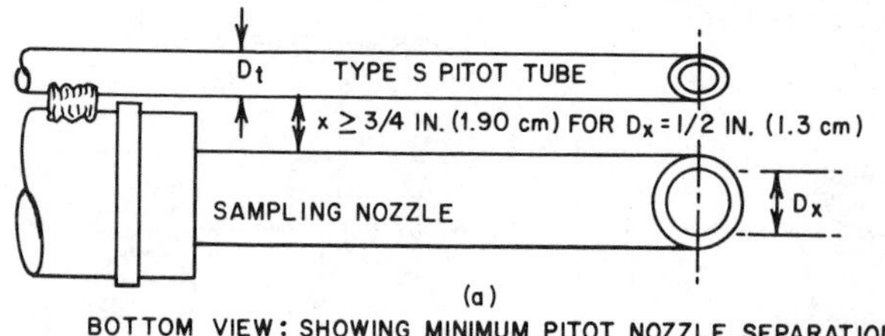

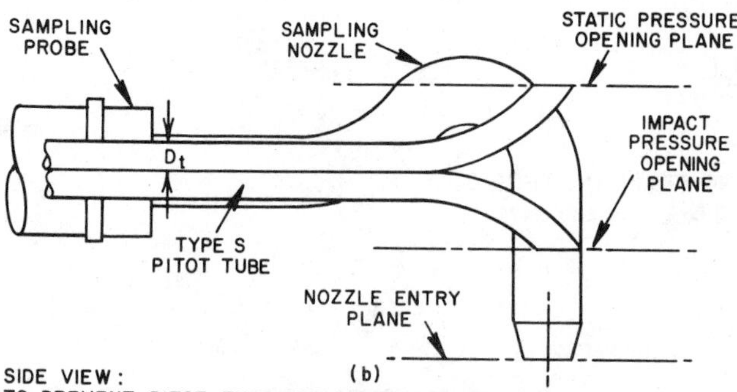

Fig. 10.3d Proper pitot tube with sampling probe nozzle configuration to prevent aerodynamic interference (A) Bottom view; minimum pitot nozzle separation; (B) Side view; to prevent pitot tube from interfering with gas flow streamlines approaching the nozzle, the impact pressure opening plane of the pitot tube shall be even with or above the nozzle entry plane

Installation

The specific points of stack for sampling are selected to ensure that the samples collected are representative of the material to be discharged or controlled. These points are determined after examination of the process or the source of emissions and its variation with time.

In general, the sampling point should be located at a distance equal to at least eight stack or duct diameters downstream and two diameters upstream from any source of flow disturbance, such as an expansion, a bend, contraction, valve, fitting or visible flame. (Note: This eight and two criterion is adopted to ensure the presence of stable, fully developed flow patterns at the test section.) For rectangular stacks, the equivalent diameter is calculated from the following equations:

$$\text{Equivalent diameter} = 2\,(\text{length} \times \text{width})/(\text{length} + \text{width}) \qquad 10.3(1)$$

Next, provisions must be made to traverse the stack. The number of traverse points is 12. If the eight and two diameter criterion is not met, the required number of traverse points depends on the sampling point distance from the nearest upstream and downstream disturbances. This number may be determined by using Figure 10.3e.

The cross-sectional layout and location of traverse points are as follows:

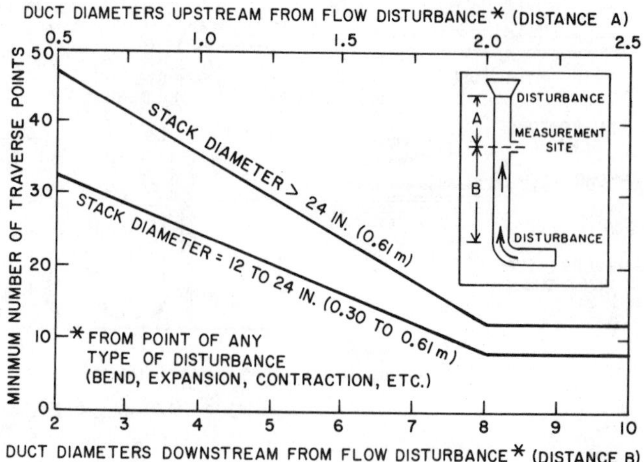

Fig. 10.3e Minimum number of traverse points for particulate traverses

1. For circular stacks, the traverse points should be located on two perpendicular diameters as shown in Figure 10.3f and Table 10.3g.
2. For rectangular stacks, the cross section is divided into as many equal rectangular areas as traverse points, such that the ratio of the length to width of the elemental area is between one and two. The traverse points are to be located at the centroid of each equal area as shown in Figure 10.3f.

Operation

The velocity head at various traverse points is measured using the pitot tube assembly shown in Figure 10.3b. The gas samples are collected at a rate proportional to the stack gas velocity and analyzed for carbon monoxide (CO), carbon dioxide (CO_2) and oxygen (O_2). The pitot tube is calibrated by measuring the velocity

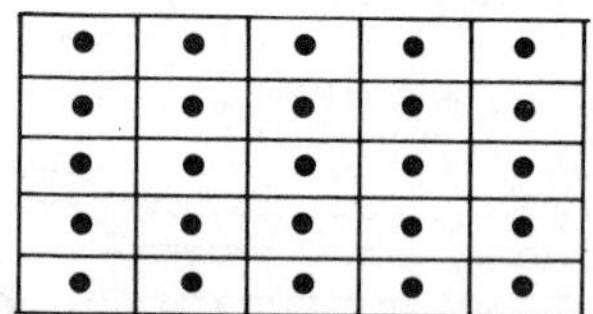

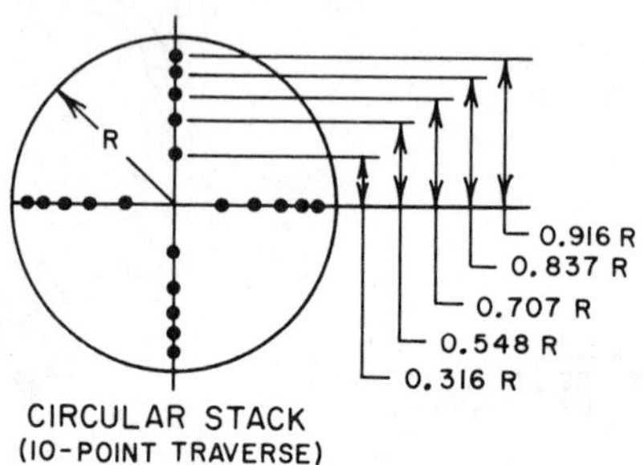

Fig. 10.3f Traverse point locations for velocity measurement or for multipoint sampling

Table 10.3g
LOCATION OF TRAVERSE POINTS
IN CIRCULAR STACKS
(Percent of stack diameter from inside wall to traverse point)

Traverse Point Number on a Diameter	*Number of Traverse Points on a Diameter*											
	2	*4*	*6*	*8*	*10*	*12*	*14*	*16*	*18*	*20*	*22*	*24*
1	14.6	6.7	4.4	3.2	2.6	2.1	1.8	1.6	1.4	1.3	1.1	1.1
2	85.4	25.0	14.6	10.5	8.2	6.7	5.7	4.9	4.4	3.9	3.5	3.2
3		75.0	29.6	19.4	14.6	11.8	9.9	8.5	7.5	6.7	6.0	5.5
4		93.3	70.4	32.3	22.6	17.7	14.6	12.5	10.9	9.7	8.7	7.9
5			85.4	67.7	34.2	25.0	20.1	16.9	14.6	12.9	11.6	10.5
6			95.6	80.6	65.8	35.6	26.9	22.0	18.8	16.5	14.6	13.2
7				89.5	77.4	64.4	36.6	28.3	23.6	20.4	18.0	16.1
8				96.8	85.4	75.0	63.4	37.5	29.6	25.0	21.8	19.4
9					91.8	82.3	73.1	62.5	38.2	30.6	26.2	23.0
10					97.4	88.2	79.9	71.7	61.8	38.8	31.5	27.2
11						93.3	85.4	78.0	70.4	61.2	39.3	32.3
12						97.9	90.1	83.1	76.4	69.4	60.7	39.8
13							94.3	87.5	81.2	75.0	68.5	60.2
14							98.2	91.5	85.4	79.6	73.8	67.7
15								95.1	89.1	83.5	78.2	72.8
16								98.4	92.5	87.1	82.0	77.0
17									95.6	90.3	85.4	80.6
18									98.6	93.3	88.4	83.9
19										96.1	91.3	86.8
20										98.7	94.0	89.5
21											96.5	92.1
22											98.9	94.5
23												96.8
24												98.9

head at some point in the flowing gas stream with both the Type S pitot tube and a standard pitot tube with a known coefficient. Other data also needed for calculation of the volumetric flow are stack temperature, stack and barometric pressures and wet-bulb and dry-bulb temperatures of the gas sample at each traverse.

Figure 10.3h gives the equations for pitot tube readings into velocity and mass flow, and a typical data sheet for stack flow measurements.[3]

Based on the range of velocity heads, a probe with a properly sized nozzle is selected to maintain isokinetic sampling of particulate matter. As shown in Figure 10.3i, a converging stream will be developed at the nozzle face if the sampling velocity is too high. Under this subisokinetic sampling condition, an excessive amount of lighter particles enters the probe. Because of the inertia effect, the heavier particles, especially those in the range of 3μ or greater, travel around the edge of the nozzle and are not collected. The result is a sample indicating an excessively high concentration of lighter particles, and the weight of the solid sample is in error on the low side. Conversely, portions of the gas stream approaching at a higher velocity are deflected if the sampling velocity is below that of the flowing gas stream. Under this superisokinetic sampling condition, the lighter particles follow the deflected stream and are not collected, while the heavier particles, because of their inertia, continue into the probe. The result is a sample indicating high concentration of heavier particles and the weight of solid sample is in error on the high side.

Isokinetic sampling requires the precise adjustment of the sampling rate with the aid of the pitot tube manometer readings and nomographs such as APTD–0576.[5] If the pressure drop across the filter in the sampling unit becomes too high, making isokinetic sampling difficult to maintain, the filter may be replaced in the midst of a sample run.

To measure the concentration of particulate matter, the sampling time for each run should be at least 60 min. and the minimum volume should be 30 dry scfm (51 m^3/hr.).[4]

Two-Module Sampling Unit

Heated Compartment

As shown in Figure 10.3a, the probe is connected to the heated compartment that contains the filter holder and other particulate-collecting devices, such as cyclone and flask. The filter holder is made of borosilicate glass, with a frit filter support and a silicone rubber gasket. The compartment is insulated and equipped with a heating system capable of maintaining a temperature around

Table 10.3h

PITOT TUBE CALCULATION SHEET

STACK VOLUME DATA

Stack No. .... Station .... Date .... Page ....

Name of Firm ....

Point	Position, in.	Reading, H, in. of H_2O	$\sqrt{H}$	Temp., t_s, °F.	Velocity, V_s, ft./sec.
1					
2					
3					
4					
5					
6					
7					
8					
9					
10					
11					
12					
13					
14					
15					
16					
	Totals				
	Average				
	Abs. temp., $T_s = t_s + 460 =$ °R.				

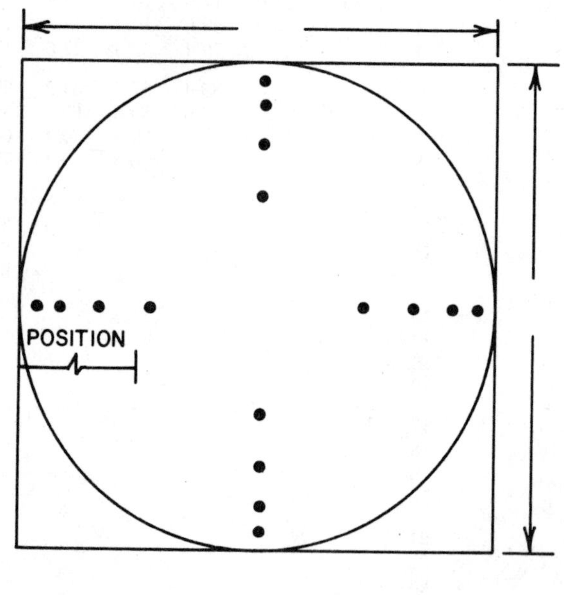

Dry bulb temp., t_d = ________________ °F.

Wet bulb temp., t_w = ________________ °F.

Absolute humidity, W = ___ lb. H_2O/lb. dry gas

Stack area, A_s = ________________ sq. ft.

Barometer, P_b = ________________ in., Hg

Stack gage pressure = ________________ in., H_2O

Stack abs. pressure, $P_s = \frac{\text{in., } H_2O}{13.6} \pm P_b$ = _____ in., Hg

Pitot correction factor, F_s = ________________

Component	Vol. fraction, dry basis ×	mol. wgt.	= wgt. fraction, dry basis
Carbon dioxide		44	=
Carbon monoxide		28	=
Oxygen		32	=
Nitrogen		28	=

Average dry gas molecular weight, M = _____

Specific gravity of stack gas, $G_s = \frac{0.62\,M\,(W+1)}{18 + MW} = \frac{0.62 \times \_\_\_ \times \_\_\_}{18 + \_\_\_\_\_} =$ ________________

(Ref. dry air at same conditions)

Velocity, $V_s = 2.9 F_s \sqrt{\frac{29.92 \times T_s}{P_s \times G_s}}\sqrt{H} = 2.9 \times \_\_\_ \sqrt{\frac{29.92 \times \_\_\_}{\_\_ \times \_\_}}\sqrt{H} =$ ________ ft./sec.

Volume = _____ ft./sec. × _____ sq. ft. × 60 _____ = _____cfm.

Standard volume = cfm. $\times \frac{530}{T_s} \times \frac{P_s}{29.92}$ = _____ $\times \frac{530}{\_\_} \times \frac{\_\_\_}{29.92}$ = _____ scfm.

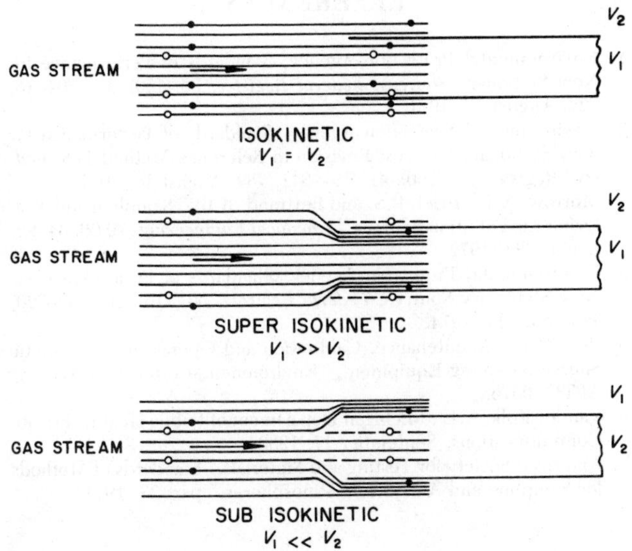

Fig. 10.3i Particle collection and sampling velocity

the filter holder during sampling at 248 ± 25°F (120 ± 14°C), or such other temperature as specified by EPA. The thermometer should measure temperature to within 5.4°F (3°C). The compartment should be provided with a circulating fan to minimize thermal gradients.

Ice-Bath Compartment

The ice-bath compartment contains the system's impingers and bubblers. The system for determining stack gas moisture content consists of four impingers connected in series as shown in Figure 10.3a. The first, third and fourth impingers are of the Greenburg-Smith design. To reduce the pressure drop, the tips are removed and replaced with a ½-in. (12.5 mm) I.D. glass tube extending to ½ in. (12.5 mm) from the bottom of the flask. The second impinger is of the Greenburg-Smith design with a standard tip. During sampling for particulates, the first and second impingers are filled with 100 ml. (3.4 oz.) of distilled and deionized water. The third impinger is left dry to separate entrained water. The last impinger is filled with 200 to 300 gm. (7 to 10.5 oz.) of precisely weighed silica gel (6 to 16 mesh) that has been dried at 350°F (177°C) for two hours to completely remove any remaining water. A thermometer, capable of measuring temperature to within 2°F (1.1°C), is placed at the outlet of the last impinger for monitoring purposes. Crushed ice should be added during the run to maintain the temperature of the gas leaving the last impinger at 60°F (16°C) or less.

Operating or Control Unit

As shown in Figure 10.3a, the control unit consists of the system's vacuum pump, valves, switches, thermometers and totalizing dry gas meter, and is connected by a vacuum line with the last Greenburg-Smith impinger. The pump intake vacuum is monitored with a vacuum gauge just after the quick disconnect. A bypass valve parallel with the vacuum pump provides fine control and permits recirculation of gases at a low-sampling rate so that the pump motor is not overloaded. Downstream from the pump and bypass valve are thermometers, dry gas meter and calibrated orifice and inclined/vertical manometers. The calibrated orifice and inclined manometer indicate the instantaneous sampling rate. The totalizing dry gas meter gives an integrated gas volume. The average of the two temperatures on each side of the dry gas meter gives the temperature at which the sample is collected. The addition of atmospheric pressure to orifice pressure gives meter pressure.

Precise measurements require that the thermometers be capable of measuring the temperature to within 5.4°F (3°C); the dry gas meter is inaccurate to within 2 percent of the volume; the barometer is inaccurate within 0.25 mmHg (torr) 0.035 kPa; and the manometer is inaccurate within 0.25 mmHg (torr) (0.035 kPa).

Umbilical Cord

The umbilical cord is an integrated multiconductor assembly containing both pneumatic and electrical conductors. It connects the two-module sampling unit to the

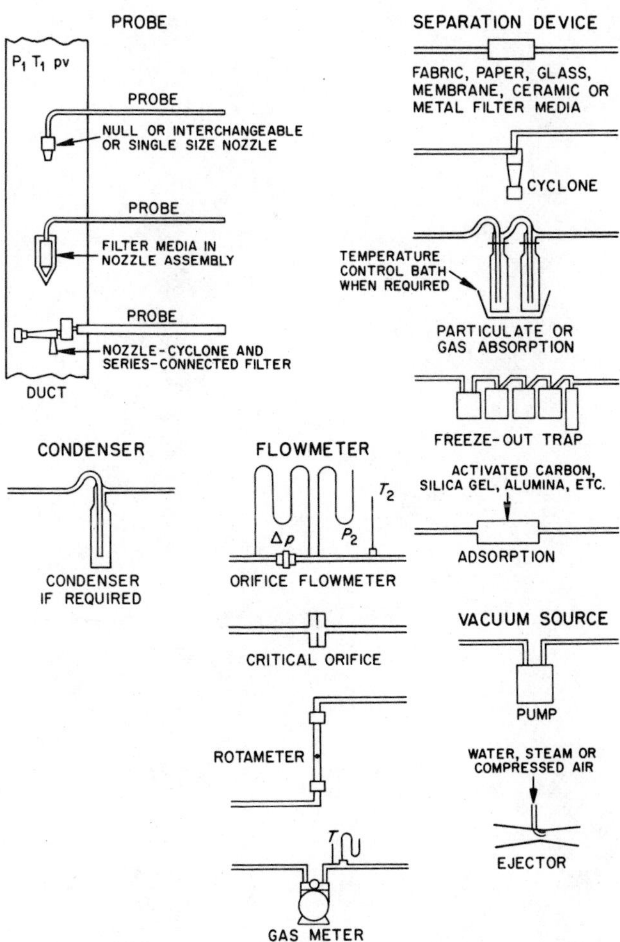

Fig. 10.3j Components of common sampling systems

control unit, as well as the pitot tube stack velocity signals to the manometers or differential pressure gauges.

Sampling for Gases and Vapors

Some commonly used components in stack-sampling systems are illustrated in Figure 10.3j. If ball-and-socket joints and compression fittings are used, any arrangement of components is readily set up for field use. The stack-sampling components are selected on the basis of the source to be sampled, the substances involved and the data needed.

A summary of sampling procedure outlines was developed by industrial hygienists[6] for specific substances. The procedural outlines serve as a starting point in assembling a stack-sampling system, after consideration has been given to the complications that might arise because of the presence of interfering substances in the gas samples. Other recommended sampling procedures for gases and vapors are given in Reference 7.

REFERENCES

1. Environmental Protection Agency, Standards of Performance for New Stationary Sources, *Federal Register,* 36 (159): 15, 704–15, 722, August 17, 1971.
2. Environmental Protection Agency, Standards of Performance for New Stationary Sources, Revision to Reference Method 1–8, *Federal Register*, 42 (160): 41, 754–841, 789, August 18, 1977.
3. Morrow, N.L., Brief, R.S. and Bertrand, R.R., "Sampling and Analyzing Air Pollution Sources," *Chemical Engineering*, 79 (2): 84–98, January 24, 1972.
4. Environmental Protection Agency, Standards of Performance for New Stationary Sources, *Federal Register,* 30 (116): 20, 790–820, 794, June 14, 1974.
5. Rom, J.J., "Maintenance, Calibration and Operation of Isokinetic Source Sampling Equipment," Environmental Protection Agency, APTD–0576.
6. Vander Kolk, A.L., Michigan Department of Public Health, Private Communications, September 17, 1980.
7. American Society for Testing and Materials, "Standards of Methods for Sampling and Analysis of Atmospheres," part 23, 1971.

Flow Sheet Symbol

SAMPLING SYSTEM

AT CHROMATOGRAPH

TO RECEIVER

10.4 CHROMATOGRAPHS FOR GAS SAMPLES (GCs)

Type of Sample:	Vapor and liquid
Standard Design Pressure:	Atmospheric
Sample Temperature	0 to 300°F (−8 to 149°C), but always less than chromatograph operating temperature
Ambient Temperature:	40 to 120°F (4 to 49°C)
Contacting Materials:	Stainless steel, Teflon, various elastomers
Auxiliary Utilities Required:	Instrument air, carrier gas; sometimes a fuel gas and "clean" air
Cost:	Laboratory units from \$4,000 to \$20,000; on-stream analyzers from \$10,000 to \$40,000, depending on application complexity and accessories
Inaccuracy:	±1% for most applications
Repeatability:	±½% for most applications
Cycle Time:	1 to 20 min. for most applications
Special Feature:	Multicomponent readout
Partial List of Suppliers:	Amscor; Analytical Instruments Corp.; Antek Instruments, Inc.; Applied Automation, Inc.; Arcas, Inc.; Beckman Instruments, Inc.; Bendix Corp., Environmental & Process Instruments Div.; Foxboro Co.; Honeywell, Inc., Process Control Div.; Perkin-Elmer Corp., Instrument Div.; Process Analyzers, Inc.; United Controls Div., Envirotech Corp.; Varian Associates

The chromatograph has become the most widely used analytical tool for the process industries. It offers extreme flexibility of applications, high-sensitivity analysis, and perhaps most important, allows multicomponent analysis. However, it functions on a semicontinuous basis; that is, a sample is periodically taken for analysis and the output is updated on a cyclic basis.

History

Ironically, the term chromatography today is a misnomer. It literally means "color writing" and refers to the technique used in the late 1800s for separating dyes and paint pigments. When the pigment solution was filtered through a fine mesh absorbent contained in a glass tube, the pigments separated into color bands or groups as they passed through the filter medium. The technique has been extended to the separation of gases, and liquids, and the term chromatography has persisted.

Principle of Operation

The process chromatograph takes a fixed volume of sample gas, or liquid in vapor form, and introduces it into the chromatographic column (Figure 10.4a). The column consists of finely granulated absorbent material

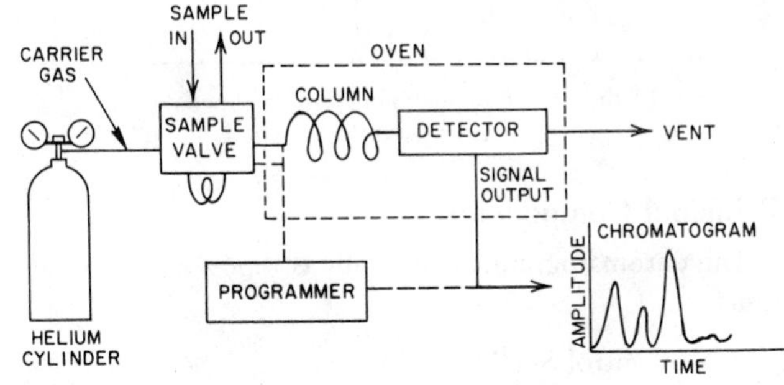

Fig. 10.4a Basic schematic of chromatograph

(or liquid-coated inert particles) packed in conventional drawn tubing. The filler for the column is selected according to the separation desired. As the sample progresses through the column, individual molecules of gas are absorbed (or adsorbed), then released at different times from the column particle surfaces.

When the column is properly selected, the components of interest elute, or emerge from the column completely separated from each other and from other components. This eluting stream passes through the detector. The relative response of the detector output is sensed by an electronic unit called an amplifier and usually presented on a recorder.

In the process chromatograph, all operating parameters are controlled and events occur on a cyclic and repetitive basis. The operating parameters that must be controlled for successful operation are 1) temperature, 2) carrier gas flow and pressure, 3) sample and column valve actuations, 4) detector sensitivity, and 5) event program.

The detector monitors the gas eluting from the column, and its output, when recorded, generates a chromatogram, or chromatographic spectrum. Such a chromatogram is illustrated in Figure 10.4b. This signal is treated in various ways (see Data Presentation).

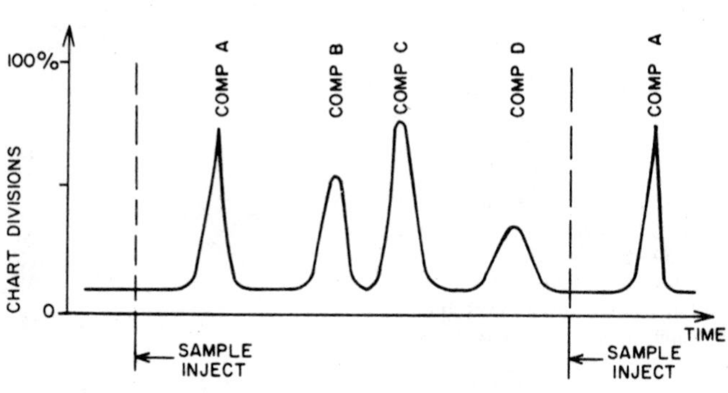

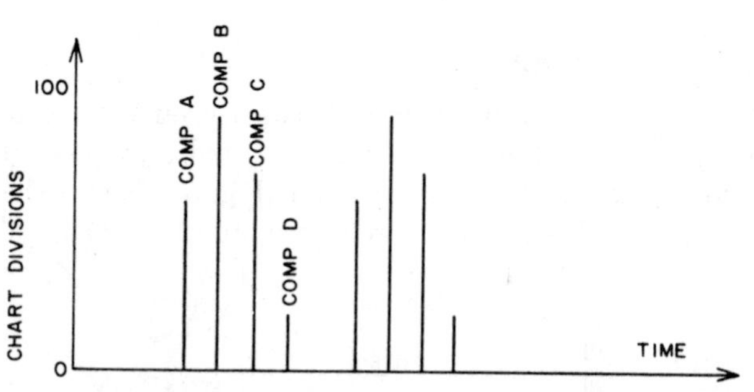

Fig. 10.4b Data presentation in form of chromatogram and bargraph

Principal Components

The chromatograph is generally composed of two sections:

- A. Control Section
  1. Programmer
  2. Stream selector (optional)
  3. Recorder
- B. Analysis Section
  1. Temperature-controlled enclosure (oven)
  2. Columns
  3. Valves
  4. Detector

Control Section

Programmer

The operation of the process chromatograph is based on the repetitive occurrence of events during an analysis. These events include:

1. Sample injection
2. Column valve operations
3. Standardization (baseline corrections)
4. Component gating (see Bargraph under Data Presentation.)

Each column valve and each component measured requires individual event control that must be variable (adjustable). For applications measuring several components and using several column-valve operations, there may be from eight to fifteen controlled events. Several programmer designs are available and will be discussed below.

Cam Timer The most prominent means of event control is the conventional cam timer. The timer consists of a bank of cams on a common shaft driven by a synchronous motor. The camshaft makes one revolution per analysis cycle. Rotation rate depends on the overall time cycle of analysis and is obtained by proper selection of motor speeds and changeable drive gears. The cams are adjustable so that dwell time of the cam follower in the cam valley can be varied, and the relative location of the cam valley on the shaft can be changed.

The cam follower is part of a microswitch assembly that opens and closes a set of contacts as the cam follower rides into and out of the cam valley. The electrical actuation of the microswitch is usually selectable for on/off or off/on operation.

The timer is then used as on/off control of the events in the cycle. This design is economical, moderately easy to adjust and above all reliable.

Optical Reader Programmer The optical reader programmer uses a light source and photocell sensors (Figure 10.4c). Between these elements is placed a film, typically mylar, on which a common lead pencil mark can be inscribed to interrupt the light to the photocell. The film is in the form of a disc with inscribed channels, one for each event. The disc is driven by a synchronous motor at a rate that allows one rotation for the analysis cycle. Motor speeds and drive gears are selected for this time cycle.

This programmer offers the advantage of rapid changes

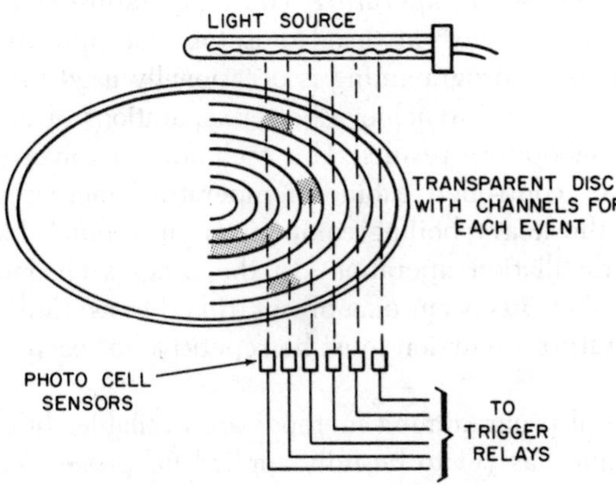

Fig. 10.4c Optical reader programmer

to programs, but is subject to source decay. The design is inherently more expensive and offers less reliability than the cam timer.

Magnetic Tape The magnetic tape programmer uses a continuous length of recording tape and a recording/play deck assembly. Pulses are recorded on the tape, which, when read by the playback heads, advance a stepping switch. The decks of the stepping switch are wired to control the required events in the analysis cycle. The cycle time is determined by the tape length.

Although an excellent programmer, this design is the most expensive and least reliable. The tape deck must be kept impeccably clean and it also wears. Minor program changes are easily implemented, but major changes require rewiring of the stepping switch.

Computer Chromatograph installations involving at least six units may be serviced by a small, special-purpose digital computer that eliminates the conventional programmer. The computer also eliminates the recorders, although at least one is required for servicing. However, the sophistication of maintenance personnel for this system is higher, sometimes requiring specialists in both technologies. Economics do not yet favor this "ultimate" programmer.

Stream Selectors

Several identical, or similar process streams can be analyzed by one chromatograph on a time-shared basis. Economics favors time-sharing, but other considerations must also be recognized. Among them are increased cycle time between analyses of a given stream, and additional complexities of the basic analysis solution and programming. The latter consideration has often been overlooked, or ignored, in the installation of one "superchromatograph" measuring many components in numerous, widely differing streams, the result being a maintenance liability.

Where time-sharing is feasible, a stream selector is generally furnished as part of the control unit; this selector consists of a stepping switch advanced once each cycle by the programmer. The preferred time to change streams is immediately following the injection of a sample because this allows the maximum purge time for the next stream. Most designs provide for bypassing, or skipping, streams that may be temporarily unimportant.

Recorders

The conventional readout of a process chromatograph uses the bargraph method of routine data presentation and chromatograms (see Figure 10.4b) for servicing and troubleshooting. These can be obtained on the same recorder by proper "mode" selection. The chromatographic recorder would normally have the following features:

Input Range:	0 to 5 mv. (0 to 1 mv. on some applications)
Inaccuracy:	±0.5% (not very important because an empirical analyzer calibration is used)
Reproducibility:	±1% (important)
Response:	1 sec. full-scale (important)
Chart width:	6 and 10 in. (150 and 250 mm) are standard
Chart speed:	1 to 5 in./min. (25 to 125 mm/min.)

Modifications are usually required in the chart advance circuit to allow the programmer to control the chart advance (see Data Presentation).

Analysis Section

Chromatographic separation and analysis are achieved in a temperature-controlled environment containing a sample inject valve, chromatographic column and detector. These components are housed in the analysis section (commonly called analyzer section, or oven). Two basic designs have evolved: the air-bath and the bell-housing (explosion-proof enclosure). The air-bath analyzer places all electrical equipment in several explosion-proof enclosures inside a large heated box. The bell-housing analyzer places all equipment inside a large conventional explosion-proof enclosure (resembling a bell). Both designs satisfy Class-I, Group-D, Division-I classification.

Air-bath housing has found wider acceptance than bell-housing, because it is easier to service.

Temperature Control

As stated previously, temperature control is very important for the chromatograph because it affects the performance of the columns and the sensitivity of the detector. Control of the air-bath analyzer involves sensing and controlling the environmental air temperature, whereas the bell-housing type requires controlling the temperature of a large heat-sink or mandrel in the center of the housing, which is electrically heated (Figure 10.4d). In either case, two types of control are possible:

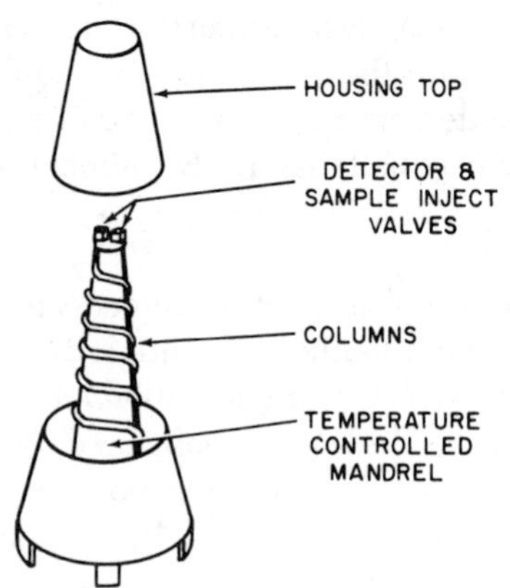

Fig. 10.4d Typical bell-housing analyzer configuration

on/off and proportional. The following discussion covers only the air-bath analyzer.

On/off mode temperature control may use a bimetallic thermostat installed through the cabinet wall into the air-bath. Large resistance heaters (500–2000 w.) are contained in an explosion-proof enclosure inside the box, and the air is passed through the heater to the analyzer case. The thermostat controls the current to the heater and thus modulates the heat input. The heated purge air is diffused into the air-bath using a distribution tube, or diffuser. It is desirable to create air turbulence in the case in order to minimize thermal gradients across the box.

On/off control permits control within 1°F (0.5°C) of the desired temperature. This is not satisfactory for most measurements and creates reproducibility errors.

Proportional Control Proportional temperature control, using compact, solid-state (SCR) controllers, is now available for most air-bath analyzers. This system utilizes a thermistor sensor located near the geometric center of the cabinet to monitor its temperature (Figure 10.4e). The controller adjusts the voltage on the resistance heater contained in an explosion-proof enclosure. Purge air is passed through the heater as in the on/off system.

These systems are generally capable of controlling between 100 and 600°F (38 and 316°C). Temperature stability of 0.1 to 0.2°F (0.05 to 0.1°C) can be expected. Air consumption is typically 2 to 4 scfm (3.4 to 6.8 m^3/hr.).

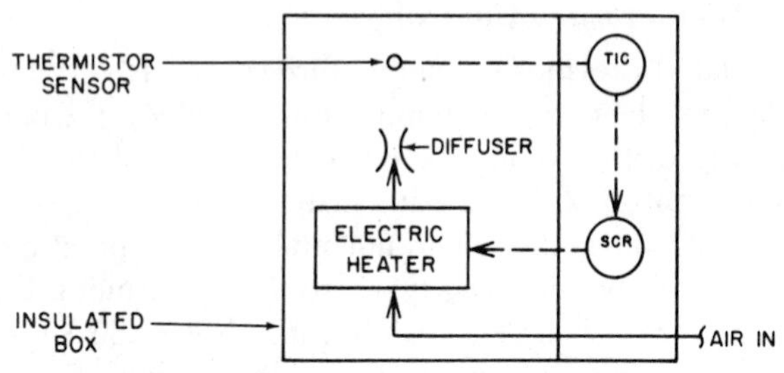

Fig. 10.4e Typical air-bath analyzer configuration

Programmed Temperature The temperature controls discussed previously are used for isothermal applications. Temperature programming is occasionally used to vary the temperature to achieve better separations or to duplicate laboratory results. This technique involves progressive, controlled cabinet temperature increases in which the heavy boiling compounds elute much like a batch distillation operation. In the process chromatograph all events occur on a preset, timed basis; thus, the temperature excursion must be repetitive for each analysis.

Several programming methods are available, but the technique has yet to be fully applied for process chromatographs. In one design, fixed resistors are added to the temperature controller input circuitry on a timed basis. Another design motorizes the set point adjustments of the temperature controller. Neither design compensates for external temperature effects.

At the completion of each cycle, the case must be cooled in preparation for the next cycle.

Columns

The chromatographic column is the working element of the chromatograph. A comprehensive discussion of the chemistry of chromatography can be found in the *Encyclopedia of Chemical Technology*.[1] The following is a general discussion of column technology for the practicing instrument engineer.

Generally speaking, there are two types of columns: partition (absorption) and adsorption. A brief classification of columns is found in Table 10.4f.

Adsorption columns use granular materials that have a large surface area and a tendency to adsorb, or retain, the component of the gas sample carried through the column. The use of this column is known as gas-solid chromatography (GSC).

Partition columns use an inert solid material of small mesh (solid support) which is coated with a high-boiling, liquid material, or substrate. A typical solid support can hold as much as 40 to 80 percent substrate material. The liquid coating absorbs (surface phenomena) the gas sample at different rates, depending on the solubility of the

Table 10.4f
TYPES OF CHROMATOGRAPHIC COLUMNS

Type	*Description*	*Example*
Partition Gas-liquid chromatography (GLS) Absorption	Liquid coating on inert solid	Carbowax on Teflon
Adsorption Gas-solid chromatography (GSC)	Highly adsorptive solid material	Charcoal, silica gel, alumina

gas in the liquid. The components are subsequently eluted selectively by the carrier gas, according to their liquid partition coefficients (Henry's law).

The partitioning agent must be suitable for the operating conditions so that it will not "bleed," or slowly elute from the column. Any loss of the liquid coating will cause the elution time of the components to change and, thus, change the programming times. For high-sensitivity work, bleeding of the partitioning agent through the detector will cause a changing baseline. The partitioning agent determines the maximum allowable operating temperature (MAOT) at which a column can be used with satisfactory results.

In either type of column, the object is to separate the sample components and allow the carrier gas to elute the component of interest to the detector as a binary mixture (i.e., mixed only with the carrier gas).

The technology of column materials is rapidly expanding for both process and laboratory chromatographs. The application, or measurement, desired dictates the proper selection. A guide to optimum selection of columns would be difficult to prepare in the face of changing technology. A short list of columns and applications is found in Table 10.4g and is intended to illustrate measurement types only. Fortunately, manufacturers offer application engineering with the equipment for a consultation fee. Many times the columns are dictated by a plant laboratory to provide directly comparable results with the laboratory—an unfortunate practice that can limit the effective use of a process unit.

The column is typically ¼, 3⁄16 or ⅛ in. (6.4, 4.8 or 3.2 mm) OD stainless steel tubing of 5 to 20 ft. (1.5 to 6 m) lengths. The tubing is generally wound around a 2 or 3 in. (50 or 75 mm) mandrel to form a coil, which facilitates handling. The bell-housing analyzer wraps the column directly around the central, temperature controlled mandrel (see Figure 10.4d). The column tubing is filled by packing the granular material into the tubing. For partition columns, the material is prepared first in the laboratory.

Some analyzers will use only one column between the sample valve and the detector; however, most applications require several columns. These columns may be different lengths of the same column-packing material which are used for different purposes, such as a backflush column (see Valves). More frequently, the columns are packed with different materials to provide supplemental separation of the sample.

Chromatographic Valves

Special valves are used in the chromatograph to introduce a fixed volume of sample to the column system and to switch column segments into and out of the carrier stream. Many different designs are available, but there are basically three general types: linear, rotary, and sliding plate (Table 10.4h). Each design can provide 6 to 10

Table 10.4g
ILLUSTRATIVE COLUMN APPLICATIONS

Application	*Substrate*	*Support*	*M.A.O.T. °F (°C)*
Light inorganic gases separates O_2 from N_2	—	5A molecular sieves	482 (250)
Light inorganic gases and hydrocarbons up to C_3	—	Silica gel	482 (250)
Light inorganic and organic gases	—	Activated charcoal	482 (250)
Hydrocarbons, boiling point separations	Squalane	Super support	302 (150)
Light inorganic gases and hydrocarbons up to C_3. Poor olefin-paraffin separation	Di-2-ethylhexyl sebacate	Silica gel	302 (150)
Selectivity for saturated hydrocarbons. Poor olefin-paraffin separation	Di-2-ethylhexyl sebacate	Super support	302 (150)
Separation of water, alcohols and amines	Tetraethylene glycol Di-methyl ether	Super support	167 (75)
Separation of mixed saturate and unsaturated hydrocarbons, H_2S, etc.	Di-2-ethylhexyl sebacated + bio-2-methoxyethyl adipate	Super support	212 (100)
Selective for aromatic hydrocarbons	Picric acid-diethylene glycol	Super support	257 (125)
Separation of high-boiling compound	DC-710 silicone oil	Super support	482 (250)
High-boiling compounds, such as fatty esters and acids	Apieyon "L"	Super support	482 (250)
Separation of esters of fatty acids	Diethylene glycol succinate	Cromasorb	392 (200)
Separation of polar compounds	Carbowax 1500	Super support	302 (150)
High-selectivity for polar compounds	B,B′-oxydipropionitrile	Cromasorb	167 (75)
Polar compounds, ammonia, alcohols, water, ketones, acromatics	Carbowax 1500	Teflon	257 (125)
General purpose	Di-isodecyl phthalate	Super support	329 (165)
General purpose	UCON Oil LB-550-X	Super support	356 (180)

Table 10.4h
CHROMATOGRAPHIC VALVES

Type	Application: Sample Valve Gas	Sample Valve Liquid	Column Switching	Temp. Rating °F (°C)	Expense	Internal Volume
Linear	√		√	248 (120)	Low	Moderate
Rotary	√	√	√	392 (200)	High	Low
Sliding plate	√	√	√	392 (200)	Moderate	Low

ports, or connections, and, by variations in their hookup configurations, these valves can be used for a multitude of flow-switching schemes (see Sample Systems). Each type valve is pneumatically operated through a solenoid, which, in turn, is actuated by switches in the control section.

Linear Valve The linear valve is a two-position linear stroke, multiport, pneumatically operated unit consisting of a body with a center bore, valve spool or stem with O-rings, and a pneumatic actuator assembly (Figure 10.4i). The linear valve stem has a smaller diameter than the valve bore so that an annular space exists between the stem and the bore. This space is interrupted by O-rings located at discrete intervals along the valve stem. The O-rings are spaced so that in each of the two valve positions, gas flow is directed between different groups of ports. The valve stem is moved by a spring-loaded pneumatic actuator and is limited in its travel by positive stops that ensure proper positioning. This valve is simple, inexpensive and useful for most gases, but is limited by the O-ring material.

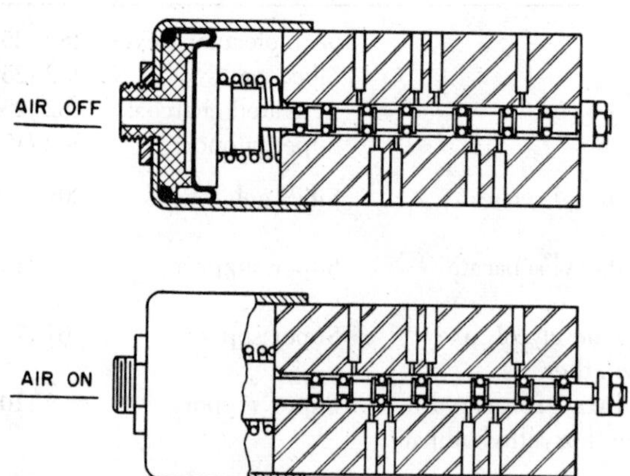

Fig. 10.4i Linear valve

Rotary Valve The rotary valve is a two-position, rotary, multiport, low-volume, pneumatically operated device consisting of two plates with precision-machined surfaces (Figure 10.4j). One plate is usually stainless steel and contains a series of small diameter openings that connect to tubing on the back side. This plate is fixed. The second surface is usually a hard elastomer, such as Rulon or Fluorgold (although a second metal block can also be used), in which small grooves have been machined to connect two adjacent ports of the lower plate. The upper plate is rotated 45° (60° for a 6-port valve) to change the flow path scheme.

The surfaces must be lapped carefully to assure mating flatness to provide a pressure-tight seal. This seal is maintained by a loading spring above the moving plate. The use of Rulon provides a lubricated surface to reduce rotation drag and allows even wear for longer service before relapping the surfaces.

The actuator may be single or dual action (dual action requires opposing pneumatic loading), but both require a mechanism, such as a gear rack and pinion, to convert linear piston motion to rotary motion.

The rotary valve offers less internal volume than the linear valve, but requires greater maintenance due to the wear on the mating surfaces. The rotary valve is more expensive than the others.

Sliding Plate Valve The sliding plate valve is a two-position, linear, multiport, low-volume, pneumatically operated device consisting of two plates with precision-machined surfaces (Figure 10.4k). One plate is usually stainless steel and contains a series of small diameter openings that connect to tubing on the back side. This plate is stationary. The second surface is usually a hard elastomer, such as Rulon, or Fluorgold, in which small grooves have been machined to connect two adjacent ports of the fixed plate. The movable plate, or slide, with the machined passages is connected to a pneumatic actuator that provides linear movement between two positions, and thus changes the path of the flow.

Like the rotary valve, the surfaces must be lapped

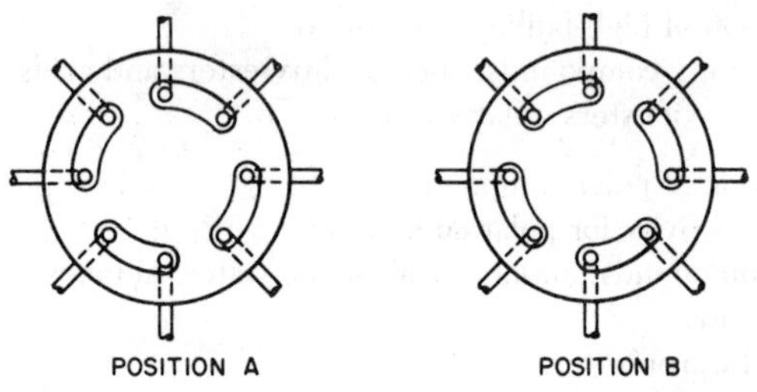

Fig. 10.4j Rotary valve

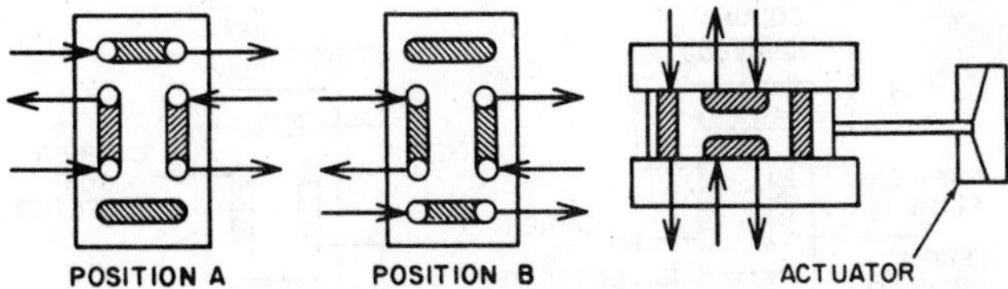

Fig. 10.4k Sliding plate valve

carefully to provide a pressure-tight seal. The seal is maintained by a loading spring above the moving plate. The use of Rulon provides a lubricated surface to reduce surface drag and enhances even wear for longer service before relapping is needed.

The sliding plate valve offers low internal volume, but requires higher maintenance for the wear surfaces. The surface-to-surface seal is more prone to leakage than that of the linear valve. Cost is moderate, and the valve can be used in temperatures up to 400°F. (204°C).

Valving Arrangements and Uses

Valves are used in the chromatograph to introduce a fixed volume of sample to the column system and to switch column segments into and out of the carrier stream. The latter permits use of programming techniques that can shorten cycle times, prolong column life, alter elution sequence and permit regrouping of selected components for a "total-components" reading. Each of the valves discussed previously can be used for these purposes. The following description of the flow-scheme arrangement and its uses is illustrated with a 6-port sliding plate valve because this type is most prominently used and its schematic is most easily depicted.

Sample Inject Valve—Gas Sample The principal function of a sample inject valve in a process chromatograph is to introduce a precise, fixed volume of sample on a repetitive basis. This fixed volume, or sample size, must remain absolutely constant for each sampling. This is achieved by use of a length of tubing between two ports of the sample valve. The diameter and length of tubing can be varied to give the sample volume required for the application. The tubing forms the sample loop and the volume is related as sample size in terms of cubic centimeters or milliliters. Sample sizes typically range from 0.2 to 50 cc.

Figure 10.4l illustrates the tubing arrangement of the sliding plate valve used for a gas sample. The valve action is rapid and the two flows are interrupted for a fraction of a second during valve position changes. The carrier flow may appear to change slightly for large sample loops. This occurs because the sample loop is normally at atmospheric pressure, and the sample volume is compressed as it is purged by the carrier from the loop into the pressurized column system. This will appear as a baseline fluctuation in the detector output.

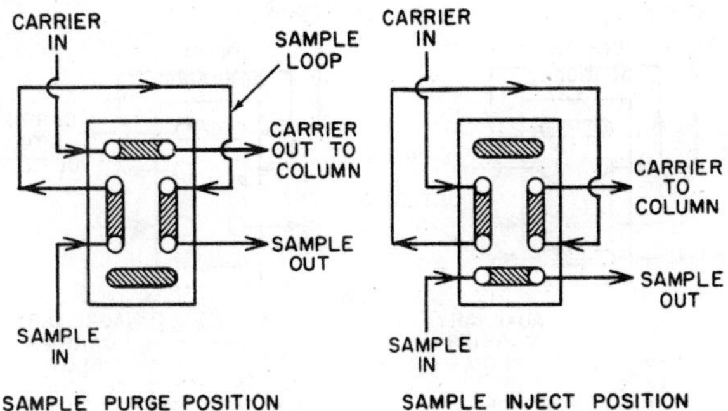

Fig. 10.4 l Gas sample inject valve

The sample flow remains unimpeded so that sample gas is kept purging to the valve. The valve is left in the "inject position" long enough for the carrier to sweep the sample loop. The valve is returned to the "purge position" when the baseline upset will not interfere with the analyses, but before the next sample injection to assure a representative sample in the loop.

Sample Inject Valve—Liquid Sample Liquid samples are frequently analyzed in the process chromatograph, but the sample must be rapidly vaporized by the valve and carrier system. Liquid sample sizes are usually very small, typically between 0.5 and 10.0 μl. Sample loops are not generally used, but a connecting groove or hole in the valve slider is sized to give the volume required (Figure 10.4m).

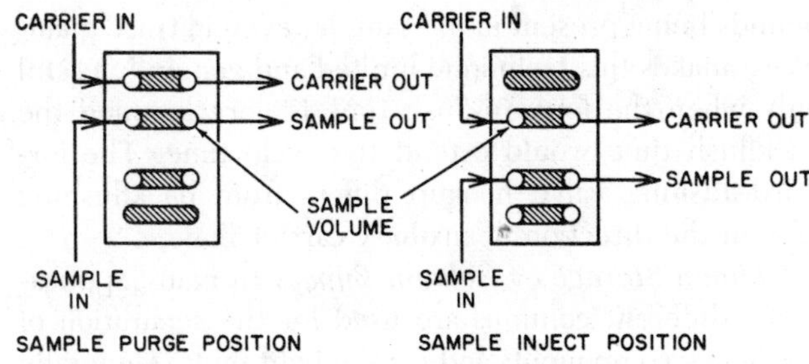

Fig. 10.4m Liquid sample inject valve

Some liquid sample valves incorporate heaters that raise the temperature of the sample valve to aid vaporization. In some applications, the valve is mounted outside the analyzer case, or in a different temperature chamber, to eliminate vapor bubble formation by a low boiler in the sample. Increasing the sample pressure may be adequate to suppress the bubble formation and can be used for noncompressible liquids. Obviously, bubbles in the sample cavity will create gross errors by reducing the effective sample injected to the column system.

Precut and Backflush Valving At times, some analyses use columns that are affected by components of the process stream, such as moisture, oxygen and so on. To avoid rapid degradation of these columns, a precut column is used to separate contaminating components from

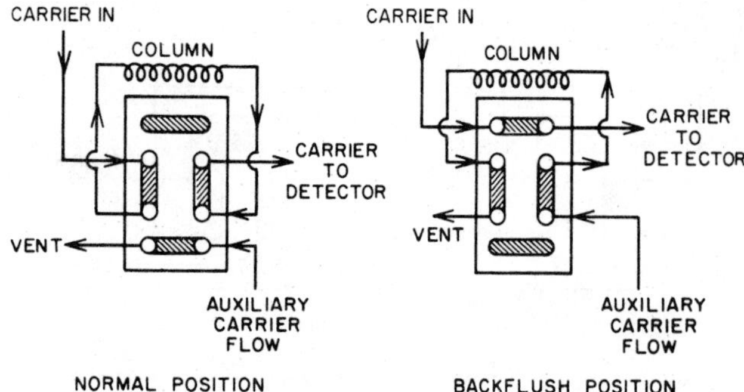

Fig. 10.4n Precut and backflush valve

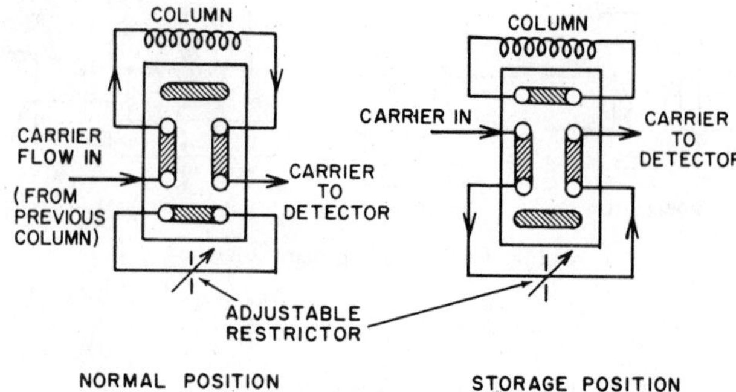

Fig. 10.4o Column storage or column bypass valve

those of interest, allowing the desired components to elute into the second column. The "undesirables" are then backflushed or purged to the vent system by an auxiliary carrier supply (Figure 10.4n).

This technique may also be used to reduce analysis time by backflushing heavy components which are not of interest and which would require excessive cycle time to elute from the entire column system.

Backflushing time must be at least equal to the precut time (at the same carrier flow rate) to assure venting of all components, but a safety factor of between two and five should be allowed, depending on the quantity and nature of the compounds. Forward-flushing of the precut column can be effectively used if the character of the sample is definitely known and the forward-flush time is, therefore, established. The risk of unpredicted compounds being present in the sample, even in trace quantities, makes this technique limited and generally useful only when the forward-flush time is very low and the backflush time would extend the cycle time. The forward-flushing valve hookup differs from backflushing only in the direction of auxiliary carrier flow.

Column Storage or Column Bypass In many applications, different columns are used for the separation of the heavy components and for the light ends. Generally speaking, a heavy compound is retained longer by a column than a lighter one, and excessive time cycles would result if the heavies were passed through the light column. The valve arrangement of Figure 10.4o illustrates how a column can be removed from the flow system for storage.

The columns are in series flow arrangement at the time of sample injection. After the unresolved light components have eluted from the first column into the second, but before any heavy components leave the first column, the second column is sealed off and bypassed. The heavies pass directly to the detector, and the storage column is then placed back in the flow system for the light components. The restrictor is used in the carrier bypass to simulate the flow restriction of the column and, thus, minimize flow differences or upsets between the two positions.

This valving arrangement is also used to provide a means of removing a column from the flow path to study the character of the separation from a previous column. This is particularly helpful when troubleshooting a suspected fouled column.

Heart Cut Technique The technique of heart cutting is frequently used in the analysis of trace components that are not readily separated from a large composition peak. Because large peaks tend to "tail" into smaller peaks which follow, separation is difficult: a longer column improves separation, but broadens both peaks, which reduces the detector output, and, thus, sensitivity.

The heart cut technique reduces the larger component in the analysis and produces a relatively sharp peak for the detector. This is accomplished by selecting a first column that produces separation of the large and small components, even though rough or incomplete. The elution of the large component is vented until the smaller one appears with the trailing edge of the large component. The valve (Figure 10.4p) is actuated to divert the flow to the second column until most of the smaller component is in that column, regardless of the larger amount of the component remaining. The valve is then repositioned, and the second column completes the separation of the two components. The heart cut technique reduces the ratio of large components to small ones for the final separation by taking a "heart cut" of a partially separated mixture.

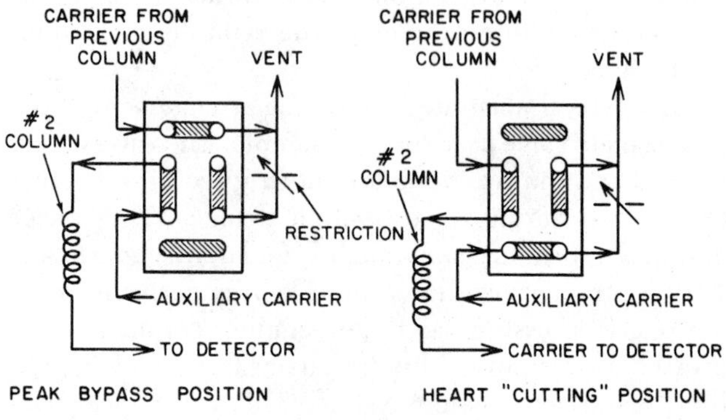

Fig. 10.4p Heart cut valve

Backflush to Measure—Regrouping Some applications are satisfied with separate analysis of light components and a total readout of heavier components. The total heavies analysis shortens the time cycle and is achieved by collecting the heavies in a column and regrouping them by backflushing the column through the detector. Exact regrouping is tricky and the calibration standard for the total heavies presents problems because each component gives different responses to the detector. Figure 10.4q illustrates the valve arrangement for this technique.

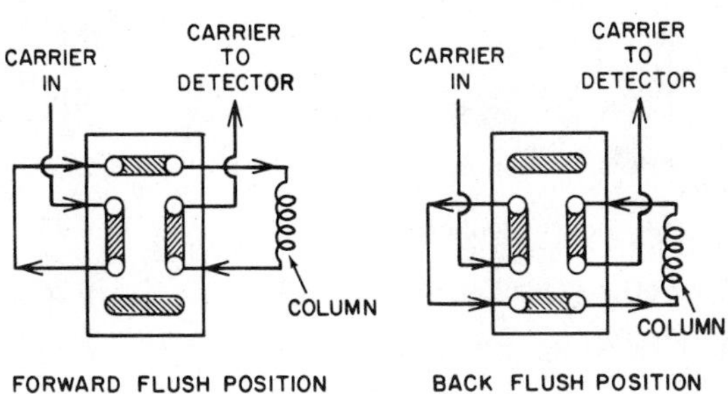

Fig. 10.4q Backflush to measure

All of the above valving arrangements include the most frequently used schemes, but they are by no means all-inclusive of the ingenious arrangements possible with the multiport valve and chromatographic technology. A typical flow scheme using several valves is depicted in Figure 10.4r.

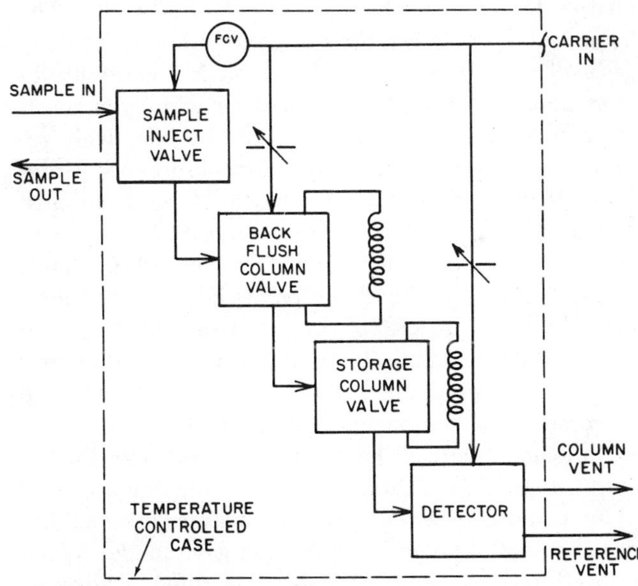

Fig. 10.4r Typical process chromatograph flow system

Detectors

The chromatograph can use several types of detectors (Table 10.4s), depending on the sensitivity and selectivity required for an application.

Table 10.4s
DETECTORS FOR PROCESS CHROMATOGRAPHS

Type	*Sensitivity*	*Cost*	*Maintenance*
Thermal conductivity	Moderate	Low	Low
Flame ionization	Very high	High	Moderate
Gas density balance	Moderate	Moderate	Low
Flame photometric	Very high	High	Moderate
Differential pressure	Low	Low	Low

Thermal Conductivity Detector This detector is discussed in detail in Section 10.16. The thermal conductivity (TC) detector has been, and continues to be, the primary detector for process gas chromatography. It is reliable, simple, easy to maintain, relatively inexpensive and has a universal response. Because the chromatographic column ideally elutes the components to be measured as a binary mixture, that is, a mixture of the component and carrier, the thermal conductivity detector is ideal when the separations are good, and the sensitivities not demanding.

Selection of the carrier gas is very important when using thermal conductivity detectors because they measure the differential conductivity between the carrier gas and the eluted component. Because most components have low conductivities, a carrier of high conductivity is usually selected. Hydrogen, with the highest conductivity, is widely used, but helium with only a slightly lower conductivity is more frequently utilized because it involves no explosion hazard. (However, helium is not readily available in some countries.) The additional expense of this carrier is considered worth the increased safety. Nitrogen, argon and carbon dioxide have also been used as carrier gases.

Flame Ionization Detector The flame ionization detector is the most versatile and sensitive detector in process chromatography today. This detector measures the ion current developed in a flame due to the ionization of organic compounds. No response is found for inorganic compounds, a fact that limits its use, while reducing the separation requirements of organic from inorganic compounds.

The source of ionization in the detector is a flame burning in a hydrogen-air mixture. Carrier gas and hydrogen mix in the detector jet and are burned in the air atmosphere (Figure 10.4t). A collector is positioned above the jet in the proximity of the flame. The collector is kept at a positive electrical potential (75–150 v. above the normally grounded jet) to assure ion flow to the collector.

Nitrogen is commonly used as a carrier gas because a favorable signal-to-noise ratio results from the dilution of hydrogen in the flame, but other inorganic carriers can also be used. A flame of clean hydrogen, nitrogen

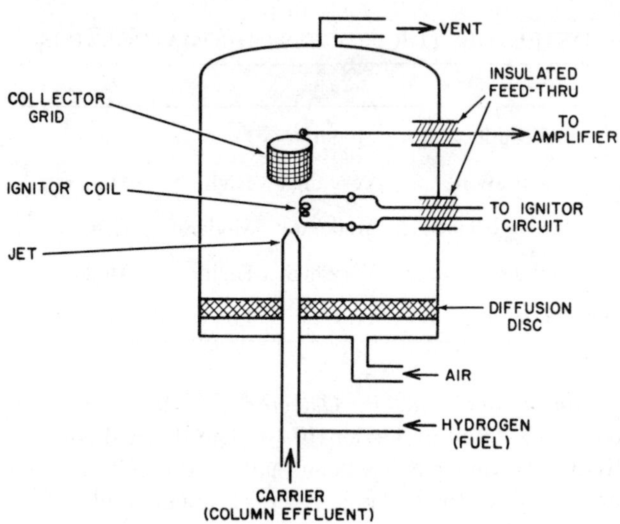

Fig. 10.4t Flame ionization detector chamber

and air produces very few ions. This background ion current is constant and can be readily zeroed, balanced. When an organic sample enters the flame, ionization occurs, releasing large ion flows to the collector. This ion flow, or current, is amplified by the use of electrometers to a useable level for recording.

The current levels involved are low (1×10^{-12} A. typical) and the source impedance very high, requiring high-gain, low-noise electrometers. These are frequently mounted very close to the detector to reduce connecting cable lengths.

Several divergent theories exist to explain ionization behavior, but the actual process of ionization is not well understood. For simple molecules, the response of the detector is nearly proportional to the carbon content of the compound. Response to other compounds is not easily predictable. Relative response is generally related to carbon number (the number of carbon atoms in the molecule), but actual response factors have been collected and published as "effective carbon numbers" (Table 10.4u). These factors applied to the carbon number of a given compound will approximate the relative sensitivity of the detector to the compound. Process designs can measure 10 ppm methane (full-scale) under most conditions; sensitivity to other compounds can be extrapolated from this.

This detector is extremely linear, with a dynamic range of at least 10^5. The flame ionization detector is not as temperature sensitive as the thermal conductivity detector, and some designs can operate up to 550°F (288°C). Higher temperature operation reduces water condensation, a product of the hydrogen-air-burning process. The insensitivity of this detector to certain compounds frequently reduces column selection problems, particularly in samples containing gross water concentrations.

Figure 10.4t depicts a typical flame ionization detector

Table 10.4u
FLAME IONIZATION DETECTOR CONTRIBUTIONS TO EFFECTIVE CARBON NUMBER

Atom	*Type*	*Effective Carbon Number Contribution*
C	Aliphatic	1.0
C	Aromatic	1.0
C	Olefins	0.95
C	Acelylinic	1.30
C	Carbonyl	0.0
C	Nitride	0.3
O	Ether	−1.0
O	Primary alcohol	−0.6
O	Secondary alcohol	−0.75
O	Tertiary alcohol, esters	−0.25
Cl	Two or more on single aliphatic C	−0.12 ea
Cl	On olefinic C	+0.05
N	In amines	Similar to O in corresponding alcohols

chamber. Various designs are employed for collector-jet configurations, coaxial cable feed-throughs and the ignitor coil. This coil is used to ignite the hydrogen-air mixture by resistance heating of a coil or glow plug. The flame ionization detector requires the additional auxiliary gases of hydrogen and air. Clean air is a necessity for eliminating the use of instrument air with even traces of compressor oils unless these can be effectively removed with molecular sieves or other absorbents. Moreover, the flame detector presents a real hazard and must be effectively housed in explosion-proof enclosures, with appropriate flash arrestors used in the fuel and vent lines.

In some designs the flame points downward. Therefore, the solid particles or droplets that form are swept away and do not fall back into the flame to create noise. A small vent maintains sufficient gas velocity to sweep the products of combustion out of the detector.

Although discussed here solely as a detector for the chromatograph, the flame ionization detector is also used as an analyzer in its own right to measure the total hydrocarbon content of a continuous gas sample. When utilized in this manner, the device is called a hydrocarbon analyzer and will be discussed in more detail in Chapter XI. This detector is being used more widely in process chromatographs due to its extreme sensitivity, linearity and selectivity. On the other hand, it increases the initial cost of the analyzer and involves higher operating cost because more auxiliary gases are required.

Photoionization detectors are similar to flame ionization units, but employ ultraviolet light sources rather than combustion to energize molecules; these detectors are suitable for organic compounds.

Gas Density Balance Detector The gas density balance detector (Figure 10.4v) is occasionally used in process chromatographs when samples that may be detrimental to the filaments, or thermistors, of the thermal conductivity detector are to be measured. The detector consists of three flow passages. The carrier from the column is connected to the vertical center passage.

The passage arrangements create a "pneumatic Wheatstone bridge" effect because the carrier entering the center passage can affect the relative flow in the horizontal passages. If the entering carrier is heavier than the reference gas, the carrier will tend to have greater flow through the lower passage. Consequently, the reference gas will be forced to flow predominantly in the upper passage. The resulting flow differential is sensed by the two detector elements. These elements can be hot wire filaments, or thermistors, and functionally resemble the thermal conductivity detector system using a Wheatstone bridge. The increased flow at the top will produce increased cooling of detector A simultaneously with decreased cooling of element B as a result of less reference gas flow.

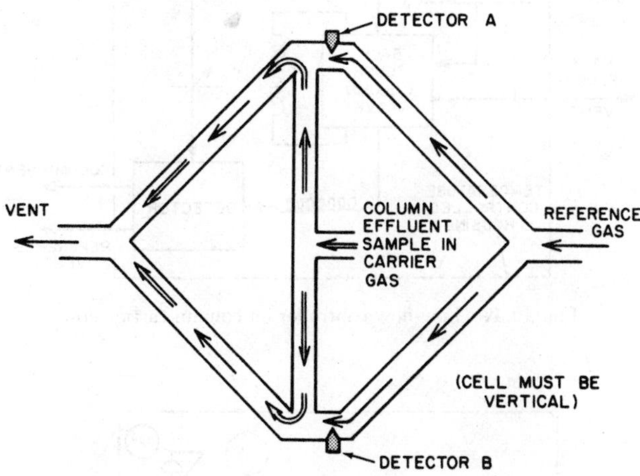

Fig. 10.4v Gas density balance detector

It is evident from the diagram that the column effluent does not contact the filaments, thus preventing possible damage due to oxidation, coatings, and so on. The reference gas can be chosen independently of the carrier gas, but in most instances the two are the same. The reference gas is normally selected to give a large density difference between it and the compounds to be analyzed.

This detector is slightly more expensive than the thermal conductivity detector with generally comparable sensitivities. It is used for applications requiring sample isolation from the thermal conductivity detector elements.

Flame Photometric Detector Other detectors are available based on spectroscopic absorption characteristics of different compounds. Selective waveband spectrometers in the infrared or ultraviolet ranges are used in fixed single-band, multiple waveband or scanning variations. Flame photometric detectors are also based on spectroscopic principles, but measure the characteristic wavelength emissions of molecules energized through controlled combustion.

The need for sensitive response to sulfur- or phosphorus-containing compounds has prompted the development of the flame photometric detector (FPD) for process applications. This detector senses sulfur at the parts per billion (ppb) level.

Differential Pressure Detector This chromatograph is also called a pneumatic composition transmitter (PCT). It takes a sample from the process stream and introduces it into carrier gas, usually helium, when the signal processor sends a demand signal to the valve.

The chromatograph column separates the gas into individual components and feeds the components to a simple orifice-capillary detector. Differential pressure across the orifice is proportional to density. A pneumatic amplifier in the detector subtracts the differential pressures of an eluted component across a 0.1-mm (0.004-in.) diameter jeweled orifice and of the carrier gas instructed by its program. The difference between the signal height and the baseline is amplified and transmitted as the output signal.

This unit, illustrated in Figure 10.4w, was developed for the more common applications in which sensitivity can be traded for increased reliability and lower installed cost. The device contains no electric components, which contributes to its increased safety and simple maintenance.

Miscellaneous Detectors The electron capture detector (ECD) responds to molecular structures that contain oxygen or chlorine; ECD sensitivity varies widely, depending on the number of oxygen or chlorine atoms and the molecular configuration.

Also available are gas chromatographs that interface mass spectrometers for sophisticated detection requirements, mainly for elemental analyses.

Another detector with recent limited usage in the chromatograph is the electrolytic moisture cell. It is used to measure the eluting moisture after separation from the background components that ordinarily would ruin the cell. This application is sometimes called the "hydrochromatograph."

Other more specialized chromatographic detector applications include the following:

Detector	*Application*
Catalytic ionization	Branched compounds
Electrolytic conductivity	Halides and nitrogen compounds

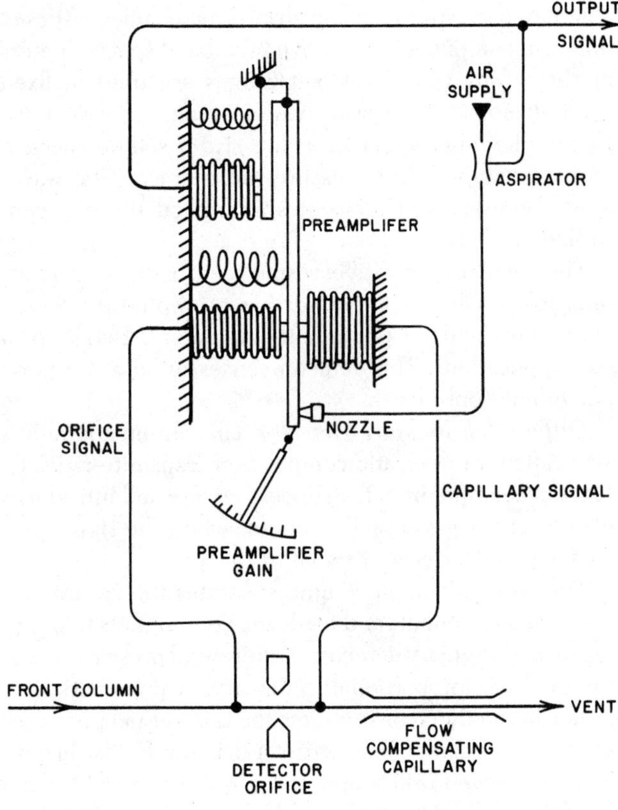

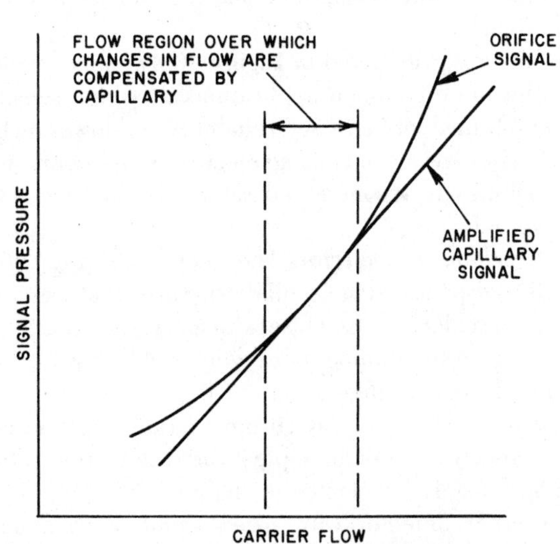

Fig. 10.4w Differential pressure analyzer

Microcoulometric	Halides, sulfur and nitrogen compounds
Reaction coulometric	Organic substances
Thermionic	Phosphorous, sulfur and nitrogen compounds

Carrier Flow Control

One of the parameters that must be closely controlled in the process chromatograph is the carrier flow rate. This is important for two reasons: some detectors are flow sensitive (thermal conductivity), and elution times are critical and must be reliably reproduced in the process chromatograph.

Two basic approaches are widely used for column flow control. The first uses a low-flow controller (Figure 10.4x) referenced upstream to a relatively constant supply pressure. This requires better pressure regulation than the conventional cylinder regulator provides, and a quality, small capacity, single-stage regulator is normally used between the cylinder regulators and the analyzer. This system provides excellent flow control, but pressure upsets caused by valve actuations may require a lengthy recovery period.

The pressure flow control system of Figure 10.4y reduces the recovery time of these pressure upsets. The carrier gas pressure is regulated inside the case and the flow rate is metered by use of an adjustable restrictor (valve) between the regulator and the columns. When

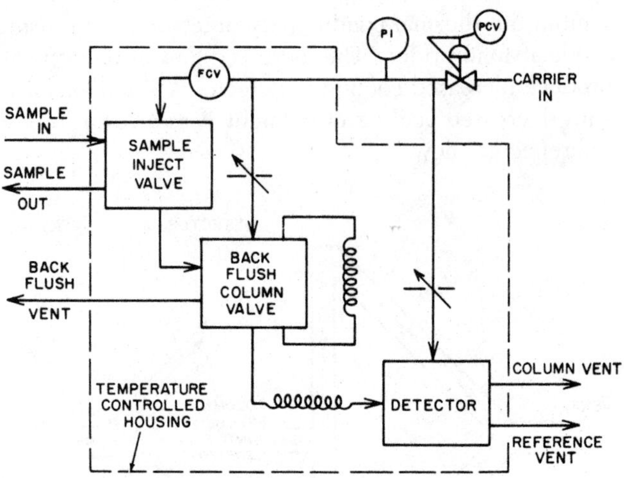

Fig. 10.4x Low-flow controller on column carrier flow

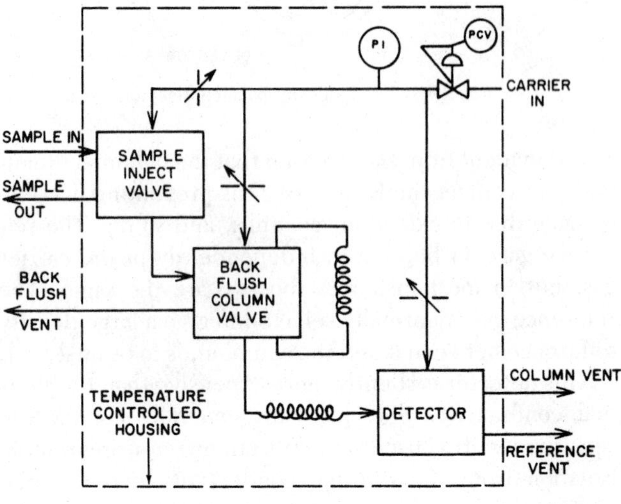

Fig. 10.4y Pressure control of column carrier flow

the pressure in the carrier flow system suddenly drops, as it does when the backflush column is returned from the vent position, the pressure regulator quickly compensates with additional carrier gas until the pressure gradient along the flow system is reestablished. Because this will occur during each analysis cycle, the elution times are reproducible. This system does provide quicker recovery, but overall flow control is not as good as with the low-flow controller. However, arguments can be advanced for both systems.

Data Presentation

The chromatograph's nature of discrete sampling and discontinuous readout of multicomponent values presents special problems in the recording of the data. The following is a brief discussion of the various readouts provided for the process chromatograph.

Spectrum or Chromatogram

The detector output for a hypothetical analysis might appear as in Figure 10.4b. This represents the detector response to a series of components emerging from the column and is commonly called the spectrum, or chromatogram. This type of presentation is conventionally used in the laboratory where the area under the component peak is measured, corrected for detector response to that component, and related to the entire analysis area to determine the composition value of the component. This presentation consumes recording paper rapidly and generates reams of paper when run continuously. Obviously, it is unsatisfactory for routine operator interpretation, so other presentations have been developed for routine recording.

The spectrum is useful to the service man in the initial programming of the analyzer (determining the event times) and in the field as a troubleshooting tool to determine column stability, elution time changes, detector sensitivity changes, baseline drift and so on.

Bargraph

Figure 10.4b also illustrates the use of bargraph data reduction. In this system, a barline is generated with a length proportional to the composition value. This is achieved by stopping the chart drive and activating the recorder servosystem only during the time interval of component elution, or component gating. This allows the recorder pen to transverse the scale to the maximum value and return to zero, thus inscribing a bar on the chart. The chart is advanced approximately ¼ in. (6 mm) between bars and ½ to ¾ in. (12 to 19 mm) between analysis sets. If multistream operation exists, the stream selector will impose a wide bar of varying height in this space to identify the stream whose analysis follows. Component identification is by appearance sequence; that is, the third bar of this illustration is component C.

It may be well to point out here that each component range is calibrated by individual range attenuators so that the height of the peak on the spectrum will probably be different from the bargraph height, unless, by coincidence, the bar-range attenuator and the spectrum attenuator are of the same value (see Calibration).

Bargraph presentation is an economical method of data reduction that normally presents the previous 2-to-3-hours analysis information on the visible recorder chart for rapid interpretation.

Trend Recording

It is frequently convenient to trend record (Figure 10.4z) the components of interest, particularly if process control action derived from the analysis results is desired. Various pneumatic and electronic systems are available to store the peak value in a memory system that outputs to the trend recorder. The memory channel is updated once each cycle when a new analysis value is obtained. Because the basic signal available from the chromatograph is continuously changing, peak-pickers are employed to track the signal to its maximum, or peak, value and remain at that value. (Many designs use capacitance charging techniques with one-way gating for this.) The peak-picker output is transferred to the appropriate memory channel for trend recording and the peak-picker is reset for the next peak.

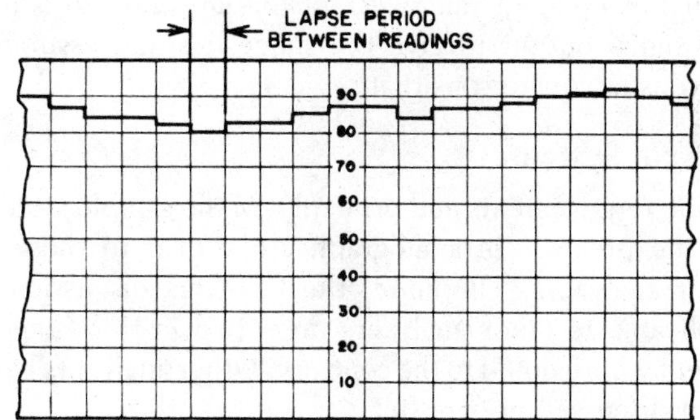

Fig. 10.4z Component trend recording on strip chart

Trend recording simplifies operator interpretation of data results, but requires additional hardware expense, for example peak-pickers and additional recorders are required.

Trend recording is also basic to the use of the chromatograph for process control. Once the signal is placed in memory, that is, a continuous output obtained, the signal can be used in a control loop, but rapid corrective action from the controller is not advisable since the "effective response" of the reading is slow. Many authorities recommend using the output for control only when the process system response is at least twice that of the analyzer. Others recommend use of sampled data techniques for control.

Integration of Peaks

Integration of the area under the peak of the chromatogram offers more exacting analyses than the peak height method of the bargraph, particularly when peak shapes are not symmetrical. This accuracy is not normally required for process applications, but can be obtained by using integrators that measure the area electronically and output an analog or a digital signal proportional to the area.

Digital Presentation

Digital presentation is the most convenient manner of reading and interpreting analysis data, but it is also the most expensive. One system uses a digital indicator printer that receives analog inputs from memory channels (as for trend recording) and prints out the analyses results in an easily used format.

Of course, computers are being widely employed for this purpose. A central process computer, or a dedicated special-purpose computer, can readily receive the peak-picker memory output and log analyses as required.

Calibration

As with most process analyzers, chromatograph calibration is empirical, using known standard samples. These samples may be synthesized, or mixed, in a laboratory, or may be actual process samples that have been carefully analyzed by a quantitative laboratory procedure. Individual component range attenuators are used to adjust the signal output to give the range desired, assuming there is adequate sensitivity.

Sample System

The requirements and principles of the sample system for the process chromatograph are similar to those of other analyzers. Therefore, the following discussion is applicable to other analyzers described in this chapter except where noted to the contrary. Other analyzers have been discussed previously.

Analyzer results will be of little value unless the analyzed sample is representative of the process stream. There is rarely one "best" design for a sampling problem because many suitable designs can be satisfactorily used. Designs will differ as a result of personal preferences, location needs for certain equipment or sampling philosophy. Certain considerations must be weighed, so that the resulting design will be the one best suited to the application. The design will be based on a sound understanding of fluid flow, the process chemistry, and the analyzer's requirements.

The analyzer sample system is generally considered to be the entire sample-handling system that continuously and automatically obtains a representative sample from the process stream, transports it, introduces it to the analyzer and disposes of the waste sample. It must do this without altering the integrity of the samples, although it may condition the sample to prepare it for the analyzer. If a multistream operation is used, the sample system must switch between the streams in the introduction of the sample to the analyzer.

The major functions of the sample system include:

1. Obtaining a representative sample
2. Sample conditioning
3. Sample transport
4. Stream switching
5. Sample disposal

Obtaining a Representative Sample

The sample tap must be located in the process line to give a homogenous, representative sample that can be conditioned and used by the analyzer. Where possible, the tap location should be selected to give adequate pressure for transporting the sample without use of additional pumps. Many times, it is possible to obtain the sample after a filtration or drying stage that may not affect the representative nature of the sample but may reduce external conditioning or other sample-handling problems.

A representative sample must have a composition representative of the process stream but may in fact differ from those conditions by selectively removing troublesome elements, such as particulate matter, which will adversely affect the sample system and/or the analyzer. Composition representation must be maintained within the system accuracy desired (see Sample Conditioning).

It is usually advisable to obtain the sample from the center part of the process pipe to avoid possible wall effects that could affect the sample composition or introduce more dirt or solid particles into the sample system. Many probe designs have been used. Some are simply pipe nipples welded through the pipe wall to protrude several inches into the pipe. Other designs use tubing probes that extend into the pipe through ball or gate valves and which can be removed through a packing gland during process operation.

If possible, the probe opening should be directed toward the downstream flow to reduce impingement of entrained solids (and liquids in gas systems).

Sample Conditioning

Conditioning of the sample to allow easier handling, or to prepare it for the analyzer is always required to some degree. This may be only a filter in the simplest case or it may be sophisticated scrubbing and dryer systems. The conditioning must not adversely affect the representative nature of the sample and, if composition representation is affected (as in a moisture removal system for high quantities of water), the effect must be continuous, stable and compensated for in the analysis results.

Sample conditioning may take place at the sample point, at the analyzer, or at various intermediate points.

The common conditioning elements include:

1. *Filters* to remove particulate matter from the sample. Most analyzers require samples with less than 10 μ particulate matter
2. *Vaporizers* at the sample point if a liquid process sample can be better handled by the analyzer as a gas at a controllable temperature and pressure. Both steam and electrical designs are available.
3. *Condensers* if a gaseous process sample can be better analyzed as a liquid at a controllable temperature and pressure. Condensers are frequently used to condense "low boilers" of a vapor stream that might otherwise condense later in the system and present problems. These are available in water, air and refrigerant designs but mechanical-packaged units are also provided.
4. *Entrainment separator* to remove large volumes of entrained solids or liquids from vapor streams. Centrifugal separators and reverse flow separators are frequently considered.
5. *Coalescer* to remove water droplets from liquid hydrocarbons (immiscible liquids).
6. *Scrubbers* to wash or scrub the sample with water (or other selected medium) to remove solid particles and condensibles, or to provide a fixed moisture level in the sample.
7. *Bubble remover* to remove vapor bubbles from liquid streams.
8. *Pump* to provide adequate flow rate if process pressure is not adequate.
9. *Pressure regulator* to regulate system pressure.
10. *Stream tracing* to maintain sample line at elevated temperature.

Table 10.4aa gives the information needed for proper evaluation of sample system requirements.

Table 10.4aa
DATA FOR EVALUATING
SAMPLE SYSTEM REQUIREMENTS

Process pressure	
Process temperature	
Liquid	*Vapor*
Bubble point @ °F at 0 PSIG	Dew point @ °F at 0 PSIG
Viscosity	Molecular weight
Specific gravity	Specific gravity
Sample composition: high, low, normal values of all components (include trace components)	
Materials of construction compatible with sample	
Sample return point pressure	
Sample line length	
Response required	
Sample pressure requirements for analyzer	
Sample temperature requirements for analyzer	
Sample flow rate required by analyzer	
Sample toxicity	

Sample Transport

The sample must be transported from the sample tap to the analyzer, a distance that may vary from 5 to 500 ft. (1.5 to 150 m), although 200 ft. (60 m) is usually the maximum desired length. The design of this sample tubing is too frequently taken lightly, or ignored entirely.

Line sizing is a function of line length, response time desired and flow rate. Line lengths should be minimized, but physical limitations on where the sample taps and the analyzers must be located sometimes take this matter out of the designer's province.

Response time is a function of flow rate and internal volume of the system. Much of this internal volume exists in the tubing run; but valves, fittings and other components must also be considered.

Flow rate of the sample will depend on the differential pressure available, on the fluid gravity and viscosity and on the tubing size. Compressibility of gases must also be considered.

For gas samples, the following equation can be used to determine the transport time in sample tubing:

$$t = \left(\frac{VL}{F}\right)\left(\frac{P + 15}{15}\right)\left(\frac{530}{T + 400}\right) \qquad 10.4(1)$$

where

t = transport time (min.)
V = internal volume of tubing cc/ft.
L = tubing length (ft.)
P = system pressure (avg.) PSIG
T = system temperature (avg.) °F
F = Flow rate, scc/min.

Note: For conversion to SI units see Appendix A.1.

The overall purge time, including fittings, regulators, and so on, is more difficult to predict because the internal volumes are not well known, and the flow patterns may include poorly purged volumes that are not cleanly swept and may require additional time to purge completely. Needless to say, these volumes should be avoided. Most designs employ flow rates between 2000 cc per minute and 2 scfm (3.4 m^3/hr.), or about 60 L. per minute.

Stream Switching

Multistream operation of analyzers is economic if the stream numbers are kept reasonable and the samples are similar in nature. This time-sharing approach requires stream-switching designs that prevent mixing of the samples.

Most switching designs employ 2- or 3-way solenoid valves that are actuated to introduce the desired stream. Unfortunately, solenoid valves are prone to leakage and system designs should provide for this eventuality. Because it is desirable to keep all sample streams purging regardless of stream analysis status, bypasses are used to maintain constant flow in the sample transport system. Figures 10.4bb and 10.4cc show some commonly used sample-switching arrangements. Each sample should have approximately the same line pressure before the first valve.

Although solenoid valves are widely utilized to switch the process streams directly, the use of miniature, pneumatically operated 2- and 3-way valves is receiving more attention. These valves introduce additional expense to the system because solenoid pilots are still required to actuate them, in order to control the air loading of the pneumatic valve. However, they allow more freedom in component layout, because running of conduit in the sample area is eliminated. The pneumatic valves are generally more leak-proof than the direct solenoids and more tolerant of steam tracing.

Stream-switching systems that require steam tracing are frequently assembled in a heat-insulated cabinet. This provides better access to the components for maintenance.

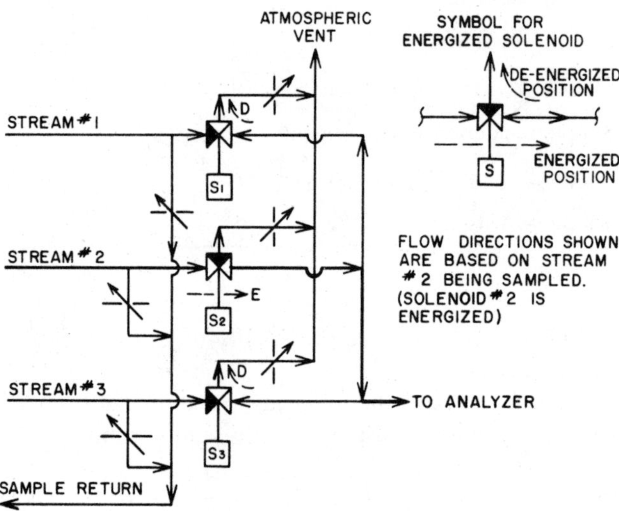

Fig. 10.4bb Small solenoids with back purge

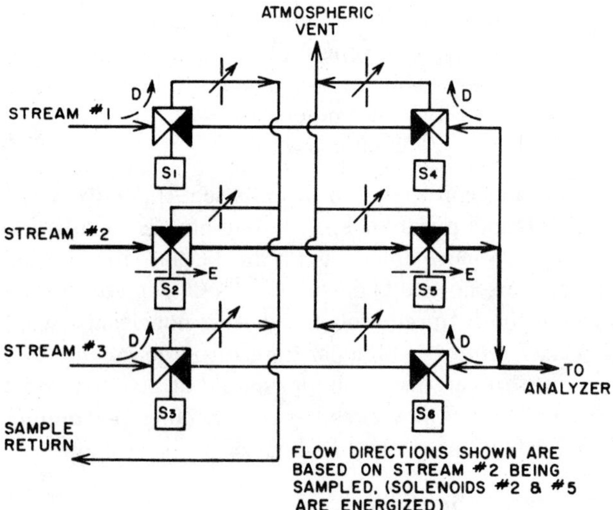

Fig. 10.4cc Dual solenoid sampling system with block, bleed and back purge

Sample Disposal

A sample that is bypassed at the analyzer, or stream-switching station, may be returned to the process if an appropriate low-pressure return point is available. The expense of installing the return line must be justified by the value of the bypassed sample; otherwise, it should be vented to the atmosphere or drained. However, if the sample would present a pollution problem, other methods of disposition must be arranged. Gases are frequently flared and liquids treated in waste treatment ponds.

If the sample discharged from the analyzer cannot be vented to the atmosphere, the flare or treatment pond must be considered. Samples that must be disposed of in a flare system must be maintained above the flare header pressure (normally 2 PSIG, or 13.8 kPa). For gas samples this can present problems. Most analyzers are pressure sensitive and perform better when operating at atmospheric pressure because pressure control system variations are usually greater than barometric variations. Absolute pressure regulators are available but offer little improvement in applications. The analyzer must be calibrated at the operating pressure of the vent system.

The column vent of the chromatograph consists mostly of carrier gas with traces of sample components. Unless the sample is very toxic or obnoxious, the column may be vented to atmosphere, particularly when the reference vent and backflush vent are also disposed of through the same header.

Atmospheric vent lines should be sized to assure that the pressure of the analyzer does not vary more than 1 in. H_2O (0.26 kPa) with the vent flow variations.

Summary

Sample system designs require the same consideration as process designs and attention to them will lead to a satisfactory sample-handling system. However, even the best designed system will require some "cut and try" after installation to balance flows, pressures, and so on. Process conditions do not always agree with the original design basis and changes may be in order.

If the service man's problems are considered in the

original design, startup problems will be less and long-term maintenance reduced.

Sample system costs may equal, or even *exceed*, the analyzer cost.

Installation

Installation requirements of the chromatograph are similar to those of other analyzers. Thus, the following discussion is applicable to other analyzers in a general sense. Special requirements for the chromatograph will be noted as such.

Today's analyzers are certainly more rugged than their forerunners; yet the installations are becoming more protective than ever before. Experience has proven that the analyzer performs best when housed in a weather protective environment with relatively small temperature variations. Yet, analyzers should be located as near the sample tap(s) as practical to reduce transport tubing. This dilemma has led to the erection of walk-in meter houses within the process unit to house the analyzer and sample switching stations.

Analyzer environment should not exceed 100°F (38°C). Many warm climate locations, such as the U.S. Gulf Coast, have found air-conditioned housing justified by improved analyzer performance.

It is usually feasible to separate the analysis and electronic sections of most analyzers so that the control sections may be installed in the control room. Increased maintenance costs are frequently encountered with this arrangement because two service men are often required to service the installation.

The chromatograph requires a carrier gas supply and, if the flame ionization detector is used, fuel gas and a clean air supply. These are normally provided from compressed gas cylinders. These cylinders can be located adjacent to the analyzer, or building, but since the heavy cylinders must be carted to and fro, easy access to them should be available. If more convenient, the cylinders can be located at a walkway and tubing run to the analyzer. If several chromatographs using the same carrier are installed together, a header system can be utilized but the number of cylinders should not be reduced because carrier consumption is not changed and more frequent cylinder replacement would otherwise be required.

Serviceability should not be overlooked. Many analyzers require access to both front and rear and a crowded installation may reduce maintenance efficiency.

Chromatograph Summary

The process chromatograph is perhaps the most useful analytical process tool available today. Multicomponent, multistream analysis is readily achieved and, while the initial cost is greater than that of most analyzers, the cost per component or readout is frequently less than that of the alternatives. Both liquids and vapors can be analyzed. Data presentation is not straightforward, but adaptations are available at nominal expense to give conventional component trend recording.

REFERENCE

1. Kirk-Othmer, *Encyclopedia of Chemical Technology*, 2nd ed., vol. 5. New York, John Wiley & Sons, Inc., pp. 413–49.

BIBLIOGRAPHY

Demming, P.L., "Process Chromatographs with Integral Microprocessors Upgrade Composition-Based Control," *InTech*, July, 1980.

Frost & Sullivan, Inc., On Stream Process Analyzer Market, Report No. 669; Frost & Sullivan, Inc., New York, August, 1979.

Griffin, D.E. and Webb, P.U., "Process Chromatographs and Computers in Optimizing Control Systems," *InTech*, July, 1979.

Jutila, J.M., "Multicomponent On-Stream Analyzers for Process Monitoring and Control," *InTech*, July, 1979.

Jutila, J.M., "Guide to Selecting Gas and Liquid Chromatographs," *InTech*, August, 1980.

Karasek, F.W., "Detection Limits in Instrumental Analysis," *Research/Development*, October, 1979.

Martin, F.D., "Developments in Process Gas Chromatography," *InTech*, January, 1977.

McCoy, R.D., "Adding Capabilities to Process Chromatography with Microprocessor-Based Programmers," ISA Spring Conference (Houston, TX), May, 1978.

Mowery, R.A., "The Use of Process Liquid Chromatography in the Analysis of Industrial Streams," ISA Analysis Instrumentation Symposium (Charleston, WV), May , 1977.

Oliver, B.M., "The Role of Micro-Electronics in Instrumentation and Control," *Scientific American*, September, 1977.

Topham, W.H., "Simplified Chromatographs for On-Line Control," *InTech*, December, 1970.

Villalobos, R; Porter, R; Leblanc, R. and Hearn, R., "New Concepts in Microprocessor-Based Process Chromatographs," 33rd Annual Texas A&M Symposium, January, 1978.

Wise, S.A. and May, W.E., "Unusual Experimental Detectors for LC," *Research/Development*, October, 1977.

Worthy, W., "Analytical Chemistry," *Chemical and Engineering News*, July 31, 1978.

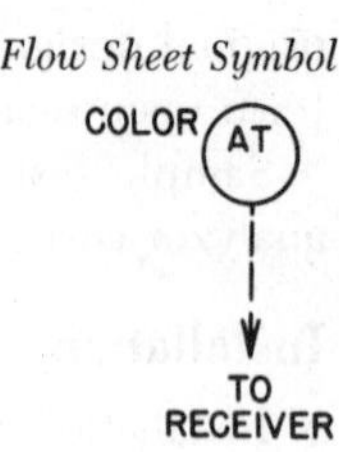

10.5 COLORIMETERS

Type of Sample:	Spectrophotometric: liquids and gases; tristimulus: film, paper, fabric, granular solids
Analyzer Sample Pressure:	Subatmospheric to 150 PSIG (1,035 kPa)
Materials of Construction:	Cell of all standard materials; windows of quartz or glass
Cell Length:	0.5 mm to 40 in. (1,000 mm)
Cost:	\$6,000 to \$20,000 for spectrophotometric types; \$30,000 to \$90,000 for tristimulus types
Inaccuracy:	±2% based on one-dimensional standard; ±1% on tristimulus standard
Drift:	±1% per 24 hr.
Ambient Temperature:	60 to 115°F (16 to 46°C)
Partial List of Suppliers:	Du Pont Co., Analytical Instruments Div.; Gardner Laboratory Div.; GCA/Precision Scientific Group; Hunter Associates Laboratory, Inc.; Macbeth, Div. of Kollmorgen Corp.; Magnuson Engineers Inc.; Teledyne Analytical Instruments

Color measurements involve that part of the electromagnetic spectrum that is visible or sensed by the human eye. This region is approximately between 400 and 700 $m\mu$. In the visible spectrum there is a definite relationship between wavelength and color (Table 10.5a).

Precise and accurate color measurements require multidimensional or coordinate systems using the tristimulus method. However, most industrial and process standards involve color-matching in which color standards are compared by eye, or with the assistance of a spectrophotometer. This spectrophotometric method is used for most colorimeters in liquid service and is essentially a one-dimensional color standard measurement.

Table 10.5a
COLOR AND WAVELENGTH ASSOCIATION

Approx. Wavelength ($m\mu$)	Associated Color
400–450	Violet
450–500	Blue
500–570	Green
570–590	Yellow
590–610	Orange
610–700	Red

Spectrophotometric Analyzers

Instrument designs for process color measurements using one-dimensional standards are similar to those discussed in Section 10.18. Interference filters are selected for desired wavelengths as determined from the spectral relationship curves. Detector photoconductors are chosen to give spectral response in the visible region.

Some UV colorimeter designs are capable of monitoring between 150 and 1,000 $m\mu$ and are frequently called photometric analyzers.

Several one-dimensional color standards are used in the process industries, including the American Public Health Association (APHA) color standards (platinum-cobalt system) and the Saybolt and ASTM color scales.

The APHA standards involve a series of known concentrations of platinum-cobalt solutions. The solutions are defined by the weight concentration of metallic platinum per L. of solution; that is, 1 mg. of platinum per L. of solution (standardized combination of potassium chloroplatimate and cobaltous chloride) is the APHA color unit. The laboratory compares the solution to be measured with the standard solutions in Nessler tubes. The spectral relationship of this standard system is shown in Figure 10.5b.

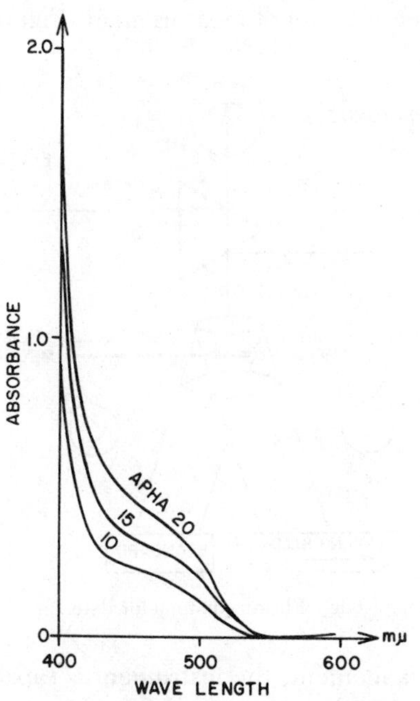

Fig. 10.5b Approximate spectral relationship of APHA color standards

The Saybolt color scale is widely used in the petroleum industry and is related to ASTM procedure D156—"Test for Color of Petroleum Products." This system uses standard glass discs that serve as color comparison standards. The fluid for color characterizing (normally oil) is added to the measuring flask or tube until the colors match, a light-colored sample requiring greater depth than a darker sample to give a comparison color. The system standard is shown in Table 10.5c, and the spectral relationship in Figure 10.5d.

The ASTM color scale is based on ASTM method D–1500, in which there is a general correlation between the "lightness" of oil and the extent of refining. Crude oil is very dark, red-brown-black, whereas refined oil is nearly water white. The standard, then, roughly approximates various stages of refining. The spectral relationship is seen in Figure 10.5e.

Tristimulus Method

Reflectance color measurements utilize the reflected light from a sample to monitor the surface color of the sample. This measurement can be done with a one-dimensional color standard such as those discussed above, but in applications requiring better color definition, a

Table 10.5c
SAYBOLT COLOR SCALE

Color Standard Number	*Depth of Oil, in. (mm)*	*Color Number*
½	20 (508)	+30
	16 (406)	+28
	12 (305)	+26
1	20 (508)	+25
	16 (406)	+23
	10.75 (273.05)	+20
	8.25 (209.55)	+18
	6.25 (158.75)	+16
2	10.50 (266.7)	+15
	9.00 (228.60)	+13
	7.25 (184.15)	+10
	5.75 (146.05)	+5
	4.50 (114.30)	0
	3.50 (88.90)	−5
	3.00 (76.20)	−9
	2.50 (63.50)	−13
	2.125 (53.975)	−16

tristimulus method must be employed. This method is a relatively complicated system of specifying the continuous reflectance spectral curve into three numbers. Generally, these correspond to amber, green and blue spectral responses and for the Commission International del'Eclairage (CIE) system the standard curves are approximately those of Figure 10.5f. A comprehensive discussion of the CIE tristimulus expression values $\overline{X}$, $\overline{Y}$, and $\overline{Z}$ can be found in Kirk-Othmer.[1] Briefly, this system

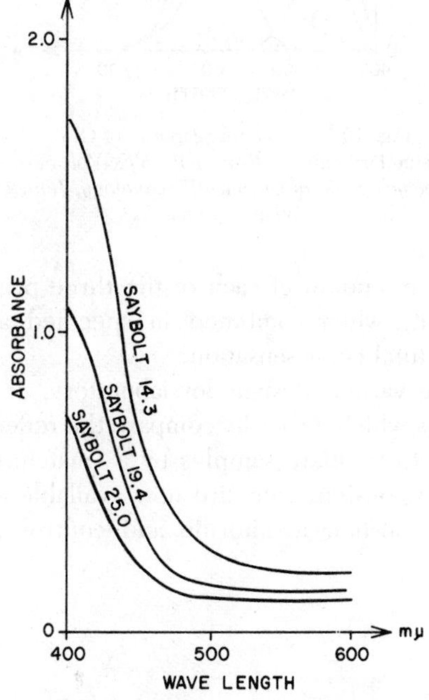

Fig. 10.5d Approximate spectral relationship of Saybolt color scales

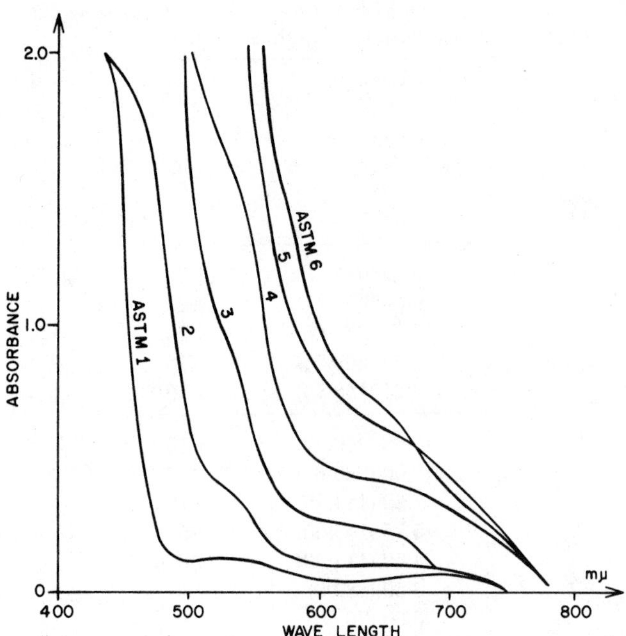

Fig. 10.5e Approximate spectral relationship of ASTM color standards

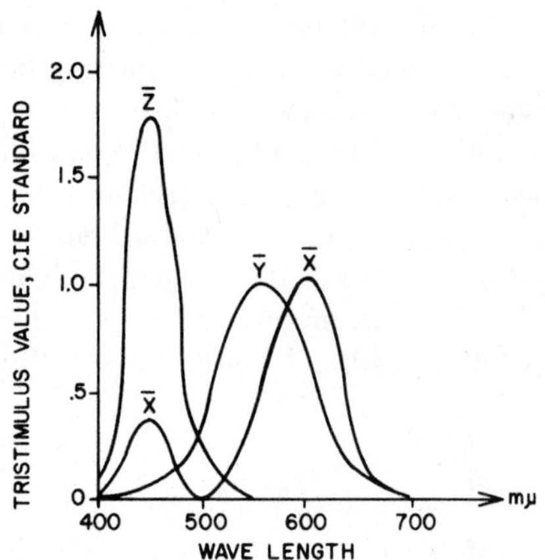

Fig. 10.5f Spectral response of CIE standard values. (*From page 802, Vol. 5, Encyclopedia of Chemical Technology, John Wiley & Sons*)

expresses the amount of each of the three primary responses that, when combined in specified amounts, produces a total color sensation.

There are various designs for laboratory, or off-line, colorimeters which generally compare the reflectance of the sample to standard samples by a "matching" technique. A few on-line units are now available and offer continuous, on-line monitoring and control of color. These systems require complicated computational units to determine the tristimulus values.

One design consists of an illumination system and optical detector positioned above the continuously moving sample. The detector employs the flicker photometer measurement principle. This involves a continuous, rapid, sequential comparison between a sample and a standard by a single photomultiplier through a common filter, for each channel of measurement (Figure 10.5g).

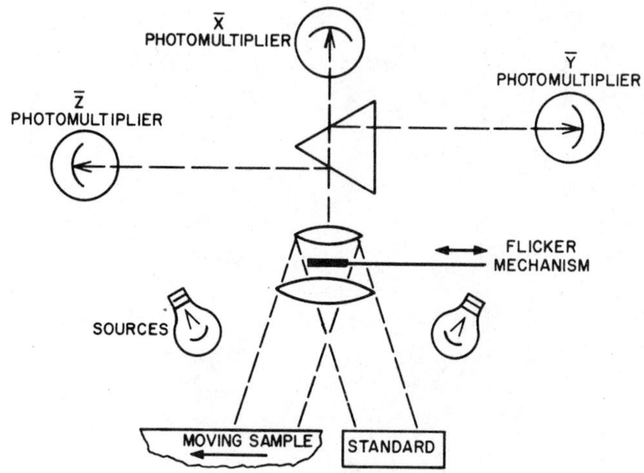

Fig. 10.5g Flicker photometer detector

With this arrangement, the instrument is capable of directly measuring the percentage difference in reflected energy between the sample and standard. The flicker photometer reduces drift and other instabilities caused by changes in illumination and photosensor aging because this measurement is a ratio determination.

REFERENCE

1. Kirk-Othmer, *Encyclopedia of Chemical Technology,* 2nd ed., vol. 5. New York, John Wiley & Sons, Inc., pp. 801–810.

BIBLIOGRAPHY

"Coloring of Plastics," Society of Plastics Engineers, Technical Conference, June, 1966.

Hardy, A.D., *Handbook of Colorimetry.* The Technology Press, 1936.

Jutila, J.M., "Multicomponent On-Stream Analyzers for Process Monitoring and Control," *InTech,* July, 1979.

Molvar, A.E., "Instrumentation and Automation Experiences in Wastewater-Treatment Facilities," EPA-600/2-76-198, October, 1976.

Flow Sheet Symbol

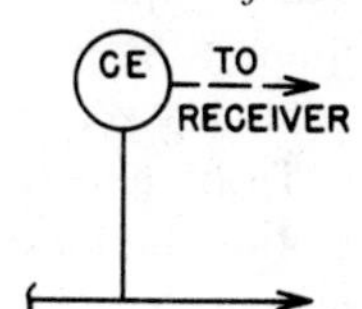

10.6 CONDUCTIVITY

Standard Design Pressure:	To 500 PSIG (3.5 MPa)
Standard Design Temperature:	To 300°F (149°C)
Element Materials:	Glass, epoxy, stainless steel cells; platinum, nickel, carbon electrodes
Cost:	$1800 to $2500
Range:	0 to 0.5 micromho (μmho) minimum; 0 to 2 mho (1 mho = 1 siemens [S])
Inaccuracy:	±½% full scale
Partial List of Suppliers:	Aquatronics Inc.; Beckman Instruments, Inc.; Delta Scientific, National Sonics Div., Envivotech Corp.; Honeywell Inc., Process Control Div.; Leeds & Northrup, a unit of General Signal; Ohmart Corp.; Raytheon Co.; Uniloc Div. of Rosemount, Inc.

Conductivity measurement is the determination of the specific conductance of a solution defined as the reciprocal of the resistance of a 1-cm (1.25 in.) cube of solution. The units of specific resistance and conductance are ohm-cm and mho/cm. The specific conductance of distilled water at room temperature is lower than 1 μmho. Table 10.6a shows the range of conductivity encountered in aqueous solutions.

Table 10.6a
CONDUCTIVITY OF AQUEOUS SOLUTIONS

Solution	*Specific Conductance mho/cm*	*Specific Resistance ohm-cm*
Ultrapure water	$.05 \times 10^{-6}$	20×10^{6}
Distilled water	1.0×10^{-6}	1×10^{6}
Rain water	50×10^{-6}	20×10^{3}
.05% salt solution	1×10^{-3}	1×10^{3}
Sea water	20×10^{-3}	50
30% sulfuric acid	1.0	1.0

Conduction in metals differs from conduction in solutions. In the latter, current is carried by ions, not by free electrons as is the case with metals. Conductivity also differs from pH because all ions present contribute to the conductivity, whereas pH is affected only by hydrogen ions.

The conductivity cell is simple in structure, consisting of two metal or carbon electrodes firmly spaced within an insulated chamber that serves to isolate a portion of the sample fluid for measurement. Thus, the measurement is independent of the sample volume and adjacent metallic surfaces. Selection of conductivity cells must be based on electrical, chemical and mechanical considerations.

Practical limits for measurement of electrolytic resistance are in the order of 50 to 100,000 ohms, where inaccuracies of the order of ½ percent can be tolerated. The first consideration, then, is to select a cell with the proper cell constant, or cell factor. Cell constant is the ratio of specific conductance of the solution to the conductance measured by the cell, and it is this constant, in conjunction with the solution conductance, that determines the resistance seen by the measuring circuit. When measuring high-electrolytic resistances, series capacitance is negligible in comparison to the measurement. However, shunt capacitances, such as long lead wires, do affect the measurement, and the effect increases with source frequency. At low-electrolytic resistance values, the effect of shunt capacitance becomes negligible and series capacitance becomes a factor. Because capacitive reactance varies inversely with frequency, it is desirable to use high-source frequencies when measuring low resistances.

To be effective, the measuring cell must be resistant to the chemical action of the process fluid. The choice of materials is complicated by the fact that an AC potential is impressed across the electrodes and materials that are normally quite unaffected by the process fluid tend to dissolve. Reduction of the source voltage and use of noble metal electrodes are usually the best methods to combat this problem.

High-flowing velocities do not affect the measurement, providing that the cell is not physically damaged. However, at low-fluid velocities, the cell should be selected and installed to ensure proper circulation of liquid between the electrodes.

Where fluid velocities are high and solids are present, the cell must be installed to avoid direct impingement of the flow on the electrodes. At low-fluid velocities, the cell should be installed in the upward flow, with the open end of the cell facing downward. The cell should never be installed in a manner that would allow solids to settle in the cell chamber.

The Measuring Circuit

The determination of conductivity involves the measurement of the resistance of a column of solution. The measuring circuit most commonly used is the ac Wheatstone bridge as shown in Figure 10.6b, where S represents a galvanometer, earphones, oscilloscope, electron tube or some other ac voltage-sensing device. The ac source may be the low-voltage side of a 60-Hz. transformer or an oscillator for higher source frequencies. Rs is the standard arm of the bridge and is usually variable either for range changes or to compensate the temperature coefficient of the solution. This coefficient takes the form $Ro/Rt = 1 + at + bt^2$ for many electrolytes where t is temperature, Ro and Rt the solution resistance at reference and operating temperature respectively and a and b are constants. R_1 and R_2 are end resistors whose function is to determine the range of calibration for the bridge. R_3 is a calibrated slide wire potentiometer. Rm is the resistance of the electrolyte between the two cell electrodes. A condition for balance of the Wheatstone bridge is that $(R_1 + Rx)/(R_2 + Ry)$ equals Rs/Rm and this condition is indicated by zero current through S.

Solution conductivity varies with concentration and temperature. The function of a temperature compensator is to nullify the temperature effect on solution conductivity, so that the measurement will always read with respect to some reference temperature regardless of process temperature. The practical method of performing the compensation is to introduce a resistor (R_S in Figure 10.6b), which will change at the same rate as the sample, into one arm of the bridge. Manual temperature compensation can be achieved with a variable resistor calibrated in temperature and requiring operator adjustment. Automatic compensation can be performed by:

1. A reference conductivity cell sampling a solution of typical composition and subject to the same temperature as the measuring cell.
2. A thermistor in contact with the process fluid and resistor network.

Electrodeless System

For measurement of slurries and very conductive or corrosive solutions, the electrodeless system can be utilized.

The electrodeless conductivity system measures the resistance of a closed loop of solution. The cell consists of two torroidal windings on a nonconducting pipe, or metallic nonconducting lined pipe. One winding is connected to a transmitter that provides a stable, audio frequency excitation voltage. The second winding is connected to a receiver that measures its output voltage. The equivalent electrical circuit, shown in Figure 10.6c, consists of two transformers. The first has the input torroid as its primary and the solution loop as its secondary. The second has the solution loop as its primary and the output torroid as its secondary. At constant input voltage,

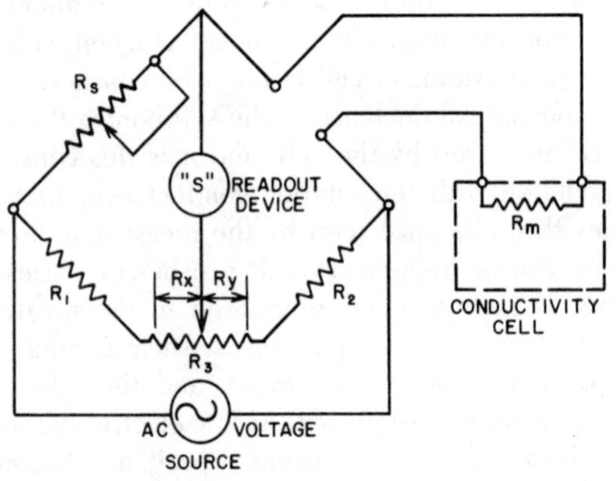

Fig. 10.6b Conductivity measuring circuit

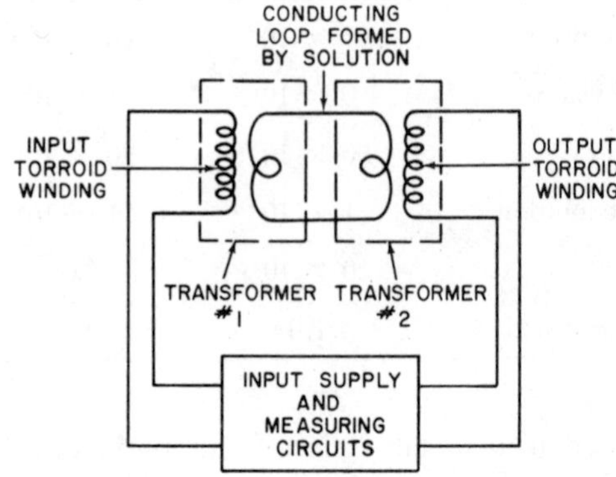

Fig. 10.6c Equivalent electrical circuit of an electrodeless conductivity system

the output is a function of solution conductivity only. A number of configurations are possible with this unit. Figure 10.6d shows an assembly of a piping spool piece for installation as shown in Figure 10.6e. The torroids can also be furnished as potted assembly for submersion in the sample liquid.

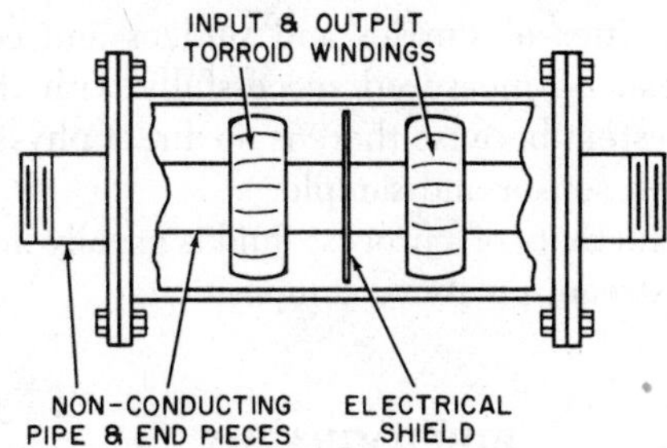

Fig. 10.6d Electrodeless conductivity "cell"

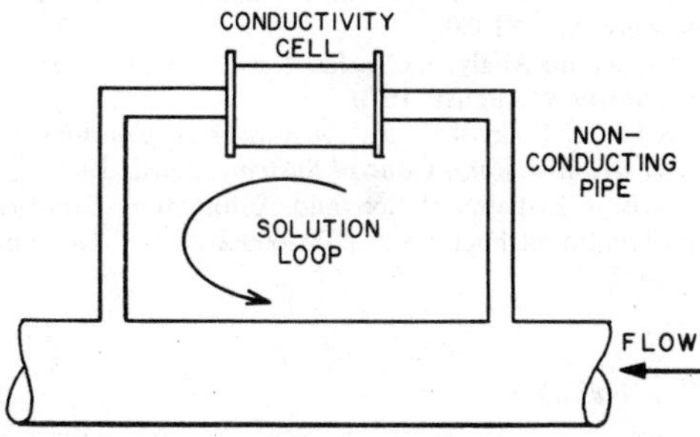

Fig. 10.6e Installation of electrodeless "cell"

Selection of Cell Constant

As noted previously, the cell constant must be selected to place the electrolytic resistance seen by the measuring circuit in the range of 50 to 100,000 ohms. A cell of low-cell constant has large electrodes relatively close together and is suitable for high-resistance measurements. Conversely, a cell of high-cell constant has smaller electrodes spaced farther apart and is suitable for low-resistance measurements.

Example: Fluid conductivity range 10 to 100 μmho per cm:

$$R_H = \frac{1}{10 \times 10^{-6}} \times K$$

$$R_L = \frac{1}{100 \times 10^{-6}} K$$

where R_H and R_L represent the electrolytic resistance at 10 μmho and 100 μmho respectively, and K is the cell constant. The range limits of the measuring circuit require that:

$$R_H < 100{,}000 \text{ and } R_L > 50$$
$$10^5\ K < 100{,}000 \text{ and } 10^4\ K > 50$$

Therefore, any cell with a constant $5 \times 10^{-3} < K < 1$ can be used.

Conductivity Measurements

Conductivity is an indirect measurement of total dissolved, ionized solids in water. Other names for dissolved solids are dissolved salts or dissolved mineral constituents. Conductivity measurements provide an estimate of the total dissolved solids in the water. When there is a preponderance of one salt, such as sodium chloride, in the water, the conductivity measurement affords a reasonably accurate measurement of this material. This is true if the concentration of the salt versus conductivity values is known.

In water and wastewater monitoring, conductivity measurement is a very valuable tool, although it is an inferential measurement. In a heat exchanger, such as a cooling tower or boiler, it measures the concentration of water and, by directly controlling blowdown or bleed-off and, thus, makeup water addition, it maintains the concentration at the desired level by dilution. Conductivity measurement can detect spills of acid or alkali or any other ionized dissolved contaminant but cannot differentiate between them. In some wastewater-monitoring applications, the ability to differentiate may not be important because inadvertent spills are always of the same material or can be easily traced on conductivity alarm.

Differential conductivity, the measurement of conductivity of water prior to and again after possible contamination, is also a valuable water quality management tool (Figure 10.6f), especially when the conductivity of the uncontaminated water is subject to variation. The potentiometric technique is particularly valuable here because of its stability. The outputs from the two analyzers are subtracted, and the difference constitutes the alarm or control signal. When the conductivity of uncontaminated water is not subject to variation, such as in a major city supply or in a well supply, conductivity measurement after the process is sufficient. Figure 10.6g illustrates an automatic diversion system in which water reuse applications are dependent on the degree of contamination after primary use.

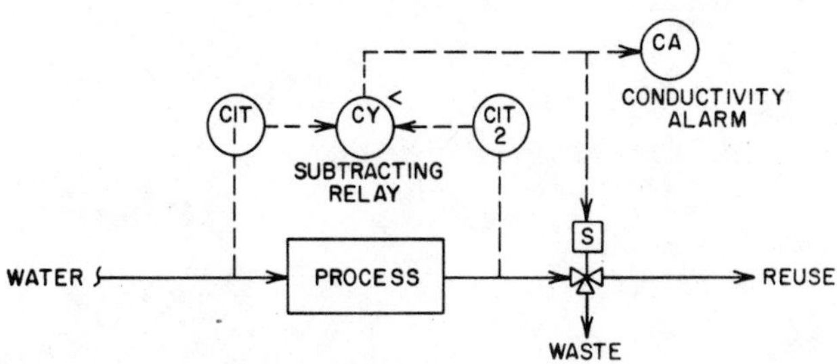

Fig. 10.6f Differential conductivity control

Summary

Conductivity measurement is not particularly difficult to perform on clean, noncorrosive fluids with conductivities above a few μmho. For such applications, deter-

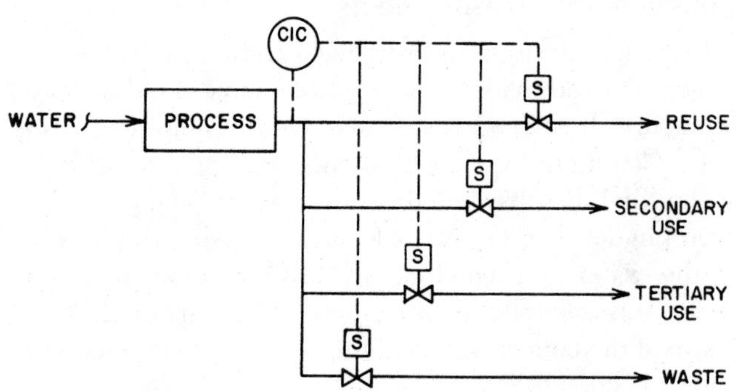

Fig. 10.6g Water reuse control by conductivity

mination of cell constant and source frequency compatible with the measuring circuit constitute the critical elements of selection. A large number of probe designs for installation in atmospheric and pressure vessels or pipes is available so that installation is not a limitation.

For samples at high temperature or pressure, the probe should be installed in a sampling chamber downstream from a sample cooler and pressure-reducing valve. If a cooler is required, it should be sized to bring the sample temperature as close as possible to cooling water temperature. Since cooling water temperature varies mainly with the seasons, short-term inaccuracies due to sample temperature fluctuations will be minimized.

At very low conductivities, polarization of the electrodes can cause inaccuracies because some of the sample is electrolyzed, thus introducing a measurement error. Platinized electrodes should be avoided for such applications because the platinum aggravates the problem.

Conductivities of slurries and sludges and corrosive materials can be measured successfully with the electrodeless system because there is no direct physical contact between sensor and sample.

The conductivity of a process fluid is usually measured to test for stream purity or composition.

BIBLIOGRAPHY

"Conductivity of Caustic Processes," *InTech*, April, 1980.

Cooper, R.L., "Industrial Wastewater Sampling," *Instruments and Control Systems*, May, 1980.

Hall, J., "On-Line Analyzers Tackle New Demands," *Instruments and Control Systems*, August, 1979.

Kidder, R.J. and Rosenthal, R., "Advances In Conductivity Measurement," *Instruments and Control Systems*, April, 1965.

Molvar, A.E., "Instrumentation and Automation Experiences in Wastewater-Treatment Facilities," EPA-600/2-76-198, October, 1976.

Flow Sheet Symbol

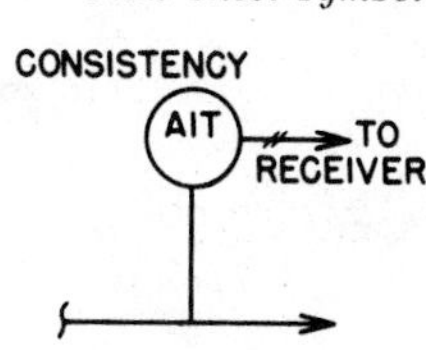

10.7 CONSISTENCY

Standard Design Pressure:	100 PSIG (690 kPa)
Standard Design Temperature:	Up to 250°F (120°C)
Element Materials:	316 stainless steel
Cost:	$2000 to $4500; lower prices for fixed sensors
Inaccuracy:	Empirical calibration
Range:	1.75 to 8.00% consistency
Partial List of Suppliers:	Automation Products Inc.; Brabender, C.W., Instruments Inc.; Brookfield Engineering Labs, Inc.; Custom Scientific Instruments Inc.; De Zurik, a unit of General Signal; Eur-Control USA Inc.; Ionics Inc.; Ronan Engineering Co.

Consistency measuring instruments detect consistency of the process fluid as shear forces acting on the sensing element. There are two basic types of consistency detectors: the fixed and the rotary. In the latter, the shear force is reflected as the torque required to maintain a rotary sensor at constant speed, as the imbalance of a strain gauge resistance bridge, or as a turning moment. The instruments are calibrated in-line; thus the output is not in terms of bone dry consistency, but rather some arbitrary, reproducible value. Fixed sensors depend on the process flow for measurement, and for such instruments the output is affected by the velocity of the flow. The sensor contour is designed to minimize flow effects on the output over the operating flow range. On the other hand, rotating sensors do not depend on process flow for a measurement. While these units are also sensitive to flow velocity variations, they can be used over wider flow ranges. In addition, rotary motion of the sensor produces some self-cleaning action when the fixed sensors depend solely on a properly designed contour to prevent material hang-up.

Strain Gauge Detectors

This sensor transmitter functions as a resistance bridge strain gauge. The bridge elements are bonded to the inner wall of a hollow cylinder that is inserted into the process. The shear force acting in the cylinder, due to the consistency of the process fluid, causes an imbalance of the resistance bridge. The amount of imbalance is proportional to the shear force and the consistency of the process fluid. The resistance bridge is powered from a recorder that also contains the ac potentiometer electronics.

The sensor is mounted through a threaded bushing furnished with the unit. Flowing velocity must be between 0.5 and 5.0 ft. (0.15 and 1.5 m) per sec. for repeatability around 0.1 percent of bone dry consistency.

Force Balance Detector

The sensing element of this instrument is a blade, specially shaped to minimize the effects of velocity. As shown in Figure 10.7a, material flowing past the blade, which is positioned along the line of flow, creates a shear force. Velocity of the process produces two drag forces, F_1 and F_2, whose resultant F_3 acts through the fulcrum. The moment arm of F_3 is, therefore, zero and the effect of velocity on the measurement is negligible over a range of 0.75 to 5 ft. (0.23 to 1.5 m) per sec. Changes in consistency are transmitted through the blade to the force bar causing small changes in the relationship between flapper and nozzle. Therefore, the relay unit output pressure changes until the force due to the feedback unit balances the shear force.

The instrument can be mounted on any line 4 in. (100 mm) or larger. Mounting is through a 2-in. (50 mm) flange supplied with the instrument.

A variation of the design described above utilizes a shaped float inserted through a pipeline tee as shown in

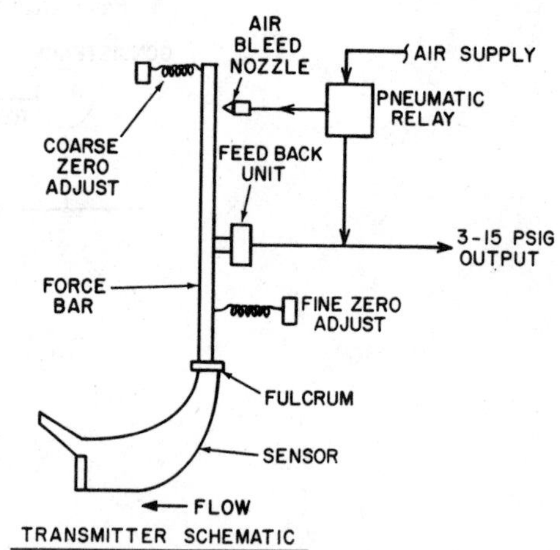

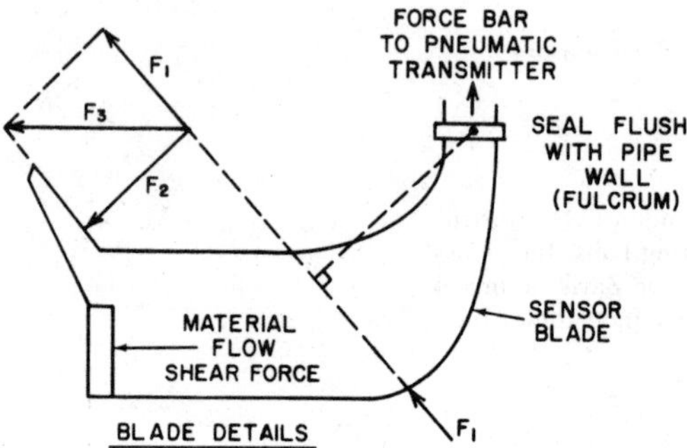

Fig. 10.7a Stationary blade sensor and transmitter schematic

Figure 10.7b. The shear forces acting on the float are transmitted to the force bar of a pneumatic transmitter mounted on top of the tee. As shown in the diagram, the unit can only be installed in a vertical pipe, with 5 pipe diameters of straight run required on the upstream side. The minimum line size is 6 in. (150 mm).

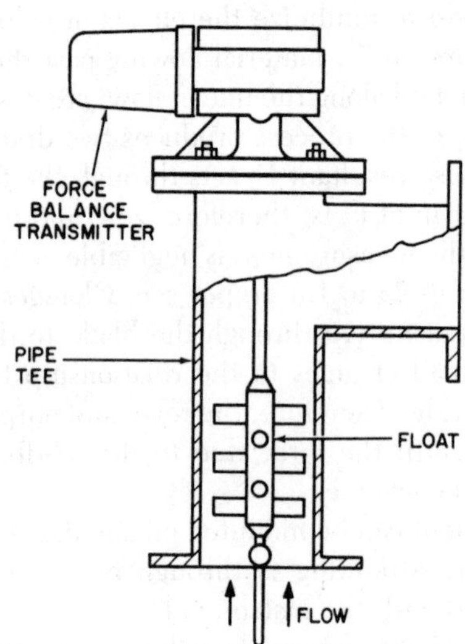

Fig. 10.7b Float sensor

Rotating Sensor

This unit consists of a motor driven, ribbed disc immersed in the process fluid. The disc is rotated at constant speed, and variations in torque output by the motor are sensed by a torque arm. The motor is suspended by flexure bearings and anchored to the torque arm which senses motor reaction torque. The tip of the torque arm is located between two air nozzles so that minute movements of the arm, caused by torque variations, are reflected as changes in two air output pressures (Figure 10.7c). The nozzle pressures are fed back to bellows that react to the torque arm movement by exerting an opposing force until equilibrium is reached between increased nozzle pressure and the force exerted by the torque arm.

While this unit is less sensitive to flow changes than the strain gauge and force balance types, problems are introduced by the shaft seal required for this design. The torque variations must reflect only consistency changes and, therefore, shaft friction variations are detrimental to the measurement. The shaft seal requires water as a seal fluid at a few PSI above the maximum process pressure. For in-line installations, the unit is mounted in a pipe spool piece with removable cover on one side for inspection of the sensors. This type of installation requires a 12-in. (0.3 m) pipeline as a minimum. Other units are available for open installation in chambers or pans.

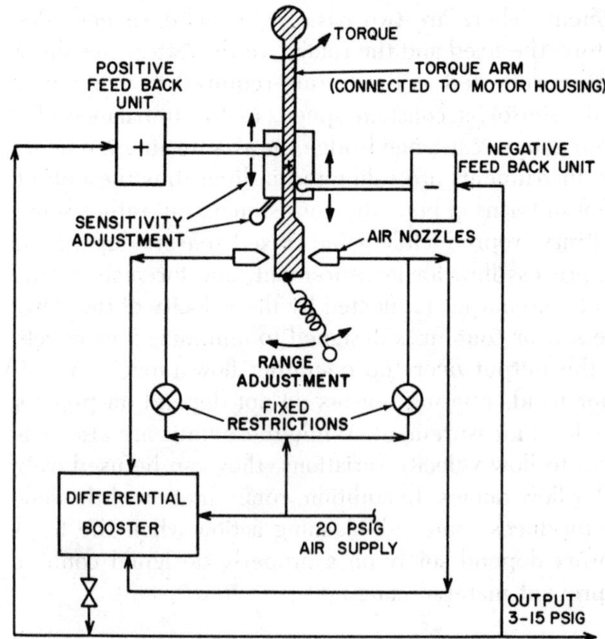

Fig. 10.7c Air schematic of rotating sensor

Level Detector

For consistency measurement of thin paper stocks, the device shown in Figure 10.7d can be used. The two chambers are separated by a screen disc. The sample

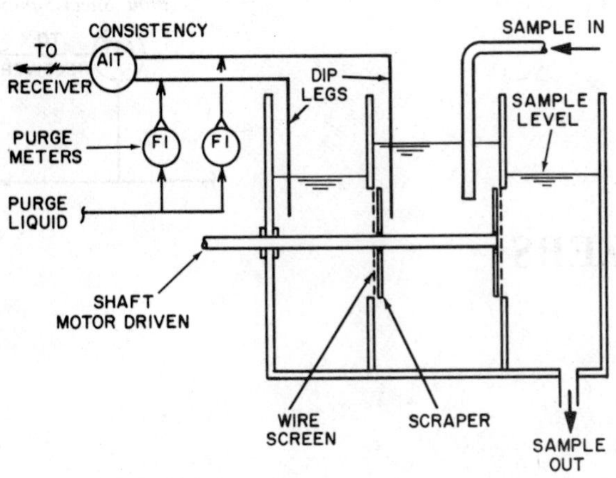

Fig. 10.7d Level consistency measurement

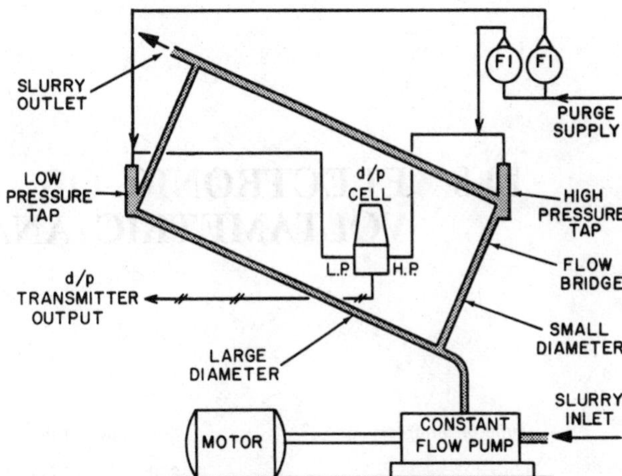

Fig. 10.7e Flow bridge consistency measurement

drains against the disc, which is kept clean by motor-driven pressure foil elements. The flow rate through the screen varies with consistency, which causes a difference in level across the screen. This level difference is measured by a transmitter as consistency.

Flow Bridge

The flow bridge, the hydraulic analog of the Wheatstone bridge, can also be used to measure consistency. A pump supplies a constant flow of sample to the flow bridge, and a d/p cell measures the differential pressure across the two legs of the bridge (Figure 10.7e). Pressure drop variations caused by sample viscosity are cancelled by the bridge arrangement, and the differential pressure across the bridge is a function of consistency only. Instrument air, or some other suitable fluid, can be used to purge the d/p cell legs. Besides the need for purging this system, two disadvantages are that it cannot be installed in-line and the flow bridge must discharge to atmosphere.

Summary

While convenient from an installation standpoint, in-line instruments are all sensitive to flow variations. Fixed sensors are more likely to be plagued by material buildup, particularly if the sample contains fibers. Rotating sensors are self-cleaning because sensor motion will tend to spin off any material; however, variations in shaft seal friction can be troublesome. The flow bridge method of consistency measurement is applicable to a wide range of materials with better accuracy than the other instruments. Since the instrument is not installed in-line, the process flow need not be shut down for maintenance on the instrument.

BIBLIOGRAPHY

Hall, J., "On-line Analyzers Tackle New Demands," *Instruments and Control Systems*, August, 1979.

Jutila, J.M., "Multicomponent On-Stream Analyzers for Process Monitoring and Control," *InTech*, July, 1979.

Molvar, A.E., "Instrumentation and Automation Experiences in Wastewater-Treatment Facilities," EPA-600/2-76-198, October, 1976.

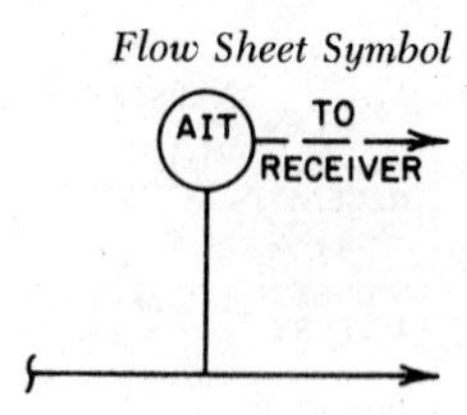

10.8 ELECTRONIC OR VOLTAMETRIC ANALYZERS

Type of Sample:	Solid and liquid
Cost:	$2000 to $10,000
Inaccuracy:	0.1 to 10.0 ppt
Partial List of Suppliers:	Beckman Instruments, Inc.; Cambridge Instrument Co. Inc.; Envirotech Corp., United Controls Div.; Fischer & Porter Co.; Foxboro Analytical; Great Lakes Instruments, Inc.; Kent Process Control Inc.; Uniloc Div. of Rosemount, Inc.; Yellow Springs Instrument Co. Inc.

Voltametric analyzers monitor the current, voltage or charge that results from the redox reaction of the analyte in question. This section presents a brief review of voltametry, including a discussion of the concepts and properties of electrochemical probes, amperometry, polarography and coulometry. The distinctions among voltametric methods are emphasized, and a summary table is provided. This table lists the common voltametric methods and gives a representative sampling of analyzable materials for each method.

Voltametry

Successful application of voltametry for monitoring and controlling process streams requires an understanding of the significant voltametric variables. The fundamental requirements for voltametry instrumentation are two electrodes, provisions for a current and/or voltage supply and monitors for the respective voltage and/or current responses as a function of time. The characteristics of the response curves indicate the nature of the electrochemical analysis occurring in the solution between the electrodes. Figure 10.8a illustrates a surface that relates the three variables (current, voltage and time) which are of interest in an electrochemical analysis. If a constant current is supplied to the electrodes, the progress of the electrochemical reaction is observed by monitoring the change in potential as a function of reaction time. This procedure is labeled chronopotentiometry, and a response curve would be similar to curve A in Figure 10.8a. If the electrochemical process occurs at a constant potential, reaction response is monitored as a changing current as a function of time. This type of analysis is called constant potential voltametry, or amperometry. A typical response curve is labeled B in Figure 10.8a. Finally, monitoring reaction current as a function of applied potential is called polarography, and a typical polarographic curve is represented by C, which is a diagonal cut through the response surface, in Figure 10.8a.

The exact shape of a particular response curve in Figure 10.8a depends on the mass transport phenomena associated with the analyte movement to the working

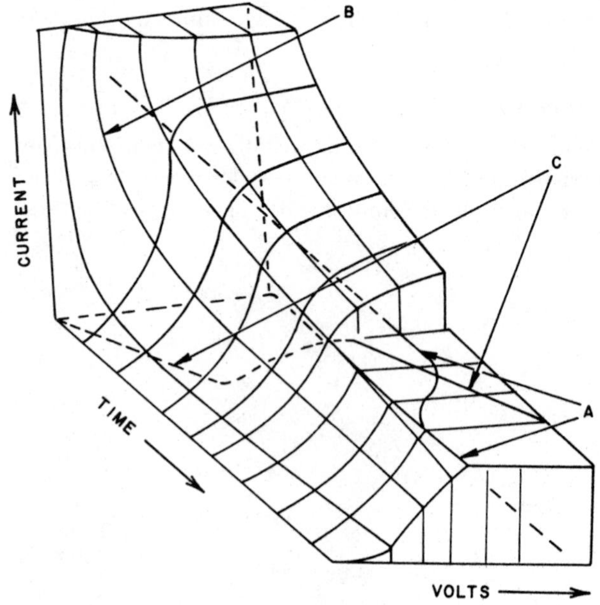

Fig. 10.8a Response surface for voltametry, showing curve A, the process of chronopotentiometry; curve B, amperometry; curve C, the polarographic process

electrode. (The working electrode is the electrode in which the electrochemical process of interest occurs. Unlike the other electrode, the reference electrode, the working electrode is made from a metal, such as mercury, platinum, palladium or gold.) Analyte diffusion to the working electrode is the usual transport mechanism required if the analytical expressions for the curves of Figure 10.8a are to be obtained. If diffusion mass transport conditions are met, the equation for Fick's first law of diffusion,

$$\frac{1}{A} dN(x,t)/dt = f(x,t) = \frac{\delta c(x,t)}{\delta x} \qquad 10.8(1)$$

and the equation for Fick's second law of diffusion,

$$\delta c(x,t)/\delta t = D\delta^2 c(x,t)/\delta x^2 \qquad 10.8(2)$$

suggest the relationship for analyte concentration with reaction time, t; distance from the working electrode, x; electrode surface area, A; moles of particles that pass a reference boundary, dN(v,t)/dt; particle flux, f(x,t); and diffusion coefficient, D. To obtain a solution to Equation 10.8(2), initial and boundary conditions are specified. These conditions reflect the experimental reality for the various forms of voltametry. For example, the boundary condition, $C(0, t) = 0$ for all $t > 0$, is imposed on electrode reactions with sufficient potential because the diffusion of the analyte to the electrode is considerably slower than the actual electrode reaction itself. An initial condition of interest, $C(X,0) = C°$, where C° is the analyte concentration before any electrode reaction occurs, is quite reasonable to expect for homogenous solutions. The use of these two constraints, together with standard methods for solving second-order homogenous differential equations leads to a solution to Equation 10.8(2):

$$c(x, t) = C° \operatorname{erf}(x/2D^{1/2}t^{1/2}) \qquad 10.8(3)$$

Equation 10.8(3) provides an analyte concentration profile in terms of the diffusion coefficient, reaction time and distance from the working electrode. This equation is differentiated with respect to electrode distance, combined with Faraday's law and the Nernst equation (see Sections 10.10 and 10.13) to produce the expression,

$$E = E° + (RT/nF) \ln(f_o/f_r(D_r/D_o)^{1/2}) + (RT/nF) \ln[(K\text{-}it^{1/2})/it^{1/2})] \qquad 10.8(4)$$

for the electrochemical surface presented in Figure 10.8a.[1] Equation 10.8(4) expresses the potential of the working electrode, E, in terms of E^0, the standard thermodynamic potential; R, the gas constant; T, the absolute temperature; F, Faraday's constant; n, the number of electrons transferred per mole of analyte; f, the activity coefficients of oxidized and reduced analyte forms; i, the current; and finally K, a constant dependent upon particular analysis parameters other than voltage current or time.

Electrochemical Probes

Electrochemical probes are essentially electrochemical cells that have been designed to respond to the presence of a particular analyte. Like electrochemical cells, electrochemical probes can be classified as either galvanic or electrolytic. Both types of probes require a pair of electrodes, usually metallic, immersed in a supporting electrolyte so that current passes through the electrodes as chemical reduction occurs at the surface of one electrode, the cathode. Most electrochemical probes reduce chemical interferences by using a thin analyte-permeable membrane to separate the cell components, that is, the two electrodes and the supporting electrolyte, from the solution to be analyzed.

The fundamental difference between an electrolytic and a galvanic probe is whether or not a potential is applied across the electrodes. Galvanic probes are constructed in such a manner that the presence of the analyte provides a flow of electrons in the cell. Because the electrochemical reaction occurs spontaneously, no external potential need be applied across the electrodes. However, galvanic probes have a major disadvantage in that the electrodes are destroyed as the reaction continues. This limitation restricts their usefulness to monitoring situations in which replacement of spent probes is not a critical restraint.

Electrolytic probes are principally used as monitoring devices for dissolved oxygen (see Section 11.8). Unlike the galvanic probe, a potential is applied across the probe electrodes. The choice of potentials is controlled by the analyte to be determined. The current from the probe is obtained at constant potential as a function of time. The analysis is an example of a constant potential voltametry, and a typical response curve is illustrated in Figure 10.8a. The exact expression for this curve is derived from Equation 10.8(4) when

$$K = nFC°(D_o/\pi)^{1/2} \qquad 10.8(5)$$

and is expressed as

$$i = (nFC°/1 + e^{\phi})(D_o/\pi t)^{1/2} \qquad 10.8(6)$$

where

$$\phi = (nF/RT)(E - E°) - \ln(f_o/f_r)(D_r/D_o) \qquad 10.8(7)$$

Thus, the usefulness of the probe is based on the fact that at a constant potential, the current is proportional to the analyte concentration.

Figure 10.8b shows a configuration for an electrolytic probe. The probe consists of an inner and outer element. The inner element contains the cathode and anode with wires leading to the power supply and current-monitoring circuit. The anode is wrapped about a plastic shaft in a way that insulates the cathode from the anode. The outer element fits over the inner element and provides a space for the electrolyte. The membrane is mounted

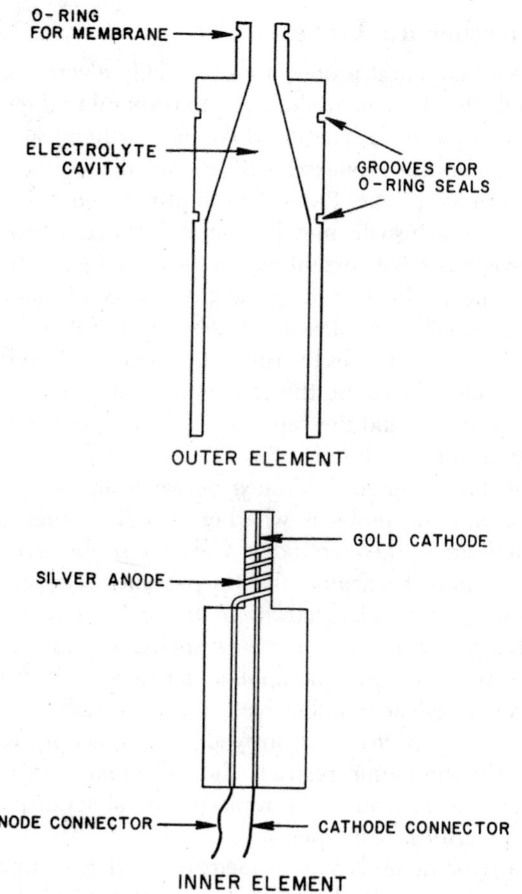

Fig. 10.8b Configuration of an electrolytic probe. The inner element contains the anode and cathode, while the outer element isolates the electrolyte from the sample

on the outside of this outer element and secured with an O-ring. The spiral anode configuration provides a large anode surface to promote large current flow, while the plastic anode support post is designed to keep the surface of the gold cathode a considerable distance from the anode surface. This is done to discourage reaction products from migrating to the opposite electrodes and generating unwanted response current.

Amperometry

If, when the reagent is coulometrically generated, the current that passes through a polarographic cell at the appropriate reduction potential is measured as a function of reagent volume or time, the process is called amperometry. Three types of response curves can be expected (Figure 10.8c). Curve (a) is the result of the analyte reacting at the electrode while the reagent does not. The curve indicates that the diffusion current is detected as long as the analyte remains. Once the reagent removes all the analyte, the current stops. This "dead-stop" endpoint technique is often used for amperometric determinations of dissolved oxygen. In curve (b), the analyte is not reduced at the working electrode, but the reagent is. In this situation, there is no current flow until the analyte is completely destroyed and the reagent is allowed to react at the electrode surface. Finally, curve (c) indicates the response when both analyte and reagent are electrode reactive. The increase after the equivalents point indicates that the current is the result of excess reagent reacting at the electrode surface.

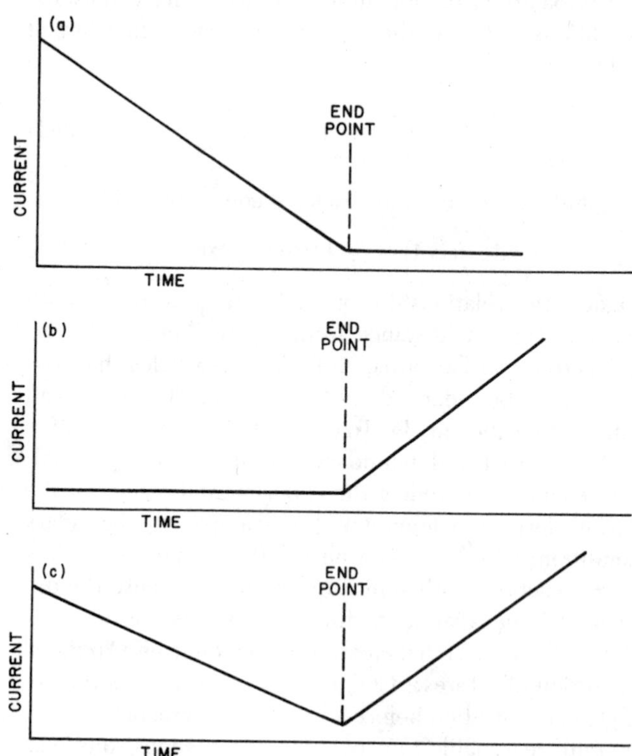

Fig. 10.8c Response curves for polarographic cell system showing curve A when analyte reacts at electrode; curve B when reagent reacts at electrode; curve C when analyte and reagent react at electrode

If the polarographic cell is replaced by a two-solid electrode system with a small applied potential (0.1 v − 0.2 v), the response curves of interest are determined by the number of reversible reactions occurring in the cell. The set of reversible half-reactions of interest are symbolized as

$$R_r \rightleftharpoons R_o + e^- \qquad 10.8(8)$$

and

$$A_o + e^- \rightleftharpoons A_r \qquad 10.8(9)$$

where A_o and R_o are the oxidized forms, and A_r and R_r are the reduced forms of analyte and reagent, respectively. The fundamental concept of interest is the fact that the applied potential will support current flow only when both forms represented in Equation 10.8(8) or Equation 10.8(9) are present in the test solution. Figure 10.8d represents the possible response curves that illustrate this concept.

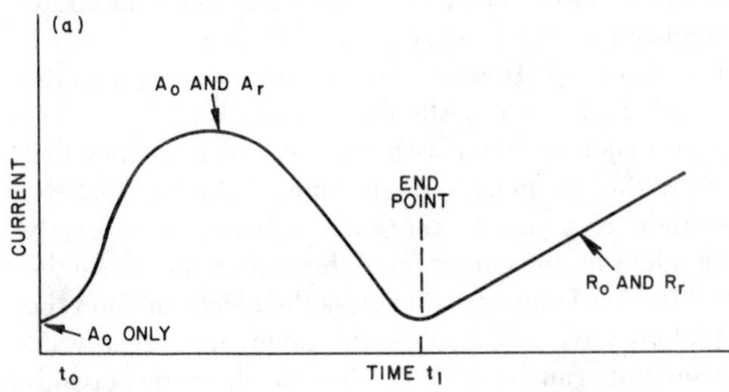

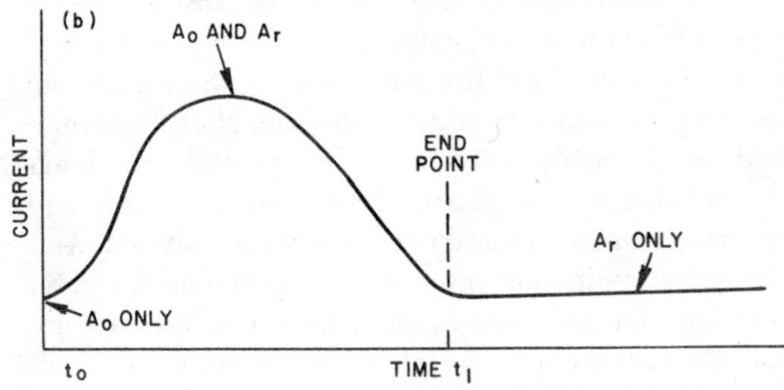

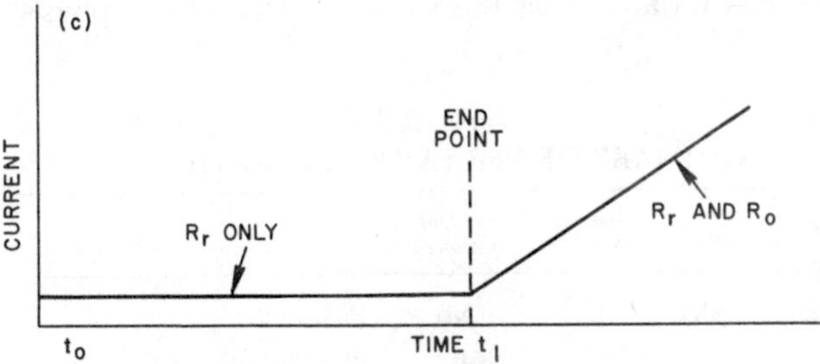

Fig. 10.8d Response curves for two solid electrode system, showing curve A when analyte and reagent are reversible; curve B when analyte only is reversible; curve C when reagent only is reversible

In curve (a) of Figure 10.8d, the absence of current at t_o is due to the lack of A_r. For values of t so that $t_o < t < t_1$, the current response is the result of varying concentrations of A_o and A_r as R_r is added. At $t = t_1$, the current returns to zero because of the absence of A_o and R_r. Finally, for $t > t_1$, the continuous addition of R_r, coupled with the R_o present from the redox reaction with A_o, results in a continuously increasing current flow. Curve (b) indicates the response when only the analyte reaction is reversible. The curve from $t_o < t \leq t_1$ is identical to the same portion of curve (a) for the same reasons. However, for $t > t_1$, no current is detected because the reagent half-reaction is irreversible. Curve (c) shows the results when the analyte reaction is irreversible but the reagent half-reaction is reversible. Current is not detected until R_r remains in the solution. This case is often called a "dead-start" titration because the current does not begin until all the analyte has been removed.

Polarography

Polarography was developed in 1922 by Nobel Prize winner Jaroslav Heyrovsky. The analysis is performed by measuring the current flow as a function of electrode potential as the latter is systematically varied. The voltage variation pattern and the voltage sweep rate are defined by the type of polarography performed and the choice of working electrode material.

Figure 10.8e shows a typical polarographic response curve. For this curve, the K term in Equation 10.8(4) is defined as

$$K = nFC^\circ(7/3D_o)^{1/2} \qquad 10.8(10)$$

and Equation 10.8(7) is rearranged to

$$(nF/KT)(E - E^\circ - \ln(f_o/f_r(D_o/D_r)^{1/2}) = \ln(i_d - i/i) \qquad 10.8(11)$$

where i_d is the diffusion current. The curve shows an S-shaped current response as the potential at the cathode, the working electrode where analyte reduction occurs, becomes more negative. Current limited by diffusion occurs at cathodic potentials beyond the potential required to initial reduction. Thus, i_d is sufficient to reduce the diffusion-supplied analyte concentration to zero.

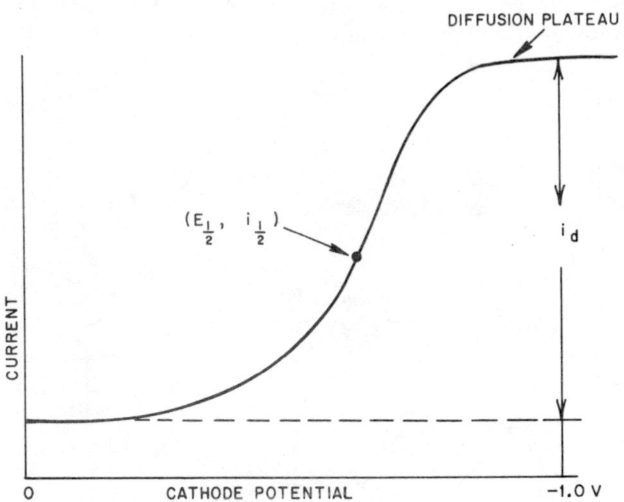

Fig. 10.8e Polarogram showing typical current vs cathode potential response. The diffusion current, i_d, the half-wave potential, $E_{1/2}$ and the half-wave current, $i_{1/2}$, are all illustrated

The value of a polarographic determination lies in the fact that the potential observed at the point $i_{1/2} = i_d/2$ is unique to the analyte in the solution. This occurs because the potential at $i_{1/2}$ is independent of the reactant concentrations but directly related to the standard potential for the analyte half-reaction. This inflection point potential is defined as the half-wave potential, $E_{1/2}$, and is a common parameter for qualitative identification of electroactive components in a solution. When the reaction is fast and reversible and the diffusion coefficients of the oxidized and reduced forms of the analyte are virtually identical, $E_{1/2}$ becomes E°.

Coulometry

Because the coulomb provides a direct connection between reaction current and analyte concentration during a redox process, coulometry offers an efficient method of determining analyte concentration. Two types of coulometry are possible: constant current and controlled potential. Constant current coulometry depends on analyte reduction to support the specified current flow. When the supply of material is insufficient to carry this current, the electrode potential drifts cathodic until another reaction begins. The result of this process is a potential time relationship similar to the results obtained in chronopotentiometry, and the charge is simply the product of the constant current with electrolysis time.

By contrast, controlled potential coulometry is conducted by maintaining a constant electrode potential at the expense of current flow. The number of coulombs must be determined by the integration of the reaction current as a function of electrolysis time. Figure 10.8f illustrates the current-time relationship for controlled potential coulometry. The choice of integration time is important because the current is a result of a faradic component and a capacitative component. As a result of this dual contribution, current integration is not started until the capacitative-charging contribution is minimal.

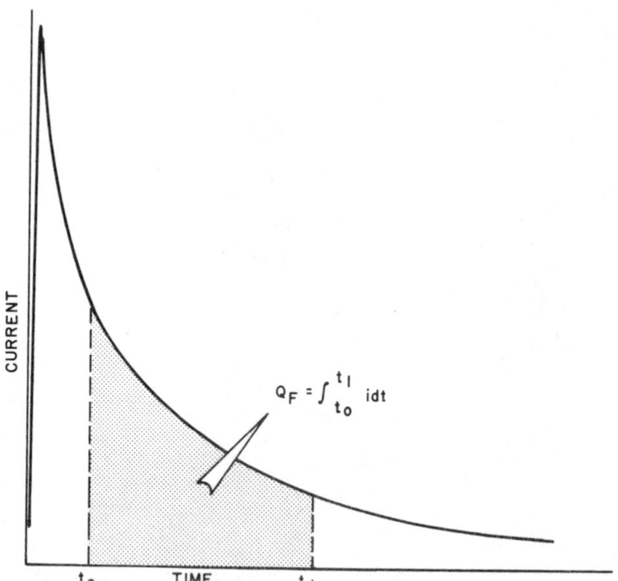

Fig. 10.8f Response curve for controlled potential coulometry, showing the relationship among current, i; time, t; and coulombs, Q_F, in the faradic portion of the curve

Summary

Although instruments are available for all types of voltametric analysis, currently only amperometry and coulometry can be performed on-line in a process stream. This situation will change as voltametry becomes an extremely important way to obtain continuous process analysis and control. However, there are problems with the methods that the user should be aware of.

One major problem with the electrolyte probes that employ thin permeable membranes is the fact that the electrode response to the analyte becomes indirectly dependent on the condition of the membrane. Once the potential has been applied across the electrodes and the depletion wave has reached the inner membrane wall, the amount of analyte that reaches the electrode depends on the diffusion of the analyte through the membrane. This situation is usually unacceptable because the membrane condition is not stable. The membrane itself is subject to failure and the outer wall of the membrane is subject to various conditions that can alter its permeability to the analyte. These conditions make the long-term reliable use of electrolytic probes impossible and constant maintenance and recalibration are advised. Amperometric endpoints pose another problem. To date, provisions for the various endpoints possible are not available and care must be taken when determining analysis endpoints. Finally, instrumentation for coulometric methods requires that the working electrode be physi-

Table 10.8g
SUMMARY OF VOLTAMETRIC METHODS

Method	*Measured Variable(s)*	*On-Line Instrument*	*Determinations**
Polarography	i,v†	No	Pesticides mercaptans thiosulfates chlorinated organics toxic heavy metals
Stripping	i,v	No	Heavy metals selenides halogens thioamides trace metals
Amperometry	i,t,c	Yes	Sulfates halides phenols aromatic amines olefins
Controlled potential coulometry	$\int i(t)$	Yes	Precious metals alloys dissolved oxygen
Differential pulse polarography	i,v	No	Aromatic hydrocarbons aromatic amines phenol ammonium salts carboxylic anhydrides N-nitrosamines

*Partial list
†i = current, v = applied potential, c = concentration, i (t) = current as functions of time

cally separated from the auxiliary electrode so that the reaction products do not migrate to the opposite electrode. This migration is also restricted by using sintered glass discs between the electrodes to restrict ion migration but permit electrical conduction through the cell.

Table 10.8g provides a summary of voltametric methods. The applications suggested are offered as guidelines for analyses available on current instrumentation.

REFERENCE

1. Reinmuth, *Analytical Chemistry*, 32: 1509, 1960.

BIBLIOGRAPHY

Bond, A. and C. Anterford, D., "Comparative Study of a Wide Variety of Polarographic Techniques with Multifunctional Instruments," *Analytical Chemistry*, 44:721, 1972.

Delahay, P., *New Instrumental Methods in Electrochemistry*. New York, John Wiley & Sons, Inc., Interscience, 1954.

Flato, J., "The Renaissance in Polarographic and Voltametric Analysis," *Analytical Chemistry*, 44:75A, 1972.

Lingane, J., *Electroanalytical Chemistry*, 2nd ed., New York, John Wiley & Sons, Inc., Interscience, 1958.

Flow Sheet Symbol

AT
IR
TO RECEIVER

10.9 INFRARED ANALYZERS

Type of Sample:	Gas and liquid
Standard Design Pressure:	Atmospheric pressure to 150 PSIG (1.3 MPa); 1000 PSIG (7 MPa) for special measurements
Standard Design Temperature:	32 to 392°F (0 to 200°C)
Materials of Construction:	Cell body of all standard materials; window materials of quartz, sodium chloride, calcium fluoride, barium fluoride, sapphire, zinc selenide
Cell Length:	0.1 mm (0.004 in.) for liquids to 20 m (66 ft.) for gases
Cost:	$2200 to $12,000
Inaccuracy:	±2% of span
Drift:	±1% per 24 hr.
Partial List of Suppliers:	Anacon, Inc.; Anarad, Inc.; Beckman Instruments, Inc.; Bendix Corp.; Foxboro Analytical; Horiba Instruments, Inc.; Infrared Industries, Inc.; Mine Safety Appliances Co.; Technicon; Teledyne Analytical Instruments

Principles of Infrared Analysis

Infrared (IR) absorption (or reflection for solids) is a technique that can be used successfully for continuous chemical analysis of a process. The IR region of the electromagnetic spectrum is generally considered to cover wavelengths from 0.8 to 1000 μm. These limits, expressed in frequency terms (cm^{-1}, wave numbers or the number of waves per cm) are 12,500 cm^{-1} to 10 cm^{-1}.

The region of the IR most used by process analyzers is broken into two parts: the near IR (12,500 to 4000 cm^{-1}) and the mid IR (4000 to 650 cm^{-1}). Except for a small overlap region, sources and detectors that are needed in the near IR will not work in the mid IR and vice versa. Most laboratory IR spectrophotometers work from 4000 cm^{-1} to between 650 and 200 cm^{-1}, depending on the model.

Infrared radiation interacts with all molecules (except the homonuclear diatomics oxygen (O_2), nitrogen (N_2), hydrogen (H_2), chlorine (Cl_2), etc.), by exciting molecular vibrations and rotations (Figure 10.9a). The oscillating electric field of the IR wave interacts with the electric dipole of the molecule, and when the IR frequency matches the natural frequency of the molecule, some of the IR power is absorbed. The pattern of wavelengths, or frequencies, absorbed identifies the molecules in the sample. The strength of absorption at particular frequencies is a measure of their concentration. Analytical laboratory IR is largely concerned with identification, or qualitative analysis, while process IR is concerned with quantitative analysis. Some typical IR spectra are shown in Figure 10.9b.

Particular groups of atoms tend to absorb at the same frequency with very little influence from the rest of the molecule. These group frequencies are a great help in identifying molecules from the IR spectra (Figure 10.9c). On the other hand, similar molecules, such as a series of homologous hydrocarbons, have very similar IR spectra. Infrared analysis is, therefore, most straightforward when the component molecules of the sample have significantly different atomic groupings. A mixture of aliphatic hydrocarbons would be better analyzed by another technique, such as gas chromatography. The part of the spectrum offering the best discrimination between molecules is between 7 and 15 μm, the so-called fingerprint region.

The starting point for quantitative analysis is the Beer-

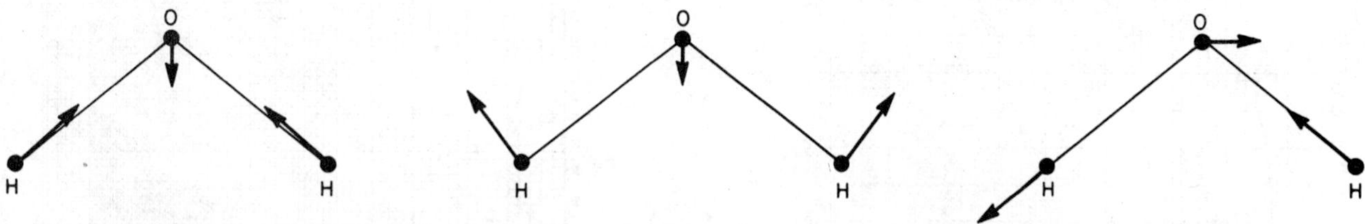

Fig. 10.9a The three fundamental vibrations of the water molecule: symmetric stretch, bend, and antisymmetric stretch; the amplitudes of the vibrations have been exaggerated for clarity

Lambert law, which relates the amount of light absorbed to the sample's concentration and path length.

$$A = abc = \log_{10} \frac{I_0}{I} \qquad 10.9(1)$$

where:

- A = absorbance
- I = IR power-reaching detector with sample in beam
- I_0 = IR power-reaching detector with no sample in beam
- a = absorption coefficient of pure component of interest at analytical wavelength; the units depend on those chosen for b and c
- b = sample path length
- c = concentration of sample component

The law states that concentration is directly proportional to absorbance at a given wavelength and path length and a specified temperature and pressure.

Calibration plots of A vs C can be made up using known samples and used to analyze unknown ones (Figure 10.9d). The Beer-Lambert law is also helpful in choosing the optimum sample path length for accurate analysis. (In some cases this linear relationship is not observed. See Linearity at end of this section.)

Lab IR Spectrophotometers

Grating Spectrophotometers

The standard laboratory IR instrument is a double-beam, optical null spectrophotometer. Diffraction gratings have completely replaced prisms for separating the IR beam into its component wavelengths. The heated "black body" source emits at all IR wavelengths and the thermal detector responds roughly equally at all wavelengths (Figure 10.9e).

The complete spectrum is scanned by rotating the grating and measuring the light intensity passing through its exit slit. Sample cells are placed in the sample beam, and as the spectrum is scanned, the instrument drives an attenuator into the reference beam until the detector sees equal energy from both beams. These spectrophotometers are ideally suited for qualitative analysis. In certain cases manufacturers have added computer data handling to make possible simple quantitative analyses.

Filter Spectrometers

A circular variable filter (CVF) selects the wavelength to be measured. Microcomputer-controlled, single-beam analyzers working from 4000 to 690 cm^{-1} have been designed for the quantitative analysis of mixtures. Measurements are made at each analytical wavelength in sequence, and the output is presented as a list of concentrations for mixtures of up to 10 components. Narrow IR bandpass filters consist of multiple thin layers of dielectrics of alternating refractive index deposited on a transparent substrate. They pass a band of wavelengths while rejecting all others. The width of the bandpass is, typically, one to a few percent of the center wavelength. Spectral resolution is low when compared to that obtained with a grating instrument, but signal-to-noise ratios are higher. A circular variable filter is made of dielectric layers of continuously varying thickness so that the wavelength selected depends on the angular position of the CVF wheel.

Fourier Transform Spectrometers

A quite different and very-high-performance approach to IR analysis makes use of a Michelson interferometer. Instead of separating different wavelengths for measurement, the complete spectrum is encoded as an interferogram in a few seconds of measurement time, and the spectrum computed by fast fourier transform. Commercial instruments for laboratory use contain a minicomputer and cost between \$30,000 and \$120,000.

Tunable Lasers

On the face of it, this is the spectroscopist's dream come true. Laser diodes are available to cover the 4000 to 400 cm^{-1} region. A given diode may cover 200 cm^{-1} with small discontinuities about every cm^{-1} in its tuning. The devices (PbSSe or PbSnTe) have very high spectral resolution and good power output but require liquid N_2 temperature or below plus other complex support equipment. They are for the research spectroscopist but have been used in isolated cases of process analysis that could not be done by more standard methods.

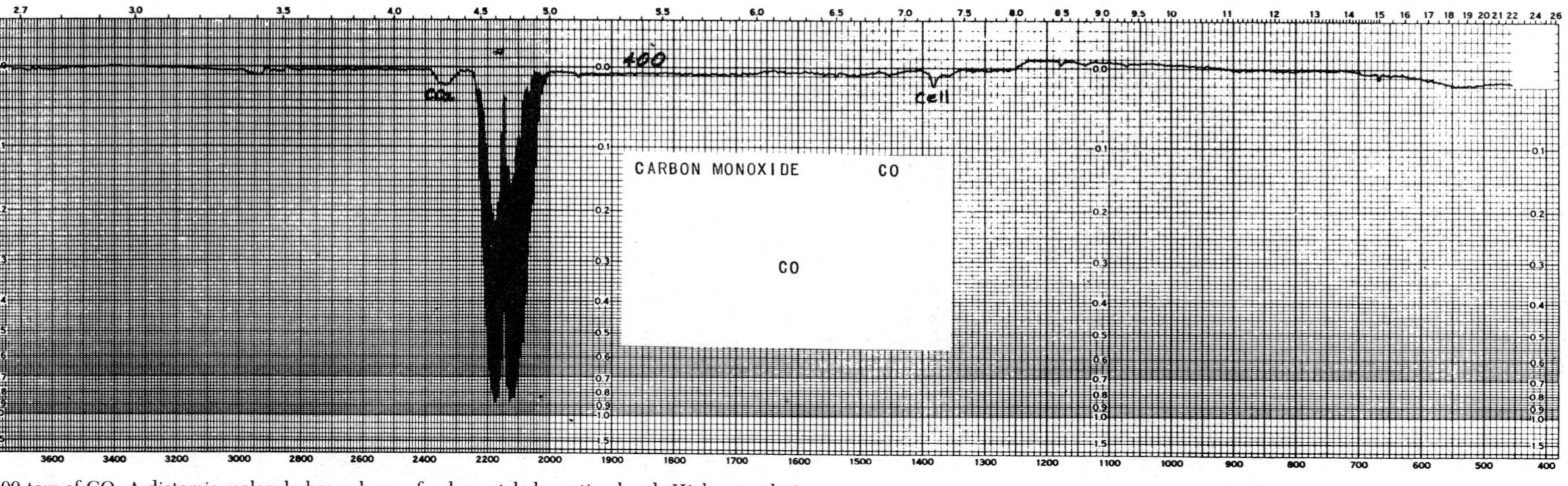

400 torr of CO. A diatomic molecule has only one fundamental absorption band. Higher resolution shows the band to consist of many sharp lines spaced about 4 cm^{-1} apart (rotational fine structure).

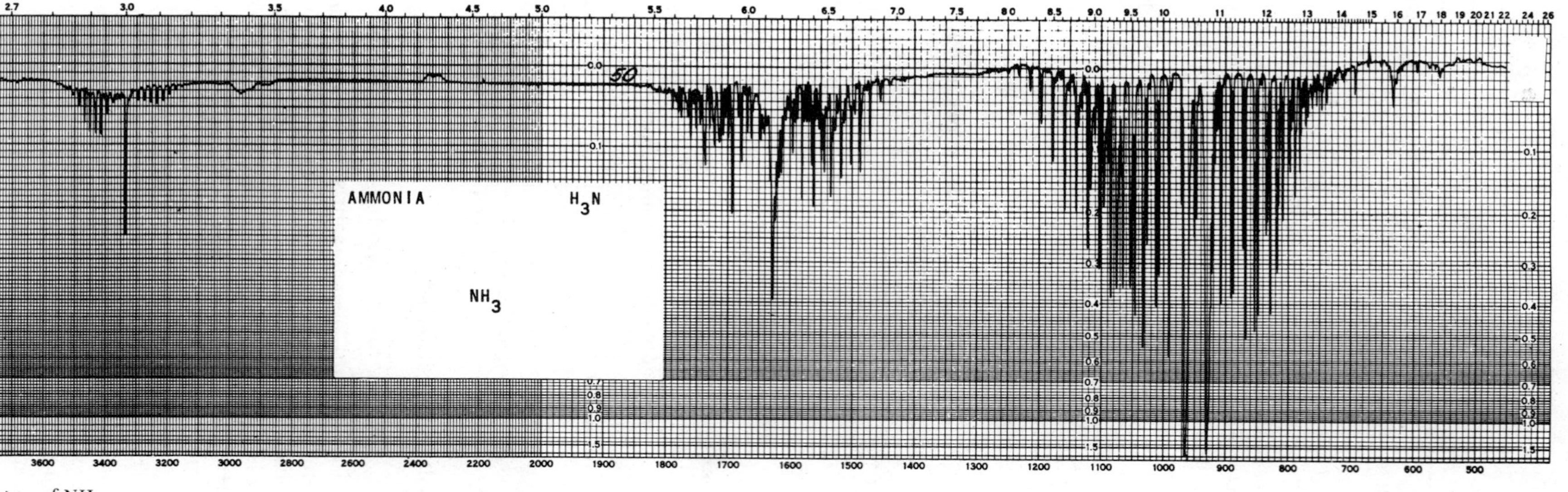

50 torr of NH_3.

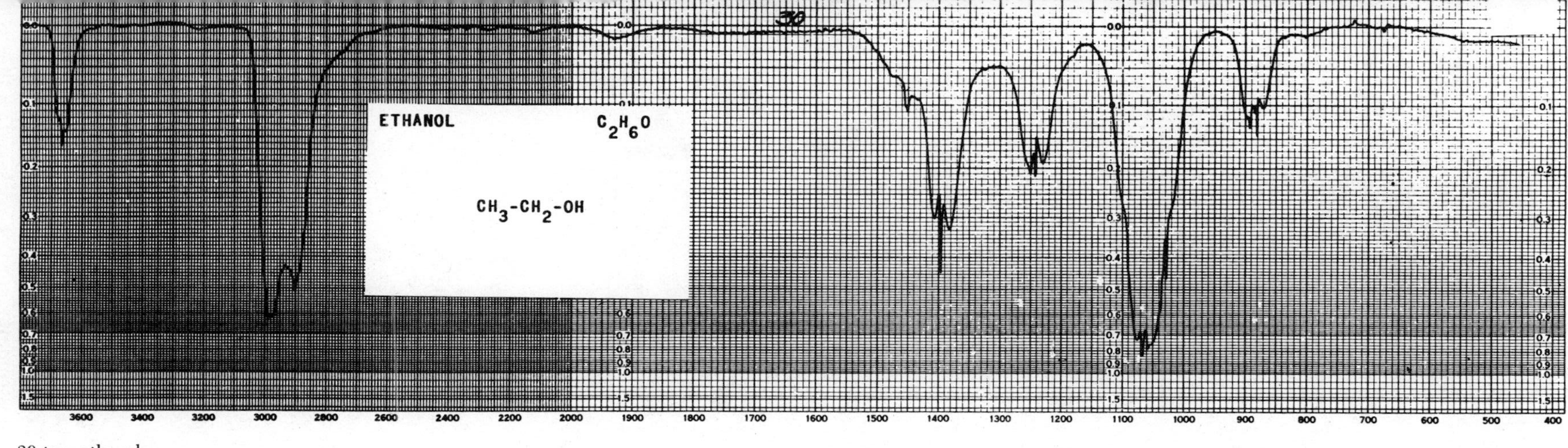

30 torr ethanol.

ETHYL CHLORIDE C_2H_5Cl

CH_3-CH_2-Cl

100 torr ethyl chloride.

Fig. 10.9b Examples of IR spectra recorded using a laboratory double-beam spectrometer (courtesy of Dow Chemical Co.). All spectra are gas phase using a 2-in. (5-cm) cell with N_2 added to give a total pressure of 600 mmHg (torr)

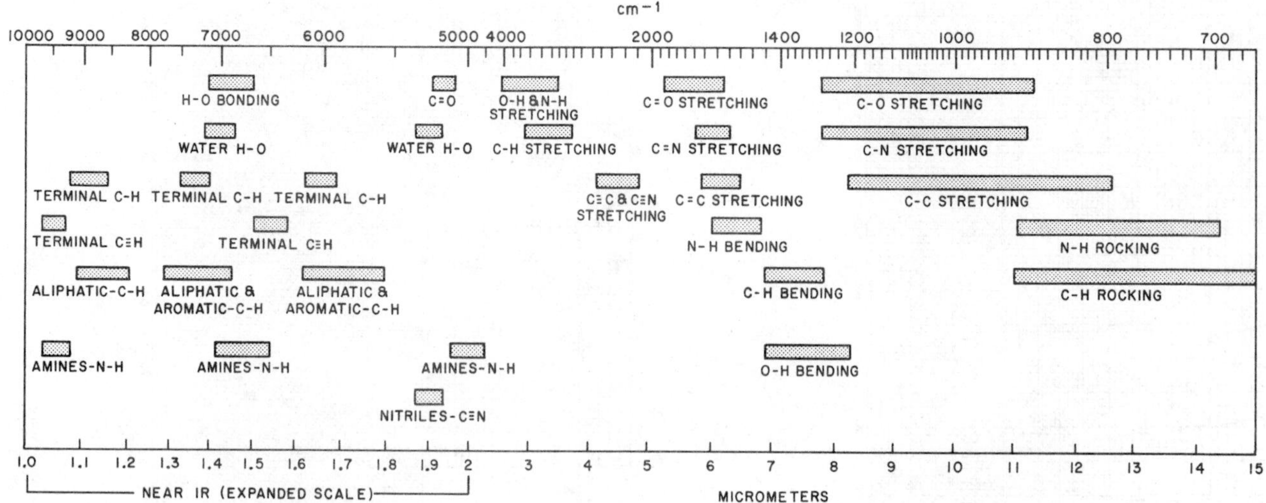

Fig. 10.9c Functional group frequency chart. Fundamental vibrations absorb in the mid IR; overtones and combination bands are 10 to 10,000 times weaker and absorb in the near IR

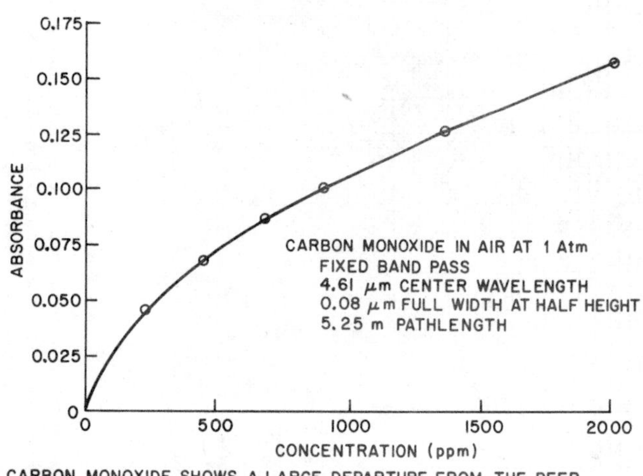

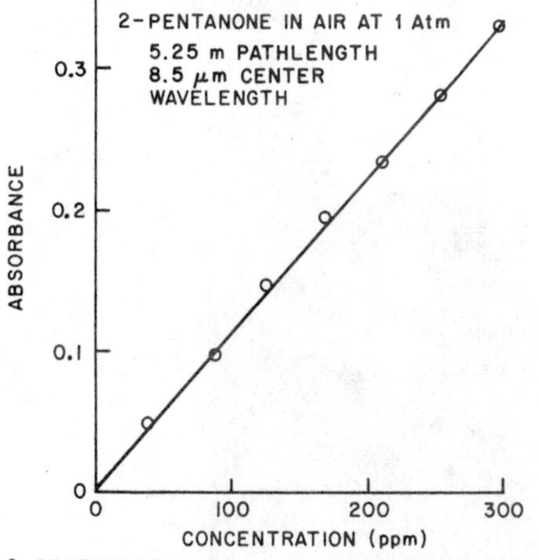

Fig. 10.9d Calibration curves for a typical filter analyzer. (Top) Carbon monoxide shows a large departure from the Beer-Lambert law; (Bottom) 2-pentenone gives a linear calibration plot

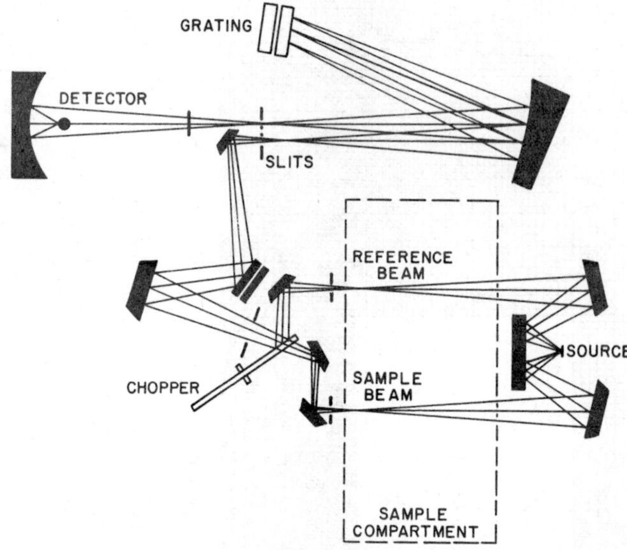

Fig. 10.9e A widely used laboratory double-beam optical null spectrophotometer

Application of Lab Instrument Technology to Process Measurement

Grating instruments have not been developed for process use. The optical system is too complex and contains several meters of internal optical path that must be purged of absorbing atmospheric species. The optical null principle is unsuited to quantitative analysis. Circular variable filter analyzers are being used for continuous workroom air monitoring (see below) and can be foreseen for process monitoring. Fourier transform technology has many advantages for continuous multicomponent analysis. To find a place as a process analyzer, Fourier transform instruments must be packaged for the industrial environment and reduced in cost. At present tunable diode lasers are too exotic to find widespread use for process analysis.

Continuous Monitoring and Process Analysis

Single-Component Analyzers: Nondispersive Infrared (NDIR)

The most commonly used process IR analyzers have no analog in laboratory instruments. They are the nondispersive infrared analyzers and used almost exclusively for gas analysis. These analyzers are invariably double-beam and use gas-selective Luft detectors filled with the gas to be analyzed (Figure 10.9f). In the usual positive filtering mode, light from single or dual sources is chopped and passed in phase through the sample and reference cells to the detector. Sample molecules attenuate the wavelengths absorbed by the detector, and the difference in power between sample and reference beams is sensed as a change in capacitance. This is amplified and displayed to give an output corresponding to concentration.

Selectivity is improved by adding a narrow bandpass filter, which selects the wavelengths of interest, and filter cells filled with interfering species, which remove undesirable wavelengths from both beams. The NDIR technique is most sensitive and selective for small molecules whose spectral fine structure is resolved under ambient conditions. The spectral resolution of these instruments is effectively set by the width of the absorption lines and in many applications is much better than the typical laboratory grating spectrophotometer.

The Luft detector has excellent sensitivity but must be temperature controlled and protected from external vibration. These analyzers need routine calibration for span and zero.

Some important variants of this basic NDIR method should be mentioned:

1. A second detector placed behind the first can be sensitized to measure and compensate for an interferant.
2. Detector chambers can be placed one behind the other with the chopper alternately delivering each beam to both. A different pressure of sensitizing gas in each detector chamber provides a built-in reference wavelength.
3. The combined beams, alternately admitted to the same chamber of the detector, cause heating and gas flow in and out of the chamber. A sensitive flowmeter gives an output proportional to sample concentration.
4. In a technique called cross-flow modulation, the mechanical chopper is replaced by a valve that exchanges sample and reference gas between the two beams at a frequency of about 1 Hz. High sensitivity and zero stability are claimed for this technique, but reference gas must be supplied continuously, flow rates and temperatures must be carefully regulated, and the response time is slow.

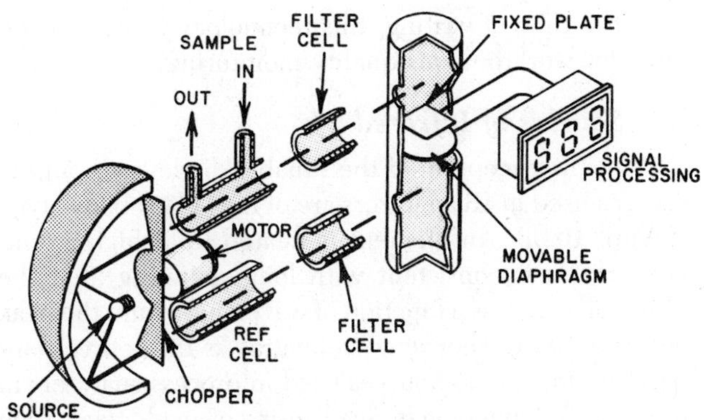

Fig. 10.9f Typical nondispersive IR analyzer using positive filtering. Both chambers of the detector are filled with a sample of the gas to be analyzed

Gas Filter Correlation Spectrometers (GFC)

These instruments (Figure 10.9g) use a nonspecific thermal detector with a specifying reference gas cell. This is an example of negative filtering. The technique is most useful when high specificity is required and it is not practical to have a reference beam. The IR beam passes alternately through a cell containing a fixed quantity of the gas to be analyzed and through a similar cell filled with a zero gas, such as N_2. The combined beam passes through the sample, for example, across a smoke

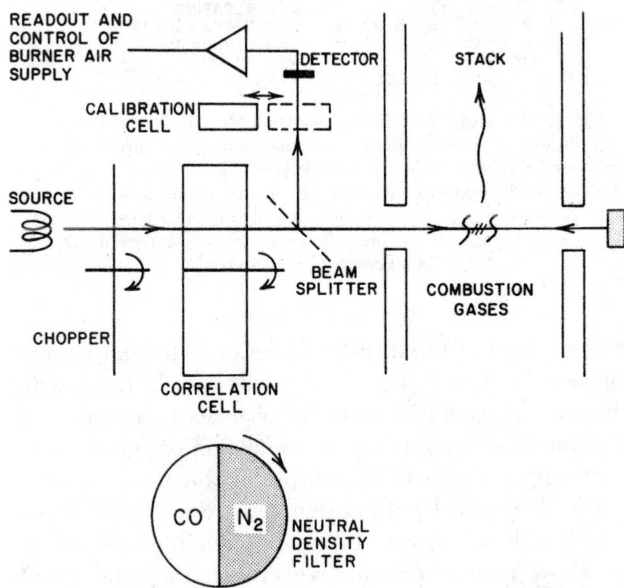

Fig. 10.9g The gas filter correlation technique is an example of negative-filtering NDIR. The diagram shows how changes can be sensed in stack CO in the 0 to 1000 ppm range and used for burner control

stack, and on to a thermal detector. This system, which is similar to NDIR, works well for analyzing small gas phase molecules whose spectra have well-resolved, rotational fine-structure in the presence of strong interference from other species. One example is analysis of carbon monoxide (CO) at the 100 ppm level in combustion gases containing large concentrations of carbon dioxide (CO_2) and water (H_2O). For accurate data, the sample must be removed from the optical path periodically to check for zero drift. A sealed calibration cell can be used to check span.

It is possible to extend the GFC method to several gases with nonoverlapping spectra by placing a gas mixture in the correlation cell and inserting different narrow bandpass filters sequentially into the beam.

Filter Analyzers

Narrow bandpass filters are used with nonspecific thermal detectors. Various configurations exist:

1. single-beam, dual-wavelength (analytical and reference) (Figure 10.9h).
2. dual-beam, single-wavelength.
3. dual-beam, dual-wavelength.

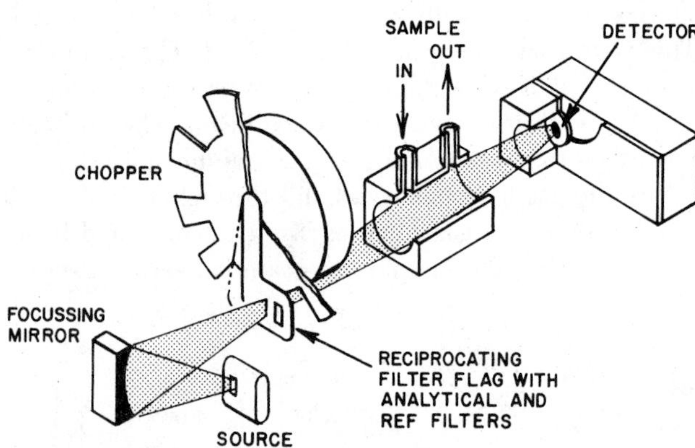

Fig. 10.9h Optical schematic of a single-beam, dual-wavelength filter analyzer. In an alternate arrangement, larger filters can be mounted directly on the rotating chopper wheel. The principle can be extended to multiple wavelengths. If the reciprocating filter flag is replaced by a microcomputer-controlled circular variable filter wheel, the optical schematic becomes that of a programmable multicomponent analyzer

The intent of the dual-beam design is to compensate for source output changes or detector and electronic gain changes. It cannot account for changes in sample cell transmission, such as dirt accumulation on windows or scattering particles in the sample stream.

The dual-wavelength design aims to use a reference wavelength not absorbed by the sample but affected by the above-mentioned sources of error to the same extent as the analytical wavelength. The single-beam design is optically and mechanically simpler and lends itself easily to interfacing with a wide variety of sample cells.

Multiple Component Process Analyzers

Multiple Fixed Filter Analyzer

When the components to be analyzed have absorption bands that do not overlap in the spectrum, the single-component analyzers can simply be expanded to measure multiple wavelengths. There are NDIR analyzers that use two detectors filled with two different gases, such as carbon monoxide and a hydrocarbon. The IR passes through first one detector and then the other. Filter analyzers are more easily expandable to multiple wavelengths enabling a wide variety of process and quality control analyses to be made. Filters can be mounted on a rotating wheel or inserted sequentially into the beam by a cam mechanism.

These instruments typically use analog signal processing. It is possible, but awkward, to compensate for spectral interferences by cross-coupling between the different wavelength outputs. Digital signal processing simplifies this problem, but at present is not widely used in multiple wavelength fixed filter analyzers.

Programmed Circular Variable Filter Analyzer

These microcomputer-controlled analyzers, derived directly from the CVF laboratory analyzers mentioned above, are simple and rugged in optical design. The same basic analyzer can be programmed for any number of analyses up to a maximum of about 10 components. The programmable CVF offers a major advantage over using fixed filters. It avoids the whole problem of filter selection and filter wavelength manufacturing tolerances and the prohibitive costs of obtaining fixed filters for one-of-a-kind analyses.

The built-in microcomputer makes interference compensation relatively straightforward. Instrument calibration coefficients and analytical wavelengths are obtained using known mixtures of the components of interest and placed in read only memory (ROM).

The microcomputer enables self-diagnostic and self-checking features to give early warning of trouble and simplify analyzer service.

At the time of writing, this technology is most developed for workroom air quality monitoring.

Sources of Infrared

With the exception of the tunable diode laser, all the sources used in IR analyzers are of the "black body" type (Figure 10.9i). An element is heated to as high a temperature as is consistent with long-operating life. The radiation varies as a function of wavelength, and the peak output shifts to shorter wavelengths as the source temperature increases. Sources used in process analyzers in the mid IR depend on ohmic heating of an element, such as Nichrome wire, either exposed or embedded in a ceramic matrix. The metal oxide surface layer or the

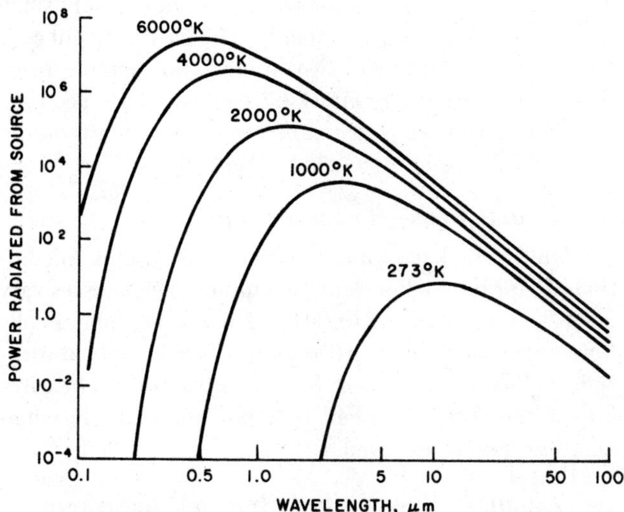

Fig. 10.9i Radiated power in w./m²/str/μm from an ideal black body as a function of wavelength and temperature. Practical sources will have lower output for a given temperature to the extent that their emissivity is less than unity. The sun's radiation is close to a black body at 6,000°K (5727°C) with peak output in the visible. A Nichrome wire source typically runs a little over 1,000°K (727°C) with a peak output near 2.5μm. Increasing the temperature increased output much more in the near IR than at longer wavelengths.

ceramic have good emissivity in the mid IR. In the near IR the much hotter tungsten halogen lamp with a quartz envelope is an excellent source. At wavelengths longer than 4μm, the output drops due to envelope absorption, but even if it did not, the emissivity of tungsten decreases at longer wavelengths.

Infrared Detectors

In all cases the IR energy is modulated so that an ac signal is detected and synchronous demodulation can be used to narrow the noise band width. Beam chopping or modulation is best done ahead of the sample. This avoids errors due to the detector sensing emission from hot samples. The error effect can be large, particularly at the longer wavelengths beyond 5μm.

NDIR Detectors

These are the gas-filled capacitive microphones already described. They use low-modulation frequencies, 10 Hz or below, and are affected by external noise and vibration. They operate in the mid IR and are very sensitive.

Thermal Detectors

These are used in the mid IR by filter analyzers.

The pyroelectric detector is usually a wafer of $LiTaO_3$, about 20μm thick, electroded on each side. It behaves electrically like a capacitor, and a current flows out of it if the temperature changes. It must be used in a chopped beam of IR, most commonly near 50 Hz. The minimum detectable signal is about 10^{-9} w. in a 1 Hz bandwidth and it is somewhat less sensitive than a typical NDIR detector. Pyroelectric detectors must be mounted and sealed to avoid microphonic and acoustic interference (Figure 10.9j).

The sensitivity of the evaporated thermopile is best at lower-chopping frequencies (below 10 Hz). It is very rugged and less microphonic than the pyroelectric detector but less sensitive.

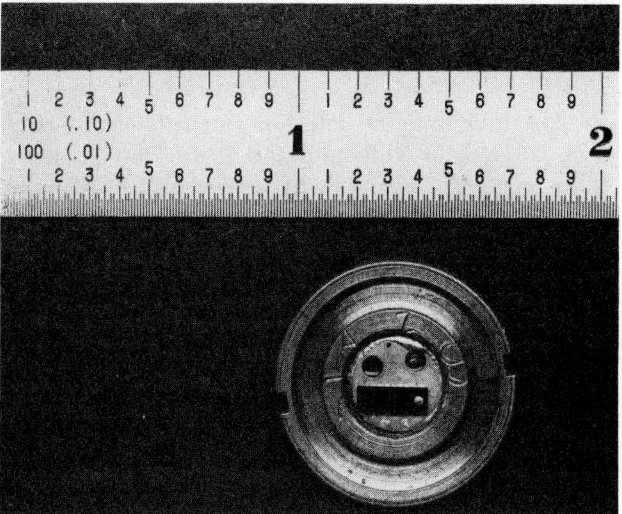

Fig. 10.9j Photograph of a pyroelectric thermal detector alongside an inch scale; when in use, an IR-transmitting window, or lens, seals the air space in front of the detector.

Photoconductive Detectors

In the near IR, photon detectors, such as PbS or PbSe, can be used. They are two or three orders of magnitude more sensitive than the thermal detectors and operate best at higher chopping frequencies. The PbSe sensitivity extends almost to 5μm, if thermoelectrically cooled, and can be used for such common analyses as CH/CO/CO_2. Response is strongly temperature-dependent in contrast with the above-mentioned thermal devices.

Sample Interface

Chosing the Best Cell Path Length

There is an optimum cell path length for analyzing a particular sample. Too short a path length means that low sample absorbance gives a weak signal in comparison to instrument noise. Too long a path length results in very little energy reaching the detector. Simple theory, based on the detector/preamplifier being the chief source of noise, unaffected by the IR power reaching the detector, predicts that the most accurate concentration measurements can be made when the sample transmits 1/e of the incident beam. This corresponds to a transmittance (T) of 36 percent, or an absorbance (A) of 0.43A. However, any absorbance between 0.1 and 1.0 A. (80 to 10 percent T.) will give good results. Measurements at higher absorbances tend to minimize the effect of certain kinds of electronic drift, while work at lower absorbances minimizes nonlinearities in the analyzer's response (see Calibration). Therefore, the cell path length should be chosen to put the sample absorbance in this desirable range. When analyzing trace gases, the absorbance is often near the lowest detection limit of the analyzer, even when the longest available path length is used, and that is the best that can be done.

Gas Cells

Figure 10.9k represents some gas cells used for process analysis. The common type transmits the IR beam straight through to the detector. The path lengths range from 0.1 mm to 50.0 cm (0.004 to 19.5 in.). Some of these long cells, especially those used on NDIR analyzers, are internally gold-coated to act as a light pipe. The single-beam analyzers can interface with a wider range of cell type, such as multiple reflection cells with path lengths adjustable between 0.75 and 40.0 m (30.0 in and 130.0 ft.). These long path cells are especially valuable for analyzing trace contaminants in the air. In choosing a gas cell, one should consider the volume if only small quantities of sample are available. The most efficient cells have a high path length-to-volume ratio. Sample pressure and temperature must be controlled for accurate results because the instrument response depends on the number of sample molecules in the cell.

Liquid Cells—Transmission

Liquid cells have much shorter path lengths to compensate for the higher density samples. Thicknesses vary from 0.1 mm to 5.0 cm (0.004 to 2.0 in.), the longer cells being used in the near IR region where sample absorption coefficients are lower. Sample streams must be carefully filtered to avoid cell plugging, particularly when very short cells are used.

Liquid Cells—Multiple Internal Reflection (MIR)

This technique, also called attenuated total reflection, is a way of avoiding the problems of thin transmission cells. The IR beam makes multiple internal reflections

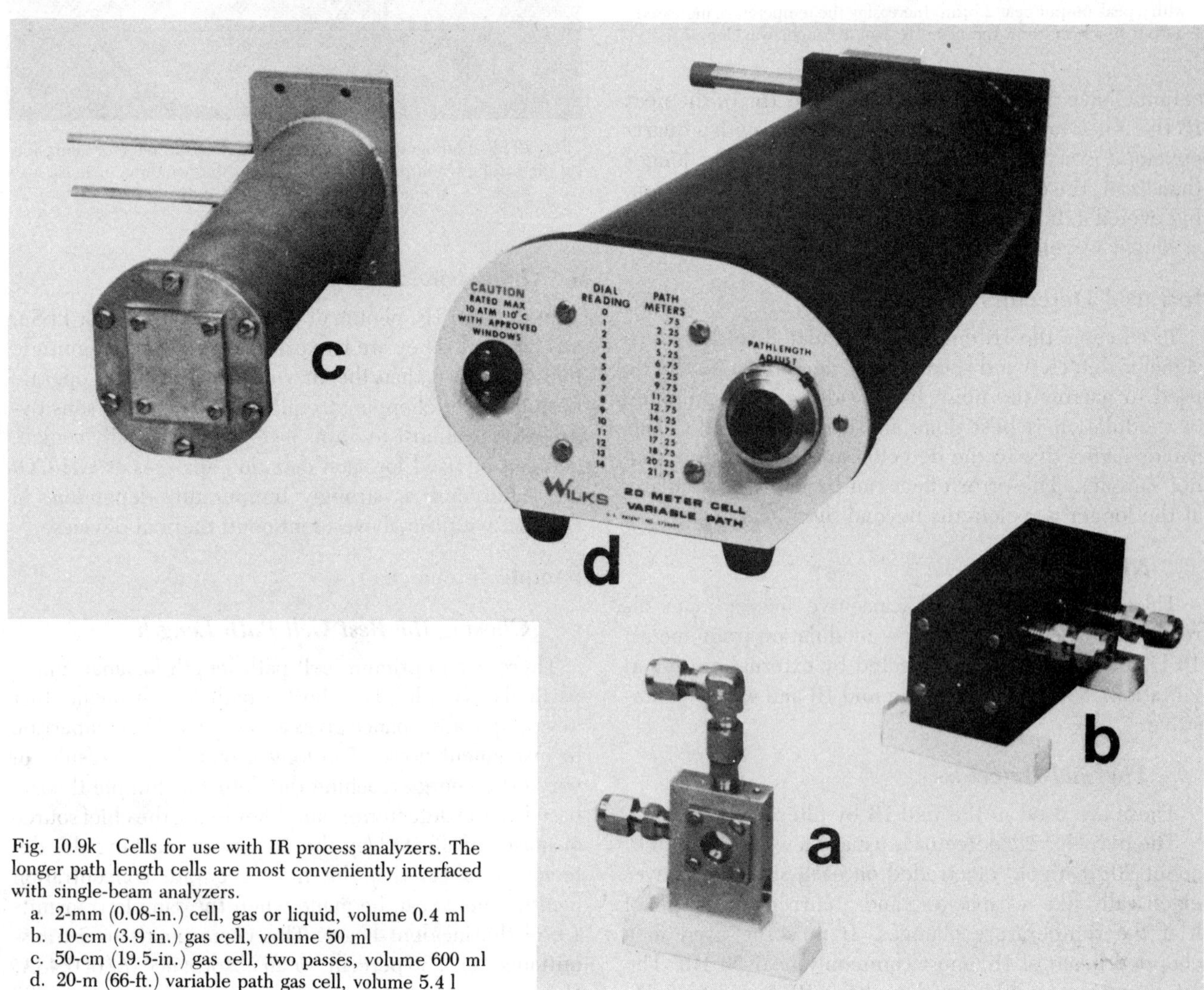

Fig. 10.9k Cells for use with IR process analyzers. The longer path length cells are most conveniently interfaced with single-beam analyzers.
a. 2-mm (0.08-in.) cell, gas or liquid, volume 0.4 ml
b. 10-cm (3.9 in.) gas cell, volume 50 ml
c. 50-cm (19.5-in.) gas cell, two passes, volume 600 ml
d. 20-m (66-ft.) variable path gas cell, volume 5.4 l

at the surface of a high-refractive index crystal wetted by the sample liquid (Figure 10.9l). The effective path length of the beam through the sample depends on the angle of incidence, the refractive indices of sample and crystal and on the number of reflections. The method is being used increasingly in the mid-IR with the sample being brought to the MIR cell via a sample loop from the process stream. Additionally, MIR has long been discussed as an excellent way to make noninvasive analyses of a process stream or reaction vessel. However, application in-line has been slow because of the problems of mounting an analyzer directly on a process stream, keeping the process-wetted surface of the crystal free of contaminating films and the difficulty of servicing the crystal without shutting down the process.

Solid Samples

Sample composition can be determined by analyzing the spectra of IR diffusely reflected from the sample surface. The largest application here is moisture measurement in, for example, paper as it is being made, or feed stocks, such as wood chips (see Section 11.15). Filter analyzers working on this principle are used for analyzing grain and other food products for such components as protein, oil and moisture. Because the percentage of IR energy which reaches the detector is so low in this diffuse reflectance sampling mode, the technique has only been used successfully in the near IR. Here the combination of tungsten halogen lamp and PbS detector provides the necessary sensitivity.

Calibration: Sources of Analyzer Drift

All IR analyzers require initial calibration with known samples. In general, the strategies of reference wavelengths and reference beams do not reduce sources of zero drift completely, and once in service, the analyzer zero must be checked and reset on a routine basis.

With constant temperature and pressure samples, an IR analyzer is inherently span stable and, in principle, requires only that the cell path length be constant. However, depending on the analyzer design and operating environment, span may require periodic checking. In general, the frequency of span calibration depends on the degree of thermostatting of the analyzer and on the application. In NDIR and GFC analyzers the characteristics of the gas-filled cells vary with temperature. Filter analyzers, as well as the NDIR and GFC types contain narrow bandpass filters that change wavelength with ambient temperature. For span stability, these filters must be temperature controlled. If high electronic scale expansion or gain is required because the cell path length is not optimized for the job (as is frequently the case in the more sensitive analysis ranges), long-term stability is degraded.

Zero and span samples can be plumbed to a calibration port for use when required. Gas mixtures in cylinders are often used but should themselves be checked as they are subject to change with age. For gas analyses, the component of interest can sometimes be removed by filtering or catalytic oxidation to provide the zero gas. The zero check is frequently automated using a timer or

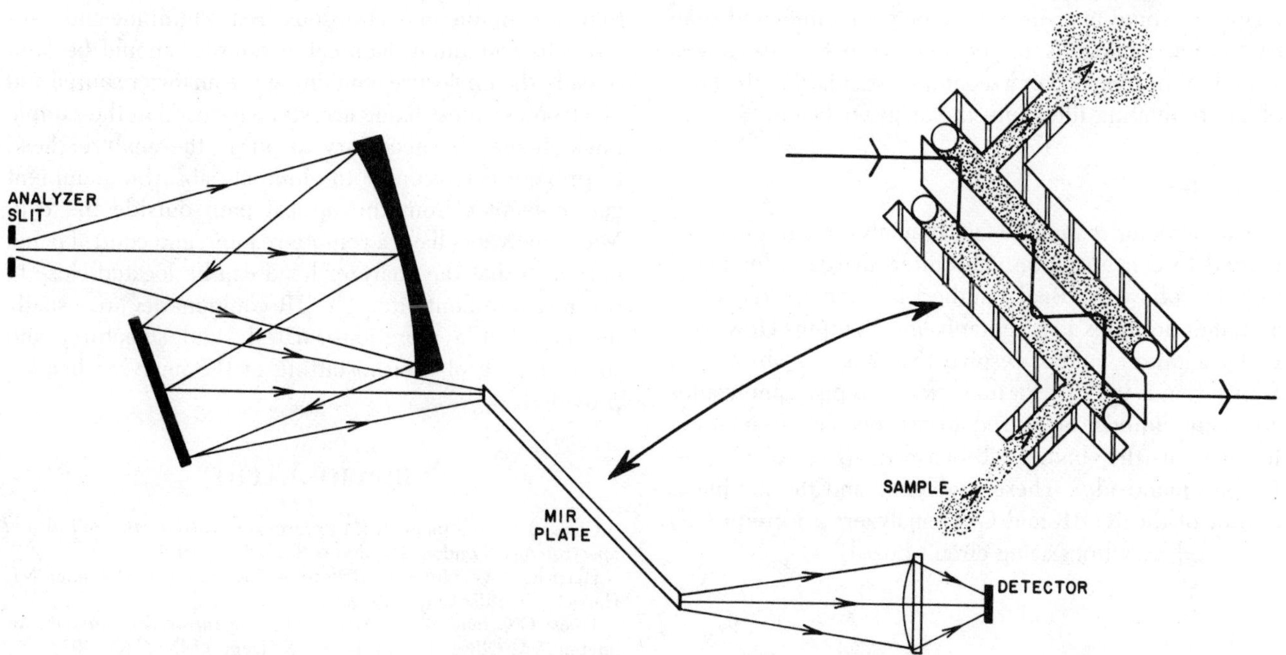

Fig. 10.9 l Optical schematic of an MIR cell suitable for use in a process sample loop interfaced with a single-beam filter analyzer. Materials used for the MIR crystal include sapphire, silicon, germanium or zinc selenide

Table 10.9m
IR ANALYZER APPLICATIONS SUMMARY

Analyzer	Carbon Monoxide	Carbon Dioxide	Organic Vapors: Simple Molecules	Organic Vapors: Complex Molecules	Organic Liquids	Solids (Reflection)	Comments
Nondispersive infrared (NDIR)	√	√	√				Single-component analysis: ethylene, CO, acetylene, methane, etc.
Mid IR filter	√	√	√	√	√		Single-component analysis: same as above, including ammonia, vinyl chloride, carbon tetrachloride, methyl ethyl ketone, ethylene dichloride, etc.
Near IR filter					√	√	Single-component analysis: ethylene dichloride, water, phenol, methyl alcohol, etc. Moisture in solids.
Correlation spectrometer	√						Stack analysis, single-component gas analysis
Multiple filter—near IR						√	Multiple components for cereal, meat and paper analysis
Multiple filter—mid IR	√	√	√	√	√		Automotive exhaust analysis ($CO, CO_2, -CH$). Multiple components for milk analysis, multiple components of gases using a programmable circular variable filter

is built into the memory of a microprocessor-controlled analyzer. Some instruments employ an automated span calibration analogous to the auto zero by inserting a sealed calibration cell or a secondary standard in the form of an attenuating filter into the analyzer beam.

Linearity

The detector response of an IR analyzer is not linearly related to concentration. The filter analyzers for many samples follow the Beer-Lambert law and a logarithmic amplifier provides an acceptably linear output. However, if the analyzer cannot resolve the sample's absorption features, as is the case for most small gas phase molecules and some liquids, measured absorbance fails to increase linearly at the higher absorbances (Figure 10.9d—Carbon monoxide). These problems, and the nonlinear output of the NDIR and GFC analyzers are frequently corrected by a linearizing circuit board.

Packaging

There is some uniformity of packaging among manufacturers of NDIR analyzers for panel mounting. At present there are no intrinsically safe IR analyzers, and they must be packaged in a purged or explosion-proof housing to be acceptable in a hazardous area. Plumbing and sample cells containing flammable samples should be kept outside the enclosure containing the analyzer source and electronics unless flame arresters are used in the sample lines. It may be necessary to purge the analyzer head to prevent corrosion or to eliminate absorbing ambient gas molecules from the optical path outside the cell. Many analyzers have a remote readout and control panel option so that the analyzer head can be located close to the measurement site. The IR components are usually mounted on a vibration-isolated rigid structure, and some degree of thermostatting of the analyzer head is provided.

BIBLIOGRAPHY

Cook, B.W. and Jones, K., *A Programmed Introduction to Infrared Spectroscopy*. London, Heyden & Sons, Ltd., 1972.

Harrick, N.G., *Internal Reflection Spectroscopy*. Ossining, NY, Harrick Scientific Corp., 1979.

Tissis, G.G. and Wolfe, W.L., ed., *The Infrared Handbook*. Arlington, VA, Office of Naval Research, Dept. of the Navy, 1978.

Van der Maas, J.H., *Basic Infrared Spectroscopy*, 2nd ed. London, Heyden & Sons, Ltd., 1972.

Willis, H.A., *Advances in Infrared and Ramon Spectroscopy*, vol. 2. London, Heyden & Sons, Ltd., 1976, ch. 3.

10.10 ION SELECTIVE ELECTRODES

Type of Electrode:	Glass, solid-state, silicon rubber matrix, liquid-ion exchange
Standard Design Pressure:	Generally dictated by electrode holder; 0 PSIG for liquid-ion exchange; 0 to 100 PSIG (0 to 0.69 MPa) for most electrode types and over 100 PSIG (over 0.69 MPa) for solid-state designs
Standard Design Temperature:	32 to 122°F (0 to 50°C) for liquid-ion exchange; 23 to 176°F (−5 to 80°C) for most others, with 212°F (100°C) intermittent exposure being permissable
Range:	From low ppm to concentrated solutions
Inaccuracy:	For direct measurements in process applications it is ±1.0 mv.; this in percentage-relative terms means ±5% for monovalent ions and ±10% for divalent ions; for endpoint detection or batch control ±0.25% or better is possible; for expanded scale commercial amplifiers better than ±1% of full-scale
Cost:	Similar to those of pH installations (Section 10.14); electrodes, $100 to $700; systems, $2000 to $7000
Partial List of Suppliers:	Beckman Instruments, Inc.; Cambridge Instrument Co.; Electrofact, Div. of Control Data Corp.; Foxboro Co.; Honeywell Inc.; Ionics Inc.; Kent Process Control, Inc.; Kernco Instrument Co.; Leeds & Northrup, a unit of General Signal; Orion Research Inc.; Uniloc Div. of Rosemount, Inc.; Van London Co. Inc.

Ion selective electrodes comprise a class of primary elements used to obtain information related to the chemical composition of a process solution. They are electrochemical transducers that generate a millivolt potential when immersed in a conducting solution containing free or unassociated ions to which the electrodes are responsive. The magnitude of the potential is a function of the logarithm of the activity of the measured ion (*not* the total concentration of that ion) as expressed by the Nernst equation (Equation 10.10[5]). The familiar pH electrode for measuring hydrogen ion activity is the best known of the ion selective electrodes and was the first one to be made commercially available (see Section 10.14 for pH measurement). With few exceptions—notably the silver-billet electrode for halide measurements and the sodium-glass electrode—the pH electrode was the only satisfactory electrode available to the process industry prior to 1966. Currently more than two dozen electrodes are suitable for industrial use. Table 10.10a lists the electrodes for which process applications have been reported.

Advantages and Disadvantages

Compared to other composition-measuring techniques, such as photometric, titrimetric, chromotographic or automated-classic analysis, the ion selective electrode measurement has an impressive list of advantages. An electrode measurement is simple, rapid, nondestructive, direct and continuous and, therefore, easily applied to closed loop process control. In this respect it is similar to using a thermocouple for temperature control. Electrodes can also be used in opaque solutions and viscous slurries. In addition, the electrodes measure the free or active-ionic species in the process, under process conditions, and, consequently, the status of a process reaction.

Table 10.10a
ION SELECTIVE ELECTRODES FOR PROCESS APPLICATIONS

Electrode Class	Electrode Designed For	Process Application
Glass	Hydrogen (pH)	See Section 10.14
	Sodium	Pulp and paper, food industry, ion exchange unit break-through
	Ammonium	Chemical industry
	Potassium	Chemical industry, food industry
	Carbon dioxide*	Food industry, chemical industry
	Ammonia*	Chemical industry
Solid-state or silicon rubber matrix	Fluoride	Potable water monitoring, HF leaks in petroleum industry, chemical industry, aluminum smelting
	Chloride Bromide Iodide	Food industry, pharmaceutical and photographic industries
	Cyanide	Waste treatment
	Lead Copper Cadmium	Plating solutions, industrial wastes and refining
	Sulfate† Phosphate†	Water pollution, pharmaceutical industry, chemical industry
Liquid-ion exchange	Calcium/water hardness	Power industry, water treatment, chemical industry
	Divalent ions	Metals industry, plating industry, waste treatment

*Based on permeable-membrane-enclosed pH electrode.
†Not yet proven in industrial applications.

However, there are several disadvantages. One is that the electrodes do not measure the total concentration of ions, the parameter that is often requested. The reason is that prior to the introduction of electrodes, concentration information was the only information available from the chemist due to his "classic" measurement techniques. Control laboratory chemists and process engineers are not accustomed to thinking in terms of activity, even when making pH measurements (Section 10.14). This habit may well disappear as this new ion selective technique becomes more popular.

However, there are times when concentration is a desirable measurement, for example, material-balance calculations or pollution control. The knowledge of material balance allows a prediction as to where a process reaction will be in the future. This information is necessary if a process is to be controlled by introducing changes that will nullify those predicted. In pollution control, it is generally accepted that many ions, even in the combined state, are detrimental to life forms. Fluorides, cyanides and sulfides, to name a few, are deleterious to fish and humans in many combined forms. They are, however, not detected by ion selective electrodes in the combined state. Consequently, pollution control agencies usually require concentration information. The electrodes can be used for this purpose if they are calibrated with solutions matching the process. If this is not satisfactory, an electrode can be used for on-line control, and separate grab samples can be analyzed by other procedures to obtain the information needed to comply with regulations.

Another disadvantage derives from a misunderstanding about precision and accuracy. Many classic analytical techniques are inaccurate to ± 0.1 percent relative error. Ion selective electrodes are inaccurate to ± 4 to 8 percent relative error (see System Accuracy). In terms of pH, this is equivalent to a measurement to ± 0.02 pH units—ordinarily regarded as a satisfactory measurement. When used with some degree of understanding, ion selective electrodes can supply satisfactory composition information and afford closed loop control that was previously unattainable. When in doubt, the user should consult with electrode manufacturers or his own analytical chemists.

Types of Electrodes

Glass

Ion selective electrodes are classified according to the type of sensing membrane employed. Glass electrodes are constructed from specially formulated glass and respond to ions by an ion exchange of mobile ions within the membrane structure. The membrane is fused to a glass body so that the outer surface makes contact with the sample or process stream, while the inner surface makes contact with an internal filling solution containing a constant activity of the ion for which the membrane is sensitive (Figure 10.10b). A stable electrical contact is made with the internal solution by a silver wire coated

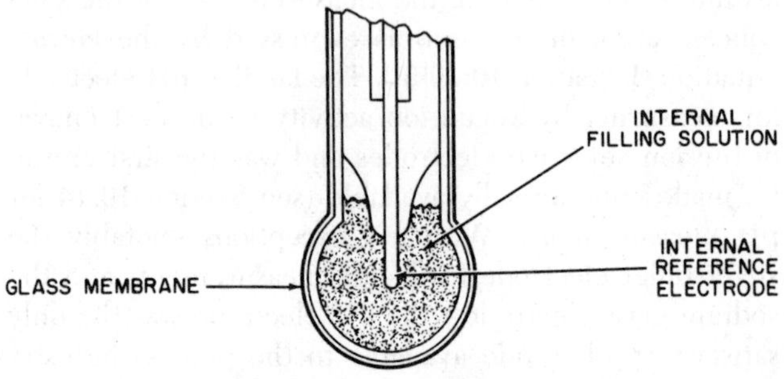

Fig. 10.10b Conventional glass pH electrode

with silver chloride. Other internal contacts have been used (mercury-mercurous chloride or thallium amalgam-thallous chloride), but the silver-silver chloride is the most popular. The internal filling solution must contain a constant chloride-ion activity and be saturated with silver chloride so that a stable potential is maintained at the metal salt-solution interface (Figure 10.10c). In the conventional glass electrode the internal solution is buffered at a pH of 7 and contains a chloride level similar to that in the external reference electrode (for details of reference electrodes see Section 10.14). Other glass electrodes in process use are the sodium, ammonium and potassium ion electrodes.

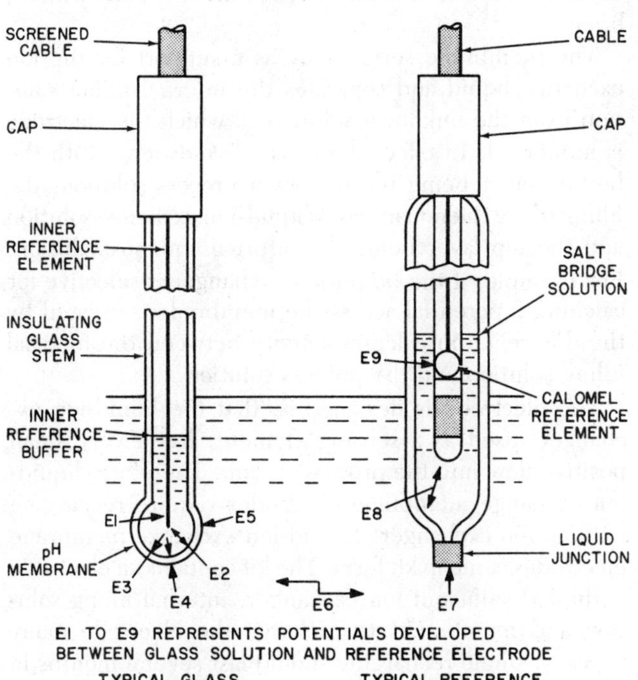

Fig. 10.10c pH electrodes

In addition to the construction already described, the sodium electrode can also be prepared by slicing a thin section from a rod of sodium-sensitive glass and cementing it to an epoxy body (Figure 10.10d). This eliminates the familiar glass body of the pH electrode. Epoxy

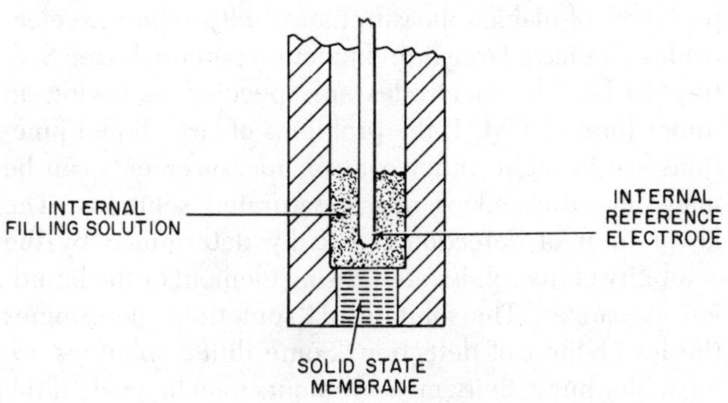

Fig. 10.10d Solid-state membrane electrode

construction is not yet available for pH measurement due to difficulties inherent in cementing pH glasses to epoxy. A carbon dioxide and an ammonia electrode can be made from a pH electrode by covering the membrane with a permeable membrane sac filled with pH buffer. The respective gas in solution will selectively diffuse in or out of the permeable membrane, causing a pH change. The latter is dependent on the activity of the gas in the process solution.

Solid-State

Solid-state electrodes are made of crystalline membranes and there are scrupulous requirements for the size and charge of the mobile ions within the membrane. The composition of the membrane varies as a function of the required measurement. For instance, the fluoride electrode has a single crystal of lanthanum fluoride for a sensing membrane. The silver and sulfide membranes are pressed pellets of insoluble silver sulfide. The solubility of silver sulfide prevents the coexistence of silver and sulfide ions, except in extremely small amounts, and the electrode can be used to measure either of these ions. Like the sodium electrode, these membranes are sealed in epoxy bodies (Figure 10.10d). Table 10.10e lists some of the commercially available solid-state electrodes and the composition of their sensing membranes.

Table 10.10e
SOLID-STATE ELECTRODES AND THEIR MEMBRANE COMPOSITION

Electrode	*Membrane*	*Form*
Fluoride	LaF_3	Single crystal
Silver sulfide	Ag_2S	Pressed pellet
Chloride, bromide or iodide	AgX* $AgX\text{-}Ag_2S$	Single crystal Pressed pellet
Cyanide	$AgI\text{-}Ag_2S$	Pressed pellet

*X = Cl. Br or I.

Some pressed pellets and the single crystalline silver-salt membranes are capable of having a metal deposited on the surface and an electrical lead connected to the metal deposit (Figure 10.10f). A solid connector permits the use of the electrodes in any position without breaking electrical continuity. Also there are no internal solutions to deteriorate with time or temperature. Figure 10.10g shows a conventional silver wire or silver billet electrode that behaves identically to its corresponding solid-state electrode. However, small imperfections in the silver-halide coating expose free silver metal to the process solution, thereby developing variable oxidation-reduction potentials, and these electrodes have not found wide use in industrial applications. When placed in clean con-

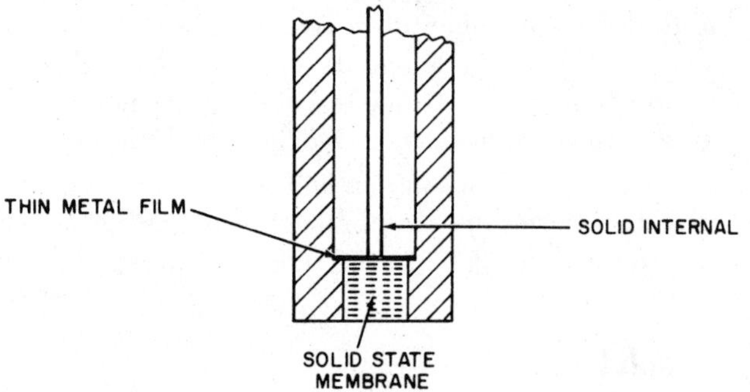

Fig. 10.10f Solid-state membrane electrodes with solid intervals

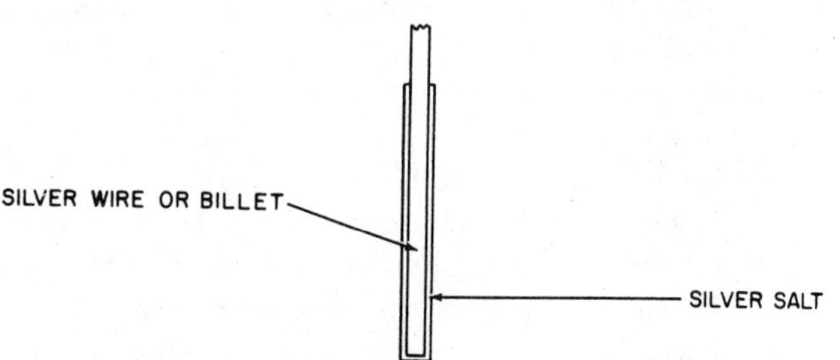

Fig. 10.10g Conventional silver-silver salt electrode

trolled environments, such as the inner-filling solution of ion selective electrodes, they produce stable reference potentials.

Silicon Rubber Matrix

Many of the solid-state electrodes, especially those made from insoluble precipitates, can be constructed by supporting the precipitate in a silicon rubber matrix and sealing the membrane in a glass body (not shown). These electrodes were developed by Professor Pungor in Hungary[1] and have not been used very much in the United States. However, as the state of the art advances and as new electrodes are proven, especially sulfate and phosphate, these electrodes will become more popular.

Liquid-Ion Exchange

There are many ions for which no glass or crystalline membrane can be found that is suitable for process measurements. Fortunately chemistry is a versatile field and by using techniques familiar in ion exchange and solvent extraction technology, electrodes can be built for some of these ions. An inert hydrophobic membrane, such as a treated filter paper, can be made selective to certain ions by saturating it with an organic ion exchange material dissolved in an organic solvent. This feature requires a construction of the electrode as shown in Figure 10.10h, which is a cross section of the tip of a liquid-ion exchange electrode. This electrode has two filling solutions, an internal aqueous filling solution in which the silver-silver chloride reference electrode is immersed,

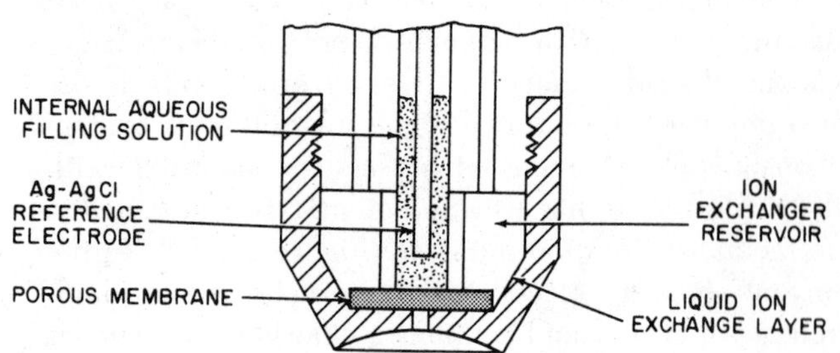

Fig. 10.10h Divalent cation electrode tip in cross section

and an ion exchange reservoir of a nonaqueous water-immiscible solution, which "wicks" into the porous membrane.

The membrane serves only as a support for the ion exchange liquid and separates the internal filling solution from the unknown solution in which the electrode is immersed. In effect, there is a "sandwich," with the bottom layer being the unknown process solution, the filling being the nonaqueous liquid-ion exchange solution and the top layer being the internal aqueous solution. For example, if the liquid-ion exchanger is selective for calcium, a potential across the membrane is created by the difference in calcium activity between the internal filling solution and the process solution.

The electrode is designed so that the liquid-ion exchanger, used as a sensing element, has a very small positive flow into the process stream. Therefore, liquid-ion exchange membrane electrodes require recharging with an ion exchanger. Liquid-ion exchange membrane electrodes come in kit form. The kit contains an electrode body and sufficient ion exchanger, internal filling solution and membranes to recharge the electrode many times. A single recharging should last several months in a properly designed system. Unlike the solid-state or glass electrodes, liquid membrane electrodes cannot be used in nonaqueous solutions because they would dissolve the liquid ion exchanger. The body of the electrode is a chemical-resistant plastic.

Measurement Range

The upper limit of detection for ion selective electrodes is the saturated solution. However, due to the problems of making measurements with reference electrodes that have large liquid junction potentials (see Section 10.14), the electrodes are specified as having an upper limit of 1 M. If the problems of large liquid junctions are brought under control, measurements can be made in saturated or nearly saturated solutions. The lower limit of detection is usually determined by the solubility of the solid-state-sensing element or the liquid-ion exchanger. The solution pH sometimes determines the lower limit of detection. Some dilute solutions are unstable, but activity measurements may be made if the solution is buffered with respect to the ion being measured, that is, if the free ion is in equilibrium with a relatively large excess of complexed ion. This is the case when free silver is measured in photographic emulsions or sulfide, cyanide or fluoride in acid solutions.

Interferences

All ion selective electrodes are similar in principle of operation and use. They differ only in the details of the process by which the ion to be measured moves across the membrane and by which other ions are kept away. Therefore, a discussion of electrode interferences will have to be in terms of the membrane materials.

The glass electrodes and the liquid-ion exchange electrodes both function by an exchange of mobile ions within the membrane, and ion exchange processes are not specific. Reactions will occur among many ions with similar chemical properties, such as the alkali metals, alkaline earths or transition elements. Thus, a number of ions may produce a potential when a given ion selective electrode is immersed in a solution. Even the pH glass electrode will respond to sodium ions at a very high pH (low hydrogen ion activity). Fortunately an empirical relationship can predict electrode interferences, and a list of selectivity ratios for the interfering ions can be obtained by consulting the manufacturers' specifications or the chemical literature. Selectivity constants will be described in connection with Equation 10.10(3).

Solid-state and silicon rubber matrix electrodes are made of crystalline materials, and interferences resulting from ions moving into the solid membrane are not to be expected. Interference is usually by a chemical reaction with the membrane. One which is observed with the silver-halide membranes (for chloride, bromide, iodide and cyanide activity measurements) involves reaction with an ion in the sample solution, such as sulfide, to form a more insoluble silver salt. As already mentioned, specific details of electrode side reactions can be found in the manufacturers' specifications and chemical literature.

A true interference is one that produces an electrode response that can be interpreted as a measure of the ion of interest. For example, the hydroxyl ion, OH^{1-}, causes a response with the fluoride electrode at fluoride levels below 10 ppm. Also the hydrogen ion, H^{1+}, creates a positive interference with the sodium ion electrode. Often an ion will be regarded as interfering if it reduces the activity of the ion of interest through chemical reaction. It is true that this reaction (complexation, precipitation, oxidation-reduction and hydrolysis) results in an activity of the ion which differs from the concentration of the ion by an amount greater than that caused by ionic interactions. However, the electrode is still measuring the true activity of the ion in the solution.

An example of solution interference will illustrate this point. Silver ion in the presence of ammonia forms a stable silver-ammonia complex that is not measured by the silver electrode. Only the free, uncombined silver ion is measured. The total silver ion may be obtained from calculations involving the formation constant of the silver-ammonia complex and the fact that the total silver is equal to the free silver plus the combined silver. Alternately a calibration curve can be drawn relating to the total silver (from analysis or sample preparation) to the measured activity. The ammonia is *not* an electrode interference.

Most of the confusion stems from the fact that analytical measurements have been in terms of concentration without regard to the actual form of the material in solution, and electrode measurements often disagree with the laboratory analyst's results. However, the electrode reflects what is actually taking place in the solution at the time of measurement. This may be far more important in process applications than the more classic information. With some of the techniques suggested, the two measurements are often reconciled.

Calibration Solutions

Calibration solutions for ion selective systems are not normally buffered to resist changes as are the standard solutions for pH systems. They are affected by dilution, evaporation or contamination with foreign matter from the process fluid and air oxidation. Thus, more care must be taken in preparing and handling these solutions than is generally needed in a typical pH application. Attention should be paid to eliminate carry-over from one test solution to another or from distilled water rinses.

Calibration solutions should be prepared by a competent laboratory in accordance with accepted principles of analytical chemistry. Many common chemical standards are available as stock solutions from laboratory supply houses. Generally, only solutions at a reasonably high-concentration level (greater than 0.01 M. or 100 ppm) should be made for storage. Serial dilutions of these stock solutions should be made at the time of use because very dilute solutions are particularly likely to lose some of their ions by absorption on the walls of the storage vessels. Use of high-grade plastic storage bottles is recommended.

Table 10.10i lists some solutions frequently used to check the performance of ion selective measuring systems. When the ionic background is held constant (sulfide, chloride, cyanide and pH), the potential difference between two of these solutions is Nernstian (see Table 10.10j). For others, the potential differences should be normalized to decade changes in activity. To achieve the utmost in accuracy and meaningful measurements, the ion selective measuring system should be standardized, and thus optimized, in a solution carefully chosen to be chemically similar to the process solution at the point of prime interest. This solution should be at a stable temperature near the actual process temperature (±3.6°F ±2°C). A grab sample of the process solution

Table 10.10i
CALIBRATING SOLUTIONS FOR ION SELECTIVE ELECTRODES

Electrode	*Chemical Composition*	*Ionic Concentration*	*Approximate Ion Activity*	*Approximate emf vs 1.0 M. KCl, AgCl, Ag Reference at 77°F (25°C)*
Hydrogen (pH)	0.05 M. KH phthalate	4.008 pH buffer at 25°C	$10^{-4.008}$ M. H^{1+}/L.	+143 (+178)*
	0.025 M. KH_2PO_4 + 0.025 M. + Na_2HPO_4	6.86 pH buffer at 25°C	$10^{-6.86}$ M. H^{1+}/L.	−35 (0)
	0.01M borax	9.18 pH buffer at 25°C	$10^{-9.18}$ M. H^{1+}/L.	−149 (−114)*
Fluoride	22.10 mg NaF/L.	10.0 mg F^{1-}/L.	9.8 mg F^{1-}/L.	−59
	2.21 mg NaF/L.	1.0 mg F^{1-}/L.	1.0 mg F^{1-}/L.	0.0
Chloride	1.00×10^{-2} M. KCl in 1.00 M. KNO_3	1.00×10^{-2} M. Cl^{1-}	0.61×10^{-2} M. Cl^{1-}	+118
	1.00×10^{-3} M. KCl in 1.00 M. KNO_3	1.00×10^{-3} M. Cl^{1-}	0.61×10^{-3} M. Cl^{1-}	+177
Silver	1.00×10^{-2} M. $AgNO_3$	1.00×10^{-2} M. Ag^{1+}	0.90×10^{-2} M. Ag^{1+}	+443
	1.00×10^{-3} M. $AgNO_3$	1.00×10^{-3} M. Ag^{1+}	0.96×10^{-3} M. Ag^{1+}	+385
Sulfide	1.00×10^{-1} M. Na_2S in 1.00 M. NaOH	1×10^{-1} M. S^{2-}	0.15×10^{-1} M. S^{2-}	−860
	1.00×10^{-3} M. Na_2S in 1.00 M. NaOH	1×10^{-3} M. S^{2-}	0.15×10^{-3} M. S^{2-}	−800
Cyanide	1.00×10^{-3} M. NaCN in 1.00×10^{-1} M. NaOH	1.00×10^{-3} M. CN^{1-}	0.76×10^{-3} M. CN^{1-}	−192
	1.00×10^{-4} M. NaCN in 1.00×10^{-1} M. NaOH	1.0×10^{-4} M. CN^{1-}	0.76×10^{-4} M. CN^{1-}	−133
Water hardness	1.00×10^{-2} M. $CaCl_2$	1.00×10^{-2} M. Ca^{2+} or 1000 mg/L. as $CaCO_3$	0.55×10^{-2} M. Ca^{2+} or 550 mg/L. as $CaCO_3$	+34
	1.00×10^{-4} M. $CaCl_2$	1.00×10^{-4} M. Ca^{2+} or 10 mg/L. as $CaCO_3$	0.92×10^{-4} M. Ca^{2+} or 9.2 mg/L. as $CaCO_3$	−18

*vs 4 M. KCl, AgCl, Ag reference electrode at 25°C.

checked in the laboratory may be the best and most convenient standard to use.

The Nernst Equation

The potential developed across an ion selective membrane is related to the ionic activity as shown by the Nernst equation:

$$E = \frac{2.3\,RT}{nF} \log \frac{a_1}{a_{int}} \qquad 10.10(1)$$

where

E = the potential developed across the membrane

a_1 = the activity of the measured ion in the sample or process

a_{int} = the activity of the same ion in the internal solution

$2.3\,RT/nF$ = the Nernst slope, or slope of the calibration curve, and is a function of the absolute temperature T and the charge on the ion being measured n

R = the gas law constant.

Table 10.10j shows how the Nernst slope changes with temperature and the charge on the ion. When the ratio of the two activities is unity, the potential across the membrane is zero.

Equation 10.10(1) assumes that the membrane has identical selectivity properties on both sides. If for some reason this is not true, the equation is written

Table 10.10j
NERNST SLOPES

Electrode Temperature		*Mv. per Decade of Activity**	
°C	°F	n = ±1	n = ±2
0	32	54.19	27.10
10	50	56.17	28.08
20	68	58.17	29.08
25	77	59.16	29.58
30	86	60.15	29.58
40	104	62.15	29.58
50	122	64.12	32.03
60	140	66.10	33.05
70	158	68.09	34.04
80	176	70.07	35.04
90	194	72.05	36.02
100	212	74.04	37.02

*The slopes are positive for cations and negative for anions.

$$E = E_{asy} + \frac{2.3\ RT}{nF} \log \frac{a_1}{a_{int}} \qquad 10.10(2)$$

where E_{asy} is the asymmetry potential and amounts to a few millivolts. This equation is simplified by the fact that a_{int} is fixed by the internal structure of the electrode, giving

$$E = E^{\circ\prime} + \frac{2.3\ RT}{nF} \log a_1 \qquad 10.10(3)$$

where $E^{\circ\prime}$ is a new constant.

In actuality, the potential of a single electrode cannot be measured by itself. It can be measured only in conjunction with a reference electrode and a high-input impedance voltmeter (Figure 10.10k). The latter is necessary to prevent current from flowing through the electrode, an action that would tend to cause electrochemical reactions in the solution phase around the membrane. The potential read on the voltmeter is equal to the algebraic sum of the potentials developed within the system. That is, the observed meter potential is the sum of the potentials developed by the measuring electrode, E, the reference electrode, E_{ref}, and a small but important liquid-junction potential, E_j.

$$E_{meter} = E - E_{ref} + E_j \qquad 10.10(4)$$

Under normal operating conditions, the reference electrode is assumed to be constant, as is the liquid-junction potential. However, this is not always the case. Substituting Equation 10.10(3) into Equation 10.10(4) and combining constant terms, including E_{ref} and E_j, gives the general form of the Nernst equation:

$$E_{meter} = E^\circ + \frac{2.3\ RT}{nF} \log a_1 \qquad 10.10(5)$$

where E° is a constant for a given electrode system at a specific temperature. It depends on the choice of reference electrode and includes the liquid-junction potential.

The Nernst equation for the electrode pair can be written as an instrument input-output equation:

$$\text{Output} = A + B \log (\text{input}), \qquad 10.10(6)$$

where the output is a millivolt signal to a meter, A is a zero adjustment and B is a span or slope adjustment around a temperature-independent, or isopotential, point. The input to the electrodes is the composition of the solution in terms of activity. Equation 10.10(5) states that the output of an electrode pair is linear with respect to the logarithm of the activity of the ion being measured (Figure 10.10l). The slope of the curve relating E_{meter} to log a_1 is 59.16 mv. (at 25°C for n = 1) or 29.58 mv. (at 25°C for n = 2). Ignoring the effects of chemical reactions that would tie up ions, the activity of the ions is related to the analytical concentration, C, as follows:

$$a = \gamma C \qquad 10.10(7)$$

where γ is the activity coefficient and is a measure of the interaction among ions in solution. It can be thought of

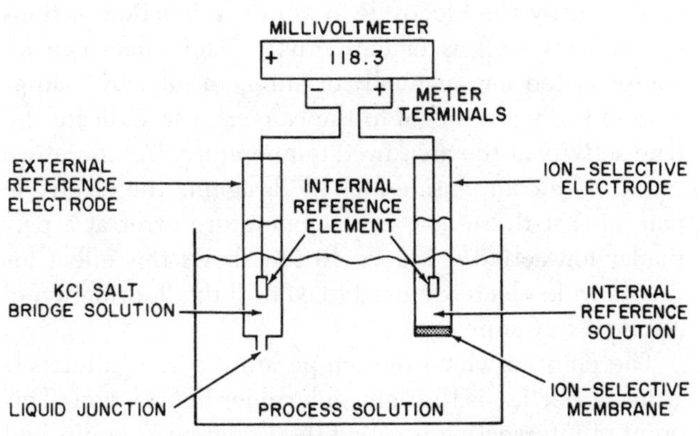

Fig. 10.10k Ion selective electrode measuring system

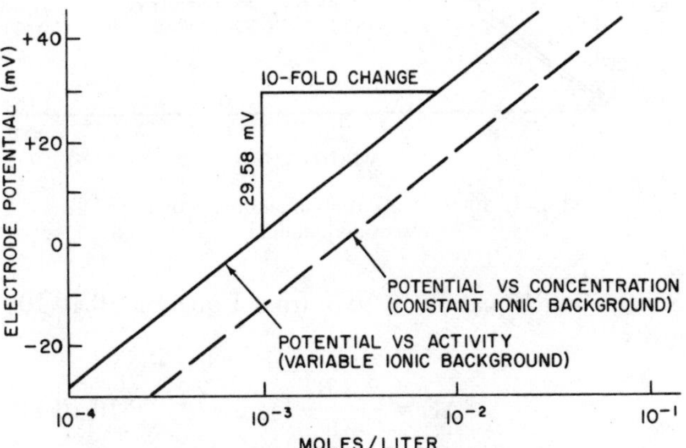

Fig. 10.10 l Electrode potential for calcium chloride solutions as a function of concentration and activity

as an empirical factor to explain the difference between the actual behavior of ions in solution and the ideal behavior. At zero ion concentration, that is, no ionic interaction, γ is taken as unity, and the activity is equal to concentration. As the concentration increases, γ decreases at first, passes to a minimum value and then rises, often to values greater than unity in very concentrated solutions.[2] The activity coefficient is constant when the ionic composition of the solution is constant. Substituting Equation 10.10(7) into Equation 10.10(5) gives

$$E_{meter} = E^\circ + \frac{2.3\ RT}{nF} \log (\gamma C) \qquad 10.10(8)$$

or, at constant total ionic conditions,

$$E_{meter} = E^\circ + \text{constant} + \frac{2.3\ RT}{nF} \log C \qquad 10.10(9)$$

where the constant term is $\frac{RT}{nF} \log \gamma$. Equation 10.10(9) is linear with respect to the concentration term (Figure 10.10l). However, if the ionic background of a solution varies, as in the preparation of a series of standards by dilution, the activity coefficient is no longer constant and Equation 10.10(8) is nonlinear (Figure 10.10m).

Equation 10.10(5) can predict the change in potential to be expected from a given change in activity or concentration (constant activity coefficient). For instance, if the activity changed twofold (100 percent change), then

$$E_{meter} = E^\circ + \frac{2.3\ RT}{nF} \log 2a_1 \qquad 10.10(10)$$

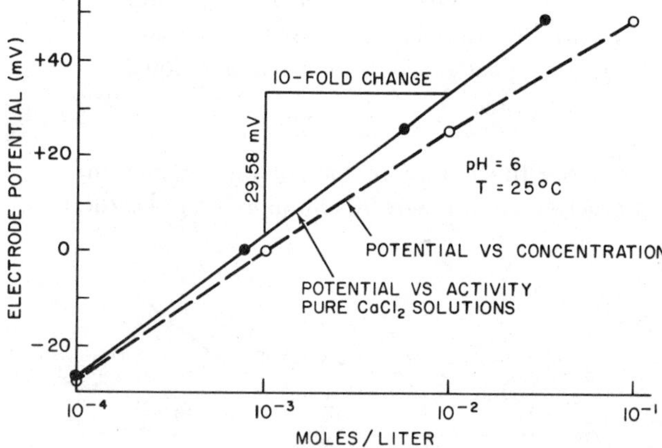

Fig. 10.10m Concentration and ion activity vs electrode potential

Subtracting Equation 10.10(5) from Equation 10.10(10) gives

$$\Delta E = \frac{2.3\ RT}{nF} \log 2 \qquad 10.10(11)$$

or 18 mv. at 25°C for n = 1. A similar argument would show that for the same sample an 18 mv. decrease would be observed if the initial activity was cut in half (50 percent change).

This change is not dependent on the magnitude of a_1 and is the same whether measuring fluoride at the ppm level, chloride in 4 percent salt solutions or pH of a 15 percent sulfuric acid solution. Table 10.10n lists the changes in potential to be expected for up to a tenfold change in activity. Column 1 shows the ratio of the original activity a_1 to the final activity a_2. The last column shows the equivalent pH change if the $[H^{1+}]$ were being measured. The data indicate that for precise measurements of small changes in activity it is necessary to use an expanded scale meter. For example, a span of 60 mv. would allow a tenfold change to be measured, using the full scale of the measuring instrument.

Table 10.10n
CHANGES IN METER POTENTIAL FOR CHANGES OF ACTIVITY
(See Equation 10.10(11))

a_1/a_2	ΔE in mv (77°F; 25°C) n = 1*	ΔE in mv (77°F; 25°C) n = 2*	Equivalent pH change (n = 1)
.1	−60†	−30†	−1.0
.25	−36	−18	−0.6
.5	−18	−9	−0.3
.79	−6	−3	−0.1
1.00	0	0	0.0
1.26	+6	+3	+0.1
2.00	+18	+9	+0.3
4.00	+36	+18	+0.6
10.00	+60	+30	+1.0

*Data are for positive ions. For negative ions, sign should be reversed.
†Values rounded off from 59.16 and 29.58 (see Table 10.10j).

Temperature Effects

There are three temperature effects on ion selective measurements, including the T term in the Nernst equation (see Table 10.10j), the thermal characteristics of the electrodes and the thermal characteristic of the solution. The T term in Equation 10.10(5) states that the potential produced by the electrode system is a function of temperature as well as of ion activity. This effect can be compensated for, manually or automatically, by manipulating the input signal to the converter to indicate the true activity at the measured temperature. Temperature can also be compensated for by designing the electrode pair so that there is a zero temperature error at a particular ion activity. Figure 10.10o shows this effect for the fluoride electrode used to control the fluoridation of public water supplies.

The point at which the temperature curves intersect, 1 mg. F per L., is the control level for fluoridation. This point of intersection is called the isopotential point, and temperature effects are negligible on either side for a 10

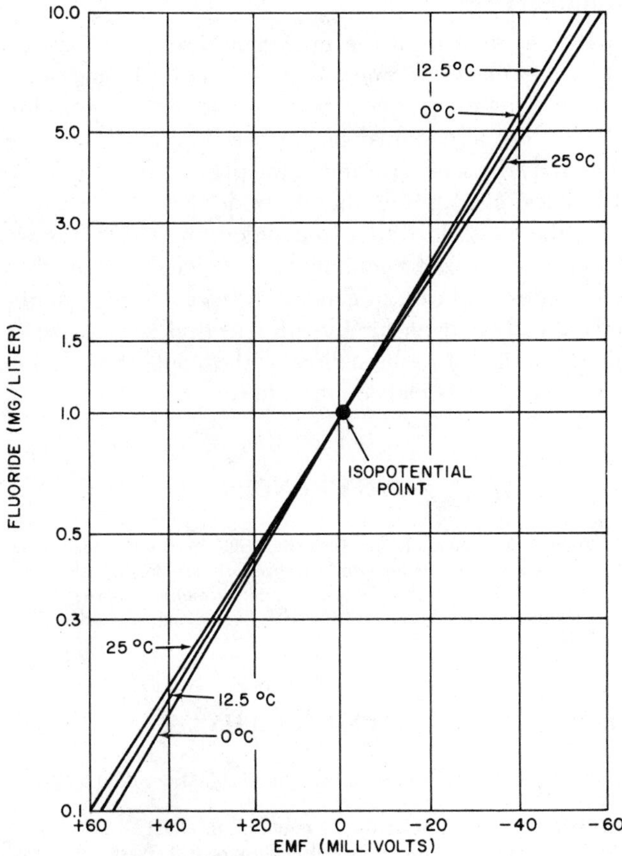

Fig. 10.10o Isopotential point for fluoride electrode

to 15°C (18 to 27°F) change in temperature. The isopotential point for pH-measuring systems is normally about pH 7 and has a potential value of 0 mv. The activity coordinate of the isopotential points of the solid-state electrodes is fixed during manufacture, whereas the millivolt coordinate is also dependent on the choice of reference electrode. However, due to the construction of the liquid-membrane electrode, the isopotential points can be changed to fit the process.

The second effect associated with temperature is created by the different internal thermal characteristics of the measuring and reference electrodes. This effect can be minimized if the internal elements of the measuring electrode are matched to the reference electrode. Most commercial pH systems employ matched internal elements for both electrodes. The temperature effect on the chemistry of solutions is the third factor that can create an apparent error in measurement. This effect is difficult to quantitate but can be offset by calibrating the system with preanalyzed process samples at the temperature of the process measurement. It should be noted that this is not a system error. The electrode indicates the true activity as a function of temperature changes. As long as the status of the process is what is required, the solution temperature effect is not important because the true activity is the quantity desired. This effect is usually not compensated for by the measuring instrumentation.

It is important to remember that when electrodes are changed from process samples to standard samples that are at different temperature, or when sudden, wide variations in process temperature are encountered, time is required for a new state of thermal equilibrium to be reached (approximately one-half hour for a 10 or 15°C change). During this time, the potential of the electrode system will drift. The duration of the drift depends on the particular electrode system and the magnitude of the temperature change. Therefore, it is important to avoid changes in temperature during calibration or to allow thermal equilibrium to be established.

System Accuracy

The accuracy of measurements derived from an analytical system is a composite of all contributing variables. These variables for ion selective measuring systems are the measuring electrode, the reference electrode, including the liquid-junction potential, the selective ion potential converter, the recorder and the temperature and solution errors. The relationship between overall emf errors (ΔE) and ionic activity (a) may be derived from the Nernst equation 10.10(5) by substituting the values of the thermodynamic constants R and F

$$E = E^\circ + \frac{0.1984T}{n} \log a \qquad 10.10(12)$$

and taking the first derivative of Equation 10.10(12) with respect to activity

$$\Delta E = (0.2568/n)\,(\Delta a/a)\ \text{(at 25°C)} \qquad 10.10(13)$$

$$\Delta E = (0.2568/n)\,(RE) \qquad 10.10(14)$$

where RE represents relative error in the activity.

A plot of Equation 10.10(14) is given in Figure 10.10p. The *relative* error in measuring activity is dependent only on the *absolute* error in the emf and is independent of the activity range and of the size of the sample being

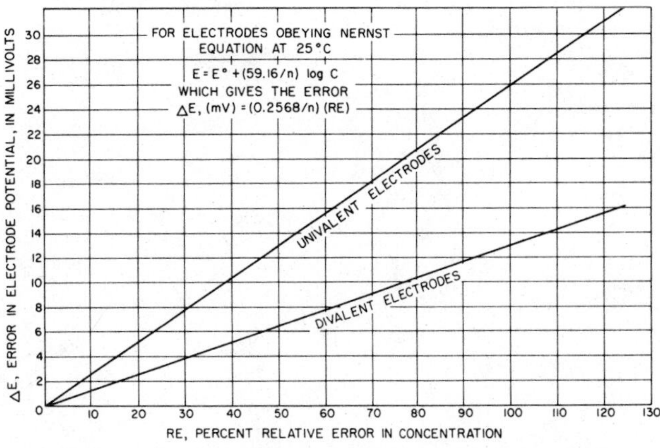

Fig. 10.10p Theoretical error in potential as a function of relative error in concentration

measured. This is similar to an equal percentage valve in which equal incremental changes in valve opening (electrode potential) produce equal percentage changes in flow (RE in activity) for all valve openings—assuming constant differential pressure (constant temperature). Being a logarithmic device, an electrode gives a constant precision throughout its dynamic range. Concentrated solutions can be analyzed with the same accuracy as dilute solutions.

Laboratory measuring instruments for ion selective electrode measurements with an inaccuracy of ±0.1 mv. have become commercially available. It is possible to make laboratory pH measurements with a relative inaccuracy of ±0.002 pH units (equivalent to ±0.12 mv.). Similarly under carefully controlled conditions, ion selective electrodes may be made repeatable within 0.1 mv. The accuracy attained to date in process instruments has been limited by the reference rather than by the ion selective electrodes.

Ion selective measurement systems for process applications are repeatable to ±1 mv. For an electrode responding to univalent ions, an overall error of 1 mv. corresponds to a 3.9 percent error in activity; for an electrode responding to divalent ions, the relative error is 7.8 percent per mv. This means that they are roughly 5 percent (of value in activity) devices when measuring univalent ions, including H^{1+}, in acidic or basic solutions or 10 percent (of value in activity) devices when making divalent ion measurements. These figures apply only to direct electrode measurements. When electrodes are used as endpoint detectors in titrations or batch reactions or in differential systems, relative errors of 0.1 percent are possible.

Summary

It is evident from the preceding discussion that ion selective electrode measurements and pH measurements using the glass electrode are not only identical in theory, but also similar in practice. The electrodes are generally the same size and fit the pH holder assemblies. Electrical insulation of the electrodes is as important as with the glass electrode. In addition, the electrodes are subject to the same fouling by oils and slimes as glass electrodes, and can, in general, be cleaned using methods already proven for the pH electrodes. Because of these similarities, discussion of the application of ion selective electrodes is omitted here.

REFERENCES

1. Pungor, E., Havas, J. and Toth, K., "Silicone Rubber Membrane Electrodes," *Instruments and Control Systems*, 38:105, 1965.
2. Frankenthal, R.P. In Meites, L., ed., *Handbook of Analytical Chemistry*. New York, McGraw-Hill Book Company, 1963, p. 1, Table 1-8.

BIBLIOGRAPHY

Butler, J.W., *Ionic Equilibrium*. Reading, MA, Addison-Wesley, 1964.

Light, T.S., "Ion Selective Electrodes." In Durst, R.A., ed., National Bureau of Standards Special Publication 314; Washington, DC, 1969, ch. 10.

Ringbom, A., *Complexation in Analytical Chemistry*. New York, John Wiley & Sons, Inc., Interscience, 1963.

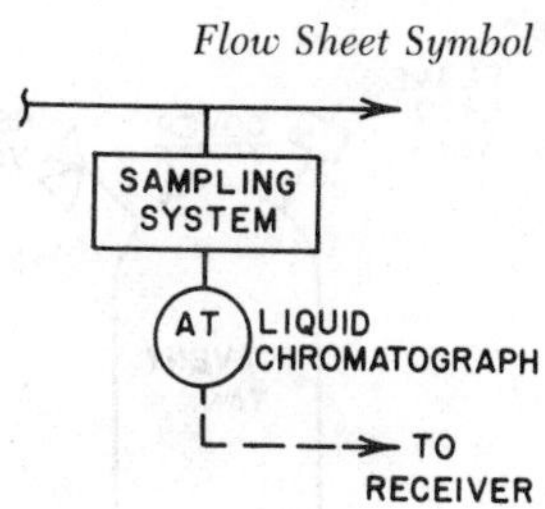

10.11 LIQUID CHROMATOGRAPHS

Type of Sample:	Liquid
Sample Pressure:	5 to 1,000 PSI (35 to 7,000 kPa)
Sample Temperature:	60 to 300°F (16 to 149°C)
Ambient Temperature:	0 to 122°F (−18 to 50°C)
Contacting Materials:	Stainless steel, Teflon standard; all conventional materials available
Utilities Required:	Electrical power, carrier solvent, air at 100 PSI (700 kPa)
Cost:	$25,000 to $75,000, depending on application
Repeatability:	± ½% for most applications
Cycle Time:	3 to 20 min. for most applications
Special Features:	Multicomponent readout, molecular weight readout
Partial List of Suppliers:	Applied Automation, Inc.; Beckman Instruments, Inc.; DuPont Co.; Hewlett-Packard Co.; Perkin-Elmer Corp., Instrument Div.; Spectra-Physics, Auto Lab Div.; Varian Associates; Waters Associates, Inc.

Comparison with Gas Chromatographs

In many ways the liquid chromatograph is similar to the gas chromatograph. The basic difference is that the carrier is a liquid instead of a gas; all instrument changes result from this difference. In fact, only components in direct contact with the carrier are significantly different. These components will be discussed here. Section 10.4 described the gas chromatograph and should be referred to for a basic understanding of the liquid chromatograph. In addition, there are current books on laboratory liquid chromatography that explain the principles of liquid chromatography and, thus, make good references for those interested in process instruments.[1,2,3]

The operating principle is basically the same as that of the gas chromatograph. A typical carrier flow diagram of the liquid chromatograph is illustrated in Figure 10.11 a. The sample valve injects a measured volume of sample into the controlled flow of carrier liquid which transports it through the columns and into the detector. Interaction between the stationary column-packing material and the flowing liquid carrier causes the sample components of interest to move through the column at different velocities, providing the separation. For each sample component, the detector provides an electrical signal proportional to its concentration in the carrier. The electrical signal is recorded as a chromatogram or otherwise displayed in any of the standard chromatographic forms.

The instrument construction is also similar to that of the gas chromatograph. The liquid chromatograph is divided into two parts: the analysis section and the control section. The analysis section is located near the process stream and is often contained in an instrument house. This section includes 1) the chromatographic oven containing the valves, columns and detector; 2) the carrier supply; 3) an electronics compartment containing the circuitry for the detector, temperature control, valve operators and local data handling; and 4) the sample preparation system, which may be in a separate temperature-controlled oven.

The control section is often located in the control room, and includes the programmer to control the instrument and process the data; and the data display, which may include a strip chart recorder, a digital display or a digital computer for further data processing. The programmer discussed here includes stream selection, which may be a physically separate unit.

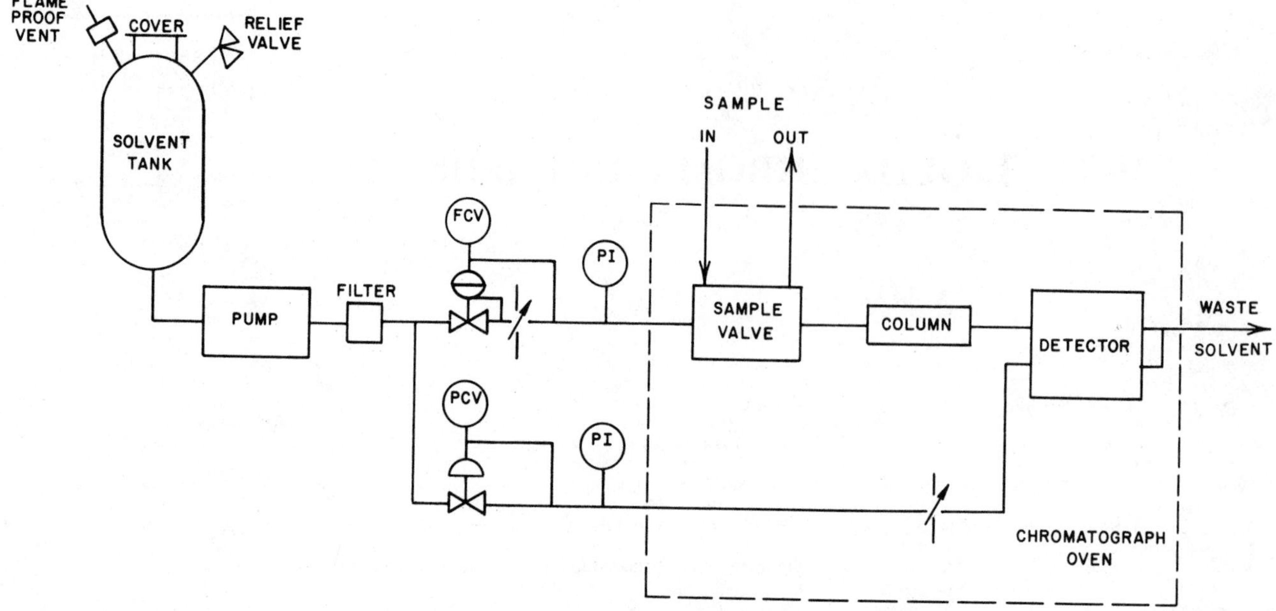

Fig. 10.11a Carrier flow diagram of the liquid chromatograph

However, the liquid chromatograph operating conditions are different from those of the gas chromatograph. In particular, the carrier pressure is typically 1,000 PSIG (7,000 kPa) at a flow rate of 1 ml per min. and the chromatographic cycle time is typically 3 to 20 minutes. Other differences will be apparent in the following discussion.

Components

Carrier Supply

The carrier supply is made up of 1) the liquid carrier; 2) a reservoir to hold it; 3) a carrier pump; 4) pressure and flow controls; and 5) a filter and gauges.

The carrier is either a single component solvent or a blend of solvents. The selection and uniformity of solvents and solvent blends is critical to the liquid chromatograph, affecting both component separation and detector response.

A typical carrier reservoir is a 15-gal. (57 L.) stainless steel tank equipped with a flame-proof vent and a relief valve for flammable solvents. Some solvents are affected by oxygen or moisture in the air and an inert gas, such as nitrogen, is either bubbled through, or blanketed over, the carrier in the tank.

There are two types of carrier pumps; air driven and motor driven. Both are reciprocating pumps containing check valves to divert the liquid through the pump. Air-driven pumps are powered by a large-diameter air piston connected to a small-diameter liquid piston. The pump recycles at the end of each stroke. The air-driven pump provides a pressure, relatively independent of flow, that is nominally some multiple of the air pressure. Motor-driven pumps are powered with an electric motor connected to an eccentric that drives one or more liquid pistons. The motor-driven pump provides a flow, relatively independent of pressure, that is controlled by motor speed or piston stroke length.

Pressure and flow control depends on the type of pump used. Air-driven pumps (see Figure 10.11a) provide a pressure that is reduced to the desired value with a diaphragm pressure regulator. Flow control is provided with a diaphragm differential pressure regulator that controls a preset pressure difference across a fixed restrictor. This system has the advantages of flexibility and pulseless control but requires numerous components. Carrier control with motor-driven pumps is generally limited to flow control. Two methods are in use. First, the motor speed or pump stroke is manually set to the desired flow. This has the advantage of simplicity, but the control is often erratic. Second, the flow is measured and a feedback signal returned to the pump. This improves flow control but is more complex and expensive.

Valves

Liquid chromatographic valves operate on the same principles as other chromatographic valves. Air-operated diaphragm valves, plug valves and rotary valves are used (see Section 10.4). These valves are strengthened and use higher seal forces for the high-operating pressures. They also have smaller internal flow passages to reduce sample-carrier mixing in the valve.

Columns

Liquid columns are packed in a ¼- or ⅜-in. (6 or 9.6 mm) OD straight stainless steel tube 2 to 12 in. (50 to 305 mm) long. The total column length in an instrument rarely exceeds 4 ft. (1.2 m) and is more often 1 ft. (0.3 m) or less. The columns are filled with a particulate packing that is typically in the 5-to-10-μ range. There are four types of liquid columns classified by principle of separation: liquid-liquid columns, liquid-solid columns, size exclusion columns and ion exchange columns. Table 10.11b lists a few typical column applications.

Table 10.11b
TYPICAL COLUMN APPLICATIONS

Column Type	*Applications*
Liquid-liquid, normal phase	Phenols Esters Pigments Cooking oils
Liquid-liquid, reverse phase	Pesticides Herbicides Organic acids Water pollutants
Liquid-solid	Plasticcizers Antioxidants Polycyclic aromatics Organic peroxides
Size exclusion	Polymers Resins Carbohydrates Hydrocarbons
Ion exchange	Inorganic ions Dies Detergents Sugars

Liquid-liquid columns (liquid-partition columns) are filled with solid support, usually silica, coated with a liquid stationary phase. To prevent loss of the liquid, it is usually chemically bonded to the support. The separation depends on the relative solubility of the sample components in the stationary and mobile, or carrier, phases. Liquid-liquid separations are subdivided into normal and reverse phase separations. In normal phase separation, the stationary phase is more polar (in the chromatographic sense, not dipole moment) than the mobile phase. While in reverse phase separation, the stationary phase is less polar than the mobile phase. When sample components can be separated with either type, the components elute from the column in reverse order when changing from the normal to the reverse phase. To improve separation, the carrier can be changed during the analysis by changing the ratio of a binary mixture of solvents. This is called solvent programming or gradient elution and is comparable to temperature programming in gas chromatography. Its use is limited almost exclusively to laboratory analysis.

Liquid-solid columns (liquid-adsorption columns) are filled with an adsorbant. Silica is the most common but alumina and even charcoal are occasionally used. Sample components are retained by the adsorption sites on the packing. The mobile phase competes with the sample for adsorption sites and displaces the sample so it can move through the column. As with liquid-liquid columns, the mobile phase is chosen or blended to provide the desired separation of sample components. Solvent programming can also be used.

Size exclusion columns (steric exclusion columns or gel permeation columns) are filled with a porous solid, such as silica, or a porous polymer, such as cross-linked polystyrene. The pore size in the column varies uniformly over a specific range, for example, 300 to 600 Å. While moving through the column in the mobile phase, a sample component with a chain length (or more correctly hydrodynamic volume) of 400 Å. can diffuse into any of the pores that are 400 Å. and larger. On the other hand, a sample component with a chain length of 500 Å. can diffuse into only those pores 500 Å. and larger. Since it enters fewer of the stationary pores, the 500-Å. component moves through the column faster than the 400-Å component. Thus, long-chain or high-molecular-weight compounds elute from the column first. The mobile phase has no direct effect on the separation process in these columns.

Ion exchange columns are filled with a cross-linked polystyrene resin containing charge-bearing functional groups on its surface. The stationary phase is called anion exchange if the functional groups are positively charged, and cation exchange if negatively charged. For the sample ions to be retained, they must be of opposite charge to the functional groups. The mobile phase is generally water containing a fixed quantity of ions of the same charge as the sample and buffered to a specific pH. The sample ions are separated, depending on how strongly they interact with the functional groups on the stationary phase in competition with the ions in the mobile phase. The quantity, the type of ion and the pH in the mobile phase control the separation.

Detectors

All detectors used in the liquid chromatograph are simplified, small internal-volume versions of other analyzers (see Sections 10.15 and 10.19). The most common are optical absorbance; refractive index and dielectric constant.

The optical absorbance detector measures the absorption of a fixed wavelength in the ultraviolet or visible spectrum. It is highly sensitive to many absorbing sample compounds and relatively insensitive to external effects, such as temperature, flow and carrier composition. It can only measure absorbing compounds in a nonabsorbing carrier.

The refractive index detector measures the difference between the refractive index of the sample compounds and the carrier. With the proper choice of carrier it is sensitive to all sample compounds, but the sensitivity is generally lower than that of the optical absorbance detector. It is also quite sensitive to temperature and carrier composition variations.

The dielectric constant detector measures the difference between the dielectric constant of the sample compounds and the carrier. Because for compounds with no dipole moment, refractive index and dielectric constant are related, the advantages and disadvantages of this detector and the refractive index detector are similar. However, if, for example, the sample compounds have a dipole moment and the carrier does not, the dielectric constant detector has higher sensitivity and more uniform response than the refractive index detector.

Application

The liquid chromatograph extends the advantages of the chromatograph to nonvolatile and thermally unstable samples, as well as polymers and inorganic salts. The prime advantages are the ability to both qualitatively and quantatively analyze multicomponent streams with the versatility to analyze a wide range of samples. When other instruments provide a satisfactory analysis, the liquid chromatograph is rarely used because it is more complex, more expensive, sometimes less sensitive and requires a longer analysis time. There are, however, many sample streams that can be analyzed satisfactorily only with the liquid chromatograph.

REFERENCES

1. Snyder, L.R. and Kirkland, J.J., *Introduction to Modern Liquid Chromatography*. New York, John Wiley & Sons, Inc., 1974.
2. Simpson, C.F., ed., *Practical High Performance Liquid Chromatography*. London, Heyden & Son, Ltd., 1976.
3. Hamilton, R.J. and Sewell, P.A., *Introduction to High Performance Liquid Chromatography*. London, Halsted, 1977.

Flow Sheet Symbol

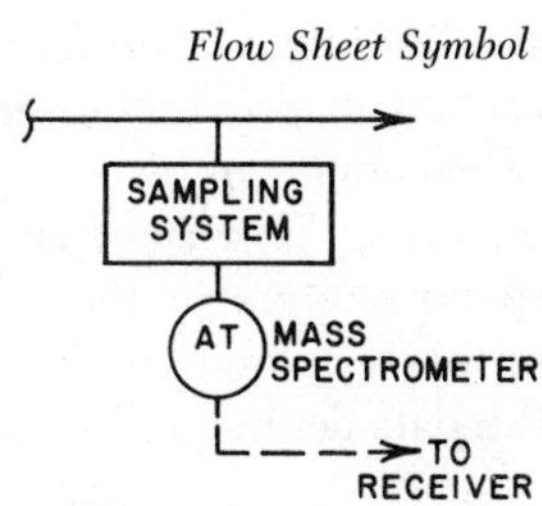

10.12 MASS SPECTROMETERS

Type of Sample:	Vapor
Standard Design Pressure:	Atmospheric
Sample Temperature:	122 to 424°F (50 to 200°C); sufficient to provide vapor sample with no condensation
Ambient Temperature:	68 to 77°F (20 to 25°C)
Contacting Materials:	Inlet materials of construction designed to be compatible with sample
Auxilary Utilities Required:	Sometimes cooling water
Cost:	\$40,000 to \$150,000 depending on application complexity and accessories
Inaccuracy:	±0.5% for most applications
Repeatibility:	±0.2% for most applications
Cycle Time:	3 to 5 sec. per stream, depending upon applications
Special Features:	Multicomponent readout
Partial List of Suppliers:	Applied Research Laboratories; CVC Products, Inc.; Extranuclear Laboratories, Inc.; Perkin-Elmer Corp., Aerospace Division; Varian Associates, Industrial Instruments Group; Veeco Instruments, Inc.

The mass spectrometer is becoming widely used as an analytical tool for the process industries. It provides multicomponent analysis with the accuracy, reliability and cycle time necessary for closed loop computer control of a process. For single-stream applications, it can function as a continuous multicomponent analyzer.

Principle of Operation

All mass spectrometers have the common property of being able to separate ions of a particular atomic or molecular species by their mass-to-charge ratios. Therefore, in its simplest form, a mass spectrometer is a vacuum device in which ions are created, separated and detected. As a prerequisite, this also implies that collisions between the ions and molecules be minimized. This is attained by making the mean free path very large in comparison to the internal dimensions of the device, thereby dictating pressures of the order of 10^{-4} to 10^{-6} mmHg (torr). Consequently, a mass spectrometer must have a vacuum envelope, an inlet leak and an evacuation pump.

A very small gas sample is introduced through the inlet leak into the ion source. Ions are produced by collision of rapidly moving electrons with molecules of the gas to be analyzed. The electrons ionize the neutral gas molecules by knocking out one or more of the orbital electrons. The resulting ions are pulled out of the gas stream as fast as they are formed by means of electrostatic fields created by the ion source electrodes and are accelerated and focused into an ion beam. The beam is directed into the resolving portion of the analyzer. The remaining nonionized molecues travel through the analyzer as a result of the internal pressure differentials until they encounter the evacuation pump.

Process mass spectrometers offer performance, versatility and flexibility, which in many cases exceed that of gas chromatography. While the instruments may be more expensive, they offer shorter cycle time, which may be necessary for closed loop computer control of a process. For example, typical instruments can sample multiple streams for concentrations of 8 to 12 components in less than 5 seconds. In rapid analysis, sample conditioning requirements become an important consideration in preventing delays. Because of the analytical speeds that can be obtained, the data acquisition system is also an important factor. The increased use of mass spectrometers in the process industry is the result of properly combining the sample stream, the analyzer and the data acquisition system into a reliable working unit.

Sample Ionization

The sample system and sample conditioning are similar to those used for gas chromatographs (see Section 10.4). Vapor samples are drawn by suction into the instrument through an inlet leak. There are several types of inlet leaks. A sintered metal leak is commonly used because it allows a uniform molecular flow into the analyzer based on the differential pressure across the leak. Thin diaphragms made from Teflon may be used for corrosive service, and ultrafine metering valves can be used. Also available are automatic metering valves that maintain a sample flow based on the pressure in the ionization chamber. While a variety of inlet leaks are available, one must be chosen that is compatible with the sample and analyzer and gives stable flow.

Several methods are available for producing ionic species from the sample gas. Electron bombardment is the most popular but uses filaments that require periodic replacement. Other methods, such as spark generation, optical and chemical-ionization methods are also used. Over the last few years atmospheric ionization mass spectrometry has been shown to provide unusually high sensitivity to a wide range of compounds and promises to open up a wide range of new applications for mass spectrometry. Because most instruments currently used in process monitoring use heated filaments, filament life and ease of replacement are of primary concern. Most process mass spectrometers have dual filament assemblies that allow switching from one to the other without any downtime. Then at some convenient time, the instrument can be shut down and the filaments replaced. Figure 10.12a shows the operation of a heated filament ion source.

Types of Analyzers

There are three types of process mass spectrometers; magnetic sector, quadrupole filter and time-of-flight. The type of instrument used defines the method used to differentiate between ions produced from the sample gas. Discriminations among the same ions from different compounds is possible after special signal processing. In the process industry, the requirements of a dedicated, online analytical instrument might be stated as follows: the instrument is expected to perform a specific analytical task, on a continuous basis, for extended periods. It must exhibit long-term stability and accuracy while operating over its environmental range. The design of the instrument must stress simplicity, automatic operation, reliability and require a minimum of operation attention. Further, it must be serviced and maintained by personnel with moderate technical ability. The key consideration is the performance of a specific analytical task.

Magnetic Sector Instruments

There are two types of magnetic sector instruments: fixed magnetic sector and electromagnetic-focusing. The fixed magnetic sector instrument utilizes a permanent magnet to produce a magnetic field at right angles to the direction of motion of the ion beam. This causes the ions to bend in a circular trajectory proportional to their mass-to-charge (M/E) ratios. This concept is shown in Figure 10.12b. Using this approach with multiple collectors obviates the need for scanning, which vastly simplifies the electronic system and does not require an external programmer to select the measurements. For each application the collectors are located at a predetermined point in the instrument focal plane. For example, if an instrument was required to measure the composition of the atmosphere, the collectors would be placed along the focal plane to intercept the ion beams for nitrogen at M/E 28, oxygen at M/E 32 and argon at M/E 40. With these three measurements, the output of the instrument could continuously monitor the atmosphere, without scanning or data multiplexing.

In 1969, Perkin-Elmer was granted U.S. Patent No. 3,648,047 for a "closed loop control" concept that effectively removes all the major measurement errors in a multiple collector mass spectrometer. Figure 10.12c details the configuration used to compensate for errors that originate in the ion source or changes in conductance of the inlet leak. In the simple atmosphere example, the three outputs would be scaled with summing resistors to create equal sensitivities for each measurement and electronically added together to create an analog of the sampled atmosphere. This sum is then compared to a fixed reference, and if properly calibrated, no error is created by this comparison. If a hypothetical drift is introduced, an error is created and fed back to each channel through the gain adjust elements. The gain gives an equal percentage change in each channel and drives the error to zero. The effectiveness of this technique for stabilizing calibration over extended periods has been proven in process applications.

The electromagnetic-focusing instrument utilizes either changes in the accelerating voltage or magnetic field to focus the desired ionic species onto a single collector. When accelerating voltage is used, the ions are focused and accelerated as a beam into a magnetic field at right angles to the direction of motion, and the acceleration voltage is varied to select the ions of a particular type. The ion beam entering the magnetic field contains ions of essentially equal energy. The momentum of each ion depends upon its mass. Because the radius of curvature of the ions in the magnetic field is different

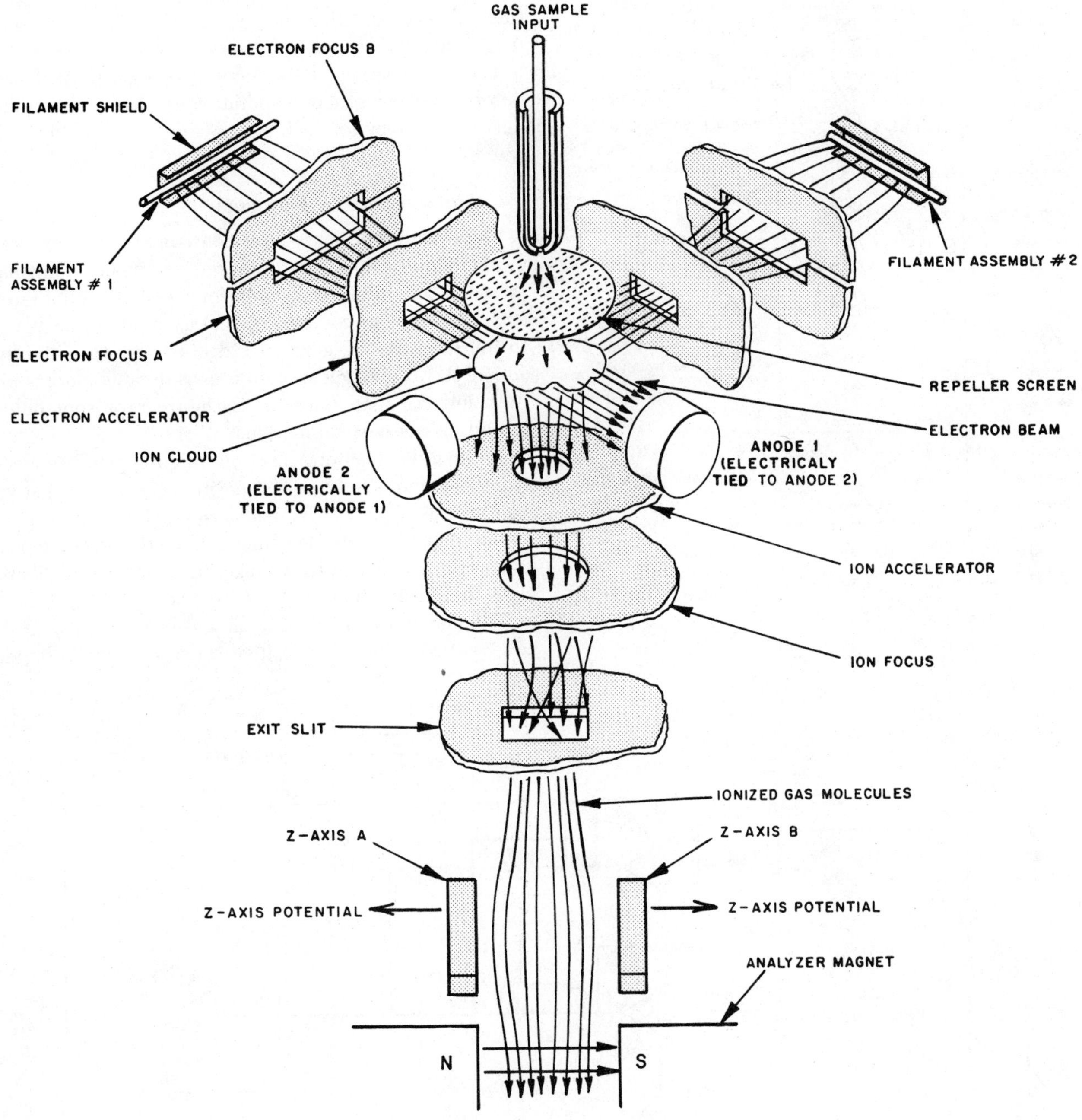

Fig. 10.12a Ion source operation

for ions of different momentum, only ions corresponding to one particular mass are focused on the collector. This signal is amplified and displayed as a signal proportional to the partial pressure of the gas type of that mass. For any particular mass selected, the product of mass and acceleration voltage is a constant.

An alternate to selecting masses by changing the accelerating voltage is to vary the magnetic field. The principal advantage of electromagnetic scanning is the wider mass range achieved. Although the mass is not directly proportional to the magnetic field, circuits have been designed to provide linear spacing of masses. When changing the magnetic field, the accelerating voltage must be held constant at an accelerating voltage corresponding to the highest mass of interest. When the masses are selected electromagnetically, the relative heights of different mass peaks do not occur in the same proportion as they do when the accelerating voltage is changed. In the latter case, the low-mass peaks are enhanced in both magnitude and resolution because of the more favorable accelerating voltage at which they appear. When the magnetic field is varied at an acceleration

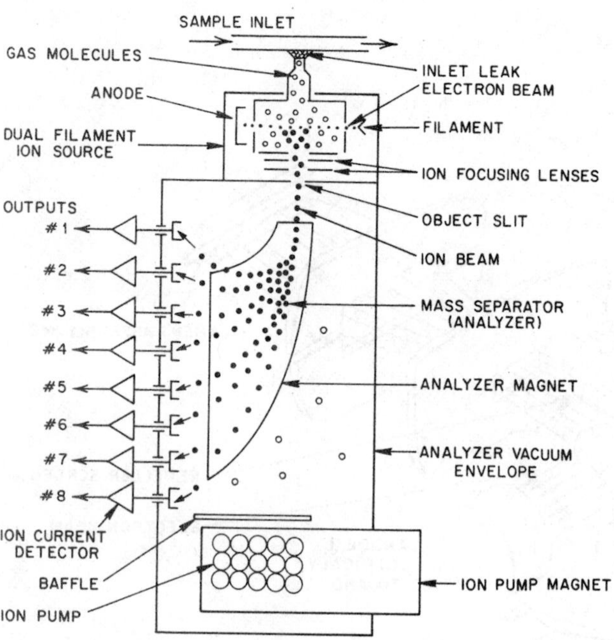

Fig. 10.12b Mass spectrometer operation

voltage corresponding to the highest mass of interest, this enhancement does not occur. Therefore, both the magnitude and resolution of mass peaks will be less.

The composition of the sample being analyzed will determine the method of operation needed to obtain the desired accuracy for each of the components to be analyzed.

Quadrupole Instruments

The quadrupole mass filter consists of four high-precision cylindrical rods located precisely in an orthogonal array. As an ion moves at uniform speed along the filter axis, it undergoes complex oscillatory motion transverse to the filter axis. Figure 10.12d illustrates the filtering action. Actual particle motion is considerably more complex and the path lines in the figure represent outer boundaries of particle motion.

In the filter, diametrically opposing rod pairs are electronically connected to form a positive pair and a negative pair of rods. The positive pair has an RF voltage superimposed on a positive dc voltage, while the negative pair has a negative dc and RF voltage that is 180° out of phase with the positive pair.

For a given set of voltages, all ions below a given mass are neutralized on the positive set of rods while larger

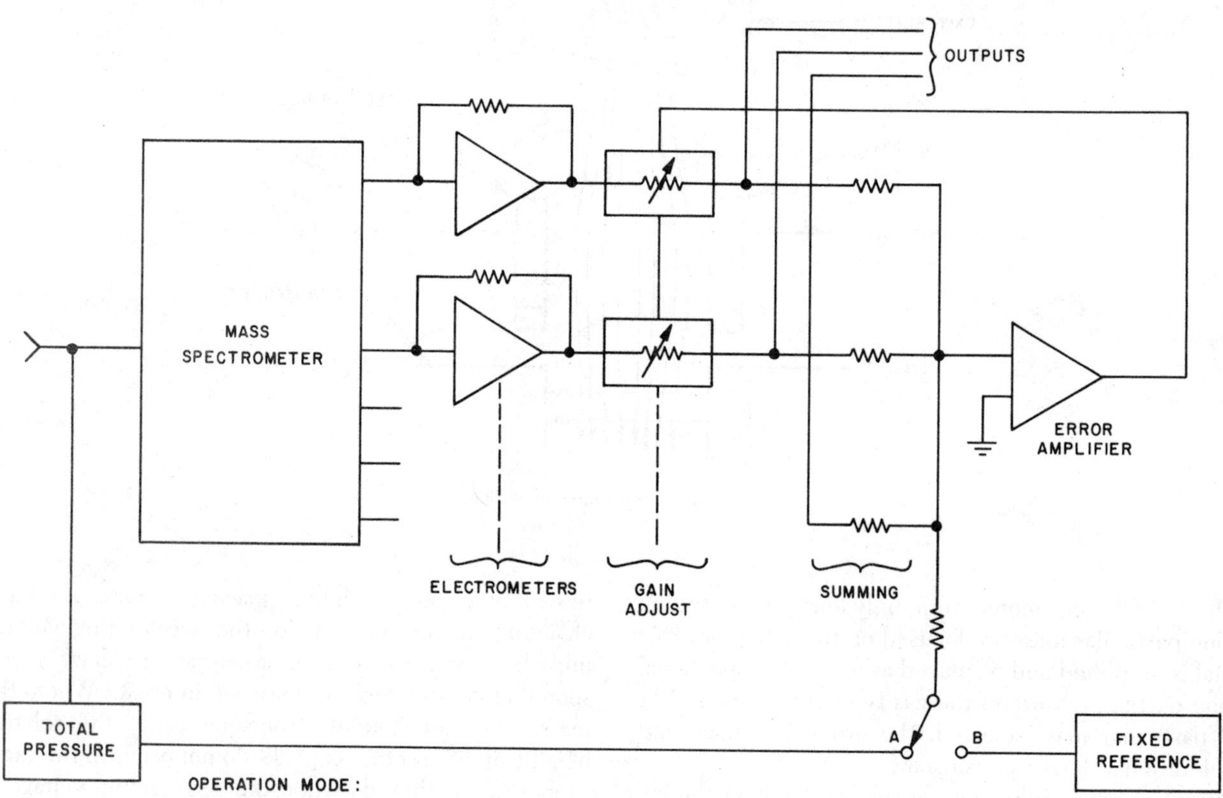

Fig. 10.12c Span calibration technique

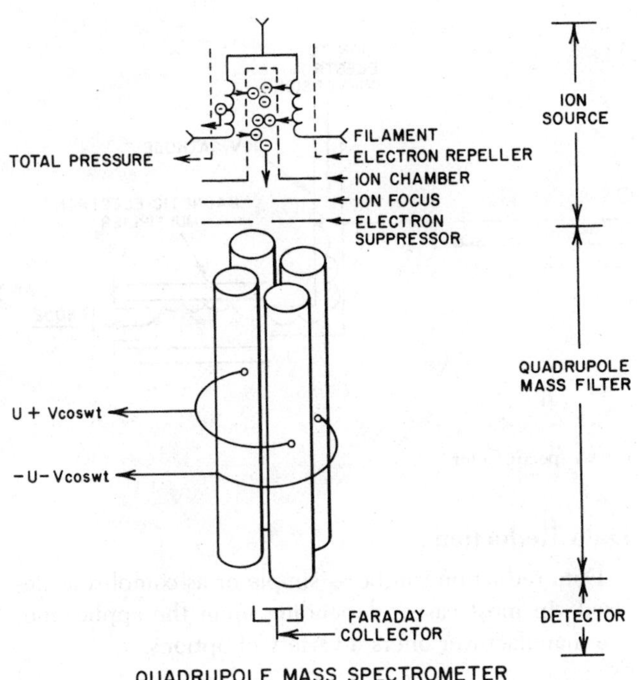

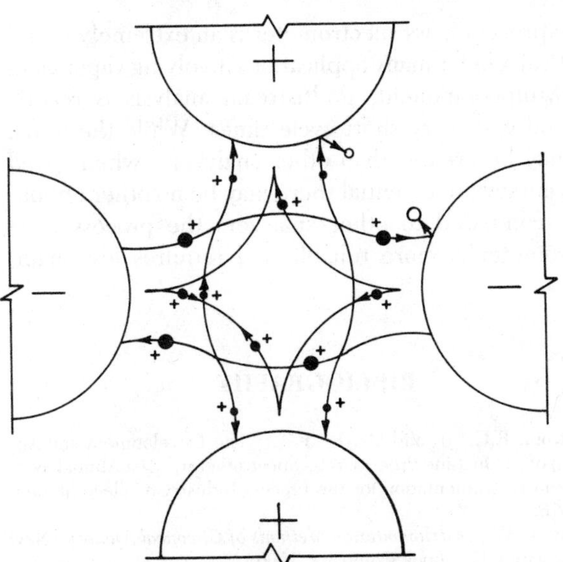

Fig. 10.12d (Top) Quadrupole mass spectrometer; (Bottom) Quadrupole mass filter

ions are passed. The negative set of rods passes low-mass ions and neutralizes ions above a given mass. By adjusting the RF to dc ratio, the passbands can be made to overlap so that only ions of a very specific mass can get through while all others are neutralized by one set of rods or the other. Under these high-resolution conditions, the passed ions barely miss the rod pairs at the extremes of their transverse oscillatory motion. A unique property of the quadrupole mass filter is that the mass which is passed is directly proportional to the voltages applied to the rods. Thus, by changing the voltages, the masses of interest are allowed to pass through to the collector where they are measured.

Time-of-Flight Instruments

In time-of-flight mass spectrometers, the ionized sample is subjected to a negative-polarity-accelerating field and allowed to pass through a drift region. The lighter ions travel faster than the heavier ions, producing mass separation by time-of-flight. It can be shown mathematically that the time-of-flight, T, for an ion of mass, M, and charge, E, is:

$$T = K\,(M/E)^{1/2}$$

where K is a constant depending upon physical dimensions.

To accomplish this type of mass separation, the ionized sample is introduced in discrete pulses of from 20,000 to 35,000 pulses per second. On each cycle the ionized molecules enter the flight tube (drift region) where the separation of different mass numbers is accomplished. The ions are then converted to electrons and amplified using a magnetic electron multiplier. The amplified signal is detected by the anode and fed to a suitable data acquisition system. Figure 10.12e illustrates this concept.

High-Vacuum Systems

Vacuum sensors associated with the high-vacuum systems were discussed in Section 5.14.

To maintain a pressure necessary for proper operation of the analyzer, one of three high-vacuum systems is employed, either a diffusion pump, an ion pump or a turbomolecular pump. In many instances the manufacturer has already included a specific pump in the process mass spectrometer package. However, because this is a vital part of the analyzer, the choice of a high-vacuum system should be given serious consideration.

For most process applications, the ion pump is most desirable. It is essentially a passive device that pumps by a combination of chemical gethering and physical burial. Figure 10.12f shows a typical pump. An electric field is established between the anode structure and the two cathodes and a magnetic field is created by a permanent magnet. This creates a trapped electron cloud that continuously ionizes the neutral gas in the pump. The ions are accelerated out of the anode and impinge upon the cathode surface where they sputter cathode material. The tantalum and titanium cathode materials coat the pump structure. These two materials are very active getters and chemically combine with most of the ions present to remove them from the vacuum system. Inerts are removed by actual burial under the sputtered material or by direct implantation.

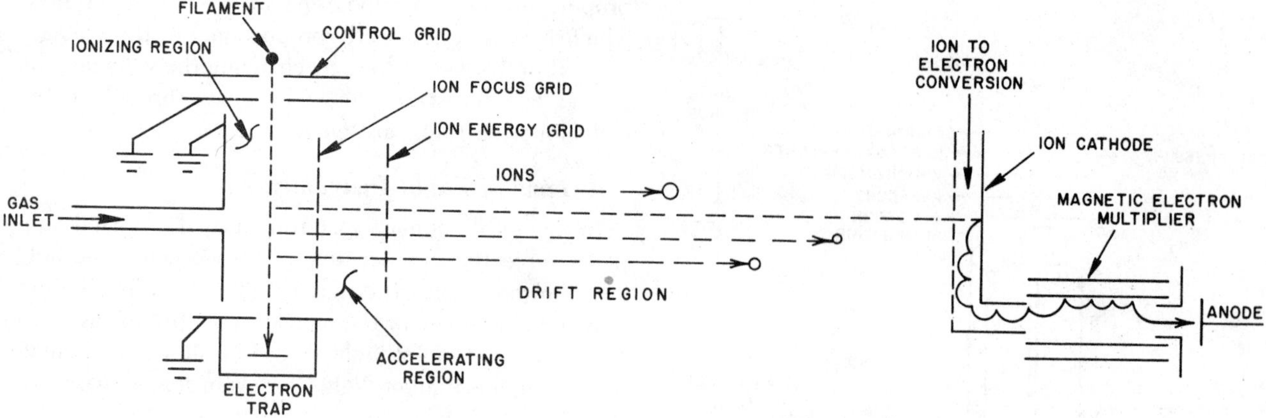

Fig. 10.12e Time-of-flight mass spectrometer

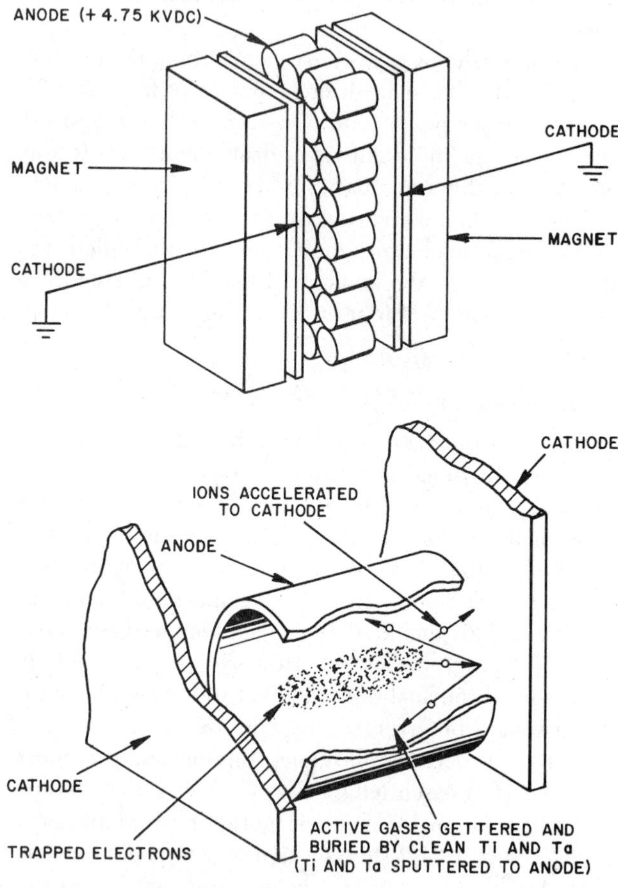

Fig. 10.12f Ion pump operation

Data Reduction

Data reduction can be as simple or as complex as desired. In most cases, depending upon the application, the manufacturer offers a variety of options.

Summary

The process mass spectrometer is an extremely useful analytical tool for many applications involving vapor analysis. Multicomponent, multistream analysis is readily achieved with very short cycle times. While the initial cost may be greater than other analyzers, when speed and accuracy are essential there may be no other option. When compared to other analyzers the process mass spectrometer is more reliable and requires less maintenance.

BIBLIOGRAPHY

Ahlstrom, R.C., Jr. and Shaver, F.A., "The Development and Application of an On-Line Process Mass Spectrometer," 31st Annual Symposium on Instrumentation for the Process Industry (College Station, TX), 1976.

Ewing, G.W., *Instrumentation Methods of Chemical Analysis*. New York, McGraw-Hill Book Company, 1960.

Gross, M.L., ed., *High Performance Mass Spectrometry: Chemical Applications*. Washington, D.C., American Chemical Society, 1978.

Robertson, A.J.B., *Mass Spectrometry*. New York, John Wiley & Sons, Inc., 1954.

10.13 OXIDATION-REDUCTION POTENTIAL (ORP) DETECTORS

Standard Design Pressures:	Vacuum to 150 PSIG (1MPa)
Standard Design Temperature:	Generally 23 to 212°F (−5 to 100°C); consult manufacturers for higher temperature applications
Materials of Construction:	Electrode hardware: stainless steel, Monel, Hastelloy, titanium, PVC, CPVC, polyethylene, polypropylene, epoxy, polyphenylene sulfide, Teflon, various elastomer materials
Inaccuracy:	System inaccuracy difficult to define; typically, amplifiers stable to 2 mv. units or better
Range:	Variable; typically any span of 200, 500 or 1,000 mv. between ±2,000 mv.
Costs:	Highly variable, depending on features ($1,000 to $3,000 per transmitter loop)
Partial List of Suppliers:	Beckman Instruments, Inc.; Electrofact, Div. of Control Data Corp.; Foxboro Co.; Great Lakes Instruments, Inc.; Honeywell, Inc.; Leeds & Northrup, a unit of General Signal; Milton Roy Co.; Orion Research Inc.; Photovolt Corp.; Rexnord Inc.; TBI Inc.; Uniloc Div. of Rosemount, Inc.

Introduction

The measurement, and control, of oxidation-reduction potential (ORP) is applied in an increasing number of applications in both processing and wastewater treatment. The following is a partial list of the more important applications:

1. In ore-leaching where the metal sought is leached from the ore and converted to the desired oxidation state for further processing.
2. When toxic cyanides are oxidized to harmless reaction products in an oxidation-reduction reaction as part of the process to remove toxic heavy metals.
3. In the pulp and paper industry for bleaching of pulp with a variety of oxidants under ORP measurement and control.
4. When hexavalent chromium is reduced to the trivalent oxidation state as part of the process for removal of toxic chromium from metal finishing wastewaters and from cooling tower blowdown waters.
5. In the manufacture of chlorine bleaches.
6. In sanitary wastewater treatment to measure and control the addition of an oxidant for odor control.

Principles of ORP Measurement

Oxidation-reduction reactions are reactions in which the substances involved gain or lose electrons and show different electron configurations before and after the reaction. Oxidation is the overall process by which a specie in a chemical reaction loses one or more electrons and increases its state of oxidation. An oxidant is a substance capable of oxidizing a chemical specie; it acquires the electron or electrons lost by the specie and is itself reduced in the overall process. Reduction is the overall process in which a specie in a chemical reaction gains one or more electrons and decreases its state of oxidation. A reductant is a substance capable of reducing a chemical specie; it loses the electrons gained by the specie and is itself oxidized in the overall process.

An ORP reaction, then, involves an electron exchange

that is capable of doing work. This capability is expressed in terms of potential for a half-cell, or electron, reaction. The potentials listed in Table 10.13a are for standard conditions, that is, where reactants and products are at unit activity. Voltages in this table are referenced to the standard hydrogen electrode (SHE) which is assigned the value of 0.000 v.

Table 10.13a
REDUCTION POTENTIALS OF SOLUTION IN ORP MEASUREMENT

Reaction	*E°, Volts*
$O_3 + 2H_3O^+ + 2e = O_2 + 3H_2O$	+2.070
$Cr_2O_7^{2-} + 14H_3O^+ + 6e = 2Cr^{3+} + 21\ H_2O$	+1.330
$C1O^- + H_2O + 2e = cl^- + 2OH^-$	+0.890
$Fe^{3+} + le = Fe^{2+}$	+0.770
Ag/AgCl electrode 4 M. KCl	+0.199
$2\ H_3O^+ + 2e = H_2 + 2H_2O$	0.000
$Zn^{2+} + 2e^- = Zn$	−0.763
$CNO^- + H_2O + 2e = CN^- + 2OH^-$	−0.970
$Na^+ + le = Na$	−2.711

Note that the Table 10.13a reactions are written as reductions, which is now the almost universally used convention. In this section, the term ox/red is used to indicate the oxidized form on the left side of the equation, the reduced form on the right. For example, the standard potential for ferric iron, Fe^{3+}, being reduced to ferrous, Fe^{2+}, is written $E°Ox/Red = +0.770$ v.

It will be noted that $E_{Red/Ox} = -E_{Ox/Red}$, which simply means that polarity is reversed when reaction is written as an oxidation reaction. For example, $Fe^{2+} = Fe^{3+} + e^-$. $E°_{Red/Ox} = -0.770$ v.

It is customary in dealing with oxidation-reduction reactions to write the two half-cell reactions that make up the overall reaction. These are written so that known reactants are on the left and known products on the right. The following is a general equation in which the hydronium ion participates in the reaction, and, therefore, potentials are pH-dependent.

Half-reaction:

$$aOx_2 + yH_3O^+ + n_2e \rightleftharpoons bRed_2 + wH_2O \quad E_{Ox_2/Red_2}$$

Half-reaction:

$$cRed_1 \rightleftharpoons dOx_1 + n_1e \quad E_{Red_1/Ox_1}$$

Overall reaction:

$$n_1aOx_2 + n_2cRed_1 + n_1yH_3O^+ \rightleftharpoons n_1bRed_2 + n_2dOx_1 + n_1wH_2O$$

$$E_{overall} = E_{Ox_2/Red_2} + E_{Red_1/Ox_1} \qquad 10.13(1)$$

At any point in the reaction the solution or cell potential is given by:

$$E_{cell} = E_{Ox_2/Red_2} = E_{Ox_1/Red_1} \qquad 10.13(2)$$

which, at 77°F (25°C):

$$E_{Ox_2/Red_2} = E°_{Ox_2/Red_2} + \frac{0.059 \log}{n_2}\frac{[Ox_2]^a[H_3O^+]^y}{[Red_2]^b[H_2O]^w} \qquad 10.13(3)$$

and

$$E_{Ox_1/Red_1} = E°_{Ox_1/Red_1} + \frac{0.059}{n_1}\log\frac{[Ox_1]^d}{[Red_1]^c} \qquad 10.13(4)$$

Setting Equations 10.13(3) and 10.13(4) to equal each other and rearranging we get:

$$(n_1 + n_2)E_{cell} = n_2E°\ Ox_2/Red_2 + n_1E°_{Ox_1/Red_1} + 0.059 \log \frac{[Ox_2]^a[Ox_1]^d}{[Red_2]^b[Red_1]^c} + 0.059 \log H_3O^{+y} \qquad 10.13(5)$$

Two other relationships hold:

at any point in the reaction after the start:

$$Red_2 = \frac{n_1b}{n_2d}[Ox_1] \qquad 10.13(6)$$

and at the equivalence point:

$$Ox_2 = \frac{n_1a}{n_2c}[Red_1] \qquad 10.13(7)$$

Substitution of the ratio relationship Ox_2^a/Red_2^b or Ox_1^d/Red_1^c, as derived from Equations 10.13(6) and 10.13(7) into Equation 10.13(5), will provide the cell potential for the equivalence point. It can be an interesting exercise to see how this equivalence point potential changes when differing values of H_3O^+, that is, pH, are substituted into Equation 10.13(5). The profound effects that pH can have on equivalence and control point potentials will become readily apparent. When hydronium, H_3O^+, or hydroxyl, OH^-, ions participate in ORP reactions, close control of pH may become equally as important as close control of ORP.

Note again Equations 10.13(2), 10.13(3) and 10.13(4) which state that

$$E°_{Ox_2/Red_2} + \frac{0.059}{n_2}\log\frac{[Ox_2]^a[H_3O]^y}{[Red_2]^b[H_2O]^w} = E°_{Ox_2/Red_2} - E°_{Ox_1/Red_1} + \underline{.059} \log\frac{[Ox_1]^d}{[Red_1]^c}$$

and

$$\frac{0.059}{n_1n_2}\log\frac{[Red_2]^{n_1b}[Ox_1]^{n_2d}[H_2O]^{n_1w}}{[Ox_2]^{n_1a}[Red_1]^{n_2c}[H_3O^+]^{n_1y}} = E°_{Ox_2/Red_2} - E°_{Ox_1/Red_1}$$

The equilibrium constant is given by:

$$\frac{[Red_2]^{n_1b}[Ox_1]^{n_2d}}{[Ox_2]^{n_1a}[Red_1]^{n_2c}} = 10^{\left(\frac{E^\circ_{Ox_2/Red_2} - E^\circ_{Ox_1/Red_1}}{.059/n_1n_2}\right)} \times [H_3O^+]^{n_1y} \qquad 10.13(8)$$

Again, using the relationships of Equations 10.13(6) and 10.13(7) substituted into Equation 10.13(8), it is possible to determine the degree of completion for the reaction of the specie of interest at the equivalence point for various pH levels.

As an example, consider a common industrial process in which hexavalent chromium is reduced with ferrous sulfate solution. The half-reactions are:

$$Cr_2O_7^{2-} + 14\,H_3O^+ + 6e^- = 2\,Cr^{3+} + 21\,H_2O$$
$$E^\circ_{Ox_2/Red_2} = 1.330$$

$$Fe^{2+} = Fe^{3+} + 1\,e^-$$

$$Cr_2O_7^{2-} + 6\,Fe^{2+} + 14\,H_3O^+ = 2Cr^{3+} + 6\,Fe^{3+} + 21\,H_2O$$
$$E_{Red_1/Ox_1} = -0.770$$

For this reaction we have the following:

$Ox_1 = Fe^{3+}$ $Red_1 = Fe^{2+}$

$n_1 = 1 \quad a = 1 \quad b = 2 \quad y = 14$

$Ox_2 = Cr_2O_7^{2-}$ $Red_2 = Cr^{3+}$

$n_2 = 6 \quad c = 1 \quad d = 1 \quad w = 21$

Assume pH = 0, that is, $[H_3O^+] = 1$
and at equivalence point $[Cr^{3+}] = 10^{-3}$.

The equivalence point is calculated from Equation 10.13(5):

$$(n_1 + n_2)\,E_{cell} = n_2\,E^\circ_{Ox_2/Red_2} + n_1E^\circ_{Ox_1/Red_1} + 0.059 \log \frac{[Ox_2]^a\,[Ox_1]^d}{[Red_2]^b\,[Red_1]^c} + 0.059 \log\,[H_3O]^y.$$

Using the ratio relationships determined from Equations 10.13(6) and 10.13(7), we get

$$7\,E_{cell} = 6(1.330) + 0.770 + .059 \log \frac{1}{2[Cr^{3+}]} + 0.059 \log\,[H_3O^+]^{14}$$

and for varying levels of H_3O^+, we get the following equivalence point potentials:

for $H_3O^+ = 1$ $\quad E_{cell} = 1.273$ v.,
for $H_3O^+ = .01$ $\quad E_{cell} = 1.037$ v.,
for $H_3O^+ = 10^{-4}$ $\quad E_{cell} = 0.793$ v.

Using Equation 10.13(8) along with Equations 10.13(6) and 10.13(7), we can determine the degree of completion at the equivalence point. This equation reduces to:

$$\frac{[Cr^{3+}]^8}{[CR_2O_7]^7} = \frac{3^6}{6^6}\,10^{56.9}\,[H_3O^+]^{14}$$

$$[Cr_2O_7] = (10^{-24} \times 6.4 \times 10^{-58.9} \times [H_3O^+]^{14})^{1/7}$$

and the following values are calculated:

pH = 0, $[H_3O^+] = 1$ $\quad [Cr_2O_7^=] = 4.9 \times 10^{-12}$,
pH = 2, $[H_3O^+] = 0.01$ $\quad [Cr_2O_7^=] = 4.3 \times 10^{-8}$,
pH = 4, $[H_3O^+] = 10^{-4}$ $\quad [Cr_2O_7^=] = 4.9 \times 10^{-4}$.

It becomes quite obvious that the reaction is not complete at pH 4.0, but is complete at pH 2.0, assuming that a concentration of 10^{-6} M. represents completion.

In industrial and laboratory work, ORP cell potentials are measured, not against a SHE but against an Ag/AgCl, 4 M. KCl reference, $E^\circ_{Ox/red} = +0.199$ v., or a saturated calomel electrode (SCE) whose $E^\circ = 0.244$ v. To convert to potential measured, which we will designate as $E_{AgCl\ ref}$, we use

$$E_{meas} = E_{cell} - E_{AgCl} \qquad 10.13(9)$$

For example, the E° for the cell

$$Fe^{3+} + e^- = Fe^{2+} \quad E_{Ox/Red} = 0.770 \text{ vs SHE}$$

and the potential measured vs the silver-silver chloride reference is

$$E_{meas/AgCl\ ref} = +.770\text{ v} - .199\text{ v} = 0.571\text{ v}$$

Absolute potentials in ORP measurement are of relatively little significance. Most equipment manufacturers use slightly modified pH analyzers for ORP measurement. These instruments normally have the standardized adjustment of the parent pH meter. Furthermore, in general, reverse of polarity is achieved merely by reversing inputs. Broad range instruments, that is, those with sufficient range to measure all conceivable redox potential ranges are of little value in process ORP measurements. A relatively narrow range instrument, such as one capable of measuring 500 or 100 mv., is better suited because almost all ORP measurements are used for control purposes. Most chemical reaction systems involving electron exchange are controlled near the equivalence point with a controlled excess of added reagent to ensure driving the reaction to completion. Thus, most ORP reactions will be controlled on the steep portion of the titration curve.

Equipment for ORP Measurement

The basic instrumentation for ORP measurement closely parallels that for pH measurement. In fact, many instrument suppliers use slightly modified pH analyzers, the main modification being a changed sensitivity and a mv. scale in place of a pH scale. The electrode hardware, that is, the equipment used to install the electrodes in the process stream, is generally the same as that used in pH systems (Figures 10.13b & c).

There are two major differences between an ORP system and a pH system. One of these differences is the

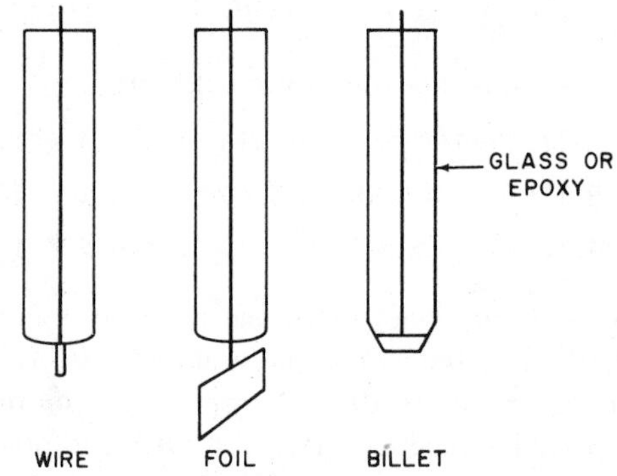

Fig. 10.13b Type of metallic ORP electrodes

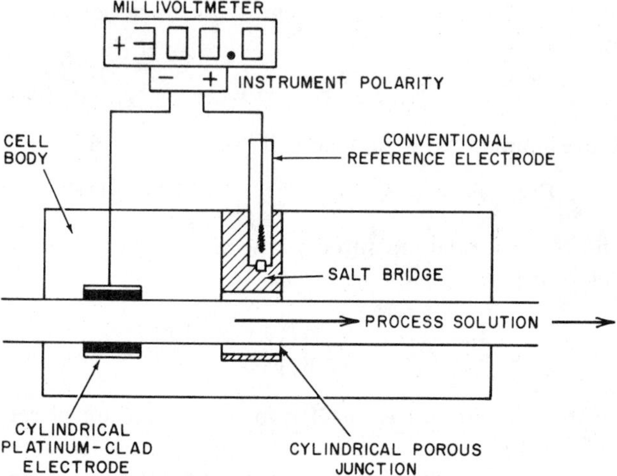

Fig. 10.13c Cylindrical ORP electrode cell

sensing electrode, which is normally a noble metal, typically platinum or gold, although other metals and carbon have been used on occasion. The second major difference is in temperature compensation. Process pH systems are typically temperature compensated, whereas ORP systems are almost never temperature compensated.

The basic thermodynamics apply to pH and ORP as expressed by the classical Nernst equation. For oxidation-reduction half-cell reactions, this may be represented as follows:

$$E_{cell} = E^{\circ}_{Ox/Red} + \frac{2.303\ RT}{nF} \log \left[\frac{Ox}{Red}\right] \qquad 10.13(10)$$

where E° represents potential under standard conditions of unit activity referred to the SHE, R is the gas constant, 1.986 cal. per mol. degree; F is the Faraday; T is temperature in °K, and n is the number of electrons exchanged in the reaction. Note even in the very abbreviated listing of Table 10.13a how the n values change from reaction to reaction. This, plus the fact that a given ORP reaction may encompass side reactions, makes it quite clear why it is difficult, if not impossible, to temperature compensate an ORP reaction. In the Nernstian representation of pH, n always equals 1.

In Equation 10.13(10), the standard potential $E^{\circ}_{Ox/Red}$ is found in tables in handbooks and in the value relative to the standard hydrogen electrode. Therefore, E_{cell} for prevailing concentration is also relative to SHE.

E_{meas} is the value that is read on the meter. If E_{cell} is known from calculation or from actual measurement, its potential value for other systems can be readily converted by

$$E \text{ vs SHE} = E \text{ vs SCE} + E_{SCE/Red}$$

or

$$E \text{ vs SCE} = E \text{ vs SHE} - E_{SCE/Red}$$

or

$$E \text{ vs Ag/AgCl Ref} = E \text{ vs SHE} - E_{Ag/AgCl/Red}$$

Practical Application of ORP

It is certainly possible by use of the equations developed above to determine the equivalence potential for an ORP control reaction. For a reaction involving H_3O^+ or OH^- ions, equation 10.13(5) shows the rather profound effects of pH on the equivalence point potential. Equation 10.13(8) can be used to calculate the degree of completion of the reaction, and, again, this can show the effects of pH. A reaction that pointedly illustrates these effects is the reduction of hexavalent chromium with sulfur dioxide or bisulfite.

However, many prefer a more empirical approach. For example, the chromium reaction typically takes place at a pH of about 2.0 to 2.5. At this pH there is a smell of sulfur dioxide when the sulfite ion is in slight excess. An experienced operator might adjust his control point potential to attain a slight odor of sulfur dioxide and then make further adjustments based on laboratory analysis for hexavalent chromium. It is certainly possible to set up a system based on calculation when reactants and products are known. However, only the very innocent would proceed without analytical verification of results.

Care of an ORP System

Maintenance of an ORP measuring and control system is generally compared to that of a similar pH system. However, because of the lack of standards analogous to pH buffers, there is sometimes a tendency to short-cut maintenance. The single most pressing problem with ORP is the sensing noble metal electrode. It is subject not only to coating, but also to poisoning, both of which may result in sluggish or inaccurate measurement of potential. This in turn may result in an improper demand for reagent by the holding of a control point not representative of the desired reagent excess. Jones[1] reported on the use of standards using quinhydrone-saturated pH buffers to establish known potentials as a check on the condition and response of the electrode system. Recommended treatment for either a change of span or a shift in potential was cleaning with aqua regia.

ORP Control

Most applications of process ORP instruments involve control. Because the measurement of ORP is quite similar to that of pH, it follows that many of control considerations for pH hold also for ORP. If, for example, an oxidizing solution, such as dichromate in acid media, is titrated with a reducing solution of ferrous ion, a titration curve that is quite similar to that of an acid-base titration can be generated. However, a more common practice is the reduction of hexavalent chromium with gaseous sulfur dioxide. Whereas it would be quite simple to catch samples of an acid waste and to titrate these samples with a standard sodium hydroxide solution, doing a titration of hexavalent chromium with sulfur dioxide would present some problems. A more expedient way of determining reagent demand is by catching representative samples and having them analyzed chemically to determine hexavalent chromium. The reagent demand of sulfur dioxide can then be determined from the balanced oxidation-reduction overall reaction. From these data a number can be assigned to the trivalent chromium concentration at the equivalence point and the operating pH level assigned. From Equations 10.13(5) and (8) equivalence point potential and unreacted hexavalent chromium at the equivalence point can be determined.

Unlike acid-base neutralizations in which the reaction is virtually instantaneous, ORP reactions are frequently time dependent. The reaction vessel might be sized accordingly to provide the required residence time. A case in point is the oxidation of cyanide with available chlorine. The reaction between free cyanide, CN^-, with the hypochlorite ion, OCl^-, is quite rapid. However, in many cases where cyanide oxidation is employed, part of the cyanide may exist as metal-cyanide complex ions. The fraction of free cyanide ions existing in equilibrium with complexed cyanide ions is extremely small. For some metal cyanide complexes, many hours of reaction time may be necessary to completely destroy the cyanide.

Considerable emphasis was put upon the fact that in ORP reactions involving hydronium, H_3O^+, or hydroxide, OH^-, ions, the measured potential at the equivalence point and the degree of completion of the reaction are pH dependent. In this case, pH control becomes equally important as ORP control. A constant potential is not indicative of an excess of reactant if pH is variable.

The considerations that will be discussed in Section 10.14 on pH generally also apply to ORP control. Of special importance is the provision of facilities to equalize the reagent demand by use of equalizing tanks or bleed-in tanks used to store and slowly emit strong solutions to the reaction vessel.

REFERENCE

1. Jones, R.H., "Oxidation Reduction Potential Measurement," *JISA*, November, 1966.

BIBLIOGRAPHY

Dick, J.G., *Analytical Chemistry*. New York, McGraw-Hill Book Company, 1973.

Latimer, W.M., *Oxidation Potentials*. Englewood Cliffs, NJ, Prentice-Hall, 1952.

Milazzo, G. and Caroli, S., *Tables of Standard Electrode Potentials*. New York, John Wiley & Sons, Inc., 1978.

Pauling, L., *General Chemistry*. San Francisco, CA, W.H. Freeman & Co., 1970.

Flow Sheet Symbol

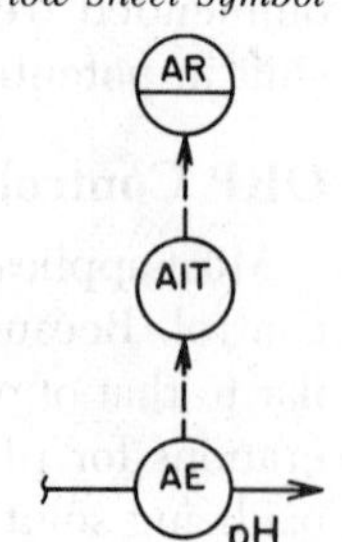

10.14 pH SENSORS

Standard Design Pressures:	Vacuum to 150 PSIG (1 MPa)
Standard Design Temperature:	Generally 23 to 212°F (−5 to 100°C); consult manufacturers for higher temperature applications
Materials of Construction:	Electrode hardware: stainless steel, Monel, Hastelloy, titanium, PVC, CPVC, polyethylene, polypropylene, epoxy, polyphenylene sulfide, Teflon, various elastomer materials
Inaccuracy:	System inaccuracy difficult to define; typical amplifiers stable to 0.02 pH units or better
Range:	0 to 14 pH; any 2-, 5-, 10-unit span with 0 to 14 range
Costs:	Highly variable, depending on features ($1000 to $3000 per transmitter loop)
Partial List of Suppliers:	Beckman Instruments, Inc., Process Instrument and Control Group; Electrofact, Div. of Control Data Corp.; Foxboro Analytical; Great Lakes Instruments, Inc.; Hach Chemical Co.; Horiba Instruments, Inc.; Lakewood Instruments; Leeds & Northrup; Orion Industrial, Div. of Orion Research Inc.; Photovolt Corp.; Rexnord Inc.; TBI Inc.; Uniloc Div. of Rosemount, Inc.

Introduction

The measurement and control of pH is applied to practically all processing areas that use water. Some of the reasons for this widespread use of continuous industrial pH measurement are listed below.

1. In many chemical processes, pH measurement and control is essential to attaining maximum yield, for example, in the manufacture of certain antibiotics made by fermentation.
2. Product quality is frequently dependent on pH measurement and control. For example, in the manufacture of chemical fertilizers close pH control is necessary to prevent an excess of one reactant in the product.
3. Corrosion control is a function of pH in certain processes. The most notable example is high-pressure boilers in which water is maintained sufficiently high in pH to avoid attack on metals used in construction.
4. Water pollution control is a principal application of pH measurement and control for a number of reasons, such as prevention of corrosive attack on sewerage systems; prevention of the release of toxic gases and protection of the treatment plant from major upsets that result from out-of-limits pH levels.

Theoretical Review

Water is a very versatile solvent and has many remarkable properties. One is that water molecules themselves will dissociate into hydronium ions and hydroxide ions in equimolar concentrations according to the equation:

$$2\,H_2O \leftrightarrows H_3O^+ + OH^-$$

and at 77°F (25°C)

$$[H_3O^+] = [OH^-] = 1 \times 10^{-7},$$

where the brackets represent concentration in terms of mols. per L.

It should be noted that when:

$$[H_3O^+] = [OH^-]$$

a condition called neutrality exists. At other temperatures, concentration of the two ions will be equal but of different values, for example:

$$0° \quad [N_3O^+] = [OH^-] = 0.339 \times 10^{-7},$$
$$40° \quad [H_3O^+] = [OH^-] = 1.71 \times {}^{-7},$$
$$60° \quad [H_3O^+] = [OH] = 3.10 \times 10^{-7}.$$

Water has a second unique property in that most ionic substances will dissociate in it to some degree, depending on the specie. If, for example, an acid is dissolved in water, it will dissociate to give an excess of hydronium ions over hydroxide ions and the solution is said to be acidic. Conversely, a base dissolved in water will dissociate to some degree, resulting in an excess of hydroxide ions, and the solution is said to be basic, or alkaline.

The degree of dissociation of an acid or base will vary with the specie. For example, a strong acid, such as hydrochloric acid, will dissociate nearly 100 percent; almost all of the acid molecules will dissociate to form hydrogen ions and chloride ions:

$$HCl + H_2O \rightarrow H_3O^+ + Cl^-$$

A weak acid, such as acetic acid, will dissociate only slightly. The reaction may be represented as follows:

$$HAc + H_2O \rightarrow H_3O^+ + Ac^- + HAc$$

with only a slight fraction of the acetic acid forming hydronium and acetate ions, and the remainder remaining undissociated.

It should be apparent that if the quantities of hydrochloric and acetic acid are added to water to yield two solutions of equal molar concentrations, the actual hydronium ion concentrations will be quite different for the two solutions. Yet it is not just the concentration of hydronium ions that determines the pH of a solution, but rather the concentration of active hydronium ions. Activity is affected by such things as concentration, temperature and other ions. The pH of an aqueous solution is defined as the negative logarithm of the active hydronium ion concentration or:

$$pH = -\log_{10}[\alpha H_3O^+]$$

The pH scale covers a range of 0.0 to 14.0, although it is possible to have pH values slightly beyond those limits. If we substitute into this equation concentration of active hydrogen ions at 77°F (25°C) in neutral water, we find that the pH under those defined conditions is 7.0. Other pH values of neutral water are: at 32°F (0°C)—7.47; 104°F (40°C)—6.77; and at 140°F (60°C)—6.51.

It should be readily apparent that the active hydronium ion concentration in a solution of pH 6.0 is 10 times greater than that of a solution whose pH is 7.0. A solution whose pH is 2.0 has 100,000 times the active hydrogen ions of a solution of pH 7.0. The logarithmic nature of pH with regard to active hydronium ion concentration is essential to the understanding and application of pH-measuring and control instrumentation (Figure 10.14a).

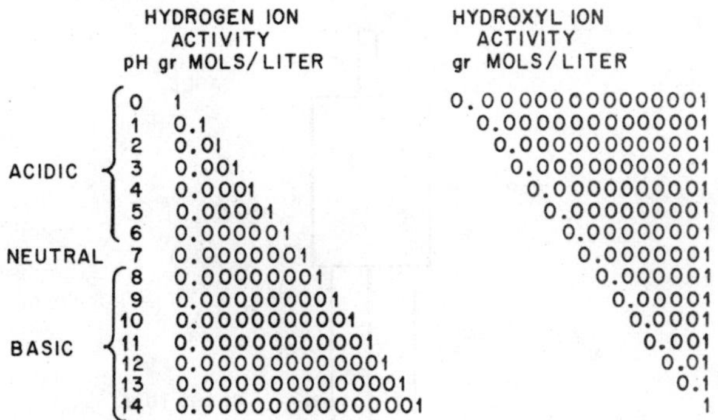

	pH	HYDROGEN ION ACTIVITY gr MOLS/LITER	HYDROXYL ION ACTIVITY gr MOLS/LITER
ACIDIC	0	1	0.00000000000001
	1	0.1	0.0000000000001
	2	0.01	0.000000000001
	3	0.001	0.00000000001
	4	0.0001	0.0000000001
	5	0.00001	0.000000001
	6	0.000001	0.00000001
NEUTRAL	7	0.0000001	0.0000001
BASIC	8	0.00000001	0.000001
	9	0.000000001	0.00001
	10	0.0000000001	0.0001
	11	0.00000000001	0.001
	12	0.000000000001	0.01
	13	0.0000000000001	0.1
	14	0.00000000000001	1

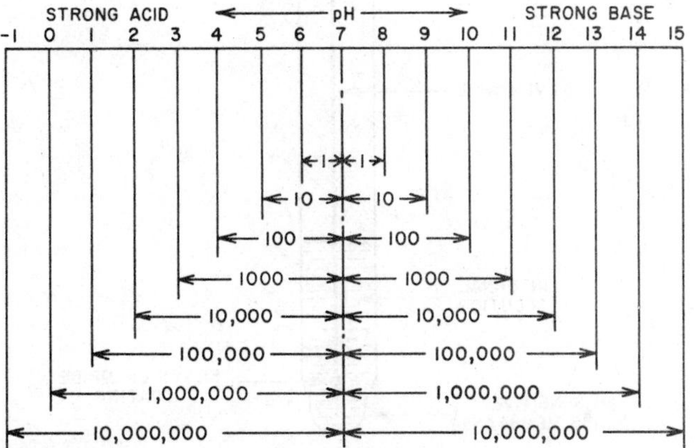

GRAPH OF REAGENT DEMAND. REAGENT ADDITION UNITS ARE 10^{-6} MOLS/LITER.

Fig. 10.14a The logarithmic nature of pH

It should also be noted that the actual concentration of an acid or base in an aqueous solution gives no direct indication of the pH of that solution. Conversely, the pH of a solution yields no direct indication of the total acidity of a solution. Total acidity is defined as an aqueous solution's capacity to don hydronium ions to a standard base (hydronium ion acceptor).

Measurement of pH

Although it is possible to measure pH by color change of certain pH sensitive dyes, by the use of certain metal-metal oxide electrodes and by other techniques, by far the most widely used method employs a pH-sensitive glass membrane electrode, a reference electrode and a high input impedance potentiometer. A process pH system also will generally incorporate automatic temperature compensation and hardware for the installation of the electrode system into the process stream.

Glass Membrane Electrode

The glass electrode, shown in Figure 10.14b, in generalized form consists of a pH-sensitive glass membrane fused to a glass tube. A silver wire coated with silver chloride protrudes into the thin-walled glass bulb. The bulb contains an internal filling solution, generally a buffered solution containing chloride ions.

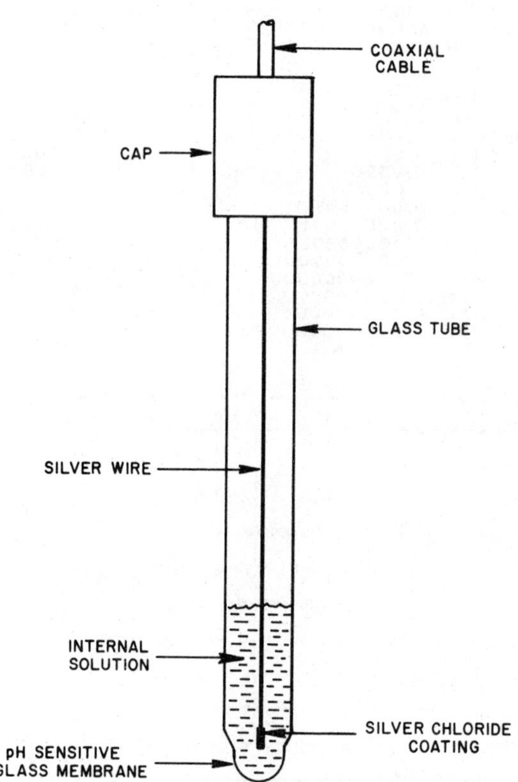

Fig. 10.14b pH glass membrane electrode

For a glass membrane electrode to function, both surfaces of the membrane must be hydrated. Hydration occurs by absorption of water by the membrane interface in contact with the solution. There is also an exchange of univalent cations of the glass for hydrogen ions from the solution. At both surfaces of the membrane there will be a boundary potential that will be a function of the hydronium ion activity in the solution at the interface. Because hydronium ion activity for the internal filling solution is constant, the potential will be a function of the hydronium ion activity of the external solution. This potential is:

$$E = k - .059 \log [\alpha H_3O^+] \text{ at } 25°C$$

where k includes such factors as the activity of hydronium ions in the internal solution and the asymmetry potential of the membrane. Asymmetry potential arises from differences in the characteristics of the two surfaces of the glass membrane and results from such things as mechanical stresses, deterioration of the external surface by dissolution of the glass, chemical attack on the glass, dirty surfaces and so on. Thus, the potential E will not be the same from electrode to electrode or for a given electrode with relation to time for a constant concentration of active hydrogen ions. For this and other reasons, a pH system must be standardized or adjusted periodically for its entire useful life.

Reference Electrode

A reference electrode is shown in generalized form in Figure 10.14c. It consists of a tube, into which protrudes a silver wire coated with silver chloride in contact with a potassium chloride solution, typically saturated. Electrical contact is made through a liquid junction between the internal potassium chloride filling solution and the process solution. In recent years "nonflowing" electrodes (a misnomer) have become extensively used. These electrodes function with a minimal interchange or intrusion of process solution into the electrode body and in a majority of applications, the intruding liquid will not cause any significant potential change in the reference potential.

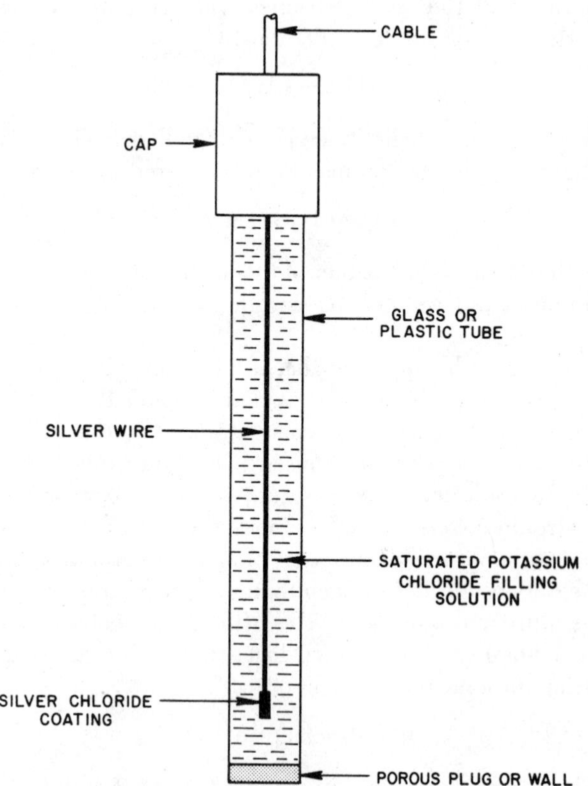

Fig. 10.14c Reference electrode

The function of a reference electrode is to complete the electrical circuit with the glass membrane electrode. It must provide a stable potential that is relatively invariable, despite changes in either the chemical composition or the physical properties of the process stream.

Thermocompensation

The measured potential in a pH system is:

$$E = k - \frac{RT}{nF} \log_{10} [\alpha H_3O^+]$$

where k is some constant covering a number of factors. In pH measurement absolute potentials are meaningless, the change in potential per concentration unit being the important parameter. In the above equation, R/nF is a constant where

R = the gas constant
F = the Faraday
n = 1.

It can be calculated that potential changes at a rate of 0.188 mv./°C. The slope, mv./pH for several temperatures is given below:

Temp. °C	*mv./pH*
0	54.20
25	59.14
50	64.08
75	69.02
100	74.00

Commercial pH analyzers generally provide manual or automatic temperature compensation or both. In general, this is done by changing the gain or sensitivity of the instrument. For example, an automatic temperature compensator consists of an element in contact with the process stream which changes in resistance with temperature; a manual thermocompensator is a calibrated variable resistor.

It must be pointed out that aqueous solutions, even buffer solutions, change in pH with temperature. This is especially true of highly alkaline solutions that exhibit quite dramatic changes in pH with temperature, whereas acid solutions change only to a minor degree. These changes are very real and no attempt is made, or should be made, to affect them with thermocompensation.

Other Electrode Considerations

As discussed previously, a glass membrane electrode must be hydrated before it can be functional. To function there must be the substitution of hydronium ions for sodium ions at the membrane interface. Therefore, there will be difficulties in the measurement of pH in any solution where the activity of water is considerably less than unity. This was first noted in strong acid solutions and the phenomenon became known as "acid error." This is a misnomer. The same errors may occur in strong salt solutions or high concentrations of nonaqueous solvents.

A second phenomenon of greater consequence occurs in highly alkaline solutions containing relatively high concentrations of certain monovalent cations, particularly sodium. This has become known as the "alkaline error" or "sodium ion error." Special formulations of glass have been developed that tend to minimize sodium ion error, and these electrodes should always be used when measuring pH values in excess of 11.0.

In general, so-called acid error results in a reading higher in pH than the true value. Alkaline or sodium ion error gives a reading lower in pH than the true value.

Electrode Station Hardware

Whereas the electrodes and thermocompensator typically used in process pH measurement are constructed with glass bodies and, thus, are chemically resistant to the great majority of process streams, less chemically resistant plastics and metals are used to install electrodes in the process. Some of the materials that have been used are glass, stainless steel, Monel, Hastelloy, titanium, PVC, CPVC, Ryton®, Kynar®, epoxy and other materials. Typically, two conditions exist:

1. Measurement is made in a tank, flume, ditch, or trough, and electrodes are installed via a submersion assembly. The term "submersion assembly" is meant to be analogous to immersion assembly, dip cell and so on.
2. Measurement is made in a flowing stream contained in a pipe.

Electrodes in flowing applications are frequently exposed to relatively high pressures. Figures 10.14d and e show typical submersion and flow chamber installations.

As previously mentioned, the so-called nonflowing junction reference electrode has replaced the flowing junction electrode in many industrial pH applications. Flowing junction electrodes were supplied with elevated

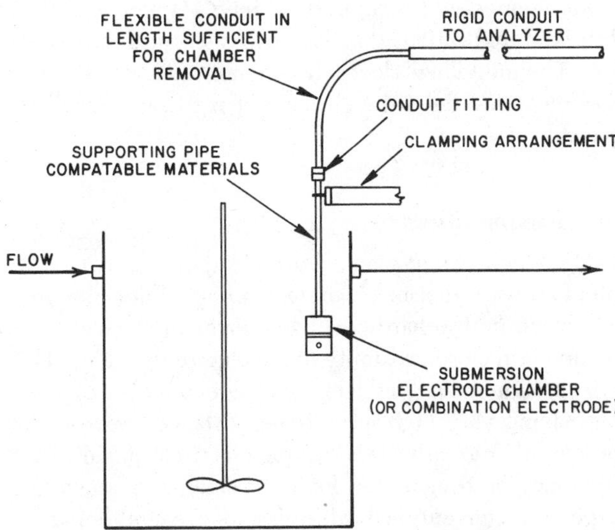

Fig. 10.14d Typical submersion assembly installation in open tank

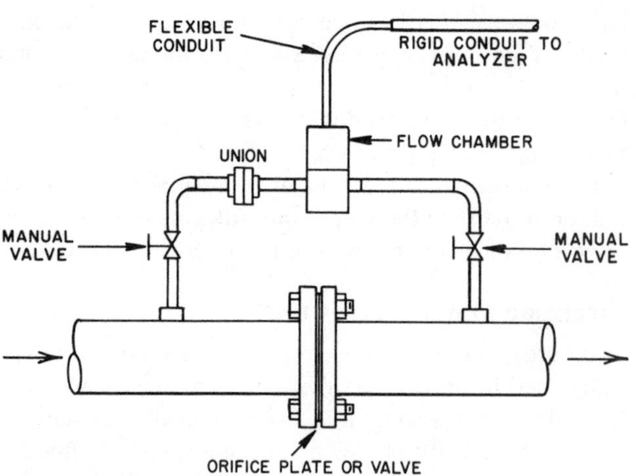

Fig. 10.14e Typical installation for pH measurement under pressure in a pipeline

electrode reservoirs for submersion applications, and flow cells with controlled gas pressure to provide a differential pressure across the liquid junction. Nonflowing junction reference electrodes do not require pressurization. Therefore, installation is generally less complicated, and less electrode maintenance is required.

In addition to the typical flow and submersion applications, pH equipment manufacturers provide systems in which electrodes may be inserted into or removed from a flowing process stream under pressure without shutting down the flow. This allows for frequent manual cleaning, or even replacement of electrodes without disturbing the process.

Another innovation is the development of process combination electrodes and throwaway electrodes. Typically, these electrodes are designed so that the reference and glass electrodes and thermocompensator are potted into a pipe nipple. For submersion applications the electrode connects to a supporting pipe via a pipe fitting. For flow applications the electrode is screwed into a pipe tee. The individual electrodes are not replaceable; when one electrode fails, the entire unit is discarded and replaced.

Preamplifiers

Many equipment manufacturers supply electrode station hardware that incorporates a preamplifier mounted either within the hardware itself, or in an electrical box mounted in close proximity to the electrode holder. This system eliminates long lengths of shielded electrode cable having very high insulation resistance. Instead the preamplifier greatly amplifies the electrode signal which then may be transmitted long distances over much less expensive conventional cable. Use of preamplifiers considerably reduces problems of electrical pickup and interferences.

Amplifiers

The amplifier receives the signal from the preamplifier and converts it to a readout. Analog and digital readout systems are available from a number of manufacturers. Some of these amplifiers provide for scale expansion that improves the readout capabilities of the instrument. Typically, scale expansion can be provided to 10.0, 5.0 or 2.0 pH units.

Amplifiers provide a variety of voltage and current outputs to drive recording and control equipment. Output capabilities vary between instruments, but some of the common outputs are: 0–10 mv., 0–100 mv., 0–1 v., 0–5 v., 0–1 ma., 0–20 ma., 4–20 ma., 10–50 ma. Current outputs may be isolated or nonisolated.

In addition to various recorder outputs, most pH analyzers can be provided with alarm contacts. These may be low-wattage relays capable of triggering audio/visual alarms or relays capable of operating relatively large metering pumps or solenoid valves.

Analyzers are available for panel mounting, surface mounting and pipe or pedestal mounting. A variety of cases are available for anything from general purpose to the most stringent hazardous area electrical classifications.

Specialized Equipment

Coating of electrodes by either solids or immiscible organic liquids has long been a deterrent to low-maintenance application of pH in many processes. Hence, a number of manufacturers offer ultrasonic cleaners that adapt to flow or submersion assemblies. A problem with ultrasonic cleaners is that cleaning is attempted with the same solution, or suspension, that caused the coating. It is difficult to predict with any certainty whether ultrasonic cleaning will be effective in any given application.

Applications of pH-measuring Equipment

As mentioned in the introduction, the applications of process pH are varied and diverse. The purpose here is not to describe these applications, but rather to describe some of the problems associated with process pH and, where possible, suggest solutions to those problems.

High Temperature

A number of important applications exist in which high temperatures, and generally high pressures, are involved. These high temperatures generally result in very short electrode life, if for no other reason than that of the solubility of the pH-responsive glass membrane. One example of high-temperature exposure with measurement at a low temperature (77 to 95°F, 25 to 35°C) is the pharmaceutical fermenter in which the electrode system is exposed to low-pressure steam for sterilization between batches. Several manufacturers have developed specialized equipment for this application. For years the chemical fertilizer industry has manufactured products, such as ammonium sulfate, by the reaction of ammonia and sulfuric acid. This reaction takes place at about 248 to 266°F (120 to 130°C), and the product is so concentrated that cooling of the solution results in crystallization. The industry has developed a technique whereby a sample is continuously diluted with water, and the pH of this more dilute solution is correlated with the pH of the undiluted solution at process temperature.

While there are many examples of high-temperature measurement within present equipment capabilities, there is increasing demand to go to significantly higher temperatures and pressures.

Low Temperature

While the measurement of pH at low temperatures is much less widespread than measurement at elevated temperatures, there are, nontheless, important measurements at temperatures below freezing. These measured solutions generally involve either soluble salts or dissolved organics that lower the freezing point. Aside from the obvious problems of the freezing of internal filling solutions within the electrodes, there can be a problem of slow response or sluggishness in the electrode system. The resistance of the glass electrode doubles for about every 45°F (7°C) temperature drop. As in the high temperature application noted above, there has been work done on correlating pH at process temperature with process solution diluted with a neutral salt solution at 77°F (25°C).

Abrasion

As noted previously, asymmetry potential is attributed to differences in the two surfaces of the glass membrane. There are many applications involving suspended solids in which there is continuous abrasion of the outer surface of the glass electrode. This in turn affects the membrane surface, and the result is a relatively rapid change in asymmetry potential manifested as a measurement drift. In such circumstances, electrode life may be very short. The problem is most severe in certain ore treatment processes where heavy agitation is required to keep solids in suspension. Another application involving severe abrasion is in coal washing. Many different approaches have been used to alleviate the abrasion problem. In some cases, a cylindrical guard around the electrodes will reduce velocity at the electrodes and improve service life. Attention must be given to response time in control applications. Some users have used cyclones to remove solids while others have used self-cleaning bypass filters. Successful measurement requires either the reduction of velocity or the removal of the abrasive solids.

Etching

Certain chemicals will attack glass, most notably fluorides. The fluoride ion in acidic solution will form a complex with the hydronium ion and this, not the fluoride ion, is what etches glass. Therefore, the glass electrode is not recommended for pH measurement in acid fluoride solutions. However, in many applications involving fluorides, the requirement is the neutralization of acids containing fluoride or even treatment for the removal of fluorides. For example, a mixture of nitric and hydrofluoric acid is widely used in etching silicone wafers and in pickling stainless steel. In the neutralization of these spent acids, the pH should be maintained as high as possible within acceptable limits. Furthermore, the control system should be sufficiently sophisticated to maintain the pH within close proximity of the control point, avoiding excursions into an acidic condition. In addition to fluorides, strong alkali will etch glass quite rapidly, particularly if hot.

Coating

Many industrial process pH measurements are hampered by coating of the sensing glass membrane electrode. As mentioned previously, many manufacturers supply ultrasonic cleaners. In addition, a few suppliers have developed mechanical cleaning devices, such as a rotating brush. Users plagued with coating problems have devised methods of intermittently washing the glass membrane by an intermittent jet of water or organic solvent.

Coating of the glass electrode will cause slow response to pH change, possibly even total insensitivity to change. This can be especially disastrous in control applications leading to complete instability of the control system. Unfortunately, it is not uncommon that the foulest stream in the plant, that is, the wastewater effluent, requires controlled neutralization. Equally unfortunate is the fact that manual cleaning by trained instrument technicians is the most effective means of keeping electrodes clean and functional.

A variety of suggestions on cleaning of pH electrodes have been made by manufacturers. This author suggests a somewhat harsher cleaning technique than is commonly practiced. Many coatings can be dissolved in strong nitric acid (50 percent v/v reagent grade nitric acid in water) applied to the electrode by a squeeze bottle. A solvent, such as naphtha, methyl ethyl ketone or ethyl alcohol, is effective on organic coatings and should also be applied by a squeeze bottle. The electrode should be scrubbed only if it is necessary to remove a difficult coating and then with a soft tissue wetted with solvent. This procedure will frequently result in the electrode taking on a static charge that may be slow to dissipate and result in minor errors. There is also the possibility of at least partial dehydration of the electrode by the solvents, but rehydration will occur rapidly when the electrodes are reimmersed in water.

Whenever electrodes are cleaned, the cleaning should be followed immediately by the rinsing of the electrodes with water. Then the system should be standardized in at least one buffer solution near the pH level of the process stream. Occasionally the system should be standardized in two or more buffers to ascertain the response of the electrode system. Table 10.14f gives an application oriented summary of the available cleaning systems.

Reference Electrode

Possibly the weakest link in the pH-measuring system is the reference electrode, and this has been given extensive attention by equipment manufacturers. As noted previously, one function of the reference electrode is to complete the electrical circuit, and this must be done with ions carrying current through a liquid junction. It was also noted that the potential of the reference electrode is actually the sum of the half-cell potential and the so-called junction potential. Changes to either of these potentials will affect the accuracy of pH measurement.

As also noted, a variety of reference electrodes are available from manufacturers with considerable emphasis on the so-called nonflowing electrodes. Despite the terminology, there must be liquid junction. This junction must bring the internal filling solution into contact with the solution being measured. The filling solution is typically saturated potassium chloride. Silver chloride in the presence of high concentrations of chloride ion will form a soluble complex ion, $AgCl_2^-$. Substances that will react with silver or chloride ions to form insoluble compounds will generally plug the liquid junction. To overcome this problem, some manufacturers offer what is termed a double-junction electrode. A generalized representation of this electrode is shown in Figure 10.14g.

In addition to reactive substances plugging the liquid junction, nonreactive solids and liquids may change junction resistance by coating the junction. In either case, the electrode will tend to become more resistant because there is less area of contact between the internal and measured solution across the junction. This may result in highly unstable measurement or, in some cases, a shift in measured pH.

There are tests the experienced instrument technician can make to determine if an anomalous condition is due to the reference electrode. He can test the junction resistance. In addition to the reference electrode under test, he needs the internal half-cell from another glass or reference electrode. (This is the silver-chloride-coated silver wire.) These are immersed in a small beaker of saturated potassium chloride (or filling solution for the reference electrode) and the resistance between the electrode and the silver wire measured with an ohmmeter. As a rule of thumb, if this measured resistance is greater than about 40–50,000 ohms, the reference electrode should be replaced.

A second test would be to expose the questionable electrode to successively more dilute solutions of potassium chloride along with a new reference electrode, the two reference electrodes being connected to the input terminals of the pH analyzer. Start with saturated potassium chloride noting the meter reading which may be set to some onstream value. Then dilute 10 ml of this solution to 100 ml with pure water and repeat the measurement without making any adjustment to the instrument controls. Take 10 ml of this diluted solution and further dilute to 100 ml with pure water. At least four or five dilutions should be possible without a significant change in reading on the pH analyzer for a "good" reference electrode. It should be noted that the new reference electrode is considered "good" for this test.

Flowing junction reference electrodes could frequently be restored by immersing the tip of the electrode into saturated potassium chloride and boiling for 15 to 20 min. This procedure is not generally effective with nonflowing junction electrodes.

Table 10.14f
SELECTION OF pH PROBE CLEANERS

Type of Application	*Type of Cleaner* Brush	*Rotary Scraper*	*Acid*	*Base*	*Emulsifier*	*Hydro-Dynamic*	*Ultrasonic*
Oils, fats		√			√		√
Resins (wood, pulp)				√			√
Emulsions of latex		√					
Fibers (paper, textile)						√	
Solid suspensions						√	
Crystalline precipitations (carbonates)	√	√	√				
Amorphous precipitations (hydroxides)	√	√	√				√

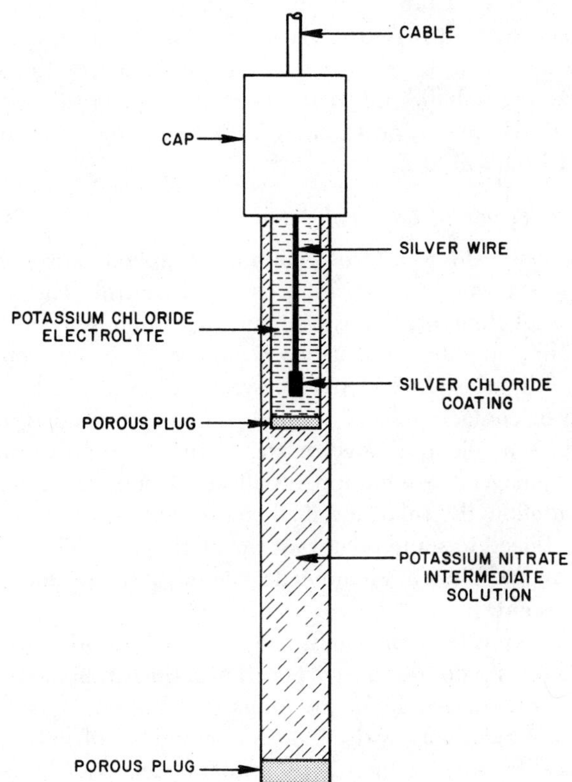

Fig. 10.14g Double-junction reference electrode

Standardization

It should be apparent that the absolute potential of an electrode system consisting of a glass membrane and reference electrode is quite meaningless. Therefore, it is necessary to standardize a pH-measuring system against solutions of known pH values. These solutions are called buffers and are supplied by equipment manufacturers and laboratory supply houses. Buffers are generally traceable to Bureau of Standards formulations.

From the above discussions, it is quite clear that many factors are involved in accurate pH measurement, not the least of which is the process stream being measured. Therefore, the frequency of standardization for any given application is difficult to predict and is better ascertained by experience. When a new system is started up, daily standardization may be recommended. It may be found that success standardizations require little or no adjustment of the instrument; in that case, the frequency of standardization is decreased.

In standardizing a pH system, it is best to follow manufacturers instructions. However, it should be emphasized that electrodes must be clean or cleaned if necessary. Second, when electrodes are exposed to a buffer, care must be exercised to make certain that all elements are in temperature equilibrium before adjustments are made. Third, before exposing electrodes to fresh buffer for making final adjustment, the electrodes and electrode hardware should be rinsed thoroughly with water followed by a rinsing with buffer. One of the most common causes of pH errors in measurement is use of contaminated buffers.

Nonaqueous pH Measurement

By definition, pH is the negative logarithm of hydronium ion activity in an aqueous solution. It is sometimes desired to make measurements in solutions which, while not strictly nonaqueous, are organics with low water content. The solution may also be low conductivity, as well as having a tendency to dehydrate glass electrodes. The important consideration in such measurements is that repeatable readings be attained for repeatable conditions. In general, such measurements are best made with a relatively fast-leaking reference electrode. Therefore, flowing junction electrodes may be preferred over the nonflowing type. In any event, some preliminary work is indicated before specifying or purchasing instrumentation, such as making measurements under laboratory conditions while simulating process conditions by heating or stirring the solution being measured.

High-purity Water

Boiler water for central station power plants is closely controlled in pH because of the great potential for corrosion. Typically, this is done with ammonia to a level of about pH 8.5 to 9.0. Despite that pH level, the water normally contains less than 0.5 ppm total dissolved solids, and its conductivity is typically less than 1 μmho. Several conditions are required to make this measurement successful: isolation of the electrode chamber from the structure, shielding of the electrodes, reduced flow rate and the proper choice of reference electrode.

It should be noted that pH is the measurement of active hydronium ions and that activity is affected by the presence of other ions. It should also be recalled that high-purity water is totally unbuffered, and that in the range of measurement, small changes in hydronium ion activity result in relatively large changes in pH. Therefore, the use of a reference electrode with a fast leak rate or an electrolyte solution interchange that causes any appreciable change in conductivity may significantly alter the activity of the measured hydronium ion and give pH readings not representative of the true pH of the stream.

Reduced flow rate is also necessary for the most accurate pH measurement, mainly because of the stability of the reference electrode junction. This also applies in solutions where high resistance is due to high organic solvent content.

pH Control—An Overview

While it is true that there are many installations in which pH is measured and recorded to prove compliance with water pollution regulations, pH is measured primarily for the purpose of control. In industrial acid-base neutralizations, the usual intent is to control to the degree necessary to meet requirements of some regulatory agency. These limits are sometimes quite wide, at least in terms of pH units. On the other hand, for the control of industrial processes, the requirement most often is for control within quite narrow ranges. Yet in some instances this may be easier than controlling to a relatively broad range.

As noted previously, pH is a logarithmic function, and a change of one pH unit from some level represents either a tenfold increase or decrease in active hydronium ions. Also, pH value has no direct relationship to the total acidity or total alkalinity of a solution. By definition, total acidity is the quantitative capacity of an aqueous solution to react with hydroxide ions, and total alkalinity is the quantitative capacity of an aqueous solution to react with hydronium ions.

A titration is the addition of a standard acid or base to a carefully measured quantity of aqueous solution, while pH values, with the quantity of reagent added, are noted. A titration curve is a plot of these data. Figure 10.14h shows a titration curve of a strong acid, hydrochloric (HCl), titrated with a strong base sodium hydroxide (NaOH). The titration curve is invaluable in considering pH control problems. It will be noted that at the extremes of the pH scale, from 0 to about 3.0 and again from about pH 11.0 to 14.0, there is relatively little change in pH even for relatively large additions of sodium hydroxide titrant. However, near the equivalence point (pH 7.0 in this case) the addition of 0.05 ml of standard NaOH (about 1 drop) changes pH by nearly 2.5 pH units

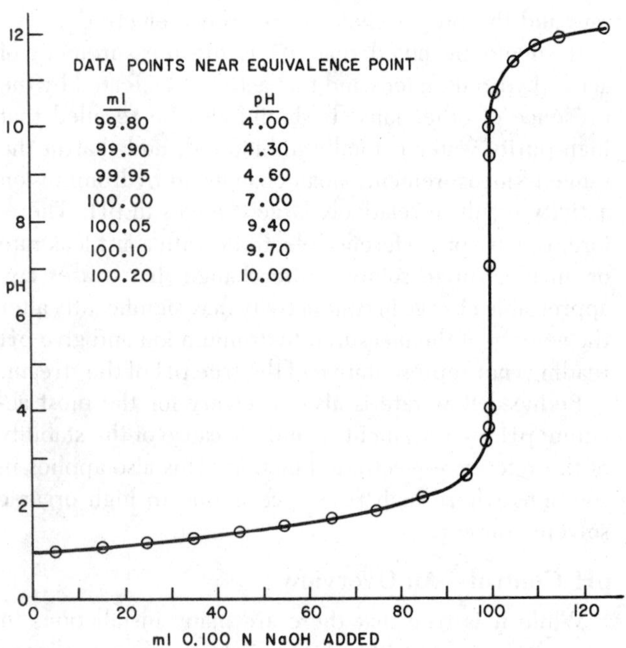

Fig. 10.14h Strong acid/strong base titration; 100ml 0.100 N HCl titrated with 0.100 N NaOH

in Figure 10.14h. It is quite obvious that it would be much easier to control pH to 2.0 ±0.1, than it would be to control to pH 7.0 ±3.0. In this example the addition of only 0.4 ml transverses the range of pH 3.0 to 10.0, whereas approximately 5 ml is required to change pH from 1.9 to 2.1.

Types of Control

Two-position Control A commonly applied type of control is termed two-position or on/off control. The final control element, that is, the unit controlling reagent addition, operates only at two positions: either fully open or fully closed. The pH analyzers are equipped with alarm contacts which, through relays, are used to energize a solenoid valve or start a pump. Sophisticated pH analyzers are equipped with deadband adjustments that allow the relay and final control element to remain energized to some point past the alarm point. Thus, the relay and control element do not "chatter" around the set point.

Two-position, or on/off, control is best applied when the control point is on a flat portion of the titration curve, (see Figure 10.14e) or the limits of pH to be controlled to are relatively wide. Control should be of only one variable, such as the neutralization of acid with base *or* of base with acid but *not* both. Relatively long holdup time in the neutralization tank (on the order of 10 minutes) should be provided. As in all acid-base neutralizations, thorough mixing is required.

Multimode Control In most pH control applications, multimode control using the conventional control modes of proportional band and reset (or integral action) and sometimes rate or derivative action is applied. The modulated output signal is used to control a motor valve or set the stroke length of a pump to automatically and continuously add reagent on demand to neutralize an acid or a base to the pH control point. Just as in on/off control, a relatively large holdup volume in the reaction vessel is desirable and as advantageous as good mixing. Controllers are available with special features for pH control; manufacturers of equipment should be consulted for details. A modification of conventional multimode control called by such various names as "pulse duration controller," "time adjusting controller" or "pulsed proportional controller" are available from several manufacturers of pH analyzers and controllers. These systems use a final control element that is fully on or fully off, but the time of the on cycle is proportional to the error signal between the set point and the controlled variable.

Special Considerations

Rangeability of Final Control Element The nonlinear nature of the pH function imposes very difficult burdens on the final control element. Take for example an industrial plant that is regenerating a demineralizer several times a day. During regeneration, the wastewater to be neutralized may be quite strongly acidic. When the demineralizer is not being regenerated, plant wastes may only be slightly acidic but outside the limits acceptable to the control agency. Of course, the control element must be sized for the maximum instantaneous load. Yet it must be equally efficient when the wastewater load is minimal. The rangeability of the element is the ratio of its maximum deliverability to its minimum. This may be several orders of magnitude in a pH control system.

Precision of the Final Control Element By examining the titration curve of the wastewater requiring neutralization, the instrument engineer can calculate the rate of change in pH per unit volume of reagent added. Knowing the limits that must be met, he can calculate the quantity of reagent that will transverse those pH limits. From titrations of representative samples and from flow rate data, the instrument engineer can determine the maximum reagent demand and from that data size a control valve or metering pump. He can also determine the precision of delivery required for maintaining pH control within specified limits. That value is:

$$\text{Precision \%} = \frac{\text{volume reagent to transverse control limits}}{\text{maximum delivery of control element}} \times 100$$

It immediately becomes obvious that when the control point is on the steep portion of the titration curve, and when the load is sometimes high, the requirements on the final control element are indeed demanding.

Elements of a pH Control System

A pH control system is not merely the instrumentation involved. It also includes tanks, pumps, mixers, valves and other equipment, because these elements affect such things as holdup time, dead time, and so on. Because this discussion is mainly concerned with pH measurement for control purposes, the following comments are very generalized.

It is usually desirable to minimize "dead time," or transport lag, which may be defined as the time interval between the beginning of some action and the first indication of a result of that action. In a pH control system, this is equated to the addition of reagent and response as a pH change. Typically reagent is added at a single point and pH measurement is made at a single point. Volume of the reaction vessel and degree of mixing significantly affect dead time, but, frequently, so does location of the pH-sensing electrode system. As a rule of thumb, it is better to locate the electrode system directly in the reaction vessel than in a flow chamber external to the vessel. Electrode systems must be carefully maintained in a control system. While reaction time and response to pH change should be nearly instantaneous in a well-mixed tank involving neutralization of a soluble acid with a soluble base, a coated electrode can increase this response time to minutes with disastrous effects on pH control.

It is normally recommended that the pumping rate of the mixer should be a minimum of one turnover per minute and preferably about three. Any attempt to control pH in a poorly mixed reaction tank (as, for example, by baffling or air sparging) is likely to end in failure.

To obtain satisfactory control, particularly in wastewater treatment applications, it is very important to dampen extreme load conditions. Examples of situations in which short-term heavy load conditions may occur are the intermittent batch dumping of pickle liquors in galvanizing steel or etch solutions in silicone wafer etching, or regeneration of water demineralizers. A mixed tank, sump- or basin-sized to generously accommodate the largest batch dump, can materially dampen surge loads on the neutralization system and, thereby, permit more optimum sizing of the reagent adding final control element.

In almost any waste or acid-base neutralization there will be a variety of reagent-demanding chemicals going to waste. This will result in a variety of problems for the pH controller, as well as frequent upsets to the system. In all likelihood the control settings will not be optimum for all conditions of the waste stream. A tank downstream of the reaction vessel will serve to attenuate the variations in the reaction tank. Recording of pH in this tank is desirable over the reaction tank for recording of pH for a control agency.

Analysis of pH is the most widely applied of analytical measurements. Being a nonlinear function pH is difficult to control. Yet it is being successfully controlled in thousands of installations. Acid-base neutralizations and wastewater treatment represent only two of the more difficult applications.

pH Control—Details

In the entire spectrum of industrial process control, the pH control problem is the most demanding. In order to cope with the chemistry of the process, the designer must undertake an exhaustive investigation of the characteristics of the material to be treated to determine the measurable ion load on the process. It is important to know how much the pH of the treatable material varies, the frequency with which the load changes and the representative titration curves at various ion loadings. The titration data obtained will reflect the difficulty of the control problem and will show whether the control problem becomes more or less severe with ion loading. Ion loading is high when the solution pH is far from neutrality.

Figure 10.14i illustrates the titration characteristics of a strong acid-strong base system with and without buffering. A strong acid or base is a compound that is completely dissociated in aqueous solutions. A buffered aqueous solution is one that resists a change in its hydrogen ion concentration when either an acid or alkali is added to the solution. The titration curves of Figure 10.14i were obtained experimentally, using the reagents indicated. The buffered solutions were prepared[1] with hydrochloric acid and potassium acid phthalate ($KHC_8H_4O_4$).

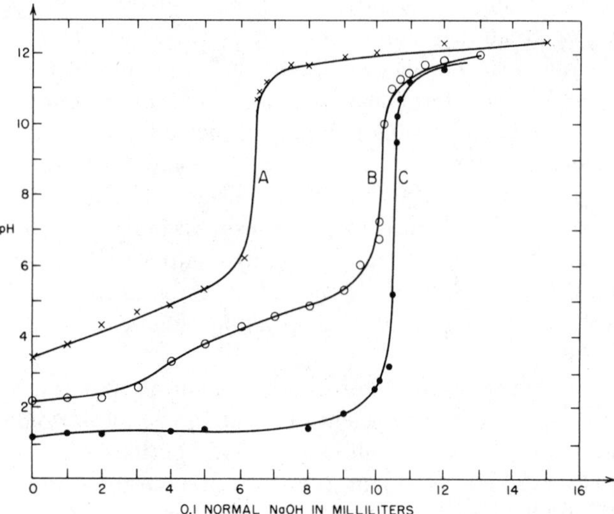

Fig. 10.14i Typical acid-base titration curves. *Key:* A = 9.9 ml. HCl + 50 ml. of 0.1N $KHC_8H_4O_4$ per 100 ml. solution; B = 6.7 ml. of 0.1N HCl + 50 ml. of 0.1N $KHC_8H_4O_4$ per 100 ml. solution; C = 100 ml. of 0.1N HCl

The problem of controlling pH is aptly demonstrated by these titrations. The control point is usually 7.0. In this range the slopes of the titration curves are steep; that is, there is a large change in pH for a small change in reagent addition. The slope is especially steep for curve C, which represents a strong acid-strong base system without buffering, and becomes less severe with buffering. This characteristic results in a high process gain (the process being sensitive) about the setpoint (pH of 7) which must be countered in an automatic control system with a low controller gain (high proportional band in the control instrument) if control loop stability is to be maintained. However, if a load upset drives the pH measurement below 4 or above 10 (for curve C), the process gain is very low (it is no longer sensitive) and small changes in pH result from large changes in reagent addition. In an automatic control system, this requires a controller with high gain if the system is to be responsive. This is one of the problems that must be resolved in connection with most pH installations.

In Figure 10.14j a strong acid has been added to water to achieve a pH of 6 and a pH of 2 (Curves A and B). The reagent flow (10 percent NaOH) requirement for each solution is plotted on separate scales (lower and upper x axes). Assuming a control specification of pH 7.0 ± 0.5, a reagent flow of ±28 percent variation can be tolerated when the pH of the inlet material is 6.0. When

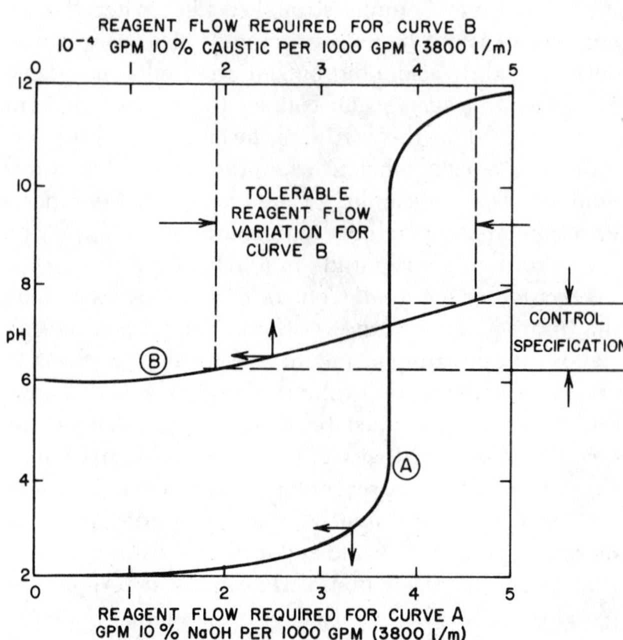

Fig. 10.14j Relationship of accuracy and rangeability to ion loading[2]

the pH is 2.0, the tolerable reagent flow variation is ±0.0028 percent, and the problem of pH control is 10,000 times more difficult. This example also points up the other problem associated with pH control: reagent flow rangeability. If faced with a waste treatment problem as shown by this example, the reagent system must be capable of a 10,000 to 1 flow turndown.

Reagent Addition Requirements

Reagent addition requirements can be handled in many ways, depending on the process loads (flow of wastewater to be neutralized) into the neutralization facility and the variation of the hydrogen and/or hydroxyl ion concentration in that flow. It should be recognized at the outset that because of the logarithmic nature of the pH measurement, a pH change of one unit is a tenfold change in load, whereas a 100 to 300 gpm (379 to 1137 l/m) change in flow (assuming no change in pH) is only a three-fold change. Thus, the consequences of flow variations in waste streams can be relatively minor in comparison to ion concentration variations.

A metering device or a control valve can be used to deliver reagents to the process under automatic control. Metering devices are very accurate; however, delivery rangeability capability is limited to about 10 to 1. This means that for pH variations greater than ±0.5 they will cycle or be inadequate because, as noted above, a pH change of one means a ten-fold change in reagent requirement. When pH load variations are minor (0.2 or less) and flow variations are less than 5 to 1, metering pumps are adequate delivery devices.

Control valves, like metering pumps, have limited

rangeability. In this category there are two types of internal plug forms usually considered for throttling service: the linear and the equal percentage types. The linear valve has a flow delivery turndown of about 15 to 1. The equal percentage* valve has a flow turndown of from about 35 to 50 to 1, depending on the design, and some have been reported to have a rangeability as high as 100 to 1.

Valve Sequencing A technique for increasing valve rangeability[3] involves the sequencing of a pair of equal percentage valves (Figure 10.14k) so as to achieve an overall rangeability approaching the product of the individual valve rangeabilities, for example, 50 × 50 = 2500. The loss of rangeability is mainly due to the amount of overlap between valves. A plot of the performance characteristics of this pair of valves is shown in Figure 10.14l. The valve positioner of the smaller sequenced valve is calibrated for full stroke over 0 to 52 percent controller output signal (closed at 0 percent; fully open at 52 percent). The positioner of the larger sequenced valve is calibrated for full stroke over the range of 48 to 100 percent of controller output. Transfer between the valves (as the controller output changes) is accomplished by a pair of three-way solenoid valves that are actuated by a pressure switch monitoring the controller output air signal pressure. Since only one valve at a time is operating, while the other valve is closed, the characteristic of the pair is equal percentage, as the semilog plot of Figure 10.14l illustrates. If the smaller valve were permitted to remain open when the larger valve came into service (or vice versa), the valve characteristic curve would have a discontinuity at the transfer point that would result in an unstable control system. There is a small flow transient at the transfer point but the characteristic curve is maintained.

The sequencing approach for achieving a wide range

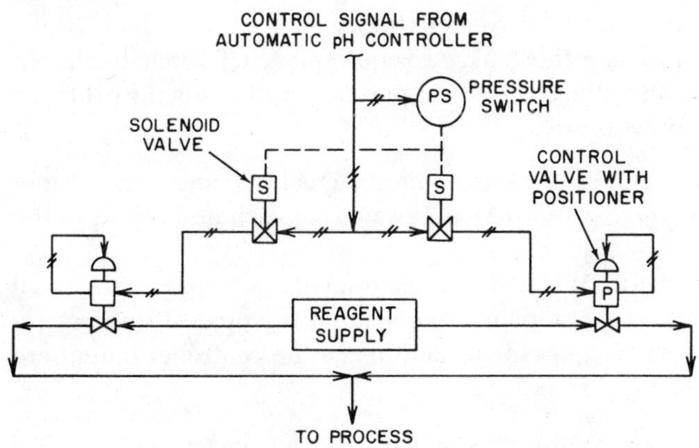

Fig. 10.14k Sequencing valve arrangement

*The term equal percentage means that the valve will produce a change in the flow rate corresponding to a unit change in lift (valve plug movement) which is a fixed percentage of the flow rate at that point.

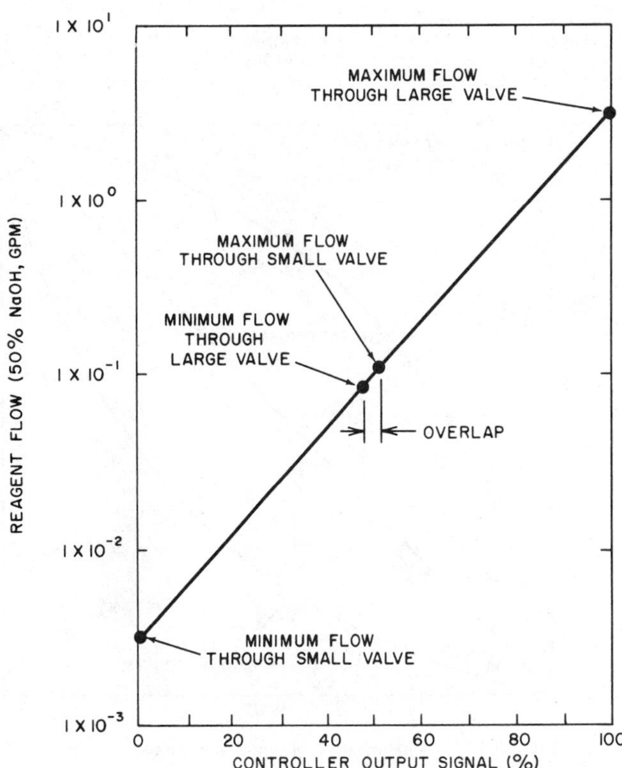

Fig. 10.14 l Reagent flow using sequenced valves

reagent delivery capability does not work satisfactorily with valves or devices having linear characteristics because the transfer point usually occurs at about 10 percent of controller output signal. This means that the larger of the two sequenced valves is essentially doing all the work.

Valve Linearization The drawback of sequencing is a high gain contribution to the overall loop gain, especially at high reagent flows, a condition that is typical of the variable-gain (sensitivity is a function of valve opening) characteristics of equal percentage valves. One approach to countering this variable-gain characteristic is by a characterizer with an input-output characteristic opposite to that of the equal percentage valve(s). This approach is illustrated in Figure 10.14m. The resultant valve characteristic using this technique is approximately linear, a characteristic that is highly desirable from an automatic control point of view, since the variable gain nature of the process makes the control problem difficult enough.

Nonlinear Controller Another method of countering the high valve gain characteristic of sequenced equal percentage valves in a control loop is to change the gain characteristics of the feedback controller itself proportional to the ion load (pH) of the process. The characteristics of the controller are as shown in Figure 10.14n. The diagonal line represents the error-output relationship for the controller (in response to an error, a corrective signal is generated—the output—which eliminates the deviation from setpoint) with a 100 percent

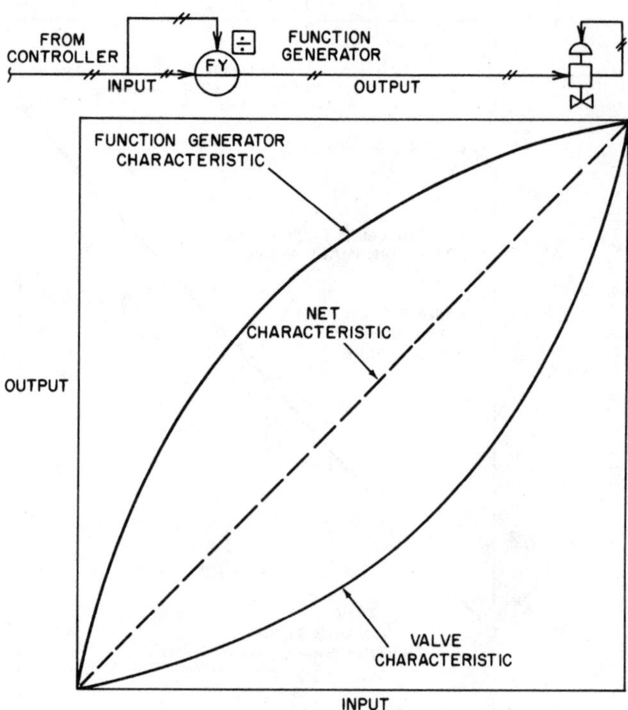

Fig. 10.14m Linearization of equal percentage valves

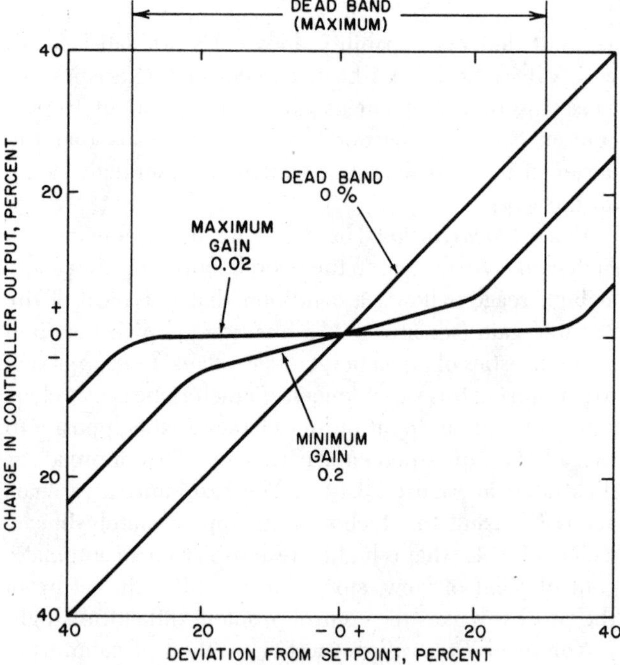

Fig. 10.14n Nonlinear controller characteristics

proportional band (gain = 1.0) without the nonlinear adjustments available with this controller. The first available adjustment is a slope adjustment that allows the proportional band to be increased (gain reduced) about the zero deviation point by a factor of 50. This means that when the gain setting of the controller is 1.0 (100 percent proportional band), the effective proportional band is 5000 percent or a gain of 0.02 (insensitive controller) at the zero deviation point. The slope adjustment is manual. The second adjustment is the error deviation range over which the slope adjustment is operative; this is referred to as "dead-band."

The dead-band is adjustable from 0 to ±30 percent error (deviation from setpoint). This latter feature allows the gain of the control loop to be adapted proportional to the ion load. If the process to be controlled resembles Figure 10.14i, a reagent flow rate or valve position signal can automatically adjust the dead band. At high ion loadings (curve A) the controller gain will be low, a desirable condition when the process valve gains are high. At lower ion loadings (curve B) the dead band can be reduced, thereby increasing the gain of the controller, a condition that is desirable when the process and valve gains are low. The effectiveness of this type of controller and the benefits achieved by adapting the control loop characteristics to those of the process have been demonstrated[4,5] on operating installations.

Control Systems

The choice of which type of control system should be used is dictated in large measure not only by the process loads, particularly ion loading, but also by the rate and the frequency with which the loads are changing. Again, the knowledge of the characteristics of the waste stream, particularly the type of information supplied by titration curves, is essential. The reagent delivery requirements deserve extensive consideration as well as the characteristics of the device used to bring the reagents into the process. For example, a reagent flow detector with a flow turndown of 50 to 1 can accommodate pH variations of 1.7 units (log of 50) for a strong acid-strong base reaction.

Batch pH Process Waste pH treatment on a batch basis is not as prevalent as the continuous approach. This is principally due to the large volumes of waste material that must be treated that would in turn dramatically increase the tankage requirements if batch treatment were utilized. Two unique characteristics of the pH batch process are[6]:

1. The measurement (actual pH) and setpoint (desired pH) are away from each other most of the time.
2. When the measurement and setpoint are equal (endpoint), the load on the process (reagent requirement) and hence the controller output are zero.

The controller characteristics for the batch pH control application should be proportional plus derivative.[12] Reset must not be used because reset windup[12] will result in overshoot of the controlled variable. In a proportional controller, the corrective action generated is proportional to the size of the error; in a reset controller, to the area under the error curve; and in a rate controller, to the rate at which the error is changing. Once the measurement goes past the setpoint, there is no way for the control system to bring it back to setpoint unless, of course, two controllers and two reagent supplies are used. In the absence of the reset control mode (proportional only), a controller is usually supplied with a 9.0 psi (62 kPa) bias spring (pneumatic controllers) so that the controller output is 9.0 psi (50 percent) when measurement and setpoint are equal. For the batch application with a proportional plus derivative controller, the bias spring must be 3.0 psi (21 kPa) so that when measurement and setpoints are equal, the controller output is zero percent.

The effect of secondary lags in the valve, process vessel and measurement are compensated for by the derivative action of the controller. For example, if reagent is added but its effect has not yet been seen by the pH electrode when measurement and set are equal, then too much reagent will have been added. With the derivative-time setting properly adjusted, the controller will shut off the reagent valve while the measurement is still away from setpoint, thereby allowing the process to come gradually to equilibrium.

Too much derivative time in the controller is preferable to too little. When there is too much, the valve will close prematurely but will open again when the measurement does not reach setpoint. Too little derivative allows the valve to remain open too long, resulting in overshooting the desired pH target.

The variable gain characteristic of the equal percentage valve is an asset to this type of control system. When the measurement is far away from setpoint, the valve will be wide open, permitting essentially unrestricted reagent flow to the process. As the measurement approaches setpoint and the valve closes, the decreasing gain of the valve counters the increasing gain of the process. Figure 10.14o illustrates the measurement-valve behavior of the batch process.

Although the installation and process design consideration for the batch process are not as severe or demanding as the continuous operation, care should be taken to insure that adequate mixing is provided, tank geometry precludes the existence of stagnant areas, reagent delivery piping between valve and process is as short as possible, and electrodes are placed in responsive locations.

Feedback Control Systems Feedback control can be used very effectively in wastewater neutralization, provided the process is not subjected to dramatic or frequent load variations. Maintained step changes in either load or setpoint can be handled effectively. Figure 10.14p illustrates a feedback control system in which the reagent flow rangeability requirements are not severe and can be handled by a single valve having either linear or equal percentage characteristics, depending on the reagent

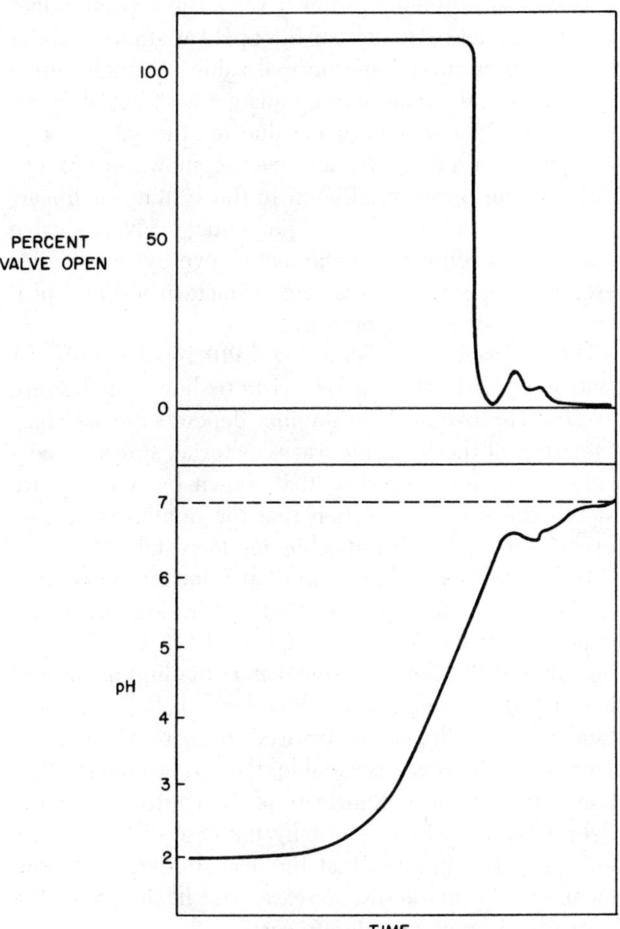

Fig. 10.14o Measurement-valve behavior for a batch process

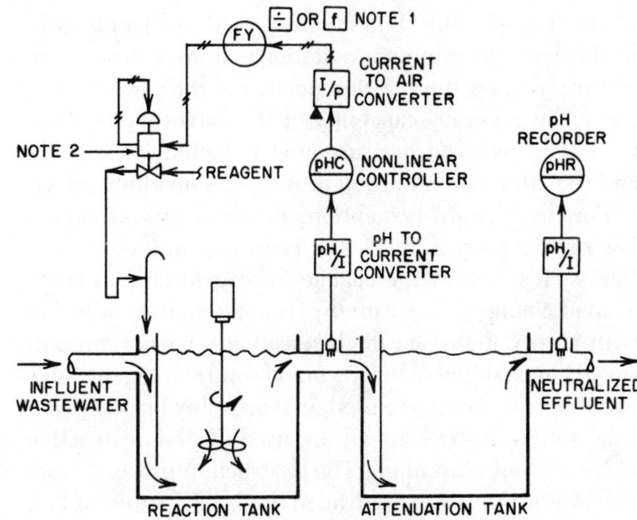

Fig. 10.14p Feedback control of pH

NOTE 1: For linearization of equal percentage characteristic, commercially available divider or function generator may be used here.

NOTE 2: Characteristics linear or equal percentage, depending on reagent delivery requirements. Positioner recommended for either choice.

delivery requirements. A linear valve can accommodate (for a strong acid-strong base) inlet pH variations of about ±0.5 units around some normal value. A single equal percentage valve can accommodate variations of about ±0.9 units about a normal value for the same set of conditions. A valve characterizer is shown in Figure 10.14p as an optional addition to the system for linearization, when using an equal percentage valve. A valve positioner is required to eliminate valve hysteresis (difference in opening and closing characteristics) and provide responsive valve movement.

The feedback controller in Figure 10.14p is a nonlinear controller with the characteristics shown in Figure 10.14n. The overall loop stability depends on the characteristics of the treatable waste material. For a process with a titration curve like that shown by A of Figure 10.14j, there is no question that the nonlinear characteristic will be helpful in achieving loop stability.

As the normal value of inlet pH increases towards neutrality (for an acid waste), the titration curve approaches that shown by B in Figure 10.14j. For this case the value of the nonlinear characteristic diminishes and therefore the nonlinearity should be dialed out, or a standard controller should be used. If the buffering characteristic of the waste is variable, there is no choice other than to adjust the nonlinearity of the controller for the severest case—which is usually the case of little or no buffering. The point is that the availability of the nonlinear feature markedly increases the flexibility of the control system at a moderate cost.

In Figure 10.14p two vessels or tanks are shown for illustration. A single vessel divided in half would also suffice. The objective is to provide a reaction section and an attenuation section. The reaction section should be as small as possible but should provide efficient backmixing with the minimum agitation cost, to permit a tight control loop on the reaction portion of the process. Because the accuracy capability of the valves will be less than required, the measurement is bound to be noisy and the attenuation portion provides a smoothing effect.

For those control conditions in which the setpoint is low or high, say 2 or 12, the process gain is very low; that is, it takes a large change in reagent flow to cause a small change in measured pH, and a linear controller with a very high gain (high sensitivity, narrow proportional band) suffices. In fact, on/off control (reagent valve is either fully open or closed) will probably be adequate. Low values of pH setpoint are used for the destruction of hexavalent chromium. The destruction proceeds rapidly when the pH is controlled at about a value of 2.0. Higher values of pH lengthen the process.

Sequenced valves can provide a wider reagent delivery capability (Figure 10.14q). The arrangement is virtually the same as that shown in Figure 10.14p except that the controller output can be switched to either valve by a pressure switch (PS), the function of which was described

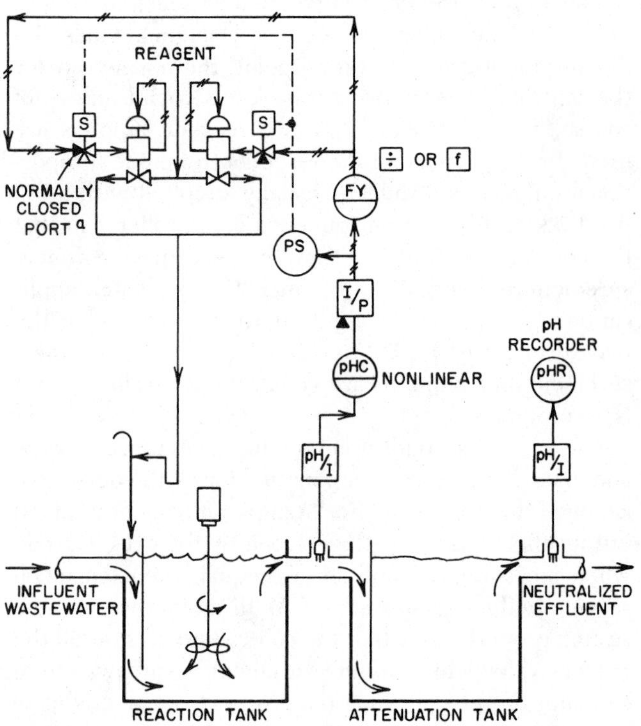

Fig. 10.14q Wide-range feedback control of pH employing sequenced valves

[a]This port is closed when the coil is de-energized due to pressure switch (PS) not having closed the electric circuit to supply current to the solenoid coil.

in Figure 10.14k. Valve positioners must be used since each valve must be calibrated to stroke over only a portion of the controller output signal range. Figure 10.14r illustrates various combinations of different pairs of sequenced valves. Table 10.14s lists the various flow rangeabilities for some valve pairs, assuming a constant pressure drop across the valves (equivalent to 9.5 feet, or 3.6 meters, of 66°Bé sulfuric acid) and assuming an individual valve rangeability of 35 to 1. The valve size coefficients (CV's) are 1.13, 0.14, 0.08 and 0.04, respectively for CV-1, 2, 3 and 4.

The overlap between each valve pair becomes smaller as the rangeability increases. The pressure switch to transfer the valves can be set anywhere in the overlap region because in this region the process loads can be satisfied by either valve.

Two-sided feedback control (Figure 10.14t) can be used when the influent may enter the system on either side of neutrality. Although only one valve for each side is shown, it would be possible to have a sequenced pair for one side of neutrality, and a single valve for the other, or a sequenced pair for both sides. Since this is a feedback control system, load changes cannot be frequent or severe in order for this system to give acceptable performance. For those applications in which load changes are frequent and severe, a combination feedforward-feedback should be considered.[7] If sequencing is employed,

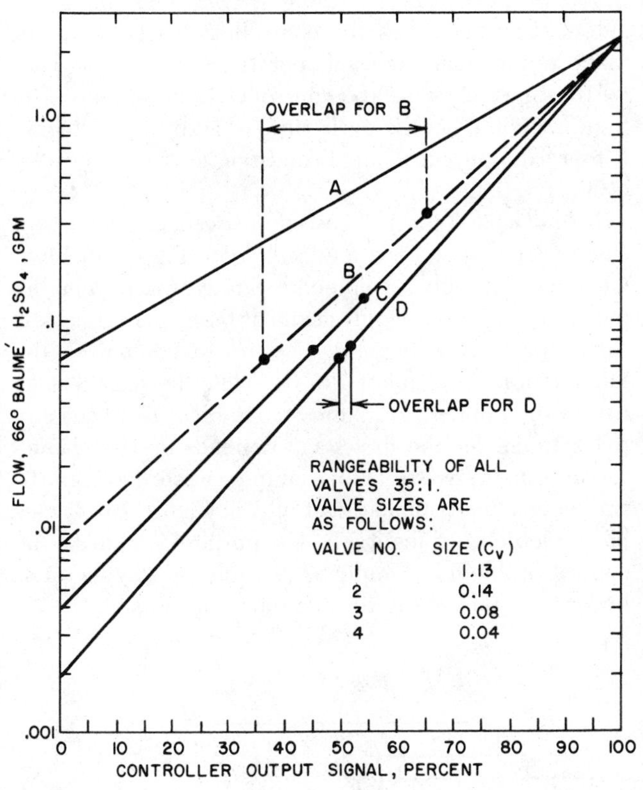

Fig. 10.14r Delivery capability for various valve pairs. *Key:* A = CV-1 alone; B = CV-1 + CV-2; C = CV-1 + CV-3; D = CV-1 + CV-4

the reagent delivery system will have a very high gain characteristic since the stroking of the pair (moving from closed to open) is accomplished with only half the controller output signal, thereby doubling the gain (making it twice as sensitive).

Table 10.14s
REAGENT DELIVERY TURNDOWN (rangeability) FOR SEQUENCED PAIRS OF EQUAL PERCENTAGE VALVES

Valve Pair	*Line on Fig. 10.14r*	*Turndown*	*Log Turndown*[a]	*Valve Positioner Calibration(s), (%)*
CV-1 (alone)	A	35 to 1	1.54	0 to 100
CV-1 + CV-2	B	275 to 1	2.44	0–63; 37–110
CV-1 + CV-3	C	570 to 1	2.76	0–58; 44–100
CV-1 + CV-4	D	1150 to 1	3.06	0–51; 50–100

[a]Signifies the approximate pH swing that valves will accommodate.

Ratio control of pH can be extremely effective when the wastewater flowrate is the major load variable, and the objective is to meet an increased waste flow with a corresponding increase of reagent. Since flow measurements may be in error and reagent concentration may vary, a means for on-line ratio adjustment must be provided. Figure 10.14u illustrates a ratio control system in which the reagent setpoint is changed in proportion to changes in wastewater flow. A feedback signal supplied by the feedback controller (pHC) also adjusts the reagent flow setpoint in proportion to a nonlinear function of the deviation between desired and actual effluent pH.

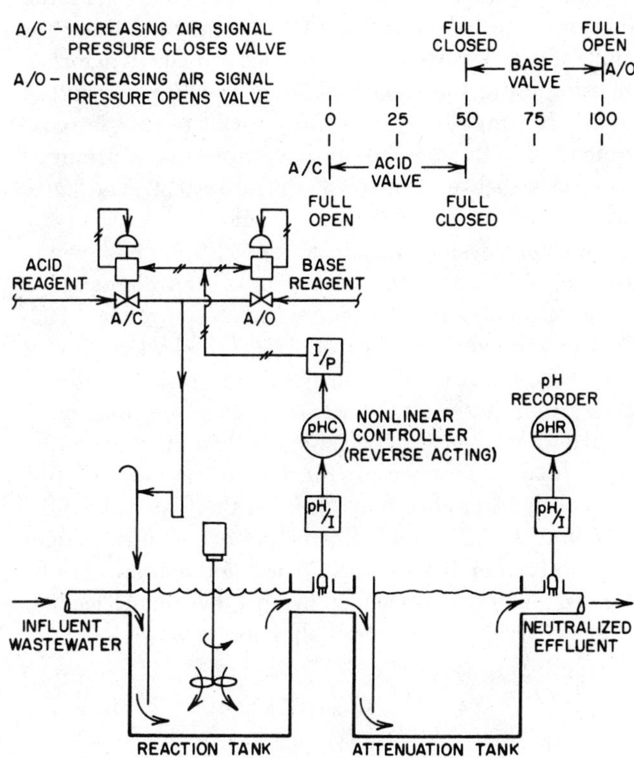

Fig. 10.14t Two-sided feedback control of pH

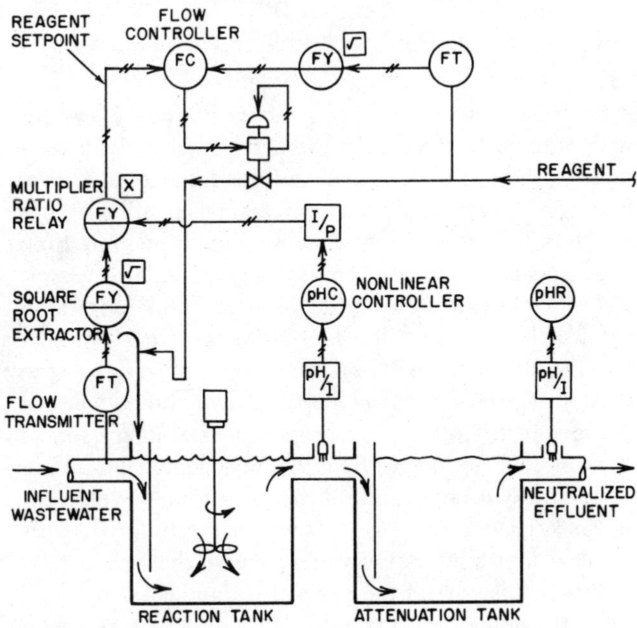

Fig. 10.14u Ratio control of pH

Cascade control (the output of one controller—the master or primary—being the setpoint of another) as applied to pH waste control systems can take two forms. In addition to the usual condition in which the output of one controller serves as the setpoint to another controller, it is also possible to have two vessels arranged in series, each with its own control system. The latter arrangement is referred to as cascaded residences.[8]

The conventional cascade control system[13] is shown in Figure 10.14v wherein the output of controller pHC-1 is the setpoint of the slave or secondary controller, pHC-2. This arrangement is particularly useful when lime is the reagent. In this instance, because of the finite reaction time between the acid and reagent, the setpoint of pHC-2 may have to be set lower than the desired pH of the final effluent because the materials are still reacting with each other after they have left the first tank. If the setpoint pHC-2 is too high, the pH of the final effluent will be greater than 7 or 8. When flocculation is to be carried out downstream of the pH treatment facility, stable pH values are extremely important.

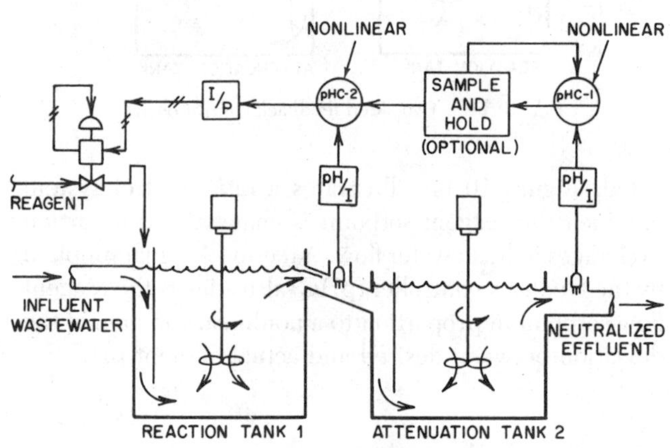

Fig. 10.14v Cascade control of pH

A delicate balance must be struck in this type of system with respect to the size of the first vessel. A long residence time in the first tank ensures long contact time between reagents, thereby producing an effluent pH which is close to the desired value, but at the same time it may result in a sluggish control loop around this vessel. For efficient cascade control, response of the inner loop (control loop around the first tank) must be fast. The outer control loop (pHC-2), sometimes referred to as the master or primary control loop, is usually tuned (control mode adjustments such as proportional band are set) so as to be less responsive than the inner loop. The tuning of pHC-2 will be a result of the dead time (a delay between a change in reagent flow and the time when its effect is first felt), capacity and process characteristics.

When this part of the process is dominated by dead time, the technique of sample data control may be useful in stabilizing the control system by a sample and hold device (Figure 10.14v). This device may be a timer that automatically switches the controller[9,10] between automatic and manual modes of operation. This can be done by the timer allowing the controller to be in automatic for a fraction (x) of the cycle time (t) and then switching it to a fixed output manual condition for the rest of the cycle $(1 - x)t$.

In the second form of cascade (cascaded residences), each vessel has its own feedback loop (Figure 10.14w). Chapman[8] strongly recommends this approach when the incoming waste is strongly acidic or basic (pH values less than 1 or greater than 13). The first stage controls the effluent at a pH of about 4 to 10, while the second stage brings the effluent to its final value of 7. The choice of pH setpoint for the first stage depends on the characteristics of the waste. For example, waste having a titration characteristic like that of A in Figure 10.14j may have a setpoint of about 3.5. The purpose is to make the control problem as simple as possible by staying on as linear a portion of the titration curve as possible.

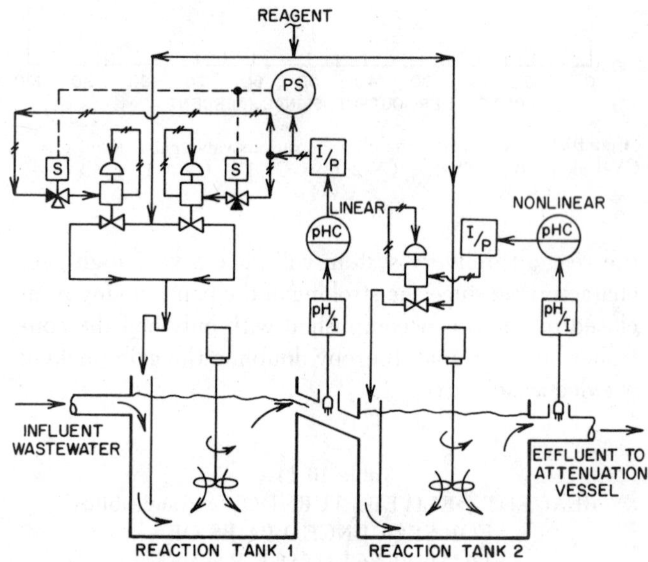

Fig. 10.14w Cascade residence control of pH

This approach is logical, considering the process gain characteristic as well as the accuracy limitation of a reagent delivery system. The remainder of the neutralization control problem is then similar to that illustrated by B in Figure 10.14j. In this manner the control system does not have to cope with the entire nonlinear characteristic of the process all at once. A sequenced pair of valves is shown in conjunction with the first stage in order to handle the pH load variations. A single valve would probably suffice for the second stage because its influent is pH controlled. Depending on the valve sizes and on the individual valve rangeabilities, this three-valve arrangement has a maximum possible reagent flow turndown of about 125,000 to 1 $(50 \times 50 \times 50)$.

Feedforward Control Systems In a feedforward control system, the consequences of upsets are anticipated and counteracted before they can influence the process. The system exerts control by correcting for changes in process load as they occur. The control system is essentially a mathematical model of the process. Ordinarily, the inclusion in the model of each and every load to which the process is subjected is neither possible nor economically justifiable. This means that a feedback control loop (usually containing the nonlinear controller for pH applications) is required in conjunction with the feedforward system.[11] The function of the feedback controller is to trim and correct for minor inaccuracies in the feedforward model.

The model for the case where a base reagent is added to an acid wastewater will be developed here. The extension to the case where an acid reagent is added to a basic waste can also be done in a similar manner.

A waste acid flowing at a rate F and normality N_a (concentration) is neutralized with a basic reagent having a normality N_b at a flowrate B:

$$BN_b = FN_a \qquad 10.14(1)$$

When computing hardware is used to perform arithmetic operations, the technique of normalizing or scaling inputs and outputs (converting into consistent engineering units) must be considered. The starting point for instrument scaling[14] is the engineering Equation 10.14(1). The next step is to write a set of normalized equations for each signal associated with the engineering equation. Normalization refers to expressing all inputs and outputs as a number between zero and one or equivalently between zero and 100 percent of instrument or transmitter input or output. The reagent and waste flows written in normalized form are

$$B = B'B_{max} \qquad 10.14(2)$$

where:

B = actual reagent flow, in engineering units
B′ = fractional output of measuring device
B_{max} = maximum value of measurable reagent flow, in engineering units

$$F = F'F_{max} \qquad 10.14(3)$$

where:

F = actual waste acid flow, in engineering units
F′ = fractional output of measuring device
F_{max} = maximum value of measurable acidic wastewater flow, in engineering units

Substituting Equations 10.14(2) and 10.14(3) into 10.14(1):

$$B' = CF'N_a \qquad 10.14(4)$$

where $C = \dfrac{F_{max}}{B_{max}N_b}$

Assuming a strong acid is being neutralized, its normality is approximately equal to its measurable hydrogen ion concentration:

$$N_a = [H^+] = 10^{-pH_i} \qquad 10.14(5)$$

where: $[H^+]$ = hydrogen ion concentration, mol. per liter, and the subscript i = inlet conditions.

Substituting Equation 10.14(5) into 10.14(4) and writing the resulting relationship in logarithmic form:

$$\log B' = \log C + \log F' - pH_i \qquad 10.14(6)$$

The log B′ portion of Equation 10.14(6) indicates that the reagent delivery device should have logarithmic characteristics; the equal percentage valve is such a device. The operating equation for the equal percentage valve is:

$$B' = R^{(X_B - 1)} \qquad 10.14(7)$$

where:

R = rangeability of the valve(s): 35:1, 50:1 and so on.
X_B = fractional control signal to valve(s), 0 to 1.0

Writing Equation 10.14(7) in logarithmic form and substituting into Equation 10.14(6) for log B′ and solving for X_B, the fractional control signal is:

$$X_B = 1.0 + \frac{\log F'}{\log R} + \frac{1}{\log R}(\log C - pH_i) \qquad 10.14(8)$$

Letting:

$$1.0 + \frac{\log F'}{\log R} = f(F') \qquad 10.14(8a)$$

where: f(F′) = function of inlet flow

$$X_B = f(F') + \frac{1}{\log R}(\log C - pH_i) \qquad 10.14(9)$$

Equation 10.14(9) is a mathematical representation of the process in which inlet flow and pH are the loads on the process. In those instances in which flow variations are less than about 3 to 1, the inclusion of flow as a load variable is *not warranted* since a flow variation of about 10 to 1 is equivalent to a pH change of one unit. If in this instance F′ is assumed to be equal to 1.0, Equation 10.14(9) becomes:

$$X_B = 1.0 + \frac{1}{\log R}(\log C - pH_i) \qquad 10.14(10)$$

Expressing pH_i on a scaled basis:

$$pH_i = S\,pH_i' \qquad 10.14(11)$$

where:

S = span of the inlet pH transmitter
pH_i = fractional transmitter output

Substituting Equation 10.14(11) into 10.14(10) and factoring:

$$X_B = 1.0 + \frac{S}{\log R}\left(\frac{\log C}{S} - pH_i'\right) \qquad 10.14(12)$$

The parenthetical portion of Equation 10.14(12) including the gain term (S/log R) is the form of a proportional controller (described by Equation 10.14[13]):

$$m_c = \frac{100}{PB}(r - c) \qquad 10.14(13)$$

where:

$\frac{100}{PB}$ = controller gain = $\frac{S}{\log R}$
PB = proportional band
r = setpoint = $\frac{\log C}{S}$
c = measurement = pH_i
m_c = controller output signal

By returning to Equation 10.14(1) and solving for the condition in which the maximum amount of ions are entering the process, Equation 10.14(14) is obtained:

$$N_{A(max)} = \frac{B_{max} N_B}{F_{max}} \qquad 10.14(14)$$

Equation 10.14(15) is developed by recalling the definition of C, applying it to the maximum acid flow condition which is being considered here, and assuming that normality and measureable $[H^+]$ are equivalent:

$$C = \frac{1}{N_{A(max)}} = \frac{1}{10^{-pH_{min}}}$$

$$\log C = pH_{min} = S\,pH'_{min} \qquad 10.14(15)$$

where: pH'_{min} = fractional equivalent of pH_{min} on a pH scale of span S.

Substituting for log C from 10.14(15) into 10.14(12):

$$X_B = 1.0 + \frac{S}{\log R}(pH'_{min} - pH_i') \qquad 10.14(16)$$

An examination of Equation 10.14(16) shows that when the inlet pH is at its minimum value ($pH'_{min} = pH_i$), Equation 10.14(16) equals 1.0, which indicates that the fractional control signal to the reagent delivery system is 100 percent corresponding to full reagent flow.

Equation 10.14(16) must now be modified to include feedback trim. Ideally, the feedback controller, which provides the trim signal, would not be required if the feedforward computation as developed were perfect. But inlet flowrate is not constant: reagent concentration may change, instrument measuring errors are present and the strong acid assumption will not be completely true 100 percent of the time. Since these loads are not likely to change suddenly or at a high frequency, feedback trim to adjust for their effect should be adequate. Ideally (when the feedforward model is perfect), the output of the feedback controller should be 50 percent or 0.5 on a scaled or normalized basis. This gives the feedback controller maximum flexibility in each direction. If the signal to the valves is to be 1.0 or 100 percent when the influent wastewater pH is at its minimum value and the output of the feedback controller is 0.5, then Equation 10.14(16) becomes:

$$X_B = 2.0 + \frac{S}{\log R}(pH'_{min} - pH_i') - 2m_b \qquad 10.14(17)$$

where: m_b = output of feedback controller, which ideally has a value of 0.5.

Equation 10.14(17) is difficult to implement with conventional analog hardware because of the 200 percent bias required; that is, the 2.0 term in equation 10.14(17). Rather than to work with the minimum value of pH, one can also work with the maximum value of influent pH corresponding to minimum reagent demand. The minimum controllable amount of reagent corresponds to a flow rate which does not yet cause the smaller sequenced valve to cycle on and off. Where basic reagent is being added to an acid waste, the maximum value of pH that can be controlled by the system is estimated from the valve rangeability and from a knowledge of the minimum expected influent pH. For example, if a pair of sequenced valves has an installed rangeability of 1500 to 1 and the minimum pH expected is 2.0, the maximum controllable influent pH value is 2.0 + log 1500 ≃ 5.2 = pH_{max}.

Equation 10.14(17) in terms of pH'_{max} is:

$$X_B = 1.0 + \frac{S}{\log R}(pH'_{max} - pH_i') - 2m_b \qquad 10.14(18)$$

The gain term (S/log R) is adjusted to give X_B a value of 1.0 when pH_i' is at the minimum expected value. For example, let the span (S) of the pH transmitter be 0 to 9.0, pH_i = 2.0 and pH_{max} = 5.2, then pH_i' = 2.0/9.0 = 0.222; pH'_{max} = 5.2/9.0 = 0.577, then:

$$1.0 = 1.0 + \frac{S}{\log R}(0.577 - 0.222) - 2(0.5)$$

$$S/\log R = 2.81$$

Since S/log R = 100/PB (Equation 10.14[13]), the proportional band setting required for this feedback controller is:

$$PB = \frac{100}{2.81} = 35.5 \text{ percent}$$

In those instances in which flow compensation is required, the 1.0 term of Equation 10.14(18) is replaced by the f(F′) expression as defined in Equation 10.14(8a). The f(F′) signal is easily generated by characterizing[3] the wastewater flow signal from a primary flow device. When the inlet pH is greater than pH_{max}, the sequenced valves are completely closed and a separate linear feedback controller can be used to operate a third valve, referred to as the trim valve. Since the reagent delivery accuracy requirement is dramatically reduced as the inlet pH approaches neutrality (Figure 10.14j), this type of control is sufficient. The trim valve is sized to deliver a maximum reagent flow, which is slightly in excess of the minimum reagent capability of the smaller sequenced valve. Figure 10.14x illustrates the reagent flow-controller output relationship for two sequenced valves in conjunction with a trim valve.

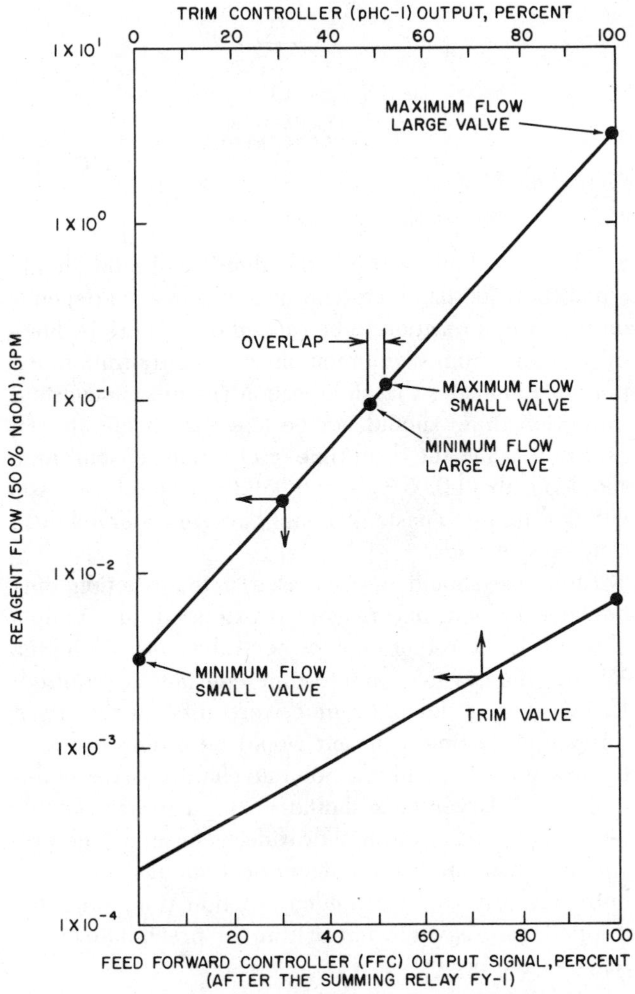

Fig. 10.14x Feedforward pH control utilizing three valves

Figure 10.14y illustrates a feedforward control system arrangement in which the flow characterization of the influent wastewater flow signal and the use of the dead band adjustment feature previously discussed are shown. A system essentially as outlined in Figure 10.14x has demonstrated the need for the combination of feedforward-feedback control because each type of control was tested individually and was found to be unsatisfactory.[4]

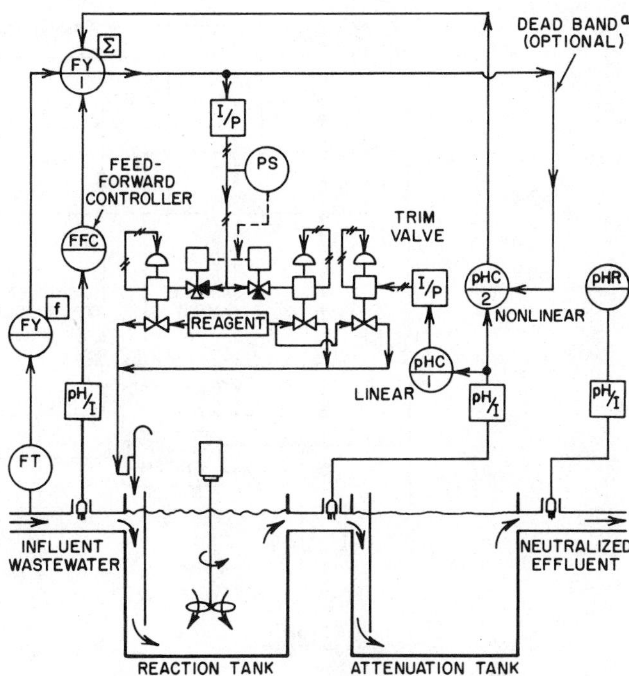

Fig. 10.14y Three-valve feedforward pH control system[4]
[a]If pHC-2 is provided with a dead band, it will be inactive when the trimming controller (pHC-1) alone is operating.

Figure 10.14z illustrates the feedforward-feedback control system arrangement for a two-sided waste neutralization task. The pH can be on either side of neutrality with severe measurement (load) variations. A logic system is also required in order to establish whether each (or neither) of the two systems is needed to be in operation. A combination of feedforward-feedback on one side of neutrality and conventional feedback on the other is also possible. The nature and characteristics of the problem to be solved will indicate the nature of the solution.

Equipment for pH and Composition Control

Composition processes, whether they be pH or "pIon" (such as pCl, pAg and so on), should be recognized as having two rather distinct aspects: one chemical, the other physical. As stated by Chapman[15] in 1952 regarding pH processes:

> Many pH control applications have become standardized to the point of catalog items while others introduce special features characteristic of their individuality. Problems of one application may be wholly physical while others may have chemical limitations and only by rigid separation and classification of all factors involved in a pH process, together with an understanding of their characteristics, can the process be most effectively instrumented.

The physical or process design considerations associated with composition control problems (pH or pIon) are

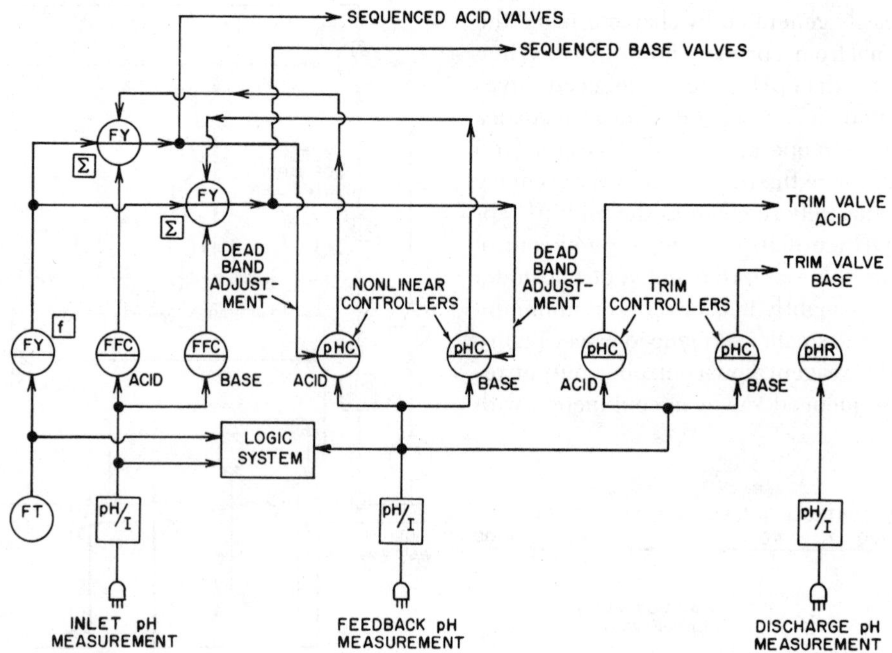

Fig. 10.14z Two-sided feedforward control of pH

tankage (size and number), baffling, agitation (how much and what type), measurement probe location(s), and reagent addition point location.

The ease or difficulty of most industrial control applications is closely related to a property of the process referred to as "dead time." Analogous terms such as "transport time," "pure delay" and "distance velocity lag" describe the same effect.[16] Dead time is defined as the time interval between the introduction of an input disturbance to a process and when a measuring device first sees the effect of that disturbance. Qualitatively, the relationship between dead time and controllability is simple: the more dead time, the more difficult the problem of control. The presence of dead time in pH or pIon processes is extremely detrimental to controllability. The major reason is the severe sensitivity of the measurement of interest at the control point. The pH process as described above illustrates this point. One of the major goals of system design is to eliminate or reduce the dead time to an absolute minimum.

Tank Size and Number

In this section the discussion will be directed at vessel(s) or tank(s) and the terms "treatment vessel(s)," "reaction vessel(s)" and "attenuation vessel(s)" will be used. "Treatment vessels" refers to all of the vessels comprising the facility. "Reaction vessels" defines the vessel in which the reaction between the waste stream and reagent takes place. "Attenuation vessel" refers to any vessel after the reaction vessel the sole purpose of which is to provide or add capacity to the treatment process.

The reaction vessel should be of cubic dimensions. If a cylindrical tank is used, the depth of liquid should equal the tank diameter. The size of the vessel depends on the rate of reaction to be carried out. There is, however, a minimum size limitation. For neutralization, experience indicates that the reaction vessel time constant (retention time) should not be less than 3 minutes (τ_1 = 3 min.), and the dead time (τ_d) to time constant ratio should be about 0.05 ($\tau_d/\tau_1 \simeq 0.05$). For example, a vessel with a 3 minute constant should have an observed dead time of 9 seconds.

Tank size should be increased if the reaction time between influent and reagent is extended, an example of which is the use of lime for neutralization. With high calcium lime, a five minute time constant is required. If, however, a dolomitic lime were used, a twenty or thirty minute time constant would be required due to the low solubility and reaction rate characteristics of this reagent. A larger tank should be used if an insoluble precipitate, such as iron hydroxide, is formed. The precipitate traps unreacted reagent or influent material, or both, and causes an extended reaction time since the trapped material must diffuse from the precipitate before reaction can occur.

The number of treatment vessels required is related to the difficulty of the control problem. The maximum or minimum pH or pIon values, or both, coupled with the degree of buffering or complexation at the control point, determine the degree of difficulty. For example, with strong acid-strong base neutralizations and with a dead time to retention time ratio of 0.05, one stirred tank will be sufficient if the influent pH is between 4 and 10. Two vessels are required, one stirred and one unstirred, if the influent pH is as low as 2 or as high as 12. If the influent pH is less than 2 or greater than 12, experience shows that three treatment vessels (two reaction and one attenuation) are required. The reaction vessels should be stirred and the third vessel should be unstirred. The unstirred vessel serves to damp the cyclic upsets that can occur in the effluent pH from the first stirred vessel. Although this description pertains to separate vessels, partitioning of an existing vessel will also serve the same purpose. Figure 10.14aa illustrates the attenuation effect.

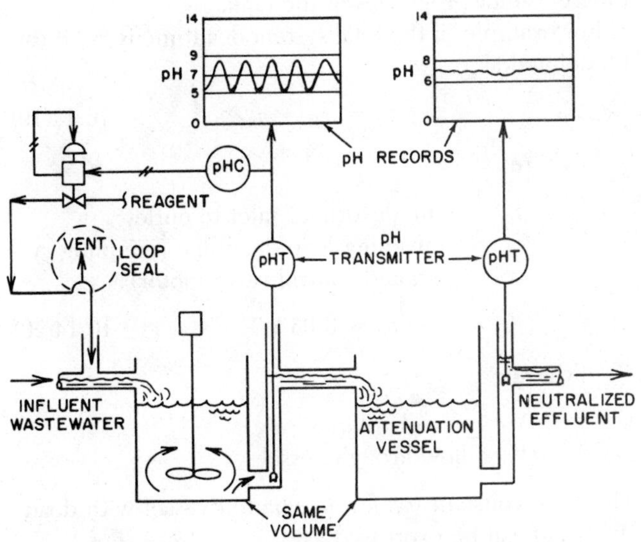

Fig. 10.14aa Effect of attenuation vessel

Tank Connection Locations

The inlet and outlet in the treatment vessel should be located at opposite sides—one high and one low—with respect to the bottom of the tank. Generally, it is most convenient to introduce the influent stream on the surface of the tank and to locate the outlet at the bottom of the vessel.

Variations in the location of the inlet and outlet can considerably change the dead time. For example, reversing the flow through the tank so that the inlet is on the bottom and the outlet at the surface causes the dead time to increase by a factor of 2 or 3. Examination of the flow patterns in the tanks (Figure 10.14bb) shows that the path from inlet to outlet can be doubled by this

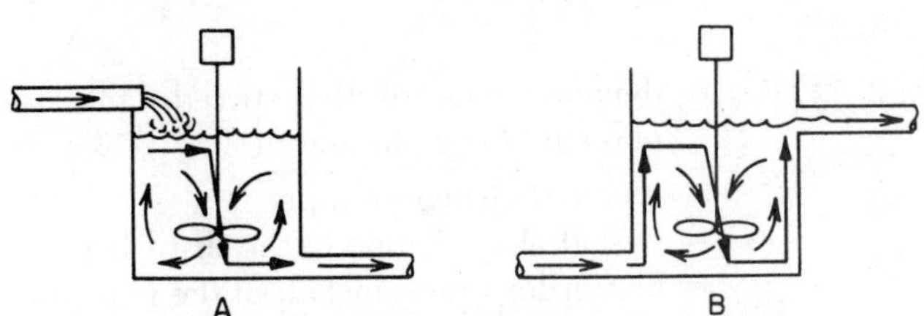

Fig. 10.14bb Flow patterns in stirred tanks. A. Recommended flowpath; B. Undesirable flowpath

change. The additional dead time is attributable to the swirl effect of the agitator, which is minimized but not eliminated by baffling.

Sensor Locations

The location of the measuring electrodes also deserves serious consideration. The general guidelines are that the locations should be responsive, and the information supplied by them timely. Submersible electrode assemblies are preferred when the measurement is used as an input to a control system. This preference is not always possible because of physical constraints. If flowthrough assemblies have to be used, the sampling time (the time required to physically transport the sample from the process to the electrodes which is essentially dead time) should be kept to a minimum. Figure 10.14aa shows a submersible assembly on the reaction vessel located as close as possible to the vessel exit. Location within the tank proper increases the measurement noise, principally because of concentration gradients. The requirements of the attenuation vessel monitoring electrodes are not as severe. Either flowthrough or submersible detectors can be used. The information supplied by these electrodes provides a clean record for the regulatory agency involved.

Equalizing Tanks

Upstream of a stirred neutralization vessel, a lagoon or holding tank can be very useful because it serves to smooth out upsets in influent pH and flow, thus allowing the use of a simple feedback system rather than a more costly feedforward control system. A lagoon can also be used to store the wastewater which is bypassed around the neutralization process in case of failure, a very important consideration if off-specification effluent causes a plant shutdown. The one thing that a lagoon cannot do is to replace a mixed vessel as part of a control system. Any attempt to control the pH of a lagoon by closed loop feedback control can only result in an effluent pH that oscillates between the influent pH and a complementary pH value on the opposite side of neutrality. The period of oscillation of such pH swings will depend on the dead time of the lagoon, but typically it will be on the order of hours.

Mixing and Agitation

The two types of mixing important to the control system are intermixing and backmixing. The reagent must be intermixed with the waste stream to furnish complete elimination of the areas of unreacted reagent or untreated waste. Adequate intermixing between influent and reagent can be readily achieved by adding the reagent at a point of small cross-sectional areas where there is some turbulence.[17] Figure 10.14aa illustrates the reagent being added in the pipeline before the influent enters the treatment facility. This is a desirable practice because it eliminates poor intermixing which can cause a noisy signal to be observed in the effluent pH[18]. A loop seal or gooseneck (Figure 10.14aa) at the point of reagent introduction is also suggested to preclude free draining of reagents contained in the line. The loop seal arrangement, particularly when long reagent transfer lines are required, allows the reagent line to remain full to the point of introduction to the process and thus eliminates a potential source of process dead time.

Backmixing is more important to close pH control than intermixing. The treated stream must be held in a vessel sufficiently long for the reagent to react and be backmixed. In general, the degree of backmixing can be defined in terms of the pumping capacity of an agitator with respect to the flow and volume of the neutralization vessel. In practice, however, this definition has limited usefulness because of variables such as agitator construction and blade pitch, baffling of the neutralization vessel and placement of inlet and outlet measuring electrodes. Experience shows that the best way to define backmixing for control purposes is by the ratio of the system dead time-to-retention time of the neutralization vessel. The retention time is the volume of the vessel divided by the flow through the vessel. A ratio of dead time-to-retention time of 0.05 is adequate for good control.

Suitable baffles or agitator positioning should be used if required in mixed neutralization vessels to avoid a whirlpool effect. The power supplied by the impeller must be used to turn the contents of the vessel over, not to whirl them about. With these effects in mind, a propeller or axial-flow impeller should be selected to direct the flow of the vessel contents toward the bottom of the tank. The flat bladed radial-flow impeller should be avoided since it generally tends to divide the vessel into two sections and increases system dead time.

Figure 10.14cc is a plot of tank size against agitator pumping capacity per unit tank volume on logarithmic coordinates. The family of curves shown for various deadtimes was developed from empirical data in tanks with capacities of 200, 1000, 10,000 and 18,000 gallons. They apply to baffled tanks of cubic dimensions with the inlet at the surface and the outlet at the bottom on the opposite side of the tank. The ratio of impeller diameter-to-tank diameter varies from 0.25 to 0.4. Square pitch propellers at an average peripheral speed of 25 fps were used in the tanks up to 1000 gallon capacity. Axial-flow turbine impellers at an average peripheral speed of 12 fps were used in the larger tanks.

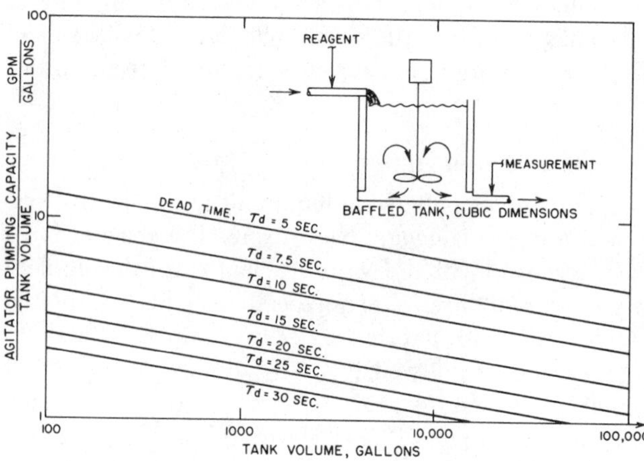

Fig. 10.14cc Dead time (τ_d) as a function of mixing intensity

Control Dynamics

The performance of a stirred tank to periodic disturbances can be evaluated by considering the deadtime and time constant properties of the tank.

For example, if the total system deadtime is τ_{dt}, it can be defined as:

$$\tau_{dt} = \tau_{d1} + \tau_{d2} \qquad 10.14(19)$$

where:

τ_{d1} = tank deadtime, inlet to outlet
τ_{d2} = remaining loop deadtime (sampling system and control valve motor)

$$\tau_{dt} = 0.05\ V/F \qquad 10.14(20)$$

where:

V = vessel volume
F = flowthrough vessel

The time constant (τ_1) for an agitated vessel with deadtime (τ_{d1}) can be expressed as[19]:

$$\tau_1 = V/F - \tau_{d1} \qquad 10.14(21)$$

Assuming that the stirred tank has the minimum 3.0 minute time constant previously mentioned and that the total deadtime is divided 80 percent to (τ_{d1}) and 20 percent to τ_{d2}, Equation 10.14(21) can be restated:

$$\tau_1 = .96\ V/F \qquad 10.14(22)$$

Expressing τ_1 in terms of deadtime by combining Equations 10.14(22) and 10.14(20):

$$\tau_1 = 19.2\ \tau_{dt} \qquad 10.14(23)$$

The dynamic gain of a stirred tank to periodic disturbances[20] is given by Equation 10.14(24):

$$G_d = \frac{\tau_o}{2\pi\ \tau_1} \qquad 10.14(24)$$

where

G_d = dynamic gain of the stirred tank = $\frac{\text{percent change in output}}{\text{percent change in input}}$
τ_o = period of oscillation of the disturbance
τ_1 = first order time constant of the tank; approximately equal to (tank volume/flow through the tank − system dead time)

To visualize the effect of dynamic gain, consider a flowing stream the pH of which falls from 7 to 4 and returns to 7 in one minute. If the stream flowed through a tank with one minute retention time (volume/flow), the spike in pH would pass through virtually unchanged and the effluent pH would closely track the influent pH. If, however, the stream flowed through a tank with 60 minutes' retention time, practically no upset would be observed in the effluent pH due to the capacity effect of the large volume.

The period of oscillation, τ_o, of a typical composition process under closed loop control with an optimally tuned (controller settings adjusted to match the process it controls) three-mode controller can be approximated as a function of the system deadtime[21]:

$$\tau_o = 4.5\ \tau_{dt} \qquad 10.14(25)$$

Substituting for τ_1 from Equation 10.14(23) and τ_o from Equation 10.14(25) into Equation 10.14(24):

$$G_d = \frac{4.5\ \tau_{dt}}{2\pi(19.2\ \tau_{dt})} = 0.0373$$

In this example the stirred tank has reduced the overall process gain by a factor of 27 (1/0.0373). Two tanks used in series reduce the process gain (slow the process down) by the product of their individual gains. Assuming a second tank identical to the first, two tanks in series would reduce the process gain by a factor of 27^2 or 729. With the stirred tank, therefore, it is possible to reduce the process gain to a controllable level. An added benefit of an increased tank capacity is to smooth out high frequency errors in reagent delivery caused by measurement noise.

This example is readily related to Figure 10.14aa in which the output of the reaction vessel is the input disturbance to the attenuation vessel. If the frequency or period (τ_o) of the input disturbance can be kept short (in the order of seconds) by virtue of a *tight* control loop around the reaction vessel, then the dynamic gain number of the attenuation vessel will be very low (0.0373 for the example), thereby increasing its attenuation capability. This results in a stable effluent pH which averages the input disturbance.

REFERENCES

1. Lange, N.A., *Lange's Handbook of Chemistry* (10th ed.) New York: McGraw-Hill, 1967.
2. Shinskey, F.G., *How Difficult is pH Control,* Publication 230A, Foxboro Co., Foxboro, Mass.
3. Shinskey, F.G., "High performance pH control systems," *Instrument Technology,* June, 1968.
4. Shinskey, F.G., and Myron, T.J., "Adaptive Feedback Applied to Feedforward pH Control," *Water and Waste Engineering,* February, 1972.
5. Special Instruction G-4430, Electronic Consotrol Model 62H Nonlinear Controller, Foxboro Co., Foxboro, Mass, August, 1970.
6. Shinskey, F.G., "End Point Control of Batch Processes," Publication 230, Foxboro Co., Foxboro, Mass.
7. Myron, T.J., "Guidelines for Effective Control System Design," *Instrument Technology,* January, 1972.
8. Chaplin, A.L., *Application of Industrial pH Control,* Pittsburgh, Pa.: Instrument Publishing Co., 1950.
9. Shinskey, F.G., *Process Control Systems,* New York: McGraw-Hill, 1967.
10. Shinskey, F.G., "Sample Data Control," Publication 230, Foxboro Co., Foxboro, Mass.
11. Lipták, B.G. (ed.). *Instrument Engineers' Handbook (Vol. II),* Philadelphia: Chilton, 1970. Section 7.6.
12. Ibid. Chapter VII.
13. Ibid. Section 7.8.
14. Ibid. Section 9.1.
15. Chapman, A.L., *Applications of Industrial pH Control.* Pittsburgh, Pa.: Instrument Publishing, 1950.
16. Shinskey, F.G., Process Control Systems. New York: McGraw-Hill, 1967.
17. Chapman, A.L., op. cit., P. 127.
18. Shinskey, F.G., and Myron, T.J., "Adaptive Feedback Applied to Feedforward pH Control," *Water and Waste Engineering,* February, 1972.
19. Shinskey, F.G., op. cit. P. 83
20. Shinskey, F.G., op. cit. P. 22.
21. Shinskey, F.G., op. cit., P. 83 and 103.

BIBLIOGRAPHY

Bates, R.G., *Electrometric pH Determinations.* New York, John Wiley & Sons, Inc., 1954.

Eisenman, G., *Glass Electrodes for Hydrogen and Other Cations.* New York, Marcel Dekker, 1967.

Eisenman, G., Bates, R.G., Mattock, G., and Friedman, S.M., *The Glass Electrode.* New York, John Wiley & Sons, Inc., Interscience, 1965.

Ives, D.J.G. and Janz, G.J., *Reference Electrodes.* New York, Academic Press, 1961.

Shinskey, F.G., *pH and pION Control in Process and Waste Streams.* New York, John Wiley & Sons, Inc., 1973.

Westcott, C.C., *pH Measurements.* New York, Academic Press, 1978.

Wilson, A., *pH Meters.* London, Kogan Page, 1970.

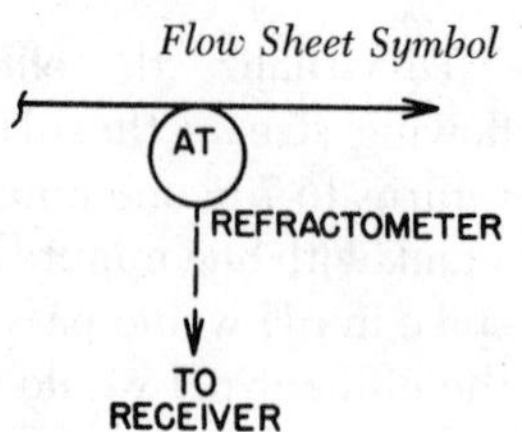

10.15 REFRACTOMETERS

Type of Sample:	Liquid and slurry
Materials of Construction:	Stainless steel and glass, standard
Inaccuracy:	±1% full-scale
Drift:	<1% full-scale per 24 hr.
Response:	10 to 30 sec
Flow:	2 to 50 cc per min. (differential)
Cost:	\$3,000 to \$9,000
Partial List of Suppliers:	American Optical Corp.; Anacon, Inc.; Bailey Controls Co.; Fisher Scientific Co.; Gaertner Scientific Co.; Lockwood & McLove, Inc.; Mine Safety Appliances Co.; Phoenix Precision Instrument Co.; Waters Associates, Inc.

A refractometer measures the refractive index (RI) of a liquid sample. Refractive index is the ratio of the velocity of light in a vacuum to the velocity of light in the material. Refraction of light (angularity change) occurs at the interface of two different mediums, except for perpendicular light incidence:

$$RI = N' = \frac{V\ (\text{Vacuum})}{V\ (\text{Material})} \qquad 10.15(1)$$

Snell's law expresses the relationship between the angle of incidence and angle of refraction when light is passed through the interface of two different materials (Figure 10.15a) as $n' = (\sin \alpha / \sin \beta)$.

The refractive index of a material is usually expressed in terms of air as a standard. A second phenomenon is observed with materials of different indices of refraction: when the angle of incidence exceeds a certain angle, the light ceases to be refracted and is totally reflected (Figure 10.15b). This angle is called the critical angle and is defined as:

$$\phi_c = \text{arc}\sin\frac{n'}{n} \qquad 10.15(2)$$

Two types of process refractometers have evolved, one based on refraction angle changes, and one based on critical angle changes (Table 10.15c). The differential refractometer measures the refraction angle changes as a function of sample RI. This is generally done by holding the incident light angle constant wherein Snell's law becomes $n = (1/\sin \beta)$.

The refractive indices of all materials vary with temperature and the temperature must be stated unless the standard reference temperature of 68°F (20°C) is applicable. Because temperature affects RI, this property must be controlled, or otherwise accounted for, in the process analyzer.

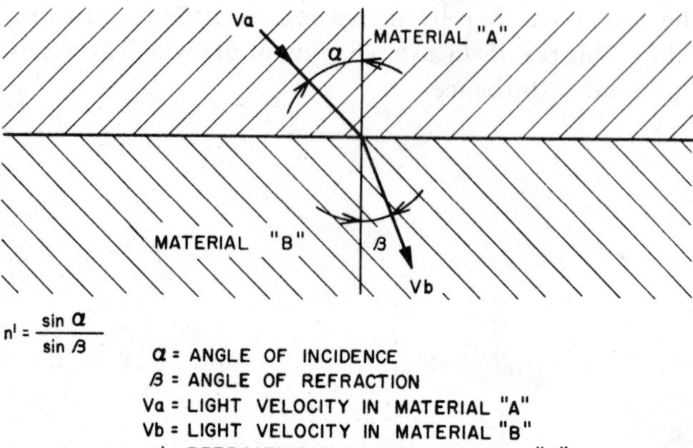

Fig. 10.15a Illustration of refraction terms

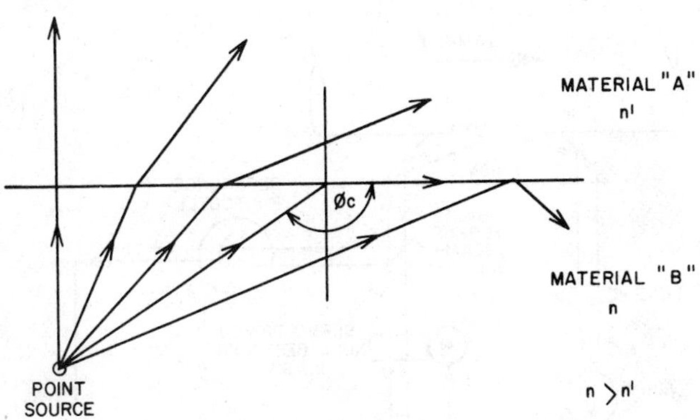

Fig. 10.15b Illustration of refraction and reflectance

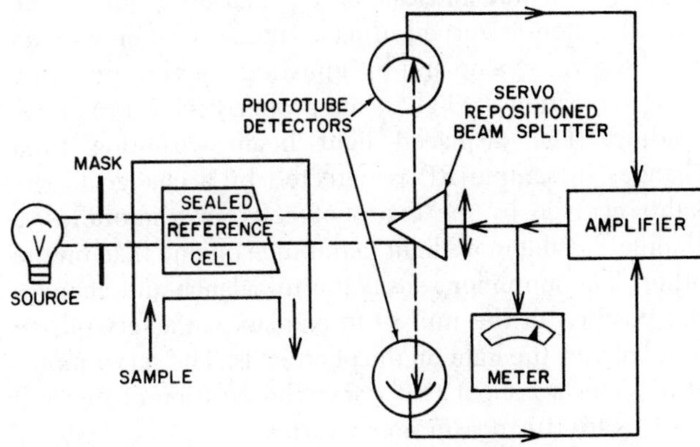

Fig. 10.15d Single-pass differential refractometer design

Table 10.15c
REFRACTOMETER COMPARISON SUMMARY

	Differential Refractometer	*Critical Angle Refractometer*
Measurement principle	Changes in angle of refraction	Changes in angle of total reflectance (critical angle)
Type sample	Clean translucent liquid	Liquids—clean or turbid, slurries
RI ranges (full-scale)	0.1 RI unit to 0.00005 RI unit	0.1 RI to 0.005 RI
Sample flow rate	2–10 cc/min.	Depends on type installation (unlimited)
Type of installation	Requires sample system	Can be used with sample system; can be installed in pipeline or vessel
Pressure rating	40 PSIG (276 kPa)	250 PSIG (1.7 MPa)
Ambient temperature	30–120°F (−1–49°C)	30–300°F (−1–149°C)

Differential Refractometer

Single-Pass Design

One elementary design illustrates the measurement technique (Figure 10.15d). This refractometer consists of a tungsten filament source, mask, sample flow cell with a sealed reference inner cell, beam splitter, and two opposed phototubes. The reference cell is filled with a solution with an RI of approximate midscale value. The light beam passes through all cell windows except one at a perpendicular angle, creating a refraction angle only at the non-normal window interface. The magnitude of the refraction angle is proportional to the difference between the RIs of the sample and of the reference solution creating a differential measurement.

The detector consists of a beam splitter and two opposed phototubes or photocells. The detector circuit can be designed so that the amplifier reads the ratio of the outputs from the two phototubes as a measure of the displacement, or refraction, of the light beam. Phototube irregularities, such as unequal surface response and unequal aging, limit the ratio system in sensitivity, linearity and range. Therefore, a null-balance system is usually employed in which the ratio output mentioned before is used to relocate the beam splitter, thereby rebalancing the light beam equally on the phototubes. The position of the beam splitter is then tracked as the measurement of RI variation.

Sensitivity and range of this design are changed by the reference cell angle and distance between the cell and detector system. Folded light path using mirrors helps maintain the unit in a reasonable package, even for a 40-inch (1 m) path length.

Two-Pass Design

The two-pass differential refractometer uses a triangular-shaped reference cell (Figure 10.15e) and a reflecting mirror. The reference cell shape presents two non-normal surfaces that interface with the sample to create two refraction angles for the forward pass and two for the return pass. The reference cell is filled with a solution

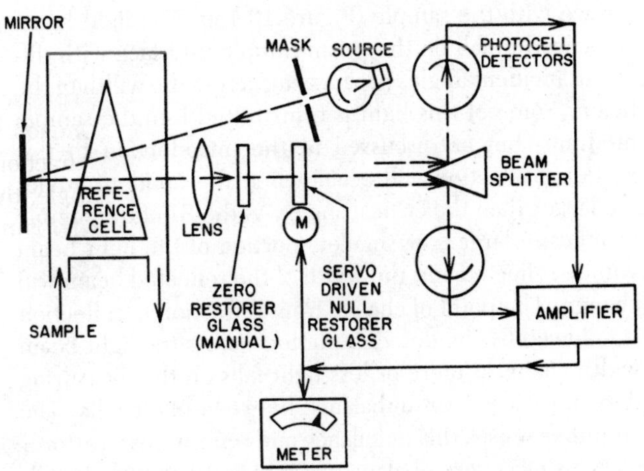

Fig. 10.15e Two-pass refractometer block diagram

with an RI about midscale of the measured range. The optical system is zeroed using a sample solution with an RI of the zero value and by adjusting the zero restoring glass to position the light beam property for a zero meter reading. The displaced light beam (resulting from changes in sample RI) is detected by a change in the light received by the two photocells, or phototubes, a change causing more light to impinge on one than on the other. The amplifier senses the unbalance and changes the position of the null restorer glass via a servo-drive to rebalance the light on the photocells. The servo-motor also drives a helipot that tracks the position of the null glass as the RI measurement varies.

As mentioned previously, sample temperature is critical in sensitive refractive index measurements. The differential refractometer generally employs a sample temperature controller to maintain the sample at a constant temperature. In some designs, the entire prism assembly is also controlled for maximum stability.

The range of this refractometer is determined by the shape of the reference cell (the angle between two refracting surfaces), the thickness of the zero restorer glass and the span attenuation in the meter circuit. This design is very sensitive, compact and flexible.

Flowing Reference Cell

The sealed reference cell of the differential refractometer can be replaced by a flowing reference cell to allow a differential measurement between two process streams. This has been used in blending operations to monitor the difference in RI continuously before and after blending. Extraction and filtration processes can use this system to monitor the difference in RI continuously before and after the extraction step.

Critical Angle Refractometer

Critical angle is the angle of incident at which light is totally reflected (see Figure 10.15b). The critical angle refractometer uses the reflected light from a prism interface with the sample (Figure 10.15f). The light beam is focused to fall on the prism-sample interface with different incident angles progressing across the width of the beam. Some of this light is transmitted into the sample medium, but as discussed in the introduction to this section, reflection will occur when the incident angles are larger than the critical angle. As the RI of the sample changes, a larger, or smaller, portion of the light beam will be reflected and the width of the reflected beam will change. The point of change from refraction to reflection is followed by the detector photocells. As the light beam width changes, more or less light falls on the measuring detector causing an unbalance between photocells. The amplifier senses this unbalance and repositions a restorer glass using a servo-balance system until the photocells are again in balance. A helipot in tandem with the null glass drive senses the position of the null glass as a measure of the critical angle.

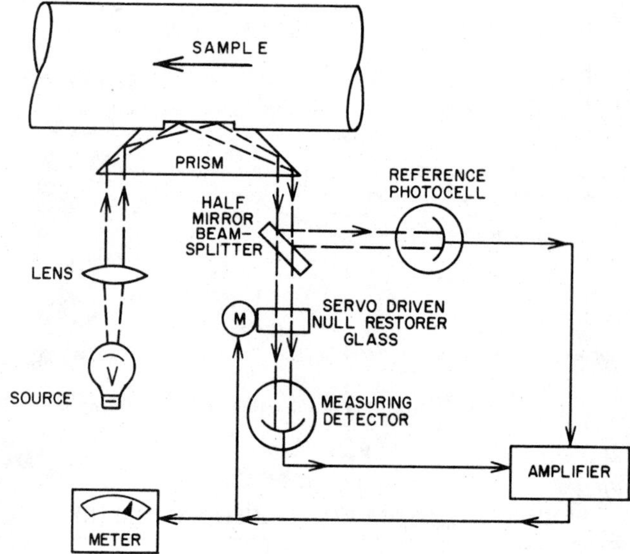

Fig. 10.15f Schematic representation for the critical angle refractometer

The range of this instrument is determined by selection of prism glass material and restorer glass thickness.

This unit is frequently installed in-line by mounting the prism assembly in a pipe and inserting this pipe section in the process line. The design also allows on-vessel mounting with the prism assembly in a flange member that can be attached to a vessel (storage tank, mixing tank, etc.). Another advantage is its potential use with slurry streams, opaque samples and viscous samples which present problems for the differential refractometer.

If temperature effects on the RI can be expected, temperature compensation can be provided using either a resistance thermometer, or a thermistor. The thermal sensor's output is utilized to counteract the thermal effects of the sample.

Limitations

The refractometer is a nonselective instrument that measures the RI of the entire sample. The RI of a mixture follows the simple mixture law:

$$N_{mixture} = C_aN_a + C_bN_b + C_cN_c + \cdots \qquad 10.15(3)$$

where

C_a, C_b, etc. are molecular concentrations of components a, b, c, etc, and N_a, N_b, etc. are the respective RI of components a, b, etc.

Thus, when used as an analyzer, the refractometer is limited to binary mixtures in the same sense as the thermal conductivity detector (Section 10.16).

Table 10.15c states the ranges applicable to both types

Table 10.15g
REFRACTIVE INDEX TABLE
All Data Based on 68°F (20°C)

Acetic acid	1.3718	Formic acid	1.3714
Acetone	1.3588	Glycerol	1.4729
Acrylic acid	1.4224	Glycol	1.4318
Amyl acetate	1.4012	Heptane	1.3876
Benzene	1.5011	Hexane	1.3749
Butyl acetate	1.3951	Hexanol	1.4135
Butyl alcohol	1.3993	Hydrazine	1.470
Butylene	1.3962	Hydrogen chloride	1.256
Carbon disulfide	1.6295	Lead tetraethyl	1.5198
Carbon tetrachloride	1.4631	Menthol	1.458
Chlorobenzene	1.5248	Methyl alcohol	1.3288
Chloroform	1.4464	Methyl ethyl ketone	1.3807
Cycloheptane	1.4440	Nitric acid	1.397
Cyclohexane	1.4262	Nonane	1.4055
Cyclohexanone	1.4503	Octane	1.3975
Cyclopentane	1.4065	Pentane	1.3575
Decane	1.41203	Perchloroethylene	1.5053
Di-ethyl benzene	1.4955	Phenol	1.5425
Di-methyl benzene	1.4972	Propanol(n)	1.3851
Di-ethyl ether	1.3497	Pronanol(iso)	1.3776
Ethyl acetate	1.3722	Styrene	1.5434
Ethyl alcohol	1.3624	Toluene	1.4969
Ethylbenzene	1.4952	Water	1.3330

of refractometers and the above relationship can be used to determine the range required for an expected concentration change. Table 10.15g gives the RI values of some common liquids.

°Brix Calibration

The sugar industry is concerned with the measurement of sucrose concentration. The weight percent concentration is called °Brix. The concentration of citrus juices and other food products is sometimes expressed in these units. Figure 10.15h shows the correlation of RI with °Brix.

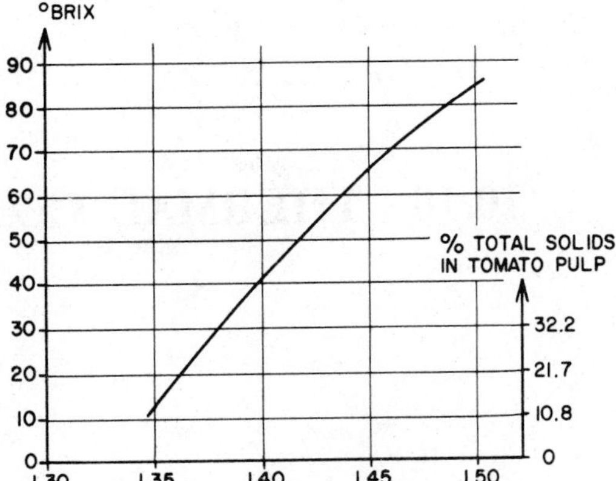

Fig. 10.15h Refractive index correlations in food industries

Summary

When used for binary mixtures, the refractometer can be a very accurate analytical tool, but trace impurities can severely limit its value. Mixtures with trace components having RI values close to one of the major components, may act as binary samples giving satisfactory results. The differential refractometer is more sensitive than the critical angle refractometer, but requires a sample handling system and a clean sample. The critical angle refractometer can monitor dirtier streams in-line.

BIBLIOGRAPHY

Frost & Sullivan, Inc., On Stream Process Analyzer Market, Report No. 669; Frost & Sullivan, Inc., New York, August, 1979.

Jutila, J.M., "Multicomponent On-Stream Analyzers for Process Monitoring and Control," *InTech*, July, 1979.

Jutila, J.M., "Guide to Selecting Gas and Liquid Chromatographs," *InTech*, August, 1980.

Maley, L.E., "Analysis by In-Line Refractometer," *ISA Journal*, October, 1962.

Wise, S.A. and May, W.E., "Unusual Experimental Detectors for LC," *Research/Development*, October, 1977.

Worthy, W., "Analytical Chemistry," *Chemical and Engineering News*, July 31, 1978.

Flow Sheet Symbol

AT
TH. CONDUCTIVITY
TO RECEIVER

10.16 THERMAL CONDUCTIVITY

Type of Sample:	Gas and vapors
Standard Design Pressure:	Atmospheric
Sample Temperature:	35 to 110°F (1.7 to 43°C)
Ambient Temperature:	30 to 100°F (−1 to 38°C)
Cell Materials:	Brass, stainless steel, Monel
Cost:	\$1000 to \$3500 for standard; \$4000 to \$9000 for high-sensitivity apparatus
Inaccuracy:	±2% full-scale for binary sample; poor for most others
Partial List of Suppliers:	Beckman Instruments, Inc.; Bendix Corp.; Condyne Instr. Inc.; DuPont Co.; Envirotech Corp.; Gow-Mac Instrument Co.; Horiba Instruments, Inc.; International Thermal Instrument Co.; Leeds & Northrup, a unit of General Signal; Mine Safety Appliances Co.; Teledyne Analytical Instruments; Uniloc Div. of Rosemount, Inc.; Varian Associates

Composition measurement using the thermal conductivity properties of gases is the simplest and earliest method of process analytical instrumentation. This technique makes use of the varying capacity of different materials to conduct heat energy from a heat source. This capacity, generally called thermal conductivity, differs for each gas and can be expressed in various unit systems such as BTU/hr./ft.²/°F/in.; w./sec./cm²/°C/cm; kiloergs/sec./cm²/°C/cm and so on. For purposes of gas analysis, comparison factors with a common gas, such as air, are generally used (Table 10.16a). In practice, continuous thermal conductivity analyzers measure a "change" in heat dissipation by comparing the change with a "reference" condition.

Early developments by the British resulted in an instrument of this type which was called a katharometer, or catharometer. The name still persists in Europe.

Principal Components

A thermal conductivity analyzer consists of three major components: measuring cell, regulated power supply and Wheatstone bridge, and case temperature control. The measuring cell consists of a relatively large mass of metal to provide a stable heat sink. Through the metal block, flow passages are drilled or formed, and a recessed cavity for inserting a heat-source-sensing element, such as a hot wire filament, is machined. The cell material must be compatible with the sample gas, yet must have a high thermal conductivity coefficient. Stainless steel is generally used.

The heat source may be either hot-wire filaments, or glass-bead thermistors. The filaments are usually tungsten or platinum alloy materials, and a wide range of thermistors may be used. These elements may be directly in the flow path, or recessed in cavities. Figure 10.16b illustrates a four-element thermal conductivity cell with recessed hot-wire filaments. Vertical mounting is preferred to prevent sagging of the wire elements. The recessed element generally has an improved noise level, but poorer response to changes. The elements are used in pairs: one in the sample stream, one in the reference gas. One or two pairs are normally used, but some cell designs include eight pairs to improve sensitivity.

The hot-wire filaments are in prevalent usage now following improved filament designs, but prior to 1965, thermal conductivity analyzers frequently utilized thermistors to achieve desired sensitivity. Thermistors are beads of metallic oxides with a thin coating, typically

Table 10.16a
THERMAL CONDUCTIVITY FACTORS

	R_0	R_{100}
Acetone	.406	.546
Acetylene	.776	.900
Air	1.000	1.000
Ammonia	.897	1.084
Argon	.709	.725
Benzene	.370	.573
Carbon dioxide	.614	.690
Carbon monoxide	.964	.962
Chlorine	.322	.381
Ethylene	.735	.919
Ethane	.807	.970
Helium	6.230	5.840
Hydrogen	7.130	6.990
Methane	1.318	1.450
Nitrogen	.996	.996
Oxygen	1.043	1.052
Pentane(n)	.520	.702
Refrigerant 12	.354	0.356
Sulfur dioxide	.344	.377

R_0, R_{100} = Thermal conductivity of gas/thermal conductivity of air @ 0°C and 100°C, respectively.

glass, over the surface. This coating tends to crack with excess heat and, when this element is used in gases with high hydrogen content, the oxides are reduced by the hydrogen and drift is experienced. Glass-bead thermistors develop frequent failures, particularly when used in high-hydrogen-containing gases. The hot-wire filaments develop surface temperatures between 400 and 750°F (204 and 400°C) and are sometimes used with a catalyst coating to promote catalytic cracking of hydrocarbons to further increase system sensitivity.

The sensing system is a basic Wheatstone bridge that uses a high-quality regulated power supply (Figure 10.16c). The power supply must be capable of delivering

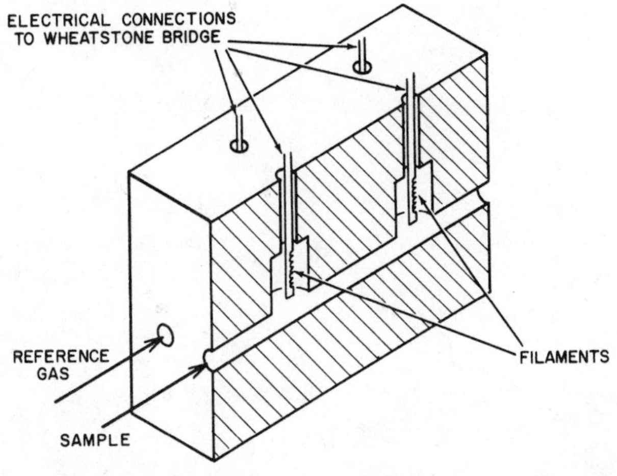

Fig. 10.16b Four-element thermal conductivity cell

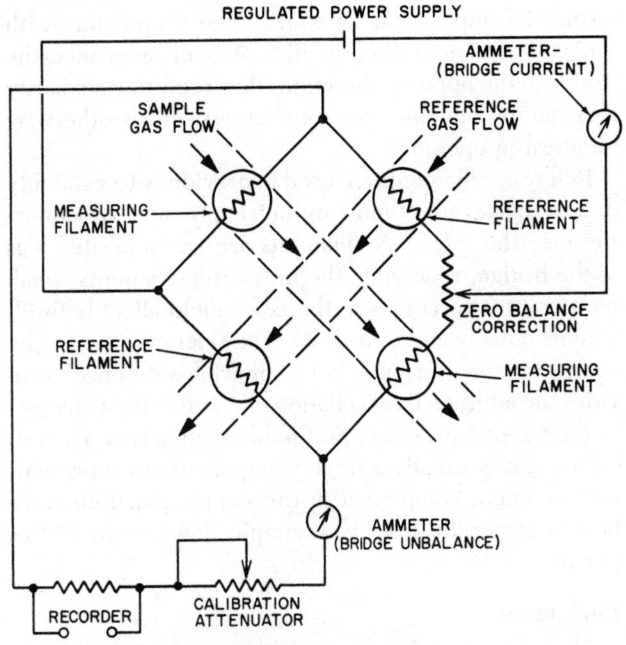

Fig. 10.16c Typical Wheatstone bridge measuring circuit for thermal conductivity analyzer

between 100 and 300 ma. Analyzer stability is primarily a function of power supply voltage regulation. In addition to the means of improving sensitivity mentioned previously, it is becoming increasingly more practical to use low-noise operational amplifiers on the bridge output. However, a low signal-to-noise ratio is required of the basic bridge output.

The case temperature control provides a constant temperature environment for the measuring cell to enhance stability. Various temperature control systems are used ranging from off/on thermal switches with bare strip heaters to the more refined ones for chromatographic ovens (see Figure 10.4e). Analyzer stability can be adversely affected by inadequate temperature control.

Operation

With the hardware involved now defined, the operation of the thermal conductivity analyzer can be described. The metered sample of 50 to 200 cc per m (flow controlled) passes through the measuring cell and across the filaments, or glass-bead thermistors, which are "hot" from current heating in the Wheatstone bridge. This resistance heating provides an elevated temperature on the surface of the filament. Heat energy is then conducted away from the filament, through the flowing gas, to the walls of the cell. The quantity of heat thus conducted away from the filament is a function of the thermal conductivity of the flowing gas. When a sample with a lower thermal conductivity than the balance (or zero standard) gas is introduced, less heat is conducted away and the filament surface temperature and its resistance increase, and cause an unbalance in the Wheatstone bridge. The degree of unbalance can be calibrated in terms of composition. Of course, the thermistor, with a negative temperature coefficient, will unbalance the bridge in the opposite direction, thus requiring a polarity reversal for the ammeter connections, but is otherwise identical in operation.

Reference filaments are used to provide better stability due to temperature and barometric pressure variations. Because the reference filaments are in "opposite" legs of the bridge, relative to the measuring filaments, small temperature variations in the cell should affect both filaments equally and cancel out. The reference gas can be sealed for this purpose, but a "flowing reference" provides the additional cancellation effects for slight changes in the venting pressure, preferably barometric. The reference gas is usually a single-component gas representing the major component in the sample gas. Reference flow is generally less than sample flow, 40 to 100 cc per m.

Packaging

The packaging of thermal conductivity analyzers varies with suppliers, but most units separate the electronics (power supply and bridge) from the temperature-controlled case, and, in some cases, the two parts can be physically separated by up to 100 ft.(30 m). In most designs, the cabinets require little more panel or wall space than conventional transmitters.

Calibration is accomplished using known samples to establish an empirical calibration.

Limitations

Although simple in detail, this analyzer has a major limitation: only binary mixtures can be accurately measured. The analyzer is nonspecific because it measures the total sample thermal conductivity and cannot distinguish what composition change causes the conductivity change—except for the binary, or binary-like mixtures. In practice, there are few industrial gas streams that are binary and require analyzing.

One notable exception to the above is the analysis of hydrogen in hydrocarbons. From Table 10.16a, it can be seen that hydrogen has a very high R_o relative to most hydrocarbon gases. The sample can be considered a hydrocarbon-hydrogen binary mixture and calibrated using an "average" background of nonhydrogen components. Accuracy in this case will depend on the true variations of the other components, but in many applications, ±5 percent full-scale inaccuracy is easily attainable.

Summary

Although the thermal conductivity analyzer is simple in design and relatively inexpensive, few true applications exist where it can be satisfactorily used. Its major application is as the detector in other analyzers, notably, the chromatograph (see Section 10.4).

BIBLIOGRAPHY

Guild, L., "Design and Performance of Thermal Conductivity Detectors," *Victoreen Issue* No. 3.

Jutila, J.M., "Multicomponent On-Stream Analyzers for Process Monitoring and Control," *InTech* July, 1979.

Kefford, F.D., "Automated Analysis as an Analytical Tool," 2nd Annual Water Technical Conference (Dallas, TX), December 2, 1974.

Tye, R.P., "The Art of Measuring Thermal Conductivity," *InTech*, March, 1969.

Flow Sheet Symbol

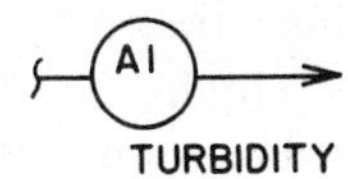

10.17 TURBIDITY

Standard Design Pressure:	To 250 PSIG (1.7 MPa)
Standard Design Temperature:	To 250°F (120°C)
Element Materials:	Stainless steel, plastics and glass
Cost:	\$2,000
Range:	0–0.5 to 0–1000 ppm; 0–0.2 to 0–10,000 JCU
Inaccuracy:	1 to 2% full-scale
Partial List of Suppliers:	Anacon, Inc.; Aquatronics Inc.; Bailey Controls Co.; Biospherics Inc.; DuPont Co.; Fisher & Porter Co.; Foxboro Co.; Gam Rad, Inc.; GCA/Precision Scientific Group; Hach Chemical Co.; Honeywell Inc.; Inter Ocean Systems Inc.; Jacoby–Tarbox Corp.; Keene Corp.; MONITEK, Inc.

The turbidity of a fluid is a measure of the amount of solids in suspension (Figure 10.17a). The units of measurement are ppm of suspended solids or Jackson candle units (JCU), which are defined as turbidity resulting from 1 ppm of fuller's earth, an inert mineral, suspended in water. The relationship between ppm and JCU depends on particle size, color and index of refraction, and correlation between the two is generally not possible. Turbidity instruments utilize a light beam projected into the sample fluid to effect a measurement. The light beam is scattered by solids in suspension, and the degree of light attenuation or the amount of scattered light can be related to turbidity. The light scattering is called the Tyndall effect and the scattered light the Tyndall light. A constant-candlepower lamp provides a light beam for measurement, and one or more photosensors appropriately placed convert the measured light intensity to an electrical signal for readout. Usually the photosensor is provided with a heater and thermostat to maintain a constant temperature because the device output is temperature sensitive.

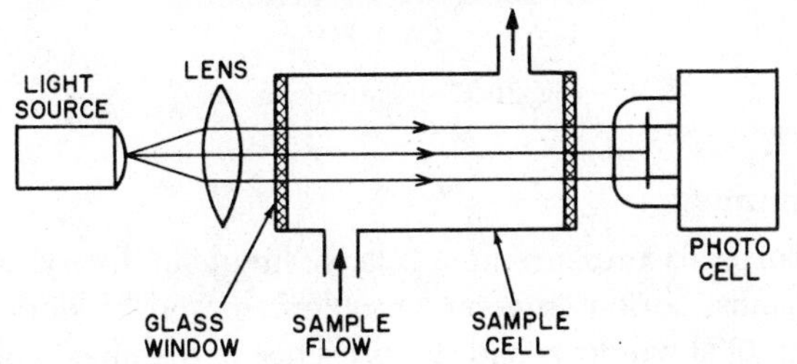

Fig. 10.17a Schematic of transmission-type turbidity meter

Instruments that measure light attenuation consist of a metallic or plastic pipe section with two glass windows located 180° apart. The light source and lenses at one window project a parallel light beam into the sample; a photosensor at the opposite window converts the attenuated light to an electric output.

Wide ranges of turbidity can be measured by matching the length of the light path to the level of turbidity. The supply voltage to the lamp must be regulated to at least ½ percent to eliminate errors due to source intensity variations because the measured light is referenced to the source. With this arrangement of lamp and sensor, the sensor output decreases with increasing turbidity and light absorbed by dissolved colors in the sample is also seen by the instrument as turbidity. Deposits formed on the flow chamber windows by the sample interfere with a proper measurement and, thus, the windows require frequent maintenance.

Instruments measuring scattered light vary somewhat in design. One type uses a flow chamber similar to the one discussed above, except that the window for the measured light is located at 90° to the window for the incident light. One window transmits light beams into the measuring chamber and the other, at right angles to the first, transmits scattered light to the photosensor. A light trap is located opposite the incident light window to eliminate reflection. With this arrangement, dissolved colors do not affect the measurement; however, instrument sensitivity is decreased by the presence of color because some light is absorbed by it (Figure 10.17b).

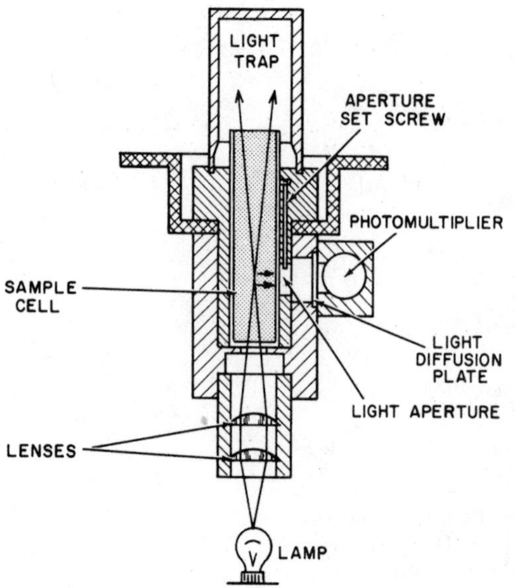

Fig. 10.17b Light-scattering turbidity meter

Variations of the basic unit include the use of two source beams and two photosensors in conjunction with two pairs of opposed windows as shown in Figure 10.17c.

Some designs utilize a separate photosensor to monitor lamp output and adjust the lamp supply voltage through a feedback circuit to maintain the light intensity constant.

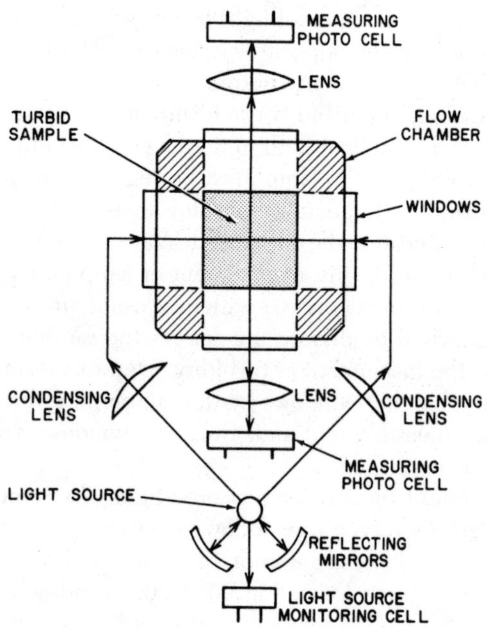

Fig. 10.17c Turbidity meter

Several units on the market do not have optical windows and are, therefore, not plagued by the problem of window deposits. Figures 10.17d and e show two such designs. The unit shown in Figure 10.17d measures surface scatter; that is, particles on the surface of the fluid scatter light in the direction of the photosensors. The design in Figure 10.17e projects the light beam into the sample and all particles in the beam path contribute to the measurement. The principal disadvantage of these designs is that they must operate at atmospheric pressure and at fairly low sample flow rates. These design features frequently exclude in-line installation of these units.

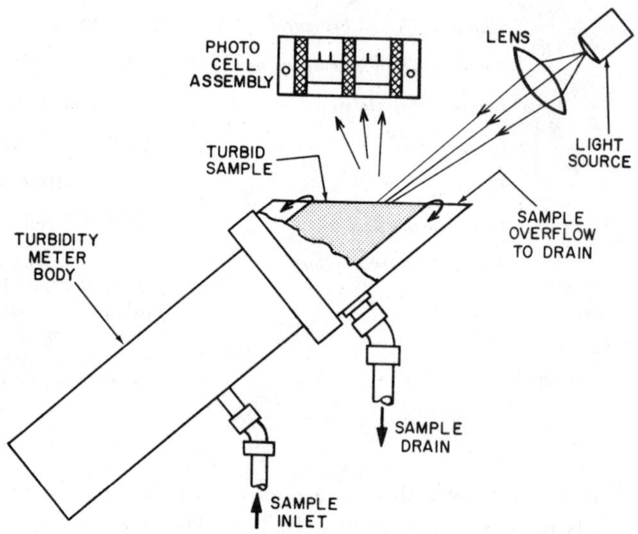

Fig. 10.17d Surface scatter turbidity meter

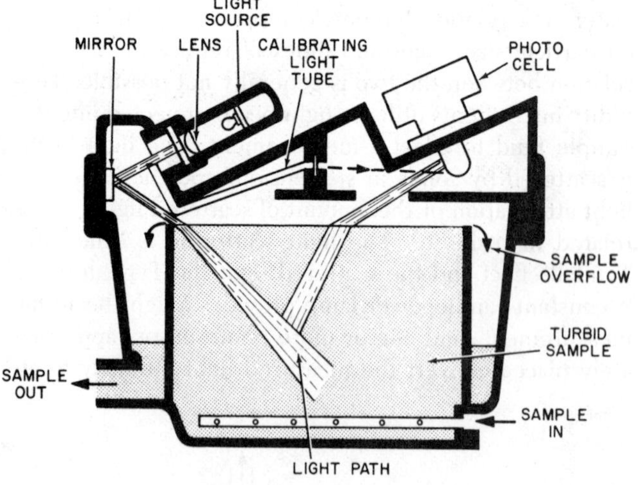

Fig. 10.17e Turbidity meter

Summary

Turbidity measurement is fairly simple in theory, with the most serious practical problems posed by deposits on optical windows and the presence of dissolved colors in the sample. The choice of a particular design will generally be governed by these factors and the materials of construction available for each type.

The same concept of utilizing light absorption is applied by instruments detecting smoke or fume density. Colorimeters were discussed in Section 10.5.

BIBLIOGRAPHY

Frost & Sullivan, Inc., On Stream Process Analyzer Market, Report No. 669; Frost & Sullivan, Inc., New York, August, 1979.

Hach, C.C., "Basic Turbidity Instrumentation," *Instruments and Control Systems*, December, 1968.

Jutila, J.M., "Multicomponent On-Stream Analyzers for Process Monitoring and Control," *InTech*, July, 1979.

Molvar, A.E., "Instrumentation and Automation Experiences in Wastewater-Treatment Facilities," EPA–600/2–76–198, October, 1976.

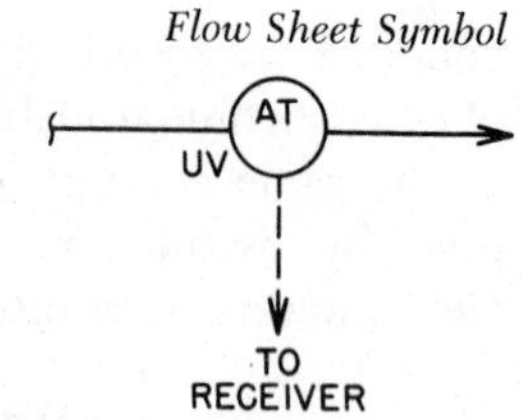

10.18 ULTRAVIOLET

Type of Sample:	Gas and liquid
Standard Design Pressure:	Atmospheric, cell can be pressurized to 150 PSIG (1 MPa) for special measurements
Standard Sample Temperatures:	30 to 200°F (−1 to 93°C)
Materials of Construction:	Quartz windows, stainless steel cell is standard—all conventional materials available
Cell Lengths:	0.25 to 2.0 in. (6 to 50 mm) standard; 0.002 to 40 in. (0.05 mm to 1 m) available
Cost:	\$7000 to \$15,000
Inaccuracy:	±1%
Partial List of Suppliers:	Anacon, Inc.; Beckman Instruments, Inc.; Du Pont Co., Analytical Instruments Div.; GCA/Precision Scientific Group; Perkins-Elmer Instrument Div.; Process Analyzers, Inc.; Teledyne Analytical Instruments; Ultra-Violet Products, Inc.

Electromagnetic Radiation Spectrum

The ultraviolet (UV) analyzer is a radiant energy (optical) analyzer that uses the ultraviolet region of the electromagnetic radiation spectrum (Figure 10.18a). This region consists of wavelengths from 100 to 400 mμ (Table 10.18b).

Chemical analysis is frequently achieved by absorption spectrophotometry in which radiation of different wavelengths is passed through a material, and the radiation transmitted, or absorbed, is measured to determine material identity and concentration. This technique is used in the ultraviolet, infrared, X-ray and similar regions, and, as applied to the laboratory, it usually surveys all the wavelengths in the IR or UV region, or at least in the region of interest, and plots the absorption vs wavelength (Figure 10.18c). The molecular structure determines the unique wavelengths at which a compound absorbs energy because the electromagnetic bonds and the energy levels associated with these bonds are affected. In the ultraviolet region, electrons experience discrete energy shifts when moving from one valence level to another.

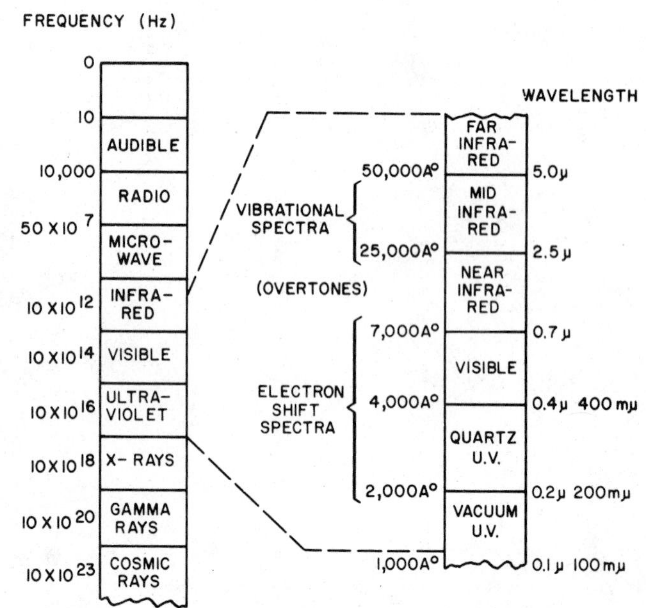

Fig. 10.18a Electromagnetic radiation spectrum

Table 10.18b
RELATIONSHIP OF WAVELENGTH TERMS

Angstrom = Å	$= 10^{-10}$ meters
Micron = μ	= 10,000Å = 10^{-6} meters
Millimicron = mμ	= 10Å = 10^{-9} meters
Meter	= 10^3mm = $10^6\mu$ = 10^9mμ = 10^{10}Å

The UV spectrum of typical absorbers (Figure 10.18c) illustrates the characteristic of the spectra shape found for absorbers in this region and is notably different from the spectra of absorbers in the IR region. For this reason, the UV analyzer can be expected to be less specific than IR; that is, in the presence of other UV-absorbing compounds, more interference can be expected from the broad characteristics of its spectra. On the other hand, UV analyzers are generally capable of more sensitivity than their IR counterparts—trace analyses are common for UV. This is due in part to the broad energy absorption and greater absorptivity of some compounds and to the availability of stronger sources that permits the use of longer cells.

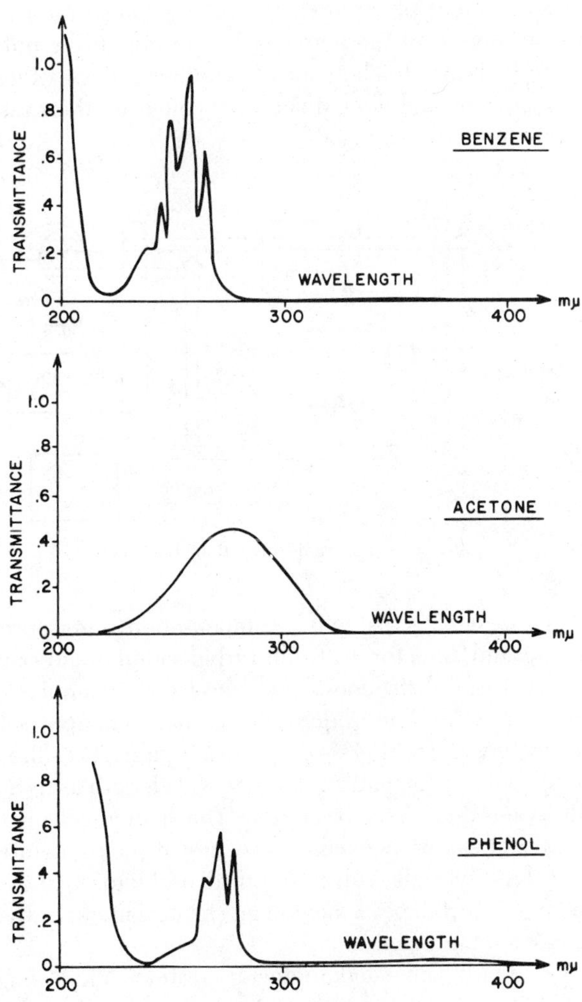

Fig. 10.18c Ultraviolet spectrums

A partial list of common gases and liquids with UV absorption are given in Table 10.18d.

Table 10.18d
PARTIAL LIST OF UV-ABSORBING COMPOUNDS

Acetic acid	Cycloheanane	Nitric acid
Acetone	Cyclohexanol	Ozone
Ammonia	1,3, Cyclopentadiene	Perchloroethane
Benzene	Ethylbenzene	Phenol
Bromine	Fluorine	Phosgene
Butadiene (1,3)	Formaldehyde	Styrene
Carbon disulfide	Formic acid	Sulfur
Carbon tetrachloride	Hydrogen peroxide	Sulfur dioxide
Chlorine	Hydrogen sulfide	Sulfuric acid
Chlorobenzene	Iodine	Toluene
Crotonaldehyde	Isoprene	Trichlorobenzene
Cumene	Mercury	Xylene (orthi, meta, and para)
Cyclohexane	Naphthalene	

There are several proven industrial designs for process UV analyzers. These basically fall into five classifications:

1. opposed-beam
2. split-beam
3. dual-beam–single-detector
4. dual-beam–dual-detector
5. flicker photometer

Opposed-beam Analyzer

This is perhaps the simplest of the designs and is limited to easy applications. The optical system (Figure 10.18e) consists of a source of two phototube detectors. The light source and detectors are so aligned that both detectors receive radiation from the same portion of the source. A sample cell and an interference filter are positioned between one detector and the source. The filter is selected to isolate the wavebands in which only the component of interest will absorb. The cell length is sized to give an absorbing path adequate for satisfactory sensitivity (i.e., high absorbers require shorter paths). A filter usually identical to the measuring filter is positioned between the source and reference detector.

The amplifier, or control circuit, compares the outputs of the two phototubes, and the difference in their outputs is related to the UV energy absorbed by the sample. The reference detector is used to provide compensation for changes in line voltage and source decay. Process analyzers have been built without this feature, measuring only the energy change at the measuring detector, but these analyzers are subject to drift and are generally unreliable.

The opposed-beam design is a simple, low-cost, moderate-accuracy instrument for simple, low-sensitivity measurements.

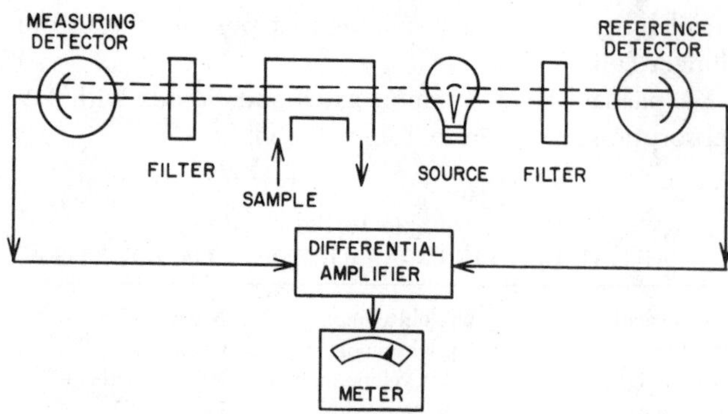

Fig. 10.18e Opposed-beam UV analyzer

Split-beam Analyzer

The split-beam UV analyzer uses a single-beam optical system relative to the sample cell (Figure 10.18f). A single sample cell is located between the source and the detectors. A beam splitter (semitransparent mirror) is used after the sample to create two paths of energy; one beam for the measuring wavelength and one beam for the reference wavelength.

Interference filters and/or broadband filters are used to isolate the desired wavelengths. The method of choosing the measuring wavelength was discussed under Electromagnetic Radiation Spectrum. The reference wavelength is selected in a region where the component of interest absorbs weakly or, preferably, not at all.

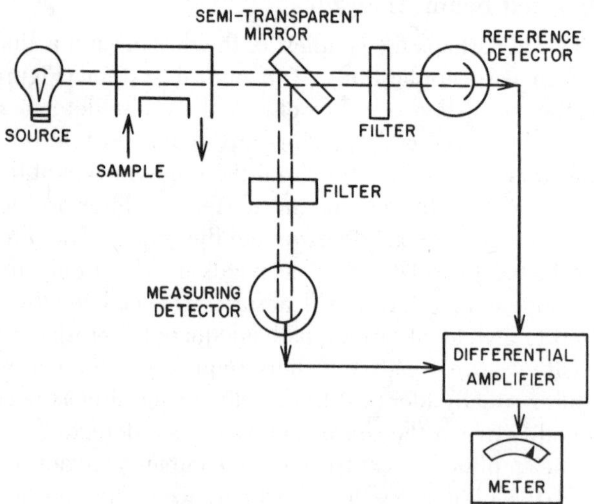

Fig. 10.18f Split-beam UV analyzer

The outputs of the phototubes are compared by the amplifier. The difference in energy absorption is measured in terms of the component of interest.

The use of the split-beam design with a reference detector offers the advantage of minimizing, or eliminating, the effect of other weak UV absorbers in the sample. Strong absorbers are rarely so eliminated. Further, source changes, sample turbidity, and cell window dirt are seen equally in both wavelengths and do not affect the measurement until the absolute energy intensity is below the sensitivity of the detector.

The split-beam system offers high sensitivity and accuracy with low drift for most applications, at a moderate cost.

Dual-beam–Single-detector Analyzer

The dual-beam-single-detector UV analyzer uses two optical paths, a single sample cell and a single photomultiplier detector (Figure 10.18g). One path includes the sample cell while the other path is used for reference because it does not pass through the sample. The paths are recombined and both pass through an interference filter that isolates the wavelengths selected for the measurement.

An interrupter or "chopper" is used to block the measuring and reference beams alternately. This creates pulses of energy through the sample cell which are 180° out of phase with the energy pulse of the reference beam. These pulses are received by the photomultiplier consecutively and are made equal in magnitude with a nonabsorber in the sample cell. When the sample contains the component of interest, the energy pulse intensity and the output of the detector for the measuring pulse are both reduced. The control amplifier receives these pulse outputs and demodulates, or converts, their ratio to a useable analog signal.

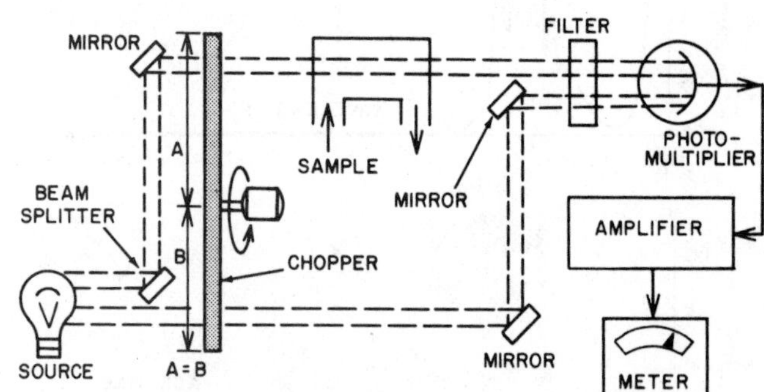

Fig. 10.18g Dual-beam–single-detector analyzer

This design gives good compensation for source changes and detector drift, but turbid samples and samples leaving a dirt deposit will cause drift. Samples of this nature sometimes require the use of two sample cells and two separate filters—one in each path. The filters create a measuring and a reference wavelength as in the split-beam–dual-detector system. The sample cells may or may not experience window coating at the same rate, which only partially solves the drift problem. Use of a "chopper" introduces a moving mechanical member that increases maintenance.

The dual-beam–single-detector system offers high sensitivity and accuracy for moderate cost, but zero drift may occur with "dirty" samples.

Dual-beam–Dual-detector Analyzer

The dual-beam–dual-detector design UV analyzer isolates the wavelength used for the measurement before the beam splitter (Figure 10.18h) and uses separate phototubes for the measuring and reference wavelengths. This design is a combination of the optics of the dual-beam–single-detector (see Figure 10.18g) and the detector of the split-beam analyzer (see Figure 10.18f). From this design, performance similar to that of the dual-beam–single-detector design can be expected; that is, high sensitivity and accuracy for a moderate cost, but "dirty" samples may create drift errors.

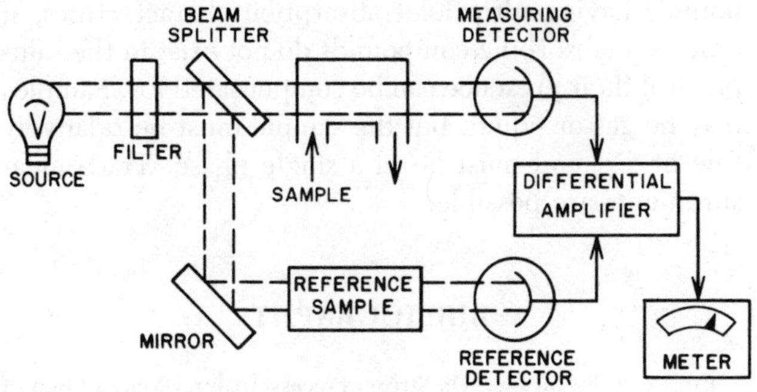

Fig. 10.18h Dual-beam–dual-detector analyzer

Flicker Photometer

The flicker photometer uses a single sample cell phototube (Figure 10.18i) and a rotating disc with two interference filters. The radiation transmitted by the sample cell alternately passes through filters that are selected to produce the measuring and reference wavelengths desired. The phototube receives these energy pulses and the amplifier calculates the change in ratio of the pulses as a measure of the component of interest.

This design minimizes effects of source decay, sample cell dirt and detector sensitivity changes by using the same optical path and phototube detector. The use of a rotating filter introduces moving parts that require maintenance. Accuracy and stability are good and cost is moderate.

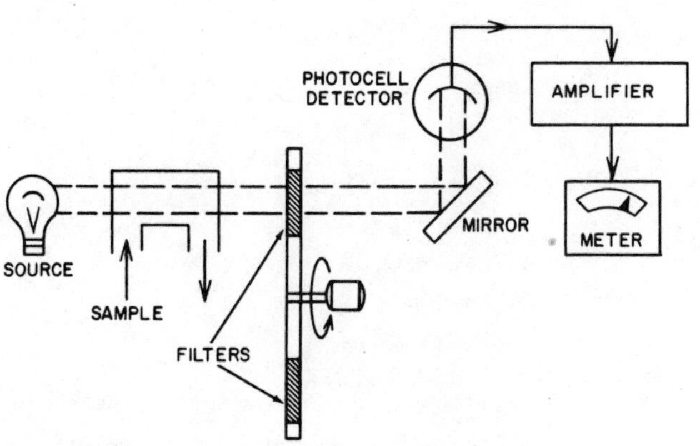

Fig. 10.18i Flicker photometer

Calibration

One unique and convenient feature the UV analyzer offers is the simple means of checking the calibration by use of a selected interference or broadband filter. Once the calibration (using known samples) is completed, a filter that absorbs at the wavelengths of interest can be used thereafter. The filter is selected to give equivalent absorbance of some percentage of full-scale when placed in the measuring beam with a nonabsorber in the sample cell. Nitrogen or water is commonly used to purge the cell during this operation.

Linearity

Beer's law of absorbance states:

$$A = abc = \ln \frac{I_0}{I} \qquad 10.18(1)$$

where:

A = absorbance
I = energy intensity leaving sample
I_0 = energy intensity entering sample
a = molar extinction coefficient or molar absorption
b = sample path length
c = absorber (sample) concentration

Ultraviolet analyzers normally require from 0.25 to 1.5 absorbance units for a full-scale measurement. When the higher absorbance units are used, the nonlinearity of the output becomes substantial. Many amplifier designs include linearizing circuits or log-amplifiers to allow a more linear analog output for ease of data presentation.

Sources

Two types of UV energy sources are used: The broad and the discrete line emission sources. The broad emission source provides energy in a broad wavelength band and narrow band filters are used to isolate the wavelengths of interest. These sources provide all wavelengths in the region but usually have a low-emission, or -energy level, at any given wavelength. Sources of this type include hydrogen, or deuterium, discharge lamps, tungsten lamps and tungsten-iodine lamps.

Discrete line sources use gas discharge lamps with narrow lines of emission. Mercury discharge lamps are prominently used due to their long life. These sources emit radiation energy at various discrete wavelengths at a high-energy level. The wavelengths that are not desired are filtered, leaving only the wavelength of interest. However, the gas discharge lamps are limited to the spectral emissions of mercury, cadmium, zinc and thallium. Therefore, all wavelengths are not available.

Selection of measuring and reference wavelengths is generally a compromise between the maximum absorbance wavelength of the component of interest and the wavelengths available from the gas discharge lamps. This usually leads to selection of a wavelength on the side of the absorption peak.

Detectors

Several types of detectors are used in process UV analyzers, including phototubes, photomultiplier tubes, and photocells.

The photoelectric effect is used in the vacuum phototube to produce a current proportional to the energy striking the tube cathode. A phototube with UV response has long life and a low-temperature coefficient.

The photomultiplier tube offers very sensitive detection of UV and visible light but large radiation energy levels will damage the light-sensitive surface. This detector has a high-temperature coefficient.

The photocell (photovoltaic) is a semiconductor light detector of the barrier layer type. A current is developed proportional to the light intensity but, unfortunately, the current output is not linear with the energy level. This may not be detrimental when used in a null-balance detection system, and the relative low cost of this device (because a voltage supply is not required) is attractive.

Packaging

Various packaging designs are available but most designs separate the amplifier/controller from the optical system. Some designs provide complete housing separation of the sample cell to provide isolation of electrical components from the flowing sample. The sample cell can then be temperature-controlled, when necessary, independent of the remaining optical system.

Both general purpose and explosion-proof housings are available.

Sample System

The criteria discussed in Section 10.2 apply generally, but an additional word of caution must be made about liquid samples. Vapor bubbles cannot be tolerated because they generate "noise" in the optical reading. If a pressurized cell is used on a gas sample, the cell pressure must be maintained constant.

Summary

The UV analyzer can be used to measure certain compounds having ultraviolet absorption characteristics, if other UV-absorbing compounds do not exist in the sample or if their presence can be compensated for. Samples may be gas or liquid, but the sample must be relatively free of dirt and must be in a single phase. Trace measurements are possible.

BIBLIOGRAPHY

Frost & Sullivan, Inc., On Stream Process Analyzer Market, Report No. 669; Frost & Sullivan, Inc., New York, August, 1979.

Jutila, J.M., "Multicomponent On-Stream Analyzers for Process Monitoring and Control," *InTech*, July, 1979.

Martin, J.M., "Ultraviolet Determination of Total Phenols," *Journal of WPCF*, 39 (1): 21–32.

Saltzman, R.S., "Process Stream Analyzers Based On Ultraviolet Absorption Changes with pH," 16th Symposium of the Analysis Division of ISA (Pittsburgh, PA), May, 1970.

10.19 WET CHEMISTRY—AUTOTITRATORS

Types of Designs:	a) Colorimetric: a1) high-precision, a2) standard accuracy; b) volumetric (autotitrator): b1) high-precision units titrating to pH endpoints, b2) standard precision units titrating to color change endpoints
Sample Requirements:	No suspended matter in sample (a1, a2)
Inaccuracy:	±2 to 3% full-scale (a2)
Cost:	\$9000 (a2, b2); \$13,500 to \$22,500 (a1, b1); reagent cost is \$2 per day per parameter
Partial List of Suppliers:	Beckman Instruments, Inc. (a1); Bran & Lubke, Inc. (a1, a2); Calgon Corp. (a1, a2, b1); Delta Scientific, National Sonics Div. (a2); Hach Chemical Co. (b2); Ionics Inc. (b1); Milton Roy Co. (a1, a2); Technicon (a1); Union Carbide Corp. (b1)

Colorimetric Analyzers

For highly accurate chemical analysis, it is necessary to have a well-trained and thoroughly experienced chemist working unhurriedly in a well-equipped laboratory.

High-precision, automatic colorimetric analyzers perform in the same manner as does the chemist. The sample and reagents are accurately metered and proportioned. Reagents are added in identical sequences with mixing and time delay steps between additions, as the chemistry and the laboratory procedure require. Where reactions are temperature dependent, a constant temperature bath becomes part of the analysis hardware. Sample preparation by distillation or dialysis can also become part of the automatic analytical procedure if required. The analysis results are about equal in inaccuracy to those obtained by an experienced analyst, but the former demonstrate better repeatability.

These analyzers contain delicate, precision electronic components that emit low-level signals requiring enormous amplification for recording and control. Therefore, it is generally advisable to make some investment in analyzer site selection and preparation. Vibration, dirt and dust should be eliminated and sudden ambient temperature variations should be avoided. An air-conditioned environment is ideal. It should also be realized that wet chemical procedures frequently liberate corrosive or deleterious gases. These must not only be vented from the analyzer cubicle to prevent corrosion of the electronic components, but must also be expelled from the temperature-conditioned room.

As colorimetry is a measure of light absorption owing to the color produced by the chemical reactions, it is essential that the sample fed to the automatic analyzer be absolutely free from suspended matter.

Reagent consumption for colorimetric analyzers of all types varies with the analysis to be performed. Generally the high-precision units use proportionately less reagent than the standard devices, but the reagents used are usually costlier and harder to prepare.

High-precision Colorimetric Analyzer

To effect accurate proportioning of sample with dilution water and reagents, most high-precision continuous analyzers use a multiple head peristaltic pump capable of accepting several plastic tubes and simultaneously pumping all liquids (Figure 10.19a). Ratio control is achieved by careful selection and control of the internal diameter of the tubes. Mixing is achieved by changing either tube diameters repeatedly downstream of the pump, thereby changing velocities, or flow directions, or both.

One analyzer introduces air through one of the pump tubes to create discrete air gaps between segments of fluid and then passes these liquid segments through glass

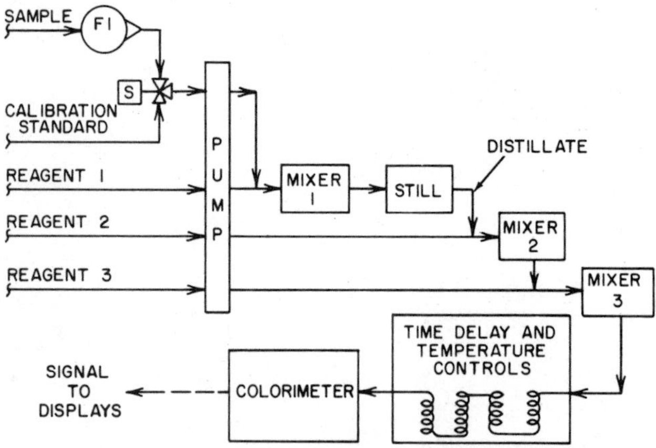

Fig. 10.19a Precision colorimetric analyzer with sample distillation as might be used for phenol or cyanide detection

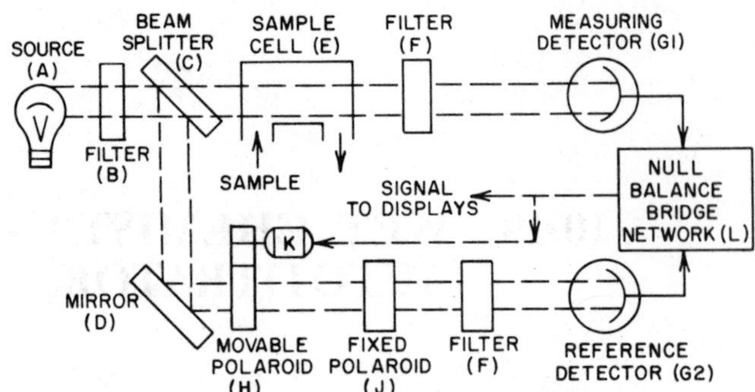

Fig. 10.19b Dual-beam colorimeter

or plastic horizontally mounted coils. The air gaps permit free fall of the liquid segments, thus achieving the mixing of sample and reagent. Retention time is provided by coils. Where temperature control is called for, these retention time coils can be made part of a constant temperature water bath that is thermostatically controlled and adjustable.

The reacted sample then passes to the colorimeter flow cell. These cells release air or gas bubbles prior to the optical measurement. Usually a single light source supplies two light beams of equal intensity. Detection wavelength for a specific test is established by the correct optical filter for the test. One light beam passes through the reacted sample to a photocell, while the second beam passes to an identical reference cell. The difference in cell outputs—the error signal—is the result of the color intensity developed by the wet chemistry steps and is later amplified and recorded.

A typical colorimeter is shown in Figure 10.19b. A single beam from light source (A) passes through an optical wavelength filter (B) and is split into two beams (C). The two beams are then deflected by mirrors (D). One beam passes through the sample flow cell (E), through an optical filter (F), which diffuses the light beam over a wide surface area of photocell (G_1). The other beam passes through a movable Polaroid lens (H), a stationary Polaroid lens (J) and an optical filter (F) and strikes an identical, matched, photocell (G_2).

With distilled water in the sample cell, maximum light is transmitted to cell G_1 and with the Polaroid lenses aligned in parallel, maximum light is also received by photocell G_2. Therefore, the outputs from the two photocells are equal and the bridge is in balance. With a colored solution in the sample cell, the light intensity striking photocell G_1 is attenuated. Output G_1 is now less than output G_2 and the error signal drives the movable Polaroid lens through drive motor (K) until the Polaroid system has decreased the light intensity striking G_2, so that the outputs of the two photocells are again equal. The distance traveled by the movable Polaroid is then a function of color intensity of the sample and can be recorded as the concentration of the constituent analyzed.

Most colorimetric analyses obey Beer's law (see Section 10.18). As such they tend to be nonlinear. Usually, however, the range requirement for a specific application is sufficiently narrow that the pertinent portion of the curve may be considered essentially linear.

Calibration

Colorimetric analyzers are generally equipped with automatic zero and standardization features. At specific intervals the reagent feed is stopped, and sample only is introduced into the sample flow cell. After a predetermined period of flushing with pure, unreacted sample, the analyzer reading should be zero. Circuitry drives the reading to zero, thus compensating for sample or flow cell discoloration and for electronic drift. Next, the sample is diverted to waste, the reagent system is reactivated and, instead of sample, a laboratory-prepared standard solution is introduced. After an interval sufficient to replace all liquids in the sample flow cell with freshly reacted standard solution, the output signal is automatically compared to the signal level expected for the standard solution strength, and adjustments are made, if needed.

The length of the optical path, that is, the depth of reacted sample penetrated by the light beam, is critical. In an analysis for trace amounts of contaminants in which color developed by the chemical reaction is expected to be weak, a long optical path is essential. The longer the light path, the less signal amplification is needed, and, therefore, the greater the accuracy of analysis. Conversely if the color intensity is high, a short optical path is needed so that the high-color intensity does not attenuate the light beam excessively. High-precision analyzers generally have interchangeable flow cells that may range from 2 to 10 in. (50 to 250 mm) in optical path length.

Simple Colorimetric Analyzers

The simpler continuous colorimetric analyzers are generally limited to analyses requiring no more than two or three reagents that can be added simultaneously. Although some of them can proportion the reagents to the sample by individual head vessels and capillaries, other designs employ multiple solution pumping heads coupled to a common drive motor unit. Each pumping unit operates in the same phase and is individually adjustable. Mixing of sample and reagents is accomplished in many ways, from mechanical stirring in the sample cell to changes in velocities and direction and to free-fall into a head vessel which, in turn, discharges into the colorimeter flow cell.

The colorimeter itself is generally unsophisticated (Figure 10.19c). A single light source (A) is used. Light is collimated by a lens (B) before it passes through the sample cell (C), either in a single or in a double pass as illustrated. It then passes through a color filter (E) to the measuring photocell (F_1). A second light beam passes directly through the color filter (E) to a second "reference" photocell (F_2). Both beams can be attenuated by shutter (G). These shutters are used for zero and full-scale adjustment. The output differential is generally displayed in concentration of constituent on a nonlinear meter scale. Recorder outputs are also available.

Ordinarily the simple analyzers are manually standardized against calibrated standard slides provided by the manufacturer. These may be in the form of one or more calibrated orifices that limit the amount of light striking the measuring photocell, or in the form of optical filters serving the same purpose. To overcome the effect of ambient temperature changes on photocell output, colorimeters may be equipped with thermostatically-controlled heating elements that provide a system relatively free from drift.

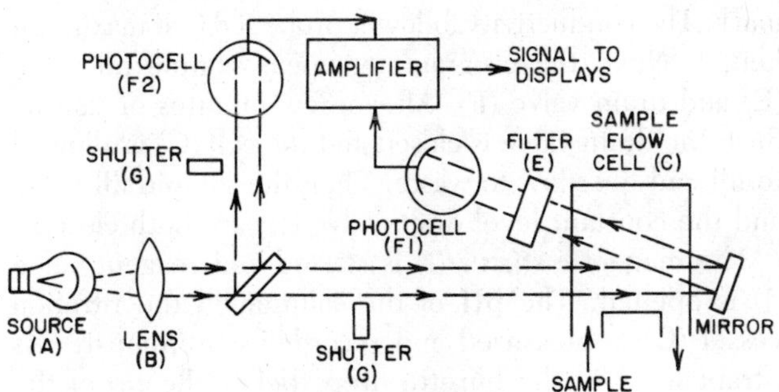

Fig. 10.19c Standard colorimeter

On/off Batch Colorimeters

Recognizing that continuous analysis may not be necessary for alarm or on/off control purposes on streams that are relatively stable, automatic batch analyzers have also been developed. These are usually single reagent devices in which the set point for alarm or control is fixed by the reagent. Hardness analysis is an example. Reagents based on the Versene method have been developed that undergo a dramatic color change from green to red when hardness exceeds the chelating properties of the reagent. Reagents exhibit this vivid color change for a wide range of water hardnesses, such as 0.75, 1.5, 5.0, 9.0 and so on up to 50 ppm or more of calcium carbonate.

The optical system (Figure 10.19d) is simple in that a photosensitive device is used which sees only the green light, a circumstance denoting that the sample analyzed is below the reagent set point. At the color change from green to red, the wavelength filter screens out all light in the red wavelength. The photocell then receives no light and an alarm signal is generated. The sample cell, illuminated from the back, is visible to the operator. Fail-safe features are inherent in this approach. Failure of the light source, the photocell, reagent or sample flow, if not visually detected by the operator, will be noted as a result of energizing the alarm circuit. With this type of analyzer it is obviously not possible to record analytical values. However, it is possible to use a simple voltmeter events recorder actuated by the alarm circuitry, that will record the time of day and duration of the alarm condition.

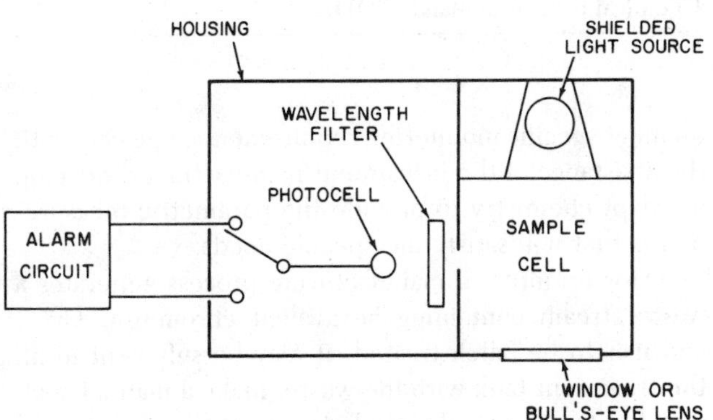

Fig. 10.19d On-off batch colorimeter

Application of Colorimetric Analyzers

Three discrete areas in the control of water pollution in which automatic colorimetric analyzers may find application include monitoring of influents to a treatment facility; monitoring of effluents from a treatment facility and the control of the actual treatment or abatement process.

Table 10.19e lists some of the more common analyses required by pollution abatement legislation, together with the maximum concentrations normally associated with standard test procedures. Other procedures and concentrations or ranges are also commercially available.

Table 10.19e
AVAILABLE AUTOMATIC COLORIMETRIC ANALYSES

Type of Analysis	Maximum Concentration (ppm)	Type of Available Unit: High-Precision Analyzer	Type of Available Unit: Standard Analyzer
Aluminum	10	√	
Ammonia	10	√	
Borate	10	√	
Carbonate	4,000	√	
Chlorate	12%	√	
Chloride	10	√	√
Chlorine	5	√	√
Chromium Cr^{6+}	5	√	√
Chromium total	5	√	
Copper	10	√	
Cyanide	5	√	√
Fluoride	5	√	√
Hardness total	500	√	√
Iron	10	√	
Magnesium	150	√	
Manganese	2%	√	√
Phenol	5	√	
Phosphate total	100	√	√
Silica	15	√	√
Sulfate	500	√	
Surfactants (anionic only)		√	
Chemical oxygen demand (COD)		√	

To meet specific monitoring requirements, one can (with the assistance of the instrument manufacturers) develop or adapt chemistry to provide the parametric measurements that will satisfy the specific needs.

If, for example, a manufacturing process generates a waste stream containing hexavalent chromium, Cr^{6+}, and it is to be batch treated, it may be sufficient to fill the treatment tank with the waste, make a manual analysis for treatment chemical dosage requirements and again manually analyze the result after treatment, prior to discharge. If this same waste is to be treated by a continuous process, it is very likely that conductivity measurement can be used more economically than can colorimetry for the control of chemical additions. The effluent from continuous treatment, however, should be monitored for residual hexavalent chrome content after dilution with other wastewaters.

If, on the other hand, cyanide waste is treated continuously by the alkaline chlorination process, it may be more economical to monitor the effluent indirectly for residual chlorine rather than for cyanide by colorimetric means. Presence of free available chlorine in excess of 5 ppm would denote absence of cyanide because these two chemicals cannot coexist.

Where colorimetric analysis is warranted, efficient sample preparation and careful analyzer site selection are essential.

Volumetric Analyzers

Much that has been said for automatic colorimetric devices is also applicable to volumetric analyzers. These are devices that automatically perform titrations either to electrometric endpoints or to their colorimetric equivalents. Again, high-precision instruments are available as well as simpler standard devices that perform valuable functions with adequate accuracies and repeatability.

The need for automatic titrations and for the development of a wide array of titrators has not yet been recognized by the instrument manufacturers. Only two distinct types of titrators are available, pH and colorimetric endpoint, for pollution control work, and these are also limited to the conventional acid-base titrations. The development of specific ion activity electrodes and of miniaturized submersible color probes using light-bundle techniques for light transmission and collection would seem to open the doors to automatic titrations of all kinds.

High-precision Volumetric Analyzer

The high-precision automatic titrator performs titrations in the exact manner used by a chemist in the laboratory. The titration beaker is flushed out with the sample and then a known volume of sample is added. An electrode assembly measures the sample pH. Under constant, gentle agitation, titrant is added until the desired, preselected endpoint is reached. The titration is stopped and the volume of titrant required to reach the endpoint is recorded. The typical analyzer of this type is depicted in simplified form in Figure 10.19f.

A multicam programmer signals the fill valve (A) to open and fill the burette (B) with reagent from the storage bottle (C) until the level in the burette reaches the full mark. The conductivity follower probe (D) is at maximum height. Next, the programmer opens sample fill valve (E) and drain valve (F). After a few minutes of sample flush the drain valve is closed and the cell (G) is allowed to fill and overflow to waste. Then the sample fill valve and the constant level drain valve (H) are both closed.

The magnetic stirrer (I) is started and reagent valve (J) is opened. The pH of the solution in the titration vessel (G) is measured and recorded continuously. As titrant level in the burette drops below the tip of the follower probe, the conductivity bridge is broken and the drive motor (K) drives the probe down until the tip of the probe is again in contact with the solution. When the pre-set endpoint has been reached, the pH analyzer (L) closes the titrant valve (J). The distance traveled by the follower probe is an indication of the volume of titrant used to complete the titration.

Titrations to two endpoints can also be made. This permits the automation of the classic phenolphthalein-

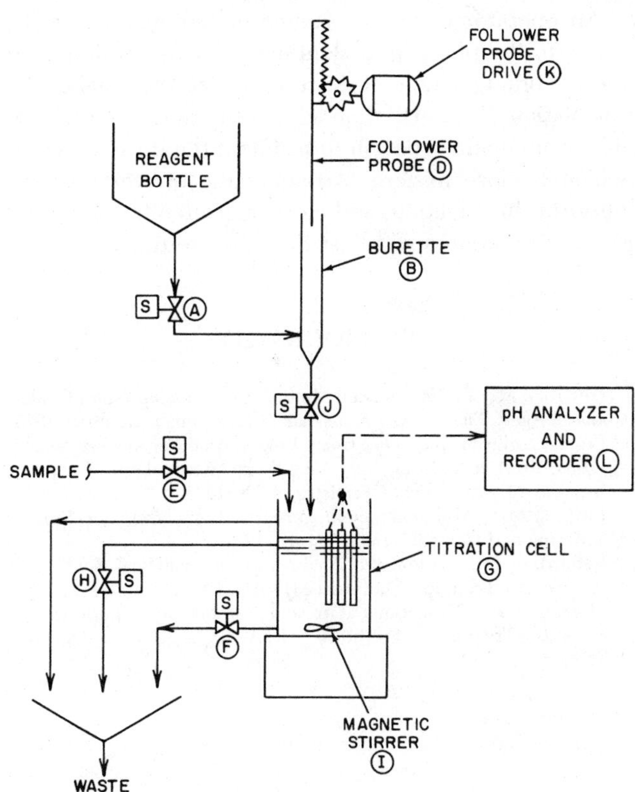

Fig. 10.19f Volumetric analyzer (autotitrator)

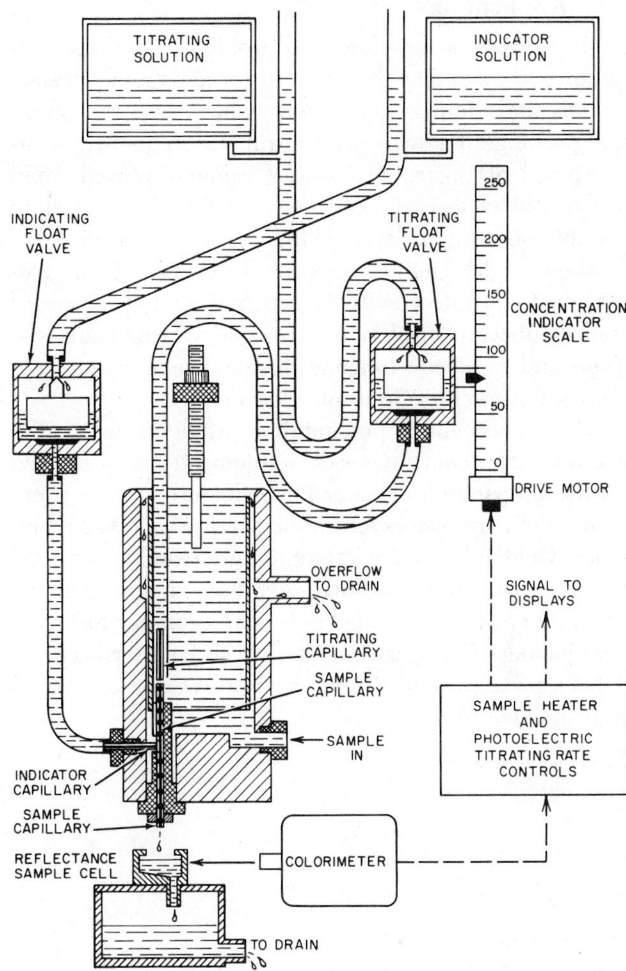

Fig. 10.19g Continuous automatic titrator using color change endpoint determination

methyl orange titration. The first endpoint is represented by the first plateau of the recorder pen and may be stored in an electronic memory device. The titration then proceeds to the second predetermined endpoint. The second plateau of the recording pen corresponds to the milliliters of reagent used for the complete titration. Control can be based on the first endpoint, the second endpoint or a computed value using both endpoints.

Whereas in the classic phenolphthalein-methyl orange titration with acid being the titrant the final solution pH is always acidic, it is also possible to titrate an acidic sample to one or two basic endpoints. In these applications it may be desirable to interpose an occasional acid rinse into the flush and fill cycle prior to titrations for electrode maintenance. This type of volumetric analyzer is not as sensitive to environmental conditions as is its colorimetric counterpart, nor does it require as clean a sample. However, it is a relatively sophisticated electronic instrument and, therefore, should be accorded an operating environment similar to a control room.

Simple Volumetric Analyzer

Illustrated in Figure 10.19g, this device follows the classic method of titrating to color change endpoints. An indicator is added to a known volume of sample, and the titrant is then admitted until the color change indicates that the desired endpoint has been reached, at which point the volume of titrant used is read as an indication of concentration. The unit illustrated uses the laminar flow principle that flow through a capillary is directly proportional to the pressure drop through the capillary. In the titrator the sample flow is maintained constant by a fixed float valve. Titrant flow can be varied by an automatic raising or lowering of the titrant float valve. A colorimeter is adjusted to sense the endpoint color and causes the float valve to drive up or down so as to maintain this color. The float valve elevation or position is indicated in concentration and can also be recorded.

Because the effectiveness of this device depends on color development and on maintaining the endpoint color, it is imperative that the sample not contain suspended matter. To perform a two-endpoint titration, two titrators are needed: one for the equivalent of the phenolphthalein endpoint, and a second for the equivalent of the methyl orange endpoint in the classic titration example cited earlier. It is then necessary to use the outputs from each titrator and to use external electronics for the calculations.

Applications

By far the most widespread problem facing the pollution control engineer today is the precipitation of heavy metals and similar applications which are, too often, lumped together with pH control. Consequently, conventional pH instrumentation is frequently used when it has little chance of success or is likely to give constant trouble because of electrode fouling. To overcome these problems, the engineer frequently chooses highly sophisticated control systems incorporating feedforward and feedback control loops, bias and variable ratio stations and electrode cleaning devices ranging from mechanical wipers to ultrasonic cleaners.

Most heavy metal precipitation problems are similar to the hot or cold lime soda softening techniques used in the preparation of water for boiler use and for preparation of hard waters for municipal drinking water supplies. Cold or hot process softeners are readily controlled by titration. Under manual conditions a flowmeter on the influent flow transmits pulses to a counter for every 100 gallons (379 liters) metered. The counter accumulates a predetermined number of pulses and activates a repeat cycle timer.

An operator collects a sample of softened water, titrates to the phenolphthalein endpoint (P reading), the methyl orange endpoint (M reading) and then makes the calculation 2P-M. He then adjusts the timer to feed sufficient lime and soda ash to maintain the desired value, which is close to zero. An automatic titrator can also perform this function, and electronic circuitry can make the computation and adjust the timer settings.

BIBLIOGRAPHY

DuCross, M.J.F., "Automated Methods for Assessing Water Quality Come of Age," *Environmental Science & Technology*, October, 1975.

Frost & Sullivan, Inc., On Stream Process Analyzer Market, Report No. 669; Frost & Sullivan, Inc., New York, August, 1979.

Jirika, A.M., *Analytical Chemistry*, 47: 8, 1975.

Jutila, J.M., "Multicomponent On-Stream Analyzers for Process Monitoring and Control," *InTech*, July, 1979.

Kefford, F.D., "Automated Analysis as an Analytical Tool," 2nd Annual Water Technical Conference (Dallas, TX), December 2, 1974.

Molvar, A.E., "Instrumentation and Automation Experiences in Wastewater-Treatment Facilities," EPA–600/2–76–198, October, 1976.

Chapter XI

ANALYZERS FOR SPECIFIC COMPONENTS AND PROPERTIES

A. Austin • L. J. Bollyky
A. Brodgesell • V. B. Cortina
A. P. Foundos • W. F. Gerdes
R. J. Gordon • C. E. Hamilton
R. A. Herrick • J. S. Jacobson
R. K. Kaminski • T. J. Kehoe
G. F. McGowan • S. Nishi
R. Nussbaum • W. H. Parth
D. J. Sibbett • A. Turk
N. S. Waner • A. Wertheimer
G. P. Whittle • I. G. Young

CONTENTS OF CHAPTER XI

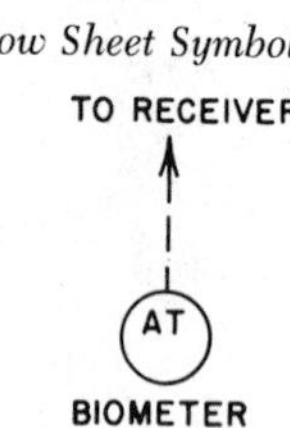

11.1 BIOMETERS

Method of Detection:	Photometric measurement of light emitted by chemical reaction
Sample Pressure:	Atmospheric
Sample Temperature:	Ambient
Sample Type:	Grab sample
Materials of Construction:	Glass
Range:	10^{-7} to 10^{-2} μg ATP per 10 ml sample of bacterial extract. Sensitivity to 10^{-7} μg per 10 μl sample. Calibratable for number of bacteria per μg ATP
Response:	Laboratory method: minutes after starting reaction
Cost:	$10,000
Supplier:	Du Pont Instruments

In a detailed study of the control parameters for the activated sludge process, measurements of great interest are biochemical oxygen demand (BOD) and chemical oxygen demand (COD), BOD and COD reduction, biological population density, and biological oxidative activity. It has been found that the amount of adenosine triphosphate (ATP) is proportionate to the viable biomass in a sample, whereas changes in ATP concentration measure the oxidative capability of the biomass.[1] Thus it is of great interest to measure the ATP content of samples in the activated sludge process as well as in rivers, lakes, and other receiving waters.

Sensitive methods for ATP analysis have been developed based on the observation that the luminescent reaction in fireflies is absolutely dependent on the presence of ATP. The in vitro light-yielding reactions are given in Equations 11.1(1) and 11.1(2).

$$LH_2 + E + ATP \xrightleftharpoons{Mg2+} E - LH_2 - AMP + PP \qquad 11.1(1)$$

$$E - LH_2 - AMP + O_2 \longrightarrow E + PRODUCT + CO_2 + AMP + H\nu \qquad 11.1(2)$$

In these reactions

LH_2 = luciferin
E = luciferase enzyme
$E - LH_2 - AMP$ = enzyme-luciferin-adenosine monophosphate complex
PP = pyrophosphate

It is seen that the yield of light quanta ($h\nu$) is in proportion to the amount of ATP present in the sample.

Luminescence Biometer

ATP assay procedures have been developed based on the reactions just described. Briefly, the procedure involves rapid killing of the live bacterial cells and immediate extraction of ATP into aqueous solution. The latter is then treated with firefly lantern extract, and the light emission of the resultant solution is measured with a photometer. The firefly lantern extract and the ATP required for calibration are commercially available. Du Pont is the only supplier who has designed a manually operated instrument for sale specifically for this measurement.

The instrument is supplied with all the required reagents. A tablet containing buffer and magnesium sulfate is dissolved in water, after which a homogeneous powder of luciferin and luciferase is added. The sample is filtered through a coarse filter to remove solid matter and the latter is discarded. The filtrate is passed through a bacterial filter to catch all the living bacteria. The bacteria on the filter are treated with butanol, which ruptures the cell walls and releases the ATP. The filtrate is made up to volume with water and a microliter aliquot is added to the prepared reagent already in a cuvette. The cuvette is then placed in the instrument for reading of its light emission. The light flash is automatically converted to ATP or microorganism concentration per milliliter, depending on how the instrument is calibrated.

REFERENCE

1. Patterson, J.W., Brezonik, P.L., and Putnam, H.D., "Sludge Activity Parameters and Their Application to Toxicity Measurements in Activated Sludge." *Proceedings*, Industrial Waste Conference, Purdue University, May 1969.

BIBLIOGRAPHY

Frost & Sullivan, Inc., *On Stream Process Analyzer Market*, Report No. 669, Frost & Sullivan, Inc., New York, August 1979.

Matzner, B.A., "Instantaneous Metering Aids Activated Sludge Plant," *Water and Wastes Engineering*, August 1976.

Molvar, A.E., "Instrumentation and Automation Experiences in Wastewater-Treatment Facilities," EPA Document 600/2-76-198, October 1976.

Sironen, E.R., "Sludge Density Control," *Journal of the Water Pollution Control Federation*, February 1970.

Flow Sheet Symbol

TO RECEIVER

AT

COD

11.2 BOD, COD, TOD

Types of Measurements: A.—Biological agency (BOD)
A1. Winkler titration, A2. Dissolved oxygen sensor, A3. Manometric methods (including on-line respirometer), A4. Coulometric (electrolysis) methods, A5. BOD for eleven samples, semiautomatic
B.—Chemical agency (COD)
B1. Oxidation with dichromate, B2. Combustion (catalytic) with carbon dioxide (including NDIR detector), B3. Combustion with oxygen
C.—Total oxygen demand (TOD)
Note: In the feature summary below, the letters A to C refer to the listed detection methods. (Some suppliers manufacture more than one type.)

Sampling Technique: Grab samples for manual methods; automatic sampling for continuous instruments (A, B)

Sample Pressure: Essentially atmospheric

Sample Temperature: 20°C (68°F) during test for biological methods (A); 150 to 1000°C (302 to 1832°F) during test for chemical methods (B)

Materials of Construction: Glass, quartz, Teflon, polyethylene, Tygon, PVC (A, B)

Ranges: 0.1 mg per liter and up (A B): Standard: 0 to 200 to 0 to 1000 mg per liter (C); higher ranges by dilution

Inaccuracy: ±2% of range at the 95% confidence level (C), 3 to 20 percent depending on method (A, B)

Response: 2 to 5 min (B, C), 2 hr to 5 days (A)

Cost: \$350 (A1), \$450 (B1), \$450 to \$2,500 (A3), \$900 to 4,500 (A2), \$9000 to 20,000 (A4), \$12,600 (B2), \$18,000 and up (C), \$45,000 (A5)

Partial List of Suppliers: Astro Ecology (C); Badger Meter Inc., Precision Products Div. (A); Calibrated Instruments Inc. (B); Consolidated Technology Inc., Environmental Products Div. (A1); Delta Scientific, National Sonics Div., Envirotech Corp. (A1,A2); Hach Chemical Co. (A3); Horiba Instruments Inc. (A2,A5); Ionics Inc. (B,C); Oceanography International Corp. (A4); PSG Industries, Inc.; Robertshaw Controls Co. (A); Technicon Industrial Systems (B); Weston Instruments, a Div. of Sangamo Weston Inc. (A2); Yellow Springs Instrument Co. (A2)

Oxygen Demand

The oxygen demand of a sample of water is the amount of elemental oxygen required to react with oxidizable or biodegradable material, dissolved or suspended in the sample. This amount is expressed as milligrams of oxygen per liter of sample. When the agent required to effect the oxidation reaction is a population of bacteria, the oxygen required is called the biochemical oxygen demand (BOD). When the oxidation is carried out with a chemical oxidizing reagent such as potassium dichromate, the oxygen equivalent is called the chemical oxygen demand (COD). Other means are also used to effect oxidation of materials in a sample of water, such as heating the sample in a furnace in the presence of oxygen: total oxygen demand (TOD); or heating in a furnace in the presence of carbon dioxide, resulting in a total carbon dioxide demand (TCO_2D) measurement.

The BOD test is perhaps the most important oxygen demand measurement for the analysis of effluents and receiving waters (streams, lakes, and rivers). Basically the BOD test measures the amount of oxygen used by microorganisms that feed on organic pollutants in the water under aerobic conditions. In this test, a bacterial culture is added to the sample under well-defined conditions and oxygen utilization is measured. Although test procedures are carefully defined,[1,2] it is difficult to obtain reproducible results, and the procedure is subject to the influence of many variables, particularly when the wastewater contains a variety of complex materials. Among the factors that contribute to variations in BOD results are the following.

The Seed

The seed is the bacterial culture which effects the oxidation of materials in the sample. If the biological seed is not acclimated to the particular wastewater, erroneous results are frequently obtained. Because different bacterial cultures are used in BOD measurements at different locations, it is not surprising that the results are inconsistent.

pH

The BOD results are also greatly affected by the pH of the sample, especially if it is lower than 6.5 or higher than 8.3. Not only is oxidation of the material itself pH dependent but also bacterial activity. In order to achieve uniform conditions, the sample should be buffered to a pH of about 7.

Temperature

Although the standard test condition calls for a temperature of 20°C (68°F), field tests often require operation at other temperatures, and consequently the results tend to vary unless temperature corrections are applied (Figure 11.2a).

Toxicity

Toxic materials in the sample, although they may be oxidizable or biodegradable, frequently have a biotoxic or biostatic effect on the biological seed. The presence of toxic materials of this type is indicated by an increase in BOD value as a specific sample is diluted for the BOD test. Consistent values may be obtained either by removing the toxic materials from the sample or by developing a seed that is compatible with the toxic materials in the sample.

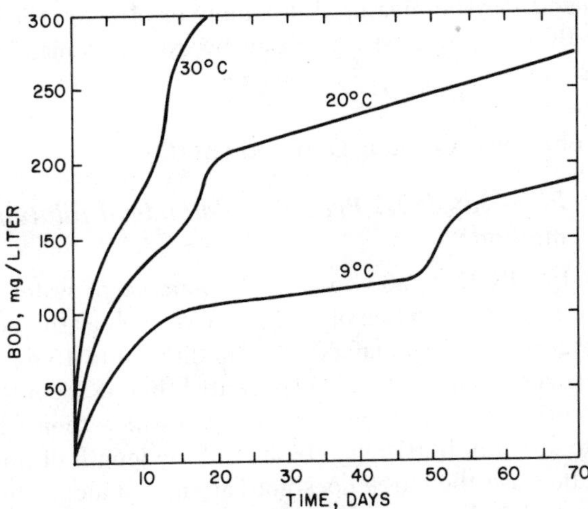

Fig. 11.2a Progress of BOD at 9°, 20°, and 30°C (48, 68, and 86°F).[3] The break in each curve corresponds to the onset of nitrification

Incubation Time

The usual incubation time is 5 days, although the time required for stabilization (complete biochemical degradation of materials in the water) may take as long as 20 or 30 days. The 5-day results may occur at a flat part of the oxygen demand versus time curve or it may occur at a steeply rising portion. Thus, depending on the type of seed and on the type of oxidizable material, divergent results can be expected for this reason alone (Figure 11.2a).

Nitrification

In the usual course of the BOD test, the oxygen consumption rises steeply at the beginning of the test owing to attack on carbohydrate materials. Another sharp increase in oxygen utilization occurs sometime during the tenth to fifteenth day in those samples containing nitrogenous materials (Figure 11.2a). Stated in another way, the rate constant for attack on nitrogenous materials is much lower than that for attack of carbonaceous materials, and the demand due to nitrification is not appreciable until most of the carbonaceous material has been destroyed.[3,4]

In view of difficulties and variabilities of the classic BOD determination, a rapid procedure that minimizes or eliminates these problems has been sought for many years. Although other procedures are used, the BOD continues to remain the universal standard method, supported by the force of tradition and the weight of legal authority in many jurisdictions. Thus those who are concerned with estimating the pollution load of effluent waters must be thoroughly acquainted with the BOD test and prepared to support other methods by suitable correlation to BOD results. Therefore, although oxygen demand may be measured in a number of ways, the 5-day BOD result is what is meant by oxygen demand in most cases.

Biochemical Oxygen Demand (BOD)

Five-Day BOD Procedure (standard dilution method)

If the BOD at 20°C (68°F) of a sample of water is measured as a function of time, a curve such as the one in Figure 11.2a is obtained. For the first 10 to 15 days, the curve is approximately exponential, but at about the fifteenth day a sharp increase is noted which then falls off to a steady BOD rate. Because of the length of time and because the curve does not flatten, a standard test period of 5 days has been adopted universally for the BOD procedure. This is a laboratory procedure requiring some skill and training to obtain concordant results. The procedure is described in greater detail in the literature;[1,2] only a brief description is given here.

A measured portion of the sample to be analyzed is mixed with seeded dilution water so that, after 5 days of incubation, the dissolved oxygen in the mixture is still sufficient for biological oxidation of materials in the sample. Of course, this cannot be known beforehand; consequently a number of dilutions are run simultaneously for an unknown sample, or experience is used as a guide for well-defined samples. The seeded dilution water contains phosphate buffer (including ammonium chloride), magnesium sulfate, calcium chloride, and ferric chloride as well as a portion of seeding material. The former group of inorganic materials is frequently referred to as nutrients. The latter group is a suspension of bacteria in water, usually supernatant liquor from a domestic sewage plant.

Seeds may also be prepared from soil, developed from cultures in the laboratory, or obtained from a receiving water 2 to 5 miles downstream of the discharge. The dissolved oxygen (DO) content of the mixture is determined at the start of the test and again after 5 days of incubation at 20°C in a special BOD bottle. The DO may be determined by the Winkler titration method[5] or instrumentally with a DO membrane electrode. The difference in DO after 5 days is used to calculate the BOD of the original sample. Corrections must be applied for immediate oxygen demand (that due to inorganic reducing materials) and for the oxygen required by the bacteria themselves for sustaining life (endogenous metabolism).

There is no standard against which the accuracy of the BOD test can be measured. The precision of the method is also difficult to ascertain because of the many variables. However, the single-operator precision of the method has been tested using a standard glucose–glutamic acid solution. Using eight different types of seed materials, the single-operator precision was 11 mg per liter at a level of 223 mg per liter, or about 5 percent. It must be recognized that these results were obtained with highly skilled personnel under well-controlled laboratory conditions.

A semiautomatic instrument is designed to measure the BOD of as many as 11 samples. The samples have to be manually placed on the instrument turntable and the controls manually set. Means are provided for automatic reaeration of those samples in which the DO has fallen to low values. Measurements of the DO are made on a preset time schedule by the polarographic DO sensor. The capability of automatic reaeration when the DO is low eliminates the need for dilution, leading to improved precision in the BOD results. The instrument consists of a measuring unit (DO probe, aerator, water-sealing mechanism, unplugging mechanism, sample bottle, and turntable) and a control unit, by which all of the operations are programmed. The measuring unit is housed in a chamber maintained at 20°C. Means for storing DO data on each sample are supplied, and the BOD is calculated from the DO values as already described. Figure 11.2b illustrates this instrument.

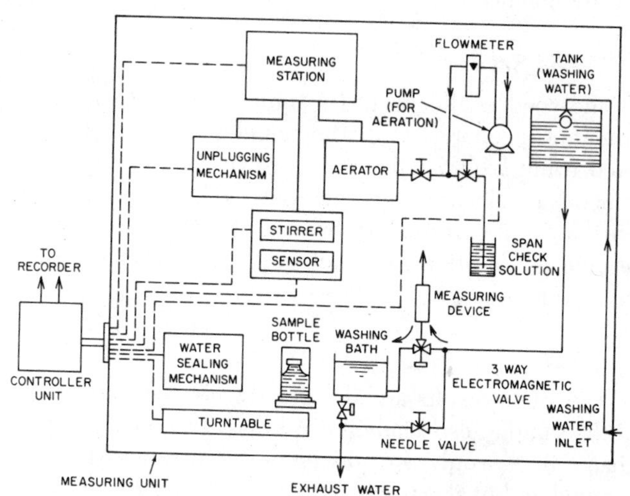

Fig. 11.2b Semiautomatic BOD instrument

Extended BOD Test

As reflected in Figure 11.2a, continuation of the BOD test beyond 5 days shows a continuing oxygen demand (OD), with a sharp increase in BOD rate at the tenth day owing to nitrification. The latter process involves biological attack on nitrogenous organic material accompanied by an increase in BOD rate. The OD continues at a uniform rate for an extended time. Knowledge of oxygen utilization of a polluted water supply is important because (1) it is a measure of the pollution load, relative to oxygen utilization by other life in the water; (2) it is a means for predicting progress of aerobic decomposition and the amount of self-purification taking place; and (3) it is a measure of the OD load removal efficiency by different treatment processes.

As a means for treatment plant control and setting the legal standards for wastewater effluents, the extended test is not used. However, it must be remembered that the 5-day BOD does not represent total OD load on a receiving water. The dynamics of oxygen removal and replenishment in a receiving water is discussed in the literature.[6]

Manometric BOD Test

In the standard dilution method, all the oxygen required must already be inside the BOD bottle, since it is sealed in a gastight manner at the initiation of the incubation period, and care is taken to prevent access of air into the sample. In the manometric procedure,[7] the seeded sample is confined in a closed system which includes an appreciable amount of air. As the oxygen in the water is depleted, it is replenished by the gas phase. A potassium hydroxide absorber within the system removes any gaseous carbon dioxide generated by bacterial action. The oxygen removed from the air phase results in a drop in pressure that is measured with a manometer. This fall is then related to the BOD of the sample.

Thus in the manometric method, the DO of the water remains at a moderately high level, close to saturation (9 mg per liter at 20°C), whereas in the standard BOD, the DO falls continuously during the 5-day incubation period to values near 1 mg per liter. Despite this marked difference in conditions of DO during incubation, results in close agreement are obtained on many samples by the two different procedures.

An apparatus is commercially available in which the BOD of five samples can be determined simultaneously by the manometric method. A measured sample of the sewage or wastewater is placed in one of the bottles of the apparatus and the bottle is connected to a closed-end mercury manometer (Figure 11.2c). Above the water sample a quantity of air is trapped. As oxygen is utilized by bacteria in the sample, it is replenished by oxygen in the air—thus lowering the air pressure. The fall in pressure is read on the mercury manometer directly in BOD units, assuming that the original air contained 21 percent oxygen. The preceding description assumes a sample that is already seeded. Of course, the method can be modified for those samples that require the addition of a bacterial culture. The procedure is carried out manually in the laboratory and, in addition to the manipulations already described, requires reading of the manometer by the laboratory technician.

The manometric method lends itself to automatic recording of the course of oxygen utilization since it is possible to monitor the pressure continuously. This has been accomplished in an automatic respirometer[8] now commercially available (Figure 11.2d). The sample, from 1 to 4 liters, is introduced into a closed system containing air. Countercurrent circulation of both air and water insures equilibrium between dissolved and gaseous oxygen. A carbon dioxide scrubber is provided in the gas-circulation line. The utilization of oxygen is detected by a recording manometer and the test is run for several hours. Published data indicate a correlation between the four-hour respirometer BOD and the standard BOD.[9] Laboratory and automatic on-line versions of this instrument are also available.

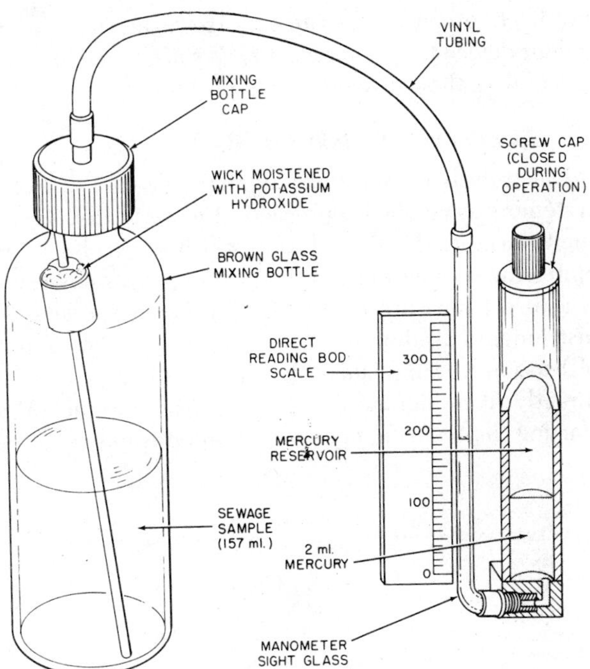

Fig. 11.2c Manometric BOD apparatus

Apart from the respirometer, there are no automatic on-line BOD detectors presently in use. It must be recognized that BOD is inherently a time-consuming process and ill suited to the requirements of process monitoring or control. The shortest period mentioned for the automatic respirometer is two hours, much too long for

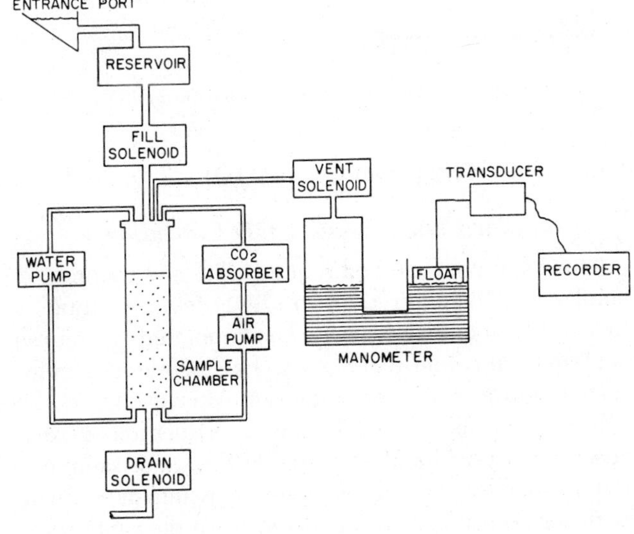

Fig. 11.2d BOD determination by automatic respirometer

an effective control instrument. However, it is an excellent device for laboratory studies since it can be made to simulate the activated sludge process.

Electrolysis System for BOD

Electrolysis of water can supply oxygen in a closed system[10,11] as incubation proceeds (Figure 11.2e). At constant current, the time during which electrolysis generates the oxygen to keep the system pressure constant is a direct measure of the OD (by Faraday's law). An instrument based on this principle permits the running of six samples simultaneously, and a readout gives BOD directly in milligrams per liter for each sample. After starting the test run, operator attention is not required.

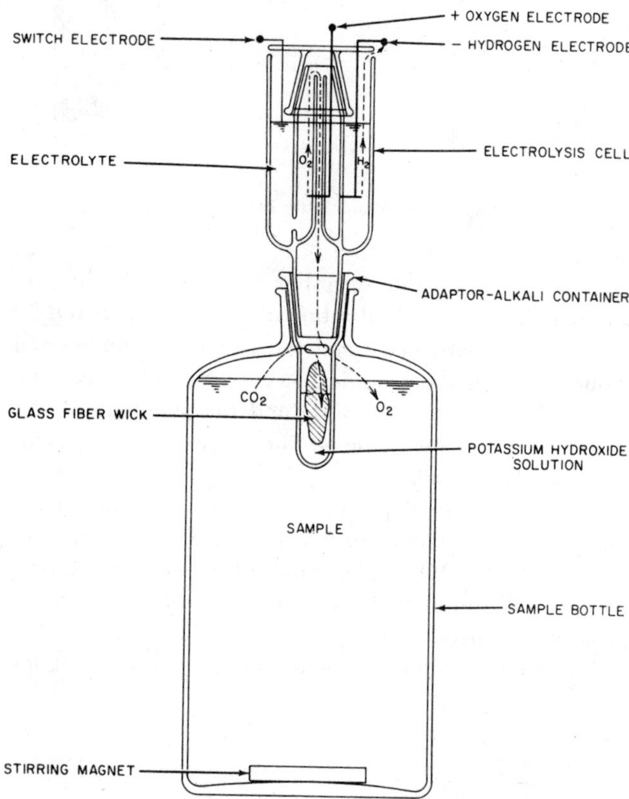

Fig. 11.2e Electrolysis system for measuring BOD

Chemical Oxygen Demand (COD)

Standard Dichromate COD Procedure

This laboratory method requires skill and training[12,13] similar to that required for the BOD test. A sample is heated to its boiling point with known amounts of sulfuric acid and potassium dichromate. Loss of water is minimized by use of a reflux condenser. After two hours the solution is cooled and the amount of dichromate that reacted with oxidizable material in the water sample is determined by titrating the excess potassium dichromate with ferrous sulfate, using ferrous 1,10-phenanthraline (ferroin) as the indicator. The dichromate consumed is calculated to oxygen equivalent for the sample and reported as milligrams of oxygen per liter of sample.

Interpretations of COD values is difficult since this method of oxidation is markedly different from the BOD method. Although ultimate BOD values can be expected to agree with COD values, a number of factors may prevent this concordance. Among these we may mention the following:

1. Many organic materials are oxidizable by dichromate but not biochemically oxidizable and vice versa. For example, pyridine, benzene, and ammonia are not attacked by the dichromate procedure.
2. A number of inorganic substances such as sulfide, sulfites, thiosulfates, nitrites, and ferrous iron are oxidized by dichromate, creating an inorganic COD which is misleading when estimating the organic content of wastewater. Although the factor of seed acclimation will give erroneously low results on the BOD tests, COD results are not dependent on acclimation.
3. Chlorides interfere with the COD analysis and their effect must be minimized in order to obtain consistent results. The standard procedure provides for only a limited amount of chlorides in the sample. Despite these limitations, the dichromate COD has been useful in control of wastewater effluents from plants concerned with caustic and chlorine, dyeing and textiles, organic and inorganic chemicals, paper, paints, plating, plastics, steel, aluminum, and ammonia.

COD Detector

The term COD usually refers to the laboratory dichromate oxidation procedure, although it has also been applied to other procedures that differ greatly from the dichromate method but which do involve chemical reaction. These methods have been embodied in instruments both for manual operation in the laboratory and for automatic operation on line. They have the distinct advantage of reducing analysis time from days (5-day BOD) and hours (dichromate, respirometer) to minutes.

In one instrument, a 20-μl water sample is manually injected into a carbon dioxide carrier stream and swept through a platinum catalyst combustion furnace where pollutants are oxidized to carbon monoxide and water, and the water is removed from the stream by a drying tube after which the reaction products receive a second platinum catalytic treatment. The concentration of carbon monoxide is then measured by an NDIR detector. The readings are converted to COD by a calibration chart. An analysis can be completed in two minutes. This instrument is available commercially for manual operation (Figure 11.2f). Data obtained on domestic sewage[14] indicate excellent correlation between this method (frequently called CO_2D) and the standard COD.

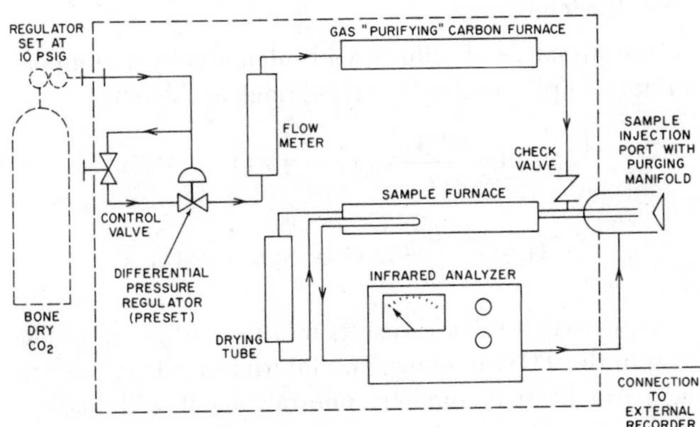

Fig 11.2f COD detection employing combustion in a carbon dioxide carrier and an NDIR sensor

Total Oxygen Demand (TOD)

The TOD method is based on the quantitative measurement of the amount of oxygen used to burn the impurities in a liquid sample. Thus it is a direct measure of the oxygen demand of the sample. Measurement is by continuous analysis[15,16] of the concentration of oxygen in a combustion process gas effluent (Figure 11.2g). The oxidizable components in a liquid sample introduced into the combustion tube are converted to their stable oxides by a reaction that disturbs the oxygen equilibrium in the carrier gas stream. The momentary depletion in the oxygen concentration in the carrier gas is detected by an oxygen detector and recorded as a negative oxygen peak on a potentiometric recorder. The TOD for the sample is obtained by comparing the recorded peak heights to the peak heights of the standard TOD calibration solutions, e.g., potassium acid phthalate (KHP).

Prepurified nitrogen from a cylinder passes through a fixed length of tube permeable to oxygen, into the combustion chamber, the gas scrubber, and then into the oxygen detector. The base-line oxygen concentration, picked up as the nitrogen passes through the temperature-controlled permeation tube, can be varied to accommodate different TOD ranges by changing the nitrogen flow rate.

The combustion chamber is a length of Vicor tubing or quartz tube containing platinum catalyst mounted in an electric furnace and held at a temperature of 900°C

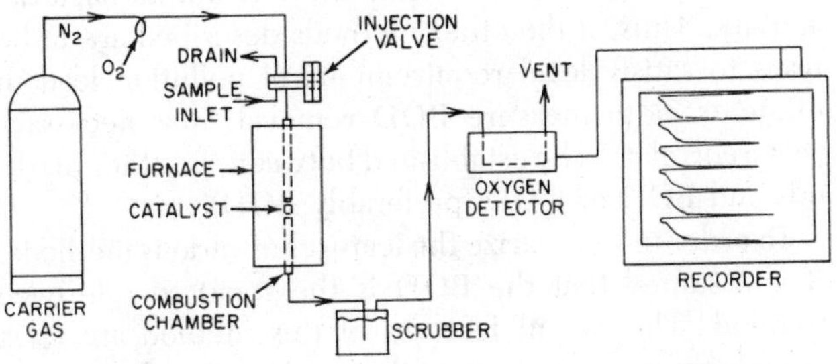

Fig. 11.2g Basic components of a TOD analyzer

(1652°F). The aqueous sample is injected into this chamber and the combustible components are oxidized.

Sample Valves

Two basic types of injection valves are in use today, the sliding plate and the rotary sampling valve. Manual injection can be accomplished through a silicone rubber septum, if desired.

Figure 11.2h shows the main features of the sliding plate valve, normally used with laboratory analyzers. Upon a signal from a cycle timer, the air actuator temporarily moves the valve to its "sample fill" position. At the same time, an air-operated aspirator pulls a 20-μl sample through the valve into the combustion tube. A stream of nitrogen moves the slug of sample into the combustion tube.

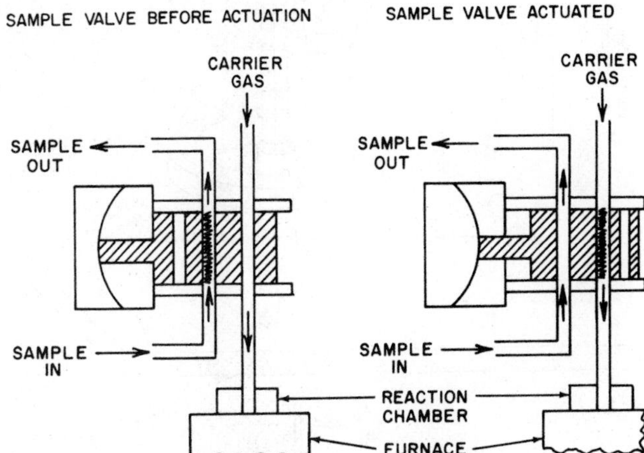

Fig. 11.2h Automatic sliding plate liquid sample injection valve

The rotary sample valve represents a more recent development and is mainly used for on-stream TOD analyzers. Figure 11.2i shows the cross section of the valve, in which a motor continuously rotates a sampling head which contains a built-in sampling syringe. For part of the time the tip of the syringe is over a trough that contains the flowing sample. Two or more cam ramps along the rotational path cause the syringe plunger to rise and fall, thus rinsing the sample chamber. Just before the syringe reaches the combustion tube, it picks up a 20-μl sample. As it rotates over the combustion tube, it discharges the sample.

Oxygen Detector

The oxygen sensor is a platinum-lead fuel cell that generates a current in proportion to the oxygen content of the carrier gas passing through it. Before entering the cell, the gas is scrubbed in a potassium hydroxide (KOH) solution both to remove acid gases and other harmful combustion products and to humidify the gas. The oxygen cell and the scrubber are located in a temperature-controlled compartment. The fuel cell output is moni-

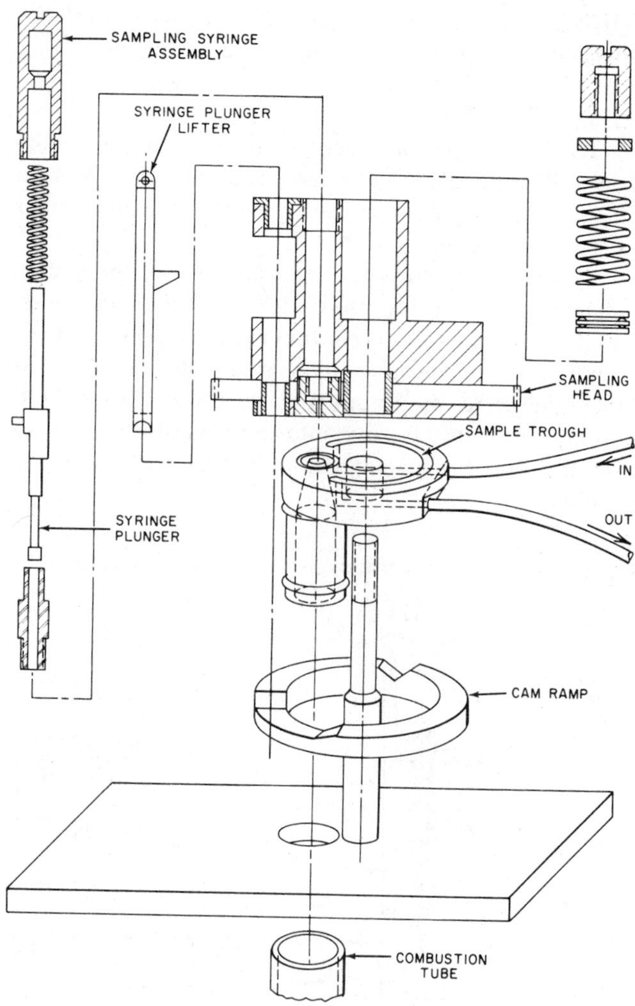

Fig. 11.2i Rotary sample valve

tored on a potentiometric recorder; the recorder system also includes an automatic zeroing circuit which maintains a constant base line. The recorded peaks are linearly proportionate to the reduced concentration of oxygen in the carrier gas as a result of the sample's TOD.

Calibration

Analysis is by comparison of peak heights to a standard calibration curve. To prepare this curve, known TOD concentrations of a primary standard, potassium acid phthalate (KHP), are prepared in deionized water. Standard solutions are stable for several weeks at room temperature. Water solutions of other pure organic compounds can also be used as standards. Several analyses at each calibrating concentration are made, and the resulting data are plotted on a parts per million (ppm) TOD versus peak height curve. Within the normal instrument range, the plot is a straight line; consequently, single-concentration calibration of the recorder chart in milligrams per liter TOD can be made.

Interferences

Nitrate salts and sulfuric acid will normally decompose under sample combustion conditions as follows:

$$2NaNO_3 \xrightarrow[Pt]{900°C} Na_2O + 1\tfrac{1}{2}O_2 + 2NO$$

$$H_2SO_4 \xrightarrow[Pt]{900°C} H_2O + SO_2 + \tfrac{1}{2}O_2$$

This oxygen release results in a proportionate reduction in the TOD reading. The interference from sulfuric acid can be overcome by neutralizing it with sodium hydroxide. Dissolved oxygen in the aqueous sample also becomes an unknown source of oxygen to the combustion reaction and lowers the TOD readings. But unless special precautions are taken, the standard solutions are also at or near saturation, thus automatically minimizing this potential interference. If absolute TOD levels are required, the oxygen equivalent of the interference should be added to the TOD reading.

Oxygen-saturated or air-saturated samples with very low TOD pose a special problem which can be circumvented by spiking the sample with a known concentration of a standard solution. The actual TOD will then be the analysis value minus the spiking concentration. Heavy metal ions give long-term interferences, usually by eventual reduction of the catalyst efficiency. Replacement of the combustion tube and a thorough cleaning of the catalyst is the only remedy.

Application Suggestions

Table 11.2j lists typical applications for the TOD analyzer. Instruments equipped with the rotary sampling valve lend themselves nicely to multistream analysis, since no carryover of the previous sample is present. If the streams have varying levels of TOD, each stream must be diluted to the basic range of the analyzer by appropriate automatic dilution techniques. On the laboratory analyzers, it is best to sample deionized water when the instrument is in a standby condition.

Correlation Between BOD, COD, and TOD

Many regulatory agencies recognize as the basis for pollution control only the BOD or COD measurements of pollution load. The reason for this is that they are concerned with the pollution load on receiving waters, which is related to lowering of the DO due to bacterial activity. Thus, if the other methods described are to be used to satisfy legal requirements of pollution load in effluents or to measure BOD removal, it is necessary that a correlation be established between the other methods and BOD or COD (preferably BOD).

In order to summarize the features of various methods, it is assumed that the BOD is the standard reference method. The salient features of this method are (1) a measurement of property of the sample, i.e., the amount

Table 11.2j
APPLICATION SUGGESTIONS FOR TOD ANALYZERS

Application	*Purpose*
Waste treatment plant influent	Determines loading of plant
Primary, secondary, and plant effluents	Determines efficiency of treatment and TOD load on the receiving waterway
Enforcement programs	Determines pollution levels
Industrial process control	Determines process efficiency and leaks by continuous monitoring
Research and development	Evaluates waste treatment processes
Stream surveillance	Monitors fresh, estuarine and marine water quality
Municipal or industrial water treatment plants	Monitors quality of influent water prior to treatment
Boiler feed and high-purity water monitoring	Measures TOD with great sensitivity

of oxygen required for bacterial oxidation of bacterial food in the water, the BOD; (2) dependence of the oxygen demand on the nature of the food as well as on its quantity; and (3) dependence of the oxygen demand on the nature and amount of the bacteria.

Variations in OD due to variations in the amount (pounds per gallon) of food in the wastewater are expected; we are less able to deal with variations in OD when the amount of food is constant but changes occur in its BOD requirements. The same observations apply to the bacterial seed. Thus variation in OD due to variation in the number of bacteria or their activity as well as those due to the changes in the nature of their foodstuffs leads to a "systematic" or "bias" error in BOD measurements which cannot be predicted or corrected for. The "standard reference" method is therefore inherently variable—one subject to analytical error. Researchers in an interlaboratory comparative study[17] employing a synthetic waste found standard deviations around the mean of ±20 percent for BOD and ±10 percent for COD.

Another extensive study[18] concluded that (1) a reliable statistical correlation between BOD and COD of a wastewater and its corresponding TOC (Section 11.30) or TOD can frequently be achieved, particularly when the organic strength is high and the diversity in dissolved organic constituents is low; (2) the relationship is best described by a least squares regression with the degree of fit expressed by the correlation coefficient. This applies to the characterization of individual chemical-processing and oil-refining wastewaters, not to all types of samples across the board; (3) the observed correspondence COD-TOD was better than COD-BOD for the wastewaters mentioned. Generally it was difficult to correlate BOD with TOD, particularly when the wastewater contained low concentrations of complex organic materials; and (4) the BOD-COD or BOD-TOC ratios of an untreated wastewater are indicative of the biological treatment possible with the particular wastewater. As these ratios increased, higher treatment efficiencies by biological methods in terms of organic removal were noted.

Several papers have indicated high correlation between BOD and other methods. This can be achieved when the nature of the pollutant is constant and only its amount changes. For complex and varying mixtures, it is difficult or impossible to obtain good correlations. An interesting example is given in the work of Nelson et al.,[19] who discuss a pyrolytic method combined with FID detection. Values from the new method agreed with BOD values within ±15 percent for BOD values greater than 100 ppm on raw sewage and primary effluent. However, discrepancies of several hundred percent were found when the BOD was 20 ppm or less. These poor results can be attributed to marked variation in biodegradability of carbonaceous products in the secondary effluent compared with products before treatment, as well as to the very small amount of total material left.

REFERENCES

1. *American Public Health Association: Standard Methods for the Examination of Water and Wastewater*, (12th ed.) APHA, New York, 1965.
2. *Industrial Water; Atmospheric Analysis, Book of ASTM Standards, Part 23,* Method, D 2329-68. ASTM, Philadelphia, 1969.
3. Gair, G.M. and Geyer, J.C., *Water Supply and Waste-Water Disposal*, Wiley, New York, 1954.
4. Ford, D.L., Eller, J.M. and Gloyna, E.F. "Analytical Parameters of Petrochemical and Refinery Waste Waters," *Journal of the Water Pollution Control Federation*, Vol. 43, 1971, p. 1713.
5. Ref. 1, p. 405; Method D 1589-60, p. 406.
6. Ref. 3, p. 835.
7. Tool, H.R. "Manometric Measurement of BOD." *Wat. Sew. Works* 114: 211, 1967.
8. Arthur, R.M., "An Automated BOD Respirometer." *Proceedings* 19th Industrial Waste Conference, Purdue University, May 1964, Part 2, p. 628.
9. Arthur, R.M. and Hursta, W.N. "Short-term BOD Using the Automatic Respirometer," *Proceedings*, 23d Industrial Waste Conference, Purdue University, May 1968, Part 1, p. 242.
10. Clark, J.W. New Mexico State University, Engineering Experiment Station, Bill No. 11, 1959.
11. Young, J.C. and Baumann, E.R. "*Demonstration of the Electrolysis Method for Measuring BOD*." Engineering Research Institute, Iowa State University, Progress Report No. 1, March 1970; Progress Report No. 2, March 1971.
12. Ref. 1, p. 510.
13. Ref. 2, Method D 1252-67, p. 243.
14. Stenger, V.A. and Van Hall, C.E., "Rapid Method for Determination of Chemical Oxygen Demand." *Analytical Chemistry*, Vol. 39, 1967, p. 206.
15. Goldstein, A.L., Katz, W., Meller, F.H., and Murdoch, D.M., *Total Oxygen Demand—A New Automatic Instrumental Method for Measuring Pollution and Loading on Oxidation Process*, Ionics Inc., Watertown, Mass., 1968.
16. Clifford, D.E., *Automatic Measurement of Total Oxygen Demand*.

Proceedings, 23d Industrial Waste Conference, Purdue University, May, 1968.

17. Ballinger, D.G., and Lishka, R.J., "Reliability and Precision of BOD and COD Determinations," *Journal of the Water Pollution Control Federation*, Vol. 34, 1962, p. 470.
18. Ford, D.L., Eller, J.M. and Gloyna, E.F., "Analytical Parameters of Petrochemical and Refinery Waste Waters," *Journal of the Water Pollution Control Federation*, Vol. 43, 1971, p. 1712.
19. Nelson, K.H., Lysyj, I., and Nagano, J., "Pyrolytic Analyzer Detects Organic Matter in Wastewaters." *Water Sewage Works*, Vol. 14, January 1970.

Flow Sheet Symbol

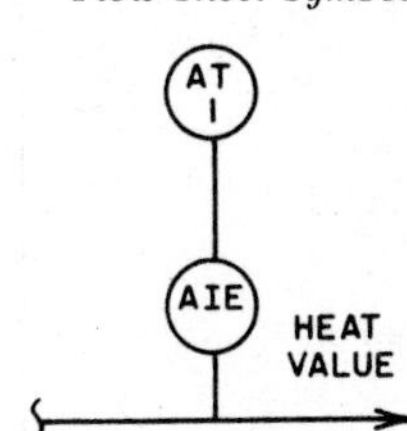

11.3 CALORIMETERS—HEAT VALUE MEASUREMENTS OF GASEOUS FUELS

Type of Designs: A.—Direct measurement by burning of fuel gas
B.—Inferential by calculation from composition and/or physical analysis

Applications: 1—Custody transfer
2—Process monitoring and control
3—Blending and mixing of fuel gases

Operation: a.—Continuous
b.—Cyclic

Performance: (1) Controlled environment
(2) Varying ambient
(3) High speed of response
(4) Inaccuracy ±0.5% of full scale or better
(5) Inaccuracy ±1.0% of full scale or better
(6) Inaccuracy ±2.0% of full scale or better

Area Classification: (a) General purpose
(b) Explosion proof

Costs: Under $10,000 [A, 2/3, a, (3)/(5)/(6)]
$10,000—$15,000 [A, 2, a, (3)/(5)/(6), (b)]
$15,000–$30,000 [A,B, 1/2 a/b, (1)/(2)/(4), (a)/(b)]

Partial List of Suppliers: Applied Automation [A, 1, a, (1)/(4), (a)]; Beckman Instruments Inc. [A, 1, a, (1)/(4), (a)]; Bendix Corp. [A, 1, a, (1)/(4), (a)]; Cutler-Hammer [A, 1, a, (1)/(4), (a)]; Fluid Data [A, 1/2/3, a/b, (2)/(3)/(4)/(5), (a), (b)]; Sigma [A, 2, a, (a) (1), (1)/(5), (a)]; Union Carbide Corp. [A, 2/3, a, 2/7, (a)]

Calorimeters are instruments that measure the heat value or energy content of gaseous fuels. There are two broad categories of this type of instrument: those that can be considered true calorimeters, which burn the gas and directly measure the heating value, and inferential calorimeters, which analyze the composition of the gas or measure a physical parameter to determine the heat value. The specific need for such a measurement most often determines the type to be used and further determines which specific type of calibration—gross calorific value, net calorific value, or Wobbe index—is most suitable for process monitoring and/or control, or for accurate measurement in sale or purchase of gaseous fuel energy.

Heat Value Terminology and Units

Basic terms and definitions used in gas calorimetry may be found in the references listed at the end of this section. The following is a summary of the measurement calibration techniques relevant to the above areas of application, as outlined in Table 11.3a.

1. Gross Calorific Value—The defined empirical measurement of the heat value or energy per unit volume at standard conditions and expressed in terms of British thermal unit per standard cubic feet (BTU/SCF) or kilocalorie per cubic Newton meters (Kcal/N·m³) or other equivalent units.
2. Net Calorific Value—The measurement of the actual available energy per unit volume at standard conditions, which is always less than the gross calorific value by an amount equal to the latent heat of vaporization of the water formed during combustion.
3. Wobbe Index—The measurement of the combined factors of net calorific value and specific gravity defined as:

Table 11.3a
SUMMARY OF CALORIMETER FEATURES AND SPECIFICATIONS

Type (Suppliers)	Type of Reading	Type				Application					Operation					Performance				
				Area Class																
		Direct	*Inferential*	*General Purpose*	*Ex-Proof*	*1*	*2*	*3*	*4*	*5*	*Continuously*	*Cyclic*	*Standard Sample*	*Empirical Calibration*	*Ambient Limits, °F (°C)*	*Local Readout*	*Remote Transmitter*	*Range Btu Full Scale*	*Accuracy ± % of full scale*	*Speed of Response (90%)*
Water ΔT (ACCU-CAL/Reineke)	Gross Calorific Value	√		√		√				√		√		√	72–77 (22–25)	√	√	120–3300	0.5	3 min
Air ΔT (Cutler-Hammer)	Gross Calorific Value	√		√		√				√	√		√		72–77 (22–25)	√	√	120–3600	0.5	15 min
Gas Chromatograph	Gross Calorific Value		√	√	√	√	√		√	√	√		√		0–100 (−18–38)		√	any	0.5	10 min
Adiabatic Flame Temperature (Therm-Titrator)	Gross Calorific Value	√		√		√				√	√		√		N/A	√	√	N/A	0.5	N/A
Airflow Calorimeter (FLO-CAL/Reineke)	Net Calorific Value	√		√	√		√	√	√		√		√		50–90 (10–32)	√	√	130–3300	1.0	8 sec
Gas Chromatograph	Net Calorific Value		√	√	√							√	√		0–128 (−18–53)		√	any	0.5	10 min
Expansion-tube Calorimeter (Sigma)	Net Calorific Value	√		√				√	√		√		√		N/A	√	√	120–3300	1.0	3.5 min
Specific Gravity	Net Calorific Value		√	√	√		√	√			√				0–128 (−18–53)	√	√	varies	2.0	N/A
Process Chromatograph (Thermeter)	Net Calorific Value	√		√				√			√		√		60–90 (16–32)	√	√	150–3600	2.0	4.5 min
Thermopile Calorimeter (Union)	Net Calorific Value	√	√	√			√		√		√		√		N/A		√	150–3300	2.0	55 sec

Table 11.3a
SUMMARY OF CALORIMETER FEATURES AND SPECIFICATIONS, *Continued*

Type (Suppliers)	*Type of Reading*	*Type*		*Area Class*		*Application*					*Operation*					*Performance*				
		Direct	*Inferential*	*General Purpose*	*Ex-Proof*	*1*	*2*	*3*	*4*	*5*	*Continuously*	*Cyclic*	*Standard Sample*	*Empirical Calibration*	*Ambient Limits, °F (°C)*	*Local Readout*	*Remote Transmitter*	*Range Btu Full Scale*	*Accuracy ± % of full scale*	*Speed of Response (90%)*
Airflow Calorimeter (FLO-CAL/Reineke)	Wobbe Index	√		√	√		√	√	√		√		√		50–110 (10–43)	√	√	130–3300	.75	8 sec
Gas Chromatograph			√	√	√	√	√		√	√		√	√		0–120 (−18–49)		√	any	0.5	10 min
Expansion-tube Calorimeter (Sigma)		√		√				√	√		√		√		N/A	√	√	120–3300	1.0	3.5 min
Thermopile Calorimeter (Union)		√		√			√	√	√		√		√		N/A		√	150–3300	2.0	55 sec

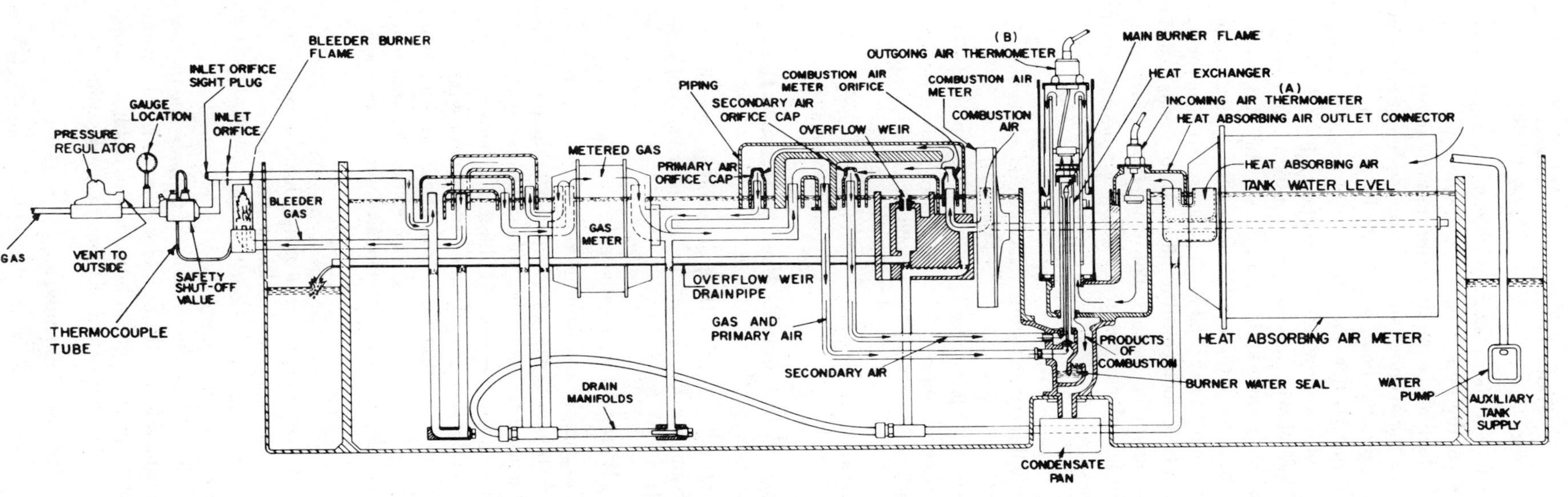

Fig. 11.3b Air ΔT calorimeter

$$\text{Wobbe Index} = \frac{\text{Calorific Value}}{\sqrt{\text{Specific Gravity}}}$$

The Wobbe index accounts for composition variations in terms of their effect on the heat value and specific gravity, which affect the flow rate through an orifice. In essence, the Wobbe index is a measurement of the available potential heat, and it can be used in conjunction with the gas flow measurement to produce a measurement of heat flow rate (see Section 2.3).

The following are basic terms used in gas calorimetry that one should know in order to assure proper specification of the requirements and sound application of the instrument. The ASTM Standards listed in the references fully document this information. Some basic units and conversion factors, however, as well as typical gas calorific values, are given below for information purposes.

English Units	*Metric Units*
British thermal unit—Btu	Kilocalorie or megajoule—Kcal or Mj
Standard temperature—°F	Standard temperature—°C
Standard pressure:—inches of mercury (in. Hg)	Standard pressure—millimeters of mercury (mmHg)
Standard cubic feet—SCF	Normal cubic meter—cubic Newton meter ($N \cdot m^3$)

Types of Instruments Used to Measure Heat Value

The following is a brief description of each type of instrument used to measure the heat value of gaseous fuels.

Water ΔT Calorimeter

Gas sample and combustion air are saturated and a fixed volume of gas is measured. The sample is burned and its heat transferred to a flow of water. A thermopile measures the temperature difference between the cold water inlet and heated water outlet to provide a signal which is compensated for barometric pressure to measure directly the gross calorific value. The unit may be empirically calibrated or standard samples may be used.

Air ΔT Calorimeter

The measurement is accomplished by continuously imparting all the combustion heat of a metered quantity of gas to a metered quantity of air (see Figure 11.3b). The temperature rise of the air is measured and related directly to gross calorific value of the gas. The thermometer is modified so the heat-absorbing air is not separated from the products of combustion, thus resulting in a more accurate measurement of the net calorific value.

Airflow Calorimeter

Variations in the heat released from continuous burning of a fuel gas sample are offset by a continuous, varying airflow to control the temperature of the products of combustion (see Figure 11.3c). Thus, the airflow is correlated to the heat value of gas as in the Wobbe index. With the addition of a constant-volume gas metering pump, compensating for specific gravity variations, the instrument can be calibrated for the net calorific value.

Process Chromatograph

A conventional chromatograph is used to analyze gas composition (see Section 10.4), and a microprocessor control section calculates the heating value and specific gravity of the gas from empirical data held in memory by the microprocessor. This information is used to calibrate for the gross or net calorific value or for the Wobbe index.

Expansion-Tube Calorimeter

The gas sample is delivered through a precision regulator system, which is relatively independent of specific gravity and atmospheric pressure, to a burner where the gas is burned at the base of a differential expansion tube unit that responds to the temperature of the products of combustion and excess air. The differential signal is calibrated as the net calorific value. Modification of the regulator to respond to specific gravity changes allows the calibration to the Wobbe index.

Adiabatic Flame Temperature

The adiabatic flame temperature is based on the principle that the gross calorific value of a fuel gas is proportionate to the ratio of air to a fuel which maximizes the adiabatic flame temperature of the mixture. Flows are controlled and two burners are used to obtain the mathematical derivative of the flame temperature composition and thus calibrate for gross calorific value.

Thermopile Calorimeter

The thermopile calorimeter measures the temperature of a thermopile which senses the hot products of combustion mixed with a constant volume of air supplied by a fan. The sample to the burner has an orifice bypass with a lag to bleed gas for specific gravity compensation; thus, the measurement is in terms of net calorific value. To measure the Wobbe index, the bleed is blocked, and the sample goes through the burner.

Areas of Application

There are five general areas of application:

1. Custody Transfer—Sale or purchase of fuel gas with accuracy is the primary consideration.
2. Process Monitoring and/or Control—To effect on-line manual or automatic control of process or efficient burning of the gas by using heat value measurement as one of the measured variables. Such applications include feedforward control of fuel gas-fired heaters or boilers, stove-firing con-

trol, vaporizer control, and synthetic natural gas (SNG) reactor control. In these applications as well as others, measurement of heat value is important as a gauge of reaction results, as a process measurement necessary for efficient control of energy consumption, or to limit the flow of heat to an energy-sensitive process. This application requires a high speed of response as well as reliability to achieve effective control.

3. Blending and Mixing of Fuel Gas—To obtain a uniform quality gas or to consume waste or by-product gases by blending them into the main supply of fuel gas. Blending is often used to achieve a desired quality; for example, propane/air and blast furnace gas/coke oven gas systems. However, when by-product or waste gases are to be injected into a stream and they vary significantly in quantity because of process conditions, then monitoring of the mixture is necessary to make proper use of the fuel. Again, speed of response and reliability are essential.

4. Processing of Gas and liquified natural gas (LNG) Operations—LNG must be vaporized and conditioned for efficient consumption. Coal gasification is a developing industry that will markedly increase available supplies. Coke oven and blast furnace gases are the main fuels used by the steel industry, and refinery gas is a major by-product which is used in petroleum refining. All these fuel gases are produced in specific processing

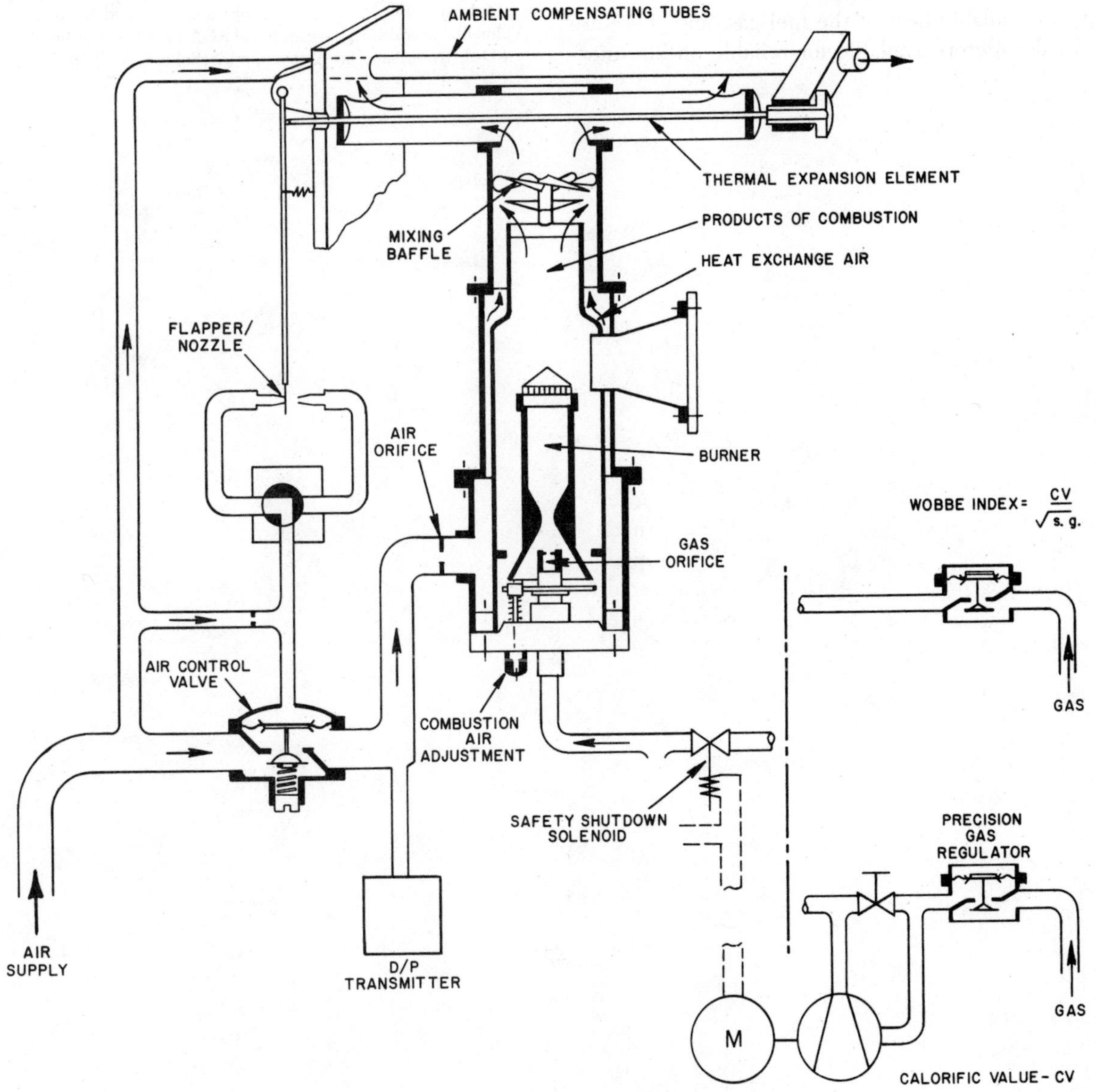

Fig. 11.3c Airflow calorimeter

operations which require monitoring and/or conditioning for efficient use.

5. Compliance Recording—For government regulated energy transfer by pipeline and utility distributors to the consumers. Accuracy and traceability are essential criteria.

Sample Conditioning

Sample conditioning of fuel gases is normally minimal. However, in many industrial applications fuel gases are generated as by-products of other processes, and these "off-gases" can be extremely dirty and require sample conditioning. Refer to Section 10.2 for a complete description of sample system considerations.

Conclusion

Gaseous fuel energy is a costly commodity that is being consumed today with much more care and efficiency than previously. The key to its efficient consumption is measuring the available heat of the fuel gas, and there are several calorimeters available for reliable on-line measurement, some of which have sufficiently fast speed of response to be used in direct-control applications.

BIBLIOGRAPHY

ASTM Standard D 900-55(70) *Calorific Value of Gaseous Fuels by the Water Flow Calorimeter*, American Society for Testing and Materials, Philadelphia, 1970.

ASTM Standard D 1826-77, *Test for Calorific Value of Gases in Natural Gas Range by Continuous Recording Calorimeter*, American Society for Testing and Materials, Philadelphia, 1977.

Armstrong, G.T., "Standard Combustion Data for the Fuel Gas Industry," AGA 1972 Distribution Conference 72-D-76, American Gas Association, Arlington, Va., 1972.

Foundos, A.P., "Measuring Heat Release Rate from Fuel Gases," *Instrumentation Technology*, Instrument Society of America, 1977.

Lange, N.A., and Forker, G.M., *Handbook of Chemistry*, 10th ed., McGraw Hill, New York, 1967, pp. 842–43, and Columbia Gas System Data.

McCoy, R., "BTU Determination by Process Chromatograph," Applied Automation, Inc., a subsidiary of Phillips Petroleum.

Melrose, D.C., "Comparison of Calculated and Measured Heating Value of Natural Gas," presented at AGA 1972 Distribution Conference 72-D-2, American Gas Association, Arlington, Va., 1972.

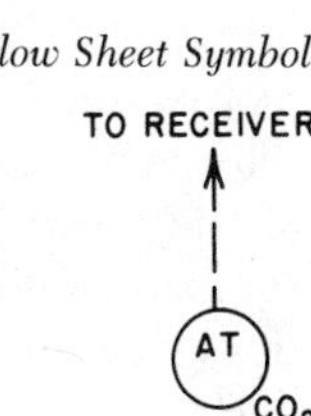

11.4 CARBON DIOXIDE

Method of Detection:	Infrared
Sample Pressure:	Up to 15 PSIG (104 kPa) but atmospheric or near atmospheric is normal
Sample Temperature:	Up to approximately 120°F (49°C). Not a consideration when freeze-out trap is used
Sample Flow Rate:	Generally less than 0.5 acfm (2.35×10^{-4} m^3/s); typically 1 to 2 liters per minute
Inaccuracy:	Can be as high as ±0.2 ppm; typically ±1% of full-scale range
Ranges:	From ppm to 100%
Response:	Determined by cell volume and sampling rate. Typically less than 30 sec
Cost:	\$6000–\$11,000 without recorder, depending on sample conditioning accessories
Partial List of Suppliers:	Anacon, Inc.; Beckman Instruments Inc.; Envirotech Corp., United Controls Div.; Foxboro Analytical; Infrared Industries Inc.; Mine Safety Appliances Co.; Perkin-Elmer, Instrument Div.; Teledyne Analytical Instruments

The measurement of carbon dioxide in the ambient air is a primary concern of the geophysicist rather than of the air pollution control engineer. The precise measurement of the carbon dioxide content of the atmosphere is of significant concern in determining long-term changes in the composition of the atmosphere.[1] The measurement techniques used by geophysicists are highly precise compared to the techniques for air pollution evaluations.

In air pollution, carbon dioxide is hardly ever measured in the ambient air. It is measured at emission points since some combustion equipment regulations are stated in terms of allowable pollutant discharges corrected to 50 percent excess air.

Ambient Air Measurement

The precise knowledge of ambient air concentrations is necessary. An increase in global carbon dioxide concentration of only 1 percent, or about 3 ppm, has significant consequences. The instruments used to measure atmospheric carbon dioxide concentrations are, of necessity, highly precise.

The universal instrument for this purpose is the nondispersive infrared analyzer (Section 10.9). Since the absorption bands of water and carbon dioxide overlap somewhat, a freeze-out trap (−80°C or −112°F) is used in the sample preparation system. When using prepared calibration standard mixtures of carbon dioxide in nitrogen, an accuracy of ±0.2 ppm is attainable. At normal atmospheric levels of approximately 314 ppm, this is the equivalent of ±0.06 percent accuracy.

Source Measurement

Measurement of the carbon dioxide, carbon monoxide and oxygen concentration of flue gases from boilers has been done for many years.[2] These measurements allow precise setting of boiler operating variables for maximum fuel economy.

Before the use of the infrared analyzer became accepted practice, mechanical instruments were used to continuously determine the carbon dioxide content of flue gases. Their operation was based on the reduction in gas volume resulting from the absorption of carbon dioxide in a strong alkaline solution. This is the principle of the Orsat analyzer (Section 11.21), still used as the standard method for manual determination of combustion gas composition.

For air pollution testing purposes, the carbon dioxide content of the flue gas is determined only during the few hours of the test. The usual procedure is to slowly withdraw a low-volume integrated sample into a plastic bag over the duration of the test. The bag sample is then analyzed manually using an Orsat analyzer or instrumentally using a nondispersive infrared analyzer.

REFERENCES

1. Lipták, B.G., ed., *Environmental Enginners' Handbook,* Vol. II, Section 2.1, Chilton, Radnor, Pa., 1974.
2. Lipták, B.G., ed., *Instrument Engineers' Handbook,* Vol. II, Section 10.6, Chilton, Radnor, Pa., 1969.

BIBLIOGRAPHY

Frost & Sullivan, Inc., *On Stream Process Analyzer Market,* Report No. 669, Frost & Sullivan, Inc., New York, August 1979.

Kusuda, T., "Intermittent Ventilation for Energy Conservation," ASHRAE Symposium 20, Paper No. 4, August 1975.

Flow Sheet Symbol

TO RECEIVER

AT

CO

11.5 CARBON MONOXIDE

Type of Detection Method:	A—Nondispersive infrared B—Mercury vapor C—Gas chromatography D—Electrochemical analyzers E—Catalytic analysis
Reference Method:	A
Range, ppm:	0 to 25 (A); 0 to 50 (A,B); 0 to 100 (A); 0 to 200 (C)
Sensitivity, ppm:	0.05 (B); 0.1 (C); 0.5 to 1.0 (A); 1.0 (D)
Response Time, minutes:	1 to 5 (A)
Cost:	\$5,600 to \$11,500 (A); \$6,700 (B); \$18,000 (C)
Partial List of Suppliers:	Bacharach Instrument Co. (B); Beckman Instruments Inc. (A,C); Bendix Corp., Process Instruments Div. (A); Calibrated Instruments Inc. (A); Mine Safety Appliances Co. (A,C); Teledyne Analytical Instruments (A); Tracor Inc. (C); Union Carbide Corp. (C)

Calibration Techniques

Methods used for analysis of carbon monoxide at atmospheric levels (Table 11.5a) require calibration with gas mixtures of known concentration. Such mixtures may be prepared by volumetric dilution pure carbon monoxide with carbon monoxide–free nitrogen or helium. If the volumes used in dilution are accurately known, the concentration may be calculated. A known small volume of pure carbon monoxide is placed in an evacuated tank of known volume and the tank is then filled with the diluent gas. The pressures of the carbon monoxide volume and the final mixture must be known (or else known to be equal). Smaller samples may be prepared in plastic bags by injecting pure carbon monoxide from a gas syringe into a stream of diluent gas metered accurately into the bag with a flow device such as a wet test meter.

Cylinder nitrogen commonly has small amounts of carbon monoxide present and must not be assumed to be pure without verification. Pure helium is reliably free of carbon monoxide. A useful procedure is to zero the detector on helium, then read the content of the available

Table 11.5a
CARBON MONOXIDE ANALYZERS

Detection Method	*Range, ppm*	*Sensitivity, ppm*	*Advantages*	*Disadvantages*
Nondispensive infrared (NDIR)	0–25, 50, 100	0.5–1	U.S. reference method, accurate, stable, dry gases	Sensitive, water and CO_2 interferences (correctable), zero gas problems
Mercury vapor (hot HgO + CO releasing Hg vapor)	0–50	0.05	Sturdy, accurate, dry gases, high sensitivity	Interferences by water and other gases
Gas chromatography (reduction of CO to CH_4, flame ionization detection)	0–200	0.1	Accurate, high sensitivity, also read CH_4, dry gases	Complex and expensive

cylinder nitrogen and make the corresponding correction. This permits using the less expensive nitrogen for most calibrations.

A reference gas mixture whose carbon monoxide content is not accurately known may be analyzed gravimetrically or volumetrically by various methods, but these techniques are usually less convenient and often require relatively large amounts of gas.

Nondispersive Infrared Analyzers

Nondispersive infrared (NDIR) analysis[1] is the reference method for the U.S. National Air Quality Standard for carbon monoxide. It allows continuous analysis based on the capacity of carbon monoxide to absorb infrared radiation. A schematic diagram of a typical NDIR analyzer is shown in Figure 11.5b. Infrared radiation from a hot filament is chopped to pass alternately through sample and reference cells to be absorbed in the detector cell divided by a pressure-sensitive diaphragm. If the sample contains carbon monoxide, it will absorb part of the radiation, causing that half of the detector to exert less pressure on the diaphragm, whose distortion is converted to an electrical signal for rectification and amplification.

Sample airflow is continuous at or sometimes above atmospheric pressure. The cell is commonly 0.5 m long. Measuring ranges usually extend from a minimum of 0.5 to 1 ppm up to a full-scale range of 25 to 100 ppm. Response times are in the range of less than 1 to 5 minutes. Although NDIR response is nonlinear, it is assumed to be linear over the limited calibration range in use. Some instruments correct for nonlinearity in the output amplifier. Instruments can be operated by nontechnical personnel.

Carbon dioxide and water vapor in the sample interfere with the measurement. Filter cells filled with these gases, or optical filters when placed in front of the cells, can minimize effects of normal atmospheric levels of these interfering gases. The control of water vapor interference by its removal with desiccants (e.g., silica gel) or a refrigerator condensor is preferable in many cases, even at the cost of some increase in response time.

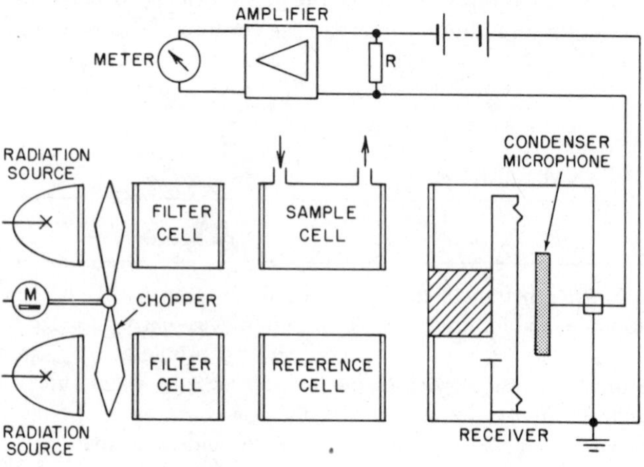

Fig. 11.5b Nondispersive infrared carbon monoxide analyzer

Mercury Vapor Analyzer

Carbon monoxide is oxidized by hot mercuric oxide[2] as follows:

$$CO + HgO(s) \xrightarrow{(210°C)} CO_2 + Hg(g) \quad 11.5(1)$$

The mercury vapor released may be measured photometrically. An analyzer based on this principle (Figure 11.5c) can be operated continuously. It has higher sensitivity than does the NDIR type, but it also suffers from some interferences. Detection levels go down to 0.025 ppm and changes of a tenth of that can be observed. Oxygenated hydrocarbons, olefins, and hydrogen interfere with the measurement, but all are normally at much lower concentrations in air than carbon monoxide. Water also interferes and should be removed by a dryer. This instrument has found particular use in nonurban measurements where carbon monoxide levels are very low.

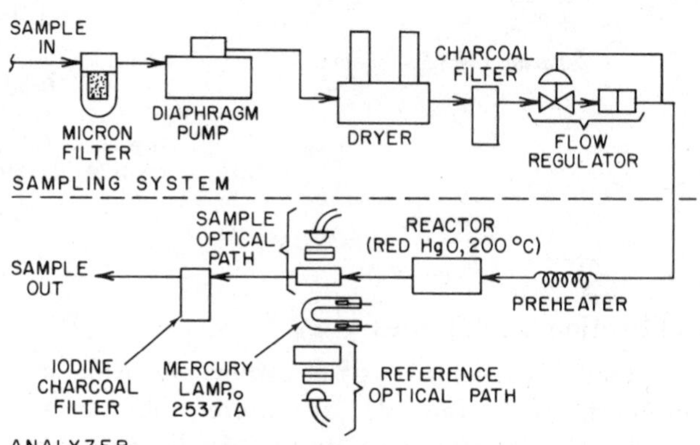

Fig. 11.5c Mercury vapor carbon monoxide analyzer

Gas Chromatographic Analyzer

An automated gas chromatograph[3] is the heart of a system to measure methane and carbon monoxide with high precision and specificity (Figure 11.5d). A precolumn holds back carbon dioxide, water, and hydrocarbons other than methane from reaching the molecular sieve separation column. After separation, a catalytic nickel reactor converts carbon monoxide to methane, which is detected by flame ionization. The system permits determination of both methane and carbon monoxide about once every 5 minutes. The output is linear for both components and can be read from 0.1 to 200 ppm. The instrument is relatively complex and expensive, however, and requires technically trained operators.

Electrochemical Analyzer

A galvanic cell for continuous carbon monoxide analysis[1] is based on its reaction with iodine pentoxide:

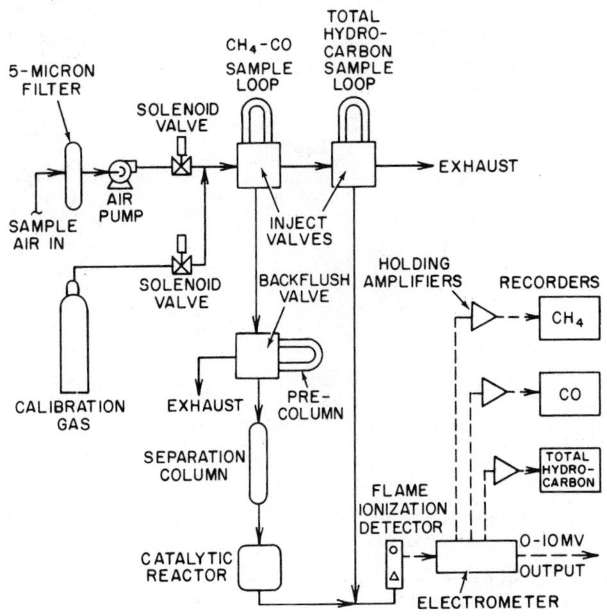

Fig. 11.5d Methane and carbon monoxide analyzer

$$5\,CO + I_2O_5 \xrightarrow{(150°C)} 5\,CO_2 + I_2 \qquad 11.5(2)$$

The iodine liberated is absorbed by an electrolyte and reaches the cathode of a galvanic cell where it is reduced. The resulting current is measured by a galvanometer. Interferences by mercaptans, hydrogen sulfide, hydrogen, olefins, and acetylenes may be minimized by sampling through an absorption tube of mercuric sulfate on silica gel. Water vapor interference is removed by a drying column.

The same reaction is used in a coulometric method with a modified Hersch-type cell. The iodine is passed into the cell and the current flow is measured by an electrometer. The interferences are the same as for the galvanic analyzer.

The minimum detectable concentrations are 1 ppm with good precision if flow rates and temperature are controlled. Careful column preparation is required and the response time is relatively slow.

Catalytic Analysis

The catalyst Hopcalite will oxidize carbon monoxide to carbon dioxide.[1] The resultant temperature rise may be recorded continuously as a measure of carbon monoxide concentration. The catalyst temperature and residence times must be controlled to avoid interference by hydrocarbons. The method is not suitable for most air monitoring applications because of low sensitivity.

Spot Sampling Methods

When only intermittent analyses are required, it is convenient to collect samples in the field for later analysis in the laboratory. Rigid glass bulbs or stainless steel tanks may be evacuated, then simply opened briefly to collect the air sample. Plastic bags may be filled by means of a small air pump. The samples may be analyzed later by various means, including use of a continuous analyzer at some other location. The samples may be analyzed in a central laboratory by an infrared spectrophotometer with a long-path gas cell or by suitable gas chromatographic apparatus.

Some colorimetric methods are available for carbon monoxide analysis, although, in general, the sensitivity and precision are low for atmospheric work. An NBS colorimetric indicating gel, if freshly prepared, will give accuracies of 5 to 10 percent, with detectability down to 0.1 ppm. The technique is simple but lengthy and tedious, with interferences by oxidizing and reducing gases.

Conclusions

For simple, not-too-precise measurement of short-term carbon monoxide levels, there are gel tube methods. For continuous and precision methods, more sophisticated equipment is necessary. The method in most widespread use currently is the nondispersive infrared analysis; but, in addition, there are now available two new types of analyzers. If it is necessary to also measure methane, the combination gas chromatographic methane–carbon monoxide analyzer, though expensive, is worthy of consideration. The other new type, the mercury vapor analyzer, is not yet thoroughly tested but seems particularly suitable where the carbon monoxide levels are low.

REFERENCES

1. "Air Quality Criteria for Carbon Monoxide," U.S. Department of Health, Education, and Welfare, National Air Pollution Control Administration Publication No. AP-62, March 1970, Chap., 5.
2. Robbins, R.C., Borg, K.M., and Robinson, E., pp. 106–110, *Journal of the Air Pollution Control Association*, vol. 18, 1968.
3. Stevens, R.K., O'Keefe, A.E., and Ortman, G.C., "A Gas Chromatographic Approach to the Semicontinuous Monitoring of Carbon Monoxide and Methane," 156th National Meeting of the American Chemical Society, Atlantic City, September 1968.

BIBLIOGRAPHY

Bay, H.W., "Electrochemical Technique for the Measurement of Carbon Monoxide," *Analytical Chemistry*, October 1974.

Dailey, W.V., "A Novel NDIR Analyzer for NO, SO_2 and CO Analysis," *Analysis Instrumentation*, Vol. 15, ISA, 1977.

Flow Sheet Symbol

TO RECEIVER

AT

Cl

11.6 CHLORINE

Methods of Detection:	A—Colorimetric, (A1) visual and (A2) spectrophotometric B—Amperometric
Sampling:	Methods A and B may be used for batch or grab sampling or in automatic continuous analyzers
Sample Pressure:	Generally atmospheric or near atmospheric. For continuous analyzer, water pressure is reduced to atmospheric
Sample Temperature:	All methods are generally limited to the range of 32° to 120°F (0° to 49°C), with method B employing automatic temperature compensation within this range. Method A may or may not require precise temperature control, depending on the reagents employed
Sample Size or Flow Rate:	In method A1, grab samples as small as 5 ml are sufficient. For method A2, flow rates of 10 to 75 ml per minute are generally specified. In method B, flow rates of 100 to 750 ml per minute are required
Materials of Construction:	Enclosures for units are available in fiberglass, styrene, urethane-painted steel, vinyl-covered aluminum, and other corrosion-resistant construction, suitable for modular or control panel installation. Wetted parts are constructed of PVC, Teflon, Lucite, Polyethylene, or glass. In method B, gold or platinum measuring and copper reference electrodes generally are employed
Readout:	All designs have indicating meters or output signals for recorders, or both. High-low alarm actuation or chlorinator controls are also available
Specificity:	Both methods A and B determine either free or total residual chlorine
Interferences:	For method A, interfering substances may include other oxidants, e.g., manganese, nitrite, and chlorine dioxide, as well as turbidity and color. In method B, nitrogen trichloride and chlorine dioxide may interfere with free chlorine determinations
Inaccuracy:	Generally ±2 to ±5% of full scale for ranges up to 20 ppm
Ranges:	For method A, available ranges include 0 to 1 ppm free chlorine, 0 to 3 ppm, and 0 to 10 ppm total chlorine. For method B, 0 to 1, 0 to 2, 0 to 5, 0 to 10, and 0 to 20 ppm ranges are available
Response to Chlorine Concentration Change: (continuous analyzer)	Three minutes or more for method A2, generally less than 10 seconds for B
Cost:	Between $750 and $8,000. The lower prices apply to method A2, and the higher prices to method B. Visual test kits are generally prices from $25 to $90

Partial List of Suppliers: Beckman Instruments Inc. (B); Capital Controls Co. (B); Fischer & Porter Co. (B); Hach Chemical Co. (A); La Motte Chemical Products Co. (A); Process Analyzers Inc. (A); Technicon Industrial Systems (A2); Teledyne Analytical Instruments (A); Uniloc Div. of Rosemount, Inc. (B); Wallace & Tiernan Div., Pennwalt Corp. (B)

Standard laboratory methods for determining aqueous chlorine concentrations involve iodometry in which free iodine liberated from potassium iodide is titrated with sodium thiosulfate, using starch as an indicator. All active forms of chlorine will liberate free iodine, limiting the iodometric methods to the determination of total active chlorine. Other titrimetric procedures employing indicators other than starch have been developed for the determination and differentiation of the active chlorine forms. The amperometric titration method is based on polarographic principles and allows determination of the various forms of active chlorine. Various colorimetric procedures are available in which a colorless indicator is oxidized to a colored product, the color intensity of which is proportionate to the chlorine concentration. Many of the colorimetric procedures allow the determination of the various forms of chlorine. Colorimetric procedures are adaptable to field determination by the use of test kits containing the necessary reagents, color standards for visual estimation, and portable battery-operated colorimeters. The amperometric titration and colorimetric methods have also been adapted to automated continuous analyzers.

Colorimetric

Figure 11.6a illustrates the basic components of a chlorine colorimetric analyzer. The water sample is introduced into a head regulator where the sample flow rate is regulated by an overflow arrangement and a capillary delivery tube. The sample is passed into a channel and mixed with the colorimetric reagent metered through another capillary tube from a storage container. The treated sample flows over a sample heater, if required, and into the sample cell which fills and overflows at a predetermined rate. If chlorine is present in the water, a characteristic color develops with an intensity in proportion to the amount of chlorine present. A filtered photocell develops an output signal which is in proportion to the reduction in intensity of the transmitted light through the sample. Periodic standardization

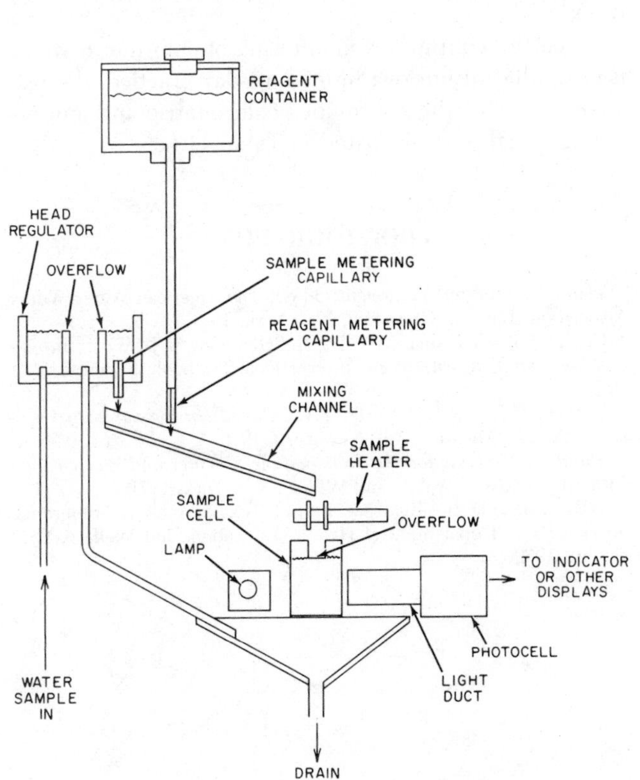

Fig. 11.6a Chlorine colorimetric analyzer

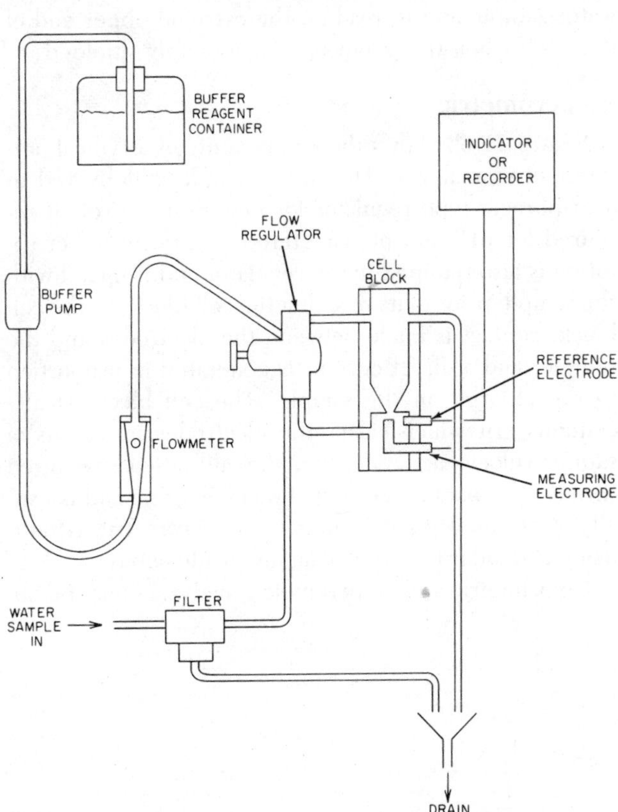

Fig. 11.6b Chlorine amperometric analyzer

Table 11.6c
RELATIVE MERITS OF COLORIMETRIC AND AMPEROMETRIC ANALYZERS

Consideration	*Colorimetric*	*Amperometric*
Type of sample	Better suited for clarified natural or treated waters than for highly turbid or colored waters and wastewaters	Turbidity and color generally not a problem, applicable to both treated water and wastewater
Interference	Interfering ions should be absent, oxidized manganese compounds produce serious interference	Copper and silver ions may interfere by plating out on electrodes
Sample temperature	Temperature control may or may not be required, depending on reagent employed	Manual or automatic temperature compensation required
Speed of response	Generally 3 or more minutes required to detect a change in chlorine concentration	Chlorine concentration change detected in 10 seconds or less
Calibration	Analyzer precalibrated; periodic standardization requires only simple manipulations	Periodic calibration required by separate analytical technique
Reagents required	External reagent solution required	External buffer may be required for varying sample pH
Maintenance	Cell staining may require periodic cleaning	Electrodes may require periodic cleaning
Stability	Drift compensated for by relatively simple standardization step	Drift not a problem when electrodes are kept clean
Initial cost	Generally less expensive	Generally more expensive

of the analyzer is required and is done by adjustment of the indicator to read 0 ppm chlorine on an untreated water sample and to read on the extreme upper end of the scale when the photocell is completely shielded.

Amperometric

Figure 11.6b shows the components of a typical amperometric analyzer. The water sample is delivered to a diaphragm-type regulator for flow rate control. If required for pH control, a metered amount of buffer solution is also pumped to the regulator and is mixed with the sample prior to its reaching the cell block. In the cell block, contact is made between the electrodes and the sample, and a direct current is generated in proportion to the chlorine in the sample. The cell block usually contains grit which cleans the electrodes by means of sample velocity agitation. Periodic calibration is required through a separate determination of chlorine and is usually performed on a laboratory amperometric titrator using a standardized phenylarsine oxide solution.

Colorimetric and amperometric analyzers may be obtained with diverse accessories including alarm actuation and external control functions. Automatic temperature compensation is also available for the amperometric analyzer.

Since the continuous monitoring of chlorine in water is of significant interest in modern-day practice, the relative merits of the continuous colorimetric and amperometric analyzers are listed in Table 11.6c.

BIBLIOGRAPHY

Water Quality and Treatment, 3d ed, The American Water Works Association, Inc., McGraw-Hill, New York, 1971.

Clarke, J.W., Viessman, W., Jr., and Hammer, M.J., *Water Supply and Pollution Control*, 2d ed, International Textbook, Scranton, Pa., 1971.

Sawyer, C.N., and McCarty, P.L., *Chemistry for Sanitary Engineers*, 2d ed., McGraw-Hill, New York, 1970.

Standard Methods for the Examination of Water and Wastewater, 13th ed., APHA, AWWA, and WPCF, New York, 1971.

Water Chlorine (Residual) No. 1, Study No. 35. Analytical Reference Service, U.S. Department of Health, Education, and Welfare, Cincinnati, 1969.

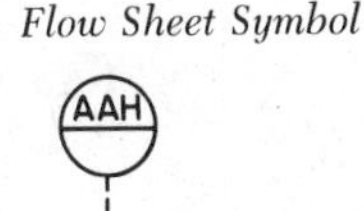

11.7 COMBUSTIBLES

Range:	Factory calibrated for 0–20% to 0–100% lower explosive limit
Inaccuracy:	±5% of full scale
Material for Wetted Parts:	Stainless steel
Cost Range:	\$3400–\$4500 per single loop
Partial List of Suppliers:	Bacharach Instrument Co.; Bailey Control Co. Div., Babcock & Wilcox; Beckman Instruments; Bendix Corp., Environmental & Process Instrument Div.; Hays-Republic Div.; Milton Roy Co.; Mine Safety Appliances Co.; Teledyne Analytical Instruments; Thermox Instruments, Inc.

Maintaining safety of personnel and property is of utmost importance in modern plant design. This section will describe the principles of operation and application of those instruments designed to detect the presence and measure the concentration of combustible gases and vapors on a continuous basis.

Many physical and chemical methods have been developed to detect combustible gases and vapors. These include measurement of changes in pressure, volume, or temperature of a sample when it is burned; reaction of a sample to a catalyst, such as electrical resistance or luminosity; thermal conductivity of a sample; infrared absorption; and gas ionization.

Of the above methods, the most widely used is catalytic combustion, where a change in resistance or temperature of the sensing elements due to catalytic combustion of the flammable gases is used to detect the concentration of combustibles.

Measuring circuits include the Wheatstone bridge for resistance and null balance potentiometers with thermocouples for temperature measurements.

In addition to the discussion of the measuring means, complete loops consisting of measuring, readout, and alarm devices and their applicability are covered in the following paragraphs.

For each combustible gas or vapor, there is a particular mixture with air that contains just enough oxygen to sustain combustion. Some combustible gas mixtures ignite more easily than others. Further, mixtures of the same constituents in different proportion require different amounts of spark energy to cause combustion.

Definitions

Lower Explosive Limit (LEL)—The leanest mixture of gas or vapor in air where, once ignition occurs, the gas or vapor will continue to burn after the source of ignition has been removed.

Upper Explosive Limit (UEL)—The richest mixture in which a flame will continue to burn after the source of ignition has been removed.

Flash Point—The lowest temperature at which a flammable liquid gives off enough vapors to form a flammable or ignitable mixture with air near the surface of the liquid or within the container used. Many hazardous liquids have flash points at or below room temperatures, and they are normally covered by a layer of flammable vapors which will ignite in the presence of a source of ignition.

Vaporization rate varies for each liquid depending on its vapor pressure. This rate increases with increased temperatures. Flammable liquids are, therefore, more hazardous at high temperatures.

PROPERTIES OF SOME FLAMMABLE LIQUIDS AND GASES

Name	Explosive Limits in Air, % by volume		Flash Point °F/°C (Closed Cup)
	Lower	Upper	
Benzene	1.4	7.1	12/−11
Carbon disulfide	1.0	50.0	−22/−30
Ethyl alcohol	3.5	19.0	55/13
Ethylene	2.7	34.0	Gas
N-hexane	1.2	7.5	−7/−22
Hydrogen	4.1	74.2	Gas
Propane	2.2	9.6	Gas

As seen from the table, explosive ranges of flammable liquids and gases are extremely wide. In detecting the presence of such vapors or gases, their lower explosive limits (LEL) are of interest, and in order to maintain safety, flammable gas and vapor concentrations are to be kept below those limits. The reasons for selecting LEL as a measure of safety are as follows. A concentration build-up of flammables above the upper explosive limit must have necessarily passed through the hazardous explosive range; and second, in bringing the concentration back down to a safe level below the lower explosive limit, the concentration must pass again through the hazardous explosive range. Since air is the diluent and is almost always present, all concentrations above the lower explosive limit are dangerous.

Gas Detection by Catalytic Combustion on a Heated Filament

The basis of most currently used flammable detectors or analyzers is the catalytic combustion technique. When a catalytically treated, fine, uniform, homogeneous platinum filament is heated and mixtures of flammable gases or vapors in air come in contact with its surface, combustion is induced at a temperature considerably below the normal ignition temperature of the particular combustible. The heat generated by the combustion on the catalytic filament is measured by sensing the change of temperature using thermocouples or by measuring the change of resistance of the filament.

Whether change of temperature or resistance is measured, it is convenient to use two filaments. One is constantly subjected to the sample (detector filament), the other is hermetically sealed in an inert atmosphere (reference filament). The reference filament is not activated with catalyst, but its temperature resistance characteristics are similar to the detector filament. The active detector filament is mounted in a measuring chamber which is relatively large with respect to the diameter of the filament, permitting a relatively large volume of sample to pass through the instrument. This insures that the sample in contact with the filament is always representative of current process conditions yet allows only a small portion of the sample to come in contact with the sensor, thereby increasing its useful life.

Measuring Circuits in Flammable Detectors

Thermocouple Detector

Two thermocouples are used. One thermocouple is bonded to the reference filament, the other to the detector filament. The two thermocouples are connected in series opposition, so that a differential electromotive force (emf) is developed and applied at the terminals of the potentiometric circuit (see Figure 11.7a). When a combustible gas or vapor is admitted to the measuring chamber, combustion increases the temperature of the detector filament, resulting in an increased emf for the thermocouple bonded to it. Temperature of the reference filament remains constant since no combustible sample comes in contact with it. Potentiometric indicating, recording, or alarming instruments respond to the resultant differential emf.

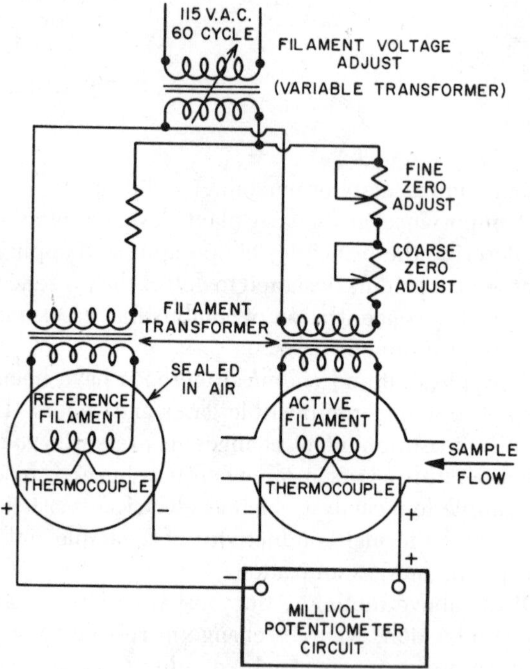

Fig. 11.7a Thermocouple detector

Wheatstone Bridge Detector

Wheatstone bridge is used for resistance measurement. Its operation is based on the comparison of an unknown resistance to a resistor of known value as shown in Figure 11.7b.

$R_1 = R_2 =$ constant
$R_3 =$ adjustable reference
$R_4 =$ resistance to be measured (by comparing to R_3)

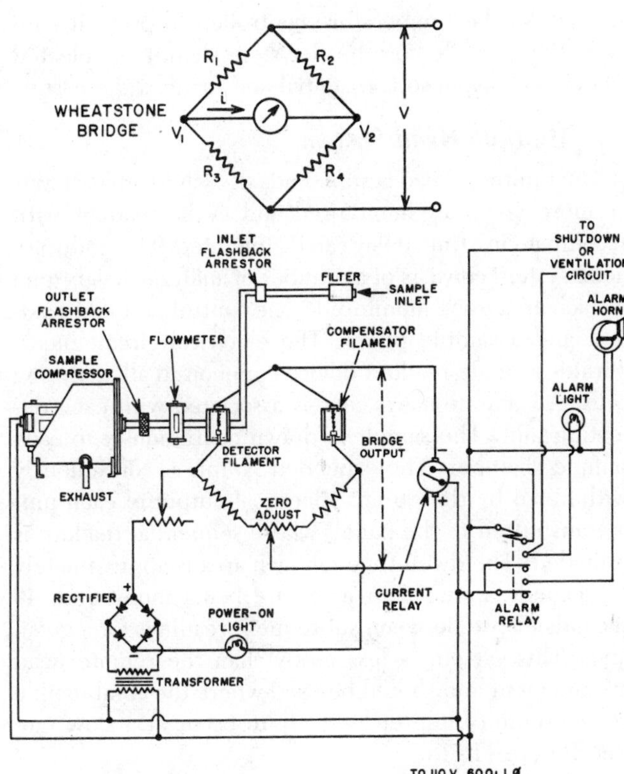

Fig. 11.7b Wheatstone bridge detector with accessories

For current I to be zero,

$$V_1 = V_2$$

$$V_1 = \frac{R_3}{R_1 + R_3}V$$

$$V_2 = \frac{R_4}{R_2 + R_4}V$$

$$\frac{R_3}{R_1 + R_3} = \frac{R_4}{R_2 + R_4}$$

$$R_3R_2 + R_3R_4 = R_4R_1 + R_4R_3$$

$$R_3R_2 = R_4R_1$$

$$R_3 = R_4$$

In the combustibles, detector R_3 is the reference filament and R_4 the detector filament. If the sample contains no combustibles, the bridge circuit remains in balance. If, however, there are combustibles in the sample, ignition will occur with consequent heating of the filament. The change of resistance of the detector filament due to heating will result in unbalancing the bridge in proportion to the amount of additional heating caused by the combustible material in the sample. The output voltage of the bridge, which is in proportion to the concentration of combustibles in the sample, is used to operate indicating or recording instruments, or to actuate alarms.

Diffusion Head Analyzer

The diffusion head analyzer differs from the two previously described detectors only in the means of delivering the sample to the measuring cell. The diffusion head analyzer does not require a sampling pump or sample flow control as all the other systems do. Rather, this system depends upon sample movement to the analyzing head by diffusion, density difference, convection, or similar motivations.

Sampling System

Considerable emphasis should be placed on sampling system design. The sample admitted to the analyzing cell should be wholly representative of that present in the monitored area, and it should be free of particulate matter or moisture.

In applications where the sample is at excessively high or low temperatures, it is advisable to use a sample conditioner. This is particularly important if the sample is hot and humid and tends to cool while passing through the sampling line. Cooling would result in condensation, which in turn could block the sample line or introduce a time lag in analyzer response. The sampling system should permit transport of the sample to the analyzer cell at the proper rate and minimum transportation time lag.

Since the vapors of all flammmable liquids are heavier than air, detection of such vapors requires that the probes be located near ground. In dealing with gases, their molecular weight (heavier or lighter than air) will decide whether sampling probes should be near the ground or at the ceiling of the monitored area. This may seem trivial, but successful detection of combustibles depends on careful consideration of probe location.

Accessories

An important precaution in the operation of combustible detectors is to avoid propagation of flame when sampled air containing an explosive mixture of gas is ignited on the detector filament.

Flashback arrestors of coiled copper screen are provided at the inlets and outlets of filament chambers. These screens prevent energy liberated by combustion from propagating to the outside.

Samples containing hydrogen or acetylene, with concentration of oxygen in excess of that found in normal air, have high rates of flame propagation. Standard flame arrestors cannot dissipate the energy liberated by combustion of such mixtures; therefore, special flame arrestors have to be used.

To insure safe operation of the detectors, alarms are provided to warn in case of: (a) filament failure, (b) power failure, (c) alarm relay failure, and (d) low sample flow (not available for diffusion head).

To insure that an adequate amount of sample passes through the measuring chamber, flow meters (rotameter) and needle valves are provided for all except the diffusion head type of units.

Selection of Complete Loops and Installation

Depending on the plant layout, speed of response required, and economic considerations, three basic systems can be considered: remote head (continuous measurement, continuous readout); multiple head (continuous measurement, sequential readout); and tube-sampling system (sequential measurement, continuous readout).

Remote Head System

While representing the highest initial cost, the remote head system offers the maximum application flexibility. As shown in Figure 11.7c, this system typically consists of a number of locally mounted analyzer heads (suitable for hazardous areas), and an equal number of panel-mounted control and readout devices. Normally, the maximum number of areas monitored from one central panel is eight, but this limitation is dictated only by the capacity of the aspirator and the physical size of the panel. Because the analyzer heads are located in the areas to be monitored, the speed of response is rapid. Samples are continuously drawn and the electrical signal corresponding to the measured combustible concentration is transmitted to the control unit instantaneously. Expended sample is continuously withdrawn from the analyzer head through the tubing to the aspirator and is exhausted. Since the analyzer head is in the monitored area, it can be temperature controlled to prevent condensation. The remote head system should be selected where fast response is essential and justifies the cost.

Multiple Head System

The multiple-head system finds use where at least four or more areas are monitored and cyclic readout with accompanying time delay can be tolerated. The multiple-head system consists of a number of analyzer heads (one in each area to be monitored), one control unit and readout, and a sample pump. The electrical circuit incorporates a single readout device common to all analyzing cells. A separate alarm unit is associated with each detecting unit. The sample is drawn continuously to each sample chamber. The expended sample is continuously withdrawn by the pump. Electrical output of each unit is transmitted to the panel, where sequential readout is provided. The dwell time for each area is approximately 10 seconds; i.e., if four areas are being monitored, 40 seconds elapse between subsequent readings for a given area. This system is less costly than the remote head arrangement and should be used where the combustible concentration build-up is likely to occur at a slow rate (see Figure 11.7d).

Tube Sampling System

The tube sampling system consists of one analyzer head, one readout device, and a sample pump. This is, therefore, the least expensive arrangement. Samples

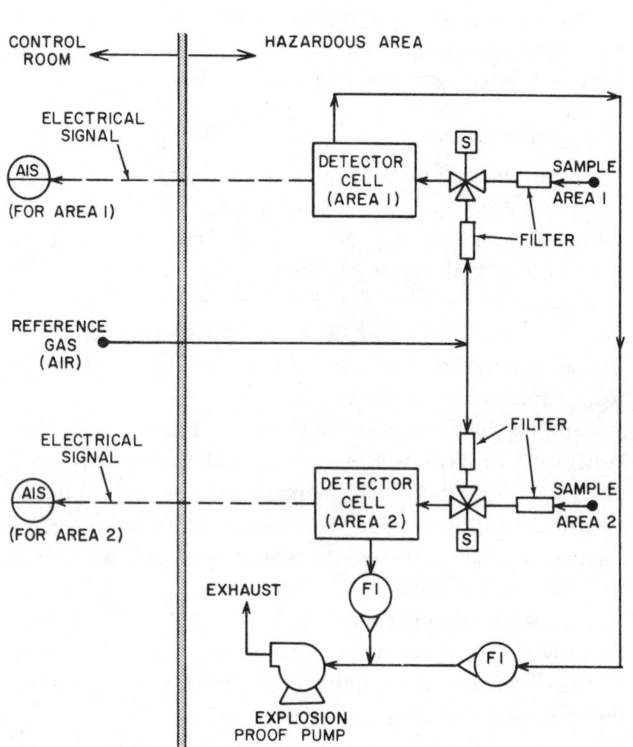

Fig. 11.7c Remote head system

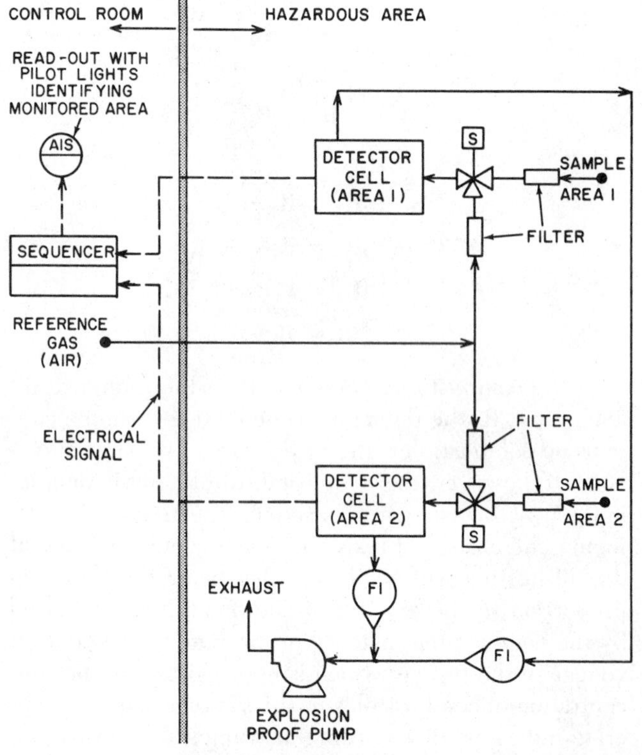

Fig. 11.7d Multiple head system

from different areas are admitted to the common analyzer head sequentially. The electrical signal is then transmitted to the readout device. A sample selector unit, consisting of time-sequenced solenoid valves, is arranged to admit one sample to the detector and connect all other sample lines to the sample pump. Sample is drawn continuously through each line. The sample selector is located at the analyzer head; thus lag time between successive analysis and delay due to sample travel is minimized since fresh sample is always present at the sample selector. One possible means of improving on the system is to use separate pumps for the sample analyzed and for those bypassing the detector.

A clean gas purge is provided after each analysis to prevent an erroneous reading caused by residual carryover in this type of system (see Figure 11.7e).

Tube sampling arrangements should be used only where true gases and vapors with boiling points well below ambient temperatures are to be detected to eliminated the problems associated with condensation in the sample tubes.

Tube sampling systems usually have a 30-second dwell time per point. Therefore, they should be considered where such slow response can be tolerated.

For additional safety, readout devices can be calibrated with full-scale ranges as low as 0 to 20 percent LEL.

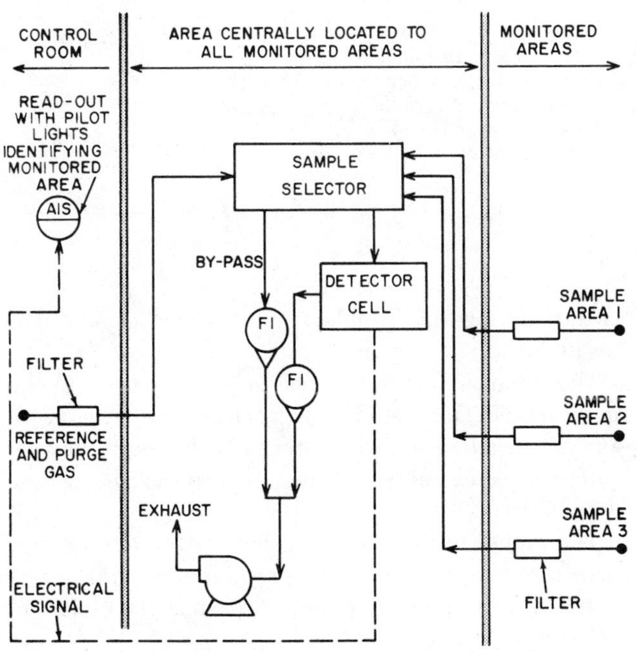

Fig. 11.7e Tube sampling system

Alarm switches contained in the measuring circuit are used to actuate alarms, start ventilation, shut down sparking devices, and so on.

These systems are found in coating ovens, solvent recovery, and soybean extraction plants, to mention a few typical applications.

Conclusions

Comparing the Wheatstone and the thermocouple cells, the following should be considered.

Whereas Wheatstone bridge cells use a fine helical filament, the thermocouple cell uses a heavy straight filament with a much longer useful life.

Further, the evaporation of the exposed filament results in a constant change of base resistance of the filament. In the Wheatstone bridge circuit, this change of base resistance produces a shifting of zero and requires frequent rebalancing of the bridge.

The temperature change measured by the thermocouple is independent of filament deterioration. Thus, for the thermocouple detector, the zero drift is reduced to a negligible amount even over long periods of time. Therefore, the thermocouple detector is superior to the Wheatstone bridge–type detector.

In the diffusion head–type analyzer, the use of sample pump or aspirator is eliminated. Dispensing with any moving part increases reliability. Therefore, the use of a diffusion head analyzer is recommended wherever clean, dry sample is to be analyzed.

Large amounts of particulate matter, moisture, and dust can and will cause plugging which is difficult to detect in diffusion head analyzers since they cannot be furnished with low sample flow alarms.

In addition to the above considerations, plant layout, required speed of response, rate of gas build-up, and economy should be some of the parameters used in selecting the flammable detectors.

BIBLIOGRAPHY

Baucke, C. G., "Application Considerations for Catalytic Combustible Gas Detectors," *Analysis Instrumentation*, Vol. 12, ISA, 1974 (ISA-AID74416).

Burgess, D., "The Flammability Limits of Lean Fuel-Air Mixtures," *Analysis Instrumentation*, Vol. 12, ISA, 1974 (ISA-AID74414).

Callahan, J., "Performance Standards for Combustible Gas Detectors," *Instrumentation Technology*, December 1981.

Dailey, W. V., "Monitoring Toxic and Flammable Hazards," *InTech*, February 1973, pp. 23–28.

Johanson, K. A., "Gas Detectors by the Acre," *InTech*, August 1974, pp. 33–37.

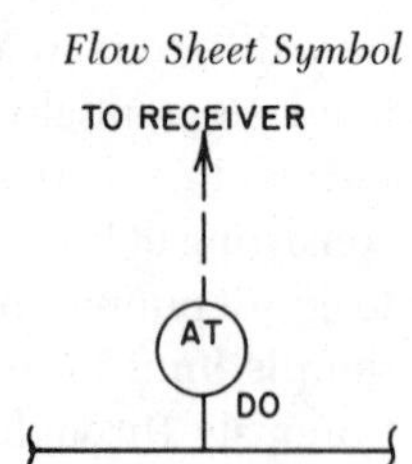

11.8 DISSOLVED OXYGEN

Methods of Detection:	A.—Galvanic cell B.—Polarographic cell C.—Multiple-anode cell D.—Thallium cell and other detectors
Sample Pressure:	Up to about 50 PSIG (345 kPa) maximum
Sample Temperature:	To about 175°F (79.5°C) maximum
Sample Flow Rate:	Should exceed 1 fps (0.3 m/s)
Materials of Construction:	Corrosion resistant
Inaccuracy:	Generally ranges from ±1% to ±2% full scale
Ranges:	0–5 to 0–20 ppm usually, and can also be calibrated in oxygen partial pressure units
Response:	90% in 60 sec. or less
Cost:	Some portable units cost less than $500. Most process-type transmitters are over $1500, not including sample systems
Partial List of Suppliers:	Beckman Instruments Inc.; Cambridge Instrument Company, Inc.; Delta Scientific, National Sonics Div., Envirotech Corp.; Hays-Republic Div. Milton Roy Co.; Honeywell Inc., Process Control Div.; Ionics Inc.; Uniloc Div. of Rosemount Inc.; Yellow Springs Instrument Co.

Oxygen Measurement in Liquid Samples

All oxygen detectors designed to handle gaseous samples can also measure the oxygen concentration in liquid streams, if components are provided to remove the gaseous oxygen from the liquid. This approach has the major advantage that only a clean gas mixture contacts the oxygen detector.

Electrochemical cells of the galvanic and polarographic types can measure both gaseous and dissolved oxygen. Because the galvanic and polarographic instruments come in a wide variety of models with numerous options, generalizations must be made.

Galvanic Cell

The theory of operation for the dissolved oxygen galvanic cell is the same as for gaseous samples (Section 11.21). With the exception of the considerations related to the "drying-out" problem, all design considerations are identical.

The main area of difference is in the nature of the applications. The majority of the dissolved oxygen sensors are installed in dirty water, and therefore they require special cleaners, agitators, and specialized sample systems.

The ranges of the galvanic-cell dissolved oxygen analyzer can be as low as 0 to 20 ppb, for applications such as the measurement of dissolved oxygen (DO) content in boiler feedwater.

All galvanic cells consist of an electrolyte and two electrodes (Figure 11.8a). The oxygen content of the electrolyte is brought into equilibrium with that of the sample. The electrodes are polarized by an applied voltage which causes electrochemical reactions to take place when oxygen contacts the electrodes. In this reaction, the cathode reduces the oxygen into hydroxide, thus

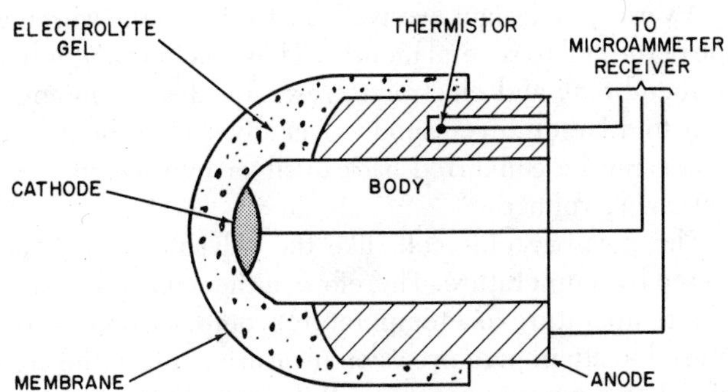

Fig. 11.8a Probe-type galvanic cell oxygen detector

releasing four electrons for each molecule of oxygen. These electrons cause a current flow through the electrolyte, the magnitude of which is in proportion to the oxygen concentration in the electrolyte.

The most common electrode materials are silver or lead, and the most frequently used electrolyte is potassium hydroxide (KOH). The cathode must be noble (silver or gold) in order for the cathode potential to reduce molecular oxygen when the cell circuit is closed. The anode is selected to be a base metal (lead, cadmium, zinc, or silver) with good stability and without any tendency toward passivation. The electrolyte [KOH, potassium chloride (KCl), or potassium bicarbonate ($KHCO_3$)] is selected so that it will not dissolve the anode at a high rate when the cell circuit is open.

In the case of a lead anode, the cell reactions can be expressed as follows:

$$\text{Cathode: } 0_2 + 2H_20 + 4 \text{ electrons} \longrightarrow 4\,(OH-)$$

$$\text{Anode: } 2Pb \longrightarrow 2(Pb^{++}) + 4 \text{ electrons}$$

The galvanic-cell designs are subject to various degrees to contamination by background gases in the process stream. As a very general rule, the following background gases can be considered harmless: argon, butane, carbon monoxide, ethane, ethylene, helium, hydrogen, methane, nitrogen, and propane. However, the following gases are likely to contaminate the cell: chlorine and other halogens, high concentrations of carbon dioxide, hydrogen sulfide, and sulfur dioxide.

Special cells have been developed to minimize the effect of the background gases. In cases where an acid gas (such as CO_2) which would neutralize a potassium hydroxide electrolyte solution is present in the background, a potassium bicarbonate electrolyte can be considered. Special cells are also available for measurement of oxygen in acetylene and fuel gases.

In the flow-through cell designs, sampling systems are usually required to bring the process stream to the analyzer and to filter it, scrub it with caustic, or treat it in other ways in preparation for the measurement. The probe-type membrane design does not require a sampling system if it can be located in a representative process area where the pressure, temperature, and velocity of the process stream are compatible with the cell's mechanical and chemical design.

Probe Design

In this design (Figure 11.8a) the electrodes are wetted by an electrolytic solution which is retained by a membrane (usually Teflon) that acts as a selective diffusion layer, allowing oxygen to diffuse into the sensor while keeping foreign matter out. The sensor is usually mounted in a thermostatically-controlled housing; therefore, the thermistor compensates for minor temperature variations.

The ion current established in the electrolyte can be expressed by the following equation:

$$i_x = \frac{nFAPmCs}{L} \qquad 11.8(1)$$

where:

i_x = the ion current
n = the number of electrons involved in the electrode reaction
A = the area of the cathode surface
F = Faraday's constant
Pm = the permeability coefficient of the membrane
L = the thickness of the membrane
Cs = oxygen concentration

Equation 11.8(1) shows the relationship of certain cell components. The Nernst equation [Equation 11.21(1)] also applies to the galvanic cell and explains why electrode potentials are a function of the absolute temperature. In addition, the ionic activity also varies with temperature, thus causing additional temperature sensitivity.

The characteristics of the membrane are critical to performance. The ideal membrane would be inert, stable, strong, permeable to oxygen, and impermeable to other ions and water molecules. In most cases a compromise solution is accepted.

Flow-Through Design

In these cells the process sample stream is bubbled through the electrolyte. The oxygen concentration of the electrolyte is therefore in equilibrium with the sample oxygen content, and the resulting ion current between electrodes is representative of this concentration.

In some trace analyzer designs, the cathode is made out of a porous metal and the sample gas passes through this electrode, immersed in the electrolyte. The oxygen reduction tends to be complete within the pores of this electrode (Figure 11.8b).

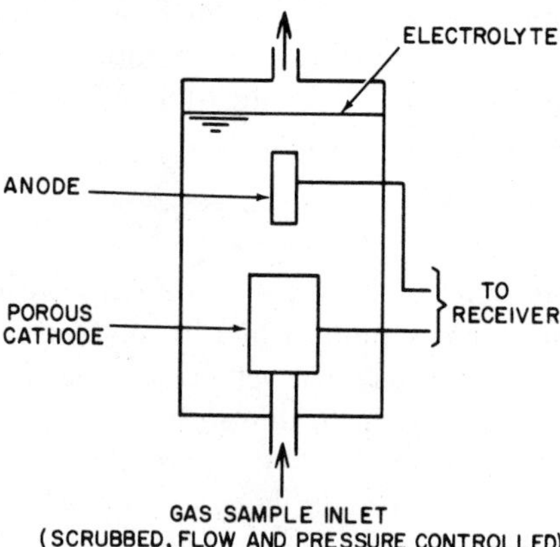

Fig. 11.8b Flow-through trace oxygen analyzer cell

Sampling systems are usually provided with these types of cells, consisting of (but not limited to) filtering and scrubbing components and flow, pressure, and temperature regulators.

Polarographic Cell

The basic polarographic cell illustrated on Figure 11.8c is very similar to the galvanic cell shown in Figure 11.8a. However, the polarographic cell has *two* noble-metal electrodes and requires a polarizing voltage to reduce the oxygen.

The dissolved oxygen in the sample diffuses through the membrane into the electrolyte which usually is an aqueous KCl solution. If there is a constant polarizing voltage (usually 0.8 V supplied by a mercury battery) across the electrodes, the oxygen is reduced at the cathode and the resulting current flow is directly proportionate to the oxygen content of the electrolyte. The oxidation–reduction reactions, in the case of a gold–silver cell with KCl electrolyte, are as follows:

at the gold cathode: $O_2 + 2H_2O + 4 \text{ electrons} \longrightarrow 4(OH^-)$

and at the silver anode: $4Ag + 4(Cl^-) \longrightarrow 4 AgCl + 4 \text{ electrons}$

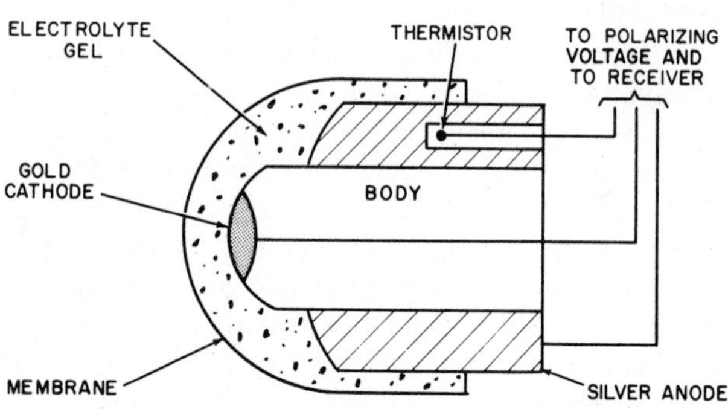

Fig. 11.8c Probe-type polarographic cell oxygen detector

A variety of designs are available to extend the working life of the cell to several months. These include a number of membrane and electrolyte reservoir designs in addition to a larger silver anode. The silver chloride at the anode can be converted back to silver by reversing the polarizing voltage.

The polarographic cell, like the galvanic cells, is affected by temperature. Therefore, either controlled sample temperature or temperature compensation is required to attain high precision measurements in the area of ±1 to 2 percent accuracy. If the sample temperature is allowed to vary between 32 and 110°F (0 and 43°C), the measurement accuracy will drop to approximately ±6 percent in some designs.

Both the galvanic and the polarographic cells require a minimum sample flow velocity. This is necessary to eliminate stagnant layers of sample over the membrane, which otherwise would interfere with the continuous transfer of oxygen into the cell. The higher sample velocities are also beneficial because of their scrubbing action. Some suppliers provide a combination cell and pump unit where the flow of 5 fps (1.5 m/s) velocity is directed against the membrane for maximum cleaning effect.

Multiple-Anode

The multiple-anode detector (Figure 11.8d) has three electrodes. Two are interspaced (+ and − in the figure) and covered with electrolyte. Oxygen is consumed at many cathodes but is generated at many anodes.

$$2H_2O \longrightarrow O_2 + 4H^+ + 4e^-$$

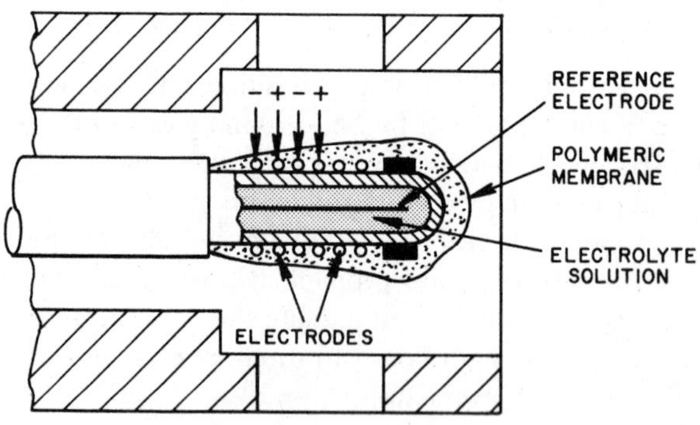

Fig. 11.8d Multiple-anode oxygen detector

Because a balance exists in this cell, there is no net generation of products. A number of advantages are claimed for this technique: fouling affects the response time only, there is no deterioration of electrodes, and membrane replacement is not required.

Thallium Cell and Other Detectors

The thallium cells are somewhat unique in their operating principle and cannot be classified into the category of either galvanic or polarographic cells. At the same time, they are of the electrochemical type.

One thallium-electrode cell design is somewhat similar in appearance to the unit illustrated on Figure 11.8c, except that it has no membrane or electrolyte. This cell has a thallium outer-ring electrode and an inner reference electrode. When oxygen contacts the thallium, the following reaction takes place:

$$4Tl + O_2 + 2H_2O \longrightarrow 4Tl\,(OH) \longrightarrow 4(Tl^+) + + 4(OH^-)$$

The potential developed by the cell is a function of the thallous ion concentration at the face of the electrode, and the ion concentration is in proportion to the concentration of dissolved oxygen.

Another cell involving the reaction between thallium and oxygen is the thallium differential conductivity analyzer. This sensor detects the amount of thallous hydroxide formed by the measurement of conductivity. The sample conductivity is sensed both before and after the reaction and the difference in conductivities is detected as a measure of dissolved oxygen content. In order to eliminate temperature gradient problems, the actual instrument has both a sample and a reference flow stream. They are maintained at the same rate and they flow through a similar flow path. Due to the high sensitivity of this measurement, accuracies of ± 0.5 ppb can be obtained over the minimum range of 0 to 10 ppb. The response speed of this sensor is relatively slow, meaning that the 95 percent response can be expected in about three minutes.

Another method of dissolved oxygen detection involves wet-chemistry analyzers. These devices operate intermittently by taking a small sample and adding reagents to it. The reagents develop a color in the sample if a certain component is present. This technique can be applied to dissolved oxygen detection—besides many other applications—by determining the concentration of the unknown component colorimetrically.

BIBLIOGRAPHY

Bond, A., and Anterford, C. D., "Comparative Study of a Wide Variety of Polarographic Techniques with Multifunctional Instruments," *Analytical Chemistry*, Vol. 44, 1972, p. 721.

Delahay, P., *New Instrumental Methods in Electrochemistry*, Wiley Interscience, New York, 1954.

DuCross, M. J. F., "Automated Methods for Assessing Water Quality Come of Age," *Environmental Science and Technology*, October 1975.

Flato, J., "The Renaissance in Polarographic and Voltametric Analysis," *Analytical Chemistry*, Vol. 44, 1972, p. 75A.

Frost & Sullivan, Inc., *On Stream Process Analyzer Market*, Report No. 669, Frost & Sullivan, Inc., New York, August 1979.

Lingane, J., *Electroanalytical Chemistry*, 2d ed., Wiley Interscience, New York, 1958.

Molvar, A. E., "Instrumentation and Automation Experiences in Wastewater-Treatment Facilities," EPA Document 600/2–76–198, October 1976.

Shinskey, F. G., *pH and pION Control in Process and Waste Streams*, John Wiley, New York, 1973.

Flow Sheet Symbol

TO RECEIVER

AT

FLUORIDE

11.9 FLUORIDE

Method of Sampling:	A—Automatic sequential collection with impingers or bubblers B—Automatic sequential collection with dual tape samplers C—Automatic sequential collection with coated glass tubes
Type of Analysis:	D—Automatic sequential microdistillation and photometric analysis E—Potentiometry
Sampling and Analysis:	F—Automatic sequential collection on a coated glass tube and photometric analysis G—Automatic sequential collection on coated glass tubes and potentiometric analysis
Airflow Rate:	Varies with device. Generally between 7 and 30 l/m
Sample Duration:	Varies with device, concentration of fluoride in air, sensitivity of analytical method, and purpose of monitoring program. Usual range for different methods is from 0.5 to 24 hr
Materials of Construction:	Surfaces of equipment that contact air or liquid containing fluoride are usually composed of stainless steel, Teflon, epoxy, polyethylene, or polypropylene
Concentrations Measured:	Usual range of interest for ambient air monitoring is less than 1 $\mu g/m^3$ to more than 50 $\mu g/m^3$. Lower concentrations usually require longer sample durations; higher concentrations may require slower airflow rates or sample dilution
Sensitivity:	Depends on magnitude of blank values, sensitivity of analytical method, collection rate, and length of sampling period. Limit of detection should be three to five times lower than lowest concentration of interest
Calibration:	Procedure varies with method and equipment. Frequently, the sampling step is not included in the calibration procedure and errors due to incomplete sampling or contamination may go undetected
Inaccuracy:	In most cases, information is insufficient for reliable estimates of inaccuracy. Probable range of absolute error: ±10 to ±50%; with percentages higher when the concentration is below 1 $\mu g/m^3$. Inaccuracy highly dependent on type and frequency of calibration
Precision:	Range of relative errors for repetitive measurements of selected methods under laboratory conditions: ±5 to ±21%. May be less precise at low concentrations and under outdoor conditions
Costs:	$550 to $5000 (A,B,C); $2200 (manual) to $22,500 (automated) (D,E); $10,000 to $21,000 (F,G)
Partial List of Suppliers:	Beckman Instruments, Inc. (E); Bendix Corp. (A,B); Coleman Environmental & Pollution Control Equipment Co., Inc. (E); Corning Glass Works (E); Hemeon Associates (C); Leigh Instruments, Inc. (C,F); Orion Research, Inc. (E); PSG Industries, Inc. (A); Research Appliance Co. (A,B,C); Technicon Corp. (D)

Table 11.9a
ORIENTATION TABLE FOR ATMOSPHERIC FLUORIDE MEASUREMENT METHODS

Sampling Methods Device	References	*Material Collected for Analysis* Wet	Dry	Gas	Particles	*Sampling, Duration, hr*	*Automatic Sequential Collection Equipment Available*	*Appropriate Sample Treatment Methods*[h] Ashing or Combustion	Acid Distillation[g]	Micro-diffusion	Ion Exchange	Extraction or Elution	*Preferred Analytical Methods* Titrimetry	Colorimetry[g]	Potentiometry	*Separate Measurement of Gaseous and Particular Fluorides*
A. Impinger	6	√		√	√	U 6–24	√		√	√	√		√	√	√	A. After F
B. Bubbler	7	√		√	P	U 6–24	√		√	√	√		√	√	√	B. After F
C. Scrubber	8	√		√	P	0.5–2.5[f]	√				√			√	√	C. After F
D. Aluminum tube, impingers	9	√	√	√[a]	√	U 6–24			√	√	√		√	√	√	D. Impinger in parallel
E. Glass fiber filter	9, 10		√	√	√	U 24–72		√	√	√			√	√	√	E. After F
F. Acid-treated or heated membrane or cellulose filter	11–13		√	[b]	P[d]	0.5–4	√	√	√	√	√	√		√	√	F. Before A, B, C, E, G, H or I
G. Alkali-treated membrane or cellulose filter	9, 12, 13		√	√	P[d]	0.5–4	√	√	√	√	√	√		√	√	G. After F
H. Coated glass tube	9, 12, 13		√	√		U 6–24	√		√	√	√	√		√	√	H. Before F
I. Coated silver beads	14, 15, 16		√	√	[e]			√	√	√	√	√	√	√	√	I. Before F
J. Limed filter paper	17–20		√	√[c]	√[c]	U 1 month										

[a]By difference.
[b]Significant quantities of gaseous fluoride may be collected depending on the type and treatment of the filter and amount of dust accumulated.
[c]Fluoride collected statically by absorption and deposition.
[d]Collection efficiency for particles dependent on flow rate, type of filter, porosity, and amount of dust accumulated.
[e]Larger particles will be collected unless excluded by using a Herpertz cap (Ixfeld, 1971).
[f]Functions as dosimeter.
[g]Equipment for automatic sequential analysis is available.
[h]Sample treatment methods may be needed to concentrate fluoride or to avoid interference with analytical measurements. In many cases, fluoride analyses are performed without concentrating or purifying the samples.
NOTE: The information in this table is for general guidance only. The reader should consult references for specific information.
P = partial; U = usually

There are two major operations in every atmospheric fluoride monitoring system: collection and analysis. Each monitoring system must possess fluoride sensing devices which are compatible with the available methods of collection.[1,2]

Selection of Analysis Methods

The circumstances and requirements of the monitoring program and the purposes for which the data are obtained dictate the choice of detection methods.[3] Air quality criteria for fluorides should also be examined[4] and the need for separating gaseous and particulate atmospheric components determined.[5] When selecting the methods of analysis, it should be kept in mind that the experience and capability of the supervisor, engineer, technician, or analyst are extremely important because monitoring methods are very susceptible to both positive and negative biases. Comparative testing and measurement of accuracy, precision, and reliability of methods of monitoring atmospheric fluorides lag far behind the rate of development and change in techniques and equipment.

Equipment

Details about individual sampling, analytical and sample treatment methods are provided in Table 11.9a.[21,22] Included in the table are the references which should be consulted for more detailed information.

Types of Monitoring Systems

There are a great many combinations of sampling and analysis methods currently in use. Basic equipment can be purchased and assembled into monitoring systems to satisfy particular requirements.[23] The fluorometric analyzer developed some years ago[24] is not commercially available. Wet chemical methods of analysis, such as spectrophotometry, colorimetry, and titrimetry, are used extensively.[7] The proliferation of methods, without thorough comparative testing, is highly undesirable because it leads to wide variations in results among laboratories. Reference should be made to the published standardized methods[6,21,25] and approved procedures should be followed exclusively.

Source Monitoring

Sampling fluoride near the source of emissions, such as in stack gases, usually requires special techniques. Higher concentrations of fluoride, particulate matter, and water vapor are encountered and higher temperatures and reactivity of the effluent gases must be overcome. An automatic dosimeter instrument for measurement of gaseous fluorides has been developed for stack monitoring.[26] In a recently developed method, atmospheric fluorides react to form silicon tetrafluoride before collection of particulate and gaseous fluoride and analysis is performed by manual methods.[27] This method has been modified to provide potentiometric analysis and calibration with hydrogen fluoride permeation tubes.[28]

Additional Research and Development

Published data on atmospheric concentrations of fluorides and factors affecting fluoride distribution in the atmosphere are scanty. The development of new instrumentation has taken place with inadequate knowledge of concentrations and fluctuations in concentration with time and distance from sources and with insufficient comparative testing of methods on known fluoride atmospheres.

REFERENCES

1. Hochheiser, S., Burmann, F.J., and Morgan, G.B., "Atmospheric Surveillance: The Current State of Air Monitoring Technology," *Environmental Science & Technology*, Vol. 5, 1971, pp. 678–684.
2. Hendrickson, E.R., "Air Sampling and Quantity Measurement," in A. C. Stern, ed., *Air Pollution*, Vol. II, 2d ed., Academic Press, New York, 1968.
3. Jacobson, J.S., Weinstein, L.H., and Farrah, G.H., *A Review of Methods for Monitoring the Fluoride Content of Air, Proceedings of the Environmental Control Symposium*, 1972 Annual AIME Meeting, Metallurgical Society of AIME, in press.
4. McCune, D.C., "On the Establishment of Air Quality Criteria, with Reference to the Effects of Atmospheric Fluorine on Vegetation," *Air Quality Monograph*, No. 69-3, American Petroleum Institute, New York, 1969.
5. National Academy of Sciences, Committee on Biologic Effects of Atmospheric Pollutants, *Fluorides*, Washington, D.C., 1971, pp. 51–65.
6. ASTM, *Standard Method of Test for Inorganic Fluoride in the Atmosphere*, D 1606-60, *Book of ASTM Standards, part 23*, American Society for Testing and Materials, Philadelphia, 1971, pp. 443–453.
7. Farrah, G.H., "Manual Procedures for the Estimation of Atmospheric Fluorides," *Journal of the Air Pollution Control Association*, Vol. 17, 1967, pp. 738–741.
8. Adams, D.F. and Koppe, R.K., "Automatic Atmospheric Fluoride Pollutant Analyzer," *Analytical Chemistry*, Vol. 31, 1959, pp. 1249–1254.
9. Pack, M.R., Hill, A.C., Thomas, M.D., and Transtrum, L.G., *Determination of Gaseous and Particulate Inorganic Fluorides in the Atmosphere*, ASTM Special Technical Publication No. 281, American Society for Testing and Materials, Philadelphia, 1959, pp. 27–44.
10. Pack, M.R. and Hill, A.C., "Further Evaluation of Glass Fiber Filters for Sampling Hydrogen Fluoride," *Journal of the Air Pollution Control Association*, Vol. 15, 1965, pp. 166–167.
11. Habel, K., "The Separation of Gaseous and Solid Fluorine Compounds During Air-Quality Measurements," *Staub* (English translation), Vol. 28, No. 7, 1968, pp. 26–31.
12. Mandl, R.H., Weinstein, L.H., Weiskopf, G.J., and Major, J.L., "Separation and Collection of Gaseous and Particulate Fluorides," Paper CD-25A, Second International Clean Air Congress, Washington, D.C., 1970.
13. Weinstein, L.H., and Mandl, R.H., "The Separation and Collection of Gaseous and Particulate Fluorides," VDI Berichte, Nr. 164, 1971, pp. 53–63.
14. Buck, M., and Stratmann, H., "Ein Verfahren zur Bestimmun sehr geringer Konzentrationen von Fluor-Ionen in der Atmosphäre," *Brennstoff-Chemie Bd.*, Vol. 46, 1965, pp. 231–235.
15. Svoboda, K., and Ixfeld, H., "Probenahme und automatische Analyse gasförmiger Fluor-Immissionen," *Staub*, Bd. 31, 1971, pp. 1–8.
16. Robinson, E. "Determining Fluoride Air Concentrations by Exposing Limed Filter Paper," *American Industrial Hygiene Association Quarterly*, Vol. 18, 1957, pp. 145–148.

17. Adams, D.F. "A Quantitative Study of the Limed-Filter Paper Technique for Fluorine Air Pollution Studies," *International Journal of Air and Water Pollution*, Vol. 4, 1961, pp. 247–255.

18. Wilson, W.L., Campbell, M.W., Eddy, L.D., and Poppe, W.H., "Calibration of Limed Filter Paper for Measuring Short-Term Hydrogen Fluoride Dosages: The Effect of Temperature, Humidity, Wind Speed, and Dose," *American Industrial Hygiene Association Journal*, Vol. 28, 1967, pp. 254–259.

19. Mukai, K., and Ishida, H. "The Alkaline Filter Paper Method for Surveying Fluorides in the Atmosphere," Paper No. A70-10, TMS-AIME Annual Meeting, Denver, Colorado, 1970.

20. Ixfeld, H., "Fluor-Immissionsmessung in Lande Nordrhein-Westfalen," VDI Berichte, Nr. 164, 1971, pp. 71–75.

21. Intersociety Committee on Methods for Ambient Air Sampling and Analysis, "Tentative Method of Analysis for Fluoride Content of the Atmosphere and Plant Tissues (Manual and Semiautomated Methods)," 12204-01-68T and 12204-02-68T, Health Laboratory Science, Vol. 6, 1969, pp. 64–91 and 94–101.

22. ASTM, "Analysis of Fluoride in Ambient Air Samples with the Selective Ion Electrode," Committee D-22, in press.

23. West, P.W., Lyles, G.R., and Miller, J.L., "Spectrophotometric Determination of Atmospheric Fluorides," *Environmental Science and Technology*, Vol. 4, 1970, pp. 487–491.

24. Thompson, C.R., Zielenski, L.F., and Ivie, J.O., "A Simplified Fluorometric Fluoride Analyzer," *Atmospheric Environment*, Vol. 1, 1967, pp. 253–259.

25. Association of Official Analytical Chemists, *Official Methods of Analysis*, 11th ed., "Fluorine," Sections 25.029–25.035, 1970.

26. Adams, D.F., "An Automatic Hydrogen Fluoride Recorder Proposed for Industrial Hygiene and Stack Monitoring," *Analytical Chemistry*, Vol. 32, 1960, pp. 1312–1316.

27. Dorsey, J.A., and Kemnitz, D.A., "A Source Sampling Technique for Particulate and Gaseous Fluorides," *Journal of the Air Pollution Control Association*, Vol. 18, 1968, pp. 12–14.

28. Elfers, L.A., and Decker, C.E., "Determination of Fluoride in Air and Stack Gas Samples by Use of an Ion Specific Electrode," *Analytical Chemistry*, Vol. 40, 1968, pp. 1658–1661.

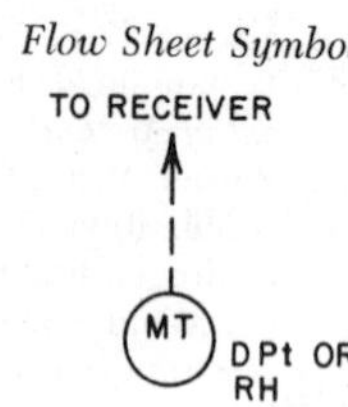

11.10 HUMIDITY AND DEW POINT SENSORS

Type:	A—Chilled mirror type B—Others
Design Pressure:	To 300 PSIG (2 MPa) (A) To 600 PSIG (4 MPa) (B)
Design Temperature:	−40 to +200°F (−40° to +93°C) (A) −25 to +170°F (−32° to +77°C) (B)
Element Material:	Metallic mirror, glass, anodized aluminum (A) Hair, glass, stainless steel (B)
Cost:	\$1000–\$5000 (A) \$110–\$2200 (B)
Inaccuracy:	±0.3 to 0.8°F (±0.2 to 0.4°C) dew point (A) ±5% relative humidity (B)
Range:	−100 to +200°F (−73° to +93°C) dew point (A) 5–100% relative humidity (B)
Partial List of Suppliers:	EG&G, Environmental Equipment Division; General Eastern Instruments Corp.; MBW; Sulzer (A); Foxboro Co.; General Eastern Instruments Corp.; Honeywell, Industrial Div.; Palmer Instruments, Inc.; Taylor Instrument Companies; Vap-Air Div., Vapor Corp. (B)

Definitions

Absorption—The taking in of a fluid to fill the cavities in a solid.

Adsorption—The adhesion of a fluid in extremely thin layers to the surfaces of a solid.

Dew point—Saturation temperature of a gas–water vapor mixture.

Hygrometer—An apparatus that measures humidity.

Hygroscopic material—A material with great affinity for moisture.

Partial pressure—In a mixture of gases, the partial pressure of one component is the pressure of that component if it alone occupied the entire volume at the temperature of the mixture.

Relative humidity—The ratio of the mole fraction of moisture in a gas mixture to the mole fraction of moisture in a saturated mixture at the same temperature and pressure. Or the ratio of the amount of moisture in a gas mixture to the amount of moisture in a saturated mixture at equal volume, temperature, and pressure.

Saturated solution—A solution which has reached the limit of solubility.

Saturation pressure—The pressure of a fluid when condensation (or vaporization) takes place at a given temperature. (The temperature is the saturation temperature.)

Specific humidity—The ratio of the mass of water vapor to the mass of dry gas in a given volume.

Wet and Dry Bulb Hygrometers

When water changes phase from liquid to vapor, an amount of heat equal to the latent heat of vaporization must be supplied from the environment for each unit mass of water evaporated. The vaporization, in addition to its dependence on heat available, is also a function of the degree of saturation or relative humidity of the atmosphere surrounding the water. The wet and dry bulb hygrometer takes advantage of this fact to effect a relative humidity measurement.

The instrument consists of two temperature-sensitive elements exposed to the atmosphere whose moisture level is to be measured. One of the two elements, the wet bulb, is wrapped with a wick soaked in water; the other element, the dry bulb, is left bare. Water evaporating from the wick lowers its temperature and that

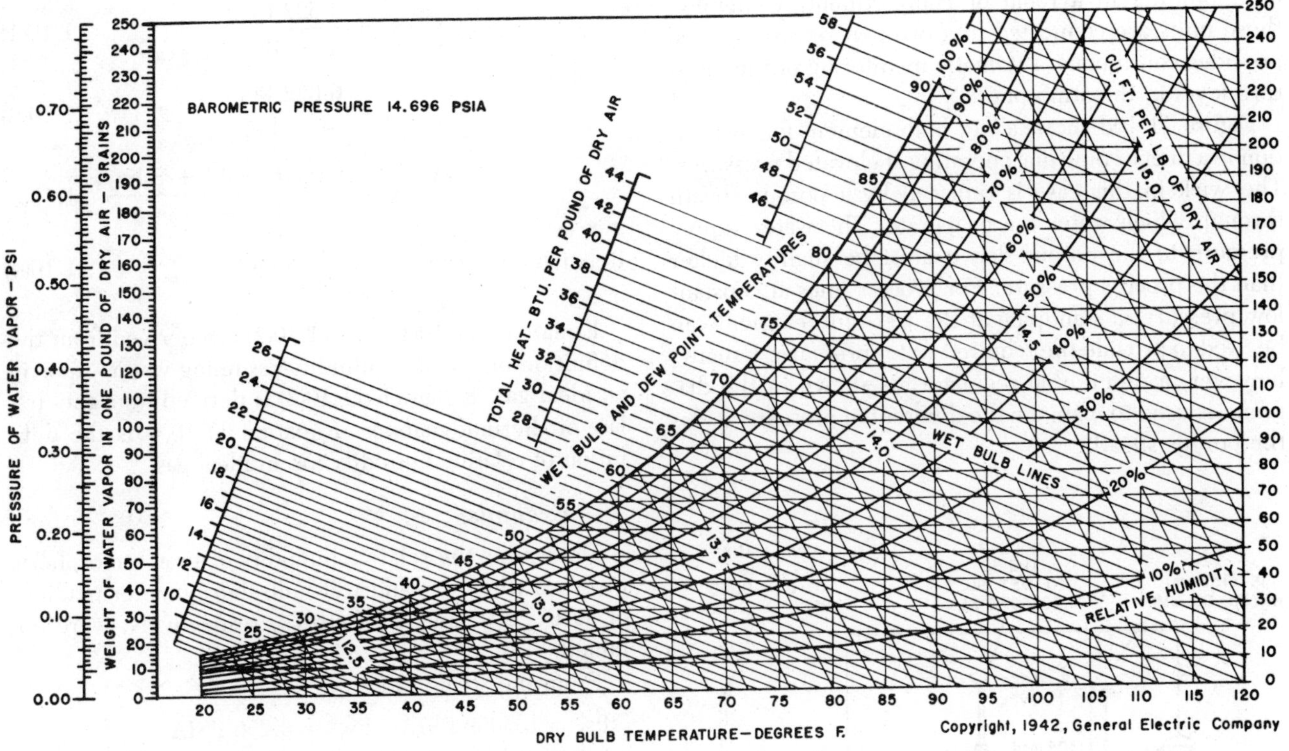

Fig. 11.10a Psychrometric chart. See Appendix A.1 for SI units

of the temperature element. The temperature values indicated by the two thermometers are related to the relative humidity of the sample atmosphere. Relative humidity can be calculated as shown on the following pages using the wet and dry bulb temperature readings, or relative humidity can be read from a psychrometric chart such as the one shown in Figure 11.10a. For a dependable measurement, the sample velocity should be in excess of 10 ft. per second (3 m/s). The sensing elements should therefore be mounted where there is adequate circulation, or forced circulation must be provided by means of a fan.

Figure 11.10b shows a sling psychrometer consisting of wet and dry bulb thermometers and mounting arranged so that the unit can be whirled manually when a reading is to be made.

Figure 11.10c shows a wet and dry bulb assembly for mounting in a duct. The wick is kept wet by a water reservoir mounted externally, or a continuous water supply can be piped to the unit. A similar device is available

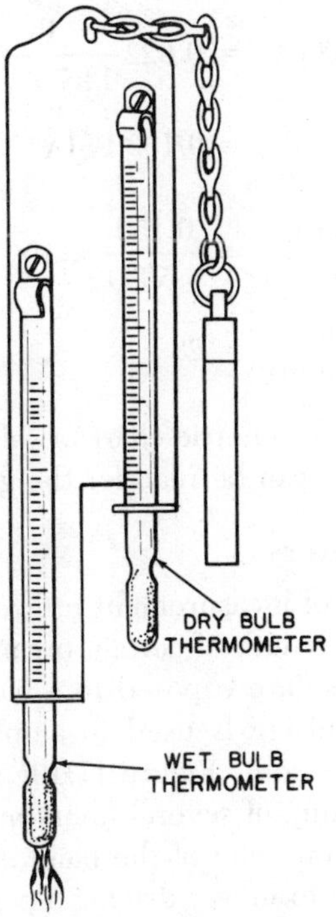

DRY BULB THERMOMETER

WET BULB THERMOMETER

Fig. 11.10b Sling psychrometer

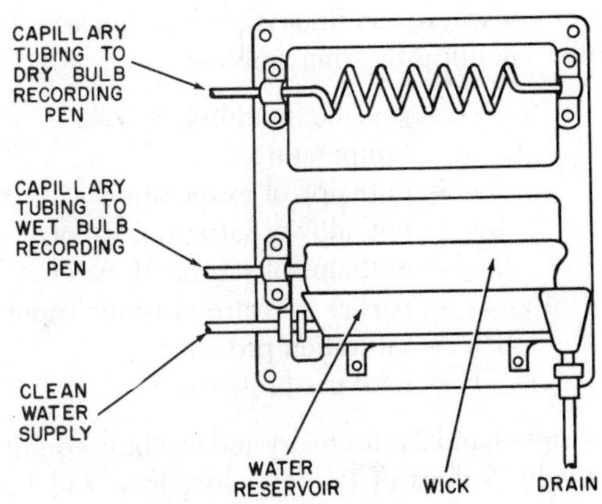

Fig. 11.10c Wet-dry bulb hygrometer using filled system temperature sensors arranged for duct mounting

for mounting in a room or compartment. Generally, these units are connected to a two-pen recorder which provides continuous, automatic monitoring of the moisture level in the atmosphere.

Figure 11.10d shows a wet bulb element for installation in a duct or pipe line operating at elevated pressures. The wick in this case is replaced by a porous sheath through which water can evaporate. The water supply pressure to the porous sleeve must be slightly higher than the process pressure. For process lines at constant low pressure, a standpipe at the wet bulb element can be used to provide the required water pressure. Usually, however, a differential regulator, as shown in the diagram, is used to maintain the differential pressure across the porous sheath.

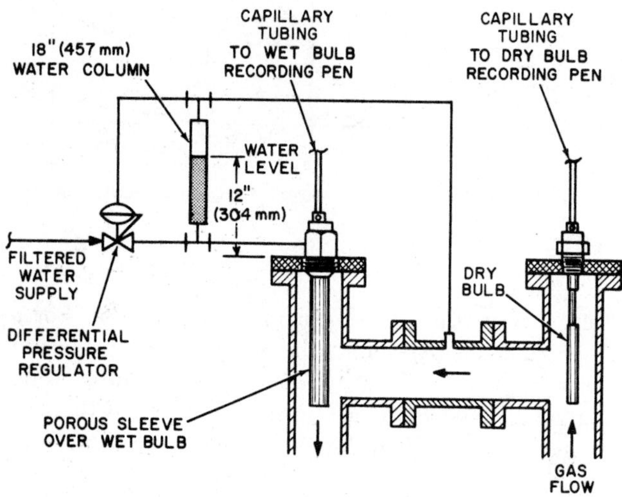

Fig. 11.10d Wet-dry bulb hygrometer using filled system temperature sensors arranged for pressure pipe-mounting

Calculation of Relative Humidity

Relative humidity can be calculated directly from wet and dry bulb thermometer readings in conjunction with a steam table; however, the calculation is valid for air at ambient conditions only and should not be applied to other gases or where conditions of temperature and pressure differ significantly from ambient.[1]

S_1, S_2 = specific humidity
t_1, t_2 = temperature
he_2 = enthalpy of evaporation
hv_1 = enthalpy of saturated vapor
hw_2 = enthalpy of saturated water
Pv_1, v_2 = partial pressure of water vapor
Ps_1 = saturation pressure
P = total gas pressure

Subscripts 1 and 2 refer to dry and wet bulb conditions, respectively. Values of Ps_1, Pv_2, hv_1, hw_2, and he_2 are obtained from a steam table.

$$S_2 = \frac{0.622\ Pv_2}{P - Pv_2} \qquad 11.10(1)$$

$$S_1 = \frac{0.622\ Pv_1}{P - Pv_1} \qquad 11.10(2)$$

$$S_1 = \frac{0.24\ (t_2 - t_1) + S_2 he_2}{hv_1 - hw_2} \qquad 11.10(3)$$

$$\text{Relative Humidity (\%)} = \frac{Pv_1}{Ps_1} \times 100 \qquad 11.10(4)$$

Equations 11.10(1) and 11.10(2) are derived from the definition of specific humidity assuming water vapor in an ideal gas. Equation 11.10(3) is derived from the first law of thermodynamics. Equation 11.10(4) is the definition of relative humidity for an ideal gas.

Example:

Determine the relative humidity of air at atmospheric pressure when dry and wet bulb thermometer readings of 80°F (27°C) and 60°F (16°C), respectively, were obtained.

From the steam table:

Ps_1 = 0.5069 PSIA Pv_2 = 0.256 PSIA
hv_1 = 1096.6 BTu/pound mole hw_2 = 28.06 BTu/pound mole
he_2 = 1059.9 BTu/pound mole

$$S_2 = 0.622 \frac{0.256}{14.7 - 0.256} = 0.011 \qquad 11.10(1)$$

$$S_1 = \frac{0.24\ (60 - 80) + 0.011\ (1059.9)}{1096.6 - 28.06} = $$

$$= 0.00642 \qquad 11.10(3)$$

$$0.00642 = 0.622 \frac{Pv_1}{14.7 - Pv_1} \qquad 11.10(2)$$

$$Pv_1 = 0.150 \text{ PSIA}$$

$$\text{Relative Humidity} = \frac{0.150}{0.5069} \times 100 \qquad 11.10(4)$$

$$\text{Relative Humidity} = 28\%$$

From the psychrometric chart, a value of 30 percent relative humidity can be read for the given data.

Hair Hygrometers

The principle of measurement utilized by this device is the change of length of certain organic and synthetic fibers when these are exposed to a moist atmosphere. A mechanical linkage is used to amplify the element movement for readout. Figure 11.10e shows a hair hygrometer consisting of several fibers mounted together to form a band; one end of the band is fixed while the other end is free to move, driving a pen arm through a

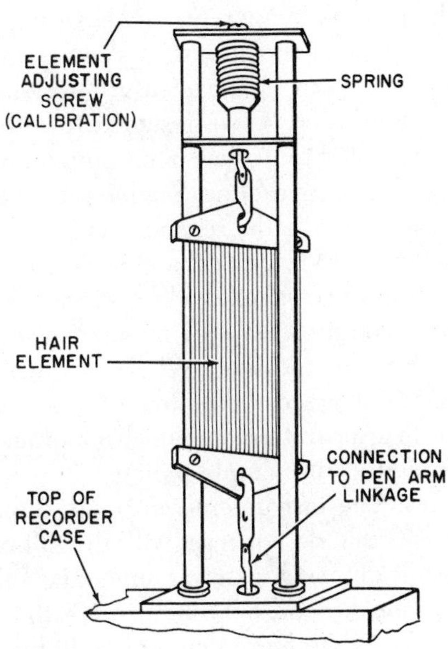

Fig. 11.10e Hair hygrometer element

linkage. Hair hygrometers, similar to wet and dry bulb hygrometers, require good circulation of the measured gas. In ducts there is generally adequate gas velocity to insure a dependable measurement; however, if the instrument is to be mounted in a room, the location must be carefully chosen. The instrument should not be mounted near doors or other openings where it will be exposed to spurious drafts; flush mounting on a panel should be avoided because the atmosphere in the back of the panel is stagnant. The hair element can be mounted on top or on the back of the instrument case, depending on the installation. The element can also be mounted on an extension in the back of the instrument so that the sensing portion is in the room or compartment where relative humidity is to be measured while the readout device is surface mounted on the wall outside. Recorders are generally available as two-pen instruments with the second pen recording temperature.

Figure 11.10f shows a pneumatic transmitter with the fiber element connected to a nozzle-baffle bleed to vary air output pressure as a function of relative humidity. The transmitting units can be installed in a room or directly in a duct.

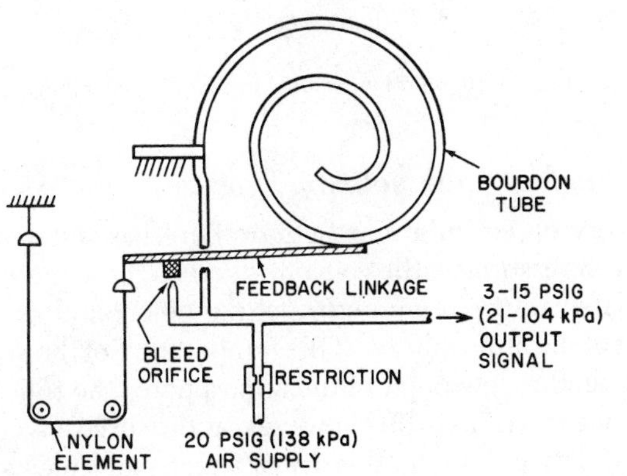

Fig. 11.10f Schematic arrangement of hygrometer transmitter

Dew-Point Hygrometers

Solution Conductivity Type

All solids condense and absorb minute quantities of water on their surfaces. If the solid is soluble, a layer of saturated solution is formed whose vapor pressure is lower than that of pure water at the same temperature. Some solids form saturated solutions in water whose vapor pressure is less than the partial pressure of water vapor in the atmosphere contacting the solution. When that is the case, water vapor will continue to condense until the vapor pressure of the solution equals the partial pressure of the water vapor. On the other hand, water vapor will evaporate from the solution should the water vapor partial pressure decrease below the vapor pressure of the solution. Under steady-state conditions, solution vapor pressure equals water vapor partial pressure, and knowledge of the former can be used to determine the moisture content of the atmosphere.

The sensing element consists of a thin-walled, hollow-metal socket wrapped with tape impregnated with salt crystals (lithium chloride). Two wires are wrapped over the tape and connected to a regulated, alternating current voltage source; the electric circuit between the wires is completed by the salt crystals. When the sensing element is exposed to the sample atmosphere, water condensing on the crystals forms an ionic solution which permits an electric current to flow between the wires. This current in turn heats the solution and raises its vapor pressure. As more water condenses, more current flows, resulting in a further increase of the solution vapor pressure until equilibrium is reached. On decreasing moisture content of the measured gas, water evaporates from the element, decreasing the current flow and resulting in a new equilibrium at a lower solution vapor pressure. A temperature sensor inside the hollow socket is used to detect sensor temperature and to provide a signal for readout.

The output calibration can be in terms of dew-point temperature water vapor pressure or specific humidity. Values of specific humidity must be corrected for pressure variations of the sample. The correction factor is (Pc − Ps)/(P − Ps), where Pc is the sample pressure the instrument is calibrated for, Ps is the water saturation pressure at the sample dew point, and P is the actual process pressure. Relative humidity is obtained by the use of a chart converting sample operating and dew-point temperature readings to percent relative humidity. Therefore, a sample temperature measurement is required if moisture level in terms of relative humidity is desired (Figure 11.10g).

Installation: The instrument does not require a high-

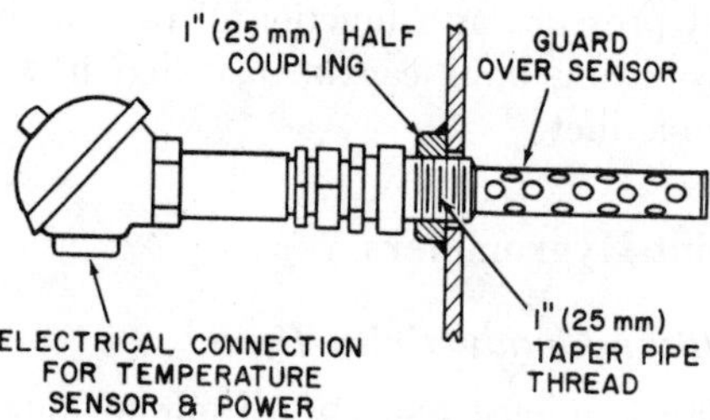

Fig. 11.10g Solution conductivity probe with resistance temperature element

velocity sample to perform satisfactorily; in fact, sample velocity in excess of 1 ft. (0.3 m) per second will result in a poor measurement and shortened element life. The reason is evident when one considers the measurement principle involved, namely, the change of solution vapor pressure and temperature on the sensing element by heating to establish equilibrium with the sample. At high sample velocities, convective heat losses swamp the measurement signal and operation at higher power causes a reduction of element life. The sensing element must, therefore, be located in a relatively quiescent zone or must be protected from direct impingement of the sample. In ducts, a sheet metal hood installed over the element and open on the downstream side is adequate. In pipelines, the element can be installed through the side outlet of a tee or can be mounted separately in a sampling chamber piped to the process line. The latter installation is preferred for samples under pressure since the element can be serviced without shutting down the pipeline (Figure 11.10h).

Sampling System: A sampling system is required when (a) the sample temperature is above the operating limit of the sensing element [operating limit approximately 200°F (93.3°C)], or (b) the process line cannot be shut down for maintenance of the element.

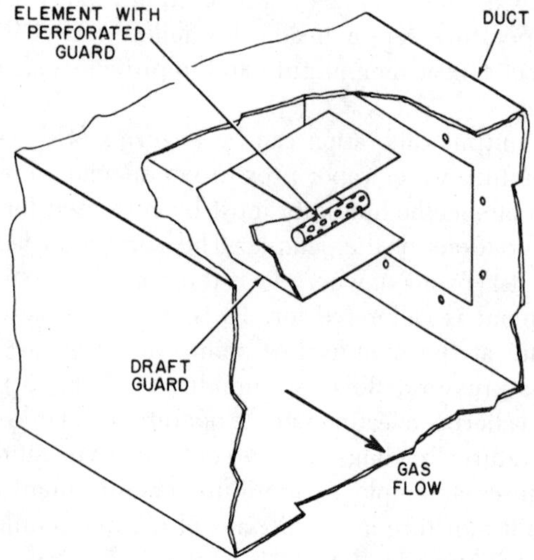

Fig. 11.10h Duct installation with draft guard

When the process temperature is above the operating temperature limit of the sensor, a sample cooler must be installed upstream of the measuring element. The sampling system must be so designed that the sample will never be cooled to its dew point under any possible process or ambient conditions. Sample tubing should be small in diameter while the sensing element is mounted in a larger diameter sampling chamber, as shown in Figure 11.10i, to reduce sample velocity at the element.

The sensor requires periodic maintenance and must be removable from the line. Whenever such removal would result in a process shutdown, the sensor should be installed in a separate sampling chamber and shut-off valves provided in the sample piping.

Limitations: The instrument can be used on all non-corrosive gases that do not react with the salt crystals or water. Specifically, such gases as ammonia, sulfur dioxide, sulfur trioxide, and chlorine react with water and this instrument is therefore not applicable for such service. The instrument is not suitable for samples containing diethylene glycol or triethanolamine.

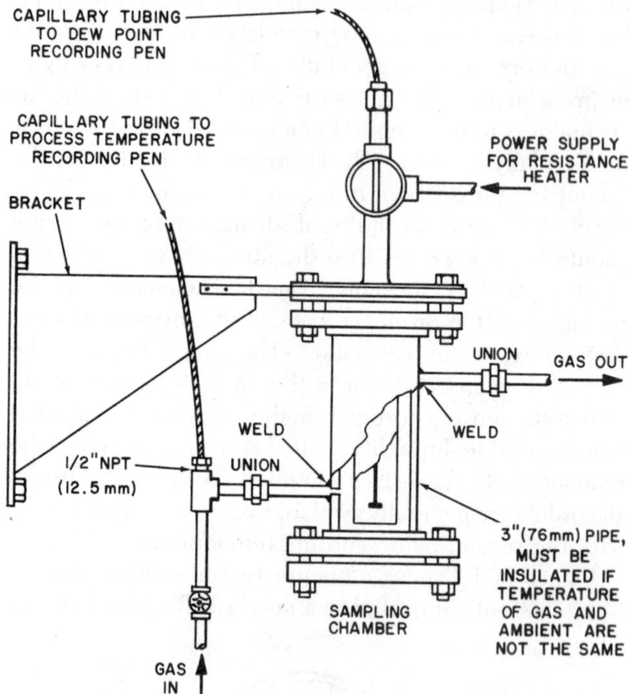

Fig. 11.10i Probe installed in sampling chamber

Surface Conductivity Type

Every object in a moist atmosphere has water molecules on its surface; the concentration on these molecules is related to the temperature of the object and the dew point of the atmosphere. If the temperature of the surface is above the dew point of the atmosphere, the thin layer of molecules is invisible; however, as the surface is cooled to the dew point, the density of water molecules at the surface becomes so great that water condenses on the surface and dew can be seen. At surface temperatures above the dew point, the moisture density at the surface can be detected electrically although the water vapor is not visible to the eye. This water vapor will permit a current to flow on the surface of even an excellent insulator. This current flow is a function of the surface material and moisture density at the surface. For a given material and fixed applied potential, the current will increase logarithmically as the surface temperature decreases to the dew point of the atmosphere. Below the dew point, the logarithmic relationship does not hold, and thus surface conductivity can be related to a dew point. Of course, dew formation can also be observed optically; however, visual observation is more difficult since dew will form at or below the dew-point temperature while conductivity is a continuous function of temperature.

The measuring element consists of a highly polished inert surface inlaid with an intermeshed gold grid and a thermocouple imbedded in the surface. A fixed potential is maintained across the gold grids, and the current flow is compared to the reference current flow at dew point. This signal is amplified and used to modulate a cooler so that the surface is maintained at the dew point of the sample. The cooler is a bismuth-telluride crystal that pumps heat away from the sensor when electric power is supplied to it.

From the thermocouple imbedded in the surface, the output is normally presented as dew-point temperature; however, readout in terms of relative or specific humidity is also possible (Figures 11.10j and k).

Sampling System: A sampling system is required where the sample gas is outside the operational limits of the instrument. At high temperatures, a sample cooler is required to bring the sample temperature below the operating limit of the sensor. The sample cooler, however, must be chosen so that the sample will never be cooled to its dew point under any possible operating or ambient condition. At high sample pressures, a pressure-reducing valve is required. For measurement of stagnant atmospheres, low-pressure samples, or closed systems, a pump may be required to draw the sample over the sensor.

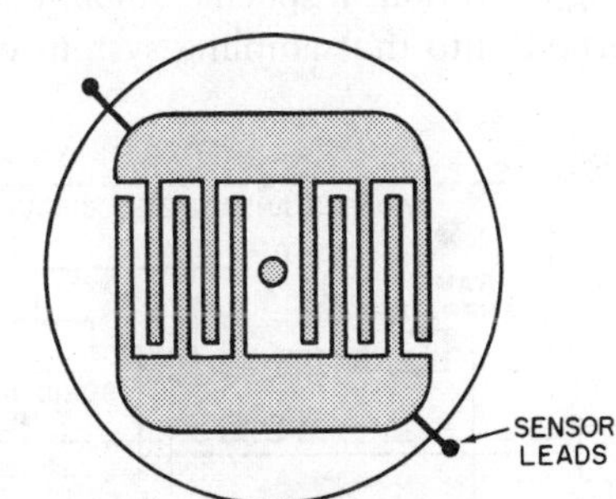

Fig. 11.10j Sensing surface

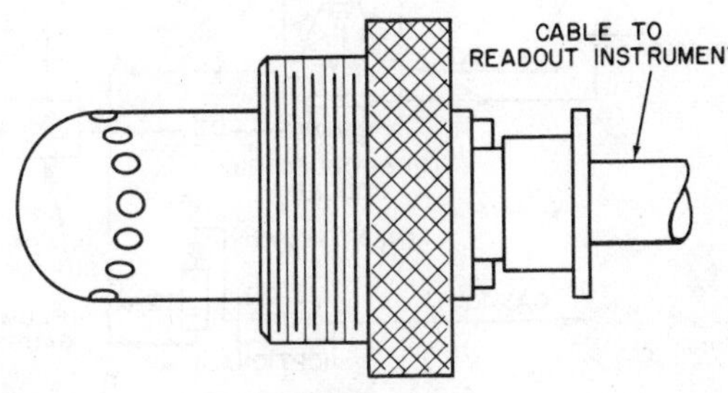

Fig. 11.10k Sensor

Limitations: The instrument can be used with all the common atmospheric gases such as nitrogen, oxygen, and carbon dioxide. The instrument is not suitable for samples containing components that react with water such as chlorine or sulfur dioxide and components with a dew point higher than that of water.

Chilled Mirror Type

The optical, chilled mirror dew-point technique is a fundamental measurement of the humidity of gas, because the saturation temperature determines the saturation partial pressure of the water vapor. These relationships have been experimentally and theoretically determined and tabulated by Goff and Gratch,[2] Keenan and Keyes,[3] and in the Smithsonian tables,[4] and they are constantly being refined. The Smithsonian tables, in particular, are used by the National Bureau of Standards as the reference documents for humidity sensor calibration. They list the saturation vapor pressures over plane surfaces of pure water and pure ice corresponding to the equivalent dew points or frost points. Once the water vapor partial pressures are determined, all definitions encountered in humidity as outlined earlier can be conveniently expressed.

The equilibrium partial pressure illustration, shown in Figure 11.10l, indicates that when the saturation temperature, or dew point, occurs and the gas mixture is saturated with respect to water or ice, the rate of water molecules leaving the atmosphere and condensing on the chilled surface is the same as the rate of water molecules leaving the chilled surface and reentering the atmosphere. At equilibrium saturation, the water vapor partial pressure of the condensate is equal to the water vapor partial pressure of the gas atmosphere. To establish this dynamic equilibrium at the mirror surface, it is necessary to precisely cool and control the mirror at the saturation temperature. A temperature element is then placed in

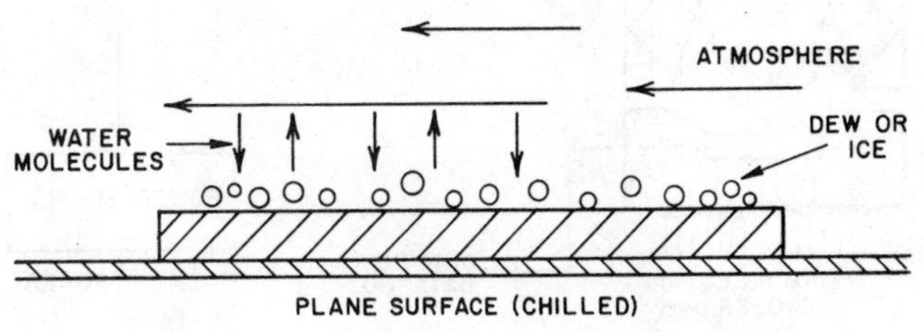

Fig. 11.10 l Equilibrium partial pressure illustration

thermal contact with the mirror, and the mirror temperature is utilized directly as the dew-point or saturation temperature.

Historically, the cooling of the mirror surface has been accomplished with acetone and dry ice, liquid CO_2, mechanical refrigeration, and, more recently, thermoelectric heat pumps. Detection of the condensation has been observed visually, and the equilibrium cooling is manually controlled. Newer versions utilize optical phototransistor detection designs to control automatically the surface at the dew point or frost point. The temperature instrumentation has included the entire spectrum from glass bulb thermometers to all types of electrical temperature elements.

The manually cooled, visually observed hygrometer is commonly known as the dew cup. It is a relatively inexpensive technique, and when operated by an experienced and skilled technician, it is quite accurate. However, it does suffer from some limitations.

1. It is not a continuous measurement.
2. It is operator dependent, and therefore readings may vary from operator to operator.
3. Versions using expendable coolants require replacement supplies.

All of these difficulties are overcome by the thermoelectrically-cooled, optically-observed dew-point hygrometer. An instrument of this type is shown in Figure 11.10m. The mirror surface is chilled to the dew point by a thermoelectric cooler while a continuous sample of the atmosphere gas is passed over the mirror. The mirror is illuminated by a light source and observed by a photodetector bridge network. As condensate forms on the mirror, the change in reflectance is detected by a reduction in the direct reflected light level received by the photodetector because of the light-scattering effect of the individual dew molecules. This light reduction forces the optical bridge toward a balance point, reduces the input error signal to the amplifier, and proportionally controls the drive from the power supply to the thermoelectric cooler. This continuously maintains the mirror at a temperature at which a constant-thickness dew layer is maintained. Independently, embedded within the mirror, a temperature-measuring element measures the dew-point temperature directly.

An additional feature of the optical hygrometer which renders it suitable for long periods of unattended operation is the automatic standardization (balance) feature, as shown in Figure 11.10n. Since moisture on the mirror surface can coexist in equilibrium with the gas adjacent to the mirror only at one unique temperature (i.e., the dew-point temperature), the measurement of the mirror temperature is a fundamental and primary measurement of dew-point temperature. However, the control loop is sensitive to change in reflectance from the mirror surface and may therefore be susceptible to changing base-line reflectance as a function of contamination. For this reason, the instruments are equipped with a balance mode, whereby the optical bridge output may be readjusted for changing dry mirror reflectance characteristics, on a time-sequential basis.

Limitations: The limitations of the optical dew-point hygrometer are (1) the sampling system, (2) the depression capability of the instrument, and (3) the absolute dew point being monitored. In the sampling system, considerations of leakage, pressure and temperature gradients, and moisture absorption/desorption characteristics must be considered.

The problem of leakage is relative; that is, if the dew point being measured is close to the ambient dew point, leakage into the system may not bias the reading substantially. If the system is pressurized above atmospheric pressure so as to create a leakage out of rather than into the system, the error introduced will be less.

The temperature stability of the sampling system components is also quite important. For any given equilibrium sampling condition, a specific amount of moisture will be absorbed onto the sampling system wetted sur-

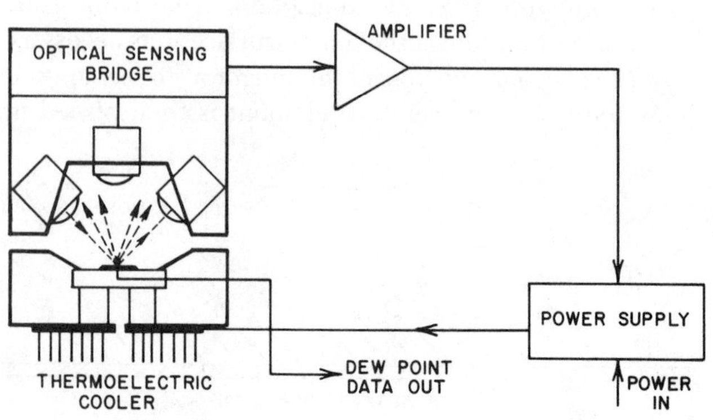

11.10m Principle of operation

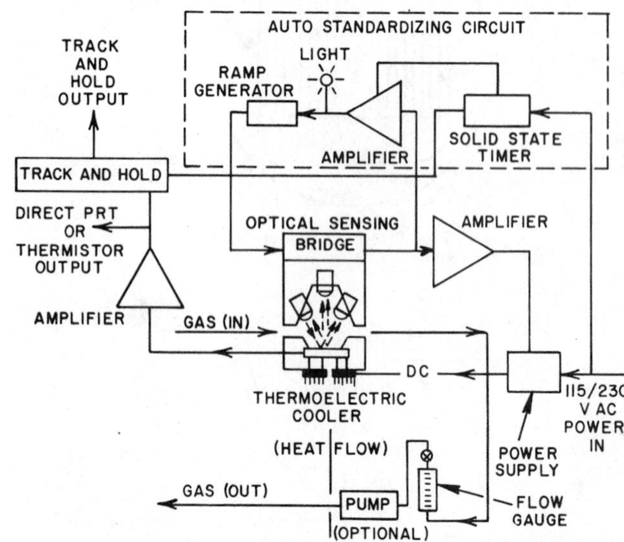

Fig. 11.10n Automatic standardization (balance) circuit

faces, so control of the sample line temperature may be necessary to insure an equilibrium condition.

Of equal importance is the effect that material absorption/desorption characteristics have on overall system response. Stainless steel and nickel alloy tubing are the best nonhygroscopic materials and should be used for dew points from 0 to −100°F (−17 to −73°C). Copper, aluminum alloys, Teflon, and polypropylene are suitable above −20°F (−29°C) dew point. Most other plastic and rubber tubing is unacceptable in all ranges.

The depression capability of the instrument is a function of the size and efficiency of the thermoelectric cooler. Recent advances in the technology of the thermoelectric coolers have made possible for the first time a practical instrument utilizing three serial stages of thermoelectric cooling, capable of depressions in excess of 160°F (71°C). Thus, newer instruments are capable of monitoring virtually all dew-point ranges encountered in current industrial processes.

The final limitation on the optical hygrometer is the absolute amount of moisture available in a gas sample. At −100°F (−73°C), or 1.5 ppm water vapor, the time required to stabilize on a frost point is dependent on the sample flow rate to the hygrometer. This type of instrument requires a very small flow rate to provide a continuous monitoring capability and is not sensitive to flow within a specified range. However, at very low frost points, it is advisable to increase the sample flow rate to provide more samples because of the very low percentage of water vapor available.

Moisture Indicators

Sometimes it is desirable to detect only a change in the amount of moisture in the sample, while the exact concentration is of only secondary importance. Such may be the case on the outlet of drying beds where moisture breakthrough will indicate the need for regeneration. Certain salt crystals indicate the presence of moisture by a change in color, and these are the most economical methods for detecting changes of moisture level. The crystals are packed into a small chamber with a glass window for observation of the color. A sample side stream is piped through the chamber, and increase of moisture in the sample is indicated to the operator by a color change.

Conclusion

The instruments discussed above are generally applicable to such easy-to-handle fluids as the atmospheric gases. Operating temperature and pressure are usually limited to atmospheric conditions, although with some designs the operating limits can be extended. The principal feature common to all the above instruments is the fact that the standard output form is not related directly to moisture content but to a parameter of moisture content. Accuracy varies widely depending on design; the greatest accuracy can be obtained with the conductivity-type hygrometer and the least accuracy with the moisture indicator. Generally, a sampling system is required only for special applications.

REFERENCES

1. Van Wylen, G.G., *Thermodynamics*, John Wiley and Sons, New York, 1965.
2. Goff, J.A., and Gratch, S., "Low Pressure Properties of Water from −160 Degrees to 212 Degrees F.," *Transactions*, ASHVE, Vol. 52, 1946, p. 95.
3. Keenan, J.H., and Keyes, F.G., *Thermodynamic Properties of Steam*, John Wiley and Sons, New York, November 1959.
4. List, Robert J., *Smithsonian Meteorological Tables*, 6th rev. ed., Publication No. 4014, Smithsonian Institution, Washington, D.C., Tables 93 through 97.

BIBLIOGRAPHY

Shinskey, F.G., "Humidity, Dew Point and Wet-Bulb Temperature," *Instruments and Control Systems*, July 1975, p. 43, and August 1975, p. 15.

Wiederhold, P.R., "Humidity Measurements," *Instrumentation Technology*, August 1975, pp. 45–50.

Wiederhold, P.R., "Which Humidity Sensor," *Instruments and Control Systems*, June 1978, pp. 31–35.

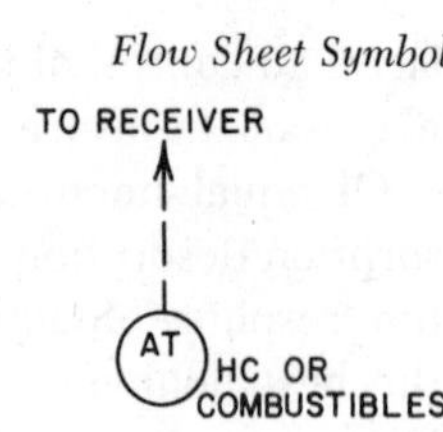

11.11 HYDROCARBONS

Type of Measurement:	A1—Total hydrocarbon by flame ionization B1—Methane by chromatography B2—Methane by flame ionization B3—Methane by mass spectrometry C1—Hydrocarbon classes by flame ionization C2—Hydrocarbon classes by mass spectrometry C3—Hydrocarbon classes by infrared D1—Individual hydrocarbons by chromatography
Reference Method:	GC with flame ionization for nonmethanes
Sensitivity:	0.1 ppm (A1, B2, C1)
Costs:	\$4500 to \$11,000 (A1); \$6800 to \$11,000 (B2, C1); \$11,000 and up (B3); \$22,500 and up (B1, C3, D1); \$45,000 and up (C2)
Partial List of Suppliers:	Beckman Instruments, Inc. (A1, B1, B2, C1, D); Bendix Corp., Process Instrument Div., (A1, B2, C1); Davis Instrument Mfg. Co., Inc. (A1, B2, C1); Foxboro Analytical (C3); Infrared Industries (C3); Mine Safety Appliances Co. (A1, B1, B2, C1, D1); Tracom, Inc., Instruments Group (B1, D1); Union Carbide Corp. (B1, D1)

Unlike most gaseous air pollutants, the hydrocarbons comprise a class containing scores of individual components (Table 11.11a).[1] The most abundant hydrocarbon, methane, is nonreactive in photochemical reactions and otherwise not a hazardous air pollutant. The remaining hydrocarbons vary widely in their reactivity. For this reason, the type of hydrocarbon analysis to be used will depend on the purpose for which data are needed. To obtain highly detailed analyses for numerous hydrocarbons requires expensive and sophisticated equipment, whereas a single measurement for total hydrocarbons is simpler and somewhat less expensive. There are also

Table 11.11a
ATMOSPHERIC HYDROCARBON ANALYZERS

Hydrocarbon Type	*Method*	*Limitations and Interferences*
Total (as carbon)	Flame ionization detection (FID)	Some response to carbon-containing nonhydrocarbons
Methane	Gas chromatography	Expensive equipment (can also be used for carbon monoxide)
	"Methane-only" subtractive column and FID	Column preparation fussy
	Mass spectrometry	Freeze-out required, expensive
Aromatics, olefins, paraffins	Subtractive columns and FID	Column preparation fussy
	Mass spectrometry	Freeze-out required, expensive, data reduction requirements large
	Infrared spectrometry	Freeze-out required, expensive, not total class coverage
Individuals	Gas chromatography	Expensive, data reduction requirements large

methods of intermediate complexity which permit determination of methane separately from other hydrocarbons.

Calibration Methods

Only dynamic calibration is completely satisfactory. In this procedure, known, realistic concentrations of the pollutant in air are passed into the analytical system, preferably in exactly the same way as unknown air samples are collected. This may be difficult to accomplish because of the low concentrations and possible reactivity or condensability of some pollutants.

Standard calibration mixtures of one or several of the common hydrocarbons in air are available commercially in high-pressure gas cylinders. These should always be checked against a reference standard, or at least against a previously calibrated analyzer. If much calibration is necessary, it may be useful to set up a gas-handling and dilution system to prepare standard mixtures. One fairly simple technique for hydrocarbons, such as butane (which can be liquefied at moderate pressure), uses permeation tubes. These are sealed tubes of a specific plastic, partially filled with liquid hydrocarbon. The hydrocarbon will effuse through the walls of a given tube at a rate depending only on temperature, as long as some liquid remains. To generate a known concentration, air is passed at a known flow rate over the tube in a temperature controlled vessel. The tube may be calibrated gravimetrically.

In another technique, a known amount of hydrocarbon is added by syringe or by crushing an ampoule, either in a large rigid vessel of known air volume or into the metered flow of air passing into a plastic bag. The bag must be of inert plastic; it has the advantage of collapsing as the sample is withdrawn, whereas a rigid vessel must be large in relation to the sample size in order to avoid substantial pressure differentials.

Flame Ionization Detectors

Flame ionization detection (FID) is the most suitable means of analysis for most hydrocarbons[2,3] at the levels found in polluted air. It may be used alone for the measurement of total hydrocarbons or as a detector after separation by a column device such as a gas chromatograph. In FID analyzers, a sensitive electrometer detects the increase in ion intensity in a hydrogen flame when air containing organic compounds is introduced. The response is approximately in proportion to the number of organically bound carbon atoms in the sample, so the detector is basically a carbon atom counter. Carbon atoms bound to oxygen, nitrogen, or halogens, however, give reduced response. There is no response to nitrogen, carbon monoxide, carbon dioxide, or water vapor. The results are usually expressed in terms of the calibration gas used, e.g., parts per million (ppm) carbon as methane. Response is rapid and can be as sensitive as 0.1 ppm. The response to various hydrocarbons is not perfectly uniform, and such variations should be taken into account in interpreting FID data. The FID analyzer is a reference method for total hydrocarbons in U.S. Air Quality Standards. Basically, it is an air metering device attached to a flame ionization detector (Figure 11.11b).

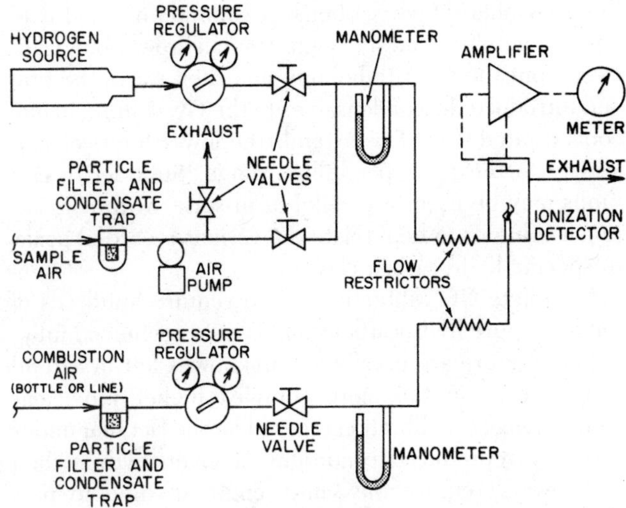

Fig. 11.11b Hydrocarbon analyzer, hydrogen flame ionization detector

Gas Chromatography

The only practical method for general analysis of specific hydrocarbons is gas chromatography (GC).[4] In GC, the basic apparatus (Figure 11.11c) is a carefully prepared tubular column of a finely divided solid with provisions for passing a steady flow of a carrier gas (often helium) through it. The column is temperature controlled, and the packing usually supports or is coated by a nonvolatile liquid phase. Preceding the column there is means for sample introduction (and sometimes sample splitting) and following the column is a detector. A high-sensitivity column useful for atmospheric work usually employs a flame ionization detector. There are a great many optional variations in the makeup of a GC instrument, involving sample valves, precolumns, columns, programmed temperature changes, stream splitters, and detectors. The output may be recorded on a strip-chart

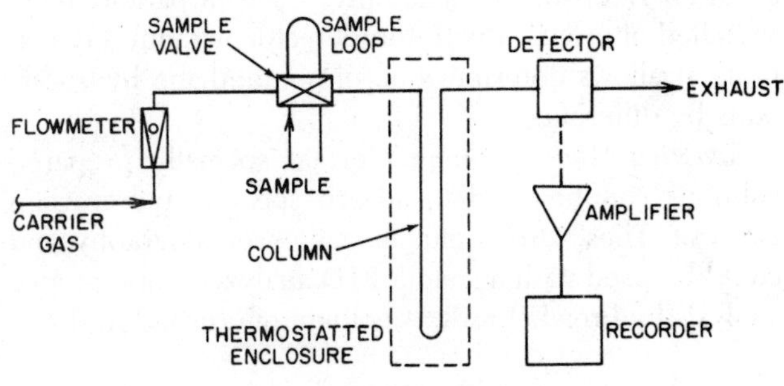

Fig. 11.11c Gas chromatograph

(possibly with a peak integrator) or converted to digital form for computer handling. In full-scale analyses for scores of hydrocarbons, data reduction from analog strip-chart recordings becomes very tedious and time-consuming. In that event, the addition of computerized readout may be worth the increase in cost.[4]

Supplemental techniques increase the utility of GC for atmospheric hydrocarbons. A common method is to pass a sample of air through a small freeze-out trap, sweep out the air with helium, and then warm the trap and introduce the condensables to the GC column in one concentrated slug. This extends the lower limits of sensitivity to the parts per billion range. Subtractive columns may be used in parallel or in series with conventional columns to help relate the various recorded peaks to specific hydrocarbon classes.

Complete GC calibration would require hundreds of different pure hydrocarbons and is never achieved fully, although there are usually few unknowns left in careful work except in the more complex higher molecular weight ranges. Calibration is based on the fact that under carefully reproduced conditions, a given hydrocarbon will always require the same length of time to pass through the column to the detector. Ambiguities arising where components overlap may be resolved by change in column packing or sometimes by the use of subtractive columns. An elaborate but useful technique is to attach another analytical instrument such as an infrared or mass spectrometer to the outlet of the detector for peak identification.

In Section 11.5 on carbon monoxide sensors, an instrument is described for combined analysis of methane and carbon monoxide. This is a simple GC, set up to determine methane, with a programmed valving arrangement and a catalytic carbon monoxide–to–methane converter. The result is a readout of two peaks, one for methane directly and the other for methane formed by reduction of carbon monoxide.

Subtractive Columns

Nonmethane Hydrocarbons: A column of adsorptive charcoal may be treated with methane just until breakthrough occurs; that is, no more methane will be adsorbed, but other hydrocarbons are still retained.[5] This column can then be used with a flame ionization detector as an analyzer for methane only. Used in parallel or in switched alternation with the detector without any column, it allows determination of nonmethane hydrocarbons by difference.

Reactive Hydrocarbons: Certain specially prepared columns can be used to adsorb specific hydrocarbon classes.[6] These are useful in gas chromatography and could be used with a simple FID analyzer alone to give analysis by broad classes. A column of crushed firebrick supporting mercuric sulfate will adsorb olefins and acetylenes, and one of palladium sulfate will then adsorb aromatic hydrocarbons (except benzene). A combination of these subtractive columns has been applied successfully to automobile exhaust analysis and shows promise for use in atmospheric work.

Spectrometric Methods

Either infrared or mass spectrometry may be used with considerable advantage for individual hydrocarbon determination, although the sensitivity of both methods would require concentration such as by a freeze-out trap. Both types of instruments are complex and expensive. Since the calibration requirements are about equal to those needed for gas chromatography, the data interpretation is generally more complicated and the extent of coverage is less; these methods have been widely supplanted by gas chromatography for atmospheric hydrocarbon detection work. These instruments may be useful for analysis of a particular type or class of hydrocarbon and might make it unnecessary to set up a gas chromatographic capability in some cases.

Assessment

For total atmospheric hydrocarbon measurement, the flame ionization analyzer is the most generally accepted detector. It requires technical attention and compressed gases but is reliable and accurate. The methods for specific classes add complexity and reduce precision and reliability but are not much more expensive than the flame ionization analyzer itself. (The spectrometric methods, unless already in use for other reasons, are not competitive.) If individual hydrocarbon determination is required, there is no good alternative* to gas chromatography (frequently augmented with a freeze-out step), even though the data reduction is time-consuming (if manual) or expensive (if computerized), and calibration requirements are demanding.[4]

*NDIR sensors can only identify hydrocarbon classes, whereas two-color IR detectors are able to identify individual hydrocarbon species. In this design, sensitivity is traded for specificity.

REFERENCES

1. Altshuller, A.P., in A.C. Stern, ed., *Air Pollution*, 2d ed., Academic Press, New York, 1968, Chap. 18.
2. Andreatch, A.J. and Feinland, R., *Analytical Chemistry*, Vol. 32, 1960, pp. 1021–1024.
3. Lipták, B.G., ed., *Instrument Engineers' Handbook* Suppl. 1, Sec. 1.15, Chilton, Philadelphia, 1972.
4. Ibid., Vol. I, Sec. 8.1.
5. Ortman, G.C., *Analytical Chemistry*, Vol. 38, 1966, pp. 644–646.
6. Klosterman, D.L. and Sigsby, J.E., Jr., *Environmental Science Technology*, Vol. 1, 1967, pp. 309–314.

Flow Sheet Symbol

TO RECEIVER

AT Hg

11.12 MERCURY IN AIR

Methods of Detection:	A—Ultraviolet, flameless atomic absorption B—Atomic absorption spectrophotometer
Typical Ambient Concentrations:	1 to 200 ng/m^3
Costs:	\$1000 to \$7000 (A—sample analyzers); \$7000 to \$70,000 (A—integrated sampler-detector); \$14,000 to \$40,000 (B)
Partial List of Suppliers:	Bacharach Instrument Company (A); Bausch and Lomb, Inc. (B); Beckman Instruments, Inc. (A,B); Bendix Corp. (B); Carle Instruments Inc. (B); Fisher Scientific Co. (B); Geomet Inc. (A); Laboratory Data Control, Div. of Milton Roy Co. (A); Olin Corp. (A); Perkin-Elmer, Instrument Div. (A,B); Varian Associates (B)

Air saturated with mercury will contain between 4 and 100 mg/m^3 in the temperature range between 45 and 110°F (7 and 43°C). Since saturation is not approached under normal circumstances with moving air, sensitive analytical and sampling methods are needed when data on actual ambient levels of mercury are required. For practical use, the methods employed must be relatively insensitive to interferences and be otherwise appropriate for the collection of data under a wide range of operating conditions. For regulatory purposes, related to emissions, three or four levels of sensitivity are of importance: 100, 50, 10, and perhaps 5 μg/m^3 of mercury in the air. Accuracy within $\pm$10 percent at these levels is required. However, ambient air monitoring requires the ability to determine levels in the 1 to 200 ng/m^3 range.

Sample Collection and Concentration

Collection of mercury samples for accurate determination of concentrations found at ambient conditions or in industrial plants requires that (1) the size of the sample taken must be adequate to satisfy the sensitivity threshold requirements of the analytical technique utilized; (2) the rate of sampling must be controlled so that efficient collection is achieved; (3) the sampling interval selected must be appropriate for the required monitoring function; (4) the collection device or procedure must be compatible in capacity and chemical reactivity of material components with the levels and properties of mercury and mercury compounds to be sampled; (5) effects of potential interferences must be avoided; (6) elapsed time between collection and analysis must be minimized in order to avoid losses; and (7) it must be possible to evaluate all unknown factors in collection and analysis by calibration procedures.

In the sampling of mercury and its compounds from the air, two basic procedures are available. These procedures utilize either the chemical reactivity of the sampled components in various aqueous media or the extraordinarily efficient extraction (amalgamation) of mercury from air by the noble metals. In the latter case, reduction of mercury compounds to elemental mercury is required before the sampling process is carried out. Currently, a wet chemical procedure for sampling is recommended by the Environmental Protection Agency; it requires considerable labor and upkeep to perform effectively.

Impinger Collection Methods

The Environmental Protection Agency has published two methods for sampling and analyzing mercury in gaseous and particulate emissions in conjunction with publication of National Emission Standards for Hazardous Air Pollutants.[1] The methods are similar with the exception that particulates must be sampled isokinetically, whereas gaseous samples are passed through filters to remove particles. This latter procedure requires equilibration of the filter prior to generation of accurate data at low mercury levels.

In principle, samples are drawn through acidic aqueous iodine monochloride solutions utilizing stringent controls over flow rates, acidity, and other process variables. This procedure gives excellent recovery of organic mercury compounds and of mercury vapor. Samples obtained with this procedure may be stored.

Figure 11.12a shows the sampling train utilized for particulate and vapor samples. It consists of (1) a Pyrex-lined, heated probe section containing a thermocouple for temperature measurement and a pair of pitot tubes and a manometer for gas flow measurement; (2) five impingers of the Smith-Greenburg type placed in an ice bath; (3) a filter for particle removal prior to passage of the gases through the two final absorber stations; and (4) miscellaneous components for control and calibration of the sample airflow. Figure 11.12b shows a simpler technique used for sampling vapors without particulates. In this method, particulates are removed by a filter at the inlet into the absorption train. For sampling particulates, the initial rate of gas flow is 2 scfh (.04 m^3/h). Flow should be maintained at a rate proportionate to the flow through the stack sampled.

Samples obtained by this procedure may be analyzed by a procedure utilizing a flameless atomic absorption spectrophotometer which has been accepted by the Environmental Protection Agency for investigations.

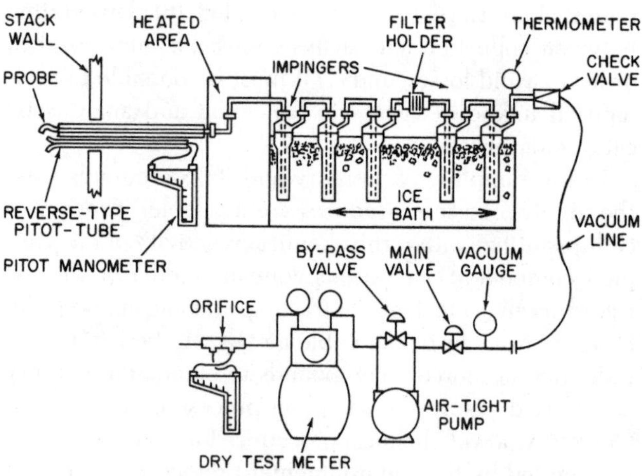

Fig. 11.12a Particulate-sampling train

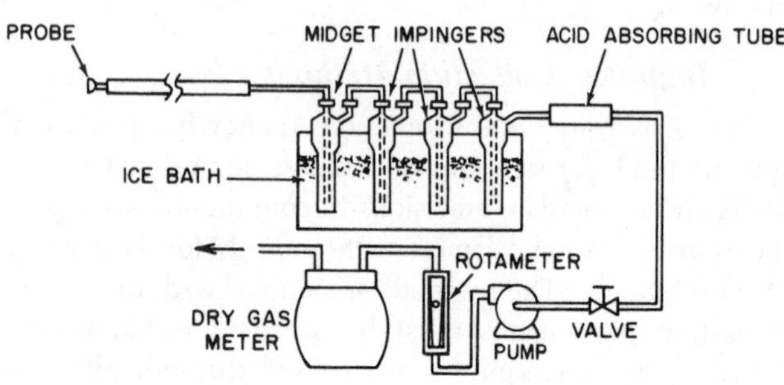

Fig. 11.12b Vapor-sampling train

A number of other collection procedures employing impingers have been examined. These utilized water, ethyl alcohol, isopropyl alcohol, potassium permanganate–sulfuric acid, potassium permanganate–nitric acid, or iodine–potassium iodide. However, for effective collection of alkyl mercury compounds, acidic iodine monochloride solutions are preferable.

Amalgamation on Wettable Metals

A number of less laborious methods of sampling mercury vapor, compounds, and particulates have been developed which utilize absorption (amalgamation) of vapor on silver and gold and on alloys of these materials. In order to sample particulates and compounds of mercury, the trapping procedure must be preceded by a decomposition step. A catalytic process which achieves this objective is described later. Examples of procedures for collecting elemental mercury vapor include Figure 11.12c, which shows a simple absorption tube filled with gold or silver wool. The tube may be sized to satisfy any specific parameters of the sampled gas stream. Fine wire may also be used as packing for flow rates under 50 liters per minute, and mercury vapor removal can be made quantitative. Gold materials are required for gases containing hydrogen sulfide or sulfur dioxide. When a simple absorption tube of this type is employed, the mercury vapor collected may be released for analysis in a flameless atomic absorption spectrophotometer (AAS) by utilizing an induction furnace. In this procedure, extensively used by the U.S. Geological Survey, a continuously flowing airstream is used to pass the mercury vapor pulse which results during heating by the furnace into the AAS cell.

Fig. 11.12c Absorption tube mercury collection device

A versatile method of sampling gas streams, employed by the Canadian Department of the Environment, uses a packing which consists of finely woven silver mesh (Figure 11.12d). One version of the sampling tube is 5 in. (125 mm) high and utilizes a 1/4-in. (6 mm) I.D. Pyrex tubing. Separation between the inner and outer tube is 3/64 in. (0.1 mm). In order for this collection procedure to function effectively, relatively low airflow rates through the wire mesh must be used. Sampling rates of 1 to 5 liters per minute have been employed, but long sampling periods are required to make measurements at low air concentrations.

Figure 11.12e shows schematically a collection device contained in an integrated mercury analyzer. In the collection section of this device, air at rates from 1 to 200 liters per minute is passed through a grid containing 160 ft. (48m) of 22-gauge wire arranged in a carefully spaced annulus. Mercury vapor is desorbed by direct passage of electrical current through the wire. Analysis is achieved by the flameless atomic absorption method in a photometer.

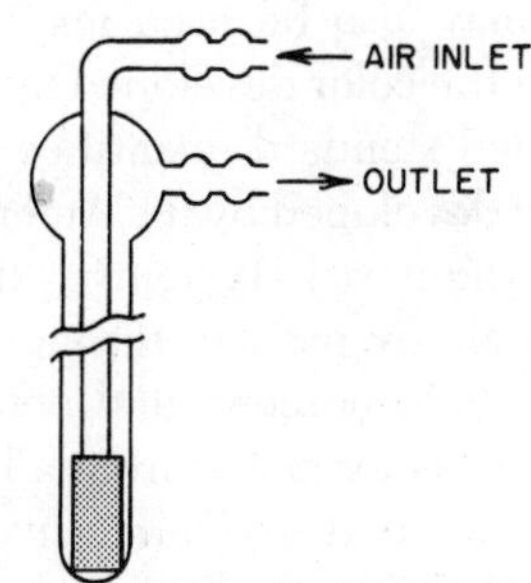

Fig. 11.12d Mercury gas sampling tube

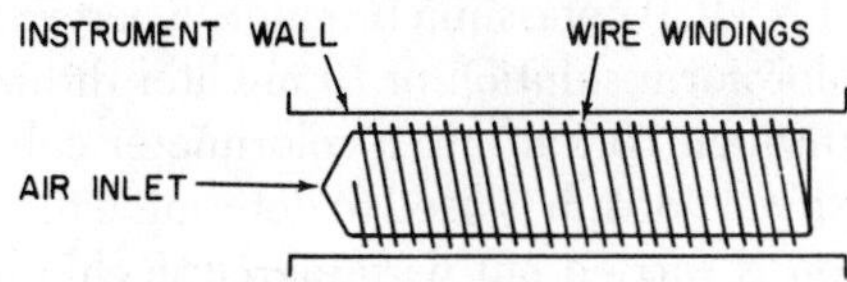

Fig. 11.12e Mercury collection device

Activated Absorption

Use of a silver on charcoal absorbent is highly efficient and may be used to remove mercury vapor from many gas streams. Extraction of the mercury vapor as a concentrated pulse for analysis, however, is relatively difficult. Induction heating is required to maximize the vapor concentration. Trapping is also possible with iodine-activated charcoal and mineral wool. Regeneration of these materials for repeated use presents difficulties not yet resolved.

Conversion of Mercury Vapors

In order to analyze for total mercury including particulates and vapor, reduction of compounds such as mercuric chloride, mercuric sulfide, dimethyl and diethyl mercury, pesticides, and fungicides to the elemental form must be achieved. Three techniques are in use: (1) thermal decomposition at 1500 to 2000°F (816 to 1093°C); (2) catalytic reduction or disproportionation, usually at 1200 to 1300°F (648 to 704°C) over materials such as cupric oxide; and (3) chemical conversion in solutions such as acidic iodine monochloride. Commercial equipment is available which may be used in conjunction with sampling and/or analysis instrumentation.

Methods of Detection

No direct analytical method is capable of attaining accurate measurements in the range required for ambient air without the use of a concentrator, discussed earlier.

Ultraviolet Light Absorption

A wide variety of methods have been developed to measure airborne mercury-containing materials in the atmosphere based on the strong absorption of ultraviolet light at 253.7 nm by elemental mercury vapor. Since many other compounds absorb light in this range, although much higher concentrations are required than with mercury, some method of separating interferences is needed when low levels of mercury are measured.

Atomic Absorption Spectrophotometry: Mercury is conveniently analyzed by atomic absorption techniques because the elemental vapor exists in atomic form under normal ambient conditions. Thus, when a cloud of mercury vapor is irradiated at an appropriate wavelength, mercury atoms absorb a portion of the energy of the beam. In the standard atomic absorption technique, liquid samples are aspirated into an acetylene air flame where the atomic mercury is released and the absorption of light at 253.7 nm is measured by a spectrophotomultiplier (Figure 11.12f). The sensitivity of 0.01 μg of mercury per milliliter of sample is readily attained by this procedure. A continuous series of calibration checks are required for maximal precision and accuracy.

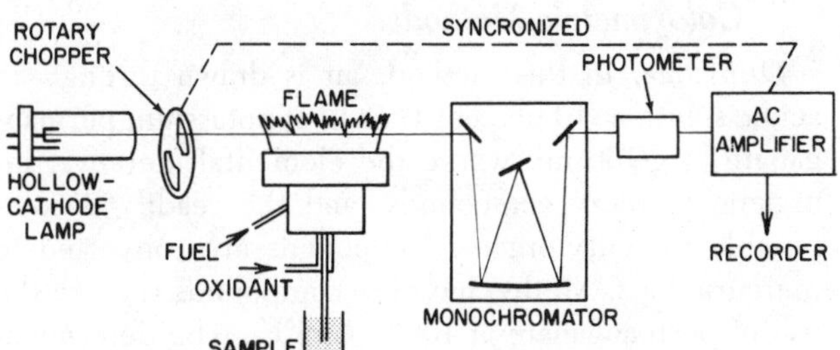

Fig. 11.12f Atomic absorption spectrophotometer

Flameless Atomic Absorption Spectroscopy: The burner-aspirator and flame indicated in the schematic of the conventional AAS (Figure 11.12f) may be replaced by a quartz-windowed glass cell and the sample may be generated by reducing an aqueous sample of mercury compounds to elemental mercury. Figure 11.12g shows the apparatus which may be used in conjunction with an AAS. In this procedure, a sample collected by utilizing a separate collection train is pipetted into the gas washing bottle. To this is added a solution of a reducing agent such as hydroxylamine sulfate or stannous chloride and the bottle is closed. With clean air supplied at a rate of 1.3 liters per minute, aeration is commenced and con-

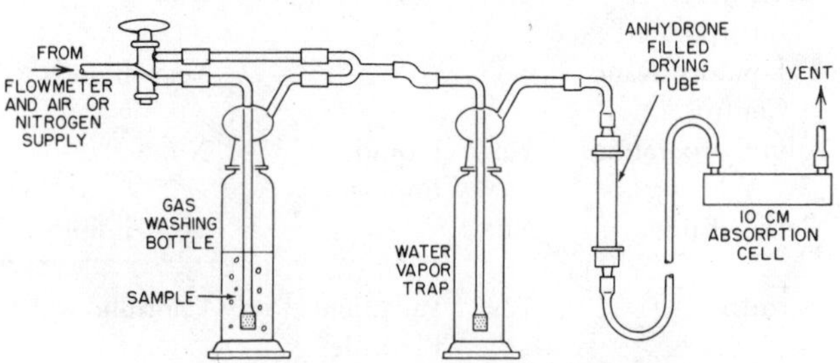

Fig. 11.12g Apparatus for flameless atomic absorption

tinued until a maximum peak height readout is obtained on the recorder. By calibration against acidic mercuric chloride standard solutions, the peak heights obtained with unknown samples can be compared with the calibration curve for direct determination of the mercuric ion concentration. This value may be converted to the gas phase concentration of mercury.

At least seven types of commercial units are available which use the principle of flameless atomic absorption spectroscopy for determination of mercury. Properties of these instruments are shown in Table 11.12h. Affirmative designations in Column 4 indicate that samples must be taken by use of a liquid sampling train and transported to the instrument.

Other devices based on the use of absorption of radiation at 253.7 nm have been developed by Barringer Research and by Williston and co-workers (Cordero Mining Company).[2] The Barringer method takes advantage of the pressure-broadening of the mercury emission lines to minimize interferences due to other compounds.

Colorimetric Methods

Dithizone: In this method, air is drawn through an acidic solution (10 percent H_2SO_4) of potassium permanganate (4 g/100 ml) where the elemental mercury, inorganic mercury compounds, and the readily decomposable mercury-organic compounds are converted to mercuric ions. Methyl mercury compounds require the use of permanganate at 100°C (212°F). The determination of mercury may be concluded colorimetrically whence the sample is titrated with dithizone and compared with the colors developed by standards. A spectrophotometer may also be used for quantitative comparison between the color developed by test samples and those of controlled standard quantities of mercury.

In the method developed by the American Conference of Government Industrial Hygienists, the sample is collected from the air by passage through an aqueous solution of iodine (0.25 percent) and potassium iodide (3 percent). Mercury is extracted from a buffered solution (ammonium citrate, hydroxylamine hydrochloride, and phenol red at pH 8.5) with dithizone in chloroform (20 mg dithizone/5 ml) until all color is removed. After washing the chloroform phase, the mercury is extracted into acidic (0.1 N HCl) potassium bromide (8 percent), shaken with a chloroform solution of 10 mg/liter dithizone, and filtered through cotton into a colorimeter cell. The optical density at 485 nm is used to determine the mercury. Calibration is carried out with mercuric chloride.

The dithizone method is relatively insensitive, requiring the use of large samples (about 10 μg are required) when the levels of mercury are low. It also requires considerable care on the part of the chemist to avoid the loss of organic mercury compounds.

Selenium Sulfide and Others: Active selenium sulfide applied to paper as a coating may be used to detect the presence of elemental mercury vapor. The paper develops a black color on exposure to air containing mer-

Table 11.12h
ESTIMATED PERFORMANCE PARAMETERS OF COMMERCIALLY AVAILABLE MERCURY MONITORS
(utilizing flameless atomic absorption photometers)

Manufacturer	*Integral Collector*	*Collection Process*	*Discrete Samples Required*	*Sensitivity Range*	*Total Hg*	*Installation*	*Price Range, $*
Antipollution Technology Corporation (Thermatron)	No	—	Yes—liquid	10 $\mu g/m^3$	With extra train	Portable	2250
Bacharach Instrument Company	No	—	No	0.1 mg/m^3	No	Portable	1100
Beckman Instruments	No	—	No	0.005–0.1 and 0.03–3.0 mg/m^3	No	Portable	1800–2700
GEOMET, Incorporated	Yes	Amalgamation	No	1–1,000,000 ng/m^3	Yes	Portable	15,750–22,500
Laboratory Data Control	No	—	Yes—liquid	1–5000 ng/m^3	With extra train	Portable	4000
Olin Corporation	Yes	Liquid Impinger	No	1 ng/m^3	Yes	Fixed	45,000–67,500
Perkin-Elmer	No	—	Yes—liquid	10 ng/m^3	With extra train	Fixed and portable	8500–33,000
Scrintrex	Yes	Palladium Fluoride Filter	Continuous	0.5 ng/m^3	No	Portable	7300–40,000

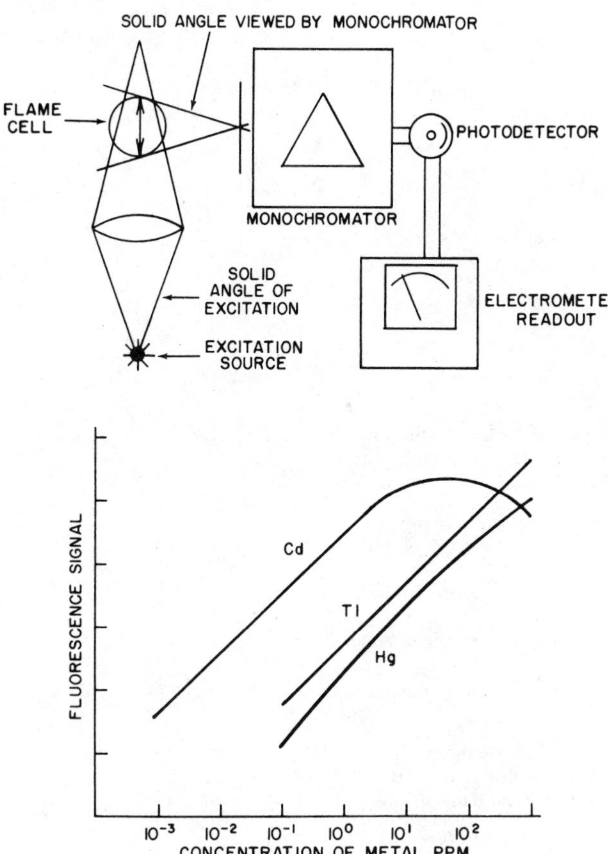

Fig. 11.12i Atomic fluorescence spectrophotometer

cury. The degree of blackening is a function of the concentration of mercury and the time of exposure. Quantitation may be made with a simple densitometer.

Mercury can also be detected with copper, iodide paper, selenium paper, and gold chloride on silica gel. These methods appear to have a limiting sensitivity in the 500 to 1000 $\mu g/m^3$ range with an accuracy of ± 5 percent.

Gas Chromatographic Methods for Mercury Organics

Organic mercurials can be determined at levels of approximately 1 ppb by gas chromatographic methods. Samples must be extracted into benzene and injected into a gas chromatographic column (Carbowax 20M on Chromasorb W) operating at 180°C (356°F) which separates methyl mercury, ethyl mercury, methoxyethyl mercury, phenyl mercury, and dimethyl mercury. Quantitation is obtained with an electron-capture detector. Dimethyl mercury requires conversion into methyl mercury halide for analysis, however. Precision is reported to be about 12 percent with 20 percent reliability.

Atomic Fluorescence Spectroscopy

Samples collected by any of the liquid impinger techniques (Figures 11.12a and b) may be analyzed by use of atomic fluorescence flame spectrophotometric methods. Figure 11.12i is a schematic diagram of a typical fluorescence spectrophotometer and of some typical calibration data. A mercury concentration of 0.1 ppm in a collection fluid is the detection limit of this method with available radiation sources.

Other Analytical Procedures

A radiochemical procedure for the determination of mercury vapor has been developed based on the isotopic exchange which takes place when air samples containing mercury are passed through solutions of Hg^{203} mercuric acetate. The method potentially has good sensitivity but requires precautions which are normal in handling radioisotopes.

Mercury which has been extracted from the air may be determined utilizing a neutron activation procedure. This method can have a precision and accuracy in the range of ± 2 percent. Analytical services are available from Gulf General Atomics and the AEC Idaho Falls Laboratory.

REFERENCES

1. *Federal Register*, Vol. 36, No. 234, December 7, 1971, p. 23,250.
2. Williston, S.H. and Morris, M.H., U.S. Patent No. 3,173,016, March 9, 1965; and Williston, S.H., U.S. Patent No. 3,178,572, April 13, 1965.

BIBLIOGRAPHY

Ewing, G.W., *Instrumental Methods of Chemical Analysis*, McGraw-Hill, New York, 1969, p. 177 (Fig. 3).

Guilbault, G.G., ed., *Fluorescence*, Marcel Dekker, New York, 1967, pp. 581 and 618.

Leithe, W., *The Analysis of Air Pollutants*, Ann Arbor Science Publishers, Ann Arbor, 1970.

Vaughn, W.W., "A Simple Mercury Vapor Detection for Geochemical Prospecting," Geological Survey Circular No. 540, U.S. Department of the Interior, 1967.

Flow Sheet Symbol

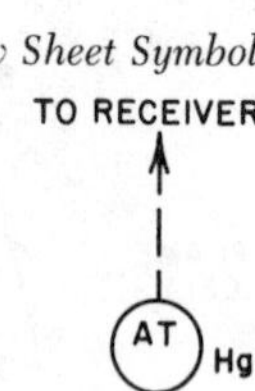

11.13 MERCURY IN WATER

Pretreatment of Sample:	Wet oxidation is used for the detection of the total mercury level, and solvent extraction is used for the detection of organic mercury
Method of Detection:	A—Colorimetric detection of total mercury level B—Atomic absorption spectrophotometry of total mercury level C—Gas chromatograph with electron capture detector for organic mercury D—Thin-layer chromatographic detection of organic mercury E—Ultraviolet detector preceded by wet chemistry package
Ranges:	The detection ranges for A and D are approximately 0.02 to 0.5 ppm of mercury in water, and for B and C approximately 0.1 to 10 ppb of mercury in water
Cost:	\$700–2500 (A); \$11,000–22,500 (C); \$16,000–34,000 (B); \$34,000–56,000 (E)
Partial List of Suppliers:	Beckman Instrument Inc. (A, B and C); Coleman Systems (A and B); Hach Chemical Co. (A); Olin Corp. (E); Packard Instrument Co. Inc. (C); Perkin-Elmer Corp. (B and C); and Philips Electronic Instruments (C)

Total Mercury Detection

Sample Treatment

The total mercury level is the sum of organic mercury and inorganic mercury in a sample. Mercury is present in the environment in organic and inorganic compounds, either dissolved in water or adsorbed on particulate matter or sediments.

To analyze a variety of mercury compounds by the commonly available methods, it is necessary to digest the sample in order to effect decomposition of the accompanying substances and thereby convert the mercury into inorganic mercury (Hg^{2+}). Digestion is carried out by heating the sample with a mineral acid and a strong oxidizing agent. The combinations in common use are H_2SO_4–$KMnO_4$; HNO_3–$KMnO_4$; H_2SO_4–H_2O_2; H_2SO_4–HNO_3; and HCl–$KClO_3$. The combination of H_2SO_4–$KMnO_4$ or HNO_3–$KMnO_4$ is best suited for analysis of water. The concentration range most often used is 1 to 5 (N) for the acid and 1 to 5 percent for $KMnO_4$. Heating is either by a water bath or by direct flame. The digestion must be carried out in the presence of excess $KMnO_4$, and therefore fresh addition must be made whenever the color of the permanganate fades (the digestion being complete when the color of $KMnO_4$ no longer fades).

Samples rich in organic matter, such as biological materials, can best be digested by heating with H_2SO_4–HNO_3 until white fumes (sulfuric acid) evolve. The digestion is complete when the solution becomes clear. When carbonization of organic matters occurs, the digestion is continued with fresh addition of nitric acid. An apparatus such as that shown in Figure 11.13a prevents the small losses due to vaporization of mercury during the digestion.

Since upon completion of the digestion an excess of the oxidizing agent is present in the digested solution, it becomes necessary to reduce the remaining oxidizing agent. Hydroxylamine is best suited for the reduction of NO_2 or $KMnO_4$. Hydroxylamine alone, however, can-

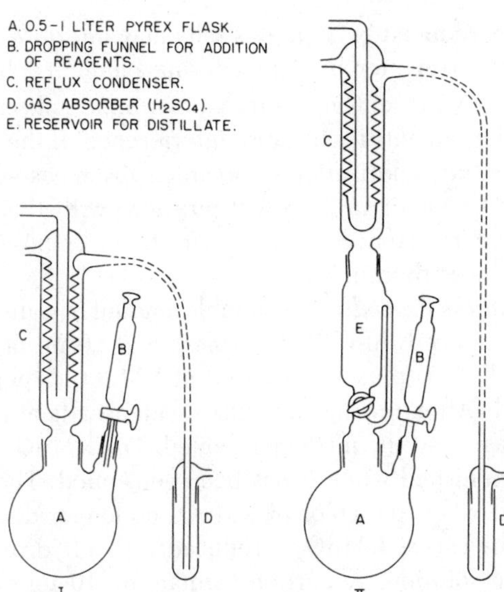

Fig. 11.13a Digestion flasks for detection of (I) mercury in water and (II) in organic materials

not reduce nitrous acid completely with nitric acid as a digesting liquor, and simultaneous addition of urea is advisable. The samples thus digested are then analyzed for mercury in one of the detectors described in this section.

Colorimetric Detection[1-3]

Dithizone is widely used in the colorimetric analysis of mercury, where the absorption of visible light is detected as a measure of mercury concentration. When an aqueous solution of mercury with a wide pH range (0 to 13) is shaken with a solution of dithizone in chloroform, carbon tetrachloride or benzene, a mercury complex is formed which dissolves in the organic layer. Dithizone is designated as H_2Dz, the mercuric (Hg^{2+}) dithizonate is represented as Hg $(HDz)_2$ and the mercurous (Hg^{1+}) dithizonate as Hg (HDz).

A solution of Hg $(HDz)_2$ in carbon tetrachloride shows a maximum absorption at the wavelength of 485 to 490 nm. Mercury forms Hg(HDz) in the absence of water or a secondary dithizonate HgDz in aqueous alkaline solutions which are deficient in dithizone. This compound gives a violet color.

Dithizone reacts with silver (Ag), copper (Cu), gold (Au), palladium (Pd), and platinum (Pt) in addition to Hg under acidic conditions, but it is generally sufficient to consider interference of Cu alone in environmental analysis. Metals such as lead (Pb), cadmium (Cd), zinc (Zn), nickel (Ni), cobalt (Co), and iron (Fe) are not extracted from an acidic solution. Moreover, tin (Sn) and bismuth (Bi) are not extracted from a strongly acidic solution unless they are present in large quantities.

Interference of Cu may be eliminated by one of the following methods:

1. When a small flow of a dithizone in carbon tetrachloride solution is added to a weakly acidic solution containing Hg and Cu, extraction of Hg is completed before that of Cu begins. Separation of Hg and Cu is thus possible.
2. Extraction of Hg and Cu can be carried out simultaneously with excess dithizone and then treated with a masking agent such as KBr, KI, and $Na_2S_2O_3$ to decompose only Hg $(HDz)_2$ and to transfer Hg^{2+} into the aqueous layer. Thereafter Hg is again extracted with dithizone.
3. Ethylenediaminetetraacetic acid (EDTA) in the form of the disodium salt can be added as a masking agent of Cu, and Hg alone is extracted with dithizone. The extraction is carried out at a pH of 2.5 or higher, preferably 5.5, at which level Cu-EDTA is stable.

A relatively large quantity of Cl^{1-} also interferes with the extraction of Hg^{2+} under strongly acidic conditions. For instance, the extraction of Hg can be done without difficulty when up to 1 mole (M) of Cl^- is in 50 ml of 1 N sulfuric acid, but the extraction becomes incomplete as Cl^{1-} exceeds 2.5 M. Under neutral to alkaline conditions, some interference is also observed when NH_4^{1+} is present in large quantities.

Analysis Procedure: The procedure generally followed in the colorimetric determination of mercury by dithizone is as follows: The sample solution is extracted repeatedly with excess dithizone, and Hg^{2+} is captured in the solvent layer as Hg $(HDz)_2$. Contaminants are also removed in this step. The Hg $(HDz)_2$ is then decomposed by a suitable masking or oxidizing agent, and the Hg^{2+} is liberated and transferred into the aqueous layer. Here it is reacted with a given excess of dithizone under specified conditions, and its concentration is measured colorimetrically.

The extraction immediately preceding the colorimetric determination is carried out under acidic conditions. The color developed is a mixture of orange (due to Hg $[HDz]_2$) and green (due to the excess of dithizone). The absorbance of such mixed colors can be measured directly, or measurements may be made at 485 to 490 nm, which is the maximum absorption wavelength of Hg $(HDz)_2$. Higher sensitivity and accuracy can be obtained at 605 to 620 nm, which is the maximum absorption wavelength of dithizone.

With the single color method, mercury is extracted with a solution of dithizone. The excess dithizone is removed by shaking it with dilute aqueous ammonia. The absorbance is measured at 485 to 490 nm. With this method, changes in the concentration of dithizone do not affect the results. Hg $(HDz)_2$ gradually fades in the light. This tendency becomes pronounced when impure dithizone is used, whereas a solution of high-grade dithizone remains stable for several hours in a lighted room.

Atomic Absorption Spectrophotometry

In atomic absorption spectrophotometry, a solution containing mercury is introduced directly into a flame. This technique does not give high sensitivity in the analysis of mercury, and it is difficult to detect mercury at a concentration below 0.2 ppm. Mercury can be analyzed with high sensitivity if the aqueous solution is first reduced to mercury vapor and then sent to the absorption cell by aeration where it is analyzed by atomic absorption spectrophotometry without the use of flame. Stannous salts are best suited for reducing mercury in an aqueous solution. Mercuric (Hg^{2+}) ions are reduced by stannous salts to metallic mercury (Hg^0) according to Equation 11.13(1).

$$Hg^{2+} + Sn^{2+} \longrightarrow Hg^0 + Sn^{4+} \qquad 11.13(1)$$

The metallic mercury thus formed is vaporized by aeration and is sent to the absorption cell (with quartz windows) where its absorbance is measured at a wavelength of 253.7 nm.

Figure 11.13b illustrates both the closed and the open system of measurement. Beer's law holds for the mercury levels of 0.1 to 10 μg involved in this method. An atomic absorption spectrophotometer is best suited for this measurement, but simpler apparatuses are also commercially available.

The presence of large quantities of metal ions which can be reduced by stannous salts may interfere with the reduction of mercury. Interfering elements, however, are rarely present in ordinary environmental samples in sufficient quantities to cause interference. If the aeration causes the vaporization of organic substances which absorb ultraviolet rays, it will interfere with the analysis of mercury. Hence it is necessary to decompose organic substances thoroughly.

Analysis Procedure: A suitable amount of water sample is introduced into a 500-ml flask and 20 ml of sulfuric acid (1 + 1) plus 15 to 20 ml of $KMnO_4$ (6 percent) is added. After mixing, a reflux condenser is attached to the flask, and the mixture is boiled. The $KMnO_4$ solution is replenished when it has been consumed. The flask is cooled when the color of $KMnO_4$ no longer disappears, and the excess $KMnO_4$ is reduced with a hydroxylamine sulfate solution. A portion (containing 10 μg or less of mercury) is transferred into a flask such as the one in Figure 11.13b. Two milliliters of sulfuric acid (1 + 1) and 2 ml of stannous sulfate (10 percent in 2 N H_2SO_4) are added and aeration is started at a rate of 2 liters per minute.

The vapor thus formed is driven into the vapor phase and circulated in a closed circuit. The ultraviolet absorption at 253.7 nm is detected in the optical cell of an atomic absorption spectrophotometer. A typical record of the UV absorbance is shown in Figure 11.13c. A calibration curve is prepared with standard solutions containing known amounts of mercury to assist in the quantitative analysis. The relative standard deviation for this method is reportedly between 2 and 10 percent.

Figure 11.13d illustrates the components of an automatic mercury monitor with an ultraviolet detector.

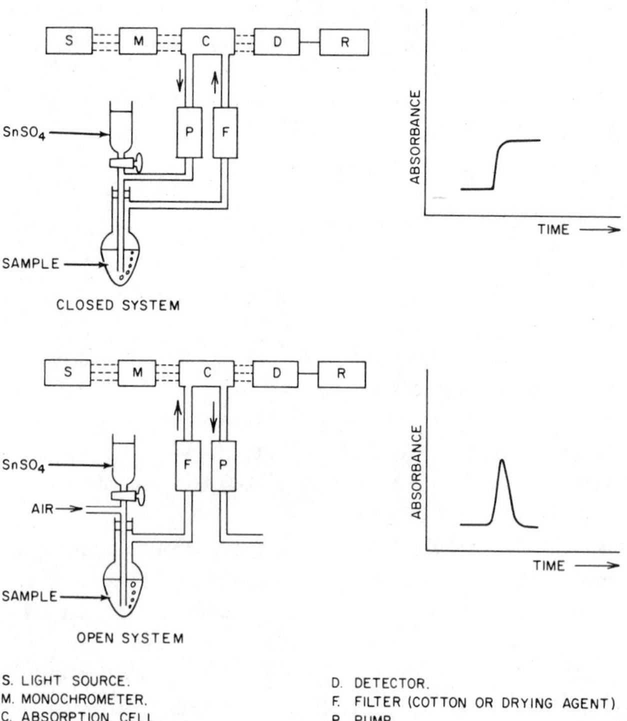

Fig. 11.13b Flameless atomic absorption spectrophotometer with reduction-aeration hardware

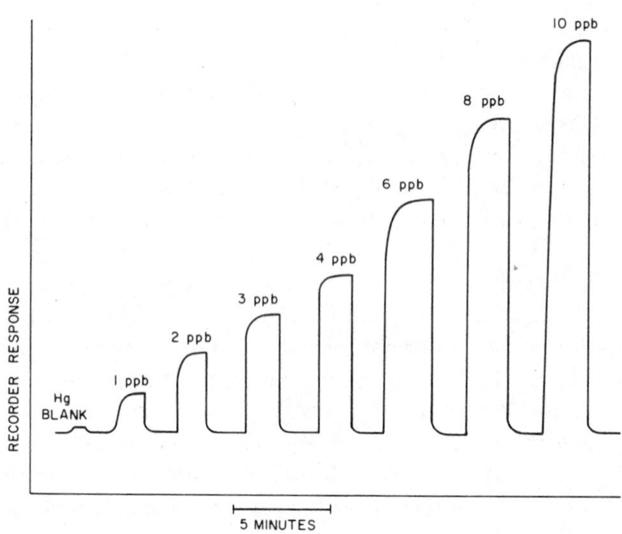

Fig. 11.13c UV absorption record of vapor sample produced by aeration

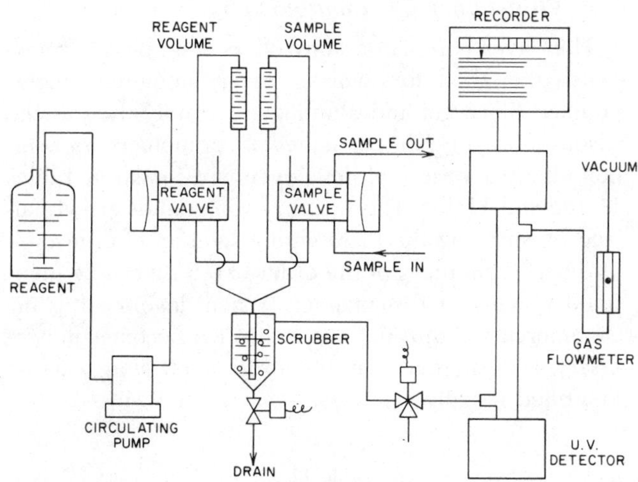

Fig. 11.13d Continuous mercury monitor

Organic Mercury Detection

Sample Treatment

In organomercury compounds of the RHgX type, R = organic group, such as alkyl and phenyl, and X is an electronegative group such as halogen and hydroxyl. Those carrying methyl, ethyl, and propyl groups are known to be the causes of Minamata disease.* When these compounds are present in water, it is first necessary to extract them with a suitable solvent.

The solubility of a compound of the RHgX type in organic solvents varies. If X is a halogen, the compound is soluble in aromatic hydrocarbons such as benzene and toluene. If X is an ion such as SO_4^{2-} OH^-, or $CH_3CO_2^{1-}$, the solubility in hydrocarbons is extremely low. Therefore it is necessary first to convert X to a halogen, and then the RHgX compound can be extracted with an organic solvent such as benzene and toluene. The aqueous solution is made acidic by addition of hydrochloric acid before extraction. The lower alkyl mercury compounds are moderately soluble in water, and a relatively large amount of solvent is necessary to effect quantitative extraction.

The organomercury compounds extracted in the organic solvent are generally contaminated and must be cleaned. The cleaning of an RHgX-type compound can be performed effectively by backextraction with an aqueous solution of a sulfur-containing compound such as cysteine. Organomercury compounds react with a sulfhydryl compound according to Equation 11.13(2) and move from the organic solvent phase to the aqueous phase

$$RHgX + R'SH \longrightarrow RHgSR' + HX \qquad 11.13(2)$$

*A severe neurological disorder resulting from poisoning by organic mercury and leading to severe permanent neurological and mental disabilities or death.

where both R and R′ are organic groups and X is an electronegative group. Many of the organic compounds remain in the organic solvent phase and the purpose of cleaning is thereby accomplished.

The transfer of the organomercury compounds from the organic solvent phase to the aqueous phase is quantitative. Therefore, when the backextraction is carried out by using a smaller volume of the aqueous solution against a known volume of the organic solvent, concentration of the organomercury compounds may be determined simultaneously with cleaning up.

The organomercury compounds extracted back into the aqueous layer are again liberated as RHgCl by addition of hydrochloric acid to the aqueous solution according to Equation 11.13(3).

$$RHgSR' + HCl \longrightarrow RHgCl + R'SH \qquad 11.13(3)$$

The RHgCl liberated is extracted with a small amount of benzene and analyzed by gas chromatographic techniques.

Gas Chromatography

Lower alkylmercury and phenylmercury compounds vaporize upon heating and can be analyzed by gas chromatography. The use of an electron capture detector gives high sensitivity and is best suited for analysis of traces of organic mercury. For example, methylmercury can be detected to a level of 1×10^{-11} g.

Organomercury compounds are highly reactive with metals; consequently it is not desirable to use metal tubing as column materials. A glass column is preferable. Also, a polar substance as a liquid phase in the column yields better results. The liquid phases most frequently used are polyethylene glycol, polydiethylene glycol succinate, and polybutanediol succinate.

The amount of the liquid phase to be coated on the support is preferably 5 to 10 percent for analysis of alkylmercury and about 2 percent for analysis of phenylmercury. An increase in the coating amount causes an increase in bleeding gas and a decrease in the standing current of the detector, with resultant lower sensitivity when a high column temperature is used. The size of the liquid phase is reduced for high sensitivity analysis.

The use of the subtractive technique is recommended for simplified identification of organomercury compounds by gas chromatography. When an organic solvent, such as benzene or toluene, which contains organomercury compounds, is mixed with the aqueous solution of a bivalent sulfur compound, such as $Na_2S_2O_3$ or cysteine, the organomercury compounds disappear from the organic solvent. If the gas chromatograms before and after this treatment are compared, it will be noticed that the peaks corresponding to the organomercury compounds disappeared or diminished markedly after treatment.

Analysis Procedure: Ten milliliters of concentrated hydrochloric acid is added to 500 ml of the sample water containing methylmercury compounds. The resultant solution is mixed with 100 ml of benzene, and the mixture is allowed to settle. The aqueous layer is separated and is extracted again with 100 ml of fresh benzene. This is repeated three times, the aqueous layer is discarded and the combined benzene layer is washed with distilled water. The benzene layer is separated and is extracted back with 10 ml of a 0.1 percent aqueous solution of 1-cysteine. The aqueous layer is also separated and 1 ml of concentrated hydrochloric acid and 2 ml of benzene are added. After mixing, the benzene layer is separated and dried over a small amount of anhydrous sodium sulfate. A 10-μl sample is analyzed by gas chromatography.

The column recommended contains 5 percent polydiethylene glycol succinate on Chromosorb W, 60 to 80 mesh, packed in glass tubing, 1 m long and with a 3-mm inside diameter. Column temperature is set at 130°C (266°F) and the flow rate of carrier gas is maintained at 60 ml per minute. Methylmercury chloride is eluted in 3 to 5 minutes under these conditions, and concentrations of 1 μg per liter or less can be detected by this method.

Table 11.13e
R_f VALUE OF ORGANOMERCURY COMPOUNDS

Compound	Developer* A	B	C	D	E
Methylmercuric chloride CH_3HgCl	0.35	0.59	0.29	—	0.42
Methylmercuric iodide CH_3HgI	0.04	0.03	—	—	—
Ethylmercuric chloride C_2H_5HgCl	0.41	0.74	0.50	0.72	0.46
Ethylmercuric phosphate $(C_2H_5Hg)_2HPO_4$	0	0	0.49	0.67	—
Methoxyethylmercuric chloride $CH_3OC_2H_4HgCl$	0.23	0.49	0.46	0.74	—
Phenylmercuric chloride C_6H_5HgCl	0.51	0.73	0.56	0.76	—
Phenylmercuric iodide C_6H_5HgI	0.74	0.84	0.92	0.86	—
Phenylmercuric acetate $C_6H_5HgCH_3CO_2$	0.39	0.64	0.86	0.83	0.48

*Key:
Adsorbent layer: silica gel
A. n-phexane:acetone (85:15)
B. n-hexane:acetone (70:30)
C. butyl alcohol saturated with water
D. isopropyl alcohol:water (90:10)
E. chloroform

Thin-Layer Chromatography

Thin-layer chromatography offers a simple and inexpensive method for analysis of organomercury compounds. Silica gel and alumina are mainly used as the adsorbent layer. The R_f values of organomercury compounds for a variety of developers are shown in Tables 11.13e and 11.13f. The mercury compounds are visualized by spraying the plate with a solution of dithizone. Ordinarily, mercury of the order of 0.5 μg can be identified visually in this manner. When developed as organomercury dithizonate, visualization becomes unnecessary and mercury on the order of 0.1 μg can be identified visually.

Table 11.13f
R_f VALUE OF ORGANOMERCURY DITHIOZONATES[4]

Compound	Developer* 1	2	3	4	5	6
Methylmercury dithizonate $CH_3Hg(HDz)$†	0.64	0.48	0.57	0.77	0.89	0.86
Ethylmercury dithizonate $C_2H_5Hg(HDz)$	0.64	0.51	0.62	0.78	0.91	0.87
Methoxyethylmercury dithizonate $CH_3OC_2H_4Hg(HDz)$	0.32	0.16	0.25	0.44	0.58	0.49
Ethoxyethylmercury dithizonate $C_2H_5OC_2H_4Hg(HDz)$	0.44	0.23	0.34	0.55	0.71	0.67
Phenylmercury dithizonate $C_6H_5Hg(HDz)$	0.48	0.34	0.46	0.62	0.72	0.69
Mercury dithizonate $Hg(HDz)_2$	0.19	0.09	0.17	0.28	0.19	0.15

*Key:
1. hexane:acetone (9:1)
2. hexane:acetone (19:1)
3. hexane:acetone (93:7)
4. petroleum ether:acetone (9:1)
5. hexane:acetone (19:1)
6. petroleum ether:acetone (19:1)
Adsorbent layer
1–4 silica gel
5–6 alumina
†HDz = abbreviation for dithizonate ligand.

REFERENCES

1. Iwantscheff, G., "Das Dithizon und seine Anwendung in der Mikro- und Spurenanalyse." *Weinheim Verlag Chemie*, 1958.
2. Sandell, E.B., *Colorimetric Determination of Traces of Metals*, Interscience, New York, 1965.
3. Kolthoff, L.M., *Treatise on Analytical Chemistry*, Vol. 3, Part II, Interscience, New York, 1961.
4. Tatton, J.O.G., and Wagstaffe, P.J., "Identification and Determination of Organomercurial Fungicide Residues by Thin-Layer and Gas Chromatography," *Journal of Chromatography*, Vol. 44, 1969, p. 284.

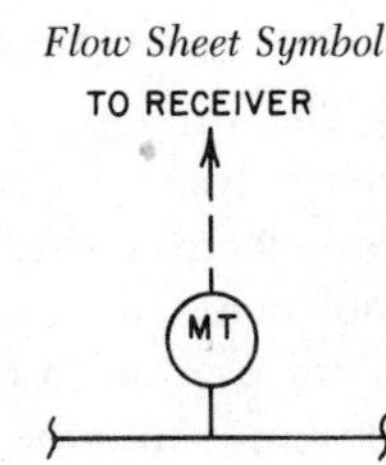

11.14 MOISTURE IN GAS OR LIQUID

Design Pressure:	To 3000 PSIG (21 MPa)
Design Temperature:	150°F (66°C)
Element Material:	Rhodium, glass, stainless steel
Cost:	\$5000–\$12,000 for piezoelectric, infrared, and microwave types. \$3000–\$7000 for all others
Inaccuracy:	1–5%
Range:	0–10 ppm to 0–90%
Partial List of Suppliers:	American Instrument Co., Industrial Products Div.; Anacon Inc/Aero Vac; Bacharach Instrument Co.; Beckman Instruments, Inc., Cedar Grove Operations; Beckman Instruments, Inc., California; Dupont Co. Analytical, Instruments Div.; DuPont Company; General Eastern Instruments Corp.; Ionics Incorporated; Kay Ray Inc.; Lockwood & McLorie Inc.; Luft Instruments Inc.; Mine Safety Appliances Co.; Moisture Control Systems Inc.; Moisture Register Co.; Moisture Systems Corp.; Omega Controls Corp.; Panametrics Inc.; Uniloc Div. of Rosemount Inc.; Weathertronics; Weschler Electric Corp.; Yellow Springs Instrument Co.; Yokogawa Corp. of America

Sections 11.10, 11.14, and 11.15 deal with the various devices for moisture detection utilizing the physical properties of water and the laws of physics and chemistry to effect a measurement. Devices discussed in Section 11.10 are applicable to gas samples only, and they are generally used for room humidity measurements on air conditioning systems, etc. Instruments covered in this section can be used to monitor moisture in both liquid and gas samples; however, a special sampling system is required with some types to make them suitable for liquid samples. These instruments are used on samples operating at elevated pressures. Section 11.15 covers instruments for monitoring moisture in solids such as flour, plastics, powders, and paper.

Sampling systems have been discussed in Section 10.4; however, special considerations applicable to moisture analyzers will be covered at the beginning of the appropriate section or in the discussion of a particular design.

Sampling Systems

For general requirements of analyzer sampling systems, the reader is asked to refer to Section 10.4; however, there are some sampling system features peculiar to moisture analyzers, and these are covered below and in the detailed discussion of each design. The function of any sampling system is to deliver a clean, representative sample to the measuring element at the required pressure, temperature, and flow rate. Measurement of moisture, although simple in principle, is complicated by the fact that all materials adsorb and desorb moisture. At moisture levels in the percent range, moisture absorption is not a serious problem; however, at moisture levels in the low parts per million (ppm) ranges, sampling system materials must be selected so as to minimize their contribution to the moisture level in the sample. Materials which are least likely to contribute to the moisture in the sample are stainless steels, Teflon, Viton, Kel-F, nickel and nickel-plated materials, and cadmium and cadmium-plated materials. Materials that should be avoided are copper and its alloys, rubber, neoprene, and elastomers.

It is important to apply the material considerations to all parts of the sampling system, including internals of valves and filters.

Electrolytic Hygrometer

The principle of measurement utilized involves the electrolysis of water into oxygen and hydrogen. Since two electrons are required for electrolysis of each water molecule, the electrolysis current is a measure of the water present in the sample. If the volumetric flow rate of sample into the electrolysis cell is controlled at a fixed value, then the electrolysis current is a function of water concentration in the sample. This relationship is illustrated in the following example.

Determine the water concentration in a sample when the sample flow rate is 0.0257 ft.^{3}/sec. (0.00073 m^3/s) at 100°F (37.8°C) and 10 PSIG (68.9 kPa). The electrolysis current is measured at 320 μA.

$$\text{Electrolysis Current} = 320 \times 10^{-6} \text{ coulomb per second}$$
$$1 \text{ coulomb} = 6.25 \times 10^{18} \text{ electrons}$$

Therefore, the electrolysis current is equal to the electron flow of 2×10^{15} electrons per second. Since two electrons of charge are required to electrolyze one water molecule, the water flow into the cell is equal to 10^{15} molecules per second. If, for purposes of this example, the water present in the sample is considered an ideal gas, the volume of water entering the sample can be calculated using the ideal gas law.

$$V_2 = \frac{P_1}{P_2}\frac{T_2}{T_1}V_1 \quad V_1 = 379\text{ N/A} \quad A = 6.02 \times 10^{23}$$

Where V, P, and T are the volume flow rate, absolute pressure, and temperature of the moisture. A is Avogadro's number, and N is the flow of moisture in molecules per second. Avogadro's law states that equal volumes of different ideal gases at the same pressure and temperature contain the same number of molecules. Subscripts 1 and 2 refer to standard and sample conditions, respectively.

$$V_1 = \frac{379 \times 10^{15}}{6.02 \times 10^{23}} = 6.3 \times 10^{-7} \text{ ft.}^3\text{/sec}$$

$$V_2 = 6.3 \times 10^{-7} \times \frac{560}{520} \times \frac{14.7}{24.7}$$

$$V_2 = 4.04 \times 10^{-7} \text{ ft.}^3\text{/sec}$$

Since the sample flow is controlled at 0.0257 ft.^{3}/sec (0.00073 m^3/s), the volume concentration of moisture is $(4.04 \times 10^{-7})/0.0257 = 1.57 \times 10^{-5}$ ft.^{3} of moisture per cubic foot of sample. Expressed in parts per million (ppm), the moisture content is 15.7 ppm. The example above is intended to illustrate the fact that the hygrometer cell sees only a mass flow of water—number of water molecules electrolyzed per unit time. In order to obtain an output in the form of moisture content, the sample flow rate must be a known, constant value. Furthermore, the accuracy of the output can never exceed the accuracy with which the sample flow is controlled.

The commercial electrolytic hygrometer cell consists of a small chamber containing two noble metal electrodes which support a thin layer of desiccant. Moisture in the sample is absorbed by the desiccant and electrolyzed by means of a voltage-regulated power supply connected to the electrodes. Units are available for use in nonhazardous areas with the sample flow control, electrolysis cell, and electronics packaged as a single unit. When used in hazardous areas, the cell and flow controller are housed in an explosion-proof conduit and the electronic circuitry is remote mounted (see Figure 11.14a).

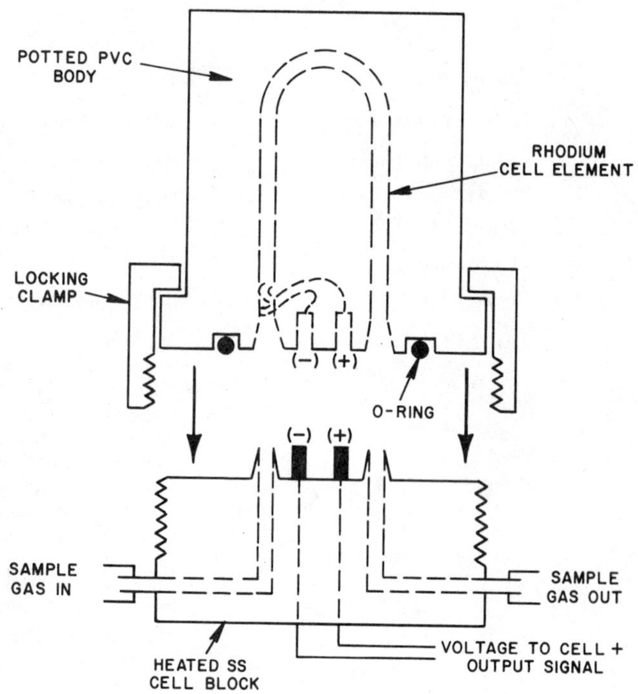

Fig. 11.14a Electrolytic hygrometer cell

Recombination Effect

Although this instrument will operate satisfactorily with a variety of samples, a phenomenon called the recombination effect introduces large errors at low moisture levels in hydrogen-rich or oxygen-rich samples. Recombination is the reverting to water of the electrolysis products; it introduces an error into the measurement when the recombined oxygen and hydrogen are reelectrolyzed. Apparently all electrodes catalyze this reaction, although some electrode materials do so more than others. The use of rhodium as the electrode material has been found to minimize recombination.

When monitoring very low moisture levels in oxygen-rich or hydrogen-rich atmospheres, the recombination produces a large error even with the best choice of electrode materials. For such an application, two sensors are used; one measures at sample flow rate X and the other at sample flow rate 2X. Since the error due to recombination is a constant, subtraction of the two sensor outputs yields a signal that is independent of recombination.

Cell Limitations

The cell will perform satisfactorily with a variety of samples; however, there are a number of sample materials which will cause problems or should not be monitored with this instrument. Gases that cause problems are hydrogen and oxygen. These gases have been discussed under the recombination effect above. Gases for which this instrument is unsuitable are unsaturated monomers, alcohols, amines, ammonia, hydrogen fluoride, and $CHClF_2$ (Freon) refrigerant.

Alcohols are seen by the cell as water, amines, and ammonia react with the desiccant, and hydrogen fluoride corrodes the internals of the cell. The data collected on $CHClF_2$ refrigerant indicates an anomaly although the reason for this is not fully understood. Unsaturated hydrocarbons such as butadiene or monomers with a strong tendency to polymerize cannot be monitored since the cell will be quickly coated with polymer. Generally, the instrument should not be used with samples whose components may deposit in the cell. When a cell becomes contaminated, it will show a memory for polarity; that is, the outputs under forward and reverse flow through the cell will not be equal.

A cell will lose sensitivity when exposed to moisture levels of a few parts per million over a period of weeks. This sensitivity loss is due to the elution of desiccant with the sample. However, this process occurs over a long period of time and the cell can be recoated fairly easily in the field during periodic maintenance.

Liquid Samples

The electrolytic hygrometer cell can be exposed only to gases. When the process sample is in the liquid state, there are two methods available to effect a measurement using the electrolytic hygrometer.

1. *Sample Vaporization*—If the liquid sample has sufficiently high vapor pressure, the sample can be vaporized, and the measurement is then made as a gas sample. The sample is vaporized by means of a vaporizing regulator located as close as possible to the sample take-off point to minimize lag. The remainder of the sampling system is the same as for gas samples.

2. *Sample Stripping*—If the sample cannot be vaporized, this method can be used to obtain a gas sample at the cell. The moisture is stripped from the liquid sample in a falling film column. Liquid sample is continuously metered into the top of the column where it descends as a thin film. Dry nitrogen, metered into the bottom of the column, ascends and removes moisture from the descending film of liquid sample. A filter located at the top of the column removes any droplets entrained in the nitrogen before the gas is passed through the measuring cell. A drain valve at the bottom of the column facilitates removal of stripped sample. The nitrogen is dried before entering the column by passing through an electrolytic dryer, similar to the measuring cell, where moisture is removed by electrolysis. The supply nitrogen moisture level should be fairly low, less than 500 ppm, to avoid overloading the dryer (see Figure 11.14b).

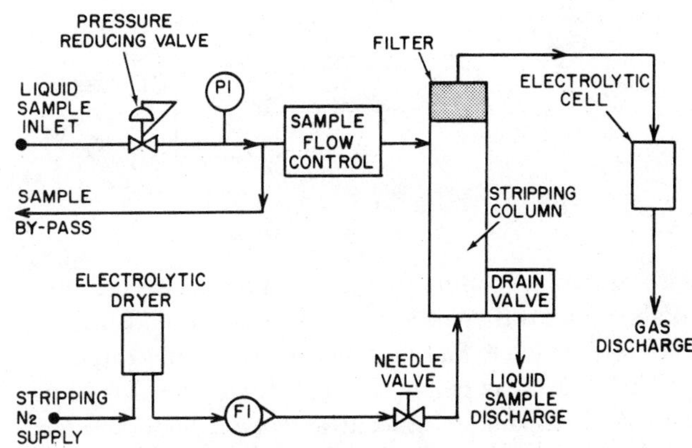

Fig. 11.14b Liquid sample stripping

Sampling System

As pointed out previously, the electrolytic hygrometer is sensitive only to the mass flow of water into the cell; the concentration of moisture is inferred from the known sample flow rate. Therefore, the accuracy of the readout is dependent on the accuracy of the sample flow control, and the accuracy of the moisture measurement can never exceed that of the sample flow control. This fact should be kept in mind when selecting a sample flow controller.

Capacitance Hygrometer

The principle of measurement utilized is the change of capacitance associated with a change of the sample dielectric constant between capacitor plates. The value of capacitance is a function of plate area, plate spacing, and the dielectric constant of the material between the plates. The dielectric constant of a material has a unique value for each substance and is related to the polar characteristics of the molecules. This property can, therefore, be used to detect the presence of a specific substance in a pipeline stream. Figure 11.14c shows an air capacitor connected to a battery. Figure 11.14d shows the same capacitor with the air between the plates displaced by mica. The capacitance of Figure 11.14d is 6 times that of Figure 11.14c because the dielectric constant, K, of mica is 6 and the K of air is 1. If the mica were inserted gradually between the plates, then the capacitance would

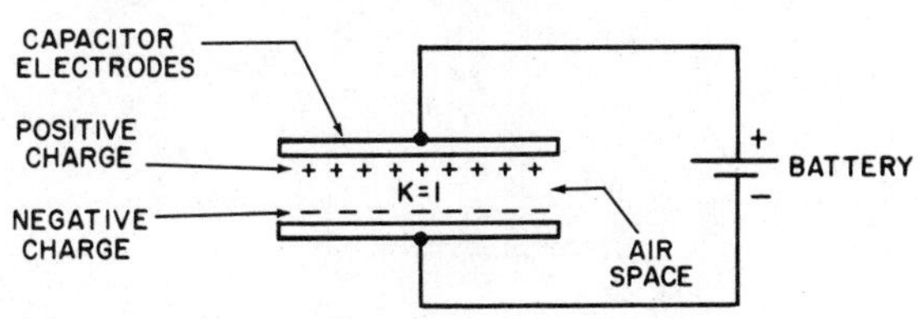

Fig. 11.14c Air capacitor

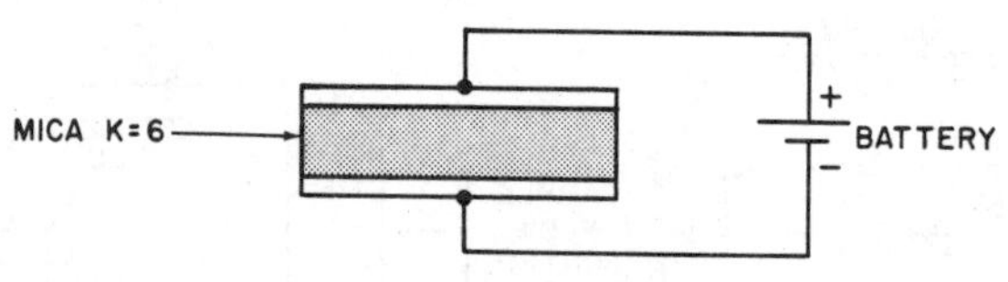

Fig. 11.14d Mica capacitor

increase correspondingly from minimum to maximum. Water, with a dielectric constant of 80, would increase the capacitance to 80 times the value of the air capacitor.

The measuring cell of the hygrometer consists of two electrically insulated concentric metal cylinders to form the measuring capacitor plates. The annulus between the cylinders is filled with alumina desiccant. Two porous metal discs support the cylinders and retain the desiccant in the annulus. The sample is allowed to flow through the annulus and in the process water is absorbed or desorbed by the desiccant which remains in equilibrium with the sample in terms of percent saturation. The moisture content of the sample is thus amplified by the desiccant because the saturation level of the desiccant is very much higher than that of the sample. The measuring capacitor is part of an electrical circuit which includes a reference capacitor. This circuit is powered by a 15-kilocycle, fixed-amplitude sine wave. The measuring and reference capacitors are switched alternately into the circuit, in such a way that its output voltage is a function of connected capacitance. The output signal is a 15-kilocycle sine wave whose amplitude varies with the measuring and reference capacitors as they are switched into the circuit. This difference in amplitude is related to the measured capacitance which in turn is a function of the moisture content of the sample (see Figure 11.14e).

The electronic chassis is normally mounted remote from the measuring cell and sampling system which should be close to the sample take-off. In addition to an

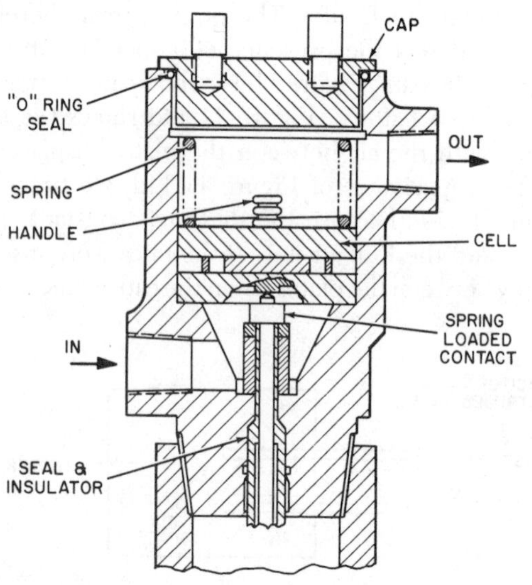

Fig. 11.14e Capacitance-type measuring cell

integral indicator and output signals of standard milliampere and millivolt ranges, the electronic chassis can be provided with multiple range switching and alarm contacts.

Sampling System

It was pointed out previously that the desiccant in the measuring cell is in equilibrium with the sample in terms of percent saturation. For this reason it is necessary to maintain the sample at constant temperature. For example, propane at 100°F (38°C) will dissolve 300 ppm of water; at 70°F (21°C) it will dissolve 150 ppm. With a moisture level of 15 ppm, the propane would be 5 percent saturated at 100°F and 10 percent saturated at 70°F. Therefore, the output signal would change 100 percent with no change in moisture content if temperature were allowed to vary from 100 to 70°F. The sample is maintained at constant temperature by passing it through a coil immersed in a constant temperature bath immediately upstream of the measuring cell. Temperature control of the sample is recommended even if the process stream is temperature controlled since ambient conditions will affect the sample temperature.

Limitations

The instrument is not suitable for polar materials such as alcohols because these become conductive at the 15-kilocycle operating frequency and short the measuring capacitor. Data on instruments operating at higher frequencies are not available.

Free sulfur and iron oxide concentrate on the desiccant and short the element.

At sample viscosities above 500 Saybolt Seconds Universal (SSU), moisture cannot contact the desiccant, and the measurement becomes meaningless.

The life of the desiccant is a function of solids in the stream which pass through the filter. While the desiccant can be replaced very readily, the short life of the desiccant excludes the use of this instrument on sample streams containing large quantities of fines.

Impedance Hygrometer

This instrument measures the water content of a sample by means of a probe whose electrical impedance is a function of the vapor pressure of moisture in the fluid. Impedance is the apparent opposition to the flow of alternating current. The probe consists of an aluminum strip, which is anodized to form a porous layer of aluminum oxide. A thin coat of gold is applied over the aluminum oxide. Leads from the gold and aluminum electrodes of the probe connect the sensing element to the measuring circuitry (see Figure 11.14f).

Water vapor penetrates the gold layer and equilibrates on the aluminum oxide. The number of molecules adsorbed to the aluminum oxide is a function of the water vapor pressure in the sample. Each water molecule ad-

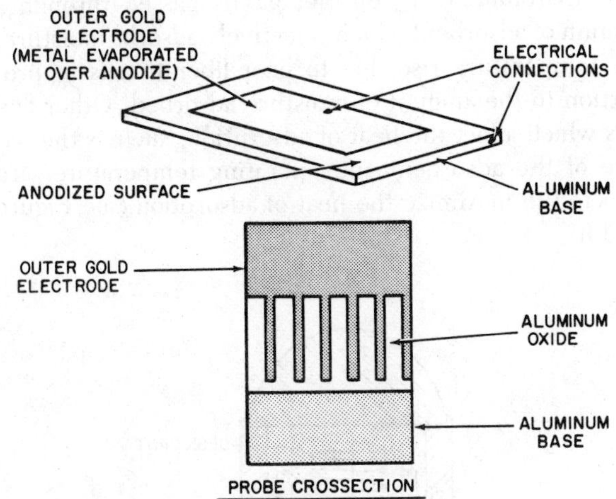

Fig. 11.14f Impedance-measuring sensor

sorbed contributes a distinct increment to the total conductivity of the aluminum oxide. The total probe impedance, the reciprocal of probe conductivity, is thus a measure of water vapor pressure in the sample. Water vapor pressure of a gas sample uniquely determines the dew-point temperature and moisture content of the sample. The output is normally calibrated to these units since they are more convenient to use than vapor pressure. In the case of certain liquid samples, the moisture content can be measured through the application of Henry's law.

In terms pertinent to this measurement, Henry's law states that at constant temperature the mass of water vapor dissolved in a given volume of liquid is in direct proportion to the partial pressure of water vapor in the sample. Henry's law can be restated in simpler form to read that the weight concentration of moisture in the sample is equal to the partial pressure of water vapor times a constant. However, Henry's law holds only for liquids with moderate solubility for water vapor such as pure hydrocarbons. The instrument cannot be used on liquid samples with high solubility for water such as the alcohols because the relationship expressed by Henry's law does not hold.

Installation

The probe is designed to be inserted directly into the process stream. However, where danger of explosion exists, the probe is inserted into a small sample chamber connected to the process line with flashback arrestors located at the sample inlet and outlet. For liquid samples, temperature must be controlled and a sampling system is required. Several probes can be connected to a single readout, and each probe can be monitored manually or by means of an automatic scanner. A millivolt output signal is available for recording or other functions; single or multiple-range measurement is standard (see Figures 11.14g and h).

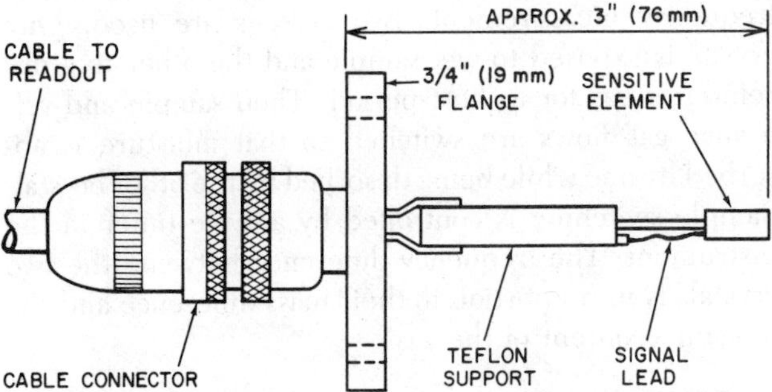

Fig. 11.14g Flanged probe

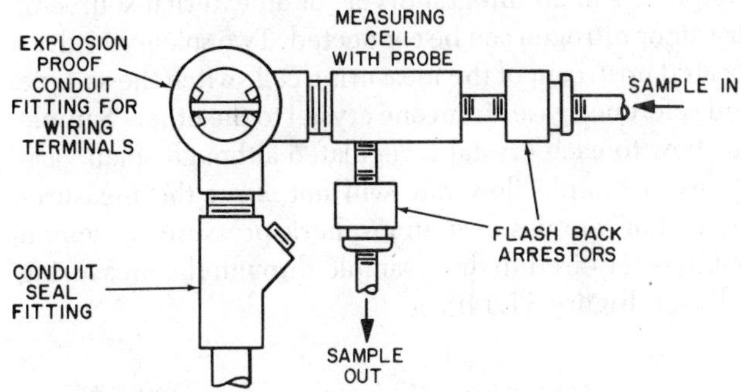

Fig. 11.14h Impedance hygrometer installation for hazardous areas

Sampling System

Normally a sampling system is not required on gas samples. On liquids, sample temperature must be held constant, as mentioned above.

Limitations

The instrument can be used on all gases not corrosive to the probe and which will not polymerize spontaneously on contact with the probe materials. On liquids, the instrument is limited to those fluids with a moderate solubility for water vapor; thus the instrument is not suitable for measurement in polar liquids such as the alcohols.

Piezoelectric Hygrometer

Piezoelectric quartz crystals have a number of uses: in communications, to control frequencies and, in industry, to measure temperature and thickness of metal films, and to generate ultrasonic waves. In moisture measurement, advantage is taken of the oscillating crystal's sensitivity to deposits of foreign material on its surface. Commercially available crystals will show a frequency change of 2000 cycles per second (cps) per microgram of material deposited. For moisture measurement, the quartz crystals are coated with a hygroscopic material and exposed to the sample; water from the sample is absorbed by the crystal coating, thus increasing the total mass and decreasing the oscillating frequency of the crystal. In order to measure changes of decreasing moisture concentration and to simplify the frequency measurement, two crystals are used. One crystal is exposed to wet sample and the other to a dry reference gas for a short period. Then sample and reference gas flows are switched so that moisture is absorbed by one while being desorbed by the other crystal. Sample switching is controlled by a cycle timer in the instrument. The frequency difference between the two crystals is in proportion to their mass difference and the moisture content of the gas.

Sampling System

Dry reference gas is obtained by drying part of the sample gas in an integral dryer, or an external source of dry air or nitrogen can be connected. Two solenoid valves located upstream of the measuring cell switch the sample and reference gases from one crystal to the other. Sample gas flow to each crystal is regulated although small variations of sample flow rate will not affect the measurement. For samples near atmospheric pressure, a vacuum pump is required to draw sample through the measuring cell (see Figure 11.14i).

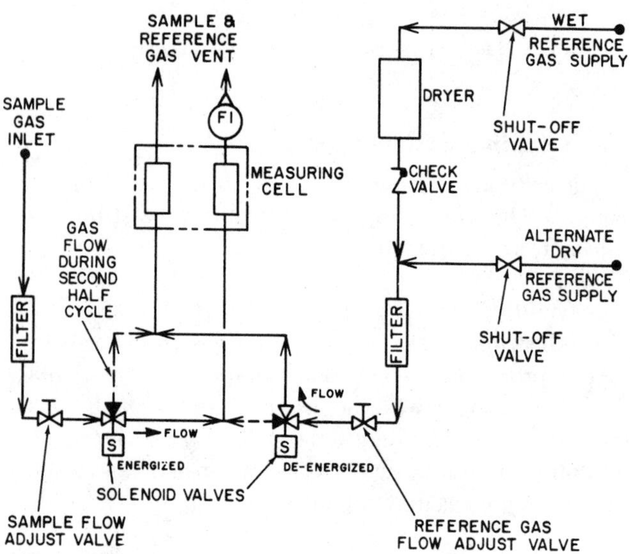

Fig. 11.14i Piezoelectric hygrometer sampling system

Limitations

The instrument is not capable of measuring moisture in liquid phase samples. For this reason the instrument is applicable to only those liquid samples which can be completely vaporized within the operating limits of the instrument. In addition, monomers with a strong tendency to polymerize, such as butadiene and styrene, may coat the crystals and prevent proper operation.

Heat of Adsorption Hygrometer

The process of adsorption and desorption involves an exchange of energy. During adsorption, energy is released to and during desorption energy is removed from the environment. When wet gas is passed through a column of adsorbent which selectively adsorbs moisture, the temperature rise due to heat liberation is in proportion to the amount of moisture adsorbed. Other factors which affect the heat of adsorption, such as the nature of the adsorbent and operating temperature, are selected to maximize the heat of adsorption (see Figure 11.14j).

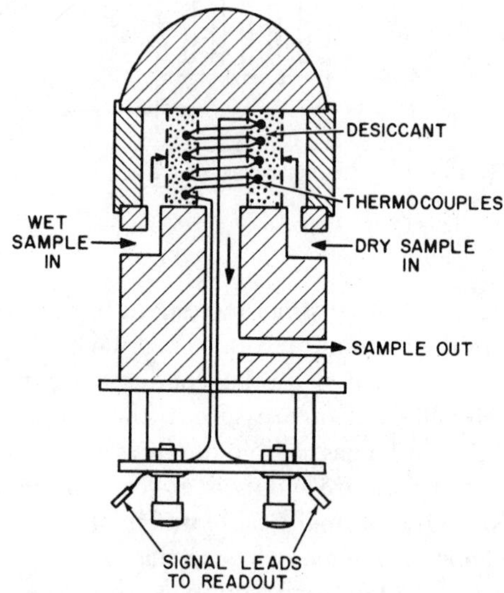

Fig. 11.14j Measuring cell for the heat of an adsorption-type hygrometer

The sensing element consists of two desiccant columns each containing a number of thermocouples connected in series. This assembly is housed in a temperature-controlled, thermally-insulated housing. In operation, wet sample gas is passed through one column and dry reference gas through the other. Since adsorption and desorption occur simultaneously, the net thermocouple output voltage represents the algebraic sum of the heat gained by one column and lost by the other. The reference gas and sample gas streams are switched on a time-cycle basis to maintain dynamic conditions necessary for measurement in the sensing element. Sample flow is closely regulated because moisture concentration is inferred from the known sample flow rate (see Figure 11.14k).

Sampling System

The sampling system contains:

a. A dryer to remove all moisture from the sample gas so that it can be used as the reference.
b. Solenoid valves and cycle timer to switch reference and sample gas between the two desiccant columns.
c. Flow regulators to hold sample and reference gas flows constant.

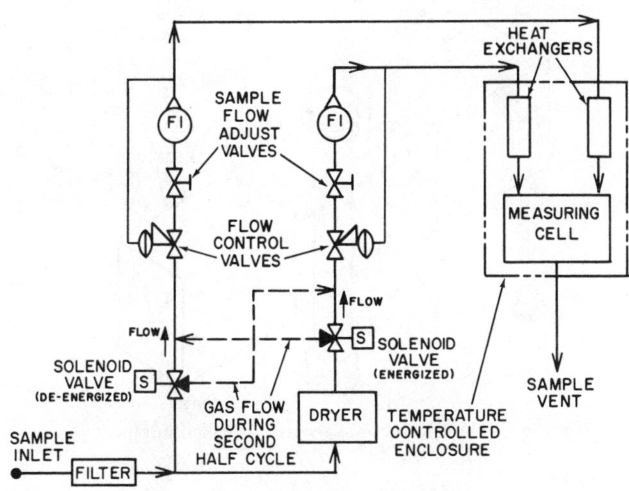

Fig. 11.14k Sampling system for the heat of an adsorption hygrometer

d. Heat exchangers within the insulated enclosure to heat both reference and sample gas to the operating temperature of the sensing element. The sample flow to the instrument is split into two equal flow rates at the sampling system; one half is the measured sample and the other half is dried for use as the reference gas. The timer actuates solenoid valves in the sample and reference gas lines so that each desiccant column is exposed alternately to reference and to sample gas.

Limitations

The sensing element cannot be exposed to liquid samples; however, the moisture content of liquid streams can be measured by one of two methods. Liquids with low boiling points can be vaporized at the sample take-off point by means of a vaporizing regulator, and the measurement is then performed on a gas sample. Or the moisture can be stripped from the liquid sample with dry gas, and from the moisture content of this gas the moisture concentration in the liquid sample can be inferred. Sample stripping is described in conjunction with electrolytic hygrometers.

Infrared Absorption Hygrometer

Water absorbs electromagnetic radiation in the infrared (IR) region of the spectrum. Specifically, infrared radiation of 1.4 and 1.93 micron wavelength is absorbed strongly by water. By measuring the attenuation (decrease of light intensity) of a beam of this wavelength as it passes through a sample, the moisture content of the sample can be determined. However, other factors such as reflection and dispersion of the radiant energy will contribute to the attenuation. Therefore, it is necessary to either calibrate these factors out of the measurement, or to use a reference wavelength which is not absorbed by moisture but is affected by all other factors to the same extent as the measuring wavelength. The difference in attenuation of the measurement and reference wavelength is then a function of moisture content only.

The sensing element consists of three groups of components: an IR radiation source, sample cell, and radiation pickup. The radiation source consists of a lamp, filters to pass the measuring and reference wavelengths, and optics to direct the beam through the glass sample cell. The radiation pickup consists of optics to collect the transmitted radiation and a photocell to convert the electromagnetic energy to an electric current. The measuring and reference wavelengths are allowed to impinge alternately on the photocell, so that two sets of current pulses are produced. These pulses are converted into two direct current (dc) signal levels whose ratio represents the moisture content of the sample (see Figure 11.14l). For more details on IR analyzers, refer to Section 10.9.

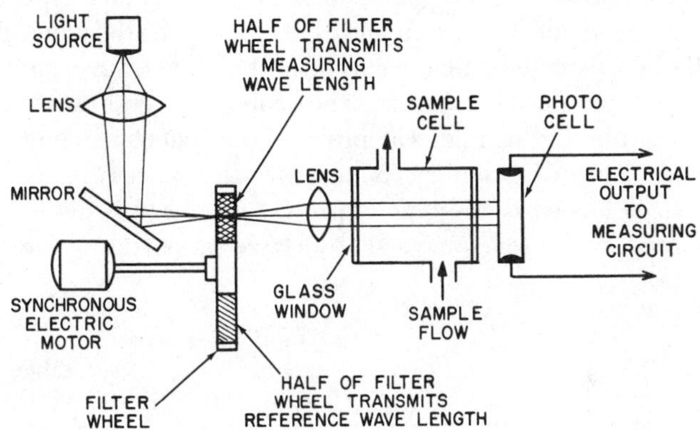

Fig. 11.14 l Schematic representation of the infrared moisture detector

Sampling System

There are no special sampling system requirements for this instrument. Selection of hydrophobic (lacking affinity for water) materials for sampling system components is not critical since the instrument can measure only relatively high moisture content, where moisture contributed to the sample by the sampling system is insignificant in its effect on the measurement.

Limitations

There are several general limitations to the field of application of this instrument. The sample fluid must not be corrosive to the glass sample cell, and it must have some minimum transparency for the measurement and reference wavelengths. In addition, there cannot be any other components present in the sample which will selectively absorb either the measuring or reference wavelength.

Microwave Absorption Hygrometer

The principle of operation of this instrument is the same as that of the infrared absorption hygrometer; namely, selective absorption of electromagnetic energy by moisture in the sample. However, in this case, radiant energy in the microwave length is used. A transmitter provides a microwave beam which can either be transmitted through or reflected by the sample material. The receiving unit senses wave attenuation and phase shift and provides a readout. The unit senses the mass of moisture in the beam path so that the readout is normally in terms of the mass of moisture per unit volume. However, with calibration, readout in weight percent can be achieved.

Several probe configurations are possible depending on the application. Figure 11.14m shows the transmitting and receiving probes welded into a section of pipe. Internally, Teflon contacts the sample. This type of arrangement can be used to advantage on slurries and pastes since there are no obstructions to flow. The units described above measure the attenuation by moisture of a microwave beam in the sample. The reflectance-type unit contains the source and detector in one housing. Rather than detecting the transmitted microwave energy, this design measures the amount of energy reflected by the sample. The probe of the reflectance unit does, however, require contact with the process material. Selection of the type of probe to be used is determined by the reflective and absorptive properties of the sample at operating conditions, and by the physical installation. An advantage of the microwave hygrometer over the infrared unit is that there are no optical windows which may coat and destroy the measurement.

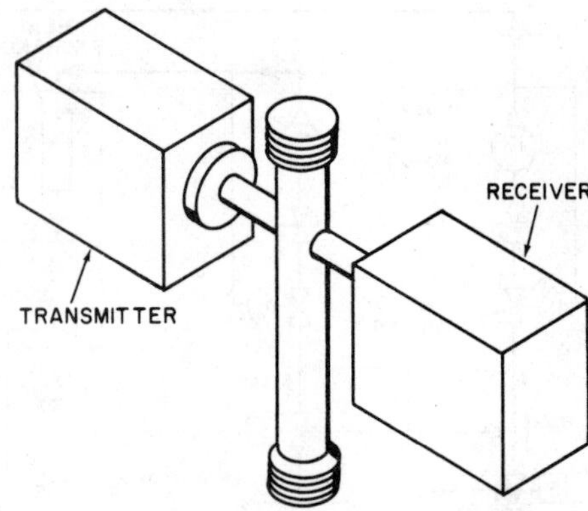

Fig. 11.14m Microwave sensor installed in pipe

Limitations

The unit is not suitable for moisture measurement in gases.

Conclusions

Table 11.14n summarizes some of the important features of the various types of instruments discussed.

Table 11.14n
SUMMARY OF MOISTURE DETECTOR FEATURES

Type	*Range*	*Sample Phase*	*Sample System Required*	*Remarks*
Electrolytic hygrometer*	0–10 to 0–1,000 ppm	Clean gas. Special sampling for liquids	Yes	Sample flow must be constant
Change of capacitance	0–10 to 0–1,000 ppm	Clean gas or liquid	Yes	Sample temperature must be constant
Impedance type*	0–20,000 ppm	Clean gas or liquid	For Liquids	Sample temperature of liquids must be constant
Piezoelectric type	0–5 to 0–25,000 ppm	Clean gas only	Yes	
Heat of adsorption type	0–10 to 0–5,000 ppm	Clean gas or liquid. Special sampling for liquids	Yes	Sample flow must be constant
Infrared absorption	0–0.05 to 0–50%	Liquids and slurries	Yes	
Microwave absorption	0–1 to 0–90%	Liquids, slurries and pastes	No	

*Generally more economical than other types.

BIBLIOGRAPHY

Belkim, H.M., "Factor Affecting Response Characteristics of Electrolytic Instruments for Detecting Moisture," AID/ISA Symposium, Pittsburgh, May 1970.

Cucchiara, O., "The Measurement of Dissolved Water in Organic Liquids Using a Hygrometer," *Analysis Instrumentation*, Vol. 15, ISA, 1977.

Frost & Sullivan, Inc., *On Stream Process Analyzer Market*, Report No. 669, Frost & Sullivan, Inc., New York, August 1979.

Jutila, J.M., "Multicomponent On-Stream Analyzers for Process Monitoring and Control," *InTech*, July 1979.

Mator, R.J., "Trace Moisture Analyzers and Their Calibration," *Analysis Instrumentation*, Vol. 12, ISA, 1975.

Flow Sheet Symbol

TO RECEIVER

MT

11.15 MOISTURE IN SOLIDS

Design Pressure:	Atmospheric
Design Temperature:	Up to 300°F (149°C)
Materials:	Stainless steel for contacting types
Cost:	\$4500–\$11,000 for capacitance, conductance or IR types; \$22,000–\$45,000 for neutron gauges
Inaccuracy:	½–2% of full scale
Range:	2 to 100% of saturation
Partial List of Suppliers:	Anacon Inc.; Foxboro Co.; Kay-Ray, Inc.; LND Inc.; Moisture Register Co.; Ohmart Corp.; Technology Mkt.; Texas Nuclear Div., Ramsey Engineering Co.

The instruments discussed in the following section are specific to moisture measurement in solids, while instruments covered in the previous section were applicable with few exceptions to both liquids and gases. Instruments for continuous moisture measurement in solids are specialized not because of a unique operating principle, but because of the difficulty in handling solids samples. Generally, instruments which infer moisture content from such properties as dielectric constant and conductivity are not as reliable as those that measure moisture directly because these properties are also affected by factors other than moisture. The dielectric constant and the conductivity of solids are also a function of particle size and packing density, parameters very difficult to predict and compensate for.

Nuclear Moisture Gauge

The moderation of neutrons by hydrogen atoms is used as the basis for moisture detection in this instrument. Neutrons are subatomic particles emitted by certain radioactive materials in the process of atomic decay. These particles are electrically neutral and have a mass approximately equal to the mass of a hydrogen atom. Neutrons, because they lack an electric charge, are not deflected by the negative and positive charges associated with atoms, but are only deflected or reflected on impact with the atomic nucleus. Such an impact involves a change of momentum of the neutron and of the impacted nucleus. Since energy and momentum must be conserved, the energy of the neutron after impact is decreased by an amount equal to the energy transferred to the impacted nucleus. The neutron energy after impact is a function of the impacted atom mass. This fact is illustrated below. Let m_1 and m_2 equal the mass of the neutron and the impacted atom, respectively, V_1 and V_3 the neutron velocities before and after impact, and V_2 the velocity of the impacted nucleus which is initially at rest.

$$m_1V_1 = m_2V_2 + m_1V_3 \text{ (conservation of momentum)} \qquad 11.15(1)$$

$$\tfrac{1}{2} m_1V_1^2 = \tfrac{1}{2} m_2V_2^2 + \tfrac{1}{2} m_1V_3^2 \text{(conservation of energy)} \qquad 11.15(2)$$

From Equation 11.15(1)

$$V_2 = \frac{m_1(V_1 - V_3)}{m_2}$$

Substituting V_2 into Equation 11.15(2) and simplifying yields

$$V_3 = \frac{-V_1(m_2 - m_1)}{(m_1 + m_2)}$$

If the mass of the impacted nucleus is equal to the mass of the neutron, all of the neutron's kinetic energy is transferred to the nucleus and V_3 equals zero.

The hydrogen atom, because it is most nearly equal

in mass to the neutron, is the most efficient energy absorber or moderator of neutrons. The number of reflected or moderated neutrons is then a measure of the number of hydrogen atoms present in the sample which relates to its moisture content. However, density also affects the amount of neutrons reflected. A nuclear density gauge is, therefore, included as part of the moisture measurement. The output of the density gauge is used to compensate the neutron moisture measurement to make it independent of sample density variations.

The measuring system contains a high-energy neutron source of plutonium-beryllium or americium-beryllium and a low-energy neutron detector. The density measurement is made by a gamma-ray source and detector arranged to measure reflected or transmitted gamma rays. The moisture and density measurement signals are fed to an electronic chassis where they are scaled to engineering units and presented for readout as a compensated moisture concentration. The sensing unit can be mounted over conveyors, as shown in Figure 11.15a, on bins or on pipelines. The sample volume should be fairly large, 16 in. (400 mm) wide by 2 in. (50 mm) thick as a minimum for conveyors, or 8 in. (200 mm) diameter minimum pipe size. On conveyors where the solids bed may be of nonuniform thickness, a scraper plow or a roller should be installed ahead of the measuring head. When measuring moisture of material in a bin or chute, care must be taken to prevent material build-up on the vessel wall. For solids with a tendency to cake, the vessel wall should be lined with a nonsticking plastic such as Teflon, or a bin vibrator can be installed.

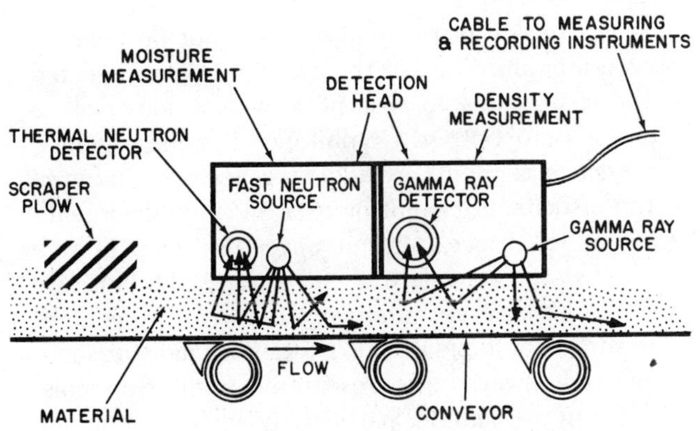

Fig. 11.15a Nuclear moisture measurement

Limitations

Since the instrument is sensitive to all hydrogen atoms, chemically bonded hydrogen in the sample is also seen as water. Only if the quantity of bonded hydrogen in the sample is constant can this portion of the signal be eliminated from the measurement by calibration. However, the amount of hydrogen present as free water must be nearly of the same order of magnitude as the bonded hydrogen in order to obtain a detectable signal change over the moisture variations.

Capacitance Moisture Gauge

The principle involved in the operation of this unit is the same as that described for the change of the capacitance-type hygrometer discussed in Section 11.14. The great difference in dielectric constants of the dry sample and water is utilized to perform moisture detection. Changes of moisture content affect the dielectric constant of the sample, which in turn affects the value of measured capacitance. The variations in measured capacitance are read out as changes in moisture level.

The measuring probe may consist of stainless steel electrodes imbedded in plastic or other insulators to form the plates of a capacitor (Figure 11.15b). The probe is allowed to contact the sample lightly, which is effectively the capacitor dielectric (the nonconducting material between capacitor plates).

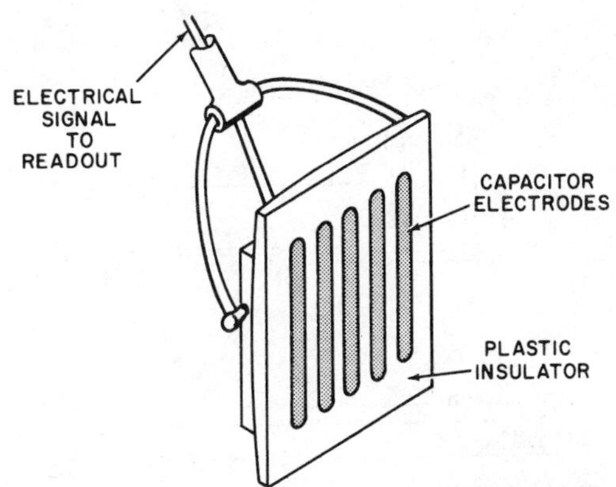

Fig. 11.15b Capacitance-measuring head for moisture detection in solids

Capacitance can be measured by two methods. One utilizes a capacitance bridge analagous to the Wheatstone bridge for resistances. The bridge is excited by a radio frequency alternating current (ac) signal, and the capacitive reactance of the measuring capacitor is continuously balanced by the reactance of a known capacitor through a servomotor. (Capacitive reactance is that part of the impedance of an ac circuit, which is due to capacitance.) Alternately, the measuring capacitor can be connected in parallel with a known inductance to form a parallel resonant circuit. This circuit is powered by a constant frequency source, and the output voltage becomes a function of connected capacitance. At resonance, this output is a maximum and decreases nonlinearly for larger and smaller values of capacitance. The nonlinearity can be minimized by appropriate choice of circuit parameters and by narrowing the operation of the instrument to a limited portion of the resonance curve.

Another type of capacitive element is designed for moisture measurement of powders in conveyors. The capacitor plates are mounted at right angles to each other, as shown in Figure 11.15c. The horizontal electrode rides on the powder surface, and the vertical electrode is submerged. Other variations include gravity-fed and force-fed measuring cells, as shown in Figures 11.15d and e.

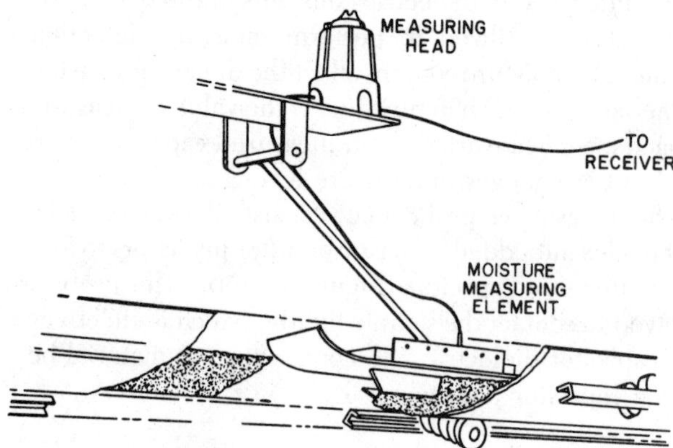

Fig. 11.15c Ski-type probe for solids moisture detection

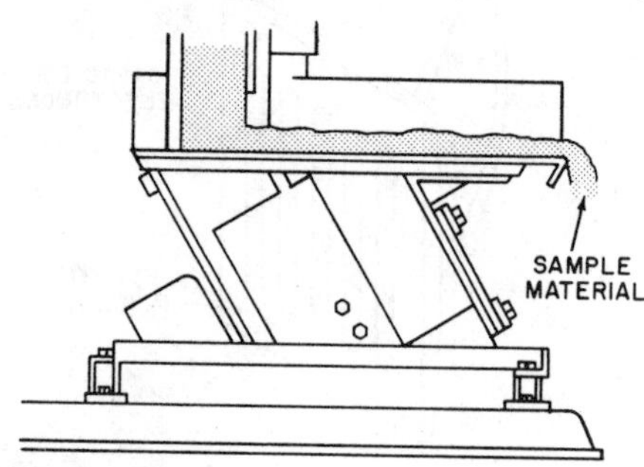

Fig. 11.15d Gravity-fed measuring cell

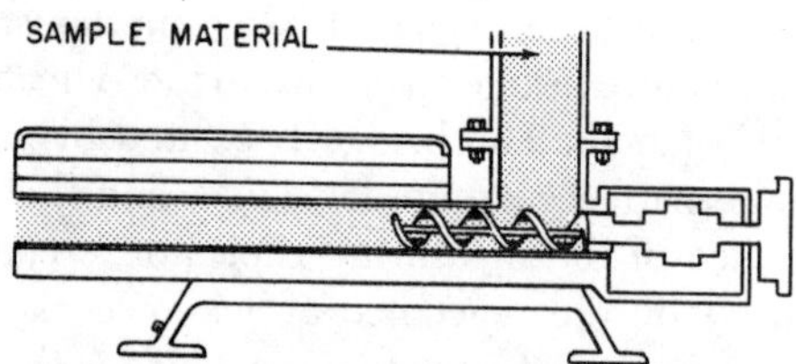

Fig. 11.15e Force-fed measuring cell

Limitations

All capacitance measurements of solids experience the same problem, namely, the measured capacitance is not a function of moisture content alone, but also of such factors as particle size, packing, and material density.

Infrared Absorption Moisture Analyzer

The principle of measurement utilized in this analyzer is the selective absorption of infrared energy by water. However, for moisture measurement in solids, the measuring and reference beams are reflected off, rather than transmitted through, the sample.

The sensing element consists of a light source, filters to pass the measuring and reference wavelengths, optics to direct the light beam onto the sample and to collect the reflected light, and a photocell to convert the reflected light into electrical current (Figure 11.15f). The two current pulse outputs of the photocell, corresponding to the reflected measurement and reference wavelengths, are converted to two dc levels whose ratio is read out as moisture content. The sensing unit is designed for mounting with the light aperture a few inches from the sample.

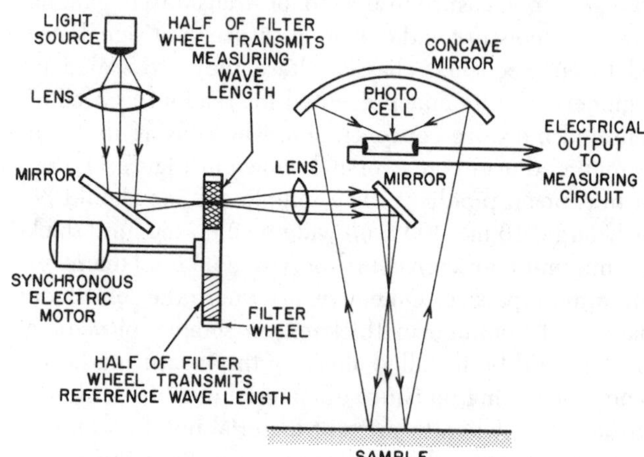

Fig. 11.15f Infrared moisture measurement (reflectance)

Limitations

The reflectivity of the sample must not be below a certain minimum value, or the amount of light reflected and the change due to absorption will be too small to produce a useful output. Graphite, coal, metal powders, and inorganic pigments are all poor reflectors of infrared and the instrument cannot be used in conjunction with these samples. Since infrared waves do not penetrate very far below the surface of the sample, the moisture measurement is representative of the moisture content on the surface of the material. The surface moisture content may not always be representative of the true moisture level in the sample, particularly where the sample is exposed to the ambient atmosphere.

Resistance Moisture Gauge

The electrical resistance of nonconducting solids is influenced by the moisture content of the sample. Instruments that take advantage of this fact use a balanced bridge circuit to detect resistance changes of the process. Measuring electrodes forming a part of the bridge circuit are permitted to contact the sample. A regulated voltage applied to the electrodes maintains a small current flow through the material. Changes in the moisture level of the sample result in a change of electrode current and bridge imbalance. The amount of imbalance is related to the moisture content. Actually the capacitive reactance due to moisture level changes is also detected by the probe. However, this portion of the measurement is continuously balanced by a reference capacitor driven by a servomotor.

Limitations

The measurement is strongly influenced by contact pressure between sample and electrodes, operating temperature, packing density, and particle size. In addition, the values of resistance measurement by the instrument are very high; therefore, good insulation is required at the electrodes to prevent leakage currents from introducing an error into the measurement.

Indirect Methods of Moisture Measurement

Indirect methods of solids moisture measurement involve the detection of moisture in the atmosphere in the proximity of the solid. The atmosphere near the solid is in equilibrium with the moisture content of the sample. Therefore, several of the instruments discussed in Sections 11.10 and 11.14 can be used for this type of measurement. In particular, the change of impedance–type hygrometer is best suited since the sensing element is small and can easily be mounted very close to the solid surface (see Figure 11.15g). However, for this measurement to be effective, the element must be ideally situated where no drafts or convection currents occur because air movement will influence the moisture measurement.

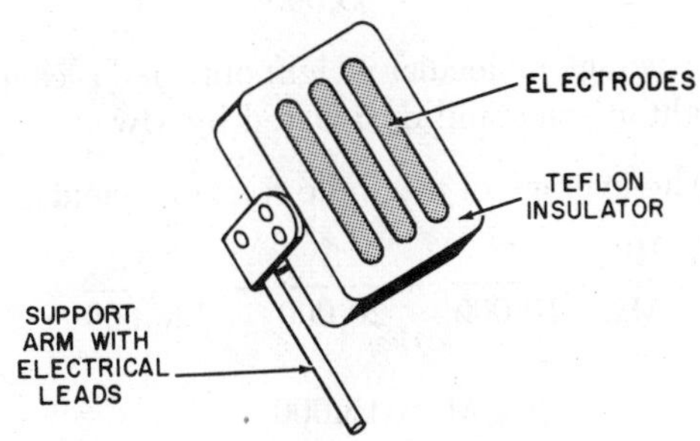

Fig. 11.15g Resistance measuring head

Discontinuous Methods

There are several other methods for moisture determination which involve discontinuous measurement of moisture in the sample. An amount of sample material can be weighed before and after thorough drying in an oven. The loss of weight is then due to the loss of moisture. Or the atmosphere in the drying oven can be monitored by an electrolytic hygrometer whose output integrated over the drying time represents the moisture content of the sample. Still another method utilizes calcium carbide to convert the moisture in a preweighed sample to acetylene. The amount of acetylene evolved is indicative of the moisture content of the sample. None of these methods provides a continuous output and each involves a considerable lag between the time of sampling and the time a measurement is completed.

Conclusions

Of the methods available for moisture measurement in solids, the nuclear moisture gauge represents the most refined and accurate technique. Other methods, however, can also be used, particularly where accuracy is not too great a factor or where the parameters affecting the reading, such as temperature and particle size, are controlled. Capacitance-type instruments can be used effectively on sheet paper, cardboard, and other materials where such factors as particle size and packing density do not enter into the measurement.

BIBLIOGRAPHY

Botsco, R., "Microwave Moisture Measurement," *Instruments and Control Systems*, May 1970.

Frost & Sullivan, Inc., *On Stream Process Analyzer Market*, Report No. 669, Frost & Sullivan, Inc., New York, August 1979.

Gardner, R.C., "Moisture/Basis Weight Infrared Gage for Paper," *InTech*, January 1968.

Jutila, J.M., "Multicomponent On-Stream Analyzers for Process Monitoring and Control," *InTech*, July 1979.

Flow Sheet Symbol

TO RECEIVER

AT Mw

11.16 MOLECULAR WEIGHT

Design Pressure:	Atmospheric
Design Temperature:	300°F (150°C)
Element Material:	Glass, Kel-F, gel-cellophane, stainless steel
Cost:	\$2200–\$34,000, \$67,500 and higher for electron microscope
Inaccuracy:	±5–10%
Range:	Molecular weight of 50 and greater
Partial List of Suppliers:	Advanced Instruments, Inc.; Anacon, Inc.; Bausch & Lomb, Inc.; Hewlett-Packard Co.; PSG Industries, Inc.; Siemens Corp.; Waters Associates, Inc.

This section deals with the various methods for molecular weight determination of polymers and other liquids.

Polymers consist of large molecules formed by the bonding of relatively simple similar parts. The size and shape of the polymer molecules are determined by the number of basic building blocks and how these are linked together. For example, the basic blocks might be arranged in chains of various lengths, in chains with branches, or two chains might be linked in several places. Thus, while the molecular weight of the basic building blocks might be known, polymers are a mixture of molecular species covering a wide range of sizes, and no definite molecular weight can be assigned to them. However, an average molecular weight or a molecular weight distribution can be determined, and these are useful in predicting the physical properties of the polymer. Among the methods available for average molecular weight determination are osmometry, light scattering, viscosimetry, gel-permeation chromatography, end group determination, electron microscopy, centrifuge sedimentation, and diffusion.

Each of these methods yields an average that must be defined, since the averages obtained are not all identical, as is illustrated by the following example:

Consider 10 pounds of polymer consisting of the following proportions:

5 lbs of molecular weight 10,000
3 lbs of molecular weight 20,000
2 lbs of molecular weight 120,000

a. The average of the weight fractions yields

$$10\,\bar{M} = 5 \times 10{,}000 + 3 \times 20{,}000 + 2 \times 120{,}000$$

$$\bar{M} = 35{,}000$$

The average molecular weight obtained is called the weight average and designated by Mw.

b. The average of the mole fractions yields

$$\frac{10}{\bar{M}} = \frac{5}{10{,}000} + \frac{3}{20{,}000} + \frac{2}{120{,}000}$$

$$\bar{M} = 15{,}000$$

This average is called the number average molecular weight and is designated by Mn.

From the above example, it is evident that the averages obtained are not identical; the weight average emphasizes the large molecules and the number average the small ones. For this reason, an average molecular weight must always be defined.

In general, Mw and Mn are defined as

$$\mathrm{Mw} = \frac{\Sigma \mathrm{WiMi}}{\Sigma \mathrm{Wi}} = \frac{\Sigma \mathrm{niMi}^2}{\Sigma \mathrm{niMi}}$$

$$\mathrm{Mn} = \frac{\Sigma \mathrm{Wi}}{\Sigma \mathrm{ni}} = \frac{\Sigma \mathrm{niMi}}{\Sigma \mathrm{ni}}$$

Where ni represents the number of moles, Wi the

weight of the fraction, and Mi the molecular weight of molecules of size i.

Additional averages can be defined as

$$Mz = \frac{\Sigma niMi^3}{\Sigma niMi^2}$$

and

$$Mv = \left(\frac{\Sigma niMi^{a+1}}{\Sigma niMi}\right)^{1/a}$$

where a is defined by the equation

$$[\eta] = K(Mv)^a$$

$[\eta]$ is the intrinsic viscosity of the polymer and K is a constant. The definition of intrinsic viscosity is given later in the section on viscometers.

Osmometers—Membrane Type

Osmotic pressure is defined as that pressure which must be applied to the solution to stop osmosis from the pure solvent. Osmotic pressure, π, is related to solution concentration and molecular weight of the solute as given by the equation

$$\pi = \frac{RT}{M}c + Bc^2 + Cc^3 \qquad 11.16(1)$$

where R is the ideal gas constant, T is absolute temperature, c is the solution concentration, and B and C are virial coefficients.

The first term of the equation is the van't Hoff relationship for ideal solutions. However, polymer solutions are anything but ideal, and, therefore, at least the second term of Equation 11.16(1) can never be neglected. Since there are two unknowns, M and B, in the equation, at least two measurements are required. Equation 11.16(1) is rewritten as

$$\pi/c = \frac{RT}{M} + Bc + Cc^2 \qquad 11.16(2)$$

The measured values of π are plotted as π/c versus c. The measurement points are linearly extrapolated to c = 0 where the terms containing the unknown virial coefficients go to zero. The value of M can then be calculated from the relationship

$$M = \frac{RT}{(\pi/c)_{c=0}} \qquad 11.16(3)$$

The molecular weight M thus obtained in the number average molecular weight Mn.

The membrane osmometer (Figure 11.16a) consists of two compartments separated by a semipermeable membrane. A pure solvent and a dilute polymer solution are introduced into the two compartments, respectively. The membrane acts as a filter to permit passage of solvent molecules, but not polymer. Migration of solvent molecules into the solution results in a change in pressure

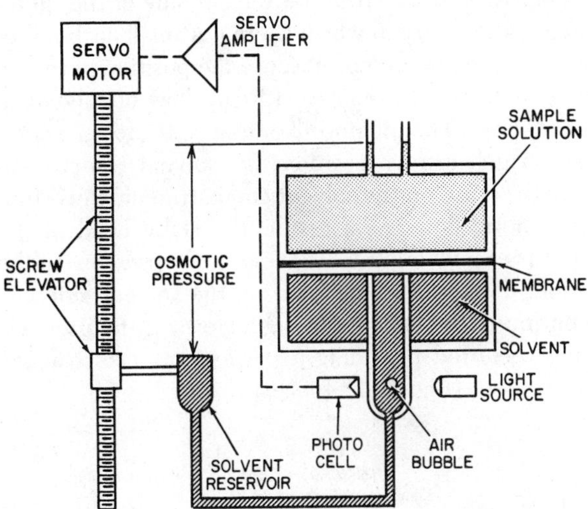

Fig. 11.16a Membrane-type automatic osmometer

between the two compartments until osmotic pressure is reached. Osmotic pressure, defined as that pressure which must be applied to the solution to stop osmosis from the pure solvent, is used to obtain the number average molecular weight. Static osmometers are provided with a capillary connection to the solution compartment so that osmotic pressure can be read as hydrostatic head. The principal disadvantage of the static osmometer is the long time required to reach equilibrium—1 to 2 hours and longer. Automatic osmometers (Figure 11.16b) are available which determine osmotic pressure dynamically; that is, the rate of solvent flow through the membrane is measured as a function of externally applied pressure. One type of automatic osmometer has a solvent reservoir mounted on a screw

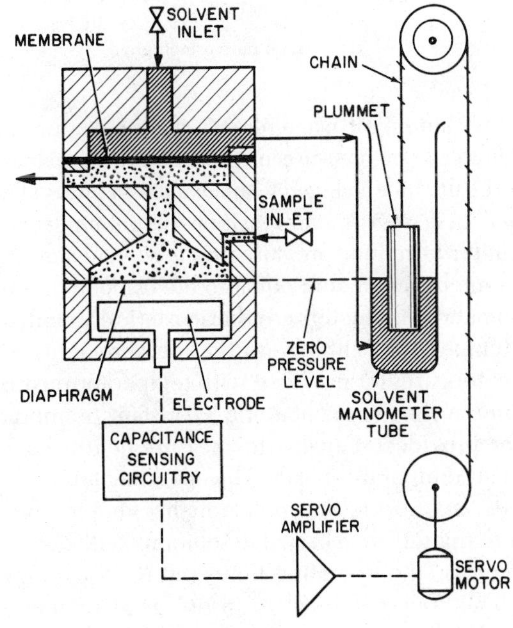

Fig. 11.16b Automatic osmometer

elevator and connected to the solvent side of the membrane by a capillary in which there is a small air bubble. A light source and photocell sense the position of the air bubble as it tends to move with any flow of solvent in the capillary. The photocell output activates a servomotor, which in turn positions the solvent reservoir to reduce pressure on the solvent compartment, thus stopping osmotic flow. As a result, the static head of the solvent is exactly opposed to the osmotic pressure of the solution. The hydrostatic head on the solvent side can be continuously displayed in digital form or the pressure can be recorded providing a permanent record of the measurement (Figure 11.16c).

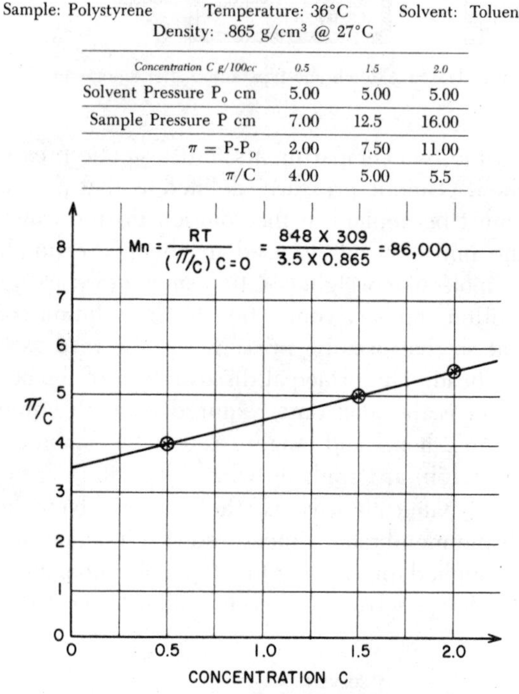

Sample: Polystyrene Temperature: 36°C Solvent: Toluene
Density: .865 g/cm³ @ 27°C

Concentration C g/100cc	0.5	1.5	2.0
Solvent Pressure P_0 cm	5.00	5.00	5.00
Sample Pressure P cm	7.00	12.5	16.00
$\pi = P - P_0$	2.00	7.50	11.00
π/C	4.00	5.00	5.5

Fig. 11.16c Membrane osmotogram

Another automatic osmometer utilizes a diaphragm to sense increase of pressure in the solvent compartment which results in a change of capacitance in an oscillator. A signal to a servomotor positions a plummet in the manometer to reduce pressure on the solvent side of the membrane by an amount equal to the osmotic pressure.

Automatic osmometers are thermostatically controlled since temperature enters into the measurement. Static osmometers are not provided with temperature control, and some means of maintaining a constant temperature must be provided. Usually the measurement is made in a constant temperature bath. The operating temperature depends on the melting point of the sample and the lowest temperature to keep it in solution. With automatic osmometers, the operating temperature is a minimum 10°F (6°C) above ambient to insure good temperature control. However, for a given sample, the lowest possible operating temperature is selected to insure longer membrane life and to inhibit possible corrosion.

Osmometers—Vapor Pressure Type

The vapor pressure of any solvent is lowered by the addition of solute. If two containers of pure solvent and solution are placed in a closed system, pure solvent will evaporate and condense in the solution. The resultant temperature difference, due to the latent heat of the solvent, can be related to the molecular weight of the solute. In practice, the temperature difference is read as the output voltage of a resistance bridge using thermistor or other high-accuracy temperature sensors (see Figure 11.16d).

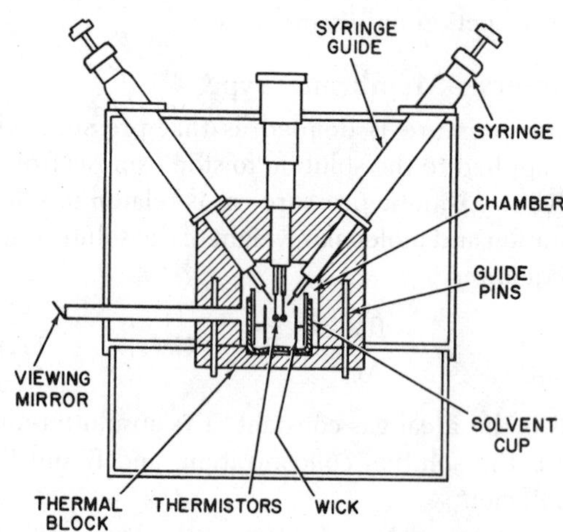

Fig. 11.16d Vapor-pressure osmometer

The relationship between bridge output voltage ΔV, solution concentration c, and average molecular weight Mn is given by

$$\left(\frac{\Delta V}{c}\right)_{c=o} = \frac{K}{Mn} \qquad 11.16(4)$$

where K is the calibration constant for the solvent–thermistor–temperature combination.

Although vapor pressure lowering is only a function of the amount of solute in solution and solute molecular weight, ΔV is a relative quantity depending on the solvent, operating temperature, and temperature-sensing element used. The instrument must, therefore, be calibrated with a known solution. In practice, several values of ΔV are obtained at different concentrations and these are plotted on a graph of ΔV/c versus c (see Figure 11.16e). The graph is extrapolated linearly to c = o and the value of K determined. Once the value of K is known, Mn can be determined in the same manner for any polymer in solution.

The vapor pressure osmometer consists of a thermostated chamber saturated with solvent vapor. Two thermistor beads are suspended in the chamber, and syringe

Sample: Fatty Acid Temperature: 35°C Solvent: Toluene
Calibration Factor K = 350

Concentration C gm/liter	5	7.5	15
1. Output ΔV	1.40	2.15	4.35
2. Output ΔV	1.40	2.16	4.35
3. Output ΔV	1.41	2.14	4.35
ΔV Average	1.40	2.15	4.35
ΔV/C	0.280	0.287	0.290

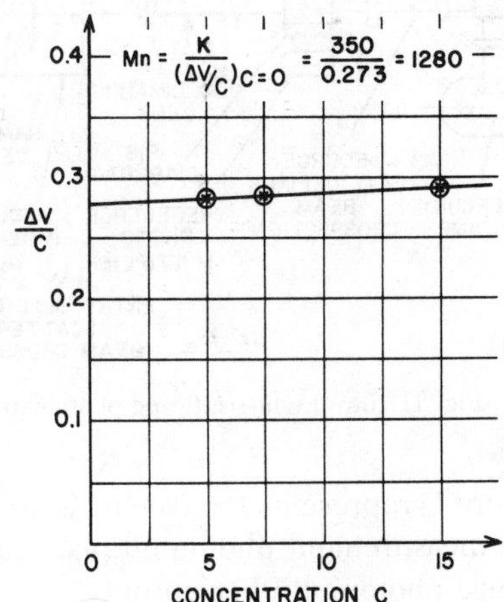

Fig. 11.16e Vapor-pressure osmotogram

guides are built into the chamber. Using syringes, a drop of solvent and a drop of solution are placed on the reference and measuring thermistor, respectively. Solvent condenses on the solution drop, thus warming the measuring thermistor and decreasing its electrical resistance. The thermistors are part of a resistance bridge whose output is in proportion to the temperature difference between measuring and reference thermistor. The resistance bridge output ΔV is plotted on a graph of ΔV/c versus c.

The instrument is limited to low molecular weight polymers because of the small difference in temperature involved. Typically, for a 1 percent solution of 50,000 molecular weight polymer, the temperature difference is in the order of 0.001°F (0.0006°C).

Light-Scattering Photometer

Polymer molecules in a suitable solvent cause scattering of the incident light. From the intensity and distribution angle of the scattered light, the weight average molecular weight, Mw, can be determined. When the molecules are small in comparison to the wavelength used, the scattered light has the same intensity for all angles of observation except for its polarization. For such molecules, a single observation at 90° to the incident light beam is sufficient. If molecular size approaches the wavelength of light used, the light is scattered by portions of the molecule which are widely separated. The light will undergo destructive interference, reducing the intensity of scattering. Destructive interference is a function of angle; that is, interference will be zero in the direction of the incident light and will increase with the angle of observation. The Zimm plot (see Figure 11.16f) utilizes data taken over as large an angular range as possible and extrapolated to zero angle. The ordinate intercept at zero angle and zero concentration is 1/Mw. The ordinate of the Zimm plot is Kc/R_θ and the abscissa is $\sin^2(\theta/2) + kc$ where

$$K = \frac{2\pi^2 n^2 \left(\frac{dn}{dc}\right)^2}{A\lambda^4} \qquad 11.16(5)$$

where

n = index of refraction of the solution
A = Avogadro's number
λ = wavelength of light used
θ = angle of observation
R_B = Rayleigh ratio at the wavelength used for the fluid the sample cell is immersed in (ratio of intensities of scattered light to incident light)

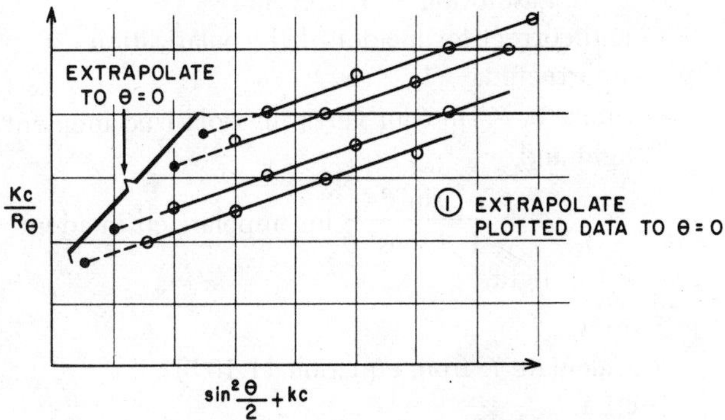

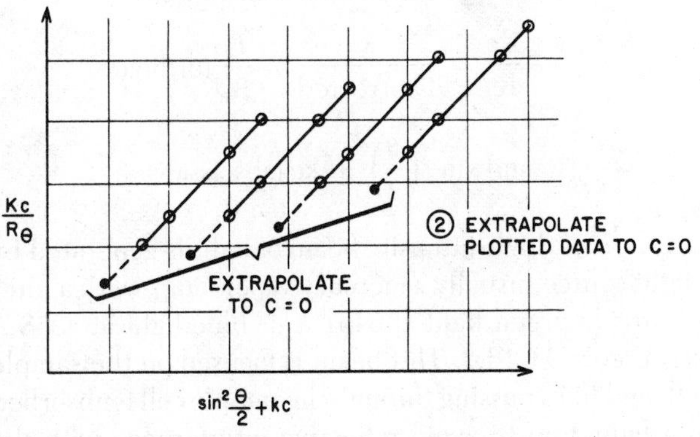

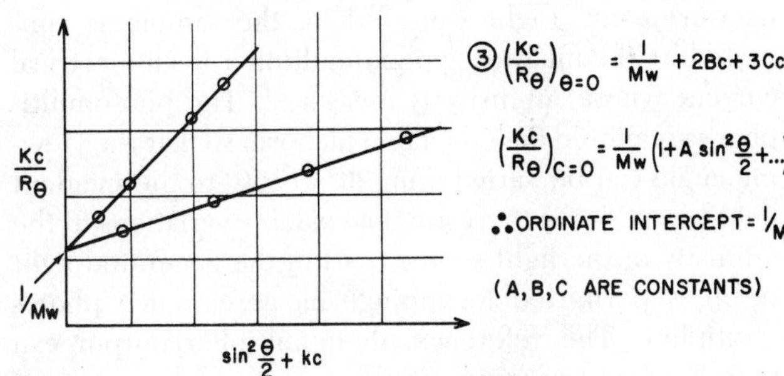

Fig. 11.16f Zimm plot construction

c = solution concentration
k = arbitrary constant chosen to give a convenient spread between data points
$\frac{dn}{dc}$ = change of refractive index with concentration

To obtain the Zimm plot, four concentrations are usually prepared along with pure solvent and the data are collected at a number of angles (10 or more). The scattered intensity for a reference substance (I_S) is measured at 90°. The change in refractive index with concentration (dn/dc) is obtained with a differential refractometer.

Step 1

$I_B = I_S \times$ calibration factor (supplied with instrument)

If the scattered light intensity (I_B) of the material the cell is immersed in is measured, I_B is obtained directly.

Step 2

Correct each measurement for solvent scattering
$I_\theta = I_\theta$ solution $- I_\theta$ solvent
and correct for incident light polarization.
I corrected $- I_\theta \times \alpha$
where $\alpha = \sin \theta$ for vertically polarized incident light and

$$\alpha = \frac{\sin \theta}{1 + \cos^2 \theta} \text{ for unpolarized incident light.}$$

Step 3

Calculate K from equation 11.16(5)

Step 4

Calculate the Zimm coordinates from

$$\frac{Kc}{R_\theta} = \frac{K}{\text{I corrected}} \times \frac{I_B}{R_B} \text{ (ordinate)}$$

$$\text{and } \sin^2\left(\frac{\theta}{2}\right) + kc \text{ (abscissa)}$$

A narrow, high-intensity beam of light is generated by a light source, usually a mercury vapor lamp, with a filter to provide green light at 5461 Å or blue light at 4358 Å (see Figure 11.16g). This beam is focused on the sample cell, and light passing through the sample cell is absorbed in a light trap to avoid reflective interference with the measurement. Light scattered by the sample is converted by the measuring photomultiplier to an electrical current whose intensity is measured. The photomultiplier is mounted on a rotatable platform so that the viewing angle can be varied from 30° to 150° to the incident light beam. Since there are unavoidable variations in the intensity of the light source, part of the generated light beam is permitted to impinge on a reference photomultiplier. The reference photomultiplier output can then be used to compensate the measurement automatically or compensation can be performed as follows: $I_\theta = i_\theta / i_S$ where I_θ represents the true measurement intensity, i_θ the measurement photomultiplier output, and i_S the reference photomultiplier output.

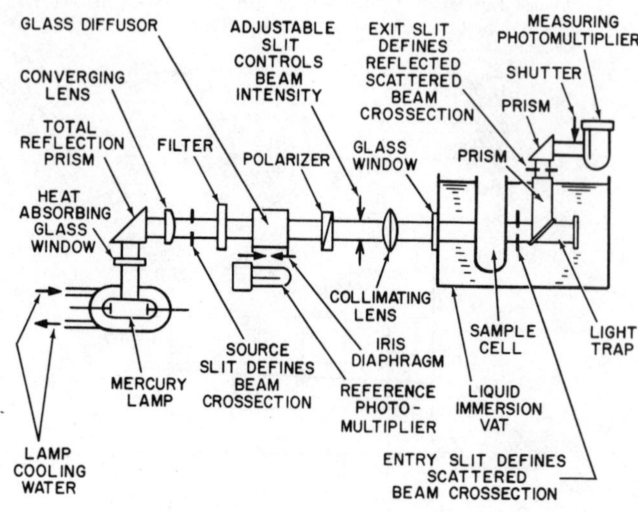

Fig. 11.16g Light-scattering photometer

Variations to the basic instrument include an immersion vat for the sample cell. This vat is filled with a fluid whose refractive index is the same as that of glass to eliminate refraction at the sample cell.

The sample for this instrument must be very carefully de-dusted by filtering to remove all suspended particles. The same holds true for the fluid in the sample vat if such is used.

Viscometer

In order to determine molecular weight of a polymer by this method, the intrinsic viscosity or limiting viscosity must be determined (see Figure 11.16h). Intrinsic viscosity $[\eta]$ is defined by the relationship

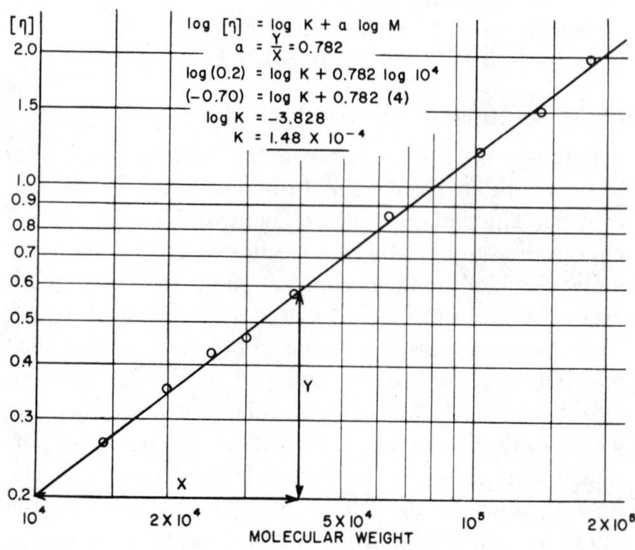

Fig. 11.16h Viscometer calibration for K and a

$$[\eta] = \lim_{c \to 0} \frac{\eta - \eta_0}{\eta_0 c} = \lim_{c \to 0} \frac{\eta_{sp}}{c} \qquad 11.16(6)$$

$$\eta_{sp} = \frac{\eta - \eta_0}{\eta_0} \qquad 11.16(7)$$

where η is the viscosity of the solution, η_0 that of the solvent, and c is the solution concentration in grams per milliliter or grams per deciliter.

Another definition of intrinsic viscosity is:

$$[\eta] = \lim_{c \to 0} \frac{\ln \eta/\eta_0}{c} \qquad 11.16(8)$$

Intrinsic viscosity is related to molecular weight as expressed by the Mark-Houwink equation:

$$[\eta] = KM^a \text{ or } \log [\eta] = \log K + a \log M \qquad 11.16(9)$$

where K and a are constants for a given polymer-solvent system at the temperature of the viscosity measurement.

Viscosity measurements are made of a well-fractionated or monodisperse polymer whose molecular weight is known or has been measured by some other method (see Figure 11.16i). The intrinsic viscosity is calculated by plotting $(\eta - \eta_0)/\eta_0$ versus c and extrapolating linearly to $c = 0$. If the plot of $(\eta - \eta_0)/\eta_0$ versus c does not fall on a straight line, log η/η_0 can be plotted versus c for better linearity.

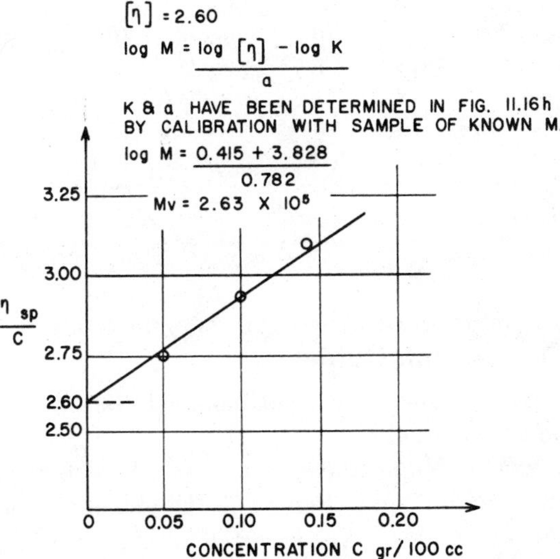

Fig. 11.16i Determination of intrinsic viscosity and average molecular weight

The values of intrinsic viscosity $[\eta]$ obtained are plotted as a function of the known molecular weight on log-log paper and the constants a and K are evaluated. Once K and a are known, Mv for a polydisperse sample can be calculated. The limitation of this method is that the empirical relationship $[\eta] = KM^a$ is valid only for linear polymers.

The various apparatus for viscosity measurement are discussed in Chapter VII.

Gel-Permeation Chromatography

This technique is based on chromatographic separation of molecules by size. A solvent stream is split and polymer sample is added to one half of the stream. The solution is directed into a column packed with a rigid, cross-linked styrene gel. As the polymer moves into the column, the smaller molecules diffuse into the gel pores while the larger ones cannot penetrate, and thus follow a shorter path. Molecules are, therefore, eluted from the column in order of size with the smallest molecules eluting last (see Figure 11.16j).

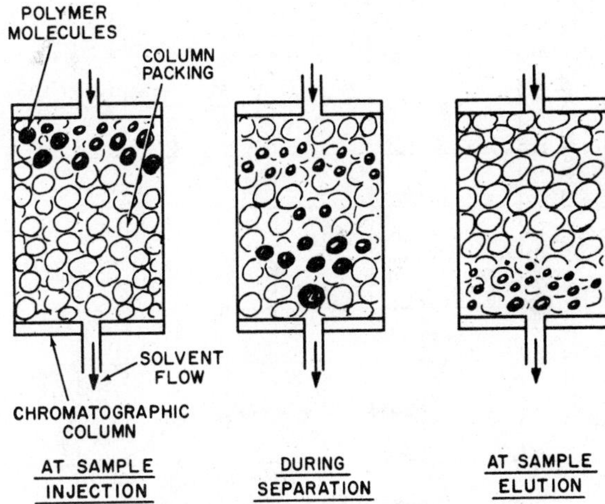

Fig. 11.16j Three stages in the chromatographic separation of polymers

Solution and solvent are passed through the measuring and reference cell of a differential refractometer where the difference in refractive index between the sample and solvent is measured (see Figure 11.16k). The output curve represents the relative abundance of molecules of a particular size from which a molecular weight distribution can be plotted and average molecular weights determined. The differential refractometer utilizes a collimated light beam which is passed through the reference and sample solutions and then reflected back through both solutions by a mirror. The beam is split by another mirror, so that equal amounts of light fall on two photo

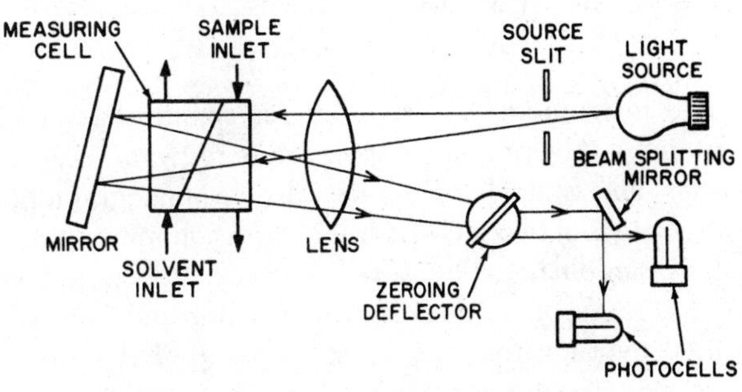

Fig. 11.16k Schematic of refractometer used with gel-permeation chromatograph

sensors. The photo sensor and two resistors form a resistance bridge whose output is in proportion to the difference in the amount of light falling on the photocells. Changes in refractive index of the solution cause the light beam to shift where now unequal amounts of light fall on the two photo sensors. These refractive index variations are related directly to solution concentration. Thus, elution time and resistance bridge output determine the relative abundance of molecules of a particular size. (See Figures 11.16l,m, and o for the method used to determine molecular weight from the output curve.)

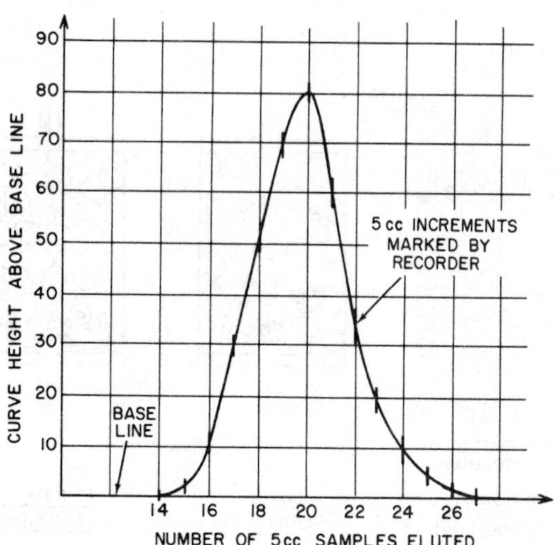

Fig. 11.16 l Chromatograph output distribution curve

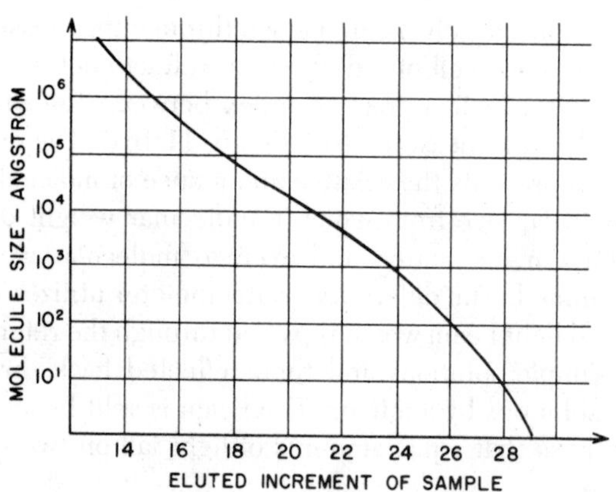

Fig. 11.16m Calibration curve furnished with instrument

The instrument consists of a free-standing assembly containing the chromatograph portion refractometer, recorder, and control electronics. The chromatograph portion consists of two sets of columns with column switching valves, sample injection valves, and solvent loop. Solvent from a reservoir is degassed in a heater and flow controlled by a positive displacement pump. The total solvent flow is split into equal parts; one half is passed through the sample loop, the other half serves as the reference. Both the sample and solvent are passed through two column banks, respectively, before entering the refractometer cell. The solvent is then returned to the reservoir and the sample is discharged to a sample collector. Each 5-ml increment of sample volume discharged is marked automatically on the recorder trace to indicate elution time.

The electronics consist of several subassemblies, including refractometer controls, automatic sample injection control, and the main control assembly.

Table 11.16n
DETERMINATION OF MOLECULAR WEIGHT FROM CHROMATOGRAPH OUTPUT

1	2	3	4	5	6	7
Sample Count	*Height Above Base Line*	*Cumulative Height*	*Cumulative Weight Percent*	*Chain Length (angstrom)*	*Number of Particles*	*Col. 2 × Col. 5*
14	0	374	100.0	3,000,000	0	0
15	1	374	100.0	1,000,000	.0000	1,000,000
16	10	373	99.8	450,000	.0000	4,500,000
17	30	363	97.0	200,000	.0002	6,000,000
18	50	333	89.0	80,000	.0006	4,000,000
19	70	283	75.7	30,000	.0023	2,100,000
20	80	213	57.0	15,000	.0053	1,200,000
21	60	133	35.5	9,000	.0067	540,000
22	35	73	19.5	5,000	.007	175,000
23	20	38	10.2	2,000	.010	40,000
24	10	18	2.7	900	.011	9,000
25	5	8	1.3	300	.017	1,500
26	2	3	0.5	140	.014	280
27	1	1	0.3	30	.033	30
Totals	374				.1071	19,565,810

Determination of Molecular Weight from Chromatograph Output

Prepare a table similar to Table 11.16n, where the columns are as follows:

Column 1: Number each successive 5-ml fraction on the distribution curve (Figure 11.16l) obtained from the chromatograph.

Column 2: Measure the height from the base line to each count on the distribution curve. The base line is drawn by the operator as shown in the diagram.

Column 3: Add successive heights to obtain cumulative height as shown.

Column 4: Normalize Column 3 (0–100) by taking the individual cumulative height and dividing by the total cumulative height and multiplying the result by 100.

$$\left(\frac{\text{Col 3}}{374} \times 100\right)$$

Column 5: Tabulate the chain lengths from the calibration curve of count number versus chain length in Figure 11.16m.

Column 6: Divide Column 2 by Column 5.

Column 7: Multiply Column 2 by Column 5.

The number average molecular weight is obtained by dividing Column 2 by Column 6 totals.

$$\frac{374}{0.107} = 3500 \text{ Å}$$

The weight average molecular weight is obtained by dividing Column 7 by Column 2 totals.

$$\frac{19{,}565{,}810}{374} = 52{,}300 \text{ Å}$$

Figure 11.16o is obtained by plotting Column 4 versus Column 5.

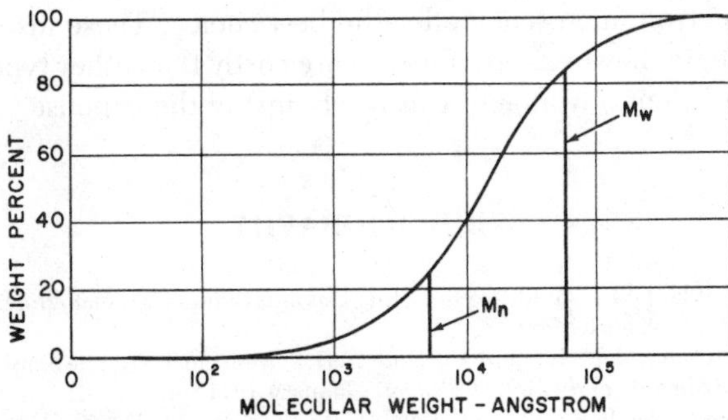

Fig. 11.16o Weight percent cumulative plot

End Group Determination

Molecular weight determination by functional group analysis requires that the polymer molecule contain a known number of distinguishable groups. Such groups are carboxyl and hydroxyl groups or amino end groups. In linear polymers, the number average molecular weight is

$$Mn = \frac{2m}{x_e} \qquad 11.16(10)$$

Where m is the number of pounds of sample and x_e the number of moles of end groups. However, this method can only be used where the number of end groups is known, as is the case with linear polymers. The number of end groups can be determined chemically or by means of an infrared spectrometer by calibration with a known sample. This method fails at molecular weights above approximately 30,000 because the fraction of end groups becomes too small to be detected.

Electron Microscope

The optical microscope is limited in resolving power to objects larger than on half the wavelength of the shortest visible light, or approximately 8000 Å. Electrons, however, not only behave as particles but also as waves, with a wavelength of about 0.5 Å. Consequently, a beam of electrons can be used to detect particles much smaller than those visible by means of the optical microscope. In the electron microscope, a heated filament provides a stream of electrons which passes through a magnetic coil acting as a condensing lens. Upon emerging from the object, the electrons pass through two additional magnetic coils which act analogous to the objective and projector lenses of the optical microscope. The focused electron beam impinges on a photographic film where a much enlarged image is produced. The photographs can be further enlarged without distortion to the point where the magnification is 100,000 diameters. The lower limit of resolution obtainable is in the order of 15 to 20 Å.

The magnifying power of the electron microscope is sufficient to permit visual observation and measurement of polymer molecules. From a photograph of the polymer sample, the size of the molecules can be measured and plotted as a molecular size distribution. Essentially, this distribution curve will be the same as the output curve of the gel-permeation chromatograph; however, the chromatograph automatically counts the number of molecules within each size range, while in the case of the electron microscope, the grouping and counting of molecules must be performed by the operator.

Since the microscope shows only molecular size, the makeup of the polymer molecule—the composition of its building blocks and distance between adjacent blocks—must be known before molecular weight can be calculated from the measured molecular size. One advantage of the electron microscope is that the molecular structure can be observed along with molecular size. Normally, the structure must be determined by independent measurements.

Sedimentation and Diffusion

Sedimentation and diffusion measurements of polymer solutions both involve the frictional properties of the polymer molecules. Therefore, both are closely related and both are needed to interpret the data obtained from the velocity ultracentrifuge. Not all polymers are suitable for analysis by means of the ultracentrifuge; for measurement success, the polymer and solvent must differ in refractive index, the solvent must have a low viscosity, and the polymer must be soluble near room temperature. Finally, mixed solvents must be avoided because of complex corrections that must be made to the measurement.

There are two different types of ultracentrifuge measurements. In the sedimentation velocity method, the centrifuge is operated at rotational speeds which produce centrifugal fields several hundred thousand times the acceleration of gravity. The polymer molecules move in the instrument under the influence of the centrifugal forces against the opposition of frictional forces. The rate of sedimentation is related to the ratio of molecular weight to the frictional force. The frictional force is determined independently by diffusion or viscosity measurements. The fact that a second, independent measurement is required is one of the principal disadvantages of the sedimentation velocity method, since a viscosity measurement alone can also be used to determine the average molecular weight.

In the sedimentation equilibrium method, the centrifuge is operated at lower speeds for periods of days or weeks. Under constant conditions, the polymer will be distributed in the cell solely as a function of its molecular weight. If the measurements are made with thermodynamically ideal solutions, the weight average molecular weight and higher averages can be obtained.

Instrument Construction

The ultracentrifuge consists of an aluminum alloy rotor several inches in diameter and a solution cell mounted within the rotor near its periphery. The rotor is mounted in an evacuated chamber provided with windows for observation of the solution in the cell. The rotor may be driven by air or oil turbine or electric motor. Concentration of polymer along the solution cell is measured as changes of refractive index or light absorption along the length of the cell.

Conclusions

Of the methods available for molecular weight determination of polymers, none is ideally suited for in-line measurement, but gel permeation and viscometry can be applied with automated sampling systems for laboratory and industrial installations. None of these methods produces a direct output in terms of molecular weight. Sedimentation and diffusion methods are not very useful for process applications because of the time required to obtain a measurement. Gel-permeation chromatography, while its analysis time is in the order of 2 to 3 hours, provides a complete molecular weight distribution in addition to molecular weight averages. Automatic membrane osmometers are relatively fast, but their useful range is between 10,000 and 300,000 molecular weight. At the lower end, special membranes can be used to extend the range to about 5000 molecular weight. Vapor pressure osmometers complement the useful range of the membrane osmometer at the lower end of the scale. The operating range of these instruments is up to approximately 20,000 molecular weight. The analysis time and range of the viscometer are comparable to the osmometers; however, its usefulness is limited to linear polymers. In terms of accuracy, range of application, and analysis speed, the light-scattering photometer and the electron microscope offer the best choice. These instruments, however, are much more costly than other types, and their advantages can rarely justify the expense.

BIBLIOGRAPHY

Jutila, J.M., "Guide for Selecting Gas and Liquid Chromatographs," *InTech*, August 1980.

LeRoy, M.J., Jr., and Gorland, S.H., "Molecular Weight Sensor," *Instruments and Control Systems*, January 1971, pp. 81–82.

Mowery, R.A., Jr., and Roof, L.B., "On-line Liquid Chromatography," *Instrumentation Technology*, June 1976, pp. 43–47.

Mowery, R.A., Jr., ""The Use of Process Liquid Chromatography in the Analysis of Industrial Streams," ISA Analysis Instrumentation Symposium, Charleston, W.Va., May 1977.

Flow Sheet Symbol

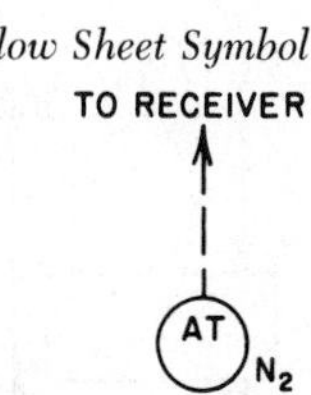

11.17 NITROGEN

Nitrogen Analysis Methods:	1.—Ammonia, (1A). direct Nessler, (1B). direct phenate, (1C). distillation-titrimetric, (1D). distillation-Nessler, (1E). distillation-phenate, (1F). microcoulometric, (1G). gas chromatographic 2.—Nitrite, (2A). sulfanilic acid, (2B). m-phenylene diamine 3.—Nitrate, (3A). phenoldisulfonic acid, (3B). ultraviolet absorption, (3C). brucine, (3D). reduction, (3E). specific ion electrode 4.—Total, (4A). Kjeldahl, (4B). reductive pyrolysis-microcoulometric detection, (4C). reductive pyrolysis-gas chromatographic detection, (4D). automated Kjeldahl
Partial List of Suppliers:	Beckman Instruments, Inc. (3B); Envirotech Corp., Dohrmann Div. (1F, 4B); Orion Research, Inc. (3E); Technicon Industrial Systems (4D); Varian Associates (1G, 4C)

Environmental Significance of Nitrogen Analyses

Although the air is 79 percent nitrogen, the supply of nitrogen to plants and animals is limited by its availability in the usable chemical compounds. Atmospheric nitrogen is an inert gas which relatively few organisms can convert to usable forms such as ammonia, nitrate, and nitrite. Until the advent of large-scale industrial fixation processes for nitrogen, the major problem with nitrogen was its limited availability. The amount of nitrogen fertilizers produced since 1950 has increased many-fold to an annual world output of about 30 million tons (27.2 million tonnes) in 1968, and it is estimated that production may exceed 100 million tons (90.7 million tonnes) by the year 2000.

Since not all of the nitrogen fertilizer is retained in the soil long enough for its intended use, several environmental problems have arisen in recent years. Nitrogen compounds in streams and lakes result in an increased growth of algae and other aquatic plants, which in turn die, and their decomposition depletes the available oxygen for oxygen-dependent organisms. The rapid eutrophication of lakes is at least partly owing to excess nitrogen compounds in the runoff.

Another process, termed nitrification, in which ammonium ion is oxidized to nitrate or nitrite, can also create problems in the aquatic environment by depleting the available oxygen. Ammonium ion in normally alkaline waters is in equilibrium with free ammonia, which is toxic to fish and other aquatic forms. Microorganisms decompose organic nitrogen compounds by a process called ammonification, which converts organic nitrogen compounds such as amino acids to ammonium ions. The ammonium ions created add to the problems described.

Another process which is part of the nitrogen cycle, denitrification, converts the nitrates or nitrites produced by nitrification to inert molecular nitrogen. Figure 11.17a shows the nitrogen cycle. Nitrate concentration, if higher than 45 mg per liter in drinking water, causes an illness in infants called methemoglobinemia. The problems caused by nitrogen compounds in water have made it necessary to analyze for nitrogen in its many forms or oxidation states. The oxidation states of nitrogen are listed in Table 11.17b.

Ammonia Nitrogen[1,2]

The direct Nessler method is sensitive under optimum conditions to about 0.02 mg per liter ammonia nitrogen. Due to interferences such as hardness, iron, acetone, organic amines, and aldehydes, most water samples require a preliminary cleanup distillation. The Nessler reagent produces a yellow to brown color which can be compared visually to standards or can be read against a

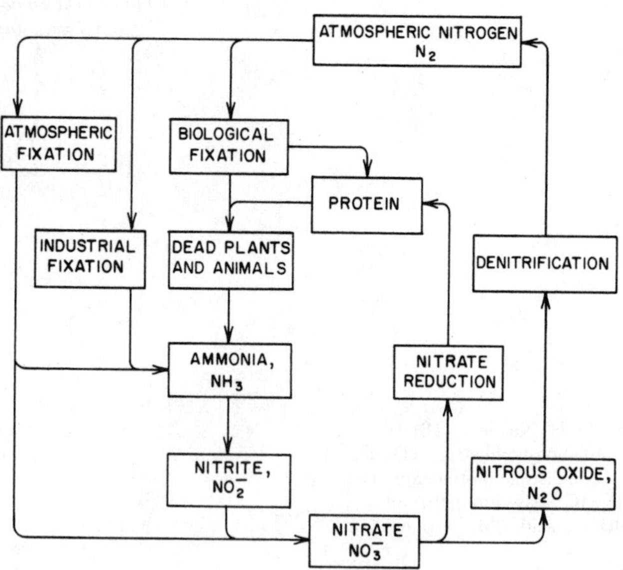

Fig. 11.17a Nitrogen cycle

Table 11.17b
OXIDATION STATES OF NITROGEN

Oxidation State	Name	Symbol
+5	nitrate ion	NO_3^{1-}
+3	nitrite ion	NO_2^{1-}
0	nitrogen	N_2
−1	hydroxyl amine	NH_2OH
−3	ammonia	NH_3

standard curve photometrically between the wavelengths of 400 and 500 nm (1 nm = 1 mμ = 10 Å), dependent on the photometer light pathlength chosen and on the concentration of the sample.

The direct phenate method, which uses the reaction of ammonia, hypochlorite, and phenol catalyzed by manganous ion to produce an intense blue color (indophenol), is subject to interferences from acidity and alkalinity. The blue color has a maximum absorbance at the wavelength of about 630 nm. Because of interferences cleanup distillation is desirable if subsequent determinations are to be made for other forms of nitrogen.

The average relative standard deviation for the sample concentration range of 0.2 to 1.5 mg per liter ammonia nitrogen by the above methods has been reported as 20.3 percent for the direct Nessler and 27.0 percent for the direct phenate methods. After cleanup distillation, both methods have an average standard deviation of 13 percent. If the sample is distilled from a phosphate buffered at a pH of 7.4 into a standard boric or sulfuric acid absorbent solution, the excess acid can be titrated with a standard base to determine the ammonia nitrogen. The relative standard deviation for this method, however, averages about 40 percent.

For certain water samples, it may be useful to distinguish the "free" ammonia nitrogen, distilled from a neutral solution (pH 7.4), the "fixed" ammonia nitrogen, distilled from solution at pH 10.5 or greater, and the "albuminoid" nitrogen obtained by conversion with permanganate and alkaline distillation. Ammonia concentration can be rapidly determined by microcoulometry or by gas chromatography.

Nitrite

Nitrite, an intermediate stage in the nitrogen cycle, can result from the decomposition of natural protein. Another source of nitrite ion in water is its use as a corrosion inhibitor.

Nitrite analyses by standard methods are based on the diazotization of sulfanilic acid with nitrite in strong acid followed by coupling with alpha naphthyl amine hydrochloride. The purple color formed can be read on a photometer at a wavelength of 520 nm or can be compared visually to standards. The best concentration range for the use of this method is below 2 mg per liter, but higher concentrations can also be determined by appropriate dilution with distilled water.

An alternative method for the higher concentration ranges uses the yellow-brown color of the reactant of nitrite with metaphenylene diamine in acid solution. Both these methods are pH dependent but otherwise suffer few significant interferences except for chlorides when present in concentrations greater than 15,000 mg per liter. Many diazotization reactions have been proposed and used to eliminate or minimize these interferences. A critical review by Sawicki[3] compares 52 methods for photometric nitrite analyses.

Nitrate

Phenol disulfonic acid reacts with nitrate to produce a yellow color which can be measured at the maximum wavelength absorption between 410 to 480 nm, depending on sample concentration range and on colorimeter cell pathlength. Chlorides—even at 10 mg per liter—represent a severe negative interference. Silver ion precipitation of chlorides has been used, but excess silver ion causes a brown precipitate which interferes with the measurement. Nitrites, at concentrations greater than 0.2 mg per liter, are a positive interference which can be removed with sodium azide.

The ultraviolet absorption at 220 nm wavelength can be used to measure nitrate. Chlorides represent no interference, but organic matter, which absorbs at or near the same wavelength, causes a positive variable interference. Hexavalent chromium and nitrite ions also interfere. Empirical correction for the normal interferences has been used successfully.

In the brucine method, nitrate reacts with brucine sulfate in a glacial acetic and dilute sulfuric acid mixture. The rate and intensity of color change with time and with temperature, so that simultaneous development of the color to its maximum is also necessary for the calibration standards. At concentrations below 1 mg per liter, results have significant negative bias. Chlorides, above about 1000 mg per liter, and nitrites both interfere with the measurement.

Nitrate has been analyzed by reduction to ammonia or nitrite, which is then analyzed by one of the several methods previously described. Several different metals and alloys have also been used to determine nitrate and nitrite, but ferric and fluoride ions interfere with the analysis. An infrared method has also been reported for determining both nitrate and nitrite simultaneously, but it requires extensive concentration and cleanup procedures.

A specific ion electrode for nitrate is available which is sensitive to less than 1 mg per liter nitrate ion, but interferences due to the normal ranges of chloride and bicarbonate ion activities make the results far from acceptable. However, for water systems in which these interferences are absent or for higher nitrate (>10 mg per liter) concentrations, the electrode is rapid and convenient. The electrode's liquid ion exchanger deteriorates with time, causing a significant error if it is not recalibrated frequently and rejuvenated at least monthly. Although Cl^{1-} and HCO_3^{1-} ions can be removed by precipitation, the procedure is difficult and time-consuming. Automated systems have been developed for the several reduction and determinative colorimetric reactions.

Total Nitrogen

The total or Kjeldahl nitrogen standard method determines free ammonia and organically-bound nitrogen in the −3 valence state but does not determine nitrites, azides, nitro, nitroso, oximes, or nitrates. Organic nitrogen is determined by subtracting the separately determined free ammonia nitrogen from the total nitrogen. Several methods based on reductive pyrolysis determine total nitrogen by microcoulometry or gas chromatography. An automated Kjeldahl has also been developed.

REFERENCES

1. *Standard Methods*, 13th ed. American Society for Testing and Materials, Philadelphia, 1971, pp. 221–247.
2. *1971 Annual Book of ASTM Standards, Part 23*, American Society for Testing and Materials, Philadelphia, 1971; D 992-71; D 1254-67; D 1426-71.
3. Sawicki, E., Stanley, T.W., Pfaff, J., and Amico, A.D., *Talanta*, Vol. 10, 641, 1963.

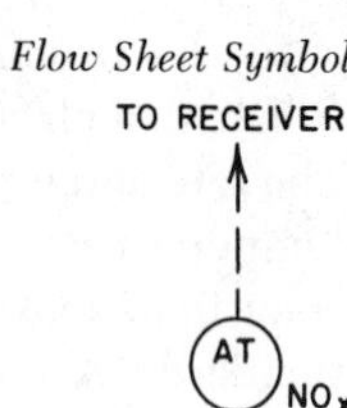

11.18 NITROGEN OXIDES

Method of Analysis:	A—Colorimetric B—Coulometric C—Chemiluminescent D—Gas Chromatography E—Electrochemical
Reference Method:	Applied to integrated samples collected in alkaline solution (A)
Costs:	\$4500 to \$11,200 (A, B, E); over \$22,500 (C, D)
Partial List of Suppliers:	AeroChem Research Labs, Inc. (C); Atlas Electric Devices Co. (A); Beckman Instruments, Inc. (A, B); Dynasciences Corp., Instrument Systems Div. (E); Environmetrics, Inc. (E); Intertech Corp. (E); Mast Development Co. (B); Monitor Labs, Inc. (A); PSG Industries, Inc. (A); Scientific Industries, Inc. (A); Technicon Corp. (A); Wilkens-Anderson Co. (A)

Calibration Methods

Dynamic calibration of a gas analyzer is always a better check on the reliability of the system than static calibration, but it is also much more difficult. Nitric oxide (NO) is oxidized to nitrogen dioxide (NO_2) by air, especially rapidly at high concentrations, and NO_2 condenses and dimerizes at high concentrations. For these reasons, dynamic calibration for NO and NO_2 requires great care. In colorimetric or coulometric analysis, the NO is in fact not measured directly at all but is measured after oxidation to NO_2, so that calibration is made only for NO_2.

Dynamic calibration for NO_2 requires preparation of a sample of inert gas containing a known concentration of NO_2. This may be accomplished by gas-dilution techniques (with special precautions in handling NO_2), gravimetrically, electrolytically, or by use of a permeation tube. The permeation tube (as used with hydrocarbons) is described in Section 11.11. NO_2 permeation tubes are moisture-sensitive, however. They should be protected from moisture in storage and only dry diluent gases should be used. If these precautions are observed, the permeation tube is probably the most convenient method of dynamic calibration for NO_2.

Static calibration is carried out with standard solutions of nitrite. Since this procedure does not involve the components of the gas sampling line, it is not as complete a test as is dynamic calibration, but is much simpler. The stoichiometric ratio of nitrite to NO_2 is usually not unity but is a constant value under controlled conditions. In the Griess-Saltzman method (although there has been controversy), the consensus is that 0.72 mole of nitrite gives the color of 1 mole of NO_2; and in the Jacobs-Hochheiser method, this factor has been found to be 0.63.

NO–NO_2 Combination Analysis

In continuous analysis, it is customary to determine both NO and NO_2 as NO_2 (Table 11.18a). The NO is oxidized to NO_2 by means of potassium permanganate or dichromate or by chromium trioxide, in various formulations. The efficiencies of conversion seem to depend on the length of service and, for the chromium oxidizers, on humidity. Aqueous permanganate seems to be the best choice even though it may not be completely efficient (hence NO may be underestimated).

In series analysis for NO and NO_2 (Figure 11.18b), the air passes through an NO_2 analyzer for measurement and removal of NO_2, then through an oxidizer to convert NO to NO_2, and finally through a second NO_2 analyzer. The second analyzer gives a measure of NO concentration. In one type of parallel analysis, two equal airstreams are analyzed for NO_2, one of them after passage through an

Table 11.18a
NITROGEN OXIDE ANALYZERS

General Method	Type	Advantages	Disadvantages
Colorimetric	Griess–Saltzman	Precise, thoroughly tested, widely used, continuous analysis	Short life of collected sample, sensitive reagents, NO oxidation required
	Jacobs–Hochheiser	Precise, stable after collection	Not adapted to continuous analysis, sensitive reagents, NO oxidation required
Coulometric		Simple apparatus, continuous analysis	Sensitive to other oxidants, NO oxidation required
Chemi-luminescent		Dry gases only, sensitive photometry, continuous analysis	Requires ozone generator NO_2 catalytic reduction
Gas chromatography		Specific, frequent analysis	Not a developed instrument, expensive and complex
Electrochemical		Simple apparatus, continuous analysis	Sensitivity not high, NO oxidation required

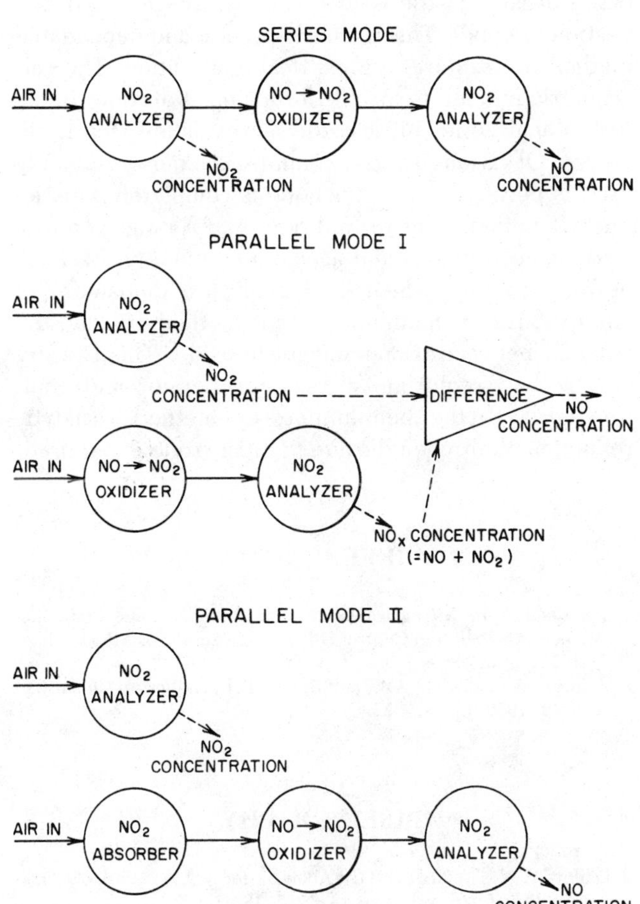

Fig. 11.18b Nitrogen oxides analyzer modes

oxidizer. The latter gives total oxides of nitrogen (NO + NO_2) from which NO is found by difference from the other parallel analyzer. In a second type of parallel analysis, one stream is analyzed directly for NO_2; the second is scrubbed free of NO_2 by passage through Ascarite, followed by oxidation of NO to NO_2 and NO_2 analysis.

Colorimetric Determination

There are two important colorimetric methods[1] for NO_2 determination, the Griess-Saltzman and the Jacobs-Hochheiser. The first is used in most continuous colorimetric NO_2 analyzers. It is based on the reaction of NO_2 with sulfanilic acid to form a diazonium salt which couples with N-(1-napthyl)-ethylenediamine dihydrochloride to form a deeply colored azo dye. Air is passed into the reagent solution for not over 30 minutes, time is allowed for development, and the color is measured at 550 nm. The range of concentrations measurable is 0.02 to 0.75 ppm. In manual use, the color is developed for 15 minutes and should be read within 1 hour (on a colorimeter or spectrophotometer). In a continuous analyzer (Figure 11.18c), the gas and liquid flow rates are adjusted for optimum response and the developed color is read in a flow cell photometrically, using a 550-nm filter. Response times are usually 5 to 15 minutes.

The Jacobs-Hochheiser method is the standard reference method for U.S. National Air Quality Standards. The reason for this is that the Air Quality Standard is an annual average; and this method allows collection of up to 24-hour integrated samples and delays in analysis of at least 2 weeks, whereas Griess-Saltzman samples must

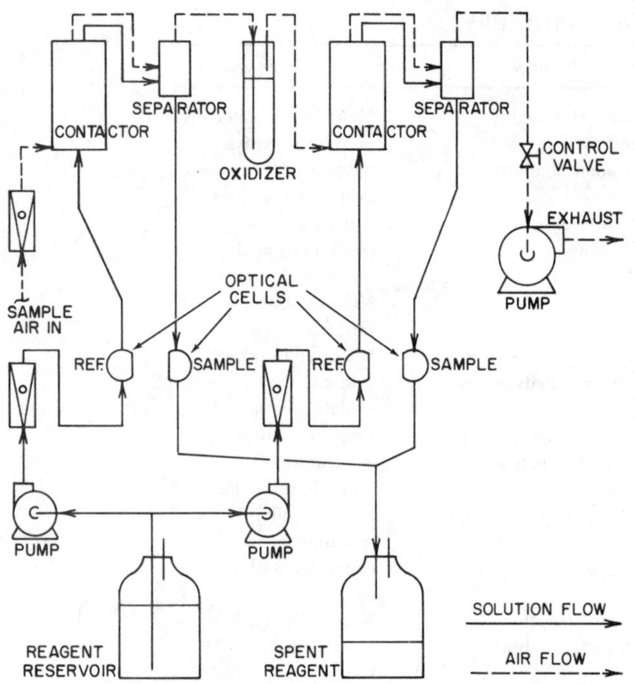

Fig. 11.18c Colorimetric nitrogen oxides analyzer

be quickly analyzed. In the Jacobs-Hochheiser method, the air is passed through aqueous sodium hydroxide, converting NO_2 to nitrite ion. The solution is freed of possible sulfur dioxide by treatment with hydrogen peroxide and acidified. The rest of the procedure is the same as for the Griess-Saltzman method except that sulfanilamide is used rather than sulfanilic acid. Efficiencies found with this procedure in the U.S. National Air Surveillance Network are approximately 35 percent.

Chemiluminescent Determination

A recent development in nitrogen oxides[2] determination makes use of the fact that nitric oxide (NO) reacts with ozone to form nitrogen dioxide (NO_2) with chemiluminescence:

$$NO + O_3 \longrightarrow NO_2 + hv \text{ (light } 0.6 - 3\ \mu) \qquad 11.18(1)$$

The light emitted can be measured photometrically as an indication of the extent of reaction. With excess ozone, the emission is in proportion to the amount of nitric oxide. Instruments based on this operating principle are available commercially. For determination of NO_2 (which reacts with ozone rather slowly), some analyzers include a catalytic converter to reduce NO_2 to NO, which can be determined directly. The apparatus is shown schematically in Figure 11.18d.

Gas Chromatography

There have been limited reports of methods for analysis of oxides of nitrogen by gas chromatography, but no commercial instrument has been marketed. Such methods involve fairly short columns, as a rule, with electron capture detectors. These are the same types of instruments used for analysis of peroxyacyl nitrates.

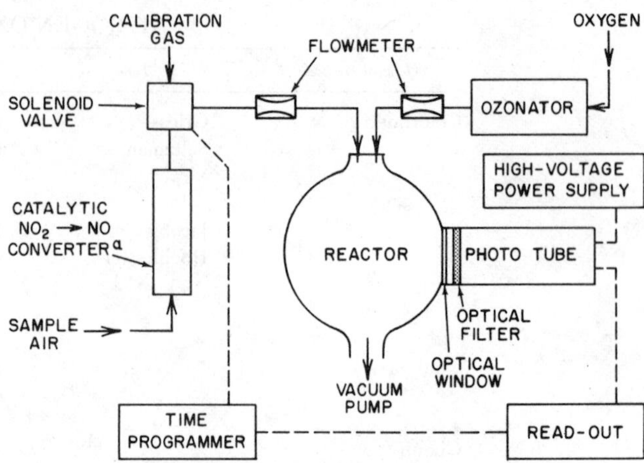

Fig. 11.18d Chemiluminescence nitric oxide analyzer

Assessment

The conventional method in widest use is colorimetric, based usually on the Griess-Saltzman reagent (a diazotization method). This is a fairly precise and dependable method but requires a great deal of attention. The colorimetric method is specific for nitrogen dioxide. In order to analyze for nitric oxide, an oxidation step is required. Of various oxidizer columns, the most commonly used is permanganate, but none is completely satisfactory. It is important to calibrate over a range of nitric oxide concentrations and at various humidity levels. A newer technique which is probably less demanding of attention is the chemiluminescent method utilizing the reaction between ozone and nitric oxide. This is a dry gas method requiring an ozone generator and compressed gas. In the chemiluminescent method, a catalytic reduction of nitrogen dioxide to nitric oxide is required.

REFERENCES

1. *Air Quality for Nitrogen Oxides*, U.S. Environmental Protection Agency, Air Pollution Control Office Publication No. AP-84, Chap. 5, 1971.
2. Fontijn, A., Sabadell, A.J., and Ronco, R.J., *Analytical Chemistry*, Vol. 42, 1970, pp. 575–579.

BIBLIOGRAPHY

Dailey, W.V., "A Novel NDIR Analyzer for NO, SO_2 and CO Analysis," *Analysis Instrumentation*, Vol. 15, ISA, 1977.

Turner, G.S., "Design and Performance of an Ambient Level NO/NO_2/NO_x Monitor," *Analysis Instrumentation*, Vol. 12, ISA, 1974.

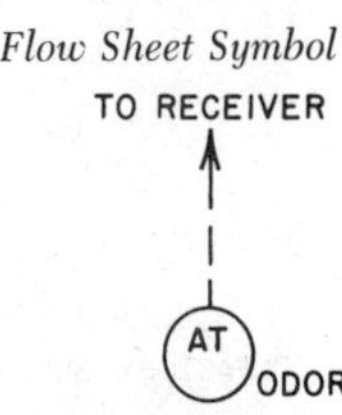

11.19 ODOR

Methods of Detection:	A—Organoleptic B—Chemical/instrumental
Sensitivity of Detection:	0.01 ppm by instrumental methods, 0.0002 ppm by the human olfactory system
Partial List of Suppliers:	Analytical Instrument Developments Inc.; Beckman Instruments, Inc.; Carle Instruments, Inc.; Gow-Mac Instrument Co.; Hewlett-Packard Co., Hewlett-Packard, Avondale Div.; Leeds & Northrup Co.; Mine Safety Appliances Company; Perkin-Elmer Corp.; Tracor Inc.; Varian Associates Aerograph Instruments Div.; Wemco Instrumentation Co.

This section compares organoleptic and chemical/instrumental methods for odor measurement and indicates the general shortcomings of the chemical/instrumental methods in comparison with the reference organoleptic method.

The organoleptic methods, which utilize the human olfactory system, are completely subjective. However, there are techniques available that can convert subjective measurements into useful objective results. Chemical/instrumental methods generally suffer from two major shortcomings:

1. Sensitivity. The human olfactory system is generally three orders of magnitude more sensitive than currently available chemical/instrumental methods. It can detect and identify odors present in quantities to which commercially available instrumentation and chemical methods are completely insensitive.
2. Flexibility. The human olfactory system is capable of detecting and identifying a wide variety of chemical structures and giving different responses to different materials. Commercially available instrumentation and chemical methods are generally restricted to particular chemical structures and give a similar response to all compounds with that structure.

The most successful instrumental methods for the measurement of odor are those using a gas chromatograph. A gas chromatography method has been developed using a flame photometric detector which can determine the concentrations of sulfur dioxide, hydrogen sulfide, and other odorous gases produced by kraft paper mills down to about 0.01 ppm.[1] Success in detecting these sulfur-containing compounds using a coulometric cell with platinum electrodes as the detector has also been reported.[2] Levels as low as 0.1 ppm were detected by this method. The threshold levels detected by the human nose[3] are as low as 0.00021 ppm (trimethylamine). Thus, it is apparent that except for special situations, the human nose, with its attendant sociological, psychological, and physiological complications, is the only odor sensor available for general odor measurement.

The Human Olfactory System

The human olfactory system actually involves more than the nose. As a stream of air is drawn in through the nostrils, it is warmed and filtered by passing over the three baffle-shaped turbinate bones in the upper part of the nose (Figure 11.19a). Some of the air swirls past the olfactory receptors located high up in the nasal passages just below the brain. These odor receptors consist of hairlike filaments attached to the end of the fibers of the olfactory nerve and the trigeminal nerve endings (Figure 11.19b). On being stimulated by odorous materials, these receptors send signals to the olfactory bulb, where they are relayed to higher centers of the brain. In these higher centers, the signals are integrated and interpreted in terms of the character and intensity of the odor.

Because of the extreme sensitivity of the human nose, the concentrations of odorants in air can be extremely

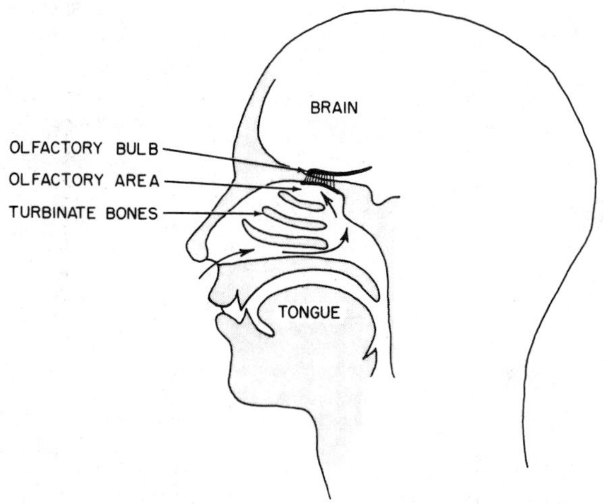

Fig. 11.19a The human olfactory system

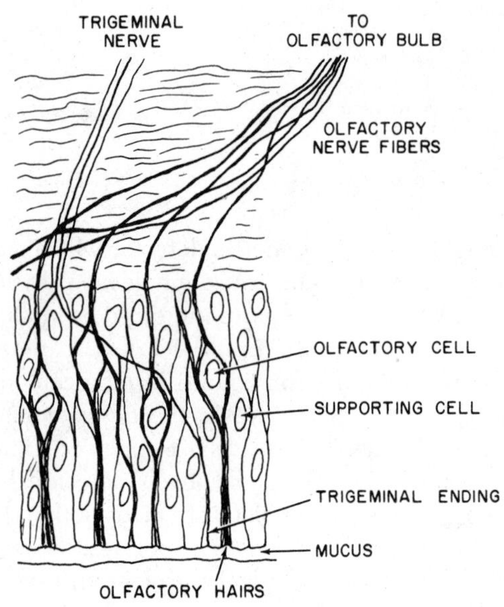

Fig. 11.19b Section of an olfactory epithelium

low yet produce a strong response. These low concentrations present a problem in the manner by which samples are presented to the human sensor. Improper handling of an ambient air sample or improper preparation of an odorant standard can result in erroneous results because of adsorption of the odorants on the walls, incomplete mixing of an odorant with dilution air, or impure dilution air. Errors can also result from the type of system used to bring the sample in contact with the olfactory sensors. The use of syringes or other means in which the nose is not immersed in the sample yields lower apparent intensities than are obtained by fully exposing the nose.[4] The use of an odor room can also yield misleading results because of natural odors generated by the body and adsorption by clothing. The best approach appears to be the "sniff box" shown in Figure 11.19c. The human sensor is kept in an air-conditioned room

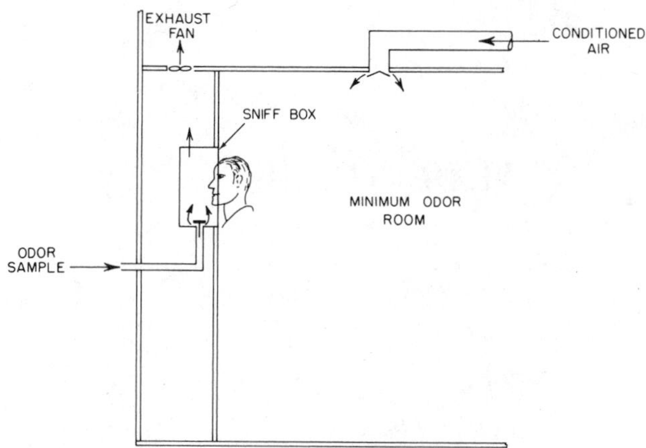

Fig. 11.19c Diagram of sniff box odor test system

into which charcoal-filtered air is blown to maintain a slight positive pressure. This prevents the sample in the sniff box and other contaminated air from entering the room. For measurement, the whole face is placed in the sample stream within the sniff box.

Odor Panels

The use of humans as measurement instruments introduces a variability that is difficult to control. This variability is due to moods, biases, and other vagaries associated with this complicated "sensor." Each human being is a unique creation, with differences in intelligence, persistence, sensitivity, experience, and interests. These differences make the response of an individual an unreliable indicator in itself. The use of a panel of individuals for the measurement of odor levels statistically eliminates the unreliability of the individual.

An odor panel generally consists of 5 to 10 people. The people who compose the odor panel need not have unusual olfactory abilities. However, they should be able to distinguish among odors of different intensities, discriminate among different odor qualities, and communicate the perceived sensations in terms of reference standards. They should also be emotionally receptive to making quantitative and discriminatory judgments without expressing their preference.

Training of an Odor Panel

The primary tasks of an odor panel are the following:

1. Judge the relative intensity of an odor at different dilutions.
2. Discriminate among the different odor qualities.
3. Combine the intensity and quality of an odor to give a composite profile that can be communicated to the scientist in charge of the test.

The panelists are expected to follow instructions but also to render independent judgments reflecting their own sensations.

With these tasks in mind, three tests can be used to train the panelists, in addition to exposing them to various levels of odors similar to those to be measured. These are the triangle test, the intensity rating test, and the multicomponent identification test. In the triangle test, three samples are presented to the panelist at the same time. Two are identical and the third is different in either intensity or quality. The panelist must select the odd sample.

The intensity rating test consists of a series of perhaps 20 dilutions of an odorant in an odorless medium. One sample is removed from the series and the panelist is asked to determine the sample's proper place in the series according to its odor intensity.

The last test is the multicomponent odor identification test in which three mixtures are presented to the panelist. These mixtures contain, in sequence, two, three, and four odors out of a possible total of eight known standards. The panelist is told how many components to look for and is asked to identify each of them.

This group of three tests develops a panelist's ability to distinguish between different odor intensities, discriminate among odor qualities, and communicate their sensations in terms of predetermined standards. After the group has some experience working together under a competent test director in actual measurement situations, highly accurate measurements of odor levels can be made.

REFERENCES

1. Stevens, R.K., O'Keefe, A.E., Mulick, J.D., and Krost, K.J. "Gas Chromatography of Reactive Sulfur Gases in Air at the Parts Per Billion Level, 1. Direct Chromatographic Analysis," National Air Pollution Control Administration, Cincinnati, Ohio, 1969.
2. Applebury, T.E., and Schaur, M.J., "Analysis of Kraft Pulp Mill Gases by Process Gas Chromatography," Department of Chemical Engineering, Montana State University, Bozeman, Montana, 1968.
3. Leonardos, G., "The Profile Approach to Odor Measurement, Proceedings of Mid-Atlantic States Section, Air Pollution Control Association Semiannual Technical Conference on Odors: Their Detection, Measurement and Control," Rutgers University, New Brunswick, New Jersey, May 13, 1970.
4. Reckner, L.R. and Squires, R.E., "Diesel Exhaust Odor Measurement Using Human Panels," SAE Paper 680444, 1968.

BIBLIOGRAPHY

Amoore, J.E., Johnston, J.W., Jr., and Rubin, M., "The Stereochemical Theory of Odor," *Scientific American*, Vol. 210, No. 2, 1964, p. 42.

Sullivan, R.J., "Preliminary Air Pollution Survey of Odorous Compounds," A literature review prepared under contract no. PH 22-68-25, National Air Pollution Control Administration Publication No. APTD 69-42, 1969.

Flow Sheet Symbol

TO RECEIVER

AT OIL

11.20 OIL IN OR ON WATER

Types of Detectors:	A—Reflected light sensing on-off oil slick detector B—Capacitor plate–type proportional oil thickness detector C—Oil-in-water detection by ultraviolet (UV) irradiation, causing visible radiation to be emitted
Partial List of Suppliers:	Bailey Controls Co.; Bull & Roberts Inc.; General Electric Co.; Foxboro Co.; PSG Industries Inc.

Oils floating on water form a mechanical barrier between the air and the water, preventing oxygenation and killing oxygen-producing vegetation on the banks of streams. By coating the gills of fish, these materials prevent breathing and cause the fish to suffocate. Outfalls from ships and municipal and industrial waste treatment plants must therefore be monitored and oil removal controlled to prevent oil-bearing wastes from entering natural waters. Continuous monitors are available for the detection of any hydrocarbon floating on the surface of water.

Oil in the water is equally undesirable. It contributes to the biochemical oxygen demand (BOD) and can also be toxic to aquatic biota, to the fish food in water, and possibly to the fish themselves. Optical methods of detection when used for both types of contamination require regular, conscientious maintenance for continuous, reliable performance. The capacitance approach for monitoring oil film thickness on water appears to require less maintenance but is limited to detection of floating oil. Each application must be evaluated separately in the light of the limited capabilities of presently available instrumentation.

On-Off Oil-on-Water Detector

This device (an application of nephelometry) is intended to detect a visible oil (hydrocarbon) slick on fresh or salt water. It consists of two parts: a sensing head and a controller. The sensing head, in an explosionproof housing supported on pontoons, floats on the body of water. An S-shaped baffle directs flowing water past the sensing head. A beam of light is focused through a lens onto the surface of the water. Reflected light is refocused by a second lens onto a photocell. In the absence of oil on the water, minimum reflection of light occurs. In the presence of floating oil, the reflected light intensity substantially increases.

The measurement is based on the differential between the reflected light photocell output and a reference photocell measuring the output of the light source itself. Alarm functions and an output signal proportional to reflected light intensity are available from the controller.

Oil Thickness on Water Detector

The previously described device measures the presence or absence of oil floating on water. A new approach permits oil layer thickness measurement. It consists of a floating sensing head connected by shielded cable to a remote controller. The sensor measures the thickness of an oil layer on water by capacitance measurement (Figure 11.20a).

The operating principle follows the series capacitor formula given in Equation 11.20(1).

$$\frac{1}{C} = \frac{t}{\epsilon A} = \frac{t\ oil}{\epsilon_{oil} A} + \frac{t\ water}{\epsilon_{water} A} \qquad 11.20(1)$$

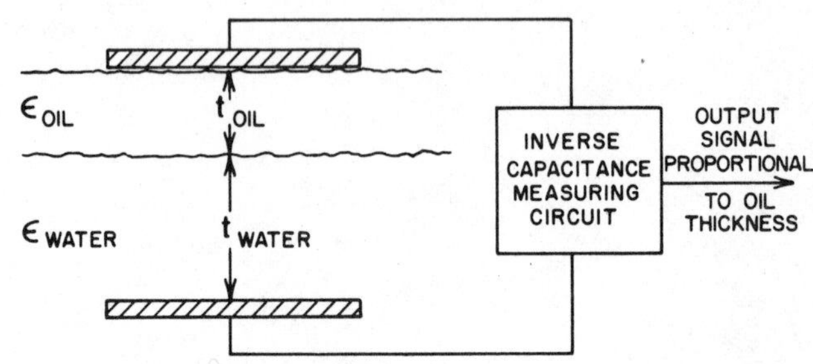

Fig. 11.20a Parallel plate capacitor detecting the thickness of an oil layer on water

where

A = effective area of one capacitor plate
t = thickness
ϵ = dielectric constant

Because ϵ_{oil} = 1.9 to 2.1, whereas ϵ_{water} = 80, Equation 11.20(1) can be simplified by eliminating the second term, as shown in Equation 11.20(2).

$$\frac{1}{C} = \frac{t\ oil}{\epsilon_{oil}A} \qquad 11.20(2)$$

Thus the inverse capacitance is in proportion to the oil thickness. The circuit generates a direct current (dc) voltage in proportion to the inverse capacitance, which is in direct proportion to the oil thickness and is available for remote transmission. The sensor depends on the large differential in dielectric constants between oil and water for its operation. It is claimed that the sensor is not confused by emulsified sludge, which has a large dielectric constant, or by oily froth, which cannot pass under the float.

Oil-in-Water Detector

By irradiating a contaminated water sample stream with ultraviolet (UV) waves at a peak intensity of 365 nm, visible radiation is emitted by the oil contaminant. This can be measured by a photocell. Visible radiation increases with increasing concentrations of the fluorescent substance. The relationship between concentration and visible radiation emitted is substantially linear in low concentrations (below 15×10^{-6}). In higher concentrations some nonlinearity is experienced as a result of a saturation effect.

The most common method of measurement is to pass a sample through the sensing head in an upflow direction (Figure 11.20b). The head is equipped with two windows set at right angles so as to minimize the intensity of direct radiation from the source striking the photocell and also to reduce the effect of multiple scattering of visible radiation. Optical filters at the incident and emergent windows are used (not shown) to reduce this effect to a negligible level.

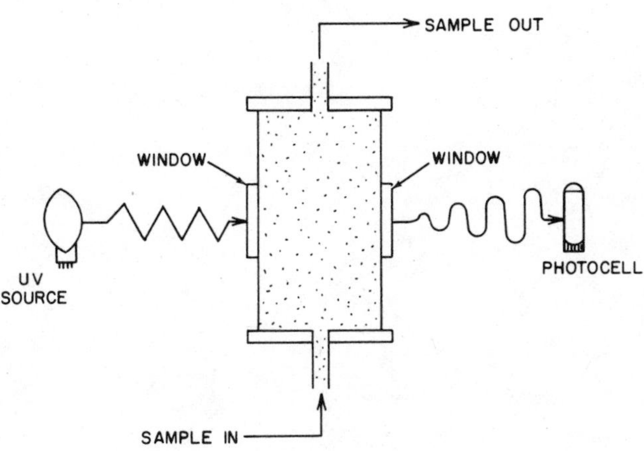

Fig. 11.20b Oil-in-water detector

For the detection of oil concentration in water, a "falling stream" type of detector is also available. With this device the sample stream is "shaped" into a rectangle (Figure 11.20c) and falls through the viewing field of the ultraviolet beam and the photocell. Efficient optical filtration is important to overcome the unavoidable effects of direct reflection of incident radiation from the surface of the shaped stream.

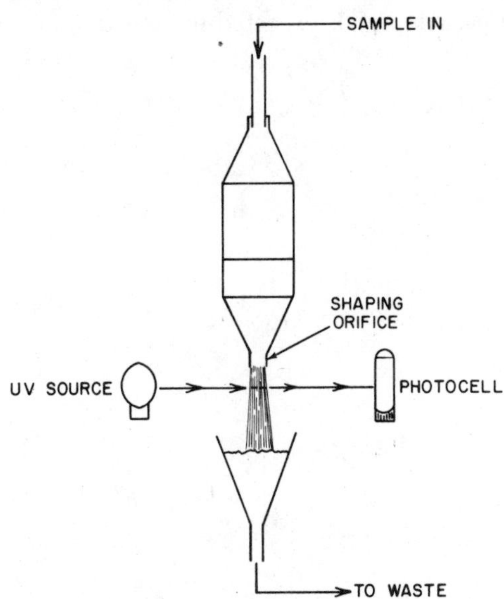

Fig. 11.20c Falling stream oil-in-water detector

Relative Merits

The on-off oil-on-water detector is said to be capable of measuring as little as a few drops of petroleum floating on the surface of water, thus making it possible to detect those oil pollution levels that are visible to the human eye. It therefore serves a useful purpose as an alarm device downstream of plant outfalls and especially during and immediately after oil loading and unloading operations from tankers and tank trucks. This device presents the maintenance problems usually associated with optical measurements, in that windows must be kept clean. The air column between the water surface and the window does reduce fouling due to splashing, but window cleanliness must be maintained for maximum sensitivity.

The oil thickness device (Figure 11.20a), being nonoptical, would appear to require less maintenance. Because both devices can provide indication of absence or presence of oil slicks, they might also find application as efficiency monitors after oil removal has been completed. Floating on the surface of wastewater storage sumps or lagoons, the output of the oil thickness monitor might be used to start and stop oil reclamation equipment. It does not appear that these devices can be calibrated for a specific oil fraction, but they will respond to any floating hydrocarbon.

The oil-in-water devices are optical, and even the falling stream type requires clean optical windows although it is less subject to fouling than the sample chamber type. They must be calibrated to a specific oil and would therefore appear to be subject to interference by other fractions. They were originally developed to monitor boiler feedwater and condensate for the presence of engine oil contamination introduced by steam-driven feedwater pumps. These devices detect the presence of a specific hydrocarbon fraction in well-segregated waste streams. Where particle size is expected to exceed 5μm, sample preparation prior to UV analysis is necessary. Use of a high shear mixer such as a blender has been found to produce a well-dispersed suspension suitable for measurement.

BIBLIOGRAPHY

Crawford, H.M., "Monitors Detect Oil in Water," API Division of Refining Conference, Houston, May 1966.

"Instrumentation and Automation Experiences in Wastewater-Treatment Facilities," EPA-600/2-76-198, Environmental Protection Agency, Washington, D.C., October 1976.

Karasek, F.W., "Detection Limits in Instrumental Analysis," *Research/Development*, October 1979.

Flow Sheet Symbol

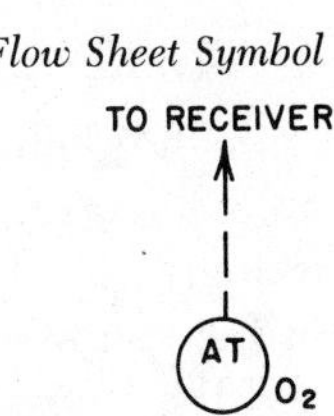

11.21 OXYGEN IN GASES

Methods of Detection:	A—Deflection-type paramagnetic B—Thermal-type paramagnetic C—Dual-gas paramagnetic D—Catalytic combustion E—Electrochemical cells (low temperature) F—Electrochemical cells (high temperature) G—Electrochemical (high temperature current-mode)
Sample Pressure:	Generally near atmospheric
Sample Temperature:	Usually under 200°F (93°C), except for F and G, which can handle 1000°F (538°C) and higher
Sample Flow Rates:	Generally low
Materials of Construction:	Many combinations of materials, most of which are suitable for corrosive service
Inaccuracy:	Generally ±1 to 2%
Range:	From less than 1% to 100% oxygen, often multiple-range instruments
Response:	Reach 90% in less than 60 sec, except F and G, which have millisecond response times
Cost:	Installed units with sample system included usually cost $3000 to $10,000
Partial List of Suppliers:	Bacharach Instrument Co. (E); Bailey Controls Co. (D,F); Beckman Instruments, Inc. (A,E); Cleveland Controls Inc. (B); Delphi Industries (E); Hays-Republic, Div. of Milton Roy Co. (B,E); Leeds & Northrup (B,F); Mine Safety Appliances Co. (B,E); Siemens Corp. (C); Thermolab Corp. (F); Westinghouse Electric Corp. (F,G)

Oxygen is vital to a large variety of well-known industrial and life processes that involve oxidation and combustion. Many industries use pure oxygen or inert gases, and both of these applications usually require analysis for various concentrations of gaseous oxygen.

Paramagnetic O_2 Detectors

Oxygen has a strong affinity for a magnetic field. This almost unique property of paramagnetism is shared by such gases as nitrous oxide and nitric oxide, which have susceptibilities of about 45 and 6 percent that of oxygen, respectively. These gases are not normally encountered during oxygen analysis.

Some gases are repelled by a magnetic field. This diamagnetic effect is exhibited by methane, ethane, ethylene, carbon monoxide and dioxide, hydrogen, and argon. Of this group, methane exhibits the greatest negative susceptibility, about −1 percent with reference to the positive susceptibility of oxygen.

Three basic types of instruments exploit the paramagnetic property of gaseous oxygen. The deflection design requires that the paramagnetic property be constant to permit measurement of the change in gas concentration. The thermal design depends on the paramagnetic effect decreasing as the temperature of the paramagnetic gas (oxygen) increases. In the reference-gas design, two gases with different oxygen contents are combined in a magnetic field and a differential pressure is generated.

Deflection Type

In the deflection analyzer (Figure 11.21a), the magnetic force acts on a test body that is free to rotate about a single axis. The force is in proportion to the difference in the volume magnetic susceptibilities of the test body and the gas around the body. The highly paramagnetic oxygen concentrates the magnetic field, and the resultant imbalance force on the test body is a linear function of the oxygen concentration. When the test body begins to swing out of the magnetic field, the mirror also rotates thus upsetting the light balance. A corrective imbalance in the electrostatic force on the test body results. This imbalance force is opposite and almost equal to the magnetic force, which is a function of the oxygen concentration.

This delicate sensor is mounted on shock absorbers. Dirt in the samples is likely to cause major difficulties; therefore, the sampling system components should insure the cleanliness of the stream.

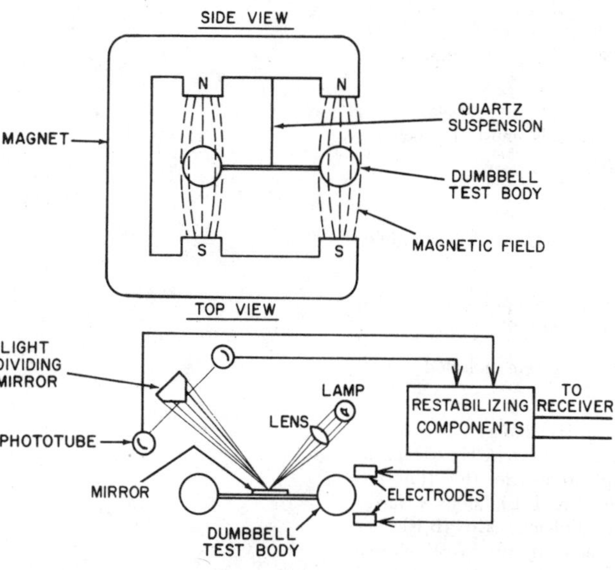

Fig. 11.21a Deflection paramagnetic oxygen analyzer

Thermal Type

A flow-through type of "ring" element is illustrated in Figure 11.21b. In other designs, the sample gas diffuses into dead-end cavities. In the case of the element illustrated, the paramagnetic oxygen content of the sample—after it enters the ring—is attracted by the magnetic field in the horizontal tube where resistors heat the gases. These resistors are connected in a Wheatstone bridge circuit to detect the resistance variations resulting from changes in flow rate. The oxygen in the heated sample loses much of its paramagnetism, thus attracting cooler oxygen from the incoming sample, which then displaces the hot, nonmagnetic oxygen. This action produces a convection current commonly called a "magnetic wind." The "wind" flow rate is a function of oxygen concentration

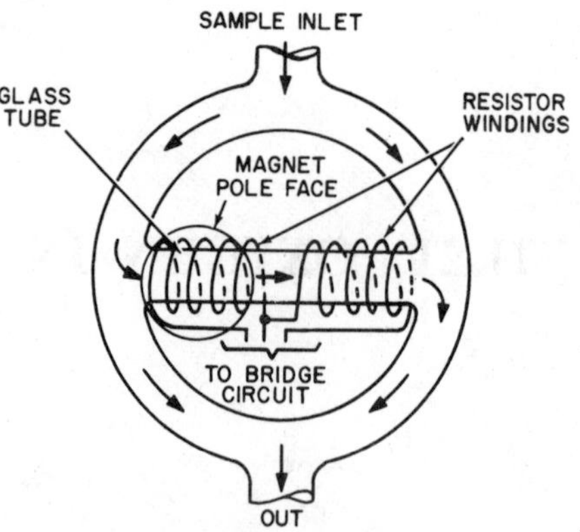

Fig. 11.21b Measuring element in a thermal paramagnetic O_2 analyzer

and is detected by resistors. The gas flow cools the left-hand winding and heats the right-hand winding, and the resulting temperature difference unbalances the bridge.

Because the heating and cooling of the resistors is not only a function of the flow rate but also of the composition and pressure, errors can be introduced by variations in these properties. Diamagnetic materials can introduce errors by affecting the magnetic wind.

Errors can also occur as a result of a change in sample pressure, because the magnetic susceptibility of oxygen varies as the square of the static pressure. Some manufacturers offer pressure compensation in the form of special cells with compensating resistors that are also in the Wheatstone bridge circuit. Others require precise pressure regulation, within a few inches H_2O. Variations in barometric pressure can cause measurement errors of up to ±2 percent on narrow ranges, because the effect of atmospheric pressure change is about 0.02 percent O_2 per inch H_2O (250 Pa).

Dual Gas Type

Figure 11.21c shows a dual-gas cell. Two gases with different oxygen contents are brought together, producing a differential pressure. The reference gas can be 100 percent oxygen, nitrogen, or air. The reference gas passes through two ducts, one of which meets the sample gas in the magnetic field. Since both ducts are connected, the pressure—in proportion to the oxygen content of the gas sample—produces a flow that can be measured. All wetted parts can be made of stainless steel or tantalum.

Catalytic Combustion–Type O_2 Detectors

This approach to oxygen analysis is very similar to the techniques outlined in Section 11.7 in connection with combustible gas detection, and the reader is referred to that section for more details.

The analysis for oxygen content in the sample is accomplished by oxidizing a fuel and measuring the amount of heat generated. The sensor consists of a measuring cell and a reference cell, with a filament in each. The

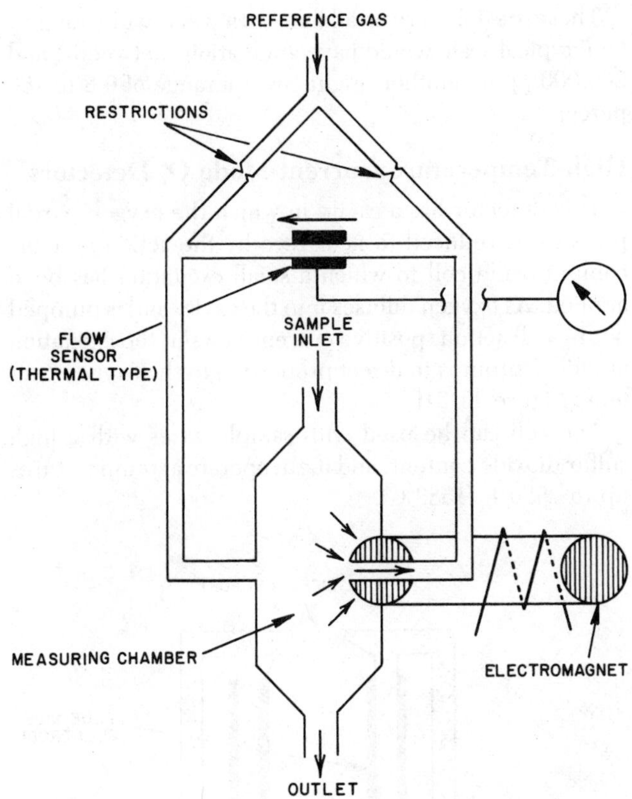

Fig. 11.21c Measuring element in a dual-gas paramagnetic O_2 analyzer

measuring filament is provided with a catalytic surface to oxidize the fuel while the reference filament serves only to compensate for variations in sample temperature and thermal conductivity. In some designs, the unit is thermostatically maintained at constant temperature (see Figure 11.21d). The filaments are connected in a bridge system.

First the sample and fuel are mixed, and then they enter both the measuring and the reference cells. In the measuring cell, the fuel is burned in the presence of the noble metal catalyst filaments. The resulting filament

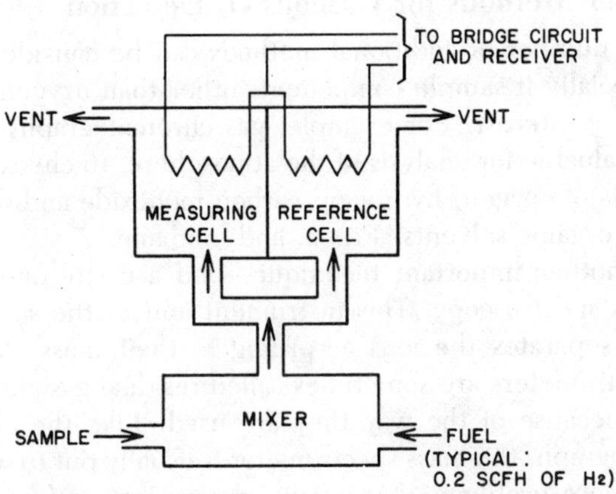

Fig. 11.21d Catalytic combustion-type oxygen analyzer

temperature is sensed by detecting its resistance, which is a measure of the oxygen content of the gas sample.

Electrochemical O_2 Detectors

The electrochemical detector category contains gaseous O_2 detectors that are similar to the catalytic combustion–type analyzer, but these high-temperature detectors are labeled "electrochemical" because they involve ionic activity in a cell. (This type is sometimes also called a fuel cell or a zirconium oxide probe.)

The second type of electrochemical detector is the galvanic cell. The galvanic cells contain two dissimilar electrodes and an electrolyte, so they are more recognizably "electrochemical." Furthermore, these cells are self-generating—that is, they produce a current that can be read on a microammeter. Many companies offer the galvanic cell for both gaseous and dissolved oxygen (DO).

The polarographic cells are also offered for the measurement of gaseous oxygen, and their performance and range are similar to those of the galvanic cells. Similarly to the galvanic cell its major application is in the measurement of DO, which is discussed in detail in Section 11.8.

All three types of electrochemical detectors have two points in common: (1) they measure the partial pressure of oxygen, and (2) they require temperature control, or compensation, for the greatest accuracy.

High-Temperature Electrochemical O_2 Detectors

The operation of these oxygen analyzers involves the ionization of oxygen in both a sample and a known reference stream. The ion concentration in each stream will be a function of the partial pressure of oxygen in that stream, and if an electrolyte separates the two streams there will be a detectable potential difference between the electrolyte surfaces. By placing electrodes on both surfaces of the electrolyte, the open circuit voltage detected will be related to the ratio of the oxygen partial pressures on the two sides of the electrolyte. The cell reactions at the two electrodes can be expressed as:

$$O_2 + 4 \text{ electrons} \longrightarrow 2(O^{--}) \text{ (at the anode)}$$

$$2(O^{--}) \longrightarrow O_2 + 4 \text{ electrons (at the cathode)}$$

As described by the cell reactions above, oxygen molecules on the anode side of the cell (higher oxygen concentration side) gain electrons to become ions, which enter the electrolyte. Simultaneously, oxygen molecules are formed by the reverse process at the cathode (low oxygen concentration side). The open-circuit voltage relates to oxygen partial pressure by the following equation, which is also referred to as the Nernst equation:

$$E = \frac{RT}{nF} \ln \left(\frac{O_2 \text{ partial pressure in reference gas}}{O_2 \text{ partial pressure in sample gas}} \right) \quad 11.21(1)$$

where

E = the open circuit voltage developed
R = the universal gas constant
T = absolute temperature
n = the number of electrons in the electrode reaction, and
F = Faraday's constant

From the above equation it can be seen that the detector output signal will change by the same amount when the oxygen content in the sample varies from 1 to 10 ppm or from 10 to 100 ppm. In other words, there is a logarithmic relationship between output signal and oxygen content. If the reference gas is ambient air, then the maximum detectable sample concentration is 21 percent, which is the O_2 concentration in air. It should also be noted that the output is in direct proportion to the absolute temperature of the cell, and therefore accurate control of this temperature is important.

Figure 11.21e illustrates one of the available element designs. Here a heater maintains the analysis zone at about 1500°F (816°C), causing ionization of oxygen in both the flowing sample and the stationary reference gas, which is usually ambient air. The platinum electrodes on the inside and outside surfaces detect the oxygen ion concentration on the sample and reference sides of the solid electrolyte. This measurement is logarithmically related to the concentration of oxygen in the sample, if the reference gas oxygen content is constant.

This analyzer should be considered only for those installations where the sample contains no combustibles, or preferably for those where the sample consists of only oxygen and inert gases. This is an important criteria because at the tube operating temperatures involved (above 1000°F or 538°C), the combustible components would oxidize, thus reducing the oxygen partial pressure in the sample and thereby introducing a measurement error.

Some versions of this design are available as transmitters while others are marketed with integral indicators for both portable and permanent installations.

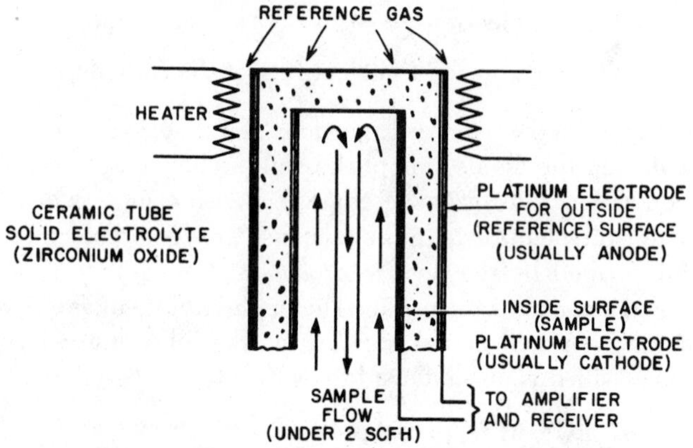

Fig. 11.21e High-temperature electrochemical oxygen detector

These designs are available with very wide ranges. One typical scale would have graduations between 1 and 200,000 ppm; another might cover a range of 0.8 to 150 percent.

High-Temperature Current-Mode O_2 Detectors

This detector has a cavity in which the oxygen partial pressure is reduced to near zero by the action of a zirconium oxide cell to which a small excitation has been applied. As oxygen diffuses into the cavity and is pumped out by cell action, positive current flows in the excitation circuit. Output is in direct proportion to the oxygen content (Figure 11.21f).

The cell can be used with sample gases with a high sulfur dioxide content, and it can operate at temperatures up to 2880°F (1582°C).

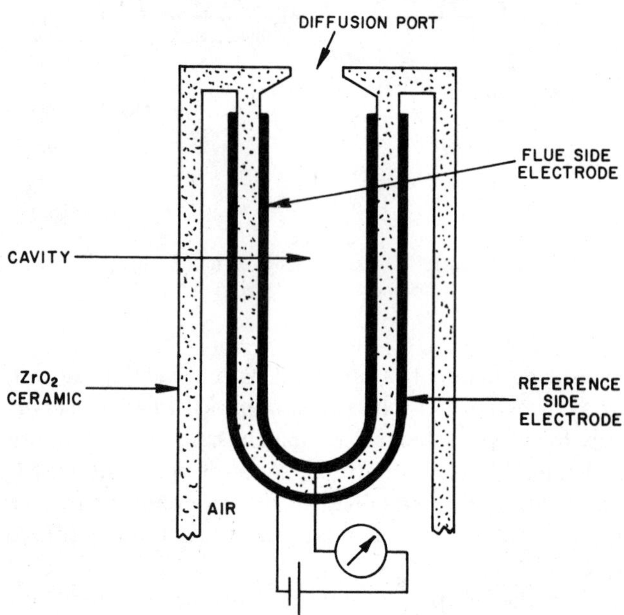

Fig. 11.21f High-temperature diffusion-limited current-mode cell

Other Methods for Gaseous O_2 Detection

A number of additional methods can be considered, especially if sample components other than oxygen are also of interest. For example, gas chromatography can be valuable for analysis of the atmosphere, to check the levels of oxygen, hydrogen, carbon monoxide and dioxide, organic solvents, Freon, and methane.

Another important technique—and a costly one—is mass spectroscopy. This instrument ionizes the sample and separates the ions according to their mass. Mass spectrometers are sometimes called residual gas analyzers because of the way they are used. Like the chromatograph, the mass spectrometer is usually put to work on those problems that involve determination (qualitative and quantitative) of more than one component in the sample.

Chemical analyzers of the manual Orsat type are available for a large variety of ranges from 0–20 percent to 100 percent oxygen. Orsat gas analyzers have been designed for specific applications, such as combustion and blast furnace testing, or checking the purity of medical and industrial oxygen.

BIBLIOGRAPHY

Kolmer, J.W., and Neuberger, E.D., "Programmed O_2 Set Point Optimizes Combustion Control," *Instruments and Control Systems*, December 1977, pp. 33–38.

Kraf, D., "Establishing Cost/Benefit and Payback for a Flue Gas Analyzer," *Pollution Engineering*, April 1979, pp. 44–46.

Nelson, R.L., "New Developments in Closed Loop Combustion Control Using Flue Gas Analysis," ISA Preprint ISBN 87664-471-X, 1980.

Flow Sheet Symbol

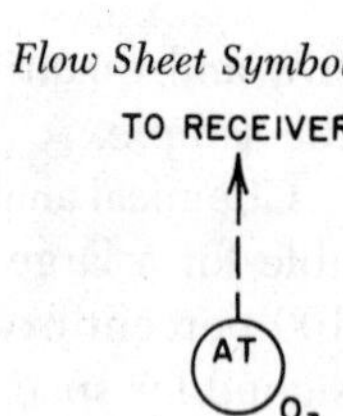

11.22 OZONE IN AIR

Methods of Detection:	A—Chemiluminescence B—Ultraviolet spectroscopy C—Coulometric D—Thermal conductivity E—Colorimetric
Reference Method:	(A) Ethylene reaction of gas phase O_3, calibrated against neutral potassium iodide method
Sample Flow Rate Liters per Minute:	0.14 (C), 0.33 (D), 1.0 (A,B), 7.0 (B)
Interferences:	Oxidizing agents interfere with C
Ranges:	0.9 to 30 ppt (D), 1 ppb to 10 ppm (A), 0.01 to 1 ppm (C), 0.01 to 10 ppm (B)
Sensitivity:	1 ppb (A), 0.01 ppm (C), 0.1 ppm (B)
Reproducibility or Inaccuracy:	2% (A,D), 2 to 3% (B), 4 to 5% (C)
Cost of Sensor Without Recorder:	\$1800 to \$6800 (C,D); \$4500 to \$9500 (D); \$7800 to \$10,000 (A)
Partial List of Suppliers:	Atlas Electric Devices Co. (C); Beckman Instruments, Inc. (C,D,E); Bendix Corp. (A,D); Dasibi Environmental Corp. (B); Mast Development Co. (C); Meloy Laboratories Inc. (A); Monitor Labs Inc. (E); Ozone Research and Equipment Corp. (C); Technicon Corp. (E)

Ozone is an unstable triatomic allotrope of oxygen which naturally occurs in the atmosphere below 0.1 ppm concentration at sea level and at substantially higher concentrations at high altitudes.[1] It is produced by electric storms and irradiation of atmospheric oxygen by sunlight. Pollutants present in the atmosphere can increase ozone production, leading to dangerously high ozone concentrations. Ozone concentrations close to 1 ppm have been recorded in Los Angeles on muggy days. Ozone is a powerful oxidizing agent, if present in high concentrations, and it attacks the pulmonary tracts.

Ozone is also used to control pollutants in both air and water. Its most desirable features are that (1) it is capable of destroying most pollutants by oxidation; (2) the oxidation products are usually nontoxic and biodegradable; (3) the unreacted excess ozone decomposes to oxygen, leaving no undesirable residue.

This section reviews those detectors which measure the ozone concentration in air or gas. For the measurement of ozone in water, see Section 11.23. Gas-phase ozone detectors are used to monitor atmospheric pollution and to detect leaks in ozone generator installations.

The highly specific and very sensitive chemiluminescence detectors were introduced in the last few years. The operating principles of the other sensors have been used for other measurements, and the detectors have been on the market for many years.

Chemiluminescence Ozone Detectors

This measurement is based on the chemiluminescent reaction between ozone and ethylene. The air sample and ethylene are continuously drawn into a reaction chamber by a pump. In this chamber, a photomultiplier is mounted on a flat window. The chemiluminescent light emission measured by the photomultiplier tube is in proportion to the ozone concentration in the sample. The signal from the photomultiplier is amplified and displayed on a meter or recorded.

The photomultiplier tube assembly and reaction chamber are enclosed in a thermoelectrically cooled housing. A low (about 10°C or 50°F) constant temperature is maintained to avoid long-term drift.

These detectors are designed for long-term, unattended operation in weather and pollution monitoring stations. The ethylene consumption is low: 15 to 20 cc per minute. The ozone measurement is specific. Oxidizing gases or other usual components of the atmosphere do not interfere with it. Chemiluminescence sensors are the most sensitive ozone detectors available.

Ultraviolet Ozone Detectors

These instruments take advantage of the strong ultraviolet (UV) absorption of ozone at the wavelength of 254 nm.

The drawn air sample is divided into two portions. From one portion, the ozone is removed by a manganese dioxide (MnO_2) catalyst filter. Then the two portions are alternately introduced into an air absorption chamber (Figure 11.22a). A UV light beam is directed through the absorption chamber into a detector. The difference between the signals from the ozone-free filtered and unfiltered air samples is displayed. The unit is calibrated for ozone on the basis of Beer's law.[2]

These detectors are somewhat less sensitive than the chemiluminescent type. They measure ozone specifically, and the normally occurring components in the atmosphere do not interfere with the measurement.

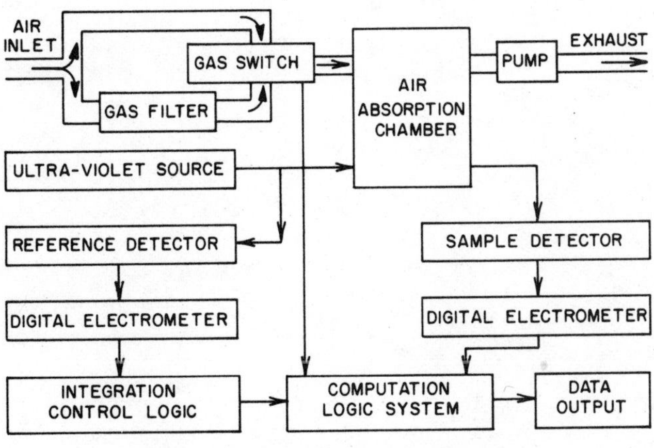

Fig. 11.22a Ultraviolet ozone detector

Coulometric Ozone Detectors

Ozone can generate iodine from potassium iodide (KI) according to the reaction in Equation 11.22(1) and the iodine can be measured coulometrically.

$$O_3 + 2\,KI + H_2O \longrightarrow O_2 + 2\,KOH + 2\,I \qquad 11.22(1)$$

The air sample is passed through a sensing cell over a thin film of potassium iodide solution which is continuously pumped through the cell (Figure 11.22b). The solution film containing the freshly produced iodine is passed over some electrodes. The current produced at the electrode is in proportion to the iodine content and ultimately to the ozone concentration.

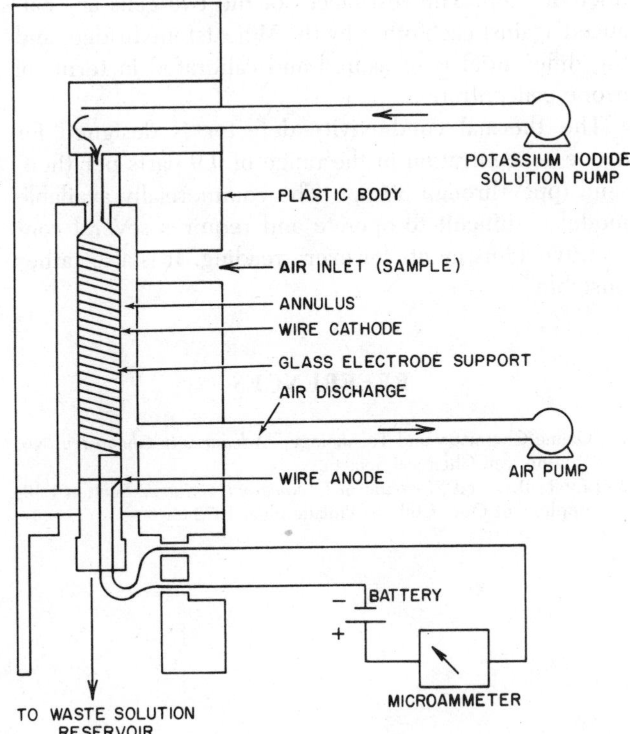

Fig. 11.22b Coulometric oxidant (ozone) analyzer

These detectors are not specific to ozone because they measure total oxidants. There are filters available to decrease interference by other oxidizing agents. Their most satisfactory application is to measure total oxidants, or ozone when other oxidizing agents are absent.

Thermal Conductivity Ozone Detector

The measurements of thermal conductivity can be correlated with the type and concentration of the gas sample (Section 10.16).

This ozone detector functions as indicated in Figure 11.22c. The gas sample enters the meter and passes through the first thermal conductivity cell and then through a heater that decomposes the ozone. The ozone-free gas is then introduced into the second thermal con-

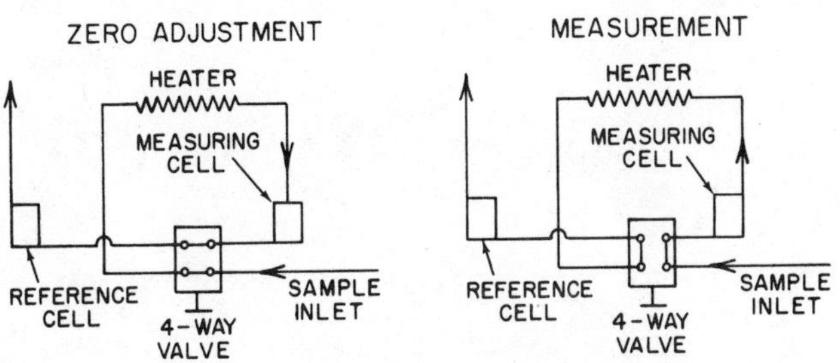

Fig. 11.22c Thermal conductivity ozone detector

ductivity cell. The resistances of the two cells are balanced against each other by the Wheatstone bridge, and the differential is measured and calibrated in terms of ozone concentration.

The thermal conductivity detector is designed for ozone concentration in the range of 0.9 parts per thousand (ppt) through 30 ppt. The commercially available model is difficult to operate and requires several consecutive adjustments for every reading. It is also rather unstable.

REFERENCES

1. "Ozone Chemistry and Technology," *Advances in Chemistry*, No. 21, American Chemical Society.
2. Lipták, B.G., ed., *Instrument Engineers' Handbook*, Section 1.10, Supplement One, Chilton, Philadelphia, 1972.

BIBLIOGRAPHY

"Air Quality Criteria for Petrochemical Oxidants," National Air Pollution Control Administration Publication No. AP-63, 1970.

Karasek, F.W., "Detection Limits in Instrumental Analysis," *Research/Development*, October 1979.

"Ozone Chemistry and Technology," *Advances in Chemistry*, No. 21, American Chemical Society.

Flow Sheet Symbol

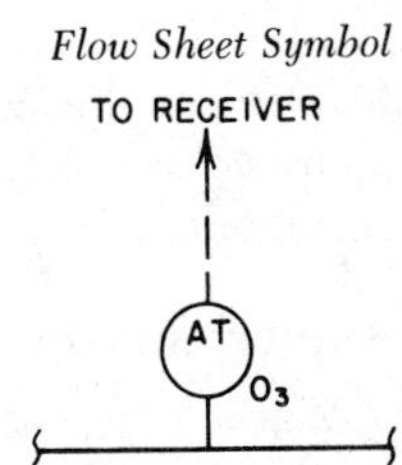

11.23 OZONE IN WATER

Methods of Detection:	A—Amperometric B—Amperometric using potassium iodide feed C—Oxidation-reduction potential (ORP) type D—Air purge of water sample followed by gas phase detection of ozone E—Colorimetric
Sample Requirements:	Detectors A and B have well-defined sample requirements. Detectors C through E are not specifically designed for dissolved ozone measurement in water and have no strict sample requirements
Sample Flow Rate:	100 to 300 cc per minute (A,B)
Sample Temperature:	40 to 120°F (4 to 49°C) (A,B)
Sample pH:	5.0 to 9.0 (A,B)
Sample Pressure:	8 to 100 PSIG (55 to 690 KPa) (A,B)
Limitations:	Sample should be substantially free of suspended solids (A,B); suspended solids and color seriously interfere with measurement (E)
Materials of Construction:	All materials in contact with the water sample are made of glass, PVC, Teflon, or aluminum coated with black-baked vinyl finish. Other ozone-resistant materials that could be used include silicone rubber, Viton, and stainless steel
Sensitivity:	Types A and B will respond to 0.01 ppm ozone concentration. Types D and E can be adapted to produce similar sensitivities
Ranges:	The complete ozone concentration range allowed by solubility can be measured. 0 to 1 ppm (A, B, D, E); 0 to 7 ppm (A, B, C, D, E)
Cost:	\$3400 to \$4500 (B, D), \$4500 to \$6800 (A), including automatic temperature compensation and recorder
Partial List of Suppliers:	Beckman Instruments, Inc. (E); Capital Controls Co. (A,B); Dasibi Environmental Corp. (D); Envirotech Corp., United Controls Div. (A); Fischer and Porter Co. (A); Mast Development Co. (D); Meloy Laboratories Inc. (C)

Ozone is an unstable, triatomic allotrope of oxygen which is normally present in the atmosphere in less than 0.1 ppm concentration at sea level and in substantially higher concentrations at higher altitudes. For industrial applications ozone is usually generated on site from air or oxygen in an electric discharge field process. As a powerful oxidizing agent, it is used for the control of oxidizable pollutants both in water and in air. Most common applications include odor, taste, sterilization, removal of refractory organics, and removal of chemical oxygen demand (COD) and biochemical oxygen demand (BOD).

This section reviews the equipment and methods for the measurement of ozone dissolved in water. For ozone detectors in the gas phase, see Section 11.22.

Detectors designed specifically to measure ozone have become available recently. In general, chlorine detectors can be modified to measure ozone, gas-phase ozone detectors can be adapted for water measurements, and many color reactions can be used for colorimetric measurements of dissolved ozone.

Amperometric

Water is fed at a constant flow rate into the cell assembly by a flow regulator or overflow device. The cell consists of two concentrically-mounted metal electrodes, usually copper and gold. An electrode scrubber plate is constantly rotated by a motor to clean the electrode surface and make the signal output independent of slight flow variations. The cell functions as a wet cell, except that the current varies linearly with the ozone concentration. In the presence of water only, the two metal electrodes become polarized, but essentially no current flows in the circuit. When ozone is present in the water, the gold measuring electrode is depolarized and the copper electrode is oxidized, thus completing the circuit and allowing current to flow (Figure 11.23a).

A built-in thermistor automatically compensates the signal for sample temperature changes in the 40 to 120°F (4 to 49°C) range. Rotating nonabrasive pellets keep the electrode surfaces clean. Pneumatic, electric, and electronic controls maintain ozone concentration at the desired level precisely and automatically.

The water sample must be clean and reasonably free from suspended particles. Solids that are likely to settle out must be filtered. Most chemicals usually present in water do not interfere; exceptions, however, are aluminum above 0.1 ppm and copper above 0.3 ppm. Chlorine and other strong oxidants also interfere with the analysis.

Amperometric detectors are rugged, reliable, well-tested, and available at medium and higher prices. For applications in which the pH and ozone concentrations fluctuate over a broad range, the amperometric detector can be modified to accept one or two chemical feeds. A buffer solution or a buffer and a potassium iodide solution can be injected.

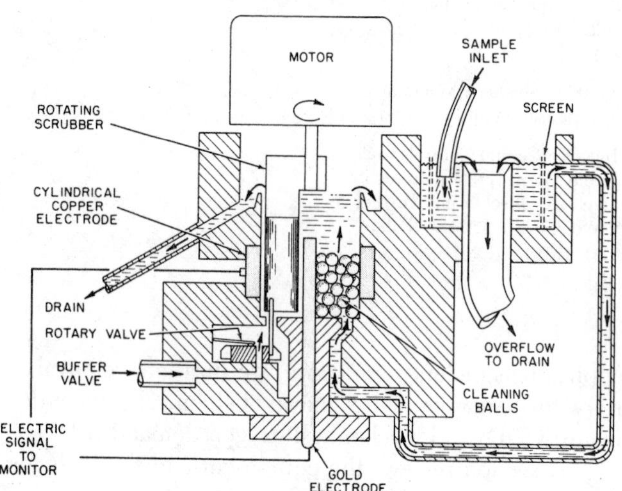

Fig. 11.23a Amperometric ozone detector

ORP

Oxidation-reduction potential (ORP) measurement is a qualitative method (see Section 10.13). Gold or platinum electrodes indicate reliably the presence of ozone in aqueous solution. The ORP method can also detect excess ozone in chemical reactions. The end point of oxidation reactions that consume ozone rapidly can be indicated by this method, and these reactions include the oxidation of cyanides and the regeneration of photographic-grade bleach.[1]

Air-Purge Gas Phase

This method relies on the adaptation of one of the gas-phase ozone detectors reviewed in Section 11.22 for the measurement of dissolved ozone. Ozone is relatively insoluble in water and can be readily removed by purging the solution with nitrogen or clean air. When a small, known volume of water containing ozone is purged with air, the purge gas can be collected and analyzed with a gas-phase ozone detector. Alternatively, the purge gas may be passed through a potassium iodide solution, in which the ozone converts a proportionate amount of iodide into iodine (which then can be measured by standard iodine or chlorine detection methods).

Colorimetric

Ozone detection is based on a color-producing or color-fading reaction of ozone with an organic dye such as crystal violet or N-phenyl-2-naphthylamine. The color is measured spectroscopically, and the ozone concentration can be calculated from Beer's law[2] (see also Section 10.5).

REFERENCES

1. Selm, R.P., "Ozone Oxidation of Aqueous Cyanide Waste Solutions." *Advances in Chemistry* No. 21, American Chemical Society, Washington, D.C., 1959.
2. "Ozone Chemistry and Technology," *Advances in Chemistry*, American Chemical Society, Washington, D.C., 1959.

BIBLIOGRAPHY

"Instrumentation and Automation Experiences in Wastewater-Treatment Facilities," EPA-600/2-76-198, Environmental Protection Agency, Washington, D.C., October 1976.

Karasek, F.W., "Detection Limits in Instrumental Analysis," *Research/Development*, October 1979.

"Ozone Chemistry and Technology," *Advances in Chemistry*, No. 21, American Chemical Society, Washington, D.C., 1959.

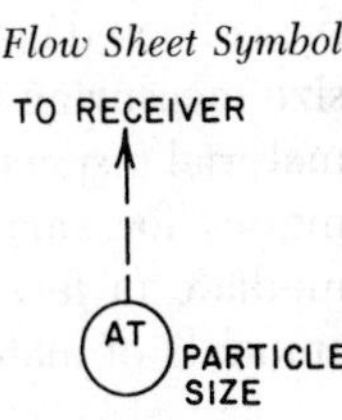

11.24 PARTICLE SIZE AND DISTRIBUTION

Measurement Technique:	A—Optical imaging B—Electron imaging C—Image analysis (with A or B) D—Optical diffraction and scattering E—Electrical resistance change F—Sieving G—Sedimentation (photo or X-ray) H—Ultrasonic attenuation I—Bulk property—absorption, permeability
Cost (in $1000s):	A—4 to 12 B—20 to 100 C—40 to 80 D—10 to 40 E—8 to 12 F—1 to 5 G—5 to 16 H—35 to 45 I—1 to 5
Partial List of Suppliers:	Alpine American Corp. (F); Autometrics (H); Bausch & Lomb Co. (A,C); Buckbee-Mears Co. (F); Cambridge Instrument Co. (B,C); Climet Instrument Co. (D); Cilas (France) (D); Coulter Electronics, Inc. (E); Fisher Scientific (F,I); Goring-Kerr (U.K.) (G); International Scientific Instruments (B); Joyce Loebl (F); Leeds & Northrup Co. (D); Leitz (A); Micromeritics Instrument Corp. (G,I); Pacific Scientific (D); Particle Data, Inc. (E)

Accurate determination of particle size and distribution is critical in many industrial processes, such as grinding, agglomeration, crystallization, and emulsification. The objectives may include improving the final product quality, as in ceramics, paints, and pigments; improving the final product performance, as in abrasives and catalysts; improving the process efficiency, as in crystallization and wastewater treatment; or minimizing the energy consumption, as in grinding of ore for subsequent processing. In many processes, more than one of these objectives may be involved.

This section discusses measurement of particles between approximately 0.01 and 1000 μm, with the emphasis on determining the quality of material in a process rather than the detection of contaminants in an otherwise particle-free stream. Performance characteristics are quoted which are attainable from standard commercial instruments. Not all measurement techniques are discussed in detail.

Particle sizing techniques differ with respect to size range, resolution, concentration, and measurement principle. Choosing the right measurement principle for the parameter of interest is important, but irrespective of the method used, the reported particle size is often dependent on other factors such as shape or physical properties. Selecting the most useful method of measuring size depends on understanding the characteristics for which the final product or process will be judged.

Accurate size measurement also depends on obtaining a representative sample of the bulk material. Bias toward smaller or larger sizes can occur if sampling is done at the wrong point or time in the process, or if a sample is extracted from a settled heap. The value of size measurement is limited by how faithfully the sample represents the process.

The amount of material used in the analysis depends on the measurement technique. This may present a second-order sampling problem if only a small fraction of the collected material is inspected. One way of classifying size measuring techniques is according to the amount of material inspected. The next three sections discuss techniques for small (a few hundredths of a gram), intermediate (a few grams), and large (hundreds of grams) amounts of material, respectively.

Techniques for Small Sample Quantities

Optical Microscopy

Optical microscopy is a standard method for particle size analysis which is often used to calibrate other particle-sizing instruments. Size determination is based on the operator's criteria and judgment, based on parameters such as the longest dimension, the diameter of a sphere of equivalent cross section, and so on. A large number of particles must be counted to convert a number distribution to a weight distribution, especially when there is a large range of sizes. Size resolution is limited by the wavelength of light, which restricts the practical use to particles larger than about 1 μm. The depth of the field, or the region in which an object is in sharp focus, decreases as the magnification is increased. Size errors can arise from out-of-focus imagery, especially with spherical particles.

Electron Microscopy

A fine electron beam is scanned across a sample, producing secondary electrons and other emissions from the material. The emissions are detected and displayed on a cathode ray tube to form an image, as in a TV picture. Operator interaction and judgment are required to determine the size. The size resolution is limited by the energy of the electron beam, and 0.006 to 0.01 μm resolution is typical for modestly priced equipment. Depth of field of the image is about 300 times greater than in optical microscopy. However, a common source of error is the evaluation of too few fields of particles.

Image Analyzers

Automating the analysis of the image of a particle field reduces much of the tedium of counting large numbers of particles. Sophisticated electronic routines, augmented by an editing pointer, provide for rapid error-free analysis of complex shapes. The input to an image analyzer is a high-resolution TV camera, so this technique must be combined with another device, such as an optical or electron microscope.

Electrical Sensing Zone (the Coulter Principle)

A particle distribution is suspended in an electrolyte and the particles are drawn through a small orifice, which is immersed between two electrodes. Each particle, in traversing the orifice, displaces its own volume of the electrolyte, producing momentary resistance change in proportion to the volume displaced. The pulses are then amplified, scaled, and separated into counting bins. Some level of training and skill is required for good reproducibility, but high resolution can be achieved for narrow distributions. The size range is determined by the orifice diameter. For very broad distributions, blockage of the orifice can be troublesome, and different orifice sizes must be used to cover the full range. The practical lower particle size limit is around 0.5 μm.

Optical Scattering (Single Particle Analysis)

When a light beam interacts with a particle, light is scattered or diffracted into other directions. The angular distribution and amount of scattered light depends on the particle size. The scattered-light pulses from individual particles are detected as they pass a probe beam, and these pulses are assigned to counting bins according to size. The size range is limited by the dynamic range of the electronics, and the size accuracy is influenced by composition for particles less than a few micrometers. The lowest size commonly measured is about 1 μm for water-suspended material and a few tenths of 1 μm for air-suspended material. Low signals for small particles and coincidence at high sample concentrations are the principal limitations.

Techniques for Intermediate Loadings

Light Scattering (Multiple Particle Analysis)

When many particles are present in the probe beam at the same time, a composite scattering pattern is generated. Several algorithms have been developed to extract the size distribution information from this scattering pattern. By using low-angle scattering and optical analog filtering, the Microtrac approach generates a signal in proportion to the volume of material present in the light beam. The volume-size histogram is calculated through a microprocessor program. By adding signals collected at high angles for particles less than 1 μm, this technique can accommodate a size range from 0.1 to 1000 μm. Resolution is lower than for single particle analysis, and the accuracy is limited by composition effects for particles below a few micrometers, but no special training or skill is needed to make the measurement. Material can be suspended in either air or liquid. Other instruments exist in this loading category using related principles of forward scattering but are limited to particles larger than 1 μm.

Sedimentation (Photo and X-ray)

Photosedimentation combines gravitational settling with optical attenuation. Sorting of particles is based on hydrodynamic properties, and sizing is based on the cumulative particulate cross section as a function of settling time. Care must be taken to avoid too rapid settling, convection currents, and improper or incomplete correction for optical scattering variations as a function of size and composition. Photosedimentation is limited to the size range of 1 to 100 μm. By replacing the optical beam with a well collimated X-ray beam, this technique can be extended to particles a few tenths of 1 μm in size. Centrifugal sedimentation is also useful in this submicrometer size range.

Techniques for High Loading

Sieving

This common method is best suited to material that is predominantly coarser than 75 μm. The sample is shaken through a series of containers whose bases have regularly-spaced uniform-size openings. The primary measure is the mass of material retained in each sieve, and the distribution is most often plotted as a cumulative fraction versus the nominal sieve aperture.

Several sieve series exist: Tyler and the American Society for Testing and Materials (ASTM) in the United States, and British and German standards in Europe. Fixed ratios of sizes of square root or fourth root of 2 are common.

For particles smaller than about 75 μm sieving becomes more difficult. Problems include the increased cost of producing sieves with uniform apertures and greater sophistication of the mechanics of the sieving process. Wet, sonic, and air-jet sieving have been used for this range, and with care, particles down to a few micrometers can be handled.

Optical Methods

Transmissometry (across a sample stream), turbidimetry (light scattered at 90° to the beam), and backscatter are often used to measure sample loading in a process stream, but since all three of these methods also respond

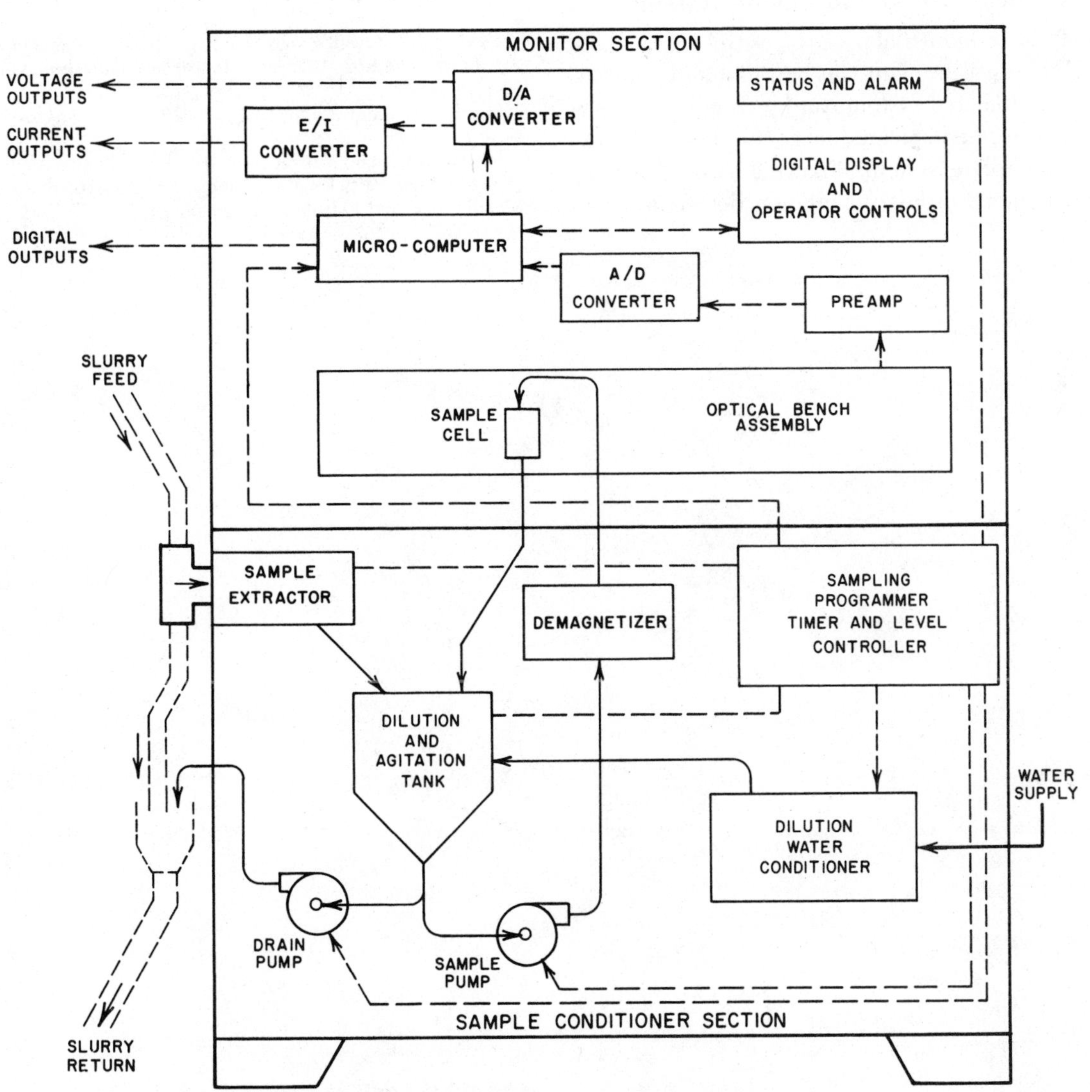

Fig. 11.24a Block diagram of complete system for outline process control using an optical particle analyzer

to size and composition changes, they are best suited for monitoring changes in a process.

Ultrasonic Attenuation

When ultrasonic energy is transmitted through a slurry, the amount of energy absorbed is dependent on the particle size, concentration, and frequency. In one approach, two pairs of sensors, each consisting of a transducer and a receiver, operate at two different frequencies. For a given material a correlation is established between the size distribution as measured by a referee method and the attenuation of the higher frequency beam. Corrections due to loading changes are made based on the lower frequency signal. This technique is restricted to particles between 25 to 600 μm, but high concentrations can be accommodated, which makes this approach attractive for on-line control. Entrained air bubbles can cause errors, so they must be avoided or eliminated.

On-Line Particle Size Measurement Systems

Size analysis for industrial process control is especially valuable if the measurement can be rapid and continuous. A fully automated system can be complex, even if the size measurement technique is simple. As an example of a complete system, Figure 11.24a illustrates critical components of instrumentation designed to extract a sample, measure the size through an optical multiple-particle analyzer, and report a specific parameter of the distribution for feedback to a control circuit.

Supply water is drawn by a fill pump into a mixing chamber through a debubbling circuit which extracts air bubbles from the entering water. A circulating pump then agitates the water in the mixing chamber and circulates the water through the sample cell in the measurement region of the laser beam. When the slurry extraction is actuated, a particulate sample is drawn and deposited into the mixing chamber. The circulating water carries the particles of the slurry through the sample cell. At the end of the measurement cycle, the mixing chamber is drained, rinsed, and refilled in preparation for the next sample.

BIBLIOGRAPHY

Allen, T., *Particle Size Measurement,* Chapman and Hall, London, 1975.

Davies, R., "Recent Progress in Rapid-Response and On-Line Methods for Particle Size Analysis," *American Laboratory,* Vol. 10, April 1978, p. 97.

Herdan, G., and Smith, M.L., *Small Particle Statistics,* Elsevier, New York, 1953.

Sresty, G.C., and Venkateswar, R., "Particle Size Analysis—A Review, in Fine Particles Processing," American Institute of Mining, Metallurgical and Petroleum Engineers, Inc., New York, 1980.

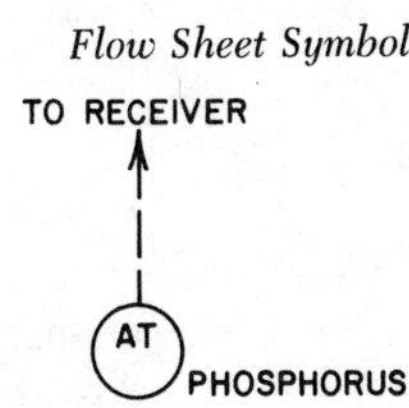

11.25 PHOSPHORUS

Methods of Detection:	A—Colorimetric B—Flame photometric C—Chromatographic
Standard Design Pressure:	Atmospheric
Materials of Construction:	Most units are available with wetted parts of stainless steel, glass, or Teflon
Inaccuracy:	±2% full scale in most cases
Ranges:	0 to 0.01 to 0 to 100 ppm
Response:	Several seconds to 15 minutes
Cost:	\$6750 to \$22,500 exclusive of sample handling system
Partial List of Suppliers:	Meloy Laboratories, Inc.; Raytheon Company; Technicon Industrial Systems; Tracor, Inc.
Special Filters:	Vane type by Raytheon Co. and "Auto-Kleen" edge type by Cuno Engineering Corp. at \$1000 or more depending on size

The principal application of a phosphorus analyzer is the control of phosphate removal in a sewage treatment plant. By knowing the phosphorus content of raw sewage and its flow, the optimum quantities of chemical additives required can be determined. In addition, a measurement of the phosphorus remaining after treatment may be desired. Other applications are those specific to an industrial process, e.g., the control of phosphate addition to high-pressure boiler water as a corrosion inhibitor, or the measurement of elemental phosphorus in the effluent from a plant that extracts phosphorus from ore.

Phosphorus occurs in wastewater almost entirely in the form of phosphates including orthophosphates, condensed phosphates (pyrophosphate, metaphosphate, and polyphosphate) and organically-bound phosphates. The diverse methods of detection do not all respond to the total phosphorus present.

Colorimetric Analysis

The simplest colorimetric procedures are to determine soluble orthophosphate. In the commonly used aminonaphtholsulfonic acid method,[1] ammonium molybdate reacts with a dilute phosphorus solution to produce molybdophosphoric acid, which is reduced to the intensely colored complex, molybdenum blue, by the combination of aminonaphtholsulfonic acid and sulfite reducing agents.

The stannous chloride method,[1] although slightly more sensitive, is similar to the method just described except for the substitution of stannous chloride for aminonaphtholsulfonic acid as the reducing agent.

A continuous phosphorus analyzer consists of a sample temperature controller, a multiple peristaltic pump for reagent and sample metering, a mixing and time-delay section, reagent storage, a colorimeter, and an electronic readout section. Referring to Figure 11.25a, the water sample is brought to a constant temperature to insure uniform sample reaction and rapid response time. The sample stream is degassed in the constant head tank and divided into two paths: one for the reference sample, and the other for the reaction sample. A multiple peristaltic pump meters the sample streams as well as the addition of ammonium molybdate and aminonaphtholsulfonic acid solutions.

The reaction and reference samples pass through separate delay coils. In one coil, which gives about 5 minutes delay, the reaction sample and the reagents complete the color reaction; in the other delay coil the reference sample is experiencing the same delay in addition to the

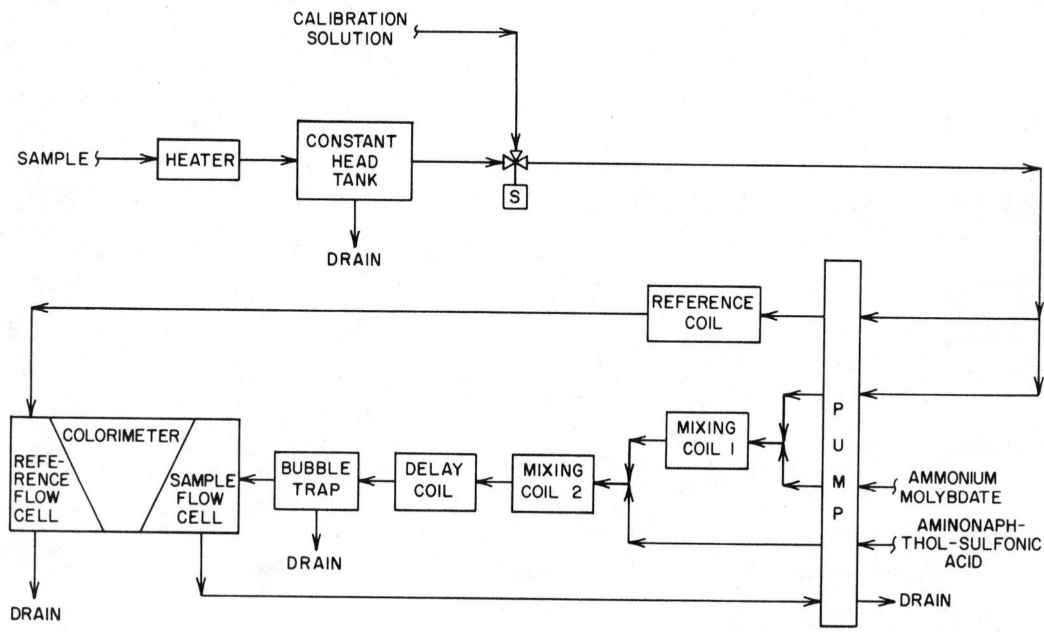

Fig. 11.25a Orthophosphate analyzer

time that the reaction sample spends in the mixing coils. The sample and reference streams from the delay coils are fed into the dual flow cells of a dual-beam colorimeter. A bubble trap ahead of the colorimeter removes any bubbles formed in the analyzer. The photodetectors sense the difference in color intensity between the reaction and reference samples. In this case, the molybdenum blue complex is measured at a wavelength of 6900 Å. The electronic section amplifies the colorimeter output and linearizes it, if necessary. The use of a dual-beam colorimeter automatically compensates for inherent color or turbidity variations in the sample.

Total inorganic phosphate can be measured by first hydrolyzing the sample with sulfuric acid at 95°C (203°F). This converts the phosphates in the meta, pyro, and polyforms to orthophosphates. Total phosphates (inorganic plus organic) can be determined by an additional step consisting of oxidizing organic compounds (e.g., by boiling in potassium persulfate) to split off the phosphate moiety which is then available for reaction. Instruments that use a single-beam colorimeter in either a continuous or batch-type analyzer are also available.

Flame Photometric Analysis

This analyzer detects the photometric flame emission of phosphorus compounds in a hydrogen-air flame. The method, developed by Draegerwerk in Germany, was first applied to the detection of phosphorus compounds in air and later was used as a detector for gas chromatography.[2]

The detector in Figure 11.25b contains a burner with separate delivery tubes for hydrogen and the air sample. The presence of phosphorus in the hydrogen-rich (reducing) flame produces a strong luminescent emission between the wavelengths of 4850 and 5650 Å. A narrow bandpass interference filter provides spectral isolation of this emission, which is at maximum at 5260 Å. The hydrogen and air are burned in a hollow tip that shields the flame from direct view of the mirror and photomultiplier tube. When phosphorus is present, the emission occurs above the shielded flame, and the light is transmitted directly and by way of the mirror through the filter to the photomultiplier tube. The shield offers specificity from carbon dioxide and hydrocarbons with flame emission at 5260 Å. The output from the photomultiplier tube is linear over several decades.

The detector can measure phosphorus in water by the addition of a nebulizer, which injects a mist of sample water into a clean air stream at a constant rate. The output is linear in the 1 to 100 ppm range.

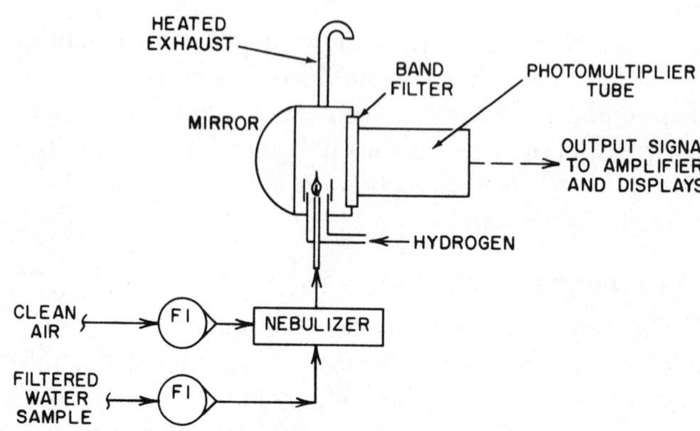

Fig. 11.25b Flame photometric detector of phosphorus in water

Gas-Liquid Chromatography

Where elemental phosphorus discharges into water, it has been found that a concentration of a few parts per billion is lethal to fish. Laboratory techniques[3] possessing this sensitivity have been developed. The phosphorus is partially isolated by extraction in a suitable solvent such as benzene or isooctane. A portion of the extract is injected into a chromatograph utilizing a flame photometric detector. Mud and tissue samples can also be analyzed rapidly by this method.

Sample Handling Systems

The successful application of the phosphorus analyzers just described depends on reliable delivery of a well-filtered sample. The system shown in Figure 11.25c is capable of handling raw sewage. The primary filter is of the motor-driven vane type with alternate stationary and rotating disks, and the clearance between the plates determines the degree of filtration.

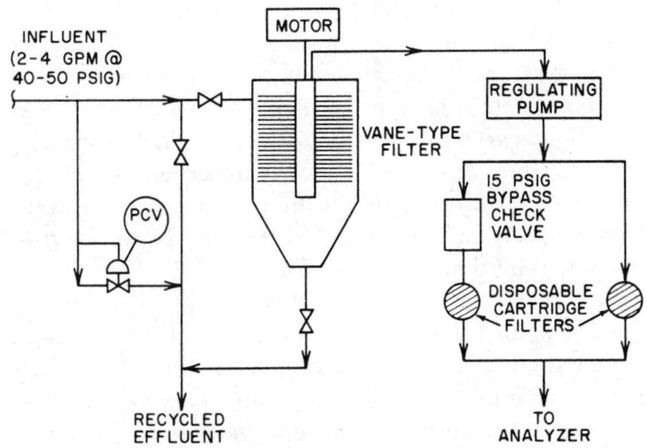

Fig. 11.25c Raw sewage sample-handling system

The second stage of filtering consists of two disposable cartridge filters. A check valve diverts the flow to the second filter when the pressure drop across the first indicates that a change is necessary. The regulating pump between the filter stages insures a constant flow rate in difficult determinations. Provision for backflushing the system is also advisable in some analyses.

As much attention should be given to the selection of a sample handling system as to the analyzer itself. Provisions for convenient calibration are also an essential part of a complete installation.

REFERENCES

1. *Standard Methods for the Examination of Water and Wastewater*, 12th ed., American Public Health Association, New York, 1965.
2. Brody, S.S., and Chaney, J.E., *Journal of Gas Chromatography*, Vol. 6, 1966, p. 42.
3. Addison, R.F., and Ackman, R.G., *Journal of Chromatography*, Vol. 47, 1970, p. 421.

BIBLIOGRAPHY

Addison, R.F., *Chromatograph*, Vol. 14, 1979, p. 421.

Frost & Sullivan, Inc., "On-Stream Process Analyzer Market," Report No. 669, Frost & Sullivan, Inc., New York, August 1979.

"Instrumentation and Automation Experiences in Wastewater-Treatment Facilities," EPA-600/2-76-198, Environmental Protection Agency, Washington, D.C., October 1976.

Jutila, J.M., "Multicomponent On-Stream Analyzers for Process Monitoring and Control," *InTech*, July 1979.

Sandford, J., "On-Stream Analyzers," *Instruments and Control Systems*, March 1977.

Standard Methods for the Examination of Water and Wastewater, 12th ed., American Public Health Association, New York, 1965.

Flow Sheet Symbol

TO RECEIVER

AT FLASH POINT OR BOILING POINT, ETC.

11.26 PHYSICAL PROPERTIES

This section deals with onstream analyzers which measure a physical property of a process stream.

Prior to the introduction of onstream analyzers, laboratory analyses were made on grab samples at intervals, and the results were reported to the process unit operator, permitting set point adjustments of parameters such as flow, temperature, pressure, and level. Continuous onstream plant analyzers offer many advantages including the following:

1. Continuous measurement of stream, eliminating high time lags.
2. Reduction of errors caused by unrepresentative samples, or by changes in sample composition due to sample handling.
3. Elimination of human errors characteristic of nonautomated laboratory procedures.
4. Ability to recognize process trends, thus permitting the direct control of a given process variable by closed-loop control.
5. Cost reductions resulting from the minimization of laboratory analyses.
6. Closer control with narrower tolerances of final product specifications, and reduction in quality "give-away."
7. Feasibility of establishing in-line blending systems, resulting in economies from elimination of tankage and from increasing system flexibility and quality control.
8. Ability to provide continuous inputs to computerized process control systems for plant optimization.
9. Direct measurement of process variables rather than detection of properties by inference.

Distillation Analyzers

Laboratory Measurements

Distillation analyzers were introduced to provide data on the volatility characteristics of process streams and separation efficiency of distillation units. ASTM Methods D 86—IP-123 and D 216—IP-191 are the currently accepted laboratory standards for determining the boiling characteristics of petroleum products distilled at atmospheric pressure. Both methods employ batch techniques and approach a single plate distillation process without reflux. The petroleum products analyzed are complex mixtures of many components, and a low level of fractionation is achieved. True boiling point distillations, in columns with 15 to 100 theoretical plates, and at reflux ratios of 5 to 1 or more, produce greater separation of components. The apparatus and procedure for true boiling point determinations are not standard, are complex, take longer to perform, and are not as widely used.

Distillation curves for a few hydrocarbons are shown in Figure 11.26a together with a comparison of curves generated by ASTM method D 86—IP-123 and by true boiling point determinations for kerosene.

ASTM Method D 86—IP-123 A 100-ml sample is heated in an Engler flask at a prescribed rate. Packing is not used and some refluxing occurs due to condensation (Figure 11.26b). The vapors produced flow through a condenser immersed in an ice-water bath, and the distillate is collected in a graduated cylinder. The initial boiling point temperature is defined as the reading of the thermometer at the instant when the first drop of condensate falls from the lip of the condenser tube. As the higher boiling fractions vaporize, condense, and col-

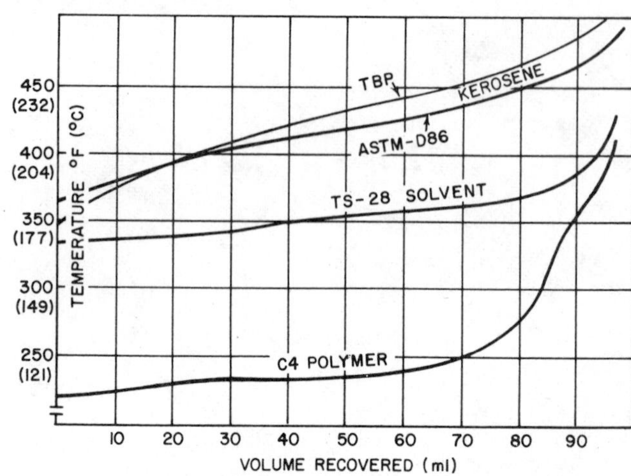

Fig. 11.26a Distillation curves

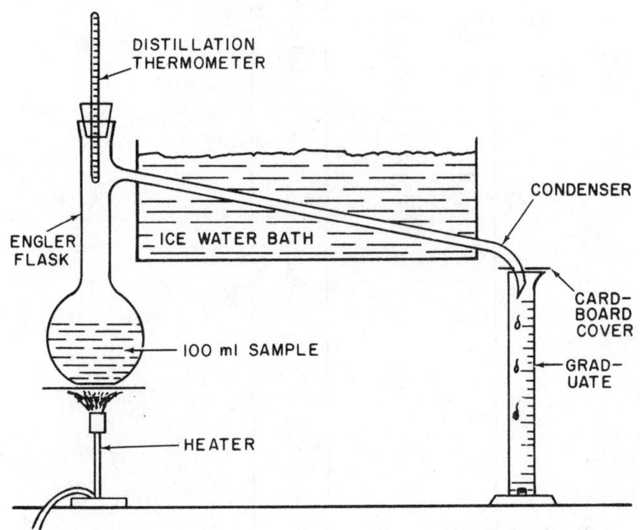

Fig. 11.26b Apparatus for ASTM D 86 distillation test

lect in the graduate, corresponding temperature readings are recorded to permit the plot of a curve of temperature versus percent of sample recovered. The end point or final boiling point is defined as the maximum thermometer reading observed during the test; it usually occurs when all the liquid has been boiled off from the bottom of the flask.

Usually, the percentage recovered does not equal the 100-ml sample charge, partly due to the inability of the apparatus to condense the lightest fractions. A curve of temperature against percent evaporation is determined by adding the percent of light ends lost to each of the recorded percentages recovered.

The precision of this method is a function of the rate temperature change. Repeatability ranges from 1 to 5°F (0.6 to 2.8°C), and the reproducibility from 4 to 13°F (2.2 to 7.2°C).

ASTM Method D 1160 This method provides for measurements under vacuums, ranging from 1 mmHg (133 Pa) absolute to atmospheric, to a maximum liquid temperature of 750°F (399°C). Results are not comparable with other ASTM distillation tests, although they may be converted to corresponding vapor temperatures at 760 mm by reference to Maxwell and Bonnell vapor pressure charts.

The sample must be moisture-free and is equivalent to a volume of 200 ml at the temperature of the receiver in which the condensed overheads are collected. The sample charge is boiled at a rate which produces a recovered distillate of approximately 4 to 8 ml per minute. The overhead vapors are condensed in a jacketed condenser and receiver in which the circulating coolant is maintained within ±5°F (±3°C) in the range of 90 to 170°F (32 to 77°C). Measurement of the vapor temperature is by a special Kovar-tipped thermocouple located at the sidearm of the boiling flask leading to the condenser. Temperatures for the 5, 10, 20, 30, 40, 50, 60, 70, 80, 90, and 95 percent recovered volumes are reported unless the liquid temperature, measured by a mercury in glass thermometer positioned in the boiling flask, reaches a value of 750°F (399°C). In these cases the test is terminated and the percentage recovered is also reported.

Depending on the percentage recovered and the operating pressure, repeatability varies from 8 to 10°F (4.4 to 5.6°C) and reproducibility from 15 to 30°F (8.4 to 17°C).

This method requires considerably greater skill to obtain optimum results and is more complicated than the more popular ASTM Method D 86—IP-123.

Plant Distillation Analyzers

Continuous plant distillation analyzers may be correlated with the results obtained by the ASTM laboratory methods but are not an exact duplication. In the laboratory (batch technique) the temperature (of a given percent evaporated value) is read off a rapidly rising vapor temperature as the sample is evaporated, whereas in the plant the analyzer measures the temperature of a continuous process, in which an equilibrium has been established for the given percent evaporated to be monitored.

Packed-Column End-Point Analyzer As shown in Figure 11.26c, the analyzer is a miniature process unit. A conditioned and pressure-regulated sample is delivered

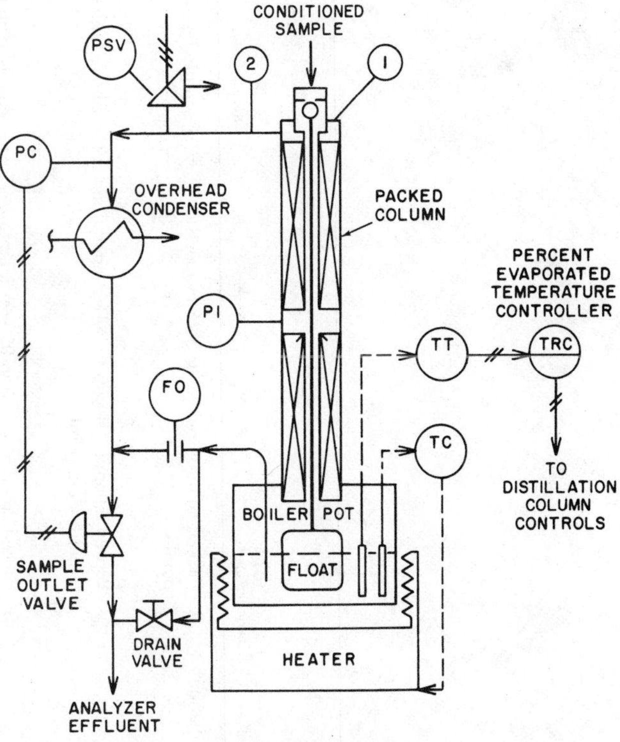

Fig. 11.26c Packed column end-point analyzer. In an initial boiling point analyzer, the temperature is detected in location 1 and a restriction orifice is added in location 2

ONSTREAM ANALYZERS

Analyzer Types		Range	Repeatability	Sample Condition: Flow Rate GPH (LPH)	Sample Condition: Pressure PSIG (kPa)	Sample Condition: Temperature °F (°C)	Response or Cycle Time Minutes	Approximate Cost in $1000s	Suppliers
Distillation	1. Packed column end point	over 30% evaporation	±1% sample boiling range	2–4 (7.6–15.1)	60 to 250 (414 to 1724)	180°F (82°C) maximum (lower than IBP)	16	21	PSG Industries
	2. Packed column initial boiling point	up to 30% evaporation	±1% sample boiling range	2–4 (7.6–15.1)	60 to 250 (414 to 1724)	180 (82) maximum (lower than IBP)	10–20	21	PSG Industries
	3. Packed column vacuum dist.	5 to 95% evaporation	±1% sample boiling range	2.3 (8.7)	50 to 250 (345 to 1724)	Flow related	16	35	PSG Industries
	4. Dist. analyzer horizontal still	1 to 99% evaporation	ASTM to 90%	0.2 (0.75) (analyzer)	10 to 400 (69 to 2758)	40 (22) lower than IBP			PSG Industries
	5. Automatic distillation	0 to 100%	ASTM	Batch	—	—	15 typ.		PSG Industries
Vapor pressure	1. Air saturated (Reid)	0–20 Reid vapor pressure (RVP)	±0.1 RVP	2 (7.6)	10 to 100 (69 to 690)	50–110 (10–43)	63% in 1 min	21	PSG Industries
	2. Dynamic	0–20 to 0–200 PSIA (0–138 to 0-1380 kPa)	Equal ASTM D 323	10–50 (38–189)	75 to 500 (517 to 3447)	70–120 (21–49)	0.75	19	PSG Industries
Vapor/liquid ratio V/L	Continuous	10–30 V/L to 150°F (66°C)	±0.5 V/L	2–4 (7.6–15.1)	10 to 100 (69 to 690)	Normal blending range	3	45	Ethyl Corp.

ONSTREAM ANALYZERS *continued*

Analyzer Types		Range	Repeatability	Sample Condition: Flow Rate GPH (LPH)	Sample Condition: Pressure PSIG (kPa)	Sample Condition: Temperature °F (°C)	Response or Cycle Time Minutes	Approximate Cost in $1000s	Suppliers
Pour point	General pour pointer	−75 to +50°F (−59 to +10°C)	±5°F (±2.8°C)	2 (7.6)	Below 20 (138)	20 (11) above pp	5 to 15	20	PSG Industries
Cloud point	Optical type	−13 to +59°F (−25 to +15°C)	±1°F (±0.6°C)	3–5 (11.3–18.9)	250 (1724) max.	20 (11) above cp	2	15	PSG Industries
Freezing point	1. Aviation fuels	−85 to +14°F (−65 to −10°C)	±1°F (±0.6°C)	3–5 (11.3–18.9)	250 (1724) max	50 to 122 (10 to 50)	3	20	PSG Industries
	2. Hydrocarbon parity	−42 to 212°F (−41 to +100°C)	ASTM	4–8 (15.1–30.2)	5 to 50 (34 to 345)	20 (11) above fp	12 to 18	on appl.	PSG Industries
Flash point	1. Low range	50 to 250°F (10 to 121°C)	±2°F (±1°C)	0.5 (1.9)		30 (17) below fp.	1 to 7	14	PSG Industries
	2. Integrated	50 to 250°F (10 to 121°C)	±2°F (±1°C)	3.5 (13.2)	1000 (6895)	250 (139) max. above fp.	1 to 7	26	PSG Industries
	3. High range	140 to 600°F (60 to 316°C)		1.0 (3.8)	20 to 90 (138 to 621)	50 (28) below fp.			PSG Industries
Octane	1. Standard engine comparator								Du Pont Co., Analytical Instr. Div.; Ethyl Corp.
	2. Continuous reactor tube	2 octane numbers	±0.1 RON	OJ (analyzer)	Regulated to <50 (345)		5	35 60	Foxboro Co.; UOP Inc.

at a rate of approximately 4 gal (15.2 liters) per hour to the top of a packed column through an inlet valve. The inlet sample flow rate is governed by a float in a boiling pot below the column which maintains an essentially constant level in the pot. A radiant heater boils the sample in the pot under an elevated pressure which is controlled. The overheads are condensed at this pressure. The bottoms' flow from the boiler pot is metered by means of a restriction orifice upstream of the outlet control valve.

Because the orifice and the packed column are subject to the same differential pressure, the ratio of overheads to bottoms is fixed. The orifice size and heater wattage are selected for the particular sample and for the percent evaporated point to be monitored. The bubble point temperature of the bottoms can be correlated with the percent evaporated point to be monitored. As the sample's percent evaporated temperature changes, the distillation process within the analyzer correspondingly adjusts the temperature in the boiler pot.

Sample effluent may be returned to a pressurized line by an educator or receiver and pump. The analyzer's thermocouple readout is calibrated by determining the temperature differential between distillation under the elevated pressure and the results of ASTM Method D 86—IP-123 measurements carried out at atmospheric pressure. The analyzer may be used for measurements from 50 percent evaporated temperatures to roughly the end-point temperature.

Packed-Column Initial Boiling Point Analyzer The analyzer operates in much the same manner as the packed-column end-point boiling point analyzer described in Figure 11.26c. One difference is that the conditioned and pressure-regulated sample first enters a preheater coil at the top of the boiler pot and is then fed to the top of the packed column. The inlet sample flow rate is controlled by a float-actuated valve in the boiler pot to maintain a constant level, and the pot is heated at a constant rate by a radiant heater. However, a restricting orifice is located in the overhead line at the top of the packed column. Where required, a restrictor may also be used in the bottoms line.

The sample is boiled at a controlled elevated pressure. The split of distilled sample into overhead and bottoms fractions is determined by the restrictions in these lines. Their ratio is constant since they are both subject to the same differential pressure. A thermocouple in the overhead line at the top of the column measures the overhead vapor temperature and serves as the analyzer readout. By the suitable selection of restrictors and heater size, a given percent evaporated temperature between the initial boiling point and the 50-percent point can be measured.

Packed-Column Vacuum Distillation Analyzer Some hydrocarbon feed stocks have very high boiling points or may decompose if boiled at atmospheric pressure. To avoid decomposition and to reduce the boiling point temperatures, the product may be distilled at a reduced pressure. Essentially the same technique is used as that shown in Figure 11.26c, except that the column and boiler pot are operated under a controlled vacuum. A conditioned and pressure-regulated sample enters the top of the column through a metering valve controlled by a float in the reboiler pot so as to maintain a constant sample level in the pot. The sample is preheated in a heat-exchanger above the reboiler pot before entering at the top of the packed column (Figure 11.26d). A fixed wattage radiant heater boils the sample. Overheads from the column flow through a water-cooled condenser from which some of the condensate is withdrawn by a precision metering pump at a constant rate, and the remaining distillate is refluxed to the top of the column through an overflow tube.

A second metering pump withdraws the bottoms fractions at a constant rate so that the ratio of overheads to bottoms flow fixes the percent evaporated material to be measured. Both pumps are gear driven by the same motor and maintained at a constant temperature by immersion in individually heated oil baths. A thermocouple

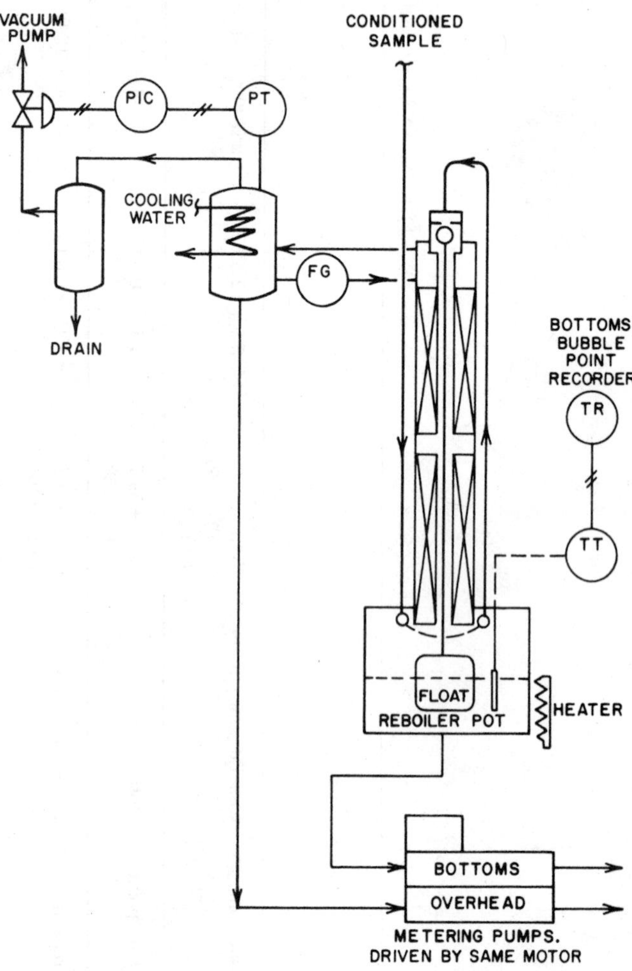

Fig. 11.26d Packed-column vacuum distillation analyzer

in the reboiler pot measures the bottoms' bubble point temperature, which may be correlated with ASTM Method D 1160.

A vacuum pump removes the noncondensable vapors from the system, a pressure controller modulates vacuum pressure, and a vacuum surge tank stabilizes the system to avoid excessive pressure fluctuations. If wide variations in product end point are expected, the heat input to the reboiler pot may be regulated by an autoformer in the heater circuit to avoid column flooding due to excessive refluxing. A sight glass in the reflux line permits observation of the reflux rate so that optimum column loading conditions may be established.

As the process boiling point temperature changes, the bottoms temperature increases or decreases accordingly.

Calibration is done by comparing analyzer readouts with ASTM D 1160 determinations for the same product drawn at the analyzer.

Horizontal Still Distillation Analyzer This analyzer, shown in Figure 11.26e, has two switch-selectable modes, the first of which permits the measurement of temperature at any preselected bottoms flow rate as a percentage of total sample flow rate through the analyzer. In the second mode, the percentage of bottoms to total sample flow at a specific temperature value is measured.

The distillation unit is an essentially horizontal still constructed with a series of baffles and weirs to promote separation of vapors and liquids at each compartment. The still is heated electrically. When the sample arrives at the last compartment, the vapor overheads and the liquid bottoms are at thermal equilibrium. A thermocouple in the bottoms section of the last compartment senses the temperature of distillation.

Vapor overheads leave the analyzer column at atmospheric pressure. They are condensed and pumped out of the analyzer system. The liquid bottoms are cooled, filtered, and then piped into a variable-speed pump system. Any escaping vapors from bottoms lines are condensed and returned to the bottoms pump. This prevents any significant losses in the volume of the bottoms product leaving the analyzer.

A differential pressure (dp) transducer continuously senses the level of the liquid bottoms at the suction side of the variable-speed pump. Heat exchangers insure that the inlet pump and the variable-speed bottoms pump operate at the same sample temperature in order to minimize errors in flow measurement.

The voltage output of the dp transducer is transduced by a voltage-to-frequency oscillator which produces the pulse train that drives the stepper-motor controlled variable-speed bottoms pump.

In one control mode, the distillation temperature is preset, and the logic control varies the speed of the bottoms pump until a constant pump suction head is established. The capacity of the sample feed pump is fixed, and the pulse rate of the bottoms pump is con-

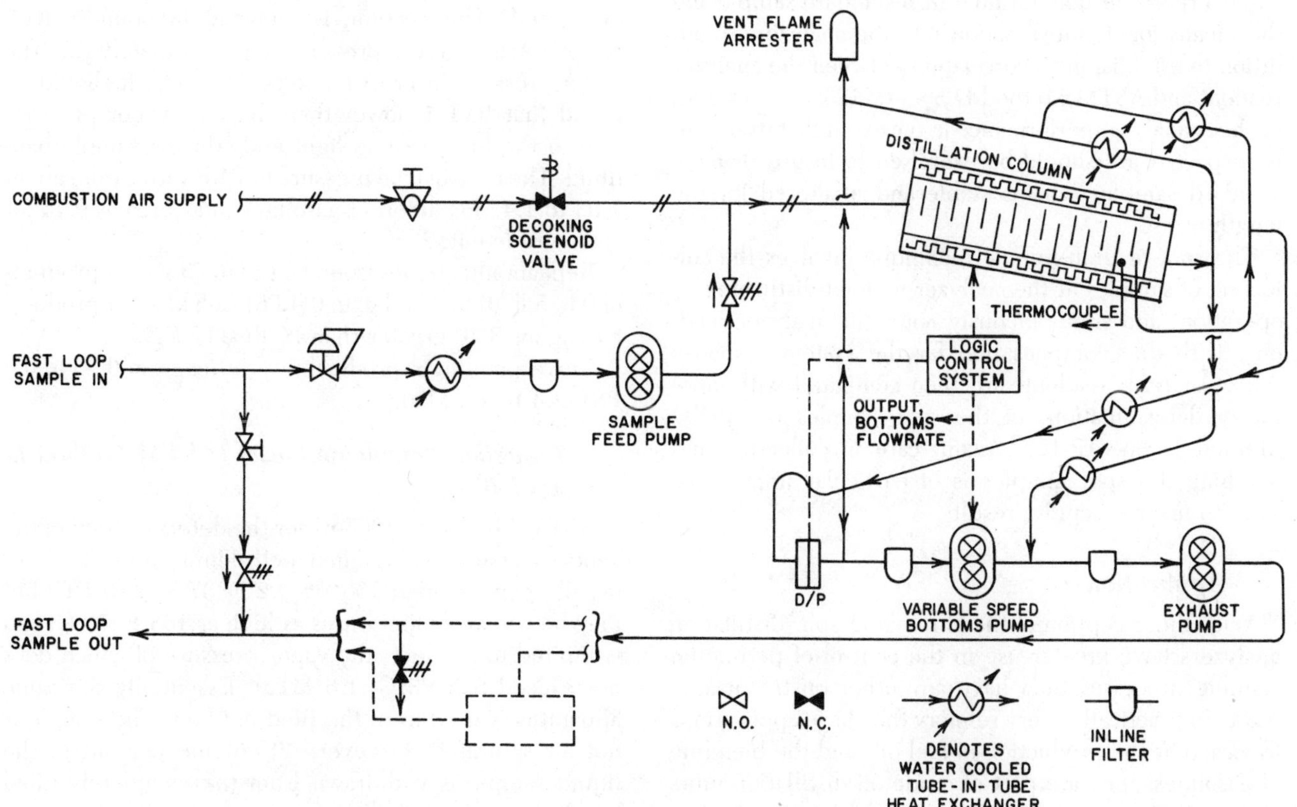

Fig. 11.26e Horizontal still distillation analyzer

verted into an equivalent flow volume, permitting the computation of the percentage distilled at temperature.

In the second control mode, the desired speed of the bottoms pump is preset, thus establishing a specific bottoms percentage. The logic control then varies the voltage to the heater until a constant pump suction head is established at the bottoms pump.

Automatic Distillation Analyzer This analyzer is an automated version of the ASTM Method D 86—IP-123 designed for plant use. It performs the same distillation test as the one illustrated in Figure 11.26b and produces a continuous temperature record with percent point lines and scribed at desired values (Figure 11.26a). Following a complete distillation, the cycle is repeated. Various and local indicators aid the operator, protect the analyzer, and avoid false data recordings due to malfunctioning.

Calibration

Techniques employed in the calibration of distillation analyzers are the standard sample method and the spot or grab sample method. Both are described in ASTM Method D 2891—70T. In the latter method, a sample is drawn off the process line and its properties determined (standardized) by replicate tests using ASTM Method D 86—IP-123. The standard sample is then introduced into the analyzer and the results are compared to the laboratory determinations. Sampling systems include a reservoir tank for holding a standard sample and the means for its introduction into the analyzer. In addition to establishing a correlation between the analyzer readout and ASTM Method D 86—IP-123, it also serves to check analyzer performance in the event that its output is suspect. Care should be exercised to insure that the standard sample remains stable and unaltered for the length of use.

The spot or grab sample technique involves the collection of samples at the analyzer without disturbing its operation, and simultaneously notes the analyzer readouts with due compensation for the system response time. Analyzer readouts are then compared with laboratory determinations of the spot samples by ASTM Method D 86—IP-123. Again, care in collecting and handling the spot samples is of particular importance so as to insure accurate results.

Application

With the exception of the chromatograph, distillation analyzers have greater use in the control of petroleum refining processes than have any other onstream analyzer. In practically every refinery they have applications as varied as the production of fuel oils and the blending of gasolines; they are part of crude oil distillation units and alkylation units; and they are used for control of catalytic reforming feed stocks and for control of the reflux ratio in gasoline fractionating towers.[1]

Vapor Pressure Analyzers

Reid Method (ASTM Method D 323-68)

The vapor pressure of petroleum products, except for liquefied petroleum gases, is usually determined by the Reid method as given by ASTM Method D 323-68. Sample handling is very critical with very volatile products, owing to the hazard of loss of light ends. For products having a vapor pressure less than 26 PSIA (179 kPa), a sample volume of at least 1 qt (0.95 liters) but not greater than 2 gal (7.5 liters) is collected and as soon as possible is placed in a bath maintained at 32 to 40°F (0° to 4°C). For higher vapor pressure products, a special bomb approximately 2 in. (50 mm) in diameter and 10 in. (250 mm) high is used.

The Reid vapor pressure (RVP) bomb consists of a liquid sample lower chamber coupled to an air chamber having a volume approximately four times that of the liquid and a Bourdon tube pressure gauge connected to the top of the air chamber.

Chilled sample is used to fill the liquid chamber and is quickly coupled to the air chamber with the attached pressure gauge. The assembly is then immersed in a water bath maintained at 100° ± 0.2°F (37.8° ± 0.1°C). After a minimum of 5 minutes, the apparatus is removed and shaken vigorously and the pressure is noted. It is then quickly reimmersed in the bath, and at intervals of not less than 2 minutes the procedure is repeated at least five times until two consecutive pressure readings are noted. This reading is reported in pounds, RVP (PSIA). Although the pressure is approximately the true vapor pressure in PSIA for some products, it should be noted that RVP is lower than the true vapor pressure due to the loss of some light ends during sample handling. The ratio of true pressure to RVP varies from about 1.03 to 1.45 for different gasolines to as high as 9.75 for some crude cuts.[2]

Repeatability varies from 0.1 PSI (0.7 kPa) for products of 0 to 5 lb (0 to 22.5 kg) to 0.4 PSI (2.8 kPa) for products having an RVP greater than 26 lb (117 Kg).

Corresponding reproducibility varies from 0.35 to 0.7 PSI (2.4 to 4.8 kPa).

Liquefied Petroleum Gases (ASTM Method D 1267-67)

ASTM Method D 1267-67 for the determination of the vapor pressure of liquefied petroleum gases (LPG) is usually performed at 100° ± 0.2°F (37.8° ± 0.1°C) but may be run at temperatures as high as 158°F (70°C) and is limited to products the vapor pressure of which does not exceed 225 PSIG (1.6 MPa). Essentially the same apparatus is used as in the Reid method. The sample is not air saturated; however, 40 volume percent of the liquid sample is withdrawn from the completely filled bomb after a prescribed purging procedure to allow space for expansion of the product when it is immersed in the test bath.

The observed pressure is reported as LPG vapor pressure in pounds per square inch gauge pressure, after corrections are made for gauge error and for standard barometric pressure at the test temperature.

Repeatability is given as 0.5 PSIG (3.5 kPa) plus 0.5 percent of the mean value for a test temperature of 100°F (37.8°C) and 1.0 PSIG (6.9 kPa) plus 1.0 percent of the mean value for a test temperature of 158°F (70°C).

Corresponding reproducibility is given as 1.0 PSIG (6.9 kPa) plus 0.5 percent of mean value and 2.0 PSIG (13.8 kPa) plus 2.0 percent of mean value.

Air-Saturated Vapor Pressure Analyzer

Sample is metered by a positive displacement pump at a rate of 100 cc per minute through a heat exchanger immersed in a constant temperature bath at 100°F (37.8°C) (Figure 11.26f). The sample is then sprayed into an air-saturation chamber, also immersed in the bath,

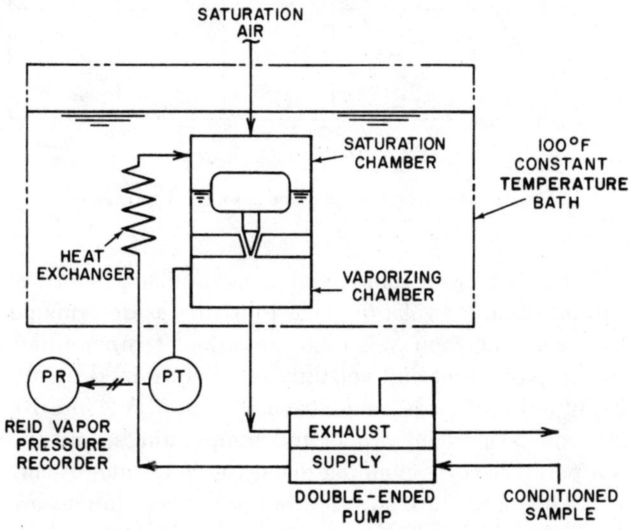

Fig. 11.26f Air-saturated vapor-pressure analyzer

where it is saturated with air. A float-controlled needle valve discharges the air-saturated sample and vapors into the vaporizing chamber and also maintains a constant level of liquid in the chamber. The sample supply pump is double ended and designed so that the exhaust end withdraws a flow of 500 cc per minute of liquid and vapor mixture from the vaporizing chamber, thus establishing the 4 to 1 liquid air ratio prescribed by the RVP (ASTM D 323-68) method. The pressure in the vapor chamber is sensed by a pressure transmitter with its 3 to 15 PSIG (21 to 104 kPa) output calibrated to represent the RVP. The exhaust pump jacket is also maintained at the 100°F (37.8°C) bath temperature.

Dynamic Vapor Pressure Analyzer

This analyzer may be calibrated to measure the vapor pressure of products covered by either ASTM Method D 323-68 or ASTM Method D 1267-67. The analysis is continuous, and the sample effluent may be returned either to a pressurized line or to an atmospheric receiver tank. Incoming sample is filtered and maintained at a constant pressure (Figure 11.26g). The sample is brought to a temperature of 100° ± 0.1°F (37.8 ± 0.06°C) in the heat exchanger, which is immersed in the constant temperature bath. The sensing device is a modified jet pump element with the suction side dead-ended into the cell of an absolute pressure transmitter. The velocity of the sample in the small-diameter nozzle causes the pressure head to approach the vapor pressure of the least volatile component in the stream. This value is lower than the RVP as determined by ASTM Method D 323-68. By reducing the efficiency, the system simulates the selective vaporization which occurs in the 4 to 1 liquid volume (vapor-liquid ratio) in the RVP test apparatus.

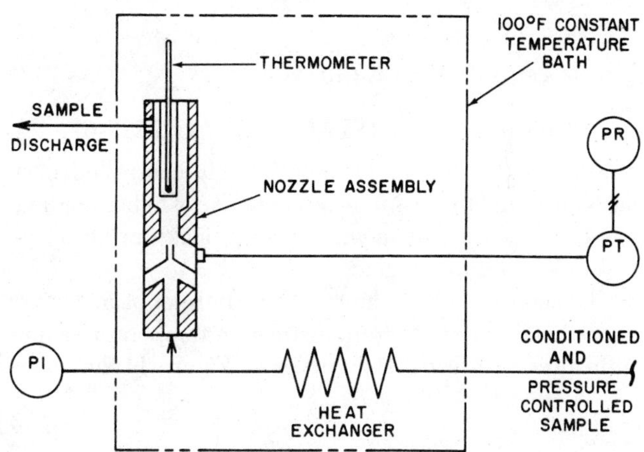

Fig. 11.26g Dynamic vapor-pressure analyzer

One of the parameters affecting the efficiency of jet pump, eductor, or aspirator is the location of the tip of the nozzle with respect to the throat of the downstream venturi. By adjusting the location of the nozzle, the operating efficiency may be adjusted so that the analyzer readout provides an essentially 1 to 1 correlation with the RVP. Figure 11.26h shows the relationship of the analyzer results against RVP for various hydrocarbons and blends.

When the analyzer has to redischarge into a pressurized line, a backpressure regulator is added to maintain a constant backpressure on the system and thereby eliminate the effects of varying return line pressure. An inlet pressure of roughly 45 to 50 PSIG (311 to 345 kPa) is required when the analyzer is discharging to atmosphere and the inlet pressure regulator is set at 40 PSIG (276 kPa). When the sample is returned to a pressurized line, the inlet pressure regulator is set at a value equal to 2.5 times the return line pressure plus 40.

Calibration

A standard is now under consideration by ASTM for a method of validating vapor pressure analyzers. The standard sample method and the spot or grab sample method are equally reliable if carefully performed. Sample collection and handling are critical in both cases as is the manner in which a stable and uniform standard sample is preserved during delivery to the analyzer.

Application

Increasing the quantity and quality of butanes and pentanes in the in-line blending of finished gasolines represents one potential application of vapor pressure analyzers. They have been used with LPG; more extensively they have been used to monitor the vapor pressure of pipeline-transported products, thereby avoiding vapor locking the pumps and minimizing safety hazards during tanker loading operations. They may also be used to detect product interfaces at pipeline receiver stations.

Vapor-Liquid (V-L) Ratio Analyzers

Volatility Test (ASTM Method D 2533-67)

Front-end volatility must be closely controlled in a gasoline blend to permit the greatest use of light blending components without incurring vapor lock during operation of a gasoline engine.

The vaporization of a fuel in the carburetor of an engine is predictable neither from distillation temperatures nor from pressure tests. The curves of Figure 11.26i show how four different fuels (with some similar properties) exhibit different volatility characteristics as determined by measuring their V-L ratios at various temperatures. In the past, front-end volatility was determined by utilizing indirect measurements such as the RVP, the 10, 20, and 50 percent evaporated temperatures, and the temperature corresponding to a given V-L ratio (usually 20). This procedure and the required computations are time-consuming, cumbersome, and inaccurate.

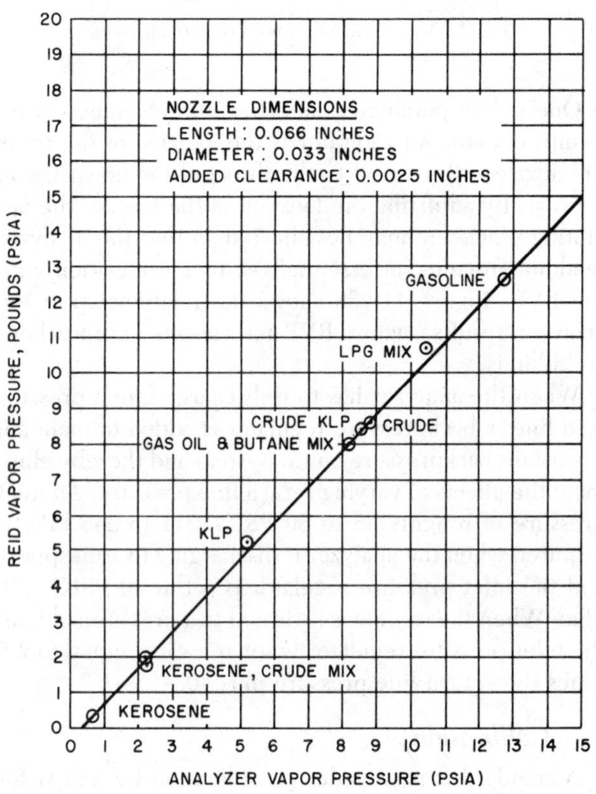

Fig. 11.26h Dynamic vapor-pressure analyzer results

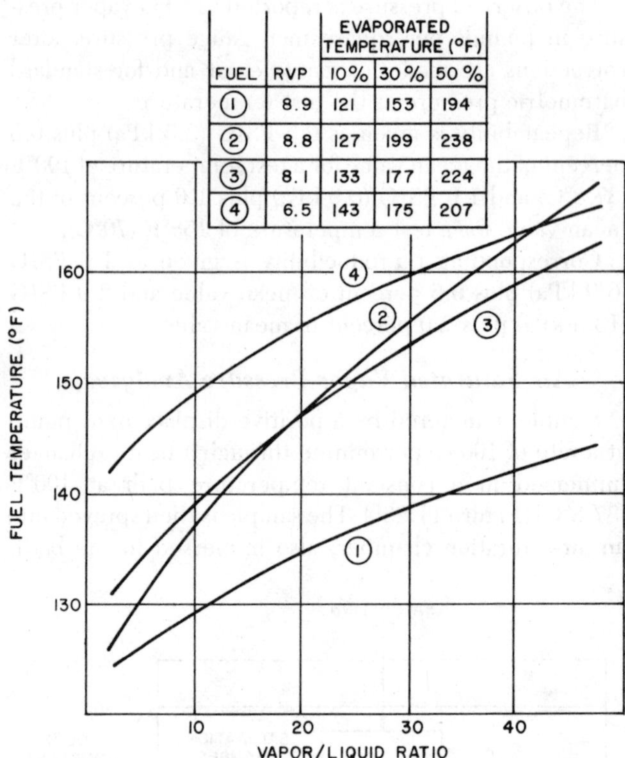

EVAPORATED TEMPERATURE (°F)
FUEL | RVP | 10% | 30% | 50%
① | 8.9 | 121 | 153 | 194
② | 8.8 | 127 | 199 | 238
③ | 8.1 | 133 | 177 | 224
④ | 6.5 | 143 | 175 | 207
FUEL TEMPERATURE (°F)
130 140 150 160
10 20 30 40
VAPOR/LIQUID RATIO

Fig. 11.26i Volatility effects on gasoline vaporization

The laboratory technique determines the V-L ratio by direct measurement for a given reference temperature. ASTM Method D 2533-67 utilizes a special buret, constructed with a stopcock at the top, 0.5 cc graduations, a short bottom arm fitted with a rubber septum, and a long bottom arm connected to a 250-ml leveling bulb by rubber tubing. The buret is filled with pure dry glycerin, and a sample of 1 cc or less is injected into the buret through the rubber septum by a hypodermic syringe.

The buret is then placed in a controlled temperature bath. As the sample vaporizes, the 250-cc leveling bulb, open at the top to the atmosphere by a drying tube, is raised or lowered to maintain the generated vapor at a pressure of 1 atmosphere absolute, which compensates for the prevailing barometric pressure. The volume of vapor is indicated on the buret's graduations, and since the liquid volume injected is known, the V-L ratio can be calculated. Repeatability is given as 4.0 percent of the mean V-L ratio and reproducibility as 1.6 plus 8.0 percent of the mean V-L ratio.

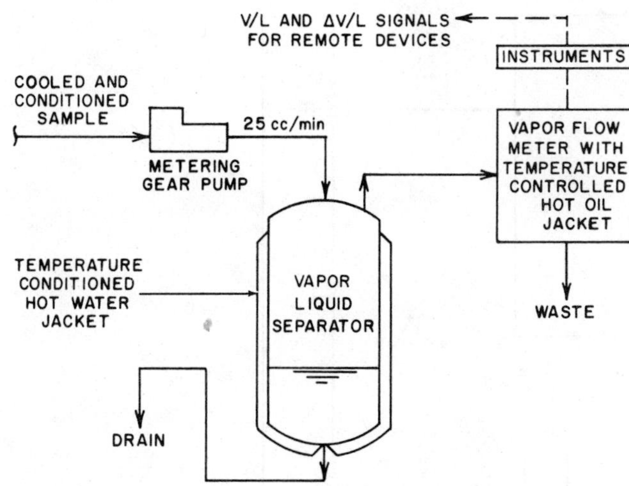

Fig. 11.26j Continuous vapor-liquid ratio analyzer

Continuous Vapor-Liquid Ratio Analyzer[3]

Figure 11.26j shows a block diagram of the analyzer. A slipstream from a sample loop is conditioned and cooled to approximately 35 to 40°F (2 to 4°C) before it enters a metering gear pump. The differential pressure across the pump is controlled to guarantee the constant delivery of liquid sample to a vapor-liquid separator at a rate of 25 cc per minute. The vapor-liquid separator is enclosed in a constant temperature bath controlled at the desired test temperature. Vaporization occurs at or near atmospheric pressure, with the residual liquid phase being discharged through a liquid seal, and the vapors are measured by a low pressure drop flow meter. The meter is maintained at 200°F (93°C) to prevent condensation and to establish a temperature reference for vapor flow measurements. With the inlet liquid flow rate fixed, the V-L ratio may be computed after compensations are made for the vapor temperature and the barometric pressure. In addition to computing the V-L ratio, the deviation of V-L from a preselected V-L set point is also displayed and provided as an output signal.

Calibration and Application

A V-L analyzer may be calibrated either by the introduction of a sample of known volatility or by the grab sample technique.

These analyzers are primarily applicable to in-line gasoline blending operations. An in-line blending system offers decided economic advantages in addition to greater flexibility, speed, and ease in switching blend formulas to meet production requirements. Although precise metering systems provide the means for implementing an in-line blending system, the final blend properties are still only implied rather than precisely known. Vapor pressure alone or its combination with distillation characteristics has been shown to be insufficient to provide efficient control of front-end volatility, because of inaccuracies and lags. The ability to maximize components like butanes and pentanes to meet seasonal and geographical requirements for volatility provides a very significant economic incentive. A V-L analyzer, in an in-line gasoline blending system, may be used to reset the set point of metering pumps. This measurement may eventually supplant vapor pressure measurements for controlling additions of butanes and pentanes to gasolines.

Pour Point Analyzers

Pour Point Tests (ASTM Method D 97-66—IP-15/67)

The standard laboratory procedure for measuring the flow characteristics of petroleum oils is given in ASTM Method D 97-66—IP-15/67. The sample must be heated without stirring to 115°F (46°C) or 15°F (8.3°C) above the expected pour point temperature, before starting the test. A thermometer immersed in a jacketed sample test jar and a cooling bath are used. The cooling rate of the sample is fixed as it is examined, at 5°F (3°C) intervals to ascertain if it will flow when the test jar is tilted. When no sample movement is detected in the tilted jar after a 5-second interval, the pour point is reported as 5°F (3°C) above the indicated temperature. This point corresponds approximately to a viscosity of about 500,000 centistokes (0.5 m^2/sec). Repeatability is given as 5°F (3°C) and reproducibility as 10°F (6°C).

Pour point analyzers were developed in an attempt not only to automate a laboratory procedure for process control, but also to improve the accuracy of such measurements. These analyzers were reasonably successful; however, with materials containing pour point depressants, some types failed to correlate with laboratory determinations.

General Pour Pointer

The analyzer (Figure 11.26k) is cyclic in operation, with one cycle consisting of five basic sequences. The analyzer is programmed so that it resets to sequence No. 1 on startup or upon interruption of any kind:

1. *Forced Drain Sequence*—Valves 4, 8, and 5 are energized, applying high-pressure air to the test cell. The electric current to the T/E module is reversed from normal polarity so that the test cell is warmed and force-drained of all residue. Valve 5 closes to terminate this sequence. Valves 4 and 8 remain energized.

2. *Fill Sequence*—Valves 6, 7, and 9 are energized. The T/E module continues to heat the sample. Depending upon product viscosity, either low or high-pressure air is applied to the test cell while sample flows into the cell and out to drain by way of the overflow. This air pressure prevents the sample from entering the sensing lines of the analyzer. Valve 6 closes to terminate this sequence.

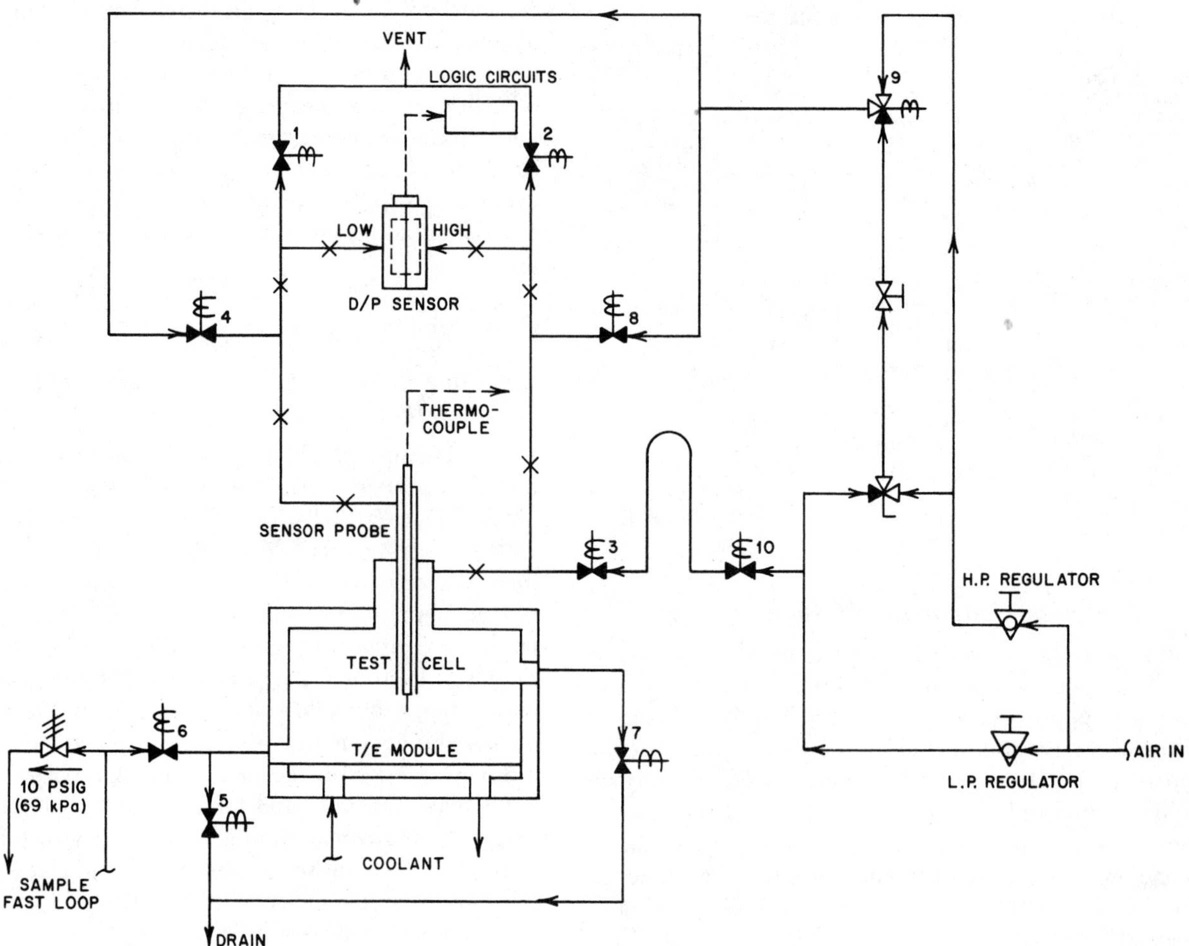

Fig. 11.26k General pour pointer

3. *Level Sequence*—Valves 4, 7, 8, and 9 remain energized, pressuring the surface of the sample and forcing surplus product out to drain. Power to the T/E module is applied at normal polarity to initiate sample cooling. At termination of this sequence, all valves deenergize, and the sensing probe is submerged below the surface of the sample.

4. *Equalizing Sequence*—Valves 1, 2, and 10 are energized. The test cell, the dp sensor, and all lines connecting these two components are vented to atmosphere through valves 1 and 2, establishing a condition of zero differential pressure in the sensing system. At the end of this sequence, all valves close. Valve 10, in closing, traps low-pressure air in the tubing between valves 3 and 10.

5. *Pulse Sequence*—Valve 3 energizes. The low-pressure air trapped between valves 3 and 10 expands into the test cell, producing a momentary pressure pulse against the surface of the liquid outside the sensor probe and against the high pressure diaphragm of the dp sensor. The output voltage of the dp sensor rises immediately, reflecting this pressure pulse.

The pressure pulse on the surface of the liquid also tends to push liquid up into the sensor probe, compressing the air inside the probe. As the sensor probe is connected to the low-pressure diaphragm of the dp sensor, this results in a reduction of the output voltage of the dp sensor.

The control logic of the analyzer has a fixed time delay built in which gives the sample time to compress the air inside the sensor probe after each pressure pulse. The output of the dp sensor is then compared to an adjustable set point voltage. If the output of the dp sensor is below set point, another equalizing sequence and another pulse sequence occur.

As the sample cools, its resistance to flow increases and its ability to compress the air inside the sensor probe decreases. This results in an increase of output voltage from the dp sensor. Eventually, this voltage exceeds set point. A peak-holding relay trips and the analyzer resets to the forced drain sequence to begin another cycle.

A thermocouple installed inside the sensor probe measures sample temperature at all times. Pour point temperature is the lowest temperature attained in each measuring cycle.

Calibration and Application

The repeatability of ASTM Method D 97 for pour point is 5°F (2.8°C) and its reproducibility is 10°F (5.6°C). A sufficient number of determinations to improve the accuracy of the ASTM results is advisable, because the process analyzer ordinarily exceeds the accuracy of the laboratory method. Either the standard or the grab sample method is suitable. The convenience of a locally available standard sample provides a rapid check on the analyzer's performance as well as serving as a means of calibration. The greater initial cost for this feature should be weighed against the delay incurred in obtaining a laboratory analysis of the spot sample.

Pour point temperature measurements are utilized more extensively in Europe than they are domestically, since the use there of furnace and fuel oils is less pronounced. When a product is sufficiently free of wax so that a cloud point determination becomes meaningless, the pour point may be used as an index of the temperature at which flow will be impeded due to semi-solidification rather than by the formation of wax crystals.

Cloud Point Analyzers

Cloud Point Tests (ASTM Method D 2500-66—IP-219/67)

This method is applicable to products with a cloud point below 120°F (49°C) which are transparent in a layer 1½ in. thick (37.5 mm).

A cylindrical, flat-bottomed, clear glass test jar 1¼ in. (31.3 mm) in diameter by 4¾ in. (118.8 mm) long, and having a scribed sample fill line 2⅛ in. (53.13 mm) above the inside bottom surface is used. A cork holds a thermometer coaxially in the test jar so that its bulb rests at the bottom.

A watertight jacket with a ¼-in. (6.25 mm) thick cork or felt pad at the bottom holds the test jar when it is immersed in a cooling water bath. The sample must be dried at a temperature at least 25°F (13.9°C) above the approximate cloud point so as to remove any moisture and minimize trace water haze formation. The test jar is fitted with a cork or felt ring approximate $^3/_{16}$-in. (4.69-mm) thick which is positioned 1 in. (25.4 mm) from the bottom to keep it centered in the jacket.

The test is begun by placing the jar vertically in a 30 to 35°F (−1.1° to 1.7°C) water bath so that the jacket projects no more than 1 in. (25.4 mm) from the bath liquid.

At intervals of 2°F (1.1°C) the jar is quickly (3 sec. at most) but gently removed and examined for formation of a wax haze. (A water haze is generally uniform throughout the sample, whereas a wax crystal haze always appears first at the bottom of the jar.)

If the cloud point is not detected when the sample reaches 50°F (10°C) the sample is transferred to a second bath which also contains a test jacket and maintained at 0 to 5°F (−17.8° to −15°C), and the test is continued. Successively lower temperature baths are used as required for low-temperature cloud point products. The temperature, expressed in increments of 2°F (1.1°C), at which a distinct wax haze is first observed is reported as the cloud point. For gas oils repeatability is 4°F (2°C) and reproducibility is 8°F (4.4°C). For other oils, both repeatability and reproducibility are 10°F (5.6°C).

Optical Cloud Point Analyzer

The cloud point analyzer (Figure 11.26l) detects the temperature at which wax crystals first appear as a liquid hydrocarbon sample is cooled.

A measuring cycle begins when a three-way solenoid valve, which normally recirculates a slipstream sample, is energized and introduces sample into the test cell. After the cell is flushed clear of all residue from any previously tested sample, the solenoid valve deenergizes, and an electrical cooling unit reduces sample temperature until the analyzer detects the presence of wax crystals. At that point the cloud point temperature is represented as the lowest temperature the sample attains, as measured by a thermocouple, and the next measuring cycle is initiated.

The detection system consists of a light beam which is directed through a polarizing lens, through the sample cell and a second polarizing lens, and into a photocell. The electrical resistance of the photocell varies with the amount of light to which it is exposed.

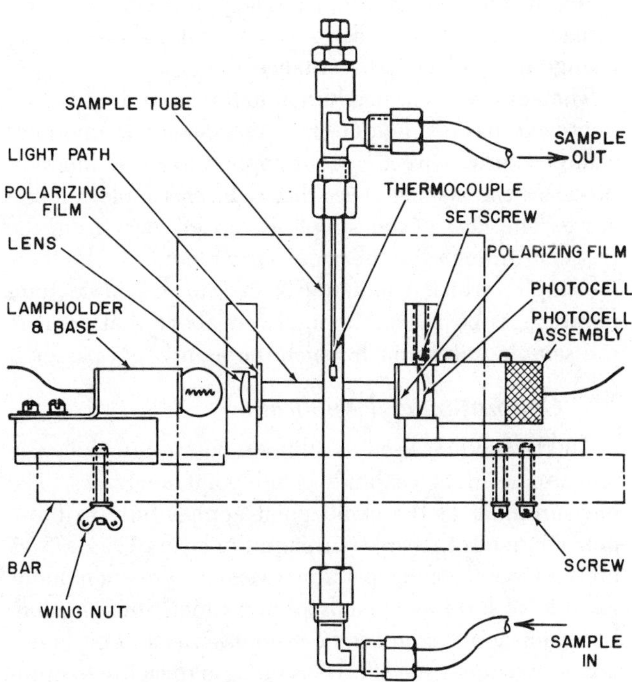

Fig. 11.26 l Optical cloud point analyzer

Polarizing lenses permit only light vibrating in a single plane to pass through them. For example, light rays may be controlled so they vibrate only in a vertical plane if the transmission axis of the lens is vertical. If two lenses are installed so that axis of light transmission of the second lens is perpendicular to that of the first, no light rays would pass through the second lens (a vertically vibrating light ray cannot pass through a horizontal lens axis). The test cell lenses are assembled in this manner, and the photocell normally senses presence of little light.

As the sample is cooled, it eventually reaches that temperature at which crystals of ice and/or wax appear. These crystals reflect some light rays and bend others so that the plane of vibration of some light rays changes as they pass through the crystals. As the angle changes at which the light rays approach the second lens, the lens is no longer capable of blocking all passage of light because some of the light rays are then vibrating in the plane in which the second lens will permit light to pass through.

Ice crystals are doubly refractive. As light rays pass through them, ice crystals will change the path of the light rays several times. However, ice crystals have very low refractive angles and, therefore, they do not seriously affect the critical direction of the light rays as they approach the second lens. Wax crystals are also doubly refractive, but have very high angles of refraction and, therefore, have a very pronounced effect of the light rays passing through them.

The photocell, which senses little light passing through a crystal-free sample, will detect small quantities of light rays passing through the lenses as a result of the presence of ice crystals and very large quantities of light as a result of the presence of wax crystals. The photocell, therefore, signals the first appearance of wax crystals by a significant change in its electrical resistance.

The analyzer cell, in addition to the lamp, lenses, and photocell, has two Peltier-effect thermoelectric modules clamped to it. Direct current applied to these modules produces the cooling effect that reduces sample temperature. The heat of the sample is carried away from the test cell through the thermoelectric modules. The thermocouple, which continuously measures sample temperature, is positioned immediately above that point in the sample line through which the light beam passes.

Calibration and Application

The techniques used to calibrate pour point analyzers are also used to calibrate cloud point analyzers. The measurement of the cloud point applies only to those petroleum oils that are transparent in layers 1½ in. (37.5 mm) thick and contain paraffin waxes or other compounds capable of forming crystals prior to total solidification. Its major application has been to gas oils and cycle oils not only to meet specifications but also to facilitate product transport and to prevent filter clogging during cold weather.

Freezing Point Analyzer

Aviation Fuel Tests (ASTM Method D 2386-67 and IP-16/68)

The freezing point as defined by this method is the temperature at which the last hydrocarbon crystal (formed during cooling) melts after the sample temperature is allowed to rise. This temperature must be within 3°C (5.4°F) of the temperature at which the appearance of hydrocarbon crystals is first observed.

A jacketed clear-glass tube is filled with either dry air or nitrogen at atmospheric pressure. The outer tube's outer diameter (OD) is 30 mm and the inner tube's inner diameter (ID) is 18 mm, with a 2-mm space between tubes. The overall length is 237 mm. The plug in the sample tube holds a total-immersion thermometer and a stirring rod formed with three spiral loops at the bottom and positioned slightly below the thermometer bulb. The tube is filled with 25 cc of fuel and placed in a clear vacuum flask (70 mm ID and 280 mm long) containing a coolant such as alcohol or solid carbon dioxide.

The sample is stirred vigorously during cooling and the temperature at which hydrocarbon crystals first appear is noted, neglecting any haze which may form at about 14°F (−10°C) owing to dissolved water in the sample. The sample tube is then removed and allowed to warm up slowly while the sample is continuously stirred. The temperature at which the last crystal disappears is reported as the freezing point temperature, provided it is within 3°C of the crystal formation point temperature. Otherwise the test must be repeated.

Repeatability is 0.7°C and reproducibility is 2.7°C.

High-Purity Hydrocarbon Test (ASTM Method D 1015-55)

This method is used in conjunction with ASTM Method D 1016-55 for determining the purity of hydrocarbons from freezing-point measurements.

The freezing point is determined from a 50-cc sample and a precision platinum resistance thermometer, calibrated by the National Bureau of Standards.

The sample is placed in a freezing tube (25 mm ID and 50 mm OD) with a silvered inner wall. A brass cylindrical sheath with an asbestos pad at the bottom holds the freezing tube in a cooling bath. The plug at the top of the freezing tube supports the precision platinum resistance thermometer, a double spiral stirring rod, and a tube with a spherical joint for admission of dry and carbon dioxide–free air. The space between inner and outer tubes is connected to a vacuum system. At the beginning of the test the bath is filled with a refrigerant suitable for the estimated freezing point temperature. Sample is introduced into the freezing tube by temporarily removing the top stopper and thermometer.

The purpose of the dry air flow into the freezing tube is to blanket the sample and prevent water vapor from entering. The sample is stirred and, when the temperature approaches the freezing point, evacuation of the jacket space is begun. Time and temperature observations are made and the vacuum on the jacket is adjusted so as to achieve a cooling rate of 2°F (1.1°C) in 1 to 3 minutes. The time (within 1 second) is recorded for a resistance thermometer change of 0.05 to 0.1 Ω. After subcooling, crystallization is induced by dipping a chilled special rod into the sample.

A cooling curve may be plotted which permits the determination of the freezing-point temperature with a sensitivity of about 0.0001°C.

Repeatability is ±0.005°C and reproducibility is ±0.015°C.

Freezing-Point Analyzer for Aviation Fuels

The freezing point analyzer essentially measures the cloud point of hydrocarbon samples at very low temperatures. The correlation between the cloud point and the freezing point of a sample is determined by laboratory analyzers, and the analyzer can be used for process measurement of either physical property.

The principle of measurement is exactly the same as that of the cloud point analyzer. The sample cell is cooled by Freon from a mechanical refrigeration unit supplied with the analyzer.

Freezing-Point Analyzer for Hydrocarbon Purity

This analyzer was initially developed to measure the purity of p-xylene but has since been applied to other organic compounds such as benzene and phthalic anhydride.

The technique follows ASTM Method D 1015-55 closely except that stirring of the sample is omitted.

A two-pen temperature recorder with programmer, limit switches, and thermistor bridge circuit is mounted in the control room. One pen (high span) triggers crystallization by jarring the sample when a predetermined supercooled state has been achieved. The second pen (narrow span) records the percent purity of the sample. The field unit contains the sample cell and related hardware.

Fresh sample is admitted at the bottom of a vertical cylindrical sample cell by a programmed sample inlet solenoid valve, purging the system for 30 seconds. Sample cooling is initiated when the inlet valve is closed, trapping a 15-cc sample in the cell. Cooling continues until 9 to 15°F (5 to 8°C) below the freezing point is achieved. The high-span pen at this point will actuate a limit switch and energize a plunger which jars the sample cell to initiate crystallization—an exothermic reaction, which raises the sample temperature to its freezing point. Since the total latent heat of fusion was not extracted from the sample during supercooling, a small volume in the center of the cell remains in the liquid phase at a temperature in equilibrium with the solidified portion. A thermistor extending coaxially down from the top of the cell displays the sample's liquid phase temperature on the narrow-span pen of the recorder. In a few minutes the heat absorbed from the surrounding air completely liquefies the sample, and the recorder chart displays the freezing-point temperature on the calibrated chart. At this point a new cycle is initiated.

When the freezing point is above ambient temperature, the same technique is used except that the sample is heated and allowed to supercool by heat loss to ambient temperature.

Calibration and Application

Calibration of the aviation fuel analyzer should be based on the comparison of analyzer readouts with multiple test results. Its principal application is in the processing of aviation fuels such as JP-4, kerosines, and similar products.

Determination of purity is needed in the production of benzene, toluene, ethylbenzene, o-xylene, p-xylene, and phthalic anhydride. These materials must be able to be supercooled and have a specified freezing-point temperature.

Calibration of hydrocarbon purity analyzers is best achieved by using a sample of known purity. A certified standard thermistor may also be used if a sample of known purity is unavailable. Comparison with the ASTM methods can also establish the purity of a sample which can then be used for analyzer calibration.

Flash Point Analyzer

Flash Point Tests (ASTM Method D 56-70, ASTM Method D 93-66—IP-34/67)

The plant analyzers are intended to correlate with ASTM Method D 56-70 (Tag closed tester) and ASTM Method D 93-66—IP-34/67 (Pensky-Martens closed tester).

ASTM Method D 56-70 is for materials with a viscosity less than 45 SSU at 100°F (37.8°C) and a flash point below 200°F (93°C). A sample of 50 cc is used at a temperature of at least 20°F (11°C) below the expected flash point. The sample cup is immersed in a bath the temperature of which may be raised at a prescribed rate. Thermometers measure the bath and sample temperature. The sample lid prevents loss of sample vapors and directs a small flame into the cup periodically. The flash point is defined as the lowest sample temperature to cause ignition of the vapor above the sample at 1 atmosphere absolute.

Repeatability is 2°F (1.1°C) for flash points below 140°F (60°C) and 3°F (1.6°C) for flash points between 140 and 199°F (60° and 92.8°C). Reproducibility is 6°F (3.3°C) for flash points below 55°F (12.8°C), 4°F (2.2°C) for flash points between 55 and 139°F 12.8° and 59°C) and 6°F (3.3°C) for flash points from 140° to 199°F (60° to 92.8°C).

The Pensky-Martens closed tester is for materials with an indicated flash point temperature as high as 700°F (371°C). Approximately 4.2 cubic inches (6.7×10^{-5} m^3) of sample are used, and the sample cup is heated directly by either a gas or electric heater at a prescribed rate. The sample cup lid is designed to support a sample stirrer, a mercury-in-glass thermometer, and an apparatus for periodically exposing the vapor above the sample to a test flame. The repeatability is 10°F (5.6°C) for materials with a flash point above 220°F (104°C) and the reproducibility is 15°F (8.3°C).

Low-Range Flash Point Analyzer

When the sample is heated to its flash point and its vapor ignited, the temperature in the vapor space increases. Sample is fed to the analyzer's heating chamber at a constant rate, and air is added at a rate of 600 cc per minute (Figure 11.26m). The air-sample mixture is heated at a controlled rate before it enters the flash cup and overflows to maintain a constant level. The vapor rises into the vapor space and is periodically exposed to a high-voltage spark. Thermocouple 1 in the flash cup measures the temperature of the air-sample mixture, and thermocouple 2 in the vapor space responds to the temperature rise when vapor ignition occurs. This shuts the heater off and causes recorder pen 1 (to which thermocouple 1 is connected) to be driven downscale. The peaks of the resulting sawtoothed record indicate the flash point temperature for each analysis cycle.

Integrated Flash Point Analyzer

This design is identical to the unit described in Figure 11.26m except that it is mounted on a frame with all the necessary accessory components piped and wired for field installation. It is particularly applicable for pipeline interface detection.

The accessories include (1) a sample conditioning system to filter and coalesce free water from the sample, regulate sample pressure, and indicate coalescer bypass flow; (2) a mechanical refrigeration system and temperature controller to cool the sample below its flash point; (3) an air compressor, filter, and flowmeter to supply combustible air; (4) a duplex positive displacement pump to provide a constant rate of sample flow to the analyzer and return analyzer effluent and coalescer bypass to the pressurized process line; and (5) block, check, relief, and backpressure valves for isolation and ability to withdraw sample for calibration.

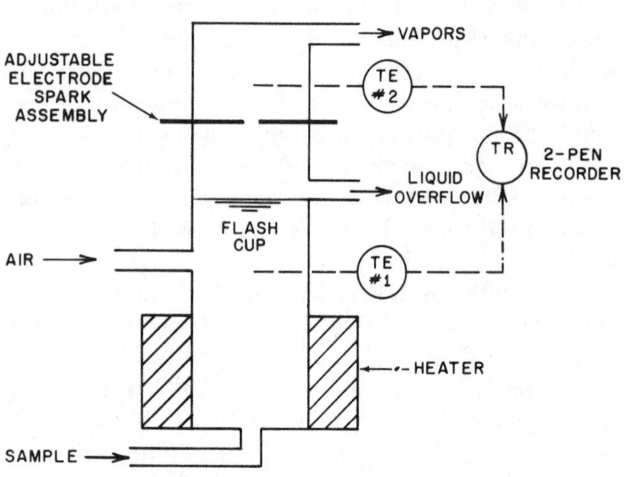

Fig. 11.26m Low-range flash point analyzer

High-Range Flash Point Analyzer

Sample from a sweepstream is metered to the system at a constant rate by one head of a duplex positive displacement pump and is preheated to a fixed temperature below the flash point as determined from the preceding analysis. It is mixed with air at a rate of 1500 cc per minute. A final heater provides the additional heat required to bring the air-sample mixture to the flash-point temperature. The liquid entering the flash chamber is returned to process by the second head of the duplex pump, and the rising vapors are exposed to a high-voltage spark every 10 seconds.

Ignition is detected by the deflection of a diaphragm caused by the combustion pressure pulse. The control circuit increases or decreases the final heater output, depending on whether or not ignition has occurred. At the same time the preheater controls are also adjusted to maintain the desired temperature differential. Flash point temperature is sensed by a thermocouple in the flash chamber liquid and displayed on a recorder chart.

Calibration and Application

Either the spot sample or standard sample method may be used. In either case, care must be exercised when a low flash point sample is used so as to prevent loss of light ends.

The analyzer can be applied to the control of vacuum distillation, dewaxing, solvent extraction and stripping, deasphalting, blending, residual fuel oil processing, and pipeline interface detection.

Octane Analyzers

Laboratory Tests

Treatment will be brief owing to the vast complexity of this subject. Basically two methods are employed in which in a standard engine an unknown fuel is compared with a standard or reference fuel. One method yields a motor octane number (MON) in which the engine is run at 900 rpm, and the second method provides a research octane number (RON) in which the engine is run at 600 rpm. The difference in octane number by these methods (or "spread") is indicative of city driving at low speeds as compared to highway engine performance. Standard fuels are based on normal heptane (zero rating) blended with isooctane (100 rating), with the octane number equal to the percentage of isooctane in the blend. The range has been extended by the addition of tetraethyl lead

(TEL) to the isooctane for ratings above 100. When the unknown fuel produces the same knocking as a standard fuel blend, it is rated equal to the octane rating of the standard blend. The RON is higher than the MON, with the spread increasing with increasing octane numbers.

Standard Engine Octane Comparator Analyzer

These analyzers serve to automate the ASTM procedure, using a standard engine with a modified carburetor fuel delivery system and standard detonation pickup and knockmeter. Comparison is made between the process stream and a prototype fuel, which serves as the standard or octane number reference point, from which an octane number difference is determined as the analyzer readout. Figure 11.26n illustrates the equipment diagram for such an analyzer.

The accuracy of the systems depends on the performance of the standard engine, which must be properly maintained for optimum system operation. Also, the prototype fuel octane number should be determined to within ±0.1 or better since it serves as the reference for stream comparison measurements.

These analyzers can trim octane "give-away" during blending to approximately 0.05 octane above specification requirements.

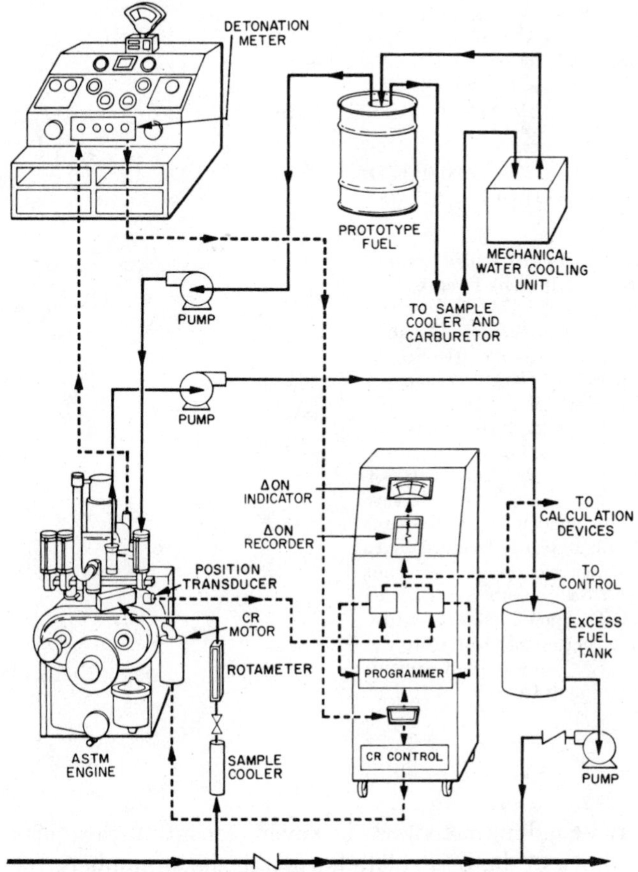

Fig. 11.26n Octane analyzer equipment

Reactor Tube Continuous Octane Analyzer

This analyzer monitors the reactions that precede engine knocking, the parameters of which may be controlled and correlated with octane number. A fuel and controlled air volume mixture is delivered at a rate of 1 cc per minute to a reactor tube maintained at an elevated temperature. Partial oxidation reactions in the tube produce a peak temperature the location of which is related to octane number. Higher octane fuels cause the peak temperature to move away from the tube inlet, whereas increasing the reactor tube pressure moves the peak closer to the tube inlet. Consequently, if the temperature peak location is fixed by varying the reactor tube pressure as fuel octane number varies, the pressure may be correlated with octane number and used as the analyzer readout. This is accomplished by locating two thermocouples in the tube, 1 in. (2.54 cm) apart and equidistant from the temperature peak. Any movement of the peak due to a change in fuel octane rating is sensed by a differential temperature controller and causes a compensating change in reactor tube pressure to restore the temperature peak location.

Cool Reactor Tube Octane Analyzer

This type of analyzer exploits the relationship between the octane of a gasoline and the temperature it attains in a partial oxidation reaction. A sample of gasoline mixed with air is introduced through a caromatographic sample valve into a reaction tube, which is controlled at a temperature of 300°C. The maximum temperature attained is recorded and translated into an octane number. The analyzer is calibrated using a sample of similar composition of a known octane number.

The outer version of the analyzer uses a pressure controller to maintain the maximum temperature at the same location in the reactor tube. The pressure required then becomes the increase of the octane of the sample.

One version of the analyzer uses a microprocessor to control the timing and data conversion functions of the analyzer. A typewriter terminal is provided for local or control room readout and for communication with a master computer.

REFERENCES

1. Waner, N.S., "Lecture Notes on Physical Property Analyzers Given at Instrument Society of America Short Courses on Process Analyzers," Instrument Society of America, Pittsburgh, Pa.
2. Nelson, W.L., "How Reid and True Vapor Pressure Vary," *Oil and Gas Journal*, June 21, 1954.
3. Huffman, H.C., Hass, R.H., O'Brien, N., Unzelman, G.H., and Jones, J.T., "An Analyzer for On-Line V/L Control," Presented at American Petroleum Institute Meeting, Houston, Texas, May 1970.
4. McLaughlin, J.H., and Bajek, W.A., "An Instrument to Continuously Monitor the Octane Quality of Gasoline," Presented at the Instrument Society of America Conference, Houston, Texas, October 1969.

Flow Sheet Symbol

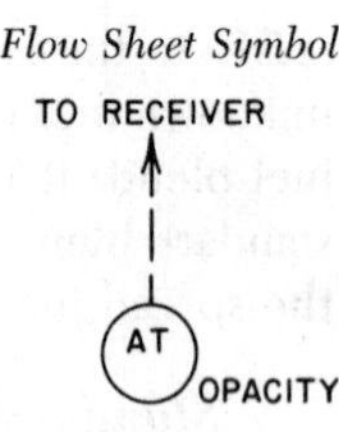

11.27 STACK/PARTICULATE MONITORS

Sampler and Monitor Types:
A—High-volume sampler
B—Dichotomous sampler
C—Tape sampler
D—Manual stack sampler
E—Piezoelectric crystal mass balance
F—Impaction devices
G—Radiometric devices
H—Charge transfer/surface ionization
I—Light attenuation/transmissometer
J—Light scattering
K—Visual observation
L—Remote sensing

Potential Applications: Visibility (I,J,K); fire/smoke detection (G,I); particle sizing (B,F,I,J)

Type of Sample and Installation: In situ (G,H,I,J,K,L); extractive (A,B,C,D,E,F,G); ambient air (A,B,C,E,F,G,I); flue/stack gas (D,F,G, H,I,J,K,L)

Reference Methods: EPA 40 CFR 50 Appendix B (A) ANSI/ASTM D1704-78 (C) EPA 40 CFR 60 Appendix A, Method 5; ANSI/ ASTM D2928-71; ANSI/ASTM D3685-78 (D) EPA 40 CFR 60 Appendix B, Perf Spec 1 (I) EPA 40 CFR 60 Appendix A, Method 9; ANSI/ASTM D3211-79 (K)

Cost: Less than \$1500 (A,F,G,H,I); \$1500–\$5000 (B,C,D, E,F,I,L); \$5000–\$15,000 (G,H,I,J,L); \$15,000–\$100,000 (J,L)

Partial List of Suppliers: Andersen Samplers, Inc. (A,B,D,F); Bailey Controls Co. (I); Beckman Instruments Inc. (B); Bendix Corp. (A,F); BGI Inc. (A,F); Contraves Goertz Corp. (I); Datatest, Inc. (I); Durag (W. Germany) (I); Dynatron Inc., (I); Extranuclear Laboratories, Inc, (H); Frieske & Hoepfner (W. Germany) (G) GCA/Precision Scientific Group (G); General Metal Works Inc. (A,F); Glasrock Products Inc., (J): Kurz Instruments Inc. (A,B); Lear Siegler, Inc. (D,F,I); Leeds & Northrup (J); Meteorology Research, Inc. (J); Mine Safety Appliances Co. (C,F); NASA/Raytheon (L); Omni-Wave Electronics Corp. (H); Photomation Inc. (I); Research Appliance Co. (A,C,D,F): Research Ventures, Inc. (L); Shell Development (K); Sick Optik Elektronik (W. Germany) (I); Sierra Instruments Inc. (A,B,F); Sierra-Misco Inc., (A,D,F); Stanford Research Inst. (L); Staplex, Co., (A,F); Thermo Electron Corp. (I); Thermo Systems, Inc. (E); Verewa (W. Germany) (G); Wallace Fisher Instrument Co. (C)

Particulate matter is the world's most ubiquitous air pollutant; it is being generated naturally by fires, volcanic action, vegetation, and wind. Manmade sources such as combustion, heating and abrasion processes, and material-handling activities, however, constitute the major source of the particulate burden in the atmosphere.

Substantial efforts have been devoted in recent years to a more precise definition of the airborne material which may be measured as particulate, depending on the measurement method, and to the size characteristics of the material. Gravity and the associated settling velocities of different size particulate selectively eliminate the larger particulate. However, the small or inhalable particulate, less than 15 microns (μm) is the most damaging to human health. Consequently, techniques have been developed to characterize and measure the distribution of particulate in relation to its size.

Particulate associated with ambient air is typically measured in terms of concentration, or weight per unit volume, although its presence may also be observed and measured in terms of visibility or light attenuation. Visibility is a much more complex parameter, however, as it involves the characteristics of the human eye. In addition to light attenuation, visibility is a function of contrast, and background luminance, which are required to establish visual range measurements of runways for airport applications.

Particulate as a by-product of combustion can also be measured in terms of concentration; however, it is often specified according to optical density or opacity, which are light-attenuation measurements. Opacity of smoke plumes is often related to the visual effects seen by a human observer.

With particulate concentrations ranging from that corresponding to a visibility of 50 mi (80 km) to the nearly opaque smoke plume arising from a coal-fired boiler without particulate emission controls, there is a wide variety of instrumental techniques available for such monitoring. These techniques have been classified above according to their applications capabilities. In-situ techniques generally designate those techniques which can be used to measure particulate in place or as it exists in the gas within its natural physical constraints. Extractive techniques alternatively include those techniques which remove a sample from the medium to be measured and transport the sample to the measuring instruments.

High-Volume Sampler

The basic instrument for determining the airborne concentration of particulate matter in the United States is the high-volume sampler (Figure 11.27a). A high-speed, multistage blower is used as the suction source. An adapter section allows the use of a flat 8 × 10-in. (203 × 254-mm) glass fiber filter. This filter has an efficiency of well over 99 percent for 0.3-μm particles. An orifice plate on the back of the blower allows the use of either manual or recording-type static pressure gauges to monitor the airflow rate through the sampler. Twenty-four hours is a common sampling time, during which period the build-up of particles increases the filter resistance and decreases the airflow rate. When the decrease is small (e.g., less than 10 percent), the initial and final readings are averaged to determine the sampling rate. If the decrease is significant, as might be expected in locations where the particulate concentration is high, it is desirable to use a pressure recorder on the instrument and determine the flow rate from the chart record. Alternatively, more precise measurements can be obtained by using instruments that incorporate a self-regulating flow controller which establishes a constant flow rate independent of filter loading.

The high-volume sampler provides an actual mass concentration of particulate matter in the air. The roof of the standardized sampling shelter has an annular opening sized so that particles larger than a nominal size of 100 μm are not drawn into the sampler. A varying airflow rate and the effect of winds make this method an approximation of a particle size cut at best; but it can furnish

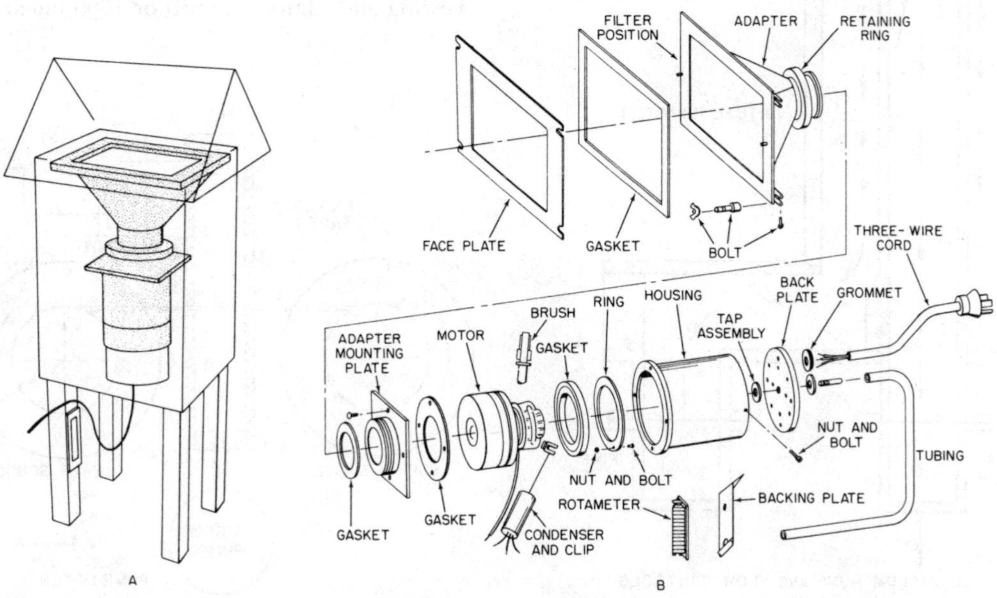

Fig. 11.27a High-volume sampler: (A) assembled sampler and shelter; (B) exploded view of typical high-volume air sampler

a large body of ambient air quality data, and air quality standards have been promulgated based on the use of this standardized instrument.

The U.S. Environmental Protection Agency (EPA) has selected as ultimate air quality goals (secondary standards) the following level of particulate matter as measured by the high-volume sampler: 60 $\mu g/m^3$, annual geometric mean and 150 $\mu g/m^3$, maximum 24-hour concentration not to be exceeded more than once per year.[1]

Dichotomous Sampler

Subsequent to the EPA's promulgation of the total suspended particulate standard, effort has been devoted to establishing the health effects of particulate in relation to particle size. As a result of these studies a revised standard is anticipated which will be size specific for inhalable particulates less than 15 μm.[2]

In concert with this regulatory activity, the dichotomous sampler has been developed to meet the size-specific requirements. The dichotomous sampler is generally designed to eliminate particulates greater than 15 μm in the inlet sampling and transport mechanism. The remaining sample stream is then segregated into two size-specific regions; hence the name dichotomous. A virtual impactor is generally used to draw off those particulates less than 2.5 μm. As a result, two sample streams are created, as shown in Figure 11.27b, which when appropriately collected on a filter medium, yield the particulate in the 15 to 2.5-μm size range, and 2.5 or below size range. Filters are typically 1.46 in. (37 mm) in diameter and may be of either glass fiber or membrane construction. Flow rates are accurately controlled to ensure the stability of the size cut points. Both automatic and manual models are available. Automatic units provide for automatic changing of filters after a preselected sampling time interval. Filters are also often analyzed using X-ray fluorescence to determine the elemental composition of the particulate.[3]

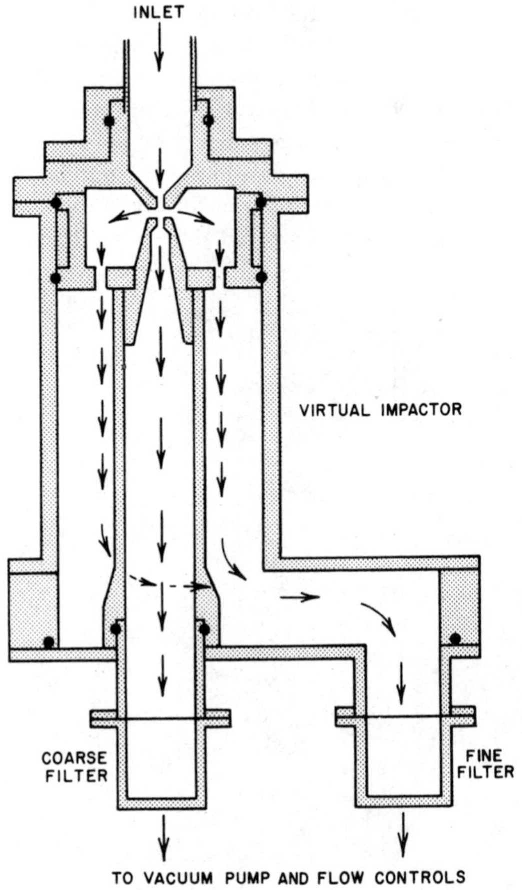

Fig. 11.27b Virtual impactor

Tape Sampler

The tape sampler (Figure 11.27c) provides an indication of the soiling properties of the sampled air. Air is drawn through a paper tape at a rate of about 0.75 actual cubic feet per minute (ACFM) for periods of 1, 2, or more hours. Filter paper and membrane tapes have been used. The soiling index (optical properties) of the filtered particulate is measured either by reflectance or transmittance. These optical measurements cannot be directly correlated with the mass concentration of particulate matter in the atmosphere, except where particle size, shape, and color are constant and a series of comparative tests are made using the high-volume sampler.

Soiling index results are expressed in coefficients of haze per 1000 linear feet (COH/1000 LF) for transmittance measurements and in reflectance units of dirt shade (RUDS) for reflectance measurements. Coefficient of haze units are defined as 100 times the optical density determined by transmittance. The reflectance unit of dirt shade is defined as 100 times the optical density determined by reflectance. RUDS are expressed either in terms of 10,000 linear feet of air (American Society for Testing and Materials unit) or 1000 linear feet of air.

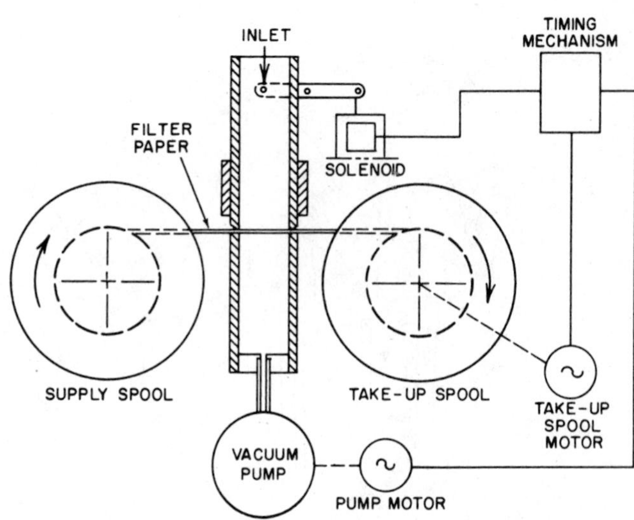

Fig. 11.27c Tape sampler

Some older tape samplers use an air pump which is quite sensitive to system resistance. It is advisable to measure the airflow rate through both a clean spot and a dirty spot to determine whether an air volume correction to an average rate is advisable. The diameter of the spot is not a variable since the results are expressed in terms of the length of the air column which is drawn through the spot.

Manual Stack Sampler

Manual stack sampling techniques have been established by both the EPA and the American Society for Testing and Materials (ASTM). The stack sampling equipment involved in the EPA technique designated as Reference Method 5 is shown in Figure 11.27d.[4] This method uses an isokinetic sampling technique whereby the velocity of the sample gas being drawn into the probe is made equal to the velocity of the flue gas at the sampling point. Correspondingly, the sample probe assembly incorporates a sampling nozzle, an S-type pitot tube, and usually a thermocouple for sensing gas temperature. In practice, the operator draws samples from a variety of points within the stack cross-sectional area to minimize the effects of particulate and velocity stratification within the gas volume of interest. The probe is inserted or withdrawn manually from various access ports on the stack to achieve the desired sampling strategy.

The probe is attached to an adjacent hot box where the particulate is separated from the gas on a filter paper and/or cyclone separator. After removal of particulate, the gas proceeds to a cold box where the gas is cooled to 32°F (0°C) by a series of ice bath–cooled impingers. The gas then flows through a heated umbilical to the control unit where the gas flow is controlled and timed. The stack gas velocity and temperature are also read out from this unit. A variety of options have been developed to make the system both easier to use and expandable to other gas measurements. One vendor has even developed an automatic microprocessor-controlled system which greatly reduces manual interaction.

EPA Method 5 is one of the most widely used techniques available today and has been used to establish the data base for particulate-emitting stationary sources.

Piezoelectric Crystal Mass Balance Measurements

Some crystals, when excited by an alternating current, exhibit a resonant frequency which is proportionate to the crystal mass. Quartz is commonly used and is readily available for this purpose. This principle is used to continuously monitor the mass concentration of particulate matter in air by trapping particulate from the impinging air stream on the sensing crystal and keeping the reference crystal clean. Thereby, the resonant frequency of the sensing crystal changes and a beat frequency develops. The particle mass is determined based on the mass versus the frequency response characteristic of the crystal. Commercially available quartz crystals show a frequency shift of 2000 Hz/μg of material deposited.

The major problem involved in this type of instrumentation involves the deposition and retention of particles on the sensing crystal. Two techniques, impaction and electrostatic precipitation, have been used commercially. Both approaches have weaknesses in collecting and retaining particles over the widesize range encountered in atmospheric aerosols.

Impaction Devices

Determination of the particle size distribution in aerosols is frequently desired. Although light-scattering devices can provide an indication of particle size distribution, most atmospheric measurements are made using

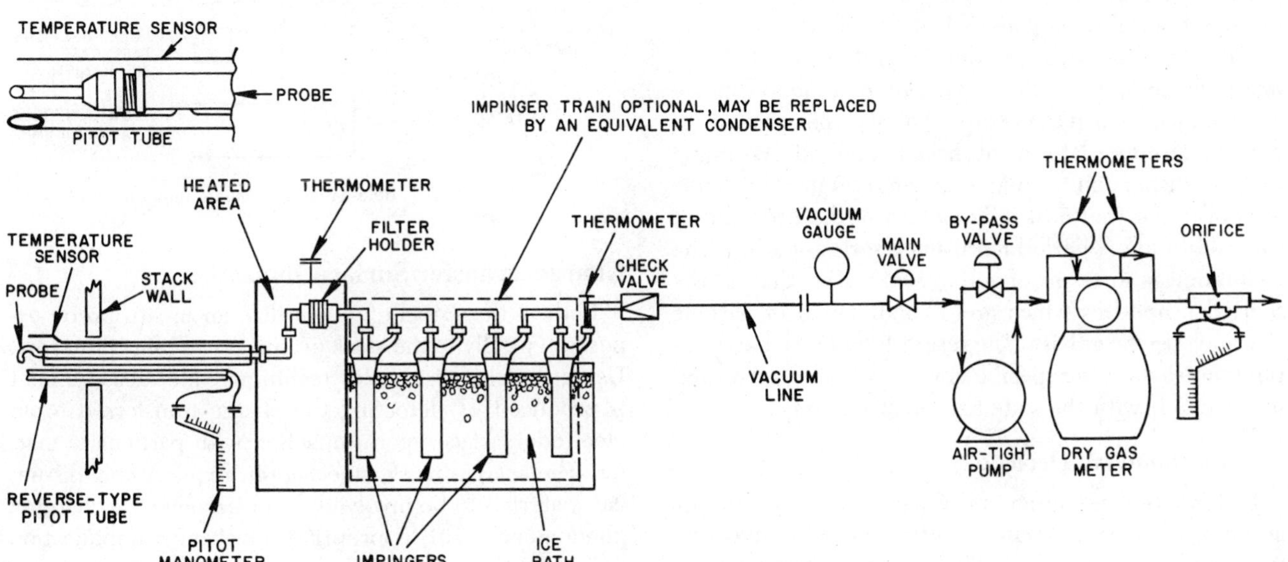

Fig. 11.27d Manual stack sampler

impaction devices. In principle, more inertial energy is required to impact small particles than large ones when they are impinged on a plane surface. This characteristic has been utilized through a series of impaction stages with succeeding stages operated at higher velocities (Figure 11.27e).

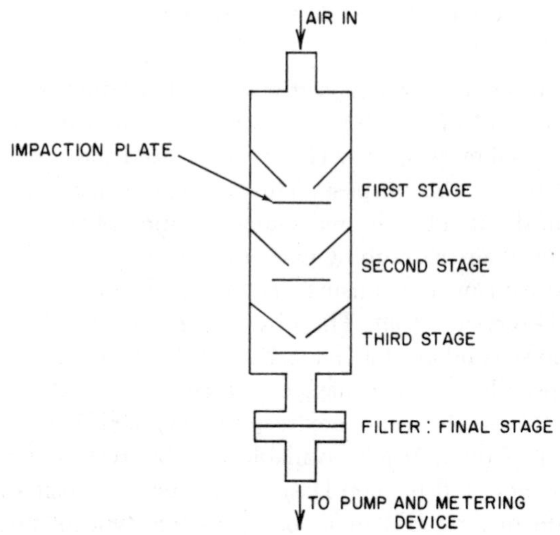

Fig. 11.27e Impaction device schematic

The upper practical size limit for impaction devices is about 10 μm. They are particularly useful in the size range of 0.5 to 7 μm, which corresponds to the approximate size range of particulate matter in the ambient air.

A cascade impactor can be added to a manual stack sampler sampling probe to evaluate stack emissions, while a fractionating sampler can be added to a high-volume sampler for ambient air analysis. In most cases, the impaction plate is removed from the sampling device and is weighed to determine the amount of sample. High plate tare weight is a potential source of inaccuracy.

Calibration of impaction devices is difficult. The normal practice involves the generation of a monodisperse aerosol of known particle size. Polystyrene spheres, as manufactured for latex paint, have been used. Assuming that the dispersion technique generates a monodisperse aerosol (often the most difficult part of the procedure), the calibration is performed using a series of sizes. The calibration is in terms of aerodynamic size, that is, the inertial properties which are a combination of particle size and specific gravity. The results from impaction sampling must state the specific gravity, whether measured or assumed, with the stated size distribution.

Radiometric Devices

Radiometric measurement of particulate is primarily accomplished using a radiation attention technique or by measuring the ionization effect of a radioactive source on ambient air. Typically, low-energy beta-type radioactive sources are used to minimize the human exposure considerations and associated regulatory problems.

In the typical beta-gauge shown in Figure 11.27f, the air to be measured is drawn through a tape filter medium where the particulate is collected. The radiometric transmission or attenuation of the filter medium is measured in the clean state, before collection, and in the dirty state, after collection. By appropriate processing of these two measurements, a measure is obtained which corresponds to the mass of collected particulate. The gas volume is usually accurately controlled so that the sample volume is precisely defined. The system is relatively insensitive to the normal chemical variation of particulate matter; however, it is especially responsive to hydrogen compounds. This technique has been widely applied to stack gas and ambient air measurements. The mechanical complexity and high costs, both the initial installed cost and maintenance costs, have limited the acceptance of these systems.

The ionizing technique has been applied to a variety of low-cost fire/smoke detectors. In this application, ambient air is exposed to a radiation source which partially ionizes the air molecules and allows a current flow through a pair of electrodes. As particulate concentration increases, the radiation is absorbed as the particulate matter decreases the ionization and, hence, the current through the electrodes.

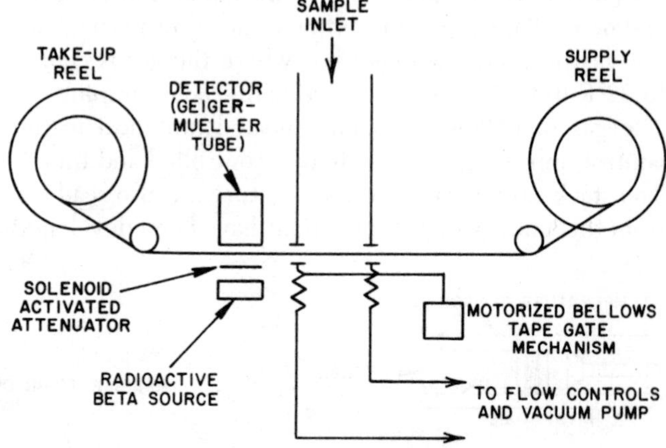

Fig. 11.27f Beta-gauge particulate monitor

Charge Transfer/Surface Ionization

Both of these techniques utilize an in-situ probe exposed typically to the stack or flue gas stream of interest. Using the charge transfer technique, the measurement is produced by detecting the charge transferred to an electrode surface upon impaction with particulate matter. The charge transfer mechanism requires only dissimilar materials to be involved, as in triboelectric charging phenomena. With appropriate electronic amplification and scaling, the output can be related to particulate concentration. This technique has not gained wide acceptance, presumably because of the difficulty in correlating charge transfer with particulate concentrations and in keeping the probe mechanism operable under the severe and diverse operating conditions experienced in stack applications.

The surface ionization technique utilizes a completely different and more consistent measurement phenomenon. In this system, airborne particulates decompose upon impaction on a hot wire with a resultant release of ions which are then collected on an electrode. By detecting both the average ion current and analyzing the pulse height and rate, this instrument is able to provide an indication corresponding to total mass concentration and the relative distribution of particle sizes, respectively. This technique has a high sensitivity to low particulate concentrations. Both techniques are essentially point measurements which provide no averaging effect including the overall stack or duct cross-sectional area.[5,6]

Light Attentuation/Transmissometer

Transmissometers are certainly the most accepted and frequently used means for the automatic and continuous monitoring of particulate in stack gases.[7] In its most basic form a transmissometer consists of a light source and a detector. By referencing all measurements of the partially obstructed optical path to the clear path measurement, the transmissometer essentially provides a measurement of light attenuation which can be expressed in units of optical density or opacity. Optical density, sometimes called extinction, varies linearly with particulate concentration, given a relatively uniform distribution of particle size and composition, and pathlength.[8] Opacity, on the other hand, varies from 0 to 100 percent—from clear to opaque conditions, respectively. Opacity can be related linearly to the Ringlemann numbers used by human observers and is also the parameter typically regulated by the EPA to insure proper operation and maintenance of particulate control equipment.

Transmissometers have become rather sophisticated as they have been developed to achieve exceptional accuracy and reliability.[9] A typical instrument is shown in Figure 11.27g. In the selection of a transmissometer, several critical design factors should be considered.[10]

Double-pass versus single-pass configurations: In a single-pass system the light source is mounted on one side of a duct and the detector on the other. Consequently, the light passes through the gas or medium of interest once. The major difficulty imposed by this configuration is in calibration—the detector cannot be easily calibrated or exposed to a known light source without adding an additional light source specifically for that purpose or using other means which are not involved in the normal measurement. Double-pass systems incorporate both a light source and a detector on the transceiver side of the stack and provide a retroreflector on the opposite side. With this configuration, the light passes through the gas of interest twice, thereby increasing its sensitivity. Further, this configuration allows all the active components to be contained in the transceiver and permits

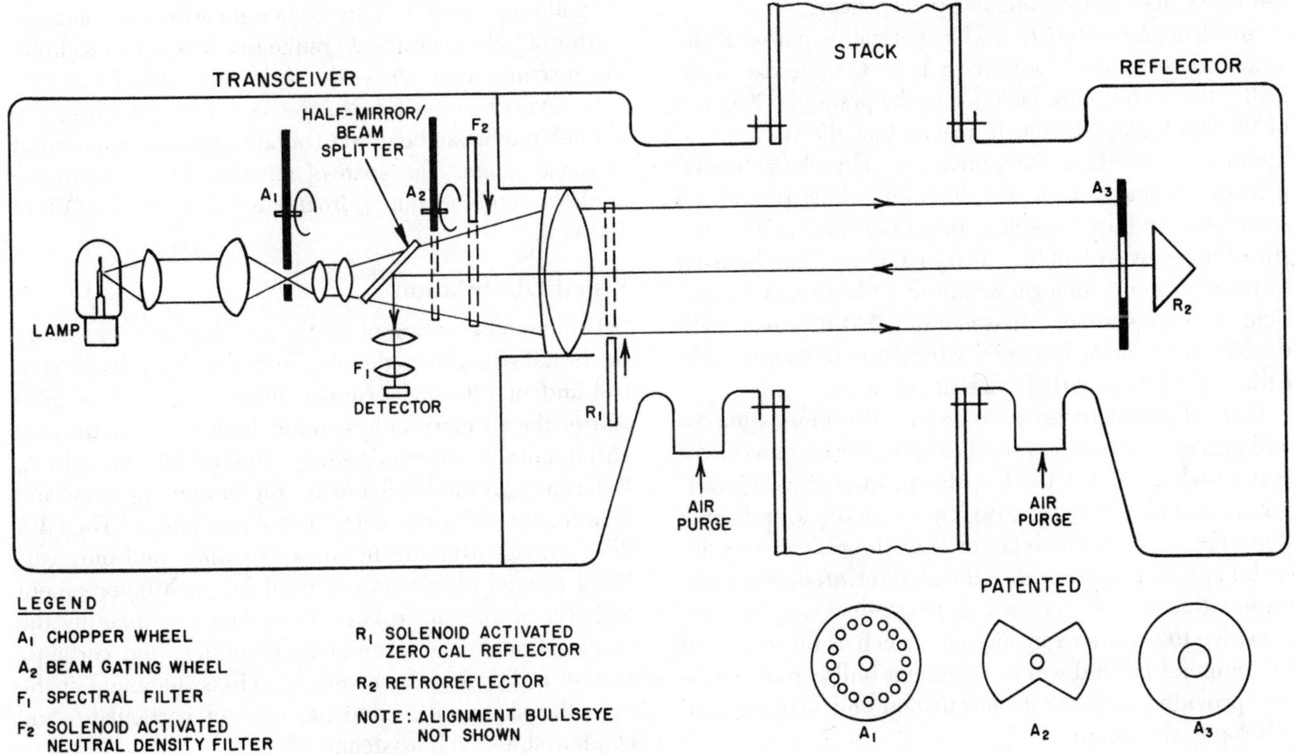

Fig. 11.27g Double-beam, double-pass transmissometer (Courtesy of Lear Siegler, Inc.)

a simple solenoid mounted reflective surface to be incorporated into the transceiver to provide a simulation or calibration check of the zero opacity condition. Internal filters can be activated sequentially to provide upscale checks of calibration.

Single, double, or dual-beam configurations: Single-beam configurations essentially measure the detected light without compensation for changes in optics, source, or detector characteristics. Obviously, this is the simplest, most inexpensive, and most inaccurate technique. Dual or double-beam configurations internally split the light emitted from the source into two beams. The measurement beam is projected through the optical medium of interest and is referenced to the second reference beam which is totally contained within the instrument. If a separate detector is provided for each beam, as is done in so-called dual-beam configurations, compensation is continuously provided for light source variations. Further, as long as the detectors for each beam are matched, only their differential errors appear in the overall measurement. If the two beams can be modulated or time-sequentially gated into a single detector, automatic compensation is provided for both light source and detector variations, which is typical of the most advanced forms of double-beam measurement systems. Regulation of lamp voltage supplies and automatic control of reference light levels further enhance the accuracy of precision instruments.

Optical divergence: Both the angles of viewing and projection should be minimized to reduce the potential interference of scattered light. Most precision units exhibit less than 5° total angular divergence.

Spectral characteristics: The spectral response of the system determines its sensitivity to particulate size. Typically, the systems are matched to the photopic response of the light-adapted human eye so that the transmissometer sees exactly as the human eye. This characteristic provides a peak response at 550 nm, which provides a somewhat uniform response for particles as small as 0.3 μm. Longer wavelengths, such as 3.39 μm, can be used to provide more uniform response with respect to particle size as desired of a mass monitor.[11] Multiwavelength devices have been designed which provide some indication of relative particle size distribution.

Optical characteristics: Typically, the more modern and precise instruments also incorporate a chopped light system which makes the instrument insensitive to variations in ambient light. Also, by providing a uniformly illuminated beam much larger than the reflector or detector and by using an autocollimated retroreflector technique, the overall system can be made essentially insensitive to variations in alignment such as those caused by temperature and wind. Alignment bullseyes are usually provided to facilitate installation and maintenance of proper alignment.

Air purge sytem: Nearly all stack-mounted systems are provided with a forced air purge system which ensures a curtain of clean air between the exposed optical surfaces and stack gas. The air filtration system, blower, and aerodynamic efficiency of the air purge system vary according to manufacturer's design characteristics.

Simplified transmissometer configurations are often used for detecting bag breaks in bag-house particulate control applications.[12] Transmissometers have also been widely used for highway visibility, runway visual range, highway tunnel, and roof vent monitoring. Exceptional accuracy and drift characteristics are required in such applications. Typically, stack-mounted transmissometers are capable of up to 75 ft. (22.5 m) separation, whereas visibility-monitoring units may have larger optics which provide pathlengths of several hundred meters.

Light Scattering

When a particle is struck, light scatters in accordance with the size, shape, and color of the particle. Light scatter can be measured as forward scatter, angular scatter, or back scatter.

Suspensions of particulate matter in the air scatter the light in all directions. Sophisticated devices are available for determining the concentration of particles in air and the size of these particles. Laser-based scattering devices have recently been designed which measure particulate concentration and/or particulate size distribution.[6] Such devices appear to be particularly useful in evaluating the behavior of particulate control equipment.

One device, the nephelometer, has had some use in air pollution studies. This device measures the angular scatter of light over a wide range in a flowing air sample. The instrument measures the scattering coefficient which is in inverse proportion to the visual range. Thus, the nephelometer can be used to obtain a point measurement of visual range. The span of measured range with the nephelometer is from approximately 1 to 100 mi (1.6 to 160 km).

Visual Observation

Human observation of smoke and dust was, of course, the initial stimulus to develop better techniques for control and monitoring. Human observation still remains one of the primary enforcement tools used by the Environmental Protection Agency (40 CFR 60 Appendix A, Reference Method 9) to ensure the proper operation and maintenance of particulate control equipment. The EPA does provide training in smoke reading, and only certified trained observers are used for such enforcement activity. Studies have been undertaken to quantify the errors of such measurement techniques, and certainly human frailties are involved as well as the angle of the sun, the sky cover, contrast, type of particulate, and condensables such as steam.

Remote Sensing

Several techniques have been developed to enable an observer to monitor smoke plume characteristics from a remote location. Telephotometers have been used manually to establish smoke plume opacity based on the luminance of the smoke plume as measured with different background characteristics. Light direction and ranging (LIDAR) has been developed as a research tool using a laser to quantify smoke plume characteristics. LIDAR is relatively expensive and complex but has proven useful in the research environment. It is normally contained in a mobile laboratory which can be driven to locations in order to provide good observation of a particular smoke plume. Aircraft have also been equipped with proper instrumentation to permit observation and measurement of smoke plume characteristics while they are flown through the plume.

Summary

Particulate-monitoring methods have been developed which achieve high standards of accuracy and reliability. Under the impetus of the EPA such methods have become widely used in the USA. Future developments are anticipated in the particle sizing area as regulatory agencies become less concerned with the visual effects of smoke, and more concerned with health effects. Visual effects, particularly visibility, will continue to be of substantial interest in relation to scenic vistas and parks as well as highway and airport conditions. Continuous particulate mass monitoring devices are not yet well accepted.

REFERENCES

1. Code of Federal Regulations, Title 40, Part 50, Subchapter C.
2. Miller, F.J., Gardner, D.E., Graham, J.A., Lee, R.E., Jr., Wilson, W.E., and Bachmann, J.D., "Size Considerations for Establishing a Standard for Inhalable Particles," *Journal of the Air Pollution Control Association*, Vol. 29, No. 6, 1979, pp. 610–615.
3. Dzubay, T.G., and Stevens, R.K., "Ambient Analysis with Dichotomous Sampler and X-Ray Fluorescence Spectrometer," *Environmental Science and Technology*, Vol. 9, No. 7, July 1975.
4. Code of Federal Regulations, Title 40, Part 60, Appendix A.
5. Shofner, F.M., Kreikebaum, G., Schmitt, H.W., and Barnhart, B.E., *In Situ, Continuous Measurement of Particulate Size Distribution and Mass Concentration Using Electro-Optical Instrumentation*, Knoxville, April 1975.
6. Farthing, W.E., and Knollenberg, R.G., *Real Time, In Situ Measurements*, APCA 80-42.5, Montreal, Quebec, 1980.
7. U.S. EPA Technology Transfer, *Handbook—Continuous Air Pollution Source Monitoring Systems*, Environmental Protection Agency, Cincinnati, June 1979.
8. Conner, W.D., Knapp, K.T., and Nader, J.S., *Applicability of Transmissometers to Opacity Measurement of Emissions Oil-Fired Power and Portland Cement Plants*, U.S. EPA-600/2-79-188, Environmental Protection Agency, September 1979.
9. Cristello, J.C., and Walther, J.E., "An Evaluation of an On-Stack Transmissometer, as a Continuous Particulate Monitor," APCA Article, 67th Annual Meeting, Denver, Colorado, 1974.
10. Lester, D.J., "Opacity Monitoring Techniques," Workshop on Sampling, Analysis and Monitoring of Stack Emissions, NTIS PB-252-748, April 1976, pp. 31–48.
11. Uthe, Edward E., "Evaluation of an Infrared Transmissometer for Monitoring Particulate Mass Concentrations of Emissions from Stationary Sources," *Journal of the Air Pollution Control Association*, 1980.
12. Saltz, Julian, and Cotler, Lee, "Opacity Monitoring Technique Predicts Baghouse Maintenance," *Pollution Engineering*, November 1978.

BIBLIOGRAPHY

Allard, Douglas, and Tombach, Ivar, *Evaluation of Visibility Measurement Methods in the Eastern United States*, APCA 80-29.3, Montreal, Quebec, 1980.

Beutner, H.P., "Monitoring of Particulate Emissions from Cement Plants," *Rock Products*, May 1974.

Brown, J.H., Cook, K.M., Ney, F.G., and Hatch, T. "Influence of Particle Size upon the Retention of Particulate Matter in the Human Lung," *American Journal of Public Health*, Vol. 40, 1950, p. 450.

Conner, W.D., "A Comparison Between In-Stack and Plume Opacity Measurements at Oil-Fired Power Plants," *Energy and the Environment—Proceedings of the Fourth National Conference*, AICHE, Dayton, Ohio, 1976, pp. 478–483.

Conner, W.D., "Measurement of the Opacity and Mass Concentration of Particulate Emissions by Transmissometry, Chemistry and Physics Laboratory," EPA-650/2-74-128, November 1974.

Conner, W.D., and Hodkinson, J.R., *Optical Properties and Visual Effects of Smoke-Stack Plumes*, 2nd printing, EPA Publication AP-30, May 1972.

Conner, W.D., Knapp, K.T., and Nader, J.S., "Applicability of Transmissometers to Opacity Measurement of Emissions Oil-Fired Power and Portland Cement Plants," EPA-600/2-79-188, Environmental Protection Agency September 1979.

Dennis, Richard, *Handbook on Aerosols*, Technical Information Center, TID-26608, 1976.

Ensor, D.S., and Pilat, M.J., "Calculation of Smoke Plume Opacity from Particulate Air Pollution Properties," *Journal of the Air Pollution Control Association*, Vol. 21, 1971, p. 496.

Farthing, William E., Hussey, David H., Smith, Wallace B., and Wilson, Rufus R., *Sampling Charged Particles with Cascade Impactors*, APCA 79-28.2, Cincinnati, 1979.

Gregory, G.L., McDougal, D.S., and Wagner, H.S., *An Air Quality Program Designed to Evaluate Remote Sensors*, APCA 80-44.5, Quebec, 1980.

Herget, William F., and Conner, William D., "Instrumental Sensing of Stationary Source Emissions," *Environmental Science and Technology*, Vol. II, No. 10, October 1977.

Hood, K.T., and Coron, A.L., *The Relationship Between Mass Emission Rate and Observed Plume Appearances from Kraft Recovery Furnaces*, 74-AP-08, Regional APCA Meeting, Boise, Idaho, November 17, 1974.

Instrumentation for Environmental Monitoring, Vol. 2, Lawrence Berkeley Laboratory, University of California, Berkeley, California.

Liu, Benjamin Y.H., Raabe, Otto G., Smith, Wallace B., Spencer, Herbert W., III, and Kuykendal, William B., "Advances in Particle Sampling and Measurement," *Environmental Science and Technology*, 1980.

Loo, B.W., Jaklevic, J.M., and Goudling, F.S., "Dichotomous Virtual Impactors for Large Scale Monitoring of Airborne Particulate Matter," Presented at the Symposium on Fine Particles, May 1975.

McCain, J.D., Cushing, K.M.,and Bird, A.N., Jr., *Field Measurements of Particle Size Distribution with Inertial Sizing Devices*, U.S. EPA-650&2-73-035, October 1973.

Miller, Frederick J., Gardner, Donald E., Graham, Judith A., Lee, Robert E., Jr., Wilson, William E., and Backmann, John D., "Size Considerations for Establishing a Standard for Inhalable Particles," *Journal of the Air Pollution Control Association*, 1979.

Molloy, R.C., "Smoke Opacity Monitoring Systems: Pollution Control and Energy Conservation," *ASHRAE Journal*, September 1976, pp. 27–32.

Montagna, John C., Smith, Gregory W., Teats, F. Gale, Vogel G. John, and Jonke, Albert A., *Evaluation of On-Line Light-Scattering Optical Particle Analyzers for Measurements at High Temperature and Pressure*, Argonne National Laboratory, Argonne, Il., ANL/CEN/FE-77-7, February 1978.

Nader, John S., Jay, Frederic, and Conner, William, *Performance Specifications for Stationary-Source Monitoring Systems for Gases and Visible Emissions*, U.S. EPA-650/2-74-013, January 1974.

NCASI, "Application of Light Transmissometry and Indicating Sodium Ion Measurement to Continuous Particulate Monitoring in the Pulping Industry," *Atmospheric Quality Improvement Technical Bulletin #79*, May 1975.

On-Stack Transmissometer Measurement of Particulate Opacity and Mass Concentration, U.S. EPA-65062-74-120, November 1974.

Osborne, Michael C., and Midgett, M. Rodney, *Training and Certification of Smoke Inspectors with Transmissometers—A Survey*, APCA 78-11.6, Houston, 1978.

Spengler, John D., Turner, William A., and Dockery, Douglas W., *Comparison of Hi-Vol, Dichotomous, and Cyclone Samplers in Four U.S. Cities*, APCA 80-43.4, Montreal, Quebec, 1980.

Stevens, R.K., and Dzubay, T.G., "Recent Developments in Air Particulate Monitoring," *Proceedings on Instrumental Methods for Air and Water Measurements and Monitoring, IEEE Transactions on Nuclear Science*, April 1975.

Wedding, T.E., McFarland, A.R., and Cermak, T.C., "Large Particle Collection Characteristics of Ambient Aerosol Samplers," *Environmental Science and Technology*, Vol. II, No. 4, April 1977, pp. 387–390.

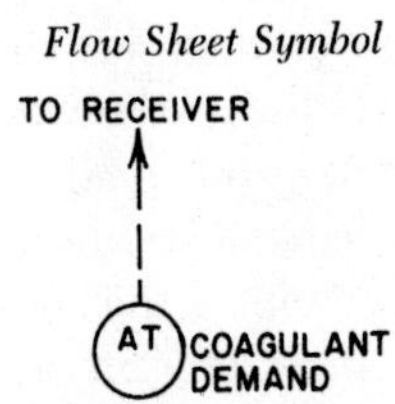

11.28 STREAMING CURRENT

Applications:	Batch operations, titrations, or continuous monitoring. The detector can control the addition of coagulation chemicals or estimate treatment demand by measuring surface charge on particles
Materials of Construction:	Stainless steel, silver, polyethylene, and Teflon
Approximate Cost:	$3,500
Partial List of Suppliers:	Komline-Sanderson Engineering Corp.; Leeds and Northrup; Panametrics, Inc.

In water or waste treatment, the streaming current detector (SCD) can estimate treatment demand or continuously control the addition of coagulation chemicals by detecting the surface charge on particles. Less directly, it can compare the relative charge-influencing ability of alternative treating chemicals, or determine the effect of pH on a chemical's ability to promote or enhance coagulation. Most effective use requires a recognition of the role of electrical charge in stabilizing suspensions.[1]

Principles of Operation

In ionic liquids, any interface with a solid or a second liquid will carry an electrical charge that originates with preferential adsorption or positioning of ions. The liquid adjacent to the surface will contain excess charges of opposite sign, called *counterions*. If a charged particle is immobilized on a filter or on the wall of a capillary, the counterions can physically be swept downstream by a stream of water. This flow of charges of predominantly one sign constitutes a current, called the streaming current. In the case of an insulating capillary, the return path is by ionic conduction through the liquid in the stream. With suitable electrodes, the return path can be arranged so as to contain apparatus for measuring the current.

Because the energy level of the streaming current is small compared to random signals from electrodes and from inadvertent thermocouples, it is advantageous to develop a reversing or alternating signal. This signal of known frequency may be distinguished for noise signals of other frequencies. Determination of polarity of the charge is by phase-sensitive rectification through the action of a switch synchronized with the instantaneous flow of liquid. This alternating design permits a pump and capillary to be combined into a single unit, a loosely fitted piston reciprocating in a dead-ended bore (Figure 11.28a). Cleaning is easy because the piston and bore can be separated by disassembling.

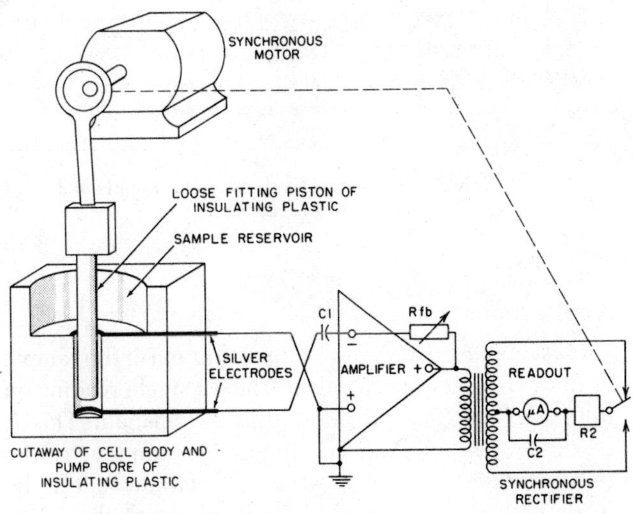

Fig. 11.28a Streaming current detector

Calibration

Although readout is in arbitrary units, reproducibility can be demonstrated by standard samples. Sensitivity is constant, but zero shift can occur owing either to inadequate cleaning or to use of strong surface-active agents. Unfortunately there is no absolute zero standard.

Colloidal suspensions are useless as routine standards because they become unstable with time. Those suspensions made stable by excess surface-active materials leave a stubborn residue on the SCD.

Often the most useful standards are two buffer solutions chosen so as to have pHs on either side of the isoelectric point of the capillary material. The isoelectric point is defined as the pH at which the material exhibits zero charge. Upscale or downscale shifts of buffer readings from previous values will indicate instrument bias caused by contamination of the surfaces; buffers near pH 4 and 5 are useful.

Dielectric constant, pH, and total ionic content all exert legitimate influences on the streaming current reading. Lowering the pH of a sample shifts its SCD reading upscale, and increasing the salt content by adding a balanced, neutral salt shifts the reading toward zero (Figure 11.28b).

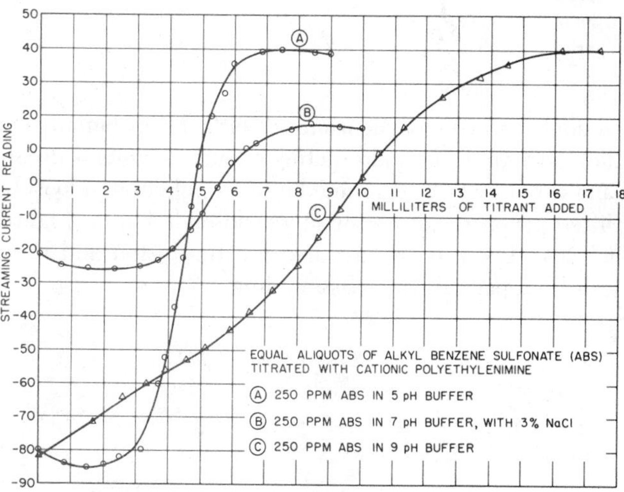

Fig. 11.28b Titration curves indicating effect of pH and salt on SCD reading

Applications

Most applications involve either titration of the sample or prior treatment in the plant, since a single reading on untreated material provides little information. This is because SCD readings are almost independent of the concentration of suspended solids. Titrations can be made with as little sample as will submerge the active part of the instrument.

To estimate the unit treatment demand of a liquid, a volumetric sample is first titrated with the contemplated treating chemical at known concentration until a zero signal is obtained. To compare alternative treating chemicals, identical samples of material are titrated with the various chemicals. When it is desired to study the effect of a change in pH on treating requirements, identical samples are titrated at various levels of Ph. This effect can be significant, with chemicals differing considerably in their tolerance of low or high pH.

In the usual treatment plant, the SCD may continuously control the feed of cationic chemicals.[2] The need for changing the rate of addition of these chemicals arises from variations in the stream flow rate, from changes in the suspended solids loading, from the unit demand of the solids, or from any combination of these factors. The main advantage of SCD control is its early response to changes. Charge neutralization occurs almost as soon as the treating chemical is dispersed in the stream, therefore, samples may be taken 1 or 2 minutes after addition of the chemical.

Sampling

Batch samples to be taken to the SCD should be adequate to permit rinsing the apparatus several times. Skimming or decanting removes sand, larger solid particles, or oil globules. Since charge is a surface phenomenon and the fines have most of the surface, little is lost by the removal of larger particles. For a continuous sample, a self-cleaning bypass filter should be used.[3] Periodic backflushing or cleaning may also be required.

Control System

For continuous control, the SCD measurement signal is fed to a two-mode controller which modulates the chemical feed pump or control valve (Figure 11.28c).

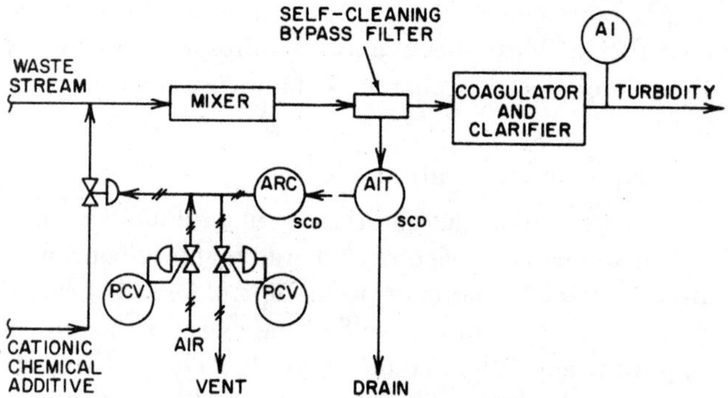

Fig. 11.28c Chemical addition control utilizing a streaming current detector

The maximum and minimum valve opening is limited as a defense against loss of sample, which can cause an open loop. Pressure regulators can serve as adjustable settings to limit the controller output to the range between minimum and maximum expected demand. The SCD controller set point determination is based on downstream turbidity measurement. If there is an existing flow proportioning control for the chemical feed, the SCD controller can influence its ratio set point in a cascade arrangement. Often more than one chemical is involved, and therefore a sequence of additions and attendant interactions must be considered.

REFERENCES

1. Riddick, T.M., "Role of the Zeta Potential in Coagulation Involving Hydrous Oxides," *TAPPI, Vol. 47, January 1964, p. 1.*
2. Gerdes, W.F., "A New Instrument, the Streaming Current Detector," *Instrument Technology*, Vol. 13, 1966, p. 12.
3. Lipták, B.G. ed., *Instrument Engineers' Handbook*, Supplement 1, Section 1.10. Chilton, Philadelphia, 1972.

BIBLIOGRAPHY

"Instrumentation and Automation Experiences in Wastewater Treatment Facilities," EPA-600/2-76-198, Environmental Protection Agency, Washington, D.C., October 1976.

Karasek, F.W., "Detection Limits in Instrumental Analysis," *Research/Development*, October 1979.

Flow Sheet Symbol

TO RECEIVER

AT SOx

11.29 SULFUR OXIDES

Methods of Detection:
A—Colorimetric
B—Conductimetric
C—Correlation spectrometry
D—Coulometric
E—Electrochemical
F—Flame photometric
G—Infrared
H—Ultraviolet

Reference Method: Pararosaniline* A

Inaccuracy: On the order of ±2% of full-scale range, limited by an inherent precision of ±0.005 ppm for ambient measurement

Ranges: Generally 0.01 to 5 ppm for ambient air sensors A, B, D, and F. With adequate dilution systems these sensors can measure up to 5000 ppm. Generally 500 to 10,000 ppm for units C, G, and H as commercially available stack gas sensors

Sample Pressure: Generally atmospheric or near atmospheric

Sample Temperatures: At sensor, limited to about 120°F (49°C) for designs A, B, D, E, and F for ambient air analysis. In stack monitoring applications, designs C, G, and H are not limited since they are not in direct contact with the sample

Sample Flow Rate: Range of 0.1 to 1.0 acfm is typical for ambient air analysis

Materials of Construction: Glass, Teflon, and stainless steel preferred in all applications due to reactivity of sulfur oxide gases. Especially avoid polyvinyl chloride for this reason

Response: Nearly instantaneous for designs C, E, and F. Designs A, B, and D rely on automated wet chemistry and the response time is determined almost entirely by the liquid capacitance of the system

Cost: Generally $3000 to $8000 not including strip chart recorder. Some portable models of designs B, D, and E may cost slightly less than this, while models of designs A, C, and F can be more expensive if specially fitted for multiple gas and/or multiple point sampling duty

Partial List of Suppliers: Atlas Electric Devices Co. (A,D); Barringer Research, Ltd. (C); Beckman Instruments, Inc. (B,D); Bendix Corp., Process Instruments Div (F); Calibrated Instruments, Inc. (B); Devco Engineering, Inc. (B); Du Pont Co. (G,H); Dynasciences, Environmental Products Div. (E); Intertech Corp. (G); ITT Barton (D); Leeds & Northrup (B,G); Litton Industries (A); Meloy Laboratories, Inc. (F); Monitor Labs, Inc. (A,B); Philips Electronic Instruments (D); Process Analyzers, Inc. (D); Technicon Industrial Systems (A); Theta Sensors, Inc. (E); Tracor, Inc. (F); Varian Associates, Aerograph Instruments Div. (F); Wilkens-Anderson Co. (A)

*The EPA study (CPA 70-101) found coulometry to be equivalent to this reference method. Of the coulometers evaluated, the Philips PW-9700 gave the best correlation.

The predominant sulfur oxide in the atmosphere is sulfur dioxide (SO_2). Some sulfur trioxide (SO_3) is formed in combustion processes, but this rapidly hydrolyzes to sulfuric acid which is considered to be a particulate matter. This section will deal with sulfur dioxide.

The measurement of sulfur dioxide in air has been performed for many years because of the simplicity of performing manual analysis based on the ability of sulfur dioxide to reduce a starch-iodine solution. The earliest instruments to detect sulfur dioxide were automated wet chemical devices which measured the increase in the conductivity of a solution after air had been intimately mixed in by some contacting procedure.

The U.S. Environmental Protection Agency (EPA) has deleted conductivity-type analyses from their approved list of ambient air sampling instruments because this technique is nonspecific for SO_2, but measures all acid gases. The EPA has designated the colorimetric technique using pararosaniline as the reference procedure, with flame photometric and coulometric detection as acceptable alternative procedures.[1]

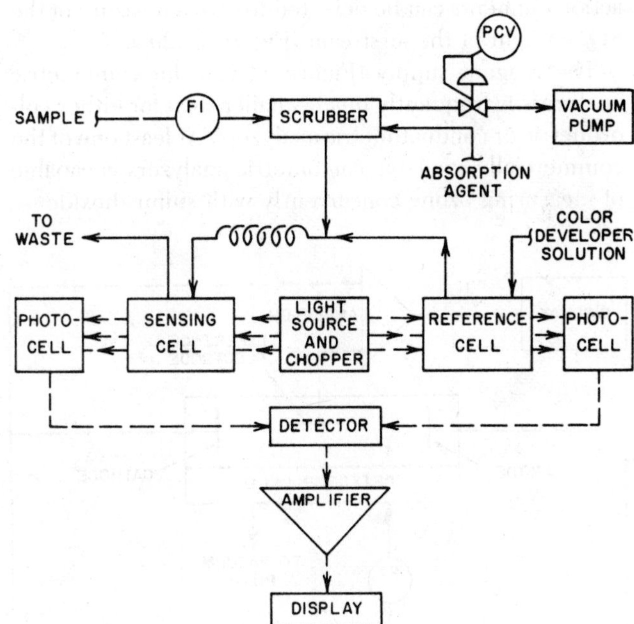

Fig. 11.29a Colorimetric analyzer for oxides of sulfur

Ambient Air Sampling

In the United States, the ultimate air quality goals (secondary standards) for sulfur dioxide are 60$\mu g/m^3$ (0.02 ppm) annual arithmetic average, 260 $\mu g/m^3$ (0.1 ppm) maximum 24-hr concentration not to be exceeded more than once a year and 1300 $\mu g/m^3$ (0.5 ppm) maximum 3-hr concentration not to be exceeded more than once a year.[2] Instrument range is generally selected for 0 to 1 ppm full-scale range except where higher levels are anticipated. There is a sensitivity problem, as few instruments are capable of accurately reading at levels of 0.01 ppm and below. Statistical procedures are usually required to determine arithmetic average concentrations.

Colorimetric Analyzers

The reference technique for determining the sulfur dioxide content of atmospheric air involves the absorption of SO_2 from the air sample in a solution of sodium tetrachloromercurate. Upon addition of formaldehyde and pararosaniline dye, a strong purple dye complex is formed. This procedure has been automated by numerous manufacturers (Figure 11.29a).

This family of analyzers requires frequent maintenance because of the complexity of the plumbing and the tendency of the dye complex to plate out on the cell windows, thus reducing its sensitivity. Properly maintained, they provide an excellent record of sulfur dioxide concentrations in air since they are based on a chemical reaction which is specific for sulfur dioxide.

Conductimetric Analyzers

There is more experience with the operation of this family of analyzers than with any other for sulfur dioxide detection. The instruments are fairly simple electronically, but the plumbing creates some maintenance problems. A basic circuit diagram is shown in Figure 11.29b. Some of the available instruments use a weak solution of sulfuric acid and hydrogen peroxide as an absorbent, while others use distilled water.

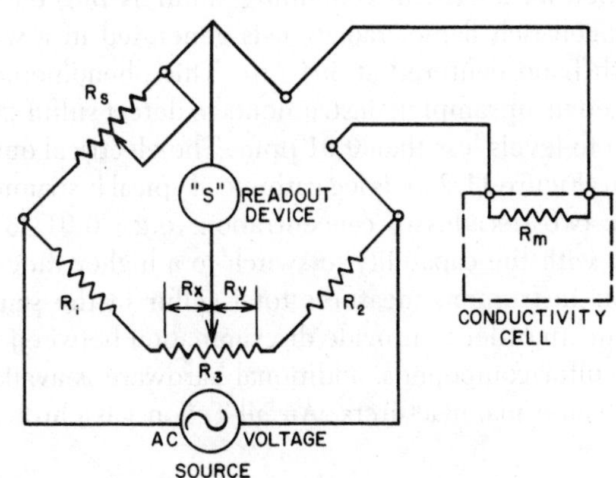

Fig. 11.29b Conductivity measuring circuit

It is important to utilize a temperature-correcting element in the conductimetric analyzer circuitry since the change in conductance of the reagent over the normal ambient temperature range is nearly as large as the conductance changes at the sulfur dioxide levels of interest.

Conductimetric analysis is not specific for sulfur dioxide. Other acid gases such as hydrogen chloride are a positive interference, while basic gases such as ammonia are a negative interference.

Coulometric Analyzers

The reducing action of sulfur dioxide will electrogenerate free bromine or iodine from a solution. These action elements can be detected to give a measure of the SO_2 content of the airstream (Figure 11.29c).

The reagent supply (Figure 11.22b) for coulometric analyzers is frequently much smaller than for either colorimetric or conductimetric analyzers. At least one of the commercially available coulometric analyzers is capable of measuring ozone concurrently with sulfur dioxide.

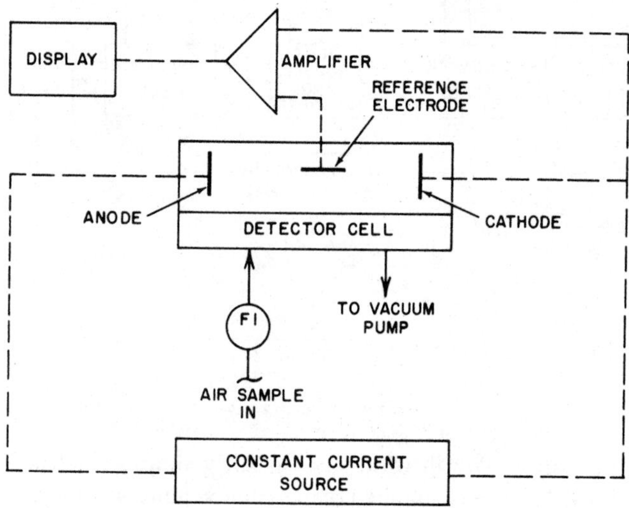

Fig. 11.29c Coulometric analyzer for oxides of sulfur

Flame Photometric Analyzers

When an airstream containing sulfur is burned in a hydrogen-rich flame, radiation is generated in a wavelength band centered at 394 mμ. This phenomenon is utilized in air sampling instruments to detect sulfur gases down to levels less than 0.01 ppm. The electrical output signal (Figure 11.29d) is logarithmic. Typical instruments cover two decades of concentration, e.g., 0.01 to 1.0 ppm, with the capability to switch to a higher range.

This instrument measures total sulfur in the sample stream. In order to provide discrimination between various sulfur compounds, additional hardware is available from some manufacturers. An all-Teflon gas chromatographic column preceding the flame photometric analyzer separates sulfur compounds quantitatively. This has proved extremely useful for the measurement of reduced sulfur compounds, e.g., hydrogen sulfide and dimethyl disulfide in the air.

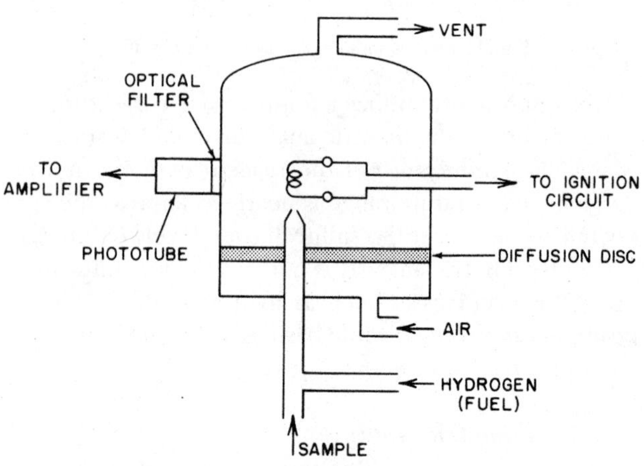

Fig. 11.29d Flame photometric analyzer

Electrochemical Analyzers

When air is passed over an element consisting of a semipermeable membrane, an electrolyte, and a voltage-sensitive pickup, certain gases may migrate selectively across the membrane and generate a signal in the electrolyte (Figure 11.29e). This principle has been used to measure many gases including sulfur dioxide.

The electrochemical principle for the measurement of air pollutants has not had extensive field use. Operational problems can develop in maintaining the proper moisture content of the membrane and the proper electrolyte strength because of the migration of water vapor.

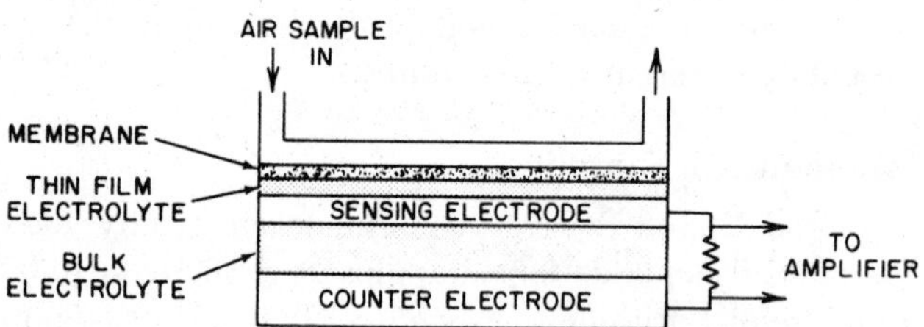

Fig. 11.29e Electrochemical gas analyzer

Source Sampling

Stack effluents contain orders of magnitude higher concentrations of sulfur dioxide then does ambient air. Few of the principles used to detect the trace amounts of sulfur dioxide in ambient air are applicable to source monitoring except by massive dilution.

The colorimetric procedure is not applicable to source monitoring at all, mainly because of the interference of nitrogen oxides. The electrochemical principle may be directly applicable to source monitoring. Instruments which are generally applicable to source monitoring (but not to ambient monitoring) are infrared (IR), ultraviolet (UV), and correlation spectrometry.

Infrared and Ultraviolet Absorption

Sulfur dioxide absorbs light energy over a broad band, which includes both the infrared and ultraviolet regions of the spectrum. Some instruments utilize a positive filtering scheme (shown in Figure 11.29f and described in detail in Section 10.9). Others utilize photometric analysis (Figure 11.29g). The lowest practical full-scale range for these instruments is 0 to 200 ppm, with the upper limit in the range of whole percentages. This adequately covers the range of concentrations encountered in practice.

The gases must be filtered before they reach the instrument to prevent a build-up of particulate matter in

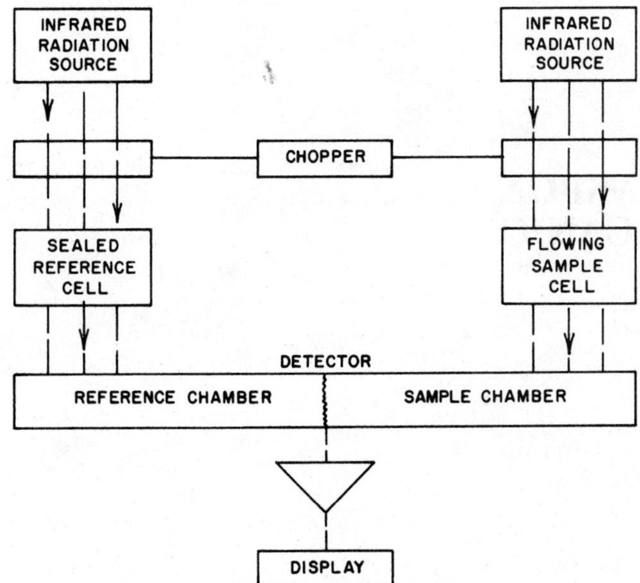

Fig. 11.29f Nondispersive infrared analyzer

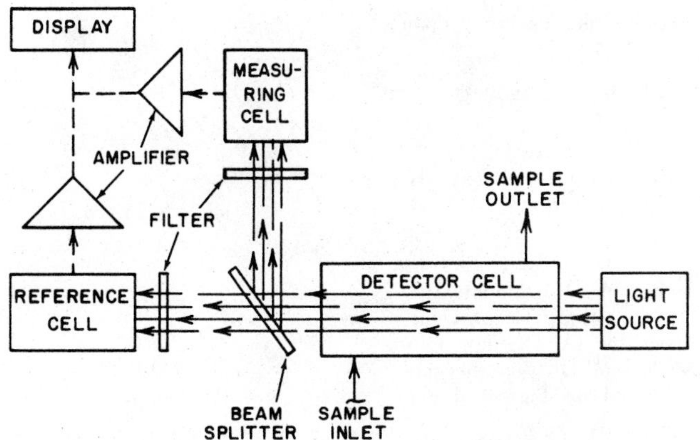

Fig. 11.29g Photometric analysis for oxides of sulfur

the cell. Likewise, the moisture content of the sample gas must be controlled to prevent condensation.

Correlation Spectrometry

The correlation spectrometer can be used at a location remote from the source to determine the sulfur dioxide content of stack gases. The instrument in Figure 11.29h is specific for SO_2 because of the photographic optical mask of the SO_2 spectrum.

In use, the natural background level of ultraviolet radiation is used to zero the instrument. When the optics are pointed across a stack plume containing sulfur dioxide, the resulting change in the ultraviolet radiation in the wavelength pattern of SO_2 is detected at the instrument. If the stack plume diameter and stack gas density are known, the sulfur dioxide concentration can thereby be determined.

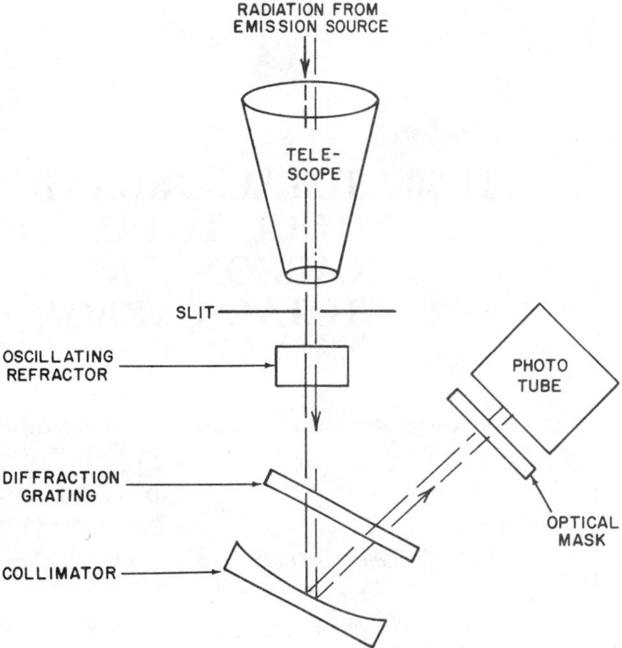

Fig. 11.29h Remote sensing spectrometer

Calibration

The standard procedure for calibrating sulfur dioxide instruments involves the preparation of known concentrations of SO_2 in air. This can be accomplished with high accuracy when permeation tubes, made of Teflon tubing and containing liquid sulfur dioxide, are maintained at constant temperature while a constant flow of air is passed over the tube. The most desirable procedure is a system calibration rather than an instrument calibration, with the known concentration introduced at the system inlet.

REFERENCES

1. *Federal Register*, Part II, Vol. 36, August 14, 1971, p. 158.
2. *Federal Register*, Part II, Vol. 36, April 30, 1971, p. 84.

BIBLIOGRAPHY

Frost & Sullivan, Inc., "On-Stream Process Analyzer Market," Report No. 669, Frost & Sullivan, Inc., New York, August 1979.

Hall, J., "On-Line Analysers," *Instruments and Control Systems*, August 1979.

Homolya, J.B., "Monitoring SO_2 Emissions from Stationary Sources," ISA Conference, Houston, October 1973.

Jutila, J.M., "Multicomponent On-Stream Analyzers for Process Monitoring and Control," *InTech*, July 1979.

Sandford, J., "On-Stream Analyzers," *Instruments and Control Systems*, March 1977.

Yaws, C.L., "Sulfur Oxides," *Chemical Engineering*, July 8, 1974.

Flow Sheet Symbol

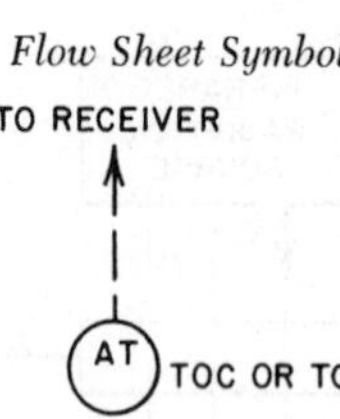

11.30 TOTAL ORGANIC CARBON (TOC), TOTAL INORGANIC CARBON (TIC), TOTAL CARBON (TC)

Methods of Detection:	A—Nondispersive infrared (NDIR) B—Flame ionization detector (FID) C—Aqueous conductivity D—Coulometry E—Colorimetry
Sample Size and Flow Rate:	10 μl to 10 ml for laboratory instruments 250 μl to 30 mm/min for on-line instruments
Materials of Construction:	Glass, quartz, Teflon, stainless steel, Hastelloy, polyethylene, polyvinyl chloride
Precision:	±2% full scale to ±5% full scale depending upon sample size and range
Ranges:	0–1 ppm (mg/liter) to 0–20,000 ppm (mg/liter)
Response and Cycle Time:	4 to 15 min
Utilities and Reagents Required:	Air, oxygen, nitrogen, hydrogen, mineral acid, oxidizing reagent, buffer, cooling water
Cost:	\$7000 to \$15,000
Partial List of Suppliers:	Astro Ecology Corp. (A), Beckman Instruments, Inc. (A); Coulometrics, Inc. (D); Envirotech Corp., Dohrmann Division (A,B); Ionics, Inc. (A); Oceanography International Corp. (A); Seiscor Division, Seismograph Service Corp. (A); Sybron Barnstead (C); Technicon Industrial Systems (E)

Significance of Measurement

The total organic carbon (TOC) found in water and wastewaters often is an index of the quality of such waters. The classical biochemical oxygen demand (BOD) and the chemical oxygen demand (COD) have long been employed as similar indices. All three methods have their strengths and weaknesses. As defined, the TOC analysis measures only the organic carbon content, and it therefore does not detect the load represented by nitrogen-based molecules. However, it is very rapid and accurate. The BOD analysis measures all molecules that exert an oxygen demand, although its readings will vary if the bioassay used is altered. Thus, the BOD, in addition to being a lengthy 5- or 7-day analysis, is often subject to significant variations. The COD analysis suffers from shortcomings in oxidation efficiencies although its analysis time is reduced.

Direct correlation between TOC, BOD, and COD usually is not possible. On the other hand, with proper interpretation, the TOC can represent a rapid and frequently accurate method of assessing the pollution levels of municipal and industrial wastes, along with various other applications.

TOC, TC, and TIC Analyzers

To arrive at the TOC content of a sample, one of two techniques must be used to eliminate the total inorganic carbon (TIC) usually present but of little or no interest to the TOC analysis. The TIC present in a water sample is usually in the form of inorganic bicarbonates and carbonates. In one technique these components are analyzed independently and then subtracted from the total carbon (TC). The TOC is then determined by the difference between TC and TIC (TC − TIC = TOC).

The other technique is to acidify the sample to a pH of 2 to 3 followed by a brief gas sparging to drive off the carbon dioxide formed by the acidification. Any carbon remaining after the sparging should be TOC. Thus, the

TC found in this sample is equal to the TOC content of the sample. One weakness in this technique is the possible loss of some volatile organic carbon (VOC) that might be present in the sample. Further techniques are available to account for such VOC.

The TOC method was first introduced in 1964 as a single-channel TC analyzer using a catalytic oxidation combustion technique followed by the analysis of the resulting carbon dioxide. The IC was either removed by acid sparging or determined by titration. A few years later a second channel was added which permitted the parallel determination of the IC in a second heated reaction chamber.

Since that earlier period, several other techniques have appeared with various changes in the methodology and detection.

Because of the rapid acceptance and usefulness of the TOC analysis as a laboratory method, on-line TOC analyzers became available in the late 1960s. Their success was limited by the relative complexity of these continuous analyzers.

By 1980 there were at least five distinctly different methodologies and means of detection to accomplish the TOC analysis.

The Catalytic Oxidation–Combustion Method by Nondispersive Infrared (NDIR)

The original and expanded method[1,2] contained a high-temperature furnace and combustion tube to catalytically oxidize all carbonaceous species (TC) to carbon dioxide (CO_2). A sample syringe is injected into the combustion tube containing a catalyst such as platinum or cobalt oxide on a suitable substrate. The temperature within the combustion tube is maintained at about 950°C (1742°F).

A continuous flow of oxygen or air carrier gas transports the resulting cloud of steam and CO_2 through a condenser and water trap into a heated NDIR analyzer. Here the CO_2 is measured in the form of a peak recorded on a suitable recorder. Peak height directly relates to CO_2 concentration present.

A second channel contains a low-temperature furnace housing a heated reactor (at 150°C or 302°F). This reactor typically contains quartz chips coated with phosphoric acid. When another portion of the same sample is injected into this channel, IC is converted to CO_2. The low temperature and absence of catalyst prevents conversion of any TOC present to CO_2. Thus an IC value is obtained which, when subtracted from the TC value, provides the TOC concentration. Figure 11.30a illustrates this method.

When the IC is very large compared with the TOC, such as in drinking waters, the difference method may not be as accurate as the acid sparge technique. For such applications, the IC must be either manually or automatically acidified and sparged from the sample prior to TOC analysis.

This general method is employed for both laboratory and on-line analysis.

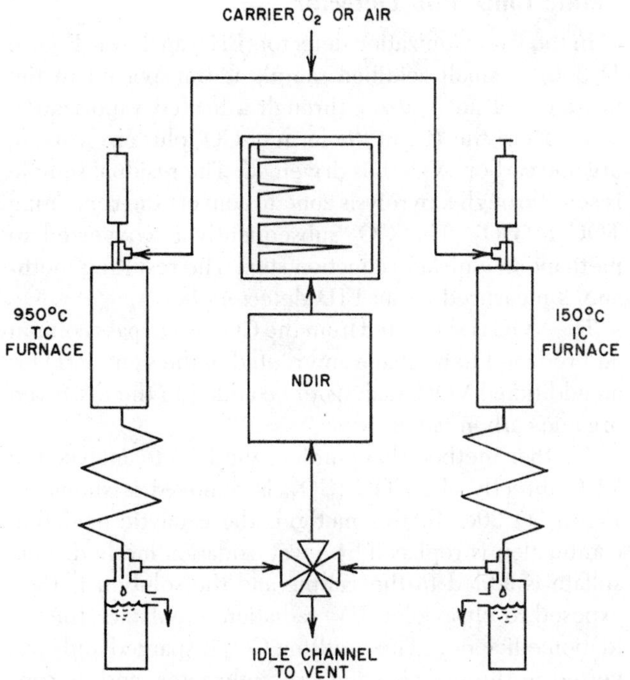

Fig. 11.30a Catalytic combustion analyzer with NDIR sensor

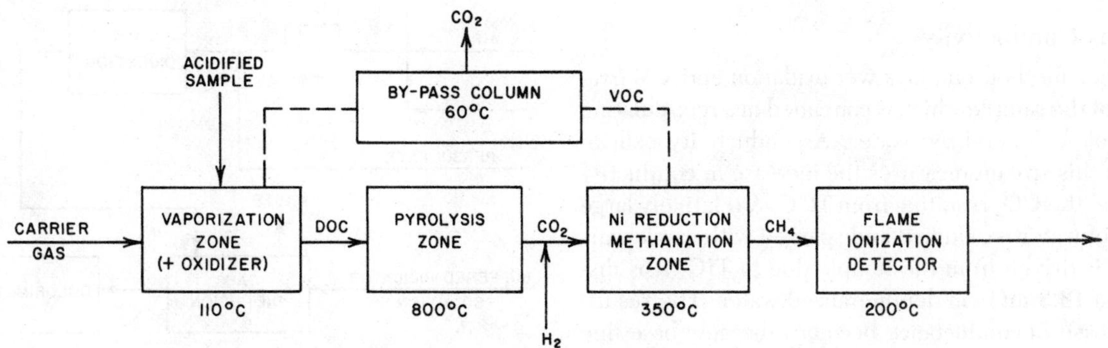

Fig. 11.30b Flame ionization detector

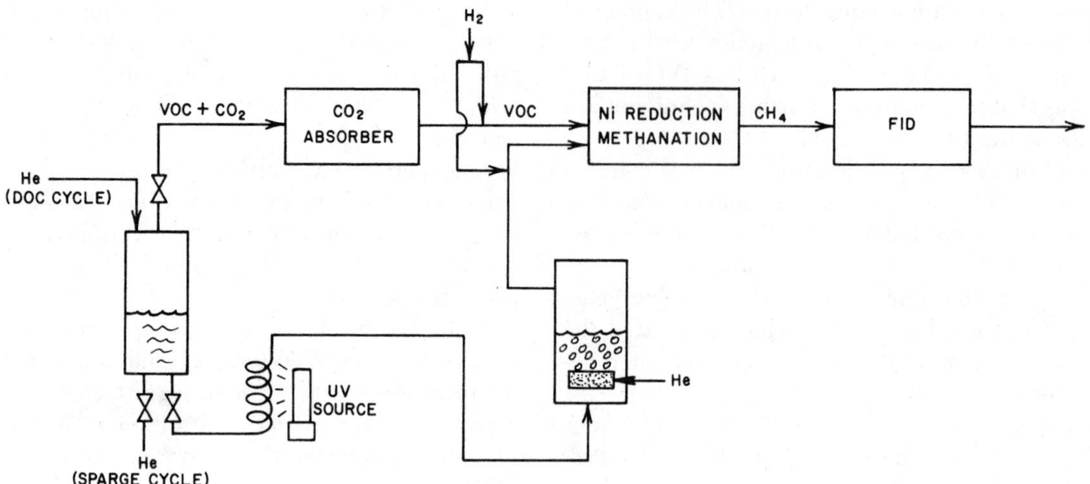

Fig. 11.30c FID analyzer with wet oxidation

Flame Ionization Detector

In the flame ionization detector (FID) analyzer (Figure 11.30b), a small acidified sample is transported in the presence of an oxidizer through a heated vaporization zone. Here the IC, in the form of CO_2 plus any volatile organic carbon (VOC), is driven off. The residual sample is sent through a pyrolysis zone to convert the remaining TOC to CO_2. The CO_2 subsequently is converted to methane in a nickel-reduction step. The resulting methane is measured by an FID detector.

The VOC is separated from the CO_2 in a bypass column and reduced to methane and routed to the same FID for an additional VOC analysis to be added to the dissolved organic carbon value.

Another method that employs the FID to analyze the VOC directly after TIC (CO_2) is removed is shown in Figure 11.30c. In this method, the catalytic oxidation combustion is replaced by a wet oxidation method. Persulfate is added to the sample and the solution is then exposed to ultraviolet (UV) radiation to enhance the oxidation efficiency. The resulting CO_2 is sparged and converted in the nickel-reduction methanator, and its concentration is measured in the FID analyzer.

This wet oxidation technique is available as an on-line analyzer.

Aqueous Conductivity

Another method employs wet oxidation and UV irradiation of the sample which is contained in a recirculating stream of demineralized water. A conductivity cell located in this stream measures the increase in conductivity due to the CO_2 resulting from TOC. A relatively large sample of water is acidified and sparged with carrier air. As CO_2 is driven from the sample due to TIC, it is dissolved in 18.3 m$\Omega \cdot$cm demineralized water. The resulting increase in conductance becomes the new base line for the next step, which entails the oxidation of the TOC remaining in the water. This is accomplished by UV radiation to oxidize the TOC to CO_2. The added CO_2 raises the conductivity to a logarithmically higher level in proportion to the TOC present. The water is then automatically demineralized as it is bypassed through an ion exchange resin bed to prepare it for the next analysis.

This method is best suited for the measurement of low TOC levels in relatively solids-free samples, such as in drinking waters. Sensitivities in the parts per billion (pbs) range are claimed.

Colorimetric Analysis

As illustrated in Figure 11.30d, the sample is acidified and sparged of TIC. An aliquot is combined with acid and persulfate and is irradiated in a UV digestor for 8 to 9 minutes. A portion of the CO_2 which is generated diffuses through a gas-permeable membrane. A weakly buffered phenolphthalein indicator solution is used as

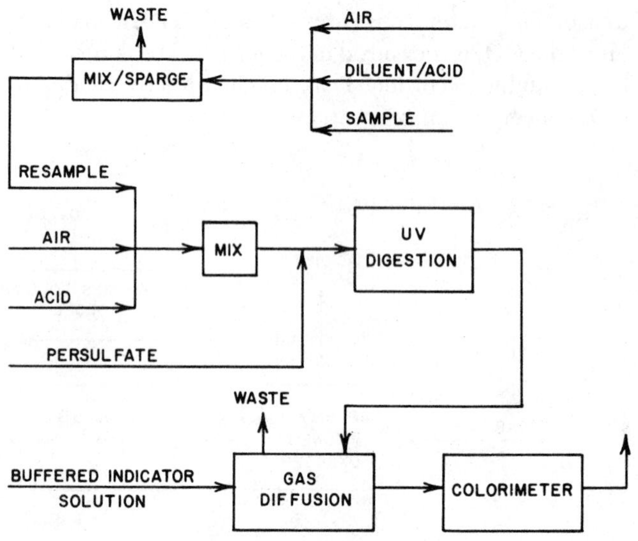

Fig. 11.30d Colorimetric analysis

the recipient stream. The color intensity of this solution decreases proportionately to the change in pH caused by the absorption of CO_2 gas. The measurement is made at a wavelength of 530 nm and is most suited for solids-free low-level measurements.

REFERENCES

1. Van Hall, C.E., Saranco, John, and Stenger, V.A., "Rapid Combustion Method for the Determination of Organic Substances in Aqueous Solutions," *Analytical Chemistry*, Vol. 35, No. 3, March 1963, pp. 315–318.
2. Van Hall, C.E., and Stenger, V.A., "An Instrumental Method for Rapid Determination of Carbonate and Total Carbon in Solutions," *Analytical Chemistry*, Vol. 39, No. 4, April 1967, pp. 503–507.

BIBLIOGRAPHY

ASTM, *Annual Book of ASTM Standards, Part 31*, American Society for Testing and Materials, Philadelphia, 1974, pp. 467–470, D 2579.

Davis, E.M., "BOD vs. COD vs. TOC vs. TOD," *Water and Wastes Engineering*, February 1971, pp. 32–38.

Handbook for Monitoring Industrial Wastewater, U.S. Environmental Protection Agency, Technology Transfer, August, 1973, pp. 5-7 to 5-11.

Helms, J.W., "Rapid Measurement of Organic Pollution by Total Organic Carbon and Comparison with Other Techniques," U.S. Department of Interior, Geological Survey, Water Resources Division, Openfile Report, May 21, 1970.

Jones, R.H., and Degaforde, A.F., "Application of a High-Sensitivity Total Organic Carbon Analyzer," *ISA Transactions*, Vol. 7, No. 4, 1968, pp. 267–272.

Kehoe, T.J., "Determining TOC in Waters," *Environmental Science and Technology*, Vol. 11, No. 2, February 1977, pp. 137–139.

Roesler, J.F., and Wise, R.H., "Variables to Be Measured in Wastewater Treatment Plant Monitoring and Control," *Journal of Water Pollution Control Federation*, Vol. 46, No 7, July, 1974, pp. 1769–1775.

Van Hall, C.E., Barth, Dennis, and Stenger, V.A., "Elimination of Carbonates from Aqueous Solutions Prior to Organic Carbon Determination," *Analytical Chemistry*, Vol. 37, No. 6, May 1965, pp. 769–771.

Williams, R.T., "The Carbonaceous Analyzer as a Water Pollution Research Tool," *Proceedings of the 21st Annual ISA Conference and Exhibit*, 1966.

Chapter XII

SYMBOLS AND TERMINOLOGY

R. F. JAKUBIK
O. P. LOVETT, JR
T. A. MAYER
C. F. MOORE
G. PLATT

CONTENTS OF CHAPTER XII

Appreciation is expressed to the Instrument Society of America for permission to abstract from their publications ANSI/ISA—S5.1—1975 (R 1981), "Instrumentation Symbols and Identifications," © Instrument Society of America, 1981, and ANSI/ISA—S5.2—1976 (R 1981), "Binary Logic Diagrams for Process Operations," © Instrument Society of America, 1981, and also for permission to reprint from their diagrams and tables, which appear here in Tables 12.1 a, b, c and 12.2 a and d.

12.1 FLOW SHEET SYMBOLS

Introduction

This section describes the major elements of a method for symbolizing and identifying instruments on flow sheets and other documents. The symbols and identifications are based on the instrument functions. This method of representation indicates the means of process measurement and control, but leaves most details of the instrumentation to be determined from specifications or other documents.

General Rules

Each instrument identification or tag number consists of a *functional* identification and a *loop* identification. A typical tag number is *PRC-8*, for a pressure recording controller, which has the functional identification *PRC* and the loop identification *8*. The tag number may be expanded to include coded information such as plant area designation, flow sheet number, etc.

Table 12.1a lists meanings of the functional identification letters. The functional identification begins with a first letter denoting a *measured or initiating variable*. *Readout or passive* functional letters follow, in any sequence, and are, in turn, followed by *output* functional letters in any sequence, except that output letter *C* (control) precedes output letter *V* (valve), e.g., *HCV*, a hand-actuated valve. Modifying letters, if used, are interposed so that they are placed immediately following the letters they modify. All identification letters are capitals for compatability with automatic printing machines.

Table 12.1.a
MEANINGS OF FUNCTIONAL INSTRUMENT-IDENTIFICATION LETTERS
Numbers in parentheses refer to notes that follow

	First Letter		*Succeeding Letters (3)*		
	Measured or Initiating Variable (4)	*Modifier*	*Readout or Passive Function*	*Output Function*	*Modifier*
A	Analysis (5)		Alarm		
B	Burner Flame		User's Choice (1)	User's Choice (1)	User's Choice (1)
C	Conductivity (Electrical)			Control (13)	
D	Density (Mass) or Specific Gravity	Differential (4)			
E	Voltage (EMF)		Primary Element		
F	Flow Rate	Ratio (Fraction) (4)			
G	Gauging (Dimensional)		Glass (9)		
H	Hand (Manually Initiated)				High (7, 15, 16)
I	Current (Electrical)		Indicate (10)		
J	Power	Scan (7)			
K	Time or Time-Schedule			Control Station	
L	Level		Light (Pilot) (11)		Low (7, 15, 16)

Table 12.1.a *Continued*
MEANINGS OF FUNCTIONAL INSTRUMENT-IDENTIFICATION LETTERS
Numbers in parentheses refer to notes that follow

	First Letter		Succeeding Letters (3)		
	Measured or Initiating Variable (4)	*Modifier*	*Readout or Passive Function*	*Output Function*	*Modifier*
M	Moisture or Humidity				Middle or Intermediate (7, 15)
N (1)	User's Choice		User's Choice	User's Choice	User's Choice
O	User's Choice (1)		Orifice (Restriction)		
P	Pressure or Vacuum		Point (Test Connection)		
Q	Quantity or Event	Integrate or Totalize (4)			
R	Radioactivity		Record or Print		
S	Speed or Frequency	Safety (8)		Switch (13)	
T	Temperature			Transmit	
U	Multivariable (6)		Multifunction (12)	Multifunction (12)	Multifunction (12)
V	Viscosity			Valve, Damper, or Louver (13)	
W	Weight or Force		Well		
X (2)	Unclassified		Unclassified	Unclassified	Unclassified
Y	User's Choice (1)			Relay or Compute (13, 14)	
Z	Position			Drive, Actuate or Unclassified Final Control Element	

Notes for Table 12.1a

1. A *user's choice* letter is intended to cover unlisted meanings that will be used repetitively in a particular project. If used, the letter may have one meaning as a first letter and another meaning as a succeeding letter. The meanings need be defined only once in a legend, or otherwise, for that project. For example, the letter N may be defined as *modulus of elasticity* as a first letter and *oscilloscope* as a succeeding letter.

2. The *unclassified* letter X is intended to cover unlisted meanings that will be used only once or to a limited extent. If used, the letter may have any number of meanings as a first letter and any number of meanings as a succeeding letter. Except for its use with distinctive symbols, it is expected that the meanings will be defined outside a tagging balloon on a flow diagram. For example, XR-2 may be a *stress recorder*, XR-3 may be a *vibration recorder*, and XX-4 may be a *stress oscilloscope*.

3. The grammatical form of the succeeding letter meanings may be modified as required. For example, *indicate* may be applied as *indicator* or *indicating, transmit* as *transmitter* or *transmitting*, etc.

4. Any first letter, if used in combination with modifying letters D (differential), F (ratio), or Q (integrate or totalize), or any combination of them, is construed to represent a new and separate measured variable, and the combination should be treated as a first-letter entity. Thus, instruments TD1 and T1 measure two different variables, namely, differential temperature and temperature. These modifying letters are used when applicable.

5. First letter A for *analysis* covers all analyses not listed in Table 12.1a and not covered by a *user's choice* letter. It is expected that the type of analysis in each instance will be defined outside a tagging balloon on a flow diagram.

   Readily recognized self-defining symbols such as pH, O_2, and CO have been used optionally in the past in place of first letter A.

6. Use of first letter U for *multivariable* in lieu of a combination of first letters is optional.

7. The use of modifying terms *high, low, middle* or *intermediate*, and *scan* is preferred, but optional.

8. The term *safety* applies only to emergency protective primary elements and emergency protective final control ele-

Table 12.1.b
RELAY FUNCTION SYMBOLS

Symbol	Function
1-0 or ON-OFF	Automatically connect, disconnect, or transfer one or more circuits, provided that this is not the first such device in a loop. (See Table 12.1a, Note 13)
Σ or ADD	Add or totalize (add and subtract), with two or more inputs.
Δ or DIFF	Subtract (with two or more inputs)
± + [−]	Bias (single input)
AVG.	Average
% or 1:3 or 2:1 (typical)	Gain or attenuate (input:output), with single input
[×]	Multiply (two or more inputs)
÷	Divide (two or more inputs)
[√] or SQ. RT.	Extract square root
x^n or $x^{1/n}$	Raise to power
f(x)	Characterize
1:1	Boost
[>] or HIGHEST (Measured Variable)	High-select. Select highest (higher) measured variable (not signal, unless so noted).
[<] or LOWEST (Measured Variable)	Low-select. Select lowest (lower) measured variable (not signal, unless so noted).
REV.	Reverse
	Convert
a. E/P or P/I (typical)	For input/output sequences of the following: Designation — Signal E — Voltage H — Hydraulic I — Current (electrical) O — Electromagnetic or sonic P — Pneumatic R — Resistance (electrical)
b. A/D or D/A	For input/output sequences of the following: A — Analog D — Digital
∫	Integrate (time integral)
D or d/dt	Derivative or rate
1/D	Inverse derivative

Note: The use of a box enclosing a symbol is optional. The box is intended to avoid confusion by setting off the symbol from other markings on a diagram.

ments. Thus, a self-actuated valve that prevents operation of a fluid system at a higher than desired pressure by bleeding fluid from the system is a back-pressure type PCV, even if the valve were not intended to be used normally. However, this valve would be a PSV if it were intended to protect against emergency conditions—i.e., conditions that are hazardous to personnel or equipment or both and that are not expected to arise normally.

The designation PSV applies to all valves intended to protect against emergency pressure conditions, regardless of whether the valve construction and mode of operation place them in the category of the safety valve, relief valve, or safety relief valve.

9. Passive function *glass* applies to instruments that provide an uncalibrated direct view of the process.
10. The term *indicate* applies only to the readout of an actual measurement. It does not apply to a scale for manual adjustment of a variable if there is no measurement input to the scale.
11. A *pilot light* that is part of an instrument loop is designated by a first letter followed by succeeding letter L. For example, a *pilot light* that indicates an expired time period may be tagged KL. However, if it is desired to tag a *pilot light* that is not part of a formal instrument loop, the *pilot light* may be designated in the same way or, alternatively, by a single letter L. For example, a running light for an electric motor may be tagged either EL, assuming that voltage is the appropriate measured variable, or XL, assuming that the light is actuated by auxiliary electric contacts of the motor starter, or simply L.

    The action of a *pilot light* may be accompanied by an audible signal.
12. Use of succeeding letter U for *multifunction* instead of a combination of other functional letters is optional.
13. A device that connects, disconnects, or transfers one or more circuits may be either a *switch*, a *relay*, an on-off *controller*, or a *control valve*, depending on the application.

    If the device manipulates a fluid process stream and is not a hand-actuated on-off block valve, it is designated as a *control valve*. For all applications other than fluid process streams, the device is designated as follows:

    A *switch*, if it is actuated by hand.

    A *switch* or an on-off *controller* if it is automatic and is the first such device in a loop. The term *switch* is generally used if the device is used for alarm, pilot light, selection, interlock or safety. The term *controller* is generally used if the device is used for normal operating control

Table 12.1.c
MISCELLANEOUS SYMBOLS

Instrument Line Symbols

All lines should be fine in relation to process piping lines.

Connection to process, or mechanical link, or instrument supply

Pneumatic signal

The pneumatic signal symbol applies to a signal using any gas as the signal medium. If a gas other than air is used, the gas is identified by a note on the signal symbol or otherwise.

Electric signal

Capillary tubing (filled system)

Hydraulic signal

Electromagnetic or sonic signal (without wiring or tubing)

Electromagnetic phenomena include heat, radio waves, nuclear radiation, and light.

Undefined signal

Power Supply Abbreviations

The following abbreviations are suggested to denote the types of power supply. These designations may also be applied for purge fluid supplies.

AS Air supply
ES Electric supply
GS Gas supply
HS Hydraulic supply
NS Nitrogen supply
SS Steam supply
WS Water supply

The power supply level may be added to the instrument supply line, e.g., AS 100 for a 100-PSIG air supply; ES 24DC for a 24-volt direct current supply, etc.

Instrument Symbol Balloons

APPROXIMATELY 7/16" DIAMETER

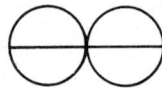

LOCALLY MOUNTED

MOUNTED ON MAIN BOARD

MOUNTED BEHIND THE BOARD

Instrument for two measured variables or single variable instrument with more than one function. Additional tangent balloons may be added as required.

A *relay,* if it is automatic and is not the first such device in a loop, i.e., if it is actuated by a *switch* or an on-off *controller*.

14. It is expected that the functions associated with the use of succeeding letter Y will be defined outside a balloon on a flow diagram when it is convenient to do so. This need not be done when the function is self-evident, as for a solenoid valve in a fluid signal line.

15. Use of modifying terms *high, low,* and *middle* or *intermediate* correspond to values of the measured variable, not of the signal, unless otherwise noted. For example, a high-level alarm derived from a reverse-acting level transmitter signal is an LAH, even though the alarm is actuated when the signal falls to a low value. The terms may be used in combinations as appropriate.

16. The terms *high* and *low,* when applied to positions of valves and other open-close devices, are defined as follows: *high* denotes that the valve is in or approaching the fully open position, and *low* denotes in or approaching the fully closed position.

The functional identification is made according to the function and not according to the construction; for example, a differential-pressure transmitter used for flow measurement is an *FT*. The first letter of the functional identification follows the measured or initiating variable and not the manipulated variable; for example, a control valve varying flow as commanded by a pressure controller is a *PV*, not an *FV*.

Each instrument loop shall have a unique identification number. This number is, in general, common to all instruments of a loop. Because each instrument should have a unique identification, suffix letters *A*, *B*, *C*, etc. shall be used to distinguish among two or more instruments of similar function in a loop, e.g., *LT-4A*, *LT-4B*, and *LT-4C*. However, multipoint recorders may more

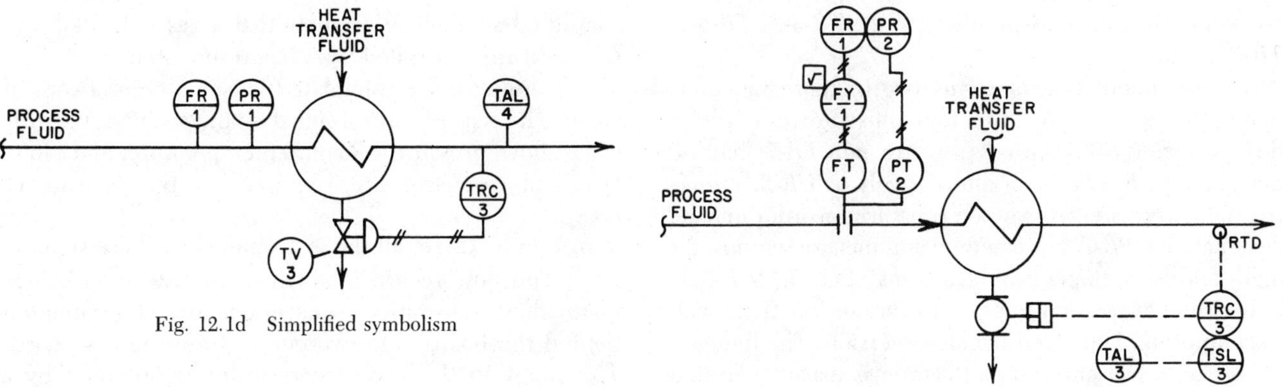

Fig. 12.1d Simplified symbolism

Fig. 12.1e Full symbolism

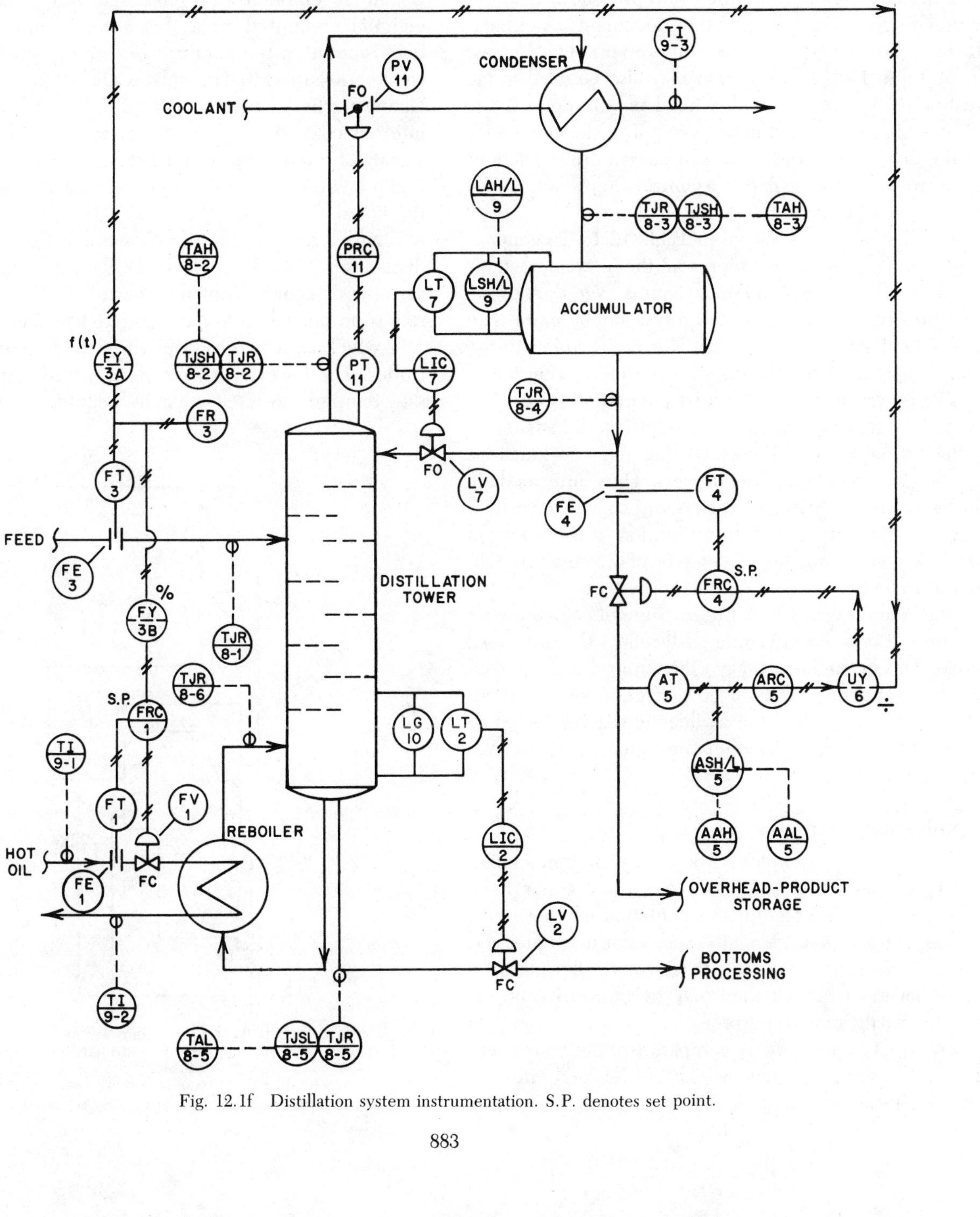

Fig. 12.1f Distillation system instrumentation. S.P. denotes set point.

conveniently use suffix numbers, e.g., *TR-5-1*, *TR-5-2*, *TR-5-3*, etc.

An instrument that performs two or more functions may be designated by all its functions. For example, a flow recorder *FR-3* with a pressure pen *PR-7* may be designated *FR-3/PR-7*, or alternatively as *UR-2*, a multivariable recorder. A two-pen pressure recorder may be designated as *PR-7/8*. A common annunciator window for high- and low-temperature alarm may be a *TAH/L-6*.

Instrument relays may perform various functions such as computing, logic, and signal conversion. The function of a relay represented on a diagram is usually clarified by placing one of the symbol designations of Table 12.1b outside the relay balloon.

Distinctive symbols are used to represent instrumentation on flow diagrams and other documents. A circular balloon represents the instrument proper in many cases, as in Table 12.1c. The balloon may also be used to tag distinctive symbols, but this need not be done if the relationship of the distinctive symbol to the remainder of the loop is apparent. For example, an orifice plate or a control valve that is part of a control loop is not usually tagged on a diagram.

Signal symbols are shown in Table 12.1c. In general, one signal line suffices to represent the interconnections between instruments on flow diagrams even though the instruments may be connected physically by more than one line. Directional arrowheads may be added to the signal lines to clarify the direction of information flow.

The representation of the instrumentation systems on a diagram may be simplified by omitting all instrumentation other than that representing the end-functions needed for operation of the process. Thus, intermediate instruments, such as transmitters and signal converters, may be eliminated. Such simplification, if used, should be done consistently for a given type of drawing throughout a project.

The sequence in which the instruments of a loop are connected on a flow diagram shall reflect the functional logic. This arrangement may differ from the actual connection sequence. Thus, a flow diagram will show instruments using electrical analog signals connected in parallel regardless of whether the signal type is voltage or current.

Application Examples

Following are examples of the symbol system applied to typical industrial processes. No attempt has been made to show complete instrumentation for these processes. Primary flow elements, such as orifice plates and control valves, are not usually tagged on the flow diagram in actual practice, but they may be tagged in the examples for illustrative purposes.

Figures 12.1d and 12.1e compare two methods of applying the symbol system, which can be used either in the full form or a simplified form. Figure 12.1d uses simplified symbolism to show that a gas is heated and temperature-controlled by a board-mounted controller. The heating fluid is modulated by some type of control valve. The type of control signal is not specified. Records of gas, flow, pressure, and outlet temperature, and a low-temperature alarm are required on the instrument board.

In Figure 12.1e, all the instruments used are symbolized. The flow record is obtained by use of an orifice plate, flow transmitter, square-root extractor mounted behind the board, and two-pen recorder on the board. The input to the pressure recorder is provided by a pressure transmitter that measures on the downstream side of the orifice plate. The signals are pneumatic. The gas outlet temperature is measured by a resistance-type element, mounted in a thermowell, connected to a board-mounted temperature recording controller, with an electric output that modulates a ball-type control valve having a cylinder-type actuator and, by implication, with internal conversion from the electric signal to a fluid signal. The temperature recording controller has an integral low-temperature switch that actuates an alarm on the board.

Figure 12.1f provides a full symbolic description of a distillation process in which the feed flow rate is measured and recorded but not controlled. The heat input rate is proportioned to the feed rate by a gain relay (FY-3B), which adjusts the set point of the hot-oil flow controller. The tower overhead is condensed, with condensate temperature controlled by maintaining a constant

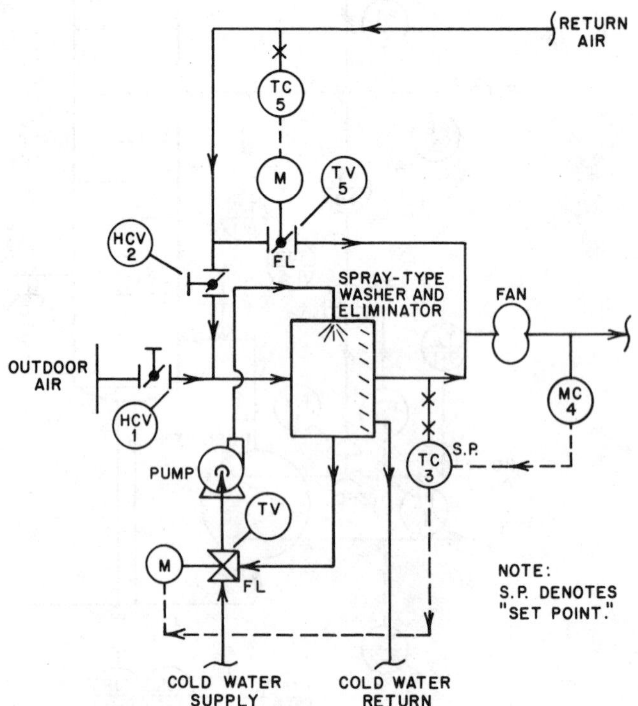

Fig. 12.1g Air cooling/humidifying system instrumentation. S.P. denotes set point.

column pressure. The overhead product drawoff rate is flow controlled. The flow controller set point is adjusted by a dividing relay (UY-6), whose inputs are the feed rate, as modified by the time-function relay (FY-3), and the output of the overhead product analysis controller. This instrument receives the product analysis from its transmitter, which also transmits it to the dual (high/low) analysis switch, which, in turn, actuates corresponding alarms. Accumulator level is maintained by throttling the tower reflux, while a separate level switch actuates a common high/low accumulator level alarm. Bottom level in the tower is controlled by modulating bottom drawoff. Local level indication is provided by a separate gauge glass. Temperature measurement at various points in the process is performed by the multipoint scanning recorder (TJR) and multipoint indicator (TI). Some of the points of TJR-8 have high or low temperature switches to actuate alarms; for example, overhead temperature is signaled by TJSH-8-2 and TAH-8-2.

Figure 12.1g illustrates an air cooling/humidifying system, in which the ratio of outdoor air to return air is manually adjusted by hand control valves. The combined streams enter a spray-type washer. Proper humidity and temperature are maintained by the humidity controller resetting the temperature controller, which, in turn, modulates the three-way control valve to vary the proportion of cold water and recirculated warm water that are pumped to the spray nozzles. Return air flow is controlled to bypass the washer in inverse proportion to its temperature.

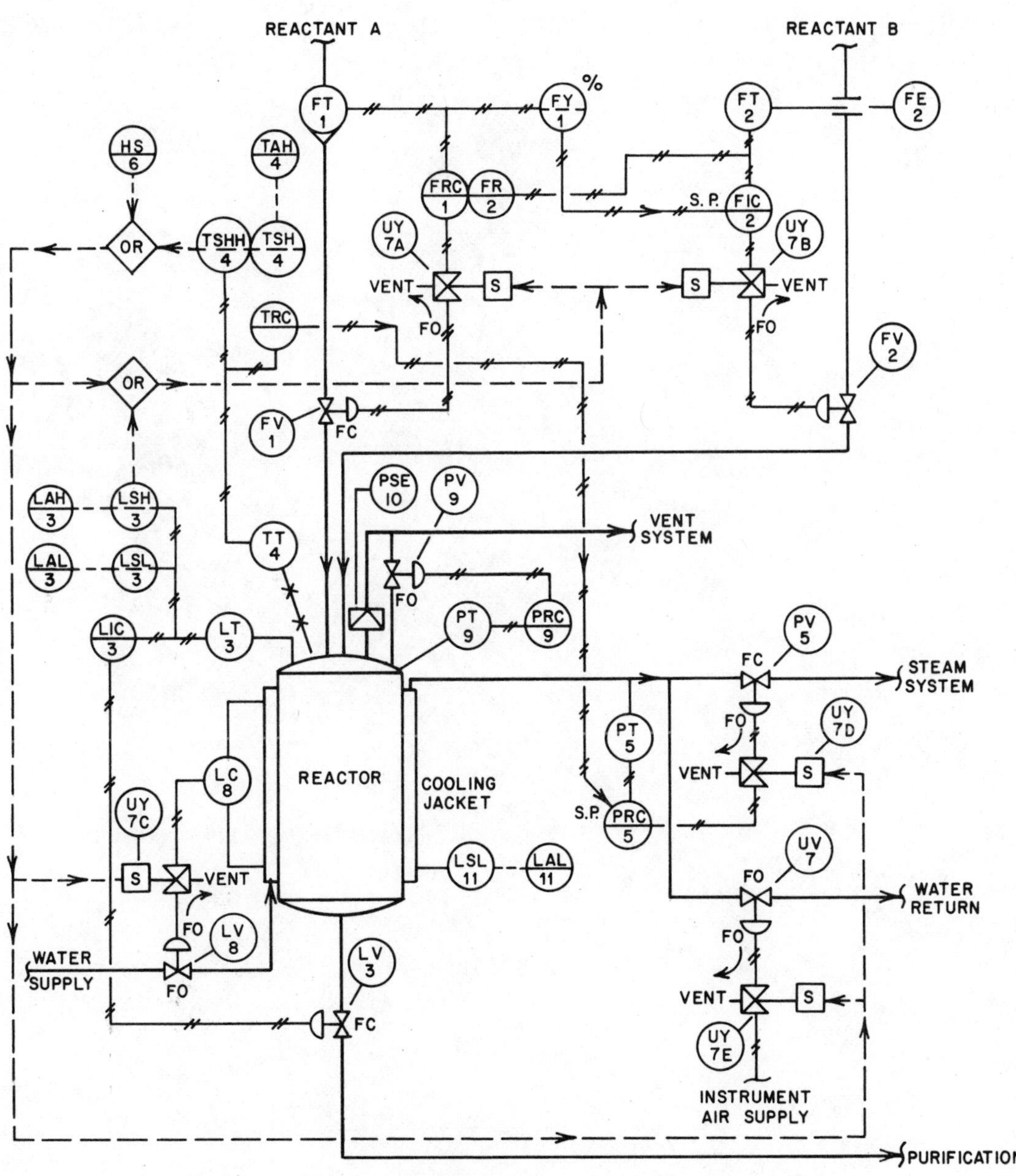

Fig. 12.1h Chemical reaction system instrumentation

Figure 12.1h illustrates the control system for a chemical reactor. The feed of reactant *A* is flow-controlled. The flows of *A* and reactant *B* are ratio-controlled, with the gain relay (FY-1) adjusting the set point of the *B* flow controller. Reactor level is kept constant by modulating the bottom drawoff. If level is high, it automatically closes the reactant feed valves through solenoid valves (UY-7A and UY-7B) and actuates a high-level alarm. A separate alarm is actuated on low reactor level. The reaction is exothermic and temperature is controlled by modulating the pressure of the coolant in the reactor jacket. This is done by the reactor temperature controller adjusting the set point of the jacket pressure controller, which controls the back-pressure of steam generated by transfer of heat to the cooling water. Reactor temperature, if high, actuates an alarm. If the temperature gets very high, it closes the *A* and *B* feed valves and the steam back-pressure valve, while it opens the water supply and return valves through interlock solenoid pilot valves. These very-high-temperature valves can also be actuated by a manual switch. A constant coolant level is maintained in the jacket by modulating the water supply, and low jacket level actuates an alarm. Reactor pressure is controlled by modulating the venting of non-condensables formed in the reaction while a rupture disc protects the reactor against hazardous overpressure.

BIBLIOGRAPHY

"Instrumentation Symbols and Identification," Instrument Society of America, American National Standard, ANSI/ISA—S5.2—1976 (R 1981).

12.2 INTERLOCK LOGIC SYMBOLS

Introduction

This section lists the symbols used to denote binary (on-off) process operations, and illustrates a typical application of the symbols to a plant process. Logic symbol diagramming is applicable to any process control system that uses switching devices to initiate normal or emergency operations. The method is primarily process-based rather than hardware-based. It describes operations in terms of the essential process functions that can be carried out by any class of hardware, whether electric, pneumatic, hydraulic, or other. The method is directed to the needs of an engineer who may have only a rudimentary knowledge of hardware circuit design but who knows what the process-sensing instruments are and how the process is supposed to operate. The hardware and circuit subfunctions needed to perform the process functions then can be detailed by the circuit designer as necessary to satisfy the instrument engineer's intentions.

Logic symbol diagrams are appropriate whenever the operating requirements of the process have to be described to operating personnel, maintenance workers, designers, or others, and it is particularly useful for group discussions. It does not require knowledge of how to read relatively complex and specialized circuit diagrams. However, where it is necessary to trace the actions of a circuit in detail, there is usually no substitute for a complete circuit diagram.

Use of Logic Symbols

A logic diagram may be more or less detailed depending on its intended use. The amount of detail in a logic diagram depends on the degree of refinement of the logic and on whether auxiliary, essentially non-logic, information is included. For example, a logic system may have two opposing inputs; a command to open and a command to close, which do not normally exist simultaneously. The logic diagram may or may not go so far as to specify the outcome if both the commands were to exist at the same time. In addition, explanatory notes may be added to the diagram to record the logic rationale. Non-logic information (reference document identification, tag numbers, terminal markings, etc.) may also be added, if desired.

The existence of a logic signal may correspond physically to either the existence or the non-existence of an instrument signal, depending on the particular type of hardware system and the circuit design philosophy that are selected. For example, a designer may choose a high-flow alarm actuated by an electric switch whose contacts open on high flow; on the other hand, the high-flow alarm may be designed to be actuated by an electric switch whose contacts close on high flow. Thus, the high-flow condition may be represented physically by the absence of an electric signal or by the presence of the electric signal. The logic diagram does not attempt to relate the logic signal to an instrument signal of any specific kind.

The flow of information is represented by lines that interconnect logic statements. The normal direction of flow is from left to right, or top to bottom. Arrowheads may be added to the flow lines wherever needed for clarity, and must be added to lines whose flow is not in a normal direction.

A summary of the status of an operating system may be put in the diagram wherever it is deemed useful, because a specified binary condition is sometimes unclear when it involves a device that does not have only two specific alternative states. For example, if it is stated that a valve is not closed, this could mean either (a) that the valve is open fully, or (b) that the valve is simply not closed: it may be in any position from almost closed to wide open. To aid accurate communication between writer and reader of the logic diagram, the diagram should be interpreted literally. Therefore, possibility (b) is the correct one.

If a valve is an open-close valve, it is necessary to do one of the following to avoid misunderstanding:

1. Develop the logic diagram in such a way that it says exactly what is intended. If the valve is intended to be open, then it should be so stated and not be stated as being not closed.
2. Have a separate note specifying that the valve always assumes either the fully closed or the fully open position.

By contrast, a device such as a motor-driven pump is

Table 12.2.a
LOGIC SYMBOLS

Function	Symbol	Definition and Truth Table	Example
INPUT	(Statement of input) ⊢ Statement may be preceded by instrument balloon with tag number	Logic sequence input	Start chemical injection by pushbutton: Start injection manually ⊢
OUTPUT	(Statement of output) ⊢ Statement may be followed by instrument balloon with tag number	Logic sequence output	The logic sequence causes drawoff to cease: Stop drawoff ⊢
AND	A, B, C → [A] → D	Output *D* exists only if and while inputs *A*, *B*, and *C* exist A B | C: 0 1 0 0 | 0 0 0 1 | 0 0 1 0 | 0 0 1 1 | 0 D	Operate pump if feed tank level is high and provided that discharge valve is open: FIELD TANK LEVEL HIGH, DISCHARGE VALVE OPEN → [A] → OPERATE PUMP
OR	A, B, C → [OR] → D	Output *D* exists only if and while one or more of inputs *A*, *B*, and *C* exist A B | C: 0 1 0 0 | 0 D 0 1 | D D 1 0 | D D 1 1 | D D	Stop compressor if cooling water pressure is low or bearing temperature is high: WATER PRESSURE LOW, BEARING TEMPERATURE HIGH → [OR] → STOP COMPRESSOR
QUALIFIED OR	A, B, C → [*] → D *Insert numerical quantity (see "Definition").	Output *D* exists only if and while a specified number of inputs *A*, *B*, and *C* exist. Mathematical symbols shall be used, as appropriate, in specifying the number, e.g., ≥3, to denote *3 or more*.	Operate feeder if and while two and only two mills are in service: MILL N, MILL E, MILL S, MILL W → [=2] → OPERATE FEEDER
NOT	A → ○ → B	Output *B* exists only if and while input *A* does not exist.	Close valve if and while pressure is not high: PRESSURE HIGH → ○ → CLOSE VALVE
FLIP-FLOP MEMORY	A → [S] → C B → [R] → D* *If output *D* is not used, it shall not be shown.	*S* denotes *set memory* and *R* denotes *reset memory*. Output *C* exists as soon as *A* exists. *C* continues to exist, regardless of the subsequent state of *A*, until the memory is reset, i.e., terminated by *B* existing. *C* remains terminated, regardless of the subsequent state of *B*, until *A* causes the memory to be set.	If standby pump operation is initiated, the pump shall operate, even on loss of the logic power supply, until the process sequence is terminated. The pump shall operate if START and STOP commands exist simultaneously.

Table 12.2.a *Continued*
LOGIC SYMBOLS

Function	*Symbol*	*Definition and Truth Table*	*Example*
		Output D, if used, exists when C does not exist, and D does not exist when C exists. *Input-Override Option* If inputs A and B exist simultaneously, and if A is then required to override B, then S should be encircled, i.e., Ⓢ; if B is to override A, then R should be encircled. *Loss-Of-Power-Supply Option* Required action of the memory on loss of power shall be symbolized by modifying the *set* letter, S, as follows: *Modified Symbol* — *Required Memory Action On Loss Of Power* S (unchanged) — Was not considered by logic designer LS — Memory lost MS — Memory maintained NS — Is not significant, no preference The use of a logic feedback to symbolize a memory that is lost in the event of loss of power is deprecated and not recommended. Thus, the following shall not be used: 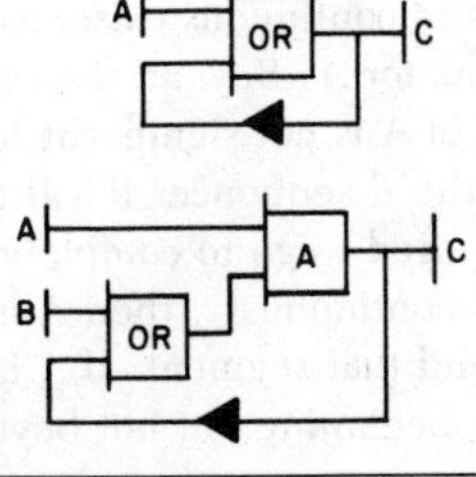	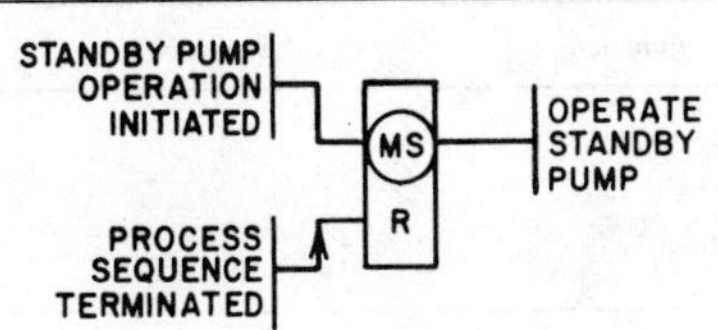
TIME ELEMENT	 *Insert symbolism for specific functions and time interval (see "Definitions").	*Basic Method* This uses the following specific symbols: *Symbol* — *Meaning* DI — *Delay Initiation* of output. The continuous existence of input A for a specified time causes output B to exist when the time expires. B terminates when A terminates. DT — *Delay Termination* of output. The existence of input A causes output B to exist immediately. B terminates when A has terminated and has not again existed for a specified time. PO — *Pulse Output*. The existence of input A causes output B to exist	If vessel purge fails even momentarily, operate evacuation pump for 3 minutes and then stop the pump. 

Table 12.2.a *Continued*
LOGIC SYMBOLS

Function	Symbol	Definition and Truth Table	Example
		immediately. *B* exists for a specified time, regardless of the state of *A*, and then terminates.	
		Generalized Method This is suitable for all time functions. The following illustrations are typical but not all-inclusive.	
	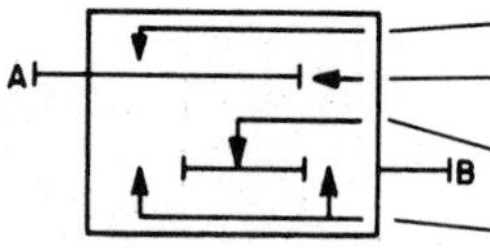 A, B (time element box)	Input logic state exists. Input logic state does not exist. Output logic state exists. Output logic state does not exist. The time at which the logic input *A* is initiated is represented by the left-hand edge of the box. Passage of time is from left to right and is usually shown unscaled. The logic output *B* always begins and ends in the same state within the time element box. More than one output may be shown, if required. The timing of logic may be applied to either the existence state or the nonexistence state, as applicable.	Steam is turned on for 15 minutes beginning 6 minutes after agitator has stopped except that the steam shall be turned off if the agitator restarts. 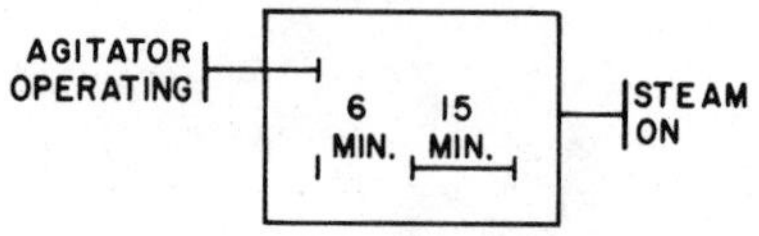 AGITATOR OPERATING; 6 MIN.; 15 MIN.; STEAM ON
	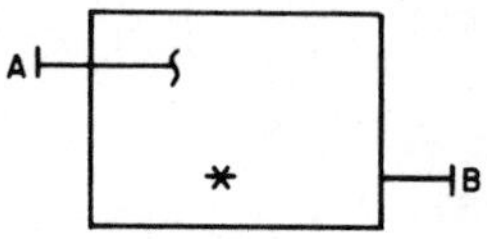 A, * , B. *Show action of output *B* as required.	The action of output *B* depends on how long input *A* is in continuous existence, up to the line break for *A*. Beyond the break in *A*, the state of *A* is not significant to the completion of the *B* sequence. If a *B* time segment is required to go to completion only if *A* exists continuously, then *A* must be drawn beyond that segment. If *A* is drawn past the beginning but not beyond the end of a time segment, then the segment will be initiated and go to completion regardless of whether *A* exists only momentarily or longer.	If pH is low continuously for ½ minute, add caustic for 3 minutes. 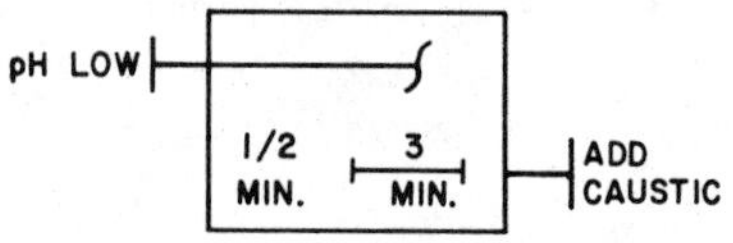 pH LOW; 1/2 MIN.; 3 MIN.; ADD CAUSTIC
SPECIAL	 A, * , B. *Insert statement of special logic requirements.	Output *B* exists with a logic relationship to input *A* as specified in the statement of special requirements.	

either operating or stopped, barring some special situations. To say that the pump is not operating usually clearly denotes that it has stopped.

The following definitions apply to devices that have open, closed, and intermediate positions. The positions stated are nominal to the extent that there are differential-gap and dead band in the instrument that senses the position of the device.

Open position: a position that is 100 percent open.

Not-open position: a position that is less than 100 percent open. A device that is not open may or may not be closed

Closed position: a position that is zero percent open

Not-closed position: a position that is more than zero percent open. A device that is not closed may or may not be open

Intermediate position: a *specified* position that is greater than zero and less than 100 percent open

Not-at-intermediate position: a position that is either above or below the *specified* intermediate position.

For a logic system having an input statement that is derived inferentially or indirectly, a condition may arise that will lead to an erroneous conclusion. For example, an assumption that flow exists because a pump motor is energized may be false because of a closed valve, a broken shaft, or other mishap. Statements based on positive measurements that a certain condition specifically exists or does not exist are generally more reliable.

A process operation may be affected by loss of the power supply—electric, pneumatic, or other—to memories and to other logic elements. In order to take such possibilities into account, it may be necessary to consider the effect of power loss to any logic component or to the entire logic system. In such cases, power supply or loss of power supply should be entered as logic inputs to a system or to individual logic elements. For memories, the power supply may be entered as a logic input or as shown in the diagrams. The effect of power supply restoration also might need to be shown. Logic diagrams do not necessarily have to cover the effect of logic power supplies on process systems, but may do so for thoroughness.

It is recommended for clarity that a single time-function symbol be used to represent each time function in its entirety. Though not incorrect, the representation of a complex or uncommon time function by the use of one time-function symbol in immediate sequence with a second time-function symbol or with a NOT symbol should be avoided.

Definitions

Table 12.2a illustrates and defines the logic symbols and some typical uses of them. The symbols shown with three inputs, *A*, *B*, and *C*, are typical for the logic functions having any number of two or more inputs. In the several truth tables, *0* denotes the non-existence of the logic input or output signal or state given at the head of the column. *1* denotes the existence of the logic input signal or state. *D* denotes the existence of the logic output signal or state as a result of appropriate logic inputs. The output states for a truth table are within a heavily outlined box.

Application to a Process

The process must have high vacuum to proceed properly. Vacuum is normally maintained by an air ejector, but in case of failure or overload of the air ejector the system pressure rises. The rise is sensed by a pressure switch (PSH), which automatically starts a vacuum pump, provided that a hand-actuated control switch (HS) for the pump motor is in the AUTOMATIC position. This switch also can be used to start and stop the pump manually. However, the pump is not permitted to start or run if the discharge temperature, as sensed by a temperature switch (TSH), is high or if the motor is overloaded and its circuit breaker is not manually reset. If high pressure is maintained for ten minutes, a high-pressure alarm (PAH) is actuated. High temperature is signalled by another alarm (TAH). Pump motor overload is signalled by the alarm (IAH). If the pump control logic circuit loses power, the pump shall stop automatically but shall not be able to be restarted until the system is reset manually.

Whenever the pump is required to operate, cooling water is automatically turned on. The water flow is controlled by an air-actuated control valve (UV), which is operated by a solenoid valve (UY) that, in turn, is operated by auxiliary contacts of the pump motor circuit breaker. The water is automatically turned off when the

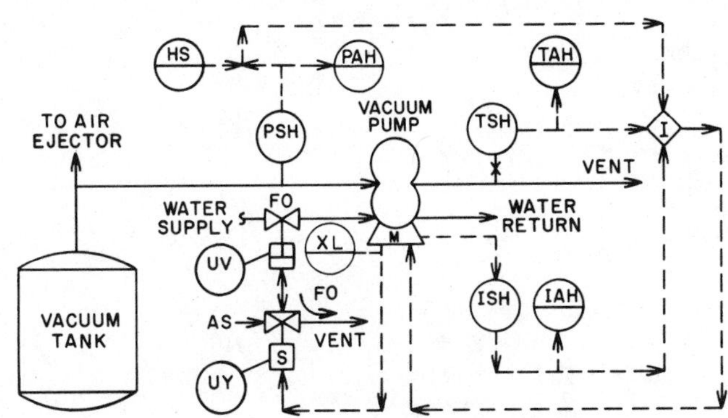

Fig. 12.2b Control system for standby vacuum pump

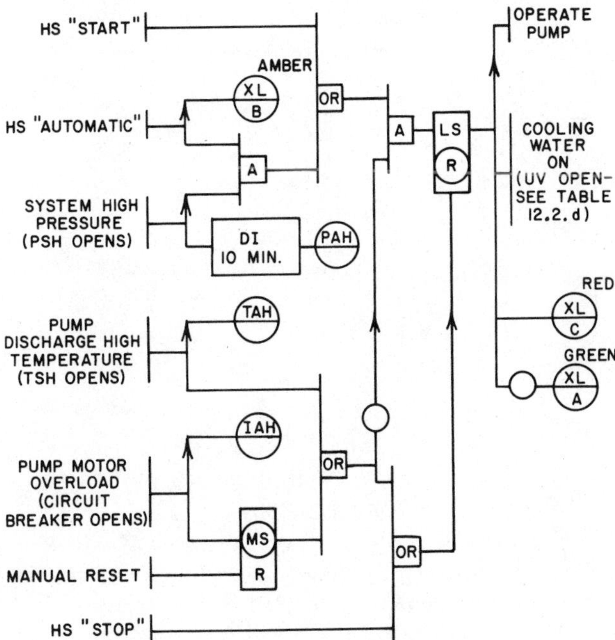

Fig. 12.2c Logic diagram for standby vacuum pump

pump is stopped. The following instruments are on the instrument board:

- HS Manual control switch for pump operation. The switch has three momentary-contact pushbuttons for *Start*, *Automatic*, and *Stop*.
- PAH Alarm actuated upon rise of pressure to abnormal value. However, this alarm is blocked for ten minutes after a pump start is required.
- TAH Alarm that is actuated if pump discharge temperature rises to abnormal value.
- XL-A Green pilot light denoting that the pump motor circuit breaker is not closed, i.e., that pump is not operating.
- XL-B Amber pilot light denoting that the pump is ready for an automatic start.
- XL-C Red pilot light denoting that the pump motor circuit breaker is closed, i.e. that pump is operating.
- IAH Alarm that is actuated upon overload of pump motor.

The required process operations are diagrammed in Figure 12.2b, while Figure 12.2c describes hardware functions.

Table 12.2.d
CONTROL VALVE OPERATING SEQUENCE

Vacuum Pump	*Motor Circuit Breaker Auxiliary Contacts*	*Solenoid Valve (UY) Coil*	*Control Valve (UV)*		*Cooling Water*
			Actuator	*Port*	
Off	Closed	Energized	Pressurized	Closed	Off
On	Open	Deenergized	Vented	Open	On

BIBLIOGRAPHY

"Binary Logic Diagrams for Process Operations," Instrument Society of America, ANSI/ISA—S5.2—1976 (R 1981).

12.3 GRAPHIC SYMBOLS

The following figures show some of the graphic symbols which may be encountered in the design, construction and operation of process control systems by the engineer, technician or operator. The symbols are used in the preparation of plan and elevation drawings for architectural and mechanical construction; in flow diagrams to indicate process flow conditions from raw material input to product output; in line, ladder and schematic diagrams to

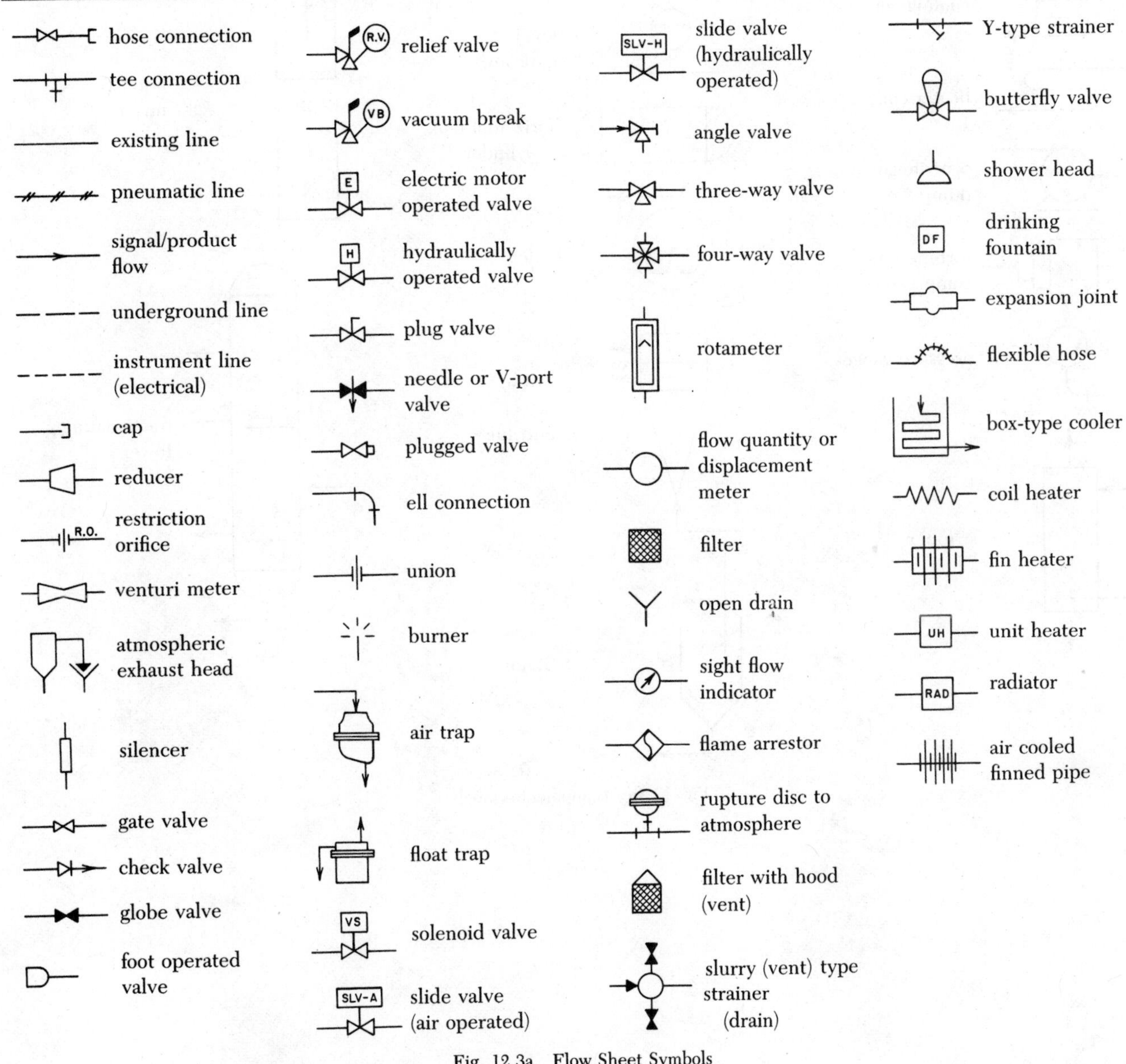

Fig. 12.3a Flow Sheet Symbols

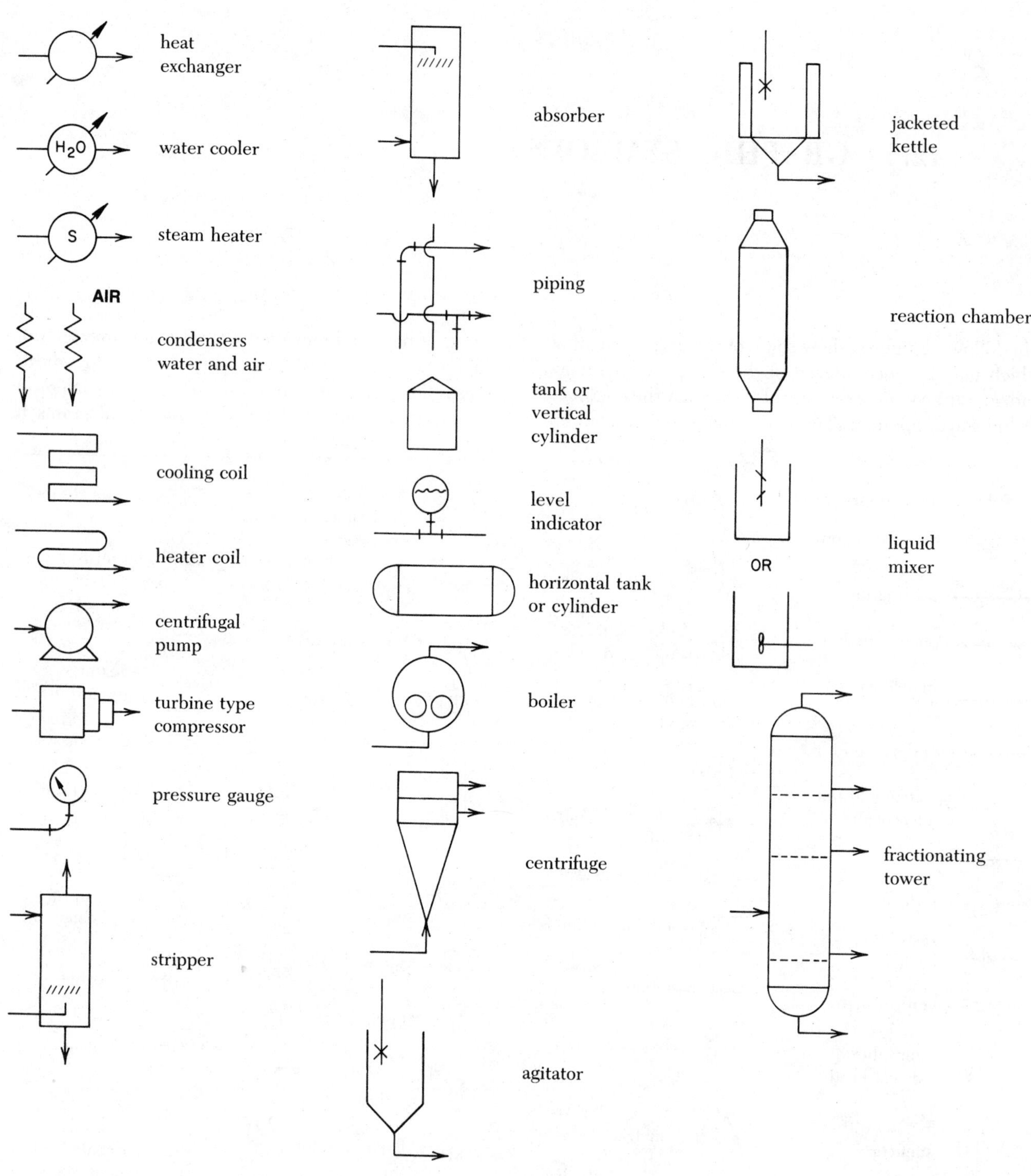

Fig. 12.3b Equipment symbols

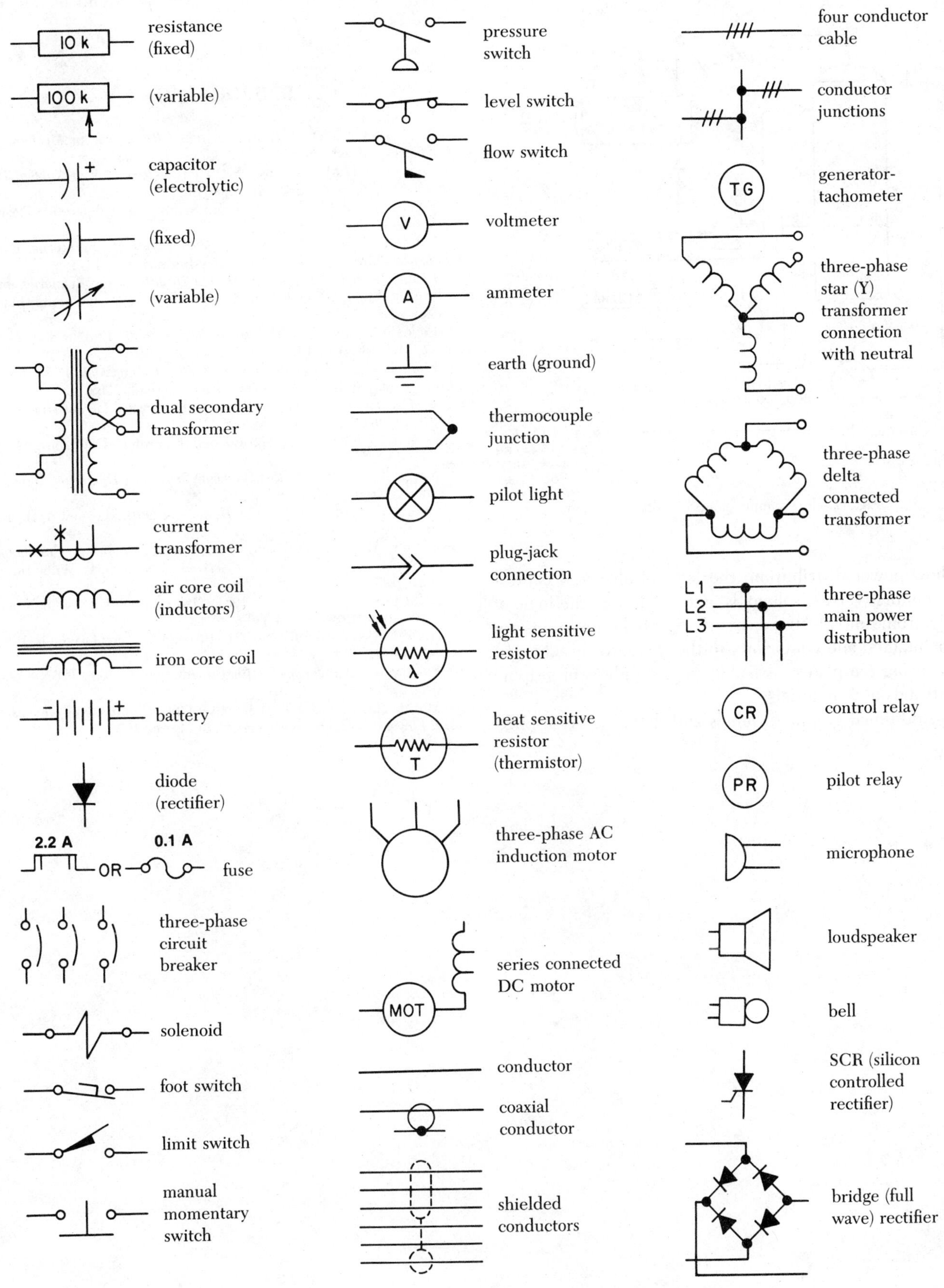

Fig. 12.3c Electrical symbols

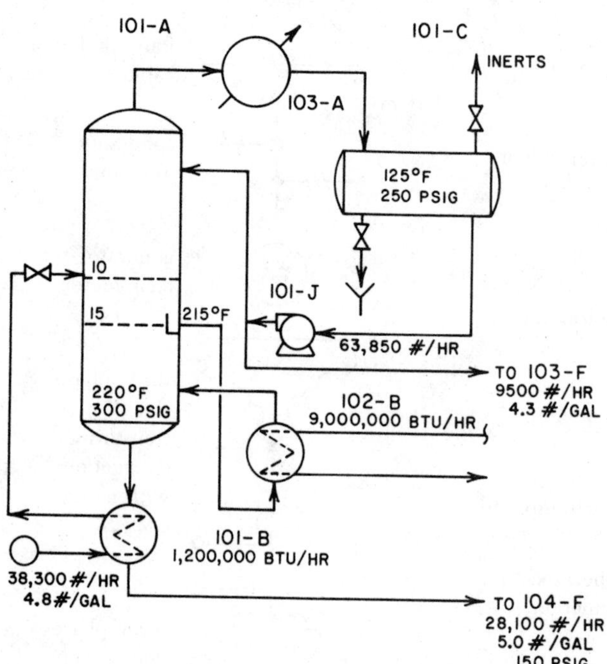

Fig. 12.3d Example of a process flow diagram

show power distribution, the logic of electrical control of motors, valves, solenoids, and so on, and the logic of electronic and electrical circuits. Symbol proportions are maintained and conformity to the standards is achieved by using templates, available from suppliers of graphic arts aids and materials.

Additional graphic symbols and their use, as applied to specific engineering fields, will be found in the selected bibliography.

BIBLIOGRAPHY

Baumeister, T., et al. *Mark's Standard Handbook for Mechanical Engineers*, 8th ed., New York, McGraw-Hill, 1967.

Considine, D.M. and Ross, S.D., eds., *Handbook of Applied Instrumentation*, New York, McGraw-Hill, 1964.

Electrical and Electronic Graphic Symbols and Reference Designations, 76-ANSI/IEEE Y32E.

Eshbach, O.W. and Souders, M., *Handbook of Engineering Fundamentals*, 3rd ed., New York, Wiley-Interscience, 1975.

Evans, Jr. F.L., *Equipment Design Handbook for Refineries and Chemical Plants*, Volumes 1 and 2. 2nd ed., Houston, Gulf Publishing, 1980.

Fink, D.C., *Standard Handbook of Electronic Engineering*, New York, McGraw-Hill, 1974.

Fink, D.C., and Carroll, J.M., *Standard Handbook of Electrical Engineering*, 10th ed., New York, McGraw-Hill, 1968.

Graphic Symbols for Architectural and Electrical Layout Drawings, ANSI Y32.9-1972.

Graphic Symbols for Electrical and Electronic Diagrams, ANSI/IEEE 315-1975.

Neumüller, R., *How to Read German Schematic Diagrams*, 2nd ed., Siemens Aktiengesellschaft, *ca*. 1975.

Oberg, E. et al., *Machinery's Handbook* 20th ed., New York, Independence Press, 1975.

Pender, H., and McIlwain, K., *Electrical Engineers Handbook: Electrical Communication and Electronics*, New York, Wiley-Interscience, 1950.

ANSI/ISA—S5.1—1975 (R 1981), Instrumentation Symbols and Identification, ISBN 87664-330-6.

ANSI/ISA—S5.4—1976 (R 1981), Instrument Loop Diagrams, ISBN 87664-331-4, 1981.

Standards and Practices for Instrumentation (6th ed.) ISBN 87664-450-7, 1980.

Weast, R.C. and Astle, M.J., eds., *CRC Handbook of Chemistry and Physics*, 59th ed., New York, CRC Press, 1979.

12.4 TERMINOLOGY AND DEFINITIONS

Control Terminology

The development of automatic control systems in the past 50 years has been equated in importance to the industrial revolution in the nineteenth century. In many respects the introduction of automatic control systems was a second industrial revolution. While the first was an extension of man's muscle, the second was an extension of his brain. In the nineteenth century we learned to harness and use various forms of natural energy; in the twentieth century we learned to make devices that could make the decisions necessary to control the various forms of energy.

Principles used in automatic control cut across virtually every scientific field and in the process created a new field. Today the basic principles of automatic control have a wide range of applications and interests, including process control, manufacturing control, aero-space control, traffic control and biomedical control.

Why Automatic Control?

The real need for automatic control in some areas is perhaps more obvious than it is in the process industries. In assembly line manufacturing facilities the need for automation is quite apparent. A machine in many cases is more suitable, both for economic and safety considerations, to perform the numerous tedious and monotonous tasks involved. It is also fairly clear to the casual observer that the control of a supersonic aircraft is much too complicated to be left entirely in the hands of a human pilot. However, in the typical process applications the reasons for control are perhaps a little less apparent.

Since most process equipment operates at a constant load, the tendency might be to suggest that the best solution to the control problem is to set all the variables which affect the process to their proper positions and forget about the process. The difficulty with this reasoning is that seldom can all the inputs to the system be fixed. Most process equipment is subject to many inputs, some of which can be manipulated (or set at a fixed value) and some which will change without regard to the operator's desires. Changes in such variables result in disturbances in the process, unless corrective action of some sort is taken.

Consider the simple direct contact water heater shown in Figure 12.4a. The heater consists of a tank from which hot water is obtained by bubbling live steam directly into the tank which is full of water. Cool water enters at the bottom of the tank and the hot water leaves at the top. A valve is available by which to regulate the flow rate of steam into the heater. In this example, if all other factors were constant the temperature of the outlet could be controlled simply by placing the steam valve at the proper setting. Note, however, that if the temperature of the inlet water changes, the outlet temperature would eventually change by the same amount unless corrective actions were taken. Other variables here besides the inlet water temperature which could disturb the process are the flow rate of the water, the steam supply pressure, the steam quality, and the ambient temperature. A change in any one of these variables would cause a change in the water outlet temperature unless some correction were made.

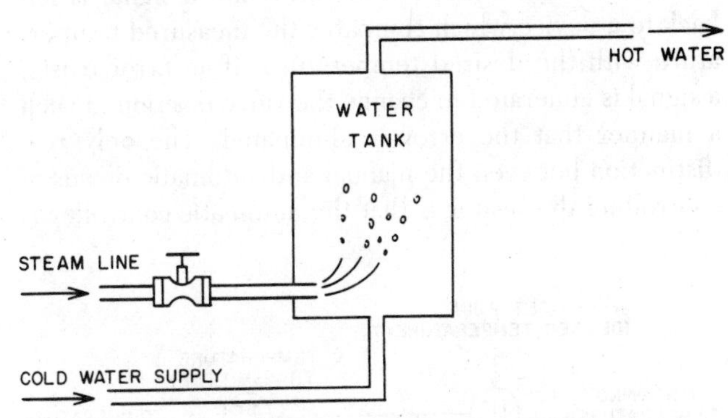

Fig. 12.4a Direct contact water heater

Feedback Control

Two concepts provide the basis for most automatic control strategies: feedback (closed-loop) control and feedforward (open-loop) control. Feedback control is the more commonly used technique of the two and is the underlying concept on which much of today's automatic control theory is based. Feedback control is a strategy designed to achieve and maintain a desired process condition by measuring the process condition, comparing the measured condition with the desired condition, and initiating corrective action based on the difference between the desired and the actual condition.

The feedback strategy is very similar to the actions of a human operator attempting to control a process manually. Consider the procedure an individual might employ in the control of the direct contact hot water heater described earlier. The operator would read the temperature indicator in the hot water line and compare its value with the temperature he desires (Figure 12.4b). If the temperature were too high, he would reduce the steam flow, and if the temperature were too low, he would increase it. Using this strategy he would manipulate the steam valve until the error was eliminated.

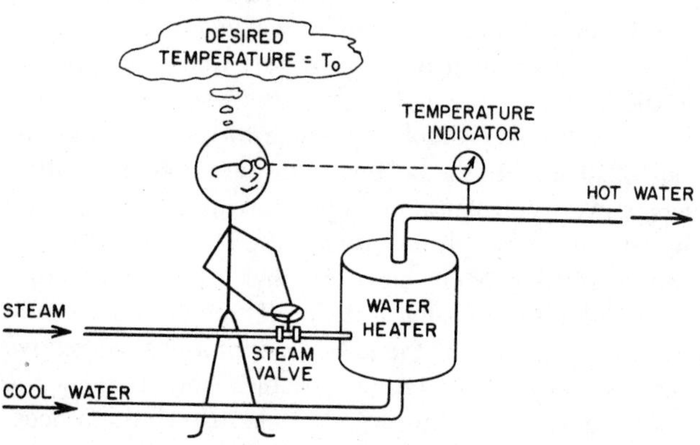

Fig. 12.4b Manual feedback control

An automatic feedback control system would operate in much the same manner (Figure 12.4c). The temperature of the hot water is measured and a signal is fed back to a device which compares the measured temperature with the desired temperature. If an error exists, a signal is generated to change the valve position in such a manner that the error is eliminated. The only real distinction between the manual and automatic means of controlling the heater is that the automatic controller is

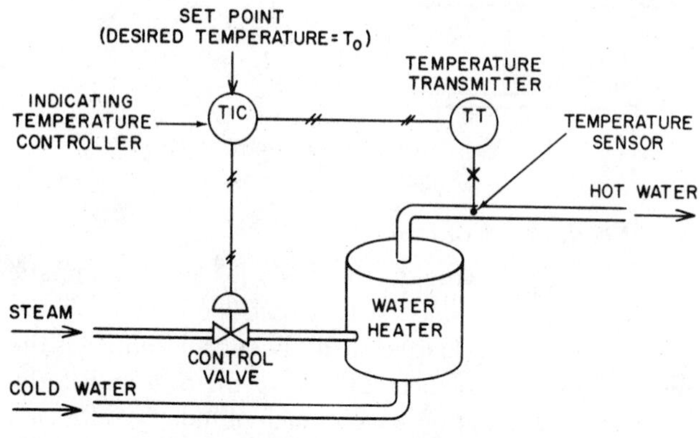

Fig. 12.4c Automatic feedback control

more accurate, consistent, and not as likely to become tired or be distracted. Otherwise, both systems contain the essential elements of a feedback control loop (Figure 12.4c and d).

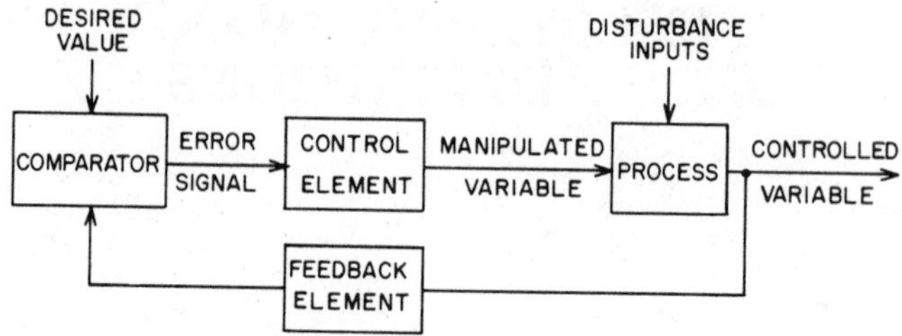

Fig. 12.4d Basic components of a feedback control loop

Feedback control has definite advantages over other techniques (such as feedforward control) in relative simplicity and potentially successful operation in the face of unknown contingencies. In general, it works well as a regulator to maintain a desired operating point by compensating for various disturbances which affect the system, and it works equally well as a servo system to initiate and follow changes demanded in the operating point.

Feedforward Control

Feedforward control is another basic technique used to compensate for uncontrolled disturbances entering the system. In this technique the control action is based on the state of a disturbance input without reference to the actual system condition. In concept, feedforward control yields much faster correction than feedback control, and in the ideal case compensation is applied in such a manner that the effect of the disturbance is never seen in the process output.

A skillful operator could use a simple feedforward strategy to compensate for changes in inlet water temperature of a direct-contact water heater. Detecting a change in inlet water temperature, he would increase or decrease the steam rate to counteract the change (Figure 12.4e). The same compensation could be made automatically with an inlet temperature detector designed to initiate the appropriate corrective adjustment in the steam valve opening.

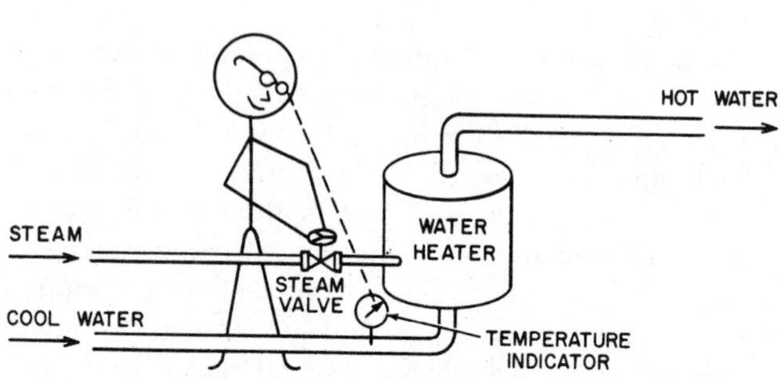

Fig. 12.4e Concept of feedforward control

The concept of feedforward control is very powerful, but unfortunately it is difficult to implement in a pure form in most process control applications. In many cases disturbances can not be accurately measured, and therefore feedforward concepts cannot be applied. Even in applications where all the inputs can be either measured or controlled, the "appropriate" action to be taken to compensate for a particular disturbance is not always obvious. In many applications, feedforward control is utilized in conjunction with feedback control in order to handle those unknown contingencies which might otherwise disturb the pure feedforward control system.

Mathematical Representation

A fundamental prerequisite of automatic control theory application is a detailed understanding of the operation of the process under control. While standard equipment and process design requires a detailed knowledge of the equipment operation for constant inputs, automatic control requires a detailed knowledge of the equipment operation when inputs are changing in time. This time-varying behavior is referred to as "process dynamics" and can be conveniently summarized in mathematical terms using differential equations (or the Laplace transfer function representation of differential equations). The application of automatic control theory presumes a knowledge of the entire control system mathematics, and without such knowledge automatic control theory is to a large extent useless.

In many cases the mathematical description (mathematical model) of the various components of the control loop can be obtained analytically, based entirely on the physics of the process components. In other cases, models can be obtained by experimental testing procedures in which the actual response of the process is analyzed in some manner to extract the desired dynamic information.

Glossary of Analog and Digital Process Control Terms

Since the birth of process control a sizable number of terms have become associated with the process control jargon. The following glossary includes terms and terminology associated mostly with analog process control. For digital terminology refer to "Computer Terminology" in this section.

Absolute Alarm An alarm caused by the detection of a variable which has exceeded its high or low limit condition.

Access The ability to place information into and retrieve information from a storage device.

Actuating Signal The variable that initiates corrective action in a control system; the same as *error signal* in feedback control.

Adaptive Control Action Control action in which the control algorithm or control parameters are changed automatically in such a way as to improve the performance of the control system.

Adder A device whose output is a representation of the sum of the inputs.

Adiabatic Occurring without transfer of heat to or from the body or system.

Ambient Pressure The pressure of the medium surrounding a device.

Ambient Temperature The temperature of the medium into which the heat of the system is dissipated.

Amplification The ratio of the signal output amplitude to the signal input amplitude.

Amplifier A device whose output is an enlarged reproduction of the essential features of an input and which draws power from some external source.

Amplitude The difference between the average value of a sinusoidal variation and the maximum (or minimum) value.

Amplitude Ratio A factor expressing the ratio of the output amplitude to the input amplitude when the input is sinusoidal.

Analog (Analogue) A system whose behavior is mathematically analogous to that of some other system and therefore has tbe same dynamic equations.

Analog Backup An alternative means to maintain control over the process in the event of a failure in the primary control system; the backup consists of conventional analog instrumentation.

Analog Controller A small special-purpose analog computer used to operate on continuous process signals such as voltages, pressures or currents to determine necessary control action. These controllers are distinguished from digital controllers operating on signals with discrete numerical values at discrete intervals of time.

AND A logic operator having the property that, if P, Q, and R are all logical statements, then the AND of P, Q, and R is true if P, Q, and R are all true, and it is false if any of the three statements are false.

AND Gate A gate that implements the logic AND operation.

Anticipatory Action Same as *Rate Action*.

ASCII American Standard Code for Information Interchange, an eight-level code intended to provide information code compatibility between digital devices of U.S. manufacturers.

Attenuation A decrease in the strength of a signal between two points or between two frequencies.

Auctioneering Device A device which automatically selects either the highest or the lowest input signal from among several input signals.

Automatic Control System Any operable combination of one or more automatic controllers connected in closed loops with one or more processes.

Automatic Controller A device which measures the

value of a variable, quantity, or condition and operates to correct, or limit, deviation of this measured value from a selected reference.

Backlash A nonlinearity typically associated with the slack in a gear train.

Backup Provision of alternative means of operation in case of a failure of the primary means of operation.

Bandwidth The range of frequencies within which performance of a component is accurate, usually extending from zero frequency to some cutoff frequency.

Bang-Bang Control The same as two-position control.

Boolean Algebra Pertaining to the operations of formal logic.

Break Point In frequency response plots, the intersection of the asymptotes of the magnitude (or amplitude) ratio plot. Also called break frequency and corner frequency.

Calculating Action The coupling of primary feedback variables with one another and/or other variables to form a computable function from which the control action is taken.

Capacitance The amount of energy or material which must be added to a closed system to cause unit change in potential; hence the partial derivative of the content with respect to potential.

Capacity A measure of the maximum quantity of energy or material which can be stored within a given piece of equipment or system.

Cascade A series of stages in which the output of one stage is the input of the next.

Cascade Control Automatic control involving cascading of controllers such that one controller manipulates the set-point input of the other controller instead of manipulating a process variable directly.

Characteristic Function A polynomial that characterizes the transient response of a system and is the denominator of the system's overall transfer function.

Closed Loop A signal path which consists of a forward path, a feedback path, and a summing point connected so as to form a closed circuit.

Combination Control Closed loops connected in combination with the loops being coupled through primary feedback or through any other controller elements.

Command Signal The set point or the reference input to a control system.

Comparator The portion of the control elements that determines the feedback error (difference between the reference input and the feedback variable) on which the controller acts.

Compensator A component or circuit added to a system to improve the characteristics of its response; hence, in many cases the controller.

Constant Value Control Automatic control of a desired constant value.

Continuous Action Control action performed continuously (analog control).

Continuous Process A process in which for extended periods of time uninterrupted flows of fluids enter and products leave a system, as opposed to a batch process, in which the input and output flows are intermittent and periodic.

Control Accuracy The degree of correspondence between the controlled variable and the desired value of the variable.

Control Action The response of a control device to an actuating signal for the purpose of decreasing the control error.

Control Algorithm The mathematical representation of the control action to be performed.

Control Elements The portion of the control system which relates the error signal to the manipulated variable; the portion of the control system which implements the algorithm.

Control Input Same as *Set Point*.

Control Mode A specific type of control action such as proportional, integral, or derivative.

Control Point The desired value of the variable under control.

Control Ratio The response of the control variable to a change in set point.

Control System A system in which deliberate guidance is employed to execute a planned set of control functions.

Controlled System The body, process, or machine which determines the relationship between the control variable and the manipulated variable.

Controlled Variable That quantity or condition of the control object which is to be directly measured and controlled.

Controller A device which operates automatically to regulate a controlled variable.

Corner Frequency Same as *Break Point*.

Correction A value to be added to a measured value to compensate for an error.

Correction Time The time required for a controlled variable to reach and stay within a band about the control point following a change of the set point or operating conditions.

Corrective Action The variation of the manipulated variable produced by the controller.

Coupled-Control-Element Combination One in which two or more controller outputs are combined to operate one manipulated variable.

Critical Damping The smallest degree of damping under which a system can operate without overshooting the desired value after a step change in input.

Critical Gain A value of system gain beyond which the system is unstable.

Critical Point In stability analysis by the Nyquist method, the point $s = -1$.

Curve Fitting The representation of a curve by a mathematical expression. It usually involves determining the "best" optimum expression by some regression technique.

Cutoff Frequency The frequency of sinusoidal forcing beyond which the amplitude ratio of the response is below some specified lower limit.

Cybernetics The theory of control and communication in both machines and animals.

Cycling A periodic change (oscillation) in the controlled variable.

Damping That property of a system which causes dissipation of energy and hence causes decay in the amplitude of oscillations.

Damping Coefficient In the characteristic function of a system, the parameter which characterizes the nature of damping of the transient response. Also called damping factor and damping ratio.

Dashpot A damping device, usually consisting of a cylinder and a piston in which relative motion of either displaces a fluid such as air or oil, resulting in friction.

Dead Band The range of values through which the measured variable can be changed without initiating a response.

Deadtime The fixed interval of time between the start of an input to a component and the beginning of response to the input.

Dead Zone See *Dead Band*.

Decade Range of frequencies of which the highest is ten times the lowest.

Decibel In frequency response terminology, a quantitative comparison of the magnitudes of the input and the output sine waves; the number of decibels is 20 times the $\log_{10}$ of the amplitude ratio.

Decimal Numbering System The numbering system using a base of 10.

Decoder A circuit which responds to specific coded signals and rejects others.

Delay Same as *Deadtime*.

Delay Time The time elapsing from the time input changes to the time the output responds to the input.

Derivative Action A controller mode in which there is a continuous linear relationship between the controller output and the derivative of the error signal.

Derivative Time The time difference by which the output of a proportional-derivative controller leads the input when the input changes linearly with time.

Desired Value The value of the controlled variable which is desired; the same as *Set Point*.

Deviation The difference between the actual value of the controlled variable and the value of the controlled variable corresponding to the set point.

Differential Action Same as *Derivative Action*.

Differential Gap In a two-position control system, an adjustment which determines the smallest range of values through which the controlled variable must pass in order to change the output signal of the controller from maximum to a minimum.

Differentiator A device, usually of the analog type, whose output is proportional to the derivative of the input signal.

Digit One of a definite set of characters which are used as coefficients of power of the base in the positional notation of numbers.

Digital Pertaining to data in the form of digits; discrete data as contrasted to continuous analog data.

Digital Backup An alternate method of controlling a process which employs a spare digital process control computer.

Direct Acting Operation of a final control element directly proportional to the control output.

Directly Controlled Variable That process variable whose value is measured to originate a feedback signal on which the control action will be taken.

Distance-Velocity Lag Same as *Deadtime*.

Distortion An undesired change in wave shape.

Disturbance An input signal other than the set point which directly affects the output of the process.

Downtime The time interval during which a system is not productive.

Drift Undesired change in instrument indication with respect to time from an initial value corresponding to a state when the measured variable and ambient conditions are constant.

Driven Response Same as *Forced Response*.

Droop An offset, particularly downward (upward is called rise).

Dynamic Error A measured error caused by variations in time of a quantity being measured, additive to static error.

Dynamic Gain The amplitude ratio of the steady-state output and input signals of an element or system when the input is sinusoidal. The dynamic gain typically changes with the frequency of the sinusoidal signal. A record of the variations in dynamic gain vs frequency is called the frequency response of the system.

Dynamic Response The behavior of the output of a device in time with respect to variations in the inputs in time.

Error Rate Control Same as *Proportional-Plus-Derivative Control*.

Error Ratio The response of the closed-loop system actuating signal to a change in set point.

Error Signal The signal resulting from subtracting the feedback signal from the reference signal. The error signal is the input to that part of the controller which contains the algorithm.

Error Squared The technique of using the square of the error on which to make the control calculation so as to produce a non-linear correction.

Exclusive OR A logic operator having the property such that, if P and Q are logical statements, then the exclusive OR of P and Q is true if either, but not both, statements are true, and false if both are true or both are false.

Exponential Stage A system whose transient response to a step input is an exponential decay; hence, a first-order linear system.

Feedback The signal to the controller representing the condition of the controlled variable.

Feedback Control Action Control action in which a measured variable is compared with the reference value to produce an actuating error signal which is acted upon to attempt to reduce the error.

Feedback Elements That portion of the controller which establishes the relationship between the primary feedback and the actual controlled variable.

Feedforward Control A control system in which corrective action is based on measurement of disturbance inputs into the process.

Filter A transducer whose frequency-response characteristics are such that input signals within a certain range are transmitted while other signals are not transmitted.

Final Control Element That portion of the control loop which directly changes the value of the manipulated variable.

First-Order Lag, or *First-Order Delay* A system whose dynamic behavior is described by a first-order linear differential equation.

Flip-Flop A circuit containing active elements capable of assuming either one of two stable positions at a given time.

Floating Control A mode of control in which the manipulated variable is changed proportional to the integral of the error. A change continuously occurs as long as an error exists.

Floating Rate Same as *Reset Rate*.

Floating Speed In single- or multi-speed controller action, the rate of motion of the final control element. It is commonly expressed in percent of full range motion per minute.

Forced Response That part of a system's output which is a direct result of an input forcing function, and remains after the transient has died out.

Forcing A change in an input to a system in a specific manner beginning with the system initially at steady state.

Forward Controlling Elements Those elements in the control system which change a variable in response to the actuating error signal.

Frequency Response The effect of input frequency on the amplitude ratio and phase shift of a system's or element's output for a sinusoidal input. The frequency response can be directly related to the differential equation which describes the system.

Fundamental Natural Frequency The lowest of a set of natural frequencies.

Gain Margin Related to the magnitude of a system response to a sinusoidal input at the frequency for which its phase angle is −180 degrees. A measure of the degree of stability a system will have under feedback control.

Harmonic A sinusoidal quantity having a frequency which is an integral multiple of some fundamental frequency to which it is related.

Head Pressure resulting from gravitational forces on liquids; measured in terms of the depth below a surface of the liquid.

High Limiting Control Action Control action in which the output never exceeds a predetermined high limit.

High Order Delay A system characterized by a high order differential equation.

Hunting The undesirable motion of an automatic control system in which the controlled variable swings or oscillates about the desired value without seeming to approach it.

Hysteresis A non-linearity usually attributed to flexibility and loose fits in linkage and to backlash in gear trains.

Idealized System An imaginary system whose controlled variable has a stipulated relationship to specified set points.

Impulse A theoretical signal which is a pulse signal of infinite magnitude and infinitesimal duration. In practical applications, a sharp increase or decrease in a variable followed immediately by a return to the original value.

Inclusive OR Same as *OR*.

Indirectly Controlled Variable A variable which is not monitored directly by the control system but is related to, and influenced by, the variable which is under direct control.

Initial Error The transient error appearing immediately after a step function input has been initiated.

Input A variable that is dependent only on conditions outside the system.

Input Element The portion of the control system which provides the reference input to the comparator in response to the set point.

Instantaneous Sampling The process of obtaining a sequence of instantaneous values of a variable continuous in time.

Integral Action A controller mode in which there is a continuous linear relationship between the integral of the error signal and the output signal of the controller.

Integral Action Limiter A device which limits the value of the output signal due to integral action to minimize the effect of reset wind-up.

Integral Control See *Integral Action*.

Integral (Reset) Controller A controller which produces integral control action only.

Integral Time The proportionality constant in the equation relating the controller output to the error for integral action. It is the time required to produce a change in controller output equal to the change in error input.

Interacting Two or more consecutive transfer stages whose effective transfer function is not the product of the individual stages.

Interacting Control Control action produced by an algorithm whose various terms are not independent.

Lag Any deviation from instantaneously complete response to an input signal. It is usually associated with lags due to resistances and capacitances in the system; however sometimes it is used synonymously with *Delay* and *Dead Time*.

Laplace Transform A mathematical transformation in which differential equations can be handled much like algebraic equations. The Laplace transform of a variable f(t) is defined as follows:

$$F(s) = \int_0^\infty f(t)e^{-st}dt \qquad 12.4(1)$$

Limit Cycle A sustained oscillatory response of a feedback system in which the amplitude of oscillation is limited. The cause is usually the presence of some nonlinearity such as saturation or hysteresis.

Limiting A condition in which the system's response is restricted to a value less than that for a linear response.

Linear Forcing A forcing function which is linear in time, such as $f(t) = A + Bt$, where A and B are constants and t is time.

Linear Programming A procedure for maximizing or minimizing some variable (such as cost or profit). The profit or cost function is written as a linear function of a number of variables which are subject to a number of constraints in the form of linear equalities. The procedure is easily adaptable to computer solution and used quite often as a strategy in supervisory control.

Linearize To substitute a linear function for a non-linear one, which gives approximately the same relationships over a small range.

Line-Out Time Same as *Settling Time*.

Live Zone A zone in the operating cycle of a machine or system during which corrective action can be initiated.

Load, or *Load Variable* An outside influence on an automatic control system other than the set point whose effect on the system must be compensated for by the control system.

Load Error Same as *Offset*.

Log A periodic summary of process operation data.

Log-Modulus Plot A rectangular plot of the logarithm of the amplitude ratio vs phase angle of frequency response data; a Nichols plot.

Loop A series of stages forming a closed path.

Loop Gain (Closed Loop) The control system gain relating a change in the controlled variable to a change in set point with the feedback element included.

Loop Gain (Open Loop) The gain of the control system relating a change controlled variable to a change in set point with the feedback element removed.

Low Limiting Control Action Control action in which the output is never less than a predetermined low limit.

Low-Pass Filter A wave filter having a transmission band extending from zero frequency up to some cutoff frequency.

Lumping In the derivation of a model, an assumption that the effects of two or more aspects of the system can be considered as a single quantity.

Magnitude Ratio In steady-state sinusoidal forcing, the ratio of the amplitude of the output signal to the amplitude of the input signal. The magnitude ratio is usually distinguished from the amplitude ratio in that it is normalized by the system gain such that:

$$\text{magnitude ratio} = \frac{\text{amplitude ratio}}{\text{system gain}} \qquad 12.4(2)$$

Manipulated Variable The process variable that is changed by the controller to eliminate error.

Manual Backup An alternative means to maintain process control in the event of a failure in the primary control system which uses manual adjustment of final control elements.

Mathematical Model A mathematical representation of a process, device, or system derived from either analytical considerations, experimental investigations, or both.

Matrix A two-dimensional rectangular array of quantities.

Measured Signal The electrical, mechanical, pneumatic or other variable which is related to some process variable such as flow rate, temperature and level.

Mode The classification of a controller by the manner in which the manipulated variable responds to the error signal. Some common modes are proportional, integral, and derivative.

Model A conceptual approximation of a physical element or system of elements; used in the prediction of the behavior of the system; usually mathematical in nature.

Modulation The process by which some characteristic of one wave is varied in accordance with some characteristic of another wave.

Multi-Element Control System A control system utilizing input signals derived from two or more process variables which are used jointly in determining the action of the control system.

Multi-Velocity Action Control action in which the velocity of the actuating variable takes one of several predetermined velocities, each corresponding to a definite range of the actuating signal.

Natural Frequency The frequency of oscillation that a system would have if the damping were reduced to zero.

Natural Response Same as *Source-Free Response*.

Neutral Zone A range of error values that gives rise to a value of zero for the controller output.

Noise An unwanted fluctuation in a variable which tends to obscure its information content.

Non-Interacting Control System A multi-element control system designed to eliminate effectively the interaction between various process loops so that adjustments can be made in one controlled variable without disturbances being introduced in the other controlled variables.

Non-Self-Operating Control Control in which energy required to operate the actuating unit is supplied by an external source.

NOR A logic operator having the property such that, if P, Q, and R are logic statements, then the NOR of P, Q, R is true if all statements are false, and is false if at least one statement is true.

Normalize To shift the representation of a variable or quantity so that the representation lies in a prescribed range.

NOT A logic operator having the property such that, if P is a logical statement, then the NOT of P is the opposite statement. If P is true, the NOT of P is false; if P is false, the NOT of P is true.

Octave A span of frequencies of which the highest is twice the lowest.

Offset The steady-state deviation of the controlled variables from the set point, usually caused by a disturbance or load change in a system employing a proportional controller.

On-Off Control A special type two-position control in which the manipulated variable has only one of two possible values: on or off.

Open Loop Refers to a feedback control system operating with the feedback loop disconnected.

Open-Loop Gain The ratio of the change in the feedback variable to the change in the set point with the feedback element disconnected.

Open-Loop Transfer Function The ratio of the transformation of the output to the transformation of the input if the feedback were disconnected.

Operating Conditions Conditions such as ambient temperature and pressure to which a device or system is subjected besides the measured variable.

Optimization A procedure whereby the optimum value of a variable, design, program, etc. is found or achieved. The optimum value is determined from a minimization (or maximization) of a criterion function such as cost (or profit).

Optimizing Control Action Control action that automatically seeks the optimum value of a specific variable or parameter rather than maintaining it at some set value.

OR A logic operator having the property such that, if P, Q, and R are logical statements, then the OR of P, Q, and R is true if at least one statement is true, and is false if all statements are false.

OR Gate A gate that implements the OR logical operator.

Oscillation A period change in the controlled variable.

Output The variable that is chosen to describe the condition of a system; the dependent variable in a dynamic equation.

Over-Damped A second-order or higher system which is damped to such a degree that the transient response has no tendency to oscillate or overshoot.

Overshoot The maximum amount by which a process output exceeds its desired value (or steady-state value) following a step change in input.

Overshoot Time The time required for a transient error to reach the overshoot point in the response of an automatic control system.

Parallel Cascade Action The regulation of the set points of two or more automatic controllers using other continuous controllers.

Parameter A constant coefficient in an equation that is determined by the physical properties of the sysem.

Peak-to-Peak Amplitude The difference between the extremes of a quantity.

Performance Operator Same as *Transfer Function*.

Period The length of time between consecutively recurring conditions; the reciprocal of frequency.

Perturbation A disturbance or input forcing, usually of small magnitude, introduced to test a system's response.

Phase The condition of a sinusoidally varying function at any particular moment in time. It may be expressed in angular form with reference angle taken as zero at the time when the function is at its average value and increasing.

Phase Angle For two functions varying sinusoidally with the same frequency, that part of the cycle which one signal has reached when the other is at zero phase.

Phase Margin A measure of the degree of stability a system will have under feedback control computed as the difference between -180 degrees and the phase angle of the system's frequency response when the magnitude ratio is unity.

Phase Shift The lag or lead that occurs when a sinusoidal signal passes through an element or control system.

Pneumatic Controller A conventional process controller whose inputs, computations, and output are all pneumatic.

Pole A real or complex value of the dummy Laplace

variable for which the value of the transfer function is infinite; hence, a root of the denominator of the transfer function.

Position Constant Same as *Steady-State Gain*.

Position Error Same as *Offset*.

Predictive Control A control scheme that involves the measurement of changes in load variables and taking corrective action before the system is disturbed. Same as *Feedforward Control*.

Primary Element That portion of the measuring device which first senses a change in the controlled variable.

Primary Feedback The signal fed back to the controller which is directly related to the controlled variable.

Process Those components of a system that are not directly related to control function; hence, the system being controlled.

Process Control System An automatic control system in which the controlled variable is associated with a process state.

Program Control A control system in which the set point varies with time according to a predetermined program.

Proportional Action A control action in which the output of the controller is proportional to the error.

Proportional Band The proportionality constant in the equation relating the controller output to the error. Physically, it is the error in percent of instrument span required to cause a unit change in controller output. It is the reciprocal of the proportional gain.

Proportional Gain, or *Proportional Sensitivity* The ratio of the change in output due to proportional control action to the change in error input.

Proportional-Plus-Derivative Control A control action which is a linear combination of proportional action and derivative (rate) action.

Proportional-Plus-Integral Control A control action which is a linear combination of proportional and integral (reset) control.

Proportional-Plus-Integral Plus-Derivative Control A control action which is a linear combination of proportional, integral, and derivative control.

Proportional-Speed Floating Control Same as *Integral Control*.

Pulse A variation for a short duration of a quantity whose value is normally constant.

Pure Lag Same as *Deadtime*.

Ramp Forcing Same as *Linear Forcing*.

Ramp Response The total time response resulting from a ramp input (an input with a constant rate of change other than zero).

Ramp Response Time The time interval by which an output lags a ramp input.

Rangeability The ratio of maximum flow to minimum controllable flow in a final control element.

Rate Action The controller mode in which there is a linear relationship between the controller output and the derivative of the error signal.

Rate Response Same as *Derivative Action*.

Rate Time The proportionality constant in the equation relating the controller output to the error for rate control. Physically, it is the time required for a unit change in the controller output when the derivative of the error with respect to time is unity.

Ratio Controller A controller that maintains a fixed ratio between two or more variables.

Real-Time Lag Same as *Lag*.

Reference Input The variable signal with which the feedback variable is compared in the computation of an error. It is related to the set point usually by a constant.

Regulatory Control Control with the primary object of maintaining the controlled variable constant in spite of external disturbances.

Relative Damping For an underdamped system, a number which is the actual damping factor divided by the critical damping factor.

Remote Control A system for control of remotely located devices.

Reset An actual or effective change in set point to eliminate an offset or static error. It can be accomplished automatically by the "integral" or "reset" mode.

Reset Rate Inverse of *Integral Time*.

Reset Wind-up The undesirable performance of the integral mode in the presence of saturation of the actuating element.

Resonance Peak A maximum occurring in the output amplitude in frequency response studies.

Response The effect on a system's output caused by a particular change in an input.

Response Time The time required for an output to increase from one specified percentage of its final value to another, based on a step input.

Reverse Acting Controller A controller in which the absolute value of the controller output decreases as the absolute value of the control error increases.

Rise An offset, particularly upward; opposite of droop.

Rise Time Same as *Response Time*.

Sampling Action Process variable sampled and control action taken at intervals.

Sampling Period The time between the intermittent observations in a sampled data system.

Saturation A nonlinearity which results from physical limitations on the maximum and minimum value an element will transmit.

Scale To change a quantity by a factor in order to bring its range within specified limits.

Scan Sequential interrogation of devices for data from process sensors for the purpose of control or data logging.

Self-Operated Controller A controller in which the energy necessary to operate the final control element is derived from the controlled process medium.

Self-Regulation The inherent characteristic of a process which comes to a steady-state value without the aid of an automatic control scheme.

Self-Tuning The technique of automatically updating the controller tuning parameters based on changing process conditions.

Sensitivity The ratio of change of output to change of input.

Servocontrol Control in which the principal objective is to follow a reference value which varies with respect to time.

Servomechanism A feedback control system in which the controlled variable is a mechanical position.

Servo Operation A control system operation whose primary objective is to follow a reference value which varies with respect to time.

Set Point The desired value of the controlled variable.

Setting Time The time required for the absolute value of the difference between the output of the process and the desired value to become and remain less than a specified amount, following the application of a step input.

Signal Information being transferred from one device or element to another. It can be accomplished by mechanical, electrical, pneumatic, hydraulic, or digital means.

Signal Transducer A device which converts one standardized transmission signal to another.

Signal-to-Noise Ratio Ratio of the signal amplitude to noise amplitude.

Simulation Using an analog or digital computer in such a manner as to represent a physical system in which information provided to the computer represents process variables. Information produced by the computer represents the results which would be obtained by the process.

Single-Velocity Action Control action in which the actuating variable changes with a constant velocity when the actuating signal is within a particular region.

Sinusoidal Change A signal having sinusoidal or cyclic characteristics.

Source-Free Response A system's natural response as it relaxes from a state of stored internal energy to an equilibrium condition.

Stability A property of a physical system in which the natural response is positively damped so that in time the response reaches some finite steady-state value or at least reaches a limit cycle which is bounded.

Stable System A system whose response to a bounded input is also bounded.

Stage, or *Transfer Stage* Some part of a larger system which is sufficiently independent of the other part so that a separate transfer function can be written for it.

Static Error A measurement error effective when the measurement is made at steady-state conditions.

Static Gain The ratio of an output change to an input change at steady-state conditions.

Steady State The condition of a system when the transient response has died out. It is implied that all properties are constant with time; however, steady-state conditions can occur even though the output is changing, such as a sine or ramp function.

Steady-State Error In a control system, the same as *Offset*.

Steady-State Gain A proportionality constant in a transfer function not containing integration.

Step Change An instantaneous change, from one value to another value, resembling a step.

Step Response The time response of an element or system to a step change in input from one operating level to another.

Successive Approximation An analog-to-digital conversion technique in which increasingly larger or smaller known voltages are compared with the unknown voltages. The logic decision in each comparison generates the binary representation of the voltage.

Superposed Action Two or more control actions superposed.

Supervisory Control Action Control action in which the direct control loops operate independently subject to periodic updating of set points of the individual controllers.

System All the materials and mechanisms contained within arbitrarily defined boundaries.

System Analysis The definition of a control problem and the development of the solution.

Three-Mode Controller A controller containing three modes of control, typically, proportional, integral and derivative.

Three-Position Controller A multi-position controller having three distinct values of output.

Throttling Band Same as *Proportional Band*.

Time Constant The time required of the output of a first-order system to reach 63.2 percent of a complete response to a step input.

Time Proportioning Controller A controller whose output consists of periodic pulses, the duration of which is varied according to the error signal.

Time Schedule Controller A controller in which the set point is varied automatically according to some predetermined time schedule.

Time Sharing Pertaining to the interleaved use of the time of a device.

Transducer An element or device which receives energy (information) from one system and retransmits it, often in a different form, to another. Generally, any device that transmits, amplifies, or changes a signal.

Transfer Function A mathematical representation of the dynamics of a system using Laplace transform notation. The transfer function is a ratio of the transform of the output of a system to the transform of the input of a system.

Transform To change, according to some standard formula, a function of a certain variable into a function of another variable, for example, to take the Laplace transformation of a function, resulting in the transformation of a time domain to a Laplace domain.

Transient Response That part of the output of a system which is related to the natural response of a system and eventually disappears if the forcing continues unchanged.

Transmittance Same as *Transfer Function*.

Transportation Lag Same as *Deadtime*.

Truth Table A table that describes a logic function by listing all possible combinations of input values and indicating, for each combination, the true output values.

Tuning The adjustment of the control parameters (gain, reset, rate, and so on) to give the desired response.

Two-Position Control A system of regulation in which the manipulated variable has only two discrete values of output.

Type-One Servo A system under servo control that contains one integration in the control loop.

Undamped Said of a system capable of oscillator transient response of constant amplitude.

Under-Damped Said of a system capable of oscillator response which diminishes in time.

Unsteady State The condition of a system undergoing a state of transient change.

Update To modify a system, program, strategy, etc. according to current information.

Velocity Constant The proportionality constant in the transfer function of a system that contains one integration; hence, the gain of a type-one servo.

Velocity-Limiting Control Action Control action in which the rate of change of a specified variable will not exceed a predetermined limit.

Zero A real or complex value of an independent variable that makes the function of that variable equal to zero.

Computer Terminology

Absolute Coding Coding written in language acceptable to a computer without further modification. Same as *Machine Language*.

Access Time (1) The time interval between the instant at which data are called for from a storage device and the instant delivery is completed, i.e., the read time. (2) The time interval between the instant at which data are requested to be stored and the instant at which storage is completed, i.e., the write time.

Accumulator A register in the arithmetic unit which stores operands and in which arithmetical results are formed. See also *Register*.

Accuracy The degree of freedom from error; the degree of conformity to truth or to a rule. Accuracy is contrasted with precision, e.g., four-place numerals are less precise than six-place numerals, but a properly computed four-place numeral might be more accurate than an improperly computed six-place numeral.

Adaptive Control Action Control action whereby automatic means are used to change the type and/or influence of control parameters in such a way as to improve the performance of the control system.

Address A label, such as an integer or other set of characters, which identifies a register, location, or device in which data are stored.

Absolute Address Actual location in storage of a particular unit of data; address that the control unit can interpret directly. Also, the label assigned by the engineer to a particular storage location in the computer.

Relative Address A label used to identify a word in a routine or subroutine with respect to its relative position in that routine or subroutine. A relative address is translated into an absolute address by addition of some specific starting address for the subroutine within the main routine.

Relativization A means by which the next instruction address and the operand address are given relative addresses when written. The relative addresses are translated automatically to absolute addresses during execution of the program.

Symbolic Address A label assigned to a selected word in a routine for the convenience of the programmer. The symbol used is independent of the location of a word within a routine. It identifies the field of data to be operated on or the operation to be used rather than its storage location.

Alarm An audible or visible signal that indicates an abnormal or out-of-limits condition in the plant or control system

ALGOL ALGOrithm Language

Allocation The assignment of blocks of data to specified blocks of storage.

Analog The representation of numerical quantities by means of physical variables such as voltage, current, resistance, rotation, etc. Contrasted with *Digital*.

Analog Backup An alternate method of process control by conventional analog instrumentation in the event of a failure in the computer system.

Analog Computer See items under *Computer*.

Analog Input Module A device which converts analog

input signals from process instrumentation into a digital code for transmission to the computer.

Annotation An added descriptive comment or explanatory note.

Argument The known reference factor necessary to find the desired item (functional) in a table.

Arithmetic Unit That portion of the hardware of an automatic digital computer in which arithmetical and logical operations are performed.

ASCII American Standard Code for Information Interchange. An eight-level code intended to provide information code compatibility between digital devices of U.S. manufacture.

Assembler A program which converts symbolic language to machine language by substitution of absolute operation codes for symbolic operation codes and absolute or relocatable addresses for symbolic addresses.

Asynchronous Computer A computer in which performance of the next command is started by a signal denoting that the previous command has been completed. Contrasted with *Synchronous Computer*, which is characterized by a fixed cycle for the execution of operations.

Attenuation (1) A decrease in signal magnitude between two points, or between two frequencies. (2) The reciprocal of gain, when the gain is less than one.

Auctioneering Device A device which automatically selects either the highest or the lowest input signal from among two or more input signals.

Automatic Data-Processing System A system that uses minimum manual operations in processing data.

Automatic Programming The process of using a computer to perform some stages of the work involved in preparing a program.

Availability The total amount of time that a computer is properly operating.

Background Program A program of no particular urgency with regard to time that may be preempted by a program of greater urgency and priority. Contrasted with *Foreground Program*.

Backup Provision of alternative means of operation in case of a failure of the primary means of operation.

Batch Processing Collection of data over a period of time to be sorted and processed as a group during a particular machine run.

Baud A unit of signaling speed equal to the number of discrete conditions or signal events per second. For example, one baud equals one half-dot cycle per second in Morse code, one bit per second in a train of binary signals, and one three-bit value per second in a train of signals each of which can assume one of eight different states.

Benchmark Problem A sample problem used to evaluate the performance of computers relative to each other.

Binary (1) Pertaining to a characteristic or property involving a selection, choice, or condition in which there are two possibilities. (2) Pertaining to the numeration system with a radix of two.

Binary Coded Decimal (BCD) Pertaining to a decimal notation in which the individual decimal digits are each represented by a group of binary digits. In the 8-4-2-1 binary coded decimal notation, the number 23 is represented as 0010 0011 whereas in binary notation 23 is represented as 10111.

Bit A binary digit; hence, a unit of data in binary notation. In the binary numbering system, only two marks (0 and 1) are used. The number 10111 contains five bits.

Block A group of consecutive machine words considered or transferred as a unit, particularly with reference to input and output.

Bootstrap A technique or device designed to bring itself into a desired state by means of its own action, e.g., a machine routine whose first few instructions are sufficient to bring the rest of itself into the computer from an input device.

Breakpoint A point in a program at which a computer may be made to stop automatically for a check on the progress of the routine.

Conditional Breakpoint A breakpoint at which the routine may be continued as coded if desired conditions are satisfied.

Buffer A storage device used to compensate for a difference in rate of flow of data, or time of occurrence of events, when transmitting data from one device to another.

Bug An error or malfunction in a program or hardware.

Bulk Memory An auxiliary memory device with storage capacity greatly in excess of working (core) memory; for example, disc file, drum.

Bus One or more conductors used for transmitting signals or power.

Byte A sequence of adjacent binary digits operated upon as a unit and usually shorter than a word.

Calibrate (1) To ascertain, usually by comparison with a standard, the locations at which scale or chart graduations should be placed to correspond to a series of values of the quantity which the instrument is to measure, receive, or transmit. (2) To adjust the output of a device to bring it to a desired value, within a specified tolerance, for a particular value of the input. (3) To ascertain the error in the output of a device by checking it against a standard.

Call To transfer control to a specified subroutine.

Calling Sequence A specified arrangement of instructions and data necessary to set up a given subroutine.

Capacity In computer terminology, the quantity of information that can be contained in a storage device, defined in terms of the basic information size, such as words or characters.

Central Processor That portion of any computer system

that performs the actual computation. It normally consists of the arithmetic and control units and working memory.

Channel A path along which data, particularly a series of digits or characters, may flow or be stored.

Check A means of verifying the accuracy of data transmitted, manipulated, or stored by any unit or device in a computer.

Built-in Check (Automatic Check) Any check constructed in hardware.

Duplication Check A check that requires identical results of two independent performances of the same operation.

Mathematical Check A check making use of mathematical identities or other properties, frequently with some degree of discrepancy being acceptable; e.g., checking multiplication by verifying that $A \cdot B = B \cdot A$, checking a tabulated function by the difference method.

Parity Check A summation check in which the binary digits, in a character or word, are added (modulo 2) and the sum checked against a single, previously computed parity digit; e.g., a check that tests whether the number of ones is odd or even.

Programmed Check A mathematical check inserted into the operating program.

Redundant Check A check that attaches one or more extra digits to a word according to rules, so that if any digit changes, the malfunction or mistake can be detected.

Summation Check A redundant check in which groups of digits are summed, usually without regard for overflow, and that sum checked against a previously computed sum to verify accuracy.

Transfer Check A check on transmitted data by temporarily storing, retransmitting, and comparing.

Twin Check A continuous duplication check achieved by duplication of hardware.

Clear To replace information in a storage device by the character zero.

Clock Frequency Master frequency of periodic pulses which schedule the operation of the computer.

Clock Rate The rate at which a word or characters of a word (bits) are transferred from one internal computer element to another. Clock rate is expressed in cycles (if a parallel-operation machine—words; if a serial-operation machine—bits) per second.

Closed Loop A signal path that includes a forward path, a feedback path, and a summing point and forms a closed circuit.

Closed Subroutine A limited-use subroutine that can be stored at one place and can be connected to a routine by linkages at one or more locations. Contrasted with *Open Subroutine*.

Code (noun) A system of rules for using a set of symbols to represent data.

Computer Code; Machine Code The code that the computer hardware was built to interpret and execute.

Instruction Code The symbols, names, and operation descriptions for all instructions represented by computer code.

Numerical Code A code in which the symbols used are all numerals.

Pseudo Code An arbitrary code, independent of the hardware of a computer and designed for convenience in programming, that must be translated into computer code if it is to direct the computer.

Code (verb) To express a program in a code that a specific computer was built or programmed to interpret and execute.

Coding The act of preparing in code or pseudo code a list of the successive computer operations required to solve a specific problem. Also, the list itself.

Command A pulse, signal, or set of signals initiating one step in the performance of a computer operation. A command is one part of an instruction.

Common Mode Interference A form of interference which appears between any measuring circuit terminals and ground.

Common Mode Rejection The ability of a circuit to discriminate against common mode voltage, usually expressed as a ratio or in decibels.

Comparator A device for comparing two different transcriptions of the same information to verify agreement or determine disagreement. A circuit that compares two signals and indicates agreement or disagreement; a signal may be given, indicating whether they are equal or unequal.

Compiler A program which translates a problem-oriented language to a machine-oriented language such as *Fortran or Algol*. A compiler, as contrasted with an assembler, can substitute subroutines as well as single machine instructions for certain symbolic inputs. A compiler which translates directly from source language to machine language is known as a single-pass compiler. A compiler that generates an interim object language which requires further translation or modification is known as a multiple-pass compiler.

Computer Any device capable of accepting data, applying prescribed processes to them, and supplying the results of these processes. The word "computer" in this glossary usually refers to a stored-program digital computer.

Analog Computer A computer which calculates by using physical analogs of the variables. Usually a one-to-one correspondence exists between each numerical variable occurring in the problem and a varying physical measurement in the analog computer. The physical quantities in an analog computer are varied continuously instead of in discrete steps as in the digital computer.

Digital Computer A computer capable of accepting and operating on only the representations of real numbers, or other characters coded numerically.

Stored Program Computer A digital computer capable of performing sequences of internally stored instructions, as opposed to calculators on which the sequence is impressed manually. Such computers usually possess the further ability to operate upon the instructions themselves, and to alter the sequence of instructions in accordance with results already calculated.

Contact Sense Module A device which monitors and converts program-specified groups of field switch contacts into digital codes for input to the computer. Inputs are scanned by the computer at programmed intervals.

Control Action Of a controller or a controlling system, the nature of the change of the output effected by the input. The output may be a signal or the value of a manipulated variable. The input may be the control loop feedback signal when the set point is constant, an actuating error signal, or the output of another controller.

Control Algorithm A mathematical representation of the control action to be performed.

Control Computer A computer which, by means of inputs from and outputs to a process, directly controls the operation of elements in that process.

Control Counter; Program Counter; Instruction Counter; Control Register A counter built into the control unit and used for sequencing instructions to be executed. It normally contains the address of the next instruction to be performed.

Control Logic The sequence of steps or events necessary to perform a particular function. Each step or event is defined to be either a single arithmetic expression or a single Boolean expression.

Control Output Module A device which stores computer commands and translates them into signals which can be used for control purposes. It can generate digital outputs to control on-off devices or to pulse set-point stations, or it can generate analog outputs—voltage or current—to operate valves and other process control devices.

Control Panel (Automatic) A panel of indicator lights and switches on which are displayed a particular sequence of routines, and from which an operator can control the operation of these routines.

Control Panel (Maintenance) A panel of indicator lights and switches on which are displayed a particular sequence of routines, and from which a maintenance engineer can check the operation of the hardware.

Control Panel (Operator's Request) A panel consisting of indicator lights and switches by which an operator can request the computer to perform particular functions.

Control Panel (Programming) A panel consisting of indicator lights and switches by which a programmer can enter or change routines in the computer.

Control System A system in which deliberate guidance or manipulation is used to achieve a prescribed value of a variable.

Control Unit Portion of the hardware of an automatic digital computer that directs sequence of operations, interprets coded instructions, and initiates proper commands to computer circuits to execute instructions.

Controller A device which operates automatically to regulate a controlled variable.

Converter A device for transfering data from one storage medium to another, as from punched cards to magnetic tape.

Core Memory A high-speed random-access storage device utilizing matrix arrays of ferrite cores, usually used as the computer's working memory.

Core Resident A term pertaining to certain pivotal programs permanently stored in core memory for frequent execution.

Counter A device or memory location whose contents can be successively incremented or decremented.

Cycle Time The basic unit of computer speed, usually the time required for a read and write operation in core memory.

Data Break An automatic input/output channel which provides external equipment with direct access to core memory.

Data Display Module A device which stores computer output and translates this output into signals which are distributed to a program-determined group of lights, annunciators, and numerical indicators in operator consoles and remote stations.

ddc Same as *Direct Digital Control*.

Dead Band The range through which an input can be varied without initiating response.

Deadtime The interval of time between initiation of an input change or stimulus and the start of the resulting response.

Debug To test a computer program to find whether it works properly and to trace and correct any errors.

Debug On-Line To *debug* a computer performing on-line functions and utilizing another routine which has previously been checked out.

Diagnostic Routine A program designed to locate malfunctions in computer hardware or software.

Digital Pertaining to data in the form of digits. Contrasted with *Analog*.

Digital Backup An alternative method of digital process control initiated by use of special purpose digital logic in the event of a failure in the computer system.

Digital Computer See items under *Computer*.

Direct Digital Control (ddc) Control performed by a

digital device which establishes the signal to the final controlling element. Examples of possible digital (D) and analog (A) combinations for this definition are:

	Feedback Elements →	Controller →	Final Controlling Element
1.	D	D	D
2.	A	D	D
3.	A	D	A
4.	D	D	A

Disc A flat circular plate with a magnetic surface on which data can be stored by selective magnetization of portions of the flat surface.

Disturbance An undesired change in a variable applied to a system which tends to affect adversely the value of a controlled variable.

Downtime Time when a computer is not operating.

Driver A small program or routine that handles the control of an external peripheral device or executes other programs.

Drum A right circular cylinder with a magnetic surface on which data can be stored by selective magnetization of portions of the curved surface.

Dummy An artificial address, instruction, or other unit of information inserted solely to fulfill prescribed conditions (such as word length or block length) without affecting operations.

Dump To copy the contents of all or part of a storage, usually from an internal storage into an external storage.

Error The difference between the indication and the true value of the measured signal. A positive error denotes that the indication of the instrument is greater than the true value.

$$\text{Error} = \text{Indication} - \text{True Value}$$

Error Signal In a closed loop, the signal resulting from subtracting a particular return signal from its corresponding input signal.

Error Squared The technique of introducing the square of the error in the error term of a linear algorithm so as to produce a non-linear correction.

Exception-Principle System An information system or data-processing system which reports on situations only when actual results differ from planned results. When results occur within a "normal range," they are not reported.

Executive Program A program that controls the execution of all other programs in the computer based on established hardware and software priorities and real time or demand requirements.

Extract To obtain certain digits from a machine word as may be specified. Or, to replace contents of specific columns of another machine word, depending on the instruction. Or, to remove from a set of items of information all those items that meet some arbitrary condition.

Field A set of one or more characters constituting a unit of data. Compare *Word*. (A field need not correspond in length to a word.)

Final Controlling Element That forward controlling element which directly changes the value of the manipulated variable.

Fixed Heads Pertaining to the use of stationary, rigidly mounted, reading and writing heads on bulk memory devices.

Flip-Flop A circuit or device containing active elements, capable of assuming either one of two stable states at a given time.

Foreground Program A time-dependent program initiated via an outside request whose urgency preempts operation of a background program.

FORTRAN *(FORmula TRANslating system)* A procedure-oriented language designed for solution of arithmetic and logical programs.

Gate Circuit An electronic circuit with one or more inputs and one output, with the property that a pulse goes to the output line only if some specified combination of pulses occurs on the input lines. Gate circuits constitute much of the hardware by means of which logical operations are built into a computer.

Guard Bit A bit contained in each word or groups of words of memory which indicates to computer hardware or software whether the content of that memory location can be altered by a program.

Half Duplex In communications, pertaining to an alternative, one way at a time, independent transmission.

Hardware Physical equipment; for example, mechanical, magnetic, electrical or electronic devices. Contrast with *Software*.

Hardware Priority Interrupt The hardware implementation of priority interrupt functions.

Head A device that reads, records, or erases data on a storage medium; for example, a small electromagnet used to read, write, or erase data on a magnetic drum or tape, or the set of perforating, reading, or marking devices used for punching, reading, or printing on paper tape.

Index Register A register to which an arbitrary integer, usually one, is added or subtracted upon the execution of each machine instruction. The register may be reset to zero or an arbitrary number. Used with indexable instructions to get "effective" instruction addressed during execution. Also called "cycle counter" and "B-box."

Indirect Address An address that specifies a storage location that contains either a direct address or another indirect address.

Initialize To set counters, switches, and addresses to

zero or other starting values at the beginning of, or at prescribed points in, a computer routine.

In-Line Processing The processing of data without sorting or any other prior treatment other than storage.

Input (1) The data to be processed. (2) The state or sequence of states occurring on a specified input channel. (3) The device or collective set of devices used for bringing data into another device. (4) A channel for impressing a state on a device or logic element. (5) The process of transferring data from an external storage to an internal storage.

Instruction Counter Same as *Control Counter*.

Interacting Control Control action produced by an algorithm whose various terms are interdependent.

Interface Logic necessary to provide electrical and communication compatibility between two devices.

Interrupt *See Priority Interrupt, Hardware Priority Interrupt*, and *Software Priority Interrupt*.

Key Punch A typewriter-like machine for recording data by cutting holes or notches in cards.

Linear Programming The analysis or solution of problems in which the linear function of a number of variables is to be maximized or minimized when those variables are subject to a number of constraints in the form of linear inequalities.

Linkage In programming, coding that connects two separately coded routines.

Logical Operation An operation in which a decision affecting the future sequence of instructions is automatically made by the computer. The decision is based upon comparisons between all or some of the characters in an arithmetic register and their counterparts in any other register on a less than, equal to, or greater than basis; or between certain characters in arithmetic registers and built-in standards.

Machine Language Same as *Absolute Coding*.

Magnetic Core Storage A storage device consisting of magnetically permeable binary cells arrayed in a two-dimensional matrix. (A large storage unit contains many such matrices.) Each cell (core) is wire wound and can be polarized in either of two directions for the storage of one binary digit. The direction of polarization can be sensed by a wire running through the core.

Magnetic Disc Storage A storage device consisting of magnetically coated discs accessible to a reading and writing arm in much the manner of an automatic record player. Binary data are stored on the surface of each disc as small, magnetized spots arranged in circular tracks around the disc. The arm is moved mechanically to the desired disc and then to the desired track on that disc. Data from a given track are read or written sequentially as the disc rotates.

Magnetic Drum Storage A storage device consisting of a rotating cylinder surfaced with a magnetic coating. Binary data are stored as small, magnetized spots arranged in close tracks around the surface. A magnetic reading and writing head is associated with each track so that the desired track can be selected by electric switching. Data from a given track are read or written sequentially as the drum rotates.

Manipulated Variable A quantity or condition which is varied as a function of the actuating signal so as to change the value of the directly controlled variable. In any practical control system, there may be more than one manipulated variable. Accordingly, when using the term, it is necessary to state which manipulated variable is being discussed. In process control work, the one immediately preceding the directly controlled system is usually intended.

Masking An operation that replaces characters in the accumulator with characters from a specified storage location that correspond to the "ones" in the mask which is in a specified storage location or register.

Measured Signal The electrical, mechanical, pneumatic, or other variable applied to the input of a device. It is the analog of the measured variable produced by a transducer (when such is used). In a thermocouple-thermometer system, for example, the measured signal is an emf which is the electrical analog of the temperature applied to the thermocouple. In a flowmeter, the measured signal may be a differential pressure which is the analog of the rate of flow through the orifice. In an electric tachometer system, the measured signal may be a voltage which is the electrical analog of the speed of rotation of the part coupled to the tachometer generator.

Measured Variable The physical quantity, property, or condition which is to be measured. It is sometimes referred to as the measurand.

Memory The capacity of a computer to receive and store data subject to recall. Loosely, any device that can store data.

Memory Protect A technique of protecting the contents of sections of memory from alteration by inhibiting the execution of any memory modification instruction upon detection of the presence of a guard bit associated with the accessed memory location. Memory modification instructions accessing protected memory are usually executed as a no-operation, and a memory protect violation program interrupt is generated.

Merge To produce a single sequence of items, ordered according to some rule (that is, arranged in some orderly sequence), from two or more sequences previously ordered according to the same rule, without changing the items in size, structure, or total number. Merging is a special kind of collating.

Microprogramming A programming capability wherein several instruction operations can be combined in one instruction for greater speed and efficient use of memory.

Mnemonic An alphanumeric designation, easy to remember and commonly used to designate a memory location or computer operation; for example, START might represent the location of the first instruction in a routine.

Multiplex The process of transferring data from several storage devices operating at relatively low transfer rates to one storage device operating at a high transfer rate in such a manner that the high-speed device is not obliged to "wait" for the low-speed units.

Multiprogramming The interleaved or time-shared execution of two or more programs by a computer.

Noise An unwanted component of a signal or variable which obscures its information content.

Normalize To shift the representation of a quantity so that the representation lies in a prescribed range.

Object Program The coding which is the output of an automatic code translation program such as an assembler or compiler.

Off Line (1) Pertaining to equipment or programs not under the direct control of the central processor. (2) Pertaining to a computer that is not actively monitoring or controlling a process or operation, or pertaining to a computer operation performed while the computer is not monitoring or controlling a process or operation.

On Line (1) Pertaining to equipment or programs under direct control of a central processor. (2) Pertaining to a computer that is actively monitoring or controlling a process or operation, or pertaining to a computer operation performed while the computer is monitoring or controlling a process or operation.

On-Line Equipment Equipment for which the transfer of data to or from the unit is under direction of the control unit of the computer.

Open Subroutine A general-use subroutine that must be relocated and inserted into a routine at each place it is used. Contrasted with *Closed Subroutine*.

Operand That which is operated upon. An operand is usually identified by an address part of an instruction.

Operating Conditions Conditions (such as ambient temperature, ambient pressure, vibration, etc.) to which a device is subjected, but not including the variable measured by the device.

Operation Code The part of a computer instruction which specifies the operation to be performed.

Operator's Console Equipment which provides for manual intervention and for monitoring computer operation.

Optimize To establish control parameters so as to make control as effective as possible.

Output Process of transferring data from internal storage of a computer to some other storage device.

Pack (1) To include several discrete items of information in one unit of information. (2) To relocate programs and data to make efficient use of available storage capacity.

Page Addressing A memory addressing technique utilized with certain computers whose addressing capability is limited to less than the total memory capacity available. Using page addressing, memory is divided into segments (pages), each of which can be addressed by the available addressing capability.

Parameter A controllable or variable characteristic of a system or device, temporarily regarded as a constant, the respective values of which serve to distinguish the various specific states of the system or device.

Parity Check See items under *Check*.

Patch A section of coding inserted into a routine (uaually by explicitly transferring control from the routine to the patch and back again) to correct a mistake or alter the routine.

Peripheral Equipment Equipment used for entering data into or receiving data from a computer.

Power Consumption The maximum wattage used by a device within its operating range during steady-state signal condition. For a power factor other than unity, power consumption is the maximum volt-amperes used under the condition stated above.

Priority Interrupt The temporary suspension of a program currently being executed in order to execute a program of higher priority. Priority interrupt functions usually include distinguishing the highest priority interrupt active, remembering lower priority interrupts which are active, selectively enabling or disabling priority interrupts, executing a jump instruction to a specific memory location(s), and storing the program counter register in a specific location(s). See *Hardware Priority Interrupt* and *Software Priority Interrupt*.

Priority Interrupt Module A device which monitors a number of priority-designated field contacts and immediately notifies the computer when any of these external priority requests have been generated. It assures servicing of urgent interrupt requests on the basis of programmer-assigned priorities when requests occur simultaneously.

Process The collective functions performed in and by the equipment in which a variable or variables is or are controlled. "Equipment" in this definition does not include automatic control equipment. The process may also be referred to as the controlled system.

Process Control Loop A system of control devices linked together to control one phase of a process.

Program A plan for the automatic solution of a problem. A complete program includes plans for the transcription of data, coding for the computer, and plans for the absorption of the result into the system. The list of coded instructions is called a "routine."

Program Counter Same as *Control Counter*.

Programmer A person who prepares computer operation procedures by means of flow charts and coding.

Proportional Control Action Control action in which there is a continuous linear relation between the output and the input. This condition applies when both the output and input are within their normal operating ranges and when operation is at a frequency below a limiting value.

Pulse A significant and sudden change of short duration in the level of an electrical variable, usually voltage.

Pulse-Counting Module A device which counts and stores a number of high- or low-speed pulse channels and transmits their status to the computer upon command.

Random Access (1) Pertaining to the process of obtaining data from, or placing data into, storage where the time required for such access is independent of the location of the data most recently obtained or placed in storage. (2) Pertaining to a storage device in which the access time is effectively independent of the location of the data.

Ratio Controller A controller that maintains a predetermined ratio between two or more variables.

Read (1) To copy, usually from one form of storage to another, particularly from external or secondary storage to internal storage. (2) To sense the meaning of arrangements of hardware.

Real Time Operation Processing data in synchronism with a physical process so that results of data processing are useful to the physical operation.

Recursive Pertaining to the use of a subroutine iteratively in the solution of a problem.

Register A device for the temporary storage of one or more words to facilitate arithmetical, logical, or transferral operations. Examples are the accumulator and the address, index, instruction, and M-Q registers.

Relative Address A label used to identify the location of data in a program by reference to its position with respect to some other location in that program. Relative addresses are translated into absolute addresses by the addition of the reference address.

Relocate In programming, to move a routine from one portion of storage to another and to adjust the necessary address references so that the routine, in its new location, can be executed.

Resolution The least interval between two adjacent discrete details which can be distinguished one from the other.

Routine A set of coded instructions arranged in proper sequence to direct the computer to perform a desired operation or series of operations. See also *Subroutine*.

Diagnostic Routine A specific routine designed to locate either a malfunction in the computer or a mistake in coding.

Executive Routine; Master Routine A routine designed to process and control other routines. A routine used in realizing *Automatic Programming*.

General Routine A routine expressed in computer coding designed to solve a class of problems, specializing to a specific problem when appropriate parametric values are supplied.

Generator A general routine that accepts a set of parameters and causes the computer to compute a specific routine for further use. Among other things, the parameters may specify the input-output devices to use, designate subroutines, or describe the form of a record.

Interpretive Routine; Interpretation An executive routine which, during the course of data-handling operations, translates a stored pseudo code program into a machine code and at once performs the indicated operations by means of subroutines.

Postmortem Routine A routine which, either automatically or on demand, prints data concerning contents of registers and storage locations when the routine is stopped in order to assist in locating a mistake in coding.

Rerun Routine; Rollback Routine A routine designed to be used in the wake of a computer malfunction or a coding or operating mistake to reconstitute a routine from the last previous rerun point.

Sampling, Analog The process by which the computer selects individual analog input signals from the process, converts tham to an equivalent binary form, and stores the data in memory.

Scale Factor A number used as a multiplier, so chosen that it causes a set of quantities to fall within a given range of values. To scale the values 856,432, −95, and −182 between −1 + 1, a scale factor of $^1/_{1000}$ would be suitable.

Scanning The action of comparing input variables to determine a particular action.

Scanning Limits The action of comparing input variables against either prestored or calculated high and/or low limits to determine if an alarm condition is present.

Search, Binary A technique for finding a particular item in an ordered set of items by repeatedly dividing in half the portion of the ordered set containing the sought-for item until only the sought-for item remains.

Self-Tuning The technique of automatic modification of control algorithm constants based upon process conditions.

Sequence Monitor Computer monitoring of the step-by-step actions that should be taken by the operator during a startup and/or shutdown of a power unit. As a minimum, the computer would check that certain milestones had been reached in the operation of the unit. The maximum coverage would have the computer check that each required step is performed, that the correct sequence is followed, and that every checked point falls within its prescribed limits. Should an incorrect action or result occur, the computer would record the fault and signal the operator.

Sequential Control The manner of control of a computer in which instructions to it are set up in a sequence and are fed in that sequence to the computer during solution of a problem.

Service Routine; Utility Routine A routine in general support of the operation of a computer; for example, an input-output, diagnostic, tracing, or monitoring routine.

Servomechanism (1) A feedback control system in which at least one of the system signals represents mechanical motion. (2) Any feedback control system.

Set Point (Command) An input variable which sets the desired value of the controlled variable. The input variable may be manually set, automatically set or programmed. It is expressed in the same units as the controlled variable.

Set-Point Control A control technique in which the computer supplies a calculated set point to a conventional analog instrumentation control loop.

Simulation A pseudo-experimental analysis of an operating system by means of mathematical or physical models which operate in a time-sequential manner similar to the system itself.

Smooth To apply procedures that decrease or eliminate rapid fluctuations in data.

Software (1) The collection of programs and routines associated with a computer, such as compilers and library routines. (2) All the documents associated with a computer, such as manuals and circuit diagrams. Contrasted with *Hardware*.

Software Priority Interrupt The programmed implementation of priority interrupt functions. See *Priority Interrupt* and *Hardware Priority Interrupt*.

Source Language A program language used as an input to a translation program such as an assembler or compiler.

Steady State A condition, such as value, rate, periodicity, or amplitude, exhibiting only negligible change over an arbitrarily long period of time.

Storage Same as *Memory*.

Suboptimization The process of fulfilling or optimizing a chosen objective which is an integral part of a broader objective. Usually the broad objective and lower-level objective are different.

Subroutine A series of computer instructions which perform a specific task for another routine. It is distinguishable from a routine in that it requires, as one of its parameters, a location specifying where in the main program to return to after its function has been accomplished.

Successive Approximation An analog-to-digital conversion technique in which increasingly larger or smaller known voltages are compared with the unknown voltage. The equality decision made in each iteration ultimately forms the binary representation of the analog value.

Supervisory A process computer application wherein the computer performs higher-level process calculations but does not actuate final elements, such as valves. Contrasted with *Direct Digital Control*. For example, the computer may handle mathematical models of the process, or may perform process calculations and relay the results to controllers for valve actuation.

Supervisory Control Control action in which the control loops operate independently, subject to intermittent corrective action; for example, set-point changes from an external source.

Symbolic Coding Any coding in which symbols other than actual binary machine language are used.

Synchronous Computer A computer in which each event, or the performance of each operation, starts as a result of a signal generated by a clock.

System A collection of hardware and software organized in such a way as to achieve an operational objective.

Systems Analysis The definition of a control problem and the development of a solution to the control problem.

Systems Engineering The implementation of a hardware and software system resulting from analysis of a control problem.

Table A block of information in memory which is used as data by a program.

Termination Rack An equipment rack containing field wiring terminals and associated signal conditioning equipment. It provides the termination interface between a computer control system and field-mounted instrumentation.

Three-Position Controller A multi-position controller having three discrete values of output.

Time Sharing Pertaining to the interleaved use of the time of a device.

Track The portion of a moving storage medium, such as a drum, tape, or disc, that is accessible to a given reading head position.

Transducer An element or device which receives information in the form of one physical quantity and converts it to information in the form of the same or other physical quantity.

Transmitter A transducer which responds to a measured variable by means of a sensing element and converts it to a standardized transmission signal which is a function only of the measurement.

Tuning The adjustment of control constants in algorithms or analog controllers to produce the desired control effect.

Utility Routine Same as *Service Routine*.

Valve Output Module A device that translates the computer's output data into analog signals suitable to position control valves or other devices.

Velocity Limit A limit which the rate of change of a specified variable cannot exceed.

Verify (1) To check, usually with an automatic machine, one typing or recording of data against another in order to minimize the number of human errors in the data transcription. (2) In preparing data for a computer, to make certain that data prepared are correct.

Watchdog Timer An electronic interval timer that generates a priority interrupt unless periodically recycled by a computer. It is used to detect program stall or hardware failure conditions.

Word A sequence of bits or characters treated as a unit and capable of being stored in one computer location.

Word Time The data transfer rate (words per second) between a device and the computer.

Write (1) To copy information usually from internal to external storage. (2) To transfer information to an output medium. (3) To record information in a register, location, or other storage device or medium.

Valve Terminology*

General Terms

Valve A pressure-dissipating device designed to modify flow of fluids in pipes.

Control Valve A valve designed to modify flow of fluids in pipes and used for control purposes via an actuator responding to an external signal.

Regulator A valve with an actuator responding to the condition of the fluids in the body.

Hand Valve A valve with a manual actuator.

Actuator The portion of a valve which responds to the applied signal and causes the motion resulting in modification of fluid flow.

Valve Body The portion of the valve containing the flowing fluid and the device which modifies the flow of fluids through it.

Terms Relating to the Valve Body

Valve Body Assembly An assembly of a body, bonnet assembly, bottom flange and trim elements. The trim includes a valve plug which opens, shuts or partially obstructs one or more ports.

Valve Body A housing for internal valve parts having inlet and outlet flow connections (Figures 12.4f, g, h and i).

Several common body arrangements are employed, as follows: Single-ported means one port and one valve plug (Figure 12.4g). *Double-ported means two ports and one valve plug* (Figure 12.4f).

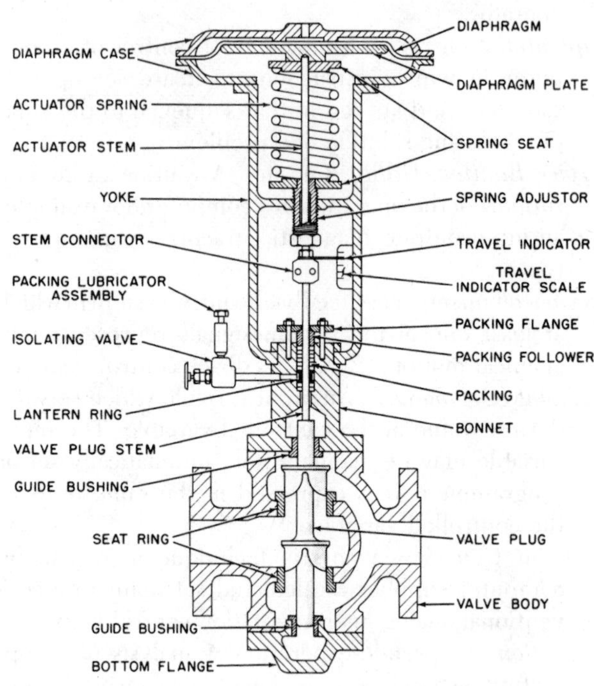

Fig. 12.4f Diaphragm-actuated control valve

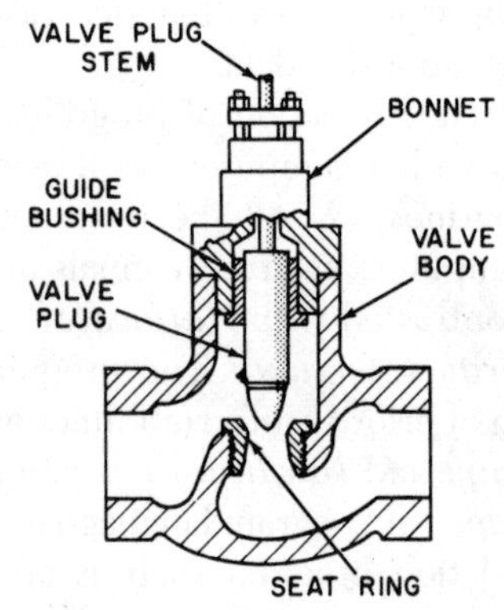

Fig. 12.4g Single-ported control valve

Two-way means two flow connections: one inlet and one outlet.

Three-way means three flow connections, two of which may be inlets with one outlet (for converging or mixing flows) (Figure 12.4i), *or one inlet and two outlets (for diverging or diverting flows)* (Figure 12.4h).

Bonnet Assembly An assembly including the part through which a valve plug stem moves and a means for sealing against leakage along the stem. It usually provides a means for mounting the actuator (Figures 12.4k, l, m and n).

*Some of the following terminology was used verbatim from Diaphragm Actuated Control Valve Terminology (ASME Standard 112) with the permission of the publisher, The American Society of Mechanical Engineers, United Engineering Center, New York, New York.

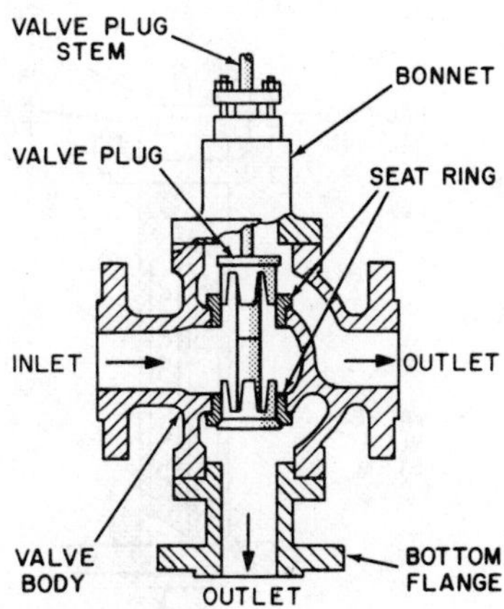

Fig. 12.4h Three-way diverting valve

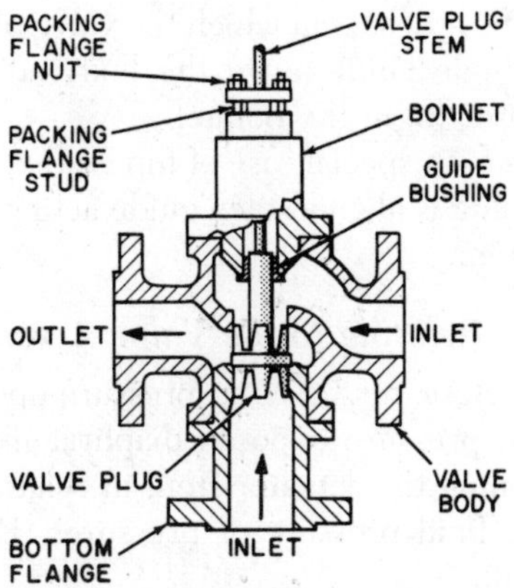

Fig. 12.4i Three-way mixing valve

Sealing against leakage may be accomplished by means of packing or a bellows. A bonnet assembly may include a packing lubricator assembly with or without isolating valve. Radiation fins or an extension bonnet may be used to maintain a temperature differential between the valve body and sealing means.

Bonnet The major part of the bonnet assembly, excluding the sealing means.

Radiation Fin Bonnet A bonnet with fins to reduce heat transfer between the valve body and packing box assembly (Figure 12.4l).

Extension Bonnet A bonnet with an extension between the packing box assembly and bonnet flange (Figure 12.4m).

Bellows Seal A seal which uses a bellows for sealing against leakage around the valve plug stem (Figure 12.4n).

Packing Box Assembly The part of the bonnet assembly used to seal against leakage around the valve plug stem, including various combinations of all or part of the parts shown in Figures 12.4f, k, and l.

Isolating Valve A hand-operated valve between the packing lubricator assembly and the packing box assembly to shut off the fluid pressure from the lubricator assembly (Figure 12.4f).

Bottom Flange A part which closes a valve body opening opposite the bonnet assembly or in a three-way valve may provide an additional flow connection.

It may include a guide bushing and, in a three-way valve, may also include a seat (Figures 12.4f, h and i).

Seat Ring A separate piece inserted in a valve body to form a valve body port (Figures 12.4f, g and h).

Seat That portion of a seat ring or valve body which a valve plug contacts for closure (Figures 12.4f, g, h and i).

Valve Plug A moveable part which provides a variable restriction in a port (Figures 12.4f, g, h and i).

Because of desired characteristics and for functional reasons, there are many forms of valve plugs, ported and contoured, a few of which are illustrated in Figure 12.4j.

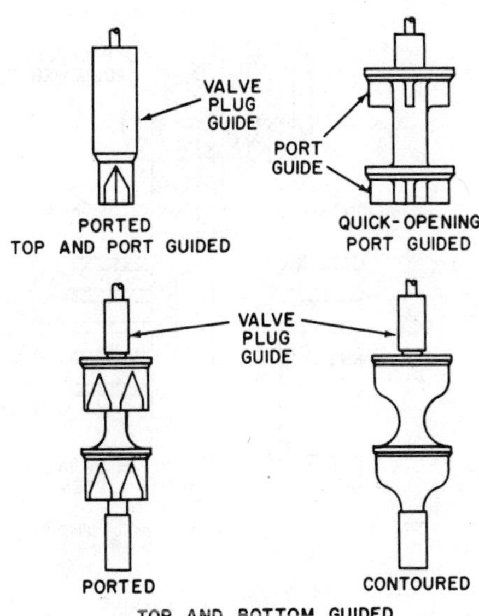

Fig. 12.4j Valve plugs

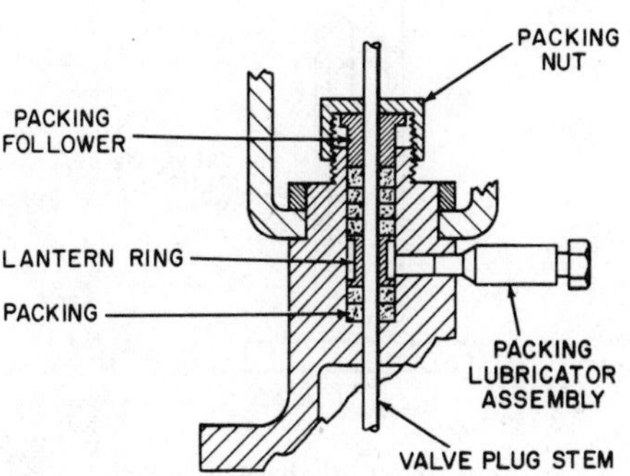

Fig. 12.4k Bonnet assembly

Valve Plug Guide That portion of a valve plug which aligns its movement in either a seat ring, bonnet, bottom flange or any two of these.

Typical examples are shown in Figure 12.4j.

Valve Plug Stem A rod extending through the bonnet assembly to permit positioning the valve plug (Figures 12.4f, g, h and i).

Guide Bushing A bushing in a bonnet, bottom flange, or body to align the movement of a valve plug with a seat ring (Figures 12.4f, g and i).

Guiding of a valve plug may be accomplished by an integral part of a bonnet or bottom flange, or by a seat ring or seat ring extension.

Top and Port Guided Design A design in which the valve plug is aligned by a guide in (a) the bonnet or body and (b) in the body port (Figure 12.4j).

Port Guided A design in which the valve plug is aligned by the body port or ports only (Figure 12.4j).

Top and Bottom Guided A design in which the valve plug is aligned by guides (a) in the body or (b) in the bonnet and bottom flange (Figure 12.4j).

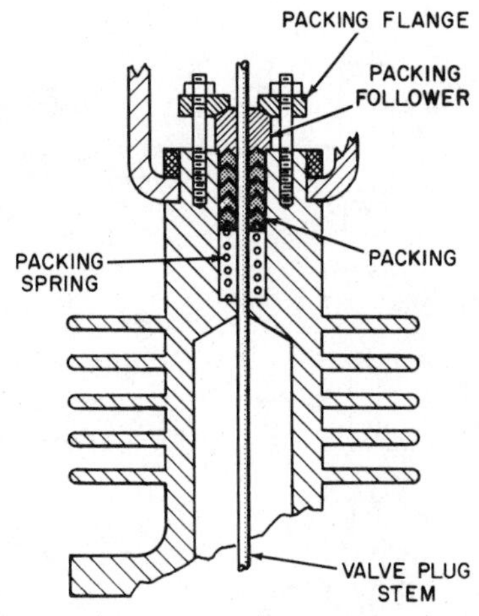

Fig. 12.4 l Radiation fin bonnet assembly

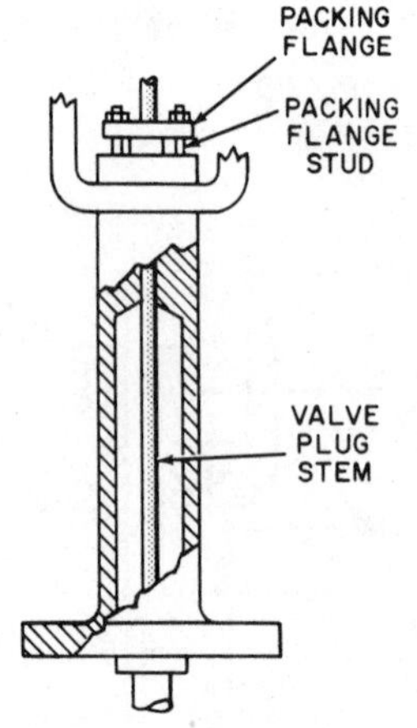

Fig. 12.4m Extension bonnet assembly

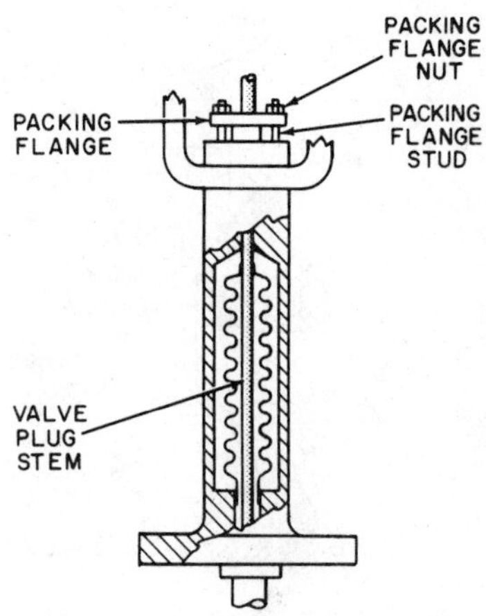

Fig. 12.4n Bellows seal assembly

Top Guided A design in which the valve plug is aligned by a single guide (a) in the body adjacent to the bonnet or (b) in the bonnet.

Stem Guided A special case of top guided in which the valve plug is aligned by a guide acting on the valve plug stem.

Terms Relating to the Valve Actuator

Diaphragm Actuator A fluid pressure operated spring or fluid pressure opposed diaphragm assembly for positioning the actuator stem in relation to the operating fluid pressure or pressures (Figures 12.4f, o and p).

Diaphragm A flexible pressure-responsive element which transmits force to the diaphragm plate and actuator stem (Figures 12.4f, o and p).

Diaphragm Plate A plate concentric with the diaphragm for transmitting force to the actuator stem (Figures 12.4f, o and p).

Diaphragm Case A housing, consisting of top and bottom sections, used for supporting a diaphragm and establishing one or two pressure chambers (Figures 12.4f, o and p).

Actuator Stem A rod-like extension of the diaphragm

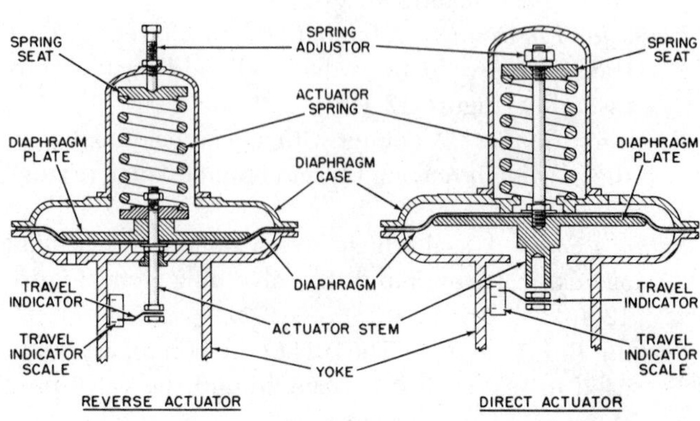

Fig. 12.4o Diaphragm actuators

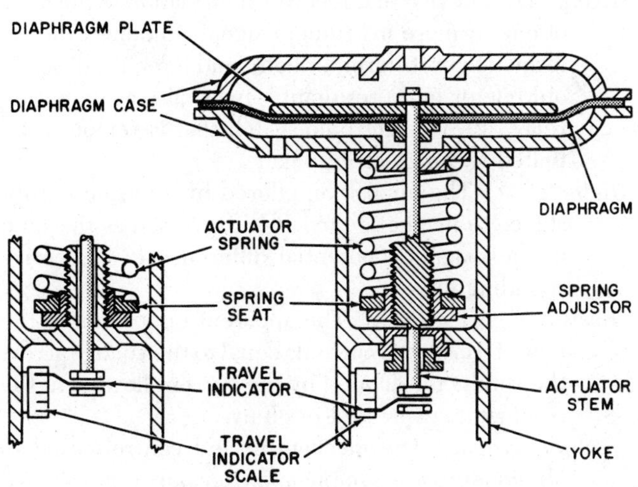

Fig. 12.4p Reverse actuator

plate to permit convenient external connection (Figures 12.4f, o and p).

Yoke A structure which supports the diaphragm case assembly rigidly on the bonnet assembly (Figures 12.4f, o and p).

Direct Actuator A diaphragm actuator in which the actuator stem extends with increasing diaphragm pressure (Figures 12.4f and o).

Reverse Actuator A diaphragm actuator in which the actuator stem retracts with increasing diaphragm pressure (Figures 12.4o and p).

Miscellaneous Definitions

For performance related definitions, see Section 1.1.

Alarm A device which signals the existence of an abnormal condition by means of an audible or visible change, or both, intended to attract attention.

Analyzer Unattended instrumentation which continuously monitors a process stream.

Binary sample A sample composed of two components whose combined concentration is 90 percent or higher.

Board A structure that has a group of instruments mounted on it. The board may consist of one or more component panels, cubicles, desks, drawers, or racks.

Board-mounted Implies an instrument which is installed on a board and which is accessible to the operator during normal use.

Capacitance (Electrical) The property of being able to collect and store a quantity of electrical energy in an insulator. The magnitude is determined by how much of the energy (charge) can be stored for a given potential difference across the terminals of the device (capacitor).

Capacitive reactance That part of the impedance of an AC circuit which is due to capacitance. It equals the reciprocal of the product of the angular frequency ($2\pi f$) of the current times the capacitance, and is expressed in ohms.

Compensated temperature range The specified limits of temperature within which the sensor is to maintain span and zero balance.

Computing relay A device that performs one or more calculations or logical functions or both, and sends out one or more resultant signals.

Conductance The ability of a substance to conduct electricity, measured by the ratio of the current through to the voltage drop across the substance, expressed in siemens.

Control station A manual loading location that also provides switching between manual and automatic control modes of a control loop. It is also known as an auto-manual station and an auto-selector station.

Converter A device that receives information in the form of an instrument signal, alters the form of the information, and sends out a resultant signal. A converter is a special form of a relay. A converter is also referred to as a transducer, although transducer is a general term and its use in connection with signal conversion is not recommended.

Elute To wash out or remove by dissolving.

Flash point The temperature at which combustion of a flammable vapor-air mixture will occur.

Function The purpose of, or action performed by, a device.

Identification The sequence of letters or digits, or both, used to designate an individual instrument or loop.

Immersion length The length from the free end of the bulb or well to the point of immersion in the medium.

Inductance The property of a conductor in which a varying current produces a varying magnetic field that induces an electromotive force across the conductor (self-inductance) or in a nearby circuit (mutual-inductance). Both are expressed in henrys.

Inductive reactance That part of the impedance of an AC circuit which is due to inductance. It equals the product of the angular frequency ($2\pi f$) of the current times the self-inductance and is expressed in ohms.

Local The location of an instrument that is neither on nor behind a board. Local instruments are commonly in the vicinity of a primary element or a final control element.

Local board A board that is not a central or main board. Local boards are commonly in the vicinity of plant subsystems.

Loop A combination of one or more interconnected instruments arranged to measure or control a process variable, or both.

Magnetostriction The phenomenon that ferro-magnetic materials show a small deformation under the influence of a magnetic field and, conversely, that the magnetic properties are affected when the materials are strained.

Manual loading station A device having a manually adjustable output which is used to actuate one or more remote devices.

Measurement The determination of the existence or magnitude of a variable. Measuring instruments include all devices used directly or indirectly for this purpose.

Measurement signal The resultant signal, observed on a readout device, from a change in a process variable when converted to a change in a different variable; for example, in a change from pressure to current, current is the measurement signal.

Piezoelectric effect The phenomenon that some crystals are slightly deformed when voltage gradients are applied in specific directions and, conversely, when they are deformed, an electric charge is obtained.

Pilot light A light that indicates which of a number of normal conditions of a system or device exist. It is unlike an alarm light, which indicates an abnormal condition. The pilot light is also known as a monitor light.

Process variable Any variable property of a process.

Reactance The opposition of inductance and capacitance to the flow of AC current. It equals the product of the sine of the angular phase difference between current and voltage and the ratio of the effective voltage to the effective current. It is expressed in ohms.

Readability The ability of an observer to distinguish between nearly equal quantities.

Refractive index (RI) The ratio of the velocity of light in a vacuum to that in the material for which the RI data are given. Refraction of light (angularity change) occurs at the interface of two different materials.

Relay (1) A device that receives information in the form of one or more instrument signals, modifies the information or its form, or both, and if required, sends out one or more resultant signals; also a computing relay. (2) An electronic switch that is remotely actuated by an electric signal.

Reluctance The opposition offered by a magnetic substance to magnetic flux. It is expressed as the ratio of the magnetic potential difference to the corresponding flux.

Resistance (Electrical) The apparent opposition of the flow of AC or DC current, equal to the voltage across the conductor divided by the current flowing in the conductor, expressed in ohms.

Sample system The mechanism and controls used to obtain a process sample and transport it to the analyzer in a condition suitable for analysis, but without affecting the integrity of the sample.

Scan To sample each of a number of inputs intermittently. A scanning device also may provide record or alarm functions.

Switch A device that connects, disconnects or transfers one or more circuits and is not designated as a controller, a relay, or a control valve.

Telemetry The practice of transmitting and receiving the measurement of a variable for readout or other uses. The term is most commonly applied to electric signal systems.

Test point A process connection to which no instrument is permanently connected, but which is intended for temporary, intermittent, or future connection of an instrument.

Thermistor A device composed of a material in which resistance varies with temperature.

Upstream Inlet side of the instrument.

APPENDIX

CONTENTS OF APPENDIX

SOURCES OF INFORMATION AND CREDIT LINES FOR THE APPENDIX

A.1 Pages in this section were reprinted, with format change only, with permission, from the Annual Book of ASTM Standards, E-380-76 Copyright, American Society for Testing and Materials, 1916 Race Street, Philadelphia, PA 19103.

A.2 Table A.2e reproduced by permission of Fisher Controls Company. All other tables in this section reproduced by permission of Jerguson Gauge and Valve Company.

A.3 Reproduced by permission of Chem Flow Corporation.

A.4 Reproduced by permission of the Foxboro Company. Data principally from Table 3, Directory of Materials for the Construction of Chemical Equipment, by permission from Chemical Engineers' Handbook, John H. Perry, Editor; Copyright 1950, McGraw-Hill Book Co., Inc.

A.7 Reproduced by permission of Jerguson Gauge and Valve Company.

A.8 Compiled by T.A. Mayer.

A.1 INTERNATIONAL SYSTEM OF UNITS

The decimal system of units was conceived in the 16th century when there was a great confusion and jumble of units of weights and measures. It was not until 1790, however, that the French National Assembly requested the French Academy of Sciences to work out a system of units suitable for adoption by the entire world. This system, based on the metre as a unit of length and the gram as a unit of mass, was adopted as a practical measure to benefit industry and commerce. Physicists soon realized its advantages and it was adopted also in scientific and technical circles. The importance of the regulation of weights and measures was recognized in Article 1, Section 8, when the United States Constitution was written in 1787, but the metric system was not legalized in this country until 1866. In 1893, the international metre and kilogram became the fundamental standards of length and mass in the United States, both for metric and customary weights and measures.[1]

Table A.1a
INTERNATIONAL SYSTEM OF UNITS

Quantity	*Unit*	*SI Symbol*	*Formula*
BASE UNITS			
length	metre	m	—
mass	kilogram	kg	—
time	second	s	—
electric current	ampere	A	—
thermodynamic temperature	kelvin	K	—
amount of substance	mole	mol	—
luminous intensity	candela	cd	—
SUPPLEMENTARY UNITS:			
plane angle	radian	rad	—
solid angle	steradian	sr	—
DERIVED UNITS:			
acceleration	metre per second squared	—	m/s^2
activity (of a radioactive source)	disintegration per second	—	(disintegration)/s
angular acceleration	radian per second squared	—	rad/s^2
angular velocity	radian per second	—	rad/s
area	square metre	—	m^2
density	kilogram per cubic metre	—	kg/m^3
electric capacitance	farad	F	A · s/V
electrical conductance	siemens	S	A/V
electric field strength	volt per metre	—	V/m
electric inductance	henry	H	V · s/A
electric potential difference	volt	V	W/A
electric resistance	ohm	Ω	V/A
electromotive force	volt	V	W/A
energy	joule	J	N · m
entropy	joule per kelvin	—	J/K
force	newton	N	$kg \cdot m/s^2$
frequency	hertz	Hz	(cycle)/s
illuminance	lux	lx	lm/m^2
luminance	candela per square metre	—	cd/m^2
luminous flux	lumen	lm	cd · sr
magnetic field strength	ampere per metre	—	A/m
magnetic flux	weber	Wb	V · s
magnetic flux density	tesla	T	Wb/m^2
magnetomotive force	ampere	A	—
DERIVED UNITS:			
power	watt	W	J/s
pressure	pascal	Pa	N/m^2
quantity of electricity	coulomb	C	A · s
quantity of heat	joule	J	N · m
radiant intensity	watt per steradian	—	W/sr
specific heat	joule per kilogram-kelvin	—	J/kg · K
stress	pascal	Pa	N/m^2
thermal conductivity	watt per metre-kelvin	—	W/m · K
velocity	metre per second	—	m/s
viscosity, dynamic	pascal-second	—	Pa · s
viscosity, kinematic	square metre per second	—	m^2/s
voltage	volt	V	W/A
volume	cubic metre	—	m^3
wavenumber	reciprocal metre	—	(wave)/m
work	joule	J	N · m

The following tables of conversion factors are intended to serve two purposes:

1. To express the definitions of miscellaneous units of measure as exact numerical multiples of coherent "metric" units. Relationships that are exact in terms of the base unit are followed by an asterisk. Relationships that are not followed by an asterisk are either the results of physical measurements or are only approximate.
2. To provide multiplying factors for converting expressions of measurements given by numbers and miscellaneous units to corresponding new numbers and metric units.

Conversion factors are presented for ready adaptation to computer readout and electronic data transmission. The factors are written as a number equal to or greater than one and less than ten with six or less decimal places. This number is followed by the letter E (for exponent), a plus or minus symbol, and two digits which indicate the power of 10 by which the number must be multiplied to obtain the correct value. For example:

$$3.523\,907 \text{ E}{-}02 \text{ is } 3.523\,907 \times 10^{-2}$$

or

$$0.035\,239\,07$$

Similarly:

$$3.386\,389 \text{ E}{+}03 \text{ is } 3.386\,389 \times 10^{3}$$

or

$$3\,386.389$$

An asterisk (*) after the sixth decimal place indicates that the conversion factor is exact and that all subsequent digits are zero.

When a figure is to be rounded to fewer digits than the total number available, the procedure should be as follows:

1. When the first digit discarded is less than 5, the last digit retained should not be changed. For example, 3.463 25, if rounded to four digits, would be 3.463; if rounded to three digits, 3.46.
2. When the first digit discarded is greater than 5, or if it is a 5 followed by at least one digit other than 0, the last figure retained should be increased by one unit. For example 8.376 52, if rounded to four digits, would be 8.377; if rounded to three digits, 8.38.
3. When the first digit discarded is exactly 5, followed only by zeros, the last digit retained should be rounded upward if it is an odd number, but no adjustment made if it is an even number. For example, 4.365, when rounded to three digits, becomes 4.36. The number 4.355 would also round to the same value, 4.36, if rounded to three digits.

Where less than six decimal places are shown, more precision is not warranted.

Table A.1b

ALPHABETICAL LIST OF UNITS

(Symbols of SI units given in parentheses)

To convert from	To	Multiply by
A		
abampere	ampere (A)	1.000 000*E+01
abcoulomb	coulomb (C)	1.000 000*E+01
abfarad	farad (F)	1.000 000*E+09
abhenry	henry (H)	1.000 000*E−09
abmho	siemens (S)	1.000 000*E+09
abohm	ohm (Ω)	1.000 000*E−09
abvolt	volt (V)	1.000 000*E−08
acre foot (U.S. survey)[1]	metre³ (m^3)	1.233 489 E+03
acre (U.S. survey)[1]	metre² (m^2)	4.046 873 E+03
ampere hour	coulomb (C)	3.600 000*E+03
are	metre² (m^2)	1.000 000*E+02
angstrom	metre (m)	1.000 000*E−10
astronomical unit	metre (m)	1.495 979 E+11
atmosphere (standard)	pascal (Pa)	1.013 250*E+05
atmosphere (technical = 1 kgf/cm²)	pascal (Pa)	9.806 650*E+04
B		
bar	pascal (Pa)	1.000 000*E+05
barn	metre² (m^2)	1.000 000*E−28
barrel (for petroleum, 42 gal)	metre³ (m^3)	1.589 873 E−01
board foot	metre³ (m^3)	2.359 737 E−03
British thermal unit (International Table)[2]	joule (J)	1.055 056 E+03
British thermal unit (mean)	joule (J)	1.055 87 E+03
British thermal unit (thermochemical)	joule (J)	1.054 350 E+03
British thermal unit (39°F)	joule (J)	1.059 67 E+03
British thermal unit (59°F)	joule (J)	1.054 80 E+03
British thermal unit (60°F)	joule (J)	1.054 68 E+03
Btu (International Table)·ft/h·ft²·°F (*k*, thermal conductivity)	watt per metre kelvin (W/m·K)	1.730 735 E+00
Btu (thermochemical)·ft/h·ft²·°F (*k*, thermal conductivity)	watt per metre kelvin (W/m·K)	1.729 577 E+00
Btu (International Table)·in/h·ft²·°F (*k*, thermal conductivity)	watt per meter kelvin (W/m·K)	1.442 279 E−01
Btu (thermochemical)·in/h·ft²·°F (*k*, thermal conductivity)	watt per metre kelvin (W/m·K)	1.441 314 E−01
Btu (International Table)·in/s·ft²·°F (*k*, thermal conductivity)	watt per metre kelvin (W/m·K)	5.192 204 E+02
Btu (thermochemical)·in/s·ft²·°F (*k*, thermal conductivity)	watt per metre kelvin (W/m·K)	5.188 732 E+02
Btu (International Table)/h	watt (W)	2.930 711 E−01
Btu (International Table)/s	watt (W)	1.055 056 E+03
Btu (thermochemical)/h	watt (W)	2.928 751 E−01
Btu (thermochemical)/min	watt (W)	1.757 250 E+01
Btu (thermochemical)/s	watt (W)	1.054 350 E+03
Btu (International Table)/ft²	joule per metre² (J/m²)	1.135 653 E+04
Btu (thermochemical)/ft²	joule per metre² (J/m²)	1.134 893 E+04
Btu (thermochemical)/ft²·h	watt per metre² (W/m²)	3.152 481 E+00
Btu (thermochemical)/ft²·min	watt per metre² (W/m²)	1.891 489 E+02
Btu (thermochemical)/ft²·s	watt per metre² (W/m²)	1.134 893 E+04
Btu (thermochemical)/in²·s	watt per metre² (W/m²)	1.634 246 E+06
Btu (International Table)/h·ft²·°F (*C*, thermal conductance)	watt per metre² kelvin (W/m²·K)	5.678 263 E+00
Btu (thermochemical)/h·ft²·°F (*C*, thermal conductance)	watt per metre² kelvin (W/m²·K)	5.674 466 E+00

1. Since 1893 the U.S. basis of length measurement has been derived from metric standards. In 1959 a small refinement was made in the definition of the yard to resolve discrepancies both in this country and abroad, which changed its length from 3600/3937 m to 0.9144 m exactly. This resulted in the new value being shorter by two parts in a million.

   At the same time it was decided that any data in feet drived from and published as a result of geodetic surveys within the U.S. would remain with the old standard (1 ft. = 1200/3937 m) until further decision. This foot is named the U.S. survey foot.

   As a result all U.S. land measurements in U.S. customary units will relate to the metre by the old standard. All the conversion factors in these tables for units referenced to this footnote are based on the U.S. survey foot, rather than the international foot.

   Conversion factors for the land measures given below may be determined from the following relationships:

   1 league = 3 miles (exactly)
   1 rod = 16½ feet (exactly)
   1 section = 1 square mile (exactly)
   1 township = 36 square miles (exactly)
   1 chain = 66 feet (exactly)

2. This value was adopted in 1956. Some of the older International Tables use the value 1.055 04 E+03. The exact conversion factor is 1.055 055 852 62*E+03.

3. The SI unit of thermodynamic temperature is the kelvin (K), and this unit is properly used for expressing thermodynamic temperature and temperature intervals. Wide use is also made of the degree Celsius (°C), which is the SI unit for expressing Celsius temperature and temperature intervals. The Celsius scale (formerly called centigrade) is related directly to thermodynamic temperature (kelvins) as follows:
   1. The temperature interval one degree Celsius equals one kelvin exactly.
   2. Celsius temperature (t) is related to thermodynamic temperature (T) by the equation $t = T - T_0$, where $T_0 = 273.15$ K by definition.

4. This is sometimes called the moment of inertia of a plane section about a specified axis.

Table A.1b
ALPHABETICAL LIST OF UNITS *continued*
(Symbols of SI units given in parentheses)

To convert from	To	Multiply by
Btu (International Table)/s·ft²·°F	watt per metre² kelvin (W/m²·K)	2.044 175 E+04
Btu (thermochemical)/s·ft²·°F	watt per metre² kelvin (W/m²·K)	2.042 808 E+04
Btu (International Table)/lb	joule per kilogram (J/kg)	2.326 000*E+03
Btu (thermochemical)/lb	joule per kilogram (J/kg)	2.324 444 E+03
Btu (International Table)/lb.°F (*c*, heat capacity)	joule per kilogram kelvin (J/kg·K)	4.186 800*E+03
Btu (thermochemical)/lb·°F (*c*, heat capacity)	joule per kilogram kelvin (J/kg·K)	4.184 000 E+03
bushel (U.S.)	metre³ (m³)	3.523 907 E−02
C		
caliber (inch)	metre (m)	2.540 000*E−02
calorie (International Table)	joule (J)	4.186 800*E+00
calorie (mean)	joule (J)	4.190 02 E+00
calorie (thermochemical)	joule (J)	4.184 000*E+00
calorie (15°C)	joule (J)	4.185 80 E+00
calorie (20°C)	joule (J)	4.181 90 E+00
calorie (kilogram, International Table)	joule (J)	4.186 800*E+03
calorie (kilogram, mean)	joule (J)	4.190 02 E+03
calorie (kilogram, thermochemical)	joule (J)	4.184 000*E+03
cal (thermochemical)/cm²	joule per metre² (J/m²)	4.184 000*E+04
cal (International Table)/g	joule per kilogram (J/kg)	4.186 800*E+03
cal (thermochemical)/g	joule per kilogram (J/kg)	4.184 000*E+03
cal (International Table)/g·°C	joule per kilogram kelvin (J/kg·K)	4.186 800*E+03
cal (thermochemical)/g·°C	joule per kilogram kelvin (J/kg·K)	4.184 000*E+03
cal (thermochemical)/min	watt (W)	6.973 333 E−02
cal (thermochemical)/s	watt (W)	4.184 000*E+00
cal (thermochemical)/cm²·min	watt per metre² (W/m²)	6.973 333 E+02
cal (thermochemical)/cm²·s	watt per metre² (W/m²)	4.184 000*E+04
cal (thermochemical)/cm·s·°C	watt per metre kelvin (W/m·K)	4.184 000*E+02
carat (metric)	kilogram (kg)	2.000 000*E−04
centimetre of mercury (0°C)	pascal (Pa)	1.333 22 E+03
centimetre of water (4°C)	pascal (Pa)	9.806 38 E+01
centipoise	pascal second (Pa·s)	1.000 000*E−03
centistokes	metre² per second (m²/s)	1.000 000*E−06
circular mil	metre² (m²)	5.067 075 E−10
clo	kelvin metre² per watt (K·m²/W)	2.003 712 E−01
cup	metre³ (m³)	2.365 882 E−04
curie	becquerel (Bq)	3.700 000*E+10
D		
day (mean solar)	second (s)	8.640 000 E+04
day (sidereal)	second (s)	8.616 409 E+04
degree (angle)	radian (rad)	1.745 329 E−02
degree Celsius	Kelvin (K)	$t_K = t_{°C} + 273.15$
degree centigrade	[see footnote 3]	
degree Fahrenheit	degree Celsius	$t_{°C} = (t_{°F} - 32)/1.8$
degree Fahrenheit	kelvin (K)	$t_K = (t_{°F} + 459.67)/1.8$
degree Rankine	kelvin (K)	$t_K = t_{°R}/1.8$
°F·h·ft²/Btu (International Table) (*R*, thermal resistance)	kelvin metre² per watt (K·m²/W)	1.761 102 E−01
°F·h·ft²/Btu (thermochemical) (*R*, thermal resistance)	kelvin metre² per watt (K·m²/W)	1.762 280 E−01
denier	kilogram per metre (kg/m)	1.111 111 E−07
dyne	newton (N)	1.000 000*E−05
dyne·cm	newton metre (N·m)	1.000 000*E−07
dyne/cm²	pascal (Pa)	1.000 000*E−01
E		
electronvolt	joule (J)	1.602 19 E−19
EMU of capacitance	farad (F)	1.000 000*E+09
EMU of current	ampere (A)	1.000 000*E+01
EMU of electric potential	volt (V)	1.000 000*E−08
EMU of inductance	henry (H)	1.000 000*E−09
EMU of resistance	ohm (Ω)	1.000 000*E−09
ESU of capacitance	farad (F)	1.112 650 E−12
ESU of current	ampere (A)	3.335 6 E−10
ESU of electric potential	volt (V)	2.997 9 E+02
ESU of inductance	henry (H)	8.987 554 E+11
ESU of resistance	ohm (Ω)	8.987 554 E+11
erg	joule (J)	1.000 000*E−07
erg/(cm²·s)	watt per metre² (W/m²)	1.000 000*E−03
erg/s	watt (W)	1.000 000*E−07
F		
faraday (based on carbon-12)	coulomb (C)	9.648 70 E+04
faraday (chemical)	coulomb (C)	9.649 57 E+04
faraday (physical)	coulomb (C)	9.652 19 E+04

Table A.1b

ALPHABETICAL LIST OF UNITS *continued*

(Symbols of SI units given in parentheses)

To convert from	To	Multiply by
fathom	metre (m)	1.828 8 E+00
fermi (femtometer)	metre (m)	1.000 000*E−15
fluid ounce (U.S.)	metre^3 (m^3)	2.957 353 E−05
foot	metre (m)	3.048 000*E−01
foot (U.S. survey)[1]	metre (m)	3.048 006 E−01
foot of water (39.2°F)	pascal (Pa)	2.988 98 E+03
ft^2	metre^2 (m^2)	9.290 304*E−02
ft^2/h (thermal diffusivity)	metre^2 per second (m^2/s)	2.580 640*E−05
ft^2/s	metre^2 per second (m^2/s)	9.290 304*E−02
ft^3 (volume; section modulus)	metre^3 (m^3)	2.831 685 E−02
ft^3/min	metre^3 per second (m^3/s)	4.719 474 E−04
ft^3/s	metre^3 per second (m^3/s)	2.831 685 E−02
ft^4 (moment of section)[4]	metre^4 (m^4)	8.630 975 E−03
ft/h	metre per second (m/s)	8.466 667 E−05
ft/min	metre per second (m/s)	5.080 000*E−03
ft/s	metre per second (m/s)	3.048 000*E−01
ft/s^2	metre per second^2(m/s^2)	3.048 000*E−01
footcandle	lux (lx)	1.076 391 E+01
footlambert	candela per metre^2 (cd/m^2)	3.426 259 E+00
ft·lbf	joule (J)	1.355 818 E+00
ft·lbf/h	watt (W)	3.766 161 E−04
ft·lbf/min	watt (W)	2.259 697 E−02
ft·lbf/s	watt (W)	1.355 818 E+00
ft·poundal	joule (J)	4.214 011 E−02
free fall, standard (g)	metre per second^2 (m/s^2)	9.806 650*E+00
G		
gal	metre per second^2 (m/s^2)	1.000 000*E−02
gallon (Canadian liquid)	metre^3 (m^3)	4.546 090 E−03
gallon (U.K. liquid)	metre^3 (m^3)	4.546 092 E−03
gallon (U.S. dry)	metre^3 (m^3)	4.404 884 E−03
gallon (U.S. liquid)	metre^3 (m^3)	3.785 412 E−03
gallon (U.S. liquid) per day	metre^3 per second (m^3/s)	4.381 264 E−08
gallon (U.S. liquid) per minute	metre^3 per second (m^3/s)	6.309 020 E−05
gallon (U.S. liquid) per hp·h (SFC, specific fuel consumption)	metre^3 per joule (m^3/J)	1.410 089 E−09
gamma	tesla (T)	1.000 000*E−09
gauss	tesla (T)	1.000 000*E−04
gilbert	ampere (A)	7.957 747 E−01
gill (U.K.)	metre^3 (m^3)	1.420 654 E−04
gill (U.S.)	metre^3 (m^3)	1.182 941 E−04
grad	degree (angular)	9.000 000*E−01
grad	radian (rad)	1.570 796 E−02

To convert from	To	Multiply by
grain (1/7000 lb avoirdupois)	kilogram (kg)	6.479 891*E−05
grain (lb avoirdupois/7000)/gal (U.S. liquid)	kilogram per metre^3 (kg/m^3)	1.711 806 E−02
gram	kilogram (kg)	1.000 000*E−03
g/cm^3	kilogram per metre^3 (kg/m^3)	1.000 000*E+03
gram-force/cm^2	pascal (Pa)	9.806 650*E+01
H		
hectare	metre^2 (m^2)	1.000 000*E+04
horsepower (550 ft·lbf/s)	watt (W)	7.456 999 E+02
horsepower (boiler)	watt (W)	9.809 50 E+03
horsepower (electric)	watt (W)	7.460 000*E+02
horsepower (metric)	watt (W)	7.354 99 E+02
horsepower (water)	watt (W)	7.460 43 E+02
horsepower (U.K.)	watt (W)	7.457 0 E+02
hour (mean solar)	second (s)	3.600 000 E+03
hour (sidereal)	second (s)	3.590 170 E+03
hundredweight (long)	kilogram (kg)	5.080 235 E+01
hundredweight (short)	kilogram (kg)	4.535 924 E+01
I		
inch	metre (m)	2.540 000*E−02
inch of mercury (32°F)	pascal (Pa)	3.386 38 E+03
inch of mercury (60°F)	pascal (Pa)	3.376 85 E+03
inch of water (39.2°F)	pascal (Pa)	2.490 82 E+02
inch of water (60°F)	pascal (Pa)	2.488 4 E+02
in^2	metre^2 (m^2)	6.451 600*E−04
in^3 (volume; section modulus)[5]	metre^3 (m^3)	1.638 706 E−05
in^3/min	metre^3 per second (m^3/s)	2.731 177 E−07
in^4 (moment of section)[4]	metre^4 (m^4)	4.162 314 E−07
in/s	metre per second (m/s)	2.540 000*E−02
in/s^2	metre per second^2 (m/s^2)	2.540 000*E−02
K		
kayser	1 per metre (1/m)	1.000 000*E+02
kelvin	degree Celsius	$t_{°C} = t_K - 273.15$
kilocalorie (International Table)	joule (J)	4.186 800*E+03
kilocalorie (mean)	joule (J)	4.190 02 E+03
kilocalorie (thermochemical)	joule (J)	4.184 000*E+03
kilocalorie (thermochemical)/min	watt (W)	6.973 333 E+01

5. The exact conversion factor is 1.638 706 4*E−05.

Table A.1b
ALPHABETICAL LIST OF UNITS *continued*
(Symbols of SI units given in parentheses)

To convert from	*To*	*Multiply by*
kilocalorie (thermochemical)/s	watt (W)	4.184 000*E+03
kilogram-force (kgf)	newton (N)	9.806 650*E+00
kgf·m	newton metre (N·m)	9.806 650*E+00
kgf·s²/m (mass)	kilogram (kg)	9.806 650*E+00
kgf/cm²	pascal (Pa)	9.806 650*E+04
kgf/m²	pascal (Pa)	9.806 650*E+00
kgf/mm²	pascal (Pa)	9.806 650*E+06
km/h	metre per second (m/s)	2.777 778 E−01
kilopond	newton (N)	9.806 650*E+00
kW·h	joule (J)	3.600 000*E+06
kip (1000 lbf)	newton (N)	4.448 222 E+03
kip/in² (ksi)	pascal (Pa)	6.894 757 E+06
knot (international)	metre per second (m/s)	5.144 444 E−01
L		
lambert	candela per metre² (cd/m²)	$1/\pi$ *E+04
lambert	candela per metre² (cd/m²)	3.183 099 E+03
langley	joule per metre² (J/m²)	4.184 000*E+04
league	metre (m)	[see footnote 1]
light year	metre (m)	9.460 55 E+15
liter[6]	metre³ (m³)	1.000 000*E−03
M		
maxwell	weber (Wb)	1.000 000*E−08
mho	siemens (S)	1.000 000*E+00
microinch	metre (m)	2.540 000*E−08
micron	metre (m)	1.000 000*E−06
mil	metre (m)	2.540 000*E−05
mile (international)	metre (m)	1.609 344*E+03
mile (statute)	metre (m)	1.609 3 E+03
mile (U.S. survey)[1]	metre (m)	1.609 347 E+03
mile (international nautical)	metre (m)	1.852 000*E+03
mile (U.K. nautical)	metre (m)	1.853 184*E+03
mile (U.S. nautical)	metre (m)	1.852 000*E+03
mi² (international)	metre² (m²)	2.589 988 E+06
mi² (U.S. survey)[1]	metre² (m²)	2.589 998 E+06
mi/h (international)	metre per second (m/s)	4.470 400*E−01
mi/h (international)	kilometre per hour (km/h)	1.609 344*E+00
mi/min (international)	metre per second (m/s)	2.682 240*E+01
mi/s (international)	metre per second (m/s)	1.609 344*E+03
millibar	pascal (Pa)	1.000 000*E+02
millimetre of mercury (0°C)	pascal (Pa)	1.333 22 E+02

6. In 1964 the General Conference on Weights and Measures adopted the name litre as a special name for decimetre. Prior to this decision the litre differed slightly (previous value, 1.000028 dm³) and in expression of precision volume measurement this fact must be kept in mind.

To convert from	*To*	*Multiply by*
minute (angle)	radian (rad)	2.908 882 E−04
minute (mean solar)	second (s)	6.000 000 E+01
minute (sidereal)	second (s)	5.983 617 E+01
month (mean calendar)	second (s)	2.628 000 E+06
O		
oersted	ampere per metre (A/m)	7.957 747 E+01
ohm centimetre	ohm metre (Ω·m)	1.000 000*E−02
ohm circular-mil per foot	ohm millimetre² per metre (Ω·mm²/m)	1.662 426 E−03
ounce (avoirdupois)	kilogram (kg)	2.834 952 E−02
ounce (troy or apothecary)	kilogram (kg)	3.110 348 E−02
ounce (U.K. fluid)	metre³ (m³)	2.841 307 E−05
ounce (U.S. fluid)	metre³ (m³)	2.957 353 E−05
ounce-force	newton (N)	2.780 139 E−01
ozf·in	newton metre (N·m)	7.061 552 E−03
oz (avoirdupois)/gal (U.K. liquid)	kilogram per metre³ (kg/m³)	6.236 021 E+00
oz (avoirdupois)/gal (U.S. liquid)	kilogram per metre³ (kg/m³)	7.489 152 E+00
oz (avoirdupois)/in³	kilogram per metre³ (kg/m³)	1.729 994 E+03
oz (avoirdupois)/ft²	kilogram per metre² (kg/m²)	3.051 517 E−01
oz (avoirdupois)/yd²	kilogram per metre² (kg/m²)	3.390 575 E−02
P		
parsec	metre (m)	3.085 678 E+16
peck (U.S.)	metre³ (m³)	8.809 768 E−03
pennyweight	kilogram (kg)	1.555 174 E−03
perm (0°C)	kilogram per pascal second metre² (kg/Pa·s·m²)	5.721 35 E−11
perm (23°C)	kilogram per pascal second metre² (kg/Pa·s·m²)	5.745 25 E−11
perm·in (0°C)	kilogram per pascal second metre (kg/Pa·s·m)	1.453 22 E−12
perm·in (23°C)	kilogram per pascal second metre (kg/Pa·s·m)	1.459 29 E−12
phot	lumen per metre² (lm/m²)	1.000 000*E+04
pica (printer's)	metre (m)	4.217 518 E−03
pint (U.S. dry)	metre³ (m³)	5.506 105 E−04
pint (U.S. liquid)	metre³ (m³)	4.731 765 E−04
point (printer's)	metre (m)	3.514 598*E−04
poise (absolute viscosity)	pascal second (Pa·s)	1.000 000*E−01
pound (lb avoirdupois)[7]	kilogram (kg)	4.535 924 E−01

7. The exact conversion factor is 4.535 923 7*E−01.

Table A.1b

ALPHABETICAL LIST OF UNITS *continued*

(Symbols of SI units given in parentheses)

To convert from	*To*	*Multiply by*
pound (troy or apothecary)	kilogram (kg)	3.732 417 E − 01
lb·ft^2 (moment of inertia)	kilogram metre^2 (kg·m^2)	4.214 011 E − 02
lb·in^2 (moment of inertia)	kilogram metre^2 (kg·m^2)	2.926 397 E − 04
lb/ft·h	pascal second (Pa·s)	4.133 789 E − 04
lb/ft·s	pascal second (Pa·s)	1.488 164 E + 00
lb/ft^2	kilogram per metre^2 (kg/m^2)	4.882 428 E + 00
lb/ft^3	kilogram per metre^3 (kg/m^3)	1.601 846 E + 01
lb/gal (U.K. liquid)	kilogram per metre^3 (kg/m^3)	9.977 633 E + 01
lb/gal (U.S. liquid)	kilogram per metre^3 (kg/m^3)	1.198 264 E + 02
lb/h	kilogram per second (kg/s)	1.259 979 E − 04
lb/hp·h (SFC, specific fuel consumption)	kilogram per joule (kg/J)	1.689 659 E − 07
lb/in^3	kilogram per metre^3 (kg/m^3)	2.767 990 E + 04
lb/min	kilogram per second (kg/s)	7.559 873 E − 03
lb/s	kilogram per second (kg/s)	4.535 924 E − 01
lb/yd^3	kilogram per metre^3 (kg/m^3)	5.932 764 E − 01
poundal	newton (N)	1.382 550 E − 01
poundal/ft^2	pascal (Pa)	1.488 164 E + 00
poundal·s/ft^2	pascal second (Pa·s)	1.488 164 E + 00
pound-force (lbf)[8]	newton (N)	4.448 222 E + 00
lbf·ft	newton metre (N·m)	1.355 818 E + 00
lbf·ft/in	newton metre per metre (N·m/m)	5.337 866 E + 01
lbf·in	newton metre (N·m)	1.129 848 E − 01
lbf·in/in	newton metre per metre (N·m/m)	4.448 222 E + 00
lbf·s/ft^2	pascal second (Pa·s)	4.788 026 E + 01
lbf·s/in^2	pascal second (Pa·s)	6.894 757 E + 03
lbf/ft	newton per metre (N/m)	1.459 390 E + 01
lbf/ft^2	pascal (Pa)	4.788 026 E + 01
lbf/in	newton per metre (N/m)	1.751 268 E + 02
lbf/in^2 (psi)	pascal (Pa)	6.894 757 E + 03
lbf/lb (thrust/weight [mass] ratio)	newton per kilogram (N/kg)	9.806 650 E + 00
Q		
quart (U.S. dry)	metre^3 (m^3)	1.101 221 E − 03
quart (U.S. liquid)	metre^3 (m^3)	9.463 529 E − 04
R		
rad (radiation dose absorbed)	gray (Gy)	1.000 000*E − 02
rhe	1 per pascal second (1/Pa·s)	1.000 000*E + 01
rod	metre (m)	[see footnote 1]

To convert from	*To*	*Multiply by*
roentgen	coulomb per kilogram (C/kg)	2.58 E − 04
S		
second (angle)	radian (rad)	4.848 137 E − 06
second (sidereal)	second (s)	9.972 696 E − 01
section	metre^2 (m^2)	[see footnote 1]
shake	second (s)	1.000 000*E − 08
slug	kilogram (kg)	1.459 390 E + 01
slug/ft·s	pascal second (Pa·s)	4.788 026 E + 01
slug/ft^3	kilogram per metre^3 (kg/m^3)	5.153 788 E + 02
statampere	ampere (A)	3.335 640 E − 10
statcoulomb	coulomb (C)	3.335 640 E − 10
statfarad	farad (F)	1.112 650 E − 12
stathenry	henry (H)	8.987 554 E + 11
statmho	siemens (S)	1.112 650 E − 12
statohm	ohm (Ω)	8.987 665 E + 11
statvolt	volt (V)	2.997 925 E + 02
stere	metre^3 (m^3)	1.000 000*E + 00
stilb	candela per metre^2 (cd/m^2)	1.000 000*E + 04
stokes (kinematic viscosity)	metre^2 per second (m^2/s)	1.000 000*E − 04
T		
tablespoon	metre^3 (m^3)	1.478 676 E − 05
teaspoon	metre^3 (m^3)	4.928 922 E − 06
tex	kilogram per metre (kg/m)	1.000 000*E − 06
therm	joule (J)	1.055 056 E + 08
ton (assay)	kilogram (kg)	2.916 667 E − 02
ton (long, 2240 lb)	kilogram (kg)	1.016 047 E + 03
ton (metric)	kilogram (kg)	1.000 000*E + 03
ton (nuclear equivalent of TNT)	joule (J)	4.184 E + 09[9]
ton (refrigeration)	watt (W)	3.516 800 E + 03
ton (register)	metre^3 (m^3)	2.831 685 E + 00
ton (short, 2000 lb)	kilogram (kg)	9.071 847 E + 02
ton (long)/yd^3	kilogram per metre^3 (kg/m^3)	1.328 939 E + 03
ton (short)/yd^3	kilogram per metre^3 (kg/m^3)	1.186 553 E + 03
ton (short)/h	kilogram per second (kg/s)	2.519 958 E − 01
ton-force (2000 lbf)	newton (N)	8.896 444 E + 03
tonne	kilogram (kg)	1.000 000*E + 03
torr (mm Hg, 0°C)	pascal (Pa)	1.333 22 E + 02
township	metre^2 (m^2)	[see footnote 1]

8. The exact conversion factor is 4.448 221 615 260 5*E + 00.

9. Defined (not measured) value.

Table A.1b

ALPHABETICAL LIST OF UNITS *continued*

(Symbols of SI units given in parentheses)

To convert from	*To*	*Multiply by*	*To convert from*	*To*	*Multiply by*
U			Y		
unit pole	weber (Wb)	1.256 637 E − 07	yard	metre (m)	9.144 000*E − 01
W			yd^2	$metre^2$ (m^2)	8.361 274 E − 01
W·h	joule (J)	3.600 000*E + 03	yd^3	$metre^3$ (m^3)	7.645 549 E − 01
W·s	joule (J)	1.000 000*E + 00	yd^3/min	$metre^3$ per second (m^3/s)	1.274 258 E − 02
W/cm^2	watt per $metre^2$ (W/m^2)	1.000 000*E + 04	year (365 days)	second (s)	3.153 600 E + 07
W/in^2	watt per $metre^2$ (W/m^2)	1.550 003 E + 03	year (sidereal)	second (s)	3.155 815 E + 07
			year (tropical)	second (s)	3.155 693 E + 07

A.2 ENGINEERING CONVERSION FACTORS

Table A.2a
CONVERSION FACTORS

To Convert	*Into*	*Multiply by*
A		
abcoulomb	statcoulombs	2.998×10^{10}
acre	sq. chain (Gunters)	10
acre	rods	160
acre	sq. links (Gunters)	1×10^{5}
acre	hectare or sq. hectometer	.4047
acres	sq. ft.	43,560.0
acres	sq. meters	4,047.
acres	sq. miles	1.562×10^{-3}
acres	sq. yards	4,840.
acre-feet	cu. ft.	43,560.0
acre-feet	gallons	3.259×10^{5}
amperes/sq. cm	amps/sq. in.	6.452
amperes/sq. cm	amps/sq. meter	10^{4}
amperes/sq. in.	amps/sq. cm	0.1550
amperes/sq. in.	amps/sq. meter	1,550.0
amperes/sq. meter	amps/sq. cm	10^{-4}
amperes/sq. meter	amps/sq. in.	6.452×10^{-4}
ampere-hours	coulombs	3,600.0
ampere-hours	faradays	0.03731
ampere-turns	gilberts	1.257
ampere-turns/cm	amp-turns/in.	2.540
ampere-turns/cm	amp-turns/meter	100.0
ampere-turns/cm	gilberts/cm	1.257
ampere-turns/in.	amp-turns/cm	0.3937
ampere-turns/in.	amp-turns/meter	39.37
ampere-turns/in.	gilberts/cm	0.4950
ampere-turns/meter	amp/turns/cm	0.01
ampere-turns/meter	amp-turns/in.	0.0254
ampere-turns/meter	gilberts/cm	0.01257
Angstrom unit	in.	$3{,}937 \times 10^{-9}$
Angstrom unit	meter	1×10^{-10}
Angstrom unit	micron or (mu)	1×10^{-4}
are	acre (US)	.02471
ares	sq. yards	119.60
ares	acres	0.02471
ares	sq. meters	100.0
Astronomical Unit	kilometers	1.495×10^{8}
atmospheres	ton/sq. in.	.007348
atmospheres	cms of mercury	76.0
atmospheres	ft. of water (at 4°C)	33.90
atmospheres	in. of mercury (at 0°C)	29.92
atmospheres	kgs/sq. cm	1.0333
atmospheres	kgs/sq. meter	10,332.
atmospheres	pounds/sq. in.	14.70
atmospheres	tons/sq. ft.	1.058
B		
barrels (U.S., dry)	cu. in.	7056.
barrels (U.S., dry	quarts (dry)	105.0
barrels (U.S., liquid)	gallons	31.5
barrels (oil)	gallons (oil)	42.0
bars	atmospheres	0.9869
bars	dynes/sq. cm	10^{6}
bars	kgs/sq. meter	1.020×10^{4}
bars	pounds/sq. ft.	2,089.
bars	pounds/sq. in.	14.50
baryl	dyne/sq. cm	1.000
bolt (US cloth)	meters	36.576
Btu	liter–atmosphere	10.409
Btu	ergs	1.0550×10^{10}
Btu	foot-lbs	778.3
Btu	gram-calories	252.0
Btu	horsepower-hrs	3.931×10^{-4}
Btu	joules	1,054.8
Btu	kilogram-calories	0.2520
Btu	kilogram-meters	107.5
Btu	kilowatt-hrs	2.928×10^{-4}
Btu/hr	foot-pounds/sec	0.2162
Btu/hr	gram-cal/sec	0.0700
Btu/hr	horsepower-hrs	3.929×10^{-4}

Table A.2a
CONVERSION FACTORS *continued*

To Convert	Into	Multiply by
Btu/hr	watts	0.2931
Btu/min	foot-lbs/sec	12.96
Btu/min	horsepower	0.02356
Btu/min	kilowatts	0.01757
Btu/min	watts	17.57
Btu/sq. ft/min	watts/sq. in.	0.1221
bucket (Br. dry)	cu. cm.	1.818×10^4
bushels	cu. ft.	1.2445
bushels	cu. in.	2,150.4
bushels	cu. meters	0.03524
bushels	liters	35.24
bushels	pecks	4.0
bushels	pints (dry)	64.0
bushels	quarts (dry)	32.0
C		
calories, gram (mean)	Btu (mean)	3.9685×10^{-3}
candle/sq. cm	lamberts	3.142
candle/sq. in.	lamberts	.4870
centares (centiares)	sq. meters	1.0
Centigrade	Fahrenheit	$(C° \times 9/5) + 32$
centigrams	grams	0.01
centiliter	ounce fluid (US)	.3382
centiliter	cu. in.	.6103
centiliter	drams	2.705
centiliters	liters	0.01
centimeters	feet	3.281×10^{-2}
centimeters	inches	0.3937
centimeters	kilometers	10^{-5}
centimeters	meters	0.01
centimeters	miles	6.214×10^{-6}
centimeters	millimeters	10.0
centimeters	mils	393.7
centimeters	yards	1.094×10^{-2}
centimeter-dynes	cm-grams	1.020×10^{-3}
centimeter-dynes	meter-kgs	1.020×10^{-8}
centimeter-dynes	pound-feet	7.376×10^{-8}
centimeter-grams	cm-dynes	980.7
centimeter-grams	meter-kgs	10^{-5}
centimeter-grams	pound-feet	7.233×10^{-5}
centimeters of mercury	atmospheres	0.01316
centimeters of mercury	feet of water	0.4461
centimeters of mercury	kgs/sq meter	136.0
centimeters of mercury	pounds/sq. ft.	27.85
centimeters of mercury	pounds/sq. in.	0.1934
centimeters/sec	feet/min	1.1969
centimeters/sec	feet/sec	0.03281
centimeters/sec	kilometers/hr	0.036
centimeters/sec	knots	0.1943
centimeters/sec	meters/min	0.6
centimeters/sec	miles/hr	0.02237
centimeters/sec	miles/min	3.728×10^{-4}
centimeters/sec/sec	feet/sec/sec	0.03281
centimeters/sec/sec	kms/hr/sec	0.036
centimeters/sec/sec	meters/sec/sec	0.01
centimeters/sec/sec	miles/hr/sec	0.02237
chain	inches	792.00
chain	meters	20.12
chains (surveyors' or Gunter's)	yards	22.00
circular mils	sq. cms	5.067×10^{-6}
circular mils	sq. mils	0.7854
circumference	radians	6.283
circular mils	sq. inches	7.854×10^{-7}
cords	cord feet	8
cord feet	cu. ft.	16
coulomb	statcoulombs	2.998×10^9
coulombs	faradays	1.036×10^{-5}
coulombs/sq. cm	coulombs/sq. in.	64.52
coulombs/sq. cm	coulombs/sq. meter	10^4
coulombs/sq. in.	coulombs/sq. cm	0.1550
coulombs/sq. in.	coulombs/sq. meter	1,550.
coulombs/sq. meter	coulombs/sq. cm	10^{-4}
coulombs/sq. meter	coulombs/sq. in.	6.452×10^{-4}
cubic centimeters	cu. ft.	3.531×10^{-5}
cubic centimeters	cu. in.	0.06102
cubic centimeters	cu. meters	10^{-6}
cubic centimeters	cu. yards	1.308×10^{-6}
cubic centimeters	gallons (U.S. liq.)	2.642×10^{-4}
cubic centimeters	liters	0.001
cubic centimeters	pints (U.S. liq.)	2.113×10^{-3}
cubic centimeters	quarts (U.S. liq.)	1.057×10^{-3}
cubic feet	bushels (dry)	0.8036
cubic feet	cu. cms	28,320.0
cubic feet	cu. in.	1,728.0
cubic feet	cu. meters	0.02832
cubic feet	cu. yards	0.03704
cubic feet	gallons (U.S. liq.)	7.48052
cubic feet	liters	28.32
cubic feet	pints (U.S. liq.)	59.84
cubic feet	quarts (U.S. liq.)	29.92
cubic feet/min	cu. cms/sec	472.0
cubic feet/min	gallons/sec	0.1247
cubic feet/min	liters/sec	0.4720
cubic feet/min	pounds of water/min	62.43
cubic feet/sec	million gals/day	0.646317

Table A.2a
CONVERSION FACTORS *continued*

To Convert	Into	Multiply by
cubic feet/sec	gallons/min	448.831
cubic inches	cu. cms	16.39
cubic inches	cu. feet	5.787×10^{-4}
cubic inches	cu. meters	1.639×10^{-5}
cubic inches	cu. yards	2.143×10^{-5}
cubic inches	gallons	4.329×10^{-3}
cubic inches	liters	0.01639
cubic inches	mil-feet	1.061×10^{5}
cubic inches	pints (U.S. liq.)	0.03463
cubic inches	quarts (U.S. liq.)	0.01732
cubic meters	bushels (dry)	28.38
cubic meters	cu. cms	10^{6}
cubic meters	cu. ft.	35.31
cubic meters	cu. in.	61,023.0
cubic meters	cu. yards	1.308
cubic meters	gallons (U.S. liq.)	264.2
cubic meters	liters	1,000.0
cubic meters	pints (U.S. liq.)	2,113.0
cubic meters	quarts (U.S. liq.)	1,057.
cubic yards	cu. cms	7.646×10^{5}
cubic yards	cu. ft.	27.0
cubic yards	cu. in.	46,656.0
cubic yards	cu. meters	0.7646
cubic yards	gallons (U.S. liq.)	202.0
cubic yards	liters	764.6
cubic yards	pints (U.S. liq.)	1,615.9
cubic yards	quarts (U.S. liq.)	807.9
cubic yards/min	cu. ft/sec	0.45
cubic yards/min	gallons/sec	3.367
cubic yards/min	liters/sec	12.74
D		
dalton	gram	1.650×10^{-24}
days	seconds	86,400.0
decigrams	grams	0.1
deciliters	liters	0.1
decimeters	meters	0.1
degrees (angle)	quadrants	0.01111
degrees (angle)	radians	0.01745
degrees (angle)	seconds	3,600.0
degrees/sec	radians/sec	0.01745
degrees/sec	revolutions/min	0.1667
degrees/sec	revolutions/sec	2.778×10^{-3}
dekagrams	grams	10.0
dekaliters	liters	10.0
dekameters	meters	10.0
drams (apothecaries' or troy)	ounces (avoidupois)	0.1371429
drams (apothecaries' or troy)	ounces (troy)	0.125
drams (U.S., fluid or apothecaries)	cu. cm	3.6967
drams	grams	1.7718
drams	grains	27.3437
drams	ounces	0.0625
dyne/cm	erg/sq. millimeter	.01
dyne/sq. cm	atmospheres	9.869×10^{-7}
dyne/sq. cm	inch of mercury at 0°C	2.953×10^{-5}
dyne/sq. cm	inch of water at 4°C	4.015×10^{-4}
dynes	grams	1.020×10^{-3}
dynes	joules/cm	10^{-7}
dynes	joules/meter (newtons)	10^{-5}
dynes	kilograms	1.020×10^{-6}
dynes	poundals	7.233×10^{-5}
dynes	pounds	2.248×10^{-6}
dynes/sq. cm	bars	10^{-6}
E		
ell	cm	114.30
ell	in.	45
em, pica	in.	.167
em, pica	cm	.4233
erg/sec	dyne–cm/sec	1.000
ergs	Btu	9.480×10^{-11}
ergs	dyne-centimeters	1.0
ergs	foot-pounds	7.367×10^{-8}
ergs	gram-calories	0.2389×10^{-7}
ergs	gram-cms	1.020×10^{-3}
ergs	horsepower-hrs	3.7250×10^{-14}
ergs	joules	10^{-7}
ergs	kg-calories	2.389×10^{-11}
ergs	kg-meters	1.020×10^{-8}
ergs	kilowatt-hrs	0.2778×10^{-13}
ergs	watt-hours	0.2778×10^{-10}
ergs/sec	Btu/min	$5,688 \times 10^{-9}$
ergs/sec	ft-lbs/min	4.427×10^{-6}
ergs/sec	ft-lbs/sec	7.3756×10^{-8}
ergs/sec	horsepower	1.341×10^{-10}
ergs/sec	kg-calories/min	1.433×10^{-9}
ergs/sec	kilowatts	10^{-10}
F		
farads	microfarads	10^{6}
faraday/sec	ampere (absolute)	9.6500×10^{4}
faradays	ampere-hours	26.80
faradays	coulombs	9.649×10^{4}
fathom	meter	1.828804
fathoms	feet	6.0
feet	centimeters	30.48
feet	kilometers	3.048×10^{-4}
feet	meters	0.3048

Table A.2a
CONVERSION FACTORS *continued*

To Convert	Into	Multiply by
feet	miles (naut.)	1.645×10^{-4}
feet	miles (stat.)	1.894×10^{-4}
feet	millimeters	304.8
feet	mils	1.2×10^{4}
feet of water	atmospheres	0.02950
feet of water	in. of mercury	0.8826
feet of water	kgs/sq. cm	0.03048
feet of water	kgs/sq. meter	304.8
feet of water	pounds/sq. ft.	62.43
feet of water	pounds/sq. in.	0.4335
feet/min	cms/sec	0.5080
feet/min	ft/sec	0.01667
feet/min	kms/hr	0.01829
feet/min	meters/min	0.3048
feet/min	miles/hr	0.01136
feet/sec	cms/sec	30.48
feet/sec	kms/hr	1.097
feet/sec	knots	0.5921
feet/sec	meters/min	18.29
feet/sec	miles/hr	0.6818
feet/sec	miles/min	0.01136
feet/sec/sec	cms/sec/sec	30.48
feet/sec/sec	kms/hr/sec	1.097
feet/sec/sec	meters/sec/sec	0.3048
feet/sec/sec	miles/hr/sec	0.6818
feet/100 feet	per cent grade	1.0
foot–candle	lumen/sq. meter	10.764
foot-pounds	Btu	1.286×10^{-3}
foot-pounds	ergs	1.356×10^{7}
foot-pounds	gram-calories	0.3238
foot-pounds	hp-hrs	5.050×10^{-7}
foot-pounds	joules	1.356
foot-pounds	kg-calories	3.24×10^{-4}
foot-pounds	kg-meters	0.1383
foot-pounds	kilowatt-hrs	3.766×10^{-7}
foot-pounds/min	Btu/min	1.286×10^{-3}
foot-pounds/min	foot-pounds/sec	0.01667
foot-pounds/min	horsepower	3.030×10^{-5}
foot-pounds/min	kg-calories/min	3.24×10^{-4}
foot-pounds/min	kilowatts	2.260×10^{-5}
foot-pounds/sec	Btu/hr	4.6263
foot-pounds/sec	Btu/min	0.07717
foot-pounds/sec	horsepower	1.818×10^{-3}
foot-pounds/sec	kg-calories/min	0.01945
foot-pounds/sec	kilowatts	1.356×10^{-3}
furlongs	miles (U.S.)	0.125
furlongs	rods	40.0
furlongs	feet	660.0
	G	
gallons	cu. cms	3,785.0
gallons	cu. ft.	0.1337
gallons	cu. in.	231.0
gallons	cu. meters	3.785×10^{-3}
gallons	cu. yards	4.951×10^{-3}
gallons	liters	3.785
gallons (liq. Br. Imp.)	gallons (U.S. liq.)	1.20095
gallons (U.S.)	gallons (Imp.)	0.83267
gallons of water	pounds of water	8.3453
gallons/min	cu. ft/sec	2.228×10^{-3}
gallons/min	liters/sec	0.06308
gallons/min	cu. ft/hr	8.0208
gausses	lines/sq. in.	6.452
gausses	webers/sq. cm.	10^{-8}
gausses	webers/sq. in.	6.452×10^{-8}
gausses	webers/sq. meter	10^{-4}
gilberts	ampere-turns	0.7958
gilberts/cm	amp-turns/cm	0.7958
gilberts/cm	amp-turns/in.	2.021
gilberts/cm	amp-turns/meter	79.58
gills (British)	cu. cm	142.07
gills	liters	0.1183
gills	pints (liq.)	0.25
grade	radian	.01571
grains	drams (avoirdupois)	0.03657143
grains (troy)	grains (avdp)	1.0
grains (troy)	grams	0.06480
grains (troy)	ounces (avdp)	2.0833×10^{-3}
grains (troy)	pennyweight (troy)	0.04167
grains/U.S. gal	parts/million	17.118
grains/U.S. gal	pouns/million gal	142.86
grains/Imp. gal	parts/million	14.286
grams	dynes	980.7
grams	grains	15.43
grams	joules/cm	9.807×10^{-5}
grams	joules/meter (newtons)	9.807×10^{-3}
grams	kilograms	0.001
grams	milligrams	1,000.
grams	ounces (avdp)	0.03527
grams	ounces (troy)	0.03215
grams	poundals	0.07093
grams	pounds	2.205×10^{-3}
grams/cm	pounds/inch	5.600×10^{-3}
grams/cu. cm	pounds/cu. ft.	62.43
grams/cu. cm	pounds/cu. in.	0.03613
grams/cu. cm	pounds/mil-foot	3.405×10^{-7}
grams/liter	grains/gal	58.417
grams/liter	pounds/1,000 gal	8.345
grams/liter	pounds/cu. ft.	0.062427
grams/liter	parts/million	1,000.0

Table A.2a
CONVERSION FACTORS *continued*

To Convert	Into	Multiply by
grams/sq. cm	pounds/sq. ft.	2.0481
gram-calories	Btu	3.9683×10^{-3}
gram-calories	ergs	4.1868×10^{7}
gram-calories	foot-pounds	3.0880
gram-calories	horsepower-hrs	1.5596×10^{-6}
gram-calories	kilowatt-hrs	1.1630×10^{-6}
gram-calories	watt-hrs	1.1630×10^{-3}
gram-calories/sec	Btu/hr	14.286
gram-centimeters	Btu	9.297×10^{-8}
gram-centimeters	ergs	980.7
gram-centimeters	joules	9.807×10^{-5}
gram-centimeters	kg-cal	2.343×10^{-8}
gram-centimeters	kg-meters	10^{-5}
H		
hand	cm	10.16
hectares	acres	2.471
hectares	sq. feet	1.076×10^{5}
hectograms	grams	100.0
hectoliters	liters	100.0
hectometers	meters	100.0
hectowatts	watts	100.0
henries	millihenries	1,000.0
hogsheads (British)	cu. ft.	10.114
hogsheads (U.S.)	cu. ft.	8.42184
hogsheads (U.S.)	gallons (U.S.)	63
horsepower	Btu/min	42.44
horsepower	ft-lbs/min	33,000.
horsepower	ft-lbs/sec	550.0
horsepower (metric) (542.5 ft lb/sec)	horsepower (550 ft-lb/sec)	0.9863
horsepower (550 ft lb/sec)	horsepower (metric) (542.5 ft-lb/sec)	1.014
horsepower	kg-calories/min	10.68
horsepower	kilowatts	0.7457
horsepower	watts	745.7
horsepower (boiler)	Btu/hr	33.479
horsepower (boiler)	kilowatts	9.803
horsepower-hrs	Btu	2,547.
horsepower-hrs	ergs	2.6845×10^{13}
horsepower-hrs	ft-lbs	1.98×10^{6}
horsepower-hrs	gram-calories	641,190.
horsepower-hrs	joules	2.684×10^{6}
horsepower-hrs	kg-calories	641.1
horsepower-hrs	kg-meters	2.737×10^{5}
horsepower-hrs	kilowatt-hrs	0.7457
hours	days	4.167×10^{-2}
hours	weeks	5.952×10^{-3}
hundredweights (long)	pounds	112
hundredweights (long)	tons (long)	0.05
hundredweights (short)	ounces (avoirdupois)	1,600
hundredweights (short)	pounds	100
hundredweights (short)	tons (metric)	0.0453592
hundredweights (short)	tons (long)	0.0446429
I		
inches	centimeters	2.540
inches	meters	2.540×10^{-2}
inches	miles	1.578×10^{-5}
inches	millimeters	25.40
inches	mils	1,000.0
inches	yards	2.778×10^{-2}
inches of mercury	atmospheres	0.03342
inches of mercury	feet of water	1.133
inches of mercury	kgs/sq. cm	0.03453
inches of mercury	kgs/sq. meter	345.3
inches of mercury	pounds/sq. ft.	70.73
inches of mercury	pounds/sq. in.	0.4912
inches of water (at 4°C)	atmospheres	2.458×10^{-3}
inches of water (at 4°C)	inches of mercury	0.07355
inches of water (at 4°C)	kgs/sq. cm	2.540×10^{-3}
inches of water (at 4°C)	ounces/sq. in.	0.5781
inches of water (at 4°C)	pounds/sq. ft.	5.204
inches of water (at 4°C)	pounds/sq. in.	0.03613
International ampere	ampere (absolute)	.9998
International volt	volts (absolute)	1.0003
J		
joules	Btu	9.480×10^{-4}
joules	ergs	10^{7}
joules	foot-pounds	0.7376
joules	kg-calories	2.389×10^{-4}
joules	kg-meters	0.1020
joules	watt-hrs	2.778×10^{-4}
joules/cm	grams	1.020×10^{4}
joules/cm	dynes	10^{7}
joules/cm	joules/meter (newtons)	100.0
joules/cm	poundals	723.3
joules/cm	pounds	22.48

Table A.2a
CONVERSION FACTORS *continued*

To Convert	Into	Multiply by
K		
kilograms	dynes	980,665.
kilograms	grams	1,000.0
kilograms	joules/cm	0.09807
kilograms	joules/meter (newtons)	9.807
kilograms	poundals	70.93
kilograms	pounds	2.205
kilograms	tons (long)	9.842×10^{-4}
kilograms	tons (short)	1.102×10^{-3}
kilograms/cu. meter	grams/cu. cm	0.001
kilograms/cu. meter	pounds/cu. ft.	0.06243
kilograms/cu. meter	pounds/cu. in.	3.613×10^{-5}
kilograms/cu. meter	pounds/mil-foot	3.405×10^{-10}
kilograms/meter	pounds/ft.	0.6720
kilogram/sq. cm	dynes	980,665
kilograms/sq. cm	atmospheres	0.9678
kilograms/sq. cm	feet of water	32.81
kilograms/sq. cm	inches of mercury	28.96
kilograms/sq. cm	pounds/sq. ft.	2,048.
kilograms/sq. cm	pounds/sq. in.	14.22
kilograms/sq. meter	atmospheres	9.678×10^{-5}
kilograms/sq. meter	bars	98.07×10^{-6}
kilograms/sq. meter	feet of water	3.281×10^{-3}
kilograms/sq. meter	inches of mercury	2.896×10^{-3}
kilograms/sq. meter	pounds/sq. ft.	0.2048
kilograms/sq. meter	pounds/sq. in.	1.422×10^{-3}
kilograms/sq. mm	kgs/sq. meter	10^{6}
kilogram-calories	Btu	3.968
kilogram-calories	foot-pounds	3,088.
kilogram-calories	hp-hrs	1.560×10^{-3}
kilogram-calories	joules	4,186.
kilogram-calories	kg-meters	426.9
kilogram-calories	kilojoules	4.186
kilogram-calories	kilowatt-hrs	1.163×10^{-3}
kilogram meters	Btu	9.294×10^{-3}
kilogram meters	ergs	9.804×10^{7}
kilogram meters	foot-pounds	7.233
kilogram meters	joules	9.804
kilogram meters	kg-calories	2.342×10^{-3}
kilogram meters	kilowatt-hrs	2.723×10^{-6}
kilolines	maxwells	1,000.0
kiloliters	liters	1,000.0
kilometers	centimeters	10^{5}
kilometers	feet	3,281.
kilometers	inches	3.937×10^{4}
kilometers	meters	1,000.0
kilometers	miles	0.6214
kilometers	millimeters	10^{6}
kilometers	yards	1,094.
kilometers/hr	cms/sec	27.78
kilometers/hr	ft/min	54.68
kilometers/hr	ft/sec	0.9113
kilometers/hr	knots	0.5396
kilometers/hr	meters/min	16.67
kilometers/hr	miles/hr	0.6214
kilometers/hr/sec	cms/sec/sec	27.78
kilometers/hr/sec	ft/sec/sec	0.9113
kilometers/hr/sec	meters/sec/sec	0.2778
kilometers/hr/sec	miles/hr/sec	0.6214
kilowatts	Btu/min	56.92
kilowatts	ft-lbs/min	4.426×10^{4}
kilowatts	ft-lbs/sec	737.6
kilowatts	horsepower	1.341
kilowatts	kg-calories/min	14.34
kilowatts	watts	1,000.0
kilowatt-hrs	Btu	3,413.
kilowatt-hrs	ergs	3.600×10^{13}
kilowatt-hrs	ft-lbs	2.655×10^{6}
kilowatt-hrs	gram-calories	859,850.
kilowatt-hrs	horsepower-hrs	1,341
kilowatt-hrs	joules	3.6×10^{6}
kilowatt-hrs	kg-calories	860.5
kilowatt-hrs	kg-meters	3.671×10^{5}
kilowatt-hrs	pounds of water evaporated from and at 212°F.	3.53
kilowatt-hrs	pounds of water raised from 62° to 212°F	22.75
knots	ft/hr	6,080.
knots	kilometers/hr	1.8532
knots	nautical miles/hr	1.0
knots	statute miles/hr	1.151
knots	yards/hr	2,027.
knots	ft/sec	1.689
L		
league	miles (approx.)	3.0
light year	miles	5.9×10^{12}
light year	kilometers	9.46091×10^{12}
lines/sq. cm	gausses	1.0
lines/sq. in.	gausses	0.1550
lines/sq. in.	webers/sq. cm	1.550×10^{-9}
lines/sq. in.	webers/sq. in.	10^{-8}
lines/sq. in.	webers/sq. meter	1.550×10^{-5}
links (engineer's)	inches	12.0
links (surveyor's)	inches	7.92
liters	bushels (U.S. dry)	0.02833
liters	cu. cm	1,000.0
liters	cu. ft.	0.03531
liters	cu. in.	61.02
liters	cu. meters	0.001

Table A.2a
CONVERSION FACTORS *continued*

To Convert	Into	Multiply by
liters	cu. yards	1.308×10^{-3}
liters	gallons (U.S. liq.)	0.2642
liters	pints (U.S. liq.)	2.113
liters	quarts (U.S. liq.)	1.057
liters/min	cu. ft/sec	5.886×10^{-4}
liters/min	gals/sec	4.403×10^{-3}
lumens/sq. ft.	foot-candles	1.0
lumen	spherical candle power	.07958
lumen	watt	.001496
lumen/sq. ft.	lumen/sq. meter	10.76
lux	foot-candles	0.0929
M		
maxwells	kilolines	0.001
maxwells	webers	10^{-8}
megalines	maxwells	10^{6}
megohms	microhms	10^{12}
megohms	ohms	10^{6}
meters	centimeters	100.0
meters	feet	3.281
meters	inches	39.37
meters	kilometers	0.001
meters	miles (naut.)	5.396×10^{-4}
meters	miles (stat.)	6.214×10^{-4}
meters	millimeters	1,000.0
meters	yards	1.094
meters	varas	1.179
meters/min	cms/sec	1.667
meters/min	ft/min	3.281
meters/min	ft/sec	0.05468
meters/min	kms/hr	0.06
meters/min	knots	0.03238
meters/min	miles/hr	0.03728
meters/sec	ft/min	196.8
meters/sec	ft/sec	3.281
meters/sec	kilometers/hr	3.6
meters/sec	kilometers/min	0.06
meters/sec	miles/hr	2.237
meters/sec	miles/min	0.03728
meters/sec/sec	cms/sec/sec	100.0
meters/sec/sec	ft/sec/sec	3.281
meters/sec/sec	kms/hr/sec	3.6
meters/sec/sec	miles/hr/sec	2.237
meter-kilograms	cm-dynes	9.807×10^{7}
meter-kilograms	cm-grams	10^{5}
meter-kilograms	pound-feet	7.233
microfarad	farads	10^{-6}
micrograms	grams	10^{-6}
microhms	megohms	10^{-12}
microhms	ohms	10^{-6}
microliters	liters	10^{-6}
microns	meters	1×10^{-6}
miles (naut.)	miles (statute)	1.1516
miles (naut.)	yards	2,027.
miles (statute)	centimeters	1.609×10^{5}
miles (statute)	feet	5,280.
miles (statute)	inches	6.336×10^{4}
miles (statute)	kilometers	1.609
miles (statute)	meters	1,609.
miles (statute)	miles (naut.)	0.8684
miles (statute)	yards	1,760.
miles/hr	cms/sec	44.70
miles/hr	ft/min	88.
miles/hr	ft/sec	1.467
miles/hr	kms/hr	1.609
miles/hr	kms/min	0.02682
miles/hr	knots	0.8684
miles/hr	meters/min	26.82
miles/hr	miles/min	0.1667
miles/hr/sec	cm/sec/sec	44.70
miles/hr/sec	ft/sec/sec	1.467
miles/hr/sec	kms/hr/sec	1.609
miles/hr/sec	meters/sec/sec	0.4470
miles/min	cms/sec	2,682.
miles/min	ft/sec	88.
miles/min	kms/min	1.609
miles/min	knots/min	0.8684
miles/min	miles/hr	60.0
mil-feet	cu. in.	9.425×10^{-6}
milliers	kilograms	1,000.
millimicrons	meters	1×10^{-9}
milligrams	grains	0.01543236
milligrams	grams	0.001
milligrams/liter	parts/million	1.0
millihenries	henries	0.001
milliliters	liters	0.001
millimeters	centimeters	0.1
millimeters	feet	3.281×10^{-3}
millimeters	inches	0.03937
millimeters	kilometers	10^{-6}
millimeters	meters	0.001
millimeters	miles	6.214×10^{-7}
millimeters	mils	39.37
millimeters	yards	1.094×10^{-3}
million gals/day	cu. ft/sec	1.54723
mils	centimeters	2.540×10^{-3}
mils	feet	8.333×10^{-5}
mils	inches	0.001
mils	kilometers	2.540×10^{-8}
mils	yards	2.778×10^{-5}
miner's inches	cu. ft/min	1.5
minims (British)	cu. cm	0.059192

Table A.2a
CONVERSION FACTORS *continued*

To Convert	Into	Multiply by
minims (U.S., fluid)	cu. cm	0.061612
minutes (angles)	degrees	0.01667
miles (naut.)	feet	6,080.27
miles (naut.)	kilometers	1.853
miles (naut.)	meters	1,853.
minutes (angles)	quadrants	1.852×10^{-4}
minutes (angles)	radians	2.909×10^{-4}
minutes (angles)	seconds	60.0
myriagrams	kilograms	10.0
myriameters	kilometers	10.0
myriawatts	kilowatts	10.0
	N	
nepers	decibels	8.686
Newton	dynes	1 × 105
	O	
ohm (International)	ohm (absolute)	1.0005
ohms	megohms	10^{-6}
ohms	microhms	10^{6}
ounces	drams	16.0
ounces	grains	437.5
ounces	grams	28.349527
ounces	pounds	0.0625
ounces	ounces (troy)	0.9115
ounces	tons (long)	2.790×10^{-5}
ounces	tons (metric)	2.835×10^{-5}
ounces (fluid)	cu. in.	1.805
ounces (fluid)	liters	0.02957
ounces (troy)	grains	480.0
ounces (troy)	grams	31.103481
ounces (troy)	ounces (avdp.)	1.09714
ounces (troy)	pennyweights (troy)	20.0
ounces (troy)	pounds (troy)	0.08333
ounce/sq. in.	dynes/sq. cm	4309
ounces/sq. in.	pounds/sq. in.	0.0625
	P	
parsec	miles	19×10^{12}
parsec	kilometers	3.084×10^{13}
parts/million	grains/U.S. gal	0.0584
parts/million	grains/Imp. gal	0.07016
parts/million	pounds/million gal	8.345
pecks (British)	cu. in.	554.6
pecks (British)	liters	9.091901
pecks (U.S.)	bushels	0.25
pecks (U.S.)	cu. in.	537.605
pecks (U.S.)	liters	8.809582
pecks (U.S.)	quarts (dry)	8
pennyweights (troy)	grains	24.0

To Convert	Into	Multiply by
pennyweights (troy)	ounces (troy)	0.05
pennyweights (troy)	grams	1.55517
pennyweights (troy)	pounds (troy)	4.1667×10^{-3}
pints (dry)	cu. in.	33.60
pints (liq.)	cu. cms	473.2
pints (liq.)	cu. ft.	0.01671
pints (liq.)	cu. in.	28.87
pints (liq.)	cu. meters	4.732×10^{-4}
pints (liq.)	cu. yards	6.189×10^{-4}
pints (liq.)	gallons	0.125
pints (liq.)	liters	0.4732
pints (liq.)	quarts (liq.)	0.5
Planck's quantum	erg/sec	6.624×10^{-27}
poise	gram/cm sec	1.00
pounds (avoirdupois)	ounces (troy)	14.5833
poundals	dynes	13,826.
poundals	grams	14.10
poundals	joules/cm	1.383×10^{-3}
poundals	joules/meter (newtons)	0.1383
poundals	kilograms	0.01410
poundals	pounds	0.03108
pounds	drams	256.
pounds	dynes	44.4823×10^{4}
pounds	grains	7,000.
pounds	grams	453.5924
pounds	joules/cm	0.04448
pounds	joules/meter (newtons)	4.448
pounds	kilograms	0.4536
pounds	ounces	16.0
pounds	ounces (troy)	14.5833
pounds	poundals	32.17
pounds	pounds (troy)	1.21528
pounds	tons (short)	0.0005
pounds (troy)	grains	5,760.
pounds (troy)	grams	373.24177
pounds (troy)	ounces (avdp.)	13.1657
pounds (troy)	ounces (troy)	12.0
pounds (troy)	pennyweights (troy)	240.0
pounds (troy)	pounds (avdp.)	0.822857
pounds (troy)	tons (long)	3.6735×10^{-4}
pounds (troy)	tons (metric)	3.7324×10^{-4}
pounds (troy)	tons (short)	4.1143×10^{-4}
pounds of water	cu. ft.	0.01602
pounds of water	cu. in.	27.68
pounds of water	gallons	0.1198
pounds of water/min	cu. ft/sec	2.670×10^{-4}
pound-feet	cm-dynes	1.356×10^{7}
pound-feet	cm-grams	13,825.
pound-feet	meter-kgs	0.1383

Table A.2a
CONVERSION FACTORS *continued*

To Convert	*Into*	*Multiply by*
pounds/cu. ft.	grams/cu. cm	0.01602
pounds/cu. ft.	kgs/cu. meter	16.02
pounds/cu. ft.	pounds/cu. in.	5.787×10^{-4}
pounds/cu. ft.	pounds/mil-foot	5.456×10^{-9}
pounds/cu. in.	gms/cu. cm	27.68
pounds/cu. in.	kgs/cu. meter	2.768×10^{4}
pounds/cu. in.	pounds/cu. ft.	1,728.
pounds/cu. in.	pounds/mil-foot	9.425×10^{-6}
pounds/ft.	kgs/meter	1.488
pounds/in.	gms/cm	178.6
pounds/mil-foot	gms/cu. cm	2.306×10^{6}
pounds/sq. ft.	atmospheres	4.725×10^{-4}
pounds/sq. ft.	feet of water	0.01602
pounds/sq. ft.	inches of mercury	0.01414
pounds/sq. ft.	kgs/sq. meter	4.882
pounds/sq. ft.	pounds/sq. in.	6.944×10^{-3}
pounds/sq. in.	atmospheres	0.06804
pounds/sq. in.	feet of water	2.307
pounds/sq. in.	inches of mercury	2.036
pounds/sq. in.	kgs/sq. meter	703.1
pounds/sq. in.	pounds/sq. ft.	144.0

Q

To Convert	*Into*	*Multiply by*
quadrants (angle)	degrees	90.0
quadrants (angle)	minutes	5,400.0
quadrants (angle)	radians	1.571
quadrants (angle)	seconds	3.24×10^{5}
quarts (dry)	cu. in.	67.20
quarts (liq.)	cu. cms	946.4
quarts (liq.)	cu. ft.	0.03342
quarts (liq.)	cu. in.	57.75
quarts (liq.)	cu. meters	9.464×10^{-4}
quarts (liq.)	cu. yards	1.238×10^{-3}
quarts (liq.)	gallons	0.25
quarts (liq.)	liters	0.9463

R

To Convert	*Into*	*Multiply by*
radians	degrees	57.30
radians	minutes	3,438.
radians	quadrants	0.6366
radians	seconds	2.063×10^{5}
radians/sec	degrees/sec	57.30
radians/sec	revolutions/min	9.549
radians/sec	revolutions/sec	0.1592
radians/sec/sec	revs/min/min	573.0
radians/sec/sec	revs/min/sec	9.549
radians/sec/sec	revs/sec/sec	0.1592
revolutions	degrees	360.0
revolutions	quadrants	4.0
revolutions	radians	6.283
revolutions/min	degrees/sec	6.0
revolutions/min	radians/sec	0.1047
revolutions/min	revs/sec	0.01667
revolutions/min/min	radians/sec/sec	1.745×10^{-3}
revolutions/min/min	revs/min/sec	0.01667
revolutions/min/min	revs/sec/sec	2.778×10^{-4}
revolutions/sec	degrees/sec	360.0
revolutions/sec	radians/sec	6.283
revolutions/sec	revs/min	60.0
revolutions/sec/sec	radians/sec/sec	6.283
revolutions/sec/sec	revs/min/min	3,600.0
revolutions/sec/sec	revs/min/sec	60.0
rod	chain (Gunters)	.25
rod	meters	5.029
rods (Surveyors' meas.)	yards	5.5
rods	feet	16.5

S

To Convert	*Into*	*Multiply by*
scruples	grains	20
seconds (angle)	degrees	2.778×10^{-4}
seconds (angle)	minutes	0.01667
seconds (angle)	quadrants	3.087×10^{-6}
seconds (angle)	radians	4.848×10^{-6}
slug	kilogram	14.59
slug	pounds	32.17
sphere	steradians	12.57
square centimeters	circular mils	1.973×10^{5}
square centimeters	sq. ft.	1.076×10^{-3}
square centimeters	sq. in.	0.1550
square centimeters	sq. meters	0.0001
square centimeters	sq. miles	3.861×10^{-11}
square centimeters	sq. millimeters	100.0
square centimeters	sq. yards	1.196×10^{-4}
square feet	acres	2.296×10^{-5}
square feet	circular mils	1.833×10^{8}
square feet	sq. cms	929.0
square feet	sq. in.	144.0
square feet	sq. meters	0.09290
square feet	sq. miles	3.587×10^{-8}
square feet	sq. millimeters	9.290×10^{4}
square feet	sq. yards	0.1111
square inches	circular mils	1.273×10^{6}
square inches	sq. cms	6.452
square inches	sq. ft.	6.944×10^{-3}
square inches	sq. millimeters	645.2
square inches	sq. mils	10^{6}
square inches	sq. yards	7.716×10^{-4}
square kilometers	acres	247.1
square kilometers	sq. cms	10^{10}
square kilometers	sq. ft.	10.76×10^{6}
square kilometers	sq. in.	1.550×10^{9}
square kilometers	sq. meters	10^{6}
square kilometers	sq. miles	0.3861
square kilometers	sq. yards	1.196×10^{6}

Table A.2a
CONVERSION FACTORS *continued*

To Convert	*Into*	*Multiply by*
square meters	acres	2.471×10^{-4}
square meters	sq. cms	10^{4}
square meters	sq. ft.	10.76
square meters	sq. in.	1,550.
square meters	sq. miles	3.861×10^{-7}
square meters	sq. millimeters	10^{6}
square meters	sq. yards	1.196
square miles	acres	640.0
square miles	sq. ft.	27.88×10^{6}
square miles	sq. kms	2.590
square miles	sq. meters	2.590×10^{6}
square miles	sq. yards	3.098×10^{6}
square millimeters	circular mils	1,973.
square millimeters	sq. cms	0.01
square millimeters	sq. ft.	1.076×10^{-5}
square millimeters	sq. in.	1.550×10^{-3}
square mils	circular mils	1.273
square mils	sq. cms	6.452×10^{-6}
square mils	sq. in.	10^{-6}
square yards	acres	2.066×10^{-4}
square yards	sq. cms	8,361.
square yards	sq. ft.	9.0
square yards	sq. in.	1,296.
square yards	sq. meters	0.8361
square yards	sq. miles	3.228×10^{-7}
square yards	sq. millimeters	8.361×10^{5}
T		
temperature (°C) +273	absolute temperature (°K)	1.0
temperature (°C) +17.78	temperature (°F)	1.8
temperature (°F) + 460	absolute temperature (°R)	1.0
temperature (°F) −32	temperature (°C)	5/9
tons (long)	kilograms	1,016.
tons (long)	pounds	2,240.
tons (long)	tons (short)	1.120
tons (metric)	kilograms	1,000.
tons (metric)	pounds	2,205
tons (short)	kilograms	907.1848
tons (short)	ounces	32,000.
tons (short)	ounces (troy)	29,166.66
tons (short)	pounds	2,000.
tons (short)	pounds (troy)	2,430.56
tons (short)	tons (long)	0.89287
tons (short)	tons (metric)	0.9078
tons (short)/sq. ft.	kgs/sq. meter	9,765.
tons (short)/sq. ft.	pounds/sq. in.	2,000.
tons of water/24 hrs	pounds of water/hr	83.333
tons of water/24 hrs	gallons/min	0.16643
tons of water/24 hrs	cu. ft/hr	1.3349
V		
volt/inch	volt/cm	.39370
volt (absolute)	statvolts	.003336
W		
watts	Btu/hr	3.4129
watts	Btu/min	0.05688
watts	ergs/sec	107.
watts	ft-lbs/min	44.27
watts	ft-lbs/sec	0.7378
watts	horsepower	1.341×10^{-3}
watts	horsepower (metric)	1.360×10^{-3}
watts	kg-calories/min	0.01433
watts	kilowatts	0.001
watts (abs.)	Btu (mean)/min	0.056884
watts (abs.)	joules/sec	1.
watt-hours	Btu	3.413
watt-hours	ergs	3.60×10^{10}
watt-hours	foot-pounds	2,656.
watt-hours	gram-calories	859.85
watt-hours	horsepower-hrs	1.341×10^{-3}
watt-hours	kilogram-calories	0.8605
watt-hours	kilogram-meters	367.2
watt-hours	kilowatt-hrs	0.001
watt (International)	watt (absolute)	1.0002
webers	maxwells	10^{8}
webers	kilolines	10^{5}
webers/sq. in.	gausses	1.550×10^{7}
webers/sq. in.	lines/sq. in.	10^{8}
webers/sq. in.	webers/sq. cm	0.1550
webers/sq. in.	webers/sq. meter	1,550.
webers/sq. meter	gausses	10^{4}
webers/sq. meter	lines/sq. in.	6.452×10^{4}
webers/sq. meter	webers/sq. cm	10^{-4}
webers/sq. meter	webers/sq. in.	6.452×10^{-4}
Y		
yards	centimeters	91.44
yards	kilometers	9.144×10^{-4}
yards	meters	0.9144
yards	miles (naut.)	4.934×10^{-4}
yards	miles (stat.)	5.682×10^{-4}
yards	millimeters	914.4

Table A.2b
PRESSURE CONVERSION

1 pound per square inch = .0703 kilograms per square centimeter

Lbs. Per Sq. In.	Kgs Per Sq. Cm	Lbs. Per Sq. In.	Kgs Per Sq. Cm	Lbs. Per Sq. In.	Kgs Per Sq. Cm	Lbs. Per Sq. In.	Kgs Per Sq. Cm
1.00	.0703	2.25	.1582	4.50	.3164	8.25	.5800
1.10	.0773	2.30	.1617	4.75	.3339	8.50	.5976
1.20	.0844	2.40	.1687	5.00	.3515	8.75	.6151
1.25	.0879	2.50	.1758	5.25	.3691	9.00	.6327
1.30	.0914	2.60	.1828	5.50	.3867	9.25	.6503
1.40	.0984	2.70	.1898	5.75	.4042	9.50	.6679
1.50	.1055	2.75	.1933	6.00	.4218	9.75	.6854
1.60	.1125	2.80	.1969	6.25	.4394	10.00	.7030
1.70	.1195	2.90	.2039	6.50	.4570		
1.75	.1230	3.00	.2109	6.75	.4746		
1.80	.1265	3.25	.2285	7.00	.4921		
1.90	.1336	3.50	.2461	7.25	.5097		
2.00	.1406	3.75	.2636	7.50	.5273		
2.10	.1476	4.00	.2812	7.75	.5448		
2.20	.1547	4.25	.2988	8.00	.5624		

1 kilogram per square centimeter = 14.223 pounds per square inch

Kgs Per Sq. Cm	Lbs. Per Sq. In.	Kgs Per Sq. Cm	Lbs. Per Sq. In.	Kgs Per Sq. Cm	Lbs. Per Sq. In.	Kgs Per Sq. Cm	Lbs. Per Sq. In.
1.0	14.22	2.5	35.56	4.0	56.89	7.5	106.67
1.1	15.65	2.6	36.98	4.1	58.31	8.0	113.78
1.2	17.07	2.7	38.40	4.2	59.74	8.5	120.90
1.3	18.49	2.8	39.82	4.3	61.16	9.0	128.01
1.4	19.91	2.9	41.25	4.4	62.58	9.5	135.12
1.5	21.33	3.0	42.67	4.5	64.00	10.0	142.23
1.6	22.76	3.1	44.09	4.6	65.43		
1.7	24.18	3.2	45.51	4.7	66.85		
1.8	25.60	3.3	46.94	4.8	68.27		
1.9	27.02	3.4	48.36	4.9	69.69		
2.0	28.45	3.5	49.78	5.0	71.12		
2.1	29.87	3.6	51.20	5.5	78.23		
2.2	31.29	3.7	52.63	6.0	85.34		
2.3	32.71	3.8	54.05	6.5	92.45		
2.4	34.14	3.9	55.47	7.0	99.56		

Table A.2c
TEMPERATURE CONVERSION

Degrees—Fahrenheit to Centigrade

$-°C = 5/9\,(°F + 32)$

$+°C = 5/9\,(°F - 32)$

°F	0	25	50	75
−200	−128.9	−142.8	−156.7	−170.6
−100	−73.3	−87.2	−101.1	−115.0
−0	−17.8	−31.7	−45.6	−59.4
+0	−17.8	−3.9	+10.0	+23.9
100	+37.8	+51.7	65.6	79.4
200	93.3	107.2	121.1	135.0
300	148.9	162.8	176.7	190.6
400	204.4	218.3	232.2	246.1
500	260.0	273.9	287.8	301.7
600	315.6	329.4	343.3	357.2
700	371.1	385.0	398.9	412.8
800	426.7	440.6	454.4	468.3
900	482.2	496.1	510.0	523.9
1000	538.0	551.7	565.6	579.4
1100	593.2	607.2	621.1	635.0

Degrees—Centigrade to Fahrenheit

$-°F = (9/5 \times °C) - 32$

$+°F = (9/5 \times °C) + 32$

°C	0	25	50	75
−200	−328	−373	−418	
−100	−148	−193	−238	−283
−0	+32	−13	−58	−103
+0	+32	+77	+122	+167
100	212	257	302	347
200	392	437	482	527
300	572	617	662	707
400	752	797	842	887
500	932	977	1022	1067
600	1112	1157	1202	1247
700	1292	1337	1382	1427

Table A.2d
VISCOSITY CONVERSION

Kinematic Viscosity Centisokes = K	*Seconds Saybolt Universal*	*Seconds Saybolt Furol*	*Seconds Redwood*	*Seconds Redwood Admiralty*	*Degrees Engler*	*Degrees Barbey*
1.00	31	—	29.1	—	1.00	6200
2.56	35	—	32.1	—	1.16	2420
4.30	40	—	36.2	5.10	1.31	1440
5.90	45	—	40.3	5.52	1.46	1050
7.40	50	—	44.3	5.83	1.58	838
8.83	55	—	48.5	6.35	1.73	702
10.20	60	—	52.3	6.77	1.88	618
11.53	65	—	56.7	7.17	2.03	538
12.83	70	12.95	60.9	7.60	2.17	483
14.10	75	13.33	65.0	8.00	2.31	440
15.35	80	13.70	69.2	8.44	2.45	404
16.58	85	14.10	73.3	8.86	2.59	374
17.80	90	14.44	77.6	9.30	2.73	348
19.00	95	14.85	81.5	9.70	2.88	326
20.20	100	15.24	85.6	10.12	3.02	307
31.80	150	19.3	128	14.48	4.48	195
43.10	200	23.5	170	18.90	5.92	144
54.30	250	28.0	212	23.45	7.35	114
65.40	300	32.5	254	28.0	8.79	95
76.50	350	35.1	296	32.5	10.25	81
87.60	400	41.9	338	37.1	11.70	70.8
98.60	450	46.8	381	41.7	13.15	62.9
110.	500	51.6	423	46.2	14.60	56.4
121.	550	56.6	465	50.8	16.05	51.3
132.	600	61.4	508	55.4	17.50	47.0
143.	650	66.2	550	60.1	19.00	43.4
154.	700	71.1	592	64.6	20.45	40.3
165.	750	76.0	635	69.2	21.90	37.6
176.	800	81.0	677	73.8	23.35	35.2
187.	850	86.0	719	78.4	24.80	33.2
198	900	91.0	762	83.0	26.30	31.3
209	950	95.8	804	87.6	27.70	29.7
220	1000	100.7	846	92.2	29.20	28.2
330	1500	150	1270	138.2	43.80	18.7
440	2000	200	1690	184.2	58.40	14.1
550	2500	250	2120	230	73.00	11.3
660	3000	300	2540	276	87.60	9.4
770	3500	350	2960	322	100.20	8.05
880	4000	400	3380	368	117.00	7.05
990	4500	450	3810	414	131.50	6.26
1100	5000	500	4230	461	146.00	5.64
1210	5500	550	4650	507	160.50	5.13
1320	6000	600	5080	553	175.00	4.70
1430	6500	650	5500	559	190.00	4.34
1540	7000	700	5920	645	204.50	4.03
1650	7500	750	6350	691	219.00	3.76
1760	8000	800	6770	737	233.50	3.52
1870	8500	850	7190	783	248.00	3.32
1980	9000	900	7620	829	263.00	3.13
2090	9500	950	8040	875	277.00	2.97
2200	10000	1000	8460	921	292.00	2.82

The viscosity is oftern expressed in terms of viscosimeters other than the Saybolt Universal. The formulas for the various viscosimeters are given opposite.

If viscosity is given at any two temperatures, the viscosity at any other temperature can be obtained by plotting the viscosity against temperature in degrees Fahrenheit on special Log paper. The points for a given oil lie in a straight line.

$$\text{Kinematic viscosity} = \frac{\text{absolute viscosity}}{\text{specific gravity}}$$

$$\text{Redwood } K = .26t - \frac{180}{t} \text{ (British)}$$

$$\text{Redwood Admiralty } K = 2.7t - \frac{20}{t} \text{ (British)}$$

$$\text{Saybolt Universal } K = .22t - \frac{195}{t} \text{ (American)}$$

$$\text{Saybolt Furol } K = 2.2\,t - \frac{184}{t} \text{ (American)}$$

$$\text{Engler } K = .147t - \frac{374}{t} \text{ (German)}$$

Table A.2e
VISCOSITY CONVERSION CHART

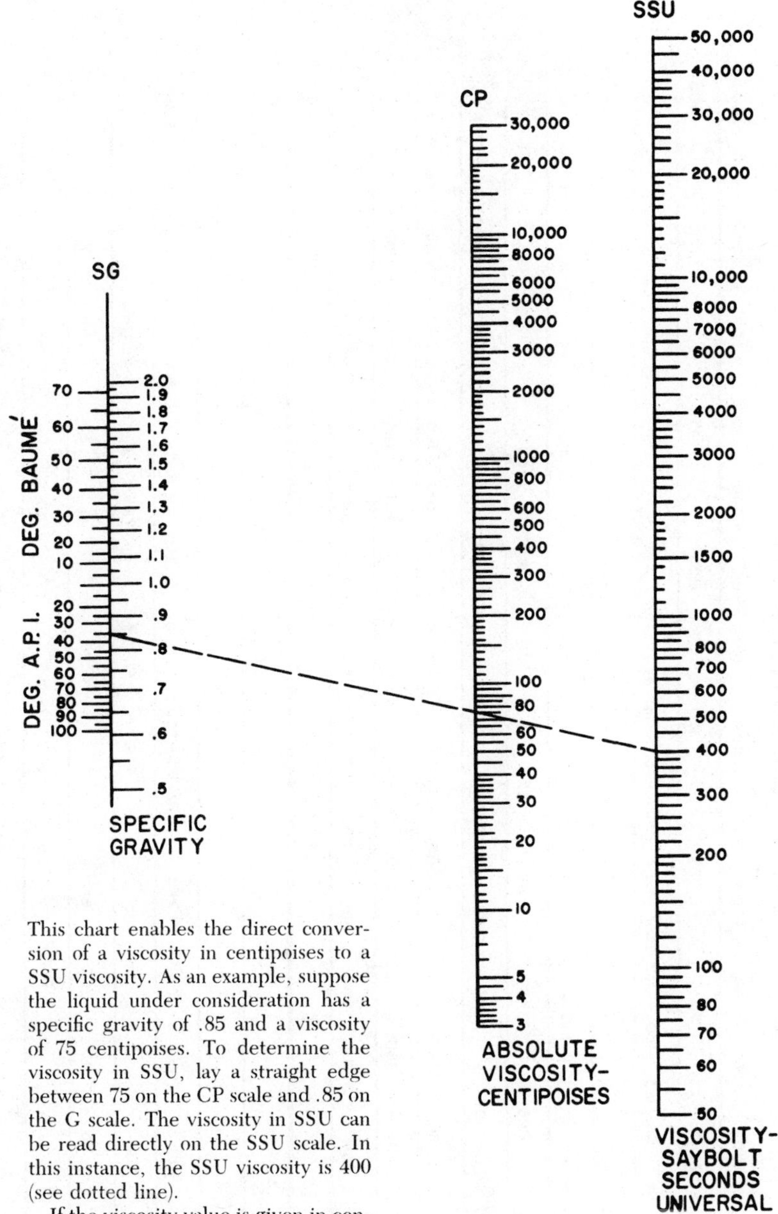

This chart enables the direct conversion of a viscosity in centipoises to a SSU viscosity. As an example, suppose the liquid under consideration has a specific gravity of .85 and a viscosity of 75 centipoises. To determine the viscosity in SSU, lay a straight edge between 75 on the CP scale and .85 on the G scale. The viscosity in SSU can be read directly on the SSU scale. In this instance, the SSU viscosity is 400 (see dotted line).

If the viscosity value is given in centistokes (kinematic viscosity), it can be used directly on the viscosity correction nomograph. The relationship between the absolute viscosity and the kinematic viscosity is expressed by the following formula:

$$\text{Centipoises} = \text{Centistokes} \times \begin{matrix}\text{Specific}\\ \text{Gravity}\end{matrix}$$

Table A.2f
APPROXIMATE VISCOSITY CONVERSION CHART

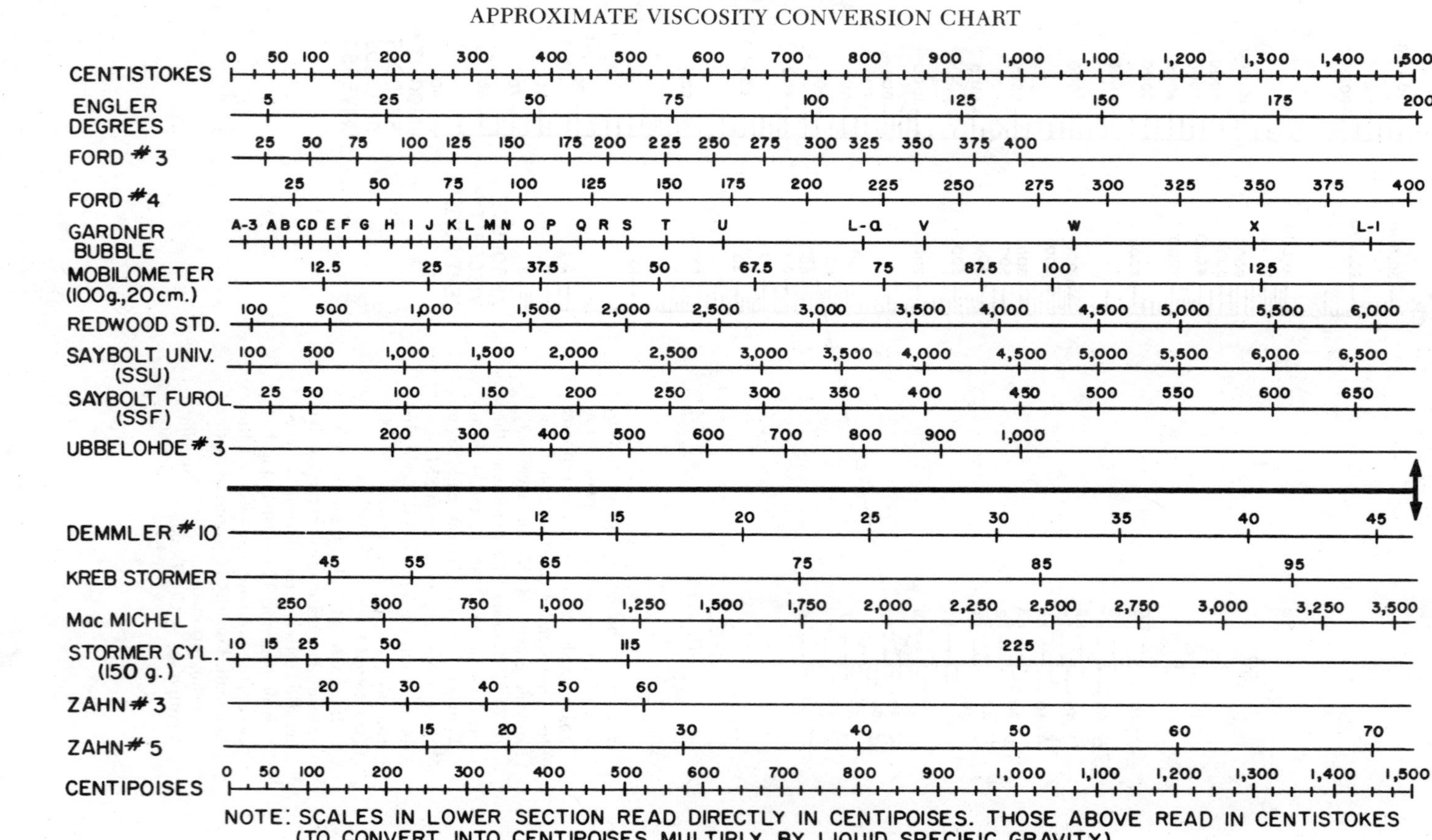

Table A.2g
PRESSURE HEAD CONVERSION

The center column, marked "Known," may be used either for head in feet or for pressure in pounds per square inch—when used as head, the corresponding pressure is found in the column at the left designated as "Pressure Wanted"; when used as pressure, the corresponding head is found in the column at the right designated as "Head Wanted." For example: a 10-foot head has a pressure of 4.33 pounds per square inch, and a 10-pound pressure has a head of 23.094 feet. By moving the decimal place, quantities larger than 100 may be used.

Pressure Wanted Lb. per Sq. In.	*Known Pressure or Head*	*Head Wanted Ft. H_2O*	*Pressure Wanted Lb. per Sq. In.*	*Known Pressure or Head*	*Head Wanted Ft. H_2O*	*Pressure Wanted Lb. per Sq. In.*	*Known Pressure or Head*	*Head Wanted Ft. H_2O*	*Pressure Wanted Lb. per Sq. In.*	*Known Pressure or Head*	*Head Wanted Ft. H_2O*
0.433	1	2.309	11.259	26	60.044	22.084	51	117.779	32.909	76	175.514
0.866	2	4.619	11.692	27	62.354	22.517	52	120.089	33.342	77	177.824
1.299	3	6.928	12.125	28	64.663	22.950	53	122.398	33.775	78	180.133
1.732	4	9.238	12.558	29	66.973	23.383	54	124.708	34.208	79	182.433
2.165	5	11.547	12.991	30	69.282	23.816	55	127.017	34.642	80	184.752
2.598	6	13.856	13.424	31	71.591	24.249	56	129.326	35.075	81	187.061
3.031	7	16.166	13.857	32	73.901	24.682	57	131.636	35.508	82	189.371
3.464	8	18.475	14.290	33	76.210	25.115	58	133.945	35.941	83	191.680
3.897	9	20.785	14.723	34	78.520	25.548	59	136.255	36.374	84	193.990
4.330	10	23.094	15.156	35	80.829	25.981	60	138.564	36.807	85	196.299
4.763	11	25.403	15.589	36	83.138	26.414	61	140.873	37.240	86	198.608
5.196	12	27.713	16.022	37	85.448	26.847	62	143.183	37.673	87	200.918
5.629	13	30.022	16.455	38	87.757	27.280	63	145.492	38.106	88	203.227
6.062	14	32.332	16.888	39	90.067	27.713	64	147.802	38.539	89	205.537
6.495	15	34.641	17.321	40	92.376	28.146	65	150.111	38.972	90	207.846
6.928	16	36.950	17.754	41	94.685	28.579	66	152.420	39.405	91	210.155
7.361	17	39.260	18.187	42	96.995	29.012	67	154.730	39.838	92	212.465
7.794	18	41.569	18.620	43	99.304	29.445	68	157.039	40.271	93	214.774
8.227	19	43.879	19.053	44	101.614	29.878	69	159.349	40.704	94	217.084
8.660	20	46.188	19.486	45	103.923	30.311	70	161.658	41.137	95	219.393
9.093	21	48.497	19.919	46	106.232	30.744	71	163.967	41.570	96	221.702
9.526	22	50.807	20.352	47	108.542	31.177	72	166.277	42.003	97	224.012
9.959	23	53.116	20.785	48	110.851	31.610	73	168.586	42.436	98	226.321
10.392	24	55.426	21.218	49	113.161	32.043	74	170.896	42.869	99	228.631
10.825	25	57.735	21.651	50	115.470	32.476	75	173.205	43.302	100	230.940

Table A.2h
WEIGHT CONVERSION

1 Kilogram = 2.2046 pounds 1 pound = 0.4536 Kilograms

Lbs.	Kgs/Lbs.	Kgs	Lbs.	Kgs/Lbs.	Kgs	Lbs.	Kgs/Lbs.	Kgs	Lbs.	Kgs/Lbs.	Kgs	Lbs.	Kgs/Lbs.	Kgs
2.205	1.0	.454	6.614	3.0	1.361	11.023	5.0	2.268	15.432	7.0	3.175	19.841	9.0	4.082
2.425	1.1	.499	6.834	3.1	1.406	11.243	5.1	2.313	15.653	7.1	3.221	20.062	9.1	4.128
2.646	1.2	.544	7.055	3.2	1.452	11.464	5.2	2.359	15.873	7.2	3.266	20.282	9.2	4.173
2.866	1.3	.590	7.275	3.3	1.497	11.684	5.3	2.404	16.094	7.3	3.311	20.503	9.3	4.218
3.086	1.4	.635	7.496	3.4	1.542	11.905	5.4	2.449	16.314	7.4	3.357	20.723	9.4	4.264
3.307	1.5	.680	7.716	3.5	1.588	12.125	5.5	2.495	16.535	7.5	3.402	20.944	9.5	4.309
3.527	1.6	.726	7.937	3.6	1.633	12.346	5.6	2.540	16.755	7.6	3.447	21.164	9.6	4.355
3.748	1.7	.771	8.157	3.7	1.678	12.566	5.7	2.586	16.975	7.7	3.493	21.385	9.7	4.400
3.968	1.8	.816	8.377	3.8	1.724	12.787	5.8	2.631	17.196	7.8	3.538	21.605	9.8	4.445
4.189	1.9	.862	8.598	3.9	1.769	13.007	5.9	2.676	17.416	7.9	3.583	21.826	9.9	4.491
4.409	2.0	.907	8.818	4.0	1.814	13.228	6.0	2.722	17.637	8.0	3.629	22.046	10.0	4.536
4.630	2.1	.953	9.039	4.1	1.860	13.448	6.1	2.767	17.857	8.1	3.674			
4.850	2.2	.998	9.259	4.2	1.905	13.669	6.2	2.812	18.078	8.2	3.720			
5.071	2.3	1.043	9.480	4.3	1.950	13.889	6.3	2.858	18.298	8.3	3.765			
5.291	2.4	1.089	9.700	4.4	1.996	14.109	6.4	2.903	18.519	8.4	3.810			
5.512	2.5	1.134	9.921	4.5	2.041	14.330	6.5	2.948	18.739	8.5	3.856			
5.732	2.6	1.179	10.141	4.6	2.087	14.550	6.6	2.994	18.960	8.6	3.901			
5.952	2.7	1.225	10.362	4.7	2.132	14.771	6.7	3.039	19.180	8.7	3.946			
6.173	2.8	1.270	10.582	4.8	2.177	14.991	6.8	3.084	19.400	8.8	3.992			
6.393	2.9	1.315	10.803	4.9	2.223	15.212	6.9	3.130	19.621	8.9	4.037			

Table A.2i
LINEAR CONVERSION

INCHES TO MILLIMETERS

1 inch = 25.4 millimeters

Inches	Millimeters	Inches	Millimeters	Inches	Millimeters
1/16	1.6	2.0	50.8	4.5	114.3
1/8	3.2	2.1	53.3	4.6	116.8
3/16	4.8	2.2	55.9	4.7	119.4
1/4	6.4	2.3	58.4	4.8	121.9
5/16	7.9	2.4	61.0	4.9	124.5
3/8	9.5	2.5	63.5	5.0	127.0
7/16	11.1	2.6	66.0	5.5	139.7
1/2	12.7	2.7	68.6	6.0	152.4
9/16	14.3	2.8	71.1	6.5	165.1
5/8	15.9	2.9	73.7	7.0	177.8
11/16	17.5	3.0	76.2	7.5	190.5
3/4	19.1	3.1	78.7	8.0	203.2
13/16	20.6	3.2	81.3	8.5	215.9
7/8	22.2	3.3	83.8	9.0	228.6
15/16	23.8	3.4	86.4	9.5	241.3
1.0	25.4	3.5	88.9	10.0	254.0
1.1	27.9	3.6	91.4		
1.2	30.5	3.7	94.0		
1.3	33.0	3.8	96.5		
1.4	35.6	3.9	99.1		
1.5	38.1	4.0	101.6		
1.6	40.6	4.1	104.1		
1.7	43.2	4.2	106.7		
1.8	45.7	4.3	109.2		
1.9	48.3	4.4	111.8		

MILLIMETERS TO INCHES

1 millimeter = .03937 inches

Millimeters	Inches	Millimeters	Inches	Millimeters	Inches	Millimeters	Inches
1.0	.0394	3.5	.1378	6.0	.2362	8.5	.3346
1.1	.0433	3.6	.1417	6.1	.2402	8.6	.3386
1.2	.0472	3.7	.1457	6.2	.2441	8.7	.3425
1.3	.0512	3.8	.1496	6.3	.2480	8.8	.3465
1.4	.0551	3.9	.1535	6.4	.2520	8.9	.3504
1.5	.0591	4.0	.1575	6.5	.2559	9.0	.3543
1.6	.0630	4.1	.1614	6.6	.2598	9.1	.3583
1.7	.0669	4.2	.1654	6.7	.2638	9.2	.3622
1.8	.0709	4.3	.1693	6.8	.2677	9.3	.3661
1.9	.0748	4.4	.1732	6.9	.2717	9.4	.3701
2.0	.0787	4.5	.1772	7.0	.2756	9.5	.3740
2.1	.0827	4.6	.1811	7.1	.2795	9.6	.3780
2.2	.0866	4.7	.1850	7.2	.2835	9.7	.3819
2.3	.0906	4.8	.1890	7.3	.2874	9.8	.3858
2.4	.0945	4.9	.1929	7.4	.2913	9.9	.3898
2.5	.0984	5.0	.1969	7.5	.2953	10.0	.3937
2.6	.1024	5.1	.2008	7.6	.2992		
2.7	.1063	5.2	.2047	7.7	.3031		
2.8	.1102	5.3	.2087	7.8	.3071		
2.9	.1142	5.4	.2126	7.9	.3110		
3.0	.1181	5.5	.2165	8.0	.3150		
3.1	.1220	5.6	.2205	8.1	.3189		
3.2	.1260	5.7	.2244	8.2	.3228		
3.3	.1299	5.8	.2283	8.3	.3268		
3.4	.1339	5.9	.2323	8.4	.3307		

By moving the decimal place, conversions for figures larger than 10 may be obtained.

Table A.2j
UNITS OF POWER

1 kw
= 1.3415 hp
= 738 ft-lb°/sec
= 44,268 ft-lb/min
= 2,656,100 ft-lb/hr
= .948 Btu/sec
= 56.9 Btu/min
= 3413 Btu/hr

1 hp
= .7455 kw
= 550 ft-lb/sec
= 33,000 ft-lb/min
= 1,980,000 ft-lb/hr
= .707 Btu/sec
= .424 Btu/min
= 2544 Btu/hr

1 ft-lb/sec
= .001355 kw
= .001818 hp
= 60 ft-lb/min
= 3600 ft-lb/hr
= .001284 Btu/sec
= .0771 Btu/min
= 4.62 Btu/hr

1 ft-lb/min
= .00002259 kw
= .0000303 hp
= .01667 ft-lb/sec
= 60 ft-lb hr
= .00002141 Btu/sec
= .001284 Btu/min
= .0771 Btu/hr

1 ft-lb/hr
= .000000376 kw
= .000000505 hp
= .000278 ft-lb/sec
= .01667 ft-lb/min
= .000000357 Btu/sec
= .00002141 Btu/min
= .001284 Btu/hr

1 Btu/sec
= 1.055 kw
= 1.416 hp
= 778 ft-lb/sec
= 46,700 ft-lb/min
= 2,802,000 ft-lb/hr
= 60 Btu/min
= 3600 Btu/hr

1 Btu/min
= .01759 kw
= .02359 hp
= 12.98 ft-lb/sec
= 778 ft-lb/min
= 46,700 ft-lb/hr
= .01667 Btu/sec
= 60 Btu/hr

1 Btu/hr
= .0002931 kw
= .0003932 hp
= .2163 ft-lb/sec
= 12.98 ft-lb/min
= 778 ft-lb/hr
= .0002778 Btu/sec
= .01667 Btu/min

°Ft-lb means foot pound, the work done in moving against one pound force a distance of one foot.

Table A.2k
UNITS OF WORK, ENERGY, HEAT

1 Btu
= 9340 in. lb
= 778.3 ft-lb
= .0002938 kwhr°
= .0003931 hphr

1 in. lb
= .0001070 Btu
= .0833 ft-lb
= .00000003137 kwhr
= .0000000421 hphr

1 ft-lb
= .001284 Btu
= 12 in. lb
= .000000376 kwhr
= .000000505 hphr

1 kwhr
= 3413 Btu
= 31,873,000 in. lb
= 2,656,100 ft-lb
= 1.342 hphr

1 hphr
= 2544 Btu
= 23,760,000 in. lb
= 1,980,000 ft-lb
= 0.7455 kwhr

°1 kilowatthour = 3413 Btu and 1 Btu = 778.3 ft-lb.

Table A.2l
UNITS OF VOLUME

1 cu. in.
= .00433 gal
= .00579 cu. ft.
= .0000214 cu. yd

1 gal
= 231 cu. in.
= .1337 cu. ft.
= .00495 cu. yd
= .00000307 acre ft.°

1 cu. ft.
= 1728 cu. in.
= 7.48 gal
= .0370 cu. yd
= .0000230 acre ft.

1 cu. yd
= 46,656 cu. in.
= 202.0 gal
= 27 cu. ft.
= .000620 acre ft.

1 acre ft.
= 325,800 gal
= 43,560 cu. ft.
= 1613 cu. yd

°Acre ft. of water is the volume in 1 ft. of depth covering 1 acre

Table A.2m
UNITS OF TIME

1 sec
= .01667 min
= .0002778 hr
= .00001157 days
= .0000003805 mo°
= .0000000317 yr

1 min
= 60 sec
= .01667 hr
= .000694 days
= .0000228 mo
= .000001903 yr

1 hr
= 3600 sec
= 60 min
= .0417 days
= .001370 mo
= .0001142 yr

1 day
= 86,400 sec
= 1440 min
= 24 hr
= .0329 mo
= .002740 yr

1 mo
= 2,628,000 sec
= 43,800 min
= 730 hr
= 30.4 days
= .0833 yr

1 yr
= 31,536,000 sec
= 525,600 min
= 8760 hr
= 365 days
= 12 mo

°Month used is exactly 1/12 year.

Table A.2n
UNITS OF VELOCITY

fps
= 60 fpm
= 3600 fph
= .01136 mpm
= .682 mph

1 fpm
= .01667 fps
= 60 fph
= .0001894 mpm
= .01136 mph

1 fph
= .002778 fps
= .01667 fpm
= .00000316 mpm
= .0001894 mph

1 mpm
= 88 fps
= 5280 fpm
= 316,800 fph
= 60 mph

1 mph
= 1.467 fps
= 88 fpm
= 5280 fph
= .01667 mpm

f = ft; hours; m = miles or minutes; p = per; s = seconds

Table A.2o
UNITS OF PRESSURE

1 in. water°
= .0833 ft. water
= .0735 in. Hg
= .577 oz/sq. in.
= 83.1 oz/sq. ft.
= .0361 lb/sq. in.
= 5.20 lb/sq. ft

1 ft. water
= 12 in. water
= .882 in. Hg
= 6.93 oz/sq. in.
= 998 oz/sq. ft.
= .433 lb/sq. in.
= 62.4 lb/sq. ft.

1 in. Hg
= 13.61 in. water
= 1.131 ft. water
= 7.84 oz/sq. in.
= 1129 oz/sq. ft.
= .491 lb/sq. in.
= 70.5 lb/sq. ft.

1 oz/sq. in.
= 1.732 in. water
= .1443 ft. water
= .1276 in. Hg
= 144 oz/sq. ft.
= .0625 lb/sq. in.
= 9 lb/sq. ft.

1 oz/sq. ft.
= .01203 in. water
= .001002 ft. water
= .000886 in. Hg
= .00694 oz/sq. in.
= .000434 lb/sq. in.
= .0625 lb/sq. ft.

1 lb/sq. in.
= 27.71 in. water
= 2.31 ft. water
= 2.04 in. Hg
= 16 oz/sq. in.
= 2304 oz/sq. ft.
= 144 lb/sq. ft.

1 lb/sq. ft.
= .1924 in. water
= .01604 ft. water
= .01418 in. Hg
= .1111 oz/sq. in.
= 16 oz/sq. ft.
= .00694 lb/sq. in.

°In. water means inches of water at 60°F.
In. Hg means inches head of mercury at 32°F.

Table A.2p
UNITS OF AREA

1 cir mil°
= .000000785 sq. in.

1 sq. in.
= 1,273,200 cir mils
= .00694 sq. ft.
= .000772 sq. yd

1 sq. ft
= 144 sq. in.
= .01111 sq. yd
= .00002296 acres

1 sq. yd
= 1296 sq. in.
= 9 sq. ft.
= .0002066 acres

1 acre
= 43,560 sq. ft.
= 4840 sq. yd

°A cir (circular) mil is the area of a circle of 1/1000 in. dia. Thus, a round rod of 1-in. dia. has an area of 1,000,000 cir mils.

Table A.2q
UNITS OF LENGTH

1 in.
= .0833 ft.
= .0277 yd
= .0000158 miles

1 ft.
= 12 in.
= .333 yd
= .000189 miles

1 yd
= 36 in.
= 3 ft.
= .000568 miles

1 mile
= 63360 in.
= 5280 ft.
= 1760 yd

Table A.2r
UNITS OF WEIGHT

1 gr
= .00229 oz°
= .0001429 lb
= .0000000714 tons

1 oz
= 438 gr
= .0625 lb
= .00003125 tons

1 lb
= 7000 gr
= 16 oz
= .000500 tons

1 ton
= 14,000,000 gr
= 32,000 oz
= 2000 lb

°Avoirdupois oz and lb and short ton of 2000 lb.

Table A.2s
UNITS OF DENSITY

1 lb/cu. in.
= 1728 lb/cu. ft.
= 0.864 tons°/cu. ft.
= 23.3 tons/cu. yd
= 231 lb/gal

1 lb/cu. ft.
= .000579 lb/cu. in.
= .000500 tons/cu. ft.
= .0135 tons/cu. yd
= .1337 lb/gal

1 ton/cu. ft.
= 1.157 lb/cu. in.
= 2000 lb/cu. ft.
= 27 tons/cu. yd
= 267 lb/gal

1 ton/cu. yd
= .0429 lb/cu. in.
= 74.1 lb/cu. ft
= .0370 tons/cu. ft.
= 9.90 lb/gal

1 lb/gal
= .00433 lb/cu. in.
= 7.48 lb/cu. ft.
= .00374 tons/cu. ft.
= .1010 tons /cu. yd

°Tons are short = 2000 lb.

Table A.2t
MASS FLOW RATES

1 lb/sec
= 60 lb/min
= 3600 lb/hr
= 86,400 lb/day
= 2,628,000 lb/mo°
= 31,536,000 lb/yr

1 lb/min
= .01667 lb/sec
= 60 lb/hr
= 1440 lb/day
= 43,800 lb/mo
= 525,600 lb/yr

1 lb/hr
= .0002778 lb/sec
= .01667 lb/min
= 24 lb/day
= 730 lb/mo
= 8760 lb/yr

1 lb/day
= .00001157 lb/sec
= .000694 lb/min
= .0417 lb/hr
= 30.4 lb/mo
= 365 lb/yr

1 lb/mo
= .000000381 lb/sec
= .0000228 lb/min
= .001370 lb/hr
= .0329 lb/day
= 12 lb/yr

1 lb/yr
= .0000000317 lb/sec
= .000001903 lb/min
= .0001142 lb/hr
= .002740 lb/day
= .0833 lb/mo

°Month used is exactly 1/12 year = 30.4 days

Table A.2u
VOLUME FLOW RATES

1 cu. ft/sec
= 60 cu. ft/min
= 3600 cu. ft/hr
= 7.48 gal/sec
= 448.8 gal/min
= 26,930 gal/hr

1 cu. ft/min
= .01667 cu. ft/sec
= 60 cu. ft/hr
= .1247 gal/sec
= 7.48 gal/min
= 448.8 gal/hr

1 cu. ft/hr
= .0002778 cu. ft/sec
= .01667 cu. ft/min
= .002078 gal/sec
= .1247 gal/min
= 7.48 gal/hr

1 gal/sec
= .1337 cu. ft/sec
= 8.02 cu. ft/min
= 481 cu. ft/hr
= 60 gal/min
= 3600 gal/hr

1 gal/min
= .002228 cu. ft/sec
= .1337 cu. ft/min
= 8.02 cu. ft/hr
= .01667 gal/sec
= 60 gal/hr

1 gal/hr
= .0000371 cu. ft/sec
= .002228 cu. ft/min
= .1337 cu. ft/hr
= .0002778 gal/sec
= .01667 gal/min

A.3 CHEMICAL RESISTANCE OF MATERIALS

Table A.3
CHEMICAL RESISTANCE OF MATERIALS

MATERIALS
× — Very Good Service
+ — Moderate Service
− — Limited or Variable Service
○ — Unsatisfactory
Blank — No Information

CHEMICALS
Solids Assumed in Sol'n.
Room Temperatures Assumed
Unless Otherwise Stated

METALS

Chemical	Carbon Steel; Fe	Cast Iron & Ductile Iron; Fe	304 Stainless Steel; Fe, 18Cr, 8Ni	316 Stainless Steel; Fe, 16Cr, 10Ni, 2Mo	347 Stainless Steel; Fe, 17Cr, 9Ni; (Cx10)Cb	Ni-Resist Iron; Fe, 14Ni, 2Cr, 2Si	Durimet 20; Carpenter 20; Fe, 4Cu, 20Cr, 29Ni, 2Mo, 1Si	Worthite; 3Mo, 2Cu, Fe, 20Cr, 24Ni, 3Si	Duriron; Fe, 14Si; Durichlor, Fe, 14Si, 3Mo*	Copper; Brass; Bronzes; Everdur	Aluminum; Al (and Alloys)	Lead; Pb	Monel; 67Ni, 30Cu, 1.4 Fe	Nickel; Ni	Inconel; 76Ni, 15Cr, 8Fe
Acetic Acid, 100%, CH_3COOH	○	○	+	×	−	○	×	×	×	○	×	○	−	+	−
Acetic Acid, Dilute	○	○	×	×	−	○	×	×	×	×	−	−	−	○	+
Acetic Anhydride, $(CH_3CO)_2O$	−	+	×	×	×	−	×	×	×	−	−	○	−	−	+
Acetone, CH_3COCH_3	×		×	×	×	×	×	×	×	×	×	×	×	×	×
Acetyl Chloride, CH_3COCl	○		−	×			×		×	○			+		
Aluminum Chloride, $AlCl_3$	○	○	○	○	○	−	×	○	×	−	○	○	○	−	−
Aluminum Hydroxide, $Al(OH)_3$			×	×			×	×	+	×	×		+	+	
Aluminum Sulfate, $Al_2(SO_4)_3$	○		−	−	×	−	×	×	×	−	×	×	−	−	−
Alums, Conc., $Al_2(SO_4)_3 \cdot K_2SO_4$, etc.	○		○	−	−		+		×	−	−	×			
Alums, Dilute	○	○	×	−	−	−	×	×	×	−	×	+	+	+	
Amines, various	×	×	×	×	×		×	×	×	−	−				
Ammonia (Gas), Moist, NH_3	×		×	×	×	×	×		×	○	○	−	−	−	−
Ammonium Carbonate, $(NH_4)_2CO_3$	+	+	+	+	×		×		×	−	−		+	×	+
Ammonium Chloride, NH_4Cl	○	○	○	−	−	×	×	×	×	○	○	−	+	+	−
Ammonium Hydroxide, NH_4OH	×	×	×	×	×	×	×		+	○	−	+	○	○	×
Ammonium Nitrate, NH_4NO_3	−		×	×	×	−	×	×	×	○	○		×		−
Ammonium Persulfate, $(NH_4)_2S_2O_8$	○		×	×	×		×	×	×	−	−		○	○	−
Ammonium Phosphate, $(NH_4)_1H_2PO_4$	○		×	×		−	×		×	○					
Ammonium Phosphate, $(NH_4)_2HPO_4$	−		×	×		×	×		×	−	−				
Ammonium Phosphate, $(NH_4)_3PO_4$	×	−	×	×	×	×	×	×	×	○	−	×	×	×	×
Ammonium Sulfate, $(NH_4)_2SO_4$	×		−	−	+	×	×	×	×	−	−	−	+	+	×
Amyl Acetate, $C_5H_{11}COOCH_3$	−		×	×	×	×	×	×	×	−	×		+	+	+
Amyl Alcohol, $C_5H_{11}OH$	×		×	×			×		×	×	×				

Chemical	Hastelloy B; Ni, 26Mo, 4Fe	Hastelloy C; Ni, 16Mo, 4Fe, 14Cr, 4W	Hastelloy D; Ni, 8Si, 3Cu	Chlorimet 3; 3Fe, 1Si, 60Ni, 18Mo, 18Cr	Chlorimet 2; 63Ni, 32Mo, 3Fe, 1Si	Stellite; Co, 28Cr, 4W	Zirconium; Zr	Tantalum; Ta	Silver; Ag	Platinum; Pt	Dowmetal; (Mg alloys)	Titanium; Ti	Molybdenum; Mo
Acetic Acid, 100%, CH_3COOH	×	×	×	×	×	×	×	×	×	×		×	
Acetic Acid, Dilute	×	×	+	×	×	×	×	×	×	×		×	×
Acetic Anhydride, $(CH_3CO)_2O$	+	×	+	×	×	×		×	×	×		×	
Acetone, CH_3COCH_3	×	×	×	×	×		×		×			×	
Acetyl Chloride, CH_3COCl				×	×								
Aluminum Chloride, $AlCl_3$	×	+	×	×	×	−	×	×	×	×		×	
Aluminum Hydroxide, $Al(OH)_3$				×	×			×					
Aluminum Sulfate, $Al_2(SO_4)_3$	×	×	×	×	×	×	×	×	×	×		×	
Alums, Conc., $Al_2(SO_4)_3 \cdot K_2SO_4$, etc.				×	×			×		×			
Alums, Dilute	+	+		×	×		×	×	×	×			
Amines, various	×	×	×	×	×	×		−	×	×			
Ammonia (Gas), Moist, NH_3	×	×	×	×	×	×		−	×	×	×	−	
Ammonium Carbonate, $(NH_4)_2CO_3$	+	+	+	×	×			×					
Ammonium Chloride, NH_4Cl	+	+	×	×	×	−	×	×	×	×		×	+
Ammonium Hydroxide, NH_4OH	+	+	+	×	×	×	×	−	×	×	×	×	−
Ammonium Nitrate, NH_4NO_3	○	×	○	×	×			×					
Ammonium Persulfate, $(NH_4)_2S_2O_8$	○	×	○	×	×			×					
Ammonium Phosphate, $(NH_4)_1H_2PO_4$				×	×								
Ammonium Phosphate, $(NH_4)_2HPO_4$				×	×								
Ammonium Phosphate, $(NH_4)_3PO_4$	×	×	×	×	×	×		×					
Ammonium Sulfate, $(NH_4)_2SO_4$	×	×	×	×	×	×	×	×	×			×	
Amyl Acetate, $C_5H_{11}COOCH_3$	×	×	×	×	×			×					
Amyl Alcohol, $C_5H_{11}OH$				×	×								

CARBONS & CERAMICS; RUBBERS

Chemical	Carbon & Graphite	Glass, "Pyrex" brand	Silicaware	Silicate Cements	Chemical Stoneware	Transite (asbestos & cement)	Chemical Porcelain	Concrete-Unbonded	Concrete-Mortar Bonded	Hard Rubber (Natural)	Soft Rubber (Natural)	Neoprene	Butadiene Derivatives	Nitrile Rubber (Chemigum)
Acetic Acid, 100%, CH_3COOH	×	×	×	×	×		×			×	−	○	○	○
Acetic Acid, Dilute	×	×	×		×		×			×	−	○	×	○
Acetic Anhydride, $(CH_3CO)_2O$	×	×	×	×	×		×			×	−	○	−	○
Acetone, CH_3COCH_3	×	×	×		×	×	×			○	○	○	×	−
Acetyl Chloride, CH_3COCl		×					×							
Aluminum Chloride, $AlCl_3$	×	×	×	×	×		×			×	×	×	×	×
Aluminum Hydroxide, $Al(OH)_3$		×					×							
Aluminum Sulfate, $Al_2(SO_4)_3$	×	×	×		×	×	×			×	×	×	×	−
Alums, Conc., $Al_2(SO_4)_3 \cdot K_2SO_4$, etc.	×	×					×					×		
Alums, Dilute	×	×	×	×	×	×	×			×	×	×	×	×
Amines, various	×	×			×		×			−				
Ammonia (Gas), Moist, NH_3	×	×	×	○	×	×	×			−	−	×	×	
Ammonium Carbonate, $(NH_4)_2CO_3$		×		×			×			×				○
Ammonium Chloride, NH_4Cl	×	×	×	×	×	×	×	×		×	×	×	×	−
Ammonium Hydroxide, NH_4OH	×	−			×		×			−	×	−		×
Ammonium Nitrate, NH_4NO_3	×	×	×		×		×	×		−	−	−	×	
Ammonium Persulfate, $(NH_4)_2S_2O_8$		×												
Ammonium Phosphate, $(NH_4)_1H_2PO_4$		×					×							
Ammonium Phosphate, $(NH_4)_2HPO_4$		×					×							
Ammonium Phosphate, $(NH_4)_3PO_4$	×	×			×		×			×				
Ammonium Sulfate, $(NH_4)_2SO_4$	×	×	×		×		×			×	×	×	×	
Amyl Acetate, $C_5H_{11}COOCH_3$	×	×	×		×		×	×		○	○	○	○	
Amyl Alcohol, $C_5H_{11}OH$	×	×	×		×	×	×			−	−	×	×	

THERMOPLASTICS

Chemical	Viton	Asphaltic, Bitumastic	Cellulose Acetate	Cellulose Acetatebutyrate	Ethyl Cellulose (Ethocel)	Cellulose Nitrate	Acrylic (Lucite, Plexiglas)	Coumarone Resins	Polyethylene	Polyvinyl Chloride, Rigid or Unplasticized	Tygon (P. V. C. & Copolymers)	Saran (Vinyl chloride, vinyldene chloride)	Kel-F (Polytrifluorochloroethylene)	Teflon (Polytetrafluoroethylene)	Uscolite CP (Styrene-acrylonitrile-butadiene)
Acetic Acid, 100%, CH_3COOH	○	○	○		×	○	○		○	○	○	−	×	×	○
Acetic Acid, Dilute		×	×	×	×	×	×	×	×	−	−	×	×	×	−
Acetic Anhydride, $(CH_3CO)_2O$	○	○	○		○	○	○			○	○	−		×	
Acetone, CH_3COCH_3	○	○	○	○	○	○	○	○	○	○		○		×	○
Acetyl Chloride, CH_3COCl													×	×	
Aluminum Chloride, $AlCl_3$	×	×				×			×	×	×	×	×	×	×
Aluminum Hydroxide, $Al(OH)_3$	×								×	×			×	×	
Aluminum Sulfate, $Al_2(SO_4)_3$	×	×				×	×	×	×	×	×	×	×	×	×
Alums, Conc., $Al_2(SO_4)_3 \cdot K_2SO_4$, etc.									×	×	×	×	×	×	×
Alums, Dilute		×				×	×	×	×	×	×	×	×	×	×
Amines, various	○									○	−		×	×	
Ammonia (Gas), Moist, NH_3	○		−	×	×		×		×	−		○	×	×	−
Ammonium Carbonate, $(NH_4)_2CO_3$		×							×	×		×	×	×	×
Ammonium Chloride, NH_4Cl	×	×	×	×		×	×	×	×	×	×	×	×	×	×
Ammonium Hydroxide, NH_4OH	+	−					×		×	×	−	○	×	×	×
Ammonium Nitrate, NH_4NO_3	+	×	×			×	×		×	×		×	×	×	×
Ammonium Persulfate, $(NH_4)_2S_2O_8$		×								×	×		×	×	
Ammonium Phosphate, $(NH_4)_1H_2PO_4$										×			×	×	×
Ammonium Phosphate, $(NH_4)_2HPO_4$										×			×	×	×
Ammonium Phosphate, $(NH_4)_3PO_4$									×	×	×		×	×	×
Ammonium Sulfate, $(NH_4)_2SO_4$		×				×	×	×		×	×	×	×	×	×
Amyl Acetate, $C_5H_{11}COOCH_3$	○	○	○	○	○	○	○	○	−	○		○	×	×	
Amyl Alcohol, $C_5H_{11}OH$	×	○	×	○	○	○	×	−	−	×		×	×	×	×

THERMOSETTING PLASTICS

Chemical	Penton (Chlorinated Polyether)	Shellac Compounds	Organic Polysulfides	Polystyrene (Styron)	Vinylidene Chlorides	Vinyl Chloride Acetates	Cast Phenol Formaldehyde	Haveg 41 (Phenolic w. asbestos)	Heresite (Phenol formaldehyde)	Molded Phenolformald. (Durez)	Phenol Furfural Plastics	Urea Formaldehyde	Casein Plastics	Epoxy Resins	Furane Resins (Haveg 61, Duralon)	Silicone Resins	Permanite (Furan, Glass Fiber)	Nylon (Adipic Acid—Hexameth, Diamine)	Durcon 6 (Modified Epoxy)
Acetic Acid, 100%, CH_3COOH	+		×	○	○	○	−	−	×	×	−			−	×	○	−	−	×
Acetic Acid, Dilute	×	×	×	○	−	−	×	×	×	×	×		×	×	×	×	×	○	×
Acetic Anhydride, $(CH_3CO)_2O$	+			○			−	○	×	×	×				○		−		×
Acetone, CH_3COCH_3	+	○	×	−	○	○	−	○		−	−	×	×	×			×	×	
Acetyl Chloride, CH_3COCl																			×
Aluminum Chloride, $AlCl_3$	×	×		×	×	×	×	×	×	×	×			×	×		×	×	×
Aluminum Hydroxide, $Al(OH)_3$	×																	×	×
Aluminum Sulfate, $Al_2(SO_4)_3$	×	×	×	×	×	×	−	×	×	×			×		×		×	×	×
Alums, Conc., $Al_2(SO_4)_3 \cdot K_2SO_4$, etc.	×							×	×					×					×
Alums, Dilute	×	×	×	×	×	×	−	×	×	×			×	×	×				×
Amines, various								×	×									×	×
Ammonia (Gas), Moist, NH_3	×	×		○	×	×	×			×	−			×	×	−			×
Ammonium Carbonate, $(NH_4)_2CO_3$	×			×						×	×			×	×			×	×
Ammonium Chloride, NH_4Cl	×	×		×	×	×	×	×	×	×	×		×	×	×		×	○	+**
Ammonium Hydroxide, NH_4OH	×			×	○			×	×	○	○			×	×	−	×	×	×
Ammonium Nitrate, NH_4NO_3	×	−		×	×	×	×			×	×		×	×				×	×
Ammonium Persulfate, $(NH_4)_2S_2O_8$	×													×	×				×
Ammonium Phosphate, $(NH_4)_1H_2PO_4$	×																		×
Ammonium Phosphate, $(NH_4)_2HPO_4$	×																		×
Ammonium Phosphate, $(NH_4)_3PO_4$	×							×	×										×
Ammonium Sulfate, $(NH_4)_2SO_4$	×	×	×		×	×	×	×	×	×	×		×	×			×		○
Amyl Acetate, $C_5H_{11}COOCH_3$	×	○		○	○	×	−			−	−		×	×			×	×	×
Amyl Alcohol, $C_5H_{11}OH$	×	○		×	−	−	×			×	×		×	×				×	×

WOODS

Chemical	Cypress	Fir	Maple	Oak	Pine	Redwood
Acetic Acid, 100%, CH_3COOH	−	−	×	×	×	−
Acetic Acid, Dilute	×	×	×	×	×	×
Acetic Anhydride, $(CH_3CO)_2O$						
Acetone, CH_3COCH_3						
Acetyl Chloride, CH_3COCl						
Aluminum Chloride, $AlCl_3$						
Aluminum Hydroxide, $Al(OH)_3$						
Aluminum Sulfate, $Al_2(SO_4)_3$	×	×	×	−	×	×
Alums, Conc., $Al_2(SO_4)_3 \cdot K_2SO_4$, etc.						
Alums, Dilute	×	×	×	−	×	×
Amines, various						
Ammonia (Gas), Moist, NH_3						
Ammonium Carbonate, $(NH_4)_2CO_3$						
Ammonium Chloride, NH_4Cl						
Ammonium Hydroxide, NH_4OH						
Ammonium Nitrate, NH_4NO_3						
Ammonium Persulfate, $(NH_4)_2S_2O_8$						
Ammonium Phosphate, $(NH_4)_1H_2PO_4$						
Ammonium Phosphate, $(NH_4)_2HPO_4$						
Ammonium Phosphate, $(NH_4)_3PO_4$						
Ammonium Sulfate, $(NH_4)_2SO_4$						
Amyl Acetate, $C_5H_{11}COOCH_3$						
Amyl Alcohol, $C_5H_{11}OH$						

* NOTE: Duriron is as shown. Durichlor is also satisfactory on chlorides and HCl.

** Durcon 5 would be the preferred formula.

Table A.3
CHEMICAL RESISTANCE OF MATERIALS *continued*

MATERIALS
× — Very Good Service
+ — Moderate Service
– — Limited or Variable Service
○ — Unsatisfactory
Blank — No Information

CHEMICALS
Solids Assumed in Sol'n.
Room Temperatures Assumed
Unless Otherwise Stated

METALS

Chemicals	Carbon Steel; Fe	Cast Iron & Ductile Iron; Fe	304 Stainless Steel; Fe, 18Cr, 8Ni	316 Stainless Steel; Fe, 16Cr, 10Ni, 2Mo	347 Stainless Steel; Fe, 17Cr, 9Ni; (Cx10)Cb	Ni-Resist Iron; Fe, 14Ni, 2Cr, 2Si	Durimet 20; Carpenter 20; Fe; 4Cu, 20Cr, 29Ni, 2Mo, 1Si	Worthite; 3Mo, 2Cu, Fe, 20Cr, 24Ni, 3Si	Duriron; Fe, 14Si; Durichlor, Fe, 14Si, 3Mo*	Copper; Brass; Bronzes; Everdur	Aluminum; Al (and Alloys)	Lead; Pb	Monel; 67Ni, 30Cu, 1.4 Fe	Nickel; Ni	Inconel; 76Ni, 15Cr, 8Fe	Hastelloy B; Ni, 26Mo, 4Fe	Hastelloy C; Ni, 16Mo, 4Fe, 14Cr, 4W	Hastelloy D; Ni, 8Si, 3Cu	Chlorimet 3; 3Fe, 1Si, 60Ni, 18Mo, 18Cr	Chlorimet 2; 63Ni, 32Mo, 3Fe, 1Si	Stellite; Co, 28Cr, 4W	Zirconium; Zr	Tantalum; Ta	Silver; Ag	Platinum; Pt	Dowmetal; (Mg alloys)	Titanium; Ti	Molybdenum; Mo
Amyl Chloride, $C_5H_{11}Cl$	×		–	–	–		×	–	×	×	○		–	+	–	×	×	×	×	×			×					
Antimony Trichloride, $SbCl_3$	○		○	○	○				×	–	○	×	×	×	×	×	×		×	×			×					
Arsenic Acid, $HAsO_8$	○		–	–		×	×	×	×	–		–			–				×	×			×	×				
Barium Carbonate, $BaCO_3$			×	×	×		×	×	×	×	○		+	×	×	×	×	×	×	×		×	×				×	
Barium Hydroxide, $Ba(OH)_2$	×			×		×	×	×	+	○	○		×	×		+	+	+	×	×		×	×				+	
Barium Sulfide, BaS	○		×	×		×	×	×	×	○	○		–						×	×				○				
Benzaldehyde, C_6H_5 CHO	×		×	×			×		×		–	○							×	×								
Benzene, C_6H_6	+	+	×	×	×	×	×	×	×	×	×	+	+	+	×	+	+	+	×	×	×		×	×	×	×		
Benzoic Acid, C_6H_5 COOH	○		×	×	+		×	×	×	+	×	○	+	+	+	×	×	×	×	×			×	×				
Borax, $Na_2B_4O_7$	×	–	×	×	×	×	×		×	–	–	×	×	×					×	×				○				
Boric Acid, H_3BO_3	○	○	+	+	+	–	×	×	×	+	+	+	+	+	×	×	×	×	×	×	×		×	×	×		×	
Bromine, Wet, Br_2	○		○	○	○	○	○	○	○	○	○	–	○	○	○	○	×	○	×	○	–		×	○	–		○	
Butanol, C_4H_9OH	×		×	×			×		×	×	×		×	×	×	×	×	×	×	×								
Butyl Acetate, $C_4H_9COOCH_3$	–			×			×	×	×	×	×	×	+			×	×	×	×	×			×			–		
Butyric Acid, C_3H_7 COOH	○	○	○	×	○		×	×	×	–	–	○	+	+	+	×	×	×	×	×			×				×	
Calcium Bisulfate, $CaHSO_4$	○		○	×		○	×			○	○		○	○					×	×								
Calcium Bisulfite, Ca HSO_3	×		○	×		○	×	×	○	–	–	×	○						×	×							×	
Calcium Carbonate, $CaCO_3$	×		×	×		×	×	×	×	×	○	×	×	×	×	×	×	×	×	×			×				×	
Calcium Chlorate, $CaClO_3$			–	×	○		+	–	+	–			+	+	+		+	+	+	+			×					
Calcium Chloride, $CaCl_2$	+	+	–	–	–	–	+	×	×	+	–	○	+	+	+	+	×	+	×	×	–	×	×				×	
Calcium Hydroxide, Ca $(OH)_2$	×		×	×	×	×	×	×	+	+	–		×	×	×	×	×	×	×	×			×				×	
Calcium Hypochlorite, Ca $(OCl)_2$	–	○	○	–	–	–	–	–	×	○	○	○	○	○	○	○	–	○	–	○	○	×	×		×		×	
Calcium Sulfate, Ca SO_4	–		×	×			×	×	×	×	–		+			+	+	+	×	×			×					
Carbon Dioxide (Dry), CO_2	×		×	×	×	×	×		×	×	×	×	×	×	×	×	×		×	×				×				×
Carbon Dioxide (Wet or H_2CO_3)	–	–	×	×	×	–	×	×	×	–	–	–	–	×	×	×	×	×	×	×	×		×	×				
Carbon Disulfide, CS_2	×	+	+	+	×	×	×	×	×	–	–	+	+	+	+				×	×			×	–				

CARBONS & CERAMICS; RUBBERS

Chemicals	Carbon & Graphite	Glass, "Pyrex" brand	Silicaware	Silicate Cements	Chemical Stoneware	Transite (asbestos & cement)	Chemical Porcelain	Concrete-Unbonded	Concrete-Mortar Bonded	Hard-Rubber (Natural)	Soft Rubber (Natural)	Neoprene	Butadiene Derivatives	Nitrile Rubber (Chemigum)
Amyl Chloride, $C_5H_{11}Cl$	×	×	×		×		×			–	○	–	×	
Antimony Trichloride, $SbCl_3$		×	×		×		×						×	
Arsenic Acid, $HAsO_8$	×	×					×					×		
Barium Carbonate, $BaCO_3$		×					×							
Barium Hydroxide, $Ba(OH)_2$		×					×							
Barium Sulfide, BaS		–					×			×				
Benzaldehyde, C_6H_5 CHO	×	×	×		×	×	×	×		○	○	○	×	
Benzene, C_6H_6	×	×		×	×		×			○	○	○		○
Benzoic Acid, C_6H_5 COOH		×					×							
Borax, $Na_2B_4O_7$	×	×	×		×	×	×	×		×	×	×	×	
Boric Acid, H_3BO_3	×	×		×	×		×			×	×	×		
Bromine, Wet, Br_2	○	×	×		×		×	×		–	○		–	
Butanol, C_4H_9OH		×					×			–	–	–		
Butyl Acetate, $C_4H_9COOCH_3$	×	×	×		×	×	×	×	×	○	○	○	○	
Butyric Acid, C_3H_7 COOH		×					×			×	–	○		
Calcium Bisulfate, $CaHSO_4$		×					×							
Calcium Bisulfite, Ca HSO_3		×	×		×		×			×	×	×	×	
Calcium Carbonate, $CaCO_3$		×					×			×	×	–		
Calcium Chlorate, $CaClO_3$	×	×					×							
Calcium Chloride, $CaCl_2$	×	×	×	×	×	×	×	×		×	×	×	×	×
Calcium Hydroxide, Ca $(OH)_2$		×					×			×	×	×		
Calcium Hypochlorite, Ca $(OCl)_2$	×	×	×	○	×		×	×		×	×	○	×	–
Calcium Sulfate, Ca SO_4		×					×			×	×	×		
Carbon Dioxide (Dry), CO_2		×					×			×	×	×		
Carbon Dioxide (Wet or H_2CO_3)	×	×	×		×	×	×	×	×	×	×	×	×	
Carbon Disulfide, CS_2	×	×	×	×	×		×			○	○	○	×	

THERMOPLASTICS

Chemicals	Viton	Asphaltic, Bitumastic	Cellulose Acetate	Cellulose Acetatebutyrate	Ethyl Cellulose (Ethocel)	Cellulose Nitrate	Acrylic (Lucite, Plexiglas)	Coumarone Resins	Polyethylene	Polyvinyl Chloride, Rigid or Unplasticized	Tygon (P. V. C. & Copolymers)	Saran (Vinyl chloride, vinyldene chloride)	Kel-F (Polytrifluorochloroethylene)	Teflon (Polytetrafluoroethylene)	Uscolite CP (Styrene-acrylonitrile-butadiene)	Penton (Chlorinated Polyether)	Shellac Compounds	Organic Polysulfides	Polystyrene (Styron)	Vinylidene Chlorides
Amyl Chloride, $C_5H_{11}Cl$	–	○								○		×	×	×		×				×
Antimony Trichloride, $SbCl_3$	×				–					×		×	×	×	×	×				
Arsenic Acid, $HAsO_8$	×								×	–			×	×		×				
Barium Carbonate, $BaCO_3$	×									×			×	×	×	×				
Barium Hydroxide, $Ba(OH)_2$	×	×							×	×			×	×	×	×				
Barium Sulfide, BaS	×								×	–	×	×	×	×	×	×				
Benzaldehyde, C_6H_5 CHO		○							○	○		–	×	×		–	○			×
Benzene, C_6H_6	–	○	–	○	○		○		–	○	○	–	×	×		+			○	
Benzoic Acid, C_6H_5 COOH		×								×	×	×	×	×		×				
Borax, $Na_2B_4O_7$	×	×	×	×	×	×	×	×		×			×	×	×	×	×		×	×
Boric Acid, H_3BO_3	×									–	×	×	×	×	×	×				
Bromine, Wet, Br_2	–	○							○	○	–	○	×	×		–			–	
Butanol, C_4H_9OH							×			×		×	×	×	×	×				
Butyl Acetate, $C_4H_9COOCH_3$	○	○	○	○	○	○	○			○		○	×	×		+			○	○
Butyric Acid, C_3H_7 COOH	○	○								○		–	×	×		×				
Calcium Bisulfate, $CaHSO_4$													×	×						
Calcium Bisulfite, Ca HSO_3										×		×	×	×	×	×				
Calcium Carbonate, $CaCO_3$	×									×		×	×	×		×				
Calcium Chlorate, $CaClO_3$		×								×			×	×	×	×				
Calcium Chloride, $CaCl_2$	×	×	×	×	×	×	×	×	×	×	×	×	×	×	×	×	×	×	×	×
Calcium Hydroxide, Ca $(OH)_2$	×	×								×		×	×	×	×	×				
Calcium Hypochlorite, Ca $(OCl)_2$		–							×	×	×	×	×	×	×	×		×	×	
Calcium Sulfate, Ca SO_4	×	×								×		×	×	×	×	×				
Carbon Dioxide (Dry), CO_2		×								×		×		×	×	×				
Carbon Dioxide (Wet or H_2CO_3)		×	×	×	×	×	×	×	×	×	×	×		×	×	×	×	×	×	×
Carbon Disulfide, CS_2	×	○	○	○	○	×	×		○	○		–	×	×		–	–		○	○

THERMOSETTING PLASTICS; WOODS

Chemicals	Vinyl Chloride Acetates	Cast Phenol Formaldehyde	Haveg 41 (Phenolic w. asbestos)	Heresite (Phenol formaldehyde)	Molded Phenolformald, (Durez)	Phenol Furfural Plastics	Urea Formaldehyde	Casein Plastics	Epoxy Resins	Furane Resins (Haveg 61, Duralon)	Silicone Resins	Permanite (Furan, Class Fiber)	Nylon (Adipic Acid—Hexameth, Diamine)	Durcon 6 (Modified Epoxy)	Cypress	Fir	Maple	Oak	Pine	Redwood
Amyl Chloride, $C_5H_{11}Cl$	×	–			–	–		×	×					×						
Antimony Trichloride, $SbCl_3$									×				○	×						
Arsenic Acid, $HAsO_8$													×	×						
Barium Carbonate, $BaCO_3$									×				×	×						
Barium Hydroxide, $Ba(OH)_2$									×				×	×						
Barium Sulfide, BaS													×	×						
Benzaldehyde, C_6H_5 CHO		×			×	×			–					×	×					
Benzene, C_6H_6	○		–	×	×	×			×	×		×	×	×**						
Benzoic Acid, C_6H_5 COOH									×				○	+**						
Borax, $Na_2B_4O_7$	×	×			×	×							×	×	×	–	–	–	×	–
Boric Acid, H_3BO_3			×	×	×	×				×		×	×	×						
Bromine, Wet, Br_2			○	×	–	–	–		○	○		○	○	×**						
Butanol, C_4H_9OH									–			×		×						
Butyl Acetate, $C_4H_9COOCH_3$	○	–			–	–	○	×	×					×						
Butyric Acid, C_3H_7 COOH									–				○	×						
Calcium Bisulfate, $CaHSO_4$														×						
Calcium Bisulfite, Ca HSO_3		×			×	×								×						
Calcium Carbonate, $CaCO_3$													×	×						
Calcium Chlorate, $CaClO_3$									×					×						
Calcium Chloride, $CaCl_2$	×	–	×	×	–	–			×	×		×	○	×	×	×	×	×	×	–
Calcium Hydroxide, Ca $(OH)_2$									×			×	×	×						
Calcium Hypochlorite, Ca $(OCl)_2$	×	○	–	–	○	○			×	×			○	×	○	○	○	○	–	○
Calcium Sulfate, Ca SO_4									×				×	×						
Carbon Dioxide (Dry), CO_2									×					○						
Carbon Dioxide (Wet or H_2CO_3)	×	×	×	×	×	×	×	×						×	×	×	×	×	×	×
Carbon Disulfide, CS_2		×			×	×			○	×		×	×	×						

*NOTE: Duriron is as shown. Durichlor is also satisfactory on chlorides and HCl.
**Durcon 5 would be the preferred formula.

Table A.3
CHEMICAL RESISTANCE OF MATERIALS *continued*

MATERIALS
× — Very Good Service
+ — Moderate Service
– — Limited or Variable Service
O — Unsatisfactory
Blank — No Information

CHEMICALS
Solids Assumed in Sol'n.
Room Temperatures Assumed
Unless Otherwise Stated

METALS

CHEMICALS	Carbon Steel; Fe	Cast Iron & Ductile Iron; Fe	304 Stainless Steel; Fe, 18Cr, 8Ni	316 Stainless Steel; Fe, 16Cr, 10Ni, 2Mo	347 Stainless Steel; Fe, 17Cr, 9Ni; (Cx10)Cb	Ni-Resist Iron; Fe, 14Ni, 2Cr, 2Si	Durimet 20; Carpenter 20; Fe; 4Cu, 20Cr, 29Ni, 2Mo, 1Si	Worthite; 3Mo, 2Cu, Fe, 20Cr, 24Ni, 3Si	Duriron; Fe, 14Si; Durichlor, Fe, 14Si, 3Mo*	Copper; Brass; Bronzes; Everdur	Aluminum; Al (and Alloys)	Lead; Pb	Monel; 67Ni, 30Cu, 1.4 Fe	Nickel; Ni	Inconel; 76Ni, 15Cr, 8Fe	Hastelloy B; Ni, 26Mo, 4Fe	Hastelloy C; Ni, 16Mo, 4Fe, 14Cr, 4W	Hastelloy D; Ni, 8Si, 3Cu	Chlorimet 3; 3Fe, 1Si, 60Ni, 18Mo, 18Cr	Chlorimet 2; 63Ni, 32Mo, 3Fe, 1Si	Stellite; Co, 28Cr, 4W	Zirconium; Zr	Tantalum; Ta	Silver; Ag	Platinum; Pt	Dowmetal; (Mg alloys)	Titanium; Ti	Molybdenum; Mo
Carbon Tetrachloride (Moist) CCl_4	–	–	O	–	–	–	×	×	×	–	–	+	+	+	×	+	×	+	×	×	×	×	×	×			×	
Chloracetic acid, $ClCH_2CO_2H$	O	O	O	O	O		–	O	×	O	O	O	–	–	+	×	×	–	×	×	–		×	×	×		×	
Chloric Acid, $HClO_3$			O	O																				O				
Chlorine (Dry), Cl_2	×	×	×	×	×		×	×	×	×	O	×	×	×	×	×	×	×	×	×			×	×	×		O	+
Chlorine (Wet), Cl_2	O	O	O	O	O	O	O	O	–	O	O	–	O	–	O	O	×	O	×	O		O	×	×	×		–	O
Chlorobenzene, C_6H_5Cl	×		×	×	×		×	×	×	+			–	–	–				×	×			×					
Chloroform, $CHCl_3$	×		×	×	×		×	×	×	×	–	×	×	×	×	×	×	×	×	×			×				×	
Chromic Acid, CrO_3 solution	–	+	O	–	O	O	+	×	×	O	+	+	O	O	–	O	+	O	+	O	×	×	×	O	×	–	×	+
Copper Chloride, $CuCl_2$	O		O	O	O		O	×	O	O	O		O	O	O	×	×	×	–	O		–	×	–			×	
Copper Cyanide, $Cu(CN)_2$			×	×	×		×	×	×	O	O		O	O	O	×	×	×	×	×			×					
Copper Nitrate, $Cu(NO_3)_2$	O		×	×	–		×	×	×	O	O		O	O	O	O	×	O	×	×			×	O				
Copper Sulfate, $CuSO_4$	O	O	×	×	×	–	×	×	×	O	O	–	–	–	–	×	×	×	×	×	×	×	×	–	×			
Cresylic acid	O		×	×	×		×	×	×	O	–		–	×	–	–	–	–	×	×								
Dichlorethane, $C_2H_4Cl_2$	O		×	×	×		×		×				×	×	×	×	×		×	×								
Diethylamine, $(C_2H_5)_2NH$	×			×			×												×	×								
Diphenyl, $C_6H_5C_6H_5$	×		×				×		×					×					×	×								
Ethers, Various	×		×	×	×		×	×	×	×	×	×	+	+	+	×	×	×	×	×								
Ethyl Acetate, $C_2H_5COOCH_3$	×		×	×		×	×	×	×	×	×		×	×	×	×	×	×	×	×			×					
Ethyl Alcohol, C_2H_5OH	+	+	×	×	×	×	×	×	×	+	–	+	+	+	+	×	×	+	×	×		×	×			×		
Ethyl Chloride, C_2H_5Cl	×		×	×	×	O	×	–	×	–	–		+	+	×	×	×	×	×	×			×					
Ethylene Chlorohydrin, $Cl(C_2H_4)OH$	×		×				×		×			×							×	×								
Ethylene Dichloride, $C_2H_4Cl_2$	×		×	×	×		×		×	–	O	×	+	+	–	×			×	×			×				×	
Ethylene Glycol, CH_2OHCH_2OH	–	+	×	×	×	×	×	×	×	+	×	+	+	+	+				×	×								
Ethylene Oxide, CH_2OCH_2	×		×				×		×										×	×								
Fatty Acids, Various	O	O	×	×	×	–	×	×	×	–	×	O	+	×	×	×	×	×	×	×	×		×	×	×			
Ferric Chloride, $FeCl_3$	O	O	O	O	O	O	O	O	–	O	O	O	O	O	O	O	+	O	–	O	–	O	×	O			×	O
Ferric Nitrate, $Fe(NO_3)_3$	O		×	×	×		×	×	×	O	O		O	O	O	O	+	O	×				×	O				
Ferric Sulfate, $Fe_2(SO_4)_3$	O		–	×	×	O	×	×	×	O	O	×	O	O	O	O	+	O	×		×		×	O			×	

CARBONS & CERAMICS; RUBBERS

CHEMICALS	Carbon & Graphite	Glass, "Pyrex" brand	Silicaware	Silicate Cements	Chemical Stoneware	Transite (asbestos & cement)	Chemical Porcelain	Concrete-Unbonded	Concrete-Mortar Bonded	Hard Rubber (Natural)	Soft Rubber (Natural)	Neoprene	Butadiene Derivatives	Nitrile Rubber (Chemigum)
Carbon Tetrachloride (Moist) CCl_4	×	×	×	×	×	×	×	×		O	O	O	×	–
Chloracetic acid, $ClCH_2CO_2H$		×			×		×			–	–	–	×	
Chloric Acid, $HClO_3$		×												
Chlorine (Dry), Cl_2	×	×	×		×		×			×	–	O	–	
Chlorine (Wet), Cl_2	–	×	×		×		×			×	×	O	×	
Chlorobenzene, C_6H_5Cl		×	×		×		×			O	O	O		
Chloroform, $CHCl_3$	×	×	×		×		×			O	O	O		
Chromic Acid, CrO_3 solution	O	×	×	×	×		×			O	O	O	O	O
Copper Chloride, $CuCl_2$		×	×		×		×			–	×	×	×	
Copper Cyanide, $Cu(CN)_2$		+					×							
Copper Nitrate, $Cu(NO_3)_2$		×					×			×		×		
Copper Sulfate, $CuSO_4$	×	×	×	×	×		×			×	×	×	×	O
Cresylic acid		×					×			O	O	O		
Dichlorethane, $C_2H_4Cl_2$		×					×							
Diethylamine, $(C_2H_5)_2NH$	×	×					×					O		
Diphenyl, $C_6H_5C_6H_5$		×					×							
Ethers, Various	×	×	×		×	×	×	×	×	O	O	O	O	
Ethyl Acetate, $C_2H_5COOCH_3$	×	×	×		×	×	×	×	×	O	O	O	O	
Ethyl Alcohol, C_2H_5OH	×	×	×	×	×	×	×		×	×	×	×	×	×
Ethyl Chloride, C_2H_5Cl	×	×					×			O		–		
Ethylene Chlorohydrin, $Cl(C_2H_4)OH$	–	×	×		×		×			–	–	–	–	
Ethylene Dichloride, $C_2H_4Cl_2$	×	×	×		×		×			O	O	O	O	
Ethylene Glycol, CH_2OHCH_2OH	×	×	×	×	×		×			–	–	–	×	×
Ethylene Oxide, CH_2OCH_2		×					×							
Fatty Acids, Various	×	×	×	×	×	×	×	×	×	O	O	O	×	
Ferric Chloride, $FeCl_3$	×	×	×	×	×		×	×		×	×	–	×	
Ferric Nitrate, $Fe(NO_3)_3$		×					×			O		×		
Ferric Sulfate, $Fe_2(SO_4)_3$	–	×	×		×		×			×	×	×	×	

THERMOPLASTICS

CHEMICALS	Viton	Asphaltic, Bitumastic	Cellulose Acetate	Cellulose Acetatebutyrate	Ethyl Cellulose (Ethocel)	Cellulose Nitrate	Acrylic (Lucite, Plexiglas)	Coumarone Resins	Polyethylene	Polyvinyl Chloride, Rigid or Unplasticized	Tygon (P. V. C. & Copolymers)	Saran (Vinyl chloride, vinylidene chloride)	Kel-F (Polytrifluorochloroethylene)	Teflon (Polytetrafluoroethylene)	Uscolite CP (Styrene-acrylonitrile-butadiene)
Carbon Tetrachloride (Moist) CCl_4	×	O	×	–		×	×		O	–	O	–	×	×	
Chloracetic acid, $ClCH_2CO_2H$	×	O			–					×	O	–	×	×	
Chloric Acid, $HClO_3$													×	×	
Chlorine (Dry), Cl_2	×	O	–	–					×	–	×	–	×	×	×
Chlorine (Wet), Cl_2		–	–	–					×	–	×	–	×	×	×
Chlorobenzene, C_6H_5Cl	–	O					–		O	O		–	×	×	
Chloroform, $CHCl_3$	–	O	O	O	O		O		O	O		–	×	×	
Chromic Acid, CrO_3 solution	–	O	O	O	O		O	O	×	–	×	×	×	×	–
Copper Chloride, $CuCl_2$	×	×							×	×		×	×	×	×
Copper Cyanide, $Cu(CN)_2$		×								×				×	×
Copper Nitrate, $Cu(NO_3)_2$	×								×	×		×	×	×	×
Copper Sulfate, $CuSO_4$	×	×	×	×		×	O		×	×	×	×	×	×	×
Cresylic acid	–									–			+	×	
Dichlorethane, $C_2H_4Cl_2$													–	×	
Diethylamine, $(C_2H_5)_2NH$										O			×	×	
Diphenyl, $C_6H_5C_6H_5$													×	×	
Ethers, Various	O	O	O	O	O	O	O		O	O		–	×	×	
Ethyl Acetate, $C_2H_5COOCH_3$	O	O	O	O	O	O	O		O	O		–	×	×	
Ethyl Alcohol, C_2H_5OH	×	×	×	–	–		–		O	×	–	×	×	×	×
Ethyl Chloride, C_2H_5Cl	×	O								O		–	×	×	
Ethylene Chlorohydrin, $Cl(C_2H_4)OH$		O	O	O						O		O	×	×	
Ethylene Dichloride, $C_2H_4Cl_2$	–	O	O				O		–	O		–	×	×	
Ethylene Glycol, CH_2OHCH_2OH	×	×	×	×		×	×		O	×	–		×	×	×
Ethylene Oxide, CH_2OCH_2	–									O		O		×	
Fatty Acids, Various	×	O	×	×		×			O	–	×	×	×	×	×
Ferric Chloride, $FeCl_3$	×	×			–	×			×	×	×	–	×	×	×
Ferric Nitrate, $Fe(NO_3)_3$	×	×								×		×	×	×	×
Ferric Sulfate, $Fe_2(SO_4)_3$	×	×			–	×				×	×	×	×	×	×

THERMOSETTING PLASTICS; WOODS

CHEMICALS	Penton (Chlorinated Polyether)	Shellac Compounds	Organic Polysulfides	Polystyrene (Styron)	Vinylidene Chlorides	Vinyl Chloride Acetates	Cast Phenol Formaldehyde	Haveg 41 (Phenolic w. asbestos)	Heresite (Phenol formaldehyde)	Molded Phenolformald. (Durez)	Phenol Furfural Plastics	Urea Formaldehyde	Casein Plastics	Epoxy Resins	Furane Resins (Haveg 61, Duralon)	Silicone Resins	Permanite (Furan, Glass Fiber)	Nylon (Adipic Acid—Hexameth. Diamine)	Durcon 6 (Modified Epoxy)	Cypress	Fir	Maple	Oak	Pine	Redwood
Carbon Tetrachloride (Moist) CCl_4	×	O	–	O	×	×	×	×	×	×	×			×	×		×	×	×						
Chloracetic acid, $ClCH_2CO_2H$	+		×	×				×	×	×	×							O	×						
Chloric Acid, $HClO_3$																			O						
Chlorine (Dry), Cl_2	+			×	×	×	–	–	–	–	–			–		–	–	O	O						
Chlorine (Wet), Cl_2	+			×	×	×	–	–		–	–			O		–	–	O	+						
Chlorobenzene, C_6H_5Cl	+		O	O	O			–						–	×		×		–**						
Chloroform, $CHCl_3$	+													×				×	×**						
Chromic Acid, CrO_3 solution	–			O	×		O	O	×	O	O			O	O		O	O	×						
Copper Chloride, $CuCl_2$	×	–						×						×	×			O	×						
Copper Cyanide, $Cu(CN)_2$	×													×					×						
Copper Nitrate, $Cu(NO_3)_2$	×																	×	×						
Copper Sulfate, $CuSO_4$	×	×		×			×	×	×	×	×			×	×		×	O	×						
Cresylic acid	×																	O	×						
Dichlorethane, $C_2H_4Cl_2$																			×						
Diethylamine, $(C_2H_5)_2NH$	×																		×						
Diphenyl, $C_6H_5C_6H_5$																			×						
Ethers, Various	×			O	O	O	–	–		–	–		×	–	×		×		×						
Ethyl Acetate, $C_2H_5COOCH_3$	+			O	O	O	–	O		–	–		×	×	×		×	×	×						
Ethyl Alcohol, C_2H_5OH	×	O	–	–	×	×		×		×	×	×		×	–			×	×						
Ethyl Chloride, C_2H_5Cl	×													×				×	×						
Ethylene Chlorohydrin, $Cl(C_2H_4)OH$	×				–		–			–	–			×					×	×					×
Ethylene Dichloride, $C_2H_4Cl_2$	+			O	O	O		×						×			×	×	×						
Ethylene Glycol, CH_2OHCH_2OH	×			×	–	–	–	×		–	–			×	×			×	×				×		
Ethylene Oxide, CH_2OCH_2	×																	×	×						
Fatty Acids, Various	×			O	×	–	–	×	×	×	–	×		×	×	–	×	×	×	×	×	×	×	×	–
Ferric Chloride, $FeCl_3$	×	×		×	×	×	×	×	×	×	×	×		×	×			O	+**						
Ferric Nitrate, $Fe(NO_3)_3$	×							×						×	×		×		×						
Ferric Sulfate, $Fe_2(SO_4)_3$	×	×		×	×	×	×	×	×	×	×	×		×	×		×	×	×						

* NOTE: Duriron is as shown. Durichlor is also satisfactory on chlorides and HCl.
** Durcon 5 would be the preferred formula.

Table A.3
CHEMICAL RESISTANCE OF MATERIALS *continued*

MATERIALS
× — Very Good Service
\+ — Moderate Service
– — Limited or Variable Service
O — Unsatisfactory
Blank — No Information

CHEMICALS
Solids Assumed in Sol'n. Room Temperatures Assumed Unless Otherwise Stated

Chemicals	Carbon Steel; Fe	Cast Iron & Ductile Iron; Fe	304 Stainless Steel; Fe, 18Cr, 8Ni	316 Stainless Steel; Fe, 16Cr, 10Ni, 2Mo	347 Stainless Steel; Fe, 17Cr, 9Ni; (Cx10)Cb	Ni-Resist Iron; Fe, 14Ni, 2Cr, 2Si	Durimet 20; Carpenter 20; Fe; 4Cu, 20Cr, 29Ni, 2Mo, 1Si	Worthite; 3Mo, 2Cu, Fe, 20Cr, 24Ni, 3Si	Duriron; Fe; 14Si; Durichlor, Fe, 14Si, 3Mo*	Copper; Brass; Bronzes; Everdur	Aluminum; Al (and Alloys)	Lead; Pb	Monel; 67Ni, 30Cu, 1.4 Fe	Nickel; Ni	Inconel; 76Ni, 15Cr, 8Fe	Hastelloy B; Ni, 26Mo, 4Fe	Hastelloy C; Ni, 16Mo, 4Fe, 14Cr, 4W	Hastelloy D; Ni, 8Si, 3Cu	Chlorimet 3; 3Fe, 1Si, 60Ni, 18Mo, 18Cr	Chlorimet 2; 63Ni, 32Mo, 3Fe, 1Si	Stellite; Co, 28Cr, 4W	Zirconium; Zr	Tantalum; Ta	Silver; Ag	Platinum; Pt	Doumetal; (Mg alloys)	Titanium; Ti	Molybdenum; Mo	Carbon & Graphite	Glass, "Pyrex" brand	Silicaware	Silicate Cements	Chemical Stoneware	Transite (asbestos & cement)	Chemical Porcelain	Concrete-Unbonded	Concrete-Mortar Bonded	Hard Rubber (Natural)	Soft Rubber (Natural)	Neoprene	Butadiene Derivatives	Nitrile Rubber (Chemigum)	Viton	Asphaltic, Bitumastic	Cellulose Acetate	Cellulose Acetatebutyrate	Ethyl Cellulose (Ethocel)	Cellulose Nitrate	Acrylic (Lucite, Plexiglas)	Coumarone Resins	Polyethylene	Polyvinyl Chloride, Rigid or Unplasticized	Tygon (P. V. C. & Copolymers)	Saran (Vinyl chloride, vinyldene chloride)	Kel-F (Polytrifluorochloroethylene)	Teflon (Polytetrafluoroethylene)	Uscolite CP (Styrene-acrylonitrile-butadiene)	Penton (Chlorinated Polyether)	Shellac Compounds	Organic Polysulfides	Polystyrene (Styron)	Vinylidene Chlorides	Vinyl Chloride Acetates	Cast Phenol Formaldehyde	Haveg 41 (Phenolic w. asbestos)	Heresite (Phenol formaldehyde)	Molded Phenolformald, (Durez)	Phenol Furfural Plastics	Urea Formaldehyde	Casein Plastics	Epoxy Resins	Furane Resins (Haveg 61, Duralon)	Silicone Resins	Permanite (Furan, Glass Fiber)	Nylon (Adipic Acid—Hexameth, Diamine)	Durcon 6 (Modified Epoxy)	Cypress	Fir	Maple	Oak	Pine	Redwood
	METALS																												CARBONS & CERAMICS									RUBBERS						THERMOPLASTICS																			THERMOSETTING PLASTICS														WOODS					
Ferrous Chloride, Fe Cl_2	O	O	O	O	O	×	–	O	+	O	O	O	O	O	O	–	+	+	+	–			×	O					×	×	×	×	×		×			×	×	×	×		×	×				×				×		–	×	×	×	×	×		×	×	×	×	×		×	×	×		×	×			O	×						
Ferrous Sulfate, $FeSO_4$	O	O	+	O	×	–	×	×	×	–	+	–	O	O	O	+	+	+	×	×	–		×	O			×		×	×	×	×	×		×			×	×	×	×	×	×	×				×			×	×	×	–	×	×	×	×	×		×	×	×	×	×	×	×	×	×		×	×		×	O	+						
Fluorine, F_2			O	O			–		O	O	O	–	O	×	×	+	×		+				O	×			O	O	O	O					O									O							–	–				–	O	O							O						O	O			O							
Formaldehyde, CH_2O	–	–	–	–	–	–	×	×	×	×	–	O	–	×	–	×	×	×	×	×	×		×	×		–	×		×	×	×		×		×	×		–	–	–	×		O	×	O	×		×	×	×	×	–	–	×	×	×	–	×	×			×	×		×	×			×	×	×				×	×						
Formic Acid, HCOOH	O	O	+	+	–		×	×	×	O	–	O	+	+	–	–	×	–	×	×	×	×	×				O	+	×	×	×	×	×		×			–	O	–	×	×	+	O			–		O		×	–	×	–	×	×	–	×	–		×	×	×	×	×	×	×	×	×	×	×	×	×			O	×						
Fuel Oil	×		–	×	×	×	×		×	×	×		×	×	+	×	×	×	×	×			×							×					×											×					O	×			×	×					–	×			×											+**						
Gallic Acid, $(OH)_3 C_6H_2COOH$	O		×	×	×		×		×	×	×		–	+	+	×	×	×	×	×										×					×								+									×	×		×	×		×																			×	×						
Gasoline, Refined	×		×	×	×	×	×		×	×	×		+	+	+	×	×	×	×	×			×						×	×	×		×		×			–	O	–	×		×	O	×	×	–	×	×			–	–	×	×	×	–	×			–	–	–	–	×		–		–			×	×	×		×	×	×						
Glycerol, CH_2 OH CHOH CH_2 OH	+		+	+	+	×	×		×	×	×	–	+	×	×	×	×	×	×	×				×			×		×	×	×		×		×	×		×	×	–	–	×	×	×	×	×		×	×		×	×	×	×	×	×	×	×			×	–	–	–			–	–			×			×	×	×						
Hydrobromic Acid, HBr	O		O	O	O		O	O	–	–	O	O	O	O	O	×	–	–	–	+			×				×		×	×	×		×		×			–	×	–	×		+	O	×						×	×	–	×	×	×	–	×	×	×	×	×	×	×	×		×			×	×			×	–		×	O	×						
Hydrochloric Acid, (Conc.), HCl	O	O	O	O	O	O	O	O	+	O	O	O	O	O	O	+	–	O	+	×	×	×	×	×	×		O	×	×	×	×	×	×		×			×	×	–	×	O	–	×	O	O	O	O	×	×	×	×	×	×	×	×	×	×	×	O	O	O		–	×	×	–	–			×	–	–	×	O	×	O	–	O	O	O	O
Hydrochloric Acid, (Dilute)	O	O	O	O	O	–	O	O	×	O	O	–	–	–	–	+	–	–	×	×	–	×	×	×	×		–		×	×	×		×		×			×	×	–	×	–		×	×	×	–	×	×	×	×	×	×	×	×	×	×	×	×	×	×	×	×	×	×	×	–	–		×	×		–	×	O	×	×	×	×	×	×	–
Hydrochloric Acid (Dry Gas)	×		–	×			×		×		O	–	–	–	–	×	×		×	×		×	×						×	×					×									O								×			×	×	×															×					×						
Hydrocyanic Acid, (Conc.), HCN	O		×	×	×	×	×	×	×	O	×	–	×	–	–	×	×	×	×	×			×							×					×									×								×			×	×	–															×					×						
Hydrocyanic Acid, (Dilute & Gas)			O	O																			×							×					×			–		–												×		×		×		–														×					×						
Hydrofluoric Acid, (Conc.), HF	–	O	O	O	O	O	+	–	O	O	O	O	+	–	–	+	+	+	×	×	–		O	×	×	×	O		O	O			O		O			O	O	O		O							×		×	–	O	O	×	×		+			×			–	O	O	–	–		–	×	O	O	–		×						
Hydrofluoric Acid, (Dilute)	O	O	O	O	O	–	+	O	O	–	O	+	+	+	–	+	+	+	×	×	–		O		×	O	O		×	O		O	O		O			×	×	–		×	+	O					×		×	–	–	–	×	×	×	×			×				O	–	×	×			×	×	O	×	O	×						
Hydrofluosilicic Acid, H_2SiF_6	O		O	O	O	O	+		O	O	–		–	–	–	–	–	–	×	×			O						×	–					O					×													–			×		×																	×		×						
Hydrocarbons (Aliphatic)	×	×	×	×	×	×	×		×	×	×		×	×	×	×	×	×	×	×	×		×	×	×	×		+	×	×	×		×		×			O	O	×	×					–	–	×	×				–		×	×			–	–	×	–	–	–	×	×	–	–	×	×						+**						
Hydrocarbons (Aromatic)	×	×	×	×	×	×	×		×	×	×		×	×	×	×	×	×	×	×	×		×	×	×	×			×	×	×		×		×			O	O		O				×	–	–		–				O		+	×			O	O	O	–	–	–	×	×	–	–		×						×						
Hydrogen Gas, H_2	×						×		×	×	×	×	×	×					×	×				×				×		×					×								×	×											×	×		×															×				×	×						
Hydrogen Peroxide (Conc.), $H_2 O_2$	O	O	+	O	–	O	+	–	+	O	×	–	–	–		–	×	–	×	×	×		×	O	×		×			×	×		×		×			–	–	O	×			–	–				×			×	–		×	×	–	+		O	O	O		–	×	×	×	–							O	×						
Hydrogen Peroxide (Dilute)	O	O	+	+			×	×	×	O	×	O	+	+	+	×	×	–	×	×	×	×	×	×	×		–		×	×	×	×	×		×			–	–	O	×	×		–	–				×		×	×	×	×	×	×	×			O	O	–		×	×	×	–	–			O	O			O	×						
Hydrogen Sulfide (Dry) H_2 S	×		–	×		–	×		×	O	O	×	–	×	+	–	×	–	×	+			×	O				+		×	×		×		×	×		×	×		×		×	×								×			×	×	×	×		–		×		–	×		×	–			×	×		×	O	×						
Hydrogen Sulfide (Wet)	–		–	×	–	–	×	×	+	O	O	O	O	×	+	–	+	–	×	+				O						×					×								×									×			×	×		×																			×						
Iodine, I_4 Wet	O	O	O	O	O		O		O	O	O	O	O	O	–	O	–	O	+	O	–		×	–	×	–			O	×	–				×	×		–	–	O	–		×	O	–						–	O		–	×	×								O	×					O	O		O	O	×						
Iodoform, CHI_3	O		O	×	×		×		×		×								×	×										×					×								–												×																				×	×						
Kerosene	×		×	×	×		×		×	×	×		×	×	×	×	×	×	×	×			×						×	×					×			O	O	O			×						×			×		×	×	×	×	×																		×	×						
Ketones, Various	×		×				×		×		–	×							×	×									–	×					×			O		O			O	O		O			–		O	O	O	O	+	×	O	+			O	–			×						–			×	×	×						
Lactic Acid, CH_3 CHOHCOOH	O	O	–	–	×	–	×	×	×	–	–	O	–	–	–	×	×	×	×	×	×	×	×	×	×		×	×	×	×	×		×		×			×	–	×			+		–	×	–				×	×	×	×	×	×	–	×			×				×	×					×	×		×	–	×						×
Lead Acetate, $Pb(CH_3 COO)_2$	O		×	×	+		×	×	×	–	O		×	+		×	×	×	×	×			×				×			×					×									×								×			×	×	×	×							×						×	×		×		×						

* NOTE: Duriron is as shown. Durichlor is also satisfactory on chlorides and HCl.

** Durcon 5 would be the preferred formula.

Table A.3

CHEMICAL RESISTANCE OF MATERIALS *continued*

MATERIALS × — Very Good Service + — Moderate Service – — Limited or Variable Service ○ — Unsatisfactory Blank — No Information *CHEMICALS* *Solids Assumed in Sol'n. Room Temperatures Assumed Unless Otherwise Stated*	METALS[*]: *Carbon Steel; Fe / Cast Iron & Ductile Iron; Fe / 304 Stainless Steel; Fe, 18Cr, 8Ni / 316 Stainless Steel; Fe, 16Cr, 10Ni, 2Mo / 347 Stainless Steel; Fe, 17Cr, 9Ni; (Cx10)Cb*	*Ni-Resist Iron; Fe, 14Ni, 2Cr, 2Si / Durimet 20; Carpenter 20; Fe; 4Cu, 20Cr, 29Ni, 2Mo, 1Si / Worthite; 3Mo, 2Cu, Fe, 20Cr, 24Ni, 3Si / Duriron; Fe, 14Si; Durichlor, Fe, 14Si, 3Mo[*] / Copper; Brass; Bronzes; Everdur*	*Aluminum; Al (and Alloys) / Lead; Pb / Monel; 67Ni, 30Cu, 1.4 Fe / Nickel; Ni / Inconel; 76Ni, 15Cr, 8Fe*	*Hastelloy B; Ni, 26Mo, 4Fe / Hastelloy C; Ni, 16Mo, 4Fe, 14Cr, 4W / Hastelloy D; Ni, 8Si, 3Cu / Chlorimet 3; 3Fe, 1Si, 60Ni, 18Mo, 18Cr / Chlorimet 2; 63Ni, 32Mo, 3Fe, 1Si*	*Stellite; Co, 28Cr, 4W / Zirconium; Zr / Tantalum; Ta / Silver; Ag / Platinum; Pt*	*Dowmetal; (Mg alloys) / Titanium; Ti / Molybdenum; Mo*	CARBONS & CERAMICS: *Carbon & Graphite / Glass, "Pyrex" brand / Silicaware / Silicate Cements*	*Chemical Stoneware / Transite (asbestos & cement) / Chemical Porcelain / Concrete-Unbonded / Concrete-Mortar Bonded*	RUBBERS: *Hard Rubber (Natural) / Soft Rubber (Natural) / Neoprene / Butadiene Derivatives / Nitrile Rubber (Chemigum)*	*Viton / Asphaltic, Bitumastic / Cellulose Acetate / Cellulose Acetatebutyrate / Ethyl Cellulose (Ethocel)*	THERMOPLASTICS: *Cellulose Nitrate / Acrylic (Lucite, Plexiglas) / Coumarone Resins / Polyethylene / Polyvinyl Chloride, Rigid or Unplasticized*	*Tygon (P. V. C. & Copolymers) / Saran (Vinyl chloride, vinyldene chloride) / Kel-F (Polytrifluorochloroethylene) / Teflon (Polytetrafluoroethylene) / Uscolite CP (Styrene-acrylonitrile-butadiene)*	*Penton (Chlorinated Polyether) / Shellac Compounds / Organic Polysulfides / Polystyrene (Styron) / Vinylidene Chlorides*	THERMOSETTING PLASTICS: *Vinyl Chloride Acetates / Cast Phenol Formaldehyde / Haveg 41 (Phenolic w. asbestos) / Heresite (Phenol formaldehyde) / Molded Phenolformald, (Durez)*	*Phenol Furfural Plastics / Urea Formaldehyde / Casein Plastics / Epoxy Resins / Furane Resins (Haveg 61, Duralon)*	*Silicone Resins / Permanite (Furan, Class Fiber) / Nylon (Adipic Acid—Hexameth, Diamine) / Durcon 6 (Modified Epoxy)*	*Cypress*	WOODS: *Fir / Maple / Oak / Pine / Redwood*
Magnesium Chloride, $Mg\,Cl_2$	– – – + –	× × × × +	– ○ + + +	× × × × ×	– × ×	×	× × ×	× × × × ×	× × × ×	× × × ×	× × × ×	× × × × ×	× × × × ×	× × × × ×	× × × ×	× ×		
Magnesium Hydroxide, $Mg\,(OH)_2$	× × × +	× × + –	○ × × ×	× × × × ×	× ×	×	×	×	× ×	×	×	× × × ×	×			× ×		
Magnesium Sulfate, $Mg\,SO_4$	+ ○ + + +	× × × × +	× + + + +	× + × × ×	× × ×	×	× × ×	× × × × ×	× × × ×	× × × ×	× × × ×	× × × ×	× × × × ×	× × ×	× × × ×	× ×		
Maleic Acid; $CO_2\,H\,C_2\,H_2CO_2\,H$	○ × ×	× × ×		× × ×	×		×	×	×	+ ○	–	– × × ×	×		×	× +[**]		
Malic Acid, $CO_2\,H\,CH_2\,CHOH\,CO_2\,H$	○ × × ×	+ × ○	– + + +	× × × ×	×		× ×	× ×	× × × ×	–	×	× ×	× ×			× ×		
Mercuric Chloride, $Hg\,Cl_2$	○ ○ ○ ○	– – + ○	○ – – +	○ × ○ + ○	× × ○	×	×	×	× –	× × ×	–	× × × × ×	×		×	×		
Mercury, Hg	× × × ×	× × ○	○ ○ + + ×	× × × × ×	○	–	×	×		× ×	–	× ×	×		×	× ×		
Methanol, (Conc.), CH_3OH	+ + + + +	× × × × +	– + + + +	× × + × ×	× × ×		× ×	×	– × – ×	– ×	○ – ×	× × × ×	× ×	– ×	× × ×	× ×		
Methanol, (Dilute)	–	× × –	–	× × – × ×	×		×	×	× ×		×	× × ×	× ×	×	× ×	× ×		
Methyl Chloride, $CH_3\,Cl$	○ – – +	× × × ×	× × × ×	× ×	×		×	×	○ ○	– ○	○	– × ×	×		×	× ×		
Naphtha, Petroleum	× × × ×	× × × ×	× + + +	× × × × ×			× × ×	× ×	– ○ × ×	+ ○	× × ×	× ×	× – –	– – –	– × ×	× ×		
Nickel Chloride, $Ni\,Cl_2$	○ – –	+ ○ + –	○ – ×	○ × ○ × ×	× ×	×	× ×	×	× ×	× ×	× ×	× × × × ×	×		× ×	×		
Nickel Sulfate, $Ni\,SO_4$	○ ○ – –	× × × +	– + + – –	○ + ○ × ×	× ×		× ×	×	× × ×	× ×	× ×	× × × × ×	× ×	×	× × ×	× ×		
Nitrating Acid (>15% H_2SO_4)	× × – – –	× × × ○	○ ○ ○	○ × ○ × ×	× – × ×	–	○ ×	× ×	– –	○		○ × ×		○ ×	○	×[**]		
Nitrating Acid (<15% $H_2\,SO_4$)	○ – – –	× × × ○	○ ○ ○	○ × ○ × ×	× – ○ ×	×	○ ×	× ×	– –	○		– × ×		– ×		×		
Nitrating Acid (<15% HNO_3)	○ – ○ –	× × ○	○ ○ ○	○ × ○ × ○	× ○ – ○ ×	–	○ ×	× ×	– –	○		– × ×		– ×		×		
Nitrating Acid (<1% Acid)	○ – × –	× × ○	○ ○ ○	○ × ○ ×	× ○ – ○ ×		○ ×	× ×	– –	○		– × ×		– ×		+[**]		
Nitric Acid (Conc.), HNO_3	○ ○ + + ×	× × × ○	– – ○ ○ –	○ + ○ × ○	× × × ○ ×	× +	○ × × ×	× ×	○ ○ ○ ○ ○	○ ○	○ ○	○ – × × ○	○ ○ ○	○ ○ ○ × ○	○ ○ ○	– ○ ○ ×		
Nitric Acid, Dilute	○ ○ × × ×	○ × × × ○	○ ○ ○ ○ ○	○ × ○ × ○	× × × ○ ×	× ○	– × ×	× ×	– ○ ○ × ○	– – –	× – ×	– × × × ×	+ × –	– – – × –	– ○	– ○ ×		
Nitrobenzene, $C_6\,N_5\,NO_2$	× × ×	× × ×	○	× ×	×		×	×	○ ○ ○	○ ○	– ○	– × ×	+		–	×		
Nitrous Acid, HNO_2	○ × × –	× × × ×	× ○	×	×		× ×	× ×	– ○ ○ ×	×		× ×	× ×			×		
Oleic Acid, $C_8\,H_{17}\,CH{:}CH(CH_2)_7CO_2H$	– – – × ×	– × × × –	× ○ × × ×	× × × × ×	× × × ×		× × ×	× × × × ×	○ ○ ○ ×	+ ○ × ×	× ○ ×	× × × × ×	× × ×	× – × × ×	– × ×	× ×	×	× × × –
Oxalic Acid, $CO_2\,H\,CO_2\,H$	○ ○ – – –	– × × × +	– ○ + – ×	+ + + × ×	× × × × ×	– +	× × × ×	× × × ×	× × – ×	– ×	× ×	× × × × ×	× ×	× × × ×	× × ×	× ○ ×		
Phenol (Conc.), $C_6\,H_5\,OH$	+ + – × –	× × × × ○	– – + + ×	+ + + × ×	× × × × ×	× –	× × ×	× × × ×	– ○ – ×	× ○ ○ ○ –	× –	– – × × ○	+ ○	– – × –	– ×	× ○ ×		
Phenol (Dilute)	+ + × × ×	× × –	× – + + –	+ + + × ×	×		×	×	– –	–	×	– – × × ○	○	×	× – ×	×		
Phosphoric Acid (100%), H_3PO_4	○ ○ – – ×	× ×	○ + – ○ –	+ × × × ×	× × × + ×	×	+ – ×	× × × × ×	× × × ×	× × ○	× ×	× × × ×	×	× × ×	–	– – – ×		
Phosphoric Acid (>45% Hot)	○ ○ ○ ○ ×	○ × × × ○	○ + ○ ○ ○	– – – × ×	× – × ×		× – ×	○	– – – ×	+	– ×	× × × ×		× × ○	○ × ×	○ ×		

[*] NOTE: Duriron is as shown. Durichlor is also satisfactory on chlorides and HCl.

[**] Durcon 5 would be the preferred formula.

Table A.3

CHEMICAL RESISTANCE OF MATERIALS *continued*

MATERIALS
× — Very Good Service
\+ — Moderate Service
— — Limited or Variable Service
○ — Unsatisfactory
Blank — No Information

CHEMICALS
Solids Assumed in Sol'n.
Room Temperatures Assumed
Unless Otherwise Stated

METALS

CHEMICALS	Carbon Steel; Fe	Cast Iron & Ductile Iron; Fe	304 Stainless Steel; Fe, 18Cr, 8Ni	316 Stainless Steel; Fe, 16Cr, 10Ni, 2Mo	347 Stainless Steel; Fe, 17Cr, 9Ni; (Cx10)Cb	Ni-Resist Iron; Fe, 14Ni, 2Cr, 2Si	Durimet 20; Carpenter 20; Fe; 4Cu, 20Cr, 29Ni, 2Mo, 1Si	Worthite; 3Mo, 2Cu, Fe, 20Cr, 24Ni, 3Si	Duriron; Fe, 14Si; Durichlor, Fe, 14Si, 3Mo*	Copper; Brass; Bronzes; Everdur	Aluminum; Al (and Alloys)	Lead; Pb	Monel; 67Ni, 30Cu, 1.4 Fe	Nickel; Ni	Inconel; 76Ni, 15Cr, 8Fe
Phosphoric Acid (>45% Cold)	○	○	—	×	×	—	×	—	×	○	○	+	—	—	—
Phosphoric Acid (<45% Cold)	○	○	—	×	×	—	×	—	×	○	—	+	—	—	—
Phosphoric Anhydride, Dry or Moist	○		×	×			×		×						
Phosphoric Anhydride Molten, P_2O_5	×		×	×			×		×	○	○		○	○	
Phthalic Anhydride, $C_6H_4(CO)_2O$	×		×				×	×	×						
Picric Acid, Sol'n., $HOC_6H_2(NO_3)_3$			×	×	+		×	×	×	○	—	—	○	○	○
Potassium Bromide, KBr	○		—	—			×		×	×	—	×	+	×	
Potassium Carbonate, K_2CO_3	—		×	×		—	×	×	×	—	○	×	×	×	+
Potassium Chlorate, $KClO_3$	×		×	×			×	—	×	—	+		+	+	+
Potassium Chloride, KCl	×		—	—		×	×	×	×	—	—		+	+	
Potassium Cyanide, KCN	×		×	×		×	×		×	○	○		×		
Potassium Dichromate, $K_2Cr_2O_7$			×	×	—		×	×	×	×	×		—	—	+
Potassium Ferrocyanide, $K_4Fe(CN)_6$	○		×	×			×	×	×	×	—	×	×	×	×
Potassium Hydroxide, KOH	+	—	+	+	—	—	×	×	+	○	○	○	×	×	×
Potassium Nitrate, KNO_3	×		×	×	×		×	×	×	×	×	—	+	+	+
Potassium Permanganate, $KMnO_4$	+	+	+	+	+		×	×	×	×	×	○	—	+	+
Potassium Sulfate, K_2SO_4	×		+	+	+	×	×	×	×	+	×	×	+	+	+
Potassium Sulfide, K_2S	—		×	×		—	×		×						
Pyrogallol, $C_6H_3(OH)_3$			×	×			×		×	×				+	
Silver Nitrate, $AgNO_3$			×	×			×	×	×	○	○	○			+
Sodium, Molten 210°-400°F.	×	×	×	×	×		×	×	×	×	×		×	×	×
Sodium Acetate, $NaCH_3COO$	○		—	×			×	×	×	×	—	×	+		
Sodium Bicarbonate, $NaHCO_3$	—		×	×	×	×	×	×	×	—	×	×	×	×	×
Sodium Bisulfate, $NaHSO_4$	○	○	—	×	×		×	×	×	—	○	×	+		
Sodium Bisulfite, $NaHSO_3$	○		×	×			×	×	—	—	×	×			

CHEMICALS	Hastelloy B; Ni, 26Mo, 4Fe	Hastelloy C; Ni, 16Mo, 4Fe, 14Cr, 4W	Hastelloy D; Ni, 8Si, 3Cu	Chlorimet 3; 3Fe, 1Si, 60Ni, 18Mo, 18Cr	Chlorimet 2; 63Ni, 32Mo, 3Fe, 1Si	Stellite; Co, 28Cr, 4W	Zirconium; Zr	Tantalum; Ta	Silver; Ag	Platinum; Pt	Dowmetal; (Mg alloys)	Titanium; Ti	Molybdenum; Mo
Phosphoric Acid (>45% Cold)	+	×	+	×	×	×	×	×	×	×		—	
Phosphoric Acid (<45% Cold)	+	×	+	×	×	×	×	×	×	×		×	
Phosphoric Anhydride, Dry or Moist				×	×								
Phosphoric Anhydride Molten, P_2O_5				×	×								
Phthalic Anhydride, $C_6H_4(CO)_2O$	×	×		×	×			×					
Picric Acid, Sol'n., $HOC_6H_2(NO_3)_3$	×	×	×	×	×								
Potassium Bromide, KBr	×	×	×	×	×			×	○			×	
Potassium Carbonate, K_2CO_3	×	×	×	×	×			—					
Potassium Chlorate, $KClO_3$	○	×	○	×	○			×					
Potassium Chloride, KCl	×	×	×	×	×			×				×	
Potassium Cyanide, KCN	×	×	×	×	×			×	○	×			
Potassium Dichromate, $K_2Cr_2O_7$	○	×	○	×	×		×	×	×			×	
Potassium Ferrocyanide, $K_4Fe(CN)_6$	×	×	×	×	×			×	○				
Potassium Hydroxide, KOH	+	+	+	×	×		×	—	×			—	+
Potassium Nitrate, KNO_3	○	×	○	×	×			×	○				
Potassium Permanganate, $KMnO_4$	○	×	○	×	○			×				×	
Potassium Sulfate, K_2SO_4	+	+	+	×	×			×				×	
Potassium Sulfide, K_2S				×	×				○				
Pyrogallol, $C_6H_3(OH)_3$	×	×	×	×	×								
Silver Nitrate, $AgNO_3$	+	+	+	×	×		×	×					
Sodium, Molten 210°-400°F.	×	×		×	×		×	×			—		×
Sodium Acetate, $NaCH_3COO$	×	×	×	×	×							×	
Sodium Bicarbonate, $NaHCO_3$	×	×	×	×	×		×	×	×				
Sodium Bisulfate, $NaHSO_4$	×	×	×	×	×	×		—		×		×	
Sodium Bisulfite, $NaHSO_3$				×	×								

CARBONS & CERAMICS — RUBBERS

CHEMICALS	Carbon & Graphite	Glass, "Pyrex" brand	Silicaware	Silicate Cements	Chemical Stoneware	Transite (asbestos & cement)	Chemical Porcelain	Concrete-Unbonded	Concrete-Mortar Bonded	Hard Rubber (Natural)	Soft Rubber (Natural)	Neoprene	Butadiene Derivatives	Nitrile Rubber (Chemigum)	Viton	Asphaltic, Bitumastic	Cellulose Acetate	Cellulose Acetatebutyrate	Ethyl Cellulose (Ethocel)
Phosphoric Acid (>45% Cold)	×	×			—		—			—	—	—				×			
Phosphoric Acid (<45% Cold)	×	×			—		—			×	×	—							
Phosphoric Anhydride, Dry or Moist	×	×																	
Phosphoric Anhydride Molten, P_2O_5		○																	
Phthalic Anhydride, $C_6H_4(CO)_2O$		×					×			×	×					×			
Picric Acid, Sol'n., $HOC_6H_2(NO_3)_3$		×					×					—				○			
Potassium Bromide, KBr		×					×			×		×							
Potassium Carbonate, K_2CO_3		—					×			×		×			×	×			
Potassium Chlorate, $KClO_3$		×					×								×				
Potassium Chloride, KCl		×					×			×		×			×	×			
Potassium Cyanide, KCN		—					×			×		×			×				
Potassium Dichromate, $K_2Cr_2O_7$		×					×			—	×	×				×			
Potassium Ferrocyanide, $K_4Fe(CN)_6$		×					×									×			
Potassium Hydroxide, KOH		—		○			—			×	×	×		○	×	×			
Potassium Nitrate, KNO_3		×					×			×		×				×			
Potassium Permanganate, $KMnO_4$		×					×			—	○	○			×	○			
Potassium Sulfate, K_2SO_4		×		×			×			×		×		×	×	×			
Potassium Sulfide, K_2S		—					×												
Pyrogallol, $C_6H_3(OH)_3$		×					×												
Silver Nitrate, $AgNO_3$		×					×			×	×				×				
Sodium, Molten 210°-400°F.		—					○										○		
Sodium Acetate, $NaCH_3COO$		×					×					×			○	×			
Sodium Bicarbonate, $NaHCO_3$	×	×	×		×	×	×	×	×	×	×	×	×			×	×	×	
Sodium Bisulfate, $NaHSO_4$	×	×	×		×		×			×	×	×	×						
Sodium Bisulfite, $NaHSO_3$	×	×	×		×		×			×	×	—	×			×			

THERMOPLASTICS — THERMOSETTING PLASTICS

CHEMICALS	Cellulose Nitrate	Acrylic (Lucite, Plexiglas)	Coumarone Resins	Polyethylene	Polyvinyl Chloride, Rigid or Unplasticized	Tygon (P. V. C. & Copolymers)	Saran (Vinyl chloride, vinylidene chloride)	Kel-F (Polytrifluorochloroethylene)	Teflon (Polytetrafluoroethylene)	Uscolite CP (Styrene-acrylonitrile-butadiene)	Penton (Chlorinated Polyether)	Shellac Compounds	Organic Polysulfides	Polystyrene (Styron)	Vinylidene Chlorides
Phosphoric Acid (>45% Cold)				×	×	×	×		×	×	×			×	
Phosphoric Acid (<45% Cold)				×	×	×	×		×	×	×				
Phosphoric Anhydride, Dry or Moist				×					×						
Phosphoric Anhydride Molten, P_2O_5									×						
Phthalic Anhydride, $C_6H_4(CO)_2O$							×		×						
Picric Acid, Sol'n., $HOC_6H_2(NO_3)_3$					○	×	○		×		×				
Potassium Bromide, KBr					×		×		×	×	×				
Potassium Carbonate, K_2CO_3				×	—		×		×	×	×				
Potassium Chlorate, $KClO_3$					×		×		×	×	×				
Potassium Chloride, KCl				×	×		×		×	×	×				
Potassium Cyanide, KCN					×	×	×		×	×	×				
Potassium Dichromate, $K_2Cr_2O_7$				×	×	×	×		×	×	×				
Potassium Ferrocyanide, $K_4Fe(CN)_6$					×		×		×		×				
Potassium Hydroxide, KOH				×	×		—		×	×	×			×	
Potassium Nitrate, KNO_3				×	×		×		×	×	×				
Potassium Permanganate, $KMnO_4$				×	×		—		×	×	×			×	
Potassium Sulfate, K_2SO_4					×		×		×	×	×			×	
Potassium Sulfide, K_2S					×				×	×					
Pyrogallol, $C_6H_3(OH)_3$		×				×			×						
Silver Nitrate, $AgNO_3$		×		×	×		×		×	×	×				
Sodium, Molten 210°-400°F.				○	○	○	○		—		○			○	○
Sodium Acetate, $NaCH_3COO$					×		×		×		×				
Sodium Bicarbonate, $NaHCO_3$	×	×			×		×		×	×	×	×		×	×
Sodium Bisulfate, $NaHSO_4$					×	×	×		×		×			×	×
Sodium Bisulfite, $NaHSO_3$				×	×	×	×		×	×	×				

THERMOSETTING PLASTICS (continued) — WOODS

CHEMICALS	Vinyl Chloride Acetates	Cast Phenol Formaldehyde	Haveg 41 (Phenolic w. asbestos)	Heresite (Phenol formaldehyde)	Molded Phenolformald, (Durez)	Phenol Furfural Plastics	Urea Formaldehyde	Casein Plastics	Epoxy Resins	Furane Resins (Haveg 61, Duralon)	Silicone Resins	Permanite (Furan, Glass Fiber)	Nylon (Adipic Acid—Hexameth, Diamine)	Durcon 6 (Modified Epoxy)	Cypress	Fir	Maple	Oak	Pine	Redwood
Phosphoric Acid (>45% Cold)			×	×					×		—			×						
Phosphoric Acid (<45% Cold)			×	×							—	×	○	×						
Phosphoric Anhydride, Dry or Moist														×						
Phosphoric Anhydride Molten, P_2O_5														×						
Phthalic Anhydride, $C_6H_4(CO)_2O$									×					×						
Picric Acid, Sol'n., $HOC_6H_2(NO_3)_3$														×						
Potassium Bromide, KBr														×						
Potassium Carbonate, K_2CO_3									×				×	×						
Potassium Chlorate, $KClO_3$													○	×						
Potassium Chloride, KCl			×						×	×		×	×	×						
Potassium Cyanide, KCN													×	×						
Potassium Dichromate, $K_2Cr_2O_7$									×					×**						
Potassium Ferrocyanide, $K_4Fe(CN)_6$									×				×	×						
Potassium Hydroxide, KOH			○		○	○			×	×		×	×	×						
Potassium Nitrate, KNO_3									×					×						
Potassium Permanganate, $KMnO_4$					×	×			×	×			○	×						
Potassium Sulfate, K_2SO_4			×		×	×			×	×		×	×	×						
Potassium Sulfide, K_2S													×	×						
Pyrogallol, $C_6H_3(OH)_3$														×**						
Silver Nitrate, $AgNO_3$		—			—				×				×	×						
Sodium, Molten 210°-400°F.			○	○										×						
Sodium Acetate, $NaCH_3COO$									×				—	×						
Sodium Bicarbonate, $NaHCO_3$	×	×	×		×	×			×	×		×	×	×						
Sodium Bisulfate, $NaHSO_4$	×	—	×	×	×	—		×	×	×		×		×						
Sodium Bisulfite, $NaHSO_3$					×			×	×				○	×**	×	×	×	×	×	

* NOTE: Duriron is as shown. Durichlor is also satisfactory on chlorides and HCl.

** Durcon 5 would be the preferred formula.

Table A.3
CHEMICAL RESISTANCE OF MATERIALS *continued*

MATERIALS
× — Very Good Service
+ — Moderate Service
– — Limited or Variable Service
○ — Unsatisfactory
Blank — No Information

CHEMICALS
Solids Assumed in Sol'n.
Room Temperatures Assumed
Unless Otherwise Stated

METALS

CHEMICALS	Carbon Steel; Fe	Cast Iron & Ductile Iron; Fe	304 Stainless Steel; Fe, 18Cr, 8Ni	316 Stainless Steel; Fe, 16Cr, 10Ni, 2Mo	347 Stainless Steel; Fe, 17Cr, 9Ni; (Cx10)Cb	Ni-Resist Iron; Fe, 14Ni, 2Cr, 2Si	Durimet 20; Carpenter 20; Fe; 4Cu, 20Cr, 29Ni, 2Mo, 1Si	Worthite; 3Mo, 2Cu, Fe, 20Cr, 24Ni, 3Si	Duriron; Fe, 14Si; Durichlor, Fe, 14Si, 3Mo*	Copper; Brass; Bronzes; Everdur	Aluminum; Al (and Alloys)	Lead; Pb	Monel; 67Ni, 30Cu, 1.4 Fe	Nickel; Ni	Inconel; 76Ni, 15Cr, 8Fe	Hastelloy B; Ni, 26Mo, 4Fe	Hastelloy C; Ni, 16Mo, 4Fe, 14Cr, 4W	Hastelloy D; Ni, 8Si, 3Cu	Chlorimet 3; 3Fe, 1Si, 60Ni, 18Mo, 18Cr	Chlorimet 2; 63Ni, 32Mo, 3Fe, 1Si	Stellite; Co, 28Cr, 4W	Zirconium; Zr	Tantalum; Ta	Silver; Ag	Platinum; Pt	Doumetal; (Mg alloys)	Titanium; Ti	Molybdenum; Mo
Sodium Borate $Na BO_2$			×	×			×		×	×	–		+			×	×	×	×	×			×					
Sodium Carbonate, $Na_2 CO_3$	+	+	+	+	×	×	×	×	×	+	○	–	+	+	–	+	×	×	×	×	×		–	×	×	×	×	
Sodium Chlorate, $Na ClO_3$	–		×	×			–	–	×	+	+				+	○	×	○	×	○			×				×	
Sodium Chloride, $NaCl$	+	+	–	–	–	–	×	×	+	+	–	+	+	+	–	+	+	–	×	×	–	×	×	×			×	
Sodium Cyanide, $NaCN$	×		×	×			×	×	×	○	○	×	×						×	×				○			×	
Sodium Fluoride, NaF	○		–	–			+	–	○	–	–		+			×	×	×	×	×		×	○				×	
Sodium Hydroxide, (Conc.), $NaOH$	–	–	–	–	×	×	×	×	–	–	○	○	×	×	×	×	+	×	×	×	×	×	○	×	×		×	
Sodium Hydroxide, (Dilute)	×	+	+	+	×		×	×	+	×	○	○	–	–	–	×	+	×	×	×	×	×	–		×	–	×	+
Sodium Hydrosulfite							×		×	–	×		○	○	○	–	×	–	×	×								
Sodium Hypochlorite, $NaOCl$	○	○	–	–	–	–	–	–	×	○	○	○	○	○	○	○	–	○	+	○	–		–	×	×		×	
Sodium Hyposulfate	–		×	×			×		×	○	○		×						×	×				○				
Sodium Nitrate, $Na NO_3$	×	+	×	×	×	–	×	×	×	+	×	○	+	+	×	○	+	○	×	○	×		–	–	×			
Sodium Peroxide, $Na_2 O_2$	–		×	×	+		×		×	–	–		×	+	+	+	×	+	×	×			○	○				
Sodium Phosphate, (Tri) $Na_3 PO_4$	×		×	×	+		×		×	–	○		+	+		×	×	×	×	×			×	×			×	
Sodium Silicate, Na_2SiO_3	×		×	×	+	×	×		×	–	–		+	×	+	×	×	×	×	×			×					
Sodium Sulfate, Na_2SO_4	×	×	–	–	×	×	×	×	×	×	×	–	+	×	×	×	×	×	×	×	–		–	×			×	
Sodium Sulfide, Na_2S	+	×	–	×	×	×	×	×	×	○	○	–	+	+	+	×	×	×	×	×	×		–	○			×	
Sodium Sulfite, $Na_2 SO_3$	+	×	–	–	–	×	×	×	○	–	–	×	–	–	–	○	×	○	×	×	×		–				×	
Stannic Chloride, $Sn Cl_4$	○		○	○	○		×	–	×	○	○	×	○	○	○	○	×	○	–	○			×				×	
Stannous Chloride, $Sn Cl_2$	○		○	–			×	–	×	○	○		○	○	○	×	×	×	×	×		×	×				×	
Stearic Acid, $CH_3 (CH_2)_{16} COOH$	–	–	×	×	×	–	×	×	×	–	–	–	×	×	×	×	×	×	×	×	×		×	×			×	
Sulfur, Molten, S	–	×	–	–	–	–	×	×	×	○	×	○	–	–	×	○	○	○	×	–	○		×	○			×	+
Sulfur Chloride, (Wet), $S_2 Cl_2$	–		○	–		×	–	–	×	○	○		–	–					×					○				
Sulfur Dioxide (Dry), SO_2	×	×	×	×	–	×	×	×	×	×	×	×	○	○	○	○	×	○	×	–	×		×	○		×		
Sulfur Dioxide (Wet)	○		–	×			×		–	○		×	○	○	○	○	×	○	×				×				+	
Sulfur Trioxide, SO_3	×		×	×		×	×	×	–	–	×		○	○	○				×				○	○				
Sulfuric Acid (Fuming to 98%)	–	–	○	○		–	×	×	○	○	–	○	○	○	○	×	○	×	×	×		○	○	○			○	
Sulfuric Acid (Hot Conc.) $H_2 SO_4$	○		○	○			+		×	○	○	–	○	○	○	○	○	×	+	+	○	○	–				○	

CARBONS & CERAMICS

CHEMICALS	Carbon & Graphite	Glass, "Pyrex" brand	Silicaware	Silicate Cements	Chemical Stoneware	Transite (asbestos & cement)	Chemical Porcelain	Concrete-Unbonded	Concrete-Mortar Bonded
Sodium Borate $Na BO_2$		×	×			×	×	×	
Sodium Carbonate, $Na_2 CO_3$	×	–	–	○	×	×	×	×	
Sodium Chlorate, $Na ClO_3$		×	×		×	×	×	×	×
Sodium Chloride, $NaCl$	×	×		×	×		×		
Sodium Cyanide, $NaCN$		–					×		
Sodium Fluoride, NaF		–					×		
Sodium Hydroxide, (Conc.), $NaOH$	×	–	○	○	○		○		
Sodium Hydroxide, (Dilute)	×	–	○	○	–		–		
Sodium Hydrosulfite		×	×		×		×		
Sodium Hypochlorite, $NaOCl$	○	×			×		×		
Sodium Hyposulfate		×					×		
Sodium Nitrate, $Na NO_3$	×	×	×	×	×		×		
Sodium Peroxide, $Na_2 O_2$		–					×		
Sodium Phosphate, (Tri) $Na_3 PO_4$		–	–		×	×	×		
Sodium Silicate, Na_2SiO_3		–					×	×	
Sodium Sulfate, Na_2SO_4	×	×			×		×		
Sodium Sulfide, Na_2S	×	–	–		×	×	×		
Sodium Sulfite, $Na_2 SO_3$	×	×	×		×		×		
Stannic Chloride, $Sn Cl_4$		×					×		
Stannous Chloride, $Sn Cl_2$		×					×		
Stearic Acid, $CH_3 (CH_2)_{16} COOH$	×	×			×		×		
Sulfur, Molten, S	–	×				–		–	
Sulfur Chloride, (Wet), $S_2 Cl_2$		×					×		
Sulfur Dioxide (Dry), SO_2	×	×	×		×		×		
Sulfur Dioxide (Wet)	×	×	×				×		
Sulfur Trioxide, SO_3		×			×		×		
Sulfuric Acid (Fuming to 98%)	○	×	×	×					
Sulfuric Acid (Hot Conc.) $H_2 SO_4$		×							

RUBBERS

CHEMICALS	Hard Rubber (Natural)	Soft Rubber (Natural)	Neoprene	Butadiene Derivatives	Nitrile Rubber (Chemigum)
Sodium Borate $Na BO_2$	×	×	×	×	
Sodium Carbonate, $Na_2 CO_3$	×	×	×	×	×
Sodium Chlorate, $Na ClO_3$	×	×	×	×	
Sodium Chloride, $NaCl$	×	×	×		×
Sodium Cyanide, $NaCN$	×	×	×		
Sodium Fluoride, NaF	×	×			
Sodium Hydroxide, (Conc.), $NaOH$	×	×	×	×	×
Sodium Hydroxide, (Dilute)	×	×	×	×	×
Sodium Hydrosulfite	×	–		×	
Sodium Hypochlorite, $NaOCl$	×	–	○	×	
Sodium Hyposulfate					
Sodium Nitrate, $Na NO_3$	×	–	×	×	×
Sodium Peroxide, $Na_2 O_2$					
Sodium Phosphate, (Tri) $Na_3 PO_4$	×	×	–	×	
Sodium Silicate, Na_2SiO_3	×	×	×	×	
Sodium Sulfate, Na_2SO_4	×	×	×		
Sodium Sulfide, Na_2S	×	–	×	×	
Sodium Sulfite, $Na_2 SO_3$	×	×	○	×	
Stannic Chloride, $Sn Cl_4$					
Stannous Chloride, $Sn Cl_2$	×	×	–		
Stearic Acid, $CH_3 (CH_2)_{16} COOH$	–		○		
Sulfur, Molten, S					
Sulfur Chloride, (Wet), $S_2 Cl_2$					
Sulfur Dioxide (Dry), SO_2	–	–	○	–	
Sulfur Dioxide (Wet)	–		○		
Sulfur Trioxide, SO_3	×	×	○	×	
Sulfuric Acid (Fuming to 98%)			○		○
Sulfuric Acid (Hot Conc.) $H_2 SO_4$	○	○	○		

THERMOPLASTICS

CHEMICALS	Viton	Asphaltic, Bitumastic	Cellulose Acetate	Cellulose Acetatebutyrate	Ethyl Cellulose (Ethocel)	Cellulose Nitrate	Acrylic (Lucite, Plexiglas)	Coumarone Resins	Polyethylene	Polyvinyl Chloride, Rigid or Unplasticized	Tygon (P. V. C. & Copolymers)	Saran (Vinyl chloride, vinylidene chloride)	Kel-F (Polytrifluorochloroethylene)	Teflon (Polytetrafluoroethylene)	Uscolite CP (Styrene-acrylonitrile-butadiene)
Sodium Borate $Na BO_2$	×	×	×	×		×	×			×	×			×	×
Sodium Carbonate, $Na_2 CO_3$	×	×	–	×		–	○		×	×	×	×	×	×	×
Sodium Chlorate, $Na ClO_3$	×	×	×	×		×	×		×	×		×		×	×
Sodium Chloride, $NaCl$	×	×	×	×			×		×	×	×	×	×	×	×
Sodium Cyanide, $NaCN$	×	×								×	×	–		×	×
Sodium Fluoride, NaF	×									×		×		×	×
Sodium Hydroxide, (Conc.), $NaOH$	–	×	○	–	×	○	×		×	×	–	○	×	×	×
Sodium Hydroxide, (Dilute)		×	–	×	×	–	×		×	×	×	–	×	×	×
Sodium Hydrosulfite														×	
Sodium Hypochlorite, $NaOCl$	–	–	–			×				×	–	×		×	×
Sodium Hyposulfate														×	
Sodium Nitrate, $Na NO_3$	×	×	×	×		×				×	×	×		×	×
Sodium Peroxide, $Na_2 O_2$										×				×	×
Sodium Phosphate, (Tri) $Na_3 PO_4$		×			×		×			×	×			×	×
Sodium Silicate, Na_2SiO_3	×	×								×				×	×
Sodium Sulfate, Na_2SO_4	×	×							×	×	×	×		×	×
Sodium Sulfide, Na_2S	×	×							×	×	×	×		×	×
Sodium Sulfite, $Na_2 SO_3$	×	×								×	×	×		×	×
Stannic Chloride, $Sn Cl_4$	×	○								×	×		×	×	×
Stannous Chloride, $Sn Cl_2$		×							×	–		×		×	×
Stearic Acid, $CH_3 (CH_2)_{16} COOH$	×	○							×	×	×	×		×	×
Sulfur, Molten, S										○	○			×	
Sulfur Chloride, (Wet), $S_2 Cl_2$														×	
Sulfur Dioxide (Dry), SO_2		×			–					×	–	–		×	
Sulfur Dioxide (Wet)	×									–		–		×	
Sulfur Trioxide, SO_3	×	○			○					×				×	
Sulfuric Acid (Fuming to 98%)		×	○	○	○	○	○			○		–		×	○
Sulfuric Acid (Hot Conc.) $H_2 SO_4$	○													×	○

THERMOSETTING PLASTICS

CHEMICALS	Penton (Chlorinated Polyether)	Shellac Compounds	Organic Polysulfides	Polystyrene (Styron)	Vinylidene Chlorides	Vinyl Chloride Acetates	Cast Phenol Formaldehyde	Haveg 41 (Phenolic w. asbestos)	Heresite (Phenol formaldehyde)	Molded Phenolformald, (Durez)	Phenol Furfural Plastics	Urea Formaldehyde	Casein Plastics	Epoxy Resins	Furane Resins (Haveg 61, Duralon)	Silicone Resins	Permanite (Furan, Glass Fiber)	Nylon (Adipic Acid—Hexameth, Diamine)	Durcon 6 (Modified Epoxy)
Sodium Borate $Na BO_2$		×		×	×	×	–			–	–							×	×**
Sodium Carbonate, $Na_2 CO_3$	×			×			×	×	×	×	×		○	×	×		×	×	×**
Sodium Chlorate, $Na ClO_3$	×	×	×	×	×	×	×			×	×	×	×					○	×
Sodium Chloride, $NaCl$	×			×				×	×	×	×			×	×	×	×	×	×
Sodium Cyanide, $NaCN$	×													×				–	×
Sodium Fluoride, NaF	×																	×	×
Sodium Hydroxide, (Conc.), $NaOH$	×			×			○	○	○	○	○	○		×	×	–	×	×	×
Sodium Hydroxide, (Dilute)	×		×	×	×	×	–	○	–	○	○	×		×	×	–		×	×
Sodium Hydrosulfite							×			×	×								×
Sodium Hypochlorite, $NaOCl$	×			×	×	×	–	○	×	–	–			○	○		○	○	
Sodium Hyposulfate																			
Sodium Nitrate, $Na NO_3$	×			×			×	×	×	×	×			×	×			○	×
Sodium Peroxide, $Na_2 O_2$																			×
Sodium Phosphate, (Tri) $Na_3 PO_4$				–			–	–		×	–			×	×		×	×	×
Sodium Silicate, Na_2SiO_3				×			×			×	×							×	×
Sodium Sulfate, Na_2SO_4	×							×	×					×	×		×	×	×
Sodium Sulfide, Na_2S				×				×	×				–	×				×	+**
Sodium Sulfite, $Na_2 SO_3$	×						–	×	×	–	–			×	×		×	×	×
Stannic Chloride, $Sn Cl_4$	×							×									×	○	×
Stannous Chloride, $Sn Cl_2$	×													×					×
Stearic Acid, $CH_3 (CH_2)_{16} COOH$	×							×	×					×				×	×
Sulfur, Molten, S	○							×	○										×
Sulfur Chloride, (Wet), $S_2 Cl_2$																			×
Sulfur Dioxide (Dry), SO_2	×				×	×		×						×				–	×
Sulfur Dioxide (Wet)	×																	○	×
Sulfur Trioxide, SO_3					×	×								×				×	×
Sulfuric Acid (Fuming to 98%)	○			○			○			○	○			○	○	○			×
Sulfuric Acid (Hot Conc.) $H_2 SO_4$	–							○						○	○				×

WOODS

CHEMICALS	Cypress	Fir	Maple	Oak	Pine	Redwood
Sodium Borate $Na BO_2$						
Sodium Carbonate, $Na_2 CO_3$	+	+	+	×	+	+
Sodium Chlorate, $Na ClO_3$	×	×	×	×	×	
Sodium Chloride, $NaCl$	×	×	×	×	+	+
Sodium Cyanide, $NaCN$						
Sodium Fluoride, NaF						
Sodium Hydroxide, (Conc.), $NaOH$	–	–	○	○	○	
Sodium Hydroxide, (Dilute)	–	–	○	○	–	–
Sodium Hydrosulfite						
Sodium Hypochlorite, $NaOCl$						
Sodium Hyposulfate						
Sodium Nitrate, $Na NO_3$						
Sodium Peroxide, $Na_2 O_2$						
Sodium Phosphate, (Tri) $Na_3 PO_4$						
Sodium Silicate, Na_2SiO_3						
Sodium Sulfate, Na_2SO_4						
Sodium Sulfide, Na_2S	–	–	○	○	–	–
Sodium Sulfite, $Na_2 SO_3$						
Stannic Chloride, $Sn Cl_4$						
Stannous Chloride, $Sn Cl_2$						
Stearic Acid, $CH_3 (CH_2)_{16} COOH$						
Sulfur, Molten, S						
Sulfur Chloride, (Wet), $S_2 Cl_2$						
Sulfur Dioxide (Dry), SO_2						
Sulfur Dioxide (Wet)						
Sulfur Trioxide, SO_3						
Sulfuric Acid (Fuming to 98%)	○	○	○	○	○	○
Sulfuric Acid (Hot Conc.) $H_2 SO_4$	○	○	○	○	○	○

* NOTE: Duriron is as shown. Durichlor is also satisfactory on chlorides and HCl.

** Durcon 5 would be the preferred formula.

Table A.3

CHEMICAL RESISTANCE OF MATERIALS *continued*

MATERIALS
× — Very Good Service
+ — Moderate Service
− — Limited or Variable Service
○ — Unsatisfactory
Blank — No Information

CHEMICALS
Solids Assumed in Sol'n. Room Temperatures Assumed Unless Otherwise Stated

METALS

Chemicals	*Carbon Steel; Fe*	*Cast Iron & Ductile Iron; Fe*	*304 Stainless Steel; Fe, 18Cr, 8Ni*	*316 Stainless Steel; Fe, 16Cr, 10Ni, 2Mo*	*347 Stainless Steel; Fe, 17Cr, 9Ni; (Cx10)Cb*	*Ni-Resist Iron; Fe, 14Ni, 2Cr, 2Si*	*Durimet 20; Carpenter 20; Fe; 4Cu, 20Cr, 29Ni, 2Mo, 1Si*	*Worthite; 3Mo, 2Cu, Fe, 20Cr, 24Ni, 3Si*	*Duriron; Fe, 14Si; Durichlor, Fe, 14Si, 3Mo**	*Copper; Brass; Bronzes; Everdur*	*Aluminum; Al (and Alloys)*	*Lead; Pb*	*Monel; 67Ni, 30Cu, 1.4 Fe*	*Nickel; Ni*	*Inconel; 76Ni, 15Cr, 8Fe*
Sulfuric Acid (Cold Conc.)	+	+	×	×	−		×	×	×	○	−	×	+		
Sulfuric Acid (75%–95%)	+	−	○	○	−	−	×	×	×	○	○	×	○	○	−
Sulfuric Acid (10%–75%)	○	○	○	○	−	−	×	×	×	○	○	×	−	○	−
Sulfuric Acid (<10%)	○	○	○	−	−	−	−	×	×	○	−	×	−	−	−
Sulfurous Acid, H_2SO_3	○	○	−	×	−	○	×	×	○	○	−	−	○	○	○
Sulfuryl Chloride, SO_2Cl_2							+		×						
Tannic Acid	−		×	×			×	×	×	−	×		+	+	+
Tartaric Acid, $(CHOH\,COOH)_2$	−		×	×			×	×	×	−	○	×	+	+	
Toluene, $CH_3C_6H_5$	×		×				×		×	×	×	×	×		
Trichlorethylene, Dry, Cl_2CCHCl	−	−	−	×	−	−	×	×	×	−	−		×	×	×
Water, Fresh, H_2O	×		×	×			×		×	×	×		×	×	
Water, Distilled Lab.	○		×	×		○	×		×	×	○		○	−	
Zinc Chloride, $ZnCl_2$	○	○	○	−	○	−	×	×	×	○	○	−	+	+	
Zinc Sulfate, $ZnSO_4$	○	○	+	+	+	×	×	×	×	+	−	+	+	+	+

METALS (continued)

Chemicals	*Hastelloy B; Ni, 26Mo, 4Fe*	*Hastelloy C; Ni, 16Mo, 4Fe, 14Cr, 4W*	*Hastelloy D; Ni, 8Si, 3Cu*	*Chlorimet 3; 3Fe, 1Si, 60Ni, 18Mo, 18Cr*	*Chlorimet 2; 63Ni, 32Mo, 3Fe, 1Si*	*Stellite; Co, 28Cr, 4W*	*Zirconium; Zr*	*Tantalum; Ta*	*Silver; Ag*	*Platinum; Pt*	*Dowmetal; (Mg alloys)*	*Titanium; Ti*	*Molybdenum; Mo*
Sulfuric Acid (Cold Conc.)	×	×	×	×	×	−	○	−		×		○	−
Sulfuric Acid (75%–95%)	×	×	−	×	×	−	−	−	○	×		○	
Sulfuric Acid (10%–75%)	×	×	×	×	×	−	×	−	×	×		−	×
Sulfuric Acid (<10%)	×	×	×	×	×	−	×	×	×	×		×	
Sulfurous Acid, H_2SO_3	○	+	+	×	×	×	×	×	×			×	
Sulfuryl Chloride, SO_2Cl_2				+	+			×					
Tannic Acid	×	×	×	×	×		×	×	×			×	
Tartaric Acid, $(CHOH\,COOH)_2$	×	×	×	×	×		×	×	×			×	+
Toluene, $CH_3C_6H_5$				×	×		×						
Trichlorethylene, Dry, Cl_2CCHCl	×	×	×	×	×	×		×			×	×	
Water, Fresh, H_2O				×	×			×	×				×
Water, Distilled Lab.				×	×				×				
Zinc Chloride, $ZnCl_2$	+	+	+	×	×		×	×	×			×	
Zinc Sulfate, $ZnSO_4$	+	+	+	×	×			×	×			×	

CARBONS & CERAMICS

Chemicals	*Carbon & Graphite*	*Glass, "Pyrex" brand*	*Silicaware*	*Silicate Cements*	*Chemical Stoneware*	*Transite (asbestos & cement)*	*Chemical Porcelain*	*Concrete-Unbonded*	*Concrete-Mortar Bonded*
Sulfuric Acid (Cold Conc.)	○	×					×		
Sulfuric Acid (75%–95%)	×	×			×		×		
Sulfuric Acid (10%–75%)	×	×	×		×		×		
Sulfuric Acid (<10%)	×	×	×		×		×		
Sulfurous Acid, H_2SO_3	×	×	×	×	×		×		
Sulfuryl Chloride, SO_2Cl_2		×	×		×		×		
Tannic Acid	×	×			×		×	×	
Tartaric Acid, $(CHOH\,COOH)_2$	×	×	×		×		×		
Toluene, $CH_3C_6H_5$	×	×	×		×		×		
Trichlorethylene, Dry, Cl_2CCHCl	×	×			×		×		
Water, Fresh, H_2O	×	×	×		×	×	×	×	×
Water, Distilled Lab.		+					×		
Zinc Chloride, $ZnCl_2$	×	×		×			×	×	
Zinc Sulfate, $ZnSO_4$	×	×					×		

RUBBERS

Chemicals	*Hard Rubber (Natural)*	*Soft Rubber (Natural)*	*Neoprene*	*Butadiene Derivatives*	*Nitrile Rubber (Chemigum)*	*Viton*	*Asphaltic, Bitumastic*	*Cellulose Acetate*	*Cellulose Acetatebutyrate*	*Ethyl Cellulose (Ethocel)*
Sulfuric Acid (Cold Conc.)	○	○	○			+				
Sulfuric Acid (75%–95%)			○		○		×			
Sulfuric Acid (10%–75%)	×	×	−	×	−	×	×	×	×	○
Sulfuric Acid (<10%)	×	×	×	×	×	×	×	×	×	−
Sulfurous Acid, H_2SO_3	−	−	×				×			
Sulfuryl Chloride, SO_2Cl_2										
Tannic Acid	×	−	×				×			−
Tartaric Acid, $(CHOH\,COOH)_2$	×	×	×			×	×	×	×	
Toluene, $CH_3C_6H_5$	○	○	○	○		−		−	−	○
Trichlorethylene, Dry, Cl_2CCHCl	○	○	○			−	○			
Water, Fresh, H_2O	×	×	×	×		×	×	×	×	×
Water, Distilled Lab.	×		×							
Zinc Chloride, $ZnCl_2$	×	×	×	×	×	×	×			×
Zinc Sulfate, $ZnSO_4$	×	×	×			×	×			

THERMOPLASTICS

Chemicals	*Cellulose Nitrate*	*Acrylic (Lucite, Plexiglas)*	*Coumarone Resins*	*Polyethylene*	*Polyvinyl Chloride, Rigid or Unplasticized*	*Tygon (P. V. C. & Copolymers)*	*Saran (Vinyl chloride, vinyldene chloride)*	*Kel-F (Polytrifluorochloroethylene)*	*Teflon (Polytetrafluoroethylene)*	*Uscolite CP (Styrene-acrylonitrile-butadiene)*
Sulfuric Acid (Cold Conc.)		−			×	○	−		×	○
Sulfuric Acid (75%–95%)				×	×	○	−		×	○
Sulfuric Acid (10%–75%)	○	×		×	×	−	×	×	×	−
Sulfuric Acid (<10%)	−	×		×	×	×	×	−	×	×
Sulfurous Acid, H_2SO_3				○	×	−	−		×	×
Sulfuryl Chloride, SO_2Cl_2				○					×	
Tannic Acid				×	×		×		×	×
Tartaric Acid, $(CHOH\,COOH)_2$					×		×		×	×
Toluene, $CH_3C_6H_5$		○		−	○	○	−	−	×	
Trichlorethylene, Dry, Cl_2CCHCl		○		○	○	○	−	○	×	
Water, Fresh, H_2O	×	×	×		×		×		×	×
Water, Distilled Lab.		×			×				×	×
Zinc Chloride, $ZnCl_2$				×	×		×		×	×
Zinc Sulfate, $ZnSO_4$				×	×	×	×		×	×

THERMOSETTING PLASTICS

Chemicals	*Penton (Chlorinated Polyether)*	*Shellac Compounds*	*Organic Polysulfides*	*Polystyrene (Styron)*	*Vinylidene Chlorides*	*Vinyl Chloride Acetates*	*Cast Phenol Formaldehyde*	*Haveg 41 (Phenolic w. asbestos)*	*Heresite (Phenol formaldehyde)*	*Molded Phenolformald. (Durez)*
Sulfuric Acid (Cold Conc.)	−							−		
Sulfuric Acid (75%–95%)	+			−				−		−
Sulfuric Acid (10%–75%)	×	×	×	×	×	×	−	×		−
Sulfuric Acid (<10%)	×	×	×	×	×	×	−	×		×
Sulfurous Acid, H_2SO_3	×			×			×	×		×
Sulfuryl Chloride, SO_2Cl_2										×
Tannic Acid	×			×			×	×		×
Tartaric Acid, $(CHOH\,COOH)_2$	×						−	×		−
Toluene, $CH_3C_6H_5$	+	○	○	○	−	−	−	−		−
Trichlorethylene, Dry, Cl_2CCHCl	+							×		
Water, Fresh, H_2O	×	×	×	×	×	×	×			×
Water, Distilled Lab.	×									
Zinc Chloride, $ZnCl_2$	×			×			−	×		×
Zinc Sulfate, $ZnSO_4$	×			×				×		×

THERMOSETTING PLASTICS (continued) and WOODS

Chemicals	*Phenol Furfural Plastics*	*Urea Formaldehyde*	*Casein Plastics*	*Epoxy Resins*	*Furane Resins (Haveg 61, Duralon)*	*Silicone Resins*	*Permanite (Furan, Glass Fiber)*	*Nylon (Adipic Acid—Hexameth. Diamine)*	*Durcon 6 (Modified Epoxy)*	*Cypress*	*Fir*	*Maple*	*Oak*	*Pine*	*Redwood*
Sulfuric Acid (Cold Conc.)				○			○	○	×	○	○	○	○	○	○
Sulfuric Acid (75%–95%)	−			−	−	−			×						
Sulfuric Acid (10%–75%)	−		○	×	×	−		−	×	−	−	−	−	−	−
Sulfuric Acid (<10%)	×	×	−		×	−	×	−	×	−	−	×	−	−	−
Sulfurous Acid, H_2SO_3	×		×	×	×	−	×	○	+**				×	×	
Sulfuryl Chloride, SO_2Cl_2									×						
Tannic Acid	×	×	×	×	×		×		×				×	×	
Tartaric Acid, $(CHOH\,COOH)_2$	−	×		×	×		×	×	×						
Toluene, $CH_3C_6H_5$	−		×		×		×	×	×	−	−	×	×	×	
Trichlorethylene, Dry, Cl_2CCHCl				×				×	×						
Water, Fresh, H_2O	×	×	○						×	×	×	×	×	×	×
Water, Distilled Lab.									×						
Zinc Chloride, $ZnCl_2$	−			×	×		×	−	×						
Zinc Sulfate, $ZnSO_4$	×			×	×		×	×	×						

* NOTE: Duriron is as shown. Durichlor is also satisfactory on chlorides and HCl.
** Durcon 5 would be the preferred formula.

A.4 COMPOSITION OF METALLIC AND OTHER MATERIALS

Table A.4
COMPOSITION OF METALLIC AND OTHER MATERIALS

No.	Material	Manufacturer	Composition or Description
			METALS
17	Aluminum		
19	Aloyco-20	Alloy Steel Products Co.	Fe; 19–21 Cr; 28–30 Ni; 4.0–4.5 Cu; 2.5–3.0 Mo; 1.5 max. Si; 0.65–0.85 Mn; 0.07 max. C
19a	720 Alloy	General Plate	20 Mn; 20 Ni; Cu
54–60	Brass		Various commercial grades ranging 60–65 Cu; 35–40 Zn, 0.5–3.0 Pb
63	Brass, red		85 Cu; 15 Zn
66	Bronze, comm.		90 Cu; 10 Zn
73	Bronze, phosphor, 5% A		94.8–95.5 Cu; 4.3–5 Sn; P
74	Bronze, phosphor, 8% C		Cu; 7–9 Sn; 0.03–0.25 P
75	Bronze, phosphor 10% D		89.5–90 Cu; 10–10.5 Sn; P
76	Bronze, phosphor, spec. free cutting		88 Cu; 4 Zn; 4 Sn; 4 Pb
81	CA-FA20	Cooper Alloy	Fe; 19–21 Cr; 28–30 Ni; 3.5 Mo; 4–4.5 Cu; 0.07 max. C
82	CA-MM	Cooper Alloy	67 Ni; 30 Cu; 1.4 Fe; 0.1 Si; 0.15 C
86	Cast Iron		Ordinary unalloyed cast iron
88	Chlorimet 2	Duriron Co.	63 Ni; 32 Mo; 3 max. Fe; 0.15 max C; 1 Si; 1 Mn
89	Chlorimet 3	Duriron Co.	60 Ni; 18 Mo; 18 Cr; 2 Fe; 0.07 max. C; 1 Si; 1Mn
111	Copper		99.9+ Cu
112	Copper, Be		97.5 Cu; 2.15 Be; 0.35 Ni
119	Corrosiron	Pacific Fdry.	Fe; 14.5 Si
140	Durichlor	Duriron Co.	Fe; 0.85 C; 14.5 Si; 3 Mo; 0.35 Mn
141	Durimet 20	Duriron Co.	Fe; 20 Cr; 29 Ni; 0.07 max. C; 2 Mo; 4 Cu; 1 Si
142	Durimet T	Duriron Co.	Fe; 19 Cr; 22 Ni; 0.07 max. C; 2 Mo; 1 Cu; 1 Si
143	Duriron	Duriron Co.	Fe; 0.80 C; 14.5 Si; 0.35 Mn
148	Everdur 1000	Amer. Brass	94.9 Cu; 4 Si; 1.1 Mn
149	Everdur 1010	Amer. Brass	95.8 Cu; 3.1 Si; 1.1 Mn
150	Everdur 1015	Amer. Brass	98.25 Cu; 1.5 Si; 0.25 Mn

Table A.4
COMPOSITION OF METALLIC AND OTHER MATERIALS *continued*

No.	Material	Manufacturer	Composition or Description
156	Gold		99.99 Au
156a	Green gold		75% Au; 25% Ag
159	Hastelloy A	Haynes Stellite	Ni; 17–21 Mo; 17–21 Fe
160	Hastelloy B	Haynes Stellite	Ni; 24–32 Mo; 3–7 Fe; 0.02–0.12 C
161	Hastelloy C	Haynes Stellite	Ni; 14–19 Mo; 4–8 Fe; 0.04–0.15 C; 12–16 Cr; 3–5.5 W
162	Hastelloy D	Haynes Stellite	Ni; 8–11 Si; 2–5 Cu; 1 max. Al
163	Stellite 1	Haynes Stellite	Co; 28–34 Cr; 11–15 W
165	Stellite 6	Haynes Stellite	Co; 25–31 Cr; 3–6 W
184	Inconel	Int'l Nickel	79.5 Ni; 13 Cr; 6.5 Fe; 0.08 C; 0.2 Cu; 0.25 Mn
191	Lead		99.9+ Pb
192	Lead, antimonial		94 Pb; 6 Sb
193	Lead, antimonial		Pb; 4–12 Sb
196	Lead, chemical		99.93 Pb; 0.06 Cu
200	Lead, Te		99.88 Pb; 0.045 Te; 0.06 Cu
216	Monel	Int'l Nickel	67 Ni; 30 Cu; 1.4 Fe; 0.1 Si; 0.15 C
219	Muntz Metal		60 Cu; 40 Zn
224	Nickel	Int'l Nickel	99.4 Ni; 0.2 Mn; 0.1 Cu; 0.15 Fe; 0.05 Si
224a	Z-Nickel	Int'l Nickel	95+ Ni
226	Nickel-Silver 18% A		65 Cu; 18 Ni; 17 Zn
227	Nickel-Silver 18% B		55 Cu; 18 Ni; 27 Zn
227a	Ni-Span	Int'l Nickel	Ni, Ti, Cr, C, Mn, Si, Al
229	Ni-Hard	Int'l Nickel	Fe; 3.4 C; 1.5 Cr; 4.5 Ni; 0.6 Si
231	Ni-Resist	Int'l Nickel	Fe; 2.8 C; 14 or 20 Ni; 6 Cu (optional); 2 Cr; 2 Si
240	Platinum		99.99 Pt
268	Silver		99.9+ Ag
275	S.S. 301		Fe; 16–18 Cr; 6–8 Ni; 0.08–0.15 C
276	S.S. 302		Fe; 17–19 Cr; 8–10 Ni; 0.08–0.15 C
278	S.S. 303		Fe; 17–19 Cr; 8–10 Ni; 0.15 max C; 0.07 min. P, S, Se; 0.6 max. Zr, Mo; 2 max. Mn
279	S.S. 304		Fe; 18–20 Cr; 8–11 Ni; 0.08 max. C; 2 max. Mn
282	S.S. 310		Fe; 24–26 Cr; 19–22 Ni; 0.25 max. C
283	S.S. 316		Fe; 16–18 Cr; 10–14 Ni; 0.1 max C; 1.75–2.75 Mo
284	S.S. 317		Fe; 17.5–20 Cr; 10–14 Ni; 0.1 max. C; 3–4 Mo
285	S.S. 321		Fe; 17–19 Cr; 8–11 Ni; Ti, 5xC min.
286	S.S. 347		Fe; 17–19 Cr; 9–12 Ni; Cb, 10xC min.
287	S.S. 403		Fe; 11.5–13 Cr; 0.15 max. C
290	S.S. 410		Fe; 11.5–13.5 Cr; 0.15 max. C

Table A.4
COMPOSITION OF METALLIC AND OTHER MATERIALS *continued*

No.	Material	Manufacturer	Composition or Description
292	S.S. 416		Fe; 12–14 Cr; 0.15 max. C: 0.07 min. P, S, Se; 0.6 max. Zr, Mo
295	S.S. 430		Fe; 14–18 Cr; 0.12 max. C
303	S.S. 446		Fe; 23–27 Cr; 0.35 max. C; 0.25 max. N
360a	Steel		Plain Carbon Steel
368	Tantalum	Fansteel	99.9+ Ta
390	Worthite	Worthington Pump	Fe; 20 Cr; 24 Ni; 0.07 max. C; 3.25 Si; 3 Mo; 1.75 Cu; 0.5 Mn
			CARBON AND GRAPHITE
401	Karbate (carbon)	National Carbon	Impervious Carbon
402	Karbate (graphite)	National Carbon	Impervious Graphite
			CERAMICS
611	Lapp Porcelain	Lapp Insulator Co.	Chemical porcelain
614	Pfaudler Glass Lining	Pfaudler Co.	Glass-lined steel equipment
615	Plate Glass		Polished plate glass, falt or bent
616	Pyrex	Corning Glass Wks.	Glass
			PLASTICS
700	Ace Saran	American Hard Rubber	Vinylidene chloride
710	Geon	B. F. Goodrich	Polyvinyl chloride
711	Haveg 41	Haveg Corp.	Phenolic-asbestos
712	Haveg 43	Haveg Corp.	Phenolic-graphite
713	Haveg 60	Haveg Corp.	Furan-asbestos
714	Haveg 63	Haveg Corp.	Furan-graphite
715	Heresite M 66	Heresite & Chem. Co.	Transparent molding powder
716	Heresite MF 66	Heresite & Chem. Co.	Black molding powder
717a	Kel-F	M. W. Kellogg	Polymerized trifluoroethylene
718	Koroseal	B. F. Goodrich	Plasticized polyvinyl chloride
731	Nylon FM-101	E. I. du Pont	Injection, compression and extrusion moldings (tubing, sheeting, wire covering, gasketing)
731a	Plastisol		Polyvinyl chloride
735	Polythene	E. I. du Pont	Polyethylene
740	Saran	Dow Chemical	Vinyl chloride-vinylidene chloride copolymer
740a	Sirvene	Chicago Rawhide	Synthetic rubber
742	Teflon	E. I. du Pont	Polymerized Tetrafluoroethylene
746	Tygon	U.S. Stoneware	Synthetic compounds
			RUBBER
800	Ace Hard Rubber	American Hard Rubber	Vulcanized rubber
805	Butyl (GR-I)	Stanco Distributors	Solid copolymer of isobutylene and isoprene

Table A.4
COMPOSITION OF METALLIC AND OTHER MATERIALS *continued*

No.	Material	Manufacturer	Composition or Description
820	Hycar (GR-A)	B. F. Goodrich	Nitrile Type Synthetic Rubber
829	Neoprene	E. I. du Pont	Polymer of chloroprene
836	Natural (soft)		
837	Natural (hard)		
838	GR-S (soft)		
839	GR-S (hard)		
853	Thiokol (GR-P)	Thiokol Corp.	

A.5 STEAM AND WATER TABLES

Table A.5a

DRY SATURATED STEAM: TEMPERATURE TABLE*

		Specific Volume, ft^3/lbm			Enthalpy; Btu/lbm			Entropy, Btu/lbm R		
Temp., °F/°C t	Abs. Press., PSIA P	Sat. Liquid v_f	Evap. v_{fg}	Sat. Vapor v_g	Sat. Liquid h_f	Evap. h_{fg}	Sat. Vapor h_g	Sat. Liquid s_f	Evap. s_{fg}	Sat. Vapor s_g
32/0	0.08854	0.01602	3306	3306	0.00	1075.8	1075.8	0.0000	2.1877	2.1877
35/1.7	0.09995	0.01602	2947	2947	3.02	1074.1	1077.1	0.0061	2.1709	2.1770
40.4.4	0.12170	0.01602	2444	2444	8.05	1071.3	1079.3	0.0162	2.1435	2.1597
45/7.2	0.14752	0.01602	2036.4	2036.4	13.06	1068.4	1081.5	0.0262	2.1167	2.1429
50/10	0.17811	0.01603	1703.2	1703.2	18.07	1065.6	1083.7	0.0361	2.0903	2.1264
60/15.6	0.2563	0.01604	1206.6	1206.7	28.06	1059.9	1088.0	0.0555	2.0393	2.0948
70/21.1	0.3631	0.01606	867.8	867.9	38.04	1054.3	1092.3	0.0745	1.9902	2.0647
80/26.7	0.5069	0.01608	633.1	633.1	48.02	1048.6	1096.6	0.0932	1.9428	2.0360
90/32.2	0.6982	0.01610	468.0	468.0	57.99	1042.9	1100.9	0.1115	1.8972	2.0087
100/37.8	0.9492	0.01613	350.3	350.4	67.97	1037.2	1105.2	0.1295	1.8531	1.9826
110/43	1.2748	0.01617	265.3	265.4	77.94	1031.6	1109.5	0.1471	1.8106	1.9577
120/49	1.6924	0.01620	203.25	203.27	87.92	1025.8	1113.7	0.1645	1.7694	1.9339
130/54	2.2225	0.01625	157.32	157.34	97.90	1020.0	1117.9	0.1816	1.7296	1.9112
140/60	2.8886	0.01629	122.99	123.01	107.89	1014.1	1122.0	0.1984	1.6910	1.8894
150/66	3.718	0.01634	97.06	97.07	117.89	1008.2	1126.1	0.2149	1.6537	1.8685
160/71	4.741	0.01639	77.27	77.29	127.89	1002.3	1130.2	0.2311	1.6174	1.8485
170/77	5.992	0.01645	62.04	62.06	137.90	996.3	1134.2	0.2472	1.5822	1.8293
180/82	7.510	0.01651	50.21	50.23	147.92	990.2	1138.1	0.2630	1.5480	1.8109
190/88	9.339	0.01657	40.94	40.96	157.95	984.1	1142.0	0.2785	1.5147	1.7932
200/93	11.526	0.01663	33.62	33.64	167.99	977.9	1145.9	0.2938	1.4824	1.7762
210/99	14.123	0.01670	27.80	27.82	178.05	971.6	1149.7	0.3090	1.4508	1.7598
212/100	14.696	0.01672	26.78	26.80	180.07	970.3	1150.4	0.3120	1.4446	1.7566
220/104	17.186	0.01677	23.13	23.15	188.13	965.2	1153.4	0.3239	1.4201	1.7440
230/110	20.780	0.01684	19.365	19.382	198.23	958.8	1157.0	0.3387	1.3901	1.7288
240/116	24.969	0.01692	16.306	16.323	208.34	952.2	1160.5	0.3531	1.3609	1.7140
250/121	29.825	0.01700	13.804	13.821	218.48	945.5	1164.0	0.3675	1.3323	1.6998
260/127	35.429	0.01709	11.746	11.763	228.64	938.7	1167.3	0.3817	1.3043	1.6860
270/132	41.858	0.01717	10.044	10.061	238.84	931.8	1170.6	0.3958	1.2769	1.6727
280/138	49.203	0.01726	8.628	8.645	249.06	924.7	1173.8	0.4096	1.2501	1.6597
290/143	57.556	0.01735	7.444	7.461	259.31	917.5	1176.8	0.4234	1.2238	1.6472
300/149	67.013	0.01745	6.449	6.466	269.59	910.1	1179.7	0.4369	1.1980	1.6350
310/154	77.68	0.01755	5.609	5.626	279.92	902.6	1182.5	0.4504	1.1727	1.6231
320/160	89.66	0.01765	4.896	4.914	290.28	894.9	1185.2	0.4637	1.1478	1.6115
330/166	103.06	0.01776	4.289	4.307	300.68	887.0	1187.7	0.4769	1.1233	1.6002
340/171	118.01	0.01787	3.770	3.788	311.13	879.0	1190.1	0.4900	1.0992	1.5891
350/177	134.63	0.01799	3.324	3.342	321.63	870.7	1192.3	0.5029	1.0754	1.5783
360/182	153.04	0.01811	2.939	2.957	332.18	862.2	1194.4	0.5158	1.0519	1.5677
370/188	173.37	0.01823	2.606	2.625	342.79	853.5	1196.3	0.5286	1.0287	1.5573
380/193	195.77	0.01836	2.317	2.335	353.45	844.6	1198.1	0.5413	1.0059	1.5471
390/199	220.37	0.01850	2.0651	2.0836	364.17	835.4	1199.6	0.5539	0.9832	1.5371
400/204	247.31	0.01864	1.8447	1.8633	374.97	826.0	1201.0	0.5664	0.9608	1.5272
410/210	276.75	0.01878	1.6512	1.6700	385.83	816.3	1202.1	0.5788	0.9386	1.5174
420/216	308.83	0.01894	1.4811	1.5000	396.77	806.3	1203.1	0.5912	0.9166	1.5078
430/221	343.72	0.01910	1.3308	1.3499	407.79	796.0	1203.8	0.6035	0.8947	1.4982
440/227	381.59	0.01926	1.1979	1.2171	418.90	785.4	1204.3	0.6158	0.8730	1.4887

Table A.5a
DRY SATURATED STEAM: TEMPERATURE TABLE* *continued*

Temp., °F/°C t	Abs. Press., PSIA P	Specific Volume, ft³/lbm Sat. Liquid v_f	Evap. v_{fg}	Sat. Vapor v_g	Enthalpy; Btu/lbm Sat. Liquid h_f	Evap. h_{fg}	Sat. Vapor h_g	Entropy, Btu/lbm R Sat. Liquid s_f	Evap. s_{fg}	Sat. Vapor s_g
450/232	422.6	0.0194	1.0799	1.0993	430.1	774.5	1204.6	0.6280	0.8513	1.4793
460/238	466.9	0.0196	0.9748	0.9944	441.4	763.2	1204.6	0.6402	0.8298	1.4700
470/243	514.7	0.0198	0.8811	0.9009	452.8	751.5	1204.3	0.6523	0.8083	1.4606
480/249	566.1	0.0200	0.7972	0.8172	464.4	739.4	1203.7	0.6645	0.7868	1.4513
490/254	621.4	0.0202	0.7221	0.7423	476.0	726.8	1202.8	0.6766	0.7653	1.4419
500/260	680.8	0.0204	0.6545	0.6749	487.8	713.9	1201.7	0.6887	0.7438	1.4325
520/271	812.4	0.0209	0.5385	0.5594	511.9	686.4	1198.2	0.7130	0.7006	1.4136
540/282	962.5	0.0215	0.4434	0.4649	536.6	656.6	1193.2	0.7374	0.6568	1.3942
560/293	1133.1	0.0221	0.3647	0.3868	562.2	624.2	1186.4	0.7621	0.6121	1.3742
580/304	1325.8	0.0228	0.2989	0.3217	588.9	588.4	1177.3	0.7872	0.5659	1.3532
600/316	1542.9	0.0236	0.2432	0.2668	617.0	548.5	1165.5	0.8131	0.5176	1.3307
620/327	1786.6	0.0247	0.1955	0.2201	646.7	503.6	1150.3	0.8398	0.4664	1.3062
640/338	2059.7	0.0260	0.1538	0.1798	678.6	452.0	1130.5	0.8679	0.4110	1.2789
660/349	2365.4	0.0278	0.1165	0.1442	714.2	390.2	1104.4	0.8987	0.3485	1.2472
680/360	2708.1	0.0305	0.0810	0.1115	757.3	309.9	1067.2	0.9351	0.2719	1.2071
700/371	3093.7	0.0369	0.0392	0.0761	823.3	172.1	995.4	0.9905	0.1484	1.1389
705.4/374.1	3206.2	0.0503	0	0.0503	902.7	0	902.7	1.0580	0	1.0580

*Abridged from *Thermodynamic Properties of Steam*, by Joseph H. Keenan and Frederick G. Keyes. Copyright 1936, by Joseph H. Keenan and Frederick G. Keyes. Published by John Wiley & Sons, Inc., New York.

Table A.5b
PROPERTIES OF SUPERHEATED STEAM*

*Abs. Press., PSIA** (Sat. Temp.)*		*Temperature, °F/°C*												
		200/93	220/104	300/149	350/177	400/204	450/232	500/260	550/288	600/316	700/371	800/427	900/482	1000/538
	v	392.6	404.5	452.3	482.2	512.0	541.8	571.6	601.4	631.2	690.8	750.4	809.9	869.5
1	*h*	1150.4	1159.5	1195.8	1218.7	1241.7	1264.9	1288.3	1312.0	1335.7	1383.8	1432.8	1482.7	1533.5
(101.74)	*s*	2.0512	2.0647	2.1153	2.1444	2.1720	2.1983	2.2233	2.2468	2.2702	2.3137	2.3542	2.3923	2.4283
	v	78.16	80.59	90.25	96.26	102.26	108.24	114.22	120.19	126.16	138.10	150.03	161.95	173.87
5	*h*	1148.8	1158.1	1195.0	1218.1	1241.2	1264.5	1288.0	1311.7	1335.4	1383.6	1432.7	1482.6	1533.4
(162.24)	*s*	1.8718	1.8857	1.9370	1.9664	1.9942	2.0205	2.0456	2.0692	2.0927	2.1361	2.1767	2.2148	2.2509
	v	38.85	40.09	45.00	48.03	51.04	54.05	57.05	60.04	63.03	69.01	74.98	80.95	86.92
10	*h*	1146.6	1156.2	1193.9	1217.2	1240.6	1264.0	1287.5	1311.3	1335.1	1383.4	1432.5	1482.4	1533.2
(193.21)	*s*	1.7927	1.8071	1.8595	1.8892	1.9172	1.9436	1.9689	1.9924	2.0160	2.0596	2.1002	2.1383	2.1744
	v		27.15	30.53	32.62	34.68	36.73	38.78	40.82	42.86	46.94	51.00	55.07	59.13
14.696	*h*		1154.4	1192.8	1216.4	1239.9	1263.5	1287.1	1310.9	1334.8	1383.2	1432.3	1482.3	1533.1
(212.00)	*s*		1.7624	1.8160	1.8460	1.8743	1.9008	1.9261	1.9498	1.9734	2.0170	2.0576	2.0958	2.1319
	v			22.36	23.91	25.43	26.95	28.46	29.97	31.47	34.47	37.46	40.45	43.44
20	*h*			1191.6	1215.6	1239.2	1262.9	1286.6	1310.5	1334.4	1382.9	1432.1	1482.1	1533.0
(227.96)	*s*			1.7808	1.8112	1.8396	1.8664	1.8918	1.9160	1.9392	1.9829	2.0235	2.0618	2.0978
	v			11.040	11.843	12.628	13.401	14.168	14.93	15.688	17.198	18.702	20.20	21.70
40	*h*			1186.8	1211.9	1236.5	1260.7	1284.8	1308.9	1333.1	1381.9	1431.3	1481.4	1532.4
(267.25)	*s*			1.6994	1.7314	1.7608	1.7881	1.8140	1.8384	1.8619	1.9058	1.9467	1.9850	2.0214
	v			7.259	7.818	8.357	8.884	9.403	9.916	10.427	11.441	12.449	13.452	14.454
60	*h*			1181.6	1208.2	1233.6	1258.5	1283.0	1307.4	1331.8	1380.9	1430.5	1480.8	1531.9
(292.71)	*s*			1.6492	1.6830	1.7135	1.7416	1.7678	1.7926	1.8162	1.8605	1.9015	1.9400	1.9762
	v				5.803	6.220	6.624	7.020	7.410	7.797	8.562	9.322	10.077	10.830
80	*h*				1204.3	1230.7	1256.1	1281.1	1305.8	1330.5	1379.9	1429.7	1480.1	1531.3
(312.03)	*s*				1.6475	1.6791	1.7078	1.7346	1.7598	1.7836	1.8281	1.8694	1.9079	1.9442
	v				4.592	4.937	5.268	5.589	5.905	6.218	6.835	7.446	8.052	8.656
100	*h*				1200.1	1227.6	1253.7	1279.1	1304.2	1329.1	1378.9	1428.9	1479.5	1530.8
(327.81)	*s*				1.6188	1.6518	1.6813	1.7085	1.7339	1.7581	1.8029	1.8443	1.8829	1.9193
	v				3.783	4.081	4.363	4.636	4.902	5.165	5.683	6.195	6.702	7.207
120	*h*				1195.7	1224.4	1251.3	1277.2	1302.5	1327.7	1377.8	1428.1	1478.8	1530.2
(341.25)	*s*				1.5944	1.6287	1.6591	1.6869	1.7127	1.7370	1.7822	1.8237	1.8625	1.8990
	v					3.468	3.715	3.954	4.186	4.413	4.861	5.301	5.738	6.172
140	*h*					1221.1	1248.7	1275.2	1300.9	1326.4	1376.8	1427.3	1478.2	1529.7
(353.02)	*s*					1.6087	1.6399	1.6683	1.6945	1.7190	1.7645	1.8063	1.8451	1.8817
	v					3.008	3.230	3.443	3.648	3.849	4.244	4.631	5.015	5.396
160	*h*					1217.6	1246.1	1273.1	1299.3	1325.0	1375.7	1426.4	1477.5	1529.1
(363.53)	*s*					1.5908	1.6230	1.6519	1.6785	1.7033	1.7491	1.7911	1.8301	1.8667
	v					2.649	2.852	3.044	3.229	3.411	3.764	4.110	4.452	4.792
180	*h*					1214.0	1243.5	1271.0	1297.6	1323.5	1374.7	1425.6	1476.8	1528.6
(373.06)	*s*					1.5745	1.6077	1.6373	1.6642	1.6894	1.7355	1.7776	1.8167	1.8534
	v					2.361	2.549	2.726	2.895	3.060	3.380	3.693	4.002	4.309
200	*h*					1210.3	1240.7	1268.9	1295.8	1322.1	1373.6	1424.8	1476.2	1528.0
(381.79)	*s*					1.5594	1.5937	1.6240	1.6513	1.6767	1.7232	1.7655	1.8048	1.8415

*Abridged from *Thermodynamic Properties of Steam*, by Joseph H. Keenan and Frederick G. Keyes.
Copyright 1936, by Joseph H. Keenan and Frederick G. Keyes, Published by John Wiley & Sons, Inc., New York.
**For SI units see Section A.1.

Table A.5b
PROPERTIES OF SUPERHEATED STEAM *continued*

Abs. Press., PSIA** (Sat. Temp.)		Temperature, °F/°C 200/93	220/104	300/149	350/177	400/204	450/232	500/260	550/288	600/316	700/371	800/427	900/482	1000/538
	v					2.125	2.301	2.465	2.621	2.772	3.066	3.352	3.634	3.913
220	*h*					1206.5	1237.9	1266.7	1294.1	1320.7	1372.6	1424.0	1475.5	1527.5
(389.86)	*s*					1.5453	1.5808	1.6117	1.6395	1.6652	1.7120	1.7545	1.7939	1.8308
	v					1.9276	2.094	2.247	2.393	2.533	2.804	3.068	3.327	3.584
240	*h*					1202.5	1234.9	1264.5	1292.4	1319.2	1371.5	1423.2	1474.8	1526.9
(397.37)	*s*					1.5319	1.5686	1.6003	1.6286	1.6546	1.7017	1.7444	1.7839	1.8209
	v						1.9183	2.063	2.199	2.330	2.582	2.827	3.067	3.305
260	*h*						1232.0	1262.3	1290.5	1317.7	1370.4	1422.3	1474.2	1526.3
(404.42)	*s*						1.5573	1.5897	1.6184	1.6447	1.6922	1.7352	1.7748	1.8118
	v						1.7674	1.9047	2.033	2.156	2.392	2.621	2.845	3.066
280	*h*						1228.9	1260.0	1288.7	1316.2	1369.4	1421.5	1473.5	1525.8
(411.05)	*s*						1.5464	1.5796	1.6087	1.6354	1.6834	1.7265	1.7662	1.8033
	v						1.6364	1.7675	1.8891	2.005	2.227	2.442	2.652	2.859
300	*h*						1225.8	1257.6	1286.8	1314.7	1368.3	1420.6	1472.8	1525.2
(417.33)	*s*						1.5360	1.5701	1.5998	1.6268	1.6751	1.7184	1.7582	1.7954
	v						1.3734	1.4923	1.6010	1.7036	1.8980	2.084	2.266	2.445
350	*h*						1217.7	1251.5	1282.1	1310.9	1365.5	1418.5	1471.1	1523.8
(431.72)	*s*						1.5119	1.5481	1.5792	1.6070	1.6563	1.7002	1.7403	1.7777
	v						1.1744	1.2851	1.3843	1.4770	1.6508	1.8161	1.9767	2.134
400	*h*						1208.8	1245.1	1277.2	1306.9	1362.7	1416.4	1469.4	1522.4
(444.59)	*s*						1.4892	1.5281	1.5607	1.5894	1.6398	1.6842	1.7247	1.7623

Abs. Press., PSIA** (Sat. Temp.)		Temperature, °F/°C 500/260	550/288	600/316	620/327	640/338	660/349	680/360	700/371	800/427	900/482	1000/538	1200/649	1400/760	1600/871
	v	1.1231	1.2155	1.3005	1.3332	1.3652	1.3967	1.4278	1.4584	1.6074	1.7516	1.8928	2.170	2.443	2.714
450	*h*	1238.4	1272.0	1302.8	1314.6	1326.2	1337.5	1348.8	1359.9	1414.3	1467.7	1521.0	1628.6	1738.7	1851.9
(456.28)	*s*	1.5095	1.5437	1.5735	1.5845	1.5951	1.6054	1.6153	1.6250	1.6699	1.7108	1.7486	1.8177	1.8803	1.9381
	v	0.9927	1.0800	1.1591	1.1893	1.2188	1.2478	1.2763	1.3044	1.4405	1.5715	1.6996	1.9504	2.197	2.442
500	*h*	1231.3	1266.8	1298.6	1310.7	1322.6	1334.2	1345.7	1357.0	1412.1	1466.0	1519.6	1627.6	1737.9	1851.3
(467.01)	*s*	1.4919	1.5280	1.5588	1.5701	1.5810	1.5915	1.6016	1.6115	1.6571	1.6982	1.7363	1.8056	1.8683	1.9262
	v	0.8852	0.9686	1.0431	1.0714	1.0989	1.1259	1.1523	1.1783	1.3038	1.4241	1.5414	1.7706	1.9957	2.219
550	*h*	1223.7	1261.2	1294.3	1306.8	1318.9	1330.8	1342.5	1354.0	1409.9	1464.3	1518.2	1626.6	1737.1	1850.6
(476.94)	*s*	1.4751	1.5131	1.5451	1.5568	1.5680	1.5787	1.5890	1.5991	1.6452	1.6868	1.7250	1.7946	1.8575	1.9155
	v	0.7947	0.8753	0.9463	0.9729	0.9988	1.0241	1.0489	1.0732	1.1899	1.3013	1.4096	1.6208	1.8279	2.033
600	*h*	1215.7	1255.5	1289.9	1302.7	1315.2	1327.4	1339.3	1351.1	1407.7	1462.5	1516.7	1625.5	1736.3	1850.0
(486.21)	*s*	1.4586	1.4990	1.5323	1.5443	1.5558	1.5667	1.5773	1.5875	1.6343	1.6762	1.7147	1.7846	1.8476	1.9056
	v		0.7277	0.7934	0.8177	0.8411	0.8639	0.8860	0.9077	1.0108	1.1082	1.2024	1.3853	1.5641	1.7405
700	*h*		1243.2	1280.6	1294.3	1307.5	1320.3	1332.8	1345.0	1403.2	1459.0	1513.9	1623.5	1734.8	1848.8
(503.10)	*s*		1.4722	1.5084	1.5212	1.5333	1.5449	1.5559	1.5665	1.6147	1.6573	1.6963	1.7666	1.8299	1.8881
	v		0.6154	0.6779	0.7006	0.7223	0.7433	0.7635	0.7833	0.8763	0.9633	1.0470	1.2088	1.3662	1.5214
800	*h*		1229.8	1270.7	1285.4	1299.4	1312.9	1325.9	1338.6	1398.6	1455.4	1511.0	1621.4	1733.2	1847.5
(518.23)	*s*		1.4467	1.4863	1.5000	1.5129	1.5250	1.5366	1.5476	1.5972	1.6407	1.6801	1.7510	1.8146	1.8729
	v		0.5264	0.5873	0.6089	0.6294	0.6491	0.6680	0.6863	0.7716	0.8506	0.9262	1.0714	1.2124	1.3509
900	*h*		1215.0	1260.1	1275.9	1290.9	1305.1	1318.8	1332.1	1393.9	1451.8	1508.1	1619.3	1731.6	1846.3
(531.98)	*s*		1.4216	1.4653	1.4800	1.4938	1.5066	1.5187	1.5303	1.5814	1.6257	1.6656	1.7371	1.8009	1.8595

Table A.5.b
PROPERTIES OF SUPERHEATED STEAM *continued*

Abs. Press., PSIA** (Sat. Temp.)		Temperature, °F/°C													
		500/260	550/288	600/316	620/327	640/338	660/349	680/360	700/371	800/427	900/482	1000/538	1200/649	1400/760	1600/871
	v		0.4533	0.5140	0.5350	0.5546	0.5733	0.5912	0.6084	0.6878	0.7604	0.8294	0.9615	1.0893	1.2146
1000	h		1198.3	1248.8	1265.9	1281.9	1297.0	1311.4	1325.3	1389.2	1448.2	1505.1	1617.3	1730.0	1845.0
(544.61)	s		1.3961	1.4450	1.4610	1.4757	1.4893	1.5021	1.5141	1.5670	1.6121	1.6525	1.7245	1.7886	1.8474
	v			0.4532	0.4738	0.4929	0.5110	0.5281	0.5445	0.6191	0.6866	0.7503	0.8716	0.9885	1.1031
1100	h			1236.7	1255.3	1272.4	1288.5	1303.7	1318.3	1384.3	1444.5	1502.2	1615.2	1728.4	1843.8
(556.31)	s			1.4251	1.4425	1.4583	1.4728	1.4862	1.4989	1.5535	1.5995	1.6405	1.7130	1.7775	1.8363
	v			0.4016	0.4222	0.4410	0.4586	0.4752	0.4909	0.5617	0.6250	0.6843	0.7967	0.9046	1.0101
1200	h			1223.5	1243.9	1262.4	1279.6	1295.7	1311.0	1379.3	1440.7	1499.2	1613.1	1726.9	1842.5
(567.22)	s			1.4052	1.4243	1.4413	1.4568	1.4710	1.4843	1.5409	1.5879	1.6293	1.7025	1.7672	1.8263
	v			0.3174	0.3390	0.3580	0.3753	0.3912	0.4062	0.4714	0.5281	0.5805	0.6789	0.7727	0.8640
1400	h			1193.0	1218.4	1240.4	1260.3	1278.5	1295.5	1369.1	1433.1	1493.2	1608.9	1723.7	1840.0
(587.10)	s			1.3639	1.3877	1.4079	1.4258	1.4419	1.4567	1.5177	1.5666	1.6093	1.6836	1.7489	1.8083
	v				0.2733	0.2936	0.3112	0.3271	0.3417	0.4034	0.4553	0.5027	0.5906	0.6738	0.7545
1600	h				1187.8	1215.2	1238.7	1259.6	1278.7	1358.4	1425.3	1487.0	1604.6	1720.5	1837.5
(604.90)	s				1.3489	1.3741	1.3952	1.4137	1.4303	1.4964	1.5476	1.5914	1.6669	1.7328	1.7926
	v					0.2407	0.2597	0.2760	0.2907	0.3502	0.3986	0.4421	0.5218	0.5968	0.6693
1800	h					1185.1	1214.0	1238.5	1260.3	1347.2	1417.4	1480.8	1600.4	1717.3	1835.0
(621.03)	s					1.3377	1.3638	1.3855	1.4044	1.4765	1.5301	1.5752	1.6520	1.7185	1.7786
	v					0.1936	0.2161	0.2337	0.2489	0.3074	0.3532	0.3935	0.4668	0.5352	0.6011
2000	h					1145.6	1184.9	1214.8	1240.0	1335.5	1409.2	1474.5	1596.1	1714.1	1832.5
(635.82)	s					1.2945	1.3300	1.3564	1.3783	1.4576	1.5139	1.5603	1.6384	1.7055	1.7660
	v							0.1484	0.1686	0.2294	0.2710	0.3061	0.3678	0.4244	0.4784
2500	h							1132.3	1176.8	1303.6	1387.8	1458.4	1585.3	1706.1	1826.2
(668.13)	s							1.2687	1.3073	1.4127	1.4772	1.5273	1.6088	1.6775	1.7389
	v								0.0984	0.1760	0.2159	0.2476	0.3018	0.3505	0.3966
3000	h								1060.7	1267.2	1365.0	1441.8	1574.3	1698.0	1819.9
(695.36)	s								1.1966	1.3690	1.4439	1.4984	1.5837	1.6540	1.7163
	v									0.1583	0.1981	0.2288	0.2806	0.3267	0.3703
3206.2	h									1250.5	1355.2	1434.7	1569.8	1694.6	1817.2
(705.40)	s									1.3508	1.4309	1.4874	1.5742	1.6452	1.7080
	v								0.0306	0.1364	0.1762	0.2058	0.2546	0.2977	0.3381
3500	h								780.5	1224.9	1340.7	1424.5	1563.3	1689.8	1813.6
	s								0.9515	1.3241	1.4127	1.4723	1.5615	1.6336	1.6968
	v								0.0287	0.1052	0.1462	0.1743	0.2192	0.2581	0.2943
4000	h								763.8	1174.8	1314.4	1406.8	1552.1	1681.7	1807.2
	s								0.9347	1.2757	1.3827	1.4482	1.5417	1.6154	1.6795
	v								0.0276	0.0798	0.1226	0.1500	0.1917	0.2273	0.2602
4500	h								753.5	1113.9	1286.5	1388.4	1540.8	1673.5	1800.9
	s								0.9235	1.2204	1.3529	1.4253	1.5235	1.5990	1.6640
	v								0.0268	0.0593	0.1036	0.1303	0.1696	0.2027	0.2329
5000	h								746.4	1047.1	1256.5	1369.5	1529.5	1665.3	1794.5
	s								0.9152	1.1622	1.3231	1.4034	1.5066	1.5839	1.6499
	v								0.0262	0.0463	0.0880	0.1143	0.1516	0.1825	0.2106
5500	h								741.3	985.0	1224.1	1349.3	1518.2	1657.0	1788.1
	s								0.9090	1.1093	1.2930	1.3821	1.4908	1.5699	1.6369

Table A.5c
PROPERTIES OF WATER
AT VARIOUS TEMPERATURES FROM 40 TO 540°F (4.4 to 282.2°C)

Temp. °F	*Temp.* °C	*Specific Volume** ft^3/lb	*Specific Gravity*	*Weight** (lb/ft^3)	*Vapor Pressure** PSIA
40	4.4	.01602	1.0013	62.42	0.1217
50	10.0	.01603	1.0006	62.38	0.1781
60	15.6	.01604	1.0000	62.34	0.2563
70	21.1	.01606	0.9987	62.27	0.3631
80	26.7	.01608	0.9975	62.19	0.5069
90	32.2	.01610	0.9963	62.11	0.6982
100	37.8	.01613	0.9944	62.00	0.9492
120	48.9	.01620	0.9901	61.73	1.692
140	60.0	.01629	0.9846	61.39	2.889
160	71.1	.01639	0.9786	61.01	4.741
180	82.2	.01651	0.9715	60.57	7.510
200	93.3	.01663	0.9645	60.13	11.526
212	100.0	.01672	0.9593	59.81	14.696
220	104.4	.01677	0.9565	59.63	17.186
240	115.6	.01692	0.9480	59.10	24.97
260	126.7	.01709	0.9386	58.51	35.43
280	137.8	.01726	0.9293	58.00	49.20
300	148.9	.01745	0.9192	57.31	67.01
320	160.0	.01765	0.9088	56.66	89.66
340	171.1	.01787	0.8976	55.96	118.01
360	182.2	.01811	0.8857	55.22	153.04
380	193.3	.01836	0.8736	54.47	195.77
400	204.4	.01864	0.8605	53.65	247.31
420	215.6	.01894	0.8469	52.80	308.83
440	226.7	.01926	0.8328	51.92	381.59
460	237.8	.0196	0.8183	51.02	466.9
480	248.9	.0200	0.8020	50.00	566.1
500	260.0	.0204	0.7863	49.02	680.8
520	271.1	.0209	0.7674	47.85	812.4
540	282.2	.0215	0.7460	46.51	962.5

*For SI units see Section A.1
Computed from Keenan & Keyes Steam Table.

A.6 FRICTION LOSS IN PIPES

Friction loss moduli for laminar flow are shown by the 45-degree lines in the upper left-hand portion of each chart. Moduli for turbulent flow are shown by the steeper curves in the lower right hand portion. Both of these regions represent stable states of flow. A diagonal line separates the regions of laminar and turbulent flow and represents the critical zone, a region in which it is difficult to predict the state of flow and hence, the friction loss. The critical zone usually represents a region of unstable flow. The critical zone line gives approximate moduli on the high side for this region of unstable flow.

The bottom scale of each chart represents flow in gallons per minute, GPM. An auxiliary top scale shows the average velocity in the pipe in feet per second. Read vertically from the GPM scale to find the corresponding velocity in feet per second. The vertical scales, labeled "Friction Loss Modulus for 100 Feet of Pipe," represent values of the ratio.

$$M = \frac{\Delta p}{SG}$$

where

M = Friction loss modulus for 100 feet (30m) of pipe
Δp = Pressure loss in pounds per square inch per 100 feet of pipe
SG = Specific gravity of fluid at 60°F (15.6°C)

The loss due to pipe friction may be obtained as follows:

$$\Delta p = M \times SG$$

To use the charts, proceed as follows:

a. Select the chart for the size of pipe in question.
b. Follow the vertical line representing the flow in GPM to its intersection with the desired viscosity curve, and read the modulus at the left.
c. If the vertical line representing the flow in GPM does not intersect the viscosity line in either turbulent or laminar flow, use the intersection with the critical zone line.
d. Compute the friction loss in pressure drop from the equation above.

Fig. A.6b
FRICTION LOSS MODULUS FOR 100 FEET OF 1½" STEEL PIPE*

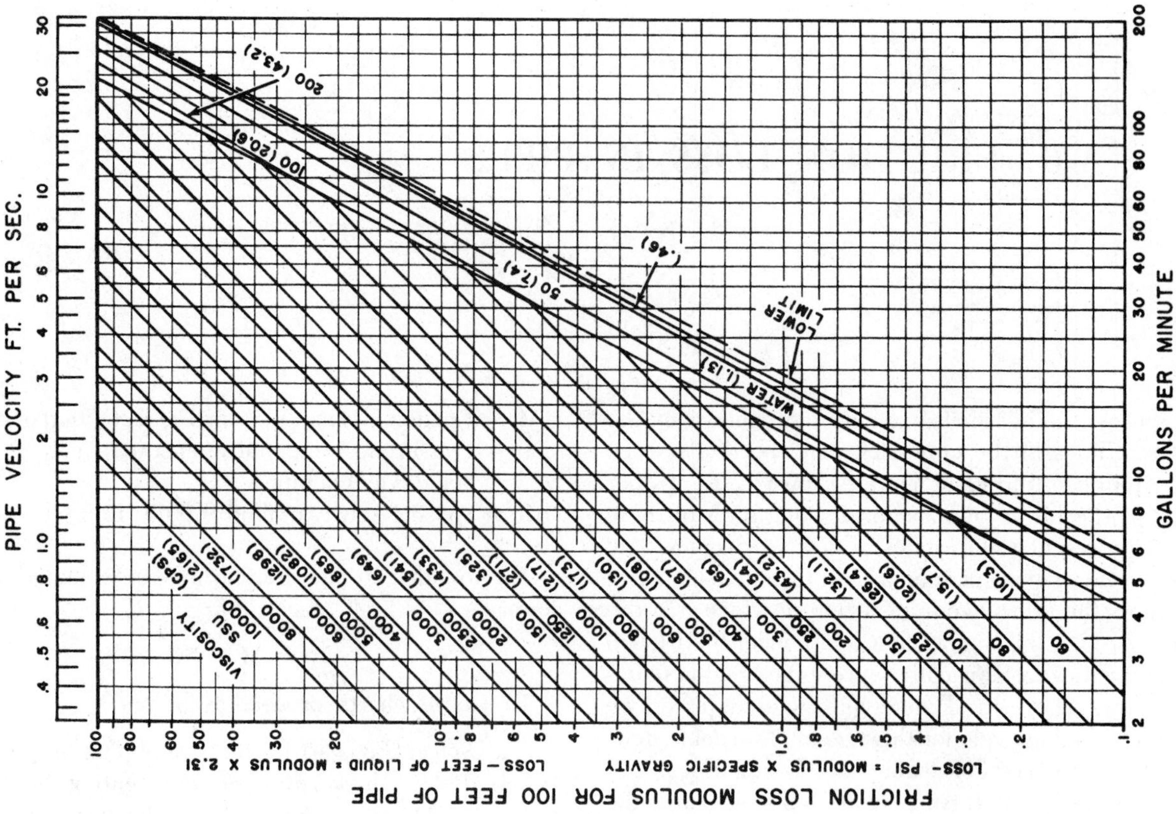

1½" STEEL PIPE (1.610" I. D.)
PIPE VELOCITY FT. PER SEC.

GALLONS PER MINUTE

FRICTION LOSS MODULUS FOR 100 FEET OF PIPE
LOSS- PSI = MODULUS X SPECIFIC GRAVITY LOSS—FEET OF LIQUID = MODULUS X 2.31

Loss—Lbs. per sq. in. = modulus × Specific Gravity
Loss—Feet of liquid = modulus × 2.31

*For SI units see Section A.1

Fig. A.6a
FRICTION LOSS MODULUS FOR 100 FEET OF 1" STEEL PIPE*

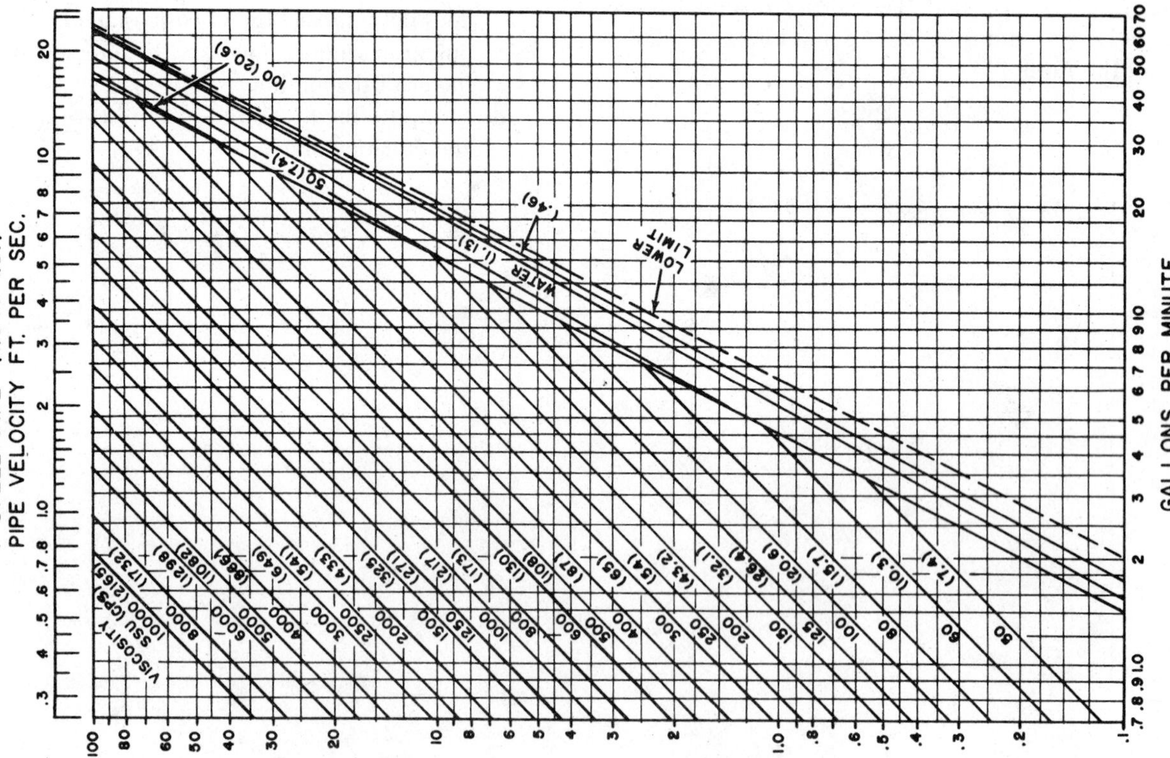

1" STEEL PIPE (1.049" I.D.)
PIPE VELOCITY FT. PER SEC.

GALLONS PER MINUTE

FRICTION LOSS MODULUS FOR 100 FEET OF PIPE
LOSS – PSI = MODULUS X SPECIFIC GRAVITY LOSS—FEET OF LIQUID = MODULUS X 2.31

Loss—Lbs. per sq. in. = modulus × Specific Gravity
Loss—Feet of liquid = modulus × 2.31

*For SI units see Section A.1

Fig. A.6c
FRICTION LOSS MODULUS FOR 100 FEET OF 2" STEEL PIPE*

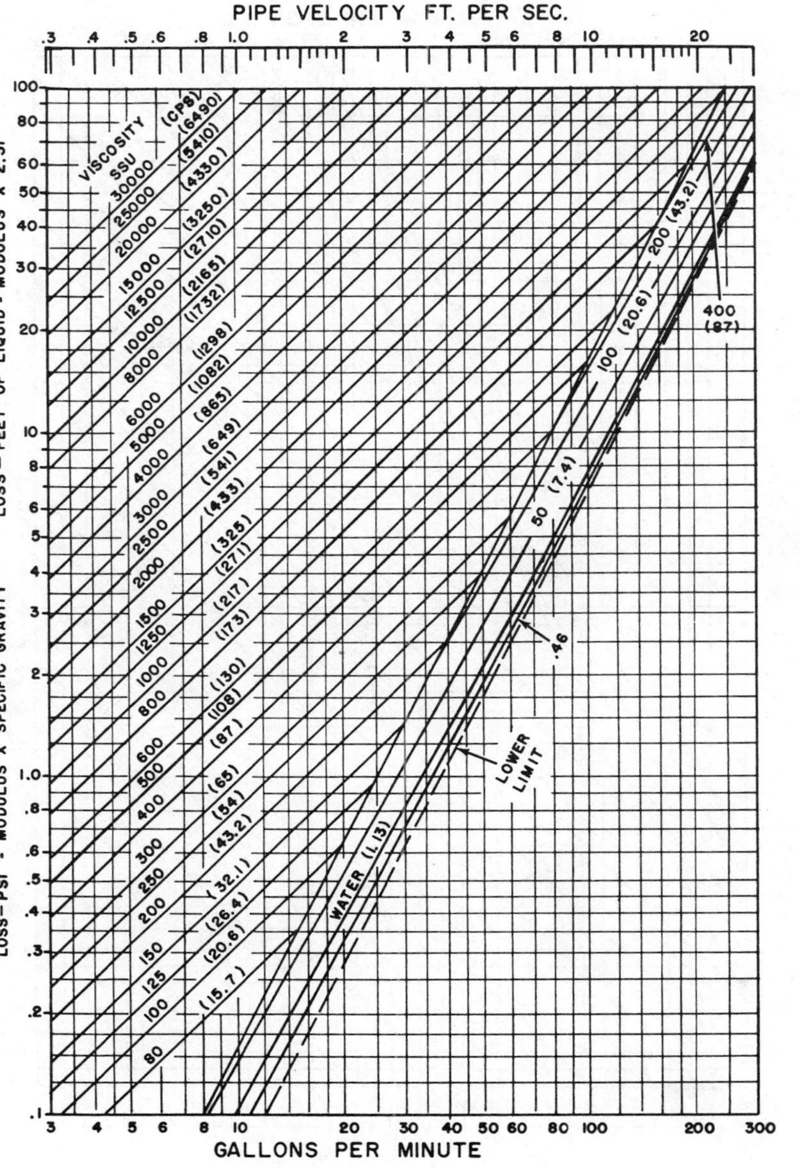

Loss—Lbs. per sq. in. = modulus × Specific Gravity
Loss—Feet of liquid = modulus × 2.31

*For SI units see Section A.1

Fig. A.6d
FRICTION LOSS MODULUS FOR 100 FEET OF 3" STEEL PIPE*

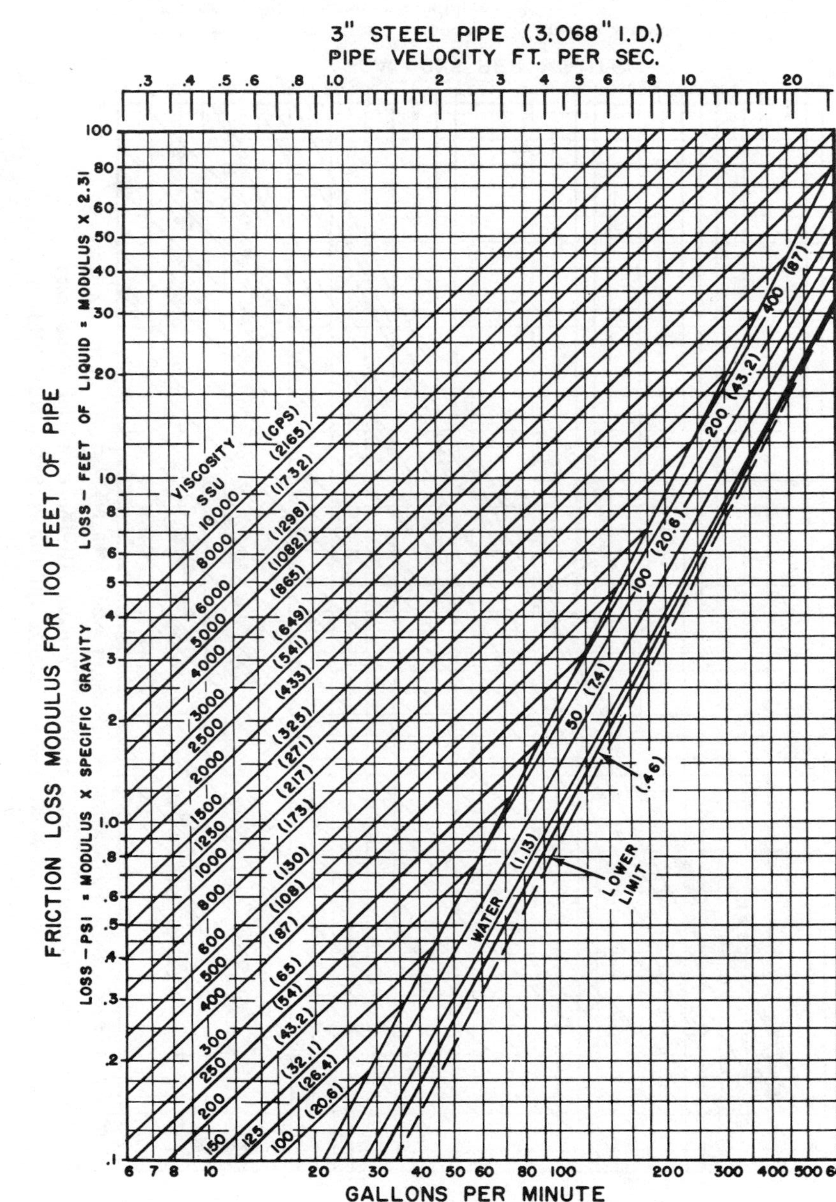

Loss—Lbs. per sq. in. = modulus × Specific Gravity
Loss—Feet of liquid = modulus × 2.31

*For SI units see Section A.1

Fig. A.6e
FRICTION LOSS MODULUS FOR 100 FEET OF 4″ STEEL PIPE*

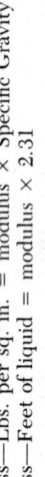

Fig. A.6f
FRICTION LOSS MODULUS FOR 100 FEET OF 6″ STEEL PIPE*

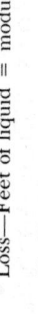

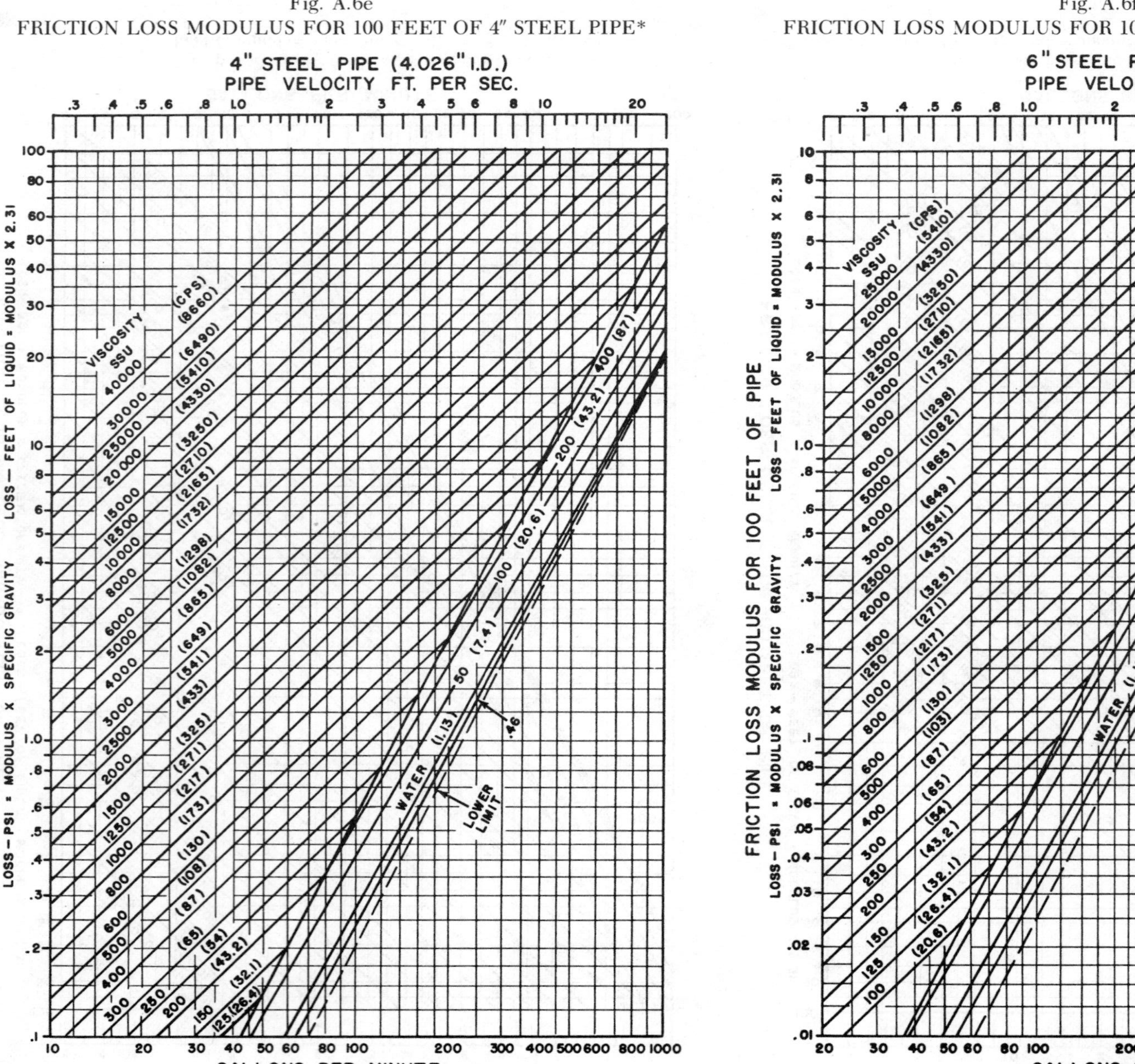

Loss—Lbs. per sq. in. = modulus × Specific Gravity
Loss—Feet of liquid = modulus × 2.31

*For SI units see Section A.1

Loss—Lbs. per sq. in. = modulus × Specific Gravity
Loss—Feet of liquid = modulus × 2.31

*For SI units see Section A.1

A.7 TANK VOLUMES

Table A.7a
CAPACITY OF ROUND TANKS*
Per Foot of Depth

Diam.	*Gals.*	*Area Sq. Ft.*	*Diam.*	*Gals.*	*Area Sq. Ft.*	*Diam.*	*Gals.*	*Area Sq. Ft.*	*Diam.*	*Gals.*	*Area Sq. Ft.*
1′	5.87	.785	4′	94.00	12.566	11′	710.90	95.03	22′	2843.60	380.13
1′1″	6.89	.922	4′1″	97.96	13.095	11′3″	743.58	99.40	22′3″	2908.60	388.82
1′2″	8.00	1.069	4′2″	102.00	13.635	11′6″	776.99	103.87	22′6″	2974.30	397.61
1′3″	9.18	1.227	4′3″	106.12	14.186	11′9″	811.14	108.43	22′9″	3040.80	406.49
1′4″	10.44	1.396	4′4″	110.32	14.748	12′	846.03	113.10	23′	3108.00	415.48
1′5″	11.79	1.576	4′5″	114.61	15.321	12′3″	881.65	117.86	23′3″	3175.90	424.56
1′6″	13.22	1.767	4′6″	118.97	15.90	12′6″	918.00	122.72	23′6″	3244.60	433.74
1′7″	14.73	1.969	4′7″	123.42	16.50	12′9″	955.09	127.68	23′9″	3314.00	443.01
1′8″	16.32	2.182	4′8″	127.95	17.10	13′	992.91	132.73	24′	3384.10	452.39
1′9″	17.99	2.405	4′9″	132.56	17.72	13′3″	1031.50	137.89	24′3″	3455.00	461.86
1′10″	19.75	2.640	4′10″	137.25	18.35	13′6″	1070.80	142.14	24′6″	3526.60	471.44
1′11″	21.58	2.885	4′11″	142.02	18.99	13′9″	1110.80	148.49	24′9″	3598.90	481.11
2′	23.50	3.142	5′	146.76	19.62	14′	1151.50	153.94	25′	3672.00	490.87
2′1″	25.50	3.409	5′3″	161.86	21.64	14′3″	1193.00	159.48	25′3″	3745.80	500.74
2′2″	27.58	3.687	5′6″	177.66	23.75	14′6″	1235.30	165.13	25′6″	3820.30	510.71
2′3″	29.74	3.976	5′9″	194.27	25.97	14′9″	1278.20	170.87	25′9″	3895.60	520.77
2′4″	31.99	4.276	6′	211.51	28.27	15′	1321.90	176.71	26′	3971.60	530.93
2′5″	34.31	4.587	6′3″	229.50	30.68	15′3″	1366.40	182.65	26′3″	4048.40	541.19
2′6″	36.72	4.909	6′6″	248.23	35.18	15′6″	1411.50	188.69	26′6″	4125.90	551.55
2′7″	39.21	5.241	6′9″	267.69	35.78	15′9″	1457.40	194.83	26′9″	4204.10	562.00
2′8″	41.78	5.585	7′	287.88	38.48	16′	1504.10	201.06	27′	4283.00	572.66
2′9″	44.43	5.940	7′3″	308.81	41.28	16′3″	1551.40	207.39	27′3″	4362.70	583.21
2′10″	47.16	6.305	7′6″	330.48	44.18	16′6″	1599.50	213.82	27′6″	4443.10	593.96
2′11″	49.98	6.681	7′9″	352.88	47.17	16′9″	1648.40	220.35	27′9″	4524.30	604.81
3′	52.88	7.069	8′	376.01	50.27	17′	1697.21	226.87	28′	4606.20	615.75
3′1″	55.86	7.467	8′3″	399.80	53.46	17′6″	1798.51	240.41	28′3″	4688.80	626.80
3′2″	58.92	7.876	8′6″	424.48	56.75	18′	1902.72	254.34	28′6″	4772.10	637.94
3′3″	62.06	8.296	8′9″	449.82	60.13	18′6″	2009.92	268.67	28′9″	4856.20	649.18
3′4″	65.28	8.727	9′	475.89	63.62	19′	2120.90	283.53	29′	4941.00	660.52
3′5″	68.58	9.168	9′3″	502.70	67.20	19′6″	2234.00	298.65	29′3″	5026.60	671.96
3′6″	71.97	9.621	9′6″	530.24	70.88	20′	2350.10	314.16	29′6″	5112.90	683.49
3′7″	75.44	10.085	9′9″	558.51	74.66	20′6″	2469.10	330.06	29′9″	5199.90	695.13
3′8″	78.99	10.559	10′	587.52	78.54	21′	2591.00	346.36	30′	5287.70	706.86
3′9″	82.62	11.045	10′3″	617.26	82.52	21′3″	2653.00	354.66	30′3″	5376.20	718.69
3′10″	86.33	11.541	10′6″	640.74	86.59	21′6″	2715.80	363.05	30′6″	5465.40	730.62
3′11″	90.13	12.048	10′9″	678.95	90.76	21′9″	2779.30	371.54	30′9″	5555.40	742.64

To find the capacity of tanks greater than shown above, find a tank of one-half the size desired, and multiply its capacity by four, or find one one-third the size desired and multiply its capacity by 9.
*For SI units see Section A.1

Table A.7b
CAPACITY OF PARTIALLY FILLED HORIZONTAL TANKS*

Diam. (in Feet)	Gallons Per Foot of Length When Tank is Filled								
	1/10	1/5	3/10	2/5	1/2	3/5	7/10	4/5	9/10
1	.3	.8	1.4	2.1	2.9	3.6	4.3	4.9	5.5
2	1.2	3.3	5.9	8.8	11.7	14.7	17.5	20.6	22.2
3	2.7	7.5	13.6	19.8	26.4	33.0	39.4	45.2	50.1
4	4.9	13.4	23.8	35.0	47.0	59.0	70.2	80.5	89.0
5	7.6	20.0	37.0	55.0	73.0	92.0	110.0	126.0	139.0
6	11.0	30.0	53.0	78.0	106.0	133.0	158.0	182.0	201.0
7	15.0	41.0	73.0	107.0	144.0	181.0	215.0	247.0	272.0
8	19.0	52.0	96.0	140.0	188.0	235.0	281.0	322.0	356.0
9	25.0	67.0	112.0	178.0	238.0	298.0	352.0	408.0	450.0
10	30.0	83.0	149.0	219.0	294.0	368.0	440.0	504.0	556.0
11	37.0	101.0	179.0	265.0	356.0	445.0	531.0	610.0	672.0
12	44.0	120.0	214.0	315.0	423.0	530.0	632.0	741.0	800.0
13	51.0	141.0	250.0	370.0	496.0	621.0	740.0	850.0	940.0
14	60.0	164.0	291.0	430.0	576.0	722.0	862.0	989.0	1084.0
15	68.0	188.0	334.0	494.0	661.0	829.0	988.0	1134.0	1253.0

*For SI units see Section A.1

Table A.7c
CAPACITIES OF VARIOUS CYLINDERS IN U.S. GALLONS*

Diam. (In Inches)	Length of Cylinder 1″	1′	5′	6′	7′	8′	9′	10′	11′	12′	13′	14′	15′	16′	17′	18′	20′	22′	24′
1		0.04	0.20	0.24	0.28	0.32	0.36	0.40	0.44	0.48	0.52	0.56	0.60	0.64	0.68	0.72	0.80	0.88	0.96
2	0.01	0.16	0.80	0.96	1.12	1.28	1.44	1.60	1.76	1.92	2.08	2.24	2.40	2.56	2.72	2.88	3.20	3.52	3.84
3	0.03	0.37	1.84	2.20	2.56	2.92	3.30	3.68	4.04	4.40	4.76	5.12	5.48	5.84	6.22	6.60	7.36	8.08	8.80
4	0.05	0.65	3.26	3.92	4.58	5.24	5.88	6.52	7.18	7.84	8.50	9.16	9.82	10.5	11.1	11.8	13.0	14.4	15.7
5	0.08	1.02	5.10	6.12	7.14	8.16	9.18	10.2	11.2	12.2	13.3	14.3	15.3	16.3	17.3	18.4	20.4	22.4	24.4
6	0.12	1.47	7.34	8.80	10.3	11.8	13.2	14.7	16.1	17.6	19.1	20.6	22.0	23.6	25.0	26.4	29.4	32.2	35.2
7	0.17	2.00	10.0	12.0	14.0	16.0	18.0	20.0	22.0	24.0	26.0	28.0	30.0	32.0	34.0	36.0	40.0	44.0	48.0
8	0.22	2.61	13.0	15.6	18.2	20.8	23.4	26.0	28.6	31.2	33.8	36.4	39.0	41.6	44.2	46.8	52.0	57.2	62.4
9	0.28	3.31	16.5	19.8	23.1	26.4	29.8	33.0	36.4	39.6	43.0	46.2	49.6	52.8	56.2	60.0	66.0	72.4	79.2
10	0.34	4.08	20.4	24.4	28.4	32.6	36.8	40.8	44.8	48.8	52.8	56.8	61.0	65.2	69.4	73.6	81.6	89.6	97.6
11	0.41	4.94	24.6	29.6	34.6	39.4	44.4	49.2	54.2	59.2	64.2	69.2	74.0	78.8	83.8	88.8	98.4	104.	118.
12	0.49	5.88	29.4	35.2	41.0	46.8	52.8	58.8	64.6	70.4	76.2	82.0	87.8	93.6	99.6	106.	118.	129.	141.
13	0.57	6.90	34.6	41.6	48.6	55.2	62.2	69.2	76.2	83.2	90.2	97.2	104.	110.	117.	124.	138.	152.	166.
14	0.67	8.00	40.0	48.0	56.0	64.0	72.0	80.0	88.0	96.0	104.	112.	120.	128.	136.	144.	160.	176.	192.
15	0.77	9.18	46.0	55.2	64.4	73.6	82.8	92.0	101.	110.	120.	129.	138.	147.	156.	166.	184.	202.	220.
16	0.87	10.4	52.0	62.4	72.8	83.2	93.6	104.	114.	125.	135.	146.	156.	166.	177.	187.	208.	229.	250.
17	0.98	11.8	59.0	70.8	81.6	94.4	106.	118.	130.	142.	153.	163.	177.	189.	201.	212.	236.	260.	283.
18	1.10	13.2	66.0	79.2	92.4	106.	119.	132.	145.	158.	172.	185.	198.	211.	224.	240.	264.	290.	317.
19	1.23	14.7	73.6	88.4	103.	118.	132.	147.	162.	177.	192.	206.	221.	235.	250.	265.	294.	324.	354.
20	1.36	16.8	81.6	98.0	114.	130.	147.	163.	180.	196.	212.	229.	245.	261.	277.	294.	326.	359.	392.
21	1.50	18.0	90.0	108.	126.	144.	162.	180.	198.	216.	238.	252.	270.	288.	306.	324.	360.	396.	432.
22	1.65	19.8	99.0	119.	139.	158.	178.	198.	218.	238.	257.	277.	297.	317.	337.	356.	396.	436.	476.
23	1.80	21.6	108.	130.	151.	173.	194.	216.	238.	259.	281.	302.	324.	346.	367.	389.	432.	476.	518.
24	1.96	23.5	118.	141.	165.	188.	212.	235.	259.	282.	306.	330.	353.	376.	400.	424.	470.	518.	564.
25	2.12	25.5	128.	153.	179.	204.	230.	255.	281.	306.	332.	358.	383.	408.	434.	460.	510.	562.	612.
26	2.30	27.6	138.	166.	193.	221.	248.	276.	304.	331.	359.	386.	414.	442.	470.	496.	552.	608.	662.
27	2.48	29.7	148.	178.	208.	238.	267.	297.	326.	356.	386.	416.	426.	476.	504.	534.	594.	652.	712.
28	2.67	32.0	160.	192.	224.	256.	288.	320.	352.	384.	416.	448.	480.	512.	544	576.	640.	704.	768.
29	2.86	34.3	171.	206.	240.	274.	309.	343.	377.	412.	446.	480.	514.	548.	584.	618.	686.	754.	824.
30	3.06	36.7	183.	220.	257.	294.	330.	367.	404.	440.	476.	514.	550.	588.	624.	660.	734.	808.	880.
32	3.48	41.8	209.	251.	293.	334.	376.	418.	460.	502.	544.	586.	628.	668.	710.	752.	836.	920.	1004.
34	3.93	47.2	236.	283.	330.	378.	424.	472.	520.	566.	614.	660.	708.	756.	802.	848.	944.	1040.	1132.
36	4.41	52.9	264.	317.	370.	422.	476.	528.	582.	634.	688.	740.	793.	844.	898.	952.	1056.	1164.	1268.

*For SI units see Section A.1.

A.8 ABBREVIATIONS

ENGINEERING ABBREVIATIONS

A

a	acceleration
A	(1) area; (2) ampere, symbol for basic SI unit of electric current; also amp
Å	Ångstrom (= 10^{-10} m)
abs	absolute (*e.g.*, value)
AC, ac, or a-c	alternating current; also ∿
ACFM	volumetric flow at actual conditions in cubic feet per minute (= 4.719×10^{-4} m^3/s)
AF or a-f	audio frequency
alt	altitude
amp	ampere; also A, *q.v.*
°API	API degrees of liquid density
asym	asymmetrical; not symmetrical
atm	atmosphere (= 101.325kPa)
aux	auxiliary

B

°Ba	Balling degrees of liquid density
bar	(1) barometer; (2) unit of atmospheric pressure measurement (= 100kPa)
bbl	barrels (= 0.1589 m^3)
°Bé	Baumé degrees of liquid density
bhp or b.h.p.	braking horsepower (= 746 W)
°Bk	Barkometer degrees of liquid density
blk	black (wiring code color for AC "hot" conductor)
bp or b.p.	boiling point
Bq	becquerel, symbol for derived SI unit of radioactivity, joules per kilogram, J/kg
°Br	Brix degrees of liquid density
BTU	British thermal unit (= 1054 J)

C

c	(1) velocity of light in vacuum (3×10^8 m/s); (2) centi, prefix meaning 0.01
°C	Celsius degrees of temperature
ca.	*circa:* about, approximately
cal	calorie (gram, = 4.184 J); also g-cal
cc	cubic centimeter (= 10^{-6} m^3)
cd	candela, symbol for basic SI unit of luminous intensity
CFM or cfm	cubic foot per minute (= 4.719×10^{-4} m^3/s)
Ci	curie (= 3.7×10^{10} Bq)
cm	centimeter (= 0.01 m)
cpm	cycles per minute; counts per minute
cps	(1) cycles per second (= Hz); (2) counts per second; (3) centipoises (= 0.001 Pa·s)
cos	cosine, trigonometric function
cp or c.p.	(1) candle power; (2) circular pitch; (3) center of pressure

D

d	(1) derivative; (2) differential as in dx/dt; (3) deci, prefix meaning 0.1; (4) depth; (5) day
D	diameter; also dia and ø
DC or dc	direct current
deg	degree; also ° (π/180 rad)
dia	diameter; also D and ø
DO	dissolved oxygen
d/p cell	differential pressure transmitter (a Foxboro trademark)
DPDT	double pole double throw (switch)

E

e	(1) error; (2) base of natural (Naperian) logarithm; (3) exponential function; also exp (-x) as in e^{-x}
E	(1) electric potential in volts; (2) scientific notation as in 1.5E − 03 = 1.5 × 10^{-3}
e.g.	*exempli gratia:* for example
E.L.	elastic limit
emf	(1) electromotive force (volts); (2) electromotive potential (volts)
EMI	electromagnetic interference
EQ or eq	equation
exp	exponential function as in exp (− at) = e^{-at}; also e

F

f	frequency; also freq
F	farad, symbol for derived SI unit of capacitance, ampere · second per volt, A·s/V
°F	Fahrenheit degrees [$t_{°C} = (t_{°F} - 32)/1.8$]
fhp	fractional horsepower (*e.g.*, ¼HP motor)
Fig.	figure
fl.	fluid
fl.oz.	fluid ounces (= 2.957×10^{-5} m³)
fp or f.p.	freezing point
FPM or fpm	feet per minute (= 5.08×10^{-3} m/s)
fps	feet per second (= 0.3048 m/s)
FS or fs	full scale

G

g	acceleration due to gravity (= 9.806 m/s²)
G	giga, prefix meaning 10^9
gal.	gallons (= 3.785×10^{-3} m³)
g-cal	gramcalorie, *q.v.*; also cal
G-M	Geiger-Mueller tube, for radiation monitoring
gph	gallons per hour (= 3.785×10^{-3} m³/s)
GPM or gpm	gallons per minute (= 6.309×10^{-5} m³/s)
grn	green (wiring code color for grounded conductor)
Gy	gray, symbol for derived SI unit of absorbed dose, joules per kilogram, J/kg

H

h	(1) height; (2) hour
H	(1) humidity expressed as pounds of moisture per pound of dry air; (2) henry, symbol of derived SI unit of inductance, volt·second per ampere, V·s/A
hhv	higher heating value
hor.	horizontal
Hz	hertz, symbol for derived SI unit of frequency, one per second (1/s)
HP or hp	horsepower (U.S. equivalent is 746 W)

I

ibid.	*ibidem:* in the same place
ID	inside diameter
i.e.	*id est:* that is
in.	inch (= 2.54×10^{-2} m)
in-lb	inch-pound (= 0.113 N·m)
IR	infrared

J

J	joule, symbol for derived SI unit of energy, heat or work, newton-meter, N·m

K

k	kilo, prefix meaning 1000
K	kelvin, symbol for SI unit of temperature
k-cal	kilogram-calories (= 4184 J)
kg	kilogram, symbol for basic SI unit of mass
kg-m	kilogram-meter (torque, = 9.806 N·m)
kip	thousand pounds (= 4448 N)
kPa	kilopascals
kVA	kilovolt-amperes
kW	kilowatts
kWh	kilowatt-hours (= 3.6×10^6J)

L

l	liter (= 0.001 m³)
L	(1) length; (2) inductance, expressed in henrys
lat	latitude
lb	pound (= 0.4535 kg)
LEL	lower explosive limit
lin.	linear
lim. or lim	limit
liq.	liquid
lm	lumen, symbol for derived SI unit of luminous flux, candela·steradian, cd·sr
ln	Naperian (natural) logarithm to base e
log or $\log_{10}$	logarithm to base 10; common logarithm
long.	longitude
LP	liquified petroleum or propane
LVDT	linear variable differential transformer
lx	lux, symbol for derived SI unit of illuminance, lumen per square meter, lm/m²

M

m	(1) meter, symbol for basic SI unit of length; (2) milli, prefix meaning 10^{-3}; (3) minute (temporal); also min
M	(1) thousand (in commerce only); Mach number; (2) molecular weight; mole; (3) mega, prefix meaning 10^6
mA	milliamperes (= 0.001 A)
max	maximum
mCi	millicuries (= 0.001 Ci)
m.c.p.	mean candle power
med.	medium or median
m.e.p.	mean effective pressure
mfg	manufacturer or manufacturing
mg	milligrams (= 0.001 kg)

mho — unit of conductance, replaced by siemens, S, *q.v.*
MI — melt index
mi — miles (= 1.609 km)
micro — prefix = 10^{-6}; also μ (mu) or μm and sometimes u, as in ug or μg, both meaning microgram (= 10^{-9} kg)
micron — micrometer (= 10^{-6}m)
min — (1) minutes (temporal); also m; (2) minimum
ml — milliliters (= 0.001 l)
mm — millimeters (= 0.001 m)
mmf — magnetomotive force in amperes
mol — mole, symbol for basic SI unit for amount of substance
mol. — molecules
mp or m.p. — melting point
mph — miles per hour (1.609 km/h)
mR — milliroentgens (= 0.001 R)
mrd — millirads (= 0.001 rd)
ms — milliseconds (= 0.001 s)
MTBF — mean time between failures
MW — megawatts (= 10^6 W)

N

N — newton, symbol for derived SI unit of force, kilogram-meter per second squared, kg·m/s²
N_0 — Avogadro's number (= 6.023×10^{23} mol^{-1})
n — (1) nano, prefix meaning 10^{-9}; (2) refractive index
NDIR — non-dispersive infrared
NDT — non-destructive testing

O

OD — outside diameter
ohm — unit of electrical resistance; also Ω (omega)
or — orange (typical wiring code color)
oz — ounce (= 0.0283 kg)

P

p — (1) pressure; (2) pico, prefix meaning 10^{-12}
Pa — pascal, symbol for derived SI unit of stress and pressure, newtons per square meter, N/m²
PB — proportional band
pct — percent; also %
pf — picofarad (= 10^{-12} F)
PF or p.f. — power factor
pH — acidity index (logarithm of hydrogen ion concentration)
ppb — parts per billion
ppm — parts per million
precip — precipitate or precipitated
psi — pounds per square inch (= 6.894 kPa)
psia — absolute pressure in pounds per square inch
psid — differential pressure in pounds per square inch
PSIG or psig — above atmospheric (gauge) pressure in pounds per square inch
pt — (1) point; (2) part; (3) pint (= 4.732×10^{-4} m³)

Q

q — (1) rate of flow; (2) electric charge in coulombs, C
Q — quantity of heat in joules, J
°Q — Quevenne degrees of liquid density
qt — quart (9.463×10^{-4} m³)
q.v. — *quod vide:* which see

R

r — radius; also rad
R — (1) resistance, electrical, ohms; (2) resistance, thermal, meter-kelvin per watt, m·K/W; (3) gas constant (= 8.317×10^7 erg·mol^{-1}·°C^{-1}); (4) roentgen, symbol for accepted unit of exposure to x and gamma radiation, (= 2.58×10^{-4} C/kg)
rad — (1) radius; also r; (2) radian, symbol for SI unit of plane angle measurement
rd — rad, symbol for accepted SI unit of absorbed radiation dose, (= 0.01 Gy)
rem — measure of absorbed radiation dose by living tissue (*r*oentgen *e*quivalent *m*an)
rev — revolution, cycle
Re — Reynolds number
RF or rf — radio frequency
RFI — radio frequency interference
RH — relative humidity
RI — refractive index
RMS or rms — square root of the mean of the square
RPM or rpm — revolutions per minute
rps — revolutions per second
RTD — resistance temperature detector

S

s — second, symbol for basic SI unit of time; also sec
S — siemens, symbol for derived SI unit of conductance, amperes per volt, A/V
sat. — saturated
SCR — silicon controlled rectifier
SCFM — standard cubic feet per minute (air flow at 1 atm and 70°F)
sec — seconds; also s
sin — sine, trigonometric function
SG — specific gravity; also sp.gr.

SPDT single pole double pole throw (switch)
sp.gr. specific gravity; also SG
sq square; also □
sr steradian, symbol for SI unit of solid angle measurement
SSU Saybolt universal seconds
std. standard
SWG standard (British) wire gauge

T

t (1) ton (metric, = 1000 kg); (2) time; (3) thickness;
T (1) temperature; (2) tera, prefix meaning 10^{12}; (3) period (= 1/Hz, in seconds); (4) tesla, symbol for derived SI unit of magnetic flux density, webers per square meter, Wb/m^2
$T_{1/2}$ half life
tan tangent, trigonometric function
T.S. tensile strength
°Tw Twadell degrees of liquid density

U

u prefix = 10^{-6} when the Greek letter μ is not available
UEL upper explosive limit
UHF ultra high frequency
UHSDS ultra high speed deluge system
UPS uninterruptable power supply
UPV unfired pressure vessel

V

v velocity
V volts, symbol for derived SI units of voltage, electric potential difference and electromotive force, watts per ampere, W/A
vert. vertical
VHF very high frequency
vs. versus

W

w (1) width; (2) mass flow rate
W (1) watt, symbol for derived SI unit of power, joules per second, J/s; (2) weight; also wt
w. water
Wb weber, symbol for derived SI unit of magnetic flux, volt·seconds, V·s
wh white (wiring code color for AC neutral conductor)
wt weight; also W

X

X reactance in ohms
x-ray electromagnetic radiation

Y

yd yard (= 0.914 m)
yr year

Z

Z (1) atomic number (proton number); (2) electrical impedance (complex) expressed in ohms

Notes

1. Whenever the abbreviated form of a unit might lead to confusion, the word should be written in full.
2. The values of SI equivalents were rounded to 3 decimal places.
3. The words meter and liter are used in their accepted spelling form instead of those in the standards, namely, metre and litre, respectively.

SOCIETIES AND ORGANIZATIONS

ACS American Chemical Society
AGA American Gas Association
APHA American Public Health Association
API American Petroleum Institute
ARI Air Conditioning and Refrigeration Institute
ASA American Standards Association
ASCE American Society of Civil Engineers
ASME The American Society of Mechanical Engineers
ASRE American Society of Refrigeration Engineers
ASTM American Society for Testing and Materials
AWG American Wire Gauge
BSI British Standards Institution
BWG Birmingham Wire Gauge
CI Cast Iron
CIL Canadian Industries Limited Flow Index
CSA Canadian Standards Association
DIN Deutsche Institut fuer Normung
DOE Department of Energy
FCI Fluid Control Institute
FIA Fluid Insurance Association
FM Factory Mutual
FPA Fire Protection Association
ICE Institute of Civil Engineers
IEC International Electrotechnical Commission
IEEE Institute of Electrical and Electronics Engineers
IPTS International Practical Temperature Scale
ISA Instrument Society of America
ISO International Organization for Standardization
ISTM International Society for Testing Materials
LPGA National LP-Gas Association
MCA Manufacturing Chemists' Association

NBFU	National Board of Fire Underwriters
NBS	National Bureau of Standards
NEMA	National Electrical Manufacturers Association
NFPA	National Fire Protection Association
NSC	National Safety Council
NSPE	National Society of Professional Engineers
NRC	Nuclear Regulatory Commission
OTS	Office of Technical Services

MISCELLANEOUS LETTER SYMBOLS

α (alpha)	(1) geometric angle; (2) radiation particle (Helium atom); (3) linear expansion coefficient
β (beta)	radiation particle (electron)
γ (gamma)	(1) electromagnetic radiation; (2) surface tension; also σ (sigma)
Δ (delta)	difference, change
ϵ (epsilon)	(1) emmissivity; (2) linear strain, relative elongation $\epsilon \times \Delta l/l_o$
η (eta)	(1) efficiency; (2) viscosity (absolute); also μ
θ (theta)	thermal resistance
λ (lambda)	(1) thermal conductivity; (2) wavelength
μ (mu)	(1) viscosity (absolute); also η; (2) linear attenuation coefficient; (3) prefix, micro = 10^{-6}; also u; (4) mμ:millimicron (10^{-9}m)
μm	micron (10^{-6}m)
ν (nu)	viscosity, kinematic
π (pi)	(1) surface pressure; (2) constant = 3.1416. . .
ρ (rho)	(1) density; (2) resistivity
σ (sigma)	(1) surface tension; also γ; (2) conductivity; (3) normal stress; (4) nuclear capture cross section
Σ (sigma)	summation
τ (tau)	(1) time delay; (2) shear stress; (3) time constant
ω (omega)	angular velocity expressed in radians per second
Ω (omega)	ohm
ϕ	diameter; also dia and D
∿	alternating current
′	minute, angular or temporal
″	second, angular
$\perp$	perpendicular to, normal to
∥	parallel
%	percent; also pct

GREEK ALPHABET

A, α	alpha	I, ι	iota	P, ρ	rho
B, β	beta	K, κ	kappa	Σ, σ	sigma
Γ, γ	gamma	Λ, λ	lambda	T, τ	tau
Δ, δ	delta	M, μ	mu	Υ, υ	upsilon
E, ϵ	epsilon	N, ν	nu	Φ, ϕ	phi
Z, ζ	zeta	Ξ, ξ	xi	X, χ	chi
H, η	eta	O, o	omicron	Ψ, ψ	psi
Θ, θ	theta	Π, π	pi	Ω, ω	omega

REFERENCES

1. Standard for Metric Practice, ASTM E380-79, American Society for Testing and Materials, Philadelphia, Pa., 1980.

BIBLIOGRAPHY

Baumeister, T. *et al.*, *Mark's Standard Handbook for Mechanical Engineers* (8th ed.,) New York: McGraw-Hill, 1967.

Considine, D.M. and Ross, S.D., ed., *Handbook of Applied Instrumentation*, New York: McGraw-Hill, 1964.

Definition of Terms Relating to Gamma and X-Radiography, ASTM E586, 1980.

Definition of Terms Relating to Thermal Analysis, ASTM E473, 1980.

ISA Standards and Practices for Instrumentation (6th ed.), 1980, ISBN 87664-450-7.

ISA Standards S5.1 (ANSI Y32.20, appproved January 1975) Instrumentation Symbols and Identification, ISBN 87644-330-6.

———S5.2 Binary Logic Diagrams for Process Operations, 1976, ISBN 87664-331-4.

———S5.4 Instrument Loop Diagrams, 1976, ISBN 87664-332-2.

———S51.1 Process Instrumentation Terminology, 1976, ISBN 87664-390-4

———MC96.1 American National Standard for Temperature Measurement Thermocouples, 1975, ISBN 87664-374-7.

Keller, J.J., *Metric System Guide, Conversion Tables*, Vol. 3, Neenah: J.J. Keller, 1974.

Oberg, E. *et al.*, *Machinery Handbook* (20th ed.) New York: Independence Press, 1975.

Pender, H. and McIlwain, K., *Electrical Engineers Handbook: Communications Electronics*, New York: Wiley-Interscience, 1950.

Standard for Metric Practice, ASTM E380-76, (ANSI Z210.1-1976 and IEEE 268-1976)

Weast, R.C. and Astle, M.J., eds., *CRC Handbook of Chemistry and Physics*, (59th ed.,) New York: CRC Press, 1979.

A.9 BIBLIOGRAPHY FOR INSTRUMENT ENGINEERS

Adams, L. F., *Engineering Measurements and Instrumentation*, The English Universities Press, Ltd., London, 1975.

Adiutori, E. F., *The New Heat Transfer*, Ventuno, Cincinnati, Ohio, 1974.

Athans, M. and Falb, P., *Optimal Control: An Introduction to the Theory and Its Applications*, McGraw-Hill Book Co., New York, 1966.

Bates, R. G., *Determination of pH—Theory and Practice*, John Wiley & Sons, Inc., New York, 1964.

Beveridge, G. S. G. and Schechter, R. S., "Optimization: Theory and Practice," *McGraw-Hill Series in Chemical Engineering*, McGraw-Hill Book Co., New York, 1970.

Buckley, P. S., *Techniques of Process Control*, John Wiley & Sons, Inc., New York, 1964.

Butler, J. N., *Ionic Equilibrium: A Mathematical Approach*, Addison-Wesley Publishing Co., Inc., Reading, Mass., 1964.

Caldwell, W. I., Coon, G. A. and Zoss, L. M., *Frequency Response for Process Control*, McGraw-Hill Book Co., New York, 1959.

Combs, C. F. (ed.), *Basic Electronic Instrument Handbook*, McGraw-Hill Book Co., New York, 1972.

Considine, D. M., *Process Instruments and Controls Handbook*, McGraw-Hill Book Co., New York, 1957.

Daniels, F., *Outlines of Physical Chemistry*, John Wiley & Sons, Inc., New York, 1948.

Diefenderfer, A. J., *Principles of Electronic Instrumentation*, W.B. Saunders Co., Philadelphia, 1972.

Doebelin, E. O., *Measurement Systems—Application and Design*, revised edition, McGraw-Hill Book Co., New York, 1975.

Eckman, D. P., *Automatic Process Control*, John Wiley & Sons, Inc., New York, 1958.

E.E.U.A., *Installation of Instrumentation and Process Control Systems*, Handbook No. 34, Constable and Co., Ltd., London, 1973.

Evans, F. L., Jr., *Equipment Design Handbook for Refineries and Chemical Plants*, Gulf Publishing Co., Houston, Texas, 1971.

Gibson, J. E., *Nonlinear Automatic Control*, McGraw-Hill Book Co., New York, 1963.

Gregory, B. A., *An Introduction to Electrical Instrumentation*, Macmillan, Inc., New York, 1973.

Harriott, P., *Process Control*, McGraw-Hill Book Co., New York, 1964.

Herrick, C. N., *Instrumentation and Measurement for Electronics*, McGraw-Hill Book Co., New York, 1972.

Hutchison, J. W., *ISA Handbook of Control Valves*, Instrument Society of America, Pittsburgh, Pa., 1971.

Lipták, B. G., *Environmental Engineers' Handbook*, Volumes I–III, Chilton Book Co., Radnor, Pa., 1974.

———, *Instrument Engineers' Handbook*, Chilton Book Co., Radnor, Pa., Vol. I, 1969, Vol. II, 1970, Supplement, 1972.

———, *Instrumentation in the Processing Industries*, Chilton Book Co., Radnor, Pa., 1973.

Johnson, A. J. and Auth, G. H., *Fuels and Combustion Handbook*, McGraw-Hill Book Co., New York, 1951.

Jones, B. E., *Instrumentation, Measurement and Feedback*, McGraw-Hill Book Co., New York, 1977.

Jones, E. B., *Instrument Technology*, Butterworth, Inc., Woburn, Mass., Vol. 1 (revised), 1965, Vol. 2, 1956, Vol. 3, 1974.

Kwakernaak, H. and Swan, R., *Linear Optimal Control Systems*, John Wiley & Sons, Inc., New York, 1972.

Levenspiel, O., *Chemical Reaction Engineering*, John Wiley & Sons, Inc., New York, 1962.

Lupfer, D. E. and Johnson, M. L., "Automatic Control of Distillation Columns to Achieve Optimum Operation," *ISA Trans.*, Instrument Society of America, Pittsburgh, Pa., April, 1974.

Meditch, J. S., *Stochastic Optimal Linear Estimation and Control*, McGraw-Hill Book Co., New York, 1969.

Murrill, P. W., *Automatic Control of Processes*, International Textbook Co., Scranton, Pa., 1967.

Oliver, B. M. and Cage, J. M., *Electronic Measurements and Instrumentation*, McGraw-Hill Book Co., New York, 1971.

Parsons, W. A., *Chemical Treatment of Sewage and Industrial Wastes*, National Lime Association, Washington, D.C., 1965.

Perry, J. H., *Chemical Engineers' Handbook*, fourth edition, McGraw-Hill Book Co., New York, 1963.

Shinskey, F. G., *Distillation Control,* McGraw-Hill Book Co., New York, 1977.

———, *Energy Conservation Through Control,* Academic Press, Inc., New York, 1978.

———, *pH and pIon Control,* John Wiley & Sons, Inc., New York, 1973.

———, *Process Control Systems,* McGraw-Hill Book Co., New York, 1979.

Smith, C. L., *Digital Computer Process Control,* International Textbook Co., Scranton, Pa., 1972.

Spink, L. K., *Principles and Practices of Flowmeter Engineering,* ninth edition, The Foxboro Company, Foxboro, Mass., 1967.

Spitzer, F. and Howarth, B., *Principles of Modern Instrumentation,* Holt, Rinehart and Winston, New York, 1972.

Tagg, C. F., *Electrical Indicating Instruments,* Butterworth, Inc., Woburn, Mass., 1974.

Technical Data Book—Petroleum Refining, American Petroleum Institute, Washington, D.C., 1970.

Treybal, R. E., *Mass-Transfer Operations,* McGraw-Hill Book Co., New York, 1955.

Van Winkle, M., *Distillation,* McGraw-Hill Book Co., New York, 1967.

Wightman, E. J., *Instrumentation in Process Control,* Butterworth, Inc., Woburn, Mass., 1972.

Wolf, S., *Guide to Electronic Measurements and Laboratory Practice,* Prentice-Hall, Inc., Englewood Cliffs, N.J., 1973.

INDEX

Note: See also Alphabetical List of Units, pp. 925–930; Engineering Conversion Factors, pp. 922ff.; Glossary of Analog and Digital Process Control Terms, pp. 899–907; Glossary of Computer Terminology, pp. 907–919; Glossary of Control Terms, pp. 899–987.